HANDBOOK OF PHYSIOLOGY

SECTION 6: The Gastrointestinal System, VOLUME III

PUBLICATIONS COMMITTEE

PAUL C. JOHNSON, *Chairman*

FRANCOIS ABBOUD
JOHN S. COOK
MELVIN J. FREGLY
STEPHEN H. WHITE

Elizabeth Cowley and Susie P. Mann,
Editorial Staff

Laurie S. Chambers, *Production Manager*

Diana Witt, *Indexer*

HANDBOOK OF PHYSIOLOGY

A critical, comprehensive presentation of physiological knowledge and concepts

---

SECTION 6: # The Gastrointestinal System

Formerly SECTION 6: Alimentary Canal

VOLUME III.
Salivary, Gastric, Pancreatic, and Hepatobiliary Secretion

Section Editor: STANLEY G. SCHULTZ

Volume Editor: JOHN G. FORTE

Executive Editor: BRENDA B. RAUNER

American Physiological Society, BETHESDA, MARYLAND, 1989

© *Copyright 1967 (Volume I), American Physiological Society*

© *Copyright 1967 (Volume II), American Physiological Society*

© *Copyright 1968 (Volume III), American Physiological Society*

© *Copyright 1968 (Volume IV), American Physiological Society*

© *Copyright 1968 (Volume V), American Physiological Society*

© *Copyright 1989, American Physiological Society*

Library of Congress Catalog Card Number 89–279

International Standard Book Number 0-19-520816-1

Printed in the United States of America by Waverly Press, Inc., Baltimore, Maryland 21202

Distributed by Oxford University Press, New York, New York 10016

Preface to *The Gastrointestinal System*

The first edition of the *Handbook of Physiology*, section 6, *Alimentary Canal*, comprised five volumes published between 1967 and 1968. This completely revised, second edition of section 6, *The Gastrointestinal System*, consists of four volumes: *Motility and Circulation* (Jackie D. Wood, Volume Editor), *Neural and Endocrine Biology* (Gabriel M. Makhlouf, Volume Editor), *Salivary, Gastric, Pancreatic, and Hepatobiliary Secretion* (John G. Forte, Volume Editor), and *Intestinal Absorption and Secretion* (Michael Field and Raymond A. Frizzell, Volume Editors).

A comparison of the contents of these two editions leads to the inescapable conclusion that the past two decades represent a golden age in the long and venerable history of gastrointestinal physiology. The advent and the application of powerful technologies have led not only to an explosion of information but also to an explosion of comprehension. Many previously enigmatic phenomena are now explicable at the cellular and often at the molecular level of biological organization. No area of this diverse discipline has been left untouched by this avalanche of knowledge.

I extend my heartfelt thanks to our many authors and reviewers for their contributions to this undertaking, and my gratitude to the volume editors, without whom this project could not have been executed, can never be duly expressed. Finally, many thanks are due Laurie Chambers, Production Manager, and her staff for their expert redactory work.

STANLEY G. SCHULTZ
Section Editor

Preface

When this volume on secretion in the gastrointestinal tract was first being developed, the edition published in 1967 was consulted as a starting point. It was clear that in the intervening years the secretory cell itself had matured and our knowledge of the cell physiology of secretion had enormously expanded. The present volume reflects that expansion with heavy emphasis on the cellular basis of secretion, in addition to chapters that review broader, more integrative aspects of secretion.

The book is intended for students of physiology at all levels, from the graduate student to the active researcher in gastrointestinal physiology. Each chapter was developed to highlight some important aspect of secretion. Chapters are grouped by traditional anatomical location, but it is apparent that there is considerable overlap in the functional activities of secretory processes, and general principles begin to emerge.

One of the most exciting areas in the physiology of secretion is the mode of regulation. Although there is great diversity in the nature of the secretory product and there are obvious differences in detailed biochemical pathways, common principles operate at the cellular level to turn a system on or off. Secretagogues interact with specific receptors, activating complex transducing machinery and producing cellular (second) messengers. These second messengers then direct the scenario of intracellular and membrane changes leading to secretion. Even the actual effector machinery for secretion can be generalized to a relatively few common principles. Most secretory mechanisms are activated by the manipulation of membranes between cytoplasmic and surface pools, by the modulation of membrane proteins that serve as channels or carriers, or by both. These include processes as diverse as macromolecular secretion and the output of voluminous isotonic salt solution.

Improvements in technology, novel methods, and creative applications have led to the many discoveries and advancements highlighted throughout this volume. Microelectrodes have extended electrophysiological analyses of secretion from transepithelial measurements to the specification of voltage and resistance at individual cell surfaces and have also provided information on intracellular ionic activities. Moreover, application of patch-clamp technology has created whole new concepts on the functional activity of single ionic channels and their role in steady state and activation processes. Particularly noteworthy has been the ingenious deployment of a whole variety of fluorescent probes to monitor ionic activities within cells and cell organelles. New probes are still being developed, but the measurements of cytosolic levels of Ca^{2+} and $H^+(pH)$ have already provided enormous insight into activation and regulatory processes of secretory cells. Advances in the separation of glands and secretory cells have led to more specific assignment of function, and such preparations have been particularly useful for in vitro analysis of cellular activation processes. Subcellular fractionation has provided biochemical perspective to epithelial cell polarity, detailing apical, basolateral, and cytoplasmic membrane domains as well as the interaction and flow of proteins among them. Moreover, studies on preparations of specific cellular organelles have unequivocally established functional mechanisms for a number of transport systems. Application of immunological techniques has promoted numerous advancements, such as cytological localization of receptors and transport proteins and quantitative immunochemistry of intracellular messengers, and is currently proving a valuable aid in the study of protein sorting and transfer associated with secretion.

The authors of this volume have skillfully incorporated the technological advances and the excitement of research into their essays on physiological function. Many chapters also offer some prospectus toward imminent and future discoveries. I am grateful to the authors for their expertise in developing and clearly expounding their assigned area and for their patience in seeing the work brought to print.

JOHN G. FORTE
Volume Editor

Contents

Salivary

1. Fluid and electrolyte secretion by salivary glands
   D. I. COOK
   J. A. YOUNG 1
2. Electrophysiology of salivary and pancreatic acinar cells
   O. H. PETERSEN
   Y. MARUYAMA 25
3. Calcium signaling system in salivary glands
   JAMES W. PUTNEY, JR. 51
4. Cellular regulation of amylase secretion by the parotid gland
   TERRY N. SPEARMAN
   FRED R. BUTCHER 63
5. Salivary mucin secretion
   DAVID O. QUISSELL
   LAWRENCE A. TABAK 79
6. Functional differentiation of salivary glands
   LESLIE S. CUTLER 93
7. Secretory membranes and the exocrine storage compartment
   RICHARD S. CAMERON
   PETER ARVAN
   J. DAVID CASTLE 107

Stomach

8. Neural and hormonal control of gastric secretion
   BASIL I. HIRSCHOWITZ 127
9. Inhibition of gastric acid secretion
   STANISLAW J. KONTUREK 159
10. Electrophysiology of gastric ion transport
    JEFFREY R. DEMAREST
    TERRY E. MACHEN 185
11. Cell biology of hydrochloric acid secretion
    JOHN G. FORTE
    ANDREW SOLL 207
12. Biochemistry of gastric acid secretion: H^+-K^+-ATPase
    G. SACHS
    J. KAUNITZ
    J. MENDLEIN
    B. WALLMARK 229
13. Intracellular activation events for parietal cell hydrochloric acid secretion
    CATHERINE S. CHEW 255
14. Cellular basis of pepsinogen secretion
    STEPHEN J. HERSEY 267
15. The gastric mucosal barrier
    BARRY H. HIRST 279
16. Secretion of bicarbonate by gastric and duodenal mucosa
    GUNNAR FLEMSTRÖM
    ANDREW GARNER 309
17. Immunological probes of gastrointestinal secretion
    ADAM SMOLKA 327
18. Functional development of stomach
    CHI-CHUAN TSENG
    LEONARD R. JOHNSON 345
19. Gastrointestinal mucus
    ADRIAN ALLEN 359

Pancreas

20. Pancreatic secretion of electrolytes and water
    R. M. CASE
    B. E. ARGENT 383
21. Cellular regulation of pancreatic secretion
    JOHN A. WILLIAMS
    DANIEL B. BURNHAM
    SETH R. HOOTMAN 419
22. Signaling transduction in hormone- and neurotransmitter-induced enzyme secretion from the exocrine pancreas
    IRENE SCHULZ 443
23. Regulation of digestive reactions by the pancreas
    STEPHEN S. ROTHMAN 465
24. Cellular compartmentation and protein processing in the exocrine pancreas
    GEORGE A. SCHEELE
    HORST F. KERN 477
25. Selective regulation of gene expression in the exocrine pancreas
    GEORGE A. SCHEELE
    HORST F. KERN 499
26. Long-term regulation of pancreatic function studied in vitro
    CRAIG D. LOGSDON 515
27. Structural and secretory polarity in the pancreatic acinar cell
    AMY CHANG
    JAMES D. JAMIESON 531

Hepatobiliary

28. Overview of bile secretion
    ALAN F. HOFMANN 549

29. Enterohepatic circulation of bile acids
    ALAN F. HOFMANN 567
30. Cellular mechanisms of hepatic fluid and electrolyte transport
    REBECCA W. VAN DYKE
    JOHN R. LAKE
    BRUCE F. SCHARSCHMIDT 597
31. Physical chemistry of bile
    DONNA J. CABRAL
    DONALD M. SMALL 621
32. Pathways and function of biliary protein secretion
    ALBERT L. JONES
    SUSAN JO BURWEN 663
33. Hepatocyte lysosomes in intracellular digestion and biliary secretion
    NICHOLAS F. LaRUSSO 677
34. Hepatic transport of organic solutes
    E. L. FORKER 693
35. Transport and metabolism in the hepatobiliary system
    DIRK K. F. MEIJER 717

Index 759

CHAPTER 1

Fluid and electrolyte secretion by salivary glands

D. I. COOK
J. A. YOUNG

Physiology Department, University of Sydney, Sydney, New South Wales, Australia

CHAPTER CONTENTS

Anatomical Nomenclature
Two-Stage Hypothesis
Mechanism of Formation of Primary Fluid
    Secretion models
        Luminal sodium pump model
        Potassium secretion model
        Basolateral Na^+-K^+-$2Cl^-$ symport model
        Bicarbonate secretion model
        Double-antiport (Na^+-H^+ and Cl^--HCO_3^-) model
    Elements of secretory mechanism
        Basolateral membrane
        Cytosol
        Luminal membrane
        Paracellular pathway
    Overview of secretion control
Ductal Electrolyte Transport
    Elements of absorptive mechanism
        Luminal membrane
        Basolateral membrane
    Plasticity of ductal transport properties
    Water permeability
    Control of ductal transport
    Overview of ductal electrolyte transport

WHEN SALIVARY SECRETION was last reviewed in the 1967 edition of the *Handbook* section on the alimentary canal (11, 96, 148), there was already fairly general agreement as to the truth of Thaysen's (163) two-stage model for salivary secretion, although most of the evidence supporting it was indirect. Soon afterward, however, this deficiency was remedied by the development of gland micropuncture (87, 99, 177) and microperfusion (174) techniques. In the ensuing decade, research efforts were largely concentrated on studying duct function (179), and on extending micropuncture studies to a wider range of glands and species (179), although in addition Petersen (123) performed interesting electrophysiological studies on the secretory cells following up Lundberg's (84) pioneer work. Nevertheless, understanding of electrolyte secretory processes generally lagged far behind understanding of absorption until 1977, when Silva et al. (155) introduced their model for secretion by the rectal salt gland of the shark. Since then our knowledge of secretion has grown rapidly.

ANATOMICAL NOMENCLATURE

Because of the structural diversity exhibited by the salivary glands, the literature on the morphology of these glands is large and complex (178). The terms used in this chapter, however, can be found in any standard textbook, and problems due to anatomical diversity do not arise because most recent physiological studies have been carried out on only a small group of glands: the parotid glands of the rat and mouse and the mandibular glands of the rat, rabbit, and cat.

All the major salivary glands consist principally of secretory end-pieces (also often called acini) gathered together in clusters at the extremities of a converging duct tree. Parotid end-pieces contain only one secretory cell type, but two or more cell types can be distinguished in mandibular and sublingual glands of many species largely on the basis of the histochemical staining properties of their secretion granules, although distinct physiological roles for each cell type have not been defined. Secretory end-pieces are usually supported by myoepithelial cells.

The most peripheral elements of the duct system are the intercalated ducts, which are short narrow tubes that connect the end-pieces to the intralobular ducts, often called striated ducts because of the extensively infolded appearance of their basolateral membranes. Although striated ducts always contain a few secretion granules, in some glands, particularly the mandibular glands of male mice, the intralobular ducts become densely packed with secretion granules and the basal striations become inconspicuous as the animal grows to maturity. They are then called granular ducts. Intralobular ducts drain into a system of extralobular or excretory ducts, which terminate in the main excretory duct. They have basolateral infoldings that are even more marked than in the intralobular ducts.

Most major salivary glands have a dual parasympathetic and sympathetic secretomotor innervation. Preganglionic parasympathetic fibers arise in the nucleus reticularis parvocellularis. Fibers destined for the parotid gland may pass via the glossopharyngeal nerve to the otic ganglion, or they may go via a buccal branch of the trigeminal nerve; fibers destined for the mandibular and sublingual glands pass mostly in the facial nerve and the chorda tympani, but some run in the lingual branch of the trigeminal nerve (178). Postganglionic parasympathetic fibers may arise from anatomically distinct extraglandular ganglia but often arise from ganglion cells within the gland substance. Postganglionic sympathetic fibers arise in the superior cervical ganglion. Sensory fibers reaching the mandibular glands arise from cells lying in the proximal vagal and geniculate ganglia and in the mandibular zone of the trigeminal ganglion (20). Parotid sensory fibers arise from cells in the trigeminal and otic ganglia (153).

TWO-STAGE HYPOTHESIS

When the two-stage hypothesis was first enunciated, Thaysen et al. (164) stated that the salivary glands form a primary secretion of constant Na^+ concentration and that this is changed by a process of Na^+ reabsorption as the saliva passes along the gland ducts, leading to the formation of a hypotonic final saliva. In 1960 Thaysen (163) modified his hypothesis to state *1*) that the concentrations of Na^+ and K^+ in the primary secretion were independent of secretory rate and that their sum was approximately equal to their sum in plasma water and *2*) that the ducts reabsorb Na^+ by a mechanism having a limited maximum capacity and either that they secrete some K^+ in exchange for Na^+ or that K^+ is concentrated secondarily to water reabsorption in a region of the duct system having a low permeability to the cation.

In the light of many subsequent experiments, the hypothesis can now be reformulated as follows.

Salivary glands produce a primary fluid in the end-pieces and perhaps also in the intercalated ducts. This fluid is isotonic or slightly hypertonic and has an electrolyte composition resembling but not identical to that of a plasma ultrafiltrate, and its composition changes only slightly when secretion rate is increased. The primary secretion is the sole source of the fluid in the final saliva; i.e., no fluid is secreted by the ducts.

The primary saliva is modified during its passage along the intralobular (striated) and extralobular (excretory) ducts. These ducts have a low permeability to water and reabsorb Na^+ actively (and Cl^- passively) so that the saliva becomes hypotonic. In addition, they secrete K^+ and HCO_3^- actively, and in a few glands they reabsorb Cl^- in exchange for HCO_3^-.

Since 1960 more than 30 studies on the relation between salivary flow rate and electrolyte concentration have been published (170, 179), and all these studies have revealed excretion patterns for the common electrolytes that can be explained in terms of the two-stage hypothesis. More direct evidence for the hypothesis, however, comes from micropuncture experiments. Martinez et al. (99), who were the first to perform such experiments, sampled fluid from the intercalated ducts of rat mandibular glands and found that it was isotonic, with an Na^+ concentration of ~146 mM and a Cl^- concentration of 121 mM. Subsequently, Young and collaborators (175, 177) showed that the primary fluid K^+ concentration was 7–8 mM and that cholinergic and β-adrenergic stimulation caused only small changes in the fluid's ionic composition. Since that time, numerous investigations have been performed in various mammalian salivary glands: rat parotid gland (87), rat sublingual gland (94), mouse mandibular and parotid glands (92), rabbit mandibular and parotid glands (91), cat mandibular and sublingual glands (67, 68), ferret mandibular and parotid glands (85), and sheep parotid gland (22).

The findings from these studies have been remarkably consistent. The primary fluid osmolality is ~290–310 mosmol/kg of water [in immature animals it is hypertonic: ~354 mosmol/kg of water (61)], the Na^+ concentration ranges between 125 and 160 mM, the K^+ concentration between 4 and 15 mM, and the Cl^- concentration between 100 and 120 mM. [Much lower primary Na^+ concentrations have been reported in the sheep parotid gland (22), but, because the solute responsible for making the primary fluid isotonic has not been identified, the findings are difficult to interpret.] Primary fluid HCO_3^- concentrations have not been measured but, from the known concentrations of Na^+, K^+, and Cl^-, it can be inferred that they are lower than or just equal to that in the interstitial fluid.

Micropuncture studies on exocrine glands are open to criticism on two technical grounds. *1*) In most glands the end-piece lumen is too narrow to allow for the direct insertion of the tip of a micropuncture pipette, so that "primary" fluid has to be sampled from the intercalated ducts into which the end-pieces drain. Should these ducts themselves transport electrolytes, they might well alter the composition of the end-piece fluid. The fact, however, that fluid obtained by micropuncture of the cat sublingual gland, in which the end-pieces can be punctured directly (67), is similar in composition to that in other salivary glands serves to assure us that the intercalated ducts are not causing any major change in the composition of the primary fluid. (They could, of course, be participating in the secretion of the primary fluid.) *2*) In micropuncture experiments, the tip of the puncture pipette is not usually visible when it is in place, so that the possibility of sample contamination with interstitial fluid cannot be excluded. Neither of these criticisms constitutes a fatal objection to the employment of micropuncture, but caution needs to be exercised in interpreting the results.

Another way to test the two-stage hypothesis is to look for the means of inhibiting the postulated ductal Na^+ reabsorptive mechanism without inhibiting the fluid secretion thought to take place in the end-pieces. This has been achieved by injection of transport inhibitors retrograde up the duct system in volumes sufficient to fill the ducts but not large enough to inhibit fluid secretion. In this way, as predicted by the two-stage hypothesis, it has proved possible to abolish the gland's capacity to form hypotonic, low-Na$^+$ saliva without reducing its capacity to secrete fluid at normal rates (11, 86, 90, 91, 100, 152). A similar result can be obtained without the need for retrograde ductal injection by vascular perfusion of isolated glands with solutions containing amiloride, which can enter the saliva and gain access to the luminal surface of the duct epithelium (117).

MECHANISM OF FORMATION OF PRIMARY FLUID

The classic demonstrations of Ludwig and Brettel (see ref. 179) that salivary glands can secrete in the absence of a blood pressure and that secretion can overcome outflow resistances generating pressures greater than the normal blood pressure established long ago that secretion must involve active solute transport. Such experiments, however, do not exclude the possibility that pressure filtration can contribute to normal secretion. In some exocrine glands, filtration appears to make such a contribution, at least under in vitro conditions. For instance, the isolated perfused rat pancreas secretes at a basal rate of ~0.5 $\mu l \cdot g^{-1} \cdot min^{-1}$ in the absence of any recognized stimulus. This basal secretion is not abolished by complete anoxia, by dinitrophenol, or by iodoacetate (or all three together), nor is it abolished by autonomic antagonists or transport blockers such as ouabain, furosemide, or amiloride (36, 83, 151). On the other hand, it depends on the vascular perfusion pressure and is abolished altogether if perfusion is stopped (172). In contrast, most salivary glands do not secrete appreciably at rest, and when they do the process can be inhibited by metabolic and transport blockers; furthermore, salivary secretion is not much affected by variations in perfusion pressure (172). It thus seems that, although filtration has a role to play in rat pancreatic secretion, it has no role in mandibular gland secretion.

Although most salivary glands do not appear to secrete at rest, some, such as the sheep parotid gland and the cat sublingual gland, secrete quite markedly in the absence of any recognized stimulus (175). This so-called spontaneous secretion can be defined as a process in which secretion takes place in the absence of neural or hormonal secretomotor activity, but, although the definition is sound, it is not very useful in practice. On the one hand, in addition to acetylcholine and norepinephrine, secretomotor nerves may release transmitters for which blockers have not yet been discovered, so it is difficult to be sure that one can block all exogenous secretomotor stimuli. On the other hand, because salivary ducts can reabsorb fluid, it is difficult to be certain that a gland not delivering any fluid to the exterior is truly quiescent—it may be secreting spontaneously at a rate low enough for complete reabsorption of the primary fluid to occur in the ducts. It seems quite possible that all glands secrete spontaneously to some extent and that a conspicuous flow of spontaneous secretion is due in part to inactive ducts; in both the cat sublingual gland and the sheep parotid gland the ducts absorb very little water (22, 67). If spontaneous secretion is a more or less universal occurrence in secretory epithelia, it would not be surprising if it should prove to depend on a transport mechanism different from that supporting stimulated secretion; this possibility needs to be borne in mind whenever one is called on to interpret results obtained from unstimulated gland cells.

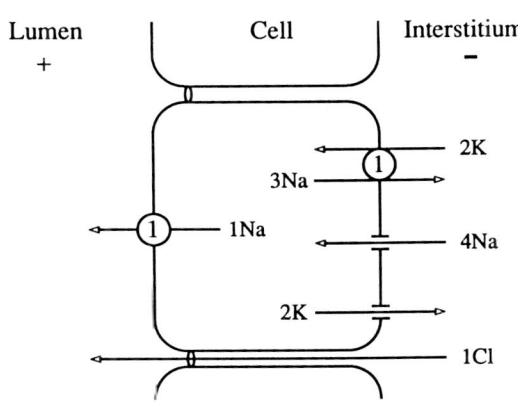

FIG. 1. Model showing secretory system dependent on active Na$^+$ transport pump in luminal plasma membrane. In this example, Na$^+$ is actively transported from cytosol across luminal membrane, but a system based on a luminal membrane Cl$^-$ pump, or even a NaCl pump, is also feasible. Cl$^-$ is shown moving passively across tight junctions, but it would be equally possible for the passively transported ion to take a transcellular route. Basolateral Na$^+$-K$^+$-ATPase, which exchanges 3Na$^+$ for 2K$^+$, can be seen to be working against luminal membrane pump.

Secretion Models

LUMINAL Na$^+$ PUMP MODEL. Until 8–10 years ago, it was widely believed that secretion depended principally on active Na$^+$ secretion (122). In such a case, it is necessary for there to be an Na$^+$-transporting ATPase in the luminal cell membrane capable of driving Na$^+$ from cytosol to lumen against a concentration gradient. Such a model is workable, but it is potentially wasteful of energy because Na$^+$ has to be pumped from cytosol to lumen against a gradient that costs energy to establish in the first instance (Fig. 1). Despite vigorous support in the early 1970s (122), the model has fallen into disfavor because efforts to demonstrate a luminal Na$^+$ pump in salivary glands have been unsuccessful, at least until recently.

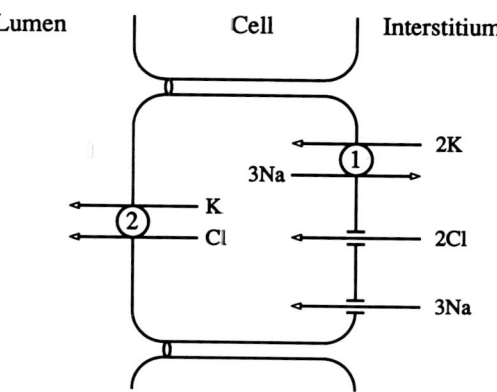

FIG. 2. Model showing secretory system dependent on active transport of K^+ by basolateral Na^+-K^+-ATPase. Potassium is concentrated in cytosol by Na^+-K^+-ATPase and enters saliva passively across luminal membrane. In this example, K^+ is shown entering saliva via a K^+-Cl^- symport, but it might as readily take a conductive pathway, and Cl^- could then take either a transcellular or a paracellular route. Number 2 on the K^+-Cl^- symport indicates that 2 molecules of KCl are secreted for each cycle of Na^+-K^+-ATPase.

POTASSIUM SECRETION MODEL. The simplest secretion model is one based on active K^+ transport (Fig. 2). This cation is normally concentrated in the cytosol by the basolateral Na^+-K^+-ATPase. If a pathway for it to cross the luminal membrane exists, then net secretion of the cation occurs, provided the current loop can be closed by the parallel movement of an anion. This anion might move paracellularly across the tight junctions, if they are anion selective, or it might move transcellularly, if suitable carriers or channels exist in the basolateral and luminal membranes. Until recently there was no compelling reason to postulate such a mechanism in salivary glands, although secretory mechanisms of this kind do operate in other tissues, e.g., gastric oxyntic cells, and K^+ concentrations in primary saliva are greater than in plasma (179). If salivary secretion depended exclusively on K^+ transport, however, the primary saliva would have much higher K^+ concentrations than those observed in micropuncture samples unless the K^+ secreted in the primary saliva is exchanged for interstitial Na at a point somewhere between the site of its initial secretion and the site of fluid collection by the micropuncture pipette (10). Recently, patch-clamp studies of the luminal membranes of cultured salivary cells have revealed the presence of plentiful, nonselective cation channels through which K^+ can pass (26). Consequently, it now appears that some K^+ secretion must occur, at least in cultured salivary cells, although it is not known whether the process is important in fresh tissue. Stimulation of secretion, however, is invariably accompanied by a large efflux of K^+ from the secretory cells across the basolateral membrane (11, 127, 129), an event that would tend to counteract any secretory process based on K^+ secretion.

BASOLATERAL Na^+-K^+-$2Cl^-$ SYMPORT MODEL. It was only in 1977 that it became clear how an anion could be pumped in the secretory direction without the need to invoke a special ATPase to do the work. In that year, Silva et al. (155), working on the rectal salt gland of the shark, had the insight to grasp how a modification of a widely accepted model for intestinal absorption might lead instead to secretion. They realized that the direction of solute transport in the rectal gland depended on the membrane location, basolateral instead of luminal, of a furosemide-sensitive cotransport protein, later found to have Na^+-K^+-$2Cl^-$ stoichiometry (Fig. 3). According to this model (see mathematical description of model in APPENDIX, p. 16), the cell Na^+ pump creates an Na^+ gradient and, as the ion runs down its gradient into the cell via the Na^+-K^+-$2Cl^-$ symport, some of its energy is transduced into the concentration of Cl^- in the cytosol above electrochemical equilibrium. Because the luminal membrane of the secretory cell possesses an anion-selective channel (57), Cl^- is then able to flow from the cytosol to the lumen. In this model, for each six Cl^- entering the cytosol, three Na^+ must also enter, but these are not secreted into the lumen with the Cl^-; instead they are returned to the interstitium by the Na^+ pump at the cost of one ATP molecule. Because the Na^+ pump drives out only one net positive charge for every six Cl^- entering the lumen, an exit pathway from the cell for five additional positive charges is needed; this current is carried by K^+, three having entered via the symport and two via the Na^+ pump. The principal pathway by which K^+ leaves, at least in the stimulated gland, is a regulated K^+ channel in the basolateral membrane. Consequently, the secretion process leads to net current flow across the cell, carried by Cl^- at the luminal membrane and by K^+ and the Na^+ pump (in a ratio of 5:1) at the basolateral membrane. For secretion to occur, the current loop must be completed;

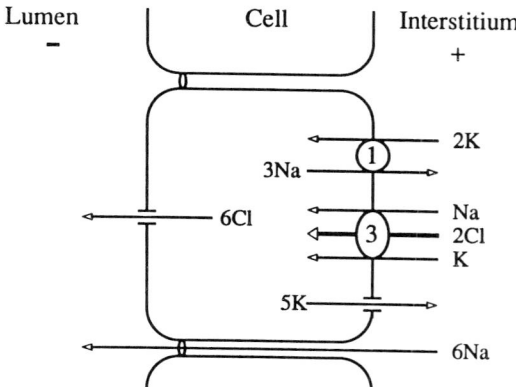

FIG. 3. Model showing secretory system dependent on active transport of Cl^- by basolateral Na^+-K^+-$2Cl^-$ symport. Cl^- is concentrated in cytosol and enters saliva passively via a conductive pathway in luminal membrane. Na^+ enters saliva passively across tight junctions. Circuit is completed by current flow across basolateral membrane, carried by K^+ through K^+ channels and via electrogenic Na^+-K^+-ATPase. Number 3 on symport indicates that 3 cycles of symport deliver $6Cl^-$ into cytosol for each cycle of Na^+-K^+-ATPase.

this is achieved by passage of Na^+ from interstitium to lumen between the secretory cells across cation-selective tight junctions (55).

BICARBONATE SECRETION MODEL. Not all exocrine glands secrete a Cl^--rich primary fluid; the pancreatic ducts, for instance, secrete HCO_3^--rich fluids. To account for active HCO_3^- secretion, which can occur in salivary glands, it is necessary only to invoke a simple modification of the Cl^--secretory model just described (Fig. 4). Instead of, or as well as, an Na^+-K^+-$2Cl^-$ symport in the basolateral membrane, an Na^+-H^+ antiport is required. This antiport can use Na^+ gradient energy to drive protons out of the cytosol and thereby raise the cytosol pH (and HCO_3^- concentration) above electrochemical equilibrium. Provided the luminal membrane has a suitable HCO_3^- channel, secretion based on HCO_3^- transport will then occur, analogous to that based on Cl^- transport, except that only three HCO_3^- will enter the lumen for each ATP molecule hydrolyzed, and the HCO_3^- current will be balanced by a basolateral cation current carried by K ions and the Na^+ pump in a ratio of 2:1.

DOUBLE-ANTIPORT (Na^+-H^+ AND Cl^--HCO_3^-) MODEL. Few salivary glands secrete fluids having very high concentrations of HCO_3^- (171) and those that do achieve this result by ductal rather than end-piece secretion of the anion. However, a simple modification of the Na^+-H^+ antiport system, which can lead to secretion of Cl^--rich rather than HCO_3^--rich fluids, appears to operate in many absorptive epithelia (78) and in some secretory epithelia as well (117, 151). This system involves the parallel operation of Na^+-H^+ and Cl^--HCO_3^- antiports in the luminal membranes of

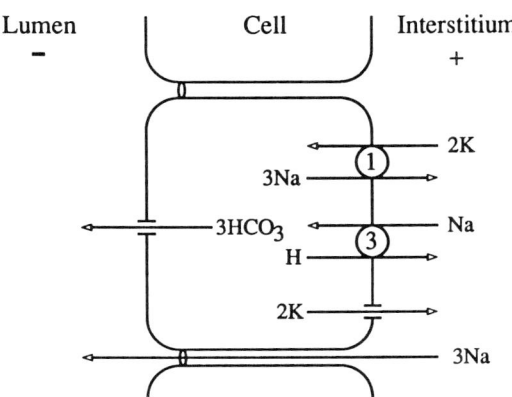

FIG. 4. Model showing secretory system dependent on active transport of HCO_3^- by basolateral Na^+-H^+ antiport. Protons are expelled from cytosol by antiport leading to intracellular accumulation of OH^-, which reacts with CO_2 to form HCO_3^-. HCO_3^- enters saliva passively via a conductive pathway in luminal membrane and Na^+ enters passively across tight junctions. Circuit is completed by current flow across basolateral membrane, carried by K^+ through K^+ channels and via electrogenic Na^+-K^+-ATPase. Number 3 on antiport indicates that 3 cycles of antiport generate $3HCO_3^-$ in cytosol for each cycle of Na^+-K^+-ATPase.

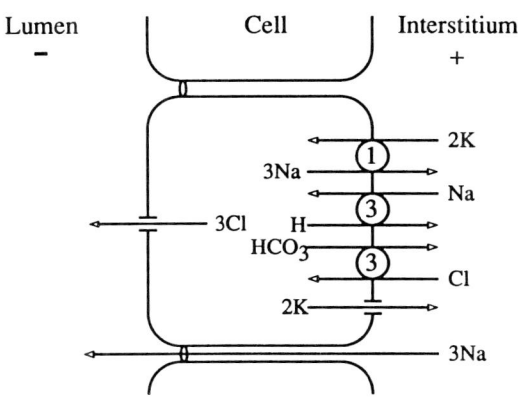

FIG. 5. Model showing secretory system dependent on active transport of Cl^- by paired basolateral Na^+-H^+ and Cl^--HCO_3^- antiports. HCO_3^- is first concentrated in cytosol by Na^+-H^+ antiport, and Cl^--HCO_3^- antiport utilizes the HCO_3^- gradient energy to concentrate Cl^- in cytosol. Cl^- enters saliva passively via a conductive pathway in luminal membrane and Na^+ enters passively across tight junctions. Circuit is completed by current flow across basolateral membrane, carried by K^+ ions through K^+ channels and via electrogenic Na^+-K^+-ATPase. Number 3 on each antiport indicates that 3 cycles of each antiport deliver 3 Cl^- ions into cytosol for each cycle of Na^+-K^+-ATPase.

absorptive cells and the basolateral membranes of secretory cells (Fig. 5). The Na^+-H^+ antiport, energized by the dissipative flux of Na^+, drives protons out of the cytosol so that intracellular pH rises above equilibrium. In turn, HCO_3^- efflux across the basolateral membrane causes Cl^- to be concentrated in the cytosol. The end result is the same as if an Na^+-Cl^- symport were present, the only difference being that use of antiport blocking drugs like 4-acetamido-4'-isothiocyanostilbene-2-2'-disulfonic acid (SITS) and amiloride can unmask the component parts of the double antiport system. Recently Knickelbein et al. (78) have argued that the two antiports may be functionally coupled to one another by membrane-bound carbonic anhydrase that catalyzes the reaction of OH^- and CO_2 and passes the HCO_3^- directly to the Cl^--HCO_3^- antiport.

Elements of Secretory Mechanism

Secretion involves a number of distinct cellular components, and control of the secretion rate might be exerted on any one of them. It is not necessary that a cell have only one secretion mechanism, however; several could be present and might be controlled independently of one another.

BASOLATERAL MEMBRANE. *Sodium pump.* With the exception of the apical Na^+ pump model, all the other secretion models just listed could depend for their energy supply solely on an Na^+-K^+-ATPase in the basolateral membrane of the secretory cell, although there is no reason why the enzyme should not be present in the apical as well as the basolateral membranes. Salivary glands are richly supplied with ouabain-sensitive Na$^+$-K$^+$-ATPase (178), although most of it appears to be associated with the absorptive ducts rather than the secretory cells (8). Two methods have commonly been applied to salivary glands for the localization of the enzyme: the autoradiographic localization of binding sites for ^{3}H-labeled ouabain and the cytochemical demonstration of ouabain-sensitive, K$^+$-dependent acylphosphatase (178). Poulsen's group (8), using the ouabain-binding method, has demonstrated that the ATPase is located in the basolateral membranes of seromucous secretory cells from cat mandibular and sublingual glands, and studies using the cytochemical method have revealed a similar localization of the enzyme in the mandibular glands of the dog, the rat, and the mouse (178). More recent studies employing a localization technique based on the binding of ferritin antibody complexes to the plasma membranes of two types of secretory epithelial cell, canine hepatocytes (106) and rat parotid end-piece cells (23), suggest that there may be considerable amounts of Na$^+$-K$^+$-ATPase in the luminal as well as the basolateral membranes.

Direct experimental evidence to support the notion that the ATPase is necessary for secretion includes the demonstration that salivary secretion is inhibited both by ouabain (21, 36, 62, 126) and by removal of extracellular K$^+$ (21, 62, 82). In some secretory tissues, such as rat pancreatic acini, it has proved possible to disassociate the effect on secretion of direct inhibition of the pump current by ouabain from the indirect effect of pump inhibition on the cell Na$^+$ gradient, but this has not yet been demonstrated in a salivary gland (36). It has been possible in salivary glands to demonstrate that the Na$^+$ pump is stimulated directly during secretion (101, 136, 160), but this fact does not help us to distinguish among the various transport models that have been put forward.

It is necessary to ask whether the rates at which salivary glands consume ATP are compatible with the amount of work done during stimulated secretion. Metabolic studies suggest that unstimulated salivary glands consume ATP at a rate of $\sim$0.1 μmol$\cdot$s$^{-1}\cdot$g^{-1} wet tissue and that this increases by a factor of 5 during maximum stimulation (79, 96, 110, 157). Making allowance for the large amount of ATPase activity in salivary ducts and the consumption of ATP associated with other cell activities, such as protein synthesis, it appears likely that the ATP consumption associated directly with fluid secretory activity is not more than $\sim$0.2–0.3 μmol$\cdot$s$^{-1}\cdot$g^{-1}. Now, when maximally stimulated, salivary glands secrete Cl$^-$ at a rate of $\sim$0.6 μmol$\cdot$s$^{-1}\cdot$g^{-1} (179). Consequently, if we assume that a gland secretes exclusively by means of a symport having Na$^+$-K$^+$-2Cl$^-$ stoichiometry and that the Na$^+$ pump consumes one molecule of ATP for every three Na$^+$ expelled, we can calculate that, when maximally stimulated, the gland utilizes ATP for secretion at a rate of at least 0.1 μmol$\cdot$s$^{-1}\cdot$g^{-1}. For an Na$^+$-Cl$^-$ symport or an Na$^+$-H$^+$ antiport, the ATP consumption would be 0.2 μmol$\cdot$s$^{-1}\cdot$g^{-1}, and, for a K$^+$-based secretory system, it would be 0.3 μmol$\cdot$s$^{-1}\cdot$g^{-1}. For a system based on a luminal ATPase, working in opposition to a basolateral ATPase, the consumption would be higher still, although it is not possible to predict the rate without more information; it could be as much as 0.4 μmol$\cdot$s$^{-1}\cdot$g^{-1} or even more. Although these calculations are too approximate to be considered conclusive, it can be said that although there appears to be more than enough energy available to support an anion-based secretion model, one based on a luminal ATPase would probably be beyond the work capacity of the gland. Calculations similar to these have been made by Poulsen and associates (34, 131), who estimated ATP requirements for secretion by the cat mandibular gland by measuring the rate at which the gland takes up K$^+$ from the interstitium after a period of intense secretomotor stimulation.

Potassium conductance. Because it is relatively easy to impale secretory cells with microelectrodes from the interstitial surface, it is not surprising that there have been numerous studies published on the resistance properties of the basolateral membranes of secretory cells in all three major salivary glands in several species (123, 179). Unfortunately, although it is simple to impale salivary secretory cells, it is extremely difficult to place the tip of a microelectrode in the lumen of a secretory end-piece, so difficult in fact that only two successful studies of this kind have been performed in salivary glands (64, 84). Consequently, in all the other conventional microelectrode studies published to the present time, interpretation of the results has been based on the untested assumption that the total resistance of the luminal membrane was high enough to prevent appreciable current flow across it. However, even for a cell in which the surface area of the luminal membrane is only $\sim$5% of the total area of plasma membrane [typical for salivary cells (178)], the specific resistance of the luminal membrane would only have to be about one-fifth that of the basolateral membrane for this assumption to be seriously wrong. That misleading conclusions do result from this difficulty has been confirmed by recent patch-clamp studies.

Nevertheless, much useful information has emerged from these microelectrode studies. The resting potential across the basolateral membranes of salivary secretory cells appears to lie between -50 and -70 mV, varying according to the species and the gland studied. The highest reported values (-73 mV) come from the rat parotid gland (49) and the lowest reliable values (-55 mV) from the mandibular and sublingual glands of the mouse (114, 166). In all salivary cells studied, the input resistance has been found to be quite low, $\sim$2–5 M$\Omega\cdot$cm^2. Although this is partly attributable to a marked degree of electrical coupling between the cells comprising an end-piece (123, 136), it also reflects the high conductance of the plasma membranes themselves. By calculating the area of plasma membrane across which the injected current flows (obtained from measurement of input capacitance and an estimate of the specific capacitance of the plasma membrane) the specific resistance of the plasma membranes can be estimated from the input resistance. From the results of Roberts and Petersen (137) for the mouse parotid gland, we calculate the resistance to be ~5 $k\Omega \cdot cm^2$. This value, which is already low, is nevertheless likely to be an overestimate, because salivary end-pieces must function as finite, two-dimensional cables (66), thereby exaggerating the extent of electrical coupling between neighboring cells.

Conventional microelectrode studies on mouse and rat parotid glands have shown that the low resistance of the plasma membrane is due largely to the presence of a K^+ conductance (136) that subsequent patch-clamp studies confirm is located in the basolateral membrane (47, 48, 103). Acetylcholine stimulation has been found to increase the K^+ conductance markedly (136), the change being associated with the well-known K^+ efflux encountered during glandular stimulation (11, 127, 129). A careful analysis of the time course of the acetylcholine-induced changes in membrane properties led Wakui and Nishiyama (166) to deduce that there were two separate K^+ conductances in the basolateral membrane, both of which were activated by gland stimulation with acetylcholine. Activation of one of these, which could be blocked with tetraethylammonium, appeared to lead to membrane hyperpolarization, whereas activation of the other led to a Na^+-dependent depolarization.

These predictions have recently received direct confirmation from patch-clamp studies (48, 103, 104, 124) of rat and mouse parotid and mouse mandibular secretory cells. These studies have demonstrated three cation-selective channels in the basolateral membrane and have failed to reveal any anion-selective channels. The most conspicuous and the best studied of the cation channels is the so-called maxi-K^+ channel. This channel is highly selective for K^+, discriminating even against Rb^+, and, when bathed symmetrically in solutions containing high concentrations of K^+ (145 mM), it has a conductance of 245 pS and a K^+ permeability of 0.46×10^{-12} cm^3/s. When bathed on the cytoplasmic surface with a K^+-rich fluid and on the interstitial surface with a plasmalike solution containing K^+ in a concentration of 4.5 mM, the channel is strongly rectifying and has a reversal potential of -90 mV and a conductance of only 35 pS at normal membrane potentials, although the K^+ permeability is unchanged. The channel is strikingly voltage sensitive when the cytoplasmic surface is exposed to Ca^{2+} in the concentration range of 10^{-9} to 10^{-8} M, but, when the Ca^{2+} concentration approaches 10^{-7} M, the channel is open almost continuously and, of course, no longer exhibits voltage sensitivity. In cell-attached patches, the channel is activated by acetylcholine, presumably mediated by a rise in intracellular Ca^{2+} activity (118, 119). It closely resembles a maxi-K^+ channel in the pig pancreas and, similarly, can be blocked by tetraethylammonium ions (65, 104). Using ensemble noise analysis on rat parotid secretory cells, Maruyama et al. (104) concluded that the maxi-K^+ channel exists in three states, one open and two closed, and that there is an average of 76 copies of the channel per secretory cell. A second K^+-selective channel of lower conductance than the maxi-K^+ channel has also been observed in mouse salivary glands by Gallacher and Morris (48), who think it may be a substate of the maxi-K^+ channel, but its properties are as yet unknown.

A third cation channel, which does not discriminate between Na^+ and K^+ and has a much lower conductance (30–35 pS when bathed symmetrically in Na^+-rich solutions), has also been reported in mouse parotid and mandibular glands. Like a similar channel in mouse pancreatic acinar cells (105), this nonspecific cation channel is Ca^{2+} but not voltage activated. It undergoes rapid inactivation in excised membrane patches, suggesting that its activation requires at least one other essential intracellular constituent in addition to Ca^{2+}.

If activation of K^+ channels is an essential step in stimulus-secretion coupling, one would predict that K^+ channel blockers should inhibit secretion. Recently it has been demonstrated that secretion is indeed blocked both by tetraethylammonium and by Ba^{2+}, agents known to block maxi-K^+ channels in other tissues (36, 63).

Basolateral Na^+-K^+-$2Cl^-$ symport. Much evidence has accumulated to show that a symport in the luminal membrane linking Na^+ and Cl^- fluxes is an important element in the process of salt transport by a number of absorptive epithelia, including the *Necturus* gallbladder (80) and the thick ascending limb of the loop of Henle in the kidney (9, 54). This symport is blocked by diuretics of the sulfamoyl-anthranilic acid group such as furosemide (142). In the loop of Henle (54), as in Ehrlich ascites cells where the observation was first made (53), the symport actually couples the fluxes of Na^+, K^+, and Cl^- with a stoichiometry of Na^+-K^+-$2Cl^-$, but not all NaCl symports necessarily have such a stoichiometry (161) and it is not certain that all those having the Na^+-K^+-$2Cl^-$ stoichiometry are sensitive to furosemide (35).

As mentioned, the first suggestion that an Na^+-Cl^- cotransporter might operate in secretory epithelia emerged from studies on the shark rectal gland (38, 155). Since publication of these studies, there have been numerous studies carried out to test the validity of the symport model in the shark rectal gland, and it can now be said that the model as outlined has overwhelming experimental support (38, 42, 43, 55–58, 60, 69, 155). The evidence includes careful biochemical studies on membrane vesicles (38, 69) and electrophysiological studies on the cell membranes and tight junctions with the determination of the cytosol ion activities (55–57).

Since the loop diuretics first became available, numerous reports have been published revealing that they can block transport in many secretory tissues including tracheal mucosa (167), corneal epithelium (13), lacrimal glands (28, 162), descending colon (44), ciliary body (138), teleost operculum (30), sweat glands (141), pancreatic acini (151), pancreatic ducts (19), mandibular glands (15, 18, 115, 117, 131), and parotid glands (113, 130). In many of these studies it has been postulated that secretion depends on a symport-based mechanism solely because secretion can be blocked by furosemide, although it is only in the dog tracheal mucosa (154, 167–169) and the shark rectal gland that the evidence for operation of the Na^+-K^+-$2Cl^-$ symport model is really compelling. Doubtless many other furosemide-sensitive secretory epithelia also utilize Na^+-Cl^- cotransport mechanisms, but other possibilities, such as secretion based on paired Na^+-H^+ and Cl^--HCO_3^- antiports, cannot be excluded without experimental justification: furosemide is not specific for the Na^+-K^+-$2Cl^-$ and Na^+-Cl^- symports but can inhibit carbonic anhydrase and block Cl^--HCO_3^- antiports and anion channels as well (12, 37, 51, 78).

In salivary glands the evidence for operation of a symport model is good but not yet conclusive, and it seems probable that at least some glands possess alternative secretory mechanisms, because there are salivary glands that can secrete in the presence of furosemide or if deprived of Cl^- and other anions accepted by the Na^+-K^+-$2Cl^-$ symport (18, 117). Because the very existence of a secretory anion symport has recently been challenged (37, 102), it seems necessary to specify what evidence there is for its existence in salivary as distinct from other glands. *1*) Secretion by mandibular glands of the cat, rat, and rabbit can be blocked reversibly by furosemide and bumetanide (15, 18, 97, 98, 115, 117, 131), although, as in other exocrine tissues, the concentration of furosemide needed to cause complete inhibition (10^{-4}–10^{-3} M) is higher than that required in absorptive epithelia (142) and other cell types (120). This observation has been used somewhat tenuously to argue that the drugs may not be acting on a symport in secretory tissues (37), but such an inference is unjustified because it assumes that all Na^+-K^+-$2Cl^-$ symports must have the same kinetic properties as those described in kidney tubules and red blood cells. *2*) Salivary secretion is inhibited by removal of extracellular Na^+, K^+, or Cl^- (18, 62, 84, 98, 112, 125). *3*) Removal of extracellular Cl^- inhibits poststimulus reuptake of K^+ (121). For a simple Na^+-Cl^- symport, this would result from inhibition of Na entry to the cytosol with subsequent blockage of the Na^+ pump; for an Na^+-K^+-$2Cl^-$ symport, K^+ uptake by the symport as well as the Na^+ pump would be blocked. *4*) Secretion by the mandibular gland shows a marked preference for Cl^- and Br^- over $CH_3CH_2COO^-$, I^-, and NO_3^- (15, 18, 98, 116, 117), which is precisely the preference shown by the Na^+-K^+-$2Cl^-$ symport in other tissues (35). *5*) Intracellular Cl^- activity is above equilibrium in salivary secretory cells both at rest and during stimulation (81, 84, 97, 108, 130, 140, 147), indicating that the cells possess an active anion-concentrating mechanism. *6*) Cytosol Cl^- uptake by rat parotid cells is furosemide sensitive (97, 113, 130). *7*) Metabolic studies on the perfused rabbit mandibular gland (157) show that, whereas the metabolic cost of secreting six Cl^- is the hydrolysis of one molecule of ATP, twice as much ATP is consumed for the absorption of six Cl^- by the salivary ducts. This indicates that salivary end-pieces secrete by a mechanism coupling the flow of twice as many Cl^- to the flow of Na^+, as is true in the ducts.

Recently, Marty (37) has argued, on the basis of experiments on rat lacimal glands, that furosemide acts not by blocking a symport but by blocking Cl^- channels. There is no reason to doubt that furosemide blocks Cl^- channels [although furosemide does not block Cl^- efflux in stimulated rat parotid cells (114, 131)], but this gives no grounds for rejecting the symport hypothesis; as mentioned, it is known that furosemide can block Cl^--HCO_3^- antiports (12, 51, 78) in addition to Na^+-K^+-$2Cl^-$ symports, so it should cause no surprise that it might also block an anion channel. Unfortunately no direct study on ion fluxes in vesicles prepared from the basolateral membranes of salivary cells has yet been performed, and, until this defect is remedied, the question of whether salivary glands have a typical symport or some other means of coupling anion uptake to an energy source cannot be considered as settled. It is difficult, however, to avoid postulating some kind of basolateral anion transporter—the questions to be addressed have more to do with the stoichiometry and ion affinities of the transporter than with its existence.

A teleological question that arises is what advantage an Na^+-K^+-$2Cl^-$ symport offers over other Cl^- secretory mechanisms such as an Na^+-Cl^- symport, a K^+-Cl^- symport, or some mechanism based on the presence of independent Na^+ and Cl^- channels in the basolateral membrane. The difference between a Na^+-Cl^- and an Na^+-K^+-$2Cl^-$ symport is clearly one of economy. In the former, three Cl^- enter the cytosol (and can be secreted) accompanied by three Na^+, and one molecule of ATP is hydrolyzed as the Na^+ are expelled again, whereas in the latter case, six Cl^- are secreted for each ATP molecule hydrolyzed. Given this, why not have a Na^+-$2K^+$-$3Cl^-$ symport? The answer is that the net driving gradient would then be insufficient to cause Cl^- uptake to take place.

Basolateral Na^+-H^+ and Cl^--HCO_3^- antiports. Recently it has become apparent that both the rat and rabbit mandibular glands are capable of secreting even when extracellular Cl^- is replaced with a nonsecreted ion such as isethionate or when the gland is exposed to a supramaximal dose of furosemide (18, 115–117). Under either of these circumstances, the maximum secretory response of the gland to acetylcholine is reduced to ~30% of normal, but the salivary HCO_3^- concentration rises from less than 20 mM to 70–80 mM. These observations suggest that the mandibular gland possesses a basolateral Na^+-H^+ antiport in addition to a furosemide-sensitive Na^+-K^+-$2Cl^-$ symport, an antiport whose action is unmasked by inhibition of the symport. This conclusion is supported by the observation that the residual secretion process can be blocked by amiloride, a recognized Na^+-H^+ antiport blocker, and by methazolamide, an inhibitor of carbonic anhydrase. Other observations, however, suggest that the system is more complex than this. On the one hand, secretion in the normal gland is not much affected by removal of extracellular HCO_3^- or by carbonic anhydrase inhibition, and, on the other hand, secretion is actually stimulated by administration of the disulfonic stilbene SITS, a recognized inhibitor of Cl^--HCO_3^- exchange (117). The simplest explanation of these findings is that the cell possesses both Na^+-H^+ and Cl^--HCO_3^- antiports that normally support electroneutral entry of NaCl into the cytosol and that furosemide, in addition to blocking Na^+-K^+-$2Cl^-$ symports, can also block Cl^--HCO_3^- antiports (12), leaving only the Na^+-H^+ antiport to drive secretion. Support for this model comes from experiments showing that furosemide can block anion self-exchange in dispersed rat mandibular secretory cells (51) and that cytosol pH rises from 7.1 to ~7.5 after gland treatments with furosemide. If we accept that the system does operate in rat and rabbit mandibular glands, it is appropriate to ask whether it is still necessary to postulate the presence of an Na^+-K^+-$2Cl^-$ symport as well. Two observations suggest that both systems must be present in the rabbit mandibular gland (117). *1*) Supramaximal doses of furosemide do not inhibit secretion by more than 80%–90% unless extracellular HCO_3^- is removed as well, and *2*) replacement of extracellular Cl^- with Br^- actually enhances secretion. Bromide is known to be well transported by Na^+-K^+-$2Cl^-$ symports in other tissues, but it is reported to be inhibitory of double antiport systems (78). A recent study on the rat mandibular gland suggests that the carriers in the basolateral membrane in that gland are active in the proportions $8(Na^+$-K^+-$Cl^-):5(Na^+$-$H^+):3(Cl^-$-$HCO_3^-)$, so that the gland secretes a mixture of Cl^- and HCO_3^- in the proportions 19:2 (128).

CYTOSOL. *Sodium, potassium, and chloride ions.* Until recently our knowledge of salivary gland electrolyte content was based entirely on the chemical analysis of tissue homogenates (84, 147). Although of limited value, useful qualitative results have emerged from such studies, and subsequent experiments based on the use of ion-selective electrodes have not revealed any major errors in the conclusions drawn from the earlier studies. As with chemical analyses of many other tissues, analysis of salivary glands reveals that the cells have a high K^+ and a low Na^+ concentration and a Cl^- concentration high enough to make it appear likely that this ion, too, is subject to active transport. Similar conclusions can be drawn from a recent study based on use of the electron microprobe (140).

Use of liquid-resin ion-selective microelectrodes has provided more accurate assessments of the intracellular activities of Cl^- and K^+ in the mandibular glands of the dog, mouse, and rabbit [Table 1; (81, 108, 109, 132)]. These studies show that K^+ and Cl^- are concentrated in the cytosol above electrochemical equilibrium both in resting and in stimulated glands and that stimulation causes their activities to fall 20%–30% and the ionic electrochemical potentials across the basolateral membrane to fall 30%–50%. Unfortunately, no information on intracellular Na^+ activities in salivary glands is yet available, although, because intracellular Na^+ concentrations certainly rise during stimulation (84, 140, 147), it is clear that the conjugate driving force across the basolateral symport, whatever its stoichiometry, must have decreased.

The pattern of change in cytosolic ionic activities is not identical in all exocrine glands, however. In the mouse lacrimal gland and the dog tracheal mucosa, for instance, although cytosol Cl^- activity falls during stimulation, the electrochemical potential for the ion across the basolateral membrane actually increases (139, 154), and in the dog mandibular gland, if one may extrapolate from electron microprobe studies, the Cl^- activity itself may actually rise during stimulation (140).

The best-studied exocrine gland is the shark rectal gland, in which the activities of Na^+, K^+, and Cl^- in the cytosol, as well as in the luminal and interstitial compartments, have been measured [Table 2; (58)]. In that gland, K^+ is close to equilibrium at rest, and its intracellular activity and basolateral electrochemical potential increase slightly during stimulation. As in salivary glands, however, Na^+ and Cl^- activities

TABLE 1. *Electrolyte Activities and Membrane Voltages in Mandibular Glands*

	Interstitium	Cytoplasm		Ref.
		Resting	Stimulated	
Potassium activity, mM				
Dog	3.4	89 (−46)	73 (−23)	109
Mouse	3.4	116 (−48)	81 (−24)	132
Chloride activity, mM				
Dog	94	27 (−7)		108
Rabbit	88	42 (−40)	34 (−28)	81
Membrane voltage, mV				
Dog		−40.2	−58.0	109
Rabbit		−59.2	−53.6	81
Mouse		−45.7	−59.8	166

Numbers in parentheses indicate electrochemical potential difference for each ion between cytosol and interstitium.

TABLE 2. *Electrolyte Activities and Membrane Voltages in Cells of Secretory Tubules of Rectal Salt Gland of Shark*

	Interstitium and Lumen	Cytosol	
		Resting	Stimulated
Electrolyte activity, mM			
Sodium	197[a]	11 (159)[b]	29.5 (122)
Potassium	3.1	123 (−7)	128 (−20)
Chloride	196	48 (−51)	41 (−35)
Potential difference, mV			
Basal		−86[c]	−74
Luminal		84	73
$\Delta\mu$ (Basal)*		50	32
$\Delta\mu_{Cl}$ (Luminal)†		−135	−108

Both epithelial surfaces were bathed in the same solution. Numbers in parentheses indicate electrochemical potentials of each ionic species with respect to interstitium. * Conjugate electrochemical driving force across basolateral symport having stoichiometry of Na^+-K^+-$2Cl^-$. † Electrochemical potential for chloride across luminal membrane. [Adapted from Greger et al. (58).]

and their basolateral electrochemical potentials fall during stimulation, although still remaining quite high. Similarly, the conjugate driving force for the three ions across the basolateral Na^+-K^+-$2Cl^-$ symport falls during stimulation, as does the electrochemical potential for Cl^- across the luminal membrane.

The available information on salivary cytosolic ion activities is fully in accord with a secretion model based on an Na^+-K^+-$2Cl^-$ symport. Two points require further comment, however: *1*) the significance of the observed decrease in the driving force for secretion during stimulation and *2*) the reason why cytosolic Cl^- activity appears to rise during stimulation in the dog mandibular gland, whereas it falls in all other glands so far studied. The fact that glandular secretion is increased by stimulation, at a time when the conjugate driving force across the symport and the Cl^- potential across the luminal membrane decrease, indicates that the properties of the symport and the channel must have changed during stimulation (i.e., they are regulated). For the anion channel there is evidence in the lacrimal gland to suggest that it may be activated by Ca^{2+} (102), and in the shark rectal gland by cAMP (57), but there is no evidence yet to indicate what might activate the basolateral symport.

Provided a model is accepted in which both the uptake of Cl^- across the basolateral membrane of the secretory cell and its subsequent exit from the cell across the luminal membrane are subject to cellular control, there is no difficulty in explaining why cytosolic Cl^- activity might rise during stimulation in some glands but fall in others. This would reflect the relative extent to which each regulatory protein has been activated. Indeed, it might be anticipated that cytosol Cl^- activity in the initial phase of stimulation, when non-steady-state conditions would prevail, might first rise and then fall if the basolateral symport and the luminal channel were not activated synchronously.

Bicarbonate and pH. No studies with pH-sensitive ion-selective electrodes have yet been performed on salivary secretory cells. Nevertheless, much can be learned simply from the study of the distribution of pH probes such as the weak acid 5,5-dimethyloxazolidine-2,4-dione (DMO). In incubated fragments of unstimulated rat mandibular glands the cell pH measured in this way is ∼7.2, and in the intact, perfused gland it is 7.12 (128). Similar and even higher pH values have been reported from studies with proton and phosphorus nuclear magnetic resonance (111, 172). It is thus clear that the secretory cell spends metabolic energy to eliminate protons from the cytosol and to maintain its pH (and HCO_3^- concentration) above equilibrium.

In view of the postulated role of Na^+-H^+ exchange in salivary secretion, Pirani et al. (128) investigated how cytosol pH changed during stimulation and what happened to pH when the gland perfusion conditions were altered to cause secretion of a HCO_3^--rich saliva. A submaximal dose of acetylcholine (0.3 μM) caused cytosol pH to fall from the resting level of 7.12 to a plateau value of 6.81 after 60–90 min continuous stimulation, whereas a supramaximal dose of acetylcholine (1.0 μM) caused cytosol pH to rise initially to ∼7.60, after which it fell slowly to ∼7.45 after 60 min. When the gland was treated with furosemide, the resting pH did not alter, but stimulation then caused pH to rise initially to above 7.45, after which it declined slowly. Replacement of extracellular Cl^- with gluconate caused the resting pH to rise to above 7.3, and stimulation caused a further rise to ∼7.5, again followed by a slow decline. The findings suggest that stimulation activates an Na^+-H^+ antiport, the presence of which is normally masked by the action of a Cl^--HCO_3^- antiport that can be blocked by gluconate substitution or by furosemide. It is not clear, however, why pH falls during continuous stimulation.

LUMINAL MEMBRANE. As mentioned, for technical reasons only two rather preliminary studies on the properties of the luminal membrane in salivary glands have been published (64, 84). The more detailed of these was performed by Lundberg (84) on the cat sublingual gland. He found that the voltage-divider ratio of luminal to basolateral membranes of the secretory cells was close to unity, which, given that the luminal membrane contributes only ∼5% to the total area of the plasma membrane, indicates that it has a much lower specific resistance than the basolateral membrane. Lundberg also found that stimulation decreased this resistance further by ∼20%.

Additional information about the luminal membranes can be obtained indirectly by comparing the membrane properties of the whole secretory cell (obtained from microelectrode studies on superfused tissue fragments or dispersed cell clusters) with what has been learned specifically about the basolateral membranes from patch-clamp studies. The intact cell has a modest Cl⁻ conductance (a transference number of 0.08 as compared with 0.25 for Na⁺ and 0.67 for K⁺), and the electrical response of the cell to stimulation is grossly altered by replacement of extracellular Cl⁻ with impermeant anions such as SO_4^{2-} (123, 136). In contrast, studies on cell-attached and excised membrane patches have failed to reveal any evidence of a Cl⁻ conductance in the basolateral membrane (48, 103, 104, 124), so it would seem that the Cl⁻ conductance must be in the luminal membrane and that it must be increased in size by stimulation.

Greger and associates (55–58) performed much more detailed studies on the luminal membranes of the cells of perfused secretory tubules from the shark rectal gland, an organ in which access to both surfaces of the secretory cell is technically somewhat easier. Like salivary end-pieces, the secretory tubules of the shark rectal gland seem to be of the leaky type, and, as mentioned, secretion is driven by a basolateral Na⁺-K⁺-2Cl⁻ symport. Greger and co-workers' studies show that the predominant electrogenic ionic pathway across the basolateral membrane of this cell is a K⁺ conductance, and across the luminal membrane is a Cl⁻ conductance, and that neither membrane possesses an appreciable Na⁺ conductance. He found that the primary event after stimulation is a >10-fold increase in the luminal Cl⁻ conductance, which patch-clamp studies show is due to activation (or insertion) of Cl⁻ channels in the luminal membrane. The channel, which has a slope conductance of 10–50 pS when bathed symmetrically in Cl⁻-rich solutions, is selective for Cl⁻ over Br⁻, is impermeable to gluconate, I⁻, and SO_4^{2-}, and can be blocked by diphenylamine-2,2′-carboxylate and its derivatives (53a, 57a).

Useful studies are also available on the luminal membranes of two flat secretory tissues, the tracheal mucosa (154, 167–169) and the corneal epithelium (70, 134), in both of which, of course, there is no problem of access to overcome. Like the salivary glands and the shark rectal gland, both these tissues secrete by mechanisms involving furosemide-sensitive symports, and in both the luminal plasma membranes contain a Cl⁻ conductance that is activated during secretomotor stimulation. For the tracheal mucosa, patch-clamp studies of the luminal membrane (of cultured cells) are also available (169). These demonstrate the presence of a Cl⁻ channel having a conductance of ~29 pS when the patch is bathed symmetrically in Cl⁻-rich solutions. Like the luminal channel seen in the shark rectal gland, the tracheal channel can be blocked by diphenylamine-2,2′-carboxylate, but it differs from that channel in exhibiting strong outward rectification in the absence of any ion gradients. In cell-attached but not excised patches, the channel is activated by isoproterenol; it is not activated by Ca^{2+}, nor is it strongly voltage gated.

This information on the luminal membranes of the shark rectal gland, the tracheal mucosa, and the corneal epithelium encourages us to predict that the Cl⁻

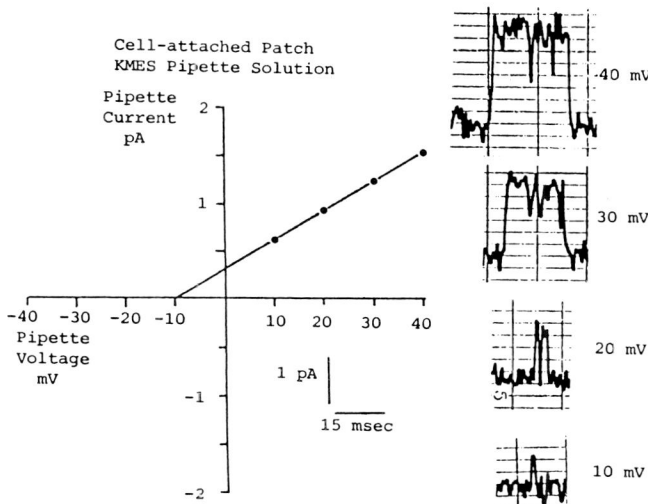

FIG. 6. Current-voltage relation for channel from luminal membrane of cultured cell from continuous mouse mandibular cell line. Recordings are from a cell-attached patch in which pipette contained K⁺ (140 mM) and Cl⁻ (25 mM). Reversal potential is close to −10 mV and slope conductance is 25 pS. *Right*, representative channel events at each pipette potential employed. [Data from Cook et al. (26).]

current induced by stimulation in dispersed salivary (and lacrimal) secretory cells will also prove to be due to the activation of a luminal Cl⁻ channel; however, a direct examination of the luminal membrane is necessary to prove the point. One may overcome the problem of access to the luminal membrane of salivary secretory cells by using tissue culture of either cell lines or freshly dispersed cells (26), because epithelial cells tend to grow in flat monolayers with the luminal surface uppermost, regardless of their original morphology. The most conspicuous channel that we have found (26) in the luminal membrane of cultured mouse mandibular cells is a cation channel, nondiscriminatory between Na⁺ and K⁺, with a slope conductance of ~25 pS (Fig. 6). Like similar channels in many other tissues, it is Ca^{2+} activated. In excised patches it is not voltage activated, although it is voltage sensitive in cell-attached patches, perhaps because voltage changes liberate Ca^{2+} from an intracellular store.

PARACELLULAR PATHWAY. *Lateral spaces.* Although solutes taking a paracellular route across an epithelium pass through the lateral space and cross the tight junctions that link the cells up to form an epithelial sheet, it would be wrong to think of the lateral space as belonging exclusively to the paracellular pathway; much of any substance taking a transcellular route must also pass through the lateral space, because its walls make up the larger part of the total area of the basolateral plasma membrane. This can best be appreciated by an examination of the relative membrane areas of a salivary secretory end-piece, calculated from morphometric information (173). A typical salivary gland, such as the rat mandibular gland, contains ~3

$\times 10^6$ end-pieces/g gland tissue. In a single end-piece, the total area of the basolateral membrane is $\sim 0.11 \times 10^6$ μm^2, of which the basal membrane contributes 23% and the lateral membranes contribute 76%; the luminal membrane area is only $\sim 0.02 \times 10^6$ μm^2 and the tight junctional area only $\sim 0.1\%$ of this. For a gland weighing 1 g, this would mean a total basolateral membrane area of $\sim 3,000$ cm^2, a luminal membrane area of 60 cm^2, and a tight junctional area of 0.05 cm^2.

Because the bulk of all transported solutes must pass through the lateral space, it is obvious that its shape and dimensions, and the degree to which it is distended or occluded in any given circumstance, will have an important influence on transepithelial transport. Unfortunately, morphological information on the geometry of these spaces in an exocrine gland is not available, so we cannot attempt a quantitative analysis of the resistances of the spaces to current flow or to fluxes of solutes and solvent. It seems quite likely, however, that the properties of these spaces will be similar in all water-transporting epithelia, absorptive as well as secretory, in particular including *Necturus* gallbladder, for which good morphological data are available (158). In that epithelium it is now generally accepted that the bulk of the transported water must pass from lumen to lateral space across the plasma membranes of the cell rather than across the tight junctions, the area of which is too small to permit much water flow even if the junctions are freely permeable to water (133, 159). For water to take a transcellular route, the plasma membranes must have a high enough water permeability: this appears to be true in salivary glands (27) just as it is in *Necturus* gallbladder (159).

Tight junctions. In contrast to water, inorganic ions cannot cross the plasma membrane at all unless appropriate membrane channels or carriers exist. In many secretion models, it is postulated that secreted Na^+ enters the saliva paracellularly across the tight junctions. For this to occur, it is necessary *1*) that the specific resistance of the junctions is low enough to permit the necessary current flow and *2*) that the junctions are sufficiently cation selective to ensure that the transcellular current will be matched by a secretory flow of Na^+ rather than an absorptive flow of the newly secreted anions. The electrical leakiness of the junctions can readily be established by comparing the transepithelial resistance of the secretory epithelium with the resistance of the plasma membranes. Thus, Lundberg's study (84) on the cat sublingual gland showed a transepithelial resistance of ~ 15 $\Omega \cdot cm^2$, whereas experiments from Petersen's group (123, 137) showed that the plasma membranes of salivary secretory cells have resistances in excess of 5 $k\Omega \cdot cm^2$. Similarly, in the shark rectal gland, Greger and associates (55, 56, 58) found that the epithelial resistance is 15–30 $\Omega \cdot cm^2$, whereas the plasma membrane resistance is at least 50 times greater. In the frog corneal epithelium (134) and the dog tracheal mucosa (154), the epithelial resistances are considerably higher (i.e., the epithelia are "tight"), but the secretory ion fluxes are correspondingly smaller. In all four cases, the same conclusion can be drawn—the tight junctions have resistances low enough to carry the necessary Na^+ current.

Some morphological investigations support the argument that salivary secretory end-pieces are leaky. Thus, the tight junctions are permeable to lanthanum ions (178) and even to horseradish peroxidase if the glands are first exposed to high concentrations of autonomic agonists (52), and freeze-fracture studies show that the junctions are composed of only a few ridges, a pattern suggestive of leakiness (29).

No studies on the permselectivity properties of tight junctions in salivary glands have appeared in the literature, although it is to be expected that they will prove to be cation selective, because the occurrence of anion-selective junctions in epithelia is extremely rare (41). Greger and Schlatter (55) have studied the junctional selectivity in the shark rectal gland and confirmed that they are cation selective.

Overview of Secretion Control

In theory, control of salivary secretion might be exerted by regulation of any of the transport elements already discussed. The most obvious control point would be the Na^+ pump itself. It is to be expected that the activity of the Na^+ pump will increase during stimulation, but this of itself does not indicate any direct regulation of the pump, because an increase in pump rate would in any case result from the rise in intracellular Na^+ that accompanies stimulation. There is, nevertheless, evidence for a more direct stimulation of the pump (160), although what intracellular mediators are involved in the process are unknown.

The control point about which we have the most information is the basolateral maxi-K^+ channel, which patch-clamp studies show is activated by increases in cytosolic free Ca^{2+} and by membrane depolarization (48, 124). Petersen and Maruyama (124) and Gallacher and Petersen (50) suggested that activation of the K^+ channels might be the primary event in initiating secretion. They argue that the drop in cytosolic K^+ and the concomitant rise in extracellular K^+ that accompany activation of the K^+ channels would reduce the K^+ driving force opposing Na^+ and Cl^- entry into the cytosol via the Na^+-K^+-$2Cl^-$ symport. It is arguable whether the drop in the K^+ driving force, which certainly does occur, could be sufficient alone to increase secretion rate to the extent actually encountered during stimulation, but, even if this were possible, the concomitant rise in cytosolic Na^+ that must accompany the fall in K^+ would counteract the effect. Nevertheless, K^+ channel activation is a necessary concomitant of any secretion process involving a large increase in transcellular current flow, and salivary secretion is acutely sensitive to exposure to K^+ channel blockers (36).

There is evidence in three glands to suggest that the luminal membrane anion channel is directly regulated: in lacrimal and salivary glands by Ca^{2+} (40, 64a, 102) and in the shark rectal gland by cAMP (57). Interestingly, however, histochemical studies for adenylate cyclase using adenylylimidodiphosphate as substrate show a much stronger reaction on the luminal than on the basolateral membranes of salivary secretory cells (135). This suggests that channel activity in the luminal membrane is controlled by cAMP even in cells utilizing Ca^{2+} and phosphoinositides as their major intracellular messengers for stimulus-secretion coupling, although a recent study (17) shows that forskolin has no additional effect on the secretion rate of the cholinergically stimulated rabbit mandibular gland. Also interesting is a recent report that cAMP may control the permeability of intercellular tight junctions (107).

As mentioned, in glands where there is sufficient information to make the necessary calculations (Table 2), it appears that the conjugate driving force for Na^+, K^+, and Cl^- across the basolateral membrane falls during stimulation. This clearly indicates that the symport itself must be regulated, although nothing is yet known about the mechanism.

Greger et al. (58) examined the question of the control of secretion in the shark rectal gland very carefully. Their studies showed three points of regulation: the luminal anion channel, the basolateral K^+ channel, and the basolateral symport. They favored the view that activation of the luminal Cl^- channel is the primary event, but more detailed studies on the initial, non-steady-state phase of secretion will be needed before this point can be settled.

DUCTAL ELECTROLYTE TRANSPORT

As indicated, the two-stage hypothesis requires that salivary ducts modify the composition of the primary saliva by reabsorption and secretion of electrolytes. The truth of this proposition has been established indirectly from a comparison of the composition of the primary saliva (obtained by micropuncture) with that of final saliva and directly, in experiments to be described, by duct microperfusion. Unfortunately, however, when Thaysen (163) and subsequent authors referred to ductal transport activity, their conclusions were mainly in terms of the intralobular (or striated) ducts, histologically speaking the most conspicuous ductal structures, whereas for technical reasons microperfusion experiments have been carried out exclusively on the extralobular (or excretory) ducts. Consequently our understanding of salivary duct transport is based on the assumption that the intralobular and extralobular ducts have similar properties. This is not an unreasonable assumption, because the ducts normally do have similar histological appearances, although there are enough differences, such as the presence of numerous secretion granules in the intralobular ducts and the presence of "dark" cells in the extralobular ducts, to make it appear unlikely that their functional properties are identical (178).

The evidence that the two ductal orders are at least similar in their transport properties is as follows. *1*) Micropuncture sampling from the most peripheral segments of the extralobular ducts shows that the saliva is already profoundly hypotonic, suggesting that this change takes place in the striated ducts (177). *2*) Retrograde injection of amiloride, poly-L-lysine, or ouabain is only able to abolish the gland's capacity to produce hypotonic saliva without altering its secretion rate if the inhibitors are injected into the intralobular as well as the extralobular ducts (86, 90, 91, 100, 152). *3*) Direct micropuncture of the intralobular ducts has been accomplished in the rat sublingual gland, and the aspirated saliva was found to be hypotonic (94). *4*) Histologically, both striated and excretory ducts undergo hypertrophy during salt deprivation and involution during salt loading (178, 179), which suggests that they both play a role in salt absorption. *5*) Even the most generous estimate of the surface area of the excretory ducts suggests that alone they could not account for the Na^+ absorptive activity of the stimulated gland. *6*) Perfusion studies on the main excretory duct revealed a pattern of electrolyte transport behavior that qualitatively can account for the conversion of the primary fluid into the final saliva; that is, there is no evidence to indicate that the intralobular ducts perform any electrolyte transport function not also performed by the extralobular ducts.

It thus seems reasonable to assume that the results of perfusion studies on the extralobular ducts can be extended qualitatively to the entire duct tree (although there is no ground for assuming that the two duct segments have absolutely identical transport properties). This is fortunate, because the extralobular ducts, particularly the main excretory duct, are readily accessible and have been studied extensively in both the rat and the rabbit. The main duct of the rat mandibular gland, which reabsorbs Na^+ and secretes K^+ and HCO_3^-, has been preferred for in vivo studies, whereas the main duct of the rabbit mandibular gland, which absorbs Na^+ powerfully but secretes little K^+ and HCO_3^-, has been preferred for in vitro studies.

Elements of Absorptive Mechanism

In Figure 7 the available information on salivary duct electrolyte transport is summarized in terms of a model that incorporates many of the features proposed by Knauf and Lübcke (74) for the rat mandibular main duct, except with respect to the location of the

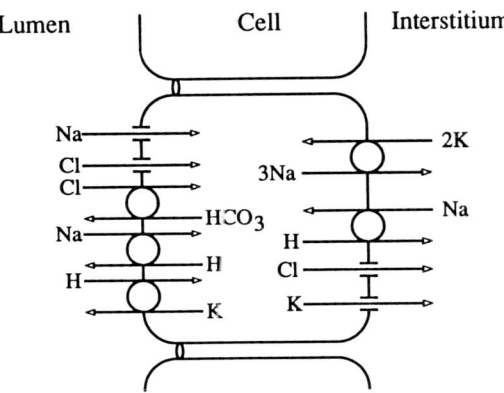

FIG. 7. Model for electrolyte transport by excretory duct epithelium of rat mandibular gland. Na^+ enters cytosol from lumen either via conductive pathway or a nonconductive Na^+-H^+ antiport and is expelled to interstitium by Na^+ pump. K^+ is transported from interstitium to cytosol by Na^+ pump and enters saliva via a luminal K^+-H^+ antiport. Cl^- is absorbed across conductive pathways in both plasma membranes, as well as via a Cl^--HCO_3^- antiport in luminal membrane. HCO_3^- is concentrated in cytosol by basolateral and luminal Na^+-H^+ antiports and can be secreted into lumen by luminal Cl^--HCO_3^- and K^+-H^+ antiports. [Adapted from Knauf et al. (74).]

anion conductances. No fewer than five elements are required in the luminal membrane, and four more are required in the basolateral membrane.

LUMINAL MEMBRANE. *Na^+ channel.* In both rat and rabbit ducts there is a large Na^+ conductance in the luminal membrane that shows a typical Nernstian relation between voltage and luminal Na^+ concentration when the shunting effect of Cl^- is eliminated (39, 71, 174). The conductance is highly selective for Na^+ and Li^+ over other cations (71, 174) and it can be blocked reversibly by amiloride and its analogues (5, 73, 77, 144).

Na^+-H^+ antiport. In the rat duct, Na^+ can also cross the luminal membrane via an Na^+-H^+ antiport (73, 74, 76, 77), although the concomitant absorption of HCO_3^- that should result from Na^+ absorption by such a mechanism is normally concealed by the parallel operation of a K^+-H^+ antiport that secretes K^+ and HCO_3^-. Like the luminal Na^+ conductance, the Na^+-H^+ antiport can be blocked by amiloride. In the rabbit duct there is a similar albeit less conspicuous Na^+-H^+ antiport (6).

Anion channel. To preserve electroneutrality, the component of salivary Na^+ that is reabsorbed via the luminal Na^+ channel must either be exchanged for secreted K^+ or reabsorbed in parallel with Cl^-. Because the duct has an extremely low K^+ conductance (143, 174), the latter of these possibilities eventuates. In the rabbit duct, this process appears to be mediated via a Cl^- conductance in the luminal membrane. The duct, which in all other respects clearly belongs to the tight class of epithelia (3, 45), has a very low transepithelial resistance of ~10 $\Omega \cdot cm^2$ when bathed in fluids containing Cl^-, but its resistance is much greater when Cl^- is replaced by SO_4^{2-}. The duct epithelium develops symmetrical Cl^- diffusion potentials when the anion gradient across it is reversed (46, 156). This might at first suggest that the Cl^- shunt is located in the tight junctions, if it were not that the anion diffusion barrier has a so-called intermediate-field anion selectivity sequence ($Br^- \geq Cl^- > I^- > F^-$), a sequence that is characteristic of the anion conductance apparent in muscle cell plasma membranes (46). Consequently most workers now agree that the conductance must be in the luminal plasma membrane (2, 3, 46) and that the confusing symmetry of the anion diffusion potentials results from the presence of a similar conductance in the basolateral membrane (2, 72). The anion selectivity properties of the rat duct have not been studied in as much detail, but the duct is clearly of the tight class and ductal Cl^- absorption in this species appears also to be transcellular and not paracellular.

K^+-H^+ antiport. The rat duct secretes K^+ at rates of up to 2.5 $nmol \cdot cm^{-2} \cdot s^{-1}$ and can develop an electrochemical gradient of ~7.5 kJ/mol. To achieve this extremely steep gradient, the duct has to secrete K^+ actively across an epithelium that is very impermeable to K^+. The site of the permeability barrier is the luminal plasma membrane (73, 77, 143, 174), not the basolateral membrane, which is quite permeable to K^+ (72, 179). The energy source for this secretion is the basolateral Na^+-K^+-ATPase, and K^+ crosses the luminal membrane down an electrochemical gradient. In view of the very low K^+ conductance of the luminal membrane, an electroneutral carrier mechanism for K^+ is required; this appears to be a K^+-H^+ antiport, because, whenever Na^+ absorption is halted (e.g., by use of amiloride), there is a close correspondence between the rates of K^+ and HCO_3^- secretion (73, 74, 76, 77). Under appropriate circumstances it can be shown that the rat duct can develop a limiting HCO_3^- gradient of 6.1 kJ/mol (179), but normally, when both the Na^+-H^+ and the K^+-H^+ antiports are operating, the duct secretes very little HCO_3^-, and ductal transport of Na^+ and K^+ behaves as if the ion fluxes were coupled by a one-for-one exchanger. The position in the rabbit duct is similar to that in the rat except that the rabbit duct's transport capacities for K^+ and HCO_3^- are much lower (25, 71, 93, 156, 176, 179).

Cl^--HCO_3^- antiport. When Na^+ transport in the rat duct is completely blocked by amiloride, exactly equal secretory net fluxes of K^+ and HCO_3^- are unmasked (73, 76, 77). In the rabbit duct, however, the same maneuver causes the HCO_3^- secretory flux to rise to more than twice that of K^+ (4, 5, 25). This is probably due to the presence of a Cl^--HCO_3^- exchanger in the luminal membrane in the rabbit duct, because the disulfonic stilbene SITS can inhibit ductal HCO_3^- secretion if applied to the luminal membrane (6). The presence of such an antiport in the luminal membrane would also explain why Na^+ absorption in the rabbit is so much more sensitive to changes in luminal

HCO_3^- concentration than could be expected from an epithelium in which nonelectrogenic Na^+ reabsorption is so inconspicuous (6, 93, 179). A luminal Cl^--HCO_3^- antiport seems also to be a prominent feature of the duct system of the sheep parotid gland, where the ducts can secrete HCO_3^- in exchange for Cl^- even when no net transport of Na^+ or K^+ is taking place (22).

BASOLATERAL MEMBRANE. *Na^+-K^+-ATPase.* Ouabain abolishes the active transport potential of the rabbit main excretory duct rapidly and reversibly when applied from the interstitial but not the luminal surface of the epithelium. The effect of the glycoside is antagonized competitively by K^+ in the interstitial bathing solution (1, 72), and removal of K^+ from the interstitial but not the luminal fluid reduces the active transport potential (1, 73). The ductal absorptive mechanism, which is completely inhibited at 0°C, has an apparent activation energy similar to that of transport Na^+-K^+-ATPases in other tissues (1), and rapid cooling experiments show that the mechanism is electrogenic (1, 7). In the rat duct the Na^+ transport mechanism obeys saturation kinetics with a maximum velocity (V_{max}) of ~5 nmol·cm^{-2}·s^{-1} and an apparent Michaelis-Menten constant (K_m) value of 27 mM (39). In the rabbit duct, the V_{max} is 15.4 nmol·cm^{-2}·s^{-1} and the apparent K_m is 4 mM (4). The pump normally operates at near maximum rates so that it is the V_{max} of the pump, not the Na^+ permeability of the luminal membrane, that is the rate-limiting step in ductal Na^+ absorption (4).

K^+ and Cl^- conductances. Although the basolateral membrane is highly selective both for K^+ (72, 176) and Cl^- (2, 72), unfortunately nothing is known about the kinetics of the channels involved or their selectivities.

Na^+-H^+ antiport. Knauf et al. (74) pointed out that it seems to be necessary to postulate the presence of an Na^+-H^+ antiport in the basolateral membrane to account for the observation that both the rat and the rabbit ducts can continue to secrete K^+ and HCO_3^-, even when all Na^+ reabsorption is blocked (e.g., during duct perfusion with Na^+-free solutions or with solutions containing amiloride). Under these circumstances, some pathway for Na^+ entry into the cytosol from the interstitium is required for the Na^+ pump to be able to continue to replenish cytosolic K^+ as it is secreted. Because at the same time HCO_3^- is secreted in equal amounts to K^+, the simplest mechanism that can account for these observations would be a basolateral Na^+-H^+ antiport.

Plasticity of Ductal Transport Properties

Knauf et al. (76) have drawn attention to the fact that, as in the distal nephron, the excretory ducts of the rat mandibular gland have two major cell types, a light and a dark cell, and that the proportion of dark cells increases during metabolic acidosis. From these experiments the authors concluded that the dark cells secrete acid and the light cells secrete HCO_3^-. Because the change in the proportions of the two cell types is a slowly developing phenomenon, these results seem likely to be due to some adaptive change evoked by a sustained alteration in cytosol pH. Because the duct cells have both an acid secretory (Na^+-H^+) and an HCO_3^- secretory (K^+-H^+) antiport in the luminal membrane, it seems probable that the adaptation involves a change in the relative proportions of the two antiport types present. This might involve relocation of Na^+-H^+ antiports from the basolateral membrane, where they would drive HCO_3^- secretion, to the luminal membrane, where they drive acid secretion (HCO_3^- absorption). Such a plasticity of transport function has recently been demonstrated in the distal nephron, an epithelium that responds similarly to the salivary ducts during sustained changes in acid-base status (150).

Water Permeability

Salivary ducts have water permeabilities as low as those seen in the mammalian distal nephron under conditions of maximum water diuresis (179): in the rat mandibular duct the osmotic filtration permeability is 6×10^{-3} cm/s (179) and in the rabbit duct it is 1×10^{-3} cm/s (165a). Nevertheless the ducts do have a finite water permeability, and significant amounts of water are reabsorbed at low secretory rates, when fluid-duct contact time is high and the transepithelial osmotic gradient is steep. This is very conspicuous in mandibular glands, where salivary urea concentrations (a rough index of ductal water reabsorption) can rise to more than three times the primary fluid concentration when salivary secretory rates are very low (176) but quite inconspicuous in parotid glands (88, 91, 92).

Control of Ductal Transport

Although it is often assumed that the salivary ducts transport electrolytes at constant rates, an analysis of salivary electrolyte excretion patterns shows that ductal electrolyte transport maxima do change during autonomic stimulation (179). This inference has received direct confirmation from duct microperfusion studies in the rat, which show that cholinergic agonists in very low concentrations inhibit Na^+ absorption and depolarize the duct potential when applied from the interstitial but not the luminal surface of the epithelium (32, 39, 93, 95, 176). Studies on the perfused rabbit mandibular gland, however, suggest the opposite action for acetylcholine (14, 16). In the rat duct, cholinergic agonists can also alter K^+ and HCO_3^- secretion rates (95, 178). Similar effects are also apparent during parasympathetic nerve stimulation (146).

Adrenergic agonists and sympathetic nerve stimulation also influence ductal electrolyte transport (31, 32, 95, 145, 149), α-agonists acting like cholinergic substances and β-agonists acting to stimulate Na⁺ transport. It appears likely that the point at which autonomic control is exerted is the Na⁺-H⁺ antiport in the luminal membrane and that both Ca^{2+} (24) and cAMP (32) have roles as intracellular mediators. A most interesting observation is that cAMP can mimic the action of acetylcholine and α-adrenergic agonists, but only when applied from the luminal surface of the epithelium (32). Ductal transport may also be under the control of nonadrenergic, noncholinergic neurons, because several biologically active peptides, including physalaemin, which stimulates Na⁺ absorption, and vasoactive intestinal peptide (VIP) and glucose-dependent insulin-releasing peptide (GIP), which inhibit it, can alter ductal transport when administered in physiological concentrations (33).

In contrast to the secretory end-pieces, salivary ducts appear also to be under the control of adrenal steroids, because the rates of ductal Na⁺ and K⁺ transport in the intact gland are directly related to body mineralocorticoid status (22, 89, 179). This contention has received direct confirmation from perfusion studies on the rat mandibular main duct (59, 75, 76) in which it has been shown that adrenalectomy or administration of spironolactone reduces Na⁺ and K⁺ transport. Knauf and associates (59, 75) argue convincingly that the principal effect of aldosterone is to increase the activity of the Na⁺-H⁺ antiport in the luminal membrane.

Overview of Ductal Electrolyte Transport

The fundamental energy source for ductal transport is the basolateral Na⁺-K⁺-ATPase, which maintains a low Na⁺ and a high K⁺ concentration in the cytosol (Fig. 7). The Na⁺ gradient is dissipated principally by influx of Na⁺ across the luminal membrane, but also partly across the basolateral membrane via an Na⁺-H⁺ antiport, thereby leading to the accumulation of HCO_3^- in the cytosol.

At the luminal membrane, Na⁺ enters the cell, either via the conductive channel or via the luminal Na⁺-H⁺ antiport, and, having entered, is pumped to the interstitium by the Na⁺-K⁺-ATPase. That component of absorbed Na⁺ entering the cytosol electrogenically is accompanied by a parallel Cl⁻ flux that passes through the cytosol to the interstitium via luminal and basolateral anion conductances. Additional Cl⁻ may enter the cytosol via a luminal Cl⁻-HCO_3^- antiport. The remainder of the absorbed Na⁺ enters the cytosol via a luminal Na⁺-H⁺ antiport, thereby promoting HCO_3^- uptake across the luminal membrane.

Potassium is concentrated in the cytosol by the basolateral Na⁺-K⁺-ATPase. Some of it is recycled across the basolateral membrane via a K⁺ conductance, but a considerable amount enters the lumen via a luminal membrane K⁺-H⁺ antiport, simultaneously promoting HCO_3^- secretion. When the luminal fluid is Na⁺ rich and the Na⁺-H⁺ and K⁺-H⁺ antiports operate at similar rates, there appears to be a simple one-for-one exchange of Na⁺ for K⁺ with little net movement of HCO_3^-, but when the luminal Na⁺ concentration falls and the Na⁺-H⁺ antiport is slowed down, the HCO_3^- secretory action of the K⁺-H⁺ antiport is unmasked.

The intercellular tight junctions appear not to be appreciably permeable to ions (i.e., the epithelium is tight) and, as in other epithelia, their small cross-sectional area severely restricts water flow across them regardless of their water permeability. Because the luminal fluid is hypotonic, there is a considerable driving force available to promote some water reabsorption, but whether this takes place across the tight junctions or across the cell membranes is unknown.

APPENDIX

Analytical Solution of Transport Model Specified in Figure 3

Any diagram of a cell secretion model, such as that depicted in Figure 3, is really a shorthand method of writing a set of simultaneous equations to describe the transport system. One can solve these equations for the steady state to make predictions about features of the system for which information is lacking. In most cases, it is convenient to obtain solutions by using computer-based search routines, but analytical solutions are also possible in many cases. In this appendix we provide an analytical solution of the model specified in Figure 3.

We assume an epithelium with a basolateral membrane permeable to Na⁺, K⁺, and Cl⁻ but conductive only to K⁺, a luminal membrane permeable and conductive only to Cl⁻, and tight junctions permeable and conductive only to Na⁺. An Na⁺-K⁺-ATPase located in the basolateral membrane is assumed to obey simple Michaelis-Menten kinetics dependent on cytosol Na⁺ concentration and to have a fixed coupling ratio linking the K⁺ and Na⁺ fluxes. Also in the basolateral membrane is an Na⁺-K⁺-2Cl⁻ symport that functions at a fixed rate for any given steady state. The interstitial fluid (b) contains Na⁺, K⁺, and Cl⁻, and the cytosol (c) contains Na⁺, K⁺, Cl⁻, and impermeable solute with a fixed charge (r). Because both the tight junctions (tj) and the luminal membrane (l) are impermeable to K⁺, the luminal fluid (l) can contain only Na⁺ and Cl⁻. We solve the model for the steady state by imposing mass balance and Kirchhoff's law.

Definitions and dimensions of symbols used in the following equations:

$P^l_{H_2O}$	luminal membrane water permeability, cm/s
$P^{bl}_{H_2O}$	basolateral membrane water permeability, cm/s
O_b	osmolality of interstitial fluid, $\mu mol/cm^3$
O_c	osmolality of cytosol, $\mu mol/cm^3$
O_l	osmolality of luminal fluid, $\mu mol/cm^3$
Na_b	interstitial fluid Na⁺ concentration, $\mu mol/cm^3$
Na_c	cytosol Na⁺ concentration, $\mu mol/cm^3$
Na_l	luminal fluid Na⁺ concentration, $\mu mol/cm^3$
K_b	interstitial fluid K⁺ concentration, $\mu mol/cm^3$
K_c	cytosol K⁺ concentration, $\mu mol/cm^3$
Cl_b	interstitial fluid Cl⁻ concentration, $\mu mol/cm^3$

Cl_c — cytosol Cl^- concentration, $\mu mol/cm^3$
Cl_l — luminal fluid Cl^- concentration, $\mu mol/cm^3$
Pr_c — cytosol impermeant solute concentration, $\mu mol/cm^3$
r — charge on each molecule of impermeant intracellular solute
Q — K^+/Na^+ coupling ratio for the Na^+-K^+-ATPase
J_{max}^{Na} — transport maximum of the Na^+-K^+-ATPase, $\mu mol \cdot s^{-1} \cdot cm^{-2}$
K_m^{Na} — Michaelis constant of the Na^+-K^+-ATPase, $\mu mol/cm^3$
J_{pump}^{Na} — transport rate of the Na^+-K^+-ATPase for a given Na_c, $\mu mol \cdot s^{-1} \cdot cm^{-2}$
E_{bl} — potential difference across basolateral membrane with respect to the interstitium, mV
E_l — potential difference across luminal membrane with respect to the cytosol, mV
E_{tj} — potential difference across tight junctions with respect to the interstitium, mV
J_{Cl} — Cl^- flux across the Na^+-K^+-$2Cl^-$ symport, $\mu mol \cdot s^{-1} \cdot cm^{-2}$
g_{Na}^{tj} — Na^+ conductance of tight junctions, mS/cm^2
g_K^{bl} — K^+ conductance of basolateral membrane, mS/cm^2
g_{Cl}^l — Cl^- conductance of luminal membrane, mS/cm^2

R, T, and F have their usual meanings and dimensions.

Given the osmolality of the secreted fluid, O_l, we define the luminal concentrations of Na^+ and Cl^-

$$Na_l = Cl_l = \frac{O_l}{2} \quad (1)$$

Because a Na^+ flux of equal size to J_{Cl} must be flowing across the tight junctions to maintain the electroneutrality of the lumen, we may write

$$J_{Cl} = g_{Na}^{tj}\left[\frac{RT}{F^2}\ln\left(\frac{Na_b}{Na_l}\right) - \frac{E_{tj}}{F}\right] \quad (2)$$

Equation 2 can be used to express E_{tj} in terms of known quantities Na_b, Na_l, and J_{Cl}

$$E_{tj} = \frac{RT}{F}\ln\left(\frac{Na_b}{Na_l}\right) - \frac{J_{Cl}F}{g_{Na}^{tj}} \quad (3)$$

The cytosol composition is determined by the following constraints

$$Na_c + K_c + Cl_c + Pr_c = O_c \quad (4)$$

where

$$O_c = \left(\frac{P_{H_2O}^{bl}}{P_{H_2O}^l + P_{H_2O}^{bl}}\right)O_b + \left(\frac{P_{H_2O}^l}{P_{H_2O}^l + P_{H_2O}^{bl}}\right)O_l \quad (5)$$

and

$$Na_c + K_c - Cl_c - rPr_c = 0 \quad (6)$$

Equation 5 states that the sum of the concentrations of the osmotically active cell constituents, O_c, as defined in Equation 4, is fixed by the requirement that water flux across the basolateral membrane must equal water flux across the luminal membrane. Equation 6 is the electroneutrality condition for the cell compartment.

Because Na^+ must enter the cell at a rate $J_{Cl}/2$ via the Na^+-K^+-$2Cl^-$ symport, it must be pumped out of the cell by the Na^+-K^+-ATPase at the same rate. Given that the Na^+-K^+-ATPase obeys simple Michaelis-Menten kinetics, we can write

$$J_{pump}^{Na} = \frac{J_{Cl}}{2} = \frac{J_{max}^{Na} Na_c}{K_m^{Na} + Na_c} \quad (7a)$$

and the cell Na^+ concentration will be given by

$$Na_c = \frac{J_{Cl} K_m^{Na}}{2J_{max}^{Na} - J_{Cl}} \quad (7b)$$

We may use Equations 4 and 6 to eliminate Pr_c, giving

$$K_c(1+r) - Cl_c(1-r) = rO_c - Na_c(1+r) \quad (8)$$

where O_c and Na_c are known quantities. Solving Equation 8 for K_c, we have

$$K_c = \frac{rO_c - Na_c(1+r) + Cl_c(1-r)}{(1+r)} \quad (9)$$

Cl^- exit across the luminal membrane, which must equal the Cl^- influx across the basolateral membrane, J_{Cl}, can be described by

$$J_{Cl} = g_{Cl}^l\left[\frac{RT}{F^2}\ln\left(\frac{Cl_c}{Cl_l}\right) + \frac{E_l}{F}\right] \quad (10)$$

and the exit of K^+ across the basolateral membrane, which is the sum of the K^+ flux due to the Na^+-K^+-$2Cl^-$ symport, $J_{Cl}/2$, and that due to the Na^+-K^+-ATPase, $QJ_{Cl}/2$, is described by

$$-\left(\frac{J_{Cl}}{2} + \frac{QJ_{Cl}}{2}\right) = -\left(\frac{(1+Q)}{2}\right)J_{Cl}$$
$$= g_K^{bl}\left[\frac{RT}{F^2}\ln\left(\frac{K_b}{K_c}\right) - \frac{E_{bl}}{F}\right] \quad (11)$$

From Kirchhoff's law we know that

$$E_l + E_{bl} = E_{tj} \quad (12)$$

Substituting Equations 9 and 12 into 11 we get

$$-\left(\frac{(1+Q)}{2}\right)J_{Cl} = g_K^{bl}\left[\frac{RT}{F^2}\ln\left(\frac{K_b(1+r)}{rO_c - Na_c(1+r) + Cl_c(1-r)}\right) + \frac{E_l - E_{tj}}{F}\right] \quad (13)$$

Eliminating E_l by substituting Equation 10 into 13 and rearranging, gives

$$Cl_c[rO_c - Na_c(1 + r) + Cl_c(1 - r)]$$
$$= K_b Cl_l(1 + r)\exp\left[\frac{F^2}{RT}\left(\frac{J_{Cl}}{g_{Cl}^l} + \frac{(1 + Q)}{2}\frac{J_{Cl}}{g_K^{bl}} - \frac{E_{tj}}{F}\right)\right] \quad (14)$$

There are two cases: if $r = 1$, then

$$Cl_c = \frac{2K_b Cl_l}{(O_c - 2Na_c)}$$
$$\cdot \exp\left[\frac{F^2}{RT}\left(\frac{J_{Cl}}{g_{Cl}^l} + \frac{(1 + Q)}{2}\frac{J_{Cl}}{g_K^{bl}} - \frac{E_{tj}}{F}\right)\right] \quad (15a)$$

and, if $r \neq 1$, then

$$Cl_c = \frac{Na_c(1+r) - rO_c}{2(1-r)} + \frac{(Na_c(1+r) - rO_c)^2}{2(1-r)} + \frac{\left\{4(1-r)K_b Cl_l(1+r)\exp\left[\frac{F^2}{RT}\left(\frac{J_{Cl}}{g_{Cl}^l} + \frac{(1+Q)}{2}\frac{J_{Cl}}{g_K^{bl}} - \frac{E_{tj}}{F}\right)\right]\right\}^{1/2}}{2(1-r)} \quad (15b)$$

Having found Cl_c, we can now find K_c

$$K_c = \frac{rO_c - Na_c(1 + r) + Cl_c(1 - r)}{(1 + r)} \quad (16)$$

and E_l

$$E_l = \frac{J_{Cl}F}{g_{Cl}^l} - \frac{RT}{F}\ln\left(\frac{Cl_c}{Cl_l}\right) \quad (17)$$

and E_{bl}

$$E_{bl} = E_{tj} - E_l \quad (18)$$

and Pr_c

$$Pr_c = O_c - Na_c - K_c - Cl_c \quad (19)$$

The order in which these equations should be solved depends on which starting values are known and which are to be found. As an example, we start here with realistic values for the ion concentrations in the interstitial and luminal fluids, the Na$^+$ concentration in the cytosol, the net rate of Cl$^-$ secretion, the conductances of the basolateral, luminal, and tight junctional membranes, and the average charge per molecule of fixed intracellular solute.

The values used are

	Unstimulated	Stimulated	Dimension
Na_b	150	150	$\mu mol/cm^3$
Na_c	25	57	$\mu mol/cm^3$
Na_l	160	160	$\mu mol/cm^3$
Cl_b	155	155	$\mu mol/cm^3$
Cl_l	160	160	$\mu mol/cm^3$
K_b	5	5	$\mu mol/cm^3$
g_K^{bl}	0.25	65	mS/cm^2
g_{Cl}^l	10	75	mS/cm^2
g_{Na}^{tj}	100	100	mS/cm^2
J_{Cl}	0.000125	0.0125	$\mu mol \cdot cm^{-2} \cdot s^{-1}$
r	0.7	0.7	

Solving Equations 3, 15b, and 16–19, we obtain

E_{tj}	−1.8	−13.8	mV
Cl_c	36	53	$\mu mol/cm^3$
K_c	111	82	$\mu mol/cm^3$
E_l	40.8	45.5	mV
E_{bl}	−42.6	−59.3	mV
Pr_c	143	123	$\mu mol/cm^3$

In this example we see that the predicted transepithelial voltage E_{tj} rises from −1.8 mV at rest to −13.8 mV during stimulation and that the basolateral membrane voltage E_{bl} rises from −42.6 to −59.3 mV. At the same time, the cytosol Cl$^-$ concentration Cl_c rises from 36 to 53 $\mu mol/cm^3$ and the K$^+$ concentration K_c^+ falls from 111 to 82 $\mu mol/cm^3$. These changes are similar to those actually observed in the dog mandibular gland (64, 108, 109, 140), indicating that we have chosen starting values appropriate for that gland. The drop of 14% in the concentration of impermeant intracellular solute Pr_c indicates that a modest degree of cell swelling must have been induced by stimulation.

This analysis may also be used to aid in the interpretation of the significance of changes in intracellular Cl$^-$ activity recorded with liquid-resin, ion-selective electrodes. To do so, let us examine the relation of the intracellular Cl$^-$ concentration Cl_c to the rate of active uptake of Cl$^-$ J_{Cl} and to the Cl$^-$ conductance of the luminal membrane g_{Cl}^l.

From Equations 2, 10, and 12, we can derive the equation

$$\frac{RT}{F}\ln\left(\frac{Cl_c Na_b}{Cl_l Na_l}\right) - E_{bl} = J_{Cl}F\left(\frac{1}{g_{Na}^{tj}} + \frac{1}{g_{Cl}^l}\right) \quad (20)$$

From this equation it is evident that the intracellular Cl$^-$ concentration is not determined solely by the rate of active uptake of Cl$^-$ and the luminal membrane Cl$^-$ conductance, but also by the Na$^+$ conductance of the tight junctions and the Na$^+$ and Cl$^-$ concentrations of the luminal fluid. It should also be noted that substitution of appropriate values in Equation 20 yields not J_{Cl}/g_{Cl}^l but $[J_{Cl}(1/g_{Cl}^l + 1/g_{Na}^{tj})]$, so that, if g_{Na}^{tj} is very small compared with g_{Cl}^l, even quite large changes in g_{Cl}^l will not result in any appreciable change in Cl_c and may therefore be overlooked.

REFERENCES

1. AUGUSTUS, J. Evidence for electrogenic sodium pumping in the ductal epithelium of rabbit salivary gland and its relationship with Na-K-ATPase. *Biochim. Biophys. Acta* 419: 63–75, 1976.
2. AUGUSTUS, J., J. BIJMAN, AND C. H. VAN OS. Electrical resistance of rabbit submaxillary main duct: a tight epithelium with leaky cell membranes. *J. Membr. Biol.* 43: 203–226, 1978.
3. AUGUSTUS, J., J. BIJMAN, C. H. VAN OS, AND J. F. G. SLEGERS. High conductance in an epithelial membrane not due to extracellular shunting. *Nature Lond.* 268: 657–658, 1977.
4. BIJMAN, J. Transport Parameters of the Main Duct of the Rabbit Mandibular Salivary Gland. Nijmegen, The Netherlands: Univ. of Nijmegen, 1982. PhD thesis.
5. BIJMAN, J., D. I. COOK, AND C. H. VAN OS. Effect of amiloride

on electrolyte transport parameters of the main duct of the rabbit mandibular salivary gland. *Pfluegers Arch.* 398: 96–102, 1983.
6. BIJMAN, J., J. F. G. SLEGERS, AND C. H. VAN OS. Mechanism for HCO_3^- stimulation of NaCl reabsorption in rabbit submaxillary main duct epithelium. In: *Hydrogen Ion Transport in Epithelia*, edited by I. Schulz, G. Sachs, J. G. Forte, and K. J. Ullrich. Amsterdam: Elsevier/North-Holland, 1980, p. 259–264.
7. BIJMAN, J., AND C. H. VAN OS. Temperature dependence of transepithelial potential in isolated rabbit mandibular main duct. In: *Electrolyte and Water Transport Across Gastrointestinal Epithelia*, edited by R. M. Case, A. Garner, L. A. Turnberg, and J. A. Young. New York: Raven, 1982, p. 173–176.
8. BUNDGAARD, M., M. MØLLER, AND J. H. POULSEN. Localization of sodium pump sites in cat salivary glands. *J. Physiol. Lond.* 273: 339–353, 1977.
9. BURG, M., L. STONER, J. CARDINAL, AND N. GREEN. Furosemide effect on isolated perfused tubules. *Am. J. Physiol.* 225: 119–124, 1973.
10. BURGEN, A. S. V. The secretion of potassium in saliva. *J. Physiol. Lond.* 132: 20–39, 1956.
11. BURGEN, A. S. V. Secretory processes in salivary glands. In: *Handbook of Physiology. Alimentary Canal*, edited by C. F. Code. Washington, DC: Am. Physiol. Soc., 1967, sect. 6, vol. II, p. 561–579.
12. CABANTCHIK, Z. I., P. KNAUF, AND A. ROTHSTEIN. The anion transport system of the red blood cells: the role of membrane protein evaluated by the use of "probes." *Biochim. Biophys. Acta* 515: 239–302, 1978.
13. CANDIA, O. A., H. F. SCHOEN, L. LOW, AND S. M. PODOS. Chloride transport inhibition by piretanide and MK-196 in bullfrog corneal epithelium. *Am. J. Physiol.* 240 (*Renal Fluid Electrolyte Physiol.* 9): F25–F29, 1981.
14. CASE, R. M., T. A. ANSAH, S. DHO, A. HOWORTH, AND B. E. ARGENT. Second messenger interactions in pancreas and salivary gland. *Biomed. Res.* 7, Suppl. 2: 95–103, 1986.
15. CASE, R. M., A. D. CONIGRAVE, E. J. FAVALORO, I. NOVAK, C. H. THOMPSON, AND J. A. YOUNG. The role of buffer anions and protons in secretion by the rabbit mandibular salivary gland. *J. Physiol. Lond.* 322: 273–286, 1982.
16. CASE, R. M., A. D. CONIGRAVE, I. NOVAK, AND J. A. YOUNG. Electrolytes and protein secretion by the perfused rabbit mandibular gland stimulated with acetylcholine or catecholamines. *J. Physiol. Lond.* 300: 467–487, 1980.
17. CASE, R. M., AND A. HOWORTH. The influence of forskolin on electrolyte transport processes in the perfused rabbit mandibular salivary gland (Abstract). *J. Physiol. Lond.* 361: 25P, 1985.
18. CASE, R. M., M. HUNTER, I. NOVAK, AND J. A. YOUNG. The anionic basis of fluid secretion by the rabbit mandibular gland. *J. Physiol. Lond.* 349: 619–630, 1984.
19. CASE, R. M., AND T. SCRATCHERD. The secretion of alkali metal ions by the perfused cat pancreas as influenced by the composition and osmolality of the external environment and by inhibitors of metabolism and Na^+,K^+-ATPase activity. *J. Physiol. Lond.* 242: 415–428, 1974.
20. CHIBUZO, G. A., AND J. F. CUMMINGS. Motor and sensory centers for the innervation of mandibular and sublingual salivary glands: a horseradish peroxidase study in the dog. *Brain Res.* 189: 301–313, 1980.
21. COMPTON, J., J. R. MARTINEZ, A. M. MARTINEZ, AND J. A. YOUNG. Fluid and electrolyte secretion from the isolated, perfused submandibular and sublingual glands of the rat. *Arch. Oral Biol.* 26: 555–561, 1981.
22. COMPTON, J. S., J. NELSON, R. D. WRIGHT, AND J. A. YOUNG. A micropuncture investigation of electrolyte transport in the parotid glands of sodium-replete and sodium-depleted sheep. *J. Physiol. Lond.* 309: 429–446, 1980.
23. CONTEAS, C. N., A. A. MCDONOUGH, T. R. KOZLOWSKI, C. B. HENSLEY, R. L. WOOD, AND A. K. MIRCHEFF. Mapping subcellular distribution of Na^+-K^+-ATPase in rat parotid gland. *Am. J. Physiol.* 250 (*Cell Physiol.* 19): C430–C441, 1986.
24. COOK, D. I. A Microperfusion Investigation of the Effects of Calcium and Calcium Mediated Drugs upon Electrolyte Transport by the Isolated Main Excretory Duct of the Rabbit Submaxillary Gland. Sydney, Australia: Univ. of Sydney, 1979. BSc (Med) thesis.
25. COOK, D. I., J. BIJMAN, AND C. H. VAN OS. The action of amiloride on the rabbit mandibular duct perfused in vitro. In: *Electrolyte and Water Transport Across Gastrointestinal Epithelia*, edited by R. M. Case, A. Garner, L. A. Turnberg, and J. A. Young. New York: Raven, 1982, p. 167–171.
26. COOK, D. I., S. P. TOWNER, AND J. A. YOUNG. Patch-clamp studies of cultured salivary cells. *Biomed. Res.* 7, Suppl. 2: 203–207, 1986.
27. COPE, I. C., P. W. KUCHEL, AND J. A. YOUNG. Water permeability of rat mandibular glands measured using ^1H NMR spectroscopy (Abstract). *Proc. Aust. Physiol. Pharmacol. Soc.* 15: 197P, 1984.
28. DARTT, D. A., M. MØLLER, AND J. H. POULSEN. Lacrimal gland electrolyte and water secretion in the rabbit: localization and role of $(Na^+ + K^+)$-activated ATPase. *J. Physiol. Lond.* 321: 557–569, 1981.
29. DE CAMILLI, P., D. PELUCHETTI, AND J. MELDOLESI. Dynamic changes of the luminal plasmalemma in stimulated parotid acinar cells. *J. Cell Biol.* 70: 59–74, 1976.
30. DEGNAN, K. J., K. J. KARNAKY, JR., AND J. A. ZADUNAISKY. Active chloride transport in the in vitro opercular skin of a teleost (*Fundulus heteroclitus*), a gill-like epithelium rich in chloride cells. *J. Physiol. Lond.* 251: 155–191, 1977.
31. DENNISS, A. R., L. H. SCHNEYER, C. SUCANTHAPREE, AND J. A. YOUNG. Action of adrenergic agonists on isolated excretory ducts of submandibular glands. *Am. J. Physiol.* 235 (*Renal Fluid Electrolyte Physiol.* 4): F548–F556, 1978.
32. DENNISS, A. R., AND J. A. YOUNG. The action of neurotransmitter hormones and analogues and cyclic nucleotides and theophylline on electrolyte transport by the excretory duct of the rabbit mandibular gland. *Pfluegers Arch.* 357: 77–89, 1975.
33. DENNISS, A. R., AND J. A. YOUNG. Modification of salivary duct electrolyte transport in rat and rabbit by physalaemin, VIP, GIP, and other enterohormones. *Pfluegers Arch.* 376: 73–80, 1978.
34. DICH-NIELSEN, J. O., L. P. LAUGESEN, AND J. H. POULSEN. Submandibular salivary secretion in the cat and associated potassium movements: dependence on temperature and perfusate flow rate. *Pfluegers Arch.* 403: 440–445, 1985.
35. ELLORY, J. C., P. B. DUNHAM, P. J. LOGUE, AND G. W. STEWART. Anion-dependent cation transport in erythrocytes. *Philos. Trans. R. Soc. Lond. B Biol. Sci.* 299: 483–495, 1982.
36. EVANS, L. A. R., D. P. PIRANI, D. I. COOK, AND J. A. YOUNG. Intraepithelial current flow in rat pancreatic secretory epithelia. *Pfluegers Arch.* 407, Suppl. 2: S107–S111, 1986.
37. EVANS, M. G., A. MARTY, Y. P. TAN, AND A. TRAUTMANN. Blockage of Ca-activated Cl conductance by furosemide in rat lacrimal glands. *Pfluegers Arch.* 406: 65–68, 1986.
38. EVELOFF, J., R. KINNE, E. KINNE-SAFFRAN, H. MURER, P. SILVA, F. H. EPSTEIN, J. STOFF, AND W. B. KINTER. Coupled sodium and chloride transport into plasma membrane vesicles prepared from dogfish rectal gland. *Pfluegers Arch.* 378: 87–92, 1978.
39. FIELD, M. J., AND J. A. YOUNG. Kinetics of Na transport in the rat submaxillary main duct perfused in vitro. *Pfluegers Arch.* 345: 207–220, 1973.
40. FINDLAY, I., AND O. H. PETERSEN. Acetylcholine stimulates a Ca^{2+}-dependent Cl^- conductance in mouse lacrimal acinar cells. *Pfluegers Arch.* 403: 328–330, 1985.
41. FINN, A. L., AND J. BRIGHT. The paracellular pathway in toad urinary bladder: permselectivity and kinetics of opening. *J. Membr. Biol.* 44: 67–83, 1978.
42. FORREST, J. N., J. L. BOYER, T. A. ARDITO, AND H. V. MURDAUGH. Structure of tight junctions during Cl secretion in the perfused rectal gland of the dogfish shark. *Am. J. Physiol.* 242 (*Cell Physiol.* 11): C388–C392, 1980.
43. FORREST, J. N., F. WANG, AND W. BEYENBACH. Perfusion of isolated tubules of the shark rectal gland. Electrical characteristics and response to hormones. *J. Clin. Invest.* 72: 1163–

1167, 1983.
44. FRIZZELL, R. A., M. FIELD, AND S. G. SCHULTZ. Sodium-coupled chloride transport by epithelial tissues. *Am. J. Physiol.* 236 (*Renal Fluid Electrolyte Physiol.* 5): F1–F8, 1979.
45. FRÖMTER, E., AND J. M. DIAMOND. Route of passive ion permeation in epithelia. *Nature Lond.* 235: 9–13, 1972.
46. FRÖMTER, E., B. GEBLER, K. SCHOPOW, AND H. POCKRANDT-HEMSTEDT. Cation and anion permeability of rabbit submaxillary main duct. In: *Secretory Mechanisms of Exocrine Glands*, edited by N. A. Thorn and O. H. Petersen. Copenhagen: Munksgaard, 1974, p. 496–513.
47. GALLACHER, D. V., Y. MARUYAMA, AND O. H. PETERSEN. Patch-clamp study of rubidium and potassium conductances in single cation channels from mammalian exocrine acini. *Pfluegers Arch.* 401: 361–367, 1984.
48. GALLACHER, D. V., AND A. P. MORRIS. A patch-clamp study of potassium currents in resting and acetylcholine stimulated mouse submandibular acinar cells. *J. Physiol. Lond.* 373: 379–395, 1986.
49. GALLACHER, D. V., AND O. H. PETERSEN. Electrophysiology of mouse parotid acini: effects of electrical field stimulation and iontophoresis of neurotransmitters. *J. Physiol. Lond.* 305: 43–57, 1980.
50. GALLACHER, D. V., AND O. H. PETERSEN. Stimulus-secretion coupling in mammalian salivary glands. In: *Gastrointestinal Physiology IV*, edited by J. A. Young, Baltimore, MD: University Park, 1983, vol. 28, p. 1–52. (Int. Rev. Physiol. Ser.)
51. GARD, G., J. A. CHRISTIE, D. I. COOK, AND J. A. YOUNG. The action of loop diuretics on the submandibular anion self-exchange antiport (Abstract). *Proc. Aust. Physiol. Pharmacol. Soc.* 16: 25P, 1985.
52. GARRETT, J. R., AND P. A. PARSONS. Movement of horseradish peroxidase in rabbit submandibular glands after ductal injection. *Histochem. J.* 8: 177–189, 1976.
53. GECK, P., C. PIETRZYK, B. C. BURCKHARDT, B. PFEIFFER, AND E. HEINZ. Electrically silent cotransport of Na, K, Cl in Ehrlich cells. *Biochim. Biophys. Acta* 600: 432–447, 1980.
53a. GÖGELEIN, H., E. SCHLATTER, AND R. GREGER. The "small" conductance chloride channel in the luminal membrane of the rectal gland of the dogfish (*Squalus acanthias*). *Pfluegers Arch.* 409: 122–125, 1987.
54. GREGER, R., AND E. SCHLATTER. Cellular mechanisms of the action of loop diuretics on the thick ascending limb of Henle's loop. *Klin. Wochenschr.* 61: 1019–1027, 1983.
55. GREGER, R., AND E. SCHLATTER. Mechanism of NaCl secretion in rectal gland tubules of spiny dogfish (*Squalus acanthias*). I. Experiments in isolated in vitro perfused rectal gland tubules. *Pfluegers Arch.* 402: 63–75, 1984.
56. GREGER, R., AND E. SCHLATTER. Mechanism of NaCl secretion in rectal gland tubules of spiny dogfish (*Squalus acanthias*). II. Effect of inhibitors. *Pfluegers Arch.* 402: 364–375, 1984.
57. GREGER, R., E. SCHLATTER, AND H. GÖGELEIN. Cl^--channels in the apical cell membrane of the rectal gland "induced" by cAMP. *Pfluegers Arch.* 406: 446–448, 1985.
57a. GREGER, R., E. SCHLATTER, AND H. GÖGELEIN. Chloride channels in the apical membrane of the rectal gland of the dogfish (*Squalus acanthias*). Properties of the "larger" conductance channel *Pfluegers Arch.* 409: 114–121, 1987.
58. GREGER, R., E. SCHLATTER, F. WANG, AND J. N. FORREST. Mechanism of NaCl secretion in rectal gland tubules of spiny dogfish (*Squalus acanthias*). III. Effect of stimulation of secretion by cyclic AMP. *Pfluegers Arch.* 402: 376–384, 1984.
59. GRUBER, W. D., H. KNAUF, AND E. FRÖMTER. The action of aldosterone on Na^+ and K^+ transport in the rat submaxillary main duct. *Pfluegers Arch.* 344: 33–49, 1973.
60. HANNAFIN, J., E. KINNE-SAFFRAN, D. FRIEDMAN, AND R. KINNE. Presence of a sodium-potassium chloride cotransport system in the rectal gland of *Squalus acanthias*. *J. Membr. Biol.* 785: 73–83, 1983.
61. HOLZGREVE, H., J. R. MARTINEZ, AND A. VOGEL. Micropuncture and histologic study of submaxillary glands of young rats. *Pfluegers Arch. Gesamte Physiol. Menchen. Tiere* 290: 134–143, 1966.
62. HUNTER, M., AND R. M. CASE. The role of the sodium gradient in the formation of saliva by the rabbit mandibular gland. *J. Physiol. Lond.* In press.
63. HUNTER, M., P. A. SMITH, AND R. M. CASE. The dependence of fluid secretion by mandibular salivary gland and pancreas on extracellular calcium. *Cell Calcium* 4: 307–317, 1983.
64. IMAI, Y. Study of the secretion mechanism of the submaxillary gland of the dog. Part 2. Effect of exchanging ions in the perfusate on salivary secretion and secretory potential, with special reference to the ionic distribution in the gland. *J. Physiol. Soc. Jpn.* 27: 313–324, 1965.
64a. IWATSUKI, N., Y. MARUYAMA, O. MATSUMOTO, AND A. NISHIYAMA. Activation of Ca^{2+}-dependent Cl^- and K^+ conductances in rat and mouse parotid acinar cells. *Jpn. J. Physiol.* 35: 933–944, 1985.
65. IWATSUKI, N., AND O. H. PETERSEN. Action of tetraethylammonium on calcium-activated potassium channels in pig pancreatic acinar cells studied by patch-clamp single-channel and whole-cell current recording. *J. Membr. Biol.* 86: 139–144, 1985.
66. JACK, J. J. B., D. NOBLE, AND R. W. TSIEN. *Electric Current Flow in Excitable Cells*. Oxford, UK: Oxford Univ. Press, 1983, p. 83–97.
67. KALADELFOS, G., AND J. A. YOUNG. Micropuncture and cannulation study of water and electrolyte excretion in the isotonic-secreting cat sublingual salivary gland. *Pfluegers Arch.* 341: 143–154, 1973.
68. KALADELFOS, G., AND J. A. YOUNG. Water and electrolyte excretion in the cat submaxillary gland studied using micropuncture and duct cannulation techniques. *Aust. J. Exp. Biol. Med. Sci.* 52: 67–79, 1974.
69. KINNE, R., B. KOENIG, J. HANNAFIN, E. KINNE-SAFFRAN, D. M. SCOTT, AND K. ZIEROLD. The use of membrane vesicles to study the NaCl/KCl cotransporter involved in active transepithelial chloride transport. *Pfluegers Arch.* 405, Suppl. S101–S105, 1985.
70. KLYCE, S. D., AND R. K. WONG. Site and mode of adrenaline action on chloride transport across the rabbit corneal epithelium. *J. Physiol. Lond.* 266: 777–799, 1977.
71. KNAUF, H. The isolated salivary duct as a model for electrolyte transport studies. *Pfluegers Arch.* 333: 82–94, 1972.
72. KNAUF, H., AND E. FRÖMTER. Studies on the origin of the transepithelial electrical potential difference in salivary duct epithelium. In: *Electrophysiology of Epithelial Cells*, edited by G. Giebisch. New York: Schattauer, 1971, p. 187–199.
73. KNAUF, H., AND R. LÜBCKE. Evidence for Na^+ independent active secretion of K^+ and HCO_3^- by rat salivary duct epithelium. *Pfluegers Arch.* 361: 55–60, 1975.
74. KNAUF, H., R. LÜBCKE, W. KREUTZ, AND G. SACHS. Interrelationships of ion transport in rat submaxillary duct epithelium. *Am. J. Physiol.* 242 (*Renal Fluid Electrolyte Physiol.* 11): F132–F139, 1982.
75. KNAUF, H., K. PANDER, AND S. FRANCK. The effect of spironolactone on transport of Na^+, K^+, and H^+. A microperfusion study in rat main submaxillary duct. *Eur. J. Clin. Invest.* 6: 17–20, 1976.
76. KNAUF, H., P. RÖTTGER, U. WAIS, AND K. BAUMANN. On the regulatory handling of Na^+, K^+, and H^+ transport by the rat salivary duct epithelium. *Fortschr. Zool.* 23: 307–321, 1975.
77. KNAUF, H., U. WAIS, R. LÜBCKE, AND G. ALBIEZ. On the mechanism of action of triamterene. Effects on transport of Na^+, K^+, and H^+/HCO_3^- ions. *Eur. J. Clin. Invest.* 6: 43–50, 1976.
78. KNICKELBEIN, R., P. S. ARONSON, C. M. SCHRON, J. SEIFTER, AND J. W. DOBBINS. Sodium and chloride transport across rabbit ileal brush border. II. Evidence for $Cl-HCO_3$ exchange and mechanism of coupling. *Am. J. Physiol.* 249 (*Gastrointest. Liver Physiol.* 18): G236–G245, 1985.
79. KOMABAYASHI, T., K. NAKANO, T. IZAWA, T. NAKAMURA, AND M. TSUBOI. Effects of Ca^{2+} and calmodulin antagonists

on the oxygen uptake induced by acetylcholine or substance-P in rat submandibular gland slices. *Jpn. J. Pharmacol.* 36: 441–447, 1984.
80. LARSON, M., AND K. R. SPRING. Bumetanide inhibition of NaCl transport by *Necturus* gallbladder. *J. Membr. Biol.* 74: 123–129, 1983.
81. LAU, K. R., AND R. M. CASE. Chloride activity in mandibular salivary gland cells during secretion. *Biomed. Res.* 7, Suppl. 2: 181–184, 1986.
82. LAUGESEN, L. P., J. O. D. NIELSEN, AND J. H. POULSEN. Partial dissociation between salivary secretion and active potassium transport in the perfused cat submandibular gland. *Pfluegers Arch.* 365: 167–173, 1976.
83. LINGARD, J. M., AND J. A. YOUNG. β-Adrenergic control of exocrine secretion by perfused rat pancreas in vitro. *Am. J. Physiol.* 245 (*Gastrointest. Liver Physiol.* 8): G690–G696, 1983.
84. LUNDBERG, A. Electrophysiology of salivary glands. *Physiol. Rev.* 38: 21–40, 1958.
85. MANGOS, J. A., R. L. BOYD, G. M. LOUGHLIN, A. COCKRELL, AND R. FUCCI. Secretion of monovalent ions and water in ferret salivary glands: a micropuncture study. *J. Dent. Res.* 60: 733–737, 1981.
86. MANGOS, J. A., AND G. BRAUN. Excretion of total solute, sodium and potassium in the saliva of the rat parotid gland. *Pfluegers Arch. Gesamte Physiol. Menschen Tiere* 290: 184–192, 1966.
87. MANGOS, J. A., G. BRAUN, AND K. F. HAMANN. Micropuncture study of sodium and potassium excretion in the rat parotid saliva. *Pfluegers Arch. Gesamte Physiol. Menschen Tiere* 291: 99–106, 1966.
88. MANGOS, J. A., N. MARAGOS, AND N. R. MCSHERRY. Micropuncture and microperfusion study of glucose excretion in rat parotid saliva. *Am. J. Physiol.* 224: 1260–1264, 1973.
89. MANGOS, J. A., AND N. R. MCSHERRY. Micropuncture study of sodium and potassium excretion in rat parotid saliva: role of aldosterone. *Proc. Soc. Exp. Biol. Med.* 132: 797–801, 1969.
90. MANGOS, J. A., N. R. MCSHERRY, AND S. N. ARVANITAKIS. Autonomic regulation of secretion and transductal fluxes of ions in the rat parotid. *Am. J. Physiol.* 225: 683–688, 1973.
91. MANGOS, J. A., N. R. MCSHERRY, K. IRWIN, AND R. HONG. Handling of water and electrolytes by rabbit parotid and submaxillary glands. *Am. J. Physiol.* 225: 450–455, 1973.
92. MANGOS, J. A., N. R. MCSHERRY, S. NOUSIA-ARVANITAKIS, AND K. IRWIN. Secretion and transductal fluxes of ions in exocrine glands of the mouse. *Am. J. Physiol.* 225: 18–24, 1973.
93. MARTIN, C. J., E. FRÖMTER, E. GEBLER, B. KNAUF, AND J. A. YOUNG. The effects of carbachol on water and electrolyte fluxes and transepithelial electrical potential differences of the rabbit submaxillary main duct perfused in vitro. *Pfluegers Arch.* 341: 131–142, 1973.
94. MARTIN, C. J., AND J. A. YOUNG. Electrolyte concentrations in primary and final saliva of the rat sublingual gland studied by micropuncture and catheterization techniques. *Pfluegers Arch.* 324: 344–360, 1971.
95. MARTIN, C. J., AND J. A. YOUNG. A microperfusion investigation of the effects of a sympathomimetic and a parasympathomimetic drug on water and electrolyte fluxes in the main duct of the rat submaxillary gland. *Pfluegers Arch.* 327: 303–323, 1971.
96. MARTIN, K. Metabolism of salivary glands. In: *Handbook of Physiology. Alimentary Canal*, edited by C. G. Code. Washington, DC: Am. Physiol. Soc., 1967, sect. 6, vol. II, p. 581–593.
97. MARTINEZ, J. R., AND N. CASSITY. ^{36}Cl fluxes in dispersed rat submandibular acini: effects of acetylcholine and transport inhibitors. *Pfluegers Arch.* 403: 50–54, 1985.
98. MARTINEZ, J. R., AND N. CASSITY. Cl$^-$ requirement for saliva secretion in the isolated, perfused rat submandibular gland. *Am. J. Physiol.* 249 (*Gastrointest. Liver Physiol.* 18): G464–G469, 1985.
99. MARTINEZ, J. R., H. HOLZGREVE, AND A. FRICK. Micropuncture study of submaxillary glands of adult rats. *Pfluegers Arch. Gesamte Physiol. Menschen Tiere* 290: 124–133, 1966.

100. MARTINEZ, J. R., AND A. M. MARTINEZ. Submandibular secretion in the dog after intraluminal injections of amiloride. *Arch. Oral Biol.* 19: 57–60, 1974.
101. MARTINEZ, J. R., A. M. MARTINEZ, AND C. COOPER. cGMP stimulates active K$^+$ uptake in rat submandibular slices. *Experientia* 39: 362–364, 1983.
102. MARTY, A., Y. P. TAN, AND A. TRAUTMANN. Three types of calcium-dependent channels in rat lacrimal glands. *J. Physiol. Lond.* 357: 293–325, 1984.
103. MARUYAMA, Y., D. V. GALLACHER, AND O. H. PETERSEN. Voltage and Ca$^+$-activated K$^+$ channel in basolateral acinar cell membranes in mammalian salivary glands. *Nature Lond.* 302: 827–829, 1983.
104. MARUYAMA, Y., A. NISHIYAMA, T. IZUMI, N. HOSHIMIYA, AND O. H. PETERSEN. Ensemble noise and current relaxation analysis of K$^+$ current in single isolated salivary acinar cells from rat. *Pfluegers Arch.* 406: 69–72, 1986.
105. MARUYAMA, Y., AND O. H. PETERSEN. Single calcium-dependent cation channels in mouse pancreatic acinar cells. *J. Membr. Biol.* 81: 83–87, 1984.
106. MATSUURA, S., S. ETO, K. KATO, AND Y. TASHIRO. Ferritin immunoelectron microscopic localization of 5'-nucleotidase on rat liver cell surface. *J. Cell Biol.* 99: 166–173, 1984.
107. MAZARIEGOS, M. R., AND A. R. HAND. Regulation of tight junctional permeability in the rat parotid gland by autonomic agonists. *J. Dent. Res.* 63: 1102–1107, 1984.
108. MORI, H., M. MURAKAMI, T. NAKAHARI, AND Y. IMAI. Intracellular Cl$^-$ activity of canine submandibular gland cells: an in vitro observation. *Jpn. J. Physiol.* 33: 869–873, 1983.
109. MORI, H., T. NAKAHARI, AND Y. IMAI. Intracellular K$^-$ activity in canine submandibular gland cells in resting [sic] and its change during stimulation. *Jpn. J. Physiol.* 34: 1077–1088, 1984.
110. MURAKAMI, M. Heat production, blood flow, O$_2$ uptake and CO$_2$ output in the secretory process of the dog submandibular gland. *J. Physiol. Soc. Jpn.* 43: 135–147, 1981.
111. MURAKAMI, M., Y. IMAI, Y. SEO, T. MORIMOTO, K. SHIGA, AND H. WATARI. Phosphorus nuclear magnetic resonance of perfused salivary gland. *Biochim. Biophys. Acta* 762: 19–24, 1983.
112. MURAKAMI, M., I. NOVAK, AND J. A. YOUNG. Choline evokes fluid secretion by perfused rat mandibular gland without densitization. *Am. J. Physiol.* 251 (*Gastrointest. Liver Physiol.* 14): G84–G89, 1986.
113. NAUNTOFTE, B., AND J. H. POULSEN. Chloride transport in rat parotid acini: furosemide-sensitive uptake and calcium-dependent release. *J. Physiol. Lond.* 357: 61P, 1984.
114. NISHIYAMA, A., K. KATOH, AND S. SAITOH. Effect of neural stimulation on acinar cell membrane potentials in isolated pancreas and salivary gland segments. *Membr. Biochem.* 3: 49–66, 1980.
115. NOVAK, I., C. DAVE, AND J. A. YOUNG. The anionic basis of secretion by rat and rabbit mandibular glands. In: *Secretion: Mechanisms and Control*, edited by R. M. Case, J. M. Lingard, and J. A. Young. Manchester, UK: Manchester Univ. Press, 1984, p. 77–80.
116. NOVAK, I., AND J. A. YOUNG. The effect of transport inhibitors and anion replacement on salivary secretion. In: *Secretion: Mechanisms and Control*, edited by R. M. Case, J. M. Lingard, and J. A. Young. Manchester, UK: Manchester Univ. Press, 1984, p. 81–87.
117. NOVAK, I., AND J. A. YOUNG. Two independent anion transport systems in rabbit mandibular salivary glands. *Pfluegers Arch.* 407: 649–656, 1986.
118. O'DOHERTY, J., R. J. STARK, S. J. CRANE, AND K. L. BRUGGE. Changes in cytosolic calcium during cholinergic and adrenergic stimulation of the parotid salivary glands. *Pfluegers Arch.* 398: 241–246, 1983.
119. O'DOHERTY, J., S. J. YOUMANS, W. MCD. ARMSTRONG, AND R. J. STARK. Calcium regulation during stimulus-secretion coupling: continuous measurement of intracellular calcium activities. *Science Wash. DC* 209: 510–513, 1980.

120. PALFREY, H. C., P. W. FEIT, AND P. GREENGARD. cAMP-stimulated cation cotransport in avian erythrocytes: inhibition by "loop" diuretics. *Am. J. Physiol.* 238 (*Cell Physiol.* 7): C139–C148, 1980.
121. PETERSEN, O. H. Some factors influencing stimulation induced release of potassium from the cat submandibular gland to fluid perfused through the gland. *J. Physiol. Lond.* 208: 431–447, 1970.
122. PETERSEN, O. H. Electrophysiology of mammalian gland cells. *Physiol. Rev.* 56: 535–577, 1976.
123. PETERSEN, O. H. *The Electrophysiology of Gland Cells*. London: Academic, 1980.
124. PETERSEN, O. H., AND Y. MARUYAMA. Calcium-activated potassium channels and their role in secretion. *Nature Lond.* 307: 693–696, 1984.
125. PETERSEN, O. H., AND J. H. POULSEN. The effects of varying the extracellular potassium concentration on the secretory rate and on resting and secretory potentials in perfused cat submandibular gland. *Acta Physiol. Scand.* 70: 293–298, 1967.
126. PETERSEN, O. H., AND J. H. POULSEN. Inhibition of salivary secretion and secretory potentials by g-strophanthin, dinitrophenol, and cyanide. *Acta Physiol. Scand.* 71: 194–202, 1967.
127. PETERSEN, O. H., AND J. H. POULSEN. The secretion of sodium and potassium in cat submandibular saliva after start of stimulation. *Acta Physiol. Scand.* 73: 93–100, 1968.
128. PIRANI, D. C., L. A. R. EVANS, D. I. COOK, AND J. A. YOUNG. Intracellular pH in the rat mandibular gland: the role of Na-H and Cl-HCO_3 antiports in secretion. *Pfluegers Arch.* 408: 178–184, 1987.
129. POULSEN, J. H., AND S. W. BLEDSOE. Salivary gland K^+ transport: in vivo studies with K^+-specific microelectrodes. *Am. J. Physiol.* 240 (*Endocrinol. Metab. Gastrointest. Physiol.* 3): E79–E83, 1978.
130. POULSEN, J. H., AND L. Ø. KRISTENSEN. Is stimulation-induced uptake of sodium in rat parotid acinar cells mediated by a sodium/chloride co-transport system? In: *Electrolyte and Water Transport Across Gastrointestinal Epithelia*, edited by R. M. Case, A. Garner, L. A. Turnberg, and J. A. Young. New York: Raven, 1982, p. 199–208.
131. POULSEN, J. H., L. P. LAUGESEN, AND J. O. D. NIELSEN. Evidence supporting that basolaterally located Na^+-K^+-ATPase and a co-transport system for sodium and chloride are key elements in secretion of primary saliva. In: *Electrolyte and Water Transport Across Gastrointestinal Epithelia*, edited by R. M. Case, A. Garner, L. A. Turnberg, and J. A. Young. New York: Raven, 1982, p. 157–159.
132. POULSEN, J. H., AND B. OAKLEY. Intracellular potassium ion activity in resting and stimulated mouse pancreas and submandibular gland. *Proc. R. Soc. Lond. B Biol. Sci.* 204: 99–104, 1979.
133. RECTOR, F. C., AND C. A. BERRY. Role of the paracellular pathway in reabsorption of solutes and water by proximal convoluted tubule of the mammalian kidney. In: *The Paracellular Pathway*, edited by S. E. Bradley and E. F. Purcell. New York: Josiah Macy, Jr., Found., 1982, p. 135–158.
134. REUSS, L., P. REINACH, S. A. WEINMAN, AND T. P. GRADY. Intracellular ion activities and Cl^- transport mechanisms in bullfrog corneal epithelium. *Am. J. Physiol.* 244 (*Cell Physiol.* 13): C336–C347, 1983.
135. REVIS, N. W., AND J. P. DURHAM. Adenylate cyclase activity in the parotid gland of the mouse after isoproterenol stimulation. *J. Histochem. Cytochem.* 27: 1317–1321, 1979.
136. ROBERTS, M. L., N. IWATSUKI, AND O. H. PETERSEN. Parotid acinar cells: ionic dependence of acetylcholine-evoked membrane potential changes. *Pfluegers Arch.* 376: 159–167, 1978.
137. ROBERTS, M. L., AND O. H. PETERSEN. Membrane potential and resistance changes induced in salivary gland acinar cells by microiontophoretic application of acetylcholine and adrenergic agonists. *J. Membr. Biol.* 39: 297–312, 1978.
138. SAITO, Y., K. ITCI, K. HORIUCHI, AND T. WATANABE. Mode of action of furosemide on the chloride-dependent short-circuit current across the ciliary body epithelium of toad eyes. *J. Membr. Biol.* 53: 85–93, 1980.
139. SAITO, Y., T. OZAWA, H. HAYASHI, AND A. NISHIYAMA. Acetylcholine-induced change in intracellular Cl^- activity of the mouse lacrimal acinar cells. *Pfluegers Arch.* 405: 108–111, 1985.
140. SASAKI, S., I. NAKAGAKI, H. MORI, AND Y. IMAI. Intracellular calcium store and transport of elements in acinar cells of the salivary gland determined by electron probe X-ray microanalysis. *Jpn. J. Physiol.* 33: 69–83, 1983.
141. SATO, K. The physiology, pharmacology and biochemistry of eccrine sweat glands. *Rev. Physiol. Biochem. Pharmacol.* 79: 51–131, 1977.
142. SCHLATTER, E., R. GREGER, AND C. WEIDTKE. Effect of "high ceiling" diuretics on active salt transport in the cortical thick ascending limb of Henle's loop of rabbit kidney. Correlation of chemical structure and inhibitory potency. *Pfluegers Arch.* 396: 210–217, 1983.
143. SCHNEYER, L. H. Secretion of potassium by perfused excretory duct of rat submaxillary gland. *Am. J. Physiol.* 217: 1324–1329, 1969.
144. SCHNEYER, L. H. Amiloride inhibition of ion transport in perfused excretory duct of rat submaxillary gland. *Am. J. Physiol.* 219: 1050–1055, 1970.
145. SCHNEYER, L. H. Sympathetic control of Na, K transport in perfused submaxillary main duct of rat. *Am. J. Physiol.* 230: 341–345, 1976.
146. SCHNEYER, L. H. Parasympathetic control of Na, K transport in perfused submaxillary duct of the rat. *Am. J. Physiol.* 233 (*Renal Fluid Electrolyte Physiol.* 2): F22–F28, 1977.
147. SCHNEYER, L. H., AND C. A. SCHNEYER. Electrolyte and inulin spaces of rat salivary glands and pancreas. *Am. J. Physiol.* 199: 649–652, 1960.
148. SCHNEYER, L. H., AND C. A. SCHNEYER. Inorganic composition of saliva. In: *Handbook of Physiology. Alimentary Canal*, edited by C. F. Code. Washington, DC: Am. Physiol. Soc., 1967, sect. 6, vol II, p. 497–530.
149. SCHNEYER, L. H., AND T. THAVORNTHON. Isoproterenol-induced stimulation of sodium absorption in perfused salivary duct. *Am. J. Physiol.* 224: 136–139, 1973.
150. SCHWARTZ, G. J., J. BARASCH, AND Q. AL-AWQATI. Plasticity of functional epithelial polarity. *Nature Lond.* 318: 368–371, 1985.
151. SEOW, F. K. T., J. M. LINGARD, AND J. A. YOUNG. Anionic basis of fluid secretion by rat pancreatic acini in vitro. *Am. J. Physiol.* 250 (*Gastrointest. Liver Physiol.* 13): G140–G148, 1986.
152. SEWELL, W., AND J. A. YOUNG. Studies on the effects of poly-L-lysine on electrolyte transport by rat salivary ducts (Abstract). *Proc. Aust. Physiol. Pharmacol. Soc.* 4: 80P, 1973.
153. SHARKEY, K. A., AND D. TEMPLETON. Substance P in the rat parotid gland: evidence for a dual origin from the otic and trigeminal ganglia. *Brain Res.* 304: 392–396, 1984.
154. SHOROFSKY, S. R., M. FIELD, AND H. A. FOZZARD. Mechanism of Cl secretion in canine trachea: changes in intracellular chloride activity with secretion. *J. Membr. Biol.* 81: 1–8, 1984.
155. SILVA, P., J. STOFF, M. FIELD, L. FINE, J. N. FORREST, AND F. H. EPSTEIN. Mechanism of active chloride secretion by shark rectal gland: role of Na-K-ATPase in chloride transport. *Am. J. Physiol.* 233 (*Renal Fluid Electrolyte Physiol.* 2): F298–F306, 1977.
156. SLEGERS, J. F. G., W. M. MOONS, P. P. IDZERDA, AND W. M. STANDHOUDERS. The contribution of a Cl-shunt to the transmucosal potential of the rabbit submaxillary duct. *J. Membr. Biol.* 25: 213–236, 1975.
157. SMAJE, L. H., J. H. POULSEN, AND H. H. USSING. Evidence from O_2 uptake measurements for Na^+-K^+-$2Cl^-$ cotransport in the rabbit submandibular gland. *Pfluegers Arch.* 406: 492–496, 1986.
158. SPRING, K. R., AND A. HOPE. Size and shape of the lateral intercellular spaces in a living epithelium. *Science Wash. DC* 200: 54–58, 1978.
159. SPRING, K. R., AND A. HOPE. Fluid transport and the dimensions of cells and interspaces of living *Necturus* gallbladder. *J.*

Gen. Physiol. 73: 287–305, 1979.
160. STEWART, D. J., M. LAANSOO, AND A. K. SEN. Release of ^{86}Rb from rat submandibular gland slices: relation to sodium pump activity. *Am. J. Physiol.* 246 (*Gastrointest. Liver Physiol.* 9): G484–G491, 1984.
161. STOKES, J. B. Sodium chloride absorption by the urinary bladder of the winter flounder. A thiazide-sensitive, electrically neutral transport system. *J. Clin. Invest.* 74: 7–16, 1984.
162. SUZUKI, K., AND O. H. PETERSEN. The effect of Na$^+$ and Cl$^-$ removal and of loop diuretics on acetylcholine-evoked membrane potential changes in mouse lacrimal acinar cells. *Q. J. Exp. Physiol.* 70: 437–446, 1985.
163. THAYSEN, J. H. Handling of alkali metals by exocrine glands other than the kidney. In: *Handbook of Experimental Pharmacology. The Alkali Metal Ions in Biology*, edited by H. H. Ussing, P. Kruhöffer, J. H. Thaysen, and N. A. Thorn. Berlin: Springer-Verlag, 1960, vol. 13, p. 424–463.
164. THAYSEN, J. H., N. A. THORN, AND I. L. SCHWARTZ. Excretion of sodium, potassium, chloride and carbon dioxide in human parotid saliva. *Am. J. Physiol.* 178: 155–159, 1954.
165. TRAUTMANN, A., AND A. MARTY. Activation of Ca-dependent K channels by carbamylcholine in rat lacrimal glands. *Proc. Natl. Acad. Sci. USA* 81: 611–615, 1984.
165a. VAN OS, C. H., G. WIEDNER, J. F. G. SLEGERS, J. BIJMAN, AND E. M. WRIGHT. Hydraulic and electrical conductivities of rabbit gallbladder and submaxillary main duct epithelium. In: *Water Transport Across Epithelia Barriers, Gradients and Mechanisms*, edited by H. H. Ussing, N. Bindslev, N. A. Lassen, and O. Sten-Knudsen. Copenhagen: Munksgaard, 1981, p. 178–189.
166. WAKUI, M., AND A. NISHIYAMA. Ionic dependence of acetylcholine equilibrium potential of acinar cells in mouse submaxillary gland. *Pfluegers Arch.* 386: 261–267, 1980.
167. WELSH, M. J. Inhibition of chloride secretion by furosemide in canine tracheal epithelium. *J. Membr. Biol.* 71: 219–226, 1983.
168. WELSH, M. J. Anthracene-9-carboxylic acid inhibits an apical membrane chloride conductance in canine tracheal epithelium. *J. Membr. Biol.* 78: 61–71, 1984.
169. WELSH, M. J. Single apical membrane anion channels in primary cultures of canine tracheal epithelium. *Pfluegers Arch.* 407, Suppl. 2: S116–S122, 1986.
170. YOUNG, J. A. Salivary secretion of inorganic electrolytes. In: *Gastrointestinal Physiology III*, edited by R. K. Crane. Baltimore, MD: University Park, 1979, vol. 19, p. 1–58. (Int. Rev. Physiol. Ser.)
171. YOUNG, J. A., R. M. CASE, A. D. CONIGRAVE, AND I. NOVAK. Transport of bicarbonate and other anions in salivary secretion. *Ann. NY Acad. Sci.* 341: 172–190, 1980.
172. YOUNG, J. A., B. E. CHAPMAN, D. I. COOK, A. P. HEALEY, P. W. KUCHEL, J. M. LINGARD, M. NICOL, I. NOVAK, AND F. SEOW. Salivary secretory mechanism: recent advances and concepts. In: *Fluid and Electrolyte Transport in Exocrine Glands in Cystic Fibrosis*, edited by P. M. Quinton, J. R. Martinez, and U. Hopfer. San Francisco, CA: San Francisco Press, 1982, p. 102–124.
173. YOUNG, J. A., D. I. COOK, E. W. VAN LENNEP, AND M. ROBERTS. Secretion by the major salivary glands. In: *Physiology of the Gastrointestinal Tract* (2nd ed.), edited by L. R. Johnson. New York: Raven, 1987.
174. YOUNG, J. A., E. FRÖMTER, E. SCHÖGEL, AND K. F. HAMANN. A microperfusion investigation of sodium resorption and potassium secretion by the main secretory duct of the rat submaxillary gland. *Pfluegers Arch. Gesamte Physiol. Menschen Tiere* 295: 157–172, 1967.
175. YOUNG, J. A., AND C. J. MARTIN. The effect of a sympatho- and a parasympathomimetic drug on the electrolyte concentrations of primary and final saliva of the rat submaxillary gland. *Pfluegers Arch.* 327: 285–302, 1971.
176. YOUNG, J. A., C. J. MARTIN, M. ASZ, AND F. D. WEBER. A microperfusion investigation of bicarbonate secretion by the rat submaxillary gland. The action of a parasympathomimetic drug on electrolyte transport. *Pfluegers Arch.* 319: 185–199, 1970.
177. YOUNG, J. A., AND E. SCHÖGEL. Micropuncture investigation of sodium and potassium excretion in rat submaxillary saliva. *Pfluegers Arch. Gesamte Physiol. Menschen Tiere* 291: 85–98, 1966.
178. YOUNG, J. A., AND E. W. VAN LENNEP. *The Morphology of Salivary Glands*. London: Academic, 1978.
179. YOUNG, J. A., AND E. W. VAN LENNEP. Transport in salivary and salt glands. I: Salivary glands. In: *Membrane Transport in Biology. Transport Organs*, edited by G. Giebisch, D. C. Tosteson, and H. H. Ussing. Berlin: Springer-Verlag, 1979, vol. 4B, p. 563–674.

CHAPTER 2

Electrophysiology of salivary and pancreatic acinar cells

O. H. PETERSEN | Medical Research Council Secretory Control Research Group, The Physiological Laboratory, University of Liverpool, Liverpool, United Kingdom

Y. MARUYAMA | Department of Physiology, Faculty of Medicine, University of Tokyo, Tokyo, Japan

CHAPTER CONTENTS

Electrophysiological Methods in Study of Acinar Cells
   Intracellular microelectrodes
   Patch-clamp methods
      Noise analysis
      Capacitance measurements
Cell-to-Cell Communication
Resting Membrane Properties
   Resting membrane potential
   Electrodiffusional control of membrane potential
Potassium Channels in Resting Membrane
   Single-channel recording from excised membrane patches
      Calcium and voltage activation
      Pharmacological properties of high-conductance calcium- and voltage-activated potassium channel
   Single-channel current recording from intact acinar cells
   Whole-cell current recording from single acinar cells
      Number of channels per cell
Electrogenic Pumps
   Sodium-potassium pump
   Sodium–amino acid cotransport
Membrane Effects of Stimulants
   Activation of potassium channels
      Single-channel current recording in cell-attached configuration
      Activation of whole-cell current by secretagogues
   Activation of chloride channels
   Activation of nonselective cation channels
   Stimulant-evoked membrane potential changes
   Stimulation-evoked changes in membrane capacitance
   Source of calcium used as messenger for stimulant-evoked membrane changes
Importance of Electrogenic Processes for Acinar Cell Function
   Potassium channels and fluid secretion
      Acinar cell potassium balance
      Acinar cell model
   Nonselective cation channels and fluid secretion
   Potassium channels and protein secretion

THE FIRST DESCRIPTION of membrane electrical properties in gland cells was made by Lundberg (53–55), who studied intracellular potentials in cat salivary glands. Lundberg mainly investigated acinar cells and found low resting membrane potentials (about −20 to −30 mV) with marked hyperpolarizations in response to electrical stimulation of parasympathetic or sympathetic nerves. Stimulant-evoked membrane hyperpolarization was thought to be due to direct activation of an electrogenic Cl^- pump localized in the basolateral plasma membrane (54, 56–58), and it was proposed that this active Cl^- transport was directly responsible for the formation of the Cl^--rich primary secretion (54, 57, 58). Since then it has become clear that the stimulant-evoked hyperpolarization of the salivary acinar cell membrane is not due to an electrogenic Cl^- pump but occurs as a consequence of K^+ channel activation (88, 89). Direct studies of single-channel currents in basolateral acinar cell membranes, from a variety of exocrine glands and many different species, have confirmed the K^+ channel hypothesis and given much detailed and quantitative information about the activation of these pores by intracellular Ca^{2+} (93, 97). In addition to the Ca^{2+}-activated K^+ channels, the basolateral acinar cell membrane contains the ATP-driven Na^+-K^+ pump (4, 84–86, 89), and there is also evidence for the presence of a Na^+-K^+-$2Cl^-$ cotransport mechanism (74, 92, 97, 119). It is now apparent that these three separate transport proteins in the basolateral acinar cell membrane can function together as an electrogenic Cl^- pump (Fig. 1). Therefore, although the original Cl^- pump hypothesis (54, 58) was incorrect as well as based on evidence that could not be supported by subsequent experimental results (88), it now appears that an operational Cl^- pump does exist and is important for the formation of the primary acinar secretion. Potassium plays a crucial role in this process because the rate of Cl^- uptake must be directly linked to the rate of transport in the cycle of K^+ release and uptake (97), and the point of regulation has been clearly identified at the molecular

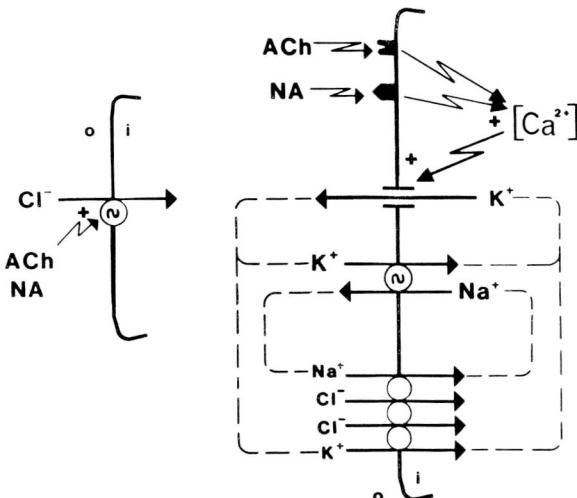

FIG. 1. Two simple models explaining stimulant-evoked Cl⁻ uptake and membrane hyperpolarization in exocrine acinar cells. *Left*: model originally proposed by Lundberg (54) for salivary glands. Autonomic neurotransmitters activate directly an electrogenic active Cl⁻ transport. *Right*: model proposed by Petersen and Maruyama (97, 119). Stimulant secretagogues [acetylcholine (ACh) and norepinephrine (NA) in salivary glands, other neurotransmitters or hormones in other glands; see ref. 89] evoke an increase in intracellular Ca^{2+} concentration ($[Ca^{2+}]_i$) that activates K^+ channels. Exit of K^+ allows K^+ reuptake through Na^+-K^+ pump and Na^+-K^+-$2Cl^-$ cotransporter. In steady-state stimulated condition all these processes constitute operational Cl⁻ pump that is electrogenic. i, Inner surface of plasma membrane; o, outer surface of plasma membrane. [From Petersen (93).]

level as the Ca^{2+}-activated K^+ channel (Fig. 1). In addition to the basolateral K^+ channels, there are also Ca^{2+}-activated Cl⁻ channels in acinar cells (19, 59), and it seems likely that these are localized in the luminal membranes (92, 119). The Ca^{2+}-activated K^+ and Cl⁻ channels dominate the electrical properties of exocrine acinar cells and are featured prominently in this chapter. Important electrogenic Na^+–amino acid cotransport systems (40, 41, 51) are also described, and no account of acinar cell electrophysiology would be complete without a discussion of cell-to-cell coupling (91).

Electrophysiology is important when considering ion transport across membranes, because transport depends on both chemical and electrical gradients as well as on the precise characteristics of the transmembrane pathways. Electrophysiological methods have been instrumental in mapping different types of ion channels, electrogenic pumps, and cotransport processes. Intracellular microelectrode recordings in association with local micropipette applications of hormones, neurotransmitters, and amino acids have contributed unique information about transport events that could not have been obtained in any other way. New recording configurations and techniques have enabled a molecular approach to channel function, and these patch-clamp methods (27, 76, 99) have also radically changed our ideas about preferred cell types for electrical studies. In the 1950s, the success of the voltage-clamp experiments of Hodgkin, Katz, and Huxley on the squid axon initiated an era in which the study of large cells provided the most useful information about the properties of the cell membrane (see ref. 28). In the 1980s the patch-clamp experiments, with their many different recording configurations, have been best suited to small mammalian cells, and studies of exocrine acinar cells have been particularly successful in clarifying aspects of cellular physiology that could not have been investigated directly with classic intracellular microelectrode techniques.

ELECTROPHYSIOLOGICAL METHODS IN STUDY OF ACINAR CELLS

Intracellular Microelectrodes

The introduction of the glass microelectrode to gland cell electrophysiology more than 30 years ago (53) was of great importance. Intracellular recordings made it possible for the first time to obtain information on the time course of cellular activation evoked by nerve stimulation or hormones as well as on the current flow across cell membranes during rest and activity. By simultaneously recording with several microelectrodes from different acinar cells, it became possible to obtain information on intracellular communication pathways and their regulation. Comparison of the effects of externally applied stimulants with the effects of internally applied substances allowed analysis of intracellular messenger actions. Figure 2 summarizes the different types of electrophysiological experiments that have been carried out with conventional intracellular microelectrodes. Although there is enormous scope for microelectrode studies, there are also serious limitations; most of these are a result of the leak introduced into the plasma membrane by the insertion into a cell of a glass microelectrode (89). Even though conventional microelectrode studies have been relatively successful in defining macroscopic transmembrane ionic currents (66, 98), they are by their very nature incapable of giving direct information about the underlying microscopic events.

Patch-Clamp Methods

Electrical currents in cells are mediated by a class of proteins in the plasma membrane called ion channels or pores. In intact cells the problem is to detect single-channel currents in the presence of background electrical noise.

Conventional intracellular microelectrode methods for current measurement are associated with a background noise of at least 100 pA, whereas the current flowing when a single channel opens is only a small fraction of this background noise. Neher and Sakmann (76) solved this problem by the patch-clamp method. Instead of inserting a microelectrode into a

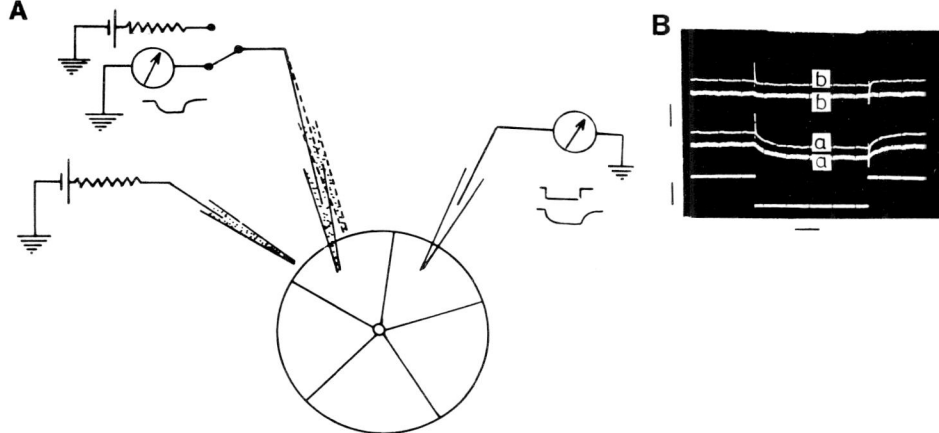

FIG. 2. Intracellular recording with 2 separate microelectrodes from neighboring acinar cells. *A*: recording configuration. Both intracellular electrodes can be used for membrane potential recording and for current injection. If electrode is filled with secretagogue (*shaded electrodes*), current injection of appropriate polarity can be used for local agonist application. *B*: membrane potential recordings from mouse pancreas. *Lower trace*, rectangular hyperpolarizing current pulse (10^{-9} A) of 100 ms duration. *Upper traces*, effects on membrane potential in injected cell as well as its neighbor during rest (*a*) and during acetylcholine application (*b*). Vertical calibration, 10 mV and 10^{-9} A; horizontal calibration, 20 ms. [From Petersen and Ueda (104).]

cell, they pressed a micropipette tip onto the surface of a cell, effectively isolating a patch of membrane. The intrinsic noise increases with the area of the membrane under study, and by isolating a small area (1–10 μm^2), such low extraneous noise levels are attained that the picoampere currents flowing through single channels can be measured directly (27).

The seal between the tip of the micropipette and the outer surface of the cell membrane has, under suitable conditions (fine-polished and clean micropipette tip and clean membrane surface), a high electrical resistance [on the order of gigaohms (10^9)] and is mechanically surprisingly stable. The discovery in 1980 of this high-resistance seal by Neher, Sakmann, and their co-workers (27) as well as by Horn and Patlak (31) was very important, as it made entirely new types of experiments possible (Fig. 3). The electrically isolated membrane patch can be pulled off the cell (excised) in such a way that the inside of the plasma membrane faces the bath solution (inside out) or alternatively so that the inside faces the solution in the micropipette (outside out). By breaking the patch membrane in the cell-attached recording conformation, the solution in the pipette interior gains direct access to the cell interior, and cell dialysis is carried out under conditions where the currents across the whole of the cell membrane can be measured. Equilibration of the cell interior with the bath solution can be done while single-channel currents are recorded by making holes in the plasma membrane outside the isolated patch area with the help of detergents such as saponin or digitonin (68).

NOISE ANALYSIS. By employing the patch-clamp whole-cell current recording configuration, it is possible to use two types of analytical methods to estimate the number of channels in single acinar cells or small communicating acinar cell clusters. Steady-state noise analysis, including spectral analysis, was first introduced by Katz and Miledi (46) and subsequently applied to the study of motor end plate currents by Anderson and Stevens (1), whereas ensemble noise analysis (transient noise analysis) introduced by Sigworth (113) was applied first to the study of voltage-operated Na^+ channels in myelinated nerve fibers. Both methods are quite applicable to exocrine gland acinar cells in situations where one type of channel with only one conducting state dominates the membrane. An example of the type of record obtained in ensemble noise analysis is shown in Figure 4.

The analysis is based on the following relations

$$I = niP \qquad (1)$$

$$\sigma^2 = i^2 nP(1 - P) \qquad (2)$$

$$\sigma^2/I = i - I/n \qquad (3)$$

where i is the single-channel current, n is the number of channels per cell, P is the open-state probability, and I and σ^2 are the mean current and current variance, respectively. It is possible to calculate i and n from plots of σ^2/I as a function of I, and P can then be obtained from Equation 1 (62).

CAPACITANCE MEASUREMENTS. An entirely new approach to the study of exocytotic and endocytotic

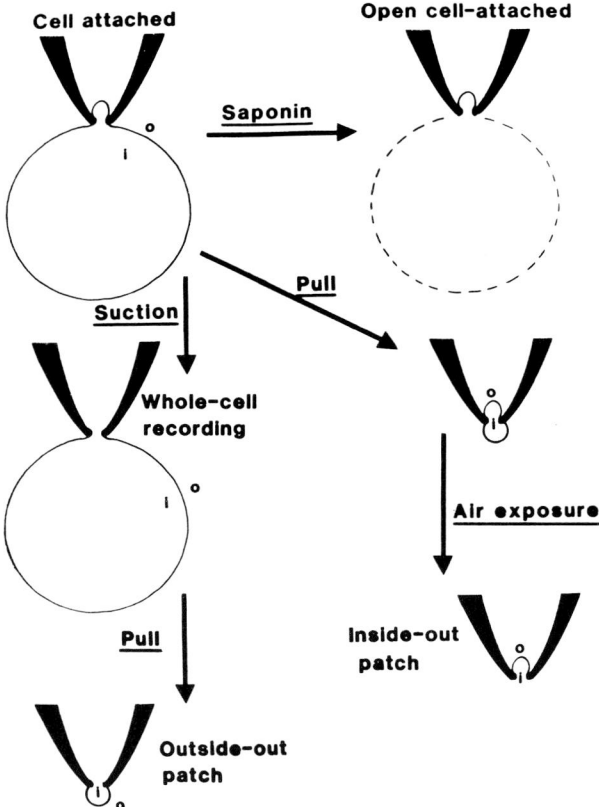

FIG. 3. Different patch-clamp recording configurations and procedures used to establish them. Patch-clamp experiments start in cell-attached configuration (*top left*). Tip of fire-polished recording pipette is gently brought into contact with cell surface and slight suction is applied to pipette resulting in high-resistance seal between tip of glass pipette and cell membrane. Detergent saponin (or digitonin) can be introduced briefly into bath to permeabilize cell membrane (outside isolated patch area), allowing equilibration between cell interior (i) and exterior (o) (open cell-attached configuration). Starting from original cell-attached situation, patch membrane can be mechanically pulled off leading to formation of closed membrane vesicle in pipette tip. Outer surface of vesicle can be disrupted by passing pipette tip briefly through air-water interface of bath, leading to excised inside-out membrane configuration. If short pulse of suction is applied to pipette in cell-attached mode, membrane patch can be broken and direct continuity between pipette and cell interior established. Currents flowing across entire cell membrane can now be recorded (whole-cell recording). If pipette is then pulled away from cell, excised membrane fragments will reseal so that excised outside-out membrane patch is obtained. [From Petersen and Petersen (99).]

processes in secretory cells was introduced by Neher and Marty (75) using the patch-clamp whole-cell recording configuration for capacitance measurements. Fusion of granules with the plasma membrane and subsequent retrieval from the cell membrane (80) should be reflected in small changes in membrane capacitance (a measure of the surface area of lipid bilayer membrane). Assuming that the diameter of a fused vesicle is ~0.5 μm, the increase in membrane capacitance is expected to be ~8 fF. The cell membrane can be regarded as an arrangement of membrane capacitance C in parallel with membrane conductance G, and the resultant oscillatory current response of the cell membrane perturbed by the application of a sine-wave voltage would be the sum of two trigonometric waves. The wave component responsible for G would be in phase with the command (reference) sine wave, whereas that for C would be shifted by $\pi/2$ from the command sine wave (cosine wave). The separation of the two components is achieved by a lock-in amplifier that locks the phase angle either at 0 or at $\pi/2$ with respect to the phase of a command sine wave and detects the magnitudes of the responding currents (75).

CELL-TO-CELL COMMUNICATION

Gland cells are linked together by cell-to-cell channels, which have an effective bore size of ~16–20 Å that allows passage of inorganic ions, metabolites, nucleotides, and high-energy phosphates but not macromolecules (52, 89). In exocrine glands, electrical communication between adjacent acinar cells was first shown in 1975 (87) in experiments where two microelectrodes were inserted into different acinar cells that were close to each other. As shown in Figure 2, two separate but simultaneous microelectrode recordings of membrane potential from two coupled mouse pancreatic acinar cells can then be made. Current injection through one of the intracellular microelectrodes causes a polarization of both cells to the same extent, thereby demonstrating the very low electrical resistance between the two cells.

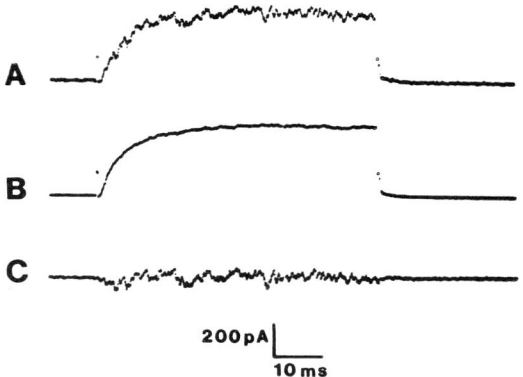

FIG. 4. Measurement of current fluctuation with help of patch-clamp whole-cell recording configuration. Recording obtained from single isolated rat parotid acinar cell dialyzed with K^+-rich and Ca^{2+}-free, ethylene glycol-bis(β-aminoethylether)-N,N'-tetraacetic acid (EGTA)-containing solution. Holding potential was −60 mV. *A*: single trace of whole-cell current elicited by 80 mV depolarizing voltage step. *B*: mean current elicited by 80 mV depolarizing voltage steps obtained by averaging 20 single sweeps. *C*: current fluctuation obtained by subtracting *trace B* from *trace A*. [From Maruyama et al. (62).]

The electrical communication network in the acinar tissue has been investigated by simultaneous intracellular recording with two microelectrodes and direct microscopic control of the localizations of the electrode tips (35). All cells within one acinus are closely electrically coupled, whereas cells in what may appear to be two different acini are sometimes coupled but in other cases are completely electrically isolated from each other. Electrical coupling is almost an all-or-nothing phenomenon, and this can most easily be explained by assuming the existence of electrical acinar units containing cells closely linked together by cell-to-cell channels but completely electrically isolated from other acinar units. On the basis of the spatial spread of electrical current pulses in the tissue as well as measurements of the total membrane capacitance in the acinar units, it can be estimated that several hundred cells make up one unit (35).

An alternative method used to map the communication network in the pancreatic acinar tissue is the observation of fluorescent tracer movements. Direct visualization of cell-to-cell communication was first achieved by demonstrating transfer of fluorescein ($M_r = 332$) as well as Procion yellow ($M_r = 697$) from the injected acinar cell to neighboring cells in the living tissue (38). Later the cell-to-cell passage of the intensely fluorescing dye Lucifer yellow ($M_r = 457$) was studied. Sustained intra-acinar microionophoretic injection of Lucifer yellow revealed a finite limit to the extent of dye spread (17, 18). In two preparations of mouse pancreas for which complete sets of serial sections could be obtained, the dye-coupled intercommunicating acinar units consisted of 110 and 230 individual acinar cells (18), which is in good agreement with the electrical studies. Lucifer yellow could only be detected in acinar cells; duct cells did not contain the dye.

Large doses of pancreatic secretagogues like acetylcholine (ACh), cholecystokinin (CCK), and bombesin can evoke electrical uncoupling of cells within functional acinar units; such effects are completely and rapidly reversible (32, 35, 37, 104). Acetylcholine-evoked electrical uncoupling restricts the passage of Lucifer yellow between the acinar cells (17). Intracellular Ca^{2+} injection can electrically isolate the injected cell (32). Intracellular acidification, brought about by exposure of the cells to CO_2 (39), can evoke electrical uncoupling, which is also fully reversible.

Gap junction particles contain the channels that directly connect the cytoplasm of adjacent cells (52). Large areas closely packed with gap junction particles have been demonstrated by electron microscopy of freeze-fractured, glutaraldehyde-fixed samples of acinar tissue (71), again confirming the tight coupling of adjacent cells within acinar units.

Recent direct studies of intercellular channels in the lacrimal acinar tissue have added new information about their electrical characteristics. Neyton and Trautmann (77) studied coupled acinar cell pairs using two patch electrodes in the whole-cell recording conformation so that the voltage and current were controlled in both cells. The signal-to-noise ratio was such that quantal junctional conductance changes caused by the opening and closing of single junctional channels could be observed. The fully open channel has a conductance of ~120 pS, but states of lower conductance probably due to partial openings or obstructions can often be found. Occasionally larger conductance jumps were observed, possibly indicating groups of channels cooperating. The junctional channels discriminate poorly between cations and anions, but the cations are slightly more permeable.

The existence of extensive cell-to-cell coupling has important consequences for the function of the exocrine acinar tissue both with regard to the resting membrane properties and the effects of stimulation (91).

RESTING MEMBRANE PROPERTIES

Resting Membrane Potential

The resting membrane potential is measured as the electrical potential difference between the tip of a glass microelectrode inserted into a cell and a reference electrode in contact with the extracellular fluid space. A leak is created in the plasma membrane around the shaft of the inserted microelectrode; thus the success of the measurement depends on the ability of the plasma membrane to seal that leak (89). Consequently as microelectrode techniques have improved, the magnitude of gland cell resting potentials has increased. Exocrine acinar cells have resting membrane potentials in the range of −30 to −70 mV (89). Petersen (89) has given a detailed discussion of the evaluation of and the criteria for correct membrane potential measurements.

Electrodiffusional Control of Membrane Potential

The membrane potential is sensitive to changes in the extracellular K^+ concentration ($[K^+]_o$). Those gland cells having high membrane potentials (salivary glands) are the most sensitive to changes in $[K^+]_o$. In the parotid gland the slope of the linear curve relating membrane potential to log $[K^+]_o$ is ~50 mV/10-fold increase in $[K^+]_o$ (for values above $[K^+]_o = 10$ mM) (83). This is high but not as high as the theoretically expected 61 mV/10-fold increase in $[K^+]_o$ for a K^+-selective membrane. Even the relatively high resting potential of −70 mV for the parotid (83) is clearly less than the calculated K^+ equilibrium potential (E_{K^+})

$$E_{K^+} = 61.5 \log(3.3/116) \text{ mV} = -95 \text{ mV} \quad (4)$$

where 116 mM is the intracellular K^+ activity as measured directly in superfused mouse salivary segments by Poulsen and Oakley (107) and 3.3 mM is the extracellular K^+ activity (at a concentration of 4.7 mM, assuming an activity coefficient of 0.7). The membrane permeability is thus dominated by K^+ but other ions do contribute.

Although there is a relatively high resting Cl^- conductance (G_{Cl^-}) in some gland cell membranes (26), there are many more glands in which G_{Cl^-} appears to be relatively low. In all salivary and lacrimal acinar cells the intracellular Cl^- concentration is substan-

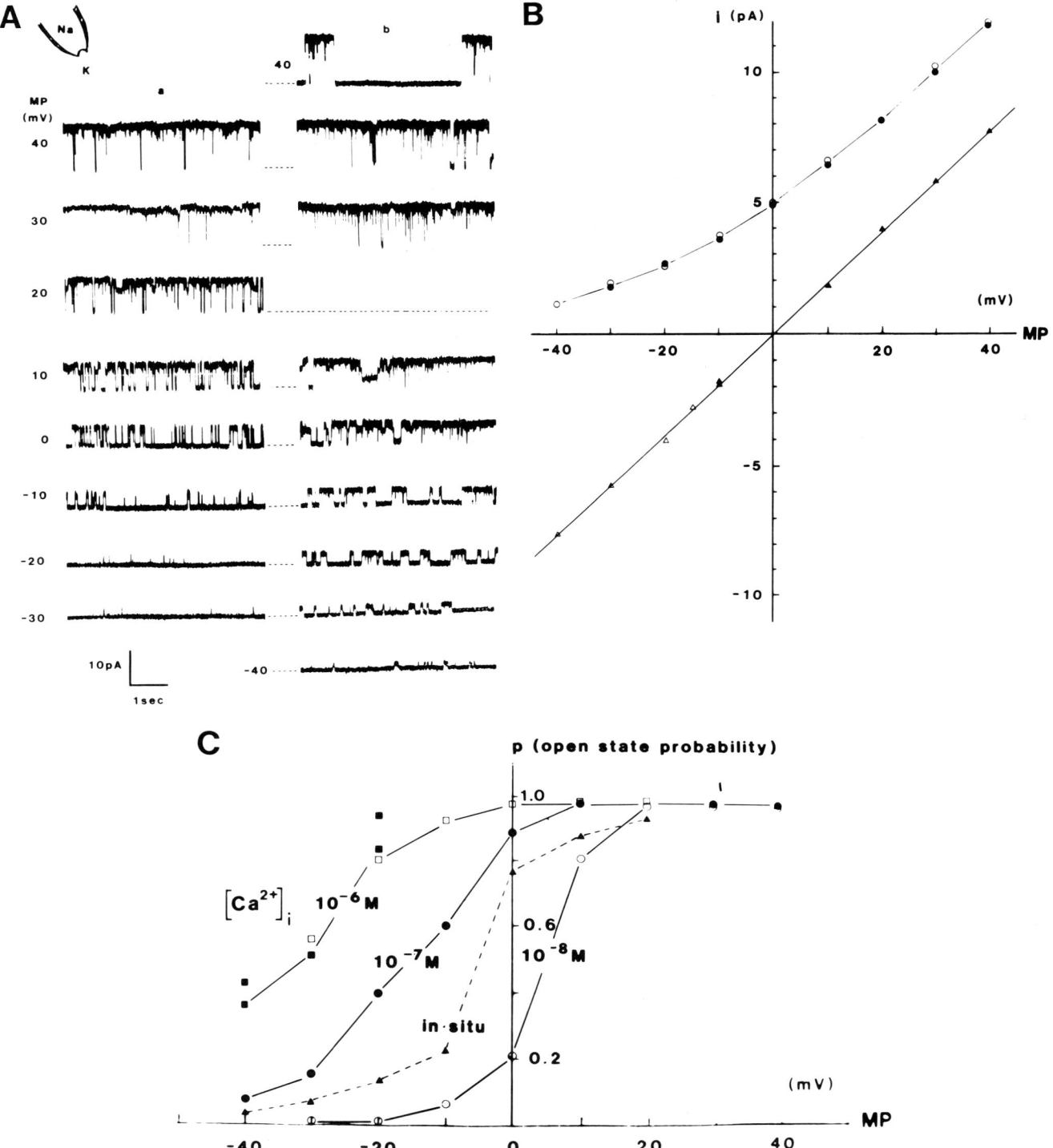

tially higher than expected on the basis of passive distribution across the membrane according to the electrical potential difference (89, 110). This is due to the operation of a furosemide-sensitive Na^+-K^+-$2Cl^-$ cotransport system that, with the help of the large electrochemical gradient favoring Na^+ uptake, accumulates Cl^- (74, 97). The resting G_{Cl^-} of the basolateral membrane appears to be very low since removal of external Cl^- does not result in an immediate membrane potential change as expected if there were opportunities for Cl^- to cross the membrane through channels (89).

POTASSIUM CHANNELS IN RESTING MEMBRANE

Single-Channel Recording From Excised Membrane Patches

CALCIUM AND VOLTAGE ACTIVATION. The properties of K^+ channels can most easily be assessed with the patch-clamp technique by analyzing single-channel current activity in excised patches of plasma membrane exposed to solutions of known composition. Figure 5A shows results obtained from an excised acinar membrane patch where the inside of the plasma membranes faces the K^+-rich (intracellular) bathing solution and the outside faces the Na^+-rich (extracellular) pipette solution. *Column a* shows the currents obtained when the Ca^{2+} concentration in the bath ($[Ca^{2+}]_i$) is 10^{-8} M. At negative membrane potentials the current steps are small and infrequent, whereas at positive potentials the channel is open much of the time, although interrupted by closings revealing the large single-channel current. Figure 5B shows the single-channel current plotted as a function of the membrane potential. The null potential has not been attained, but it is more negative than -40 mV and because K^+ is the only ion with a negative equilibrium (Nernst) potential in this experimental situation, the channel must be K^+ selective. Unitary activity can also be seen in excised patches exposed to symmetrical K^+ solutions, and the single-channel current-voltage relationship (Fig. 5E) is in this case linear (the slope corresponding to a conductance of ~200 pS) with a null potential of 0 mV. In symmetrical Na^+-rich, K^+-free solutions no unitary activity can be observed. The relationship between the probability of the channel being open (P) and the membrane potential is shown in Figure 5C. Although channel openings occur at a membrane potential of -30 mV (Fig. 5A), Figure 5C shows that at $[Ca^{2+}]_i = 10^{-8}$ M, P is very low indeed. The relationship between P and membrane potential is represented by a sigmoid curve, and the membrane potential at which the channel is open half the time is about $+5$ mV. Increasing $[Ca^{2+}]_i$ to 10^{-7} M activates the channel. As seen in *column b* of Figure 5A, the channel spends much more time in the open state at the negative membrane potentials than at the lower Ca^{2+} concentration, as is also seen in Figure 5C. However, the amplitude of the single-channel currents is unaffected by the change in $[Ca^{2+}]_i$ (Fig. 5A, B). In Figure 5C, the open-state probability curve obtained in the intact cell is situated between those obtained at $[Ca^{2+}]_i = 10^{-8}$ M and 10^{-7} M in the excised patch, indicating that $[Ca^{2+}]_i$ in the resting pig pancreatic acinar cell lies somewhere between these two values.

Results very similar to those in Figure 5 were first obtained in patch-clamp experiments on submandibular and parotid glands from rats and mice (61). The presence of this type of Ca^{2+}- and voltage-activated high-conductance K^+ channel has subsequently been directly demonstrated in acinar cells from rat and mouse lacrimal glands (14, 121) and in human pancreas (94) as well as human submandibular and parotid glands (63).

In addition to the high-conductance K^+ channel (200–300 pS), there are also smaller K^+-selective channels with unit conductances of ~50 pS and 25 pS, respectively. These channels, which have been described in salivary glands (61) as well as in human and guinea pig pancreas (94, 120), have much the same properties with respect to Ca^{2+} and voltage activation as the high-conductance channel. In the pig and human pancreas (69, 94), the 50-pS channel is

FIG. 5. Single-channel current recording from excised (inside-out) pig pancreatic acinar membrane patches. A: single-channel current traces from inside-out membrane patch exposed to extracellular solution on outside (pipette) and intracellular solution on inside (bath) (see *cartoon*). a: Ca^{2+} concentration in bath ($[Ca^{2+}]_i$) was 10^{-8} M; b: in same patch, $[Ca^{2+}]_i$ was adjusted to 10^{-7} M. Membrane potentials (MP) at which individual traces were obtained are indicated. *Horizontal dashed lines*, current level when channel is closed. Upward deflections represent outward current. *Upper 2 traces*, response at membrane potential of 40 mV. Although channel was generally open, there were times when channel went from open state into state of prolonged closure. B: single-channel current-voltage (i/V) relationships from excised membrane patches. *Circles*, experiment shown in A (normal Na^+/K^+ gradients). *Closed circles*, $[Ca^{2+}]_i = 10^{-8}$ M; *open circles*, $[Ca^{2+}]_i = 10^{-7}$ M. *Triangles*, different experiment in which intracellular solution was used on both sides of membrane (K^+/K^+). *Closed triangles*, $[Ca^{2+}]_i = 10^{-8}$ M; *open triangles*, $[Ca^{2+}]_i = 10^{-7}$ M. C: relationship between open-state probability (P) of K^+ channel and membrane potential (MP) at 3 different levels of $[Ca^{2+}]_i$: *open circles*, 10^{-8} M; *closed circles*, 10^{-7} M; and *open and closed squares*, 10^{-6} M. *Open and closed circles*, experiment shown in A. *Triangles* (dashed line), experiment on in situ (cell-attached) patch. In intact cell $[Ca^{2+}]_i$ was apparently between 10^{-7} M and 10^{-8} M. [From Maruyama et al. (69). Reprinted by permission from *Nature*, copyright 1983, Macmillan Journals Limited.]

less frequently observed than the high-conductance pore shown in Figure 5, whereas the guinea pig pancreatic acinar cells are electrically dominated by the low-conductance channel (20–30 pS) (120).

PHARMACOLOGICAL PROPERTIES OF HIGH-CONDUCTANCE CALCIUM- AND VOLTAGE-ACTIVATED POTASSIUM CHANNEL. In experiments on excised inside-out or outside-out patches from pig pancreatic acinar cell membranes it has been possible to investigate the effects of various blocking agents. Barium is of particular interest, because it is generally used to reduce K^+ conductance in epithelial tissues (10, 124). Barium (0.5–5 mM) acting on the outside of the plasma membrane evoked prolonged periods of channel closure but had little effect on the amplitude of the single-channel currents, whereas Ba^{2+} in much lower concentrations (6×10^{-7}–3×10^{-5} M) acting on the membrane inside had both excitatory and inhibitory effects on channel opening depending on the membrane potential. The effects of Ba^{2+} from the inside are qualitatively similar to those evoked by Ca^{2+}, but Ba^{2+} is less potent than Ca^{2+} in promoting channel opening and more potent in evoking channel closure at positive potentials (43).

Tetraethylammonium (TEA, 1–5 mM) acting from the outside of the cell membrane evokes a drastic reduction in the channel open-state probability and also markedly reduces the single-channel conductance. In contrast, TEA in a concentration as high as 10 mM has virtually no effect when added to a solution that is exclusively in contact with the membrane inside (42). Because other K^+ channels (e.g., ATP-sensitive inward rectifier channels in pancreatic endocrine cells) are insensitive to external TEA in low concentrations [<10 mM; (15, 16)], this drug may be particularly useful as a relatively specific blocking agent for the high-conductance channel.

Quinine or its stereoisomer quinidine (200–500 μM), which has been used to reduce stimulant-evoked K^+ efflux from salivary glands (48), reduces K^+ current flow through membrane patches (acting from both inside and outside) by chopping the single-channel events so that what in the control situation appears as a real unitary current trace is transformed by the drug into a rapidly fluctuating noise pattern (43). The effects of quinine or quinidine are not, however, specific for the Ca^{2+}- and voltage-activated high-conductance channel. In an insulin-secreting cell line it has been shown that quinine, in as low a concentration as 100 μM, totally blocks a Ca^{2+}-independent inward rectifier K^+ channel but has much less effect on the Ca^{2+}-activated high-conductance channel (16).

Single-Channel Current Recording From Intact Acinar Cells

It is unfortunately difficult to investigate single-channel currents at physiological membrane poten-

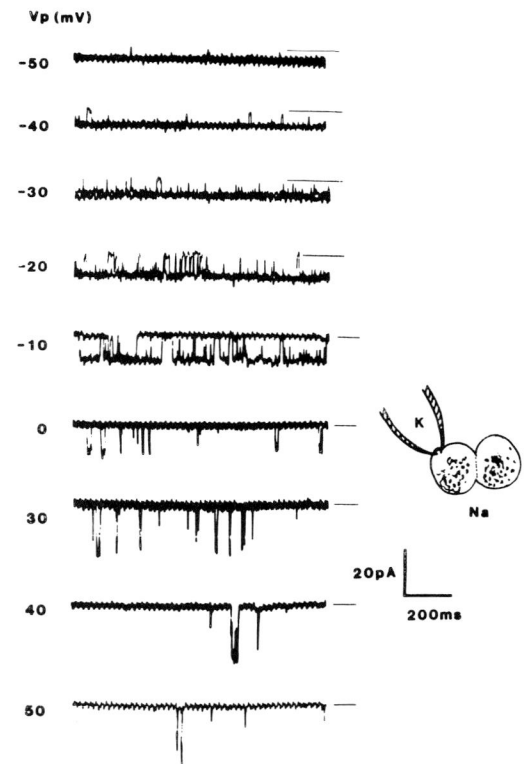

FIG. 6. Single-channel current recording in cell-attached configuration (isolated pig pancreatic acinar cell). Cartoon shows recording configuration. Recording pipette is filled with intracellular solution (high K^+ concentration), whereas bath fluid is extracellular solution (high Na^+ concentration). Traces, single-channel currents through K^+ channel in isolated patch area at different pipette potentials (V_p). [From Petersen (93).]

tials (i.e., about −60 mV) under conditions with quasi-physiological ion gradients, because the driving force for K^+ outflow [the difference between the K^+ equilibrium (Nernst) potential and the membrane potential] is so small that the single-channel currents are barely detectable. However, when the K^+ concentration in the recording pipette is raised to intracellular levels, as in the experiment represented by Figure 6, clear channel openings (inward current steps) can be observed at the normal resting potential [pipette potential (V_p) = 0 mV], although the probability of channel opening is low. The driving force for these inward currents must be the intracellular negativity, as there is no K^+ gradient over the isolated patch membrane. When V_p is made positive, the inward-going single-channel currents have larger amplitudes, and they also become less frequent, as the probability of channel opening is dependent on the membrane potential. Making V_p negative reduces the single-channel current amplitude, and the channel now spends more time in the open state. The single-channel current (i) depends only on the transmembrane potential of the isolated patch membrane ($V_m - V_p$) (as there is no concentration gradient for K^+) and the single-channel conductance (g)

$$i = g(V_m - V_p) \qquad (5)$$

Therefore i becomes zero when the pipette potential V_p equals the intracellular potential V_m. As seen in Figure 6, i is very small at $V_p = -50$ mV. There is a roughly linear relationship between i and V_p (corresponding to a single-channel conductance of ~200 pS) with the null potential having a value of −50 to −60 mV in all experiments. It is therefore reasonable to conclude that the resting potential in the isolated pig acinar cells is about −60 mV.

Whole-Cell Current Recording From Single Acinar Cells

Single-channel recording is a good technique for characterizing ion channels in the plasma membrane, as it allows direct observation of the properties of individual molecules. However, to obtain a quantitatively valid picture of the total ion conductance in the cell membrane, whole-cell current recording is very useful. Starting out from the recording situation shown in Figure 6 (cell-attached configuration) it is possible to break the isolated membrane patch and establish continuity between the pipette interior and the cell interior (see also Fig. 3). This allows a much better seal between pipette and cell membrane than can be achieved with a conventional intracellular microelectrode. In addition there is the advantage of being able to equilibrate the cell interior with the pipette solution and in this way control the intracellular ionic composition (99). Figure 7 shows examples of the type of voltage-clamp experiment that can be carried out using this technique. The membrane potential can be clamped close to the resting level and the currents associated with depolarizing and hyperpolarizing voltage jumps recorded. As seen in Figure 7, the whole plasma membrane acts as a rectifier. Depolarizing voltage jumps are associated with large outward currents, whereas hyperpolarizing jumps of the same magnitude generate very little inward current. This extreme rectifier property of the acinar cell membrane is due to the fact that the K^+ channels are voltage gated. Depolarization opens, whereas hyper-

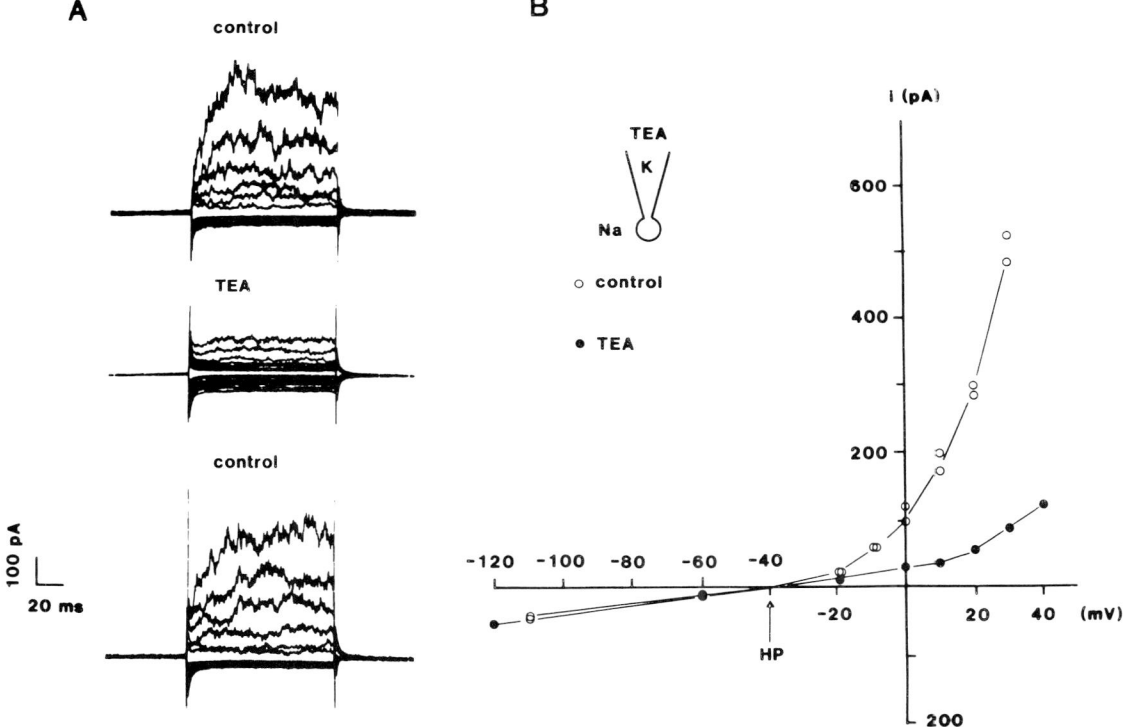

FIG. 7. Effects of intracellular and extracellular tetraethylammonium (TEA) on whole-cell currents in pig pancreatic acinar cell evoked by depolarizing or hyperpolarizing voltage steps from holding potential (HP) of −40 mV. Pipette contained intracellular K^+-rich solution to which 2 mM TEA had been added. Bath was filled with extracellular Na^+-rich solution. A: whole-cell currents before, during presence of TEA (2 mM) in bath, and after return to control situation. Voltage steps of ±20 to 70 mV (in one case 80 mV) were applied. B: whole-cell current-voltage relationship in control situations before and after TEA application and in presence of 2 mM TEA in bath. [From Iwatsuki and Petersen (42).]

polarization closes, the channels. The K$^+$ conductance of the plasma membrane therefore acts as a very efficient machine for preventing depolarization, as a depolarizing influence will activate the K$^+$ conductance and in a normal cell, which does not operate under voltage-clamp conditions, evoke repolarization.

NUMBER OF CHANNELS PER CELL. The whole-cell voltage-clamp currents can be regarded as the sum of many small elementary currents flowing through individual K$^+$ channels of the type characterized in single-channel studies. There are a number of strong arguments that support this conclusion. *1*) The whole-cell K$^+$ current is Ca^{2+} activated, and the Ca^{2+} sensitivity is similar to that described for the single channels (Fig. 8). *2*) The activation of the whole-cell currents occurs within the same voltage range as described for single-channel currents (67, 69). *3*) The Ca^{2+}-activated K$^+$ currents can be blocked by a low concentration (~2 mM) of TEA acting specifically from the outside of the plasma membrane (Fig. 7). *4*) Ensemble fluctuation analysis of the whole-cell currents (see Fig. 4) indicates that the outward current can be entirely accounted for by a relatively small number (50–100) of high-conductance channels (62, 121).

On the basis of these four arguments, the whole-cell current in the acinar cells is the sum of the elementary currents flowing through individual K$^+$ channels when they are open. It follows then that the relationship between the whole-cell current (I) and the single-channel current (i) is given by

$$I = nPi \qquad (6)$$

where n is the total number of operational K$^+$ channels and P the open-state probability. At positive membrane potentials where P is close to 1 (see Fig. 5)

$$n = I/i \qquad (7)$$

In the resting pig acinar cell this approach has indicated that the total number of operational channels is quite small, only ~50/cell (69), which is in agreement with the ensemble fluctuation analysis on salivary [30–108 channels/cell; (62)] and lacrimal cells [50–150 channels/cell; (121)].

This small number of high-conductance channels would cause problems if not for the extensive cell-to-cell coupling in the acinar units. The open-state probability in the resting cells at the spontaneous membrane potential is ~0.002 (22, 118). If we assume 100 channels per cell, on average only one-fifth of a channel will be open in a single cell. Because, as was seen in Figure 6, channels open and close frequently, a single cell would usually be in a position of having no channels open, whereas in other periods 1 and sometimes 2 channels would be conducting. This situation would lead to a very unstable membrane potential if the cells were electrically isolated from their neighbors. However, as discussed in CELL-TO-CELL COMMUNICATION, p. 28, several hundred cells form a func-

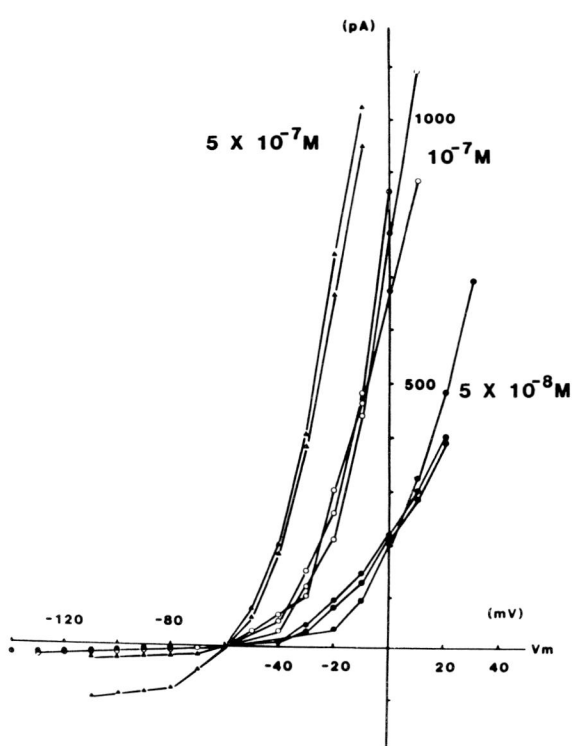

FIG. 8. Whole-cell currents in isolated pig pancreas acinar cell at different intracellular ionized Ca^{2+} concentrations (strongly buffered Ca^{2+}-EGTA solution) plotted as function of membrane potential. [From Maruyama and Petersen (67).]

tional unit; thus on the average ~50–100 channels will be open in the resting situation in 1 unit. With this number of channels the effects of openings and closings of single channels will be modest. The sudden closure of 1 channel, for example, would reduce the overall conductance by only 1%–2%, whereas in an isolated cell this event would have caused a 100% loss of conductance. It would appear therefore that cell-to-cell junctional channels are essential for maintaining a stable resting potential in cells with small numbers of K$^+$ channels, such as the exocrine acinar cells (91). Many cells sharing their small number of K$^+$ channels can therefore overcome the threat of an unstable and unpredictable membrane potential arising from the stochastic nature of channel behavior.

ELECTROGENIC PUMPS

The resting membrane potential is due to the large transmembrane K$^+$ gradient and the existence of resting K$^+$ channels. The transmembrane K$^+$ gradient is established and maintained by the ATP-driven Na$^+$-K$^+$ pump. The transmembrane Na$^+$ gradient, which results from the operation of the Na$^+$-K$^+$ pump together with the membrane potential, is used as the driving force for concentrative uptake of a number of amino acids (Fig. 9). In the steady-state situation, K$^+$ recirculates via the Na$^+$-K$^+$ pump and the K$^+$ channel,

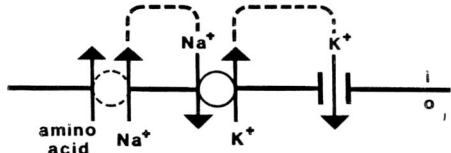

FIG. 9. Simplified diagram for transport of amino acids and cations across basolateral plasma membrane of acinar cells. Na⁺–amino acid cotransport system is coupled to Na⁺-K⁺ pump that is closely linked to the K⁺ channels. i, Inner surface of plasma membrane; o, outer surface of plasma membrane. [From Singh and Petersen (115).]

whereas Na⁺ recirculates via the Na⁺-K⁺ pump and the Na⁺–amino acid cotransporter. All three transport processes illustrated in Figure 9 contribute directly to the membrane potential (electrogenic) or generate current under voltage-clamp conditions (rheogenic).

Sodium-Potassium Pump

The Na⁺-K⁺ pump is electrogenic (or rheogenic), because it extrudes 3 Na⁺ in exchange for 2 K⁺. In the resting acinar cell, however, the direct pump contribution to the resting potential is small (a few millivolts), as shown when the pump is suddenly arrested by acute application of a maximal dose of ouabain (86). However, when the pump is working at maximal intensity in Na⁺-loaded cells, ouabain application may result in a depolarization of 20–30 mV (86). The direct electrogenic character of the pump is best seen when previously blocked Na⁺-K⁺ pumps are suddenly and maximally reactivated. This can be achieved by exposing pancreatic cells to K⁺-free solutions for a long period (e.g., 1 h) and thereafter readmitting K⁺ to the extracellular solution. In the absence of external K⁺ the pump cannot work, and the cells will therefore accumulate Na⁺ and lose K⁺. The internal Na⁺ site of the pump will be fully activated, and in the moment when extracellular K⁺ is readmitted maximal pump operation occurs (86, 107).

Sodium–Amino Acid Cotransport

Sodium gradient–driven amino acid transport is an important mechanism for cellular uptake, and such a process has been directly demonstrated for L-alanine in membrane vesicles isolated from the cat pancreas (122). Iwatsuki and Petersen (40, 41) first demonstrated that neutral amino acids, such as L-alanine and L-valine, when added to the bathing fluid of isolated pancreatic segments, evoked marked membrane depolarization and some increase in membrane conductance. The depolarization evoked by L-alanine is stereospecific and is half-maximal at a concentration of 1.2 mM in the superfusion fluid. The amplitude of the L-alanine–evoked depolarization is markedly reduced when the tissue is exposed to nominally Na⁺-free solutions, and this effect is fully reversible (40).

These results indicate that when L-alanine is present in the extracellular fluid a Na⁺ conductance pathway is opened, which is the result expected from the known presence of the Na⁺–L-alanine cotransport system.

It has been possible to study the L-alanine–evoked current under voltage-clamp conditions, maintaining control also over the intracellular ionic concentrations, by using the patch-clamp method for whole-cell current recording from isolated cells (see Fig. 3). The time course of transmembrane current after external addition and removal of 1 mM L-alanine in the presence of 145 mM NaCl on both sides of the membrane is shown in Figure 10. Addition of 1 mM methylaminoisobutyric acid (MeAIB) resulted in a somewhat smaller current. In the presence of 20 mM MeAIB, L-alanine was virtually ineffective, indicating that L-alanine and MeAIB compete for the same transport site. Because N-methylated amino acids only bind to the "A" system (9), this finding indicates that L-alanine is also transported by this system.

The experiment in Figure 10, *bottom*, was carried out under conditions similar to those in the experiment in Figure 10, *top*, but the cells were exposed for a longer time to external L-alanine. In this case the current rose to a maximum and thereafter declined to zero in the continued presence of the amino acid. When the external medium was changed to an L-alanine–free solution, a transient outward current was seen. In the first part of the experiment, inward L-alanine movement occurs until internal and external concentrations have become approximately equal (at least in a compartment close to the plasma membrane). After removal of external L-alanine the gradient is reversed, and L-alanine moves out leading to an outward-directed Na⁺ current.

In patch-clamp whole-cell current recording experiments, the current under "zero-trans" conditions (i.e., with varying finite concentrations of Na⁺ and L-ala-

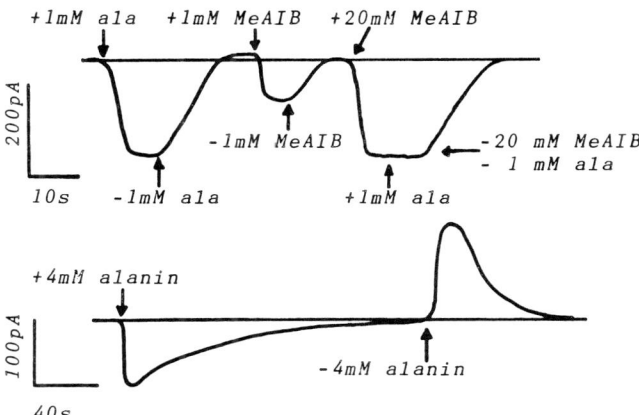

FIG. 10. Whole-cell recording from isolated mouse pancreatic acinar cell with Na⁺-rich solutions on both sides of membrane. Transmembrane current at membrane voltage of 0 mV after external addition and removal of L-alanine (ala) or N-methylaminoisobutyric acid (MeAIB). Inward currents are plotted downward. [From Jauch et al. (44).]

nine on the outside and zero concentrations on the inside) exhibits Michaelis-Menten behavior. The half-saturation concentration of L-alanine at the external site decreases from 18 mM at $[Na^+]_o = 5$ mM to 2.9 mM at $[Na^+]_o = 150$ mM. The half-saturation concentration of Na^+ decreases from 63 mM at $[L\text{-}Ala]_o = 1$ mM to 14 mM at $[L\text{-}Ala]_o = 10$ mM. From reversal potential measurements in the presence of various transmembrane L-alanine and Na^+ concentration gradients, a Na^+/L-alanine stoichiometric ratio of 1:1 could be inferred (45).

All the neutral amino acids evoke similar electrical effects, and the null potentials for the actions of L-alanine, L-phenylalanine, L-valine, L-proline, and glycine are the same (40, 41, 51). Interaction experiments demonstrate mutual inhibition of depolarization responses. For example, the depolarizing action of L-proline is much reduced in the presence of L-alanine even when the depolarizing effect of L-alanine has been canceled by direct injection of hyperpolarizing current (40, 51).

The action of the basic amino acid L-arginine is different from that of the neutral amino acids. The maximal depolarization evoked by supramaximal L-arginine concentrations is much smaller (~4 mV) than that observed for L-alanine (~18 mV), and the null or reversal potential for L-arginine ($E_{L\text{-}Arg}$) is much less positive than that for L-alanine ($E_{L\text{-}Ala}$). In experiments in which $E_{L\text{-}Ala}$ and $E_{L\text{-}Arg}$ were both measured with the same electrodes in the same acinar units, the values were +30 mV for L-alanine and +7 mV for L-arginine (51). L-Arginine is also unable to inhibit L-alanine–evoked depolarizations, and similarly L-alanine fails to reduce L-arginine–evoked membrane depolarization, although in the same experiments mutual inhibition between the different neutral amino acids can be readily demonstrated (51). Although it is clear that L-arginine opens up a membrane pathway with a different permeability from the one activated by the neutral amino acids, there is no positive evidence for the ionic mechanisms underlying the L-arginine–evoked depolarization. In addition to the amino acids already mentioned, a number of other amino acids such as the neutral asparagine, glutamine, hydroxyproline, isoleucine, leucine, serine, tryptophan, tyrosine, and methionine; the basic histidine, lysine, and ornithine; and the acidic aspartic acid, glutamic acid, cysteine, and cystine all evoke membrane depolarization in varying degrees. Detailed dose-response curves and ionic mechanisms have, however, not been investigated (90).

MEMBRANE EFFECTS OF STIMULANTS

The original work of Lundberg (53–58) indicated that secretagogues evoke hyperpolarization of the acinar cell membrane. Later a more varied picture emerged that also included depolarizing and multiphasic responses in some gland preparations (88). In

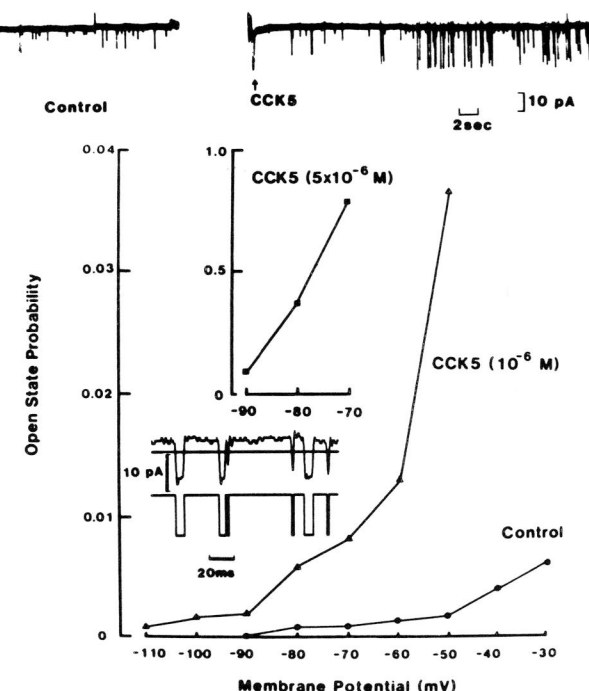

FIG. 11. Single-channel current recording from basolateral membrane patch in pig pancreatic acinar cell. All records were obtained in cell-attached configuration. Recording pipette was filled with K^+-rich (intracellular) solution containing 2.5 mM Ca^{2+}, whereas bath was filled with Na^+-rich (extracellular) solution. Resting membrane potential was about −60 mV. *Upper traces*, recording made on slow time base. In control situation there are only a few inward current steps. Application of 10^{-6} M cholecystokinin-5 (CCK-5) to bath evokes, after latency of ~18 s, a clear and sustained increase in frequency of channel openings (inward current steps). Single-channel current amplitude is also enhanced because of membrane hyperpolarization (to about −80 mV). *Lower graphs*, plots of channel open-state probability as function of membrane potential in control situation and during stimulation with 10^{-6} M and 5×10^{-6} M CCK-5. *Inset*, single-channel current recorded on fast time base together with idealized current trace obtained from computerized threshold analysis of digitized data (*bottom trace*). *Inset trace* was obtained in presence of CCK-5 (10^{-6} M) at membrane potential of −50 mV. [From Suzuki et al. (118).]

the majority of cases it would now appear that the electrical responses to secretagogues can be explained by a combination of K^+ channel and Cl^- channel activation, although in some preparations nonselective cation channels play an important role. In this section evidence is presented for secretagogue activation of the various ion channels, and membrane potential changes in the intact cells are explained on the basis of the known channel properties.

Activation of Potassium Channels

SINGLE-CHANNEL CURRENT RECORDING IN CELL-ATTACHED CONFIGURATION. The most direct evidence for stimulation of Ca^{2+}- and voltage-activated high-conductance K^+ channels by secretagogues is from single-channel current recording experiments using the cell-attached configuration. Figure 11 shows such an experiment on an isolated pig pancreatic acinar cell cluster. In the control period before stimulation there were only infrequent and short-lasting channel openings. Within 10–20 s after the start of stimulation with the cholecystokinin-gastrin peptide CCK-5 (hormone is added to the bath so there is no direct contact between the hormone and the channel from which recording is made), a marked increase in the frequency of channel openings was observed, and this higher frequency of channel opening (activation) was maintained as long as the hormonal stimulation continued. The long latency may relate to the known restrictions of intracellular Ca^{2+} diffusion (2). Figure 11 also shows the relationship between the channel open-state probability and the membrane potential in the control period and in the presence of two different CCK-5 concentrations. There is a marked CCK-5–evoked dose-dependent increase in open-state probability and also a marked voltage sensitivity. Thus CCK-5 has acted to increase the open-state probability of the Ca^{2+}- and voltage-activated high-conductance K^+ channel. The evidence for the channel classification is as follows (118). *1*) The relationship between single-channel amplitude and change in membrane potential is identical before and after CCK-5 stimulation with a slope corresponding to a unit conductance of 200–250 pS. *2*) The channel is voltage dependent with depolarization increasing the open-state probability in the absence as well as the presence of CCK-5. *3*) In a low concentration (5 mM), TEA acting from the outside of the membrane blocks channel openings both before and after stimulation. *4*) The CCK-5–evoked increase in channel open-state probability is only sustained in the presence of Ca^{2+} on the outside of the membrane patch from which the channel recording is made.

Activation of K^+ channels by ACh (10^{-5} M) has been demonstrated in submandibular acinar cells (22); in lacrimal acinar cells, Marty et al. (59) have shown activation of K^+ channels by bath application of the Ca^{2+} ionophore A23187 (0.2 µM).

ACTIVATION OF WHOLE-CELL CURRENT BY SECRETAGOGUES. Stimulation with CCK enhances the voltage-activated outward K^+ current in pig pancreatic acinar cells as illustrated in Figure 12. The acinar intracellular compartment is equilibrated with the K^+-rich pipette solution containing 0.5 mM EGTA and no added Ca^{2+}. Thus there is a stable resting situation in which depolarizing voltage pulses, from the −40 mV holding potential, evoke large outward K^+ currents, whereas hyperpolarizing voltage pulses are associated with very little inward current. Stimulation with 10^{-6} M CCK-5 or 2×10^{-10} M CCK-8 (67) causes a sustained increase in the voltage-activated outward currents and a small increase in the inward current. Application of TEA (5 mM) to the bath reduces the outward currents drastically (below the control level) but has no effect on the inward current. Subsequent removal of the stimulant hormone reduces the small inward current.

The effect of CCK on the pancreatic acinar cells [Fig. 12; (67)] or the similar effect of ACh on lacrimal acinar cells (14, 19, 121) must be due to intracellular Ca^{2+} activating the high-conductance K^+ channels, and the evidence for this can be summarized as follows. *1*) The stimulant-evoked increase in outward K^+ current is blocked by equilibrating the cell interior with a high concentration (10 mM) of the Ca^{2+} chelator ethylene glycol-bis(β-aminoethylether)-N,N'-tetraacetic acid (EGTA) (67). *2*) The sustained stimulant-evoked increase in outward K^+ current is abolished by removal of external Ca^{2+} (67). *3*) Ensemble fluctuation analysis suggests that the additional outward K^+ current evoked by stimulation is due to the opening of high-conductance K^+ channels (121).

Activation of Chloride Channels

In addition to the large increase in outward current evoked by CCK in pancreatic acinar cells, there is also a small increase in inward current (Fig. 12). The inward current has not been studied in detail in pancreatic or salivary acinar cells, but in the lacrimal acinar cells, stimulation with ACh also evokes two currents—an outward and an inward current. The outward current is carried by K^+, whereas the inward current is carried by Cl^-, as it is not seen when extra- and intracellular Cl^- is replaced by SO_4^{2-}. This inward current is also activated by Ca^{2+}, as a high intracellular EGTA concentration abolishes both the outward and inward current responses (19). Therefore ACh activates K^+ and Cl^- conductance pathways, and these effects are mediated by an increase in $[Ca^{2+}]_i$. The precise localization of the Cl^- conductance pathway is unknown, but, as discussed in ACINAR CELL MODEL, p. 45, it is most likely confined to the luminal plasma membrane.

There are no patch-clamp studies of Ca^{2+}-activated Cl^- currents in salivary or pancreatic acinar cells, but the striking similarity between electrophysiological responses in lacrimal acinar cells and salivary acinar cells (89) suggests that the lacrimal gland may be a useful model for the discussion of exocrine acinar cell Cl^- currents.

Marty and co-workers (11, 12, 59) studied Cl^- currents and their response to carbachol and the Ca^{2+} ionophore A23187 with the patch-clamp technique (whole-cell recording) in isolated rat lacrimal acinar cells. All K^+ currents were blocked by using K^+-free internal solutions containing Cs^+ or Na^+ and/or by using TEA as a blocking agent for K^+ channels. When internal or external Cl^- concentrations were changed by partial or full replacement with glutamate or isethionate, the stimulant-evoked current always reversed at the calculated values for the Cl^- equilibrium potential [E_{Cl^-}; (59)].

The stimulant-evoked Cl^- current was accompanied by a small noise increase (59). The relationship between noise variance and mean current provides an estimate of the elementary (single-channel) current,

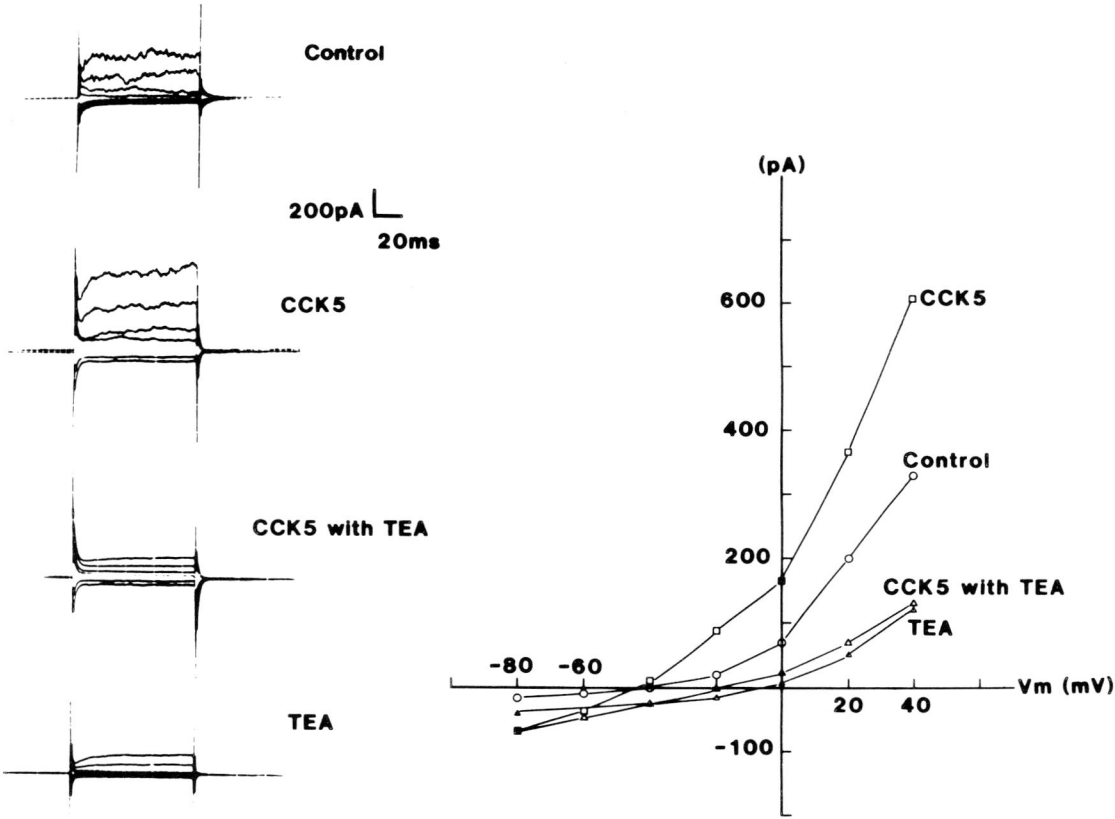

FIG. 12. Whole-cell voltage-clamp current recordings from single pig pancreatic acinar cell. Bath contained extracellular Na$^+$-rich solution and pipette was filled with intracellular K$^+$-rich solution without Ca^{2+} (containing 0.5 mM EGTA). Currents associated with depolarizing voltage steps are shown as upward deflections (outward current) and with hyperpolarizing steps as downward deflections. Holding potential was −40 mV, and 90 ms voltage steps to −20, 0, +20, and +40, as well as −60, −80, and in one case −100 mV, were applied. *Control*, currents recorded before stimulation; *CCK-5*, 3 min after start of continued exposure to 10^{-6} M CCK-5; *CCK-5 with TEA*, 3 min after addition of 5 mM TEA still in presence of CCK-5; *TEA*, 3 min after discontinuation of CCK-5 stimulation but still in presence of TEA. Relationship between steady-state currents and membrane potential in different experimental situations obtained from displayed current traces is shown in graph. [From Suzuki et al. (118).]

if it can be assumed that the total response results from the opening of a homogeneous population of channels that are independent of each other. Such an analysis indicates that the Cl$^-$ channel activated by the Ca^{2+} ionophore A23187 or by carbachol has a unit conductance of ~1–2 pS (59). The Cl$^-$ channel density was also calculated and found to be ~5,000–20,000 channels/cell (59). Thus the Ca^{2+}-activated Cl$^-$ channels are far more numerous than the Ca^{2+}-activated high-conductance K$^+$ channels for which the density is only 50–150 channels/cell (62, 69, 121).

In single-channel current recording studies on cell-attached patches (59), small inward current steps were evoked by adding the Ca^{2+} ionophore A23187 to the bath solution. True single-channel events could rarely be identified, but noise analysis indicated that the channels responsible for the inward current (presumably carried by Cl$^-$) had a unit conductance of ~1–2 pS in agreement with the noise analysis of the whole-cell currents (59).

Further studies on the Ca^{2+} activation of the Cl$^-$ currents have been carried out with the whole-cell voltage-clamp technique, where the cell interior was equilibrated with well-buffered Ca^{2+} solutions so that the precise relationship between [Ca^{2+}]$_i$ and the Cl$^-$ current could be evaluated (11). The Cl$^-$ conductance was negligible at [Ca^{2+}]$_i$ = 10^{-7} M and only increased slightly up to 5 × 10^{-7} M but thereafter increased steeply to become fully activated at 2 × 10^{-6} M (11). This is in marked contrast to the K$^+$ current, which is highly activated by changing [Ca^{2+}]$_i$ from 5 × 10^{-8} M to 10^{-7} M (see Fig. 8).

The selectivity of the Cl$^-$ conductance pathway was also investigated in anion substitution experiments, and the following permeability sequence was obtained: I$^-$ > NO$_3^-$ > Br$^-$ > Cl$^-$ > F$^-$ > isethionate, methanesulfonate > glutamate (11). Finally, it has been shown that the diuretic agent furosemide (1 mM) can markedly reduce the Ca^{2+}-activated Cl$^-$ conductance without entirely abolishing it (12).

Activation of Nonselective Cation Channels

Single-channel current recording with the patch-clamp technique has helped to identify and characterize a Ca^{2+}-activated channel in the basolateral membrane of mouse and rat pancreatic acinar cells that can be activated by secretagogues such as CCK and ACh. In these cells there are no detectable single-channel currents in the resting situation, but local application of CCK or ACh to the area surrounding the membrane patch isolated by the recording micropipette evokes clear unitary current steps (Fig. 13), and these effects can be blocked by the appropriate antagonist (65). In this type of experiment the agonist cannot gain access to the isolated membrane patch itself, and the only plausible explanation for the ability of the agonist to open channels in an area to which it has no access is that an intracellular messenger mediates the action (65, 95). Because unitary current activity of the same type can also be observed after excision of the membrane patch and exposure of the inner aspect of the membrane to a high Ca^{2+} concentration (Fig. 13), it seems likely that the effects of CCK and ACh in opening channels are indeed mediated in the intact cell by an increase in $[Ca^{2+}]_i$.

The unitary currents shown in Figure 13 go through nonselective cation channels. In excised membrane patches the magnitude of single-channel currents and the pattern of channel opening and closing are independent of the presence or absence of Cl^- (64, 95). In a situation with quasi-physiological cation gradients, i.e., a Na^+-rich extracellular solution on the outside

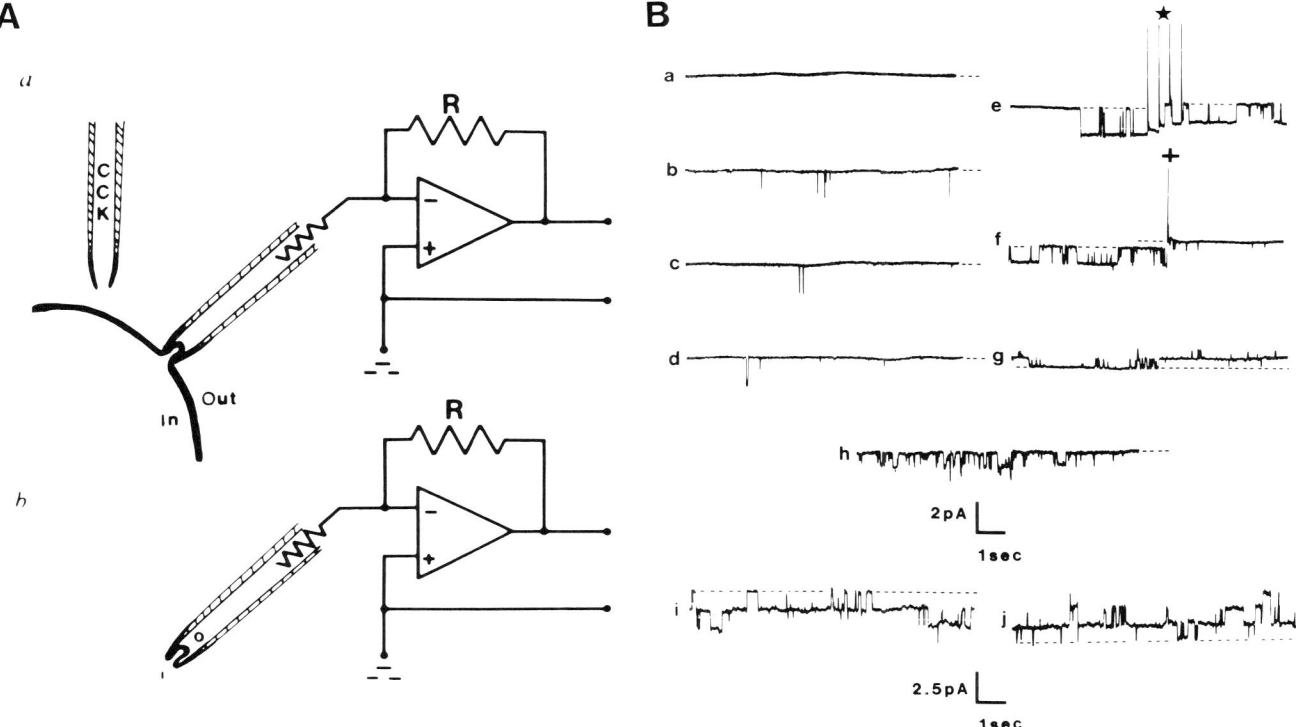

FIG. 13. Activation of single-channel current in mouse pancreatic acinar cell by local CCK application. A: schematic of experimental system. a: Recording from in situ membrane patch. Recording pipette was always filled with extracellular bath solution. No agonist was present in pipette solution. *CCK pipette* contained the octapeptide CCK-8 in concentration of 5 µM. Spontaneous diffusion of CCK from tip of micropipette was used as means of stimulation. b: After in situ recording patch-clamp pipette was withdrawn from acinus to obtain excised inside-out membrane patch. Single-channel recordings were made in symmetrical saline solutions or bath fluid was changed to one having an intracellular composition. B: a–h: consecutive traces from same membrane patch. a: Recording situation as described in A but with CCK pipette far away from acinus under investigation; pipette potential +40 mV. b–d: Three continuous traces obtained 20, 30, and 40 s, respectively, after tip of CCK pipette had been brought close to patch pipette; pipette potential +40 mV. e: 50 s after start of CCK stimulation. *Star*, potential of patch pipette was changed (in steps of 5 mV) from +40 to +20 mV. f: Pipette potential +20 mV. *Plus sign*, potential was changed to 0. g: Pipette potential −40 mV. h: currents from same patch after excision so now inside out [recording situation as shown in A (b)]. Symmetrical extracellular saline solutions; pipette potential +50 mV. i, j: Currents recorded from another excised (inside-out) membrane patch with intracellular solution (high K^+, low Na^+) in bath and normal extracellular solution (high Na^+, low K^+) in pipette. Pipette potential +60 mV in i and −60 mV in j. *Dashed lines*, current level when all channels are closed. Downward deflections represent current from extracellular to intracellular side of membrane. [From Maruyama and Petersen (65). Reprinted by permission from *Nature*, copyright 1982, Macmillan Journals Limited.]

and a K^+-rich intracellular solution on the inside of the membrane patch, the amplitude of the single-channel currents is independent of the polarity of the membrane potential, and the null potential is 0 mV. The single-channel current-voltage relationship is linear, with a slope corresponding to a single-channel conductance of ~30–35 pS. The conductance and the null potential are the same in symmetrical Na^+/Na^+, K^+/K^+, or asymmetrical Na^+/K^+ solutions, demonstrating that the channel does not discriminate between these two ions (64, 95). If all Na^+ or K^+ is replaced by a divalent cation, for example Ca^{2+}, no unitary current activity can be observed (95, 96). The channel is therefore a nonselective monovalent cation channel, but this does not exclude the existence of a small Ca^{2+} leak (96). The conductance and the null potential of the channel activated by CCK or ACh in the intact cell match exactly those of the Ca^{2+}-activated channel in the excised patch (65).

The Ca^{2+} sensitivity of the channel has been investigated in excised inside-out patches. In a steady-state situation, micromolar Ca^{2+} concentrations in the solution bathing the membrane inside are needed to observe any channel openings, and for near-maximal activation, concentrations >100 µM are required, indicating that the nonselective channel in the mouse pancreas is relatively insensitive to internal Ca^{2+} (64, 68). This conclusion, however, may not necessarily be correct, because very marked channel activation can be observed when an isolated patch membrane is excised and the inner membrane aspect is acutely exposed to a solution having $[Ca^{2+}]_i = 5 \times 10^{-7}$ M. This activation, however, is only transient. Either the channels become desensitized or some essential factor loosely attached to the cytoplasmic side of the membrane is washed out after the excision. Several minutes after the excision, 10 µM Ca^{2+} on the inside of the membrane only causes modest channel activation compared with the effect of 5×10^{-7} M immediately after excision. Further studies employing saponin to make the plasma membrane of intact cells leaky (open cell-attached configuration), therefore allowing equilibration between extra- and intracellular fluid, have shown that single-channel activity can be evoked by exposure to solutions with $[Ca^{2+}]_i = 10^{-7}$ M (68). The Ca^{2+}-activated nonselective cation channel in the intact mouse and rat acinar cell may therefore be controlled by variations in $[Ca^{2+}]_i$ in the range of 5×10^{-8}–10^{-6} M, as in the case of the larger voltage-gated K^+ channel.

In mouse and rat pancreatic acinar cells it has been estimated that there may be ~500 channels/cell (95), which is a somewhat higher density than that estimated for the high-conductance K^+ channel [50–150 per cell; (62, 69)] but considerably smaller than the number of Cl^- channels/cell [5,000–20,000; (59)].

Even though the first and most detailed studies of the Ca^{2+}-activated nonselective cation channel have been on mouse and rat pancreatic acinar cells (64, 65), this same channel has also been reported to exist in salivary acinar cells (61) and lacrimal acinar cells (59). From the studies on lacrimal acinar cells, Marty et al. (59) have concluded that activation of the nonselective cation channel only occurs when $[Ca^{2+}]_i$ attains values >1 µM.

In those glands where the basolateral acinar cell membrane is dominated by the high-conductance K^+ channels (pig pancreas and all salivary glands), there is no indication that stimulation with hormones or neurotransmitters evokes the opening of nonselective cation channels. In cell-attached single-channel recording experiments of the type shown in Figure 11, there was only activation of high-conductance K^+ channels (22, 118). In the lacrimal gland, activation of nonselective channels has been demonstrated in similar experiments but the Ca^{2+} ionophore A23187 rather than physiological stimulation was employed (59), and consequently $[Ca^{2+}]_i$ may have been raised to much higher levels than those attained with neurotransmitters or hormones. In the mouse and rat pancreatic acinar cells, where stimulation with CCK has been shown to activate the nonselective cation channel (65), the high-conductance K^+ channel would appear to be totally absent (97).

Stimulant-Evoked Membrane Potential Changes

A bewildering variety of membrane potential changes in response to physiological types of stimulation have been recorded with conventional intracellular microelectrodes in acinar cells from different exocrine glands and different species. They range from simple hyperpolarizations in pig pancreatic acinar cells and cockroach salivary gland cells (24, 25, 81) to mixed and complex depolarization-hyperpolarization responses in mammalian salivary glands (23, 108, 109), multiphasic hyperpolarizations in lacrimal glands (36, 82), and simple depolarizations in mouse and rat pancreatic acinar cells (33, 34).

There is now little doubt that the simple hyperpolarizing responses are due to the opening of K^+ conductance pathways. By inserting two microelectrodes into neighboring communicating acinar cells (see Fig. 2), it is possible to use one electrode for measurement of membrane potential and the other for passing current. Employing a protocol with repetitive injection of hyperpolarizing current pulses, records can be obtained where effectively the stimulant-evoked membrane potential change is monitored simultaneously starting from two different levels of resting potential (Fig. 14). From such records the null potential for the transmitter action can be worked out. In the cockroach salivary gland acinar cells the null potential for the action of the stimulant transmitter dopamine (E_{dop}) corresponds to the K^+ equilibrium potential (E_{K^+}) over a range of extracellular K^+ concentrations. Similarly in the pig pancreatic acinar cells, the null potential for the action of pentagastrin was found to correspond

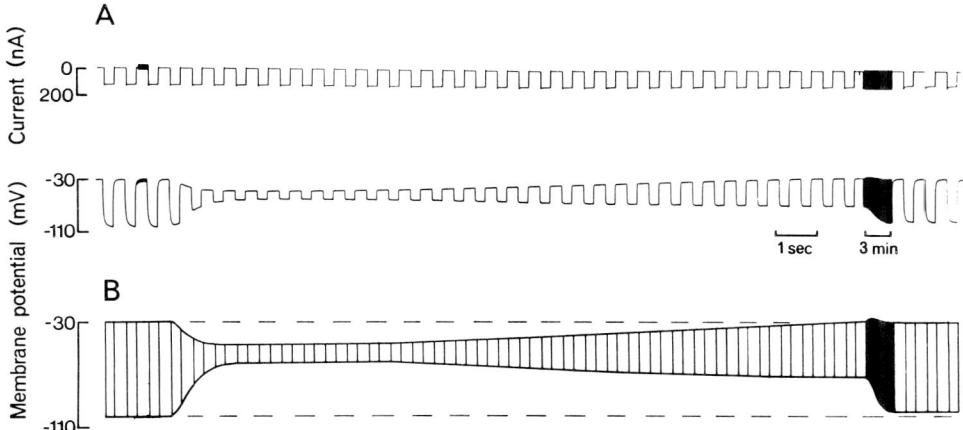

FIG. 14. Conductance change during response to nerve stimulation using two intracellular microelectrodes in neighboring cells of cockroach salivary gland acinus (see Fig. 2). *A: upper trace*, current pulses; *lower trace*, electrotonic potentials. Period of stimulation indicated by thickening of traces due to stimulus artifacts. Extracellular K⁺ concentration = 10 mM. *B*: upper and lower envelopes of voltage trace in *A*. [From Ginsborg et al. (25).]

almost exactly to E_{K^+}, both at $[K^+]_o = 4.7$ mV and at $[K^+]_o = 47$ mV (69). In these cases therefore stimulation of acinar cells with a hormone or a neurotransmitter has selectively increased the membrane K⁺ conductance, and in the pig pancreatic acinar cells, where extensive patch-clamp work has been carried out, this can be fully explained by activation of high-conductance K⁺ channels in the basolateral membrane (see Fig. 11). There is also good agreement between the microelectrode data and the patch-clamp data with regard to the involvement of Ca^{2+}. In patch-clamp experiments it was shown that external Ca^{2+} is needed for the sustained stimulant-evoked increase in outward K⁺ current from isolated acinar cells (67). Figure 15 demonstrates that this is also the case in classic microelectrode recordings of membrane potential changes in response to sustained ACh stimulation. In the presence of external Ca^{2+}, ACh evokes a sustained hyperpolarization, whereas the first ACh response after external Ca^{2+} deprivation only results in a transient hyperpolarization, and thereafter no ACh responses can be obtained.

In the salivary gland acinar cells, ACh and epinephrine evoke a marked reduction in input resistance (Fig. 16). Recording with two separate indwelling microelectrodes showed that the resistance reduction occurs to the same extent in two neighboring cells, directly demonstrating that it is due to a decrease in plasma membrane resistance (108). In many cases the resistance decrease is so dramatic that the electrotonic potential changes set up by current injection virtually disappear (Fig. 16). After the initial membrane resistance reduction there is a hyperpolarization not associated with any marked resistance change. The null potentials for the initial action of ACh (E_{ACh}) and epinephrine (E_{epi}) are the same, with a value of about −60 to −65 mV in both the mouse and rat parotid

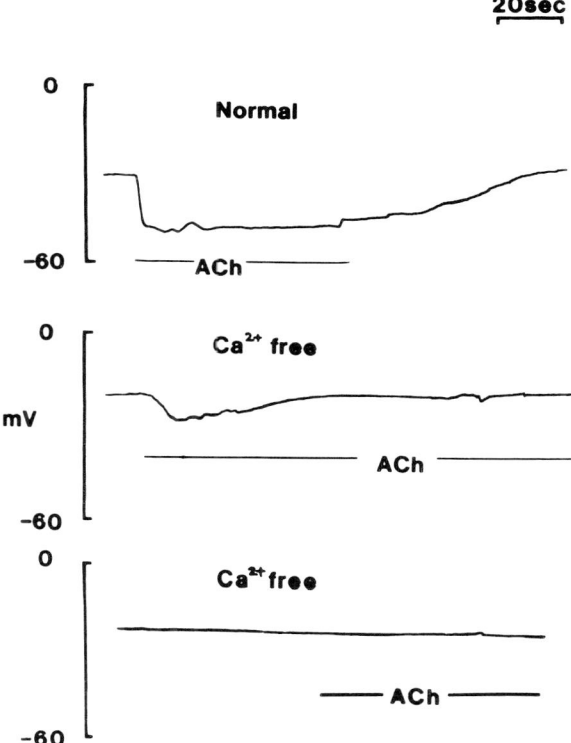

FIG. 15. Intracellular microelectrode recording from pig pancreatic acinar cell. Effects of external Ca^{2+} removal on ACh-evoked membrane potential changes. *Upper trace*, recording during control conditions with normal extracellular Ca^{2+} levels in superfusion fluid. *Lower 2 traces*, part of continuous recording made 20 min after switch to superfusion with Ca^{2+}-free solution containing Ca^{2+}-chelating agent EGTA (10^{-4} M). *Horizontal bars*, period of 5×10^{-7} M ACh superfusion. [From Pearson et al. (81).]

glands (108, 109). This value is less negative than in the pig pancreatic acinar cells [about −85 mV (69)] but nevertheless indicates a marked permeability of

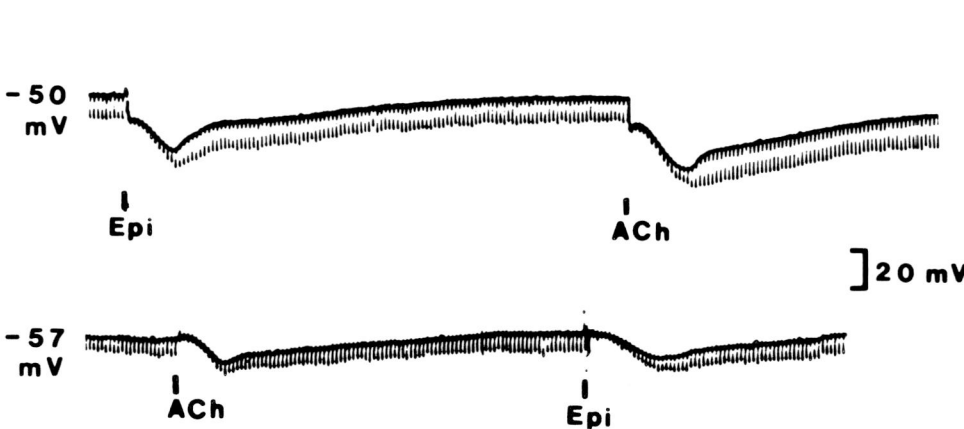

FIG. 16. Effects of acetylcholine (ACh) and epinephrine (Epi), applied by microionophoresis, on membrane potential and resistance of mouse parotid acinar cells. Resting potential of each cell is written to left of its potential recording. Pulses on potential record are produced by passage of rectangular current pulses (2 nA, 100 ms, 1 s^{-1}) through recording microelectrode. Shorter intervals on time marker trace (*top*) represent 1 s. [From Roberts and Petersen (109).]

the receptor-activated channels to K^+. The value for E_{ACh} is roughly midway between those for E_{Cl^-} and E_{K^+}. In the absence of Cl^-, E_{ACh} attained a mean value of about -80 mV. In individual cases, values of -90 mV were observed (108). Furthermore, during continued exposure of acinar cells to Cl^--free solution, the slope of the linear curve representing the membrane potential as a function of $[K^+]_o$ during ACh stimulation is close to the theoretically expected value for a K^+-selective electrode (78). Because E_{ACh} under normal ionic conditions is less negative than E_{K^+}, it would appear that stimulation has opened up a Cl^- conductance pathway in addition to the K^+ channels.

Although the initial stimulant-evoked potential change can be adequately explained by the opening of K^+ and Cl^- channels in the plasma membrane, in good agreement with the patch-clamp data, this is not so for the delayed hyperpolarization (Fig. 16). It has proved impossible to reverse the secondary hyperpolarization by making the membrane potential very negative prior to a stimulus (108, 109). Furthermore the secondary hyperpolarization is specifically blocked by ouabain and exposure to Na^+-free or K^+-free perfusion solutions (108). All these findings can be explained most easily by the secondary hyperpolarization being due to activation of the electrogenic Na^+-K^+ pump. The opening of K^+ channels in the basolateral plasma membrane and Cl^- channels in the luminal membrane enables Na^+-K^+-$2Cl^-$ cotransport into the acinar cell through the basolateral membrane (see IMPORTANCE OF ELECTROGENIC PROCESSES FOR ACINAR CELL FUNCTION, p. 44). It is likely therefore that there will be an increase in $[Na^+]_i$ and that the Na^+-K^+ pump will be activated by this stimulant-evoked increase in intracellular Na^+ concentration.

All the pancreatic secretagogues (ACh, CCK-gastrin peptides, and bombesin peptides) evoking hyperpolar-

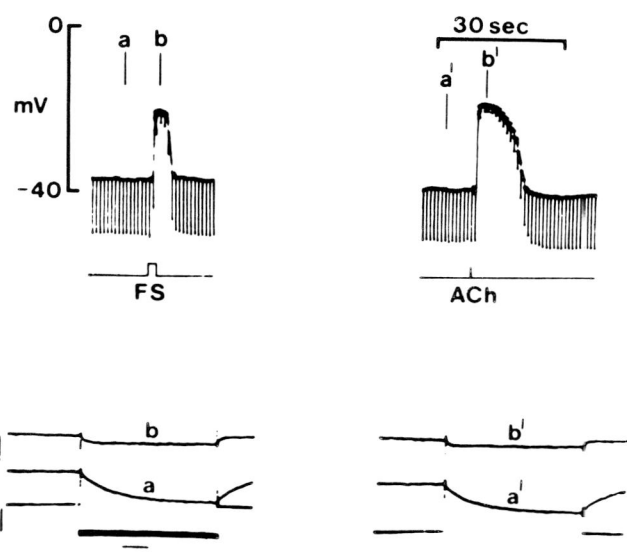

FIG. 17. Comparison between effect of nerve stimulation (FS) and local acetylcholine (ACh) application on rat pancreatic acinar cell membrane potential and resistance. *Upper traces*, pen recordings; *lower traces*, oscilloscope photographs taken at times indicated in pen recording. Calibration in oscilloscope photographs: vertical, 10 mV and 1 nA; horizontal, 20 ms. [From Petersen (90).]

ization in pig acinar cells (Fig. 15) evoke depolarization in the mouse and rat [Fig. 17; (37, 89, 90)]. The ionic mechanism underlying the stimulant-evoked depolarization and conductance increase has been investigated, and important information about the characteristics of the conductance pathway opened by stimulation has been revealed in two microelectrode experiments in which E_{ACh} has been measured under different ionic conditions. In addition to increasing the membrane permeability to Na^+ and K^+, ACh also evokes a marked increase in Cl^- permeability. Iwatsuki and Petersen (34) showed that the resistance of the ACh-controlled ionic pathways doubled after Cl⁻ removal (Cl⁻ replaced by sulfate). The reversal potential for the ACh-evoked macroscopic current was about 0 to +10 mV in the absence of Cl⁻, whereas it was about −15 to −20 mV under control conditions (34, 66, 100). On the basis of these measurements and the intracellular ionic activities determined by ion-selective electrodes, Petersen et al. (98) calculated that the ratio of Na^+ and K^+ permeability (P_{Na^+}/P_{K^+}) of the ACh-controlled pathway is close to unity but that P_{Cl^-} is about five times higher than P_{Na^+} or P_{K^+}. Thus in the absence of Cl⁻, the ACh-evoked macroscopic current could at least qualitatively be explained entirely in terms of the sum of microscopic currents through the Ca^{2+}-activated nonselective cation channels (95). The anionic selectivity sequence of the ACh-activated Cl⁻ conductance pathway in mouse pancreatic acinar cells, as determined from changes in E_{ACh} with various anionic substitutions (100), is very similar to that from patch-clamp whole-cell current studies in rat lacrimal acinar cells (11), but there are no patch-clamp studies of Cl⁻ channels in mouse and rat pancreatic acinar cells.

Stimulation-Evoked Changes in Membrane Capacitance

The final step of enzyme discharge from the luminal membrane in exocrine gland acinar cells is commonly known to occur by exocytosis and endocytosis, which should lead to a transient increase and decrease of membrane surface area and therefore membrane capacitance.

Measurement of small capacitance changes with stimulation has been attempted in single acinar cells from rat pancreas (60) in which an increase in cytosolic Ca^{2+} undoubtedly enhances exocytotic enzyme discharge (112). Whole-cell dialysis with a quasi-intracellular solution containing EGTA evoked no significant fluctuation of membrane capacitance, whereas dialysis with a solution containing 5×10^{-7} M Ca^{2+} (ATP and Mg^{2+} were also present) induced steplike capacitance changes without any obvious correlation with changes in membrane conductance (Fig. 18). The mean amplitude of on-steps (probably responsible for exocytosis) was 9 ± 3 fF ($n = 50$, where n is the number of experiments) and that of off-steps (responsible for endocytosis) was 10 ± 4 fF ($n = 46$), corresponding to vesicle spherical diameters of ~0.5 μm. Similar steplike changes have also been reported in chromaffin cells (75) and mast cells (13) and correspond to the images of exocytosis and endocytosis. Although capacitance measurements using the phase-sensitive detection method in combination with whole-cell dialysis seem to be applicable to exocrine acinar cells, whole-cell dialysis can easily wash out essential substances needed for various physiological responses involving exocytotic and endocytotic proc-

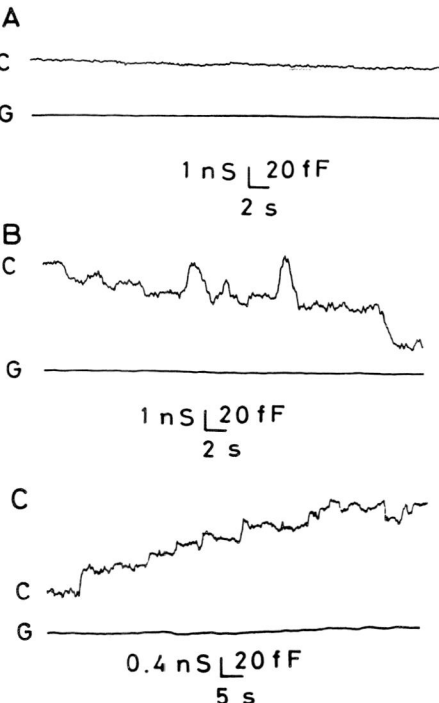

FIG. 18. Patch-clamp whole-cell recording of Ca^{2+}-evoked capacitance changes in isolated rat pancreatic acinar cells. Cell was dialyzed with K^+-glutamate solution containing Mg^{2+} and ATP, and Ca^{2+}-EGTA buffered solutions were used to attain specified free Ca^{2+} concentrations inside cell. In each of 3 records (*A*, *B*, and *C*), trace labeled *C* represents capacitance, whereas trace labeled *G* represents conductance. *A*: Ca^{2+} concentration in cell $<10^{-9}$ M (nominally Ca^{2+}-free solution, 1 mM EGTA); *B* and *C*: Ca^{2+} concentration in cell = 5×10^{-7} M (1 mM Ca^{2+}, 1.2 mM EGTA). [From Maruyama (60).]

esses. Searches for the substances needed for maintaining the natural internal cell environment in acini during dialysis remain for future investigations.

Source of Calcium Used as Messenger for Stimulant-Evoked Membrane Changes

There is evidence from conventional intracellular microelectrode studies that there are two pools of Ca^{2+} involved in channel activation. In the absence of extracellular Ca^{2+}, transient stimulant-evoked potential changes can easily be observed, but to sustain the stimulant-evoked electrical responses, extracellular Ca^{2+} is required (50, 81). These conclusions have been confirmed in patch-clamp single-channel and whole-cell current recording experiments (67, 118). The first phase of stimulation that is independent of external Ca^{2+} nevertheless relies on release of intracellular Ca^{2+} (49). Evidence for stimulant-evoked release of intracellular Ca^{2+} has come from studies using the fluorescent probe chlortetracycline (7, 8) as well as $^{45}Ca^{2+}$ flux measurements (70, 112). It has also been shown that this Ca^{2+} release causes an increase in $[Ca^{2+}]_i$ (79). The mechanism by which the intracellular Ca^{2+} release occurs is now relatively clear. Hormone-receptor activation of phospholipase C splits phosphatidylinositol bisphosphate into inositol trisphosphate (IP_3) and diacylglycerol (3, 29). Then IP_3 acts on the endoplasmic reticulum membrane by activating a Ca^{2+} pathway, allowing release of Ca^{2+} into the cytosol (72, 117). The sustained phase of stimulation depends on extracellular Ca^{2+} (50, 81, 104), and it has been shown that pancreatic secretagogues increase unidirectional Ca^{2+} flux into acinar cells (47). The mechanism underlying this Ca^{2+} uptake remains obscure, but single-channel current studies demonstrating an external Ca^{2+} requirement for sustained CCK-evoked K^+ channel opening at a site distinct from the location of hormone-receptor interaction may indicate a messenger-mediated Ca^{2+} uptake rather than receptor-operated Ca^{2+} channels (118).

IMPORTANCE OF ELECTROGENIC PROCESSES FOR ACINAR CELL FUNCTION

The two main acinar cell functions are enzyme and fluid secretion. A role for ion channels and electrogenic pumps in fluid secretion is more immediately apparent than in enzyme secretion and is therefore discussed first.

Potassium Channels and Fluid Secretion

The acinar cells in the pancreas and in the salivary glands secrete an isotonic NaCl-rich fluid (89, 111, 125). Pancreatic acinar fluid secretion has only been studied in detail in the rat where there is a small basal secretion, and all the stimulants (ACh, CCK, bombesin) acting on cellular Ca^{2+} metabolism evoke a marked increase in fluid output (105, 123). The stimulant-evoked fluid secretion, as well as the enzyme secretion, is Ca^{2+} dependent but, unlike the secretin-evoked fluid secretion, is independent of the presence of CO_2 and HCO_3^- in the perfusion fluid (105, 123). This is very similar to the earlier findings made in the salivary glands (85).

ACINAR CELL POTASSIUM BALANCE. Acinar fluid secretion is clearly dependent on the operation of the Na^+-K^+ pump (74). Thus it is useful to approach the problem of the relationship between the stimulant-evoked membrane conductance changes and electrolyte and fluid secretion by considering the handling of K^+ by the acinar cells. Figure 19 shows the changes in the K^+ concentration measured in the venous outflow from a salivary gland and in the saliva after the start of parasympathetic nerve stimulation. After the start of stimulation, K^+ is clearly lost from the cells to both the saliva and the blood in the initial phase, and there is a reduction of $[K^+]_i$ (107). During continued stimulation, the initial net K^+ release is followed by a slight net reuptake [Fig. 19; (5, 6, 106)]. This same pattern can also be observed in superfused gland fragments. In experiments carried out on isolated pieces

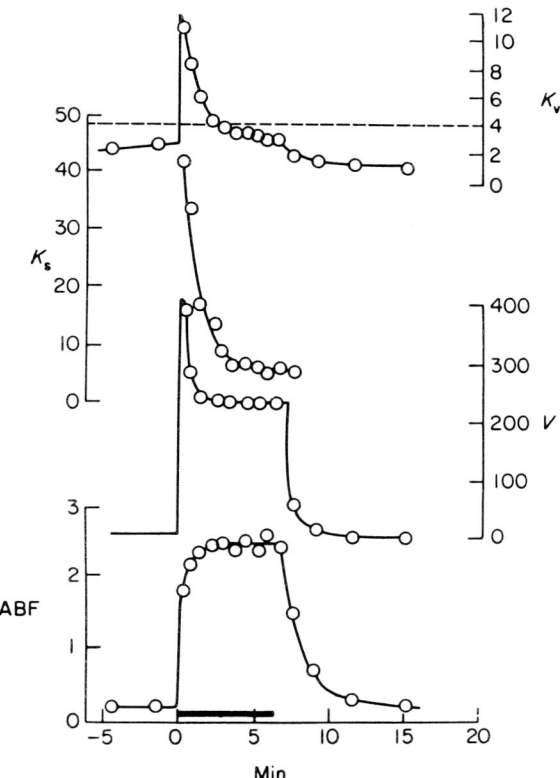

FIG. 19. Dog submaxillary gland in vivo. Submaxillary gland circulation was isolated so that venous blood draining from gland could be collected. Following rest period, chorda tympani (parasympathetic nerves) was stimulated at 20 Hz. K_v, venous plasma K^+ concentration (mM); K_s, salivary K^+ concentration (mM); V, saliva flow rate in $mg \cdot g^{-1} \cdot min^{-1}$; ABF, arterial blood flow through gland in $ml \cdot g^{-1} \cdot min^{-1}$. At beginning of stimulation K^+ concentration in saliva was >40 mM, whereas in steady state it was only 6 mM. At beginning of stimulation concentration of K^+ in venous plasma reached >11 mM and later settled down to 2.6 mM. [From Burgen (5).]

of human pancreas, the initial K^+ release is followed by reuptake during sustained ACh stimulation. Unidirectional outfluxes, using Rb^+ as a tracer, show that ACh does evoke a small sustained increase in ^{86}Rb outflux. Therefore at the time during stimulation when there is net K^+ reuptake into the acinar cells there is still a stimulant-evoked increase in K^+ permeability (94). This conclusion is in agreement with results obtained on lacrimal acinar cells where sustained repetitive electrical nerve stimulation or continued exposure to ACh evokes a sustained increase in acinar membrane conductance (82). The normal K^+ concentration in the venous effluent, which is reached again after ~10-15 min of continuous stimulation, therefore reflects a balance between stimulant-evoked passive K^+ loss and active reuptake.

The stimulant-evoked K^+ release can be explained by the Ca^{2+} activation of K^+ channels (61). However, in order to explain the K^+ release it is not sufficient to have an exit pathway; there must also be an electrochemical gradient favoring outflux. This is achieved by the simultaneous activation of K^+ and Cl^- conductance pathways (19).

After cessation of stimulation there is marked net reuptake of K^+ (Fig. 19). The K^+ uptake takes place against an electrochemical gradient (89). The ubiquitous mechanism for cellular K^+ uptake is the ATP-driven Na^+-K^+ pump. The Na^+-K^+ pump sites in acinar cells have been localized to the basolateral cell membrane (4). It has been known for many years that digitalis glycosides, such as ouabain (G-strophanthin), specifically block the Na^+-K^+ pump acting from the outside of the membrane. Ouabain inhibits the K^+ uptake following stimulant-evoked release and also causes a net release when given in the absence of stimulation (84, 85). This is not, however, the whole story. Reuptake of K^+ after stimulant-evoked release can also be immediately and reversibly abolished by replacing extracellular Na^+ with Li^+ or by replacing Cl^- with SO_4^{2-} or NO_3^- (84, 101). The K^+ uptake is therefore acutely dependent on extracellular Na^+ and Cl^-. It is now clear that, in addition to the presence of the ATP-driven Na^+-K^+ pump, there is a Na^+-K^+-$2Cl^-$ carrier mechanism in which the "downhill" movement of Na^+ along its electrochemical gradient (created by the Na^+-K^+ pump) is coupled to the "uphill" movement of Cl^- and K^+ (74, 97).

ACINAR CELL MODEL. Figure 20 summarizes the findings in salivary acinar cells and the pancreatic acinar cells from pigs and humans. Three different aspects are emphasized, and for clarity these are shown separately in the three cells depicted. In the upper cell, stimulant neurotransmitters or hormones act on the acinar cells by releasing intracellular Ca^{2+}, evoking an increase in $[Ca^{2+}]_i$ (79). The released Ca^{2+} activates K^+ channels in the basolateral plasma membrane. The Ca^{2+}-activated Cl^- channels are not nearly so well characterized. It is assumed that they are present in the luminal cell membrane because the lumen becomes more negative with respect to the interstitial fluid after stimulation (54, 55, 58).

In the middle cell of Figure 20 the most important transport events are shown. In the basolateral plasma membrane there are three transport proteins: the Ca^{2+}- and voltage-activated K^+ channel, the Na^+-K^+-$2Cl^-$ cotransporter, and the energy-requiring Na^+-K^+ pump. In the steady secreting state, K^+ recirculates via the channel, cotransporter, and pump; Na^+ recirculates via the pump and cotransporter. The only net transport is that of Cl^- uptake. Chloride leaves the cell via the luminal Cl^- channel. The lumen negativity allows Na^+ to move between the cells through the narrow intercellular spaces and the so-called tight junctions placed at their luminal end. These are generally cation selective (89), and the Na^+ flux dominates simply because Na^+ is the main cation in the extracellular fluid. The net result of all these transport events is transcellular NaCl transport. Water flows

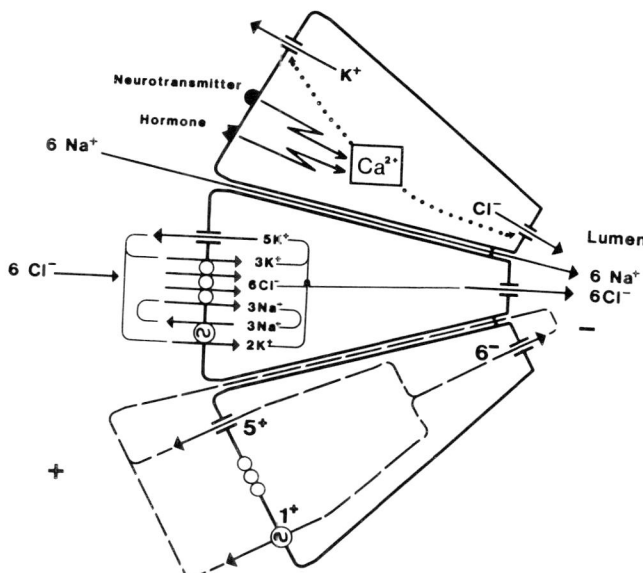

FIG. 20. Semiquantitative model of transport events in exocrine acini. Three different aspects are, for clarity, represented by three separate cells. *Upper cell*, intracellular Ca^{2+} release by transmitter (e.g., acetylcholine) or hormone (e.g., epinephrine) and Ca^{2+} activation of K^+ and Cl^- conductance pathways. *Middle cell*, relation between transport rates via different routes in steady-state stimulation situation. *Lower cell*, overall electrical circuit. [From Suzuki and Petersen (119).]

along by osmosis both through the cells and between the cells. The result of all these transport events is the formation of an isotonic NaCl-rich fluid (111, 125). Many aspects of this model are similar to the one originally proposed by Field and co-workers (114) to account for Cl^- secretion by the shark rectal gland, but the novel feature is that the rate of Cl^- uptake is directly linked to the rate of transport in the cycle of K^+ release and uptake (97).

In the bottom cell of Figure 20 the overall electrical circuit is presented. The outward current across the basolateral membrane is mostly due to exit through K^+ channels but also partly to the imbalance of ion movements through the Na^+-K^+ pump, and this total outward current must be precisely matched by the inward current through the luminal Cl^- channels. To complete the circuit, current must pass between the cells and through the tight junction. This paracellular current is carried by Na^+ moving from the interstitial fluid into the acinar lumen. The low transcellular specific resistance (much lower than the specific transmembrane resistance), which has been found in many epithelia (20), was demonstrated for the first time in salivary acini (56). It is now generally recognized that this is explained by the leakiness of the tight junctions (20, 89).

The model (Fig. 20) can clearly explain acinar NaCl secretion, but can it also explain all aspects of the K^+ concentration changes in secreted juice and blood in the initial 10–15 min following start of stimulation?

The K^+ release to the blood side is of course due to Ca^{2+}-activated K^+ channels but, as indicated in Figure 20, K^+ release must be accompanied by something else, namely Cl^- release (19, 74, 110). Calcium activation of K^+ and Cl^- channels allows cellular KCl release, and water must follow for osmotic reasons, causing the acinar cells to shrink (6). Most of the outward K^+ current across the basolateral membrane will be directed into the lumen to accompany the inward Cl^- current (outward Cl^- flux); thus most of the K^+ release is secreted into the acinar lumen. In this initial phase a major part of the paracellular cation flow is made up of K^+. The release of cellular KCl creates a more favorable overall electrochemical gradient for Na^+-K^+-$2Cl^-$ uptake via the cotransporter, and the Na^+ inflow raises $[Na^+]_i$, which stimulates the Na^+-K^+ pump (30). Both transport proteins help to reaccumulate K^+, and in the secreting steady state this exactly balances the K^+ lost via the channels. When stimulation ceases, $[Ca^{2+}]_i$ decreases, and both K^+ and Cl^- channels close. There is now net KCl reuptake (mediated by pump and cotransporter), followed by water uptake, and the cells swell back to their original size. The overall electrochemical gradient for Na^+-K^+-$2Cl^-$ cotransport is soon less favorable for uptake, and when Na^+ inflow is reduced, the activity of the Na^+-K^+ pump is also brought back to the prestimulation level, and the gland is again in a true resting state.

Nonselective Cation Channels and Fluid Secretion

The acinar transport model shown in Figure 20 would seem to account very well for the main findings in most exocrine gland preparations, but there are alternative models, and the results from at least the mouse and rat pancreatic acinar cells do not fit in at all with this pattern. It may be convenient to consider K^+ transport in the mouse pancreatic acinar cells, because this differs somewhat from that described for the salivary glands and the human pancreas.

Acetylcholine can evoke marked K^+ release from mouse pancreatic segments. This extrusion of K^+ is insensitive to ouabain or the specific blocker of Ca^{2+}-activated channels, TEA, but is sensitive to loop diuretics such as furosemide, piretanide, and bumetanide. It is also dependent on Na^+, Cl^-, and to a smaller extent Ca^{2+} (103). In perfused salivary glands the immediate effect of replacing extracellular Cl^- by NO_3^- or SO_4^{2-} is an increase in the K^+ concentration in the venous effluent, and readmission of the normal Cl^--containing solution results in an immediate reduction in the venous K^+ concentration (84). In mouse pancreatic segments Cl^- replacement by NO_3^- or SO_4^{2-} results in an immediate reduction in the effluent K^+ concentration, which returns to the control level on readmission of the control solution (103). These differences between results obtained on salivary glands and the mouse pancreas indicate that the mechanisms underlying cellular K^+ transport may be different, which is in agreement with the absence of Ca^{2+}- and voltage-activated K^+-selective channels in the mouse and rat pancreas (97, 103). Blockade of such K^+ channels by Ba^{2+} and TEA virtually abolishes ACh-evoked fluid secretion in perfused rat salivary glands, whereas secretion from the perfused rat pancreas is only affected to a minor extent (10). The K^+ transport results in the mouse pancreas can most easily be explained by assuming that the Na^+-K^+-$2Cl^-$ cotransporter moves the ions from the cell interior to the exterior, i.e., in the direction opposite to that suggested for the salivary glands or the pig pancreas. However, it is not possible to arrive at final conclusions about the localization of transport events in polarized cells on the basis of studies on superfused tissue fragments in which both basolateral and luminal membranes are in contact with the bath solution.

The only ion transport process that has been precisely localized in the mouse pancreatic acinar cell is the Ca^{2+}-activated nonselective cation channel that is in the basolateral membrane (64, 65). Although this channel is permeable to both Na^+ and K^+, Na^+ influx will dominate as the membrane potential is much closer to the K^+ equilibrium potential than the Na^+ equilibrium potential (95). This channel, which is activated via internal Ca^{2+} by CCK or ACh, could play a crucial role in the activation of the Na^+-K^+-$2Cl^-$ cotransporter providing the Na^+ necessary for extrusion (95, 103).

Potassium Channels and Protein Secretion

The role of ion channels in protein secretion is quite uncertain. To manufacture and secrete protein it is necessary for the acinar cells to have well-developed Na^+-dependent amino acid uptake systems, and for these systems to work, both the Na^+-K^+ pump and the resting K^+ channels are essential since the resting potential is providing a major driving force for the "downhill" Na^+ uptake linked to the "uphill" amino acid transport. The Na^+-K^+ pump obviously creates the inwardly directed Na^+ gradient that is utilized to move neutral amino acids into the cell (see Fig. 9). Operation of the Na^+–amino acid cotransporter depolarizes the cell and tends to increase $[Na^+]_i$. The depolarization activates voltage-activated K^+ channels in those cells where they are present and partially repolarizes the cell and helps to maintain the driving force for Na^+–amino acid cotransport. The increase in $[Na^+]_i$ activates the Na^+-K^+ pump, and the K^+ that is accumulated when Na^+ is extruded exits via the K^+ channel.

Major issues relating to the function of ion channels in the plasma membrane of the acinar cells remain to be settled. The role of the gap junctional channels is also not entirely clear. Because the number of K^+ channels in individual acinar cells is very small (62, 69, 121), the junctional channels clearly have an important function in allowing the acinus as a whole to share the existing channels and therefore establish a stable resting potential (91). In the pig pancreas and most of the other mammalian exocrine glands (except the mouse and rat pancreas) where the K$^+$ channels are voltage sensitive, the junctional channels are of great importance during nerve stimulation in cases where every acinar cell does not receive its own innervation. Any potential change evoked by stimulation of one cell will virtually instantaneously be transmitted to all cells in the unit and therefore affect the open-state probability of all K$^+$ channels. An increase in [Ca^{2+}]$_i$ opens channels but also allows K$^+$ exit and therefore hyperpolarization that tends to close channels. Thus K$^+$ flux through the conductance pathway is kept in good control, and the cell-to-cell junctional channels will ensure a uniform regulation throughout the acinar network.

A model for exocytosis based on the opening of Ca^{2+}-activated K$^+$ channels in the vesicle membranes has been proposed (116). The opening of these channels coupled with anion transport across the vesicle membranes would result in an influx of K$^+$ and anions increasing the osmotic pressure of the vesicles. For vesicles situated very close to the cell membrane this could lead to fusion with the plasma membrane and exocytosis of the contents. The model predicts a number of characteristic features of secretion (116) but has not yet been critically tested.

In general specific knowledge is lacking about channel properties of the vesicle membranes in exocrine acinar cells, and much work is needed on the properties of these membranes as well as the rest of the luminal cell membranes before a reasonably complete picture can be had of the electrophysiology of the acinar cells.

We thank Beverley Houghton for great efficiency in preparing this manuscript in its final form.

Work cited in this chapter was supported by the Medical Research Council, United Kingdom, and the Wellcome Trust, United Kingdom, as well as the Ministry of Education, Japan.

REFERENCES

1. ANDERSON, C. R., AND C. F. STEVENS. Voltage-clamp analysis of acetylcholine produced end-plate current fluctuations of frog neuromuscular junction. *J. Physiol. Lond.* 235: 655–691, 1973.
2. BAKER, P. F. The regulation of intracellular calcium in giant axons of *Loligo* and *Myxicola. Ann. NY Acad. Sci.* 307: 250–268, 1978.
3. BERRIDGE, M. J., AND R. F. IRVINE. Inositol trisphosphate, a novel second messenger in cellular signal transduction. *Nature Lond.* 312: 315–321, 1984.
4. BUNDGAARD, M., M. MØLLER, AND J. H. POULSEN. Localization of sodium pump sites in cat salivary glands. *J. Physiol. Lond.* 273: 339–353, 1977.
5. BURGEN, A. S. V. The secretion of potassium in saliva. *J. Physiol. Lond.* 132: 20–39, 1956.
6. BURGEN, A. S. V., AND N. G. EMMELIN. *Physiology of the Salivary Glands.* London: Arnold, 1961.
7. CHANDLER, D. E., AND J. A. WILLIAMS. Intracellular divalent cation release in pancreatic acinar cells during stimulus-secretion coupling. I. Use of chlorotetracycline as a fluorescent probe. *J. Cell Biol.* 76: 371–385, 1978.
8. CHANDLER, D. E., AND J. A. WILLIAMS. Intracellular divalent cation release in pancreatic acinar cells during stimulus-secretion coupling. II. Subcellular localization of the fluorescent probe chlorotetracycline. *J. Cell Biol.* 76: 386–399, 1978.
9. CHRISTENSEN, H. N. Organic ion transport during seven decades. The amino acids. *Biochim. Biophys. Acta* 779: 255–269, 1984.
10. EVANS, L. A. R., D. PIRANI, D. I. COOK, AND J. A. YOUNG. Intraepithelial current flow in rat pancreatic secretory epithelia. *Pfluegers Arch.* 407, Suppl. 2: S107–S111, 1986.
11. EVANS, M. G., AND A. MARTY. Calcium-dependent chloride currents in isolated cells from rat lacrimal glands. *J. Physiol. Lond.* 378: 437–460, 1986.
12. EVANS, M. G., A. MARTY, AND Y. TAN. Blockage of a Ca-activated Cl conductance by furosemide in rat lacrimal glands. *Pfluegers Arch.* 406: 65–68, 1986.
13. FERNANDEZ, J. H., E. NEHER, AND B. D. GOMPERTS. Capacitance measurements reveal stepwise fusion events in degranulating mast cells. *Nature Lond.* 312: 453–455, 1984.
14. FINDLAY, I. A patch-clamp study of potassium channels and whole-cell currents in acinar cells of the mouse lacrimal gland. *J. Physiol. Lond.* 350: 179–195, 1984.
15. FINDLAY, I., M. J. DUNNE, AND O. H. PETERSEN. ATP-sensitive inward rectifier and voltage- and calcium-activated K$^+$ channels in cultured pancreatic islet cells. *J. Membr. Biol.* 88: 165–172, 1985.
16. FINDLAY, I., M. J. DUNNE, S. ULLRICH, C. B. WOLLHEIM, AND O. H. PETERSEN. Quinine inhibits Ca^{2+}-independent K$^+$ channels whereas tetraethylammonium inhibits Ca^{2+}-activated K$^+$ channels in insulin-secreting cells. *FEBS Lett.* 185: 4–8, 1985.
17. FINDLAY, I., AND O. H. PETERSEN. Acetylcholine-evoked uncoupling restricts the passage of Lucifer Yellow between pancreatic acinar cells. *Cell Tissue Res.* 225: 633–638, 1982.
18. FINDLAY, I., AND O. H. PETERSEN. The extent of dye-coupling between exocrine acinar cells of the mouse pancreas. *Cell Tissue Res.* 232: 121–127, 1983.
19. FINDLAY, I., AND O. H. PETERSEN. Acetylcholine stimulates a Ca^{2+}-dependent Cl$^-$ conductance in mouse lacrimal acinar cells. *Pfluegers Arch.* 403: 328–330, 1985.
20. FRÖMTER, E., AND J. DIAMOND. Route of passive ion permeation in epithelia. *Nat. New Biol.* 235: 9–13, 1972.
21. GALLACHER, D. V., Y. MARUYAMA, AND O. H. PETERSEN. Patch-clamp study of rubidium and potassium conductances in single cation channels from mammalian exocrine acini. *Pfluegers Arch.* 401: 361–367, 1984.
22. GALLACHER, D. V., AND A. P. MORRIS. A patch-clamp study of K$^+$ currents in resting and acetylcholine stimulated mouse submandibular acinar cells. *J. Physiol. Lond.* 373: 379–395, 1986.
23. GALLACHER, D. V., AND O. H. PETERSEN. Electrophysiology of parotid acini: effects of electrical field stimulation and ionophoresis of neurotransmitters. *J. Physiol. Lond.* 305: 43–57, 1980.
24. GINSBORG, B. L., AND C. R. HOUSE. Stimulus-response coupling in gland cells. *Annu. Rev. Biophys. Bioeng.* 9: 55–80, 1980.
25. GINSBORG, B. L., C. R. HOUSE, AND E. M. SILINSKY. Conductance changes associated with the secretory potential in the cockroach salivary gland. *J. Physiol. Lond.* 236: 723–731, 1974.
26. GRAF, J., AND O. H. PETERSEN. Cell membrane potential and resistance in liver. *J. Physiol. Lond.* 284: 105–126, 1978.
27. HAMILL, O. P., A. MARTY, E. NEHER, B. SAKMANN, AND F. J. SIGWORTH. Improved patch-clamp technique for high-res-

olution current recording from cells and cell-free membrane patches. *Pfluegers Arch.* 391: 85–100, 1981.
28. HODGKIN, A. L. *The Conduction of the Nervous Impulse.* Liverpool, UK: Liverpool Univ. Press, 1964.
29. HOKIN, L. E. Receptors and phosphoinositide-generated second messengers. *Annu. Rev. Biochem.* 54: 205–235, 1985.
30. HOOTMAN, S. R., D. L. OCHS, AND J. A. WILLIAMS. Intracellular mediators of Na^+-K^+ pump activity in guinea pig pancreatic acinar cells. *Am. J. Physiol.* 249 (*Gastrointest. Liver Physiol.* 12): G470–G478, 1985.
31. HORN, R., AND J. B. PATLAK. Single-channel currents from excised patches of muscle membrane. *Proc. Natl. Acad. Sci. USA* 77: 6930–6934, 1980.
32. IWATSUKI, N., AND O. H. PETERSEN. Acetylcholine-like effects of intracellular calcium application in pancreatic acinar cells. *Nature Lond.* 268: 147–149, 1977.
33. IWATSUKI, N., AND O. H. PETERSEN. Pancreatic acinar cells: localization of acetylcholine receptors and the importance of chloride and calcium for acetylcholine-evoked depolarization. *J. Physiol. Lond.* 269: 723–733, 1977.
34. IWATSUKI, N., AND O. H. PETERSEN. Pancreatic acinar cells: the acetylcholine equilibrium potential and its ionic dependency. *J. Physiol. Lond.* 269: 735–751, 1977.
35. IWATSUKI, N., AND O. H. PETERSEN. Electrical coupling and uncoupling of exocrine acinar cells. *J. Cell Biol.* 79: 533–545, 1978.
36. IWATSUKI, N., AND O. H. PETERSEN. Intracellular Ca^{2+} injection causes membrane hyperpolarization and conductance increase in lacrimal acinar cells. *Pfluegers Arch.* 377: 185–187, 1978.
37. IWATSUKI, N., AND O. H. PETERSEN. In vitro action of bombesin on amylase secretion, membrane potential and membrane resistance in rat and mouse pancreatic acinar cells. A comparison with other secretagogues. *J. Clin. Invest.* 61: 41–46, 1978.
38. IWATSUKI, N., AND O. H. PETERSEN. Direct visualization of cell to cell coupling: transfer of fluorescent probes in living mammalian pancreatic acini. *Pfluegers Arch.* 380: 277–281, 1979.
39. IWATSUKI, N., AND O. H. PETERSEN. Pancreatic acinar cells: the effect of CO_2, NH_4Cl and acetylcholine on intercellular communication. *J. Physiol. Lond.* 291: 317–326, 1979.
40. IWATSUKI, N., AND O. H. PETERSEN. Amino acid-evoked membrane potential and resistance changes in pancreatic acinar cells. *Pfluegers Arch.* 386: 153–159, 1980.
41. IWATSUKI, N., AND O. H. PETERSEN. Amino acids evoke short-latency membrane conductance increase in pancreatic acinar cells. *Nature Lond.* 283: 492–494, 1980.
42. IWATSUKI, N., AND O. H. PETERSEN. Action of tetraethylammonium on calcium-activated potassium channels in pig pancreatic acinar cells studied by patch-clamp single-channel and whole-cell current recording. *J. Membr. Biol.* 86: 139–144, 1985.
43. IWATSUKI, N., AND O. H. PETERSEN. Inhibition of Ca^{2+}-activated K^+ channels in pig pancreatic acinar cells by Ba^{2+}, Ca^{2+}, quinine and quinidine. *Biochim. Biophys. Acta* 819: 249–257, 1985.
44. JAUCH, P., Y. MARUYAMA, O. H. PETERSEN, H. A. KOLB, AND P. LÄUGER. Electrophysiological study of the alanine-sodium cotransporter in pancreatic acinar cells. In: *Ion Gradient-Coupled Transport,* edited by F. Alvardo and C. H. Van Os. Amsterdam: Elsevier, 1986, p. 241–244. (INSERM Symp. Ser., vol. 26.)
45. JAUCH, P., O. H. PETERSEN, AND P. LÄUGER. Electrogenic properties of the sodium-alanine cotransporter in pancreatic acinar cells: I. Tight-seal whole-cell recordings. *J. Membr. Biol.* 94: 99–115, 1986.
46. KATZ, B., AND R. MILEDI. The statistical nature of the acetylcholine potential and its molecular components. *J. Physiol. Lond.* 224: 665–699, 1972.
47. KONDO, S., AND I. SCHULZ. Calcium ion uptake in isolated pancreas cells induced by secretagogues. *Biochim. Biophys. Acta* 419: 76–92, 1976.
48. KURTZER, R. J., AND M. L. ROBERTS. Calcium-dependent K^+ efflux from rat submandibular gland. The effects of trifluoroperazine and quinidine. *Biochim. Biophys. Acta* 693: 479–484, 1982.
49. LAUGIER, R., AND O. H. PETERSEN. Effects of intracellular EGTA injection on stimulant-evoked membrane potential and resistance changes in pancreatic acinar cells. *Pfluegers Arch.* 386: 147–152, 1980.
50. LAUGIER, R., AND O. H. PETERSEN. Pancreatic acinar cells: electrophysiological evidence for stimulant-evoked increase in membrane calcium permeability in the mouse. *J. Physiol. Lond.* 303: 61–72, 1980.
51. LAUGIER, R., AND O. H. PETERSEN. Two different types of electrogenic amino acid action on pancreatic acinar cells. *Biochim. Biophys. Acta* 641: 216–221, 1981.
52. LOEWENSTEIN, W. R. Junctional intercellular communication: the cell-to-cell membrane channel. *Physiol. Rev.* 61: 829–913, 1981.
53. LUNDBERG, A. The electrophysiology of the submaxillary gland of the cat. *Acta Physiol. Scand.* 35: 1–25, 1955.
54. LUNDBERG, A. Secretory potentials and secretion in the sublingual gland of the cat. *Nature Lond.* 177: 1080–1081, 1956.
55. LUNDBERG, A. Secretory potentials in the sublingual gland of the cat. *Acta Physiol. Scand.* 40: 21–34, 1957.
56. LUNDBERG, A. The mechanism of establishment of secretory potentials in sublingual gland cells. *Acta Physiol. Scand.* 40: 35–58, 1957.
57. LUNDBERG, A. Anionic dependence of secretion and secretory potentials in sublingual gland cells. *Acta Physiol. Scand.* 40: 101–112, 1957.
58. LUNDBERG, A. Electrophysiology of salivary glands. *Physiol. Rev.* 38: 21–40, 1958.
59. MARTY, A., Y. P. TAN, AND A. TRAUTMANN. Three types of calcium-dependent channel in rat lacrimal glands. *J. Physiol. Lond.* 357: 293–325, 1984.
60. MARUYAMA, Y. Ca^{2+}-induced excess capacitance fluctuation studied by phase-sensitive detection method in exocrine pancreatic acinar cells. *Pfluegers Arch.* 407: 561–563, 1986.
61. MARUYAMA, Y., D. V. GALLACHER, AND O. H. PETERSEN. Voltage and Ca^{2+}-activated K^+ channel in baso-lateral acinar cell membranes of mammalian salivary glands. *Nature Lond.* 302: 827–829, 1983.
62. MARUYAMA, Y., A. NISHIYAMA, T. IZUMI, N. HOSHIMIYA, AND O. H. PETERSEN. Ensemble noise and current relaxation analysis of K^+ current in single isolated salivary acinar cells from rat. *Pfluegers Arch.* 406: 69–72, 1986.
63. MARUYAMA, Y., A. NISHIYAMA, AND T. TESHIMA. Two types of cation channels in the basolateral cell membrane of human salivary gland acinar cells. *Jpn. J. Physiol.* 36: 219–223, 1986.
64. MARUYAMA, Y., AND O. H. PETERSEN. Single-channel currents in isolated patches of plasma membrane from basal surface of pancreatic acini. *Nature Lond.* 299: 159–161, 1982.
65. MARUYAMA, Y., AND O. H. PETERSEN. Cholecystokinin activation of single-channel currents is mediated by internal messenger in pancreatic acinar cells. *Nature Lond.* 300: 61–63, 1982.
66. MARUYAMA, Y., AND O. H. PETERSEN. Voltage clamp study of stimulant-evoked currents in mouse pancreatic acinar cells. *Pfluegers Arch.* 399: 54–62, 1983.
67. MARUYAMA, Y., AND O. H. PETERSEN. Control of K^+ conductance by cholecystokinin and Ca^{2+} in single pancreatic acinar cells studied by the patch-clamp technique. *J. Membr. Biol.* 79: 293–300, 1984.
68. MARUYAMA, Y., AND O. H. PETERSEN. Single calcium-dependent cation channels in mouse pancreatic acinar cells. *J. Membr. Biol.* 81: 83–87, 1984.
69. MARUYAMA, Y., O. H. PETERSEN, P. FLANAGAN, AND G. T. PEARSON. Quantification of Ca^{2+}-activated K^+ channels under hormonal control in pig pancreas acinar cells. *Nature Lond.*

305: 228–232, 1983.
70. MATTHEWS, E. K., O. H. PETERSEN, AND J. A. WILLIAMS. Pancreatic acinar cells: acetylcholine-induced membrane depolarization, calcium efflux and amylase release. *J. Physiol. Lond.* 234: 689–701, 1973.
71. MEDA, P., I. FINDLAY, E. KOLOD, L. ORCI, AND O. H. PETERSEN. Short and reversible uncoupling evokes little change in the gap junctions of pancreatic acinar cells. *J. Ultrastruct. Res.* 83: 69–84, 1983.
72. MUALLEM, S., M. SCHOEFFIELD, S. PANDOL, AND G. SACHS. Inositol trisphosphate modification of ion transport in rough endoplasmic reticulum. *Proc. Natl. Acad. Sci. USA* 82: 4433–4437, 1985.
73. NAGEL, W. Inhibition of potassium conductance by barium in frog skin epithelium. *Biochim. Biophys. Acta* 552: 346–357, 1979.
74. NAUNTOFTE, B., AND J. H. POULSEN. Effects of Ca^{2+} and furosemide on Cl^- transport and O_2 uptake in rat parotid acini. *Am. J. Physiol.* 251 (*Cell Physiol.* 20): C175–C185, 1986.
75. NEHER, E., AND A. MARTY. Discrete changes of cell membrane capacitance observed under conditions of enhanced secretion in bovine adrenal chromaffin cells. *Proc. Natl. Acad. Sci. USA* 79: 6712–6716, 1982.
76. NEHER, E., AND B. SAKMANN. Single-channel currents recorded from membrane of denervated frog muscle fibres. *Nature Lond.* 260: 799–802, 1976.
77. NEYTON, J., AND A. TRAUTMANN. Single-channel currents of an intercellular junction. *Nature Lond.* 317: 331–335, 1985.
78. NISHIYAMA, A., AND O. H. PETERSEN. Membrane potential and resistance measurement in acinar cells from salivary glands in vitro: effect of acetylcholine. *J. Physiol. Lond.* 242: 173–188, 1974.
79. OCHS, D. L., J. I. KORENBROT, AND J. A. WILLIAMS. Relationship between free cytosolic calcium and amylase release by pancreatic acini. *Am. J. Physiol.* 249 (*Gastrointest. Liver Physiol.* 12): G389–G398, 1985.
80. PALADE, G. Intracellular aspects of the process of protein synthesis. *Science Wash. DC* 189: 347–358, 1975.
81. PEARSON, G. T., P. M. FLANAGAN, AND O. H. PETERSEN. Neural and hormonal control of membrane conductance in the pig pancreatic acinar cell. *Am. J. Physiol.* 247 (*Gastrointest. Liver Physiol.* 10): G520–G526, 1984.
82. PEARSON, G. T., AND O. H. PETERSEN. Nervous control of membrane conductance in mouse lacrimal acinar cells. *Pfluegers Arch.* 400: 51–59, 1984.
83. PEDERSEN, G. L., AND O. H. PETERSEN. Membrane potential measurement in parotid acinar cells. *J. Physiol. Lond.* 234: 217–227, 1973.
84. PETERSEN, O. H. Some factors influencing stimulation-induced release of potassium from the cat submandibular gland to fluid perfused through the gland. *J. Physiol. Lond.* 208: 431–447, 1970.
85. PETERSEN, O. H. Formation of saliva and potassium transport in the perfused cat submandibular gland. *J. Physiol. Lond.* 216: 129–142, 1971.
86. PETERSEN, O. H. Electrogenic sodium pump in pancreatic acinar cells. *Proc. R. Soc. Lond. B Biol. Sci.* 184: 115–119, 1973.
87. PETERSEN, O. H. Electrical coupling between pancreatic acinar cells. *J. Physiol. Lond.* 250: 2P–4P, 1975.
88. PETERSEN, O. H. Electrophysiology of mammalian gland cells. *Physiol. Rev.* 56: 535–577, 1976.
89. PETERSEN, O. H. *Electrophysiology of Gland Cells.* New York: Academic, 1981.
90. PETERSEN, O. H. Stimulus-excitation coupling in plasma membranes of pancreatic acinar cells. *Biochim. Biophys. Acta* 694: 163–184, 1982.
91. PETERSEN, O. H. Importance of electrical cell-cell communication in secretory epithelia. In: *Gap Junctions*, edited by M. V. Bennett and D. C. Spray. Cold Spring Harbor, NY: Cold Spring Harbor, 1985, p. 315–324.
92. PETERSEN, O. H. Potassium channels and fluid secretion. *News Physiol. Sci.* 1: 92–95, 1986.
93. PETERSEN, O. H. Calcium-activated potassium channels and fluid secretion by exocrine glands. *Am. J. Physiol.* 251 (*Gastrointest. Liver Physiol.* 14): G1–G13, 1986.
94. PETERSEN, O. H., I. FINDLAY, N. IWATSUKI, J. SINGH, D. V. GALLACHER, C. M. FULLER, G. T. PEARSON, M. J. DUNNE, AND A. P. MORRIS. Human pancreatic acinar cells: studies of stimulus-secretion coupling. *Gastroenterology* 89: 109–117, 1985.
95. PETERSEN, O. H., AND Y. MARUYAMA. Cholecystokinin and acetylcholine activation of single-channel currents via second messenger in pancreatic acinar cells. In: *Single-Channel Recording*, edited by B. Sakmann and E. Neher. New York: Plenum, 1983, p. 425–435.
96. PETERSEN, O. H., AND Y. MARUYAMA. What is the mechanism of the calcium influx to pancreatic acinar cells evoked by secretagogues? *Pfluegers Arch.* 396: 82–84, 1983.
97. PETERSEN, O. H., AND Y. MARUYAMA. Calcium-activated potassium channels and their role in secretion. *Nature Lond.* 307: 693–696, 1984.
98. PETERSEN, O. H., Y. MARUYAMA, J. GRAF, R. LAUGIER, A. NISHIYAMA, AND G. T. PEARSON. Ionic currents across pancreatic acinar cell membranes and their role in fluid secretion. *Philos. Trans. R. Soc. Lond. B Biol. Sci.* 296: 151–166, 1981.
99. PETERSEN, O. H., AND C. C. H. PETERSEN. The patch-clamp technique: recording ionic currents through single pores in the cell membrane. *News Physiol. Sci.* 1: 5–8, 1986.
100. PETERSEN, O. H., AND H. G. PHILPOTT. Pancreatic acinar cells: the anion selectivity of the acetylcholine-opened chloride pathway. *J. Physiol. Lond.* 306: 481–492, 1980.
101. PETERSEN, O. H., AND J. H. POULSEN. Secretory potentials, potassium transport and secretion in the cat submandibular gland during perfusion with sulphate Locke's solution. *Experientia* 24: 919–920, 1968.
102. PETERSEN, O. H., AND J. SINGH. Amino acid-evoked membrane current in voltage-clamped mouse pancreatic acini. *J. Physiol. Lond.* 319: 99P–100P, 1981.
103. PETERSEN, O. H., AND J. SINGH. Acetylcholine-evoked potassium release in the mouse pancreas. *J. Physiol. Lond.* 365: 319–329, 1985.
104. PETERSEN, O. H., AND N. UEDA. Pancreatic acinar cells: the role of calcium in stimulus-secretion coupling. *J. Physiol. Lond.* 254: 583–606, 1976.
105. PETERSEN, O. H., AND N. UEDA. Secretion of fluid and amylase in the perfused rat pancreas. *J. Physiol. Lond.* 264: 819–835, 1977.
106. POULSEN, J. H., AND S. W. BLEDSOE. Salivary gland K^+ transport: in vivo studies with K^+-specific microelectrodes. *Am. J. Physiol.* 234 (*Endocrinol. Metab. Gastrointest. Physiol.* 3): E79–E83, 1978.
107. POULSEN, J. H., AND B. OAKLEY II. Intracellular potassium ion activity in resting and stimulated mouse pancreas and submandibular gland. *Proc. R. Soc. Lond. B Biol. Sci.* 204: 99–104, 1979.
108. ROBERTS, M. L., N. IWATSUKI, AND O. H. PETERSEN. Parotid acinar cells: ionic dependence of acetylcholine-evoked membrane potential changes. *Pfluegers Arch.* 376: 159–167, 1978.
109. ROBERTS, M. L., AND O. H. PETERSEN. Membrane potential and resistance changes induced in salivary gland acinar cells by microiontophoretic application of acetylcholine and adrenergic agonists. *J. Membr. Biol.* 30: 297–312, 1978.
110. SAITO, Y., T. OZAWA, H. HAYASHI, AND A. NISHIYAMA. Acetylcholine-induced change in intracellular Cl^- activity of the mouse lacrimal acinar cells. *Pfluegers Arch.* 405: 108–111, 1985.
111. SCHULZ, I. Electrolyte and fluid secretion in the exocrine pancreas. In: *Physiology of the Gastrointestinal Tract* (2nd ed.), edited by L. R. Johnson. New York: Raven, 1987, p. 1147–1171.
112. SCHULZ, I., AND H. H. STOLZE. The exocrine pancreas: the role of secretagogues, cyclic nucleotides and calcium in enzyme secretion. *Annu. Rev. Physiol.* 42: 127–156, 1980.

113. SIGWORTH, F. J. The variance of sodium current fluctuations at the node of Ranvier. *J. Physiol. Lond.* 307: 97–129, 1980.
114. SILVA, P., J. STOFF, M. FIELD, L. FINE, J. N. FORREST, AND F. H. EPSTEIN. Mechanism of active chloride secretion by shark rectal gland: role of Na-K-ATPase in chloride transport. *Am. J. Physiol.* 233 (*Renal Fluid Electrolyte Physiol.* 2): F298–F306, 1977.
115. SINGH, J., AND O. H. PETERSEN. The effects of L-alanine and acetylcholine on membrane potential, $^{45}Ca^{2+}$ and $^{86}Rb^+$ efflux and amylase secretion in the isolated mouse pancreas. *Q. J. Exp. Physiol. Cogn. Med. Sci.* 69: 531–540, 1984.
116. STANLEY, E. F., AND G. EHRENSTEIN. A model for exocytosis based on the opening of calcium-activated potassium channels in vesicles. *Life Sci.* 37: 1985–1995, 1985.
117. STREB, H., E. BAYERDORFFER, W. HAASE, R. F. IRVINE, AND I. SCHULZ. Effect of inositol-1,4,5-trisphosphate on isolated subcellular fractions of rat pancreas. *J. Membr. Biol.* 81: 241–253, 1984.
118. SUZUKI, K., C. C. H. PETERSEN, AND O. H. PETERSEN. Hormonal activation of single K^+ channels via internal messenger in isolated pancreatic acinar cells. *FEBS Lett.* 192: 307–312, 1985.
119. SUZUKI, K., AND O. H. PETERSEN. The effects of Na^+ and Cl^- removal and of loop diuretics on acetylcholine-evoked membrane potential changes in mouse lacrimal acinar cells. *Q. J. Exp. Physiol. Cogn. Med. Sci.* 70: 437–445, 1985.
120. SUZUKI, K., AND O. H. PETERSEN. Patch-clamp studies of K^+ channels in guinea-pig pancreatic acinar cells (Abstract). *J. Physiol. Lond.* 378: 62P, 1986.
121. TRAUTMANN, A., AND A. MARTY. Activation of Ca-dependent K channels by carbamoylcholine in rat lacrimal glands. *Proc. Natl. Acad. Sci. USA* 81: 611–615, 1984.
122. TYRAKOWSKI, T., S. MILUTINOVIC, AND I. SCHULZ. Studies on isolated subcellular components of cat pancreas. III. Alanine-sodium co-transport in isolated plasma membrane vesicles. *J. Membr. Biol.* 38: 333–346, 1978.
123. UEDA, N., AND O. H. PETERSEN. The dependence of caerulein-evoked pancreatic fluid secretion on the extracellular calcium concentration. *Pfluegers Arch.* 370: 179–183, 1977.
124. WELSH, M. J. Barium inhibition of basolateral membrane potassium conductance in tracheal epithelium. *Am. J. Physiol.* 244 (*Renal Fluid Electrolyte Physiol.* 13): F639–F645, 1983.
125. YOUNG, J. A., AND E. W. VAN LENNEP. Transport in salivary and salt glands. In: *Membrane Transport in Biology. Transport Organs*, edited by G. Giebisch. Berlin: Springer-Verlag, 1979, p. 563–692.

CHAPTER 3

Calcium signaling system in salivary glands

JAMES W. PUTNEY, JR. | Calcium Regulation Section, National Institute of Environmental Health Sciences, Research Triangle Park, North Carolina

CHAPTER CONTENTS

Stimulus-Permeability Coupling in Salivary Glands
Phosphoinositides and Salivary Receptor Mechanisms
    Phosphoinositide turnover in salivary glands
    Pathways of phosphoinositide turnover in salivary glands
    Phosphoinositides and intracellular calcium release
    Mechanism of action of inositol-1,4,5-trisphosphate
    Calcium entry
    Capacitative calcium entry in parotid gland
Conclusions

SINCE RINGER (75) discovered the essential function of Ca^{2+} for cardiac muscle contraction, appreciation of the diverse biological roles of this divalent cation has grown considerably. One of the more recently studied biological actions of Ca^{2+} is its role as a second messenger for hormones and neurotransmitters that cause changes in membrane permeability in exocrine glands. These permeability changes probably are involved in the mediation of transepithelial water flow in these glands. By analogy with the terms *excitation-contraction coupling* (80) and *stimulus-secretion coupling* (78), the term *stimulus-permeability coupling* has been used to describe the sequence of events and the role of Ca^{2+} in the translation of agonist-receptor interaction into membrane permeability responses (65, 73).

Historically this concept developed primarily from research in three somewhat unrelated cell types: the erythrocyte, the vertebrate and invertebrate nerve cell, and the acinar cell of the parotid salivary gland. Studies in the erythrocyte originated with the observation that metabolic poisoning of these cells, which would elevate $[Ca^{2+}]$ by compromising active Ca^{2+} extrusion, inevitably led to an increase in membrane K^+ permeability (24, 33, 95). By using chelating agents, Gardos (30) demonstrated the Ca^{2+} requirement for this reaction. Accordingly the general phenomenon of a Ca^{2+}-mediated increase in K^+ permeability is often described as the Gardos effect. It was some 10 years later, however, when Whittam (95) first proposed that the intracellular concentration of ionized Ca^{2+} controlled the K^+ permeability in the erythrocyte. This contention has been largely confirmed by use of divalent Ca ionophores and resealed ghost preparations (47).

The second major area in which the role of Ca^{2+} in controlling K^+ permeability was developed emerged from studies of intracellular Ca^{2+} injections in nerve cells. Godfraind and co-workers (31, 32) observed that the uncoupling agent dinitrophenol increased the K^+ conductance of cortical neurons. This effect could be mimicked in spinal motoneurons by the direct injection of Ca^{2+} intracellularly (45), suggesting that, as for the erythrocyte, an increase in intracellular Ca^{2+} probably mediates the effect of the cellular poison. In similar experiments, Meech (53, 54) observed an increase in K^+ conductance on injection of Ca^{2+} into *Aplysia* nerve cells.

The first demonstration of a requirement for Ca^{2+} in membrane permeability regulation through receptors came from studies of rat parotid tissue by Schramm, Selinger, and others (5–7, 81–83). Muscarinic-cholinergic or α-adrenergic agonists caused a sizable loss of K^+ from the parotid salivary gland acinar cells, and extracellular Ca^{2+} was found to be an absolute requirement for this reaction (82). The effect could also be mimicked by the divalent Ca ionophore A23187 (83). Accordingly these investigators proposed that receptor activation led to a stimulation of Ca^{2+} influx, and the resulting increase in cytosolic Ca^{2+} in some manner increased membrane K^+ permeability. This hypothesis forms the basic framework of the general hypothesis of Ca^{2+} and stimulus-permeability coupling: receptor occupation → Ca^{2+} mobilization → activation of Ca^{2+}-dependent permeabilities.

In this chapter the various steps in this sequence are analyzed in the salivary glands and to a lesser extent in other mammalian exocrine systems that have been sufficiently characterized to permit meaningful comparisons and generalizations. In addition to the action of Ca^{2+} on membrane permeability, the mechanisms of Ca^{2+} mobilization by receptors are also discussed.

This specific subject was covered in an extensive review in 1978 (65); therefore emphasis is placed on the most recent developments in this area. In addition there have been many recent reviews dealing with topics related to those discussed here (8, 10, 12, 14, 20, 22, 27, 36, 48, 66, 67, 78).

STIMULUS-PERMEABILITY COUPLING IN SALIVARY GLANDS

As mentioned previously, in studies of K^+ release from the rat parotid gland, Schramm and Selinger (81) were the first to demonstrate that Ca^{2+} could serve as a second messenger for receptor-mediated alterations in membrane permeability. Thus these investigators and others found that slices of rat parotid gland would release 20%–40% of intracellular K^+ into the medium on incubation with α-adrenergic or muscarinic-cholinergic agonists (6, 7, 81). Extracellular Ca^{2+} was required for this effect (72, 81), and the response could be mimicked by the divalent Ca ionophore A23187 (83). On the basis of these results, it was proposed that muscarinic and α-adrenergic receptor activation stimulates Ca^{2+} influx into the acinar cell, and the resulting elevation in cytosolic Ca^{2+} in some manner activates the K^+-release mechanism (83). Several investigators demonstrated that stimulation of these same receptor pathways leads to an enhanced cellular Ca^{2+} uptake (42, 44, 56, 63, 72). Subsequently, it was established that a peptide receptor, activated by substance P and congeners, similarly initiates Ca^{2+}-dependent K^+ efflux (79) and also stimulates $^{45}Ca^{2+}$ uptake (72). Similar investigations have shown that comparable mechanisms operate in the submaxillary (13, 51, 52, 74, 86) and the sublingual (69) glands.

Electrophysiological studies of salivary gland cells have also detected receptor-mediated alterations in membrane permeability. Thus, Pedersen and Peterson (59, 60) found that acetylcholine or epinephrine caused a transient hyperpolarization of parotid acinar cells and concluded that the effect was probably due to an increased K^+ conductance. In contrast to results obtained by others measuring net K^+ release, however, they found that hyperpolarization was unaffected by Ca^{2+} omission or by chelating agents.

This discrepancy was resolved when unidirectional isotope-flux techniques were employed to examine K^+ movements in the parotid gland. Thus carbachol, phenylephrine, and substance P were found to stimulate unidirectional efflux of $^{42}K^+$ or $^{86}Rb^+$ from parotid gland slices or cells (18, 50, 62). This response was biphasic; an initial four- to fivefold increase in $^{86}Rb^+$ efflux fell after 4–6 min to a lesser but significantly elevated (about twofold) rate [Fig. 1; (62)]. When Ca^{2+} was omitted from the medium, the initial transient increase was little affected, but the later sustained phase of the response was completely

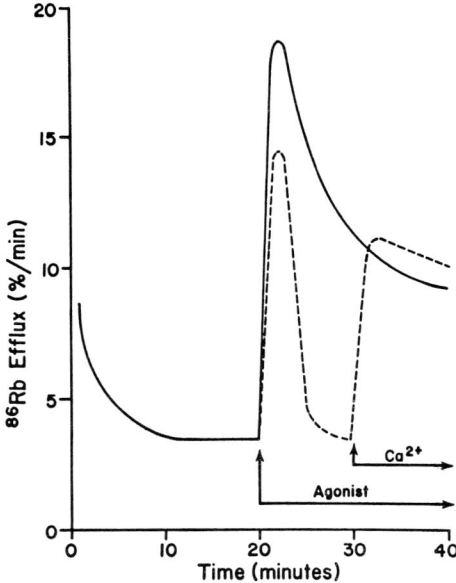

FIG. 1. Efflux of ^{86}Rb from salivary gland. Ordinate values represent apparent first-order rate coefficients (×100%). Agonist (i.e., acetylcholine) is present from 20 to 40 min. Solid line, Ca^{2+} present throughout. Dotted line, Ca^{2+} absent but restored from 30 to 40 min. [From Putney (65).]

blocked. Thus the transient hyperpolarization observed in the electrophysiological studies probably results from the short-lived rapid increase in K^+ permeability that does not require Ca^{2+} in the medium, whereas net K^+ loss from the tissue requires the longer-lasting sustained phase of the response that is Ca^{2+} dependent (65).

The biphasic nature of the $^{86}Rb^+$-efflux response in the parotid gland has been useful in disclosing the mechanism of action of pharmacological antagonists. Neomycin, La^{3+}, Co^{2+}, and Ni^{2+} were found to block selectively the Ca^{2+}-dependent phase of the $^{86}Rb^+$-efflux response with little or no effect on the transient phase (50). The local anesthetics procaine and tetracaine and the putative Ca^{2+} antagonist D 600 (methoxyverapamil) blocked both phases of the response equally, suggesting an alternative mode of action for these compounds (50). Because the responses to carbachol and phenylephrine but not to substance P and A23187 were blocked (50, 56, 57), an early step in the reaction sequence would seem to be affected. In fact procaine blocks muscarinic receptor binding in the parotid gland (71), and D 600, lidocaine, and tetracaine have been shown to block α-adrenergic and muscarinic receptors in brain homogenates (28). These results suggest caution in extrapolating demonstrated mechanisms of drug action from one tissue, such as smooth muscle, to nonexcitable tissues, such as the salivary glands.

Although the transient phase of the $^{86}Rb^+$ efflux apparently does not result from Ca^{2+} influx, there is considerable evidence that this phase of the response may be mediated by Ca^{2+} released from some intracellular store. Thus, after the stimulation of a transient release of $^{86}Rb^+$ by an agonist in the absence of Ca^{2+}, a second challenge will not produce a second response unless the tissue is at least temporarily returned to a Ca^{2+}-containing medium (64). This relationship holds whether or not the second agonist acts on the same receptor or a different receptor from the first agonist. Also, sizable release of $^{45}Ca^{2+}$ occurs from tissue treated with carbachol or substance P, and this release exceeds the Ca^{2+} influx occurring during the same interval (20, 65, 66). Finally, prolonged (60-min) incubation of dispersed parotid cells in a medium without added Ca^{2+} and containing a chelating agent prevented a subsequent transient response (18). Readdition of Ca^{2+} to the medium for 2–4 min restored the capacity of the cells to respond normally (18). These results indicate that in the parotid gland, receptor activation leads to a release of Ca^{2+} from some cellular store into the cytoplasm and the resulting increase in intracellular ionized Ca^{2+} activates the K^+ permeability pathway. Moreover the data indicate that each of the three receptors (muscarinic, α-adrenergic, and substance P) regulates the same cellular Ca^{2+} pool (64).

Similar conclusions were reached concerning the regulation of Ca^{2+} entry by these receptors. Thus carbachol, epinephrine, and substance P, when added in combination, did not summate in stimulating the Ca^{2+}-dependent phase of $^{86}Rb^+$ efflux, despite the fact that a limiting (1.0 mM) Ca^{2+} concentration was employed (50). Direct measurement of $^{45}Ca^{2+}$ influx produced a similar result (72). Thus the conclusion is that the three receptors, in addition to mobilizing a common pool of cellular Ca^{2+}, also activate a common entry mechanism.

Recently it has become possible to examine more directly changes in cytosolic $[Ca^{2+}]$ in cells by using the penetrating intracellular Ca^{2+} indicator quin 2 (91). Takemura (90) has used this technique to measure cytosolic free $[Ca^{2+}]$ in isolated rat parotid acinar cells. Resting $[Ca^{2+}]$ in these cells was estimated to be 163 nM and was approximately doubled by 10 μM or 100 μM carbachol (90). In the presence of extracellular Ca^{2+}, this increase in intracellular $[Ca^{2+}]$ was maintained for the duration of the measurement (~10 min). In the absence of extracellular Ca^{2+}, an increase in cytosolic $[Ca^{2+}]$ due to carbachol was still observed, but the response was diminished in magnitude and was transient, lasting only 3–4 min (90).

Although the release of K^+ has been traditionally the simplest ionic flux to quantitate, a sizable increase in Na^+ influx also occurs because of receptor activation in the salivary glands. Thus in more recent experiments when Petersen and co-workers (29, 76, 77) applied agonists to parotid cells by microiontophoresis, often depolarization rather than hyperpolarization was observed. The reversal potential for this effect was about −60 mV. Ion-substitution experiments led to the conclusion that membrane permeability to Na^+ and K^+ was increased. In support of this, uptake of $^{22}Na^+$ by parotid slices or dispersed cells is markedly enhanced by carbachol, epinephrine, or substance P (46, 70). This uptake of $^{22}Na^+$ was dependent on the extracellular $[Ca^{2+}]$, and the response to agonists could be mimicked by the divalent Ca ionophore A23187 (46), completing the parallel with the Ca^{2+}-activated K^+ efflux.

Receptor activation also leads to an enhanced Na^+-K^+ pump activity, as indicated by the ability of carbachol to stimulate uptake of $^{86}Rb^+$ by parotid gland slices (70). This effect could be blocked by omission of Ca^{2+}, by decreasing external Na^+, or by ouabain. The conclusion was that receptor stimulation, by a Ca^{2+}-dependent process, causes Na^+ influx, and the resulting elevation in intracellular $[Na^+]$ activates the Na^+-K^+ pump (70). Consistent with this interpretation are the results of electrophysiological studies carried out by Roberts et al. (76). These investigators found that the initial rapid potential change due to acetylcholine was generally followed by a delayed hyperpolarization. This hyperpolarization was relatively insensitive to alterations in membrane potential, required the presence of Na^+ and K^+, and could be blocked by ouabain (76).

The ionic events after receptor activation in the parotid gland can therefore be summarized as follows: *1)* receptor activation leads to a mobilization of Ca^{2+} (both release and influx); *2)* the increase in intracellular $[Ca^{2+}]$ in some manner activates permeabilities to Na^+ and K^+ with resulting Na^+ influx and K^+ efflux; and *3)* the increase in intracellular $[Na^+]$ activates the Na^+-K^+ pump. The manner in which these phenomena interact to effect the transepithelial flow of water is unknown, but a general premise of exocrine gland function is that fluid secretion is a functional consequence of modified monovalent ion fluxes or transport (65). The essential ingredients of increased passive ionic permeability and increased active transport have been demonstrated, however, and these events are all seemingly initiated by Ca^{2+} as a second messenger. The physiological significance of this mechanism is supported by reports that water secretion by intact salivary glands is a Ca^{2+}-dependent process (cf. ref. 65).

Although not the central theme of this chapter, some mention should be made of the roles of the cyclic nucleotides in the salivary glands. Stimulation of cAMP formation through β-adrenergic receptors in the parotid gland leads to a considerable rate of exocytosis of α-amylase (20, 81), but little or no effects on membrane permeability have been noted (41). On the other hand, muscarinic and α-adrenergic agonists increase the parotid levels of cGMP (17, 19, 21). Submicromolar concentrations of 8-bromo cGMP stimulated K^+ efflux from dispersed parotid cells, and K^+ efflux due to carbachol could be potentiated by a phosphodiesterase inhibitor (49). On the basis of these observations, it was concluded that cGMP may be a mediator of the muscarinic receptor in the parotid gland (49). Observations from several other laboratories do not support this contention. First, Butcher and associates (19, 21) found that the phosphodiesterase inhibitor methylisobutylxanthine enhanced cGMP formation by submaximal concentrations of phenylephrine or carbachol but did not affect the stimulation of K^+ release. Second, other investigators have failed to reproduce the effect of 8-bromo cGMP on K^+ efflux (cf. ref. 20). Third, substance P can produce all of the ionic effects of muscarinic stimuli but does not increase tissue levels of cGMP (79).

PHOSPHOINOSITIDES AND SALIVARY RECEPTOR MECHANISMS

Phosphoinositide Turnover in Salivary Glands

Hokin and Hokin (38) first reported an enhanced turnover of phosphatidylinositol (PI) and its precursor phosphatidic acid (PA) (measured as ^{32}P incorporation from inorganic phosphate) with cholinergic stimulation of exocrine pancreas. Hokin and Sherwin (37) subsequently described a similar phenomenon for the salivary glands. The possibility initially considered was that this reaction might be involved directly in the secretory process, but this idea was soon discounted when it was shown that omission of Ca^{2+} blocked pancreatic secretion due to acetylcholine but did not block PI turnover (35). It was later shown that for the parotid gland, β-adrenoceptor activation, which is the most efficient pathway for inducing enzyme secretion, did not induce PI turnover (58); however, activation of receptors associated with Ca^{2+} mobilization (muscarinic, α-adrenergic, substance P) caused a substantial effect (58, 94).

Based on his findings with the parotid gland, and the results of others, in 1975 Michell (55) suggested that turnover of inositol lipids might couple receptor activation to cellular Ca^{2+} mobilization. He based this idea on three circumstantial points of evidence (55): *1*) the PI effect was almost invariably associated with Ca^{2+}-mobilizing receptors and never, for example, with receptors linked to adenylate cyclase; *2*) the PI effect was relatively resistant to Ca^{2+} depletion; and *3*) the PI effect was not activated by Ca ionophores. These observations suggested to Michell that the PI effect was not a consequence of cellular Ca^{2+} mobilization and thus might precede Ca^{2+} mobilization in the stimulus-response coupling pathway (55).

The actions of receptors in regulating PI turnover in dispersed parotid acinar cells were investigated by Weiss and Putney (94). Methacholine (a muscarinic-cholinergic agonist), epinephrine (with propranolol, an α-agonist), and substance P each substantially increased cellular PI labeling (by $^{32}PO_4$ added to the incubation medium). The maximum effects of the three agonists differed markedly, however. The rank order was methacholine > epinephrine > substance P, in proportions of ~1.0:0.4:0.25. One possible explanation for such a discrepancy is that it reflects the relative numbers of the three kinds of receptors present. However, this suggestion was essentially ruled out by the results of experiments in which the agonists were applied in combination. Thus methacholine and epinephrine, as well as methacholine and substance P, when applied in combination gave responses intermediate between the responses obtained with each agonist alone; i.e., epinephrine partially inhibited the response to methacholine. Furthermore, when both epinephrine and substance P were applied, the response was no greater than that obtained with epinephrine alone (94). These results indicate that, as previously shown for Ca^{2+} mobilization, the three receptors regulated a common step (probably a phospholipase C) in the PI turnover pathway. In addition they indicate that the differences in maximum response are not due to differences in receptor number but probably reflect differences in the efficiency with which the different receptors, when activated, in turn can activate the appropriate PI-metabolizing enzyme(s).

Pathways of Phosphoinositide Turnover in Salivary Glands

When chemical determinations of PI and PA have been made (instead of or in addition to radiochemical labeling measurements), almost invariably receptor activation has been found to decrease PI and increase PA levels (55). Accordingly, the initial reaction involved in the PI effect was suggested to be the breakdown of PI to diacylglycerol (DG) and inositol phosphate by a PI-specific phospholipase C or phosphodiesterase (55).

Most cell types are capable of synthesizing phosphorylated derivatives of PI, the polyphosphoinositides [Fig. 2; (55)]. Specific kinases add monoester phosphates first to position 4 on the inositol ring to form phosphatidylinositol-4-phosphate (PIP) and then to position 5 to form phosphatidylinositol-4,5-bisphosphate (PIP_2). Specific phosphomonoesterases also exist that sequentially remove position-5 phosphates and then position-4 phosphates. These monoester phosphates turn over rapidly through a futile cycle that is fueled by and dependent on a constant level of ATP. Abdel-Latif et al. (1) first demonstrated an agonist-mediated decline in polyphosphoinositides in iris smooth muscle. Kirk and co-workers (43) observed a similar phenomenon in vasopressin-stimulated hepatocytes and suggested that the initial reaction involved in receptor-regulated PI turnover might be the phosphodiesteratic (phospholipase C) breakdown of PIP_2, PIP, or both, rather than PI as was originally supposed.

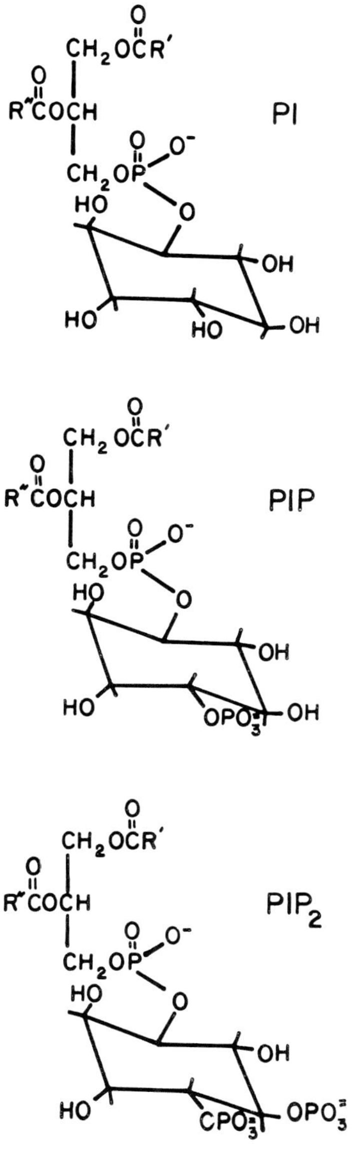

FIG. 2. Structures of phosphatidylinositol (PI), phosphatidylinositol-4-phosphate (PIP), and phosphatidylinositol-4,5-bisphosphate (PIP_2).

formation of ^{3}H-labeled inositol monophosphate (IP), inositol bisphosphate (IP_2), and inositol trisphosphate (IP_3). Both groups found that in the initial few seconds after receptor activation, only IP_3 and IP_2 levels increased (Fig. 3). Furthermore, when turnover of the inositol phosphates was analyzed under quasi-steady-state conditions, IP_2 breakdown could quantitatively account for the rate of formation of IP. However, breakdown of IP_3 could only account for about one-half of the rate of appearance of IP_2. Thus it was concluded that in intact cells, no direct degradation of PI occurred but that both PIP_2 and PIP were hydrolyzed to DG and their respective inositol phosphates (3, 26).

An assumption in the original analysis by both groups was that IP_3 behaves as a kinetically homogeneous pool. However, if there were a small pool of IP_3 with a relatively rapid turnover rate, this could account for the extra IP_2 formation that Downes and Wusteman (26) as well as Aub and Putney (3) attributed to PIP hydrolysis. Recent evidence suggests that just such a pool of IP_3 does exist. Irvine et al. (40) have found that the [^{3}H]IP_3 formed in stimulated parotid cells is actually a mixture of two isomers, 1,3,4-IP_3 and 1,4,5-IP_3. The 1,4,5-IP_3 is quantitatively the minor fraction, but its turnover rate appears to be much faster than that of 1,3,4-IP_3 (39). Similar findings have been reported for the hepatocyte by Burgess

Studies of parotid acinar cells also disclosed a rapid, Ca^{2+}-independent fall in radiolabeled PIP_2 after muscarinic, α-adrenergic, or peptidergic (substance P) receptor stimulation (93). When techniques were developed for labeling with [^{3}H]inositol and extracting and separating [^{3}H]inositol phosphates (10), it was possible to carry out kinetic studies to determine the primary substrates involved in phosphoinositide turnover in the parotid gland. Studies carried out by Downes and Wusteman (26) and by Aub and Putney (3) reached similar conclusions. These investigators prelabeled the inositol lipids of parotid cells by incubating with [^{3}H]inositol and examined the rates of

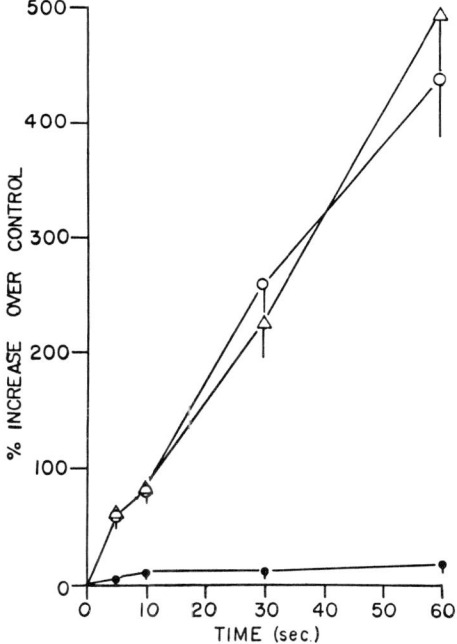

FIG. 3. Formation of [^{3}H]inositol phosphates in parotid acinar cells during initial 60 s after addition of 10^{-4} M methacholine. *Closed circles*, inositol monophosphate; *open circles*, inositol bisphosphate; *open triangles*, inositol trisphosphate. Values are means ± SE of 4 experiments. Control values did not change appreciably over 60-s period. [From Aub and Putney (3).]

et al. (16). Thus the direct breakdown of PIP is now less certain, and it is possible that in the parotid gland the PI cycle is initiated by a single enzymatic reaction, the phosphodiesteratic breakdown of PIP_2.

The biochemical picture of inositol phosphate metabolism has become considerably complicated by the recent disclosure of the existence of inositol tetrakisphosphate (IP_4), inositol pentakisphosphate (IP_5), and inositol hexakisphosphate (IP_6; phytic acid) in mammalian cells (4, 34). In brain slices, muscarinic receptor activation caused a rapid, parallel increase in IP_3 and IP_4 (4). In *Calliphora* salivary glands, serotonin increased both isomers of IP_3 and IP_4 but not IP_5 or IP_6 (34). The structure of IP_4 was elucidated by Batty et al. (4) and found to be 1,3,4,5-IP_4. Because a highly active 5-phosphomonoesterase exists that degrades 1,4,5-IP_3 to 1,4-IP_2 (87), it is likely that this same enzyme degrades IP_4 to 1,3,4-IP_3. The sources of IP_4, IP_5, and IP_6 have not been elucidated to date, but a likely possibility is that they are formed by phosphorylation of soluble inositol phosphates.

Phosphoinositides and Intracellular Calcium Release

Berridge (9) first suggested that phosphoinositide hydrolysis might couple receptor activation to Ca^{2+} mobilization through one or more of the soluble inositol phosphates acting as a second messenger to release sequestered Ca^{2+} from an intracellular pool. Streb et al. (89) tested this idea utilizing 1,4,5-IP_3 prepared from human erythrocytes in studies with permeable pancreatic acini. By a mechanism apparently quite similar to that for the salivary glands, the pancreas responds to muscarinic-cholinergic stimuli with a biphasic mobilization of Ca^{2+} (internal release followed by influx) and a rapid breakdown of polyphosphoinositides (68). When the plasma membranes of isolated acini were made permeable by a low-Ca^{2+} technique, the intracellular organelles of the permeable acini (in the presence of ATP) sequestered Ca^{2+} and reduced the $[Ca^{2+}]$ of the medium to ~400 nM, as determined by a Ca^{2+} electrode (89). When 1,4,5-IP_3 was added, the medium $[Ca^{2+}]$ increased rapidly, indicating release of some of the sequestered Ca^{2+} (89). The concentration of 1,4,5-IP_3 that produced a half-maximal effect was ~0.4 μM. The uptake of Ca^{2+} and its subsequent release were inhibited by vanadate but not by mitochondrial poisons, indicating that the IP_3-sensitive Ca^{2+} pool is probably a component of the endoplasmic reticulum (89). Burgess et al. (15) obtained essentially similar results using guinea pig hepatocytes made permeable with saponin and by measuring $^{45}Ca^{2+}$ exchange of the permeable cells. Again, 1,4,5-IP_3 caused release of Ca^{2+} from an internal pool, apparently a component of the endoplasmic reticulum (15). Subsequently, a number of investigators confirmed this ability of 1,4,5-IP_3 to release internal Ca^{2+} in a variety of systems (cf. ref. 12).

Recently this observation has been confirmed for the rat parotid salivary gland (2). In these experiments, parotid cells were made permeable with saponin and suspended in a medium containing the fluorescent Ca^{2+} indicator quin 2 so that the $[Ca^{2+}]$ surrounding the permeable cells could be monitored spectrophotometrically. In the presence of oligomycin (to eliminate mitochondrial Ca^{2+} uptake) and ATP, the permeable parotid cells lowered the medium $[Ca^{2+}]$ to ~200 nM, which is similar to the value for resting cytosolic $[Ca^{2+}]$ in the parotid gland determined with the quin 2 technique (90). Addition of 1,4,5-IP_3 caused a rapid increase in quin 2 fluorescence (Fig. 4), indicating release of Ca^{2+} from a nonmitochondrial pool. As shown previously in other systems, 2,4,5-IP_3 and 4,5-IP_2 had some activity in releasing Ca^{2+}, whereas 1,4-IP_2, 1-IP, and inositol had no effect. These results, taken with those just described, suggest that in the parotid salivary gland (and presumably in other salivary glands) receptor occupation leads to the phosphodiesteratic breakdown of PIP_2, which liberates 1,4,5-IP_3. The 1,4,5-IP_3 then signals the release of sequestered Ca^{2+} from a component of the endoplasmic reticulum, which results in a transient elevation in cytosolic Ca^{2+} and a transient increase in K^+ flux that is independent of extracellular $[Ca^{2+}]$.

Little is known of the biological activity of the other recently discovered inositol polyphosphates previously described. In the few systems thus far examined, IP_5 and IP_6 do not appear to increase in response to receptor activation. The levels of IP_4 are rapidly increased in the brain and in the *Calliphora* salivary gland, and presumably its biological activity or lack

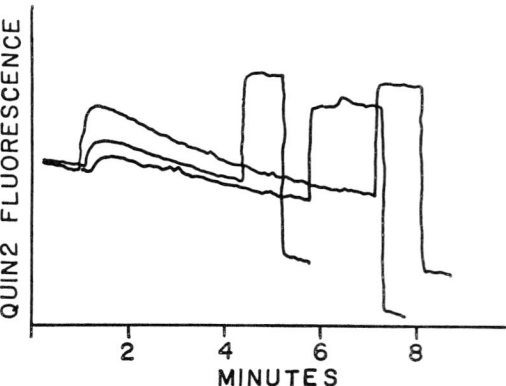

FIG. 4. Time course of inositol trisphosphate (IP_3)-stimulated Ca^{2+} mobilization. Isolated parotid acinar cells were permeabilized with saponin. $CaCl_2$ was added and cells were allowed to reequilibrate, after which inositol-1,4,5-trisphosphate (1,4,5-IP_3) was added ($t = 1$ min). Change in quin 2 fluorescence was monitored as an indicator of changing free $[Ca^{2+}]$. Excess Ca^{2+} was added after response was completed, followed by alkaline ethylene glycol-bis(β-aminoethylether)-N,N'-tetraacetic acid (EGTA) for calibration purposes. Representative traces for 3 concentrations of IP_3 are superimposed; 0.3 μM, 1 μM, and 3 μM IP_3 produced similarly graded increases in fluorescence. [From Aub (2).]

thereof will soon be reported. Indirect evidence has been presented that suggests that 1,3,4-IP$_3$ probably has no role in the action of Ca^{2+}-mobilizing agonists (16).

Mechanism of Action of 1,4,5-IP$_3$

In addition to 1,4,5-IP$_3$, certain other inositol phosphates have been tested for their relative abilities to release Ca^{2+}. Burgess et al. (15) found that 2,4,5-IP$_3$ and 4,5-IP$_2$ could also release Ca^{2+} from permeable guinea pig hepatocytes, but they were less potent than 1,4,5-IP$_3$. No Ca^{2+}-releasing activity was detected for 1,4-IP$_2$, 1-IP, cyclic 1,2-IP, or free inositol. On the basis of apparent structural requirements for Ca^{2+}-releasing activity, it was suggested that 1,4,5-IP$_3$ acts by binding to a specific site on the endoplasmic reticulum, an IP$_3$ receptor (15).

Recently, Spät et al. (85) have obtained more direct biochemical evidence for the existence of such a receptor. These investigators prepared ^{32}P-labeled 1,4,5-IP$_3$ with high specific radioactivity ($\sim$40 Ci/mmol) and were able to demonstrate specific and saturable binding of this radioligand to permeable hepatocytes and neutrophils. For the hepatocytes, competition experiments indicated that occupancy of this saturable site by 1,4,5-IP$_3$ and 2,4,5-IP$_3$ was well correlated with their relative abilities to release Ca^{2+}. In addition, occupancy of the site by 1,4,5-IP$_3$ in rabbit neutrophils occurred at lower concentrations than in the hepatocytes, which is consistent with the finding that the neutrophil also responds to lower concentrations of 1,4,5-IP$_3$ for releasing Ca^{2+}. Collectively these findings suggest that the binding site detected by Spät et al. (85) is in all likelihood the physiological receptor for 1,4,5-IP$_3$. For the guinea pig hepatocytes the cellular concentration of these receptors was estimated to be ~200 fmol/mg protein (85).

Calcium Entry

As just discussed, in salivary glands and indeed in a wide variety of receptor-regulated cell types, Ca^{2+} mobilization occurs in two phases: an initial release of Ca^{2+} from an internal pool followed by an accelerated entry of Ca^{2+} from the extracellular space. A number of hypotheses have been advanced to explain the regulation of Ca^{2+} entry by receptors, and the merits and disadvantages of some of these have been discussed in recent reviews (66, 67). More recently a hypothesis has been developed that provides a mechanism through which 1,4,5-IP$_3$ may mediate both phases of Ca^{2+} mobilization.

Because solid evidence suggests that 1,4,5-IP$_3$ releases Ca^{2+} by increasing the permeability of the endoplasmic reticulum to Ca^{2+}, it is not unreasonable to propose that 1,4,5-IP$_3$ might have a similar action at the plasma membrane. Recent data obtained from subcellular fractionation studies of exocrine pancreas argue against this idea, however (88). When IP$_3$-induced Ca^{2+} release was assayed in various subcellular fractions, a strong positive correlation (0.83–0.89) was found between this activity and biochemical markers for the endoplasmic reticulum, and a strong negative correlation (-0.81 to -0.77) was obtained for plasma membrane markers. Because the plasma membrane fractions presumably contained vesicles capable of ATP-dependent Ca^{2+} sequestration, an effect of IP$_3$ on plasma membrane permeability would have been detected as Ca^{2+} release from those vesicles.

Ueda et al. (92) recently examined the effects of 1,4,5-IP$_3$ on plasma membrane vesicles prepared from rat brain synaptosomes and on a microsomal endoplasmic reticulum preparation from N1E-115 neuronal cells. Both types of vesicles were capable of ATP-dependent ^{45}Ca^{2+} uptake; however, 1,4,5-IP$_3$ only released ^{45}Ca^{2+} from the endoplasmic reticulum vesicles.

On the other hand, recent electrophysiological studies by Slack et al. (84), in which 1,4,5-IP$_3$ was injected into sea urchin eggs, suggest that in intact cells, 1,4,5-IP$_3$ may mediate Ca^{2+} entry. If these findings can be generalized to other systems, they would suggest that 1,4,5-IP$_3$ may mediate Ca^{2+} entry in cells but not by acting directly on the plasma membrane. The model proposed in the next section provides such a mechanism.

Capacitative Calcium Entry in Parotid Gland

In 1977 it was suggested that in the parotid gland, Ca^{2+} release and entry were tightly linked processes, with the receptor-regulated Ca^{2+} pool being a binding site in the Ca^{2+} channel; i.e., an intermediate or capacitative step between extracellular Ca^{2+} and the cytosol (64). The major flaw in this model is probably the conception of the Ca^{2+} pool as a binding site. Present evidence suggests that the IP$_3$-sensitive pool is an ATP-dependent vesicular pool (15, 89). Its persistence in preparations whose plasma membranes have been treated with detergents (saponin, digitonin) and in the subcellular fractionation studies previously discussed suggests that the pool is distinct from the plasma membrane. However, an overwhelming body of circumstantial evidence still favors a close functional association between the pool and the plasma membrane. In intact cells the pool depends almost entirely on extracellular Ca^{2+} to maintain or restore its [Ca^{2+}] under conditions such that more than adequate cellular [Ca^{2+}] is available in other cellular pools (61, 64); refilling of the pool from the extracellular space occurs rapidly and without an apparent elevation in cytosolic [Ca^{2+}] (61, 64). The putative signal for release of Ca^{2+} from the pool (IP$_3$) is believed to be formed from PIP$_2$ at the plasma membrane (10, 12), and it has been shown that the enzyme that degrades IP$_3$ (a 5-phosphomonoesterase) is primarily located in the plasma membrane (87). Thus, because

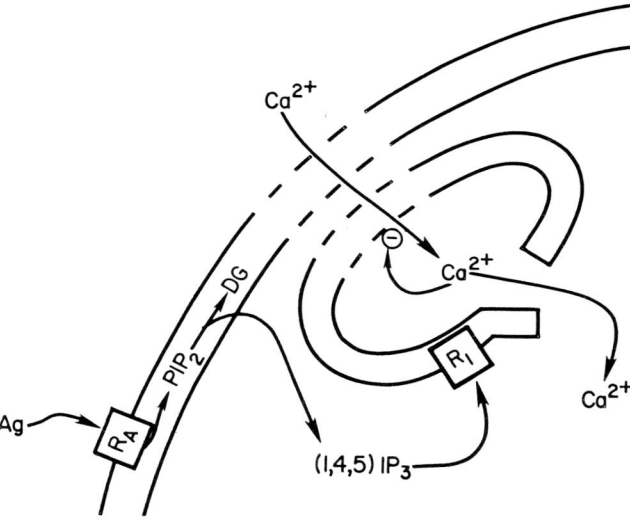

FIG. 5. Capacitative model for receptor regulation of Ca^{2+} release and entry. Agonist (Ag) binding to its receptor (R_A) leads to breakdown of phosphatidylinositol-4,5-bisphosphate (PIP_2) into diglyceride (DG) and inositol-1,4,5-trisphosphate (1,4,5-IP_3). 1,4,5-IP_3 binds to a receptor (R_I) on the receptor-regulated pool, presumably an endoplasmic reticulum component, which activates Ca^{2+} discharge to the cytosol and the Ca^{2+}-release phase of the response. Decrease in Ca^{2+} content of the pool relieves an inhibitory constraint on a direct pathway for Ca^{2+} to enter the pool from the extracellular space. With continued presence of 1,4,5-IP_3, Ca^{2+} continues down its concentration gradient to cytosol, resulting in a sustained Ca^{2+}-entry phase of the response. [From Putney (67).]

the pool is able to obtain Ca^{2+} from the extracellular space so rapidly, sustained Ca^{2+} entry to the cytosol could simply reflect the rapid refilling of a pool that is open to the cytosol because of the action of IP_3.

An important point to consider now is the question: What regulates Ca^{2+} entry into the receptor-regulated pool? It has previously been pointed out that refilling of the pool from outside of the cell after withdrawing an agonist occurs much more rapidly than the exchange of a loaded pool with $^{45}Ca^{2+}$. In the presence of Ca^{2+} chelators, the pool is quite stable; i.e., Ca^{2+} flux from a loaded pool to the extracellular space is much slower than Ca^{2+} flux from the extracellular space into an empty pool (61, 64). Finally, the rapid reloading of the pool has been shown to be independent of the time after the removal of the agonist (67). This latter finding indicates that signals generated by the agonist are probably not the major determinants of Ca^{2+} entry into the pool. Rather it suggests that it is the [Ca^{2+}] of the pool that determines its interaction with the extracellular space. More simply stated, an empty pool is open to the outside, whereas a loaded pool is not. The rapid entry of extracellular Ca^{2+} into the pool can only be demonstrated experimentally when the agonist has been withdrawn so that the loading of the pool can be assessed. However, it seems more than reasonable to suggest that as long as IP_3 is present in the cytosol and a low Ca^{2+} content of the pool is maintained, then Ca^{2+} entry into the pool will also be maintained. The result would be a sustained phase of Ca^{2+} entry. The appeal of this idea is that both phases of Ca^{2+} mobilization can be explained by the action of a single messenger substance, 1,4,5-IP_3.

The salient features of the proposed capacitative model are shown schematically in Figure 5. One shortcoming in terms of evidence for this hypothesis is that no clear morphological association between the endoplasmic reticulum and plasma membrane has been described for the cell types on which these arguments are based, largely nonexcitable cells. However, in smooth muscle, associations or couplings between sarcoplasmic reticulum and plasma membrane have been described (25). Furthermore, evidence for direct and rapid entry of extracellular Ca^{2+} into an agonist-regulated pool in arterial smooth muscle has been reported by Casteels and Droogmans (23), and these investigators have also suggested and provided evidence that such a pathway could contribute to receptor-regulated Ca^{2+} entry.

CONCLUSIONS

In salivary glands, a Ca^{2+}-messenger system exists that links surface membrane receptors to the control of epithelial ion fluxes that are probably important for fluid secretion. Receptor activation leads to a biphasic Ca^{2+} mobilization, composed of a component of intracellular Ca^{2+} release and a component of accelerated Ca^{2+} entry from the extracellular space. The resulting elevation in cytosolic Ca^{2+} leads to the opening of Ca^{2+}-sensitive monovalent ion channels and a subsequent efflux of K^+ and entry of Na^+. The elevation in cytosolic Na^+ secondarily activates the Na^+-K^+ pump.

Considerable evidence suggests that the mechanism by which receptors are coupled to Ca^{2+} mobilization involves the hydrolysis of surface membrane PIP_2, liberating soluble 1,4,5-IP_3. The 1,4,5-IP_3 binds to a specific receptor on the endoplasmic reticulum that signals Ca^{2+} release. The Ca^{2+}-release process may secondarily activate the Ca^{2+}-entry phase of cellular Ca^{2+} mobilization.

REFERENCES

1. ABDEL-LATIF, A. A., R. A. AKHTAR, AND J. N. HAWTHORNE. Acetylcholine increases the breakdown of triphosphoinositide of rabbit iris muscle prelabelled with [^{32}P]phosphate. *Biochem. J.* 162: 61–73, 1977.

2. AUB, D. L. Role of Inositol Phosphates in Stimulus-Response Coupling. Richmond: Virginia Commonwealth University, 1985. PhD thesis.

3. AUB, D. L., AND J. W. PUTNEY, JR. Metabolism of inositol

phosphates in parotid cells: implications for the pathways of the phosphoinositide effect and for the possible messenger role of inositol trisphosphate. *Life Sci.* 23: 1347–1355, 1984.

4. BATTY, I. R., S. R. NAHORSKI, AND R. F. IRVINE. Rapid formation of inositol 1,3,4,5-tetrakisphosphate following muscarinic receptor stimulation of rat cerebral cortical slices. *Biochem. J.* 232: 211–215, 1985.

5. BATZRI, S., A. AMSTERDAM, Z. SELINGER, I. OHAD, AND M. SCHRAMM. Epinephrine-induced vacuole formation in parotid gland cells and its independence of the secretory process. *Proc. Natl. Acad. Sci. USA* 68: 121–123, 1971.

6. BATZRI, S., Z. SELINGER, AND M. SCHRAMM. Potassium ion release and enzyme secretion: adrenergic regulation by α- and β-receptors. *Science Wash. DC* 174: 1029–1031, 1971.

7. BATZRI, S., Z. SELINGER, M. SCHRAMM, AND M. R. ROBINOVITCH. Potassium release mediated by the epinephrine α-receptor in rat parotid slices. Properties and relation to enzyme secretion. *J. Biol. Chem.* 248: 361–368, 1973.

8. BERRIDGE, M. J. Relationship between calcium and the cyclic nucleotides in ion secretion. In: *Mechanisms of Intestinal Secretion*, edited by H. J. Binder. New York: Liss, 1979, p. 65–81. (Kroc Found. Ser., vol. 12.)

9. BERRIDGE, M. J. Rapid accumulation of inositol trisphosphate reveals that agonists hydrolyze polyphosphoinositides instead of phosphatidylinositol. *Biochem. J.* 212: 849–858, 1983.

10. BERRIDGE, M. J. Inositol trisphosphate and diacylglycerol as second messengers. *Biochem. J.* 220: 345–360, 1984.

11. BERRIDGE, M. J., R. M. C. DAWSON, C. P. DOWNES, J. P. HESLOP, AND R. F. IRVINE. Changes in the levels of inositol phosphates following agonist-dependent hydrolysis of membrane phosphoinositides. *Biochem. J.* 212: 473–482, 1983.

12. BERRIDGE, M. J., AND R. F. IRVINE. Inositol trisphosphate, a novel second messenger in cellular signal transduction. *Nature Lond.* 312: 315–321, 1984.

13. BOGART, B. I., AND J. PICARELLI. Agonist-induced secretions and potassium release from rat submandibular gland slices. *Am. J. Physiol.* 235 (*Cell Physiol.* 4): C256–C268, 1978.

14. BOLTON, T. B. Mechanisms of action of transmitters and other substances on smooth muscle. *Physiol. Rev.* 59: 606–718, 1979.

15. BURGESS, G. M., R. F. IRVINE, M. J. BERRIDGE, J. S. McKINNEY, AND J. W. PUTNEY, JR. Actions of inositol phosphates on Ca^{2+} pools in guinea-pig hepatocytes. *Biochem. J.* 224: 741–746, 1984.

16. BURGESS, G. M., J. S. McKINNEY, R. F. IRVINE, AND J. W. PUTNEY, JR. Inositol 1,4,5-trisphosphate and inositol 1,3,4-trisphosphate formation in Ca^{2+}-mobilizing hormone-activated cells. *Biochem. J.* 232: 237–243, 1985.

17. BUTCHER, F. R. The role of calcium and cyclic nucleotides in α-amylase release from slices of rat parotid: studies with the divalent cation ionophore A-23187. *Metabolism* 24: 409–418, 1975.

18. BUTCHER, F. R. The calcium requirement for stimulation of rubidium efflux from isolated rat parotid acinar cells by carbachol. *Life Sci.* 24: 1979–1982, 1979.

19. BUTCHER, F. R., P. A. McBRIDE, AND L. RUDICH. Cholinergic regulation of cyclic nucleotide levels, amylase release and K^+ efflux from rat parotid glands. *Mol. Cell. Endocrinol.* 5: 243–254, 1976.

20. BUTCHER, F. R., AND J. W. PUTNEY, JR. Regulation of parotid gland function by cyclic nucleotides and calcium. In: *Advances in Cyclic Nucleotide Research*, edited by P. Greengard and A. Robison. New York: Raven, 1980, vol. 13, p. 215–249.

21. BUTCHER, F. R., L. RUDICH, C. EMLER, AND M. NEMEROVSKI. Adrenergic regulation of cyclic nucleotide levels, amylase release, and potassium efflux in rat parotid gland. *Mol. Pharmacol.* 12: 862–870, 1976.

22. CASE, R. M. Synthesis, intracellular transport and discharge of exportable proteins in the pancreatic acinar cell and other cells. *Biol. Rev.* 53: 211–354, 1978.

23. CASTEELS, R., AND G. DROOGMANS. Exchange characteristics of the noradrenaline-sensitive calcium store in vascular smooth muscle cells of rabbit ear artery. *J. Physiol. Lond.* 317: 263–279, 1981.

24. DAWSON, H. The effect of some metabolic poisons on the permeability of the rabbit erythrocyte to potassium. *J. Cell. Comp. Physiol.* 18: 173–185, 1941.

25. DEVINE, C. E., A. V. SOMLYO, AND A. P. SOMLYO. Sarcoplasmic reticulum and excitation-contraction coupling in mammalian smooth muscle. *J. Cell Biol.* 52: 690–718, 1972.

26. DOWNES, C. P., AND M. M. WUSTEMAN. Breakdown of polyphosphoinositides and not phosphatidylinositol accounts for muscarinic agonist-stimulated inositol phospholipid metabolism in rat parotid glands. *Biochem. J.* 216: 633–640, 1983.

27. EXTON, J. H. Mechanisms involved in α-adrenergic phenomena. *Am. J. Physiol.* 248 (*Endocrinol. Metab.* 11): E633–E647, 1985.

28. FAIRHURST, A. S., M. L. WHITTAKER, AND F. J. EHLERT. Interactions of D-600 (methoxyverapamil) and local anesthetics with rat brain α-adrenergic and muscarinic receptors. *Biochem. Pharmacol.* 29: 155–162, 1980.

29. GALLACHER, D. V., AND O. H. PETERSEN. Substance P increases membrane conductance in parotid acinar cells. *Nature Lond.* 283: 393–395, 1980.

30. GARDOS, G. The function of calcium in the potassium permeability of human erythrocytes. *Biochim. Biophys. Acta* 30: 653–654, 1958.

31. GODFRAIND, J. M., H. KAWAMURA, K. KRNJEVIĆ, AND R. PUMAIN. Actions of dinitrophenol and some other metabolic inhibitors on cortical neurones. *J. Physiol. Lond.* 215: 199–222, 1971.

32. GODFRAIND, J. M., K. KRNJEVIĆ, AND R. PUMAIN. Unexpected features of the action of dinitrophenol on cortical neurones. *Nature Lond.* 228: 562–564, 1970.

33. HENRIQUES, V., AND S. L. ORSKOV. Untersuchungen über die Schwankungen des Kationengehalts der roten Blutkörperchen. II. Änderung des Kaliumgehaltz der Blutkörperchen bei Bleivergiftung. *Skand. Arch. Physiol.* 74: 78–85, 1936.

34. HESLOP, J. P., R. F. IRVINE, A. H. TASHJIAN, AND M. J. BERRIDGE. Inositol tetrakis- and pentakisphosphates in GH_4 cells. *J. Exp. Biol.* 119: 395–401, 1985.

35. HOKIN, L. E. Effects of calcium omission on acetylcholine-stimulated amylase secretion and phospholipid synthesis in pigeon pancreas slices. *Biochim. Biophys. Acta* 115: 219–221, 1966.

36. HOKIN, L. E. Receptors and phosphoinositide-generated second messengers. *Annu. Rev. Biochem.* 54: 205–235, 1985.

37. HOKIN, L. E., AND A. L. SHERWIN. Protein secretion and phosphate turnover in the phospholipids in salivary glands in vitro. *J. Physiol. Lond.* 135: 18–29, 1957.

38. HOKIN, M. R., AND L. E. HOKIN. Effects of acetylcholine on phospholipids in the pancreas. *J. Biol. Chem.* 209: 549–558, 1954.

39. IRVINE, R. F., E. A. ANGGÅRD, A. J. LETCHER, AND C. P. DOWNES. Metabolism of inositol 1,4,5-trisphosphate and inositol 1,3,4-trisphosphate in rat parotid glands. *Biochem. J.* 229: 505–511, 1985.

40. IRVINE, R. F., A. J. LETCHER, D. J. LANDER, AND C. P. DOWNES. Inositol trisphosphates in carbachol-stimulated rat parotid glands. *Biochem. J.* 223: 237–243, 1984.

41. IWATSUKI, N., AND O. H. PETERSEN. Does exocytosis influence membrane resistance in mouse parotid acini? *J. Physiol. Lond.* 292: 81P–82P, 1979.

42. KANAGASUNTHERAM, P., AND P. J. RANDLE. Calcium metabolism and amylase release in rat parotid acinar cells. *Biochem. J.* 160: 547–564, 1976.

43. KIRK, C. J., J. A. CREBA, C. P. DOWNES, AND R. H. MICHELL. Hormone-stimulated metabolism of inositol lipids and its relationship to hepatic receptor function. *Biochem. Soc. Trans.* 9: 377–379, 1981.

44. KOELZ, H. R., S. KONDO, A. L. BLUM, AND I. SCHULZ. Calcium ion uptake induced by cholinergic and α-adrenergic stimulation in isolated cells of rat salivary glands. *Pflugers Arch.* 370: 37–44, 1977.

45. KRNJEVIĆ, K., AND A. LISIEWICZ. Injections of calcium ions

into spinal motoneurones. *J. Physiol. Lond.* 225: 363–390, 1972.
46. LANDIS, C. A., AND J. W. PUTNEY, JR. Calcium and receptor regulation of radiosodium uptake by dispersed rat parotid acinar cells. *J. Physiol. Lond.* 297: 369–377, 1979.
47. LEW, V. L., AND H. G. FERREIRA. Ca transport and the properties of a Ca-sensitive K channel in red cell membranes. *Cur. Top. Membr. Transp.* 10: 217–277, 1978.
48. LEW, V. L. (editor). Ca-activated K^+ channels. *Cell Calcium* 4: 321–517, 1983.
49. MANGOS, J. A. Role of cGMP in isolated rat parotid acinar cells. *J. Dent. Res.* 57: 989–994, 1978.
50. MARIER, S. H., J. W. PUTNEY, JR., AND C. M. VAN DE WALLE. Control of calcium channels by membrane receptors in the rat parotid gland. *J. Physiol. Lond.* 279: 141–151, 1978.
51. MARTINEZ, J. R., AND D. O. QUISSEL. Potassium release from the rat submaxillary gland in vitro. II. Induction by parasympathomimetic secretagogues. *J. Pharmacol. Exp. Ther.* 199: 518–525, 1976.
52. MARTINEZ, J. R., D. O. QUISSEL, AND M. GILES. Potassium release from the rat submaxillary gland in vitro. I. Induction by catecholamines. *J. Pharmacol. Exp. Ther.* 198: 385–394, 1976.
53. MEECH, R. W. Intracellular calcium and the control of membrane permeability. In: *Calcium in Biological Systems*, edited by C. J. Duncan. Cambridge, UK: Cambridge Univ. Press, 1976, p. 161–191.
54. MEECH, R. W. Calcium-dependent potassium activation in nervous tissues. *Annu. Rev. Biophys. Bioeng.* 7: 1–18, 1978.
55. MICHELL, R. H. Inositol phospholipids and cell surface receptor function. *Biochim. Biophys. Acta* 415: 81–147, 1975.
56. MILLER, B. E., AND D. L. NELSON. Calcium fluxes in isolated acinar cells from rat parotid: the effect of adrenergic and cholinergic stimulation. *J. Biol. Chem.* 252: 3629–3636, 1977.
57. MILLER, B., D. L. NELSON, AND F. R. BUTCHER. Tetracaine blocks the responses of isolated acinar cells from rat parotid to carbachol but not to substance P. *Biochim. Biophys. Acta* 587: 446–454, 1979.
58. ORON, Y., M. LOWE, AND Z. SELINGER. Involvement of the α-adrenergic receptor in the phospholipid effect in rat parotid. *FEBS Lett.* 34: 198–200, 1973.
59. PEDERSEN, G. L., AND O. H. PETERSEN. Membrane potential measurement in parotid acinar cells. *J. Physiol. Lond.* 234: 217–227, 1973.
60. PETERSEN, O. H., AND G. L. PEDERSEN. Membrane effects mediated by alpha- and beta-adrenoceptors in mouse parotid acinar cells. *J. Membr. Biol.* 16: 353–362, 1974.
61. POGGIOLI, J., AND J. W. PUTNEY, JR. Net calcium fluxes in rat parotid acinar cells: Evidence for a hormone-sensitive calcium pool in or near the plasma membrane. *Pfluegers Arch.* 392: 239–243, 1982.
62. PUTNEY, J. W., JR. Biphasic modulation of potassium release in rat parotid gland by carbachol and phenylephrine. *J. Pharmacol. Exp. Ther.* 198: 375–384, 1976.
63. PUTNEY, J. W., JR. Stimulation of ^{45}Ca influx in rat parotid gland by carbachol. *J. Pharmacol. Exp. Ther.* 199: 526–537, 1976.
64. PUTNEY, J. W., JR. Muscarinic, α-adrenergic and peptide receptors regulate the same calcium influx sites in the parotid gland. *J. Physiol. Lond.* 268: 139–149, 1977.
65. PUTNEY, J. W., JR. Stimulus-permeability coupling: role of calcium in the receptor regulation of membrane permeability. *Pharmacol. Rev.* 30: 209–245, 1978.
66. PUTNEY, J. W., JR. Identification of cellular activation mechanisms associated with salivary secretion. *Annu. Rev. Physiol.* 48: 75–88, 1986.
67. PUTNEY, J. W., JR. A model for receptor-regulated calcium entry. *Cell Calcium* 7: 1–12, 1986.
68. PUTNEY, J. W., JR., G. M. BURGESS, S. P. HALENDA, J. S. MCKINNEY, AND R. P. RUBIN. Effects of secretagogues on [^{32}P]phosphatidylinositol 4,5-bisphosphate metabolism in the exocrine pancreas. *Biochem. J.* 212: 483–488, 1983.
69. PUTNEY, J. W., JR., B. A. LESLIE, AND S. H. MARIER. Calcium and the control of potassium efflux in the sublingual gland. *Am. J. Physiol.* 235 (*Cell Physiol.* 4): C128–C135, 1978.
70. PUTNEY, J. W., JR., AND R. J. PAROD. Calcium-mediated effects of carbachol on cation pumping and Na uptake in rat parotid gland slices. *J. Pharmacol. Exp. Ther.* 205: 449–458, 1978.
71. PUTNEY, J. W., JR., AND C. M. VAN DE WALLE. The relationship between muscarinic receptor binding and ion movements in the rat parotid gland. *J. Physiol. Lond.* 299: 521–531, 1980.
72. PUTNEY, J. W., JR., C. M. VAN DE WALLE, AND B. A. LESLIE. Receptor control of calcium influx in parotid acinar cells. *Mol. Pharmacol.* 14: 1046–1053, 1978.
73. PUTNEY, J. W., JR., AND S. J. WEISS. Relationship between receptors, calcium channels and responses in exocrine gland cells. In: *Methods in Cell Biology. Basic Mechanisms of Cellular Secretion*, edited by D. Prescott and A. R. Hand. New York: Academic, 1981, vol. 23, p. 503–511.
74. QUISSELL, D. O. Secretory response of dispersed rat submandibular cells. I. Potassium release. *Am. J. Physiol.* 238 (*Cell Physiol.* 7): C90–C98, 1980.
75. RINGER, S. A further contribution regarding the influence of the different constituents of the blood on the contraction of the heart. *J. Physiol. Lond.* 4: 29–42, 1883.
76. ROBERTS, M. L., N. IWATSUKI, AND O. H. PETERSEN. Parotid acinar cells: ionic dependence of acetylcholine-evoked membrane potential changes. *Pfluegers Arch.* 376: 159–167, 1978.
77. ROBERTS, M. L., AND O. H. PETERSEN. Membrane potential and resistance changes induced in salivary gland acinar cells by microiontophoretic application of acetylcholine and adrenergic agonists. *J. Membr. Biol.* 39: 297–312, 1978.
78. RUBIN, R. P. *Calcium and Cellular Secretion.* New York: Plenum, 1982.
79. RUDICH, L., AND F. R. BUTCHER. Effect of substance P and eledoisin on K^+ efflux, amylase release and cyclic nucleotide levels in slices of rat parotid gland. *Biochim. Biophys. Acta* 444: 704–711, 1976.
80. SANDOW, A. Excitation-contraction coupling in skeletal muscle. *Pharmacol. Rev.* 17: 265–320, 1965.
81. SCHRAMM, M., AND Z. SELINGER. The functions of cyclic AMP and calcium as alternative second messengers in parotid gland and pancreas. *J. Cyclic Nucleotide Res.* 1: 181–192, 1975.
82. SELINGER, Z., S. BATZRI, S. EIMERL, AND M. SCHRAMM. Calcium and energy requirements for K^+ release mediated by the epinephrine α-receptor in rat parotid slices. *J. Biol. Chem.* 248: 369–372, 1973.
83. SELINGER, Z., S. EIMERL, AND M. SCHRAMM. A calcium ionophore simulating the action of epinephrine on the α-adrenergic receptor. *Proc. Natl. Acad. Sci. USA* 71: 128–131, 1974.
84. SLACK, B. E., J. E. BELL, AND D. J. BENOS. Inositol-1,4,5-trisphosphate injection mimics fertilization potentials in sea urchin eggs. *Am. J. Physiol.* 250 (*Cell Physiol.* 19): C340–C344, 1986.
85. SPÄT, A., P. G. BRADFORD, J. S. MCKINNEY, R. P. RUBIN, AND J. W. PUTNEY, JR. A saturable receptor for [^{32}P]inositol-(1,4,5)trisphosphate in hepatocytes and neutrophils. *Nature Lond.* 319: 514–516, 1986.
86. SPEARMAN, T. N., AND E. T. PRITCHARD. Potassium release from submandibular salivary gland in vitro. *Biochim. Biophys. Acta* 466: 198–207, 1977.
87. STOREY, D. J., S. B. SHEARS, C. J. KIRK, AND R. H. MICHELL. Stepwise enzymatic dephosphorylation of inositol-1,4,5-trisphosphate to inositol in liver. *Nature Lond.* 312: 374–376, 1984.
88. STREB, H., E. BAYERDÖRFFER, W. HAASE, R. F. IRVINE, AND I. SCHULTZ. Effect of inositol-1,4,5-trisphosphate on isolated subcellular fractions of rat pancreas. *J. Membr. Biol.* 81: 241–253, 1984.
89. STREB, H., R. F. IRVINE, M. J. BERRIDGE, AND I. SCHULZ. Release of Ca^{2+} from a nonmitochondrial intracellular store in pancreatic acinar cells by inositol-1,4,5-trisphosphate. *Nature Lond.* 306: 67–69, 1983.
90. TAKEMURA, H. Changes in cytosolic free calcium concentration in isolated rat parotid cells by cholinergic and α-adrenergic agonists. *Biochem. Biophys. Res. Commun.* 131: 1048–1055, 1985.

91. TSIEN, R. Y., T. POZZAN, AND T. J. RINK. Calcium homeostasis in intact lymphocytes: cytoplasmic free calcium monitored with a new, intracellularly trapped fluorescent indicator. *J. Cell Biol.* 94: 325–334, 1982.
92. UEDA, T., S. H. CHURCH, M. W. NOEL, AND D. L. GILL. Influence of inositol 1,4,5-trisphosphate and guanine nucleotides on intracellular calcium release within the N1E-115 neuronal cell line. *J. Biol. Chem.* 216: 3184–3192, 1986.
93. WEISS, S. J., J. S. MCKINNEY, AND J. W. PUTNEY, JR. Receptor-mediated net breakdown of phosphatidylinositol 4,5-bisphosphate in parotid acinar cells. *Biochem. J.* 206: 555–560, 1982.
94. WEISS, S. J., AND J. W. PUTNEY, JR. The relationship of phosphatidylinositol turnover to receptors and calcium-ion channels in rat parotid acinar cells. *Biochem. J.* 194: 463–468, 1981.
95. WHITTAM, R. Control of membrane permeability to potassium in red blood cells (Abstract) *Nature Lond.* 219: 610, 1968.
96. WILBRANDT, W. A relation between the permeability of the red cell and its metabolism. *Trans. Faraday Soc.* 33: 956–959, 1937.

CHAPTER 4

Cellular regulation of amylase secretion by the parotid gland

TERRY N. SPEARMAN
FRED R. BUTCHER

Department of Biochemistry, West Virginia University School of Medicine, Morgantown, West Virginia

CHAPTER CONTENTS

Composition of Rat Parotid Saliva
Ultrastructure of Rat Parotid Gland
Neurotransmitter Receptors
Parotid Slice System
Cyclic Adenosine 5′-Monophosphate
   Evidence for involvement in amylase release
   Mechanism of action
Calcium
   Evidence for involvement in amylase release
      Amylase release in calcium-free medium
      Effects on amylase release of treatments known to elevate intracellular calcium
      Calcium fluxes
      Effects of β-agonists on intracellular calcium levels
   Source of calcium mobilized by β-adrenergic agonists
   Mechanism of calcium mobilization
   Mechanism of action
Endogenous Protein Phosphorylation
Speculations

THREE ANATOMICALLY DISTINCT PAIRS of major salivary glands exist—the parotid, the submandibular (also referred to as the submaxillary), and the sublingual—in addition to a number of minor mucous glands. Although the mechanisms of macromolecule secretion appear similar in the different glands and in glands from various species, differences do exist, especially in the proteins secreted. The rat parotid gland is the primary focus of this chapter, but where indicated, results from other species or glands are discussed.

Mammalian salivary glands are innervated by both the sympathetic and parasympathetic branches of the autonomic nervous system (83). Generally stimulation of the parasympathetic nervous supply to the glands leads to a copious flow of dilute saliva, whereas stimulation of the sympathetic nervous supply leads to a much smaller volume of a viscous secretion (116). Parasympathetic nerve terminals in salivary glands use acetylcholine as a neurotransmitter, whereas norepinephrine is released by sympathetic terminals. In addition nerve fibers staining for substance P (58) and vasoactive intestinal peptide (VIP) (140) have been identified by immunohistochemistry, and both these substances have been detected in salivary gland homogenates by radioimmunoassay (22, 23, 140). There is evidence suggesting both function as neurotransmitters. Electric field stimulation of rat parotid gland in vitro has been used to induce secretion by liberating endogenous neurotransmitter from nerve endings within the gland (50). A portion of the resulting secretory response was resistant to inhibition by antagonists of acetylcholine and norepinephrine but was inhibited by an antagonist of substance P (50). Both substance P and VIP exert effects on parotid slices in vitro similar to those produced by other neurotransmitters. In the rat parotid gland, nerve terminals that release substance P and VIP appear to be components of the parasympathetic branch of the autonomic nervous system rather than an independent neural pathway. Parasympathetic denervation of the gland decreases the glandular level of substance P and VIP (22) and produces supersensitivity of the gland to substance P (45), whereas sympathetic denervation is essentially without effect. Lowered levels of specific neurotransmitters in the gland and supersensitivity to those neurotransmitters are well-known effects of severing the nerves that release these neurotransmitters at their synapses.

COMPOSITION OF RAT PAROTID SALIVA

Anionic and cationic electrophoresis of rat parotid saliva has revealed 21–22 distinct bands of protein (111). Four of these were shown to be isozymes of α-amylase, two were shown to be DNase, and one was RNase. Enzyme activity has not been identified for the others, including two proline-rich proteins (75). Purified parotid secretory granules appear to contain a protein profile identical with that of parotid saliva (75) and in addition contain Ca^{2+} (140) and L-ascorbic acid (138). Both Ca^{2+} (66, 139) and ascorbate (138) have been found in parotid saliva, suggesting they are secreted concomitantly with protein. Calcium has been found in secretory granules from several different secretory cell types; in parotid secretory granules, Ca^{2+} has been proposed to be involved in the packaging of proteins (139). Ascorbate has been found in secretory granules from the adrenal medulla and the pituitary gland, where it is believed to function as a cofactor in catecholamine synthesis and neuropeptide α-amidation, respectively. Its function in the parotid gland is unknown, but its presence in secretory granules from a cell where catecholamine and neuropeptides are not known to be synthesized may suggest it has a more general role than previously suspected. Amylase, DNase, and RNase have been shown to be secreted in parallel from parotid regardless of the type of stimulation (14, 15, 66). Thus in studies of parotid secretion usually only one enzyme is monitored. Because of the ease and sensitivity of its assay, α-amylase release is normally used as an indicator of protein secretion.

ULTRASTRUCTURE OF RAT PAROTID GLAND

The structure of the rat parotid gland is fairly typical of exocrine glands. The parotid contains two major cell types [acinar cells and duct cells (83)] with acinar cells occupying >85% of the volume of the gland. The acinar cells form secretory end pieces at the termini of the branching ductal system. Each acinus is composed of several acinar cells grouped around a central lumen continuous with the ductal system. Each acinar cell is pyramidal in shape. The nucleus occupies the basal end of the cell with the majority of the mitochondria and large quantities of endoplasmic reticulum (119). The apical portion of the cell contains large numbers of secretory granules. The Golgi apparatus is usually located between the nucleus and the secretory granules.

It is widely accepted that macromolecule secretion occurs through the process of exocytosis: the secretory granules containing the exportable protein fuse to the plasmalemma, and the fused secretory granule membrane–plasma membrane thin to a single bilayer and subsequently rupture at their point of contact. Thus the contents of the granule are released to the external milieu (lumen of ductal system) while preserving the integrity of the cell. Compound fusion has also been observed; i.e., a secretory granule will fuse to the plasmalemma and rupture, then a second granule will fuse to the first. On stimulation of the rat parotid gland, release of amylase was correlated with a depletion of secretory granules coupled with an increase in the size of the ductal lumen (4), which is consistent with the addition of secretory granule membrane to the apical plasmalemma as a consequence of exocytosis. This chapter reviews the mechanism and regulation of this process.

NEUROTRANSMITTER RECEPTORS

The existence of multiple neurotransmitter receptors in the rat parotid gland has been confirmed by binding studies with specific receptor agonists and antagonists. Thus the β-adrenergic receptor has been characterized by the use of the β-adrenergic antagonists $(-)$-$[^3H]$dihydroalprenalol (7, 24, 54) and $[^{125}I]$-iodohydroxybenzyl pindolol (89). Two studies reported linear Scatchard plots (7, 54), indicating a single binding site with no cooperativity among sites, whereas a third (24) reported concave upward plots, indicating either negative cooperativity or multiple sites of different affinity for the ligand. Experiments examining the ability of a number of β-adrenergic agonists and antagonists to compete with dihydroalprenolol binding (7) or to elicit or block amylase release (33, 34, 49) indicate that the parotid contains mainly the β_1-subtype. Similarly, α-adrenergic receptors have been characterized by the binding of $[^3H]$-dihydroergocryptine (36, 62, 130), $[^3H]$prazosin (62), and $[^3H]$WB-4101 (54). The number of α-adrenergic receptors per cell has been estimated to be ~15,000 (130), divided approximately equally between α_1- and α_2-subtypes (62). Norepinephrine, the neurotransmitter released in vivo on stimulation of the sympathetic branch of the autonomic nervous system, does not discriminate between α_1- and α_2-, β_1- and β_2-, or α- and β-adrenergic receptors.

The parotid muscarinic cholinergic receptor has been characterized by the use of $[^3H]$quinuclidinyl benzilate (54, 59, 89, 105). The receptor number per cell has been calculated as 21,400 (59) or 23,000 (105). A receptor for substance P has been demonstrated by ^{125}I-labeled N(1)acylated substance P (86), substance P labeled with $[^{125}I]$-labeled Bolton-Hunter reagent (87), and $[^{125}I]$-labeled physalaemin (107). Both substance P and physalaemin bind to the same receptor, as evidenced by the specific displacement of labeled substance P by physalaemin (86). The number of peptide receptors per cell has been estimated to be ~215 (107). Recently rat parotid acinar cells have also been shown to possess receptors for VIP (38, 60); ^{125}I-labeled VIP demonstrated 41,000 high-affinity sites per cell (60).

Dopamine has recently been reported to produce effects on the rat parotid gland (1, 131), but specific receptors for this substance have yet to be demonstrated and its mechanism of action is not clear. Its effects were blocked by β-adrenergic antagonists (1, 131), and levels of dopamine 10-fold higher were required to induce comparable effects in glands from rats treated with reserpine to deplete endogenous catecholamine stores (131). Thus at least a portion of its effects apparently are due to release of endogenous neurotransmitter in the tissue rather than a direct effect on acinar cells.

PAROTID SLICE SYSTEM

Major advances in understanding the regulation of parotid secretion began when Schramm and co-workers demonstrated that the rat parotid gland would secrete in vitro in response to neurotransmitters under appropriate conditions (15). These investigators concluded that amylase release is primarily regulated by the β-adrenergic receptor (10) and mediated by cAMP (16) and that rat parotid acinar cells possess α-adrenergic, β-adrenergic, and cholinergic receptors coexisting on the same cell (118). These conclusions serve as the basis for many of the studies reviewed here.

Most of the research discussed has been performed with parotid slices. However, enzymatically dissociated acinar cell preparations have been employed in some studies and yield essentially equivalent results. Unless the data pertain solely to slices or dissociated cells, the type of preparation used in each study is not specified.

CYCLIC ADENOSINE 5'-MONOPHOSPHATE

Evidence for Involvement in Amylase Release

Parotid acinar cells contain adenylate cyclase activity (5, 11, 24, 28, 117, 127) that is stimulated by β-adrenergic agonists (5, 11, 24, 28, 117). β-Adrenergic agonists (28, 30, 42, 49, 53, 54, 84, 143) and VIP (38, 60, 120), which also induces amylase release (38, 60, 120), elevate parotid cAMP levels. Derivatives of cAMP able to traverse cellular membranes such as $N^6,O^{2'}$-dibutyryl adenosine 3':5'-cyclic monophosphate (Bt$_2$cAMP) induce amylase release (8, 16, 31, 44, 67, 81, 88, 145). Phosphodiesterase inhibitors also induce amylase release and potentiate the effects of submaximal β-adrenergic stimulation (16, 49, 88, 143, 145) and the effects of VIP (120). Furthermore methods of activating adenylate cyclase not involving the β-adrenergic receptor also induce amylase release. Thus cholera toxin, believed to activate adenylate cyclase by a mechanism involving ADP-ribosylation of the GTP-binding regulatory subunit (36, 52), induces release (127), as does forskolin (42, 145), which activates adenylate cyclase by a mechanism that is not completely understood (119, 128).

There is, however, a wealth of data showing that the relationship between cAMP levels and amylase release is not simple. Low levels of β-adrenergic agonists induce measurable amylase release without producing detectable alterations in cAMP levels (28, 143). Ethyleneglycol-bis(β-aminoethylether)-N,N'-tetraacetic acid (EGTA) has been reported to inhibit the cAMP response to epinephrine without significantly inhibiting protein secretion (53). Theophylline, a phosphodiesterase inhibitor, induces protein secretion without affecting cAMP levels and potentiates the effects of the β-agonist isoproterenol without enhancing its effects on cAMP accumulation (3). Cyclic AMP accumulation has been reported to be higher with 0.2 μM norepinephrine plus 0.5 mM 3-isobutyl-1-methylxanthine (a phosphodiesterase inhibitor) than with 1 μM norepinephrine alone, although 1 μM norepinephrine was more effective in inducing amylase release (143). Similarly, less amylase was reported to be released by increased cAMP levels stimulated by forskolin than by similar increases in cAMP produced by β-agonists (145). Some β-adrenergic agonists have been reported to induce substantial amylase release without affecting cAMP levels (35, 126, 143). Different investigators have explained these results in different ways. Importantly, as pointed out by Afari et al. (3), experiments using phosphodiesterase inhibitors must be interpreted with caution, because there is a real possibility that these drugs have biological effects in addition to inhibiting the hydrolysis of cAMP. One possible explanation of the ability of β-adrenergic agonists to elicit amylase release without significantly elevating cAMP levels is that the exocytotic machinery responds to smaller increases in the cyclic nucleotide levels than can be reliably measured. It may also be that increases occur in a highly localized site in the cell; i.e., the production, utilization, and hydrolysis of cAMP occur in a very limited area and the increases in total cellular cAMP seen at higher concentrations of some β-agonists represent overshoot. Note that cytochemical techniques have located parotid adenylate cyclase on the luminal plasma membrane (77), i.e., on the membrane to which the secretory granules fuse. At least one of the β-adrenergic agonists that induces amylase release without elevating cAMP levels has been shown to elevate cAMP levels in a dose-dependent manner when parotid slices are incubated in the presence of a phosphodiesterase inhibitor. This agonist also increases the phosphorylation of endogenous proteins (in the absence of the inhibitor) whose phosphorylation is stimulated by cAMP in broken-cell preparations (126). These results suggest that this agonist induces release by a cAMP-dependent mechanism even though, by itself, it does not detectably increase intracellular cAMP levels.

More difficult to explain are instances where increases in cAMP levels are observed in the absence of significant amylase release. In one study terbutaline, a β_2-agonist, induced a substantial increase in cAMP levels accompanied by only a modest release of amylase, whereas the β_1-agonist prenalterol induced amylase release without significantly elevating cAMP levels (35). Similarly, cholera toxin induced a significant increase in cAMP levels 30 min before significantly increasing amylase release (127). If very small increases in cAMP produced by low concentrations of one β-agonist are sufficient to elicit significant amylase release, then comparable or larger increases in cAMP levels produced by other β-agonists, cholera toxin, or phosphodiesterase inhibitors would be expected to elicit comparable amylase release. One explanation for this discrepancy is that these other mechanisms of elevating cAMP might not increase the levels at the appropriate site. As previously discussed in NEUROTRANSMITTER RECEPTORS, p. 64, amylase release appears to be primarily regulated by the β_1-adrenergic receptor. However, terbutaline, believed to be relatively specific for the β_2-receptor, produces effects on rat parotid gland such as increases in cAMP levels in vitro (35) and increases in gland weight when injected in vivo (56). Therefore parotid acinar cells may possess distinct β_2-receptors. If the production, utilization for induction of amylase release, and degradation of cAMP occur in a highly localized area in the immediate vicinity of the β_1-receptor, then the cAMP produced by stimulation of the β_2-receptor may not increase cAMP levels at its site of utilization as effectively as that produced by stimulation of the β_1-receptor. Similarly, agents that activate adenylate cyclase independently of the β-receptor (e.g., cholera toxin, forskolin) would activate enzyme not associated with this specific site of utilization. A similar situation could occur with respect to elevations in cAMP induced by phosphodiesterase inhibitors if the hypothetical β_1-receptor–associated phosphodiesterase activity was not the sole site of cAMP degradation in parotid acinar cells. Although this hypothesis may account for some of the discrepancy between cAMP levels and exocytosis, it is not able to account for all the discrepancies reported to date. Clearly other factors are involved in the regulation of amylase release in addition to cAMP.

Mechanism of Action

Cyclic AMP is believed to exert its effects in eucaryotic cells by activating cAMP-dependent protein kinase (PK) and hence increasing the phosphorylation state of cAMP-dependent PK substrate proteins, thereby altering their properties or activity (if enzymes). Rat parotid glands have been shown to contain cAMP-dependent PK (31, 141), and β-adrenergic agonists (12, 13, 125, 144) and cholera toxin (127) have been reported to increase the ratio of activated to nonactivated cAMP-dependent PK in vitro. Tolbutamide, an inhibitor of cAMP-dependent PK, has been reported to inhibit amylase release induced by isoproterenol (73), norepinephrine (144), and Bt$_2$cAMP (81). However, results obtained with this compound must be interpreted with caution, because it has been reported that a concentration inhibitory to amylase release did not alter the ability of norepinephrine to elevate the ratio of activated to nonactivated cAMP-dependent PK (144). This may indicate that tolbutamide inhibits amylase release by a mechanism unrelated to cAMP-dependent PK inhibition or alternatively that tolbutamide does not interfere with the dissociation of cAMP-dependent PK holoenzyme but does inhibit the activity of the free catalytic subunit. If this latter theory is correct, the lack of inhibition could result from a dilution of tolbutamide, during homogenization of the tissue, to a level below that required for inhibition.

CALCIUM

Evidence for Involvement in Amylase Release

Several independent lines of investigation have been pursued to reveal and elucidate the role of Ca^{2+} in amylase release from the rat parotid gland. The main ones are *1*) examining the ability of β-adrenergic agonists and cyclic nucleotide derivatives to induce amylase release in Ca^{2+}-free media, *2*) determining whether treatments thought to affect intracellular Ca^{2+} levels stimulate amylase release themselves or modify the release induced by β-adrenergic agonists or cyclic nucleotide derivatives, *3*) examining the effects of β-adrenergic agonists and cyclic nucleotide derivatives on $^{45}Ca^{2+}$ fluxes, and *4*) measuring the effects of β-adrenergic agonists on intracellular Ca^{2+} levels with a fluorescent Ca^{2+} indicator.

AMYLASE RELEASE IN CALCIUM-FREE MEDIUM. Experiments on the induction of amylase release in Ca^{2+}-free media by β-adrenergic agonists and cyclic nucleotide derivatives have produced a great deal of conflicting data, much of which can be explained by differences in experimental protocols.

Epinephrine reportedly elicits more amylase release in the absence of extracellular Ca^{2+} than in its presence (9, 67). Batzri and Selinger (9) attributed this to the fact that the K^+-release response due to epinephrine activating the α-adrenergic receptor does not occur in the absence of extracellular Ca^{2+}, thereby conserving cellular ATP for exocytosis. However, a similar augmentation of amylase release in Ca^{2+}-free medium, compared with Ca^{2+}-containing medium, was found with isoproterenol (2), which activates only the β-adrenergic receptor. Most studies report little or no effect on amylase release when the tissue is stimulated at relatively short periods of time after extracellular Ca^{2+} removal or replacement with the Ca^{2+} chelator EGTA (40, 53, 84). However, most note a partial inhibition when the tissue is preincubated with EGTA in the absence of extracellular Ca^{2+} for a relatively long time before stimulation, whether epinephrine (40, 53), isoproterenol (94, 108), VIP (120), or Bt$_2$cAMP (108, 110, 122) is employed as the secretagogue. Selinger and Naim (122) reported that the readdition of extracellular Ca^{2+} restored the response to Bt$_2$cAMP, although Quissell et al. (110) obtained an opposite result.

Butcher (26), in determining the length of time parotid tissue needs to be preincubated in Ca^{2+}-free medium containing EGTA to affect amylase release, noted an inhibition of secretion immediately on removing Ca^{2+}, provided release was measured very soon after isoproterenol addition. When slices were stimulated in Ca^{2+}-free media they did not secrete for a short period of time, then began to release amylase at a rate similar to that for tissues stimulated in the presence of Ca^{2+}. Preincubating the tissue in Ca^{2+}-free media with EGTA for progressively longer times increased the length of lag period between agonist addition and the commencement of secretion (26). Argent and Arkle (5) reported similar data, except they found that longer preincubations with EGTA also began to lower the maximal rate of amylase release. Butcher (26) noted differences in the response to Bt_2cAMP and isoproterenol when extracellular Ca^{2+} was withdrawn; an immediate inhibition of amylase release was observed with both agents but the rate of release recovered to that obtained in the presence of Ca^{2+} only with isoproterenol. Because Bt_2cAMP was previously found to be a more effective inducer of amylase secretion than monobutyryl cAMP, whereas the monobutyryl derivative was a more effective activator of rat parotid cAMP-dependent PK (31), Butcher (26) proposed that Bt_2cAMP is intracellularly converted to a more active species in a Ca^{2+}-dependent reaction. However, data shown by Quissell et al. (110) indicate that under his conditions Bt_2cAMP induces the same rate of exocytosis in the absence and presence of extracelullar Ca^{2+} once the lag period seen in the absence of Ca^{2+} is over.

Butcher (26) also noted differences in the Ca^{2+} sensitivity of Bt_2cAMP, monobutyryl cAMP, and other cAMP derivatives; only the effects of Bt_2cAMP were influenced by the lack of extracellular Ca^{2+}. Putney (102) subsequently demonstrated that the effects of monobutyryl cAMP could also be inhibited by incubation in Ca^{2+}-free media but that a more extensive depletion of intracellular Ca^{2+} stores was required to demonstrate inhibition using the monobutyryl derivative than was required for the dibutyryl derivative.

Note that β-adrenergic agonists do not elevate cAMP levels to the same extent in Ca^{2+}-free media as they do in the presence of extracellular Ca^{2+} (5, 26, 53, 108). This may result from a direct effect on adenylate cyclase, since this enzyme has been reported to be sensitive to the concentration of this divalent cation in broken-cell preparations; maximal cyclase activity was observed at 10^{-7} M Ca^{2+}, with both higher and lower levels decreasing activity (5). However, it is unlikely that inhibition of cyclase is responsible for the decreased amylase release observed. β-Agonists appear to elevate cAMP to levels in excess of that required to produce maximal amylase release, and although β-agonists do not elevate cAMP levels in Ca^{2+}-free media to the same extent they do in the presence of extracellular Ca^{2+}, the increase in cAMP levels in the absence of extracellular Ca^{2+} is still substantial. If the inhibition of amylase release is due solely to inhibition of adenylate cyclase, secretion induced by cAMP derivatives should be unaffected by preincubation of the tissue in Ca^{2+}-free media. As discussed above, this is not the case.

In summary, although the literature contains conflicting data on some points, studies of the effects of Ca^{2+} removal on amylase release show that, although extracellular Ca^{2+} does not directly participate in exocytosis, Ca^{2+} derived from an intracellular pool is required at a step in amylase secretion after cAMP production.

EFFECTS ON AMYLASE RELEASE OF TREATMENTS KNOWN TO ELEVATE INTRACELLULAR CALCIUM. Although their major function is thought to be inducing the formation of the fluid component of saliva (see the chapter by Putney in this *Handbook*), α-adrenergic (18, 28, 30, 54, 67, 84, 143, 145) and cholinergic (29, 54, 57, 61, 67, 84, 112) agonists and substance P and related peptides (42, 50, 60, 85, 113) have all been reported to induce limited amylase release. Extracellular Ca^{2+} is required for these agents to exert this effect (42, 61, 67, 84, 113). They are believed to act by both mobilizing intracellular Ca^{2+} pools and increasing Ca^{2+} influx into the acinar cells. The Ca^{2+} ionophore A23187 has also been shown to induce limited amylase release (25, 67, 112).

Low concentrations of α-adrenergic (94, 136) and cholinergic (124, 133) agonists potentiate the effects of isoproterenol on amylase release; i.e., the release produced by the combination is greater than the sum of the two separately. The potentiation seen with carbachol, a cholinergic agonist, is not due to potentiation of cAMP accumulation (124, 133) or activation of cAMP-dependent PK (124). Potentiation may be due to an elevation of intracellular Ca^{2+} to a level more favorable for some step of the exocytotic process after cAMP-dependent PK activation. Although one study showed that α-adrenergic agonists were capable of inducing potentiation in the absence of extracellular Ca^{2+}, prolonged incubation in Ca^{2+}-free media abolished the effect (133). This may indicate that the intracellular Ca^{2+} pool mobilized by cholinergic and β-adrenergic agonists contains sufficient stores to elevate the intracellular level to that optimal for exocytosis in the absence of Ca^{2+} influx.

In contrast to the potentiation seen at low concentrations, high concentrations of carbachol inhibit isoproterenol-induced amylase release (124, 132). Although carbachol has been reported to inhibit the ability of isoproterenol to increase cAMP levels (96), apparently by inhibiting adenylate cyclase (95), this effect occurs predominately at isoproterenol concentrations higher than those used in the studies demonstrating cholinergic inhibition of amylase release (96). Under conditions where amylase release was inhibited, cAMP levels were the same with the combination of isoproterenol and carbachol as they were with isoproterenol alone (132, 133). Surprisingly, activation of cAMP-dependent PK was even greater with the combination of isoproterenol and the high concentration of carbachol than it was with isoproterenol alone (124). Further evidence suggesting that high concentrations of carbachol do not inhibit isoproterenol-induced amylase release by inhibiting cAMP production is the finding that Bt_2cAMP-induced release is also inhibited (132). Extracellular Ca^{2+} is required for carbachol to exhibit its inhibitory effect on isoproterenol-induced amylase release (132), whereas extracellular Ca^{2+} is not required for inhibition of adenylate cyclase (95). Although the mechanism of the inhibition of secretion is not clear, it is possible that the postulated Ca^{2+}-sensitive step in exocytosis after cAMP-dependent PK activation has a biphasic Ca^{2+} sensitivity, so that Ca^{2+} levels both too low and too high are suboptimal for amylase secretion. According to this hypothesis, low concentrations of carbachol increase cytoplasmic Ca^{2+} concentrations to the optimal level and high concentrations produce further increases to inhibitory concentrations. Studies of the effects of combinations of isoproterenol and carbachol must be interpreted with caution. Yoshimura et al. (146) recently reported some very complex effects of combinations of isoproterenol and cholinergic agonists on parotid cAMP levels. Some combinations produced an inhibition of cAMP levels (compared with isoproterenol alone) at early time points, no effect at intermediate times, and an augmentation at later times. The pattern of response was markedly influenced by both the concentrations of isoproterenol and agonist and the specific cholinergic agonist used. Unfortunately none of the conditions examined in that study corresponded exactly to those used to examine the effects of combinations of isoproterenol and carbachol on amylase release. It is therefore possible, although unlikely, that the levels of cAMP were altered by carbachol in the amylase-release experiments but that by coincidence the cAMP levels were measured at times when the inhibitory and stimulatory effects exactly canceled each other.

CALCIUM FLUXES. *Uptake.* Dormer and Ashcroft (40) found epinephrine increased $^{45}Ca^{2+}$ uptake by rat parotid slices and suggested that this uptake was important in the regulation of amylase release. However, most later studies with specific α-adrenergic and β-adrenergic agonists and antagonists have concluded that although α-adrenergic agonists do increase $^{45}Ca^{2+}$ uptake, β-agonists (67, 78, 108) and Bt_2cAMP (67, 78, 82) do not. Although some studies have described an effect of β-agonists (92, 106) or Bt_2cAMP (92), this has been suggested to occur via pinocytosis resulting from the retrieval of membrane added to the luminal plasmalemma during exocytosis rather than a result of a direct effect of β-agonists on Ca^{2+} influx (106). This conclusion is based on the finding that β-agonists stimulated uptake of the impermeant species [^{14}C]-sucrose as well as $^{45}Ca^{2+}$ (106).

Efflux. Essentially all reports on the effects of β-adrenergic agonists and cyclic nucleotide derivatives on the washout of $^{45}Ca^{2+}$ from preloaded tissue agree that these agents stimulate Ca^{2+} efflux (5, 27, 67, 102, 108), although a negative result with Bt_2cAMP was reported in one study (82). The efflux has been observed in both the absence (5, 67) and presence (5, 27, 102, 108) of extracellular Ca^{2+}, indicating that the increased efflux was not secondary to an increased uptake of unlabeled Ca^{2+}. Vasoactive intestinal peptide has also been reported to stimulate efflux of $^{45}Ca^{2+}$ from preloaded tissue (120). Butcher (27) demonstrated that the $^{45}Ca^{2+}$ efflux stimulated by isoproterenol or monobutyryl cAMP originated from an intracellular Ca^{2+} pool different from that mobilized by α-adrenergic and cholinergic agonists and substance P. Isoproterenol- and carbachol-induced $^{45}Ca^{2+}$ effluxes were additive; furthermore induction of $^{45}Ca^{2+}$ efflux with carbachol did not prevent additional efflux upon subsequent stimulation with isoproterenol. This is unlike the situation with carbachol and phenylephrine (α-adrenergic agonist), or carbachol and substance P, where stimulation of $^{45}Ca^{2+}$ efflux with one prevents a response to subsequent stimulation with the other, presumably because both induce efflux from the same intracellular pool (27).

EFFECTS OF β-AGONISTS ON INTRACELLULAR CALCIUM LEVELS. Takemura (134) examined the effect of isoproterenol on intracellular Ca^{2+} levels by loading parotid acinar cells with the fluorescent Ca^{2+} indicator quin 2. A concentration of isoproterenol optimal for amylase release (1 μM) produced a small (~5%) but statistically significant increase in the intracellular free Ca^{2+} level. Concentrations of isoproterenol supramaximal with respect to amylase secretion produced greater increases in intracellular Ca^{2+} levels. These findings indicate that the increased efflux of $^{45}Ca^{2+}$ induced by β-agonists and cAMP derivatives is secondary to an increased intracellular Ca^{2+} level due to the mobilization of an intracellular Ca^{2+} pool. An alternate interpretation of the $^{45}Ca^{2+}$ efflux data, ruled out by the results obtained with quin 2, is that cAMP directly stimulates pumping of Ca^{2+} out of the cell, resulting in lowered cytoplasmic Ca^{2+} levels upon β-adrenergic stimulation. Takemura (134) found that resuspending quin 2–loaded cells in Ca^{2+}-free media containing 0.1 mM EGTA lowered the basal intracellular Ca^{2+} levels by about half and that isoproterenol was subsequently incapable of increasing quin 2 fluorescence. This finding seems to indicate that the extracellular media, rather than an intracellular pool, provides the Ca^{2+} responsible for increasing the cytoplasmic level upon isoproterenol stimulation. However, this conclusion is at odds with the results of $^{45}Ca^{2+}$-flux experiments. It is possible that in the absence of extracellular Ca^{2+} the internal stores are depleted to the point where the amount of Ca^{2+} they are able to release upon isoproterenol stimulation is reduced to a level below that necessary for detection by quin 2 but still sufficient to induce amylase release or to be detected in $^{45}Ca^{2+}$-efflux experiments. An additional possibility is that in the absence of extracellular Ca^{2+}, Ca^{2+} released from the intracellular pool remains highly localized, so that an increase in the cytoplasmic level at some critical site is not detected against the background of unaffected levels in the rest of the cell. Another option is that the Ca^{2+} released from the intracellular store is immediately bound by a Ca^{2+}-binding protein component of the exocytotic machinery, so that the cytoplasmic free-Ca^{2+} concentration is not substantially elevated. The increases in quin 2 fluorescence seen in the presence of extracellular Ca^{2+}, when the intracellular stores are filled to capacity and thus able to release greater quantities than possible in the absence of extracellular Ca^{2+} (when stores are partially depleted), would thus represent overshoot according to this hypothesis. Alternatively, the samples stimulated with isoproterenol in Ca^{2+}-free media may not have been monitored long enough to detect an increase in quin 2 fluorescence. In the absence of extracellular Ca^{2+} there is a time lag between isoproterenol addition and the commencement of amylase release, and the length of the lag period increases proportionately to the length of time the cells have been maintained in Ca^{2+}-free media (5, 26). A similar lag period may exist between isoproterenol addition and the increase in intracellular Ca^{2+} levels. It is not apparent from the report of Takemura (134) how long the cells were exposed to Ca^{2+}-free media before stimulation nor how long they were monitored for increases in intracellular Ca^{2+} levels after stimulation.

Source of Calcium Mobilized by β-Adrenergic Agonists

A number of studies have examined the ability of subcellular fractions prepared from rat parotid gland homogenates to sequester Ca^{2+} (19, 40, 55, 68, 69, 101, 123, 135). Interpretation of some of these studies is complicated by the fact that these fractions were only partially characterized as to the relative amounts of the various subcellular organelles they contain. Clearly, however, several different subcellular organelles can transport or sequester Ca^{2+}. These include mitochondria (40, 67), endoplasmic reticulum (68, 69, 101), and basolateral plasma membrane (135).

Dormer and Ashcroft (40) reported that epinephrine added to parotid tissue simultaneously with $^{45}Ca^{2+}$ increased the amount of $^{45}Ca^{2+}$ associated with the mitochondrial fraction. When epinephrine was added to parotid tissue preloaded with $^{45}Ca^{2+}$, and the tissue then subsequently transferred to Ca^{2+}-free medium or medium containing unlabeled Ca^{2+}, a reduction in the amount of $^{45}Ca^{2+}$ associated with both a mitochondrial and a microsomal fraction was noted compared with unstimulated tissue (40). However, the use of epinephrine in these studies makes it impossible to determine whether these effects are due to the activation of the α- or β-adrenergic receptor. Kanagasuntheram and Randle (67) reported that both isoproterenol and Bt_2cAMP decreased the amount of $^{45}Ca^{2+}$ associated with the mitochondrial fraction when added simultaneously with the $^{45}Ca^{2+}$ and that isoproterenol decreased the amount of label in this fraction when $^{45}Ca^{2+}$-preloaded tissue was used. In a cytochemical study, Sampson et al. (114) reported that isoproterenol decreased the amount of Ca^{2+} associated both with mitochondria and the plasma membrane.

Mitochondria have been proposed to act as hormone-sensitive Ca^{2+} stores in a number of different tissues (3a, 93a, 141a). Experiments with isolated mitochondria have shown them capable of sequestering or releasing Ca^{2+}, depending on the Ca^{2+} concentration of their suspension medium, but it appears that the Ca^{2+} concentration at which neither net uptake nor net release occurs (i.e., the Ca^{2+} concentration mitochondria would tend to maintain in cytoplasm) is higher than the measured cytoplasmic level (39). A convincing mechanism by which hormones could mobilize Ca^{2+} from mitochondria has yet to be demonstrated; an initial report that cAMP promoted Ca^{2+} efflux from mitochondria (20) could not be confirmed (21, 115). The theory that mitochondria have an active role in regulating extramitochondrial Ca^{2+} levels has recently been questioned (39). It has been suggested that observations of decreased Ca^{2+} levels in mitochondria isolated from hormone-stimulated tissue may result from Ca^{2+} redistribution during subcellular fractionation or to other organelles contaminating mitochondrial preparations (39). Alternatively, hormones may promote cytoplasmic redistribution of Ca^{2+} in addition to mobilizing intracellular stores, so that some areas of the cell experience a decrease in cytoplasmic levels, whereas others experience an increase. If a localized decrease were to occur in the vicinity of the mitochondria (basal end of parotid cell) a passive efflux of Ca^{2+} from mitochondria could occur.

In summary, several different parotid subcellular organelles appear capable of sequestering Ca^{2+}, but the relative importance of each in releasing Ca^{2+} in response to β-adrenergic stimulation is not clear.

Mechanism of Calcium Mobilization

The mechanism by which β-adrenergic agonists mobilize intracellular Ca^{2+} is not understood at present. Dreux et al. (42) suggested that β-adrenergic agonists mobilize Ca^{2+} independently of their activation of adenylate cyclase. This hypothesis is based on their observation that both isoproterenol and forskolin increase cAMP levels equally well in the absence and presence of extracellular Ca^{2+}, whereas forskolin requires extracellular Ca^{2+} to induce amylase release but isoproterenol does not. In contrast, Putney et al. (108) suggested that it is the elevated cAMP level resulting from the stimulation of adenylate cyclase that mobilizes intracellular Ca^{2+}. The possibility that cAMP promotes Ca^{2+} release from the endoplasmic reticulum by stimulating the phosphorylation of an endogenous endoplasmic reticulum–associated protein is discussed in ENDOGENOUS PROTEIN PHOSPHORYLATION, this page.

Inositol trisphosphate mediates the effects of many neurotransmitters and hormones on intracellular Ca^{2+} pool mobilization in a wide variety of cells, including rat parotid acinar cells. Certain specific neurotransmitter receptors appear to be functionally coupled to phospholipase C through a GTP-binding protein such that neurotransmitters activate the lipase, which subsequently cleaves inositol trisphosphate from phosphatidylinositol bisphosphate [see the chapter by Putney in this *Handbook*; (18)]. Although α-adrenergic and muscarinic cholinergic agonists and substance P evidently mobilize intracellular Ca^{2+} by this mechanism in the rat parotid, the same does not appear to be true for β-adrenergic agonists. Unlike the former agonists, β-agonists do not increase the incorporation of ^{32}P-labeled inorganic phosphate ($^{32}P_i$) into phosphatidylinositol in parotid minces (91, 97, 98). Increased phosphatidylinositol labeling (phospholipid effect) is a well-studied phenomenon whose discovery preceded and led to the elucidation of the role of inositol trisphosphate, which is involved in the resynthesis of phosphatidylinositol bisphosphate after its breakdown by phospholipase C.

Mechanism of Action

Little is known regarding how Ca^{2+} might function to promote amylase release. Calmodulin is thought to mediate many of the intracellular actions of Ca^{2+} in other systems. Rat parotid gland contains calmodulin (32, 70, 80) and calmodulin-stimulated phosphodiesterase (80, 142). Phenothiazines inhibit calmodulin-sensitive reactions and are commonly used to demonstrate an involvement of calmodulin in biological processes, but the use of these drugs to investigate calmodulin involvement in amylase release from the parotid gland has yielded conflicting data. Kanagasuntheram and Teo (70) found that trifluoperazine inhibited amylase release induced by isoproterenol, phenylephrine, carbachol, and Bt_2cAMP but not that induced by substance P. The inability to inhibit substance P–induced release was of concern since substance P is believed to act by a mechanism similar to phenylephrine and carbachol. In addition phenothiazines act as neurotransmitter antagonists in some systems. However, trifluoperazine, at concentrations inhibitory to isoproterenol-induced amylase release, did not inhibit the ability of isoproterenol to increase glucose utilization, suggesting that the phenothiazine did not act as a β-adrenergic antagonist at the concentration employed. Argent et al. (6), using a higher concentration, found isoproterenol-induced release was inhibited, whereas Bt_2cAMP-induced release was not. Spearman and Butcher (125) were unable to demonstrate an inhibition of Bt_2cAMP-stimulated amylase release at any concentration of several different phenothiazines. Only a nonphenothiazine calmodulin antagonist (W-7) and a purported intracellular Ca^{2+} antagonist (TMB-8) showed inhibitory effects. The antagonist W-7 inhibited only slightly at relatively high concentrations, which also elevated basal release, suggesting nonspecific effects, and TMB-8 markedly lowered parotid ATP levels, indicating it was toxic to parotid cells (125). Thus a role for calmodulin in parotid exocytosis must be considered unproved. The lack of a specific inhibitor of the intracellular actions of Ca^{2+} in parotid function complicates attempts to elucidate its role.

Putney et al. (104) recently reported that a phorbol ester, 4α-phorbol dibutyrate, induces amylase release from rat parotid. Because phorbol esters activate PK C, a Ca^{2+}-dependent, phospholipid-sensitive protein kinase, the ability of the phorbol ester to induce release may reflect a role for this enzyme in the regulation of exocytosis. However, PK C from rat brain has been shown to be inhibited by chlorpromazine (93), a phenothiazine Spearman and Butcher (125) found did not inhibit Bt_2cAMP-induced amylase release. Further study is required to determine if this enzyme participates in mediating amylase release. In summary, although Ca^{2+} appears to be an important regulator of amylase release from rat gland, its mechanism of action is not yet clear.

ENDOGENOUS PROTEIN PHOSPHORYLATION

Cyclic AMP is thought to exert its biological effects by activating cAMP-dependent PK, thus increasing the phosphorylation of endogenous cAMP-dependent PK substrate proteins. In addition Ca^{2+}-stimulated protein kinases (PK C, Ca^{2+}-calmodulin–stimulated protein kinases) and at least one Ca^{2+}-stimulated phosphoprotein phosphatase are known to exist in other tissues. Investigators have examined the effects of secretagogues on the phosphorylation of endogenous parotid proteins to gain insight into the mechanisms by which cAMP and Ca^{2+} regulate exocytosis and to identify proteins involved in the exocytotic process at steps after increases in second messenger levels. These experiments are performed by preincubating parotid slices or dissociated acinar cells with $^{32}P_i$ to label the endogenous ATP pool, stimulating an aliquot of the prelabeled tissue with secretagogues, preparing homogenates of the control and stimulated labeled tissue, resolving the labeled proteins by electrophoresis, preparing autoradiograms from the electrophoresis gels, and comparing the autoradiograms from control and stimulated tissue to find proteins whose degree of phosphorylation is affected by secretagogues.

Investigators have reported differences in the number and molecular masses of proteins seen to be influenced by secretagogues, although they are probably seeing many of the same proteins. The number of stimulus-affected proteins generally increases in proportion to the resolving power of the techniques used to examine them, especially those proteins either weakly labeled or present in low amounts. Resolution can be increased either before electrophoresis (purifying specific subcellular fractions before assay) or during electrophoresis (with two-dimensional electrophoresis as opposed to one-dimensional electrophoresis or with two-dimensional electrophoresis and a narrow-range ampholine mix in the first direction as opposed to a wide-range mix). Poorly labeled proteins or those low in abundance are examined by increasing the amount of $^{32}P_i$ used to label the tissue, increasing the autoradiography exposure time, or both. In practice, examining minor proteins requires high resolution in addition to highly labeled samples and long exposures because more highly labeled proteins mask the poorly labeled ones that run near them during electrophoresis.

Thus the best studied stimulus-affected phosphoproteins would be expected to be those that are both highly labeled and easily resolved from other endogenous phosphoproteins. In the rat parotid the easiest to detect appears to be a protein of ~31 kilodaltons (kDa). An endogenous stimulus-affected protein of approximately this molecular mass has been described by Jahn et al. [35 kDa (65)], Kanamori and Hayakawa [30 kDa (71)], Baum et al. [27 kDa (13)], Spearman et al. [31 kDa (126)], and Freedman and Jamieson [29 kDa (46)]. A protein of similar molecular mass has also been described in mouse parotid gland (63), rat lacrimal gland (64), rabbit parotid gland [34 kDa (41)], rat submandibular salivary gland [34 kDa (109)], and rat pancreas [29 kDa (46)]. Because this protein occurred in more than one secretory tissue and its altered phosphorylation appeared to be stimulated only by those neurotransmitters that induced protein secretion in each respective tissue, Jahn and Söling (64) initially speculated that it might have a general role in mediating exocytosis. However, the protein has subsequently been demonstrated to be the S6 ribosomal protein (47), making such a role unlikely.

This illustrates one of the disadvantages of this experimental approach. Neurotransmitters activate a wide range of responses in the rat parotid gland in addition to stimulating exocytosis. These include effects on DNA synthesis (137), protein synthesis (90) and posttranslational modification (79), oxygen consumption (103), glucose oxidation (51), and amino acid uptake (76). There is no reliable method to ascertain that stimulus-affected phosphoproteins are involved in exocytosis as opposed to these other effects. Phosphorylation of the S6 protein has been observed in a variety of cell types, and it has been reported that ribosomal subunits containing phosphorylated S6 are preferentially utilized in the formation of the initiation complex during protein synthesis (43).

Although phosphorylation of the ribosomal S6 protein is most likely not involved in exocytosis, information regarding its phosphorylation is still able to shed some light on the interaction of cAMP and Ca^{2+} as second messengers in parotid. In rat parotid slices, the phosphorylation of the protein is increased by both isoproterenol and carbachol (63); furthermore in homogenates the phosphorylation of this protein is stimulated by both cAMP (48, 72, 126) and Ca^{2+} (126). In guinea pig parotid slices, stimulation with either isoproterenol or carbachol resulted in the same pattern of S6 labeling as revealed by tryptic fingerprinting (99). When purified guinea pig parotid ribosomes were phosphorylated in vitro with purified cAMP-dependent PK, calmodulin-dependent PK, and PK C, and then tryptic fingerprinting was performed, only PK C was able to reproduce the complete pattern seen when slices were stimulated with secretagogues (100). This implies that, even with isoproterenol, PK C rather than cAMP-dependent PK phosphorylates S6 in situ. Together with Putney's finding that a phorbol ester induces amylase release from rat parotid slices, these data suggest that PK C may have a significant role in mediating the effects of β-agonists in parotid gland.

The stimulus-affected protein most easily detected after the S6 protein is one of ~20. 5 kDa (128) [20.4 kDa, Jahn et al. (65); 14 kDa, Baum et al. (13); 19 kDa, Kanamori et al. (74)]. This protein can be detected in whole homogenates run on one-dimensional electrophoresis gels, although under these conditions a major nonphosphorylated protein distorts its location so that it appears to have a lower molecular mass (~19 kDa). The phosphorylation of this protein is increased by isoproterenol (13, 65, 128) and Bt_2cAMP (65) but not by carbachol (Fig. 1). In parotid homogenates its phosphorylation has been reported to be stimulated by cAMP (111, 126). This protein has been localized to the rough endoplasmic reticulum (111, 128) and has been suggested to function similarly to phospholamban in heart sarcoplasmic reticulum in the sequestration of Ca^{2+} or mobilization of a Ca^{2+} pool (111). Differences have been reported in the sensitivity of the rat parotid protein and dog heart phospholamban to heating in 4% sodium dodecyl sulfate, indicating that the two are not identical (111). We are unaware of any studies showing that the ability of parotid endoplasmic reticulum vesicles to transport Ca^{2+} or to hydrolyze ATP in a Ca^{2+}-dependent manner is correlated with the phosphorylation state of the 20.5-kDa protein. Such a study should confirm or refute this hypothesis.

The phosphorylation of a protein of ~22.5 kDa [25.7 kDa (65)] is increased by isoproterenol (65, 128) and by Bt_2cAMP (65). Carbachol does not increase the phosphorylation of this protein (Fig. 1). This protein is weakly labeled by $^{32}P_i$ compared with the proteins discussed previously, and it can not be readily seen in parotid whole homogenates fractionated on one-dimensional electrophoresis gels. Spearman et al. (128),

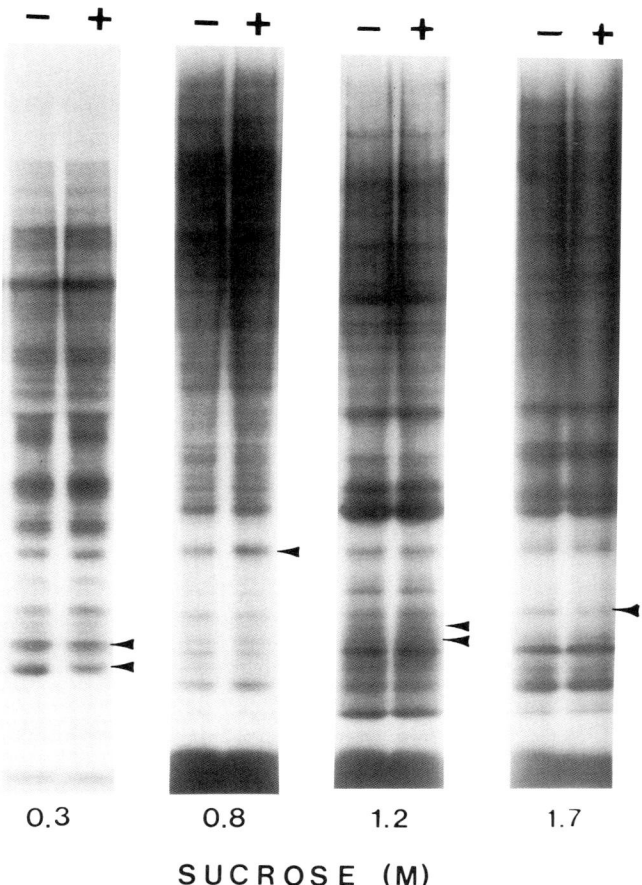

FIG. 1. Effect of carbachol on phosphorylation of endogenous parotid phosphoproteins. Autoradiograms of selected sucrose gradient fractions from carbachol-stimulated (+) and nonstimulated (−) rat parotid gland slices; carbachol concentration = 10 μM. *Arrowheads* indicate (from *top*) 18-kDa and 16-kDa proteins whose phosphorylation is decreased by carbachol (0.3 M sucrose tracks), 31-kDa protein whose phosphorylation is increased by carbachol (0.8 M sucrose), 22.5-kDa and 20.5-kDa proteins whose phosphorylation is unaffected by carbachol (1.2 M sucrose), and 24-kDa protein whose phosphorylation is decreased by carbachol (1.7 M sucrose). (See ref. 128 for experimental details.)

are cytoplasmic (128) and the phosphorylation of both is decreased by carbachol (Fig. 1) in addition to isoproterenol (128).

As discussed above, a finding that a secretagogue using analytical sucrose gradient fractionation, concluded that this protein was also localized to the endoplasmic reticulum. Autoradiograms of an endoplasmic reticulum fraction prepared by Plewe et al. (101) do not show this protein, although these investigators had reported it earlier in crude particulate fractions (65).

A somewhat surprising finding is that isoproterenol decreases the phosphorylation of other specific endogenous parotid phosphoproteins. One protein of ~16 kDa [13.6 kDa, Baum et al. (13); 17 kDa, Kanamori et al. (74)] is easily detected when whole parotid homogenates are fractionated on one-dimensional electrophoresis gels (13, 74, 126), whereas the other (~18 kDa) is more weakly labeled and is seen only when homogenates are first partially prepurified or two-dimensional electrophoresis is employed. Both

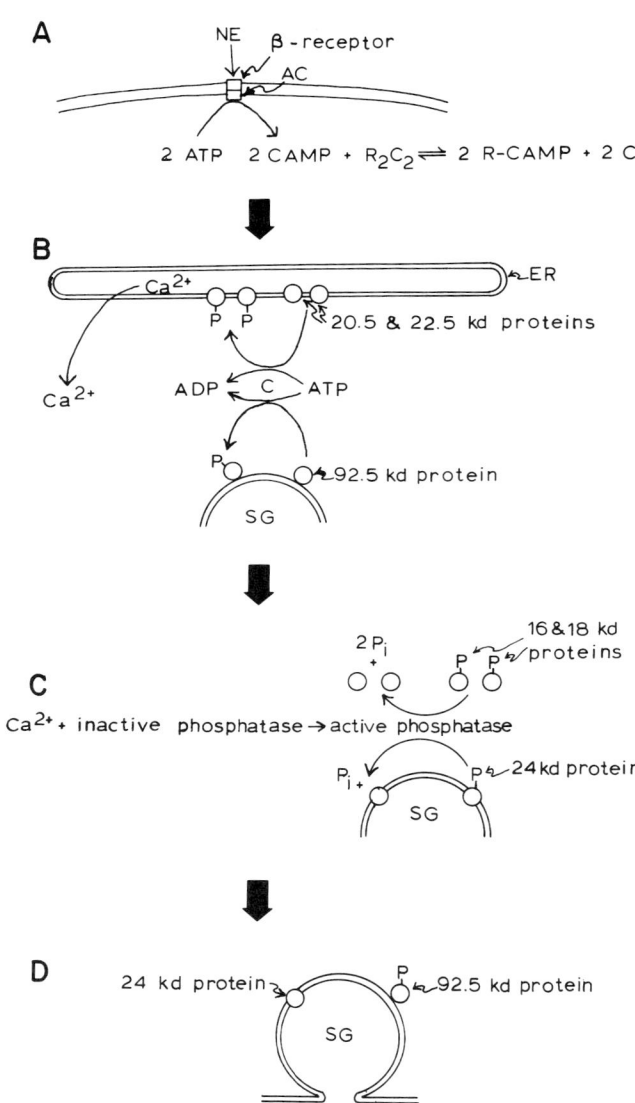

FIG. 2. Diagram of working hypothesis of mechanism of β-agonist–induced amylase secretion from rat parotid gland. *A*: norepinephrine (NE) binds to β-receptor, activating adenylate cyclase (AC), resulting in increased production of cAMP that dissociates cAMP-dependent protein kinase (R_2C_2), liberating free catalytic subunit (C). *B*: catalytic subunit (C) phosphorylates 92.5-kDa protein associated with secretory granule (SG) membrane and 20.5-kDa and 22.5-kDa proteins associated with endoplasmic reticulum (ER). Phosphorylation of one or both ER-associated proteins results in release of Ca^{2+} from cisterna of ER to cytoplasm by unknown mechanism. *C*: mobilized Ca^{2+} activates a phosphoprotein phosphatase that dephosphorylates 16-kDa and 18-kDa cytoplasmic proteins (function unknown) and 24-kDa protein associated with SG membrane. *D*: as a result of phosphorylation of 92.5-kDa SG membrane–associated protein and dephosphorylation of 24-kDa SG membrane–associated protein, SG fuses to plasma membrane (PM) and ruptures at point of contact, releasing contents into ductal lumen.

influences the phosphorylation of a specific phosphoprotein does not guarantee that protein a role in stimulus-secretion coupling. However, stimulus-affected phosphoproteins associated with secretory granules would appear better candidates than proteins associated with other structures. It is of interest that stimulus-affected phosphoproteins associated with this organelle have been reported (128). Demonstrating this fact is a formidable task, however, because the secretory granule is poorly labeled when parotid slices are incubated with $^{32}P_i$. It is therefore necessary to highly purify granules before electrophoresis, and it is never certain that stimulus-affected phosphoproteins in the granule fraction are not due to very low levels of contaminating organelles. However, we believe that secretory granule membranes have at least two proteins associated with them whose phosphorylation is influenced by secretagogues. One is a protein of ~92.5 kDa whose phosphorylation is increased by isoproterenol. The second is a protein of ~24 kDa whose phosphorylation is decreased by both isoproterenol and carbachol. We had initially reported that these proteins were not extracted from the granule membrane by high-salt washes (128), but subsequent experimentation has shown that this was in error for the 92.5-kDa protein. We have noticed substantial experiment-to-experiment variability in the degree of radiolabeling of the 24-kDa protein in unstimulated controls, and thus it is frequently necessary to resolve purified secretory granules by two-dimensional electrophoresis to demonstrate effects of secretagogues. We have not been able to attribute this variability to any specific experimental condition. Preliminary evidence with broken-cell preparations suggests that the phosphorylation of the 92.5-kDa protein is stimulated by cAMP. Phosphorylation of the 24-kDa protein has only been observed in intact cells to date. We have also obtained preliminary evidence that purified secretory granules have an endogenous protein kinase activity associated with them but have not characterized it nor determined whether it phosphorylates those proteins whose degree of phosphorylation is influenced by secretagogues in the intact cell.

SPECULATIONS

It is clear that exocytosis is regulated by cAMP and Ca^{2+} in the rat parotid gland, but the mechanisms by which they do so are still subject to speculation. Two main schemes have been put forward. 1) β-Agonists increase cAMP levels, which then mobilize an intracellular Ca^{2+} pool; the elevated intracellular Ca^{2+} level then stimulates exocytosis by an unknown mechanism. 2) β-Agonists increase cAMP levels, which in turn mobilize an intracellular Ca^{2+} pool; the cAMP and Ca^{2+} then function in concert to stimulate exocytosis by unknown mechanisms.

We currently favor the second scheme. Although the mechanism by which cAMP mobilizes Ca^{2+} is unknown, it may involve phosphorylation of the stimulus-affected protein(s) associated with the endoplasmic reticulum (Fig. 2). This liberated Ca^{2+} may activate a phosphoprotein phosphatase, which then dephosphorylates both the 24-kDa protein integral to the secretory granule membrane and the two cytoplasmic stimulus-affected proteins. In addition to promoting Ca^{2+} mobilization, cAMP would also increase the phosphorylation of the 92.5-kDa protein loosely associated with the secretory granule membrane. The altered phosphorylation of the secretory granule membrane proteins is proposed to lead to exocytosis, perhaps by affecting the interaction of the secretory granule with the cytoskeleton or the luminal plasma membrane. According to this hypothesis, both increased phosphorylation of the 92.5-kDa protein and decreased phosphorylation of the 24-kDa protein are required for exocytosis to proceed at an optimal rate. This could explain why amylase release is inhibited by Ca^{2+}-depleted tissues (insufficient Ca^{2+} remains to dephosphorylate the 24-kDa protein) and why agonists promoting Ca^{2+} influx are poorer inducers of amylase release than β-agonists (they would not promote phosphorylation of the 92.5-kDa protein). However, the relationship between altered secretory granule membrane phosphorylation and exocytosis is by no means established. It is only known that both occur as a result of neurotransmitter stimulation, not that one causes the other. A substantial amount of research is required to fully understand exocytosis and its regulation in the rat parotid gland. Our purpose in formulating this hypothesis is to synthesize the known facts into the simplest mechanism able to explain the data and to develop a set of predictions resulting from the hypothesis that can be experimentally tested. Determining the biochemical events underlying a physical event that occurs only in intact cells is not a trivial problem. It is likely that a satisfactory understanding of the process will result only from the efforts of many people using a variety of experimental approaches.

REFERENCES

1. ABE, K., AND C. DAWES. Dopamine-induced secretion of protein and of some electrolytes by rat submandibular and parotid glands. *Arch. Oral Biol.* 27: 635–643, 1982.
2. AFARI, G., A. TENENHOUSE, AND J. KLEIN. The effect of ouabain and of Ca^{2+} deprivation on isoproterenol- and DBcAMP-stimulated protein secretion from the superfused rat parotid gland. *Can. J. Physiol. Pharmacol.* 55: 419–426, 1977.
3. AFARI, G., A. TENENHOUSE, AND C. VACHON. The effects of theophylline and 4-(ε-butoxy-4-methoxybenzyl)-2-imidazolidinone (RO 20-1724) on protein secretion from rat parotid gland. *Br. J Pharmacol.* 77: 405–411, 1982.
3a. AKERMAN, K. E. O., AND D. G. NICHOLLS. Physiological and bioenergetic aspects of mitochondrial calcium transport. *Rev.*

Physiol. Biochem. Pharmacol. 95: 149–201, 1983.
4. AMSTERDAM, A., I. OHAD, AND M. SCHRAMM. Dynamic changes in the ultrastructure of the acinar cell of the rat parotid gland during the secretory cycle. *J. Cell Biol.* 41: 753–773, 1969.
5. ARGENT, B. E., AND S. ARKLE. Mechanism of action of extracellular calcium on isoprenaline-evoked amylase secretion from isolated rat parotid glands. *J. Physiol. Lond.* 369: 337–353, 1985.
6. ARGENT, B. E., S. ARKLE, P. D. PICKFORD, AND P. S. SCHOFIELD. The effects of trifluoperazine on amylase secretion by the in vitro rat parotid gland (Abstract). *J. Physiol. Lond.* 354: 38P, 1984.
7. AU, D. K., C. C. MALBON, AND F. R. BUTCHER. Identification and characterization of beta-adrenergic receptors in rat parotid membranes. *Biochim. Biophys. Acta* 500: 361–371, 1977.
8. BADAD, H., R. BEN-ZVI, A. BDOLAH, AND M. SCHRAMM. The mechanism of enzyme secretion by the cell. 4. Effects of inducers, substrates and inhibitors on amylase secretion by rat parotid slices. *Eur. J. Biochem.* 1: 96–101, 1967.
9. BATZRI, S., AND Z. SELINGER. Enzyme secretion mediated by the epinephrine β-receptor in rat parotid slices. Factors governing efficiency of the process. *J. Biol. Chem.* 248: 356–360, 1973.
10. BATZRI, S., Z. SELINGER, AND M. SCHRAMM. Potassium ion release and enzyme secretion: adrenergic regulation by α- and β-receptors. *Science Wash. DC* 174: 1029–1031, 1971.
11. BATZRI, S., Z. SELINGER, M. SCHRAMM, AND M. R. ROBINOVITCH. Potassium release mediated by the epinephrine α-receptor in rat parotid slices. Properties and relation to enzyme secretion. *J. Biol. Chem.* 248: 361–368, 1973.
12. BAUM, B. J., F. T. COLPO, AND C. R. FILBURN. Characterization and relationship to exocrine secretion of rat parotid gland cyclic AMP-dependent protein kinase. *Arch. Oral Biol.* 26: 333–337, 1981.
13. BAUM, B. J., J. M. FREIBERG, H. ITO, G. S. ROTH, AND C. R. FILBURN. β-Adrenergic regulation of protein phosphorylation and its relationship to exocrine secretion in dispersed rat parotid gland acinar cells. *J. Biol. Chem.* 256: 9731–9736, 1981.
14. BDOLAH, A., R. BEN-ZVI, AND M. SCHRAMM. The mechanism of enzyme secretion by the cell. II. Secretion of amylase and other proteins by slices of rat parotid gland. *Arch. Biochem. Biophys.* 104: 58–66, 1964.
15. BDOLAH, A., AND M. SCHRAMM. Factors controlling the process of enzyme secretion by the rat parotid slice. *Biochem. Biophys. Res. Commun.* 8: 266–270, 1962.
16. BDOLAH, A., AND M. SCHRAMM. The function of 3'5' cyclic AMP in enzyme secretion. *Biochem. Biophys. Res. Commun.* 18: 452–454, 1965.
17. BERRIDGE, M. J., AND R. F. IRVINE. Inositol trisphosphate, a novel second messenger in cellular signal transduction. *Nature Lond.* 312: 315–321, 1984.
18. BODNER, L., M. T. HOOPES, M. GEE, H. ITO, G. S. ROTH, AND B. J. BAUM. Multiple transduction mechanisms are likely involved in calcium-mediated exocrine secretory events in rat parotid cells. *J. Biol. Chem.* 258: 2774–2777, 1983.
19. BONIS, D., AND B. ROSSIGNOL. Effect of sodium and potassium on ATP-dependent Ca^{2+} uptake in rat parotid microsomes. *FEBS Lett.* 137: 63–66, 1982.
20. BORLE, A. B. Cyclic AMP stimulation of calcium efflux from kidney, liver, and heart mitochondria. *J. Membr. Biol.* 16: 221–236, 1974.
21. BORLE, A. B. Cyclic AMP stimulation of calcium efflux from mitochondria: a negative report. *J. Membr. Biol.* 29: 209–210, 1976.
22. BRODIN, E., R. EKMAN, R. EKSTROM, R. HAKANSON, AND F. SUNDLER. Effect of denervation on substance P and vasoactive intestinal peptide in rat salivary glands (Abstract). *J. Physiol. Lond.* 348: 67P, 1983.
23. BRODIN, E., AND G. NILSSON. Concentration of substance P-like immunoreactivity (SPLI) in tissues of dog, rat and mouse. *Acta Physiol. Scand.* 112: 305–312, 1981.
24. BURKE, G. T., AND T. BARKA. Beta-adrenergic receptors and adenylate cyclase in hypertrophic and hyperplastic rat salivary glands. *Biochim. Biophys. Acta* 539: 54–61, 1978.
25. BUTCHER, F. R. The role of calcium and cyclic nucleotides in α-amylase release from slices of rat parotid: studies with the divalent cation ionophore A-23187. *Metabolism* 24: 409–418, 1975.
26. BUTCHER, F. R. Calcium and cyclic nucleotides in the regulation of secretion from the rat parotid by autonomic agonists. In: *Advances in Cyclic Nucleotide Research*, edited by W. J. George and L. J. Ignarro. New York: Raven, 1978, vol. 9, p. 707–721.
27. BUTCHER, F. R. Regulation of calcium efflux from isolated rat parotid cells. *Biochim. Biophys. Acta* 630: 254–260, 1980.
28. BUTCHER, F. R., J. A. GOLDMAN, AND M. NEMEROVSKI. Effect of adrenergic agents on α-amylase release and adenosine 3'-5'-monophosphate accumulation in rat parotid tissue slices. *Biochim. Biophys. Acta* 392: 82–94, 1975.
29. BUTCHER, F. R., P. A. MCBRIDE, AND L. RUDICH. Cholinergic regulation of cyclic nucleotide levels, amylase release, and K^+ efflux from rat parotid glands. *Mol. Cell. Endocrinol.* 5: 243–254, 1976.
30. BUTCHER, F. R., L. RUDICH, C. EMLER, AND M. NEMEROVSKI. Adrenergic regulation of cyclic nucleotide levels, amylase release and potassium efflux in rat parotid gland. *Mol. Pharmacol.* 12: 862–870, 1976.
31. BUTCHER, F. R., M. THAYER, AND J. A. GOLDMAN. Effect of adenosine 3',5'-cyclic monophosphate derivatives on α-amylase release, protein kinase and cyclic nucleotide phosphodiesterase activity from rat parotid tissue. *Biochim. Biophys. Acta* 421: 289–295, 1976.
32. CAMPOS GONZALES, R., J. F. WHITFIELD, A. L. BOYNTON, J. P. MACMANUS, AND R. M. RIXON. Prereplicative changes in the soluble calmodulin of isoproterenol-activated rat parotid glands. *J. Cell. Physiol.* 118: 257–261, 1984.
33. CARLSÖÖ, B., A. DANIELSSON, AND R. HENRIKSSON. Effects of a new selective $β_1$-adrenoceptor agonist on amylase secretion from the rat parotid gland. *Br. J. Pharmacol.* 62: 364–366, 1978.
34. CARLSÖÖ, B., A. DANIELSSON, R. HENRIKSSON, AND L. A. IDAHL. Characterization of the rat parotid β-adrenoceptor. *Br. J. Pharmacol.* 72: 271–276, 1981.
35. CARLSÖÖ, B., A. DANIELSSON, R. HENRIKSSON, AND L. A. IDAHL. Dissociation of β-adrenoceptor-induced effects on amylase secretion and cyclic adenosine 3',5'-monophosphate accumulation. *Br. J. Pharmacol.* 75: 633–638, 1982.
36. CASSEL, D., AND T. PFEUFFER. Mechanism of cholera toxin action: covalent modification of the guanyl nucleotide-binding protein of the adenylate cyclase system. *Proc. Natl. Acad. Sci. USA* 75: 2669–2673, 1978.
37. DAVIS, J. N., AND W. MAURY. Clonidine and related imidazolines are postsynaptic alpha adrenergic antagonists in dispersed rat parotid cells. *J. Pharmacol. Exp. Ther.* 207: 425–430, 1978.
38. DEHAYE, J. P., J. CHRISTOPHE, F. ERNST, P. POLOCZEK, AND P. VAN BOGAERT. Binding in vitro of vasoactive intestinal peptide on isolated acini of rat parotid glands. *Arch. Oral Biol.* 30: 827–832, 1985.
39. DENTON, R. M., AND J. G. MCCORMACK. Ca^{2+} transport by mammalian mitochondria and its role in hormone action. *Am. J. Physiol.* 249 (*Endocrinol. Metab.* 12): E543–E554, 1985.
40. DORMER, R. L., AND S. J. ASHCROFT. Studies on the role of calcium ions in the stimulation by adrenaline of amylase release from rat parotid. *Biochem. J.* 144: 543–550, 1974.
41. DOWD, F. J., E. L. WATSON, B. HORIO, Y.-S. LAU, AND K. PARK. Phosphorylation of rabbit parotid microsomal protein occurs only with β-adrenergic stimulation. *Biochem. Biophys. Res. Commun.* 101: 281–288, 1981.
42. DREUX, C., V. IMHOFF, AND B. ROSSIGNOL. Substance P as a modulator of β-adrenergic regulation of protein secretion in rat parotid gland. In: *Regulatory Peptides in Digestive, Nervous and Endocrine Systems*, edited by M. J. M. Lewin and S.

Bonfils. Amsterdam: Elsevier, 1985, p. 157–160. (INSERM Symp. 25.)
43. DUNCAN, R., AND E. H. MCCONKEY. Preferential utilization of phosphorylated 40-S ribosomal subunits during initiation complex formation. *Eur. J. Biochem.* 123: 535–538, 1982.
44. ECKSTEIN, F., S. EIMERL, AND M. SCHRAMM. Adenosine 3',5' cyclic phosphorothioate: an efficient inducer of amylase secretion in rat parotid slices. *FEBS Lett.* 64: 92–94, 1976.
45. EKSTRÖM, J., AND C. WAHLESTEDT. Supersensitivity to substance P and physalaemin in rat salivary glands after denervation or decentralization. *Acta Physiol. Scand.* 115: 437–446, 1982.
46. FREEDMAN, S. D., AND J. D. JAMIESON. Hormone-induced protein phosphorylation. I. Relationship between secretagogue action and endogenous protein phosphorylation in intact cells from the exocrine pancreas and parotid. *J. Cell Biol.* 95: 903–908, 1982.
47. FREEDMAN, S. D., AND J. D. JAMIESON. Hormone-induced protein phosphorylation. II. Localization to the ribosomal fraction from rat exocrine pancreas and parotid of a 29,000-dalton protein phosphorylated in situ in response to secretagogues. *J. Cell Biol.* 95: 909–917, 1982.
48. FREEDMAN, S. D., AND J. D. JAMIESON. Hormone-induced protein phosphorylation. III. Regulation of the phosphorylation of the secretagogue-responsive 29,000-dalton protein by both Ca^{2+} and cAMP in vitro. *J. Cell Biol.* 95: 918–923, 1982.
49. FULLER, C. M., AND D. V. GALLACHER. β-Adrenergic receptor mechanisms in rat parotid glands: activation by nerve stimulation and 3-isobutyl-1-methylxanthine. *J. Physiol. Lond.* 356: 335–348, 1984.
50. GALLACHER, D. V. Substance P is a functional neurotransmitter in the rat parotid gland. *J. Physiol. Lond.* 342: 438–498, 1983.
51. GEE, M. V., B. J. BAUM, AND G. S. ROTH. Stimulation of parotid cell glucose oxidation. Role of alpha$_1$-adrenergic receptors and calcium mobilization. *Biochem. Pharmacol.* 32: 3351–3354, 1983.
52. GILL, D. M., AND R. MEREN. ADP-ribosylation of membrane proteins catalyzed by cholera toxin: basis of the activation of adenylate cyclase. *Proc. Natl. Acad. Sci. USA* 75: 3050–3054, 1978.
54. HARFIELD, B., AND A. TENENHOUSE. Effect of EGTA on protein release and cyclic AMP accumulation in rat parotid gland. *Can. J. Physiol. Pharmacol.* 51: 997–1001, 1971.
54. HATA, F., H. ISHIDA, K. KAGAWA, E. KONDO, S. KONDO, AND Y. NOGUCHI. β-Adrenoceptor alterations coupled with secretory response in rat parotid tissue. *J. Physiol. Lond.* 341: 185–196, 1983.
55. HAYAKAWA, M., H. AOKI, N. TERAO, Y. ABIKO, AND H. TAKIGUCHI. Vitamin D-mediated decrease of Ca^{2+}-pump activity in the rat parotid gland. *Int. J. Biochem.* 15: 1175–1178, 1983.
56. HENRIKSSON, R. β_1- And β_2-adrenoceptor agonists have different effects on rat parotid acinar cells. *Am. J. Physiol.* 242 (*Gastrointest. Liver Physiol.* 5): G481–G485, 1982.
57. HERMAN, G., S. BUSSON, L. OVTRACHT, C. MAURS, AND B. ROSSIGNOL. Regulation of protein discharge in two exocrine glands: rat parotid and exorbital lacrimal glands. Analogies between cholinergic (muscarinic) and α-adrenergic stimulation and importance of extracellular calcium. *Biol. Cell* 31: 255–262, 1978.
58. HÖKFELT, T., O. JOHANNSON, J.-O. KELLERTH, A. LJUNGDAHL, G. NILSSON, A. NYGARDS, AND B. PERNOW. Immunohistochemical distribution of substance P. In: *Substance P*, edited by U. S. von Euler and B. Pernow. New York: Raven, 1977, p. 117–145. (Nobel Symp. Ser. 37.)
59. HOOTMAN, S. R., T. M. PICADO-LEONARD, AND D. B. BURNHAM. Muscarinic acetylcholine receptor structure in acinar cells of mammalian exocrine glands. *J. Biol. Chem.* 260: 4186–4194, 1985.
60. INOUE, Y., K. KAKU, T. KANEKO, N. YANAIHARA, AND T. KANNO. Vasoactive intestinal peptide binding to specific receptors on rat parotid acinar cells induces amylase secretion accompanied by intracellular accumulation of cyclic adenosine 3'-5'-monophosphate. *Endocrinology* 116: 686–692, 1985.
61. ISHIDA, H., N. MIKI, AND H. YOSHIDA. Role of Ca^{2+} in the secretion of amylase from the parotid gland. *Jpn. J. Pharmacol.* 21: 227–238, 1971.
62. ITO, H., M. T. HOOPES, B. J. BAUM, AND G. S. ROTH. K$^+$ release from rat parotid cells: an α_1-adrenergic mediated event. *Biochem. Pharmacol.* 31: 567–573, 1982.
63. JAHN, R., AND H. D. SÖLING. Phosphorylation of the same specific protein during amylase release evoked by β-adrenergic or cholinergic agonists in rat and mouse parotid glands. *Proc. Natl. Acad. Sci. USA* 78: 6903–6906, 1981.
64. JAHN, R., AND H. D. SÖLING. Protein phosphorylation during secretion in the rat lacrimal gland. A general role of EC-protein in stimulus-secretion coupling in exocrine organs? *FEBS Lett.* 131: 28–30, 1981.
65. JAHN, R., C. UNGER, AND H. D. SÖLING. Specific protein phosphorylation during stimulation of amylase secretion by β-agonists or dibutyryl adenosine 3',5'-monophosphate in the rat parotid gland. *Eur. J. Biochem.* 112: 345–352, 1980.
66. KANAGASUNTHERAM, P., AND S. C. LIM. Parallel secretion of secretory proteins and calcium by the rat parotid gland. *J. Physiol. Lond.* 312: 445–454, 1981.
67. KANAGASUNTHERAM, P., AND P. J. RANDLE. Calcium metabolism and amylase release in rat parotid acinar cells. *Biochem. J.* 160: 547–564, 1976.
68. KANAGASUNTHERAM, P., AND T. S. TEO. Calmodulin-sensitive ATP-dependent calcium transport by the rat parotid endoplasmic reticulum. *FEBS Lett.* 141: 233–236, 1982.
69. KANAGASUNTHERAM, P., AND T. S. TEO. Parotid microsomal Ca^{2+} transport. Subcellular localization and characterization. *Biochem. J.* 208: 789–794, 1982.
70. KANAGASUNTHERAM, P., AND T. S. TEO. Does calmodulin mediate stimulus-secretion coupling in the parotid gland? Studies using trifluoperazine. *Biochem. Int.* 7: 511–518, 1983.
71. KANAMORI, T., AND T. HAYAKAWA. Cyclic AMP-dependent ^{32}P incorporation into a protein in rat parotid slices. *Biochem. Int.* 1: 395–402, 1980.
72. KANAMORI, T., AND T. HAYAKAWA. Phosphorylation of the rat parotid Mr = 30,000 protein by cyclic AMP-dependent protein kinase in a cell-free system. *Biochem. Int.* 4: 39–46, 1982.
73. KANAMORI, T., T. HAYAKAWA, AND T. NAGATSU. Adenosine 3',5'-monophosphate-dependent protein kinase and amylase secretion from rat parotid gland. *Biochem. Biophys. Res. Commun.* 57: 394–398, 1974.
74. KANAMORI, T., T. HAYAKAWA, AND T. NAGATSU. Involvement of β_1-adrenergic receptors in regulation of the phosphorylation state of rat parotid gland proteins. *Biomed. Res.* 5: 77–82, 1984.
75. KELLER, P. J., M. ROBINOVITCH, J. IVERSON, AND D. L. KAUFFMAN. The protein composition of rat parotid saliva and secretory granules. *Biochim. Biophys. Acta* 379: 562–570, 1975.
76. KERYER, G., AND B. ROSSIGNOL. Effects of carbachol on extracellular Na-dependent AIB uptake in rat parotid gland. *Am. J. Physiol.* 239 (*Gastrointest. Liver Physiol.* 2): G183–G189, 1980.
77. KIM, S.-K. The cytochemical localization of adenylate cyclase activity in mucous and serous cells of the salivary gland. *J. Supramol. Struct.* 4: 185–197, 1976.
78. KOELZ, H. R., S. KONDO, A. L. BLUM, AND I. SCHULZ. Calcium ion uptake induced by cholinergic and α-adrenergic stimulation in isolated cells of rat salivary glands. *Pfluegers Arch.* 370: 37–44, 1977.
79. KOUSVELARI, E. E., S. R. GRANT, D. K. BANERJEE, M. J. NEWBY, AND B. J. BAUM. Cyclic AMP mediates β-adrenergic-induced increases in N-linked protein glycosylation in rat parotid acinar cells. *Biochem. J.* 222: 17–24, 1984.
80. KU, K. Y., AND F. R. BUTCHER. Detection of a calmodulin-sensitive cyclic nucleotide phosphodiesterase in rat parotid gland. *Biochim. Biophys. Acta* 631: 70–78, 1980.
81. KUSEK, J. C. Amylase release from rat parotid glands. I.

General characteristics. *Biochim. Biophys. Acta* 583: 295–308, 1979.
82. KUSEK, J. C. Amylase release from rat parotid glands. II. Calcium kinetics. *Biochim. Biophys. Acta* 583: 309–319, 1979.
83. LEESON, C. R. Structure of salivary glands. In: *Handbook of Physiology. Alimentary Canal*, edited by C. F. Code. Washington, DC: Am. Physiol. Soc., 1967, sect. 6, vol. II, chapt. 32, p. 463–495.
84. LESLIE, B. A., J. W. PUTNEY, JR., AND J. M. SHERMAN. α-Adrenergic, β-adrenergic and cholinergic mechanisms for amylase secretion by rat parotid gland in vitro. *J. Physiol. Lond.* 260: 351–370.
85. LIANG, T., AND M. A. CASCIERI. Substance P stimulation of amylase release by isolated parotid cells and inhibition of substance P induction of salivation by vasoactive peptides. *Mol. Cell. Endocrinol.* 15: 151–162, 1979.
86. LIANG, T., AND M. A. CASCIERI. Specific binding of an immunoreactive and biologically active ^{125}I-labeled N(1) acylated substance P derivative to parotid cells. *Biochem. Biophys. Res. Commun.* 96: 1793–1799, 1980.
87. LIANG, T., AND M. A. CASCIERI. Substance P receptor on parotid cell membranes. *J. Neurosci.* 1: 1133–1141, 1981.
88. LINDSAY, R. H., T. UEHA, B. S. HULSEY, AND R. W. HANSON. Relationship of chemically initiated enzyme secretion to metabolism in rat parotid in vitro. *Am. J. Physiol.* 221: 80–85, 1971.
89. LUDFORD, J. M., AND B. R. TALAMO. β-Adrenergic and muscarinic receptors in developing rat parotid glands. Selective effect of neonatal sympathetic denervation. *J. Biol. Chem.* 255: 4619–4627, 1980.
90. McPHERSON, M. A., AND C. N. HALES. Control of amylase biosynthesis and release in the parotid gland of the rat. *Biochem. J.* 176: 855–863, 1978.
91. MICHELL, R. H., AND L. M. JONES. Enhanced phosphatidylinositol labelling in rat parotid fragments exposed to α-adrenergic stimulation. *Biochem. J.* 138: 47–52, 1974.
92. MILLER, B. E., AND D. L. NELSON. Calcium fluxes in isolated acinar cells from rat parotid. Effect of adrenergic and cholinergic stimulation. *J. Biol. Chem.* 252: 3629–3636, 1977.
93. MORI, T., Y. TAKAI, R. MINAKUCHI, B. YU, AND Y. NISHIZUKA. Inhibitory action of chlorpromazine, dibucaine, and other phospholipid-interacting drugs on calcium-activated, phospholipid-dependent protein kinase. *J. Biol. Chem.* 255: 8378–8380, 1980.
93a. NICHOLLS, D. G., AND M. CROMPTON. Mitochondrial calcium transport. *FEBS Lett.* 111: 261–268, 1980.
94. OHSHIKA, H., H. TAKEMURA, J. ENDO, S. HATTA, AND M. TANAKA. Stimulating effect of α-adrenoceptor agonists on isoproterenol-induced amylase release in rat parotid tissue. *Jpn. J. Pharmacol.* 31: 1021–1027, 1981.
95. ORON, Y., S. CREACY, J. KELLOGG, AND J. LARNER. Stable cholinergic-muscarinic and α-adrenergic inhibition of rat parotid adenylate cyclase. *J. Cyclic Nucleotide Res.* 6: 105–120, 1980.
96. ORON, Y., J. KELLOGG, AND J. LARNER. Alpha adrenergic and cholinergic-muscarinic regulation of adenosine cyclic 3′,5′-monophosphate levels in the rat parotid. *Mol. Pharmacol.* 14: 1018–1030, 1978.
97. ORON, Y., M. LOWE, AND Z. SELINGER. Involvement of the α-adrenergic receptor in the phospholipid effect in rat parotid. *FEBS Lett.* 34: 198–200, 1973.
98. ORON, Y., M. LOWE, AND Z. SELINGER. Incorporation of inorganic [^{32}P]phosphate into rat parotid phosphatidylinositol. Induction through activation of alpha adrenergic and cholinergic receptors and relation to K$^+$ release. *Mol. Pharmacol.* 11: 79–86, 1975.
99. PADEL, U., J. KRUPPA, R. JAHN, AND H. D. SÖLING. Phosphopeptide patterns of the ribosomal protein S6 following stimulation of guinea pig parotid glands by secretagogues involving either cAMP or calcium as second messenger. *FEBS Lett.* 159: 112–118, 1983.
100. PADEL, U., AND H. D. SÖLING. Phosphorylation of the ribosomal protein S6 during agonist-induced exocytosis in exocrine glands is catalyzed by calcium-phospholipid-dependent protein kinase (protein kinase C). Experiments with guinea pig parotid glands. *Eur. J. Biochem.* 151: 1–10, 1985.
101. PLEWE, G., R. JAHN, A. IMMELMANN, C. BODE, AND H. D. SÖLING. Specific phosphorylation of a protein in calcium accumulating endoplasmic reticulum from rat parotid glands following stimulation by agonists involving cAMP as second messenger. *FEBS Lett.* 166: 96–103, 1984.
102. PUTNEY, J. W., JR. On the role of cellular calcium in the response of the parotid to dibutyryl and monobutyryl cyclic AMP. *Life Sci.* 22: 631–638, 1978.
103. PUTNEY, J. W., JR. Oxygen consumption in the parotid gland. *Life Sci.* 22: 1731–1736, 1978.
104. PUTNEY, J. W., JR., J. S. McKINNEY, D. L. AUB, AND B. A. LESLIE. Phorbol ester-induced protein secretion in rat parotid gland. Relationship to the role of inositol lipid breakdown and protein kinase C activation in stimulus-secretion coupling. *Mol. Pharmacol.* 26: 261–266, 1984.
105. PUTNEY, J. W., JR., AND C. M. VAN DE WALLE. The relationship between muscarinic receptor binding and ion movements in rat parotid cells. *J. Physiol. Lond.* 299: 521–531, 1980.
106. PUTNEY, J. W., JR., C. M. VAN DE WALLE, AND B. A. LESLIE. Receptor control of calcium influx in parotid acinar cells. *Mol. Pharmacol.* 14: 1046–1053, 1978.
107. PUTNEY, J. W., JR., C. M. VAN DE WALLE, AND C. S. WHEELER. Binding of ^{125}I-physalaemin to rat parotid acinar cells. *J. Physiol. Lond.* 301: 205–212, 1980.
108. PUTNEY, J. W., JR., S. J. WEISS, B. A. LESLIE, AND S. H. MARIER. Is calcium the final mediator of exocytosis in the rat parotid gland? *J. Pharmacol. Exp. Ther.* 203: 144–155, 1977.
109. QUISSELL, D. O., L. M. DEISHER, AND K. A. BARZEN. Role of protein phosphorylation in regulating rat submandibular mucin secretion. *Am. J. Physiol.* 245 (*Gastrointest. Liver Physiol.* 8): G44–G53, 1983.
110. QUISSELL, D. O., J. L. LAFFERTY, AND K. A. BARZEN. Dispersed rat parotid cells: role of calcium and cAMP in the regulation of amylase release. *J. Dent. Res.* 62: 131–134, 1983.
111. ROBINOVITCH, M. R., AND L. M. SREEBNY. Separation and identification of some of the protein components of rat parotid saliva. *Arch. Oral Biol.* 14: 935–939, 1969.
112. ROSSIGNOL, B., G. HERMAN, A. M. CHAMBAUT, AND G. KERYER. The calcium ionophore A23187 as a probe for studying the role of Ca^{2+} ions in the mediation of carbachol effects in rat salivary glands: protein secretion and metabolism of phospholipids and glycogen. *FEBS Lett.* 43: 241–246, 1974.
113. RUDICH, L., AND F. R. BUTCHER. Effect of substance P and eledoisin on K$^+$ efflux, amylase release and cyclic nucleotide levels in slices of rat parotid gland. *Biochim. Biophys. Acta* 444: 704–711, 1976.
114. SAMPSON, H. W., D. J. KIESSEL, L. MACKENZIE-GRAHAM, AND I. PISCOPO. A cytochemical study of the effect of cholinergic and β-adrenergic stimulation on calcium fluxes of rat parotid gland. *Histochemistry* 79: 193–203, 1983.
115. SCARPA, A., K. MALMSTROM, M. CHIESI, AND E. CARAFOLI. On the problem of the release of mitochondrial calcium by cyclic AMP. *J. Membr. Biol.* 29: 205–208, 1976.
116. SCHNEYER, L. H., J. A. YOUNG, AND C. A. SCHNEYER. Salivary secretion of electrolytes. *Physiol. Rev.* 52: 720–777, 1972.
117. SCHRAMM, M., AND E. NAIM. Adenyl cyclase of rat parotid gland. Activation by fluoride and norepinephrine. *J. Biol. Chem.* 245: 3225–3231, 1970.
118. SCHRAMM, M., AND Z. SELINGER. The function of α- and β-adrenergic receptors and a cholinergic receptor in the secretory cell of rat parotid gland. In: *Advances in Cytopharmacology*, edited by B. Ceccarelli, F. Clementi, and J. Meldolesi. New York: Raven, 1974, vol. 2, p. 29–32.
119. SCOTT, B. L., AND D. C. PEASE. Electron microscopy of the salivary gland and lacrimal glands of the rat. *Am. J. Anat.* 104: 115–161, 1959.
120. SCOTT, J., AND B. J. BAUM. Involvement of cyclic AMP and calcium in exocrine protein secretion induced by vasoactive

intestinal polypeptide in rat parotid cells. *Biochim. Biophys. Acta* 847: 255–262, 1985.
121. SEAMON, K., AND J. W. DALY. Activation of adenylate cyclase by the diterpene forskolin does not require the guanine nucleotide regulatory protein. *J. Biol. Chem.* 256: 9799–9801, 1981.
122. SELINGER, Z., AND E. NAIM. The effect of calcium on amylase secretion by rat parotid slices. *Biochim. Biophys. Acta* 203: 335–337, 1970.
123. SELINGER, Z., E. NAIM, AND M. LASSER. ATP-dependent calcium uptake by microsomal preparations from rat parotid and submaxillary glands. *Biochim. Biophys. Acta* 203: 326–334, 1970.
124. SPEARMAN, T. N., AND F. R. BUTCHER. Rat parotid gland protein kinase activation. Relationship to enzyme secretion. *Mol. Pharmacol.* 21: 121–127, 1982.
125. SPEARMAN, T. N., AND F. R. BUTCHER. The effect of calmodulin antagonists on amylase release from the rat parotid gland in vitro. *Pfluegers Arch.* 397: 220–224, 1983.
126. SPEARMAN, T. N., J. P. DURHAM, AND F. R. BUTCHER. The role of cyclic AMP in the regulation of exocytosis in the rat parotid gland: evidence obtained with the isoproterenol analog PI-39. *J. Cyclic Nucleotide Res.* 8: 225–234, 1982.
127. SPEARMAN, T. N., J. P. DURHAM, AND F. R. BUTCHER. Cyclic AMP in the regulation of exocytosis in the rat parotid gland. Evidence obtained with cholera toxin. *Biochim. Biophys. Acta* 759: 117–124, 1983.
128. SPEARMAN, T. N., K. P. HURLEY, R. OLIVAS, R. G. ULRICH, AND F. R. BUTCHER. Subcellular location of stimulus-affected endogenous phosphoproteins in the rat parotid gland. *J. Cell Biol.* 99: 1354–1363, 1984.
129. STENGEL, D., L. GUENET, M. DESMIER, P. INSEL, AND J. HANOUNE. Forskolin requires more than the catalytic unit to activate adenylate cyclase. *Mol. Cell. Endocrinol.* 28: 681–690, 1982.
130. STRITTMATTER, W. J., J. N. DAVIS, AND R. J. LEFKOWITZ. α-Adrenergic receptors in rat parotid cells. I. Correlation of [^{3}H]dihydroergocryptine binding and catecholamine-stimulated potassium efflux. *J. Biol. Chem.* 252: 5472–5477, 1977.
131. SUNDSTRÖM, S., B. CARLSÖÖ, A. DANIELSSON, AND R. HENRIKSSON. Differences in dopamine- and noradrenaline-induced amylase release from the rat parotid gland. *Eur. J. Pharmacol.* 109: 355–361, 1985.
132. TAKEMURA, H. Inhibitory effect of carbachol on isoproterenol-induced amylase release from isolated rat parotid cells. *Jpn. J. Pharmacol.* 35: 9–17, 1984.
133. TAKEMURA, H. Potentiation of amylase release from isolated rat parotid cells—studies on the combination of isoproterenol and a low dose of carbachol. *Jpn. J. Pharmacol.* 36: 107–109, 1984.
134. TAKEMURA, H. Changes in cytosolic free calcium concentration in isolated rat parotid cells by cholinergic and β-adrenergic agonists. *Biochem. Biophys. Res. Commun.* 131: 1048–1055, 1985.
135. TAKUMA, T., B. L. KUYATT, AND B. J. BAUM. Calcium transport mechanisms in basolateral plasma membrane-enriched vesicles from rat parotid gland. *Biochem. J.* 227: 239–245, 1985.
136. TEMPLETON. D. Augmented amylase release from rat parotid gland slices, in vitro. *Pfluegers Arch.* 384: 287–289, 1980.
137. TSANG, B. K., R. H. RIXON, AND J. F. WHITFIELD. A possible role for cyclic AMP in the initiation of DNA synthesis by isoproterenol-activated parotid gland cells. *J. Cell. Physiol.* 102: 19–26, 1980.
138. VON ZASTROW, M., T. R. TRITTON, AND J. D. CASTLE. Identification of L-ascorbic acid in secretion granules of the rat parotid gland. *J. Biol. Chem.* 259: 11746–11750, 1984.
139. WALLACH, D., AND M. SCHRAMM. Calcium and the exportable protein in the rat parotid gland. Parallel subcellular distribution and concomitant secretion. *Eur. J. Biochem.* 21: 433–437, 1971.
140. WHARTON, J., J. M. POLACK, M. G. BRYANT, S. VAN NOORDEN, S. R. BLOOM, AND A. G. E. PEARSE. Vasoactive intestinal polypeptide (VIP)-like immunoreactivity in salivary glands. *Life Sci.* 25: 273–280, 1979.
141. WILCHEK, M., Y. SALOMON, M. LOWE, AND Z. SELINGER. Conversion of protein kinase to a cyclic AMP independent form by affinity chromatography on N^6-caproyl 3′,5′-cyclic adenosine monophosphate–Sepharose. *Biochem. Biophys. Res. Commun.* 45: 1177–1184, 1971.
141a. WILLIAMSON, J. R., R. H. COOPER, AND J. B. HOEK. Role of calcium in the hormonal regulation of liver metabolism. *Biochim. Biophys. Acta* 639: 243–295, 1981.
142. YOKOYAMA, N., M. ABE, AND S. FURUYAMA. Cyclic nucleotide phosphodiesterase in rat parotid gland. *Can. J. Physiol. Pharmacol.* 59: 293–299, 1981.
143. YOSHIMURA, K., E. NEZU, AND A. CHIBA. Stimulation of α-amylase release and cyclic AMP accumulation by catecholamine in rat parotid slices in vitro. *Jpn. J. Physiol.* 32: 121–135, 1982.
144. YOSHIMURA, K., E. NEZU, AND T. YONEYAMA. Stimulation of cyclic AMP-dependent protein kinase by catecholamines and its relationship to α-amylase release in rat parotid gland. *Jpn. J. Physiol.* 32: 699–716, 1982.
145. YOSHIMURA, K., E. NEZU, AND T. YONEYAMA. Mechanism of regulation of amylase release by α- and β-adrenergic agonists in rat parotid tissue. *Jpn. J. Physiol.* 34: 665–667, 1984.
146. YOSHIMURA, K., E. NEZU, AND T. YONEYAMA. Augmentation of isoproterenol-stimulated tissue cyclic AMP level by cholinergic agonists in rat parotid gland. *Jpn. J. Physiol.* 35: 765–781, 1985.

CHAPTER 5

Salivary mucin secretion

DAVID O. QUISSELL | Oral Research Center, School of Dentistry, University of Colorado Health Sciences Center, Denver, Colorado

LAWRENCE A. TABAK | Department of Dental Research and Biochemistry, School of Medicine and Dentistry, University of Rochester, Rochester, New York

CHAPTER CONTENTS

Chemical Composition and Structure of Salivary Mucins
Physiological Alteration of Mucin Structure
Biosynthesis of Mucins
Submandibular Gland
    Submandibular mucin secretion
    Membrane receptors and their role in mucin secretion
    Other possible effectors of mucin secretion
    Signal transduction
    Cellular mediators of mucin secretion
      Rat submandibular mucin secretion
    Importance of calcium in submandibular mucin secretion
    Summary
Sublingual Gland
Minor Salivary Glands

EPITHELIAL SURFACES ARE PROTECTED by a slimy, viscoelastic coat termed *mucus*. This adherent layer, which consists of glycoproteins, proteins, and lipids, forms a hydrated gellike structure that serves as the first line of defense against mechanical, chemical, or microbial insult to the underlying tissue. Of particular functional significance are the high-molecular-weight mucin glycoproteins (mucins) that are believed to be largely responsible for the unique physiochemical and rheological properties of the mucous coat.

Compelling evidence suggests that salivary mucins play a significant role in the nonimmune protection of the oral cavity (119). In particular, mucins coat the hard and soft structures of the mouth, serving to lubricate and hydrate these tissues as well as provide a selective permeability barrier to withstand exogenous insults. In addition, salivary mucins modulate the oral microflora by favoring attachment and subsequent proliferation of certain microorganisms and/or by promoting the clearance of others.

This chapter reviews the current knowledge of the molecular events involved in mucin structure, function, and secretion. Because space restrictions prohibit a complete scientific review, the major emphasis of this chapter is on the more recent studies in the area of salivary gland cell biology, cellular physiology, pharmacology, and biochemistry rather than the more traditional in vivo studies. These more recent studies have provided important new information on mucin structure, physiological effectors of mucin structure and function, and the molecular events involved in the regulation of salivary mucin secretion. However, much additional information is required before a complete comprehension of these important physiological components can be obtained.

CHEMICAL COMPOSITION AND STRUCTURE OF SALIVARY MUCINS

Mucins are high-molecular-weight glycoproteins that are characterized by common structural features including a peptide core (apomucin), which is enriched in serine, threonine, proline, glutamic acid, and glycine and carbohydrate side chains, termed *oligosaccharides*, which are linked by *O*-glycosides to threonine or serine. Complexing between mucin subunits and nonmucin species results in the formation of elaborate mucin suprastructures that may be stabilized by both noncovalent and covalent forces (41, 53, 119). It is this suprastructure that is responsible for the unique rheological properties characteristic of these molecules.

Mucin oligosaccharides may contain galactose (Gal), fucose (Fuc), *N*-acetylgalactosamine (GalNAc), *N*-acetylglucosamine (GlcNAc), and sialic acids. Sulfate, if present, is covalently linked to either Gal or GlcNAc. In addition, the presence of a small number of *N*-linked oligosaccharides is possible (27, 82). The sequence of carbohydrates found in the oligosaccharides of many salivary mucins has been elucidated (for review, see refs. 41, 53, 119). These include neutral sulfated and sialic acid–containing chains (Table 1).

The high content of carbohydrate makes it likely that the glycosylated peptide is present as a random

TABLE 1. *Representative Oligosaccharide Structures of Salivary Mucins*

Type	Structure	Ref.
Armadillo submandibular	SAα2,6GalNAcα1-OThr(Ser)	132
Cow submandibular		8
Sheep submandibular		8
Dog submandibular	Fucα1,2Galβ1,3GalNAcα1-OThr(Ser)	65
Human submandibular-sublingual		96
Human submandibular-sublingual	SAα2,3Galβ1,3GalNAcα1-OThr(Ser)	96
Pig submandibular	GalNAcα1,3(Fucα1,2)Galβ1,3(SAα2,6)GalNAcα1-OThr(Ser)	15, 27
Rat submandibular		78
Dog submandibular	Fucα1,2Galβ1,3(3,4,6)GlcNAc3,4-SO$_4$$\beta$1,6Gal$\beta$1,3GalNAc$\alpha$1-OThr(Ser)	65
Monkey submandibular-sublingual	Fucα1,2Galβ1,6(Galβ1,4GlcNAcβ1,3)GalNAcα1-OThr(Ser)	121
Rat sublingual	GlcNAcβ1,3(SAα2,6)Galβ1,4GlcNAcβ1,3(SAα2,6)Galβ1,4GlcNAcβ1,3(SAα2,6)Gal β1,4GlcNAcβ1,3(SAα2,6)Gal β1,4GlcNAcβ1,3(SAα2,6)GalNAcα1-OThr(Ser)	112
Human submandibular-sublingual	Fucα1,2Galβ1,4(Fucα1,3)GlcNAcβ1,6(SAα2,3Galβ1,3)GalNAcα1-OThr(Ser)	96

coil with little secondary structure (16). Regions of the peptide backbone that are not glycosylated are said to be "naked" and hence are more susceptible to proteolysis (74). Compositional studies have indicated that the naked regions of mucins are enriched in both hydrophobic amino acids and cysteine residues (30, 109). These findings have been corroborated by fluorescence spectroscopic measurements of hydrophobic binding sites in several mucins, including rat submandibular gland (113, 120) and MG1 (high-molecular-weight mucin) from human submandibular-sublingual saliva (M. J. Levine, unpublished observations).

Available evidence suggests that the distribution and extent of these naked regions play an important role in determining the nature of mucin suprastructure. Two distinct patterns have been reported for salivary mucins. The first category is characterized by a somewhat uniform distribution of oligosaccharides about the entire length of the peptide core. Mucins of this type tend to be smaller in molecular weight than has been classically assumed ($M_r \sim 100,000-200,000$), although they may form suprastructures via noncovalent interactions. This pattern is typified by ovine submandibular gland mucin (55), for which the most detailed information is available. The mucin is formed by the noncovalent aggregation of glycoprotein subunits, each having a molecular weight of ~154,000, of which 58,000 is contributed by the apomucin. Glycosylated serine and threonine appear in clusters of three to nine residues along the protein core, in which every other residue is serine or threonine. The segments are interspersed with nonglycosylated regions of four to seven amino acids that are devoid of serine and threonine. Relatively small molecular weight mucins having no covalent subunit structure have also been purified from mouse submandibular glands (26, 100), rat submandibular glands (36, 69, 120), and human submandibular-sublingual saliva (88). In contrast, mucins displaying a more biased distribution of oligosaccharides (and hence longer stretches of naked regions) are characterized by a covalent stabilization of their suprastructure via disulfide bonds. This pattern has been found for mucins isolated from monkey extraparotid saliva (54), porcine submandibular glands (109), and most recently, human submandibular-sublingual saliva (64).

Considerable evidence indicates that mucus-secreting cells from salivary glands are able to synthesize both types of mucins and that each molecule plays a distinct functional role. Classic studies of submandibular gland mucins have repeatedly pointed to the presence of two mucins [major and minor (45)] based on affinity for hydroxyapatite (the major inorganic constituent of tooth enamel). Extensive studies with mucins purified from human submandibular-sublingual saliva have indicated that MG2, a low-molecular-weight species characterized by uniformly distributed oligosaccharides (64, 88), interacts specifically with the oral bacteria *Streptococcus mitis* and *Streptococcus sanguis* (63, 79, 117). In contrast, MG1, a high-molecular-weight species characterized by disulfide-linked subunits and a biased distribution of oligosaccharides (64), displays physicochemical properties that enable it to effectively coat the hard and soft tissues of the oral cavity (119). Recent in vitro studies have supported this view; MG1 has a greater affinity for hydroxyapatite than MG2 (118). Collectively, the data suggest that MG1 serves a tissue-coating function, thereby protecting the tooth, whereas MG2 fails to adsorb in appreciable quantity to the tooth surface, making it available to interact with specific bacteria and effect their clearance from the oral cavity (64).

PHYSIOLOGICAL ALTERATION OF MUCIN STRUCTURE

Heterogeneity of mucins has been attributed largely to incomplete posttranslational modification (44) or degradation occurring concomitant with isolation procedures. This implies that the control of mucin biosynthesis is largely indirect, that is, at the glycosyltransferase level. An alternative possibility is that the heterogeneity of mucins is controlled in part by the expression of unique apomucins (42), which are then subjected to the indirect controls of posttranslational modification (98).

Alteration of mucin composition, in response to hormones and neurotransmitters, has been suggested by analysis of either submandibular gland extracts or secretions. For example, pilocarpine stimulation of canine submandibular glands leads to a decrease in the ratio of sialic acid–fucose in the secretion (29). Because these carbohydrates occupy terminal positions on oligosaccharide units, it was suggested that they might compete to terminate a growing carbohydrate chain. Adrenalectomy or castration of male rats results in a decreased sialic acid content in the submandibular glands (21, 23), whereas spaying female rats had no measurable effect (22). Chronic isoproterenol administration leads to an increased sialic acid content in the submandibular glands of male rats (20). However, recent findings have demonstrated that such treatment leads to the induction of N-linked glycoproteins (56, 76) in rat submandibular glands that may account in part for the changes in sialic acid content observed. These data point to the need to chemically characterize specific molecules from the secretion or glandular extract under study. In this regard, recent work using a radioimmunoassay (RIA) to detect mucins has indicated that the concentration of mucins per gland wet weight increases in isoproterenol-treated mice. However, carbohydrate analysis of mucins purified from single glands indicated less heterogeneity in sialic acid and GalNAc content relative to control animals (P. A. Denny and P. C. Denny, unpublished observations).

Variation in mucin expression and structure in response to normal physiological events has been suggested by several investigations. Mucin is detected by RIA in developing mouse submandibular glands by the 14th day of gestation. In the interval between the 18th day of gestation and birth, the concentration of mucin increases 500-fold. The level obtained by birth remains constant through adulthood (P. A. Denny and P. C. Denny, unpublished observations). In vitro experiments with dispersed cell aggregates prepared from rat submandibular glands suggest that there is a decreased level of mucin synthesis in cells derived from aged animals (7).

Collectively these observations suggest that the structure of mucin glycoproteins is subject to neurotransmitter, hormonal, and age-related controls. However, little is known about the regulatory mechanisms that effect these controls.

BIOSYNTHESIS OF MUCINS

The biosynthesis of mucin glycoproteins is an extremely complex process consisting of multiple steps, each of which is subject to a wide range of cellular controls. Thus the ultimate structure (and hence function) of a mucin is related to the aggregate expression of these multiple control points.

A high proportion of genes in eucaryotes are now known to be members of multigene families, including salivary molecules such as the proline-rich proteins (67). In view of the structural and functional data that point to the expression of multiple mucins in salivary glands summarized in the previous section, it would not be surprising if mucins were found to be members of multigene families. Although there has been a preliminary report in which putative salivary gland mucin mRNAs have been characterized (81), thus far there are no data available that directly address the issue of transcriptional or translational control of apomucin expression.

By analogy to other secretory proteins, it is presumed (but not yet demonstrated) that the preproapomucin contains a hydrophobic amino terminal sequence (signal sequence) that mediates the transport of the nascent polypeptide across the endoplasmic reticulum membrane (129). During translocation, the signal peptide is removed by proteolytic processing, an event that has been shown to be a prerequisite for the correct conformational development of several exocrine proteins (103). In addition, defective secretory proteins may be eliminated by intracellular degradation that occurs, in part, within the endoplasmic reticulum (10).

Available evidence indicates that O-glycosylation is a posttranslational event that occurs in the Golgi apparatus (48, 58, 104) and requires no lipid-bound intermediates (3, 50). Rather, nucleotide sugars are added stepwise to the growing oligosaccharide chain under the direction of specific glycosyltransferases. Thus an initial control point of O-glycosylation involves the availability of the sugar nucleotide pool. Sugar nucleotides are produced largely in the cytoplasm of the cell (49, 103, 114), the exception being CMP sialic acid, which forms in the nucleus (17, 59, 126). The translocation of UDP galactose into the lumen of rat submandibular gland Golgi apparatus occurs via facilitated transport (6). Recently it has been proposed that this process is effected by antiport proteins that span the Golgi membrane (14). The activated sugar binds via the nucleotide to the antiport protein and is then subsequently translocated across the membrane. Following glycosylation, the remaining nucleoside monophosphate binds to the luminal domain of the antiport protein for equimolar exchange with the next activated sugar to be translocated.

The substrate specificity of the enzyme UDP-GalNAc:polypeptide α-N-acetylgalactosaminyltransferase, which has been partially characterized from the submandibular glands of several different species, represents a major control point of O-glycosylation. Although specific amino acid sequences have been proposed as generalized acceptors for O-glycosylation, it has become apparent that there is no consensus signal (2, 12). Perhaps the only common feature to emerge is an apparent requirement of proline residues to be near the glycosylation site. This suggests that the determinant for O-glycosylation is conformational in nature, rendering the substrate more accessible to the glycosyltransferase (2, 55).

The elongation of the mucin oligosaccharides is controlled by the strict substrate specificity (98, 104) and the topographical distribution (from cis- to trans-Golgi) of the glycosyltransferases (33). In this process, the product of the first glycosyltransferase reaction becomes the substrate for the second glycosyltransferase reaction, et cetera. Thus the degree of substrate specificity displayed by each enzyme dictates the intrinsic fidelity of the overall process. Also implicit in this model is the prediction that there is a separate transferase for each sugar added to the mucin. Studies using highly purified glycosyltransferases (9, 131) have borne out the concept of "one linkage–one glycosyltransferase." Because in many instances the prior activity of one transferase severely inhibits the subsequent action of another transferase, regulation of the process is dictated not only by which transferases are present but also by the order in which they are permitted to act. Accordingly, the lectin from Helix pomatia, which interacts with terminal GalNAc, has recently been shown to bind to sites in the cis- and trans-Golgi of intestinal goblet cells (99). No labeling was detected in the medial compartment of the Golgi, however. These data suggest that core O-glycosylation takes place in the cis-Golgi cisternae; extension (which would render the core GalNAc residue cryptic) takes place in the medial compartment and termination takes place with GalNAc in the trans-Golgi cisternae. It is not clear, however, if these proposed compartments correspond to those defined for processing N-linked oligosaccharide chains (33).

Toward the later stages of posttranslational modification, mucins may undergo acylation (68) in which fatty acids are covalently attached to the peptide core. Recent evidence suggests that a rat submandibular gland mucin is acylated near the end of subunit processing but prior to assembly into high-molecular-weight mucin polymer (111). The functional consequences of the covalent attachment of fatty acids to a salivary mucin are not yet appreciated. However, mucin-bound fatty acids have been demonstrated to retard hydrogen ion diffusion (102) and contribute significantly to the viscosity of stomach mucin (80) in in vitro studies. Quite clearly, these properties could influence the tissue-coating function of salivary mucins in the oral cavity (119).

Mucins complex with a wide range of peptides and proteins prior to or concomitant with packaging in secretory granules. Studies with mucins from rat (120) and ovine submandibular glands (55), as well as human submandibular-sublingual saliva (88), have shown that small peptides are released after modification of the ξ-amino group of lysine. These peptides may be analogous to link proteins that have been described for goblet cell mucins (37, 70). In addition, a wide range of proteins with protective qualities enter into heterotypic associations with mucins, including antiproteases (106, 110), lysozyme (19, 110), and secretory IgA (88, 116). These findings lead to the speculation that mucins serve to concentrate protective molecules at the mucosal interface and the inhospitable external environment (34, 119). Moreover, these interactions may represent "prepackaged" associations, which are formulated to express a specific biological function.

SUBMANDIBULAR GLAND

Submandibular Mucin Secretion

Exocytosis in the submandibular salivary gland is regulated by the autonomic nervous system (107). In the rat, stimulation of the β-adrenergic receptors resulted in the production of a thick mucous secretion, whereas α-adrenergic or cholinergic receptor stimulation resulted in a rapid rate of saliva production with a low macromolecular content. Canine submandibular mucin secretion also appears to be predominantly regulated via β-adrenergic receptors (61). Unfortunately, the actual molecular basis of exocytosis and its regulation is still poorly understood. The stimulus-secretion coupling mechanism involved in the control of submandibular mucin release has not been as extensively studied as in other exocrine or endocrine cells.

In the past, secretory studies involved both in vivo rat submandibular glands (107, 125) and the use of in vitro rat submandibular gland slices (11, 73). Cell dispersion procedures, which have also been developed in recent years, have eliminated complications due to the presence of neural tissue and possible secondary release of endogenous neurotransmitters, removing other extracellular elements that impose diffusional barriers and eliminating the existence of pericellular compartments that may alter the secretory response of the cells. The cell dispersion technique results in the isolation of functionally intact submandibular cells with most of the cells still in their normal acinar-intercalated duct complexes (40, 95). Although straightforward in design, the cell dispersion procedure requires careful monitoring of the pH of the medium, oxygenation of the cells, and the use of only highly purified dispersion enzymes during the dissociation procedures. If these conditions are not controlled, the resultant population of submandibular cells will demonstrate a diminished secretory response and loss of receptor sensitivity.

The in vitro evaluation of the regulation of mucin secretion has involved the use of three different analytical approaches. The extent of ^{14}C-radiolabeled mucin release (40, 75, 89, 91, 94), the extent of sialic acid secretion (11, 28), and recently an RIA procedure has been developed to evaluate mucin secretion (28, 38). Each experimental approach has certain advantages and disadvantages. The use of ^{14}C-glucosamine for radiolabeling the submandibular mucins has been the procedure most widely utilized. The radiolabel was incorporated into amino sugars with a relative radiospecific activity ratio for glucosamine, galactosamine, and sialic acid of 3.4, 1.1, and 1.0, respectively (94). Although the radiolabel was predominantly incorporated into the mucin molecules, some of the radiolabel was also incorporated into other cellular glycoproteins (40). The sialic acid assay procedure can be tedious, and rat submandibular mucins have low sialic acid content (39, 120). Although the recently developed RIA procedure shows great promise, the usefulness is predicated on the definitive characteristics of the antibody specificity (28, 38). Environmental or genetic variations in the actual composition of the mucins in the submandibular secretory granules could affect the relative extent of release of a particular mucin independent of receptor stimulation. In addition, if more than one mucin molecule exists within the submandibular secretory granule, which appears to be true, then differences in the secretory response of the cells, independent of the actual rate of mucin release, could occur when comparing the above assay procedures.

Membrane Receptors and Their Role in Mucin Secretion

Table 2 presents a summary in which the radiolabeling procedure was used to evaluate the effects of various secretagogues on rat submandibular mucin secretion. Those reported studies in which the submandibular cells were not oxygenated or in which ascorbic acid was not included in the incubation medium to prevent the nonenzymatic oxidation of the catecholamines were not included in Table 2.

As shown in Table 2, rat submandibular mucin secretion required β-adrenergic receptor stimulation to elicit a significant secretory response. Pure cholinergic or pure α-adrenergic receptor stimulation was unable to elicit a significant release of radiolabeled mucin material. The essential requirement for β-adrenergic receptor stimulation to elicit mucin secretion was further supported by the fact that the Ca^{2+} ionophore A23187 was unable to elicit mucin release in the absence of β-adrenergic receptor stimulation or

TABLE 2. *Effect of Various Secretagogues on Rat Submandibular Mucin Secretion*

Secretagogue	Mucin Secretion,* %	Ref.
Unstimulated	9.36	89, 91
Carbamylcholine, 10^{-6} M	9.96	89
Phenylephrine†, 5×10^{-6} M	8.87	89
Isoproterenol, 10^{-5} M	24.6	89, 91
Norepinephrine, 10^{-5} M	47.0	89, 91
Isoproterenol, 10^{-5} M, and carbamylcholine, 10^{-5} M	33.7	89
A23187, 10^{-6} M	6.90	89
Bt$_2$cAMP, 10^{-3} M	17.4	89
A23187, 10^{-6} M, and Bt$_2$cAMP, 10^{-3} M	31.1	89

Values are means ± SD (<10%) of at least 4 separate experiments. * (−)-Propranolol, 10^{-5} M, was included in culture medium. † Release of ^{14}C-radiolabeled trichloroacetic acid and phosphotungstic acid precipitable material after secretagogue stimulation. Extent of mucin secretion was normalized by expressing amount of radiolabeled material released as percent of total cellular precipitable ^{14}C disintegrations per minute present at beginning of each study.   [Adapted from Quissell et al. (89, 91)].

Bt$_2$cAMP stimulation. These data in Table 2 also document the fact that α-adrenergic or cholinergic receptor stimulation augments the β-adrenergic receptor-mediated response. The augmentation of mucin secretion by either of these two receptors required the presence of extracellular Ca^{2+} (89, 91). The Ca^{2+} ionophore A23187 was also able to augment mucin release during Bt$_2$cAMP stimulation, and extracellular Ca^{2+} was required for augmentation.

The availability of highly specific radioligands has permitted the pharmacological identification and characterization of the membrane receptors in submandibular glands. The α_1- and β_1-adrenergic and M$_1$ muscarinic cholinergic receptors were found to be the predominant receptor subtypes in the adult animals (5, 13, 18, 87). Although receptor density can vary because of sex and age, most studies presented similar results for the various radioligands tested. With the specific muscarinic ligand [^{3}H]quinuclidinyl benzilate, kinetic analysis indicated a single population of high-affinity binding sites with an apparent dissociation constant of 87–450 pM and a receptor density of 214–257 pmol/g of membrane protein. With [^{3}H]dihydroalprenolol in the characterization of the β-adrenergic receptors, kinetic analysis indicated an apparent dissociation constant of 800 pM, with receptor density of 226 pmol/g of protein. A radioligand for α_1-receptor binding studies, [^{3}H]prazosin, indicated that the adult submandibular gland had an α_1-receptor density of 116 pmol/g of protein, with an apparent dissociation constant of 430 pM.

Other Possible Effectors of Mucin Secretion

The major salivary glands of the rat are innervated with nerve fibers that contain vasoactive intestinal peptide (VIP) and substance P (43, 66, 130). The VIP probably coexists with acetylcholine in the parasympathetic nerve fibers, and both acetylcholine and VIP are released from postganglionic nerve fibers. When injected intravenously, VIP was found to elicit saliva production in rat submandibular glands (35). The effect of VIP on saliva production appeared to be direct and occurred in the presence of autonomic receptor blockers. Saliva elicited by VIP injection had a lower total protein content than that elicited by isoproterenol, a β-adrenergic agonist. The physiological significance of VIP in the modulation of mucin secretion is unknown. Substance P is present in peripheral nerve fibers in many tissues, including the salivary glands. The neuropeptide may have an important role in modulating salivary secretion, but its mechanism of action is unknown. Intravenous infusion of the neuropeptide into rats induced changes in saliva flow and the volume of whole saliva secreted (72). Substance P receptors have been identified in rat submandibular tissue. Using a substance P analogue [tyrosine^1, norleucine^{11}]-substance P iodinated by the chloramine-T method, a single class of noninteracting binding sites was identified with an affinity of 9.26 nM and maximum binding capacity of 15.7 pmol/g of protein (4).

Signal Transduction

As shown in Table 2, the predominant receptor involved in the direct regulation of rat submandibular mucin is the β_1-adrenergic receptor. The activation of this receptor system results in a dramatic increase in intracellular cAMP. The signal transduction process appears to involve the selective activation of adenylate cyclase. Unfortunately, little information is currently available regarding the biochemical and enzymatic characteristics of submandibular adenylate cyclase and its coupling to the β-adrenergic receptor. Various agonists and antagonists have been tested for their ability to activate adenylate cyclase activity (5, 46). As expected, those agonists that have β-adrenergic receptor binding activity also showed an increased adenylate cyclase activity in crude membrane preparations, and those β-adrenergic receptor antagonists that are able to block β-receptor specific binding activity also blocked the stimulation of adenylate cyclase activity. Fluoride ion was able to stimulate adenylate cyclase activity in the absence of β-adrenergic receptor stimulation (5, 46).

The mechanism by which β-adrenergic receptor stimulation leads to mucin secretion is not well understood, but given the fact that β-adrenergic receptor stimulation was required and the fact that adenylate cyclase activity was modulated via β-adrenergic receptor stimulation, cAMP metabolism appears to play a central role in regulating submandibular mucin secretion. As shown in Figure 1, total rat submandibular cellular cAMP levels increased dramatically following β-adrenergic receptor stimulation. Within the first 30

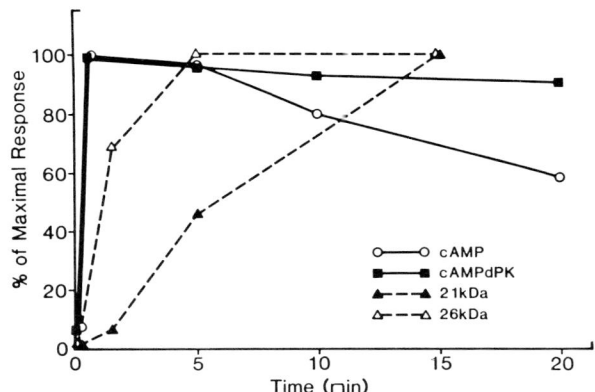

FIG. 1. Time course of effect of β-adrenergic receptor stimulation on rat submandibular cellular cAMP levels, on cellular cAMP-dependent protein kinase activity, and on endogenous protein phosphorylation of 2 β-adrenergic–dependent, integral membrane phosphoproteins. [Adapted from Quissell et al. (90, 91, 93).]

s, the cAMP content of the cells was increased 10-fold over basal levels. Analogues of cAMP, which are known to activate cAMP-dependent protein kinase, can also elicit a secretory response in the absence of β-adrenergic receptor stimulation (91).

Cellular Mediators of Mucin Secretion

Cyclic AMP is thought to mediate most, if not all, of its effects by the activation of cAMP-dependent protein kinase. As shown in Figure 1, concomitant with the increase in intracellular cAMP levels after β-adrenergic receptor stimulation, rat submandibular cAMP-dependent protein kinase activity also increased, with a maximal response obtained within 30 s after β-adrenergic receptor stimulation. The activation of cAMP-dependent protein kinase remained fairly constant throughout the entire secretory period, unlike intracellular cAMP levels. The extent of cAMP-dependent protein kinase activity closely correlated with β-adrenergic receptor–stimulated mucin release (Fig. 2). Recent studies have suggested that cAMP-dependent protein kinase activation may have an important regulatory role in mucin secretion (90). In addition, termination of β-adrenergic receptor stimulation caused a rapid decrease in endogenous cAMP-dependent protein kinase activity. As expected, cholinergic or pure α-adrenergic receptor stimulation, which was unable to elicit significant mucin secretion in rat submandibular cells, had no effect on the endogenous cAMP-dependent protein kinase activity. Combined α- and β-adrenergic receptor stimulation also had no inhibitory or potentiating effect on β-adrenergic receptor–mediated cAMP-dependent protein kinase activity.

Cyclic AMP–dependent protein kinase exists as two separate isozymes, type I and type II, which can be separated and identified by diethylaminoethyl-cellulose (DEAE-cellulose) chromatography and by sodium

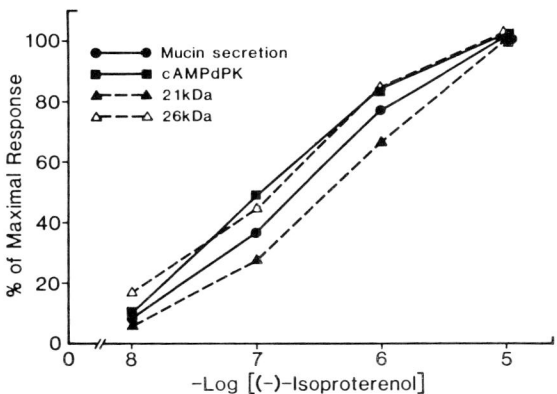

FIG. 2. Concentration dependence of isoproterenol stimulation on mucin secretion, cellular cAMP-dependent protein kinase activation, and on endogenous protein phosphorylation of 2 integral membrane phosphoproteins. [Adapted from Quissell et al. (89, 90, 93).]

dodecyl sulfate–polyacrylamide gel electrophoresis (SDS-PAGE). The mechanism of activation by cAMP involves the binding of cAMP to the regulatory subunit with subsequent dissociation of the catalytic subunit. The DEAE-cellulose chromatography of cytosolic rat submandibular cAMP-dependent protein kinase indicated that ~70% of total glandular cAMP-dependent protein kinase activity was of the type II subtype (32, 90).

A cAMP-dependent metabolic pathway has been implicated in the stimulus-secretion coupling mechanism of rat submandibular mucin secretion. This suggests that protein phosphorylation and dephosphorylation may be one of the actual molecular mechanisms by which mucin secretion is regulated. Recent studies support such a possible mechanism. Beta-adrenergic receptor stimulation or Bt$_2$cAMP stimulation has been associated with the enhanced phosphorylation of three submandibular protein subunits: 34,000-, 26,000-, and 21,000-mol-wt phosphoproteins (92, 93). The extent of phosphorylation of each of these three proteins correlated with the β-adrenergic–dependent dose response for mucin secretion (Fig. 2) and with time (Fig. 1) after stimulation. Termination of the stimulus resulted in a decrease in the extent of phosphorylation of each of the phosphoproteins (92, 93). Pure cholinergic or pure α-adrenergic receptor stimulation had no effect on the extent of phosphorylation for each of the phosphoproteins. Subsequent studies indicated that the 34,000-mol-wt phosphoprotein was protein S6 of the ribosome (D. O. Quissell, unpublished observations) and that the 26,000- and 21,000-mol-wt substances were integral membrane phosphoproteins (92). As shown in Figure 2, the dose-response relationship for the activation of cAMP-dependent protein kinase closely correlated with the β-adrenergic dose-dependent phosphorylation of the two integral membrane phosphoproteins. These studies suggest that, if protein phosphorylation is one of the actual cellular molecular mechanisms involved in regulating exocytosis, then the phosphate turnover rate (phosphorylation and dephosphorylation rate) should be similar to the rate of activation and inactivation of the cAMP-dependent protein kinase (92), which is presumably involved in the actual phosphotransferase reaction. Of these, only the 26,000-mol-wt phosphoproteins appeared to display the appropriate rate of phosphorylation (Fig. 1) and dephosphorylation (92), which was compatible with a direct role for the phosphoprotein in regulating mucin secretion. Phosphate turnover studies indicated that the in situ dephosphorylation rate for the 26,000 phosphoprotein was rather rapid ($t_{1/2}$ = 5–6 min), whereas the dephosphorylation rates for the 34,000- and 21,000-mol-wt phosphoproteins were much slower ($t_{1/2} >$ 20 min) (92, 93).

RAT SUBMANDIBULAR MUCIN SECRETION. There are five cAMP-related events that occur during rat submandibular mucin secretion.

1. Mucin secretion requires β-adrenergic receptor stimulation.
2. Beta-adrenergic receptor stimulation causes a rapid increase in endogenous cAMP levels.
3. Beta-adrenergic receptor stimulation results in rapid activation of cAMP-dependent protein kinase.
4. The rate of phosphorylation and dephosphorylation of the 26,000-mol-wt integral membrane phosphoprotein correlates with rate of cAMP-dependent protein kinase activation and inactivation.
5. The extent of phosphorylation of the 26,000-mol-wt integral membrane phosphoprotein correlates with the dose-response relationship for β-adrenergic receptor–mediated mucin secretion and cAMP-dependent protein kinase activation.

The information just summarized presents the salient features relating to the importance of the cAMP-mediated events during rat submandibular mucin secretion. The current hypothesis based on these data suggest that there exists within the rat submandibular acinar cell membrane fraction a "stimulus-coupled regulatory complex for exocytosis" in which the 26,000-mol-wt phosphoprotein is a subunit. The 26,000-mol-wt phosphoprotein subunit transfers to the enzyme complex the cAMP-mediated signal, and thus the 26,000-mol-wt phosphoprotein is the actual regulatory site of cAMP-mediated exocytosis, with the extent of 26,000-mol-wt phosphorylation determining the actual rate of mucin secretion.

Importance of Calcium in Submandibular Mucin Secretion

An important role for Ca^{2+} in the exocytotic process of salivary glands was first indicated by the studies of Douglas and Poisner (31). Perfusion of a cat submandibular gland with Ca^{2+}-free incubation medium inhibited the production of saliva after adrenergic stimulation. The precise role of Ca^{2+} in regulating secretion is not known, however, nor is its role(s) in exocytosis fully understood. Dispersed rat submandibular acinar cells pretreated with ethylene glycol bis(β-amino ethyl ether)tetraacetic acid (EGTA) to deplete their intracellular stores of Ca^{2+} were unable to elicit mucin secretion after adrenergic stimulation, whereas removal of extracellular Ca^{2+} diminished but did not totally inhibit mucin secretion (91).

The regulation of intracellular free Ca^{2+} concentrations by submandibular acinar cells before, during, and after receptor stimulation has not been characterized. Several subcellular organelles have demonstrated an ability to bind or concentrate Ca^{2+} (1, 84, 108). Microsomal fractions isolated from rat submandibular glands accumulate Ca^{2+} in an ATP-dependent manner (1, 84, 108, 124). The rate of accumulation was concentration dependent with a Michaelis-Menten constant (K_m) of 20-25 μM (124) and a V_{max} of 12 mM $Ca^{2+} \cdot mg^{-1}$ protein $\cdot min^{-1}$. The presence of the divalent cations Sr^{2+} and Mn^{2+} inhibited Ca^{2+} accumulation. Chlorpromazine, trifluoperazine, and 4,4'-diisothiocyanostilbene-2,2'-disulfonate all inhibited Ca^{2+} accumulation, but mitochondrial transport inhibitors, ruthenium red, and carbonyl cyanide m-chlorophenylhydrazone had no effect on Ca^{2+} transport. Marker enzyme analysis suggested that a specialized region of the endoplasmic reticulum may be involved. Indomethacin (101, 122) and pyridoxal 5'-phosphate inhibited microsomal Ca^{2+}-ATPase activity, whereas aspirin had no effect on enzyme activity (86). The Ca^{2+} concentration at half-maximal activity was 82 nM and ATPase activity reached a steady level at $\sim$200 pmol phosphate $\cdot min^{-1} \cdot \mu g^{-1}$ protein (57). The Ca^{2+}-ATPase activity required micromolar concentrations of Mg^{2+} for maximal activity. Because cAMP has been shown to increase Ca^{2+} binding in a plasma membrane fraction isolated from the rat submandibular gland, a role for cAMP in the modulation of cellular Ca^{2+} metabolism during β-adrenergic receptor–mediated exocytosis may exist.

The steady-state level of cellular free Ca^{2+} is probably modulated by the release or uptake of Ca^{2+} from intracellular stores such as the endoplasmic reticulum and mitochondrion, the influx of Ca^{2+} from the external milieu through various plasma membrane channels, and the removal of Ca^{2+} via plasma membrane Ca^{2+}-ATPase. The Ca^{2+} present in the secretory granules may not be metabolically active, but it is probably released into the lumen during exocytosis. Recent studies have shown that α-adrenergic and cholinergic agonists stimulate Ca^{2+} uptake into rat submandibular cells (62, 75), whereas β-adrenergic receptor stimulation only increased Ca^{2+} uptake after prolonged receptor stimulation (in excess of 10 min). The Ca^{2+} efflux from acini was also detected after isoproterenol stimulation but, as with the uptake studies, only after prolonged stimulation. If cAMP is involved in the release of Ca^{2+} from intracellular stores during exocytosis, the magnitude of the release was not sufficient to cause an initial detectable efflux of Ca^{2+} from the cells.

Although an extracellular influx of Ca^{2+} may not be an absolute requirement for submandibular mucin secretion, it may still have an important biological role in vivo. As shown in Table 2, α-adrenergic receptor activation in conjunction with β-adrenergic receptor stimulation dramatically augmented mucin secretion. In a similar manner cholinergic receptor stimulation also augmented β-adrenergic receptor–mediated mucin secretion. These studies were further supported by the observation that the presence of A23187, a divalent Ca^{2+} ionophore, was not sufficient to elicit mucin secretion by itself but was able to dramatically increase the rate of mucin release elicited by Bt_2cAMP. Current experimental evidence does not indicate a direct regulatory role for Ca^{2+} (91) in the signal transduction process for rat submandibular mucin secretion. However, cellular Ca^{2+}-depletion studies clearly document a critical role for Ca^{2+} in the actual process of exocytosis. Therefore the increase in intracellular free Ca^{2+} concentration during secretion may not act as an effector of mucin secretion, but instead intracellular Ca^{2+} stores may be important for providing the necessary free intracellular Ca^{2+} for the various Ca^{2+}-dependent cellular processes involved in exocytosis, such as modulation of the cellular cytoskeleton elements and membrane fusion. The augmentation observed during α-adrenergic or cholinergic stimulation may be due to the enhanced availability of intracellular free Ca^{2+} levels for those Ca^{2+}-dependent processes involved in exocytosis.

Summary

As presented in Figure 3, both cellular cAMP and Ca^{2+}-mediated events appear to be involved during rat submandibular mucin secretion. Cellular cAMP appears to be the actual cellular effector of mucin secretion, with Ca^{2+} playing an essential mechanistic role in the secretory process. However, much additional information must be obtained before those regulatory events and cellular processes that control rat submandibular mucin secretion can eventually be understood at the molecular level. Additional structural and biochemical data are required before we will be able to fully understand the complex cellular process of submandibular mucin secretion.

SUBLINGUAL GLAND

The rodent sublingual gland is a compound tubuloacinar gland that consists of acinar units and ducts. In each acinus, the mucin-containing acinar cells predominate (60). The blind end of the tubule is capped by a demilune that contains serous cells. The sublingual glands are predominantly innervated by the parasympathetic nervous system when only a few sympathetic nerve fibers are present (83).

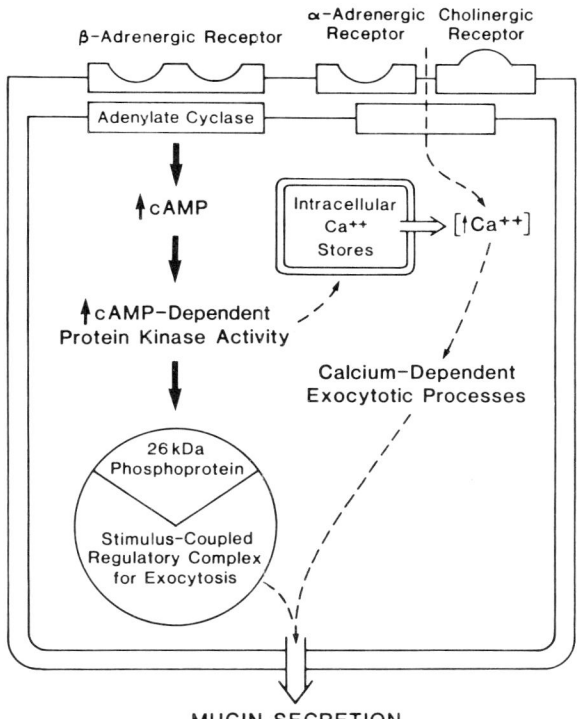

FIG. 3. Summary of cellular molecular events that appear to be involved during rat submandibular mucin secretion. [Adapted from Quissell et al. (90–93).]

Mucin secretion from mouse sublingual glands was elicited by acetylcholine injection, whereas norepinephrine and epinephrine injections induced only a slow rate of mucin secretion (128). Extracellular Ca^{2+} was required to elicit a secretory response. Although cAMP was present in the gland, cAMP levels were unaffected by isoproterenol stimulation. Adenylate cyclase was not stimulated in glandular homogenates by adrenergic agonists. In vivo secretory studies also document the insensitivity of rat sublingual gland secretion to isoproterenol stimulation (85). Thus in the sublingual gland both mucin secretion and fluid secretion appear to be controlled via the parasympathetic nervous system.

The autonomic receptors of the rat sublingual gland have been characterized with appropriate radioligands (71). Both muscarinic and α_2-adrenergic receptors were present in significant amounts, whereas the α_1- and β-adrenergic receptors were present at a low density. The [^{3}H]quinuclidinyl benzilate–binding and [^{3}H]clonidine-binding studies gave B_{max} of 18.4 and 9.9 pmol/g tissue, respectively; [^{3}H]prazosin and [^{3}H]-dihydroalprenolol gave B_{max} of 3.2 and 3.6 pmol/g tissue, respectively.

The signal transduction mechanism and the intracellular mediators of sublingual mucin secretion have not been identified. Given the predominant role of the muscarinic receptors in mediating sublingual mucin secretion and the lack of a role for the sympathetic nervous system, the intracellular events involved in the initiation and modulation of sublingual mucin secretion may be very similar to that of the pancreas rather than the parotid or submandibular salivary gland; that is, cellular free Ca^{2+} may be the primary intracellular regulator of sublingual mucin secretion rather than cAMP.

MINOR SALIVARY GLANDS

The minor salivary glands located throughout most of the oral cavity are of the mucous or seromucous type, except for the serous glands of von Ebner (36, 123), and are named according to their location in the oral cavity (97). The labial glands are located throughout the submucosa of the lip. The buccal glands are located in the buccal mucosa, and the heaviest concentration of buccal glands are between the first molars and the lips. The palatal glands are in the soft palate, uvula, and posterior hard palate. No salivary glands are found in the anterior hard palate. The glossopalatal glands are found in the glossopalatal fold of the oropharynx. The lingual glands are made up of three distinct subgroups. The serous glands of von Ebner are associated with the vallate and foliate papillae. Posterior, deep, and lateral to the von Ebner's glands are located a group of lingual mucous glands, and located in the submucosa ventral to the anterior tip of the tongue are the glands of Blandin-Nuhn.

The term *minor* salivary gland is a physiological misnomer if one considers the importance of these glands in the maintenance of normal oral health. Although the minor salivary glands contribute only 10% of the total saliva volume produced in humans (24), their secretory products make immediate and intimate contact with the oral mucosa and enamel surfaces of the teeth. During resting secretion, such as during sleep (119), the minor salivary gland secretions have a major role in maintaining normal oral health. It has been estimated that the minor salivary glands produce up to 70% of the mucins found in the human whole saliva (77).

Gustatory stimuli or mechanical stimuli result in the enhanced secretion from the lower lip mucous gland in humans (115). Pilocarpine-stimulated labial and palatine secretions from human subjects indicated an increased secretion of mucosubstances (51). Sour lemon-drop stimulation indicated an increase in the mucosubstance-containing secretions from the lip mucous glands of young adults (25). In the rat, pilocarpine injection stimulated the secretion of ^{35}S-labeled sulfated glycoproteins from minor salivary glands (47). In vitro glandular culture procedures documented the incorporation of $^{35}SO_4$ and [^{3}H]lysine into the secretory products of the minor mucous glands of the monkey *Macaca irus* (52). These studies confirm the previous histochemical studies (123) indicating that the minor mucous salivary glands are also involved in the synthesis and secretion of sulfated mucosubstances.

We thank Wanda Valentine for her assistance in the preparation of the manuscript. We thank our colleagues who generously provided us with their prepublications. L. A. Tabak wishes to thank Michael J. Levine for his helpful discussions.

Original research reported here from the authors' laboratories was supported by National Institutes of Health Grant AM-33835, Cystic Fibrosis Foundation Grant XG0905 (D. O. Q.) and DE-06970 (L. A. Tabak). L. A. Tabak is the recipient of Research Career Development Award DE-00132.

ADDENDUM

Since the original preparation of this review, several exceptional developments have occurred.

The structural properties of the porcine submandibular gland apomucin have recently been reported (33a). Enzymatically deglycosylated mucin was analyzed by gel filtration and circular dichroism. It was concluded that the apoprotein is devoid of secondary structure and serves as a scaffold for the sugar side chains.

The conformation of ovine and porcine submandibular gland mucins has also been studied by light scattering (109a, 109b) and natural abundance ^{13}C NMR spectroscopy (42a, 42b). The rigidity of the protein core of each mucin is substantially greater than that observed for nonglycosylated polypeptide random coils (33a, 109a). Interestingly the peptide core of porcine submandibular gland mucin exhibits an internal segmented flexibility similar to that of the ovine submandibular gland mucin (42a). Because the porcine mucin is characterized by longer oliosaccharide chains (see Table 1), it appears that these prosthetic groups have no significant effect on peptide core mobility compared with the simpler side chains that typify the ovine mucin. Furthermore, the blood group active side chain of the porcine glycoprotein is relatively inflexible.

The effects of various secretagogues on the mucin concentration of mouse submandibular gland saliva have been determined by radioimmune assay (28a). Submandibular gland secretions, elicited with isoproterenol, contained the highest concentrations of protein and mucin. Although the nature of the secretagogue employed is a significant determinant of mucin concentration, total salivary protein, and flow rate, little evidence for the coordination of these factors has been obtained.

REFERENCES

1. ALONSO, G. J., P. M. BAZERQUE, D. M. ARSIGO, AND O. R. TUMILASCI. Adenosine triphosphate-dependent calcium uptake by rat submaxillary gland microsomes. *J. Gen. Physiol.* 58: 340–350, 1971.
2. AUBERT, J. P., G. BISERTE, AND M. H. LOUCHEUX-LEFEBURE. Carbohydrate-peptide linkage in glycoproteins. *Arch. Biochem. Biophys.* 175: 410–418, 1976.
3. BABCZINSKI, P. Evidence against the participation of lipid intermediates in the in vitro biosynthesis of serine-(threonine)-N-acetyl-D-galactosamine linkages in submaxillary mucin. *FEBS Lett.* 117: 207–211, 1980.
4. BAHOUTH, S. W., J. M. STEWART, AND J. M. MUSACCHIO. Specific binding of a ^{125}I-labeled substance P analog to rat submaxillary gland. *J. Pharmacol. Exp. Ther.* 230: 116–123, 1984.
5. BARKA, T., AND H. VAN DER NOEN. Adenylate cyclase activity in rat submandibular gland during postnatal development. *Life Sci.* 14: 267–280, 1974.
6. BARTHELSON, R., AND S. ROTH. Topology of UDP-galactose cleavage in relation to N-acetyl-lactosamine formation in Golgi vesicles. *Biochem. J.* 225: 67–75, 1985.
7. BAUM, B. J., B. L. KUYATT, AND S. HUMPHREYS. Protein production and processing in young adult and aged rat submandibular gland cells in vitro. *Mech. Ageing Dev.* 23: 123–136, 1983.
8. BERTOLINI, M., AND W. PIGMAN. The existence of oligosaccharides in bovine and ovine submaxillary mucins. *Carbohydr. Res.* 14: 53–63, 1970.
9. BEYER, T. A., J. I. REARICK, J. C. PAULSON, J.-P. PRIEELS, J. E. SADLER, AND R. L. HILL. Biosynthesis of mammalian glycoproteins. Glycosylation pathways in the synthesis of the nonreducing terminal sequences. *J. Biol. Chem.* 254: 12531–12541, 1979.
10. BIENKOWSKI, R. S. Intracellular degradation of newly synthesized protein. *Biochem. J.* 214: 1–10, 1983.
11. BOGART, B. I., AND J. PISCARELLI. Agonist-induced secretions and potassium release from rat submandibular gland slices. *Am. J. Physiol.* 235 (*Cell Physiol.* 4): C256–C268, 1978.
12. BRIAND, J. P., S. P. ANDREWS, E. CAHIL, N. A. CONWAY, AND J. D. YOUNG. Investigation of the requirements for O-glycosylation by bovine submaxillary gland UDP-N-acetylgalactosamine: polypeptide N-acetylgalactosamine transferase using synthetic peptide substrates. *J. Biol. Chem.* 256: 12205–12207, 1981.
13. BYLUND, D. B., J. R. MARTINEZ, AND D. L. PIERCE. Regulation of autonomic receptors in rat submandibular gland. *Mol. Pharmacol.* 21: 27–35, 1982.
14. CAPASSO, J. M., AND C. B. HIRSCHBERG. Mechanisms of glycosylation and sulfation in the Golgi apparatus: evidence for nucleotide sugar/nucleoside monophosphate and nucleotide sulfate/nucleoside monophosphate antiports in the Golgi apparatus membrane. *Proc. Natl. Acad. Sci. USA* 81: 7051–7055, 1984.
15. CARLSON, D. M. Structures and immunochemical properties of oligosaccharides isolated from pig submaxillary mucins. *J. Biol. Chem.* 243: 616–626, 1968.
16. CLAMP, J. R., A. ALLEN, R. A. GIBBONS, AND G. P. ROBERTS. Chemical aspects of mucus. *Br. Med. Bull.* 34: 25–41, 1978.
17. COATES, S. W., T. GURNEY, L. W. SOMMERS, M. YEH, AND C. B. HIRSCHBERG. Subcellular localization of sugar nucleotide synthetases. *J. Biol. Chem.* 255: 9225–9229, 1980.
18. COSTA, L. G., AND S. D. MURPHY. Characterization of muscarinic cholinergic receptors in the submandibular gland of the rat. *J. Auton. Nerv. Syst.* 13: 287–301, 1985.
19. CREETH, J. M., J. L. BRIDGE, AND J. R. HORTON. An interaction between lysozyme and mucus glycoproteins. *Biochem. J.* 181: 717–724, 1979.
20. CURBELO, H. M., J. J. DEVALLE, A. B. HOURSSAY, C. H. GAMPER, AND A. A. TOCCI. Effects of isoproterenol upon the sialic acid content of salivary glands in the rat. *J. Oral Ther. Pharmacol.* 4: 431–438, 1968.
21. CURBELO, H. M., C. H. GAMPER, J. A. KOFOED, AND A. A. TOCCI. Effects of testosterone propionate on sialic acid content of submaxillary and retrolingual glands in the rat. *J. Dent. Res.* 53: 1164–1166, 1974.
22. CURBELO, H. M., A. B. HOUSSAY, C. H. GAMPER, J. A. KOFOED, AND A. A. TOCCI. Effects of oestrogens upon the sialic acid in the submaxillary and sublingual glands in the rat. *Arch. Oral Biol.* 19: 421–423, 1974.
23. CURBELO, H. M., A. B. HOUSSAY, C. H. GAMPER, AND O.

SANCHO. Influence of the adrenal glands on the sialic acid content of submaxillary and retrolingual glands in the rat. *J. Dent. Res.* 52: 1265–1267, 1973.
24. DAWES, C., AND C. M. WOOD. The contribution of oral minor mucous gland secretions to the volume of whole saliva in man. *Arch. Oral Biol.* 18: 337–342, 1973.
25. DAWES, C., AND C. M. WOOD. The composition of human lip mucous gland secretions. *Arch. Oral Biol.* 18: 343–350, 1973.
26. DENNY, P. A., AND P. C. DENNY. Purification and biochemical characterization of a mouse submandibular sialomucin. *Carbohydr. Res.* 87: 265–274, 1980.
27. DENNY, P. A., AND P. C. DENNY. A mouse submandibular sialomucin containing both N- and O-glycosylic linkages. *Carbohydr. Res.* 110: 305–314, 1982.
28. DENNY, P. C., AND P. A. DENNY. Diurnal variation of sialomucin concentration in female mouse submandibular glands measured by radioimmunoassay. *Arch. Oral Biol.* 29: 1033–1040, 1984.
28a. DENNY, P. C., P. A. DENNY, AMD M. S. YIM. The effects of various secretagogues on the mucin content of pure submandibular salivas. *J. Dent. Res.* 66: 1011–1015, 1987.
29. DISCHE, Z., C. M. BURGHER, A. DANILCHENKO, AND C. ROTHCHILD. Variations in the composition of glycoproteins of the submaxillary and parotid saliva of dog in relation to intensity of secretory stimulus. *Arch. Biochem. Biophys.* 135: 1–9, 1969.
30. DONALD, A. S. R. The products of pronase digestion of purified blood group-specific glycoproteins. *Biochem. Biophys. Acta* 317: 420–436, 1973.
31. DOUGLAS, W. W., AND A. M. POISNER. The influence of calcium on the secretory response of the submaxillary gland to acetylcholine or to noradrenaline. *J. Physiol. Lond.* 165: 528–541, 1963.
32. DOWD, F., P. CHEUNG, J. WARREN, T. FAERBER, AND D. TRAUB. Comparison of cyclic AMP-dependent protein kinases from salivary glands of four species. *J. Dent. Res.* 64: 1199–1203, 1985.
33. DUNPHY, W. G., AND J. E. ROTHMAN. Compartmental organization of the Golgi stack. *Cell* 42: 13–21, 1985.
33a. ECKHARDT, A. E., S. CANDACE, J. L. ABERNETHY, A. TOUMADJE, W. C. JOHNSON, JR., AND R. L. HILL. Structural properties of porcine submaxillary gland apomucin. *J. Biol. Chem.* 262: 11339–11344, 1987.
34. EDWARDS, P. A. W. Is mucus a selective barrier to macromolecules. *Br. Med. Bull.* 34: 55–56, 1978.
35. EKSTROM, J., B. MANSSON, AND G. TOBIN. Vasoactive intestinal peptide evoked secretion of fluid and protein from rat salivary glands and the development of supersensitivity. *Acta Physiol. Scand.* 119: 169–175, 1983.
36. EVERSOLE, L. R. The histochemistry of mucosubstances in human minor salivary glands. *Arch. Oral Biol.* 17: 1225–1239, 1972.
37. FAHIM, R. E. F., G. G. FORSTNER, AND J. T. FORSTNER. Heterogeneity of rat goblet-cell mucin before and after reduction. *Biochem. J.* 209: 117–124, 1983.
38. FLEMING, N., S. BHANDAL, AND J. F. FORSTNER. Radioimmunoassay of rat submandibular salivary gland mucin. *Arch. Oral Biol.* 28: 423–429, 1983.
39. FLEMING, N., M. BRENT, R. ARELLANO, AND J. F. FORSTNER. Purification and immunofluorescent localization of rat submandibular mucin. *Biochem. J.* 205: 225–233, 1982.
40. FLEMING, N., M. TEITELMAN, AND J. M. STURGESS. The secretory response in dissociated acini from the rat submandibular gland. *J. Morphol.* 163: 219–230, 1980.
41. FORSTNER, J. F. Intestinal mucins in health and disease. *Digestion* 17: 234–263, 1978.
42. GALLAGHER, J. T., AND A. P. CORNFIELD. Mucin-type glycoproteins: new perspectives on their structure and synthesis. *Trends Biochem. Sci.* 3: 38–41, 1978.
42a. GERKEN, T. A., AND D. G. DEARBORN. Carbon-13 NMR studies of native and modified ovine submaxillary mucin. *Biochemistry* 23: 1485–1497, 1984.
42b. GERKEN, T. A., AND N. JENTOFT. Structure and dynamics of porcine submaxillary mucin as determined by natural abundance carbon-13 NMR spectroscopy. *Biochemistry* 26: 14689–14699, 1987.
43. GOEDERT, M., J. I. NAGY, AND P. C. EMSON. The origin of substance P in the rat submandibular gland and its major duct. *Brain Res.* 252: 327–333, 1982.
44. GOTTSCHALK, A. Biosynthesis of glycoproteins and its relationship to heterogeneity. *Nature Lond.* 222: 452–454, 1969.
45. GOTTSCHALK, A., A. S. BHARGAVA, AND V. L. N. MURTY. Ovine submaxillary mucins. In: *Glycoproteins*, edited by A. Gottschalk. New York: Elsevier, 1972, vol. 5, p. 810–817.
46. GRAND, R. J., D. A. CHONG, AND S. H. RYAN. Postnatal development of adenylate cyclase in rat salivary glands: patterns of hormonal sensitivity. *Am. J. Physiol.* 228: 608–612, 1975.
47. GREEN, D. R. J., AND G. EMBERY. Partial chemical characterization and biological activities of sulphated glycoproteins isolated from in vivo pilocarpine-stimulated secretions of rat minor salivary glands. *Arch. Oral Biol.* 29: 859–863, 1984.
48. HANOVER, J. A., AND W. J. LENNARZ. Transmembrane assembly of membrane and secretory glycoproteins. *Arch. Biochem. Biophys.* 211: 1–19, 1981.
49. HANOVER, J. A., AND W. J. LENNARZ. Transmembrane assembly of N-linked glycoproteins. *J. Biol. Chem.* 257: 2787–2794, 1982.
50. HANOVER, J. A., W. J. LENNARZ, AND J. D. YOUNG. Synthesis of N- and O-linked glycopeptides in oviduct membrane preparations. *J. Biol. Chem.* 255: 6713–6716, 1980.
51. HENSTEN-PETTERSEN, A. Some chemical characteristics of human minor salivary gland secretions. *Acta Odontol. Scand.* 34: 13–22, 1976.
52. HENSTEN-PETTERSEN, A., AND J. JACOBSEN. In vitro production of sulphated mucosubstances by the labial and palatine glands of the monkey *Macaca irus*. *Arch. Oral Biol.* 20: 111–114, 1975.
53. HERP, A., A. M. WU, AND J. MOSCHERA. Current concepts of the structure and nature of mammalian salivary mucous glycoproteins. *Mol. Cell. Biochem.* 23: 27–44, 1979.
54. HERZBERG, M. C., M. J. LEVINE, S. A. ELLISON, AND L. A. TABAK. Purification and characterization of a monkey salivary mucin. *J. Biol. Chem.* 254: 1487–1494, 1979.
55. HILL, H. D., M. SCHWYZER, H. M. STEINMAN, AND R. L. HILL. Ovine submaxillary mucin. Primary structure and peptide substrates of UDP-N-acetylgalactosamine: mucin transferase. *J. Biol. Chem.* 252: 3799–3804, 1977.
56. HUMPHREYS-BEHER, M. G. Isoprenaline-induced changes in rat parotid and submandibular glands are age- and dose-dependent. *Biochem. J.* 221: 15–20, 1984.
57. HURLEY, T. W., AND J. R. MARTINEZ. Characterization of the kinetic and regulatory properties of high-affinity Ca^{2+}-ATPase activity in acinar preparations of rat submandibular salivary glands. *Arch. Oral Biol.* 30: 587–594, 1985.
58. JOHNSON, D. C., AND P. G. SPEAR. O-linked oligosaccharides are acquired by herpes simplex virus glycoproteins in the Golgi apparatus. *Cell* 32: 987–997, 1983.
59. KEAN, E. L. Nuclear cytidine 5'-monophosphosialic acid synthetase. *J. Biol. Chem.* 245: 2301–2308, 1970.
60. KIM, S. K., C. E. NASJLETI, AND S. S. HAN. The secretion processes in mucous and serous secretory cells of the rat sublingual gland. *J. Ultrastruct. Res.* 38: 371–389, 1972.
61. KINJO, K., T. NISHIKAWA, AND A. TSUJIMOTO. Role of the autonomic nervous system in regulating salivary mucin secretion by the canine submandibular gland in vivo. *Arch. Oral Biol.* 28: 97–98, 1983.
62. KOELZ, H. R., S. KONDO, A. L. BLUM, AND I. SCHULZ. Calcium ion uptake induced by cholinergic and alpha-adrenergic stimulation in isolated cells of rat salivary glands. *Pfluegers Arch.* 370: 37–44, 1977.
63. LEVINE, M. J., M. C. HERZBERG, M. S. LEVINE, S. A. ELLISON, M. W. STINSON, H. C. LI, AND T. VAN DYKE. Specificity of

salivary-bacterial interactions: role of terminal sialic acid residues in the interaction of salivary glycoproteins with *Streptococcus sanguis* and *Streptococcus mutans*. *Infect. Immun.* 19: 107–115, 1978.

64. LEVINE, M. J., L. A. TABAK, M. S. REDDY, AND I. D. MANDEL. Nature of salivary pellicles in microbial adherence: role of salivary mucins. In: *Molecular Basis of Oral Microbial Adhesion*, edited by S. E. Mergenhagen and B. Rosan. Washington, DC: Am. Soc. Microbiol., 1985, p. 125–130.

65. LOMBART, C. G., AND R. J. WINZLER. Isolation and characterization of oligosaccharides from canine submaxillary mucin. *Eur. J. Biochem.* 49: 77–86, 1974.

66. LUNDBERG, J. M., A. ANGGARD, J. FAHRENBURG, T. HOKFELT, AND V. MUTT. Vasoactive intestinal polypeptide in cholinergic neurons of exocrine glands: functional significance of coexisting transmitters for vasodilatation and secretion. *Proc. Natl. Acad. Sci. USA* 77: 1651–1655, 1980.

67. MAEDA, N., H. S. KIM, E. A. AZEN, AND O. SMITHIES. Differential RNA splicing and post-translational cleavages in the human salivary proline-rich protein gene system. *J. Biol. Chem.* 260: 11123–11130, 1985.

68. MAGEE, A. I., AND M. J. SCHLESINGER. Fatty acid acylation of eucaryotic cell membrane proteins. *Biochim. Biophys. Acta* 694: 279–289, 1982.

69. MALINOWSKI, C. E., AND A. HERP. Purification and partial characterization of rat submaxillary mucin. *Comp. Biochem. Physiol. B Comp. Biochem.* 69: 605–609, 1981.

70. MANTLE, M., G. G. FORSTNER, AND J. F. FORSTNER. Antigenic and structural features of goblet-cell mucin of human small intestine. *Biochem. J.* 217: 159–167, 1984.

71. MARTINEZ, J. R., D. B. BYLUND, AND J. CAMDEN. Characterization of autonomic receptors in the rat sublingual gland by biochemical and radioligand assays. *Naunyn-Schmiedeberg's Arch. Pharmacol.* 318: 313–318, 1982.

72. MARTINEZ, J. R., AND A. M. MARTINEZ. Stimulatory and inhibitory effects of substance P on rat submandibular secretion. *J. Dent. Res.* 60: 1031–1038, 1981.

73. MARTINEZ, J. R., D. O. QUISSELL, AND M. GILES. Potassium release from the rat submaxillary gland in vitro. I. Induction by catecholamines. *J. Pharmacol. Exp. Ther.* 198: 385–394, 1976.

74. MASSON, P. L. Carbohydrate component of cervical mucus. In: *Cervical Mucus in Human Reproduction*, edited by A. Epstein. Copenhagen: Scriptor, 1972, p. 82–92.

75. MCPHERSON, M. A., AND R. L. DORMER. Mucin release and calcium fluxes in isolated rat submandibular acini. *Biochem. J.* 224: 473–481, 1984.

76. MEHANSHO, H., AND D. M. CARLSON. Induction of protein and glycoprotein synthesis in rat submandibular glands by isoproterenol. *J. Biol. Chem.* 258: 6616–6620, 1983.

77. MILNE, R. W., AND C. DAWES. Relative contributions of different salivary glands to the blood group activity of whole saliva in humans. *Vox Sang.* 25: 298–307, 1973.

78. MONTE, L. D., L. A. TABAK, M. S. REDDY, M. J. LEVINE, B. L. KUYATT, AND B. J. BAUM. Structures of the O-glycosidic units of a mucin glycoprotein from rat submandibular glands (Abstract). *J. Dent. Res.* 62: 202, 1983.

79. MURRAY, P. A., M. J. LEVINE, L. A. TABAK, AND M. S. REDDY. Specificity of salivary-bacterial interactions. II. Evidence for a lectin on *Streptococcus sanguis* with specificity for a NeuAcα2,3Galβ1,3GalNac sequence. *Biochem. Biophys. Res. Commun.* 106: 390–396, 1984.

80. MURTY, V. L. N., J. SAROSIEK, A. SLOMIANY, AND B. L. SLOMIANY. Effect of lipids and proteins on the viscosity of gastric mucus glycoprotein. *Biochem. Biophys. Res. Commun.* 121: 521–529, 1984.

81. NICHOLS, R., D. M. CARLSON, AND J. E. DIXON. Molecular characterization of salivary gland mucin mRNAs. In: *Glycoconjugates: Proceedings of the VIII International Symposium*, edited by E. A. Davidson, J. C. Williams, and N. M. DiFerrante. New York: Praeger, 1985, vol. 2, p. 503–504.

82. NIEUW AMERONGEN, A. V., C. H. ODERKERK, P. A. ROUKEMA, J. H. WOLF, J. J. W. LISMAN, AND B. OVERDIJK. Murine submandibular mucin (MSM): a mucin carrying N- and O-glycosylically bound carbohydrate-chains. *Carbohydr. Res.* 115: C1–C5, 1983.

83. NORBERG, K. A., AND L. OLSON. Adrenergic innervation of the salivary glands in the rat. *Histochemie* 68: 183–189, 1965.

84. NYJAR, M. A., AND E. T. PRITCHARD. Calcium binding by a plasma membrane fraction isolated from rat submandibular glands. *Biochim. Biophys. Acta* 323: 391–395, 1971.

85. OHLIN, P., AND C. PEREC. Salivary secretion of the major sublingual gland of rats. *Experientia* 21: 408–409, 1965.

86. OKZEKI, H., N. TERAO, M. HAYAKAWA, AND H. TAKIGUCHI. Effect of pyridoxal 5'-triphosphate on Ca^{2+}-stimulated adenosine triphosphatase activity in microsomes of rat submandibular gland in vitro. *Int. J. Biochem.* 15: 603–607, 1983.

87. POINTON, S. E., AND S. P. BANERJEE. Alpha- and beta-adrenergic receptors of the rat salivary gland. *Biochim. Biophys. Acta* 584: 231–241, 1979.

88. PRAKOBPHOL, A., M. J. LEVINE, L. A. TABAK, AND M. S. REDDY. Purification of a low molecular weight mucin-type glycoprotein from human submandibular-sublingual saliva. *Carbohydr. Res.* 108: 111–122, 1982.

89. QUISSELL, D. O., AND K. A. BARZEN. Secretory response of dispersed rat submandibular cells. II. Mucin secretion. *Am. J. Physiol.* 238 (*Cell Physiol.* 7): C99–C106, 1980.

90. QUISSELL, D. O., K. A. BARZEN, AND L. M. DEISHER. Role of cyclic AMP-dependent protein kinase activation in regulating rat submandibular mucin secretion. *Biochim. Biophys. Acta* 762: 215–220, 1983.

91. QUISSELL, D. O., K. A. BARZEN, AND J. L. LAFFERTY. Role of calcium and cAMP in the regulation of rat submandibular mucin secretion. *Am. J. Physiol.* 241 (*Cell Physiol.* 10): C76–C85, 1981.

92. QUISSELL, D. O., L. M. DEISHER, AND K. A. BARZEN. Role of protein phosphorylation in regulating rat submandibular mucin secretion. *Am. J. Physiol.* 245 (*Gastrointest. Liver Physiol.* 8): G44–G53, 1983.

93. QUISSELL, D. O., L. M. DEISHER, AND K. A. BARZEN. The rate-determining step in cAMP-mediated exocytosis in the rat parotid and submandibular glands appear to involve analogous 26-kDa integral membrane phosphoproteins. *Proc. Natl. Acad. Sci. USA* 82: 3237–3241, 1985.

94. QUISSELL, D. O., T. P. MAWHINNEY, K. A. BARZEN, AND L. M. DEISHER. Comparison of the incorporation of D[2-3H(N)]-mannose and D-[1-^{14}C]-glucosamine into glycoproteins of dispersed rat submandibular salivary gland cells. *Arch. Oral Biol.* 28: 827–831, 1983.

95. QUISSELL, D. O., AND R. S. REDMAN. Functional characteristics of dispersed rat submandibular cells. *Proc. Natl. Acad. Sci. USA* 76: 2789–2793, 1979.

96. REDDY, M. S., M. J. LEVINE, AND A. PRAKOBPHOL. Oligosaccharide structures of the low-molecular-weight salivary mucin from a normal individual and one with cystic fibrosis. *J. Dent. Res.* 64: 33–36, 1985.

97. REDMAN, R. S. The importance of the minor salivary glands. *Northwest Dent.* 53: 19–23, 1974.

98. ROSEMAN, S. The synthesis of complex carbohydrates by multiglycosyltransferase systems and their potential function in intercellular adhesion. *Chem. Phys. Lipids* 5: 270–297, 1970.

99. ROTH, J. Cytochemical localization of terminal N-acetyl-D-galactosamine residues in cellular compartments of intestinal goblet cells: implications for the topology of O-glycosylation. *J. Cell Biol.* 98: 399–406, 1984.

100. ROUKEMA, P. A., C. H. ODERKERK, AND M. S. SALKINOJA-SALONEN. The murine sublingual and submandibular mucins. Their isolation and characterization. *Biochim. Biophys. Acta* 428: 432–440, 1976.

101. SAITO, H., R. MATSUKAWA, AND H. SOKE. Correlation between Ca^{2+}-stimulated adenosine triphosphatase activity and Ca^{2+} uptake in microsomal fraction of rat submandibular

gland. *Int. J. Biochem.* 17: 723–726, 1985.
102. SAROSIEK, J., A. SLOMIANY, A. TAKAGI, AND B. L. SLOMIANY. Hydrogen ion diffusion in dog gastric mucus glycoprotein: effect of associated lipids and covalently bound fatty acids. *Biochem. Biophys. Res. Commun.* 118: 523–531, 1984.
103. SCHACHTER, H., AND S. ROSEMAN. Mammalian glycosyltransferases. Their role in the synthesis and function of complex carbohydrates and glycolipids. In: *The Biochemistry of Glycoproteins and Proteoglycans,* edited by W. J. Lennarz. New York: Plenum 1980, p. 85–160.
104. SCHACHTER, H., AND D. WILLIAMS. Biosynthesis of mucus glycoproteins. *Adv. Exp. Med. Biol.* 144: 3–28, 1982.
105. SCHEELE, G., AND R. JACOBY. Proteolytic processing of presecretory proteins is required for development of biological activities in pancreatic exocrine proteins. *J. Biol. Chem.* 258: 2005–2009, 1983.
106. SCHILL, W. B., O. WALLNER, H. SCHIESSLER, AND H. FRITZ. Immunofluorescent localization of the acid-stable proteinase inhibitor (antileukoprotease) of human cervical mucus. *Experientia* 34: 509–510, 1978.
107. SCHNEYER, L. H., J. A. YOUNG, AND C. A. SCHNEYER. Salivary secretions of electrolytes. *Physiol. Rev.* 52: 720–777, 1972.
108. SELINGER, Z., E. NAIM, AND M. LASSAR. ATP-dependent calcium uptake by microsomal preparations from rat parotid and submaxillary glands. *Biochim. Biophys. Acta* 203: 326–334, 1970.
109. SHOGREN, R. L., A. M. JAMIESON, J. BLACKWELL, AND N. JENTOFT. The thermal depolymerization of porcine submaxillary mucin. *J. Biol. Chem.* 259: 14657–14662, 1984.
109a.SHOGREN, R. L., A. M. JAMIESON, J. BLACKWELL, AND N. JENTOFT. Comformation of mucous glycoproteins in aqueous solvents. *Biopolymers* 25: 1505–1517, 1986.
109b.SHOGREN, R. L., N. JENTOFT, T. A. GERKEN, A. M. JAMIESON, AND J. BLACKWELL. Light-scattering studies of fractionated ovine submaxillary mucins. *Carbohydr. Res.* 160: 317–327, 1987.
110. SHOMERS, J. P., L. A. TABAK, M. J. LEVINE, I. D. MANDEL, AND S. A. ELLISON. The isolation of a family of cysteine-containing phosphoproteins from human submandibular-sublingual saliva. *J. Dent. Res.* 61: 973–977, 1982.
111. SLOMIANY, B. L., V. L. N. MURTY, A. TAKAGI, H. TSUDKADA, M. KOSMALA, AND A. SLOMIANY. Fatty acid acylation of salivary mucin in rat submandibular glands. *Arch. Biochem. Biophys.* 242: 402–410, 1985.
112. SLOMIANY, A., AND B. L. SLOMIANY. Structures of the acidic oligosaccharides isolated from rat sublingual glycoprotein. *J. Biol. Chem.* 253: 7301–7306, 1978.
113. SMITH, B. F., AND J. T. LAMONT. Hydrophobic binding properties of bovine gallbladder mucin. *J. Biol. Chem.* 259: 12170–12177, 1984.
114. SNIDER, M. D., L. A. SULTZMAN, AND P. W. ROBBINS. Transmembrane localization of oligosaccharide-lipid synthesis in microsomal vesicles. *Cell* 21: 385–392, 1980.
115. SPEIRS, R. L. Secretion of saliva by human lip mucous glands and parotid glands in response to gustatory stimuli and chewing. *Arch. Oral Biol.* 29: 945–948, 1984.
116. SPOHN, M., AND I. McCOLL. Studies on gastric mucosal IgA: separation of immunoglobulin rich fraction from gastric mucoproteins. *Biochem. Biophys. Res. Commun.* 79: 837–842, 1977.
117. STINSON, M. W., M. J. LEVINE, J. M. CAVESE, A. PRAKOBPHOL, P. A. MURRAY, L. A. TABAK, AND M. S. REDDY. Adherence of *Streptococcus sanguis* to salivary mucin bound to glass. *J. Dent. Res.* 61: 390–393, 1982.
118. TABAK, L. A., M. J. LEVINE, N. K. JAIN, A. R. BRYAN, R. E. COHEN, L. D. MONTE, A. ZAWACKI, G. H. NANCOLLAS, A. SLOMIANY, AND B. L. SLOMIANY. Adsorption of human salivary mucins to hydroxyapatite. *Arch. Oral Biol.* 30: 423–427, 1985.
119. TABAK, L. A., M. J. LEVINE, I. D. MANDEL, AND S. A. ELLISON. Role of salivary mucins in the protection of the oral cavity. *J. Oral Pathol.* 11: 1–17, 1982.
120. TABAK, L. A., M. MIRELS, L. D. MONTE, A. L. RIDALL, M. J. LEVINE, R. E. LOOMIS, F. LINDAUER, M. S. REDDY, AND B. J. BAUM. Isolation and characterization of a mucin-glycoprotein from rat submandibular glands. *Arch. Biochem. Biophys.* 242: 383–392, 1985.
121. TABAK, L. A., M. S. REDDY, AND M. J. LEVINE. Characterization of a pentasaccharide in salivary mucin from the stumptail monkey *Macaca arctoides. Arch. Oral Biol.* 27: 297–303, 1982.
122. TAKIGUCHI, H., K. ITO, R. OTSUKA, AND Y. ABIKO. Effect of indomethacin on Ca^{2+}-stimulated adenosine triphosphatase in the microsomes of rat submandibular gland. *J. Dent. Res.* 58: 1714–1716, 1979.
123. TANDLER, B., C. R. DENNING, I. D. MANDEL, AND A. H. KUTSCHER. Ultrastructure of human labial salivary glands. I. Acinar secretory cells. *J. Morphol.* 127: 383–408, 1969.
124. TERMAN, B. I., AND T. E. GUNTER. Characterization of the submandibular gland microsomal calcium transport system. *Biochim. Biophys. Acta* 730: 151–160, 1983.
125. THORN, N. A., AND O. H. PETERSON (editors). *Secretory Mechanisms of Exocrine Glands.* Copenhagen: Munksgaard, 1974.
126. VAN DEN EIJNDEN, D. H. The subcellular localization of cytidine 5'-monophospho *N*-acetylneuraminic acid synthetase in calf brain. *J. Neurochem.* 21: 949–958, 1973.
127. VAN HALBEEK, H., L. DORLAND, J. HAVERKAMP, G. A. VELDINK, J. F. G. VLIEGENTHART, B. FOURNET, G. RICART, J. MONTREUIL, W. D. GATHMAN, AND D. AMINOFF. Structure determination of oligosaccharides isolated from A+, H+, and A–H–hog-submaxillary-gland mucin glycoproteins, by 360 MHz 1H-NMR spectroscopy, permethylation analysis and mass spectrometry. *Eur. J. Biochem.* 118: 487–495, 1981.
128. VREUGDENHIL, A. P., AND P. A. ROUKEMA. Comparison of the secretory processes in the parotid and sublingual glands of the mouse. I. Regulation of the secretory processes. *Biochim. Biophys. Acta* 413: 79–94, 1975.
129. WALTER, P., R. GILMORE, AND G. GLOBEL. Protein translocation across the endoplasmic reticulum. *Cell* 38: 5–8, 1984.
130. WHARTON, J., J. M. POLAK, M. G. BRYANT, S. VAN NOORDEN, S. R. BLOOM, AND A. G. E. PEARSE. Vasoactive intestinal peptide (VIP)-like immunoreactivity in salivary glands. *Life Sci.* 25: 273–280, 1979.
131. WILLIAMS, D., G. LONGMORE, K. L. MATTA, AND H. SCHACHTER. Mucin synthesis. II. Substrate specificity and product identification studies on canine submaxillary gland UDP-GlcNAc:Gal beta 1-3GalNAc(GlcNAc leads to GalNAc)beta 6-N-acetylglucosaminyltransferase. *J. Biol. Chem.* 255: 11253–11261, 1980.
132. WU, A. M., A. SLOMIANY, A. HERP, AND B. L. SLOMIANY. Structural studies on the carbohydrate units of armadillo submandibular glycoprotein. *Biochim. Biophys. Acta* 578: 297–304, 1979.

CHAPTER 6

Functional differentiation of salivary glands

LESLIE S. CUTLER | *Department of Oral Diagnosis, University of Connecticut School of Dental Medicine, Farmington, Connecticut*

CHAPTER CONTENTS

Morphogenesis
  Descriptive embryology
  Developmental regulation
Cytodifferentiation
  Descriptive embryology
  Developmental regulation
Stimulus-Secretion Coupling Systems
  Cyclic adenosine 5'-monophosphate–generating system
  Cell surface receptors
  Physiological coupling with nerves
Summary

FUNCTIONAL DIFFERENTIATION of salivary glands is a complex process that involves dependent and independent developmental events. These events are highly regulated and temporally and spatially coordinated. Functional differentiation begins during fetal development when specific cells of the oral epithelium are induced to produce unique secretory proteins and mucins that are specific to the glands. This primary or initial induction is referred to as the *determination* step. After this molecular determination step, the salivary gland rudiment undergoes primary morphogenetic development during which the gland-specific branching architecture is established. Once the initial arborized morphogenetic pattern has been established, cells within the developing rudiment are induced to undergo cytodifferentiation. This process involves amplification of the synthesis of salivary gland–specific exportable proteins. In addition the cellular architecture is modified to one best suited for the specific function of the cell. There is a further refinement of glandular structure as the parenchymal cells assume their functional form and are organized into compartments (acinar or ductal) with specific functions (secretory or conducting) within the gland. The intracellular signaling mechanisms as well as the intracellular response systems must differentiate so that the developing exocrine cells can secrete the exportable proteins they are synthesizing. Finally, receptors that transduce external stimuli to the intracellular effector systems must develop at the cell surface to make the cells competent to respond to physiological activators of secretion. Ultimately, complete functional differentiation requires that the secretory cells are anatomically coupled to the sympathetic nerves that activate the secretory process.

Salivary glands are typical exocrine secretory organs; their functional epithelial parenchymal structure is characterized by an acinar or secretory compartment and various ductal compartments (intercalated, striated, major excretory) that modify the primary acinar secretion while conducting it from the acinus into the oral cavity. The gross, microscopic, and fine structure of salivary glands from numerous vertebrate and invertebrate species has been described (for reviews, see refs. 64, 103). The regulation of salivary gland secretion is complex and is described in the chapters by Cook and Young; Petersen and Maruyama; Putney; Spearman and Butcher; Quissel and Tabak; and Cameron, Arvan, and Castle in this *Handbook*. In brief, protein secretion by salivary gland acini is mediated by sympathetic nerve stimulation of β-adrenergic receptors on the acinar cell surface. Receptor stimulation leads to the activation of adenylate cyclase and an increase in the intracellular cAMP concentration. Cyclic AMP accumulation initiates a cascade of subsequent events in the cytoplasm that ultimately results in exocytosis by the acinar cells.

This chapter focuses on the developmental integration of the processes of glandular morphogenesis and cellular cytodifferentiation with the maturation of the stimulus-secretion coupling systems that regulate protein secretion by salivary gland acinar cells. There are several descriptive studies on salivary gland development in a variety of species (20, 30, 33, 34, 41, 56, 58, 65, 74–76, 78, 101–103). However, studies on mice and rats have provided most of the available knowledge on the development of salivary gland structure and function and therefore are the basis for this chapter.

This chapter describes the general aspects of the structural development of the salivary glands, the roles of epithelial-mesenchymal interactions and extracellular matrix in the regulation of glandular morphogenesis and acinar cell cytodifferentiation, the differentiation of the structure and function of acinar cells, and the maturation of the stimulus-secretion coupling systems (cell surface receptors, intracellular mediator systems) that regulate salivary gland secretion. The functional linkage that develops between the surface receptors of the acinar cells and the neural components that regulate glandular activity in vivo is also discussed in *Physiological Coupling With Nerves*, p. 101.

MORPHOGENESIS

Descriptive Embryology

Although the major salivary glands (parotid, submandibular, and sublingual) develop at different times and at different rates, they all follow the same basic sequence of morphogenetic development shown in Figure 1. Salivary glands are epithelial in origin; thus the first evidence of salivary gland development is a thickening and protrusion of the oral epithelium into the underlying connective tissue. This glandular anlage is surrounded by a condensation of mesenchymal cells forming a discrete, thick capsule that separates the anlage from the adjacent connective tissue. The anlage continues to grow into the underlying connective tissue, giving rise to an elongated stalk that terminates in one or more knoblike clusters of cells or end buds. Highly regulated differential growth from various regions of the end bud leads to the initial branching of the gland. Further differential growth at the tips of these branches gives rise to the specific arborized pattern unique to salivary glands (9–13, 61). At this early point in development the primary morphogenetic branching pattern of the salivary gland has been established. Continued differential growth at the ends of the newly forming branches leads to glandular enlargement and lobule formation within the parenchyma. Because this process is highly regulated, the unique branching pattern is maintained (6, 11, 13, 15, 48–51, 73, 90, 91).

The early salivary gland rudiment does not have a lumen (33, 74, 76) and thus differs from the early pancreatic rudiment, which does from inception (72a). The events in the formation of the lumen of the developing salivary gland are shown in Figure 2. The more central duct cells and end buds of the early branching rudiment form junctional complexes (tight, close, and desmosomal junctions) at their presumptive luminal aspect, and their cell nuclei shift basally.

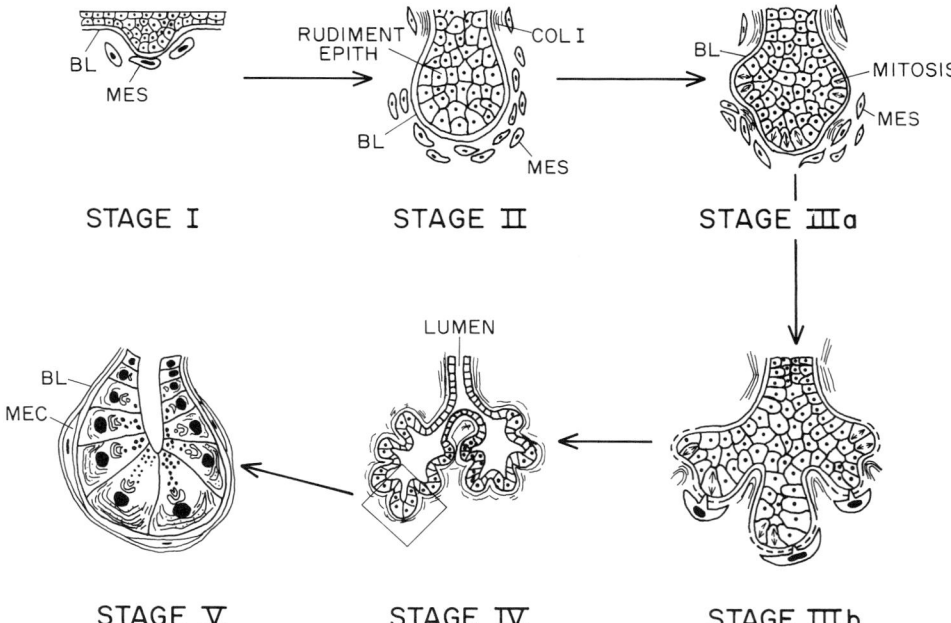

FIG. 1. Diagram of major stages in development of salivary glands. *Stage I*: salivary epithelium begins to grow down from oral epithelium into underlying mesenchyme (MES). Epithelium is separated from mesenchyme by intact basal lamina (BL). *Stage II*: epithelium continues to proliferate and grow. Mesenchyme condenses to form a discrete capsule. Collagen type I is deposited in proximity to epithelial basal lamina. *Stages IIIa and IIIb*: specific morphogenetic branching pattern is established through differential cell division at growing ends of branch buds and through changes in synthesis and degradation of basal lamina. Collagen type I stabilizes lamina in clefts but lamina becomes discontinuous at growing ends. Direct epithelial-mesenchymal interactions occur at growing ends of early branches. *Stage IV*: branching continues. Cells become polarized and amplify secretory protein synthesis. *Stage V*: cells initiate packaging of secretory proteins into zymogen granules. Myoepithelial cells (MEC) differentiate at periphery of developing acini to aid in raising intraluminal pressure to facilitate movement of secretory products out of acinus and through ductal system.

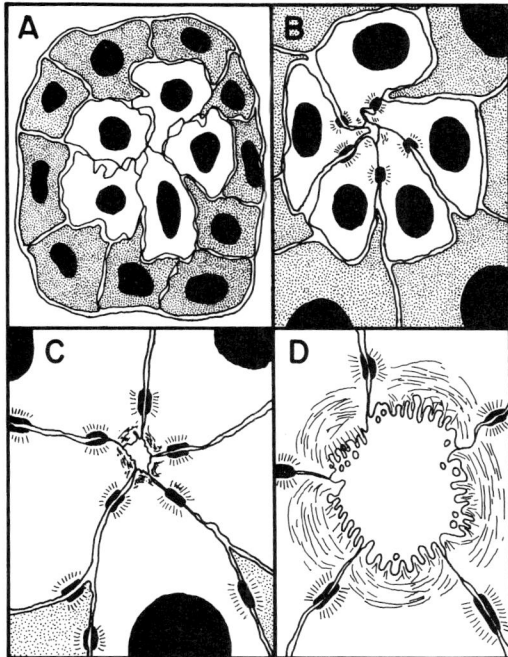

FIG. 2. Diagram of formation of acinar lumen in developing salivary gland. A: growing end buds that give rise to acini are solid cellular structures. B: central cells in end bud begin to elongate and form junctional complexes at presumptive luminal end (differentiation of apical domain of cells). C: cytofilament aggregates and myosin ATPase activity appear concentrated beneath membranes at presumptive apical end of cells. D: apparent contraction of apical cytofilaments gives rise to end bud lumen.

Cytofilament aggregates become prominent in the cytoplasm adjacent to the newly formed junctional complexes. This change in structure is the first evidence of the apical-basal polarity characteristic of the acinar and ductal cell types. Myosin ATPase activity can be seen in association with these "apical" cytofilament bundles. Apparent contraction of the apical cytoplasm, mediated by the cytofilaments, is the final step in lumen formation. Until a lumen forms to conduct the secretory products out of the gland and into the oral cavity, neither the secretory nor the transport functions of the salivary glands can be operational (35b).

During the development of the initial branching pattern the parenchymal cells show very little structural or biochemical evidence of cellular specialization or any significant synthesis of specific secretory proteins [Fig. 3; (33, 34, 74, 76, 78, 97–99)]. However, small amounts of salivary gland–specific secretory proteins can be cytochemically localized in the endoplasmic reticulum and Golgi apparatus of a few cells in the earliest rudiment (25, 97), indicating that the cells have been "determined" but that a high level of salivary gland–specific secretory protein synthesis is not yet underway. Thus the major morphogenetic events in salivary gland development precede the major events in the cytodifferentiation and specialization of specific cell types within the parenchyma (25–27, 30, 31, 33, 34, 74–76, 78, 97–99), as is exemplified by the rat parotid gland, in which morphogenesis occurs well before birth but cellular cytodifferentiation is not seen until 1 day after birth (44, 45, 53, 76–78, 82–85).

Developmental Regulation

Interaction between the epithelial and mesenchymal components of developing organ rudiments is a fundamental phenomenon in the morphogenesis and cytodifferentiation of many tissues. There are epithelial-mesenchymal interactions that are required for the induction and maintenance of salivary gland morphogenesis (9–13, 15, 22, 27, 48–51, 61–63). These interactions appear to be mediated by diffusible molecules that are associated with the extracellular matrix at the epithelial-mesenchymal interface (9–12, 22, 48–51, 88, 89, 95, 96). Furthermore these interactions occur continuously during the entire period the gland is growing. The pioneering work of Borghese (15) and Grobstein and Cohen (48–51) demonstrated the morphogenetic dependence of the mouse submandibular gland epithelium on its capsular mesenchyme by separating the epithelial portion of the rudiment from its surrounding mesenchyme and placing it in tissue culture. Upon culturing, the epithelium involuted and failed to undergo further morphogenesis. When the epithelium was separated from and then recombined with its mesenchyme, morphogenesis continued in a normal fashion. Grobstein (50) performed in vitro transfilter experiments demonstrating that the induction of branching morphogenesis appeared dependent on a diffusible macromolecule that subsequent studies suggested was collagen (51).

Several studies suggested that the driving forces behind submandibular gland morphogenesis were selective epithelial cell proliferation and changes in cell shape (12, 13, 90, 91). Bernfield and his co-workers (5, 6, 9–12, 22) demonstrated that the branching morphology of the mouse submandibular gland was maintained by its basal lamina, which was shown to be derived solely from the salivary epithelium (6, 9, 12, 22, 57). In the early branching rudiment the basal lamina was continuous and collagen fibers were seen in the clefts between newly growing buds or branches. In contrast, the lamina was either thinned or showed breaks at the distal ends of the growing buds. Ultrastructural studies of the epithelial-mesenchymal interface indicated that the basal lamina was not uniform and often showed discontinuities that permitted direct epithelial-mesenchymal cell interaction (22–24, 26, 31). Histochemical and immunohistochemical studies (6, 9, 11) indicated that glycosaminoglycans (proteoglycans) and collagen type I were concentrated in the clefts and sparse at the distal ends of the buds (see Fig. 1). Biochemical studies (5, 22, 37) indicated that the types of glycosaminoglycans synthesized var-

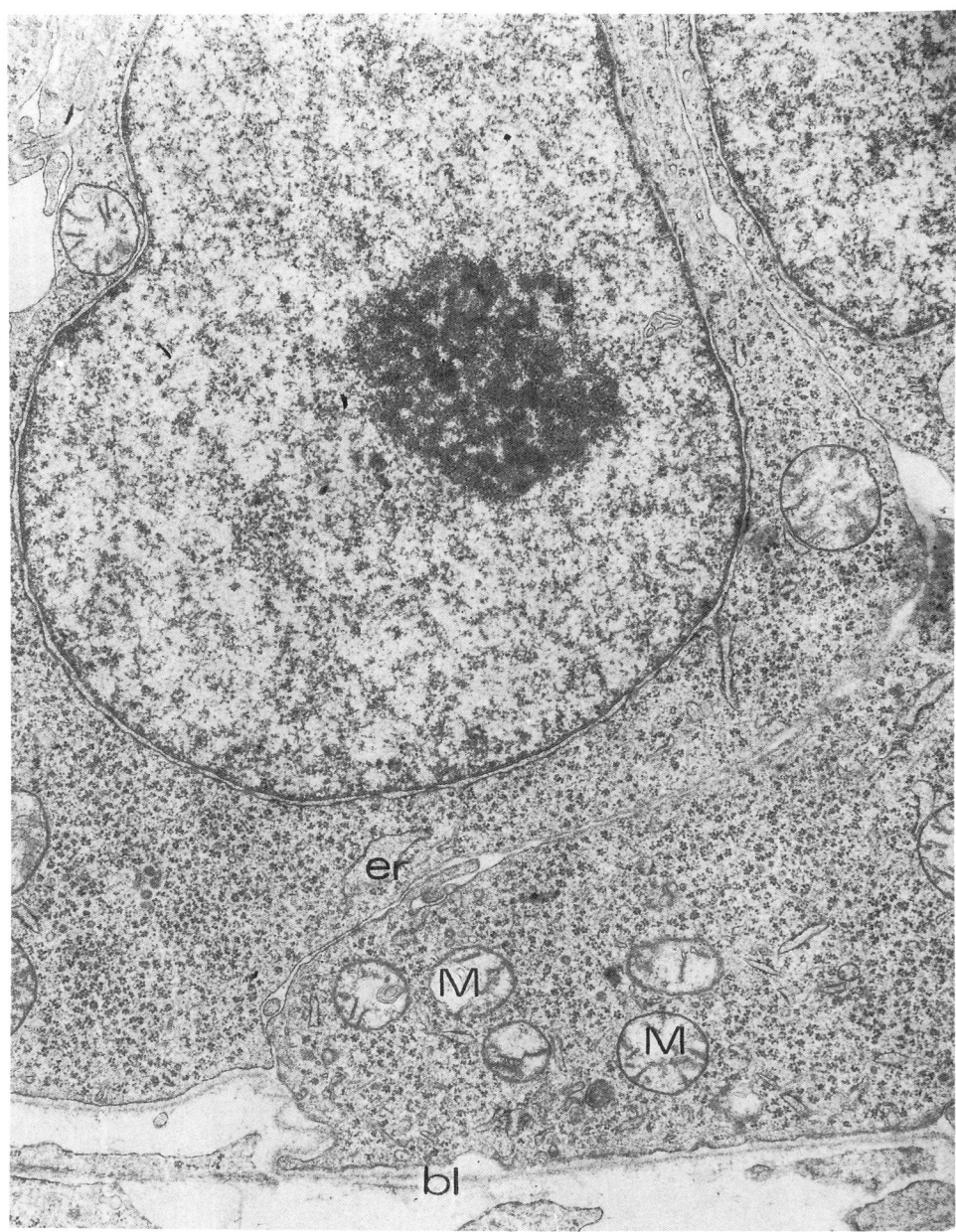

FIG. 3. Electron micrograph showing basal portion of typical salivary gland epithelial cell at stage II or III. This particular cell is from developing rat submandibular gland at stage IIIa. Cytoplasm of cell is filled with free polyribosomes. A few strands of endoplasmic reticulum (er) and few scattered mitochondria (M) are in cytoplasm. Cells rest on intact basal lamina (bl). × 10,000.

ied as the morphogenetic branching pattern was initiated, became established, and was stabilized.

Mesenchyme appears to influence epithelial branching by modifying the basement membrane, in particular glycosaminoglycans, permitting changes in epithelial structure to proceed (5, 6, 10, 12, 87). Direct measurements indicate that glycosaminoglycan synthesis and degradation rates in the clefts, at the tips of the growing buds, and along the stalk of the early branching submandibular gland rudiment differ. Glycosaminoglycan synthesis is at a steady state along the stalk, which is not undergoing much morphogenetic change. The clefts show a dramatic accumulation of glycosaminoglycans due to both a high rate of glycosaminoglycan synthesis and a reduced rate of degradation (10). At the growing ends of the rudiment buds (branch points), glycosaminoglycan synthesis is low and degradation rates are relatively high. Both autoradiographic and histochemical studies indicate regional differences in the rate of basal lamina glycosaminoglycan turnover (6, 22). The increased degradation of the basement membrane at the distal end of the growing branches appears to be mediated by the surrounding mesenchyme, which contains a developmentally regulated neutral hyaluronidase that degrades hyaluronic acid and chondroitin sulfate (5, 10, 12, 87). In synchrony with the appearance of the mesenchymal hyaluronidase, the pattern of basement membrane glycosaminoglycans synthesized by the developing submandibular gland rudiment changes. When the early rudiment is beginning to branch, the predominant glycosaminoglycans synthesized by the salivary gland epithelium are hyaluronic acid and chondroitin sulfate, but very little heparan sulfate is produced. As branching continues and becomes more pronounced the pattern of glycosaminoglycan synthesis shifts to an increase in production of heparan sulfate and a decrease in production of hyaluronic acid and chondroitin sulfate (5, 22, 37). In the fully differentiated state, heparan sulfate is the predominant glycosaminoglycan synthesized and only small amounts of hyaluronic acid and virtually no chondroitin sulfate are produced (35a, 37).

Type I collagen seems to be responsible for stabilizing the basal lamina against degradation. The collagen apparently immobilizes heparan sulfate and reduces the breakdown rate of heparan sulfate–rich proteoglycans synthesized by the epithelial cells (12). Collagen deposition within the cleft areas, in coordination with the changing pattern of glycosaminoglycan synthesis, appears to be the mechanism regulating assembly of the basement membrane and stabilizing the differentiated branching pattern. More direct evidence for the role of basement membrane glycosaminoglycans in salivary gland morphogenesis comes from studies that demonstrated that specific inhibition of the terminal steps in proteoglycan synthesis and deposition of synthesized sulfated glycosaminoglycans inhibit branching morphogenesis of the developing submandibular gland (88, 95, 96). This inhibition can be reversed by removing the agents that blocked glycosaminoglycan processing, thus allowing normal proteoglycan synthesis and deposition in the basal lamina to proceed.

Lawson (61–63) demonstrated that the specific branching pattern expressed by salivary epithelium could be directed by varying the origin of the inducing mesenchyme. When salivary epithelium was combined with lung mesenchyme, a lung branching pattern was observed. Similar findings were made when other non-salivary mesenchyme was used as the inducing tissue. However, the synthesis of salivary gland–specific proteins by the cells within the epithelium was neither altered by the type of inducing mesenchyme nor by the final morphogenetic pattern established. In all cases the cells synthesized salivary gland-specific secretory proteins independent of the branching pattern displayed. These studies were among the first to demonstrate a dissociation between the processes of morphogenesis and cytodifferentiation.

CYTODIFFERENTIATION

Descriptive Embryology

In the development of salivary glands and of virtually all exocrine glands, the initial phases of morphogenesis precede the onset of significant cell-specific protein synthesis. Low levels of synthesis can be detected by cytochemical techniques in developing salivary glands just before branching morphogenesis begins (25, 98), although there are no signs of cellular cytodifferentiation [e.g., no significant amounts of rough endoplasmic reticulum, no secretory granules, no apical-basal polarization of organelles; (Fig. 3)]. Morphogenesis at the rudiment level is relatively advanced before fine structural signs of cellular differentiation become apparent (25, 30, 33, 34, 74–78, 97–99).

Cytodifferentiation of specific cell types within the salivary gland rudiment appears to be initiated as the morphogenetic branching pattern is being established. In the rat submandibular gland this occurs on the 16th day of gestation when the initial 4–12 branches of the rudiment form. Although the precise point of induction has not been determined, cytodifferentiation in the rat parotid gland does not proceed until several days after birth, even though the gland develops at about the same prenatal time as the submandibular gland (76–78). There are several cellular changes associated with the induction of cytodifferentiation; the initial change is cell shape. The cells that are going to become acinar exocrine cells change from cuboidal to columnar and the nuclei migrate basally. As the cells change shape there is an amplification of the synthesis of cell-specific secretory proteins. This amplification often causes the few strands of endoplasmic reticulum in the cytoplasm to swell (Fig. 4) because the Golgi complex apparently is not fully capable of processing and packaging the significantly increased amounts of newly synthesized secretory proteins. With time the Golgi is able to process the newly produced secretory proteins at a rate commensurate with synthesis. Increasing amounts of endoplasmic reticulum appear in the cytoplasm that together with the Golgi become distributed in the cells in the typical exocrine pattern, the Golgi in a supranuclear position and the endoplasmic reticulum in the basal cytoplasm. As this process proceeds, condensing vacuoles and secretory granules are then observed in the apical cytoplasm [Fig. 5; (25, 33, 74, 89)]. The basic cellular changes related to acinar cell cytodifferentiation are complete at this point, although there are subtle changes in the nature of the secretory granules and in the organization of the cytoplasmic organelles that occur as development continues and that are unique to each gland (see refs. 1–4, 8, 20, 21, 41–43, 56, 59, 65, 66, 72, 100). A comparison of the timing of the events in the morphogenesis and differentiation of the parotid, sublingual, and submandibular glands of the rat is shown in Table 1.

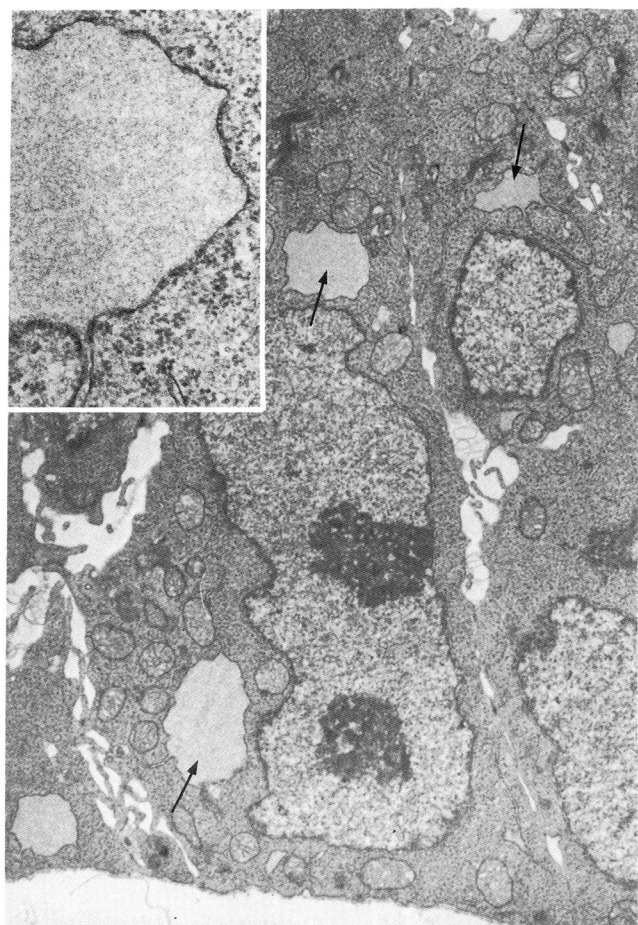

cytoplasm has filled with cytofilaments (21, 30, 67, 75).

Differentiation of the striated and extralobular ducts also parallels the development of the acinar cells. Development of the striated ducts is characterized by the extension of many fingerlike projections from the basal and basolateral aspects of the duct cells. Numerous mitochondria appear in the cell cytoplasm, become situated in the basal portion of the cell, and are associated with the cytoplasmic extensions. This configuration increases the cell surface area and provides an energy resource to facilitate rapid movements of water and ions in a manner analogous to that of the kidney tubules (34, 70, 73). In the rodent submandibular gland there is a special secretory segment of the striated duct, the granular convoluted tubule. This

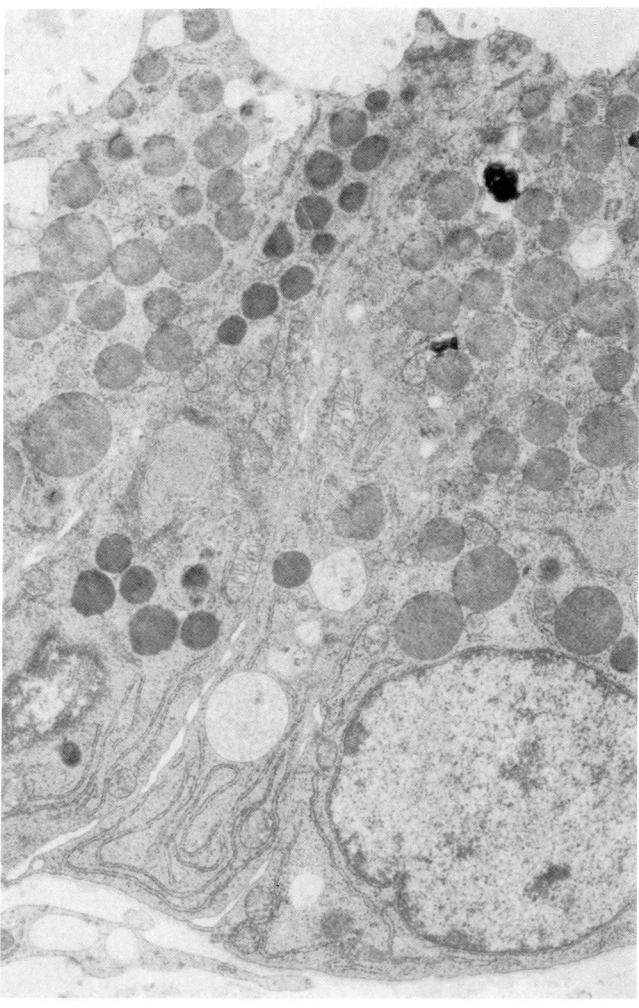

FIG. 4. Electron micrograph of differentiating submandibular gland acinar cell (stage IV) shortly after direct epithelial-mesenchymal interactions seen in stage III have taken place. Cell has assumed columnar shape and started to amplify secretory protein synthesis. Cytoplasm is dominated by free polyribosomes with few mitochondria and scattered strands of endoplasmic reticulum. Endoplasmic reticulum is filled with newly synthesized secretory product, causing cisternae to be widely dilated (arrows). ×4,250. Inset: high-magnification electron micrograph of dilated ER cisterna. Ribosomes studding cisternal membrane are clearly visible, as is flocculent material within cisternal lumen. × 30,000.

Myoepithelial cells (epithelial cells with a structure and function analogous to smooth muscle) surround the acini and intercalated ducts of all the salivary glands. Myoepithelial cell contraction at the time of secretion raises the pressure within the acinar lumen and aids in the flow of saliva from the acinus into the ductal system. These cells develop from undifferentiated epithelial cells situated at the periphery of the end buds in the developing salivary glands and differentiate in parallel with the acinar cells. Myoepithelial cell cytodifferentiation is characterized by a change in shape from cuboidal to stellate and an intracellular accumulation of cytofilaments and contractile proteins. The cells do not produce secretory proteins or undergo significant amounts of cell division once their

FIG. 5. Electron micrograph of several differentiated acinar cells (stage V). Cells are from developing submandibular gland acinus at 21 days in utero. Cells show typical apical-basal polarity of organelles. Nucleus and strands of endoplasmic reticulum are in basal cytoplasm; Golgi, condensing vacuoles, and secretory granules are in more apical cytoplasm. × 4,050.

TABLE 1. *Morphogenesis and Cytodifferentiation of Acinar Secretory Cells of Rat Salivary Glands*

	Submandibular Gland	Parotid Gland	Sublingual Gland
Morphogenesis			
Anlage appears	14	15	14
Branching pattern established	15–16	16–17	15–16
Lobular pattern established	18	19	17
Cytodifferentiation			
Initial secretory protein synthesis	14–15	17–18	
Amplification of secretory protein synthesis	16	1*	17
Polarization of secretory cells	16–17	1*	16–17
Secretory granules in cytoplasm	18	1*	17
Myoepithelial cells differentiate	18		17–18
Striated ducts develop	18–21	20–21	18–21

Numbers, developmental age in days in utero. * Developmental age in days after birth. Data from refs. 1–4, 25, 33, 74, 76–78, 97–99.

duct segment produces a variety of alkaline protease enzymes and a number of hormonelike substances, such as nerve growth factor and epidermal growth factor, all of which are secreted in an exocrine fashion (32). This duct segment develops in the postnatal period (see refs. 34, 47, 56, 92), and its maturation depends on circulating levels of thyroxine and testosterone (47).

Developmental Regulation

Short-term epithelial-mesenchymal interactions, involving direct contact between the salivary epithelium and the surrounding mesenchyme (Fig. 6), play a significant role in initiating the amplification of gland–specific protein synthesis and the subsequent structural changes at the cellular level that ultimately give rise to cytodifferentiation typical of exocrine secretory cells [see Fig. 5; (23, 24, 26, 27, 31)]. These contacts occur at the time the characteristic branching pattern is established and appear to be limited in time and space, only involving interaction between the mesenchymal and epithelial cells at the ends of the first few growing branches. Direct contacts of this type are not seen before or after this developmental stage. If submandibular gland epithelial cells are isolated prior to the occurrence of these direct epithelial-mesenchymal interactions and cultured without their mesenchyme, the developing epithelial cells fail to undergo cytodifferentiation. However, if the cells are isolated after the contacts have occurred and then cultured without their mesenchyme, normal acinar cell cytodifferentiation proceeds (27). The mesenchyme is apparently not required for continued cytodifferentiation once it has been induced, as the process proceeds even in the absence of mesenchyme. Furthermore

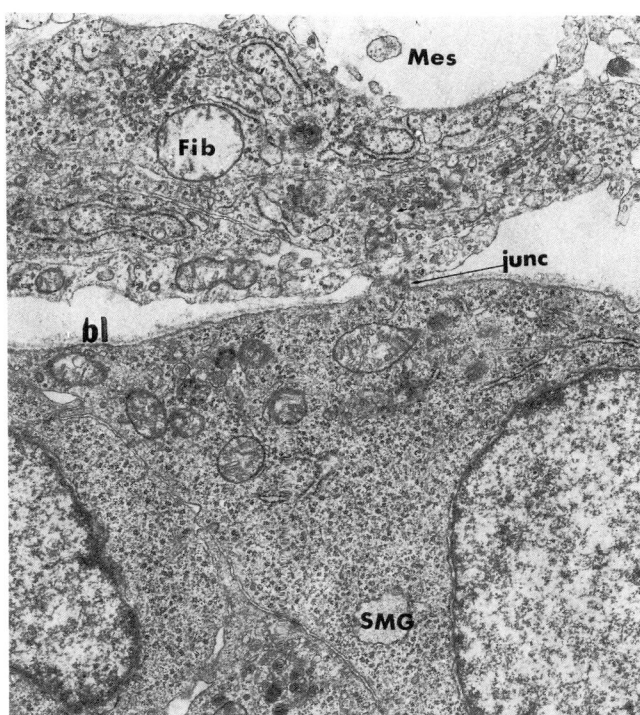

FIG. 6. Electron micrograph of direct epithelial-mesenchymal contact (*junc*) between submandibular gland end bud cell (*SMG*) at stage IIIb and mesenchymal (*MES*) fibroblast (*Fib*). Epithelial cell is extending projection through basal lamina (*bl*) to contact fibroblast. × 7,100.

cytodifferentiation occurs even though the developing epithelial cells are not organized into acini or other glandular configurations (27). These experiments demonstrate that salivary gland rudiment morphogenesis and cytodifferentiation are partially coupled but remain independently regulated processes. The early phases of glandular morphogenesis (rudiment downgrowth and initiation of primary branching) appear to be essential for cytodifferentiation to proceed. However, once the transient direct epithelial-mesenchymal interactions required for cytodifferentiation have occurred, secretory cell maturation can proceed in the absence of continued glandular morphogenesis or mesenchymal influences (27). However, amplification of cell-specific protein synthesis occurs even if the cells do not undergo apical-basal polarization or develop typical exocrine cell architecture. The level of protein synthesis by cells that have not undergone structural cytodifferentiation is apparently below that of cells that display both structural and biosynthetic differentiation. Thus, although cellular morphogenesis and cytodifferentiation are independent processes, full expression of differentiated potential is closely linked to the most efficient cell shape for the task.

Amplification of the synthesis of cell-specific secretory proteins, structural maturation of the cells into typical exocrine cells, and the packaging of exportable proteins into secretory granules do not a priori indicate that the cell is capable of secreting the newly produced proteins in response to a physiological stimulus. There is substantial evidence that the biochemical and structural maturation of the secretory cell is independent of the development of the stimulus-secretion coupling systems that regulate the physiological release of the secretory material (16, 18, 35, 39, 44, 45, 68, 69).

STIMULUS-SECRETION COUPLING SYSTEMS

Cyclic Adenosine 5′-Monophosphate–Generating System

Protein secretion in salivary glands is mediated by cAMP, which is generated in response to β-adrenergic stimulation of receptors at the secretory cell surface (see the chapter by Spearman and Butcher in this *Handbook*). Cyclic AMP is produced by the catalytic action of the membrane-bound enzyme adenylate cyclase on ATP. The complete adenylate cyclase complex or system is composed of three subunits: *1*) the catalytic subunit of the enzyme, *2*) a guanylnucleotide/fluoride (G/F) regulatory protein or subunit, and *3*) a receptor subunit. The receptor subunit transduces extracellular signals to the cytoplasmic side of the cell membrane by spanning the plasmalemma to connect, via the G/F regulatory subunit, with the catalytic subunit of adenylate cyclase. During the earliest phases of submandibular gland development (stages I–III; see Fig. 1) the amount of membrane-associated adenylate cyclase is quite low (38). As cytodifferentiation of secretory cells begins (stage IV) the amount of adenylate cyclase at the cell surface rises rapidly and approaches adult levels as the cells assume typical exocrine cell structure [stage V; (7, 36, 38)]. In the submandibular gland the development of the G/F regulatory subunit closely parallels the appearance of adenylate cyclase. It is possible to stimulate the adenylate cyclase activity of embryonic submandibular gland secretory cells (stage V) more than 15-fold with appropriate concentrations of fluoride or guanylnucleotides (35). A somewhat different pattern is observed in the developing parotid gland, where the G/F protein appears to develop independently from the catalytic unit of adenylate cyclase. In the developing parotid gland, G/F activation of adenylate cyclase increases during the first few weeks of postnatal development as the number of β-adrenergic receptors increases on the secretory cells (69).

Both the developing submandibular gland (prenatal) and parotid gland (early neonatal) synthesize and package secretory proteins in a typical exocrine fashion. Adenylate cyclase is present on the surface membranes of the secretory cells at this time, and the cells can be induced to release their secretory proteins in a typical exocrine fashion in tissue culture experiments by the administration of dibutyryl cAMP (35, 44, 45). These experiments indicate that as the cells initiate accelerated synthesis and packaging of secretory proteins, develop typical exocrine structure (apical-basal polarization), and develop the cAMP-generating system, they are capable of exocrine secretion in response to the intracellular mediator of the exocrine process (cAMP). However, neither the prenatal or early neonatal submandibular gland nor the early neonatal parotid gland is able to secrete its exportable proteins in response to stimulation with saturating doses of the adenylate cyclase–activating β-adrenergic agonist, L-isoproterenol (35, 44, 45). In addition at this stage of development the adenylate cyclase activity from both the submandibular and parotid glands is refractile to isoproterenol stimulation (35, 44, 45).

Cell Surface Receptors

The failure of adenylate cyclase activity from the prenatal and early postnatal submandibular gland to respond to hormonal activation is due to a paucity of β-adrenergic receptors at the secretory cell surface (18, 35). The failure of parotid gland adenylate cyclase to respond to stimulation is due to both the absence of β-adrenergic receptors and to low levels of the G/F regulatory protein (68, 69). In the developing submandibular gland, adenylate cyclase becomes responsive to hormone activation at 5–6 days of age. This onset of hormone responsiveness is directly related to an increase in the number of β-adrenergic receptors at the cell surface at this time (18, 35). The adenylate cyclase activity of the parotid gland becomes responsive to β-adrenergic stimulation between 14 and 18 days after birth (69). Like the submandibular gland, this responsiveness is due in part to an increased number of β-adrenergic receptors and to the development of the G/F regulatory subunit of the adenylate cyclase enzyme complex. Furthermore, secretion studies, both in vitro and in vivo (using chemical or electrophysiological stimulation), suggest that the normal physiological pattern of stimulus-secretion coupling becomes operational in the submandibular and parotid glands with the appearance of increased numbers of β-adrenergic receptors at the secretory cell surface (16, 35, 44, 45).

Muscarinic cholinergic and α-adrenergic responses regulate the water and ion movements that accompany β-adrenergic stimulation of exocrine protein secretion. Muscarinic cholinergic receptors develop in both the submandibular and parotid glands in a fashion analogous to that of β-adrenergic receptors (18, 68). However, the number of muscarinic receptors in the submandibular gland is relatively high at the time of birth and is sufficient to permit apparently normal movements of ions and water by both ductal and secretory cells (16, 53, 71, 82).

In contrast with the relatively low numbers of β-adrenergic and muscarinic cholinergic receptors observed in the neonatal submandibular and parotid glands, the total number of α-adrenergic receptors is slightly higher in these developing glands than in the adult glands (18, 35). However, the α-adrenergic subtypes change dramatically during postnatal differentiation. α_1-Adrenergic receptors are barely detectable in the submandibular gland at birth but increase to adult levels by 3 wk of age. The number of α_2-adrenergic binding sites is very high at birth, but these receptors are virtually nondetectable in the submandibular gland by 6 wk of age when the animals reach puberty (18). Thus, although the number of α-adrenergic receptors is high at birth, there is a developmental change in receptor subtype, but the functional significance of this change is not known.

A summary of the development of adenylate cyclase and autonomic receptors in the parotid and submandibular glands of the rat is shown in Table 2.

Physiological Coupling With Nerves

Both electrophysiological and catecholamine fluorescence data indicate that there are no adrenergic nerve terminals in the submandibular gland during the prenatal and early postnatal period (prior to 5 days after birth). However, catecholamine-containing nerve processes appear in the submandibular gland at 5-6 days after birth, the same time that the β-adrenergic receptor increase occurs (16). Electrophysiological studies show that the sympathetic nervous system becomes functionally linked to the exocrine process in the submandibular gland at this same time (16, 35). Conversely, the cholinergic nerves that innervate the salivary glands are in place and functionally connected at the time of birth (16, 85).

Treatment of neonatal rats (5 days of age and older) with the β-adrenergic agonist isoproterenol for 5 days leads to a rapid, premature cytodifferentiation of the acinar cell compartment of the submandibular gland (19, 80, 86, 94). However, this effect is not seen if the animals are treated with isoproterenol from birth to 5 days of age, suggesting that changes occur in the gland about this time. These observations combined with those on receptor development suggest that changes occur in the rat submandibular gland at ~4-6 days of age that lead to the physiological coupling of the stimulus-secretion system of the acinar cells. These cellular and physiological changes are correlated with the ingrowth of catecholamine-containing nerve processes. This temporal and spatial correlation suggests that the ingrowth of the nerves might be responsible for the induction of the increased numbers of β-adrenergic receptors that appear at the cell surface as well as have a role in the regulation of the postnatal maturation of the gland.

The relationship (cause and effect?) between the ingrowth of the sympathetic nerves that stimulate exocytosis and the appearance of β-adrenergic receptors on the submandibular and parotid gland secretory cell surfaces and the development of cellular responsiveness to neurohormonal secretory stimulation has been investigated. Experiments that examined the effect of isoproterenol-induced premature acinar cell development revealed a concomitant increase in the number of β-adrenergic receptors (39).

Conversely, the effect of chemical sympathectomy with reserpine or 6-hydroxydopamine on the appearance of β-adrenergic receptors in developing neonates revealed that the increase in the number of receptors observed in both the submandibular and parotid glands followed the same pattern seen in untreated, control neonates (39, 69). These data indicate that the increase in the number of β-adrenergic receptors seen during early postnatal development of the salivary glands is a specifically programmed step that is closely associated with the degree of maturation attained by the secretory cells. The increase in receptor number

TABLE 2. *Development of Adenylate Cyclase Activity and Autonomic Receptors in Rat Salivary Glands*

	Submandibular Gland	Parotid Gland
Adenylate Cyclase		
First measured	15	
Activity begins to rise	16	
Adult levels reached	18	Birth
Fluoride activation		
First observed	18	Birth
Maximum activation	21	18-20*
Guanylnucleotide activation		
First observed	18	Birth
Maximum activation	21	18-20*
Autonomic Receptors		
β-Adrenergic		
First measured	21	Birth
Number begins to increase	6*	15*
Adult levels reached	21*	20-25*
α-Adrenergic		
Total	High at birth. Increase for ~2 wk after birth.	
α_1	Low at birth. Adult levels reached by 3-4 wk after birth.	
α_2	High at birth. Increase for 2 wk, then decrease to almost zero by 6 wk after birth.	
Muscarinic	Present at birth. Increase to adult levels by 2 wk after birth.	Present at birth. Increase to adult levels between 10 and 25 days after birth.

Numbers, developmental age in days in utero. *Developmental age in days after birth. Data from refs. 18, 35, 39, 44, 45, 68, 69.

is not related to or caused by nerve process ingrowth and is not dependent on the neurosecretory hormone or chemical inducers associated with the nerve process itself. Similarly, development of competence to secrete in response to neurohormonal stimuli, at the level of the cell, is linked to the appearance of the β-adrenergic receptors on the surface of the secretory cell but is independent from the ingrowth or presence of catecholamine-containing nerves (39, 69).

Once the stimulus-secretion systems have become functionally linked to the autonomic nervous system there appears to be a trophic effect of these nerves on the continued growth and development of the salivary glands (14, 17, 29, 40, 52, 54, 55, 60, 81–85). The continued presence of sympathetic innervation during the neonatal phase of secretory cell development appears to be required for the proper differentiation of the specific β-adrenergic receptor subtype (40). In the adult parotid gland a β_1-adrenergic response is responsible for eliciting amylase secretion. However, if the superior cervical ganglion is surgically removed from neonatal rats, the β_1-adrenergic receptor subtype does not develop in the parotid gland; rather the β_2-adrenergic receptor subtype develops and regulates parotid secretion (40, 54). Furthermore the autonomic nervous system plays a role in the postnatal regulation of amylase development, secretory function development, the number of cells, and the size of the rat parotid gland (14, 17, 55, 60, 81–85). Neonatal sympathectomy results in a reduction in cell size, cell number, and amylase content of the rat parotid gland. There is evidence that the trophic factors regulating postnatal development of the parotid gland may be associated with the sympathetic nerve process itself rather than the neurosecretory hormone. Postnatal development of the parotid gland was depressed by surgical sympathectomy but not by a chemical sympathectomy procedure that only eliminated catecholamines (55). Conversely, postnatal development of the submandibular gland appears to be less dependent than the parotid gland on the continued presence of an intact sympathetic nervous system. Neonatal sympathectomy, surgical or chemical, delays but does not permanently or substantially alter submandibular gland development (46, 93).

SUMMARY

The complete functional differentiation of a salivary gland is a process that requires multiple steps. The specific cells within the oral cavity of the embryo must first undergo induction and determination to become "salivary" in nature. The factors involved in this primary induction and determination of the salivary glands are not known. Once induction and determination have occurred the cells begin to synthesize very small amounts of salivary gland–specific proteins. Induction and determination are followed by the development of the salivary gland–specific structure or branching pattern by the growing rudiment. The establishment and nature of the branching pattern are determined by the mesenchyme that surrounds the salivary epithelium. The regulation and maintenance of the branching pattern are controlled by the composition of the epithelial basal lamina combined with interactions with the investing mesenchyme. The basement membrane composition is determined by the synthesis of glycosaminoglycans (proteoglycans) by the epithelium, the selective degradation by mesenchymal neutral hyaluronidases, and the stabilization of specific lamina components by mesenchyme-derived type I collagen. Once the morphogenetic pattern has been established, cytodifferentiation of the specific cell types within the gland proceeds. Cytodifferentiation appears to be induced by direct cell-cell interactions between cells of the investing mesenchyme and cells of the end buds of the initial 4–12 branches of the developing rudiment. Cytodifferentiation is characterized by the amplification of the synthesis of salivary gland–specific secretory proteins or mucins and changes in cellular structure that are consistent with efficient exocrine protein synthesis, packaging, and secretion (apical-basal polarity of function).

Secretion of salivary proteins is mediated by cAMP that is generated by the catalytic action of the cell surface enzyme adenylate cyclase on ATP. Adenylate cyclase activity appears at the secretory cell surface as cytodifferentiation commences. Enzyme activity reaches adult levels as the secretory cells begin to package their exportable proteins into secretory granules. Although these embryonic and early neonatal secretory cells are synthesizing and packaging proteins for export and the enzyme system for generating cAMP is present, the cells do not respond to externally applied hormones known to induce secretion. This failure to respond is due in part to the absence of β-adrenergic receptors at the cell surface of the secretory cells. In addition in the developing parotid gland the guanylnucleotide regulatory component of adenylate cyclase does not appear until the β-adrenergic receptors appear at the cell surface. In the submandibular gland the ingrowth of catecholamine-containing nerve processes is coordinated with but does not regulate the appearance of the β-adrenergic receptors on the cell surface. Thus the ingrowth of the nerves and the appearance of the receptors complete the maturation of the stimulus-secretion coupling system and permit normal physiological functioning of the gland.

REFERENCES

1. Ball, W. D. Development of the rat salivary glands. I. Accumulation of parotid gland DNAse activity. *J. Exp. Zool.* 178: 331–342, 1971.
2. Ball, W. D. Development of the rat salivary glands. III.

Mesenchymal specificity in the morphogenesis of the embryonic submaxillary and sublingual glands of the rat. *J. Exp. Zool.* 188: 277–288, 1974.
3. BALL, W. D. Development of the rat salivary glands. IV. Amylase and ribonuclease activity during embryonic and neonatal development of the parotid and submaxillary glands. *Dev. Biol.* 41: 267–277, 1974.
4. BALL, W. D., AND R. S. REDMAN. Two independently regulated secretory systems within the acini of the submandibular gland of the perinatal rat. *Eur. J. Cell Biol.* 33: 112–122, 1984.
5. BANERJEE, S. D., AND M. R. BERNFIELD. Developmentally regulated neutral hyaluronidase activity during epithelial-mesenchymal interaction (Abstract). *J. Cell Biol.* 83: 469a, 1979.
6. BANERJEE, S. D., R. H. COHN, AND M. R. BERNFIELD. Basal lamina of embryonic salivary epithelia. Production by the epithelium and role in maintaining lobular morphology. *J. Cell Biol.* 73: 445–463, 1977.
7. BARKA, T., AND H. VAN DER NOEN. Adenylate cyclase activity in rat submandibular gland during postnatal development. *Life Sci.* 14: 267–280, 1974.
8. BARRETT, M. L., AND W. D. BALL. Development of the rat salivary glands. II. The identification of a trypsinlike protease in the submaxillary gland. *Dev. Biol.* 36: 195–201, 1974.
9. BERNFIELD, M. R., AND S. D. BANERJEE. Acid mucopolysaccharide (glycosaminoglycan) at the epithelial-mesenchymal interface of mouse embryo salivary glands. *J. Cell Biol.* 52: 664–673, 1972.
10. BERNFIELD, M., AND S. D. BANERJEE. The turnover of basal lamina glycosaminoglycan correlates with epithelial morphogenesis. *Dev. Biol.* 90: 291–305, 1982.
11. BERNFIELD, M. R., S. D. BANERJEE, AND R. H. COHN. Dependence of salivary epithelial morphology and branching morphogenesis upon acid mucopolysaccharide-protein (proteoglycan) at the epithelial surface. *J. Cell Biol.* 52: 674–689, 1972.
12. BERNFIELD, M., S. D. BANERJEE, J. KODA, AND A. C. RAPRAEGER. Remodelling of the basement membrane: morphogenesis and maturation. In: *Basement Membranes and Cell Movement*, edited by R. Porter and J. Whelan. London: Pitman, 1986, p. 179–196. (Ciba Found. Symp.)
13. BERNFIELD, M. R., AND N. K. WESSELLS. Intra- and extracellular control of epithelial morphogenesis. *Dev. Biol. Suppl.* 4: 195–249, 1970.
14. BLOOM, G. D., B. CARLSÖÖ, A. DANIELSSON, S. HELLSTRÖM, AND R. HENRIKSSON. Trophic effect of the sympathetic nervous system on the early development of the rat parotid gland: a quantitative ultrastructural study. *Anat. Rec.* 201: 645–654, 1981.
15. BORGHESE, E. Explantation experiments on the influence of the connective tissue capsule on the development of the epithelial part of the submandibular gland of *Mus musculus*. *J. Anat.* 84: 303–318, 1950.
16. BOTTARO, B., AND L. S. CUTLER. An electrophysiological study of the postnatal development of the autonomic innervation of the rat submandibular salivary gland. *Arch. Oral Biol.* 29: 237–242, 1984.
17. BRENNER, G. M., AND R. G. WULF. Adrenergic beta receptors mediating submandibular salivary gland hypertrophy in the rat. *J. Pharmacol. Exp. Ther.* 218: 608–612, 1981.
18. BYLUND, D. B., J. R. MARTINEZ, J. CAMDEN, AND S. B. JONES. Autonomic receptors in the developing submandibular glands of neonatal rats. *Arch. Oral Biol.* 27: 945–950, 1982.
19. CHANG, W. W. L., AND T. BARKA. Stimulation of acinar cell proliferation by isoproterenol in the postnatal rat submandibular gland. *Anat. Rec.* 178: 203–210, 1974.
20. CHAUDHRY, A. P., L. S. CUTLER, J. A. SCHMUTZ, G. M. YAMANE, M. SUNDERRAJ, AND L. K. PIERRI. Development of the hamster submandibular gland I. The acinar intercalated duct complex. *J. Submicrosc. Cytol.* 17: 555–567, 1985.
21. CHAUDHRY, A. P., J. A. SCHMUTZ, L. S. CUTLER, AND M. SUNDERRAJ. Prenatal and postnatal histogenesis of myoepithelium in hamster submandibular gland. *J. Submicrosc. Cytol.* 15: 787–798, 1983.

22. COHN, R. H., S. D. BANERJEE, AND M. R. BERNFIELD. Basal lamina of embryonic salivary epithelia. Nature of glycosaminoglycan and organization of extracellular materials. *J. Cell Biol.* 73: 464–478, 1977.
23. COUGHLIN, M. D. Early development of parasympathetic nerves in the mouse submandibular gland. *Dev. Biol.* 43: 123–139, 1975.
24. COUGHLIN, M. D. Target organ stimulation of parasympathetic nerve growth in the developing mouse submandibular gland. *Dev. Biol.* 43: 140–158, 1975.
25. CUTLER, L. S. Morphogenesis and Functional Differentiation of the Rat Submandibular Gland in Vivo and in Vitro: Ultrastructural and Biochemical Studies. Buffalo: State Univ. of New York at Buffalo, 1973. PhD thesis.
26. CUTLER, L. S. Intercellular contacts at the epithelial-mesenchymal interface of the developing rat submandibular gland in vitro. *J. Embryol. Exp. Morphol.* 39: 71–77, 1977.
27. CUTLER, L. S. The dependent and independent relationships between cytodifferentiation and morphogenesis in developing salivary gland secretory cells. *Anat. Rec.* 196: 341–347, 1980.
29. CUTLER, L. S., J. BOCCUZZI, L. YAEGER, B. BOTTARO, C. P. CHRISTIAN, AND J. R. MARTINEZ. Effects of reserpine treatment on β-adrenergic/adenylate cyclase modulated secretion and resynthesis by the rat submandibular gland. *Virchows Arch. Pathol. Anat. Physiol. Klin. Med.* 392: 185–198, 1981.
30. CUTLER, L. S., AND A. P. CHAUDHRY. Differentiation of the myoepithelial cells of the rat submandibular gland in vivo and in vitro: an ultrastructural study. *J. Morphol.* 140: 343–354, 1973.
31. CUTLER, L. S., AND A. P. CHAUDHRY. Intercellular contacts at the epithelial-mesenchymal interface during the prenatal development of the rat submandibular gland. *Dev. Biol.* 33: 229–240, 1973.
32. CUTLER, L. S., AND A. P. CHAUDHRY. Release and restoration of the secretory granules of the convoluted granular tubules of the rat submandibular gland. *Anat. Rec.* 176: 405–420, 1973.
33. CUTLER, L. S., AND A. P. CHAUDHRY. Cytodifferentiation of the acinar cells of the rat submandibular gland. *Dev. Biol.* 41: 31–41, 1974.
34. CUTLER, L. S., AND A. P. CHAUDHRY. Cytodifferentiation of striated duct cells and secretory cells of the convoluted granular tubules of the rat submandibular gland. *Am. J. Anat.* 143: 201–217, 1975.
35. CUTLER, L. S., C. P. CHRISTIAN, AND B. BOTTARO. Development of stimulus-secretion coupling in salivary glands. In: *Methods in Cell Biology. Basic Mechanisms of Cellular Secretion*, edited by A. R. Hand and C. Oliver. New York: Academic, 1981, vol. 23, p. 531–545.
35a. CUTLER, L. S., C. P. CHRISTIAN, AND J. K. RENDELL. Glycosaminoglycan synthesis by adult rat submandibular salivary-gland secretory units. *Arch. Oral Biol.* 32: 413–420, 1987.
35b. CUTLER, L. S., AND B. A. MOORADIAN. Lumen formation during development of the rat submandibular gland. *J. Dent. Res.* 66: 1562–1669, 1987.
36. CUTLER, L. S., B. A. MOORADIAN, AND C. P. CHRISTIAN. Concurrent cytochemical localization of adenylate cyclase and peroxidase in the developing submandibular gland. *J. Histochem. Cytochem.* 25: 1207–1212, 1977.
37. CUTLER, L. S., AND J. K. RENDELL. Glycosaminoglycan synthesis by developing and adult submandibular gland secretory units (Abstract). *J. Dent. Res.* 63: 302, 1984.
38. CUTLER, L. S., AND S. B. RODAN. Biochemical and cytochemical studies on adenylate cyclase activity in the developing rat submandibular gland: differentiation of the acinar secretory compartment. *J. Embryol. Exp. Morphol.* 36: 291–303, 1976.
39. CUTLER, L. S., C. SCHNEYER, AND C. CHRISTIAN. The influence of the sympathetic nervous system on the development of β-adrenergic receptors in the rat submandibular salivary gland. *Arch. Oral Biol.* 30: 341–344, 1985.
40. DANIELSSON, A., R. HENRIKSSON, P. LINDSTRÖM, AND J. SEHLIN. The importance of an intact sympathetic innervation for the differentiation of the β-adrenoceptor subtypes in the rat parotid gland. *Acta Physiol. Scand.* 115: 377–379, 1982.

41. DEVI. N. S., AND F. JACOBY. The submaxillary gland of the golden hamster and its post-natal development. *J. Anat.* 100: 269–285, 1966.
42. DVOŘÁK, M. The secretory cells of the submaxillary gland in the perinatal period of development in the rat. *Z. Zellforsch. Mikrosk. Anat.* 99: 346–356, 1969.
43. GERSTNER, R., H. FLON, AND T. O. BUTCHER. Onset and type of salivary secretion in fetal rats (Abstract). *J. Dent. Res.* 42: 1250, 1963.
44. GRAND, R. J., D. A. CHONG, AND S. J. RYAN. Postnatal development of adenylate cyclase in rat salivary glands: patterns of hormonal sensitivity. *Am. J. Physiol.* 228: 608–612, 1975.
45. GRAND, R. J., AND M. I. SHAY. Development of secretory function in rat parotid gland. *Pediatr. Res.* 12: 100–104, 1975.
46. GRESIK, E. W., AND T. BARKA. The effect of neonatal sympathectomy on the response of the rat submandibular gland to isoproterenol. *J. Pharmacol. Exp. Ther.* 200: 101–106, 1977.
47. GRESIK, E. W., AND T. BARKA. Precocious development of granular convoluted tubules in the mouse submandibular gland induced by thyroxine or by thyroxine and testosterone. *Am. J. Anat.* 159: 177–185, 1980.
48. GROBSTEIN, C. Analysis in vitro of the early organization of the rudiment the mouse submandibular gland. *J. Morphol.* 93: 19–44, 1953.
49. GROBSTEIN, C. Epithelio-mesenchymal specificity in the morphogenesis of mouse submandibular rudiments in vitro. *J. Exp. Zool.* 124: 383–413, 1953.
50. GROBSTEIN, C. Morphogenetic interaction between embryonic mouse tissues separated by a membrane filter. *Nature Lond.* 172: 869–871, 1953.
51. GROBSTEIN, C., AND J. COHEN. Collagenase: effect on morphogenesis of embryonic salivary epithelium in vitro. *Science Wash. DC* 150: 626–628, 1965.
52. HALL, H. D. Neural regulation of compensatory enlargement of the parotid gland of the rat. *Cell Tissue Res.* 187: 147–151, 1978.
53. HALL, H. D., AND C. A. SCHNEYER. Physiological activity and regulation of growth of developing parotid. *Proc. Soc. Exp. Biol. Med.* 131: 1288–1291, 1969.
54. HENRIKSSON, R., B. CARLSÖÖ, A. DANIELSSON, S. HELLSTRÖM, AND L. A. IDAHL. Effects of neonatal sympathetic denervation on amylase secretion in the adult rat parotid gland: difference in β_1- and β_2-adrenoceptor response. *Eur. J. Pharmacol.* 78: 195–200, 1982.
55. HENRIKSSON, R., B. CARLSÖÖ, A. DANIELSSON, S. SUNDSTRÖM, AND G. JÖNSSON. Developmental influences of the sympathetic nervous system on rat parotid gland. *J. Neurol. Sci.* 71: 183–191, 1986.
56. JACOBY, F., AND C. R. LEESON. The post-natal development of the rat submaxillary gland. *J. Anat.* 93: 201–216, 1959.
57. KALLMAN, F., AND C. GROBSTEIN. Localization of glucosamine-incorporating materials at epithelial surfaces during salivary epithelio-mesenchymal interactions in vitro. *Dev. Biol.* 14: 52–67, 1966.
58. KAWAKATSU, K., M. MORI, M. FUKUDA, S. KAGAWA, AND K. OSANAI. Histochemical studies of the developing submaxillary gland in the human fetus. *J. Osaka Dent. Univ.* 3: 3–9, 1963.
59. KIM, S. K., S. S. HAN, AND C. E. NASJLETI. The fine structure of secretory granules in submandibular glands of the rat during early postnatal development. *Anat. Rec.* 168: 463–467, 1970.
60. KLEIN, R. M. Alteration of neonatal rat parotid gland acinar cell proliferation by guanethidin-induced sympathectomy. *Cell Tissue Kinet.* 12: 411–423, 1979.
61. LAWSON, K. A. Morphogenesis and functional differentiation of rat parotid gland in vivo and in vitro. *J. Embryol. Exp. Morphol.* 24: 411–424, 1970.
62. LAWSON, K. A. The role of mesenchyme in the morphogenesis and functional differentiation of rat salivary epithelium. *J. Embryol. Exp. Morphol.* 27: 497–513, 1972.
63. LAWSON, K. A. Mesenchyme specificity in rodent salivary gland development: the response of salivary epithelium to lung mesenchyme in vitro. *J. Embryol. Exp. Morphol.* 32: 469–493, 1974.
64. LEESON, C. R. Structure of salivary glands. In: *Handbook of Physiology. Alimentary Canal. Secretion*, edited by C. F. Code. Washington, DC: Am. Physiol. Soc., 1967, sect. 6, vol. II, chapt. 32, p. 463–495.
65. LEESON, C. R., AND W. G. BOOTH. Histological, histochemical, and electron microscopic observations on the postnatal development of the major sublingual gland of the rat. *J. Dent. Res.* 40: 838–845, 1961.
66. LEESON, C. R., AND F. JACOBY. An electron microscopic study of the rat submaxillary gland during its post-natal development and in the adult. *J. Anat.* 93: 287–295, 1959.
67. LINE, S. E., AND F. L. ARCHER. The postnatal development of myoepithelial cells in the rat submandibular gland. An immuno-histochemical study. *Virchows Arch. B Cell Pathol.* 10: 253–262, 1972.
68. LUDFORD, J. M., AND B. A. TALAMO. β-Adrenergic and muscarinic receptors in developing rat parotid glands. *J. Biol. Chem.* 255: 4619–4627, 1980.
69. LUDFORD, J. M., AND B. A. TALAMO. Independent regulation of β-adrenergic receptor and nucleotide binding proteins of adenylate cyclase. *J. Biol. Chem.* 258: 4831–4838, 1983.
70. MARTINEZ, J. R., AND J. CAMDEN. Ouabain-sensitive K^+ uptake in submandibular salivary gland slices of rats of different post-natal ages. *Arch. Oral Biol.* 28: 1109–1114, 1983.
71. MARTINEZ, J. R., AND J. CAMDEN. Volume and composition of pilocarpine- and isoproterenol-stimulated submandibular saliva of early postnatal rats. *J. Dent. Res.* 62: 543–547, 1983.
72. MOORADIAN, B. A., AND L. S. CUTLER. Developmental distribution of microperoxisomes in the rat submandibular gland. *J. Histochem. Cytochem.* 26: 989–999, 1978.
72a. PICTET, R. L., W. R. CLARK, R. H. WILLIAMS, AND W. J. RUTTER. An ultrastructural analysis of the developing embryonic pancreas. *Dev. Biol.* 29: 436–467, 1972.
73. PLASMEYER, J. H. J. The action of parathyroid hormone on salivary glands in organ culture. I. Some aspects of the organ culture of mouse and rat embryonic salivary glands with special references to the duct system. *Proc. K. Ned. Akad. Wet. Series C Biol. Med. Sci.* 73: 75–92, 1969.
74. REDMAN, R. S., AND W. D. BALL. Cytodifferentiation of secretory cells in the sublingual gland of the prenatal rat: a histological, histochemical and ultrastructural study. *Am. J. Anat.* 153: 367–389, 1978.
75. REDMAN, R. S., AND W. D. BALL. Differentiation of myoepithelial cells in the developing rat sublingual gland. *Am. J. Anat.* 156: 543–565, 1979.
76. REDMAN, R. S., AND L. M. SREEBNY. The prenatal phase of morphosis of the rat parotid gland. *Anat. Rec.* 168: 127–138, 1970.
77. REDMAN, R. S., AND L. M. SREEBNY. Proliferative behavior of differentiating rat parotid gland. *J. Cell Biol.* 46: 81–87, 1970.
78. REDMAN, R. S., AND L. M. SREEBNY. Morphologic and biochemical observations on the development of the rat parotid gland. *Dev. Biol.* 25: 248–279, 1971.
79. RUFO, M. B., AND T. BARKA. Cell differentiation in the terminal tubule of fetal rat submandibular gland in organ culture. *Anat. Rec.* 184: 301–310, 1976.
80. SCHNEYER, C. A. Premature increases in amylase of postnatal rat parotid with chronic isoproterenol. *Proc. Soc. Exp. Biol. Med.* 155: 440–444, 1977.
81. SCHNEYER, C. A., AND H. D. HALL. Effects of denervation on development of function and structure of immature rat parotid. *Am. J. Physiol.* 212: 871–876, 1967.
82. SCHNEYER, C. A., AND H. D. HALL. Time course and autonomic regulation of development of secretory function of rat parotid. *Am. J. Physiol.* 214: 808–813, 1968.
83. SCHNEYER, C. A., AND H. D. HALL. Autonomic regulation of postnatal changes in cell number and size of rat parotid. *Am.*

J. Physiol. 219: 1268–1272, 1970.
84. SCHNEYER, C. A., AND H. D. HALL. Autonomic regulation of changes in rat parotid amylase during postnatal development. Am. J. Physiol. 223: 172–175, 1972.
85. SCHNEYER, C. A., AND H. D. HALL. Effects of removal of superior cervical ganglion or auriculotemporal nerve on course of postnatal change in rat parotid amylase. Proc. Soc. Exp. Biol. Med. 140: 911–915, 1972.
86. SCHNEYER, C. A., AND J. M. SHACKLEFORD. Accelerated development of salivary glands of early postnatal rats following isoproterenol. Proc. Soc. Exp. Biol. Med. 112: 320–324, 1963.
87. SMITH, R. L., AND M. BERNFIELD. Mesenchymal cells degrade epithelial basal lamina glycosaminoglycan. Dev. Biol. 94: 378–390, 1982.
88. SPOONER, B. S., K. BASSETT, AND B. STOKES. Sulfated glycosaminoglycan deposition and processing at the basal epithelial surface in branching and β-D-xyloside-inhibited embryonic salivary glands. Dev. Biol. 109: 177–183, 1985.
89. SPOONER, B. S., AND J. M. FAUBION. Collagen involvement in branching morphogenesis of embryonic lung and salivary gland. Dev. Biol. 77: 84–102, 1980.
90. SPOONER, B. S., AND N. K. WESSELS. Effects of cytochalasin B upon microfilaments involved in morphogenesis of salivary epithelium. Proc. Nat. Acad. Sci. USA 66: 360–364, 1970.
91. SPOONER, B. S., AND N. K. WESSELS. An analysis of salivary gland morphogenesis: role of cytoplasmic microfilaments and microtubules. Dev. Biol. 27: 38–54, 1972.
92. SRINIVASAN, R., AND W. W. L. CHANG. The development of the granular convoluted duct in the rat submandibular gland. Anat. Rec. 182: 29–40, 1975.
93. SRINIVASAN, R., AND W. W. L. CHANG. Effect of neonatal sympathectomy on the postnatal differentiation of the submandibular gland of the rat. Cell Tissue Res. 180: 99–109, 1977.
94. SRINIVASAN, R., W. W. L. CHANG, H. VAN DER NOEN, AND T. BARKA. The effect of isoproterenol on the postnatal differentiation and growth of the rat submandibular gland. Anat. Rec. 177: 243–254, 1973.
95. THOMPSON, H. A., AND B. S. SPOONER. Inhibition of branching morphogenesis and alteration of glycosaminoglycan biosynthesis in salivary glands treated with β-D-xyloside. Dev. Biol. 89: 417–424, 1982.
96. THOMPSON, H. A., AND B. S. SPOONER. Proteoglycan and glycosaminoglycan synthesis in embryonic mouse salivary glands: effects of β-D-xyloside, an inhibitor of branching morphogenesis. J. Cell Biol. 96: 1443–1450, 1983.
97. YAMASHINA, S., AND T. BARKA. Localization of peroxidase activity in the developing submandibular gland of normal and isoproterenol-treated rats. J. Histochem. Cytochem. 20: 855–872, 1972.
98. YAMASHINA, S., AND T. BARKA. Development of endogenous peroxidase in fetal rat submandibular gland. J. Histochem. Cytochem. 21: 42–50, 1973.
99. YAMASHINA, S., AND T. BARKA. Peroxidase activity in the developing rat submandibular gland. Lab. Invest. 31: 82–89, 1974.
100. YAMASHINA, S., AND V. MIZUHIRA. Postnatal development of acinar cells in rat submandibular gland as revealed by electron microscopic staining for carbohydrates. Am. J. Anat. 146: 211–235, 1976.
101. YOHRO, T. Development of secretory units of mouse submandibular gland. Z. Zellforsch. Mikrosk. Anat. 110: 173–184, 1970.
102. YOHRO, T. Nerve terminals and cellular junctions in young and adult mouse submandibular glands. J. Anat. 108: 409–417, 1971.
103. YOUNG, J. A., AND E. W. VAN LENNEP. The Morphology of Salivary Glands. New York: Academic, 1978.

CHAPTER 7

Secretory membranes and the exocrine storage compartment

RICHARD S. CAMERON | Departments of Cell Biology and Internal Medicine,
PETER ARVAN | Yale University School of Medicine,
J. DAVID CASTLE | New Haven, Connecticut

CHAPTER CONTENTS

Granule Formation and Storage
Granule Isolation and Preparation of Membrane Subfractions
    Secretory granules
    Membrane subfractions
Composition of Exocrine Granule Membranes
    Membrane lipids
    Membrane proteins
        Enzyme activities
        Common antigens in secretory membranes
        Polypeptides characterized by gel electrophoresis
Storage Function and Biophysical Properties of
    Secretion Granules
    Granule packaging, stability, and osmotic behavior
    Ion permeation of granule membranes
    Intragranular pH and buffering capacity
    Hydrogen-ATPase activity
Interactions With Cytoplasmic Proteins
    Possible roles of microtubules
    Microfilamentous networks and secretory organelle interactions
    Other polypeptides binding to secretory membranes
    Stimulus-enhanced phosphorylation of membrane and
        associated proteins
Exocytosis
Recycling, Turnover, and Sorting From Other Vesicular Traffic
    Fate of reinternalized secretory membrane
    Secretory membrane as a plasmalemmal protein carrier
    Secretory membrane and sorting sites

AN IMPORTANT INTERACTION between eucaryotic cells and the surrounding environment is the bidirectional exchange of macromolecules. Both export and import activities involve the segregation and movement of transported macromolecules through a specific sequence of membrane-bounded compartments. For export the series of distinct compartments includes the endoplasmic reticulum, various subdivisions of the Golgi complex, and the plasma membrane, which are all functionally interconnected by vesicular carriers that transport products vectorially from site of synthesis to site of release (141). The final carriers involved in export at the cell surface consist of secretory membrane, whose structure and function are discussed in this chapter. For import, vesicular carriers derived from the plasma membrane provide for the vectorial delivery of internalized macromolecules to the endosome-lysosome system (186). Thus the plasma membrane is simultaneously the destination of secretory membrane and the origin of membrane used in import.

Export and import activities are especially important in the epithelial layer of multicellular organisms, which mediates exchanges between the external world and the internal environment of the organism. The epithelium is a two-dimensional functional mosaic where certain regions are highly specialized for amplification of export (exocrine-secretion) or import (nutrient-uptake) activities. The individual cells in these regions are polarized; not only is the plasma membrane divided into compositionally distinct domains (apical and basolateral) that communicate selectively with the two different extracellular environments but export and import in these cells are organized around the surface domain that is in direct contact with the external world. This feature distinguishes exocrine tissues from endocrine cells, whose intracellular compartments are otherwise similar in structure and function (59).

Cells have at least two routes along which they direct secretory proteins from post-Golgi compartments to the cell surface. The ubiquitous pathway [termed *constitutive* (99)] involves continuous dispatching of vesicular carriers from Golgi/post-Golgi compartments, usually with little concentration of secretory products over levels observed in preceding compartments. The constitutive pathway may be branched in polarized epithelial cells to accommodate selective delivery of products to the distinct surface domains (103, 118, 150, 159) and may be involved in the maintenance and growth of these domains as well as in the production and turnover of components in the immediate extracellular environment.

The second route [termed *regulated* (194)] is a major

export pathway in exocrine as well as endocrine secretory cells; its rate is controlled by integration of various stimuli of neural and endocrine origin. Because protein synthesis is relatively slow to respond to these stimulatory factors (14, 112), regulation in these cells entails the mobilization and discharge of already synthesized products that have been maintained in intracellular stores [stimulus-secretion coupling (50)]. The stored products are found at high concentrations in secretion granules, whose distinctive structural features are condensed cores bounded by secretory membrane. The size of the granule population reflects multiple factors, which include the ultimate function of the secretion and the extent and duration of secretory stimulation.

Cells specialized for regulated export characteristically show an extreme level of differentiation, involving commitment to secretory function at the expense of activities related to growth and replication. Examples of this are the exocrine (acinar) cells of the parotid or pancreatic glands (Fig. 1), which are the primary experimental systems described in this chapter. Their

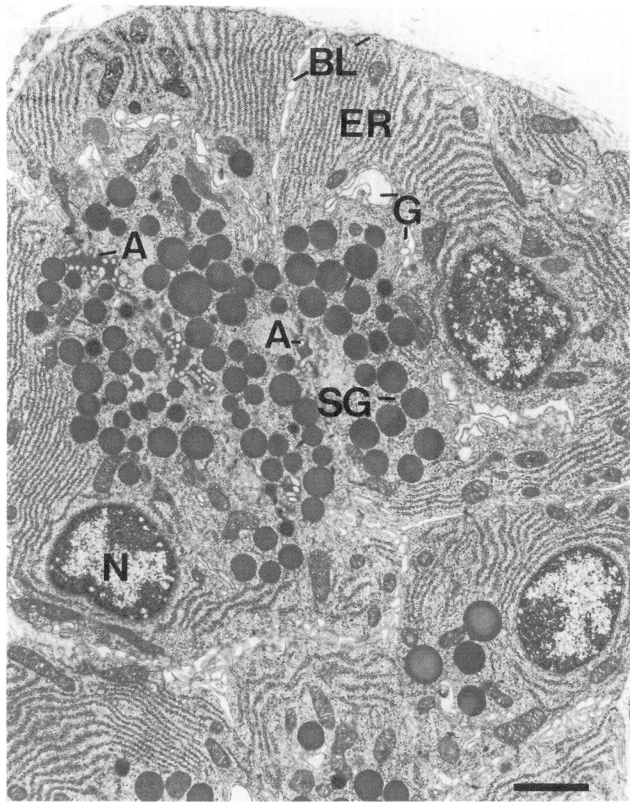

FIG. 1. Low-power electron micrograph of a rat parotid acinus consisting of functionally polarized secretory cells. Basolateral (*BL*) surfaces extend inward from the periphery and are separated from apical surfaces (*A*) by junctional complexes. Basal rough endoplasmic reticulum (*ER*) and centrally located extensive Golgi complexes (*G*) are evident. Secretory storage granules (*SG*; 1 μm diam) are collected near cell apices. Apical lumina are filled with discharged secretion. *N*, nucleus. Bar, 2 μm.

accentuation of internal storage and a stimulus-dependent secretory pathway has been characterized extensively (1, 30, 93). Furthermore, these tissues have been unusually favorable as sources of highly purified populations of isolated secretion granules and their membranes through cell fractionation experiments (21, 24, 31, 44). Consequently, recent compositional and biophysical studies, reviewed and compared with other systems (see COMPOSITION OF EXOCRINE GRANULE MEMBRANES, p. 109, and STORAGE FUNCTION AND BIOPHYSICAL PROPERTIES OF SECRETION GRANULES, p. 114), form a basis for identifying the parameters that distinguish segregation, packaging, and discharge activities, thereby contributing to a general scheme for regulated secretion.

GRANULE FORMATION AND STORAGE

For protein storage in secretion granules, concentration of products is generally first evident morphologically in the trans-most (acid phosphatase–positive) cisterna of the Golgi complex (77, 83, 136). Certain exceptions to this location, including condensation as intracisternal granules in the endoplasmic reticulum (140, 193) or at the periphery of other Golgi saccules (55, 59), seem to specify that packaging may be both dynamic, with respect to compartment, and opportunistic, in that it can commence under the appropriate conditions regardless of location.

In unstimulated exocrine cells, storage granule formation has been traced autoradiographically (Fig. 2) and, depending on the cell type, can be an extended process lasting >1 h (30, 93, 210). Its onset normally occurs by deposition and packaging of secretory content in a condensing vacuole associated with the trans-most Golgi cisterna (27, 136). Over and above the increasing internal concentration [by an estimated twofold (12)], the maturation process involves a net [apparently compositionally selective (137, 197)] reduction in vacuolar surface area [by as much as 30% (207)]. An immature storage granule that can be recognized in parotid acinar cells as a kinetic intermediate of the final homogeneously dense mature storage granule (30) may not be intimately associated with the Golgi complex. Its maturation entails a progressive increase in the density of packaged secretion, possibly involving continued (vesicle-mediated) input of proteins destined for storage (104), in addition to concentration. Thus the storage function initially observed in the Golgi may extend more peripherally into still-forming granules.

Granules within an individual cell are thought to contain approximately the same ratio of different secretory proteins (12, 105); however, there may be heterogeneities in posttranslational age of different polypeptides within individual granules, resulting from nonparallel intracellular transport after synthesis (170). Furthermore, distinct transcriptional or

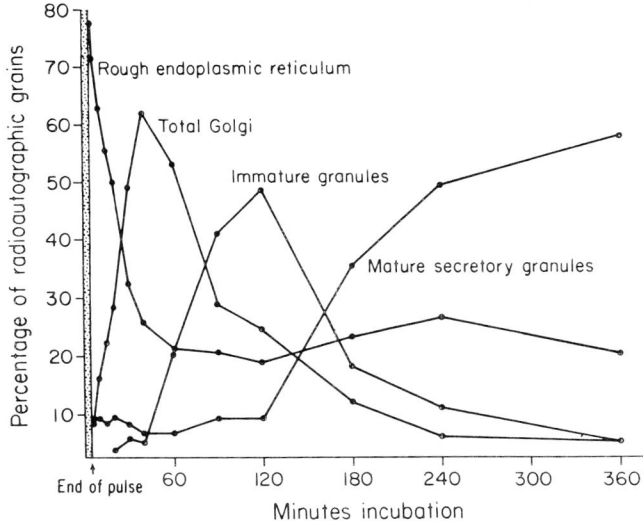

FIG. 2. Summary of an autoradiographic study of intracellular transport of newly synthesized secretory protein in rabbit parotid tissue. Tissue lobules were pulse labeled in vitro with [^{3}H]leucine and subjected to chase incubation for the indicated times. Distributions of autoradiographic grains determined from electron-macrograph autoradiograms are expressed as percent of total that were quantitated at each time point. Precursor-product kinetic relationships between rough endoplasmic reticulum, Golgi complex, immature secretion granules, and mature secretion granules are apparent. [From Castle et al. (30).]

adsorbed secretory content. A two-step procedure has often been used for this purpose. After lysis in hypotonic (2, 31) or isotonic salt solutions [25; in some cases at alkaline pH (121)], membranes are separated from residual organelle contaminants, unlysed granules, and most soluble content polypeptides by sucrose density-gradient centrifugation (25, 31, 121). Subsequent removal of extractable contaminants has been achieved by 37°C incubation of isolated membrane subfractions in saline solution (205) or treatment with 0.25 M NaBr (163); however, thorough extraction of these species generally requires a more drastic treatment. Exposure to a combination of saponin and 0.3 M Na_2SO_4 (25, 32) or to 100 mM Na_2CO_3, pH > 11 (65, 89), has proven to be effective, although changes in membrane bilayer integrity ranging from subtle to substantial are evident. Consequently the catalog of proteins established for secretion granule membranes purified by the more extensive procedures should be regarded as a subset containing integral, but possibly excluding peripheral, membrane polypeptides.

Examples of a highly purified parotid secretion granule fraction obtained in hyperosmotic sucrose-Ficoll and a saponin-sulfate–treated membrane subfraction having negligible residual organellar or soluble protein contamination (25) are shown in Figure 3.

translational programs from one cell to another, or in the same cell at different times, may contribute to different ratios of particular secretory products undergoing storage (129, 169).

GRANULE ISOLATION AND PREPARATION OF MEMBRANE SUBFRACTIONS

Secretory Granules

Because exocrine secretory granules are large (0.5–1.5 μm diam) and quite dense, their isolation by a variety of cell fractionation procedures is readily achieved. Repeated differential centrifugation after homogenization in an isosmotic buffered sucrose solution is convenient and rapid. However, the granule fractions obtained are contaminated considerably by mitochondria and to a lesser degree by lysosomes and plasmalemmal elements (2). More highly purified preparations suitable for isolating granule membrane subfractions require use of isopycnic density centrifugation procedures (4, 21, 25, 31, 44).

Membrane Subfractions

Even with highly purified granule fractions as starting material, the study of the polypeptide composition of granule membranes has required thorough removal of residual contaminating organelle membrane and

COMPOSITION OF EXOCRINE GRANULE MEMBRANES

Membrane Lipids

Membranes constituting the secretory storage compartment have been found to possess a very high phospholipid-protein weight ratio: 4.4 (mg/mg) for rat parotid (25); 2.5–2.6 (mg/mg) for rat pancreas (18); 5.3 (mg/mg) for synaptic vesicles (48); and 2.2 (mg/mg) for bovine adrenal chromaffin granules (213). Such ratios imply that 60%–80% of the membrane dry weight is lipid. This range can be contrasted with that observed in membrane fractions from cellular compartments of higher functional diversity where values ≤1.0 (mg/mg) are characteristically observed (89, 98, 107). The low buoyant density of granule membranes observed during centrifugation (25) as well as their paucity of intramembranous particles [viewed by freeze fracture (42)] are consistent with these biochemical findings.

Although analysis of exocrine granule membrane phospholipid composition is limited, Table 1 indicates overall compositional similarities to more thoroughly studied endocrine and neural granule membranes. Thus, considered as a group, secretory membranes are composed predominantly of phosphatidylcholine (30%–40%) and phosphatidylethanolamine (20%–30%), with phosphatidylserine and phosphatidylinositol (together 10%–15%) as minor constituents

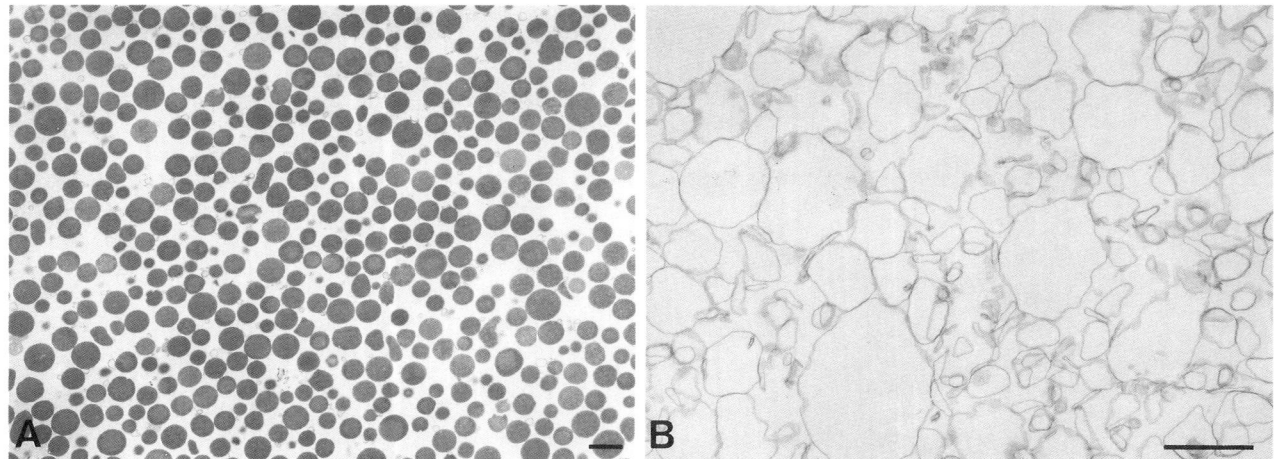

FIG. 3. Representative electron micrographs of a purified parotid secretion granule (A) and granule membrane fraction (B). With detailed consideration of the extent of contamination by residual mitochondria and incompletely removed secretory polypeptides as a basis, it is possible to estimate that 95% of the protein associated with the purified secretion granule membrane is bona fide granule membrane protein. Bars, 1 μm. [From Cameron and Castle (25).]

TABLE 1. *Representative Lipid Composition of Isolated Fractions of Secretory Membrane*

Lipid	Adrenal Chromaffin	Synaptic Vesicles		Anterior Pituitary		Parotid		Exocrine Pancreas	
	Bovine	Marine electric organ		Rat	Bovine	Rat	Rabbit	Guinea pig	Pig
Phosphatidylcholine*	25–27	48	41	36	41	33	52	31†	35
Lysophosphatidylcholine	11–16	~0		5	~0	6	7		13
Sphingomyelin	13–15	4	12	21	22	14	9	24	17
Phosphatidylethanolamine	32–34	31	36	17	26	26	21	32	18
Lysophosphatidylethanolamine	4–6	3							
Phosphatidylserine	5–8	9	7	12	6	9	8‡	5‡	16
Phosphatidylinositol	2–3	2	4	7	3	10			2
Cholesterol/phospholipid§	0.53	0.50	0.63					0.55	0.30

* Values are percentage of total lipid phosphorus recovered or mole %. † Percent total phosphatidylcholine plus lysophosphatidylcholine. ‡ Percent total phosphatidylserine plus phosphatidylinositol. § Molar ratio. Data for chromaffin granules from Dreyfus et al. (52) and De Oliveira Filgueiras et al. (47); data for synaptic vesicles from Michaelson et al. (123) and Deutsch and Kelly (48); data for pituitary granules from Meldolesi et al. (120) and De Oliveira Filgueiras et al. (47); data for parotid granules from Williams et al. (211) and Meldolesi et al. (120); data for exocrine pancreas from Meldolesi et al. (121) and LeBel and Beattie (110).

among the glycerolipids. In contrast to other membranes comprising the secretory pathway (but analogous to plasma membranes), granule membranes are enriched in sphingomyelin (10%–20%) and cholesterol (0.5–0.6 mol/mol phospholipid) and exhibit an increased degree of phospholipid fatty acid saturation. Variations have been observed in compositions reported for the same tissue and may reflect either residual organelle contamination, e.g., mitochondrial elements, as gauged by the presence of cardiolipin [<2% (47, 120)] or inadequate resolution of individual phospholipid species (phosphatidylserine and phosphatidylinositol) and incomplete recoveries. The unusually large amount of lysophosphatidylcholine observed in pancreatic zymogen granules has been attributed to artifactual phosphatidylcholine degradation by pancreatic neutral lipase (120); however, the even higher amount observed in adrenal chromaffin granule membranes (15) is considered to be real but of unexplained significance (46).

Membrane Proteins

The protein components (integral) of exocrine granule membranes are estimated to comprise 0.5%–1.0% of the total protein present in exocrine secretory granules. This percentage is below that generally specified for endocrine granules [20%–25% (213)], which have higher membrane surface-volume ratios and where membrane transport systems are prominent granule activities. Low membrane protein content not only reemphasizes the functional specialization of the storage compartment but also signifies in practical terms that extreme care must be taken in fractionation studies to guarantee insignificant contamination. Thus far analyses of the composition of integral polypeptides in granule membranes have included 1) tests for particular enzyme activities either in situ (by histochemistry) or in vitro after organelle isolation; 2) examination of the distribution of antigenic determinants by immunochemical or immunocytochemical techniques, either between granules and other cellular organelles or between different granule populations; and *3*) examination of isolated fractions by polyacrylamide gel electrophoresis (PAGE).

ENZYME ACTIVITIES. Exocrine secretory granule membranes are relatively poor in known membrane enzyme activities. Accordingly, activities characteristic of endoplasmic reticulum (either NADH- or NADPH–cytochrome *c* reductases), Golgi (galactosyltransferase, thiamine pyrophosphatase/nucleoside-diphosphatase), Golgi-associated transtubular network (acid CMPase), lysosomal content (acid β-glycerophosphatase and β-*N*-acetylglucosaminidase), and plasma membrane (5′-nucleotidase, adenylate cyclase, Na^+-K^+-ATPase) are undetectable or present in exceedingly low quantities in purified exocrine granule fractions and their membrane subfractions. Other membrane components, including electron-transport devices (54, 168), nucleotide translocators (e.g., 185, 206), and ATP-driven, inward-directed H^+ pumps (166), which are prominent constituents of endocrine and neural secretory granules, are thought to be present in at most low amounts in mature exocrine storage granules (6).

Currently the only enzyme activities known to be associated with exocrine granule membranes are γ-glutamyltransferase [parotid, pancreas, submandibular, and lacrimal glands (24, 25)], which is present in many epithelial cells as an apical plasma membrane dimeric ectoenzyme (29); phosphatidylinositol kinase [parotid gland (139)], which has also been found at varying levels in other intracellular membranes, including those of the nucleus, lysosomes, Golgi, plasmalemma, and coated vesicles (26, 34, 179) and in endocrine granules as well (151); ATP phosphohydrolase, an enzyme of unknown function that has been characterized, so far, only in pancreatic zymogen granule membranes (110); and cAMP-dependent protein kinase, found associated with pancreatic (21) and, potentially, parotid (178) granule membranes [and chromaffin granule membranes (198)]. The latter activity is of interest particularly with regard to its potential involvement in second-messenger–mediated regulation of secretory function and will be considered in INTERACTIONS WITH CYTOPLASMIC PROTEINS, p. 116.

COMMON ANTIGENS IN SECRETORY MEMBRANES. With limited knowledge of the function of secretory membrane components, exploratory approaches have focused on identifying elements of compositional overlap in several granule types that may merit further study. One form of such an analysis has involved the examination of membrane antigenic determinants. A polyclonal, polyspecific antibody to parotid granule membrane has been prepared that binds to most of the polypeptides present in the catalog of radioiodinated membrane components (24). Although the complexity of such a probe limits its utility for exploring individual polypeptide species in detail, its use has led to three striking insights concerning the general composition of storage membranes. First, indirect immunofluorescence staining of parotid tissue indicates very clearly that antigens are largely restricted to granule membranes (Fig. 4*A*). Thus, as for other endomembrane compartments in eucaryotic cells [e.g., rough endoplasmic reticulum, Golgi, and lysosome (115, 196)], the storage membrane is at least partly endowed with a specific composition, which must involve some degree of selective segregation of membrane components during granule formation (197). Second, the same kind of study reveals selective granule membrane staining in other exocrine glandular tissue (Fig. 4*B*), which has led to a more refined comparative analysis of secretory membranes reviewed next. Finally, when the parotid granule membrane antiserum (in particular, a subset of immunoglobulins selected by adsorption for interaction with the cytoplasmic aspect of the membrane) has been used to stain endocrine (epithelial and nonepithelial) and neural tissues, organelles involved in secretion (or cell regions highly enriched in such organelles) were selectively fluorescent (Fig. 4*C–F*). Such positive results contrast with findings obtained with antibodies (either polyclonal-polyspecific or monoclonal) developed against neural or endocrine secretory membranes (which may contain a more varied spectrum of antigens than exocrine granules because of the presence of amine translocation and other membrane transport machinery), where staining of exocrine tissues was not observed (20, 117). Because complications arising from any highly antigenic oligosaccharides have been avoided by the immunoselection strategy with adsorption on intact (aldehyde-fixed) granules, these observations indicate that most vesicular carrier membranes involved in export function may bear similar, cytosolically oriented determinants.

POLYPEPTIDES CHARACTERIZED BY GEL ELECTROPHORESIS. For secretory membranes from both exocrine and endocrine storage granules, one-dimensional [sodium dodecyl sulfate (SDS)] and two-dimensional, isoelectric-focusing SDS-PAGE analyses have shown that the overall polypeptide composition is less complex than that of membranes of other cellular organelles, especially those comprising distinct compartments along the secretory pathway (89, 106). With radioiodinated preparations, the spectrum of major granule membrane polypeptides consists of 25–30 discrete species. Nearly all of these polypeptides have isoelectric points between pH 5 and 7, and they fall roughly into three sets distinguished by molecular weight: *1*) a relatively high-molecular-weight set (80,000 to >110,000) that is enriched in glycosylated species; *2*) a set of intermediate molecular weight (40,000–65,000); and *3*) a prominent low-molecular-weight set (20,000–35,000). These features are illustrated in Figure 5A for storage granule membranes prepared from rat parotid.

Comparison of this standard profile for parotid membranes with those for granule membrane fractions obtained from different exocrine glands (pancreas, lacrimal, and submandibular) reveals a striking degree of compositional overlap (24). As shown in Figure 5B, between 10 and 15 distinct polypeptides of a pancreatic zymogen granule membrane preparation exhibit identical isoelectric points and apparent molecular weights as well as overall mobility patterns relative to one another. The major overlap involves

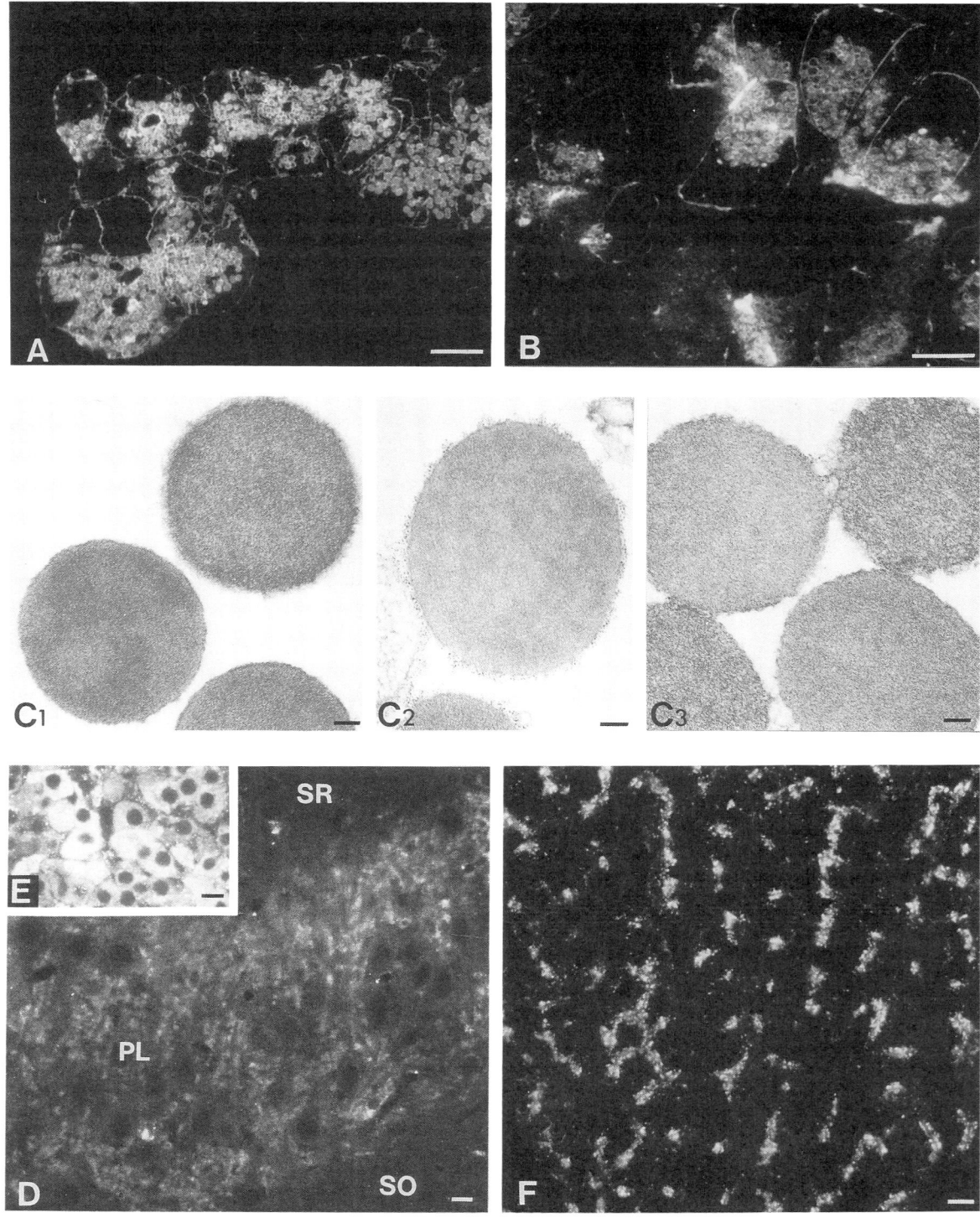

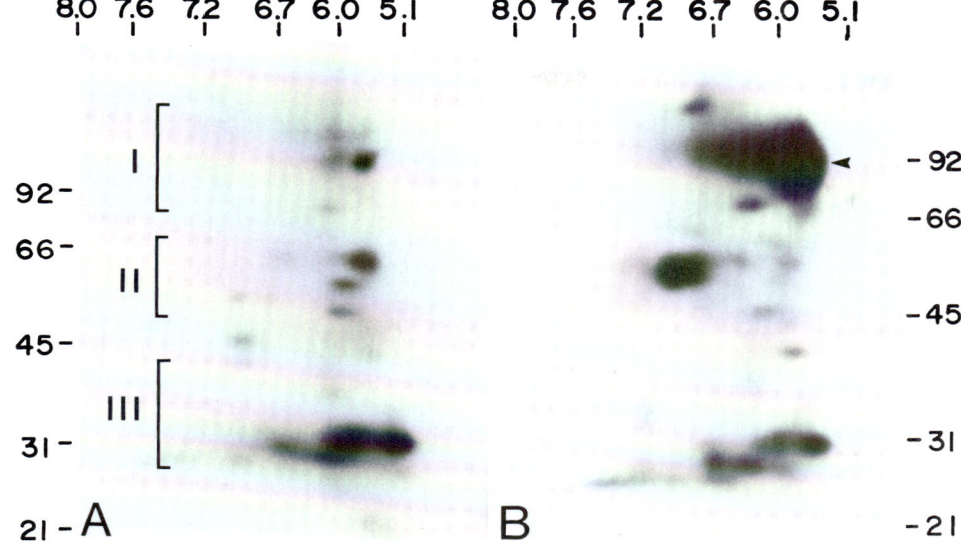

FIG. 5. Comparative two-dimensional sodium dodecyl sulfate–polyacrylamide gel electrophoretic analyses of purified exocrine secretory granule membrane polypeptides. Autoradiographic patterns obtained by parallel electrophoresis of radioiodinated (chloramine T) parotid and pancreas secretory granule membrane polypeptides. A: parotid pattern; 3 groups of polypeptides distinguished in the text on the basis of molecular weight are indicated by brackets (I–III). B: *solid arrow*, position of pancreatic zymogen granule membrane protein GP-2. Horizontal values, pH values; vertical values, molecular weight $\times 10^{-3}$. [From Cameron et al. (24).]

set 3, the prominent group of lower-molecular-weight polypeptides (although apparent identities can be observed within set 2 and, to a lesser extent, within set 1), and may represent one of the most constant features among the membranes of different granule types investigated so far.

The identities for selected parotid and pancreatic species have been confirmed by comparative two-dimensional peptide mapping of tryptic and chymotryptic digests of radioiodinated polypeptides (24). Furthermore, this approach has been extended over the entire molecular weight range of parotid membrane polypeptides to provide useful information concerning specific primary structures of individual species (these data argue against proteolysis as a major cause for the previously noted enrichment in low-molecular-weight species). Thus even though acinar cells in these exocrine glands package and store widely differing spectra of secretory proteins, they use a storage membrane of very similar composition.

Because the immunofluorescence studies presented in COMMON ANTIGENS IN SECRETORY MEMBRANES, p.

FIG. 4. Immunocytochemical demonstration of antigen distribution with heterologous antiserum to parotid secretory granule membranes. *A, B*: bars, 10 μm. Indirect immunofluorescent staining of 0.5-μm-thick frozen sections of parotid (*A*) and exocrine pancreas (*B*). Immunoreactivity is concentrated over secretion granule membrane profiles that are identified on the basis of their apical cytoplasmic location, size, and frequency. Staining is not detected in the basal cytoplasm, which contains abundant rough endoplasmic reticulum and mitochondria. *C1–3*: bar, 0.1 μm. Indirect immunolabeling with ferritin conjugates on isolated, aldehyde-fixed parotid secretion granules. *C2*, ferritin decoration reveals that the granule membrane antiserum recognizes determinants on the cytosolic aspect of the granule membrane. No ferritin decorates secretory granules when preimmune serum (*C1*) or antiserum specific for content polypeptides (proline-rich polypeptides (*C3*)) is substituted for the membrane antiserum (*C2*). *D–F*: bar, 10 μm. Indirect immunofluorescent staining of endocrine (epithelial and nonepithelial) and neural tissues with parotid granule membrane antibodies selected by adsorption for determinants on the cytoplasmic surface. For recovery of antibodies bound to intact parotid granules, granule suspensions were treated in 200 mM glycine (pH 2.8), and desorbed antibodies were recovered in supernate after sedimentation of granules through 0.75 M sucrose. Soluble IgG fractions served subsequently as the primary antiserum. *D*: 6-μm-thick frozen section of the CA3 region of rat hippocampus. *SR*, stratum radiatum; *PL*, pyramidal cell layer; *SO*, stratum oriens. Immunolabeling observed throughout the neuropile is particularly prominent adjacent to the pyramidal cell layer. Cell bodies and dendrites are unstained. *E*: 2-μm-thick frozen section of rat adrenal medulla. Chromaffin cells show a diffuse (or in some cases punctate) fluorescent pattern in the cytoplasm that is assumed to mark chromaffin granules. *F*: 4-μm-thick frozen section of rat liver. Label is associated with the bile canalicular surface and associated pericanalicular cellular organelles. [*A* and *B* from Cameron et al. (24).]

111, indicated antigenic overlap between exocrine and endocrine secretory membranes, the two-dimensional electrophoretic analysis has been extended to membranes of adrenal medullary chromaffin granules and to the membranes of Golgi vesicles of hepatocytes [Golgi fraction 1 (GF_1); (55)] that serve to convey very low density lipoproteins and other constituents to the blood plasma. For the chromaffin membranes, the pattern of compositional overlap is noticeable but limited, probably because these membranes are enriched in machinery involved in biogenic amine translocation and metabolism. However, for liver Golgi vesicles, the compositional similarity to exocrine storage granule membranes is quite striking (Fig. 6). In this case the two-dimensional patterns resemble one another, but the analogous polypeptides, especially the 20,000- to 35,000-M_r polypeptides, do not appear to have identical mobilities. The structural similarities between individual species from the two membrane sources have been substantiated by observations of considerable peptide overlap in comparative analyses of chymotryptic or tryptic digests (R. S. Cameron, unpublished observations). Thus the current picture suggests that the export carriers as disparate as regulated exocrine and constitutive endocrine secretion possess a subset of analogous polypeptide components that are likely to share domains of structural identity.

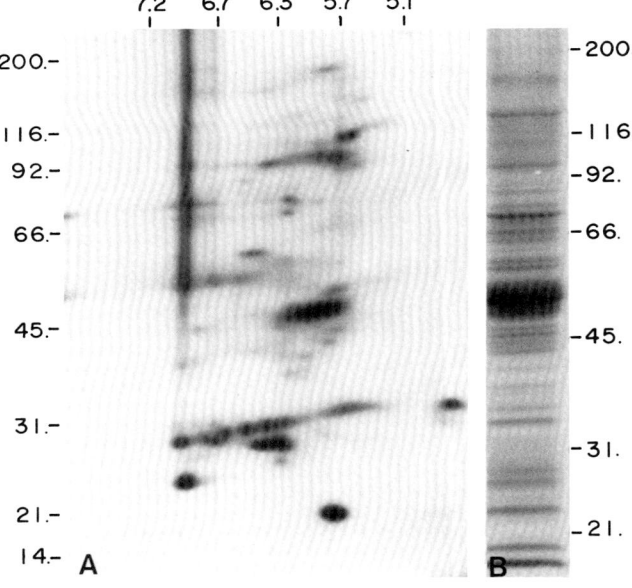

FIG. 6. Two-dimensional electrophoretic analysis of rat liver Golgi secretory vesicle membranes. Autoradiogram of radioiodinated (chloramine T) membrane polypeptides from a sodium carbonate–treated Golgi fraction 1 (GF1) fraction (A). Analogous polypeptides, especially 30,000-M_r polypeptides, have similar (but not identical) mobilities. Even though this group of polypeptides are prominent radiolabeled species, they may not be such major constituents of the Golgi vesicle membrane as signified by their minor contributions to a Coomassie blue–stained polypeptide profile (B). Horizontal values, pH values; vertical values, molecular weight × 10^{-3}. [A from Ehrenreich et al. (55).]

STORAGE FUNCTION AND BIOPHYSICAL PROPERTIES OF SECRETION GRANULES

Exocrine secretions consist of heterogeneous macromolecules diluted to varying degrees by fluid and electrolytes. Macromolecular and fluid secretions often derive from distinct sources that are under separate regulatory control (57, 66). Thus exportable proteins are stored within acinar cell granules at extremely high concentrations—on the order of several hundred milligrams per milliliter of internal aqueous space. Similar concentrations are achieved in adrenal medullary chromaffin granules (132). The latter endocrine granules have been studied extensively in relation to the storage function and constitute a useful model for the comparative analysis to properties of exocrine granules.

Granule Packaging, Stability, and Osmotic Behavior

Several observations suggest that intragranular packaging may vary between granule types. First, the spectrum of macromolecules stored in endocrine granules is considerably more limited than in exocrine granules, and accumulation of biogenic amines is a supplementary activity to macromolecular storage solely in certain endocrine granules. Second, there is apparently a large range of solute concentrations among different granule types. Chromaffin granules contain ~750 mM total solutes, including 550 mM catecholamine, 120 mM ATP, ~20 mM ascorbate, 20 mM divalent cations, low (less than ~10 mM) Na^+ and K^+, and 3.3 mM protein (as calculated from a measured internal aqueous space of ~5 μl/mg protein, corresponding to 200 mg/ml protein of average M_r ~ 60,000) (see ref. 132 for a review). By contrast, exocrine granules, particularly isolated parotid granules, contain (as far as is known) less than ~50 mM solutes, including no catecholamines, <10 μM ATP (6), ≤3 mM ascorbate (216), ~10 mM divalent ions (204), <10 mM Na^+ and K^+ (6), and ~12.5 mM protein [as calculated from a measured internal aqueous space of 2 μl/mg protein (5), corresponding to 500 mg/ml protein of average M_r 40,000].

Third, there is no evidence for insoluble complexes of sequestered secretory products in chromaffin granules, even though internal solute concentration is almost three times higher than the osmolality of the cytoplasm. In fact, ^{31}P nuclear magnetic resonance (of intragranular ATP) and other biophysical studies suggest that the chromaffin granule interior is a highly nonideal solution (reviewed in ref. 132). By contrast, stable protein-rich cores have been reported in demembranated mast cell granules (199) and prolactin-containing granules (70). Also, electrostatic complexation [possibly involving sulfated proteoglycans (158)] has been invoked as a basis for the aggregation of pancreatic zymogen granule content (95) and may contribute to the ordered substructure of stored salivary secretions (193).

Despite these major compositional differences in content, osmotic forces are an important consideration in relation to the stability of both endocrine and exocrine granules (132, 174). Isolated granules [chromaffin granules (28, 132), neurosecretory granules from posterior pituitary (167, 173), and parotid granules (5)] behave as osmometers, changing the volume of their internal aqueous space in response to changes in external osmotic strength; in most cases internal osmotic pressure will cause granule rupture in substantially hyposmotic media.

To minimize the osmotic activity of highly concentrated macromolecules containing multiple charged residues (Donnan effect), granules may employ mechanisms such as greatly restricting selected ionic permeabilities, as in amine-storing granules (28), or storing a range of differently charged molecules with a relatively low aggregate net fixed charge, as in exocrine granules (5). In the latter case, protein-protein interactions and possible complexation with selected ions (e.g., H^+, Ca^{2+}) may be used to achieve optimal packaging. What is difficult to rationalize in the case of parotid granules is the simultaneous evidence for low internal solute concentration and negligible net fixed internal charge, yet osmotically driven lysis under mildly hypotonic conditions. Perhaps the granule membrane permeability to selected ions is not so restricted (which potentially could contribute to both granule lysis and an underestimation of the intragranular solute content).

Ion Permeation of Granule Membranes

Studies of both isolated exocrine granules and adrenal chromaffin granules have shown that granule lysis in isosmotic media is generally higher in ion-containing media than in sucrose and that the choice of anion is a major factor governing the extent of release of content (5, 43, 132). Lysis increases with the lyotropic series of anions (5, 43, 49); in addition, comparable evidence for graded anion permeabilities was obtained for parotid granules (6) by measuring inside-negative diffusion potentials with tracer concentrations of $^{86}Rb^+$ as a probe in the presence of valinomycin. Thus granule membranes in general probably contain a selective anion transporter (146).

Cation permeabilities are less well studied in secretion granules. At least two reports show that chromaffin granules (and their vesiculated membranes) are only slightly permeated by Na^+, K^+, H^+, Ca^{2+}, and Mg^{2+} (28, 152). Pancreas (43) and parotid (6) exocrine granules do not show differences in their levels of lysis when Na^+ and K^+ salts are compared; however, diffusion potential measurements have not been carried out with these or most other cations. Hydrogen diffusion potentials, however, have been examined in parotid granules and have been found to be quite low (6). Evidently more detailed information concerning cation permeation of exocrine granules is needed, especially in relation to the stability properties discussed here.

Intragranular pH and Buffering Capacity

While exocrine and neuroendocrine granules share certain osmotic and ionic permeability characteristics, the existing data indicate a difference in their respective internal pH values. An acidic interior (pH ≤6.2) has been described for chromaffin, platelet, mast cell, endocrine pancreatic, anterior pituitary, and neurosecretory granules (28). Notably, neither the absence of ATP nor the presence of protonophores significantly changes the chromaffin granule pH (28). Evidently it is strongly buffered by acidic granule contents (primarily ATP and secondarily the chromogranins) with a buffer capacity of 70–210 mM H^+ per pH unit (132).

In contrast, the intragranular pH of rat parotid secretion granules is nearly neutral (~6.8) (5). This value is correlated with a much lower level of internal ATP (6) and a pI of the major stored secretory protein (α-amylase) of 6.8. The intragranular buffer capacity is 85–140 mM H^+ per pH unit (5); however, the identity of content species that are responsible for this high level of buffering is uncertain. Major contributions to intragranular pH by secretory protein are suggested indirectly through experimental manipulation of the composition of stored protein. In rats, repeated injection of isoproterenol, which alters the spectrum of parotid secretory proteins to an overall composition that is enriched in considerably more basic (pI >10) species, causes a rise in intragranular pH to ~7.7 (4). The measurements made in vitro on both parotid granule preparations are supported by microscopic observations of intact cells treated with acridine orange. This fluorescent weak base does not concentrate in parotid granules in situ, signifying an intragranular pH roughly comparable with that of the cytosol (4, 5).

The presence of a large buffering capacity may promote packaging and stability of the granule content, regardless of the internal pH value. Thus concentration of stored products could result from the concerted operation of electrogenic H^+ uptake (driven by H^+-ATPase; see next section) with internal H^+ buffering and electrically dissipative efflux of permeant cations leading to concomitant extrusion of water. Such considerations are supported indirectly by converse observations of chromaffin and parotid granule lysis elicited by exogenously added monovalent cations in the presence of exchange ionophores (6, 67). In Na^+- or K^+-containing media, nigericin- (or monensin-) treated chromaffin granules undergo rapid lysis by a countertransport mechanism that is independent of the species of anion present (28, 67, 152). The extent of lysis matches or exceeds that observed with K^+-Cl^- cotransport facilitated by valinomycin and is suppressible by using either acidic or hyperosmotic media (67). Thus the interpretation is that there are major osmotic consequences when buffered (bound) internal H^+ is exchanged for osmotically active (unbound) cations.

Hydrogen-ATPase Activity

All endocrine granules examined to date have an inwardly directed transmembrane H^+ gradient maintained by Mg-ATP–driven H^+ pump. For intact granules or isolated membrane vesicles devoid of content, the H^+ pump can create an internal acidification or inside-positive membrane potential, depending, respectively, on whether permeant or nonpermeant anions are present in the medium (132, 166). When isolated parotid granules were examined directly for H^+-pump activity, they were found not to acidify nor to generate more than 6-mV change in transmembrane potential (6). This relative lack of H^+-ATPase in the parotid granule fraction suggests an important difference between mature exocrine and endocrine granules. This difference may largely reflect the absence of amine storage capabilities in mature exocrine granules, since electrogenic H^+-pump activity appears to play a central role in promoting uptake and storage of biogenic amines (132, 166).

Although our initial investigations and those of De Lisle and Hopfer (43) have failed to detect H^+-translocating ATPase activity in pancreatic zymogen granules, the near absence of H^+ pumping is not an invariant property of the exocrine storage compartment. The change in parotid granule content that follows chronic isoproterenol administration (see previous section) is accompanied by appearance of ATP-dependent acidification in granules (4). The role of this activity is not understood but may reflect a sustained action in macromolecular storage. In relation to granule origins in the trans-Golgi in which H^+-ATPase activity may be present (3, 74), the conditional nature of exocrine granule H^+ pumping raises the possibility for further exploration that retention of activity in post-Golgi compartments can be differential.

INTERACTIONS WITH CYTOPLASMIC PROTEINS

A variety of studies have examined the roles of microtubule and microfilament systems in mediating secretory organelle translocation and proximity to the site of discharge. The organelles (or their membranes) have been used as an affinity matrix for identifying the binding of selected cytosolic components. Although this discussion emphasizes results obtained in exocrine secretory tissues, certain considerations rely on studies from a broader range of cell types.

Possible Roles of Microtubules

Observations by electron microscopy and indirect immunofluorescence (with tubulin antibodies) show microtubule organizing centers in the centrosphere region and in close association with the site of origin of secretory membranes in the Golgi complex (161). From this point microtubules are characteristically concentrated along the axis of directed secretion. Particularly in neuronal axons (180) and endocrine cells (116), a role in conveying packaged products to the site of discharge has been suggested. Recent studies have provided not only support for specific, ATP-mediated association between microtubules and either insulin-containing granules or large neuronal vesicles in vitro (109, 191) but also direct evidence for bidirectional microtubule-based organelle movement in extruded neuronal axoplasm (109, 126, 176, 200, 201). In the latter case, structural bridges between vesicles and microtubules have been identified (71, 126) that are similar to those observed in axons in situ (81).

In cells that are specialized for directed discharge in response to external stimuli, microtubules do not appear to expedite the mobilization or discharge of stored secretion. In neurons this observation is supported by the absence of microtubules at neurotransmitter storage sites in synaptic terminals (149). However, the conclusion can be generalized to cells of less extreme spatial polarity, since a substantial body of evidence indicates that microtubule-disruptive drugs such as colchicine do not appreciably affect discharge after secretory stimulation in exocrine or endocrine cells (16, 22, 108, 144, 209). By contrast, constitutive vesicular transport is either defocused (160) or, in large part, inhibited in response to colchicine administration (56, 156, 157, 195). If newly synthesized secretory proteins are examined selectively in exocrine and endocrine cells, colchicine treatment alters their ultimate fate, as indicated by inhibition of their discharge (16, 22, 144, 209). This behavior is correlated with a major disorganization of the intracellular transport pathway, particularly at the level of the Golgi complex (144, 145) but possibly also involving perturbation of exit from the endoplasmic reticulum (22, 156). Thus these studies provide indirect evidence for a possible general role of microtubules in organizing and expediting relocation of vesicular carriers in the cascade of interactions that comprise the secretory pathway.

Microfilamentous Networks and Secretory Organelle Interactions

The evidence for participation of microfilamentous structures in secretion is fairly limited, although particular attention to their possible role stems from the analogy between excitation-contraction coupling in muscle and stimulus-secretion coupling in cells that retain internal granule stores (187). Thus a number of studies have shown that structural perturbants of actin filament organization, such as the cytochalasins, can influence (enhance at low dose and inhibit at high dose) stimulus-dependent secretion (10, 23) and that actin and myosin are closely apposed to exocrine granules in situ (11). Furthermore, isolated secretion granules, particularly those from endocrine cells, are able to bind F-actin (63). The nature and specificity of actin binding are not well characterized; it may (7) or may not (63) involve membrane-associated α-actinin.

Two recent studies addressing more directly the functional role of actin suggest that actin-based filaments merit more serious consideration as possible mediators of organelle translocation and secretion. First, actin-containing fibers have been shown to support rapid and oriented particle movement in fractured *Nitella* preparations (184). Second, actin mutants of yeast (133) exhibit the phenotype of previously described *sec* mutants (134), characterized by intracellular accumulation of fully glycosylated invertase in presumptive secretory vesicles as a result of an apparent disorganization of the budding site and export machinery.

In exocrine secretory systems, the terminal web situated beneath the apical plasma membrane appears to present a physical barrier to the interaction of the membranes of storage granules and the apical cell surface. This organization is particularly clear in the parotid gland and pancreas, where immunofluorescence studies have identified the terminal web as the principal site for the concentration of actin, myosin, and tropomyosin (51). Thus reorganization or penetration of this barrier appears to constitute an essential step for granule discharge. A similar suggestion has been made for adrenal chromaffin cells, where α-fodrin was found to undergo rearrangement on external stimulation (148).

The present evidence appears to favor a major role for microtubules in maintaining the overall spatial organization of membrane-bounded structures comprising the secretory pathway in polarized cells. Actin-based microfilaments, on the other hand, may play a more prominent role in organelle translocation in actively growing or motile cells, where well-defined functional polarity is either transient or variable. In established polarized systems, microfilaments may be more intimately concerned with regulating the discharge process by controlling contact between the fusion partners.

Other Polypeptides Binding to Secretory Membranes

Because second-messenger systems have been implicated as regulators of stimulus-secretion coupling (66, 100, 101, 122, 142, 154, 175), investigators have examined associations between secretory membranes and cytosolic regulatory proteins. Thus calmodulin (45, 70, 87), protein kinase C–like activity (17, 21, 190), phosphatidylinositol-specific phospholipase C (37), and the regulatory subunit(s) and activities of cAMP-dependent protein kinases (21, 111, 198) have all been reported to be present in various secretion granule fractions.

Other studies have identified a variety of cytosolic proteins that interact with granule membranes in a calcium-specific manner. This set includes a group of 22 "chromobindins" with varying Ca^{2+} requirements for membrane association (38) and 4 polypeptides that require calmodulin for membrane interaction (68). Thus far no immunocytochemical studies have provided a convincing, high-resolution localization of any of this group of proteins to the secretory machinery in situ, and it is unclear whether the capacities of certain proteins for calcium-specific self-association [synexin (39); calelectrin (202)] and binding to phospholipid vesicles [synexin (86); calpactin (73)] are relevant to secretory processes.

For the most part these investigations, particularly of interactions with chromaffin granule membranes, have been duplicated to only a limited extent in other secretory systems. Surface interactions potentially related to the regulation of exocrine secretion (as well as granule formation and secretory membrane retrieval) remain an area for investigation.

Stimulus-Enhanced Phosphorylation of Membrane and Associated Proteins

Examinations of secretagogue-dependent protein phosphorylation reactions involving polypeptides that are integral or associated components of secretory membranes have been quite limited. Three recent reports used the in situ phosphorylation approach on stimulated (versus unstimulated) cells. The first report involves a 29,000-M_r pancreatic zymogen granule membrane protein (147), apparently the same species that exhibits Ca^{2+}-specific phosphorylation in vitro (21) and potentially one of the common membrane polypeptides shared by all exocrine secretion granules studied so far (Fig. 5). This species may be related to the phorbol ester–stimulatable 29,000-M_r membrane phosphoprotein of insulinoma cells (17). The second report involves K^+ or cholinergic agonist-induced phosphorylation of 27,000-M_r polypeptide (chromobindin 9) that is known to interact with chromaffin granules in a Ca^{2+}-dependent manner (124) and may be related to calpactin I, the calcium-, actin-, and phospholipid-binding protein (73). In contrast, a third study has suggested that a 24,000-M_r parotid granule membrane polypeptide exhibits isoproterenol-dependent decreased phosphorylation in situ (183).

Probably the most extensively studied regulatory phosphorylation reactions at the level of secretory membranes involve synapsin I, the extrinsic protein associated with synaptic vesicles (41). Synapsin I is a substrate for both calmodulin- and cAMP-dependent protein kinases (90); its association with vesicle membranes decreases significantly with calmodulin kinase–mediated phosphorylation (91); its phosphorylation in situ can be altered by stimuli that evoke discharge of neurotransmitters (131); and microinjection of calmodulin kinase II into neurons seems to facilitate neurotransmitter release (114). Although these properties, in combination, form an attractive model for mechanisms of secretory regulation involving stimulus-dependent changes in restrictions to exocytotic fusion, two factors should guide any generalizations to other cell types. First, synaptic vesicles may be questionable as representatives of what has been defined as secretory membrane. They function primarily as a locally recycling neurotransmitter storage compartment within the nerve terminal without return to the Golgi complex. It may be more appropriate to view them as analogous to endocytic membrane that is restrained from access to the cell surface, thereby creating an internal reserve (97) primed for stimulated surface fusion. Possible exocrine analogies for such local, conditional recycling may include vesicles of gastric parietal cells (62) and kidney and urinary bladder epithelia (96, 177). In these cases, stimulus-dependent insertion of membrane components into the cell surface is used to regulate the extracellular ionic environment rather than to export secretory macromolecules.

Second, within neurons, synapsin is not associated significantly with peptide-storing granules (130); release for these granules may be less focal than for synaptic vesicles, and separate mechanisms may control their membrane interactions. Even though synapsin is not a general component of nonneuronal secretory cells, the recent observation of structural homologies between synapsin and the erythrocyte peripheral membrane protein 4.1 (8) suggests that it may be worthwhile to search for functional equivalents of synapsin in different kinds of secretory cells as a first step in attempts to generalize from neuronal models of regulated secretion.

EXOCYTOSIS

During exocytosis, secretory vesicles selectively fuse with the plasma membrane, thereby externalizing their contents without apparent loss of other cellular components. Very little is known about the mechanism of membrane interaction and fusion, except that in nearly all cases it requires energy (60, 94) in the form of ATP (100) and, where internal storage is involved, its rate is modulated by second-messenger systems. In cell types where internal storage is unusually large and direct access to the cell surface is restricted, secretory stimulation can result in compound exocytosis, in which the surface of a discharged granule subsequently serves as a fusion site for other granules. Thus secretory release proceeds inwardly as a cascade, without relocation of each granule to the periphery. Compound exocytosis has been observed in a large variety of cell types (e.g., refs. 1, 92, 162), and its occurrence in some cases is too rapid to involve lateral diffusion of plasma membrane effector molecules to the newly formed granule discharge site (33). Thus the granule membrane itself appears to contain all the necessary membrane fusogens whose activation toward other storage granules is contingent on becoming continuous with the cell surface.

The efficiency and selectivity of exocytotic events are reminiscent of other recognition-related membrane processes involving receptors, such as the entry of animal viruses into cells. In this case separate polypeptides (172) or separate domains of a single polypeptide (212) are responsible for binding versus fusion, and activation of the fusion domain is conditional, requiring proteolysis and altered pH (208). No counterparts that might act from the cytoplasmic aspects of secretory membranes have been discovered, although recent circumstantial evidence suggests that proteolysis might participate in facilitating exocytosis (36). If secretion does proceed by a receptor-mediated mechanism, then the capacity for compound exocytosis implies that the granule membrane probably contains the information for both recognition and membrane fusion.

In considering the membrane-membrane interactions involved in exocytosis, at least three potential processes requiring energy input can be identified. One is the reduction of distance separating the prospective fusion partners. This may involve relocation of vesicles over substantial distances. Additionally it may be necessary to induce cytoskeletal rearrangements to achieve access to the cell surface (148). Furthermore, rather extensive studies using phospholipid bilayer membranes have identified the water of hydration of phospholipid headgroups as a major energetic barrier to the intimate membrane interaction that is required for fusion (155). This may be an important consideration because the compositional analyses (see *Membrane Lipids*, p. 109) have identified a high lipid content of secretory membranes.

A second potential requirement for energy utilization in effecting exocytosis could be to facilitate the phosphorylation of a regulatory molecule, recognition protein, or secretory membrane fusogen. In the latter instance, this mechanism could conceivably entail stabilization of a fusogenic polypeptide conformation or activation of enzyme machinery that generates lipid fusogens (e.g., phospholipase C or A_2). Either of these processes might promote a reduction in the energy barrier of hydration, but there is no precedent for operation of either mechanism in fusion events.

A third site for possible involvement of energy in the discharge process is in the production of transmembrane osmotic gradients that in certain systems (fusion of phospholipid vesicles with planar bilayers, discharge of cortical granules in in vitro preparations from sea urchin oocytes) drive membrane fusion (35, 125, 217). The osmotic gradient increases surface free energy that may be required to initiate lipid rearrangement in the region of bilayer contact. Although mounting evidence argues against the involvement of ATP-driven H^+ pumping in promoting exocytosis (6, 9, 85), ATP may exert a regulatory role on the permeability of other ions that may increase internal osmotic activity.

Finally, although experimental availability has resulted in an initial focus on secretory membranes involved in storage and stimulus-dependent release, possible compositional differences or interactions that distinguish membranes involved in constitutive secretion must be addressed. According to the hypothesis that all secretory mechanisms are variations on a common theme, membranes under stimulatory control selectively could be exposed to additional factors that are responsible for preventing constitutive surface interactions. This inhibition could be released by second-messenger activation. Alternatively, these storage membranes could have decreased concentrations of discharge-promoting components in relation to membranes that operate constitutively. In this case the modulatory action of second-messenger systems would involve either increasing the concentration of the necessary components or changing the sensitivity such that the levels present become sufficient for evoking release.

RECYCLING, TURNOVER, AND SORTING FROM OTHER VESICULAR TRAFFIC

Discharge from secretory cells results in export of storage volume but retention of the bounding storage membrane. The continuity established between the secretory membrane and the cell surface is only transient, since massive export of stored secretion is not accompanied by a sustained increase in cell surface area (1, 141). The secretory membrane that returns to the cytoplasm must be characterized by a higher surface-volume ratio than that of the original storage container; small vesicles and/or flattened sacs have been implicated in this action (80, 113). In many cases the principal mechanism for this compensatory membrane reinternalization appears to be analogous structurally to receptor-mediated endocytosis, as signified by the increased incidence of coated pits appearing on the former storage membrane after discharge (13, 50, 80). Both freeze-fracture studies (42) and immunofluorescence localization of granule membrane-specific antigens (153) suggest that the internalization is selective for the return of secretory membrane that normally resides within the cell.

Stereological studies carried out on isoproterenol-stimulated parotid tissue indicate that the rate of secretory membrane retrieval depends on the presence of extracellular calcium (102).

Fate of Reinternalized Secretory Membrane

The idea that secretory membrane is conserved and reutilized in multiple rounds of both storage and discharge (see ref. 141 for a review) has been supported by observations suggesting a considerably slower rate of synthesis of granule membrane polypeptides relative to the exported secretion (32) and a prolonged turnover time for membrane components (119, 214). However, these earlier studies, particularly those involving exocrine tissues, were limited by uncertainties concerning the purity of the fractions used for investigation. Reevaluation with fractions of documented purity (25) is essential to make legitimate comparisons to cell surface membranes undergoing endocytosis, where extensive reutilization is required (186).

The pathways taken by reinternalized granule membrane have been followed qualitatively with probes that either are fluid-phase markers or exhibit nonspecific adsorption to the discharged secretory membranes. A range of results has been obtained that to some extent stems from properties (e.g., transport kinetics, size of secretory membrane reserves, and storage capacity) of the particular cell type under investigation but also appears to reflect in part the chemical nature, quantity, and valence of the exogenous probe used. Accordingly, dextran, a fluid-phase marker, identifies a reinternalization pathway that rapidly and efficiently relocates the probe exclusively to all levels of the Golgi complex in the exocrine pancreas (80) but only partially to Golgi (favoring the transcisternae) and partially to lysosomes in the parotid (79). Furthermore, cationized ferritin (which adsorbs to membranes) relocates to both Golgi and lysosomes in parotid, exocrine pancreatic, and pituitary cells (58, 79, 113), whereas anionic ferritin and horseradish peroxidase primarily mark pathways to lysosomes in the same cell types (58) but are delivered to the Golgi complex elsewhere (75). More recently, specific membrane antibody probes have been used to trace internalization in secretory cells. In adrenal chromaffin cells the pathway appears to be divided between routes leading to lysosomes and to the Golgi complex (143), whereas in constitutively secreting myeloma cells, nearly all vesicular traffic is directed to the Golgi and may even favor an initial cis and medial cisternal location (215). Thus the limited view so far suggests that reutilization of secretory membrane is variable and may be nearly complete in cases where the major export function is constitutive.

Most of the evidence obtained by examining vesicular traffic patterns with exogenous probes suggests that it is prudent to retain a distinction between surface membrane that is transiently relocated (either inward during the endocytic cycle or to a distinct region of the cell surface during transcytosis) and secretory membrane that is reinternalized after discharge. Indeed, neither endocytosed asialoglycoproteins (targeted to lysosomes) nor secretory immunoglobulin A's (IgAs) (targeted to the bile canaliculus) in hepatocytes trace a route to the Golgi (88, 203). Similarly, nonspecific probes undergoing enhanced basolateral endocytosis after secretory stimulation of parotid or pancreatic acinar cells do not appear to gain access to the exocrine secretory pathway (137).

In contrast to this general picture suggesting minimal interactions between the secretory and endocytic pathways, three apparent exceptions may clarify the nature of limited or conditional crossover. First, radioactive ligands for selected cell surface receptors have been identified in both endocrine storage granules [a gonadotropin-releasing hormone analogue in pituitary gonadotroph granules (53)] and exocrine granules [radioiodinated insulin in pancreatic zymogen granules (40)]. Unfortunately the pathway and distribution of receptors have not been traced, so it is not possible to judge to what extent the granule compartment constitutes a bona fide receptor depot. The second type of study suggests the colocalization of a secretory protein, albumin, and an endocytic marker, asialoglycoprotein receptor, after ammonium chloride treatment of hepatoma cells (188). Although insufficient data are presented to judge what conditions lead to colocalization, it is possible that intracompartmental pH is an important determinant of the extent of crossover. Finally, recycling of transferrin receptors and other glycoproteins through a mannosidase I–containing (presumably Golgi) compartment has been identified in erythroleukemia cells (181). Because this process is slow, it is possibly tracing retrieval of secretory membrane involved in cell surface biogenesis and repair. Alternatively, it may signify that there is a continuous limited crossover of import and export pathways.

Secretory Membrane as a Plasmalemmal Protein Carrier

Because secretory membranes make regular or intermittent contact with the cell surface, they potentially could serve as carriers for vectorial delivery of newly synthesized plasmalemmal constituents. This issue is significant quantitatively in growing cells between division, in infected or transformed cells that are producing viral membrane proteins, and in secretory cells that produce a class of short-lived surface macromolecules. In the first group the use of shared carriers for secretory and plasmalemmal proteins seems probable, because certain yeast *sec* mutants having the abnormal phenotype of accumulated secretory vesicles exhibit a defect in the appearance of both secretory invertase and cell surface polypeptides (135). In the second group comprising virally infected or transformed cells, many studies indicate that the post-Golgi transits of constitutively secreted proteins and viral membrane glycoproteins to the cell surface occur at similar rates (see ref. 99 for a review). Furthermore, these species may, in part, occupy common post-Golgi carriers (189). In contrast, secretory proteins that are packaged in mature storage granules are segregated to a very large extent from either of these species in mature storage granule subfractions (76, 127).

The final group includes normal differentiated cells, particularly epithelial cells with multiple targeting pathways for membrane and secretory proteins. In the case of hepatocytes and lactating mammary cells, the membrane receptor for secretory immunoglobulins is synthesized rapidly and continuously and targeted to the basolateral plasma membrane where it binds IgA and subsequently undergoes transcytosis and proteolysis at the cell apex (88, 128, 182). For both cell types the secretory products (which are concentrated but not stored) and membrane polypeptides probably travel to the cell surface in distinct vesicular carriers (182, 192).

In the case of exocrine acinar cells, γ-glutamyltransferase (particularly structurally related high-molecular-weight polypeptides) and a pancreas-specific glycoprotein (GP-2) are common granule- and apical cell surface components (24, 25, 29, 171) that exhibit high rates of biosynthesis. Initial studies have suggested that the appearance of newly synthesized glycoproteins at both cellular locations is nearly simultaneous (78; R. S. Cameron, A. K. Ma, and J. D. Castle, unpublished observations), and so far it is uncertain whether the species in storage granule membranes can serve as precursors to forms present at the cell surface. Evidently, further explorations of post-Golgi pathways for membrane proteins in these cells are warranted to delineate traffic patterns for both rapidly synthesized and more stable polypeptide components of recycling membranes.

Secretory Membrane and Sorting Sites

Post-Golgi targeting of secretory proteins may not be confined to a single cellular site, and, especially in cells where storage is a major function, the operations of concentration-packaging and sorting may occur in the same compartment(s). Certain activities directly involved in (or supporting) terminal glycosylation exist not only in the trans-Golgi cisternae and trans-tubular network (165) but also may conditionally extend into the immature granule compartment (137, 164). Similarly, activities involved in posttranslational processing of precursor polypeptides mostly appear to be nonintegral membrane or even soluble species (61, 64, 72). Thus they can be transported along with their substrates and can act either at the level of Golgi cisternae or, as has been elegantly shown (19, 138), within the maturing granule compartment. Apparently these operations are designed to occur over a range of functionally continuous late- and post-Golgi compartments, and to the extent that vesicular traffic from these compartments is sustained, a possible role in membrane and secretory sorting ought to be considered. Consequently, the branch point for distinct export pathways may not be exclusive or focal and may differ from cell type to cell type.

The final impression from these considerations is that secretory membrane has certain unique characteristics related to secretory function. However, the extent of refinement of its composition and the location within the distal transport pathway in different cell types are functions of the degree of specialization for internal storage and the kinetics of processing and packaging in relation to release. The generally intracellular location of this vesicular carrier system distinguishes it from the functionally (and possibly compositionally) analogous, but oppositely oriented, endocytic carrier system. In the latter case, similar variations in steady-state location and compositional refinement according to specific cell function may be anticipated.

We are grateful to Dr. Anna Castle for her scientific and editorial advice and for assistance in preparing the manuscript.

Research on exocrine glands was supported by National Institutes of Health Grants GM-26524 and AM-29868.

REFERENCES

1. AMSTERDAM, A., I. OHAD, AND M. SCHRAMM. Dynamic changes in the ultrastructure of the acinar cell of the rat parotid gland during the secretory cycle. *J. Cell Biol.* 41: 753–773, 1969.
2. AMSTERDAM, A., M. SCHRAMM, I. OHAD, Y. SALOMON, AND Z. SELINGER. Concomitant synthesis of membrane protein and exportable protein of the secretory granule in rat parotid gland. *J. Cell Biol.* 50: 187–200, 1971.
3. ANDERSON, R. G. W., AND R. K. PATHAK. Vesicles and cisternae in the *trans* Golgi apparatus of human fibroblasts are acidic compartments. *Cell* 40: 635–643, 1985.
4. ARVAN, P., AND J. D. CASTLE. Isolated secretion granules from parotid glands of chronically stimulated rats possess an alkaline internal pH and inward-directed H^+ pump activity. *J. Cell Biol.* 103: 1257–1267, 1986.
5. ARVAN, P., G. RUDNICK, AND J. D. CASTLE. Osmotic properties and internal pH of isolated rat parotid secretory granules. *J. Biol. Chem.* 259: 13567–13572, 1984.
6. ARVAN, P., G. RUDNICK, AND J. D. CASTLE. Relative lack of H^+ translocase activity in isolated parotid secretory granules. *J. Biol. Chem.* 260: 14945–14952, 1985.
7. BADER, M. F., AND D. AUNIS. The 97-kd α-actinin-like protein in chromaffin granule membranes from adrenal medulla: evidence for localization on the cytoplasmic surface and for binding to actin filaments. *Neuroscience* 8: 165–181, 1983.
8. BAINES, A. J., AND V. BENNETT. Synapsin I is a spectrin-binding protein immunologically related to erythrocyte protein 4.1. *Nature Lond.* 315: 410–413, 1985.
9. BAKER, P. F., AND D. E. KNIGHT. Chemiosmotic hypothesis of exocytosis: a critique. *Biosci. Rep.* 4: 285–298, 1984.
10. BAUDUIN, H., C. STOCK, D. VINCENT, AND J. F. GRENIER. Microfilamentous system and secretion of enzyme in the exocrine pancreas. Effect of cytochalasin B. *J. Cell Biol.* 66: 165–181, 1975.
11. BENDAYAN, M. Ultrastructural localization of cytoskeletal proteins in pancreatic secretory cells. *Can. J. Biochem. Cell Biol.* 63: 680–690, 1985.
12. BENDAYAN, M., J. ROTH, A. PERRELET, AND L. ORCI. Quantitative immunocytochemical localization of pancreatic secretory proteins in subcellular compartments of the rat acinar cell. *J. Histochem. Cytochem.* 28: 149–160, 1980.
13. BENEDECZKY, I., AND A. D. SMITH. Ultrastructural studies on the adrenal medulla of golden hamster: origin and fate of secretory granules. *Z. Zellforsch. Mikrosk. Anat.* 124: 367–386, 1972.
14. BIEGER, W., J. SEYBOLD, AND H. F. KERN. Studies in intracellular transport of secretory proteins in the rat exocrine pancreas. V. Kinetic studies on accelerated transport following caerulein infusion. *Cell Tissue Res.* 170: 203–219, 1976.
15. BLASCHKO, H., H. FIREMARK, A. D. SMITH, AND H. WINKLER. Lipids of the adrenal medulla. Lysolecithin, a characteristic constituent of chromaffin granules. *Biochem. J.* 104: 545–549, 1967.
16. BOYD, A. E., III, W. E. BOLTON, AND B. R. BRINKLEY. Microtubules and beta cell function: effect of colchicine on microtubules and insulin secretion in vitro by mouse beta cells. *J. Cell Biol.* 92: 425–434, 1982.
17. BROCKLEHURST, K. W., AND J. C. HUTTON. Involvement of protein kinase C in the phosphorylation of an insulin-granule membrane protein. *Biochem. J.* 220: 283–290, 1984.
18. BROCKMEYER, T. F. Isolation and Subfractionation of a Zymogen Granule Fraction From the Rat Pancreas. New Haven, CT: Yale Univ., 1981. PhD thesis.
19. BROWNSTEIN, M. J., J. T. RUSSELL, AND H. GAINER. Synthesis, transport, and release of posterior pituitary hormones. *Science Wash. DC* 207: 373–378, 1980.
20. BUCKLEY, K., AND R. B. KELLY. Identification of a transmembrane glycoprotein specific for secretory vesicles of neural and endocrine cells. *J. Cell Biol.* 100: 1284–1294, 1985.
21. BURNHAM, D. B., P. MUNOWITZ, N. THORN, AND J. A. WILLIAMS. Protein kinase activity associated with pancreatic zymogen granules. *Biochem. J.* 227: 743–751, 1985.
22. BUSSON-MABILOT, S., A.-M. CHAMBAUT-GUERIN, L. OVTRACHT, P. MILLER, AND B. ROSSIGNOL. Microtubules and protein secretion in rat lacrimal glands: localization of short-term effects of colchicine on the secretory process. *J. Cell Biol.* 95: 105–117, 1982.
23. BUTCHER, F. R., AND R. H. GOLDMAN. Effect of cytochalasin B and colchicine on α-amylase release from rat parotid tissue slices. Dependence of the effect on $N^5,O^{2'}$-dibutyryl adenosine $3',5'$-cyclic monophosphate concentration. *J. Cell Biol.* 60: 519–523, 1974.
24. CAMERON, R. S., P. L. CAMERON, AND J. D. CASTLE. A common spectrum of polypeptides occurs in secretion granule membranes of different exocrine glands. *J. Cell Biol.* 103: 1299–1313, 1986.
25. CAMERON, R. S., AND J. D. CASTLE. Isolation and compositional analysis of secretion granules and their membrane subfraction from the rat parotid gland. *J. Membr. Biol.* 79: 127–144, 1984.
26. CAMPBELL, C. R., J. B. FISHMAN, AND R. E. FINE. Coated vesicles contain a phosphatidylinositol kinase. *J. Biol. Chem.* 260: 10948–10951, 1985.
27. CARO, L. G., AND G. E. PALADE. Protein synthesis, storage, and discharge in the pancreatic acinar cell. A radioautographic study. *J. Cell Biol.* 20: 473–495, 1964.
28. CARTY, S. E., R. G. JOHNSON, AND A. SCARPA. H^+-translocating ATPase and other membrane enzymes involved in the accumulation and storage of biological amines in chromaffin granules. In: *The Enzymes of Biological Membranes. Membrane Transport* (2nd ed.), edited by A. N. Martonosi. New York: Plenum, 1985, vol. 3, p. 449–495.
29. CASTLE, J. D., R. S. CAMERON, P. L. PATTERSON, AND A. K. MA. Identification of high molecular weight antigens structurally related to gamma-glutamyl transferase in epithelial tissues. *J. Membr. Biol.* 87: 13–26, 1985.
30. CASTLE, J. D., J. D. JAMIESON, AND G. E. PALADE. Radioautographic analysis of the secretory process in the parotid acinar cell of the rabbit. *J. Cell Biol.* 53: 290–311, 1972.
31. CASTLE, J. D., J. D. JAMIESON, AND G. E. PALADE. Secretion granules of the rabbit parotid gland. Isolation, subfractionation, and characterization of the membrane and content subfractions. *J. Cell Biol.* 64: 182–210, 1975.
32. CASTLE, J. D., AND G. E. PALADE. Secretion granules of the

rabbit parotid. Selective removal of secretory contaminants from granule membranes. *J. Cell Biol.* 76: 323–340, 1978.
33. CHANDLER, D. F., AND J. E. HEUSER. Arrest of membrane fusion events in mast cells by quick-freezing. *J. Cell Biol.* 86: 666–674, 1980.
34. COCKROFT, S., J. A. TAYLOR, AND J. D. JUDAH. Subcellular localization of inositol lipid kinases in rat liver. *Biochim. Biophys. Acta* 845: 163–170, 1985.
35. COHEN, F. S., M. H. AKABAS, AND A. FINKELSTEIN. Osmotic swelling of phospholipid vesicles causes them to fuse with a planar phospholipid bilayer membrane. *Science Wash. DC* 217: 458–460, 1982.
36. COUCH, C. B., AND W. J. STRITTMATTER. Rat myoblast fusion requires metalloprotease activity. *Cell* 32: 257–265, 1983.
37. CREUTZ, C. E., L. G. DOWLING, E. M. KYGER, AND R. C. FRANSON. Phosphatidylinositol-specific phospholipase C activity of chromaffin granule-binding proteins. *J. Biol. Chem.* 260: 7171–7173, 1985.
38. CREUTZ, C. E., L. G. DOWLING, J. J. SANDO, C. VILLAR-PILASI, J. H. WHIPPLE, AND W. J. ZAKS. Characterization of chromobindins. Soluble proteins that bind to the chromaffin granule membrane in the presence of Ca^{2+}. *J. Biol. Chem.* 258: 14664–14674, 1983.
39. CREUTZ, C. E., C. J. PAZOLES, AND H. B. POLLARD. Identification and purification of an adrenal medulla protein (synexin) that causes calcium-dependent aggregation of isolated chromaffin granules. *J. Biol. Chem.* 253: 2858–2866, 1978.
40. CRUZ, J., B. I. POSNER, AND J. J. M. BERGERON. Receptor-mediated endocytosis of [^{125}I]insulin into pancreatic acinar cells in vivo. *Endocrinology* 115: 1996–2008, 1984.
41. DE CAMILLI, P., R. S. CAMERON, AND P. GREENGARD. Synapsin I (protein I), a nerve terminal-specific phosphoprotein. I. Its general distribution in synapses of the central and peripheral nervous system demonstrated by immunofluorescence in frozen and plastic sections. *J. Cell Biol.* 96: 1337–1354, 1983.
42. DE CAMILLI, P., D. PELUCHETTI, AND J. MELDOLESI. Dynamic changes of the luminal plasmalemma in stimulated parotid acinar cells. A freeze fracture study. *J. Cell Biol.* 70: 59–74, 1976.
43. DE LISLE, R. C., AND U. HOPFER. Electrolyte permeabilities of pancreatic zymogen granules: implications for pancreatic secretion. *Am. J. Physiol.* 250 (*Gastrointest. Liver Physiol.* 13): G489–G496, 1986.
44. DE LISLE, R. C., I. SCHULZ, T. TYRAKOWSKI, W. HAASE, AND U. HOPFER. Isolation of stable pancreatic zymogen granules. *Am. J. Physiol.* 246 (*Gastrointest. Liver Physiol.* 9): G411–G418, 1984.
45. DE LORENZO, R. J., S. D. FREEDMAN, W. B. YOHE, AND S. C. MAURER. Stimulation of Ca^{2+}-dependent neurotransmitter release and presynaptic nerve terminal protein phosphorylation by calmodulin and a calmodulin-like protein isolated from synaptic vesicles. *Proc. Natl. Acad. Sci. USA* 76: 1838–1842, 1979.
46. DE OLIVEIRA FILGUEIRAS, O. M., A. M. H. P. VAN DEN BESSELAAR, AND H. VAN DEN BOSCH. Localization of lysophosphatidylcholine in bovine chromaffin granules. *Biochim. Biophys. Acta* 558: 73–84, 1979.
47. DE OLIVEIRA FILGUEIRAS, O. M., H. VAN DEN BOSCH, R. G. JOHNSON, S. E. CARTY, AND A. SCARPA. Phospholipid composition of some amine storage granules. *FEBS Lett.* 129: 309–313, 1981.
48. DEUTSCH, J. W., AND R. B. KELLY. Lipids of synaptic vesicles: relevance to the mechanism of membrane fusion. *Biochemistry* 20: 378–385, 1981.
49. DOLAIS-KITABGI, J., AND R. L. PERLMAN. The stimulation of catecholamine release from chromaffin granules by valinomycin. *Mol. Pharmacol.* 11: 745–750, 1975.
50. DOUGLAS, W. W. Involvement of calcium in exocytosis and the exocytosis-vesiculation sequence. Calcium and cell regulation. *Biochem. Soc. Symp.* 39: 1–28, 1974.
51. DRENCKHAHN, D., AND H. G. MANNHERZ. Distribution of actin and the actin-associated proteins myosin, tropomyosin, alpha-actinin, vinculin, and villin in rat and bovine exocrine glands. *Eur. J. Cell Biol.* 30: 167–176, 1983.
52. DREYFUS, H., D. AUNIS, S. HARTH, AND P. MANDEL. Gangliosides and phospholipids of the membranes from bovine adrenal medullary chromaffin granules. *Biochim. Biophys. Acta* 489: 89–97, 1977.
53. DUELLO, T., T. M. NETT, AND M. G. FARQUHAR. Fate of a gonadotropin-releasing hormone agonist internalized by rat pituitary gonadotrophs. *Endocrinology* 112: 1–10, 1983.
54. DUONG, L. T., P. J. FLEMING, AND J. T. RUSSELL. An identical cytochrome b_{561} is present in bovine adrenal chromaffin vesicles and posterior pituitary neurosecretory vesicles. *J. Biol. Chem.* 259: 4885–4889, 1984.
55. EHRENREICH, J. H., J. J. M. BERGERON, P. SIEKEVITZ, AND G. E. PALADE. Golgi fractions prepared from rat liver homogenates. I. Isolation procedure and morphological characterization. *J. Cell Biol.* 59: 45–72, 1973.
56. EHRLICH, H. P., R. ROSS, AND P. BORNSTEIN. Effects of antimicrotubular agents on the secretion of collagen. A biochemical and morphological study. *J. Cell Biol.* 62: 390–405, 1974.
57. EMMELIN, N. Nervous control of mammalian salivary glands. *Philos. Trans. R. Soc. Lond. B Biol. Sci.* 296: 27–35, 1981.
58. FARQUHAR, M. G. Recovery of surface membrane in anterior pituitary cells. Variations in traffic detected with anionic and cationic ferritin. *J. Cell Biol.* 77: R35–R42, 1978.
59. FARQUHAR, M. G., J. J. REID, AND L. W. DANIELL. Intracellular transport and packaging of prolactin: a quantitative electron microscopic autoradiographic study of mammotrophs dissociated from rat pituitaries. *Endocrinology* 102: 296–311, 1978.
60. FEINSTEIN, H., AND M. SCHRAMM. Energy production in rat parotid gland. *Eur. J. Biochem.* 13: 158–163, 1970.
61. FLETCHER, D. J., J. P. QUIGLEY, G. E. BAUER, AND B. D. NOE. Characterization of proinsulin- and proglucagon-converting activities in isolated islet secretory granules. *J. Cell Biol.* 90: 312–322, 1981.
62. FORTE, T. M., T. E. MACHEN, AND J. G. FORTE. Ultrastructural changes in oxyntic cells associated with secretory function: a membrane-recycling hypothesis. *Gastroenterology* 73: 941–955, 1977.
63. FOWLER, V. M., AND H. B. POLLARD. Chromaffin granule membrane-F-actin interactions are calcium sensitive. *Nature Lond.* 295: 336–339, 1982.
64. FRICKER, L. D., AND S. H. SNYDER. Enkephalin convertase: purification and characterization of a specific enkephalin-synthesizing carboxypeptidase localized to adrenal chromaffin granules. *Proc. Natl. Acad. Sci. USA* 79: 3886–3890, 1982.
65. FUJIKI, Y., A. L. HUBBARD, S. FOWLER, AND P. B. LAZAROW. Isolation of intracellular membranes by means of sodium carbonate treatment: application to endoplasmic reticulum. *J. Cell Biol.* 93: 97–102, 1982.
66. GARDNER, J. D., AND R. T. JENSEN. Regulation of pancreatic enzyme secretion in vitro. In: *Physiology of the Gastrointestinal Tract*, edited by L. R. Johnson. New York: Raven, 1981, vol. 2, p. 831–869.
67. GEISOW, M. J., AND R. D. BURGOYNE. Cation-dependent lysis of chromaffin granules—an alternative hypothesis for osmotically driven exocytosis. *Cell Biol. Int. Rep.* 6: 353–359, 1982.
68. GEISOW, M. J., AND R. D. BURGOYNE. Recruitment of cytosolic proteins to a secretory granule membrane depends on Ca^{2+}-calmodulin. *Nature Lond.* 301: 432–435, 1983.
69. GEISOW, M. J., R. D. BURGOYNE, AND A. HARRIS. Interaction of calmodulin with adrenal chromaffin granule membranes. *FEBS Lett.* 143: 69–72, 1982.
70. GIANNATTASIO, G., A. ZANINI, AND J. MELDOLESI. Molecular organization of rat prolactin granules. I. In vitro stability of intact and "membraneless" granules. *J. Cell Biol.* 64: 246–251, 1975.

71. GILBERT, S. P., R. D. ALLEN, AND R. D. SLOBODA. Translocation of vesicles from squid axoplasm on flagellar microtubules. *Nature Lond.* 315: 245–248, 1985.
72. GLEMBOTSKI, C. C., B. A. EIPPER, AND R. E. MAINS. Characterization of a peptide α-amidation activity from rat anterior pituitary. *J. Biol. Chem.* 259: 6385–6392, 1984.
73. GLENNEY, J. Phospholipid-dependent Ca^{2+} binding by the 36-kDa tyrosine kinase substrate (calpactin) and its 33-kDa core. *J. Biol. Chem.* 261: 7247–7252, 1986.
74. GLICKMAN, J., K. CROEN, S. KELLY, AND Q. AL-AWQATI. Golgi membranes contain an electrogenic H^+ pump in parallel to a chloride conductance. *J. Cell Biol.* 97: 1303–1308, 1983.
75. GONATAS, N. K., A. STIEBER, W. F. HICKEY, S. H. HERBERT, AND J. O. GONATAS. Endosomes and Golgi vesicles in adsorptive and fluid phase endocytosis. *J. Cell Biol.* 99: 1379–1390, 1984.
76. GUMBINER, B., AND R. B. KELLY. Secretory granules of an anterior pituitary cell line, AtT-20, contain only mature forms of corticotropin and β-lipotropin. *Proc. Natl. Acad. Sci. USA* 78: 318–322, 1981.
77. HAND, A. R. Morphology and cytochemistry of the Golgi apparatus of the rat salivary gland acinar cells. *Am. J. Anat.* 130: 141–157, 1971.
78. HAVINGA, J. R., G. J. A. M. STROUS, AND C. POORT. Biosynthesis of the major glycoprotein associated with zymogen-granule membranes in the pancreas. *Eur. J. Biochem.* 133: 449–454, 1983.
79. HERZOG, V., AND M. G. FARQUHAR. Luminal membrane retrieved after exocytosis reaches most Golgi cisternae in secretory cells. *Proc. Natl. Acad. Sci. USA* 74: 5073–5077, 1977.
80. HERZOG, V., AND H. REGGIO. Pathways of endocytosis from luminal plasma membrane in rat exocrine pancreas. *Eur. J. Cell Biol.* 21: 141–150, 1980.
81. HIRAKAWA, N. Cross-linker system between neurofilaments, microtubules, and membranous organelles in frog axons revealed by the quick-freeze, deep-etching method. *J. Cell Biol.* 94: 129–142, 1982.
82. HOKIN, L. E. Isolation of the zymogen granules of dog pancreas and a study of their properties. *Biochim. Biophys. Acta* 18: 379–388, 1955.
83. HOLTZMAN, E., AND R. DOMINITZ. Cytochemical studies of lysosomes, Golgi apparatus and endoplasmic reticulum in secretion and protein uptake by adrenal medulla cells of the rat. *J. Histochem. Cytochem.* 16: 320–336, 1968.
84. HOLZ, R. W. Measurement of membrane potential of chromaffin granules by the accumulation of triphenylmethylphosphonium cation. *J. Biol. Chem.* 254: 6703–6709, 1979.
85. HOLZ, R. W., R. A. SENTER, AND R. R. SHARP. Evidence that H^+ electrochemical gradient across membranes of chromaffin granules is not involved in exocytosis. *J. Biol. Chem.* 258: 7506–7513, 1983.
86. HONG, K., N. DUZGUNES, AND D. PAPAHADJOPOULOS. Role of synexin in membrane fusion. *J. Biol. Chem.* 256: 3641–3644, 1981.
87. HOOPER, J. E., AND R. B. KELLY. Calmodulin is tightly associated with synaptic vesicles independent of calcium. *J. Biol. Chem.* 259: 148–153, 1984.
88. HOPPE, C. A., T. P. CONNOLLY, AND A. L. HUBBARD. Transcellular transport of polymeric IgA in rat hepatocyte: biochemical and morphological characterization of the transport pathway. *J. Cell Biol.* 101: 2113–2123, 1985.
89. HOWELL, K. E., AND G. E. PALADE. Hepatic Golgi fractions resolved into membrane and content polypeptides. *J. Cell Biol.* 92: 822–832, 1982.
90. HUTTNER, W. B., L. J. DEGENNARO, AND P. GREENGARD. Differential phosphorylation of multiple sites in purified protein I by cyclic AMP-dependent and calcium-dependent protein kinases. *J. Biol. Chem.* 256: 1482–1488, 1981.
91. HUTTNER, W. B., N. SCHIEBLER, P. GREENGARD, AND P. DE CAMILLI. Synapsin I (protein I), a nerve terminal-specific phosphoprotein. III. Its association with synaptic vesicles studied in a highly purified synaptic vesicle preparation. *J. Cell Biol.* 96: 1374–1388, 1983.
92. ISHIKAWA, A. Fine structural changes in response to hormonal stimulation of the perfused canine pancreas. *J. Cell Biol.* 24: 369–385, 1965.
93. JAMIESON, J. D., AND G. E. PALADE. Intracellular transport of secretory proteins in the pancreatic exocrine cells. I. Role of the peripheral elements of the Golgi complex. *J. Cell Biol.* 34: 577–596, 1967.
94. JAMIESON, J. D., AND G. E. PALADE. Intracellular transport of secretory proteins in the pancreatic exocrine cell. IV. Metabolic requirements. *J. Cell Biol.* 39: 589–603, 1968.
95. JAMIESON, J. D., AND G. E. PALADE. Condensing vacuole conversion and zymogen granule discharge in pancreatic exocrine cells: metabolic studies. *J. Cell Biol.* 48: 503–522, 1971.
96. KACHADORIAN, W. A., J. MULLER, AND A. FINKELSTEIN. Role of osmotic forces in exocytosis: studies of ADH-induced fusion in toad urinary bladder. *J. Cell Biol.* 91: 584–588, 1981.
97. KAPLAN, J., D. M. WARD, AND H. S. WILEY. Phenylarsine oxide-induced increase in alveolar macrophage surface receptors: evidence for fusion of internal receptor pools with the cell surface. *J. Cell Biol.* 101: 121–129, 1985.
98. KEENAN, T. W., AND D. J. MORRE. Phospholipid class and fatty acid composition of Golgi apparatus isolated from rat liver and comparison with other cell fractions. *Biochemistry* 9: 19–25, 1970.
99. KELLY, R. B. Pathways of protein secretion in eukaryotes. *Science Wash. DC* 230: 25–32, 1985.
100. KNIGHT, D. E., AND P. F. BAKER. Calcium-dependence of catecholamine release from bovine adrenal medullary cells after exposure to intense electron fields. *J. Membr. Biol.* 68: 107–140, 1982.
101. KNIGHT, D. E., AND M. C. SCRUTTON. Cyclic nucleotides control a system which regulates Ca^{2+} sensitivity of platelet secretion. *Nature Lond.* 309: 66–68, 1984.
102. KOIKE, H., AND J. MELDOLESI. Post-stimulation retrieval of luminal surface membrane in parotid acinar cells is calcium-dependent. *Exp. Cell Res.* 134: 377–388, 1981.
103. KONDOR-KOCH, C., R. BRAVO, S. D. FULLER, D. CUTLER, AND M. GAROFF. Exocytotic pathways exist to both the apical and the basolateral cell surface of the polarized epithelial cell MDCK. *Cell* 43: 297–306, 1985.
104. KOUSVELARI, E., F. G. OPPENHEIM, AND L. S. CUTLER. Ultrastructural localization of salivary acidic proline-rich proteins from Macaca fascicularis. *J. Histochem. Cytochem.* 30: 274–278, 1982.
105. KRAEHENBUHL, J.-P., L. RACINE, AND J. D. JAMIESON. Cytochemical localization of secretory proteins in bovine pancreatic exocrine cells. *J. Cell Biol.* 72: 406–423, 1977.
106. KREIBICH, G., P. DEBEY, AND D. D. SABATINI. Selective release of content from microsomal vesicles without membrane disassembly. I. Permeability changes induced by low detergent concentrations. *J. Cell Biol.* 58: 436–462, 1973.
107. KREMMER, T., M. H. WISHER, AND W. H. EVANS. The lipid composition of plasma membrane subfractions originating from the three major functional domains of the rat hepatocyte cell surface. *Biochim. Biophys. Acta* 455: 655–664, 1976.
108. LAGUNOFF, D., AND E. Y. CHI. Effect of colchicine on rat mast cells. *J. Cell Biol.* 71: 182–195, 1976.
109. LASEK, R. J., AND S. T. BRADY. Attachment of transported vesicles to microtubules in axoplasm is facilitated by AMP-PNP. *Nature Lond.* 316: 645–647, 1985.
110. LEBEL, D., AND M. BEATTIE. The integral and peripheral proteins of the zymogen granule membrane. *Biochim. Biophys. Acta* 769: 611–621, 1984.
111. LEWIS, D. S., AND R. A. RONZIO. An assessment of the role of protein kinase and zymogen granule phosphorylation during secretion by the rat exocrine pancreas. *Biochim. Biophys. Acta* 583: 422–433, 1979.
112. LILLIE, J. H., AND S. S. HAN. Secretory protein synthesis in the stimulated parotid gland. Temporal dissociation of the

112. maximal response from secretion. *J. Cell Biol.* 59: 708–721, 1973.
113. LIVNE, E., AND C. OLIVER. Internalization of cationized ferritin by isolated pancreatic acinar cells. *J. Histochem. Cytochem.* 34: 167–176, 1986.
114. LLINÁS, R., T. L. MCGUINNESS, C. S. LEONARD, M. SUGIMORI, AND P. GREENGARD. Intraterminal injection of synapsin I or calcium/calmodulin-dependent protein kinase II alters neurotransmitter release at the squid giant synapse. *Proc. Natl. Acad. Sci. USA* 82: 3035–3039, 1985.
115. LOUVARD, D., H. REGGIO, AND G. WARREN. Antibodies to the Golgi complex and the rough endoplasmic reticulum. *J. Cell Biol.* 92: 92–107, 1982.
116. MALAISSE, W. J., F. MALAISSE-LAGE, M. O. WALKER, AND P. E. LACY. The stimulus-secretion coupling of glucose-induced insulin release. V. The participation of a microtubular-microfilamentous system. *Diabetes* 20: 257–265, 1971.
117. MATLIN, K., AND K. SIMONS. Sorting of a plasma membrane glycoprotein occurs before it reaches the cell surface in cultured epithelial cells. *J. Cell Biol.* 99: 2131–2139, 1984.
118. MATTHEW, W. D., L. TSAVALER, AND L. F. REICHARDT. Identification of a synaptic vesicle-specific membrane protein with a wide distribution in neuronal and neurosecretory tissue. *J. Cell Biol.* 91: 257–269, 1981.
119. MELDOLESI, J. Dynamics of cytoplasmic membranes in guinea pig pancreatic acinar cells. I. Synthesis and turnover of membrane proteins. *J. Cell Biol.* 61: 1–13, 1974.
120. MELDOLESI, J., P. DE CAMILLI, AND D. PELUCHETTI. The membrane of secretory granules: structure, composition and turnover. In: *Secretory Mechanisms of Exocrine Glands*, edited by N. A. Thorn and O. H. Peterson. Copenhagen: Munksgaard, 1974, p. 137–148.
121. MELDOLESI, J., J. D. JAMIESON, AND G. E. PALADE. Composition of cellular membranes in the pancreas of the guinea pig. Isolation of membrane fractions. *J. Cell Biol.* 49: 109–129, 1971.
122. MERRITT, J. E., AND R. P. RUBIN. Pancreatic amylase secretion and cytoplasmic free calcium. *Biochem. J.* 230: 151–159, 1985.
123. MICHAELSON, D. M., G. BARKAI, AND Y. BARENHOLZ. Asymmetry of lipid organization in cholinergic synaptic vesicle membranes. *Biochem. J.* 211: 155–162, 1983.
124. MICHENER, M., W. B. DAWSON, AND C. E. CREUTZ. Phosphorylation of a chromaffin granule-binding protein in stimulated chromaffin cells. *J. Biol. Chem.* 261: 6548–6555, 1986.
125. MILLER, C., P. ARVAN, J. N. TELFORD, AND E. RACKER. Ca^{2+}-induced fusion of proteoliposomes. Dependence on transmembrane osmotic gradient. *J. Membr. Biol.* 30: 271–282, 1976.
126. MILLER, R. H., AND R. J. LASEK. Cross-bridges mediate anterograde and retrograde vesicle transport along microtubules in squid axoplasm. *J. Cell Biol.* 101: 2181–2193, 1985.
127. MOORE, H.-P. H., AND R. B. KELLY. Secretory protein targeting in a pituitary cell line: differential transport of foreign secretory proteins to distinct secretory pathways. *J. Cell Biol.* 101: 1773–1781, 1985.
128. MOSTOV, K. E., AND N. E. SIMISTER. Transcytosis. *Cell* 43: 389–390, 1985.
129. MROZ, E. A., AND C. LECHENE. Pancreatic zymogen granules differ markedly in protein composition. *Science Wash. DC* 232: 871–874, 1986.
130. NAVONNE, F., P. GREENGARD, AND P. DE CAMILLI. Synapsin I in nerve terminals: selective association with small synaptic vesicles. *Science Wash. DC* 226: 1209–1215, 1985.
131. NESTLER, E. J., AND P. GREENGARD. Nerve impulses increase the phosphorylation state of Protein I in rabbit superior cervical ganglion. *Nature Lond.* 296: 452–454, 1982.
132. NJUS, D., J. KNOTH, AND M. ZALLAKIAN. Proton-linked transport in chromaffin granules. *Curr. Top. Bioenerg.* 11: 107–147, 1981.
133. NOVICK, P., AND D. BOTSTEIN. Phenotypic analysis of temperature-sensitive yeast actin mutants. *Cell* 40: 405–416, 1985.
134. NOVICK, P., C. FIELD, AND R. SCHEKMAN. Identification of 23 complementation groups required for post-translational events in the yeast secretory pathway. *Cell* 21: 205–215, 1980.
135. NOVICK, P., AND R. SCHEKMAN. Export of major cell surface proteins is blocked in yeast secretory mutants. *J. Cell Biol.* 96: 541–547, 1983.
136. NOVIKOFF, A. B., M. MORI, N. QUINTANA, AND A. YAM. Studies of the secretory process in the mammalian exocrine pancreas. I. The condensing vacuole. *J. Cell Biol.* 75: 148–165, 1977.
137. OLIVER, C., AND A. R. HAND. Enzyme modulation of the Golgi apparatus and GERL: a cytochemical study of parotid acinar cells. *J. Histochem. Cytochem.* 31: 1041–1048, 1983.
138. ORCI, L., M. RAVAZZOLA, M. ARHERDT, O. MADSEN, J. D. VASSALI, AND A. PERRELET. Direct identification of prohormone conversion in insulin-secretory cells. *Cell* 42: 203–219, 1985.
139. ORON, Y., Y. SHARONI, H. LEFKOVITZ, AND Z. SELINGER. Phosphatidylinositol kinase and diphosphoinositide kinase in the rat parotid gland. In: *Cyclitols and Phosphoinositides*, edited by W. W. Wells and F. Eisenberg. New York: Academic, 1978, p. 383–397.
140. PALADE, G. E. Intracisternal granules in the exocrine cells of the pancreas. *J. Biophys. Biochem. Cytol.* 2: 417–425, 1956.
141. PALADE, G. E. Intracellular aspects of the process of protein synthesis. *Science Wash. DC* 189: 347–358, 1975.
142. PANDOL, S. J., M. S. SCHOEFFIELD, G. SACHS, AND S. MUALLEM. Role of free cytosolic calcium in secretagogue-stimulated amylase release from dispersed acini from guinea pig pancreas. *J. Biol. Chem.* 260: 10081–10086, 1985.
143. PATZAK, A., AND H. WINKLER. Exocytotic exposure and recycling of membrane antigens of chromaffin granules: ultrastructural evaluation after immunolabeling. *J. Cell Biol.* 102: 510–515, 1986.
144. PATZELT, C., D. BROWN, AND B. JEANRENAUD. Inhibitory effect of colchicine on amylase secretion by the rat parotid glands. Possible localization in the Golgi area. *J. Cell Biol.* 73: 578–593, 1977.
145. PAVELKA, M., AND A. ELLINGER. Effect of colchicine on the Golgi complex of rat pancreatic acinar cells. *J. Cell Biol.* 97: 737–748, 1983.
146. PAZOLES, C., AND H. B. POLLARD. Evidence for stimulation of anion transport in ATP-evoked transmitter release from isolated secretory vesicles. *J. Biol. Chem.* 253: 3962–3969, 1978.
147. PEIFFER, A., C. GAGNON, AND S. HEISLER. Phosphorylation of zymogen granule proteins in intact rat pancreatic acinar cells. *Biochem. Biophys. Res. Commun.* 122: 413–419, 1984.
148. PERRIN, D., AND D. AUNIS. Reorganization of α-fodrin induced by stimulation in secretory cells. *Nature Lond.* 315: 589–592, 1985.
149. PETERS, A., S. L. PALAY, AND H. F. WEBSTER. Synapses. In: *The Fine Structure of the Nervous System: The Neurons and Supporting Cells*. Philadelphia, PA: Saunders, 1976, p. 156–157.
150. PFEIFFER, S., S. D. FULLER, AND K. SIMONS. Intracellular sorting and basolateral appearance of the G protein of vesicular stomatitis virus in Madin-Darby canine kidney cells. *J. Cell Biol.* 101: 470–476, 1985.
151. PHILLIPS, J. H. Phosphatidylinositol kinase. A component of the chromaffin-granule membrane. *Biochem. J.* 136: 579–587, 1973.
152. PHILLIPS, J. H. Passive ion permeability of the chromaffin-granule membrane. *Biochem. J.* 168: 289–297, 1977.
153. PHILLIPS, J. H., K. BURRIDGE, S. P. WILSON, AND N. KIRSHNER. Visualization of the exocytosis/endocytosis secretory cycle in cultured adrenal chromaffin cells. *J. Cell Biol.* 97: 1906–1917, 1983.
154. PUTNEY, J. W., J. S. MCKINNEY, D. L. AUB, AND B. A. LESLIE. Phorbol ester-induced protein secretion in rat parotid gland. *Mol. Pharmacol.* 26: 261–266, 1984.
155. RAND, R. Interacting phospholipid bilayers: measured forces

and induced structural changes. *Annu. Rev. Biophys. Bioeng.* 10: 277–314, 1981.
156. REAVEN, E. P., AND G. M. REAVEN. Evidence that microtubules play a permissive role in hepatocyte very low density lipoprotein secretion. *J. Cell Biol.* 84: 28–39, 1980.
157. REDMAN, C. M., D. BANERJEE, K. HOWELL, AND G. E. PALADE. Colchicine inhibition of plasma protein release from rat hepatocytes. *J. Cell Biol.* 66: 42–59, 1975.
158. REGGIO, H., AND J. C. DAGORN. Ionic interaction between bovine chymotrypsinogen A and chondroitin sulfate A, B, C. A possible novel model for molecular aggregation in zymogen granules. *J. Cell Biol.* 78: 951–954, 1978.
159. RODRIGUEZ-BOULAN, E., D. E. MISEK, D. V. SALAS, P. J. I. SALAS, AND E. BARD. Protein sorting in the secretory pathway. *Curr. Top. Membr. Transp.* 24: 251–294, 1985.
160. ROGALSKI, A. A., J. E. BERGMANN, AND S. J. SINGER. Effect of microtubule assembly states on the intracellular processing and surface expression of an integral protein of the plasma membrane. *J. Cell Biol.* 99: 1101–1109, 1984.
161. ROGALSKI, A. A., AND S. J. SINGER. Associations of elements of the Golgi apparatus with microtubules. *J. Cell Biol.* 99: 1092–1100, 1984.
162. ROHLICH, P., P. ANDERSON, AND B. UVNAS. Electron microscope observations on compound 48/80-induced degranulation in rat mast cells. Evidence for sequential exocytosis of storage granules. *J. Cell Biol.* 51: 465–483, 1971.
163. RONZIO, R. A., K. E. KRONQUIST, D. S. LEWIS, R. J. MACDONALD, S. H. MOHRLOK, AND J. J. O'DONNELL, JR. Glycoprotein synthesis in the adult rat pancreas. IV. Subcellular distribution of membrane glycoproteins. *Biochim. Biophys. Acta* 508: 65–84, 1978.
164. ROTH, J. Subcellular organization of glycosylation: results from in situ immuno electron microscopy. In: *Glycoconjugates*, edited by E. A. Davidson, J. C. Williams, and N. M. Di Ferrante. New York: Praeger, 1985, p. 497–498. (Proc. 8th Intern. Symp.)
165. ROTH, J., D. J. TAATJES, J. M. LUCOCQ, J. WEINSTEIN, AND J. C. PAULSON. Demonstration of an extensive trans-tubular network continuous with the Golgi apparatus stack that may function in glycosylation. *Cell* 43: 287–295, 1985.
166. RUDNICK, G. ATP-driven H^+ pumping into intracellular organelles. *Annu. Rev. Physiol.* 48: 403–413, 1986.
167. RUSSELL, J. T., AND R. W. HOLZ. Measurement of ΔpH and membrane potential in isolated neurosecretory vesicles from bovine neurohypophyses. *J. Biol. Chem.* 256: 5950–5953, 1981.
168. RUSSELL, J. T., M. LEVINE, AND D. NJUS. Electron transfer across posterior pituitary neurosecretory vesicle membranes. *J. Biol. Chem.* 260: 226–231, 1985.
169. SCHEELE, G. A. Regulation of gene expression in the exocrine pancreas. In: *The Exocrine Pancreas: Biology, Pathology, and Diseases*, edited by V. L. W. Go. New York: Raven, 1986, p. 60–65.
170. SCHEELE, G., AND A. TARTAKOFF. Exit of nonglycosylated secretory proteins from the rough endoplasmic reticulum is asynchronous in the exocrine pancreas. *J. Biol. Chem.* 260: 926–931, 1985.
171. SCHEFFER, R. C. T., C. POORT, AND J. W. SLOT. Fate of the major zymogen granule membrane-associated glycoproteins from rat pancreas. A biochemical and immunocytochemical study. *Eur. J. Cell Biol.* 23: 122–128, 1980.
172. SCHEID, A., AND P. W. CHOPPIN. Identification of biological activities of paramyxovirus glycoproteins. Activation of cell fusion, hemolysis, and infectivity by proteolytic cleavage of an inactive precursor protein of Sendai virus. *Virology* 57: 475–490, 1974.
173. SCHERMAN, D., AND J. NORDMAN. Internal pH of isolated newly formed and aged neurohypophyseal granules. *Proc. Natl. Acad. Sci. USA* 79: 476–479, 1982.
174. SCHRAMM, M., AND D. DANON. The mechanism of enzyme secretion by the cell. I. Storage of amylase in the zymogen granules of the rat parotid gland. *Biochim. Biophys. Acta* 50: 102–112, 1961.
175. SCHRAMM, M., AND Z. SELINGER. The functions of cyclic AMP and calcium as alternative second messengers in parotid gland and pancreas. *J. Cyclic Nucleotide Res.* 1: 181–192, 1975.
176. SCHROER, T. A., S. T. BRADY, AND R. B. KELLY. Fast axonal transport of foreign synaptic vesicles in squid axoplasm. *J. Cell Biol.* 101: 568–572, 1985.
177. SCHWARTZ, G. H., AND Q. AL-AWQATI. Regulation of transepithelial H^+ transport by exocytosis and endocytosis. *Annu. Rev. Physiol.* 48: 153–162, 1986.
178. SCHWOCH, G., S. M. LOHMAN, U. WALTER, AND U. JUNG. Determination of cyclic AMP-dependent protein kinase subunits by an immunoassay reveals a different subcellular distribution of the enzyme in rat parotid than does determination of the enzyme activity. *J. Cyclic Nucleotide Protein Phosphorylation Res.* 10: 247–258, 1985.
179. SMITH, C. D., AND W. W. WELLS. Phosphorylation of rat liver nuclear envelopes. *J. Biol. Chem.* 258: 9368–9373, 1983.
180. SMITH, D. S., U. JARLFORS, AND R. BERANEK. The organization of synaptic axoplasm in the lamprey central nervous system. *J. Cell Biol.* 46: 199–219, 1970.
181. SNIDER, M. D., AND ROGERS, O. C. Membrane traffic in animal cells: cellular glycoproteins return to the site of Golgi mannosidase I. *J. Cell Biol.* 103: 265–276, 1986.
182. SOLARI, R., AND J.-P. KRAEHENBUHL. Biosynthesis of the IgA antibody receptor: a model for the transepithelial sorting of a membrane glycoprotein. *Cell* 36: 61–71, 1984.
183. SPEARMAN, T. N., K. P. HURLEY, R. OLIVAS, R. G. ULRICH, AND F. R. BUTCHER. Subcellular localization of stimulus-affected endogenous phosphoproteins in the rat parotid gland. *J. Cell Biol.* 99: 1354–1363, 1984.
184. SPUDICH, J. A., S. J. KRON, AND M. P. SHEETZ. Movement of myosin-coated beads on oriented filaments reconstituted from purified actin. *Nature Lond.* 315: 584–586, 1985.
185. STADLER, H., AND E. M. FENWICK. Cholinergic synaptic vesicles from *Torpedo marmorata* contain an atractyloside-binding protein related to the mitochondrial ADP/ATP carrier. *Eur. J. Biochem.* 136: 377–382, 1983.
186. STEINMAN, R. M., I. S. MELLMAN, W. A. MULLER, AND Z. A. COHN. Endocytosis and the recycling of plasma membrane. *J. Cell Biol.* 96: 1–27, 1983.
187. STOSSEL, T. P. Actin filaments and secretion. The macrophage model. *Methods Cell Biol.* 23: 215–245, 1981.
188. STROUS, G. J., A. D. MAINE, J. E. ZIJDERHAND-BLEEKEMOLEN, J. W. SLOT, AND A. L. SCHWARTZ. Effect of lysosomotropic amines on the secretory pathway and on the recycling of the asialoglycoprotein receptor in human hepatoma cells. *J. Cell Biol.* 101: 531–539, 1985.
189. STROUS, G. J. A. M., R. WILLEMSEN, P. VAN KECKHOF, J. W. SLOT, H. J. GEUZE, AND H. F. LODISH. Vesicular stomatitis virus glycoprotein, albumin, and transferrin are transported to the cell surface via the same Golgi vesicles. *J. Cell Biol.* 97: 1815–1822, 1983.
190. SUMMERS, T. A., AND C. E. CREUTZ. Phosphorylation of a chromaffin granule-binding protein by protein kinase C. *J. Biol. Chem.* 260: 2437–2443, 1985.
191. SUPRANANT, K. A., AND W. L. DENTLER. Association between endocrine pancreatic secretory granules and in vitro-assembled microtubules is dependent upon microtubule-associated proteins. *J. Cell Biol.* 93: 164–174, 1982.
192. SZTUL, E. S., K. E. HOWELL, AND G. E. PALADE. Biogenesis of the polymeric IgA receptor in rat hepatocytes. II. Localization of its intracellular forms by cell fractionation studies. *J. Cell Biol.* 100: 1255–1261, 1985.
193. TANDLER, B., AND R. A. ERLANDSON. Ultrastructure of baboon parotid gland. *Anat. Rec.* 184: 115–132, 1975.
194. TARTAKOFF, A., P. VASSALI, AND M. DÉTRAZ. Comparative studies of intracellular transport of secretory proteins. *J. Cell Biol.* 79: 694–707, 1978.
195. TARTAKOFF, A. M., P. VASSALLI, AND M. DÉTRAZ. Plasma cell immunoglobulin secretion. Arrest is accompanied by al-

terations of the Golgi complex. *J. Exp. Med.* 146: 1332–1345, 1977.
196. TOUGARD, C., D. LOUVARD, R. PICART, AND A. TIXIER-VIDAL. The rough endoplasmic reticulum and the Golgi apparatus visualized using specific antibodies in normal and tumoral prolactin cells in culture. *J. Cell Biol.* 96: 1197–1207, 1983.
197. TOUGARD, C., D. LOUVARD, R. PICART, AND A. TIXIER-VIDAL. Immunocytochemical identification of membrane compartments involved in secretory process and endocytosis in prolactin cells. In: *Prolactin. Basic and Clinical Correlates*, edited by R. M. Macleod, M. O. Thorner, and U. Scapagnini. Padua, Italy: Livania, 1985, vol. 1, p. 37–44. (Fidia Res. Ser.)
198. TREIMAN, M., W. WEBER, AND M. GRATZL. 3′,5′-Cyclic adenosine monophosphate- and Ca^{2+}-calmodulin-dependent endogenous protein phosphorylation activity in membranes of the bovine chromaffin secretory vesicles: identification of two phosphorylated components as tyrosine hydroxylase and protein kinase regulatory subunit Type II. *J. Neurochem.* 40: 661–669, 1983.
199. UVNAS, B. Histamine storage and release. *Federation Proc.* 33: 2172–2176, 1974.
200. VALE, R. D., B. J. SCHNAPP, T. MITCHISON, E. STEUER, T. S. REESE, AND M. P. SHEETZ. Different axoplasmic proteins generate movement in opposite directions along microtubules in vitro. *Cell* 43: 623–632, 1985.
201. VALE, R. D., B. J. SCHNAPP, T. S. REESE, AND M. P. SHEETZ. Movement of organelles along filaments dissociated from the axoplasm of the squid giant axon. *Cell* 40: 449–454, 1985.
202. WALKER, J. H., J. OBROCKI, AND T. C. SUDHOF. Calelectrin, a calcium-dependent membrane-binding protein associated with secretory granules in *Torpedo* cholinergic electromotor nerve endings and rat adrenal medulla. *J. Neurochem.* 41: 139–145, 1983.
203. WALL, D. A., AND A. L. HUBBARD. Receptor-mediated endocytosis of asialoglycoproteins by rat liver hepatocytes: biochemical characterization of the endosomal compartments. *J. Cell Biol.* 101: 2104–2112, 1985.
204. WALLACH, D. The secretory granule of the parotid gland. In: *The Secretory Granule*, edited by A. Poisner and J. M. Trifaro. New York: Elsevier, 1982, p. 247–276.
205. WALLACH, D., N. KIRSCHNER, AND M. SCHRAMM. Non-parallel transport of membrane proteins and content proteins during assembly of the secretory granule in rat parotid gland. *Biochim. Biophys. Acta* 375: 87–105, 1975.
206. WEBER, A., E. W. WESTHEAD, AND H. WINKLER. Specificity and properties of the nucleotide carrier in chromaffin granules from bovine adrenal medulla. *Biochem. J.* 210: 789–794, 1983.
207. WEINSTOCK, M., AND C. P. LEBLOND. Elaboration of the matrix glycoprotein of enamel by the secretory ameloblasts of the rat incisor as revealed by radioautography after galactose-^{3}H injection. *J. Cell Biol.* 51: 26–51, 1971.
208. WHITE, J. M., M. KIELIEN, AND A. HELENIUS. Membrane fusion proteins of envelopal viruses. *Q. Rev. Biophys.* 16: 151–195, 1983.
209. WILLIAMS, J. A. Effects of antimitotic agents on ultrastructure and intracellular transport of protein in pancreatic acini. *Methods Cell Biol.* 23: 247–258, 1981.
210. WILLIAMS, M. A., AND G. H. COPE. A system for the study of zymogen granule genesis in rabbit parotid gland tissue in vitro. *J. Anat.* 116: 431–444, 1973.
211. WILLIAMS, M. A., M. K. PRATTEN, J. W. TURNER, AND G. M. COPE. Comparative studies on the nature of purified cytomembranes of the rabbit parotid gland. *Histochem. J.* 11: 19–50, 1979.
212. WILSON, I. A., J. J. SKEHEL, AND D. C. WILEY. Structure of the haemagglutinin membrane glycoprotein of influenza virus at 3 Å resolution. *Nature Lond.* 289: 366–373, 1981.
213. WINKLER, H. The composition of adrenal chromaffin granules: an assessment of controversial results. *Neuroscience* 1: 65–80, 1976.
214. WINKLER, H. The biogenesis of the adrenal chromaffin granules. *Neuroscience* 2: 657–683, 1977.
215. WOODS, J. W., M. DORIAUX, AND M. G. FARQUHAR. Transferrin receptors recycle to cis and middle as well as trans Golgi cisternae in Ig-secretory myeloma cells. *J. Cell Biol.* 103: 277–286, 1986.
216. ZASTROW, M. V., T. R. TRITTON, AND J. D. CASTLE. Identification of L-ascorbic acid in secretion granules of the rat parotid gland. *J. Biol. Chem.* 259: 11746–11750, 1984.
217. ZIMMERBERG, J., C. SARDET, AND D. EPEL. Exocytosis of sea urchin egg cortical vesicles in vitro is retarded by hyperosmotic sucrose: kinetics of fusion monitored by quantitative light-scattering microscopy. *J. Cell Biol.* 101: 2398–2410, 1985.

CHAPTER 8

Neural and hormonal control of gastric secretion

BASIL I. HIRSCHOWITZ | Division of Gastroenterology, Department of Medicine, University of Alabama at Birmingham, Birmingham, Alabama

CHAPTER CONTENTS

Central Controls
   Central vagal complex
   Functional anatomy of the vagal complex
   Brain peptides relating to gastric secretion
   Vagal excitation by glucoprivation
     Glucose analogues: 2-deoxy-D-glucose and 3-O-methylglucose
     Other glucose analogues
   Non-glucose-related chemical stimuli of the central vagus
     Mechanism of nerve cell stimulation
     Central antagonists of glucoprivic stimulation
     Cortical and cephalic phase vagal stimulation
     Electrical stimulation of the gastric vagus
   Vagal afferents
     Vagovagal reflexes
     Macromolecules in afferent neurons
Peripheral Controls of Gastric Secretion
   Enteric nervous system
   Chemical transmitters of neural effects
     Acetylcholine
     Gastrin
     Other peptides
     Histamine
   Trophic effects
     Fundic mucosa
     Antral mucosa
   Antrofundal interactions
     Antrofundal reflexes
     Fundoantral reflexes
     Antrectomy
   Vagotomy
     Truncal vagotomy
     Regional vagotomy
   Intestinal phase of gastric secretion
     Stimulation
     Inhibition

GASTRIC SECRETION IS PART of a complex integrated system by which higher organisms feed, digest, and regulate their nutrition. The system controls and integrates feeding behavior; swallowing; salivary, gastric, and pancreatic secretion; gastrointestinal motility, blood flow, and absorption; and, ultimately, blood sugar and caloric homeostasis. Gastric function is controlled and integrated at several hierarchical levels that share or overlap many components of the digestive complex. This chapter discusses gastric acid and pepsin secretion and only touches on the other components as they are relevant. Secretion of mucus is discussed in the chapter by Allen and bicarbonate secretion is discussed in the chapter by Flemström and Garner in this *Handbook*. Intrinsic factor secretion is briefly discussed in this chapter.

Compartmentalization of the controls of gastric secretion into cephalic, gastric, and intestinal phases is no longer useful. These phases overlap, and each affects gastric secretion by interactions between neural and hormonal mechanisms (19, 80), which are discussed in this chapter.

CENTRAL CONTROLS

Central Vagal Complex

The central nervous system (CNS) controls secretion and motility of the stomach via the vagus nerves (19) through fibers originating in the medulla from neurons of the dorsomotor nucleus of the vagus (DMNV), as well as the nucleus ambiguus (NA) and the nucleus tractus solitarius (NTS).

The vagus nerves are complex mixed brachiomeric nerves containing afferent fibers from *1*) around and in the ear, *2*) the pharynx, larynx, trachea, esophagus, and thoracic and abdominal viscera, and *3*) taste buds in the region of the epiglottis. Efferent fibers are distributed to *1*) parasympathetic ganglia of thoracic and abdominal viscera and *2*) the striated muscles of the larynx and pharynx. Visceral efferents comprise only ~3% of the nearly 60,000 fibers that constitute the abdominal vagi in humans (137).

The neurons of various abdominal vagal efferents have been localized by selective retrograde tracing using the fluorescent dye True Blue (67) or horseradish peroxidase (86, 277). In rats, each DMNV contains a column occupying the medial two-thirds or more of the nucleus corresponding to one of the gastric vagus branches: left DMNV, anterior gastric and hepatic; right DMNV, posterior gastric. The lateral one-third of the DMNV forms the celiac nerves (67). In cats and monkeys, gastric vagal efferents originate in both the DMNV and the medial NTS (86, 277). Pancreatic (262), oropharyngeal, cardiovascular, and lung parasympathetic efferent neurons coexist with those of the abdominal viscera.

The NA contains cell bodies of preganglionic parasympathetic neurons of the vagus, as well as motor neurons supplying branchiometric skeletal muscles via the vagus. The NA cells labeled by retrograde markers injected around the pylorus (86, 196) can be electrically activated to stimulate gastroduodenal motility (and slow the heart) without affecting gastric secretion (198).

Primary visceral sensory information is relayed to the DMNV via the tractus solitarius and the NTS that surrounds it. General somatic afferent fibers of the vagus nerve arise from cells of the superior ganglion of the vagus nerve, located in or immediately beneath the jugular foramen, whereas both general and special visceral afferent fibers originate in the larger inferior vagal ganglion (nodose ganglion). The fibers entering the solitary tract bifurcate into short ascending and longer descending components. Descending vagal components in the tractus solitarius gradually diminish in number as collaterals and terminals are given off to the NTS. Some vagal visceral fibers descend caudal to the obex, where the NTS of the two sides merge to form the commissural vagal nucleus.

The rostral or lateral part of the NTS receives mainly taste fibers and is also known as the gustatory nucleus; the caudal or medial NTS receives mainly general visceral afferents. Gastrointestinal afferents end in the parvocellular subnucleus (86, 277). The NTS connects with the NA, the thalamic ventral posteromedial nucleus concerned with gustatory sensations, the hypoglossal and salivary nuclei, the DMNV, and the paraventricular parvocellular nucleus of the hypothalamus (255). The NTS is also closely linked with the respiratory and vasomotor control centers.

A cross section of the medulla (Fig. 1A) shows the relative location of the nuclei comprising the central vagal complex. Major connections are illustrated in Figure 1B. For more detailed descriptions of the medullary nuclei, see Carpenter (27). The anatomy of the distal vagal innervation of the stomach is described in detail in standard textbooks of anatomy.

Functional Anatomy of the Vagal Complex

Kerr and Preshaw (155) first proved that the DMNV was the critical relay for stimulation of gastric secretion by showing that destruction of one DMNV with contralateral vagotomy eliminated the gastric response to insulin hypoglycemia. Electrical stimulation of the DMNV in cats leads to vigorous, prolonged gastric secretion (242, 275). Andrews et al. (4) stimulated the DMNV and recorded discharge patterns in single efferent neurons, half of which entered the stomach and half of which entered the celiac and other branches of the abdominal vagus. They found no evidence of fiber branching (4, 67).

The DMNV neurons are not directly stimulated by a lack of glucose (154), but there are a number of sites from which the DMNV can be stimulated indirectly either electrically or by glucose deprivation. Shiraishi (242) has recorded responses in 68% of neurons of the DMNV from electrical stimulation of the lateral hypothalamus; only 9% of these neurons in turn stimulated gastric secretory efferents (4). Gastric secretion is stimulated when the lateral hypothalamus is deprived of glucose by local application of the glucose analogue 2-deoxy-D-glucose (2-DG) (33, 151, 152). It also responds to direct application of pentagastrin (260). Local anesthesia or destruction of the lateral hypothalamus blocks both feeding behavior and stimulation of secretion resulting from systemically injected 3-O-methylglucose (3-O-MG) (33) or insulin (152, 217), confirming the importance of the lateral hypothalamus in control of gastric secretion. The connections between the hypothalamus and the vagal nuclei have been traced by ter Horst et al. (262) and by Sawchenko and Swanson (232).

Destruction of the ventromedian hypothalamus (VMH) causes a large increase in basal acid and pepsin secretion in rats (220), suggesting that the VMH restrains the lateral hypothalamus, one major site from which both feeding (20) and gastric secretion are stimulated. Ishikawa et al. (145) found that in most but not all rats tested, electrical stimulation of the VMH partly inhibited 2-DG–stimulated acid secretion.

Bilateral lesions of the medial forebrain bundle (MFB) at the level of the middle and posterior hypothalamus, immediately adjacent to the lateral hypothalamic nucleus, also blocked the gastric secretory response to systemically administered 2-DG (152). Various "upstream" lesions failed to interrupt the 2-DG response, ruling out the possibility that the effects of the lesions of the MFB resulted from interruption of impulses passing through the MFB (152). Confirmation of the direct involvement of neurons in the MFB as a cytoglucopenic signal source came from microinjections of 2-DG into highly localized areas of the MFB, with resulting intense prolonged gastric secretory responses (151). Presumably the area of the MFB is coexistent with the lateral hypothalamic nucleus similarly stimulated by 2-DG (32, 33, 99). By contrast, destruction of the caudal two-thirds of the globus pallidus blocked both 2-DG–induced gastric secretion and feeding in cats (151), but microinjection of 2-DG did not stimulate secretion. The implication is that hypothalamic signals destined for the DMNV involve transfer via neurons in the globus pallidus (151), with probable relay in the brain stem reticular formation (27).

Intense and prolonged gastric secretion follows local injection of 2-DG, but not saline, into the medial and lateral components of both right and left NTS, especially the rostral ends (153, 154). In decerebrate cats (153), injection of 2-DG into the liver caused gastric secretion, probably via the NTS in the medulla (152, 154, 243).

The glucose-dependent trigger sites for activation of the DMNV to stimulate gastric secretion were confirmed by autoradiographic topography of the preferential uptake of $[^{14}C]$2-DG in brains of rats made hypoglycemic by insulin (150). Newer techniques, such as positron emission tomography (212), are likely to provide insights into the dynamic localized cerebral functions associated with gastric secretion and feeding in intact animals and humans.

Brain Peptides Relating to Gastric Secretion

There is a remarkable concordance between the peptides found in the gut in both endocrine and neuronal tissue and those represented in the brain (1, 42–44, 50, 68). At least two dozen peptides have been identified in brain (50, 110, 222, 223) by immunostaining, but Dockray and Sharkey (44) caution that even with the most rigorous controls, it is not always possible to make a definitive identification by immunohistochemistry. Peptides and other neurotransmitters in the neural structures associated with the DMNV have been much less well studied than those associated with the hypothalamus and pituitary (233). In the medulla, thyrotropin-releasing hormone (TRH) is localized to the DMNV, NTS, and NA (161).

To bypass the blood-brain barrier, a number of peptides and other compounds (1, 189, 201, 221, 257–259, 260, 265) have been injected by the intracerebroventricular route to study effects on gastric secretion, appetite, thirst, behavior, and other events via direct effect on sensitive structures adjacent to the cerebral ventricles. Gastric secretion has been stimulated via the vagus by TRH (259), sulfated cholecystokinin (CCK) (146), and short-chain analogues of somatostatin given intracerebroventricularly but not systemically. Somatostatin-14, acetylcholine (ACh), neurotensin, vasoactive intestinal polypeptide (VIP), and angiotensin (257) were ineffective. Despite the fact that feeding and gastric secretion are both stimulated by glucoprivation, the action of peptides that promote feeding does not correspond very well with their effects on secretion (188, 189, 257); e.g., CCK acting centrally or peripherally inhibits feeding (9, 213) but stimulates gastric secretion.

Stimulation by TRH is inhibited by central catecholamine depletion (257) but not by dopamine blockade. Noradrenergic circuitry may be involved in DMNV afferent and efferent circuits (232). Gastric secretion is inhibited by intracerebroventricular epi-

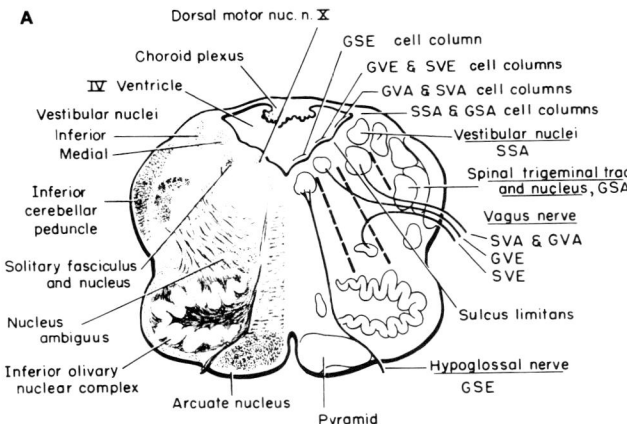

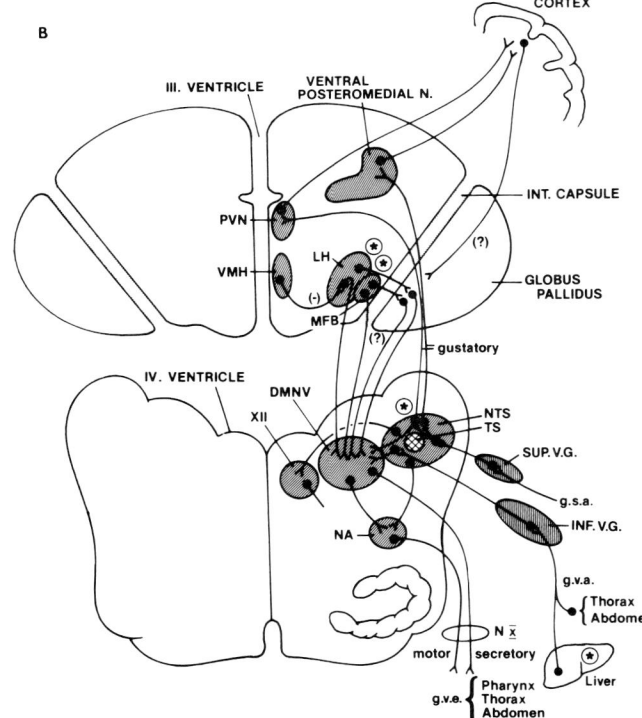

FIG. 1. *A*: schematic transverse section of the medulla showing its basic features. Cell columns related to functional components of the cranial nerve are indicated on the *right*. Functional components of cranial nerves are both general and special. Functional components of the vagus nerve are shown in relation to particular nuclei. *Heavy dashes* separate nuclei of the various cell columns on the *right. B*: schematic illustration of the connections of the central vagal complex. DMNV, dorsomotor nucleus of the vagus; gsa, general somatic afferents; gse, general somatic efferents; gva, general visceral afferents; gve, general vagal efferents; INF VG, inferior vagal ganglion; LH, lateral hypothalamus; MFB, median forebrain bundle; NA, nucleus ambiguus; NTS, nucleus tractus solitarius; NX, vagus nerve; PVN, paraventricular nucleus; SSA, special somatic afferents; SUP VG, superior vagal ganglion; SVA, special visceral afferent; SVE, special visceral efferents; TS, tractus solitarius; VMH, ventromedian hypothalamus. Sites stimulated by glucoprivation: LH, MFB, NTS, NA. [*A* from Carpenter (27).]

nephrine, norepinephrine, opioids, and most potently (50 times greater than opioids) by the amphibian peptide bombesin (201, 257), even after bilateral vagotomy (259). The mammalian counterpart of bombesin, gastrin-releasing peptide (GRP), is less effective. Bombesin, which inhibits secretion, and TRH, which stimulates secretion, produce several other profound effects: epinephrine release is stimulated, causing an increase in circulating glucagon and glucose; body temperature is lowered; behavior is altered; and feeding is depressed. It is thus evident that intracerebroventricular injection may produce a number of nonspecific effects, since all structures surrounding the ventricles are equally accessible to the compound being injected.

Vagal Excitation by Glucoprivation

Simici, Popesco, and Diculesco (246) first reported stimulation of gastric secretion during insulin-induced hypoglycemia 60 years ago. La Barre and de Cespédès (162) showed stimulation of gastric acid secretion by cross-circulating hypoglycemic blood into the head of a recipient dog connected to its stomach by only the vagus nerves. The effects of hypoglycemia can be blocked irreversibly by bilateral vagotomy and reversibly by cooling the vagal trunks to 4°C–8°C (74), by injection of local anesthetic into the nerve trunks (30), and by systemically given atropine (19, 20). The gastric acid response in dogs to insulin at doses >0.3 U/kg are bimodal, because insulin directly and potently inhibits stimulated acid secretion, but not pepsin secretion, by a mechanism that involves K^+ (101, 103).

GLUCOSE ANALOGUES: 2-DEOXY-D-GLUCOSE AND 3-O-METHYLGLUCOSE. The brain can also be deprived of glucose by intravenous administration of nonmetabolized glucose analogues (2-DG) and 3-O-MG (32, 45, 97, 98, 133). Unlike insulin (101, 103), these agents do not inhibit acid secretion (133). The glucose analogue 2-DG is a competitive inhibitor of glucose transport and phosphohexose isomerase activity, forming 2-DG-6-phosphate and blocking the further uptake and metabolism of glucose, thus effectively causing cytoglucopenia [Fig. 2; (133, 265)]. Threshold doses of 2-DG in most species exceed 25 mg/kg, and maximal stimulation of gastric secretion is seen at 100–200 mg/kg. Rather than inhibiting phosphohexose isomerase activity, 3-O-MG competes for cellular glucose transport. Thus much larger doses (~1 g/kg) of 3-O-MG than of 2-DG are required to activate glucose-dependent neural centers (97, 100, 133). The gastric response to a single dose of 2-DG is prolonged (often >4 h) and not readily reversed by subsequent injection of glucose (45, 100, 133). Gastric secretion can be sustained without fading for 24 h or more by continuous 2-DG infusion (48). In large doses, 3-O-MG augments the effects of 2-DG (133), but prior injection of a low dose of 3-O-MG, insufficient by itself to cause gastric secretion, may by dilution prevent the effect of 2-DG (100). Glucose-dependent cerebral centers have been further identified by activation through localized microinjection of 2-DG (33, 153). Effects of systemically injected 2-DG or 3-O-MG have been shown in humans (45), dogs (48, 133), cats (151), rats (32, 97, 100), chickens (23), and sheep (104), among others, whereas insulin hypoglycemia does not stimulate the vagus in sheep (136) and goats (96) at glucose levels as low as 10 mg/100 ml. In sheep brains, the high levels of hexokinase (211) may explain the tolerance of hypoglycemia and intolerance to 2-DG, manifested by convulsions (141).

Vagal excitation stimulates equally acid, pepsin (Fig. 3), and gastric antral motility (78), and all are equally and rapidly inhibited by atropine (78) or interruption of vagal trunk transmission (18, 74). Glucoprivation stimulates not only gastric secretion via cholinergic pathways and the release of gastrin (116, 123) but also causes the pancreas to secrete (184) and the stomach, gallbladder, and gut to contract; produces hunger and thirst (45, 265); and stimulates sympa-

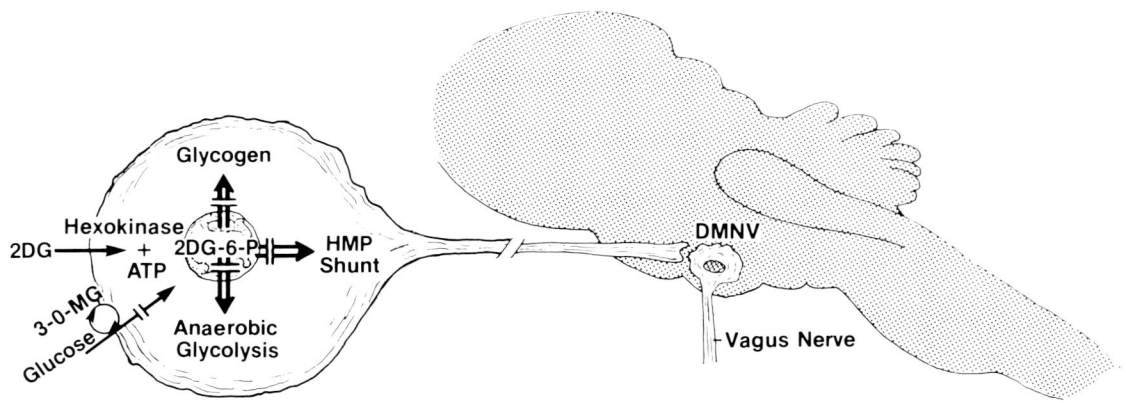

FIG. 2. Mechanism of action of glucose analogues in causing cerebral glucoprivation at sites that act via the dorsomotor nucleus of the vagus (DMNV). 2-DG, 2-deoxy-D-glucose; 2DG-6-P, 2-deoxy-D-glucose-6-phosphate; 3-O-MG, 3-O-methylglucose; HMP, hexose monophosphate.

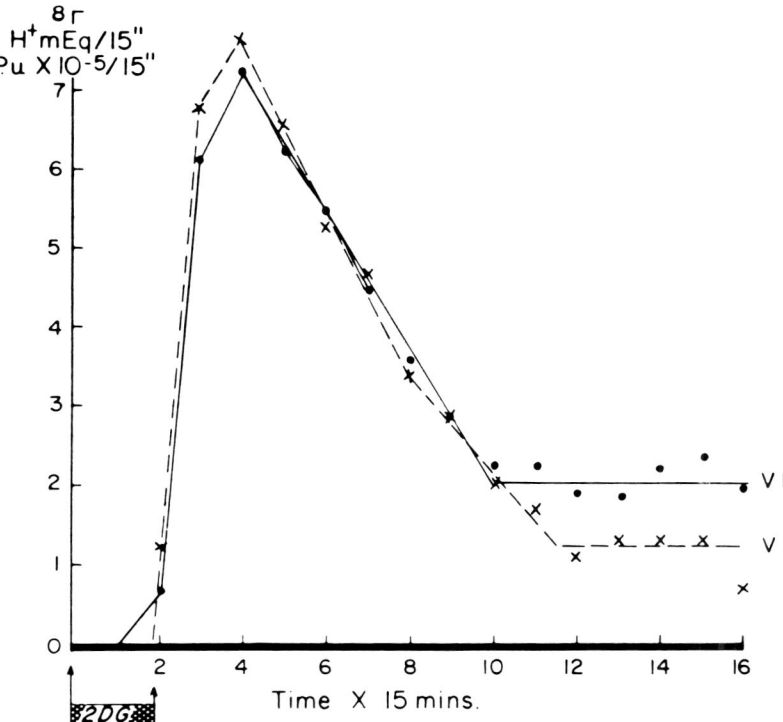

FIG. 3. Acid and pepsin outputs in gastric fistula dogs with intact vagi given 2-DG, 100 μg/kg iv over 30 min. VH, acid output; VP, pepsin output. [From Hirschowitz and Sachs (133).]

thetic adrenomedullary discharge (97, 265), anterior pituitary hormones [especially growth hormones and renin and vasopressin release (265)], and directly and indirectly the many homeostatic responses required to restore the glucose equilibrium. Serum [K^+] falls and [Na^+] rises (45, 133). With 2-DG, serum glucose (45) and plasma free fatty acids increase because of adrenergic activation (97, 99, 265). Insulin secretion fails to rise (265), presumably because after 2-DG injection the β-cells recognize and respond to the cellular glucoprivation rather than to the secondary hyperglycemia.

Vagal effects with various stimuli can be shown to be graded. Thus gastrin release and the gastric secretory (123) and motor (78) responses to 2-DG are dose dependent. Levels of pepsin output in dogs (Fig. 4) and acid output in humans (186) given insulin can be related to blood glucose. Electrical stimulation of the vagus can produce graded responses by varying either the voltage (73) or stimulus frequency (250). Thus the vagus does not act as an all-or-none switch but more like a tuning device. However, it is not normal for blood glucose concentrations to fall to levels generally associated with experimental maximal activation of the vagus. It is not known how physiological changes of glucose content or metabolism in glucose-dependent neurons normally modulate the gastric and other mechanisms invoked by other stimuli of the DMNV and feeding centers.

OTHER GLUCOSE ANALOGUES. In experiments largely relating to control of feeding behavior, other glucose

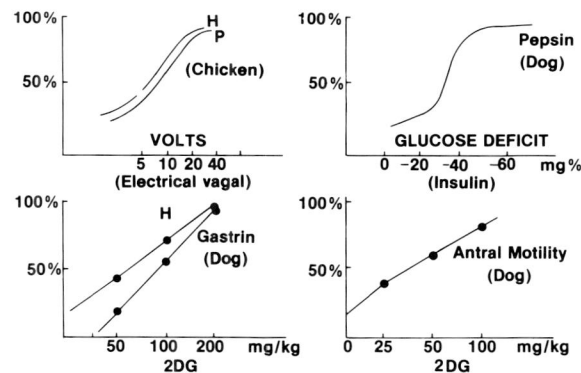

FIG. 4. Responses to vagal stimulation are dose or stimulus dependent. Acid (H) and pepsin (P) related to electrical stimulation of vagal trunks (70); pepsin secretion related to the level of blood sugar (101); gastrin release and acid secretion (114) and antral motility (74) related to dose of 2-deoxy-D-glucose. [From Hirschowitz (106).]

analogues, such as 5-thioglucose and gold thioglucose, produced localized lesions in the hypothalamus. These lesions result in hyperphagia, polydipsia, obesity, and increased basal gastric secretion (40). Lesions produced by 5-thioglucose or gold thioglucose are most prominent in the ventromedian hypothalamus (VMH) and are similar to the effects of electrical destruction of this area in the study by Ridley and Brooks (220). The VMH is thought to modulate or restrain the stimulatory center of the lateral hypothalamus by pathways common to appetite and gastric secretion.

Intracerebroventricular injections of 2-DG or 5-thioglucose stimulate feeding (221) by a noradrenergic pathway (14). It is not clear why the VMH preferentially accumulates the thioglucose analogues, since in the hypoglycemic state the lateral hypothalamus rather than the VMH preferentially takes up [^{14}C]2-DG (150). Thioglucose analogues have not been similarly tested under conditions of altered blood glucose.

Non-Glucose-Related Chemical Stimuli of the Central Vagus

Most non-glucose-related compounds that stimulate the secretory vagus are also neural inhibitors. These include systemically injected 5($\gamma\gamma$-dimethylallyl)-5-ethylbarbituric acid (6), alcohol [the effect of which is eliminated by bilateral vagotomy (131)], and GABA (γ-aminobutyric acid). The GABA$_B$ agonist, baclofen, a lipophilic derivative of GABA [β(p-chlorophenyl)-γ-aminobutyric acid], is a potent gastric secretory stimulus when given parenterally or intracerebroventricularly (75, 76). Although the increase in vagal spike activity (from ~30 Hz to 250 Hz) induced by baclofen in rats was not altered by atropine, stimulation of gastric secretion was eliminated (76) by either bilateral vagotomy or a large dose of atropine. The stimulant action of another GABA agonist, muscimol (given intracerebroventricularly), was blocked by the GABA$_A$ antagonist bicuculline and by atropine (170). The exact location of the central action of the GABA agonists is not known. Although GABA stimulates gastric secretion, application of the GABA antagonist bicuculline to previously identified neurons in the DMNV stimulates antral motility (272). This suggests that gastric secretion and antral motility are affected oppositely by GABA, whereas glucoprivation stimulates both together.

MECHANISM OF NERVE CELL STIMULATION. The cellular mechanism that might excite neuronal activity through deprivation of an essential energy substrate, such as glucose, is unknown. Possibly the neurons of the sites sensitive to the lack of glucose exist in a suppressed state, maintained by energy- or substrate-dependent, endogenous ionic (e.g., K$^+$) currents, analogous to the M-current described by Brown (21, 22) that is depolarized by ACh in ACh-sensitive cells. If the analogy holds, glucose deprivation would activate the cell by altering the frequency and pattern of cell firing. The mechanisms remain to be studied.

The prolonged response to baclofen (75, 76) and to glucoprivation and the difficulty of reversing the vagal stimulation by elevating blood sugar after insulin hypoglycemia (102) or 2-DG injection (97, 133) indicate that the mechanism of stimulation may not depend directly on glucose concentration but on the presence of a slowly cleared metabolite or a slowly corrected cell deficit accumulated as a result of glucose lack. It is unlikely that central vagal stimulation is entirely glucose related, since many other apparently unrelated compounds also stimulate the gastric vagus.

CENTRAL ANTAGONISTS OF GLUCOPRIVIC STIMULATION. *Sugars.* Pretreatment of the intact animal with glucose or mannose (133) or subthreshold (diluting) amounts of 3-*O*-MG (98, 100) prevents or delays 2-DG–stimulated gastric secretion, whereas fructose, galactose, and xylose, which are not substrates for brain metabolism, are ineffective (133). Whether anomers of glucose would all be equally effective substrates is not known (230).

Anticholinergics. Although antimuscarinic inhibition of vagally stimulated gastric secretion by atropine (78) might be central or peripheral, the effect of pirenzepine, a specific antagonist of a subtype of muscarinic ACh receptors with similar potency (112) indicates that the muscarinic link is peripheral rather than central. Pirenzepine, a quaternary compound, does not cross the blood-brain barrier. Moreover, although the GABA$_B$ vagal stimulation of the stomach by baclofen (76) is atropine sensitive, the vagal spike activity induced by baclofen is not. Thus the central vagal excitation by GABA$_B$ or glucoprivation is not cholinergically mediated.

GABA. Thirlby et al. (264) report that a GABA precursor (progabide) inhibits insulin hypoglycemic (i.e., vagal) stimulation of acid, gastrin, pancreatic polypeptide, and somatostatin release in dogs. The antagonist effect of this GABA precursor contrasts sharply with the stimulant effect of the GABA$_A$ and GABA$_B$ agonists, muscimol (170) and baclofen (75, 76).

Opiates. Morphine, a μ-agonist, augments the 2-DG response in dogs, whereas naloxone, a μ-antagonist, inhibits both 2-DG stimulation and the morphine effect (3). Met-enkephalin, a δ-opiate agonist that does not cross the blood-brain barrier, inhibits secretion after intravenous injection, presumably at a peripheral site. The evidence, albeit indirect, implicates an opiate pathway in the neuronal activation by glucoprivation or transmission from glucose-dependent neurons to the DMNV.

Dopamine. Bromocriptine reduced 2-DG–stimulated acid secretion in either conscious or anesthetized rats (176). Apomorphine, another dopamine agonist, reduced 2-DG effects only in anesthetized rats. Maximum inhibitory effects were 68% with bromocriptine, 68% with apomorphine, and 93% with atropine. Neither bromocriptine nor apomorphine exhibited peripheral gastric effects. Dopamine antagonists haloperidol, sulpiride, or metoclopramide partly blocked the inhibitory effects of bromocriptine in both conscious and anesthetized rats, though these agents themselves have no effect on gastric secretion. The significance of these data awaits further topographic and neurophysiological verification.

Anesthetics. Injection of a local anesthetic (lignocaine) in the lateral hypothalamus blocks the stimulant effect of locally applied 2-DG or systemically injected 3-O-MG (33). General anesthesia with pentobarbital sodium greatly reduces the gastric secretory response to insulin hypoglycemia in dogs (207). Studies of central vagal stimulation in anesthetized animals should be interpreted cautiously.

CORTICAL AND CEPHALIC PHASE VAGAL STIMULATION. Apart from glucoprivation and various chemicals given intracerebroventricularly that appear to act on hypothalamic or medullary structures near the third and fourth ventricles, there are signals that originate more distally and act on the DMNV. These include signals from the cerebral cortex, the limbic system, and (more peripherally) the liver, as well as responses to conditions requiring additional caloric input, such as lactation or exposure to cold (19).

The stimulation of gastric secretion by electrical stimulation of various cortical structures is largely inconclusive. Decortication and decerebration experiments distorted or delayed the response to hypoglycemia but did not prevent gastric stimulation (19). Nevertheless, since the time Pavlov first studied sham feeding and described conditioned reflexes, it has been clear that a strong cortical component was part of the normally integrated process of food-related stimulation of gastric secretion (55, 219). Gastric secretion and gastrin release can both be stimulated in humans by intensive discussion of food (57, 187) or suggestions of pleasurable eating under hypnosis (47). The olfactory and visual effects alone of watching and smelling food being prepared provide lesser input than concentrated discussion about food for 30 min (57). The combined multiple gustatory signals from sham feeding are about twice as potent as thoughts or discussion of food alone (56, 61).

Sham feeding. In sham feeding, food is chewed and swallowed but prevented from reaching the stomach by esophageal occlusion using a balloon (252) or by external diversion through a cervical esophagostomy, as first practiced by Pavlov (16). In modified sham feeding (MSF), food, usually meat and bread, is chewed and expectorated (16, 158). Though chewing flavored gum provides nearly as much stimulation as MSF (93), chewing a rubber or plastic tube is ineffective (158, 219), suggesting involvement of gustatory afferents. Sham feeding may stimulate gastric secretion to 40%–50% of the maximum obtainable with pentagastrin or histamine (16, 58, 59, 94, 158, 219) and to 60%–80% of the output attained by insulin hypoglycemia (254); however, the pepsin response relative to acid may be lower than that obtained with hypoglycemia (135).

Sham feeding stimulates gastric secretion solely via the vagus through a muscarinic step, since the response can be completely blocked by atropine (60, 94, 158). More than one afferent is involved in sham feeding; i.e., the mental input (56, 57, 221) is supplemented by olfactory, taste, and tactile sensations from the oropharynx. The various signals are summated and integrated, probably in the limbic system (153, 267). The signals converge on the DMNV and probably the NA via the MFB and descending reticular pathways.

Hunger alone is insufficient to initiate gastric secretion in the fasting gastric fistula dog. Even after 24–48 h of fasting, dogs trained to stand in a Pavlov stand secrete no gastric juice for observation periods of up to 6 additional hours (B. I. Hirschowitz, unpublished observations), even if food can be seen or smelled. However, when food is chewed, swallowed, and diverted out of a cervical esophagostomy, the dogs begin to secrete actively, and this secretion long outlasts (1–2 h) the period of sham feeding (19, 135, 158, 196). This suggests that although hunger may prime or sensitize the system, the cephalic phase of gastric secretion in animals represents the integration of several input signals triggered by actual food contact with the mouth. The oral signal converges with others ultimately on the gustatory center to stimulate salivary flow, initiate swallowing reflexes, and, via the DMNV and NA, stimulate the gastric vagus. The difference between hunger in the dog and thoughts or discussion of food in humans poses the question of how animals think about food.

Although the influence of reduced blood glucose levels on appetite and the cephalic phase of gastric secretion is well defined, the question of whether elevated glucose levels influence central vagal control was addressed by Moore and Crespin (187), who showed no inhibition of sham feeding responses in human volunteers given graded intravenous infusions of glucose to increase blood sugar concentrations. Only intraduodenally administered glucose inhibited secretion. Similar inhibition may be seen with fat and acid in the gut. The site (gut or liver) or mechanism of this effect of glucose is not known, but reduction in the gastric acid response to MSF does not appear to be due directly to the increase in circulating and hence cerebral glucose concentration.

Modified sham feeding also reportedly stimulates HCO_3^- secretion from the human stomach. Using direct techniques, Forssell et al. (66) found a stimulated value of 0.7 mmol/h, or ~150% of basal, while Feldman (53), using an indirect method, reported 6 mmol/h during MSF (see also the chapter by Flemström and Garner in this *Handbook*).

In humans, the magnitude and duration of response to MSF are independent of the duration (from 5 to 30 min) of the stimulus (158). The prolonged response indicates that a complex central reflex mechanism is entrained and sustained without external reinforcement. No studies have been done to truncate the response centrally; the distal limb can be rapidly blocked by atropine (78) or vagal interruption (19, 82).

In the majority of studies in humans, serum gastrin levels are unchanged by MSF (2, 16, 161, 253). In others, a small rise has been reported (60, 62, 158),

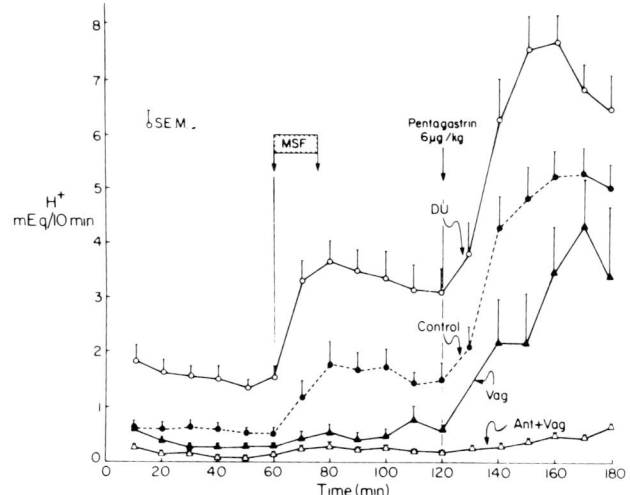
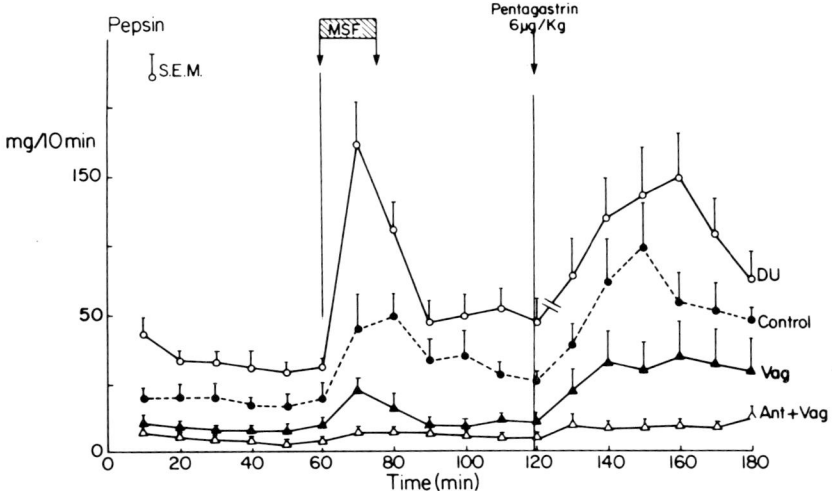

FIG. 5. Secretion of acid and pepsin in the basal state (60 min) during and after 15-min modified sham feeding (MSF) (60–120 min) and after pentagastrin 6 μg/kg subcutaneously in 4 groups of subjects. Duodenal ulcer (DU) (n = 60), normal controls (n = 40), duodenal ulcer after fundic vagotomy (Vag) (n = 20), and duodenal ulcer after truncal vagotomy and antrectomy (Ant) (n = 16). n, Number of subjects.

but to levels that would not sustain a prolonged increase in gastric secretion. In this respect, MSF is different from the more potent vagal stimulation by insulin and by 2-DG, both of which promote release of gastrin (104, 114, 123, 252).

Modified sham feeding has been used as a test for vagal intactness in surgically treated duodenal ulcer patients (6, 15, 58, 59, 62, 156, 160). Though it may not be quite as sensitive (145), the test is now widely used, because it avoids the possible risks of insulin hypoglycemia (101) or 2-DG (45). Figure 5 illustrates the gastric acid and pepsin responses to MSF in normals and in patients with duodenal ulcer; the responses in duodenal ulcer patients are eliminated by adequate vagotomy of the fundic area, leaving the antrum innervated. If gastrin were released by MSF, it is not in sufficient quantity to stimulate acid or pepsin secretion. However, antrectomy with vagotomy caused a much greater decrease in response than vagotomy alone (Fig. 5). Antrectomy also greatly reduced the fundic response to pentagastrin, indicating that the antrectomy effect on MSF is not solely due to loss of any gastrin released by MSF.

ELECTRICAL STIMULATION OF THE GASTRIC VAGUS. Vagal stimulation of the stomach is most directly accomplished by electrical stimulation of the vagus nerve trunks (91). If the nerve is not damaged, secretion of acid may be stimulated for hours. Such experiments are obviously not physiological, since the animals are anesthetized (207) and electrical parameters used do not necessarily mimic endogenous electrical excitatory signals (83). Also, because one or both vagus nerves are transected, physiological modulating reflexes are inoperative.

In cats and dogs, graded responses of acid (and pepsin) secretion have been obtained by varying frequency [0.5–8 Hz, 4–5 V, 4–5 ms impulses (17, 250)] and in chickens by varying voltage [20–50 V, 1 ms, 10 Hz (72)]. Andrews et al. (4) have shown that central electrical stimulation of brain stem neurons produces impulses bilaterally at a rate of firing of 0.5–5 Hz, suggesting that the optimum frequencies found by most investigators in the range of 4–10 Hz may be appropriate for efferent secretory signals, compared with a rate of 30–50 Hz for afferent impulses and for motor stimulation from central nuclei (196). At low frequencies, only mucin secretion is stimulated, whereas at higher frequencies, up to 30 Hz, acid and pepsin predominate (250), indicating that the vagus may be able to independently stimulate different elements of gastric function: motility, sphincter function, gastrin release, and (separately) the secretion of mucin (261, 273), HCO_3^- (52), pepsin, acid, and chloride. Above 30 Hz and up to 120 Hz, acid secretion was inhibited (83). However, Goto et al. (76) reported frequencies of up to 250 Hz in vagi of rats stimulated by a GABA agonist. The origin of such signals and the precise mechanisms whereby they are discriminated are not known.

Although the antrum is important for the full gastric response to stimulation of the vagi (267), the fundus can be stimulated directly by the vagus, even with gastrin release blocked by antral cocainization (248) or the antrum selectively vagally denervated (239, 261) or removed (80).

Vagal Afferents

The more than 95% of the fibers in the abdominal vagal trunks that are afferent may be preferentially devoted to processing signals for intestinal motor reflexes (190, 199). From recordings of single vagal fibers, the sensory system is activated by both chemical (142, 143) and mechanical stimuli. These signals can also be detected in the DMNV (7, 90, 274). Mechanoreceptors are divided into slow (i.e., steady-state or tone sensors) and rapid, representing rate of change that presumably determines anticipatory motor changes. In the stomach both distension (7) and contraction increase the rate of firing (199). In cats, unmyelinated C-fibers (the predominant type in the vagus) with conduction velocities of ~1–5 m/s (143) responded with no delay to distension at pressures of 2–25 mmHg in the esophagus, stomach, or small gut. The impulses, which arise from the muscle layers and not the mucosa or submucosa, lasted only as long as the stimulus was applied and varied from 30 to 50 Hz. Slowly adapting mucosal chemoreceptors, especially abundant in the antral mucosa, activated another set of C-fiber afferents when stimulated by 0.1 N NaOH or 0.1 N HCl, threshold pH >8.0 or <3.0. Individual fibers responded to acid or alkali but never to both. The area innervated by a single axon was as large as 1.5–4 cm^2, with the most sensitive area ~5 mm^2. The receptors are not affected by osmotic extremes nor by stretch but are stimulated by light stroking (35, 142). The responses of chemoreceptors were not considered to be injury currents and may be the same as the mechanoreceptors but with a longer stimulation threshold (227).

VAGOVAGAL REFLEXES. Distension of the innervated stomach produces secretion from the fundus as well as gastrin release via long-loop reflexes (82). After mechanically stimulating vagal afferents, Davison and Grundy (36) recorded activity from efferent vagal fibers in chloralose-anesthetized rats. Distension of the stomach increased the rate of firing in one group of fibers, while another group responded in a reciprocal manner with a decreased spontaneous rate of firing. Vagotomy eliminated the reflexes. The destination of the efferent signals (excitatory or inhibitory) could not be determined, nor were the physiological responses recorded. It is probable that the reflexes involve motor rather than secretory efferents (190).

Electrical stimulation (5 V, 2 ms, 10 Hz; 10 s on/10 s off) of the proximal end of the cut left cervical vagus caused about half as much gastric secretion as unilateral right efferent vagus stimulation in pentobarbital sodium–anesthetized rats (224). The response was diminished by ~40%–50% after electrolytic lesions of the ipsilateral (left) paraventricular nucleus of the lateral hypothalamus. The results were interpreted as being due to the interruption of the predominantly excitatory influence of the paraventricular nucleus on the NTS and hence on the DMNV. These results and those of Harper et al. (91), using the central cut end of one abdominal vagus rather than the cervical vagus, describe long-loop vagovagal secretory reflexes, as well as contralateral reflexes.

The liver provides yet another source of afferent signals that can stimulate gastric secretion via the vagus. Both glucoprivation of the liver (153, 243) and electrical stimulation of the central end of the hepatic branch of the vagus (222) produce impulses via the left vagus to the NTS (225) and result in gastric secretion (230). In hypoglycemic rats, portal injection of glucose reduced gastric secretion (79, 230). The β-D-glucose anomer was more effective than the α-D-glucose anomer (220). The extent to which these signals are physiological has not been established. Infusion of glucose into the liver produces uncertain brain stem signals that cannot be distinguished from signals resulting from distension of the stomach (7).

MACROMOLECULES IN AFFERENT NEURONS. The afferent neurons of the abdominal vagus have their nerve cell bodies in the inferior (nodose) vagal ganglion, peripheral processes originating in the gastric wall and central processes projecting to the medullary nuclei described in *Central Vagal Complex*, p. 127. The presence and antidromic direction of transport of a number of peptides and proteins present a view of the afferent nerves as being more than electrical conduits from the periphery to the CNS.

Visceral afferents involve both vagal and spinal neurons (44). Identification by immunohistochemistry, radioimmunoassay, nerve ligature and toxin block of axonal transport, retrograde marker tracing, and more sophisticated measurements of physiological functions have shown that visceral afferents synthesize and transport a number of biologically active peptides, all of which occur as well in the brain and in nerves and endocrine cells elsewhere in the body (49). Each belongs to a family of related peptides with similar patterns of biological activity. Moreover, as with many transmitters, each peptide may have a wide spectrum of biological activity, depending on its site of release (1). Physiological interpretations of topographic location should thus be made with appropriate caution.

Nerve cell bodies in the nodose ganglion of the vagus contain peptides that stain with antibodies to substance P, CCK-gastrin, somatostatin, VIP, and possibly calcitonin gene-related peptide. These are transported distally in afferent fibers, since they accumulate central to vagal trunk ligatures. Substance P, somatostatin-14, G-17, and CCK-8 have also been found in vagal nerve extracts (43, 72, 184, 215). Substance P is the most abundant (50–250 pmol/g) of vagal peptides that also include CCK, somatostatin, and VIP (1–10 pmol/g). Most of the substance P–containing afferent neurons of the nodose ganglion of the vagus involve the thoracic organs rather than the upper gastrointestinal tract, though a few substance P–containing fibers can be traced by double labeling from the anterior wall of the rat stomach to the ventral end of the nodose ganglion (44). Peptides and proteins synthesized by cell bodies of afferent neurons are also transported to the NTS as well as to the periphery. Because axonal flow is bidirectional, transported substances accumulate on both sides of a ligature on the vagus nerve trunk; double ligature and immobilization of axonal flow (including axonal flow of the dye True Blue) by colchicine or capsaicin (241) suggest that ~70% of peptides or proteins are stationary.

Small-diameter vagal axons also contain and transport several important receptors, three of which have been studied in some detail: muscarinic ACh receptors (280), opioid receptors (278), and CCK receptors (279). A proportion of these receptors, like the various messenger peptides, is transported in both directions at rates roughly two times faster in the centrifugal than the centripetal direction (279, 280). Although the transport rates for peptides are given as 0.6–4 mm/h (43), those for receptors are estimated to be ~0.3 mm/h (279). When corrected for the proportion of axonally flowing receptors that are immobile, the rates may be more than twice as high and may be classed as fast axonal transport. Although muscarinic and opioid receptors are located in afferent or sensory fibers (278), Zarbin et al. (280) have suggested that CCK receptors also occur in efferent nerves and originate in the brain stem nuclei. Moreover, centripetally moving muscarinic receptors differ from those flowing to the periphery. The latter exhibit high affinity for carbachol, while those flowing to the cell body and those stationary along the nerve trunk exhibit low affinity and are less sensitive to modulation by a guanine nucleotide, GppNHp (279). These probably represent different modulated forms of a single receptor molecule.

Different biologically active mobile peptides and receptors may coexist in afferent neurons with conventional transmitter substances (49, 271). Although the flow of signals in the afferent neurons is toward the CNS, the bulk of peptides are transported to the periphery, where they are released and may contribute to axonal reflexes. Moreover, peptides released distally produce responses in secretion, circulation, smooth muscle, and gut mucosa, either directly or indirectly through action on ganglia or release of active intermediates (e.g., histamine) (44). From these data it would appear that afferent neurons of the vagus may also be effectors. The afferent vagal neurons may represent yet another level of integration and control of gastric function between the enteric nervous system (ENS) and CNS.

PERIPHERAL CONTROLS OF GASTRIC SECRETION

Both the traditional gastric and intestinal phases of gastric secretion fall largely under the heading of peripheral controls.

The gastric phase describes gastric secretion resulting from the presence of food in the stomach. Responses result from distension, which activates antrofundal and local neuronal reflexes and long-loop vagovagal mechanisms. Other responses depend on chemical stimuli: food products after digestion stimulate gastrin and probably other peptides, and secreted acid completes the feedback loop by inhibiting gastrin release. The intestinal phase comprising stimulation of gastric secretion by amino acids and inhibition by fat and acid in the intestine is much less well defined.

Enteric Nervous System

The notion that the stomach is "denervated" by surgical vagotomy in the treatment of peptide (duodenal) ulcer has obscured the fact that the ENS is not only interposed between the CNS and the gut as a relay station but is an independent integrating neural network, with structural, chemical, and functional properties analogous to those of the CNS. The ganglia of the gut are even more developed than the autonomic prevertebral ganglia (274) and more closely resemble CNS organization in the compact organization of neurons and glial cells. A blood-ganglion barrier exists and blood vessels do not penetrate ganglia (274). The endogenous independent function of vessels and muscles and exocrine, endocrine, and absorptive cells are further tuned by precisely balanced input from the ENS via neurotransmitters and endocrine and paracrine neuroeffectors.

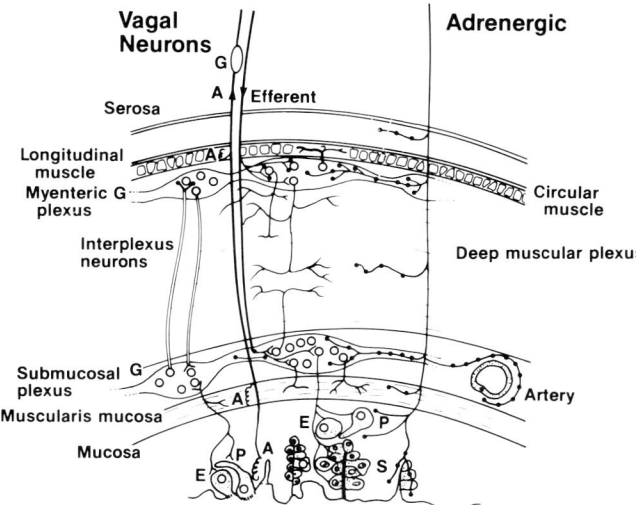

FIG. 6. Schematic drawing of autonomic innervation of the stomach by vagal and adrenergic neurons. A, afferent; E, endocrine (shown as both open and closed cells); G, ganglia; P, paracrine; S, secretory cells.

The ENS of the upper GI tract is connected to the CNS by two pathways: *1*) the parasympathetic via the vagus and *2*) the sympathetic via the celiac ganglion and the splanchnic nerves (77). Of the two, the vagus is functionally more important to gastric secretion. The myenteric and submucosal plexuses of the stomach and gut wall (Fig. 6) are each composed of a network of ganglia densely connected to each other by nervous cell processes, which also connect the two plexuses (71). In turn, the ganglia are connected to the extraenteric nervous structures by both afferent and efferent axons. The afferent vagal fibers terminate in cell bodies in the nodose ganglion and thence connect to the brain stem.

The efferent fibers of the vagus, representing some 2,000 of the 60,000 fibers in the human subdiaphragmatic vagus (137), originate in the DMNV and NA and terminate on the estimated 10^7 ganglia of the gastrointestinal tract via axosomatic, axoaxonal, and axodendritic synapses (210). Given the large numerical disparity and the sometimes marginal or subtle alteration or loss of gastrointestinal function and only fractional loss of gastric ENS neurons by ultrastructural studies (195) after external denervation, there must be a large amplification of any central signal via the intramural neural networks. There are fewer ganglia in the gastric submucosal than in the intestinal submucosal plexus or either gastric or gut myenteric plexus (210). Although some adrenergic fibers from the splanchnic nerves apparently bypass the intramural ganglia to innervate directly blood vessels (71, 274), as illustrated by fluorescent staining (210) and electron microscopy (195), the vagal efferents apparently all terminate in ganglia.

Although electrical nerve events and muscle contraction can be measured in spatial and temporal relationships (274), the relationships between nerves and exocrine or endocrine secretion or absorption, which are much slower events than contraction, are less amenable to isolation and measurement at the cellular level. This makes it difficult to know which nerves subserve which function or whether any neuron or combination is multifunctional. The coincidence of periodic gastric and pancreatic secretion with periodic motor activity, persisting after vagal interruption and eliminated by atropine or pentolinium, demonstrates at least one area of overlapping motor and secretion activity of the ENS (177).

Modern histochemical techniques have demonstrated a much more complex chemical picture of the ENS than was generally held until quite recently. Beyond the classification into cholinergic and adrenergic neurons, the same intramural neurons may also contain other transmitters or one or more of the peptides VIP, substance P, gastrin, CCK, somatostatin, bombesin, neurotensin, and enkephalin (44, 49, 215, 271), many of uncertain function (238). Their counterparts and others yet to be discovered exist in the CNS, where their precise functions are not well defined (50). Although some neuroenteric peptides have known hormonal functions, some also have paracrine functions, and their location and coexistence with classical transmitters in neurons indicate a neurocrine or transmitter role for peptides as well (43, 49, 143, 271).

The circular muscle is a syncytial system with inherent contractility in which the ENS plays a largely modulatory and integrative role to produce peristalsis and propulsion by a mixture of carefully regulated excitatory and inhibitory signals. The vagus appears to serve as the overall regulator of the integration provided by the ENS, since for a time after vagotomy, electrical and motor activity becomes chaotic (71, 227, 274). Although gut function apparently recovers or adapts after vagotomy, the ENS probably then responds within narrower limits. The best recognized disturbances after truncal abdominal vagotomy are a decreased sensitivity of the secretory mucosa to noncholinergic stimuli, delayed gastric emptying, and diarrhea. Nervous control of intestinal absorption and secretion have not been well enough defined to describe a clear role for the vagus.

Gastric secretion, on the other hand, although using many and perhaps most of the same neuronal structures and transmitters, is organized differently from muscle at the effector level. Secretion is controlled at several interacting levels—nervous, hormonal, and paracrine (including cell-to-cell communication)—by interstitial tissue environment and by luminal factors (106). Nervous control of secretion by the vagus is probably entirely mediated by the postganglionic and intramural neural networks. Postganglionic cholinergic nerve terminals have been shown in proximity to some, but not all, parietal cells (192, 193, 195, 210); taken with the pharmacologic evidence of atropine sensitivity (112), both the ganglionic and postganglionic stimulation (e.g., with bethanechol) of acid and pepsin secretion are muscarinic. Moreover, parietal cells (231, 250) and peptic cells (95, 244) can be stimulated directly by cholinergic agents.

Although ACh is clearly the transmitter of the efferent vagal fibers, the peptides gastrin, CCK, GRP, substance P, gastric inhibitory peptide (GIP), VIP, and perhaps others probably occur largely in afferent vagal neurons or in postganglionic enteric neurons (49). The peptides and transmitters in postganglionic fibers may act through neurocrine mechanisms directly on effector cells or through stimulating and inhibiting paracrine or endocrine cells, such as gastrin, somatostatin, and histamine-secreting cells [Figs. 6, 7, 13; (231)].

Chemical Transmitters of Neural Effects

The principal chemicals involved in vagal stimulation of gastric secretion are ACh, gastrin, the bombesinlike GRP, histamine, and the inhibitor somatostatin.

ACETYLCHOLINE. Vagal stimulation of gastric secretion and motility is mediated at one or more critical steps by ACh. This conclusion follows from the complete suppression of vagally stimulated gastric secretion and motility (78) by atropine at doses of 10–20 µg/kg (113). Inhibition by the muscarinic antagonist pirenzepine (74, 112), a quaternary compound that does not cross the blood-brain barrier, indicates that the cholinergic step is peripheral and not central. Inhibition by atropine of vagal stimulation is seen in all species and is independent of the vagal stimulant: 2-DG (112), hypoglycemia (18, 19, 134), and sham feeding (94, 129) are equally susceptible to atropine inhibition. The sensitivity to atropine, reflected by an ED_{50} of ~1 nM/kg (112), clearly defines vagal stimulation of gastric secretion as muscarinic (8). Bethanechol stimulation of gastric secretion in dogs is equally sensitive to atropine (112).

The susceptibility of vagal stimulation to atropine provides a basis for examining basal (fasting) secretion. In dogs there is no basal secretion, whereas in humans basal secretion is present and is generally increased in patients with duodenal ulcer. The susceptibility of basal secretion in both normals and duodenal ulcer patients to atropine with ED_{50} in the range of ~1 nM/kg (111, 129) shows that basal secretion is sustained by a muscarinic mechanism. That the vagus is in turn largely responsible for basal secretion is shown by the virtual elimination of basal secretion by vagotomy (58, 160, 207).

Considering the marked sensitivity to atropine, it is worth stressing that many studies (including some of my own) report the use of very large doses of atropine, some as high as 200 µg/kg or almost 300 nM/kg. Doses in excess of 25 µg/kg probably include many nonspecific peripheral and central effects and should not be relied on to define muscarinic receptor specific events.

The vagus stimulates both acid and pepsin secretion via muscarinic pathways, and both are equally susceptible to atropine. However, histamine H_2 antagonists inhibit only the acid stimulated by the vagus or by cholinomimetics (126). This suggests that ACh acts either through or with histamine to stimulate parietal cells in intact stomachs, even though there is evidence for direct cholinergic stimulation of isolated parietal cells through specific cholinergic receptors that act on the cell by pathways distinct from those of histamine (231). Moreover, cholinergic receptors have been localized on parietal cells by electron-microscopic autoradiography with [^{3}H]-labeled 3-quinuclidinyl benzilate (QNB) (191, 192). Stimulation of peptic cells by ACh apparently occurs without any further intermediation (95), and during histamine infusion in dogs, where acid is maximally stimulated and pepsin not, superimposition of vagal excitation by 2-DG (133) or insulin (101) results in a strong stimulation of pepsin secretion without change in acid secretion (Fig. 8). The added pepsin stimulation is completely inhibited by atropine (133). Electrical vagal stimulation also increases pepsin secretion under similar circumstances (248).

Although the role of ACh in vagal stimulation may be defined in part through the use of antagonists, cholinergic agonists have provided conflicting evidence regarding the role of ACh. Despite the evidence from antagonists indicating cholinergic pathways for

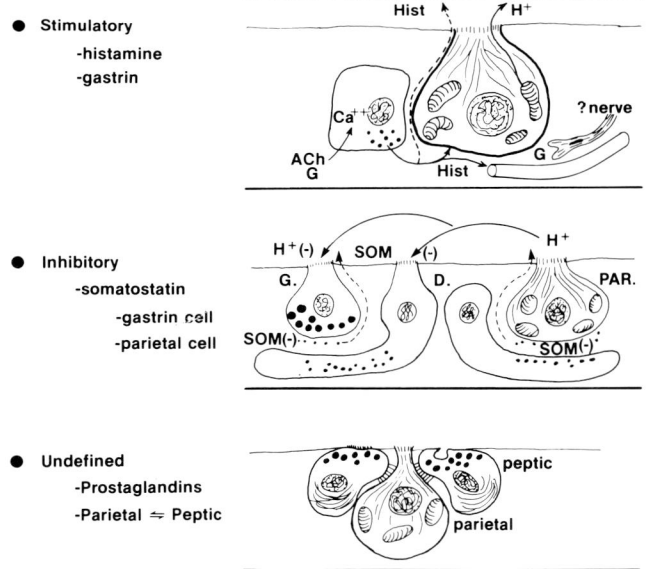

FIG. 7. Paracrine controls of gastric secretion include possible stimulatory release of histamine (HIST) from a tissue histamine cell by acetylcholine (ACh) and perhaps gastrin (G). Inhibitory effects of somatostatin may involve both gastrin and parietal (PAR) cells. Undefined paracrine effects include unknown parietal–peptic cell interactions via gap junctions. *Middle panel*, both open (antral) and closed (fundus) G and D cells are shown. D, D cells; SOM, somatostatin; (−), inhibition.

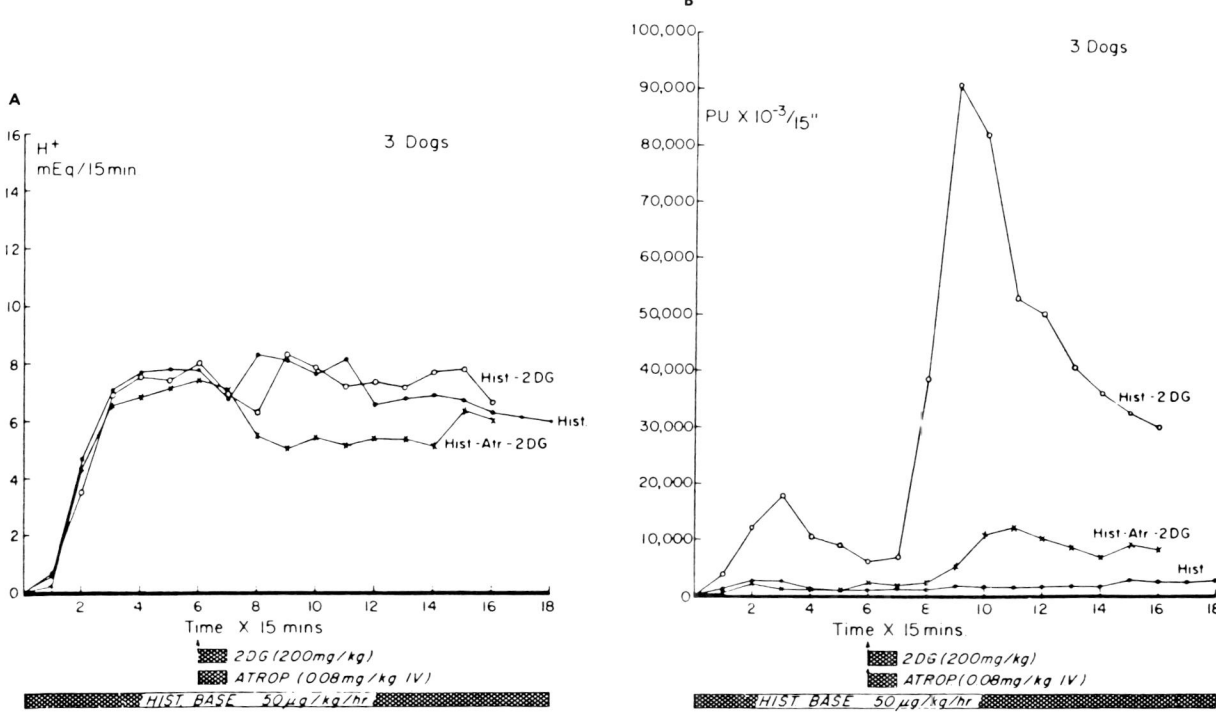

FIG. 8. Effects of an intravenous injection of 2-DG on acid and pepsin output during a continuous infusion of histamine. In another experiment, 2-DG was given with atropine, 80 µg/kg. PU, peptic unit. [From Hirschowitz and Sachs (133).]

vagal stimulation of secretion in humans, cholinomimetics are weak and erratic stimuli of both acid and pepsin secretion (61, 63, 118, 226) and do not stimulate gastrin release (61, 118). In the dog, however, bethanechol is a potent stimulus of gastric secretion (80, 112, 121, 126). The species difference remains unexplained.

Fundic vagotomy in dogs results in a decreased response to stimulation by gastrin or histamine (120, 121). This effect can be reversed by a background infusion of a subthreshold dose (10 $\mu g \cdot kg^{-1} \cdot h^{-1}$) of bethanechol. In humans bethanechol may also restore pentagastrin- (118) or histamine-stimulated (203) acid and pepsin to near prevagotomy levels. The exact mechanism whereby fundic vagotomy decreases sensitivity and cholinergic agents reverse the effect is not known. Because atropine also shifts the histamine (104, 134) and pentagastrin (122) dose-response curves to the right, it must be assumed that the pseudocompetitive inhibitory effects of vagotomy and of atropine are due to a lack of ACh at the cell level.

Contrasting with decreased sensitivity after vagotomy, the administration of bethanechol as background in both dogs and rats potentiates stimulation by histamine and gastrin of gastric acid and pepsin secretion. The effects of bethanechol are twofold: 1) an increase of maximum output of acid and 2) a leftward shift of the dose-response curve of both histamine (119, 120) and pentagastrin (121). Both effects are proportional to the dose of bethanechol and are equally effective before and after vagotomy (120). In humans bethanechol does not potentiate either gastrin or histamine stimulation of acid but does elevate pepsin output when given in addition to either secretagogue (117, 226).

Because bethanechol acts at postganglionic rather than ganglionic sites (24, 70), these effects of cholinergics are more likely to be cellular than via the ENS. Also, activation of the ENS by vagal stimulation during histamine infusion in dogs does not potentiate stimulation of acid secretion, while stimulating pepsin through muscarinic pathways (Fig. 8).

A cellular site of potentiation by cholinergics can also be inferred from effects demonstrable in isolated gastric glands (9) and parietal cells (231, 250). It is therefore likely that the pseudocompetitive effects of vagotomy and atropine, the reversal of the vagotomy effect by bethanechol, and potentiation by bethanechol of histamine or pentagastrin effects (in dogs) all occur through actions on the same mechanism. Such a mechanism most likely involves events downstream in the calcium-dependent messenger pathways.

The role of ACh in the release of gastrin is discussed next.

GASTRIN. *Physiological effects.* Gastrin is important in both central and peripheral stimulation of gastric secretion. Central stimulation is augmented by syn-

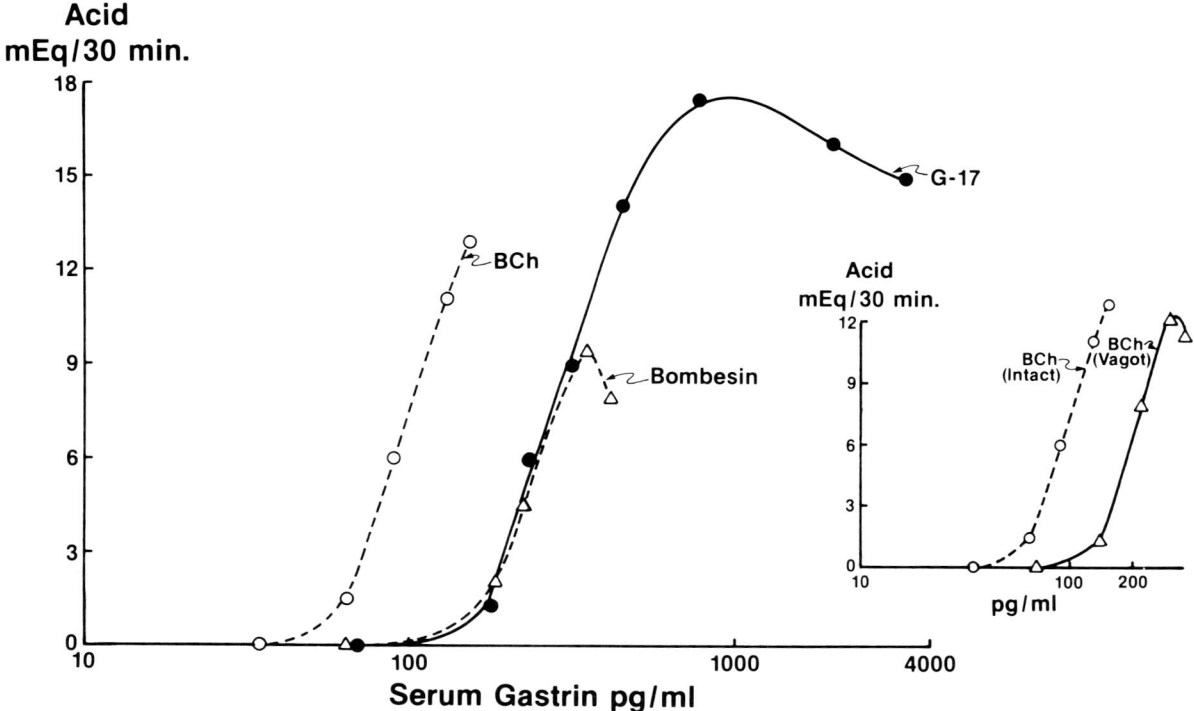

FIG. 9. Acid output in fistula dogs related to measured concentrations of serum gastrin. Dogs were stimulated by graded doses (20–160 $\mu g \cdot kg^{-1} \cdot h^{-1}$) of bethanechol (BCh), bombesin (0.1–2 $\mu g \cdot kg^{-1} \cdot h^{-1}$), and gastrin G-17 (0.05–5 $\mu g \cdot kg^{-1} \cdot h^{-1}$). *Inset*: bethanechol dose response in 3 dogs before (intact) and after fundic vagotomy (Vagot).

ergistic interaction between gastrin and ACh; in peripheral stimulation, gastrin is the principal and perhaps the only pathway for stimulation by food.

Gastrin stimulates gastric secretion solely by a hormonal mechanism (269) rather than by a paracrine action or an intragastric portal circulation. Peripheral venous levels, though lower than those in portal venous blood (12), represent the effective concentrations responsible for stimulation of gastric secretion (64, 125, 130, 269), since gastrin reaches the fundus only through the systemic circulation. The role of circulating gastrin on gastric secretion and its interrelation with other stimuli are best understood when secretion is related to blood levels of gastrin. Because gastrin and ACh act synergistically, the curve relating acid secretion to serum gastrin during vagal or bethanechol stimulation is much steeper and to the left of that resulting from exogenous or endogenous gastrin acting alone (Fig. 9). By the synergism between ACh and gastrin, the effects of the relatively small amounts of gastrin released by the vagus could be greatly amplified to produce the vigorous secretory response seen with vagal excitation. With bethanechol stimulation after fundic vagotomy, the curve relating acid to gastrin concentrations is shifted to the right, because more gastrin is released by bethanechol, whereas the fundus is less sensitive to stimuli (Fig. 9, *insert*).

Release of gastrin. Gastrin cells (G cells) produce both G-34 and G-17. Both are found in blood in equal proportions with small amounts of G-14 (42, 269), whereas in antral mucosa G-34 comprises only 10% of total gastrin. The higher proportion of G-34 in blood than in antrum results from the slower metabolic clearance of G-34 (269, 270). For physiological purposes, however, a general antibody with specificity for total COOH-terminal activity and roughly equal affinity for G-17 and G-34 expressed in terms of G-17 standard or equivalent is adequate (270). Gastrin is released into the circulation from G cells, which are almost solely confined to the antrum; a much smaller number of G cells is located in the duodenum (except in humans, where the number is higher), whereas the fundic mucosa contains over 100 times fewer G cells than the antrum. Release of gastrin from G cells may be stimulated in several ways—nervous, paracrine, and luminal; the principal stimuli are ACh, bombesin-like GRP, and protein food products. Gastrin release is modulated or inhibited by somatostatin and by luminal acid.

Vagal and cholinergic regulation of gastrin release. Electrical stimulation of the vagus and excitation of the vagus through cytoglucopenia promote gastrin release (216). Gastrin release depends on the strength of the vagal signal. Insulin hypoglycemia (104, 252) and 2-DG (see Fig. 4) elevate serum gastrin dose-dependently. In dogs sham feeding produces a modest increase in serum gastrin (261), whereas in humans MSF, a weaker stimulus of secretion than hypoglycemia, causes no significant increase (2, 16, 219) or at best a very small increase [<5–10 pg/ml; (56, 62)].

Vagal stimulation of gastrin release has been studied in two different ways. One depends on the effects of selective vagotomy in intact animals or humans, and the other depends on experiments with isolated perfused stomachs.

Vagotomy effects. Vagal release of gastrin depends solely on the nerve fibers innervating the antrum and is eliminated by selective antral vagotomy [Fig. 10; (38, 112, 261)], thus defining the source of vagally released gastrin as well as the pathway for stimulation. Selective fundic vagotomy, leaving the antrum innervated, results in a greatly augmented vagal release of gastrin [Fig. 11; (115)]. Bethanechol (113) and bombesin (114) stimulation are also augmented by fundic vagotomy. This effect has been ascribed to the removal by fundic vagotomy of an inhibitor of gastrin release (39, 116, 123). The augmented vagal or bombesin stimulated release of gastrin can be reversed by a background infusion of bethanechol at a dose (10 μg·kg^{-1}·h^{-1}) that is subthreshold for acid secretion [Fig. 12; (113, 121)]. In an apparent paradox, the anticholinergic drug atropine also inhibits the augmented gastrin response to vagal stimulation [Fig. 11; (123)] but fails to antagonize the increased bombesin response (116). Thus the disinhibiting effect of fundic denervation on gastrin release is not confined to neural release of gastrin. Because resection of the fundic mucosa also results in augmentation of gastrin release (37, 251), inhibition and disinhibition are probably not direct vagal or neural phenomena but apparently involve cholinergic mechanisms and the fundic mucosa.

Bethanechol had a biphasic effect on gastrin release in intact dogs. Where low doses (20 μg·kg^{-1}·h^{-1}) reduced basal gastrin levels, a higher dose (120 μg·kg^{-1}·h^{-1}) stimulated gastrin release. After fundic vagotomy, both doses stimulated gastrin release. Stimulation by bethanechol was competitively antagonized by atropine (114). Atropine also largely inhibited the augmented vagal release of gastrin after fundic vagotomy, leaving a small residual response. However, atropine augments (2-DG) vagal gastrin release in intact dogs, suggesting that vagal control of gastrin release is largely but not completely muscarinic and partly nonmuscarinic and is the algebraic result of stimulation and inhibition (Fig. 13).

Acute elevation of serum gastrin 24 h after truncal vagotomy in dogs (138) suggests that gastrin release is under the immediate control of the ENS and that inhibition is under control of the vagus. Vagal excitation during bombesin stimulation of dogs with antral vagotomy and fundic vagus intact did not inhibit gastrin release; thus the vagus does not actively stimulate an inhibitor from the fundus (B. I. Hirschowitz, unpublished observations). The nature of the fundic-cholinergic inhibitor of gastrin release thus remains to be discovered.

Enteric nervous system and paracrine regulation of gastrin release and gastric secretion. In further examining the role of the ENS in regulation of gastrin release, major consideration must be given to both bombesin and somatostatin.

Bombesin, a 14–amino acid peptide isolated from the skin of the frog *Bombina bombina*, is a potent gastrin-releasing peptide with additional effects on other endocrine, exocrine, muscle, and neural cells. A family of structurally related peptides of amphibian origin has been isolated and found to have similar actions (51). A synthetic compound comprising the terminal 9 amino acids also actively releases gastrin (115–117). A 27–amino acid peptide, GRP, isolated from mammalian stomach and intestine (181), shares 9 of the 10 COOH-terminal amino acids with bombesin and most of its actions. Immunohistochemical

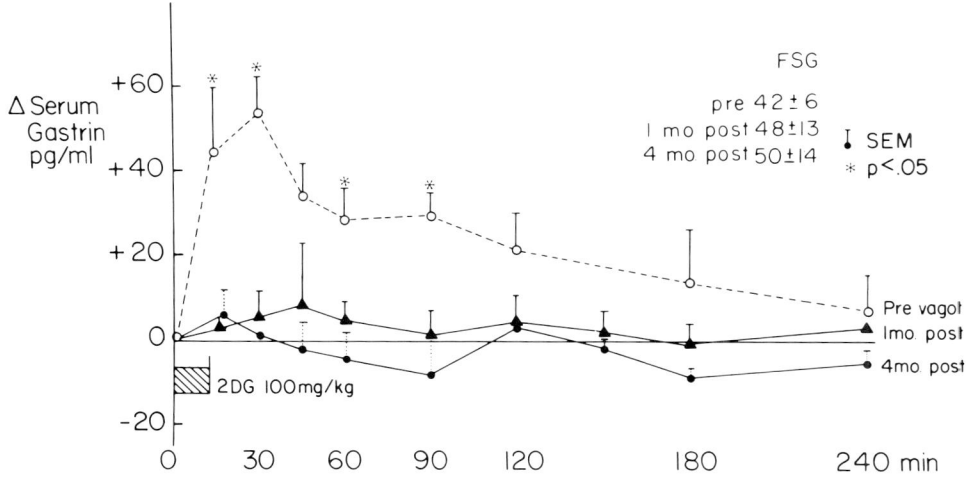

FIG. 10. Change in serum gastrin in the 4 h after intravenous injection of 2-DG, 100 mg/kg, in 4 dogs before and 1 and 4 mo after antral vagotomy. FSG, fasting serum gastrin.

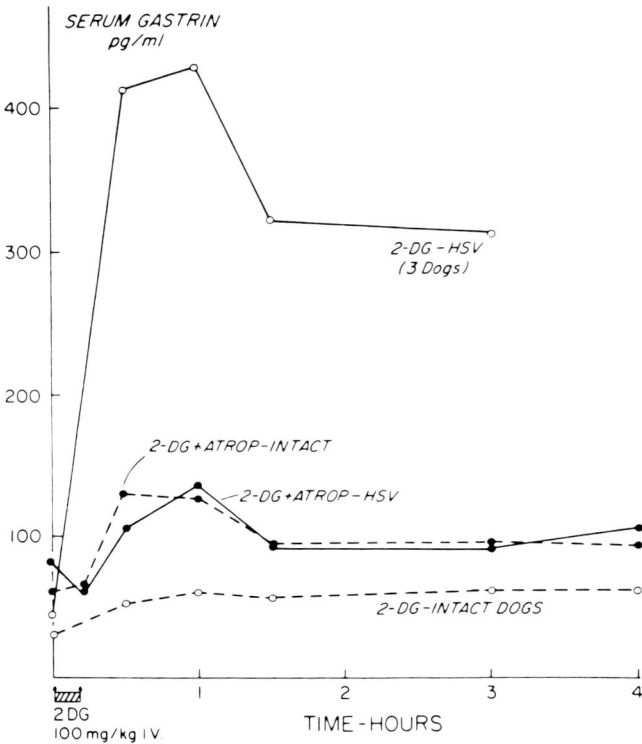

FIG. 11. Effect of atropine on serum gastrin after an injection of 2-DG, 100 µg/kg, given intravenously over a 10-min period with or without atropine. Studies were performed in conscious fistula dogs before (intact) and after highly selective (fundic) vagotomy (HSV).

staining has localized bombesinlike peptides in both myenteric and mucosal plexus neurons in the stomach (140) and in the brain (49), where it has a number of very potent effects (257).

Gastrin-releasing peptide acts as a major intermediate for vagal release of gastrin (46, 140, 193) and probably represents the atropine-resistant nicotinic component of vagal gastrin release [Fig. 11; (139, 140)]. No role has been defined in gastric secretion for the GRP cells that are found in abundance in the fundus (139, 140), since bombesin does not stimulate acid secretion directly and is ineffective after antrectomy (10). Assays have not been sensitive enough to define a possible hormonal role for circulating GRP. Bombesin directly stimulates smooth muscle contraction, as well as mammalian pancreatic cells (147) and amphibian peptic cells (244). The pancreatic and peptic effects are antagonized by a specific receptor antagonist, the substance P analogue (DArg^1,DPro^2,DTrp^{7,9},Leu^{11}-substance P) (147, 244). This antagonist apparently did not block the release of gastrin by bombesin in intact dogs (201), though the doses used fell short of the ED$_{50}$ (~10^{-6} M) calculated from in vitro studies (147, 244).

With the antrum in place, bombesin stimulates gastric acid and pepsin only through release of gastrin, with dose-response curves relating acid output to serum gastrin curves that are superimposable on those derived from exogenously infused gastrin G-17 [see Fig. 9; (125, 130, 269)]. However, bombesin acting via endogenous gastrin does not stimulate either acid or

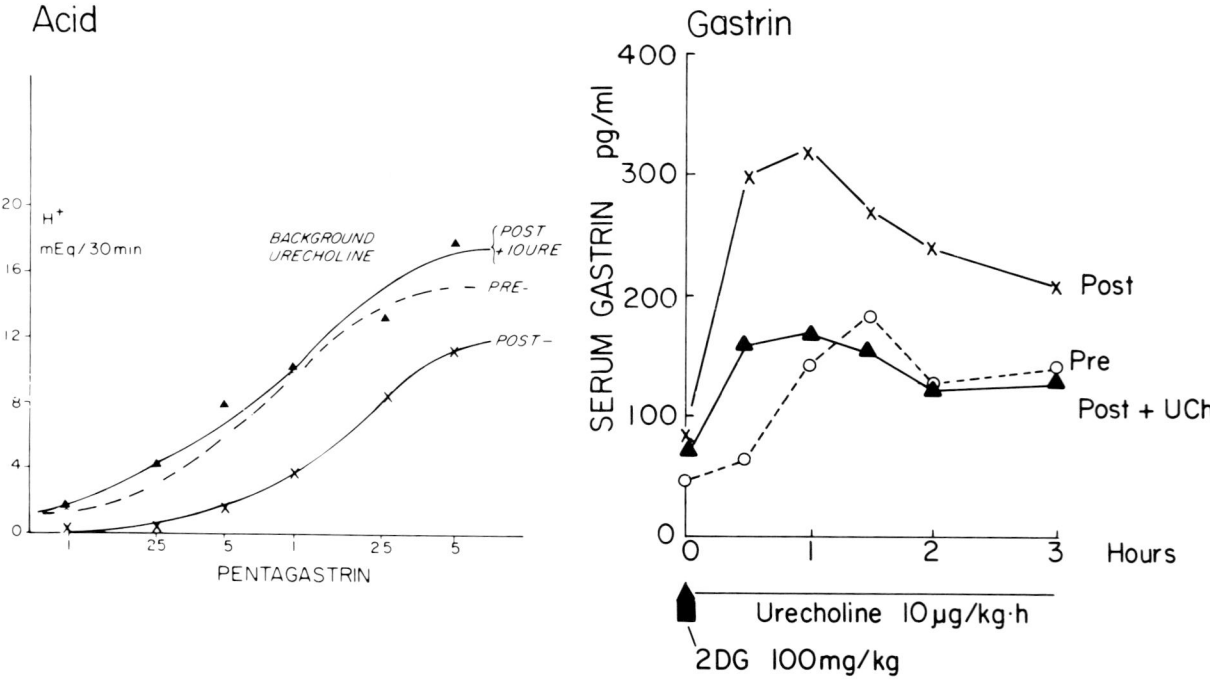

FIG. 12. Gastric acid response to graded doses of pentagastrin (*left panel*) and serum gastrin response to 2-DG (*right panel*) in 3 dogs before and after fundic vagotomy. After fundic vagotomy, bethanechol (Urecholine), 10 µg·kg^{-1}·h^{-1}, was given as background (Post + UCh). [From Hirschowitz (106).]

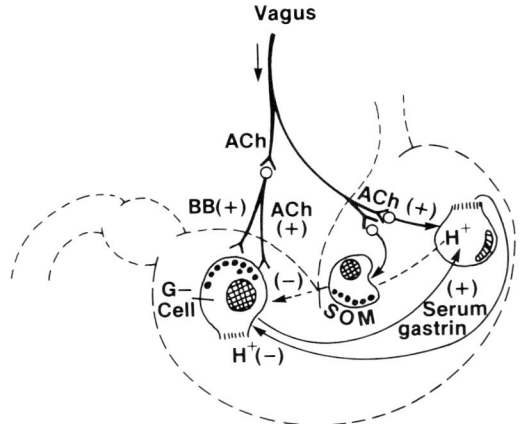

FIG. 13. Major factors controlling release of gastrin by the vagus. Negative feedback inhibiting gastrin release may act directly via acid in the lumen on the open end of the gastrin cell (G cell); both H+ and acetylcholine (ACh) may act by releasing somatostatin (SOM). BB, bombesin; (+), stimulation; (−), inhibition. Interpretation from results of studies in intact animals.

pepsin output to more than 60%–80% of gastrin maximum, because at higher doses bombesin produces a concomitant inhibition of acid and pepsin secretion (125, 127, 130, 243). The failure to reach full response (e.g., see Fig. 9) may be due to the corelease at high bombesin doses of somatostatin in the fundus from D cells, which are liberally distributed in proximity to parietal and chief cells (140, 165, 166).

Somatostatin-secreting D cells are widely distributed in both brain and gut. Somatostatin exhibits remarkably wide effects: inhibiting transport, exocrine, endocrine, neural, immune system, and contractile cells and antagonizing a broad range of stimuli. Nevertheless, somatostatin is not a universal inhibitor. It is at times quite stimulus specific, being relatively ineffective against histamine-stimulated gastric secretion [see the chapter by Konturek in this *Handbook*; (110, 231)] but very potent against cholinergic stimulation. By its strategic location and extensions of the D cell body, the somatostatin-secreting D cell maintains paracrine contact with both antral G cells and the secretory cells of the fundus (140, 165, 166). Somatostatin is therefore well placed to modulate both gastrin release and gastric secretion (see Fig. 7). In addition, somatostatin can inhibit the release of ACh at synapses (84) and is thus capable of acting at yet another site concerned with neurohormonal control of secretion.

Antral and fundic D cells differ in one important respect. Antral D cells, like G cells, are of the open type (69), with one surface exposed to the lumen, whereas D cells of the fundus do not open on the lumen (165). This difference has two possible implications: *1*) the D cell of the antrum, but not the fundus, could sense signals from the lumen, e.g., acid (139); and *2*) somatostatin may have a function in the antral lumen. No such function has thus far been defined.

Neuroparacrine interactions. The intermediary and interacting roles of ACh, GRP, and somatostatin in vagal release of gastrin have also been studied by experiments with in situ perfused stomachs in rats (28–30, 46, 180, 182, 205, 228, 230, 236) and pigs (139, 140), in which venous effluent is collected for peptide measurements. The rat models do not measure acid or pepsin secretion; the actions of the measured peptides on secretion are thus implied. Where acid is measured, as in the pig stomach model (139, 140), circulating hormones are not involved. These studies suggest that the vagus stimulates gastrin release directly via ACh, as well as by releasing GRP from postganglionic neurons stimulated through nonmuscarinic (nicotinic) pathways (237); the vagus simultaneously inhibits somatostatin release from D cells by a muscarinic step (140, 237). Somatostatin is also released by a noncholinergic pathway (probably by GRP); cholinergic vagal inhibition of D cells predominates and somatostatin decreases with vagal stimulation (193, 236). If muscarinic transmission is blocked by atropine, vagal stimulation gives a lower gastrin response and a positive somatostatin response (89, 236).

Five ways in which the D cells could be involved by paracrine mechanisms in gastrin release have been deduced from in vitro studies. *1*) Somatostatin limits basal gastrin release, and removal of somatostatin [e.g., by somatostatin antibody (28, 140, 229)] results in gastrin release. *2*) Stimulation of the target G cell by GRP or ACh could be accompanied by simultaneous suppression of the D cell by ACh; such was the case with electrical vagal trunk stimulation (140, 236), electrical field stimulation of the antrum (228), and cholinergic agonist stimulation (229). Coadministration of somatostatin antibody with methacholine or bombesin resulted in higher stimulated gastrin release (28, 46). This mechanism does not explain inhibition of gastrin release in vivo by low doses of bethanechol (114) nor stimulation by low doses of atropine (111, 129). *3*) Agents may stimulate both the G cell and the D cell; e.g., bombesin and the gastrin it stimulates may both release somatostatin and, through this feedback loop, may thus limit the excessive release of gastrin via antral D cells (46) and restrain the secretion of acid and pepsin via fundic D cells (130). *4*) The D cell may be stimulated alone and thus suppress gastrin release. Such a mechanism may mediate inhibition of stimulated gastrin release by low doses of bethanechol after fundic vagotomy; it may also mediate the effect of acid in the antrum (139, 165, 166), possibly acting via the open end of antral D cells (69, 139), though the evidence is not conclusive. *5*) Because somatostatin appears in gastric juice, a role for somatostatin in inhibition of vagally induced gastrin release by secretion into the lumen and action on the open end of the G cell is possible but remains to be demonstrated.

Not all the observations in vitro and in vivo fit the models (236, 237). Thus Pederson et al. (205) have shown that most of the somatostatin recovered from perfusion of isolated rat stomachs arose from the fundus. They also report that the acute gastrin increase after vagotomy was not accompanied or explained by a change in somatostatin release. Moreover, neither electrical nerve trunk (140) nor field stimulation (228) exactly simulate the normal neural signals. Electrical stimulation of either type is nonselective, so that both stimulatory and inhibitory neurons may be activated together (275). Electrical field stimulation of the antrum releases gastrin (228) and stimulates acid secretion from the mouse and rat stomach in vitro (5), but in the absence of an effective vascular circulation, the stimulation of acid cannot be ascribed to gastrin and must occur via neural or neuroparacrine pathways.

Also, the neuroendocrine model derived from studies with isolated perfused stomachs (236, 237) does not explain the many observations in intact animals concerning vagal release of gastrin and the effects of bethanechol and atropine from which the model in Figure 13 is derived. When better methods become available for study of paracrine transmitters and events, it may be possible to define the role of these and other messengers in the regulation of gastrin and gastric secretion by the ENS.

Gastric phase. Stimulation of secretion during the gastric phase of gastric secretion is almost exclusively due to gastrin released by food. Distension and afferent vagal impulses involved in normal eating by themselves probably contribute little in the intact stomach, though some gastrin is released by nonnutrient liquids in the dog (107) and human (234). Graded secretory responses to intragastric peptone are mediated by proportional elevations of circulating gastrin (11, 163, 218).

Like the antral D cell, the G cell has one surface, covered by microvilli, that is exposed to the gastrin lumen—the so-called open-end endocrine cell of Fujita and Kobayashi (69). Food products within the lumen acting directly on the open end of the G cell stimulate the cell to secrete gastrin. It is likely that the components of digested food that stimulate the G cell (172) are protein products or amines, though whether they act on cell surface recognition sites or intracellularly is not known. Calcium (169) and ACh (80, 261, 267) in the lumen also stimulate gastrin release, probably through direct effects on the G cell via the luminally exposed open end. Their effects on antral D cells are unknown.

Vagal denervation prolongs the gastrin response to food (116), due to a combination of delayed gastric emptying and delayed acidification resulting from reduced acid secretion. In the long run, gastric stasis and hypoacidity would produce G cell hyperplasia, leading to a permanently elevated serum gastrin. In the dog, atropine dose responsively inhibits food-stimulated gastrin release irrespective of vagal innervation and independently of pH of the contents (116). Similar findings were reported in humans (235).

The release of gastrin by food (F) and by bombesin (B) is affected oppositely ($\downarrow$, inhibited; 0, no effect) by each of the following: atropine (B0, F$\downarrow$), bethanechol (B$\downarrow$, F0), somatostatin (B$\downarrow$, F0), prostaglandin E$_2$ (B$\downarrow$, F0), and antral acidification (B0, F$\downarrow$), indicating different pathways for food and bombesin stimulation of the G cell (127, 128).

Inhibition by acid of gastrin release. Acidification of the luminal open-end surface (69) of the G cell below pH 3.0 partly and below 2.0 absolutely inhibits gastrin release stimulated by food, ACh, calcium in the lumen, or vagal excitation (Fig. 14) but does not inhibit gastrin release stimulated by bombesin (116, 127, 128). Feedback through acidification of gastric contents thus terminates the gastric phase of gastrin-dependent gastric secretion. Acid inhibition is unaffected by vagal denervation (39, 46, 80). Atropine alone may increase fasting serum gastrin (37, 111, 129, 228); combining acidification and atropine nullifies the effect of both (62) but does not explain either. The concept that somatostatin mediates the effects of acid through a neural cholinergic ENS reflex blocked by atropine (62) is inconsistent with the inhibition of somatostatin release by cholinergics (228, 236, 237). It has also been suggested that acid in the lumen might release somatostatin by acting on the open end of the antral D cell exposed to the gastric lumen (139, 140). Because somatostatin at a dose (0.5 $\mu g \cdot kg^{-1} \cdot h^{-1}$) sufficient to inhibit bombesin stimulation did not inhibit food-stimulated gastrin release and acidification-inhibited food [but not bombesin (127, 128)] stimulation (Fig. 14), it is unlikely that somatostatin mediates the inhibitory effect of luminal acid on gastrin release. On balance, it is most probable that acid inhibits the G cell by a direct effect exerted through its surface exposed to the antral lumen.

OTHER PEPTIDES. Few of the other 20 or more peptides that have been identified in neural or gastric tissue and could possibly play a role in vagal stimulation of secretion have been identified as physiologically important (238, 269, 270). These include pancreatic polypeptide, which increases most prominently with vagal stimulation and is extremely sensitive to atropine inhibition (239) but for which no gastric effect has been described. Gastric inhibitory peptide may be involved by potentiating somatostatin release (140, 183). Others are discussed in the chapter by Konturek in this *Handbook*.

It is perhaps too early to identify specific neurohormonal modulatory or transmitter roles for the other gut peptides and neuropeptides in the control of gastric secretion, either centrally or peripherally. Those that are relatively abundant in the vagus and enteric nerves and for which a role might be surmised include substance P, VIP, CCK, and enkephalin.

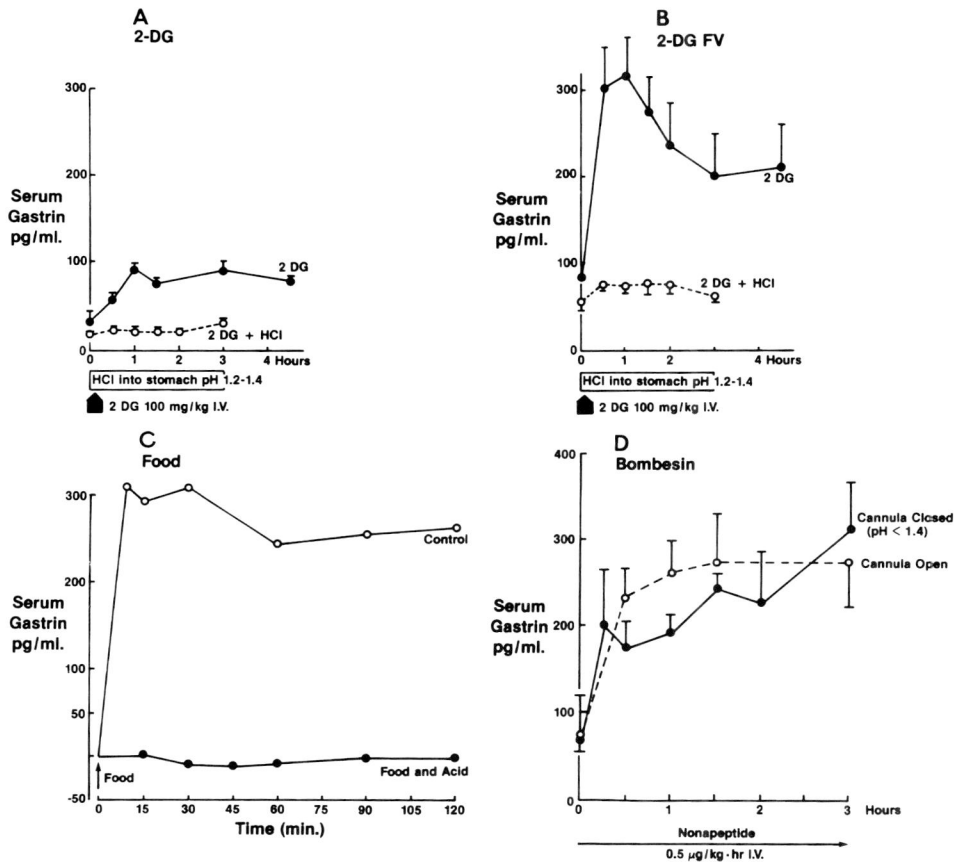

FIG. 14. Acidification of antrum to pH <1.4 in fistula dogs inhibited gastrin release by 2-DG with intact vagi (A), 2-DG after fundic vagotomy (B), a meal of 350-g meat (C), but not bombesin nonapeptide (D).

HISTAMINE. Despite intense research on gastric secretion in general and on histamine in particular, in the dozen years since the development of H_2 receptor antagonists, the connection between the vagus and histamine in the stimulation of the stomach remains circumstantial (31, 231).

It is not possible to reconcile the evidence implicating histamine in the mediation of vagal stimulation of gastric secretion: acid and pepsin secretion have to be separately considered (103), and there are significant species differences in the metabolism and effects of histamine (231). The evidence to be considered involves stimulation by histamine and its analogues of acid and pepsin 1) in various species in the intact state and 2) in isolated stomachs, isolated mucosa, isolated glands, and gastric cells. Another line of evidence derives from the inhibition by H_2 antagonists of the action of histamine and nonhistamine stimuli (vagal, cholinergic, gastrin) on acid and pepsin secretion in various models. The synergism between histamine and other agonists must also be considered. Further evidence is circumstantial: e.g., the location of histamine-containing cells (231, 250) and neurons (89) in the gastric mucosa and the synthesis and release of histamine from the stomach under various conditions (231). Nervous control of histamine release is deduced from proximity of postganglionic fibers to mast cells in gastric mucosa of rat, guinea pig, and rabbit (217).

Briefly, histamine stimulates acid secretion in all species and in all experimental preparations, including isolated stomachs, mucosae, glands, and cells (13, 105). This effect is competitively and specifically inhibited by histamine H_2-receptor antagonists, the action of which simulates agonist withdrawal (124). Moreover, H_2 antagonists inhibit stimulation of acid secretion in intact animals by stimuli other than histamine (see the chapter by Konturek in this *Handbook*), including vagal, cholinergic (126), gastrin, food, and even caffeine in humans (103). This effect implies that all stimuli of acid secretion in intact animals act through or depend on histamine. From such evidence in intact animals one might conclude that histamine is a critical intermediary in acid stimulation by the vagus and all agonists; i.e., all release histamine to stimulate the parietal cell. Alternatively histamine could be an essential costimulus and the occupation of the H_2 receptor by an agonist would be obligatory for stimulation via any other receptor. Neither is likely to be generally true and the facts still beg a rational explanation.

Experiments in vitro present a somewhat different picture of histamine as the "final common pathway." In isolated mouse stomachs (5), although H_2 antagonists block ganglionic (electrical field or carbachol) stimulation of acid, they do not block postganglionic (bethanechol) cholinergic stimulation. Moreover, in isolated parietal cells neither cholinergic (13, 250) nor gastrin stimulation (13, 230, 231) is blocked by H_2 antagonists. Histamine intermediation thus occurs in intact stomachs but not at the cellular level, where nonhistamine stimuli do bypass H_2 antagonists.

The peptic cell presents a different picture. There are major species differences in the action of histamine on pepsinogen secretion, and H_2-receptor antagonists antagonize only the action of histamine on such cells (87). Histamine via H_2 receptors stimulates pepsinogen secretion in humans (87), monkeys, and pigs, among others, but has a biphasic effect on pepsinogen secretion in dogs and cats (103). In dogs histamine at low doses stimulates and at high doses inhibits pepsinogen secretion, both apparently via H_2 receptors but presumably of different affinity (132). Histamine does not stimulate isolated peptic cells (95). Moreover, histamine H_2 antagonists do not block the stimulation of peptic cells in dogs by cholinergic (126) or vagal stimulation (105). Because the vagus stimulates acid and pepsin equally well, it is unlikely that histamine mediates vagal stimulation, at least of the peptic cell. Gastrin also stimulates the peptic cell in the dog much more effectively than does histamine, making it unlikely that gastrin acts via the intermediation of histamine. Thus histamine does not appear to be necessary for peptic cell secretion by either the vagus or gastrin.

Intrinsic factor (IF) secretion presents another aspect to the role of histamine in gastric secretion. A glycoprotein required for efficient absorption of cyanocobalamin (vitamin B_{12}), IF is unusual in that it is synthesized and secreted by parietal cells in human, monkey, sheep, cat, guinea pig, and rabbit gastric mucosa (240) but by the peptic cells in rats and mice (233). As expected, IF secretion is stimulated by the same agents that stimulate the parent cell. In parietal cells that synthesize IF in small amounts in the rough endoplasmic reticulum, IF is localized to tubulovesicles and moves rapidly to the microvilli with the structural changes that follow stimulation. Intrinsic factor is secreted early and for only a short period before acid secretion is fully established, the cell stores being small and rapidly depleted. Histamine (171) and dibutyryl cAMP (Bt_2cAMP) (11) stimulate parietal cell IF, and the effect of histamine is blocked by H_2 antagonists. Atropine and omeprazole do not block IF secretion, whereas vagotomy decreases it. In the rat, carbachol stimulates IF from the peptic cell and this effect is antagonized by atropine. It may be anticipated that vagal stimulation would promote IF secretion in the context of its localization in specific cells.

The interactions between cholinergic and histamine effects—pseudocompetitive inhibition by vagotomy and atropine and potentiation by cholinomimetics—are discussed in ACETYLCHOLINE, p. 138.

Circumstantial evidence for a role for histamine in the gastric mucosa that is derived from the presence of histamine-containing cells (210, 231, 250) or the increase in histamine or histamine-forming capacity, as in the rat (89), does not extend to all species. Vagal stimulation in dogs does not increase venous histamine nor does it alter histidine decarboxylase or N-methyltransferase (173). Pentagastrin likewise does not release histamine from the canine stomach in vivo or in vitro (214).

Therefore no single model clearly defines a specific role for histamine in vagal stimulation of gastric secretion. Species differences and the apparent discrepancies between in vivo and in vitro effects of histamine and H_2 antagonists need to be reconciled before a definitive picture can be drawn of the place of histamine in the regulation of gastric secretion.

Trophic Effects

FUNDIC MUCOSA. Hyperplasia of gastric secretory mucosa is well recognized in cases of gastrinoma (252). Because gastrinomas are neoplastic and are outside the acid stream, they produce gastrin in an unrestrained way, resulting in parietal cell hyperplasia, with a much increased maximum acid output. An increase in maximum acid output, which reflects an enlarged parietal cell mass (26, 179), is also seen in many patients with duodenal ulcer (108, 252). The trophic action of gastrin on the stomach is independent of stimulation of secretion (149), since DNA synthesis is increased even while acid secretion is inhibited by an H_2-receptor antagonist. Somatostatin, however, prevents gastrin-induced hyperplasia of the fundus (149, 167), as well as inhibiting gastrin-stimulated secretion [see the chapter by Konturek in this *Handbook*; (127)].

Based on the premise that physiological stimuli are all able to produce hypertrophy or hyperplasia in the target organs, a possible trophic role for the vagus was investigated by chronically stimulating the vagus through sham feeding of dogs 7 h/day for 6 wk. In 6 dogs, maximal acid output with pentagastrin from the innervated gastric fistula, but not the denervated pouch, increased in each dog between 10% and 35% (27 ± 4%) (263). All dogs recovered their base-line maximum acid output 6 wk after the sham-feeding period. The data suggest that the vagus caused hyperplasia of parietal cells directly and not through hypergastrinemia, since serum gastrin remained unchanged, basal secretion was not increased, and secretion from the denervated pouch did not change. In another study in cats, anterior hypothalamic nuclei were subjected to electrical stimulation 20 h/day, resulting in increased numbers of parietal, peptic, and mucous neck cells within a few days; these changes were not seen in vagotomized cats (204). Both studies need to be confirmed.

Atrophy after vagotomy has been even more difficult to establish; maximal acid output in rats (268) and in humans (203) falls immediately with vagal transection, a change too soon and too extreme to ascribe to atrophy, and the decrease is not progressive (160). Unilateral vagotomy in rats has been reported to produce ipsilateral reduction in weight and height of oxyntic mucosa (88), but parietal and peptic cells were not counted or measured. Ultrastructural changes in parietal cells after vagotomy are quite subtle (92) and insufficient to explain the large and rapid decrease in acid output after vagotomy. By contrast, in dogs after vagotomy there is no decrease in maximal or supramaximal acid output and thus no atrophy (120) but rather a state of functional cholinergic deficiency (120, 121).

ANTRAL MUCOSA. Experimentally, G cell hyperplasia can be produced by bombesin, which stimulates G cell secretion and is also trophic to G cells (167). No state is known in which there is a chronic excess of bombesin. In the absence of acid, antral G cells increase in number and hypergastrinemia results. Acid secretion may be absent as a result of parietal cell atrophy (as in pernicious anemia), as a result of gastritis with atrophy, or as a result of prolonged inhibition of acid secretion by omeprazole. In cases where the antrum has been surgically separated from the fundus and thus from the acid stream (excluded antrum), a lack of feedback inhibition by acid results in G cell hyperplasia. Experimentally, antral pouches transplanted to the colon cause a marked increase in gastric acid secretion (194), presumably due to a lack of acid feedback and unrestrained gastrin secretion. The G cell populations have not been counted in such transplanted pouches. The mechanisms that cause both G cell secretion and hyperplasia are not known.

Inhibition of G cell function by acid also apparently exerts an antitrophic effect, since recovery of acid secretion after discontinuation of omeprazole treatment leads to involution of hypergastrinemia and G cell hyperplasia. The mechanisms of either the inhibitory or the antitrophic effects of acid are not known, but they may have a common cellular pathway. Clearly, however, acid is important in the regulation of antral G cell function and population kinetics.

Antrofundal Interactions

The extensive autonomic nervous network in the gastric wall provides a mechanism for the functional integration between anatomic regions of the stomach. Further integration is provided by circulating hormones. Communication between antrum and fundus by a portal intramural circulation has not been established.

ANTROFUNDAL REFLEXES. Distension of separated, vagally innervated antral pouches stimulates acid secretion from the innervated fundus via two mechanisms: *1*) gastrin release into the circulation and *2*) vagovagal reflexes (3, 5, 80). Each can be separately eliminated, the former by luminal acidification and the latter by vagotomy. Both together eliminate the antrofundal reflex. Antral distension is much less effective if the antrum is kept in continuity with the rest of the stomach (172), indicating that the balance between the antrum and fundus is modulated largely via the ENS.

FUNDOANTRAL REFLEXES. In the other direction, distension of an innervated fundic (Pavlov) pouch stimulates the release into the circulation of gastrin (39) from an innervated antral pouch, resulting in turn in acid secretion from the Pavlov pouch. The antral response is eliminated by antral acidification as well as by vagotomy (80). The pathway is presumed to be via long vagovagal reflexes. Distension of the innervated fundus after antral resection also causes acid and pepsin secretion (80–82), presumably via local ENS reflexes. However, distension of the fundus in an intact stomach produces little release of gastrin (82), suggesting a modulating effect of the ENS on vagovagal reflexes.

ANTRECTOMY. A functional antrum is critical to the full response to electrical or central vagal stimulation (80–82, 267). Either antral acidification or topical anesthesia, e.g., cocainization of the antrum, eliminates the gastrin response to most stimuli (80). Antrectomy sharply reduces the response of the innervated remaining fundus to vagal stimulation, by either sham feeding or insulin hypoglycemia (25, 248), and the fundic responses to the nonvagal stimuli, histamine and gastrin, are markedly depressed as well (see Fig. 5). The mechanism of this effect is unknown, though Sjodin (248) reported that infusions of gastrin or pentagastrin restored the response to sham feeding in antrectomized dogs. This finding has not been reported by others. Even separation of the antrum from the body of the stomach reduces the fundic secretory response to stimuli (208). The effect may be due to disruption of the neural network integrating the two regions of the stomach.

Vagotomy

Because vagotomy has been used for treating duodenal ulcer (15, 54, 174), the role of the vagus in disease and function of the stomach is a matter of considerable interest in medicine.

TRUNCAL VAGOTOMY. If the vagal trunks are interrupted during central vagal stimulation in dogs, gastric secretion ceases almost immediately, indicating that the ENS requires continuous input from the vagus and is not entrained by the vagus for prolonged action. Moreover, if the vagi are cut in rats during stimulation by the combined infusion of histamine, gastrin, and carbachol (268), gastric secretion falls immediately by 90%, an effect that cannot be explained as due to an immediate transmitter deficiency. Acute vagal denervation may induce a transient chaotic state in the ENS by removal of modulatory or oscillatory control and result in dominance of inhibitory effects.

After bilateral truncal vagotomy in dogs, there is neither gastrin released nor acid or pepsin secreted in response to 2-DG central vagal excitation (Figs. 15 and 16). Moreover, in such animals, superimposed 2-DG injection does not inhibit secretion stimulated by histamine or pentagastrin infusion (B. I. Hirschowitz, unpublished observations). Thus *1)* there is no evidence for a "vagogastrone" (81, 157); *2)* there are no extravagal pathways for central stimulation or inhi-

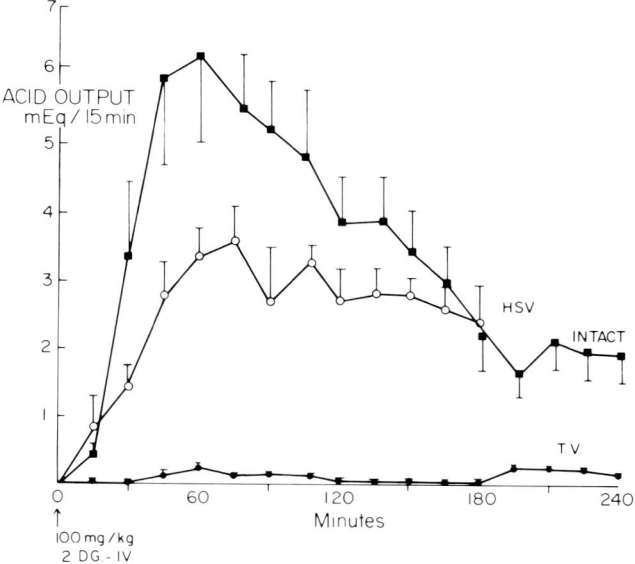

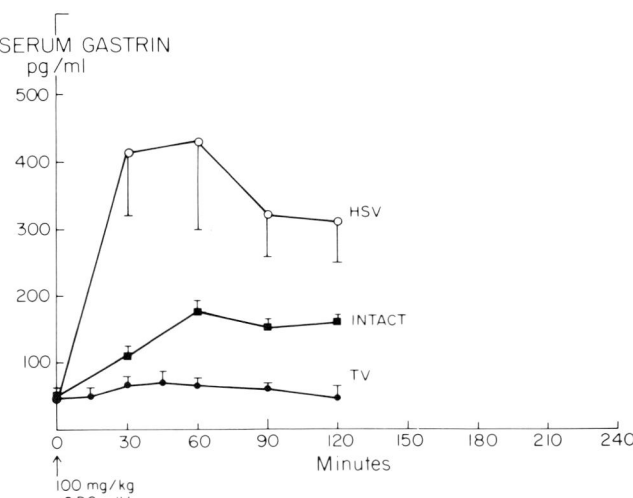

FIG. 16. Acid output and gastrin release with 2-DG in dogs before (intact) and after highly selective (fundic) vagotomy (HSV). In the same dogs, subsequent truncal vagotomy (TV) eliminated both acid and gastrin responses. [From Hirschowitz (106).]

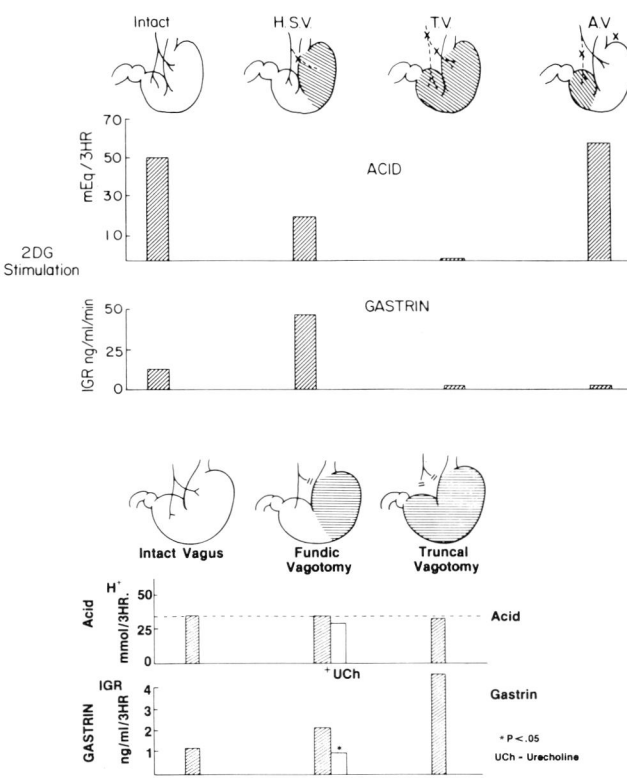

FIG. 15. *Top*: summary of the integrated acid and gastrin output in response to intravenous 2-DG stimulation in dogs with intact vagi and after fundic, truncal, and antral vagotomy. HSV, highly selective (fundic) vagotomy; TV, truncal vagotomy; AV, antral vagotomy; IGR, integrated gastrin response. *Shading*, vagally denervated parts of the stomach. *Bottom*: acid and gastrin responses to 3-h intravenous infusion of bombesin-14 in 3 dogs before and after fundic and subsequent truncal vagotomy. In the fundic vagotomy experiments, the effect of a background of bethanechol [Urecholine (UCh), 10 $\mu g \cdot kg^{-1} \cdot h^{-1}$] on acid and gastrin are shown. [From Hirschowitz (106).]

bition of gastrin release or gastric secretion (16); and *3)* the lack of response is not due to concomitant inhibition, e.g., by the strong adrenergic stimulation that would be unaffected by gastric vagotomy or the other actions of 2-DG–induced cytoglucopenia, such as growth hormone, vasopressin (209), or renin release (265) and hyperglycemia and hypokalemia (133). In similar experiments with insulin instead of 2-DG, acid secretion is inhibited by a direct action of insulin (101, 103) and not through central vagal excitation.

In humans, bilateral truncal vagotomy produces a decrease in both sensitivity and maximum response to nonvagal stimuli (203). Basal secretion and vagal responses tend to remain depressed after complete vagotomy (160), but partial recovery of the response to histamine or pentagastrin is seen in many vagotomized subjects. Bethanechol has been reported by some (119, 203) but not others (63) to restore histamine or gastrin secretory responses to prevagotomy levels. Recovery of secretory (174) and, in many instances, motor function after partial or complete gastric vagotomy has remained unexplained. In rats, recovery of acid secretion within 4–6 wk has been shown to coincide with regeneration of fundic vagus nerves cut close to the serosa (148) and remaining in proximity to the target organ. In the absence of regeneration, partial functional recovery may represent adaptation of the intramural and myenteric nervous plexuses to assume some of the controlling functions of the full hierarchical system—brain, vagus nerves and ganglia, and peripheral autonomic system. Restoration by bethanechol of sensitivity of the fundus to gastrin and histamine and the reversal of hypersensitivity of the antral G cells indicate that the vagus normally exerts its modulation of the ENS via ACh.

REGIONAL VAGOTOMY. *Fundus.* Selective vagal denervation of the fundus, leaving the antrum innervated, immediately eliminates the gastric acid and pepsin response to vagal excitation [Fig. 17; (85, 120)]. The sensitivity, but not the maximum response, of the fundus to histamine and gastrin is depressed; i.e., curves shift to the right (pseudocompetitive inhibition) (see Fig. 12). This effect is reversed by subthreshold doses of bethanechol (114, 121, 123), which also reversed the increased gastrin release by 2-DG (see Fig. 12). There was also an increased gastrin response to bombesin (see Fig. 15), i.e., the effect was not purely neural, but this effect too was reversed by bethanechol (113, 114).

After a variable period, usually 2–4 mo, the vagal responses to 2-DG begin to recover but reach a plateau well below prevagotomy levels (121). Acid secretion recovers to ~60%, whereas pepsin secretion recovers to <15%. However, both acid and peptic cells remain fully responsive to direct stimulation by bethanechol (Fig. 17).

Selective antral vagotomy. An experimental operation performed only in animals, antral vagotomy provides an excellent model for studying vagal effects on the fundus (37, 112, 261). After selective antral vagotomy, the vagally stimulated release of gastrin by 2-DG (see Fig. 10) or insulin hypoglycemia (37, 261) is eliminated. However, the fundic acid and pepsin secretory responses to vagal stimulation by 2-DG or insulin hypoglycemia (261) remain unimpaired, perhaps because of increased sensitivity of the fundus to

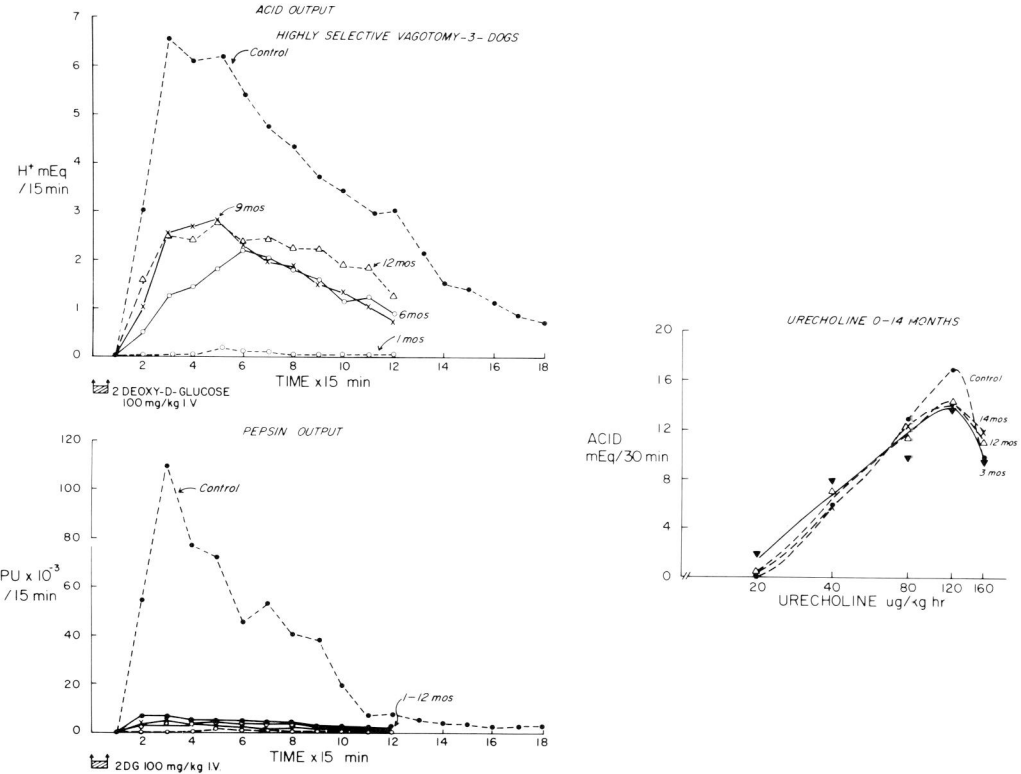

FIG. 17. Effect of fundic (highly selective) vagotomy on acid and pepsin responses to 2-DG, 100 µg/kg, before and 1, 6, 9, and 12 mo later. *Right panel*, acid responses to graded doses of bethanechol (Urecholine) before and up to 14 mo after vagotomy. [Adapted from Hirschowitz and Hutchison (121).]

cholinergic stimulation. By contrast, the fundic response to pentagastrin was lower (Fig. 18).

With the loss of gastrin release after antral vagotomy, it is possible to analyze direct vagal stimulation of the fundus. After antral vagotomy, the secretory response to 2-DG was highly sensitive to atropine (112) with $ED_{50} \sim 1$ nM/kg. This showed that the direct vagal stimulation of the fundus is muscarinic at pre- or postganglionic sites or both.

Intestinal Phase of Gastric Secretion

STIMULATION. Two physiological phenomena, one stimulatory, one inhibitory, draw attention to an intestinal effect on gastric secretion. Amino acids in the duodenum and jejunum stimulate gastric acid secretion. Experimental data used to explain this finding are contradictory. In one study (168), amino acids were equally effective in stimulating acid by intravenous or intraduodenal administration but not by the intrajejunal route, though all three produced the same serum levels of amino acids and none increased serum gastrin. In another study (276), intraduodenal liver extract stimulated gastrin release and thus acid and pepsin. The response was eliminated by antral acidification or antrectomy. However, antral defunctioning also greatly reduces the effect of infused G-17 (64) or pentagastrin (see Fig. 5). A third group (178) reported secretion after enteral amino acids in dogs only if they had had portocaval transpositions, i.e., portal blood bypassing the liver. Because unshunted dogs had the same blood levels but no secretion, a putative hormone mechanism was proposed. In another report (164) the whole effect was ascribed to circulating amino acids. Others have been more sure of the existence of a hormone that is not gastrin and have named it intestinal phase hormone (197); such a hormone may be responsible for stimulating gastric secretion in the presence of a portocaval shunt. In summary, several groups have conflicting data as to whether intrajejunally delivered amino acids do or do not stimulate gastrin secretion or gastrin release and whether the same effect is seen with or without portocaval shunt. Whether circulating amino acids could explain all the findings is not settled. The question awaits accumulation of verifiable and duplicated results.

INHIBITION. The second aspect of the intestinal phase concerns inhibition of secretion by intraduodenal or intestinal instillation of fat, acid, and hyperosmolar solutions. There is less dispute about inhibition than there is about stimulation. The inhibition occurs in vagotomized stomachs as well and is thus not mediated by nervous pathways. The effect has been ascribed to hormone action via a putative "enterogastrone." Candidate enterogastrone hormones dance on and off the stage with the predictability of aspiring princes in *Grimms' Fairy Tales*. At one time or an-

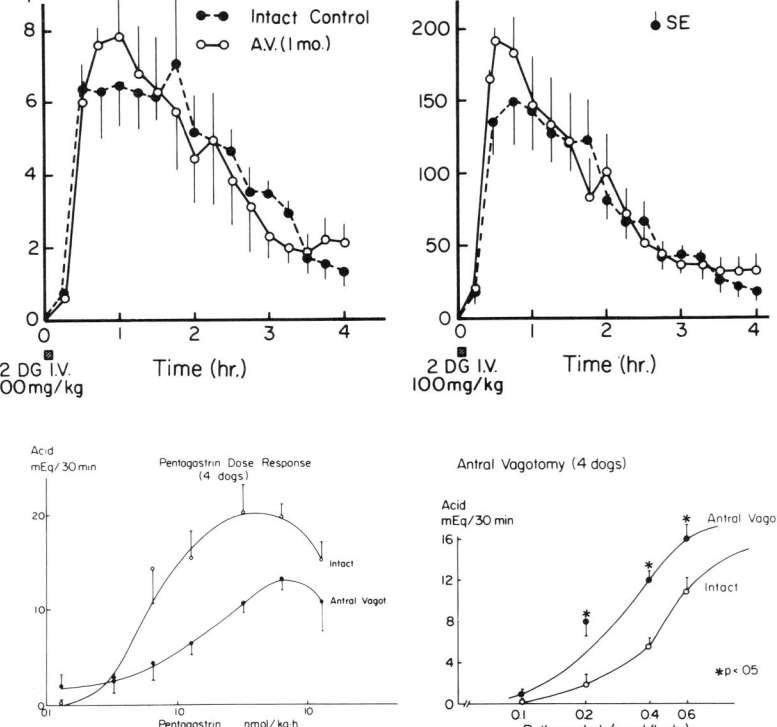

FIG. 18. Acid and pepsin output in response to 100 mg/kg 2-DG (*top*) remained substantially unaltered by antral vagotomy (AV) in 4 dogs. Response to pentagastrin (*bottom left*) was reduced; response to bethanechol (*lower right*) was increased by antral vagotomy.

other, secretin, CCK, neurotensin, somatostatin, VIP, and enteroglucagon (among others) have been thought to cause the observed inhibition that might reduce the response to various stimuli by 10%–90%. The most recent candidate, peptide-YY (PYY), is an intestinal and ileocolonic peptide that inhibits sham feeding by 90% and is released by fat in sufficient amounts to produce this effect (200), perhaps by preventing ACh release from the vagus. The hypothetical enterogastrone(s) remains to be discovered.

The author is indebted to Jo Ann Stone for typing the manuscript; to Mary Helen Stringer, Nancy Clemons, and Jack G. Smith, Jr., of the Lister Hill Library, University of Alabama at Birmingham, for checking the references; to Dr. Jerry Brown for helpful discussion and criticism; and to Drs. Graham Dockray and Mark Feldman for making available unpublished information.

REFERENCES

1. ALTMAN, J. Tuning in to neurotransmitters. *Nature Lond.* 315: 537, 1985.
2. ANAGNOSTIDES, A., V. S. CHADWICK, A. C. SELDEN, AND P. N. MATON. Sham feeding and pancreatic secretion. *Gastroenterology* 87: 109–114, 1984.
3. ANDERSON, W., E. MOLINA, J. RENTZ, AND B. I. HIRSCHOWITZ. Analysis of the 2-deoxy-D-glucose-induced vagal stimulation of gastric secretion and gastrin release in dogs using methionine-enkephalin, morphine and naloxone. *J. Pharmacol. Exp. Ther.* 222: 617–622, 1982.
4. ANDREWS, P. L. R., H. L. DUTHIE, I. F. FUSSEY, AND A. MELLERSH. Brainstem neurones sending efferent fibres to the abdominal viscera of the dog. *J. Physiol. Lond.* 266: 90–91, 1977.
5. ANGUS, J. A., AND J. W. BLACK. The interaction of choline esters, vagal stimulation and H_2-receptor blockade on acid secretion in vitro. *Eur. J. Pharmacol.* 80: 217–224, 1982.
6. ANTIA, F., C. E. ROSIERE, C. ROBERTSON, AND M. I. GROSSMAN. Effect of vagotomy on gastric secretion and emptying time in dogs. *Am. J. Physiol.* 166: 470–479, 1951.
7. APPIA, F., W. R. EWART, B. S. PITTAM, AND D. L. WINGATE. Convergence of sensory information from abdominal viscera in the rat brain stem. *Am. J. Physiol.* 251 (*Gastrointest. Liver Physiol.* 14): G169–G175, 1986.
8. ARUNLAKSHANA, O., AND H. O. SCHILD. Some quantitative uses of drug antagonists. *Br. J. Pharmacol.* 14: 48–58, 1959.
9. BAILE, C. A., C. L. MCLAUGHLIN, AND M. A. DELLA-FERA. Role of cholecystokinin and opioid peptides in control of food intake. *Physiol. Rev.* 66: 172–234, 1986.
10. BASSO, N., G. IMPROTA, P. MELCHIORRI, AND N. SOPRANZI. Gastrin release by bombesin in the antral pouch dog. *Rend. Gastroenterol.* 6: 95–98, 1974.
11. BATZRI, S., J. W. HARMON, M. D. WALKER, W. F. THOMPSON, AND R. TOLES. Secretion of intrinsic factor from dispersed mucosal cells isolated from guinea pig and rabbit stomach. *Biochim. Biophys. Acta* 720: 217–221, 1982.
12. BECKER, H. D., D. D. REEDER, AND J. C. THOMPSON. Direct measurement of vagal release of gastrin. *Surgery St. Louis* 75: 101–106, 1974.
13. BERGLINDH, T. The mammalian gastric parietal cell in vitro. *Annu. Rev. Physiol.* 46: 377–392, 1984.
14. BERTHOUD, H. R., AND G. J. MOGENSON. Ingestive behavior after intracerebral and intracerebroventricular infusions of glucose and 2-deoxy-D-glucose. *Am. J. Physiol.* 233 (*Regulatory Integrative Comp. Physiol.* 2): R127–R133, 1977.
15. BLAIR, A. J., C. T. RICHARDSON, J. H. WALSH, P. CHEW, AND M. FELDMAN. Effect of parietal cell vagotomy on acid secretory responsiveness to circulating gastrin in humans. Relationship to postprandial serum gastrin concentration. *Gastroenterology* 90: 1001–1007, 1986.
16. BONFILS, S., M. MIGNON, AND C. ROZE. Vagal control of gastric secretion. *Int. Rev. Physiol.* 19: 61–106, 1979.
17. BRAUER, R. F. Electrically-induced vagal stimulation of gastric acid secretion. In: *Nerves and the Gut*, edited by F. P. Brooks and P. W. Evers. Thorofare, NJ: Slack, 1977, p. 86–95.
18. BREMER, A. Contribution a l'etude des mecanismes de stimulation de la secretion acid de l'estomac. *Acta Gastro-Enterol. Belg.* 22: 467–526, 1959.
19. BROOKS, F. P. Central neural control of gastric secretion. In: *Handbook of Physiology. Alimentary Canal*, edited by C. F. Code. Washington, DC: Am. Physiol. Soc., sect. 6, vol. 2, chapt. 45, 1967, p. 805–826.
20. BROOKS, F. P., AND P. W. EVERS (editors). *Nerves and the Gut*. Thorofare, NJ: Slack, 1977, p. 541.
21. BROWN, D. A. Muscarinic excitation of sympathetic and central neurons. *Trends Pharmacol. Sci.* 1: 32–34, 1984.
22. BROWN, D. A. Neuropharmacology. Acetylcholine and brain cells. *Nature Lond.* 319: 358–359, 1986.
23. BURHOL, P. G., AND B. I. HIRSCHOWITZ. Intravenous injection of 2-deoxy-D-glucose in gastric fistula chickens. *Scand. J. Gastroenterol. Suppl.* 11: 35–40, 1971.
24. BURLEIGH, D. E. Selectivity of bethanechol on muscarinic receptors. *J. Pharm. Pharmacol.* 30: 398–399, 1978.
25. CABOCIO, J. L. F., M. M. WOLFE, M. P. HOCKING, J. E. MCGUIGAN, AND E. R. WOODWARD. Effect of antrectomy on the nervous phase of gastric secretion in the dog. *Am. J. Surg.* 142: 324–327, 1981.
26. CARD, W. I., AND I. N. MARKS. The relationship between acid output of the stomach following "maximal" histamine stimulation and the parietal cell mass. *Clin. Sci. Lond.* 19: 147–163, 1960.
27. CARPENTER, M. B. *Core Text of Neuroanatomy* (3rd ed). Baltimore, MD: Williams & Wilkins, 1985.
28. CHIBA, T., S. KADOWAKI, T. TAMINATO, K. CHIHARA, Y. SEINO, S. MATSUKURA, AND T. FUJITA. Effect of antisomatostatin γ-globulin on gastrin release in rats. *Gastroenterology* 81: 321–326, 1981.
29. CHIBA, T., T. TAMINATO, S. KADOWAKI, H. ABE, K. CHIHARA, Y. SEINO, S. MATSUKURA, AND T. FUJITA. Effects of glucagon, secretin and vasoactive intestinal polypeptide on gastric somatostatin and gastrin release from isolated perfused rat stomach. *Gastroenterology* 79: 67–71, 1980.
30. CHIBA, T., T. TAMINATO, S. KADOWAKI, Y. INOUE, K. MORI, Y. SEINO, H. ABE, K. CHIHARA, S. MATSUKURA, T. FUJITA, AND Y. GOTO. Effects of various gastrointestinal peptides on gastric somatostatin release. *Endocrinology* 106: 145–149, 1980.
31. CODE, C. F. Reflections on histamine, gastric secretion and the H_2 receptor. *N. Engl. J. Med.* 296: 1459–1462, 1977.
32. COLIN-JONES, D. G., AND R. L. HIMSWORTH. The secretion of gastric acid in response to a lack of metabolizable glucose. *J. Physiol. Lond.* 202: 97–109, 1969.
33. COLIN-JONES, D. G., AND R. L. HIMSWORTH. The location of chemoreceptor controlling gastric acid secretion during hypoglycemia. *J. Physiol. Lond.* 206: 397–409, 1970.
34. COLTURI, T. J., R. H. UNGER, AND M. FELDMAN. Role of circulating somatostatin in regulation of gastric acid secretion, gastrin release, and islet cell function: studies in healthy subjects and duodenal ulcer patients. *J. Clin. Invest.* 74: 417–423, 1984.
35. DAVISON, J. S. Response of single vagal afferent fibres to mechanical and chemical stimulation of the gastric and duodenal mucosa in cats. *Q. J. Exp. Physiol.* 57: 405–416, 1972.
36. DAVISON, J. S., AND D. GRUNDY. Modulation of single vagal

efferent fibre discharge by gastrointestinal afferents in the rat. *J. Physiol. Lond.* 284: 69–82, 1978.
37. DEBAS, H. T., J. HOLLINSHEAD, A. SEAL, P. SOON-SHIONG, AND J. H. WALSH. Vagal control of gastrin release in the dog: pathways for stimulation and inhibition. *Surgery St. Louis* 95: 34–37, 1984.
38. DEBAS, H. T., S. J. KONTUREK, J. H. WALSH, AND M. I. GROSSMAN. Proof of a pyloro-oxyntic reflex for stimulation of acid secretion. *Gastroenterology* 66: 526–532, 1974.
39. DEBAS, H. T., J. H. WALSH, AND M. I. GROSSMAN. Evidence for oxyntopyloric reflex for release of antral gastrin. *Gastroenterology* 68: 687–690, 1975.
40. DEBONS, A. F., L. SILVER, E. P. CRONKITE, H. A. JOHNSON, G. BRECHER, D. TENZER, AND I. L. SCHWARTZ. Localization of gold in mouse brain in relation to gold thioglucose obesity. *Am. J. Physiol.* 202: 743–750, 1962.
41. DE GRAEF, J., AND M. C. WOUSSEN-COLLE. Effects of sham feeding, bethanechol, and bombesin on somatostatin release in dogs. *Am. J. Physiol.* 248 (*Gastrointest. Liver Physiol.* 11): G1–G7, 1985.
42. DOCKRAY, G. J. Comparative biochemistry and physiology of gut hormones. *Annu. Rev. Physiol.* 41: 83–95, 1979.
43. DOCKRAY, G. J., R. A. GREGORY, H. J. TRACY, AND W. Y. ZHU. Transport of cholecystokinin-octapeptide-like immunoreactivity toward the gut in afferent vagal fibers in cat and dog. *J. Physiol. Lond.* 314: 501–511, 1981.
44. DOCKRAY, G. J., AND K. A. SHARKEY. Neurochemistry of visceral afferent neurones. In: *Progress in Brain Research.* Amsterdam: Elsevier, 1986, vol. 67, p. 133–148.
45. DUKE, W. W., B. I. HIRSCHOWITZ, AND G. SACHS. Vagal stimulation of gastric secretion in man by 2-deoxy-D-glucose. *Lancet* 2: 871–876, 1965.
46. DUVAL, J. W., B. SAFFOURI, G. C. WEIR, J. H. WALSH, A. ARIMURA, AND G. M. MAKHLOUF. Stimulation of gastrin and somatostatin secretion from the isolated rat stomach by bombesin. *Am. J. Physiol.* 241 (*Gastrointest. Liver Physiol.* 4): G242–G247, 1981.
47. EICHHORN, R., AND J. TRACKTIR. The effect of hypnosis upon gastric secretion. *Gastroenterology* 29: 417–421, 1955.
48. EISENBERG, M. M., R. C. CHAWLA, AND K. SUGAWARA. Sustained gastric secretion in response to 2-deoxy-D-glucose. *Gastroenterology* 59: 174–179, 1970.
49. EKBLAD, E., M. EKELUND, H. GRAFFNER, R. HAKANSON, AND F. SUNDLER. Peptide containing nerve fibers in stomach wall of rat and mouse. *Gastroenterology* 89: 73–85, 1985.
50. EMSON, P. C. (editor). *Chemical Neuroanatomy.* New York: Raven, 1983.
51. ERSPARMER, V., AND P. MELCHIORRI. Active polypeptides of the amphibian skin and their synthetic analogues. *Pure Appl. Chem.* 35: 463–494, 1973.
52. FANDRIKS, L., AND D. DELBRO. Neural stimulation of gastric bicarbonate secretion in the cat. An involvement of vagal axon reflexes and substance P. *Acta Physiol. Scand.* 118: 301–304, 1983.
53. FELDMAN, M. Gastric H^+ and HCO_3^- secretion in response to sham feeding in humans. *Am. J. Physiol.* 248 (*Gastrointest. Liver Physiol.* 11): G188–G191, 1985.
54. FELDMAN, M., R. M. DICKERMAN, R. N. MCCLELLAND, K. A. COOPER, J. H. WALSH, AND C. T. RICHARDSON. Effect of selective proximal vagotomy on food-stimulated gastric acid secretion and gastrin release in patients with duodenal ulcer. *Gastroenterology* 76: 926–931, 1979.
55. FELDMAN, M., AND C. T. RICHARDSON. Gastric acid secretion in humans. In: *Physiology of the Gastrointestinal Tract*, edited by L. R. Johnson. New York: Raven, 1981, p. 693–707.
56. FELDMAN, M., AND C. T. RICHARDSON. "Partial" sham feeding releases gastrin in normal human subjects. *Scand. J. Gastroenterol.* 16: 13–16, 1981.
57. FELDMAN, M., AND C. T. RICHARDSON. Role of thought, sight, smell, and taste of food in the cephalic phase of gastric acid secretion in humans. *Gastroenterology* 90: 428–433, 1986.
58. FELDMAN, M., C. T. RICHARDSON, AND J. S. FORDTRAN. Experience with sham feeding as a test for vagotomy. *Gastroenterology* 79: 792–795, 1980.
59. FELDMAN, M., C. T. RICHARDSON, AND J. S. FORDTRAN. Effect of sham feeding on gastric acid secretion in healthy subjects and duodenal ulcer patients: evidence for increased basal tone in some ulcer patients. *Gastroenterology* 79: 796–800, 1980.
60. FELDMAN, M., C. T. RICHARDSON, I. L. TAYLOR, AND J. H. WALSH. Effect of atropine on vagal release of gastrin and pancreatic polypeptide. *J. Clin. Invest.* 63: 294–298, 1979.
61. FELDMAN, M., AND L. SCHILLER. Effect of bethanechol (urecholine) on gastric acid and nonparietal secretion in normal subjects and duodenal ulcer patients. Comparison with atropine, pentagastrin, and histamine. *Gastroenterology* 83: 262–265, 1982.
62. FELDMAN, M., AND J. H. WALSH. Acid inhibition of sham feeding-stimulated gastrin release and gastric acid secretion: effect of atropine. *Gastroenterology* 78: 772–776, 1980.
63. FELDMAN, M., AND J. H. WALSH. Effect of bethanechol on gastric acid secretion and serum gastrin concentration after proximal gastric vagotomy. *Ann. Surg.* 196: 14–17, 1982.
64. FELDMAN, M., J. H. WALSH, H. C. WONG, AND C. T. RICHARDSON. Role of gastrin heptadecapeptide in the acid secretory response to amino acids in man. *J. Clin. Invest.* 61: 308–313, 1978.
65. FOKINA, A., S. J. KONTUREK, N. KWIECIEN, AND T. RADECKI. Role of gastric antrum in gastric and intestinal phases of gastric secretion in dogs. *J. Physiol. Lond.* 295: 229–239, 1979.
66. FORSSELL, H., B. STENQUIST, AND L. OLBE. Vagal stimulation of human gastric bicarbonate secretion. *Gastroenterology* 89: 581–586, 1985.
67. FOX, E. A., AND T. L. POWLEY. Longitudinal columnar organization within the dorsal motor nucleus represents separate branches of the abdominal vagus. *Brain Res.* 341: 269–282, 1985.
68. FUJITA, T., T. KANNO, AND N. YANAIHARA. *Brain-Gut Axis.* Tokyo: Biomedical Res. Found., 1983, p. 333.
69. FUJITA, T., AND S. KOBAYASHI. Structure and function of gut endocrine cells. *Int. Rev. Cytol.* 6: 187–233, 1977.
70. FURCHGOTT, R. F., AND P. BURSZTYN. Comparison of dissociation constants and of relative efficacies of selected agonists acting on parasympathetic receptors. *Ann. NY Acad. Sci.* 144: 882–899, 1967.
71. GABELLA, G. Structure of muscles and nerves in the gastrointestinal tract. In: *Physiology of the Gastrointestinal Tract*, edited by L. R. Johnson. New York: Raven, 1981, p. 197–241.
72. GAMSE, R., F. LEMBECK, AND A. C. CUELLO. Substance P in the vagus nerve: immunochemical and immunohistochemical evidence for axoplasmic transport. *Naunyn-Schmiedeberg's Arch. Pharmacol.* 306: 37–44, 1979.
73. GIBSON, R. G., H. W. COLVIN, JR., AND B. I. HIRSCHOWITZ. Kinetics of gastric response in chickens to graded electrical vagal stimulation. *Proc. Soc. Exp. Biol. Med.* 145: 1058–1060, 1974.
74. GLEYSTEEN, J. J., M. J. ESSER, AND A. L. MYRVIK. Reversible truncal vagotomy in conscious dogs. *Gastroenterology* 85: 578–583, 1983.
75. GOTO, Y., AND H. DEBAS. GABA-mimetic effect on gastric acid secretion: possible significance in central mechanisms. *Dig. Dis. Sci.* 28: 56–59, 1983.
76. GOTO, Y., Y. TACHE, H. DEBAS, AND D. NOVIN. Gastric acid and vagus nerve response to GABA agonist baclofen. *Life Sci.* 36: 2471–2475, 1985.
77. GOYAL, R. K. Neurology of the gut. In: *Gastrointestinal Disease: Pathophysiology, Diagnosis, Management* (3rd ed.), edited by M. H. Sleisenger and J. S. Fordtran. Philadelphia: Saunders, 1983, p. 97–115.
78. GRAHAME, G. R., J. M. GARRETT, AND B. I. HIRSCHOWITZ. 2-Deoxyglucose and histamine effects on gastric motility and secretion in the dog. *Am. J. Physiol.* 215: 243–248, 1968.

79. GRANNEMAN, J., AND M. I. FRIEDMAN. Hepatic modulation of insulin-induced gastric acid secretion and EMG activity in rats. *Am. J. Physiol.* 238 (*Regulatory Integrative Comp. Physiol.* 7): R346–R352, 1980.
80. GROSSMAN, M. I. Neural and hormonal stimulation of gastric secretion of acid. In: *Handbook of Physiology. Alimentary Canal*, edited by C. F. Code. Washington, DC: Am. Physiol. Soc., 1967, sect. 6, vol. 2, chapt. 47, p. 835–863.
81. GROSSMAN, M. I. Vagal stimulation and inhibition of acid secretion and gastrin release: which aspects are cholinergic? In: *Gastrin and the Vagus*, edited by J. F. Rehfeld and E. Amdrup. New York: Academic, 1979, p. 103–113.
82. GROSSMAN, M. I. Regulation of gastric acid secretion. In: *Physiology of the Gastrointestinal Tract*, edited by L. R. Johnson. New York: Raven, 1981, p. 659–671.
83. GRUNDY, D., AND T. SCRATCHERD. Effect of stimulation of the vagus nerve in bursts on gastric acid secretion and motility in the anaesthetized ferret. *J. Physiol. Lond.* 333: 451–461, 1982.
84. GUILLEMIN, R. Somatostatin inhibits release of acetylcholine induced electrically in the myenteric plexus. *Endocrinology* 99: 1653–1654, 1976.
85. GULDVOG, I., D. GEDDE-DAHL, AND A. BERSTAD. "Non-active" pepsin secretion compared with stimulated secretion by bethanechol, histamine, pentagastrin, and 2-deoxy-D-glucose. *Scand. J. Gastroenterol.* 15: 939–948, 1980.
86. GWYN, D. G., R. A. LESLIE, AND D. A. HOPKINS. Observations on the afferent and efferent organization of the vagus nerve and the innervation of the stomach in the squirrel monkey. *J. Comp. Neurol.* 239: 163–175, 1985.
87. HAGGSTROM, G. D., AND B. I. HIRSCHOWITZ. Histamine H_1 and H_2 effects on gastric acid and pepsin, heart rate and blood pressure in humans. *J. Pharmacol. Exp. Ther.* 231: 120–123, 1984.
88. HÅKANSON, R., S. VALLGREN, M. EKELUND, J. F. REHFELD, AND F. SUNDLER. The vagus exerts trophic control of the stomach in the rat. *Gastroenterology* 86: 28–32, 1984.
89. HÅKANSON, R., C. WAHLESTEDT, L. WESTLIN, S. VALLGREN, AND SUNDLER, F. Neuronal histamine in the gut wall releasable by gastrin and cholecystokinin. *Neurosci. Lett.* 42: 305–310, 1983.
90. HARDING, R., AND B. F. LEEK. Central projections of gastric afferent vagal inputs. *J. Physiol. Lond.* 228: 73–90, 1973.
91. HARPER, A. A., C. KIDD, AND T. SCRATCHERD. Vago-vagal reflex effects on gastric and pancreatic secretion and gastrointestinal motility. *J. Physiol. Lond.* 148: 417–436, 1959.
92. HELANDER, H. F., S. O. SVENSSON, AND S. EMAS. Effects of vagotomy on the structure and function of the parietal cells in cats (Abstract). *Acta Hepato-Gastroenterol.* 24: 483, 1977.
93. HELMAN, C. A. Chewing gum is equally effective as modified sham feeding in stimulating gastric acid secretion (Abstract). *Gastroenterology* 84: 1185, 1983.
94. HELMAN, C. A., AND B. I. HIRSCHOWITZ. Pirenzepine and atropine inhibition of modified sham feeding stimulated gastric acid and pepsin secretion (Abstract). *Gastroenterology* 90: 1456, 1986.
95. HERSEY, S. J., S. H. NORRIS, AND A. J. GIBERT. Cellular control of pepsinogen secretion. *Annu. Rev. Physiol.* 46: 393–402, 1984.
96. HILL, K. J. The effect of insulin on the secretion of gastric juice in the goat. *Q. J. Exp. Physiol.* 37: 143–147, 1952.
97. HIMSWORTH, R. L. Compensatory reactions to a lack of metabolizable glucose. *J. Physiol. Lond.* 198: 451–465, 1968.
98. HIMSWORTH, R. L. Interference with the metabolism of glucose by a non-metabolizable hexose (3-methylglucose). *J. Physiol. Lond.* 198: 467–477, 1968.
99. HIMSWORTH, R. L. Hypothalamic control of adrenaline secretion in response to insufficient glucose. *J. Physiol. Lond.* 206: 411–417, 1970.
100. HIMSWORTH, R. L., AND D. G. COLIN-JONES. Factors which determine the gastric secretory response to 2-deoxy-D-glucose. *Gut* 10: 1015–1018, 1969.
101. HIRSCHOWITZ, B. I. Quantitation of inhibition of gastric electrolyte secretion by insulin in the dog. *Am. J. Dig. Dis.* 11: 173–182, 1966.
102. HIRSCHOWITZ, B. I. Continuing gastric secretion after insulin hypoglycemia despite glucose injection. *Am. J. Dig. Dis.* 12: 19–25, 1967.
103. HIRSCHOWITZ, B. I. The secretion of pepsinogen. In: *Handbook of Physiology. Alimentary Canal*, edited by C. F. Code. Washington, DC: Am. Physiol. Soc., 1967, sect. 6, vol. 2, chapt. 50, p. 889–918.
104. HIRSCHOWITZ, B. I. The vagus and gastric secretion. In: *Nerves and the Gut*, edited by F. P. Brooks and P. W. Evers. Thorofare, NJ: Slack, 1977, p. 96–118.
105. HIRSCHOWITZ, B. I. H-2 histamine receptors. *Annu. Rev. Pharmacol. Toxicol.* 19: 203–244, 1979.
106. HIRSCHOWITZ, B. I. Controls of gastric secretion: a roadmap to the choice of treatment for duodenal ulcer. Stuart Lecture. *Am. J. Gastroenterol.* 77: 281–293, 1982.
107. HIRSCHOWITZ, B. I. Gastrin release in fistula dogs with solid compared to nutrient and non-nutrient liquid meals. *Dig. Dis. Sci.* 28: 705–711, 1983.
108. HIRSCHOWITZ, B. I. Apparent and intrinsic sensitivity to pentagastrin acid and pepsin secretion in peptic ulcer. *Gastroenterology* 86: 843–851, 1984.
109. HIRSCHOWITZ, B. I. Desensitization within and between secretagogues of canine gastric acid and pepsin secretion. *J. Pharmacol. Exp. Ther.* 236: 14–23, 1986.
110. HIRSCHOWITZ, B. I. Selective inhibition by somatostatin-14 (SOM) of gastric secretory stimulants in the conscious fistula dog (Abstract). *Gastroenterology* 90: 1461, 1986.
111. HIRSCHOWITZ, B. I., M. DANILEWITZ, AND E. MOLINA. Inhibition of basal acid, chloride and pepsin secretion in duodenal ulcer by graded doses of ranitidine and atropine with studies of pharmacokinetics of ranitidine. *Gastroenterology* 82: 1314–1326, 1982.
112. HIRSCHOWITZ, B. I., J. FONG, AND E. MOLINA. Effects of pirenzepine and atropine on vagal and cholinergic gastric secretion and gastrin release and on heart rate in the dog. *J. Pharmacol. Exp. Ther.* 225: 263–268, 1983.
113. HIRSCHOWITZ, B. I., AND R. G. GIBSON. Cholinergic stimulation and suppression of gastrin release in gastric fistula dogs. *Am. J. Physiol.* 235 (*Endocrinol. Metab. Gastrointest. Physiol.* 4): E720–E725, 1978.
114. HIRSCHOWITZ, B. I., AND R. G. GIBSON. Stimulation of gastrin release and gastric secretion: effect of bombesin and a nonapeptide in fistula dogs with and without fundic vagotomy. *Digestion* 18: 227–239, 1978.
115. HIRSCHOWITZ, B. I., AND R. G. GIBSON. Augmented vagal release of antral gastrin by 2-deoxyglucose after fundic vagotomy in dogs. *Am. J. Physiol.* 236 (*Endocrinol. Metab. Gastrointest. Physiol.* 5): E173–E179, 1979.
116. HIRSCHOWITZ, B. I., R. G. GIBSON, AND E. MOLINA. Atropine suppresses gastrin release by food in intact and vagotomized dogs. *Gastroenterology* 81: 838–843, 1981.
117. HIRSCHOWITZ, B. I., R. GIBSON, L. WRIGHT, AND G. HUTCHISON. Changed cholinergic receptor characteristics after vagotomy in gastric fistula dogs. *Am. J. Physiol.* 230: 105–109, 1976.
118. HIRSCHOWITZ, B. I., AND C. A. HELMAN. Effects of fundic vagotomy and cholinergic replacement on pentagastrin dose-responsive gastric acid and pepsin secretion in man. *Gut* 23: 675–682, 1982.
119. HIRSCHOWITZ, B. I., AND G. A. HUTCHISON. A working hypothesis for urecholine effects on histamine stimulation of gastric secretion. *Scand. J. Gastroenterol.* 8: 569–576, 1973.
120. HIRSCHOWITZ, B. I., AND G. A. HUTCHISON. Effects of vagotomy on urecholine-modified histamine dose responses in dogs. *Am. J. Physiol.* 228: 1313–1318, 1975.
121. HIRSCHOWITZ, B. I., AND G. A HUTCHISON. Long-term effects of highly selective vagotomy (HSV) in dogs on acid and pepsin

secretion. *Am. J. Dig. Dis.* 22: 81–95, 1977.
122. HIRSCHOWITZ, B. I., AND G. A. HUTCHISON. Kinetics of atropine inhibition of pentagastrin-stimulated H⁺, electrolyte, and pepsin secretion in the dog. *Am. J. Dig. Dis.* 22: 99–107, 1977.
123. HIRSCHOWITZ, B. I., AND E. MOLINA. Gastrin release after truncal vagotomy in fistula dogs: hypersensitivity to bombesin but not bethanechol. *Peptides Fayetteville* 1: 217–222, 1980.
124. HIRSCHOWITZ, B. I., AND E. MOLINA. Effect of cimetidine on histamine-stimulated gastric acid and electrolytes in dogs. *Am. J. Physiol.* 244 (*Gastrointest. Liver Physiol.* 7): G416–G420, 1983.
125. HIRSCHOWITZ, B. I., AND E. MOLINA. Relation of gastric acid and pepsin secretion to serum gastrin levels in dogs given bombesin and gastrin-17. *Am. J. Physiol.* 244: (*Gastrointest. Liver Physiol.* 7): G546–G551, 1983.
126. HIRSCHOWITZ, B. I., AND E. MOLINA. Effects of four H$_2$ histamine antagonists on bethanechol-stimulated acid and pepsin secretion in the dog. *J. Pharmacol. Exp. Ther.* 224: 341–345, 1983.
127. HIRSCHOWITZ, B. I., AND E. MOLINA. Somatostatin, prostaglandin E$_2$ and atropine inhibition of the gastric actions of bombesin in the dog. *Peptides Fayetteville* 5: 29–34, 1984.
128. HIRSCHOWITZ, B. I., AND E. MOLINA. Analysis of food stimulation of gastrin release in dogs by a panel of inhibitors. *Peptides Fayetteville* 5: 35–40, 1984.
129. HIRSCHOWITZ, B. I., E. MOLINA, L. OU TIM, AND C. HELMAN. Effects of very low doses of atropine on basal acid and pepsin secretion, gastrin, and heart rate in normals and DU. *Dig. Dis. Sci.* 29: 790–796, 1984.
130. HIRSCHOWITZ, B. I., L. OU TIM, C. A. HELMAN, AND E. MOLINA. Bombesin and G-17 dose responses in duodenal ulcer and controls. *Dig. Dis. Sci.* 30: 1092–1103, 1985.
131. HIRSCHOWITZ, B. I., H. M. POLLARD, S. W. HARTWELL, JR., AND J. LONDON. The action of ethyl alcohol on gastric secretion. *Gastroenterology* 30: 244–253, 1956.
132. HIRSCHOWITZ, B. I., J. RENTZ, AND E. MOLINA. Histamine H-2 receptor stimulation and inhibition of pepsin secretion in the dog. *J. Pharmacol. Exp. Ther.* 218: 676–680, 1981.
133. HIRSCHOWITZ, B. I., AND G. SACHS. Vagal gastric secretory stimulation by 2-deoxy-D-glucose. *Am. J. Physiol.* 209: 452–460, 1965.
134. HIRSCHOWITZ, B. I., AND G. SACHS. Atropine inhibition of insulin-, histamine-, and pentagastrin-stimulated gastric electrolyte and pepsin secretion in the dog. *Gastroenterology* 56: 693–702, 1969.
135. HIRSCHOWITZ, B. I., S. SCHENKER, AND J. D. BOYETT. A highly active gastric secretagogue extracted from a metastasis of a Zollinger-Ellison tumor. *Am. J. Dig. Dis.* 8: 499–508, 1963.
136. HITCHCOCK, M. W. S., M. J. KARVONEN, AND A. T. PHILLIPSON. The effect of insulin on the acidity of the abomasal contents of lamb. *Acta Physiol. Scand.* 16: 33–34, 1948.
137. HOFFMAN, H. H., AND H. N. SCHNITZLEIN. The number of nerve fibers in the vagus nerve of man. *Anat. Rec.* 139: 429–436, 1961.
138. HOLLINSHEAD, J. W., H. T. DEBAS, T. YAMADA, J. ELASHOFF, B. OSADCHEY, AND J. H. WALSH. Hypergastrinemia develops within 24 hours of truncal vagotomy in dogs. *Gastroenterology* 88: 35–40, 1985.
139. HOLST, J. J., S. L. JENSEN, S. KNUHTSEN, O. V. NIELSEN, AND J. F. REHFELD. Effect of vagus, gastric inhibitory polypeptide, and HCl on gastrin and somatostatin release from perfused pig antrum. *Am. J. Physiol.* 244 (*Gastrointest. Liver Physiol.* 7): G515–G522, 1983.
140. HOLST, J. J., S. KNUHTSEN, S. L. JENSEN, J. FAHRENKRUG, I.-I. LARSSON, AND O. V. NIELSEN. Interrelation of nerves and hormones in stomach and pancreas. *Scand. J. Gastroenterol.* 82: 85–99, 1983.
141. HOUPT, T. R. Stimulation of food intake in ruminants by 2-deoxy-D-glucose and insulin. *Am. J. Physiol.* 227: 161–167, 1974.
142. IGGO, A. Gastro-intestinal tension receptors with unmyelinated afferent fibres in the vagus of the cat. *Q. J. Exp. Physiol.* 42: 130–143, 1957.
143. IGGO, A. Gastric mucosal chemoreceptors with vagal afferent fibres in the cat. *Q. J. Exp. Physiol.* 42: 398–409, 1957.
144. ISENBERG, J. I. "Vagal tone" in duodenal ulcer and sham feeding as a test for completeness of vagotomy: another point of view. *Gastroenterology* 79: 952–953, 1980.
145. ISHIKAWA, T., M. NAGATA, AND Y. OSUMI. Dual effects of electrical stimulation of ventromedial hypothalamic neurons on gastric acid secretion in rats. *Am. J. Physiol.* 245 (*Gastrointest. Liver Physiol.* 8): G265–G269, 1983.
146. ISHIKAWA, T., Y. OSUMI, AND T. NAKAGAWA. Cholecystokinin intracerebroventricularly applied stimulates gastric acid secretion. *Brain Res.* 333: 197–199, 1985.
147. JENSEN, R. T., S. W. JONES, K. FOLKERS, AND J. D. GARDNER. A synthetic peptide that is a bombesin receptor antagonist. *Nature Lond.* 309: 61–63, 1984.
148. JOFFE, S. N., A. CROCKET, M. CHEN, AND K. BRACKETT. In vitro and in vivo technique for assessing vagus nerve regeneration after parietal cell vagotomy in the rat. *J. Auton. Nerv. Syst.* 9: 27–51, 1983.
149. JOHNSON, L. R. Regulation of gastrointestinal growth. In: *Physiology of the Gastrointestinal Tract*, edited by L. R. Johnson. New York: Raven, 1981, p. 169–196.
150. KADEKARO, M., H. SAVAKI, AND L. SOKOLOFF. Metabolic mapping of neural pathways involved in gastrosecretory response to insulin hypoglycaemia in the rat. *J. Physiol. Lond.* 300: 393–407, 1980.
151. KADEKARO, M., C. TIMO-IARIA, AND L. E. VALLE. Neural systems responsible for the gastric secretion provoked by 2-deoxy-D-glucose cytoglucopoenia. *J. Physiol. Lond.* 252: 565–584, 1975.
152. KADEKARO, M., C. TIMO-IARIA, L. E. VALLE, AND L. P. VELHA. Site of action of 2-deoxy-D-glucose mediating gastric secretion in the cat. *J. Physiol. Lond.* 221: 1–13, 1972.
153. KADEKARO, M., C. TIMO-IARIA, AND M. DE L. VICENTINI. Control of gastric secretion by the central nervous system. In: *Nerves and the Gut*, edited by F. P. Brooks and P. W. Evers. Thorofare, NJ: Slack, 1977.
154. KADEKARO, M., C. TIMO-IARIA, AND M. DE L. VICENTINI. Gastric secretion provoked by functional cytoglucopoenia in the nuclei of the solitary tract in the cat. *J. Physiol. Lond.* 299: 397–407, 1980.
155. KERR, F. W., AND R. M. PRESHAW. Secretomotor function of the dorsal motor nucleus of the vagus. *J. Physiol. Lond.* 205: 405–415, 1969.
156. KNUTSON, U., AND L. OLBE. Gastric acid response to sham feeding in the duodenal ulcer subject. *Scand. J. Gastroenterol.* 8: 513–522, 1973.
158. KONTUREK, S. J., N. KWIECIEN, W. OBTULOWICZ, E. MIKOS, E. SITO, J. OLESKY, AND T. POPIELA. Cephalic phase of gastric secretion in healthy subjects and duodenal ulcer patients: role of vagal innervation. *Gut* 20: 875–881, 1979.
159. KOWALEWSKI, K. Effect of vagotomy and/or antrectomy on gastric secretion stimulated by intravenous infusion of bethanechol chloride. Study on rats with gastric fistula. *Am. J. Dig. Dis.* 16: 19–26, 1971.
160. KRONBORG, O. Completeness of vagotomy: anatomy, pathophysiology and consequences. *Scand. J. Gastroenterol.* 16: 577–580, 1981.
161. KUBEK, M. J., M. A. REA, Z. I. HODES, AND M. H. APRISON. Quantitation and characterization of thyrotropin-releasing hormone in vagal nuclei and other regions of the medulla oblongata of the rat. *J. Neurochem.* 40: 1307–1313, 1983.
162. LA BARRE, J., AND C. DE CESPÉDÈS. Rôle du systèmes nerveux central dans l'hypersécrétion gastrique consécutive a l'administration d'insuline. *C. R. Seances Soc. Biol. Fil.* 106: 1249–1251, 1931.
163. LAM, S. K., J. I. ISENBERG, M. I. GROSSMAN, W. H. LANE, AND J. H. WALSH. Gastric acid secretion is abnormally sen-

sitive to endogenous gastrin released after peptone test meals in duodenal ulcer patients. *J. Clin. Invest.* 65: 555–562, 1980.
164. LANDOR, J. H., A. L. GOUGH, V. S. RAI, AND M. K. LIM. Amino acids as possible mediators of the intestinal phase of gastric secretion. *Surg. Gynecol. Obstet.* 150: 203–207, 1980.
165. LARSSON, L.-I. Peptide secretory pathways in GI tract: cytochemical contributions to regulatory physiology of the gut. *Am. J. Physiol.* 239 (*Gastrointest. Liver Physiol.* 2): G237–G246, 1980.
166. LARSSON, L.-I., N. GOLTERMANN, L. DE MAGISTRIS, J. F. REHFELD, AND T. W. SCHWARTZ. Somatostatin cell processes as pathways for paracrine secretion. *Science Wash. DC* 205: 1393–1395, 1979.
167. LEHY, T. Trophic effect of some regulatory peptides on gastric exocrine and endocrine cells of the rat. *Scand. J. Gastroenterol. Suppl.* 101: 27–30, 1984.
168. LENZ, H. J., D. L. HOGAN, AND J. I. ISENBERG. Intestinal phase of gastric acid secretion in humans with and without portacaval shunt. *Gastroenterology* 89: 791–796, 1985.
169. LEVANT, J. A., J. H. WALSH, AND J. I. ISENBERG. Stimulation of gastric secretion and gastrin release by single oral doses of calcium carbonate in man. *N. Engl. J. Med.* 289: 555–558, 1973.
170. LEVINE, A. S., J. E. MORLEY, J. KNEIP, M. GRACE, AND S. E. SILVIS. Muscimol induces gastric acid secretion after central administration. *Brain Res.* 229: 270–274, 1981.
171. LEVINE, J. S., P. K. NAKANE, AND R. H. ALLEN. Immunocytochemical localization of human intrinsic factor: the nonstimulated stomach. *Gastroenterology* 79: 493–502, 1980.
172. LICHTENBERGER, L. M. Importance of food in the regulation of gastrin release and formation. *Am. J. Physiol.* 243 (*Gastrointest. Liver Physiol.* 6): G429–G441, 1982.
173. LORENZ, W., K. THON, H. BARTH, E. NEUGEBAUER, H.-J. REIMANN, AND J. KUSCHE. Metabolism and function of gastric histamine in health and disease. In: *Receptors and the Upper GI Tract*, edited by B. I. Hirschowitz and J. G. Spenney. New York: Advanced Therapeutics Communications, 1983, p. 148–173.
174. LYNDON, P. J., M. J. GREENALL, R. B. SMITH, J. C. GOLIGHER, AND D. JOHNSTON. Serial insulin tests over a five-year period after highly selective vagotomy for duodenal ulcer. *Gastroenterology* 69: 1188–1195, 1975.
175. MACLEAN, D. B., AND S. F. LEWIS. De novo synthesis and axoplasmic transport of [^{35}S]methionine-substance P in explants of nodose ganglion/vagus nerve. *Brain Res.* 310: 325–335, 1984.
176. MAEDA-HAGIWARA, M., AND K. WATANABE. Bromocriptine inhibits 2-deoxy-D-glucose-stimulated gastric acid secretion in the rat. *Eur. J. Pharmacol.* 90: 11–17, 1983.
177. MAGEE, D. F., AND S. NARUSE. Neural control of periodic secretion of the pancreas and the stomach in fasting dogs. *J. Physiol. Lond.* 344: 153–160, 1983.
178. MARIANO, E. C., S. DEAK, M. T. REDDELL, AND J. H. LANDOR. Mechanisms of protein activation of the intestinal phase of gastric secretion. *Surgery St. Louis* 95: 492–496, 1984.
179. MARKS, I. N., S. A. KOMAROV, AND H. SHAY. Maximal acid secretory response to histamine and its relation to parietal cell mass in the dog. *Am. J. Physiol.* 199: 579–588, 1960.
180. MARTINDALE, R., G. L. KAUFFMAN, S. LEVIN, J. H. WALSH, AND T. YAMADA. Differential regulation of gastrin and somatostatin secretion from isolated perfused rat stomachs. *Gastroenterology* 83: 240–244, 1982.
181. MCDONALD, T. J., H. JÖRNVALL, G. NILSSON, M. VAGNE, M. GHATEI, S. R. BLOOM, AND V. MUTT. Characterization of a gastrin-releasing peptide from porcine non-antral gastric tissue. *Biochem. Biophys. Res. Commun.* 90: 227–233, 1979.
182. MCINTOSH, C. H. S., Y. N. KWOK, T. MORDHORST, E. NISHIMURA, R. A. PEDERSON, AND J. C. BROWN. Enkephalinergic control of somatostatin secretion from the perfused rat stomach. *Can. J. Physiol. Pharmacol.* 61: 657–663, 1983.
183. MCINTOSH, C. H. S., R. A. PEDERSON, H. KOOP, AND J. C. BROWN. Gastric inhibitory polypeptide stimulated secretion of somatostatinlike immunoreactivity from the stomach: inhibition by acetylcholine or vagal stimulation. *Can. J. Physiol. Pharmacol.* 59: 468–472, 1981.
184. MEYER, J. H. Control of pancreatic exocrine secretion. In: *Physiology of the Gastrointestinal Tract*, edited by L. R. Johnson. New York: Raven, 1981, p. 821–829.
185. MOGHIMZADEH, E., R. EKMAN, R. HÅKANSON, N. YANAIHARA, AND F. SUNDLER. Neuronal gastrin-releasing peptide in mammalian gut and pancreas. *Neuroscience* 10: 553–563, 1983.
186. MOORE, J. G. The relationship of gastric acid secretion to plasma glucose in five men. *Scand. J. Gastroenterol.* 15: 625–632, 1980.
187. MOORE, J. G., AND F. CRESPIN. Influence of glucose on cephalic-vagal-simulated gastric acid secretion in man. *Dig. Dis. Sci.* 25: 117–122, 1980.
188. MORLEY, J. E., AND A. S. LEVINE. The pharmacology of eating behavior. *Annu. Rev. Pharmacol. Toxicol.* 25: 127–146, 1985.
189. MORLEY, J. E., A. S. LEVINE, AND S. E. SILVIS. Mini-review: central regulation of gastric acid secretion: the role of neuropeptides. *Life Sci. Part II Biochem. Gen. Mol. Biol.* 31: 399–410, 1982.
190. MORRISON, T. F. B. Afferent innervation of the gastrointestinal tract. In: *Nerves and the Gut*, edited by F. P. Brooks and P. W. Evers. Thorofare, NJ: Slack, 1977, p. 297–326.
191. NAKAMURA, M., M. ODA, N. WATANABE, N. TSUKADA, Y. YONEI, AND M. TSUCHIYA. Evidence for direct parasympathetic innervation of parietal cells in the rat glandular stomach. Histochemical and electron microscopic cytochemical study. *Okajimas Folia Anat. Jpn.* 59: 167–180, 1982.
192. NAKAMURA, M., M. ODA, Y. YONEI, N. TSUKADA, N. WATANABE, H. KOMATSU, AND M. TSUCHIYA. Demonstration of the localization of muscarinic acetylcholine receptors in the gastric mucosa. *Acta Histochem. Cytochem.* 17: 297–309, 1984.
193. NISHI, S., Y. SEINO, J. TAKEMURA, H. ISHIDA, M. SENO, T. CHIBA, C. YANAIHARA, N. YANAIHARA, AND H. IMURA. Vagal regulation of GRP, gastric somatostatin, and gastrin secretion in vitro. *Am. J. Physiol.* 248 (*Endocrinol. Metab.* 11): E425–E431, 1985.
194. OBERHELMAN, H. A., JR., E. R. WOODWARD, J. M. ZUBIRAN, AND L. R. DRAGSTEDT. Physiology of the gastric antrum. *Am. J. Physiol.* 169: 738–748, 1952.
195. OKI, M., AND E. E. DANIEL. Effects of vagotomy on the ultrastructure of the nerves of dog stomach. *Gastroenterology* 73: 1029–1040, 1977.
196. OLBE, L., AND U. KNUTSON. Gastric acid response to sham feeding in the dog and the duodenal ulcer patient. *Acta Hepato-Gastroenterol.* 23: 455–458, 1976.
197. ORLOFF, M. J., P. V. HYDE, L. D. KOSTA, R. C. GUILLEMIN, AND R. H. BELL, JR. The intestinal phase hormone. *World J. Surg.* 3: 523–538, 1979.
198. PAGANI, F. D., W. P. NORMAN, D. K. KASBEKAR, AND R. A. GILLIS. Effects of stimulation of nucleus ambiguus complex on gastroduodenal function. *Am. J. Physiol.* 246 (*Gastrointest. Liver Physiol.* 9): G253–G262, 1984.
199. PAINTAL, A. S. Vagal sensory receptors and their reflex effects. *Physiol. Rev.* 53: 159–227, 1973.
200. PAPPAS, T. N., H. T. DEBAS, AND I. L. TAYLOR. Enterogastrone-like effect of peptide γγ is vagally mediated in the dog. *J. Clin. Invest.* 77: 49–53, 1986.
201. PAPPAS, T., D. HAMEL, H. DEBAS, J. H. WALSH, AND Y. TACHE. Cerebroventricular bombesin inhibits gastric acid secretion in dogs. *Gastroenterology* 89: 43–48, 1985.
202. PAPPAS, T., D. HAMEL, H. DEBAS, J. WALSH, AND Y. TACHE. Spantide: failure to antagonize bombesin-induced stimulation of gastrin in dogs. *Peptides Fayetteville* 6: 1001–1003, 1985.
203. PAYNE, R. A., AND A. W. KAY. The effect of vagotomy on the maximal acid secretory response to histamine in man. *Clin. Sci. Lond.* 22: 373–382, 1972.
204. PEARL, J. M., W. P. RITCHIE, JR., R. B. GILSDORF, J. P.

DELANY, AND A. S. LEONARD. Hypothalamic stimulation and feline gastric mucosal cellular populations. *J. Am. Med. Assoc.* 195: 281-284, 1966.

205. PEDERSON, R. A., C. H. MCINTOSH, M. K. MUELLER, AND J. C. BROWN. Absence of a relationship between immunoreactive-gastrin and somatostatin-like immunoreactivity secretion in the perfused rat stomach. *Regul. Pept.* 2: 53-60, 1981.

206. POLAK, J. M., AND S. R. BLOOM. Organization of the gut peptidergic innervation. In: *Gut Hormones*, edited by S. R. Bloom and J. M. Polak. New York: Churchill Livingstone, 1981, p. 487-494.

207. POWELL, D. W., AND B. I. HIRSCHOWITZ. Sodium pentobarbital depression of histamine- or insulin-stimulated gastric secretion. *Am. J. Physiol.* 212: 1001-1006, 1967.

208. PRESHAW, R. M. Influence of the antrum on the acid response to distension of the body of the stomach in dogs. *Can. J. Physiol. Pharmacol.* 48: 661-669, 1970.

209. PUURUNEN, J., AND J. LEPPALUOTO. Centrally administered PGE_2 inhibits gastric secretion in the rat by releasing vasopressin. *Eur. J. Pharmacol.* 104: 145-150, 1984.

210. RADKE, R., W. STACH, AND R. WEISS. Innervation of the gastric wall related to acid secretion: a light and electron microscopy study on rats, rabbits and guinea pigs. *Acta Biol. Med. Ger.* 39: 687-696, 1980.

211. RAGGI, F., AND D. S. KRONFELD. Higher glucose affinity of hexokinase in sheep brain than in rat brain. *Nature Lond.* 209: 1353-1354, 1966.

212. RAICHLE, M. E. Positron emission tomography. Progress in brain imaging. *Nature Lond.* 317: 574-576, 1985.

213. RAYBOULD, H. E., R. J. GAYTON, AND G. J. DOCKRAY. CNS effects of circulating CCK8: involvement of brainstem neurones responding to gastric distension. *Brain Res.* 342: 187-190, 1985.

214. REDFERN, J. S., R. THIRLBY, M. FELDMAN, AND C. T. RICHARDSON. Effect of pentagastrin on gastric mucosal histamine in dogs. *Am. J. Physiol.* 248 (*Gastrointest. Liver Physiol.* 11): G369-G375, 1985.

215. REHFELD, J. F. Gastrin and cholecystokinin in the vagus. *J. Auton. Nerv. Syst.* 9: 113-118, 1983.

216. REHFELD, J. F., AND E. AMDRUP (editors). *Gastrins and the Vagus*. New York: Academic, 1979, p. 315.

217. RICHARDSON, C. T., C. C. BARNETT, AND M. FELDMAN. Gastric acid secretory responsiveness to food in duodenal ulcer (DU) (Abstract). *Gastroenterology* 86: 1219, 1984.

218. RICHARDSON, C. T., M. N. PETERS, M. FELDMAN, R. N. MCCLELLAND, J. H. WALSH, K. A. COOPER, G. WILLEFORD, R. M. DICKERMAN, AND J. S. FORDTRAN. Treatment of Zollinger-Ellison syndrome with exploratory laparotomy, proximal gastric vagotomy, and H_2-receptor antagonists. A prospective study. *Gastroenterology* 89: 357-367, 1985.

219. RICHARDSON, C. T., J. H. WALSH, K. A. COOPER, M. FELDMAN, AND J. S. FORDTRAN. Studies on the role of cephalic-vagal stimulation in the acid secretory response to eating in normal human subjects. *J. Clin. Invest.* 60: 435-441, 1977.

220. RIDLEY, P. T., AND F. P. BROOKS. Alterations in gastric secretion following hypothalamic lesions producing hyperphagia. *Am. J. Physiol.* 209: 319-323, 1965.

221. RITTER, R. C., P. G. SLUSSER, AND S. STONE. Glucoreceptors controlling feeding and blood glucose: location in hindbrain. *Science Wash. DC* 213: 451-452, 1981.

222. ROBERTS, G. W., T. J. CROW, AND J. M. POLAK. Neuropeptides in the brain. In: *Gut Hormones* (2nd ed.), edited by S. R. Bloom and J. M. Polak. New York: Churchill Livingstone, 1981, p. 457-463.

223. ROBERTS, G. W., J. M. POLAK, AND T. J. CROW. Peptide circuitry of the limbic system. In: *Psychopharmacology of the Limbic System*, edited by M. R. Trimble and E. Zarifian. Oxford, UK: Oxford Univ. Press, 1984, p. 226-243.

224. ROGERS, R. C., AND G. E. HERMANN. Vagal afferent stimulation-evoked gastric secretion suppressed by paraventricular nucleus lesion. *J. Auton. Nerv. Syst.* 13: 191-199, 1985.

225. ROGERS, R. C., P. J. KAHRILAS, AND G. E. HERMANN. Projection of the hepatic branch of the splanchnic nerve to the brainstem of the rat. *J. Auton. Nerv. Syst.* 11: 223-225, 1984.

226. ROLAND, M., A. BERSTAD, AND I. LIAVAG. Effect of urecholine and carbacholine on pentagastrin-stimulated gastric secretion after proximal gastric vagotomy in duodenal ulcer patients. *Scand. J. Gastroenterol.* 10: 315-319, 1975.

227. ROMAN, C., AND J. GONELLA. Extrinsic control of digestive tract motility. In: *Physiology of the Gastrointestinal Tract*, edited by L. R. Johnson. New York: Raven, 1981, p. 289-333.

228. SAFFOURI, B., J. W. DUVAL, AND G. M. MAKHLOUF. Stimulation of gastrin secretion in vitro by intralumenal chemicals: regulation by intramural cholinergic and non-cholinergic neurons. *Gastroenterology* 87: 557-561, 1984.

229. SAFFOURI, B., G. C. WEIR, K. N. BITAR, AND G. M. MAKHLOUF. Gastrin and somatostatin secretion by perfused rat stomach: functional linkage of antral peptides. *Am. J. Physiol.* 238 (*Gastrointest. Liver Physiol.* 1): G495-G501, 1980.

230. SAKAGUCHI, T. Alterations in gastric acid secretion following hepatic portal injections of D-glucose and its anomers. *J. Auton. Nerv. Syst.* 5: 337-344, 1982.

231. SANDERS, M. J., AND A. H. SOLL. Characterization of receptors regulating secretory function in the fundic mucosa. *Annu. Rev. Physiol.* 48: 89-101, 1986.

232. SAWCHENKO, P. E., AND L. R. SWANSON. Central noradrenergic pathways for the integration of hypothalamic neuroendocrine and autonomic responses. *Science Wash. DC* 214: 685-687, 1981.

233. SCHEPP, W., S. E. MIEDERER, AND H. J. RUOFF. Intrinsic factor secretion from isolated gastric mucosal cells of rat and man—two different patterns of secretagogue control. *Agents Actions* 14: 522-528, 1984.

234. SCHILLER, L. R., J. H. WALSH, AND M. FELDMAN. Distension-induced gastrin release: effects of luminal acidification and intravenous atropine. *Gastroenterology* 78: 912-917, 1980.

235. SCHILLER, L. R., J. H. WALSH, AND M. FELDMAN. Effect of atropine on gastrin release stimulated by an amino acid meal in humans. *Gastroenterology* 83: 267-272, 1982.

236. SCHUBERT, M. L., K. N. BITAR, AND G. M. MAKHLOUF. Regulation of gastrin and somatostatin secretion by cholinergic and noncholinergic intramural neurons. *Am. J. Physiol.* 243 (*Gastrointest. Liver Physiol.* 6): G442-G447, 1982.

237. SCHUBERT, M. L., AND G. M. MAKHLOUF. Regulation of gastrin and somatostatin secretion by intramural neurons: effect of nicotinic receptor stimulation by dimethyl-phenylpiperazinium. *Gastroenterology* 83: 626-632, 1982.

238. SCHULTZBERG, M. The peripheral nervous system. In: *Chemical Neuroanatomy*, edited by P. C. Emson. New York: Raven, 1983, p. 1-52.

239. SCHWARTZ, T. W. Pancreatic polypeptide: a hormone under vagal control. *Gastroenterology* 85: 1411-1425, 1983.

240. SERFILIPPI, D., AND R. M. DONALDSON, JR. Production and secretion of intrinsic factor by isolated rabbit gastric mucosa. *Am. J. Physiol.* 251 (*Gastrointest. Liver Physiol.* 14): G287-G292, 1986.

241. SHARKEY, K. A., R. G. WILLIAMS, AND G. J. DOCKRAY. Sensory substance P innervation of the stomach and pancreas: demonstration of capsaicin-sensitive sensory neurons in the rat by combined immunohistochemistry and retrograde tracing. *Gastroenterology* 87: 914-921, 1984.

242. SHIRAISHI, T. Effects of lateral hypothalamic stimulation on medulla oblongata and gastric vagal neural responses. *Brain Res. Bull.* 5: 245-250, 1980.

243. SHIRAISHI, T., AND T. TAKAHASHI. Glucose sensing cells and gastric acid secretion. *J. Physiol. Soc. Jpn.* 32: 422, 1970.

244. SHIRAKAWA, T., AND B. I. HIRSCHOWITZ. Interaction between stimuli and their antagonists on frog esophageal peptic glands. *Am. J. Physiol.* 249 (*Gastrointest. Liver Physiol.* 12): G668-G673, 1985.

245. SILVERMAN, A.-J., AND G. E. PICKARD. The hypothalamus. In: *Chemical Neuroanatomy*, edited by P. C. Emson. New York:

Raven, 1983, p. 295–336.
246. SIMICI, D., M. POPESCO, AND G. DICULESCO. L'action de l'insuline sur la sécrétion de l'estomac a l'état normal et pathologique. *Arch. Mal. Appar. Dig. Mal. Nutr.* 17: 28–43, 1927.
247. SJODIN, L. Potentiation of the gastric secretory response to sham feeding in dogs by infusion of gastrin and pentagastrin. *Acta Physiol. Scand.* 85: 24–32, 1972.
248. SJODIN, L. Electrical stimulation of nerves and secretion from digestive glands. In: *Nerves and the Gut*, edited by F. P. Brooks and P. W. Čvers. Thorofare, NJ: Slack, 1977, p. 1–13.
249. SMOLEN, A. J., AND R. C. TRUEX. The dorsal motor nucleus of the vagus nerve of the cat. *Anat. Rec.* 189: 555–565, 1977.
250. SOLL, A. H. Receptors modulating acid secretion. In: *Receptors and the Upper GI Tract*, edited by B. I. Hirschowitz and J. G. Spenney. New York: Advanced Therapeutics Communications, 1983, p. 101–115.
251. SOON-SHIONG, P., AND H. T. DEBAS. Fundic inhibition of acid secretion and gastrin release in the dog. *Gastroenterology* 79: 867–872, 1980.
252. STADIL, F., AND J. G. STAGE. Gastrinomas as model for duodenal ulcer disease. In: *Gastrin and the Vagus*, edited by J. F. Rehfeld and E. Amdrup. New York: Academic, 1979, p. 199–210.
253. STENQUIST, B., U. KNUTSON, AND L. OLBE. Gastric acid responses to adequate and modified sham feeding and to insulin hypoglycemia in duodenal ulcer patients. *Scand. J. Gastroenterol.* 13: 357–362, 1978.
254. STENQUIST, B., G. NILSSON, J. F. REHFELD, AND L. OLBE. Plasma gastrin concentrations following sham feeding in duodenal ulcer patients. *Scand. J. Gastroenterol.* 14: 305–311, 1979.
255. SWANSON, L. W., AND H. G. KUYPERS. The paraventricular nucleus of the hypothalamus. *J. Comp. Neurol.* 194: 555–570, 1980.
256. SZURSZEWSKI, J. H. Physiology of mammalian prevertebral ganglia. *Annu. Rev. Physiol.* 43: 53–68, 1981.
257. TACHE, Y., D. LESIEGE, W. VALE, AND R. COLLU. Gastric hypersecretion by intracisternal TRH: dissociation from hypophysiotropic activity and role of central catecholamine. *Eur. J. Pharmacol.* 107: 149–155, 1985.
258. TACHE, Y., W. VALE, AND M. BROWN. Thyrotropin releasing hormone—CNS action to stimulate gastric acid secretion. *Nature Lond.* 287: 149–151, 1980.
259. TACHE, Y., W. VALE, J. RIVIER, AND M. BROWN. Brain regulation of gastric secretion: influence of neuropeptides. *Proc. Natl. Acad. Sci. USA* 77: 5515–5519, 1980.
260. TEPPERMAN, B. L., AND M. S. EVERED. Gastrin injected into the lateral hypothalamus stimulates secretion of gastric acid in rats. *Science Wash. DC* 209: 1142–1143, 1980.
261. TEPPERMAN, B. L., J. H. WALSH, AND R. M. PRESHAW. Effect of antral denervation on gastrin release by sham feeding and insulin hypoglycemia in dog. *Gastroenterology* 63: 973–980, 1972.
262. TER HORST, G. J., P. G. LUITEN, AND F. KUIPERS. Descending pathways from hypothalamus to dorsal motor vagus and ambiguus nuclei in the rat. *J. Auton. Nerv. Syst.* 11: 59–75, 1984.
263. THIRLBY, R. C., AND M. FELDMAN. Effect of chronic sham feeding on maximal gastric acid secretion in the dog. *J. Clin. Invest.* 73: 566–569, 1984.
264. THIRLBY, R. C., M. H. STEVENS, A. J. BLAIR, I. L. CRAWFORD, I. A. TAYLOR, AND M. FELDMAN. Effect of a GABA agonist, progabide, on gastric acid secretion and serum gastrin (G), pancreatic polypeptide (PP), and somatostatin-like immunoreactivity (SLI) during insulin-hypoglycemia in dogs (Abstract). *Gastroenterology* 90: 1663, 1986.
265. THOMPSON, D. A., R. G. CAMPBELL, U. LILAVIVAT, S. L. WELLE, AND G. L. ROBERTSON. Increased thirst and plasma arginine vasopressin levels during 2-deoxy-D-glucose-induced glucoprivation in humans. *J. Clin. Invest.* 67: 1083–1093, 1981.
266. TRIMBLE, M. R., AND E. ZARIFIAN (editors). *Psychopharmacology of the Limbic System*. Oxford, UK: Oxford Univ. Press, 1984.
267. UVNAS, B. The part played by the pyloric region in the cephalic phase of gastric secretion. *Acta Physiol. Scand. Suppl.* XIII: 5–86, 1942.
268. VALLGREN, S., M. EKELUND, AND R. HÅKANSON. Mechanism of inhibition of gastric acid secretion by vagal denervation in the rat. *Acta Physiol. Scand.* 119: 77–80, 1983.
269. WALSH, J. H. Evidence for the hormonal role of gastrin in regulation of gastric acid secretion. In: *Gut Peptides: Secretion, Function and Clinical Aspects*, edited by A. Miyoshi. Amsterdam: Elsevier, 1979, p. 137–144.
270. WALSH, J. H. Endocrine cells of the digestive system. In: *Physiology of the Gastrointestinal Tract*, edited by L. R. Johnson. New York: Raven, 1981, p. 59–144.
271. WHITE, J. D., K. D. STEWART, J. E. KRAUSE, AND J. F. MCKELVY. Biochemistry of peptide-secreting neurons. *Physiol. Rev.* 65: 553–606, 1985.
272. WILLIFORD, D. J., H. S. ORMSBEE III, W. NORMAN, J. W. HARMON, T. Q. GARVEY III, J. A. DIMICCO, AND R. A. GILLIS. Hindbrain GABA receptors influence parasympathetic outflow to the stomach. *Science Wash. DC* 214: 193–194, 1981.
273. WISE, L., AND W. F. BALLINGER. Effect of vagal stimulation and inhibition on gastric mucus and sulfated aminopolysaccharide secretion. *Ann. Surg.* 174: 976–982, 1971.
274. WOOD, J. D. Physiology of the enteric nervous system. In: *Physiology of the Gastrointestinal Tract*, edited by L. R. Johnson. New York: Raven, 1981, p. 1–37.
275. WYRWICKA, W., AND R. GARCIA. Effect of electrical stimulation of the dorsal nucleus of the vagus nerve on gastric acid secretion in cats. *Exp. Neurol.* 65: 315–325, 1979.
276. YAGI, Y., A. MISUMI, AND A. MURAKAMI. Role of antral mucosa in intestinal phase of gastric secretion in dog. *Gastroenterol. Jpn.* 19: 24–33, 1984.
277. YAMAMOTO, T., H. SATOMI, H. ISE, AND K. TAKAHASHI. Evidence of the dual innervation of the cat stomach by the vagal dorsal motor and medial solitary nuclei as demonstrated by the horseradish peroxidase method. *Brain Res.* 122: 125–131, 1977.
278. YOUNG, W. S., J. K. WAMSLEY, M. A. ZARBIN, AND M. J. KUHAR. Opioid receptors undergo axonal flow. *Science Wash. DC* 210: 76–78, 1980.
279. ZARBIN, M. A., J. K. WAMSLEY, R. B. INNIS, AND M. J. KUHAR. Cholecystokinin receptors: presence and axonal flow in rat vagus nerve. *Life Sci. Part I Physiol. Pharmacol.* 29: 697–705, 1981.
280. ZARBIN, M. A., J. K. WAMSLEY, AND M. J. KUHAR. Axonal transport of muscarinic cholinergic receptors in rat vagus nerve: high and low affinity agonist receptors move in opposite directions and differ in nucleotide sensitivity. *J. Neurosci.* 2: 934–941, 1982.

CHAPTER 9

Inhibition of gastric acid secretion

STANISŁAW J. KONTUREK | *Institute of Physiology, Academy of Medicine, Krakow, Poland*

CHAPTER CONTENTS

Cephalic Inhibition of Gastric Secretion
  Cephalic phase
  Inhibition of cephalic phase
    Effects of vagotomy and antrectomy
    Vagus-dependent inhibitor—vagogastrone
    Somatostatin as vagogastrone
    Pharmacological inhibition of cephalic phase
Inhibition Arising From Stomach
  Gastric phase
    Effects of vagotomy and antrectomy
    Inhibition arising from oxyntic mucosa
      Direct inhibition by luminal acidification
      Local inhibitors in oxyntic mucosa
      Somatostatin and prostaglandins as local inhibitors
    Inhibition arising from antral mucosa
      Effects of luminal acidification
      Roles of somatostatin and prostaglandins
      Inhibitory pylorooxyntic reflex
  Pharmacological inhibition of gastric phase
Intestinal Inhibition of Gastric Secretion
  Intestinal phase
  Intestinal inhibition by acid
    Secretin versus bulbogastrone
    Somatostatin as bulbogastrone
    Secretin as physiological inhibitor
    Role of cholecystokinin
  Intestinal inhibition by fat
    Contribution of cholecystokinin and secretin
    Gastric inhibitory peptide as enterogastrone
    Role of neurotensin
    Peptide YY and ileocolonic inhibition
  Intestinal inhibition by hyperosmolar solutions
Gastric Inhibition by Urogastrone and Epidermal Growth Factor
Gastric Inhibition by Neuropeptides
  Gastric inhibitory effects of neuropeptides administered intracerebrally
    Bombesin-like peptides
    Opioid peptides
    Corticotropin-releasing factor
    Calcitonin gene-related peptide
    Other peptides
  Gastric inhibitory effects of neuropeptides administered intravenously
    Somatostatin
    Vasoactive intestinal peptide
    Thyrotropin-releasing hormone
    Corticotropin-releasing factor
    Opioid peptides
Sympathetic Nervous System and Gastric Inhibition

THE OBSERVATION THAT the secretion of HCl by gastric glands is markedly increased after the ingestion of a meal and virtually abolished during the interdigestive period formed the concept that gastric secretion depends on the interplay of many stimulatory and inhibitory influences that arise within the central nervous system and the gastrointestinal tract. Through interactions between nervous and hormonal stimuli, gastric secretion is controlled by complex feedback mechanisms. The inhibitory influences on gastric secretion are usually classified as central, gastric, and intestinal in origin, although under normal conditions inhibitory, like stimulatory, processes occur and interact continuously. With the discovery of functional links between the brain and the gut (brain-gut axis), a number of neuropeptides acting within the brain or in the gastrointestinal tract have been implicated in the peptidergic control of gastric secretion. In addition to the physiological control of gastric secretion, a number of pharmacological agents are now available that suppress the secretory activity of the parietal cells either at their receptor sites (e.g., anticholinergics or histamine H_2-receptor blockers) or at nonreceptor sites (e.g., inhibitors of cAMP formation or inhibitors of the proton pump).

The inhibition covers a whole series of situations ranging from severe and irreversible loss of the response to stimulants, through more or less quickly reversible decrease, to a rebound increase in the secretory process immediately after the cessation of the inhibitory process. The inhibition with rebound hypersecretion may sometimes be wholly or partly compensated; therefore, the term *inhibition* always requires explicit definition, description, and analysis.

CEPHALIC INHIBITION OF GASTRIC SECRETION

Cephalic Phase

The cephalic phase of gastric secretion refers to the stimulation by factors acting in the region of cranial nerve distribution. It can be elicited physiologically by sham feeding or pharmacologically by glucopenic agents such as insulin or 2-deoxy-D-glucose (2-DG). Factors evoking conditioned reflexes established by pairing unconditioned stimuli such as food with indifferent stimuli such as light or sound can also be effective.

The cephalic phase is assumed to be mediated entirely by the vagal nerves, and the vagal effects on the stomach are attributable to three major components: *1*) direct vagal stimulation of the parietal cells, *2*) vagal release of gastrin and possibly also histamine, and *3*) vagal sensitization of the parietal cells to other secretagogues (60). Because the maximal secretory response to the physiological vagal stimulation such as sham feeding reaches in humans only ~50% (89, 116, 180) and in dogs ~70%–90% (109, 121) of the maximal response to pentagastrin or histamine, it has been postulated that vagal excitation includes both the stimulatory and inhibitory influences on gastric secretion (60).

Inhibition of Cephalic Phase

The cephalic stimulation of gastric secretion induced by sham feeding depends on the "desire for food" and so the inhibition of appetite by emotional and environmental influences depresses or abolishes gastric secretory response. For the same reason, glucose introduced intraduodenally during the cephalic phase induced by sham feeding or insulin hypoglycemia was found to suppress the acid secretory response, possibly through the action on glucose-sensitive receptors in the hypothalamus (151). The exclusion of the cephalic phase, by introducing food directly into the stomach of the unwary subject or by vagotomy, results in a reduction of postprandial acid secretion and the prolongation of the whole digestive process. Certain substances such as thyrotropin-releasing hormone (TRH), which are known to reduce the food intake (152), were also found to suppress gastric acid secretion induced by sham feeding (125).

If the cephalic phase is initiated by conditioned reflexes, the resulting gastric secretory response declines and is finally extinguished unless it is periodically reinforced (162). The disappearance of unreinforced conditioned reflex stimulation of gastric secretion was considered by Pavlov to originate from its active central inhibition. Gastric secretion can also be inhibited by suggestion under hypnosis, indicating an important role of cerebral cortex in the control of vagal activity.

EFFECTS OF VAGOTOMY AND ANTRECTOMY. Truncal vagotomy in dogs permanently abolished the gastric acid response to all types of cephalic stimulation (60). Some recovery of the gastric response to vagal excitation was observed in dogs a few months after selective gastric vagotomy, possibly due to the regeneration of vagal fibers (67). Also in humans, in the early period after vagotomy, responses to all forms of vagal excitation were completely abolished, but then the secretory responses were partly recovered (56, 83).

Removal of the main source of gastrin by antrectomy (155) or antral mucosectomy (121) in dogs reduced acid response to sham feeding 70%–90%, despite intact vagal innervation of the oxyntic mucosa. Because exogenous gastrin in the subthreshold dose completely restored the secretory response to sham feeding to its preantrectomy level, it has been concluded that vagally released gastrin plays a crucial role in the cephalic stimulation of gastric secretion, possibly due to the potentiating effect on direct vagal stimulation of parietal cells. In accordance with these findings, acidification of the antral mucosa to suppress the release of gastrin (155) or vagal antral denervation to abolish vagal release of gastrin (213) completely inhibited the cephalic-phase stimulation of acid secretion from the vagally innervated oxyntic mucosa in dogs.

In addition to the stimulatory fibers, vagal nerves may also possess the inhibitory fibers for gastrin release. This has been documented by the simple observation that immediately after vagotomy there is a marked rise in basal and postprandial serum gastrin levels (70). The postvagotomy hypergastrinemia is a common finding that cannot be attributed to the alteration in the plasma somatostatin level and likely results from withdrawal of the usual tonic vagal inhibitory influence on gastrin release (Fig. 1). In humans, the cephalic stimulation of gastric secretion was not accompanied by any significant alterations in plasma gastrin and appears to be less dependent on gastrin release. Accordingly, antrectomy resulted in a smaller reduction in cephalic stimulation of gastric secretion, and exogenous gastrin failed to restore the secretory response (89). Vagotomy in humans, as in dogs, resulted in a marked increase in serum gastrin (46, 47, 216, 218), indicating that vagal nerves exert a tonic inhibitory effect on gastrin release.

VAGUS-DEPENDENT INHIBITOR—VAGOGASTRONE. Interestingly, after vagotomy of the oxyntic mucosa in dogs and humans, when the vagal innervation of antral mucosa is intact, sham feeding produced greater gastrin release than it did before vagotomy, but no acid response occurred (46). The failure of the denervated oxyntic mucosa to respond to sham feeding, despite the excessive gastrin release, was initially explained by the reduced sensitivity of the vagally denervated oxyntic glands to gastrin (60). This explanation seemed unlikely because truncal vagotomy performed on dogs with vagally denervated fundic pouches greatly increased the acid response of these pouches to gastrin so that the maximal output induced by gastrin reached that evoked by histamine (39). Because truncal vagotomy did not affect maximal response to histamine but enhanced that to gastrin in these animals, it was proposed that the activity of intact vagal nerves releases an inhibitor that decreases gastrin-stimulated secretion from the vagally denervated oxyntic mucosa.

The existence of such a vagus-dependent inhibitor was suspected previously when several authors reported that sham feeding in dogs, despite excessive gastrin release, provoked little or no acid secretion from the vagally denervated Heidenhain pouches. The hypothetical inhibitor substance has been named *vagogastrone* (60). Further evidence supporting the vagogastrone concept was provided by Preshaw (172), who found that gastrin-stimulated secretion from the Heidenhain pouch was actively inhibited when the animals were sham fed. Preshaw hypothesized that the inhibition results from the action of an inhibitor released by excited vagal nerves. This finding was confirmed by Sjodin (190), who also observed that the inhibition disappeared after truncal vagotomy. Because the inhibition was eliminated by resection of the antrum and duodenum (191), it has been postulated that the release of vagogastrone requires vagal integrity and is confined to the antroduodenal region.

SOMATOSTATIN AS VAGOGASTRONE. Vagogastrone has not yet been isolated or identified, but recently several regulatory peptides were found to be released during vagal excitation, and some of them have been supposed to act as vagus-dependent inhibitors of gastric secretion. One of these is somatostatin, which was found to be released during vagal excitation by sham feeding (30, 114), 2-DG (77), or insulin hypoglycemia (70, 230) and to inhibit gastric secretion, particularly in response to gastrin (123). More detailed studies of De Graef and Woussen-Colle (30) indicated that somatostatin release is largely under vagal-cholinergic control (Fig. 2), and the amounts of hormone released after sham feeding might be sufficient for the inhibition of gastrin-stimulated gastric secretion. Thus somatostatin seems to fulfill the criteria for vagogastrone. It remains to be determined, however, whether vagotomy is capable of abolishing the release of somatostatin as it is with vagogastrone. Recently, Hollinshead et al. (70) reported that indeed vagotomy in dogs abolished plasma somatostatin response to insulin hypoglycemia and caused a marked elevation of serum gastrin, but this has been interpreted as showing that vagus exerts a tonic inhibitory influence on gastrin release independently of its effect on somatostatin. Other studies suggested that the increase in the plasma level of somatostatin after vagal excitation such as insulin hypoglycemia (230) is not due to direct vagal excitation of the D cells but probably results from the increase in gastric acid secretion and subsequent acid stimulation of these cells, because the inhibition of acid secretion by cimetidine prevented this insulin-induced somatostatin release.

PHARMACOLOGICAL INHIBITION OF CEPHALIC PHASE. The major component of vagally stimulated gastric secretion is direct cholinergic activation of the parietal cells with the muscarinic action of acetylcholine as a postganglionic neuroeffector modulator. As expected, anticholinergic drugs reduce dose dependently gastric acid responses induced by sham feeding or insulin hypoglycemia in humans (45, 120) and in animals (109). Because anticholinergics tend to increase plasma gastrin, their gastric inhibitory action is mainly due to direct suppression of vagal-cholinergic stimulation of the parietal cells. Recent studies with selective anticholinergics such as pirenzepine (45, 120) suggest the existence of subtypes of muscarinic receptors for acetylcholine (M_1 and M_2 receptors). Isolated parietal cells appear to be predominantly equipped with M_2 receptors of low affinity to pirenzepine so that atropine is a more potent inhibitor of acid formation than pirenzepine is. However, in in vivo studies pirenzepine was found to be relatively more potent, suggesting that the high-affinity M_1 receptors for this drug must be present somewhere in the gastric vagal pathway at a site remote from the parietal cells, such as the enteric ganglia.

Data of Figure 3 also show that, in addition to

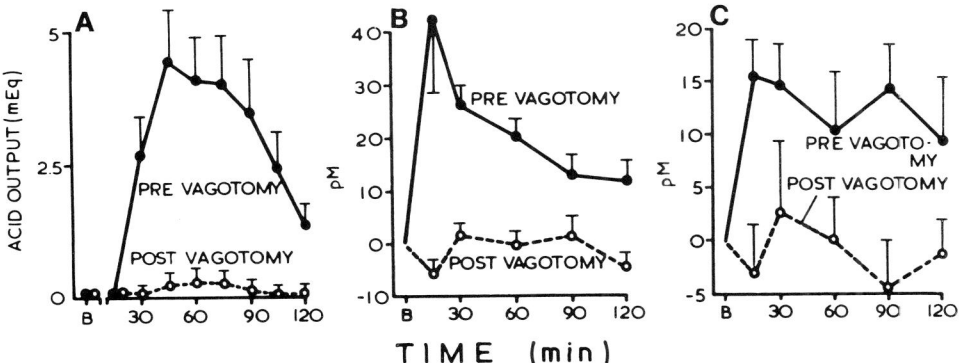

FIG. 1. *A*: gastric fistula acid output; *B*: plasma gastrin; *C*: somatostatin-like immunoreactivity increments over basal levels in response to insulin hypoglycemia before and >3 days after truncal vagotomy in dogs. [From Hollinshead et al. (70). Copyright 1985 by The American Gastroenterological Association.]

anticholinergics, other inhibitors of parietal cells such as H$_2$-receptor antagonists (e.g., cimetidine or ranitidine), agents interfering with cAMP production (e.g., prostaglandins), and drugs blocking the proton pump (e.g., omeprazole) suppress cephalic stimulation of gastric secretion (8, 93). These agents are highly effective inhibitors of all types of vagal excitation of gastric acid secretion, though their primary mode of action is unrelated to vagal-cholinergic stimulation of the oxyntic glands.

This apparent nonspecificity of the effects of various inhibitors on gastric secretion could be explained by the fact that in vivo stimulation of the parietal cells, such as that occurring after vagal excitation, results from the action of not only acetylcholine released from the nerve terminals but also from the action of other secretagogues, particularly histamine. These secretagogues activate the parietal cells by means of at least three different membrane receptors, resulting in the potentiating interaction on acid stimulation (193). In the in vitro isolated parietal cells, an individual antagonist to any of these receptors blocks the parietal cell response only to the respective agonist. When these cells are stimulated by several different secretagogues and the potentiating interaction occurs, both the anticholinergics and H$_2$ blockers are effective inhibitors and so display an apparent nonspecificity reminiscent of that found in vivo. Anticholinergics withdraw the cholinergic enhancement, whereas H$_2$ blockers abolish the histaminic enhancement of the interaction (194).

Vagal stimulation such as that induced by sham

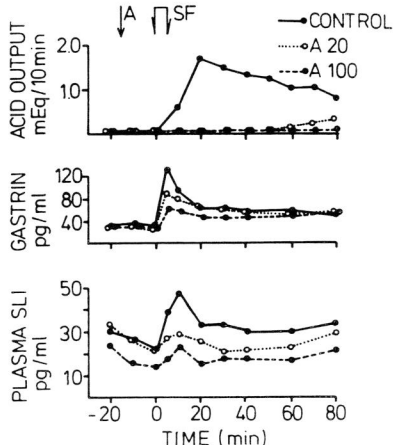

FIG. 2. Mean acid output from gastric fistula and plasma gastrin and somatostatin-like immunoreactivity, SLI, before and after sham feeding, SF, (10 min) in dogs without and with intravenous injection of atropine, A, at dose of 20 or 100 µg/kg. [From De Graef and Woussen-Colle (30).]

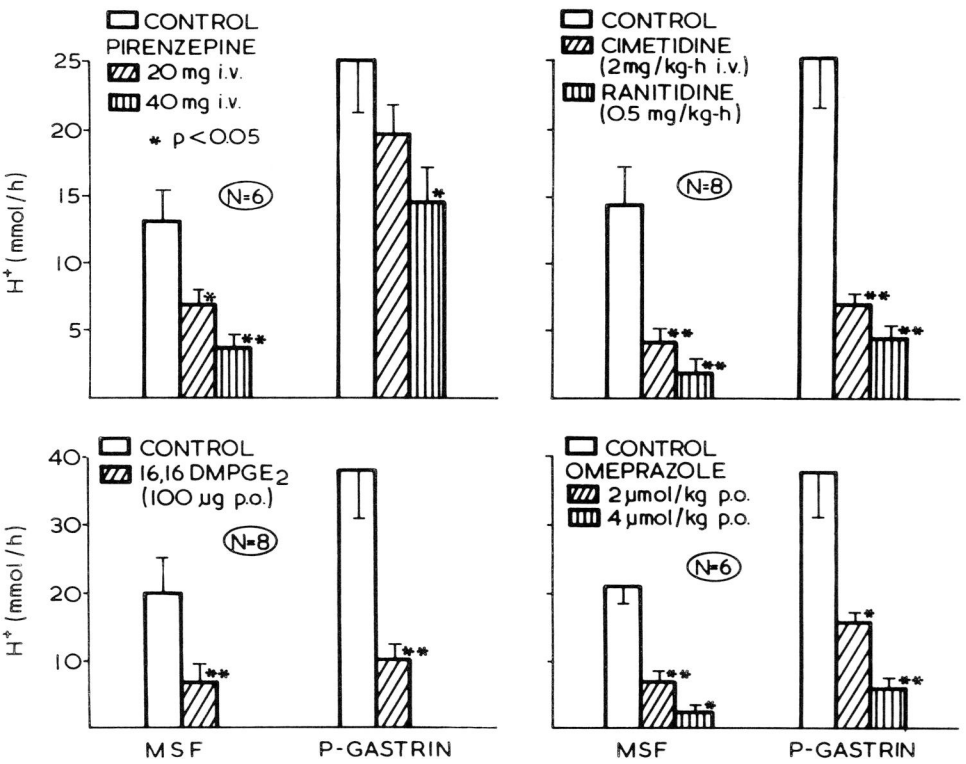

FIG. 3. Mean ± SE acid response to modified sham feeding, MSF, and pentagastrin infusion (2 µg·kg^{-1}·h^{-1}) in duodenal ulcer patients (n = 6–8) without and with administration of pirenzepine, cimetidine or ranitidine, 16,16-dimethyl prostaglandin E$_2$ (16,16 DMPGE$_2$), or omeprazole. *Single asterisk*, $P < 0.05$; *double asterisk*, $P < 0.01$. [From Konturek (93).]

feeding or insulin hypoglycemia appears to activate all three major pathways (cholinergic, histaminergic, and gastrinergic) of the parietal cell stimulation, but the histamine–H_2-receptor pathway may be dominant because H_2 blockers are capable of inhibiting a greater portion of vagally stimulated acid secretion than are anticholinergics (61, 93). The residual secretion observed after H_2 blockade may represent the interaction between remaining cholinergic and gastrinergic inputs at the parietal cells (61).

Because prostaglandins specifically inhibit histamine-stimulated cAMP production in the isolated parietal cells (194), they have been expected to suppress histamine-induced gastric secretion to an extent similar to that of H_2 blockers. However, both natural PGE_2 and its stable analogues were found to suppress not only histaminic but also vagal stimulation of acid secretion (79, 93). Prostaglandin analogues appear to be particularly effective inhibitors when applied directly on the gastric mucosa, possibly because of their local action on the gastric glands and the suppression of the release of gastrin.

The inhibition of the proton pump, which is distal to cAMP and is probably the terminal stage of secretory pathway, provides a means to completely block the secretory activity of the parietal cells (8). Indeed, benzimidazole derivatives such as omeprazole, which are specific and irreversible blockers of the H^+-K^+-ATPase of the parietal cells, have been found to suppress all known modes of the gastric acid stimulation, including that induced by vagal excitation (8, 50, 103, 115). This gastric inhibition by omeprazole occurred despite the increase of serum gastrin level, which was probably due to the pronounced reduction of intragastric acidity and the removal of acid-induced suppression of gastrin release.

INHIBITION ARISING FROM STOMACH

Gastric Phase

The gastric phase may be defined as that fraction of the secretory response that occurs when a food stimulus acts on receptors within the stomach. It begins soon after the food contacts the gastric mucosa and continues for several hours afterwards. With a newly devised technique of measuring gastric acid secretion in the presence of food in the intact stomach, known as intragastric titration (53), it has been established that the postprandial acid secretory rate reaches a gastrin or histamine maximum and depends on the distension of the stomach and the action of the chemical constituents of food on the oxyntic glands and the G cells (53, 60).

In dogs, the distension of the oxyntic or antral mucosa initiates both long (vagovagal) and short (intramural) reflexes with efferents to both the oxyntic glands and the G cells. These reflexes are believed to be cholinergic, because they can be inhibited by a sufficient dose of atropine (60).

Food in the stomach stimulates gastric secretion also by its chemical ingredients, particularly the products of protein digestion such as peptones and amino acids. This chemical stimulation includes direct action on the G cells to release gastrin and possibly also direct excitation of the oxyntic glands to secrete acid.

EFFECTS OF VAGOTOMY AND ANTRECTOMY. Gastric-phase stimulation of acid secretion can be reduced in part by vagotomy because of vagovagal reflexes activated by the distension of the stomach, but this is accompanied by an increase in serum gastrin level. This action seems to indicate that vagal nerves exhibit an inhibitory effect on the release of gastrin (57, 76, 216, 218). Antrectomy also reduces the postprandial gastric acid secretion but causes a marked reduction in the postprandial gastrin release. It also diminishes by ~50% the maximal secretory response to exogenous stimulants such as histamine or pentagastrin. Adding vagotomy to this procedure eliminated 80% of the response to histamine (218). It seems that the removal of the endogenous source of gastrin and the subsequent absence of its trophic action is the main cause of the reduction in postantrectomy diminution of acid secretory capacity.

Inhibition Arising From Oxyntic Mucosa

In the intact stomach with full vagal innervation and undisturbed release of gastrin, the effectiveness of mechanical, chemical, or vagal-cholinergic stimulation is greatly reduced when the gastric content is acidified. The mechanism by which an acidified meal in the stomach inhibits the secretory activity of the oxyntic glands has not been fully elucidated, but both the oxyntic and the antral mucosa have been proposed as the source of the inhibition.

DIRECT INHIBITION BY LUMINAL ACIDIFICATION. The possibility of the gastric inhibitory mechanism arising from the oxyntic gland area was first suggested by Debas and Grossman (26), who reported that bathing the oxyntic mucosa with a peptide solution stimulates acid secretion by a local mechanism sensitive to the luminal pH. It has been confirmed (19) that peptic digests in contact with oxyntic mucosa stimulate acid secretion that might be inhibited pH dependently with the reduction of the luminal pH below 4.0. The concept of direct chemical stimulation of oxyntic mucosa has been refuted by Spenney (198), who indicated that a solution of luminal peptone at a high pH may act as a sink for the passage of gaseous CO_2 from the blood into the gastric lumen, where it might be mistaken for secreted HCl. Because luminal peptones and amino acid solutions were also effective stimulants at lower pH (19, 27, 217), it remains to be established whether secretory response to topical peptic digests is an artifact, as suggested by Spenney, or whether it represents a genuine stimulation of HCl secretion.

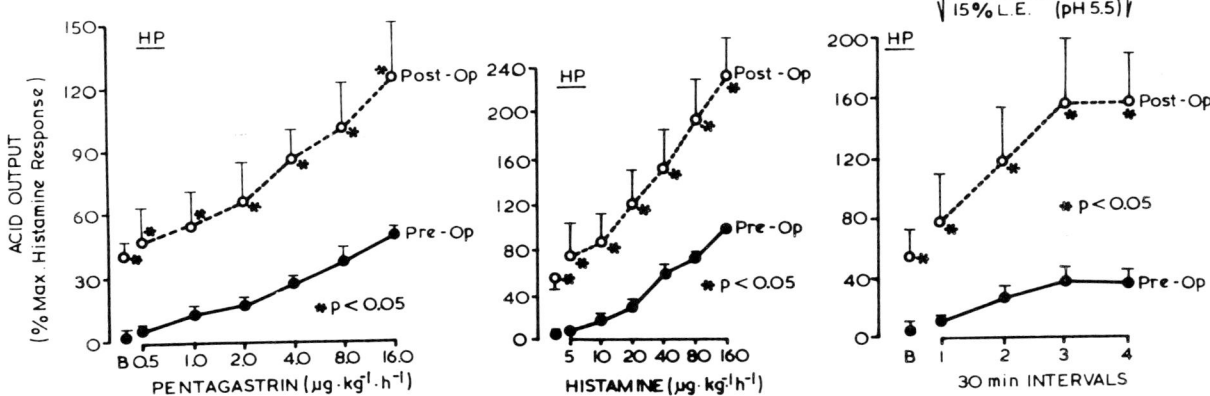

FIG. 4. Mean acid outputs from Heidenhain pouch, HP, in response to graded doses of intravenous pentagastrin, histamine, or 15% liver extract meal, LE, before, pre-op, and after, post-op, excision of fundic mucosa. [From Soon-Shiong and Debas (197). Copyright 1980 by The American Gastroenterological Association.]

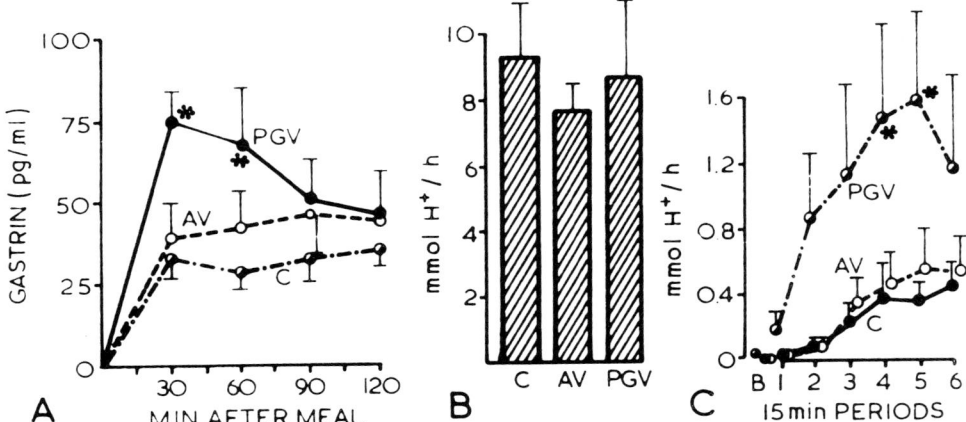

FIG. 5. Responses to 300-ml liquid meal of 15% liver extract in control stage, C, after truncal vagotomy, AV, and after proximal gastric vagotomy, PGV. A: plasma gastrin increment over basal aftermeal ($n = 6$ dogs); B: first 4-h gastric fistula acid output in response to meal ($n = 6$ dogs); C: Heidenhain pouch acid output in response to meal in three stages ($n = 3$ dogs). [From Debas et al. (27).]

The possibility that luminal acidification of the meal in contact with the oxyntic mucosa suppresses the secretory activity of the oxyntic glands has been criticized by Carter and Grossman (11), who reinvestigated the effects of low luminal pH using validated intrapouch titration and volume markers. They recognized larger errors using intragastric titration at low pH and were unable to confirm previous observations (19, 27) that the luminal pH affects gastric acid secretion by direct action on the oxyntic mucosa.

LOCAL INHIBITORS IN OXYNTIC MUCOSA. Strong support for the existence of a local inhibitor of acid secretion in the oxyntic mucosa has been provided by Soon-Shiong and Debas (197). They found that the excision of the oxyntic mucosa from the main stomach to abolish its secretory capacity resulted in an excessive acid secretion from the Heidenhain pouch both under basal conditions and after stimulation with histamine, gastrin, or food (Fig. 4). Although excessive resting acid secretion and increased sensitivity of the oxyntic glands to various secretagogues might reflect an accompanying hypergastrinemia, it has been proposed that these were due to the removal of the inhibitor of the oxyntic glands and of the G cells. The active principle has not yet been identified, but preliminary results obtained with purified oxyntic mucosa extracts indicate that this oxyntic chalone is a low-molecular-weight peptide (26). The mechanism releasing the oxyntic inhibitor has not been elucidated, but Debas and colleagues (27) believe that such an inhibitor could be released by a gastric meal through the vagovagal, cholinergic pathway to the proximal stomach. The release of the inhibitor requires at least partial integrity of vagal fibers supplying both the antral and the oxyntic mucosa because vagal denervation of these mucosal areas eliminated the inhibitory influence [Fig. 5; (27)]. The humoral nature of the inhibitor is supported by the fact that the antral distension in dogs with vagally innervated main stomachs also inhibited acid secretion from the vagally denervated Heidenhain pouches in the same dogs. The inhibitor originating from the oxyntic mucosa may also have been responsible for the tonic suppression of gastrin release, because the denervation of the oxyntic mucosa led to a marked hypergastrinemia.

SOMATOSTATIN AND PROSTAGLANDINS AS LOCAL INHIBITORS. Other candidates for the local inhibitors of acid secretion in the oxyntic mucosa include somatostatin and prostaglandins. Somatostatin immunoreactivity has been identified in the fundic mucosa and found to be released by luminal acidification (187). Because this peptide is known to suppress all modes of gastric acid secretion and gastric release (101), it might be regarded as a local feedback inhibitor of gastric secretion.

Prostaglandins of E series, which are known to inhibit gastric secretion (79, 118) and to be released by mucosal acidification (79), may also contribute to the inhibitory mechanism operating in the oxyntic mucosa in response to topical acidification. The possibility of the involvement of local mucosal prostaglandin in the control of acid secretion is supported by several studies showing that the suppression of prostaglandin formation by inhibitors of cyclooxygenase such as indomethacin resulted in an increase in gastric secretion in both in vivo studies on dogs (119), monkeys (36), and humans (140) and in vitro experiments on isolated gastric glands (139). Topically applied prostaglandins were effective local inhibitors of acid secretion from the oxyntic mucosa (64). Naturally released prostaglandins in the stomach of the brooding frog *Rheobatrachus silus* were found to be capable of complete suppression of acid secretion until the tadpoles had completed development and emerged as juvenile frogs by way of the female's mouth (219). These studies indicate that endogenous prostaglandins generated in the stomach may be involved in the control of acid secretion as local feedback inhibitors.

Another natural releaser of mucosal prostaglandins might be the hyperosmolar solutions that may occur after ingesting certain mixed meals (63). The inhibitory effect of hyperosmolar solution was found to be restricted to the oxyntic mucosa without the mediation of intramural or extragastric nervous reflexes or gastrin release. Because prostaglandins are known to be released by hyperosmolar solution (88), it is possible that they are mediators of local inhibition of acid secretion by direct action on the oxyntic glands (121).

Inhibition Arising From Antral Mucosa

The vital role of the antral mucosa in gastric inhibitory mechanisms was first recognized by Woodward et al. (234), who found that luminal acidification of the antrum markedly diminished the gastric acid response of the vagally innervated or denervated portion of the stomach to a liver extract meal in the antrum or to antral distension. Then the inhibitory function of the antral gland mucosa was confirmed by numerous workers, and its mechanism was studied in relation to the cephalic, gastric, and intestinal phases of gastric secretion (60, 231).

EFFECTS OF LUMINAL ACIDIFICATION. In dogs, antral acidification was found to inhibit gastric acid secretion and gastrin release by sham feeding (155), by gastric distension (28), and by bathing the antral mucosa with chemicals such as liver extract meal (234). Debas et al. (28) found that gastrin release by direct irrigation of the antral mucosa with solutions of varying pH depends on the degree of the distension of the antrum. With increased antral distension, the acidification of the irrigating solution from pH 4.0 to pH 1.0 resulted in a pH-dependent inhibition of both gastrin release and the accompanying gastric acid secretion. Antral distension at higher pH (pH 7.0–10.0) failed to increase gastrin release or gastric acid secretion. Thus, although the acidification of antral mucosa suppressed the action of gastrin releasers, mucosal alkalinization per se did not act as a stimulus for gastrin release (Fig. 6).

In humans, the acidification of the gastric content also suppressed gastrin release, and this was associated with an inhibition of gastric acid secretion (97, 217, 229). Luminal acidification was found to abolish completely the gastrin response to modified sham feeding (48). Because atropine partly prevented acid inhibition of sham feeding–stimulated gastrin release, it has been speculated that a cholinergic pathway is involved in the inhibition (48). This suggestion requires verification because sham feeding in humans

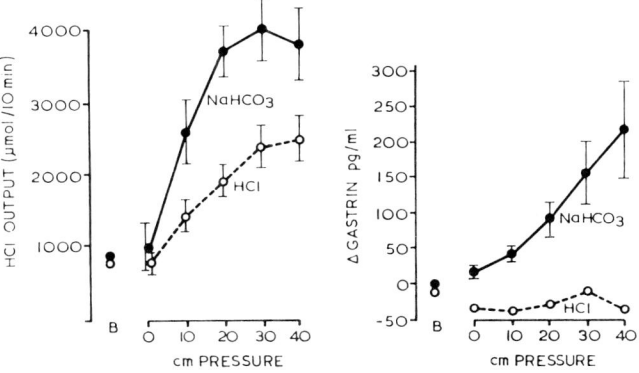

FIG. 6. Effects of distension of vagally innervated antral pouch with 100 mM NaHCO₃ or 100 mM HCl on acid output from vagally innervated gastric fistula and on increments in serum gastrin concentrations. Acidification of antral pouch abolished gastrin release but only moderately reduced acid output, indicating that pylorooxyntic reflex was operating. Vagal denervation of antral pouch abolished the effect. [From Debas et al. (28).]

appears to cause little or no increase in gastrin release (116, 201).

Several studies (97, 217, 229) using amino acid or peptone meal in humans found that luminal acidification to pH 2.5 reduced gastrin release and inhibited gastric acid response to these meals. At higher luminal pH (pH 5.5), it was found that gastric acid secretion and serum gastrin level were increased both in normal subjects and in duodenal ulcer (DU) patients. With a decrease in luminal pH, acid secretion and serum gastrin declined both in normal subjects and DU patients. Nevertheless, the effect of this lower pH in DU patients was less pronounced and resulted in a significantly greater acid output and higher gastrin levels in the patients with ulcer disease. These observations have been cited as evidence of abnormality in the antral autoregulation of acid secretion and gastrin release in DU patients (229). Other studies (97, 217) failed to confirm the existence of such a defect in the antral regulation of gastric secretion and gastrin release in DU patients, but the suppression of gastric secretory functions by luminal acidification is now generally accepted (Fig. 7). An exception appears to be the gastric secretory response to gastric distension, which was found to be relatively resistant to luminal acidification to pH 2.5 (185).

The mechanism by which luminal acid inhibits gastrin release has not yet been explained, but acid may directly suppress the secretory activity of the G cells acting on the microvilli on the luminal surface of these cells. Some have believed that acid in the antral mucosa causes the release of an inhibitory hormone, antral chalone or antrogastrone, that is supposed to suppress the activity of the G cells and the oxyntic glands, but such an inhibitor has not been identified (215).

ROLES OF SOMATOSTATIN AND PROSTAGLANDINS. With the identification of somatostatin in the D cells located in the close vicinity of the G cells of the antral mucosa (136), it has been suggested that this peptide might be responsible for the acid-induced inhibition of gastric secretion and gastrin release. Animal experiments have shown that antral acidification results in an increase in plasma levels as well as luminal release of somatostatin (187, 221). It has been suggested that somatostatin could be identical with antral chalone. Its major action would be the suppression of gastrin release due to its local or paracrine effects on the G cells. The failure of the antral acidification to inhibit gastric acid secretin induced by exogenous gastrin has been explained either by the release into the bloodstream of insufficient amounts of somatostatin for acid inhibition or by concurrent reflex stimulation of the oxyntic glands, thereby preventing the inhibitory effects of somatostatin. However, studies in humans (20) have shown that the postprandial increment in plasma somatostatin may not be sufficient to suppress gastrin release, though it may inhibit gastric acid

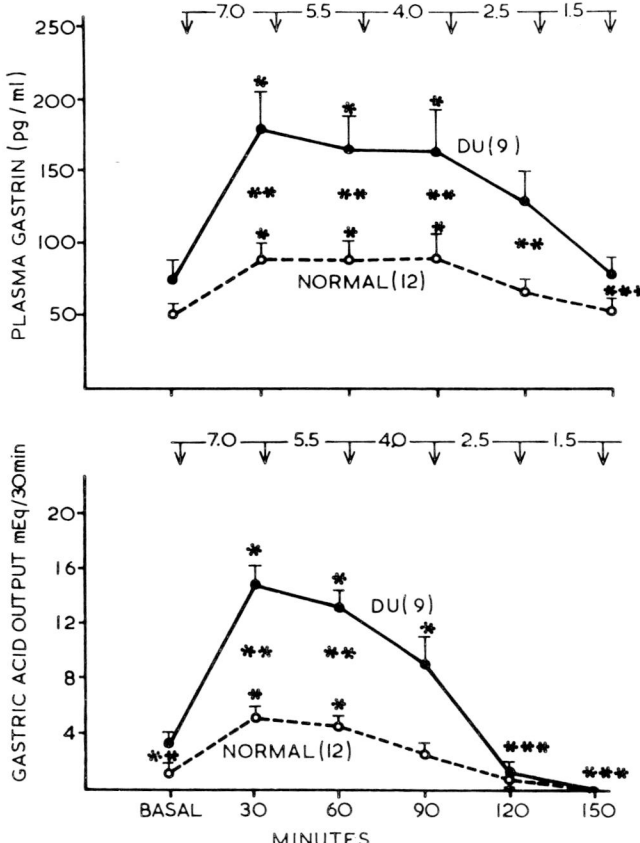

FIG. 7. Mean serum gastrin level (*top*) and gastric acid secretion (*bottom*) in response to amino acid meal of varying pH in duodenal ulcer, DU, ($n = 9$) and normal subjects ($n = 12$). *Single asterisk*, significant increase above basal levels; *double asterisk*, significant difference between duodenal ulcer and normal subjects; *triple asterisk*, significantly less than peak. [From Thompson and Swierczek (217).]

secretion. It is uncertain whether plasma somatostatin reflects true hormone secretion or is just an overflow of somatostatin released primarily into the interstitium to act locally on the G cells.

More recently, Ligumsky et al. (141) suggested that prostaglandins may mediate the gastric inhibitory effects of somatostatin because this peptide enhanced PG release from the isolated rat stomach and indomethacin blocked somatostatin-induced inhibition of bethanechol-stimulated acid secretion in rats. Similarly, Befritis et al. (6) reported that in humans, antral acidification inhibited basal and vagally stimulated gastric acid secretion and gastrin release. Indomethacin reduced the inhibition of the peak acid output and the suppression of plasma gastrin levels induced by antral acidification. As in studies with rats, studies in humans suggest that the pH-dependent inhibitory regulation of gastric acid secretion and gastrin release is mediated by locally produced prostaglandin.

There are some doubts about the validity of the concept of somatostatin-prostaglandin mediation of gastric inhibition by antral acidification, because other studies showed that somatostatin was an effective gastric acid inhibitor in rats (148), cats (2), and humans (149) despite the suppression of prostaglandin release by indomethacin. Thus neither in animals nor in humans is there any evidence to support the hypothesis that somatostatin inhibition is mediated by prostaglandin.

INHIBITORY PYLOROOXYNTIC REFLEX. As already mentioned, studies showed that the distension of the antrum inhibits pentagastrin-stimulated gastric acid secretion in dogs with at least partial preservation of vagal innervation of the oxyntic mucosa (28). This inhibitory effect has been attributed to the release of an inhibitor in the vagally innervated oxyntic mucosa. An alternative explanation is that antral inhibition of gastric secretion is of an inhibitory neuroreflex character. Williams and Forest (233) demonstrated that antral inhibition requires nervous connections between the antral and oxyntic mucosa. It has been suggested that proximal selective vagotomy may be useful for the treatment of DU partly because of the hypothesis that the innervated antrum acts to suppress gastric secretion (69). The inhibition of gastric secretion by selective antral distension in humans also seems to be mediated by an inhibitory reflex (7). Interestingly, DU patients failed to respond with gastric inhibition to antral distension, suggesting the existence of a defect in the control of gastric secretion in these patients. Thus the antral mucosa may be a source of both stimulatory and inhibitory effects mediated by hormonal or neural pathways; which of these predominates depends on the condition under which the study is carried out.

Pharmacological Inhibition of Gastric Phase

Several drugs are now available that can inhibit the postprandial gastric acid secretion by direct action on the parietal cells (93). Anticholinergics and tricyclic agents such as trimipramine and pirenzepine mildly reduce (by 30%–50%) acid response to a meal (45, 93). Anticholinergics tend to increase, whereas tricyclic antidepressants and pirenzepine reduce the postprandial serum gastrin levels, but this may depend on the dose used. As shown by Hirschowitz et al. (65), atropine at lower doses (1–20 μg/kg) reduced dose dependently the postprandial serum gastrin in intact and vagotomized animals, suggesting the existence of a cholinergic pathway leading to gastrin release by food. Current evidence suggests that the mechanism of acid secretory inhibition by nonselective (atropine) or selective (pirenzepine or telenzepine) antimuscarinic agents is independent of changes in serum gastrin levels (45). These agents may act directly on the parietal cells by blocking their muscarinic (M_1 and M_2) receptors and at the site remote from the parietal cells such as enteric ganglia (pirenzepine) where the high-affinity receptors have been detected. It is also possible that they block muscarinic-mediated release of an acid secretagogue such as histamine and the inhibitory muscarinic pathway of an inhibitor of acid secretion such as somatostatin. Both the histamine-containing mast cells and the somatostatin-releasing D cells have been identified in the close vicinity of the parietal cells and found to be innervated by the cholinergic nerves.

Histamine H_2-receptor antagonists such as cimetidine, ranitidine, or famotidine are much more potent gastric acid inhibitors that in short-term treatment do not affect serum gastrin levels but after prolonged therapy may raise serum gastrin (93). They strongly inhibit basal, nocturnal, and postprandial gastric acid secretion, with ranitidine and famotidine being several times more potent on a molar basis than cimetidine (92, 150, 153). Although the half-lives of these drugs are short, they are capable of reducing gastric secretion for long periods over 24 h during treatment in humans with usual therapeutic doses. Substantial suppression of acid secretion (to ~65%–80% of control maximal pentagastrin value) was accompanied by moderate increase in serum gastrin level. After withdrawal of a 6-wk treatment, acid secretion returned to pretreatment levels within ~1 wk. The fact that H_2 blockers are equally effective inhibitors of histamine- and meal-induced gastric secretion and suppress this secretion despite increasing serum gastrin levels emphasizes the important role of histamine in the postprandial stimulation of gastric acid secretion (92).

Benzimidazole derivatives such as omeprazole are the most thorough inhibitors of gastric secretion induced by a meal and other secretagogues (gastrin, histamine, or urecholine) (103). Omeprazole was found to have ~5–10 times greater potency than that of cimetidine and longer duration of action due to its unique, irreversible blockade of the H^+-K^+-ATPase of parietal cells (103, 115, 137). Repeated once daily, the dose of omeprazole exhibited a strong antisecretory activity accompanied by severalfold increase in basal and postprandial serum gastrin levels (189). After stopping treatment, there was a rapid return of serum gastrin and gastric acid to the pretreatment level without any rebound hyperacidity. However, test animals treated for 2 yr with omeprazole showed the hyperplasia of the parietal cells and endocrine enterochromaffin-like cells with formation of micronodules or carcinoid proliferation. These effects have been attributed to the profound inhibition of acid secretion and subsequent hypergastrinemia (137).

Methylated PGE_2 analogues such as arbaprostil, misoprostil, or enprostil are also effective gastric acid inhibitors, but unlike other potent gastric inhibitors they tend to reduce serum gastrin levels, particularly when given orally (64, 93, 117, 118). They are very potent inhibitors of the postprandial gastric secretion because of direct suppression of the parietal cell activity and the reduction of gastrin release. Because the blockade of PG generation by indomethacin or aspirin fails to affect the postprandial gastric secretion, it is unlikely that endogenous PGs are involved in the control of gastric secretion (64, 117).

The explosive growth of new potent gastric acid inhibitors such as highly selective antimuscarinic agents, new generation of potent H_2-receptor antagonists, proton pump blockers, and methylated prostaglandin analogues, virtually free of serious adverse reactions, greatly improved the prospect of the therapy of gastric acid–pepsin disorders (45, 47, 93, 179).

INTESTINAL INHIBITION OF GASTRIC SECRETION

Intestinal Phase

Unlike the first two phases of gastric secretion that are referred to as "turning on" this secretion, the intestinal phase is considered to be "turning it off." Gastric secretion in this phase is maintained for some time, mainly by intestinal gastrin and enterooxyntin as well as by neural reflexes activated by peptic digests entering the intestines and their absorption products in the circulation. Secretion tends to decline as the stimulatory mechanisms are extinguished and the inhibitory mechanisms become effective (60, 91).

Three substances that under physiological conditions may enter the duodenum and inhibit gastric acid secretion are acid, fat, and hyperosmolar solutions.

Intestinal Inhibition by Acid

After Sokolov's (192) demonstration that acid instilled into the duodenum inhibits gastric acid secretion, a considerable number of reports on this subject have accumulated, but its mechanism has not yet been fully explained.

SECRETIN VERSUS BULBOGASTRONE. It is generally accepted that large amounts of unbuffered acid solution irrigating the intestinal mucosa inhibit gastric acid secretion probably through releasing secretin from the intestinal S cells (80, 95). The concept of secretin mediation in gastric inhibition induced by duodenal acid was originally proposed by Greenlee et al. (57), and this supposition was supported by showing that *1*) acid in the intestine and exogenous secretin inhibit gastrin- but not histamine-stimulated gastric secretion (81); *2*) gradually increasing rates of duodenal acidification and graded doses of intravenous secretin produce graded increases in pancreatic bicarbonate secretion and parallel inhibition of gastrin-induced gastric secretion (106); and *3*) the kinetics of gastric inhibition by duodenal acid and by exogenous secretin are both noncompetitive (81).

With the advent of reliable secretin radioimmunoassay, the ability of duodenal acid to release secretin in amounts sufficient to inhibit gastrin-stimulated gastric secretion has been confirmed (15, 16). There is, however, a controversy whether secretin is the only enterogastrone released by an acidified meal entering the duodenum, as previously supposed, or even whether secretin is a physiological inhibitor of the postprandial gastric secretion (80).

Earlier studies on the acidification or surgical removal of the consecutive portions of the duodenum from the exposure to the gastric acid stream revealed that the strongest inhibition is confined to the upper portion of the duodenum, particularly to the duodenal bulb (3, 106, 108, 127). Bulbar acidification was effective in the inhibition of all types of gastric secretion from the vagally innervated or denervated fundic pouches in which either exogenous or endogenous gastrin was used as a stimulant (3, 112, 127). Andersson, Nilsson, and colleagues (3, 154) suggested that a humoral, nonsecretin factor, bulbogastrone, is involved in the mediation of bulbar inhibition because this inhibition was demonstrated after denervation or transplantation of the bulb and was not accompanied by any changes in pancreatic secretion. The postbulbar segment had no inhibitory effect when exposed to acid solution. In contrast, Konturek and Grossman (108) obtained evidence for the involvement of the entire duodenum in the acid-induced gastric inhibition, whereas acid in the jejunum had a stimulatory effect and in the ileum was without influence (107). The most potent inhibition was obtained after bulbar acidification, particularly when both the duodenal bulb and the stomach were fully innervated (112). After the separation of the duodenal bulb from the pylorus or denervation of the fundic pouch the bulbar inhibitory effect declined, suggesting that there may be an enterogastric reflex, either intramural or vagovagal, triggered by acid in the bulb (112). This concept does not exclude the possibility that a humoral agent such as bulbogastrone is involved in the bulbar inhibition of gastric secretion. Secretin is, however, unlikely to be involved in this inhibition because an acidification of the bulb results only in a negligible pancreatic bicarbonate secretion or secretin release (122, 127). Furthermore, exogenous secretin, given in a dose producing a similar increment in plasma secretin to that obtained by bulbar acidification, failed to affect gastric secretion (112, 127). Thus, gastric inhibition by bulbar acidification cannot be ascribed to the release and action of endogenous secretion (Fig. 8).

SOMATOSTATIN AS BULBOGASTRONE. Because somatostatin can be released by acid in the duodenum (187) and exerts more potent inhibitory influence on gastrin-induced than on histamine-induced gastric secretion (123), it has been presumed that it may play the role of bulbogastrone released by bulbar acidification (221). Recent studies in humans have confirmed that duodenal acidification increases plasma somatostatin only when large amounts of acid are

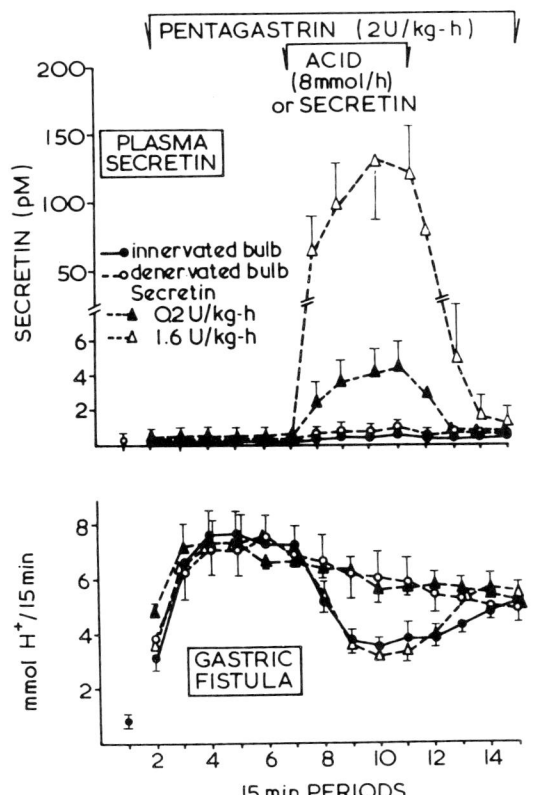

FIG. 8. Effect of acidification of duodenal bulb before and after separation from pylorus or of exogenous secretin infused in small (0.2 U·kg⁻¹·h⁻¹) and large (1.6 U·kg⁻¹·h⁻¹) doses on gastric acid secretion from gastric fistula and on plasma secretin levels. [Adapted from Konturek and Johnson (112).]

used (142). Duodenal acidification at lower, more physiological rates caused only negligible increments in plasma somatostatin that were unlikely to suppress gastric acid secretion. The possibility of the interaction of both secretin and somatostatin released in small amounts by duodenal acid has been ruled out by the demonstration that no potentiation or even additive effects on gastric functions occurred between the two peptides administered in small doses (133). Further studies are needed to clarify whether bulbar acidification inhibits gastric secretion by hormonal, neural, or neurohormonal mechanisms.

SECRETIN AS PHYSIOLOGICAL INHIBITOR. Although bulbar acidification does not seem to inhibit gastric secretion by releasing secretin, the involvement of this hormone in the postprandial control of gastric secretion cannot be ruled out. The question remains whether short-lived and small increases in plasma secretin that correspond to brief periods of proximal duodenal acidification (182) occurring postprandially could influence gastric secretory functions. Chey et al. (15) attempted to settle the controversy over the significance of the inhibitory action of secretin by using antisecretin serum to neutralize the circulating hormone after feeding in dogs. After almost complete neutralization of secretin, there was a small but significant rise both in gastric acid secretion from the gastric pouches and in serum gastrin levels, suggesting that secretin released postprandially may regulate gastric secretory functions and therefore acts as an enterogastrone released by acid in the duodenum (15) This was supported by the demonstration that exogenous secretin given in physiological doses inhibited gastric responses to exogenous gastrin. Unfortunately, the plasma secretin level in tests with exogenous secretin was not measured, so it is unknown whether the doses of hormone used corresponded to the amounts of endogenous hormone released postprandially. Similar studies with antisecretin serum (105) in dogs confirmed that secretin immunoneutralization abolished the pancreatic bicarbonate responses to acidified meal or to exogenous secretin infused in doses mimicking the increments in plasma hormone levels observed after a meal. However, neither gastric acid nor serum gastrin responses to acidified meals were altered by the removal of circulating secretin at all but the lowest pH (pH 2.5), when excessive duodenal acidification was performed. Exogenous secretin given in doses reproducing the physiological increments in the plasma hormone was ineffective in the inhibition of gastric acid secretion or gastrin release (Fig. 9). These results are consistent with studies in humans (87), in whom a peptone meal acidified from pH 7.0 to pH 2.5 resulted in the inhibition of gastric acid secretion and the suppression of serum gastrin level, whereas exogenous secretin given in physiological doses reproducing the postprandial increases in plasma secretin were ineffective. Thus little evidence was found to support the notion that endogenous secretin is involved in the postprandial gastric acid inhibition or gastrin release.

ROLE OF CHOLECYSTOKININ. Another candidate for an inhibitor of gastric secretion by acid in the duodenum is cholecystokinin (CCK). It has long been suspected and recently confirmed by radioimmunoassay (13) that duodenal acidification releases CCK that in some species such as the human or dog (but not the rat or cat) exerts an inhibitory action on gastrin-stimulated gastric acid secretion. This effect seems to be unrelated to the contamination of CCK preparation by other peptides such as gastric inhibitory peptide (GIP) because highly purified CCK preparation retained full inhibitory potency on gastric secretion (145). Thus CCK fulfills the requirements implicit in the definition of an enterogastrone released by acid, but whether it is released by acid in amounts sufficient to exert its inhibitory action on gastric secretion is uncertain. It is also unknown whether it acts directly on the parietal cells or by releasing other inhibitors such as somatostatin in the gastric mucosa (195).

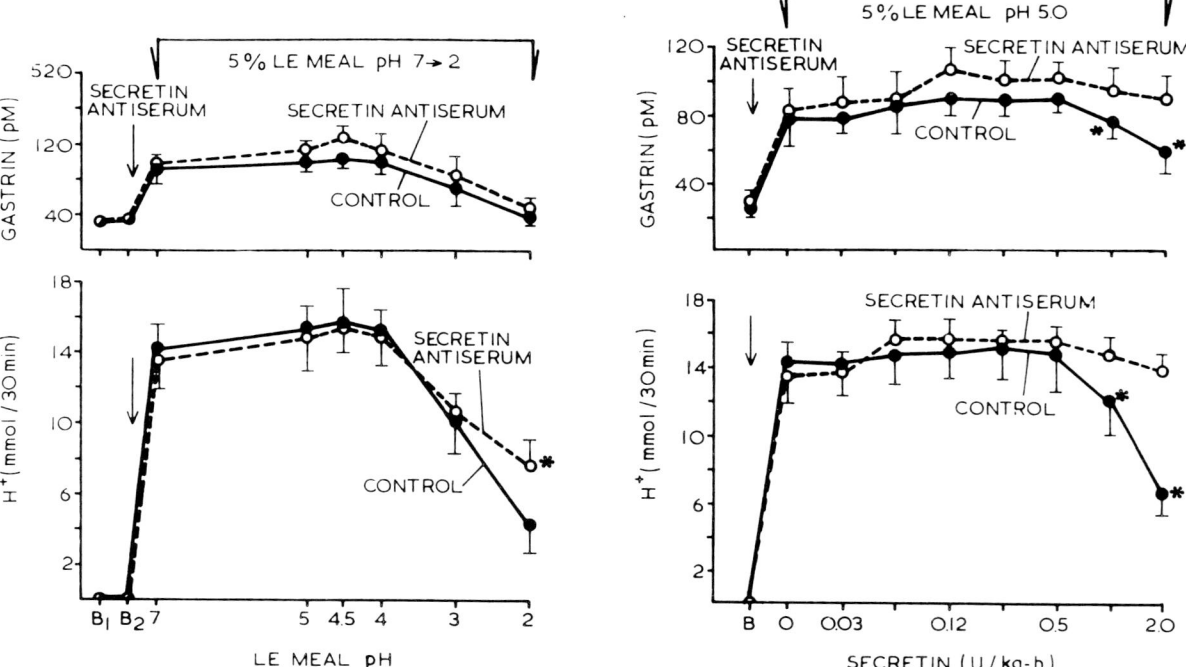

FIG. 9. Effects of immunoneutralization of circulating secretin on serum gastrin and gastric acid responses to 5% liver extract, LE, meals in stomach adjusted to pHs varying from 7.0 to 2.0 or to 5% liver extract meal of pH 7.0 combined with intravenous infusion of graded doses of secretin (0.03 – 2.0 $U \cdot kg^{-1} \cdot h^{-1}$). Mean ± SE of 10 tests on 5 dogs with gastric and pancreatic fistulas. Secretin antiserum almost completely removed circulating secretin both in tests with meals of varying pH and exogenous secretin. *Single asterisk*, signifcantly different from control. [From Konturek et al. (105).]

Intestinal Inhibition by Fat

Fat in the small intestine has long been recognized as a potent inhibitor of gastric secretion, and the term *enterogastrone* was originally used by Kosaka and Lim (128) to describe the fat-released inhibitor of gastric secretion. Fat appears to inhibit gastrin-, histamine-, and food-induced gastric acid secretion in various species, including humans (91, 113).

For fat to exert its inhibitory action, it must be present in the intestine in an absorbable form, and only the fatty acids with 10 or more carbons in the chain are effective. Konturek and Grossman (107) used perfused loops prepared from different levels of the intestine to show that inhibition of gastric secretion by micellar fat mixture could be produced from all levels of the small bowel but most strongly from the jejunum. The absorption of fatty acids corresponded to the degree of inhibition so that the greatest absorption occurred in the jejunal loop, the most potent site of inhibition.

The mechanism of fat-induced inhibition of gastric secretion is apparently mediated mainly by a humoral mechanism, because this inhibition persisted after the extrinsic nerves to the stomach and intestine had been severed. Among the peptides from the intestinal mucosa that have been isolated and sequenced, the likely candidates for enterogastrone released by fat include CCK, secretin, GIP, neurotensin, and peptide YY (PYY).

CONTRIBUTION OF CHOLECYSTOKININ AND SECRETIN. Fat in the duodenum and jejunum is a potent releaser of CCK from the intestinal endocrine I cells (94), and CCK in turn is known to inhibit gastrin-stimulated gastric secretion in dogs and humans (21, 145). Fat in the duodenum is a potent inhibitor of histamine-induced secretion from the Heidenhain pouch, an effect that cannot be reproduced by exogenous CCK, regardless of the dose used (82). It may be concluded therefore that fat in the duodenum causes the release of enterogastrone, which is distinct and separate from CCK.

Fat in the intestine is also known to release small amounts of secretin (44) that may be implicated in the stimulation of pancreatic secretion, but because secretin is ineffective against histamine-induced gastric secretion, it is unlikely to contribute to the fat-induced gastric inhibition.

GASTRIC INHIBITORY PEPTIDE AS ENTEROGASTRONE. The GIP identified in the endocrine cells (K cells) present in the duodenum and to a lesser extent in the jejunum of dogs and humans (169) was originally reported to be released by fat and to inhibit strongly gastric acid secretion from the Heidenhain pouch in response not only to gastrin but also to histamine (9, 163). The inhibition of this secretion by exogenous GIP and intestinal fat was reported to be well correlated with the increments in plasma GIP concentrations, indicating that GIP satisfied most completely the criteria for enterogastrone described by Kosaka and Lim (128). However, GIP does not appear to affect acid secretion from the vagally innervated stomach stimulated by gastrin, peptone meal, or histamine (23, 113), whereas fat in the intestine inhibits secretion from the innervated stomach even more effectively than from the denervated Heidenhain pouches (30, 113). Creutzfeldt et al. (23) observed that although duodenal instillation of fat strongly inhibited meal-induced gastric secretion and increased plasma GIP levels, there was no close correlation between the changes in plasma GIP levels and the inhibition of gastric secretion. The same authors (113) found that exogenous GIP infused intravenously in doses raising plasma GIP levels several times higher than those recorded with duodenal fat was virtually without any major effect on acid secretion from the vagally innervated portion of the stomach stimulated by pentagastrin, meal, or histamine (30, 113). A small but significant inhibition of acid secretion was observed only from the Heidenhain pouch stimulated by pentagastrin, but not histamine, when the dose of exogenous GIP increased the plasma peptide level several times higher than that induced by duodenal fat (Fig. 10). These findings are consistent with the observations in humans (144, 224) that exogenous GIP did not cause any inhibition of gastric secretion and that fat was equally effective as a gastric inhibitor whether given orally or intravenously, its inhibitory effects being completely unrelated to the plasma GIP concentration. These results in dogs and humans may show that GIP is of no physiological importance in the control of gastric acid secretion, and fat ingestion may suppress gastric secretion at least in part by its digestion products.

ROLE OF NEUROTENSIN. Neurotensin was identified in the nerves of the gut and the endocrine N cells found mainly in the mucosa of the ileum (10). Because plasma levels of neurotensin increase within a few minutes after feeding, not long enough for the chyme to reach the ileum, it is possible that the proximal part of the gut plays an important role in the postprandial release of this peptide (42, 110).

Exogenous neurotensin was reported to result in a potent inhibition of gastric secretion in dogs stimulated by gastrin, sham feeding or ordinary feeding, but not by histamine (110). The inhibitory effect (but not the release) of neurotensin was eliminated by vagal denervation of the oxyntic mucosa, indicating that the peptide does not act directly on the oxyntic glands. In

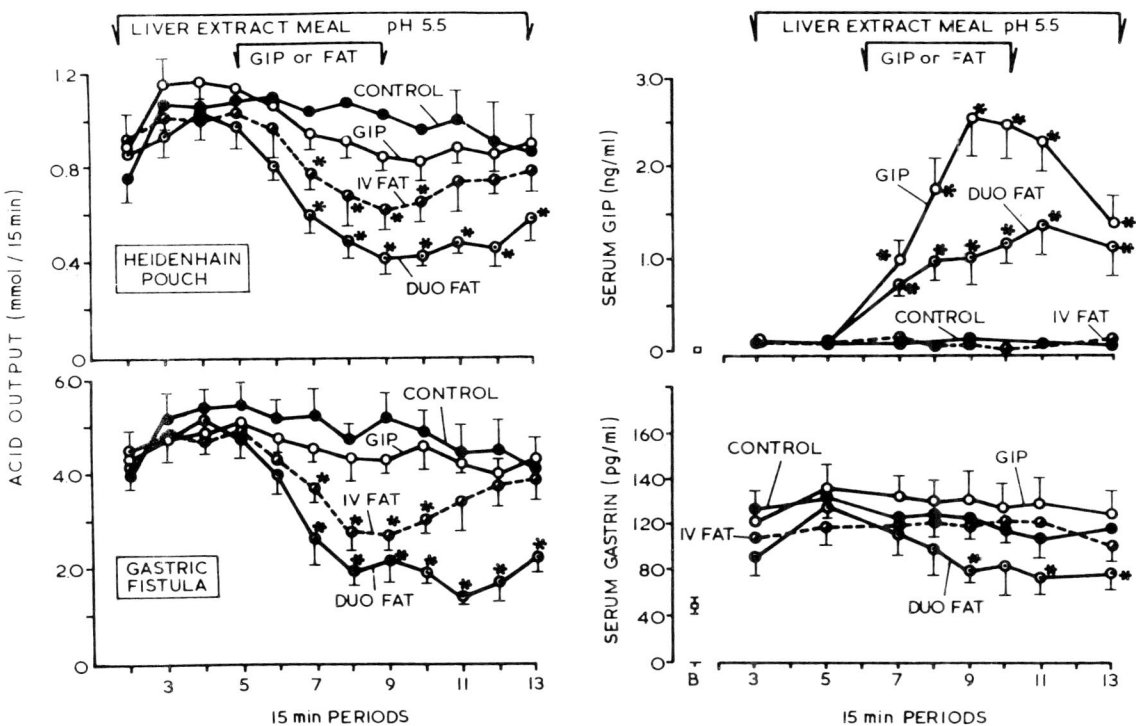

FIG. 10. Effects of intravenous infusion of exogenous gastric inhibiting peptide, GIP, (1.0 µg·kg^{-1}·h^{-1}) or fat administered intraduodenally or intravenously on gastric acid secretion from Heidenhain pouch and gastric fistula stimulated by liver extract meal of pH 5.5 in main stomach (*left*) and serum GIP and gastrin levels (*right*) in dogs. Mean ± SE of 6 tests on 6 dogs. *Single asterisk*, significantly different from control. [From Konturek et al. (113).]

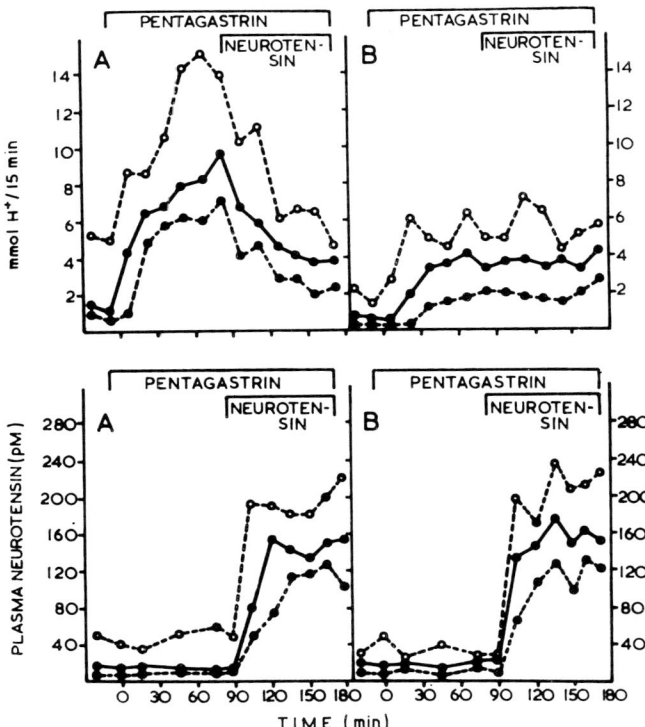

FIG. 11. Gastric acid and plasma neurotensin concentration during infusion of pentagastrin (150 ng·kg^{-1}·h^{-1}) and neurotensin (500 ng·kg^{-1}·h^{-1}) in 7 duodenal ulcer patients before (A) and after (B) parietal cell vagotomy. [From Olsen et al. (156).]

humans, neurotensin was also found to be released by duodenal oleate and to inhibit gastrin or meal-stimulated acid secretion in the vagally innervated but not denervated stomach [Fig. 11; (84, 85)]. These results suggest that neurotensin may contribute to the fat-induced gastric secretory inhibition but that vagal innervation is required for its inhibitory action. The fact that neurotensin loses its inhibitory activity after vagal denervation of the stomach (156) and is ineffective against histamine-induced secretion militates against its primary role as an enterogastrone released by fat in the intestine.

PEPTIDE YY AND ILEOCOLONIC INHIBITION. Recently, Tatemoto (211) isolated a new peptide from hog intestine, PYY, which was then localized by immunocytochemistry in the endocrine-type cells of the distal ileum and colon (143). Peptide YY has a structure similar to pancreatic polypeptide (PP) but, unlike PP, it was found to cause a potent inhibition of pentagastrin- and meal-induced gastric secretion in dogs (159). Because ileal and colonic perfusion with micellar fat, sodium oleate, or liver extract has long been known to inhibit gastric acid secretion in dogs (107, 188) and because the intestinal instillation of these substances has recently been shown to increase the plasma levels of PYY, it has been suggested that this newly discovered peptide may be another enterogastrone-like substance released by nutrients in the distal portion of the gut (159). Adrian et al (1) reported recently that PYY in humans also inhibited gastrin-induced gastric acid secretion while raising plasma PYY to the levels observed postprandially. Further studies correlating gastric inhibition with plasma PYY increments by various substances in the intestine will evaluate the possible physiological role of PYY in the control of gastric secretion.

INTESTINAL INHIBITION BY HYPEROSMOLAR SOLUTIONS. The intestinal inhibition of gastric secretion by hyperosmolar solutions was established long ago (91) and ascribed to a humoral factor, because it could be demonstrated after the denervation of the stomach and the intestine (91). Konturek and Grossman (107) found that 20% glucose solution instilled into the canine duodenum inhibited the Heidenhain pouch response to exogenous gastrin. The degree of inhibition depended on the type of solute employed and the region of intestine perfused. Inhibition was greater with glucose than with saline solution and could be obtained only from the duodenum.

The mechanism by which a hyperosmolar solution in the small intestine inhibits gastric secretion is unknown. Intraduodenal glucose is a potent stimulant of GIP release (23), but because GIP is a poor inhibitor of gastric acid secretion (113), it is unlikely to be the major factor in glucose-induced gastric inhibition.

In humans, intrajejunal hypertonic glucose solution was found to be a potent inhibitor of gastrin or meal-stimulated gastric secretion. This inhibition was accompanied by a rise in plasma glucagon and enteroglucagon (72), the former being a known inhibitor of gastric secretion not only in pharmacological but also in physiological doses (18, 95). Recently Petersen et al. (166) reported that hypertonic glucose (but not hypertonic saline) instilled intrajejunally in healthy volunteers reduced acid secretion dose dependently, and this inhibition was closely related to the increase in plasma enteroglucagon. Because glicentin, a highly purified form of the major component of enteroglucagon, was found to be a potent inhibitor of gastric acid secretion in rats (86), it has been speculated that glicentin might be directly responsible for the glucose-specific inhibitory mechanism located in the gut (166).

GASTRIC INHIBITION BY UROGASTRONE AND EPIDERMAL GROWTH FACTOR

Historically, urogastrone was discovered as a consequence of the beneficial effect of pregnancy on healing peptic ulceration. It was supposed that urine extract contains two different substances, one inhibiting gastric acid secretion and another promoting ulcer healing (181). With the isolation of pure urogastrone from the urine, it was established that both gastric inhibitory and mitogenic properties reside in one molecule of urogastrone, which was found to be a mixture of the 52- and 53-amino acid peptides. The structurally similar 53-amino acid peptide epidermal growth factor (EGF) was isolated from the mouse submandibular glands (58) and was identified by immunocytochemistry in the submandibular salivary glands and Brunner's glands of cats and humans (54).

Urogastrone was reported to inhibit gastric acid secretion in gastrinoma patients (38), normal subjects (37), and duodenal ulcer patients (90). The inhibition affected acid secretion but not pepsin output and was not accompanied by any change in serum gastrin levels.

Experimental studies in dogs revealed that either urogastrone or EGF was a highly effective inhibitor of acid secretion from the vagally innervated and denervated portions of the stomach stimulated by histamine, pentagastrin, or liver extract meal [Fig. 12; (102)]. The inhibitory effect appears to be short lived, due to the short half-life of the peptide, but not to be followed by any rebound hypersecretion after urogastrone or EGF withdrawal (102).

The mechanism of the inhibitory action and the possible physiological importance of urogastrone or EGF has been meagerly investigated. Studies in vitro on the isolated rabbit gastric glands (31, 102, 174) or isolated guinea pig gastric mucosa (51) showed that EGF acts directly on the parietal cells and inhibits histamine- and cAMP-induced stimulation of acid formation. It has been proposed that EGF interferes with acid formation at steps beyond the histamine H_2 receptors, possibly between the cyclic nucleotides and the proton pump of the parietal cells.

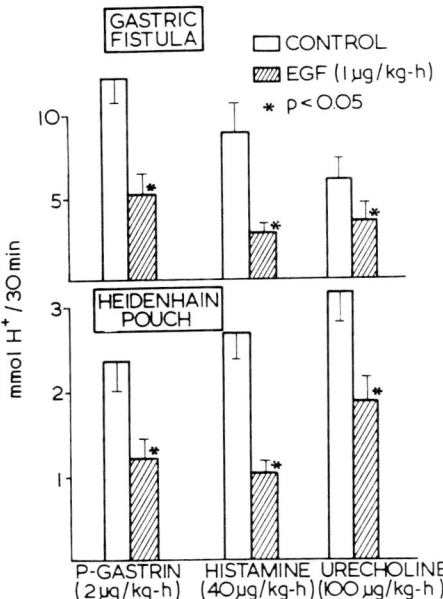

FIG. 12. Effect of epidermal growth factor (EGF) (1.0 $\mu g \cdot kg^{-1} \cdot h^{-1}$) on acid secretion from gastric fistula and Heidenhain pouches stimulated by pentagastrin (2.0 $\mu g \cdot kg^{-1} \cdot h^{-1}$), histamine (40 $\mu g \cdot kg^{-1} \cdot h^{-1}$), or urecholine (100 $\mu g \cdot kg^{-1} \cdot h^{-1}$). *Single asterisk*, significantly different from control. [From Konturek at al. (102).]

The EGF is remarkably resistant to acid-pepsin digestion and so theoretically could act as a luminal inhibitor of gastric acid secretion. However, studies in vivo with intragastric administration of the peptide (32) or its topical application on the luminal surface of the gastric mucosa (51) did not show that EGF retains any inhibitory activity unless applied in a large dose. It has not been ruled out that EGF or its active fragments might be released directly into the bloodstream or absorbed from the gut into the portal circulation and to act systemically, but the evidence for such mechanisms has not been described.

Interestingly, luminal EGF exhibits mucosa growth promoting (32) and gastroprotective effects against various ulcerogens (100). Probably the major action of EGF is the maintenance of the mucosal growth and mucosal integrity rather than the inhibition of gastric acid secretion.

GASTRIC INHIBITION BY NEUROPEPTIDES

Over the past decade a number of peptides have been identified both in the brain and in the peripheral nervous system as well as in the gastroenteropancreatic endocrine system (35, 202). The list of the peptides identified both in the brain and the gut (so-called brain-gut axis) is constantly growing, and the most important include bombesin-like peptides [bombesin or gastrin-releasing peptide (GRP)], somatostatin, opioid peptides, TRC, corticotropin-releasing factor (CRF), vasoactive intestinal peptide (VIP), substance P, neurotensin, and so forth.

Gastric Inhibitory Effects of Neuropeptides Administered Intracerebrally

Among the peptides that were found to inhibit gastric acid secretion after intracerebral administration are bombesin-like peptides, CRF, and opioid peptides. Some other peptides such as TRH (208), gastrin (212), and somatostatin (161, 210) have been reported to elicit brain stimulation of gastric secretion, probably by excitation of the vagus-dependent mechanism.

The physiological relevance of the pharmacological effects of neuropeptides administered intracerebrally is unknown. All of them have been identified by immunocytochemistry in the neurons of the brain and the gut, but some of them (e.g., bombesin-like peptides) were found in the highest concentration in the brain (40). There is no unifying concept to explain the significance of their effects on gastric secretion, but most of them appear to have opposite actions when given centrally or peripherally. It has been suggested that this may provide a basis for the neurohormonal regulation of gastric secretory functions (161, 209).

BOMBESIN-LIKE PEPTIDES. Bombesin-like peptides given intravenously are known to cause potent stimulation of gastric acid secretion (225). The major

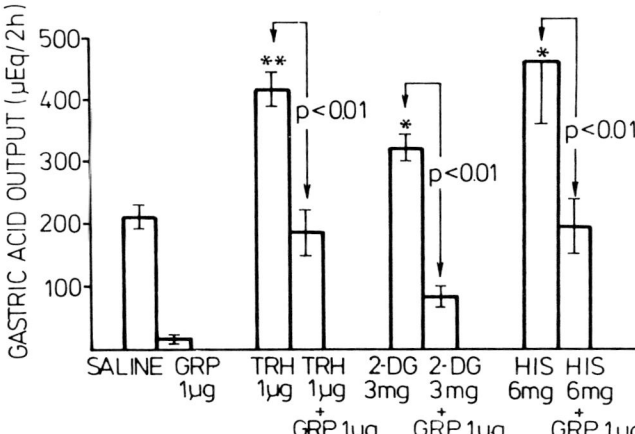

FIG. 13. Inhibitory effects of gastrin-releasing peptide (GRP) on gastric acid output in pylorus-ligated rats stimulated by intracisternal thyrotropin-releasing hormone (TRH), 2-deoxy-D-glucose (2-DG), or histamine (HIS) without and with intracisternal administration of GRP. [From Taché et al. (205). Copyright 1981 by The American Gastroenterological Association.]

mechanism of their peripheral action appears to be the release of gastrin (75), because antrectomy (75) or antral mucosectomy (52) greatly reduced their stimulatory action. On the contrary, the intracerebral administration of bombesin-like peptides caused dose-dependent inhibition of basal and TRH or 2-DG–stimulated gastric acid secretion in rats (207) and pentagastrin-induced secretion in dogs [Fig. 13; (160)].

Because vagal tone has a crucial role in the gastric stimulation in pylorus-ligated rats and because secretagogues such as TRH or 2-DG require intact vagal innervation, it was believed that bombesin-like peptides given intracerebrally may act by decreasing vagal activity (209). However, vagotomy caused only partial inhibition, whereas GRP injected intracysternally abolished gastric acid secretion completely (207). Furthermore, the inhibitory action of this peptide administered intracysternally was found to be unaffected by subdiaphragmatic vagotomy in rats (206). Studies of bombesin microinfusion into specific hypothalamic nuclei of intact rats or injection into the cysterna magna of midbrain-transected rats demonstrated that the peptide can trigger inhibition of gastric secretion from both forebrain and midbrain structures. The neural pathway mediating bombesin action requires an intact spinal cord (206) and probably involves the activation of the sympathetic system (135, 196).

Another mechanism of the central action of bombesin on gastric secretion might be a modification of the release of gastrin. However, centrally acting bombesin appears to increase rather than decrease the serum gastrin level in pylorus-ligated rats (205, 207) and in fasted or fed dogs (225) despite the suppression of gastric acid secretion. Although the hypergastrinemic response to central bombesin might be an indirect effect due to gastric acid inhibition, it is not excluded that it may result from the direct effect of the peptide on the brain or simply from the leakage of bombesin from the brain to the peripheral blood stream. The former possibility is supported by the finding that hypergastrinemia induced by the central action of bombesin may be blocked by lesion of the lateral hypothalamic area.

OPIOID PEPTIDES. Opioid peptides such as β-endorphin or dynorphin given intracisternally also decreased gastric acid secretion in pylorus-ligated rats (209) and prevented stress-induced gastric lesions (209). Similar results were obtained with β-endorphins and D-Ala-Met-enkephalin injected intracerebroventricularly, which caused a dose-dependent inhibition of basal and TRH-induced gastric secretion. Dermorphine, an opioid peptide isolated from the amphibian skin (41), was found to suppress gastric acid response to distension or insulin but not to histamine, the inhibitory effect being antagonized by naloxone.

CORTICOTROPIN-RELEASING FACTOR. In rats (204) CRF injected intracerebroventricularly caused a potent inhibition of gastric acid secretion. The inhibitory effect did not require vagal integrity because it persisted after vagotomy. The effect was blocked by adrenalectomy, suggesting that CRF acts within the brain, at least in part, through an adrenal mechanism (203, 204).

CALCITONIN GENE-RELATED PEPTIDE. Calcitonin gene-related peptide (CGRP) and its mRNA have been identified in the brain areas implicated in the central control of gastrointestinal functions. When administered intravenously, CGRP was found to inhibit basal and pentagastrin-stimulated acid secretion in rats and humans (130). When given intracerebroventricularly, CGRP and calcitonin (138) itself strongly inhibited gastric acid response to pentagastrin, histamine, and bethanechol. The inhibitory effect was not altered by adrenalectomy or adrenergic blockage but was abolished by vagotomy, suggesting that CGRP effects on the stomach are mediated by vagal nerves.

OTHER PEPTIDES. Interestingly, other peptides, which are known inhibitors of gastric acid secretion after peripheral administration such as somatostatin, substance P, VIP, or CCK have little or no effect on that secretion after intracerebral administration. Of these peptides, somatostatin and VIP were found to inhibit vagal activity, yet they had little influence on gastric acid secretion (196, 209).

Gastric Inhibitory Effects of Neuropeptides Administered Intravenously

When given intravenously many of the neuropeptides, with the exception of GRP, exert an inhibitory effect on gastric acid secretion. The possible physiological role of these neuropeptides in the control of gastric acid secretion is unknown because neither the factors responsible for their release nor the amounts of peptides released have been determined. Our present knowledge of neuropeptides is based mainly on pharmacological studies of their intravenous administration, tissue or plasma detection by radioimmunoassay or immunocytochemistry, and the blockade of their receptors in the target cells.

SOMATOSTATIN. Originally isolated from the hypothalamus, somatostatin was subsequently identified in nerves and cell bodies in the brain and peripheral nervous system and in the endocrine-paracrine D cells of the gastrointestinal mucosa and the pancreas (22, 136). Larsson et al. (136) demonstrated that in the gastric mucosa, many somatostatin-containing cells have processes ending on the G cells in the antrum and the parietal cells in the oxyntic mucosa. This provides an anatomical explanation for the local (paracrine) release of somatostatin in the vicinity of gastrin-producing and parietal cells. Arnold et al. (4, 5) found that in humans the radio of the D cells to G cells depends on the gastric acid secretory status. In patients with gastric hyperchlorhydria such as gastrinoma this ratio tended to decrease, whereas in states of reduced acid secretion such as after vagotomy it increased.

Somatostatin isolated from the brain was characterized as cyclic tetradecapeptide (SS-14), but more recently larger forms (SS-28) have been found in the gastrointestinal mucosa and the pancreas. Many somatostatin analogues have been synthesized in an attempt to increase or prolong the biological activity of the native peptide. The high potency of these mini-somatostatins has been attributed to their reduced susceptibility to enzymatic degradation in the blood plasma (226).

The effects of somatostatin include a potent and dose-dependent inhibition of the secretion of gastric acid, intrinsic factor, and pepsin; the suppression of gastrin release; and a reduction in the mucosal blood flow (5, 123). Intravenous somatostatin is capable of inhibiting gastric acid responses to various stimulants including sham feeding, ordinary feeding, peptone or amino acid meal, pentagastrin, urecholine, and insulin (5). It produces less inhibition of histamine-stimulated secretion (123). It also causes a marked suppression of basal, meal-, and vagus-stimulated gastrin release (5, 123). The most sensitive to somatostatin inhibition appears to be the postprandial acid secretion, possibly because of the suppression of the release and action of gastrin (123). The results regarding the kinetics of somatostatin-induced gastric inhibition are controversial, and both competitive and noncompetitive inhibition of pentagastrin and histamine-induced secretion has been proposed (6, 214). Interestingly, after withdrawal of somatostatin infusion, a quick return of the secretory activity and plasma hormone level toward the preinfusion values or even a rebound increase in gastric secretory functions has been described (5).

In animals and humans, SS-28 was found to be equipotent with SS-14 as a gastric inhibitor when compared on the basis of molar doses but was 4–10 times less potent on the basis of plasma hormone concentrations (114, 125). The difference in the inhibitory activity between SS-14 and SS-28 is likely to be a reflection of the longer half-life of SS-28 in the circulation.

Short-chain cyclic analogues of somatostatin were found to have longer action and to be effective also after oral administration (101, 226), but some of them were found to be weaker inhibitors of gastric secretion than native peptides (68).

Somatostatin was found to be a rather weak inhibitor of acid formation in isolated rat parietal cells (184), probably acting by interfering with the parietal cell adenylate cyclase (55). However, Chew (14) was able to increase the sensitivity of the isolated parietal cells to the inhibitory action of somatostatin by adding reducing agents to the incubation medium. She found that somatostatin exhibits a strong and noncompetitive inhibition of acid formation in the gastric glands stimulated by histamine or gastrin. This inhibition was specific, because the peptide did not affect basal acid formation or that induced by carbachol, cAMP, or elevated extracellular K^+. The enriched parietal cells were inhibited by somatostatin only when histamine but not gastrin was used for their stimulation. This was interpreted as showing that somatostatin affects acid secretion by direct action on the parietal cells to inhibit the effect of histamine and indirectly to reduce the gastrin effect, possibly by blocking the release of histamine from the histaminocytes present in the gastric glands.

The physiological role of somatostatin in the control of gastric secretion is far from full elucidation. Little information is available on the relations of the concentrations of different molecular forms of somatostatin in the circulation to physiological events. Most important, there is no method of assessing the amounts of somatostatin released locally in the gastric mucosa. It has been reported that selective nutrients such as amino acids and/or fat in the canine stomach or the duodenum increased the plasma levels of somatostatin-like immunoreactivity (SLI) in the venous outflow from fundic, antral, and pancreatic areas (187). In contrast, peptone solution in the rat stomach suppressed the release of somatostatin while increasing gastrin secretion, both effects being mediated by intramural cholinergic and noncholinergic neurons (176, 177). The most potent stimulant of SLI release was the acidification of the gastroduodenal mucosa, suggesting that somatostatin may serve as a mediator of the gastroduodenal feedback inhibition of gastric secretion (221)

Food and selective nutrients (fat and protein) as well as insulin hypoglycemia have also been found to simulate both the release of somatostatin and gastric acid secretion in dogs and humans (12, 164, 230). It has been suggested that luminal acid is an important factor in the somatostatin release, but this is a controversial problem because other studies (142) found that only grossly supraphysiological amounts of acid in the duodenum caused a moderate release of somatostatin, so it is unlikely that gastric acid plays a major role in the postprandial- or hypoglycemia-induced release of somatostatin.

Studies on the isolated perfused rat stomach indicated that somatostatin may play a tonic inhibitory influence on gastrin release because immunoneutralization of released somatostatin by specific antiserum added to the stomach perfusate caused a prompt increase in gastrin output (177). The increase in the release of somatostatin accompanied by the decrease in gastrin was observed after adding some gut peptides (secretin, VIP, glucagon) (175) and PGE_2, which are known to inhibit gastric acid secretion. A similar reverse relationship between the release of somatostatin and gastrin was found after adding methacholine. Because the somatostatin inhibition by methacholine was reversed by atropine, it has been suggested that a separate cholinergic pathway is involved in this inhibition (178). Another atropine-resistant pathway involving peptidergic stimulation of the G cells by bombesin has been supposed to activate the release of gastrin (186). The cholinergic inhibition is said to act predominantly to inhibit paracrine secretion of somatostatin and thus to eliminate its tonic inhibitory influence on gastrin secretion and to allow the bombesin pathway to stimulate gastrin release (Fig. 14).

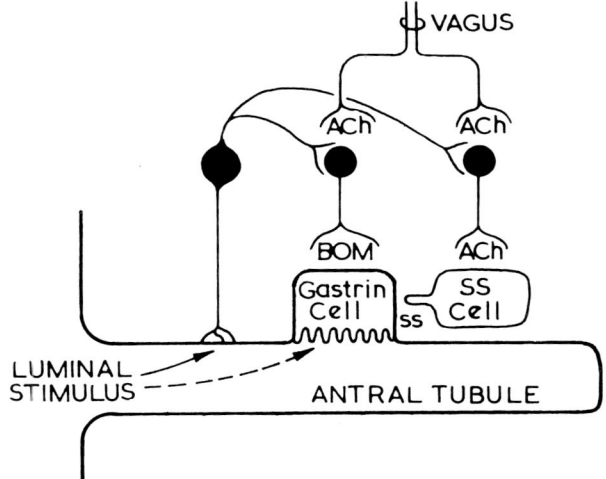

FIG. 14. Model for regulation of gastrin and somatostatin release in antrum by cholinergic and noncholinergic (bombesinergic) intramural neurons. Somatostatin (SS) cell is shown structurally and functionally coupled to gastrin cell. Sensory bipolar neuron links the regulatory cholinergic (ACh) and bombesinergic (BOM) neurons to luminal stimulus. Luminal stimulus could act directly on gastrin and somatostatin cells. [From Saffouri et al. (176). Copyright 1984 by The American Gastroenterological Association.]

This interesting concept of somatostatin-gastrin interaction has not been supported by recent observations in dogs (25), in which vagal stimulation by sham feeding increased rather than decreased the plasma somatostatin, whereas bombesin was found to raise not only plasma gastrin but also somatostatin. Unlike its action in the perfused rat stomach, atropine inhibited the release of plasma somatostatin while maintaining gastrin response to sham feeding unaltered. The reason for the discrepancy between the in vitro results obtained from the perfused rat stomach and those observed in conscious dogs is not obvious. In addition to differences in species, it is likely that the release of somatostatin may be under both cholinergic stimulatory and cholinergic inhibitory pathways and sham feeding in dogs as a physiological vagal stimulus predominantly activates the stimulatory pathway, whereas pharmacological stimulation with cholinergic agonists in rats activates mainly the inhibitory pathway.

VASOACTIVE INTESTINAL PEPTIDE. Isolated from the porcine intestine (178), VIP has been shown to be widely distributed in the central and peripheral nervous system (43, 178). The bulk of the body VIP is present in the gastrointestinal tract where the VIPergic nerves are distributed in all its layers (43). The VIP coexists with acetylcholine in the same postganglionic cholinergic nerves supplying the exocrine glands, such as pancreas, and may mediate the atropine-resistant vagal stimulation of their secretory activity (43).

Vagal stimulation by sham feeding or insulin hypoglycemia in dogs causes only a minor increase in the plasma VIP level (220), but direct electrical excitation of the vagus in pigs was found to increase both the portal and peripheral plasma VIP levels (43). This neuronal release of VIP was not affected by atropine or β-adrenergic antagonists but was abolished by hexamethonium, indicating that this release is mainly due to the stimulation of noncholinergic postganglionic vagal fibers. Ingestion of normal mixed meal did not significantly affect the plasma VIP levels in dogs (220) or humans (147). Direct duodenal instillation of HCl, fat, or ethanol in humans caused significant rise in plasma VIP concentration (183).

Vasoactive intestinal peptide is known to inhibit gastric acid response to pentagastrin and meal in dogs (104). This effect was similar to that exerted by secretin, which is structurally related to VIP, though VIP has a wider spectrum of gastric effects because, unlike secretin, it inhibits histamine-induced acid secretion. Both VIP and secretin also inhibit gastrin release, so it was suggested that their enterogastrone-like action resulted from both the direct action on the parietal cells and the inhibition of gastrin release (104, 227). Studies on the isolated perfused rat stomach confirmed that VIP suppressed gastrin release possibly because of the enhancement of somatostatin release, which then exerted its restraint on gastrin secretion (175). The inhibitory effect of VIP on histamine-induced cAMP production and acid formation in the isolated gastric glands indicates that VIP may act directly on the parietal cells (55), but this requires supraphysiological doses.

Because VIP was found to be released by duodenal acidification or fat instillation (183), it has been proposed that the peptide might be one of the enterogastrone-like substances involved in the control of gastric acid secretion. Although the effects of exogenous VIP on gastric secretion in dogs (104) are consistent with this proposal, recent studies in humans showed that exogenous VIP does not affect gastric acid or plasma gastrin responses to a meal or to pentagastrin (71). In some species such as the cat, VIP may even enhance basal secretion and the acid response to a meal and pentagastrin (112). These results in humans and cats, as opposed to those in dogs, indicate marked species differences in the gastric effect of VIP and militate against the hypothesis that VIP is an enterogastrone in the classical sense.

THYROTROPIN-RELEASING HORMONE. This hormone is a 3—amino acid peptide, originally isolated from the hypothalamus and subsequently detected throughout the body (78), particularly in the gastrointestinal tract and the islets of Langerhans.

Because of rapid enzymatic inactivation, it is difficult to measure TRH in the plasma, thus little is known about its release from the peripheral tissues. This hormone crosses the blood-brain barrier poorly, but central affects may occur after systemic administration of high doses of the peptide.

Intravenous administration of TRH resulted in an immediate and dose-dependent inhibition of gastric motor and secretory activities. The inhibition by TRH of pentagastrin-, histamine-, and insulin-induced gastric acid and pepsin secretion was observed in humans (34). Although TRH had little effect on the serum gastrin level, it showed a potent inhibitory action on gastric acid secretion in gastrinoma patients (33). In dogs, TRH was found to suppress sham feeding-induced gastric acid secretion (111) and the reduction in food intake (125, 152) without affecting the vagal release of gastrin. It also caused a moderate decrease in pentagastrin- or meal-induced gastric secretion without affecting the plasma gastrin level.

The secretory and food intake effects so far described should be considered pharmacological, because little is currently known about the possible physiological role of TRH in the control of gastric functions.

CORTICOTROPIN-RELEASING FACTOR. This 41–amino acid peptide has been identified in the neurons of the hypothalamus and in the endocrine type cells of the gastrointestinal mucosa and the pancreas (168, 223).

Corticotropin-releasing factor injected intravenously in dogs was found to inhibit dose dependently pentagastrin-induced but not histamine-induced gastric acid secretion without affecting the serum gastrin levels (98). Because the inhibitory effect of CRF was observed after vagal denervation of the oxyntic mucosa, it is obvious that in dogs the inhibitory action of CRF does not require the integrity of the vagus nor is it mediated by changes in the plasma gastrin, because neither basal nor postprandial gastrin level was affected by CRF (98).

The cellular mechanism of CRF-induced gastric inhibition is not known, but it is not due to direct action of the peptide on the parietal cells because neither basal nor stimulated acid production in the isolated rabbit gastric glands was altered by CRF (98).

OPIOID PEPTIDES. Enkephalins and opiate receptors were originally discovered in the brain tissue (73, 165). Subsequently immunological studies revealed a wide distribution of opioid peptides in the gastrointestinal tract and the pancreas both in the peripheral nerves and endocrine cells (170).

The fact that vagal nerves contain enkephalin-like immunoreactivity suggests that opioid peptides may have some role in the cephalic phase of gastric secretion. Indeed, studies in dogs (100) revealed that infusion of Met-enkephalin causes a suppression of gastric acid response to sham feeding, an effect that was partly reversed by naloxone, an opiate receptor antagonist. Because opiates are known to suppress the release of acetylcholine (74), it is likely that the inhibitory effects were due to the removal of the cholinergic background required for the stimulation of gastric secretion in response to vagal stimulation. In humans, enkephalin had no effect on vagally stimulated gastric secretion, but naloxone completely abolished basal and Met-enkephalin–stimulated gastric acid secretion (49, 157), suggesting that endogenous opioid peptides may participate in the physiological stimulation of gastric secretion. Studies in dogs also support the assumption that enkephalin may stimulate basal or pentagastrin-induced acid secretion through a mechanism involving opiate receptors and an increase in tissue histamine release (124, 129).

SYMPATHETIC NERVOUS SYSTEM AND GASTRIC INHIBITION

Studies in animals and humans indicate that catecholamines exert their potent inhibitory effect on gastric acid secretion through both α- and β-adrenoceptors, though α-blockade may also result in gastric inhibition, whereas β-blockade causes stimulation (24).

Studies in dogs showed that the inhibitory effects of catecholamines depend in part on the type of gastric stimulation. Histamine-induced secretion was more inhibited by epinephrine than by isoprenaline, whereas that induced by bethanechol, pentagastrin, or food was more affected by isoprenaline (24). This suggests that α-receptor activation is required for the inhibition of gastric secretion induced by histamine, whereas β-receptor activation is more effective for the inhibition of secretion provoked by bethanechol, pentagastrin, or food (24). Because histamine is a direct stimulant of the parietal cells, the inhibitory effects of α-stimulation might be attributed to the interference with the histamine activation of these cells. However, α-adrenoceptors are also known to mediate the reduction in the gastric mucosal blood flow (62, 134), and this effect might contribute to the observed inhibition of gastric secretion. On the other hand, isoprenaline increases the mucosal blood flow due to β-adrenoceptor stimulation but causes an inhibition of acid secretion (134), indicating that the changes in the mucosal blood flow may not be of major importance in the inhibitory action of catecholamines on gastric secretion (24).

The blockade of α-adrenoceptors was unexpectedly inhibiting, whereas the blockade of β-adrenoceptors had little effect on gastric acid secretion in dogs. The inhibitory action of α-blockade is difficult to explain, particularly because it affects gastric secretion induced by various stimulants (24).

In humans, intravenous administration of epinephrine increased the circulating gastrin level and raised basal gastric acid secretion (17, 199), whereas norepinephrine slightly increased serum gastrin but had no significant effect on acid secretion. β-Blockade with propranolol had little effect on basal or stimulated secretion but blocked the gastrin response to food and insulin (131). These results have been interpreted as showing that insulin stimulates gastric secretion through both α- and β-adrenoceptors owing to the elevation of the circulating epinephrine and norepinephrine. These catecholamines may also activate the G cells, mainly via β-adrenoceptors (24). There are no satisfactory reports to clarify the effects of α-blockade on gastric functions in humans. The studies now reported do not permit a distinction between the physiological and pharmacological influences of catecholamines and adrenoceptors on gastric acid secretion.

REFERENCES

1. ADRIAN, T. E., A. P. SAVAGE, G. R. SAGOR, J. M. ALLEN, A. J. BACARESE-HAMILTON, K. TATEMOTO, J. M. POLAK, AND S. R. BLOOM. Effect of peptide YY on gastric, pancreatic and biliary functions in humans. *Gastroenterology* 89: 494–499, 1985.
2. ALBINUS, M., A. GOMEZ-PAN, B. H. HIRST, AND B. SHAW. Evidence against prostaglandin-mediation of somatostatin-inhibition of gastric secretions. *Regul. Pept.* 10: 259–266, 1985.
3. ANDERSSON, S., G. NILSSON, AND B. UVNAS. Effect of acid in proximal and distal duodenal pouches on gastric secretory responses to gastrin and histamine. *Acta Physiol. Scand.* 71: 368–378, 1967.
4. ARNOLD, R., M. V. HÜLST, C. H. NEUHOF, S. SCHWARTING, H. D. BECKER, AND W. CREUTZFELDT. Antral gastrin-producing G-cells and somatostatin-producing D-cells in different states of gastric acid secretion. *Gut* 23: 285–291, 1982.
5. ARNOLD, R., AND P. G. LANKISCH. Somatostatin and the gastrointestinal tract. *Clin. Gastroenterol.* 9: 733–753, 1980.
6. BEFRITS, R., K. SAMUELSSON, AND C. JOHANSSON. Gastric acid inhibition by antral acidification mediated by endogenous prostaglandins. *Scand. J. Gastroenterol.* 19: 899–904, 1984.
7. BERGEGÅRDH, S., AND L. OLBE. Gastric acid response to antrum distension in man. *Scand. J. Gastroenterol.* 10: 171–176, 1975.
8. BERGLINDH, T., AND G. SACHS. Emerging strategies in ulcer therapy: pumps and receptors. *Scand. J. Gastroenterol. Suppl.* 108: 7–14, 1985.
9. BROWN, J. C., V. MUTT, AND R. A. PETERSON. Further purification of a polypeptide demonstrating enterogastrone activity. *J. Physiol. Lond.* 209: 57–64, 1970.
10. CARRAWAY, R., AND S. E. LEEMAN. Characterization of radioimmunoassayable neurotensin in the rat. Its differential distribution in the central nervous system, small intestine, and stomach. *J. Biol. Chem.* 251: 7045–7052, 1976.
11. CARTER, D. C., AND M. I. GROSSMAN. Effect of luminal pH on acid secretion from Heidenhain pouches evoked by topical and parenteral stimulants. *J. Physiol. Lond.* 281: 227–237, 1978.
12. CHAYVIALLE, J. A., M. MIYATA, P. L. RAYFORD, AND J. C. THOMPSON. Effect of test meal, intragastric nutrients and intraduodenal bile on plasma concentrations of immunoreactive somatostatin and vasoactive intestinal peptide in dogs. *Gastroenterology* 79: 844–852, 1980.
13. CHEN, Y. F., W. Y. CHEY, T.-M. CHANG, AND K. Y. LEE. Duodenal acidification releases cholecystokinin. *Am. J. Physiol.* 249 (*Gastrointest. Liver Physiol.* 12): G29–G33, 1985.
14. CHEW, C. S. Inhibitory action of somatostatin on isolated gastric glands and parietal cells. *Am. J. Physiol.* 245 (*Gastrointest. Liver Physiol.* 8): G221–G229, 1983.
15. CHEY, W. Y., M. S. KIM, K. Y. LEE, AND T. M. CHANG. Secretin is an enterogastrone in the dog. *Am. J. Physiol.* 240 (*Gastrointest. Liver Physiol.* 3): G239–G244, 1981.
16. CHEY, W. Y., AND S. J. KONTUREK. Plasma secretion and pancreatic secretion in response to liver extract meal with varied pH and exogenous secretion in the dog. *J. Physiol. Lond.* 324: 263–272, 1982.
17. CHRISTENSEN, K. C., AND F. STADIL. On beta-adrenergic contribution to the gastric acid and gastrin responses to hypoglycaemia in man. *Scand. J. Gastroenterol. Suppl.* 11: 87–92, 1976.
18. CHRISTIANSEN, J., J. J. HOLST, AND E. KALAJA. Inhibition of gastric acid secretion in man by exogenous and endogenous pancreatic glucagon. *Gastroenterology* 70: 688–692, 1976.
19. CIESZKOWSKI, M., S. J. KONTUREK, W. OBTULOWICZ, AND J. TASLER. Chemical stimulatory mechanism in gastric secretion. *J. Physiol. Lond.* 246: 143–157, 1975.
20. COLTURI, T. J., R. H. UNGER, AND M. FELDMAN. Role of circulating somatostatin in regulation of gastric acid secretion, gastrin release and islet cell function. Studies in healthy subjects and duodenal ulcer patients. *J. Clin. Invest.* 74: 417–423, 1984.
21. CORAZZIARI, E., T. E. SOLOMON, AND M. I. GROSSMAN. Effect of ninety-five percent pure cholecystokinin on gastrin stimulated acid secretion in man and dog. *Gastroenterology* 77: 91–95, 1979.
22. COSTA, M., Y. PATEL, J. B. FURNESS, AND A. ARIMURA. Evidence that some intrinsic neurons of the intestine contain somatostatin. *Neurosci. Lett.* 6: 215–222, 1977.
23. CREUTZFELDT, W., R. EBERT, U. FINKE, S. J. KONTUREK, N. KWIECIEŃ, AND T. RADECKI. Inhibition of gastric secretion by fat and hypertonic glucose in the dog: role of gastric inhibitory polypeptide. *J. Physiol. Lond.* 334: 91–101, 1983.

24. DALY, M. J. The classification of adrenoceptors and their effects on gastric acid secretion. *Scand. J. Gastroenterol. Suppl.* 89: 3–10, 1984.
25. DEBAS, H. T., Y. GOTO, J. REEVE, N. BUNNET, AND J. H. WALSH. Preliminary isolation of an inhibitor of acid secretion from canine fundic mucosa (Abstract). In: *Int. Symp. Gastrointest. Horm. Stockholm, 4th, June 20–23, 1982*, p. 56.
26. DEBAS, H. T., AND M. I. GROSSMAN. Chemicals bathing the oxyntic gland area stimulate acid secretion in dog. *Gastroenterology* 69: 651–659, 1975.
27. DEBAS, H. T., J. HOLLINSHEAD, A. SEAL, P. SOON-SHIONG, AND J. H. WALSH. Vagal control of gastrin release in the dog: pathways for stimulation and inhibition. *Surgery St. Louis* 95: 34–37, 1984.
28. DEBAS, H. T., S. J. KONTUREK, J. H. WALSH, AND M. I. GROSSMAN. Proof of a pyloro-oxyntic reflex for stimulation of acid secretion. *Gastroenterology* 66: 526–532, 1974.
29. DEBAS, H. J., AND T. YAMAGISHI. Gastric inhibitory polypeptide (GIP) is not the primary mediator of the enterogastrone action of fat in the dog. *Gastroenterology* 78: 931–936, 1980.
30. DE GRAEF, J., AND M. C. WOUSSEN-COLLE. Effects of sham feeding, bethanechol, and bombesin on somatostatin release in dogs. *Am. J. Physiol.* 248 (*Gastrointest. Liver Physiol.* 11): G1–G7, 1985.
31. DEMBINSKI, A., D. DROZDOWICZ, H. GREGORY, S. J. KONTUREK, AND Z. WARZECHA. Inhibition of acid formation by epidermal growth factor in the isolated rabbit gastric glands. *J. Physiol. Lond.* 378: 347–357, 1986.
32. DEMBINSKI, A., H. GREGORY, S. J. KONTUREK, AND M. POLANSKI. Trophic action of epidermal growth factor on the pancreas and gastroduodenal mucosa in rats. *J. Physiol. Lond.* 325: 35–42, 1982.
33. DOLVA, L. O., AND K. F. HANSSEN. Thyrotropin-releasing hormone: distribution and actions in the gastrointestinal tract. *Scand. J. Gastroenterol.* 17: 705–708, 1982.
34. DOLVA, L. O., K. F. HANSSEN, A. BERSTAD, AND H. M. M. FREY. Thyrotropin-releasing hormone inhibits the pentagastrin stimulated gastric secretion in man. A dose response study. *Clin. Endocrinol.* 10: 281–286, 1979.
35. EKBLAD, E., M. EKELUND, H. GRAFFNER, R. HÅKANSON, AND F. SUNDLER. Peptide-containing nerve fibers in the stomach wall of rat and mouse. *Gastroenterology* 89: 73–85, 1985.
36. EL-BAYAR, H., L. STEEL, E. MONTCALM, E. DANQUECHIN-DORVAL, A. DUBOIS, AND T. SHEA-DONOHUE. The role of endogenous prostaglandin in the regulation of gastric acid secretion in rhesus monkeys. *Prostaglandins* 30: 401–417, 1985.
37. ELDER, J. B., P. C. GANGULI, I. E. GILLESPIE, E. L. GERRING, AND H. GREGORY. Effect of urogastrone on gastric secretion and plasma gastrin levels in normal subjects. *Gut* 16: 887–893, 1982.
38. ELDER, J. B., P. C. GANGULI, I. E. GILLESPIE, AND H. GREGORY. Effect of urogastrone in Zollinger-Ellison syndrome. *Br. J. Surg.* 61: 916–920, 1974.
39. EMÅS, S., AND M. I. GROSSMAN. Response of Heindenhain pouch to histamine, gastrin and feeding before and after truncal vagotomy in dogs. *Scand. J. Gastroenterol.* 4: 497–503, 1969.
40. ERSPAMER, V., AND P. MELCHIORRI. Actions of amphibian skin peptides on the central nervous system and the anterior pituitary. *Neurol. Endocrinol. Perspect.* 2: 37–106, 1983.
41. ERSPAMER, V., P. MELCHIORRI, M. BROCCARDO, G. F. ERSPAMER, P. FALASCHI, G. IMPROOTA, L. NEGRI, AND T. RENDA. The brain-gut-skin triangle: new peptides. *Peptides Fayetteville* 2, Suppl. 2: 7–16, 1981.
42. EYSSELEIN, V. E. Neurotensin—what is known about its role as a hormone in the gastrointestinal tract? *Klin. Wochenschr.* 62: 523–530, 1984.
43. FAHRENKRUG, J. Vasoactive intestinal peptide. *Clin. Gastroenterol.* 9: 633–643, 1980.
44. FAICHNEY, A., W. Y. CHEY, Y. C. KIM, K. Y. LEE, M. S. KIM, AND T. M. CHANG. Effect of sodium oleate on plasma secretin concentration and pancreatic secretion in dog. *Gastroenterology* 81: 458–462, 1981.
45. FELDMAN, M. Inhibition of gastric acid secretion by selective and nonselective anticholinergics. *Gastroenterology* 86: 361–366, 1984.
46. FELDMAN, M., R. M. DICKERMAN, R. N. MCCLELLAND, K. A. COOPER, J. H. WALSH, AND C. T. RICHARDSON. Effect of selective proximal vagotomy on food-stimulated gastric acid secretion and gastrin release in patients with duodenal ulcer. *Gastroenterology* 76: 926–931, 1979.
47. FELDMAN, M., AND C. T. RICHARDSON. Gastric acid secretion in humans. In: *Physiology of the Gastrointestinal Tract*, edited by L. R. Johnson. New York: Raven, 1981, vol. 1, p. 693–707.
48. FELDMAN, M., AND J. H. WALSH. Acid inhibition of sham feeding stimulated gastrin release and gastric acid secretion: effect of atropine. *Gastroenterology* 78: 772–776, 1980.
49. FELDMAN, M., J. H. WALSH, AND I. L. TAYLOR. Effect of naloxone and morphine on gastric acid secretion and on serum gastrin and pancreatic polypeptide concentrations in humans. *Gastroenterology* 79: 294–298, 1980.
50. FELLENIUS, E., T. BERGLINDH, G. SACHS, L. OLBE, B. ELANDER, S. E. SJÖSTRAND, AND B. WALLMARK. Substituted benzimidazoles inhibit gastric secretion by blocking ($H^+ + K^+$) ATPase. *Nature Lond.* 290: 159, 1981.
51. FINKE, U., M. RUTTEN, R. A. MURPHY, AND W. SILEN. Effects of epidermal growth factor on acid secretion from guinea pig gastric mucosa: in vitro analysis. *Gastroenterology* 88: 1175–1182, 1985.
52. FOKINA, A., S. J. KONTUREK, N. KWIECIEŃ, AND T. RADECKI. Role of gastric antrum in gastric and intestinal phases of gastric secretion in dogs. *J. Physiol. Lond.* 295: 229–239, 1979.
53. FORDTRAN, J. S., AND J. H. WALSH. Gastric acid secretion rate and buffer content of the stomach after eating: results in normal subjects and in patients with duodenal ulcer. *J. Clin. Invest.* 52: 645–657, 1972.
54. GEITZ, P. H. V., M. KASPAR, S. VAN NOORDEN, J. M. POLAK, H. GREGORY, AND A. G. E. PEARSE. Immunohistochemical localization of urogastrone in human duodenal and submandibular glands. *Gut* 19: 408–413, 1978.
55. GESPACH, C., D. HUI BON HOA, AND G. ROSSELIN. Regulation by vasoactive intestinal peptide, histamine, somatostatin-14 and -28 of cyclic adenosine monophosphate levels in gastric glands isolated from the guinea pig fundus or antrum. *Endocrinology* 112: 1597–1606, 1983.
56. GREENALL, M. J., P. J. LYNDON, J. C. GOLIGHER, AND D. JOHNSTON. Long-term effect of highly selective vagotomy on basal and maximal acid output in man. *Gastroenterology* 68: 1421–1425, 1975.
57. GREENLEE, H. B., E. H. LONGHI, J. D. GUERRERO, T. S. NELSEN, A. L. EL-BEDRI, AND L. R. DRAGSTEDT. Inhibitory effect of pancreatic secretion on gastric secretion. *Am. J. Physiol.* 190: 396–402, 1975.
58. GREGORY, H. Isolation and structure of urogastrone and its relationship to epidermal growth factor. *Nature Lond.* 257: 325–327, 1975.
59. GRIJALVA, C. V., E. LINDHOLM, AND D. NOVIN. Physiological and morphological changes in the gastrointestinal tract induced by hypothalamic intervention: overview. *Brain Res. Bull.* 5, Suppl. 1: 19–31, 1980.
60. GROSSMAN, M. I. Regulation of gastric acid secretion. In: *Physiology of the Gastrointestinal Tract*, edited by L. R. Johnson. New York: Raven, 1981, vol. 1, p. 659–671.
61. GROSSMAN, M. I., AND S. J. KONTUREK. Inhibition of acid secretion in dog by metiamide, a histamine antagonist acting on H_2-receptors. *Gastroenterology* 66: 517–521, 1974.
62. HAIGH, A. L., AND W. M. STEEDMAN. The action of catecholamines and adrenergic blockade on gastric blood flow and acid secretion in the dog (Abstract). *J. Physiol. Lond.* 198: 79P–80P, 1968.
63. HARPER, A. A., J. D. REED, AND J. R. SMY. Effects of intragastric hyperosmolar solutions on gastric functions. *J. Physiol. Lond.* 209: 453–472, 1970.
64. HAWKEY, C. J., AND D. S. RAMPTON. Prostaglandins and the

gastrointestinal mucosa: are they important in the function, disease, or treatment? *Gastroenterology* 89: 1162–1188, 1985.
65. HIRSCHOWITZ, B. I., R. GIBSON, AND E. MOLINA. Atropine suppresses gastrin release by food in intact and vagotomozed dogs. *Gastroenterology* 81: 838–843, 1981.
66. HIRSCHOWITZ, B. I., AND G. A. HUTCHINSON. Long-term effects of highly selective vagotomy (HSV) in dogs on acid and pepsin secretion. *Am. J. Dig. Dis.* 22: 81–95, 1977.
67. HIRSCHOWITZ, B. I., AND G. A. HUTCHINSON. Kinetics of atropine inhibition of pentagastrin-stimulated H^+, electrolyte, and pepsin secretion in the dog. *Am. J. Dig. Dis.* 22: 99–107, 1977.
68. HIRST, B. H., E. ARILLA, D. H. COY, AND B. SHAW. Cyclic hex- and pentapeptide somatostatin analogues with reduced gastric inhibitory activity. *Peptides Fayetteville* 5: 857–860, 1984.
69. HOLLE, F. Effect of selective proximal vagotomy on parietal cells in man. In: *Vagotomy*, edited by F. Holle and S. Anderson. Berlin: Springer-Verlag, 1974, p. 24–32.
70. HOLLINSHEAD, J. W., H. T. DEBAS, T. YAMADA, J. ELASHOFF, B. OSADCHEY, AND J. H. WALSH. Hypergastrinemia develops within 24 hours of truncal vagotomy in dogs. *Gastroenterology* 88: 35–40, 1985.
71. HOLM-BENTZEN, M., J. CHRISTIANSEN, P. KIRKEGAARD, P. S. OLSEN, B. PETERSEN, AND J. FAHRENKRUG. The effect of vasoactive intestinal polypeptide on meal-stimulated gastric acid secretion in man. *Scand. J. Gastroenterol.* 18: 659–661, 1983.
72. HOLST, J. J., J. CHRISTIANSEN, AND C. KÜHL. The enteroglucagon response to intrajejunal infusion of glucose, triglycerides, and sodium chloride, and its relation to jejunal inhibition of gastric acid secretion in man. *Scand. J. Gastroenterol.* 11: 297–304, 1976.
73. HUGHES, J., T. W. SMITH, H. W. KOSTERLITZ, L. A. FOTHERGILL, B. A. MORGAN, AND H. R. MORRIES. Identification of two related pentapeptides from the brain with potent opiate agonist activity. *Nature Lond.* 258: 577–580, 1975.
74. IHAMANDAS, K., C. PINSKY, AND J. W. PHILLIS. Effects of morphine and its antagonists on release of cerebral cortical acetylcholine. *Nature Lond.* 228: 176–177, 1980.
75. IMPICCIATORE, M., H. DEBAS, J. H. WALSH, M. I. GROSSMAN, AND G. BERTACCINI. Release of gastrin and stimulation of acid secretion by bombesin in dog. *Rend. Gastroenterol.* 6: 99–101, 1974.
76. IMPICCIATORE, M., J. H. WALSH, AND M. I. GROSSMAN. Low doses of atropine enhance serum gastrin response to food in dogs. *Gastroenterology* 72: 995–996, 1977.
77. IPP, E., U. PIRAN, H. RICHTER, C. GARBEROGLIO, A. MOOSSA, AND A. H. RUBINSTEIN. Central control of peripheral circulating somatostatin in dogs: effect of 2-deoxyglucose. *Am. J. Physiol.* 243 (*Endocrinol. Metab.* 6): E213–E216, 1982.
78. JACKSON, I. M. D. Thyrotropin-releasing hormone. *N. Engl. J. Med.* 306: 145–155, 1982.
79. JOHANSSON, C. Oral PGE_2 inhibits gastric acid secretion in man. *Prostaglandins* 29: 143–152, 1985.
80. JOHNSON, L. R., AND M. I. GROSSMAN. Secretin: the enterogastrone release by acid in the duodenum. *Am. J. Physiol.* 215: 885–888, 1968.
81. JOHNSON, L. R., AND M. I. GROSSMAN. Characteristics of inhibition of gastric secretion by secretin. *Am. J. Physiol.* 217: 1401–1404, 1969.
82. JOHNSON, L. R., AND M. I. GROSSMAN. Analysis of inhibition of acid secretion by cholecystokinin in dogs. *Am. J. Physiol.* 218: 550–554, 1970.
83. JOHNSTON, D., A. R. WILKINSON, C. S. HUMPHREY, R. B. SMITH, J. C. GOLIGHER, E. KRAGELUND, AND E. AMDRUG. Serial studies on gastric secretion in patients after highly selective (parietal cell) vagotomy without a drainage procedure for duodenal ulcer. II. The insulin test after highly selective vagotomy. *Gastroenterology* 64: 12–21, 1973.
84. KIHL, B., AND L. OLBE. Fat inhibition of gastric acid secretion in duodenal ulcer patients before and after proximal gastric vagotomy. *Gut* 21: 1056–1061, 1980.
85. KIHL, B., A. RÖKAEUS, S. ROSELL, AND L. OLBE. Fat inhibition of gastric acid secretion in man and plasma concentration of neurotensin-like immunoreactivity. *Scand. J. Gastroenterol.* 16: 513–526, 1981.
86. KIRKEGAARD, P., A. J. MOODY, J. J. HOLST, F. B. LOUD, P. S. OLSEN, AND J. CHRISTIANSEN. Glicentin inhibits gastric acid secretion in rat. *Nature Lond.* 297: 156–157, 1982.
87. KLEIBEUKER, J. H., V. E. EYSSELEIN, V. E. MAXWELL, AND J. H. WALSH. Role of endogenous secretin in acid induced inhibition of human gastric functions. *J. Clin. Invest.* 73: 526–532, 1984.
88. KNAPP, H. R., O. OELZ, B. J. SWEETMAN, AND J. A. OATES. Synthesis and metabolism of prostaglandin E_2, F_2, α, and D_2 by gastrointestinal tract. Stimulation by a hypertonic environment in vitro. *Prostaglandins* 15: 751–757, 1978.
89. KNUTSON, U., AND L. OLBE. The effect of exogenous gastrin on the acid sham feeding response in antrum-bulb-resected duodenal ulcer patients. *Scand. J. Gastroenterol.* 9: 231–238, 1974.
90. KOFFMAN, C. G., J. B. ELDER, P. C. GANGULI, H. GREGORY, AND C. G. GEARY. Effect of urogastrone on gastric secretion and serum gastrin concentration in patients with duodenal ulceration. *Gut* 23: 951–956, 1982.
91. KONTUREK, S. J. Intestinal mechanisms regulating gastric secretion. In: *Progress in Gastroenterology*, edited by G. B. J. Glass. New York: Grune & Stratton, 1977, vol. 3, p. 395–416.
92. KONTUREK, S. J. Pharmacology and clinical use of rantidine. *Mt. Sinai J. Med.* 49: 370–382, 1982.
93. KONTUREK, S. J. Pharmacological control of gastric acid secretion in peptic ulcer. *Mt. Sinai J. Med.* 50: 457–467, 1983.
94. KONTUREK, S. J., W. BIELAŃSKI, J. TASLER, J. BILSKI, J. JENSEN, A. DE JONG, AND C. LAMERS. Release of secretin, CCK, and PP by regional perfusion of intestine with HCl and oleate. *Gastroenterology* 86: 1140, 1984.
95. KONTUREK, S. J., J. BIERNAT, AND T. GRZELEC. Inhibition by secretin of the gastric acid responses to meals and to pentagastrin in duodenal ulcer patients. *Gut* 14: 842–846, 1973.
96. KONTUREK, S. J., J. BIERNAT, N. KWIECIEŃ, AND J. OLEKSY. Effect of glucagon on meal-induced gastric secretion in man. *Gastroenterology* 68: 448–454, 1975.
97. KONTUREK, S. J., J. BIERNAT, AND J. OLEKSY. Serum gastrin and gastric responses to meals at various pH levels in man. *Gut* 15: 526–528, 1974.
98. KONTUREK, S. J., J. BILSKI, W. PAWLIK, P. THOR, K. CZARNOBILSKI, B. SZOKE, AND A. V. SCHALLY. Gastrointestinal secretory motor and circulatory effects of corticotropin releasing factor (CRF). *Life Sci.* 37: 1231–1240, 1985.
99. KONTUREK, S. J., J. BILSKI, J. TASLER, AND J. LASKIEWICZ. Gastroduodenal alkaline response to acid and taurocholate in conscious dogs. *Am. J. Physiol.* 247 (*Gastrointest. Liver Physiol.* 10): G149–G154, 1984.
100. KONTUREK, S. J., T. BRZOZOWSKI, I. PIASTUCKI, A. DEMBINSKI, T. RADECKI, A. DEMBINSKA-KIEC, A. ZMUDA, AND H. GREGORY. Role of mucosal prostaglandins and DNA synthesis in gastric cytoprotection by luminal epidermal growth factor. *Gut* 22: 927–932, 1981.
101. KONTUREK, S. J., M. CIESZKOWSKI, J. BILSKI, J. KONTUREK, W. BIELANSKI, AND A. V. SCHALLY. Effects of cyclic hexapeptide analog of somatostatin on pancreatic secretion in dogs. *Proc. Soc. Exp. Biol. Med.* 178: 68–72, 1985.
102. KONTUREK, S. J., M. CIESZKOWSKI, J. JAWOREK, J. KONTUREK, T. BRZOZOWSKI, AND H. GREGORY. Effects of epidermal growth factor on gastrointestinal secretions. *Am. J. Physiol.* 246 (*Gastrointest. Liver Physiol.* 9): G580–G586, 1984.
103. KONTUREK, S. J., M. CIESZKOWSKI, N. KWIECIEŃ, J. KONTUREK, J. TASLER, AND J. BILSKI. Effect of omeprazole, a substituted benzimidazole, on gastrointestinal secretions, serum gastrin, and gastric mucosal blood flow in dogs. *Gastroenterology* 86: 71–77, 1984.
104. KONTUREK, S. J., A. DEMBINSKI, P. THOR, AND R. KROL. Comparison of vasoactive intestinal peptide (VIP) and secretin

in gastric acid secretion and mucosal blood flow. *Pfluegers Arch.* 361: 175–181, 1976.
105. KONTUREK, S. J., S. DOMSCHKE, W. DOMSCHKE, L. VARGA, AND F. HALTER. Effects of secretin antibodies on gastric H$^+$ inhibition and pancreatic HCO$_3^-$ stimulation by acidified liver extract meal in dogs. *Hepato-Gastroenterol.* 33: 170–175, 1986.
106. KONTUREK, S. J., B. GABRYS, AND J. DUBIEL. Effect of exogenous and endogenous secretin on gastric and pancreatic secretion in cats. *Am. J. Physiol.* 217: 1110–1113, 1969.
107. KONTUREK, S. J., AND M. I. GROSSMAN. Localization of the mechanism for inhibition of gastric secretion. *Gastroenterology* 49: 74–78, 1965.
108. KONTUREK, S. J., AND M. I. GROSSMAN. Effect of perfusion of intestinal loops with acid, fat and dextrose on gastric secretion. *Gastroenterology* 49: 481–489, 1965.
109. KONTUREK, S. J., J. JAWOREK, W. BIELANSKI, M. CIESZKOWSKI, M. DOBRZAŃSKA, AND D. H. COY. Comparison of enkephalin and atropine in the inhibition of vagally stimulated gastric and pancreatic secretion and gastrin and pancreatic polypeptide release in dogs. *Peptides Fayetteville* 3: 601–606, 1982.
110. KONTUREK, S. J., J. JAWOREK, M. CIESZKOWSKI, W. PAWLIK, J. KANIA, AND S. R. BLOOM. Comparison of effects of neurotensin and fat on pancreatic stimulation in dogs. *Am. J. Physiol.* 244 (*Gastrointest. Liver Physiol.* 7): G590–G598, 1983.
111. KONTUREK, S. J., J. JAWOREK, M. CIESZKOWSKI, AND A. V. SCHALLY. Effect of thyrotropin-releasing hormone on gastrointestinal secretions in dogs. *Life Sci.* 29: 2289–2298, 1981.
112. KONTUREK, S. J., AND L. R. JOHNSON. Evidence for an enterogastric reflex for inhibition of acid secretion. *Gastroenterology* 61: 667–674, 1971.
113. KONTUREK, S. J., J. KONTUREK, M. CIESZKOWSKI, R. ERBERT, AND W. CREUTZFELDT. Comparison of gastric inhibitory polypeptide and intraduodenal or intravenous fat on gastric acid secretion from vagally innervated and denervated canine stomach. *Dig. Dis. Sci.* 31: 49–56, 1986.
114. KONTUREK, S. J., N. KWIECIEŃ, W. OBTULOWICZ, W. BIELAŃSKI, J. OLEKSY, AND A. V. SCHALLY. Effects of somatostatin 14 and somatostatin 28 on plasma hormonal and gastric secretory response to cephalic and gastrointestinal stimulation in man. *Scand. J. Gastroenterol.* 20: 31–38, 1985.
115. KONTUREK, S. J., N. KWIECIEŃ, W. OBTULOWICZ, B. KOPP, AND J. OLEKSY. Action of omeprazole (a benzimidazole derivative) on secretory responses to sham feeding and pentagastrin and upon serum gastrin and pancreatic polypeptide in duodenal ulcer patients. *Gut* 25: 14–18, 1984.
116. KONTUREK, S. J., N. KWIECIEŃ, W. OBTULOWICZ, E. MIKOŚ, E. SITO, J. OLEKSY, AND T. POPIELA. Cephalic phase of gastric secretion in healthy subjects and duodenal ulcer patients: role of vagal innervation. *Gut* 20: 875–881, 1978.
117. KONTUREK, S. J., N. KWIECIEŃ, W. OBTULOWICZ, AND J. OLEKSY. Prostaglandins and vagal stimulation of gastric secretion in duodenal ulcer patients. *Scand. J. Gastroenterol.* 18: 43–47, 1983.
118. KONTUREK, S. J., N. KWIECIEŃ, J. SWIERCZEK, J. OLEKSY, E. SITO, AND A. ROBERT. Comparison of methylated prostaglandin E$_2$ analogues given orally in the inhibition of gastric responses to pentagastrin and peptone meal in man. *Gastroenterology* 70: 683–687, 1976.
119. KONTUREK, S. J., E. MIKOŚ, W. PAWLIK, AND K. WALUS. Direct inhibition of gastric secretion and mucosal blood flow by arachidonic acid. *J. Physiol. Lond.* 286: 15–28, 1979.
120. KONTUREK, S. J., W. OBTULOWICZ, N. KWIECIEŃ, AND J. OLEKSY. Effects of pirenzepine and atropine on gastric secretory and plasma hormonal responses to sham feeding in patients with duodenal ulcer. *Scand. J. Gastroenterol. Suppl.* 66: 63–69, 1980.
121. KONTUREK, S. J., AND T. E. SOLOMON. Role of gastrin in cephalic phase of gastric acid and pepsin secretion in dogs. *Hepato-Gastroenterology* 32: 43, 1985.
122. KONTUREK, S. J., J. TASLER, J. BILSKI, J. KONTUREK, AND W. BIELANSKI. Studies on the localization of secretin release from canine intestine. *Digestion* 34: 207–215, 1986.
123. KONTUREK, S. J., J. TASLER, M. CIESZKOWSKI, D. H. COY, AND A. V. SCHALLY. Effect of growth hormone release—inhibiting hormone on gastric secretion, mucosal blood, flow, and serum gastrin. *Gastroenterology* 70: 737–741, 1976.
124. KONTUREK, S. J., J. TASLER, M. CIESZKOWSKI, E. MIKOŚ, D. H. COY, AND A. V. SCHALLY. Comparison of methionine-enkephalin and morphine in the stimulation of gastric acid secretion in the dog. *Gastroenterology* 78: 294–300, 1980.
125. KONTUREK, S. J., J. TASLER, J. JAWOREK, M. DOBRZANSKI, D. H. COY, AND A. V. SCHALLY. Comparison of TRH and anorexigenic peptide on food intake and gastrointestinal secretions. *Peptides Fayetteville* 2, Suppl. 2: 235–240, 1981.
126. KONTUREK, S. J., J. TASLER, J. JAWOREK, W. PAWLIK, K. M. WALUS, V. SCHUSDZIARRA, C. A. MEYERS, D. H. COY, AND A. V. SCHALLY. Gastrointestinal secretory, motor, circulatory and metabolic effects of prosomatostatin. *Proc. Natl. Acad. Sci. USA* 78: 1967–1971, 1981.
127. KONTUREK, S. J., J. TASLER, AND W. OBTULOWICZ. Duodenal mechanisms for inhibition of gastric acid secretion in the dog. *Am. J. Physiol.* 220: 918–921, 1971.
128. KOSAKA, T., AND R. K. S. LIM. Demonstration of the humoral agent in fat inhibited gastric secretion. *Proc. Soc. Exp. Biol. Med.* 27: 880–891, 1930.
129. KOSTRITSKY-PEREIRA, A., M. C. WOUSSEN-COLLE, AND J. DE GRAEF. Effect of Met-enkephalin on acid secretion from gastric fistulas and Heidenhain pouches in dogs stimulated by pentagastrin, pentagastrin plus bethanechol, or a meal. *Int. J. Tissue Reactions* 6: 167–173, 1984.
130. KRAENZLIN, M. E., J. L. CHANG, P. K. MULDERRY, M. A. GHATEI, AND S. R. BLOOM. Infusion of a novel peptide, calcitonin gene-related peptide (CGRP) in man. Pharmacokinetics and effects on gastric acid secretion and on gastrointestinal hormones. *Regul. Pept.* 10: 189–197, 1985.
131. KRONBORG, O., T. PEDERSEN, F. STADIL, AND J. F. REHFELD. The effect of beta-adrenergic blockage upon gastric acid response to peptone meal. *Scand. J. Gastroenterol.* 9: 173–176, 1974.
132. KWAK, Y. W., C. H. S. MCINTOSH, R. A. PETERSON, AND J. C. BROWN. Effect of substance P on somatostatin release from the isolated perfused rat stomach. *Gastroenterology* 88: 90–95, 1985.
133. LAFONTAINE, M., G. B. CADIÈRE, M. C. WOUSSEN-COLLE, AND J. DE GRAEF. Interaction between secretin and somatostatin on acid secretion and gastric emptying (Eng. Abstr.). *Gastroenterol. Clin. Biol.* 8: 343–346, 1984.
134. LANCIAULT, G., AND E. D. JACOBSON. The gastrointestinal circulation. *Gastroenterology* 71: 851–873, 1976.
135. LARSON, G. M., B. H. AHLMAN, C. T. BOMBECK, AND L. M. NYHUS. The effect of chemical and surgical sympathectomy on gastric secretion and innervation. *Scand. J. Gastroenterol. Suppl.* 89: 27–32, 1984.
136. LARSSON, L. I., N. GOLTERMANN, L. DE MAGISTRIS, J. REHFELD, AND T. W. SCHWARTZ. Somatostatin cell processes as pathways for paracrine secretion. *Science Wash. DC* 205: 1393–1395, 1979.
137. LARSSON, L. I., H. MATTSON, G. SUNDELL, AND E. CARLSSON. Animal pharmacodynamics of omeprazole. A survey of its pharmacological properties in vivo. *Scand. J. Gastroenterol. Suppl.* 108: 23–35, 1985.
138. LENZ, H. J., M. T. MORTRUD, J. E. RIVIER, AND M. R. BROWN. Central nervous system actions of calcitonin gene-related peptide on gastric acid secretion in the rat. *Gastroenterology* 88: 539–544, 1985.
139. LEVINE, R. A., K. R. KOHEN, E. H. SCHWARTZEL, JR., AND C. E. RAMSAY. Prostaglandin E$_2$-histamine interactions on cAMP and acid production in isolated fundic glands. *Am. J. Physiol.* 242 (*Gastrointest. Liver Physiol.* 5): G21–G26, 1982.
140. LEVINE, R. A., AND E. H. SCHWARTZEL. Effect of indomethacin on basal and histamine stimulated human gastric acid secretion. *Gut* 25: 718–722, 1984.
141. LIGUMSKY, M., Y. GOTO, H. DEBAS, AND T. YAMADA. Pros-

taglandins mediate inhibition of gastric acid secretion by somatostatin in the rat. *Science Wash. DC* 219: 301–303, 1983.

142. LUCEY, M. R., J. A. H. WASS, P. D. FAIRCLOUGH, M. O'HARE, P. KWASOWSKI, E. PENMAN, J. WEBB, AND L. H. REES. Does gastric acid release plasma somatostatin in man? *Gut* 25: 1217–1220, 1984.

143. LUNDBERG, J. M., K. TATEMOTO, I. TERENIUS, P. M. HELLSTROM, V. MUTT, AND T. HÖKFELT. Localization of peptide YY (PYY) in gastrointestinal endocrine cells and effects on intestinal blood flow and motility. *Proc. Natl. Acad. Sci. USA* 79: 4471–4475, 1972.

144. MAXWELL, V., A. SHULKES, J. C. BROWN, T. E. SOLOMON, J. H. WALSH, AND M. I. GROSSMAN. Effect of gastric inhibitory polypeptide on pentagastrin stimulated acid secretion in man. *Dig. Dis. Sci.* 25: 113–116, 1980.

145. MAYER, E. A., J. ELASHOFF, V. MUTT, AND J. H. WALSH. Reassessment of gastric inhibition by cholecystokinin and gastric inhibitory polypeptide in dogs. *Gastroenterology* 83: 1047–1050, 1982.

146. MISHER, A., AND F. P. BROOKS. Electrical stimulation of hypothalamus and gastric secretion in the albino rat. *Am. J. Physiol.* 211: 403–406, 1966.

147. MITCHELL, S. J., AND S. R. BLOOM. Measurements of fasting and postprandial plasma VIP in man. *Gut* 19: 1043–1048, 1978.

148. MOGARD, M. H., G. L. KAUFFMAN, JR., M. PEHLEVANIAN, E. GOLANSKA, J. D. ELASHOFF, AND J. H. WALSH. Prostaglandin may not mediate inhibition of gastric acid secretion by somatostatin in the rat. *Regul. Pept.* 10: 231–236, 1985.

149. MOGARD, M. H., V. MAXWELL, T. KOVACS, G. V. DEVENTED, J. D. ELASHOFF, T. YAMADA, G. L. KAUFFMAN, AND J. H. WALSH. Somatostatin inhibits gastric acid secretion after gastric mucosal prostaglandin synthesis inhibition by indomethacin in man. *Gut* 26: 1189–1191, 1985.

150. MOHAMMED, R., R. J. HOLDEN, J. B. HEARNS, B. M. MCKIBBEN, K. B. BUCHMAN, AND G. P. CREAN. Effect of eight weeks' continuous treatment with oral rantidine and cimetidine on gastric acid secretion, pepsin secretion and fasting serum gastrin. *Gut* 24: 61–66, 1983.

151. MOORE, J. G., AND F. CRESPIN. Influence of glucose on cephalic vagal stimulated gastric acid secretion in man. *Dig. Dis. Sci.* 25: 117–122, 1980.

152. MORLEY, J. E., B. A. GOSNELL, AND A. S. LEVINE. The role of peptides in feeding. *Trends Pharmacol. Sci.* 5: 468–471, 1984.

153. MÜLLER, P., B. SIMON, G. FEURLE, K. LICHTWALD, AND H. SCHMIDT-GAYK. Human acid secretion during daily administration of H_2 blockers. *Schweiz. Med. Wochenschr.* 114: 667, 1984.

154. NILSSON, G. Bulbogastrone: physiological evidence for its significance. In: *Gastrointestinal Hormones*, edited by G. B. J. Glass. New York: Raven, 1980, p. 911–928.

155. OLBE, L. Potentiation of sham feeding response in Pavlov pouch dogs by subthreshold amounts of gastrin with and without acidification of denervated antrum. *Acta Physiol. Scand.* 61: 244–254, 1964.

156. OLSEN, P. S., J. H. PEDERSEN, P. KIRKEGAARD, H. BEEN, F. STADIL, J. FAHRENKRUG, AND J. CHRISTENSEN. Neurotensin induced inhibition of gastric acid secretion in duodenal ulcer patients before and after parietal cell vagotomy. *Gut* 25: 481–485, 1984.

157. OLSON, P., P. KIRKEGAARD, B. PETERSEN, AND J. CHRISTIANSEN. Effect of naloxone on met-enkephalin-induced gastric acid secretion and serum gastrin in man. *Gut* 23: 63–65, 1982.

158. OSUMI, Y., Y. NAGASAKA, L. H. WANG FU, AND M. FUJIWARA. Inhibition of gastric acid secretion and mucosal blood flow induced by intraventricularly applied neurotensin in rats. *Life Sci.* 23: 2275–2280, 1978.

159. PAPPAS, T. N., H. T. DEBAS, Y. GOTO, AND I. L. TAYLOR. Peptide YY inhibits meal-stimulated pancreatic and gastric secretion. *Am. J. Physiol.* 248 (*Gastrointest. Liver Physiol.* 11): G118–G123, 1985.

160. PAPPAS, T. N., D. HAMEL, H. DEBAS, J. H. WALSH, AND Y. TACHÉ. Cerebroventricular bombesin inhibits gastric acid secretion in dogs. *Gastroenterology* 89: 43–48, 1985.

161. PAPPAS, T. N., Y. TACHÉ, AND H. T. DEBAS. Opposing central and peripheral actions of brain-gut peptides: a basis for regulation of gastric function. *Surgery St. Louis* 98: 183–190, 1985.

162. PAVLOV, I. P. *The Work of the Digestive Glands* (2nd ed.) (transl. by H. Thompson). London: Griffin, 1910.

163. PEDERSON, R. A., AND J. C. BROWN. Inhibition of histamine-pentagastrin- and insulin-stimulated canine gastric secretion by pure "gastric inhibitory polypeptide." *Gastroenterology* 62: 393–400, 1972.

164. PENMAN, E., J. A. H. WASS, S. MEDBAK, L. MORGAN, J. M. LEWIS, G. M. BESSER, AND L. H. REES. Response of circulating immunoreactive somatostatin to nutritional stimuli in normal subjects. *Gastroenterology* 81: 692–699, 1981.

165. PERT, C. B., AND S. H. SNYDER. Opiate receptor: demonstration in nervous tissue. *Science Wash. DC* 179: 1011–1014, 1973.

166. PETERSEN, B., J. CHRISTIANSEN, AND J. J. HOLST. A glucose-dependent mechanism in jejunum inhibits gastric acid secretion: a response mediated through enteroglucagon? *Scand. J. Gastroenterol.* 20: 193–197, 1985.

167. PETERSON, W., M. FELDMAN, J. TAYLOR, AND M. BREMER. The effect of 15 (R) 15 methyl prostaglandin E_2 on meal-stimulated gastric acid secretion, serum gastrin and pancreatic polypeptide in duodenal ulcer patients. *Dig. Dis. Sci.* 24: 281–284, 1979.

168. PETRUSZ, P., I. MERCHENTHALER, S. ORDRONNEAU, J. L. MADERDRUT, S. VIGH, AND A. V. SCHALLY. Corticotropin-releasing factor (CRF)-like immunoreactivity in the gastro-entero-pancreatic endocrine system. *Peptides Fayetteville* 5, Suppl. 3: 71–78, 1984.

169. POLAK, J. M., S. R. BLOOM, M. KUZIO, J. C. BROWN, AND A. G. E. PEARSE. Cellular localization of gastric inhibitory polypeptide in the duodenum and jejunum. *Gut* 14: 284–288, 1973.

170. POLAK, J. M., S. R. BLOOM, S. N. SULLIVAN, P. FACER, AND A. C. E. PEARSE. Enkephalin-like immunoreactivity in the human gastrointestinal tract. *Lancet* 1: 972–974, 1977.

171. PRADAYROL, L., H. JÖRNVALL, V. MUTT, AND A. RIBET. N-terminally extended somatostatin: the primary structure of somatostatin-28. *FEBS Lett.* 109: 55–58, 1980.

172. PRESHAW, R. M. Inhibition of pentagastrin stimulated gastric output by sham feeding. *Federation Proc.* 32: 410, 1973.

173. PRICHARD, P. J., D. RUBINSTEIN, D. D. JONES, F. J. DUDLEY, R. A. SMALLWOOD, W. J. LOUIS, AND N. D. YEOMANS. Double blind comparative study of omeprazole 10 mg and 30 mg daily for healing duodenal ulcers. *Br. Med. J.* 290: 601–603, 1985.

174. REICHSTEIN, B. J., C. OKAMOTO, AND J. G. FORTE. Inhibition of acid secretion by epidermal growth factor (EGF) in isolated gastric glands is secretagogue-specific (Abstract). *Federation Proc.* 43: 1072, 1984.

175. SAFFOURI, B., J. W. DUVAL, A. ARIMURA, AND G. M. MAKHLOUF. Effects of vasoactive intestinal peptide and secretin on gastrin and somatostatin secretion in the perfused rat stomach. *Gastroenterology* 86: 839–842, 1984.

176. SAFFOURI, B., W. DUVAL, AND G. M. MAKHLOUF. Stimulation of gastrin secretion in vitro by intraluminal chemicals: regulation by intramural cholinergic and noncholinergic neurons. *Gastroenterology* 87: 557–561, 1984.

177. SAFFOURI, B., G. C. WEIR, K. N. BITAR, AND G. M. MAKHLOUF. Gastrin and somatostatin secretion by perfused rat stomach: functional linkage of antral peptides. *Am. J. Physiol.* 238 (*Gastrointest. Liver Physiol.* 1): G495–G501, 1980.

178. SAID, S. I. Vasoactive intestinal peptide (VIP): current status. *Peptides Fayetteville* 5: 143–150, 1984.

179. SAID, S. I., AND V. MUTT. Polypeptide with broad biological activity: isolation from small intestine. *Science Wash. DC* 169: 1217–1218, 1970.

180. SANDVIK, A., B. K. KAUL, H. WALDUM, AND H. PETERSEN. Gastric acid and pepsin secretion in response to modified sham feeding in active and duodenal ulcer disease. *Scand. J. Gastroenterol.* 20: 602–606, 1985.

181. SANDWEISS, D. J. The immunizing effect of the anti-ulcer

factor in normal human urine (Anthelone) against the experimental gastrojejunal (peptic) ulcer in dogs. *Gastroenterology* 1: 965, 1943.
182. SCHAFFALITZKY DE MUCKADELL, O. B., AND J. FAHRENKRUG. Secretin pattern of secretin in man: regulation by gastric acid. *Gut* 19: 812–818, 1978.
183. SCHAFFALITZKY DE MUCKADELL, O. B., J. FAHRENKRUG, J. J. HOLST, AND K. B. LAURISTSEN. Release of vasoactive intestinal polypeptide (VIP) by intraduodenal stimuli. *Scand. J. Gastroenterol.* 12: 793–799, 1977.
184. SCHEPP, W., H. J. RUOFF, AND S. MAŚLINSKI. Aminopyrine accumulation of isolated parietal cells from the rat stomach. Effect of histamine and interaction with endogenous inhibitors. *Arch. Int. Pharmacodyn. Ther.* 265: 293–308, 1983.
185. SCHILLER, L. R., J. H. WALSH, AND M. FELDMAN. Distention-induced gastrin release: effects of luminal acidification and intravenous atropine. *Gastroenterology* 78: 912–917, 1980.
186. SCHUBERT, M. L., D. SAFFOURI, J. H. WALSH, AND G. M. MAKHLOUF. Inhibition of neurally mediated gastrin secretion by bombesin antiserum. *Am. J. Physiol.* 248 (*Gastrointest. Liver Physiol.* 11): G456–G462, 1985.
187. SCHUSDZIARRA, V., V. J. HARRIS, J. M. CONLON, A. ARIMURA, AND R. UNGAR. Pancreatic and gastric somatostatin release in response to intragastric and intraduodenal nutrients and HCl in the dog. *J. Clin. Invest.* 62: 509–518, 1978.
188. SEAL, A. M., AND H. T. DEBAS. Colonic inhibition of gastric acid secretion in the dog. *Gastroenterology* 79: 823–826, 1980.
189. SHARMA, B. K., R. P. WALT, R. E. POUNDER, F. A. GOMES, E. C. WOOD, AND L. H. LOGAN. Optimal dose of oral omeprazole for maximal 24 hour decrease of intragastric acidity. *Gut* 25: 957–964, 1984.
190. SJÖDIN, L. Inhibition of gastrin-stimulated canine acid secretion by sham-feeding. *Scand. J. Gastroenterol.* 10: 73–80, 1975.
191. SJÖDIN, L., AND S. ANDERSSON. Effect of resection of antrum and duodenal bulb on sham-feeding induced inhibition of canine gastric secretion. *Scand. J. Gastroenterol.* 12: 43–47, 1977.
192. SOKOLOV, A. P. Analysis of the Secretory work of the Stomach in the Dog (transl. from Russian). PhD thesis. St. Petersburg, Russia, 1950. Cited by B. P. Babkin. In: *Secretory Mechanism of the Digestive Glands*. New York: Hoeber, 1950, p. 642.
193. SOLL, A. H. Specific inhibition by prostaglandin E₂ and I₂ of histamine-stimulated [^{14}C]aminopyrine accumulation and cyclic adenosine monophosphate generation by isolated canine parietal cells. *J. Clin. Invest.* 65: 1222–1229, 1980.
194. SOLL, A. H. Physiology of isolated canine parietal cells: receptors and effectors regulating function. In: *Physiology of Gastrointestinal Tract*, edited by L. R. Johnson. New York: Raven, 1981, p. 1673.
195. SOLL, A. H., D. A. AMIRIAN, J. PARK, J. D. ELASHOFF, AND T. YAMADA. Cholecystokinin potently releases somatostatin from canine fundic mucosal cells in short-term culture. *Am. J. Physiol.* 248 (*Gastrointest. Liver Physiol.* 11): G569–G573, 1985.
196. SOMIYA, H., AND T. TONOUE. Neuropeptides as central integrators of autonomic nerve activity: effects of TRH, SRIF, VIP, and bombesin on gastric and adrenal nerves. *Regul. Pept.* 9: 47–52, 1984.
197. SOON-SHIONG, P., AND H. T. DEBAS. Fundic inhibition of acid secretion and gastrin release in the dog. *Gastroenterology* 79: 867–872, 1980.
198. SPENNEY, J. G. Physical, chemical and technical limitations to intragastric titration. *Gastroenterology* 76: 1025–1034, 1979.
199. STADIL, F., AND J. F. REHFELD. Release of gastrin by epinephrine in man. *Gastroenterology* 65: 210–215, 1973.
200. STENING, G. F., AND M. I. GROSSMAN. Gastric acid response to pentagastrin and histamine after extragastric vagotomy in dogs. *Gastroenterology* 59: 364–371, 1970.
201. STENQUIST, B., G. NILSSON, J. F. REHFELD, AND L. OLBE. Plasma gastrin concentrations following sham-feeding in duodenal ulcer patients. *Scand. J. Gastroenterol.* 14: 305–311, 1979.
202. SUNDLER, F., R. HÅKANSON, AND S. LEANDER. Peptidergic nervous systems in the gut. *Clin. Gastroenterol.* 9: 517–543, 1980.
203. TACHÉ, Y., Y. GOTO, M. GUNION, J. RIVIER, AND H. T. DEBAS. Inhibition of gastric secretion in rats and in dogs by corticotropin-releasing factor. *Gastroenterology* 86: 281–286, 1984.
204. TACHÉ, Y., Y. GOTO, M. GUNION, W. VALE, J. RIVIER, AND M. BROWN. Inhibition of gastric acid secretion in rats by intracerebral injection of corticotropin releasing factor. *Science Wash. DC* 222: 935–937, 1983.
205. TACHÉ, Y., C. GRIJALVA, M. GUNION, P. H. COOPER, J. WALSH, AND D. NOVIN. Lateral hypothalamic mediation of hypergastrinemia induced by intracisternal bombesin. *Neuroendocrinology* 39: 114–119, 1984.
206. TACHÉ, Y., AND M. GUNION. Central nervous system action of bombesin to inhibit gastric acid secretion. *Life Sci.* 37: 115–123, 1985.
207. TACHÉ, Y., W. MARKI, J. RIVIER, W. VALE, AND M. BROWN. Central nervous system inhibition of gastric secretion in the rat by gastrin-releasing peptide, a mammalian bombesin. *Gastroenterology* 81: 298–302, 1981.
208. TACHÉ, Y., W. VALE, AND M. BROWN. Thyrotropin-releasing hormone: central nervous system action to stimulate gastric acid secretion. *Nature Lond.* 287: 149–151, 1980.
209. TACHÉ, Y., W. VALE, J. RIVIER, AND M. BROWN. Brain regulation of gastric secretion in rats by neurogastrointestinal peptides. *Peptides Fayetteville* 2: 51–55, 1981.
210. TACHÉ, Y., W. VALE, J. RIVIER, AND M. BROWN. Is somatostatin or somatostatin-like peptide involved in central nervous system control of gastric secretion? *Regul. Pept.* 1: 307–315, 1981.
211. TATEMOTO, K. Isolation and characterization of peptide YY (PYY), a candidate gut hormone that inhibits pancreatic exocrine secretion. *Proc. Natl. Acad. Sci. USA* 79: 2514–2518, 1982.
212. TEPPERMAN, B. L., AND M. D. EVERED. Gastrin injected into the lateral hypothalamus stimulates secretion of gastric acid in rats. *Science Wash. DC* 209: 1142–1143, 1980.
213. TEPPERMAN, B. L., J. H. WALSH, AND R. M. PRESHAW. Effect of antral denervation on gastrin release by sham feeding and insulin hypoglycemia in dogs. *Gastroenterology* 63: 973–980, 1972.
214. THOMAS, W. E. G. The kinetics of somatostatin inhibition of gastric acid secretion. *Regul. Pept.* 1: 245–251, 1981.
215. THOMPSON, J. C. Candidate hormones of the gut. XVII. Antral chalone. *Gastroenterology* 67: 752–754, 1974.
216. THOMPSON, J. C., W. S. LOWDER, J. T. PEURIFOY, J. S. SWIERCZEK, AND P. L. RAYFORD. Effect of selective proximal vagotomy and truncal vagotomy on gastric acid and serum gastrin responses to a meal in duodenal ulcer patients. *Ann. Surg.* 188: 431–438, 1978.
217. THOMPSON, J. C., AND J. S. SWIERCZEK. Acid and endocrine responses to meals varying in pH in normal and duodenal ulcer subjects. *Ann. Surg.* 186: 541–548, 1977.
218. THOMPSON, J. C., AND I. WIENER. Evaluation of surgical treatment of duodenal ulcer: short- and long-term effects. *Clin. Gastroenterol.* 13: 569–600, 1984.
219. TYLER, M. J., D. J. SHEARMAN, R. FRANCO, P. O'BRIEN, R. F. SEAMARK, AND R. KELLY. Inhibition of gastric acid secretion in gastric brooding frog, *Rheobatrachus silus*. *Science Wash. DC* 220: 609–610, 1983.
220. UVNÄS-MOBERG, K., M. GOINY, L. E. BLOMQUIST, AND C. E. ELWIN. Release of VIP-like immunoreactivity in response to dopamine agonists, insulin hypoglycemia and feeding in conscious dogs. *Peptides Fayetteville* 399–402, 1984.
221. UVNÄS-WALLENSTEN, K., S. EFENDIC, C. JOHANSSON, L. SJODIN, AND P. D. CRANWELL. Effect of intraantral and intrabulbar pH on somatostatin-like immunoreactivity in peripheral venous blood of conscious dogs. The possible function of somatostatin as an inhibitory hormone of gastric acid secretion and its possible identity with bulbagastrone and antral chalone. *Acta Physiol. Scand.* 111: 397–408, 1981.
222. VAGNE, M., S. J. KONTUREK, AND J. A. CHAYVIALLE. Effect

223. VALE, W., J. SPIESS, C. RIVIER, AND J. RIVIER. Characterization of 41-residue ovine hypothalamic peptide that stimulates secretion of corticotropin and beta-endorphin. *Science Wash. DC* 213: 1394-1397, 1981.
224. VARNER, A. A., J. I. ISENBERG, J. D. ELASHOFF, C. B. H. W. LAMERS, V. MAXWELL, AND A. A. SHULKES. Effect of intravenous lipid on gastric acid secretion stimulated by intravenous amino acids. *Gastroenterology* 79: 873-876, 1979.
225. VARNER, A. A., I. M. MODLIN, AND J. H. WALSH. High potency of bombesin for stimulation of human gastrin release and gastric secretion. *Regul. Pept.* 1: 289-296, 1981.
226. VEBER, D. F., R. M. FREIDLINGER, D. S. PERLOW, W. J. PALEVEDA, JR., F. W. HOLLY, R. G. STRACHAN, R. F. NUTT, B. H. ARISON, C. HOMNICK, W. C. RANDALL, M. S. GLITZER, R. SAPERSTEIN, AND R. HIRSCHMANN. A potent cyclic hexapeptide analogue of somatostatin. *Nature Lond.* 295: 55-58, 1981.
227. VILLAR, H. V., H. R. FENDER, P. L. RAYFORD, AND J. C. THOMPSON. Suppression of gastrin release and gastric secretion by gastric inhibitory polypeptide (GIP) and vasoactive intestinal peptide (VIP). *Ann. Surg.* 184: 97-102, 1976.

of vasoactive intestinal peptide on gastric secretion in the cat. *Gastroenterology* 83: 250-255, 1982.

228. WALSH, J. H. Gastrointestinal hormones and peptides. In: *Physiology of the Gastrointestinal Tract*, edited by L. R. Johnson. New York: Raven, 1981, vol. 1, p. 59-144.
229. WALSH, J. H., C. T. RICHARSON, AND J. S. FORDTRAN. pH dependence of acid secretion and gastrin release in normal and ulcer patients. *J. Clin. Invest.* 57: 1125-1131, 1976.
230. WEBB, S., I. LEVY. J. A. WASS, A. LLORENS, E. PENMAN, R. CASAMITJANA, P. WU, J. GAYA, M. J. MARTINEZ, AND F. RIVIERA. Studies on the mechanisms of somatostatin release after insulin induced hypoglycemia in man. *Clin. Endocrinol.* 21: 667-675, 1984.
231. WHEELER, M. H. Inhibition of gastric secretion by pyloric antrum. *Gut* 15: 420-432, 1974.
232. WICKBOM, G., J. H. LANDOR, F. L. BUSHKIN, AND J. E. MCGUIGAN. Changes in canine gastric acid output and serum gastrin levels following massive small intestine resection. *Gastroenterology* 69: 448-452, 1975.
233. WILLIAMS, C. B., AND A. P. M. FORREST. Effect of antral acidification and acid secretion induced by meat and pentagastrin. *Gastroenterology* 57: 399-405, 1969.
234. WOODWARD, E. R., E. S. LYON, J. LANDOR, AND L. R. DRAGSTEDT. The physiology of the gastric antrum. *Gastroenterology* 27: 766-785, 1954.

CHAPTER 10

Electrophysiology of gastric ion transport

JEFFREY R. DEMAREST

TERRY E. MACHEN

Physiology-Anatomy Department, University of California, Berkeley, California

CHAPTER CONTENTS

Structure of Gastric Mucosa
Voltage, Current, Ionic Flux, and Resistance
Membrane Permeability and Transport Properties
  Cation and anion requirements for acid secretion
  Sodium absorption by surface cells and chloride secretion by oxyntic cells
  Potassium transport
  Bicarbonate secretion
  Cell membrane potentials and intracellular ion activities
    Surface cells in intact, isolated mucosa
    Isolated oxyntic cells
    Measurements from oxyntic cells in intact mucosa
    Changes in oxyntic cells during stimulation
  Equivalent circuit analysis with microelectrodes
    General comments
    Consequences of multiple cell types
    Resistance of gland lumina
    Estimating shunt resistance
  Equivalent circuit analysis from impedance measurements
  Segregation of transport functions and associated shunt pathways
  Electroneutral hydrogen and electrogenic chloride transport by oxyntic cells
Summary

THIS CHAPTER REVIEWS electrophysiological experiments used to gain insights into the mechanisms by which H^+, K^+, Na^+, Cl^-, and HCO_3^- are transported across the mucosal and serosal membranes of the different cell types of the stomach. The nature of the paracellular shunt (i.e., the transepithelial pathway between the cells through the tight junctions and lateral spaces) is also considered, because this pathway is obviously important for an epithelium that must withstand a $\geq 10^6$ pH gradient between its lumen and the blood. We also consider the role that the gross structure of the epithelium may play in its overall function: how do membrane transport and epithelial structure interact? The final section integrates the information from electrophysiological measurements into a description of the ion-transport mechanisms found in the gastric mucosa.

STRUCTURE OF GASTRIC MUCOSA

By way of introduction we review briefly the gross and microscopic structure of the gastric mucosa and then some of the ultrastructural changes occurring at the apical membrane of the oxyntic (acid-secreting) cells during the stimulation of acid secretion (see also the chapter by Forte and Soll in this *Handbook*).

The stomach is folded at many levels. At a gross level the whole epithelium is thrown into elaborate folds (rugae) that increase the nominal surface area several times. The epithelium is indented into crypts, which split further into four or five gastric glands. The second-level folding, represented by the crypts, further increases the surface area [by another factor of 13 in the frog mucosa (48)]. The total length of the glands is ~300 μm, and their diameter may vary, depending on the depth within the tissue and perhaps the secretory state of the tissue. Mucus-secreting surface epithelial cells (SEC) cover the surface of the areas of the stomach in direct contact with the luminal contents and also project part way down the crypts. Enzyme and acid secretion take place in the glands, in separate chief cells and oxyntic cells (OC) in mammals but together in a single cell type in nonmammalian vertebrates.

A third level of folding is provided within the OC. In the resting state the OC are covered with short microvilli; the apical cytoplasm is filled with membrane-bound vesicles and tubules. The incorporation of these membranes into the apical cell membrane on stimulation leads to the development of the secretory canaliculus in mammalian OC and a large increase in apical membrane area in nonmammals. The overall effect is to cause an increase in OC apical membrane area of 6- to 10-fold. The actual area of apical membrane exceeds the nominal smooth surface area by a factor of at least several hundred.

VOLTAGE, CURRENT, IONIC FLUX, AND RESISTANCE

The gastric mucosa of most vertebrate species maintains a transepithelial electrical potential difference

(V_{ms}), with the mucosal side negative to the serosal side by -10 to -60 mV (see ref. 30). This potential is generated by the active transport of monovalent ions across a transepithelial resistance that varies widely, depending on the species and the secretory state of the tissue [e.g., from 50 $\Omega \cdot cm^2$ in the actively secreting piglet gastric mucosa (38) to >1,000 $\Omega \cdot cm^2$ in the resting *Necturus* gastric mucosa (22)].

The absorption of Na^+ and secretion of Cl^- and H^+ across the short-circuited piglet gastric mucosa accounts for most of the mucosally directed negative short-circuit current, I_{sc} (38). There may be other active transport processes (the most likely candidates are K^+ and HCO_3^-) that contribute to I_{sc} and account for the small discrepancy between net ionic transport and the I_{sc}. The frog gastric mucosa exhibits quite similar transport properties to the mammalian tissue, except that Na^+ absorption is absent, or at least quantitatively very small (50).

During stimulation of the OC with histamine or cAMP, a number of dramatic changes occur: transepithelial resistance (R_T) rapidly decreases, and H^+ and Cl^- secretion both increase (92, 93) concomitant with the increase in apical membrane surface area (60). These are the most characteristic changes associated with stimulation. The changes in V_{ms} are more variable. In some species V_{ms} increases [e.g., piglet (38) or *Necturus* (22)]. In others [e.g., frog (100)] it usually decreases. The direction of change in V_{ms} reflects the relative changes in active H^+ and Cl^- transport and the drop in R_T. The V_{ms} increases in those species in which I_{sc} increases; V_{ms} decreases when I_{sc} either remains constant or decreases during stimulation. There is general agreement that the drop in R_T is associated with the activation of the OC, but it is still unclear how much can be associated with any or all of the following possibilities.

1. An electrogenic H^+ pump is activated (100). According to this hypothesis, the H^+ pump is electrogenic and exhibits a significant conductance. Activation of such a pump could account for the drop in R_T associated with the stimulation of H^+ secretion. However, a substantial body of evidence indicates that the H^+ pump in the stomach is electrically neutral [i.e., 1:1 H^+-K^+ exchange; (57, 104)], and because the resistance of other active transport pumps is high, this electrogenic pump explanation seems unlikely.

2. The resistance of the apical membrane (R_a) of OC decreases because of an increased conductance for K^+ and/or Cl^-. The surface area of the OC mucosal membrane increases 6- to 10-fold during stimulation, and this "simple" increase in membrane area alone (with no change in the conductance per unit area of the membrane) would be expected to cause a drop in R_a and thereby a drop in R_T. Another possibility is that there is both an increase in membrane area and an increase in the number or activity of K^+- and/or Cl^--conductance channels per unit area of inserted membrane. Wolosin and Forte (143) found that apical membrane vesicles isolated from stimulated OC contain K^+ and Cl^- conductances, which are absent from microsomes from resting OC; this finding supports this possibility [see the chapter by Forte and Soll in this *Handbook*; (14, 143)]. Recent impedance analysis experiments (10, 11) support the possibility that changes in membrane area alone are adequate to account for the drop in R_T. However, there are some problems with the impedance analysis technique that need to be resolved before definite conclusions can be reached (see *Equivalent Circuit Analysis From Impedance Measurements*, p. 197). It is clear, however, that the large increase in apical membrane area of OC during stimulation is at least one of the causes of the drop in R_T. Whether there are increases in both membrane area and membrane conductance (opening K^+ and/or Cl^- channels) per unit membrane area remains to be determined.

3. A drop in resistance of the basolateral membrane (R_b) of OC could also cause a drop in R_T. Until recently this possibility had not been considered. However, microelectrode experiments on OC in the intact *Necturus* gastric mucosa have shown that during histamine stimulation the ratio of apical-to-basolateral membrane resistance (R_a/R_b) of these cells did not change (22). This indicates that both R_a and R_b have decreased during stimulation.

4. Changes in the diameter of the long and narrow gastric gland lumen could cause changes in R_T. Inhibition of acid secretion with thiocyanate or omeprazole causes R_T to increase, but this increase is prevented if the osmolarity of the mucosal solution is increased (49, 77, 87, 88). A possible explanation for these data is that inhibition of water flow through the glands (by inhibiting H^+ secretion) causes the glands to collapse, and R_T increases; subsequent treatment with hyperosmotic luminal solution restores water flux, presumably opens gland lumens, and causes R_T to decrease. Thus it is further possible that during normal gastric stimulation the gland lumens open because of the increased water flow through them, and this opening contributes to the drop in R_T. Direct experimental tests of these calculations and assumptions are needed to clarify this question, which is discussed again in ESTIMATING SHUNT RESISTANCE, p. 196.

MEMBRANE PERMEABILITY AND TRANSPORT PROPERTIES

Cation and Anion Requirements for Acid Secretion

The secretion of HCl by gastric mucosa is virtually independent of the composition of the mucosal solution (38, 63). However, this secretion is critically dependent on the composition of the serosal solution. Treatment of in vitro frog gastric mucosa with Na^+-

free Ringer's solution usually, but not always (106), causes H$^+$ secretion to be reduced to zero (65). The K$^+$-free serosal solutions also cause H$^+$ and Cl$^-$ transport to decrease slowly to zero. In the intact frog gastric mucosa, the effects of K$^+$-free serosal solution can be reversed by increasing [K$^+$] in the mucosal solution (19). The effects of Na$^+$-free solutions on H$^+$ secretion have been proposed to be the result of inhibition of the serosal Na$^+$-K$^+$-ATPase and consequent reduction of cellular [K$^+$], which may reduce the activity of the H$^+$-K$^+$-ATPase (53). The K$^+$-free solutions appear to exert the additional effects of increasing cellular [Na$^+$], which is also inhibitory to the H$^+$-K$^+$-ATPase (53).

The establishment of H$^+$ gradients across the epithelium requires not only functioning cellular transport mechanisms but also intact junctional complexes. Extracellular Ca^{2+} plays an important role in the maintenance of tight junctional integrity. The threshold for this effect is between 10^{-5} and 10^{-4} M. When [Ca^{2+}] < 10^{-5} M, the tight junctions are disrupted (116), R_T decreases to zero, and transepithelial fluxes of extracellular space markers increase dramatically (50, 116).

Both Cl$^-$ and HCO$_3^-$ are also required in the serosal solution to maintain maximum rates of H$^+$ secretion, though specific experimental maneuvers can overcome these requirements. For example, when serosal Cl$^-$ is replaced with an impermeant anion such as gluconate or isethionate, H$^+$ secretion is reduced ~75% in amphibians (37, 38) and 100% in mammals (38). This Cl$^-$ requirement is not simply for maintenance of electroneutrality, because voltage clamping V_{ms} to zero does not eliminate the Cl$^-$ requirement (20, 100). In Cl$^-$-free solutions, H$^+$ secretion is increased back toward control rates when luminal [K$^+$] is increased to 79 mM (20, 101). Substitution of Cl$^-$ with other halides (24) or halidelike anions (62) allows acid secretion to proceed at varying rates. Removal of HCO$_3^-$ (and also CO$_2$ gassing) and replacement with another buffer reduce acid secretion 20% (108, 122). Normal rates of acid secretion can be restored if the pH of the serosal solution is reduced to pH ~4.5 (108, 123).

Sodium Absorption by Surface Cells and Chloride Secretion by Oxyntic Cells

The gastric mucosa secretes Cl$^-$ in excess of H$^+$, and this Cl$^-$ secretion is the major determinant of the lumen-negative V_{ms} observed in the gastric mucosa (50). Resting tissues (i.e., H$^+$ secretion = 0) secrete Cl$^-$ actively and maintain a lumen-negative V_{ms} (38, 65). Although a large portion of the I_{sc} is due to net Cl$^-$ secretion (i.e., when V_{ms} = 0), at open circuit (when there is a negative transepithelial potential) net Cl$^-$ secretion is very small (38, 68). The rate of Cl$^-$ transport is directly related to the transepithelial conductance (63). The Cl$^-$ secretion depends on the presence of Na$^+$ (65), K$^+$ (67), and HCO$^-$ (71) in the serosal solution. However, in HCO$_3^-$-free solutions, Cl$^-$ secretion can be maintained at control rates if the serosal pH is increased from pH 7.2–7.4 to pH 9.0 (71).

In addition to its secretory properties, the gastric mucosa also actively absorbs Na$^+$ at a significant rate. Active Na$^+$ absorption has now been identified in mammalian stomach both in vivo (2, 67) and in vitro (52, 67, 130) and in the in vitro lizard (43) and *Necturus* (120) gastric mucosa. Thus it appears that the frog gastric mucosa is an exception to the general rule that the gastric mucosa actively absorbs Na$^+$.

The active Na$^+$ absorption by gastric mucosa shares many properties in common with other "tight" (e.g., rabbit urinary bladder or frog skin) or "moderately tight" (e.g., rabbit colon) epithelia. *1*) Gastric Na$^+$ transport is blocked by adding amiloride (10^{-5} M) to the mucosal solution (25, 67, 130) or ouabain (56, 67, 130) to the serosal solution. *2*) Absorption of Na$^+$ is inhibited by acidifying the luminal solution below pH 6, with complete inhibition at pH 2–3 (67). From an analysis of the I_{sc} versus pH of the luminal solution, an apparent pK 4.0 was found (67).

Given the similarities between Na$^+$ transport in the stomach and that in other tight epithelia, it is interesting that Palmer (84) has shown that the amiloride-inhibitable Na$^+$ channel of toad bladder is quite permeable to H$^+$. If the Na$^+$ channels of the stomach are like those of the toad bladder, then penetration of H$^+$ through the Na$^+$ channels could be an important site of backdiffusion of H$^+$. In other words, are the SEC particularly susceptible to ulceration because of the presence of these Na$^+$ channels in the luminal membrane? Investigation of the interactions between H$^+$ and Na$^+$ in SEC could yield important information.

Although it has been well established that OC alone are responsible for acid secretion by the stomach, the assignment of responsibility for the transport of other ions to specific cell types has remained uncertain. The transport functions of the various cell types that make up the gastric mucosa have in the past been inferred from developmental studies. However, direct measurements of Na$^+$ absorption and Cl$^-$ secretion by the two principle cell types, OC and SEC of *Necturus* gastric mucosa, have recently been made with the vibrating probe (25).

The vibrating probe is basically a metal microelectrode that can be used to measure local extracellular ionic currents. The probe is vibrated in the medium directly above an identified cell, and local current flow causes the probe to measure slightly different potentials at the extremes of its excursion. This potential difference is processed with a phase-sensitive detector such as a lock-in amplifier and calibrated for the determination of current density. An example of the measurement of the current arising from an SEC in *Necturus* fundus is shown in Figure 1. The height of the probe above the epithelium and open- or short-

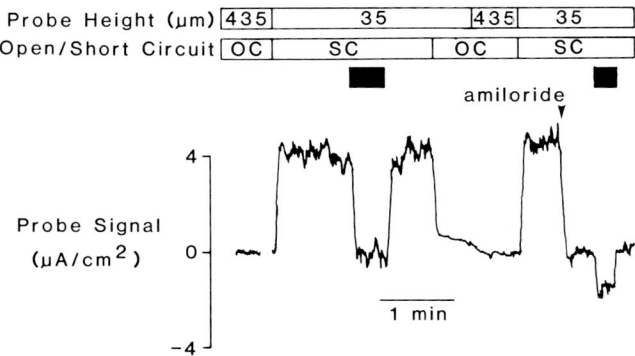

FIG. 1. Vibrating-probe measurement of current emerging from a surface epithelial cell (SEC) during control conditions and during amiloride (10^{-6} M, mucosal) treatment. Height of probe above epithelium (measured to midpoint of probe's excursion) and open-circuit (OC) or short-circuit (SC) conditions are indicated across top. Black bars, time periods when transepithelial voltage was clamped to −20 mV, mucosal negative. Resistance of this SEC (including the shunt pathway) was ~5,000 $\Omega \cdot cm^2$ (20 mV $\cdot$ 4 $\mu A^{-1} \cdot cm^{-2}$) during control conditions and 15,000 $\Omega \cdot cm^2$ (20 mV $\cdot$ 1.33 $\mu A^{-1} \cdot cm^{-2}$) during amiloride treatment. [From Demarest et al. (25).]

circuit conditions are indicated at the top of the figure. Zero probe current was determined at 435 μm above the epithelium in the short-circuited state when a net ionic current, I_{sc}, was flowing across the epithelium. The black bars in the figure indicate periods when the epithelium was clamped to −20 mV for the determination of the resistance of the surface epithelium (see ESTIMATING SHUNT RESISTANCE, p. 196). Under these conditions the probe measured a current of ~4 $\mu A/cm^2$ for this SEC, and after the addition of amiloride the probe-measured current was promptly reduced to zero, indicating that the current measured above this cell was due exclusively to Na^+ transport. Reduction of the current measured above the SEC to zero was observed only in a few cells; usually a small residual current remained after addition of amiloride. However, the amiloride-induced change in the current measured above the SEC could account for the change in the I_{sc} measured across the whole epithelium (25). Because amiloride abolishes the contribution of Na^+ transport to the I_{sc}, these results indicate that Na^+ is transported exclusively by SEC in Necturus.

In amiloride-treated tissues, where the I_{sc} equals the net Cl^- secretion, vibrating-probe measurements showed that all of the I_{sc} could be accounted for by current arising from crypts and that residual current measured over SEC in the presence of amiloride was due to contamination from nearby crypts (25). Thus all Cl^- is secreted by OC in the crypts. No similar studies have been performed on the mammalian fundus, but the experiments on primary cultures of monolayers of dog chief cells show that they are capable of net transepithelial Na^+ transport (107).

Potassium Transport

Despite abundant evidence of a role for K^+ in the H^+ secretory process, little is known about K^+ transport by the intact stomach. Most early studies (12, 44, 50, 69) indicated that transepithelial K^+ transport was purely a passive process. However, work on isolated membrane vesicles has shown that K^+ is actively transported by the H^+-K^+-ATPase (57, 104). Recent studies have also shown that inhibition of H^+ secretion by p-chloromercuribenzoate or the specific H^+-K^+-ATPase antagonist omeprazole (136, 137) caused increases in K^+ secretion. In these studies it was not determined whether the enhanced K^+ flux was active or solely an increased passive flux due to the increased V_{ms} induced by these agents.

Reenstra et al. (88), unlike the previous workers, found that the stimulated frog gastric mucosa exhibited a small but significant net K^+ absorption of 0.07 $\mu eq \cdot cm^{-2} \cdot h^{-1}$ under short-circuited conditions that was increased to 0.3 $\mu eq \cdot cm^{-2} \cdot h^{-1}$ when ouabain was added to the serosal solution. Subsequent treatment with mucosal omeprazole reduced the net K^+ flux to zero. Omeprazole treatment alone caused K^+ transport to reverse from an absorption of 0.08 $\mu eq \cdot cm^{-2} \cdot h^{-1}$ to a secretion of 0.09 $\mu eq \cdot cm^{-2} \cdot h^{-1}$. These investigators proposed that during ouabain treatment net K^+ absorption was driven by the accumulation of K^+ via the H^+-K^+-ATPase followed by the leak of K^+ out of the cells through the serosal K^+ channels; during omeprazole treatment net K^+ secretion was driven by the accumulation of K^+ by the Na^+-K^+-ATPase and the leak of K^+ through the mucosal K^+ channels.

An unexpected finding was that in omeprazole-treated tissues, net K^+ secretion was still only a small fraction of the rate of H^+ secretion. In the steady state the rate of K^+ movement from cell to lumen should equal the rate of H^+ secretion. Reenstra et al. (88) have argued that this finding means that the rate-limiting step for K^+ movement across the epithelium is the diffusion of K^+ down the gland lumen. The lack of net K^+ secretion in the control state is postulated to be due to the efficient recycling of K^+ from the gland lumen into the OC via the H^+-K^+-ATPase. Consistent with this idea is the fact that during omeprazole treatment, making the mucosal solution hyperosmotic (by adding 100 mM sucrose to the normal Ringer's solution to open gland lumina) caused net K^+ secretion to increase. However, K^+ secretion in this situation was still only 2% of control rates of H^+ secretion. An alternative interpretation is that omeprazole inhibition of the H^+-K^+ pump may cause a decrease in K^+ conductance at the apical membrane of OC (23). It will be important to determine the degree to which the gland lumen is rate limiting for the transepithelial movement of K^+ (see RESISTANCE OF GLAND LUMINA, p. 196).

Bicarbonate Secretion

It has been inferred from studies on a variety of intact animals that the stomach is capable of secreting an alkaline fluid, although the mechanism was unknown (70, 81). The presence of this alkaline secretion was first reported in the isolated amphibian antrum (35); it has been subsequently observed in fundus of several other species (33, 34, 127). It is generally accepted that the SEC are the source of alkaline secretion. This is based on the indirect argument that the secretion is present in both the antrum and the fundus, yet the antrum contains no OC. However, the antrum does contain glands, and more experiments are required to confirm the cell type that contributes HCO_3^- secretion. It has been difficult to assign specific electrophysiological responses to HCO_3^- transport because the rate of HCO_3^- transport is low and because HCO_3^- and Cl^- transport appear to be closely interrelated (127). This alkaline secretion has recently attracted a great deal of interest, and this topic is covered in detail in the chapter by Flemström and Garner in this Handbook.

Cell Membrane Potentials and Intracellular Ion Activities

SURFACE CELLS IN INTACT, ISOLATED MUCOSA. The basolateral membrane potentials (V_{cs}) of SEC in resting mucosa range between −41 and −67 mV with respect to the serosal bathing solution (Table 1). The lower values are from studies published in the 1960s and 1970s; the higher values reported in the more recent studies probably reflect improvements in microelectrode techniques. In all cases the membrane potential profile across the cells from the mucosal to serosal side has the shape of a well: both membrane potentials (V_{mc} and V_{cs}) are negative inside the cell with respect to the adjacent bathing solutions. A similar potential profile has been found in the dog gastric mucosa in vivo (142). In the *Necturus* fundus and antrum, which both actively absorb Na^+, the luminal addition of amiloride causes large hyperpolarizations of both V_{mc} and V_{cs} and greatly increases the relative resistance of the apical membrane (25, 35, 43). Despite these pronounced effects on the SEC, V_{ms} is only slightly decreased and R_T only slightly increased by amiloride. These results indicate that although there is a large Na^+ conductance in the mucosal membrane of the SEC of both fundus and antrum, the Na^+ conductive pathway makes a relatively small contribution to their transepithelial electrical properties. In the frog prolonged (50-min) removal of serosal Na^+ depolarized V_{cs} of SEC by 25 mV (16).

The V_{mc} in frog SEC does not change significantly when the mucosal concentration of K^+, $[K^+]_m$, is increased, indicating that there is no K^+ conductance at this membrane (28a). In *Necturus*, V_{cs} is depolarized by increasing serosal $[K^+]$, $[K^+]_s$, consistent with a substantial K^+ conductance at this membrane (22, 125). In the frog, intracellular $[K^+]$, $[K^+]_i$, is above electrochemical equilibrium with respect to V_{cs} (109), consistent with an active accumulation by the Na^+-K^+-ATPase located in the basolateral membrane (47).

Investigations of fundic SEC of both *Necturus* (68, 110) and frog (16) with ion-selective electrodes have shown that $[Cl^-]_i$ is also maintained at a concentration above that predicted for electrochemical equilibrium with respect to V_{cs} but approximately at equilibrium concentration with respect to V_{mc}. Thus it appears that Cl^- is actively accumulated by SEC. However, there is little unambiguous information concerning the mechanisms by which Cl^- moves across the apical and basolateral membranes of SEC. For example, lowering $[Cl^-]_m$ has been reported to cause a large reduction of $[Cl^-]_i$ of SEC in *Necturus*, consistent with a large, passive Cl^- permeability at this membrane (68). However, a more recent study has shown that low $[Cl^-]_m$ has little effect on $[Cl^-]_i$ (110). The reason for this discrepancy remains unclear. However, several studies have shown that there is no effect of low $[Cl^-]_m$ on V_{mc} (25, 110) or on the relative resistance of the apical membrane of SEC (25), indicating that there is no Cl^- conductance in the apical membranes of SEC.

Experiments to determine mechanisms by which

TABLE 1. *Electrical Properties and Intracellular Ion Activities of Surface Cells in Gastric Mucosa*

Species	V_{ms}	V_{mc}, mV	V_{cs}	R_a/R_b	R_T, $\Omega \cdot cm^2$	$[K^+]_i$, mM	$[Cl^-]_i$, mM	Ref.
Rana pipiens								
Fundus	−29	20	−49	0.8	508	89*		134, 135
Rana esculenta								
Fundus	−26	41	−67	9.7	316	99	15	16, 109
Necturus maculosus								
Fundus	−20	30	−50	1.2	701			125
	−18	34	−52	4.5	230		29	68
	−18	39	−57	2.7			26	110
	−17	33	−50	3.6	797			22
Antrum	−11	36	−47	2.8	828			35
	−7	34	−41	1.3	1,424			124
	−10	45	−55	2.8	434			42

V_{ms}, transepithelial voltage; V_{mc}, voltage across the apical membrane with respect to the mucosal solution; V_{cs}, voltage across the basolateral membrane with respect to the serosal solution; R_a/R_b, ratio of apical to basolateral membrane resistances; R_T, transepithelial resistance; $[K^+]_i$, intracellular $[K^+]$; $[Cl^-]_i$, intracellular $[Cl^-]$. * Calculated from concentration assuming an activity coefficient of 0.77. All other activities were measured with ion-selective microelectrodes.

Cl⁻ moves across the serosal membrane have also been ambiguous. In SEC of the frog gastric mucosa, V_{cs} and $[Cl^-]_i$ are initially unchanged by removing Cl⁻ from the serosal solution. However, long-term incubation (>30 min) with a Cl⁻-free solution results in a 7-mV depolarization of V_{cs} and a decrease of $[Cl^-]_i$ to 7 mM (16). Although the authors interpreted these changes as evidence for the existence of a basolateral Cl⁻ conductance, the time course of the observed change in V_{cs} was much slower than would be expected if this membrane has a significant Cl⁻ conductance. Furthermore, $[Cl^-]_i$ appears to be independent of V_{cs}, because it is unaffected by hyperpolarization of V_{cs} due to transepithelial voltage clamping or by the removal of serosal Na⁺, which eventually results in a significant depolarization of V_{cs} without affecting $[Cl^-]_i$ (16).

The effects of stimulating acid secretion by OC on the membrane potentials of *Necturus* SEC have been examined in two studies (22, 121). The results were in general agreement, and those of the former study are shown in Table 2. Stimulation caused a 7- to 10-mV increase in V_{ms}; V_{mc} depolarized, and V_{cs} hyperpolarized. These data are consistent with a purely passive electrical response by the SEC to the effects of histamine on the OC. There is no evidence for a direct effect of histamine on the electrical properties of the SEC, although recent microelectrode measurements have shown that $[Cl^-]_i$ in SEC decreases during histamine stimulation (17). This effect has been interpreted as indicating that $[Cl^-]_i$ of OC decreases during stimulation, and Cl⁻ from SEC then diffuses in the plane of the epithelium down a concentration gradient, from cell to cell, through gap junctions, into OC (17). Direct measurements of $[Cl^-]_i$ in OC are required to confirm this finding. Flemström (33) has shown that SEC of *Necturus* antrum are unaffected by histamine.

It can be concluded that SEC have a large apical membrane Na⁺ conductance and a large basolateral membrane K⁺ conductance. Both apical and basolateral membranes appear to have a low-Cl⁻ conductance. Potassium is actively accumulated, presumably by the Na⁺-K⁺-ATPase in the basolateral membrane. Chloride is also actively accumulated by SEC. The specific mechanism(s) by which Cl⁻ moves across the cell membranes is (are) unknown. During histamine stimulation of OC, changes of V_{mc} and V_{cs} in SEC occur solely because they are electrically coupled to the OC through the paracellular shunt.

ISOLATED OXYNTIC CELLS. Because of the inaccessibility of OC in the intact epithelium, most of the electrical measurements on OC have been obtained from isolated cells or gastric glands. Table 3 summarizes the membrane potential (V_m), intracellular ion activities, and pH values of these cells. The precise relationship of V_m to the apical and basolateral membrane potentials measured in cells in the intact epithelium is uncertain, although it may be close to the potential that would be measured in the intact mucosa under short-circuit conditions (i.e., a mucosa bathed with identical solutions on both sides and clamped to a transepithelial potential of 0 mV by passing current across it from an external source). However, because it is a composite of the potentials of the two cell membranes with respect to a common bathing solution, changes in V_m of an isolated cell in response to alterations of the bathing solution cannot be ascribed to a specific membrane. The V_m ranges from about −5 mV (cell interior negative), measured in cells in enzymatically isolated rabbit gastric glands, to −44 mV, measured in OC isolated from *Necturus*. Isolated *Necturus* OC depolarize by ~25 mV per decade change in external $[K^+]$ (1), which is similar to the depolarization observed in OC in the intact mucosa when the $[K^+]$ is increased only on the serosal side of the tissue (23).

Microelectrode impalements of OC and chief cells in isolated gastric glands from the rabbit have been reported by two groups (Table 3), neither of which detected any differences between the electrical potentials of the two cell types. Although Schettino et al. (111) reported a somewhat larger potential of −27 mV for cells in microdissected rabbit glands, the difference between this value and that obtained from collagenase-treated glands of −6 mV (51) may be more apparent than real, because Schettino et al. employed more restrictive criteria for the acceptability of impalements. Schettino et al. (111) reported that Cl⁻-free solutions caused V_m to depolarize transiently and then to hyperpolarize in the steady state by up to 50 mV; a 10-fold increase in solution $[K^+]$ caused V_m to depolarize by 9 mV. In contrast, Kafoglis et al. (51) reported the surprising result that V_m was unaffected by large changes of $[Cl^-]$ or $[K^+]$.

Kafoglis et al. (51) also reported ion-sensitive microelectrode measurements of $[K^+]_i$ and intracellular pH (pH_i) in gland cells, which were not identified as to cell type. Stimulation of OC with histamine caused

TABLE 2. *Effects of Histamine Stimulation on Electrical Properties of Oxyntic and Surface Cells*

	V_{ms}	V_{mc}, mV	V_{cs}	R_a/R_b	R_T, $\Omega \cdot cm^2$
Oxyntic cells					
Resting	−26.6	19.3	−45.9	1.1	854
Stimulated	−32.9	26.3	−59.2	1.1	466
Δ	−6.7	7.0	−13.3	0.0	−388
P	<0.001	<0.001	<0.001	NS	<0.001
Surface cells					
Resting	−17.2	32.9	−50.2	3.6	797
Stimulated	−26.7	26.6	−53.6	3.4	498
Δ	−9.5	−6.3	−3.4	−0.2	−299
P	<0.001	<0.02	<0.06	NS	<0.001

Significance levels are for the difference (Δ) between resting and stimulated according to the *t* test for unpaired observations. V_{ms} was measured with respect to serosal (nutrient) solution. NS, not significant. [Data from Demarest and Machen (22).]

TABLE 3. *Membrane Potentials and Intracellular Ion Activities of Isolated Oxyntic Cells*

Species	V_m, mV	$[K^+]_i$, mM	pH_i	Ref.
Necturus				
Resting	-44 ± 3	52 ± 2[a]		1
Bullfrog				
Resting			7.2 ± 0.1[b]	76
Stimulated			7.3 ± 0.1[b]	
Rat				
Resting	-20 ± 5			83
Stimulated	-26 ± 3			
Rabbit				
Resting	-6 ± 1	42 ± 2[e]	7.1 ± 0.1[c]	51
Stimulated	-4 ± 1	53 ± 3[c]	6.9 ± 0.1[c]	
Resting	-26			111
Resting			7.16 ± 0.01[d]	
Stimulated			7.21 ± 0.01[d]	e

V_m, membrane potential. [a] Calculated from the concentration assuming an activity coefficient of 0.77. [b] Measured with [^{14}C]-5,5-dimethyl-2,4-oxazolidinedione (DMO). [c] Measured with ion-selective microelectrodes. [d] Measured with the fluorescent, pH-sensitive dye bis-carboxyethylcarboxyfluorescein in single parietal cells of intact gastric glands. Stimulation was achieved by using histamine and/or dibutyryl cAMP and/or isobutylmethylxanthine and/or forskolin. pH_i increased by an average of 0.05 pH units during stimulation. [e] A. M. Paradiso and T. E. Machen, unpublished observations.

a rise in $[K^+]$ (Table 3), which was blocked by picoprazole, a benzimidazole inhibitor of the H^+-K^+-ATPase (138). The pH_i of gland cells was ~7.1 under resting conditions, and it decreased to 6.9 during histamine stimulation. In contrast, measurements of pH_i of individual OC in the same preparation with an intracellularly trapped fluorescent pH indicator have indicated that pH_i increases by ≤ 0.1 pH units after stimulation (A. M. Paradiso and T. E. Machen, unpublished observations). A similar, small histamine-induced increase in pH_i, as measured by the distribution of the weak acid [^{14}C]5,5-dimethyl-2,4-oxazolidinedione (DMO), has been reported for isolated frog OC [Table 3; (76)]. Both of the microelectrode studies on the isolated rabbit glands appear to have suffered from many technical difficulties, as evidenced by the skewed distributions of V_m toward low values (51, 111). This important and difficult work needs to be repeated with further technical refinements before conclusions can be drawn about the electrical properties of the cells of the mammalian glands.

Membrane potentials have recently been reported from OC in primary cultures of rat fundic epithelium (83). After 2-4 days of culture the OC had a resting V_m of -20 mV. The V_m hyperpolarized transiently by an average of 18 mV in response to carbachol and somewhat less in response to histamine or gastrin. The V_m was weakly depolarized by increasing $[K^+]$ in the bath, 8 mV per decade change, but during the peak response to carbachol the response increased to 20 mV depolarization per decade, implying that the relative K^+ conductance of the cell membrane was increased by stimulation. It was not possible, however, to determine if this change occurred in the apical, basolateral, or both cell membranes.

MEASUREMENTS FROM OXYNTIC CELLS IN INTACT MUCOSA. Recently, OC have been impaled across their basolateral membranes by mounting the gastric mucosa serosal side up and dissecting away the muscle and connective tissue surrounding the gastric glands (15, 22). Figure 2 shows an example of a basolateral impalement of an OC in a stimulated *Necturus* gastric mucosa. Table 2 lists mean values of the transepithelial and cellular electrical properties (22). These data show that V_{mc} in the resting OC is oriented opposite to that previously reported (121, 134) and similar to that found in SEC; both cell types exhibit well-shaped potential profiles, not "stair steps" (121, 134). The mean V_{cs} measured in OC in the resting intact mucosa is not significantly different from the mean V_m measured in isolated *Necturus* OC (Table 2).

In the isolated mucosa, changes of $[K^+]_s$ cause V_{cs} to depolarize by 26–27 mV per decade (15, 23). These data indicated that the basolateral membranes of OC have a large K^+ conductance. The patch-clamp technique (131) has also been used on the serosal membranes of *Necturus* OC, and two types of K^+ channels have been identified: one channel was relatively voltage independent but could be activated by increases in cellular [cAMP]; a second channel could be activated by depolarization or increases in intracellular Ca^{2+}. Both Ca^{2+} and cAMP increase in response to the stimulation of acid secretion by either histamine or carbachol (7, 79), and their subsequent activation of K^+ channels may account for some of the stimulation-associated hyperpolarization of V_{cs} observed in OC in the intact tissue.

The presence of other conductive pathways in the basolateral membrane of OC is less clear. In early transepithelial studies, measurements of V_{ms} during changes of serosal ion concentrations were interpreted as indicating that the OC had significant basolateral Cl^- conductance (45) but no conductance for H^+, OH^-, or HCO_3^- (99). However, direct measurements of V_{cs} of OC in *Necturus* with microelectrodes during the reduction of $[Cl^-]_s$ show that the immediate effect (from 10 s to several minutes) is a hyperpolarization (J. R. Demarest and T. E. Machen, unpublished observations). This change in V_{cs} is in the direction opposite to that expected if there were a significant Cl^- conductance in the basolateral membrane of the OC. Furthermore, although changes in $[Na^+]_s$ were originally reported to have little effect on V_{ms} (45), more recent experiments showed that V_{ms} is affected by changes in $[Na^+]_s$ and an electrogenic Na^+-Cl^- cotransport system (in which net charge is transferred by the transport of Cl^- in excess of Na^+) has been proposed (6). Finally, an investigation with microelectrodes has shown for OC in frog gastric mucosa that decreasing either $[Na^+]_s$ or $[HCO_3^-]_s$ depolarized V_{cs}, these depolarizations were blocked in the presence of

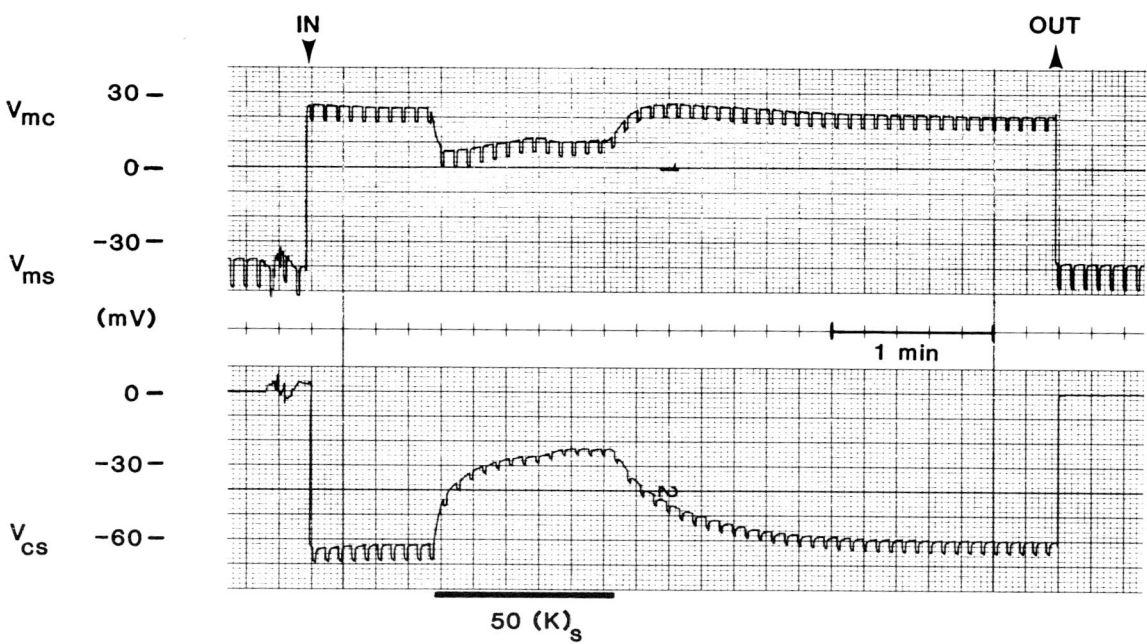

FIG. 2. Example of a basolateral impalement of an oxyntic cell (OC) in a stimulated (0.1 mM histamine) mucosa. Record starts with microelectrode in serosal bath registering a potential of 0 mV (*lower trace*). IN arrow, OC was impaled, and basolateral membrane potential (V_{cs}) of -62 mV was recorded. Upper trace, transepithelial potential (V_{ms}) of -38 mV prior to impalement and apical membrane potential (V_{mc}) (= $V_{ms} - V_{cs}$) of 24 mV after impalement. Repeated *downward deflections* in *traces* were due to transepithelial current pulses (amplitude = 20 µA/cm^2; duration = 1 s; frequency = 12/min) used to determine transepithelial resistance, $R_T = \Delta V_{ms} \cdot 20$ µA$^{-1} \cdot$ cm^{-2}, and ratio of apical to basolateral membrane resistances, $R_a/R_b = \Delta V_{mc}/\Delta V_{cs}$. At ~45 s after impalement, serosal solution [K$^+$] was raised from 5 to 50 mM (*black bar*), causing a rapid and reversible depolarization of both membrane potentials and a decrease in relative resistance of the basolateral membrane, i.e., an increase in R_a/R_b. After return to 5 mM serosal [K$^+$], microelectrode was withdrawn from cell at OUT arrow. [From Demarest and Machen (22).]

serosal 4-acetamido-4'-isothiocyanatostilbene-2,2'-disulfonic acid (SITS), and the effect of [HCO$_3^-$] was dependent on the presence of Na$^+$ in the serosal solution (15). Based on the changes in V_{cs} alone, an electrogenic Na$^+$-HCO$_3^-$ cotransport similar to that reported for the proximal tubule (3) was proposed (15). Despite the large changes observed in V_{cs}, V_{ms} was relatively unaffected, and there appeared to be no alteration in the relative resistance of the basolateral membrane, as judged by the absence of any significant effects on the ratio of the membrane resistances (15). The present picture of the basolateral membrane is thus one in which K$^+$ conductance predominates and there is no Cl$^-$ conductance. Although both Na$^+$ and HCO$_3^-$ appear to be important determinants of conductance, it is not clear whether these ions themselves are conductive. The identity of an Na$^+$-HCO$_3^-$ cotransporter also needs clarification.

The conductance properties of the apical membranes of OC in the intact epithelium are also poorly characterized, because the appropriate direct measurements have not been performed. Based on the results of experiments on isolated membrane vesicles, K$^+$ and Cl$^-$ conductances should dominate in the stimulated state (14, 143). In the intact epithelium, changes in [K$^+$]$_m$ affect V_{ms} only when Cl$^-$ has been removed from the solutions bathing both sides of the tissue (37, 100, 145). Although V_{ms} is sensitive to [Cl$^-$]$_m$, their relationship is significantly less than expected for an ideally Cl$^-$-selective membrane; changes in mucosal Na$^+$, H$^+$, and HCO$_3^-$ do not affect V_{ms} (36, 45, 46, 63).

One technique that has been applied with some success is that of fluctuation-noise analysis (145). This technique measures the small, random fluctuations in I_{sc} during conditions that are designed to alter specifically the transport of one ionic species or another. Under the appropriate conditions it is possible to analyze the so-called power spectra in terms of the number of channels and their conductance and mean open times. Zeiske et al. (145) changed Cl$^-$ transport by altering [Cl$^-$]$_s$, stimulating with histamine, and adding SITS to the serosal solution. The overall level of noise changed during these treatments in ways consistent with the presence of Cl$^-$ channels in the luminal membrane of OC. The Cl$^-$-channel activity increased during stimulation.

In experiments performed to look for K$^+$-dependent noise in the luminal membranes, the clearest results were generated in stimulated tissues incubated with Cl⁻-free Ringer's solutions. Under these conditions, data consistent with K⁺ channels were observed when large gradients of [K⁺] were established across the epithelium or when Ba^{2+} was added to the mucosal solution. Overall the data indicated the presence of both Cl⁻ and K⁺ channels in the luminal membranes of OC. Van Driessche and Zeiske (133) recently reviewed the results of the application of noise-analysis techniques to epithelial membranes.

CHANGES IN OXYNTIC CELLS DURING STIMULATION. The time course of stimulation followed continuously in single OC of *Necturus* is illustrated in Figure 3, and the steady-state electrical properties are summarized in Table 2. After exposure to histamine, there is a significant hyperpolarization of V_{cs} and decrease in R_T by 10 min and a significant hyperpolarization of V_{ms} after 20 min. In the steady state V_{mc}, V_{cs}, and V_{ms} were hyperpolarized, and R_T was significantly reduced.

Homeostatic mechanisms would be expected to be well developed in OC for the regulation of pH_i, $[K^+]_i$, and $[Cl^-]_i$. Recent experiments employing an intracellularly trapped fluorescent pH-sensitive dye have shown that there is virtually no change in pH_i of the OC in isolated rabbit gastric glands upon their transition from the resting to the stimulated state (A. M. Paradiso and T. E. Machen, unpublished observations). This indicates that pH_i is closely controlled in these cells, which are extruding large and equal quantities of H⁺ and OH⁻ (HCO_3^-) across their apical and basolateral membranes in the stimulated state. Machen and Paradiso proposed (66) that the increased flux of base across the serosal membrane that occurs in the stimulated state is the result of the activation of the Cl⁻-HCO_3^- exchanger that has been demonstrated in this membrane (77, 87).

Electrophysiological studies of *Necturus* OC in the intact epithelium have shown that the ratio of the apical to basolateral membrane resistances (R_a/R_b) is not significantly affected by stimulation, while R_T falls by ~40% (Fig. 3). This suggests that stimulation results in simultaneous decreases in R_a and R_b of OC. A decrease in R_a would be expected from the large elaboration of the apical membrane during the stimulation of H⁺ secretion and the activation of K⁺- and Cl⁻-conductance pathways. However, the decrease in R_b

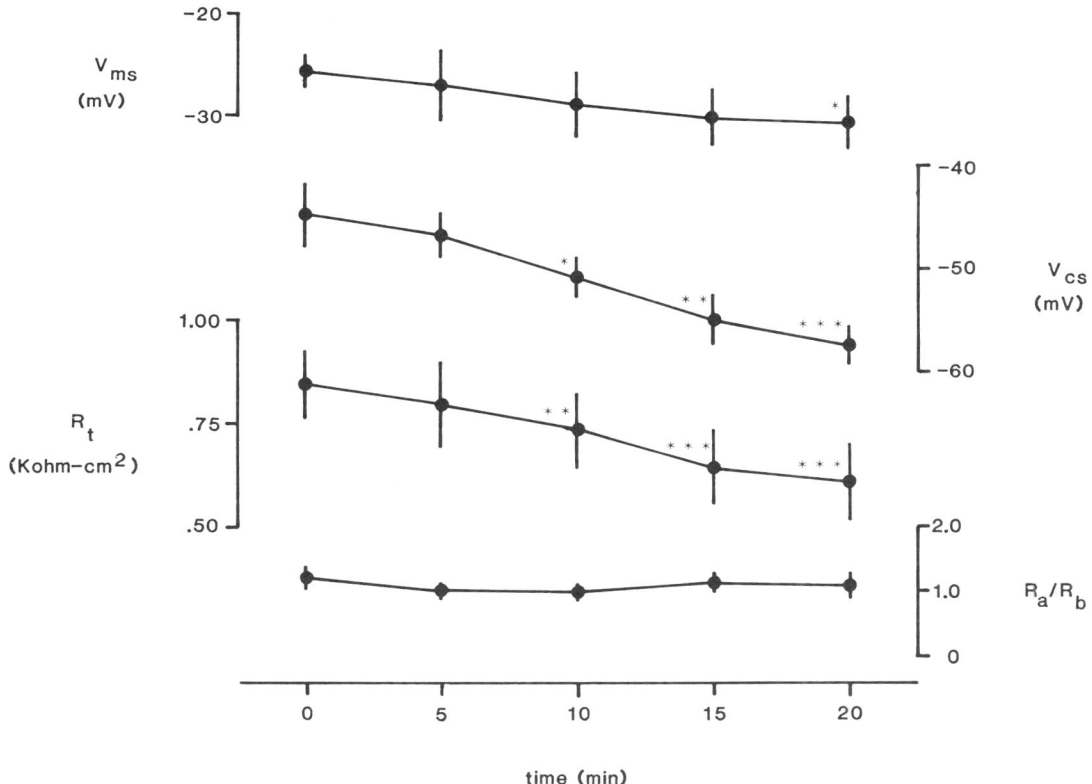

FIG. 3. Time course of stimulation followed continuously in a single OC in each of 5 mucosae. Values for time 0 were determined immediately before switching serosal superfusate to a superfusate containing 0.1 mM histamine. V_{ms} and V_{cs}, transepithelial and basolateral membrane potential differences referred to serosal solution. R_t, transepithelial resistance; R_a/R_b, ratio of apical to basolateral membrane resistances. *, $P < 0.05$; **, $P < 0.005$; ***, $P < 0.001$. [From Demarest and Machen (22).]

was unanticipated and may be related to an increase in K^+ conductance. Thus there may be increased recycling of K^+ across the apical membrane by the H^+-K^+-ATPase and across the basolateral membrane by the Na^+-K^+-ATPase under stimulated conditions. The latter may also contribute to the stimulation-induced hyperpolarization of V_{cs} of the OC. These parallel changes may reflect underlying cellular homeostatic mechanisms similar to those proposed to exist in other epithelia (18, 26, 113).

Equivalent Circuit Analysis With Microelectrodes

GENERAL COMMENTS. Any attempt at an electrophysiological equivalent circuit analysis of the gastric epithelium must take into account its structural complexity. Not only is the mucosa composed of several different transporting cell types, but in addition the epithelium has a complicated three-dimensional structure. The comparatively simple electrical circuits used to model flat-sheet epithelia composed of a single cell type with a single type of paracellular shunt pathway are simply inadequate. With this caveat in mind we proceed, however, to use just such a simple circuit to illustrate the approaches to equivalent circuit analysis of the gastric mucosa. Subsequently the complicating factors particular to the gastric mucosa are considered.

A conventional equivalent circuit for a simple epithelium is shown in Figure 4A. The apical and basolateral membranes of the cells are represented by series arrangements of membrane resistances (R_a and R_b) and Thevenin electromotive forces (E_a and E_b); a shunt pathway bypassing the cells [represented as a resistance (R_s)] completes the circuit. To simplify our discussion, only cases in which the epithelium is bathed with the same solution on both sides are considered. Therefore the electromotive force of the shunt is zero and is omitted from the circuit. According to the circuit given in Figure 4A, the transepithelial resistance (R_T) is given as (102)

$$R_T = [(R_a + R_b)R_s]/(R_a + R_b + R_s) \quad (1)$$

The R_T is measured by passing a brief pulse of current from an external source (I_{ext}) across the epithelium. The resulting deflection in the V_{ms} is divided by the magnitude of the current pulse, i.e., $R_t = \Delta V_{ms}/I_{ext}$. When determined in this way, R_T is the slope resistance of the epithelium (129).

All of the strategies used to obtain estimates of the cell membrane resistances and electromotive forces involve the measurement of the ratio of the cell membrane resistances, R_a/R_b. This ratio is determined as the ratio of the deflections in the measured cell membrane potentials, i.e., $\Delta V_{mc}/\Delta V_{cs}$, resulting from transepithelial current pulses. When current from an external source is passed across the epithelium, it is divided between the cells and shunt in proportion to their conductances. That portion of the current that enters a cell across one of its membranes must leave

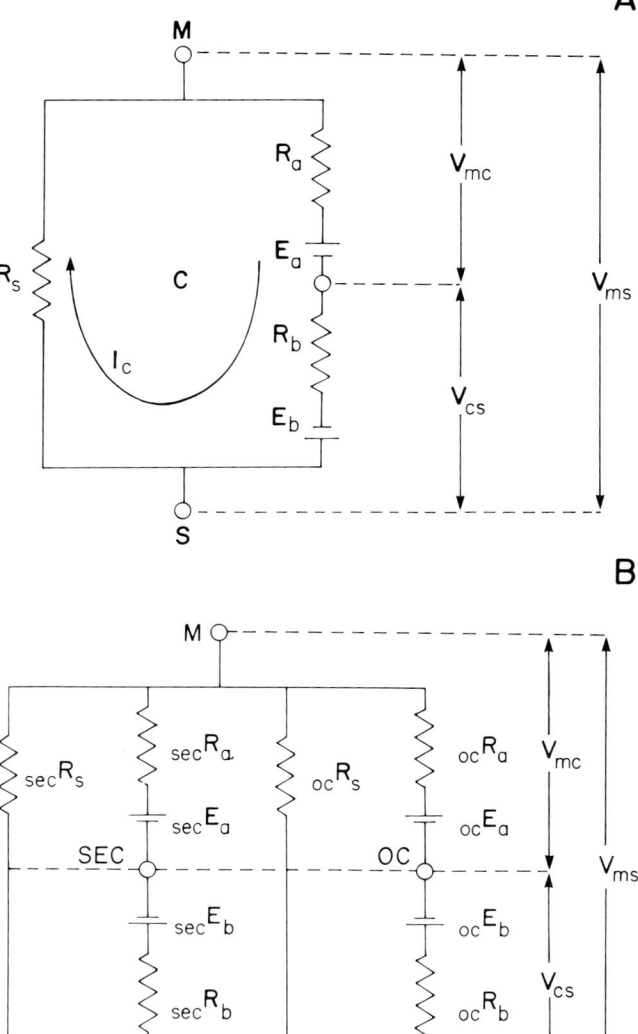

FIG. 4. Equivalent circuit models of epithelia. M, mucosal compartment; C, cellular compartment; S, serosal compartment. Apical and basolateral cell membranes are represented by electromotive forces E_a and E_b, in series with resistances R_a and R_b. R_s, shunt pathway; electromotive force across this resistance was assumed to be 0. *Horizontal dashed lines* connect transepithelial (V_{ms}), apical (V_{mc}), and basolateral (V_{cs}) membrane potentials to points in circuit from which they are measured. *Curved arrow*, current (I_c) that circulates within epithelium under open-circuit conditions. *A*: simple circuit for flat epithelium with only 1 cell type. *B*: more realistic circuit for gastric mucosa that represents 2 cell types (oc, oxyntic cells; sec, surface epithelial cells, each with a paracellular shunt).

the cell across the other membrane. Therefore no knowledge of the amount of the current traversing the cell is required. This measurement of R_a/R_b is based on the assumption that the portion of the total transepithelial current that enters a cell across one of its membranes must leave uniformly across the other membrane. The $\Delta V_{mc}/\Delta V_{cs}$ is an accurate measure of

R_a/R_b, unless the resistance of the lateral intercellular spaces makes a large contribution to the total resistance of the paracellular pathway, as in some leaky epithelia (4, 78). Under these circumstances what are known as distributed resistance effects become significant, and R_a/R_b is underestimated (4, 78).

From the measured values of R_T and R_a/R_b and an estimated value of R_s (which is discussed in ESTIMATING SHUNT RESISTANCE, p. 196), R_a and R_b can be calculated (102)

$$R_a = [(R_a/R_b)R_sR_T]/[(R_a/R_b + 1)(R_s - R_T)] \quad (2)$$

$$R_b = R_sR_T/[(R_a/R_b + 1)(R_s - R_T)] \quad (3)$$

Under open-circuit conditions an internal current (I_c in Fig. 4A) circulates continuously within the epithelial circuit. The V_{ms} can be viewed as the manifestation of the I_c that is due to active, electrogenic transcellular ion transport. The V_{ms} results from I_c passing through R_s

$$V_{ms} = I_cR_s \quad (4)$$

The conservation of charge (i.e., Kirchoff's first law) requires that current must flow in a closed loop, so I_c flows back through the cells to complete the loop and thus determines, in part, the measured cell membrane potentials. The V_{mc} and V_{cs} are the algebraic sums of the potentials due to the I_c circulating across the resistance of each membrane and the membrane electromotive forces, composed of contributions from the gradients of all the conductively permeable ion species and contributions from any electrogenic transport mechanisms (e.g., electrogenic ion pumps and exchangers) that reside in the membrane. With the calculated values of the resistances in the circuit, the values of E_a and E_b can be calculated from the measured membrane potentials by correcting for the influence of the I_c

$$E_a = V_{mc} - I_cR_a \quad (5)$$

and

$$E_b = V_{cs} - I_cR_b \quad (6)$$

The use of slope resistances rather than chord resistances results in underestimates of the actual cell membrane electromotive forces, if the current-voltage relationships of the cell membranes are nonlinear (129). However, the error is small if the membrane potentials are near the equilibrium or reversal potentials (129). To the extent that the membrane potentials in OC are dominated by K^+ conductances that would be expected to have relatively large reversal potentials (approx. −70 mV), this should not be a serious problem, because the membrane potentials are relatively large negative values with respect to the adjacent bathing solutions (Table 2). However, for the SEC where V_{mc} is 20 mV negative with respect to the secretory solution and the apical membrane conductance is dominated (at least in *Necturus* and the mammal) by an amiloride-sensitive Na^+ channel with a large positive reversal potential (e.g., 70–100 mV), the problem is potentially much more serious. No determinations have been made of the nature of the current-voltage relationships of the individual membranes of gastric epithelial cells.

With an estimated value for R_s (see ESTIMATING SHUNT RESISTANCE, p. 196), it is possible to calculate all of the circuit parameters for the equivalent circuit shown in Figure 4A from measurements of V_{ms}, R_T, V_{mc}, V_{cs}, and R_a/R_b. Because of the complicating factors introduced by multiple cell types and the heterogeneous nature of the paracellular shunt pathways of the gastric mucosa (see ESTIMATING SHUNT RESISTANCE, p. 196), no equivalent circuit analysis of the gastric mucosa has been made that gives cell membrane resistances and electromotive forces for specific cell types. Such a circuit analysis requires some method for determining the contributions of the different paracellular pathways to the overall paracellular shunt resistance. The next three sections highlight a few of the complications involved in such analyses.

CONSEQUENCES OF MULTIPLE CELL TYPES. A more realistic circuit for the gastric mucosa is shown in Figure 4B, where the SEC and OC are represented separately and are each shown with a specific shunt pathway. Although not shown in the circuit of Figure 4B, there are two components to I_c that passes through the cells.

An important aspect of this more complicated circuit (Fig. 4B) is that alteration of the transcellular resistance of one cell type will change the fraction of I_c that circulates through the other cell type and therefore alter its membrane potentials. This effect appears to be the reason that V_{mc} and V_{cs} of SEC change when OC are stimulated with histamine [see Table 2; (23)]. If it is assumed that R_s is unaffected by histamine, it is apparent that the total I_c is increased by histamine, because V_{ms} is increased. However, histamine dramatically reduces the transcellular resistance of the OC, which is reflected as a decrease in R_T. This could result in a reduction in the magnitude of current that circulates through the SEC. The prediction of the circuit shown in Figure 4B is that a decrease in the current circulating through the SEC will result in a depolarization of V_{mc} and a hyperpolarization of V_{cs} (Eqs. 5 and 6), which is observed experimentally (Table 3). Similar effects may explain the fact that changes of solution [Cl^-] give rise to changes of membrane potentials of SEC. For example, low-[Cl^-] serosal solution causes a hyperpolarization of V_m of OC in isolated rabbit gastric glands (111) and of both cell membrane potentials in the intact *Necturus* gastric mucosa (J. R. Demarest and T. E. Machen, unpublished observations). This elimination of Cl^- transport by the OC (by removing serosal Cl^-) could result in a redistribution of I_c due to increased resistance through the OC, thereby producing the depolarization of V_{cs} in SEC. These examples show that in the intact tissue, multiple factors contribute to V_{mc} and V_{cs}, and experimentally induced changes in V_{mc} and V_{cs} must be interpreted with caution.

RESISTANCE OF GLAND LUMINA. If the gland lumen represents a significant resistance (i.e., compared with the membranes of OC), then accurate measurements of $\Delta V_{mc}/\Delta V_{cs}$ ($= R_a/R_b$) for OC in intact mucosae become complicated. Because the V_{mc} is measured between the intracellular microelectrode and a macroelectrode in the secretory solution, changes in V_{mc} may represent changes across the gland luminal resistance as well as across the apical membrane of OC. Thus, OC R_a in the circuit shown in Fig. 4B would actually be the sum of the apical membrane resistance and the resistance of the gland lumen. In addition, current passed across the epithelium for the determination of R_a/R_b of the OC could be dissipated as it flows down the lumen of the gland. Such distributed effects depend on the cable properties of the gland lumen and of the gland cells (11a, 80a).

The available information on the resistance of the gland lumen is somewhat contradictory. On the one hand, morphological studies (48) have shown that when the secreting frog gastric mucosa was bathed on both sides for 50 min with hypotonic solutions (25 mosM), there was a dramatic increase in R_T from 130 to 1,200 $\Omega \cdot cm_2$. Electron microscopy showed that the OC were swollen and completely obliterated the lumina of the glands. On the basis of these and other studies [see VOLTAGE, CURRENT, IONIC FLUX, AND RESISTANCE, p. 185; (48, 88, 90, 97)], it has been suggested that the increase in R_T that accompanies the inhibition of acid secretion by a variety of inhibitors (90, 98) may be due at least partly to an increase in the resistance of the lumina due to their collapse when solute-linked volume flow is reduced (91).

On the other hand, hyperosmotic mucosal solution (which might be expected to induce serosal-to-mucosal volume flow and dilate gland lumens) has no effect on R_a/R_b measured in OC of resting Necturus gastric mucosa (22). The conclusion of this study was that luminal resistance is much smaller than the resistance across the OC in the cimetidine-treated resting mucosa and that R_a/R_b measured during the transition from this resting state to the histamine-stimulated state can safely be interpreted in terms of the membrane resistances alone.

More information about the relatize size of the gland lumina under these various conditions and correlations between volume flow and electrical resistance are needed to determine the contribution of the luminal resistance to the overall electrophysiology of gastric mucosa.

ESTIMATING SHUNT RESISTANCE. The shunt is the pathway across the epithelium through the tight junctions and the lateral intercellular spaces between the cells. This pathway may be important in the stomach's ability to maintain its barrier function. Due to the large pH gradient that can be established by the fundus, it has been generally assumed that the shunt is tight in this tissue. In epithelia isolated for in vitro study the shunt also includes some degree of damage inherent in the manipulations necessary for experimentation.

The first attempt to obtain a quantitative estimate of the shunt in the gastric mucosa with electrophysiological methods was reported by Shiba (118). In this pioneering paper the newt gastric mucosa was modeled as a two-dimensional electrical cable. A comprehensive analysis of the gastric mucosa according to the circuit in Figure 4A was reported by Spenney et al. (125). They used two independent methods: cable analysis and a method that we call the IR drop method. Rehm (94) has used similar analysis to interpret transepithelial electrical measurements.

Cable analysis is performed by measuring the spread of intracellularly injected current within the epithelium. The epithelium is modeled as a uniform conducting core of cytoplasmic fluid bounded by two thin parallel sheets of a leaky insulator, the cell membranes. The effective total resistance offered by the cell membranes to intracellularly injected current (R_z) is a function of the electrical space constant (r) for the spread of current within the epithelium (118)

$$R_z = (2\pi A r^2)/i_0 \qquad (7)$$

where r and A are constants derived from the fit of a Bessel function to a plot of the voltage deflections resulting from intracellularly injected current (i_0) as a function of the radial distance from the current injection site. For their analysis Spenney et al. (125) modeled their estimate of R_z as

$$R_z = (R_a + R_T)R_b/(R_a + R_b + R_T) \qquad (8)$$

The R_b was calculated as the positive solution of the quadratic equation

$$-(R_a/R_b)R_b^2 + [R_z + (R_a/R_b)R_z - R_T]R_b \qquad (9)$$
$$+ R_T R_z = 0$$

The R_a was calculated from the measured ratio of the cell membrane resistances (i.e., R_a/R_b) and R_s was calculated from

$$R_s = R_T(R_m + R_b)/[(R_a + R_b) - R_T] \qquad (10)$$

Spenney et al. (125) found that the ratio of R_s to the total cellular resistance was 3.8, and R_s was 3,367 $\Omega \cdot cm^2$ (referred to the nominal surface area of the epithelium without correcting for the increased surface area of the gastric glands).

The IR drop method utilizes a sudden change in the internal circulating current caused by the alteration of the R_b and/or E_b, which in turn causes an alteration of the potential across the transepithelial shunt (i.e., V_{ms}), which is proportional to R_s. The value of R_s can be calculated from the ratio of the change in V_{ms} (ΔV_{ms}) and the change in V_{cs} (ΔV_{cs}) resulting from a change in $[K^+]_s$, if the change in $[K^+]_s$ does not affect any of the circuit parameters other than those of the basolateral membrane and, in the case of an epithelium with more than one transporting cell type, if all of the cell types behave comparably in response to the change. It can be shown that (59)

$$\Delta V_{cs}/\Delta V_{ms} = R_a/R_s + 1 \quad (11)$$

$$R_s = \frac{R_T(1 + R_a/R_b)}{[(\Delta V_{cs}/\Delta V_{ms}) - 1](R_a/R_b + 1)} \quad (12)$$

This method provides an estimate of the total shunt resistance for the entire epithelium. Using it, Spenney et al. (125) estimated that R_s was 4,469 $\Omega \cdot cm^2$; the shunt conductance accounted for ~20% of the total transepithelial conductance.

Despite the apparent consistency between the findings with cable analysis and the IR drop method in *Necturus* fundus, these estimates of R_s may be in error because of technical difficulties. For example, the space constant for current spread in the plane of the epithelium obtained from the cable analysis experiments was 406 μm, but the distance between crypt openings in *Necturus* fundus averages only 158 μm (25). Therefore the cable analysis estimate of the shunt, which presumably reflects more the properties of the shunt associated with the SEC than that of the whole epithelium, may have been underestimated due to the influence of the crypts.

In the IR drop experiments it was implicitly assumed that R_a/R_b was the same for both SEC and OC, and this has proven not to be true. The SEC have a ratio of 3.5, and OC have a ratio of 1 (22). In addition a serious problem arises when comparing the results of the two methods, because the experiments were not conducted using the same conditions. The Cl$^-$-free mucosal solution used in the IR drop experiments significantly increased R_T from ~700 $\Omega \cdot cm^2$ to 950 $\Omega \cdot cm^2$ (125) (again referred to the nominal surface of the epithelium), and this seems to have resulted in an artificially high estimate of R_s for the whole epithelium, because recent experiments employing Cl$^-$-containing solutions on both sides of the epithelium indicate that R_s is ~1,300 $\Omega \cdot cm^2$, accounting for ~60% of the total transepithelial conductance under resting conditions (24).

Vibrating-probe measurements [see Fig. 1; (24)] of the specific resistance of the surface epithelium of a *Necturus* fundus indicate that the surface epithelium has a resistance that is severalfold greater than R_T (24). This agrees with the studies of O'Callaghan et al. (82) on the frog fundus that indicated that the ratio of the surface cell to glandular resistance was 2.4. In the presence of apical amiloride the surface epithelial resistance in *Necturus* rose to values as high as ~15,000 $\Omega \cdot cm^2$ (see Fig. 1). This resistance in the presence of amiloride, which makes the resistance across the SEC increase to a very high level, reflects predominantly the shunt resistance associated with these cells. Because this resistance is much higher than the total R_s for the whole epithelium of ~1,300 $\Omega \cdot cm^2$ [estimated using the IR drop method with Cl$^-$-containing solutions on both sides of the epithelium (24)], the shunt resistance associated with the OC must be ~1,400 $\Omega \cdot cm^2$ (i.e., $1/R_s$ total $- 1/R_s$ surface $= 1/R_s$ oxyntic) or <10% of that associated with the SEC. This analysis indicates that the shunt pathway of the gastric mucosa is heterogeneous, with a much higher shunt resistance associated with the SEC than with the OC.

An alternative approach that has been used to assess the magnitude of the total paracellular shunt in the gastric mucosa is to correlate passive unidirectional ion or nonelectrolyte fluxes with total transepithelial conductance. Measurements of transepithelial fluxes of Na$^+$ and mannitol (65, 130) have indicated that the paracellular conductance contributes between 14% and 30% of total tissue conductance in resting gastric fundic mucosae from rabbit (67), piglet (38), and monkey (130). This interpretation assumes that the passive fluxes of Na$^+$ and mannitol are confined to the paracellular pathway and in the case of Na$^+$ that the junctions are cation selective. In the only study in which both Na$^+$ and Cl$^-$ fluxes were measured and correlated with the mannitol flux, it was concluded that the shunt contributes 60% of tissue conductance for rat gastric fundic mucosa (21).

Equivalent Circuit Analysis From Impedance Measurements

The basic principle of impedance analysis is to examine and model the tissue's potential response during passage of a current of constant amplitude over a range of frequencies. This yields information about both membrane resistance (R) and membrane capacitance (C). Because diverse biological membranes have $C = 1 \mu F/cm^2$, knowledge of the capacitance yields information about membrane areas. This is of particular importance for measurements on the stomach, which exhibits enormous membrane folding. Impedance analysis may also yield information about membrane configuration (e.g., how folded the individual membranes may be) and about so-called distributed effects. Distributed effects arise when the series path resistance of a membrane-lined structure becomes comparable with the transmembrane impedance. A corollary here is that impedance analysis can yield information about the resistances of pathways within the epithelium, e.g., within lateral intercellular spaces and the gastric crypts. Thus impedance analysis could potentially provide unique information. However, the technique also has its drawbacks. This whole topic has been reviewed (27), and two excellent papers detailing new advances in this field have appeared (54, 55).

Three different methods and six different equivalent circuit models have been utilized for impedance analysis of gastric mucosa. The simplest model that matches in any way the actual tissue structure is the so-called lumped one-cell model (80, 101, 144), in which the apical and basolateral membranes of all cell types are represented as discrete (lumped) parallel RC circuit elements. This model is identical in form to Figure 4A. Modifications of the lumped one-cell model that account for distributed effects within the gastric pits (103) or within both the apical tubulovesicles and also intercellular spaces (10, 11) have also been used.[1]

Four major conclusions resulted from studies in which the impedance of resting and stimulated tissues were compared (10, 11). *1)* The gastric glands represent the predominant (90%) conductive pathways across the epithelium, even in the resting state. *2)* The resistances of the apical (R_a = 41 Ω) and basolateral membranes (R_b = 89 Ω) were both very low when normalized to 1 cm^2 of chamber area, but when normalized to the area (i.e., capacitance) of each individual membrane, these values were R_a = 3,500 Ω·μF and R_b = 25,000 Ω·μF. *3)* The resistances within the tubulovesicles and gastric crypts and within the lateral spaces were sometimes within 80% of the values for R_a and R_b, suggesting that under some conditions the apical and basolateral membranes represent only 50% of the total tissue resistance. *4)* Stimulation-induced drop in R_T was caused by an increase in conductance of the apical membrane, and this increase in conductance could be accounted for by an increase in the area of the apical membrane alone.

Although these findings are interesting, the quantitative conclusions may not be justified. For example, direct microelectrode measurements of R_a/R_b in OC indicated that both R_a and R_b decreased during stimulation (22). The impedance data may be incorrect for at least two reasons. *1)* The lumped one-cell model is too simplistic. *2)* The resistance of the junctions was assumed to be infinite, and this is certainly not the case. Furthermore the junctions between SEC and between OC appear to have different properties. If these discrepancies can be resolved, and if further technical developments (54, 55) can be successfully applied to impedance measurements in the gastric mucosa, impedance analysis may yield further important information about the function of the OC in the intact tissue.

Segregation of Transport Functions and Associated Shunt Pathways

The absorption of Na$^+$ and secretion of H$^+$ and Cl$^-$ in the *Necturus* gastric mucosa are performed by sep-

[1] Impedance data have also been modeled in terms of other, less realistic models; a single RC circuit (128), a single RC circuit in parallel with a series capacitance-plus resistance (32), and two RC subcircuits in parallel (80). Although these models can be made to fit the experimental data, they do not correspond to single membranes within the epithelium, so their usefulness is questionable.

TABLE 4. *Differences in Paracellular Junction Permeability Associated With Absorptive and Secretory Cells in Different Regions of Gastrointestinal Tract*

Cell	Ion Transport		Junctional Permeability		Ref.
	Na$^+$ absorption[a]	Cl$^-$ secretion[b]	Tracer permeability[c]	Freeze fracture[d]	
Salivary gland					
Duct	Yes	No	T	T	41, 42, 113
Acinar	No	Yes	L	L	122, 124
Stomach					
Surface	Yes	No	T	T	26, 76, 140
Oxyntic	No	Yes	L	L	
Ileum					
Villus	Yes[e]	No	[f]	T	73, 142
Crypt	[e]	Yes	[f]	L	
Colon					
Surface	Yes	No	T	T	62, 142
Gland	No	Yes		L	115

T, tight junctional complex; L, leaky junctional complex. [a] Apical membrane amiloride-sensitive Na$^+$ conductance. [b] Apical membrane Cl$^-$ conductance. [c] Permeability to electron-dense tracers horseradish peroxidase and/or ionic lanthanum. [d] Morphology of freeze-fractured junctional complexes. [e] Na$^+$ absorption by a variety of mechanisms is localized to villus cells. Under conditions of Na$^+$ depletion, part of the Na$^+$ absorption involves an amiloride-inhibitable electrogenic pathway. [f] Although ionic lanthanum has been reported to penetrate junctions of the ileum in at least 2 studies (72, 73), the identity of the cell type with which the junctions were associated was not determined (72).

arate cell types, the SEC and the OC. The paracellular shunt pathways associated with these two cell types appear to have different electrical resistances. A similar segregation of absorptive and secretory function and regional heterogeneity of the shunt pathway appears to occur in the epithelia throughout the gastrointestinal tract (Table 4).

There is a clear correlation between the morphological characteristics revealed by freeze-fracture replicas of the tight junctions of epithelial cells and the electrical resistance of their paracellular pathways (8, 72). The number of strands of particles, the apical-to-basolateral depth of the junctions, and the density of junction per unit surface area are important determinants of the junctional contribution to transepithelial conductance (8, 72). Although the validity of this correlation has been challenged on the basis of studies comparing the electrically leaky mammalian ileum and the tight toad urinary bladder (73), a recent elegant reexamination of this question strongly supported the correlation between resistance and structure of the tight junctions (72). (The discrepancy seems to have arisen from a lack of appreciation of the fact that the tight junctions of the ileum are different between villus cells as compared with those between glandular cells, a feature that appears to be common in other gastrointestinal epithelia.) In addition the permeation of electron-dense tracers (e.g., ionic lanthanum or horseradish peroxidase) is correlated with tight junctional resistance (see refs. in Table 4).

Electrophysiological and morphological data have been collected that show that throughout the gastrointestinal tract there is a strong correlation between the ion-transport function of the cell type and the permeability of its associated junctions within the epithelium of each region (Table 4). The data show that absorptive and secretory functions are segregated into separate cell types in many regions of the gastrointestinal tract and that leakier paracellular pathways are associated with the secretory cells and tighter paracellular pathways with the absorptive cells.

For example, the crypt cells of the rabbit colon are the site of the cAMP-induced fluid secretion that is coupled to an electrogenic Cl^--secretory process (140). Electrogenic Na^+ absorption by the colon is restricted to its surface epithelial cells (114, 140), similar to the pattern observed in the gastric mucosa (20). Accordingly the number of strands of particles observed in freeze-fracture replicas of the zonula occludens and the depth of the junctional complexes, which are indices of junctional ion permeability (8, 72), indicated that the junctions between the crypt cells of the rabbit colon are relatively leakier than those between its surface cells (61). Similar morphological evidence for the relatively leaky junctions between OC has also been reported for the dog and rat gastric mucosae (75, 139). In addition the junctions between OC, but not SEC, in the dog gastric mucosa were permeable to ionic lanthanum, which is indicative of leaky junctions (62, 75). These morphological observations correlate well with paracellular electrical resistances estimated for the *Necturus* gastric mucosa that show that a high-resistance paracellular pathway is associated with the SEC and that the paracellular pathway resistance associated with the OC is ~1 order of magnitude lower. This heterogeneity of the paracellular permeability may have important implications for the barrier function of the gastric mucosa, its pathological disruption, and the site of action of so-called cytoprotective agents, because the most permeable junctions are located between the cells responsible for the production of gastric acid and therefore must support the greatest pH gradients.

Electroneutral Hydrogen and Electrogenic Chloride Transport by Oxyntic Cells

One major unresolved question regarding H^+ secretion is whether it is electrogenic in the intact tissue. There have been three general arguments in favor of the electrogenic hypothesis. *1)* Current passage through the intact tissue elicits changes in H^+ secretion (13, 91). *2)* During stimulation of H^+ secretion, R_T decreases (92). *3)* In Cl^--free solutions the normal mucosa-negative V_{ms} inverts and becomes mucosa positive. Furthermore, during inhibition of H^+ secretion, V_{ms} becomes less positive on the mucosal side (28, 58, 98, 105, 115), and there is a straight-line relationship between the rate of H^+ secretion and V_{ms} (100). In addition, in Cl^--free media with a high $[K^+]$ on the mucosal side, passing positive current through the epithelium toward the lumen causes a 2- to 3-fold increase in H^+ secretion (20).

These data appear to argue for an electrogenic mechanism for H^+ secretion. However, it is very difficult to distinguish between an electrogenic H^+ pump and a neutral H^+-K^+ exchanger operating in parallel with K^+ and Cl^- conductances. For example, in Cl^--free solution, apparent electrogenic H^+ secretion could be provided by the conductive cell-to-lumen movement of K^+, which is then recycled back into the cell via the neutral H^+-K^+-ATPase.

In Cl^--containing solutions a model for the stimulated OC that could accommodate most of the experimental data is one in which the H^+-K^+-ATPase operates in parallel with the K^+ and Cl^- conductances

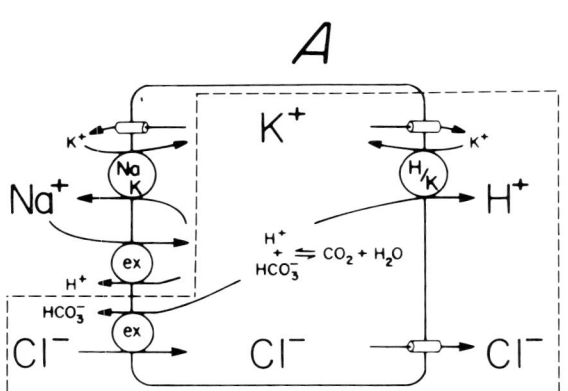
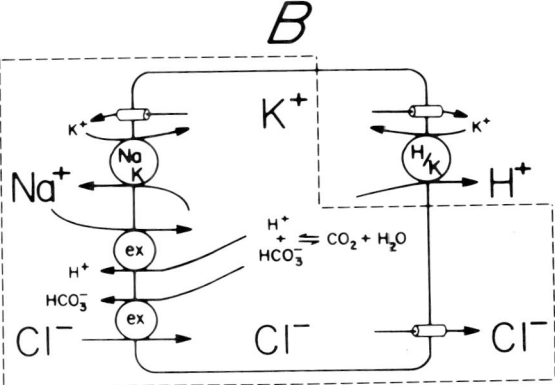

FIG. 5. Models for H^+ and Cl^- transport by secreting (*A*) and resting (*B*) OC. Apical membrane contains H^+-K^+ pump and conductive pathways for K^+ and Cl^-. Basolateral membrane contains Na^+-K^+ pump, K^+ conductance, and separate Na^+-H^+ and Cl^--HCO_3^- exchangers (ex). Mechanisms surrounded with *dotted lines* in *A* can account for HCl secretion by stimulated OC; mechanisms surrounded with *dotted lines* in *B* can account for Cl secretion by resting OC. [From Reenstra et al. (89).]

[Fig. 5A; (89)]. The K^+ and Cl^- exit the cell driven by their respective electrochemical gradients. The K^+ then recycles via the ATPase, leaving HCl in the glandular lumen. The H^+-K^+-ATPase secretes H^+, and this process leaves OH^- behind in the cytoplasm, which is converted to HCO_3^- by carbonic anhydrase. The neutral exchange of HCO_3^- for Cl^- at the serosal membrane would complete the overall process of HCl^- secretion into the lumen and HCO_3^- transport into the blood.

Active electrogenic Cl^- transport by the resting OC has been proposed (27, 64) to operate according to a model that is similar to that proposed originally by Frizzell et al. (39) to account for intestinal Cl^- secretion. Neutral, active, Na^+-dependent accumulation of Cl^- across the serosal membrane of OC could be accomplished by Na^+-H^+ and Cl^--HCO_3^- exchangers that can operate independently yet are loosely coupled through changes in pH_i [Fig. 5B; (64)]. According to this model the out-to-in gradient of $[Na^+]$ (maintained by the activity of the Na^+-K^+-ATPase) is used to drive H^+ out of the cell [via the Na^+-H^+ exchanger that has been recently demonstrated (77, 86, 87)], leading to an elevation of both cellular pH and $[HCO_3^-]$ above their passive, electrochemical equilibrium levels. The HCO_3^- gradient then serves as the driving force for uphill accumulation of Cl^- via Cl^--HCO_3^- exchange (77, 86, 87). The HCO_3^- and H^+ that exit the cell can then recombine to form CO_2 and recycle back into the cell. The net result is neutral accumulation of NaCl in exchange for H_2CO_3, which recycles across the serosal membrane as CO_2. In the short-circuited state, I_{sc} is carried across the mucosal membrane by Cl^- moving out of the cell through the Cl^- conductance and across the basolateral membrane by K^+ moving out of the cell (74).

Several aspects of the schemes shown in Figures 5A and B have been tested in the intact gastric mucosa. For example, in stimulated tissues, H^+ secretion is inhibited by omeprazole [blocks H^+-K^+-ATPase (31, 88, 126, 136, 137)], SITS [blocks Cl^--HCO_3^- exchange (T. E. Machen and J. R. Demarest, unpublished observations)], and Cl^--free serosal solution (35). Secretion of H^+ is also inhibited by K^+-free solutions, and this inhibition is reversed by adding K^+ back to the mucosal solution only (16). In resting tissues, Cl^- secretion is blocked by Na^+-free serosal (but not mucosal) solution (65, 130), ouabain [blocks Na^+-K^+-ATPase (65)], Ba^{2+} [blocks serosal K^+ channel and depolarizes both membrane potentials (71)], SITS (e.g., see ref. 136), and amiloride analogues [specific for Na^+-H^+ exchange (T. E. Machen, J. R. Demarest, and E. J. Cragoe, unpublished observations)].

Despite the apparent consistency of the available data with the models (Figs. 5A and B), questions remain. For example, little is known about the mechanism(s) by which K^+ and Cl^- leave OC in the intact tissue. It has been assumed for some time (74) that Cl^- efflux must be conductive because Cl^- secretion generates a current in the resting state. Also experiments on membrane vesicles isolated from stimulated OC have demonstrated the apparent presence of both K^+ and Cl^- conductances (14, 143). However, changing luminal $[Cl^-]$ or $[K^+]$ does not change V_{ms} in ways that are simply compatible with the presence of apical Cl^- and K^+ conductances. Microelectrode measurements of the membrane potentials and resistances of OC during changes of the mucosal $[Cl^-]$ will likely yield important information.

Finally, Rehm and his colleagues (6) have proposed that Cl^- transport by gastric mucosa is driven by a passive cotransporter that carries more Cl ions than Na ions (89). This proposal is based on measurements of V_{ms} during changes of the serosal solution of frog gastric mucosa. These types of studies need to be viewed with caution, because changes in V_{ms} (e.g., due to changes in composition of the serosal solution) can be caused by changes in V_{mc} and/or V_{cs}. Recent microelectrode experiments [J. R. Demarest and T. E. Machen, unpublished observations; (16)] have shown that decreasing $[Cl^-]$ of the serosal solution causes V_{ms} to decrease primarily because of changes at the mucosal membrane of OC; V_{cs} of OC hyperpolarizes. These microelectrode results indicate that the Cl^- conductance of the serosal membrane is negligible. It is unlikely that an electrogenic NaCl cotransporter exists in OC.

SUMMARY

A model of the transport pathways of the gastric mucosa is shown in Figure 6. There are both Na^+ absorptive and Cl^- secretory cell types, which have different paracellular junctional pathways associated with them, segregated into separate regions of the epithelium. The SEC of mammals and *Necturus* actively absorb Na^+ by the same mechanisms found in classic tight epithelia such as the frog skin and toad urinary bladder (132). The Na^+ enters the cells across the apical membrane through a conductive, amiloride-sensitive channel (that appears also to be pH sensitive) and is pumped out of the cells across the basolateral membrane by the Na^+-K^+-ATPase. The K^+ pumped into the cells in exchange for Na^+ recycles across the basolateral membrane by exiting the cells through conductive channels. Within the cells, Cl^- is accumulated against an electrochemical gradient by an Na^+-dependent process located in the basolateral membrane. However, these cells have no Cl^- conductance in either the mucosal or serosal membrane and appear to play no role in electrogenic Cl^- secretion. The SEC do, however, appear to be the sites of active HCO_3^- secretion, but the mechanisms by which this takes place are not well understood. Data currently available indicate that the junctional pathway between the SEC has the relatively high-resistance characteristic of other Na^+-absorbing epithelia.

measurements have not been made in intact cells. These K⁺ and Cl⁻ conductances may not be the only mechanisms by which these ions exit the cells into the lumen. In the resting state, Cl⁻ secretion can be explained by the combined operation of the Cl⁻-HCO$_3^-$ exchanger with the Na⁺-H⁺ exchanger, Na⁺-K⁺-ATPase and K⁺ conductance in the serosal membrane, and the Cl⁻ conductance in the mucosal membrane. Stimulation results in the incorporation into the apical membrane of intracellular membranes containing the H⁺-K⁺-ATPase and additional permeation pathways for K⁺ and Cl⁻, as well as the activation of two different K⁺-conductance channels of the serosal membrane. The secretion of HCl can be explained by the operation at the mucosal membrane of the H⁺-K⁺-ATPase and the K⁺ and Cl⁻ conductances and at the serosal membrane of the Cl⁻-HCO$_3^-$ exchanger (and Na⁺-HCO$_3^-$ cotransport?). There is likely to be extensive recycling of K⁺ by the H⁺-K⁺-ATPase at the mucosal membrane and the Na⁺-K⁺-ATPase at the serosal membrane, both of which must increase their activities with stimulation. The H⁺-K⁺-ATPase, Na⁺-H⁺, and Cl⁻-HCO$_3^-$ exchangers (and Na⁺-HCO$_3^-$ cotransport?) operate in conjunction to keep pH$_i$ constant during both resting and stimulated conditions. The mechanisms for coordinating all the transport pathways are unknown. The resistance of the junctional pathways between OC is lower than that between SEC, indicating a higher ion permeability, similar to the situation found in other Cl⁻-secreting cell types.

We thank Drs. C. William Davis, Arthur Finn, and William Reenstra for critical readings of the manuscript and Patricia LaForce and Asata Iman for help with typing the manuscript.

Work in this laboratory was supported by National Institutes of Health Grants AM-19520 and AM-17328.

REFERENCES

1. BLUM, A. L., G. T. SHAH, V. D. WIEBELHAUS, F. T. BRENNAN, H. F. HELANDER, R. CEBALLOS, AND G. SACHS. Pronase method for isolation of viable cells from *Necturus* gastric mucosa. *Gastroenterology* 61: 189-200, 1971.
2. BORNSTEIN, A. M., W. H. DENNIS, AND W. S. REHM. Movement of water, sodium, chloride and hydrogen across the resting stomach. *Am. J. Physiol.* 197: 332-336, 1959.
3. BORON, W. F., AND E. L. BOULPAEP. Intracellular pH regulation in the renal proximal tubule of the salamander. Basolateral HCO$_3^-$ transport. *J. Gen. Physiol.* 81: 53-94, 1983.
4. BOULPAEP, E. L., AND H. SACKIN. Electrical analysis of intraepithelial barriers. *Curr. Top. Membr. Transp.* 13: 169-197, 1980.
5. CABANTCHIK, Z. I., AND A. ROTHSTEIN. Membrane proteins related to anion permeability of human red blood cells. *J. Membr. Biol.* 15: 207-226, 1979.
6. CARRASQUER, G., T.-C. CHU, W. S. REHM, AND M. SCHWARTZ. Evidence for electrogenic Na-Cl symport in the in vitro frog stomach. *Am. J. Physiol.* 242 (*Gastrointest. Liver Physiol.* 5): G620-G627, 1982.
7. CHEW, C. S., AND M. BROWN. Release of intracellular Ca²⁺ and elevation of inositol triphosphate by secretagogues. *Biochim. Biophys. Acta* 888: 116-125, 1986.
8. CLAUDE, P. Morphological factors influencing transepithelial permeability: a model for the resistance of the zonula occludens. *J. Membr. Biol.* 39: 219-232, 1978.
9. CLAUDE, P., AND D. A. GOODENOUGH. Fracture faces of zonulae occludentes from "tight" and "leaky" epithelia. *J. Cell Biol.* 58: 390-400, 1973.
10. CLAUSEN, C., T. E. MACHEN, AND J. M. DIAMOND. Changes in the cell membranes of the bullfrog gastric mucosa with acid secretion. *Science Wash. DC* 217: 448-459, 1982.
11. CLAUSEN, C., T. E. MACHEN, AND J. M. DIAMOND. Use of AC impedance analysis to study membrane changes related to acid secretion in amphibian gastric mucosa. *Biophys. J.* 41: 167-178, 1983.
11a.COOK, D. I., AND E. FRÖMTER. Is the voltage divider ratio a reliable estimate of the resistance ratio of the cell membranes in tubular epithelia? *Pflügers Arch.* 403: 388-395, 1985.
12. CRANE, E. E., AND R. E. DAVIES. Transport of radioactive Na and K through the gastric mucosa. *Biochem. J.* 45: 23-24, 1949.
13. CRANE, E. E., R. E. DAVIES, AND N. M. LONGMUIR. The effect

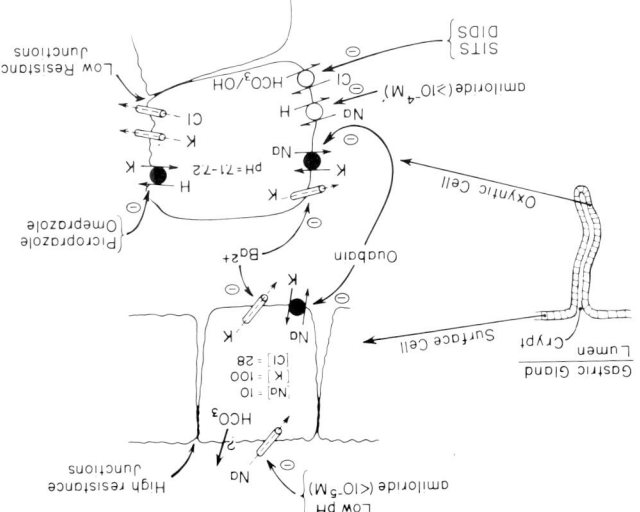

FIG. 6. Ion-transport pathways in SEC and OC. *Filled circles*, ATP-driven pumps (Na⁺-K⁺- and H⁺-K⁺-ATPases). *Open circles*, coupled exchangers (Na⁺-H⁺ and Cl⁻-HCO$_3^-$). *Open cylinders*, conductance channels (K⁺ and Cl⁻). Known intracellular concentrations and effects of inhibitors are indicated. Value for [Na⁺]$_i$ in SEC is taken from unpublished measurements by T. Zeuthen and T. E. Machen.

The OC, which secrete H⁺ and Cl⁻ in the stimulated state and Cl⁻ alone in the resting state, have parallel mechanisms for Na⁺-H⁺ and Cl⁻-HCO$_3^-$ exchange, a K⁺ conductance, and the Na⁺-K⁺-ATPase located in their basolateral membranes. A mechanism for Na⁺-HCO$_3^-$ cotransport (not shown in Fig. 6) appears to be present in frog OC. Whether such a mechanism is also present in mammalian cells remains to be determined. The basolateral membrane does not contain a Cl⁻ conductance. The apical membranes are likely to contain conductances for both K⁺ and Cl⁻, although direct

14. CUPPOLETTI, J., AND G. SACHS. Regulation of gastric acid secretion via modulation of a chloride conductance. *J. Biol. Chem.* 259: 14952-14959, 1984.

15. CURCI, S., L. DEBELLIS, AND E. FRÖMTER. Evidence for rheogenic sodium bicarbonate cotransport in the basolateral membrane of oxyntic cells of frog gastric fundus. *Pflügers Arch.* 408: 497-504, 1987.

16. CURCI, S., AND T. SCHETTINO. Effect of external sodium on intracellular chloride activity in the surface cells of frog gastric mucosa. Microelectrode studies. *Pflügers Arch.* 401: 152-159, 1984.

17. CURCI, S., T. SCHETTINO, AND E. FRÖMTER. Histamine reduces Cl⁻ activity in surface epithelial cells of frog gastric mucosa. Suggestive evidence for ionic coupling between surface epithelial and oxyntic cells. *Pflügers Arch.* 406: 204-211, 1986.

18. DAVIS, C. W., AND A. L. FINN. Cell volume regulation in frog urinary bladder. *Federation Proc.* 44: 2520-2525, 1985.

19. DAVIS, T. L., J. R. RUTLEDGE, D. C. KEESER, F. J. VAJANDAS, AND W. S. REHM. Acid secretion, potential, and resistance of frog stomach in K⁺-free solutions. *Am. J. Physiol.* 209: 146-152, 1965.

20. DAVIS, T. L., J. R. RUTLEDGE, AND W. S. REHM. Effect of potassium on secretion and potential of frog's gastric mucosa in Cl⁻-free solutions. *Am. J. Physiol.* 205: 873-877, 1963.

21. DAWSON, D. C., AND A. R. COOKE. Parallel pathways for ion transport across rat gastric mucosa: effect of ethanol. *Am. J. Physiol.* 235 (*Endocrinol. Metab. Gastrointest. Physiol.* 4): E7-E15, 1978.

22. DEMAREST, J. R., AND T. E. MACHEN. Microelectrode measurements from oxyntic cells in intact *Necturus* gastric mucosa. *Am. J. Physiol.* 249 (*Cell Physiol.* 18): C535-C540, 1985.

23. DEMAREST, J. R., AND T. E. MACHEN. Effects of omeprazole on the electrical properties of oxyntic cells (Abstract). *Federation Proc.* 45: 894, 1986.

24. DEMAREST, J. R., AND T. E. MACHEN. Different paracellular resistances are associated with surface and oxyntic cells in the gastric mucosa (Abstract). *Biophys. J.* 51: 340a, 1987.

25. DEMAREST, J. R., C. SCHEFFEY, AND T. E. MACHEN. Segregation of gastric Na and Cl transport: a vibrating probe and microelectrode study. *Am. J. Physiol.* 251 (*Cell Physiol.* 20): C643-C648, 1986.

26. DIAMOND, J. M. Transcellular cross-talk between epithelial cell membranes. *Nature Lond.* 300: 683-685, 1982.

27. DIAMOND, J. M., AND T. E. MACHEN. Impedance analysis in epithelia and the problem of gastric acid secretion. *J. Membr. Biol.* 73: 17-41, 1983.

28. DINNO, F. M., DINNO, C. H. LEE, M. SCHWARTZ, AND T. H. MACKRELL. Determination of electrogenicity of frog gastric mucosa in chloride-free solutions by using barbiturates. *Proc. Soc. Exp. Biol. Med.* 138: 479-481, 1971.

28a.DINNO, M. A., M. SCHWARTZ, G. CARRASQUER, AND W. S. REHM. Microelectrode study of effects of luminal K⁺ on surface cells of frog stomach. *Biochim. Biophys. Acta* 856: 629-633, 1986.

29. DURBIN, R. P. Anion requirements for gastric acid secretion. *J. Gen. Physiol.* 4: 735-748, 1964.

30. DURBIN, R. P. Electrical potential difference of the gastric mucosa. In: *Handbook of Physiology. Alimentary Canal*, edited by C. F. Code. Washington, DC: Am. Physiol. Soc., 1967, sect. 6, vol. II, chapt. 49, p. 879-888.

31. FELLENIUS, E., T. BERGLINDH, G. SACHS, L. OLBE, B. ELANDER, S. E. SJOSTRAND, AND B. WALLMARK. Substituted benzimidazoles inhibit gastric acid secretion by blocking (H⁺ + K⁺)-ATPase. *Nature Lond.* 290: 159-161, 1981.

32. FLEMSTRÖM, G. Na⁺ transport and impedance properties of the isolated frog gastric mucosa at different O_2 tensions. *Biochim. Biophys. Acta* 225: 35-45, 1971.

33. FLEMSTRÖM, G. Active alkalinization by amphibian gastric fundic mucosa in vitro. *Am. J. Physiol.* 233 (*Endocrinol. Metab.*

34. FLEMSTRÖM, G., AND A. GARNER. Gastroduodenal HCO_3^- transport: characteristics and proposed role in acidity regulation and mucosal protection. *Am. J. Physiol.* 242 (*Gastrointest. Liver Physiol.* 5): G183-G193, 1982.

35. FLEMSTRÖM, G., AND G. SACHS. Ion transport by amphibian antrum in vitro. I. General characteristics. *Am. J. Physiol.* 228: 1188-1198, 1975.

36. FORTE, J. G. Three components of Cl⁻ flux across isolated bullfrog gastric mucosa. *Am. J. Physiol.* 216: 167-174, 1969.

37. FORTE, J. G., P. H. ADAMS, AND R. E. DAVIES. The source of the gastric mucosal potential. *Nature Lond.* 197: 874-876, 1963.

38. FORTE, J. G., AND T. E. MACHEN. Transport and electrical phenomena in resting and secreting piglet gastric mucosa. *J. Physiol. Lond.* 244: 33-51, 1975.

39. FRIZZELL, R. A., M. FIELD, AND S. G. SCHULTZ. Sodium-coupled chloride transport by epithelial tissues. *Am. J. Physiol.* 236 (*Renal Fluid Electrolyte Physiol.* 5): F1-F8, 1979.

40. FRÖMTER, E., B. GEBLER, H. SCHOROM, AND H. POCKRANDT-HEMSTEDT. Cation and anion permeability of rabbit submaxillary main duct. In: *Secretory Mechanisms of Exocrine Glands*, edited by N. A. Thorn and O. H. Petersen. Copenhagen: Munksgaard, 1974, p. 496-509.

41. GARRATT, J. R., AND P. A. PARSONS. Preliminary observations on the permeability of submandibular glands to horseradish peroxidase in rabbits. In: *Secretory Mechanisms of Exocrine Glands*, edited by N. A. Thorn and O. H. Petersen. Copenhagen: Munksgaard, 1974, p. 487-492.

42. GRADY, T. P., AND L. V. CHEUNG. Microelectrode studies of *Necturus* antral mucosa: electrical potentials and resistances. *Am. J. Physiol.* 244 (*Gastrointest. Liver Physiol.* 7): G71-G75, 1983.

43. HANSEN, T., J. F. G. SLEGERS, AND S. L. BONTING. Gastric acid secretion in the lizard. Ionic requirements and effects of inhibitors. *Biochim. Biophys. Acta* 382: 590-609, 1975.

44. HARRIS, J. B., AND I. S. EDELMAN. Transport of potassium by the gastric mucosa of the frog. *Am. J. Physiol.* 198: 280-288, 1960.

45. HARRIS, J. B., AND I. S. EDELMAN. Chemical concentration gradients and electrical properties of gastric mucosa. *Am. J. Physiol.* 206: 769-782, 1964.

46. HEINZ, E., AND R. P. DURBIN. Studies of the chloride transport in the gastric mucosa of the frog. *J. Gen. Physiol.* 41: 101-117, 1957.

47. HELANDER, H. F., AND R. P. DURBIN. Localization of ouabain binding sites in frog gastric mucosa. *Am. J. Physiol.* 243 (*Gastrointest. Liver Physiol.* 6): G297-G303, 1982.

48. REHM, W. S. SANDERS, L. SHABOUR, AND W. S. REHM. Reversibility of effects of very hypotonic fluids on in vitro frog gastric mucosa: a functional morphological study. *Acta Physiol. Scand.* 95: 353-363, 1975.

49. HERSEY, S. J., G. SACHS, AND D. K. KASBEKER. Acid secretion by frog gastric mucosa is electroneutral. *Am. J. Physiol.* 248 (*Gastrointest. Liver Physiol.* 11): G246-G250, 1985.

50. HOGBEN, C. A. M. Active transport of chloride by isolated frog gastric epithelium. Origin of the gastric mucosal potential. *Am. J. Physiol.* 180: 641-649, 1955.

51. KAFOGLIS, K., S. J. HERSEY, AND J. F. WHITE. Microelectrode measurements of K⁺ and pH in rabbit gastric glands: effect of histamine. *Am. J. Physiol.* 246 (*Gastrointest. Liver Physiol.* 9): G433-G444, 1984.

52. KITAHARA, S., K. R. FOX, AND C. A. M. HOGBEN. Acid secretion, Na⁺ absorption, and the origin of the potential difference across isolated mammalian stomachs. *Am. J. Dig. Dis.* 14: 221-238, 1969.

53. KOELZ, H. R., G. SACHS, AND T. BERGLINDH. Cation effects on acid secretion in rabbit gastric glands. *Am. J. Physiol.* 241 (*Gastrointest. Liver Physiol.* 4): G431-G442, 1981.

54. KOTTRA, G., AND E. FRÖMTER. Rapid determination of intraepithelial resistance barriers by alternating current spectroscopy. I. Experimental procedures. *J. Membr. Biol.* 402: 409-

55. KOTTRA, G., AND E. FRÖMTER. Rapid determination of intraepithelial resistance barriers by alternating current spectroscopy. II. Test of model circuits and quantification of results. *J. Membr. Biol.* 402: 421–432, 1984.

56. KUO, Y.-J., AND L. L. SHANBOUR. Effects of cyclic AMP, ouabain and furosemide on ion transport in isolated canine gastric mucosa. *J. Physiol. Lond.* 309: 29–43, 1980.

57. LEE, C. H. BREITBART, M. BERMAN, AND J. G. FORTE. Potassium-stimulated ATPase activity and hydrogen transport in gastric microsomal vesicles. *Biochim. Biophys. Acta* 553: 107–131, 1979.

58. LEFEVRE, M. E., J. GOHMANN, JR., AND W. S. REHM. An hypothesis for discovery of inhibitors of gastric acid secretion. *Am. J. Physiol.* 207: 613–618, 1964.

59. LEWIS, S. A., AND N. K. WILLS. Electrical properties of the rabbit urinary bladder assessed using gramiciden D. *J. Membr. Biol.* 67: 45–53, 1982.

60. LOGSDON, C. D., AND T. E. MACHEN. Involvement of extracellular calcium in gastric stimulation. *Am. J. Physiol.* 241 (*Gastrointest. Liver Physiol.* 4): G365–G375, 1981.

61. LUCIANO, E., E. REALE, G. RECHKEMMER AND W. V. ENGELHARDT. Structure of the zonulae occludentes and the permeability of the epithelium to short chain fatty acids in the proximal and distal colon of the guinea pig. *J. Membr. Biol.* 82: 145–156, 1984.

62. MACHEN, T. D., ERLIJ, AND F. B. P. WOODING. Permeable junctional complexes. The movement of lanthanum across rabbit gallbladder and intestine. *J. Cell Biol.* 54: 302–312, 1972.

63. MACHEN, T. E., AND J. G. FORTE. Gastric secretion. In: *Membrane Transport in Biology*, edited by G. Giebisch, D. C. Tosteson, and H. H. Ussing. Berlin: Springer-Verlag, 1979, vol. 4B, p. 693–747.

64. MACHEN, T. E., AND J. G. FORTE. Anion secretion by gastric mucosa. In: *Chloride Transport Coupling in Biological Membranes and Epithelia*, edited by G. A. Gerencser. Amsterdam: Elsevier, 1983, p. 415–460.

65. MACHEN, T. E., AND W. L. MCLENNAN. Na^+-dependent H^+ and Cl^- transport in in vitro frog gastric mucosa. *Am. J. Physiol.* 238 (*Gastrointest. Liver Physiol.* 1): G403–G413, 1980.

66. MACHEN, T. E., AND A. M. PARADISO. Regulation of intracellular pH in the stomach. *Annu. Rev. Physiol.* 49: 19–33, 1987.

67. MACHEN, T. E., W. SILEN, AND J. G. FORTE. Na^+ transport by mammalian stomach. *Am. J. Physiol.* 234 (*Endocrinol. Metab. Gastrointest. Physiol.* 3): E228–E235, 1978.

68. MACHEN, T. E., AND T. ZEUTHEN. Cl^- transport by gastric mucosa: Cl^- activity and membrane permeability. *Phil. Trans. R. Soc. Lond. B Biol. Sci.* 299: 559–573, 1982.

69. MAKHLOUF, G. M. A model for passive transport of potassium by the stomach: evidence from in vitro studies. *Am. J. Physiol.* 227: 1285–1288, 1974.

70. MAKHLOUF, G. M., AND G. R. DUCKWORTH. Secretion and electrical activity of a unilateral in vitro gastric mucosa. *Gastroenterology* 65: 907–911, 1973.

71. MANNING, E. C., AND T. E. MACHEN. Effects of bicarbonate and pH on chloride transport by gastric mucosa. *Am. J. Physiol.* 243 (*Gastrointest. Liver Physiol.* 6): G60–G68, 1982.

72. MARCIAL, M. A., S. L. CARLSON, AND J. L. MADARA. Partitioning of paracellular conductance along the ileal crypt-villus axis: a hypothesis based on structural analysis with detailed consideration of tight junction structure-function relationships. *J. Membr. Biol.* 80: 59–70, 1984.

73. MARTINEZ-PALOMO, A., AND D. ERLIJ. Structure of tight junctions in epithelia with different permeability. *Proc. Natl. Acad. Sci. USA* 72: 4487–4491, 1975.

74. MCLENNAN, W. L., T. E. MACHEN, AND T. ZEUTHEN. Ba^{2+} inhibition of electrogenic Cl^- secretion by in vitro frog and piglet gastric mucosa. *Am. J. Physiol.* 239 (*Gastrointest. Liver Physiol.* 2): G151–G160, 1980.

75. MEYER, R. A., D. MCGINLEY, AND Z. POSALAKY. The gastric mucosal barrier: structure of intercellular junctions in the dog. *J. Ultrastruct. Res.* 86: 192–201, 1984.

76. MICHELANGELI, F. Acid secretion and intracellular pH in isolated oxyntic cells. *J. Membr. Biol.* 38: 31–50, 1978.

77. MULLEN, S., C. BURNHAM, D. BLISSARD, T. BERGLINDH, AND G. SACHS. Electrolyte transport across the basolateral membrane of the parietal cells. *J. Biol. Chem.* 260: 6641–6649, 1985.

78. NAGEL, W., J. F. GARCIA-DIAS, AND A. ESSIG. Contribution of junctional conductance to the cellular voltage-divider ratio in frog skins. *Pfluegers Arch.* 399: 336–341, 1983.

79. NEGULESCU, P. A., AND T. E. MACHEN. Intracellular Ca regulation during secretagogue-secretion of the parietal cell. *Am. J. Physiol.* 254 (*Cell Physiol.* 23): C130–C140, 1988.

80. NOYES, D. H., AND W. S. REHM. Voltage response of frog gastric mucosa to direct current. *Am. J. Physiol.* 219: 184–192, 1970.

80a. OBERLEITHNER, H., W. GUGGINO, AND G. GIEBISCH. Resistance properties of the diluting segment of the amphibian kidneys: influence of potassium adaptation. *J. Membr. Biol.* 88: 139–147, 1985.

81. OBRINK, K. J. Studies on the kinetics of the parietal secretion of the stomach. *Acta Physiol. Scand. Suppl.* 5: 1–106, 1948.

82. O'CALLAGHAN, J., S. S. SANDERS, R. L. SHOEMAKER AND W. S. REHM. Barium and K^+ on surface and tubular cell resistances of frog stomach with microelectrodes. *Am. J. Physiol.* 227: 273–280, 1974.

83. OKADA, Y., AND S. UEDA. Electrical membrane responses to secretagogues in parietal cells of the rat gastric mucosa in culture. *J. Physiol. Lond.* 354: 109–119, 1984.

84. PALMER, L. G. Ion selectivity of the apical membrane Na channel in the toad urinary bladder. *J. Membr. Biol.* 67: 91–98, 1982.

86. PARADISO, A. M., P. A. NEGULESCU, AND T. E. MACHEN. Na^+-H^+ and Cl^--OH^- (HCO_3^-) exchange in gastric glands. *Am. J. Physiol.* 250 (*Gastrointest. Liver Physiol.* 13): G524–G534, 1986.

87. PARADISO, A. M., R. Y. TSIEN, J. R. DEMAREST, AND T. E. MACHEN. Na-H and Cl-HCO_3 exchange in rabbit oxyntic cells using fluorescence microscopy. *Am. J. Physiol.* 253 (*Cell Physiol.* 22): C30–C36, 1987.

88. REENSTRA, W. W., J. D. BETTENCOURT, AND J. G. FORTE. Active K^+ absorption by the gastric mucosa: inhibition by omeprazole. *Am. J. Physiol.* 250 (*Gastrointest. Liver Physiol.* 13): G455–G460, 1986.

89. REENSTRA, W. W., J. D. BETTENCOURT, AND J. G. FORTE. Mechanisms of active Cl^- secretion by frog gastric mucosa. *Am. J. Physiol.* 252 (*Gastrointest. Liver Physiol.* 15): G543–G547, 1987.

90. REENSTRA, W. W., AND J. G. FORTE. Mechanism of inhibition of gastric acid secretion by SCN^-: interrelation of SCN^- flux and inhibition. *Am. J. Physiol.* 250 (*Gastrointest. Liver Physiol.* 13): G76–G84, 1986.

91. REHM, W. S. Effect of electric current on gastric hydrogen and chloride ion secretion. *Am. J. Physiol.* 185: 325–331, 1956.

92. REHM, W. S. Acid secretion, resistance, short-circuit current, and voltage-clamping in frog's stomach. *Am. J. Physiol.* 203: 63–72, 1962.

93. REHM, W. S. Ion permeability and electrical resistance of the frog's gastric mucosa. *Federation Proc.* 26: 1303–1313, 1967.

94. REHM, W. S. The metabolic state and the response of the potential of frog gastric mucosa to changes in external ion concentration. *J. Gen. Physiol.* 51: 250S–260S, 1968.

95. REHM, W. S., G. CARRASQUER, AND M. SCHWARTZ. Contributions of electrogenic pumps and parallel passive pathways of gastric mucosa. In: *Membrane Biophysics. Physical Methods in the Study of Epithelia*, edited by M. A. Dinno, A. B. Callahan, and T. C. Rozzell. New York: Liss, 1983, vol. II, p. 313–327.

96. REHM, W. S., G. CARRASQUER, AND M. SCHWARTZ. Effects of NaSCN and omeprazole on resistance and potential of fundus of *Rana pipiens*. *Am. J. Physiol.* 250 (*Gastrointest. Liver Physiol.* 13): G511–G517, 1986.

97. REHM, W. S., T.-C. CHU, M. SCHWARTZ, AND G. CARRASQUER. Mechanisms responsible for SCN increase in resistance of in vitro frog gastric mucosa. *Am. J. Physiol.* 245 (*Gastrointest. Liver Physiol.* 8): G143–G156, 1983.
98. REHM, W. S., AND M. E. LEFEVRE. Effect of dinitrophenol on potential, resistance, and H⁺ rate of frog stomach. *Am. J. Physiol.* 208: 922–930, 1965.
99. REHM, W. S., AND S. S. SANDERS. Implications of the neutral Cl⁻-HCO₃⁻ exchange mechanism in gastric mucosa. *Ann. NY Acad. Sci.* 264: 442–455, 1975.
100. REHM, W. S., AND S. S. SANDERS. Electrical events during activation and inhibition of gastric HCl secretion. *Gastroenterology* 73: 959–969, 1977.
101. REHM, W. S., S. SANDERS, M. G. TANT, F. M. HOFFMAN, AND J. T. TARVIN. Conductance of frog gastric mucosa under varying conditions as determined by square wave analysis. In: *Gastric Hydrogen Ion Secretion*, edited by D. K. Kasbekar, G. Sachs, and W. S. Rehm. New York: Dekker, 1977, p. 29–53.
102. REUSS, L., AND A. L. FINN. Passive electrical properties of the toad urinary bladder epithelium. Intracellular electrical coupling and transepithelial cellular and shunt conductance. *J. Gen. Physiol.* 64: 1–25, 1974.
103. RING, A., AND J. SANDBLOM. Impedance of the frog gastric mucosa: the Teorell-Wensall effect. *Ups. J. Med. Sci.* 85: 283–293, 1980.
104. SACHS, G., H. H. CHANG, E. RABON, R. SCHACKMANN, M. LEWIN, AND G. SACCOMANI. A nonelectrogenic H⁺ pump in plasma membranes of hog stomach. *J. Biol. Chem.* 251: 7690–7698.
105. SACHS, G., R. L. SHOEMAKER, AND B. I. HIRSCHOWITZ. Action of 2-deoxy-D-glucose on frog gastric mucosa. *Am. J. Physiol.* 209: 461–466, 1965.
106. SACHS, G., R. L. SHOEMAKER, AND B. I. HIRSCHOWITZ. Effects of sodium removal on acid secretion by the frog gastric mucosa. *Proc. Soc. Exp. Biol. Med.* 123: 47–52, 1986.
107. SANDERS, M. J., A. AYALON, M. ROLL, AND A. H. SOLL. The apical surface of canine chief cell monolayers resists H⁺ back diffusion. *Nature Lond.* 313: 52–54, 1985.
108. SANDERS, S. S., V. B. HAYNE, JR., AND W. S. REHM. Normal H⁺ rates in frog stomach in absence of exogenous CO₂ and a note on pH stat method. *Am. J. Physiol.* 225: 1311–1321, 1973.
109. SCHETTINO, T., AND S. CURCI. Intracellular potassium activity in epithelial cells of frog fundic mucosa. *Pfluegers Arch.* 383: 99–103, 1980.
110. SCHETTINO, T., AND S. CURCI. On the luminal membrane permeability to Cl⁻ of *Necturus* gastric surface cells. *Pfluegers Arch.* 403: 331–333, 1985.
111. SCHETTINO, T., M. KOHLER, AND E. FRÖMTER. Membrane potentials of individual cells of isolated gastric glands of rabbit. *Pfluegers Arch.* 405: 58–65, 1985.
112. SCHNEYER, L. H. Amiloride inhibition of ion transport in perfused excretory duct of rat submaxillary gland. *Am. J. Physiol.* 219: 1050–1055, 1970.
113. SCHULTZ, S. G. Homocellular regulatory mechanisms in sodium-transporting epithelia: avoidance of extinction by "flush-through." *Am. J. Physiol.* 241 (*Renal Fluid Electrolyte Physiol.* 10): F579–F590, 1981.
114. SCHULTZ, S. G., R. A. FRIZZELL, AND H. N. NELLANS. Active sodium transport and the electrophysiology of the rabbit colon. *J. Membr. Biol.* 33: 351–384, 1976.
115. SCHWARTZ, M., AND T. N. MACKRELL. A confirmation of the electroheterogeneity of frog gastric mucosa by using methexyfluorane. *Proc. Soc. Exp. Biol. Med.* 130: 1048–1051, 1969.
116. SEDAR, A. W., AND J. G. FORTE. Effects of calcium depletion on the junctional complex between oxyntic cells of gastric glands. *J. Cell Biol.* 22: 173–188, 1964.
117. SEN, P. C., L. TAGUE, AND T. K. RAY. Secretion of H⁺ and K⁺ by bullfrog gastric mucosa: characterization of K⁺ transport pathway. *Am. J. Physiol.* 239 (*Gastrointest. Liver Physiol.* 2): G485–G492, 1980.
118. SHIBA, H. Heaviside's "Bessel cable" as an electric model for flat simple epithelial cells with low resistive junctional membranes. *J. Theor. Biol.* 30: 59–68, 1970.
119. SHIMONO, M., T. YAMAMURA, AND G. FUMAGALLI. Intracellular junctions in salivary glands: freeze-fracture and tracer studies of normal rat sublingual gland. *J. Ultrastruct. Res.* 7: 286–299, 1980.
120. SHOEMAKER, R. L. Characteristics of the *Necturus* In Vitro Gastric Mucosa. Birmingham: Univ. of Alabama, 1967. PhD thesis.
121. SHOEMAKER, R. L. Micropuncture studies using the amphibian fundic gastric mucosa, in vitro. In: *Gastric Ion Transport*, edited by K. J. Obrink and G. Flemstrom. Uppsala, Sweden: *Acta Physiol. Scand.*, 1978, p. 173–180. (Special Suppl.)
122. SILEN, W., T. E. MACHEN, AND J. G. FORTE. Acid-base balance in amphibian gastric mucosa. *Am. J. Physiol.* 229: 721–730, 1975.
123. SIMONSON, J. V, A, AND H. L. BANK. Freeze-fracture and lead ion tracer evidence for a paracellular fluid secretory pathway in rat parotid glands. *Anat. Rec.* 208: 69–80, 1984.
124. SPENNEY, J. G., G. FLEMSTRÖM, R. L. SHOEMAKER, AND G. SACHS. Quantitation of conductance pathways in antral gastric mucosa. *J. Gen. Physiol.* 65: 645–662, 1975.
125. SPENNEY, J. G., R. L. SHOEMAKER AND G. SACHS. Microelectrode studies of fundic gastric mucosa: cellular coupling and shunt conductance. *J. Membr. Biol.* 19: 105–128, 1974.
126. STARLINGER, M. J., M. J. HOLLANDS, P. H. ROWE, J. B. MATTHEWS, AND W. SILEN. Chloride transport of frog gastric fundus: effects of omeprazole. *Am. J. Physiol.* 250 (*Gastrointest. Liver Physiol.* 13): G118–G126, 1986.
127. TAKEUCHI, K., A. MERHAV, AND W. SILEN. Mechanism of luminal alkalinization by bullfrog gastric mucosa. *Am. J. Physiol.* 243 (*Gastrointest. Liver Physiol.* 6): G377–G388, 1982.
128. TEORELL, T., AND R. WESSALL. Electrical impedance properties of surviving gastric mucosa of the frog. *Acta Physiol. Scand.* 10: 243–257, 1945.
129. THOMPSON, S. M. Relations between chord and slope conductances and equivalent electromotive forces. *Am. J. Physiol.* 250 (*Cell Physiol.* 19): C333–C339, 1986.
130. TRIPATHI, S., AND P. K. RANGACHARI. In vitro primate gastric mucosa: electrical characteristics. *Am. J. Physiol.* 239 (*Gastrointest. Liver Physiol.* 2): G77–G82, 1980.
131. UEDA, S., D. D. F. LOO, AND G. SACHS. Regulation of K channels in the basolateral membrane of *Necturus* oxyntic cells. *J. Membr. Biol.* 97: 31–41, 1987.
132. USSING, H. H., AND A. LEAF. Transport across multimembrane systems. In: *Membrane Transport in Biology*, edited by G. Giebisch, D. C. Tosteson, and H. H. Ussing. New York: Springer-Verlag, 1978, p. 1–26.
133. VAN DRIESSCHE, W., AND W. ZEISKE. Ionic channels in epithelial membranes. *Physiol. Rev.* 65: 833–903, 1985.
134. VILLEGAS, L. Cellular location of the electrical potential difference in the frog gastric mucosa. *Biochim. Biophys. Acta* 64: 359–367, 1962.
135. VILLEGAS, L., F. MICHELANGELI, AND L. SANANES. Asymmetrical response of oxyntic cells to direct current in nonstimulated frog gastric mucosa. *Biochim. Biophys. Acta* 219: 518–520, 1970.
136. WALLMARK, B., A. BRANDSTRÖM, AND H. LARSSON. Evidence for acid-induced transformation of omeprazole into an active inhibitor of (H⁺ + K⁺)-ATPase within the parietal cell. *Biochim. Biophys. Acta* 778: 549–558, 1984.
137. WALLMARK, B., B.-M. JARESTEN, H. LARSSON, B. RYBERG, A. BRANDSTRÖM, AND E. FELLENIUS. Differentiation among inhibitory actions of omeprazole, cimetidine, and SCN⁻ on gastric acid secretion. *Am. J. Physiol.* 245 (*Gastrointest. Liver Physiol.* 8): G64–G71, 1983.
138. WALLMARK, B., G. SACHS, S. MARDH, AND E. FELLENIUS. Inhibition of gastric (H⁺ + K⁺)-ATPase by the substituted benzimidazole, picoprazole. *Biochim. Biophys. Acta* 728: 31–38, 1983.
139. WEINSTEIN, R. S., B. F. BANNER, J. R. KUSZAK, N. J. THOMAS, AND B. U. PAULI. Ultrastructure of tight junctions in prostaglandin-exposed rat stomach. *Dig. Dis. Sci.* 31, Suppl.

2: 115S–119S, 1986.
140. WELSH, M. J., P. L. SMITH, M. FROMM, AND R. A. FRIZZELL. Crypts are the site of intestinal fluid and electrolyte secretion. *Science Wash. DC* 218: 1219–1221, 1982.
141. WILL, P. C., R. N. CORTWRIGHT, R. G. GROSECLOSE, AND U. HOPFER. Amiloride-sensitive salt and fluid absorption in small intestine of sodium-depleted rats. *Am. J. Physiol.* 248 (*Gastrointest. Liver Physiol.* 11): G133–G141, 1985.
142. WINSHIP, D. H., AND C. R. CAFLISCH. Intramucosal gastric acid concentration determined by glass microelectrode technique. *Biochim Biophys. Acta* 291: 280–286, 1973.
143. WOLOSIN, J. M., AND J. G. FORTE. Stimulation of oxyntic cell triggers K^+ and Cl^- conductances in apical H^+-K^+-ATPase membrane. *Am. J. Physiol.* 246 (*Cell Physiol.* 15): C537–C545, 1984.
144. WRIGHT, G. H. Electrical impedance, ultrastructure and ion transport in foetal gastric mucosa. *J. Physiol. Lond.* 242: 661–672, 1974.
145. ZEISKE, W., T. E. MACHEN, AND W. VAN DRIESSCHE. Cl^- and K^+-related fluctuations of ionic current through oxyntic cells in frog gastric mucosa. *Am. J. Physiol.* 245 (*Gastrointest. Liver Physiol.* 8): G797–G807, 1983.

CHAPTER 11

Cell biology of hydrochloric acid secretion

JOHN G. FORTE | Department of Physiology-Anatomy, University of California, Berkeley, California, and Medical and Research Services,

ANDREW SOLL | Wadsworth Veterans Administration Hospital Center and University of California, Los Angeles, California

CHAPTER CONTENTS

Functional Morphology of the Oxyntic Cell
  General structure
  Ultrastructure of oxyntic cells
  Membrane turnover and recycling
  Cytoskeleton of oxyntic cell
Receptors Regulating Oxyntic Cell Function
  Techniques for studying gastric glands and cell preparations
    Dispersion and isolation of gastric glands and cells
    Techniques for cell separation
    Studies of oxyntic cell function
  Studies of receptor specificity
    Histamine receptors
    Cholinergic receptors
    Gastrin receptors
    Other oxyntic cell receptors
  Receptors on nonoxyntic cells
    Histamine cells in the fundic mucosa
Cellular Basis of Hydrochloric Acid Secretion
  H^+-K^+-ATPase: the primary H^+-pump enzyme
  Membrane changes associated with hydrochloric acid secretion
  Hydrochloric acid transport at the apical cell membrane
  Transport at the basolateral membrane
  Oxyntic cell transport model
Cell-Activation Mechanisms
  Oxyntic cell activation by cyclic AMP
  Prostaglandin inhibition of oxyntic cell function
  Oxyntic cell activation by calcium-dependent mechanisms

SECRETION OF HYDROCHLORIC ACID by the stomach is a carefully regulated process involving a succession of steps between the initial stimulus and the final effluent product. These steps include the binding of secretagogue to appropriate receptor, transduction to cytoplasmic second messengers, activation of cytoplasmic regulatory proteins, transformation of membranes within the secretory cell, transport activities at the apical and basal lateral membranes, and the flow of metabolites to provide the energetic requirements. Regulation of gastric secretory activity in the intact animal is discussed in the chapters by Hirschowitz and Konturek in this *Handbook*. We deal with the cell biology of HCl secretion, discussing the membrane and cytoplasmic factors that are responsible for cellular activation and secretion of the acid juice.

FUNCTIONAL MORPHOLOGY OF THE OXYNTIC CELL

General Structure

The gastric epithelial cell that secretes HCl is called the oxyntic cell, a name derived from the Greek word *oxyntos* (to generate an acidic substance). The gastric acid-secreting cell is also called the parietal cell because of its histological location "on the walls" of the gastric gland. A tubular gastric gland isolated from the fundic region of the rabbit stomach is shown in Figure 1. The oxyntic cells tend to protrude away from the lumen of the gland; chief cells, which secrete pepsinogen, are much more centrally located around the gland lumen. Oxyntic cells are found almost exclusively in the fundus and/or corpus of gastric mucosa, and they tend to predominate within the upper and middle regions of these glands. However, there are some variations among species of oxyntic cell density within the mucosa [for an excellent review of gastric histology, see Helander (60)].

As with other cells of the gastric epithelium, oxyntic cells are renewed by differentiation of progenitor cells located in the neck region of the glands. Newly differentiated oxyntic cells progressively migrate from the neck toward the base of the gland. On the basis of morphological and histochemical criteria, it has been suggested that there are functional differences that may be related to age of the oxyntic cells: the younger, more luminally located cells are presumably more active in HCl secretion than the more basal cells (28, 62, 70). Although oxyntic cells constantly turn over, there appears to be considerable species variation in their life span. For example, in rodents the oxyntic cell population is replaced in ~3 mo (96), whereas in humans, renewal of the oxyntic cell population may take 1 yr or more (78).

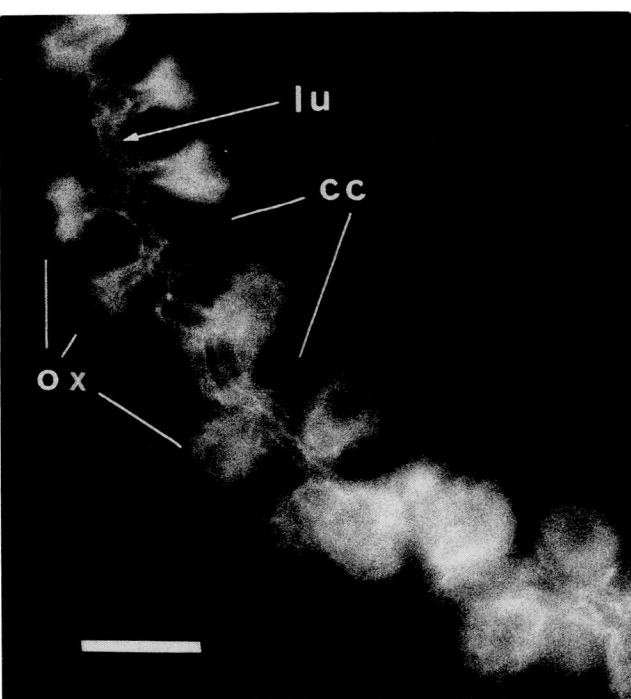

FIG. 1. Isolated gastric gland. Glands were isolated from rabbit fundic mucosa, stimulated with Bt₂cAMP, and subsequently fixed and stained with a fluorescent-labeled probe for filamentous actin (phallacidin) (144). The glands appear as a tube of cells, with the oxyntic cells (*ox*) tending to bulge outward from the wall of the gland. Under fluorescence microscopy the distribution of actin microfilaments provides additional detail of cytoskeletal organization. Oxyntic cells are well stained; distribution of the actin stain is localized to microvilli in the expanded secretory canaliculi of stimulated oxyntic cells. No stain is visible within the cytoplasm of chief cells (*cc*), the other major cell type within the gland. Thin deposits of stain define the gland lumen (*lu*); this pattern is consistent with actin microfilaments within luminal microvilli of both cell types. Bar, 20 μm.

Ultrastructure of Oxyntic Cells

The oxyntic cell has several distinctive morphological features that have functional significance (see refs. 60 and 68 for a more complete description). The nonsecreting, or resting, cell has a limited area of apical membrane in direct contact with the gland lumen, but distinctive invaginations extend into the cell from the luminal surface, forming a system of tiny canals or secretory canaliculi (Fig. 2). The entire apical surface, including the canaliculi, contains short, stubby microvilli. In thin sections one can see an abundance of tubular and vesicular membrane profiles, especially prominent in the apical pole of oxyntic cells, the system of so-called tubulovesicles. There are also numerous, large mitochondria to support the high demand for oxidative energy to secrete acid. In resting oxyntic cells of many species, the majority of the cytoplasmic volume (>60%) is occupied by tubulovesicles and mitochondria.

Oxyntic cells undergo profound morphological change when stimulated to secrete acid. Nearly 100 years ago Golgi (53) described the secretory canaliculi and noted that they were enlarged in oxyntic cells from the "digesting stomach." Electron micrographs of maximally stimulated cells reveal enlarged canalicular spaces, a greatly expanded apical membrane surface with elongated microvillar projections, and a diminution of cytoplasmic tubulovesicles (Fig. 3). Morphometric analyses have shown that the 5- to 10-fold increase in apical surface area typically seen in the transition from rest to stimulation can be quantitatively accounted for by the decrease in surface area of the cytoplasmic tubulovesicles (61, 69, 146).

Membrane Turnover and Recycling

Two hypotheses have been proposed to account for the elaborate ultrastructural transformations that accompany oxyntic cell secretory activity: the osmotic flow hypothesis and the membrane recycling hypothesis. According to the osmotic flow hypothesis (9), the compartment of tubulovesicles is always confluent with the apical cell membrane but presents the morphological appearance of a supercollapsed system of infolded membrane; osmotic flow accompanying HCl secretion forces the membrane spaces to expand, and the apical surface of stimulated cells appears in its extended form. Although most investigators agree that bulk fluid secretion represents a dilating force in the gland lumen, a number of experimental tests do not support the osmotic flow hypothesis as it relates to membrane transformations. For example, the induction of osmotic flow not directly related to cell stimulation did not produce apical surface expansion (52), and no structural continuities have been observed between the gland lumen and tubulovesicles in the resting state.

The membrane recycling hypothesis contends that tubulovesicles are a distinct membrane component that must fuse with the apical surface to provide increased area and access of the intrinsic proton pumps of the gland lumen (48). Most existing experimental data are consistent with some means of membrane recycling through fusion processes, although specific identification of fusion sites has been difficult; fusion events are only occasionally captured by ultrastructural analyses, and static electron micrographs must be cautiously interpreted in terms of dynamic processes.

Macromolecular tracers applied to the luminal side of functioning gastric mucosa are taken up from the apical surface into tubulovesicles, and this is especially apparent when the tissue returns from the stimulated to resting state (41, 112). On the other hand, when gastric mucosa were fixed prior to the exposure to tracers (macromolecules or extracellular surface stains) the tubulovesicular compartment remained free of the probes (41, 47). Thus a permanent access pathway does not exist between gland lumen and tubulovesicles, but transient access and communication between these compartments occurs during functional activity, most likely through fusion and fission.

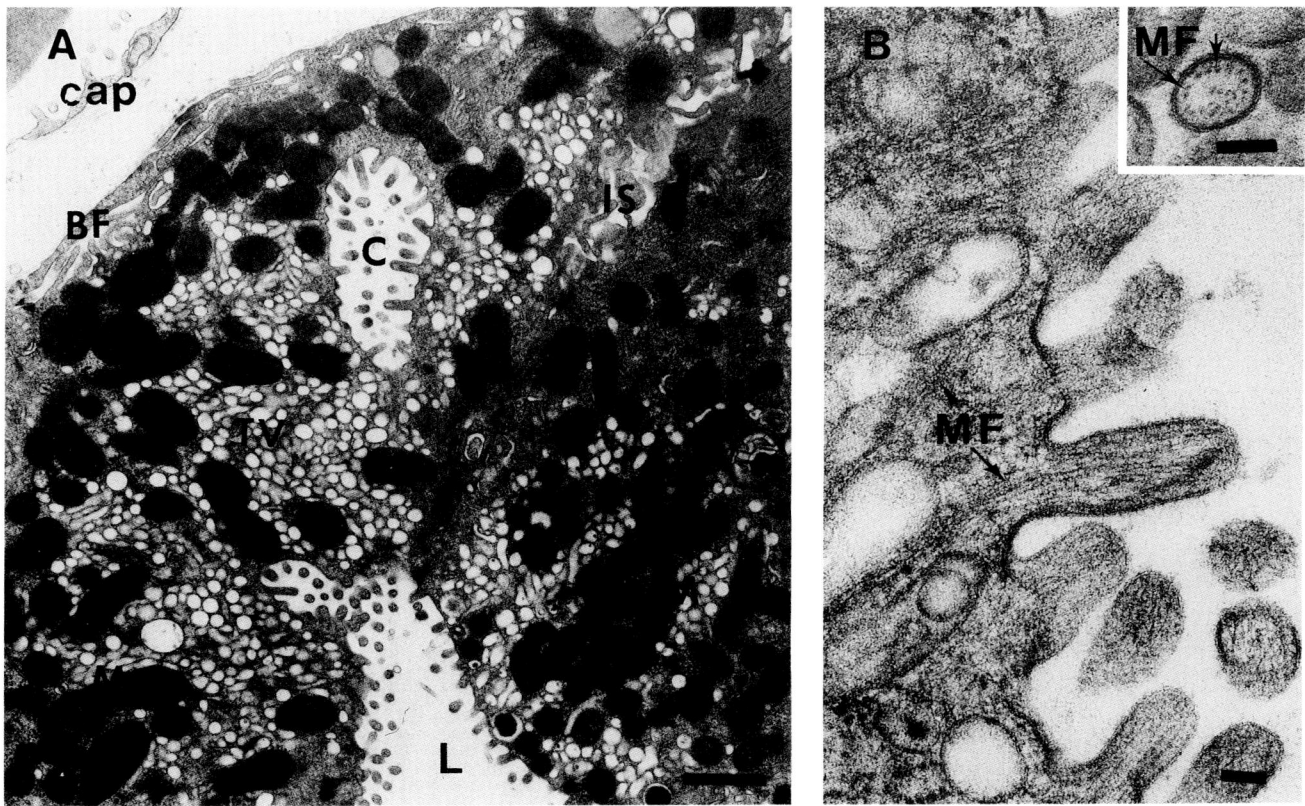

FIG. 2. Resting (nonsecreting) oxyntic cells from piglet gastric mucosa. A: low-power view showing oxyntic cells in a section cut across the gastric gland. Lumen (L) of the gland is visible, as well as intercellular spaces (IS), basolateral folds (BF), and a capillary (cap) on the interglandular space. Secretory canaliculi (C) extend from the gland lumen into the cell and contain short stubby microvilli. The cytoplasm of the oxyntic cell, especially the apical portion, contains numerous tubular and vesicular membranous profiles of the "tubulovesicles" (TV), and large mitochondria (M) are readily apparent. Bar, 1 μm. B: high-power view of apical microvilli showing organization of microfilaments (MF) in longitudinal as well as cross section (inset). Tubulovesicles can also be seen beneath the apical plasma membrane. Bars, 0.1 μm. [A from Forte and Machen (46a).]

Freeze-fracture electron microscopy has demonstrated differences in the structural organization of the apical plasma membrane of resting and stimulated oxyntic cells (15). On the other hand, substructure of tubulovesicular membranes closely resembles that seen in the apical membrane of the stimulated cells, conforming to the hypothesis that the large apical membrane expansion associated with stimulation is derived from fusion with tubulovesicles. This view is substantiated by immunocytochemistry using a monoclonal antibody to the H$^+$-pump enzyme, the H$^+$-K$^+$-ATPase (see *H$^+$-K$^+$-ATPase: the Primary H$^+$-Pump Enzyme*, p. 216). In resting cells, H$^+$-K$^+$-ATPase antibody was heavily localized to tubulovesicles with only a light staining pattern within the apical plasma membrane, whereas the apical membrane was heavily stained with antibody in stimulated cells (115).

These cytological data show that the tubulovesicle compartment is distinct from the apical plasma membrane in the resting oxyntic cell and that continuity between the two membrane compartments is achieved in the stimulated state. Withdrawal of the stimulus leads to a reversal of the process, whereby the apical surface area is diminished and tubulovesicles reform within the apical pole of the cell. Thus a scheme of membrane recycling, through fusion and fission, represents an adequate working hypothesis to account for secretion-dependent structural changes.

Secretion of gastric HCl does not occur unless the structural transformations have taken place within the oxyntic cell. However, it has been possible to induce the membrane changes associated with oxyntic cell activation without the production of HCl. For example, acid secretion can be inhibited at several sites distal to secretagogue-receptor activation without any significant change in the morphological transformation. Inhibition of acid output by such diverse modes as substrate depletion (16), uncoupling of H$^+$ transport by SCN$^-$ (16, 19, 146), or direct inhibition of the H$^+$-pump enzyme by picoprazole or omeprazole (62) does not prevent the morphological transformation induced by secretagogues. Thus membrane fusion and associated apical membrane surface expansion are necessary but not sufficient for HCl secretion. The

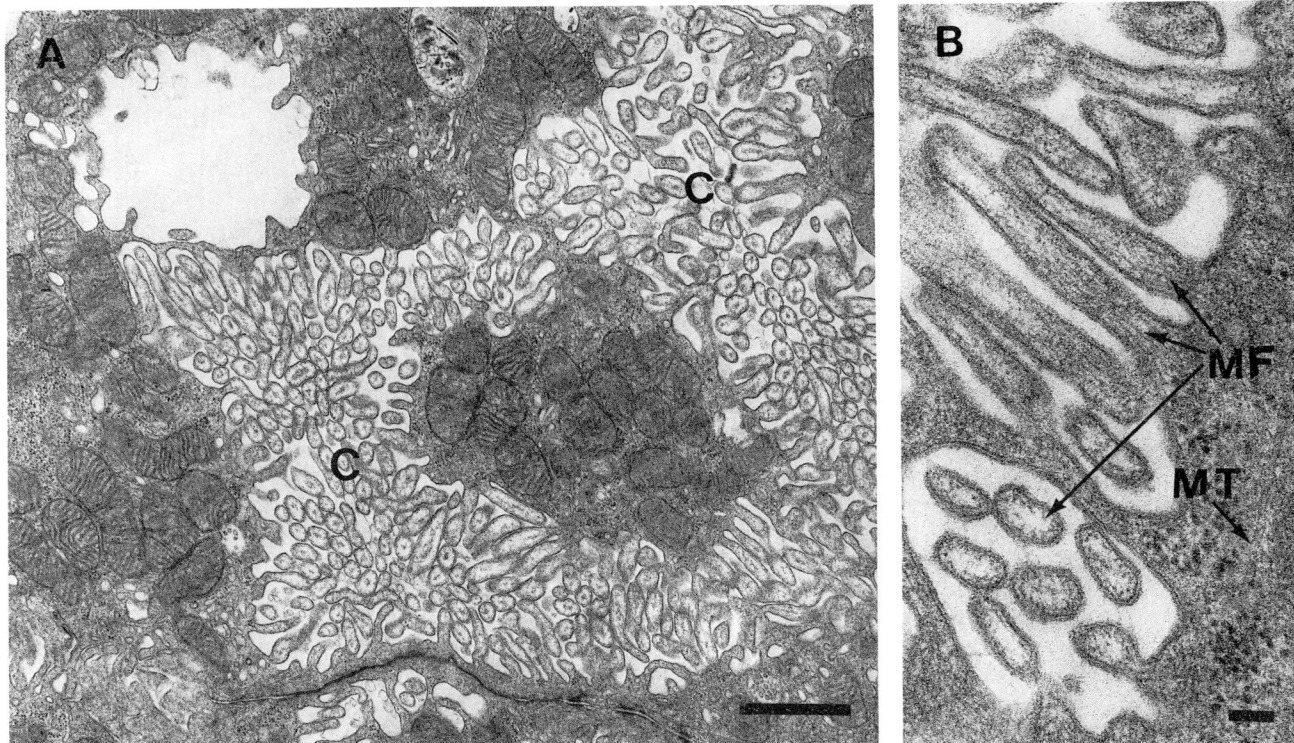

FIG. 3. Maximally stimulated oxyntic cells from piglet gastric mucosa. A: secretory canaliculi (C) are filled with long microvilli, greatly expanding the apical surface area. Only a relatively few tubulovesicles are left within the cytoplasm. Bar, 1 µm. B: high-power view of expanded apical surface in which microfilaments (MF) and microtubules (MT) can be seen. Bar, 0.1 µm. [A from Forte and Machen (46a).]

structural changes provide the external access port for the H^+ pump, and they are a fundamental step in the pathway of stimulus-secretion coupling.

Cytoskeleton of Oxyntic Cell

The forces responsible for the mass movement of membranes and organization of the apical surface during the secretory cycle of the oxyntic cell cannot be specified. Some role for microfilaments and microtubules has been implied from studies that have used inhibitors, or disruptors, of these cytostructural elements.

A system of highly organized microfilaments is a characteristic feature of gastric oxyntic cells. In Figure 1 the gastric gland microfilaments composed of polymerized actin can be seen distributed in the microvillar region throughout the apical and canalicular membrane surfaces. At the ultrastructural level, microfilaments are arranged within the periphery of the microvillus and run the entire length of the microvillus (17, 135). The peripheral arrangement of oxyntic cell microfilaments, near the microvillar plasma membrane (Figs. 2B and 3B), is distinctly different from the central core of microfilaments characteristic of intestinal microvilli. In the resting oxyntic cell, organized microfilaments extend into the cytoplasm up to 1 µm below the bases of the short microvilli. In going from rest to the secretory state there are few structural changes apparent in the organization of microfilaments, with the exception of possible sites of crossbridges (between microfilaments and microvillar membrane) and have been tentatively identified in the extended microvilli of the stimulated cell (17). On the other hand, in the initial stages of return to the resting structure after removal of secretagogues, there are striking changes of apparent disorganization of microfilamentous structure and the apparent collapse of the apical canalicular surface (48). Treatment of gastric mucosa with cytochalasin B, a microfilament-disrupting agent, causes an inhibition of H^+ secretion and the severe disorganization of microfilaments within the collapsed microvilli (17). An important difference between these two treatments is that the cytochalasin B-treated cells remain in the collapsed, disorganized state, whereas after removal of secretagogues the cells follow an orderly progression toward reestablishment of the resting structure with highly organized microfilaments aligned parallel within the microvillar walls.

A role for microtubules in oxyntic cell secretion has been suggested on the basis of studies with colchicine and vinblastine, agents that disrupt microtubules and also inhibit gastric acid secretion (71). Furthermore, when secreting amphibian gastric mucosa was treated with a histamine H_2 antagonist to block acid secretion, a time-dependent increase in free-tubulin content was observed, indicating a progressive depolymerization of microtubules in the secretion-inhibited tissue (72).

These observations implicate some role for both microfilaments and microtubules in gastric acid secretion. It is not surprising that cytoskeletal elements would be important in a process where extensive structural change and membrane rearrangement are so fundamental to functional activity. However, the specific mechanisms that regulate the state of the cytoskeletal components, their interactions with membranes, and the means for force generation remain to be determined. Characterization of the modulating elements within oxyntic cells, such as actin-binding proteins and microtubule-associated proteins, is an important area in which to extend our knowledge of the cell biology of acid secretion.

RECEPTORS REGULATING OXYNTIC CELL FUNCTION

This section focuses on the regulation of oxyntic cell function by chemical transmitters, based on information gained from studies with isolated fundic mucosal cells and gastric glands. Characterization of receptors requires a reductionistic approach for several reasons. Acid secretion in vivo is regulated by inputs from endocrine, paracrine, and neurocrine pathways; gastrin, histamine, and acetylcholine, respectively, are the major recognized transmitters mediating the effects of these pathways. Evidence indicates that the receptors for these chemotransmitters are present on other cell types, in addition to the oxyntic cell itself. Furthermore there are several other chemotransmitters that modulate acid secretion, such as somatostatin, adrenergic agents, and prostaglandins, with receptors possibly on oxyntic and/or non-oxyntic cells. The existence of these several pathways and receptors complicates receptor localization and characterization of the cellular mechanisms regulating acid secretion. Study of the regulation of acid secretory mechanisms in intact mucosa or in vivo is also complicated by the cellular heterogeneity of the fundic mucosa; not only are there oxyntic, chief, and mucous cells but many endocrine cells and cellular elements of the lamina propria. As an additional complicating factor, several of the chemotransmitters that potentially regulate acid secretion are present in the mucosa in the immediate vicinity of the oxyntic cell, including histamine, acetylcholine, somatostatin, glucagon, and prostaglandins. To simplify these problems, a variety of methods have been developed over the last decade for the isolation of cells and glands from fundic mucosa, thereby allowing the cell separation and direct characterization of the receptors and mechanisms regulating cell function. The following sections briefly review the methods used to disperse cells and glands from the fundic mucosa and the data obtained pertaining to the receptors regulating acid secretion.

Techniques for Studying Gastric Glands and Cell Preparations

DISPERSION AND ISOLATION OF GASTRIC GLANDS AND CELLS. Enzyme digestion has been used to disperse cells and glands from fundic mucosa; this method has been reviewed recently (123). Crude collagenase, which contains enzyme activities in addition to collagenase, including clostripain, proteases, and trypsin-like activity, allows preparation of gastric glands; high-pressure perfusion facilitates the recovery of glands (8). Adding a Ca^{2+}-chelation step facilitates dispersion of cells; alternatively, pronase can be added to collagenase to allow dispersion of single cells. There is some variability among species with respect to how readily mucosal cells are dispersed; rabbit mucosa is most readily dispersed, whereas pig mucosa requires more aggressive techniques. Treatment with enzymes and Ca^{2+} chelation may be deleterious to cell function, requiring cautious interpretation of findings, particularly negative ones.

TECHNIQUES FOR CELL SEPARATION. Because of the heterogeneous cell population, cell separation is mandatory for many studies. Enrichment of various cell populations has been accomplished by use of both velocity and density separation techniques. Velocity separation can be accomplished by unit gravity sedimentation (51, 82, 105), by use of an elutriator rotor (116, 119, 138), and by brief, repeated centrifugation (127, 128). Velocity separation can provide enrichment of oxyntic cells up to ~60%; chief cells and a few mucous cells are the main contaminating cells.

Density separations have been performed with a variety of media, including sucrose, Ficoll, albumin plus Ficoll, Percoll, and Nycodenz. Both step gradients and linear gradients have been used. The success of these gradients depends (as does velocity separation) on the starting preparation and on the desired end point. Step gradients are technically easier and may provide a good yield of highly enriched cells; however, these gradients may not provide resolution of complex subpopulations of mucosal cells. Density gradients are generally effective in separating the denser chief cells from the lighter oxyntic cells. The main contaminating cells in the oxyntic cell fraction from a density separation are mucous cells and the lighter small cells, which may include endocrine and endocrine-like cells and macrophages. The choice of media for density separation is important; these media may have deleterious effects on cell viability and function.

The limitations encountered with the density or velocity techniques can be reduced by combining these techniques, with their sequential application providing excellent enrichment of selected cell types. An early application of sequential velocity and density techniques permitted further separation of the canine fundic small cell elutriator fraction that contained several cell types, including mast cells, endocrine cells (with either somatostatin and glucagon), endocrine-like cells (containing serotonin and dopa decarboxylase activity), macrophages, and endothelial cells (121, 124). This sequential use of elutriation and linear density gradient allowed enrichment of mast cells to >80% and separation of the other cell types. Sequential velocity and density separation yields a high degree of enrichment of canine fundic parietal and chief cells (122). These techniques have been used in reverse order for enrichment of rabbit parietal cells, with removal of the small cells (24).

STUDIES OF OXYNTIC CELL FUNCTION. Because isolated oxyntic cells have lost their polar orientation, indirect indices of cell function need to be used. Four indices have proved useful in recent studies. The secretion of acid is a highly energy-dependent process, and both O_2 consumption and glucose oxidation provide useful quantitative measures of total cell responsiveness to stimulation. Recently the formation of $^{14}CO_2$ from [^{14}C]glucose has been successfully used to determine metabolic activity in oxyntic cells from the dog (33, 123) and in rabbit gastric glands. The oxyntic cell also undergoes a dramatic morphological transformation with stimulation, as previously discussed. With stimulation of oxyntic cells in vitro, this morphological transformation also occurs, providing evidence for the preservation of functional integrity in these dispersed cells and glands and a useful qualitative measure of cell function.

The accumulation of weak bases into membrane-bound acid spaces provides an extremely useful measure of acid sequestration by isolated cells and glands. This technique was adapted by Berglindh (10) from the studies with isolated vesicles that demonstrated weak base accumulation by pH partition into acid compartments. Radiolabeled probes such as [^{14}C]aminopyrine allow quantitation of sequestered acid in isolated oxyntic cells and glands (8, 10, 119). Fluorescent weak base probes, such as acridine orange or 9-aminoacridine, can also be quantitated with fluorometric or microscopic techniques.

Glucose oxidation and O_2 consumption optimally reflect the full spectrum of oxyntic cell function, since they are indices of the energy requirements to support H^+ secretion. Unrelated changes in cell respiration can occur, such as an uncoupling of oxidative phosphorylation, which can alter this correlation. Weak base accumulation provides the most convenient and widely utilized index of oxyntic cell function; however, weak base accumulation is a measure of the sequestered acid compartment and does not reflect the actual rate of acid secretion. Ideally patterns of cell responsiveness should be confirmed with different functional modalities, such as a combination of aminopyrine accumulation and a metabolic parameter.

Studies of Receptor Specificity

The locus and specificity of the receptors for histamine, gastrin, and acetylcholine have been the main focus for many of the studies with isolated oxyntic cells and gastric glands. The pattern of responsiveness and the apparent pharmacological properties of receptors are similar with each of the indices of cell response mentioned in the preceding section, and in general there has been good agreement between studies. However, certain findings, such as the presence of gastrin receptors on the oxyntic cell and the occurrence of direct modulation of histamine release by gastrin and acetylcholine, remain controversial. Some of these differences appear to be due to species differences, as discussed subsequently.

HISTAMINE RECEPTORS. Several lines of evidence indicate that the oxyntic cell has specific receptors for histamine. In studies with isolated oxyntic cells and gastric glands, histamine stimulates O_2 consumption, aminopyrine accumulation, and glucose oxidation. Histamine stimulates oxyntic cell function in rabbit gastric glands (10, 27) and the function of oxyntic cells dispersed from the rat (35, 36, 111), guinea pig (3), frog (83), and human (40). There are considerable species differences in the magnitudes of the response to histamine; for example, histamine produced a more pronounced enhancement of aminopyrine accumulation into rabbit than canine oxyntic cells. In general, histamine alone stimulates oxyntic cell function, with the response enhanced by use of a phosphodiesterase inhibitor, such as isobutylmethylxanthine (IBMX).

Histamine action on oxyntic cells is mediated by a pharmacologically typical H_2 receptor. The H_2 blockers competitively inhibit histamine stimulation of O_2 consumption and aminopyrine accumulation by rabbit gastric glands (6, 27) and isolated oxyntic cells from dogs, guinea pig, and rat (3, 35, 116, 119). The kinetics of inhibition are similar to those obtained with other typical H_2-receptor systems, such as guinea pig atrium and rat uterus. Anticholinergic agents in concentrations that markedly inhibit the response to cholinomimetics fail to shift the dose response to histamine in the oxyntic cell. The H_2 antagonists do inhibit histamine-stimulated cyclic AMP (cAMP) production and aminopyrine accumulation, but only at concentrations 10,000-fold greater than those necessary to inhibit typical H_1 receptors. Furthermore these oxyntic cell effects of H_1 antagonists are not specific for histamine, since the responses to carbachol and dibutyryl cAMP (Bt_2cAMP) are also blocked (27, 119). Therefore present data indicate a pharmacologically typical H_2 receptor on the oxyntic cell.

Direct study of oxyntic cell histamine receptors using radiolabeled ligands has proved difficult (14), probably in large part because of histamine uptake into oxyntic cells (11) and the presence of histamine methyltransferase in oxyntic cells (1, 5). Studies with [^{3}H]cimetidine found high-affinity sites in guinea pig tissues (104), but the specificity of these sites was not consistent with an H_2 receptor. In guinea pig oxyntic cells, [^{3}H]histamine interacted with sites that also did not have an H_2 specificity (4). Characterization of histamine H_2 receptors using these methods will probably require development of radioligands that selectively interact with H_2 receptors and not with potential uptake and degradative sites.

CHOLINERGIC RECEPTORS. Cholinergic agents stimulate oxyntic cells and gastric glands from rabbit (7), dog (116, 119), rat (36), guinea pig (3), and human. Antimuscarinic agents inhibit carbachol action on oxyntic cells (3, 6, 36, 117, 119). Atropine at concentrations between 3.2 nM and 100 nM produced a progressive, parallel rightward shift of dose response for carbachol stimulation of aminopyrine accumulation by isolated canine oxyntic cells, while cimetidine did not inhibit this cholinergic response (119). The dissociation constant determined for this atropine effect was 1 nM, which is similar to the constant found for atropine inhibition of cholinergic stimulation in other tissues. Even after enzyme dispersion, the oxyntic cell appears to retain pharmacologically typical muscarinic and H_2 receptors. Caution is necessary in studies with atropine, as is true for other drugs; atropine at concentrations >100 μM inhibits the response to histamine, Bt_2cAMP, or gastrin, effects clearly unrelated to specific blockade of the muscarinic receptor.

Direct studies of muscarinic-cholinergic receptors in the gastric mucosa have been done with [^{3}H]quinuclidinyl benzilate (QNB) and N-methylscopalamine. Pharmacological specificity of binding to an enriched rat oxyntic cell preparation (37) and an enriched canine oxyntic cell preparation (A. H. Soll, unpublished observations) was consistent with interaction at a typical muscarinic receptor site. Culp et al. (30) examined the distribution of QNB-binding sites to crude membrane fractions from elutriator-separated fractions of canine fundic mucosal cells. The QNB sites were found in high number in all of the fractions, with the number of QNB-receptor sites per milligram protein greatest in fractions maximally enriched in chief cells.

Two muscarinic receptor antagonists, pirenzepine and telenzepine, compared to atropine have a greater potency inhibiting acid secretion than altering heart rate (38, 64). These findings suggest that M_1 receptors may be a subtype of muscarinic receptors regulating acid secretion, but the localization of these putative receptors remains elusive. In studies with rat oxyntic cells (106) and rabbit gastric glands (109), high concentrations of pirenzepine were required to inhibit cholinergic-stimulated parietal cell function. In studies with canine oxyntic cells (A. H. Soll, unpublished observations), telenzepine and pirenzepine inhibited cholinergic stimulation of aminopyrine accumulation with potencies that were 10- and 80-fold less, respectively, than atropine (21). These findings indicate that the oxyntic cell possesses a muscarinic receptor with at most an intermediate affinity for M_1 antagonists.

GASTRIN RECEPTORS. Considerable controversy still surrounds the localization of gastrin receptors within the fundic mucosa. With isolated canine oxyntic cells, gastrin produced a small but definite increase in O_2 consumption (117), aminopyrine accumulation (119), and glucose oxidation. These responses were not blocked by either H_2 antagonists or anticholinergic agents, indicating interaction at a third receptor site (Fig. 4). Because this gastrin effect was found in fractions enriched in canine oxyntic cells (119, 124), direct action on oxyntic cells, rather than action mediated by histamine release, appeared likely. Proglumide, an agent shown to selectively block cholecystokinin (CCK) receptors in pancreatic acinar cells (56), inhibits gastrin, but not histamine, stimulation of canine oxyntic cells (122).

The existence of a specific gastrin receptor on canine oxyntic cells is supported by findings from studies with a biologically active ^{125}I-labeled [Leu^{15}]gastrin (122). In these studies the biological activity of the gastrin tracer was preserved with [Leu^{15}]gastrin-17 (133); alternatively, biological activity can be preserved with conditions that avoid oxidation of native gastrin, such as oxidizing with lactoperoxidase or reducing the tracer after iodination. In cell separation studies with the elutriator rotor, ^{125}I-[Leu^{15}]gastrin binding correlated with the distribution of oxyntic cells, indicating that oxyntic cells accounted for the majority of the gastrin binding to canine fundic mucosal cells. This finding was confirmed in studies in which step density gradients were performed, starting with fractions enriched in content of oxyntic or chief cells using the elutriator; with these step gradients, enrichment of the oxyntic cell and of the chief cell content in excess of 85%, respectively, was achieved. With the ^{125}I-[Leu^{15}]gastrin tracer, binding correlated positively with oxyntic cell content and negatively with the chief cell content. These data clearly indicate that the canine oxyntic cell, and probably not the chief cell, has a specific gastrin receptor. However, in the elutriator separation there was indication of gastrin binding to an additional cell type(s) that eluted in a small cell fraction, as discussed in *Receptors on Nonoxyntic Cells*, p. 214.

When canine oxyntic cells were studied, gastrin binding was found to correlate to gastrin stimulation of oxyntic cell function (122). Proglumide inhibited gastrin binding, and the inhibition of binding was proportional to inhibition of gastrin stimulation of oxyntic cell function. The octapeptide of CCK (CCK-8) inhibited gastrin binding and stimulated oxyntic cell function with a potency similar to that found for

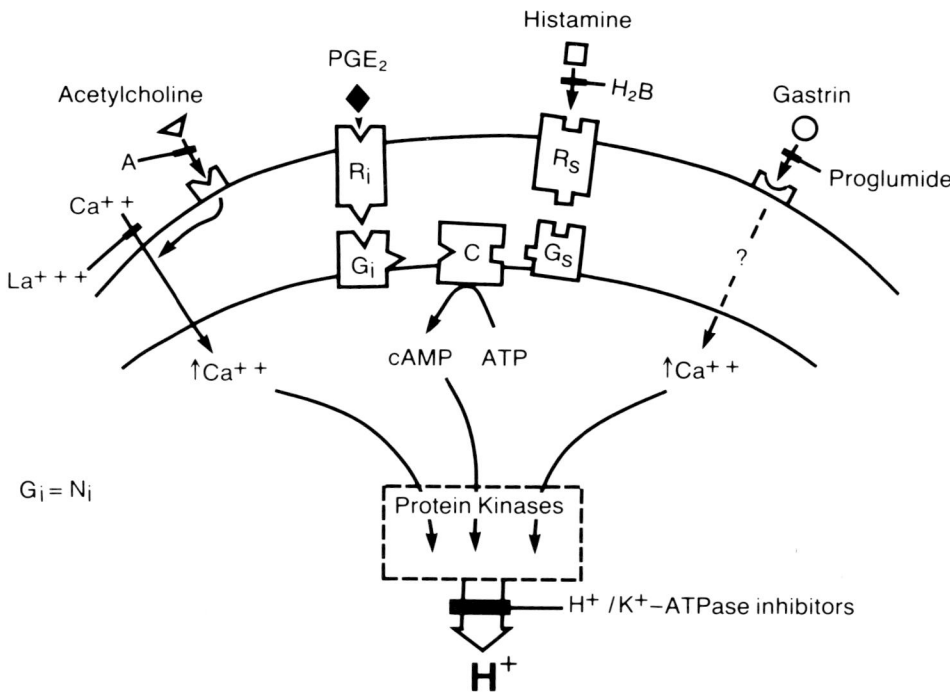

FIG. 4. Model for receptor regulation of canine parietal cell function. This revised model (116) depicts pharmacologically specific muscarinic, histamine, and gastrin receptors on the canine parietal cell. Respectively, these 3 receptors can be selectively blocked by atropine (A), H_2 histamine antagonists (H_2B), and proglumide (119, 121). Histamine activates adenylate cyclase, acting at a stimulatory receptor (R_s), which in turn activates the stimulatory GTP-binding protein G_s. There also appears to be an inhibitory receptor (R_i) for PGE_2 that is linked to the catalytic subunit (C) of adenylate cyclase via the inhibitory GTP-binding protein G_i. Acetylcholine and gastrin act in a Ca^{2+}-dependent fashion, although specific details regarding mechanisms increasing cytosolic Ca^{2+} remain uncertain.

the heptadecapeptide gastrin. Thus the oxyntic cell gastrin receptor differs from the receptor for CCK on the pancreatic acinar cell, since the acinar cell receptor has a much higher affinity for CCK than gastrin.

In studies with oxyntic cells isolated from frog, rat, and guinea pig, gastrin did not produce significant stimulation (3, 83). In rabbit results have been variable; no gastrin stimulation was found in gastric glands (6, 12), while a small response was reported with both glands and isolated cells in the presence of the sulfhydryl-reducing agent dithiothreitol (26). When the same techniques as employed with canine fundic cells (122) were used, studies with ^{125}I-$[Leu^{15}]$ gastrin did not detect specific binding to any fractions of rabbit fundic cells separated by elutriation (A. H. Soll, unpublished observations).

In contrast, in the presence of the phosphodiesterase inhibitor IBMX, nanomolar concentrations of gastrin produced clear stimulation of aminopyrine accumulation into rabbit gastric glands (11, 12, 26). The response to gastrin in the presence of IBMX was totally inhibited by cimetidine (100 μM). If it is assumed that the effects of 100 μM cimetidine solely reflect blockade of the H_2 receptor, these findings indicate either a critical permissive action of hista-

mine on the gastrin response or mediation of gastrin action by induction of histamine release. In the same nanomolar concentration range, gastrin also caused the release of histamine from rabbit gastric glands (13). The differences in the findings for gastrin stimulation between canine oxyntic cells and rabbit gastric glands do not result from a difference between glands and cells; gastrin alone also stimulated aminopyrine accumulation into isolated canine gastric glands (8) but not into isolated, enriched rabbit oxyntic cells (T. Berglindh, unpublished observations).

OTHER OXYNTIC CELL RECEPTORS. Several other agents may directly modulate oxyntic cell function, including somatostatin, adrenergic agents, epidermal growth factor (EGF), enteroglucagon, and purinergic agents, such as adenosine. However, the pattern of responses varies considerably among species and model systems and few firm conclusions can be formulated at this time regarding the physiological relevance of these actions.

Receptors on Nonoxyntic Cells

Although the focus of this chapter is the oxyntic cell, brief mention is necessary of the nonoxyntic cell

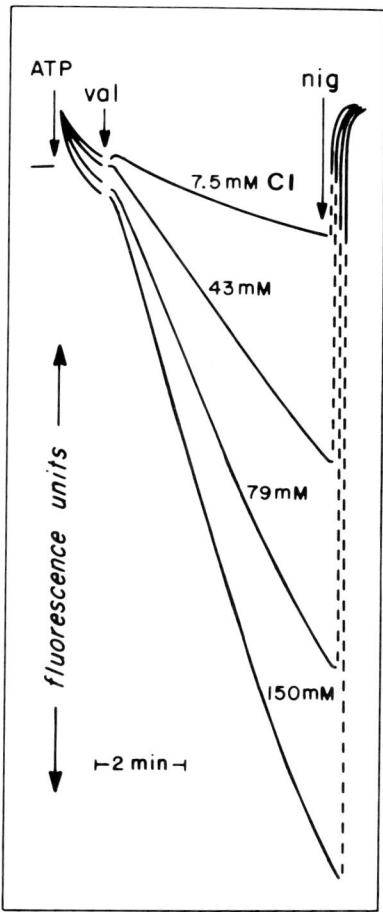

FIG. 6. H$^+$ uptake in gastric microsomal vesicles is dependent on Cl$^-$ concentration and K$^+$ ionophores. H$^+$ uptake was monitored by the acridine orange fluorescence-quenching method. Addition of ATP to a suspension of microsomal vesicles resulted in a small uptake of H$^+$ that was accelerated by the subsequent addition of the K$^+$ ionophore valinomycin (val). K$^+$ concentration was constant at 150 mM in all cases; Cl$^-$ concentration was varied as shown; isethionate was used as the balance anion. The rate of H$^+$ uptake was greatly enhanced by increasing the concentration of the permeant anion, Cl$^-$. All H$^+$ gradients were abolished by the H$^+$-K$^+$ exchange ionophore nigericin (nig). [From Lee and Forte (75).]

HO(CH$_2$)$_2$SO$_3^-$ $\gg$ carboxylates was obtained (141). This was somewhat different than the series for the basal anionic permeability of microsomal vesicles.

The basic observation that there is a change in the K$^+$- and Cl$^-$-transport characteristics of H$^+$-K$^+$-ATPase–rich vesicles associated with stimulation has now been confirmed by several other laboratories (31, 55). Cuppoletti and Sachs (31) found an increase in Cl$^-$ conductance for vesicles from histamine-stimulated rabbit stomach. From their measurements of bi-ionic diffusion potentials using a potential-sensitive cyanine dye, 3,3″-dipropyl-2,2′-thiodicarbocyanine-iodide [diSC$_3$(5)], these authors concluded that there was little or no change in vesicular conductance in going from rest to secretion. However, the recent experiments of Gunther et al. (55) carefully reexamined this question of K$^+$ conductance. They clearly demonstrated that, apart from any nonconductive isotopic exchange diffusion, there was a very large component of K$^+$ flux through a conductive pathway in membrane vesicles from stimulated stomach (i.e., SA vesicles), whereas microsomal vesicles from resting stomach were virtually devoid of a K$^+$ conductance pathway. Furthermore their demonstration that diSC$_3$(5) was an inhibitor of K$^+$ conductance in SA vesicles suggests a basis for the negative results of Cuppoletti and Sachs (31).

The search for inhibitors of the K$^+$- and Cl$^-$-transport systems in the gastric SA vesicles has begun to produce interesting results. Both the exchange fluxes and unidirectional fluxes of K$^+$ and Cl$^-$ in SA vesicles are inhibited by a number of divalent cations, including Zn^{2+}, Mn^{2+}, Co^{2+}, and Ni^{2+}. For these cations, Wolosin and Forte (143) found that the kinetics for inhibiting K$^+$ conductance were nearly identical to those for inhibiting Cl$^-$ conductance. Such similarities in kinetics of inhibition suggest some interaction or proximity between the apparently independent K$^+$ and Cl$^-$ conductance pathways and that they may even be housed within a single membrane protein or protein complex. The physiologically interesting divalent cations, Ca^{2+} and Mg^{2+}, had no significant effect on K$^+$ or Cl$^-$ conductances. The K$^+$ pathway in SA vesicles was inhibited by Ba^{2+} (which blocks K$^+$

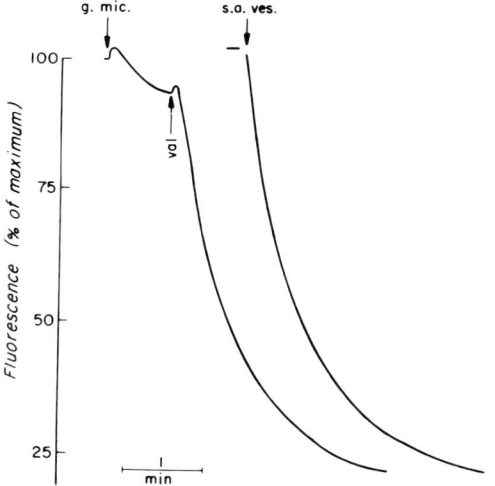

FIG. 7. Comparison of ATP-driven H$^+$ uptake in gastric microsomal vesicles and stimulation-associated vesicles. As in Figure 6, H$^+$ uptake was monitored by the acridine fluorescence-quenching method. At the indicated time, gastric microsomes (g. mic.) were added to uptake medium consisting of 150 mM KCl, 0.5 mM MgATP, and 5 μM acridine orange, with the result of very little H$^+$ uptake. Addition of valinomycin (val) increased K$^+$ permeability and K$^+$ influx, also resulting in greatly increased H$^+$ uptake and ATP turnover. When stimulation-associated vesicles (s.a. ves.) were added to the same medium there was an immediate rapid uptake of H$^+$ that was not increased by valinomycin, indicating that a pathway for rapid K$^+$ entry was present in these apical membrane vesicles from stimulated oxyntic cells.

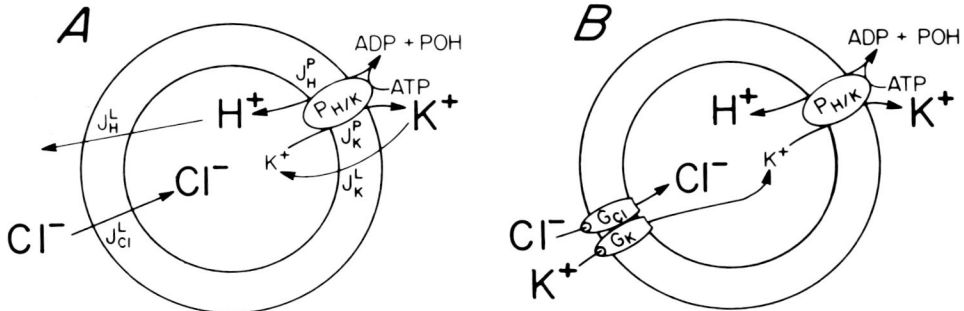

FIG. 5. Models of H$^+$ pumping and ion transport in gastric membrane vesicles. *A*: pump-leak model for ion transport in isolated microsomal vesicles from resting gastric mucosa. Microsomal model consists of an ATP-driven H$^+$-K$^+$ exchange pump and passive leak pathways for the principal ions, H$^+$, K$^+$, and Cl$^-$. *J*, ion flux, with superscripts designating the pump flux (P) or leak pathway (L). *B*: schematic representation of the mechanism of HCl accumulation in vesicles isolated from apical plasma membrane of stimulated oxyntic cells (stimulation-associated vesicles). These vesicles are significantly different from microsomes by virtue of ionic conductances for K$^+$ (G_K) and Cl$^-$ (G_{Cl}), which provide pathways for rapid entry of KCl. As for the microsomal vesicles, the ATP-driven H$^+$-K$^+$ exchange pump provides the force to accumulate intravesicular HCl.

low in the absence of exogenous ionophores; thus to account for rates of proton transport sufficient to support gastric HCl secretion, some means to accelerate K$^+$ and Cl$^-$ flux must be available. An important key to understanding how the H$^+$-K$^+$-ATPase participated in the overall HCl secretory process remained closely aligned with the membrane recycling events that accompanied secretion. Wolosin and Forte (139, 140) examined membrane fractions from stomachs that were in defined physiological states of rest or maximal HCl secretory activity. For resting stomachs, H$^+$-K$^+$-ATPase was predominantly in the low-density microsomal fraction, in conformance with much of the earlier work. On the other hand, for the maximally stimulated stomach, H$^+$-K$^+$-ATPase was greatly reduced in the microsomal fraction and much of the activity was redistributed to a fraction of larger and denser membranes, which were then called stimulation-associated (SA) vesicles. On the basis of their size, complex morphological characteristics (139), and entrainment of actin microfilaments (144), it was reasoned that the SA vesicles are derived from the elaborate apical membrane of the stimulated cell. This is in contrast to the low-density, unilamellar, microsomal vesicles, derived from tubulovesicles, that predominate in the resting oxyntic cell.

The observed redistribution of enzyme activity was consistent with a model for recycling H$^+$-K$^+$-ATPase between a "cytoplasmic" compartment and the apical plasma membrane during rest and secretion. Moreover, changes in the functional activities of H$^+$-K$^+$-ATPase–rich vesicles in different secretory states provided a new way to view the mechanism of HCl secretion and its regulation through the secretory cycle (140). The operation of the H$^+$-K$^+$-ATPase appears to be identical in both the SA vesicles and the microsomal vesicles; i.e., the pump enzyme per se is not activated when oxyntic cells are stimulated. However, the presence of an intrinsic system to promote rapid entry of K$^+$ and Cl$^-$ in SA vesicles provides the most important functional difference between the membrane types. Thus SA vesicles do not require exogenous K$^+$ ionophores for maximal rates of ATP-generated proton uptake, in contrast to microsomes isolated from resting tissue (Fig. 7). The relatively low concentrations of K$^+$ and Cl$^-$ that maximally stimulated proton-transport rates suggested that a change in permeability for both cation and anion occurred with stimulation. Pathways for rapid permeation of K$^+$ and Cl$^-$ in membranes from activated cells operate in parallel with the ATP-driven H$^+$-K$^+$ exchange pump as a sustained source for net HCl accumulation (see Fig. 4*B*).

Because of an apparent interdependence between K$^+$ and Cl$^-$ flux, Wolosin and Forte (141) first proposed that transport in SA vesicles operated via an electroneutral KCl transporter. However, more extensive analyses by these investigators have shown that observed interdependence between K$^+$ and Cl$^-$ fluxes is due to electrical coupling and that these ions move through independent conductive pathways (142). Evidence for separate pathways comes from studies of isotopic exchange, where the self-exchange fluxes of ^{86}Rb (as K$^+$ substitute) and ^{36}Cl were independent of the nature of the counterion. Using measurements of electrically coupled dissipation of H$^+$ gradients, as well as direct isotopic flux studies, Wolosin and Forte (142) concluded that both K$^+$ and Cl$^-$ conductances were increased by as much as 50-fold in H$^+$-K$^+$-ATPase–rich membranes from stimulated cells compared with the microsomal vesicles from resting ones. Like many other K$^+$ conductance pathways, the cationic conductance in SA vesicles is relatively specific for K$^+$ and its cogeners (e.g., Rb$^+$), excluding Na$^+$ and Li$^+$ (142). On the other hand, many anions will substitute for Cl$^-$ with varying efficiency. For example, with the rate of H$^+$ uptake into SA vesicles as a parameter, the series I$^-$ > Br$^-$ > Cl$^-$ > NO$^-$ >> CH$_3$SO$^-$ >

basolateral membranes. For example, H^+ in the gastric juice is derived from H_2O; thus for each H^+ secreted, an equivalent amount of base must be produced and ultimately transported from cell to plasma HCO_3^-. In the operation of the gastric proton pump, K^+ plays an essential role, and the maintenance of cellular K^+ homeostasis via the Na^+-K^+-ATPase at the basolateral membrane is another important transport function. In this section, we review the various transport systems at the apical and basolateral surfaces of the oxyntic cell and ultimately bring these together in a comprehensive picture of the formation of gastric HCl.

H^+-K^+-ATPase: the Primary H^+-Pump Enzyme

The oxyntic cell contains a unique membrane-bound ATPase that provides the driving force for HCl secretion. This gastric ATPase was first identified as a K^+-stimulated ATPase activity and an associated K^+-stimulated p-nitrophenylphosphatase activity in the microsomal fraction isolated from homogenates of oxyntic mucosa (42, 44, 49). Further purification of gastric microsomes by density-gradient centrifugation yields a relatively homogeneous population of low-density membrane vesicles, ~100–200 nm in diameter. In highly purified gastric microsomes, >80% of the total protein occurs as a single band of ~95,000–100,000 Da, which has been identified as the molecular mass of the K^+-stimulated ATPase (45, 94, 107, 115). Similarities in morphological characteristics (74), chemical constituents (43), and immunocytochemical localization of K^+-stimulated ATPase activity (115) demonstrate that the light microsomal vesicles are derived from the tubulovesicles of the intact cell and that the isolated microsomes are oriented in the same manner as the tubulovesicles, with their cytoplasmic side facing out.

Studies on isolated gastric microsomes provide a comprehensive picture of the interrelationships between ATP hydrolysis and vesicular ion transport. A site for MgATP is present on the extravesicular surface. When K^+ is present at the intravesicular surface, hydrolysis of ATP provides the driving force for the pumping of H^+ into the vesicles and the development of large gradients of up to 4.5 pH units (75, 94). Based on their observations using membrane potential probes and the measured equivalency between $^{86}Rb^+$ (as K^+ substitute) efflux and H^+ uptake, Sachs and co-workers (108, 110) concluded that the microsomal H^+ pump occurred as an ATP-driven, electroneutral H^+-K^+ exchange. Because of this closely coupled ion exchange, this gastric proton-pump enzyme is called the H^+-K^+-ATPase. The gastric H^+-K^+-ATPase shares a number of functional similarities with two other cation-transporting ATPases, the ubiquitous plasma membrane Na^+-K^+-ATPase and the Ca^{2+}-ATPase of the sarcoplasmic reticulum. All three of these ATPases are classified as $E_1 \rightarrow E_2$-ATPases in which the enzyme undergoes phosphorylation and dephosphorylation along with ligand-dependent conformational rearrangements during their catalytic cycles. The three enzymes are about the same molecular mass (~100,000 Da), share a similar amino acid profile at the active site of phosphorylation (39), and recent descriptions of their primary amino acid structure demonstrate numerous sequence homologies (113), thus underscoring the common origin of the three cation-pumping ATPases. A more detailed analysis of the biochemistry of the gastric H^+-K^+-ATPase can be found in the chapter by Sachs, Kaunitz, Mendlein, and Wallmark in this Handbook.

The gastric H^+-K^+-ATPase is capable of supporting high rates of proton transport and generating large transvesicular pH gradients, as long as there is an adequate supply of intravesicular K^+ (74, 75). Given the characteristics of H^+ transport by gastric microsomes, the pump-leak model depicted in Figure 5 was developed to account for pH-gradient formation in terms of the H^+-K^+ pump, diffusive entry of K^+ and anions, and an H^+ leak pathway. Assuming an adequate supply of ATP, pump activity depends on intravesicular $[K^+]$. Ordinarily, gastric microsomal vesicles are relatively impermeable to passive ion flux (49, 110); thus the rate of ATP hydrolysis and proton transport is limited by the rate of permeation of K^+ to the intravesicular site. Application of highly specific K^+ ionophores, such as valinomycin, greatly enhances ATPase and vesicular proton uptake, as exemplified in Figure 6 (74, 75). Figure 6 also demonstrates the anionic dependence of proton pumping. Entry of K^+ must be accompanied by a counterion, and this is ordinarily satisfied by Cl^-. In conditions where K^+ entry is not rate limited (e.g., valinomycin present), the microsomal ATPase activity and pH-gradient formation depend on the rate of anion penetration by passive diffusion with the observed permeability being $SCN^- > NO_3^- > Br^- > Cl^- > I^- \gg$ isethionate $\approx$ acetate (75).

Involvement of SCN^- represents a special case in that the initial rate of H^+ uptake can be very fast in KSCN supplemented with valinomycin, but the accumulated product, HSCN, leaks out rapidly and thus SCN^- does not support a pH gradient (75, 99). This action of SCN^- as an uncoupler of the gastric H^+ pump may be the means by which gastric acid secretion is effectively inhibited by SCN^- and other closely related pseudohalogens (76, 100).

Membrane Changes Associated With Hydrochloric Acid Secretion

Analysis of gastric microsomes in terms of H^+-K^+-ATPase and ion fluxes has provided an enormous wealth of detail to aid in the characterization of the nature of the proton pump, but additional transport systems are required to complement the activity of the pump enzyme. There is the obvious kinetic limitation that K^+ permeability of the microsomes is extremely

gastrin and acetylcholine receptors that probably play a role in the regulation of acid secretion. The preceding discussion of the receptors on the oxyntic cell is an oversimplification in that gastrin and acetylcholine receptors are also present on cells in addition to the oxyntic cell. In studies with canine mucosal cells, specific gastrin binding was localized to a small cell population, as well as to oxyntic cells (121, 122). Combined velocity and density separations were used to further separate this small cell fraction and revealed a close association between the presence of gastrin receptors and somatostatin cells (122). To further study the regulation of somatostatin release, this small cell elutriator fraction was placed in short-term culture (126); somatostatin cells adhered to the collagen substrate and after 48 h responded to treatment with a variety of agents. Gastrin and epinephrine stimulated somatostatin release, and marked potentiation was found between these agents. The β-adrenergic antagonist propranolol, but not α-adrenergic receptor antagonists, competitively inhibited this epinephrine response. Muscarinic receptors also regulated somatostatin cell function, but cholinergic agents inhibited—rather than stimulated—somatostatin release (145). This cholinergic effect was blocked by atropine acting at a typical muscarinic receptor. The affinity of this muscarinic receptor for pirenzepine was nearly 100-fold lower than for atropine; therefore neither the oxyntic cell nor somatostatin cell muscarinic receptor appeared to account for any selective action of pirenzepine on acid secretion.

The finding that gastric and acetylcholine receptors also regulate release of somatostatin, a potent inhibitor of acid secretion, complicates models for the regulation of acid secretion. Gastrin potentially short-circuits itself, activating both stimulatory and inhibitory pathways via receptors, respectively, on oxyntic and somatostatin cells. In contrast, muscarinic agents directly inhibit release of somatostatin while stimulating oxyntic cell function. These findings relate to the interdependency of cholinergic pathways and gastrin, evident by the ability of anticholinergic agents to block gastrin action. Cholinergic input may provide a balancing element, modulating gastrin action on oxyntic and somatostatin cells, attenuating gastrin stimulation of somatostatin release, and adding to gastrin stimulation of the oxyntic cell. Anticholinergic agents may inhibit acid secretion by both enhancing somatostatin release and inhibiting the oxyntic cell. This model is speculative; the physiological importance of the release of fundic somatostatin in vivo remains uncertain.

HISTAMINE CELLS IN THE FUNDIC MUCOSA. Despite the central role for histamine in the regulation of gastric acid secretion, knowledge remains incomplete regarding the identity of the cells that store histamine within the gastric mucosa and the regulation of histamine formation and release. One of the complexing factors in this area is the fact that histamine in the fundic mucosa is potentially stored in two cell types: mast cells and histamine-containing endocrine-like cells. To further complicate matters, the proportion of these two histamine cells in the fundic mucosa is highly variable among species. Any discussion of the cellular mechanisms regulating acid secretion requires mention of histamine cells, since the delivery of histamine to the oxyntic cell is probably a major point of physiological control of acid secretion.

In the dog, morphologically typical mast cells appear to fully account for the histamine content present in the population of dispersed cells (124). In short-term culture these mast cells released histamine in response to the cross-linking of IgE receptors and treatment with the Ca^{2+} ionophore A23187 (125). However, neither acetylcholine nor gastrin caused histamine release in this system. The absence of gastrin and muscarinic receptors on canine fundic mast cells contrasted with their presence on fundic mucosal somatostatin cells, as discussed in Receptors on Nonoxyntic Cells, p. 214. Mast cells also have two receptors that inhibit histamine release. Prostaglandin E_2 inhibited histamine release; concentrations >100 nM are necessary for this effect, but these levels may be achieved locally under certain circumstances. Additionally, adrenergic agents inhibit histamine release, acting at a β_2-receptor site. Previous studies had not detected endocrine-like histamine cells in dog fundic mucosa, although recent studies have indicated that a small number of such cells may be present (57).

In contrast, in certain other species such as the rat, rabbit, and frog, a significant proportion of fundic mucosal histamine is stored in endocrine-like cells. These cells have been termed *enterochromaffin-like* (ECL) cells because of their silver staining properties. Unlike the mast cell, histamine-containing ECL cells possess stimulatory acetylcholine and gastrin receptors. Studies with rabbit gastric glands indicate that gastric and acetylcholine induce the release of histamine (13, 90). The extent to which stimulation of acid secretion by gastrin and acetylcholine reflects stimulation of histamine release versus interaction with receptors on the oxyntic cell remains a subject of considerable controversy. Gastrin has one clear physiological effect on the histamine-containing ECL cell of the rat fundic mucosa; prolonged hypergastrinemia causes profound hyperplasia of these cells (73).

CELLULAR BASIS OF HYDROCHLORIC ACID SECRETION

Because HCl is the primary secretory product of the oxyntic cell, it is not surprising that well-defined systems for the transport of H^+ and Cl^- exist within the apical cell membrane. However, steady-state secretion of HCl requires the participation of several additional transport systems, both at the apical and

channels in a number of systems), with no effect on the Cl⁻ pathway (143). In contrast to their pronounced effects in SA vesicles, the divalent cations did not significantly alter ion fluxes in gastric microsomes. This latter result and observations that the conductive fluxes of Li^+, Na^+, and K^+ were similar and very low support the absence of any specific permeability pathway in microsomes. Recently several agents, in addition to the divalent cations, have been cited as inhibitors of conductive pathways in gastric SA vesicles. For inhibiting K^+ conductance the list includes quinine, tetraphenylmethylphosphonium, $diSC_3(5)$, and 1,8-bis-guanidinium-n-octane (55); for gastric vesicular, anionic conductance pathways, furosemide, 9-carboxyanthracene, and picrotoxin have been cited as inhibitors (31).

The recent studies of Takeguchi and his colleagues (2, 131, 132) have added an interesting note to the question of changing the fundamental conductance properties of the ordinarily highly impermeable microsomes from resting stomach. Cross-linking of S-S bonds by an oxidizing reagent, Cu^{2+}-o-phenanthroline, markedly increased KCl permeability in microsomal vesicles. The authors interpreted their results to mean that cross-linking caused only Cl⁻ conductance to increase, but the data are equivocal in this respect, since they show no good measurement of K^+ conductance. The general observation could conceivably have some physiological role, e.g., regulation of K^+ and Cl⁻ permeability by reversible oxidation-reduction steps, but this remains to be seen.

Hydrochloric Acid Transport at the Apical Cell Membrane

On the basis of studies with isolated vesicles and from the cytological features of the oxyntic cell, a model of transport events at the apical surface membrane is illustrated in Figure 8. Tubulovesicles of the resting cell have properties of the isolated microsomal vesicles: rich in H^+-K^+-ATPase but devoid of permeability pathways for K^+ and Cl⁻ transport. Consequently, transport by the pump, and certainly pump turnover, is very low in the resting cells. Stimulation by secretagogues leads to recruitment of tubulovesicles into the apical plasma membrane, greatly expanding that surface area and providing a rich supply of H^+-K^+-ATPase. Simultaneously there is the appearance of pathways for conductive transport of K^+ and Cl⁻ in the apical plasma membrane. The K^+ and Cl⁻ appear to move through independent electroconductive channels, but based on the similarities of inhibition by certain divalent cations, there may be some interaction between the K^+ and anionic channels.

Although the relocation of the pump enzyme during the secretory cycle is well established, the disposition of the channel protein(s) is uncertain. We can surmise several scenarios. One possibility is that K^+- and Cl⁻-transport paths are always present and active in the apical plasma membrane, and tubulovesicle fusion simply brings the pump enzyme into functional parallelism with the conductive paths. Another could be that channel protein(s) requires activation by stimulus-related cytoplasmic events (e.g., phosphorylation). Finally, it is possible that channel proteins are entrained within an entirely separate population of vesicles that must also fuse with the apical surface upon stimulation. Some support for this latter idea is available from the studies of Im et al. (67), who isolated from stimulated rat gastric mucosa a fraction of membrane vesicles that demonstrated high fluxes of K^+ and Cl⁻ but contained low H^+-K^+-ATPase activity. It is now important to define *1*) the mechanism by which the membrane proteins are activated, *2*) the location specificity of membrane fusion events, and *3*) the cytoplasmic pathways that regulate the membrane turnover.

As long as the transport systems remain aligned in the apical membrane, sustained flow of HCl is assured by the efficient ATP-dependent H^+-K^+ exchange by the pump enzyme and the consequent generation of osmotically active solutes (Fig. 8). For example, the H^+-K^+-ATPase produces, de novo, one proton and one equivalent of base from H_2O, and the bulk flow associated with this "net synthesis" of osmotic particles may be very important for the dynamics of secretion. Inhibition of acid secretion either by SCN^-, which is thought to uncouple the process and shunt protons and base back to water (100), or by omeprazole, a direct inhibitor of the H^+-K^+-ATPase (136), leads to an increase in transmucosal electrical resistance without diminishing the elaborated apical surface area of the oxyntic cells (16, 62). If the total apical surface area of the oxyntic cell has not changed, then the inhibition-related increase in transmucosal resistance must be due to a change in specific resistance of the apical membrane or to a change in resistance of some other element in the tissue. Virtually all the current that flows across the gastric epithelium passes through the glandular cells (as opposed to surface epithelial cells); consequently the current also must flow through the gland lumina as a series element (34). Although the gland lumina contain highly conductive fluid, they are long (200–500 μm) and narrow (5–10 μm diam) structures that could contribute to transtissue resistance, especially where diminished bulk flow associated with inhibition would lead to a constriction or even collapse of the tubular lumina. Luminal resistance represents a significant fraction of the transmucosal resistance in SCN^--inhibited or omeprazole-inhibited frog gastric mucosa, and hyperosmotic secretory solutions decreased the transmucosal resistance, presumably by promoting an osmotically induced "secretory flow" and expanding the luminal cross-sectional area (97, 100, 102).

When acid secretion is inhibited by removal of secretagogue or by the application of an H_2-receptor antagonist, the oxyntic cell is restored to its resting

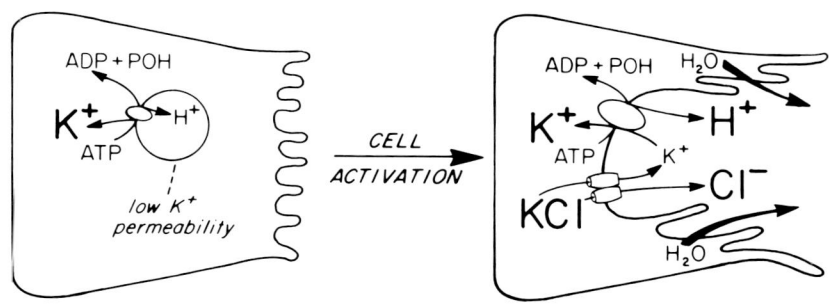

FIG. 8. Schematic representation of the oxyntic cell transport in the transformation from rest to activate HCl secretion. In the resting cell, tubulovesicles contain H^+-K^+-ATPase, but because of low membrane permeability to K^+ (and Cl^-) there is very little H^+ accumulation and virtually no ATP turnover. Cell activation brings about a fusion of tubulovesicles with the apical plasma membrane, transferring the H^+-K^+-ATPase to that surface. In addition the participation of conductive pathways (possibly activated?) for K^+ and Cl^- movement provides the means for KCl movement into the secretory canaliculi. The H^+-K^+ exchange pump recycles K^+ back into the cytoplasm with the net effect of HCl transfer and ATP turnover. Water flux into the canaliculus is osmotically driven by the net solute flux.

configuration with diminished apical membrane surface area and expanded compartment of cytoplasmic tubulovesicles (48). Cell fractionation studies during the return to rest show that H^+-K^+-ATPase activity progressively decreases in the apical cell membrane fraction with complementary increments in H^+-K^+-ATPase activity of microsomal vesicles (46). Changes in the subcellular redistribution of H^+-K^+-ATPase activity in rabbit stomach taken at various stages of stimulation and after inhibition with an H_2-receptor antagonist are illustrated in Table 1. These observations conform to the membrane recycling hypothesis, where H^+-K^+ pumps that had been recruited to the apical surface for secretory activity are taken back into a "storage" compartment during quiescent interdigestive periods.

Transport at the Basolateral Membrane

Gastric juice from maximally stimulated stomach contains high concentrations of H^+ and Cl^- (nearly isotonic HCl) and typically contains concentrations of K^+ that exceed the levels in plasma by at least twofold. The transport systems (discussed in the preceding section) in the apical cell membrane provide the means of access and power to form the ionic constituents of the juice, but in addition specific mechanisms must be present at the basolateral membrane for ion uptake and homeostasis. At a minimum the basolateral membrane must have transport systems to catalyze the elimination of base that must be produced in the cell for each mole of H^+ secreted and for the uptake of Cl^- and K^+ from the serosal fluid. The operation of an electroneutral Cl^--HCO_3^- exchanger at the basolateral membrane was first deduced by Rehm (101) on the basis of electrophysiological studies in frog gastric mucosa. This Cl^--HCO_3^- exchanger simultaneously provides a means for Cl^- uptake and elimination of base from the oxyntic cell. The electrical conductance of the basolateral membrane is dominated by a K^+ pathway. This conclusion comes from several experimental systems, including observed changes in transepithelial potential difference with

TABLE 1. H^+-K^+-ATPase is Redistributed Among Subcellular Fractions During Stimulation or Inhibition of Acid Secretion

	H^+ Secretory Rate		Total H^+-K^+-ATPase Activity in Cell Fraction, %				Redistribution Ratio
	n	µeq/min	P1	P2	P3	S3	P1/P3
Resting stomach	5	4.8 ± 2.3	12.0	8.1	78.1	1.8	0.15
Histamine stimulated	5	78.9 ± 9.0	60.4	19.0	20.7	1.2	2.92
Time after H_2 blocker							
10 min	4	37.3 ± 7.6	38.8	14.6	42.4	4.1	0.92
15 min	4	26.6 ± 6.6	35.0	21.8	39.4	3.8	0.89
20 min	3	17.5 ± 6.6	28.7	12.5	59.5	0	0.48
45 min	3	4.8 ± 2.2	22.2	14.2	61.5	2.1	0.36

Rabbits (2.1–2.6 kg) were fasted overnight and anesthetized for experimentation. A plastic cannula was secured into the stomach for collecting gastric juice; the juice was subsequently titrated to determine H^+ secretory rates. Resting animals were treated with an H_2-receptor antagonist, SKF 93479 (18). All other animals were treated for 20 min with histamine to produce maximal acid output. Some animals were then treated with 12.5 µmol SKF 93479 iv. Under conditions and/or times as indicated, stomachs were removed, and the fundic mucosa was scraped, homogenized, and separated by centrifugation into cell fractions (64a). After removing a very low speed fraction of tissue debris (40 g × 5 min), the following fractions were separated: P1, 4,000 g × 10 min; P2, 14,000 g × 10 min; P3, 49,000 g × 90 min; and S3, supernatant. K^+-stimulated p-nitrophenylphosphatase was measured as a marker of H^+-K^+-ATPase activity, and values are shown for percentage of total enzyme activity in each fraction. The redistribution ratio, P1/P3, demonstrates the large redistribution of H^+-K^+-ATPase activity among particulate fractions associated with histamine stimulation and recovery after H_2-receptor blockade. Although H^+-K^+-ATPase activity markedly changed with secretory activity, there were virtually no changes in the total protein distributed among the various fractions.

changes in serosal [K^+] (59, 129), a negative intracellular potential in keeping with the direction of the K^+ gradient (34), and a demonstrated K^+ conductance in isolated basolateral membrane vesicles (85). The basolateral membrane of the oxyntic cell is also rich in Na^+-K^+-ATPase (29), which serves as a path for K^+ uptake from the blood and sustains the cationic balance within the cell (i.e., high [K^+], low [Na^+]). Inhibition of the Na^+-K^+-ATPase by ouabain or the complete removal of Na^+ from the serosal bathing solution leads to a predictable slow loss of cellular K^+ and a consequent reduction in H^+ secretion (130); however, Na^+-pump inhibition also causes a relatively rapid and large diminution of Cl^- secretion (20, 79). From observations such as these in intact mucosa, it was reasoned that the basolateral uptake of Cl^- was catalyzed by an NaCl symport (82).

More recently it has become apparent that Na^+ uptake and Cl^- uptake are not directly coupled but move through a system of parallel exchangers. Studies with isolated gastric glands and probes to measure intracellular pH (pH_i) have shown that NaCl uptake is based on independent electroneutral Na^+-H^+ and Cl^--HCO_3^- exchangers (85, 92, 93). The Na^+-H^+ exchanger has been demonstrated by a tight coupling between pH_i and the Na^+ gradient. When oxyntic cells were first preloaded with Na^+ and then resuspended into Na^+-free media, pH_i decreased. Conversely, when cells were first loaded with acid (by an NH_4Cl preloading technique), the rate of intracellular acid release was increased by the presence of Na^+ in the extracellular medium but not by K^+ or impermeant cations. These interrelationships between pH_i and the Na^+ gradient were blocked or slowed down by amiloride, an inhibitor of Na^+-H^+ antiport activity. After blockade by amiloride, the changes in pH_i could be mimicked by the addition of monensin, an exogenous Na^+-H^+ exchange ionophore. A similar experimental approach was used to support the existence of a Cl^--HCO_3^- exchange system at the basolateral membrane (85, 92). Replacement of extracellular Cl^- by impermeable anions caused an increase in pH_i that was rapidly reversed on reintroduction of Cl^-. The Cl^--dependent changes in pH_i were abolished by treatment with known anion exchange inhibitors, stilbene disulfonates, and were then reestablished by introduction of the Cl^--OH^- exchanger, tributyl tin.

The existence of parallel Na^+-H^+ and Cl^--HCO_3^- exchange activities is a common feature in many epithelial cells, particularly in the gastrointestinal tract, although tissue-specific differences in activity and membrane localization (e.g., apical vs. basolateral) provide a range of functional attributes for the exchangers (65, 80, 103). Although physically independent, the separate systems of Na^+-H^+ and Cl^--HCO_3^- exchangers are, in practice, coupled through cellular pH, and they act cooperatively as an NaCl cotransport. For the gastric mucosa, when the exchangers are combined with the Na^+ pump and a high basolateral K^+ conductance, these systems collaborate as a mechanism to accumulate Cl^- against its electrochemical gradient.

The system of parallel exchangers also serves an important role in homeostasis of cellular pH, maintaining and adjusting disequilibria in pH_i, especially to expel excess base during acid secretory activity. In this regard, Paradiso et al. (92) made the insightful observation that when Na^+ and Cl^- were removed from the bathing solution (i.e., replaced by N-methylglucamine and N-methylgluconate, respectively), the cystolic pH of oxyntic cells exhibited biphasic changes. At first the cells rapidly alkalinized because of the Cl^- efflux with concomitant influx of base. Then there was a more gradual return of pH_i toward its starting value. Several implications are apparent in these results. 1) The results that Cl^--base exchange may be faster than the Na^+-H^+ antiport, providing an adequate mechanism for extrusion of base created by acid secretion. 2) The slower reversal of cellular alkalinization, presumably the result of Na^+ efflux with concomitant H^+ influx, suggests a possible delayed activation of the Na^+-H^+ antiport in response to a significant shift from the steady-state pH_i, as suggested for other systems (54). 3) The biphasic change in pH_i seems to rule out the existence of a pH-independent electroneutral Na^+ and Cl^- cotransport mechanism. Otherwise, Na^+ and Cl^- would exit the cell before the Cl^- gradient could elicit cellular alkalinization.

Oxyntic Cell Transport Model

On the basis of apical and basolateral membrane transport properties and direct measurements of transepithelial ion fluxes measured in chambered gastric mucosa (97, 98), a model for ion transport by the oxyntic cell has been developed. As depicted in Figure 9, the minimal model specifies the membrane location of pumps, passive exchangers, and ionic conductances and can be incorporated into the general ion-transport characteristics of the gastric epithelium. There are

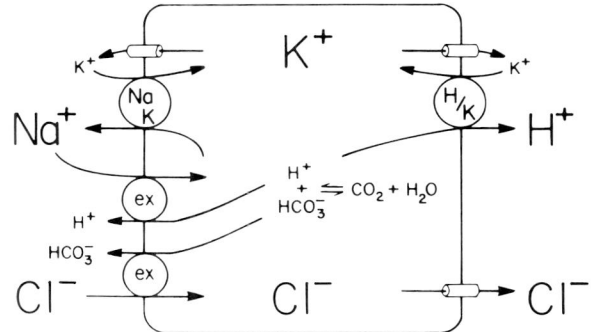

FIG. 9. Transport model of gastric oxyntic cell. Apical membrane contains H^+-K^+ pump and conductive pathways for K^+ and Cl^-. Basolateral membrane contains N^+-K^+ pump, a K^+ conductance, and separate Na^+-H^+ and Cl^--HCO_3^- exchange (ex) mechanisms. [From Reenstra et al. (98).]

two primary active transport systems, both of which are ATP-driven cation exchange pumps: the H^+-K^+-ATPase at the apical membrane and the Na^+-K^+-ATPase at the basolateral membrane. Because both pumps recycle K^+ into the cell, there is relatively little net transcellular movement of K^+, but application of H^+-K^+-ATPase inhibitors (e.g., omeprazole) leads to net K^+ secretion and application of Na^+-K^+-ATPase inhibitors (e.g., ouabain) leads to net K^+ absorption (97). For the oxyntic cell, apical membrane permeability to Na^+ is extremely low; thus the basolateral Na^+-K^+ pump and Na^+-H^+ exchange permit Na^+ recycling between cytosol and serosal medium.

The model in Figure 9 contains no primary active transport system for Cl^- and yet it provides for secondary active Cl^- transport driven by indirect coupling to either of the two cationic pumps. The H^+-K^+-ATPase drives Cl^- by providing a large K^+ gradient and electrical driving force across the apical membrane (i.e., K^+ and Cl^- leakage and H^+-K^+ exchange). Compensatory Cl^- transport across the basolateral membrane occurs through the Cl^--HCO_3^- exchanger. For this entire process, no net charge transfer occurs; i.e., Cl^- moves as net HCl secretion across the apical membrane and Cl^--HCO_3^- exchange across the basolateral membrane. Hence this mode of Cl^- transport can be called electroneutral HCl transport driven by the H^+-K^+-ATPase. Stimulation-associated changes in the apical membrane transport activities were discussed in *Membrane Changes Associated With Hydrochloric Acid Secretion*, p. 216; recent evidence suggests that there may also be an increased activity of the basolateral Cl^--HCO_3^- exchanger in histamine-stimulated glands (91). A separate means of secondary active Cl^- transport is driven by the basolateral Na^+-K^+-ATPase. Essentially, the Na^+ gradient created by the pump provides the force for accumulating Cl^- from the serosal solution into the cell against its electrochemical potential gradient. In actual fact, Na^+ and Cl^- movements occur via the parallel but closely coupled Na^+-H^+ and Cl^--HCO_3^- exchangers. When there is no pH disequilibrium, H^+ and HCO_3^- simply form and recombine on either side of the basolateral membrane. The elevated intracellular electrochemical activity of Cl^- is the basis for an apical cell membrane Cl^- diffusion potential that is capable of delivering a current of Cl^- under short-circuit conditions. This mode of Cl^- flux can be called electrogenic Cl^- transport driven by the Na^+-K^+-ATPase. Working in concert, these various oxyntic cell transport systems provide the basis for total Cl^- transport as the sum of electroneutral HCl secretion and electrogenic Cl^- secretion.

CELL-ACTIVATION MECHANISMS

The mechanisms activating oxyntic cell function have been reviewed in the chapter by Chew in this *Handbook*; selected points are discussed in the following sections. Current dogma is that after agonist binding to membrane receptors, secondary changes transduce the signal-activating cell function by two general mechanisms: *1*) the enhanced generation of cAMP and increases in cytosolic Ca^{2+} concentration in the cell and *2*) activation of protein kinase C by diacylglycerol.

Oxyntic Cell Activation by Cyclic AMP

Stimulation of oxyntic cell function by histamine but not by cholinergic agonists or gastrin is closely linked to enhanced formation of cAMP. In cell separation studies utilizing velocity and density separation techniques, the elevation of cAMP levels by histamine, as well as the activation of adenylate cyclase activity by histamine, has been correlated with the presence of oxyntic cells, suggesting that these cells are the primary locus of H_2 receptors in the fundic mucosa. The time profile and dose response for histamine activation of oxyntic cell function and cAMP production are correlated, suggesting a causal relationship (27). Moreover, gastric acid production is also stimulated by an exogenous supply of membrane-permeable derivatives of cAMP, blockers of the cAMP catabolism, and agents that directly activate adenylate cyclase, such as forskolin (22, 58, 63).

The secretagogue action of cAMP is most likely to be mediated by activation of cAMP-dependent protein kinase. Addition of cAMP to oxyntic cell homogenates increased phosphorylation activity by 10- to 20-fold (84). More directly, stimulation of whole fundic mucosa (137) or isolated oxyntic cells (23) by histamine resulted in increased activity of intracellular cAMP-dependent protein kinase. Distinct isozymes of cAMP-dependent protein kinase, so-called type I and type II, have been identified in rabbit oxyntic cell homogenates (23). The type I isoform was found exclusively in the cytosol and appeared to be selectively activated by histamine treatment. The type II isoform showed no such activation by histamine and was found both in particulate and soluble cell fractions, leading Chew (23) to suggest the possibility of complex regulatory action and compartmental restrictions.

Three recent studies have sought to identify specific phosphoproteins that may be activated in conjunction with stimulation of acid secretion: two of these studies used highly enriched preparations of oxyntic cells (25, 32), and the other used whole gastric glands (134). Each of these groups has provided evidence for proteins whose phosphorylation is increased by the histamine-cAMP pathway for cell activation. However, two major problems exist concerning these works in secretagogue-dependent changes in oxyntic cell protein phosphorylation. *1*) There is no agreement among investigators regarding the individual proteins that are phosphorylated (e.g., large variation in molecular size and putative cellular location). *2*) Functional activity cannot be specified for any of the phosphoproteins thus far identified.

A different approach was taken by Im et al. (66), who addressed questions of phosphoprotein metabolism in vitro, using a preparation of membrane vesicles derived from apical plasma membranes of stimulated oxyntic cells (i.e., SA vesicles). They reported that a protein phosphatase activity was copurified with the SA vesicles. Furthermore they presented evidence to suggest that dephosphorylation of certain (unidentified) membrane proteins, possibly through the intrinsic protein phosphatase, was responsible for abolishing the K^+ conductive pathway, thus converting the SA vesicles into a valinomycin-dependent preparation, more characteristic of microsomal vesicles from resting stomach.

The state of knowledge regarding the precise pathway and role of protein phosphorylation in cAMP-mediated oxyntic cell activation is still preliminary and evolving. Because of recent intensive and suggestive efforts, it is likely that specific sites and mechanisms for activation and regulation will soon be defined.

Prostaglandin Inhibition of Oxyntic Cell Function

Nanomolar concentrations of prostaglandins E_2 and I_2 directly inhibit oxyntic cell function (77, 111, 114, 119). Prostanoid inhibition was specific for histamine; the response to either carbachol or gastrin was not inhibited. However, when carbachol or gastrin action was enhanced by interaction with histamine, but not Bt_2cAMP, the effects were inhibited by prostaglandins, with the residual response similar to that found with either gastrin or carbachol alone. Histamine stimulation of cAMP production was also inhibited over the same nanomolar concentration range of prostaglandins with which histamine-stimulated aminopyrine accumulation was inhibited (77, 81, 111, 119). These findings indicate that prostaglandins inhibit histamine interaction with its receptor or the adenylate cyclase complex itself. The adenylate cyclase complex includes a stimulatory guanine nucleotide–binding subunit (G_s or N_s) that in turn activates the subunit that catalyzes the conversion of ATP to cAMP. The G_s is the target for cholera toxin; this toxin induces the transfer of an ADP-ribose moiety to G_s, thereby locking it into the "on" position and inducing cAMP generation. There exists a mirror image of this stimulatory limb of the cAMP pathway that mediates inhibition of cell function by a variety of transmitters, such as opiates, α-adrenergic agonists, and somatostatin. In this inhibitory limb there is another GTP-binding protein, G_i, that inhibits rather than stimulates the catalytic subunit of adenylate cyclase (see Fig. 4). Another bacterial toxin, one produced by *Bordetella pertussis*, has proved an important

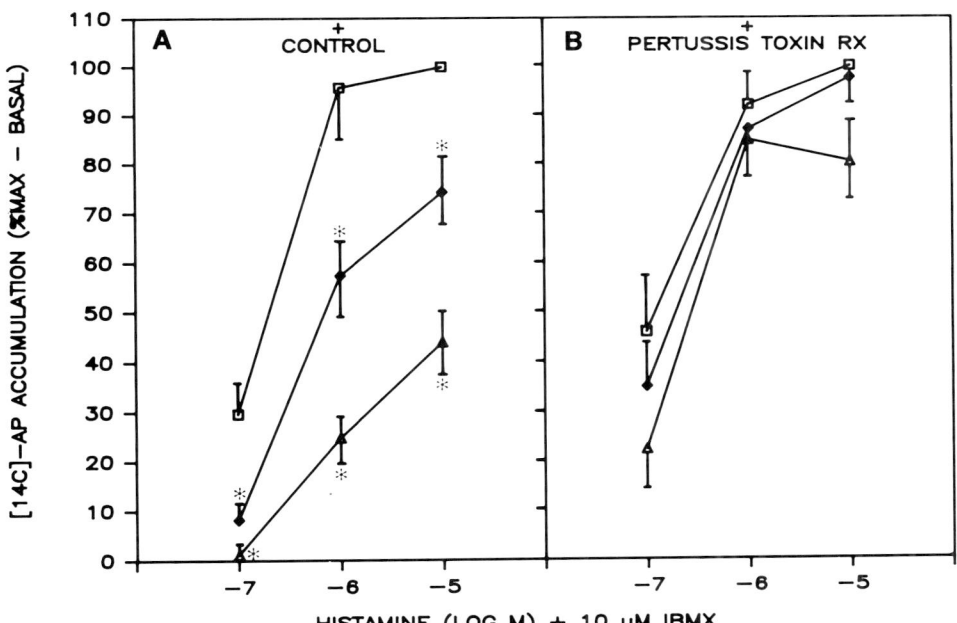

FIG. 10. Pertussis toxin attenuation of enprostil inhibition of histamine-stimulated [^{14}C]aminopyrine ([^{14}C]-AP) accumulation. Enriched parietal cells were cultured overnight in the absence (*A*) or presence (*B*) of 300 ng/ml of pertussis toxin. [^{14}C]aminopyrine accumulation was then studied in response to the indicated concentrations of histamine alone (*open squares*) or in the presence of 10 nM (*closed diamonds*) or 1 μM enprostil (*open triangles*). Data (means ± SE from 5 separate cell preparations) were expressed as percentage of the maximal response to histamine. This maximal value over basal in control cells was a [^{14}C]aminopyrine ratio of 21.3 ± 3.3, whereas in the cells treated with pertussis toxin the maximal response to histamine was 19.2 ± 4.6. IBMX, isobutylmethylxanthine.

tool for dissecting this inhibitory mechanism. In a similar fashion to cholera toxin, pertussis toxin acts by transferring an ADP-ribose group to G_i, but pertussis toxin inactivates rather than stimulates this inhibitory GTP-binding protein. Therefore pertussis toxin treatment of cells blocks G_i-mediated inhibition. Treatment of oxyntic cells with pertussis toxin markedly attenuates prostanoid inhibition of histamine-stimulated function and cAMP generation, while not impairing stimulatory responses (Fig. 10). These findings suggest that prostanoids act via G_i to turn off the oxyntic cell (21).

Oxyntic Cell Activation by Calcium-Dependent Mechanisms

For many cell systems the secretory response to cholinergic stimulation is mediated through a Ca^{2+}-activation pathway, and this also appears to be the case for the oxyntic cell, but there are some noteworthy features regarding Ca^{2+} metabolism in HCl secretion. Functional studies with canine oxyntic cells and rabbit gastric glands indicate that stimulation by cholinergic agents was dependent on the concentration of extracellular Ca^{2+} (12, 120). In both gland and cell preparations, the extent of stimulation obtained by either histamine or by manipulations leading to increased cellular levels of cAMP was insensitive to extracellular Ca^{2+} (87, 120). On the other hand, carbachol-induced acid generation was depressed in the absence of external Ca^{2+} or when Ca^{2+} entry was blocked by the inorganic Ca^{2+}-channel blocker, lanthanum. Cholinergic stimulation enhanced $^{45}Ca^{2+}$ uptake into oxyntic cells from the extracellular space (120). This Ca^{2+} uptake is probably via selective receptor-activated Ca^{2+} channels in the cell membrane, but these channels are different from the channels present in muscle, nerve, and other tissues in that they are not blocked by the organic Ca^{2+}-channel antagonists such as verapamil or nifedipine.

Studies with fluorescent probes to monitor cytosolic Ca^{2+} concentration yield a somewhat different picture. First, increases in free cytosolic Ca^{2+} were measured not only in response to carbachol but also to histamine and gastrin, albeit the cholinergic response was most profound (24, 88). Furthermore the secretagogue-dependent changes in cytosolic free Ca^{2+} signal occurred from an intracellular store of bound Ca^{2+} and were independent of extracellular Ca^{2+} unless the intracellular stores were first depleted (89). Intracellular Ca^{2+} stores were relatively easily depleted by cholinergic stimulation in a Ca^{2+}-free medium, and the stores could be replaced from extracellular sources via a lanthanum-sensitive pathway (89). These latter observations provide a basis to account for some of the differences noted by various investigators. Depending on the species used and aggressiveness in isolating the particular preparation, there may occur some variation in depletion of intracellular Ca^{2+} stores, which ultimately must be replaced from the medium. Thus the oxyntic cell conforms to the dogma that cholinergic stimulation works via a Ca^{2+}-dependent pathway. It remains to be established what this pathway (or pathways) is and to what extent it might play a role in histaminergic or cAMP-mediated cell activation. For a discussion of some of these activation pathways, the reader is referred to the chapter by Chew in this *Handbook*.

REFERENCES

1. ALBINUS, M., AND K. F. SEWING. Histamine uptake and metabolism in intact isolated parietal cells. *Agents Actions* 11: 223–227, 1981.
2. ASANO, S., M. INOIE, AND N. TAKEGUCHI. The Cl⁻ channel in hog gastric vesicles is part of the function of H,K-ATPase. *J. Biol. Chem.* 262: 13263–13268, 1987.
3. BATZRI, S., AND J. DYER. Aminopyrine uptake by guinea pig gastric mucosal cells: mediation by cyclic AMP and interaction among secretagogues. *Biochim. Biophys. Acta* 675: 416–426, 1981.
4. BATZRI, S., J. W. HARMON, AND W. F. THOMPSON. Interaction of histamine with gastric mucosal cells: effect of histamine agonists on binding and biological response. *Mol. Pharmacol.* 22: 33–40, 1982.
5. BEAVEN, M. A., A. H. SOLL, AND K. J. LEWIN. Histamine synthesis by intact mast cells from canine fundic mucosa and liver. *Gastroenterology* 82: 254–262, 1982.
6. BERGLINDH, T. Effects of common inhibitors of gastric acid secretion on secretagogue-induced respiration and aminopyrine accumulation in isolated gastric glands. *Biochim. Biophys. Acta* 464: 217–233, 1977.
7. BERGLINDH, T. Potentiation by carbachol and aminophylline of histamine- and dbcAMP-induced parietal cells activity in isolated gastric glands. *Acta Physiol. Scand.* 99: 75–84, 1977.
8. BERGLINDH, T. The mammalian gastric parietal cell in vitro. *Annu. Rev. Physiol.* 46: 377–392, 1984.
9. BERGLINDH, T., D. R. DIBONA, S. ITO, AND G. SACHS. Probes of parietal cell function. *Am. J. Physiol.* 238 (*Gastrointest. Liver Physiol.* 1): G165–G176, 1980.
10. BERGLINDH, T., H. F. HELANDER, AND K. J. ÖBRINK. Effects of secretagogues on oxygen consumption, aminopyrine accumulation and morphology in isolated gastric glands. *Acta Physiol. Scand.* 97: 401–414, 1976.
11. BERGLINDH, T., AND K. J. ÖBRINK. Histamine as a physiological stimulant of gastric parietal cells. In: *Histamine Receptors*, edited by T. Yellin. New York: Spectrum, 1979, p. 35–56.
12. BERGLINDH, T., G. SACHS, AND N. TAKEGUCHI. Ca^{2+}-dependent secretagogue stimulation in isolated rabbit gastric glands. *Am. J. Physiol.* 239 (*Gastrointest. Liver Physiol.* 2): G90–G94, 1980.
13. BERGQVIST, E., M. WALLER, L. HAMMER, AND K. J. ÖBRINK. Histamine as the secretory mediator in isolated gastric glands. In: *Hydrogen Ion Transport in Epithelia*, edited by I. Schulz, G. Sachs, J. G. Forte, and K. J. Ullrich. Amsterdam: Elsevier/North-Holland, 1980, p. 429–437.
14. BERTACCINI, G., AND G. CORUZZI. Evidence for and against heterogeneity in the histamine H_2-receptor population. *Pharmacology Basel* 23:1–13, 1981.

15. BLACK, J. A., T. M. FORTE, AND J. G. FORTE. Structure of oxyntic cell membranes during conditions of rest and secretion of HCl as revealed by freeze-fracture. *Anat. Rec.* 196: 163–172, 1980.
16. BLACK, J. A., T. M. FORTE, AND J. G. FORTE. Inhibition of HCl secretion and the effects of ultrastructure and electrical resistance in isolated piglet gastric mucosa. *Gastroenterology* 81: 509–519, 1981.
17. BLACK, J. A., T. M. FORTE, AND J. G. FORTE. The effects of microfilament disrupting agents on HCl secretion and ultrastructure of piglet oxyntic cells. *Gastroenterology* 83: 595–604, 1982.
18. BLAKEMORE, R. C., T. H. BROWN, G. J. DURAN, C. R. GANELLIN, M. E. PARSONS, A. C. RASMUSSEN, AND D. A. RAWLINGS. SKF93479, a potent and long acting histamine H_2-receptor antagonist (Abstract). *Br. J. Pharmacol.* 74: 200P, 1981.
19. CARLISLE, K. S., C. S. CHEW, AND S. J. HERSEY. Ultrastructural changes and cyclic AMP in frog gastric cells. *J. Cell Biol.* 76: 31–42, 1978.
20. CARRASQUER, G., T.-C. CHU, W. S. REHM, AND M. SCHWARTZ. Evidence for electrogenic Na-Cl symport in the in vitro frog stomach. *Am. J. Physiol.* 242 (*Gastrointest. Liver Physiol.* 5): G620–G627, 1982.
21. CHEN, M. C., D. AMIRIAN, M. TOOMEY, M. SANDERS, AND A. H. SOLL. Prostanoid inhibition of canine parietal cells: mediation by the inhibiting GTP-binding protein of adenylate cyclase. *Gastroenterology.* 94: 1121–1129, 1988.
22. CHEW, C. S. Forskolin stimulation of acid and pepsinogen secretion in isolated gastric glands. *Am. J. Physiol.* 245 (*Cell Physiol.* 14): C371–C380, 1983.
23. CHEW, C. S. Parietal cell protein kinases. *J. Biol. Chem.* 260: 7540–7550, 1985.
24. CHEW, C. S. Cholecystokinin, carbachol, gastrin, histamine, and forskolin increase $[Ca^{2+}]_i$ in gastric glands. *Am. J. Physiol.* 250 (*Gastrointest. Liver Physiol.* 13): G814–G823, 1986.
25. CHEW, C. S., AND M. R. BROWN. Histamine increases phosphorylation of 27- and 40-kDa parietal cell proteins. *Am. J. Physiol.* 253 (*Gastrointest. Liver Physiol.* 16): G823–G829, 1987.
26. CHEW, C. S., AND S. J. HERSEY. Gastrin stimulation of isolated gastric glands. *Am. J. Physiol.* 242 (*Gastrointest. Liver Physiol.* 5): G504–G512, 1982.
27. CHEW, C. S., S. J. HERSEY, G. SACHS, AND T. BERGLINDH. Histamine responsiveness of isolated gastric glands. *Am. J. Physiol* 238 (*Gastrointest. Liver Physiol.* 1): G312–G320, 1980.
28. COULTON, G. R., AND J. A. FIRTH. Cytochemical evidence for functional zonation of parietal cells within the gastric glands of the mouse. *Histochem. J.* 15: 1141–1150, 1983.
29. CULP, D. J., AND J. G. FORTE. An enriched preparation of basolateral plasma membranes from gastric glandular cells. *J. Membr. Biol.* 59: 135–142, 1981.
30. CULP, D. J., J. M. WOLOSIN, A. H. SOLL, AND J. G. FORTE. Muscarinic receptors and guanylate cyclase in mammalian gastric glandular cells. *Am. J. Physiol.* 245 (*Gastrointest. Liver Physiol.* 8): G760–G768, 1983.
31. CUPPOLETTI, J., AND G. SACHS. Regulation of gastric acid secretion via modulation of a chloride conductance. *J. Biol. Chem.* 259: 14952–14959, 1984.
32. CUPPOLETTI, J., G. SACHS, AND D. H. MALINOWSKA. Cytoskeletal proteins in gastric H^+ secretion: cAMP dependent phosphorylation, immunolocalization, and protein blotting (Abstract). *Federation Proc.* 45: 1676, 1986.
33. DAVIDSON, W. D., K. L. KLEIN, K. KUROKAWA, AND A. H. SOLL. Instantaneous and continuous measurement of ^{14}C-labeled substrate oxidation to $^{14}CO_2$ by minute tissue specimens: an ionization chamber method. *Metabolism* 30: 596–600, 1981.
34. DEMAREST, J. R., AND T. E. MACHEN. Microelectrode measurements from oxyntic cells in intact *Necturus* gastric mucosa. *Am. J. Physiol.* 249 (*Cell Physiol.* 18): C535–C540, 1985.
35. DIAL, E., W. J. THOMPSON, AND G. C. ROSENFELD. Isolated parietal cells: histamine response and pharmacology. *J. Pharmacol. Exp. Ther.* 219: 585–590, 1981.
36. ECKNAUER, R., E. DIAL, W. J. THOMPSON, L. R. JOHNSON, AND G. C. ROSENFELD. Isolated rat gastric parietal cells: cholinergic response and pharmacology. *Life Sci.* 28: 609–621, 1981.
37. ECKNAUER, R., W. J. THOMPSON, L. R. JOHNSON, AND G. C. ROSENFELD. Isolated parietal cells: [^3H]QNB binding to putative cholinergic receptors. *Am. J. Physiol.* 239 (*Gastrointest. Liver Physiol.* 2): G204–G209, 1980.
38. ELTZE, M., S. GÖNNE, R. RIEDEL, B. SCHLOTKE, C. SCHUDT, AND W. A. SIMON. Pharmacological evidence for selective inhibition of gastric acid secretion by telenzepine, a new antimuscarinic drug. *Eur. J. Pharmacol.* 112: 211–224, 1985.
39. FARLEY, R. A., AND L. D. FALLER. The amino acid sequence of an active site peptide from the H,K-ATPase of gastric mucosa. *J. Biol. Chem.* 260: 3899–3901, 1985.
40. FELLENIUS, E., B. ELANDER, B. WALLMARK, U. HAGLUND, H. F. HELANDER, AND L. OLBE. A micro-method for the study of acid secretory function in isolated human oxyntic glands from gastroscopic biopsies. *Clin. Sci* 64: 423–431, 1983.
41. FORTE, J. G., J. A. BLACK, T. M. FORTE, T. E. MACHEN, AND J. M. WOLOSIN. Ultrastructural changes related to functional activity in gastric oxyntic cells. *Am. J. Physiol.* 241 (*Gastrointest. Liver Physiol.* 4): G349–G358, 1981.
42. FORTE, J. G., T. M. FORTE, AND P. SALTMAN. K^+-stimulated phosphatase in microsomes isolated from gastric mucosa. *J. Cell. Physiol.* 69: 175–189, 1967.
43. FORTE, J. G., AND T. M. FORTE. Histochemical staining and characterization of glycoproteins in acid-secreting cells of frog stomach. *J. Cell Biol.* 47: 437–452, 1970.
44. FORTE, J. G., A. L. GANSER, R. C. BEESLEY, AND T. M. FORTE. Unique enzymes of purified microsomes from pig fundic mucosa. *Gastroenterology* 69: 175–189, 1975.
45. FORTE, J. G., A. L. GANSER, AND T. K. RAY. The K^+-stimulated ATPase from oxyntic glands of gastric mucosa. In: *Gastric Hydrogen Ion Secretion*, edited by D. K. Kasbekar, G. Sachs, and W. Rehm. New York: Dekker, 1976, p. 302–330.
46. FORTE, J. G., AND D. J. KEELING. Cellular recycling of gastric H,K-ATPase with stimulation and inhibition of HCl secretion. *FASEB J.* 2: 1275A, 1988.
46a. FORTE, J. G., AND T. E. MACHEN. Ion transport by gastric mucosa. In: *Physiology of Membrane Disorders*, edited by T. E. Andreoli, J. F. Hoffman, D. D. Fanestil, and S. G. Schultz. New York: Plenum, 1986, p. 535–558.
47. FORTE, T. M., AND J. G. FORTE. Definition of extracellular space in secreting and non-secreting cells. *J. Cell Biol.* 47: 782–786, 1970.
48. FORTE, T. M., T. E. MACHEN, AND J. G. FORTE. Ultrastructural changes in oxyntic cells associated with secretory function: a membrane recycling hypothesis. *Gastroenterology* 73: 941–955, 1977.
49. GANSER, A. L., AND J. G. FORTE. Ionophoretic stimulation of K^+-ATPase of oxyntic cell microsomes. *Biochem. Biophys. Res. Commun.* 54: 690–696, 1973.
50. GANSER, A. L., AND J. G. FORTE. K^+-stimulated ATPase in purified microsomes of bullfrog oxyntic cells. *Biochim. Biophys. Acta* 307: 169–180, 1973.
51. GESPACH, C., D. BOUHOURS, J. F. BOUHOURS, AND G. ROSSELIN. Histamine interaction on surface recognition sites of H_2-type in parietal and nonparietal cells isolated from the guinea pig stomach. *FEBS Lett.* 149: 85–90, 1982.
52. GIBERT, A. J., AND S. J. HERSEY. Morphometric analysis of parietal cell membrane transformations in isolated gastric glands. *J. Membr. Biol.* 67: 113–124, 1982.
53. GOLGI, C. Sur la fine organisation des glandes peptiques des mammifères. *Arch. Ital. Biol.* 19: 448–453, 1893.
54. GRINSTEIN, S., C. A. COHEN, AND A. ROTHSTEIN. Cytoplasmic pH regulation in thymic lymphocytes by an amiloride-sensitive Na^+/H^+ antiport. *J. Gen. Physiol.* 83: 341–369, 1984.

55. GUNTHER, R. D., S. BASSILIAN, AND E. RABON. Cation transport in vesicles from secreting rabbit stomach. *J. Biol. Chem.* 262: 13966–13972, 1987.
56. HAHNE, W. F., R. T. JENSEN, G. F. LAMP, AND J. D. GARDNER. Proglumide and benzotript: members of a different class of cholecystokinin receptor antagonists. *Proc. Natl. Acad. Sci. USA* 10: 6304–6308, 1981.
57. HAKANSON, R., G. BOTTCHER, E. EKBLAD, P. PANULA, M. SIMONSSON, M. DOHLSTEN, T. HALLBERG, AND F. SUNDLER. Histamine in endocrine cells in the stomach: a survey of several species using a panel of histamine antibodies. *Histochemistry* 86: 5–17, 1986.
58. HARRIS, J. B., AND D. ALONSO. Stimulation of the gastric mucosa by adenosine-3′,5′-monophosphate. *Federation Proc.* 24: 1368–1376, 1965.
59. HARRIS, J. B., AND I. S. EDELMAN. Chemical concentration gradients and electrical properties of gastric mucosa. *Am. J. Physiol.* 206: 769–782, 1964.
60. HELANDER, H. F. The cells of the gastric mucosa. *Int. Rev. Cytol.* 70: 217–289, 1981.
61. HELANDER, H. F., AND B. I. HIRSCHOWITZ. Quantitative ultrastructural studies of microvilli and changes in the tubulovesicular compartment of mouse parietal cells in relation to gastric acid secretion. *J. Cell Biol.* 63: 951–961, 1972.
62. HELANDER, H. F., AND G. W. SUNDELL. Ultrastructure of inhibited parietal cells in the rat. *Gastroenterology* 87: 1064–1071, 1984.
63. HIGH, W. L., AND S. J. HERSEY. Mechanism of theophylline stimulation of acid secretion by frog gastric mucosa. *Am. J. Physiol.* 226: 1408–1412, 1974.
64. HIRSCHOWITZ, B. I., J. FONG, AND E. MOLINA. Effects of pirenzepine and atropine on vagal and cholinergic gastric secretion and gastric release and on heart rate in the dog. *J. Pharmacol. Exp. Ther.* 225: 263–268, 1983.
64a. HIRST, B. H., AND J. G. FORTE. Redistribution and characterization of ($H^+ + K^+$)-ATPase membranes from resting and stimulated gastric parietal cells. *Biochem. J.* 231: 641–649, 1985.
65. HOPFER, U., AND C. M. LIEDTKE. Proton and bicarbonate transport mechanisms in the intestine. *Annu. Rev. Physiol.* 49: 51–67, 1987.
66. IM, W. B., D. P. BLAKEMAN, J. E. BLEASDALE, AND J. P. DAVIS. A protein phosphatase associated with rat heavy gastric membranes enriched with (H^+-K^+)-ATPase influences membrane K^+ transport activity. *J. Biol. Chem.* 262: 9865–9871, 1987.
67. IM, W. B., D. P. BLAKEMAN, AND J. P. DAVIS. Studies on K^+ permeability of rat gastric microsomes. *J. Biol. Chem.* 260: 9452–9460, 1985.
68. ITO, S. Functional gastric morphology. In: *Physiology of the Gastrointestinal Tract* (2nd ed.), edited by L. R. Johnson. New York: Raven, 1987, vol. 1, p. 817–851.
69. ITO, S., AND G. C. SCHOFIELD. Studies on the depletion and accumulation of microvilli and changes in the tubulovesicular compartment of mouse parietal cells in relation to gastric acid secretion. *J. Cell Biol.* 63: 364–382, 1974.
70. JACOBS, D. M., AND R. P. STURTEVANT. Circadian ultrastructural changes in rat gastric parietal cells under altered feeding regimens: a morphometric study. *Anat. Rec.* 203: 101–113, 1982.
71. KASBEKAR, D. K., AND G. S. GORDON. Effects of colchicine and vinblastine on in vitro gastric secretion. *Am. J. Physiol.* 236 (*Endocrinol. Metab. Gastrointest. Physiol.* 5): E550–E555, 1979.
72. KASBEKAR, D. K., AND H. E. STEWART. Colchicine binding activity of the frog gastric mucosa. In: *Hydrogen Ion Transport in Epithelia*, edited by I. Schulz, G. Sachs, J. G. Forte, and K. J. Ullrich. New York: Elsevier, 1980, p. 105–112.
73. LARSSON, H., E. CARLSSON, H. MATTSSON, K. LUNDELL, F. SUNDLER, G. SUNDELL, B. WALLMARK, T. WATANABE, AND R. HAKANSON. Plasma gastrin and gastric enterochromaffin-like cell activation and proliferation. *Gastroenterology* 90: 391–399, 1986.
74. LEE, H. C., H. BREITBART, M. BERMAN, AND J. G. FORTE. Potassium-stimulated ATPase activity and H^+ transport in gastric microsomal vesicles. *Biochim. Biophys. Acta* 553: 107–131, 1979.
75. LEE, H. C., AND J. G FORTE. A study of H^+ transport in gastric microsomal vesicles using fluorescent probes. *Biochim. Biophys. Acta* 508: 339–356, 1978.
76. LEFEVRE, M. E., E. J. GOHMANN, JR., AND W. S. REHM. An hypothesis for discovery of inhibitors of gastric acid secretion. *Am. J. Physiol.* 207: 613–618, 1964.
77. LEVINE, R. A., K. R. KOHEN, E. H. SCHWARTZEL, JR., AND C. E. RAMSAY. Prostaglandin E_2-histamine interactions on cAMP, cGMP, and acid production in isolated fundic glands. *Am. J. Physiol.* 242 (*Gastrointest. Liver Physiol.* 5): G21–G26, 1982.
78. LIPKIN, M. Proliferation and differentiation of gastrointestinal cells. *Physiol. Rev.* 53: 891–915, 1973.
79. MACHEN, T. E., AND W. L. MCLENNAN. Na^+-dependent H^+ and Cl^- transport in in vitro frog gastric mucosa. *Am. J. Physiol.* 238 (*Gastrointest. Liver Physiol.* 1): G403–G413, 1980.
80. MACHEN, T. E., AND A. M. PARADISO. Regulation of intracellular pH in the stomach. *Annu. Rev. Physiol.* 49: 19–33, 1987.
81. MAJOR, J. S., AND P. SCHOLES. The localization of a histamine H_2-receptor adenylate cyclase system in canine parietal cells and its inhibition by prostaglandins. *Agents Actions* 8: 324–331, 1978.
82. MCLENNAN, W. L., T. E. MACHEN, AND T. ZEUTHEN. Ba^{2+} inhibition of electrogenic Cl^- secretion in vitro frog and piglet gastric mucosa. *Am. J. Physiol.* 239 (*Gastrointest. Liver Physiol.* 2): G151–G160, 1980.
83. MICHELANGELI, F. Isolated oxyntic cells: physiological characterization. In: *Gastric Hydrogen Ion Secretion*, edited by D. K. Kasbekar, G. Sachs, and W. S. Rehm. New York: Dekker, 1976, p. 212–236.
84. MILLER, M., AND A. J. HERSEY. Cyclic nucleotide-dependent protein kinase from isolated gastric glands. In: *Hydrogen Ion Transport in Epithelia*, edited by J. G. Forte, D. G. Warnock, and F. C. Rector. New York: Wiley, 1984, p. 353–362.
85. MUALLEM, S., C. BURNHAM, D. BLISSARD, T. BERGLINDH, AND G. SACHS. Electrolyte transport arccross basolateral membrane of the parietal cells. *J. Biol. Chem.* 260: 6644–6653, 1985.
86. MUALLEM, S., AND G. SACHS. Changes in cytosolic free Ca^{2+} in isolated parietal cells. Differential effects of secretagogues. *Biochim. Biophys. Acta* 805: 181–185, 1984.
87. MUALLEM, S., AND G. SACHS. Ca^{2+} metabolism during cholinergic stimulation of acid secretion. *Am. J. Physiol.* 248 (*Gastrointest. Liver Physiol.* 11): G216–G228, 1985.
88. NEGULESCU, P. A., AND T. E. MACHEN. Intracellular Ca regulation during secretagogus stimulation of the parietal cell. *Am. J. Physiol.* 254 (*Cell Physiol.* 23): C130–C140, 1988.
89. NEGULESCU, P. A., AND T. E. MACHEN. Release and reloading of intracellular Ca stores after cholinergic stimulation of the parietal cell. *Am. J. Physiol.* 254 (*Cell Physiol.* 23): C498–C504, 1988.
90. NYLANDER, O., T. BERGLINDH, AND K. J. ÖBRINK. Prostaglandin interaction with histamine release and parietal cell activity in isolated gastric glands. *Am. J. Physiol.* 250 (*Gastrointest. Liver Physiol.* 13): G607–G616, 1986.
91. PARADISO, A. M., AND T. E. MACHEN. Histamine activates H/K-ATPase and Na/H and Cl/HCO_3 exchange in oxyntic cells (OC). *FASEB J.* 2: 1727A, 1988.
92. PARADISO, A. M., P. A. NEGULESCU, AND T. E. MACHEN. Na^+-H^+ and Cl^--OH^-(HCO_3^-) exchange in gastric glands. *Am. J. Physiol.* 250 (*Gastrointest. Liver Physiol.* 13): G524–G534, 1986.
93. PARADISO, A. M., R. Y. TSIEN, AND T. E. MACHEN. Na^+-H^+ exchange in gastric glands as measured with a cytoplasmic-trapped, fluorescent pH indicator. *Proc. Natl. Acad. Sci. USA* 81: 7436–7440, 1984.

94. RABON, E., H. CHANG, AND G. SACHS. Quantitation of hydrogen ion and potential gradients in gastric plasma membrane vesicles. *Biochemistry* 17: 3345–3353, 1978.
95. RABON, E., G. SACCOMANI, D. K. KASBEKAR, AND G. SACHS. Transport characteristics of frog gastric membranes. *Biochem. Biophys. Acta* 551: 432–447, 1987.
96. RAGINS, H., F. WINCZE, S. M. LIU, AND M. DITTENBRENNER. The origin and survival of gastric parietal cells in the mouse. *Anat. Rec.* 162: 99–110, 1968.
97. REENSTRA, W. W., J. D. BETTENCOURT, AND J. G. FORTE. Active K^+ absorption by the gastric mucosa: inhibition by omeprazole. *Am. J. Physiol.* 250 (*Gastrointest. Liver Physiol.* 13): G455–G460, 1986.
98. REENSTRA, W. W., J. D. BETTENCOURT, AND J. G. FORTE. Mechanism of active Cl^- secretion by frog gastric mucosa. *Am. J. Physiol.* 252 (*Gastrointest. Liver Physiol.* 15): G543–G547, 1987.
99. REENSTRA, W. W., AND J. G. FORTE. Action of thiocyanate on pH gradient formation by gastric microsomal vesicles. *Am. J. Physiol.* 244 (*Gastrointest. Liver Physiol.* 7): G308–G313, 1983.
100. REENSTRA, W. W., AND J. G. FORTE. Mechanism of inhibition of gastric acid secretion by SCN^-: interrelation of SCN^- flux and inhibition. *Am. J. Physiol.* 260 (*Gastrointest. Liver Physiol.* 13): G76–G84, 1986.
101. REHM, W. S. Ion permeability and electrical resistance of the frog's gastric mucosa. *Federation Proc.* 26: 1303–1313, 1967.
102. REHM, W. S., T.-C. CHU, M. SCHWARTZ, AND G. CARRASQUER. Mechanisms responsible for SCN increase in resistance of in vitro frog gastric mucosa. *Am. J. Physiol.* 245 (*Gastrointest. Liver Physiol.* 8): G143–G156, 1983.
103. REUSS, L., AND J. S. STODDARD. Role of H^+ and HCO_3^- in salt transport in gallbladder epithelium. *Annu. Rev. Physiol.* 49: 35–49, 1987.
104. RISING, T. J., D. B. NORRIS, S. E. WARRANDER, AND T. P. WOOD. High affinity ^{3}H-cimetidine binding in guinea-pig tissues. *Life Sci.* 27: 199–206, 1980.
105. ROMRELL, L. J., M. R. COPPE, D. R. MUNRO, AND S. ITO. Isolation and separation of highly enriched fractions of viable mouse gastric parietal cells by velocity sedimentation. *J. Cell Biol.* 65: 428–438, 1975.
106. ROSENFELD, G. C. Pirenzepine (LS 519): a weak inhibitor on acid secretion by isolated rat parietal cells. *Eur. J. Pharmacol.* 86: 99–101, 1983.
107. SACCOMANI, G., H. F. HELANDER, S. CRAGO, H. H. CHANG, AND G. SACHS. Characterization of gastric mucosal membranes. X. Immunological studies of gastric $(H^+ + K^+)$-ATPase. *J. Cell Biol.* 83: 271–283, 1979.
108. SACHS, G., H. H. CHANG, E. RABON, R. SCHACKMANN, M. LEWIN, AND G. SACCOMANI. A non-electrogenic H^+ pump in plasma membranes of hog stomach. *J. Biol. Chem.* 251: 7690–7698, 1976.
109. SACHS, G., D. K. KASBEKAR, AND T. BERGLINDH. The mechanism of action of pirenzepine on gastric secretion. In: *Die Behandlung des ulcus pepticum mit perenzepin*, edited by A. L. Blum, Z. Biberach, and R. H. Demeter. Berlin: Springer-Verlag, 1978, p. 18–34.
110. SCHACKMANN, R. A., A. SCHWARTZ, G. SACCOMANI, AND G. SACHS. Cation transport by gastric $H^+ + K^+$ ATPase. *J. Membr. Biol.* 32: 361–381, 1977.
111. SCHEPP, W., H.-K. HEIM, AND H.-J. RUOFF. Comparison of the effect of PGE_2 and somatostatin on histamine stimulated ^{14}C-aminopyrine uptake and cyclic AMP formation in isolated rat gastric mucosal cells. *Agents Actions* 13: 200–206, 1983.
112. SEDAR, A. W. Uptake of peroxidase into the smooth-surfaced tubular system of the gastric acid-secreting cell. *J. Cell Biol.* 43: 179–184, 1969.
113. SHULL, G. E., AND J. B. LINGREL. Molecular cloning of the rat stomach $(H^+ + K^+)$-ATPase. *J. Biol. Chem.* 261: 16788–16791, 1986.
114. SKOGLUND, M. L., A. S. NIES, AND J. G. GERBER. Inhibition of acid secretion in isolated canine parietal cells by prostaglandins. *J. Pharmacol. Exp. Ther.* 220: 371–374, 1982.
115. SMOLKA, A., H. F. HELANDER, AND G. SACHS. Monoclonal antibodies against gastric $H^+ + K^+$ ATPase. *Am. J. Physiol.* 245 (*Gastrointest. Liver Physiol.* 8): G589–G596, 1983.
116. SOLL, A. H. The actions of secretagogues on oxygen uptake by isolated mammalian parietal cells. *J. Clin. Invest.* 61: 370–380, 1978.
117. SOLL, A. H. The interaction of histamine with gastrin and carbamylcholine on oxygen uptake by isolated mammalian parietal cells. *J. Clin. Invest.* 61: 381–389, 1978.
118. SOLL, A. H. Specific inhibition by prostaglandins E_2 and I_2 of histamine-stimulated [^{14}C]aminopyrine accumulation and cyclic adenosine monophosphate generated by isolated canine parietal cells. *J. Clin. Invest.* 65: 1222–1229, 1978.
119. SOLL, A. H. Secretagogue stimulation of [^{14}C]aminopyrine accumulation by isolated canine parietal cells. *Am. J. Physiol.* 238 (*Gastrointest. Liver Physiol.* 1): G366–G375, 1980.
120. SOLL, A. H. Extracellular calcium and cholinergic stimulation of isolated canine parietal cells. *J. Clin. Invest.* 68: 270–278, 1981.
121. SOLL, A. H., D. A. AMIRIAN, L. P. THOMAS, J. PARK, J. D. ELASHOFF, M. A. BEAVAN. AND T. YAMADA. Gastrin receptors on nonparietal cells isolated from canine fundic mucosa. *Am. J. Physiol.* 247 (*Gastrointest. Liver Physiol.* 10): G715–G723, 1984.
122. SOLL, A. H., D. A. AMIRIAN, L. P. THOMAS, T. J. REEDY, AND J. D. ELASHOFF. Gastrin receptors on isolated canine parietal cells. *J. Clin. Invest.* 73: 1434–1447, 1984.
123. SOLL, A. H., AND T. BERGLINDH. Physiology of isolated gastric glands and parietal cells: receptors and effectors regulating function. In: *Physiology of the Gastrointestinal Tract* (2nd ed.), edited by L. R. Johnson. New York: Raven, 1987, vol. 1, p. 883–909.
124. SOLL, A. H., K. LEWIN, AND M. A. BEAVEN. Isolation of histamine containing cells from canine fundic mucosa. *Gastroenterology* 77: 1283–1290, 1979.
125. SOLL, A. H., M. TOOMEY, D. CULP, F. SHANAHAN, AND M. A. BEAVEN. Modulation of histamine release from canine fundic mucosal mast cells. *Am. J. Physiol.* 254 (*Gastrointest. Liver Physiol.* 17): G40–G48, 1988.
126. SOLL, A. H., T. YAMADA, J. PARK, AND L. P. THOMAS. Release of somatostatinlike immunoreactivity from canine fundic mucosal cells in primary culture. *Am. J. Physiol.* 247 (*Gastrointest. Liver Physiol.* 10): G558–G566, 1984.
127. SONNENBERG, A., W. HUNZIKER, H. R. KOELZ, J. A. FISCHER, AND A. L. BLUM. Stimulation of endogenous cyclic AMP (cAMP) in isolated gastric cells by histamine and prostaglandin. *Acta Physiol. Scand. Suppl.* 307–317, 1978.
128. SOUMARMON, A., A. M. CHERET, AND M. J. M. LEWIN. Localization of gastrin receptors in intact isolated and separated fundic rat cells. *Gastroenterology* 73: 900–903, 1978.
129. SPANGLER, S. G., AND W. S. REHM. Potential responses of nutrient membrane of frog's stomach to step changes in external K^+ and Cl^- concentrations. *Biophys. J.* 8: 1211–1227, 1968.
130. TAKEGUCHI, N., M. HATTORI, A. SANO, AND I. HORIKOSHI. Intracellular potassium ion in relation to acid secretory rate by frog gastric mucosa. *Am. J. Physiol.* 237 (*Endocrinol. Metab. Gastrointest. Physiol.* 6): E51–E55, 1979.
131. TAKEGUCHI, N., R. JOSHIMA, Y. INOUE, T. KASHIWAGURA, AND M. MORII. Effects of Cu^{2+}-o-phenanthroline on gastric $(H^+ + K^+)$-ATPase. Evidence for opening of a closed anion conductance by S-S cross-linkings. *J. Biol. Chem.* 258: 3094–3098, 1983.
132. TAKEGUCHI, N., AND Y. YAMAZAKI. Disulfide cross-linking of H,K-ATPase opens Cl^- conductance, triggering proton uptake in gastric vesicles. Studies with specific inhibitors. *J. Biol. Chem.* 261: 2560–2566, 1986.
133. TAKUCHI, R., G. R. SPEIR, AND L. R. JOHNSON. Mucosal gastric receptor. I. Assay standardization and fulfillment of receptor criteria. *Am. J. Physiol.* 237 (*Endocrinol. Metab.*

Gastrointest. Physiol. 6): E284–E294, 1979.
134. URUSHIDANI, T., D. K. HANZEL, AND J. G. FORTE. Protein phosphorylation associated with stimulation of rabbit gastric glands. *Biochim. Biophys. Acta* 930: 209–219, 1987.
135. VIAL, J. D., AND J. GARRIDO. Actin-like filaments and membrane rearrangement of oxyntic cells. *Proc. Natl. Acad. Sci. USA* 73: 4032–4036, 1976.
136. WALLMARK, B., G. SACHS, S. MARDH, AND E. FELLENIUS. Inhibition of gastric $(H^+ + K^+)$-ATPase by the substituted benzimidazole, picoprazole. *Biochim. Biophys. Acta* 728: 31–38, 1983.
137. WOLLIN, A., L. D. BARNES, Y. S. HUI, AND T. P. DOUSA. Activation of protein kinase in the guinea pig fundic gastric mucosa by histamine. *Life Sci.* 17: 1303–1306, 1974.
138. WOLLIN, A., A. H. SOLL, AND I. M. SAMLOFF. Actions of histamine, secretin, and PGE_2 on cyclic AMP production by isolated canine fundic mucosal cells. *Am. J. Physiol.* 237 (*Endocrinol. Metab. Gastrointest. Physiol.* 6): E437–E443, 1979.
139. WOLOSIN, J. M., AND J. G. FORTE. Changes in the membrane environment of the $(K^+ + H^+)$-ATPase following stimulation of the gastric oxyntic cell. *J. Biol. Chem.* 256: 3149–3152, 1981.
140. WOLOSIN, J. M., AND J. G. FORTE. Functional differences between K^+-ATPase-rich membrane isolated from resting and stimulated rabbit fundic mucosa. *FEBS Lett.* 125: 208–212, 1981.
141. WOLOSIN, J. M., AND J. G. FORTE. Kinetic properties of the KCl transport at the secreting apical membrane of the oxyntic cell. *J. Membr. Biol.* 71: 195–207, 1983.
142. WOLOSIN, J. M., AND J. G. FORTE. Stimulation of oxyntic cell triggers K^+ and Cl^- conductances in apical $H^+ + K^-$-ATPase membrane. *Am. J. Physiol.* 246 (*Cell Physiol.* 15): C537–C545, 1984.
143. WOLOSIN, J. M., AND J. G. FORTE. K^+ and Cl^- conductances in the apical membrane from secreting oxyntic cells are concurrently inhibited by divalent cations. *J. Membr. Biol.* 83: 261–272, 1985.
144. WOLOSIN, J. M., C. OKAMOTO, T. M. FORTE, AND J. G. FORTE. Actin and associated proteins in gastric epithelial cells. *Biochim. Biophys. Acta* 761: 171–182, 1983.
145. YAMADA, T., A. H. SOLL, J. PARK, AND J. ELASHOFF. Autonomic regulation of somatostatin release: studies with primary cultures of canine fundic mucosal cells. *Am. J. Physiol.* 247 (*Gastrointest. Liver Physiol.* 10): G567–G573, 1984.
146. ZALEWSKY, C. A., AND F. G. MOODY. Stereological analysis of the parietal cell during acid secretion and inhibition. *Gastroenterology* 73: 66–74, 1977.

CHAPTER 12

Biochemistry of gastric acid secretion: H^+-K^+-ATPase

G. SACHS

J. KAUNITZ

J. MENDLEIN

B. WALLMARK

Center for Ulcer Research and Education, Wadsworth Veterans Administration Hospital and University of California, Los Angeles, California

Department of Physiology, University of California School of Medicine, Los Angeles, California

Hassle AB, Molndahl, Sweden

CHAPTER CONTENTS

Parietal Cell Acid Secretion and H^+-K^+-ATPase
General Aspects of H^+-K^+-ATPase
Structure of H^+-K^+-ATPase
   Sequence
   Protein composition
   Dimensions
Transport by H^+-K^+-ATPase
   Resting vesicles
   Stimulated vesicles
   Transport modes of H^+-K^+-ATPase
   H^+-K^+-ATPase in parietal cell
Kinetics of H^+-K^+-ATPase
   Steady-state kinetics of ATPase
   Steady-state kinetics of phosphatase reaction
   Phosphorylation from ATP
   ATP-ADP exchange
   Potassium-dependent dephosphorylation
   Phosphorylation from P_i
Conformations of H^+-K^+-ATPase
   Tryptic digestion
   Fluorescence
Inhibition of H^+-K^+-ATPase
   Site-specific inhibitors
      N_3-ATP
      Vanadate
      Protonatable amines
      Omeprazole
   Group-selective reagents
Biosynthesis of H^+-K^+-ATPase
Tissue Distribution of H^+-K^+-ATPase
Model for H^+-K^+-ATPase

SINCE THE MECHANISM of gastric acid secretion was reviewed previously in this series, there has been a major reorientation of thinking in this area (13). In 1967 the major model for mechanistic investigation of acid secretion was the amphibian gastric mucosa in the Ussing chamber, and the major techniques used were electrophysiological and pH metric. Much discussion revolved around the energy source for acid secretion: whether ATP or oxidation-reduction reaction provided the driving force (116). Another area of controversy focused on the electrogenicity or electroneutrality of H^+ secretion (114, 115). The basis for active Cl^- transport was also not understood (48). In the intervening years, adequate mammalian models of acid secretion were developed, such as the rabbit gastric gland (5) and the permeable gland model (4, 82). Gastric membranes (27), containing an ATPase capable of transport of H^+ in exchange for K^+ from a variety of species (72, 130), were isolated in the resting and stimulated state (9, 53, 172). These models have resulted in a major change in emphasis in investigation into mechanisms of gastric acid secretion. The present focus is at the membrane and molecular level and involves questions such as the sequence of the H^+-K^+-ATPase (141), the K^+ and Cl^- conductance pathways (which may be involved in stimulation), and the mechanism of pump activation (35).

Most of this chapter is focused on the H^+-K^+-ATPase and what is known of its structure, reaction pathways, and transport mechanism in the mammalian parietal cell.

PARIETAL CELL ACID SECRETION AND H^+-K^+-ATPASE

The first evidence that K^+ was required in acid secretion was obtained in the isolated frog mucosa (37, 38). A similar K^+ requirement for acid secretion was shown in isolated rabbit gastric glands (65). Direct evidence that acid secretion by the parietal cell was ATP dependent was obtained in glands that had been permeabilized by electric shock (4) or by digitonin (82). The accumulation of the weak base aminopyrine, which demonstrates acid secretion, was shown to be ATP dependent and to occur despite the presence of mitochondrial inhibitors and anoxia, eliminating a role for an oxidation-reduction component in acid secretion. In vivo it was shown that a covalent inhibitor of the H^+-K^+-ATPase, omeprazole, inhibited secretion and that the prevailing acid rate was correlated with the residual activity of the H^+-K^+-ATPase isolated from the same stomach in which acid secretion had been measured (167). Thus H^+ secretion is K^+ dependent, ATP driven, and caused by the H^+-K^+-ATPase.

GENERAL ASPECTS OF H^+-K^+-ATPASE

The enzyme H^+-K^+-ATPase has been positively identified only in the gastric mucosa and only in the parietal cell of the mammalian stomach by polyclonal or monoclonal antibody and enzyme techniques (123, 147). There may be other tissues in which this enzyme or a similar enzyme occurs, but the evidence is less strong. In the parietal cell the enzyme is found in the resting state in intracellular smooth membrane structures, the tubulovesicles; the enzyme is located on the microvilli of the expanded intracellular canaliculus in the stimulated state (66, 147). Thus in the morphological transformation that occurs with stimulation of the parietal cell, there is a change in the cellular site of this enzyme but a maintained intracellular location (24).

The H^+-K^+-ATPase has been found in a variety of species, e.g., frog (25, 27, 105), dog (72), hog (76, 126), rat (50), rabbit (172), and human (119). The biochemistry of the ATPase is best characterized for hog enzyme, but the amino acid sequence is known only for the rat catalytic subunit (141), and the sequence has significant homology with the Na^+-K^+-ATPase.

Purification of the enzyme in vesicular form has shown that the major protein constituent has an apparent molecular weight (M_r) of 90,000 and 100,000 on reducing sodium dodecyl sulfate–polyacrylamide gel electrophoresis (SDS-PAGE) (96, 126, 150). The enzyme is phosphorylated by $Mg^{2+}[\gamma$-$^{32}P]ATP$, and dephosphorylated in the presence of K^+ (110, 125). Transport studies show that in hog gastric vesicles the ATPase drives an H^+-K^+ exchange (8, 130).

These data suggest that this enzyme is a member of the class of phosphorylating, transport ATPases, which contains enzymes such as the K^+-dependent ATPase of Escherichia coli (45), the H^+-ATPase of Neurospora (34) and Saccharomyces cerevisiae (140), the Ca^{2+}-ATPase of the sarcoplasmic reticulum and plasma membrane (83), and the Na^+-K^+-ATPase (142, 143). These enzymes vary in the nature of the ion transported, whether there is a counterion, whether they are electrogenic, in the number of subunits, and in amino acid sequence (25%–60% homology). In common, there is a major subunit that undergoes cyclic covalent phosphorylation of an aspartyl residue in one of the hydrophilic sectors during ATPase activity, and this region appears highly conserved (161). These enzymes are often referred to as E_1E_2 transport ATPases because the two major forms of the phosphorylated intermediate are found (E_1-P and E_2-P), arguably related to the transport function in moving an ion cis (E_1-P) to trans (E_2-P) [i.e., ion-binding site, cytosolic (cis) or luminal (trans)].

The gastric ATPase activity is monovalent cation activated. A small, basal, Mg^{2+}-dependent activity (5% of maximum) is probably due to spontaneous, alkali cation independent, turnover of the phosphoenzyme. Various monovalent cations stimulate the ATPase in the order of affinity $Tl^+ > K^+ = Rb^+ > NH_4^+ > Cs^+ > Na^+ > Li^+$ (127). The pH optimum of K^+-ATPase is ~7.1. Specific activation by protons equivalent to activation by Na^+ of the Na^+-K^+-ATPase has not been shown, because pH affects several groups in proteins, not only the transport site. It is the H^+-transport capability of the K^+-activated ATPase, rather than a specific H^+ effect on ATP turnover, that has resulted in naming the enzyme the H^+-K^+-ATPase.

STRUCTURE OF H^+-K^+-ATPASE

Understanding the mechanism of a pump is a formidable task that requires integration of physiology, biochemistry, and structure. The amino acid sequence is a first and necessary step, but the final analysis requires description of the tertiary and quaternary structure of these membrane-bound enzymes. This requires high-resolution X-ray diffraction of three-dimensional crystals in at least the two major conformations of a pump. That this is a feasible, and not hopeless, endeavor is shown by the recent success in the structural analysis at 3 Å of the photoreaction center of *Rhodopseudomonas viridis* (10) and improved methods for two-dimensional structural analysis of the Ca^{2+}-ATPase (15).

Sequence

The amino acid sequence of the catalytic subunit of the enzyme has been deduced from cDNA clones for the rat H^+-K^+-ATPase [Fig. 1; (141)]. From this sequence it is possible to calculate the probability of hydrophobic, membrane-spanning sequences, using one of various hydropathy plots (67). In addition, regions of α-helix, β-sheet, and β-turn in the hydrophilic domain can be predicted from known sequences in other, nonmembrane proteins (Fig. 2). The reliability of these probability calculations is limited (162). For example, there may be errors in such an analysis of the cholinergic receptor (109) and the β-subunit of the Na^+-K^+-ATPase (175); both an even number (142) and an odd number of spanning sequences have been postulated (94). Nevertheless a structural model can

```
  1 MGKENYELYS VELGTGPGGD MAAKMSKKKA GGGGGKKKEK LENMKKEMEM
 51 NDHQLSVSEL EQKYQTSATK GLKASLAAEL LLRDGPNALR PPRGTPEYVK
101 FGRQLAGGLQ CLMWVAAAIC LIAFAIQASE GDLTTDDNLY LALALIAVVV
151 VTGCFGYYQE FKSTNIIASF KNLVPQQATV IRDGDKFQIN ADQLVVGDLV
201 EMKGGDRVPA DIRILSAQGC KVDNSSLTGE SEPQTRSPEC THESPLETRN
251 IAFFSTMCLE GTAQGLVVST GDRTIIGRIA SLASGVENEK TPTAIEIEHF
301 VDIIAGLAIL FGATFFVVAM CIGYTFLRAM VFFMAIVVAY VPEGLLATVT
351 VCLSLTAKRL ASKNCVVKNL EAVETLGSTS VICSDKTGTL TQNRMTVSHL
401 WFDNHIHTAD TTEDQSGQTF DQSSETWRAL CRVLTLCNRA AFKSGQDAVP
451 VPKRIVIGDA SETALLKFSE LTLGNAMGYR DRFPKVCEIP FNSTNKFQLS
501 IHTLEDPRDP RHLLVMKGAP ERVLERCSSI LIKGQELPLD EQWREAFKTA
551 YLSLGGLGER VLGFCQLYLN EKDYPPGYTF DVEAMNFPSS GLCFAGLVSM
601 IDPPRVTVPD AVLKCRTAGI RVIMVTGDHP ITAKAIAASV GIISEGSETV
651 EDIAARLRMP VDQVNKKDAR ACVINGMQLK DMDPSELVEA LRTHPEMVFA
701 RTSPQQNLVI VESCQRLVAI VAVTGDGVND SPALKKADIG VAMGIAGSDA
751 AKNAADMILL DDNFASIVTG VEQGRLIFDN LKKSIAYTLT KNIPELTPYL
801 IYITVSVPLP LGCITILFIE LCTDIFPSVS LAYEKAESDI MHVRPRNPRR
851 DRLVNEPLAA YSYFQIGAIQ SFAGFADYFT AMAQEGWFPL LCVGLRPQWE
901 DHHLQDLEDS YGQEWTFGQR LYQQYTCYTV FFISIEMCQI ADVLIRKTRR
951 LSAFQQGFFR NRILVIAIVF QVCIGCFLCY CPGMPNIFNF MPIRFQWWLV
1001 PMPFGLLIFV YDEIRKLGVR CCPGSWWDQE LYY
```

FIG. 1. Amino acid sequence of rat H^+-K^+-ATPase as deduced by cDNA cloning methods. An additional translatable region that could code for a peptide of ~40 amino acids is shown below the sequence for the catalytic subunit. Whether it is synthesized or its function is unknown, so far this structure is unique for the E_1, E_2 cDNA sequences.

```
5' END : MHTTASKGRMLFMNVTMYMRRGVQIMGWSVRMAFMACFTQ
```

be derived to serve as a template for future experiments.

It is of interest to compare the hydrophilic sequences postulated in the Na^+-K^+-ATPase and the H^+-K^+-ATPase. Of the eight membrane sequences postulated, a major difference is found in the sequences toward the COOH-terminal end, where a large number of cysteines are found in the H^+-K^+-ATPase, as compared with the Na^+-K^+-ATPase. One possibility that has been suggested for the lac permease, which couples the electrochemical H^+ gradient to lactose transport in *E. coli*, is a thiol-disulfide exchange as a means of H^+ translocation across this hydrophobic peptide (85). However, recent mutations in seven of the eight cysteines of this protein have shown that transport is only partially inhibited, hence arguing against a role for thiol-disulfide exchange in proton transport by this protein. However, the membrane-located cysteines in the H^+-K^+-ATPase could play the role of a membrane-spanning H^+ pathway. Inhibition of this enzyme by a luminal SH reagent, omeprazole (79, 80), and the sensitivity of the H^+-K^+-ATPase to a hydrophobic SH reagent, tributyl tin (101), may relate to this idea.

The ion-translocation pathway across this and other pumps is unlikely to be across a single membrane-spanning sequence but will rather involve interaction between hydrophilic faces of several of these segments. Inspection of the eight postulated sequences in the H^+-K^+-ATPase suggests that three are anchoring in function, whereas five may be involved in H^+ and K^+ countertransport. It may be possible by computer

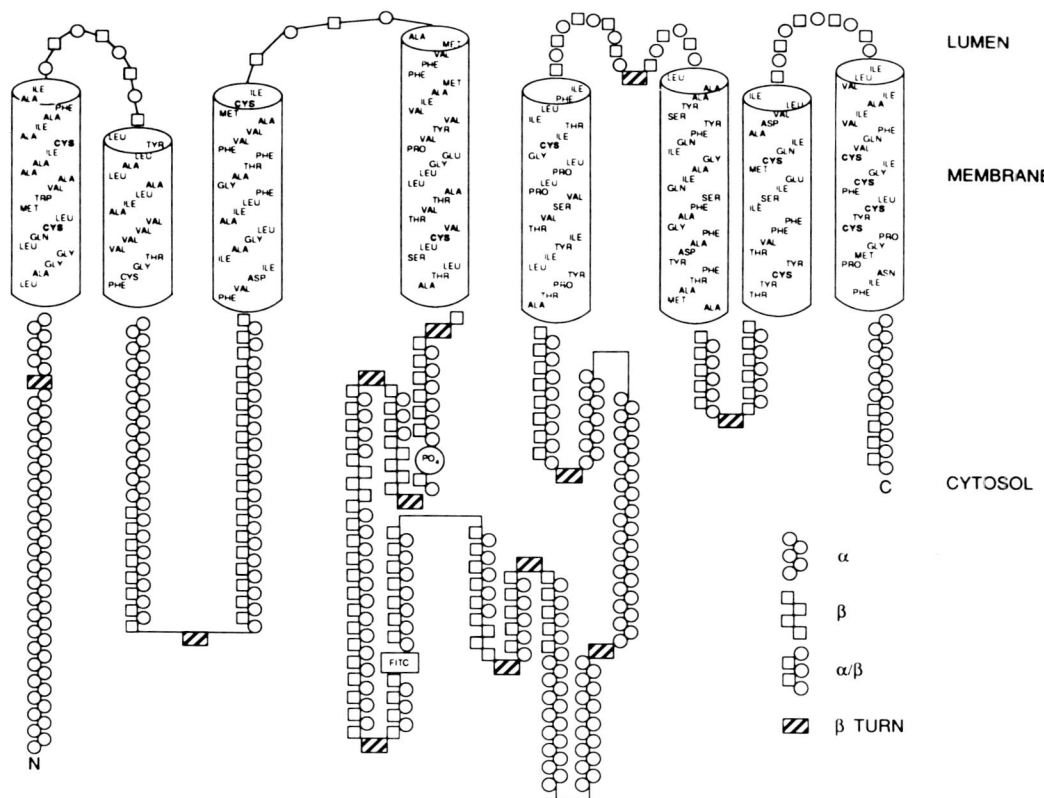

FIG. 2. Model of the secondary structure of the H^+-K^+-ATPase catalytic subunit, where various symbols denote predictions for α-helix, β-sheet, and β-turns. If no discrimination between α and β is statistically significant, a separate symbol is used. Probabilities used depend on known structures for soluble proteins and may not apply to membrane-embedded molecules (162). FITC, fluorescein isothiocyanate.

modeling of the H^+-K^+- and Na^+-K^+-ATPase hydrophobic peptides to determine the H^+, Na^+, and K^+ pathways, because the restrictions on conformation in the membrane sector are more severe than in the hydrophilic domain.

The region where ATP is thought to interact and other regions of the large hydrophilic sector between membrane sequences H_4 and H_5, containing the phosphorylation site (161) and fluorescein isothiocyanate (FITC) site (22), show the largest degree of homology with the Na^+-K^+-ATPase. As is seen later, there are also significant analogies in the kinetic steps of the two enzymes, and this may be due to the homologies in this region (presumably catalytic) between the two enzymes.

A major difference between the RNA-derived cDNA sequence of this ATPase and the other known E_1E_2 ATPases is the presence of an upstream sequence that has both start and stop codons. As illustrated in Figure 1, this sequence would code for a protein of ~40 amino acids, with a large number of methionines, a single cysteine, and four positively charged residues (141). The function of this possible sequence is unknown, although a Cl^--channel function was suggested (141). An alternative is that this sector is incorporated into the catalytic subunit of the enzyme by disulfide linkage, perhaps to provide a targeting sequence for tubulovesicular homing of the holoenzyme. This may account for the doublet appearance at 94,000 M_r of the hog enzyme.

Protein Composition

Most studies on the protein composition of the H^+-K^+-ATPase have been performed on hog gastric material. It seems that the major protein is in the 90,000- to 100,000-M_r region, often a doublet (20), and in the most highly purified material this region accounts for ~80%–90% of the material visualized on reducing SDS-PAGE. In addition, there may be a minor glycoprotein at ~80,000 M_r (25, 96, 126). These results are obtained from staining with Coomassie Blue [other minor peptides at 43,000 and 35,000 M_r are revealed with silver staining (E. Rabon, personal communication)]. No other peptide has been identified at a 1:1 stoichiometry to act in a manner equivalent to the β-subunit of the Na^+-K^+-ATPase.

Evidence that the 90,000-M_r region of the SDS-PAGE contains the catalytic subunit is substantial, although the amino acid sequence gives a molecular weight of 113,000, a characteristic discrepancy in this class of enzymes (154). Thus it is covalently labeled by [^{32}P]-ATP, with K$^+$-dependent dephosphorylation (110, 125). It is recognized by both monoclonal (147) and polyclonal (123) antibodies, which also selectively stain the intracellular membranes of the parietal cell. It is labeled by FITC, which reacts with intact ATPase in an ATP-protectable manner (57), and is labeled by [^{3}H]omeprazole, an inhibitor of the H$^+$-K$^+$-ATPase (80, 163). Reconstitution of enzyme activity requires the presence of the 94,000-M_r subunit (103). Also, the homology of the sequence of the rat H$^+$-K$^+$-ATPase to Na$^+$-K$^+$-ATPase argues for the functional role of this protein.

The role of the other proteins that might be in the same membrane region is unknown. A protein has been described as an activating factor (3), and proteins that are required for transport of K$^+$ and Cl$^-$ (171) into the lumen of the gastric tubulovesicle or canaliculus may be present, as well as proteins that determine the location of the membrane in the parietal cell (35).

In the most highly purified preparations of the hog gastric ATPases, activities as high as 300 μM P$_i$ produced per milligram per hour have been recorded (E. Rabon, personal communication). A higher activity ($\sim$500 μM P$_i \cdot$mg$^{-1} \cdot$h^{-1}) has been produced by a low-concentration SDS treatment of hog gastric microsomes (3). However, most preparations seem to have an activity of $\sim$100–150 μM$\cdot$mg$^{-1} \cdot$h^{-1}. With phosphorylation from [γ-^{32}P]ATP, 1.5 nmol E-P/mg can be isolated (168, 170). Labeling with FITC also provided 1.5 nmol/mg enzyme (57), but eosin binding was 3.5 nmol/mg (42). Labeling with [γ-^{32}P-8-N$_3$]ATP gave a value of 3 nmol/mg enzyme, but half of the label was removed by K$^+$ addition, suggesting more than one form of the label (20, 121). Vanadate binding was 3 nmol/mg protein (21), and [^{3}H]omeprazole labeling from the luminal surface of the enzyme bound at a level of 2 mol compound/1 mol E-P (80).

To relate stoichiometry to composition, on the assumption that each subunit is active, 1 mg of an 80%-pure protein of 100,000 M_r would be expected to contain 8 nmol of binding sites. Thus a large fraction of the protein at that molecular weight does not display the binding expected of various ligands. Stoichiometric arguments have also surrounded other pumps of this class, for example, the Na$^+$-K$^+$-ATPase (93). Perhaps a large fraction of the enzyme is damaged during isolation, or there is more than one class of 94,000-M_r subunit. The ability to sequence the hog gastric enzyme for $\sim$20 amino acids from the NH$_2$ terminus argues against the latter interpretation (68) and argues for homogeneity of the constituent peptides.

The functional molecular weight of the enzyme has been studied by target analysis. Electron irradiation has given a 270,000 M_r, whereas irradiation has given a molecular weight of up to 440,000 for the ATPase activity (124, 139). A reexamination of this problem, also using electron-beam irradiation in a different laboratory, has given a molecular weight for ATPase activity, p-nitrophenylphosphatase (pNPPase) activity, and K$^+$-K$^+$ exchange of $\sim$220,000 M_r. This could indicate that a homodimeric unit is required for activity (E. Rabon and E. Kempner, personal communication).

Dimensions

The general dimensions of the molecule have been obtained by 1) measurements of tannic acid– and glutaraldehyde-fixed hog gastric vesicles (128), which allow an estimate of the extension of the hydrophilic peptide outside the membrane region; 2) freeze fracture to estimate the dimensions of the membrane-spanning sector; and 3) measurements of two-dimensional crystals (108). These have allowed the general shape of the cytosolic sector to be calculated. The composite in Figure 3 illustrates the measured structures and the approximate dimensions and shape that have been obtained. The finding that the crystals, which are obtained in the E$_2$ form and have P$_2$ symmetry, may only be coincidentally dimeric does correlate, nevertheless, with the target data discussed here.

TRANSPORT BY H$^+$-K$^+$-ATPASE

Vesicles isolated from both nonstimulated and secretagogue-stimulated gastric mucosa of a variety of species have been shown to be capable of acidification of their lumen in the presence of K$^+$- and Mg^{2+}-ATP (53, 72, 131, 173). Thus this type of vesicle is inside out with respect to the orientation of acid secretion in the intact parietal cell. Vectorial H$^+$ transport can be measured by alkalinization of the medium using a pH electrode or pH metric dyes (70) or by trapping of protonatable weak bases in the vesicle space (70, 101). The trapping can be measured either by radioactivity distribution {e.g., [^{14}C]aminopyrine (50, 51)} or by optical methods using concentration-dependent chromic shifts (acridine orange) or quenching of fluorescence (e.g., 9-aminoacridine, chloroquin) (70, 101). The transport properties of vesicles isolated from resting or stimulated mucosa are distinct and are reviewed separately.

Resting Vesicles

The K$^+$ activation of ATP hydrolysis is largely dependent on the presence of ionophores when isosmolarity has been maintained to retain vesicular integrity. The degree of K$^+$ stimulation is also ionophore specific (25, 129). The cation-proton exchange ionophores, such as nigericin or gramicidin, produce as high an activity as is found in broken (lyophilized, resuspended) vesicles, whereas valinomycin also stim-

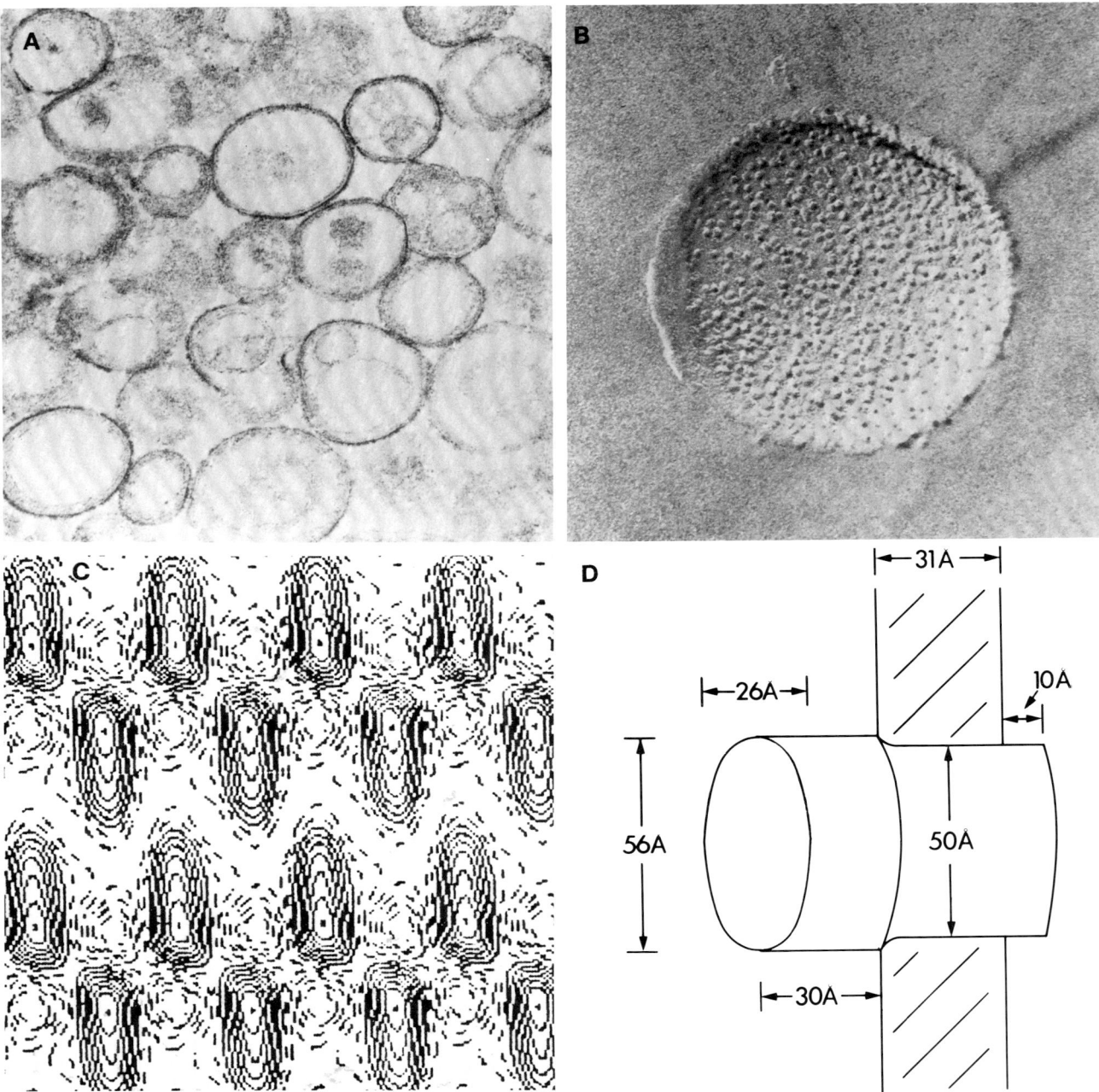

FIG. 3. Composite illustrating various structural features of hog gastric vesicle, which contains the H^+-K^+-ATPase. A: tannic acid–glutaraldehyde–fixed section of vesicles prepared by centrifugation and free-flow electrophoresis. B: freeze-fracture image of gastric vesicles showing membrane-embedded region. C: reconstruction of an optical diffraction pattern of 2-dimensional crystals of the H^+-K^+-ATPase. D: model of outline of the H^+-K^+-ATPase determined from dimensions calculated from A, B, and C. [From Sachs (127a).]

ulates the enzyme in the presence of KCl (but not K^+ sulfate). However, the stimulation by valinomycin is considerably less than that by nigericin or gramicidin (see Fig. 5, *inset*). A protonophore, such as 3,3′,4′,5-tetrachlorosalicylanilide (TCS), does not stimulate ATPase activity, suggesting that the formation of a proton or potential gradient does not limit ATP turnover.

These ionophore effects suggest that the activity of the H^+-K^+-ATPase is limited by access of K^+ to the luminal surface of the enzyme. Gramicidin and nigericin, which act as K^+-H^+ exchange ionophores, produce the maximal rate of ATP hydrolysis, as compared with the slower rate with valinomycin, which is accounted for by the slow rate of valinomycin [$K^+ \cdot Cl^-$] transport across the membrane. The activity in the presence of KCl alone is due to high ion permeability of broken membranes, as shown by the inability of such vesicles to form a proton gradient or to be inhibited by the acid-activated inhibitor, omeprazole (80).

These observations determine the conditions necessary for observing proton transport by H^+-K^+-ATPase in resting membranes. It is required to have K^+ on the luminal, or the acidifying, side of H^+-K^+-ATPase. Because K^+-H^+ exchange ionophores prevent proton-gradient formation, either the K^+-selective ionophore valinomycin (at pH 7 the selectivity ratio K^+/H^+ is ~100,000 in favor of K^+) can be used to provide internal K^+ or the vesicles can be loaded by preincubation in K^+ salts or by osmotic shock using glycerol dilution (W. B. Im, S. Bassilian, E. Rabon, and G. Sachs, unpublished observations). The development of the pH gradient is faster after K^+ salt loading than in the presence of only external KCl and valinomycin (8). This finding correlates with the slower rate of ATPase activity in the presence of valinomycin, compared with nigericin.

A loading experiment that shows the K^+ internal requirement is illustrated in Figure 4A. Increasing times of preincubation in KCl resulted in progressively larger H^+ gradients because of the higher internal KCl reached as a function of preincubation time (130). As a consequence of progressive KCl loading, the osmotic sensitivity of the proton transport was also progressively lost. This latter result can be explained if the transport reaction is an equal exchange of cytosolic H^+ for luminal K^+ (Fig. 4B).

In resting vesicles, H^+ transport has no specific anion requirement. Loading vesicles with K^+ sulfate, rather than KCl, required longer preincubation periods (130). Acid transport was still obtained with these K_2SO_4-loaded vesicles. This is in contrast to the Cl^- requirement for acid secretion by the isolated rabbit gastric gland, where aminopyrine uptake could not be demonstrated in Cl^--free, SO_4^--containing solutions (81).

When ATPase activity was measured in KCl-loaded membranes, there was a transient burst of ATPase activity, followed by a return to the level of activity seen in the absence of K^+ (Fig. 5). Internal K^+ was thus required in the vesicles, but the addition of ATP resulted in the loss of K^+ from the interior and hence loss of K^+-dependent ATPase activity. From this enzyme assay procedure it is possible to infer that K^+ is transported by the H^+-K^+-ATPase from the luminal to cytosolic face of the pump.

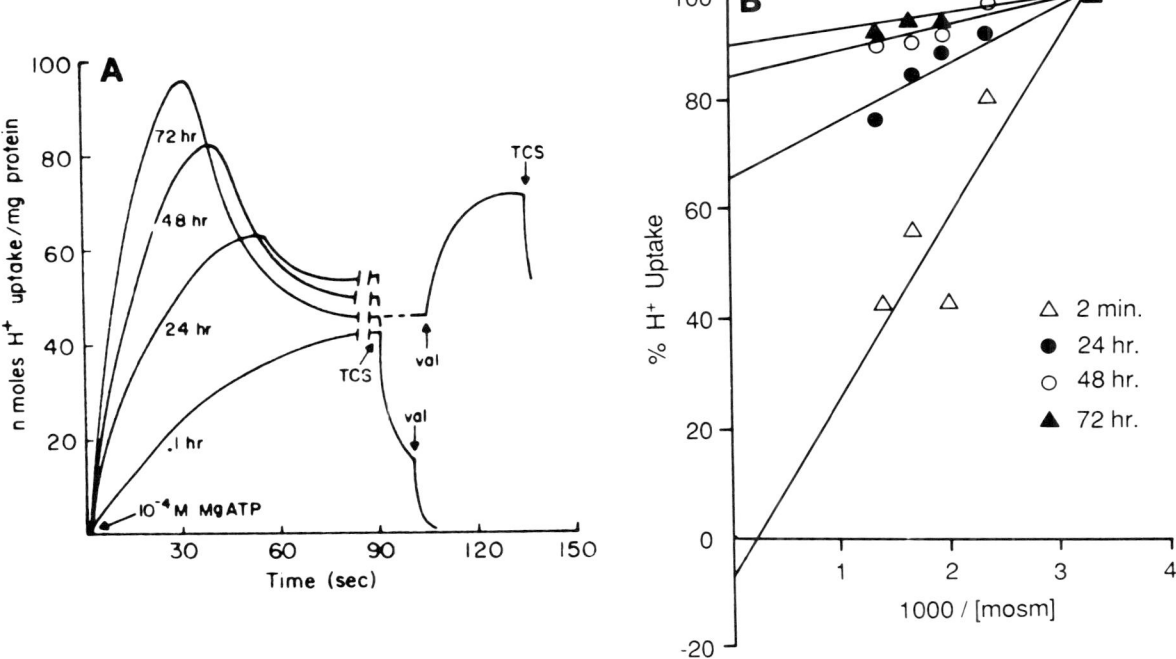

FIG. 4. A: pH metric study of proton transport by hog gastric vesicles. Vesicles were loaded in 150 mM KCl for the indicated times, and pH-gradient formation was initiated by addition of Mg^{2+}-ATP in a 5 mM glycylglycine buffer at pH 6.12. Gradient formation was monitored by alkalinization of the medium using a pH electrode. B: osmotic sensitivity of the pH gradient formed after KCl loading for different times, showing that there is a progressive loss of osmotic sensitivity, as KCl is present before the addition of Mg^{2+}-ATP. TCS, tetrachlorosalicylanilide; val, valinomycin.

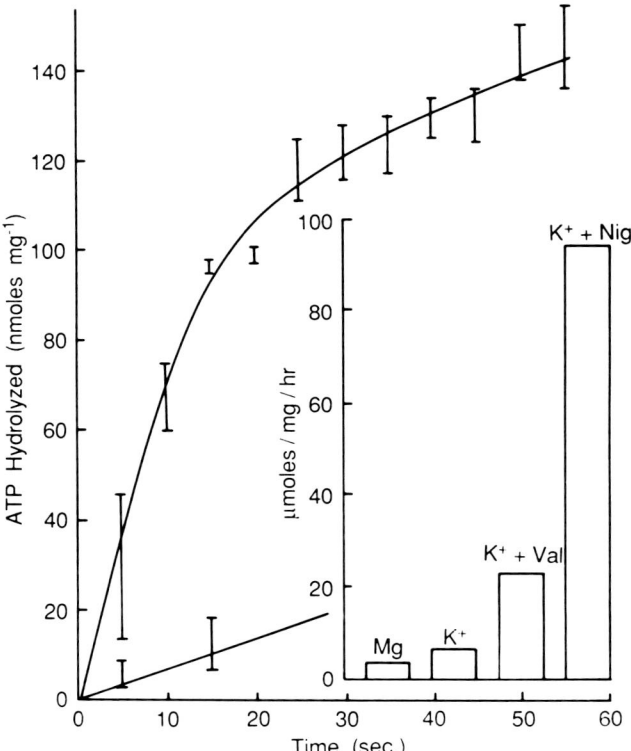

FIG. 5. ATPase activity of hog gastric vesicles that are first loaded with 150 mM KCl diluted 10-fold with choline Cl⁻, and then Mg^{2+}ATP is added. Burst of K$^+$-ATPase activity is observed, followed by a return to base-line (K$^+$-independent) ATP hydrolysis (*lower curve*). This shows that internal (luminal) K$^+$ is required for K$^+$ stimulation of ATPase activity and that K$^+$ is exported during the enzyme reaction. Calculation gives a K$^+$/ATP stoichiometry = 2. *Inset*: ATPase activity of this preparation in unequilibrated vesicles, with K$^+$, K$^+$ + valinomycin (Val), and K$^+$ + nigericin (Nig).

More direct evidence for K$^+$ transport was obtained from studies in which the vesicles were equilibrated with ^{86}Rb$^+$ in place of K$^+$ and Mg^{2+}-ATP was added directly to the equilibrated mixture. There was ATP-dependent loss of ~60% of the ^{86}Rb$^+$, and hence the pump is able to actively transport ^{86}Rb$^+$ from lumen to cytosol, thus acting as an H$^+$-K$^+$ exchange enzyme (136).

In these ^{86}Rb$^+$-loaded vesicles, a cation-cation exchange pathway was also found where the presence of external cation stimulated loss of ^{86}Rb$^+$ under isotope-gradient conditions in the selective sequence K$^+$ > Rb$^+$ > Cs$^+$ > Na$^+$, Li$^+$, but the selectivity was less distinct than that for the H$^+$-K$^+$-ATPase (136). However, this exchange appears to be a property of the ATPase, rather than due to a distinct cation-transporting peptide, because ligands of the enzyme ATP, HVO$_4^{2-}$ and Schering compound (SCH 28080), block the exchange (31, 165). In contrast, anion exchange is slow (133).

The Rb$^+$-Rb$^+$ exchange in the ATPase is electroneutral exchange only, because K$^+$ gradients outward do not generate pH gradients in the presence of the electrogenic protonophore TCS, although such gradients are formed when valinomycin is also present. Net uptake of RbCl is slow (107, 136), with a rate equivalent to that of Cl⁻-Cl⁻ exchange (102), suggesting that another limitation to KCl loading is the absence of a Cl⁻ pathway (136). Flux techniques also show that these vesicles have neither a K$^+$ nor Cl⁻ conductance (31). Other evidence for the absence of significant conductances is obtained in H$^+$-transport experiments where ATP is added to KCl-preequilibrated vesicles. The electrogenic protonophore TCS hardly affects the pH gradient, hence there is neither a K$^+$ nor Cl⁻ conductance that would allow H$^+$-gradient dissipation by electrical coupling to the TCS-dependent proton pathway (8, 130).

The H$^+$-ATP stoichiometry of the H$^+$-K$^+$-ATPase has been measured by several groups of investigators (104, 112, 145, 146). When corrected for the valinomycin-independent ATPase, which reflects broken or leaky membranes, it seems that at pH 6.12, 2 H$^+$ are transported per ATP hydrolyzed when initial rates are

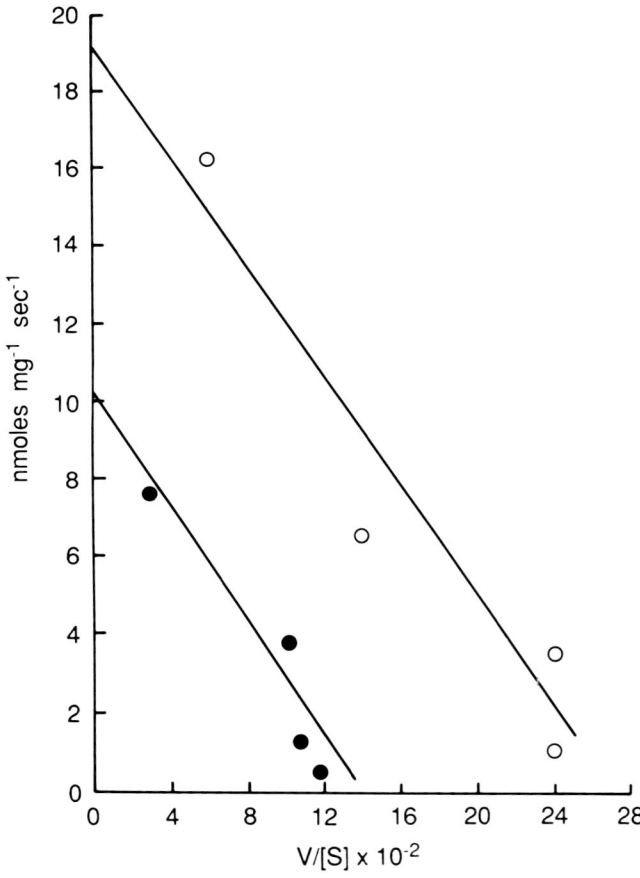

FIG. 6. H$^+$- to ATP-transport stoichiometry, measured at a medium pH of 6.12, in vesicles that were preequilibrated with KCl. Ionophore-independent ATPase activity was subtracted, because this reflects transport-incompetent enzyme activity. H$^+$-transport rate and ATPase rates were initial rates measured at various ATP concentrations. Ratio of 2 H$^+$/ATP is obtained. ○, H$^+$; ●, P$_i$.

used in the calculation (Fig. 6). Because under in vivo conditions the pH gradient formed is ~6.3 units, along with a 10-fold K^+ gradient, from the free energy of hydrolysis of ATP a stoichiometry of 1 H^+ per ATP would seem to be the maximum attainable, suggesting that the stoichiometry of H^+ per ATP varies between 2 and 1, depending on the magnitude of the gradient or cytosolic pH.

The H^+-K^+-ATPase proton and cation transport in these isolated hog gastric membranes is electroneutral, thus exchanging H^+ and K^+ with a stoichiometric ratio of unity. If the pump were electrogenic, exchanging with an H^+/K^+ ratio $\neq$ 1, then it would be anticipated that a protonophore such as TCS would accelerate ATPase activity and affect the pH gradient, contrary to experience. An uptake of lipid-permeable ions would be observed dependent on the direction of the potential. Neither lipid-permeable cations nor anions accumulate in response to the addition of ATP in KCl-equilibrated vesicles. However, if TCS is added, a negative interior potential develops because of the H^+ diffusion potential that is created, with development of a pH gradient and with consequent uptake of positively charged carbocyanine dyes, and SCN^- or ANS^- is taken up when valinomycin is added because of the development of a K^+ diffusion potential due to the inward K^+ gradient developed by the ATPase (73, 130). Moreover, if vesicle size is monitored by light scattering, from a KCl equilibrium situation, there is no change in vesicle volume until HCl leaks from the vesicle interior (107). Therefore the exchange of H^+ for K^+ is isosmotic, hence electroneutral.

These vesicle data stand in contrast to much of what has been concluded for acid secretion in the intact gastric mucosa (115), where manipulation of the potential difference (PD) affects H^+ transport rate in a direction predicted from the change in transepithelial potential (114). There is only one series of vesicle experiments, albeit in stimulated vesicles from rabbit, that is not simply reconciled with electroneutrality of the pump under all conditions. Thus, when the fluorescent oxonol anionic dye bis(1,3-dibutylbarbituric acid)-5-pentamethylene oxonol [diBAC(4)5] is used in K_2SO_4 solutions, there is an ATP-dependent quenching of fluorescence because of formation of a positive interior potential (20). The rate of dye quenching is slow and nigericin sensitive and is probably therefore not due to a primary pump potential but to a diffusion potential. The dye signal is lowered progressively as Cl^- is added to the medium, and proton transport increases. A possible interpretation is the development of a low-pH region in the vesicle membrane with access to the interior, rather than exterior, of the vesicle with the production of an

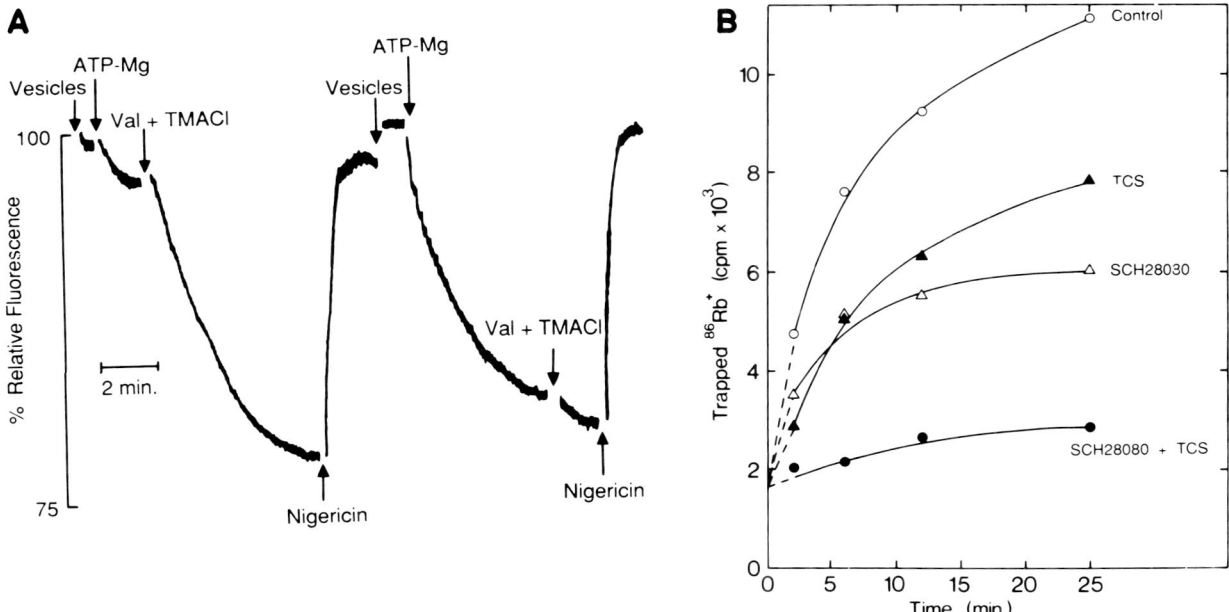

FIG. 7. A: comparison of H^+ transport in rabbit microsomes derived from an animal treated with cimetidine (resting, *left trace*) or from an animal treated with histamine (*right trace*), showing that when the vesicles were derived from a stimulated stomach, neither KCl preincubation nor ionophore was necessary for ATP-dependent H^+ transport. B: flux measurements of conductance in stimulated rabbit vesicles. Vesicles were loaded with KCl, external medium was removed, and vesicles were placed in a medium containing only $^{86}Rb^+$. Uptake of $^{86}Rb^+$ depends on the presence of the exchange-only mode of the H^+-K^+-ATPase and is inhibited by SCH 28080, a K^+-competitive antagonist of this enzyme, and on the presence of a conductance pathway where $^{86}Rb^+$ uptake is driven by the K^+ diffusion potential across this conductance and is inhibited by the electrogenic protonophore tetrachlorosalicylanilide (TCS). TMACl, tetramethylammonium chloride.

inwardly divalent H^+ diffusion potential. Previously it had been noted that the hydrophobic pH indicator dye, malachite green, also showed a slow response to ATP addition, perhaps again consistent with intramembranal acidification (101).

Stimulated Vesicles

When vesicles were isolated from secretagogue-stimulated rabbit or rat mucosa, several differences in the properties of the H^+-K^+-ATPase system were found as compared with resting tissue. There was a redistribution of enzyme activity from a light microsomal to a heavier, denser fraction of the gastric homogenate (43, 51, 53, 179). This fraction also showed proton transport properties in the presence of KCl and Mg^{2+}-ATP, without either valinomycin or KCl preincubation [Fig. 7A; (9, 173, 174)]. Shrink-reswell experiments monitoring vesicle light scattering as a function of volume, similar to the experiments in resting vesicles (107), showed the presence of a rapid KCl-uptake pathway (171, 173). The proton gradient was dissipated by TCS, showing that the ATPase in stimulated vesicles was located in membranes with either a K^+ or Cl^- conductance (173). Dye probe measurements showed the presence of a measurable Cl^- conductance (9). The $^{86}Rb^+$ uptake driven by a K^+ gradient (29) demonstrated two types of K^+ pathway in the vesicles derived from the stimulated rabbit mucosa (31). One was the exchange pathway across the H^+-K^+-ATPase, because uptake of tracer was inhibited by ~50% by the ATPase inhibitors, vanadate and SCH 28080. The remaining 50% was due to a conductance pathway, because this flux was inhibited by TCS, which would collapse the K^+ diffusion potential developed by the K^+ gradient (Fig. 7B). The cation selectivity of this conductance, measured in the presence of SCH 28080, has the sequence $K^+ > Rb^+ = Cs^+ > Na^+, Li^+$, which is unusual in that Cs^+ appears to be readily permeable. Figure 8 shows a model of the resting and stimulated pump membrane. The K^+ and Cl^- conductances are concomitantly inhibited by divalent cations (174).

Evidence has also been obtained for a Cl^- pathway in resting gastric microsomes (149) and for an ATP-dependent electrogenic H^+ pump in rat microsomes (52), as well as a K^+ pathway in the same preparations (51, 150). The relationship of these observations to the H^+-K^+-ATPase and ion pathways in the stimulated state is not clear.

These predictions for regulation of the H^+-K^+-ATPase by changes in KCl permeability properties of the H^+-K^+-ATPase–associated membranes can be tested in the permeable gastric gland or parietal cell by measurement of the H^+-K^+-ATPase–dependent or ATP-dependent benzylamine uptake (an index of acid accumulation). With stimulation there is a reduction in the valinomycin-dependent [^{14}C]benzylamine accumulation and an increase in the valinomycin-independent weak base accumulation (44). Thus there is an increase in the KCl permeability of the ATPase-associated membranes. Similarly the H^+-K^+-ATPase activity that is KCl dependent increases with stimulation, and there is a reduction in the enhancement of activity induced by NH_4^+ (which bypasses a KCl pathway because of NH_3 entry and cycling of NH_4^+ by the pump), showing a relaxation of the KCl restriction on ATP turnover (132).

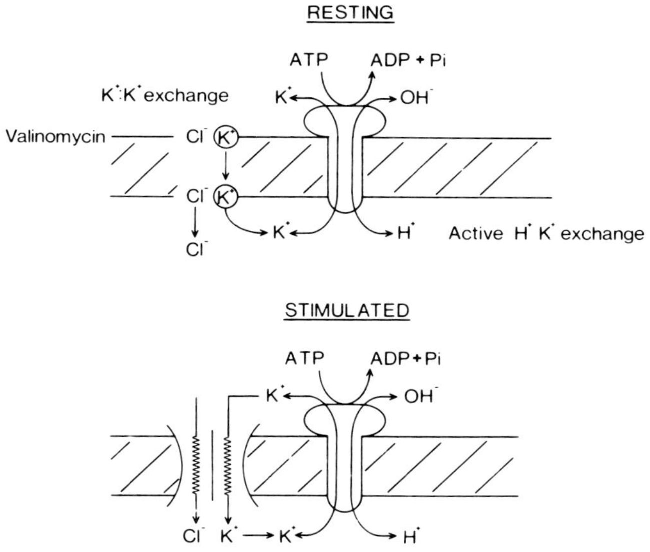

Activation of the ATPase may also depend in part on the presence of an activating protein (3), as does the activity of the plasma membrane Ca^{2+}-ATPase depend on the presence of calmodulin (87).

However, the major difference between resting and stimulated membranes containing the ATPase appears to be the KCl permeability of the membrane, because K^+ and Cl^- conductances are measurable in stimulated, but not in resting, membranes. Whether these conductance proteins are present in the resting membrane and then activated as a function of stimulation or whether they are present in a different membrane location in the cell and associate with the ATPase on stimulation is not known.

Transport Modes of H^+-K^+-ATPase

The Na^+-K^+-ATPase can allow various cation movements, such as reverse Na^+-K^+ exchange, Na^+-Na^+ exchange, K^+-K^+ exchange, uncoupled Na^+ efflux, and normal Na^+-K^+ exchange (30, 59). Cation movements are regulated by ligands of the pump. Nucleotide and Mg^{2+} with P_i, for example, stimulate K^+-K^+ exchange and are thought to deocclude K^+

FIG. 8. Conceptual model of the H^+-K^+-ATPase in a membrane derived from resting (*upper*) and stimulated (*lower*) stomach. In resting membranes, valinomycin is shown as providing a KCl pathway to allow K^+ access to the luminal face of the pump, whereas in the stimulated state the membrane contains K^+ and Cl^- conductances that allow the penetration of KCl. [From Sachs (127a).]

bound to the Na^+-K^+-ATPase (60, 99). At pH 6.1, in the absence of Na^+, the Na^+-K^+-ATPase is able to transport H^+ in exchange for K^+ (36), thus performing as an H^+-K^+-ATPase. The H^+-H^+ exchange in the H^+-K^+-ATPase would seem impossible to measure because of the high passive proton and water permeability, and reversal of this enzyme by ion gradients has not been shown. However, K^+-K^+ exchange has been measured, but in contrast to Na^+-K^+-ATPase is rapid relative to turnover in the absence of other ligands, and only inhibitory effects of ATP have been found (103). The H^+-K^+-ATPase, in the absence of phosphorylation, does not appear to occlude the K^+ ion as tightly as the sodium pump, and the E_1K and E_2K forms of the enzyme seem readily interconvertible. It has been claimed that passive H^+-K^+ exchange occurs in the absence of other ligands (144), but this mode has not been described for the Na^+ pump (i.e., passive Na^+-K^+ exchange).

H^+-K^+-ATPase in Parietal Cell

The alteration in the parietal cell on stimulation or secretion of HCl is profound, morphologically, biochemically, and functionally.

The morphological change that occurs on stimulation involves expansion of the surface of the intracellular canaliculus at the expense of the tubulovesicles. Thus, the H^+-K^+-ATPase is moved from one membrane location to another (24, 39, 147). There is considerable evidence that cytoskeletal elements are involved in this transformation, and a 120,000- and an 80,000-MW protein associated with the stimulated membranes has been shown to be phosphorylated by stimulation (35). A cellular rearrangement of carbonic anhydrase is also associated with the change in location of the H^+-K^+-ATPase (153, 159). This transformation appears to be able to be achieved with stimuli that change only intracellular Ca^{2+} levels. In the presence of cimetidine in the in vivo rat (to block the effects of endogenous histamine release and hence cAMP formation) this morphological change can be observed with carbachol and pentagastrin stimulation, which are $[Ca^{2+}]_i$ dependent (41). However, under these conditions there is little, if any, acid secretion, which shows that the $[Ca^{2+}]_i$-dependent transformation alone is not sufficient for acid secretion in this species. Cell morphology and secretion have also been dissociated by pump inhibitors such as omeprazole and SCH 28080 (H. F. Helander, personal communication).

The major biochemical change with onset of secretion that has been noted is a large increase in metabolism, as monitored by O_2 consumption in vivo (86), or CO_2 production from glucose in vitro. For example, Figure 9 shows the change in $^{14}CO_2$ production from [U-^{14}C]glucose with histamine or cAMP stimulation of rabbit gastric glands. Measurement of changes in metabolite levels have shown rather little, if any,

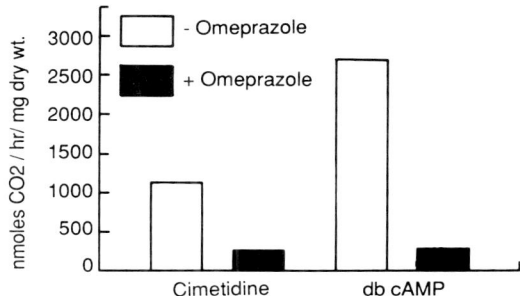

FIG. 9. Use of $^{14}CO_2$ production to demonstrate stimulation of acid secretion in cimetidine-treated rabbit gastric glands by histamine and dibutyryl cAMP (db cAMP), as well as to show the basal CO_2 production when the H^+-K^+-ATPase is inhibited by the pump blocker, omeprazole.

change in the ATP/ADP ratio or the phosphocreatine-creatine ratios that would account for the large increase in turnover of mitochondrial metabolites that has been noted (14, 134, 135). Thus the steps of metabolism that are regulated in the parietal cell by secretagogues are not known.

With stimulation the gastric mucosa secretes an isotonic fluid containing ~140 mM HCl and 20 mM KCl in many mammalian species. In humans a rate of ~200 ml/h can be measured under maximally stimulated conditions. Thus the parietal cell loses H^+, K^+, and Cl^- across the open canalicular membrane. From vesicle studies previously discussed, the mechanism of stimulation is the appearance of K^+ and Cl^- conductances in parallel to the H^+-K^+ exchange ATPase. The enzyme apparently does not recycle the K^+ perfectly, accounting for the appearance of K^+ in the secreted fluid with stimulation of HCl transport.

With H^+ pumping out of the cell there is an associated equivalent cytosolic production of OH^- by the pump, which is rapidly converted to HCO_3^- by the carbonic anhydrase that is associated with the canalicular membrane. The parietal cell has on its basal lateral surface both an Na^+-H^+ and HCO_3^--Cl^- exchange pathway (88, 95). With histamine stimulation, there is activation of the Na^+-H^+ exchange, which results in a slight rise in cell pH from ~7.1 to ~7.2, which is presumably sufficient to activate the HCO_3^--Cl^- exchange (87). This pathway is probably the major pathway for Cl^- entry across the basal surface in the stimulated cell and performs the dual function of supplying Cl^- to the apical membrane and removing the HCO_3^- produced. Activation of Na^+-H^+ exchange also supplies Na^+ to the cell, which in turn presumably increases the rate of the Na^+-K^+-ATPase on the basal surface, thus replenishing the K^+ that is lost across the apical surface. Hyperpolarization of both membranes is associated with these activities (12), and this could be accounted for by the activation of K^+ conductances on both surfaces of the cell. Both cAMP- and Ca^{2+}-activated K^+ channels have been observed by patch-clamp methods in the basal-lateral surface of *Necturus* oxyntic cells (157). The pumps and ion pathways on apical and basal surfaces of this cell appear to be coupled.

Thus the ion pathways in the stimulated parietal cell can be illustrated as in Figure 10. This is incomplete, because acid secretion is not abolished by 4,4'-diisothiocyanostilbene-2,2'-disulfonic acid (DIDS), the anion-exchange inhibitor, nor by amiloride or its analogues, which inhibit the Na^+-H^+ exchange (87). However, a combination of both types of inhibitors reduces acid secretion by >80%, arguing for a major role of the pathways suggested in the model compared with those derived from studies on intact mucosa (114–116). There is no Cl^- conductance on the basallateral surface, which was implied by changes of transtissue PD with Cl ion substitution in the frog mucosa. In intact parietal cells or basal-lateral vesicles (88) no evidence for a Cl^- conductance was found, only for electroneutral anion pathways. In oxyntic cells from *Necturus*, using patch-clamp methods, again no anion signals were obtained, only K^+ channels (157). However, with a rapid anion exchange path, Cl^- changes in the serosal medium could rapidly alter $[Cl^-]_i$ and, with an apical conductance for this ion, produce the transepithelial changes previously described. The proton pump is also illustrated as being electroneutral. Most studies on proton transport by H^+-K^+-ATPase have been performed on mammalian tissue. In one study, differences were found between frog gastric vesicles and their mammalian counterpart (105), such as SCN^- and TCS sensitivities. The presence of a K^+ conductance that is altered by stimulation has recently been shown in frog gastric mucosa (62), and this result is consistent with the model derived from studies comparing the resting and stimulated mammalian vesicle system. The electrogenicity of acid secretion in the model of Figure 10 would thus depend on the coupling of a K^+ gradient to the activity of the H^+-K^+-ATPase, which recycles the K^+ in exchange for H^+. The K^+ outward gradient would thus be proportional to the rate of the pump, namely the rate of secretion of HCl (115). Various arguments can be advanced against this K^+-gradient model, but it is difficult to counteract the case that has been made for the electroneutral nature of the isolated enzyme (130, 131).

KINETICS OF H^+-K^+-ATPASE

The usefulness of a kinetic investigation of this enzyme lies in the ability of kinetics to dissect the steps involved in coupling the scalar hydrolysis of ATP to the vectorial transport of H^+ and K^+. Much of what has been learned about the kinetics of this enzyme derives from studies on Na^+-K^+-ATPase (30, 59). However, there appear to be some major differences, which must be reflected in structural differences between the two enzymes.

In a countertransport pump it is convenient to discuss the reactions as if they were sequential, but there is no direct evidence for or against sequential or simultaneous movement of the two ions in the H^+-K^+-ATPase. For H^+ transport to occur, H^+ binding must occur at a high-affinity site or sites on the cytosolic face. With the addition of Mg^{2+} and ATP, the enzyme phosphorylates, forming an intermediate where the protons are trapped within the enzyme. In other words, with phosphorylation a cis barrier to protons leaving the enzyme toward the cytosolic face is raised. In the pump, in the absence of Mg^{2+} or nucleotide, a barrier must be present preventing passive backleak of H^+, otherwise considerable wastage of energy would happen. This barrier must be on the trans side of the H^+-binding site, which is accessed from the cytosolic face. For reasonable transport rates to occur, the site that binds protons must decrease markedly in affinity for H^+, and a barrier on the luminal side must decrease to allow proton release into the luminal solution. Thus the proton-binding site is alternatively exposed to the cytosolic and then to the luminal face, and in this sequence the binding affinity must change. Expressed in the standard nomenclature for this type of pump, there must be a conformational transition between an E_1-P and E_2-P form, whereby convention the E_1 conformer binds the ion from the cytosolic face (cis) and the E_2 conformer binds or releases ions on the luminal face. Conversely,

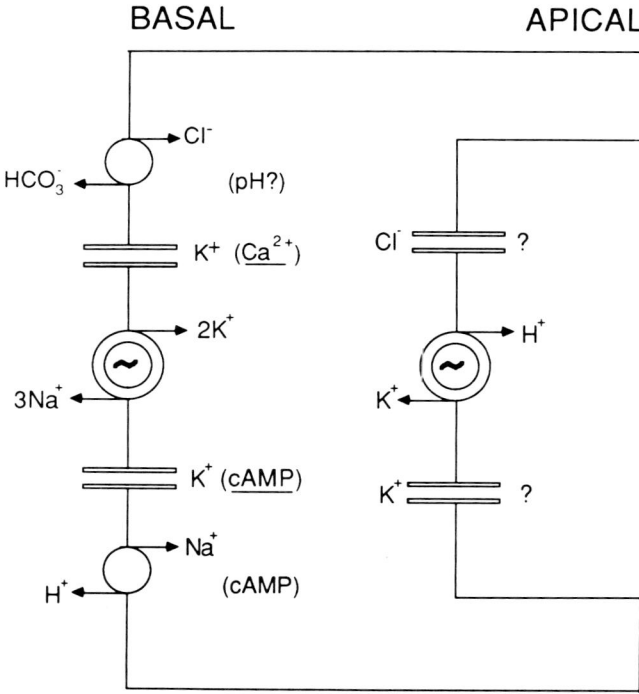

FIG. 10. Ion pathways that have been described in stimulated oxyntic cell and are related to acid secretion. Basal-lateral membrane contains the H^+-K^+-ATPase, Na^+-H^+ and HCO_3^--Cl^- exchange pathways, and cAMP- and $[Ca^{2+}]_i$-activatable K^+ conductances. Canalicular membrane contains the H^+-K^+-ATPase and K^+ and Cl^- conductances demonstrated in vesicle studies.

for K^+ movement from luminal to cytosolic face, the E_2-P form must bind luminal K^+, and with this a barrier re-forms on the luminal side of the bound K^+, the enzyme dephosphorylates, and the barrier on the cytosolic side of the bound K^+ ion disappears, along with a decrease in the affinity of the K^+ site, to allow K^+ to be transported into the cytosol. The situation where either ion is situated between the barriers during the pump cycle is often referred to as the occluded form of the pump (60, 99). This general picture of the steps (Fig. 11) of a countertransport pump can also be restated in conformational terms, where there must be several conformers corresponding to the status of the monovalent ion-binding sites, the presence or absence of divalent ion, nucleotide or inorganic phosphate. One means of investigating the different steps in the sequence is kinetic analysis of overall ATPase activity and of the phosphorylation and dephosphorylation steps.

Steady-State Kinetics of ATPase

Analysis of the enzyme under turnover conditions is complementary to measurements of the various partial reactions, such as phosphorylation from ATP and P_i and dephosphorylation of the phosphoenzyme intermediate. Interpretation of the data also depends on the purity of the preparation and on the meaning of the low stoichiometry of binding sites that are found. Because the reason for the latter is not known, the additional possible complication of multiple reaction paths in the enzyme has been ignored.

All the preparations of the H^+-K^+-ATPase thus far described have, with one exception (3), been shown to have a basal Mg^{2+}-ATPase present in the absence of activating cation. In part, it can be shown that this basal activity is a property of the H^+-K^+-ATPase, because of its inhibition by vanadate or SCH 28080,

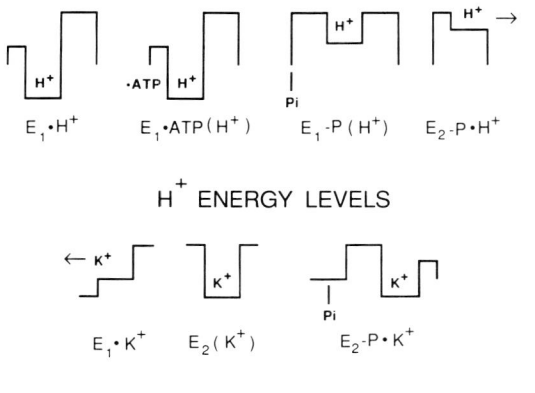

FIG. 11. Barrier-site model for the H^+-K^+-ATPase, showing the necessity for alteration of barriers in the transmembrane path, as well as the change in affinity for transported ions. Conformation where ion-binding sites face the cytosol defines the E_1 state and where ion-binding sites face the lumen defines the E_2 state of this class of enzyme.

which specifically inhibit the H^+-K^+-ATPase component of the preparation (103, 165). The component of the ATPase that is insensitive to these inhibitors must be considered to be due to contamination with other ATPases and is a function of the method of preparation.

The Mg^{2+}-ATPase activity is ~5% of the total K^+-dependent activity and can be accounted for by the spontaneous loss of phosphoenzyme that has been described (168, 170). This would be equivalent to the Na^+-ATPase activity of the Na^+-K^+-ATPase (30), which depends on the K^+-independent hydrolysis of the E_2-P form of the Na^+ pump. In fact, this can be related to uncoupled Na^+ transport in the latter system. However, H^+ transport in the absence of K^+ has not been described for the H^+-K^+-ATPase, perhaps because the equivalent uncoupled H^+ efflux is slower than the H^+ leak rate in gastric vesicles.

Other divalent cations such as Mn^{2+} can substitute for Mg^{2+}. However, Ca^{2+} appears to have an effect on the enzyme that might be the result of a significant alteration of conformation. In the case of the Na^+ pump, phosphorylation is slowed and K^+ dephosphorylation is almost abolished because of interaction with the dephosphorylation site (158). In the proton pump there is a low level of Ca^{2+}-ATPase activity. The Ca^{2+}-ATPase reaction is sensitive to SCH 28080 and thus is catalyzed by the H^+-K^+-ATPase. The Ca^{2+} slows phosphorylation, virtually abolishes the additional dephosphorylation due to K^+, and the phosphoenzyme form does not react with ADP (84). However, in contrast to the Na^+ pump, the addition of ethylene glycol-bis(β-aminoethylether)-N,N'-tetraacetic acid (EGTA) to complex the Ca^{2+} after phosphorylation does not result in a K^+-sensitive phosphoenzyme form. This implies that Ca^{2+} is tightly bound to the phosphoenzyme. Because K^+-induced dephosphorylation can be seen after the addition of diaminocyclohexane tetraacetic acid (CDTA) to enzyme phosphorylated in the presence of Mg^{2+}, the Ca^{2+} effect could mean that Mg^{2+} is also tightly bound to the enzyme. This would be consistent with the requirement for Mg^{2+} for ^{18}O exchange with P_i (19) and the apparent lack of requirement for Mg^{2+} for K^+-dependent dephosphorylation.

Measurement of the affinity of ATP for the enzyme in the absence of K^+ showed that there are two apparent Michaelis-Menten constant (K_m) values. Whether these were due to the H^+-K^+-ATPase only or to the presence of a contaminating Mg^{2+}-ATPase (75, 170) is not clear, because the selective inhibitor SCH 28080 was not used in these experiments. The interaction of Mg^{2+} and ATP with apoenzyme, in terms of steady-state kinetics, can be written as

$$E + Mg + ATP \longleftrightarrow E\text{-}P \cdot Mg \longleftrightarrow E \cdot Mg + P_i$$

The K^+ stimulation of the Mg^{2+}-ATPase activity is between 20- and 30-fold in reasonably pure vesicular preparations. The simplest explanation for this observation is that K⁺ activates the ATPase by permitting rapid breakdown of the phosphoenzyme intermediate, which is slow in the presence of Mg^{2+} alone, and a different reaction step becomes rate limiting. Variation of K⁺ concentration at constant ATP results in stimulation followed by inhibition at high K⁺/ATP ratios (132, 170). In ion-tight vesicles, in the presence of valinomycin, increasing K⁺ concentration does not inhibit the ATPase, in contrast to the lyophilized, reconstituted enzyme. The difference between these two conditions is the presence of acid on the luminal face of the enzyme. The biphasic effect of K⁺ on the ATPase can probably be accounted for by two sites for K⁺ binding. The luminal site is responsible for activation of ATPase activity by stimulating the breakdown of E-P, whereas a cytosolic site competes with ATP for initiation of phosphorylation. This is discussed later in the sections on the partial reactions of the ATPase (152, 170). In the Na⁺-K⁺-ATPase, at low ATP, the loss of K⁺ from the enzyme appears to be rate limiting (99), and this kinetic evidence resulted in postulation of an occluded form of K⁺ bound to the enzyme. This concept has been confirmed repeatedly in transport studies of reconstituted Na⁺-K⁺-ATPase (60). As has been seen, the H⁺-K⁺-ATPase in the absence of phosphorylation does not display K⁺ occlusion. Equally the sidedness of K⁺ in stimulation of dephosphorylation indicates that under transport conditions, K⁺-K⁺ exchange is blocked.

A second feature of the steady-state hydrolysis of ATP in the presence of K⁺ is a nonlinear Eadie-Hofstee plot (170) with a high and low apparent K_m for ATP. If K⁺ concentration is set at the maximal stimulatory concentration at any given ATP and the enzyme velocity is plotted as a function of ATP concentration, the data of Figure 12 are obtained. This illustrates the initial effect of ATP on activity, with a relatively low apparent K_m, and then an activating effect with a higher apparent K_m at higher ATP concentrations. Because the K⁺ concentration was set to generate the highest rate at any given ATP, the explanation for the biphasic effect of ATP is most likely to be binding of the substrate to two forms of the enzyme or to two separate sites on the enzyme. Alternatively, if the enzyme is a functional dimer, half of site reactivity or different reaction paths of the two enzyme subunits may be considered, as for the Na⁺-K⁺-ATPase (98), to explain the two apparent K_m values. Because in the primary sequence of the enzyme more than one nucleotide site is unlikely, with a sequential model for a single site, one explanation is that at low-ATP concentrations, ATP binds only to one form of unphosphorylated enzyme with high affinity and to another form at higher concentrations, with lower affinity. The binding of ATP to the low-affinity site accelerates the turnover of the enzyme. There are several lines of evidence for ATP binding to two enzyme forms. For example, nucleotide chase of E-P depends on the Mg^{2+} concentration. At high Mg^{2+}, ADP appears not to bind (43), whereas it is effective at low Mg^{2+}. The reversible inhibitor, SCH 28080, which binds to the E_2 form of the enzyme, is uncompetitive with respect to ATP (165), interpreted as due to ATP binding to the E_2 form complexed with the inhibitor. Phosphorylation of the enzyme depends on ATP binding to the K⁺-free E_1 form (170). From these steady-state data the sequence of reactions in the presence of K⁺ is

$$E_1 \cdot K + ATP + Mg^{2+} \longleftrightarrow Mg^{2+} \cdot ATP \cdot E_1 + K^+$$

$$Mg^{2+}\text{-}ATP \cdot E_1 \longleftrightarrow Mg^{2+} \cdot E_1\text{-}P + ADP$$

$$Mg^{2+} \cdot E\text{-}P + ATP \longleftrightarrow Mg^{2+}\text{-}ATP \cdot E\text{-}P$$

$$Mg^{2+}ATP \cdot E_2\text{-}P + K \longleftrightarrow Mg^{2+}ATP \cdot E_2 \cdot K + P_i$$

In these equations the two major forms of the enzyme, E_1 and E_2, are not specified in terms of which form ATP binds to in the low-affinity form.

Steady-State Kinetics of Phosphatase Reaction

The H⁺-K⁺-ATPase displays a K⁺-stimulated phosphatase reaction, whereby K⁺ activates a vanadate-sensitive pNPPase, acetyl phosphatase or carbamyl phosphatase reaction (8, 25, 132). In contrast to the ATPase in intact vesicles, there is no ionophore stimulation of this reaction nor is there any detectable proton transport associated with these activities (130, 131). The absence of transport is consistent with data on the Na⁺-K⁺-ATPase (59), but the Ca^{2+}-ATPase of the sarcoplasmic reticulum does transport using the phosphatase reaction pathway (97). The reaction pathway taken by pNPP as compared with ATP has not been studied extensively. The ATP and pNPP are competitive, which does not necessarily mean that they bind to the same site, because, for example, FITC irreversibly inhibits the ATPase with little effect on

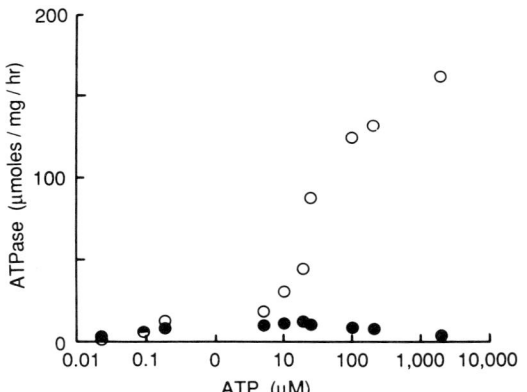

FIG. 12. Biphasic effect of ATP concentration on H⁺-K⁺-ATPase activity, where at each ATP concentration the level of K⁺ was set to give maximal activity. Data translate into a high and low apparent K_m, with maximal activity found with binding of ATP to the low-affinity state of the enzyme. ○, +K⁺; ●, −K⁺.

the pNPPase activity (57). Phospholipase A$_2$ also differentially inhibits ATPase activity, and thimerosal stimulates pNPPase while inhibiting ATPase activity (26, 120). Differential action of these inhibitors can be contrasted to the concomitant inhibition of the two reactions by vanadate, K$^+$ removal, omeprazole, or SCH 28080 (21, 79, 80, 165). In addition, the copurification and homogeneity of the preparation argue against the idea that the pNPPase is not a property of the catalytic subunit of the H$^+$-K$^+$-ATPase (111). However, the differences in inhibition of the two activities point to a difference in the binding region and possibly a different reaction pathway. For example, it is not known whether pNPP phosphorylates either the H$^+$-K$^+$- or Na$^+$-K$^+$-ATPase enzymes. The $K_{0.5}$ for K$^+$ activation of the pNPPase activity is about an order of magnitude greater than that found for K$^+$ activation of ATPase activity (5 mM and 0.45 mm, respectively), which again suggests only partial overlap of the reactions. It seems likely that the form of the enzyme responsible for pNPP hydrolysis is either the E$_2$ or E-occluded form, rather than the E$_1$ form. For example, an impermeant K$^+$ antagonist, the methyl derivative of SCH 28080, does not inhibit either the ATPase or the pNPPase in intact vesicles (B. Wallmark, personal communication). Also, the addition of K$^+$ to FITC enzyme produces the E$_2$ form of the enzyme, and pNPP is hydrolyzed under these conditions (57).

The phosphatase reaction of the H$^+$-K$^+$-ATPase can therefore be written as only involving the E$_2$ class of ATPase conformer, and in the absence of direct evidence for phosphorylation, a simple scheme is

$$Mg^{2+} \cdot E_2 + pNPP \longleftrightarrow Mg^{2+} \cdot E_2 \cdot pNPP$$

$$Mg^{2+} E_2 \cdot pNPP \longleftrightarrow Mg^{2+} E_2 + P_i + pNP$$

Phosphorylation From ATP

When the enzyme Mg^{2+} and labeled ATP are rapidly mixed in a flow-quench machine, a rapid phosphorylation occurs with a rate of ~1,400/min, which is faster than the overall turnover rate of the enzyme of ~200/min (170). If ATP and enzyme are mixed first and then Mg^{2+} added, the initial rate of phosphorylation at 20°C increased to >4,400/min (75, 170), whereby it can be concluded that ATP binding to the apoenzyme induces a form favorable for phosphorylation. In intact vesicles with K$^+$ added to the cytosolic face only, the rate of phosphorylation is slowed and accelerates with an increasing ATP/K$^+$ ratio (170). Thus the K$^+$ form of the enzyme obtained with K$^+$ addition to the cytosolic face seems to slow the rate of phosphorylation from ATP (168). The binding of Na$^+$ to the enzyme also reduces phosphorylation (168), but whether this is due to binding of Na$^+$ to the same K$^+$ site has not been established.

The rate of phosphorylation is increased with decreasing cytosolic pH, which might mean that protonation of a site on the enzyme is required for phosphorylation, and increasing protonation of this site thus accelerates phosphorylation. The effect of cytosolic K$^+$ is also decreased with decrease of medium pH, suggesting interaction of H$^+$ with the cytosolic K$^+$ site that inhibits phosphorylation. Perhaps H$^+$ and K$^+$ interact with the H$^+$-K$^+$-ATPase, as do Na$^+$ and K$^+$ with the Na$^+$-K$^+$-ATPase (77, 152).

At stoichiometric Mg^{2+} and ATP levels, a fraction of the E-P that is formed is chased by ADP, but the majority (80%) is ADP insensitive. At high ratios of Mg^{2+} to ATP, no ADP chase can be detected (43, 170). Thus an excess of Mg^{2+} may prevent effective ADP binding to the enzyme. An additional observation that has to be fitted into the effect of ADP on E-P is that the addition of CDTA to prevent further phosphorylation from labeled ATP induces a biphasic dephosphorylation in the same way that an ATP + ADP chase does (43). This may mean that removal of Mg^{2+} allows the form of enzyme that is E$_1$-P·ADP to re-form ATP spontaneously.

From the studies on phosphorylation of the H$^+$-K$^+$-ATPase, it is possible to expand the reactions in the phosphorylation sequence more explicitly than from the overall reaction.

$$E_1 \cdot K^+ + ATP \longleftrightarrow E_1 \cdot ATP + K^+$$

$$E_1 \cdot K^+ + H \longleftrightarrow E_1 \cdot H + K^+$$

$$E_1 \cdot H^+ + ATP \longleftrightarrow E_1 \cdot H^+ \cdot ATP$$

$$E_1 \cdot ATP + H^+ \longleftrightarrow E_1 \cdot ATP \cdot H^+$$

This group of reactions forms the competent initial intermediate for phosphorylation after the binding of Mg^{2+}. The order of binding appears to be random.

After the addition of Mg^{2+} or with the addition of Mg^{2+}-ATP, the reaction proceeds to the formation of phosphoenzyme

$$E_1 \cdot ATP \cdot H^+ + Mg^{2+} \longleftrightarrow Mg^{2+} \cdot E_1 \cdot P \cdot H^+ \cdot ADP$$

$$Mg^{2+} \cdot E_1 \cdot P \cdot H^+ \cdot ADP \longleftrightarrow Mg^{2+} \cdot E_1 \cdot P \cdot H^+ + ADP$$

and there is significant conversion of the phosphoenzyme to an ADP-insensitive form

$$Mg^{2+} \cdot E_1 \cdot P \cdot H^+ \longleftrightarrow Mg^{2+} \cdot E_2 \cdot P \cdot H^+$$

ATP-ADP Exchange

The H$^+$-K$^+$-ATPase catalyzes an ATP-ADP exchange reaction, due to the reversal of the phosphorylation step that forms E$_1$-P. The Mg^{2+} rate in this enzyme is relatively slow but is stimulated by K$^+$ at a high-affinity site (106). This is probably due to the rapid formation of E$_2$-P with Mg^{2+} and ATP, which is inactive in the exchange reaction, and the presence of K$^+$ allows re-formation of E$_1$-P by stimulating the dephosphorylation of E_2-P. Thus this reaction and the ADP sensitivity of part of the E-P formed are evidence for the presence of an E_1-P form of the enzyme.

The exchange reaction can be written as

$$Mg^{2+} \cdot ATP \cdot E_1 \longleftrightarrow Mg^{2+} \cdot E_1\text{-}P \cdot ADP$$
$$Mg^{2+} \cdot E_1\text{-}P \cdot ADP \longleftrightarrow Mg^{2+} \cdot E_1\text{-}P + ADP \quad \Big\} \text{ Exchange}$$
$$Mg^{2+} \cdot E_1\text{-}P \longleftrightarrow Mg^{2+} \cdot E_2\text{-}P$$
$$Mg^{2+} \cdot E_2\text{-}P + K \longleftrightarrow Mg^{2+} \cdot E_2 \cdot K^+ + P_i \quad \Big\} \; K^+ \text{ stimulation}$$
$$Mg^{2+} \cdot E_2 \cdot K^+ + ATP \longleftrightarrow Mg^{2+} \cdot ATP \cdot E_1 + K^+ \quad \text{of exchange}$$

which shows the ADP reactive form of phosphoenzyme, as from the dephosphorylation kinetics, and also the reason for stimulation by K^+, which allows re-formation of the E_1 form from E_2-P.

Potassium-Dependent Dephosphorylation

In the absence of K^+, when the enzyme is phosphorylated by labeled ATP and the phosphorylation by labeled phosphate stopped by either excess cold ATP or CDTA at 20°C, the spontaneous dephosphorylation rate is ~12/min. The addition of K^+ accelerates dephosphorylation biphasically: a fast phase that is accelerated by K^+ with a $K_{0.5}$ of ~0.8 mM and a rate of 4,400/min and a slow phase, accounting for ~20% of the E-P formed, that is only slightly increased by K^+ (170). When the E-P is formed in tight vesicles, the addition of K^+ does not accelerate dephosphorylation unless an ionophore such as gramicidin is present (75, 152), showing that luminal K^+ is required for the acceleration of dephosphorylation, as it is for stimulation of overall ATPase activity. Because the majority of the E-P formed is K^+ chased and ADP insensitive, this is defined as the E_2-P form of the enzyme. From these data and the data on phosphorylation, where 80% of the phosphoenzyme is in the E_2-P form and is produced more rapidly than the overall rate, the two forms of phosphoenzyme are competent to serve as intermediates in the overall reaction.

With the K^+-dependent breakdown of E_2-P, P_i is lost from the enzyme (as discussed in the subsequent section) and the $E_2 \cdot K$ form can rapidly convert to the $E_1 \cdot K$ form, as shown by the rapid K^+-K^+ exchange discussed earlier, thus allowing the cycle of ATP binding, phosphorylation, and transport to start again.

The sequence of reactions after the addition of K^+ to phosphoenzyme involves, therefore, the loss of P_i and the reformation of the $E_1 \cdot K^+$ form, with loss of H^+ to the luminal side of the enzyme.

$$Mg^{2+} \cdot E_2\text{-}P \cdot H^+ + K^+ \longleftrightarrow Mg^{2+} \cdot E_2\text{-}P \cdot K^+ + H^+$$
$$Mg^{2+} \cdot E_2\text{-}P \cdot K^+ \longleftrightarrow Mg^{2+} \cdot E_2 \cdot K^+ + P_i$$
$$Mg^{2+} \cdot E_2 \cdot K^+ \longleftrightarrow Mg^{2+} \cdot E_1 \cdot K^+$$

Phosphorylation From P_i

As a class these enzymes are able to phosphorylate from P_i the same aspartyl residue that is phosphorylated by ATP, in the presence of Mg^{2+}. The H^+-K^+-ATPase is also phosphorylated to a level of ~3 nmol/mg protein with a $K_{0.5}$ for P_i of ~60 μM (58). The majority of this phosphoenzyme, however, is K^+ insensitive and can be regarded as a dead-end phosphoenzyme, as can occur in the Na^+-K^+-ATPase (100).

Another means of determining whether the enzyme can react with P_i to form a covalent intermediate is to measure the exchange of ^{18}O with P_i, either from ^{18}O-H_2O incorporation into P_i or exchange out from ^{18}O-labeled P_i. These reactions have been shown to occur, require Mg^+, are accelerated by K^+, and the rate is much faster than overall. Thus the steps from E_2-P to free enzyme and P_i are rapidly reversible. An additional advantage of this approach is that the distribution of label allows the relative rate constants of the formation of E_2-P and dissociation of the noncovalent form $E \cdot P$ to $E + P_i$ to be compared. It appears that in this enzyme the rate of dissociation of $E \cdot P$ is much faster than the formation of $E_2 \cdot P$; thus there is not a form where noncovalently bound phosphate (acid labile) is tightly bound on the enzyme.

The phosphorylation of the enzyme from inorganic phosphate and the exchange of ^{18}O into or from P_i shows the reversibility of the dephosphorylation reactions and that the steps involved are

$$E_2 + Mg^{2+} + K^+ + P_i \longleftrightarrow Mg^{2+} \cdot E_2 \cdot P \cdot K^+$$
$$Mg^{2+} \cdot E_2 \cdot P \cdot K^+ \longleftrightarrow Mg^{2+} \cdot E_2\text{-}P \cdot K^+$$

The ability of these enzymes to form a phosphoenzyme from P_i shows that in the presence of Mg^{2+} these pumps operate virtually isoenergetically. It has been suggested that it is the binding of divalent cation that alters the state of the enzyme so that the various reaction steps subsequently do not have large energy barriers restricting the rate of the enzyme reaction (17).

A major contribution of these kinetic analyses is that various enzyme conformers with different reactivities toward the known ligands of the enzyme have been defined either by steady-state or transient-state kinetics. Figure 13 presents a model of the different kinetic steps that have been discussed, with the addition of the interaction of the H^+-K^+-ATPase with the K^+ competitive antagonist SCH 28080, which is discussed in PROTONATABLE AMINES, p. 246.

CONFORMATIONS OF H^+-K^+-ATPASE

Apart from the kinetic methods discussed previously, there are other general means of showing the occurrence of a conformational change induced by ligands such as K^+ and ATP. These methods often do not specify the site that changes conformation, because they depend on changes either in optical properties, binding of ligands, or the protein digestion pattern.

Tryptic Digestion

The H^+-K^+-ATPase is inactivated in a time-dependent manner when incubated with trypsin. Ligands such as ATP and ADP protected against tryptic inactivation, and two tryptic fragments were formed of 67,000 and 35,000 M_r (11, 122). The presence of K^+ also modified the digestion pattern, where two peptides of 55,000 and 42,000 M_r were formed (11, 122). Both the 67,000- and 42,000-M_r fragments contain the phosphorylation site, and thus in moving from the ATP·E_1 form to the E_2·K^+ form there is a shift of a trypsin-sensitive bond. This change in location, whereby an arginine residue becomes trypsin inaccessible in the E_2 conformation, might be due to a movement of a hydrophilic sector to a more hydrophobic location (1, 6).

Fluorescence

The intrinsic fluorescence of the H^+-K^+-ATPase is low and has not proved useful in monitoring changes in protein structure in this enzyme. However, two extrinsic probes of fluorescence, FITC (57) and eosin (42), have shown fluorescent changes as a function of the presence of other ligands, interpreted as showing conformational differences between the E_1 and E_2 forms of the enzyme.

The FITC binds irreversibly to the H^+-K^+-ATPase at Lys-517 when the enzyme is incubated at pH 9 with the probe. The uniqueness of this binding may be ascribed to the low acidic dissociation constant (pK_a) of this particular lysine or to its accessibility to the reagent. The ATP protects against this inhibition and binding. The addition of K^+ results in a rapid quench of the fluorescence, as does the addition of the K^+-competitive antagonist SCH 28080 (57, 165), whereas HVO_4^{4-} produces a slower quench, compatible with the slow inhibition by this anion (21). The low $K_{0.5}$ for K^+ argues that the formation of the E_2 form is due to binding of K^+ to the high-affinity K^+ site on the enzyme, which exists in the E_2 conformation. In the Na^+-K^+-ATPase it is possible to alternate Na^+ and K^+ to induce the E_1 and E_2 conformations, respectively (61), but the quench of fluorescein fluorescence by lower pH precludes this type of experiment in the H^+-K^+-ATPase.

Eosin binds reversibly to the H^+-K^+-ATPase and competes with ATP (11, 42). Fluorescence due to eosin binding is enhanced by Mg^{2+}, and K^+ quenches fluorescence due to the release of eosin. In this way it has been possible to follow the change in conformation between the $E_1 \cdot Mg^{2+}$ and $E_2 \cdot K^+$ forms of the enzyme and to show that the interconversion of the forms is rapid and not rate limiting for enzyme turnover, as implied by the rapid K^+-K^+ exchange (136).

The fluorescent ATP analogue TNP-ATP (2′,3′-[2,4,6-trinitrohexadienylidine]-ATP) binds to the ATPase with enhancement of fluorescence but is not a substrate for the enzyme (20). This compound binds with a stoichiometry of 3.4 nmol/mg protein with a single class of high-affinity binding sites in the presence or absence of Mg^{2+}. In the absence of Mg^{2+}, displacement of TNP-ATP occurs monophasically, whereas in the presence of Mg^{2+} the displacement is biphasic, suggesting that the presence of Mg^{2+} has resulted in the alteration of affinity of a class of nucleotide-binding sites, as suggested, for example, by the effects of Mg^{2+} on the dephosphorylation of E-P by ADP and the two apparent K_m values for ATP (43, 170). The addition of K^+ quenches the fluorescence, whereas Mg^{2+} enhances the fluorescence, again providing evidence for E_1 and E_2 forms of the enzyme (L. D. Faller, G. Sartor, and G. Sachs, unpublished observations).

INHIBITION OF H^+-K^+-ATPASE

Studies of inhibition of the H^+-K^+-ATPase yield information as to the sites involved in catalysis or transport. Protection by ligands has often been used to show that a given reagent reacts at the same site or

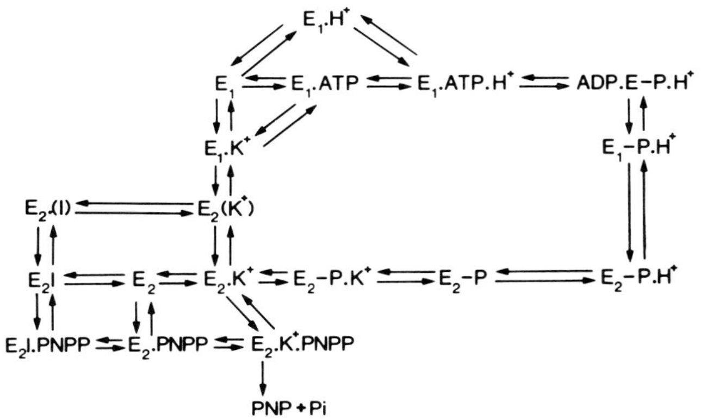

FIG. 13. Schematic of the sequence of reactions that have been shown for the H^+-K^+-ATPase, including the interaction of the enzyme with the K^+-competitive inhibitor SCH 28080. Reactions are written as reversible, but backreaction from E_2-P to E_1- has not been shown. [From Sachs (127a).]

region as the protective ligand. This may be the case, but protection or even simple competitive kinetics only prove that binding of one ligand excludes binding by the other, which could result from conformational changes and not necessarily direct competition at the same site. The reagents that have been used are, in general, amino acid reagents of varying specificity, and only in the case of FITC has the site been defined (22). Often the stoichiometry is unknown, and thus the conclusions that are derivable are limited. The inhibitors to be discussed can be classified into site-specific and group-specific inhibitors, with subclasses in each area.

Site-Specific Inhibitors

N$_3$-ATP. This photoaffinity analogue of ATP is hydrolyzed at ~40% of the rate of ATP and is able to phosphorylate the enzyme with twice the stoichiometry of ATP (121). When the ^{32}P form was used and photolysed, ~3 nmol of radioactivity was incorporated per milligram enzyme protein, but about half was lost after the addition of K$^+$ (20, 121). With the ^{32}P form, 1.5 mmol was incorporated (121). The peptide sequence to which this compound binds has not been defined but must be close to the ATP-binding region sequence of the catalytic subunit of the H$^+$-K$^+$-ATPase. The presence of both N$_3$-ATP binding and ^{32}P binding suggests that ATP can bind to the phosphoenzyme form, which is perhaps the low-affinity state of the ATP site as determined by kinetic analysis.

VANADATE. The vanadate anion inhibits the H$^+$-K$^+$-ATPase, as it does the Na$^+$-K$^+$-ATPase (4, 7). In these enzymes, in the presence of Mg^{2+}, which is also required for binding of P$_i$, vanadate forms a stable vanadyl enzyme complex, where vanadate probably binds in the region of the phosphate transferred from ATP. However, the complex is not covalent. The formation of E-P and pNPPase was inhibited monophasically, whereas inhibition of the H$^+$-K$^+$-ATPase activity was biphasic, with binding of 3 nmol of the inhibitor required for total inhibition of ATPase activity. Vanadate quenched fluorescence of FITC enzyme; hence, the complex is E$_2$·vanadate. Vanadate also inhibited the K$^+$-K$^+$ exchange reaction of ATPase but not the conductive pathway associated with the stimulated ATPase fraction (31). Thus it has proved to be a useful reagent for the H$^+$-K$^+$-ATPase.

PROTONATABLE AMINES. Various protonatable amines have been shown to inhibit the H$^+$-K$^+$-ATPase by competing with K$^+$. These compounds include nolinium bromide (92), chlorpromazine, trifluperazine, TMB 8 (54), and SCH 28080 (165). In addition, hydrophobic cations, such as tetraphenylphosphonium, also compete with K$^+$ (31). As a class these compounds do not appear to interfere with Na$^+$-K$^+$-ATPase activity, showing that the K$^+$ sites in these two enzymes have significant differences.

The best studied of these inhibitors is SCH 28080, a pyridyl-1,2-amidazole, which has been shown to inhibit gastric acid secretion in vivo (78). This compound also inhibits secretion in rabbit gastric glands and inhibits the gastric H$^+$-K$^+$-ATPase by competing with K$^+$. It is more active in the protonated form. The 50% inhibitory concentration (IC$_{50}$) in glands, where cytosolic K$^+$ is high, is the same as in the enzyme preparation; thus the cationic form (Fig. 14A) inhibits from the luminal face of the enzyme. Thus it is a luminally active, K$^+$-competitive antagonist of the H$^+$-K$^+$-ATPase. With respect to ATP or pNPP, the inhibition is essentially uncompetitive, which can be interpreted as due to the formation of a relatively stable E$_2$·SCH·ATP form. The compound apparently does not compete with K$^+$ for the E$_1$ form of the enzyme, because the rate constant of phosphorylation is unaffected by SCH 28080, nor does the drug either induce dephosphorylation of E$_2$-P or affect the K$^+$ dephosphorylation of this enzyme form. Thus it seems to be selective for the E$_2$ form and to bind from the luminal face of the K$^+$-K$^+$-ATPase in this conformation (165). In the azido form, specific inhibition is observed at pH 6.5 that is completely K$^+$, but not Na$^+$, protected, and ATP enhances the photolytic inhibition (9). This compound should, therefore, be a useful probe of the K$^+$-binding region of the H$^+$-K$^+$-ATPase. Its interaction with the H$^+$-K$^+$-ATPase is illustrated in Figure 13.

OMEPRAZOLE. Omeprazole is the best studied of this class of inhibitors of the H$^+$-K$^+$-ATPase (23), because it has achieved use in humans as a therapy for acid-related disease of the upper gastrointestinal tract (16, 74, 160). This compound is a substituted benzimidazole sulfoxide, which is inactive in neutral solutions but becomes active in acid solutions. The chemistry of the rearrangement of the prodrug, omeprazole, to the active SH-reactive species is complicated and shown in Figure 14B. In its active form, it is a cationic sulfenamide or sulfenic acid that covalently derivatives SH residues on the H$^+$-K$^+$-ATPase (163, 164, 166).

It can be shown that in gastric glands, acidity is necessary for the antisecretory activity of omeprazole, and the same phenomenon can be shown in isolated vesicles forming a proton gradient (64, 79, 80, 163). In vivo and in vitro the compound, due to its weak base properties (pK_a 4.3), accumulates in the acid space of the parietal cell and there binds almost exclusively to the catalytic subunit of the H$^+$-K$^+$-ATPase in the parietal cell (B. Wallmark and J. Fryklund, unpublished observations). A similar binding selectivity is found in gastric vesicles (80). The reaction with the enzyme derivatizes two SH groups per mole E-P when activated in the luminal space, and ~20 when added

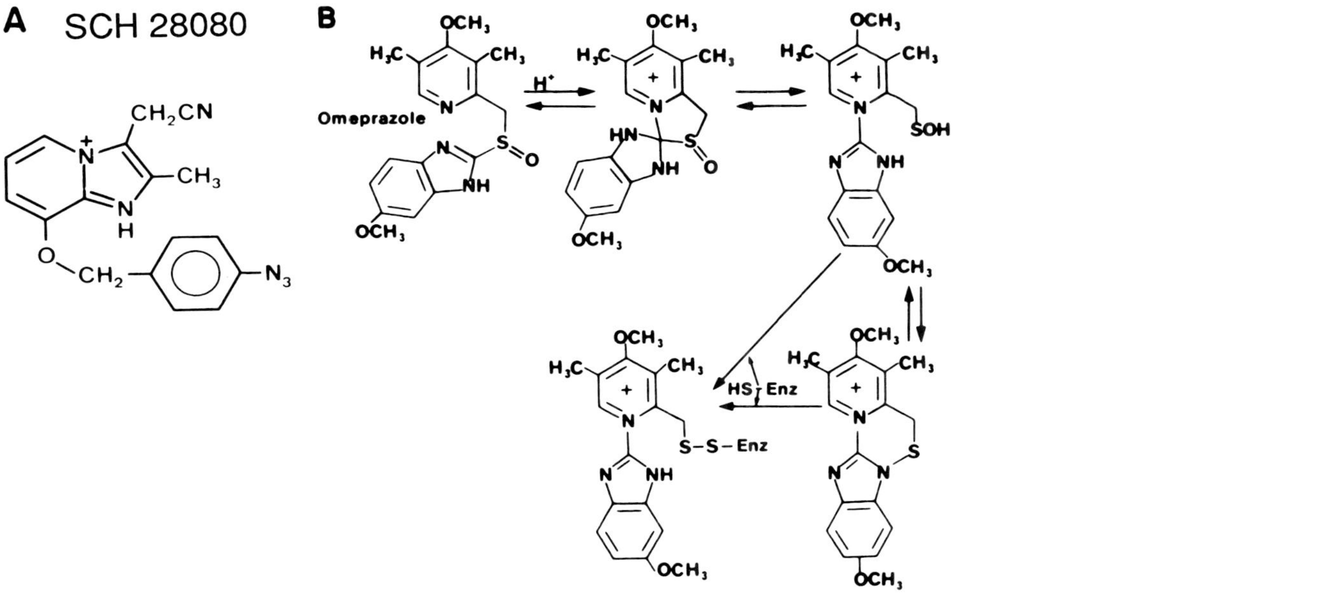

FIG. 14. *A*: formula of SCH 28080 in the active, protonated form with charge distribution shown. *B*: reactive cycle of omeprazole after protonation of the prodrug. Initial protonation on benzimidazole nitrogen is followed by slow rearrangement to the spirocompound, which then rapidly forms either the sulfenamide (the major reactive species) or the sulfenic acid, both of which are cationic SH reagents.

in the acid-activated form to lyophilized preparations of the H^+-K^+-ATPase [Fig. 15; (80)]. Thus the cationic nature of the reactive species provides a highly selective modification of the H^+-K^+-ATPase, which is reversed by SH-reducing agents in the vesicle preparation (80) and in gastric glands (55, 163). This reagent should, therefore, prove useful in defining the location and nature of the essential cysteines accessed from the luminal surface of the H^+-K^+-ATPase. Given that both SCH 28080 and omeprazole in their active form are bulky cations, it may be that the gastric H^+-K^+-ATPase has a luminal vestibule that can accommodate these cations and result in K^+ competition by SCH 28080 and SH group reaction with the omeprazole sulfenamide.

Group-Selective Reagents

A large variety of group-selective reagents have been used, but few have been characterized in terms of binding site or stoichiometry. An amino group reagent, 2-methoxy-2,4-diphenyl-3-dihydrofuranone, has been shown to inhibit the enzyme in an ATP-protectable manner (71). Butanedione, an arginine reagent, has also been shown to inhibit the H^+-K^+-ATPase and ATP, or activating cations such as K^+, Rb^+, and NH_4^+, protected against activation, suggesting movement of the arginine group into a protected location in the E_2 enzyme form (137). The histidine reagent, diethyl pyrocarbonate (DEPC), was also an inhibitor of the H^+-K^+-ATPase, but several histidines were

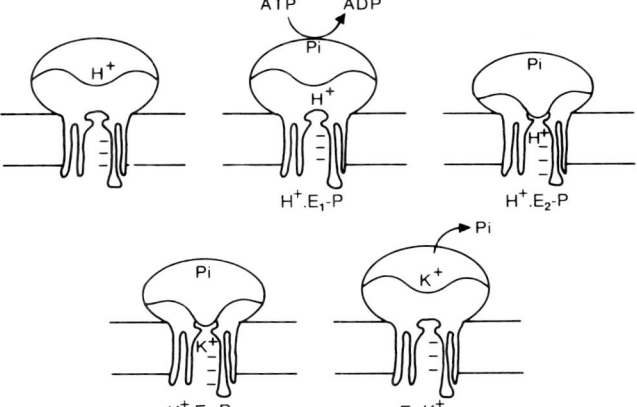

FIG. 15. Conceptual model of E_1E_2-type pump, where coupling between scalar ATP hydrolysis and vectorial ion transport is due to a mobile sector of the pump that inserts partway into the hydrophobic domain. Thus these pumps act as liner mechanochemical devices, and conformations analyzed by different means are represented by different locations of mobile pump sector.

modified by DEPC. The pK_a of the essential histidine was found to be 6.7 (117) and may be involved in the cytosolic H^+-binding region.

Carboxyl group reagents, such as dicyclohexylcarbodiimide (DCCD), inhibit H^+-K^+-ATPase (130, 131), but their region of action has not been defined. In the case of *Neurospora* ATPase, it appears that DCCD modifies a single carboxyl in the calculated hydrophobic sector (154). The carboxyl-activating reagent, ethoxyethyl dihydroquinoline (EEDQ), was shown to inhibit the H^+-K^+-ATPase by binding to a single class of carboxylic acids, and this inhibition was prevented by luminal K^+ (118). The action involved intramolecular cross-linking, and the carboxylate involved might thus be related to the sites blocked by SCH 28080 and the reactive omeprazole species.

Various SH reagents inhibit this as well as other ATPases. The enzyme was inhibited by 5,5'-dithiobis(2-nitrobenzoic acid) (DTNB), and the inhibition was reversed by dithiothreitol (138). Thimerosal (26) inhibited the H^+-K^+-ATPase and activated the pNPPase reaction of the enzyme, whereas carboxypyridine disulfide inhibited the H^+-K^+-ATPase in iontight vesicles, in an ATP-modifiable region (91).

Various reagents are therefore available to study some of the functional residues in the H^+-K^+-ATPase, both from the point of view of catalysis and transport. With the primary sequence now known, location of these sites is feasible and will form one of the new approaches to the mechanism of this enzyme.

BIOSYNTHESIS OF H^+-K^+-ATPASE

The mRNA for the H^+-K^+-ATPase may be a relatively abundant message. By in vitro translation from total gastric mRNA, however, the major peptide that is made is likely to be pepsinogen (148). With immunoprecipitation, using a monoclonal antibody against the H^+-K^+-ATPase, both total and size-fractionated mRNA from rabbit mucosa have been shown to synthesize the catalytic subunit of the H^+-K^+-ATPase (A. Smolka, P. Lorentzon, and G. Sachs, unpublished observations) and the calculated message frequency is ~0.1%. Turnover of the enzyme can be estimated in various species by the recovery of acid secretion after covalent inhibition by omeprazole. Thus, in the rat the half-life of restoration of acid secretion and of H^+-K^+-ATPase activity after omeprazole treatment in vivo is ~30 h (163, 167), and similar data seem to apply to humans. Long-term treatment in the rat results in rebound hypersecretion after cessation of omeprazole dosage, and the decay of this hypersecretion has a half-life of ~30 days, suggesting that the parietal cell has a longer half-life than its H^+-K^+-ATPase (69). The regulation of the biosynthesis of this enzyme and its targeting to tubulovesicles will benefit from a molecular biological approach to these problems.

TISSUE DISTRIBUTION OF H^+-K^+-ATPASE

By antibody techniques, the only cell in the gastric mucosa that contains measurable quantities of the H^+-K^+-ATPase is the parietal cell. This cell is arranged spirally in the mammalian gastric gland; thus it probably originates from a single class of stem cells in the neck region of the gastric gland and is thus absent in the peptic cell. Antibody techniques using a lysosomally derived antibody have shown some common region of antigenicity between the lysosomal and hog gastric mucosal H^+-K^+-ATPase (113, 155), and several monoclonal antibodies have been shown to stain the plasma membrane of the nonpigmented cell of the ciliary body of the eye (18). There may also be a similar enzyme in osteoclasts, because omeprazole, at very high doses, inhibits bone resorption by these cells (2, 156).

The only nongastric tissue where there is direct evidence for a functional H^+-K^+-ATPase is the distal colon. It was shown that a ouabain-insensitive pNPPase activity was present in colonic homogenates, as well as a K^+-sensitive, ouabain-insensitive phosphoenzyme (32, 33). Furthermore, in colonic vesicles a vanadate-sensitive, K^+-dependent, N-ethylmaleimide–insensitive proton pump was discovered (63). The function of this enzyme in this region is not clear. It may play a role in pH regulation in metabolic acidosis or in K^+ reabsorption. The similarity between the distal colon and the collecting duct of the kidney suggests that a similar enzyme might be present in the kidney, and an ouabain-insensitive pNPPase activity that is K^+ stimulated has been described in several sections of the nephron (28). The use of mRNA that is H^+-K^+-ATPase specific should provide a rapid advance in measurement of tissue distribution of this enzyme. It would be surprising if the parietal cell were the only cell type to express the H^+-K^+-ATPase gene.

MODEL FOR H^+-K^+-ATPASE

Knowledge of how biological pumps function is rudimentary and fragmented in the absence of analyzed three-dimensional crystals in the various conformations thought to be occupied by these pumps. However, there are some experiments that can suggest a working hypothesis. In analogy with manufactured pumps, several models can be thought of, such as a rotary pump, a screw mechanism, or a piston mechanism, to mention three simple examples. The piston mechanism, essentially a mechanochemical device, may at the moment be the most attractive and the most easily explored. In the case of the Ca^{2+}-ATPase, low-angle X-ray scattering has shown that the hydrophilic sector contracts with the formation of the E_2 form of the enzyme. Tighter packing of the protein may account for part of this shrinkage, but equally plausible is the idea that a segment of the hydrophilic sector associates with or inserts into the hydrophobic sector (6). This requires that the hydrophobic peptides are relatively loosely packed in the membrane, as appears to be the case for the photoreaction center of *Pseudomonas viridis* (10). Furthermore, using a photoreactive hydrophobic label (trifluoromethyl iodophenyl diazirine) in Ca^{2+}-pump vesicles of the sarcoplasmic reticulum, hydrophobic labeling of a tryptic cleavable portion of the hydrophilic sector increased in the E_2 form of the enzyme, again as if there were partial insertion of a segment of the enzyme into the hydrophobic region in the transition between the E_1 and E_2 form of the Ca^{2+}-ATPase (7). In the case of the H^+-K^+-ATPase, it is striking that relatively large cations binding from the luminal face, presumably distant from the ATP region of the enzyme, are able to inhibit various reactions of the H^+-K^+-ATPase. Thus, for example, omeprazole inhibits pNPPase and ATPase activity, proton transport, and phosphorylation, and SCH 28080 has similar effects. It would therefore be reasonable to speculate that the binding of ATP and subsequent transphosphorylation result in movement of the ion-binding site from a hydrophilic location (E_1-P form) across the hydrophilic/hydrophobic interface, with insertion between the membrane-spanning regions of the enzyme. This would correspond to the E_2 configuration of the E-P form, with the corresponding change in H^+-binding site affinity. A change of ~6 orders of magnitude is required. With the release of H^+ and binding of K^+ from the luminal face, the mobile site can return to the hydrophilic sector as the enzyme dephosphorylates, reforming the E_1 configuration. The reactivity of the omeprazole sulfenamide and the cationic SCH 28080, if this model is correct, would also argue for the presence (at least with this pump) of a relatively large vestibule facing the lumen, a vestibule that has been shown to be present in many ion channels [Fig. 15; (46)]. The combination of molecular biology and crystallography will undoubtedly change our orientation in the next 20 years as much as enzymology and vesicology changed thinking about gastric acid secretion in the last 20 years.

For half a century Warren S. Rehm has electrified the field of gastric acid secretion. His significant contributions are gratefully acknowledged.

REFERENCES

1. ANDERSEN, J. P., B. VILSEN, J. H. COLLINS, AND P. L. JØRGENSEN. Localization of E_1-E_2 conformational transitions of sarcoplasmic reticulum Ca^{2+}-ATPase by tryptic cleavage and hydrophobic labeling. *J. Membr. Biol.* 93: 85–92, 1986.
2. ANDERSON, R. E., D. M. WOODBURY, AND W. S. JEE. Humoral and ionic regulation of osteoclast acidity. *Calcif. Tissue Int.* 39: 252–258, 1986.
3. BANDOPADHYAY, S., P. K. DAS, M. V. WRIGHT, J. NANDI, D. BHATTACHARYYAY, AND T. K. RAY. Characteristics of a pure endogenous activator of the gastric H^+,K^+-ATPase system. *J. Biol. Chem.* 262: 5664–5670, 1987.
4. BERGLINDH, T., D. R. DIBONA, C. S. PACE, AND G. SACHS. ATP dependence of H^+ secretion. *J. Cell Biol.* 85: 392–401, 1980.
5. BERGLINDH, T. O., H. F. HELANDER, AND K. J. OBRINK. Effects of secretagogues on oxygen consumption, aminopyrine accumulation and morphology on isolated gastric glands. *Acta Physiol. Scand.* 97: 401–414, 1976.
6. BLASIE, J. K., L. HERBETTE, AND J. PACHENCE. Biological membrane structure as seen by X-ray and neutron diffraction techniques. *J. Membr. Biol.* 86: 1–8, 1985.
7. CANTLEY, L. C., L. G. CANTLEY, AND L. JOSEPHSON. A characterization of vanadate interactions with the (Na,K)-ATPase. *J. Biol. Chem.* 253: 7361–7368, 1978.
8. CHANG, H., G. SACCOMANI, E. RABON, R. SCHACKMANN, AND G. SACHS. Proton transport by gastric membrane vesicles. *Biochim. Biophys. Acta* 464: 313–327, 1977.
9. CUPPOLETTI, J., AND G. SACHS. Regulation of gastric acid secretion via modulation of a chloride conductance. *J. Biol. Chem.* 259: 14952–14959, 1984.
10. DEISENHOFER, J., O. EPP, K. MIKI, R. HUBER, AND H. MICHEL. Structure of the protein subunits in the photoreaction center in the photosynthetic reaction center of *Rhodopseudomonas viridis* at three angstrom resolution. *Nature Lond.* 318: 618–624, 1985.
11. DEJONG, M. Conformational Changes of H^+,K^+-ATPase. Nijmegen, Netherlands: University of Nijmegen, 1986. Dissertation.
12. DEMAREST, J. R., AND T. E. MACHEN. Microelectrode measurements form oxyntic cells in intact *Necturus* gastric mucosa. *Am. J. Physiol.* 249 (*Cell Physiol.* 18): C535–C540, 1985.
13. DURBIN, R. P. Electrical potential difference of the gastric mucosa. In: *Handbook of Physiology. Alimentary Canal. Secretion*, edited by C. F. Code. Washington, DC: Am. Physiol. Soc., 1967, sect. 6, vol. II, chapt. 49, p. 879–888.
14. DURBIN, R. P., F. MICHELANGELI, AND A. NICKEL. Active transport and ATP in frog gastric mucosa. *Biochim. Biophys. Acta* 3367: 177–189, 1974.
15. DUX, L., P. SLAWOMIR, N. MULLNER, AND A. MARTONOSI. Crystallization of Ca^{2+}-ATPase in detergent-solubilized sarcoplasmic reticulum. *J. Biol. Chem.* 262: 6439–6443, 1987.
16. ELANDER, B., E. FELLENIUS, R. LETH, L. OLBE, AND B. WALLMARK. Inhibitory action of omeprazole on acid formation in gastric glands and on H^+,K^+-ATPase isolated from human gastric mucosa. *Scand. J. Gastroenterol.* 12: 268–272, 1986.
17. EPSTEIN, M., Y. KURIKI, R. BILTONEN, AND E. RACKER. Calorimetric studies of ligand-induced modulation of calcium adenosine 5′-triphosphatase from sarcoplasmic reticulum. *Biochemistry* 19: 5565–5568, 1980.
18. FAIN, G., A. SMOLKA, M. C. CILLUFFO, M. J. FAIN, D. A. LEE, N. C. BRECHA, AND G. SACHS. Monoclonal antibodies to the H^+,K^+-ATPase of gastric mucosa selectively stained the nonpigmented cell of the rabbit ciliary body epithelium. *Invest. Ophthamol. Visual Sci.* 29: 785–794, 1988.
19. FALLER, L. D., AND G. A. EL GAVISH. Catalysis of ^{18}O exchange between inorganic phosphate and water by gastric H^+,K^+-ATPase. *Biochemistry* 23: 6584–6590, 1984.
20. FALLER, L., R. JACKSON, D. MALINOWSKA, E. MUKIDJAM, E. RABON, G. SACCOMANI, G. SACHS, AND A. SMOLKA. Mechanistic aspects of gastric (H^+ + K^+)-ATPase. *Ann. NY Acad. Sci.* 402: 146–163, 1982.
21. FALLER, L. D., E. RABON, AND G. SACHS. Vanadate binding to the gastric H^+,K^+-ATPase. *Biochemistry* 22: 4676–4685, 1983.
22. FARLEY, R. A., AND L. D. FALLER. Amino acid sequence of an active site peptide from H^+,K^+-ATPase of gastric mucosa. *J. Biol. Chem.* 260: 3899–3901, 1985.
23. FELLENIUS, E., T. BERGLINDH, G. SACHS, L. OLBE, B. ELANDER, S. E. SJÖSTRAND, AND B. WALLMARK. Substituted benzimidazoles block acid secretion by blocking the H^+,K^+-ATPase. *Nature Lond.* 290: 159–161, 1981.
24. FORTE, J. G., J. A. BLACK, T. M. FORTE, T. E. MACHEN, AND J. M. WOLOSIN. Ultrastructural changes related to functional activity in gastric oxyntic cells. *Am. J. Physiol.* 241 (*Gastrointest. Liver Physiol.* 4): G349–358, 1981.
25. FORTE, J. G., A. L. GANSER, AND A. S. TANISAWA. The K^+ stimulated ATPase system of microsomal oxyntic cells. *Ann. NY Acad. Sci.* 242: 255–267, 1974.
26. FORTE, J. G., J. L. POULTER, R. DYKSTRA, J. RIVAS, AND H. C. LEE. Specific modification of gastric K^+-stimulated ATPase

activity by thimerosal. *Biochim. Biophys. Acta* 644: 257–265, 1981.
27. GANSER, A. L., AND J. G. FORTE. K^+-ATPase activity in purified microsomes of bullfrog oxyntic cells. *Biochim. Biophys. Acta* 307: 169–180, 1973.
28. GARG, L., AND N. NARANG. Ouabain-insensitive K^+-ATPase activity in distal nephron segments (Abstract). *Federation Proc.* 46: 363, 1987.
29. GARTY, H., B. RUDY, AND S. J. D. KARLISH. A simple and sensitive procedure for measuring isotope flux through ion specific channels in heterogeneous populations of membrane vesicles. *J. Biol. Chem.* 258: 169–175, 1983.
30. GLYNN, I. M., AND S. J. D. KARLISH. The sodium pump. *Annu. Rev. Physiol.* 37: 13–55, 1975.
31. GUNTHER, R. D., S. BASSILIAN, AND E. C. RABON. Conductance properties of stimulated rabbit vesicles. *J. Biol. Chem.* 262: 13966–13972, 1987.
32. GUSTIN, M. C., AND D. P. GOODMAN. Isolation of brush border membrane from rabbit descending colon epithelium. *J. Biol. Chem.* 265: 10651–10656, 1981.
33. GUSTIN, M. C., AND D. B. GOODMAN. Characterization of the phosphorylated intermediate of K^+-ATPase of rabbit colon brush border membrane. *J. Biol. Chem.* 257: 9629–9633, 1982.
34. HAGER, K., AND C. W. SLAYMAN. Amino acid sequence of the plasma membrane ATPase of *Neurospora crassa*: deduction from genomic and cDNA sequence. *Proc. Natl. Acad. Sci. USA* 83: 7693–7697, 1986.
35. HANGEL, D. K., T. URUSHIDAKI, AND J. G. FORTE. Phosphorylation of microtubule-associated proteins in gastric glands (Abstract). *Federation Proc.* 46: 1082, 1987.
36. HARA, Y., AND M. NAKAO. ATP-dependent proton uptake by proteoliposomes reconstituted with purified Na^+,K^+-ATPase. *J. Biol. Chem.* 261: 12655–12658, 1986.
37. HARRIS, J. B., AND I. S. EDELMAN. Transport of potassium by the gastric mucosa of the frog. *Am. J. Physiol.* 198: 280–284, 1960.
38. HARRIS, J. B., H. FRANK, AND I. S. EDELMAN. Effect of potassium on ion transport and bioelectric potentials of frog gastric mucosa. *Am. J. Physiol.* 195: 499–504, 1958.
39. HELANDER, H. F., AND B. I. HIRSCHOWITZ. Quantitative ultrastructural studies on gastric parietal cell. *Gastroenterology* 63: 951–961, 1972.
40. HELANDER, H. F., C. H. RAMSAY, AND C. G. REGARDH. Localization of omeprazole and metabolites in the mouse. *Scand. J. Gastroenterol.* 108: 95–104, 1985.
41. HELANDER, H. F., AND G. SACHS. Effects of secretagogues on parietal cell morphology in rat. *IUPS Symposium, Univ. of Calgary*, in press.
42. HELMICH-DEJONG, M. L., J. P. VAN DUYNHOVEN, F. M. SCHUURMANS-STEKHOVEN, AND J. J. DE PONT. Eosin, a fluorescent marker for the high affinity ATPase of H^+,K^+-ATPase. *Biochim. Biophys. Acta* 858: 254–262, 1986.
43. HELMICH-DEJONG, M. L., S. E. VAN EMST DE VRIES, J. J. DE PONT, F. M. SCHUURMANS-STEKHOVEN, AND S. L. BONTING. Direct evidence for an ADP-sensitive phosphointermediate of the H^+,K^+-ATPase. *Biochim. Biophys. Acta* 821: 377–383, 1985.
44. HERSEY, S. J., AND L. STEINER. Stimulation of acid formation in permeable gastric glands by valinomycin. *Am. J. Physiol.* 255 (*Gastrointest. Liver Physiol.* 18): G313–G318, 1988.
45. HESSE, J. E., J. WIECZOREK, K. H. ALTENDORF, A. S. REICIN, E. DORUS, AND W. EPSTEIN. Sequence homology between two membrane transport ATPases, the Kdp-ATPase of *Escherichia coli* and the Ca^{2+}-ATPase of sarcoplasmic reticulum. *Proc. Natl. Acad. Sci. USA* 81: 4746–4750, 1984.
46. HILLE, B. *Ionic Channels of Excitable Membranes*. Boston, MA: Sinauer, 1984, p. 249.
47. HIRST, B. A., AND J. G. FORTE. Redistribution and characterization of H^+,K^+-ATPase membranes from resting and stimulated gastric parietal cells. *Biochem. J.* 231: 641–649, 1985.
48. HOGBEN, C. A. M. Biological aspects of active chloride transport. In: *Electrolytes in Biological Systems*, edited by A. M. Shanes. Washington, DC: Am. Physiol. Soc., 1955, p. 176–204.
49. IM, W. B., AND D. P. BLAKEMAN. Protective effect of EGTA on rat gastric membranes enriched in gastric H^+,K^+-ATPase. *Biochem. Biophys. Res. Commun.* 180: 635–639, 1982.
50. IM, W. B., D. P. BLAKEMAN, AND J. P. DAVIS. Studies on K^+ permeability of rat gastric microsomes. *J. Biol. Chem.* 260: 9452–9460, 1985.
51. IM, W. B., D. P. BLAKEMAN, AND J. P. DAVIS. Finding of a KCl-independent, electrogenic and ATP-driven H^+ pumping activity in rat light gastric microsomes and its effect on the membrane K^+ transport activity. *J. Biol. Chem.* 261: 11686–11692, 1986.
52. IM, W. B., D. P. BLAKEMAN, J. M. FIELDHOUSE, AND E. C. RABON. Effect of carbachol or histamine stimulation on rat gastric membranes enriched in H^+,K^+-ATPase. *Biochim. Biophys. Acta* 772: 167–175, 1984.
53. IM, W. B., D. P. BLAKEMAN, J. MENDLEIN, AND G. SACHS. Inhibition of $(H^+ + K^+)$-ATPase and H^+ accumulation in hog gastric membranes by trifluoperazine, verapamil and 8-(N,N-diethylamino)octyl-3,4,5-trimethoxybenzoate. *Biochim. Biophys. Acta* 770: 65–72, 1984.
54. IM, W. B., D. P. BLAKEMAN, AND G. SACHS. Reversal of antisecretory activity of omeprazole by sulfhydryl compounds in isolated rabbit gastric glands. *Biochim. Biophys. Acta* 845: 54–59, 1985.
55. IM, W. B., J. P. DAVIS, AND D. P. BLAKEMAN. Preparation of rat gastric microsomal heavy and light membranes enriched in $(H^+ + K^+)$-ATPase using $2H_2O$ and Percoll gradients. *Biochem. Biophys. Res. Commun.* 131: 905–911, 1985.
56. JACKSON, R. J., J. MENDLEIN, AND G. SACHS. Interaction of fluorescein isothiocyanate with the H^+,K^+-ATPase. *Biochim. Biophys. Acta* 731: 9–15, 1983.
57. JACKSON, R. J., AND G. SACCOMANI. Phosphorylation of gastric H^+,K^+-ATPase by inorganic phosphate (Abstract). *Biophys. J.* 45: 83, 1984.
58. KAPLAN, J. D. Ion movements through the sodium pump. *Annu. Rev. Physiol.* 47: 535–544, 1985.
59. KARLISH, S. J. D., W. R. LIEB, AND W. D. STEIN. Combined effects of ATP and phosphate on Rb^+ exchange mediated by Na^+,K^+-ATPase reconstituted into phospholipid vesicles. *J. Physiol. Lond.* 328: 333–350, 1982.
60. KARLISH, S. J. D., AND U. PICK. Sidedness of the effects of sodium and potassium on the conformational state of the sodium potassium pump. *J. Physiol. Lond.* 312: 505–529, 1981.
61. KASBEKAR, D. K. Potassium selectivity of frog gastric luminal membrane. *Am. J. Physiol.* 250 (*Gastrointest. Liver Physiol.* 13): G765–G772, 1986.
62. KAUNITZ, J., AND G. SACHS. Identification of a vanadate-sensitive, potassium-dependent proton pump from rabbit colon. *J. Biol. Chem.* 261: 14005–14010, 1986.
63. KEELING, D. J., C. FALLOWFIELD, AND A. H. UNDERWOOD. The specificity of omeprazole as an H^+,K^+-ATPase inhibitor depends upon the means of its activation. *Biochem. Pharmacol.* 36: 339–344, 1987.
64. KOELZ, H. R., G. SACHS, AND T. BERGLINDH. Cation effects on acid secretion in rabbit gastric glands. *Am. J. Physiol.* 241 (*Gastrointest. Liver Physiol.* 4): G431–G442, 1981.
65. KOENIG, C. S. Redistribution of gastric K^+ pNNPase in vertebrate oxyntic cells in relation to hydrochloric acid secretion: a cytochemical study. *Anat. Rec.* 210: 583–596, 1984.
66. KYTE, J., AND R. F. DOOLITTLE. A simple method for displaying the hydropathic character of a protein. *J. Mol. Biol.* 157: 105–132, 1983.
67. LANE, L. K., T. K. KIRLEY, AND W. J. BALL. Structural studies on H^+,K^+-ATPase: determination of the NH_2 terminal amino acid sequence and immunological cross reactivity with Na^+,K^+-ATPase. *Biochem. Biophys. Res. Commun.* 138: 185–192, 1986.
68. LARSSON, H., E. CARLSSON, B. RYBERG, J. FRYKLUND, AND B. WALLMARK. Rat parietal cell function after prolonged

inhibition of gastric acid secretion. *Am. J. Physiol.* 254 (*Gastrointest. Liver Physiol.* 17): G33–G39, 1988.
70. LEE, H. C., AND J. G. FORTE. A study of H^+ transport in gastric microsomal vesicles using fluorescent probes. *Biochim. Biophys. Acta* 508: 339–356, 1978.
71. LEE, H. C., AND J. G. FORTE. Chemical modification of gastric microsomal potassium-stimulated ATPase. *Biochim. Biophys. Acta* 598: 595–605, 1980.
72. LEE, J., G. SIMPSON, AND P. SCHOLES. An ATPase from dog gastric mucosa: changes of outer pH in suspensions of membrane vesicles accompanying ATP hydrolysis. *Biochem. Biophys. Res. Commun.* 60: 825–832, 1974.
73. LEWIN, M. J. M., G. SACCOMANI, R. SCHACKMANN, AND G. SACHS. The use of ANS as a probe of gastric vesicle transport. *J. Membr. Biol.* 32: 301–318, 1977.
74. LIND, T., C. CEDERBERG, G. EKENVED, U. HAGLUND, AND L. OLBE. Effect of omeprazole, a gastric proton pump inhibitor, on pentagastrin-stimulated acid secretion in man. *Gut* 24: 270–276, 1983.
75. LJUNGSTRÖM, M., AND S. MÅRDH. Kinetics of the acid pump in the stomach. Proton transport and hydrolysis of ATP and *p*-nitrophenyl phosphate by the gastric H^+,K^+-ATPase. *J. Biol. Chem.* 260: 5440–5444, 1985.
76. LJUNGSTRÖM, M., L. NORBERG, H. OLAISSON, C. WERNSTEDT, F. V. VEGA, G. ARVIDSON, AND S. MÅRDH. Characterization of proton-transporting membranes from resting pig gastric mucosa. *Biochim. Biophys. Acta* 769: 209–219, 1984.
77. LJUNGSTRÖM, M., F. V. VEGA, AND S. MÅRDH. Effects of pH on the interaction of ligands with the $(H^+ + K^+)$-ATPase purified from pig gastric mucosa. *Biochim. Biophys. Acta* 769: 220–230, 1984.
78. LONG, J. F., P. J. S. CHIU, M. J. DERELANKO, AND M. STEINBERG. Gastric antisecretory and cytoprotective activities of SCH 28080. *J. Pharmacol. Exp. Ther.* 226: 114–120, 1983.
79. LORENTZON, P., B. EKLUNDH, A. BRÄNDSTRÖM, AND B. WALLMARK. The mechanism for inhibition of gastric H^+,K^+-ATPase by omeprazole. *Biochim. Biophys. Acta* 778: 549–558, 1984.
80. LORENTZON, P., R. JACKSON, B. WALLMARK, AND G. SACHS. Inhibition of H^+,K^+-ATPase by omeprazole in isolated gastric vesicles requires proton transport. *Biochim. Biophys. Acta* 897: 41–51, 1987.
81. MALINOWSKA, D. H., J. CUPPOLETTI, AND G. SACHS. Cl^- requirement of acid secretion in isolated gastric glands. *Am. J. Physiol.* 245 (*Gastrointest. Liver Physiol.* 8): G573–G581, 1983.
82. MALINOWSKA, D. M., H. R. KOELZ, S. J. HERSEY, AND G. SACHS. Properties of the gastric proton pump in permeable, unstimulated gastric glands. *Proc. Natl. Acad. Sci. USA* 78: 5908–5912, 1981.
83. McLENNAN, D. H., C. J. BRANDL, B. KORCZAK, AND N. M. GREEN. Amino acid sequence of a Ca^{2+},Mg^{2+}-dependent ATPase from rabbit muscle sarcoplasmic reticulum, deduced from its complementary DNA sequence. *Nature Lond.* 316: 696–700, 1985.
84. MENDLEIN, J., AND G. SACHS. Ca^{2+} generated form of the H^+,K^+-ATPase (Abstract). *Federation Proc.* 41: 1125, 1982.
85. MENIK, D. R., J. A. LEE, R. J. BROOKER, T. H. WILSON, AND H. R. KABAK. Role of cysteine residues in Lac permease of *E. coli*. *Biochemistry* 26: 1132–1136, 1987.
86. MOODY, F. G. Oxygen consumption during thiocyanate inhibition of gastric acid secretion in dogs. *Am. J. Physiol.* 215: 127–131, 1968.
87. MUALLEM, S., D. BLISSARD, E. M. CRAGOE, AND G. SACHS. Role of Na^+/H^+ exchange in parietal cell function. *J. Biol. Chem.* In press.
88. MUALLEM, S., C. BURNHAM, T. BERGLINDH, D. BLISSARD, AND G. SACHS. Electrolyte transport across the basal lateral membrane of the parietal cell. *J. Biol. Chem.* 260: 6641–6653, 1985.
89. MUALLEM, S., AND S. J. D. KARLISH. Regulation of the Ca^{2+} pump by calmodulin in the intact cell. *Biochim. Biophys. Acta* 687: 329–332, 1982.
90. MUNSON, K. B., AND G. SACHS. Inactivation of H^+,K^+-ATPase by a K^+ competitive photoaffinity inhibitor. *Biochemistry* 27: 3932, 1988.
91. NANDI, J., Z. MEN-AI, AND T. K. RAY. Role of membrane-associated thiol groups in the functional regulation of gastric microsomal H^+,K^+-transporting ATPase system. *Biochem. J.* 213: 587–594, 1983.
92. NANDI, J., M. V. WRIGHT, AND T. K. RAY. Mechanism of gastric antisecretory effects of nolinium bromide. *Gastroenterology* 85: 938–945, 1983.
93. OTTOLENGHI, P., J. G. MORBY, AND J. JENSEN. Solubilization and further chromatographic purification of highly purified, membrane-bound Na^+,K^+-ATPase. *Biochem. Biophys. Res. Commun.* 135: 1008–1014, 1986.
94. OVCHINNIKOV, Y. A., N. N. MODYANOV, N. E. BROUDE, K. E. PETRUHKIN, A. V. GRISHIN, N. M. ARZAMAZOVA, N. A. ALDANOVA, G. S. MONASTYRSKAYA, AND E. S. SVERDLOV. Pig kidney Na^+,K^+-ATPase. *FEBS Lett.* 201: 237–245, 1986.
95. PARADISO, A. M., R. Y. TSIEN, AND T. E. MACHEN. Na^+/H^+ exchange in gastric glands as measured with a cytoplasmic trapped fluorescent pH indicator. *Proc. Natl. Acad. Sci. USA* 81: 7436–7440, 1984.
96. PETERS, W. H., A. M. FLEUREN-JAKOBS, J. J. SCHRIJEN, J. J. DE PONT, AND S. L. BONTING. Studies on $(H^+ + K^+)$-ATPase. V. Chemical composition and molecular weight of the catalytic subunit. *Biochim. Biophys. Acta* 688: 803–807, 1982.
97. PICK, U., AND S. BASSILIAN. The effects of ADP, phosphate and arsenate on Ca^{2+} efflux from sarcoplasmic reticulum vesicles. *Eur. J. Biochem.* 131: 393–399, 1983.
98. PLESNER, I. W., L. PLESNER, J. G. NORBY, AND I. KLODOS. The steady state kinetic mechanism of ATP hydrolysis catalyzed by membrane bound Na^+,K^+-ATPase from ox brain. *Biochim. Biophys. Acta* 643: 483–494, 1981.
99. POST, R. L., C. HEGYVARY, AND S. KUME. Activation by adenosine triphosphate in the phosphorylation kinetics of sodium and potassium ion transport adenosine triphosphatase. *J. Biol. Chem.* 247: 6530–6540, 1972.
100. POST, R. L., G. TODA, AND F. N. ROGERS. Phosphorylation by inorganic phosphate of sodium plus potassium ion transport adenosine triphosphatase. *J. Biol. Chem.* 250: 691–701, 1975.
101. RABON, E., H. CHANG, AND G. SACHS. Quantitation of hydrogen ion and potential gradients in gastric plasma membrane vesicles. *Biochemistry* 17: 3345–3353, 1978.
102. RABON, E., J. CUPPOLETTI, D. MALINOWSKA, A. SMOLKA, H. F. HELANDER, J. MENDLEIN, AND G. SACHS. Proton secretion by the gastric parietal cell. *J. Exp. Biol.* 106: 119–133, 1983.
103. RABON, E., R. D. GUNTHER, A. SOUMARMON, S. BASSILIAN, M. LEWIN, AND G. SACHS. Solubilization and reconstitution of the gastric H^+,K^+-ATPase. *J. Biol. Chem.* 260: 10200–10207, 1985.
104. RABON, E., T. McFALL, AND G. SACHS. The gastric H^+,K^+-ATPase H^+/ATP stoichiometry. *J. Biol. Chem.* 257: 6296–6299, 1982.
105. RABON, E., G. SACCOMANI, D. K. KASBEKAR, AND G. SACHS. Transport characteristics of frog gastric membranes. *Biochim. Biophys. Acta* 551: 432–437, 1979.
106. RABON, E., G. SACHS, S. MARDH, AND B. WALLMARK. ATP/ADP exchange activity of gastric H^+,K^+-ATPase. *Biochim. Biophys. Acta* 688: 515–524, 1982.
107. RABON, E., N. TAKEGUCHI, AND G. SACHS. Water and salt permeability of gastric vesicles. *J. Membr. Biol.* 53: 109–117, 1980.
108. RABON, E., M. WILKE, G. SACHS, AND G. ZAMPIGHI. Crystallization of the H^+,K^+-ATPase. *J. Biol. Chem.* 261: 1434–1439, 1986.
109. RATNAM, M., D. L. NGUYEN, J. RIVIER, P. B. SARGENT, AND J. LINDSTROM. Transmembrane topography of nicotinic acetylcholine receptor: immunochemical tests contradict theoret-

ical predictions based on hydrophobicity profiles. *Biochemistry* 25: 2633-2643, 1986.
110. RAY, T. K., AND J. G. FORTE. Studies on the phosphorylated intermediates of a K^+-stimulated ATPase from rabbit gastric mucosa. *Biochim. Biophys. Acta* 443: 451-467, 1976.
111. RAY, T. K., AND J. NANDI. K^+-stimulated *p*-nitrophenyl phosphatase is not a partial reaction of the gastric H^+,K^+-ATPase. Evidence supporting a new model for the univalent cation-transporting ATPase systems. *Biochem. J.* 233: 231-238, 1986.
112. REENSTRA, W. W., AND J. G. FORTE. H^+/ATP stoichiometry for the gastric H^+,K^+-ATPase. *J. Membr. Biol.* 61: 55-60, 1981.
113. REGGIO, H., D. BAINTON, E. HARMS, E. COUDRIER, AND D. LOUVARD. Antibodies against lysosomal membranes reveal a 100,000-mol-wt protein that cross-reacts with purified H^+,K^+-ATPase from gastric mucosa. *J. Cell Biol.* 99: 1511-1526, 1984.
114. REHM, W. S. The effect of electric current on gastric secretion and potential. *Am. J. Physiol.* 144: 115-125, 1945.
115. REHM, W. S. Electrophysiology of the gastric mucosa in Cl^--free solutions. *Federation Proc.* 24: 1387-1395, 1965.
116. REHM, W. S. *Proton Transport in Metabolic Transport*, edited by L. E. Hokin. New York: Academic, 1972, p. 188-241.
117. SACCOMANI, G., M. L. BARCELLONA, AND G. SACHS. Site-directed modifications of gastric H^+,K^+-ATPase in hydrogen ion transport. In: *Epithelia*, edited by I. Schulz, G. Sachs, J. Forte, and K. J. Ullrich. New York: Elsevier, 1980, p. 175-184.
118. SACCOMANI, G., M. L. BARCELLONA, AND G. SACHS. Reactivity of gastric $(H^+ + K^+)$-ATPase to *N*-ethoxycarbonyl-2-ethoxy-1,2-dihydroquinoline. *J. Biol. Chem.* 256: 12405-12410, 1981.
119. SACCOMANI, G., H. H. CHANG, A. A. MIHAS, S. CRAGO, AND G. SACHS. An acid transporting enzyme in human gastric mucosa. *J. Clin. Invest.* 64: 627-635, 1979.
120. SACCOMANI, G., H. H. CHANG, A. SPISNI, H. F. HELANDER, H. L. SPITZER, AND G. SACHS. Effect of phospholipase A_2 on purified gastric vesicles. *J. Supramol. Struct.* 11: 429-444, 1979.
121. SACCOMANI, G., L. COLE, AND E. MUKIDJAM. Interaction of photoaffinity label 8-azido ATP with the gastric H^+,K^+-ATPase. In: *Hydrogen Ion Transport in Epithelia*, edited by J. G. Forte, D. G. Warnock, and R. C. Rector. New York: Wiley-Interscience, 1984, p. 195-208.
122. SACCOMANI, G., D. W. DAILEY, AND G. SACHS. The action of trypsin on the gastric $(H^+ + K^+)$-ATPase. *J. Biol. Chem.* 254: 2821-2827, 1979.
123. SACCOMANI, G., H. F. HELANDER, S. CRAGO, H. H. CHANG, D. W. DAILEY, AND G. SACHS. Characterization of gastric mucosal membranes. X. Immunological studies of gastric $(H^+ + K^+)$-ATPase. *J. Cell Biol.* 83: 271-283, 1979.
124. SACCOMANI, G., G. SACHS, J. CUPPOLETTI, AND C. Y. JUNG. Target molecular weight of the gastric $(H^+ + K^+)$-ATPase functional and structural molecular size. *J. Biol. Chem.* 256: 7727-7729, 1981.
125. SACCOMANI, G., G. SHAH, J. G. SPENNEY, AND G. SACHS. Characterization of gastric mucosal membranes. VIII. Localization of peptides by iodination and phosphorylation. *J. Biol. Chem.* 250: 4802-4809, 1975.
126. SACCOMANI, G., H. B. STEWART, D. SHAW, M. LEWIN, AND G. SACHS. Characterization of gastric mucosal membranes: fractionation by zonal centrifugation and free flow electrophoresis. *Biochim. Biophys. Acta* 465: 311-330, 1977.
127. SACHS, G. H^+ transport by a non-electrogenic ATPase as a model for acid secretion. *Rev. Physiol. Biochem. Pharmacol.* 79: 133-162, 1977.
127a.SACHS, G. The gastric proton pump: the H^+,K^+-ATPase. In: *Physiology of the Gastrointestinal Tract*, edited by L. R. Johnson. New York: Raven, 1987, vol. 1, p. 865-881.
128. SACHS, G., T. BERGLINDH, E. RABON, H. B. STEWART, M. L. BARCELLONA, B. WALLMARK, AND G. SACCOMANI. Aspects of parietal cell biology: cells and vesicles. *Ann. NY Acad. Sci.* 341: 312-334, 1980.
129. SACHS, G., T. BERGLINDH, E. RABON, B. WALLMARK, M. BARCELLONA, H. B. STEWART, AND G. SACCOMANI. Interaction of K^+ with gastric parietal cells and gastric ATPase. *Ann. NY Acad. Sci.* 358: 118-137, 1980.
130. SACHS, G., H. H. CHANG, E. RABON, R. SCHACKMANN, M. LEWIN, AND G. SACCOMANI. A nonelectrogenic H^+ pump in plasma membranes of hog stomach. *J. Biol. Chem.* 251: 7690-7698, 1976.
131. SACHS, G., H. CHANG, E. RABON, R. SCHACKMANN, H. M. SARAU, AND G. SACCOMANI. Metabolic and membrane aspects of gastric H^+ transport. *Gastroenterology* 73: 931-940, 1977.
132. SACHS, G., J. CUPPOLETTI, J. KAUNITZ, AND D. MALINOWSKA. Aspects of H^+ translocating ATPases. In: *Gastrointestinal and Hepatic Secretions: Mechanisms and Control*, edited by J. S. Davison and E. A. Shaffer. Alberta, Canada: Univ. of Calgary Press, 1988.
133. SACHS, G., R. J. JACKSON, AND E. C. RABON. Use of plasma membrane vesicles. *Am. J. Physiol.* 238 (*Gastrointest. Liver Physiol.* 1): G151-G164, 1980.
134. SARAU, H. M., J. FOLEY, G. MOONSAMY, AND G. SACHS. Metabolism of dog gastric mucosa. II. Levels of glycolytic, citric acid cycle and other intermediates. *J. Biol. Chem.* 252: 8572-8581, 1977.
135. SARAU, H. M., J. FOLEY, G. MOONSAMY, V. D. WIEBELHAUS, AND G. SACHS. Metabolism of dog gastric mucosa. I. Nucleotide levels in parietal cells. *J. Biol. Chem.* 280: 83321-83329, 1975.
136. SCHACKMANN, R., A. SCHWARTZ, G. SACCOMANI, AND G. SACHS. Cation transport by the gastric H^+,K^+-ATPase. *J. Membr. Biol.* 32: 361-381, 1977.
137. SCHRIJEN, J. J., W. A. LUYBEN, J. J. DE PONT, AND S. L. BONTING. Studies on H^+,K^+-ATPase. I. Essential arginine residue in its substrate binding center. *Biochim. Biophys. Acta* 597: 331-344, 1980.
138. SCHRIJEN, J. J., W. A. VAN GRONINGEN-LUYBEN, J. J. DE PONT, AND S. L. BONTING. Studies on $(K^+ + H^+)$-ATPase. II. Role of sulfhydryl groups in its reaction mechanism. *Biochim. Biophys. Acta* 640: 473-486, 1981.
139. SCHRIJEN, J. J., W. A. VAN GRONINGEN-LUBEN, H. NAUTA, J. J. DE PONT, AND S. L. BONTING. Studies on $(K^+ + H^+)$ATPase. VI. Determination on the molecular size by radiation inactivation analysis. *Biochim. Biophys. Acta* 731: 9-15, 1983.
140. SERRANO, R., M. KIELLAND-BRANDY, AND G. R. FINK. Yeast plasma membrane ATPase is essential for growth and has homology with Na^+, K^+, and Ca^{2+} ATPase. *Nature Lond.* 319: 689-693, 1986.
141. SHULL, G. E., AND J. B. LINGREL. Molecular cloning of the gastric H^+,K^+-ATPase. *J. Biol. Chem.* 261: 16788-16791, 1986.
142. SHULL, G. E., A. SCHWARTZ, AND J. B. LINGREL. Amino acid sequence of the catalytic subunit of the Na^+,K^+-ATPase deduced from a complementary DNA. *Nature Lond.* 316: 691-695, 1985.
143. SKOU, J. C. The Na^+,K^+ activated enzyme system and its relationship to transport of sodium and potassium. *Q. Rev. Biophys.* 7: 401-434, 1975.
144. SKRABANJA, A. T., P. ASTY, A. SOUMARMON, J. JOEP, J. J. DE PONT, AND M. J. LEWIN. H^+ transport by reconstituted gastric H^+,K^+-ATPase. *Biochim. Biophys. Acta* 860: 131-136, 1986.
145. SKRABANJA, A. T., J. J. DE PONT, AND S. L. BONTING. H^+/ATP transport ratio of the H^+,K^+-ATPase of pig gastric membrane vesicles. *Biochim. Biophys. Acta* 774: 91-95, 1984.
146. SMITH, G. S., AND P. B. SCHOLES. The H^+/ATP stoichiometry of the H^+,K^+-ATPase of dog gastric microsomes. *Biochim. Biophys. Acta* 688: 803-807, 1982.
147. SMOLKA, A. K., H. F. HELANDER, AND G. SACHS. Monoclonal antibodies against the gastric H^+,K^+-ATPase. *Am. J. Physiol.* 245 (*Gastrointest. Liver Physiol.* 8): G589-G596, 1983.
148. SMOLKA, A., AND P. LORENTZON. In vitro synthesis of gastric proteins. In: *IUPS Symposium, Calgary*. Alberta, Canada: Univ. of Calgary Press, 1988.

149. SOUMARMON, A., M. ABASTADO, S. BONFILS, AND M. J. LEWIN. Chloride transport in gastric microsomes: an ATP-dependent influx sensitive to membrane potential and to protein kinase inhibitor. *J. Biol. Chem.* 255: 11662–11687, 1980.
150. SOUMARMON, A., AND M. J. LEWIN. Gastric H^+,K^+-ATPase. *Biochimie Paris* 68: 1287–1291, 1986.
151. SOUMARMON, A., P. RANGACHARI, AND M. LEWIN. Passive transport of Rb^+ by hog gastric H^+,K^+-ATPase. *J. Biol. Chem.* 259: 11861–11867, 1984.
152. STEWART, H. B., B. WALLMARK, AND G. SACHS. The interaction of H^+ and K^+ with the partial reactions of gastric H^+,K^+-ATPase. *J. Biol. Chem.* 256: 2682–2690, 1981.
153. SUGAI, N. A., AND S. ITO. Carbonic anhydrase, ultrastructural localization in marine gastric mucosa and improvements in technique. *J. Histochem. Cytochem.* 28: 511–525, 1980.
154. SUSSMAN, M. R., AND C. W. SLAYMAN. Modification of *Neurospora crassa* plasma membrane proton translocating ATPase with N,N'-dicyclocarbodiimide. *J. Biol. Chem.* 258: 1839–1843, 1983.
155. TOUGARD, C., D. LOUVARD, R. PICART, AND A. TIXIER-VIDAL. Antibodies against a lysosomal membrane antigen recognize a prelysosomal compartment involved in the endocytic pathway in cultured prolactin cells. *J. Cell Biol.* 100: 786–793, 1985.
156. TUUKKANEN, J., AND H. K. VÄÄNÄNEN. Omeprazole, a specific inhibitor of H^+-K^+-ATPase, inhibits bone resorption in vivo. *Calcif. Tissue Int.* 38: 123–125, 1986.
157. UEDA, S., D. LOO, AND G. SACHS. Regulation of K^+ channels in basal lateral membrane of *Necturus* oxyntic cells. *J. Membr. Biol.* 97: 31–41, 1987.
158. VASALLO, P. M., AND R. L. POST. Calcium ion as a probe of the monovalent cation center of sodium potassium ATPase. *J. Biol. Chem.* 261: 16957–16962, 1986.
159. VEGA, F. V., H. OLAISSON, AND S. MÅRDH. Distribution of carbonic anhydrase in cells and membranes isolated from pig gastric mucosa. *Acta Physiol. Scand.* 124: 573–579, 1985.
160. WALAN, A. Clinical perspectives of drugs inhibiting acid secretion—H^+-K^+-ATPase inhibitors. *Scand. J. Gastroenterol.* 125: 50–54, 1986.
161. WALDERHAUG, M. O., R. L. POST, G. SACCOMANI, R. T. LEONARD, AND D. P. BRISKIN. Structural relatedness of three ion-transport adenosine triphosphatases around their active sites of phosphorylation. *J. Biol. Chem.* 260: 3852–3859, 1985.
162. WALLACE, B. A. Evaluation of methods for prediction of membrane protein secondary structure. *Proc. Natl. Acad. Sci. USA* 83: 9423–9428, 1986.
163. WALLMARK, B. Mechanism of action of omeprazole. *Scand. J. Gastroenterol.* 118: 11–17, 1986.
164. WALLMARK, B., A. BRÄNDSTRÖM, AND H. LARSSON. Evidence for acid-induced transformation of omeprazole into an active inhibitor of H^+,K^+-ATPase within the parietal cell. *Biochim. Biophys. Acta* 778: 549–558, 1984.
165. WALLMARK, B., C. BRIVING, J. FRYKLUND, K. MUNSON, R. JACKSON, J. MENDLEIN, E. RABON, AND G. SACHS. Inhibition of gastric H^+,K^+-ATPase and acid secretion by SCH 28080, a substituted pyridyl (1,2α)-imidazole. *J. Biol. Chem.* 262: 2077–2084, 1987.
166. WALLMARK, B., E. CARLSSON, H. LARSSON, A. BRÄNDSTRÖM, AND P. LINDBERG. New inhibitors of gastric acid secretion: properties and design of H^+,K^+-ATPase blockers. In: *SCI-RSC Medicinal Chemistry Symposium 3rd, London*, edited by R. W. Lambert. 1985, p. 293–311.
167. WALLMARK, B., H. LARSSON, AND L. HUMBLE. The relationship between gastric acid secretion and gastric H^+,K^+-ATPase. *J. Biol. Chem.* 260: 13681–13684, 1985.
168. WALLMARK, B., AND S. MÅRDH. Phosphorylation and dephosphorylation kinetics of potassium-stimulated ATP phosphohydrolase from hog gastric mucosa. *J. Biol. Chem.* 254: 11899–11902, 1979.
169. WALLMARK, B., G. SACHS, S. MÅRDH, AND E. FELLENIUS. Inhibition of gastric H^+,K^+-ATPase by substituted benzimidazole, picoprazole. *Biochim. Biophys. Acta* 728: 31–38, 1983.
170. WALLMARK, B., H. B. STEWART, E. RABON, G. SACCOMANI, AND G. SACHS. The catalytic cycle of the gastric H^+,K^+-ATPase. *J. Biol. Chem.* 255: 5313–5319, 1980.
171. WOLOSIN, J. M. Ion transport studies with H^+K^+-ATPase-rich vesicles: implications for HCl secretion and parietal cell physiology. *Am. J. Physiol.* 248 (*Gastrointest. Liver Physiol.* 11): G595–G607, 1985.
172. WOLOSIN, J. M., AND J. G. FORTE. Changes in the membrane environment of the H^+,K^+-ATPase following stimulation of the gastric oxyntic cell. *J. Biol. Chem.* 256: 3149–3152, 1981.
173. WOLOSIN, J. M., AND J. G. FORTE. Kinetic properties of the KCl transport at the secreting apical membrane of the oxyntic cell. *J. Membr. Biol.* 71: 195–207, 1983.
174. WOLOSIN, J. M., AND J. G. FORTE. K^+ and Cl^- conductances in the apical membranes from secreting oxyntic cells are concurrently inhibited by divalent cations. *J. Membr. Biol.* 83: 261–272, 1985.
175. ZIBIRVE, R., G. H. FELDMAN, J. KUHNE, P. PORZNIK, G. WARNECKE, AND G. KOCH. Detection and localization of a cytoplasmic domain on the subunit of the Na^+,K^+-ATPase: a monoclonal antibody study. *J. Biol. Chem.* 262: 4349–4354, 1987.

CHAPTER 13

Intracellular activation events for parietal cell hydrochloric acid secretion

CATHERINE S. CHEW | Department of Physiology, Morehouse School of Medicine, Atlanta, Georgia

CHAPTER CONTENTS

Evidence Supporting a Role for Cyclic AMP in Control of Parietal Cell Hydrochloric Acid Secretion
Cyclic AMP–Dependent and Cyclic AMP–Independent Protein Kinases in Parietal Cells
Role of Calcium in Control of Acid Secretion
Future Perspectives

ALTHOUGH SECRETION of HCl by the stomach has been investigated for many years, studies designed to elucidate the intracellular mechanisms involved in the control of gastric acid secretion have only recently begun to yield significant results. Progress has been slow in this area because of the extreme cellular heterogeneity of the gastric mucosa. However, in the past 10 years two different hormonally responsive cellular models have become available for the study of parietal cell intracellular control mechanisms. These models include isolated gastric glands, which are small groups of parietal and chief cells with a few endocrine/paracrine cells and intact intercellular connections (14), and isolated enriched parietal cells, which range in purity from 50% to 99% (32, 123). As detailed in the chapters by Berglindh and by Forte and Soll in this *Handbook* and recent reviews (111, 129), glands and parietal cells respond to appropriate doses of the major acid secretory stimulants (histamine, gastrin, and acetylcholine) with increases in oxygen consumption and uptake of the weak base [^{14}C]aminopyrine. Both measurements indirectly estimate acid secretory activity, which cannot be determined directly in isolated cells because protons secreted by the proton pump [presumably an H^+- and K^+ activated ATPase (H^+-K^+-ATPase) (53, 78, 112)] are immediately neutralized by bicarbonate that is secreted at the basolateral membrane. In vivo neutralization of acid does not occur, because parietal cell apical and basolateral membranes are anatomically separated so that secreted bicarbonate is carried away in the portal circulation (alkaline tide) while HCl is retained in the gastric lumen. Although indirect measurements of parietal cell HCl secretion are less desirable than more direct measurements of acid secretion, many studies have shown that when conditions are carefully controlled, indirect measurements are reliable indicators of parietal cell acid secretory activity (11–13, 26, 27, 36–39, 123–125, 130). With the ability to measure such activity in isolated cells, it has become possible to correlate more accurately changes in parietal cell HCl secretion with activation of a variety of intracellular events.

EVIDENCE SUPPORTING A ROLE FOR CYCLIC AMP IN CONTROL OF PARIETAL CELL HYDROCHLORIDE ACID SECRETION

The discovery by Sutherland (133) that certain hormones activate cellular events via an increase in intracellular cAMP content strongly influenced early attempts to define the cellular mechanisms of action of histamine, acetylcholine, and gastrin. Between 1965 and 1969, Harris and colleagues (62, 63) published the earliest evidence that cAMP plays a role in the control of parietal cell HCl secretion. Their experiments with chambered amphibian gastric mucosae, in which acid secretion can be measured directly, demonstrated that addition of exogenous cAMP increased HCl secretion, whereas methylxanthines, which inhibit phosphodiesterase (the enzyme that degrades cellular cAMP), also increased secretion in chambered mucosae and elevated mucosal cAMP content. The increases in cAMP content and acid secretory activity induced by methylxanthines were temporally correlated and exhibited a similar dependency on the dose of methylxanthine that was administered (62, 63). Because methylxanthines have actions independent of their ability to elevate cellular cAMP content (35, 37, 104, 122), it was necessary to demonstrate that endogenous factors, such as histamine, also increased HCl secretion via cAMP-dependent mechanisms. This seemingly simple question proved to be most difficult to answer, and considerable controversy ensued over the role of cAMP in the control of gastric acid secretion.

From 1970 to 1979 many laboratories attempted to measure changes in cAMP in vivo after injection of secretagogues or to initiate HCl secretion by infusion of cAMP and cAMP analogues. Intravenous infusion of cAMP inhibited acid secretion in several species (80, 87); histamine was reported to increase mucosal cAMP content in vivo under some conditions in dog and rat gastric mucosa (19, 45, 97, 108), to have no effect on cAMP metabolism (42, 79, 86, 134), or to decrease mucosal cAMP content (1). With in vitro studies utilizing guinea pig and piglet gastric mucosae, histamine increased cAMP levels (23, 50, 72, 73). A linear relationship was also observed between stimulation of acid secretion by histamine and the phosphodiesterase inhibitors theophylline and IC 63197 and increases in cAMP (23). Other studies were performed to determine the effects of histamine, acetylcholine, and gastrin on the activity of adenylate cyclase, the enzyme that converts ATP to cAMP. Histamine activated mucosal adenylate cyclase in several species, including dog, guinea pig, rat, rabbit, human, bullfrog, and *Necturus* (18, 46, 49, 88, 95, 101, 109, 120, 132). In a few studies gastrin was reported to activate this enzyme (18, 81, 95). Negative results were also obtained with respect to histamine activation of adenylate cyclase in homogenates from dog, rat, and rabbit mucosae (42, 70, 86, 137, 138). Furthermore both histamine H_1- and H_2-receptor antagonists were reported to inhibit histamine-stimulated gastric mucosal adenylate cyclase activity (4, 9, 97, 132), whereas inhibitors of acid secretion such as prostaglandin E_1 (PGE_1) and secretin were found to activate this enzyme (7-9, 47, 70, 137, 138). During this time reviews were written both in favor of (77) and opposing (70) a role for cAMP in acid secretion.

The major reason for most of the controversy surrounding a role for cAMP as a second messenger in the control of gastric acid secretion arose because of the heterogeneity of the gastric mucosa. Parietal cells comprise only 10% of the fundic mucosa. Therefore changes in cAMP in this cell are often not readily detectable when the entire gastric mucosa is used, because agonist/antagonist elevation of cAMP in other cell types may occur. With the development of techniques to prepare hormonally responsive isolated glands and parietal cells, it has become possible to measure more directly the effects of agonists on parietal cell cAMP metabolism. Most data obtained from these cellular models support a role for cAMP in the acid secretory process. Histamine increases cAMP and activates adenylate cyclase in isolated gastric glands and enriched parietal cell preparations (7, 36, 37, 39, 83, 114, 130); histamine-induced increases in cAMP content, adenylate cyclase activity, and acid secretion (measured as accumulation of the weak base [^{14}C]-aminopyrine and increased oxygen consumption) are activated with similar 50% effective concentrations (EC_{50}). All of these histamine-stimulated activities are inhibited by the H_2-receptor antagonist cimetidine. Comparisons of histamine dose-response curves generated in the presence of varying cimetidine concentrations indicated that the affinity of cimetidine for the histamine receptor (pA_2 value) was the same for histamine-stimulated increases in cAMP, adenylate cyclase, [^{14}C]aminopyrine accumulation, and oxygen consumption (39). In contrast, inhibition of histamine-stimulated [^{14}C]aminopyrine uptake by the H_1-receptor antagonist mepyramine occurs only at concentrations well above those reported to be maximally active in tissues responding to H_1-receptor agonists. Moreover the concentrations of mepyramine required to inhibit the secretory response to histamine also inhibit the response to the cAMP analogue dibutyryl cAMP (Bt_2cAMP) (39, 129). These data indicate that H_1-receptor antagonists inhibit parietal cell HCl secretion independent of a specific effect on histamine receptors.

Initially conflicting results that indicated that the acid secretory inhibitor PGE_2 has an excitatory rather than inhibitory action of adenylate cyclase in homogenates from gastric mucosae were clarified by experiments with isolated canine and rat mucosal cell preparations in which PGE_2 activated adenylate cyclase and increased cAMP content in cells other than parietal cells (83, 126, 128, 130). Experiments with isolated enriched parietal cells and glands have also demonstrated that neither gastrin nor acetylcholine has a direct stimulatory effect on adenylate cyclase activity in parietal cells (7, 8, 37, 38, 130). Epinephrine, which inhibits gastric acid secretion in vivo, has no detectable effect on parietal cell cAMP metabolism in rabbit and dog but appears to increase parietal cAMP content and [^{14}C]aminopyrine accumulation in the rat (107, 110). Whether species differences in β-adrenergic control of parietal cell HCl secretion can explain these conflicting results remains to be determined.

Other evidence supporting a role for cAMP in acid secretion includes the observations that Bt_2cAMP, 8brcAMP, forskolin, and cholera toxin stimulate acid secretory responses in isolated parietal cells and glands (7, 12, 26, 27, 39, 65, 130, 135). Dibutyryl cAMP also stimulates HCl secretion in the chambered amphibian gastric mucosa, in which acid secretion can be measured directly (62, 143). Forskolin, a diterpene isolated from *Coleus forskohlii* (116), elevates cellular cAMP content in many different cell types by a direct activation of the catalytic subunit and/or an action on regulatory subunit(s) of this enzyme (6, 57, 131). Stimulation of gastric glands with this agent results in increased cAMP concentrations and acid secretory responses. The acid secretory response to maximal concentrations of forskolin is not inhibited by the histamine H_2-receptor blocker cimetidine (27, 65, 135); however, there is a parallel rightward shift in the forskolin dose-response curve in gastric glands on addition of cimetidine (27). These data suggest that forskolin potentiates the action of endogenous histamine present in glands. Cholera toxin has been shown to activate adenylate cyclase irreversibly by stimulating NADH-dependent ADP ribosylation of a regulatory component (G_s, where s is stimulatory activity) of this enzyme (54, 55). Cholera toxin also increases cAMP content in glands with a time course that is correlated with the increase in [^{14}C]aminopyrine accumulation (26).

Presently a limited amount of data suggests that regulatory components are involved in the control of secretagogue-induced activation of parietal cell adenylate cyclase. In other systems activity of the catalytic portion of adenylate cyclase is influenced by stimulatory and inhibitory proteins identified variously as G_s and G_i or N_s and N_i, by different authors, where s and i are stimulatory and inhibitory activity, respectively (41, 55). Both G_s and G_i possess GTPase activity and bind guanyl nucleotides. Both G_s and G_i are heterotrimers composed of three subunits, α, β, and γ. The α-subunit of G_s has a molecular weight of 45,000 or 52,000, whereas that of G_i is 41,000; the β-subunits appear identical (molecular weight 35,000–36,000), whereas the γ-subunits are peptides with molecular weights of 5,000–10,000 (55, 66, 85). Cholera and pertussis toxins stimulate ADP ribosylation of α-subunits of G_s and G_i, respectively. Similar results have been obtained with parietal cells (22). The cholera toxin experiments with glands (26) also indicate that parietal cells possess a regulatory subunit similar to G_s. Other evidence indicating the presence of G_i in parietal cells is derived from experiments in which pretreatment of glands with pertussis toxin enhanced histamine-stimulated [^{14}C]aminopyrine uptake and reduced PGE$_2$ inhibition of acid secretory responses (Fig. 1). In other cell types, pertussis toxin appears to block the action of certain inhibitory hormones by stimulating ADP ribosylation of the α-subunit of G_i (20, 76, 94). A role for regulatory subunits in the control of parietal cell adenylate cyclase activity is also supported by observations that the stable analogue of GTP, Gpp(NH)p, enhances histamine-stimulated adenylate cyclase activity in gastric cell homogenates (10).

CYCLIC AMP–DEPENDENT AND CYCLIC AMP–INDEPENDENT PROTEIN KINASES IN PARIETAL CELLS

Protein kinases activated by cAMP were first isolated from skeletal muscle by Walsh et al. (142). It is now established that cAMP-dependent protein kinases are present in all eucaryotic cells and that the holoenzyme is a tetramer composed of two catalytic and two regulatory subunits. Cyclic AMP activates the kinase by binding to the regulatory subunits, causing a dissociation of the inhibitory subunits from the

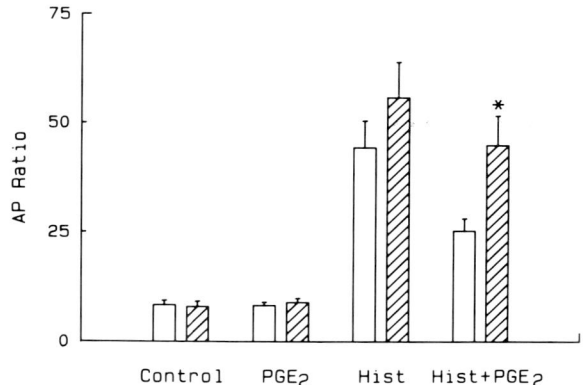

FIG. 1. Effect of islet-activating protein and pertussis toxin on PGE$_2$ inhibition of histamine-stimulated [^{14}C]aminopyrine (AP) accumulation in rabbit gastric glands. Glands were preincubated 4 h at 37°C with 250 ng/ml pertussis toxin (hatched bars) or no additions (open bars), rinsed, then stimulated in the presence of [^{14}C]aminopyrine for 45 min with 1 μM histamine ± 1 μM PGE$_2$. [^{14}C]aminopyrine accumulation was determined at the end of the 45-min incubation period. Values are means ± SE for 5 experiments. [Data from Brown and Chew (22).]

catalytic portion of the enzyme. The ratio of enzyme activity in cellular homogenates measured in the presence and absence of exogenous cAMP provides the basis for quantitation of activation of the enzyme in previously stimulated intact cells. At least two cAMP-dependent protein kinase isozymes, types I and II (identified initially by order of elution from anion-exchange columns, rate of dissociation in the presence of high-salt concentrations, and regulatory subunits of different molecular weights), exist and are found in varying concentrations in different cell types and species (50, 115). The only known action of these kinases is phosphorylation of cellular proteins (58).

A limited number of studies have been performed to define the role of cAMP-dependent protein kinases in the control of parietal cell HCl secretion. Reimann and Rapino (103) isolated both type I and type II isozymes from the soluble fraction of homogenates from rabbit gastric mucosa; Corbin et al. (43) found mainly type II isozyme in pig mucosa. In an earlier study, Ray and Forte (102) reported the presence of both soluble and membrane-bound cAMP-dependent protein kinases in rabbit gastric mucosa. Histamine activation of soluble cAMP-dependent protein kinases has been demonstrated in canine gastric mucosa (67), gastric mucosal cells from guinea pig (84), and rabbit parietal cells (28). Forskolin has been shown to activate this enzyme in rabbit gastric glands (27). A decrease in binding of 8-azido-[^{32}P]cAMP, a photoaffinity label that binds cAMP-dependent protein kinase regulatory subunits with high affinity (60), has also been observed in rabbit gastric glands and parietal cells after histamine stimulation (28, 68). Such data are compatible with observations that histamine activates the holoenzyme by increasing cellular cAMP content. Cyclic AMP binds to the regulatory subunit of the holoenzyme form of cAMP-dependent protein kinase, causing a dissociation of the enzyme into free catalytic and regulatory subunits. Endogenous cAMP bound to regulatory subunits competes with the photoaffinity label for cAMP-binding sites. The net result is decreased binding of the photoaffinity label with increased cellular cAMP content.

With histamine and forskolin, cAMP-dependent protein kinase activation has been shown to precede [^{14}C]aminopyrine accumulation and to exhibit a dose-response relationship similar to that for acid secretory response parameters observed in glands and parietal cells [Fig. 2; (27, 28)]. Recent evidence suggests that histamine may selectively activate a cAMP-dependent protein kinase type I isozyme in soluble fractions of rabbit parietal cells. Both type I and type II isozymes were detected. However, type II was present in both particulate and soluble cell fractions, whereas type I was found only in the soluble fraction (28). Similar patterns of isozyme distribution have been reported in heart, where the type II isoform was detected in both particulate and soluble fractions and type I was found only in the soluble fraction. In these studies the free catalytic subunit of cAMP-dependent protein kinase bound to the particulate fraction when NaCl was omitted from the homogenizing medium but not when NaCl (0.1 M) was included. Increasing the salt concentration has no effect on the amount of type II regulatory subunit present in the particulate fraction (43, 50). Similar results have been obtained with rabbit parietal cell homogenates (28).

It is not clear whether the salt-sensitive binding of free catalytic subunit to particulate cellular material is the result of a hormonally controlled translocation of the subunit from cytosol to some membrane site or whether this binding is artifactual due to low-salt conditions. Because the type II subunits are not displaced from particulate fractions by relatively high-salt concentrations, these subunits may be specific membrane components. Such data suggest that there may be cellular compartmentalization of cAMP-dependent protein kinase isozymes. Furthermore the activity of these isozymes may be under separate hormonal control.

In addition to cAMP-dependent protein kinases, several other protein kinases have been detected in parietal cell extracts, including Ca^{2+}-phospholipid–dependent protein kinase or protein kinase C (3, 28) and casein kinases (28). Calmodulin-dependent protein kinase activity has been measured only in basolateral membranes from rabbit gastric mucosa (93). The role, if any, of these protein kinases in the control of acid secretion remains to be defined. There is recent evidence that suggests that protein kinase C can regulate parietal cell acid secretory activity in vitro. In experiments with 12-O-tetradecanoylphorbol-13-acetate (TPA), a phorbol ester that activates protein kinase C in a number of cell types, including parietal cells (3, 28), TPA increased parietal cell oxygen consumption and stimulated a transient (albeit weak) increase in [^{14}C]aminopyrine accumulation. The stimulatory action of Bt_2cAMP and 8brcAMP (another permeable cAMP analogue) on [^{14}C]aminopyrine uptake and oxygen consumption was also potentiated by TPA. However, when TPA was added either prior to or after histamine, forskolin, or carbachol (an acetylcholine analogue), a time-dependent inhibition was observed (22). The H_2-receptor antagonist cimetidine was added with all agonists except histamine. Therefore the inhibitory effects of TPA on forskolin and carbachol were independent of potentiating interactions between these agonists and endogenous histamine. Such data suggest that if protein kinase C does play a role in the control of acid secretion, there may be both membrane-directed and postreceptor activities associated with activation of this enzyme.

In other studies with rat parietal cells, TPA inhibited both Bt_2cAMP and histamine-stimulated [^{14}C]aminopyrine uptake (2). Because a portion of the secretory response to Bt_2cAMP is inhibited by H_2-receptor antagonists, it may be that the inhibitory effect of TPA on [^{14}C]aminopyrine uptake observed to rat parietal cells was actually associated with inhibition of the histaminergic component of the response. Although these preliminary studies are interesting, much work must be done in this area before a true physiological function can be ascribed to either protein kinase C or other cAMP-independent protein kinases(s).

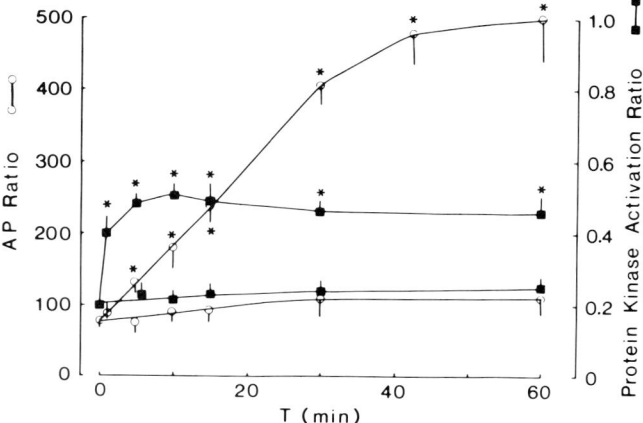

FIG. 2. Time course of activation of cAMP-dependent protein kinase and [^{14}C]aminopyrine (AP) accumulation in response to histamine. Parietal cells from same preparations were preincubated for 30 min at 37°C and then stimulated with 10^{-4} M histamine. Aliquots of cells were removed at indicated times for determination of [^{14}C]aminopyrine uptake and cAMP-dependent protein kinase activity. Values are means ± SE for 5 experiments. *Lower lines* in each group, basal values. *$P < 0.01$, significantly different from basal. [From Chew (28).]

ROLE OF CALCIUM IN CONTROL OF ACID SECRETION

There have been intermittent attempts to demonstrate a physiological role for Ca^{2+} in the control of parietal cell HCl secretion for at least 60 years (56, 90). Until recently the major test used to determine whether Ca^{2+} is necessary for the initiation of acid secretion was to measure the acid secretory responses of chambered amphibian gastric mucosae to secretagogues under conditions in which medium Ca^{2+} was reduced by addition of Ca^{2+} chelators. Prolonged exposure of mucosae to such media decreases the transmucosal electrical potential difference and increases conductance. These changes appear to be due to increased permeability of intercellular tight junctions. Because there is backflux of HCl under these conditions, it is not possible to determine whether medium Ca^{2+} removal directly affects parietal cell activity (52, 69). When medium Ca^{2+} concentrations are briefly reduced, there is an increase in tissue resistance, which has been postulated to be associated with a direct inhibitory effect on parietal cells (69, 117). Replacement of Ca^{2+} in the serosal bathing medium restores normal secretory activity; replacement of mucosal Ca^{2+} partially restores secretory function. It has been suggested that Ca^{2+} that is added to the mucosal solution prevents disruption of tight junctions but does not penetrate into mucosal cells (69).

Prolonged chelation of Ca^{2+} in the serosal bathing solution attenuates responses to all secretory stimulants, including Bt_2cAMP. When serosal Ca^{2+} is chelated for shorter times, acid secretion in response to Bt_2cAMP is partially inhibited, but there appears to be complete inhibition of responses to histamine, carbachol, and gastrin. Such data have led to the proposal that Ca^{2+} may be involved in two separate cellular events, the first associated with activation of adenylate cyclase and the second with a step beyond cAMP elevation (74, 75). In other experiments with chambered amphibian gastric mucosae, replacement of Ca^{2+} in the mucosal solution with lanthanum prevented histamine-induced HCl secretion but suppressed incompletely the response to supramaximal concentrations of histamine + Bt_2cAMP + isobutylmethylxanthine (IBMX) (82). A possible explanation for these results is that, even in the presence of lanthanum, cellular tight junctions are sufficiently leaky to allow backflux of H^+ ions. With supramaximal stimulation, sufficient HCl is secreted to be detectable despite H^+ ion backflux.

The use of isolated cellular models avoids problems associated with disruption of mucosal tight junctions. With parietal cells and gastric glands, chelation of medium Ca^{2+} significantly decreases acid secretory responses to carbachol but has only partial or no inhibitory effect on histamine-stimulated secretion (15, 29, 127). In gastric glands there is a biphasic effect of Ca^{2+} removal on [^{14}C]aminopyrine uptake in response to carbachol. The initial transient rise in [^{14}C]aminopyrine uptake is completely blocked by medium Ca^{2+} chelation, whereas the smaller steady-state increase in [^{14}C]aminopyrine accumulation is still present. Potentiating interactions between carbachol and histamine are detectable under these conditions but display a temporal delay (29). One interpretation of these results is that extracellular Ca^{2+} is required for only a portion of the acid secretory response.

It has not been possible to detect HCl secretion in isolated parietal cells with Ca^{2+} ionophores, presumably because these agents initiate a Ca^{2+}-H^+ exchange. In chambered amphibian gastric mucosae, the ionophore A23187 has been reported to stimulate HCl secretion (71). The response to A23187 was partially blocked by the H_2-receptor antagonist metiamide, which suggests that A23187 causes release of histamine from the gastric mucosa. A second component of the response did not appear to be associated with histamine release because addition of A23187 along with maximal stimulatory concentrations of histamine resulted in further stimulation of acid secretion. There is no simple explanation for the different effects of ionophores on isolated cells and mucosal preparations.

Recent evidence suggests that at least some of the effects of Ca^{2+} on cellular activities may be mediated by the heat-stable, acidic Ca^{2+}-binding protein, calmodulin (89). Attempts to define a role for this protein in the control of gastric secretion have, however, not been successful. Phenothiazine antagonists of calmodulin, such as trifluoperazine and chlorpromazine, potently inhibit acid secretion in chambered amphibian gastric mucosae and isolated parietal cells (5, 67a, 100); unfortunately, as in other cell types, these antagonists appear to interact with cellular enzymes that have no detectable Ca^{2+} dependency (67a). More definitive experiments are required before it can be determined whether calmodulin is active in the control of HCl secretion.

The Ca^{2+} fluxes in partially purified parietal cells and gastric glands have been measured with $^{45}Ca^{2+}$. In canine parietal cells enriched to an average purity of 50%, increased influx of $^{45}Ca^{2+}$ occurred on addition of the acetylcholine analogue carbachol but not histamine or gastrin. No increased efflux in response to any of the three secretagogues was detected (127). In rabbit gastric glands, which contain ~50% parietal and 50% chief cells (13), increased influx of $^{45}Ca^{2+}$ in response to carbachol was detected only under conditions in which intracellular Ca^{2+} content was decreased by preincubation of glands in phosphate-free medium containing inosine. With depletion of cellular Ca^{2+} content, carbachol increased both influx and efflux of $^{45}Ca^{2+}$ (93). More recently, changes in free intracellular Ca^{2+} ([Ca^{2+}]$_i$) in parietal cells and glands have been measured with the fluorescent intracellular Ca^{2+} indicators quin 2 and fura 2 (59, 140). With these indicators carbachol has been shown to elevate intracellular Ca^{2+} content in parietal cells from dog and rabbit (30-33, 91-93). Carbachol and cholecystokininoctapeptide (CCK-8) increased [Ca^{2+}]$_i$ in >95% pure parietal cells and in enriched chief cells. In contrast, gastrin was equipotent with CCK-8 in increasing parietal cell [Ca^{2+}]$_i$ but had no detectable effect on chief cell $[Ca^{2+}]_i$ at similar concentrations (32). These results agree with observations that gastrin and CCK-8 are equipotent in the stimulation of parietal cell [^{14}C]aminopyrine uptake but not chief cell pepsinogen release (64).

Difficulties have been encountered in attempts to define Ca^{2+} entry pathways in parietal cells through the use of Ca^{2+} channel blockers. Concentrations of Ca^{2+} blockers far in excess of those that inhibit Ca^{2+} entry into excitable cells are required to suppress [^{14}C]aminopyrine accumulation in parietal cells (29, 32, 92). Furthermore Ca^{2+} antagonists such as nicardipine and verapamil have no effect on histamine- and forskolin-stimulated parietal cell oxygen consumption at concentrations that totally block [^{14}C]-aminopyrine uptake (29). In other experiments, the Ca^{2+} channel blockers nicardipine and nifedipine had no effect on carbachol-induced changes in $[Ca^{2+}]_i$ in parietal cells at concentrations as high as 10^{-5} M (29–32). At these same concentrations nicardipine strongly inhibits [^{14}C]aminopyrine accumulation in response to both histamine and carbachol. In contrast, nifedipine has no effect on [^{14}C]aminopyrine uptake in response to either agonist (29). These data suggest that Ca^{2+} entry into parietal cells is not controlled by voltage-operated channels. Therefore results obtained with presently available Ca^{2+} channel antagonists should be interpreted with caution.

The carbachol-induced rise in $[Ca^{2+}]_i$ is prevented by 8-(N,N-diethylamino)-octyl-3,4,5-trimethoxybenzoate (TMB 8), a putative antagonist of intracellular Ca^{2+} release (40), but TMB 8 has no effect on gastrin or CCK-induced increases in intracellular Ca^{2+} in gastric glands (30). These observations and the knowl-

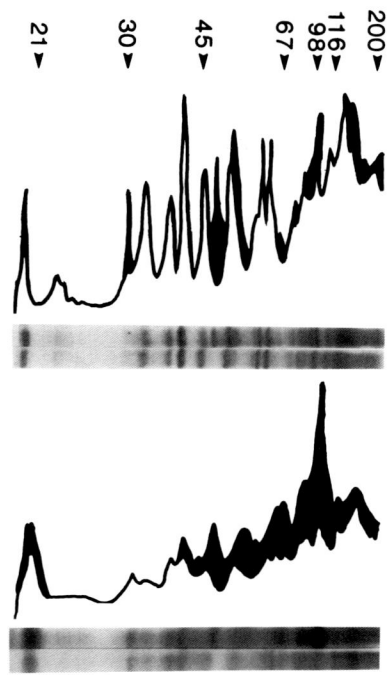

FIG. 4. Comparison of Ca^{2+}-dependent protein phosphorylation patterns in parietal and chief cells from rabbit gastric mucosa. Cells were enriched on Nycodenz gradients followed by centrifugal elutriation. Supernatants (50,000 g) of homogenates of each cell type were prepared and assayed for Ca^{2+}-dependent protein kinase activity. Phosphorylated proteins were identified with sodium dodecyl sulfate–polyacrylamide electrophoresis (SDS-PAGE) followed by autoradiography. *Top*: superimposed densitometric tracings of phosphorylated proteins from parietal cells (99% parietal). *Bottom*: chief cell (<5% parietal) protein phosphorylation. *Darkened areas*, increased phosphorylation due to Ca^{2+} addition. Autoradiographs of control (*lower lane*) and Ca^{2+}-treated (*upper lane*) homogenates are aligned with densitometric tracings for direct comparisons. Molecular weights (×10^3) are indicated at top of figure. Note differences in phosphorylation patterns in 2 cell types. (C. S. Chew and M. R. Brown, unpublished observations.)

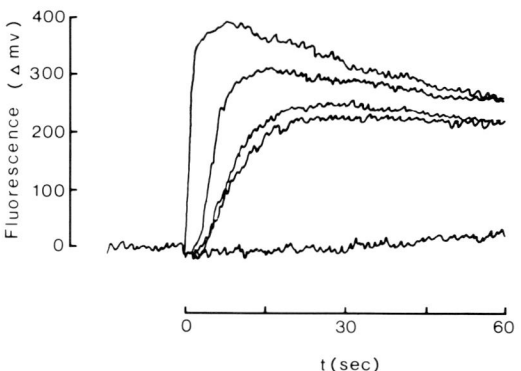

FIG. 3. Time course of changes in parietal cell free intracellular Ca^{2+} ($[Ca^{2+}]_i$) in response to several agonists that elevate cAMP content, as compared with carbachol, which does not increase cAMP. Parietal cells (97% enriched) were loaded with 2 µM fura 2 AM, the membrane-permeant acetomethylester form of the fluorescent Ca^{2+} indicator, fura 2, for 20 min at 37°C, rinsed, and then stimulated with agonists. Tracings of responses to different agonists were superimposed for direct comparisons. *Left to right*: carbachol (10^{-4} M), histamine (10^{-4} M), forskolin (10^{-4} M), isobutylmethylxanthine (10^{-4} M). Neither 8-brcAMP (5 × 10^{-4} M) nor Bt$_2$cAMP (1 mM) had any detectable effect on the fura 2 signal (*lower tracing*). [From Chew and Brown (33).]

edge that TMB 8 and several Ca^{2+} channel blockers antagonize acetylcholine-receptor binding (136, 139) support the conclusion that presently available Ca^{2+} blockers do not specifically block Ca^{2+} fluxes in parietal cells.

There are two opposing views regarding the source of carbachol-induced increases in $[Ca^{2+}]_i$. The first proposes that there is no release of intracellular Ca^{2+} upon carbachol stimulation (93). The second suggests that the initial rapid increase in parietal cell $[Ca^{2+}]_i$ is the result of agonist-induced release of Ca^{2+} from some site(s) within the cell, whereas the lower, steady-state increase in $[Ca^{2+}]_i$ is due to movement of Ca^{2+} across the cell membrane (31, 32). Recent evidence that supports the second hypothesis includes the observation that acute chelation of extracellular Ca^{2+} does not prevent the carbachol-induced increase in $[Ca^{2+}]_i$ in >95% pure parietal cells. Carbachol also increases inositol trisphosphate (IP$_3$) production in parietal cells (32). In several cell types, IP$_3$ has been shown to cause release of intracellular Ca^{2+} [see the chapters by Putney, by Schulz, and by Williams, Burnham, and Hootman in this *Handbook*; (16, 118)]. Acetylcholine-induced increases in IP_3 appear to be associated with M_2- but not M_1-type cholinergic receptors (21). Parietal cell cholinergic receptors exhibit M_2-type characteristics (106).

A surprising finding is that histamine increases parietal cell $[Ca^{2+}]_i$. This action of histamine is blocked by the H_2-receptor antagonist cimetidine but not by equimolar concentrations of H_2-receptor antagonists (30–33). Because forskolin also elevates parietal cell $[Ca^{2+}]_i$, the increase in $[Ca^{2+}]_i$ may be associated with increased cellular cAMP content. An alternative explanation is that histamine and forskolin elevate $[Ca^{2+}]_i$ via an action on G_s or some other regulatory subunit. In partially purified parietal cells, no significant effect of histamine on IP_3 levels was detected (32). With increased parietal cell enrichment, however, histamine stimulated a small increase in IP_3 concentrations (33). It remains to be determined whether forskolin has a similar effect on parietal cell inositol phosphate metabolism. Because neither Bct_2cAMP nor 8-brcAMP appears to mimic the effect of histamine $[Ca^{2+}]_i$, the second hypothesis is more tenable than the first. The increase in $[Ca^{2+}]_i$ in re-

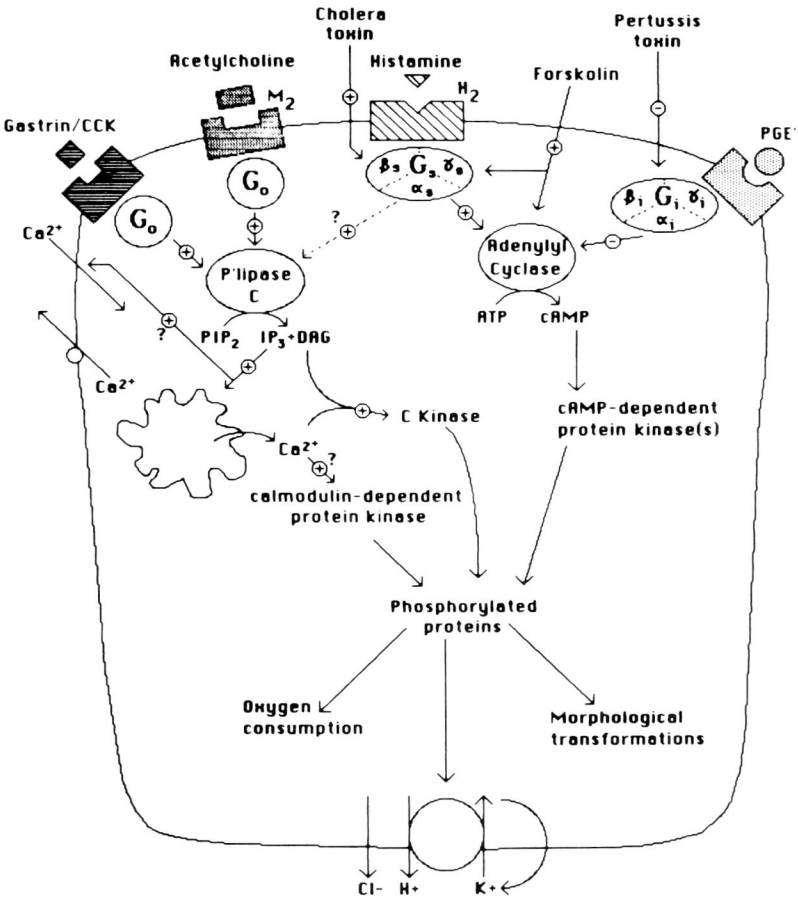

FIG. 5. Model of neural, endocrine, paracrine control mechanisms in parietal cell HCl secretion. Histamine binds H_2 receptors, causing release of α-subunits from G_s, a GTP-binding protein with stimulatory activity. Forskolin also causes release of α-subunits from G_s. The α-subunits activate adenylate cyclase and phospholipase C (P'lipase C). Result is increased cellular cAMP concentration and breakdown of phosphatidylinositol 1,4-bisphosphate (PIP_2) to form inositol 1,4,5-trisphosphate (IP_3) and diacylglycerol (DAG). Cholera toxin elevates cAMP by NAD-dependent ribosylation of G_s; prostaglandins inhibit histamine-stimulated activation of adenylate cyclase by causing release of inhibitory α-subunits from G_i, a GTP-binding protein with inhibitory activity. Release of α_i-subunits is blocked by pertussis toxin-induced NAD-dependent ribosylation of G_i. Acetylcholine and gastrin activate phospholipase C and increase IP_3 by activating other GTP-dependent regulatory components (G_o) that may also contain α-subunits (α_o). The α_o-subunits could also have actions independent of phospholipase C activation. Agonist-induced increases in $[Ca^{2+}]_i$ and cAMP lead to activation of Ca^{2+} and cAMP-dependent protein kinases, which then phosphorylate a variety of intracellular proteins. Increased protein phosphorylation may lead to morphological transformation that precede HCl secretion and increased cellular metabolism that parallels acid secretory activity. Protein phosphorylation may directly or indirectly regulate the H^+-K^+-ATPase and/or other cellular transport processes.

sponse to both histamine and forskolin is slower than that observed with carbachol stimulation (Fig. 3). These data suggest that additional intracellular events may be interposed between histamine-induced increases in $[Ca^{2+}]_i$ as compared with carbachol.

Glucagon, which increases cAMP content in hepatocytes, also increases $[Ca^{2+}]_i$ in this cell type (25, 121). As with histamine, the time course of the rise in $[Ca^{2+}]_i$ in response to glucagon is slower than that observed with agonists that do not act via cAMP (25). Both cAMP-dependent agonists, glucagon and histamine, transiently increase cellular autofluorescence in hepatocytes and gastric glands, respectively (25, 30, 31). This increase in cellular fluorescence is maximal at wavelengths that are optimal for detection of reduced NAD(P). These data suggest that at least some cAMP-dependent agonists may act at the same metabolic sites as agonists that elevate $[Ca^{2+}]_i$ but not cAMP.

FUTURE PERSPECTIVES

Present evidence suggests that cAMP plays an important role in the control of gastric acid secretion. Considerably more data must be accumulated, however, before a definitive role for Ca^{2+} in this process can be established. In many tissues, hormones that increase cellular cAMP or free Ca^{2+} concentrations also cause increased phosphorylation of cellular proteins (58). Calcium-dependent phosphorylation of proteins in membranes from rat gastric mucosa that retain H^+-K^+-ATPase activity has been detected (113). There is also Ca^{2+}-dependent phosphorylation and dephosphorylation of parietal cell cytosolic proteins (Fig. 4). Agonists such as carbachol that elevate parietal cell $[Ca^{2+}]_i$ but not cAMP content have not yet been shown to alter the phosphorylation states of parietal cell proteins. With all agonists the ability to demonstrate reproducible changes in phosphorylation has been hampered by the presence of a large number of phosphorylated proteins in unstimulated cells. Recent data suggest, however, that it may be possible to detect histamine-stimulated increases in protein phosphorylation (34, 44, 61). With improved techniques, it may become possible to detect increased phosphorylation with other agonists as well. It would be most interesting if different agonists, such as histamine and acetylcholine, overlap in their ability to phosphorylate one or more parietal cell proteins. Possible cellular functions that could be affected by protein phosphorylation include intermediary metabolism, membrane transport, and/or potentiating interactions between histamine and gastrin and histamine and acetylcholine. Parietal cells are rich in actin and contain other cytoskeletal proteins (51, 141, 147). Morphological transformations that precede HCl secretion may also be initiated by secretagogue-induced changes in cAMP and/or Ca^{2+} followed by activation of protein kinase(s) and phosphorylation of specific cytoskeletal proteins.

A more global hypothesis is that the pleiotypical responses induced by different agonists are not mediated by simple unitary increases in the second messengers cAMP and Ca^{2+}. As Rodbell (105) recently proposed, hormones could interact with receptor–GTP-binding protein (RG) regulatory complexes, causing release of α-subunits from these complexes. Released α-subunits could then be transformed into new structures by multiple modifying enzymes, with the new subunits activating different effector enzymes. Secondary signals, such as cAMP generation, would be initiated by an action of the altered α-subunits on various effector enzymes. In this model the ultimate pleiotypical cellular response depends on the number of secondary signals initiated by activated α-subunits. If, for example, there is some overlap in the activities of subunits produced by histamine and carbachol RG complexes, a potentiating interaction between these two agonists could result. Negative interrelationships between regulatory subunits of adenylate cyclase and Ca^{2+} metabolism have been described in mast cells and neutrophils (96, 98). Positive interactions may also occur. Experiments with cholera and pertussis toxins indicate that parietal cells possess both G_s and G_i regulatory subunits. Furthermore, alterations of intracellular Ca^{2+} content by histamine and forskolin but not cAMP analogues (Fig. 5) suggest that other types of G subunits may also be present in parietal cells. A model of possible activation mechanisms in parietal cells is depicted in Figure 5.

REFERENCES

1. AMER, M. S. Cyclic GMP and gastric acid secretion. *Am. J. Dig. Dis.* 19: 71–74, 1974.
2. ANDERSON, N. G., AND P. J. HANSON. Inhibition of gastric acid secretion by a phorbol ester: effect of 12-O-tetradecanoyl phorbol-13-acetate on aminopyrine accumulation by rat parietal cells. *Biochem. Biophys. Res. Commun.* 121: 566–572, 1984.
3. ANDERSON, N. G., AND P. J. HANSON. Involvement of calcium-sensitive phospholipid-dependent protein kinase in control of acid secretion by isolated rat parietal cells. *Biochem. J.* 232: 609–611, 1985.
4. ANTILLA, P., C. LUCKE, AND E. WESTERMANN. Stimulation of adenylate cyclase of guinea pig gastric mucosa by histamine, sodium fluoride and 5-guanylimidophosphate and inhibition by histamine H_1- and H_2-receptor antagonists in vitro. *Naunyn-Schmiedeberg's Arch. Pharmacol.* 296: 31–36, 1976.
5. BARBOZA, R., AND J. CHACIN. Effects of phenothiazines on acid secretion in the toad gastric mucosa. *Comp. Biochem. Physiol. C Comp. Pharmacol.* 81: 419–424, 1985.
6. BAROVSKY, K., AND G. BROOKER. Forskolin potentiation of cholera toxin-stimulated cyclic AMP accumulation in intact C6-2B cells. Evidence for enhanced G_s-C coupling. *Mol. Pharmacol.* 28: 502–507, 1985.
7. BATZRI, S., AND J. DYER. Aminopyrine uptake by guinea pig gastric mucosal cells. Mediation by cyclic AMP and interactions among secretagogues. *Biochim. Biophys. Acta* 675: 416–426, 1981.

8. BATZRI, S., AND J. D. GARDNER. Cellular cyclic AMP in dispersed mucosal cells from guinea pig stomach. *Biochim. Biophys. Acta* 541: 181–189, 1978.
9. BATZRI, S., AND J. D. GARDNER. Action of histamine on cyclic AMP in guinea pig gastric cells: inhibition by H_1- and H_2-receptor antagonists. *Mol. Pharmacol.* 16: 406–416, 1979.
10. BEARER, C. F., L. K. CHANG, G. C. ROSENFELD, AND W. J. THOMPSON. Histamine stimulation of rat gastric parietal cell adenylyl cyclase: modulation by guanine nucleotides. *Arch. Biochem. Biophys.* 207: 325–336, 1981.
11. BERGLINDH, T. Effects of common inhibitors of gastric acid secretion on secretagogue induced respiration and aminopyrine accumulation in isolated gastric glands. *Biochim. Biophys. Acta* 464: 217–233, 1977.
12. BERGLINDH, T. Potentiation by carbachol and aminophylline of histamine- and db-cAMP-induced parietal cell activity in isolated gastric glands. *Acta Physiol. Scand.* 99: 75–84, 1977.
13. BERGLINDH, T., H. F. HELANDER, AND K. J. OBRINK. Effects of secretagogues on oxygen consumption, aminopyrine accumulation and morphology in isolated gastric glands. *Acta Physiol. Scand.* 97: 401–414, 1976.
14. BERGLINDH, T., AND K. J. OBRINK. A method for preparing isolated glands from the rabbit gastric mucosa. *Acta Physiol. Scand.* 96: 150–159, 1976.
15. BERGLINDH, T., G. SACHS, AND N. TAKEGUCHI. Ca^{2+}-dependent secretagogue stimulation in isolated gastric glands. *Am. J. Physiol.* 239 (*Gastrointest. Liver Physiol.* 2): G90–G94, 1980.
16. BERRIDGE, M. J., AND R. F. IRVINE. Inositol trisphosphate, a novel second messenger in cellular signal transduction. *Nature Lond.* 312: 315–321, 1984.
18. BERSIMBAEV, R. I., S. V. ARGUTINSKAYA, AND R. I. SALGANIK. The stimulating action of gastrin pentapeptide and histamine on adenyl cyclase activity in rat stomach. *Experentia Basel* 27: 1389–1390, 1971.
19. BIECK, P. R., J. A. OATES, G. A. ROBISON, AND R. B. ADKINS. Cyclic AMP in the regulation of gastric secretion in dogs and humans. *Am. J. Physiol.* 224: 158–164, 1973.
20. BOKOH, G. M., T. KATADA, J. K. NORTHUP, E. C. HEWLETT, AND A. G. GILMAN. Identification of the predominant substrate for ADP-ribosylation by islet activating protein. *J. Biol. Chem.* 258: 2072–2075, 1983.
21. BROWN, J. H., D. GOLDSTEIN, AND S. B. MASTERS. The putative M_1 muscarinic receptor does not regulate phosphoinositide hydrolysis. *Mol. Pharmacol.* 27: 525–531, 1985.
22. BROWN, M. R., AND C. S. CHEW. Multiple effects of phorbol ester on secretory activity in rabbit gastric glands and parietal cells. *Can. J. Physiol. Pharmacol.* 65: 1840–1847, 1987.
23. CANFIELD, S. P., B. P. CURWAIN, AND J. SPENCER. A possible role for cyclic adenosine 3′,5′-monophosphate in the regulation of acid secretion in the isolated stomach of the guinea pig. *Br. J. Pharmacol.* 59: 327–332, 1977.
24. CARLISLE, K. S., C. S. CHEW, AND S. J. HERSEY. Ultrastructural changes and cyclic AMP in frog oxyntic cells. *J. Cell Biol.* 76: 31–42, 1978.
25. CHAREST, R., P. F. BLACKMORE, B. BERTHON, AND J. H. EXTON. Changes in free cytosolic Ca^{2+} in hepatocytes following α-adrenergic stimulation. *J. Biol. Chem.* 258: 8769–8773, 1983.
26. CHEW, C. S. Inhibitory action of somatostatin on isolated gastric glands and parietal cells. *Am. J. Physiol.* 245 (*Gastrointest. Liver Physiol.* 8): G221–G229, 1983.
27. CHEW, C. S. Forskolin stimulation of acid and pepsinogen secretion in isolated gastric glands. *Am. J. Physiol.* 245 (*Cell Physiol.* 14): C371–C380, 1983.
28. CHEW, C. S. Parietal cell protein kinases: selective activation of type I cAMP-dependent protein kinase by histamine. *J. Biol. Chem.* 25: 7540–7550, 1985.
29. CHEW, C. S. Differential effects of extracellular calcium removal and nonspecific effects of Ca^{2+} antagonists on acid secretory activity in isolated gastric glands. *Biochim. Biophys. Acta* 846: 370–378, 1985.
30. CHEW, C. S. Role of intracellular calcium in the action of secretagogues on isolated gastric glands. In: *Regulatory Peptides in Digestive, Nervous and Endocrine Systems*, edited by M. J. M. Lewin and S. Bonfils. Amsterdam: Elsevier, 1985, p. 81–84. (INSERM Symp. no. 25.)
31. CHEW, C. S. Cholecystokinin, carbachol, gastrin, histamine, and forskolin increase $[Ca^{2+}]_i$ in gastric glands. *Am. J. Physiol.* 250 (*Gastrointest. Liver Physiol.* 13): G814–G823, 1986.
32. CHEW, C. S., AND M. R. BROWN. Release of intracellular Ca^{2+} and elevation of inositol trisphosphate by secretagogues in parietal and chief cells. *Biochim. Biophys. Acta* 888: 116–125, 1986.
33. CHEW, C. S., AND M. R. BROWN. Release of intracellular calcium in parietal cells by the cAMP-dependent agonists, histamine and forskolin. In: *Gastrointestinal and Hepatic Secretion*, edited by J. S. Davison, in press.
34. CHEW, C. S., AND M. R. BROWN. Histamine increases phosphorylation of 27- and 40-kDa parietal cell proteins. *Am. J. Physiol.* 253 (*Gastrointest. Liver Physiol.* 16): G823–G829, 1987.
35. CHEW, C. S., AND S. J. HERSEY. Dissociation between oxyntic cell cAMP formation and HCl secretion in bullfrog gastric mucosa. *Am. J. Physiol.* 235 (*Endocrinol. Metab. Gastrointest. Physiol.* 4): E140–E149, 1978.
36. CHEW, C. S., AND S. J. HERSEY. Characteristics of histamine receptor in isolated gastric glands. In: *Hormone Receptors in Digestion and Nutrition*, edited by G. Rosselin, P. Fromageot, and S. Bonfils. Amsterdam: Elsevier/North Holland, 1979, p. 361–372.
37. CHEW, C. S., AND S. J. HERSEY. cAMP and secretagogue interactions in isolated gastric glands. In: *Nutrition—Digestion—Metabolism: Proc. 28th Int. Congr. Physiol. Sci., Budapest, 1980*, edited by T. Gati, L. G. Szollar, and G. Unguary. New York: Pergamon, 1980, vol. 12, p. 149–156. (Adv. Physiol. Sci. Ser.)
38. CHEW, C. S., AND S. J. HERSEY. Gastrin stimulation of isolated gastric glands. *Am. J. Physiol.* 242 (*Gastrointest. Liver Physiol.* 5): G504–G512, 1982.
39. CHEW, C. S., S. J. HERSEY, G. SACHS, AND T. BERGLINDH. Histamine responsiveness of isolated gastric glands. *Am. J. Physiol.* 238 (*Gastrointest. Liver Physiol.* 1): G312–G320, 1980.
40. CHIOU, C. Y., AND M. H. MALAGODI. Studies on the mechanism of action of a new Ca^{2+} antagonist, 8-(N,N-diethylamino) octyl 3,4,5-trimethoxybenzoate hydrochloride in smooth and skeletal muscle. *Br. J. Pharmacol.* 53: 279–285, 1975.
41. CODINA, J., J. HILDEBRANDT, R. IYENGAR, L. BIRNBAUMER, R. D. SEKURA, AND C. R. MANCLARK. Pertussis toxin substrate, the putative N_i component of adenylyl cyclases, is an alpha beta heterodimer regulated by guanine nucleotide and magnesium. *Proc. Natl. Acad. Sci. USA* 80: 4276–4280, 1983.
42. COOKE, A. R., T. E. CHVASTA, AND D. K. GRANNA. Histamine, pentagastrin, methylxanthines and adenyl cyclase activity in acid secretion. *Proc. Soc. Exp. Biol. Med.* 147: 674–678, 1974.
43. CORBIN, J. D., S. L. KEELY, AND C. R. PARK. The distribution and dissociation of cyclic adenosine 3′:5′-monophosphate-dependent protein kinases in adipose, cardiac, and other tissues. *J. Biol. Chem.* 250: 218–225, 1975.
44. CUPPOLETTI, J., G. SACHS, AND D. H. MALINOWSKA. Cytoskeletal proteins in gastric H^+ secretion: cAMP dependent phosphorylation, immunolocalization and protein blotting (Abstract). *Federation Proc.* 45: 1676, 1986.
45. DOMSCHKE, W., S. DOMSCHKE, M. CLASSEN, AND L. DEMLING. Histamine and cyclic 3′,5′-AMP in gastric acid secretion. *Nature Lond.* 241: 454–455, 1973.
46. DOUSA, T. P., AND C. F. CODE. Effect of histamine and its methyl derivatives on cyclic AMP metabolism in gastric mucosa and its blockade by the receptor antagonists. *J. Clin. Invest.* 53: 334–357, 1974.
47. DOUSA, T. P., AND R. R. DOZOIS. Interrelationships between histamine, prostaglandins, and cyclic AMP in gastric secretion: a hypothesis. *Gastroenterology* 73: 904–912, 1977.
48. DOZOIS, R. R., A. WOLLIN, R. D. RETTMAN, AND T. P. DOUSA. Effect of histamine on canine gastric mucosal adenylate cyclase. *Am. J. Physiol.* 232 (*Endocrinol. Metab. Gastrointest. Physiol.* 1): E35–E38, 1977.

49. EKBLAD, E. B. M., T. E. MACHEN, V. LICKO, AND M. J. RUTTEN. Histamine, cyclic AMP and the secretory response of piglet gastric mucosa. *Acta Physiol. Scand. Suppl.* 68–80, 1978.
50. FLOCKHART, D. A., AND J. D. CORBIN. Regulatory mechanisms in the control of protein kinases. *CRC Crit. Rev. Biochem.* 12: 133–186, 1982.
51. FORTE, J. G., T. M. FORTE, J. A. BLACK, C. OKAMOTO, AND J. M. WOLOSIN. Correlation of parietal cell structure and function. *J. Clin. Gastroenterol.* 5, Suppl. 1: 17–27, 1983.
52. FORTE, J. G., AND A. H. NAUSS. Effects of calcium removal on bullfrog gastric mucosa. *Am. J. Physiol.* 205: 631–637, 1963.
53. GANSER, A. L., AND J. G. FORTE. K^+-stimulated ATPase in purified microsomes of bullfrog oxyntic cells. *Biochim. Biophys. Acta* 307: 169–180, 1973.
54. GILL, D. M. Mechanism of action of cholera toxin. *Adv. Cyclic Nucleotide Res.* 8: 85–118, 1977.
55. GILMAN, A. G. G proteins and dual control of adenylate cyclase. *Cell* 36: 577–579, 1984.
56. GRAY, J. S., AND J. L. ADKISON. The effect of inorganic ions on gastric secretion in vitro. *Am. J. Physiol.* 134: 27–31, 1941.
57. GREEN, D. A., AND R. B. CLARK. Direct evidence for the role of the coupling proteins in forskolin activation of adenylate cyclase. *J. Cyclic Nucleotide Res.* 8: 337–346, 1982.
58. GREENGARD, P. Phosphorylated proteins as physiological effectors. *Science Wash. DC* 199: 146–152, 1978.
59. GRYNKIEWICZ, G., M. POENIE, AND R. Y. TSIEN. A new generation of Ca^{2+} indicators with greatly improved fluorescence properties. *J. Biol. Chem.* 260: 3440–3450, 1985.
60. HALEY, B. F. Adenosine 3′,5′-cyclic monophosphate binding sites. *Methods Enzymol.* 46: 339–346, 1977.
61. HANZEL, D. K., T. URUSIDANI, AND J. G. FORTE. Phosphorylation of microtubule-associated proteins in stimulated gastric glands (Abstract). *Federation Proc.* 46: 1082, 1987.
62. HARRIS, J. B., AND D. ALONSO. Stimulation of the gastric mucosa by adenosine 3′,5′-monophosphate. *Federation Proc.* 24: 1368–1376, 1965.
63. HARRIS, J. B., K. NIGON, AND D. ALONSO. Adenosine 3′,5′-monophosphate: intracellular mediator for methylxanthine stimulation of gastric secretion. *Gastroenterology* 47: 377–389, 1969.
64. HERSEY, S. J., D. MAY, AND D. SCHYBERG. Stimulation of pepsinogen release from isolated gastric glands by cholecystokininlike peptides. *Am. J. Physiol.* 244 (*Gastrointest. Liver Physiol.* 7): G192–G197, 1983.
65. HERSEY, S. J., A. OWIRODU, AND M. MILLER. Forskolin stimulation of acid and pepsinogen secretion by gastric glands. *Biochim. Biophys. Acta* 755: 293–299, 1983.
66. HILDEBRANDT, J. D., J. CODINA, R. RISINGER, AND L. BIRNBAUMER. Identification of a γ subunit associated with the adenylyl cyclase regulatory proteins N_s and N_i. *J. Biol. Chem.* 259: 2039–2042, 1984.
67. HOLIAN, O., C. RUIZ, C. T. BOMBECK, AND L. M. NYHUS. Action of histamine and 3-isobutyl-1-methylxanthine on cAMP activation of protein kinase in dog gastric mucosa. *Agents Actions* 13: 5–9, 1983.
67a. IM, W. B., D. P. BLAKEMAN, J. MENDLEIN, AND G. SACHS. Inhibition of $(H^+ + K^+)$-ATPase and H^+ accumulation in hog gastric membranes by trifluoperazine, verapamil and 8-(*N,N*-diethylamino)octyl-3,4,5-trimethoxybenzoate. *Biochim. Biophys. Acta* 770: 65–72, 1984.
68. JACKSON, R. J., AND G. SACHS. Identification of gastric cyclic AMP binding proteins. *Biochim. Biophys. Acta* 717: 453–458, 1982.
69. JACOBSEN, A., M. SCHWARTZ, AND W. S. REHM. Effects of removal of calcium from bathing medium on frog stomach. *Am. J. Physiol.* 209: 134–140, 1965.
70. JACOBSON, E. D., AND W. J. THOMPSON. Cyclic AMP and gastric secretion: the illusive second messenger. *Adv. Cyclic Nucleotide Res.* 7: 199–224, 1976.
71. JIRON, C., M. C. RUIZ, AND F. MICHELANGELI. Role of Ca^{++} in stimulus-secretion coupling in the gastric oxyntic cell: effect of A23187. *Cell Calcium* 2: 573–585, 1981.
72. KARPPANEN, H. O., P. J. NEUVONEN, P. R. BIECK, AND E. WESTERMANN. Effect of histamine, pentagastrin and theophylline on the production of cyclic AMP in the isolated gastric tissue of the guinea pig. *Naunyn-Schmiedeberg's Arch. Pharmacol.* 284: 15–23, 1974.
73. KARPPANEN, H. O., AND E. WESTERMANN. Increased production of cyclic AMP in isolated gastric tissue by stimulation of histamine$_2$ (H_2)-receptors. *Naunyn-Schmiedeberg's Arch. Pharmacol.* 279: 83–87, 1973.
74. KASBEKAR, D. K. Calcium-secretagogue interaction in the stimulation of gastric acid secretion. *Proc. Soc. Exp. Biol. Med.* 145: 235–239, 1974.
75. KASBEKAR, D. K., AND H. CHUGANI. Role of calcium ion in in vitro gastric acid secretion. In: *Gastric Hydrogen Ion Secretion*, edited by D. K. Kasbekar, G. Sachs, and W. S. Rehm. New York: Dekker, 1976, p. 187–211.
76. KATADA, T., AND M. UI. ADP ribosylation of the specific membrane protein of C_6 cells by islet-activating protein associated with modification of adenylate cyclase activity. *Proc. Natl. Acad. Sci. USA* 79: 3129–3133, 1982.
77. KIMBERG, D. V. Cyclic nucleotides and their role in gastrointestinal secretion. *Gastroenterology* 67: 1023–1064, 1974.
78. LEE, J., G. SIMPSON, AND P. SCHOLES. An ATPase from dog gastric mucosa: changes in outer pH in suspensions of membrane vesicles accompanying ATP hydrolysis. *Biochim. Biophys. Res. Commun.* 60: 825–832, 1974.
79. LEVINE, R. A., E. H. SCHWARTZEL, JR., S. BACHMAN, AND J. N. TALEV. Effects of secretagogues and theophylline on canine gastric mucosal cyclic nucleotides. *J. Lab. Clin. Med.* 92: 813–821, 1978.
80. LEVINE, R. A., AND D. E. WILSON. The role of cyclic AMP in gastric secretion. *Ann. NY Acad. Sci.* 185: 363–375, 1971.
81. LEWIN, M., A. SOUMARMON, J. BALI, S. BONFILS, J. P. GIRMA, J. L. MORGOT, AND P. FROMAGEOT. Interaction of ^{3}H-labelled synthetic human gastrin I with rat gastric plasma membrane. Evidence for biologically reactive gastrin receptor site. *FEBS Lett.* 66: 168–172, 1976.
82. LOGSDON, C. D., AND T. E. MACHEN. Involvement of extracellular calcium in gastric stimulation. *Am. J. Physiol.* 241 (*Gastrointest. Liver Physiol.* 4): G365–G375, 1981.
83. MAJOR, J. S., AND P. SCHOLES. The localization of histamine H_2 receptor adenylate cyclase system in canine parietal cells and its inhibition by prostaglandins. *Agents Actions* 8: 324–331, 1978.
84. MANGEAT, P., G. MARCHIS-MOUREN, A. M. CHERET, AND M. J. M. LEWIN. Specific activation of cyclic AMP-dependent protein kinase(s) by H_2-histamine agonists in isolated gastric mucosal cells from guinea pig. *Biochim. Biophys. Acta* 629: 604–608, 1980.
85. MANNING, D. R., AND A. G. GILMAN. The regulatory components of adenylate cyclase and transducin. *J. Biol. Chem.* 258: 7059–7063, 1983.
86. MAO, C. C., E. D. JACOBSON, AND L. L. SHANBOUR. Mucosal cyclic AMP and secretion in the dog stomach. *Am. J. Physiol.* 225: 893–896, 1973.
87. MAO, C. C., L. L SHANBOUR, D. S. HODGINS, AND E. D. JACOBSON. Adenosine 3′,5′-monophosphate (cyclic AMP) and secretion in the canine stomach. *Gastroenterology* 63: 427–438, 1972.
88. MCNEILL, J. H., AND S. C. VERMA. Stimulation of rat gastric adenylate cyclase by histamine and histamine analogues and blockade by burimamide. *Br. J. Pharmacol.* 52: 104–106, 1974.
89. MEANS, A. R., AND J. R. DEDMAN. Calmodulin—an intracellular calcium receptor. *Nature Lond.* 285: 73–77, 1980.
90. MOND, R. Über die elektromotorischen Kräfte der Magenschleimhaut von Frosch. *Pfluegers Arch. Gesamte Physiol. Menchen Tiere Physiol.* 215: 468–480, 1927.
91. MUALLEM, S., C. J. FIMMEL, S. J. PANDOL, AND G. SACHS. Regulation of free cytosolic Ca^{2+} in the peptic and parietal

cells of the rabbit gastric gland. *J. Biol. Chem.* 261: 2660–2667, 1986.
92. MUALLEM, S., AND G. SACHS. Changes in cytosolic free Ca^{2+} in isolated parietal cells. Differential effects of secretagogues. *Biochim. Biophys. Acta* 805: 181–185, 1984.
93. MUALLEM, S., AND G. SACHS. Ca^{2+} metabolism during cholinergic stimulation of acid secretion. *Am. J. Physiol.* 248 (*Gastrointest. Liver Physiol.* 11): G216–G228, 1985.
94. MURAYAMA, T., AND M. UI. Loss of the inhibitory function of the guanine nucleotide regulatory component of adenylate cyclase due to its ADP ribosylation by islet-activating protein, pertussis toxin, in adipocyte membranes. *J. Biol. Chem.* 285: 3319–3326, 1983.
95. NAKAJIMA, S., B. I. HIRSCHOWITZ, AND G. SACHS. Studies on adenyl cyclase in *Necturus* gastric mucosa. *Arch. Biochem. Biophys.* 143: 123–126, 1971.
96. NAKAMURA, T., AND M. UI. Simultaneous inhibitions of inositol phospholipid breakdown, arachadonic acid release, and histamine secretions in mast cells by islet-activating protein, pertussis toxin. A possible involvement of the toxin-specific substrate in the Ca^{2+}-mobilizing receptor-mediated biosignaling system. *J. Biol. Chem.* 260: 3584–3593, 1985.
97. NAURMI, S., AND Y. MAKI. Possible role of cyclic AMP in gastric acid secretion in rat. *Biochim. Biophys. Acta* 311: 90–97, 1973.
98. OKAJIMA, F., AND M. UI. ADP-ribosylation of a specific membrane protein by islet-activating protein, pertussis toxin, associated with inhibition of a chemotactic peptide-induced arachidonate acid release in neutrophils: a possible role of the toxin substrate in Ca^{2+} mobilizing biosignaling. *J. Biol. Chem.* 259: 13863–13871, 1984.
99. PERRIER, C. V., AND M. GRIESSEN. Action of H_1 and H_2 inhibitors on the response of histamine sensitive adenylyl cyclase from guinea pig mucosa. *Eur. J. Clin. Invest.* 6: 113–120, 1976.
100. RAPHAEL, N., E. B. EKBLAD, AND T. E. MACHEN. Reversible effects of phenothiazines on frog gastric stimulation. *Am. J. Physiol.* 247 (*Gastrointest. Liver Physiol.* 10): G366–G376, 1984.
101. RAY, T. K., AND J. G. FORTE. Adenyl cyclase of oxyntic cells. Its association with different cellular membranes. *Biochim. Biophys. Acta* 363: 320–339, 1971.
102. RAY, T. K., AND J. G. FORTE. Soluble and bound protein kinases of rabbit gastric secretory cells. *Biochem. Biophys. Res. Commun.* 61: 1199–1206, 1974.
103. REIMANN, E. M., AND N. G. RAPINO. Partial purification and characterization of adenosine 3′,5′monophosphate dependent protein kinase from rabbit gastric mucosa. *Biochim. Biophys. Acta* 350: 201–204, 1974.
104. ROBBERECHT, P., M. DESCHODT-LANCKMAN, AND P. DE NEEF. Hydrolysis of the cyclic 3′:5′-monophosphates of adenosine and guanosine by rat pancreas. *Eur. J. Biochem.* 41: 585–591, 1974.
105. RODBELL, M. Programmable messengers: a new theory of hormone action. *Trends Biomed. Sci.* 10: 461–464, 1985.
106. ROSENFELD, G. C. Pirenzepine (LS519): a weak inhibitor of acid secretion in rat parietal cells. *Eur. J. Pharmacol.* 86: 99–101, 1983.
107. ROSENFELD, G. C. Isolated parietal cells: adrenergic responses and pharmacology. *J. Pharmacol. Exp. Ther.* 229: 763–767, 1984.
108. RUOFF, H. J., AND K.-F. SEWING. Rat gastric mucosal cAMP following cholinergic and histamine stimulation. *Eur. J. Pharmacol.* 28: 338–343, 1974.
109. RUOFF, H. J., AND K.-F. SEWING. Adenylate cyclase of the dog gastric mucosa: stimulation by histamine and inhibition by metiamide. *Naunyn Schmiedeberg's Arch. Pharmacol.* 294: 207–208, 1976.
110. RUOFF, H.-J., M. WAGNER, C. GÜNTHER, AND S. MAŚLIŃSKI. Adrenergic stimulation of isolated rat gastric mucosal cells. Effect on adenylate cyclase and ^{14}C-aminopyrine uptake. *Naunyn-Schmiedeberg's Arch. Pharmacol.* 320: 175–181, 1982.
111. SACHS, G., AND T. BERGLINDH. Physiology of the parietal cell. In: *Physiology of the Gastrointestinal Tract* (1st ed.), edited by L. R. Johnson. New York: Raven, 1981, vol. 1, p. 567–602.
112. SACHS, G., H. H. CHANG, E. RABON, R. SCHACKMAN, M. LEWIN, AND G. SACCOMANI. A nonelectrogenic H^+ pump in plasma membranes of hog stomach. *J. Biol. Chem.* 251: 7690–7698, 1976.
113. SCHLATZ, L. J., C. BOOLS, AND E. M. REIMANN. Phosphorylation of membranes from the rat gastric mucosa. *Biochim. Biophys. Acta* 673: 539–551, 1981.
114. SCHOLES, P. A., A. COOPER, D. JONES, J. MAJOR, M. WALTERS, AND C. WILDE. Characterization of an adenylate cyclase sensitive to histamine H_2-receptor excitation in cells from dog gastric mucosa. *Agents Actions* 6: 677–682, 1976.
115. SCHWARTZ, D. A., AND C. S. RUBIN. Regulation of cAMP-dependent protein kinase subunit levels in Friend erythroleukemic cells. Effects of differentiation and treatment with 8-Br-cAMP and methylisobutylxanthine. *J. Biol. Chem.* 258: 777–784, 1983.
116. SEAMON, K. B., AND J. W. DALY. Forskolin, a unique diterpine activator of cyclic AMP generating system. *J. Cyclic Nucleotide Res.* 7: 201–224, 1981.
117. SEDAR, A. W., AND J. G. FORTE. Effects of calcium depletion on the junctional complex between oxyntic cells and gastric glands. *J. Cell Biol.* 22: 173–188, 1964.
118. SEKAR, M. C., AND C. E. HOKIN. The role of phosphoinositides in signal transduction. *J. Membr. Biol.* 89: 193–210, 1986.
119. SEWING, K.-F., AND H. HANNEMAN. Calcium channel antagonists verapamil and gallopamil are powerful inhibitors of acid secretion in isolated and enriched guinea pig parietal cells. *Pharmacology Basel* 27: 9–14, 1983.
120. SIMON, B., AND H. KATHER. Histamine-sensitive adenylate cyclase of human gastric mucosa. *Gastroenterology* 73: 429–431, 1977.
121. SISTARE, F. D., R. A. PICKING, AND R. C. HAYNES, JR. Sensitivity of the response to cytosolic calcium in quin-2-loaded rat hepatocytes to glucagon, adenine nucleosides and adenine nucleotides. *J. Biol. Chem.* 260: 12744–12747, 1985.
122. SOBEL, B. E., AND S. E. MAYER. Cyclic adenosine monophosphate and cardiac contractility. *Circ. Res.* 32: 407–413, 1973.
123. SOLL, A. H. The actions of secretagogues on oxygen uptake by isolated mammalian parietal cells. *J. Clin. Invest.* 61: 370–380, 1978.
124. SOLL, A. H. The interaction of histamine with gastrin and carbamylcholine on oxygen uptake by isolated mammalian parietal cells. *J. Clin. Invest.* 61: 381–389, 1978.
125. SOLL, A. H. Secretagogue stimulation of [^{14}C]aminopyrine accumulation by isolated canine parietal cells. *Am. J. Physiol.* 238 (*Gastrointest. Liver Physiol.* 1): G366–G375, 1980.
126. SOLL, A. H. Specific inhibition by prostaglandins E_2 and I_2 of histamine-stimulated [^{14}C]-aminopyrine accumulation and cyclic adenosine monophosphate generation by isolated canine parietal cells. *J. Clin. Invest.* 65: 1222–1229, 1980.
127. SOLL, A. H. Extracellular calcium and cholinergic stimulation of isolated canine parietal cells. *J. Clin. Invest.* 68: 270–278, 1981.
128. SOLL, A. H. Prostacyclin analogues inhibit canine parietal cell activity and cyclic AMP formation. *Prostaglandins* 21: 353–365, 1981.
129. SOLL, A. H., AND T. BERGLINDH. Physiology of isolated gastric glands and parietal cells: receptors and effectors regulating function. In: *Physiology of the Gastrointestinal Tract* (2nd ed.), edited by L. R. Johnson. New York: Raven, 1987, vol. 1, p. 883–909.
130. SOLL, A. H., AND A. WOLLIN. Histamine and cyclic AMP in isolated canine parietal cells. *Am. J. Physiol.* 237 (*Endocrinol. Metab. Gastrointest. Physiol.* 6): E444–E450, 1979.
131. STENGEL, D., L. GUENET, M. DESIMER, AND J. HANOUNE. Forskolin requires more than the catalytic unit to activate adenylate cyclase. *Mol. Cell. Endocrinol.* 28: 681–690, 1982.

132. SUNG, C. P., B. C. JENKINS, L. R. BURNS, V. HACKNEY, J. G. SPENNEY, G. SACHS, AND V. D. WIEBELHAUS. Adenyl and guanyl cyclase in rabbit gastric mucosa. *Am. J. Physiol.* 225: 1359-1363, 1973.
133. SUTHERLAND, E. W. On the biological role of cyclic AMP. *J. Am. Med. Assoc.* 214: 1281-1288, 1970.
134. TAGUE, L. L., M. S. AMER, AND E. D. JACOBSON. Histamine, cyclic AMP and gastric secretion in the dog. *Am. J. Dig. Dis.* 22: 13-15, 1977.
135. TAKAHASHI, S., K. MORIWAKI, S. HIMENO, T. KUROSHIMA, Y. SHINOMURA, S. HAMABE, M. KUROKAWA, R. SAITO, T. KITANI, AND S. TAURI. Forskolin-induced cyclic AMP production and gastric acid secretion in dispersed rabbit parietal cells: novel evidence for a major role of cyclic AMP in acid release. *Life Sci.* 33: 1400-1408, 1983.
136. TENNES, L. A., J. A. KENNEDY, AND M. L. ROBERTS. Inhibition of agonist-induced hydrolysis of phosphatidylinositol and muscarinic receptor binding by the calcium antagonist, 8-(N,N-diethylamino)-octyl-3,4,5-trimethoxybenzoate (TMB-8). *Biochem. Pharmacol.* 32: 2116-2118, 1983.
137. THOMPSON, W. J., L. K. CHANG, G. C. ROSENFELD, AND E. D. JACOBSON. Activation of rat gastric mucosal adenyl cyclase by secretory inhibitors. *Gastroenterology* 72: 251-254, 1977.
138. THOMPSON, W. J., AND E. D. JACOBSON. Comparison of the effects of secretory stimulants and inhibitors on gastric mucosal adenylyl cyclases of various species. *Proc. Soc. Exp. Biol. Med.* 154: 377-381, 1977.
139. TRIGGLE, D. J. Calcium antagonists: basic chemical and pharmacologic aspects. In: *New Perspectives on Calcium Antagonists*, edited by G. B. Weiss. Baltimore, MD: Williams & Wilkins, 1981, p. 1-18.
140. TSIEN, R. Y., T. POZZAN, AND T. J. RINK. Calcium homeostasis in intact lymphocytes: cytoplasmic free calcium monitored with a new, intracellular trapped fluorescent indicator. *J. Cell Biol.* 94: 325-334, 1982.
141. VIAL, J. D., AND J. GARRIDO. Actin-like filaments and membrane rearrangements in oxyntic cells. *Proc. Natl. Acad. Sci. USA* 73: 4032-4036, 1976.
142. WALSH, D. A., J. P. PERKINS, AND E. G. KREBS. An adenosine 3',5'-monophosphate-dependent protein kinase from rabbit skeletal muscle. *J. Biol. Chem.* 243: 3763-3765, 1968.
143. WAY, L., AND R. P. DURBIN. Inhibition of gastric acid secretion in vitro by prostaglandin E_1. *Nature Lond.* 221: 874-875, 1969.
144. WEISS, B. Techniques for measuring the interaction of drugs with calmodulin. *Methods Enzymol.* 102: 171-184, 1983.
145. WOLLIN, A., C. F. CODE, AND T. P. DOUSA. Interaction of prostaglandins and histamine with enzymes of cyclic AMP metabolism from guinea pig gastric mucosa. *J. Clin. Invest.* 57: 1548-1553, 1976.
146. WOLLIN, A., A. H. SOLL, AND I. M. SAMLOFF. Actions of histamine, secretin, and PGE_2 on cyclic AMP production by isolated canine fundic mucosal cells. *Am. J. Physiol.* 237 (*Endocrinol. Metab. Gastrointest. Physiol.* 6): E437-E443, 1979.
147. WOLOSIN, J. M., C. OKAMOTO, T. M. FORTE, AND J. G. FORTE. Actin and associated proteins in gastric epithelial cells. *Biochim. Biophys. Acta* 761: 171-182, 1983.

CHAPTER 14

Cellular basis of pepsinogen secretion

STEPHEN J. HERSEY | Department of Physiology, Emory University, Atlanta, Georgia

CHAPTER CONTENTS

General Considerations
  Basic model for pepsinogen secretion
  Experimental models for studying pepsinogen secretion
  Assay of pepsinogen
Properties of Pepsinogen
  Biochemistry
  Distribution
Secretion of Pepsinogen
  Stimulation of secretion
    Cholinergic
    Adrenergic
    Histamine
    Gastrin and cholecystokinin
    Secretin and vasoactive intestinal peptide
    Other peptides
  Intracellular mediators
    Cyclic AMP
    Calcium
  Secretory mechanism
Summary

PEPSIN HAS BEEN RECOGNIZED for over a century as the major proteolytic enzyme secreted by the stomach. The role of pepsin in protein digestion was first described by Schwann (91) as a ferment in gastric juice, which he named pepsin. By the beginning of the twentieth century, the classic studies of Langley and co-workers (59, 60) had established that pepsin was secreted by the gastric chief cells as a proenzyme, pepsinogen, and converted to the active enzyme by acid in the gastric juice. Numerous studies over the past decades have served primarily to confirm the fundamental features of Langley's model for the secretion of pepsin while adding more detail to the mechanisms that underlie this process. Substantial information has accumulated on the biochemistry of pepsin, including its synthesis and enzymatic activity. Additionally the recent development of cellular models for studying secretion of pepsinogen has provided more detailed information on the mechanisms involved in the stimulation of secretion as well as new insights on the processes responsible for the release of pepsinogen. This chapter provides a general description of the cellular events involved in the synthesis and secretion of pepsinogen with emphasis on some areas that would benefit from further investigation.

GENERAL CONSIDERATIONS

Basic Model for Pepsinogen Secretion

The major steps involved in the secretion of pepsinogen at the cellular level are depicted in Figure 1. This basic model for the function of the gastric chief cell is essentially similar to that proposed by Langley (59) and described in the previous edition of this *Handbook* (41). According to this model, the major events include synthesis of a proenzyme, pepsinogen, by the endoplasmic reticulum; concentration and storage of pepsinogen in secretory granules; and the release of pepsinogen into the lumen of the oxyntic glands, where, under the influence of acid, pepsinogen is converted to the active enzyme pepsin. The basic model also provides for the regulation of pepsinogen secretion through the action of stimulating agents. These agents act at receptors located in the basolateral plasma membrane to initiate intracellular events leading to the release of pepsinogen from the secretory granules. While the general concepts depicted by this model have been known for several decades, more recent studies, particularly those employing cellular systems, have added significant detail to the basic model. Despite these advances in the understanding of pepsinogen secretion, a number of fundamental questions remain unanswered. It is hoped that with improvements in the techniques for studying pepsinogen secretion more rapid progress can be made in understanding the mechanisms that underlie this important physiological process.

Experimental Models for Studying Pepsinogen Secretion

Advances in understanding the cellular basis of pepsinogen secretion have depended, in part, on the recent development of in vitro model systems. For most previous studies the major model system employed was the intact animal. Particularly frequent

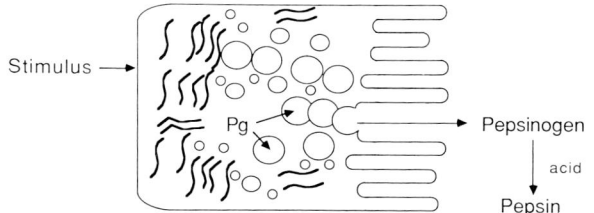

FIG. 1. Model for basic features of pepsinogen secretion by gastric chief cells. Major events include synthesis of pepsinogen (Pg) by endoplasmic reticulum, storage in secretory granules, and release of pepsinogen through apical surface on stimulation of cell. Released pepsinogen is converted to active pepsin in presence of acid.

use has been made of the conscious dog with gastric fistula or stomach pouch. The intact animal models have yielded a wealth of information and continue to be important models for studying integrated responses. The major disadvantage of the intact animal model is the complexity of potential regulatory mechanisms. Thus dose-response studies are complicated by plasma clearance rates, and responses may be indirect because of secondary activation of endogenous stimuli. For studies related to cellular mechanisms or biochemical events, the intact animal models are very limited. Investigators interested in cellular events, therefore, have sought to develop simpler model systems.

A few investigators have reported the use of isolated intact gastric mucosa. This preparation is simplified but retains its polar orientation and could be of value in studying release of pepsinogen into the serosal or blood side. Mammalian intact gastric mucosa has been studied in organ culture (106), but this seems to be a poor compromise since the separation of epithelial surfaces is lost. Recently the intact esophageal mucosa of frog has been employed as an in vitro preparation for studying pepsinogen secretion (96). Initial results indicate that this may be a highly useful preparation. The esophageal mucosa is stable in vitro and responsive to various stimuli (94). Moreover, the absence of acid-secreting cells offers some added advantage for this preparation. While the intact epithelium has the advantages of polar orientation and more extensive control over experimental variables, this type of preparation is not ideal for studies requiring multiple samples, e.g., receptor binding and intracellular composition. For these studies the more dispersed preparations of isolated gastric glands and gastric cells have proven to be useful.

Isolated gastric glands were developed initially for studies of gastric acid formation (7) and later adapted for studies of pepsinogen secretion (53, 55). This preparation has been employed to identify stimuli for secretion, to investigate stimulus-secretion coupling mechanisms, and to study pepsinogen release mechanisms. The major advantages of the preparation include responsiveness, control of experimental conditions, and rapid multiple sampling. The major disadvantage is that the gastric glands consist of a heterogeneous cell population, including parietal and mucus neck cells. While this heterogeneity is a problem for certain biochemical studies, the maintenance of an epithelial-like organization, e.g., cell polarity and tight junctions, appears to be important for maintaining secretory responses as compared with isolated, dispersed cells (87). Attempts to develop a preparation of responsive, purified chief cells recently have been more successful (77), and this preparation should be a major advantage for studies of intracellular regulatory mechanisms.

Studies on the isolation and characterization of subcellular components of the chief cell are virtually nonexistent. Preliminary studies of the isolation and partial characterization of peptic granules have been reported (72), but substantially more work is required to describe the properties of this key organelle. A potentially useful alternative to isolating subcellular components is the use of detergent-permeabilized cells (68). This preparation allows direct access to the intracellular environment while maintaining the spatial and presumably the functional organization of organelles. Further characterization of the permeable preparation is required before its full potential can be determined.

Assay of Pepsinogen

The usual method for assay of pepsinogen involves the conversion of the proenzyme to pepsin and measurement of the resulting acid protease. The classical method of Anson and Mirsky (3) using hemoglobin substrate at pH 2.0 remains a reliable, inexpensive assay. The sensitivity of the assay can be enhanced by processing the proteolytic end products through Folin phenol reagent or other procedures developed for high-sensitivity protein assays (37). For large numbers of samples this assay can be adapted for automated analysis (113). Alternative methods for increasing sensitivity, though more expensive, are to employ radiolabeled hemoglobin as substrate (53) or to use synthetic chromogenic substrates (63, 73).

Specific radioimmunoassays for pepsinogen have been developed and employed for measurements of the proenzyme in tissues, serum, and urine (84, 86). This sensitive and highly specific method has not found use for routine assays. However, for cytochemical applications or instances where a distinction between pepsinogen isoenzymes is required, this would be the method of choice.

The rate of pepsinogen secretion in vivo is expressed typically as units of peptic activity, referred to a standard preparation of porcine pepsinogen. For cellular systems in vitro, secretion is usually expressed as a percent of the initial total pepsinogen content that is released during the incubation period (55). This allows for normalization between tissues that vary in total enzyme content. The total pepsinogen content is determined on tissue samples that have been disrupted by freeze-thaw, sonication, or detergent treatment. Apart from providing a normalizing procedure, the expression of pepsinogen release as a fraction of the total content provides an important piece of information itself. Accordingly the percent of total can be used to assess the relationship between content and secretory rate or to investigate the role of pepsinogen content and synthesis.

PROPERTIES OF PEPSINOGEN

Biochemistry

Pepsinogens consist of an active pepsin plus a variable NH_2-terminal sequence of amino acids (24, 51, 110, 111). Based on the DNA sequence, it appears that pepsinogen is synthesized initially as a prepepsinogen containing a 15–amino acid signal sequence at the NH_2-terminus (99). In vitro translation of pepsinogen mRNA from rat gastric mucosa also results in a prepepsinogen containing a 16–amino acid signal sequence (48, 100). The signal sequence presumably is lost during posttranslational processing since the secreted protein lacks the signal sequence. It is not known at what stage of the synthesis-storage cycle the NH_2-terminal sequence is cleaved.

Under acidic conditions pepsinogen is converted to pepsin by the autocatalytic, i.e., intramolecular catalysis, loss of a variable NH_2-terminal sequence (51, 54). The mechanism of this activation appears to involve a pH-dependent conformation change that positions the NH_2-terminal activation peptide within the active site of the enzyme. This is followed by an intramolecular hydrolysis of the peptide and dissociation of the activation peptide (63). Before the activation steps, pepsinogen is able to bind substrates, but the binding does not result in hydrolysis. Binding of substrates, including the activation peptide and its analogues, to pepsinogen results in competitive inhibition of the intramolecular hydrolysis (58, 63). At lower pH, the activation peptide or substrate inhibitors tend to dissociate rapidly, allowing for activation of the enzyme. After the loss of the activation sequence, pepsin undergoes a further conformational change to expose an apparent binding cleft that can accommodate a substrate with about eight amino acid residues (110, 111). The active site contains two aspartyl residues that are found in separate coding sequences suggesting that the pepsinogen gene evolved by duplication of a smaller ancestral gene (111). Similar primary and tertiary structures are found in other carboxyl proteases, e.g., renin and cathepsin, suggesting that all carboxyl proteases may have evolved from a common ancestral peptide (1, 111). Once activated, pepsin may be irreversibly denatured at pH <7.2 or temperatures <65°C. In contrast, pepsinogen is resistant to denaturation by pH up to 10 or temperature up to boiling. The difference in alkali resistance led Langley to postulate the existence of a proenzyme and provides an operational definition of pepsinogens as distinct from other acid proteases. Physiologically the denaturation of pepsin by alkaline pH provides a mechanism for inactivation after passage into the intestine. What advantage is gained by inactivating pepsin is not clear since at the relatively alkaline pH of the intestine, proteolytic activity would be slight if any.

The catalytic activity of pepsin is described in general as acid-active proteolysis (82). The pH optimum for hydrolysis of hemoglobin is 1.5–2.5, but the pH optimum is quite broad and depends on the isozyme and specific substrate; e.g., milk clotting exhibits an optimum at pH 5.5. The substrate specificity of pepsin is fairly broad, but the preferred peptide bond is between two aromatic amino acids such as Phe-Trp (26). Studies of synthetic substrates indicate that hydrolysis is enhanced if the aromatic residues are flanked by hydrophobic amino acids. The presence of a flanking cationic residue, e.g., histamine, both enhances activity and shifts the pH optimum upward (26). Interestingly, if either of the aromatic amino acids is substituted by a D-enantiomer, the substrate becomes a competitive inhibitor with a K_i equal to the K_m for the L,L-substrate (26). Such observations suggest the possibility of developing specific and potent inhibitors for pepsin and other acid proteases.

Mammalian gastric pepsinogen can be separated electrophoretically into multiple bands. There are at least seven distinct pepsinogens, Pg 1–7, in order of decreasing electronegativity (81, 82, 92). The seven pepsinogens have been divided into two major groups, pepsinogen I and II (PG-I and PG-II), based on immunological reactivity using specific, non-cross-reacting antibodies. Pepsinogen I contains Pg 1–5 and PG-II contains Pg 6 and 7. The heterogeneity of pepsinogens appears to arise from several factors including variation in the number of structural genes and allelic variation at individual gene loci (107). Posttranslational modification may contribute also to the phenotypic heterogeneity (107).

In recent years, pepsinogens and pepsins from a number of species have been isolated and characterized. Amino acid sequencing and X-ray crystallographic studies (1, 2, 24, 95, 99–101) revealed substantial homologies and similar tertiary structures of pepsinogens from different species. Molecular weight estimates for pepsinogens from various species, including amphibians, fish, and mammals, range from 29 kDa to 65 kDA. Some of this heterogeneity arises from differences in the NH_2-terminal activation sequence (24, 48, 95, 110), but significant differences exist in the catalytic peptide as well (95). Despite such differences the catalytic site appears to be similar in all species. Somewhat surprisingly the immunological distinction between PG-I and PG-II, i.e., no cross-reactivity, persists across species despite the variation in size and NH_2-terminal sequences. Thus the major antigenic determinant, like the catalytic site, appears to have been conserved during evolution of the pepsinogens.

Distribution

The gastric chief cell is the major source of pepsinogen in mammals (83, 85). The chief cells are found primarily in the basal portion of the gastric glands of the body and fundus of the stomach, but the exact distribution of chief cells varies from species to species (35). Pepsinogen is found also in the mucus neck cells of fundic glands (83, 85) and in the cardiac and pyloric glands. The fundic glands contain both PG-I and PG-II. The cardiac glands in the gastric cardia and the pyloric glands of the gastric antrum contain pepsinogen, but only PG-II has been found in these regions (85). Since the cardiac and pyloric glands do not contain chief cells, the pepsinogen is thought to be associated with mucus neck cells. Pepsinogen II but not PG-I is found in the Brunner's glands of the proximal duodenum (85) and is secreted by the prostate gland into seminal fluid (61). The physiological significance of pepsinogen in nongastric tissues is unknown. Since the pH of these tissues is too high to allow activation of pepsinogen, it is unlikely that pepsinogen serves as a proteolytic enzyme.

Vertebrate species other than mammals do not possess chief cells per se. Instead, a single cell type, the oxyntic cell, is responsible for gastric secretion of both acid and pepsinogen (35). As with the mammalian chief cell, the oxyntic cells are confined to the fundus or body of the stomach, being absent from the antrum. In certain amphibia (frog and *Necturus*), glands located in the lower esophagus contain peptic-like cells that secrete pepsinogen (25, 96). The esophageal pepsinogen is unusual in having a low molecular weight but has immunological reactivity and a pH optimum characteristic of the PG-I group (95).

SECRETION OF PEPSINOGEN

Stimulation of Secretion

The initial cellular event in the secretion of pepsinogen involves the interaction of stimuli with cell surface receptors. Recent studies have identified a wide variety of agents that can stimulate pepsinogen secretion by chief cells. Some of these agents have been identified as acting through specific receptors while others act at intracellular sites or through unknown mechanisms. It seems probable that only the receptor-mediated stimuli will prove to be physiological regulators of pepsinogen secretion. The other stimuli have proven to be useful in defining some of the intracellular events of the secretory process but would be considered pharmacological agents rather than physiological regulators. For this reason, stimulation or inhibition by these agents is covered in *Intracellular Mediators*, p. 273. This section deals only with identified, receptor-mediated stimulation.

CHOLINERGIC. Acetylcholine and its analogues stimulate pepsinogen secretion in intact animals (41, 82), gastric glands (45, 46), and monolayer cultures of chief cells (87). The stimulation is inhibited by atropine indicating a muscarinic-type receptor. Recently, muscarinic receptors have been classified into two subtypes, M-1 and M-2 (32). Both subtypes have a high affinity for atropine but can be distinguished on the basis of their relative affinities for the newer muscarinic antagonist pirenzepine. Pirenzepine exhibits about equal potency with atropine against M-1–type receptors but is 20–100 times less potent than atropine against the M-2 subtype (32). Studies using intact dog (65) and isolated mouse stomach (42) indicated potent inhibition of pepsinogen secretion by pirenzepine suggesting that gastric secretion of both acid and pepsin is mediated by M-1–type receptors. In contrast, studies with isolated parietal cells (62) and gastric mucosal membranes (36) revealed only M-2–subtype receptors. Using rabbit isolated gastric glands, which permit measurement of both acid and pepsinogen secretions, we have found that pirenzepine inhibition of carbachol stimulation is less potent than atropine by ~100-fold for both acid and pepsinogen secretion (Fig. 2). These results indicate that both types of secretion are mediated at the cellular level by M-2 receptors. It may be suggested that the results with more intact preparations reflect effects on cholinergic neurons rather than direct cellular responses (70). Mediation of cholinergic stimulation of pepsinogen secretion, as well as acid secretion, by M-2 receptors rather than M-1

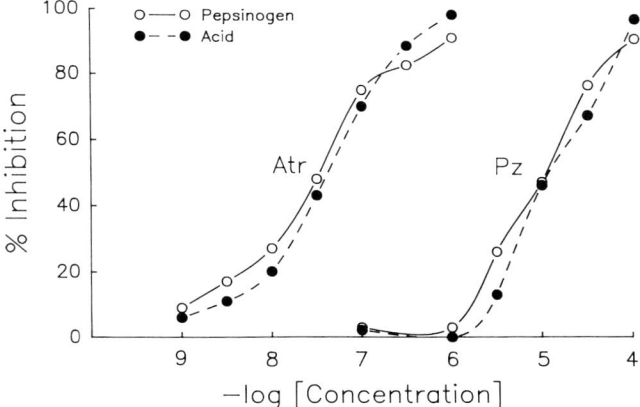

FIG. 2. Cholinergic stimulation of both acid and pepsinogen secretions are mediated by M-2–subtype receptor. Secretions of acid and pepsinogen by gastric glands are inhibited more potently by atropine (Atr) (nonselective) than by M-1–selective antagonist pirenzepine (Pz). Acid and pepsinogen secretions were stimulated by carbachol. Acid secretion was measured by accumulation of aminopyrine and pepsinogen secretion as percent of total released.

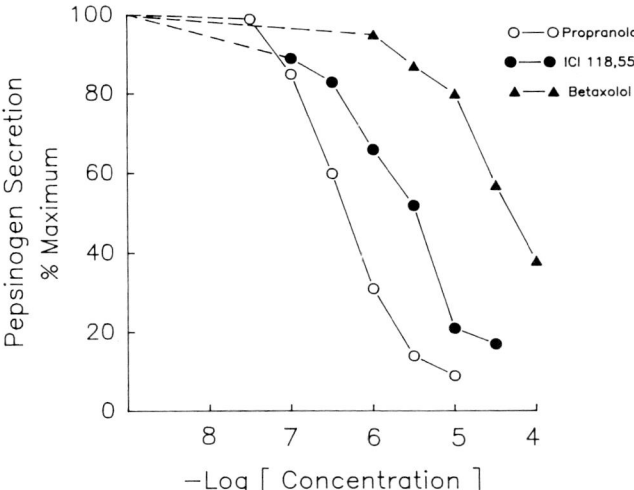

FIG. 3. Adrenergic stimulation of pepsinogen secretion occurs through β_2-adrenoceptor. Stimulation of pepsinogen secretion by isoproterenol in gastric glands is antagonized with a potency sequence of propranolol (nonselective) > ICI 118551 (β_2-selective) >> betaxolol (β_1-selective).

receptors also would be more consistent with the postulated intracellular coupling mechanism for this secretory process (see *Intracellular Mediators*, p. 273). This is indicated by the postulate that M-1 receptors are associated with inhibition of adenylate cyclase, while M-2 receptors are associated with calcium mobilization (33).

ADRENERGIC. A somewhat novel finding from in vitro studies is that the β-adrenergic agonist isoproterenol stimulates pepsinogen secretion by rabbit isolated gastric glands (55) and amphibian peptic glands (94). The adrenergic stimulation was found to be inhibited by propranolol but not by atropine or cimetidine indicating a direct action on the chief cell. Moreover the stimulation appears to be specific for pepsinogen secretion since isoproterenol does not stimulate acid secretion by the parietal cells (38). The adrenergic stimulation is suggested to be due to β_1-type receptors as evidenced by sensitivity to the selective antagonist pafenolol (98). However, our studies with gastric glands revealed that the β_2-selective antagonist ICI 118551 (11) is >100 times more potent than the β_1-selective antagonist betaxolol (12, 13) for inhibiting isoproterenol stimulation of pepsinogen secretion (Fig. 3). The difference in these findings may be due to the use of more selective antagonists. Studies using intact animals have demonstrated both adrenergic stimulation and inhibition of pepsinogen secretion, and thus the physiological significance of the in vitro results is unknown.

HISTAMINE. Pepsinogen secretion in dogs, humans, and several other species (41) is stimulated by histamine as judged by in vivo results. These responses are probably mediated by an H_2-type of histamine receptor (30, 44). The role of histamine in regulating pepsinogen secretion may be more complex than a simple direct action on chief cells. Histamine, at least in the dog, both stimulates and, at higher doses, inhibits pepsinogen secretion (44). Both actions appear to be mediated by H_2 receptors. Acidification of the gastric lumen, which might result from histamine stimulation of acid secretion, can result in secondary stimulation of pepsinogen secretion, probably via a local cholinergic reflex (49). In vitro studies have shown that histamine does not stimulate or inhibit pepsinogen secretion by rabbit isolated gastric glands (38, 55) or frog esophageal glands (96). The combined possibilities of multiple actions and species differences indicate that caution is needed in the interpretation of results concerning a direct action of histamine on chief cells. Although the results of in vivo experiments suggest a coupling between parietal and chief cells, possibly through local reflexes, such coupling does not appear to exist at a cellular level (38). For rabbit isolated gastric glands, the selective stimulation of acid versus pepsinogen secretions by histamine and isoproterenol (Fig. 4) demonstrates that activity of one cell type does not induce activity in the other cell type. The independence of the two cell types is further demonstrated by the observation that selective inhibition of acid formation by thiocyanate does not influence pepsinogen secretion (38). Thus, at a cellular level, acid and pepsinogen secretions appear to be regulated independently, although the two processes share some common stimuli. In the absence of clear evidence for a direct action of histamine on chief cells, it may be suggested that histamine is not a physiological regulator of pepsinogen secretion.

GASTRIN AND CHOLECYSTOKININ. Gastrin and pentagastrin stimulate pepsinogen secretion by intact an-

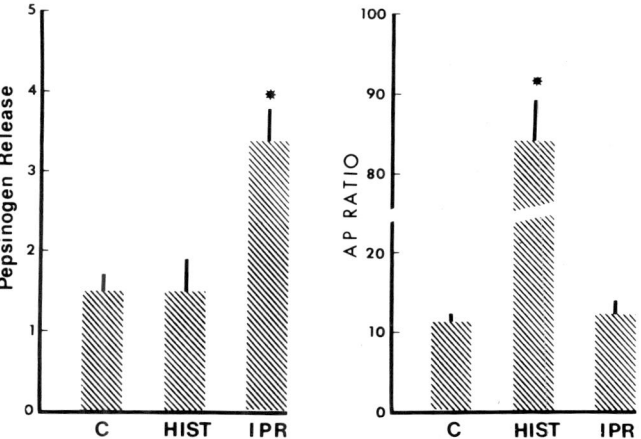

FIG. 4. Selective stimulation of acid and pepsinogen secretions by gastric glands. Isoproterenol (IPR) stimulates pepsinogen release (*left*) but not acid formation (*right*). In contrast, histamine (HIST) stimulates acid but not pepsinogen secretion. Pepsinogen secretion was measured as percent of total released, and acid secretion was measured as accumulation ratio of aminopyrine (AP). Selective action indicates independent regulation of the 2 secretory processes.

imals. The interpretation of this finding is complicated by the observation that gastrin stimulation is inhibited by atropine (35) or histamine H_2-receptor antagonists (30, 43). Thus a direct stimulation of the chief cell cannot be demonstrated. Studies with in vitro preparations indicate that gastrin is not a potent stimulus for pepsinogen secretion. Thus gastric glands isolated from rabbit secrete pepsinogen in response to gastrin but only at doses well above that required to stimulate acid secretion (37, 53), and pentagastrin does not stimulate pepsinogen secretion by either canine chief cells (87) or frog esophagus (96). Presently therefore the evidence does not favor direct activation of chief cells by gastrin as a physiological mechanism for stimulating pepsinogen secretion.

In contrast to the results with gastrin, in vitro studies have shown a potent stimulation of pepsinogen secretion by cholecystokinin-like peptides (37, 53). Stimulation by the COOH-terminal octapeptide of cholecystokinin (CCK-8) appears to be a direct action on the chief cell since it is not inhibited by atropine, propranolol, or cimetidine but is antagonized competitively by the nonpeptide antagonists dibutyryl cGMP (38) and asperlicin (L. Tang and S. Hersey, unpublished results). Potency sequences for CCK-related peptides indicate that one crucial feature for chief cell activation is sulfation of the tyrosine residue (37). This does not appear to be the case for parietal cells since gastrin I (nonsulfated) and gastrin II (sulfated) are equipotent for stimulation of acid secretion (31, 50, 52). The exact position of the sulfated tyrosine also appears to be important for stimulating pepsinogen secretion since sulfated gastrin (tyrosine at sixth position from COOH terminus) is much less potent than CCK-8 (tyrosine at seventh position from COOH terminus) and only slightly more potent than desulfated CCK (Fig. 5). These results indicate that the mammalian chief cell is able to distinguish between gastrin and CCK despite the structural similarity of the active COOH-terminal portions of these peptides.

At present, it is unclear whether the chief cell possesses a single type of receptor for CCK-like peptides or if there are distinct receptors for CCK and gastrin (18). However, it seems clear that chief cells possess different receptors for these peptides than do parietal cells. This concept is supported by two lines of evidence, selective inhibition and radiolabeled ligand binding studies. Both dibutyryl cGMP and asperlicin are specific, nonpeptide antagonists of CCK and both act as competitive inhibitors of CCK-stimulated pepsinogen secretion (38). However, neither cGMP (38) nor asperlicin inhibit acid secretion stimulated by CCK-8 or gastrin (Fig. 6). Thus these agents are selective for the chief cell receptor versus the parietal cell receptor rather than being selective for a specific ligand. Further support for different peptide receptors comes from ligand binding studies. Using gastrin I as the ligand resulted in identification of a high-affinity binding site in gastric homogenates (108, 109) and isolated gastric cells (101, 102). This "gastrin receptor" is characterized by a single class of binding sites, similar affinities for gastrin I and sulfated CCK-8, and insensitivity to asperlicin (17). Gastrin binding is associated with parietal cells, whereas chief cells exhibit little or no gastrin binding (101). On the other hand, when radiolabeled CCK-8 is used as the ligand, a complex binding curve results (75) that can be interpreted as showing at least two classes of binding sites. A low affinity site appears to be associated with CCK-8 stimulation of pepsinogen secretion. The "CCK receptor" exhibits higher affinity for CCK-8 than for gastrin I, is selective for sulfated peptides (75), and is inhibited by asperlicin (L. Tang and S. Hersey, unpublished results). Thus the chief cell receptor behaves more like the pancreatic CCK receptor than the parietal cell gastrin receptor (78, 88).

SECRETIN AND VASOACTIVE INTESTINAL PEPTIDES. Secretin has been reported to stimulate pepsinogen secretion in the intact animal (66, 103) and human (9). The dose of secretin required to elevate pepsinogen secretion appears to be within the normal range indicating that this hormone may play a physiological role. Since secretin inhibits acid secretion, the stimulation of pepsinogen output is not secondary to activation of parietal cells. Results using in vitro preparations are somewhat conflicting. Secretin stimulates pepsinogen secretion by rat gastric glands (76) and guinea pig isolated chief cells (77) but is ineffective in frog peptic glands (96) and is only a weak stimulus for monolayer cultures of canine chief cells (87) and rabbit gastric glands (55). It is not certain whether these differences are due to species variability or to differences between experimental preparations. An additional problem in regard to studying the action of secretin is the variation among preparations of the peptide. Crude preparations of secretin, often em-

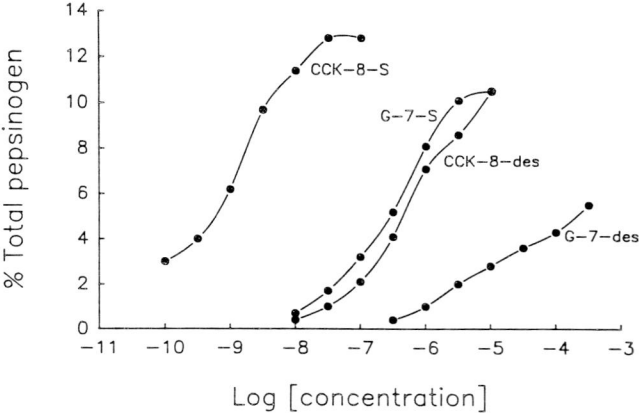

FIG. 5. Pepsinogen secretion by gastric glands shows selectivity for cholecystokinin (CCK) over peptide analogues. Potency sequence is CCK-8-S (sulfated) >> gastrin 7 (sulfated) (G-7-S) = CCK-8-des (desulfated) >> G-7-des (desulfated). Sequence indicates receptor specificity for both sulfation of tyrosine residue and position of tyrosine residue.

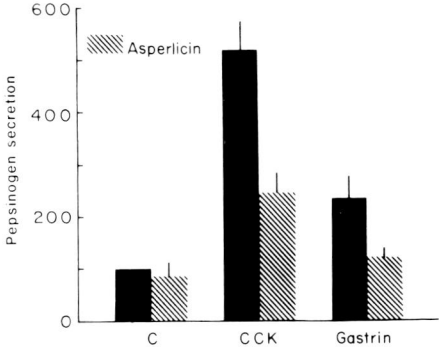
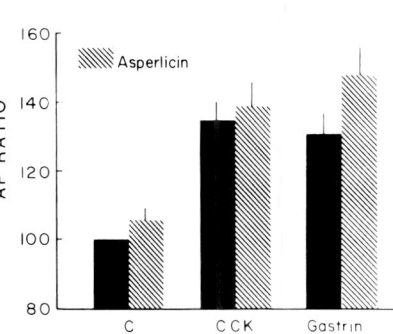

FIG. 6. Selective inhibition of peptide stimulation by asperlicin. Asperlicin inhibits pepsinogen secretion stimulated by either CCK-8 or gastrin (*left*) but does not inhibit peptide-stimulated acid secretion (*right*). Both pepsinogen secretion and acid secretion (aminopyrine accumulation) are expressed as percent of control (C). Selective inhibition indicates that acid and pepsinogen secretions are mediated by distinct receptor types.

ployed for earlier studies, are known to contain several different peptides. Such crude preparations have been found to be more effective in stimulating pepsinogen secretion than purified or synthetic secretin both in the intact animal (14) and in vitro (55). It has been suggested that the crude preparation contains an additional stimulus for pepsinogen secretion besides secretin (55), but this agent has not been identified. Based on in vitro studies (76, 77) it is likely that vasoactive intestinal peptide (VIP) shares a common receptor mechanism with secretin, but a physiological role for VIP in regulating pepsinogen secretion has not been demonstrated as yet.

OTHER PEPTIDES. Additional peptide hormones that have been implicated as modulators of pepsinogen secretion include bombesin, glucagon, and somatostatin. Bombesin, or a bombesin-like peptide, stimulates pepsinogen secretion by frog esophageal glands (94) through a direct action, but in mammals this peptide is thought to act indirectly by modulating gastrin release (46, 97). Glucagon, which is structurally similar to secretin, is reported to inhibit both acid and pepsinogen secretion (56). The interpretation of these findings is complicated since glucagon has several actions in the intact animal including elevation of blood glucose and reduction in gastric volume secretion. Appropriate in vitro experiments potentially could resolve some of these discrepancies. Somatostatin, which is found throughout the gastrointestinal tract, has been shown to inhibit pepsinogen secretion elicited by several stimuli in the intact animal (4, 47, 57, 112). The mechanism of this inhibition remains unknown, but given the localization of somatostatin within the gastric mucosa, a physiological role for this peptide might well be expected.

Intracellular Mediators

The interaction of stimuli with cell surface receptors results in a series of intracellular reactions, ultimately leading to the release of pepsinogen. These intracellular reactions are generically referred to as stimulus-secretion coupling. In the case of the gastric chief cell, stimulus-secretion coupling is thought to be mediated by two separate pathways. One pathway involves an increase of intracellular cAMP, while the alternate pathway involves a rise in cellular calcium. The precise steps involved in these pathways have not been defined. However, there is substantial evidence to indicate the existence of two pathways, each of which is coupled to specific stimuli.

CYCLIC AMP. The best evidence for an involvement of cAMP comes from the observation that cAMP and its derivatives stimulate pepsinogen secretion by all of the in vitro preparations examined (23, 38, 76, 77, 87, 96). No stimulation is observed with adenosine, AMP, or ATP indicating that an adenosine receptor is not involved in the response to the cyclic nucleotides (38, 96, 115). 8-Bromo-cIMP is a potent stimulus for pepsinogen secretion while 8-bromo-cGMP is a weak stimulus and neither cGMP nor dibutyryl cGMP stimulate (38). The potency sequence of the cyclic nucleotides is similar to the sequence for activating cAMP-dependent protein kinase (79), suggesting that these agents act through this enzyme. Cyclic AMP analogues also stimulate pepsinogen secretion by permeable gastric glands, indicating an intracellular site of action (68). In addition to the adenosine cyclic nucleotides themselves, three agents that are believed to act via increasing cellular cAMP levels (forskolin, cholera toxin, and prostaglandins) have been shown to stimulate pepsinogen secretion (6, 40, 68, 89).

Stimulation of pepsinogen secretion by adrenergic agents and secretin (or VIP) is believed to be mediated by cAMP. β-Adrenergic stimulation of various cell types involves activation of adenylate cyclase and elevation of cAMP content. Stimulation of pepsinogen secretion by isoproterenol is associated with elevation of cAMP content, and isoproterenol stimulates adenylate cyclase in gastric gland homogenates (55). Secretin was reported to increase the cAMP content of canine gastric cells, tentatively identified as an enriched population of chief cells (116). An adenylate cyclase activity sensitive to secretin (or VIP) has been identified in rat gastric glands (27). These results are consistent with the hypothesis that stimulation by adrenergic agents and secretin is mediated by activation of adenylate cyclase and an increase in intracellular cAMP. Further support for cAMP mediation of adrenergic and secretin stimulation comes from studies in which these stimuli are given in combination with cAMP derivatives (76). The combination of stimuli results in a less-than-additive response indicating that the exogenous cyclic nucleotide duplicates the rise in endogenous cAMP. In contrast, when the stimuli are combined with agents believed not to act via cAMP, the responses are additive or greater than additive. Further studies are required to firmly establish cAMP as a mediator for these stimuli, but the evidence so far is entirely consistent with such a hypothesis.

The manner in which cAMP may pass along the secretory stimulus in the chief cell remains speculative. By analogy with other systems, it would be expected that cAMP activates one or more protein kinases and that subsequent phosphorylation of key proteins activates the secretory mechanism. At present neither the specific protein kinases nor their endogenous substrates have been identified.

CALCIUM. A number of stimuli for pepsinogen secretion do not appear to be mediated by cAMP. These include cholinergic agonists and CCK-like peptides. Doses of CCK-8 or carbachol that are maximal for stimulating pepsinogen secretion do not stimulate adenylate cyclase or elevate cAMP content of gastric glands (20). Further evidence against cAMP mediation comes from the observation that carbachol and CCK responses are additive with agents that are thought to act via cAMP (76). However, responses to CCK-8 and carbachol are less than additive (37, 76). Thus CCK-8 and carbachol appear to share a common intracellular mechanism that does not involve cAMP. Since the responses to carbachol or CCK are less than additive when combined with the calcium ionophore A23187, it is postulated that these agents act via an increase in intracellular calcium (76). Direct measurements using fluorescent probes demonstrated a rise in intracellular calcium after addition of CCK-8 or carbachol to preparations of purified chief cells (19). This provides the most direct evidence for calcium mediation of these stimuli. The rise in intracellular calcium must result from a release of calcium from an intracellular pool since the responses are not blocked by removal of calcium from the medium unless there is prolonged incubation with calcium chelating agents (55).

Based on analogy with other systems, e.g., pancreas or liver, a possible mechanism for initiating calcium release would be formation of inositol trisphosphate (104, 105, 114). This would be accomplished through receptor activation of phospholipase C (phosphatidylinositol diesterase), which not only releases inositol phosphates but also produces 1,2-diacylglycerol (DAG). The DAG formed is a potent activator of protein kinase C (PK-C). Evidence for this sequence of events being involved in stimulation of pepsinogen secretion is rather preliminary, but initial results are at least consistent with such a model. Thus CCK-8 and carbachol have been shown to increase inositol phosphate levels in isolated chief cells (19). Beyond this initial observation it will be important to examine the relationships between inositol phosphates, intracellular calcium, and pepsinogen secretion to establish a sequence of events. A possible role of PK-C is suggested by the ability of phorbol esters to stimulate secretion (64, 80). These compounds are known to mimic the action of DAG in activating this kinase. While it seems likely that PK-C activation can lead to pepsinogen secretion, it is not clear whether PK-C plays a role in stimulation by secretagogues. In peptic cells from frog esophagus, phorbol esters display a synergistic interaction with A23187 and forskolin suggesting that PK-C action is independent of cAMP and calcium (64). However, in rabbit isolated gastric glands, combinations of phorbol esters with carbachol, CCK-8, or forskolin result in a less-than-additive response (S. Hersey and S. Matheravidathu, unpublished results), indicating a role for PK-C in the action of all three stimuli. Thus it seems likely that PK-C is involved in pepsinogen secretion, but the exact role has not been clarified as yet.

Secretory Mechanism

Since the original observations of Langley it has been postulated that pepsinogen is released from the peptic granules by a process now known as exocytosis. More recent studies employing electron microscopy and autoradiography have provided a qualitative description of the exocytotic process. The major criteria for demonstrating exocytosis are *1*) a loss of secretory granules accompanying secretion and *2*) the observation of exocytotic figures or fusion figures. In the gastric chief cell, a loss of peptic granules after stimulation has been observed both in vivo and in vitro. Exocytotic figures have been demonstrated by electron microscopy in chief cells of rat (34) and gastric glands from rabbit (28). In rabbit gastric glands, the granule fusion process includes both the fusion of granules with the plasma membrane and granule-to-granule fusion (28). The process involving granule-to-granule fusions is referred to as compound exocytosis and has been described for other cell types (6).

The demonstration that exocytosis occurs in chief cells provides qualitative evidence that this process is responsible for secretion of pepsinogen but does not eliminate the possible existence of alternative mechanisms. Secretion of pepsinogen in vivo occurs in two phases. After stimulation there is a rapid output of pepsinogen, which is presumed to reflect a loss of peptic granules. The rapid phase is followed by slower rate of continuous secretion, which can persist for hours. Since the continuous phase of secretion can persist after depletion of peptic granules, it has been suggested that this secretion involves a rapid release of newly synthesized pepsinogen that may not require encapsulation into granules before release (41). Secretion of pepsinogen in vitro also exhibits a biphasic pattern (28, 53, 54). The initial rapid release is associated with exocytotic figures (28), but it is not clear whether the slow phase involves exocytosis or some alternate mechanisms. Additional studies of the secretory dynamics, particularly in association with measurements of pepsinogen synthesis, may provide a firmer understanding of the secretory patterns.

The biochemical basis for the exocytotic release of pepsinogen is largely unknown. In the absence of stimuli, the peptide granules must be maintained in a state that prevents fusion with the plasma membrane or, for compound exocytosis, with each other. Events following stimulation must then alter the state of the granules to permit fusion and release of granule contents. Studies using model systems (21, 117) suggest that osmotic swelling of the granules may be an essential event in the initiation of exocytosis. A loss of electron density of peptic granules has been observed during exocytosis in chief cells (28, 35), and this could be interpreted as evidence for swelling of the granules (10). Exposure to hyperosmotic medium inhibits exocytosis in some cells (74) but stimulates secretion by chief cells (28). For chief cells, stimulation by hyperosmotic medium is blocked by ouabain or furosemide indicating that the secretion is secondary to membrane ion transport processes or changes in cell composition (29). These observations suggest that the sensitivity of the peptic granule to stimuli can be altered by changes in intracellular ion composition. This concept is supported by the observations that pepsinogen secretion is sensitive to ion composition of the incubation medium (29, 67), membrane transport inhibitors (29), and certain ionophores (67, 68). While no coherent pattern of ion specificity has emerged from these studies as yet, one might speculate that the peptic granule behaves like an independent organelle that regulates its composition to maintain a resting state. Alterations in the intracellular composition may interfere with the granule regulatory mechanisms so as to increase or decrease the sensitivity to stimuli. Investigations of the properties of peptic granules have only begun, but the initial studies indicate that the granules can maintain a pH gradient and may possess a proton transport mechanism (71). This property would be similar to other secretory granules that maintain a low intragranular pH (68, 93). The role of a pH gradient in peptic granules is unknown, but it could serve to maintain the pepsinogen in a condensed state before release (6, 10). The possible existence of other regulatory mechanisms remains to be demonstrated.

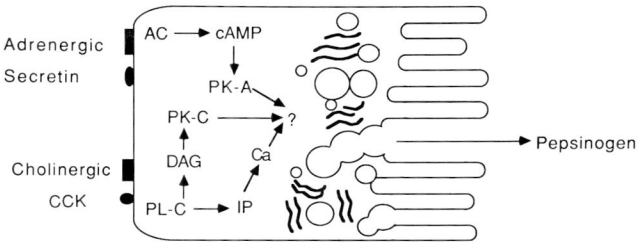

FIG. 7. Proposed model for cellular events in pepsinogen secretion by gastric chief cells. Adrenergic agonists and secretin bind to cell membrane receptors that are coupled to adenylate cyclase (AC). Subsequent increase in cAMP activates a protein kinase (PK-A). Cholinergic agonists and CCK-type peptides produce receptor-mediated activation of phospholipase C (PL-C). Phospholipase generates intracellular inositol phosphates (IP) that lead to release of Ca from intracellular stores. PL-C also generates diacylglycerol (DAG), which activates second protein kinase (PK-C). Activities of protein kinases and Ca result in secretion of pepsinogen by compound exocytotic process. Identifying mechanisms by which intracellular mediators activate exocytosis remains major challenge.

SUMMARY

The recent development of in vitro models for studying pepsinogen secretion has led to rapid progress in understanding the cellular basis of this physiological process. Although knowledge of the cellular mechanisms of secretion is far from complete, some general concepts have evolved from the in vitro investigations (Fig. 7). It is clear now that peptic cells possess a variety of receptors for stimuli, including cholinergic, adrenergic, and peptide receptors. Regulation of peptic cell activity appears to be independent of parietal cell acid secretion, at least at the cellular level. Stimulus-secretion coupling in the peptic cell involves at least two separable pathways. One pathway employs cAMP as a mediator while another pathway involves elevation of intracellular calcium. In addition, PK-C appears to be involved in the regulatory process. Once activated, the peptic cell releases pepsinogen through one or more processes including exocytosis or compound exocytosis.

Future studies will provide additional details for this general outline. Particular attention is called toward investigating the general areas of relating cellular stimuli to in vivo regulation, identifying the mechanisms of action of the intracellular mediators, and describing the basic mechanism of exocytotic release. Research efforts in these areas would both enhance knowledge of pepsinogen secretion and contribute to understanding of basic biological control mechanisms.

I am grateful to E. Christian and L. Tang for assistance in preparation of the manuscript. This work was supported in part by National Institutes of Health Grants DK-14752 and DK-36548.

REFERENCES

1. ANDREEVA, N. S. Structure of pepsin. *Mol. Biol.* 19: 185–190, 1985.
2. ANDREEVA, N. S., A. E. GUSTCHINA, A. A. FEDOROV, N. E. SHUTZKEVER, AND T. V. VOLNOVA. X-ray crystallographic studies of pepsin. *Adv. Exp. Med. Biol.* 95: 3–22, 1976.
3. ANSON, M. L., AND A. E. MIRSKY. The estimation of pepsin

with hemoglobin. *J. Gen. Physiol.* 16: 59–63, 1932.
4. BARROS, D., S. R. BLOOM, AND J. H. BARON. Inhibition by somatostatin (growth hormone release-inhibiting hormone, GH-RIH) of gastric acid and pepsin and G cell release of gastrin. *Gut* 19: 315–320, 1978.
5. BERGER, S., AND J.-P. RAUFMAN. Prostaglandin-induced pepsinogen secretion from dispersed gastric glands from guinea pig stomach. *Am. J. Physiol.* 249 (*Gastrointest. Liver Physiol.* 12): G592–G598, 1985.
6. BERGER, W., G. DOHL, AND H. P. MEISSNER. Structural and functional alterations in fused membranes of secretory granules during exocytosis in pancreatic islet cells of the mouse. *Cytobiologie* 12: 119–139, 1975.
7. BERGLINDH, T., AND K. J. OBRINK. A method for preparing isolated glands from the rabbit gastric mucosa. *Acta Physiol. Scand.* 96: 150–159, 1976.
8. BERQUIST, E., AND K. J. OBRINK. Gastrin-histamine as a normal sequence in gastric acid stimulation in the rabbit. *Upsala J. Med. Sci.* 84: 145–154, 1979.
9. BERSTAD, A., AND H. PETERSEN. Dose-response relationship of the effect of secretin on acid and pepsin secretion in man. *Scand. J. Gastroenterol.* 5: 647–654, 1970.
10. BILINSKI, M., H. PLATTNER, AND H. MATT. Secretory protein decondensation as a distinct Ca mediated event during the final steps of exocytosis in paramecium cells. *J. Cell Biol.* 88: 179–188, 1981.
11. BILSKI, A. J., S. E. HALLIDAY, J. D. FITZGERALD, AND J. L. WALE. The pharmacology of a β_2-selective adrenoceptor antagonist (ICI 118,551). *J. Cardiovasc. Pharmacol.* 5: 430–437, 1983.
12. BOJAMIC, D., J. D. JANSEN, S. R. NAHYORSKI, AND J. ZAAGSMA. Atypical characteristic of the β-adrenoceptor mediating cyclic AMP generation and lipolysis in the rat adipocyte. *Br. J. Pharmacol.* 84: 131–137, 1985.
13. BONDOT, J. P., I. CAVERO, S. FENARD, F. LEFEVRE-BORG, P. MANOURY, AND A. G. ROACH. Preliminary studies on SL75212, a new potent cardioselective β-adrenoceptor antagonist. *Br. J. Pharmacol.* 66: 445P, 1979.
14. BRAGANZA, J. M., H. T. HOWAT, AND G. H. KAY. A comparison of the pepsin stimulating effects of secretin preparations. *J. Physiol. Lond.* 258: 63–72, 1976.
15. BUSA, W. B., AND R. NUCCITELLI. Metabolic regulation via intracellular pH. *Am. J. Physiol.* 246 (*Regulatory Integrative Comp. Physiol.* 15): R409–R438, 1984.
16. CASTAGNA, M., Y. TAKAI, K. KAIBUCHI, K. SANO, J. KIKKAWA, AND Y. NISHIZUKA. Direct activation of calcium-activated, phospholipid-dependent protein kinase by tumor promoting phorbol esters. *J. Biol. Chem.* 257: 7847–7851, 1982.
17. CHANG, R. S. L., V. J. LOTTI, R. L. MONAGHAN, J. BIRNBAUM, E. O. STAPLEY, M. A. GOETZ, G. ALBERO-SCHONBERG, A. A. PATCHETT, J. M. LIESCH, O. D. HENSENS, AND J. P. SPRINGER. A potent nonpeptide cholecystokinin antagonist selective for peripheral tissues isolated from *Aspergillus alliaceus*. *Science Wash. DC* 230: 177–179, 1985.
18. CHERNER, J. A., V. E. SUTLIFF, D. M. GRYBOWSKI, R. T. JENSEN, AND J. D. GARNER. Functionally distinct receptors for cholecystokinin and gastrin on dispersed chief cells from guinea pig stomach. *Am. J. Physiol.* 254 (*Gastrointest. Liver Physiol.* 17): G151–G155, 1988.
19. CHEW, C. S., AND M. R. BROWN. Release of intracellular Ca and elevation of inositol trisphosphate by secretagogues in parietal and chief cells isolated from rabbit gastric mucosa. *Biochim. Biophys. Acta* 888: 116–125, 1986.
20. CHEW, C. S., AND S. J. HERSEY. Gastrin stimulation of isolated gastric glands. *Am. J. Physiol.* 242 (*Gastrointest. Liver Physiol.* 5): G504–G512, 1982.
21. COHEN, F. S., J. ZIMMERBERG, AND A. FINKELSTEIN. Fusion of phospholipid vesicles with planar phospholipid bilayers. II. Incorporation of a vesicular membrane marker into the planar membrane. *J. Gen. Physiol.* 75: 251–270, 1980.
22. FELDMAN, E. J., AND M. I. GROSSMAN. Liver extract and its free amino acids equally stimulate gastric acid secretion. *Am.*

J. Physiol. 239 (*Gastrointest. Liver Physiol.* 2): G493–G496, 1980.
23. FIMMEL, C. J., M. M. BERGER, AND A. L. BLUM. Dissociated response of acid and pepsin secretion to omeprazole in an in vitro perfused mouse stomach. *Am. J. Physiol.* 247 (*Gastrointest. Liver Physiol.* 10): G240–G247, 1984.
24. FOLTMAN, B., AND V. B. PEDERSEN. Comparison of primary structures of acid proteases and their zymogens. *Adv. Exp. Med. Biol.* 95: 3–22, 1976.
25. FRIEDMAN, M. H. Oesophageal and gastric secretion in the frog. *J. Cell Comp. Physiol.* 10: 37–50, 1937.
26. FRUTON, J. S. The mechanism of the catalytic action of pepsin and related acid proteinases. *Adv. Enzymol.* 44: 1–36, 1976.
27. GESPACH, C., D. BATAILLE, C. DUPONT, G. ROSSELIN, E. WUNCH, AND E. JAEGER. Evidence for a cyclic AMP system highly sensitive to secretin in gastric glands isolated from the rat fundus and antrum. *Biochim. Biophys. Acta* 630: 433–441, 1980.
28. GIBERT, A. J., AND S. J. HERSEY. Exocytosis in isolated gastric glands induced by secretagogues and hyperosmolarity. *Cell Tissue Res.* 227: 535–542, 1982.
29. GIBERT, A. J., AND S. J. HERSEY. Effect of ouabain and furosemide on pepsinogen secretion by gastric glands in vitro. *J. Cell Physiol.* 119: 220–226, 1984.
30. GIBSON, R., B. I. HIRSCHOWITZ, AND G. HUTCHISON. Actions of metiamide, an H_2-histamine receptor antagonist, on gastric H^+ and pepsin secretion in dogs. *Gastroenterology* 67: 93–99, 1977.
31. GREGORY, R. A., AND H. J. TRACEY. The constitution and properties of two gastrins extracted from hog antral mucosa. *Gut* 5: 103–117, 1964.
32. HAMMER, R., AND A. GIACHETTI. Muscarinic receptor subtypes: M1 and M2. Biochemical and functional characterization. *Life Sci.* 31: 2991–2998, 1982.
33. HARDEN, T. K., L. I. TANNER, M. W. MARTIN, N. NAKAHATA, A. R. HUGHES, J. R. HEPLER, T. EVANS, S. B. MASTERS, AND J. H. BROWN. Characteristics of two biochemical responses to stimulation of muscarinic cholinergic receptors. *Trends Pharmacol. Sci.* 7, Suppl.: 14–18, 1986.
34. HELANDER, H. F. Quantitative ultrastructural studies on rat gastric zymogen cells under different physiological and experimental conditions. *Cell Tissue Res.* 189: 287–303, 1965.
35. HELANDER, H. F. The cells of the gastric mucosa. *Int. Rev. Cytol.* 70: 217–289, 1981.
36. HERAWI, M., G. LAMBRECHT, E. MUTSCHLER, U. MOSER, AND A. PFIEFFER. Different binding properties of muscarinic M_2-receptor subtypes for agonists and antagonists in porcine gastric smooth muscle and mucosa. *Gastroenterology* 94: 630–637, 1988.
37. HERSEY, S. J., D. MAY, AND D. SCHYBERG. Stimulation of pepsinogen release from isolated gastric glands by cholecystokininlike peptides. *Am. J. Physiol.* 244 (*Gastrointest. Liver Physiol.* 7): G192–G197, 1983.
38. HERSEY, S. J., M. MILLER, D. MAY, AND S. H. NORRIS. Lack of interaction between acid and pepsinogen secretion in isolated gastric glands. *Am. J. Physiol.* 245 (*Gastrointest. Liver Physiol.* 8): G775–G779, 1983.
40. HERSEY, S. J., A. OWIRODU, AND M. MILLER. Forskolin stimulation of acid and pepsinogen secretion by gastric glands. *Biochim. Biophys. Acta* 755: 293–299, 1983.
41. HIRSCHOWITZ, B. I. Secretion of pepsinogen. In: *Handbook of Physiology. Alimentary Canal. Secretion*, edited by C. F. Code. Washington, DC: Am. Physiol. Soc., 1967, vol. II, chapt. 50, p. 889–918.
42. HIRSCHOWITZ, B. I., J. FONG, AND E. MOLINA. Effects of pirenzepine and atropine on vagal and cholinergic gastric secretion and gastrin release and on heart rate in the dog. *J. Pharmacol. Exp. Ther.* 225: 263–268, 1983.
43. HIRSCHOWITZ, B. I., AND R. G. GIBSON. Effect of cimetidine on stimulated gastric secretion and serum gastrin in the dog. *Am. J. Gastroenterol.* 70: 437–447, 1978.
44. HIRSCHOWITZ, B. I., AND G. A. HUTCHISON. Evidence for a

histamine H$_2$ receptor that inhibits pepsin secretion in the dog. *Am. J. Physiol.* 233 (*Endocrinol. Metab. Gastrointest. Physiol.* 2): E225–E228, 1977.
45. HIRSCHOWITZ, B. I., AND G. HUTCHISON. Kinetics of atropine inhibition of pentagastrin-stimulated H$^+$, electrolyte, and pepsin secretion in the dog. *Am. J. Dig. Dis.* 22: 99–107, 1977.
46. HIRSCHOWITZ, B. I., AND E. MOLINA. Relation of gastric acid and pepsin secretion to serum gastrin levels in dogs given bombesin and gastrin-17. *Am. J. Physiol.* 244 (*Gastrointest. Liver Physiol.* 7): G546–G551, 1983.
47. HIRST, B. H., J. M. CONLON, D. H. COY, J. HOLLAND, AND B. SHAW. Comparison of the gastric exocrine inhibitory activities and plasma kinetics of somatostatin-28 and somatostatin-14 in cats. *Regul. Pept.* 4: 227–237, 1982.
48. ICHIHARA, Y., K. SOGAWA, AND K. TAKAHASHI. Rat gastric prepepsinogen: in vitro synthesis and partial amino-terminal signal sequence. *J. Biochem. Tokyo* 92: 603–606, 1982.
49. JOHNSON, L. R. Regulation of pepsin secretion by topical acid in the stomach. *Am. J. Physiol.* 223: 847–850, 1972.
50. JOHNSON, L. R., G. E. STENNING, AND M. I. GROSSMAN. Effect of sulfation on the gastrointestinal actions of caerulein. *Gastroenterology* 58: 208–216, 1970.
51. KAGEYAMA, T., AND K. TAKAHASHI. Isolation of an activation intermediate and determination of the amino acid sequence of the activation segment of human pepsinogen A. *J. Biochem. Tokyo* 88: 571–582, 1980.
52. KAMINSKI, D. L., M. J. RUWART, AND M. JELLINEK. Structure-function relationships of peptide fragments of gastrin and cholecystokinin. *Am. J. Physiol.* 233 (*Endocrinol. Metab. Gastrointest. Physiol.* 2): E286–E292, 1977.
53. KASBEKAR, D. K., R. T. JENSEN, AND J. D. GARDNER. Pepsinogen secretion from dispersed glands from rabbit stomach. *Am. J. Physiol.* 244 (*Gastrointest. Liver Physiol.* 7): G392–G396, 1983.
54. KAY, J., AND C. W. DYKES. The first cleavage site in pepsinogen activation. *Adv. Exp. Med. Biol.* 95: 103–127, 1976.
55. KOELZ, H. R., S. J. HERSEY, G. SACHS, AND C. S. CHEW. Pepsinogen release from isolated gastric glands. *Am. J. Physiol.* 243 (*Gastrointest. Liver Physiol.* 6): G218–G225, 1982.
56. KONTUREK, S. J., J. BIERNAT, N. KWIECIEN, AND J. OLEKSY. Effect of glucagon on meal-induced gastric secretion in man. *Gastroenterology* 68: 448–454, 1975.
57. KONTUREK, S. J., N. KWIECIEN, W. OBTULOWICZ, W. BIELANSKI, J. OLEKSY, AND A. V. SCHALLY. Effects of somatostatin-14 and somatostatin-28 on plasma hormonal and gastric secretory responses to cephalic and gastrointestinal stimulation in man. *Scand. J. Gastroenterol.* 20: 31–38, 1985.
58. KUMAR, P. M. H., P. H. WARD, AND B. KASSELL. Chemical modification of a pepsin inhibitor from the activation peptides of pepsinogen. *Adv. Exp. Med. Biol.* 95: 211–221, 1976.
59. LANGLEY, J. N. On the histology and physiology of pepsin-forming glands. *Phil. Trans. R. Soc. Lond. Ser. B* 172: 664–711, 1881.
60. LANGLEY, J. N., AND J. S. EDKINS. Pepsinogen and pepsin. *J. Physiol. Lond.* 7: 371–415, 1886.
61. LICHTENBERGER, L. M. Importance of food in the regulation of gastrin release and formation. *Am. J. Physiol.* 243 (*Gastrointest. Liver Physiol.* 6): G429–G441, 1982.
62. MAGOUS, R., B. BAUDIERE, AND J.-P. BALIE. Muscarinic receptors in isolated gastric fundic mucosal cells. *Biochem. Pharmacol.* 34: 2269–2274, 1985.
63. MARCINISZYN, J., J. S. HUANG, J. A. HARTSUCK, AND J. TANG. Mechanisms of intramolecular activation of pepsinogen. *J. Biol. Chem.* 251: 7095–7102, 1976.
64. MATSUMOTO, H., K. KOMIYAMA, T. SHIRAKAWA, A. HELDMAN, W. ANDERSON, AND B. I. HIRSCHOWITZ. Stimuli of pepsinogen secretion from frog isolated peptic cells. *Federation Proc.* 45: 1043, 1986.
65. MULLER-LISSNER, S. A., I. SZELENYI, H. ENGLER, AND A. L. BLUM. Pirenzepine inhibits acid and pepsinogen secretion by the isolated perfused mouse stomach. *Scand. J. Gastroenterol. Suppl.* 72: 101–103, 1982.

66. NAKAJIMA, S., AND D. F. MAGEE. Influences of duodenal acidification on acid and pepsin secretion of the stomach in dogs. *Am. J. Physiol.* 218: 545–549, 1970.
67. NORRIS, S. H., AND S. J. HERSEY. Inhibition of pepsinogen secretion in isolated gastric glands by amphotericin B is secretagogue specific. *Can. J. Physiol. Pharmacol.* 62: 1518–1524, 1984.
68. NORRIS, S. H., AND S. J. HERSEY. Stimulation of pepsinogen secretion in permeable isolated gastric glands. *Am. J. Physiol.* 249 (*Gastrointest. Liver Physiol.* 12): G408–G415, 1985.
69. PACE, C. S., AND G. SACHS. Glucose induced proton uptake in secretory granules of B-cells in monolayer culture. *Am. J. Physiol.* 242 (*Cell Physiol.* 11): C382–C387, 1982.
70. PAGANI, F., A. SCHIAVONE, E. MONFERINI, R. HAMMER, AND A. GIACHETTI. Distinct muscarinic receptor subtypes (M$_1$ and M$_2$) controlling acid secretion in rodents. *Trends Pharmacol. Sci.* 5, *Suppl.*: 66–68, 1984.
71. PEERCE, B. E., A. SMOLKA, AND G. SACHS. Isolation of pepsinogen granules from rabbit gastric mucosa. *J. Biol. Chem.* 259: 9255–9262, 1984.
72. PEIKIN, S. R., C. L. COSTENBADER, AND J. D. GARDNER. Actions of derivatives of cyclic nucleotides on dispersed acini from guinea pig pancreas: discovery of a competitive antagonist of the action of cholecystokinin. *J. Biol. Chem.* 254: 5321–5327, 1979.
73. POHL, J., M. BAUDYS, AND V. KOSTKA. Chromophoric peptide substrates for activity determination of animal aspartic proteinases in the presence of their zymogens: a novel assay. *Anal. Biochem.* 133: 104–109, 1983.
74. POLLARD, H. B., C. J. PAZOLES, C. E. CRENTZ, AND O. SINDER. The chromaffin granule and possible mechanisms of exocytosis. *Int. Rev. Cytol.* 58: 159–197, 1979.
75. PRAISSMAN, M., M. E. WALDEN, AND C. PELLECCHIA. Identification and characterization of a specific receptor for cholecystokinin on isolated fundic glands from guinea pig gastric mucosa using a biologically active ^{125}I-CCK-8 probe. *J. Receptor Res.* 3: 647–665, 1983.
76. RAUFMAN, J.-P., D. K. KASBEKAR, R. T. JENSEN, AND J. D. GARDNER. Potentiation of pepsinogen secretion from dispersed glands from rat stomach. *Am. J. Physiol.* 245 (*Gastrointest. Liver Physiol.* 8): G525–G530, 1983.
77. RAUFMAN, J.-P., V. E. SUTLIFF, D. K. KASBEKAR, R. T. JENSEN, AND J. D. GARDNER. Pepsinogen secretion from dispersed chief cells from guinea pig stomach. *Am. J. Physiol.* 247 (*Gastrointest. Liver Physiol.* 10): G95–G104, 1984.
78. REHFELD, J. F. Four basic characteristics of gastrin-cholecystokinin system. *Am. J. Physiol.* 240 (*Gastrointest. Liver Physiol.* 3): G255–G266, 1981.
79. REIMANN, E. M., D. A. WALSH, AND E. G. KREBS. Purification and properties of rabbit skeletal muscle adenosine 3′,5′-monophosphate-dependent protein kinase. *J. Biol. Chem.* 246: 1986–1995, 1971.
80. SAKAMOTO, C., T. MATOZAKI, M. NAGAO, AND S. BABA. Combined effect of phorbol ester and A23187 or dibutyryl cyclic AMP on pepsinogen secretion from isolated gastric glands. *Biochem. Biophys. Res. Commun.* 131: 319–324, 1985.
81. SAMLOFF, I. M. Slow moving protease and the seven pepsinogens. Electrophoretic demonstration of the existence of eight proteolytic fractions in human gastric mucosa. *Gastroenterology* 57: 659–669, 1966.
82. SAMLOFF, I. M. Pepsinogens, pepsins, and pepsin inhibitors. *Gastroenterology* 60: 586–604, 1971.
83. SAMLOFF, I. M. Cellular localization of group I pepsinogens in human gastric mucosa by immunofluorescence. *Gastroenterology* 61: 185–188, 1971.
84. SAMLOFF, I. M. Pepsinogens I and II: purification from gastric mucosa and radioimmunoassay in serum. *Gastroenterology* 82: 26–33, 1982.
85. SAMLOFF, I. M., AND W. M. LIEBMAN. Cellular localization of the group II pepsinogens in human stomach and duodenum. *Gastroenterology* 65: 36–42, 1973.
86. SAMLOFF, I. M., AND W. M. LIEBMAN. Radioimmunoassay of

group I pepsinogens in serum. *Gastroenterology* 66: 494–502, 1984.
87. SANDERS, M. J., D. A. AMIRIAN, A. AYALON, AND A. H. SOLL. Regulation of pepsinogen release from canine chief cells in primary monolayer culture. *Am. J. Physiol.* 245 (*Gastrointest. Liver Physiol.* 8): G641–G646, 1983.
88. SANKARAN, H., I. D. GOLDFINE, C. W. DEVENEY, K.-Y. WONG, AND J. A. WILLIAMS. Binding of cholecystokinin to high affinity receptors on isolated rat pancreatic acini. *J. Biol. Chem.* 255: 1849–1853, 1980.
89. SCHAFER, D. E., AND G. N. GARSHFIELD. Cholera enterotoxin stimulates marked pepsinogen secretion by isolated gastric fundic glands. *Federation Proc.* 41: 1432, 1982.
90. SCHUURMANS STEKHOVEN, F., AND S. L. BONTING. Transport adenosine triphosphatases: properties and functions. *Physiol. Rev.* 61: 1–76, 1981.
91. SCHWANN, T. L. Ueber das wesen des verdauungsprozessen. *Poggendorf Ann. Phys. Chem.* 38: 358–364, 1836.
92. SEIJFFERS, M. J., M. D. TURNER, L. L. MILLER, AND H. L. SEGAL. Human pepsinogens and pepsins. *Gastroenterology* 48: 122–125, 1965.
93. SHERMAN, O., AND J. P. HENRY. Role of the proton electrochemical gradient monoamine transport by bovine chromaffin granules. *Biochim. Biophys. Acta* 601: 64–677, 1980.
94. SHIRAKAWA, T., AND B. I. HIRSCHOWITZ. Bombesin-induced pepsinogen secretion from frog esophagus peptic glands in vitro. *Gastroenterology* 86: 1250, 1984.
95. SHUGERMAN, R. P., B. I. HIRSCHOWITZ, A. S. BHOWN, R. E. SCHROHENLOHER, AND J. G. SPENNEY. A unique "mini" pepsinogen isolated from bullfrog esophageal glands. *J. Biol. Chem.* 257: 795–798, 1982.
96. SIMPSON, L., D. GOLDENBERG, AND B. I. HIRSCHOWITZ. Pepsinogen secretion by the frog esophagus in vitro. *Am. J. Physiol.* 238 (*Gastrointest. Liver Physiol.* 1): G79–G84, 1980.
97. SINGER, M. V., W. NIEBEL, C. LAMERS, S. BECKER, J. VESPER, W. HARTMANN, J. DIEMEL, AND H. GOEBELL. Effects of truncal vagotomy and antrectomy on bombesin-stimulated pancreatic secretion, release of gastrin, and pancreatic polypeptide in the anesthetized dog. *Dig. Dis. Sci.* 26: 871–877, 1981.
98. SKOUHO-KRISTENSEN, E., AND J. FRYKLUND. Adrenergic stimulation of pepsinogen release from rabbit isolated gastric glands. *Naunyn-Schmeideberg's Arch. Pharmacol.* 330: 37–41, 1985.
99. SOGAWA, K., Y. FUJII-KURIYAMA, Y. MIZUKAMI, Y. ISCHIHARA, AND K. TAKAHASHI. Primary structure of human pepsinogen gene. *J. Biol. Chem.* 258: 5306–5311, 1983.
100. SOGAWA, K., Y. ICHIHARA, Y. TAKAHASHI, Y. FUJII-KURIYAMA, AND M. MURAMATSU. Molecular cloning of complementary DNA to swine pepsinogen mRNA. *J. Biol. Chem.* 256: 12561–12565, 1981.
101. SOLL, A. H., D. A. AMIRIAN, L. P. THOMAS, T. J. REEDY, AND J. D. ELASHOFF. Gastrin receptors on isolated canine parietal cells. *J. Clin. Invest.* 73: 1434–1447, 1984.
102. SOUMARON, A., A. M. CHERET, AND M. J. M. LEWIN. Localization of gastrin receptors in intact isolated and separated rat fundic cells. *Gastroenterology* 73: 900–903, 1977.
103. STENING, G. F., L. R. JOHNSON, AND M. I. GROSSMAN. Effect of secretin on acid and pepsin secretion in cat and dog. *Gastroenterology* 56: 468–475, 1969.
104. STREB, H., E. BAYERDORFFER, W. HAASE, R. F. IRVINE, AND I. SCHULZ. Effect of inositol-1,4,5-triphosphate on isolated subcellular fractions of rat pancreas. *J. Membr. Biol.* 81: 241–253, 1984.
105. STREB, H., R. F. IRVINE, M. J. BERRIDGE, AND I. SCHULZ. Release of Ca from a nonmitochondrial intracellular store in pancreatic acinar cells by inositol-1,4,5-trisphosphate. *Nature Lond.* 306: 67–69, 1983.
106. SUTTON, D. R., AND R. M. DONALDSON. Synthesis and secretion of protein and pepsinogen by rabbit gastric mucosa in organ culture. *Gastroenterology* 69: 166–174, 1975.
107. TAGGART, R. T., T. K. MOHANDAS, T. B. SHOWS, AND I. B. GRAEME. Variable numbers of pepsinogen genes are located in the centromeric region of human chromosome 11 and determine the high-frequency electrophoretic polymorphism. *Proc. Natl. Acad. Sci. USA* 82: 6240–6244, 1985.
108. TAKEUCHI, K., G. R. SPEIR, AND L. R. JOHNSON. Mucosal gastrin receptor. I. Assay standardization and fulfillment of receptor criteria. *Am. J. Physiol.* 237 (*Endocrinol. Metab. Gastrointest. Physiol.* 6): E284–E294, 1979.
109. TAKEUCHI, K., G. R. SPEIR, AND L. R. JOHNSON. Mucosal gastrin receptor. IV. Binding specificity. *Am. J. Physiol.* 239 (*Gastrointest. Liver Physiol.* 2): G395–G399, 1980.
110. TANG, J. (editor). *Advances in Experimental Medicine and Biology. Acid Proteases.* New York: Plenum, 1976, vol. 95.
111. TANG, J. Evolution in the structure and function of carboxyl proteases. *Mol. Cell. Biochem.* 26: 93–109, 1979.
112. VAGNE, M., C. ROCHE, J. A. CHAYVIALLE, AND C. GESPACH. Effect of somatostatin on the secretin-induced gastric secretions of pepsin and mucus in cats. *Regul. Pept.* 3: 183–191, 1982.
113. VATIER, M. M., A. M. CHERET, AND S. BONFILS. Le dosage automatique de l'activite proteolytique du suc gastrique. *Biol. Gastro-Enterol.* 1: 15–29, 1968.
114. WILLIAMSON, J. R., R. H. COOPER, S. K. JOSEPH, AND A. P. THOMAS. Inositol triphosphate and diacylglycerol as intracellular second messengers in liver. *Am. J. Physiol.* 248 (*Cell Physiol.* 19): C203–C216, 1985.
115. WOLF, J., D. LONDOS, AND D. M. F. COOPER. Adenosine receptors and the regulation of adenylate cyclase. *Adv. Cyclic Nucleotide Res.* 14: 199–214, 1981.
116. WOLLIN, A., A. H. SOLL, AND I. M. SAMLOFF. Actions of histamine, secretin and PGE_2 on cyclic AMP production by isolated canine fundic mucosal cells. *Am. J. Physiol.* 237 (*Endocrinol. Metab. Gastrointest. Physiol.* 6): E437–E443, 1979.
117. ZIMMERBERG, J., F. S. COHEN, AND A. FINKELSTEIN. Fusion of phospholipid vesicles with planar phospholipid bilayer membranes. I. Discharge of vesicular contents across the planar membrane. *J. Gen. Physiol.* 75: 241–250, 1980.

CHAPTER 15

The gastric mucosal barrier

BARRY H. HIRST | Department of Physiological Sciences, University of Newcastle upon Tyne Medical School, Newcastle upon Tyne, United Kingdom

CHAPTER CONTENTS

Electrolyte Composition of Gastric Juice
Electrical Characteristics of Gastric Mucosa
    Comparison of fundic and antral gastric mucosa
Proton Permeability of Gastric Mucosa
Constituents of the Barrier
Mucus-Bicarbonate Layer
Epithelial Cell Layer
    Tight junctions
    Apical cell membranes
Studies With Barrier-Breaking Agents
    Aspirin and weak acids
    Bile acids
    Alcohols
    Pepsin
Gastric Membrane Composition
    Lipid constituents
    Glycosubstances
Surface Hydrophobicity
Direct Evidence for Impermeability of Apical Membranes to Protons
    Reconstituted epithelial cell monolayers
    Isolated apical plasma membrane vesicles
Regulation of Intracellular pH
Gastric Microcirculation
    Tissue acid-base balance
    Tissue oxygenation
    Oxygen radical generation
Rapid Reepithelialization of Gastric Mucosa
Cytoprotection
    Prostaglandins
    Sulfhydryl compounds and other agents
Conclusions

HOW THE STOMACH is able to carry out its normal digestive functions while not compromising its own integrity has intrigued scientists at least from the time of John Hunter (126). The acidity of gastric juice can reach 160 mM (i.e., pH 0.8), which represents a $>10^6$ concentration gradient across the mucosa. The ability of the stomach to maintain such an H^+ gradient has been termed the *gastric mucosal barrier*. This concept of a barrier has been generally extended to include the defense of the stomach against a variety of other potentially aggressive agents, including pepsin.

Gastric mucosal defense and repair mechanisms must be in balance with the aggressive agents, including acid and pepsin, for the normal functioning of the stomach. The gastric mucosal barrier is a physiological concept rather than a precise anatomical barrier. In the mechanistic sense the barrier consists of a series of physiological processes.

In this chapter the cellular mechanisms underlying barrier function, in particular the epithelial cell layer, are emphasized. The mechanisms of action of specific barrier-breaking agents are discussed, and an attempt is made to assign barrier function to specific epithelial sites. Other aspects of the barrier, including the mucus-HCO_3^- layer overlying the mucosa, and tissue acid-base balance and blood flow are also discussed in relation to the barrier, although these are covered in more detail in the chapters by Allen, by Flemström and Garner, and by Demarest and Machen in this *Handbook*. Finally, the processes of restoration of the damaged epithelial cell layer and cytoprotection are discussed.

ELECTROLYTE COMPOSITION OF GASTRIC JUICE

The composition of gastric juice is not constant but varies with flow rate. In particular, $[H^+]$ increases asymptotically with flow rate, accompanied by an inversely related decrease in $[Na^+]$ [Fig. 1; (117, 120, 122, 158, 159, 175, 239)]. At maximal rates of acid secretion the $[H^+]$ achieved is >150 mM, and the calculated primary $[H^+]$ in various mammalian species is ~140–170 mM, with the lower values from experiments in humans (87, 93, 95, 117, 120, 149, 159, 160). Compared with plasma, there is a $>10^6:1$ $[H^+]$ within the gastric lumen. The production of this H^+ gradient is achieved by the gastric H^+-K^+-ATPase localized in the parietal cells. The ability of the gastric mucosa to maintain a $>10^6$ concentration gradient for H^+ is a function of the gastric mucosal barrier.

The low permeability of the gastric mucosa to H^+ was first examined in detail by Teorell (239), who

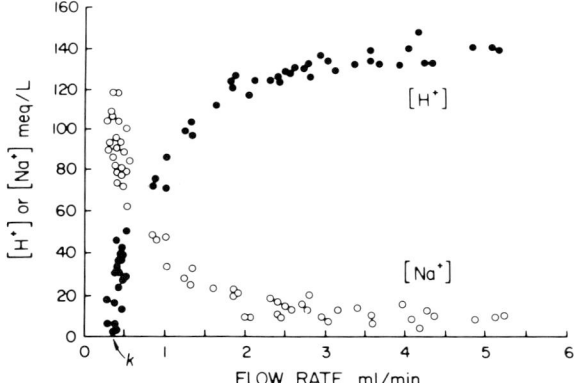

FIG. 1. Relationship between [H$^+$] and [Na$^+$] and flow rate of gastric juice in humans. Data were obtained in the unstimulated state and during stimulation with gastrin. k, Flow rate of nonparietal component of gastric juice. [From Makhlouf (158). In: *Physiology of the Gastrointestinal Tract*, © 1981, Raven Press, New York.]

described the disappearance of H$^+$ from acidic solutions instilled into the stomach. Qualitatively the H$^+$ loss was compensated by a gain in Na$^+$ from plasma. Teorell explained the varying ionic composition of gastric juice in terms of this backdiffusion of H$^+$ in exchange for Na$^+$ and found it difficult to accept that a nonacidic secretion could be the major cause of the reduction of gastric acidity. In contrast, Hollander (122) championed the concept of two components of gastric secretion, the gastric juice being a variable mixture of an isotonic solution of HCl secreted by the parietal cells, which is diluted and neutralized by an Na$^+$-rich alkaline nonparietal secretion. Potassium is a component of both the parietal and nonparietal secretions. The [K$^+$] of gastric juice is linearly related to [H$^+$], suggesting a link between these two ions, with average [K$^+$] ~10% of the [H$^+$] (158). In Hollander's original proposals the nonparietal component was seen as being produced at a constant rate, and the constancy of the Na$^+$ output in gastric juice supports this (24, 93, 159, 160). However, all these earlier studies observed the constancy of Na$^+$ output while changing the flow of gastric juice with stimulants of parietal cell secretion, e.g., histamine and gastrin. More recent data illustrate that the nonparietal component of gastric juice is also not constant. With appropriate agents, e.g., cholinomimetics and prostaglandins, the secretion of the nonparietal component, including Na$^+$ and HCO$_3^-$, can be increased (28, 65, 84, 87, 114, 221). Thus the two-component hypothesis to explain the ionic composition of gastric juice should be modified as a mixture of variable amounts of acidic parietal cell secretion and variable amounts of Na$^+$-rich alkaline nonparietal cell secretion. However, in quantitative terms the maximum capacity for parietal cell secretion far exceeds that for nonparietal secretion.

Makhlouf (158) examined these two hypotheses to explain the varying ionic composition of gastric juice and described a formal similarity between the two. Moreover, quantitatively the calculated flow of nonparietal secretion in humans is identical to the calculated permeability coefficient for H$^+$ [both ~0.35 ml·min^{-1} (159, 175)]. These values represent 6%–8% of the maximal flow rate of gastric juice. Direct quantitative evidence for nonparietal secretion is now available, and in particular much attention has been placed on the HCO$_3^-$ constituent as a component of the mucus-HCO$_3^-$ layer (see the chapter by Flemström and Garner in this *Handbook*). The maximal rate of unstimulated alkalinization (equivalent to HCO$_3^-$ secretion) of 0.25–0.55 μmol·cm^{-2}·h^{-1} in *Rana* species or *Necturus* gastric mucosa in vitro is <10% of maximal H$^+$ secretion in the same species (67). Similarly in the cat in vivo, unstimulated HCO$_3^-$ secretion of ~3 μmol·15 min^{-1} (~1 μmol·cm^{-2}·h^{-1}) from a vagally denervated (Heidenhain) pouch (222) is <2% of maximal acid secretion (~200 μmol·15 min^{-1}) in this preparation (232). Stimulation of gastric HCO$_3^-$ transport in vivo and in vitro, for example, with 16,16-dimethyl prostaglandin E$_2$ results in a maximum doubling of the secretory rate (86, 118, 222). Therefore secretion of an alkaline fluid can quantitatively account for a large proportion, if not the majority, of the rate of disappearance of H$^+$ from the gastric lumen. In addition, increased intraluminal acidity may cause a further increase in gastric HCO$_3^-$ secretion (68), although such increases have not been uniformly described (6).

ELECTRICAL CHARACTERISTICS OF GASTRIC MUCOSA

The gastric mucosa is generally assumed to be a tight epithelium, i.e., its conductance properties are dominated by transcellular rather than paracellular pathways (80). This view of the gastric mucosa is in accordance with its ability to maintain large H$^+$ and Na$^+$ gradients. The greater proportion of the total tissue conductance appears to be located in the gastric glands rather than the surface cells. Measurements in frog gastric mucosa indicate that as much as 90% of the transepithelial conductance may be provided through the tubules (31). This is not surprising when the large surface area provided by the glandular region is considered. It has been estimated that presence of gastric glands and pits increases the nominal surface area of the stomach by >10-fold (106). The surface topography of the gastric cells provides a further increase in surface area. In particular the apical membranes of all the major cell types are covered in microvilli (127). The apical surface of the parietal cells is further folded to form intracellular canaliculi (106, 127). In the unstimulated state, intracellular canaliculi are relatively modest in area, but on stimulation the surface is increased by ~10-fold because of the fusion of intracellular tubulovesicles with the apical membrane (70, 89, 106). Because parietal cells comprise ~30%–40% of the total cell population of the mucosa of the body of the stomach (127), this surface topography and its increase on stimulation are significant.

Further evidence that the major permeability pathways in the gastric mucosa originate within the glands comes from ultrastructural studies in the dog. Using a mixture of quantitative freeze-fracture and thin-section extracellular tracer electron-microscopic methods, Meyer et al. (165) have correlated tight-junction structure with epithelial permeability. They documented a difference in the arrangement of strands within the tight junctions of surface epithelial, as compared with parietal and peptic, cell junctions. In addition, the surface cell junctions were impermeable to the extracellular tracer lanthanum, whereas the gland cell junctions were permeable.

Measurements of the transepithelial resistance of the gastric mucosa, particularly when stimulated to secrete acid, yield nominal values as low as 50–100 $\Omega \cdot cm^2$ (71, 191), comparable with those generally considered to indicate a leaky epithelium (80). However, if the complexity of the membrane convolutions is included in the calculation, the value for transepithelial resistance increases to closer to 9,000 $\Omega \cdot cm^2$ (53), consistent with values observed for other tight epithelia, such as frog skin and rabbit urinary bladder. Microelectrode measurements (226) and measurement of transepithelial flux of ions and small polar nonelectrolytes, such as mannitol (157, 242), indicate that the paracellular pathway contributes only ~25% of the total tissue conductance (72). Thus, consistent with the relative impermeability of the gastric mucosa to H^+, the tissue can be classified as a tight epithelium.

The paracellular (shunt) pathways in the amphibian antrum are cation selective (17, 224). The isoelectric point (the pH at which permselectivity switches from cation to anion selectivity) is 3.0–4.4 for this amphibian tissue (17). Thus the negatively charged groups envisaged to line the paracellular pathway would appear to have a pK_a ($-\log_{10}$ of the acidic dissociation constant) of ~4–5 (185). In dog fundic mucosa, however, the isoelectric point estimated from measurements of Na^+ permeation at different pH is ~10.0 (131). The fundus would therefore be protected against paracellular H^+ permeation over the entire physiological range of pH, and the antrum at acidic pH, inasmuch as the paracellular pathways would be anion selective (185).

Comparison of Fundic and Antral Gastric Mucosa

The studies detailed in the preceding section concentrate on the fundic or body region of the stomach. In canine antral pouches the flux of Na^+ and H^+ is 14–19 times greater than across fundic mucosa (61). Consistent with this difference, the paracellular pathway contributes ~75%–80% of the transepithelial conductance in *Necturus* antral mucosa (224), considerably greater than the <25% contribution in fundic mucosa (226). Thus antral mucosa could be classified as a moderately tight epithelium, the ratio of paracellular to cellular conductance in antrum being considerably less than observed in the duodenum, gallbladder, and proximal renal tubule but similar to that in the colon (80, 185). In the rabbit gastric mucosa, however, the antrum is 2–3 times less permeable than the fundus to H^+ (139), although disappearance of H^+ is slow from both areas of the stomach, and this is consistent with the lower susceptibility of the antrum in this species to acute ulceration (138). Rabbit antrum also has a lower electrical conductance than the fundic region of the stomach (77, 79). The species differences in H^+ permeability, conductance, and susceptibility to ulceration between the antrum and fundus may point to the importance of the paracellular pathway as a barrier. These physiological differences between the antral and fundic gastric mucosa cannot be correlated with differences in the structure of the tight junctions in the two regions (165).

PROTON PERMEABILITY OF GASTRIC MUCOSA

From the preceding discussion it can be seen that the barrier function of the stomach, with respect to maintaining a $>10^6$ H^+ gradient, is well documented. Accurate assessment of the permeability coefficient for H^+ has, however, been difficult. In particular, as is the problem in obtaining accurate electrical resistance measurements in the stomach, inaccuracies in assessment of the surface area of the stomach preclude an accurate estimate of the permeability coefficient. The best estimate of the H^+ permeability coefficient, although based on macroscopic surface area, comes from the isolated frog gastric mucosa, where a value of 0.4×10^{-5} $cm \cdot s^{-1}$ was obtained (60). This value is not dissimilar to values estimated for mammalian gastric mucosa of 0.8–3.9×10^{-5} $cm \cdot s^{-1}$. These values could be reduced by two orders of magnitude if normalized to the serosal membrane area [~100 cm^2 serosal membrane per cm^2 chamber area; (72)]. To support secretion of 150 mM H^+, the proton permeability of the secretory membrane of the parietal cell must have an H^+ permeability of $<10^{-7}$ $cm \cdot s^{-1}$ (50). However, from estimates of an average vesicle radius of 0.15 μm (257) and internal buffering capacity of 20 mM/pH, the net proton/hydroxide permeability coefficient for parietal cell apical membrane vesicles at 20°C is $\sim 4 \times 10^{-4}$ $cm \cdot s^{-1}$ (J. M. Wilkes and B. H. Hirst, unpublished observations). This value is considerably greater than the proton permeability coefficients in the preceding description and may partially explain and be related to the failure of gastric microsomal membranes to generate intravesicular pH as low as in vivo, as described next.

In the frog gastric mucosa the major part of H^+ backdiffusion is in the form of molecular HCl (60).

This is similar to that reported for phospholipid bilayers at low pH (100). At neutral pH other mechanisms are necessary to account for the high observed permeability (98).

It is interesting to compare the intraluminal pH that the stomach in vivo is capable of generating and sustaining with the permeability of gastric membrane vesicles rich in H^+-K^+-ATPase, gastric microsomes. Gastric microsomes can only generate an intravesicular pH of 2.2–3.7 units if measured with intravesicular probes, such as acridine orange or aminopyrine (145, 188). From measurement of loss of H^+ from the extravesicular medium, an intravesicular pH as low as 1.1 is indicated, assuming no significant buffering capacity (188). This is close (although still 80 mM H^+ lower) to the lowest intragastric pH of 0.8. The failure to generate intravesicular pHs as low as in vivo may indicate intravesicular buffering and/or proton leaks. In planar phospholipid bilayer membranes, proton conductance may be reduced by albumin, probably by removing fatty acids, which act as endogenous protonophores (99). Similarly, albumin reduces the H^+ permeability of gastric microsomes (101). The putative fatty acid protonophores could be produced during vesicle preparation and in this respect represent an artifact.

A further explanation for the inability of gastric microsomes to generate pH gradients as large as observed in vivo stems from consideration of the degree of curvature of the membrane vesicles. Membranes with a small radius of curvature have significantly greater water permeability (179); small increases in the degree of curvature result in disproportionate increases in water flow (192). The increased water permeability is the result of reduced packing of the outer leaflet of the lipid bilayer, with increased fluidity (179). The effect will be most pronounced with diameters <40–80 nm (179) but might still increase permeability in the range of 100–200 nm, the diameter of gastric microsomes (70, 72, 145). If H^+ permeability is also increased in relation to increased curvature, this might contribute to the presumed increased H^+ permeability of the gastric microsomes.

CONSTITUENTS OF THE BARRIER

Powell (186) usefully divided the concept of epithelial barriers into two categories, extrinsic and intrinsic (Fig. 2). The extrinsic was further divided into pre-epithelial, i.e., the specialized secretions of the epithelium, and postepithelial, which includes the tissue acid-base balance and blood flow. The intrinsic barrier is the epithelium per se, including the pathways and mechanisms for H^+ permeation through the cellular junctions and cellular membranes and the mechanisms for controlling intracellular pH in the epithelial cells. A further element of the epithelial barrier is the ability of the damaged epithelium to repair itself.

MUCUS-BICARBONATE LAYER

Mucus has long been thought to subserve an important protective role in the stomach (90, 102). Mucus acts as a lubricant, both on the surface of the epithelium and, in the case of soluble mucus, within the lumen of the stomach. However, the case for further roles for mucus in mucosal defense has not been without controversy (48). More recent detailed studies on the physical and biochemical properties of gastric mucus (1–3) have helped to clarify its role in mucosal resistance, as the first line of defense (123). As first suggested by Claude Bernard (21), there is now experimental evidence that mucus acts as a permeability barrier to pepsin (2).

Of relevance to the impermeability of the gastric mucosa to H^+ is the contribution subserved by mucus. Heatley (105) found that mucus did not significantly reduce the diffusion of H^+ as compared with its diffusion in free solution. More recent studies (183, 206, 215) have demonstrated a retardation of H^+ diffusion through mucus in vitro, but this is only by up to a factor of 3- to 4-fold, to give a value of 1.75×10^{-5} cm · s^{-1} (255). This retardation of H^+ diffusion depends on the gel-forming polymeric structure of the undegraded mucous glycoprotein and may be enhanced by interaction with other proteins (206). Gas-

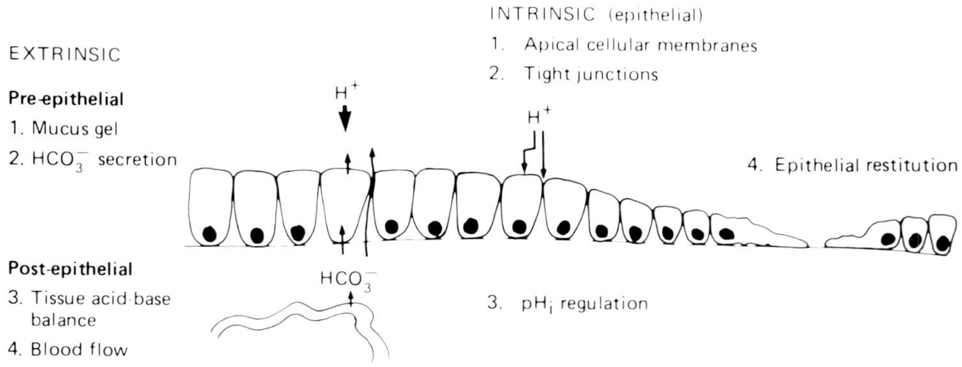

FIG. 2. Constituents of the gastric mucosal barrier. [Adapted from Powell (186).]

EXTRINSIC

Pre-epithelial
1. Mucus gel
2. HCO_3^- secretion

Post-epithelial
3. Tissue acid-base balance
4. Blood flow

INTRINSIC (epithelial)
1. Apical cellular membranes
2. Tight junctions
3. pH$_i$ regulation
4. Epithelial restitution

tric mucus, in the range pH 4–8, was reported to reduce proton permeability, as estimated from the reduced response time of a pH electrode (244). A similar reduction in response time was not observed with Na^+- or K^+-selective electrodes. In these experiments the mucus also buffered the pH response, possibly because of protein contaminants. At pH < 4 the pH electrode response time was unaffected by mucus. The response time at neutral pH was also increased by increasing the HCO_3^- concentration from 1 to 10 mM. This latter observation was explained in terms of the HCO_3^- acting as a buffer shuttle, with the HCO_3^- transporting protons along the pH gradient as carbonic acid. These authors' results remain anomalous with current views of the mucous-HCO_3^- layer.

The studies on proton permeability of mucus have been criticized (154) as neglecting turbulence effects, the contribution of the supporting membranes to resistance, and the effect of buffering. In an attempt to overcome some of these criticisms, Lucas (154) measured the diffusion of Na^+ through mucus and found it to be reduced, compared with free-solution values. The reduction in the Na^+ diffusion coefficient was at maximum three- to fourfold, being dependent on mucus concentration. Obviously there are problems in extrapolating from Na^+ diffusion to that for H^+. However, considering all the evidence, mucus per se is unlikely to provide a significant barrier to the diffusion of H^+.

The mucous layer, however, does sustain a measurable pH gradient from the lumen of the mucosa to the surface of the epithelial cells (Fig. 3). This gradient is such that, with a luminal fluid of pH 2–3, the pH at the apical surface of the mucosal cells is near neutrality (14, 200, 237, 256). In a recent study in humans, Quigley and Turnberg (187), using a relatively large flexible pH electrode, reported lower pH gradients of from 2.0, 1.8, and 3.5 (luminal) to 4.8, 5.5, and 5.4 (juxtamucosal) in the gastric fundus, body, and antrum, respectively. With acid luminal pH the antrum was not able to maintain a pH gradient, whereas juxtamucosal values in the fundus and body were relatively unaffected by luminal pH. The gastric (HCO_3^-) alkaline secretion, presumably from the surface epithelial cells, sustains this pH gradient. However, measurement of pH gradients is difficult (155) and, as was discussed in ELECTROLYTE COMPOSITION OF GASTRIC JUICE, p. 279, gastric HCO_3^- secretion is only at maximum 10% of maximal gastric acid secretory rates. This is consistent with the pH gradient being dissipated with intraluminal pH values <2 (200, 237). In addition, Engel et al. (63) calculated that the lowering of $[H^+]$ within the mucous gel layer may be as small as 5 mM compared with free intraluminal $[H^+]$, an insignificant value when compared with intraluminal acidities of ~150 mM. Machen and Paradiso (155) have argued that since low luminal pH reduces Na^+ transport to a similar extent in rabbit fundic surface epithelial cells covered by mucus and in isolated frog skin cells that are not covered by mucus nor secrete base, the mucous-HCO_3^- layer does not significantly buffer the luminal surface of the stomach.

The mucous layer subserves the first line of defense of the gastric mucosa. This appears to be most important in protecting the underlying mucosa from damage by mechanical shear forces and abrasion, chemical digestion by macromolecular enzymes (e.g., pepsin), and as a barrier to other damaging macromolecules and microbials. The role of the mucous layer in resisting H^+, particularly at pH <2, appears more controversial. Furthermore the mucous layer is not a barrier to freely diffusible small molecules that can damage the gastric mucosa and break the barrier to H^+ diffusion, such as ethanol, aspirin, and bile salts. Mucus, however, is likely to provide an important substrate, mixed with cellular debris and fibrin, under which the damaged epithelium can repair itself while protected from the effects of further luminal damaging agents [see Fig. 6; (170, 212, 249)].

There is one area of controversy concerning the mucous layer that is particularly relevant to this discussion: is the mucous layer continuous? Studies by Allen and co-workers [see the chapter by Allen in this *Handbook*; (133)], using thick unfixed sections of frog, rat, and human gastric mucosa, indicate that the mucous layer forms a continuous layer of variable thickness above the mucosa. In these unfixed thick sections, mean mucous thickness is up to 192 μm in human and 73 μm in rat, with a large variation evident; values of 5–400 μm are observed in a single human stomach (133). However, in sections of gastric mucosa histologically fixed, up to two-thirds of the gastric mucosa has no visible covering of mucus (169), although some earlier studies report a continuous layer in fixed sections (123). The lack of an observable continuous

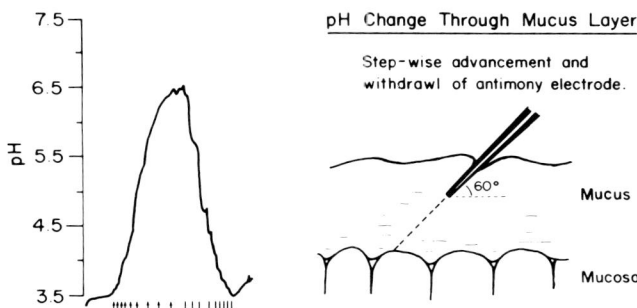

FIG. 3. Demonstration of a pH gradient above frog gastric mucosa in vitro. An antimony microelectrode was positioned at an angle of 60° to the tissue and advanced toward the mucosal surface in 40 or 80 μm steps. Although originally believed to measure the pH within the mucus, the measurements probably also include any unstirred layer above the mucus (69, 216). *Arrows*, changes in position of the electrode tip by either 40 or 80 μm. [From Takeuchi et al. (237). Copyright 1983 by The American Gastroenterology Association.]

mucous layer in some fixed sections has been argued to be an artifact of the fixation process (127, 212). More importantly for this discussion, if the mucous layer is not continuous then the mucous-HCO_3^- layer cannot be an important factor in the barrier, except after mucosal damage, when all authors agree it increases substantially (170, 212, 249). If in vivo the mucus does form a continuous layer covering the gastric mucosa, then the question arises as to how the acid and pepsin, produced deep within the gastric glands, appear in the gastric lumen. Suggestions for this include the presence of small pores within the mucous layer above the openings of the gastric glands or buildup of fluid within the gastric pits forcing, by hydrostatic pressure, a pathway through the mucous layer (69). Experimental evidence to clarify this question is required.

The roles of mucus and gastric alkaline secretions are reviewed in detail in the chapters by Flemström and Garner and by Allen in this *Handbook* (see also refs. 5, 68, 69, 85, 190).

EPITHELIAL CELL LAYER

Hollander (123) recognized the importance of the epithelial cell layer in gastric mucosal defense. Davenport (36–49), in a series of papers extending from the mid-1960s to mid-1970s, laid the foundations for our understanding of the gastric mucosal barrier. The majority of our knowledge still rests on experiments, like those described by Davenport, in which the gastric mucosa is subjected to damaging insult that breaks the barrier. These experiments have given us important insights into the essential constituents of the cellular barrier.

Tight Junctions

The luminal surface of the gastric epithelium consists of the apical membranes of the mucosal cells, connected by tight junctions. As discussed in ELECTRICAL CHARACTERISTICS OF GASTRIC MUCOSA, p. 280, the tight junctions of the fundic mucosa are less conductive than the cellular pathway for ions and thus must constitute a significant component of the gastric mucosal barrier. This is not to say, however, that barrier-breaking agents all act on the tight junctions; there is substantial evidence that the opposite is more likely.

In frog gastric mucosa in vitro, Ca^{2+} removal or chelation decreased resting potential difference and increased the flux of sucrose (73). These reversible changes were accompanied by a widening of tight junctions between oxyntic cells (211). Similarly, cation permeability is increased in vivo upon Ca^{2+} chelation, without changes in cellular structure (30). As Ca^{2+} is required for normal tight junctional characteristics, these experiments may support the importance of these complexes in barrier function. However, Ca^{2+} also regulates plasma membrane permeability to ions (179).

Hypertonic solutions of urea break the gastric mucosal barrier, i.e., increase the flux of Na^+, K^+, and H^+ across the gastric mucosa (10, 43). This effect of urea is associated with distortion of the tight junctions and formation of vacuoles within the surface epithelial cells, with little change to the apical membrane or cellular organelles at 300 mM urea [Fig. 4; (62)]. Similar increases in ion permeability are observed with intragastric instillation of hypertonic solutions of other nonelectrolytes such as glucose, sucrose, mannitol, sorbitol, and glycine (10, 104, 221). In the dog, intragastric hyperosmolal glucose, sucrose, and urea were associated with increased output of protein, although the type and source of protein were not determined (10). The output of Na^+, Cl^-, and HCO_3^- from cat gastric pouches in response to hyperosmolal mannitol solutions was accompanied by and correlated with increases in the flux of [^{14}C]thiourea from blood. The increase in [^{14}C]thiourea flux was taken to indicate an increase in paracellular permeability (221). A common mechanism, therefore, may relate the action of these relatively impermeable substances, increasing gastric mucosal permeability and thus "breaking" the barrier by reducing the resistance of paracellular pathways. The reported increases in H^+ backdiffusion observed with various agents may, therefore, rather reflect increased plasma ultrafiltration and efflux of HCO_3^- (9, 10, 39, 141). This mechanism does not account for the action of other barrier breakers such as ethanol, bile acids, and aspirin. The increases in permeability observed with hypertonic nonelectrolyte solutions are not associated with gross mucosal damage, e.g., marked lesions and hemorrhage (62).

Hypertonic solutions of urea, sucrose, mannitol, raffinose, and also NaCl and KCl result in increased electrical conductance and increased passive permeability in other "tight" epithelia, including frog skin and toad urinary bladder (22, 54, 243, 246). Although deformations (blisters; cf. Fig. 4) of the tight junctions are also observed upon mucosal hypertonicity in these epithelia (54, 246), the increased permeability and conductance are argued not to be a result of this blister formation (22). Nevertheless, in frog skin, apparently normal-looking tight junctions are permeable to lanthanum after exposure to hypertonic solutions (64).

Apical Cell Membranes

In parallel with the tight junctions, the apical cell membranes of the gastric mucosal cells provide the second layer of the barrier. Evidence for the importance of the apical cell membranes as a significant, if not the major, component of the gastric mucosal barrier has in the main come from studies with barrier-breaking agents.

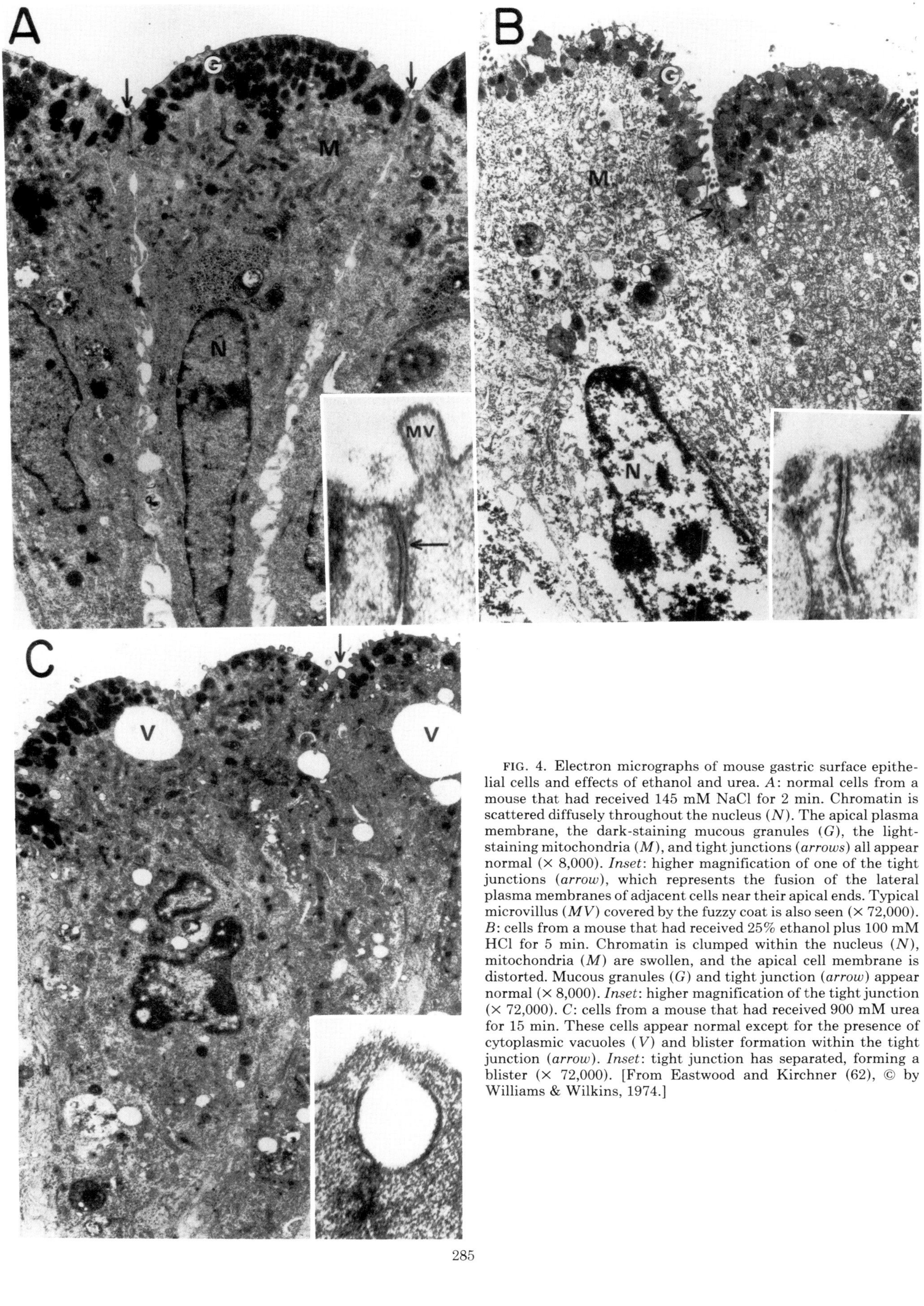

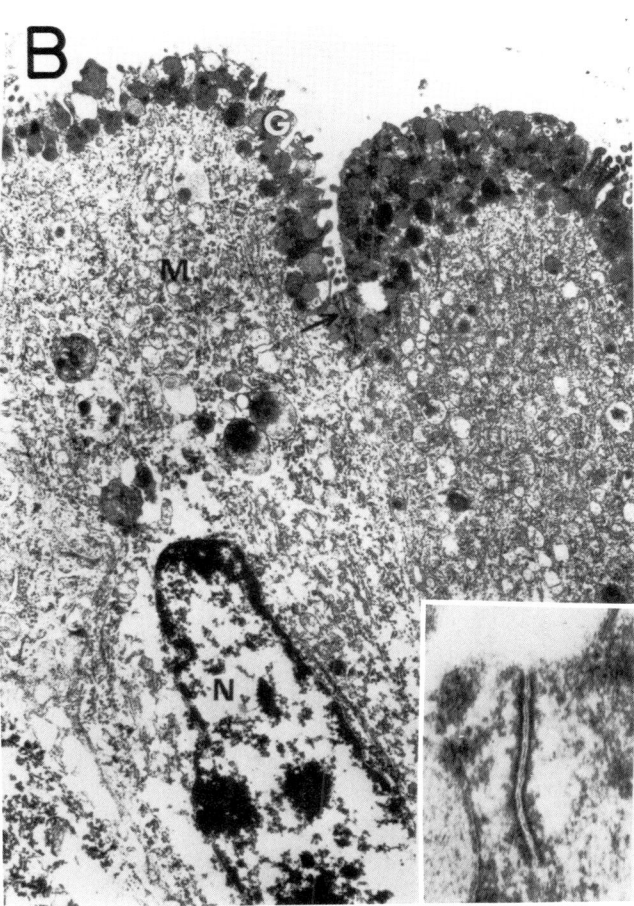

FIG. 4. Electron micrographs of mouse gastric surface epithelial cells and effects of ethanol and urea. *A*: normal cells from a mouse that had received 145 mM NaCl for 2 min. Chromatin is scattered diffusely throughout the nucleus (*N*). The apical plasma membrane, the dark-staining mucous granules (*G*), the light-staining mitochondria (*M*), and tight junctions (*arrows*) all appear normal (× 8,000). *Inset*: higher magnification of one of the tight junctions (*arrow*), which represents the fusion of the lateral plasma membranes of adjacent cells near their apical ends. Typical microvillus (*MV*) covered by the fuzzy coat is also seen (× 72,000). *B*: cells from a mouse that had received 25% ethanol plus 100 mM HCl for 5 min. Chromatin is clumped within the nucleus (*N*), mitochondria (*M*) are swollen, and the apical cell membrane is distorted. Mucous granules (*G*) and tight junction (*arrow*) appear normal (× 8,000). *Inset*: higher magnification of the tight junction (× 72,000). *C*: cells from a mouse that had received 900 mM urea for 15 min. These cells appear normal except for the presence of cytoplasmic vacuoles (*V*) and blister formation within the tight junction (*arrow*). *Inset*: tight junction has separated, forming a blister (× 72,000). [From Eastwood and Kirchner (62), © by Williams & Wilkins, 1974.]

STUDIES WITH BARRIER-BREAKING AGENTS

Physiological breaking of the gastric mucosal barrier, as indicated by increased flux of ions across the gastric mucosa (e.g., see Fig. 2) or decreased electrical resistance and a fall in transmucosal potential difference (see Fig. 7), is observed after subjecting canine gastric mucosa to insult with salicylates and weak organic acids (36-38, 42, 132), ethanol (41), eugenol (49), lysolecithin and digitonin (45), phospholipase A (45), and bile acids (43, 140, 217). These agents are usually also associated with damage to the surface epithelial cells, including desquamation and often (though by no means always) overt hemorrhage (36, 39, 42, 44, 47, 132, 140). With several of these agents, extensive damage to the surface epithelial cells has been noted in ultrastructural studies, while the tight junctions remained intact (62, 76, 113, 132).

In addition to increased ion fluxes and electrical conductance across the gastric mucosa not always being associated with overt hemorrhage, the opposite is also true (48). For example, the increased gastric mucosal blood, albumin, and fluid loss observed after gastric ischemia in the rat was not associated with increased H^+ backflux, although Na^+ and K^+ fluxes were increased. Similarly, intravenous aspirin induced gastric hemorrhage before any net changes in H^+ flux and without changes in Na^+ or K^+ flux (184). Thus overt damage to the stomach, as demonstrated by hemorrhage, can occur without net changes in ion fluxes across the mucosa, i.e., the gastric mucosal barrier is still intact. This discussion is mainly concerned with the barrier to Na^+ and H^+ and not ulceration and hemorrhage of the mucosa. Thus, for clarity, barrier-breaking agents are defined as agents that increase the flux of Na^+ and H^+ or the electrical conductance across the gastric mucosa. These effects may or may not be associated with hemorrhage or ulceration.

Aspirin and Weak Acids

The damaging weak acids, including aspirin (acetylsalicylic acid), all have a pK_a between 3 and 5. From an acidic solution they readily diffuse across the plasma membrane in an un-ionized form, followed by rapid dissociation within the cell. The trapped acid anion may then interfere with intracellular metabolism and/or osmolality, and the dissociated proton may overcome intracellular buffering (160). Thus in one form of the hypothesis the weak acids may act as protonophores. The model proposed by Martin (160) contained three compartments: *1*) the luminal acidic fluid in which the un-ionized form of the acid would predominate, *2*) the mucosal cell in which the acid would largely dissociate, and *3*) the interstitial fluid and blood (Fig. 5). Thus, assuming the apical membrane is selectively permeable to the un-ionized form of the acid, the model will accumulate very large quantities of the acid anion intracellularly. However, this model requires revision with the recognition that the external face of the apical membrane of the surface epithelial cells may be nearer to neutrality (14, 200, 237, 242). With this extra compartment added the acid would appear to be ionized at the membrane surface. The acids, including aspirin, are unlikely to act as a carrier for protons through mucus, since H^+ itself has a very fast rate of diffusion through mucus. However, aspirin has been shown to dissipate quickly the pH gradient above rat gastric mucosa (200).

The accumulation of aspirin in the gastric mucosa is greater from acidic solutions (38, 44, 83), however, which appears inconsistent with a pH gradient above the surface epithelial cells. Aspirin, apart from damaging surface epithelial cells, selectively damages parietal cells located deep within the gastric mucosa, presumably as a result of the pH-partitioning hypothesis (189). This selective damaging effect of aspirin on parietal cells is not observed with butyric acid (132), and even with aspirin the initial damage appears to be to the surface epithelial cells (113). When administered in 100 mM HCl, aspirin and salicylic acid not only break the mucosal barrier, as demonstrated by increased fluxes of Na^+ and H^+, but also produce severe hemorrhage in the dog stomach (38, 42, 45). In contrast, in lower concentrations of HCl (1-10 mM) the salicylates still result in increased Na^+ and H^+ fluxes, but without the hemorrhage (38, 44). The reduced damage does not appear to be due to reduced total absorption of salicylate (38, 44), although, when absorbed from neutral solutions, a higher proportion of the absorbed aspirin appears to be extracellular (83). Thus the salicylates appear to permeabilize the gastric mucosa to cations, and, in the presence of strongly acidic solutions, H^+ rapidly permeates the tissue to cause more overt damage. In this respect, damage to amphibian gastric mucosa by aspirin is reduced by increasing the nutrient $[HCO_3^-]$ (201).

Intravenous aspirin is also damaging to the gastric mucosa (184, 252). However, unlike the effects of

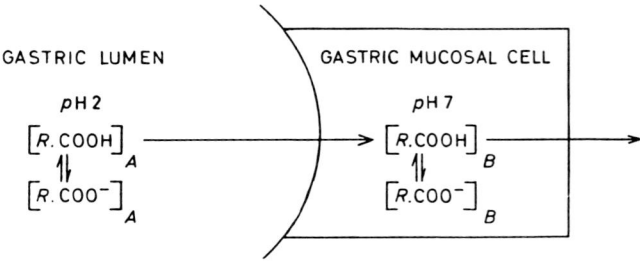

FIG. 5. Diagram illustrating the pH gradient that exists in the stomach between the gastric fluid and a mucosal cell and equilibrium between the dissociated and undissociated forms of a weak acid (e.g., aspirin) $R \cdot COOH$. *A*, concentration of the weak acid in the gastric lumen; *B*, concentration of the weak acid in the gastric mucosal cell. [From Martin (160). Reprinted by permission from *Nature*, copyright 1963, Macmillan Magazines Limited.]

topical aspirin, the overt damage is not correlated with changes in ion fluxes. In the rat (184), increased H⁺ backflux was preceded by hemorrhage, suggesting the H⁺ backflux is a consequence rather than the cause of the gastric damage.

The mechanism by which salicylates increase ionic permeability is still unclear. Although aspirin and other nonsteroidal anti-inflammatory drugs are potent inhibitors of gastric prostaglandin synthesis, this is not thought to be the primary mechanism of action for topically applied drugs (252). The mechanism appears dependent on the carboxyl group but is independent of the effects of inhibition of ion transport. Salicylates interfere with cellular metabolism. The effects of salicylates have been reviewed by Fromm (78) and Whittle (252).

Bile Acids

Luminal bile acids increase the flux of Na⁺, H⁺, and polar nonelectrolytes across the gastric mucosa, while decreasing tissue electrical resistance; i.e., bile acids break the mucosal barrier (43, 217). Ultrastructurally the initial site of action of bile acids, as exemplified by deoxycholate, is the apical membrane of the surface epithelial cell (76). Swelling of intracellular organelles eventually results in rupture of the apical membrane and, with more prolonged exposure, complete loss of the surface epithelial cells. The underlying basement membrane remains intact (76).

The effects of bile acids are dependent on pH, concentration, and the specific bile acid in question. The severity of the increased permeability of the gastric mucosa is directly related to the concentration of bile acid (58, 140, 217). Differences in damage due to conjugation and hydroxylation have also been noted (103, 194, 217). These latter effects are also related to the pH of the solutions in which they are administered. The conjugated bile acids are stronger acids than their unconjugated counterparts. At relatively acidic pH, unconjugated and dehydroxylated bile acids, with pK_a >5, precipitate, thus lowering their effective concentration. Taurocholic acid, in contrast, with a pK_a ~1.8 and therefore greater solubility at acidic pH, damages the gastric mucosa in acidic solution.

Recently interest has been shown in the mechanism by which bile acids damage the gastric mucosa. In particular Duane and co-workers (57, 58) have focused on the question of whether bile acids enter the mucosa, with subsequent cellular injury, or whether dissolution of mucosal lipids by intraluminal bile explains their action. The gastric mucosa absorbs taurocholate (40, 57), but addition of lecithin and cholesterol reduces mucosal uptake while diminishing barrier-breaking effects [Fig. 6; (55, 57)]. Damaging quantities of dissociated bile acids (i.e., bile salts at neutral pH) increase efflux of mucosal phospholipids (55, 57, 241). The phospholipids eluted with bile acids include lecithin (phosphatidylcholine), phosphatidylethanola-

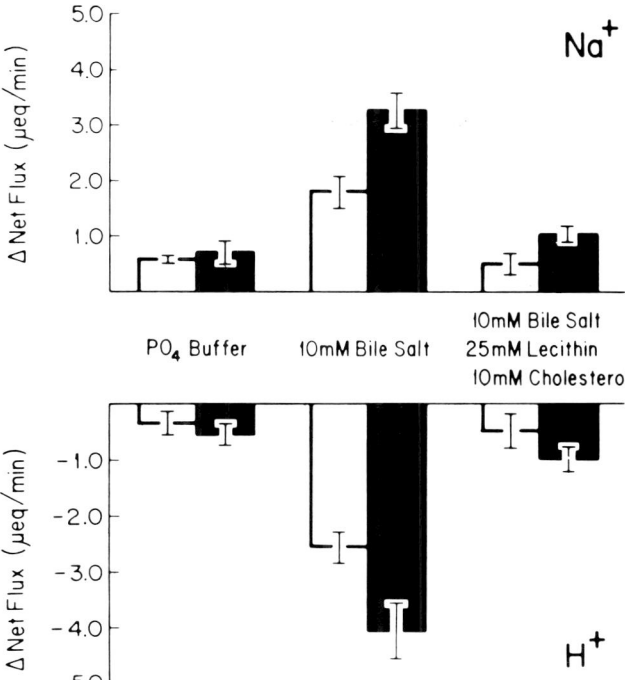

FIG. 6. Changes in net Na⁺ and H⁺ flux from the gastric lumen of the rat and the effect of conjugated bile acids. Test solutions were 0.1 M phosphate buffer (pH 7) and 10 mM mixture of conjugated bile acids in this buffer with and without added lecithin and cholesterol. *Filled bars*, test solution was continuously mixed during incubation in the stomach. *Open bars*, test solution was not mixed. Bile acids in mixed or unmixed solutions significantly increased net forward diffusion of Na⁺ and backdiffusion of H⁺. Saturating this solution with lecithin and cholesterol prevented these changes in net ion fluxes. Mixing bile acid solutions resulted in significantly greater increases in the net fluxes of both ions. Results are means ±1 SE. [From Duane et al. (55). Copyright 1986 by The American Gastroenterology Association.]

mine, phosphatidylserine, phosphatidylinositol, and sphingomyelin, with a disproportionate amount of lecithin eluted compared with that in mucosal scraping (57). The barrier-breaking effects of dissociated bile acids at neutral pH are argued to occur at concentrations above their critical micellar concentration (CMC). Thus these dissociated bile acids disrupt the gastric mucosal barrier by dissolving mucosal lipids with the formation of intraluminal micelles (57). High concentrations (100 mM) of taurocholic acid increased the polar part of the electron spin resonance spectra of 16-doxylstearic acid spin label added to unpurified rat gastric mucosal cell membranes, consistent with disintegration of the membranes and formation of micellar aggregates with taurocholic acid (96). This effect of taurocholic acid could be prevented by prostaglandin E₂.

In contrast to the dissociated bile acids, micelle formation does not appear to be important for the disruption of protonated taurocholic acid. Taurocholic acid at pH 1 disrupts the gastric mucosal barrier at concentrations below its CMC at this pH. The disruption is not prevented by saturation with lecithin and cholesterol, and in the ionized state taurocholic acid is absorbed 5 times more rapidly than at pH 7 (58). Therefore the damaging effects of bile acids at acid pH are caused by accumulation within the gastric mucosa and may in part be explained in terms similar to those for other weak acids. However, even at concentrations below their CMC, bile acids have important detergent-like qualities (108) and may accumulate in the apical plasma membranes of the surface epithelial cells, increasing their permeability to H^+.

The damage caused by dissociated bile acids is dependent on the extent of the unstirred layer overlying the gastric mucosa. Mixing of gastric contents reduces the thickness of the unstirred layer, with a concomitant increase in Na^+ and H^+ fluxes across the gastric mucosa induced by bile acids at pH 7 [Fig. 6; (55)]. The protection afforded by the unstirred layer may be to create a concentration gradient of bile acid from lumen to mucosal surface or to slow diffusion of mixed micelles away from the mucosa. Mucus is likely to be an important component of any unstirred layers, particularly in vivo with normal mixing of the luminal contents by gastric motility. Thus increased mucous thickness, for example, with prostaglandins or secretin (4, 133), may enhance mucosal resistance to damaging agents such as bile acids by increasing this unstirred layer, as has already been argued for its role in protection against acid and pepsin attack (5, 7). Whether the unstirred layer affords any protection against other damaging agents remains to be investigated.

Tepperman et al. (240) have reported that net mucosal H^+ and Na^+ fluxes in response to acid loads are low in rats until the animals are 25 days old, after which threefold increases are observed (Fig. 7). This time period relates to the period of weaning, when appearance of acid secretion and response to secretagogues are developing. The young rat stomach also demonstrates little increase in Na^+ and H^+ fluxes or protein efflux in response to taurocholic acid until 10–12 days old. These changes with age are unlikely to have simple explanations, and, for example, some of the increase in ion fluxes with age may be associated with as yet undetermined changes in gastric nonparietal secretions. However, the development of gastric mucosal permeability and the damaging effects of bile acids are interesting. The neonatal rat stomach may prove a useful model in studies of the gastric mucosal barrier.

Other molecules with detergent properties also disrupt the mucosal barrier. These include digitonin and lysolecithin (45, 56, 140). Lysolecithin is disruptive at both acidic and neutral pH and associated with increased transmucosal fluxes of Na^+ and H^+ and increased efflux of membrane phospholipids. Saturation of lysolecithin solutions with lecithin prevents the damage. Thus it has been argued that, in a similar manner to the dissociated bile acids, lysolecithin damages the gastric mucosa by extracting membrane lipids to form intraluminal micelles (56). Further work in this area is required to fully substantiate this proposed mechanism of action.

Lysolecithin is produced from biliary lecithin by the action of pancreatic phospholipase A. Human pancreatic juice (from one patient) containing phospholipase A activity had no effect on the gastric mucosal barrier. However, venom from the snake *Naja naja* but not that from *Vipera palestinae* was effective in breaking the barrier (45). Both venoms contain phospholipase A activity, particularly phospholipase A_2, and have similar differential activity in their ability to lyse erythrocytes. If the barrier-breaking activity of *Naja naja* venom is solely related to its phospholipase A_2 activity (and this has to be substantiated), then the question arises as to how this enzyme achieves access to the susceptible phospholipids at the gastric apical membrane surface, covered as they are by mucus. Mucus is permeable to vitamin B_{12} [molecular weight $(M_r) = 1,346$] but is an effective barrier to myoglobin ($M_r = 17,500$) (2). *Naja naja* phospholipase A_2 enzymes have a molecular weight of 11,000–15,000 (51), which may therefore allow them to slowly permeate mucus. Bee venom phospholipase A_2 is able to completely disrupt gastric microsomal membranes, attacking phosphatidylcholine and phosphatidylethanolamine (173).

Alcohols

Ethanol breaks the gastric mucosal barrier, as evinced by increased flux of Na^+ and H^+ and a reduced transmucosal potential difference (Fig. 8; see Fig. 13). The threshold for this gross damaging effect of ethanol is ~10% wt/vol or 2 M [Fig. 8; (34, 41, 251)]. Lower ethanol concentrations of ~2% wt/vol induce changes, i.e., increased fluidity and proton permeability, in the plasma membranes of gastric cells [see Fig. 9; (15, 16, 253, 254)]. Ethanol is rapidly absorbed across the gastric mucosa (41, 198, 251). Its absorption is unaffected by intragastric pH, and as a consequence the threshold concentration for ethanol damage is similar in acidic and neutral solutions [Fig. 8; (41)]. The ultrastructural damage caused by ethanol is similar to that described for aspirin and bile acids. The initial event is cellular damage within the surface epithelial cells [see Fig. 4; (62)]. High concentrations of ethanol are associated with very marked damage to the gastric mucosa, including overt bleeding, hemorrhagic lesions, and desquamation of the entire layer of surface mucosal cells [see Fig. 12; (47, 128, 170)], with the result that this procedure is a popular model for gastric mucosal damage.

Some of the actions of ethanol may be related to the hypertonicity of the solutions, although the damage is generally not similar to that produced by other

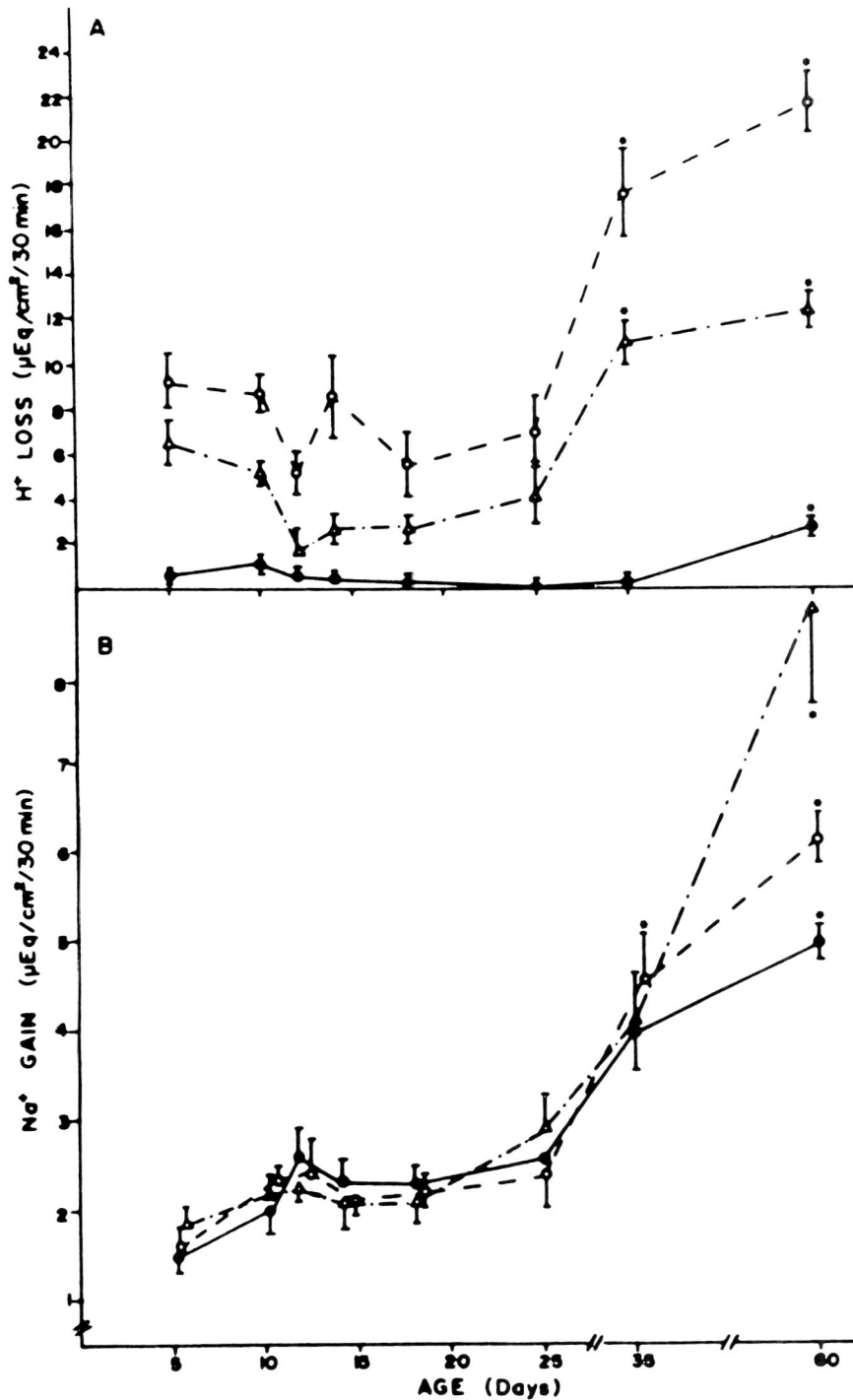

FIG. 7. Net H$^+$ loss and Na$^+$ gain in the gastric lumen of rats of various ages in response to instillation of 10 mM (●), 50 mM (△), or 150 mM (○) HCl. Net ion fluxes increase after 25 days. Results are means ± 1 SE. *, Significant difference between groups. [From Tepperman et al. (240).]

hypertonic agents such as urea [see Fig. 4; (62)]. In addition, solutions of methanol and ethylene glycol having comparable osmolality to 70% ethanol (~27 osmol) do not result in a similar extent of mucosal injury (234). The early ultrastructural damage observed with ethanol is similar to that recorded with bile acids (76) and aspirin (113), but the lack of dependence on H$^+$ for damage argues for a separate mechanism of action. Very high concentrations of ethanol (>40% vol/vol or ~8 M) dehydrate and denature mucus in vitro (20). Similar dehydration and fixation of the superficial surface epithelial cell are

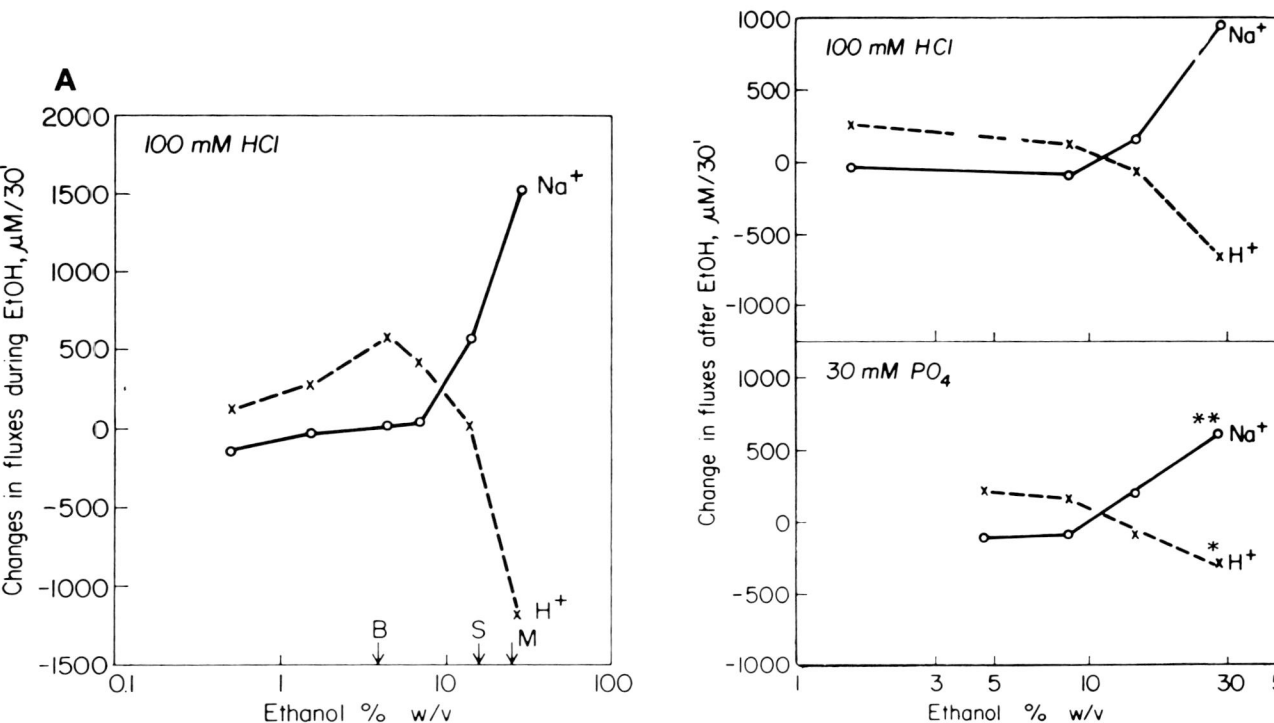

FIG. 8. Changes in fluxes of Na^+ and H^+ across canine gastric mucosa during instillation of solutions of ethanol. *A*: increasing concentrations of ethanol were instilled in 100 mM HCl plus 54 mM NaCl. Concentrations of ethanol normally found in beer (B), sherry (S), and a martini cocktail [4 parts gin to 1 part vermouth (M)] are indicated. *B*: increasing concentrations of ethanol were instilled in 100 mM HCl or 30 mM phosphate buffer (pH 7.5). Ion fluxes from phosphate buffer were significantly (**) lower than from HCl solutions. [From Davenport (41).]

likely. An important factor in ethanol-induced damage, however, is vascular injury with increased vascular permeability (47, 234). Because an essential component of the gastric mucosal barrier is the apical plasma membranes of the surface epithelial cells, paradoxically, ethanol, due to its lipid solubility, finds this no barrier at all.

Acid is not essential for ethanol damage, but the presence of H^+ exacerbates the damage caused by lower concentrations of ethanol [Fig. 8; (41, 44, 170)]. The damaging effects of a range of alcohols correlate with their oil-water partition coefficient (34, 251). These effects of alcohols may be related to their effects on plasma membrane fluidity and to subsequent changes in membrane enzyme, transport, and permeability functions (see *Lipid Constituents*, p. 291).

Pepsin

Pepsin is an important factor in damage to the gastric mucosa. As has been pointed out (172), the oft-quoted Schwartz's dictum, "No acid, no ulcer," is more correctly, "Without acid gastric juice, no peptic ulcer" (209). Thus it has been argued that Schwartz's dictum should read, "No peptic activity, no peptic ulcer" (172).

The mammalian gastrointestinal tract is resistant to prolonged infusion of H^+ (8, 208); ulceration is usually only observed with excessive concentrations of acid, producing systemic acidosis. Addition of pepsin, either in the form of native gastric juice or exogenous pepsin, results in marked ulceration of the gastric mucosa (8, 172, 208). The increased incidence of ulceration is observed with physiological concentrations of pepsin: 100–1,000 $\mu g \cdot ml^{-1}$. Similarly, in frog gastric mucosa in vitro, acid does not result in ulceration until pepsin (400 $\mu g \cdot ml^{-1}$) is added to the solution (137). In addition, inhibition of peptic activity in natural gastric juice reduces the incidence of ulceration (8). Mucus acts as a permeability barrier to pepsin (2). Nevertheless, pepsin is able to digest mucus in vitro (20), and in vivo pepsin dissolves this mucus layer overlying the mucosa, followed by subsequent damage to the mucosa (4). The reepithelialization of ethanol-damaged rat gastric mucosa is severely reduced by pepsin (5 $mg \cdot ml^{-1}$), assessed histologically and electrophysiologically [see Fig. 13; (249)]. Pepsin solubilization of mucus occurs from the luminal surface, and thus the maintenance of an effective mucous gel depends on the balance between this digestion and secretion of fresh gel.

Peptic cell monolayers, however, are resistant to

exposure to pepsin (800 μg·ml^{-1}) at pH 2.5 (205). In like manner, parietal cell apical membrane vesicles are resistant to incubation with another proteolytic enzyme, trypsin (115, 116).

GASTRIC MEMBRANE COMPOSITION

The problem of the nature of the gastric mucosal barrier is one of membrane physiology (48). The physiological properties of the luminal plasma membranes are dependent on their physical and chemical properties. Apart from specific pumps and carriers, membrane function may be modulated by passive factors, as determined by protein-lipid interactions. Understanding the role of membranes in gastric mucosal function therefore depends on characterization of the membrane constituents. Gastric microsomal membranes rich in H$^+$-K$^+$-ATPase are the most fully characterized system, but even here the information is sparse.

Lipid Constituents

Gastric microsomal membranes are characterized by a lipid composition typical of plasma membranes, containing a mixture of phospholipids, cholesterol, and glycolipids. The cholesterol-phospholipid ratio varies from 1.5–2.0 in the rabbit and pig (150, 213, 214) to just over 1.0 in the dog and frog (19, 213, 225). In rabbit isolated surface epithelial cells, the cholesterol-phospholipid ratio was lower, 0.38 (15), perhaps reflecting the total cellular lipids rather than just plasma membrane proportions. Phospholipids comprise approximately half of the total lipids (19, 213), with phosphatidylcholine and phosphatidylethanolamine the most abundant, accounting for 22%–49% each of the total phospholipid (19, 150, 213). In the rabbit, phosphatidylcholine and phosphatidylethanolamine together account for 93% of the gastric microsomal phospholipid, but only ~53%–75% in the pig and frog (150, 213). In the pig, sphingomyelin (19%–29%), and in the frog, phosphatidylinositol (24%) are the other major phospholipids (150, 213). In the dog, phosphatidylethanolamine (~30%) and phosphatidylcholine (~55%) are the most abundant phospholipids, with sphingomyelin contributing ~7.5% of the total (29, 225). Canine gastric mucosal scrapings show a similar phospholipid composition to the H$^+$-K$^+$-ATPase–rich microsomes: phosphatidylcholine (43%) and phosphatidylethanolamine (30%) are the most abundant, with lesser amounts of phosphatidylserine and phosphatidylinositol, sphingomyelin, and cardiolipin (57). Rabbit isolated surface epithelial cells and gastric glands also show a predominance of phosphatidylcholine (44%) and phosphatidylethanolamine (25%) (15, 146). In pig, rabbit, and frog gastric microsomes the predominant fatty acid constituent in phosphatidylcholine, phosphatidylinositol, and sphingomyelin was palmitoyl (16:0) (173, 213, 214), while oleoyl (18:1) was the predominant unsaturated fatty acid in all phospholipids except rabbit and pig phosphatidylcholine, where linoleoyl (18:2) was predominant. Oleoyl and linoleoyl are the most abundant (~25% each) fatty acids of phosphatidylethanolamine from pig gastric microsomes (173). In total lipid extracts from whole rat gastric mucosa, dipalmitoyl (16:0/16:0) phosphatidylcholine constituted 31% of the total phosphatidylcholine (250), whereas saturated fatty acids comprise ~40% of the phospholipid in dog gastric mucosal membranes: palmitoyl (16:0) is the most common, followed by oleoyl (18:1) and linoleoyl (18:2) (225).

The lipid composition of gastric membranes is likely to be an important modulator of passive components of membrane function. In particular the lipid components determine, to a large degree, the fluidity of the membrane. Fluidity is a concept reflecting the viscosity within the membrane. The lipid makeup of the gastric membranes described in this section hints at a rather rigid membrane structure. Cholesterol, under physiological conditions, is the main component imparting rigidity to the lipid membranes. Up to a cholesterol-phospholipid ratio of 2.0 (the highest reported for gastric membranes), cholesterol increases rigidity (215). In general terms, nonelectrolyte and electrolyte passive permeabilities of membranes are decreased with decreasing membrane fluidity and hence increased proportion of cholesterol (179).

Parietal cell apical membranes from rabbit gastric mucosa are less fluid, as determined by diphenylhexatriene fluorescence anisotropy, than gastric microsomal membranes (253, 254). The fluidity of these gastric membranes and duodenal brush-border membranes is temperature dependent and is increased with increasing concentrations of a range of alcohols. The H$^+$ permeation in these membrane vesicles can be quantified by acridine orange fluorescence quenching, and over a range of alcohol concentrations (Fig. 9) and temperatures, the rate constants for permeation of H$^+$ are correlated with the membrane fluidity (119, 253, 254). The membrane-fluidizing and H$^+$ permeation effects of the alcohols correlate with their oil-water partition coefficients, in like manner to their barrier-breaking effects (34, 250). Without alcohol treatment the order of proton permeability of these membranes is parietal cell apical < duodenal brush-border < gastric microsomal membranes (254). The membrane-fluidizing effects of ethanol have also been demonstrated with proton nuclear magnetic resonance (NMR) in gastric microsomes (16) and isolated surface epithelial cells (15).

Conclusions about the physiological significance, in terms of the gastric mucosal barrier, of the phospholipid composition of gastric membranes and their relation to membrane fluidity may be inferred from experiments on erythrocytes. Erythrocytes from different species contain varying proportions of sphingomyelin to phosphatidylcholine, and the greater the

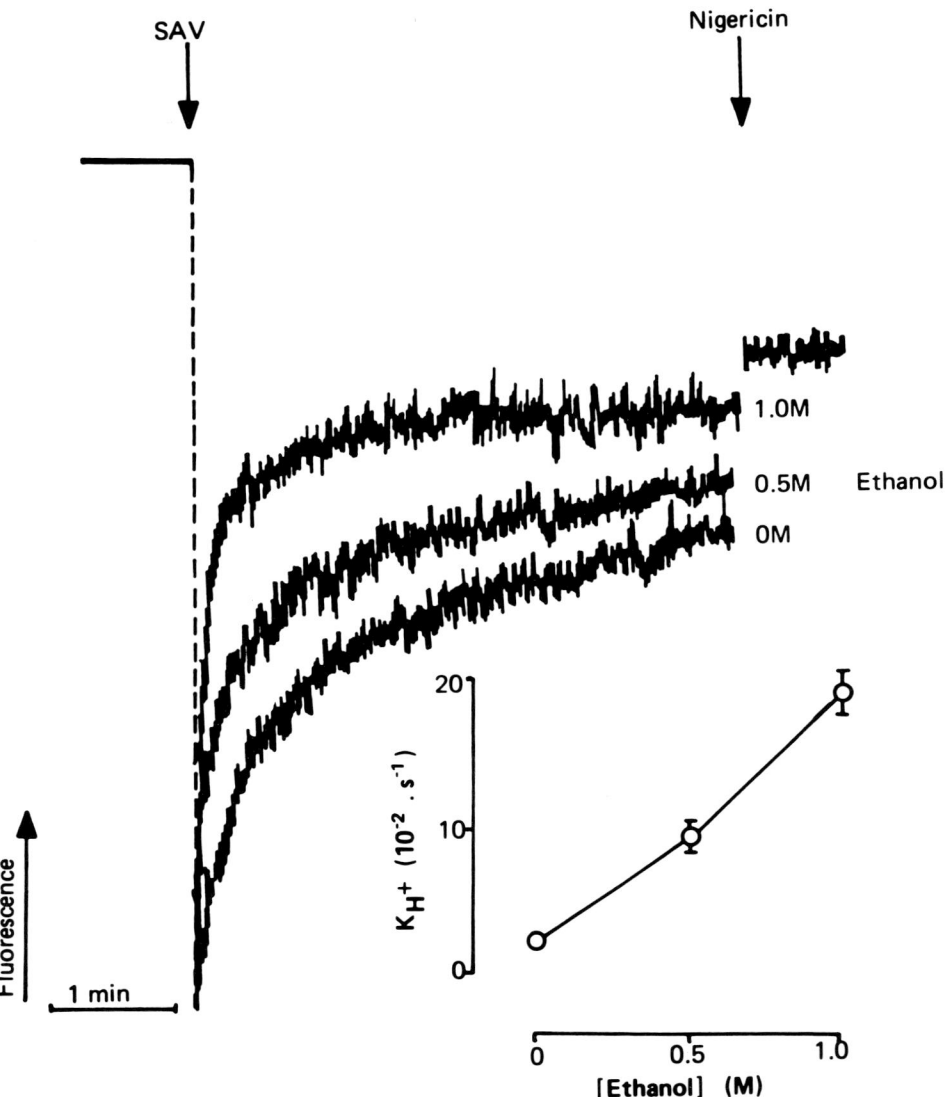

FIG. 9. Effects of ethanol on proton permeation in parietal cell isolated apical membrane vesicles. Parietal cell apical membrane vesicles [stimulation-associated vesicles (SAV)] were equilibrated in an acidic (pH 6.5) solution. H$^+$ permeation was measured by diluting the vesicles into a pH 8.0 solution containing acridine orange (*first arrow*). Vesicles were voltage-clamped with a K$^+$-valinomycin system to prevent generation of diffusion potentials. Addition of vesicles leads to a quenching of the fluorescence signal from acridine orange [λ_{ex} (excitation wavelength) 492 nm; λ_{em} (emission wavelength) 530 nm] as the dye is accumulated within the acidic vesicles. Recovery of the fluorescence gives the rate of H$^+$ permeation. Nigericin, an H$^+$-K$^+$ ionophore, was added at the *second arrow* to fully dissipate the proton gradient. Increasing concentrations of ethanol accelerated the rate of recovery of fluorescence, i.e., rate of H$^+$ permeation. *Inset*: apparent rate constant for recovery of fluorescence was analyzed by simple first-order kinetics. Rate constant for H$^+$ permeation (K_{H^+}) is plotted against ethanol concentration (mean ± 1 SE for 12–17 observations). (See refs. 254 and 119 for further details.)

fraction of sphingomyelin the lower the membrane fluidity. The erythrocyte membranes with a high fraction of sphingomyelin (and low fluidity) are less susceptible to lysis by bile salts (33, 152). An increasing sphingomyelin fraction also reduces the permeability of the erythrocyte membrane to small molecules (urea, thiourea, ethylene glycol, and glycerol) (33, 174). Decreasing the sphingomyelin fraction increases fluidity and is associated with increased susceptibility of the erythrocytes to lysis (33). These studies on a model cell membrane system suggest that reduced membrane fluidity is an important factor in the resistance of membranes to damage by bile salts. This conclusion is supported by studies on rat liver plasma membranes

(153). The microvillus membrane of the bile canaliculus has the lowest fluidity of the plasma membrane of the hepatocyte. As this membrane is normally exposed to bile salts in high concentration, its low fluidity is likely to be an important factor in resisting damage.

The importance of the lipid composition of gastric membranes and their fluidity in contributing to mucosal resistance is an area with more questions than answers.

Glycosubstances

Gastric cells show considerable staining for carbohydrate constituents. The mucous content of the surface epithelial cells is one obvious source, but parietal cells also demonstrate considerable staining with the periodic acid–Schiff reagent (88, 227). Ultrastructural studies employing phosphotungstic acid staining have further localized this staining to the tubulovesicles and apical plasmalemma of the parietal cell (210). Histochemical staining of mouse gastric mucosa with periodic acid–thiocarbohydrizide–silver proteinate (hexoses rich in vicinal hydroxyls), high iron diamine (sulfated mucosubstances), and dialyzed iron (sulfated and carboxyl-rich glycoconjugates) procedures have shown localized differences in the luminal glycocalyx from different cell types (207). Whereas the luminal surfaces of the surface epithelial and parietal cells were shown to have neutral glycoproteins, the peptic cells had carboxylated glycoproteins, and the isthmus cells had sulfated glycoproteins. Forte and Forte (74) have further localized the glycoproteins with periodic acid–silver methenamine staining to the external aspect of the plasmalemma and the internal face of the tubulovesicles. The cisternae and vesicles of the Golgi also showed heavy staining, but, in comparison, staining of the basolateral membranes was indifferent and the tight junctions were almost devoid of stain. Microsomal membrane vesicles isolated from parietal cells (equivalent to the tubulovesicular system in the intact cell) also demonstrate heavy staining for glycoproteins on their internal face (74).

As was the case for the lipid composition of gastric membranes, the microsomal membranes of the parietal cell are the best characterized in terms of their glycoconjugate composition. The carbohydrate is associated with both lipid and protein. The majority of carbohydrate in frog microsomes is associated with protein (73%). The most abundant sugars in the glycoprotein are hexoses (1.0) with molar ratios for other sugars: fucose 0.4, hexosamine 0.6–0.8, sialic acid 0.02, and uronic acid 0.04 (18, 74). The most abundant hexose is galactose (82%–95%), and N-acetylglucosamine is the only detectable hexosamine (19, 74). The glycoproteins from rabbit and pig microsomes show a similar pattern to the frog, i.e., greatest proportion of hexose (1.0) with a lesser proportion of hexosamine (0.3–0.5) and very little sialic acid (0.02), except that the proportion of fucose is much lower in the mammalian species (0.05–0.18) (18).

A more recent approach to the characterization of glycoconjugates associated with the membranes of gastric cells is the use of lectins. With fluorescein isothiocyanate (FITC)–conjugated lectins, the apical membrane of the parietal cell shows strong staining with *Helix pomata* agglutinin (HPA: specificity = GalNAc) and peanut agglutinin [PNA: β-D-Gal(1→3)-D-GalNAc] in sections of normal human gastric mucosa (66) and rabbit isolated gastric glands (177). However, with horseradish peroxidase–conjugated PNA no labeling of mouse parietal cells was observed (207). In rabbit gastric glands, wheat germ agglutinin [WGA: β-D-GlcNAc(1→4)] also stained parietal cell apical, as well as basolateral, membranes (177), although staining was absent in human tissue (66). Weak staining of rabbit parietal cell apical membranes was also observed with *Ricinus communis* agglutinin 1 (RCA 1: β-D-Gal) (177). The apical membrane of the peptic cell shows a similar staining pattern to the parietal cell: PNA staining is most marked in rabbit, human, and mouse; HPA staining is strong in rabbit but absent in human; and RCA 1 staining is present in human but not rabbit (66, 177, 207). In contrast to the marked PNA staining in the parietal and peptic cells, staining of the surface epithelial cells is very weak or absent in the human and mouse (66, 207). Mannose does not appear to be a component of the apical membrane of gastric cells, as evidenced by the lack of FITC–concanavalin agglutinin (ConA) staining (66, 177). *Ulex europeaus* agglutinin (UEA: fucose) staining is also surprisingly low in parietal and peptic cells (177).

In summary, the histochemical studies employing lectins, specific for particular carbohydrates, have confirmed and extended the biochemical analyses of apical membranes from the stomach. They indicate the glycocalyx to be a substantial component of the external face of the membrane with galactose and N-acetylgalactosamine as particularly important constituents, probably with terminal galactose. As evidenced by the lack of UEA staining, fucose appears to be a nonterminal sugar.

The gastric microsomes also contain glycolipids, constituting ~15% of the total lipid in the pig and frog, with a slightly lower (11%) value in rabbit (213). The most abundant glycolipids of frog microsomes are monoglucosylceramide and monogalactosylceramide (19). The glycolipids of guinea pig gastric mucosal scrapings are predominantly monoglucosylceramide, monogalactosylceramide, and digalactosylceramide, with digalactosylglucosylceramide also a significant component (25). When the glycolipid components of guinea pig individual epithelial cell types, isolated by pronase digestion followed by unit gravity velocity sedimentation, were analyzed, mucous and peptic cells showed roughly equal proportions of monohexosylceramide, dihexosylceramide, and free ceramide, with a smaller (~10%) proportion of trihexosylceramide. In contrast, parietal cells showed a greater proportion of dihexosylceramide (46%) (25).

In addition to these simple glycolipids, a number of complex glycosphingolipids are also found in pig gastric mucosa. This heterogeneous group of fucolipids with 12-36 sugar residues exhibit blood group (A + H) activities (220).

What of the possible function(s) of the luminal glycoprotein coat of the gastric cells? The membrane glycoproteins and glycolipids may play a role in protecting the stomach from proteolysis and the harsh acidic environment. The rich carbohydrate coat on the surface of the cells may present a mechanical barrier to the approach of the proteolytic enzymes, thus preventing hydrolysis of membrane proteins (18). The glycopeptides isolated from the parietal cell microsomes show inherent resistance to nonspecific proteolytic enzymes (19) and trypsin (176). In addition, isolated apical membranes of parietal cell are resistant to proteolytic digestion by trypsin (115, 116). On the surface epithelial cells, mucus will also subserve this function (2). The glycocalyx may also influence the lipid organization of the membrane. Removal of sialic acid residues with neuroaminidase increases the membrane fluidity of intestinal brush-border membrane vesicles (175a). The hydrophilic nature of oligosaccharides and their ability to form hydrogen bonds may also endow the membrane surface with a degree of stability, rendering it resistant to denaturation by the high concentration of H^+ found in the stomach (18). The presence of an extensive glycocalyx reduces the permeability of membranes to water (179) and may similarly reduce permeability to H^+.

SURFACE HYDROPHOBICITY

Phosphatidylcholine is the major component of lung surfactant, constituting 75% of the lipid fraction. The commonest fatty acid component of lung phosphatidylcholine is palmitate (71%), which results in ~35% of the phosphatidylcholine being disaturated, dipalmitylphosphatidylcholine in particular (134). Thus the lipid composition of lung surfactant shows remarkable similarity to the lipid composition described for gastric mucosal membranes and extracts (see *Lipid Constituents*, p. 291). Starting from these observations, Hills and Lichtenberger (112) have argued that these lipids may provide a similar surfactant function in the stomach, the phospholipids aligning as a monolayer above the surface of the gastric mucosa, thus presenting a hydrophobic or nonwettable layer to the acidic luminal contents (Fig. 10). The arguments of Hills and Lichtenberger are supported by observations on the hydrophobicity of the gastric mucosal surface, as determined by the contact angle between a drop of aqueous fluid and the gastric mucosal surface (111). With this method the contact angle measured for canine fundic mucosa was equivalent to that of relatively nonwettable substances. Other areas of the gastrointestinal tract, including the antrum, showed lower hydrophobicity. Stretching the gastric mucosa reduced the contact angle, arguably as the result of thinning or breaking up of the adsorbed phospholipid monolayer (112). Dissociated deoxycholate and acidified aspirin solutions abolished the surface hydrophobicity consistent with their gastric mucosal disrupting actions (91, 111). Moreover, 16,16-dimethyl prostaglandin E_2 increased the surface hydrophobicity and prevented the reduction caused by acidified aspirin

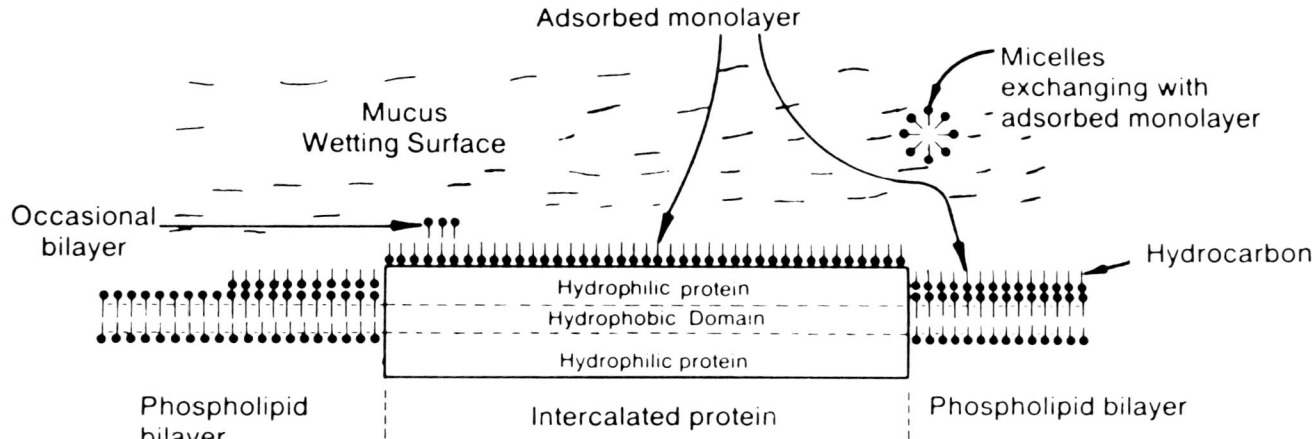

FIG. 10. Diagrammatic representation of an adsorbed monolayer of phospholipid on the gastric mucosal surface. Schematic includes *1*) intercalated protein zones in plasma membrane lipid bilayer, which might also be coated with monolayer; *2*) occasional bilayer formation; and *3*) wetting of the monolayer external surface by mucus, argued to reduce interfacial energy and provide an aqueous phase containing micelles and dissolved molecules in equilibrium with the adsorbed monolayer. Folding of intercalated proteins is argued to provide the ideal orientation of protein domains for adsorption of phospholipid. [From Hills (109).]

(148). Further circumstantial evidence for this barrier role of surface-active phospholipids includes reduced acid-induced ulcerogenesis with intragastric administration of a phospholipid mixture (147) or milk phospholipids (52), increased phospholipid secretion with exogenous prostaglandins (147), and the lubricating effect of phospholipids (29).

The experiments of Hills and Lichtenberger form the basis of a novel hypothesis. However, there are important caveats that should be placed on this hypothesis before it gains wide acceptance. The most important caveat is that all the experiments relate to a situation where the surface of the gastric mucosa forms the boundary between air and water. Thermodynamically it would be expected that, in the physiological situation, the phospholipids would form a bilayer, rather than the hypothesized monolayer. Hills (109) has argued that the phospholipids may not be deposited in their most stable form and that the transition from monolayer to bilayer may be very slow. In addition, experiments with mucus and phospholipids showed the mucus to be a powerful wetting agent, and Hills (109) argued that this effect would be great enough to provide stability to a phospholipid monolayer (Fig. 10).

Other caveats to the phospholipid monolayer hypothesis include the observation that the hydrophobicity was greatest in the fundic mucosa, with minimal hydrophobicity measured in the duodenum (111). However, the phospholipid constituents did not vary along the gastrointestinal tract (29). In the lungs, surfactant phospholipids are formed into intracellular lamellar bodies from which they are secreted; these events occur in the specialized alveolar type II cell (92). There are no reports of similar cells in the gastrointestinal tract, although preliminary electron-microscopic studies indicate the presence of lamellar bodies in parietal, mucous neck, and peptic cells (110). These observations need confirming, particularly since pentalaminar bodies are often, and only, observed in parietal cells as apical membrane is retrieved during transition from the stimulated to the resting state (75, 107). Mucus is argued to stabilize the adsorbed monolayer and to act as a reservoir for surfactant phospholipids (109). However, isolated mucus only weakly retards H^+ diffusion, suggesting only a minor role for luminal phospholipids as a barrier.

The phospholipid monolayer hypothesis for the gastric mucosal barrier requires further experimental analysis. In particular, evidence for its ability to function in the stomach under normal conditions requires substantiation. Because the plasma membranes of the gastric mucosa are constituted essentially of the same phospholipids as the proposed adsorbed monolayer, the question arises as to whether these luminal phospholipids are not just reflections of normal cellular turnover. This latter point may not, however, mitigate against any proposed functional role. Lastly, experimental evidence that an adsorbed monolayer of phospholipids presents any additional significant barrier to the diffusion of protons, above and beyond that offered by the plasma membrane bilayer, is lacking.

DIRECT EVIDENCE FOR IMPERMEABILITY OF APICAL MEMBRANES TO PROTONS

Reconstituted Epithelial Cell Monolayers

Isolated canine peptic cells have been grown in short-term culture to form reconstituted epithelial monolayers (206). The monolayers retain functional differentiation, as evidenced by stimulation of pepsinogen release by agents such as carbachol, secretin, and cyclic AMP analogues. These peptic cell monolayers are electrically tight, have a transepithelial resistance of ~1,500 $\Omega \cdot cm^2$, and form tight junctions (13). In an important study, Sanders et al. (205) demonstrated that these canine peptic cell monolayers are able to maintain electrical integrity despite a large pH gradient (Fig. 11). The transepithelial resistance increased 2.6-fold upon acidification of the apical solution to pH 2, and the monolayer maintained a pH gradient for >4 h. The barrier to H^+ was overcome at a critical pH, which was between 2 and 2.5 in 70% of monolayers and <2 in 20%. Aspirin (4 mM) caused a rapid decay in the monolayer transepithelial electrical properties when added to acidified apical solution (Fig. 11) but not when added at neutral pH. The electrical changes observed with apical acidification were fully reversible, indicating preservation of cellular integrity upon acidification. In contrast to the resistance of the apical membrane to acid, acidification of the basolateral solution to pH <5.5 resulted in a rapid decay of electrical resistance.

The resistance of these peptic cell monolayers to acid was not dependent on Na^+ or Cl^- transport (205). Treatment with ouabain or absence of basolateral HCO_3^- did not prevent the barrier function of the monolayers. Thus the resistance to H^+ of these peptic monolayers is consistent with properties of the apical membranes and tight junctions, these intact monolayers being practically impermeable to protons up to a concentration gradient of 10^5 (apical solution pH 2.5). Pepsin, 800 $\mu g \cdot ml^{-1}$ for 2 h at pH 2.5, was without effect on electrical resistance, in keeping with the peptic cells normal physiological function.

The monolayers were damaged, i.e., loss of electrical resistance, by pH <2–2.5. Normally the intraluminal pH of the gastric gland would be expected to fall below pH 1, producing essentially 150 mM HCl (see ELECTROLYTE COMPOSITION OF GASTRIC JUICE, p. 279). The failure of the peptic monolayers to withstand such low pH may indicate that other factors, such as intracellular buffering, are important for acid resistance at such low pH. Alternatively, this reconstituted system may not completely model the properties and resistance of the gastric gland in vivo.

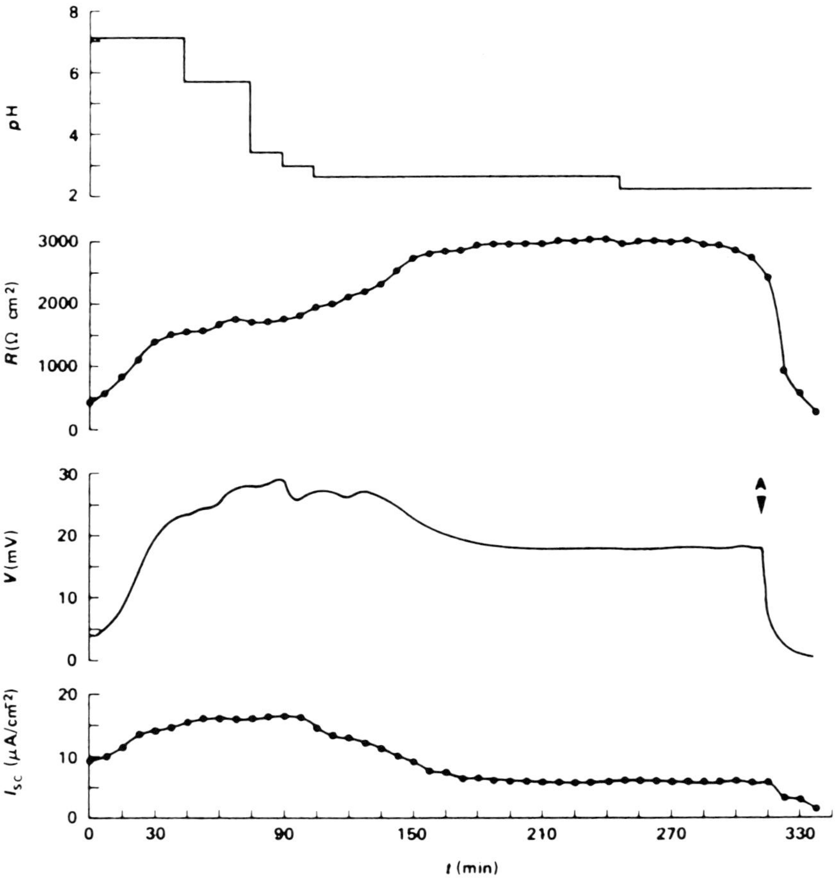

FIG. 11. Electrical response of a peptic cell monolayer to mucosal acidification. *Top panel*, apical solution pH as a function of time; a decrease reflects addition of acid. R, membrane resistance. V, potential difference. I_{sc}, short-circuit current. With reduction of apical pH to <2.5, R increased and remained stable for >3 h. Addition of 4 mM aspirin (A, *arrowhead*) caused a rapid decay in R, V, and I_{sc}. [From Sanders et al. (205). Reprinted by permission from *Nature*, copyright 1985, Macmillan Magazines Limited.]

Gastric surface epithelial cells from guinea pig (203) and fetal rabbit gastric epithelial cells (151) also form monolayers in culture. These cellular monolayers maintain resistances of ~280 $\Omega \cdot cm^2$, much lower than the values for the peptic cell monolayers (205). This difference does not correlate with reported differences in surface and glandular cell tight junction permeability to lanthanum (165). Future studies need to determine whether this difference is a reflection of differences in electrical conductance of surface and glandular cells in vivo. Intracellular pH in *Necturus* antral surface epithelial cells is independent of mucosal acidification to values as low as pH 2 (12, 155). This is consistent with the apical membrane of surface epithelial cells being extremely impermeable to H^+. However, one must apply the caveat that in *Necturus* antrum the cells will be covered by a mucous layer, and the results are equally consistent with the mucus-HCO_3^- layer hypothesis for resistance to H^+.

These reconstituted epithelial cell monolayers should yield further information about barrier properties of the gastrointestinal epithelia and their transport properties.

Isolated Apical Plasma Membrane Vesicles

Another useful model for studying directly the barrier properties of the plasma membrane in isolation is apical plasma membrane vesicles prepared from gastric epithelial cells. Vesicles isolated from parietal cell apical membranes show a resistance to low (1.9) pH (115). The rate of permeation of protons in these parietal cell apical membranes is lower than duodenal brush-border or gastric microsomal membranes (253, 254). The proton permeability is increased with increasing membrane fluidity, as observed with agents such as ethanol (see Fig. 9) and increasing temperature (254). Ethanol increases the rate of proton permeation at concentrations as low as 0.5 M, i.e., ~2% wt/vol (see Fig. 9). This is much lower than the threshold concentration of ethanol required for gross damage to the gastric mucosa (see *Alcohols*, p. 288, and Fig. 8). Thus these apical membrane vesicles are a sensitive model for study of damage to the gastric mucosal barrier. The parietal cell apical membrane vesicles are also resistant to proteolytic digestion with trypsin (115, 116). The ability to study the properties of defined membrane populations in isolation should lead to clarification of the specific barrier functions of the apical gastric mucosal membranes.

REGULATION OF INTRACELLULAR pH

The cells lining the stomach need to be able to cope with small quantities of proton that are able to penetrate the gastric mucosal barrier, whether as a result of normal permeability or enhanced by low concentrations of barrier-breaking agents. The buffering capacity of both parietal and peptic cells is 45–50 mM/pH (155, 171). The cells of the gastric mucosa also have specific mechanisms for controlling intracellular pH (pH_i).

Studies in isolated parietal cells using the trapped pH-sensitive dye dimethylcarboxyfluorescein have demonstrated these cells to possess both Na^+-H^+ and Cl^--HCO_3^- exchange mechanisms (171). With acridine orange as a pH probe, these two pH-regulating mechanisms were localized to the basolateral membrane with isolated membrane vesicles (171). Microspectrofluorimetric measurements of isolated gastric glands loaded with the pH-sensitive dye 2′,7′-bis(2-carboxyethyl)-5(6)-carboxyfluorescein (BCECF) have confirmed the presence of both Na^+-H^+ and Cl^--HCO_3^- exchange mechanisms in parietal cells (180, 181). A greater activity was observed for the anion exchanger consistent with this cell having to balance apical H^+ secretion and subsequent cytosolic alkalinization.

Peptic cells also possess both Na^+-H^+ and Cl^--HCO_3^- exchange mechanisms. However, as determined by microspectrofluorimetric observation of BCECF-loaded peptic cells in isolated gastric glands, in these cells the cation exchanger predominates (181). This is consistent with the function of peptic cells, which are periodically exposed to apical conditions of low pH and may therefore regulate pH_i by extruding H^+ using this exchange mechanism.

Isolated surface epithelial cell pH_i regulation has been studied with acridine orange. The experiments indicate these cells possess both Na^+-H^+ and Cl^--HCO_3^- exchange mechanisms (81, 178). Measurement of pH_i in BCECF-loaded surface epithelial cells indicates that the Na^+-H^+ exchanger is much more active than the Cl^--HCO_3^- exchanger (see the chapter by Demarest and Machen in this *Handbook*). The loss of polarity of these isolated cells does not allow comment on the localization of the exchangers. It might be argued that the presence of an Na^+-H^+ exchange mechanism on the apical membrane would predispose these cells to apical H^+ influx and subsequent damage. However, the Na^+-H^+ exchanger in other cell types is normally inactive at low pH (11, 155). An apical Na^+-H^+ exchanger, if active at low pH, would also be inconsistent with the data on the near impermeability of apical membranes to H^+ (see PROTON PERMEABILITY OF GASTRIC MUCOSA, p. 281). Mammalian gastric mucosa absorbs Na^+ through amiloride-sensitive apical conductive pathways (157, 242), and these channels might allow permeation of H^+ (72). However, low pH inhibits Na^+ absorption (121, 157), while the conductance of rat gastric mucosa in vitro is able to withstand a luminal pH of 2.5 (121).

The reader is referred to a recent review on regulation of pH_i in the stomach (155) and the chapter by Demarest and Machen in this *Handbook* for further discussion.

GASTRIC MICROCIRCULATION

Virchow (245) proposed that peptic ulcer disease was primarily a vascular problem, with local ischemia together with thrombosis and stasis leading to autodigestion of the gastric mucosa. Recent studies on the gastric microcirculation have centered on its buffering and metabolic roles in mucosal injury and defense.

Tissue Acid-Base Balance

Once H^+ has breached the gastric mucosal barrier it rapidly causes gross damage, including hemorrhage. This may be partly explained by access of H^+ to the exchangers normally involved in pH_i regulation on the basolateral membranes of the gastric epithelial cells. The damage initiated by this H^+ may be minimized by an adequate interstitial, and thus eventually blood, supply of buffer, particulary HCO_3^-.

Closed sacs of frog gastric mucosa ulcerate in the absence of serosal HCO_3^- (136, 137). The ulceration cannot be prevented by serosal buffers other than HCO_3^-, and the protection afforded by HCO_3^- is similar, whether CO_2 is present or absent. Acetazolamide increases the susceptibility to ulceration. In the presence of nutrient HCO_3^-, pH_i, measured by 5,5-dimethyloxazolidine-2,4-dione (DMO) distribution, was higher than in its absence (136). High concentrations of nutrient HCO_3^- are also able to attenuate the injury caused by acidic aspirin solutions (201). In vivo, increased plasma HCO_3^- is also able to reduce acid-induced ulceration (combined with hemorrhagic shock). The backdiffusion of H^+ was not altered by the HCO_3^- infusions, but the pH of the lamina propria was significantly greater (135). Increased rates of gastric acid secretion are also able to protect the mucosa against acid-induced ulceration, again accompanied by a high tissue pH (139). The anatomic arrangement of the blood vessels within the gastric mucosa will mean that the HCO_3^- extruded by the parietal cells and measurable as the alkaline tide will be made available to the surface epithelial cells (82). The blood flow to the mucosa will thus assume importance for the delivery of adequate quantities of HCO_3^- (228), as well as washing away any backflux of acid. Whether the protection afforded by increased HCO_3^- is due to increased supply for gastric alkaline secretion or for intracellular and intramucosal buffering of backdiffusing H^+ is yet to be answered. The absence of effect of HCO_3^- on barrier function of peptic cell monolayers (205) is consistent with the supply of HCO_3^- relating to these secondary barrier functions of the gastric mucosa.

Tissue Oxygenation

In addition to a role in buffering acid, the blood flow to the gastric mucosa is also important in providing adequate tissue oxygenation. Menguy and colleagues (161, 163) observed a large reduction in gastric mucosal ATP content during hemorrhagic shock in rats and rabbits. The decrease in gastric ATP concentrations was much greater than in liver or skeletal muscle, as a result of the small glycogen stores available in the gastric tissue (161). During hemorrhagic shock or complete ischemia, ATP concentrations were reduced to a far greater extent in rabbit and rat fundic compared with antral mucosa (161, 162). This difference may contribute to the greater resistance of the antral mucosa to stress ulceration. Large decreases in gastric surface epithelial cell intracellular partial pressure of oxygen (P_{O_2}) were observed during hemorrhagic shock or tourniquet ischemia, and the critical PO_2 below which gastric active Cl^- transport was inhibited was 9 mmHg (26).

However, the gastric mucosa does not ulcerate in response to reduced oxygen supply alone. Ulceration of the fundic mucosa in response to hemorrhagic shock depends on acidic luminal contents; ulceration was not observed with buffered luminal solution (164, 219). Topical application of 10% ethanol to the canine gastric mucosa results in an initial fall in surface epithelial cell potential difference, with a subsequent decrease in cellular P_{O_2} (26). In addition, in rabbit gastric mucosa in vitro, severe anoxia (gassing with 100% N_2), although reducing the transmucosal potential difference to zero, only resulted in a small increase in gastric mucosal permeability, as determined by H^+ and erythritol diffusion (23). Therefore it is unlikely that the degree of hypoxia likely to be encountered in the normal gastric mucosa plays a role in reducing the function of the gastric mucosal barrier, although it may enhance the action of other barrier-breaking agents.

Oxygen Radical Generation

Recent studies have implicated the generation of oxygen free radicals in the gastric mucosal damage observed after ischemia (129, 182). Oxygen radical scavengers are able to prevent the lesions caused upon reperfusion of ischemic gastric mucosa. During ischemia, catalyzed by xanthine oxidase, hypoxanthine accumulates as a result of ATP catabolism. Upon reperfusion of the tissue, xanthine oxidase forms the superoxide anion ($O_2^-\cdot$) and hydrogen peroxide from oxygen and hypoxanthine. The superoxide anion and hydrogen peroxide may react further to form the highly reactive and cytotoxic hydroxyl radical ($HO\cdot$) (223, 247). The damage produced by ischemia may be mimicked by exogenously generated oxygen radicals in the circulation (247) and prevented by inhibitors of xanthine oxidase (223). Endogenous sulfhydryl compounds, such as reduced glutathione, may bind these reactive free radicals and offer a possible explanation for the protective action of sulfhydryl agents (235). Gastric mucus may subserve a similar oxidant scavenger role within the gastric lumen (35, 94).

Lipid peroxidation, oxidation of the fatty acid side chains of membrane lipids by hydroperoxides, does not appear to be a major mechanism involved in gastric damage after ischemia and reperfusion (143). Similarly, lipid peroxidation does not appear to play a major role in damage induced by HCl, NaOH, or ethanol, although a role in the pathogenesis of aspirin-induced mucosal lesions may be possible (143).

RAPID REEPITHELIALIZATION OF GASTRIC MUCOSA

After the gastric epithelium has been compromised by injurious agents, its integrity and continuity are rapidly restored. The rapid regeneration of the epithelial cell layer was anticipated by Bernard (21) and recognized by other early workers, including Hollander (123). The restitution of a functional epithelial cell layer occurs within a few hours or less, much less time than required for cell division. This rapid process has been variously termed *restitution, reconstitution, reepithelialization,* or simply *rapid repair.* Reepithelialization is the process by which epithelial integrity and continuity are rapidly reestablished after injury, before cell proliferation or extensive inflammatory responses can occur (127, 216, 218). Reepithelialization occurs by migration of viable cells from adjacent areas or from the gastric pits to cover the denuded area. The process is graphically illustrated in Figure 12 and described in detail in reviews by Silen (216) and Silen and Ito (218).

Reepithelialization is observed after a short exposure to a variety of damaging agents, both in vivo and in vitro. These include hypertonic saline in frog (231) and guinea pig gastric mucosa in vitro (202) and ethanol in rat in vivo (128, 170). The reestablishment of epithelial continuity is accompanied by reestablishment of transepithelial electrical parameters (Fig. 13), i.e., reestablishment of the gastric mucosal barrier (202, 231).

During reepithelialization the damaged mucosa is protected from further damage by a fibrin-mucoid cap (see MUCUS-BICARBONATE LAYER, p. 282, and Fig. 12). The process is also accompanied by a large net alkaline secretion (231), which is likely to provide an optimal environment under the mucoid cap for reepithelialization to occur. In particular, mechanical removal or damage to the mucoid cap by mucolytic agents reduces the recovery of transmucosal potential difference (Fig. 13) associated with a reduction in the amount of histologically intact epithelium (249). The alkaline environment, in the main a plasma transudate, may be particularly important for reepithelialization, since acid is able to degrade the basal lamina, and in these circumstances reepithelialization is not observed (128, 218). Increasing nutrient HCO_3^-, even with a low luminal pH, allows reepithelialization to occur in a normal manner (218).

Reepithelialization is an important process in re-

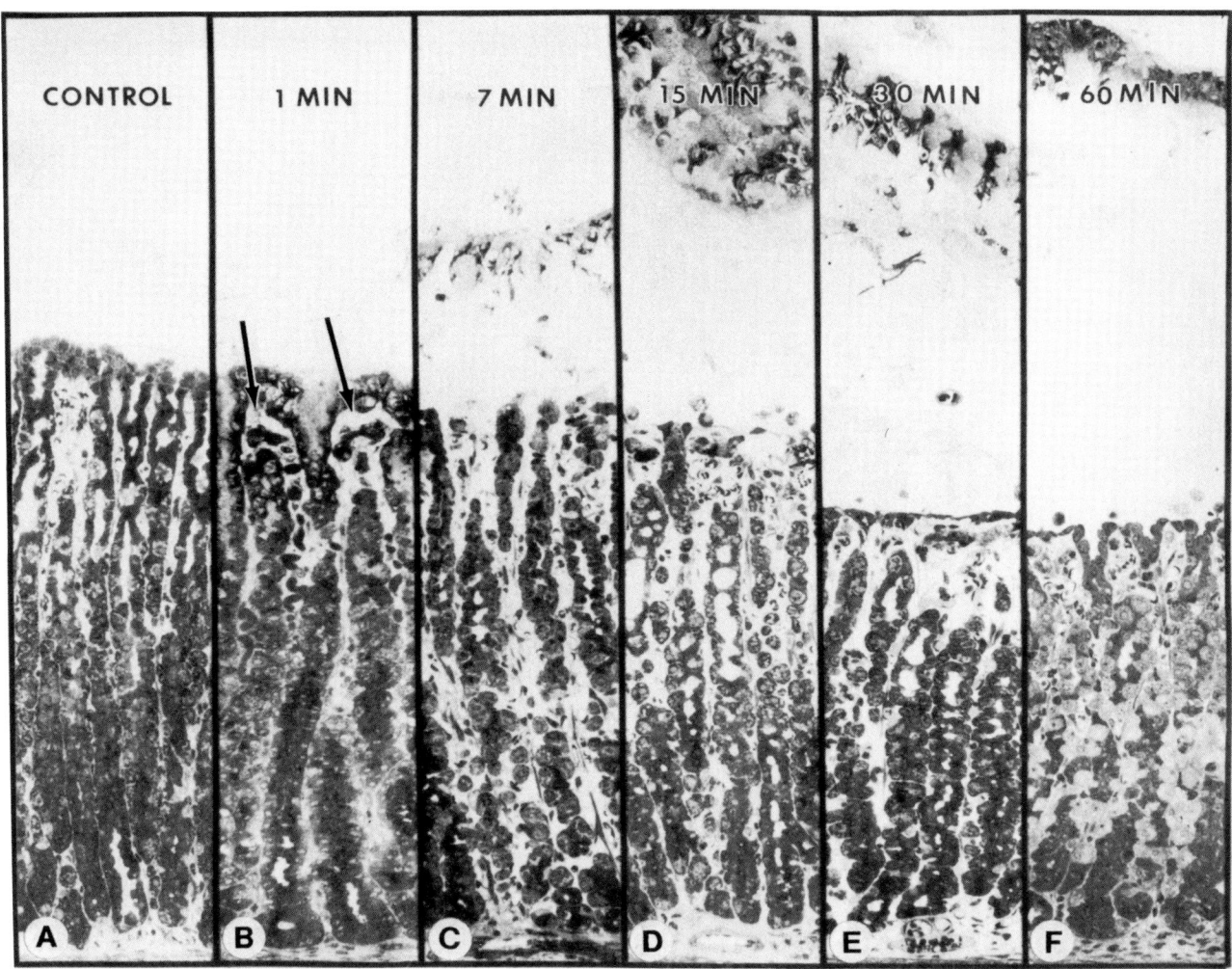

FIG. 12. Rat gastric mucosa during the process of damage by ethanol and subsequent reepithelialization. Low-power photomicrographs of semithin sections of rat gastric mucosa at various intervals after distension with absolute ethanol. A: control mucosa, luminal saline for 1 h. B: 1 min after ethanol exposure the surface epithelium is necrotic and lifting off the lamina propria (arrows). Deep pit and gland cells are intact. C: after 7 min of exposure to ethanol the surface epithelium is separated from the mucosa. D: 15-min sample shows exfoliated surface, intervening layer, and early evidence of reepithelialization. E: after 30 min, much of the luminal surface is reepithelialized. F: by 60 min the restituted surface is clearly evident. Space between dead cell layer and mucosa contains fibrin and mucus. × 162. [From Ito and Lacy (128). Copyright 1985 by The American Gastroenterology Association.]

pairing the gastric mucosal barrier after injurious insults. Breaking the mucosal barrier, at least with extreme conditions of damaging agents, is likely to be synonymous with destruction and shedding of the surface epithelial cells. The recovery of barrier function after removal of the damaging agent is the result of the restoration of epithelial integrity. Recognition of the reepithelialization process has provided evidence for anatomic site for the barrier as the surface epithelial cells. The process of reepithelialization has also been described in various *Necturus* epithelial tissues, including gallbladder, urinary bladder, and intestine (124, 125). Thus reepithelialization appears to be a fundamental property of epithelia and is likely to represent a generalized and important repair mechanism.

CYTOPROTECTION

Prostaglandins

The term *cytoprotection* has probably captivated the imaginations of more scientists and resulted in more experiments and publications than any other single area in gastroenterology in recent years. The term

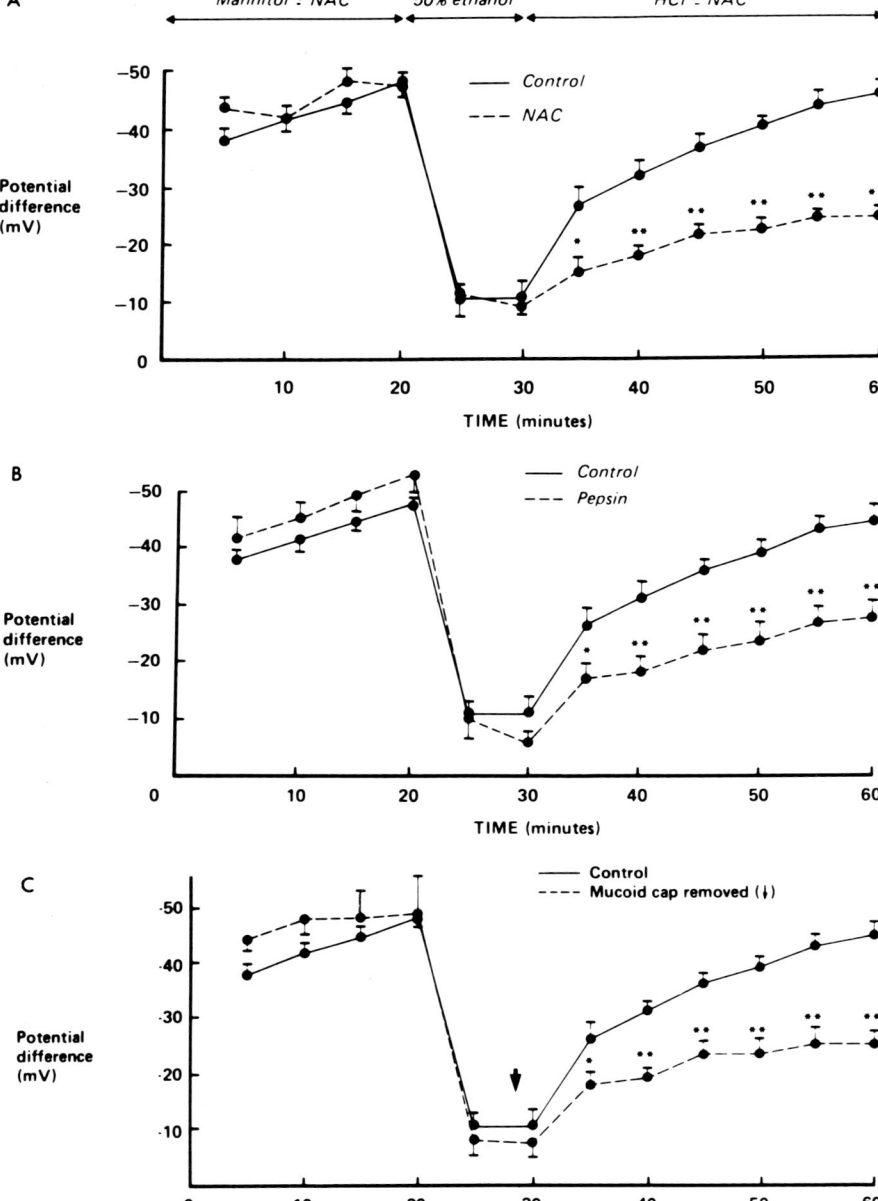

FIG. 13. Transmucosal potential difference recordings from chambered rat gastric mucosae damaged with 50% ethanol. All tissues were bathed in 0.3 M mannitol before ethanol was added. In control tissues, luminal bathing fluid was then 50 mM HCl in 0.2 M mannitol. *A*: in the test group, 5% *N*-acetylcysteine (NAC) was added to the chamber in each period except the period in which 50% ethanol was applied (21–30 min). *B*: protocol for these experiments was the same as shown at the top of *A*, except that 0.5% pepsin was added to the chamber in place of NAC. *C*: in the test group, solutions added to the chamber were identical to those in the control group. Under a stereomicroscope, the mucoid cap that formed after application of 50% ethanol was meticulously peeled off the underlying mucosa at 28 min (*arrow*). Results illustrated as mean ± 1 SE; double asterisks, significant difference from control group. [From Wallace and Whittle (249). Copyright 1986 by The American Gastroenterology Association.]

cytoprotection was first employed to describe the property of prostaglandins to protect the cells of the intestinal mucosa against the necrotizing effect of indomethacin (195). Subsequently, experiments showed a similar effect of many prostaglandins in the stomach (e.g., Fig. 14), where they prevented this mucosa from becoming inflamed and necrotic when exposed to noxious agents (196). Most outstandingly this effect of prostaglandins has been described to extend to prevention of gastric ulceration induced by such extreme challenges as 100% ethanol (Fig. 14), 0.6 M HCl, 0.2 M NaOH, 25% NaCl, and even boiling water (199). Careful histological evaluation of gastric mucosa subjected to such gross insults, however, has shown that the surface epithelial cells are almost completely destroyed (e.g., see Fig. 12), even in the presence of prostaglandins (97, 144, 248). The prostaglandins do reduce or even prevent the degree of deep mucosal necrosis caused by these damaging agents. Therefore the effect of the prostaglandins appears to be at the vascular rather than epithelial level, preventing hemorrhagic lesions (130, 234, 248). The term *cytoprotection* has thus been controversial (197).

The possible mechanism(s) of action of prostaglandins proposed for their cytoprotective actions has included *1*) prevention of gastric mucosal barrier dis-

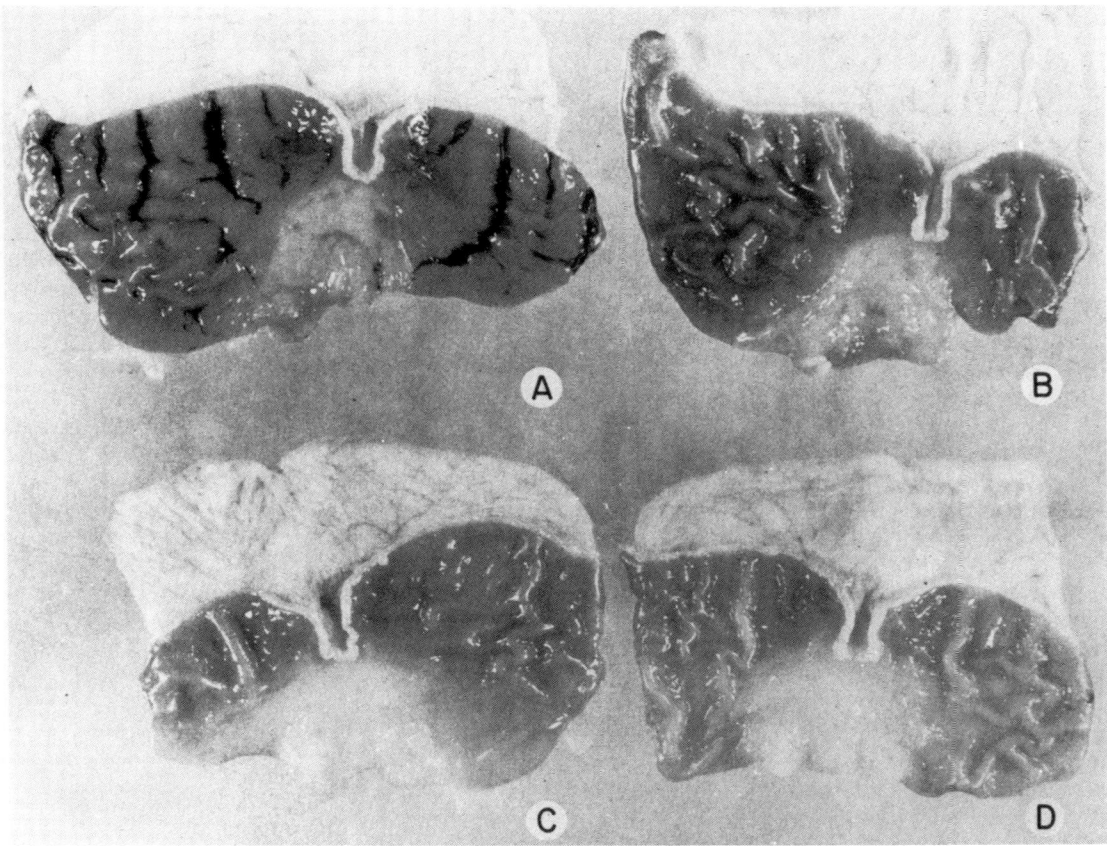

FIG. 14. Gastric cytoprotection of prostaglandins against ethanol. One ml of absolute ethanol was given orally. Rats were killed 1 h later and stomachs were removed and opened along the greater curvature. A: control vehicle was given orally 30 min before the ethanol. Multiple and severe necrotic lesions of the body of the stomach caused by ethanol are visible. B–D: a prostaglandin was administered 30 min before the ethanol. B: PGE_2 500 $\mu g \cdot kg^{-1}$ subcutaneously. C: PGE_2 150 $\mu g \cdot kg^{-1}$ orally. D: 16,16-dimethyl PGA_2 50 $\mu g \cdot kg^{-1}$ orally. These 3 prostaglandins prevented formation of visible gastric lesions due to ethanol. [From Robert et al. (199). Copyright 1979 by The American Gastroenterology Association.]

ruption; 2) stimulation of mucous secretion; 3) stimulation of gastric nonparietal, particularly alkaline, secretion; 4) enhancement of gastric mucosal blood flow; 5) stimulation of cellular transport processes; 6) stimulation of macromolecular synthesis; 7) stabilization of tissue lysosomes; 8) maintenance of gastric sulfhydryl compounds; 9) dissolution of gastric mucosal folds; and 10) stimulation of surface-active phospholipids. The varied effects of prostaglandins are reviewed by Miller (166) and in the volumes edited by Cohen (32).

There are many reports that prostaglandins are able to prevent or ameliorate the damage caused by a variety of agents to the stomach. In particular the conclusion has been reached that prostaglandins prevent disruption of the gastric mucosal barrier (166). However, with the recognition that prostaglandins do not prevent destruction of the surface epithelium, considered with the ability of the gastric mucosa to restore its continuity after damage (reepithelialization), this conclusion must be questioned (216). Prostaglandins do protect against vascular damage (234), and the enhanced blood flow, supplying buffer capacity and volume to dilute noxious luminal agents, is likely to be an important factor in their mechanism of action (130, 216). Prostaglandins do not alter the process of reepithelialization nor provide protection against the damaging effects of hypertonic NaCl in frog gastric mucosa in vitro (230) or the effects of 40% ethanol in ex vivo chambered rat gastric mucosa (248). In contrast, in the rat in vivo, prostaglandin pretreatment was reported to enhance reepithelialization and recovery of barrier function after ethanol damage (238). The foregoing reservations as to the nonvascular cytoprotective actions of prostaglandins are not meant to imply that these actions are unimportant. Indeed, if substantiated, they are likely to provide effective protective actions for specific, more physiologically relevant damaging agents than absolute ethanol and the like (32, 166).

Sulfhydryl Compounds and Other Agents

Prostaglandins are not the only group of compounds described to have cytoprotective activity. Acute gastric lesions may also be prevented or ameliorated by, among others, epidermal growth factor (142), somatostatin (236), and polyamines (168). Much recent interest has been shown in the role of sulfhydryl compounds in mediating these cytoprotective actions.

The acute mucosal injury in response to ethanol is associated with dose-dependent decreases in the gastric mucosal concentration of nonprotein sulfhydryl compounds (167, 235). The gastric mucosa contains uncommonly high concentrations of reduced glutathione, which is the major constituent of the nonprotein sulfhydryl pool (27). Several sulfhydryl-containing compounds, including cysteamine and cysteine, are able to prevent the gross damage caused by ethanol, and the protective action of these compounds may be prevented by sulfhydryl blocking agents, such as iodoacetate and N-ethylmaleimide (233, 235). The sulfhydryl blocking agents can also counteract the cytoprotective actions of prostaglandins (233, 235). Thus it may be argued that endogenous nonprotein sulfhydryls may play a role in maintaining normal gastric mucosal integrity. Protein sulfhydryl groups may subserve a similar or complementary role (59).

The role and mechanism of action of sulfhydryl compounds in maintenance of gastric mucosal integrity is complex. For example, some sulfhydryl reagents (e.g., N-acetylcysteine) are cytoprotective (233), while others (e.g., dithiothreitol) demonstrate barrier-breaking properties (46). One factor likely to be important in the cytoprotective actions of the sulfhydryls, as that of the prostaglandins, may be the maintenance of vascular integrity (233, 234). Reduced glutathione is oxidized in the synthesis of prostaglandins, and thus the role of these two cytoprotective agents may be at least in part interrelated. However, sulfhydryl compounds are involved in many aspects of cellular metabolism, and the contribution of these to direct epithelial barrier function has yet to be elucidated.

CONCLUSIONS

Our knowledge of the gastric mucosal barrier has been to a large extent founded on investigation of the mechanisms of action of specific damaging agents. With the advent of modern cell biological techniques and their application to study of the barrier, this barrier is being better defined on an anatomical as well as a physiological basis. As championed by Davenport, the epithelial cell layer, and in particular the apical cell membranes and epithelial tight junctions, is proving to be the major anatomical site of the barrier. However, other epithelial properties and constituents also subserve essential barrier functions. These importantly include the mucous layer overlying the epithelium, the adequate supply of blood with its buffering capacity, and the processes of rapid repair to damaged regions of the epithelium. Further developments in our understanding of the barrier properties of the gastric epithelium may be expected as studies employing isolated membrane vesicles and reconstituted epithelial cell monolayers in culture bear fruit.

I thank my colleagues, Drs. Nick Simmons, Heather Ballard, Lynda Sellers, and Jeff Pearson, for critical evaluation of this manuscript. The assistance of Cynthia Wood, Alison Speed, Elizabeth Hill, and Jennifer Wallis in preparing this manuscript is gratefully recorded.

This work was supported by grants from the Medical Research Council, SmithKline Foundation, and University of Newcastle upon Tyne Research Committee and carried out in collaboration with Drs. Jon Wilkes and Heather Ballard and with the technical assistance of Ken Elliott.

REFERENCES

1. ALLEN, A. The structure of gastrointestinal mucus glycoproteins and the viscous and gel-forming properties of mucus. *Br. Med. Bull.* 34: 28–33, 1978.
2. ALLEN, A. The structure and function of gastrointestinal mucus. In: *Basic Mechanisms of Gastrointestinal Mucosal Cell Injury and Protection*, edited by J. W. Harmon. Baltimore: Williams & Wilkins, 1981, p. 351–367.
3. ALLEN, A. Structure and function of gastrointestinal mucus. In: *Physiology of the Gastrointestinal Tract* (1st ed.), edited by L. R. Johnson. New York: Raven, 1981, vol. 1, p. 617–639.
4. ALLEN, A., N. J. H. CARROLL, AND B. H. HIRST. Gastric mucus in the anaesthetized rat: response to secretin, prostaglandin, ethanol and pepsin (Abstract). *J. Physiol. Lond.* 371: 135P, 1985.
5. ALLEN, A., AND A. GARNER. Mucus and bicarbonate secretion in the stomach and their possible role in mucosal protection. *Gut* 21: 249–262, 1980.
6. ALLEN, A., B. H. HIRST, AND L. A. SMEATON. Regulation of gastroduodenal HCO_3^- output by luminal acidification in the cat. *J. Physiol. Lond.* 342: 82P–83P, 1983.
7. ALLEN, A., D. A. HUTTON, A. J. LEONARD, J. P. PEARSON, AND L. A. SELLERS. The role of mucus in protection of the gastroduodenal mucosa. *Scand. J. Gastroenterol. Suppl.* 125: 71–77, 1986.
8. ALPHIN, R. S., V. A. VOKAC, R. L. GREGORY, P. M. BOLTON, AND J. W. TAWES. Role of intragastric pressure, pH, and pepsin in gastric ulceration in the rat. *Gastroenterology* 73: 495–500, 1977.
9. ALTAMIRANO, M. Alkaline secretion produced by intra-arterial acetylcholine. *J. Physiol. Lond.* 168: 787–803, 1963.
10. ALTAMIRANO, M. Action of concentrated solutions of nonelectrolytes on the dog gastric mucosa. *Am. J. Physiol.* 216: 33–40, 1969.
11. ARONSON, P. S. Kinetic properties of the plasma membrane Na^+-H^+ exchanger. *Annu. Rev. Physiol.* 47: 545–560, 1985.
12. ASHLEY, S. W., D. I. SOYBEL, AND L. Y. CHEUNG. Measurements of intracellular pH in *Necturus* antral mucosa by microelectrode technique. *Am. J. Physiol.* 250 (*Gastrointest. Liver Physiol.* 13): G625–G632, 1986.
13. AYALON, A., M. J. SANDERS, L. P. THOMAS, D. A. AMIRIAN, AND A. H. SOLL. Electrical effects of histamine on monolayers formed in culture from enriched canine gastric chief cells. *Proc. Natl. Acad. Sci. USA* 79: 7009–7013, 1982.
14. BAHARI, H. M. M., I. N. ROSS, AND L. A. TURNBERG. Dem-

onstration of a pH gradient across the mucus layer on the surface of human gastric mucosa in vitro. *Gut* 23: 513–516, 1982.
15. BAILEY, R. E., R. A. LEVINE, J. NANDI, E. H. SCHWARTZEL, JR., D. H. BEACH, P. N. BORER, AND G. C. LEVY. Effects of ethanol on gastric epithelial cell phospholipid dynamics and cellular function. *Am. J. Physiol.* 252 (*Gastrointest. Liver Physiol.* 15): G237–G243, 1987.
16. BAILEY, R. E., J. NANDI, R. A. LEVINE, T. K. RAY, P. N. BORER, AND G. C. LEVY. NMR studies of pig gastric microsomal H^+,K^+-ATPase and phospholipid dynamics: effects of ethanol perturbation. *J. Biol. Chem.* 261: 11086–11090, 1986.
17. BAJAJ, S. C., J. G. SPENNEY, AND G. SACHS. Properties of gastric antrum. III. Selectivity and modification of shunt conductance. *Gastroenterology* 72: 72–77, 1977.
18. BEESLEY, R. C., AND J. G. FORTE. Glycoproteins and glycolipids of oxyntic cell microsomes. I. Glycoproteins: carbohydrate composition, analytical and preparative fractionation. *Biochim. Biophys. Acta* 307: 372–385, 1973.
19. BEESLEY, R. C., AND J. G. FORTE. Glycoproteins and glycolipids of oxyntic cell microsomes. II. Glycopeptides and glycolipids. *Biochim. Biophys. Acta* 356: 144–155, 1974.
20. BELL, A. E., L. A. SELLERS, A. ALLEN, W. J. CUNLIFFE, E. R. MORRIS, AND S. B. ROSS-MURPHY. Properties of gastric and duodenal mucus: effect of proteolysis, disulfide reduction, bile, acid, ethanol, and hypertonicity on mucus gel structure. *Gastroenterology* 88: 269–280, 1985.
21. BERNARD, C. *Leçons de physiologie expérimentale appliquée à la médicine*. Paris: Ballière, 1856, vol. 2.
22. BINDSLEV, N., J. M. TORMEY, R. J. PIETRA, AND E. M. WRIGHT. Electrically and oncotically induced changes in permeability and structure of toad urinary bladder. *Biochim. Biophys. Acta* 332: 286–297, 1974.
23. BIRKETT, D., AND W. SILEN. Effect of severe anoxia on the permeability of gastric mucosa. *Proc. Soc. Exp. Biol. Med.* 148: 256–260, 1975.
24. BLAIR, E. L., AND A. K. YASSIN. The electrolyte content of histamine-stimulated gastric secretion in the cat (Abstract). *J. Physiol. Lond.* 159: 82P–83P, 1961.
25. BOUHOURS, J.-F., AND D. BOUHOURS. Neutral glycosphingolipids of three cell types isolated from guinea pig gastric mucosa. *Biochem. Biophys. Res. Commun.* 85: 1314–1317, 1978.
26. BOWEN, J. C., AND R. B. FAIRCHILD. Oxygen in gastric mucosal protection. In: *Mechanisms of Mucosal Protection in the Upper Gastrointestinal Tract*, edited by A. Allen, G. Flemström, A. Garner, W. Silen, and L. A. Turnberg. New York: Raven, 1984, p. 259–266.
27. BOYD, S. C., H. A. SASAME, AND M. R. BOYD. High concentrations of glutathione in glandular stomach: possible implications for carcinogenesis. *Science Wash. DC* 205: 1010–1012, 1979.
28. BUNCE, K. T., AND N. M. CLAYTON. The effects of the stable thromboxane A_2-mimetic, U46619, on gastric mucosal damage and gastric non-parietal secretion in the rat. *Br. J. Pharmacol.* 91: 23–29, 1987.
29. BUTLER, B. D., L. M. LICHTENBERGER, AND B. A. HILLS. Distribution of surfactants in the canine gastrointestinal tract and their ability to lubricate. *Am. J. Physiol.* 244 (*Gastrointest. Liver Physiol.* 7): G645–G651, 1983.
30. CHUNG, R. S. K., P. T. SUM, H. GOLDMAN, M. FIELD, AND W. SILEN. Effects of chelation of calcium on the gastric mucosal barrier. *Gastroenterology* 59: 200–207, 1970.
31. CLAUSEN, C., T. E. MACHEN, AND J. M. DIAMOND. Use of AC impedance analysis to study membrane changes related to acid secretion in amphibian gastric mucosa. *Biophys. J.* 41: 167–178, 1983.
32. COHEN, M. M. (editor). *Biological Protection With Prostglandins*. Boca Raton, FL: CRC, 1986.
33. COLEMAN, R., P. J. LOWE, AND D. BILLINGTON. Membrane lipid composition and susceptibility to bile salt damage. *Biochim. Biophys. Acta* 599: 294–300, 1980.
34. COOKE, A. R., AND M. G. KIENZLE. Studies on anti-inflammatory drugs and aliphatic alcohols on antral mucosa. *Gastroenterology* 66: 56–62, 1974.
35. CROSS, C. E., B. HALLIWELL, AND A. ALLEN. Antioxidant protection: a function of tracheobronchial and gastrointestinal mucus. *Lancet* 1: 1328–1330, 1984.
36. DAVENPORT, H. W. Gastric mucosal injury by fatty and acetylsalicylic acids. *Gastroenterology* 46: 245–253, 1964.
37. DAVENPORT, H. W. Damage to the gastric mucosa: effects of salicylates and stimulation. *Gastroenterology* 49: 189–196, 1965.
38. DAVENPORT, H. W. Potassium fluxes across the resting and stimulated gastric mucosa: injury by salicylic and acetic acids. *Gastroenterology* 49: 238–245, 1965.
39. DAVENPORT, H. W. Fluid produced by the gastric mucosa during damage by acetic and salicylic acids. *Gastroenterology* 50: 487–499, 1966.
40. DAVENPORT, H. W. Absorption of taurocholate-24-^{14}C through the canine gastric mucosa. *Proc. Soc. Exp. Biol. Med.* 125: 670–673, 1967.
41. DAVENPORT, H. W. Ethanol damage to canine oxyntic glandular mucosa. *Proc. Soc. Exp. Biol. Med.* 126: 657–662, 1967.
42. DAVENPORT, H. W. Salicylate damage to the gastric mucosal barrier. *N. Engl. J. Med.* 276: 1307–1312, 1967.
43. DAVENPORT, H. W. Destruction of the gastric mucosal barrier by detergents and urea. *Gastroenterology* 54: 175–181, 1968.
44. DAVENPORT, H. W. Gastric mucosal hemorrhage in dogs: effects of acid, aspirin, and alcohol. *Gastroenterology* 56: 439–449, 1969.
45. DAVENPORT, H. W. Effect of lysolecithin, digitonin, and phospholipase A upon the dog's gastric mucosal barrier. *Gastroenterology* 59: 505–509, 1970.
46. DAVENPORT, H. W. Protein-losing gastropathy produced by sulfhydryl reagents. *Gastroenterology* 60: 870–879, 1971.
47. DAVENPORT, H. W. Plasma protein shedding by the canine oxyntic glandular mucosa induced by topical application of snake venoms and ethanol. *Gastroenterology* 67: 264–270, 1974.
48. DAVENPORT, H. W. The gastric mucosal barrier: past, present and future. *Mayo Clin. Proc.* 50: 507–514, 1975.
49. DAVENPORT, H. W., H. A. WARNER, AND C. F. CODE. Functional significance of gastric mucosal barrier to sodium. *Gastroenterology* 47: 142–152, 1964.
50. DEAMER, D. W., AND J. W. NICHOLS. Proton-hydroxide permeability of liposomes. *Proc. Natl. Acad. Sci. USA* 80: 165–168, 1983.
51. DEEMS, R. A., AND E. A. DENNIS. Phospholipase A_2 from cobra venom (*Naja naja naja*). *Methods Enzymol.* 71: 703–710, 1981.
52. DIAL, E. J., AND L. M. LICHTENBERGER. A role for milk phospholipids in protection against gastric acid: studies in adult and suckling rats. *Gastroenterology* 87: 379–385, 1984.
53. DIAMOND, J. M., AND T. E. MACHEN. Impedance analysis in epithelia and the problem of gastric acid secretion. *J. Membr. Biol.* 72: 17–41, 1983.
54. DIBONA, D. R., AND M. M. CIVAN. Pathways for movement of ions and water across toad urinary bladder. I. Anatomic site of transepithelial shunt pathways. *J. Membr. Biol.* 12: 101–128, 1973.
55. DUANE, W. C., M. D. LEVITT, N. A. STALEY, A. P. MCHALE, D. M. WIEGAND, AND C. A. FETZER. Role of the unstirred layer in protecting the murine gastric mucosa from bile salt. *Gastroenterology* 91: 913–918, 1986.
56. DUANE, W. C., A. P. MCHALE, AND C. E. SIEVERT. Lysolecithin-lipid interactions in disruption of the canine gastric mucosal barrier. *Am. J. Physiol.* 250 (*Gastrointest. Liver Physiol.* 13): G275–G279, 1986.
57. DUANE, W. C., AND D. M. WIEGAND. Mechanism by which bile salt disrupts the gastric mucosal barrier in the dog. *J. Clin. Invest.* 66: 1044–1049, 1980.
58. DUANE, W. C., D. M. WIEGAND, AND C. E. SIEVERT. Bile acid and bile salt disrupt gastric mucosal barrier in the dog by

different mechanisms. *Am. J. Physiol.* 242 (*Gastrointest. Liver Physiol.* 5): G95–G99, 1982.
59. DUPUY, D., AND S. SZABO. Protection by metals against ethanol-induced gastric mucosal injury in the rat. Comparative biochemical and pharmacologic studies implicate protein sulfhydryls. *Gastroenterology* 91: 966–974, 1986.
60. DURBIN, R. P. Backdiffusion of H^+ in isolated frog gastric mucosa. *Am. J. Physiol.* 246 (*Gastrointest. Liver Physiol.* 9): G114–G119, 1984.
61. DYCK, W. P., J. L. WERTHER, J. RUDICK, AND H. D. JANOWITZ. Electrolyte movement across canine antral and fundic mucosa. *Gastroenterology* 56: 489–495, 1969.
62. EASTWOOD, G. L., AND J. P. KIRCHNER. Changes in the fine structure of mouse gastric epithelium produced by ethanol and urea. *Gastroenterology* 67: 71–84, 1974.
63. ENGEL, E., A. PESKOFF, G. L. KAUFFMAN, JR., AND M. I. GROSSMAN. Analysis of hydrogen ion concentration in the gastric gel mucus layer. *Am. J. Physiol.* 247 (*Gastrointest. Liver Physiol.* 10): G321–G338, 1984.
64. ERLIJ, D., AND A. MARTINEZ-PALOMO. Opening of tight junctions in frog skin by hypertonic urea solutions. *J. Membr. Biol.* 9: 229–240, 1972.
65. FELDMAN, M., AND L. SCHILLER. Effect of bethanechol (Urecholine) on gastric acid and nonparietal secretion in normal subjects and duodenal ulcer patients. *Gastroenterology* 83: 262–266, 1982.
66. FISCHER, J., P. J. KLEIN, M. VIERBUCHEN, B. SKUTTA, G. UHLENBRUCK, AND R. FISCHER. Characterization of glycoconjugates of human gastrointestinal mucosa by lectins. I. Histochemical distribution of lectin binding sites in normal alimentary tract as well as in benign and malignant gastric neoplasms. *J. Histochem. Cytochem.* 32: 681–689, 1984.
67. FLEMSTRÖM, G. Active alkalinization by amphibian gastric fundic mucosa in vitro. *Am. J. Physiol.* 233 (*Endocrinol. Metab. Gastrointest. Physiol.* 2): E1–E12, 1977.
68. FLEMSTRÖM, G., AND A. GARNER. Gastroduodenal HCO_3^- transport: characteristics and proposed role in acidity regulation and mucosal protection. *Am. J. Physiol.* 242 (*Gastrointest. Liver Physiol.* 5): G183–G193, 1982.
69. FLEMSTRÖM, G., AND L. A. TURNBERG. Gastroduodenal defence mechanisms. *Clin. Gastroenterol.* 13: 327–354, 1984.
70. FORTE, J. G., J. A. BLACK, T. M. FORTE, T. E. MACHEN, AND J. M. WOLOSIN. Ultrastructural changes related to functional activity in gastric oxyntic cells. *Am. J. Physiol.* 241 (*Gastrointest. Liver Physiol.* 4): G349–G358, 1981.
71. FORTE, J. G., AND T. E. MACHEN. Transport and electrical phenomena in resting and secreting piglet gastric mucosa. *J. Physiol. Lond.* 244: 33–51, 1975.
72. FORTE, J. G., AND T. E. MACHEN. Ion transport by gastric mucosa. In: *Physiology of Membrane Disorders* (2nd ed.), edited by T. E. Andreoli, J. F. Hoffman, D. D. Fanestil, and S. G. Schultz. New York: Plenum, 1986, p. 535–558.
73. FORTE, J. G., AND A. H. NAUSS. Effects of calcium removal on bullfrog gastric mucosa. *Am. J. Physiol.* 205: 631–637, 1963.
74. FORTE, T. M., AND J. G. FORTE. Histochemical staining and characterization of glycoproteins in acid-secreting cells of frog stomach. *J. Cell Biol.* 47: 437–452, 1970.
75. FORTE, T. M., T. E. MACHEN, AND J. G. FORTE. Ultrastructural changes in oxyntic cells associated with secretory function: a membrane-recycling hypothesis. *Gastroenterology* 73: 941–955, 1977.
76. FORTE, T. M., W. SILEN, AND J. G. FORTE. Ultrastructural lesions in gastric mucosa exposed to deoxycholate: implications toward the barrier concept. In: *Gastric Hydrogen Ion Secretion*, edited by D. K. Kasbekar, G. Sachs, and W. S. Rehm. New York: Dekker, 1976, p. 1–28.
77. FROMM, D. Ion selective effects of salicylate on antral mucosa. *Gastroenterology* 71: 743–749, 1976.
78. FROMM, D. Gastric mucosal barrier. In: *Physiology of the Gastrointestinal Tract* (1st ed.), edited by L. R. Johnson. New York: Raven, 1981, vol. 1, p. 733–748.
79. FROMM, D., J. H. SCHWARTZ, AND R. QUIJANO. Effects of salicylate and bile salt on ion transport by isolated gastric mucosa of the rabbit. *Am. J. Physiol.* 230: 319–326, 1976.
80. FRÖMTER, E., AND J. M. DIAMOND. Route of passive ion permeation in epithelia. *Nature New Biol.* 235: 9–13, 1972.
81. FURUKAWA, T., E. OLENDER, D. FROMM, AND M. KOLIS. Effects of cyclic adenosine monophosphate and prostaglandins on Na^+- and HCO_3^--induced dissipation of a proton gradient in isolated gastric mucosal surface cells of rabbits. *Gastroenterology* 89: 500–506, 1985.
82. GANNON, B., J. BROWNING, P. O'BRIEN, AND P. ROGERS. Mucosal microvascular architecture of the fundus and body of human stomach. *Gastroenterology* 86: 866–875, 1984.
83. GARNER, A. Mechanisms of action of aspirin on the gastric mucosa of the guinea pig. *Acta Physiol. Scand. Special Suppl.*: 101–110, 1978.
84. GARNER, A., AND G. FLEMSTRÖM. Gastric HCO_3^- secretion in the guinea pig. *Am. J. Physiol.* 234 (*Endocrinol. Metab. Gastrointest. Physiol.* 3): E535–E541, 1978.
85. GARNER, A., G. FLEMSTRÖM, A. ALLEN, J. R. HEYLINGS, AND S. MCQUEEN. Gastric mucosal protective mechanisms: role of epithelial bicarbonate and mucus secretions. *Scand. J. Gastroenterol. Suppl.* 101: 79–86, 1984.
86. GARNER, A., AND J. R. HEYLINGS. Stimulation of alkaline secretion in amphibian-isolated gastric mucosa by 16,16-dimethyl PGE_2 and $PGF_{2\alpha}$. *Gastroenterology* 76: 497–503, 1979.
87. GASCOIGNE, A. D., AND B. H. HIRST. Prostaglandins alter the relationship between gastric hydrogen ion concentration and flow: evidence for stimulation of non-parietal secretion in the cat. *J. Physiol. Lond.* 316: 427–438, 1981.
88. GERARD, A., R. LEV, AND G. B. J. GLASS. Histochemical study of the mucosubstances in the canine stomach. I. The resting stomach. *Am. J. Dig. Dis.* 12: 891–912, 1967.
89. GILBERT, A. J., AND S. J. HERSEY. Morphometric analysis of parietal cell membrane transformations in isolated gastric glands. *J. Membr. Biol.* 67: 113–124, 1982.
90. GLOVER, J. An Attempt to Prove that Digestion, in Man, Depends on the United Causes of Solution and Fermentation. Philadelphia: University of Pennsylvania, 1800. Dissertation.
91. GODDARD, P. J., B. A. HILLS, AND L. M. LICHTENBERGER. Does aspirin damage canine gastric muocsa by reducing its surface hydrophobicity? *Am. J. Physiol.* 252 (*Gastrointest. Liver Physiol.* 15): G421–G430, 1987.
92. GOERKE, J. Lung surfactant. *Biochim. Biophys. Acta* 344: 241–261, 1974.
93. GRAY, J. S., AND G. R. BUCHER. The composition of gastric juice as a function of the rate of secretion. *Am. J. Physiol.* 133: 542–550, 1941.
94. GRISHAM, M. B., C. VON RITTER, B. F. SMITH, J. T. LAMONT, AND D. N. GRANGER. Interaction between oxygen radicals and gastric mucin. *Am. J. Physiol.* 253 (*Gastrointest. Liver Physiol.* 16): G93–G96, 1987.
95. GUDIKSEN, E. Investigations on the composition of gastric juice. *C. R. Trav. Lab. Carlsberg* 27: 145–278, 1950.
96. GÜLDÜTUNA, S., G. ZIMMER, W. KURTZ, AND U. LEUSCHNER. Prostaglandin E_2 directly protects isolated rat gastric surface cell membranes against bile salts. *Biochim. Biophys. Acta* 902: 217–222, 1987.
97. GUTH, P. H., G. PAULSEN, AND H. NAGATA. Histologic and microcirculatory changes in alcohol-induced gastric lesions in the rat: effect of prostaglandin cytoprotection. *Gastroenterology* 87: 1083–1090, 1984.
98. GUTKNECHT, J. Proton/hydroxide permeabilities of lipid bilayer membranes. In: *Hydrogen Ion Transport in Epithelia*, edited by J. G. Forte, D. G. Warnock, and F. C. Rector. New York: Wiley, 1984, p. 3–12.
99. GUTKNECHT, J. Proton/hydroxide conductance through phospholipid bilayer membranes: effects of phytanic acid. *Biochim. Biophys. Acta* 898: 97–108, 1987.
100. GUTKNECHT, J., AND J. WALTER. Transport of protons and hydrochloric acid through lipid bilayer membranes. *Biochim.*

Biophys. Acta 641: 183–188, 1981.
101. HANZEL, D., AND J. G. FORTE. Cited by Gutknecht, J. Proton/hydroxide conductance through phospholipid bilayer membranes: effects of phytanic acid. *Biochim. Biophys. Acta* 898: 97–108, 1987.
102. HARLEY, G. Contribution to our knowledge of digestion. *Br. Foreign Med. Chir. Rev.* 25: 206, 1860.
103. HARMON, J. W., T. DOONG, AND T. R. GADACZ. Bile acids are not equally damaging to the gastric mucosa. *Surgery St. Louis* 84: 79–86, 1978.
104. HARPER, A. A., J. D. REED, AND J. R. SMY. Effects of intragastric hyperosmolal solutions on gastric function. *J. Physiol. Lond.* 209: 453–472, 1970.
105. HEATLEY, N. G. Mucosubstance as a barrier to diffusion. *Gastroenterology* 37: 313–317, 1959.
106. HELANDER, H. F. The cells of the gastric mucosa. *Int. Rev. Cytol.* 70: 217–289, 1981.
107. HELANDER, H. F., AND B. I. HIRSCHOWITZ. Quantitative ultrastructural studies on gastric parietal cells. *Gastroenterology* 63: 951–961, 1972.
108. HELENIUS, A., AND K. SIMONS. Solubilization of membranes by detergents. *Biochim. Biophys. Acta* 415: 29–79, 1975.
109. HILLS, B. A. Gastric mucosal barrier: stabilization of hydrophobic lining to the stomach by mucus. *Am. J. Physiol.* 249 (*Gastrointest. Liver Physiol.* 12): G342–G349, 1985.
110. HILLS, B. A. *The Biology of Surfactant.* Cambridge, UK: Cambridge Univ. Press, 1988.
111. HILLS, B. A., B. D. BUTLER, AND L. M. LICHTENBERGER. Gastric mucosal barrier: hydrophobic lining to the lumen of the stomach. *Am. J. Physiol.* 244 (*Gastrointest. Liver Physiol.* 7): G561–G568, 1983.
112. HILLS, B. A., AND L. M. LICHTENBERGER. Gastric mucosal barrier: hydrophobicity of stretched stomach lining. *Am. J. Physiol.* 248 (*Gastrointest. Liver Physiol.* 11): G643–G647, 1985.
113. HINGSON, D. J., AND S. ITO. Effect of aspirin and related compounds on the fine structure of mouse gastric mucosa. *Gastroenterology* 61: 156–177, 1971.
114. HIRST, B. H. Gastric electrolyte composition. In: *Mechanisms of Mucosal Protection in the Upper Gastrointestinal Tract*, edited by A. Allen, G. Flemström, A. Garner, W. Silen, and L. A. Turnberg. New York: Raven, 1984, p. 103–106.
115. HIRST, B. H., H. J. BALLARD, J. M. WILKES, AND J. G. FORTE. Gastric and small intestinal membrane vesicle resistance to trypsin: implications for mucosal protection. In: *Gastrointestinal and Hepatic Secretions: Mechanisms and Control*, edited by J. S. Davison and E. A. Shaffer. Calgary, Canada: Univ. of Calgary Press, 1988, sect. II, in press.
116. HIRST, B. H., AND J. G. FORTE. Redistribution and characterization of ($H^+ + K^+$)-ATPase membranes from resting and stimulated gastric parietal cells. *Biochem. J.* 231: 641–649, 1985.
117. HIRST, B. H., L. A. LABIB, J. D. REED, AND J. G. STEPHEN. Relationship between hydrogen ion concentration and flow of gastric juice during inhibition of gastric secretion in the cat. *J. Physiol. Lond.* 306: 51–63, 1980.
118. HIRST, B. H., AND L. A. SMEATON. Gastric bicarbonate transport. In: *Biological Protection With Prostaglandins*, edited by M. M. Cohen. Boca Raton, FL: CRC, 1986, vol. 2, p. 63–75.
119. HIRST, B. H., AND J. M. WILKES. Proton permeability and membrane fluidity of rabbit duodenal brush-border membrane vesicles (Abstract). *J. Physiol. Lond.* 391: 26P, 1987.
120. HOBSLEY, M., AND W. SILEN. The relation between the rate of production of gastric juice and its electrolyte concentrations. *Clin. Sci.* 39: 61–75, 1970.
121. HOGBEN, C. A. M., AND D. R. KARAL. Further observations on the isolated rat gastric mucosa. In: *Transport Mechanisms in Epithelia*, edited by H. H. Ussing and N. A. Thorn. Copenhagen: Munksgaard, 1972, p. 236–253.
122. HOLLANDER, F. Studies on gastric secretion. IV. Variations in the chloride content of gastric juice and their significance. *J. Biol. Chem.* 97: 585–604, 1932.
123. HOLLANDER, F. The two-component mucous barrier: its activity in protecting the gastroduodenal mucosa against peptic ulceration. *Arch. Intern. Med.* 93: 107–120, 1952.
124. HUDSPETH, A. J. Establishment of tight junctions between epithelial cells. *Proc. Natl. Acad. Sci. USA* 72: 2711–2713, 1975.
125. HUDSPETH, A. J. The recovery of local transepithelial resistance following single-cell lesions. *Exp. Cell Res.* 138: 331–342, 1982.
126. HUNTER, J. On the digestion of the stomach after death. *Philos. Trans.* 62: 447–454, 1772.
127. ITO, S. Functional gastric morphology. In: *Physiology of the Gastrointestinal Tract* (2nd ed.), edited by L. R. Johnson. New York: Raven, 1987, vol. 1, p. 817–851.
128. ITO, S., AND E. R. LACY. Morphology of rat gastric mucosal damage, defense, and restitution in the presence of luminal ethanol. *Gastroenterology* 88: 250–260, 1985.
129. ITOH, M., AND P. H. GUTH. Role of oxygen-derived free radicals in hemorrhagic shock-induced gastric lesions in the rat. *Gastroenterology* 88: 1162–1167, 1985.
130. KAUFFMAN, G. L. Gastric circulation. In: *Biological Protection With Prostaglandins*, edited by M. M. Cohen. Boca Raton, FL: CRC, 1986, vol. 2, p. 39–44.
131. KAUFFMAN, G. L., AND M. R. THOMPSON. Titration of sodium channels in canine gastric mucosa. *Proc. Natl. Acad. Sci. USA* 72: 3731–3734, 1975.
132. KELLY, D. G., C. F. CODE, J. LECHAGO, J. BUGAJSKI, AND J. F. SCHLEGEL. Physiological and morphological characteristics of progressive disruption of the canine gastric mucosal barrier. *Dig. Dis. Sci.* 24: 424–441, 1979.
133. KERSS, S., A. ALLEN, AND A. GARNER. A simple method for measuring thickness of the mucus gel layer adherent to rat, frog and human gastric mucosa: influence of feeding, prostaglandin, *N*-acetylcysteine and other agents. *Clin. Sci.* 63: 187–195, 1982.
134. KING, R. J., AND J. A. CLEMENTS. Surface active materials from dog lung. II. Composition and physiological correlations. *Am. J. Physiol.* 223: 715–726, 1972.
135. KIVILAAKSO, E. High plasma HCO_3^- protects gastric mucosa against acute ulceration in the rat. *Gastroenterology* 81: 921–927, 1981.
136. KIVILAAKSO, E. Contribution of ambient HCO_3^- to mucosal protection and intracellular pH in isolated amphibian gastric mucosa. *Gastroenterology* 85: 1284–1289, 1983.
137. KIVILAAKSO, E., A. BARZILAI, R. SCHIESSEL, R. CRASS, AND W. SILEN. Ulceration of isolated amphibian gastric mucosa. *Gastroenterology* 77: 31–37, 1979.
138. KIVILAAKSO, E., A. BARZILAI, R. SCHIESSEL, D. FROMM, AND W. SILEN. Experimental ulceration of rabbit antral mucosa. *Gastroenterology* 80: 77–83, 1981.
139. KIVILAAKSO, E., D. FROMM, AND W. SILEN. Effect of the acid secretory state on intramural pH of rabbit gastric mucosa. *Gastroenterology* 75: 641–648, 1978.
140. KIVILAAKSO, E., AND W. SILEN. Pathogenesis of experimental gastric-mucosal injury. *N. Engl. J. Med.* 301: 364–369, 1979.
141. KIVILUOTO, T., J. VOIPIO, AND E. KIVILAAKSO. Is "H^+ back diffusion" following disruption of the gastric mucosal barrier in fact alkali (HCO_3^-) efflux? (Abstract). *Gastroenterology* 92: 1470, 1987.
142. KONTUREK, S. J., T. BRZOZOWSKI, I. PIASTUCKI, A. DEMBINSKI, T. RADECKI, A. DEMINSKA-KIEC, A. ZMUDA, AND H. GREGORY. Role of mucosal prostaglandins and DNA synthesis in gastric cytoprotection by luminal epidermal growth factor. *Gut* 22: 927–932, 1981.
143. KUSTERER, K., G. PIHAN, AND S. SZABO. Role of lipid peroxidation in gastric mucosal lesions induced by HCl, NaOH, or ischemia. *Am. J. Physiol.* 252 (*Gastrointest. Liver Physiol.* 15): G811–G816, 1987.
144. LACY, E. R., AND S. ITO. Microscopic analysis of ethanol damage to rat gastric mucosa after treatment with a prosta-

glandin. *Gastroenterology* 83: 619–625, 1982.
145. LEE, H.-C., H. BREITBART, M. BERMAN, AND J. G. FORTE. Potassium-stimulated ATPase activity and hydrogen transport in gastric microsomal vesicles. *Biochim. Biophys. Acta* 553: 107–131, 1979.
146. LEVINE, R. A., G. D. LEVINE, A. P. HEALEY, D. I. COOK, P. W. KUCHEL, AND J. A. YOUNG. Proton nuclear magnetic resonance spectroscopy of isolated rabbit fundic glands. *Biochim. Biophys. Acta* 804: 324–330, 1984.
147. LICHTENBERGER, L. M., L. A. GRAZIANI, E. J. DIAL, B. D. BUTLER, AND B. A. HILLS. Role of surface-active phospholipids in gastric cytoprotection. *Science Wash. DC* 219: 1327–1329, 1983.
148. LICHTENBERGER, L. M., J. E. RICHARDS, AND B. A. HILLS. Effect of 16,16-dimethyl prostaglandin E_2 on the surface hydrophobicity of aspirin-treated canine gastric mucosa. *Gastroenterology* 88: 308–314, 1985.
149. LINDE, S. Studies on the stimulation mechanism of gastric secretion. *Acta Physiol. Scand. Suppl.* 74: 1950.
150. LJUNGSTRÖM, M., L. NORBERG, H. OLAISSON, C. WERNSTEDT, F. V. VEGA, G. ARVIDSON, AND S. MÅRDH. Characterization of proton-transporting membranes from resting pig gastric mucosa. *Biochim. Biophys. Acta* 769: 209–219, 1984.
151. LOGSDON, C. D., C. A. BISBEE, M. J. RUTTEN, AND T. E. MACHEN. Fetal rabbit gastric epithelial cells cultured on floating collagen gels. *In Vitro Rockville* 18: 233–242, 1982.
152. LOWE, P. J., AND R. COLEMAN. Membrane fluidity and bile salt damage. *Biochim. Biophys. Acta* 640: 55–65, 1981.
153. LOWE, P. J., AND R. COLEMAN. Fluorescence anisotrophy from diphenylhexatriene in rat liver plasma membranes. *Biochim. Biophys. Acta* 689: 403–409, 1982.
154. LUCAS, M. Estimation of sodium chloride diffusion coefficient in gastric mucin. *Dig. Dis. Sci.* 29: 336–345, 1984.
155. MACHEN, T. E., AND A. M. PARADISO. Regulation of intracellular pH in the stomach. *Annu. Rev. Physiol.* 49: 19–33, 1987.
157. MACHEN, T. E., W. SILEN, AND J. G. FORTE. Na^+ transport by mammalian stomach. *Am. J. Physiol.* 234 (*Endocrinol. Metab. Gastrointest. Physiol.* 3): E228–E235, 1978.
158. MAKHLOUF, G. M. Electrolyte composition of gastric secretion. In: *Physiology of the Gastrointestinal Tract* (1st ed.), edited by L. R. Johnson. New York: Raven, 1981, p. 551–566.
159. MAKHLOUF, G. M., J. P. A. MCMANUS, AND W. I. CARD. A quantitative statement of the two-component hypothesis of gastric secretion. *Gastroenterology* 51: 149–171, 1966.
160. MARTIN, B. K. Accumulation of drug anions in gastric mucosal cells. *Nature Lond.* 198: 896–897, 1963.
161. MENGUY, R., L. DESBAILLETS, AND Y. F. MASTERS. Mechanism of stress ulcer: influence of hypovolemic shock on energy metabolism in the gastric mucosa. *Gastroenterology* 66: 46–55, 1974.
162. MENGUY, R., AND Y. F. MASTERS. Mechanism of stress ulcer. II. Differences between antrum, corpus, and fundus with respect to the effects of complete ischemia on gastric mucosal energy metabolism. *Gastroenterology* 66: 509–516, 1974.
163. MENGUY, R., AND Y. F. MASTERS. Mechanism of stress ulcer. III. Effects of hemorrhagic shock on energy metabolism in the mucosa of the antrum, corpus, and fundus of the rabbit stomach. *Gastroenterology* 66: 1168–1174, 1974.
164. MERSEREAU, W. A., AND E. J. HINCHEY. Effect of gastric acidity on gastric ulceration induced by hemorrhage in the rat, utilizing a gastric chamber technique. *Gastroenterology* 64: 1130–1135, 1973.
165. MEYER, R. A., D. MCGINLEY, AND Z. POSALAKY. The gastric mucosal barrier: structure of intercellular junctions in the dog. *J. Ultrastruct. Res.* 96: 192–201, 1984.
166. MILLER, T. A. Protective effects of prostaglandins against gastric mucosal damage: current knowledge and proposed mechanisms. *Am. J. Physiol.* 245 (*Gastrointest. Liver Physiol.* 8): G601–G623, 1983.
167. MILLER, T. A., D. LI, Y.-J. KUO, K. L. SCHMIDT, AND L. L. SHANBOUR. Nonprotein sulfhydryl compounds in canine gastric mucosa: effects of PGE_2 and ethanol. *Am. J. Physiol.* 249 (*Gastrointest. Liver Physiol.* 12): G137–G144, 1985.
168. MIZUI, T., AND M. DOTEUCHI. Effect of polyamines on acidified ethanol-induced gastric lesions in rats. *Jpn. J. Pharmacol.* 33: 939–945, 1983.
169. MORRIS, G. P., R. K. HARDING, AND J. L. WALLACE. A functional model for extracellular gastric mucus in the rat. *Virchows Arch. B. Cell Pathol.* 46: 239–251, 1984.
170. MORRIS, G. P., AND J. L. WALLACE. The roles of ethanol and of acid in the production of gastric mucosal erosions in rats. *Virchows Arch B. Cell Pathol.* 38: 23–38, 1981.
171. MUALLEM, S., C. BURNHAM, D. BLISSARD, T. BERGLINDH, AND G. SACHS. Electrolyte transport across the basolateral membrane of parietal cells. *J. Biol. Chem.* 260: 6641–6653, 1985.
172. NAGASHIMA, R., AND I. M. SAMLOFF. Aggressive factors. II. Pepsin. In: *Contemporary Issues in Gastroenterology. Peptic Ulcer Disease*, edited by F. P. Brooks. New York: Churchill Livingstone, 1985, vol. 3, chapt. 6, p. 181–214.
173. NANDI, J., M. V. WRIGHT, AND T. K. RAY. Effects of phospholipase A_2 on gastric microsomal H^+,K^+-ATPase system: role of "boundary lipids" and the endogenous activator protein. *Biochemistry* 22: 5814–5821, 1983.
174. NELSON, G. J. Lipid composition and metabolism of erythrocytes. In: *Blood Lipids and Lipoproteins: Quantitation, Composition, and Metabolism*, edited by G. J. Nelson. New York: Wiley-Interscience, 1972, p. 317–386.
175. NORDGREN, B. The rate of secretion and electrolyte content of normal gastric juice. *Acta Physiol. Scand. Suppl.* 202: 1963.
175a. OHYASHIKI, T., M. TAKA, AND T. MOHRI. Effect of neuroaminidase treatment on the lipid fluidity of the intestinal brush-border membranes. *Biochim. Biophys. Acta* 905: 57–64, 1987.
176. OKAMOTO, C., AND J. G. FORTE. Isolation and characterization of oxyntic cell microsomal membrane glycoproteins (Abstract). *Federation Proc.* 46: 365, 1987.
177. OKAMOTO, C., J. M. WOLOSIN, AND J. G. FORTE. FITC-conjugated lectin binding sites in glands and subcellular fractions of rabbit gastric mucosa (Abstract). *Federation Proc.* 44: 1900, 1985.
178. OLENDER, E. J., D. FROMM, T. FURUKAWA, AND M. KOLIS. H^+ disposal by rabbit gastric mucosal surface cells. *Gastroenterology* 86: 698–705, 1984.
179. OSCHMAN, J. L., B. J. WALL, AND B. L. GUPTA. Cellular basis of water transport. In: *Transport at the Cellular Level*. Cambridge, UK: Cambridge Univ. Press, 1974, p. 305–350. (Soc. Exp. Biol. Symp. Ser. 28.)
180. PARADISO, A. M., R. Y. TSIEN, J. R. DEMAREST, AND T. E. MACHEN. Na-H and Cl-HCO_3 exchange in rabbit oxyntic cells using fluorescence microscopy. *Am. J. Physiol.* 253 (*Cell Physiol.* 22): C30–C36, 1987.
181. PARADISO, A. M., R. Y. TSIEN, AND T. E. MACHEN. Digital image processing of intracellular pH in gastric oxyntic and chief cells. *Nature Lond.* 325: 447–450, 1987.
182. PERRY, M. A., S. WADHWA, D. A. PARKS, W. PICKARD, AND D. N. GRANGER. Role of oxygen radicals in ischaemia-induced lesions in the cat stomach. *Gastroenterology* 90: 362–367, 1986.
183. PFIEFFER, C. J. Experimental analysis of hydrogen ion diffusion in gastrointestinal mucus glycoprotein. *Am. J. Physiol.* 240 (*Gastrointest. Liver Physiol.* 3): G176–G182, 1981.
184. PIPKIN, G., C. A. PRICE, AND M. E. PARSONS. Effect of cimetidine on net ion fluxes across the rat gastric mucosa during mucosal damage after gastric ischaemia and after intravenous acetylsalicylic acid. *Gastroenterology* 87: 1283–1291, 1984.
185. POWELL, D. W. Barrier function of epithelia. *Am. J. Physiol.* 241 (*Gastrointest. Liver Physiol.* 4): G275–G288, 1981.
186. POWELL, D. W. Physiological concepts of epithelial barriers. In: *Mechanisms of Mucosal Protection in the Upper Gastrointestinal Tract*, edited by A. Allen, G. Flemström, A. Garner, W. Silen, and L. A. Turnberg. New York: Raven, 1984, p. 1–5.

187. QUIGLEY, E. M. M., AND L. A. TURNBERG. pH of the microclimate lining human gastric and duodenal mucosa in vivo: studies in control subjects and in duodenal ulcer patients. *Gastroenterology* 92: 1876–1884, 1987.
188. RABON, E., H. CHANG, AND G. SACHS. Quantitation of hydrogen ion and potential gradients in gastric plasma membrane vesicles. *Biochemistry* 17: 3345–3353, 1978.
189. RAINSFORD, K. D., AND K. BRUNE. Selective cytotoxic actions of aspirin on parietal cells: a principal factor in the early stages of aspirin-induced gastric damage. *Arch. Toxicol.* 40: 143–150, 1978.
190. REES, W. D. W., AND L. A. TURNGERG. Mechanisms of gastric mucosal protection: a role for the "mucus-bicarbonate" barrier. *Clin. Sci.* 62: 343–348, 1982.
191. REHM, W. S. Electrical resistance of resting and secreting stomach. *Am. J. Physiol.* 172: 689–699, 1953.
192. REHM, W. S. Acid secretion, resistance, short-circuit current, and voltage-clamping in frog's stomach. *Am. J. Physiol.* 203: 63–72, 1962.
193. RICHARDSON, I. W., V. LICKO, AND E. BARTOLI. The nature of passive flows through tightly folded membranes. The influence of microstructure. *J. Membr. Biol.* 11: 293–308, 1973.
194. RITCHIE, W. P., AND T. S. FELGER. Differing ulcerogenic potential of dihydroxy and trihydroxy bile acids in canine gastric mucosa. *Surgery St. Louis* 89: 342–347, 1981.
195. ROBERT, A. An intestinal disease produced experimentally by a prostaglandin deficiency. *Gastroenterology* 69: 1045–1047, 1975.
196. ROBERT, A. Cytoprotection by prostaglandins. *Gastroenterology* 77: 761–767, 1979.
197. ROBERT, A. Role of endogenous and exogenous prostaglandins in mucosal protection. In: *Mechanisms of Mucosal Protection in the Upper Gastrointestinal Tract*, edited by A. Allen, G. Flemström, A. Garner, W. Silen, and L. A. Turnberg. New York: Raven, 1984, p. 377–382.
198. ROBERT, A., C. LANCASTER, J. P. DAVIS, S. O. FIELD, A. J. WICKREMA SINHA, AND B. A. THORNBURGH. Cytoprotection by prostaglandin occurs in spite of penetration of absolute ethanol into the gastric mucosa. *Gastroenterology* 88: 328–333, 1985.
199. ROBERT, A., J. E. NEZAMIS, C. LANCASTER, AND A. J. HANCHAR. Cytoprotection by prostaglandins in rats: prevention of gastric necrosis produced by alcohol, HCl, NaOH, hypertonic NaCl and thermal injury. *Gastroenterology* 77: 433–443, 1979.
200. ROSS, I. N., H. M. M. BAHARI, AND L. A. TURNBERG. The pH gradient across mucus adherent to rat fundic mucosa in vivo and the effect of potential damaging agents. *Gastroenterology* 81: 713–718, 1981.
201. ROWE, P. H., R. LANGE, G. MARRONE, J. B. MATTHEWS, E. KASDON, AND W. SILEN. In vitro protection of amphibian gastric mucosa by nutrient HCO_3^- against aspirin damage. *Gastroenterology* 89: 767–778, 1985.
202. RUTTEN, M. J., AND S. ITO. Morphology and electrophysiology of guinea pig gastric mucosal repair in vitro. *Am. J. Physiol.* 244 (*Gastrointest. Liver Physiol.* 7): G171–G182, 1983.
203. RUTTEN, M., D. RATTNER, AND W. SILEN. Transepithelial transport of guinea pig gastric mucous cell monolayers. *Am. J. Physiol.* 249 (*Cell Physiol.* 18): C503–C513, 1985.
204. SANDERS, M. J., D. A. AMIRIAN, A. AYALON, AND A. H. SOLL. Regulation of pepsinogen release from canine chief cells in primary monolayer culture. *Am. J. Physiol.* 245 (*Gastrointest. Liver Physiol.* 8): G641–G646, 1983.
205. SANDERS, M. J., A. AYALON, M. ROLL, AND A. H. SOLL. The apical surface of canine chief cell monolayers resists H^+ back-diffusion. *Nature Lond.* 313: 52–54, 1985.
206. SAROSIEK, J., A. SLOMIANY, AND B. L. SLOMIANY. Retardation of hydrogen ion diffusion by gastric mucus constituents: effect of proteolysis. *Biochem. Biophys. Res. Commun.* 115: 1053–1060, 1983.
207. SATO, A., AND S. S. SPICER. Ultrastructure visualization of galactosyl residues in various alimentary epithelial cells with peanut lectin-horseradish peroxidase procedure. *Histochemistry* 73: 607–624, 1982.
208. SCHIFFRIN, M. J., AND A. A. WARREN. Some factors concerned in the production of experimental ulceration of the GI tract in cats. *Am. J. Dig. Dis.* 9: 205–209, 1942.
209. SCHWARTZ, K. Über penetrierende Magen- und jejunal Geschwüre. *Beitr. Klin. Chir.* 67: 96–128, 1910.
210. SEDAR, A. W. Electron microscopic demonstration of polysaccharides associated with acid-secreting cells of the stomach after "inert dehydration." *J. Ultrastruct. Res.* 28: 112–124, 1969.
211. SEDAR, A. W., AND J. G. FORTE. Effects of calcium depletion on the junctional complex between oxyntic cells of gastric glands. *J. Cell Biol.* 22: 173–188, 1964.
212. SELLERS, L. A., A. ALLEN, AND M. K. BENNETT. Formation of a fibrin based gelatinous coat over repairing rat gastric epithelium after acute ethanol damage: interaction with adherent mucus. *Gut* 28: 835–843, 1987.
213. SEN, P. C., AND T. K. RAY. Characterization of gastric mucosal membranes: lipid composition of purified microsomes from pig, rabbit, and frog. *Arch. Biochem. Biophys.* 198: 548–555, 1979.
214. SEN, P. C., AND T. K. RAY. Control of the potassium ion-stimulated adenosine triphosphatase of pig gastric microsomes: effects of lipid environment and the endogenous activator. *Arch. Biochem. Biophys.* 202: 8–17, 1980.
215. SHINITZKY, M. Membrane fluidity and cellular functions. In: *Physiology of Membrane Fluidity*, edited by M. Shinitzky. Boca Raton, FL: CRC, 1984, vol. 1, p. 1–51.
216. SILEN, W. Gastric mucosal defense and repair. In: *Physiology of the Gastrointestinal Tract* (2nd ed.), edited by L. R. Johnson. New York: Raven, 1987, vol. 2, p. 1055–1069.
217. SILEN, W., AND J. G. FORTE. Effects of bile salts on amphibian gastric mucosa. *Am. J. Physiol.* 228: 637–644, 1975.
218. SILEN, W., AND S. ITO. Mechanisms for rapid re-epithelialization of the gastric mucosal surface. *Annu. Rev. Physiol.* 47: 217–229, 1985.
219. SKILLMAN, J. J., S. A. GOULD, R. S. K. CHUNG, AND W. SILEN. The gastric mucosal barrier: clinical and experimental studies in critically ill and normal man, and in the rabbit. *Ann. Surg.* 172: 564–582, 1970.
220. SLOMIANY, B. L., A. SLOMIANY, AND V. L. MURTY. Partial characterization of the highly complex fucolipids from gastric mucosa. *Biochem. Biophys. Res. Commun.* 88: 1092–1097, 1979.
221. SMEATON, L. A., AND B. H. HIRST. Gastroduodenal ion outputs: prostaglandins and hyperosmolal solutions stimulate via different mechanisms. In: *Mechanisms of Mucosal Protection in the Upper Gastrointestinal Tract*, edited by A. Allen, G. Flemström, A. Garner, W. Silen, and L. A. Turnberg. New York: Raven, 1984, p. 107–111.
222. SMEATON, L. A., B. H. HIRST, A. ALLEN, AND A. GARNER. Gastric and duodenal HCO_3^- transport in vivo: influence of prostaglandins. *Am. J. Physiol.* 245 (*Gastrointest. Liver Physiol.* 8): G751–G759, 1983.
223. SMITH, S. M., M. B. GRISHAM, E. A. MANCI, D. N. GRANGER, AND P. R. KVIETYS. Gastric mucosal injury in the rat. Role of iron and xanthine oxidase. *Gastroenterology* 92: 950–956, 1987.
224. SPENNEY, J. G., G. FLEMSTRÖM, R. L. SHOEMAKER, AND G. SACHS. Quantitation of conductance pathways in antral mucosa. *J. Gen. Physiol.* 65: 645–662, 1975.
225. SPENNEY, J. G., G. SACCOMANI, H. L. SPITZER, M. TOMANA, AND G. SACHS. Characterization of gastric mucosal membranes. Composition of gastric cell membranes and polypeptide fractionation using ionic and nonionic detergents. *Arch. Biochem. Biophys.* 161: 456–471, 1974.
226. SPENNEY, J. G., R. L. SHOEMAKER, AND G. SACHS. Microelectrode studies of fundic gastric mucosa: cellular coupling and shunt conductance. *J. Membr. Biol.* 19: 105–128, 1974.
227. SPICER, S. S., AND D. C. H. SUN. Carbohydrate histochemistry of gastric epithelial secretions in dog. *Ann. NY Acad. Sci.* 140: 762–783, 1967.

228. STARLINGER, M., R. JAKESZ, J. B. MATTHEWS, C. YOON, AND R. SCHIESSEL. The relative importance of HCO_3^- and blood flow in the protection of rat gastric mucosa during shock. *Gastroenterology* 81: 732–735, 1981.
229. SUE, M. W., AND P. H. GUTH. A fluorescent in vivo microscopic method to assess surface mucosal integrity in the rat stomach: effect of ethanol and prostaglandin. *Gastroenterology* 89: 415–420, 1985.
230. SVANES, K., J. CRITCHLOW, K. TAKEUCHI, D. MAGEE, S. ITO, AND W. SILEN. Factors influencing reconstitution of frog gastric mucosa: role of prostaglandins. In: *Mechanisms of Mucosal Protection in the Upper Gastrointestinal Tract*, edited by A. Allen, G. Flemström, A. Garner, W. Silen, and L. A. Turnberg. New York: Raven, 1984, p. 33–39.
231. SVANES, K., S. ITO, K. TAKEUCHI, AND W. SILEN. Restitution of the surface epithelium of the in vitro frog gastric mucosa after damage with hyperosmolar sodium chloride. Morphological and physiological characteristics. *Gastroenterology* 82: 1409–1426, 1982.
232. SVENSSON, S.-O., AND S. EMÅS. Acid secretory responses to histamine, pentagastrin, and feeding before and after vagal denervation of fundic pouches in cats. *Scand. J. Gastroenterol.* 12: 357–362, 1977.
233. SZABO, S. Role of sulfhydryls and early vascular lesions in gastric mucosal injury. *Acta Physiol. Hung.* 64: 203–214, 1984.
234. SZABO, S., J. S. TRIER, A. BROWN, AND J. SCHNOOR. Early vascular injury and increased vascular permeability in gastric mucosal injury caused by ethanol in the rat. *Gastroenterology* 88: 228–236, 1985.
235. SZABO, S., J. S. TRIER, AND P. W. FRANKEL. Sulfhydryl compounds may mediate gastric cytoprotection. *Science Wash. DC* 214: 200–202, 1981.
236. SZABO, S., AND K. H. USADEL. Cytoprotection-organoprotection by somatostatin: gastric and hepatic lesions. *Experientia Basel* 38: 254–256, 1982.
237. TAKEUCHI, K., D. MAGEE, J. CRITCHLOW, J. MATTHEWS, AND W. SILEN. Studies of the pH gradient and thickness of frog gastric mucus gel. *Gastroenterology* 84: 331–340, 1983.
238. TARNAWSKI, A., D. HOLLANDER, J. STACHURA, W. J. KRAUSE, AND H. GERGELY. Prostaglandin protection of the gastric mucosa against alcohol injury—a dynamic time-related process. Role of the mucosal proliferative zone. *Gastroenterology* 88: 334–352, 1985.
239. TEORELL, T. Electrolyte diffusion in relation to the acidity regulation of the gastric juice. *Gastroenterology* 9: 425–443, 1947.
240. TEPPERMAN, B. L., D. B. BARR, AND M. C. PALMER. Ontogeny of gastric mucosal permeability responses to luminal H^+ and bile salt in the rat. *Am. J. Physiol.* 250 (*Gastrointest. Liver Physiol.* 13): G617–G624, 1986.
241. THOMAS, A. J., D. L. NAHRWOLD, AND R. C. ROSE. Detergent action of sodium taurocholate on rat gastric mucosa. *Biochim. Biophys. Acta* 282: 210–213, 1972.
242. TRIPATHI, S., AND P. K. RANGACHARI. In vitro primate gastric mucosa: electrical characteristics. *Am. J. Physiol.* 239 (*Gastrointest. Liver Physiol.* 2): G77–G82, 1980.
243. USSING, H. H., AND E. E. WINDHAGER. Nature of shunt-path and active sodium transport through frog skin epithelium. *Acta Physiol. Scand.* 61: 484–504, 1964.
244. VADGAMA, P., AND K. G. M. M. ALBERTI. The effect of a gastric mucus barrier on the dynamic response of a pH electrode. *Experientia Basel* 39: 573–576, 1983.
245. VIRCHOW, R. Historiches, Kritsches und Positives zur Lehre der Unterleibsaffektonen. *Virchows Arch. Pathol. Anat. Physiol. Klin. Med.* 5: 281–375, 1853.
246. WADE, J. B., J.-P. REVEL, AND V. A. DISCALA. Effect of osmotic gradients on intercellular junctions of the toad bladder. *Am. J. Physiol.* 224: 407–415, 1973.
247. WADHWA, S. S., AND M. A. PERRY. Gastric injury induced by hemorrhage, local ischemia, and oxygen radical generation. *Am. J. Physiol.* 253 (*Gastrointest. Liver Physiol.* 16): G129–G133, 1987.
248. WALLACE, J. L., G. P. MORRIS, E. J. KRAUSSE, AND S. E. GREAVES. Reduction by cytoprotective agents of ethanol-induced damage to the rat gastric mucosa: a correlated morphological and physiological study. *Can. J. Physiol. Pharmacol.* 60: 1686–1699, 1982.
249. WALLACE, J. L., AND B. J. R. WHITTLE. Role of mucus in the repair of gastric epithelial damage in the rat: inhibition of epithelial recovery by mucolytic agents. *Gastroenterology* 91: 603–611, 1986.
250. WASSEF, M. K., Y. N. LIN, AND M. I. HOROWITZ. Molecular species of phosphatidylcholine from rat gastric mucosa. *Biochim. Biophys. Acta* 573: 222–226, 1979.
251. WEISBRODT, N. W., M. KIENZLE, AND A. R. COOKE. Comparative effects of aliphatic alcohols on the gastric mucosa. *Proc. Soc. Exp. Biol. Med.* 142: 450–454, 1973.
252. WHITTLE, B. J. R. The mechanisms of gastric damage by nonsteroid anti-inflammatory drugs. In: *Biological Protection With Prostaglandins*, edited by M. M. Cohen. Boca Raton, FL: CRC, 1986, vol. 2, p. 1–27.
253. WILKES, J. M., H. J. BALLARD, AND B. H. HIRST. Correlation of proton permeation and fluidity in gastrointestinal apical membrane vesicles (Abstract). *Gastroenterology* 92: 1695, 1987.
254. WILKES, J. M., H. J. BALLARD, J. A. E. LATHAM, AND B. H. HIRST. Gastroduodenal epithelial cells: the role of the apical membrane in mucosal protection. In: *Cells, Membranes, and Disease, Including Renal*, edited by E. Reid, G. M. W. Cook, and J. P. Luzio. New York: Plenum, 1987, vol. 17, p. 243–254. (Method. Survey. Biochem. Analysis Ser.)
255. WILLIAMS, S. E., AND L. A. TURNBERG. Retardation of acid diffusion by pig gastric mucus: a potential role in mucosal protection. *Gastroenterology* 79: 299–304, 1980.
256. WILLIAMS, S. E., AND L. A. TURNBERG. Demonstration of a pH gradient across mucus adherent to rabbit gastric mucosa: evidence for a "mucus-bicarbonate" barrier. *Gut* 22: 94–96, 1981.
257. WOLOSIN, J. M., AND J. G. FORTE. K^+ and Cl^- conductances in the apical membrane from secreting oxyntic cells are concurrently inhibited by divalent cations. *J. Membr. Biol.* 83: 261–272, 1985.

CHAPTER 16

Secretion of bicarbonate by gastric and duodenal mucosa

GUNNAR FLEMSTRÖM | *Department of Physiology and Medical Biophysics, Uppsala University Medical Center, Uppsala, Sweden*

ANDREW GARNER | *Department of Bioscience, ICI Pharmaceuticals Division, Macclesfield, England*

CHAPTER CONTENTS

Methods of Measurement
    Gastric bicarbonate secretion
        Intragastric partial pressure of carbon dioxide
        Osmolality
        Direct titration of gastric bicarbonate
        Other methods of measuring gastric bicarbonate
    Duodenal bicarbonate secretion
Mechanisms of Gastroduodenal Bicarbonate Secretion
    Origin and transport of gastric bicarbonate
    Contribution of bicarbonate leakage
    Duodenal mucosal transport of bicarbonate
    Gradient in secretion along the duodenum
    Biochemical basis for duodenal bicarbonate secretion
Physiological Control of Gastroduodenal Bicarbonate Secretion
    Humoral control
    Neural influence
    Control by mucosal endogenous prostaglandins
    Relation to gastric acid secretion and blood flow
Pharmacological Modulation
    Inhibitors of gastric alkaline secretion
    Stimulation of gastric alkaline secretion by prostaglandins
    Duodenal mucosal alkaline secretion and cAMP
    Stimulation of duodenal secretion by prostaglandins
    Other drugs affecting duodenal alkaline secretion
Protective Role of Bicarbonate
    Surface pH gradient
    Possible therapeutic implications
Summary and Perspectives

THE DANISH PHYSIOLOGIST SCHIERBECK provided in 1892 the first evidence to suggest that gastric mucosa transports HCO_3^- as well as acid (145). He demonstrated that a feeding maneuver in dogs not only stimulated acid secretion but also increased gastric intraluminal partial pressure of CO_2 (P_{CO_2}) to levels considerably higher than those in blood (>100 mmHg). Studies performed over the past 10 yr have shown that gastric HCO_3^- is transported into the lumen by a metabolism-dependent process (45, 53, 54).

Alkaline secretion occurs in all species tested and is stimulated by a variety of means, including physiological stimuli such as sham feeding (33, 58, 106), gastric distension (57), and the presence of acid in the lumen (71, 82). Use of pH-sensitive microelectrodes has allowed the experimental demonstration of a standing gradient of pH in the viscoelastic mucous gel adherent to the gastric surface and provided evidence for the dependence of this gradient on mucosal HCO_3^- secretion (171, 175). The pH in the gel adjacent to the luminal cell membranes remains at neutrality when the luminal bulk solution is maintained at pH 2.0–3.0. The alkaline secretion together with the mucous gel is thus proposed as one line of mucosal defense against luminal acid.

The finding that gastric transport of HCO_3^- depends on tissue metabolism fostered interesting studies of whether surface epithelium in the duodenum possesses a similar ability. It has been found in a variety of species that duodenal mucosa devoid of Brunner's glands secretes HCO_3^- into the lumen at a greater rate than the stomach does (45, 78, 152). A pH gradient that is generated at the duodenal surface is more resistant to acid than that in the stomach (49, 103). There are some distinct differences between the mechanisms of HCO_3^- transport and the stimulatory pathways in the duodenum and those in the stomach.

This chapter presents the various methods developed to study gastric and duodenal mucosal HCO_3^- secretion and summarizes present knowledge about the mechanisms for transport and the physiological control of the secretions. Pharmacological modulation of the secretions and some of the pitfalls involved in studying the process are also described.

METHODS OF MEASUREMENT

Determination of HCO_3^- in gastric contents presents a number of difficulties. The rate of transport is low and therefore usually obscured even by resting acid secretion. Furthermore metabolism-dependent gastric HCO_3^- transport is probably caused by Cl^-/HCO_3^- exchange at the apical membrane such that alkalinization of the lumen may not be accompanied by volume flow. The duodenum presents accessibility problems, and in both stomach and duodenum, obtaining samples from intact animals or humans uncontaminated by saliva or pancreaticobiliary secretions requires careful control. A variety of surgically modified preparations and a number of approaches to assaying HCO_3^- have thus been developed. Alkaline secretion has now been demonstrated in the gastric and duodenal mucosa of all common laboratory species as well as in humans (16, 32, 46, 66, 95, 98, 131). Many of the problems associated with in vivo models can be overcome by using isolated gastric or duodenal preparations, and much of the current information on cellular mechanisms of HCO_3^- transport has been obtained using mucosal sheets. On the other hand, studies of the complex neural control of the secretion require the use of intact animals.

Gastric Bicarbonate Secretion

The HCO_3^- produced by the surface epithelial cells and H^+ produced by the parietal cells interact at the mucosal surface and in the lumen of the stomach according to Equation 1

$$H^+ + HCO_3^- \rightleftharpoons H_2CO_3 \rightleftharpoons CO_2 + H_2O \qquad (1)$$

Carbonate formation is outside the biological range [acidic dissociation constant (pK_a) = 10.4], and because protonation to form carbonic acid (H_2CO_3) is essentially instantaneous, the reaction can be simplified according to Equation 2 for which the pK_a is 6.1

$$H^+ + HCO_3^- \rightleftharpoons CO_2 + H_2O \qquad (2)$$

At the acidic pH, which normally prevails in the stomach, HCO_3^- is fairly rapidly converted to CO_2 and H_2O ($t_{1/2} \cong 14$ s in absence of carbonic anhydrase). Based on Equation 2, three approaches have been developed for determining gastric HCO_3^- production: *1*) calculation of HCO_3^- concentration from pH and P_{CO_2} measurements by using the Henderson-Hasselbalch equation, *2*) calculation of HCO_3^- concentration from the decrease in osmolality associated with the reaction, and *3*) determination of secretory rate by titration of net HCO_3^- secretion after specific inhibition of acid secretion. Both P_{CO_2} and osmolality measurements have the potential advantage of enabling simultaneous estimation of HCO_3^- and parietal H^+ secretions. This can be useful especially in view of evidence that low luminal pH is itself a stimulant of HCO_3^- transport.

INTRAGASTRIC PARTIAL PRESSURE OF CARBON DIOXIDE. Measurement of P_{CO_2} together with pH enables free HCO_3^- concentration in the luminal fluid (perfusate) to be calculated from the Henderson-Hasselbalch equation

$$pH = pK_a + \log(HCO_3^-/\alpha P_{CO_2}) \qquad (3)$$

where HCO_3^- in the form of CO_2 is calculated from P_{CO_2} and α, the solubility coefficient. The HCO_3^- in the form of CO_2 added to the free HCO_3^- gives total HCO_3^-. A more detailed account is given in the original papers describing this technique and its validation in the guinea pig (66), rat (16), cat (32), and human (56, 95, 131) stomach. The prerequisite of the method is that permeability of the gastric epithelium to CO_2 is low, so that CO_2 remains in the lumen rather than diffusing back into the mucosa.

This condition appears to hold in the stomach where the achieved luminal P_{CO_2} values far exceed venous levels, whereas artificially imposed CO_2 gradients are maintained for prolonged periods (32, 66, 99, 140). Increasing the proportion of HCO_3^- in the form of free HCO_3^- minimizes the possible loss of CO_2 and can be achieved by pretreatment with antisecretory agents (e.g., histamine H_2-receptor antagonists or omeprazole). These drugs have no effect on HCO_3^- secretion by the gastric mucosa (32, 37, 54, 58, 66). Permanent gastric pouches in conscious animals or tied-off stomachs in anesthetized animals can be used to prevent contamination by extragastric HCO_3^-. In clinical studies, however, contamination by salivary HCO_3^- and HCO_3^- in duodenogastric reflux should be assessed (56, 131).

OSMOLALITY. This approach is based on the two-component model of gastric secretion. It is assumed that HCO_3^- is secreted at a fixed concentration and thus its amount is directly related to the volume of nonparietal secretion (87). Alkaline secretion dilutes and neutralizes luminal acid secreted from the parietal cells, leading to a reduction in acidity and osmolality. Bicarbonate concentration can then be calculated by measuring volume, osmolality, and H^+ concentration.

This approach has been used by Feldman and others (34, 36, 77, 172) and gives values for gastric HCO_3^- concentration of ~100 mM and output of ~2.6 mmol/h, values ~5- to 10-fold greater than that determined from P_{CO_2} and pH (58, 95, 131). Gastric juice concentrations of HCO_3^- in humans in the same range as those obtained from P_{CO_2} and pH measurements have also been measured by direct titration in achlorhydric subjects (128) and in healthy subjects during spontaneous H^+ secretory arrest (111). Although HCO_3^- secretion estimated by osmolality was determined in subjects not treated with antisecretory agents, it seems unlikely that the then-prevailing low luminal pH induces a 10-fold stimulation in HCO_3^- output. The high calculated HCO_3^- concentration (6 times higher than plasma) could be resulting from erroneous assumptions in the calculation as well as reflecting some hypotonicity of gastric parietal cell acid secretion (42, 58, 136). It should, however, be noted that HCO_3^- secretion values in humans calculated either from osmolality or from P_{CO_2} and pH are very similar with respect to the action of stimulants and inhibitors.

DIRECT TITRATION OF GASTRIC BICARBONATE. The aim of this approach is to abolish H^+ secretion so that HCO_3^- transport can be determined by discrete or continuous back-titration. This can be achieved by pretreatment with specific inhibitors of acid secretion. In theory, inhibiting H^+ production (and hence the parietal cell alkaline tide) would reduce mucosal availability of HCO_3^- and could decrease alkaline secretion. Also, if complete abolition of H^+ secretion were not achieved, there could be a small and masked acid secretion. The latter could affect titratable (net) alkalinity with apparent stimulation of HCO_3^- secretion due to further inhibition of H^+ transport (45, 52).

Fundic mucosa from amphibians (37, 82, 167) and from mammals (52, 59, 132) have been stripped of seromuscle layers, mounted as sheets in Ussing-type in vitro chamber, and bathed on both sides by oxygenated electrolyte solutions. Amphibian tissue in particular remains viable for many hours. If the nutrient (blood) side of the membrane is bathed with an HCO_3^--containing buffer (pH 7.2–7.4), alkalinization of the luminal solution can be detected 1–2 h after addition of an acid-inhibitory drug such as the histamine H_2-receptor antagonist cimetidine (1 mM, serosal side). The unbuffered luminal fluid is gassed with O_2 to remove CO_2, thus raising its pH. Alkalinization can then be continuously monitored by the pH-stat technique in which dilute hydrochloric acid is infused to maintain pH at a predetermined end point. Use of high end points (7.4–8.0) ensures that measured alkalinization is not due to loss of H^+ ions from the luminal solution by diffusion into the mucosa ("back-diffusion"). A similar method has been used to measure HCO_3^- secretion in animals prepared with permanent fundic pouches (25, 71, 98, 108) or in whole-stomach preparations of anesthetized rats (173). When saline is recirculated through the Heidenhain pouch of cimetidine-treated dogs, pH of the perfusate rarely exceeds 7, even during continuous gassing with O_2 (71, 98). This may be caused by a cimetidine-insensitive residual acid secretion or by incomplete CO_2 removal and requires use of lower than ideal end points for back-titration (up to pH 6.0). Validity of this method has, however, been established by quantitative recovery of exogenous HCO_3^-.

The antrum, a region devoid of acid-secreting cells, has been used for direct measurement of HCO_3^- secretion (53, 59, 75, 82, 110). Sheets of antral mucosa mounted in chambers as well as antral pouches constructed from this region of dog stomach have been used. Permanent antral pouches are considerably more difficult to construct and maintain by comparison with fundic pouches, because of the smaller area of mucosa and greater thickness of the muscle layers.

OTHER METHODS OF MEASURING GASTRIC BICARBONATE. Transmucosal fluxes of HCO_3^- have been determined in vitro in the antrum by using [^{14}C]HCO_3^- (59). Nonparietal secretion in vivo has been estimated by determining flow and H^+ concentration of gastric juice and by extrapolating this relationship to zero acid output. In cats with gastric fistulae, calculated nonparietal flow was ~2 ml/h (73). Finally it is important to remember that, although measurement of gastric HCO_3^- has been based on its appearance in the lumen of the stomach, the role of this secretion is probably one of surface neutralization of H^+ ions (see PROTECTIVE ROLE OF BICARBONATE, p. 320). To assess surface neutralization, extracellular pH in the mucous gel at the apical membrane of the surface epithelium has been measured with pH-sensitive microelectrodes (103, 171).

Duodenal Bicarbonate Secretion

Spontaneous alkaline secretion by in vivo preparations or by isolated duodenal epithelium bathed with an HCO_3^--containing serosal buffer solution can be readily detected by titration of the luminal perfusate. Furthermore, mucosal HCO_3^- production or H^+ disposal can be estimated in the presence of luminal acid from measurement of P_{CO_2} or osmolality, using principles similar to those described for the stomach. A variety of preparations have been used in studies of the alkaline secretion by the duodenal mucosa. They include isolated epithelial sheets or mucosal tubes mounted in chambers (40, 152), acute duodenal segment in anesthetized animals (46, 165), permanent preparations in conscious animals (93, 110), and human duodenal segments isolated between balloons (91).

The American bullfrog (*Rana catesbeiana*) is well suited as a source of tissue for in vitro studies. The proximal duodenum, extending from pylorus to the entry of the pancreatic duct, is devoid of Brunner's glands and sufficiently large for mounting in an in vitro chamber. After stripping off external muscle layers, the epithelium can be mounted as either a tube (40) or a flat sheet (152). The use of a tube preparation should reduce edge damage but does not allow determination of electrical resistance and short-circuit current. Isolated mammalian duodenal preparations have also been tested, but their stability and viability tend to be inferior to amphibian duodenum, and the secretory rates are much lower than those observed in the same species in vivo (13, 17).

Secretion in anesthetized animals has been studied in segments of duodenum with intact blood supply and devoid of pancreatic and biliary secretions. Conscious-animal models include rats with Thiry-Vella-type open loops (93) or dogs with loops of proximal or distal duodenum closed at one end to form pouches (110). However, duodenal pouches deteriorate with time, and it is advisable to regularly assess the functional status of the mucosa. When measured by in situ pH-stat titration of luminal perfusates (46), increases in HCO_3^- secretory rate take up to 30 min to achieve a new equilibrium steady state. This partially reflects time required for elimination of CO_2 formed during reaction between secreted HCO_3^- and acid titrant, which is easily demonstrated by comparing the titration curve for HCO_3^- with that of a strong base such as NaOH. Recently duodenal mucosal HCO_3^- secretion in human volunteers has been measured in short (4-cm) segments isolated between balloons positioned under fluoroscopy in the bulbar region or in the distal portion of the duodenum (91, 148). The HCO_3^- output was determined by back-titration of perfusate effluent samples that also contained a volume marker. Additional markers were infused into the stomach and distal to the test segment in order to assess contamination by gastric and pancreatic secretions.

Very high P_{CO_2} levels have been recorded in the duodenum consistent with the interaction of (gastric) H^+ with HCO_3^- in the duodenal lumen. In the intact dog, for example, basal P_{CO_2} in the duodenal bulb was in the range of 160–265 mmHg (140). In humans P_{CO_2} 10 cm distal to the pylorus was ~300 mmHg and increased postprandially to 500 mmHg (139). Both duodenal mucosal and pancreatic HCO_3^- should have contributed to CO_2 released luminally in these experiments. Harmon et al. (78), however, reported similar high P_{CO_2} levels in tied-off segments of canine duodenum during acidification of the luminal perfusate. These latter experiments, performed in duodenal loops distal to the Brunner's gland region in which pancreaticobiliary secretions were excluded, indicate substantial HCO_3^- secretion by the mucosa. Evidence for similar removal of acid by duodenal mucosa has recently been obtained from patients devoid of pancreatic HCO_3^- secretion as a result of chronic pancreatitis (129), and comparable conclusions were reached based on measurement of the reduction in osmolality of acid solutions after instillation into pouches constructed from dog duodenum (28, 85). From these latter data it was concluded that mucosal HCO_3^- secretion accounted for 50%–70% of acid disposal by duodenal epithelium, the remainder being caused by diffusion of H^+ ions into the mucosa. However, the high permeability of duodenal mucosa to CO_2 and H_2O is likely to underestimate HCO_3^- production measured by osmolality and consequently underestimate the proportion of acid lost by neutralization.

MECHANISMS OF GASTRODUODENAL BICARBONATE SECRETION

The secretion of HCO_3^- by frog (37, 53, 167) and mammalian (52, 59, 132) gastric fundic and antral mucosa mounted in an in vitro chamber is inhibited by anoxia, cyanide, or 2,4-dinitrophenol, indicating its dependence on tissue metabolism. The transport of HCO_3^- by bullfrog duodenum in vitro is similarly reduced or abolished by inhibitors of tissue metabolism (40, 83, 152). These types of inhibitors cannot be tested in vivo. However, both gastric and duodenal mucosal transport of HCO_3^- in vivo show similar sensitivity to many (but not all) stimulants and inhibitors, suggesting that the mechanisms for transport of HCO_3^- in vivo and in vitro may be similar.

Origin and Transport of Gastric Bicarbonate

Secretion by frog fundic mucosa in vitro is stimulated by cGMP but not by cAMP (37). Cholinergic stimuli that increase alkaline secretion also elevate mucosal cGMP concentration in canine fundus and antrum (19). The gastric surface epithelial cells contain high concentrations of mucosal cGMP diesterase (160) and carbonic anhydrase activity (80, 115, 159). Acetazolamide, which inhibits this latter enzyme, decreases HCO_3^- secretion by frog fundic mucosa in vitro (37) as well as dog and rat stomach in vivo (99, 137). Failure of acetazolamide to inhibit secretion has been reported (167), possibly reflecting lower HCO_3^- secretory rates in these preparations. Alkaline secretion by frog fundic mucosa in vitro and the electrical potential generated by monolayers of guinea pig surface epithelial cells are considerably more sensitive to luminal- than to serosal-side administration of acetazolamide (33, 141). Much of the carbonic anhydrase activity of these cells is localized within the apical cytoplasmic matrix and microvillar core (159). These combined findings strongly suggest that metabolism-dependent transport of HCO_3^- is a property of the surface epithelial cells. It should also be noted that antral mucosa is composed mainly of this cell type and that gastric antral and fundic alkaline secretion display very similar properties (37).

Metabolism-dependent secretion of HCO_3^- by fundic mucosa from the European frog *Rana temporaria* is dependent on luminal but not on serosal Cl^- (39). In contrast to gastric H^+ secretion, stimulation of HCO_3^- transport is not associated with changes in the transmucosal electrical potential difference or resistance (37). This suggests that electroneutral Cl^--HCO_3^- exchange at the luminal membrane is the predominant transport process. The electrical potential generated by monolayers of surface epithelial cells is sensitive to luminal 4,4'-diisothiocyanatostilbene-2,2'-disulfonic acid (DIDS), whereas serosal application of this exchange inhibitor has no effect (141). Also, DIDS (0.6 mM) has been reported to inhibit alkaline secretion by fundic mucosa from the American bullfrog, although only after serosal-side application (167). Furthermore alkaline secretion by this preparation reportedly requires serosal (but not luminal) Cl^- and showed some dependence on serosal Na^+ and HCO_3^-. Ouabain binds to the basal membrane of the surface epithelial cells of R. temporaria fundic mucosa (81) and also inhibits prostaglandin-stimulated HCO_3^- transport by this preparation, suggesting that maintenance of a transmembrane Na^+ gradient is necessary for HCO_3^- secretion.

The gastric surface epithelial cells have been further characterized by use of micropuncture techniques and isolated cell preparations. The luminal membrane of these cells in isolated Necturus antrum and fundus displays an amiloride-sensitive Na^+ conductance (27, 53), whereas varying luminal HCO_3^- concentration has minimal effects on the luminal transmembrane electrical potential (53). The luminal membrane also displays only negligible Cl^- conductance in Necturus (27, 144) as well as in Rana esculenta mucosa (22). Sodium-dependent Cl^- uptake across the basolateral membrane has been demonstrated in frog fundus (22, 116, 117), and use of pH-sensitive microelectrodes reveals that serosal HCO_3^- enhances the intracellular pH in Necturus isolated fundus (7). Intracellular pH in isolated mammalian surface epithelial cells has been studied with the dye acridine orange (60, 127). Extracellular HCO_3^- dissipates intracellular acid. Dissipation is inhibited by 4-acetamido-4'-isothiocyanatostilbene-2,2'-disulfonic acid (SITS) and cAMP but the precise mechanism has not been established. Treatment of Necturus antrum with 16,16-dimethyl prostaglandin E_2 (16,16-dimethyl PGE_2) decreases the resistance of both the apical and the basolateral cell membrane and, in addition, increases the resistance of the paracellular shunt pathway (6).

Further studies are essential to establish the precise processes of transport of HCO_3^- and other ions by the gastric surface epithelial cell. Dependence on luminal Cl^- as well as sensitivity to SITS would suggest that HCO_3^- transport occurs by Cl^--HCO_3^- exchange at the luminal cell membrane. Dependence of transport (and intracellular pH) on serosal HCO_3^- suggests uptake of HCO_3^- at the basolateral membrane of the surface epithelial cells, whereas the inhibition of transport by acetazolamide provides evidence that HCO_3^- produced by cellular metabolism contributes to the luminal alkalinization. A recently developed method (127) enables in situ distinction between transport by surface epithelial and oxyntic cells.

Contribution of Bicarbonate Leakage

A possible mechanism of luminal alkalinization in studies of gastric HCO_3^- secretion is leakage of plasma HCO_3^- resulting from mucosal damage. This process should not be confused with cellular transport. The gastric fundus and antrum are normally tight epithelia with low-conductance paracellular pathways (156, 157). Agents that increase mucosal permeability by affecting tight junctional integrity or by simply desquamating the surface epithelium inevitably permit increased passive migration and ultrafiltration of interstitial HCO_3^- into the gastric lumen. Such increases in permeability are associated with a markedly lowered transmucosal electrical potential difference and resistance (23, 43, 63, 168); also, in whole animals, substantial amounts of serum albumin and glucose appear in the gastric lumen (23).

It is vital that permeation of interstitial HCO_3^- after mucosal damage is distinguished from metabolism-dependent secretion of HCO_3^- when investigating transport mechanisms. In terms of mucosal protection, however, diffusion of HCO_3^- may be important in facilitating mucosal repair after superficial damage (113, 122, 150). Mucosal leakage of HCO_3^- occurs after gastric intraluminal instillation of chelating agents and high concentrations of aspirin, ethanol, and taurocholic acid (2, 20, 23, 24, 43, 63, 168). At lower concentrations, these latter agents inhibit metabolism-dependent transport of HCO_3^-, which results in a net decrease of the amount of HCO_3^- appearing in the gastric lumen (43, 63, 132, 135). At higher concentrations any decrease in metabolism-dependent transport is masked by the greatly increased passive permeation of HCO_3^-. Finally it should be noted that prostaglandins, which at low concentrations stimulate metabolism-dependent secretion of HCO_3^-, may at very high luminal concentrations induce mucosal damage and HCO_3^- leakage (162).

Duodenal Mucosal Transport of Bicarbonate

Proximal duodenum isolated from the American bullfrog, a species devoid of Brunner's glands, spontaneously alkalinizes the lumen at a greater rate (~1.0 $\mu eq \cdot cm^{-1} \cdot h^{-1}$) than does gastric antrum or fundus (0.1–0.4 $\mu eq \cdot cm^{-1} \cdot h^{-1}$) from the same species (37, 40, 82, 152, 167). Secretion of HCO_3^- by the bullfrog duodenum is abolished by inhibitors of tissue metabolism and stimulated by a variety of agents, including dibutyryl cAMP (but not cGMP), prostaglandins, and the hormones glucagon and gastric inhibitory peptide (40, 47, 154). It probably occurs by two different mechanisms: 1) electroneutral Cl^--HCO_3^- exchange at the luminal membrane stimulated by the hormones and inhibited by furosemide and 2) transmucosal transport stimulated by cAMP and prostaglandins (Fig. 1). The latter process is quantitatively larger and has been studied in detail (153, 154). It is inhibited by ouabain and dependent on serosal-side Na^+. It has been suggested by studies of the transmucosal fluxes of ions across short-circuited duodenum that HCO_3^- enters the cell at the serosal (lateral) membrane by $NaHCO_3$ cotransport and that Na^+ is then recycled at this membrane because of Na^+-K^+-ATPase activity. Passive exit across the luminal membrane results in a net electrogenic secretion of HCO_3^-. This model is similar to that suggested for Cl^- secretion by more distal parts of the small intestine. In bullfrog duodenum bathed in Cl^--free solutions, addition of Cl^- to the nutrient side results in a decrease in HCO_3^- appearance on the luminal side (41). This finding suggests a nutrient membrane transport mechanism with affinity for both Cl^- and HCO_3^-. It was recently re-

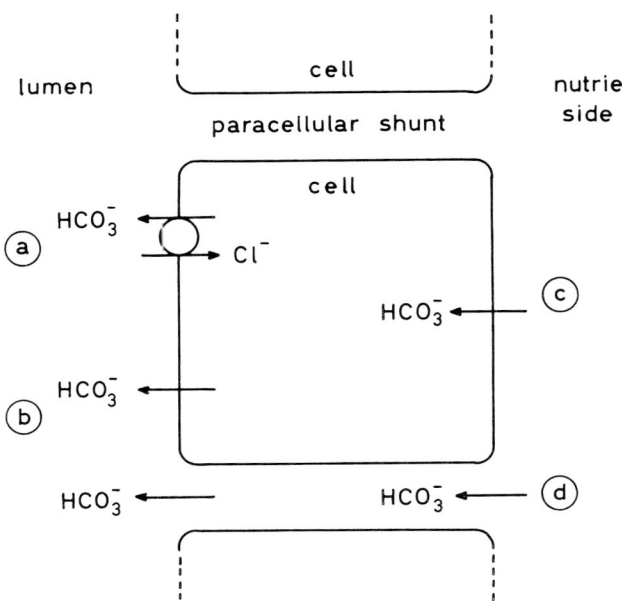

FIG. 1. Surface epithelia secretion of HCO_3^- in bullfrog proximal duodenum based on in vitro studies. *a*, Cl^--HCO_3^- exchange stimulated by glucagon and inhibited by furosemide; *b*, transport independent of luminal Cl^-, stimulated by prostaglandins and cAMP but insensitive to furosemide; *c*, anion carrier that under some conditions displays affinity for Cl^-; *d*, passive migration of HCO_3^- through shunt pathways sensitive to variations in transmucosal hydrostatic pressure. Similar inhibition of glucagon-stimulated secretion by furosemide occurs in rat duodenum in vivo (51). [From Flemström and Garner (45).]

ported (14) that purified apical membrane vesicles isolated from guinea pig duodenal enterocytes exchange HCO_3^- for Cl^- and OH^- at a high rate. The uptake of Cl^- stimulated by an outwardly directed pH gradient occurred via an electroneutral mechanism and was significantly inhibited by furosemide, DIDS, and SITS. In addition an anion-selective pathway was described with equal affinity for HCO_3^- and Cl^-.

The sensitivity of mucosal HCO_3^- secretion to prostaglandin E_2 (PGE_2), glucagon, and furosemide has been tested also in mammalian (rat) duodenum in vivo. As in vitro, glucagon- but not prostaglandin-stimulated secretion was sensitive to luminal furosemide (50). This suggests similar transport processes in vivo and in vitro. However, stimulation by the synthetic 16,16-dimethyl PGE_2 but not that by the natural compound PGE_2 in rat duodenum is associated with a marked rise in transmucosal electrical potential difference (49). Further studies of the ion-transport processes in the duodenum and their relation to eicosanoid metabolism are clearly required.

Gradient in Secretion Along the Duodenum

The rate of HCO_3^- secretion declines distally along the duodenum with proximal segments secreting at greater rates. This has been found in rats, dogs, and humans, all of which contain Brunner's glands in the proximal duodenum (90, 91, 110). However, a similar situation exists in the bullfrog (152), where these glands are absent throughout the duodenum. Furthermore secretion in both regions of the duodenum in the bullfrog in vitro as well as in the rat in vivo shows similar sensitivity to stimulation by prostaglandins and inhibition by acetazolamide. This suggests that also in mammalian proximal duodenum the surface epithelium is the main source of HCO_3^-. It cannot be excluded, however, that Brunner's glands of the serous type secrete an alkaline juice. Thus rabbits, which have numerous serous-type glands, secrete HCO_3^- at a 10- to 20-fold greater basal rate than the cats, which have very few of these glands (46, 55, 74, 125).

Biochemical Basis for Duodenal Bicarbonate Secretion

Carbonic anhydrase activity has been demonstrated in cells of the duodenal mucosa (115), and acetazolamide inhibits stimulated rates of HCO_3^- secretion in this tissue (49). The duodenal enterocyte brush border reportedly contains HCO_3^--stimulated ATPase, distinct from mitochondrial ATPase or alkaline phosphatase. It has been suggested that these enzymes play a role in the alkaline secretion. There is also a decline in enzyme activity together with the appearance of duodenal ulceration in cysteamine-treated rats (157, 161). The possible role of anion-dependent ATPase activity in epithelial transport is, however, much debated (11). It has also been recently reported that apparent activation of duodenal ATPase by exogenous HCO_3^- is caused by changes in pH and in ion strength (174).

PHYSIOLOGICAL CONTROL OF GASTRODUODENAL BICARBONATE SECRETION

It has been proposed that the secretion of HCO_3^- by the gastric mucosa constitutes one of the protective mechanisms against acid and pepsin in the stomach. Because mucosal HCO_3^- secretion in the duodenum is very probably a major mechanism of protection in this region of the gut, it is important to understand the control of these secretions. Luminal acid is a potent stimulus of alkaline secretion. Local mucosal as well as humoral and neural mechanisms influence secretion and also mediate the rise in secretion in response to acid. In addition recent studies have provided evidence that gastroduodenal mucosal HCO_3^- secretion, like gastric acid secretion, is under the control of the central nervous system (CNS).

Humoral Control

Instillation of hydrochloric acid (10–100 mM) into the gastric remnant of dogs with Heidenhain (vagally denervated) pouches increases HCO_3^- secretion by the pouch (71, 108). These studies suggested that stimulation, which was associated with a rise in plasma enteroglucagon concentration (71), is humorally mediated. The response was unaffected by pretreatment of animals with indomethacin, implying independence of mucosal prostaglandin synthesis. Definitive evidence for humoral mediation has been obtained with in vitro preparations in which two bullfrog mucosae were mounted in parallel, facing a common (buffered) serosal-side solution (82, 84). Exposure of the luminal side of fundus to pH 2 or of duodenum to pH 4 stimulated the alkaline secretion by parallel, non-acid-exposed tissue. As illustrated in Figure 2, acidification of fundus stimulates secretion by a parallel fundus; acidification of duodenum stimulates HCO_3^- secretion by both parallel duodenal and fundic mucosae. Similar stimulation of mucosal alkaline secretion by acid has been demonstrated in conscious dogs prepared with denervated fundic, antral, and duodenal pouches (108). It has also been shown in human volunteers that perfusion of the lumen of a proximal duodenal segment occluded between balloons stimulates alkaline secretion by a more distal duodenal segment (91). Mediation of the stimulation in dogs is probably humoral, although involvement of neural (sympathetic) mechanisms cannot be excluded. The response in humans may be mediated either humorally or by extrinsic or local enteric neural reflexes.

Several peptides are possible mediators of the humorally mediated rise in gastric alkaline secretion in response to luminal acid. Stimulation of gastric alkaline secretion in conscious dogs occurs in response to intravenous administration of low doses of pancreatic polypeptide and has been observed also with cholecystokinin and neurotensin (109). Gastrin and histamine have been tested and found to be without effect on gastric alkaline secretion in dogs and humans (34, 36, 109) and in frog mucosa in vitro (38, 47). The effects of a variety of peptides have been tested in the frog gastric mucosa in vitro (47), but only cholecystokinin was a stimulant in this preparation. Secretin has been tested in vitro, in animals in vivo, and in humans (26, 47, 109, 172). Although this peptide stimulates gastric mucus release, it does not affect gastric alkaline secretion.

Duodenal mucosal alkaline secretion in anesthetized guinea pigs and cats is stimulated by (pancreatic) glucagon and gastric inhibitory peptide (46), which resembles the response in bullfrog duodenum in vitro (47). Pancreatic polypeptide, cholecystokinin, and neurotensin have been tested and found to stimulate secretion in conscious dogs (109). Exposure of the gastroduodenal lumen to acid releases vasoactive intestinal peptide (VIP) and β-endorphin (9, 119), and both peptides are potent stimulants of duodenal mucosal HCO_3^- secretion in rats (48, 50, 94). Vasoactive intestinal peptide is a potent stimulant of duodenal mucosal secretion also in cats and humans (88, 94) but has no effect in bullfrog duodenum in vitro (47). Secretin has been tested in vitro, in several species in vivo, and in humans (18, 46, 47, 94, 109, 163, 170). This hormone has little or no effect on duodenal mucosal alkaline secretion despite its release by luminal acid and its ability to stimulate pancreatic HCO_3^- secretion. A study of receptors present on duodenal enterocytes would prove valuable in trying to understand the sensitivity of HCO_3^- secretion to peptides and other transmitters.

Neural Influence

Sham feeding is a stimulant of gastric HCO_3^- secretion in humans (33, 58) and of duodenal and gastric mucosal HCO_3^- secretion in conscious dogs (106). The stimulation in humans (Fig. 3) is abolished by the antimuscarinic agent benzilonium bromide but unaffected by indomethacin. Furthermore electrical stimulation of the peripheral cut vagal nerves in cats (31, 125) and rats (97) increases both gastric and duodenal HCO_3^- secretion. These responses are abolished by hexamethonium but only partially inhibited by atropine. Mediation could thus be peptidergic in part. Infusion of carbachol causes only a small (~20%) increase in duodenal HCO_3^- secretion in rats (51) and is without effect in rabbits (74). This muscarinic ag-

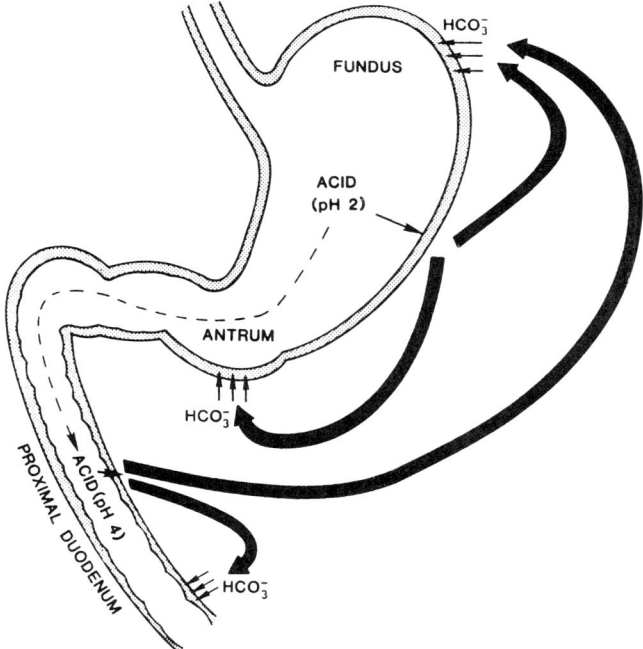

FIG. 2. Humoral mediation of rise in HCO_3^- secretion in response to luminal acid, based on experiments where isolated amphibian mucosae were mounted in parallel (82). Exposure of fundic mucosa to pH 2 releases agents that simulate secretion in parallel (non-acid-exposed) fundus and antrum. Exposure of duodenal mucosa to pH 4 releases agents that stimulate secretion in both duodenum and fundus. Experiments with fundic, antral, and duodenal pouches in conscious dog (109, 110) suggest that similar stimulation of mucosa alkaline secretions operates in mammals. [From Heylings et al. (84).]

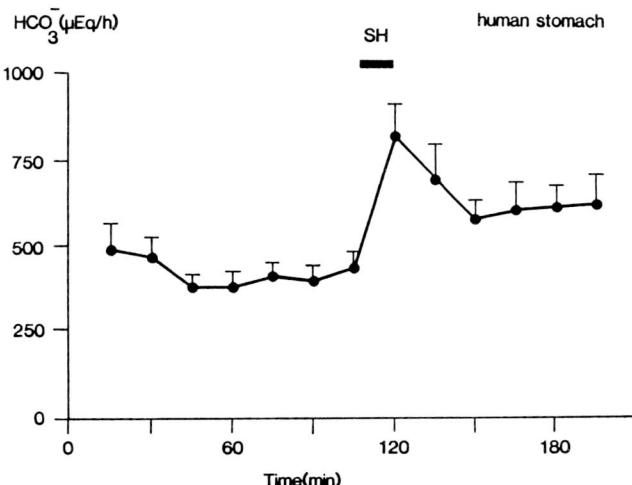

FIG. 3. Sham feeding (SH) stimulates gastric HCO_3^- secretion in healthy volunteers. Means ± SE; $n = 6$, where n is number of experiments. [From Forssell et al. (58). Copyright 1985 by The American Gastroenterological Association.]

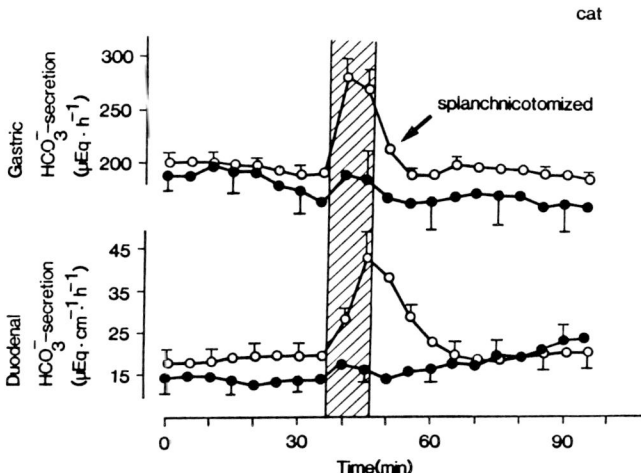

FIG. 4. Effects of bilateral vagal stimulation (shaded area, 10 Hz for 10 min) on gastric and duodenal HCO_3^- secretion in anesthetized cats with intact (closed circles, $n = 5$) and cut (open circles, $n = 6$) splanchnic nerves; n, number of experiments. Note much greater response after splanchnicotomy. [From Fändriks (30).]

onist is a relatively potent stimulant of alkaline secretion by the guinea pig and cat stomach in vivo and by frog fundic mucosa in vitro (31, 37, 66). Mucosal intrinsic neural activity influences gastric HCO_3^- secretion in humans (57). Fundic distension thus stimulates secretion, and the response is very similar in healthy volunteers and in patients after vagotomy.

Duodenal and gastric HCO_3^- secretion in vivo as well as its vagal control are under potent sympathetic influence. Infusion of the adrenergic α_2-agonist clonidine inhibits the duodenal secretion in cats and rats, and the inhibition is counteracted by the α_2-antagonist yohimbine (30, 124). The α_1-agonist phenylephrine, in contrast, stimulates the duodenal secretion in rats, whereas drugs acting at β-adrenoceptors were without effect when tested in this species (124). The vagally induced rise in both duodenal and gastric HCO_3^- secretion in cats was enhanced by splanchnicotomy and/or ligation of the adrenal glands (Fig. 4). Yohimbine and the adrenolytic agent guanethidine had similar stimulating effects (30, 31). Enhancement of the response to vagal stimulation of duodenal secretion has been demonstrated also in rats (97). Norepinephrine inhibits secretion by frog (38) and guinea pig (59) gastric mucosa in vitro, suggesting that there is a direct inhibitory effect on this epithelium. In contrast, secretion by isolated bullfrog duodenum is stimulated by epinephrine, norepinephrine, and isoprenaline. This stimulation has been shown to be abolished by β_2-antagonists (69) and to be inhibited by the α-antagonist phentolamine (47).

Neural reflexes, humoral stimuli, and mucosal prostaglandin production are all involved in mediation of the duodenal alkaline response to luminal acid. The role of neural mediation is illustrated in Figure 5. The ganglionic-blocking agent hexamethonium and the muscarinic antagonist atropine both significantly reduce the rise in secretion in response to 10 mM of luminal hydrochloric acid in rats. Vagal afferent fibers activated by luminal acid have previously been dem-

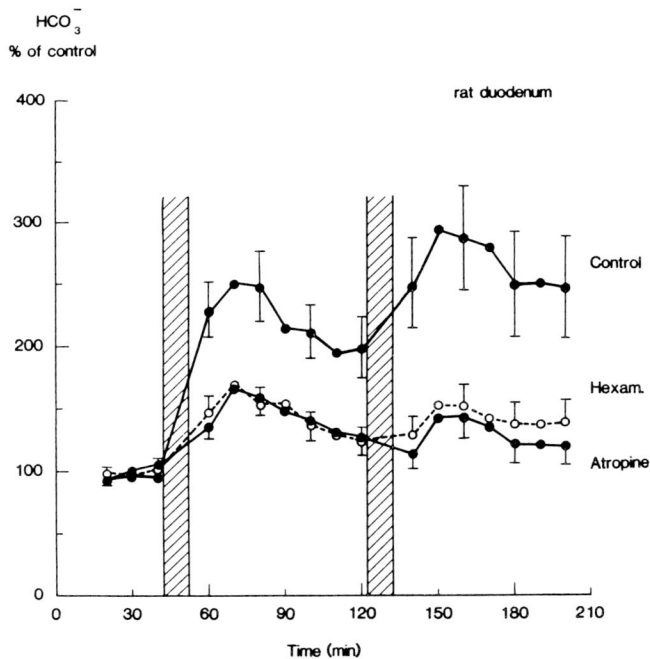

FIG. 5. Brief 10-min exposures of duodenal segments devoid of Brunner's glands in anesthetized rat to 10 mM hydrochloric acid increase the HCO_3^- secretion titrated at neutral luminal pH 7.4. Pretreatment with ganglionic blocking agent hexamethonium (10 mg/kg) or muscarinic antagonist atropine (40 µg/kg) decreases response to luminal acid. Data are means ± SE of normalized HCO_3^- secretion; $n = 8$ in all groups, where n is the number of rats.

onstrated in cat and rabbit stomach and duodenum (5).

The fact that sham feeding stimulates mucosal alkaline secretion in acid-inhibited human stomach and in canine duodenal pouch is evidence that these secretions are under the control of the CNS. Other evidence for CNS control of alkaline secretion is that injection of corticotropin-releasing factor (CRF) into the hypothalamus causes a 20-fold increase in gastric HCO_3^- content in rats (76). Furthermore intravenous injection of pirenzepine increases duodenal HCO_3^- secretion in this species (142). This tricyclic compound binds potently to muscarinic (M_1) receptors in ganglia and in some areas in the brain. The stimulation of duodenal alkaline secretion by this agent is abolished by vagotomy, which may suggest that it is exerted centrally and mediated via the vagus. Finally, acupuncture, a procedure that releases endogenous opioid peptides, is reported to stimulate gastric alkaline secretion in conscious dogs (176).

Control by Mucosal Endogenous Prostaglandins

The rise in duodenal mucosal alkaline secretion in response to hydrochloric acid in the duodenal lumen in rats, dogs, cats, and humans is inhibited by the cyclooxygenase inhibitors aspirin or indomethacin (3, 46, 49, 93, 108, 148, 164, 165) and the phospholipase A_2 inhibitor quinacrine (164), suggesting that local production of prostaglandins is involved in mediating the response. Furthermore indomethacin abolishes stimulation of secretion by the prostaglandin precursor arachidonic acid (106, 110). It has also been observed that duodenal luminal acidification in cats, dogs, and humans is associated with release of E-type prostaglandins into the lumen (46, 107, 148). Such release has been demonstrated also during acidification of the gastric lumen in humans (4). Prostaglandin mediation of duodenal mucosal HCO_3^- secretion in response to acid may be an important protective mechanism at high luminal acidities. The response to continuous exposure of the duodenal lumen to pH 2, but not that to pH 5, in rats is sensitive to intravenously administered aspirin (49). Finally, mucosal endogenous production of prostaglandins influences basal rates of secretion in the intestine (15), including the duodenum (83), and thereby may affect sensitivity to their exogenous administration. In the duodenum, for example, stimulation of HCO_3^- secretion in the anesthetized guinea pig can only be demonstrated if the animals are first pretreated with indomethacin (46).

Relation to Gastric Acid Secretion and Blood Flow

It has been demonstrated that stimulation of (parietal cell) acid secretion enhances the ability of the stomach to resist instilled hydrochloric acid. Conversely, potent inhibition of acid secretion decreases gastric mucosal resistance to instilled hydrochloric acid. Parenteral infusion of HCO_3^- but not other buffer species has a protective effect similar to that induced by stimulation of acid secretion (101, 102, 104). Recent work has shown that there are fenestrated capillaries in the rat and human gastric mucosa (61, 62) that may facilitate vascular transport of HCO_3^-, released interstitially to the surface epithelial cells from the parietal cells during acid secretion. Increased availability of HCO_3^- at the surface epithelial cell basal membrane together with stimulation of HCO_3^- secretion by neural, local, and humoral stimuli may thus all contribute to mucosal protection against luminal acid (Fig. 6).

Studies with anesthetized rabbits provide evidence that mucosal vascular supply of HCO_3^- in the duodenum is rate limiting for the mucosal alkaline secretion (147). Decreasing duodenal mucosal blood flow by vasopressin or hemorrhagic shock caused a parallel decrease in alkaline secretion. Parenteral infusion of HCO_3^- at a rate causing metabolic alkalosis increased the secretion. It has also been demonstrated in rats that mild hemorrhage reduces duodenal alkaline secretion (96). This inhibition, however, is abolished by epidural anesthesia and thus probably mediated by a sympathetic reflex. Furthermore reductions in mucosal blood flow play a considerably more important role in the duodenum than in the stomach as a means to increase susceptibility of the mucosa to acid-induced injury (114).

PHARMACOLOGICAL MODULATION

Alkaline secretion in response to drugs and other synthetic chemicals has provided valuable information

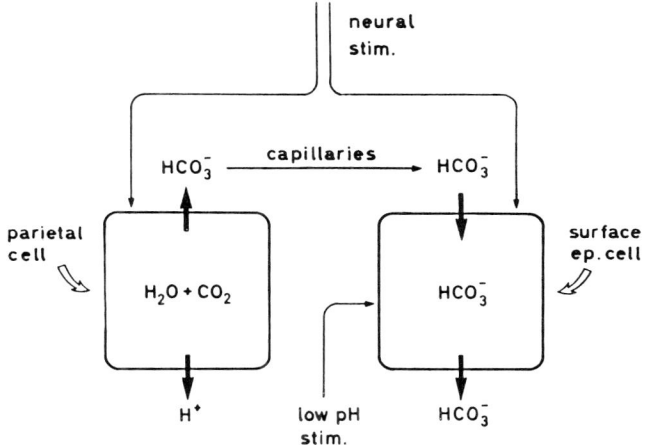

FIG. 6. Model for protection of gastric mucosa. Stimulation of the H^+ secretory process in the parietal cells increases the amount of HCO_3^- available for secretion by surface epithelial cells. Direct stimulation of HCO_3^- secretory process by neural stimuli and the subsequent effect of low luminal pH generated by H^+ secretion further increase protection. [From Flemström (43). In: *Physiology of the Gastrointestinal Tract* (2nd ed.), © 1987, Raven Press, New York.]

about the mechanism, control, and possible role of mucosal HCO_3^- transport in the stomach and duodenum. A variety of ulcerogens have been shown to inhibit HCO_3^- transport, whereas prostaglandins, which have an antiulcer action in animal models, stimulate alkalinization. These findings support the contention that HCO_3^- transport has a protective function (65, 68). Greater attention has been paid to the duodenum, where mucosal HCO_3^- secretion is probably a main protective mechanism against luminal acid.

Inhibitors of Gastric Alkaline Secretion

A number of agents known to induce gastric ulceration have been found to inhibit HCO_3^- transport when applied to the serosal side of acid-inhibited fundus in vitro (45, 54). The agents include the carbonic anhydrase inhibitor acetazolamide, ethanol, and a range of cyclooxygenase inhibitors such as aspirin, fenclofenac, and indomethacin (Fig. 7). Attenuation of gastric alkaline secretion in vitro by nonsteroidal anti-inflammatory agents is antagonized by simultaneous application of E-type prostaglandins (67). Secretion is also inhibited by bile salts such as sodium taurocholate but only when applied to the luminal side of the mucosa (133). In studies comparing different regions of the stomach, inhibitory responses in vitro are generally smaller in antrum than in fundus, which may reflect greater contribution of transmucosal diffusion of HCO_3^- to luminal alkalinization in this tissue. Inhibition of gastric HCO_3^- secretion in animals in vivo has been demonstrated after parenteral administration of acetazolamide (99, 137), low doses of aspirin (64), or the steroid prednisolone (123). It has also been reported in human volunteers given aspirin or sodium taurocholate (135).

Stimulation of Gastric Alkaline Secretion by Prostaglandins

The first report that exogenous E-type prostaglandins stimulate gastric production of an HCO_3^--rich fluid was in the canine Heidenhain pouch (10). When infused intravenously (0.6 $\mu g \cdot kg^{-1} \cdot h^{-1}$) or applied luminally (10 $\mu g/kg$) the synthetic analogue 16,16-dimethyl PGE_2 increased nonparietal volume twofold and HCO_3^- output rose fourfold over a period of 3 h. The PGE_2 itself had no effects in this model at doses 1,000-fold greater. Stimulation of alkaline secretion also occurred in amphibian gastric mucosa in vitro in response to serosal-side application of 10^{-6} M 16,16-dimethyl PGE_2, which suggests that stimulation of epithelial HCO_3^- secretion can contribute to the antiulcer activity of these agents (68). A number of studies have since reported that E-type prostaglandins enhance luminal alkalinization by the stomach in various species (Fig. 8). Many studies have been performed in dogs with denervated fundic pouches (71, 98, 110, 121), but prostaglandins are also stimulants of alkaline secretion in rats (173) and in human volunteers (34, 56, 95). However, stimulation of gastric alkaline secretion is not a universal finding. Failure of 16,16-dimethyl PGE_2 to stimulate gastric alkaline secretion was reported in bullfrog fundus in vitro (146). However, sensitivity of this preparation to other HCO_3^- stimulants was not reported. Also, the increase in luminal alkalinization by rabbit fundus in vitro was not dose related (133), whereas canine antral (but not fundic) mucosa responded to 16,16-dimethyl PGE_2 (112). In vivo, cats are relatively insensitive to synthetic E-type prostaglandins compared with other species (155), and in one study in humans (135) PGE_2 had no effect on basal gastric HCO_3^- secretion but it ameliorated the inhibition of secretion by luminal

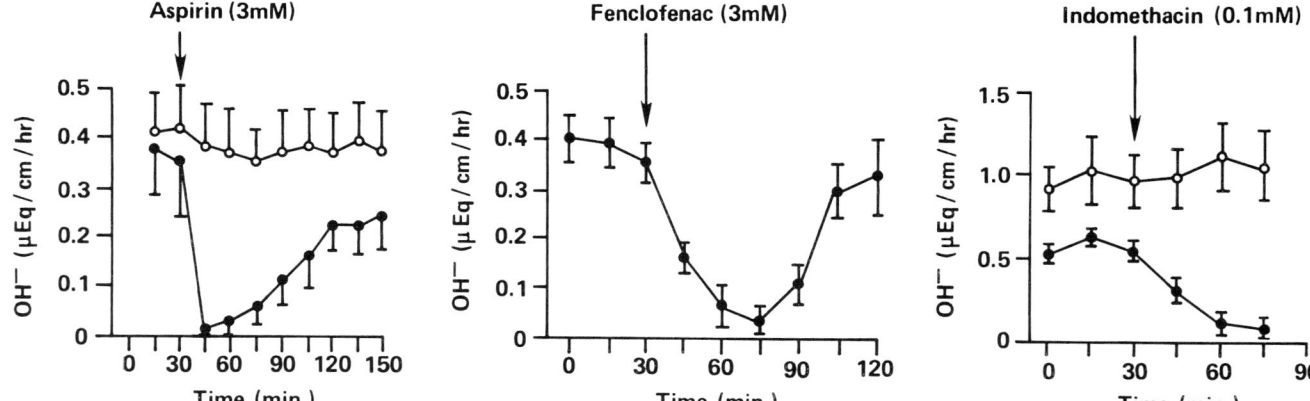

FIG. 7. Inhibitory action of various nonsteroidal, anti-inflammatory drugs on gastric fundic HCO_3^- secretion in vitro. Frog tissues were pretreated with a histamine H_2-receptor antagonist, and drugs were added to serosal side (*Closed circles*, exposed to an anti-inflammatory agent; *open circles*, exposed to histamine H_2-receptor antagonist alone. [Adapted from Garner (63; aspirin results), and from Garner et al. (67; fenclofenac and indomethacin results).]

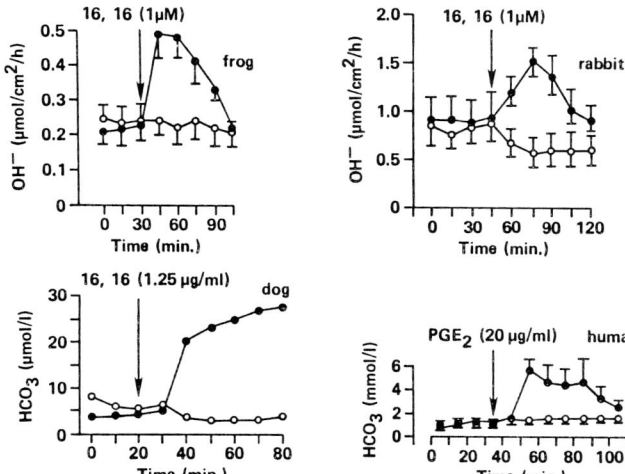

FIG. 8. Stimulation of gastric alkaline secretion by 16,16-dimethyl prostaglandin E_2 in frog fundus, rabbit fundus, and canine Heidenhain pouches and by PGE_2 in human stomach. Compounds were applied to serosal side of isolated preparations and to luminal side in vivo [Adapted from Garner and Heylings (68; frog data), Rees et al. (133; rabbit data), Kauffman et al. (98; dog data), and Johansson et al. (95; human data).]

taurocholic acid. Prostaglandins of A-, F-, and I-types also reportedly stimulate gastric alkaline secretion (110, 173), but E-types seem more effective. Synthetic analogues appear more potent, probably because of greater metabolic stability, whereas luminal administration results in maximum responses greater than those obtained after parenteral injection.

Prostaglandins also increase gastric mucus production, and there is evidence that this stimulation is mediated by cAMP (8). Studies where gastric mucus and HCO_3^- secretions have been determined simultaneously show a direct correlation between the rise in the two secretions in response to 16,16-dimethyl PGE_2 (98). Note that purified gastric mucus has little buffering capacity (1). Eicosanoids also have potent vasoactive properties, and the extent to which changes in mucosal blood flow contribute to effects of prostaglandins or indeed inhibitors of cyclooxygenase on gastric alkaline secretion in vivo must be considered in future studies. A possible link between prostaglandin and cholinergic stimulation of gastric alkaline secretion has been observed in dogs (120). The stimulation by 16,16-dimethyl PGE_2 was thus blocked by atropine and, interestingly, also by tetrodotoxin. Finally, aluminium ions have been reported to stimulate gastroduodenal HCO_3^- secretion by bullfrog mucosa (21) and an aluminium-containing compound increases secretion in the human stomach (77). It has not been examined whether stimulation by aluminium relates to effects on mucosal prostaglandin synthesis.

Duodenal Mucosal Alkaline Secretion and cAMP

Transport of HCO_3^- into the duodenum is stimulated by a wide range of pharmacological agents. Cyclic AMP appears to be one mediator of HCO_3^- transport, and so agents that elevate its cellular concentration can stimulate alkaline secretion. Addition of exogenous dibutyryl cAMP (but not dibutyryl cGMP) induces a dose-related increase in HCO_3^- secretion and in the electrical potential difference across amphibian proximal duodenum in vitro (47). Forskolin, which activates adenylate cyclase and thus raises cAMP levels in a variety of tissues, acts similarly to PGE_2 as a stimulant of HCO_3^- secretion by the duodenum (Fig. 9). Furthermore preventing breakdown of cAMP by the inhibition of phosphodiesterase is associated with stimulation of HCO_3^- secretion. Established phosphodiesterase inhibitors such as 3-isobutyl-1-methylxanthine (BMX) or 2-amino-6-methyl-5-oxo-4-n-propyl-4,5-dihydro-S-triazolo(1,5-α)-pyrimidine (ICI 63197) increase alkaline secretion in isolated duodenal preparations (47, 70). Inhibition of tissue phosphodiesterase is a common action among synthetic organic chemicals, and it is possible that this property underlies the stimulation of alkaline secretion in vitro demonstrated with a number of compounds, e.g., benzodiazepines (70).

Stimulation of Duodenal Secretion by Prostaglandins

Presence of a prostaglandin-sensitive adenylate cyclase in duodenal enterocytes suggests that prostaglandins stimulate alkaline secretion via cAMP (151). Administration of E-type prostaglandins stimulates

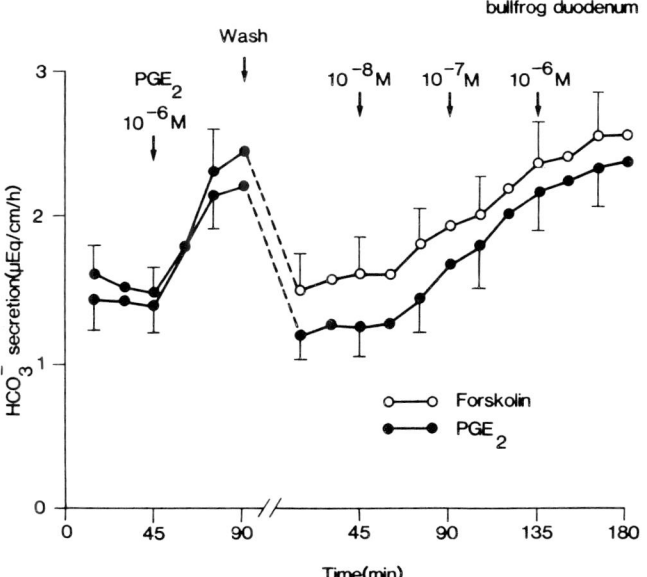

FIG. 9. In the isolated proximal duodenum, prostaglandin E_2 and forskolin (serosal side) show comparable activity as stimulants of HCO_3^- secretion, suggesting that prostaglandins act by elevating intracellular cAMP levels. Sensitivity of tissues to a standard agonist (PGE_2) was established at the beginning of experiments. Data are means ± SE; $n = 8$ for both, where n is number of experiments.

secretion in vitro, in vivo, and in human volunteers (3, 40, 46, 49, 91, 93, 154, 164). In isolated preparations PGE_1, PGE_2, and their synthetic analogues are of comparable activity, whereas in mammals in vivo the synthetic compounds appear more potent; this probably reflects their greater metabolic stability. By comparison with the stomach, the increases in secretory rate after prostaglandins are considerably greater and the response is generally sustained rather than transient. A variety of other eicosanoids also stimulate duodenal alkalinization. In the bullfrog in vitro, PGA_2, PGD_2, $PGF_{2\alpha}$, and prostacyclin are at least 10-fold less potent than PGE_2, whereas thromboxanes and leukotriene B_4 have no effect (100). Indomethacin inhibits the stimulatory action of arachidonic acid, but pretreatment enhances the net effect of PGE_2. This latter action is consistent with endogenous prostaglandin level being a major determinant of the basal secretory rate (83). The mixed cyclooxygenase-lipoxygenase inhibitor 3-amino-1-(3-trifluoromethylphenyl)-2-pyrazoline (BW755C), which increases prostaglandin production by rat gastrointestinal mucosa (12), increases duodenal mucosal alkaline secretion after intravenous administration in this species (44).

Other Drugs Affecting Duodenal Alkaline Secretion

The activity of a variety of standard pharmacological agents has been tested in the isolated bullfrog duodenum (70, 89). In addition to eicosanoids and phosphodiesterase inhibitors, compounds such as 6-hydroxydopamine, dipyridamole, 2-chloroadenosine, methallibure, testosterone, spizofurone, and a range of benzodiazepines were found to stimulate secretion when added to the serosal-side bathing solution. The response to each of these agonists was reduced on subsequent addition of the metabolic inhibitor 2,4-dinitrophenol, indicating that the increase in alkalinization was not due to mucosal damage and a subsequent increase in passive permeation of HCO_3^-. Each of these stimulants also increased the transmucosal electrical potential difference. Although they were less potent than prostaglandin itself, 6-hydroxydopamine, 2-chloroadenosine, and the benzodiazepines (diazepam, nitrozepam, and chlordiazepoxide) all produced a maximal response comparable to the tissue PGE_2 maxima. When indomethacin was present in the serosal bathing solution, only 2-chloroadenosine and the benzodiazepines retained activity, suggesting that many agonists operate by stimulating endogenous prostaglandin production. Agents that have no influence on basal secretion in frog duodenum in vitro include agonists and antagonists of histamine, γ-aminobutyric acid (GABA), serotonin, as well as monoamine oxidase and cholinesterase inhibitors, diuretics, and a range of steroids. Most CNS depressants are inactive.

Opioid peptides have been reported to stimulate duodenal alkaline secretion both in vitro and in vivo. Stimulation by morphine in frog mucosa in vitro (134) was blocked by the antagonist $(Allyl)_2$ Tyr-Gly-Gly-ψ (CH_2S)-Phe-Leu-OH (ICI 154129) that acts at δ-receptors. In contrast, naloxone, a μ-receptor antagonist, inhibits secretion stimulated by β-endorphin and morphine in anesthetized rat in vivo (48). Also in the rat in vivo, the stimulation of duodenal mucosal alkaline secretion has been observed with the tricyclic compound pirenzepine (142), whereas potent inhibitors of gastric acid secretion such as histamine H_2-receptor antagonists and the H^+-K^+-ATPase inhibitor omeprazole (52, 105) are without effect. The duodenal ulcerogens cysteamine, propionitrile, and mepirizole decrease disposal of H^+ ions by duodenal mucosa by inhibiting the rise in mucosal alkaline secretion in response to luminal acidification. The effect is thus similar to that observed with the cyclooxygenase inhibitors indomethacin and aspirin, which also are well-known duodenal ulcerogens (13, 126, 163).

PROTECTIVE ROLE OF BICARBONATE

Understanding of gastric and duodenal mucosal protection has important implications for physiology as well as etiology and possible therapy of peptic ulcer disease. Epithelial HCO_3^- secretion may serve as a first line of defense against luminal acid and pepsin, whereas stimulants of the process could provide a basis for finding therapeutical alternatives to antagonists of gastric acid secretion.

Surface pH Gradient

Surface pH microclimates occur throughout the gastrointestinal tract by virtue of H^+ or HCO_3^- transport into an unstirred boundary zone and/or because of an inherent property of the viscoelastic mucous gel adherent to the epithelial surface (130, 149). In the stomach and duodenum, HCO_3^- secreted by the mucosa passes into the overlying mucous layer, which should prevent immediate dissipation and neutralization by luminal acid. Mucous gel also reduces H^+ diffusion and may contain fixed charges. With pH-sensitive microelectrodes (antimony or glass), pH gradients have been demonstrated at the mucosal surface of the stomach and duodenum of a variety of species. At the stomach epithelial cell surface in human, rabbit, and frog mucosa in vitro and in rat and human mucosa in vivo, pH is close to neutral, when in the luminal bulk solution it is 2.0–3.0 (103, 118, 138, 166, 171). Higher acidities in the bulk solution dissipate surface alkalinity with consequent acidification of the cell surface. The characteristics of a possible pH gradient at the gastric surface have also been analyzed mathematically, using known or estimated values of HCO_3^- secretory rate, H^+ ion diffusion, H_2O bulk flow, ion-ion interaction, and ion-fixed charge interaction (29). It was concluded that a surface gradient able to resist acidities in the luminal bulk solution of ~5 mM (pH 2.3) could be generated. This acidity is somewhat lower than that observed experimentally in some studies. The stimulation of HCO_3^- transport by acid in the lumen was, however, not considered in the mathematical analysis. The surface pH gradient can thus only provide a first line against luminal acid (and as a consequence of increased pH also against pepsin) in the stomach. However, reduction of H^+ concentration as opposed to complete neutralization of H^+ may be important in terms of lowering H^+ concentration to a level where other mechanisms are able to cope. It is of interest in this context that monolayers of chief cells, which most probably do not secrete mucus or HCO_3^-, exhibit considerable luminal (but not serosal) resistance to acid (143). Recent demonstrations that acid-sensitive bacteria frequently occur in the mucous gel adjacent to gastric epithelial cell surface in humans (79, 86) represent another interesting aspect of an extracellular alkaline environment at this surface.

In the duodenum absolute rates of alkaline secretion are higher and the rise in response to stimulants, particularly luminal acidification, is greater. Furthermore, when tested in the same animal, the pH gradient is considerably more resistant to acid in the duodenum than in the stomach. A surface gradient in the duodenum is thus maintained even at pH 1.3 in the luminal bulk solution (103). These factors combined with higher ambient pH (duodenal luminal pH only rarely falls below pH 2) suggest that extracellular surface neutralization of acid can play a major role in the mechanism of mucosal protection in the duodenum.

Possible Therapeutic Implications

Inhibition of gastric acid secretion, principally with histamine H_2-receptor antagonists, has proved highly effective in the initial therapy of peptic ulceration. However, concerns have been expressed over the consequences of long-term achlorhydria, and stimulation of mucosal alkaline secretion could provide an alternative to the use of antisecretory agents in therapy of ulcer disease (65). Recent findings that rates of proximal duodenal alkaline secretion (and in particular the rise in secretion in response to luminal acid) are considerably lower than normal in patients with inactive duodenal ulcers (92) adds interest to this approach. Gastric alkaline secretion in asymptomatic duodenal ulcer patients has been reported to be close to normal (35).

SUMMARY AND PERSPECTIVES

Secretion of HCO_3^- by the gastric mucosa is increased by cholinergic stimuli, by some hormones, and by prostaglandins. It probably provides a first line of mucosal defense against acid by alkalinizing the viscoelastic mucous gel adherent to the surface. This process is distinct from the leakage/ultrafiltration of HCO_3^- that occurs across damaged gastric epithelium, although the latter is probably important in the process of mucosal repair. Further studies of the mechanisms of HCO_3^- transport across the surface epithelial cell and its proposed relation to the parietal cell alkaline tide should provide better understanding of mucosal acid-base balance. Recent findings indicate that the regulation of this secretion is under the influence of both peripheral and central nervous systems. The HCO_3^- secretion by duodenal mucosa is probably a major mechanism of mucosal defense against acid discharged from the stomach. The process of HCO_3^- transport in duodenum involves Cl^--HCO_3^- exchange as well as a conductance pathway. The relation between eicosanoids and neural influences in the control of duodenal secretion is of particular interest. Furthermore duodenal mucosal blood supply of HCO_3^- is probably rate limiting for transport of HCO_3^- into the lumen. However, little is known about the relative extent to which stimulants and inhibitors in vivo affect the secretory process per se compared with mucosal blood flow.

We wish to thank Ann-Sofie Göransson and Ailsa Walters for excellent secretarial assistance. This work was supported in part by the Swedish Medical Research Council, Grant 04X-3515.

REFERENCES

1. ALLEN, A. Structure and function of gastric mucus. In: *Physiology of the Gastrointestinal Tract*, edited by L. R. Johnson, J. Christensen, M. I. Grossman, E. D. Jacobson, and S. G. Schultz. New York: Raven, 1981, p. 617-639.
2. ALLISON, J. G., AND J. J. CULLEN. Paracellular back-diffusion of H^+ across bullfrog fundic mucosa resulting from Ca^{2+} depletion and its effects on taurocholate-induced mucosal injury. *Am. Surg.* 52: 226-232, 1986.
3. ALY, A., AND G. FLEMSTRÖM. Alkaline secretion by the stomach and intestine: effects of eicosanoids and anti-inflammatory drugs. In: *Eicosanoids and the Gastrointestinal Tract*, edited by K. Hillier. Lancaster, UK: MTP, 1987.
4. ALY, A., K. GREEN, AND C. JOHANSSON. Acid instillation increases gastric luminal prostaglandin E_2 output in man. *Acta Med. Scand.* 218: 505-510, 1985.
5. ANDREWS, C. J. H., AND W. H. H. ANDREWS. Receptors, activated by acid, in the duodenal wall of rabbits. *Q. J. Exp. Physiol. Cogn. Med. Sci.* 56: 221-230, 1971.
6. ASHLEY, S. W., D. I. SOYBEL, AND L. Y. CHEUNG. Effect of 16,16-dimethyl prostaglandin E_2 on gastric epithelial cell membrane potentials and resistances. *Surgery St. Louis* 98: 166-173, 1985.
7. ASHLEY, S. W., D. I. SOYBEL, AND L. Y. CHEUNG. Measurements of intracellular pH in *Necturus* antral mucosa by microelectrode technique. *Am. J. Physiol.* 250 (*Gastrointest. Liver Physiol.* 13): G625-G632, 1986.

8. BERSIMBAEV, R. I., M. M. TAIROV, AND R. I. SAGALNIK. Biochemical mechanisms of regulation of mucus secretion by prostaglandin E_2 in rat gastric mucosa. *Eur. J. Pharmacol.* 115: 259–266, 1985.
9. BLOOM, S. R., S. J. MITCHELL, G. R. GREENBERG, N. CHRISTOFIDES, W. DOMSCHKE, S. DOMSCHKE, P. MITZNEGG, AND L. DEMLING. Release of VIP, secretin, and motilin after duodenal acidification in man. *Acta Hepato-Gastroenterol.* 25: 365–368, 1978.
10. BOLTON, J. P., AND M. M. COHEN. Stimulation of non-parietal secretion in canine Heidenhain pouches by 16,16-dimethyl prostaglandin E_2. *Digestion* 17: 291–299, 1978.
11. BONTING, S. L. Lack of evidence that chloride transport involves an anion ATP-ase. In: *Proc. 30th Congr. Int. Union Physiol. Sci., Vancouver, BC, Canada, July 1986*, p. 333.
12. BOUGHTON-SMITH, N. K., AND B. J. R. WHITTLE. Stimulation and inhibition of prostaglandin formation in the gastric mucosa and ileum in vitro by anti-inflammatory agents. *Br. J. Pharmacol.* 78: 173–180, 1983.
13. BRIDÉN, S., G. FLEMSTRÖM, AND E. KIVILAAKSO. Cysteamine and propionitrile inhibit the rise of duodenal mucosal alkaline secretion in response to luminal acid in rats. *Gastroenterology* 88: 295–302, 1985.
14. BROWN, C. D. A., AND L. A. TURNBERG. Mechanisms of anion transport across the apical membrane of duodenal enterocytes (Abstract). *J. Physiol. Lond.* 382: 49P, 1987.
15. BUKHAVE, K., AND J. RASK-MADSEN. Saturation kinetics applied to in vitro effects of low prostaglandin E_2 and $F_{2\alpha}$-concentrations on ion transport across human jejunal mucosa. *Gastroenterology* 78: 32–42, 1980.
16. CANFIELD, S. P., AND B. P. CURWAIN. Measurement of gastric bicarbonate secretion in the presence of acid secretion in the anaesthetized rat (Abstract). *J. Physiol. Lond.* 374: 2P, 1986.
17. CANFIELD, S. P., AND J. P. N. WHITE. Duodenal alkalinization in a rat isolated duodenum preparation (Abstract). *J. Physiol. Lond.* 374: 3P, 1986.
18. CASE, R. M., A. GARNER, AND K. K. UDDIN. Simultaneous determination of duodenal and pancreatic bicarbonate transport: differential effects of secretin and prostaglandin E_2 in the cat in vivo (Abstract). *J. Physiol. Lond.* 340: 36P–37P, 1983.
19. CHEUNG, L. Y., AND W. T. NEWTON. Cyclic guanosine monophosphate response to acetylcholine stimulation of gastric alkaline secretion. *Surgery St. Louis* 86: 156–161, 1979.
20. CHUNG, R. S. K., B. S. SUM, H. GOLDMAN, M. FIELD, AND W. SILEN. Effects of chelation of calcium on the gastric mucosal barrier. *Gastroenterology* 59: 200–207, 1970.
21. CRAMPTON, J. R., L. C. GIBBONS, AND W. D. W. REES. The effect of aluminum on bicarbonate secretion by isolated amphibian gastroduodenal mucosa (Abstract). *Gut* 26: A1111, 1985.
22. CURCI, S., AND T. SCHETTINO. Effect of external sodium on intracellular chloride activity in the surface cells of frog gastric mucosa. *Pfluegers Arch.* 401: 152–159, 1984.
23. DAVENPORT, H. W. Salicylate damage to the gastric mucosal barrier. *N. Engl. J. Med.* 276: 1307–1312, 1967.
24. DAYTON, M. T., G. L. KAUFFMAN, J. F. SCHLEGEL, C. F. CODE, AND J. H. STEINBACH. Gastric bicarbonate appearance with ethanol ingestion. Mechanism and significance. *Dig. Dis. Sci.* 28: 449–455, 1983.
25. DAYTON, M. T., AND J. SCHLEGEL. Cimetidine-induced bicarbonate production in canine gastric pouches. *J. Surg. Res.* 32: 464–470, 1982.
26. DAYTON, M. T., AND J. SCHLEGEL. The effect of secretin on canine gastric mucosal HCO_3^- production. *J. Surg. Res.* 35: 319–324, 1983.
27. DEMAREST, J. R., C. SCHEFFEY, AND T. E. MACHEN. Segregation of gastric Na and Cl transport: a vibrating probe and microelectrode study. *Am. J. Physiol.* 251 (*Cell Physiol.* 20): C643–C648, 1986.
28. DORRICOTT, N. J., R. G. FIDDIAN-GREEN, AND W. SILEN. Mechanisms of acid disposal in canine duodenum. *Am. J. Physiol.* 228: 269–275, 1975.
29. ENGEL, E., A. PESKOFF, G. L. KAUFFMAN, JR., AND M. I. GROSSMAN. Analysis of hydrogen ion concentration in the gastric gel mucus layer. *Am. J. Physiol.* 247 (*Gastrointest. Liver Physiol.* 10): G321–G338, 1984.
30. FÄNDRIKS, L. Vagal and splanchnic neural influences on gastric and duodenal bicarbonate secretions. *Acta Physiol. Scand. Suppl.* 555: 1–39, 1986.
31. FÄNDRIKS, L. Sympatho-adrenergic inhibition of vagally induced gastric motility and gastroduodenal HCO_3^- secretion in the cat. *Acta Physiol. Scand.* 128: 555–562, 1986.
32. FÄNDRIKS, L., AND L. STAGE. Simultaneous measurements of gastric motility and acid-bicarbonate secretion in the anesthetized cat. *Acta Physiol. Scand.* 128: 563–573, 1986.
33. FELDMAN, M. Gastric H^+ and HCO_3^- secretion in response to sham feeding in humans. *Am. J. Physiol.* 248 (*Gastrointest. Liver Physiol.* 11): G188–G191, 1985.
34. FELDMAN, M., AND C. C. BARNETT. Gastric bicarbonate secretion in humans. Effects of pentagastrin, bethanechol, and 11,16,16-trimethyl prostaglandin E_2. *J. Clin. Invest.* 72: 295–303, 1983.
35. FELDMAN, M., AND C. C. BARNETT. Gastric bicarbonate secretion in patients with duodenal ulcer. *Gastroenterology* 88: 1205–1208, 1985.
36. FELDMAN, M., AND L. S. SCHILLER. Effect of bethanechol (urecholine) on gastric acid and non-parietal secretion in normal subjects and duodenal ulcer patients. *Gastroenterology* 83: 262–266, 1982.
37. FLEMSTRÖM, G. Active alkalinization by amphibian gastric fundic mucosa in vitro. *Am. J. Physiol.* 233 (*Endocrinol. Metab. Gastrointest. Physiol.* 2): E1–E12, 1977.
38. FLEMSTRÖM, G. Effect of catecholamines, Ca^{++} and gastrin on gastric HCO_3^- secretion. *Acta Physiol. Scand. Special Suppl.*, 81–90, 1978. (Proc. Symp. Gastric Ion Transport, Uppsala, Sweden, July, 1977.)
39. FLEMSTRÖM, G. Cl^- dependence of HCO_3^- transport in frog gastric mucosa. *Upsala J. Med. Sci.* 85: 303–310, 1980.
40. FLEMSTRÖM, G. Stimulation of HCO_3^- transport in isolated proximal bullfrog duodenum by prostaglandins. *Am. J. Physiol.* 239 (*Gastrointest. Liver Physiol.* 2): G198–G203, 1980.
41. FLEMSTRÖM, G. Properties of hormone and prostaglandin-stimulated duodenal HCO_3^- transport. In: *Electrolyte and Water Transport Across Gastrointestinal Epithelia*, edited by R. M. Case, A. Garner, L. A. Turnberg, and J. A. Young. New York: Raven, 1982, p. 85–94.
42. FLEMSTRÖM, G. Measurement of gastric bicarbonate secretion. *Gastroenterology* 88: 2000–2002, 1985.
43. FLEMSTRÖM, G. Gastric and duodenal mucosal bicarbonate secretion. In: *Physiology of the Gastrointestinal Tract* (2nd ed.), edited by L. R. Johnson, J. Christensen, M. J. Jackson, E. D. Jacobson, and J. H. Walsh. New York: Raven, 1987, vol. 2, p. 1011–1029.
44. FLEMSTRÖM, G., A. BERGMAN, AND S. BRIDÉN. Stimulation of mucosal bicarbonate secretion in rat duodenum in vivo by BW755C. *Acta Physiol. Scand.* 121: 39–43, 1984.
45. FLEMSTRÖM, G., AND A. GARNER. Gastroduodenal HCO_3^- transport: characteristics and proposed role in acidity regulation and mucosal protection. *Am. J. Physiol.* 242 (*Gastrointest. Liver Physiol.* 5): G183–G193, 1982.
46. FLEMSTRÖM, G., A. GARNER, O. NYLANDER, B. C. HURST, AND J. R. HEYLINGS. Surface epithelial HCO_3^- transport by mammalian duodenum in vivo. *Am. J. Physiol.* 243 (*Gastrointest. Liver Physiol.* 6): G348–G358, 1982.
47. FLEMSTRÖM, G., J. R. HEYLINGS, AND A. GARNER. Gastric and duodenal HCO_3^- transport in vitro: effects of hormones and local transmitters. *Am. J. Physiol.* 242 (*Gastrointest. Liver Physiol.* 5): G100–G110, 1982.
48. FLEMSTRÖM, G., G. JEDSTEDT, AND O. NYLANDER. β-Endorphin and enkephalins stimulate duodenal mucosal alkaline secretion in the rat in vivo. *Gastroenterology* 90: 368–372, 1986.

49. FLEMSTRÖM, G., AND E. KIVILAAKSO. Demonstration of a pH gradient at the luminal surface of rat duodenum in vivo and its dependence on mucosal alkaline secretion. *Gastroenterology* 84: 787–794, 1983.
50. FLEMSTRÖM, G., E. KIVILAAKSO, S. BRIDÉN, O. NYLANDER, AND G. JEDSTEDT. Gastroduodenal bicarbonate secretion in mucosal protection. Possible role of vasoactive intestinal peptide and opiates. *Dig. Dis. Sci.* 30, Suppl. 11: 63S–68S, 1985.
51. FLEMSTRÖM, G., E. KIVILAAKSO, AND A. GARNER. Bicarbonate secretion by the duodenal surface epithelium and its protective role. In: *Hydrogen Ion Transport in Epithelia*, edited by J. G. Forte, D. G. Warnock, and F. C. Rector. New York: Wiley-Interscience, 1984, p. 389–398.
52. FLEMSTRÖM, G., AND H. MATTSSON. Effects of omeprazole on gastric and duodenal bicarbonate secretion. *Scand. J. Gastroenterol.* 21, Suppl. 118: 65–67, 1986.
53. FLEMSTRÖM, G., AND G. SACHS. Ion transport by amphibian antrum in vitro. I. General characteristics. *Am. J. Physiol.* 228: 1188–1198, 1975.
54. FLEMSTRÖM, G., AND L. A. TURNBERG. Gastroduodenal defence mechanisms. *Clin. Gastroenterol.* 13: 327–354, 1984.
55. FLOREY, H. W., AND H. E. HARDING. The functions of Brunner's glands and the pyloric end of the stomach. *J. Pathol. Bacteriol.* 37: 431–453, 1933.
56. FORSSELL, H., AND L. OLBE. Continuous computerized determination of gastric bicarbonate secretion in man. *Scand. J. Gastroenterol.* 20: 767–774, 1985.
57. FORSSELL, H., AND L. OLBE. Effect of fundic distension on gastric bicarbonate secretion in man. *Scand. J. Gastroenterol.* 22: 627–633, 1987.
58. FORSSELL, H., B. STENQUIST, AND L. OLBE. Vagal stimulation of human gastric bicarbonate secretion. *Gastroenterology* 89: 581–586, 1985.
59. FROMM, D., J. H. SCHWARTZ, R. ROBERTSON, AND R. FUHRO. Ion transport across isolated antral mucosa of the rabbit. *Am. J. Physiol.* 231: 1783–1789, 1976.
60. FURUKAWA, T., E. OLENDER, D. FROMM, AND M. KOLIS. Effects of cyclic adenosine monophosphate and prostaglandins on Na^+- and HCO_3^--induced dissipation of a proton gradient in isolated gastric surface cells of rabbits. *Gastroenterology* 89: 500–506, 1985.
61. GANNON, B., J. BROWNING, AND P. O'BRIEN. The microvascular architecture of the glandular mucosa of the rat stomach. *J. Anat.* 135: 667–683, 1982.
62. GANNON, B., J. BROWNING, P. O'BRIEN, AND P. ROGERS. Mucosal microvascular architecture of the fundus and body of human stomach. *Gastroenterology* 86: 866–875, 1984.
63. GARNER, A. Effects of acetylsalicylate on alkalinization, acid secretion and electrogenic properties in the isolated gastric mucosa. *Acta Physiol. Scand.* 99: 281–291, 1977.
64. GARNER, A. Mechanisms of action of aspirin on the gastric mucosa of the guinea pig. *Acta Physiol. Scand. Special Suppl.*, 101–110, 1978. (Proc. Symp. Gastric Ion Transport, Uppsala, Sweden, July, 1977.)
65. GARNER, A. Future opportunities for drug therapy in peptic ulcer disease. *Scand. J. Gastroenterol. Suppl.* 125: 203–209, 1986.
66. GARNER, A., AND G. FLEMSTRÖM. Gastric HCO_3^- secretion in the guinea pig. *Am. J. Physiol.* 234 (*Endocrinol. Metab. Gastrointest. Physiol.* 3): E535–E541, 1978.
67. GARNER, A., G. FLEMSTRÖM, AND J. R. HEYLINGS. Effects of anti-inflammatory agents and prostaglandins on acid and bicarbonate secretions in the amphibian-isolated gastric mucosa. *Gastroenterology* 77: 457–461, 1979.
68. GARNER, A., AND J. R. HEYLINGS. Stimulation of alkaline secretion in amphibian-isolated gastric mucosa by 16,16-dimethyl PGE_2 and $PGF_{2\alpha}$: a proposed explanation for some of the cytoprotective actions of prostaglandins. *Gastroenterology* 76: 497–503, 1979.
69. GARNER, A., J. R. HEYLINGS, T. J. PETERS, AND J. M. WILKES. Adrenergic agonists stimulate HCO^- secretion by amphibian duodenum in vitro via an action on β_2-receptors (Abstract). *J. Physiol. Lond.* 354: 34P, 1984.
70. GARNER, A., J. R. HEYLINGS, AND A. M. STANIER. Stimulants of epithelial HCO_3^- transport in isolated proximal duodenum from *Rana catesbeiana* (Abstract). *J. Physiol. Lond.* 357: 131P, 1984.
71. GARNER, A., AND B. C. HURST. Alkaline secretion by the canine Heidenhain pouch in response to exogenous acid, some gastrointestinal hormones and prostaglandin. In: *Advances in Physiological Sciences. Nutrition, Digestion and Metabolism*, edited by T. Gati, L. G. Szollar, and G. Ungvary. Oxford, UK: Pergamon, 1981, vol. 12, p. 215–219.
72. GARNER, A., B. C. HURST, J. R. HEYLINGS, AND G. FLEMSTRÖM. Role of gastroduodenal HCO_3^- transport in acid disposal and mucosal protection. In: *Electrolyte and Water Transport Across Gastrointestinal Epithelia*, edited by R. M. Case, A. Garner, L. A. Turnberg, and J. A. Young. New York: Raven, 1981, p. 239–252.
73. GASCOIGNE, A. D., AND B. H. HIRST. Prostaglandins alter the relationship between gastric hydrogen ion concentration and flow: evidence for stimulation of nonparietal secretion in the cat. *J. Physiol. Lond.* 316: 427–438, 1981.
74. GRANSTAM, S. O., G. FLEMSTRÖM, AND O. NYLANDER. Bicarbonate secretion by the rabbit duodenal mucosa in vivo: effects of prostaglandins, vagal stimulation and some drugs. *Acta Physiol. Scand.* 131: 377–385, 1987.
75. GROSSMAN, M. I. The secretion of the pyloric glands of the dog. *Symp. Special Lectures, 21st Int. Congr. Physiol. Sci., Buenos Aires, August, 1959*, p. 226–228.
76. GUNION, M. W., Y. TACHE, AND G. L. KAUFFMAN. Intrahypothalamic corticotropin-releasing factor (CRF) increases gastric bicarbonate content (Abstract). *Gastroenterology* 88: 1407, 1985.
77. GUSLANDI, M. Sucralfate and gastric bicarbonate. *Pharmacology Basel* 31: 298–300, 1986.
78. HARMON, J. W., M. WOODS, AND N. J. GURLL. Different mechanisms of hydrogen ion removal in stomach and duodenum. *Am. J. Physiol.* 235 (*Endocrinol. Metab. Gastrointest. Physiol.* 4): E692–E698, 1978.
79. HAZELL, S. L., A. LEE, L. BRADY, AND W. HENNESSY. Campylobacter pyloridis and gastritis: association with intercellular spaces and adaptation to an environment of mucus as important factors in colonization of the gastric epithelium. *J. Infect. Dis.* 153: 658–663, 1986.
80. HELANDER, H. F. The cells of gastric mucosa. *Int. Rev. Cytol.* 70: 279–351, 1981.
81. HELANDER, H. F., AND R. P. DURBIN. Localization of ouabain-binding sites in frog gastric mucosa. *Am. J. Physiol.* 243 (*Gastrointest. Liver Physiol.* 6): G297–G303, 1982.
82. HEYLINGS, J. R., A. GARNER, AND G. FLEMSTRÖM. Regulation of gastroduodenal HCO_3^- transport by luminal acid in the frog in vitro. *Am. J. Physiol.* 246 (*Gastrointest. Liver Physiol.* 9): G235–G242, 1984.
83. HEYLINGS, J. R., S. E. HAMPSON, AND A. GARNER. Endogenous E-type prostaglandins in regulation of basal alkaline secretion by amphibian duodenum in vitro. *Gastroenterology* 88: 290–294, 1985.
84. HEYLINGS, J. R., B. C. HURST, AND A. GARNER. Effect of luminal acid on gastric and duodenal bicarbonate transport. *Scand. J. Gastroenterol. Suppl.* 92: 59–62, 1984.
85. HIMAL, H. S., H. W. WERTHER, M. L. CHAPMAN, H. D. JANOWITZ, AND J. RUDICK. Acid absorption in the canine duodenum. *Ann. Surg.* 185: 481–487, 1977.
86. HO, J., I. LUI, W. M. HUI, M. M. T. NG, AND S. K. LAM. A study on the correlation of duodenal-ulcer healing with campylobacter-like organisms. *J. Gastroenterol. Hepatol.* 1: 69–74, 1986.
87. HOLLANDER, H. F. The two-component mucous barrier. *Arch. Intern. Med.* 92: 107–120, 1954.
88. HURST, B. C., J. R. HEYLINGS, O. NYLANDER, K. K. UDDIN, A. GARNER, AND G. FLEMSTRÖM. Duodenal bicarbonate se-

cretion in response to glucagon and related peptides (Abstract). *Regul. Pept.* 4: 367, 1982.
89. INADA, I., H. SATOH, N. INATOMI, H. NAGAYA, AND Y. MAKI. Spizofurone, a new anti-ulcer agent, increases alkaline secretion in isolated bullfrog duodenal mucosa. *Eur. J. Pharmacol.* 124: 149–155, 1986.
90. ISENBERG, J. I., G. FLEMSTRÖM, AND C. JOHANSSON. Mucosal bicarbonate secretion is significantly greater in the proximal versus distal duodenum in the in vivo rat. In: *Mechanism of Mucosal Protection in the Upper Gastrointestinal Tract*, edited by A. Allen, G. Flemström, A. Garner, W. Silen, and L. A. Turnberg. New York: Raven, 1983, p. 175–180.
91. ISENBERG, J. I., D. L. HOGAN, M. A. KOSS, AND J. A. SELLING. Human duodenal mucosal bicarbonate secretion. Evidence for basal secretion and stimulation by hydrochloric acid and a synthetic prostaglandin E_1 analogue. *Gastroenterology* 91: 370–378, 1986.
92. ISENBERG, J. I., J. A. SELLING, D. L. HOGAN, AND M. A. KOSS. Impaired proximal duodenal mucosal bicarbonate secretion in duodenal ulcer patients. *N. Engl. J. Med.* 316: 374–379, 1987.
93. ISENBERG, J. I., B. SMEDFORS, AND C. JOHANSSON. Effect of graded doses of intraluminal H^+, prostaglandin E_2, and inhibition of endogenous prostaglandin synthesis on proximal duodenal bicarbonate secretion in unanesthetized rat. *Gastroenterology* 88: 303–307, 1985.
94. ISENBERG, J. I., B. WALLIN, C. JOHANSSON, B. SMEDFORS, V. MUTT, K. TAKEMOTO, AND S. EMÅS. Secretin, VIP, and PHI stimulate rat proximal duodenal surface epithelial bicarbonate secretion in vivo. *Regul. Pept.* 8: 315–320, 1984.
95. JOHANSSON, C., A. ALY, E. NILSSON, AND G. FLEMSTRÖM. Stimulation of gastric bicarbonate secretion by E_2 prostaglandins in man. *Adv. Prostaglandin Thromboxane Leukotriene Res.* 12: 395–401, 1983.
96. JÖNSSON, C., AND L. FÄNDRIKS. Bleeding decreases duodenal HCO_3^- secretion by a nervous mechanism. *Acta Physiol. Scand.* 127: 273–274, 1986.
97. JÖNSSON, C., O. NYLANDER, G. FLEMSTRÖM, AND L. FÄNDRIKS. Vagal stimulation of duodenal HCO_3^- secretion in anaesthetized rats. *Acta Physiol. Scand.* 128: 65–70, 1986.
98. KAUFFMAN, G. L., JR., J. J. REEVE, JR., AND M. I. GROSSMAN. Gastric bicarbonate secretion: effect of topical and intravenous 16,16-dimethyl prostaglandin E_2. *Am. J. Physiol.* 239 (*Gastrointest. Liver Physiol.* 2): G44–G48, 1980.
99. KAUFFMAN, G. L., JR., AND J. H. STEINBACH. Gastric bicarbonate secretion: effects of pH and topical 16,16-dimethyl prostaglandin E_2. *Surgery St. Louis* 89: 324–328, 1981.
100. KEOGH, J. P., A. M. STANIER, J. R. HEYLINGS, A. ALLEN, AND A. GARNER. Effects of eicosanoids on amphibian duodenal luminal alkaline secretion in vitro (Abstract). *Gut* 26: A1110, 1985.
101. KIVILAAKSO, E. High plasma HCO_3^- protects gastric mucosa against acute ulceration in the rat. *Gastroenterology* 81: 921–927, 1981.
102. KIVILAAKSO, E. Contribution of ambient HCO_3^- to mucosal protection and intracellular pH in isolated amphibian gastric mucosa. *Gastroenterology* 85: 1284–1289, 1983.
103. KIVILAAKSO, E., AND G. FLEMSTRÖM. Surface pH gradient in gastroduodenal mucosa. *Scand. J. Gastroenterol. Suppl.* 105: 50–52, 1984.
104. KIVILAAKSO, E., AND W. SILEN. Pathogenesis of experimental gastric mucosal injury. *N. Engl. J. Med.* 301: 364–369, 1979.
105. KNUTSON, L., G. FLEMSTRÖM, S. GUSTAVSSON, G. JEDSTEDT, AND G. LÖNNERHOLM. HCO_3^- secretion in rat duodenum after treatment with omeprazole and ranitidine. *Scand. J. Gastroenterol.* 22: 87–90, 1987.
106. KONTUREK, S. J., J. BILSKI, A. J. BILSKI, AND J. TASLER. Cephalic phase of gastric and duodenal alkaline secretion in conscious dogs (Abstract). *Gastroenterology* 90: 1499, 1986.
107. KONTUREK, S. J., J. BILSKI, J. TASLER, J. W. KONTUREK, H. BIELANSKI, AND A. KAMINSKA. Role of endogenous prostaglandins in duodenal alkaline response to luminal hydrochloric acid or arachidonic acid in conscious dogs. *Digestion* 34: 268–274, 1986.
108. KONTUREK, S. J., J. BILSKI, J. TASLER, AND J. LASKIEWICZ. Gastroduodenal alkaline response to acid and taurocholate in conscious dogs. *Am. J. Physiol.* 247 (*Gastrointest. Liver Physiol.* 10): G149–G154, 1984.
109. KONTUREK, S. J., J. BILSKI, J. TASLER, AND J. LASKIEWICZ. Gut hormones in stimulation of gastroduodenal alkaline secretion in conscious dogs. *Am. J. Physiol.* 248 (*Gastrointest. Liver Physiol.* 11): G687–G691, 1985.
110. KONTUREK, S. J., J. TASLER, J. BILSKI, AND J. KANIA. Prostaglandins and alkaline secretion from oxyntic, antral, and duodenal mucosa of the dog. *Am. J. Physiol.* 245 (*Gastrointest. Liver Physiol.* 8): G539–G546, 1983.
111. KRISTENSEN, M. Titration curves for gastric secretion. *Scand. J. Gastroenterol. Suppl.* 32: 1–149, 1975.
112. KUO, Y. J., L. L. SHANBOUR, AND T. A. MILLER. Effects of 16,16-dimethyl prostaglandin E_2 on alkaline secretion in isolated canine gastric mucosa. *Dig. Dis. Sci.* 12: 1121–1126, 1983.
113. LACY, E. R., AND S. ITO. Rapid epithelial restitution of the rat gastric mucosa after ethanol injury. *Lab. Invest.* 51: 573–583, 1984.
114. LEUNG, F. W., M. ITOH, K. HIRABAYASHI, AND P. H. GUTH. Role of blood flow in gastric and duodenal mucosal injury in the rat. *Gastroenterology* 88: 281–289, 1985.
115. LÖNNERHOLM, G., Ö. SELKING, AND P. J. WISTRAND. Amount and distribution of carbonic anhydrases CA I and CA II in the gastrointestinal tract. *Gastroenterology* 88: 1151–1161, 1985.
116. MACHEN, T. E., W. L. MCLENNAN, AND T. ZEUTHEN. Electrogenic Cl^- secretion by resting gastric mucosa: Na-Cl cotransport model. In: *Hydrogen Ion Transport in Epithelia*, edited by I. Schultz, G. Sachs, J. G. Forte, and K. J. Ullrich. Amsterdam: Elsevier, 1980, p. 379–390.
117. MACHEN, T. E., AND A. M. PARADISO. Regulation of intracellular pH in the stomach. *Annu. Rev. Physiol.* 49: 21–35, 1987.
118. MASAKI, H., AND M. FUJIMOTO. Microelectrode study on pH gradient across gastric mucus gel and intracellular pH of surface epithelial cells. In: *Proc. 30th Congr. Int. Union Physiol. Sci., Vancouver, BC, Canada, July 1986*, p. 373.
119. MATSUMURA, M., H. WADA, AND S. SAITO. Effect of solution of low pH on releases of β-endorphin-like immunoreactivity and ACTH-like immunoreactivity from human gastric antral mucosa. *Gastroenterol. Jpn.* 18: 210–215, 1983.
120. MILLER, T. A., J. M. HENAGAN, L. A. WATKINS, AND T. M. LOY. Prostaglandin induced bicarbonate secretion in the canine stomach: characteristics and evidence for a cholinergic mechanism. *J. Surg. Res.* 35: 105–112, 1983.
121. MILLER, T. A., B. B. KRAEMER, J. M. HENAGAN, AND C. E. FOUCAR. Topical 16,16-dimethyl prostaglandin E_2. Effects on gastric morphology, hydrogen ion loss, and bicarbonate secretion. *Dig. Dis. Sci.* 28: 641–648, 1983.
122. MORRIS, G. P., R. K. HARDING, AND J. L. WALLACE. A functional model for extracellular gastric mucus in the rat. *Virchows Arch. B Cell Pathol.* 46: 239–251, 1984.
123. NOBUHARA, Y., S. UEKI, AND K. TAKEUCHI. Influence of prednisolone on gastric alkaline response in rat stomach. A possible explanation for steroid-induced gastric lesions. *Dig. Dis. Sci.* 30: 1166–1173, 1985.
124. NYLANDER, O., AND G. FLEMSTRÖM. Effects of alpha-receptor agonists and antagonists on duodenal surface epithelial HCO_3^- secretion in the rat in vivo. *Acta Physiol. Scand.* 126: 433–441, 1986.
125. NYLANDER, O., G. FLEMSTRÖM, D. DELBRO, AND L. FÄNDRIKS. Vagal influence on gastroduodenal HCO_3^- secretion in the cat in vivo. *Am. J. Physiol.* 252 (*Gastrointest. Liver Physiol.* 15): G522–G528, 1987.
126. OHE, K., Y. OKADA, T. FUJIWARA, M. INOUE, AND A. MIYOSHI. Cysteamine-induced inhibition of acid neutralization and the increase in hydrogen ion back-diffusion in the duodenal mucosa. *Dig. Dis. Sci.* 27: 250–256, 1983.

127. OLENDER, E. J., D. FROMM, T. FURUKAWA, AND M. KOLIS. H$^+$ disposal by rabbit gastric mucosal surface cells. *Gastroenterology* 86: 698–705, 1984.
128. OSKOSDINOSSIAN, E. T., AND H. A. EL MUNSHID. Composition of the alkaline component of human gastric juice: effect of swallowed saliva and duodeno-gastric reflux. *Scand. J. Gastroenterol.* 12: 945–950, 1977.
129. OVESEN, L., F. BENDTSEN, U. TAGE-JENSEN, N. T. PEDERSEN, B. R. GRAM, AND S. RUNE. Intraluminal pH in the stomach, duodenum, and proximal duodenum in normal subjects and patients with exocrine pancreatic insufficiency. *Gastroenterology* 90: 958–962, 1986.
130. RECHKEMMER, G., M. WAHL, W. KUSCHINSKY, AND W. VON ENGELHARDT. pH-microclimate at the luminal surface of intestinal mucosa of guinea pig and rat. *Pfluegers Arch.* 407: 33–40, 1986.
131. REES, W. D. W., D. BOTHAM, AND L. A. TURNBERG. A demonstration of bicarbonate production by the normal human stomach in vivo. *Dig. Dis. Sci.* 27: 961–966, 1982.
132. REES, W. D. W., A. GARNER, L. A. TURNBERG, AND L. C. GIBBONS. Studies of acid and alkaline secretion by rabbit gastric fundus in vitro: effect of low concentrations of sodium taurocholate. *Gastroenterology* 83: 435–440, 1982.
133. REES, W. D. W., L. C. GIBBONS, AND L. A. TURNBERG. Effects of nonsteroidal anti-inflammatory drugs and prostaglandins on alkali secretion by rabbit gastric fundus in vitro. *Gut* 24: 784–789, 1983.
134. REES, W. D. W., L. C. GIBBONS, AND L. A. TURNBERG. Influence of opiates on alkali secretion by amphibian gastric and duodenal mucosa in vitro. *Gastroenterology* 90: 323–327, 1986.
135. REES, W. D. W., L. C. GIBBONS, G. WARHURST, AND L. A. TURNBERG. Studies of bicarbonate secretion by the normal human stomach in vivo—effect of aspirin, sodium taurocholate and prostaglandin E$_2$. In: *Mechanisms of Mucosal Protection in the Upper Gastrointestinal Tract*, edited by A. Allen, G. Flemström, A. Garner, W. Silen, and L. A. Turnberg. New York: Raven, 1983, p. 119–124.
136. REHM, W. S., C. F. BUTLER, S. G. SPANGLER, AND S. S. SANDERS. A model to explain uphill water transport in the mammalian stomach. *J. Theor. Biol.* 27: 433–453, 1970.
137. REICHSTEIN, B. J., AND M. M. COHEN. Effect of acetazolamide on rat gastric mucosal protection and stimulated bicarbonate secretion with 16,16-dimethyl PGE$_2$. *J. Lab. Clin. Med.* 104: 797–804, 1984.
138. ROSS, I. N., H. M. M. BAHARI, AND L. A. TURNBERG. The pH gradient across mucus adherent to rat fundic mucosa in vivo and the effect of potential damaging agents. *Gastroenterology* 81: 713–718, 1981.
139. RUNE, S. J. Acid-base parameters of duodenal contents in man. *Gastroenterology* 62: 533–539, 1972.
140. RUNE, S. J., AND F. W. HENRIKSEN. Carbon dioxide tensions in the proximal part of the canine gastrointestinal tract. *Gastroenterology* 56: 758–762, 1969.
141. RUTTEN, M., D. RATTNER, AND W. SILEN. Transepithelial transport of guinea pig gastric mucous cell monolayers. *Am. J. Physiol.* 249 (*Cell Physiol.* 18): C503–C513, 1985.
142. SÄFSTEN, B., AND G. FLEMSTRÖM. Stimulatory effect of pirenzepine on mucosal bicarbonate secretion in rat duodenum in vivo. *Acta Physiol. Scand.* 127: 267–268, 1986.
143. SANDERS, M. J., A. AYALON, M. ROLL, AND A. H. SOLL. The apical surface of canine chief cell monolayers resists H$^+$ back diffusion. *Nature Lond.* 313: 52–54, 1985.
144. SCHETTINO, R., AND S. CURCI. On the luminal membrane permeability to Cl$^-$ of *Necturus* gastric surface cells. *Pfluegers Arch.* 403: 331–353, 1985.
145. SCHIERBECK, N. P. Ueber Kohlensäure im Ventrikel. *Skand. Arch. Physiol.* 3: 437–474, 1892.
146. SCHIESSEL, R., J. G. ALLISON, A. BARZILAI, L. A. FLEISCHER, J. B. MATTHEWS, A. MERHAV, J. SIMSON, AND W. SILEN. Failure of 16,16-dimethyl-prostaglandin E$_2$ to stimulate alkaline secretion in the isolated amphibian mucosa. *Gastroenterology* 78: 1513–1519, 1980.
147. SCHIESSEL, R., M. STARLINGER, E. KOVATS, W. APPEL, W. FEIL, AND A. SIMON. Alkaline secretion of rabbit duodenum in vivo: its dependence on acid base balance and mucosal blood flow. In: *Mechanisms of Mucosal Protection in the Upper Gastrointestinal Tract*, edited by A. Allen, G. Flemström, A. Garner, W. Silen, and L. A. Turnberg. New York: Raven, 1983, p. 267–272.
148. SELLING, J. A., D. L. HOGAN, A. ALY, M. A. KOSS, AND J. I. ISENBERG. Indomethacin inhibits duodenal mucosal bicarbonate secretion and endogenous prostaglandin output in human subjects. *Ann. Intern. Med.* 106: 368–371, 1987.
149. SHIAU, Y.-F., P. FERNANDEZ, M. J. JACKSON, AND S. MCMONAGLE. Mechanisms maintaining a low-pH microclimate in the intestine. *Am. J. Physiol.* 248 (*Gastrointest. Liver Physiol.* 11): G608–G617, 1985.
150. SILEN, W., AND S. ITO. Mechanism for rapid re-epithelialization of the gastric mucosal surface. *Annu. Rev. Physiol.* 47: 217–229, 1985.
151. SIMON, B., AND H. KATHER. PGI$_2$-sensitive human adenylate cyclase in biopsy specimens of corpus, antral and duodenal mucosa. *Digestion* 20: 111–114, 1980.
152. SIMSON, J. N. L., A. MERHAV, AND W. SILEN. Alkaline secretion by amphibian duodenum. I. General characteristics. *Am. J. Physiol.* 240 (*Gastrointest. Liver Physiol.* 3): G401–G408, 1981.
153. SIMSON, J. N. L., A. MERHAV, AND W. SILEN. Alkaline secretion by amphibian duodenum. II. Short-circuit current and Na$^+$ and Cl$^-$ fluxes. *Am. J. Physiol.* 240 (*Gastrointest. Liver Physiol.* 3): G472–G479, 1981.
154. SIMSON, J. N. L., A. MERHAV, AND W. SILEN. Alkaline secretion by amphibian duodenum. III. Effects of DBcAMP, theophylline, and prostaglandins. *Am. J. Physiol.* 241 (*Gastrointest. Liver Physiol.* 4): G528–G536, 1981.
155. SMEATON, L. A., B. H. HIRST, A. ALLEN, AND A. GARNER. Gastric and duodenal HCO$_3^-$ transport in vivo: influence of prostaglandins. *Am. J. Physiol.* 245 (*Gastrointest. Liver Physiol.* 8): G751–G759, 1983.
156. SPENNEY, J. G., G. FLEMSTRÖM, R. L. SHOEMAKER, AND G. SACHS. Quantitation of conductance pathways in antral gastric mucosa. *J. Gen. Physiol.* 65: 645–662, 1975.
157. SPENNEY, J. G., R. L. SHOEMAKER, AND G. SACHS. Conductance properties of gastric fundic mucosa. *J. Membr. Biol.* 19: 105–128, 1974.
158. STIEL, D., D. J. MURRAY, AND T. J. PETERS. Activities and subcellular localizations of enzymes implicated in gastroduodenal bicarbonate secretion. *Am. J. Physiol.* 247 (*Gastrointest. Liver Physiol.* 10): G133–G139, 1984.
159. SUGAI, N., AND S. ITO. Carbonic anhydrase, ultrastructural localization in the mouse gastric mucosa and improvements in the technique. *J. Histochem. Cytochem.* 6: 511–525, 1980.
160. SUNG, C. P., V. D. WIEBELHAUS, B. C. JENKINS, P. ADLERCREUTZ, B. I. HIRSCHOWITZ, AND G. SACHS. Heterogeneity of 3′,5′-phosphodiesterase of gastric mucosa. *Am. J. Physiol.* 223: 648–650, 1972.
161. SUZUKI, S., AND N. OZAKI. Mg^{2+}-HCO$_3^-$-ATPase and carbonic anhydrase in rat intestinal mucosa. *Experientia Basel* 39: 872–873, 1983.
162. ŚWIERCZEK, J. S., AND S. J. KONTUREK. Gastric alkaline response to mucosa-damaging agents: effect of 16,16-dimethyl prostaglandin E$_2$. *Am. J. Physiol.* 241 (*Gastrointest. Liver Physiol.* 4): G509–G515, 1981.
163. TABATA, K., E. D. JACOBSON, M. H. CHEN, R. F. MURPHY, AND S. N. JOFFE. Decrease in alkaline secretion during duodenal ulceration induced by mepirizole. *Gastroenterology* 87: 396–401, 1984.
164. TAKEUCHI, K., O. FURUKAWA, H. TANAKA, AND S. OKABE. Determination of acid-neutralizing capacity in rat duodenum. Influences of 16,16-dimethyl prostaglandin E$_2$ and nonsteroidal antiinflammatory drugs. *Dig. Dis. Sci.* 31: 631–637, 1986.

165. TAKEUCHI, K., O. FURUKAWA, H. TANAKA, AND S. OKABE. A new model of duodenal ulcers induced in rats by indomethacin plus histamine. *Gastroenterology* 90: 636–645, 1986.
166. TAKEUCHI, K., D. MAGEE, J. CRITCHLOW, J. MATTHEWS, AND W. SILEN. Studies of the pH gradient and thickness of frog gastric mucus gel. *Gastroenterology* 84: 331–340, 1983.
167. TAKEUCHI, K., A. MERHAV, AND W. SILEN. Mechanism of luminal alkalinization by bullfrog fundic mucosa. *Am. J. Physiol.* 243 (*Gastrointest. Liver Physiol.* 6): G377–G388, 1982.
168. TAKEUCHI, K., AND S. OKABE. Role of luminal alkalinization in repair process of ethanol-induced mucosal damage in rat stomach. *Dig. Dis. Sci.* 28: 993–1000, 1983.
169. TANAKA, H., Y. KUWAHARA, AND S. OKABE. Augmentation of cysteamine and mepirizole-induced lesions in the rat duodenum by histamine or indomethacin. *Jpn. J. Pharmacol.* 41: 545–549, 1986.
170. THOMAS, F. J., D. L. HOGAN, G. J. KREJS, M. ALGAZI, J. W. SACKMAN, AND J. I. ISENBERG. The effect of secretin and VIP on duodenal bicarbonate secretion in humans (Abstract). *Gastroenterology* 90: 1663, 1986.
171. TURNBERG, L. A. pH gradients in the gastroduodenal mucous gel. *Gastroenterol. Clin. Biol.* 9: 13–15, 1985.
172. VON KLEIST, D., H. BECHTEL, H. D. JANISCH, F. E. BAUER, AND K. E. HAMPEL. Effects of secretin on gastric pH and alkali secretion. *Innere Med.* 12: 133–135, 1985.
173. WHITTLE, B. J. R., G. L. KAUFFMAN, AND N. G. BOUGHTON-SMITH. Stimulation of gastric alkaline secretion by stable prostacyclin analogues in rat and dog. *Eur. J. Pharmacol.* 100: 277–283, 1984.
174. WILKES, J. M., A. GARNER, AND T. J. PETERS. Studies on the localization and properties of rat duodenal HCO_3^- ATPase with special relation to alkaline phosphatase. *Biochim. Biophys. Acta* 924: 159–166, 1987.
175. WILLIAMS, S. E., AND L. A. TURNBERG. Demonstration of a pH gradient across mucus adherent to rabbit gastric mucosa: evidence for a "mucus-bicarbonate" barrier. *Gut* 22: 94–96, 1981.
176. ZHOU, L., AND W. Y. CHEY. Electric acupuncture stimulates nonparietal cell secretion in dog. *Life Sci.* 34: 2233–2238, 1982.

CHAPTER 17

Immunological probes of gastrointestinal secretion

ADAM SMOLKA | *Center for Ulcer Research and Education, Department of Medicine, University of California, Los Angeles, California*

CHAPTER CONTENTS

Antibodies as probes
   Polyclonal antisera
   Monoclonal antibodies
   Immunocytochemistry
   Immunoassay
   Immunoblotting
   Immunoprecipitation
   Immunoaffinity chromatography
Salivary Secretion
Gastric Secretion
   Hydrochloric acid
   Pepsinogen
   Intrinsic factor
Pancreatic Secretion
Hepatobiliary Secretion
Colonic Secretion
Future Directions
   Anti-idiotypes
   Immunodissection
   Reverse hemolytic plaque assay
   Antibodies and molecular biology

GASTROINTESTINAL EXOCRINE SECRETIONS are formed by polarized epithelial cells responsive to neural and hormonal stimulants reacting with basolateral membrane receptors. This interaction leads to formation of inositol triphosphate and transient elevation of cellular calcium levels, both through opening basolateral calcium channels and through releasing calcium from intracellular stores (see the chapter by Forte and Soll in this *Handbook*). These events may be accompanied by internalization of the receptor-ligand complex. The increased cytosolic calcium level is responsible for the next phase of the secretory response, the mobilization of membrane-bound secretory granules, and their fusion with the apical membrane of the secretory cells. After exocytosis, the functional protein/lipid components of the secretory membrane are endocytically recovered from the apical membrane for lysosomal degradation or for reprocessing for future use. These pathways of membrane traffic are complemented by the biosynthetic pathway, whereby secretory products, lysosomal enzymes, and membrane proteins are transported from the endoplasmic reticulum to the Golgi complex and thence to secretory granules or lysosomes.

Every step of the secretory process entails interactions between molecules that in principle can elicit immune responses in appropriate hosts and can therefore give rise to antibodies to unique specificity. Once purified, such antibodies can be used as probes of the subcellular distribution and structure of the molecular antigens eliciting their formation. To date, the major applications of these antibodies in gastrointestinal secretion have been in immunocytochemical studies defining the morphology and anatomical location of secretory cells and in radioimmunoassay of various peptide hormones. The recent flowering of immunochemical techniques nurtured by advances in antibody production, notably hybridoma technology for synthesis of monoclonal antibodies, presents investigators with new approaches to the study of gastrointestinal secretion. In this chapter, the use of antibodies as probes of salivary, gastric, pancreatic, hepatic, and colonic exocrine secretion is discussed, and their potential for further clarification of these secretory mechanisms is assessed.

ANTIBODIES AS PROBES

The repertoire of immunological techniques available to the gastrointestinal physiologist supports both analytical and preparative applications. Immunocytochemistry, radioimmunoassay, enzyme-linked immunosorbent assay, immunoblotting, and immunoprecipitation are analytical methods that localize antigens within tissues and cells, quantitate those antigens, and provide structural and metabolic information. Each of these methods exists in several forms, distinguished usually by the manner in which presence of an antigen-antibody complex is signaled. Immunodissection of cells, affinity chromatography, and fluorescence-activated cell sorting are preparative methods allowing more than 10^7 cells and milligram quantities of purified antigen to be isolated. In this section the basic principles of the analytical approaches are described. Theoretical treatments of antigen-antibody interaction and detailed accounts of the immunological methods discussed here can be found in several comprehensive sources (15, 44, 77, 78, 92, 103, 176).

Polyclonal Antisera

Production of conventional antisera has been extensively reviewed (71). The immunogenicity of small ligands, for example, peptides or steroids, is usually augmented by conjugation to large carrier proteins such as keyhole limpet hemocyanin. For reasons of cost and convenience, rabbits are commonly used as hosts; because variations in the responses of individual rabbits to a given antigen may be varied, several animals are usually immunized. Typically, 1–10 mg of antigen emulsified in a 1:1 ratio with Freund's complete adjuvant are given subcutaneously or intramuscularly into multiple sites along the animal's back. High-affinity antibodies may result from immunization with very small amounts of antigen (50 μg), with the advantage of reducing the response to copurifying contaminants of the antigen (170). An alternative approach to generation of antigen-specific antisera calls for sodium dodecyl sulfate–polyacrylamide gel electrophoresis (SDS-PAGE) of the antigen. The protein band of interest is excised from the gel, homogenized, and used directly for immunization (27, 28). The strongly denaturing action of SDS may eliminate conformational antigenic determinants; the specificity of the resulting antisera is thus restricted to sequence determinants only. This approach is particularly valuable for generation of antibodies to membrane proteins, which are of necessity often probed in SDS-solubilized form and may therefore display determinants not accessible in the native configuration.

The polyclonal response is characterized by a multitude of antibodies, directed against all the epitopes of the immunizing antigen, with a wide range of affinities and specificities. Scatchard plots are consequently curvilinear, and many assays are dominated by the higher-affinity antibodies. Inactivation of sensitive subsets of antibodies by freeze-thaw cycles, lyophilization, chemical derivatization, or changes in salt concentration tends not to affect assays because sufficient insensitive antibodies remain. In polyclonal sera, antibodies against all impurities of the antigen are present. Such nonspecific antibodies can be removed by affinity purification of the serum using the immunizing antigen bound to a solid support, if the antigen is available in pure form (105). This approach, however, does not allow isolation of antibodies to specific single epitopes on the antigen.

Monoclonal Antibodies

In contrast to conventional methods for raising antisera that result in vast heterogeneity of antibody response, hybridoma techniques introduced in 1976 by Kohler and Milstein (85) allow the generation of homogeneous antibodies of unique affinity and specificity. After conventional immunization of mice, lymphocytes are recovered from the lymph nodes or spleen and fused with murine myeloma cells (54, 62, 85). The resulting immortalized "hybridoma" cell lines continue to secrete antibodies. Cloned hybridomas secreting homogeneous antibodies of interest are identified by appropriate screening assays (92, 176). Propagation of the clones in vitro (32) or as ascites tumors in syngeneic mice yields milligram quantities of monoclonal antibody, which may be further purified chromatographically on protein A–agarose (91), on anion-exchange resins (38), or by high-performance liquid chromatography (HPLC) (33).

As immunological probes, monoclonal antibodies have several advantages over polyclonal antisera. Most important, monoclonals are pure, homogeneous reagents with unique specificities. They are available in large quantities, allowing standardization of protocols among several laboratories. Because a hybridoma-secreting antibody with desired properties (such as protein A reactivity, good cytochemical staining, or a particular specificity) can be selected at will from the multiple hybridomas generated by somatic fusion, the immunizing antigen need not be pure. Indeed, crude cell homogenates may be used as antigen in the generation of monoclonal antibodies against cell-surface receptors. The paramount requirement is that an assay is available to recognize hybridoma clones secreting antibodies of the desired specificity (92, 176).

Monoclonal antibodies show linear Scatchard plots and a wider range of affinities (10^{-6}–10^{-11} M) than do polyclonal sera. Every monoclonal antibody has a characteristic stability to denaturation and change in affinity with temperature, pH, or salt concentration (102); in extreme cases, monoclonal antibodies may precipitate at low temperature or low salt concentration. An important caveat in the use of monoclonal antibodies concerns their specificity. Because a monoclonal antibody recognizes a single antigenic determinant on its target, any modification of that determinant may prevent binding of the antibody. The sensitivity of the antigen-antibody complex to radioiodination or partial denaturation of the antigen must therefore be established, and the possibility that genetic polymorphisms or glycosylation of the antigen may affect its determinants must be recognized.

Immunocytochemistry

Immunofluorescence microscopy was introduced by Coons et al. in 1941 and was the first technique to use covalently labeled antibodies as probes of cellular structure (34). Various enzymes and radioisotopes have since joined fluorescein isothiocyanate and other fluorescent probes as labels of antibodies used in immunocytochemistry. Briefly, a tissue section is incubated with specific antibody, washed in saline, and exposed to a second, labeled antibody directed against the first antibody. As a rule, the animal species in which the primary and secondary antibodies originate should differ from one another and from the species in which presence of antigen is being probed. Thus mouse monoclonal antibody binding to antigens on a rabbit tissue section might be detected using goat antimouse immunoglobulin (Ig) covalently linked to tetramethylrhodamine isothiocyanate, with visualization of the immune complexes by fluorescence microscopy with excitation at 515 nm. Alternatively, polyclonal goat antibodies binding to the same section might be detected with sheep antigoat Ig coupled to horseradish peroxidase (HRP). In this case the site of antibody binding on the section is visualized by addition of H_2O_2 and the chromogen diaminobenzidine, which precipitates in situ. Variations on this theme providing amplification of the signal include biotinylated primary antibody and detection with avidin-labeled HRP (5) and the peroxidase-antiperoxidase system (PAP) (163). A particular advantage of peroxidase immunocytochemistry is that diaminobenzidine forms an electron-dense complex with osmium tetroxide and is thus easily visualized in electron microscopy, allowing cellular ultrastructure to be studied. Other prominent immunoelectron-microscopic techniques used immunoglobulin-ferritin conjugates directly or as probes of primary antibody (122) and colloidal gold particles with adsorbed protein A, a surface antigen of *Staphylococcus aureus* with high affinity for the Fc region of Ig molecules (134).

Because protein A figures prominently in many immunochemical procedures (vide infra), a precautionary note is in order. Protein A does not bind avian IgG, human IgG3, rat IgG2a and IgG2b, and bovine and sheep IgG1. It binds weakly to mouse, rat, and goat IgG1. Protein G, an IgG-binding protein recently isolated from human group G streptococci, does not show these species or subclass restrictions and may therefore be a preferable reagent to protein A in certain applications (18).

An essential control for specificity of the primary antiserum is absorption of the antibody with excess purified antigen, with consequent loss of staining. Controls for nonspecific binding of the second labeled antibody are substitution of preimmune serum for primary antiserum and nonspecific hybridoma antibody for a monoclonal antibody.

Immunocytochemical analysis is critically dependent on the simultaneous preservation of cellular morphology and the antigenic determinants of interest. Often, however, optimal enhancement of morphology occurs under conditions detrimental to antigenic structure, and vice versa. As a matter of course, morphology and immunogenicity should be assessed in a series of fixatives (94): cold ethanol, paraformaldehyde, glutaraldehyde, periodate-lysine-paraformaldehyde (PLP) (105), or picric acid–paraformaldehyde (Zamboni's fixative) (162).

Immunoassay

Two forms of assay have come to dominate detection of very small amounts (picograms) of antigen. Radioimmunoassay (RIA) and enzyme-linked immunosorbent assay (ELISA) are based on antigen-antibody interaction and depend on measurement of exogenous label covalently coupled to one of the partners in the reaction. In RIA, initially described by Yalow and Berson in 1960 (181), equilibrium is established between limited specific antibody, radiolabeled standard antigen, and unknown antigen, and it is followed by separation of antibody-bound and free radiolabeled antigen. Because both forms of the antigen compete for antibody-binding sites, bound counts are inversely proportional to concentration of the unknown antigen. The antigen-antibody complex remains soluble at the low reactant concentrations typical of RIA, and many strategies have evolved to separate bound and free counts, including nonspecific adsorption or precipitation of the latter by charcoal, silica, polyethylene glycol, or saturated ammonium sulfate (reviewed in ref. 29); second antibody precipitation (70); and solid-phase separations (178). Radioiodine (^{125}I) antigen-labeling methods have been extensively reviewed (35, 70), as have procedures for optimizing the sensitivity, specificity, and precision of RIAs (29, 61). Although endocrinology was the first beneficiary of RIA techniques, with rapid development of steroid and protein hormone RIAs, virtually any biomolecule to which an antibody is available can be assayed by RIA, and indeed RIAs exist for many of the exocrine gastrointestinal secretions.

In the simplest form of ELISA (43), antigen is coupled irreversibly to test tubes or to the wells of polystyrene microtitration plates (Fig. 1). Coupling may be simply hydrophobic adsorption or adsorption through activation of the surface with glutaraldehyde. Remaining protein-binding sites are blocked by bovine serum albumin (BSA), and specific antiserum is added to the wells. After appropriate incubation, free antibody is removed by washing. Anti-Ig antibody covalently coupled to an indicator enzyme, most commonly AP (alkaline phosphatase) HRP, is added. After a further incubation and wash cycle, a chromogenic substrate is added, and appearance of the reaction product is monitored visually or spectrophotometrically.

The disadvantages of this form of ELISA lie in sensitivity of the antigen to passive or covalent coupling onto polystyrene surfaces, and for very dilute antigens in protein-rich media, preemption of binding sites on the surface by irrelevant proteins. These deficiencies are corrected in double-antibody ELISA and competitive-binding ELISA [Fig. 1; (179)]. In the

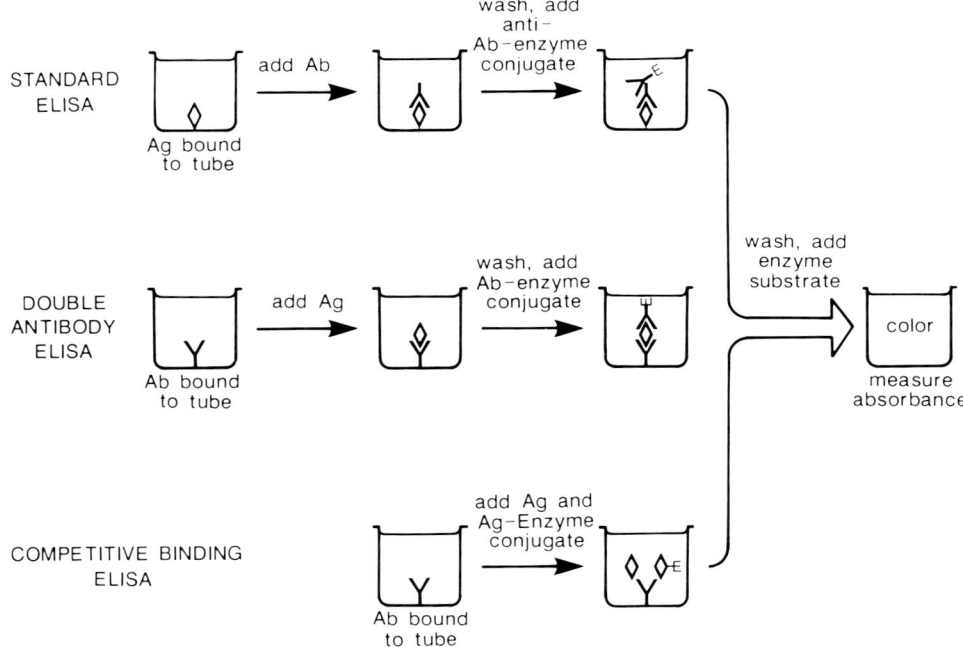

FIG. 1. Schematic diagram of 3 forms of enzyme-linked immunosorbent assay (ELISA). Ab, antibody; Ag, antigen.

former assay, primary antibody is immobilized in microtitration wells and incubated with the antigen. Primary antibody covalently coupled to HRP or AP is then added, and after incubation, washing, and addition of substrate, the color reaction is quantitated as described, the signal being proportional to antigen concentration. In the competitive-binding assay, primary antibody is again bound to wells and incubated with enzyme-linked standard antigen plus the "unknown" antigen. After washing, bound enzyme is assayed by substrate addition; in this case, the spectrophotometric signal is inversely proportional to sample concentration.

Complementing radioisotopes and enzymes as immunoassay labels are spin labels (45), luminescent compounds (121), metals (25), and fluorescent probes (65, 160). The latter probes show high sensitivity, and modulation of fluorescence in response to the environment of the label allows homogeneous assay, dispensing with the separation of reactants typical of heterogeneous assays.

Immunoblotting

Electrophoresis in conjunction with antibody binding gives rise to the immunoblotting technique (23). Cell homogenates or extracts are separated into constituent polypeptide bands by SDS-PAGE or isoelectric focusing (IEF). The band pattern is then transferred to a nitrocellulose membrane by electrophoresis, transfer being confirmed by staining part of the membrane with amido black 10B. The membrane is blocked with gelatin or BSA to minimize nonspecific binding of antibodies and then incubated with specific antibody against the antigen of interest. After washing in phosphate-buffered saline, the membrane is exposed to [^{125}I]protein A (10^6 counts·min^{-1}·ml^{-1}) or anti-Ig antibody conjugated to HRP (anti-Ig–HRP). Binding of [^{125}I]protein A to the nitrocellulose is detected by autoradiography, whereas anti-Ig–HRP binding is detected by addition of H_2O_2 and one of several chromogenic substrates. Immunoblotting of integral membrane proteins is subject to the limitation that the antigenic determinants must be stable to solubilization and denaturation, although the less stringent conditions of IEF may allow successful blotting when SDS-PAGE blotting has failed.

Immunoprecipitation

Study of synthesis and processing of components of the secretory pathway in whole cells or tissues or in cell-free translation systems often depends on immunoprecipitation for detection and quantitation of antigens (2). A requirement of this technique is that the antigen be radiolabeled; for cells, ^{125}I labeling of externally accessible cell surface proteins is possible (69, 111), whereas biosynthetic labeling of antigen, usually with [^{35}S]methionine, is used for cytoplasmic proteins in isolated cells maintained briefly in culture and for cell-free translation systems (124). Once labeled, cells are lysed and fractionated by differential centrifugation. If the antigen is membrane bound, solubilization is accomplished with 0.5% Nonidet P-40 or 1% Triton

X-100. The extract or appropriate cell fraction is preadsorbed with nonspecific antibody and anti-Ig antibody or protein A–agarose to minimize background. After incubation with specific antibody, either anti-Ig antibody or protein A–agarose is added. Heat-killed, formalin-fixed *Staphylococcus aureus* cells are also useful immune complex precipitants, again by virtue of the high-affinity binding between Fc regions of antibodies and protein A in *S. aureus* cell walls (82). The precipitated immune complexes are extensively washed and characterized by SDS-PAGE fluorography or chromatographic techniques. More complete elimination of nonspecifically adsorbed proteins may be accomplished by dissolution of the immunoprecipitate in SDS and reformation of the immune complex by addition of fresh antibody and protein A–agarose (125).

Immunoaffinity Chromatography

The specificity of antigen-antibody reactions allows many thousandfold purification of antigens by selection on chromatographic matrices bearing covalently coupled antibody (97). The coupling reaction must be mild enough to conserve the specific binding affinity of the antibody, and recovery of bound antigen must occur without compromising its biological activity. The former requirement is most often met by spontaneous coupling of antibody with CNBr-activated Sepharose 4B at pH 7.4; the resulting immunosorbent is stable to extremes of pH, detergents, and various dissociating reagents. Should steric effects hinder the antigen-antibody reaction, alternative matrices with long sidearms are available (79). Antigen-elution conditions are favored by lower antigen-antibody affinity constants (10^{-4} M to 10^{-8} M) and range from relatively mild salt dissociation (1.5 M LiCl or NaCl), through 3 M KSCN or 0.1 M glycine HCl, pH 3.0, to highly denaturing 5 M guanidine HCl. Eluted antigen should be neutralized immediately if necessary and dialyzed against suitable buffer.

SALIVARY SECRETION

Immunocytochemical methods have been prominent in mapping the distribution of secretory proteins in salivary glands. Amylase was first localized to human parotid acinar cells by Kraus and Mestecky (89) and confirmed by Korsrud and Brandtzaeg (86), who also found amylase, albeit with reduced staining intensity, in the serous acini and demilunes of the human submandibular gland. The latter study (86) found a granular cytoplasmic distribution of amylase with formaldehyde fixation and a diffuse distribution with ethanol fixation. This result was probably due to better preservation of granules with formaldehyde, emphasizing the necessity of trying several different fixatives in immunocytochemical studies. Korsrud and Brandtzaeg (86) also detected the iron-binding glycoprotein lactoferrin in parotid gland acini and, less frequently, in submandibular gland acini; in both glands, staining was most concentrated in the intercalated duct epithelium. The role of immunocytochemistry in mapping the cellular origin of secreted proteins is exemplified in Figure 2, from reference 86, which shows schematically by immunofluorescence the distribution of four proteins in human salivary glands.

Immunocytochemical localization of lysozyme in salivary glands, however, has been somewhat controversial. Lysozyme was first identified by immunofluorescence (89) and later by immunoperoxidase staining (83) in striated duct cells of the parotid gland. In contrast, Korsrud and Brandtzaeg (86) saw no lysozyme in striated duct cells; instead, with paired immunofluorescence staining, they localized lysozyme to intercalated duct cells in the parotid and submandibular glands (Fig. 2). Brandtzaeg (21) reported that formaldehyde fixation followed by pronase digestion of the tissue was necessary for lysozyme staining, and he attributed the earlier absence of lysozyme staining in intercalated ducts (83, 89) to antigen "masking."

Equally controversial is the immunocytochemical

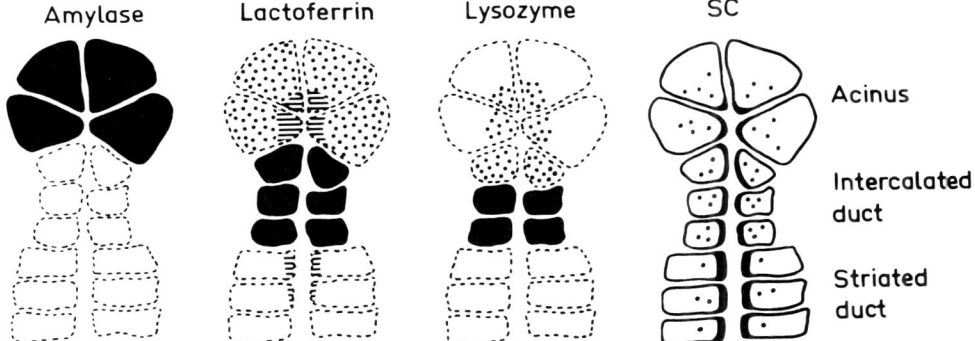

FIG. 2. Schematic diagram of distribution of various epithelial components in human major salivary glands, based on paired immunofluorescence staining. *Dotted, shaded,* and *black* areas indicate increasing occurrence and staining intensity. SC, secretory component. [From Korsrud and Brandtzaeg (86).]

localization of secretory component (SC). It has been reported in parotid acinar cells only (168), in parotid acinar and duct cells and in mucous submandibular acini (169), in serous submandibular acinar cells only (133), or in both ductal and serous acinar cells of both the parotid and submandibular gland (Fig. 2). These uncertainties cannot readily be resolved; variable fixation protocols and antisera in different laboratories make unique interpretation of immunocytochemical data difficult. Other salivary gland secretions mapped immunocytochemically include carbonic anhydrase (166), haptocorrin (119), and nerve growth factor and epidermal growth factor (115).

Immunoaffinity chromatography was used by Rajan et al. (130) to characterize proline-rich proteins in rabbit parotid saliva. The acidic proline-rich proteins of human saliva bind calcium and inhibit hydroxylapatite formation, preventing tooth dissolution and hydroxylapatite deposition (14, 63). Rajan et al. (130) bound antibodies against human acidic proline-rich protein C to CNBr-activated Sepharose 4B and applied rabbit parotid proteins collected by cannulation. A single peak of specifically bound protein was eluted with formic acid, and SDS-PAGE showed five proteins ranging from 19,000 M_r to 61,000 M_r in this peak, two of which consisted almost exclusively of glutamic acid, proline, and glycine by amino acid analysis.

Monoclonal antibodies were used by Barka et al. (8) in a study that illustrates well the application of several immunochemical techniques to salivary secretion. Spleen cells of mice immunized with rat submandibular saliva were fused with NS1 myeloma cells, giving rise to antibody-secreting hybridomas, identified by ELISA using rat salivary protein as the bound antigen. One of the purified monoclonal antibodies immunoprecipitated a 39,000-M_r protein from submandibular saliva, and reacted with a 39,000-M_r band in a salivary protein immunoblot. The antibody also immunoprecipitated a 39,000-M_r band from extracts of submandibular glands biosynthetically labeled with [^{3}H]leucine. A monoclonal antibody–Sepharose affinity column was used to purify a protein of the same relative molecular weight to near homogeneity, judged by SDS-PAGE, from rat submandibular gland extracts. Immunocytochemical analysis of rat submandibular glands showed localization of the antibody only in secretory granules of acinar cells and loss of staining with isoproterenol stimulation of salivary secretion. Despite the unknown identity and function of this salivary secretory protein, the study emphasizes the wealth of information acquired from purely immunological approaches to secretion. Further biochemical analysis of the purified protein, for example, proteolytic or antimicrobial activity, should define its physiological role in saliva, whereas monoclonal antibody screening of a rat salivary gland cDNA expression library will allow deduction of its amino acid sequence.

GASTRIC SECRETION

Hydrochloric Acid

Gastric acid secretion is mediated by a proton-translocating, K$^+$-dependent adenosine triphosphatase (H$^+$-K$^+$-ATPase) in the parietal cells of the gastric mucosa (137). Both polyclonal (136) and monoclonal antibodies (157) have been raised against this ATPase by using free-flow electrophoretically purified pig H$^+$-K$^+$-ATPase showing a single band at a molecular weight of 94,000 as the immunizing antigen. Of the five polyclonal rabbit antisera raised with this antigen (136), two antisera inhibited ATPase activity to a maximum of 80% and inhibited the K$^+$-pNPPase (p-nitrophenylphosphatase) activity by ~30%. The other three antisera were without effect on these activities. Immunocytochemically, all five antisera were localized in parietal cells of the pig gastric mucosa. Immunoelectron microscopy of rat gastric mucosa revealed one of the inhibitory antisera labeling secretory canalicular membranes of parietal cells, suggesting an apical membrane distribution of H$^+$-K$^+$-ATPase in these cells.

These results were confirmed and extended by using monoclonal antibodies (157). Hybridomas secreting antibodies against the H$^+$-K$^+$-ATPase were selected initially by ELISA, using the purified ATPase bound to polystyrene microwell plates as antigen, and then by indirect fluorescent immunocytochemical staining of pig parietal cells in ethanol-fixed, paraffin-embedded gastric mucosal sections. Immunoelectron microscopy of rabbit parietal cells with unstimulated morphology showed monoclonal antibody labeling tubulovesicles most intensely and rudimentary secretory canalicular membranes relatively weakly. In contrast, in parietal cells with stimulated morphology, the antibody labeled secretory canalicular membranes most intensely. This result was interpreted as confirming the transformation of tubulovesicles into secretory canaliculi on stimulation of acid secretion. In a later study by Smolka and Weinstein (158), cross-reactivity of anti–pig H$^+$-K$^+$-ATPase monoclonal antibodies with human parietal cells was demonstrated by light microscopy, immunoblotting confirmed that the monoclonal antibodies specifically labeled a 94,000-M_r polypeptide in gastric microsomes, and this specificity was used as the basis for immunoassay of the gastric ATPase in human biopsy specimens.

Pepsinogen

In 1966 Yasuda et al. (183) were the first to demonstrate binding of anti-pepsin antibodies to cytoplasmic granules in hog zymogen cells. Similar results were later obtained for the human gastric mucosa (67, 138, 141). On the basis of these studies, five types of peptic cells have been identified in humans: mucus neck cells and chief cells in fundic gland mucosa, pyloric gland cells in antral mucosa, cardiac gland cells in the gastric cardia, and Brunner's glands in the proximal duodenum. In addition, these studies have shown that pepsinogen I (PG I) is found only in chief cells and mucus neck cells (138), whereas pepsinogen II (PG II) is found in all five cell types. More recently, Meuwissen et al. (108) used a direct immunoperoxidase technique, in which HRP was covalently coupled to anti–PG I antibody to reveal abundant PG I and lesser amounts of PG II in fundic chief cells, whereas antral gland cells contained only PG II. In the same study, and in an earlier report by Reid et al. (131), dramatic reductions in PG I but not PG II were found in adenocarcinoma of the gastric corpus. Asymmetrical pepsinogen subclass distribution occurs also in the human gastric cardia (Fig. 3), PG II being the predominant pepsinogen in the deeper parts of the crypts.

Abnormal PG I and PG II distributions have been identified by immunofluorescence in several other gastrointestinal disorders. Varis et al. (172) showed that the clear-staining cells typical of pseudopyloric metaplasia in atrophic gastritis contain only PG II and that both pepsinogens are present in heterotopic gastric glands in the duodenum (171). Lechago et al. (93) showed PG I and PG II in chief cells and PG II in pyloric cells in ileal gastric metaplasia; both pepsinogens were found in heterotopic fundic glands in Meckel's diverticulum (96). These studies emphasize the value of immunocytochemistry in identifying functional and geographic perturbations of secretory cells in metaplastic tissue.

Some neonatal mammals, including ruminants, pigs, and cats, secrete in addition to pepsinogen another gastric proteinase called chymosin, which causes milk to coagulate (47). Edkins (41) proposed in 1898 that prochymosin was secreted by chief cells; in 1982, his proposal was confirmed by Andren et al. (3), who used indirect immunofluorescence to localize prochymosin and pepsinogen to the same cell type in bovine abomasal mucosa.

Apart from cellular and subcellular localization of

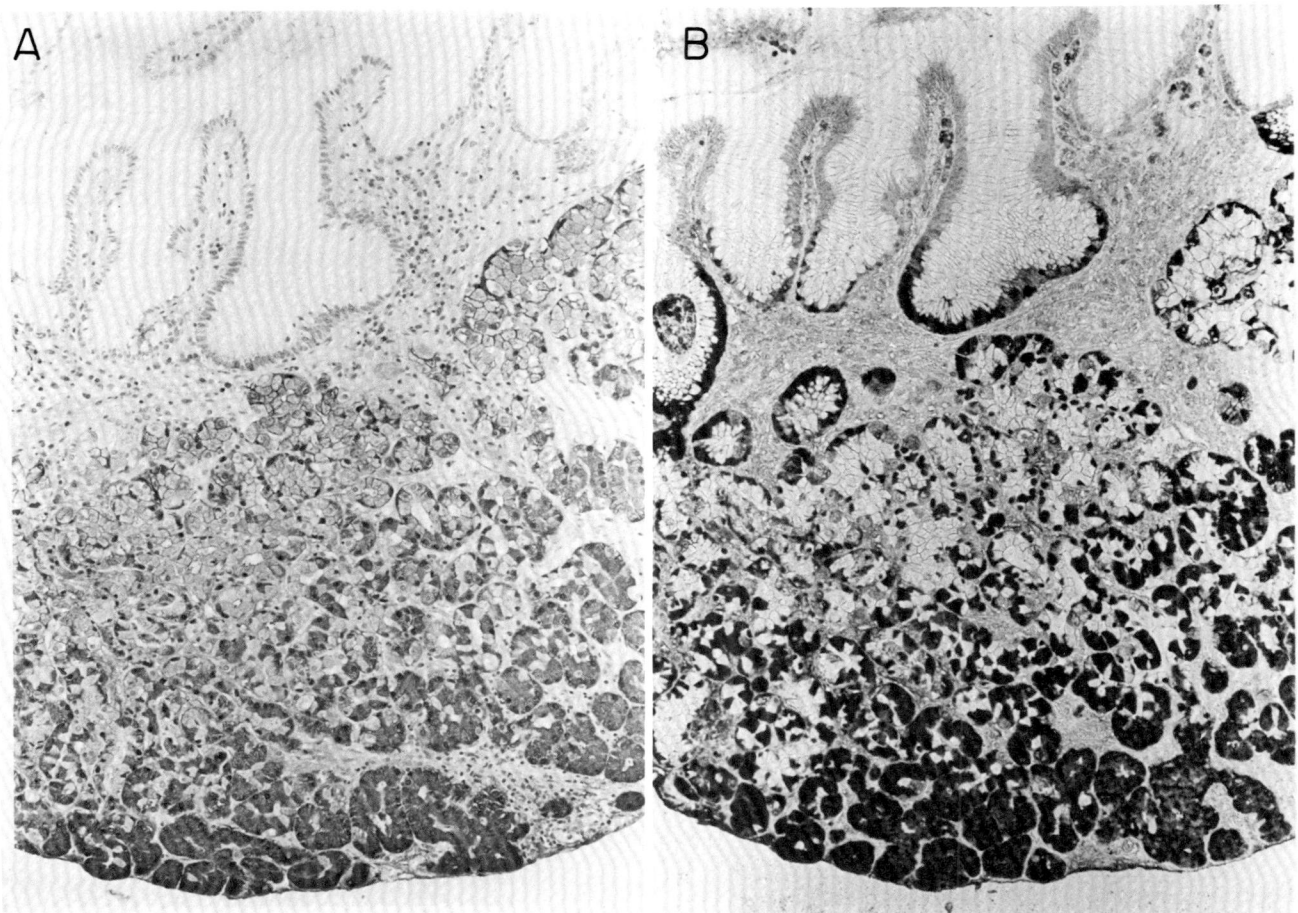

FIG. 3. Normal human gastric mucosa (cardia) stained with antibodies to (A) pepsinogen I (PG I) (× 75) and (B) pepsinogen II (PG II) (× 75) by the indirect immunoalkaline phosphatase technique. (Courtesy of Dr. Stephan Meuwissen.)

pepsinogen, antibodies to pepsinogens have been used to measure the zymogen in gastric juice and serum. Classically, serum pepsinogen has been measured by proteolytic assays, which cannot distinguish between PG I and PG II. Serum PG I levels are of interest, because they correlate well with peak acid output in the stomach (174).

Several RIAs of PG I in serum (7, 74, 139, 142, 175) and at least two RIAs of pepsin I in gastric juice (66, 173) have been reported. All these RIAs used the chloramine T method of ^{125}I incorporation into the PG I standard, polyclonal primary antisera, and the double-antibody technique for separation of bound and free counts. A hitherto unresolved problem of serum PG I RIAs concerns the source of the radiolabeled standard antigen. Measured serum PG I levels are twice as high when urinary PG I instead of gastric PG I is used as the standard. Although electrophoretic and immunochemical identity of urinary and gastric PG I has been suggested (143, 186), other factors such as differential susceptibility of PG I epitopes to iodination and presence of genetic variants could be responsible. Presumably RIAs based on monoclonal antibodies to defined epitopes on PG I would eliminate these uncertainties; to date, high-affinity monoclonals suitable for PG I RIA have not been available (16, 165).

To circumvent the use of radioisotopes intrinsic to RIA and the associated short lifetime of iodinated standards, Biemond et al. (16) developed an ELISA for serum PG I. Double-antibody sandwich and competitive binding ELISAs gave useful calibration curves with urinary PG I in the range 0.2–15 µg/l and 5–200 µg/l, respectively.

Studies of the prepepsinogens and preprochymosins of several species have depended heavily on specific antibodies to pepsinogen and chymosin. Thus Ichihara et al. (72, 73) translated total RNA from human, pig, rat, and calf gastric mucosae in wheat germ cell-free systems. Having specifically immunoprecipitated the major translation products with antipepsinogen or antichymosin antisera, Ichihara et al. (72, 73) resolved the products by SDS-PAGE, eluted them into dialysis bags, and obtained NH$_2$-terminal sequences by manual Edman degradation. Comparison of these sequences with the NH$_2$-terminal sequences of mature pepsinogen and prochymosin allowed deduction of the respective signal sequences.

Intrinsic Factor

In 1929, Castle (28) defined the principal defect in pernicious anemia to be lack of intrinsic factor (IF) in gastric juice. Subsequent studies have delineated the role of IF in intestinal absorbtion of vitamin B$_{12}$ (reviewed in ref. 39). More than 85% of patients with pernicious anemia have circulating antibodies against IF and other antigens of parietal cells (167), and these autoantibodies have been widely used to study IF secretion. Anti-IF autoantibodies show type I specificity, blocking formation of an IF-cobalamin complex (4, 146), or type II specificity, binding to the IF-cobalamin complex and preventing its attachment to the ileal receptor (140, 147). The first use of anti-IF autoantibodies in mapping of IF synthesis was by Hoedemaeker et al. (68), who localized cellular uptake of [^{57}Co]vitamin B$_{12}$ to parietal cells in the human and to chief cells in the rat and showed that this uptake could be blocked by a γ-globulin fraction of serum with anti-IF activity from pernicious anemia patients.

Immunofluorescent studies by Fisher and Taylor (46) in 1969 using autoantibodies localized IF to parietal cells in the human gastric mucosa. In elegant studies that cast light on parietal cell membrane recycling, Levine et al. (94) used peroxidase immunoelectron microscopy of normal human gastric biopsies with a monospecific anti–human IF antiserum. Intrinsic factor was found on membranes of the perinuclear envelope, rough endoplasmic reticulum (RER), Golgi, and tubulovesicles. Biopsies from pentagastrin-stimulated subjects, however, showed altered IF distributions (95). Shortly after stimulation, IF-staining tubulovesicles were found close to secretory canalicular membranes; 15 min after stimulation, IF was bound to canalicular membranes (Fig. 4), and after 1 h, 25% of the parietal cells had regained a resting morphology, with IF once again on basal RER and tubulovesicular membranes. This sequence of IF distribution is consistent with fusion of tubulovesicles with canalicular membrane after stimulation (50).

Smolka et al. (159) have prepared a panel of monoclonal antibodies against human IF. These antibodies bind human IF-cobalamin complex with affinities (K_d) of 13–116 nM, and two of them also block the binding of human IF–cobalamin complex to its ileal receptor. Three of the antibodies bind to rabbit IF–cobalamin complex, and when incubated with membrane suspensions from rabbit gastric glands biosynthetically labeled with [^{35}S]methionine and lysed with Triton X-100, these antibodies precipitated two peptides of 53,000 and 94,000 M_r, detectable by SDS–PAGE fluorography. Because the same two peptides were precipitated by anti–H$^+$-K$^+$-ATPase monoclonal antibodies, these results imply that IF and the gastric proton pump are localized in the same Triton-stable membrane in the rabbit parietal cell.

Immunoaffinity purification of IF was reported by Rothenberg in 1965 (135), although vitamin B$_{12}$–Sepharose affinity chromatography subsequently became the standard IF preparative method (1, 30). More recently, Shepherd et al. (154) purified human IF–vitamin B$_{12}$ complex from gastric juice by immunoadsorption to Sepharose CL-4B–protein A with covalently linked (177) type II antibody recovered from a pernicious anemia patient. The advantages of IF immunoaffinity purification are elimination of non-IF–vitamin B$_{12}$ binding proteins, and less stringent column elution.

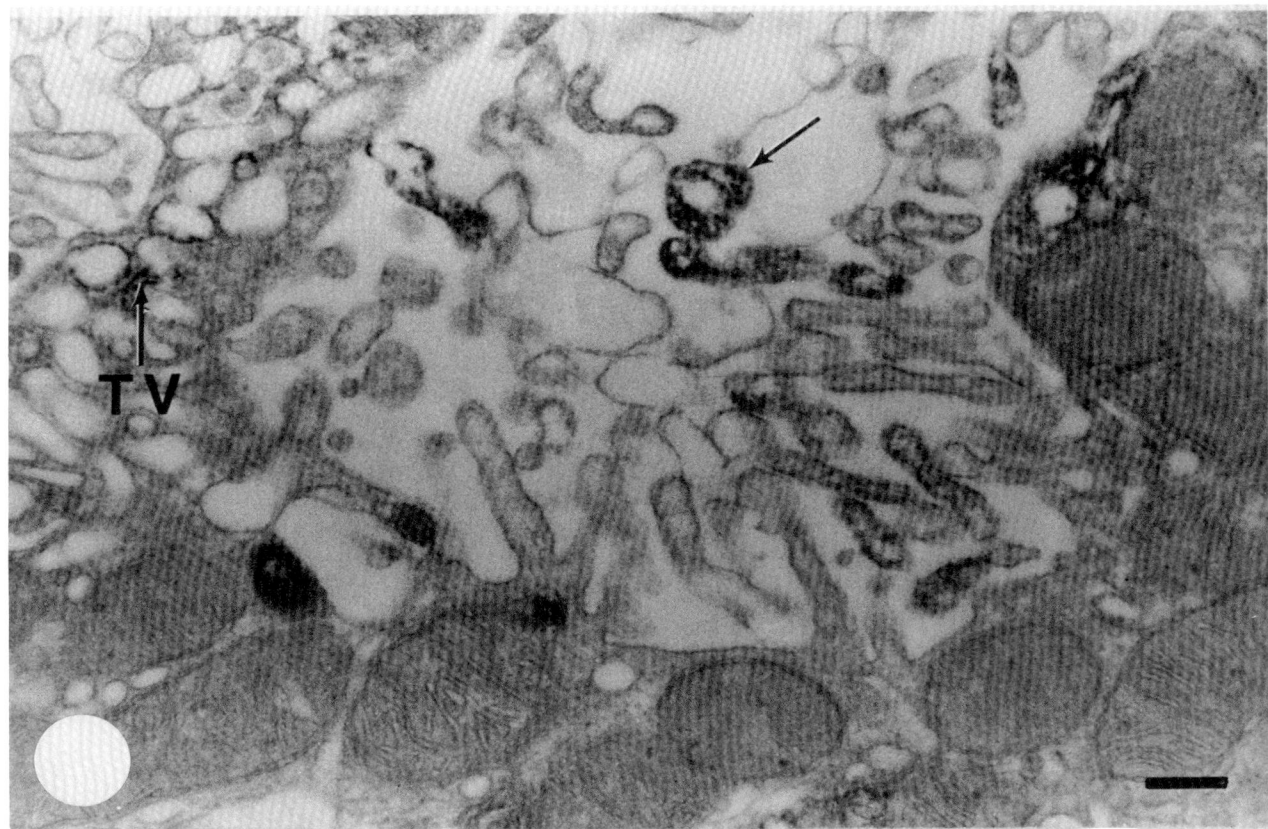

FIG. 4. Electron micrograph of secretory canalicular membrane of human parietal cell stained with antihuman intrinsic factor (IF) antibody by the indirect immunoperoxidase technique. The secretion was prepared from a gastric biopsy recovered 15 min after gastrin stimulation. Intrinsic factor is present on microvilli and on inner and outer membranes of cross-sectioned microvillus (*arrow*). TV, tubulovesicles. Bar, 0.2 µm. × 50,000. [From Levine et al. (95).]

PANCREATIC SECRETION

The exocrine pancreas secretes at least 19 enzymes (148). Because no enzyme histochemical methods have been available to study their localization, immunological methods have played a prominent role in defining their distribution and in unraveling their secretory pathways. Thus the study of pancreatic digestive enzyme distribution was among the early applications of the direct immunofluorescence technique (34). In 1954, Marshall (101) localized chymotrypsinogen, procarboxypeptidase, deoxyribonuclease (DNase), and ribonuclease (RNase) to zymogen granules in the apex of bovine pancreatic acinar cells; he proposed that neither acinar cells nor zymogen granules are specialized with respect to zymogen synthesis and storage. This hypothesis found qualitative support in the subsequent immunofluorescent studies of Ehinger (42), Yasuda and Coons (182), and Morikawa (110), which also added amylase to the list of enzymes sequestered in zymogen granules (182). Bendayan and Ito (11) localized nine enzymes (amylase, lipase, chymotrypsinogen A, trypsinogen, elastase, carboxypeptidase A and B, DNase, and RNase) to zymogen granules in the rat by using indirect immunofluorescence with specific anti–exocrine enzyme antisera. Additionally, Bendayan and Ito (11) showed a gradient of fluorescence in acinar cells, such that cells close to pancreatic islets were more intensely stained for most enzymes than cells farther away. Although the functional basis of this immunocytochemical asymmetry is unclear, the result is consistent with the division of exocrine pancreatic tissue into peri- and teleinsular regions in which acinar cells differ morphologically (64, 88) and biochemically (99).

With the introduction of pancreatic immunoelectron microscopy by Kraehenbuhl and Jamieson (87) in 1972, who used ferritin-labeled antirabbit IgG Fab fragments and rabbit antitrypsinogen antiserum, precise subcellular localization of pancreatic enzymes to the cisternae of the RER, zymogen granules, and secretory ducts became possible. Thus Geuze et al. (57) used antiamylase and antichymotrypsinogen and the immunoferritin technique with ultrathin frozen sections to identify these enzymes in the RER and the Golgi and to note an increase in their labeling intensity between these locations. The more quantitative nature of the protein A-gold technique [Fig. 5; (134)] allowed Bendayan et al. (13) to extend their light-microscopic study (11), showing that all the membrane-bound compartments implicated in the intracellular processing of secretory protein in the acinar cell contain at least nine different enzymes and that some of the enzymes undergo concentration as they migrate from the ER through the Golgi to the zymogen granules.

Whether various pancreatic enzymes migrate passively through the subcellular processing compartments or are selectively concentrated or even secreted is still controversial. Immunocytochemical evidence addressing this question has been provided by Geuze and Slot (56), who used both immunoferritin and fluorescence in rat and guinea pig exocrine pancreas. Their study showed that amylase, although prominent in the RER, did not appear to be concentrated in the intracisternal granules characteristic of pancreatic cisternal cavities (123), whereas chymotrypsinogen, only weakly reactive in the RER, was heavily concentrated in these granules. Furthermore, the increase in labeling intensity between RER and Golgi complex was more pronounced for chymotrypsinogen than for amylase.

Fluctuations of pancreatic secretory composition as a function of dietary stimuli have been well documented. For example, protein-rich diets stimulate secretion of proteolytic zymogens at the expense of amylase (127). Postuma et al. (128) used protein A-gold immunocytochemistry to study the effect of dietary soybean trypsin inhibitor (STI) in decreasing amylase relative to chymotrypsinogen (38), and they found that the majority of acinar cells contributed to the amylase/chymotrypsinogen shift after STI treatment and that the shift resulted both from decrease in amylase and parallel increase in chymotrypsinogen concentration in secretory granules. Equally significant was the correlation made by Postuma et al. (128) between immunochemical labeling intensity in their embedded sections and the antigen concentration in the various cell compartments, as measured biochemically. However, because of the multiplicity of factors affecting labeling efficiency in immunocytochemistry (penetrability, denaturation, and redistribution or extraction of antigens during fixation or subsequently), extreme caution should generally be exercised in assigning cellular concentrations to antigens on the basis of immunochemical staining intensity.

Shuttling of exocytotic granules between the Golgi apparatus and the plasma membrane of secretory cells

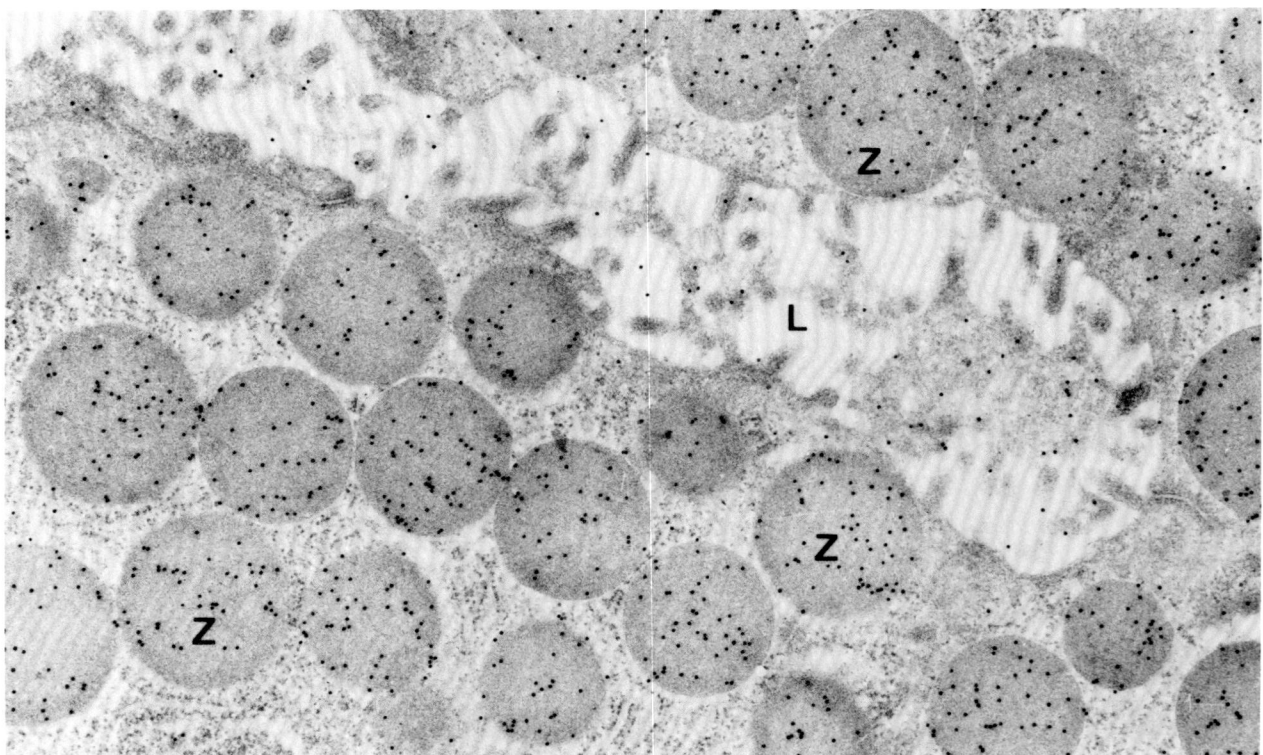

FIG. 5. Electron micrograph of rat pancreatic tissue stained with antiamylase antibody by the protein A-gold technique. Gold particles are clustered over the zymogen granules (Z) and the flocculent material in the acinar lumen (L). A few gold particles decorate the rough endoplastic reticulum. × 26,500. [From Roth et al. (134).]

has suggested possible involvement of contractile proteins, particularly actin (24). Drenckhahn et al. (40) used antibodies to localize myosin and actin in the apical area of acinar cells of rat salivary glands and exocrine pancreas at the light-microscopic level. More recently, the protein A–gold technique (134) was used to identify actin closely associated with the membranes of secretory granules and the Golgi complex of rat pancreatic acinar cells; the content of secretory granules, although not free of actin, was found to be less intensely labeled with antiactin antibody (12).

Another approach to the study of membrane recycling in pancreatic acinar cells suggested by Meldolesi (106) involves immunoelectron microscopy of zymogen granule membranes by using antibodies against one of the few proteins characteristic of these membranes. Again, caution must be exercised in deciding what is truly an integral membrane protein of zymogen granules; Scheffer et al. (149), using an antibody against GP2, a prominent zymogen granule membrane protein (22), showed that GP2 is also present in the granule content and is released to the pancreatic juice. A more general approach uses clathrin antibodies. Because clathrin is known to coat vesicles migrating between the ER, the Golgi complex, and the secretory membrane, antibodies to clathrin allow the isolation of such vesicles by immunoadsorption of secretory cell homogenates (81, 107). Subsequent separation of vesicle subpopulations should be possible over density gradients or by free-flow electrophoresis, allowing biochemical definition of vesicle membranes and content.

HEPATOBILIARY SECRETION

Of central interest in hepatic secretion are the mechanisms by which macromolecular serum components are actively and selectively transported from blood to bile, particularly the targeting of endocytic vesicles formed at the sinusoidal (basal lateral) membrane to either lysosomes or the canalicular (apical) membrane. In 1978, Mullock et al. (112) detected 16 proteins in rat bile using classic (31) and crossed (6) immunoelectrophoresis with polyclonal sera against rat serum and rat liver plasma membrane; only albumin, SC, IgA, and a bile lipoprotein were identified. Transport of IgA across hepatocytes against a 10-fold concentration gradient was shown by Jackson et al. (75), who perfused rat livers with exogenous rat monoclonal IgA and quantitated IgA in cannulated bile with a latex agglutination technique (26) and a polyclonal serum against an IgA idiotype. That IgA transport occurred in endocytic vesicles was confirmed in 1979 by Mullock et al. (113), who found exogenous [^{125}I]-IgA and endogenous IgA in vesicular components of liver homogenates, the latter IgA being quantitated by immunoelectrophoresis against antiserum to rat α-chain. Immunoelectrophoresis was also used by Mullock et al. (114) to localize SC to Golgi and sinusoidal membrane vesicles in rat liver homogenates. These findings are consistent with a model of rat IgA secretion based on interaction of serum IgA with hepatocyte sinusoidal membrane-bound SC, internalization of the IgA-SC complex in endocytic vesicles, transcellular migration and fusion of the vesicles with the canalicular membrane, and finally exocytotic secretion of IgA into the bile canaliculus.

Immunocytochemical studies of human liver have led to a somewhat different model for IgA secretion in this species. Nagura et al. (116) localized IgA and SC by the peroxidase-labeled antibody method to epithelial cells of intrahepatic and extrahepatic bile ductules. Immunoglobulin A and SC were not found in the cytoplasm or canalicular membranes of hepatocytes. Subsequently, Nagura et al. (117) argued, again on the basis of immunocytochemical staining, that human hepatic bile IgA originates in plasma cells in the biliary mucosa and is then transferred across ductular epithelial cells by SC-mediated endocytosis. Present evidence suggests therefore that cellular pathways for IgA secretion differ in the rat and the human.

Because serum concentrations of bile salts are elevated in many hepatobiliary disorders, their measurement is potentially a sensitive test of liver function. The first bile salt RIA was reported by Simmonds et al. (155) in 1973, based on rabbit antiserum to cholylglycine coupled to BSA. Procedures for setting up and validating RIAs of bile salts have been reviewed by Ross (132). The study of bile salt synthesis and metabolism has benefited greatly from RIAs for cholic (10), chenodeoxycholic (9), and tauro-β-muricholic acids (19), which are effective in isolated rat hepatocytes (48).

COLONIC SECRETION

A wealth of evidence assigns maintenance of mammalian potassium balance to the colon, where the epithelium both absorbs and secretes potassium (51). In addition, the colonic epithelium is the site of bicarbonate secretion, and it also elaborates and secretes mucus. In contrast to studies of salivary, gastric, pancreatic, and hepatobiliary exocrine secretion, relatively little use has so far been made of antibodies as immunological probes of these colonic secretory processes. Yeh and Chopra (184) used a rabbit antiserum raised against brush-border membranes of adult rat colon and indirect fluorescent labeling to verify the identity of cells propagated from dissociated colonic epithelia from suckling rats. No attempt was made to characterize the cell surface antigens eliciting the response. A K$^+$-ATPase isolated by Gustin and Goodman (59, 60) from rabbit colon showed antigenic cross-reactivity with monoclonal antibodies against the gastric proton ATPase, and immunocytochemistry of rat colonic sections with the same antibodies revealed weak labeling of surface epithelial cells (A. Smolka, unpublished observations). These results, together with biochemical data identifying a phosphorylated intermediate of the colonic ATPase (60), suggest that the two ATPases may be structurally and functionally similar. Further parallels between proton-pumping gastric parietal cells and K^+-transporting colonic epithelial cells are suggested by the presence of carbonic anhydrase (CA) in both cell types. Spicer et al. (161), using a rabbit antiserum against human erythrocyte CA and the indirect immunoperoxidase method, localized CA to superficial nongoblet cells in the human colon and appendix (Fig. 6). The same antibody gave strong immunocytochemical reactivity with gastric parietal cells, which are obliged to secrete bicarbonate basolaterally to offset proton secretion apically, and with pancreatic duct cells, gallbladder columnar cells, kidney distal and collecting tubule cells, and submandibular gland seromucous acinar cells, all of which have been shown to secrete bicarbonate. The staining of nongoblet cells for CA suggests that goblet cells themselves are not involved in ion secretion; however, the colonic cell type responsible for fluid and electrolyte secretion has not been identified.

With respect to colonic goblet cell structure and function, both immunocytochemical and RIA approaches have been provocative, underlying the need for further studies. Recent findings that kallikrein causes net chloride secretion in isolated descending colon of the cat (36, 99) have been complemented by immunocytochemical localization of the protease to cat and human colonic goblet cells (144, 145); whether kallikrein is involved in the presecretory processing of mucins or acts as a luminal kininogenase to promote secretion is unknown. For mucins, a study by Podolsky et al. (126) illustrates nicely how antibodies betray heterogeneity among otherwise indistinguishable subcellular structures. Having raised a panel of monoclonal antibodies to human colonic mucin, Podolsky et al. used indirect immunofluorescence to localize glycoproteins in human colonic goblet cells. Whereas some antibodies decorated mucin granules within goblet cells, others stained granule membranes, cytoplasm, or goblet cell apical membrane. Moreover, staining of goblet cell granule content by mucin subclass-specific antibodies showed that at least three mucin subclasses existed in mutually exclusive granules. Antibody-defined differences in granule mucin content also suggested the presence of seven distinct subpopulations of goblet cells in colonic mucosa. Qureshi et al. (129) used RIA with a rabbit antiserum to

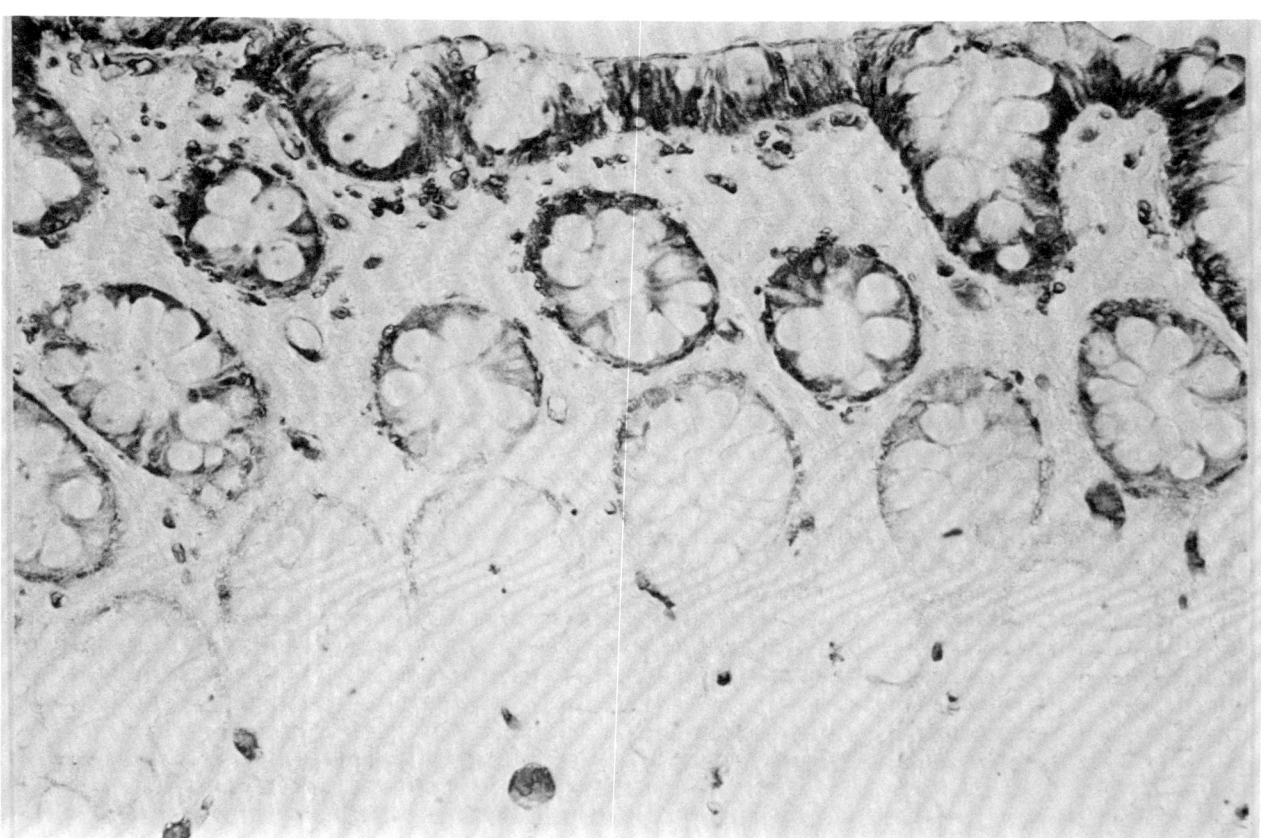

FIG. 6. Human colonic epithelium stained with anti–carbonic anhydrase antibody by the immunoperoxidase technique. Staining is limited to nongoblet columnar cells in superficial epithelium. Bar, 100 μm. × 250. [From Spicer et al. (161).]

human goblet cell mucin to show immunological identity between human ileal, rectal, and colonic mucins. The RIA allowed quantitation of mucin in the ileum and colon, could be used as a marker for mucin purification, and revealed partial cross-reactivities among mucins of different species. The same RIA allowed Forstner et al. (49) to show a dose-dependent increase in mucin secretion in rat ileum in vitro in response to cholera toxin; more recently, the same group has used solid-phase RIA of human ileal mucin and immunoblotting to identify nonglycosylated conformation-dependent antigenic sites on human mucin (100).

FUTURE DIRECTIONS

Immunohistochemical techniques, both light and electron microscopic, will undoubtedly continue to dominate morphological studies of gastrointestinal secretion, and for gastric and colonic secretion they may become more prominent as new antibodies against proton ATPases and intrinsic factor become available. Although the other immunological methods discussed here have only recently, and infrequently, been applied to questions of gastrointestinal secretion, they clearly provide cell-specific and product-specific information not accessible by other means. In addition to these, several novel immunological approaches not yet applied to gastrointestinal secretion may be equally promising.

Anti-idiotypes

Immunoglobulins share with other molecules the capacity to elicit the formation of antibodies when administered to suitable animal hosts (156). The heavy- and light-chain variable regions of immunoglobulins are sites of antigen binding and exhibit various antigenic determinants known individually as idiotopes; a complete antigen-binding site is called an idiotype. Occasionally, anti-idiotypic antibodies may have conformations similar to the original immunogen and may compete with that immunogen for the idiotype. Sege and Peterson (152) showed in 1978 that an anti-idiotypic antibody raised against antibody to retinol-binding protein bound to the retinol-binding receptor on human prealbumin. Should the immunogen be a membrane-receptor ligand, for example, a peptide hormone, the anti-idiotype may even mimic the physiological effects of the ligand. Thus Shechter et al. (150) demonstrated that anti-idiotypes formed in response to circulating anti-insulin antibodies in mice blocked binding of insulin to fat cell receptors, stimulated glucose oxidation, and inhibited lipolysis. These and similar studies of cell surface receptors, for example, β-adrenergic receptor (151), T cell antigen receptor (17), and mammalian reovirus receptor (80), suggest potential roles for anti-idiotypic antibodies as probes of secretory cell hormone receptors, and particularly in the study of potential autoimmune defects in gastrointestinal secretion.

Immunodissection

The irreversible hydrophobic adsorption of antibody molecules to polystyrene surfaces allows solid-phase immunoabsorption of specific cells to these surfaces. This technique was originally applied to the purification of lymphocyte subpopulations (180) and later to isolation of cortical collecting tubule cells (55). The method is based on brief incubation of heterogeneous single-cell suspensions in polystyrene Petri dishes coated with antibody against a surface antigen of the target cell. Subsequent washing leaves target cells bound to the dishes, where they can be left to grow in an appropriate culture medium, or removed by trypsinization for further biochemical or biophysical characterization. The requirement of this technique for milligram amounts of cell-surface–specific antibodies virtually dictates a hybridoma approach, with immunization of mice with whole cells and subsequent screening of hybridoma supernatants by immunofluorescent labeling of specific target cells in cytocentrifuge preparations. The high capacity of immunodissection (10^7 cells) may be of great value in studying gastrointestinal secretory cells, which in the intact tissue are always dispersed among other cell types and which are in any case difficult to maintain in culture. Antibody-mediated attachment of cells to a substratum as in immunodissection may overcome a major problem of gastrointestinal cell culture, although a greater advantage of immunodissection, again related to its high capacity, may be the opportunity it provides to characterize biochemically constituents of the secretory pathway, such as receptors, which are present in few copies per cell. For example, two-dimensional electrophoresis of whole-cell lysates is sensitive enough to detect modulations in the level of single peptides in response to physiological stimuli (120); such analysis of immunodissected gastric parietal cells after histamine or omeprazole administration would contribute to an understanding of the intracellular pathways of acid secretion.

Reverse Hemolytic Plaque Assay

Secretory products of single cells can be identified and quantitated by the reverse hemolytic plaque assay (RHPA) (109, 118). A mixed monolayer of secreting cells and protein A–coated red blood cells (RBC) is incubated with antibody against the secretory product and complement. The antigen-antibody complex is bound by the protein A–RBC, and complement then mediates RBC lysis, forming a clear hemolytic plaque around the secreting cell. The rate of formation of plaques is proportional to the cellular secretory rate (76), and plaque diameter is related to the amount of product secreted (53). To date, the major applications of RHPA have been in studies of hormone release from pituitary cells (reviewed in ref. 52). These studies have uncovered functional differences among hormone-secreting cells of the same type, have identified cells secreting at least two different hormones, and have clarified the ontogeny of such multiple secretors in the rat. The assay is not in principle limited to measurement of endocrine cell secretion; possible applications in gastrointestinal secretion are *1*) study of the apparent cosecretion of intrinsic factor and pepsinogen by rat gastric chief cells and *2*) further clarification of the parallel-nonparallel secretion controversy surrounding pancreatic and salivary acinar cells.

Antibodies and Molecular Biology

The first step in deducing a polypeptide primary sequence by molecular cloning methods is purification of the specific messenger RNA (mRNA) expressed in vivo and formation of a complementary DNA (cDNA) library. However, low-abundance mRNAs are often inefficiently recovered by classic mRNA purification techniques. Immunopurification of mRNA is based on antibody recognition of nascent polypeptide chains still bound to polyribosomes; detailed procedures based on solid-phase immunoadsorbents and allowing recovery of mRNAs representing only 0.02% of total mRNA have been described by Shapiro and Young (153) and Boyer et al. (20). If the cDNA library is formed by using the bacteriophage vector λgt11, whereby the cDNA is inserted into the viral β-galactosidase gene lacZ, cDNA clones of interest can be identified by expressing lacZ fusion proteins in *Escherichia coli* and probing a replica of the plaque pattern on nitrocellulose with specific antibodies. This approach has been pioneered by Young and Davis (185) and is similar to the immunoblotting technique described earlier. Should it not prove possible, however, to use an expression vector for the cDNA library, antibodies against the target protein are still indirectly valuable as probes of the library. By means of immunoaffinity chromatography or immunoprecipitation, the protein or proteolytic fragments thereof are purified to homogeneity and automatically microsequenced. From the sequence one can deduce and synthesize a family of oligodeoxynucleotide probes. After appropriate labeling, these probes can be used to screen a plasmid cDNA library (164). Finally, antibodies are indispensable probes of in vitro mRNA translation either in rabbit reticulocyte lysates (124) or in *Xenopus* oocytes (90). The importance of these assays is that proteins implicated in secretory processes can be followed biochemically throughout their biosynthetic pathway, for example, from their origin on ribosomes, through the processes of membrane targeting and insertion, and ultimately to their expression in the secretory membrane.

Studies in this chapter performed at the Center for Ulcer Research and Education, Department of Medicine, University of California at Los Angeles, were supported in part by Public Health Service Grant AM-34092.

REFERENCES

1. ALLEN, R. H., AND C. S. MEHLMAN. Isolation of gastric vitamin B_{12} binding proteins using affinity chromatography. Purification and properties of human intrinsic factor. *J. Biol. Chem.* 248: 3660–3669, 1973.
2. ANDERSON, D. J., AND G. BLOBEL. Immunoprecipitation of proteins from cell-free translations. *Methods Enzymol.* 96: 111–120, 1983.
3. ANDREN, A., B. LENNART, AND O. CLAESSON. Immunohistochemical studies on the development of prochymosin- and pepsinogen-containing cells in bovine abomasal mucosa. *J. Physiol. Lond.* 327: 247–254, 1982.
4. ARDEMAN, S., AND I. CHANARIN. A method for the assay of human gastric intrinsic factor and for the detection of titration of antibodies against intrinsic factor. *Lancet* 2: 1350–1354, 1963.
5. AVRAMEAS, S., AND T. TERNYNCK. Peroxidase-labeled antibody and Fab conjugates with enhanced intracellular penetration. *Immunochemistry* 8: 1175–1179, 1971.
6. AXELSEN, N. H., J. KROLL, AND B. WEEKE. *A Manual of Quantitative Immunoelectrophoresis.* Oslo: Universitetsforlaget, 1973.
7. AXELSSON, C. K., M. DAMKJAER-NIELSEN, AND A. M. KAPPELGAARD. Solid-phase double-antibody radioimmunoassay of pepsinogen I in serum. *Clin. Chim. Acta* 121: 309–319, 1982.
8. BARKA, T., E. W. GRESIK, AND H. VAN DER NOEN. Monoclonal antibodies against rat saliva and salivary gland antigens. *J. Histochem. Cytochem.* 33: 209–218, 1985.
9. BECKETT, G. J., J. E. T. CORRIE, AND I. W. PERCY-ROBB. The preparation of ^{125}I-labelled bile acid ligands for use in the radioimmunoassay of bile acids. *Clin. Chim. Acta* 93: 145–150, 1979.
10. BECKETT, G. J., W. M. HUNTER, AND I. W. PERCY-ROBB. Investigations into the choice of immunogen, ligand, antiserum, and assay conditions for the radioimmunoassay of conjugated cholic acid. *Clin. Chim. Acta* 88: 257–266, 1978.
11. BENDAYAN, M., AND S. ITO. Immunohistochemical localization of exocrine enzymes in normal rat pancreas. *J. Histochem. Cytochem.* 27: 1029–1034, 1979.
12. BENDAYAN, M., N. MARCEAU, A. R. BEAUDOIN, AND J. M. TRIFARO. Immunocytochemical localization of actin in the pancreatic exocrine cell. *J. Histochem. Cytochem.* 30: 1075–1078, 1982.
13. BENDAYAN, M., J. ROTH, A. PERRELET, AND L. ORCI. Quantitative immunocytochemical localization of pancreatic secretory proteins in subcellular compartments of the rat acinar cell. *J. Histochem. Cytochem.* 28: 149–160, 1980.
14. BENNICK, A., A. C. MCLAUGHLIN, A. A. GREY, AND G. MADDAPALLIMATTAM. The location and nature of calcium-binding sites in salivary acidic proline-rich phosphoproteins. *J. Biol. Chem.* 256: 4741–4746, 1981.
15. BERZOFSKY, J. A., AND I. J. BERKOWER. Antigen-antibody interactions. In: *Fundamental Immunology*, edited by W. E. Paul. New York: Raven, 1976, p. 595–644.
16. BIEMOND, I., G. PALS, J. P. GILIAMS, J. KREUNING, AND A. S. PEÑA. Enzyme linked immunosorbent assay of pepsinogen. I. *Prog. Clin. Biol. Res.* 173: 55–65, 1985.
17. BINZ, H., AND H. WIGZELL. Shared idiotypic determinants on B and T lymphocytes reactive against the same antigenic determinants. V. Biochemical and serological characteristics of naturally occurring antigen-binding T lymphocyte-derived

molecules. *Scand. J. Immunol.* 5: 559–571, 1976.
18. BJÖRCK, L., AND G. KRONVALL. Purification and some properties of streptococcal protein G, a novel IgG-binding reagent. *J. Immunol.* 133: 969–974, 1984.
19. BOTHAM, K. M., G. S. BOYD, D. WILLIAMSON, AND G. J. BECKETT. A radioimmunoassay for tauro-β-muricholic acid suitable for use with isolated rat liver cells. *FEBS Lett.* 151: 19–21, 1983.
20. BOYER, S. H., K. D. SMITH, A. N. NOYES, AND K. E. YOUNG. Adjuvants to immunological methods for mRNA purification. *J. Biol. Chem.* 258: 2068–2071, 1983.
21. BRANDTZAEG, P. Prolonged incubation time in immunohistochemistry: effects on fluorescence staining of immunoglobulins and epithelial components in ethanol and formaldehyde-fixed paraffin-embedded tissues. *J. Histochem. Cytochem.* 29: 1302–1315, 1981.
22. BROCKMEYER, T. F., AND G. E. PALADE. A major glycoprotein in zymogen granule membranes (Abstract). *J. Cell Biol.* 83: 272a, 1979.
23. BURNETTE, W. N. Western blotting: electrophoretic transfer of proteins from sodium dodecyl sulfate-polyacrylamide gels to unmodified nitrocellulose and radiographic detection with antibody and radioiodinated protein A. *Anal. Biochem.* 112: 195–203, 1981.
24. BURRIDGE, K., AND J. H. PHILLIPS. Association of actin and myosin with secretory granule membranes. *Nature Lond.* 254: 526–529, 1975.
25. CAIS, M., S. DANI, Y. EDEN, O. GANDOLFI, M. HORN, E. E. ISSACS, Y. JOSEPHY, Y. SARR, E. SLOVIN, AND L. SNARSKY. Metalloimmunoassay. *Nature Lond.* 270: 534–535, 1977.
26. CAMBASIO, C. L., A. E. LEEK, F. DE STEENWINKEL, J. BILLEN, AND P. L. MASSON. Particle counting immunoassay (PACIA). I. A general method for the determination of antibodies, antigens, and haptens. *J. Immunol. Methods* 18: 33–44, 1977.
27. CARROL, R. B., S. M. GOLDFINE, AND J. A. MEHERO. Antiserum to polyacrylamide gel-purified simian virus T40 antigen. *Virology* 87: 194–198, 1978.
28. CASTLE, W. B. Observations on the etiologic relationship of achylia gastrica to pernicious anemia. I. Effect of administration to patients with pernicious anemia of contents of normal human stomach recovered after ingestion of beef muscle. *Am. J. Med. Sci.* 178: 748–764, 1929.
29. CHARD, T. *An Introduction to Radioimmunoassay and Related Techniques.* Amsterdam: North-Holland, 1978.
30. CHRISTENSEN, J. M., E. HIPPE, H. OLESON, M. RYE, E. HABER, L. LEE, AND J. THOMSEN. Purification of human intrinsic factor by affinity chromatography. *Biochim. Biophys. Acta* 303: 319–332, 1973.
31. CLAUSEN, J. Immunochemical techniques for the identification and estimation of macromolecules. In: *Laboratory Techniques in Biochemistry and Molecular Biology*, edited by T. S. Work and E. Work. Amsterdam: North-Holland, 1969, vol. 1, p. 399–572.
32. CLEVELAND, W. L., Z. STEPLEWSKI, AND H. KOPROWSKI. Production of monoclonal antibodies in serum free medium. *J. Immunol. Methods* 39: 369–375, 1985.
33. CLEZARDIN, P., J. L. MCGREGOR, M. MANACH, H. BOUKERCHE, AND M. DECHAVANNE. One-step procedure for the rapid isolation of mouse monoclonal antibodies and their antigen binding fragments by fast protein liquid chromatography on a mono Q anion-exchange column. *J. Chromatogr.* 319: 67–77, 1985.
34. COONS, A. H., H. J. CREECH, AND R. N. JONES. Immunological properties of an antibody containing a fluorescent group. *Proc. Soc. Exp. Biol. Med.* 47: 200–202, 1941.
35. CORRIE, J. E. T., AND W. M. HUNTER. [^{125}I]-iodinated tracers for haptene-specific radioimmunoassays. *Methods Enzymol.* 73: 79–112, 1981.
36. CUTHBERT, A. W., AND H. S. MARGOLIUS. Kinins stimulate net chloride secretion by the rat colon. *Br. J. Pharmacol.* 75: 587–598, 1982.
37. DIJKHOF, J., S. R. PORT, AND C. POORT. Effect of feeding soybean flour-containing diets on the protein synthetic pattern of the rat pancreas. *J. Nutr.* 107: 1985–1995, 1977.
38. DISSANAYAKE, S., AND F. C. HAY. Isolation of pure normal IgGl from mouse serum. *Immunochemistry* 12: 101–103, 1975.
39. DONALDSON, R. M., JR. Intrinsic factor and the transport of cobalamin. In: *Physiology of the Gastrointestinal Tract* (1st ed.), edited by L. R. Johnson. New York: Raven, 1981, p. 641–658.
40. DRENCKHAHN, D., U. GROSCHEL-STEWART, AND K. UNSICKER. Immunofluorescence-microscopic demonstration of myosin and actin in salivary glands and exocrine pancreas of the rat. *Cell Tissue Res.* 183: 273–279, 1977.
41. EDKINS, J. S. Mechanism of secretion of gastric, pancreatic, and intestinal juices. In: *Textbook of Physiology*, edited by E. A. Schafer. Edinburgh: Young J. Pentland, 1898, vol. 1, p. 531–558.
42. EHINGER, B. The cytological localization of rat pancreas ribonuclease by means of fluorescent antibodies. *Histochemie* 5: 145–153, 1965.
43. ENGVALL, E., AND P. PERLMANN. Enzyme-linked immunosorbent assay (ELISA). Quantitative assay of immunoglobulin G. *Immunochemistry* 8: 871–879, 1971.
44. ERLANGER, B. F. The preparation of antigenic hapten-carrier conjugates: a survey. *Methods Enzymol.* 70: 85–104, 1980.
45. ESSER, A. F. Principles of electron-spin resonance assay and immunologic applications. *Lab. Res. Methods Biol. Med.* 4: 213–233, 1980.
46. FISHER, J. M., AND K. B. TAYLOR. The intracellular localization of Castle's intrinsic factor by an immunofluorescent technique using autoantibodies. *Immunology* 16: 779–784, 1969.
46. FOLTMAN, B., AND N. H. AXELSEN. Gastric proteinases and their zymogens. Phylogenetic and developmental aspects. In: *Trends in Enzymology: Enzyme Regulation and Mechanism of Action*, edited by P. Mildner and B. Ries. Oxford, UK: Pergamon, 1980, vol. 60, p. 271–280.
48. FORD, R. P., K. M. BOTHAM, K. E. SUCKLING, AND G. S. BOYD. Characterization of rat hepatocyte monolayers for investigation of the metabolism of bile salts. *Biochim. Biophys. Acta* 836: 185–191, 1985.
49. FORSTNER, J. F., N. W. ROOMI, R. E. F. FAHIM, AND G. G. FORSTNER. Cholera toxin stimulates secretion of immunoreactive intestinal mucin. *Am. J. Physiol.* 240 (*Gastrointest. Liver Physiol.* 3): G10–G16, 1981.
50. FORTE, T. M., T. E. MACHEN, AND J. G. FORTE. Ultrastructural changes in oxyntic cells associated with secretory function: membrane recycling hypothesis. *Gastroenterology* 73: 941–955, 1977.
51. FOSTER, E. S., J. P. HAYSLETT, AND H. J. BINDER. Mechanism of active potassium absorption and secretion in the rat colon. *Am. J. Physiol.* 246 (*Gastrointest. Liver Physiol.* 9): G611–G617, 1984.
52. FRAWLEY, L. S., F. R. BOOCKFOR, AND J. P. HOEFFLER. Focusing on hormone release. *Nature Lond.* 321: 793–794, 1986.
53. FRAWLEY, L. S., J. P. HOEFFLER, AND F. R. BOOCKFOR. Functional maturation of somatotropes in fetal rat pituitaries: analysis by reverse hemolytic plaque assay. *Endocrinology* 116: 2355–2360, 1985.
54. GALFRE, G., AND C. MILSTEIN. Preparation of monoclonal antibodies: strategies and procedures. *Methods Enzymol.* 73: 3–46, 1981.
55. GARCIA-PEREZ, A., AND W. L. SMITH. Use of monoclonal antibodies to isolate cortical collecting tubule cells: AVP induces PGE release. *Am. J. Physiol.* 244 (*Cell Physiol.* 13): C211–C220, 1983.
56. GEUZE, H. J., AND J. W. SLOT. Disproportional immunostaining patterns of two secretory proteins in guinea pig and rat exocrine pancreatic cells. An immunoferritin and fluorescence study. *Eur. J. Cell Biol.* 21: 93–100, 1980.
57. GEUZE, H. J., J. W. SLOT, AND K. T. TOKUYASU. Immunocytochemical localization of amylase and chymotrypsinogen in the exocrine pancreatic cell with special attention to the

Golgi complex. *J. Cell Biol.* 82: 697–707, 1979.
58. GRANGER, B. L., AND E. LAZARIDES. Desmin and vimentin coexist at the periphery of the myofibril z disc. *Cell* 18: 1053–1063, 1979.
59. GUSTIN, M. C., AND D. B. P. GOODMAN. Isolation of brush-border membrane from the rabbit descending colon epithelium. Partial characterization of a unique K-activated ATPase. *J. Biol. Chem.* 256: 10651–10656, 1981.
60. GUSTIN, M. C., AND D. B. P. GOODMAN. Characterization of the phosphorylated intermediate of the K-ouabain-insensitive ATPase of the rabbit colon brush-border membrane. *J. Biol. Chem.* 257: 9629–9633, 1982.
61. HALFMAN, C. J. Concentrations of binding protein and labelled analyte that are appropriate for measuring at any analyte concentration range in radioimmunoassays. *Methods Enzymol.* 74: 481–497, 1981.
62. HAMMERLING, G. J., U. HAMMERLING, AND J. F. KEARNEY. *Monoclonal Antibodies and T-Cell Hybridomas.* Amsterdam: Elsevier, 1981.
63. HAY, D. I., AND E. C. MORENO. Macromolecular inhibitors of calcium phosphate precipitation in human saliva. Their roles in providing a protective environment. In: *Saliva and Dental Caries*, edited by I. Kleinberg, S. A. Ellison, and I. D. Mandel. New York: Information Retrieval, 1979, p. 45–48.
64. HELLMAN, B., A. WALLGREN, AND B. PETERSSON. Cytological characteristics of the exocrine pancreatic cells with regards to their position in relation to the islets of Langerhans. *Acta Endocrinol.* 39: 465–473, 1962.
65. HEMMILA, I. Fluoroimmunoassays and immunofluorometric assays. *Clin. Chem.* 31: 359–370, 1985.
66. HENGELS, K. J. Determination of pepsin I by radioimmunoassay. *Prog. Clin. Biol. Res.* 173: 73–74, 1985.
67. HIRSCH-MARIE, H., F. LOISILLIER, J. P. TOUBOUL, AND P. BURTIN. Immunochemical study and cellular localization of human pepsinogens during ontogenesis and in gastric cancers. *Lab. Invest.* 34: 623–632, 1976.
68. HOEDEMAEKER, P. J., J. ABELS, J. J. WACHTERS, A. ARENDS, AND H. O. NIEWEG. Investigations about the site of production of Castle's gastric intrinsic factor. *Lab. Invest.* 13: 1394–1399, 1964.
69. HUBBARD, A. L., AND Z. A. COHN. Specific labels for cell surfaces. In: *Biochemical Analysis of Membranes*, edited by A. H. Maddy. London: Wiley, 1976, p. 427–501.
70. HUNTER, W. M. Radioimmunoassay. In: *Handbook of Experimental Immunology* (3rd ed.), edited by D. M. Weir. Oxford, UK: Blackwell, 1978, p. 14.1–14.40.
71. HURN, B. A. L., AND M. CHANTLER. Production of reagent antibodies. *Methods Enzymol.* 70: 104–142, 1980.
72. ICHIHARA, Y., K. SOGAWA, AND K. TAKAHASHI. Rat gastric prepepsinogen: in vitro synthesis and partial amino-terminal signal sequence. *J. Biochem.* 92: 603–606, 1982.
73. ICHIHARA, Y., K. SOGAWA, AND K. TAKAHASHI. Isolation of human, swine, and rat prepepsinogens and calf preprochymosin, and determination of the primary structures of their NH_2-terminal signal sequences. *J. Biochem.* 98: 483–492, 1985.
74. ICHINOSE, M., K. MIKI, C. FURIHATA, T. KAGEYAMA, H. NIWA, H. OKA, T. MATSUSHIMA, AND K. TAKAHASHI. Radioimmunoassay of serum group I and group II pepsinogens in normal controls and patients with various disorders. *Clin. Chim. Acta* 126: 183–191, 1982.
75. JACKSON, G. D. F., I. LEMAITRE-COELHO, J. P. VAERMAN, H. BAZIN, AND A. BECKERS. Rapid disappearance from serum of intravenously injected rat myeloma IgA and its secretion into bile. *Eur. J. Immunol.* 8: 123–126, 1978.
76. JERNE, N. K., C. HENRY, A. A. NORDIN, H. FUJI, A. M. C. KOROS, AND I. LEFKOVITS. Plaque forming cells: methodology and theory. *Transplant. Rev.* 18: 130–191, 1974.
77. JOHNSTONE, A., AND R. THORPE. *Immunochemistry in Practice.* Oxford, UK: Blackwell, 1982.
78. KABAT, E. A. Basic principles of antigen-antibody reactions. *Methods Enzymol.* 70: 3–49, 1980.
79. KANELLOPOULOS, J., G. ROSSI, AND H. METZGER. Preparative isolation of the cell receptor for immunoglobulin E. *J. Biol. Chem.* 254: 7691–7697, 1979.
80. KAUFFMAN, R. S., J. H. NOSEWORTHY, J. T. NEPOM, R. FINBERG, B. N. FIELDS, AND M. I. GREENE. Cell receptors for the mammalian reovirus. II. Monoclonal anti-idiotypic antibody blocks viral binding to cells. *J. Immunol.* 131: 2539–2541, 1983.
81. KEEN, J. H. Preparation of antibodies to clathrin and use in cytochemical localization. *Methods Enzymol.* 98: 359–368, 1983.
82. KESSLER, S. W. Use of protein A-bearing staphylococci for the immunoprecipitation and isolation of antigens from cells. *Methods Enzymol.* 73: 442–471, 1981.
83. KLOCKARS, M., AND S. REITAMO. Tissue distribution of lysozyme in man. *J. Histochem. Cytochem.* 23: 932–940, 1975.
84. KOHLER, G. The technique of hybridoma production. In: *Immunological Methods*, edited by I. Lefkovits and B. Pernis. New York: Academic, 1981, vol. 2, p. 285–298.
85. KOHLER, G., AND C. MILSTEIN. Derivation of specific antibody-producing tissue culture and tumor lines by cell fusion. *Eur. J. Immunol.* 6: 511–519, 1976.
86. KORSRUD, F. R., AND P. BRANDTZAEG. Characterization of epithelial elements in human major salivary glands by functional markers: localization of amylase, lactoferrin, lysozyme, secretory component, and secretory immunoglobulins by paired immunofluorescence staining. *J. Histochem. Cytochem.* 30: 657–666, 1982.
87. KRAEHENBUHL, J. P., AND J. D. JAMIESON. Solid-phase conjugation of ferritin to Fab-fragments of immunoglobulin G for use in antigen localization on thin sections. *Proc. Natl. Acad. Sci. USA* 69: 1771–1775, 1972.
88. KRAMER, M. F., AND H. T. TAN. The peri-insular acini of the pancreas of the rat. *Z. Zellforsch. Mikrosk. Anat.* 86: 163–170, 1968.
89. KRAUS, F. W., AND J. MESTECKY. Immunohistochemical localization of amylase, lysozyme, and immunoglobulins in the human parotid gland. *Arch. Oral. Biol.* 16: 781–790, 1971.
90. LANE, C. D., A. COLMAN, T. MOHUN, J. MORSER, J. CHAMPION, I. KUORIDES, R. CRAIG, S. HIGGINS, T. C. JAMES, S. W. APPLEBAUM, R. I. OHLSSON, E. PAUCHA, M. HOUGHTON, J. MATTHEWS, AND B. J. MIFLIN. The *Xenopus* oocyte as a surrogate secretory system. The specificity of protein export. *Eur. J. Biochem.* 111: 225–235, 1980.
91. LANGONE, J. J. Applications of immobilized protein A in immunological techniques. *J. Immunol. Methods* 55: 277–296, 1982.
92. LANGONE, J. J., AND H. VAN VUNAKIS. *Methods in Enzymology.* New York: Academic, 1981, vol. 73.
93. LECHAGO, J., C. BLACK, AND I. M. SAMLOFF. Immunofluorescence studies of gastric heterotopia of the small intestine in Crohn's disease. *Gastroenterology* 70: 429–432, 1976.
94. LEVINE, J. S., P. K. NAKANE, AND R. H. ALLEN. Immunocytochemical localization of human intrinsic factor: the nonstimulated stomach. *Gastroenterology* 79: 493–502, 1980.
95. LEVINE, J. S., NAKANE, P. K., AND R. H. ALLEN. Human intrinsic factor secretion: immunocytochemical demonstration of membrane-associated vesicular transport in parietal cells. *J. Cell Biol.* 90: 644–655, 1981.
96. LIEBMAN, W. M., AND Y. BUJANOVER. Groups I and II pepsinogens in Meckel's diverticulum. *J. Histochem. Cytochem.* 26: 867–868, 1978.
97. LOWE, C. R. An introduction to affinity chromatography. In: *Laboratory Techniques in Biochemistry and Molecular Biology*, edited by T. S. Work and R. H. Burdon. Amsterdam: Elsevier/North-Holland, 1979, vol. 7, pt. 2, p. 269–522.
98. MALAISSE-LAGAE, F., M. RAVAZZOLA, P. ROBBERECHT, A. VANDERMEERS, W. J. MALAISSE, AND L. ORCI. Exocrine pancreas: evidence for topographic partition of secretory function. *Science Wash. DC* 190: 795–797, 1975.
99. MANNING, D. C., S. H. SNYDER, J. F. KACHUR, R. J. MILLER, AND M. FIELD. Bradykinin receptor-mediated chloride secretion in intestinal function. *Nature Lond.* 299: 256–259, 1982.

100. MANTLE, M., G. G. FORSTNER, AND J. F. FORSTNER. Antigenic and structural features of goblet-cell mucin of human small intestine. *Biochem. J.* 217: 159–167, 1984.
101. MARSHALL, J. M., JR. Distributions of chymotrypsinogen, procarboxypeptidase, desoxyribonuclease, and ribonuclease in bovine pancreas. *Exp. Cell Res.* 6: 240–242, 1954.
102. MASON, D. W., AND A. F. WILLIAMS. The kinetics of antibody binding to membrane antigens in solution and at the cell surface. *Biochem. J.* 187: 1–20, 1980.
103. MAURER, P. H., AND H. J. CALLAHAN. Proteins and polypeptides as antigens. *Methods Enzymol.* 70: 49–70, 1980.
104. MCDONOUGH, A. A., A. HIATT, AND I. S. EDELMAN. Characteristics of antibodies to guinea pig ($Na^+ + K^+$)-adenosine triphosphatase and their use in cell-free synthesis studies. *J. Membr. Biol.* 69: 13–22, 1982.
105. MCLEAN, I. W., AND P. K. NAKANE. Periodate-lysine-paraformaldehyde fixative. A new fixative for immunoelectron microscopy. *J. Histochem. Cytochem.* 22: 1077–1083, 1974.
106. MELDOLESI, J. Membranes of pancreatic zymogen granules. *Methods Enzymol.* 98: 67–75, 1983.
107. MERISKO, E. M., M. G. FARQUHAR, AND G. E. PALADE. Coated vesicle isolation by immunoadsorption on *Staphylococcus aureus* cells. *J. Cell Biol.* 92: 846–857, 1982.
108. MEUWISSEN, S. G. M., H. MULLINK, A. BOSMA, G. PALS, J. DÉFIZE, M. FLIPSE, B. D. WESTERVELD, M. TAS, J. BRAKKÉ, J. KREUNING, A. W. ERIKSSON, AND C. J. L. M. MEYER. Immunocytochemical localization of pepsinogen I and II in the human stomach. *Prog. Clin. Biol. Res.* 173: 185–197, 1985.
109. MOLINARO, G. A., W. C. EBY, AND C. A. MOLINARO. Antigen quantitation by a reverse hemolytic assay. *Methods Enzymol.* 73: 319–338, 1981.
110. MORIKAWA, S. Studies on alkaline and acid ribonucleases in mammalian tissues: immunohistochemical localization and immunochemical properties. *J. Histochem. Cytochem.* 15: 662–673, 1967.
111. MORRISON, M. Lactoperoxidase-catalyzed iodination as a tool for the investigation of protein. *Methods Enzymol.* 70: 214–220, 1980.
112. MULLOCK, B. M., M. DOBROTA, AND R. H. HINTON. Sources of the proteins of rat bile. *Biochim. Biophys. Acta* 543: 497–507, 1978.
113. MULLOCK, B. M., R. H. HINTON, M. DOBROTA, J. PEPPARD, AND E. ORLANS. Endocytic vesicles in liver carry polymeric IgA from serum to bile. *Biochim. Biophys. Acta* 587: 381–391, 1979.
114. MULLOCK, B. M., R. H. HINTON, M. DOBROTA, J. PEPPARD, AND E. ORLANS. Distribution of secretory component in hepatocytes and its mode of transfer into bile. *Biochem. J.* 190: 819–826, 1980.
115. MURPHY, R. A., A. Y. WATSON, J. METZ, AND W. G. FORSSMANN. The mouse submandibular gland: An exocrine organ for growth factors. *J. Histochem. Cytochem.* 28: 890–902, 1980.
116. NAGURA, H., P. D. SMITH, P. K. NAKANE, AND W. R. BROWN. IgA in human bile and liver. *J. Immunol.* 126: 587–595, 1981.
117. NAGURA, H., Y. TSUTSUMI, H. HASEGAWA, K. WATANABE, P. K. NAKANE, AND W. R. BROWN. IgA plasma cells in biliary mucosa: a likely source of locally synthesized IgA in human hepatic bile. *Clin. Exp. Immunol.* 54: 671–680, 1983.
118. NEILL, J. D., AND L. S. FRAWLEY. Detection of hormone release from individual cells in mixed populations using a reverse hemolytic plaque assay. *Endocrinology* 112: 1135–1137, 1983.
119. NEXO, E., M. HANSEN, S. S. POULSEN, AND P. S. OLSEN. Characterization and immunohistochemical localization of rat salivary cobalamin-binding protein and comparison with human salivary haptocorrin. *Biochim. Biophys. Acta* 838: 264–269, 1985.
120. O'FARRELL, P. H. High resolution two-dimensional electrophoresis of proteins. *J. Biol. Chem.* 250: 4007–4021, 1975.
121. OLSSON, T., AND A. THORE. Chemiluminescence and its use in immunoassay. In: *Immunoassays for the 80s*, edited by A. Voller, A. Bartlett, and D. Bidwell. Lancaster, UK: MTP Press, 1981, p. 113–125.
122. PAINTER, R. G., K. T. TOKUYASU, AND S. J. SINGER. Immunoferritin localization of intracellular antigens: the use of ultracryotomy to obtain ultrathin sections suitable for direct immunoferritin staining. *Proc. Natl. Acad. Sci. USA* 70: 1649–1653, 1973.
123. PALADE, G. E. Intracisternal granules in the exocrine cells of the pancreas. *J. Biophys. Biochem. Cytol.* 2: 417–422, 1956.
124. PELHAM, H. R. B., AND R. J. JACKSON. An efficient mRNA-dependent translation system from reticulocyte lysates. *Eur. J. Biochem.* 67: 247–256, 1976.
125. PLATT, E. J., K. KARLSEN, A. LOPEZ-VALDIVIESO, P. W. COOK, AND G. L. FIRESTONE. Highly sensitive immunoadsorption procedure for detection of low abundance proteins. *Anal. Biochem.* 156: 126–135, 1986.
126. PODOLSKY, D. K., D. A. FOURNIER, AND K. E. LYNCH. Human colonic goblet cells. Demonstration of distinct subpopulations defined by mucin-specific monoclonal antibodies. *J. Clin. Invest.* 77: 1263–1271, 1986.
127. POORT, S. R., AND C. POORT. Effect of feeding diets of different composition on the protein synthetic pattern of rat pancreas. *J. Nutr.* 111: 1475–1479, 1981.
128. POSTUMA, G., J. W. SLOT, AND H. J. GEUZE. Immunocytochemical assays of amylase and chymotrypsinogen in rat pancreas secretory granules. Efficacy of using immunogold-labelled ultrathin cryosections to estimate relative protein concentrations. *J. Histochem. Cytochem.* 32: 1028–1034, 1984.
129. QURESHI, R., G. G. FORSTNER, AND J. F. FORSTNER. Radioimmunoassay of human intestinal goblet cell mucin. Investigation of mucus from different organs and species. *J. Clin. Invest.* 64: 1149–1156, 1979.
130. RAJAN, A. I., AND A. BENNICK. Demonstration of proline-rich proteins in rabbit parotid saliva and partial characterization of some of the proteins. *Arch. Oral Biol.* 28: 431–439, 1983.
131. REID, W. A., W. D. THOMSON, AND J. KAY. Pepsinogen in gastric carcinoma cells. *J. Clin. Pathol.* 36: 137–139, 1983.
132. ROSS, P. E. Radioimmunoassay of serum bile acids. *Methods Enzymol.* 84: 321–349, 1982.
133. ROSSEN, R. D., C. MORGAN, K. C. HSU, W. T. BUTLER, AND H. M. ROSE. Localization of 11S external secretory IgA by immunofluorescence in tissue lining the oral and respiratory passages in man. *J. Immunol.* 100: 706–717, 1968.
134. ROTH, J., M. BENDAYAN, AND L. ORCI. Ultrastructural localization of intracellular antigens by the use of protein A-gold complex. *J. Histochem. Cytochem.* 26: 1074–1081, 1978.
135. ROTHENBERG, S. P. Immunologic isolation of human intrinsic factor. *Blood* 26: 868, 1965.
136. SACCOMANI, G., H. F. HELANDER, S. CRAGO, H. H. CHANG, D. W. DAILEY, AND G. SACHS. Characterization of gastric mucosal membranes. X. Immunological studies of gastric ($H^+ + K^+$)-ATPase. *J. Cell Biol.* 83: 271–283, 1979.
137. SACHS, G. H^+ transport by a non-electrogenic gastric ATPase as a model for gastric secretion. *Rev. Physiol. Biochem. Pharmacol.* 79: 133–162, 1977.
138. SAMLOFF, I. M. Cellular localization of group I pepsinogens in human gastric mucosa by immunofluorescence. *Gastroenterology* 61: 185–188, 1971.
139. SAMLOFF, I. M. Pepsinogens I and II: purification from gastric mucosa and radioimmunoassay in serum. *Gastroenterology* 82: 26–33, 1982.
140. SAMLOFF, I. M., M. D. KLEINMAN, M. TURNER, V. SOBEL, AND G. H. JEFFRIES. Blocking and binding antibodies to intrinsic factor and parietal cell antibody in pernicious anemia. *Gastroenterology* 55: 575–583, 1968.
141. SAMLOFF, I. M., AND W. M. LIEBMAN. Cellular localization of group II pepsinogens in human stomach and duodenum by immunofluorescence. *Gastroenterology* 65: 36–42, 1973.
142. SAMLOFF, I. M., AND W. M. LIEBMAN. Radioimmunoassay of group I pepsinogens in serum. *Gastroenterology* 66: 494–502, 1974.
143. SAMLOFF, I. M., AND P. L. TOWNES. Electrophoretic heterogeneity and relationships of pepsinogens in human urine,

serum, and gastric mucosa. *Gastroenterology* 58: 462–469, 1970.
144. SCHACHTER, M., D. J. LONGRIDGE, G. D. WHEELER, J. G. MEHTA, AND Y. UCHIDA. Immunocytochemical and enzyme histochemical localization of kallikrein-like enzymes in colon, intestine, and stomach of rat and cat. *J. Histochem. Cytochem.* 34: 927–934, 1986.
145. SCHACHTER, M., M. W. PERET, A. G. BILLING, AND G. D. WHEELER. Immunolocalization of the protease kallikrein in the colon. *J. Histochem. Cytochem.* 31: 1255–1260, 1983.
146. SCHADE, S. G., J. ABELS, AND R. F. SCHILLING. Studies on antibody to intrinsic factor. *J. Clin. Invest.* 46: 615–620, 1967.
147. SCHADE, S. G., P. FEICK, M. MUCKERHEIDE, AND R. F. SCHILLING. Occurrence in gastric juice of antibody to a complex of intrinsic factor and vitamin B_{12}. *N. Engl. J. Med.* 275: 528–531, 1966.
148. SCHEELE, G., D. BARTELT, AND W. BIEGER. Characterization of human exocrine pancreatic proteins by two-dimensional isoelectric focusing/sodium dodecyl sulfate gel electrophoresis. *Gastroenterology* 80: 461–473, 1981.
149. SCHEFFER, R. C. T., G. POORT, AND J. W. SLOT. Fate of the major zymogen granule-associated glycoprotein from rat pancreas. A biochemical and immunocytochemical study. *Eur. J. Cell Biol.* 23: 122–128, 1980.
150. SCHEHTER, Y., R. MARON, D. ELIAS, AND I. R. COHEN. Autoantibodies to insulin receptor spontaneously develop as anti-idiotypes in mice immunized with insulin. *Science Wash. DC* 216: 542–545, 1982.
151. SCHRIEBER, A. B., P. O. COURAUD, C. ANDRE, B. VRAY, AND D. A. STROSBERG. Anti-alprenolol anti-idiotypic antibodies bind to β-adrenergic receptors and modulate catecholamine-sensitive adenylate cyclase. *Proc. Natl. Acad. Sci. USA* 77: 7385–7389, 1980.
152. SEGE, K., AND P. A. PETERSON. Anti-idiotypic antibodies against anti-vitamin A transport protein reacts with prealbumin. *Nature Lond.* 271: 167–168, 1978.
153. SHAPIRO, S. Z., AND J. R. YOUNG. An immunochemical method for mRNA purification. *J. Biol. Chem.* 256: 1495–1498, 1981.
154. SHEPHERD, H. A., J. D. PRIDDLE, W. J. JENKINS, AND D. P. JEWELL. The preparation of human intrinsic factor-cobalamin complex from human gastric juice by immunoadsorption. *Clin. Chim. Acta* 139: 155–165, 1984.
155. SIMMONDS, W. J., M. G. KORMAN, V. L. W. GO, AND A. F. HOFMANN. Radioimmunoassay of conjugated cholyl bile acids in serum. *Gastroenterology* 65: 705–711, 1973.
156. SIRISINHA, S., AND H. H. EISEN. Autoimmune-like antibodies to the ligand-binding sites of myeloma proteins. *Proc. Natl. Acad. Sci. USA* 68: 3130–3135, 1971.
157. SMOLKA, A., H. F. HELANDER, AND G. SACHS. Monoclonal antibodies against gastric $H^+ + K^+$ ATPase. *Am. J. Physiol.* 245 (*Gastrointest. Liver Physiol.* 8): G589–G596, 1983.
158. SMOLKA, A., AND W. M. WEINSTEIN. Immunoassay of pig and human gastric proton pump. *Gastroenterology* 90: 532–539, 1986.
159. SMOLKA, A., D. ZETTEL, P. LORENTZON, AND R. DONALDSON. Monoclonal antibodies against intrinsic factor. *Gastroenterology* 90: 1640, 1986.
160. SOINI, E., AND I. HEMMILA. Fluoroimmunoassay: present status and key problems. *Clin. Chem.* 25: 353–361, 1979.
161. SPICER, S. S., M. A. SENS, AND R. E. TASHIAN. Immunocytochemical demonstration of carbonic anhydrase in human epithelial cells. *J. Histochem. Cytochem.* 30: 864–873, 1982.
162. STEFANINI, M., C. DE MARTINO, AND L. ZAMBONI. Fixation of ejaculated spermatozoa for electron microscopy. *Nature Lond.* 216: 173–174, 1967.
163. STERNBERGER, L. A., P. H. HARDY, JR., J. J. CUCULIS, AND H. G. MEYER. The unlabeled antibody enzyme method of immunohistochemistry. Preparation and properties of soluble antigen-antibody complex (horseradish peroxidase-antihorseradish peroxidase) and its use in identification of spirochaetes. *J. Histochem. Cytochem.* 18: 315–333, 1970.

164. SUGGS, S. V., R. B. WALLACE, T. HIROSE, E. H. KAWASHIMA, AND K. ITAKURA. Use of synthetic oligonucleotides as hybridization probes: isolation of cloned cDNA sequences for human B_2-microglobulin. *Proc. Natl. Acad. Sci. USA* 78: 6613–6617, 1981.
165. TAGGART, R. T., AND I. M. SAMLOFF. Stable antibody producing murine hybridomas. *Science Wash. DC* 219: 1228–1230, 1983.
166. TAKASHI, I., AND S. ITO. Carbonic anhydrase in mouse salivary glands and saliva: a histochemical, immunohistochemical, and enzyme activity study. *J. Histochem. Cytochem.* 32: 625–635, 1984.
167. TAYLOR, K. B., I. M. ROITT, D. DONIACH, K. G. COUCHMAN, AND C. SHAPLAND. Autoimmune phenomena in pernicious anemia: gastric antibodies. *Br. Med. J.* II: 1347–1352, 1962.
168. TOMASI, T. B., JR., E. M. TAN, A. SOLOMON, AND R. A. PRENDERGAST. Characteristics of an immune system common to certain external secretions. *J. Exp. Med.* 121: 101–124, 1965.
169. TOURVILLE, D. R., R. H. ADLER, J. BIENENSTOCK, AND T. B. TOMASI. The human secretory immunoglobulin system: immunohistological localization of IgA, secretory piece, and lactoferrin in normal human tissues. *J. Exp. Med.* 129: 411–429, 1969.
170. VAITUKAITIS, J. L. Production of antisera with small doses of immunogen: multiple intradermal injections. *Methods Enzymol.* 73: 46–52, 1981.
171. VARIS, K. S. Peptic cells. *Prog. Clin. Biol. Res.* 173: 177–184, 1985.
172. VARIS, K. S., M. MARIN-SORENSEN, I. M. SAMLOFF, AND W. M. WEINSTEIN. Immunocytochemical characterization of pseudopyloric metaplasia. *Gastroenterology* 84: 1341, 1983.
173. WALDUM, H. L. Radioimmunoassay of pepsin in gastric juice. *Prog. Clin. Biol. Res.* 173: 75–78, 1985.
174. WALDUM, H. L., AND P. G. BURHOL. The effect of insulin-induced hypoglycaemia on group I pepsinogens, serum gastrin, and plasma secretin and on gastric H^+ and pepsin outputs. *Scand. J. Gastroenterol.* 15: 259–266, 1980.
175. WALDUM, H. L., B. K. STRAUME, AND P. G. BURHOL. Radioimmunoassay of group I pepsinogens (PG I) and the effect of food on serum PG I. *Scand. J. Gastroenterol.* 14: 241–247, 1979.
176. WEIR, D. M. (Editor). *Handbook of Experimental Immunology* (3rd ed.). Oxford, UK: Blackwell, 1978.
177. WERNER, S., AND M. MACHLEIDT. Isolation of precursors of cytochrome oxidase from *Neurospora crassa*; application of subunit-specific antibodies and protein A from *Staphylococcus aureus*. *Eur. J. Biochem.* 90: 99–105, 1978.
178. WIDE, L. Use of particulate immunosorbents in radioimmunoassay. *Methods Enzymol.* 73: 203–224, 1981.
179. WISDOM, G. B. Enzyme-immunoassay. *Clin. Chem.* 22: 1243–1255, 1976.
180. WYSOCKI, L. J., AND G. L. SATO. "Panning" for lymphocytes: a method of cell selection. *Proc. Natl. Acad. Sci. USA* 75: 2844–2848, 1978.
181. YALOW, R. S., AND S. A. BERSON. Immunoassay of endogenous plasma insulin in man. *J. Clin. Invest.* 39: 1157–1175, 1960.
182. YASUDA, K., AND A. H. COONS. Localization by immunofluorescence of amylase, trypsinogen, and chymotrypsinogen in the acinar cells of the pig pancreas. *J. Histochem. Cytochem.* 14: 303–313, 1966.
183. YASUDA, K., T. SUZUKI, AND K. TAKANO. Localization of pepsin in the stomach, revealed by fluorescent antibody technique. *Folia Anat. Jpn.* 42: 355–367, 1966.
184. YEH, K. Y., AND D. P. CHOPRA. Epithelial cell cultures from the colon of the suckling rat. *In Vitro* 16: 976–986, 1980.
185. YOUNG, R. A., AND R. W. DAVIS. Efficient isolation of genes using antibody probes. *Proc. Natl. Acad. Sci. USA* 80: 1194–1198, 1983.
186. ZOLLER, M., S. MATZKU, AND W. RAPP. Purification of human gastric proteases by immunoabsorbents. Pepsinogen I group. *Biochim. Biophys. Acta* 427: 708–718, 1976.

CHAPTER 18

Functional development of stomach

CHI-CHUAN TSENG
LEONARD R. JOHNSON

Department of Physiology and Cell Biology, University of Texas, Medical School of Houston, Houston, Texas

CHAPTER CONTENTS

Parietal Cells
    Structure
    Basal acid secretion
    Acid secretory response to stimuli
Chief Cells
    Structure
    Mucosal pepsinogen content
    Pepsin secretory response to stimuli
Gastrin
    Tissue levels
    Serum levels
Growth
Regulation of Gastric Development
    Diet or weaning
    Corticosterone
    Thyroxine
    Influence of sex
    Intrinsic mechanism
Summary

THERE ARE RELATIVELY few studies of the development of gastric function, and even fewer studies that have attempted to understand developmental regulation. Numerous morphological events in gastric development have been described, but quantitative data have been available only recently. This chapter summarizes most of the existing literature on the functional development of the stomach. The correlation between functional and morphological development of gastric chief and parietal cells is examined as well. Almost all studies of the development of gastric function and the mechanisms controlling that development have been on rats and mice. Existing studies in humans are almost totally descriptive (7, 9). Thus this chapter concentrates on the results of experiments done on the rodent, although whenever appropriate, related knowledge from the human and other species is included.

The stomach has three basic activities: secretion, an endocrine function, and motility. The most studied and best understood of these is gastric secretory activity. The cells of the gastric mucosa secrete hydrochloric acid, pepsinogen, intrinsic factor, mucus, water, and a few other electrolytes. Acid is produced by parietal cells in concentrations of ~160 meq/l in most species (63). Acid digests some protein, kills bacteria, and converts inactive pepsinogen into the active enzyme pepsin. Pepsin digests protein by cleaving interior peptide linkages. Intrinsic factor is secreted by the parietal cells in humans and is necessary for the adequate absorption of vitamin B_{12} (10). Mucus serves to lubricate the walls of the stomach and prevent physical damage by the ingesta.

The mucosa of the antral region produces the hormone gastrin, which with the vagal mediator acetylcholine regulates the stimulation of gastric acid secretion (90). These two substances interact with the third major stimulator of the parietal cells, histamine, to produce large increases in acid secretion in response to a meal. Gastrin also stimulates mucosal growth in the oxyntic gland portion of the stomach (40). The gastric mucosa and its nerves contain somatostatin and bombesin, respectively, two substances that may have physiological functions in regulating gastrin release (90).

A third group of activities carried out by the stomach are under the heading of motility (48). These include the ability to function as a reservoir; to mix food and fluid, thereby reducing the size of the particles; and to propel gastric contents into the duodenum at a regulated rate.

PARIETAL CELLS

Structure

The ultrastructural development of rat gastric mucosa has been described by Helander (28). In the 16-day-old fetus the gastric epithelium is composed of primitive stem cells whose primary function is to divide. The same type of cell was found in the rabbit gastric mucosa in 19- to 23-day-old fetuses (26). On the 18th day of gestation in the rat a slight differentiation could be traced in some gastric epithelial cells. By day 19 the first parietal cell was distinguished by the presence of intracellular canaliculi. At this time carbonic anhydrase was also detected (28, 88). The differentiation of parietal cells continued through the rest of the gestation period, as shown by increases in mitochondrial number and area of secretory membrane. The conclusion from most ultrastructural studies is that the gastric mucosa of most species is capable of secreting acid shortly before birth (9).

Tseng et al. (87) recently investigated the morphological development of rat parietal cells in the postnatal period quantitatively. During the first 20 postnatal days, parietal cell structure changed little compared with the increase in basal acid secretion. As shown in Figure 1, overall acid secretion increased 10-fold between days 5 and 20. During the same period there was almost no increase in the volume fraction of mitochondria and only a threefold increase in the surface density of tubulovesicles and intracellular canaliculi. Most of the latter increases occurred before the major change in acid secretion. This is consistent with data from Helander (28) that there was a doubling of the volume of parietal cell cytoplasm occupied by mitochondria during the first 10 days after birth. Those data clearly indicate that the morphological maturation of parietal cells takes place in the very early postnatal days and that the postnatal development of acid secretion is not correlated with ultrastructural changes. This latter conclusion is further supported by Ackerman and Shindledecker (2), who showed that 30-day-old rats secreted 2-3 times more acid than 100-day-old rats, although they had only one-fourth the parietal cell mass. These developmental features of parietal cell structure contrast with those of the chief cell, where the developmental increase in pepsinogen content parallels structural maturation (86).

Basal Acid Secretion

Accurate measurement of acid secretion in fetal and newborn rats is difficult because of the small volumes of fluid involved. In general, investigators have taken four different approaches to this problem. Measurement of the pH of the gastric contents has been used to detect acid secretory activity, but this tells nothing about the quantity of acid produced (21, 22). Fetal stomachs mounted in Ussing chambers are capable of basal secretion and of responding to stimuli; however, the results are difficult to relate to live animals (11, 23). Gastric juice can easily be collected from newborn rats after pylorus ligation (35, 68, 80), but this procedure does not yield true basal secretion, because pylorus ligation itself is a weak stimulus. Ackerman (1) has used continuous saline perfusion of the gastric lumen of anesthetized rats to determine acid output. Although this technique is certainly the most accurate in terms of measuring the acid secreted, it introduces anesthesia as another variable (2). Despite shortcom-

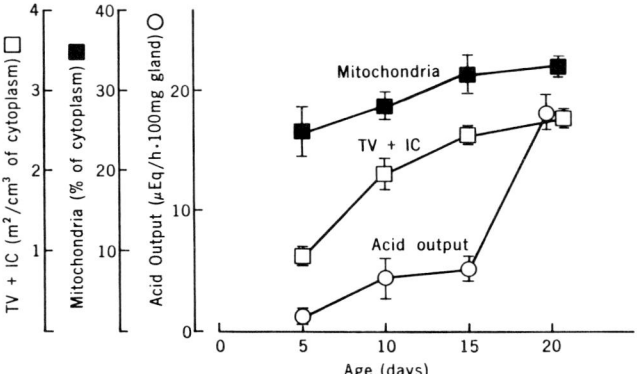

FIG. 1. Volume fraction of mitochondria and surface density of tubulovesicles (TV) and intracellular canaliculi (IC) of parietal cells as well as basal acid output of neonatal rats aged 5-20 days. [From Tseng et al. (86).]

ings in these techniques, however, a relatively clear pattern of the onset of acid secretion has emerged.

Although morphological data indicate that rat gastric mucosa is able to secrete acid before birth, there is little acid in the fetal stomach. Garzon et al. (21) found that the pH of the gastric contents of 19-day-old rat fetuses was nearly neutral. On day 20 the pH had dropped to ~6 and then fell markedly to below 4 shortly after birth. Johnson and co-workers (35, 80) found that the gastric pH of rats 5 and 10 days after birth is ~6 and does not fall to 4 until day 15. This discrepancy was addressed by Garzon et al. (23) using a modified Ussing chamber in which they measured net transepithelial H^+ flux. They found significant H^+ secretion on day 20 and 21 of gestation but no basal secretion of acid 5 days after birth. Thus there appears to be a biphasic development of the ability of the rat stomach to secrete acid basally (23).

Significant development of basal acid secretion in rats occurs during the third week of life when weaning takes place. During this time the gastric pH decreases from 4 to ~2.5 and acid secretion becomes measurable. As shown in Figure 1, neonatal rats ages 5 through 15 days secreted very little acid. By day 20 acid secretion increased significantly. Overall acid secretion in 20-day-old rats is ~10 times what it is on day 5.

From the above studies it is obvious that the rat is capable of secreting acid shortly before birth. The amount of basal acid secretion, however, remains of little consequence until the third week of life when the animal begins to eat solid food. Between days 15 and 21, gastric pH decreases ~2 log units and basal acid output increases from 3% or 4% of adult levels to ~25%. By day 25 basal acid secretion approaches 40% of the adult level, which is reached on approximately day 40 (7). The rat is a basal secretor of gastric acid, and basal secretion occupies as much as 40%-60% of the response to maximal stimulation (35, 44, 75).

In human infants basal acid secretion is also low at birth and increases dramatically in the first 24 h of life (13, 14). As shown in Figure 2, the acid output displays a gradual increase until 3 wk after birth, followed by a short decline in weeks 3 to 4, and a sustained rise continuing into the 2nd and 3rd mo of life (3, 67). By 3 mo of age acid secretion approaches the lower limit of adult levels (3, 70).

Acid Secretory Response to Stimuli

The development of responses to stimuli of acid secretion in the rat gastric mucosa begins during late gestation. In the fetal stomach, Garzon et al. (22) demonstrated that pentagastrin, gastrin (G-17), histamine, and carbachol significantly decreased the pH of the gastric contents from day 20. During the same time, administration of cimetidine completely prevented the acid response to histamine but not to carbachol or pentagastrin. Moreover atropine totally inhibited the acid secretory response to carbachol but not to either pentagastrin or histamine. Garzon et al. concluded that the fetal gastric mucosa was sensitive to the three primary stimulants of acid secretion at the time when differentiated parietal cells start secreting acid. In Ussing chambers the same group demonstrated that pentagastrin significantly increased the transepithelial H^+ fluxes in 20- and 21-day-old fetuses but failed to do so in the 5-day-old neonatal rat (23). Others have shown that newborn pig (17), dog (65), and rat (80) stomachs are insensitive to gastrin. In general, gastric mucosa from 20-day-old rat fetuses is sensitive both in vivo and in vitro to the stimulants. The quantity of acid secretion, however, is small and of no consequence until the 3rd wk of life.

There are slight but statistically insignificant increases in acid secretion in response to pentagastrin in 15- and 18-day-old rats (35, 80). A significant increase in acid secretion did not occur until day 20. To explain these results and the fact that newborn rats do not secrete acid in response to endogenously elevated serum gastrin, Takeuchi et al. (80) measured the binding of ^{125}I-labeled Leu-15-gastrin-17 to rat gastric mucosal membranes. Specific binding of gastrin to its receptor was not detected until day 20. Thus it appears that the reason for gastrin insensitivity in newborn rats is the lack of receptors for the hormone (80). Peitsch et al. (68) injected 7-day-old rats with corticosterone and found that gastrin receptors developed precociously 3 days later. By day 12 there was a significant acid secretory response to gastrin coincident with the highly significant increase in specific gastrin binding (68). These data support the hypothesis that the development of sensitivity to secretagogues may be determined by the presence or absence of receptors (80).

Sensitivity to cholinergic stimulation appears to develop earlier than sensitivity to gastrin. There is a significant acid secretory response to carbachol in 15-day-old rats (35). In the same study, animals of identical age failed to respond to histamine or gastrin. Seidel and Johnson (75) used [^{3}H]quinuclidinyl benzilate to measure the appearance of cholinergic receptors in rat gastric mucosa and to correlate it with the acid response to carbachol. In contrast with the gastrin receptor, cholinergic receptors were present at birth in concentrations comparable with those in adult rats. The affinity of the receptor for quinuclidinyl benzilate actually decreased about fivefold from birth to 40 days of age. This dose of carbachol, which was fourfold higher than that used in an earlier study (35), caused a significant increase in acid secretion in rats aged 0.5, 5, and 10 days as well as in the older animals. It therefore appears that the response to cholinergic stimuli is present from birth along with the necessary receptors. Beginning on day 18, however, there was a dramatic increase in the secretory response to cholinergic stimulation. The response on day 18 was threefold more than on day 15 and by day 25 had increased 12-fold (75).

The development of the secretory response to histamine in the rat is generally similar to that of gastrin and cholinergic stimuli, with the major increase in sensitivity appearing during the 3rd wk of life (1, 35). Ackerman (1) was unable to measure a statistically significant response to histamine in the 14-day-old

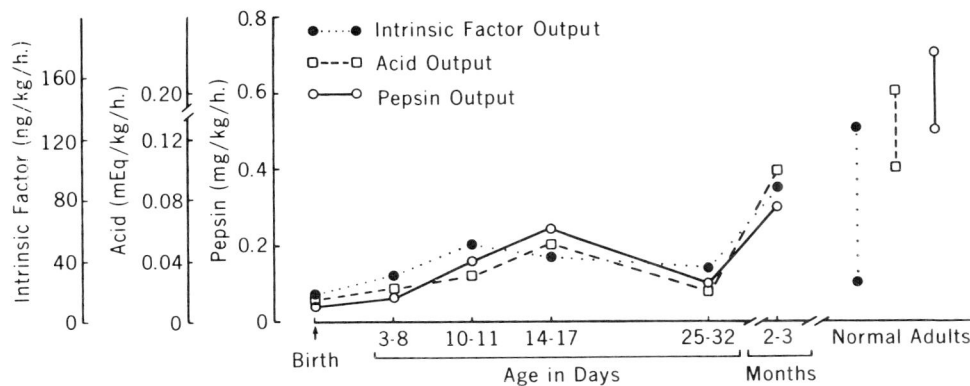

FIG. 2. Gastric secretion of acid, pepsin, and intrinsic factor in human newborns and infants. Data have been corrected for body weight. [From Agunod et al. (3).]

rat, yet animals of the same age responded well to pentagastrin and carbachol. Based on these data he concluded that histamine cannot act as the final mediator of acid secretion in the newborn rat. This conclusion was supported by the failure of cimetidine to block pentagastrin-stimulated acid secretion in 14-day-old rats (1).

On the other hand, in one study significant responses to both histamine and pentagastrin occurred on day 20 and responses to carbachol on day 15 (35). In a different study the response to pentagastrin again appeared on day 20, but the response to histamine appeared on day 15 (80). The amounts of acid secretion involved and the differences between statistically significant and insignificant responses, however, were small. In summary, in the rat it appears that sensitivity to carbachol is present the earliest and with high doses can be demonstrated from birth; responsiveness to gastrin and histamine follows between days 14 and 20. Biologically significant acid secretion in response to all stimuli is established at the end of the 3rd wk.

The newborn human infant has hypergastrinemia and hyposecretion of acid (14, 15). During the first 48 h there is no difference between basal and pentagastrin-stimulated acid secretion (15). Therefore the newborn human is insensitive to exogenous as well as endogenous gastrin. In newborn dogs there is little or no acid secretion until day 9, in spite of high serum gastrin levels, and exogenous gastrin has no effect on acid secretion or motility until day 9 (65). These data are similar to those in the rat and may be explained by the absence of gastrin receptors in these species as well (80). There are few studies of human infants involving histamine. Agunod et al. (3), however, demonstrated a response to histamine beginning on day 1. In that study the basal concentration of H^+ was increased from 3.3 meq/l to 8.1 meq/l after histamine. Acid output in response to histamine, however, was only 0.03 meq/h. One other study involving sensitivity to different secretagogues has been done in newborn pigs (17). Forte et al. (17) found that in vitro mucosa from piglets aged 1-5 days responded to histamine but not to pentagastrin or acetylcholine.

There is excellent evidence that histamine, acting as a paracrine, potentiates the effects of both cholinergic stimulants and gastrin (78). The role of histidine decarboxylase in this process is uncertain. It is only in the rat that this enzyme responds to feeding and secretagogues. Studies of whole mucosa from a variety of mammals, including the cat and dog, have not demonstrated the presence of the enzyme (4). It remains to be seen whether the enzyme can be induced in isolated histamine-containing cells from species in which the enzyme is undetectable in whole mucosa.

Histidine decarboxylase is undetectable in rat gastric mucosa until day 10 (35). Significant amounts of enzyme activity are present on days 10 and 15, but none of the three major secretagogues increases activity. By day 18 pentagastrin significantly increases enzyme activity. Neither histamine nor carbachol increases histidine decarboxylase activity in rats of any age. In fact in 25- and 40-day-old animals these two agents decrease enzyme activity (35). The timing of the appearance of histidine decarboxylase activity is similar to that found previously by Håkanson et al. (25) for mucosal histamine content.

The parallels between the development of acid secretion and histidine decarboxylase are striking. Enzyme activity is high in fetal rat gastric mucosa until the 19th day after mating and then disappears at birth on the 22nd day after mating (4). This pattern is nearly identical to that described by Garzon et al. (23) for acid secretion using the in vitro chamber preparation.

In the human the parietal cell also secretes intrinsic factor. Little is known about the development of its secretion, and all available data come from human studies. Intrinsic factor is detectable in gastric mucosa during the 14th wk of gestation and increases fourfold by week 25 (7). Intrinsic factor secretion rises slowly and can be detected by day 10–11 (3). Secretion of intrinsic factor after stimulation with betazole parallels acid production in the newborn, rising to levels ~50% of adult values by 10–14 days and reaching mature levels by 3 mo of age (Fig. 2).

CHIEF CELLS

Structure

In the fetal rat stomach, Helander (29) found the first trace of peptic activity on the 19th day of gestation. Zymogen cells could be distinguished after 20 days of gestation by the presence of large cytoplasmic granules. These granules increased in size and number during the rest of the fetal period. Zymogen cells contained a large number of secretory granules and a great deal of rough endoplasmic reticulum (RER) at birth. However, during the first 10 days after birth, zymogen cells contained only a few secretory granules and peptic activity in homogenates of gastric mucosa remained constant. Twenty days after birth, rat zymogen cells attained a normal adult appearance (29).

Similar to Helander's observation, the morphological studies of Tseng et al. (86) demonstrated that in the first 10 postnatal days zymogen cells displayed very little ultrastructural maturation. As shown in Figure 3, the volume fraction of zymogen granules and the surface density of RER exhibit little change from day 5 to day 10. A dramatic increase in the surface area of RER occurs after day 10 followed by a similar increment in zymogen granules between days 15 and 20. These developmental changes parallel mucosal pepsinogen levels. Thus, unlike parietal cells, the functional changes and the morphological development of chief cells are closely related. Furthermore the ultrastructural maturation of chief cells begins relatively late, between days 10 and 15.

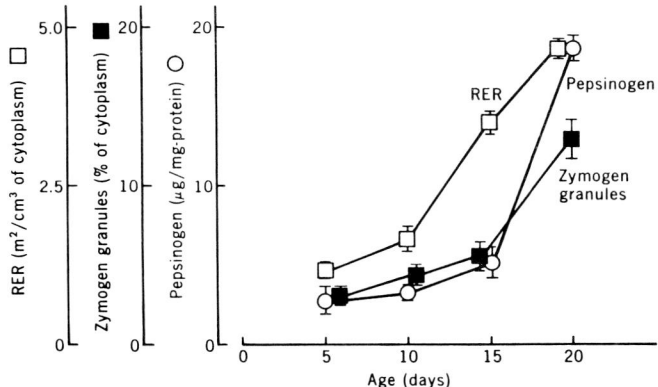

FIG. 3. Volume fraction of zymogen granules and the surface density of rough endoplasmic reticulum (RER) of chief cells as well as pepsinogen content from neonatal rats aged 5–20 days.

Mucosal Pepsinogen Content

The developmental increase in mucosal pepsinogen is identical to that for acid secretion. Tseng and Johnson (84) found that the mucosal pepsinogen content (Fig. 3) was low during the first 15 postnatal days and increased dramatically on day 20. Similar results were reported by Furihata et al. (19), who measured the mucosal pepsinogen activity in rats of various ages postnatally and found that the peptic activity remained constant for the first 17 days at ~20% of adult levels. Activity began to increase on day 18 and reached the adult level by day 30. In a subsequent study, Furihata et al. (18) demonstrated differences in molecular species of pepsinogen found in newborn and adult rats and suggested that the mucous neck cell was the source of pepsinogen in newborns, whereas the chief cell contained most of the pepsinogen in the adult. The increase in adult-type pepsinogen after day 17, observed by Furihata et al. (18), coincided with the elevation in total pepsinogen and the appearance of mature chief cells observed by Helander (29) and Tseng et al. (86). Kumegawa et al. (51) found similar results for mucosal pepsinogen activity in mice aged 10, 20, and 28 days. The activity on day 20 was fourfold that present on day 10 and then doubled between days 20 and 28. In general these studies indicate that oxyntic mucosal pepsinogen levels are low at birth and increase rapidly during the 3rd wk of life.

Furihata et al. (18), however, obtained different results in antral mucosa. They showed that peptic activity and the molecular species of pepsinogen in the rat pyloric mucosa were constant from birth and suggested that the maturation of pyloric gland cells was complete before birth.

Pepsin Secretory Response to Stimuli

Only two reports have included data on the development of pepsin secretion as opposed to tissue content of the enzyme. Basal pepsinogen secretion occurs from birth, but it is probably biologically insignificant until approximately day 15 (75). By day 15 basal secretion was equal to ~1% of adult levels. By day 20 it had increased to 5% and by day 25 to 20% of adult amounts.

A significant biological response to secretagogues also appears during the 3rd wk of life (35). Carbachol effectively increases pepsin output on day 15, whereas histamine and pentagastrin are not effective until day 20. Recently, Yahav et al. (93) studied the pepsin secretory response to various secretagogues in isolated rat gastric glands. Isolated glands from full-term neonatal and 1-day-old rats failed to respond to cholecystokinin octapeptide, carbachol, and Ca^{2+} ionophore. The same glands, however, were responsive to dibutyryl cAMP. Significant responses to these secretagogues were demonstrated in 2-day-old rats. They suggested that the pepsinogen-release mechanism is functioning on the first day after birth and that the lack of responsiveness of the gastric glands to carbachol and cholecystokinin is due to a deficiency at the receptor level. Using higher doses of carbachol, Seidel and Johnson (75) found a statistically significant peptic secretory response on the first postnatal day. From those data it appears that cholinergic regulation of gastric pepsin secretion is functional shortly after birth. However, the amount of pepsin secreted is biologically insignificant until the 3rd wk of life.

The presence of physiological amounts of H^+ in the dog stomach stimulates pepsin secretion (37, 38). The mechanism involves a mucosal receptor sensitive to H^+ that can be blocked with topical Xylocaine. Receptor activation triggers a cholinergic reflex that can be blocked by atropine (38). The finding that topical 0.1 N HCl stimulated pepsin secretion in 15-day-old rats is further evidence that the cholinergic receptor system on chief cells is fully developed by this time (35).

Thus, as is the case for acid secretion, pepsin secretion is first capable of being stimulated by cholinergic agents. Because peptic activity is dependent on acid secretion, this is not surprising. The autonomic nervous system of the gut also develops early. In the longitudinal smooth muscle of the fetal rabbit and mouse, functional cholinergic innervation and responsiveness to acetylcholine appear on the 16th–17th day of gestation (24). A recent report demonstrates that choline acetyltransferase and acetylcholinesterase activity of the rat gastric pylorus are as great 1 day after birth as 50 days after birth (27). Finally, histological studies suggest the enteric plexus is well developed in the neonate. Although it has fewer total nerve cells than in the adult, the neonatal rat plexus has a density of 64,000 nerve cells/cm^2 of the gastrointestinal surface area compared with 9,000 nerve cells for the same area in the adult (20).

In human fetuses, identifiable chief cells are first seen after 13–14 wk of gestation (73). Peptic activity has been demonstrated as early as the 16th wk in gestation (47), and 3- to 4-fold increases occurred from the 28th to 40th wk (91). As shown in Figure 2, pepsin secretion into the lumen of the stomach parallels acid secretion during the neonatal period and by 2-3 mo of age is ~50% of the adult levels. By 2 yr of age pepsin secretion has reached the adult values (69).

GASTRIN

Development of the gastrin radioimmunoassay in 1970 (66, 94) made it possible to study tissue and serum gastrin levels. Numerous investigators have reported that human infants at birth have elevated serum gastrin levels compared with adults (14, 15, 61, 71). Most investigators agree that the hypergastrinemia is of short duration—lasting only a few days (14, 15, 61). One report of human fetal antral gastrin indicates amounts ≅10% of adult levels (83).

Most studies involving the ontogeny of gastrin have been performed on the rat. Studies of developmental changes in serum gastrin have involved larger mammals—primarily sheep. In general the animal results are similar to the human results, demonstrating hypergastrinemia and low tissue levels of the hormone in newborns.

Tissue Levels

Immunoreactive gastrin first appears on the 18th day of gestation in the rat (6, 55). It gradually increases until the 18th day after birth and then increases dramatically to adult levels at day 21 (58).

Lichtenberger and Johnson (58) found that rat antral gastrin levels were ~2 μg/g tissue from birth through day 18; by day 21 there was a 10-fold increase. Takeuchi et al. (80) recently reported similar values. This abrupt change coincides with the weaning period and the major changes in gastric secretion already discussed in *Basal Acid Secretion*, p. 346. In the sheep, gastrin first appears in significant amounts in the 107-day-old fetus. The abomasal gastrin is almost totally G-17, with a small peak eluting with G-34 (57, 76). This same pattern persists through birth and at least until 6 days after birth (57). In the sheep therefore there are no changes in the molecular species of gastrin during development.

Serum Levels

In addition to humans, neonatal hypergastrinemia occurs in dogs (65), sheep (57, 76), and rats (41, 80). Malloy et al. (65) found that serum gastrin in puppies rises above adult levels shortly after birth and continues to increase through the 9th day of life. In the sheep, Shulkes and Hardy (76) detected gastrin in the serum of 101-day-old fetuses. Serum levels then rose steadily to adult levels by the 124th day of gestation and to 5 times the adult levels on the 5th day after birth. By day 15 serum gastrin had returned to adult levels. Similar results were reported by Lichtenberger et al. (57) except that hypergastrinemia remained through day 60 after birth. Takeuchi et al. (80) found markedly elevated serum gastrin levels in newborn rats that decreased abruptly (3- to 4-fold) on day 18 and reached adult levels by day 40.

A number of explanations for hypergastrinemia in newborn animals exist. Lichtenberger (56) has suggested that the high protein and calcium content of the neonate's diet plays a role, because both of these are releasers of gastrin. However, this seems unlikely because hypergastrinemia in the human disappears long before there is a change in the diet (14, 15). Furthermore, if the newborn rat is prevented from weaning, serum gastrin levels decrease at the same time as those of weaned littermates eating only solid food (80).

Gastrin release is inhibited when the pH of the antral contents drops to ~3.0 (92). As described in *Basal Acid Secretion*, p. 346, there is little gastric acid secretion and the pH of the contents remains above 4.0 until 18 days after birth, which is not low enough to significantly inhibit gastrin release. If 150 mN HCl is introduced into the stomachs of 10- and 15-day-old rats, there is a prompt and significant decrease in circulating gastrin (41). This experiment clearly demonstrates that a significant portion of postnatal hypergastrinemia is of antral origin and is due to insufficient acid to trigger the inhibitory mechanism.

When the gastric contents were acidified in the above experiment, serum gastrin did not decrease to adult values (41). This might be expected if some of the gastrin were derived from extra-antral sources. However, on a per gram basis the duodenal mucosa of the unweaned rat contains only 1/50 as much gastrin as the antrum (58). Incomplete development of the inhibitory mechanism itself could also account for this finding. A number of observations indicate that somatostatin may mediate the acid inhibition of gastrin release (53, 54, 74) and that somatostatin exerts a tonic inhibition of gastrin release (72). Somatostatin injection had no effect on serum gastrin levels in 10- and 15-day-old rats (41). Beginning on day 18, somatostatin significantly lowered serum gastrin. The antral mucosa of newborn rats, therefore, appears to lack receptors and/or the mechanism to respond to somatostatin until weaning. Furthermore, Koshimizu (50) has shown that gastric somatostatin content gradually increases in the rat until the time of weaning when there is an abrupt, large increase. Therefore there is evidence that the decrease in serum gastrin levels at the time of weaning is due to two independent mechanisms: an acid-sensitive one, triggered by the onset of significant acid secretion, that is intact before weaning and a separate mechanism, dependent on somatostatin, that is not present until weaning (41).

GROWTH

Changes in growth of the gastric mucosa can be expected to affect its functional development. Unfortunately not many studies have addressed this topic. The rate of DNA synthesis per unit of DNA in the gastric mucosa of the 5- and 10-day-old rat is ~5 times higher than in the adult (62). Between days 15 and 20 the rate decreases to adult levels. During this time, however, the mucosa grows rapidly and, as gastrin receptors appear, the mucosa becomes sensitive to the trophic action of gastrin (39, 40).

In adults normal populations of gastric mucosal cells are maintained by sustained proliferation of progenitor cells, their maturation, and exfoliation (60). Figure 4 shows that the high rate of DNA synthesis in the oxyntic gland of unweaned rats is matched by a high rate of cell loss (62). Although there was a decrease in the rate of DNA synthesis per cell at the time of weaning, there was a significant increase in DNA content and DNA synthesis for the entire oxyntic gland mucosa (62). Both oxyntic gland and antral mucosa undergo a rapid increase in weight as well as DNA, RNA, and protein content around the time of weaning (84). The ratios of protein to DNA and RNA to DNA in the oxyntic and antral gland increased gradually from days 5 through 25. These data indicate that the rapid growth of the gastric mucosa in the 3rd wk of life is associated with hypertrophy of the organ (84).

These data also suggest that, as the mucosa develops, a large proportion of cells is removed from the proliferative pool. To investigate whether these cells mature into functioning mucosal cells, Tseng and Johnson examined the relative number of chief, parietal, and undifferentiated cells over the bases of gastric glands where mature cells are located, as described by Kataoka (46). From postnatal days 15 to 20, the relative percentage of chief cells in the gastric mucosa increased from 19% to 60%, whereas parietal cells increased from 15% to 19%. At the same time undifferentiated cells decreased from 66% to 21% (C.-C. Tseng and L. R. Johnson, unpublished observations). These data suggest that although gastric mucosal cells are proliferating, they also differentiate into mature forms.

The growth rate of individual cells in oxyntic glands was studied by Jacobs and Ackerman (36) in 15-, 21-, 30-, and 100-day-old rats. They found that surface epithelial cells reached adult dimensions by day 21, whereas parietal and chief cells continued to increase in size until day 100. The numbers of each individual cell per unit area during development also showed great variation (36). Surface epithelial cells decreased in relative number over time, whereas the chief cell population increased. Parietal cells at the base of glands continued to increase in number after day 30; those in the glandular neck did not. These investigators concluded that the growth characteristics of gastric cells differ from each other. They also pointed out that surface epithelial cells, believed to have protective functions, reach mature dimensions earlier than chief and parietal cells, which produce substances that are potentially harmful to the gastric mucosa (36).

REGULATION OF GASTRIC DEVELOPMENT

The inescapable conclusion is that all the gastric mucosal functions in the rat undergo important developmental changes at approximately the same time. These occur during the 3rd wk of life coincident with the beginning of weaning. The coordination of these events suggests the presence of an overriding factor or factors that regulate that development. The most obvious candidates for this role are hormones, the change to a solid diet or the weaning process itself, and intrinsic or genetically programmed changes. Most of these have been examined in relation to small intestine development (32). However, only a few comparable studies have been carried out on gastric function.

Diet or Weaning

Throughout this chapter it has been mentioned that changes have taken place at the time of weaning. This was in no way meant to imply that these developmental changes were triggered or regulated by the act of weaning or the dietary change that occurs at that

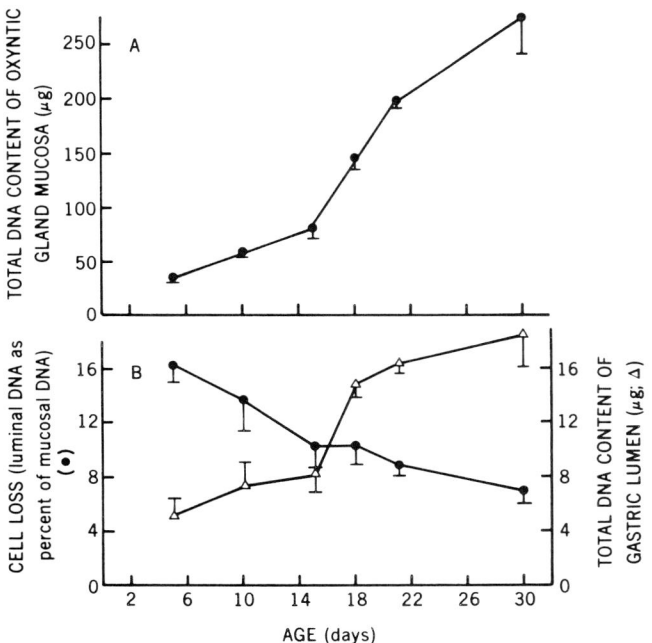

FIG. 4. Changes in DNA content in oxyntic gland mucosa (A) and in gastric lumen (B) at various stages of development. [From Majumdar and Johnson (62).]

time. Actually the evidence from other gastrointestinal tissues is that weaning plays only a minor role, if any, in modulating these changes (5, 32). Ontogenic changes in the small intestine, liver, and pancreas begin at the normal time in rats prevented from weaning.

Similar results have been obtained regarding antral and serum gastrin (58, 68, 80). As shown in Figure 5, antral gastrin levels begin to increase in unweaned rats at the same time as in normally weaned animals. The rate of increase, however, is less, and the final level of hormone reached in the tissue is significantly lower (80). Serum gastrin also falls at the same time in unweaned as it does in weaned rats and in fact is more rapid in unweaned animals. The important finding is that the changes still took place at the same time. The differences in final levels of tissue and serum gastrin in unweaned animals are not related to development but are dependent on the stimulation of gastrin release, which is significantly less in rats ingesting their calories in liquid form (42, 77). In rats separated from their mothers and weaned prematurely on day 13, there were no significant differences in antral or serum gastrin when they were killed on day 18 and compared with normally developing animals of the same age. However, a positive result in the above study would be suspect, because premature weaning causes stress and elevates corticosterone, which is the true mediator of such changes.

There are no reports of the effect of weaning on development of acid and pepsin secretion or other gastric functional changes.

Corticosterone

Glucocorticoids have received wide attention as mediators of developmental changes in rat pups because serum concentrations increase significantly at the beginning of the 3rd postnatal wk. In rat serum, concentrations of free corticosterone begin to rise on postnatal day 14 and reach peak levels between days 15 and 19 (31). This period of high serum corticosterone coincides with the start of the normal gastric development.

Glucocorticoids have been shown to affect the development and function of many small intestinal and hepatic enzymes (32). There have been a few similar studies indicating a role for glucocorticoids in gastric development. Using organ culture methods, Yeomans et al. (95) demonstrated that addition of cortisone markedly increased pit gland formation in gastric explants from 18-day-old fetuses. Cytodifferentiation was also augmented by cortisol (95). During the 1st or 2nd postnatal wk, injection of hydrocortisone (19, 51, 52, 82), ACTH (19), or corticosterone (35, 84) prematurely increases the amount of gastric mucosal pepsinogen. Adrenalectomy of pups on the 7th postnatal day significantly decreases mucosal pepsinogen levels by day 20 (85). As shown in Figure 6A, corticosterone replacement in adrenalectomized animals prevented the decrease in pepsinogen levels and actually brought pepsinogen to levels above those found in normal rats. Thyroxine administration, however, failed to do so. Those data clearly indicate that corticosterone is essential for the development of pepsinogen levels in the rat gastric mucosa.

To determine whether the above effect of corticosterone is due to the stimulation of mucosal cell growth, Tseng and Johnson (84) measured the weight as well as the DNA, RNA, and protein content of the gastric mucosa in normal and corticosterone-treated animals during development. Corticosterone acetate was given on postnatal days 7, 9, and 11, and the gastric mucosa was studied from days 10 through 25. The weight of oxyntic gland mucosa was not changed by corticosterone treatment at any day measured nor was the DNA, RNA, or protein content (84). Although these data indicate that corticosterone treatment does not change overall cell proliferation, it may change relative cell populations within the gastric mucosa. An-

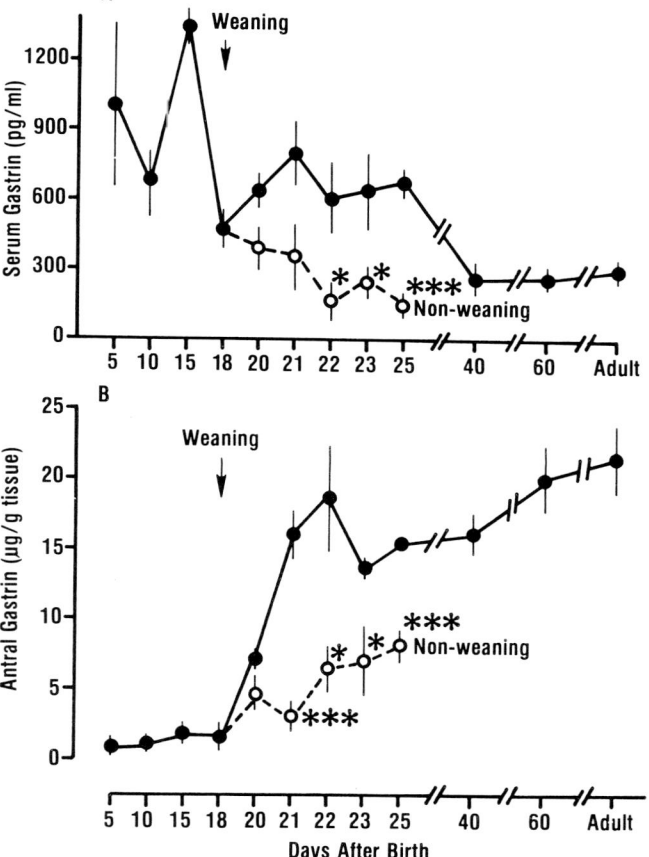

FIG. 5. Serum (A) and antral (B) gastrin levels in rats of various ages showing effect of weaning on day 18. *, $P < 0.05$; ***, $P < 0.001$. [From Takeuchi et al. (80).]

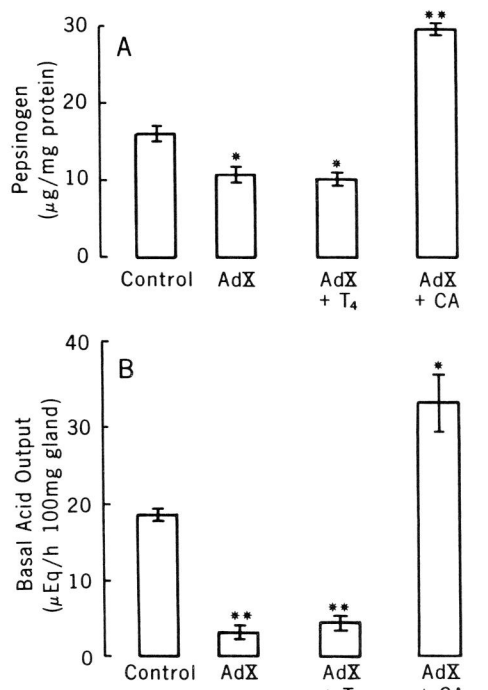

FIG. 6. Pepsinogen content (A) and basal acid secretion (B) on day 20 in control, adrenalectomized (AdX), adrenalectomized treated with thyroxine (AdX + T$_4$), and adrenalectomized treated with corticosterone (AdX + CA) rats. *, $P < 0.05$; **, $P < 0.01$. Adrenalectomy occurred on day 7 and rats were injected daily with 10 μg/g body wt of corticosterone or with 0.1 μg/g body wt of L-thyroxine. [From Tseng and Johnson (85).]

other possibility is that corticosterone causes maturational changes in existing differentiating cells.

Tseng et al. (86) recently studied the effect of corticosterone on the structural development of chief cells during the early postnatal period. Administration of corticosterone significantly increased the relative number of chief cells from 59% to 70% and decreased undifferentiated cell numbers from 22% to 9%. In those animals basophilia was increased in the basal portion of the chief cells. Under electron microscopy RER was more abundant and pepsinogen granules were in greater numbers than in controls. These changes were reflected quantitatively by an increase in the volume fraction of zymogen granules and the surface density of RER in the chief cells of corticosterone-treated animals (86). When rats were subjected to adrenalectomy, the percentage of identifiable chief cells decreased to 40% while that of undifferentiated cells increased to 49%; the size and number of pepsinogen granules decreased and the cytoplasm of chief cells contained less RER. After stereological analysis the volume fraction of pepsinogen granules and the surface density of RER were significantly reduced. Those results showed that the developmental increase in pepsinogen levels in corticosterone-treated animals is associated with the precocious maturation of chief cells, whereas the low pepsinogen levels in adrenalectomized animals were accompanied by retardation of morphological development. Furthermore corticosterone was effective in restoring the normal structural appearance of chief cells in the adrenalectomized rats, whereas thyroxine had no such effect. Therefore corticosterone plays an important part in both functional and morphological development of chief cells during the early postnatal period.

Information regarding the role of corticosterone in the development of parietal cells is scarce. Garzon et al. (21) found that triamcinolone injection in 20-day-old fetuses significantly decreased the gastric pH 3 h after injection. They suggested that this short-term effect of corticosterone was due to stimulation of the hydrogen generation process in already differentiated parietal cells rather than a corticosterone-induced architectural maturation of parietal cells. Tseng and Johnson (84) found that administration of corticosterone on postnatal days 7, 9, and 11 significantly increased basal acid secretion from days 10 through 20 and that this increase was related to the dose of corticosterone. As indicated in Figure 6B, basal acid secretion in adrenalectomized animals is low. Throxine has no effect on basal acid secretion in adrenalectomized rats, whereas corticosterone prevented the decrease and resulted in higher than normal rates of acid secretion. Those studies suggest that corticosterone also performs a significant role in the development of acid secretion.

Structurally, however, corticosterone appeared to have little effect on parietal cell development in the 20-day-old rat. In response to additional corticosterone, parietal cells of normal animals appeared to be larger and to contain more mitochondria as well as tubulovesicles and intracellular canaliculi than did controls (87). Unlike chief cells, parietal cells retained normal ultrastructure in the absence of adrenal hormones. These data indicate that the morphological development of parietal cells does not depend on corticosterone after postnatal day 7. Therefore changes in ultrastructure cannot account for the failure of acid secretion to develop in adrenalectomized animals. Tseng et al. (86) also found that adrenalectomy decreased the number of identifiable parietal cells from 18.5% to 13.6%. Although this was a significant decrease, it could not account for such drastic changes in acid secretion. As described in *Acid Secretory Response to Stimuli*, p. 347, the development of acid secretory responses to the three secretagogues was not complete until postnatal day 20 (35). Injection of corticosterone into 8-day-old rats prematurely induced acid secretion on day 12 in response to all three secretagogues. It is therefore possible that the development of receptors to the three naturally occurring stimulants may be delayed in the absence of normal levels of adrenal hormones. We have shown that the normal response of parietal cells to the three secretagogues was absent in 20-day-old adrenalectomized rats (87). Another possibility is that levels of enzymes, such as carbonic anhydrase or proton-K^+-ATPase, involved in acid secretion are functionally related to corticosterone.

Injection of 250 mg/kg corticosterone on postnatal day 7 caused the precocious appearance of gastrin receptors on day 10 (68). By day 12 gastrin binding was well developed and the stomach secreted significant amounts of acid in response to pentagastrin. In the same rats, antral levels of gastrin also developed prematurely and by day 15 had reached 80% of adult levels. Adrenalectomy delayed the normal development of both gastrin receptors and tissue levels of the hormone by ~5–7 days. After this period, however, development proceeded to normal levels. The delay in the onset of development was prevented if adrenalectomized animals were treated with corticosterone (68). For full development of receptors and tissue gastrin the rats had to be maintained on normal solid laboratory pellets as opposed to liquid diets. This dietary dependence reflects the necessity for gastrin release to maintain both tissue levels of the hormone and receptor levels. Corticosterone has been shown to precociously induce sensitivity of pepsin secretion to carbachol in 15-day-old rats (35) and to significantly increase pepsinogen secretion in response to cholecystokinin, carbachol, and Ca^{2+} ionophore A23187 in isolated gastric glands from 1-day-old rats (93).

Thyroxine

Thyroxine is also an attractive candidate for the regulation of gastric development. In rats, circulating thyroid hormone concentrations are low at birth and increase rapidly between days 4 and 16 to reach a peak of 6 μg/dl (12). The ontogenic rise of free serum thyroxine is similar to that of total thyroxine. Free thyroxine concentrations rise rapidly between days 5 and 12 to levels identical to adult concentrations (89).

Administration of thyroxine induces the precocious development of many enzymes in the rat small intestine (for review see ref. 31). However, few studies involving thyroid hormones and gastric development have been reported (51, 52, 85, 86). Most of these showed that thyroxine administration in the 1st and 2nd postnatal wk induced a premature increase in pepsinogen activity (51, 85, 86). Because thyroxine had been shown to increase glucocorticoids levels (64), it is possible that the effect of administered thyroxine may reflect a primary action of corticosterone. A study of the postnatal development of sucrase and maltase in rat small intestine has indicated that the precocious increase of both enzymes is due to the accompanying increase in serum corticosterone (8).

Daily injection of thyroxine from postnatal day 7 significantly increased pepsinogen levels of oxyntic gland mucosa (85). As shown in Figure 7A, pepsinogen levels in hypothyroid rats were significantly decreased. Thyroxine injections prevented their decrease, whereas treatment with corticosterone significantly increased pepsinogen above normal (85). The combination of thyroxine and corticosterone resulted in levels not significantly different from those seen with corticosterone alone. In 15-day-old rats the serum corticosterone concentration of thyroxine-treated pups was twice that of normal pups (85). Figure 8 demonstrates that when serum corticosterone and mucosal pepsinogen levels in normal, adrenalectomized, and thyroxine-treated animals are measured at day 15, there is a linear relationship between serum corticosterone and pepsinogen content (85). Moreover chief cells from thyroxine-injected animals were structurally similar to those from rats treated with corticosterone, as demonstrated by increased numbers of zymogen granules and increased area of RER. When pups were made hypothyroid by feeding them propylthiouracil-containing water, the morphological development of chief cells was retarded. These changes are similar to those found in adrenalectomized animals. Moreover, unlike the situation with adrenalectomy, either corticosterone or thyroxine restored the ultrastructure of the chief cells to normal in hypothyroid rats. These studies indicate that administration of

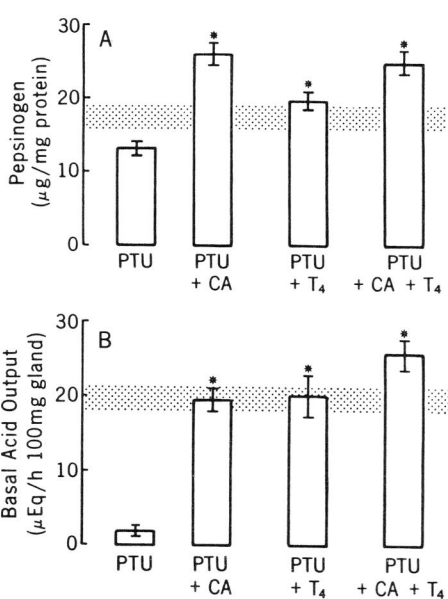

FIG. 7. Pepsinogen content (A) and basal acid secretion (B) on day 20 of propylthiouracil-induced hypothyroid rats (PTU) and of hypothyroid rats treated with corticosterone (PTU + CA), thyroxine (PTU + T_4), and both (PTU + CA + T_4). Shaded area represents normal values from control rats of same age. *, $P < 0.05$ compared with PTU group. Water containing propylthiouracil (0.001%) was given to dams and pups from day 0 through entire experimental period to induce hypothyroidism. Hypothyroid rats were injected on days 7, 9, and 11 with 200 μg/g body wt of corticosterone or daily from day 7 with 0.05 μg/g body wt of thyroxine or both. [From Tseng and Johnson (85).]

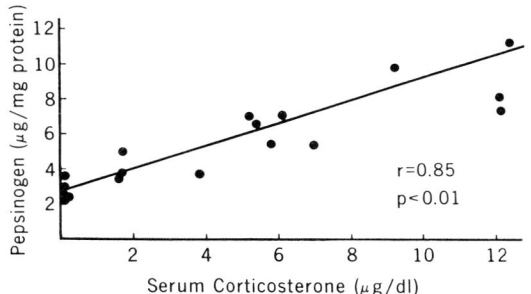

FIG. 8. Correlation between serum corticosterone levels and pepsinogen content of oxyntic gland mucosa from normal, thyroxine-treated, and adrenalectomized rats on day 15. Adrenalectomy occurred on day 7. L-Thyroxine (0.1 µg/g body wt) was given daily from day 7. [From Tseng and Johnson (85).]

thyroxine during the 1st and 2nd postnatal wk precociously induces chief cell maturation. However, the effect of thyroxine is secondary to corticosterone.

In response to daily thyroxine injections, basal acid secretion increased significantly over controls 3 days after the first injection (85). Basal acid output of hypothyroid pups was only 10% of normal (Fig. 7B), although injection of either corticosterone or thyroxine totally prevented this decrease. The combination of corticosterone and thyroxine resulted in secretory rates not significantly higher than when either was administered alone. These results also indicate that the effect of thyroxine on the development of acid secretion is due to corticosterone. Parietal cells in thyroxine-injected rats were larger and contained more mitochondria, tubulovesicles, and intracellular canaliculi than those in controls (87). In hypothyroid pups, parietal cells maintained normal ultrastructure, and either thyroxine or corticosterone injections increased mitochondria and secretory membrane surface area (87). These studies also support the conclusion that the effect of thyroxine on parietal cell development is due to corticosterone.

Administration of thyroxine from postnatal day 7 significantly decreased serum gastrin and increased antral gastrin levels on day 20 (Fig. 9). In contrast, hypothyroid rats had higher serum and lower antral gastrin levels than did normal animals.

Influence of Sex

To this point no distinction has been made between male and female rats. Male rats secrete more gastric acid (43, 79) and have higher concentrations of serum gastrin (43, 59) and gastrin receptors (43) than females. Development of receptors and serum gastrin in the two sexes is identical until day 40, which is the approximate time of puberty in the rat (43), after which it plateaus in the female but keeps increasing in the male. Both gastrin levels and receptors reach final adult values by day 60 in the male. If female rats were ovariectomized, development proceeds to male levels. Castration of males has no effect on the development of final levels of either serum gastrin or receptors. Injection of castrated males with estrogen, however, lowers these values. Serum gastrin values and hence gastrin-receptor concentrations are dependent on food intake (81). Because male rats begin to eat more than females after puberty, this explains part of the differences. However, if ovariectomized females are pair fed to normal females, they still develop significantly higher serum gastrins and mucosal gastrin-binding capacities. If ovariectomized females are adrenalectomized as well and pair fed to normal females, gastrin levels and receptors are not significantly different from those of males (43).

In summary, functional and morphological development of the chief cell is dependent on glucocorticoids up until the normal initiation of change during the 3rd wk of life. This dependence, however, is for timing only, because development, although delayed, reaches normal levels in adrenalectomized rats. In contrast, the morphological development of parietal cells does not appear to depend on corticosterone. This is not surprising, considering that structural maturation of parietal cells takes place before the developmental surge in serum corticosterones. The discrepancy between functional and morphological development of parietal cells may be due to the delayed development of receptors to the various secretagogues. Another possibility is that enzymes involved in acid secretion do not reach full activity until the 3rd postnatal wk. Injection of thyroxine during the 1st and

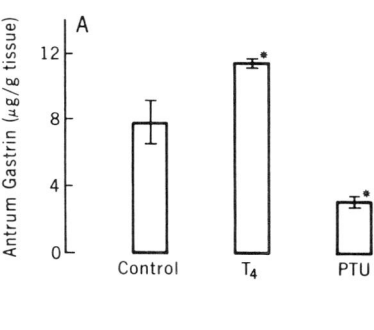

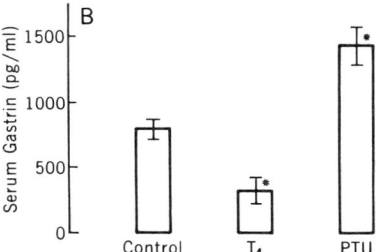

FIG. 9. Antral (A) and serum (B) gastrin levels in control, thyroxine-treated (T_4), and propylthiouracil-induced hypothyroid (PTU) rats on day 20. *, $P < 0.05$. L-Thyroxine (0.1 µg/g body wt) was given daily from day 7. Water containing propylthiouracil (0.001%) was given to dams and pups from day 0 through 20.

2nd postnatal wk induced precocious increases in acid secretion and pepsinogen levels. However, those effects were secondary to corticosterone. Normal intake of solid food is necessary for total development of gastrin levels and receptor-binding capacities. Estrogen influences development in female rats, resulting in reductions in food intake, acid secretion, serum gastrin, and gastrin receptors. These effects are due in part to direct hormonal action and in part to the indirect effects of reduced food intake.

Intrinsic Mechanism

Although the dietary and hormonal changes associated with the weaning period modulate, and in the case of corticosterone may trigger, events in gastric functional development, development can proceed in the absence of these changes. In recent studies the pepsinogen content of the oxyntic gland mucosa still increased gradually when serum corticosterone concentrations were undetectable (85). This indicates that the elimination of the developmental surge in circulating corticosterone delays but does not abolish the development of the chief cell. Moreover the pepsinogen content of normal animals reaches the same level as that of corticosterone-treated animals by day 25 (84), even in the absence of excess corticosterone. This suggests that the time course as well as the events themselves are ultimately regulated intrinsically by the genetic material of the cells involved. This hypothesis is strengthened by the observation that small intestine mucosa taken from a near-term fetal rat and implanted under the kidney capsule of a syngeneic adult host shows normal patterns of sucrase, lactase (16, 49), and β-galactosidase (49) development. The jejuno-ileal gradient of enzyme activity develops fully in isografts without exposure to luminal contents (49). Similar studies with gastric tissue have not been reported.

SUMMARY

Large increases in gastric acid and pepsin secretion, antral gastrin concentration, and decreases in serum gastrin occur during the 3rd wk of life in the neonatal rat. At the same time gastrin receptors appear and gastrin release becomes sensitive to somatostatin, indicating that absence and then appearance of specific hormone receptors may be responsible for some of the ontogenic pattern. At this time the mucosa also begins to grow rapidly, with a greater proportion of cells leaving the proliferative pool and differentiating. For the first 2-3 wk these ontogenic changes can be triggered by corticosterone. Their full expression depends on dietary changes associated with weaning. Thyroxine plays an indirect role by maintaining the normal development of corticosterone. Neither hormones, dietary changes, nor the weaning process itself is essential for development; in the absence of these, all of the changes still occur, although they may be delayed or be smaller in magnitude. These findings indicate that ontogeny is genetically programmed and that the full expression of this program depends on hormones, luminal contents, and other environmental factors.

The morphological development of chief cells parallels the increase in mucosal pepsinogen content. However, the ultrastructural maturation of parietal cells takes place earlier than the development of acid secretion.

In comparison with the small intestine, gastric ontogeny has not received adequate attention. For example, there are essentially no studies directed toward understanding changes in motility during this period. In addition little is known about the roles of epidermal growth factor, insulin, glucagon, and prostaglandin in gastric development. Studies implanting fetal tissue into adult hosts are needed to determine which gastric functions can develop in the absence of luminal stimulation and hormone changes.

The cell biology of the gastric mucosa is difficult to examine—especially that involving the cells concerned with growth and differentiation. The stem cells are dispersed throughout the tissues and are a small portion of the cell population; consequently these have never been isolated for study. In vitro culture of mucosal cells, however, is a technique that can possibly be used to examine development at cellular and molecular levels.

This work was supported by National Institutes of Health Grant DK 37260.

REFERENCES

1. ACKERMAN, S. H. Ontogeny of gastric acid secretion in the rat: evidence for multiple response system. *Science Wash. DC* 217: 75-77, 1982.
2. ACKERMAN, S. H., AND R. D. SHINDLEDECKER. Maturational increases and decreases in acid secretion in the rat. *Am. J. Physiol.* 247 (*Gastrointest. Liver Physiol.* 10): G638-G644, 1984.
3. AGUNOD, M., N. YAMAGUCHI, R. LOPEZ, A. L. LUHBY, AND G. B. J. GLASS. Correlative study of hydrochloric acid, pepsin, and intrinsic factor secretion in newborns and infants. *Am. J. Dig. Dis.* 14: 400-414, 1969.
4. AURES, D., W. D. DAVIDSON, AND R. HÅKANSON. Histidine decarboxylase in gastric mucosa of various animals. *Eur. J. Pharmacol.* 8: 100-117, 1969.
5. BOYLE, J. T., AND O. KOLDOVSKÝ. Critical role of adrenal glands in precocious increase in jejunal sucrase activity following premature weaning in rats: negligible effect of food intake. *J. Nutr.* 101: 169-177, 1980.
6. BRAATEN, J. T., M. H. GREIDER, J. E. MCGUIGAN, AND D. H. MINTZ. Gastrin in the perinatal rat pancreas and gastric antrum: immunofluorescence localization of pancreatic gastrin cells and gastrin secretion in monolayer cell cultures. *Endocrinology* 99: 684-691, 1976.
7. CHRISTIE, D. L. Development of gastric function during the first month of life. In: *Textbook of Gastroenterology and Nutri-*

tion in Infancy, edited by E. Lebenthal. New York: Raven, 1981, vol. 1, p. 109–120.

8. D'AGOSTINO, J. B., AND S. J. HENNING. Role of thyroxine in coordinate control of corticosterone and CBG in postnatal development. *Am. J. Physiol.* 242 (*Endocrinol. Metab.* 5): E33–E39, 1982.

9. DEREN, J. S. Development of structure and function in the fetal and newborn stomach. *Am. J. Clin. Nutr.* 24: 144–159, 1971.

10. DONALDSON, R. M., JR. Intrinsic factor and the transport of cobalamin. In: *Physiology of the Gastrointestinal Tract*, edited by L. R. Johnson. New York: Raven, 1981, vol. 1, p. 641–658.

11. DUCROC, R., J.-F. DESJEUX, B. GARZON, J.-P. ONOLFO, AND J.-P. GELOSO. Acid secretion in fetal rat stomach in vitro. *Am. J. Physiol.* 240 (*Gastrointest. Liver Physiol.* 3): G206–G210, 1981.

12. DUSSAULT, J. H., AND F. LABRIE. Development of the hypothalamic-pituitary-thyroid axis in the neonatal rat. *Endocrinology* 97: 1321–1324, 1975.

13. EBERS, D. W., D. I. SMITH, AND G. E. GIBBS. Gastric acidity during the first month of life. *Pediatrics* 18: 800–802, 1956.

14. EULER, A. R., W. J. BYRNE, L. M. COUSINS, M. E. AMENT, R. D. LEAKE, AND J. H. WALSH. Increased serum gastrin concentrations and gastric acid hyposecretion in the immediate newborn period. *Gastroenterology* 72: 1271–1273, 1977.

15. EULER, A. R., W. J. BYRNE, P. J. MEIS, R. D. LEAKE, AND M. E. AMENT. Basal and pentagastrin-stimulated acid secretion in newborn human infants. *Pediatr. Res.* 13: 36–37, 1979.

16. FERGUSON, A., V. P. GERSHOWITCH, AND R. I. RUSSELL. Pre- and post-weaning disaccharidase patterns in isografts of fetal mouse intestine. *Gastroenterology* 64: 292–297, 1973.

17. FORTE, J. G., T. M. FORTE, AND T. E. MACHEN. Histamine stimulated hydrogen ion secretion by in vitro piglet gastric mucosa. *J. Physiol. Lond.* 244: 15–31, 1975.

18. FURIHATA, C., Y. IWASAKI, T. SUGIMURA, M. TATEMATSU, AND M. TAKAHASHI. Differentiation of pepsinogen-producing cells in the fundic and pyloric mucosa of developing rats. *Cell Differ.* 2: 179–189, 1973.

19. FURIHATA, C., T. KAWACHI, AND T. SUGIMURA. Premature induction of pepsinogen in developing rat gastric mucosa by hormones. *Biochem. Biophys. Res. Commun.* 47: 705–711, 1972.

20. GABELLA, G. Neuron size and number in the myenteric plexus of the newborn and adult rat. *J. Anat.* 109: 81–95, 1971.

21. GARZON, B., R. DUCROC, AND J.-P. GELOSO. Ontogenesis of gastric acid secretion in fetal rat. *Pediatr. Res.* 15: 921–925, 1981.

22. GARZON, B., R. DUCROC, AND J.-P. GELOSO. Ontogenesis of gastric response to agonists and antagonists of acid secretion in fetal rat. *J. Dev. Physiol. Oxf.* 4: 195–205, 1982.

23. GARZON, B., R. DUCROC, J.-P. ONOLFO, J.-F. DESJEUX, AND J.-P. GELOSO. Biphasic development of pentagastrin sensitivity in rat stomach. *Am. J. Physiol.* 242 (*Gastrointest. Liver Physiol.* 5): G111–G115, 1982.

24. GERSHON, M. D., AND E. B. THOMPSON. The maturation of neuromuscular function in a multiple innervated structure: development of the longitudinal smooth muscle of the fetal mammalian gut and its cholinergic excitatory, adrenergic inhibitory, and nonadrenergic inhibitory innervation. *J. Physiol. Lond.* 234: 257–277, 1973.

25. HÅKANSON, R., C. OWMAN, AND N. P. SJÖBERG. Cellular stores of gastric histamine in the developing rat. *Life Sci.* 6: 2535–2543, 1967.

26. HAYDARD, A. F. The fine structure of gastric epithelial cells in the suckling rabbit with particular reference to the parietal cell. *Z. Zellforsch. Mikrosk. Anat.* 78: 474–483, 1967.

27. HEITKEMPER, M. M., AND S. F. MAROTTA. Development of neurotransmitter enzyme activity in the rat gastrointestinal tract. *Am. J. Physiol.* 244 (*Gastrointest. Liver Physiol.* 7): G58–G64, 1983.

28. HELANDER, H. F. Ultrastructure and function of gastric parietal cells in the rat during development. *Gastroenterology* 56: 35–52, 1969.

29. HELANDER, H. F. Ultrastructure and function of gastric mucoid and zymogen cells in the rat during development. *Gastroenterology* 56: 53–70, 1969.

30. HELANDER, H. F. Stereological change in rat parietal cells after vagotomy and antrectomy. *Gastroenterology* 71: 1010–1018, 1976.

31. HENNING, S. J. Plasma concentrations of total and free corticosterone during development in the rat. *Am. J. Physiol.* 235 (*Endocrinol. Metab. Gastrointest. Physiol.* 4): E451–E456, 1978.

32. HENNING, S. J. Postnatal development: coordination of feeding, digestion, and metabolism. *Am. J. Physiol.* 241 (*Gastrointest. Liver Physiol.* 4): G199–G214, 1981.

33. HENNING, S. J., AND L. L. LEEPER. Coordinate loss of glucocorticoid responsiveness by intestinal enzymes during postnatal development. *Am. J. Physiol.* 242 (*Gastrointest. Liver Physiol.* 5): G89–G94, 1982.

34. HENNING, S. J., AND J. M. SIMS. Delineation of the glucocorticoid-sensitive period of intestinal development in the rat. *Endocrinology* 104: 1158–1163, 1979.

35. IKEZAKI, M., AND L. R. JOHNSON. Development of sensitivity to different secretagogues in the rat stomach. *Am. J. Physiol.* 244 (*Gastrointest. Liver Physiol.* 7): G165–G170, 1983.

36. JACOBS, D. M., AND S. H. ACKERMAN. Differential growth rate of rat gastric mucosal cells during postnatal ontogeny. *Am. J. Physiol.* 247 (*Gastrointest. Liver Physiol.* 10): G645–G650, 1984.

37. JOHNSON, L. R. Pepsin secretion stimulated by topical hydrochloric and acetic acids. *Gastroenterology* 62: 33–38, 1971.

38. JOHNSON, L. R. Regulation of pepsin secretion by topical acid in the stomach. *Am. J. Physiol.* 223: 847–850, 1972.

39. JOHNSON, L. R. The trophic action of gastrointestinal hormones. *Gastroenterology* 70: 278–288, 1976.

40. JOHNSON, L. R. Regulation of gastrointestinal growth. In: *Physiology of the Gastrointestinal Tract*, edited by L. R. Johnson. New York: Raven, 1981, vol. 1, p. 169–196.

41. JOHNSON, L. R. Effects of somatostatin and acid on inhibition of gastrin release in newborn rats. *Endocrinology* 114: 743–746, 1984.

42. JOHNSON, L. R., AND P. D. GUTHRIE. Regulation of antral gastrin content. *Am. J. Physiol.* 245 (*Gastrointest. Liver Physiol.* 8): G725–G729, 1983.

43. JOHNSON, L. R., W. PEITSCH, AND K. TAKEUCHI. Mucosal gastrin receptor. VIII. Sex-related differences in binding. *Am. J. Physiol.* 243 (*Gastrointest. Liver Physiol.* 6): G469–G474, 1982.

44. JOHNSON, L. R., AND D. B. TUMPSON. Effect of secretin on histamine-stimulated secretion in gastric fistula rats. *Proc. Soc. Exp. Biol. Med.* 133: 125–127, 1970.

45. JUMAWAN, J., AND O. KOLDOVSKÝ. Comparison of the effect of various doses of thyroxine on jejunal disaccharidases in intact and adrenalectomized rats during the first 3 weeks of life. *Enzyme Basel* 23: 206–209, 1978.

46. KATAOKA, K. Electron microscopic observation on cell proliferation and differentiation in the gastric mucosa of the mouse. *Arch. Histol. Jpn.* 32: 251–273, 1970.

47. KEENE, M. F. L., AND E. E. HEWER. Digestive enzymes of the human fetus. *Lancet* 1: 767–769, 1929.

48. KELLY, K. A. Motility of the stomach and gastroduodenal junction. In: *Physiology of the Gastrointestinal Tract*, edited by L. R. Johnson. New York: Raven, 1981, vol. 1, p. 393–410.

49. KENDALL, K., J. JUMAWAN, AND O. KOLDOVSKÝ. Development of jujunoileal differences of activity of lactase, sucrase and acid β-galactosidase in isografts of fetal rat intestine. *Biol. Neonate* 36: 206–214, 1979.

50. KOSHIMIZU, T. The development of pancreatic and gastrointestinal somatostatin-like immunoreactivity and its relationship to feeding in neonatal rats. *Endocrinology* 112: 911–916, 1983.

51. KUMEGAWA, M., T. TAKUMA, S. HOSODA, S. KUNII, AND Y. KANDA. Precocious induction of pepsinogen in the stomach of suckling mice by hormones. *Biochim. Biophys. Acta* 543: 243–250, 1978.

52. KUMEGAWA, M., T. YAJIMA, N. MAEDA, T. TAKUMA, AND S. HOSODA. Permissive role of L-thyroxine in the induction of

stomach pepsinogen by cortisol in neonatal rats. *J. Endocrinol.* 87: 431–435, 1980.
53. LARSSON, L.-I. Peptide secretory pathways in GI tract: cytochemical contributions to regulatory physiology of the gut. *Am. J. Physiol.* 239 (*Gastrointest. Liver Physiol.* 2): G237–G246, 1980.
54. LARSSON, L.-I., N. GOLTERMANN, L. deMAGISTRIS, J. F. REHFELD, AND T. W. SCHWARTZ. Somatostatin cell processes as pathways for paracrine secretion. *Science Wash. DC* 205: 1393–1395, 1979.
55. LARSSON, L.-I., R. HÅKANSON, J. F. REHFELD, F. STADIL, AND F. SANDLER. Occurrence of neonatal development of gastrin immunoreactivity in the gastric tract of rat. *Cell Tissue Res.* 149: 275–281, 1974.
56. LICHTENBERGER, L. M. A search for the origin of neonatal hypergastrinemia. *J. Pediatr. Gastroenterol. Nutr.* 3: 161–166, 1984.
57. LICHTENBERGER, L. M., S. S. CRANDELL, P. A. PLAMA, AND F. H. MORRISS, JR. Ontogeny of tissue and serum gastrin concentrations in fetal and neonatal sheep. *Am. J. Physiol.* 241 (*Gastrointest. Liver Physiol.* 4): G235–G241, 1981.
58. LICHTENBERGER, L., AND L. R. JOHNSON. Gastrin in the ontogenic development of the small intestine. *Am. J. Physiol.* 227: 390–395, 1974.
59. LICHTENBERGER, L. M., D. M. NANCE, AND R. A. GORSKI. Sex-related differences in antral and serum gastrin levels in the rat. *Proc. Soc. Exp. Biol. Med.* 151: 785–788, 1976.
60. LIPKIN, M. Growth and development of gastrointestinal cells. *Annu. Rev. Physiol.* 47: 175–197, 1985.
61. LUCAS, A., T. E. ADRIAN, N. CHRISTOFIDES, S. R. BLOOM, AND A. AYNSLEY-GREEN. Plasma motilin, gastrin, and enteroglucagon and feeding in the human newborn. *Arch. Dis. Child.* 55: 673–677, 1980.
62. MAJUMDAR, A. P. N., AND L. R. JOHNSON. Gastric mucosal cell proliferation during development in rats and effects of pentagastrin. *Am. J. Physiol.* 242 (*Gastrointest. Liver Physiol.* 5): G135–G139, 1982.
63. MAKHLOUF, G. M. Electrolyte composition of gastric secretion. In: *Physiology of the Gastrointestinal Tract*, edited by L. R. Johnson. New York: Raven, 1981, vol. 1, p. 551–556.
64. MALINOWSKA, K. W., W. S. CHAN, P. W. NATHANELZ, AND R. N. HARDY. Plasma adrenocorticosteroid changes during thyroxine-induced accelerated maturation of the neonatal rat intestine. *Experientia Basel* 30: 61, 1974.
65. MALLOY, M. H., F. H. MORRISS, S. E. DENSON, N. W. WEISBRODT, L. M. LICHTENBERGER, AND E. W. ADCOCK III. Neonatal gastric motility in dogs: maturation and response to pentagastrin. *Am. J. Physiol.* 236 (*Endocrinol. Metab. Gastrointest. Physiol.* 5): E562–E566, 1979.
66. McGUIGAN, J. E., AND W. L. TRUDEAU. Studies with antibodies to gastrin: radioimmunoassay in human serum and physiological studies. *Gastroenterology* 58: 139–150, 1970.
67. MILLER, B. A. Observations on the gastric acidity during the first month of life. *Arch. Dis. Child.* 16: 22–30, 1941.
68. PEITSCH, W., K. TAKEUCHI, AND L. R. JOHNSON. Mucosal gastrin receptor. VI. Induction by corticosterone in newborn rats. *Am. J. Physiol.* 240 (*Gastrointest. Liver Physiol.* 3): G442–G449, 1981.
69. RODBRO, P., P. A. KRASILNIKOFF, AND V. BITSCH. Gastric secretion of pepsin in early childhood. *Scand. J. Gastroenterol.* 2: 257–260, 1967.
70. RODBRO, P., P. A. KRASILNIKOFF, AND P. M. CHRISTIANSEN. Parietal cell secretory function in early childhood. *Scand. J. Gastroenterol.* 2: 209–213, 1967.
71. ROGERS, I. M., D. C. DAVIDSON, J. LAWRENCE, J. ARDILL, AND K. D. BUCHANAN. Neonatal secretion of gastrin and glucagon. *Arch. Dis. Child.* 49: 796–801, 1974.
72. SAFFOURI, B., G. WEIR, K. BITAR, AND G. MAKHLOUF. Stimulation of gastrin secretion from the perfused rat stomach by somatostatin antiserum. *Life Sci.* 25: 1749–1753, 1979.
73. SALENIUS, P. On the ontogenesis of the human gastric epithelial cells. *Acta Anat. Suppl.* 46: 1–76, 1962.
74. SCHUSDZIARRA, V., V. HARRIS, J. M. CONLON, A. ARIMURA, AND R. UNGER. Pancreatic and gastric somatostatin release in response to intragastric and intraduodenal nutrients and HCl in the dog. *J. Clin. Invest.* 62: 509–518, 1978.
75. SEIDEL, E. R., AND L. R. JOHNSON. Ontogeny of gastric mucosal muscarinic receptor and sensitivity to carbachol. *Am. J. Physiol.* 246 (*Gastrointest. Liver Physiol.* 9): G550–G555, 1984.
76. SHULKES, A., AND K. J. HARDY. Ontogeny of circulating gastrin and pancreatic polypeptide in the foetal sheep. *Acta Endocrinol.* 100: 565–572, 1982.
77. SIRCAR, B., L. R. JOHNSON, AND L. M. LICHTENBERGER. Effect of chemically defined diets on antral and serum gastrin levels in rats. *Am. J. Physiol.* 238 (*Gastrointest. Liver Physiol.* 1): G376–G383, 1980.
78. SOLL, A. H. The interaction of histamine with gastrin and carbamylcholine on oxygen uptake by isolated mammalian parietal cells. *J. Clin. Invest.* 61: 381–389, 1978.
79. TAKEUCHI, K., S. OKABE, AND K. TAGAKI. Influence of pregnancy and lactation on the healing process of gastric and duodenal ulcers in rats. *Experientia Basel* 30: 366–368, 1974.
80. TAKEUCHI, K., W. PEITSCH, AND L. R. JOHNSON. Mucosal gastrin receptor. V. Development in newborn rats. *Am. J. Physiol.* 240 (*Gastrointest. Liver Physiol.* 3): G163–G169, 1981.
81. TAKEUCHI, K., G. R. SPEIR, AND L. R. JOHNSON. Mucosal gastrin receptor. III. Regulation by gastrin. *Am. J. Physiol.* 238 (*Gastrointest. Liver Physiol.* 1): G135–G140, 1980.
82. TATEMATSU, M., M. TAKAHASHI, H. TSUDA, M. HIROSE, C. FURIHATA, AND T. SUGIMURA. Precocious differentiation of immature chief cells in fundic mucosa of infant rats induced by hydrocortisone. *Cell Differ.* 4: 285–294, 1975.
83. TRACK, N. S., C. CREUTZFELDT, J. LITZENBERGER, C. NEUHOFF, R. ARNOLD, AND W. CREUTZFELDT. Appearance of gastrin and somatostatin in the human fetal stomach, duodenum, and pancreas. *Digestion* 19: 292–306, 1979.
84. TSENG, C.-C., AND L. R. JOHNSON. Does corticosterone affect gastric mucosal cell growth during development? *Am. J. Physiol.* 250 (*Gastrointest. Liver Physiol.* 13): G633–G638, 1986.
85. TSENG, C.-C., AND L. R. JOHNSON. Role of thyroxine in functional gastric development. *Am. J. Physiol.* 251 (*Gastrointest. Liver Physiol.* 14): G111–G116, 1986.
86. TSENG, C.-C., K. L. SCHMIDT, AND L. R. JOHNSON. Hormonal effects on development of the secretory apparatus of chief cells. *Am. J. Physiol.* 253 (*Gastrointest. Liver Physiol.* 16): G274–G283, 1987.
87. TSENG, C.-C., K. L. SCHMIDT, AND L. R. JOHNSON. Hormonal effects on development of the secretory apparatus of parietal cells. *Am. J. Physiol.* 253 (*Gastrointest. Liver Physiol* 16): G284–G289, 1987.
88. VOLLRATH, L. Über Entwicklung und Funktion der Belegzellen der Magendrüsen. *Z. Zellforsch. Mikrosk. Anat.* 50: 36–60, 1959.
89. WALKER, P., J. D. DUBOIS, AND J. H. DUSSAULT. Free thyroid hormone concentration during postnatal development in the rat. *Pediatr. Res.* 14: 247–249, 1980.
90. WALSH, J. H. Gastrointestinal hormones and peptides. In: *Physiology of the Gastrointestinal Tract*, edited by L. R. Johnson. New York: Raven, 1981, vol. 1, p. 59–145.
91. WERNER, B. Peptic and tryptic capacity of the digestive glands in newborns. *Acta Paediatr. Scand. Suppl.* 70: 1–80, 1948.
92. WOODWARD, E. R., E. S. LYON, J. LANDOR, AND L. R. DRAGSTEDT. The physiology of the gastric antrum: experimental studies on isolated antrum pouches in dogs. *Gastroenterology* 27: 766–785, 1954.
93. YAHAV, J., P. C. LEE, AND E. LEBENTHAL. Ontogeny of pepsin secretory response to secretagogues in isolated rat gastric glands. *Am. J. Physiol.* 250 (*Gastrointest. Liver Physiol.* 13): G200–G204, 1986.
94. YALOW, R. S., AND S. A. BERSON. Radioimmunoassay of gastrin. *Gastroenterology* 58: 1–14, 1970.
95. YEOMANS, N. D., J. S. TRIER, D. C. MOXEY, AND E. T. MARKEZIN. Maturation and differentiation of culture fetal stomach. Effects of corticosterone, pentagastrin and cytochalasin B. *Gastroenterology* 71: 770–777, 1976.

CHAPTER 19

Gastrointestinal mucus

ADRIAN ALLEN | *Department of Physiological Sciences, The Medical School, Framlington Place, Newcastle upon Tyne, United Kingdom*

CHAPTER CONTENTS

Adherent Mucus Gel
   In vivo continuity and thickness
   Physical and permeability properties
Isolated Mucin Glycoprotein
   Technology
   Mucin carbohydrate chain structure
   Protein core of mucins and macromolecular polymeric structure
   Gel-formation and physical properties of mucins
   Lipid and protein components of mucus secretions
Biosynthesis and Secretion of Mucus: Cellular Aspects
   Mucus-secreting cells
   Mucin biosynthesis
   Secretion of mucus
Secretion of Mucus: Measurement and Control
   Measurement
   Control of secretion of gastric mucus
   Control of secretion of intestinal and colonic mucus
Degradation of Mucus
Function of Mucus

MUCUS IS A UBIQUITOUS SECRETION from stomach to colon. It is a major organic secretion of the gut in terms of output by weight and biosynthetic requirements. Mucus contrasts with the other gastrointestinal secretions: its primary physical form is a water-insoluble gel adherent to the mucosal surfaces. This mucus gel layer marks the interface between the mucosal epithelium and the liquid luminal environment, which is teeming with nutrients, destructive hydrolases, foreign compounds, and microbial life. The adherent mucus gel thus provides a protective barrier and a stable unstirred layer between the mucosa and the lumen. Mucus also occurs in a predominantly soluble form mixed with the luminal contents.

   The foundations for the gastrointestinal mucus field were laid in the 1950s in the areas of gastroduodenal mucosal protection and blood group antigen structure. The work of Florey (67), Heatley (86), and Hollander (90) highlighted the protective roles of gastroduodenal mucus and led to the concept of a two-component barrier against acid and pepsin, comprising an alkaline mucus layer and the underlying rapidly regenerating epithelium (90). Much of the current understanding of mucus structure stems from the classic work by Kabat, Morgan, and Watkins (253, 254) on the ABH antigenic determinants carried by the soluble mucin (mucus glycoprotein components). Since then a wealth of knowledge has accumulated on the mechanisms and control of mucin secretion by the different cell types in the gastrointestinal tract and the structure of the component mucin subunits, although the tertiary polymeric structure of mucins has yet to be clearly defined. More recently, particularly in the stomach and duodenum, emphasis has been on the adherent gel that provides a protective barrier and stable unstirred layer at the enteric interface. The adherent mucus gel is emphasized in this chapter. Salivary mucus and biliary mucus, secreted primarily as a viscous soluble phase, are not considered here; the reader is referred to the chapter by Quissel and Tabak and the chapter by Hofmann in this *Handbook*.

   The viscous and gel-forming components of mucus secretions are glycoproteins, often referred to as mucins. In this chapter the term *mucus* is used for the secretion and *mucin* for its mucus glycoprotein components. [The spelling *mucus* is used throughout for both the noun and adjective (41).]

ADHERENT MUCUS GEL

In Vivo Continuity and Thickness

   The recognition that there is a distinct layer of water-insoluble mucus gel adherent to the mucosal surface in vivo and the characterization of this gel have been important developments over the last decade. This adherent mucus gel layer is subject to dehydration and shrinkage when exposed to routine histological fixation procedures (10, 197), and therefore little or no extracellular mucus is visible on sections routinely fixed and stained for light microscopy. Similarly standard preparation procedures for electron microscopy distort the adherent mucus layer to give condensed strands or fenestrated patches (32, 196). An extracellular mucus layer is observed if special techniques are employed to fix the gel matrix prior to processing. An adherent mucus layer over the proximal and distal colon of the rat (average thickness 150 µm and 16 µm) has been observed by light microscopy of sections made after freeze substitution and vapor fixation (197). When stabilized with antibodies, extracellular mucus is seen under the scanning electron microscope as a continuous cover over the mouse ileum and rat colon (32, 196).

Two techniques have been developed for measuring the thickness of the adherent mucus layer on unfixed, fresh mucosa: *1*) an indirect method that makes use of a slit lamp and pachymeter to measure gel thickness over an everted mucosa bathed with solution (26, 187) and *2*) direct observation of unfixed sections of mucosa by light microscopy (111, 149). The pachymeter is an image-splitting device that allows the measurement of optically distinct objects to an accuracy of ± 20 µm. With this method mucus thickness over the rat fundus was a mean of 166 µm for six stomachs (3–4 readings per stomach) and an average of 652 µm and 500 µm for two samples of human antrum (26).

A simple, direct, and reliable method for observing adherent mucus gel and measuring its thickness is to view it by light microscopy on unfixed sections of fresh mucosa (111, 149). The mucosa, mounted luminal surface uppermost, is sectioned by razor blades 1.6 mm apart. During the procedure the mucosa and subsequent sections are bathed in physiological saline to prevent dehydration. This technique has been applied to human and rat stomach and rat duodenum. In all cases the adherent mucus layer was an optically distinct, translucent layer readily distinguishable from the mucosa and luminal bathing solutions (Fig. 1A). The mucus layer was always continuous but of variable thickness: the median range for thickness values for gastric mucus was an average of 180 µm (individual readings 50–450 µm) and ~70–80 µm (individual readings 10–400 µm) in human and rat, respectively (15, 111, 149, 211). Success of this technique depends on the observed physical stability of the adherent mucus gel layer, of which the section thickness does not significantly change over 60 min and measurements are independent of the sectioning and mounting procedures.

In rat stomach, prostaglandins [topical; (26, 111)], carbachol (intraperitoneal), and secretin [intravenous; (5, 148)] administered in vivo result in an up to threefold significant increase in median mucus thickness. The mucus gel layer in vivo is unchanged after exposure to acid pH 2.2 for 2 h but becomes granular in appearance and discontinuous after topical pepsin [2 mg/ml, pH 2.2; (128)]. There is a reduction of mucus gel thickness to 15% of control values, with substantial discontinuities and epithelial exfoliation, after exposure of a ligated stomach to distending volumes (4 ml) of 70% ethanol for 45 s (209).

Although both direct observation of unfixed mucosae and indirect measurements by the pachymeter give qualitatively similar results for mucus thickness, the direct observations are consistently about half the pachymeter dimensions (Table 1). India ink particles, added prior to sectioning, settled on the surface of the

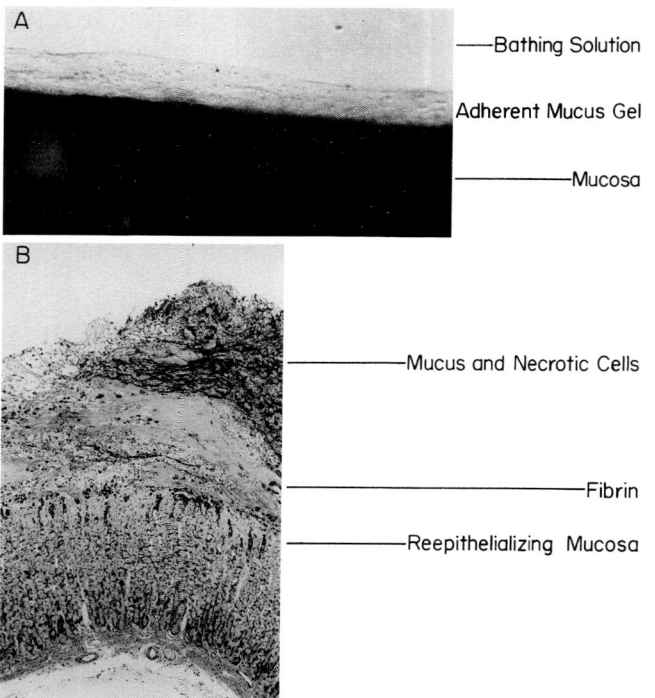

FIG. 1. *A*: adherent gastric mucus viewed under bright field on a transverse section (1.6 mm thick) of rat gastric mucosa. Three distinct phases can be seen clearly: mucosa, mucus gel layer, and bathing solution. For method, see Kerss et al. (111). *B*: reepithelialized mucosa: rat gastric mucosa after exposure to 70% (vol/vol) ethanol for 45 s and rapid wash with 0.9% (wt/vol) sodium chloride followed by 0.9% (wt/vol) sodium chloride for 1 h. Mucosa was formalin fixed, embedded in paraffin wax, and sections were stained with periodic acid–Schiff stain for mucin. ([*A* courtesy of L. A. Sellers; *B* from Sellers et al. (209).]

TABLE 1. *Comparison of Adherent Mucus Thickness and Surface pH Gradients*

	Rat		Human	Ref.
	Basal	Prostaglandin stimulated	Basal	
Stomach				
Direct, unfixed sections	77	139	180	15, 111
Pachymeter	166	399	500	26
			652	
pH gradient	695			192
	396	294		114
Duodenum				
Direct, unfixed sections	80			149
pH gradient	750	1,470		65

In all cases values are given for 16,16-dimethyl prostaglandin E_2; see references for doses and administration routes. Results are expressed as mean (µm), except from direct, unfixed sections, which are expressed as median (µm).

adherent mucus gel viewed on unfixed sections. This demonstrates that the site of the mixing boundary between the insoluble mucus gel phase and the bulk luminal fluid phase is truly represented in the direct method (111). It has been suggested that these differences in thickness dimensions between the two methods could be due to the pachymeter measuring a component of unstirred solution exterior to the adherent mucus gel (12). Also, there is an uncertainty in the pachymeter method, in that the refractive index of mucus is assumed to be the same as water. Direct observation of mucus gel on unfixed sections would appear to be the more sensitive method of the two, and it shows clearly the variable contours of the gel at its luminal surface, which are not observed by the pachymeter.

A functional method that measures the thickness of the unstirred layer at the mucosal surface is the pH gradient, and its dimensions have been equated with the thickness of the adherent mucus gel layer. These pH gradients, measured using antimony microelectrodes, progressively increase from an acid pH (e.g., pH 2) in the lumen to near neutral pH at the mucosal surface. Such surface pH gradients have been demonstrated in the rabbit and rat stomach in vivo (114, 192, 258), the amphibian stomach in vitro (239), and the rat duodenum in vivo (65). The dimensions of these pH gradients, several hundred microns (Table 1), are clearly substantially greater than the dimensions of the adherent mucus layer measured either by direct observation or by the pachymeter. The pH-gradient dimensions do, however, compare closely to those for unstirred water layers recorded at gastrointestinal mucosal surfaces by other methods under conditions of low shear (82, 130, 243). It seems reasonable to conclude that the measured pH gradients extend into the lumen in a layer of unstirred solution beyond the phase boundary of the adherent mucus gel (12). The effectiveness of such an unstirred layer outside of the gel matrix would be enhanced by the presence of a viscous soluble mucus. However, such an unstirred layer would be decreased or eliminated by shear, and under conditions of high motility in vivo the adherent mucus gel alone might be expected to provide the basis for the pH gradient. This is borne out by the large reduction in the dimensions of the unstirred layers associated with stirring (82, 243), i.e., a reduction from 500 to 63 μm in the rat intestine, a value close to that found for the adherent gastric mucus gel. A value of 650 μm for unstirred layers in rat intestine, calculated from a kinetic analysis of substrate hydrolysis, is more than one-quarter of the intestinal diameter and nearly half its volume (226). This value, which is still lower than the measured pH-gradient dimensions for the rat duodenum (Table 1), is unlikely to apply in vivo under normal conditions.

Interesting differences in the rate of action of N-acetylcysteine on decreasing the dimensions of mucus thickness and collapsing pH gradients have been observed. Direct observations of mucus thickness showed the adherent gel was unaffected by 20% N-acetylcysteine in vitro over 30 min at pH 2.2 or 7.4 and little change was seen until after a 60-min incubation (111). Physical studies on isolated mucus gel show that it takes ~30–60 min to effectively dissolve in strong solutions of thiol agents [0.2 M mercaptoethanol; (21)]. In contrast, mucus gel thickness measured by the pachymeter is reduced from a mean of 166 μm to a mean of 88 μm within 3 min after treatment with 20% N-acetylcysteine (26), whereas pH gradients at a fixed point over the rat stomach surface collapsed from pH 5.69 to 3.88 after treatment with 5% N-acetylcysteine for 15 min (193). It would appear therefore that the rapid changes after N-acetylcysteine observed by the pachymeter and with pH gradients reflect events other than a collapse in thickness of the visible adherent mucus gel layer. One possible explanation is that the rapid effect of N-acetylcysteine observed with the pachymeter and with measurement of pH gradients reflects a decrease in the viscosity of the more readily thiol-reducible soluble mucus exterior to the adherent gel.

After acute mucosal damage by alcohol and subsequent epithelial repair, in the rat animal model there is a massive release of surface mucoid gel mixed with the exfoliated, necrotic epithelium (118, 119, 157, 158). This gelatinous or mucoid coat, which can be observed by routine histochemical methods, forms rapidly (within 7 min of exposure to ethanol) and although it contains substantial numbers of necrotic cells and fibrin it has been assigned by some workers as mucus (119, 158, 215). Recent histochemical studies with peroxidase-antiperoxidase staining for fibrinogen show this mucoid coat is predominantly a fibrin gel with an exterior layer rich in mucus and necrotic cells [Fig. 1B; (10, 209)]. This fibrin-based gelatinous coat is granular in appearance, sloppier, and substantially thicker (median thickness 680 μm; range 40–1,560 μm on unfixed mucosal sections) than the thin, translucent, rigid adherent mucus gel covering the undamaged mucosa [median thickness ~200 μm after maximal topical prostaglandin stimulation (149, 211)]. The fibrin-mucoid coat is dissolved by pepsin and N-acetylcysteine but, unlike the adherent mucus gel, is resistant to ethanol, although relatively easily removed mechanically (118, 251). Studies in vivo in rat stomach demonstrate that the fibrin-mucoid coat forms after exudation of plasma from the damaged mucosa and is thought to protect the subsequent repair by reepithelialization (101, 118, 251).

To provide an effective barrier it is important that adherent mucus forms a continuous cover over the gastroduodenal mucosa in vivo. Evidence favors such a continuous cover. If special techniques to preserve the mucus gel in situ are employed, then a continuous extracellular mucus layer is observed over the mucosa with both light and electron microscopy (32, 196, 197). A continuous adherent mucus layer is always observed in unfixed sections of gastric mucosa, with minimum values of 50 µm in the human and 10 µm in rat (148, 149, 211). No discontinuities in the mucus layer are observed when mucus thickness is measured using a pachymeter (26) or in the pH gradients at the mucosal surface (114, 192, 239, 258). The continuity of the mucus layer over undamaged mucosa, however, has been questioned on the basis of a discontinuous layer seen under both light and electron microscopy (156, 157). Such an interpretation can be criticized, because no attempt was made in these studies to preserve the gel matrix in situ prior to histological processing. Glutaraldehyde used in these particular studies for fixing the mucosa has been shown with unfixed sections to cause discontinuities and an overall shrinkage in thickness of the mucus layer to a mean of 45% of control values (148). The contrast between the lack of adherent mucus on histological sections of undamaged mucosa and the thick gelatinous coat observed on sections from mucosa after ethanol damage (157) has also been seen as supporting the discontinuous mucus layer in the former case in vivo. This is now explained by the formation of an ethanol-resistant, sloppy, fibrin-base gelatinous coat 10-fold thicker and quite different from the original adherent mucus gel (209).

Physical and Permeability Properties

Mucus gel gently removed from the mucosal surface contains 90%–95% by weight water and an overall electrolytic composition similar to plasma (91). In addition to the gel-forming component, the mucus glycoprotein (mucin), such preparations contain varying amounts of protein, nucleic acid, and lipid, depending on the severity of the procedure used to remove the gel and the underlying epithelium. Also associated with the mucus gel are other gastrointestinal secretions, spent epithelial cells, bacteria, and bacterial components (2, 40, 161). The concentration of mucin glycoprotein forming the gel varies according to the source of the mucus; pig gastric mucus and pig duodenal mucus contain ~50 mg/ml glycoprotein, pig small intestinal mucus contains ~20 mg/ml glycoprotein, and pig colonic mucus contains ~30 mg/ml glycoprotein (21, 210).

Adherent gastrointestinal mucus secretions have mechanical properties characteristic of weak viscoelastic gels (20, 59, 112, 113, 150). Detailed physical studies show such mechanical properties are characteristic of adherent mucus from all regions of the pig gastrointestinal tract (stomach through to colon) and human gastric mucus (21, 210). Adherent mucus is a "true" gel; i.e., it does not dissolve on dilution and can be sedimented by centrifugation. At the same time mucus gel will reform when sectioned and slowly flow over a relatively long time scale (30–120 min), properties that enable it to form a continuous cover over the mucosal surface in vivo.

Adherent gastrointestinal mucus gel does not dissolve in solutions of pH 1–8, hypertonic salt (e.g., 2 M NaCl), or bile (21, 210). Denaturants such as high concentrations of salt, guanidinium chloride, and urea are frequently used to disperse mucus secretions. The success of such dispersion depends on the preparation; relatively uncontaminated gastroduodenal and colonic mucus gels remain unaffected (20, 21), while crude small intestinal mucus gel and preparations of respiratory and cervical mucus are dissolved by such agents (38, 112, 146, 210). The susceptibility of a given mucus preparation to denaturants could reflect, at least partly, the amount of nonmucin material present in the sample (210). Various amounts of mucin glycoprotein, 20%–70% by weight, can be solubilized from the gel by stirring or dialysis over several hours. However, to fully disrupt the noncovalent bonds between mucin molecules it is necessary to use shear (e.g., homogenization for 1 min) at levels substantially higher than those that occur in vivo (2). For the larger species of mucin glycoprotein [molecular weight (M_r) = several million] such shear procedures can cause fragmentation of the native molecule (38).

An important aspect of mucus gels from all regions of the gastrointestinal tract is their solubilization by proteolytic enzymes or, after reduction of disulfide bridges, by agents such as mercaptoethanol, dithiothreitol, or N-acetylcysteine (3, 21, 56, 213). The mucolytic effect of such procedures is due to cleavage of the protein core of the polymeric mucin, which yields smaller glycoprotein units with lower viscosity and no gel-forming properties (2, 3, 201, 227). The rate of collapse of mechanical properties of pig gastric mucus gel on reduction after incubation with 0.2 M mercaptoethanol is relatively slow, and the sample only assumes liquid properties after ~1 h (21). Whereas mercaptoethanol penetrates the interstices of the gel, pepsin does not (1), and consequently the enzyme dissolves the mucus gel at its surface, with the sample becoming progressively smaller, but what remains is still a gel. Proteolytic degradation by pepsins of adherent gastroduodenal mucus to produce soluble, degraded mucin glycoprotein is a major feature governing mucus dynamics in vivo and is discussed in DEGRADATION OF MUCUS, p. 372.

The permeability properties of mucus gels have not been precisely defined. The gel matrix of adherent mucus provides an excellent, stable, unstirred layer, although it is still >90% solution. Hydrogen ions and other small-molecular-weight ions and solutes (e.g., glucose, salicylates) and molecules as large as vitamin B (M_r 1,346) can permeate gastric mucus gel (1, 173, 257). The gastric mucus gel matrix (and likely that of other mucus gels from the gastrointestinal tract) is impermeable to proteins the size of myoglobin (M_r 17,000), and therefore it is also a diffusion barrier to proteinases such as pepsin (M_r ~35,000) (1). The measured rate of diffusion of H^+ through mucus gel is about fourfold slower than an equivalent unstirred layer of solution (173, 257). This rate of H^+ diffusion is still relatively fast, and it would be a matter of minutes for H^+ in the lumen to equilibrate across the adherent mucus layer in vivo in the absence of surface neutralization (7, 51, 52). It has been proposed that lipids associated with the adherent mucus layer significantly reduce the rate of H^+ diffusion through mucus (88, 131, 200). The ion-exchange properties of mucus have also been implicated in influencing the rate of H^+ diffusion through mucus (225, 244). However, it remains that acid will rapidly diffuse through isolated mucus gel both in vitro and in vivo, and therefore it is unlikely that lipids or the ion-exchange properties of the adherent mucus layer enable it to provide a significant diffusion barrier to H^+. There is a need for more detailed studies to determine how diffusion through the negatively charged mucus gels, from different regions of the gastrointestinal tract, is governed by the charge, size, and shape of the solute.

ISOLATED MUCIN GLYCOPROTEIN

Technology

Mucin glycoproteins are very complex molecules of large molecular size (millions of daltons), and available technology has largely determined progress in their isolation and purification, as well as elucidation of their chemical and physical structures. (For more comprehensive information, see refs. 2, 9, 38, 39, 41, 161, 167.)

Variations and combinations of three widely used methods are usually employed in the isolation and purification of mucin glycoproteins: *1*) gel filtration, *2*) equilibrium density-gradient centrifugation, and *3*) ion-exchange chromatography. These techniques make use respectively of differences in size, density, and charge to separate mucins from other macromolecular components in the secretion and to fractionate different species of mucin. Gel filtration on large-pore gels will separate mucins (excluded) from proteins with lower molecular weights (included), although a drawback can be that some protein will adhere to and fractionate with the mucin. Gels with very large pores can also be employed to separate glycoprotein subunits, or proteolytically degraded units (included), from the polymeric mucin (excluded) (138, 170). A successful and widely used method for purification of mucins employs one or more fractionations by equilibrium centrifugation in a density gradient (usually cesium chloride), which separates the mucin ($\rho = 1.45$–1.55) from lower-density lipid and protein and from higher-density nucleic acid (45, 47, 235). Ion-exchange chromatography, which is more often used to subfractionate mucins once they have been partially purified, utilizes the wide variations in negative charge on different mucin species, primarily from anionic sialic acid and ester sulfate residues (123, 179). A more important consideration in isolation of mucins is to prevent their degradation by endogenous proteinases. This is usually achieved by the addition of a spectrum of proteinase inhibitors and/or denaturants such as guanidinium chloride to the mucus secretions immediately after their collection from the animal (13, 37, 38).

Classic chemical techniques and glycosidase enzymes have been used widely to elucidate the structures of the carbohydrate side chains of mucins (34, 35, 219, 253, 254). More recently proton nuclear magnetic resonance spectroscopy and mass spectroscopy combined with high-pressure liquid chromatography have been applied to mucin structure. These powerful analytical tools use relatively small amounts of material and will no doubt surplant older classic methods (97, 237, 246, 247). The use of lectins and mucin antibodies, particularly monoclonals, has been employed as a very sensitive means for identifying specific oligosaccharide and peptide structures in tissue sections, for quantitating mucins, and for structural studies (28, 30, 31, 58, 62, 103, 137, 140, 180).

The large size of mucins, their inherent polydispersity, and their viscous, non-Newtonian behavior in solution make study of their physical properties particularly difficult. Polydispersity in size of mucins means that their molecular-weight values represent a mean around a broad distribution, which can vary widely from one glycoprotein preparation to another and lead to different interpretations of results (11, 13, 36, 37, 45, 46, 75). Analytical ultracentrifugation, viscosity determinations, and light-scattering measurements have been used to determine size and shape of mucins in solution (2, 38, 45, 75). Electron microscopy has given essential information about shape and size of mucins, estimated from measured contour lengths (83, 84, 151, 212, 216, 217). To obtain a reasonable assessment of physical properties of mucin glycoproteins a combination of methods is necessary.

Mucin Carbohydrate Chain Structure

Composition and structure distinguish mucin glycoproteins from other macromolecules and in particular other classes of glycoproteins (2, 9, 38, 41, 80, 92, 161). They are large: molecular weights of 2–44 million have been reported for isolated mucin glycoproteins. They have a high-carbohydrate content: gastrointestinal mucins comprise 70%–80% carbohydrate, 12%–25% protein, and up to ~5% ester sulfate. The carbohydrate composition consists of galactose, fucose, *N*-acetylgalactosamine, *N*-acetylglucosamine, and sialic acid. The absence of uronic acid and only trace amounts of mannose (<1%) distinguish mucin glycoproteins from the proteoglycans of connective tissue and serum glycoproteins, respectively. The oligosaccharide chains of 2–19 sugars per chain are attached covalently to serine and threonine residues of the protein core by *O*-glycosidic linkages from *N*-acetylgalactosamine at their reducing ends. In gastric mucins, for example, it can be calculated there must be at least ~100–200 chains per glycoprotein unit of 5×10^5 M_r. A useful analogy is often made between this characteristic secondary structure of mucin glycoprotein units and "bottle brushes," with the carbohydrate chains representing the bristles and the wire support the protein core (Fig. 2).

The carbohydrate side chains vary in composition, structure, and size between glycoprotein species from different secretions (Fig. 3; Table 2). The smallest completed carbohydrate chains are those from the submaxillary mucins; sheep submaxillary mucin has a disaccharide of sialic acid α-(2-6)-N-acetylgalactosamine (81); in pig submaxillary mucin the chain is a pentasaccharide (34). Pig small intestinal mucin has oligosaccharide chains each of ~6–8 sugars (139); much more complex branched chains are found in pig and human gastric mucins [up to 19 sugars per chain (207, 219, 223)] and human and rat colonic mucin [up to 12 sugars per chain (177, 178, 220)]. The order of the different sugars comprising the oligosaccharide chain sequence, the location of branch points, the several possible linkage sites between two adjacent sugars, and the α- and β-anomeric conformation of each bond combine to give many theoretical permutations in chain structure. However, common structural features that are apparent in the oligosaccharide chains of mucins as a whole are determined by the expression of specific glycosyltransferase enzymes within the secreting cell. N-acetylgalactosamine always occurs at the reducing ends of the carbohydrate chains as the linkage sugar to threonine and serine residues in the peptide core (35). The only other position where N-acetylgalactosamine occurs in the oligosaccharide chain is a terminal position at the nonreducing end of the chain in some mucins. The backbone of the oligosaccharide chain is alternating residues of β-linked galactose or N-acetylgalactosamine. In gastric mucins of the human and pig the sugar linking the chain to N-acetylgalactosamine at the reducing end is β-galactose, whereas in human and rat colonic mucins it is β-N-acetylglucosamine. Sialic acids and fucose are always found at the nonreducing ends of main or branched chains.

There are various common antigenic features between different mucins that are determined by the high degree of stereospecificity of their carbohydrate chains [Fig. 3; (97, 253, 254)]. The ABH(O) and Lewis and antigenic determinants are conferred on many mucins by the structure of the nonreducing ends of their oligosaccharide chains. The primary group A determinant is a terminal N-acetylgalactosamine, which is linked α-(1-3) to a subterminal galactose, which also is linked to an α-(1-2)–terminal fucose.

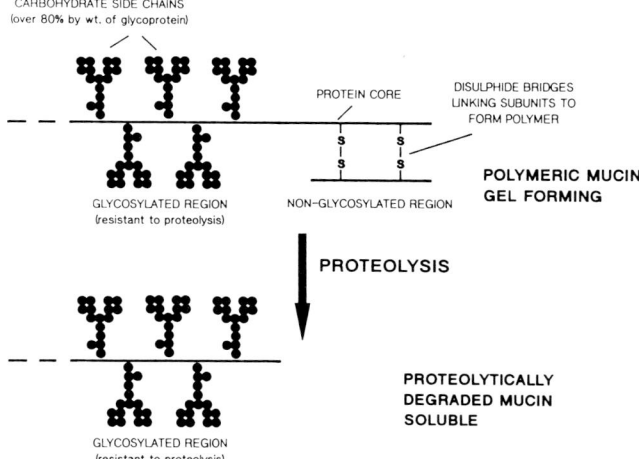

FIG. 2. Diagram of a subunit of mucus glycoprotein structure showing glycosylated "bottle brush" and nonglycosylated regions. Subunits are joined by disulfide bridges to form overall linear polymeric mucins. Detailed tertiary structure of subunits within the polymeric mucin has not been elucidated.

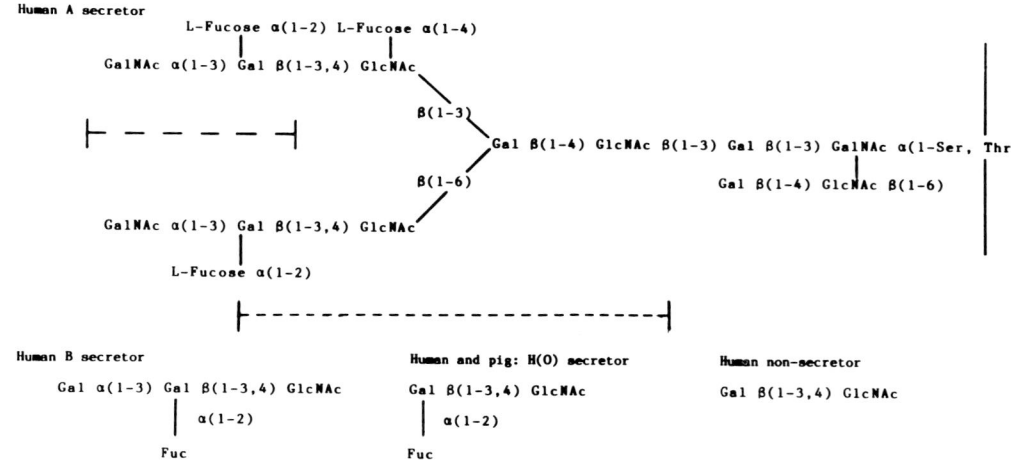

FIG. 3. Composite diagram of proposed structures for typical branched oligosaccharides in human (223) and pig (219) gastric mucin glycoproteins showing the immunodeterminant blood group A and H (— — —) and I (- - -) saccharide structures. Location of sulfate esters on the chains is not known. Fuc, fucose; Gal, galactose; GalNAc, N-acetylgalactosamine; GlcNAc, N-acetylglucosamine.

TABLE 2. *Carbohydrate and Protein Compositions of Purified Gastrointestinal Mucins*

	Carbohydrate Side Chains						Protein Core			Ref.
	GalNAc	GlcNAc	Gal	Fuc	Sialic acid	Sulfate, % by wt	Protein, % by wt	Thr/Ser/Pro, mol/100 mol protein		
								Polymeric	Degraded*	
Gastric mucins										
Human	1	1.8	2.3	1.6	0.9	7.0	17.0	57.2	ND	207
Pig	1	2.8	2.9	1.9	0.2	3.1	13.0	41.6	58.9	201
Small intestinal mucins										
Human	1	1.1	1.0	1.0	0.4	1.6	18.0	50.9	69.1	141
Pig	1	0.6	0.6	0.3	0.6	2.6	18.0	52.3	69.3	139
Rat	1	1.0	1.3	0.7	0.8	ND	16.0	55.1	ND	57
Colonic mucins										
Human	1	0.9	1.2	0.3	0.2	2.0	31.1	ND	ND	179
Pig	1	2.9	2.5	1.5	0.2	3.0	13.3	58.5	78.0	145

GalNAc, N-acetylgalactosamine; GlcNAc, N-acetylglucosamine; Gal, galactose; Fuc, fucose; Thr, threonine; Ser, serine; Pro, proline; ND, not determined. Carbohydrate data are expressed as the molar ratio to 1 mol GalNAc. *Proteolytically degraded by pronase.

Replacement of the terminal N-acetylgalactosamine with galactose converts the structure from blood group A to blood group B specificity. When neither α-N-acetylgalactosamine nor α-galactose is present, the terminal structure, fucose α-(1-2) galactose, is the H antigen determinant, which is otherwise masked in A or B antigens. The Ii antigens are located deeper into the oligosaccharide side chains; the branched core structure of repeating type 2 [galactose β-(1-4) N-acetylglucosamine β-(1-3)] constitutes the I antigen, whereas the same repeating units in a linear but unbranched chain confer blood group antigen i.

Many gastrointestinal glycoproteins are strongly negatively charged because of the presence of sialic acid and carbohydrate-bound ester sulfate residues. A contribution to the charge of the glycoprotein will also be made by the protein core. Intestinal mucins appear to carry more negative charge than gastric mucins (Table 2). *Sialic acids* is a generic name for a family of sugars based on N-acetyl- or N-glycolylneuraminic acids, both forms of which occur in mucins as well as mono-, di-, and tri-O-acetyl derivatives of N-acetylneuraminic acid (60, 206). Many sialic acid residues in human colonic mucins are O-acetylated, such as N-acetyl-7,9-di-O-acetylneuraminic acids (50, 185, 190). Sialic acids, always at the nonreducing ends of the carbohydrate chains, are removed by enzymes known as sialidases or neuraminidases, although O-acetyl substitution can make some sialic acids resistant to cleavage. Ester sulfate groups are located within the carbohydrate side chains toward the protein core, but their linkage and location have not been clearly defined. In pig gastric mucin and human tracheobronchial mucin, N-acetylglucosamine 6-sulfate and galactose 6-sulfate have been identified, respectively (147, 219). In human intestinal mucin it has been proposed that sialic acid is associated with short oligosaccharide chains, whereas sulfate occurs in longer branched chains (255).

In a given mucin preparation there is considerable heterogeneity in carbohydrate chain length and structure. In pig submaxillary mucin, five different sizes of carbohydrate chains are found, from a single N-acetylgalactosamine to the complete pentasaccharide chain (34). In human colonic mucin, 25 different oligosaccharide chains have been isolated, ranging from a disaccharide [sialic acid α-(2-6)-N-acetylgalactosamine] to a decasaccharide (177, 178). All of the carbohydrate chains in a mucin preparation have related structures, however, indicating common biosynthetic pathways. This microheterogeneity in oligosaccharide chain length is thought to arise from incomplete biosynthesis in the cell and in some instances degradation by bacterial glycosidases in the gut lumen (93, 94). It is also likely that the number of oligosaccharide chains per molecule varies, and there may also be heterogeneity in the protein core (79, 256). The consequence of this microheterogeneity is that any "pure" mucin preparation is polydisperse with respect to its physical properties of molecular weight, density, and charge. On sedimentation analysis, mucus glycoproteins sediment as a broad peak, indicative of a wide variation in sedimentation values for individual mucin molecules around a mean (11, 75). Variations in sialic acid and ester sulfate between different mucin preparations absorbed onto ion-exchange columns result in their elution of a continuous series of glycoprotein fractions with increasing salt concentrations (123, 179).

Protein Core of Mucins and Macromolecular Polymeric Structure

The polypeptide core of gastrointestinal mucin glycoproteins comprises 12%–30% by weight of the molecule (Table 2). The mucin polypeptide is characteristically high in threonine and serine, the sites of attachment of the numerous oligosaccharide side chains. Mucins also have unusually high proline content, and it has been proposed that this amino acid maintains a conformation that is most suitable for close packing of these side chains (3). It is clear from studies on a number of mucins that the polypeptide backbone consists of two distinct types of segments distinguished by their sensitivity to proteolytic digestion: *1*) those to which the oligosaccharide side chains are attached, the glycosylated regions (here the protein core is protected by the carbohydrate from proteolytic digestion) and *2*) the "naked" or nonglycosylated regions that are accessible to proteolytic enzymes and have an amino acid composition more characteristic of a typical globular protein: low in threonine, serine, and proline but with disulfide bridges (2, 54, 201). The nonglycosylated regions can be digested away by a variety of proteinases (pepsin, trypsin, papain, and pronase) to leave a digested glycosylated peptide containing most, if not all, of the carbohydrate originally present in the mucin, of large molecular weight (2.7–7.5×10^5) and resistant to further proteolytic digestion (143, 145, 201). Threonine, serine, and proline are conserved in the digested glycoprotein such that they constitute up to 70%–80% by weight of the remaining peptide core (Table 2). Calculation shows that, on average, one in every two or three amino acid residues in the glycosylated region of the peptide chain carries an oligosaccharide chain. Such a sheath of carbohydrate around the protein core protects the digested protein from further proteolytic attack (2, 248).

Mucin glycoproteins possess a polymeric structure where glycoprotein subunits are joined together, in many cases by disulfide bridges located between the nonglycosylated regions of protein core (3, 227). One of the best characterized is that for a mucin glycoprotein ($M_r \sim 2 \times 10^6$) isolated from pig gastric mucus, which on reduction is split into subunits of $M_r \sim 5 \times 10^5$ (227). Proteolytic enzymes cleave the polymeric glycoprotein into glycopeptides ($M_r \sim 5 \times 10^5$) by digesting the nonglycosylated region of the protein core, where the intersubunit disulfide bridges are located (201). A pig small intestinal mucin preparation ($M_r \sim 1.7 \times 10^6$) is split on reduction into subunits of $M_r \sim 2.4 \times 10^5$, which were not decreased in size by further proteolytic digestion (143). A mucin glycoprotein isolated from pig colonic mucus has a larger polymeric structure ($M_r \sim 15 \times 10^6$), which on reduction produces subunits ($M_r \sim 6 \times 10^6$), whereas proteolytic and further reduction yields glycopeptides of $M_r \sim 7.6 \times 10^5$ (145). Polymeric structures of glycoprotein subunits joined by disulfide bridges [first demonstrated for pig gastric mucins (227)] have also been shown for human (170) and rat (233) gastric mucins; human (140), pig (143), and rat (57) small intestinal mucins; human and pig biliary mucins (171) and mucin glycoproteins from human cervical (38, 76) and respiratory tracts (189). Salivary polymeric mucins differ from gastric mucins in that they are formed from noncovalent, ionic-dependent associations of subunits, although disulfide bridges still play an important role in their conformation (87, 208).

Although a polymeric structure based on interchain disulfide bridges is generally accepted for mucus glycoproteins, controversy exists over the size of the covalent unit and the size and arrangement of the subunits (2, 11, 13, 36–38, 46). Mucins isolated under nondenaturing conditions are $\sim 2 \times 10^6$ M_r; upon concentration these form a gel with the same mechanical properties as the native secretion and on reduction yield subunits of $M_r \sim 5 \times 10^5$ (2). More recently, glycoproteins of much larger size, up to $M_r \sim 44 \times 10^6$, have been isolated from mucus secretions using isolation procedures involving guanidine hydrochloride and proteolytic inhibitors (36, 37). These glycoproteins are heterogeneous in size and will not form a typical mucus gel. Discrepancies also exist over the subunit size in these mucins, in that different reduction procedures yield subunits of $M_r \sim 2.3 \times 10^6$ (37) and $M_r \sim 5 \times 10^5$ (13) from the same mucins. The full explanation for the size differences in the same mucin glycoprotein prepared by different methods has yet to be given. Hydrolysis of very sensitive peptide bonds during isolation can explain at least some of the differences (37), but evidence from dissociation studies also suggests that larger mucins may be noncovalent associations of glycoproteins of $M_r \sim 2 \times 10^6$ (13). The three-dimensional arrangement of the subunits in the polymeric structure is also unclear: chemical evidence based on NH_2-terminal amino acid analysis (201) suggested that subunits in pig gastric mucus may be joined centrally by disulfide bridges located in the nonglycosylated protein core (2). However, electron microscopy of gastric mucus and other mucus glycoproteins together with light-scattering studies (36, 212, 216, 217) clearly shows mucins to be long, straight-chain molecules.

Some gastrointestinal mucins possess a protein component that is integrated into their polymeric structures and separates on reduction of disulfide bridges. Pig and human gastric mucins each contain a 70,000 M_r protein (169), whereas human (141) and rat (57) intestinal mucins have larger peptides of M_r $\sim$118,000. These peptide components are part of the nonglycosylated portion of the glycoprotein and are destroyed on proteolytic digestion. Pig gastric mucin contains one $\sim$70,000-M_r peptide per glycoprotein of $M_r \sim 2 \times 10^6$, which suggests a possible structural role in polymerization (169). Another possibility is that the peptide might equally be a part of the secretory mechanism in a manner similar to that of secretory IgA. The role of these peptides joined to mucins by disulfide bridges awaits clarification.

Gel-Formation and Physical Properties of Mucins

The component that determines the viscous and gel-forming properties of mucus secretions is the polymeric mucin glycoprotein. This is demonstrated by formation in vitro of a gel, with the same mechanical

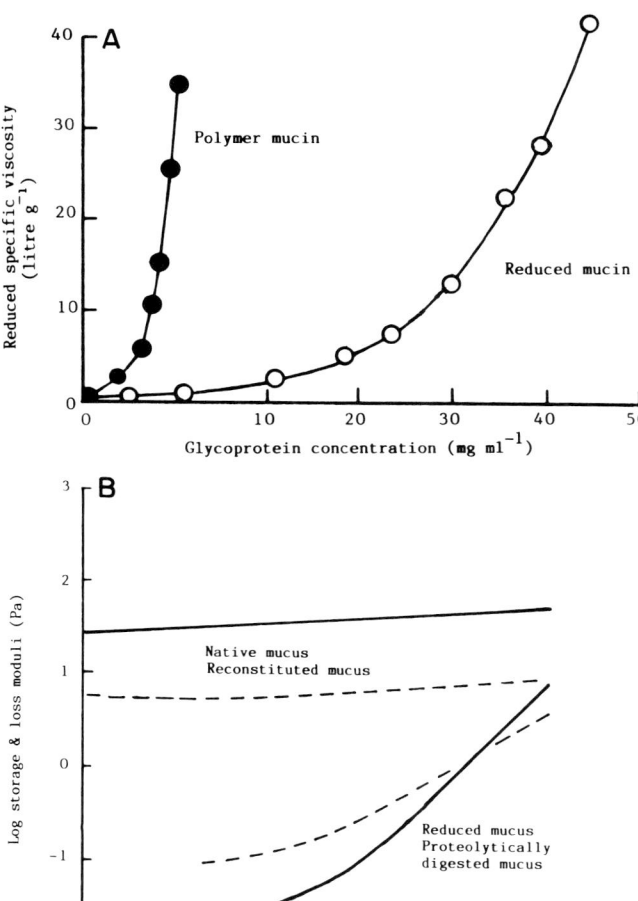

FIG. 4. Physical properties of pig gastric mucus in dilute solution (A) and in native mucus gel (B). A: dependence of solution viscosity of pig gastric mucus on glycoprotein concentration. Viscosity of the solution rises asymptotically with increased glycoprotein concentration until it assumes the properties of a gel at a glycoprotein concentration of ~50 mg/ml (14, 20). This can be explained by the large hydrated glycoprotein molecules at higher concentrations occupying the whole solution volume and their overlapping molecular domains interacting noncovalently to form the mucus-gel matrix. Concentration of glycoprotein in gastric mucus (pig and human) is ~50 mg/ml. Reduced or proteolytically degraded glycoprotein subunits have a low solution viscosity and do not form gel at high concentrations of glycoprotein. B: mechanical spectrum for pig gastric mucus showing a plot of elastic (storage modulus, ——) and viscous (loss modulus, ---) moduli against frequency of oscillatory deformations (20). Mucus gel is sandwiched between a flat plate and shallow cone and subjected to small deformations at different oscillatory frequencies without disrupting gel structure. From the magnitude and phase of stress generated in resistance to applied deformation the storage modulus G′ (elastic or solid component) and the loss modulus G″ (viscous or liquid component) can be calculated. Native gastric mucus and mucus reconstituted from isolated glycoprotein have the behavior characteristic of a weak viscoelastic gel with the elastic modulus dominant over the viscous modulus. Reduced and proteolytically digested mucus has the behavior of viscous liquid where the viscous modulus is dominant over the loss modulus for much of the frequency range accessed.

properties as the native gastrointestinal secretions, from the purified mucin at concentrations the same as those found in vivo (20, 21, 210). Gel formation in vitro can be followed by measuring solution viscosity as the mucin is concentrated (Fig. 4). The intrinsic viscosity (viscosity extrapolated to infinite dilution) of mucins is high, 320 ml/g and 400 ml/g respectively for pig gastric (228) and rat small intestinal (70) mucins in isotonic salt solution. At finite concentrations mucin glycoproteins show a substantial shear dependence of viscosity and non-Newtonian behavior. As the mucin concentration increases, the viscosity rises not linearly but asymptotically until the solution assumes viscoelastic gel-like properties and a gel is formed (~50 mg/ml, pig gastric mucin). This behavior can be understood from the known solution properties of mucins (14, 20). Hydrodynamic measurements (3, 29, 37, 84, 228) and some electron-microscope studies (83, 84) indicate that the mucin glycoproteins in dilute solution are highly expanded, hydrated, roughly spherical molecules occupying a large volume of solution. As the concentration of the mucin solution is increased, a point is reached where it occupies the whole solution volume and the probability of intermolecular interactions becomes appreciable. It is above this concentration (~20 mg/ml for pig gastric mucin; Fig. 4) that the viscosity rises sharply because of increasing intermolecular interactions as the hydrodynamic domains of the glycoprotein molecules overlap and gel formation occurs.

To study the nature of the intermolecular gel-forming interactions in mucins, it is necessary to investigate the intact gel, by techniques such as mechanical spectroscopy (20, 59, 112, 113, 150). The full extent of the intermolecular gel-forming interactions in mucins is only apparent at concentrations too high for solution viscosity measurements (20). From mechanical spectroscopy measurements of gastrointestinal mucus gels, it is clear that there are positive gel-forming interactions between the mucin molecules over and above those characteristic of a superviscous solution (e.g., concentrated hyaluronic acid) where polymers only interact by molecular entanglement (20, 21). Such positive (noncovalent) interactions between mucin molecules are reflected in the insolubility of the mucus gel, and they must be sufficiently strong to resist the osmotic forces of dilution of the hydrophilic glycoprotein interacting with its aqueous environment (155). At the same time, mucus gels will flow over a time scale of several minutes and reform when sectioned, properties explained by the making and breaking of the gel-forming interactions between the mucin molecules in the gel matrix (20). This suggests these mucin gel-forming interactions are still relatively weak when compared with the cooperative formation of strong periodic interactions of various other rigid polysaccharide gels, e.g., carrageenan, alginates, and agar (155).

Although the gel-forming interactions between the mucin molecules have yet to be explained in molecular terms, two broad structural features of the mucin molecule can be identified as requirements for gel formation: the polymeric structure and the carbohydrate side chains. That an intact polymeric mucin structure is an integral part of the gel matrix of all gastrointestinal mucus secretions is apparent from their solubilization after reduction of mucin intersubunit disulfide bridges (by mucolytic thiol agents, e.g., mercaptoethanol and N-acetylcysteine) or on proteolysis (2, 21, 227). The reduced or proteolytically digested subunits have a lower intrinsic viscosity (50–100 ml/g pig gastric mucus) and, at a given mucin concentration, markedly lower specific viscosity than the polymeric mucin. Isolated, reduced subunits do not form a gel at the same space-filling concentration as the polymeric mucin. A full understanding of how the polymeric mucin is incorporated into the gel network must await elucidation of its tertiary structure, as well as the nature of the noncovalent gel-forming interactions.

The oligosaccharide chains comprise over 70% by weight mucin and ensure these molecules interact strongly with an aqueous environment. The close packing of the oligosaccharide chains also results in stiffening of the glycoprotein molecules and may thereby favor gel formation by decreasing the entropy change associated with this process. There is some evidence that the carbohydrate side chains may be involved in gel-forming interactions. Mechanical spectroscopy studies show that noncovalent, intermolecular interactions characteristic of the mucus gel are present between the proteolytically digested subunits (20). Because such digested glycopeptides have lost all their "accessible" protein core, this implies the observed intermolecular interactions are between the carbohydrate side chains. Based on this observation, a model for mucus gel formation was proposed where the matrix is formed by interdigitation of the carbohydrate side chains of polymeric mucin molecules (20). What is clear from mechanical spectroscopy studies is that although there are wide variations in average composition and chain length of the carbohydrate side chains of mucins from different regions of the gastrointestinal tract, they all form gels with the same overall mechanical structure (20, 21, 210).

The functional significance of the negative charge on mucins is not understood. The amount of negative charge influences the viscosity of the mucin but only at very low ionic strengths. Thus the viscosity of pig gastric mucin (228), rat small intestinal mucin (70), and sheep submaxillary mucin (81) increases markedly if the total ionic content falls below $\sim$10 mM. This viscosity increase can be explained by a polyelectrolyte effect, whereby repulsion of the unshielded negative charges causes molecular expansion and increases the hydrated size of the mucin. These effects of ionic strength on viscosity are at salt concentrations <10 mM and unlikely to be significant in vivo. Various cations such as Ca^{2+}, Mg^{2+}, Fe^{2+}, Ba^{2+}, and Zn^{2+} (49, 69, 144) and also preparatory cesium chloride gradients (228) condense mucin molecules, with the exclusion of water. Calcium ions have been implicated in condensing presecreted mucin in secretory vesicles at concentrations very much in excess of that found in the secreted adherent gel (249).

Lipid and Protein Components of Mucus Secretions

Recently some attention has been paid to the role of lipids in the structure and properties of mucus secretions. Gastrointestinal mucus preparations contain a variety of lipids, including neutral lipids (mostly free fatty acids and cholesterol), glycolipids, and phospholipids, varying in amounts from 5% to 25% by weight of the preparation (260, 261). Mucins contain hydrophobic binding sites (224) and will bind lipids. Gastric mucin has been reported to contain small amounts of covalently bound, acyl-linked fatty acids, between $\sim$0.1% and 0.3% by weight of glycoprotein (221, 222). However, studies on purified intestinal mucin have failed to confirm association of either noncovalently or covalently bound lipids (142). Crude mucus will contain lipids, if nothing else, from sloughed epithelial cells and membrane lost during secretion (161, 229). Once purified it would seem mucin glycoproteins contain at the most only very small quantities of lipid.

Both proteins (40, 133) and lipids (159, 200) enhance the viscosity of mucin preparations. On the other hand, nonmucin components can decrease the gel quality of mucus (210). Whether these effects are significant in vivo is not clear, because it is very difficult from collection procedures in vitro to ensure that the adherent mucus gel is collected without removing part of the epithelium. The purified gastrointestinal mucins, essentially free of protein, lipid (<0.5% by weight), and nucleic acid, will alone reproduce the viscous and gel-forming properties of the native mucus secretion (21, 210). It has been reported that lipids can protect gastric mucin from proteolytic degradation (218); however, even without any removal of the lipid components, gastric mucus gel is readily degraded by pepsin (21).

BIOSYNTHESIS AND SECRETION OF MUCUS: CELLULAR ASPECTS

Mucus-Secreting Cells

Mucus is secreted from a variety of cell types in the gastrointestinal tract: surface epithelial cells found throughout the gut, but particularly in the stomach; goblet cells of the small and large intestine; the glands in the stomach; and Brunner's glands in the duodenum (60, 100, 129, 161). All these cells are an integral part of the continuous, simple columnar epithelium that forms the gut and comprises both the mucosal surface and the invaginated epithelium of the glands and crypts. Mucin-producing cells display polarity of secretion, junction complexes, and other specializations characteristic of gut epithelia. Mucus-secreting cells, as with other exocrine cells, have a well-developed rough endoplasmic reticulum concentrated in the basal and lateral cytoplasm. The Golgi complex of these cells is characteristically large, with 8–12 broad stacks of membrane cisternae associated with small smooth and coated vesicles together with newly formed secretory granules in the supranuclear cytoplasm. The rough endoplasmic reticulum and the Golgi complex are the site of synthesis respectively of the protein core and oligosaccharide side chains of the mucin (24, 117, 129, 161, 163, 164, 182, 194). Mucin-secreting cells are often characterized by the presence of close-packed, stored mucin granules beneath the apical membrane, a feature that also provides the distinction between epithelial mucin cells and goblet cells. In goblet cells the cuplike theca, a particularly thick filament-rich layer of cytoplasm, separates the mass of stored mucin granules from the lateral membranes and supranuclear Golgi region (161, 232). The goblet shape, which is maintained primarily by noncontractile intermediate filaments, does not collapse after accelerated mucin secretion with the loss of many granules.

Histochemical studies show differences in the carbohydrate chain structure of mucins between functional regions of the gastrointestinal tract, within the same cell population, during maturation, and in disease. Gastric mucus in the pits stains for acidic sulfomucins, whereas that at the mucosal surface is neutral staining. Neutral and acidic sialomucins predominate in the goblet cell secretions of the small intestine, whereas acidic sulfomucins are characteristic of colonic goblet cell secretions. In stomach, mucus-secreting cells lose their ability to incorporate [^{35}S]sulfate as they move upward from the pit to the crypt (117). In human rectal mucosa it is the reverse, with maximal [^{35}S]sulfate incorporation in the mature surface goblet cells in contrast to the immature cells in the crypts (162). The relation between such histochemical observations and overall mucin structure has yet to be understood; changes in the carbohydrate structures could signify new species of mucin, but, alternatively, such changes may reflect degrees in carbohydrate chain completeness within the same mucin species. There is a wealth of information on histochemical studies of mucins in health and disease; the reader is referred to specialist reviews (9, 50, 60, 61, 129).

Mucin Biosynthesis

The general features of the pathways of mucin biosynthesis are known, and the mechanisms and regulation are under active investigation. They have been extensively reviewed (43, 202–205), and only key aspects are outlined here. The mucin peptide core is synthesized in the rough endoplasmic reticulum, whereas most (if not all) of the oligosaccharide side chains are added during passage of the nascent mucin molecule through the Golgi. The oligosaccharide chains are constructed by the sequential addition of sugars one at a time from activated nucleotide sugar donors, namely UDP-galactose, UDP-N-acetylglucosamine, UDP-N-acetylgalactosamine, GDP-fucose, and CMP sialic acids. Each of the many linkages in the oligosaccharide chain requires the presence of a specific glycosyltransferase for its formation. In this manner the oligosaccharide chains are built up on the nascent protein core by the integrated specific and sequential actions of the multienzyme glycosyltransferase systems. The oligosaccharide chain is initiated by the transfer of N-acetylgalactosamine (GalNAc) from UDP-GalNAc to serine or threonine of the nascent mucin polypeptide by membrane-bound UDP-GalNAc:polypeptide α-GalNAc-transferase. Thereafter the cores of the oligosaccharide chains are built up sequentially by different β-glycosyltransferases ending in chain termination by the α-sialyl-, fucosyl-, galactosyl- or N-acetylgalactosyltransferases. Examples of the chain termination process are the ABH, Lewis, and secretory genes, each of which codes for synthesis of a specific transferase and whose presence or absence thereby dictates the terminal structures of the gastrointestinal mucin chains and their antigenicity (253, 254).

The oligosaccharide chains on any given glycoprotein molecule vary in length, number of branches, and type of end-group sugars; consequently the mechanics and the controls required for the assembly of these heterogeneous chains must be complex. Known controls include *1*) genetic controls of the synthesis of specific sugar nucleotide transferase enzymes; *2*) the differential gene expression of various transferases within the same individual that is seen both in the organ specificity of secreted mucins and in the phased assembly of different oligosaccharide chain structures during ontogenesis; and *3*) controls within the Golgi at the substrate level, where the nature of the acceptor sugar, its linkage, and the presence of branching determine whether it is recognized as a suitable substrate for sugar transfer. Competition between certain transferase enzymes for a common acceptor mucin precursor will in turn determine the ratio of differently structured oligosaccharide chains in a given molecule. In the newly synthesized mucin, chain length will vary with type of chain structure and with incompletely biosynthesized chains.

Further postribosomal processing occurs with addition of ester sulfate to mucin oligosaccharide chains in the Golgi complex (117, 164). The activated donor is 3′-phosphoadenine-5-phosphosulfate (PAPS), but little is known about the sulfotransferases involved.

How and at what site in the cell interpeptide chain disulfide bond formation takes place to join the glycoprotein subunits together to form the polymeric mucin is unknown (2). It would be expected (from analogy with other proteins and glycoproteins) to occur at an early stage in biosynthesis at the level of the rough endoplasmic reticulum. However, the possibility exists of disulfide exchange occurring at later stages of mucin biosynthesis or during packaging for secretion. In this respect an interesting analogy could be made between the 70,000- to 118,000-M_r proteins associated by disulfide bridges with gastrointestinal mucins and the transfer through the epithelial cell of secretory IgA with secretory component.

There are important differences between mucin biosynthetic pathways and those for the other major class of mammalian glycoproteins with N-linked oligosaccharides, for example, serum glycoproteins (98). Formation of N-linked oligosaccharide chains proceeds by a lipid intermediate dolichol pyrophosphate and the transfer of a preassembled oligosaccharide chain onto a peptide backbone in the rough endoplasmic reticulum. Subsequently there are a series of processing and trimming reactions in N-linked oligosaccharides that have not been observed with mucin biosynthesis.

Secretion of Mucus

Elegant electron-microscopy studies have given insight into the cellular events preceding and during exocytosis of the mucin-containing granules, although (as with other exocytotic processes) the detailed molecular events remain to be elucidated (160, 161, 163, 164, 264). Autoradiographic studies of radioactive mucin granules demonstrate that mucus-secreting cells in the gastrointestinal tract transport and secrete mucins through their life spans (24, 162–164). In unstimulated cells there is intermittent release of mucin granules from the apical surface, balanced by new granule formation from the Golgi. The average transit time for a mucin granule from the Golgi to exocytosis at the apical membrane in vivo in rat stomach is ~1 h, whereas in the rat intestine the rate is considerably slower, 4–8 h (117, 164). In mature human and rabbit goblet cells, labeled, freshly synthesized mucin granules appear at the base of the theca and move upward along the thecal margins, not mingling with mucin granules in the center (162, 232). On the other hand, labeling of central granules did occur in the immature goblet cells at the base of the crypts. This suggests that mucin synthesized early in the life of the goblet cell remains with it in mucin granules in the center of its theca during migration up the crypt and that central granules constitute a mucin reserve not routinely replaced. Colchicine blocks the movement of mucin vesicles from the Golgi to the cell apex, demonstrating a dependence of this process in microfilament polymerization (232).

Three mechanisms have been identified for the release of gastric mucus (264): *1*) a continuous exocytosis of a few granules at a time, *2*) an explosive release of mucus by apical expulsion of the older cells in the interfoveolar area after maximal stimulation by secretagogues (prostaglandin, cholinergic, or irritant mustard oil), and *3*) the relatively rare event of cell exfoliation. On stimulation of goblet cells (cholinergic or irritant mustard oil), there is an orderly sequence of events, culminating in a compound exocytosis of mucin granules contiguous with the apical membrane and rupture of the pentalamellar membranes surrounding adjacent granules. This results in rapid, progressive fusion of individual granules, thereby forming a mucus bolus that expands and exudes into the lumen. The result is an "empty" goblet cell, with a concave depression representing the thecal lining and the rest of its architecture intact. During compound exocytosis much of the secretory granule membrane is lost, together with some goblet cell cytoplasm (229).

An intriguing question is how the completed mucin molecule transversing the Golgi cisternae, where presumably it is soluble, finishes up after exocytosis as a concentrated (~50 mg/ml in stomach and duodenum) water-insoluble gel. There is no evidence for a special mechanism of polymerization after secretion, and the ability of the isolated mucin glycoprotein (when concentrated) to form a gel would argue against this (20). The secretory vesicles become condensed while they are in the supranuclear region after separation from the Golgi (199). Once in the apical mass the stored granules remain relatively uniform in size and density. Mucin in secretory granules appears more concentrated than after secretion, pointing to an expansion on exocytosis to form the adherent gel (55, 249). There is some evidence for special mechanisms enabling superconcentration of the mucin. Electron-probe analyses indicate a high-calcium content of intracellular mucus granules (252), and in vitro mucus volume is reduced by divalent cations (68, 69). A lectin-binding galactose residue has been identified in mucin granules of goblet cells in chick intestine, and it has been suggested that this may play a role in mucin concentration (18).

SECRETION OF MUCUS: MEASUREMENT AND CONTROL

Measurement

Mucus in vivo exists in three phases: presecreted mucus in intracellular vesicles, insoluble adherent mucus gel, and freely mobile (largely soluble) luminal mucus. This makes for a complex secretion. For a complete picture of the physiological state of the mucus barrier, all three phases should be assayed, because there is no proportional relationship between them. In functional terms it is the adherent gel that is important, and this is the most difficult to measure, because it cannot be sampled quantitatively without removing some of the underlying epithelium with its intracellular mucus. Luminal mucus, on the other hand, although easier to sample, can be derived from different sources: directly by secretion, from erosion of the adherent gel by proteases (of endogenous and bacterial origin), or through mechanical shear. Consequently the amount of mucin in luminal juice cannot be taken as a direct measure of mucus secretion or as an index of changes in the functional mucus gel barrier, as has frequently been assumed in the past (2, 4). Gastric luminal mucin content, for example, has been demonstrated to vary quite independently of changes in the thickness of the adherent gastric mucus barrier (5, 148, 149). A wide variety of methods have been used to assess mucus secretions in vivo (2, 161), but no one method on its own will give a complete, quantitative measure of the process.

Biosynthesis of mucus and the subsequent secretory process has been extensively investigated by several groups using incorporation of radioactive precursors into mucins (105–107, 110, 162–164, 181, 191). Radioactivity is frequently incorporated into nonmucinous glycoproteins, connective tissue proteoglycans, and gut bacterial products; therefore it is essential that radioactive material studied is shown to be only mucin glycoprotein; this has not always been the case. Radioactive mucins have been identified autoradiographically from the density of silver grains or by analysis of the isolated mucin. Morphological methods based on the characteristic apical cavitation of goblet cells and visible loss of granules from stomach surface mucus cells have been employed successfully to monitor mucus secretion, particularly when combined with radioactive methods (161, 166, 230, 264). Such methods have the advantage of circumventing the isolation of individual mucins, as well as enabling the monitoring of mucus secretion by individual cells.

The most satisfactory method for measuring changes in the functional adherent mucus gel barrier is by mucus thickness (see ADHERENT MUCUS GEL, p. 359). Thickness can be directly related to the protective barrier properties. Changes in thickness may be related quantitatively to changes in mucus gel secretion if *1*) it is assumed the degree of hydration of the mucus gel does not change under different physiological conditions and *2*) the amount of soluble mucus, from adherent gel degradation and mucus secretion into the lumen, is estimated at the same time. Physical studies indicate that the degree of hydration of the adherent mucus is unlikely to change that significantly in vivo (21). Mucus output into the lumen can be measured by the mucin content and gel fractionation (Sepharose 2B), used to determine the proportions of polymeric gel-forming mucin from smaller degraded material (138, 170).

The glycoprotein content of mucus secretions has been determined with a variety of chemical methods. Simple colorimetric assays most often employed are the anthrone estimation for neutral sugars (124), the thiobarbituric acid assay for sialic acid (53, 124), and the periodic acid–Schiff (138) and Alcian blue–binding assays (33, 124) for total glycoprotein. Although these methods are simple and quick, they are all subject to interference from other macromolecular components, e.g., proteins and nucleic acids and other glycoconjugates (connective tissue proteoglycans, serum glycoproteins, and cell membrane glycolipids). The most informative chemical method for measuring mucin content is that of gas-liquid chromatography, which displays the entire spectrum of sugars present (135, 237), but it is important to eliminate interference from other glycoproteins and glycolipids. With these estimations it is usually desirable to partially purify the mucin before analysis to remove interfering substances. Immunological assays [either radioimmunoassay (RIA) or enzyme-linked immunosorbent assay (ELISA)] that use an antibody directed against the mucin have provided a particularly sensitive and specific technique for estimating its content in secretions (71, 140). However, with all methods based on quantitation of mucus, satisfactory separation of the secreted adherent gel phase from nonsecreted intracellular reserves cannot as yet be satisfactorily achieved.

Control of Secretion of Gastric Mucus

Cholinergic pathways control the secretion of gastric mucus. Copious amounts of viscous mucus are released into the stomach in response to splanchnic or vagal stimulation or after topical application of acetylcholine (67, 104). Carbachol increases the adherent mucus thickness in the rat stomach in vivo and in the isolated frog gastric mucosa in vitro, a response attenuated by atropine (148). A large increase in both adherent gel and luminal mucus accompanies an outpouring of plasma from the dog gastric mucosa after intra-arterial acetylcholine infusion (265). Bombesin, when given intracerebrally (but not topically) in the rat, stimulates gastric mucus by an adrenal-dependent, prostaglandin-independent mechanism (238).

Secretin (intravenous) stimulates a 75% increase in adherent mucus thickness in the rat over a 2-h period (5). In the cat, intravenous secretin induced marked increases in gastric juice viscosity, up to 16-fold, associated with secretion of a large molecular-sized mucin aggregate (89). An increase in gastric juice mucin content (measured as carbohydrate) has been observed after secretin infusion in the human (16), cat (245), and dog (116). Gastrin, cholecystokinin (CCK), and histamine all increase mucin carbohydrate content in rat gastric juice (245), but in the absence of further analysis it is not possible to define its origin as secretion, peptic erosion, or both. Insulin (intravenous) in humans has been shown to induce a threefold increase in mucin content of gastric washouts, but it is degraded mucin that is paralleled by a similar rise in pepsin activity (262).

Many studies have demonstrated that prostaglandins (both topically and intravenously) stimulate increases in luminal mucus, adherent mucus gel thickness, and mucin biosynthesis (8). In the human (53, 108, 259) and rat (25, 33, 124, 149), luminal mucus glycoprotein content increased between two- and fivefold, according to the particular prostaglandin administered, its dose, route of administration, and method used to estimate the glycoprotein. The most potent prostaglandins appear to be the stable E-type analogues and the enteroperoxide. This prostaglandin-mediated stimulation of soluble mucus is independent of the inhibitory action on acid secretion (33, 108) but associated with a rise in nonparietal secretions (149). Topical prostaglandins of the E series induce an up to threefold increase in the thickness of adherent gastric mucus in rat in vivo and in frog in vivo (26, 111, 149, 211). This rise in mucus thickness is rapid (70% of the maximal response occurs within 5 min), pointing to a release of stored intracellular preformed mucus (149). The E-type prostaglandins, in several studies, stimulated the uptake of radioactive precursors (^{3}H and ^{14}C sugars and amino acids and $^{35}SO_4^{2-}$) into glycoconjugates by the rat stomach in vivo and in vitro (25, 105, 240). Incorporation of radioactive precursors, including glucosamine, by mucosa is not mucin specific, and in several studies no distinction is made between labeled mucin and other labeled cellular components. In one careful study of a perfused rat stomach, radioactivity from [^{3}H]glucosamine into mucin, identified autoradiographically, was confirmed by analysis by equilibrium centrifugation in a cesium chloride gradient (105). In the same study, however, only 8% of the incorporated radioactive serine was in mucin; much of the rest was incorporated into pepsinogen, emphasizing the care needed when interpreting studies on radioactive incorporation into mucus (106, 107).

Control of Secretion of Intestinal and Colonic Mucus

Control of goblet cell mucus secretion depends on the location of the cells, whether in crypts or in the surface epithelium. Surface goblet cells secrete mucus in response to topical irritants, for example, mustard oil, alcohol, hypertonic saline, bile salts, and enterotoxins such as those produced by *Escherichia coli* and *Vibrio cholerae* (165, 191). The secretory response is independent of autonomic innervation and may be mediated by eicosanoids. Crypt goblet cells, in contrast to surface goblet cells, secrete mucin in response to cholinergic stimulation. Increased mucin secretion after cholinergic stimulation has been demonstrated for rat intestine both in vivo and in vitro with mucosal slices or explants (160, 166, 231), a preparation of rat intestinal crypt goblet cells (174), and explants of rabbit intestinal tissue in organ culture (134). This cholinergic stimulation is attenuated by atropine (231). Because all goblet cells originate in the crypts, they must all be initially sensitive to cholinergic stimulation, but for some unknown reason, as these cells mature and migrate toward the mucosal surface, they appear to lose this sensitivity.

Many potential secretagogues, including adrenergic agents and various peptide hormones [e.g., secretin, CCK, pentagastrin, vasoactive intestinal polypeptide (VIP), and somatostatin], have no measurable short-term effects on mucin secretion, either in surface or crypt goblet cells (165, 231). Electrical field stimulation of rat intestinal sheets stimulates the goblet cells to secrete mucin, an effect partially blocked by atropine (175). Intracellular "second-messenger" pathways mediating mucin secretion from goblet cells are not yet established. A cyclic AMP (cAMP) system would not seem to be involved, because agents that increase intracellular cAMP (dibutyryl cAMP) or theophylline (a diesterase inhibitor) fail to stimulate release of preformed mucin from rat small intestinal slices (191) or colonic explants (122, 165). Cholera toxin appears to act on intestinal cells by two different mechanisms: *1*) a proved cAMP-mediated system that induces marked increases in fluid and electrolyte transport and *2*) a different, cAMP-independent system that stimulates mucin secretion (72, 191). Although neither cAMP nor β-adrenergic agents stimulate goblet cell mucin secretion, they do increase the incorporation of radioactivity into intestinal mucosal glycoproteins (73, 122), suggesting these components may be involved in longer term control of mucin production. Serotonin increased the hexose content of perfused rat colon (27) but it did not stimulate mucus release from goblet cells in rabbit intestinal biopsies (165). Because serotonin induces fluid and ion secretion from intestinal crypts, a washout from the surface into the lumen of secreted mucus could account for the serotonin-induced increase in luminal carbohydrate.

DEGRADATION OF MUCUS

Enzymatic degradation of gastrointestinal mucins can be considered in two stages (2, 9, 94, 96). *1*) Proteolysis of the mucin polymer to degraded subunits is a process that is thought to occur throughout the gut from stomach to colon, mediated by proteases in gastrointestinal secretions or of bacterial origin. Such proteolysis will erode the adherent mucus gel to produce soluble, degraded mucin (3, 4). *2*) Glycosidases remove sugars from the oligosaccharide chains of the glycopeptide in a stepwise process. These glycosidases are produced by the indigenous microflora but not by the host and are found only in the large intestine (94). Further proteolytic digestion of the peptide core once the protective oligosaccharide side chains have been removed also probably occurs.

Mucus gel secretions and isolated mucins from

stomach to colon undergo proteolytic degradation in vitro by a variety of secreted proteases (e.g., pepsin, trypsin, chymotrypsin) and bacterial proteases (e.g., pronase) (2, 3, 143, 145, 201). The products of digestion are soluble, degraded, glycopeptide subunits; resistant to further proteolysis; and still relatively large ($M_r \sim 2.5-7.5 \times 10^5$). There is evidence that proteolytic degradation of the adherent mucus gel is an ongoing process in vivo, at least by pepsin in the stomach. Analysis by gel-filtration chromatography of mucin from gastric washouts shows a predominance of degraded mucin relative to mucin polymer (170, 262). After indirect vagal stimulation via insulin-induced hypoglycemia, the content of pepsin and degraded mucin both rise about threefold in gastric washouts from duodenal ulcer patients (262). In vagotomized patients there is no increase in pepsin or the associated degraded mucin in gastric washouts after stimulation, further evidence that the insulin-induced rise in degraded mucin is due to pepsin. Pepsin dissolution of the adherent mucus barrier can be simulated in vivo in the anesthetized rat ligated stomach after infusion of a two- to threefold excess of enzyme (10, 128). In this model addition of pepsin to the HCl pH 2.2 instillate results in a significant (almost threefold) increase in the degraded mucin content, visible disruption of the adherent mucus layer, and focal mucosal erosions. Besides proteolysis, a further contributing factor to the loss of mucus gel in vivo is mechanical erosion due to the passage of solid food and the motile forces of digestion. Over the normal gastric mucosa, a dynamic balance must exist between erosion of the adherent mucus layer and secretion of new gel by the mucosa to maintain the observed continuity of the mucus barrier.

Studies point to an increased degradation by pepsin of the adherent gastric mucus barrier in peptic ulcer disease. Pepsin 1, the pepsin form raised four- to fivefold in peptic ulcer disease (241), has increased mucolytic activity compared with the major form, pepsin 3. For the same proteolytic activity (albumin substrate), pepsin 1 had two- and sixfold respectively greater mucolytic activity (gastric mucus substrate) at the optimal pH 2 and at pH 4 than did pepsin 3 (172). Similar increased mucolytic activity was found in gastric juice from duodenal ulcer patients compared with that from nonsymptomatic controls. This increased pepsin 1–mediated mucolytic activity in gastric juice is associated with impaired structure of the adherent gastric mucus gel in these patients (263). Adherent antral mucus from nonulcerated patients contains primarily polymeric mucin, mean content 67%; the rest is of equivalent molecular size to the pepsin-degraded mucin. By contrast, adherent antral mucus from gastric ulcer and duodenal ulcer patients contains significantly less polymeric mucin, mean of 35% and 50%, respectively. Rheological studies show a direct correlation between polymeric mucin content and the quality of the mechanical properties of the mucus gel (21), and therefore these studies implicate a weaker mucus barrier in peptic ulcer disease. Observation of the mucus gel on unfixed sections of gastric mucosa also shows a disrupted mucus gel over the antral mucosa of gastric ulcer patients (15). A study in rats has shown a predisposition to aspirin-induced ulceration in animals whose adherent gastric mucus had the greatest amounts of low-molecular-weight mucin relative to polymer mucin (17).

Proteolytic erosion of the mucus barrier in the intestine and colon is likely to occur as a consequence of attack by pancreatic digestive enzymes and bacterial proteases (143, 145, 168). An interesting observation is the recent demonstration of a protease with mucolytic activity in human feces that increases in ulcerative colitis patients (44, 99).

Degradation of the oligosaccharide side chains occurs stepwise, one monosaccharide at a time from the outer nonreducing end. For complete degradation of the oligosaccharide chains the action of several glycosidases is required, each enzyme with a different linkage specificity. These glycosidases are produced by the indigenous microflora and not by the host (94, 234). The initial event in chain degradation is cleavage of the terminal α-glycosidic linkages. The capacity to produce the requisite extracellular α-glycosidases for making this initial cleavage, particularly cleavage of the ABH determinants, is restricted to a few specialized strains among normal intestinal bacteria, notably *Ruminococcus* and *Bifidobacteria* species (95). These strains comprise a subset numbering about 1% of the total human fecal bacteria that appear responsible for the major degradation of mucin oligosaccharides in humans (95, 96, 154). The β-galactosidases and β-N-acetylhexosaminidases required for further degradation of the backbone of the oligosaccharide chains are produced by a wider range of gut bacteria. The fecal population density of bacteria that degrade the ABH determinants varies according to the host blood group. For example the density of bacteria that produce the α-galactosidase that cleaves the B-determinant galactose averages 50,000-fold more in B secretors than in blood group A or O secretors or nonsecretors. Bacterial sialidase and sulfatase activities are also present in fecal extracts. The driving force for production of the extracellular glycosidases by the indigenous microflora is presumably to utilize mucus as a food source.

FUNCTION OF MUCUS

The physical properties of mucus provide both a stable, protective, adherent gel barrier over the mucosal surface and a soluble viscous lubricant within the lumen. The slimy soluble mucus lubricates the passage of solids, undigested food, and feces through the gut lumen and protects the delicate epithelium from mechanical damage by the vigorous forces that attend digestion. The soluble mucus may also lubricate the surfaces of the adherent gel barrier and minimizes its erosion by mechanical forces. In contrast the adherent gel provides a stable unstirred layer at the mucosal surface that is independent of the fluid shear forces within the lumen and thereby provides a constant aqueous environment over the epithelial surfaces.

The adherent gel restricts the passage of solutes at least to a rate as slow as that which would be expected for diffusion through an unstirred layer. The passage of H^+, glucose, and other low-molecular-weight solutes has been shown to be restricted across the gastroduodenal mucus gel (1, 173, 226, 257). Diffusion of ions may be restricted by binding; for example, gastric mucus binds iron (23), colonic mucus binds sodium (78), and calcium interacts with mucus secretions from various sources (68, 69). Gastric mucus is impermeable to proteins (M_r 17,000) (1), and it would be expected that mucus gels from other regions of the tract would show similar impermeability to large molecules. Although evidence favors a continuous adherent mucus gel and its attendant restrictive permeability properties over the gastric, upper duodenal, and colonic mucosa (see *In Vivo Continuity and Thickness*, p. 359), it is not known how continuous the mucus cover is in the small intestine, the major absorptive area.

Invading microorganisms from the gut lumen must penetrate the adherent mucus gel, although there is little evidence that mucus provides a protective blanket that excluded pathogenic microorganisms (78, 161). Many microbes attached to mucosal surfaces must spend much of their existence in association with mucus and may depend on the special microenvironment it creates. For example, *Campylobacter pyloridis* colonizes the gastric epithelial surface–mucus interface and recently has been directly implicated in gastritis (183). Colonic mucus is a nutrient source, and probably a major one, for the endogenous flora [see DEGRADATION OF MUCUS, p. 372; (94, 96, 234)].

For many enteric pathogens a key element of their pathogenicity is their possession of specialized structures ("adhesins") with which they bind to specific saccharide structures at the mucosal surface. To the extent that mucins comprising the mucus layer share the same saccharide structures as the receptors on the mucosal surface, the mucus layer could compete with surface receptors for bacterial binding and thus promote bacterial associations. Thus salivary mucins bind to oral streptococci (77), whereas bovine submaxillary mucin was found to be a potent inhibitor of the binding of the enterotoxigenic *E. coli* K99 strain to erythrocytes (132). Some strains of *E. coli* bind to a protein component of mouse small intestinal mucin, resulting in their having a competitive advantage in colonization (42, 125), whereas *Shigella flexneri* adheres to guinea pig colonic mucus (102). Oral introduction of *Campylobacter jejuni* strains into germ-free mice led to rapid colonization in mucus at the luminal surface and in the crypts of cecum and colon (126).

Entamoeba histolytica has also been shown to bind to glutaraldehyde-fixed human and rat colonic mucus (184). Pig gastric mucin has been shown to bind to cholera toxin (236). The excessive secretions of mucus and fluid that can result from infections by pathogens should aid in washing out the offending organisms (22). Intestinal infection of rats by parasitic nematodes produces goblet cell hyperplasia and a mucus release that traps and expels the worms (22, 127, 152, 153). An excess production of intestinal mucus is induced by inflammation and immunological stimulation (120, 121, 161, 250). Considerably more research is needed to establish the full subtleties and implications of interactions between gastrointestinal mucus and microorganisms.

Current concepts of gastroduodenal mucosal protection against acid, pepsin, and other damaging agents in the lumen recognize many factors, including the mucus-bicarbonate barrier, a rapidly repairing epithelium, and a good vascular supply [see the chapter in this *Handbook* by Hirst; (6, 66, 85)]. The role of mucus in mucosal protection can be considered under two headings (10): *1*) protection against the natural endogenous aggressors acid, pepsin, and bile, which are secreted into the lumen, and *2*) protection against exogenous damaging agents, such as alcohol and nonsteroidal anti-inflammatory drugs (Table 3). Evidence points to adherent mucus providing an important part of the protective barrier against acid and pepsin but not exogenous damaging agents such as alcohol (10). Mucus has also been proposed to provide antioxidant protection to the underlying gastrointestinal epithelium by scavenging highly reactive oxygen-derived species (48).

Mucus does not seem to be a significant diffusion barrier to acid, although it does restrict the passage of H^+ ions compared with that through an equivalent unstirred layer of solution (see *Physical and Permeability Properties*, p. 362). The H^+ will rapidly equilibrate (in the absence of neutralizing HCO_3^-) across the relatively thin adherent mucus gel layer (median 180 µm, minimum 50 µm in humans). Evidence that this is so in vivo is seen by the formation of mucosal lesions after histamine stimulation of excess acid in humans (109), while increased backdiffusion of acid in response to mucosal damaging agents also occurs very rapidly, within 2 or 3 min (52). Mucus gels, even when mixed with substantial amounts of mucosal tissue, do not have significant buffering capacity against luminal acid in the absence of mucosal HCO_3^- (19).

The primary function of the adherent mucus gel in protection against acid is considered to be the provision of a stable unstirred layer at the mucosal surface supporting neutralization of H^+ by the epithelial HCO_3^- secretion (7, 10, 63). Mucus gel acts as a mixing barrier, preventing the small amount of HCO_3^- secretion from rapidly mixing with the large amount of acid in the luminal juice. Evidence supporting this

TABLE 3. *Role of Mucus in Gastroduodenal Protection*

Damaging Factors	Protective Action
Endogenous aggressors	
Acid	Stable unstirred layer; mixing barrier supporting surface neutralization by HCO_3^-
Pepsin	Permeability barrier
Biliary reflux	Bile permeates through mucus to epithelium; mucus structure unchanged; mucus continues to protect against acid and pepsin
Exogenous damaging agents	
Ethanol; nonsteroidal anti-inflammatory drugs; hypertonic solutions; bile salts, high concentration	Mucus permeated but remains; fibrin gel with mucus and necrotic cells protects re-epithelializing cells; mucus may act as template for fibrin formation
Severe damage resulting in lesions	None assigned
Mechanical damage from abrasion	Lubricant, particularly soluble mucus

From Allen et al. (10).

mucus-HCO_3^- barrier is *1*) the extensive characterization of gastroduodenal mucosal HCO_3^- secretion [see the chapter in this *Handbook* by Flemström and Garner; (64), *2*) the demonstration of pH gradients at the gastroduodenal mucosal surface from an acid pH in the lumen to a near neutral pH at the mucosal surface, and *3*) the observation of a continuous layer of adherent gel over the mucosal surface (see *In Vivo Continuity and Thickness*, p. 359). In the duodenum, mucosal HCO_3^- secretion would appear sufficient to neutralize acid down to the lowest pH values measured in vivo (64). In the stomach, however, maximal stimulated HCO_3^- secretion is not sufficient to neutralize acid below pH ~1.5, and the surface pH gradient is dissipated at this point (7, 193). Under these circumstances, mucosal acid-base balance is maintained in both the interstitial and intracellular compartments by perfusion with HCO_3^--rich plasma and operation of epithelial membrane Na^+-H^+ and Cl^--HCO_3^- exchanges (136). The requirement for nutrient HCO_3^- to protect the gastric mucosa against acid damage in vivo and in vitro supports the concept of neutralization of acid by plasma HCO_3^- (115, 195). Studies show the adherent mucus gel and therefore its unstirred layer are maintained down to at least pH 1.

Gastroduodenal mucus is a diffusion barrier to pepsins in the lumen, preventing them from digesting the underlying epithelium. Pepsin will dissolve the adherent gel at its luminal aspect to produce soluble degraded mucin (1, 2). However, because a continuous layer of adherent gel is observed over the undamaged mucosa in vivo, what is lost by erosion is normally replaced by secretion of new mucus. Excess pepsin (two- to threefold in the ligated rat stomach) will disrupt the mucus gel and cause focal erosions and luminal bleeding (128).

There is no mucus-HCO_3^- barrier within the gastric glands where acid concentrations of pH <1 are attained and newly secreted pepsinogen is rapidly and autocatalytically converted to pepsin. The apical membranes of these glands must be resistant to their own secretions (198). Another problem in this respect is how newly secreted pepsin and H^+ gain access through an apparently continuous mucus layer from the glands to the gastric lumen. One feasible explanation could be that the hydraulic pressure from the volume of secretion in the gland forces the acid and pepsin into the lumen through channels or bubbles in the overlying mucus gel.

The adherent mucus barrier is readily permeated by damaging agents such as bile salts, hypertonic saline, ethanol, and nonsteroidal anti-inflammatory drugs, resulting in the destruction of the underlying epithelium (10, 209). After such acute damage the epithelium is rapidly reepithelialized (101, 118, 119, 157, 158). A thick fibrin-based mucoid coat with the remaining mucus and necrotic cells forms over the regenerating epithelia after acute ethanol damage (209). Studies in vivo in rat stomach demonstrate that the fibrin-mucoid coat that forms after exudation of plasma from the damaged mucosa protects the subsequent repair by reepithelialization from toxic agents in the lumen, namely ethanol and acid (118, 251). The plasma clotting time of human blood is decreased by pig gastric mucus gel and soluble mucus glycoprotein, suggesting in vivo that the mucus layer and necrotic cells might act as a template for fibrinogen-fibrin conversion (209). It remains to be seen whether such protective fibrin-mucoid coats are formed during mucosal repair after acute damage by agents other than ethanol and under less severe conditions. Prostaglandins in laboratory animals attenuate acute mucosal hemorrhagic damage induced by exogenous agents, e.g., alcohol and aspirin, a phenomenon sometimes referred to as cytoprotection (186, 188). This effect is not mediated by the prostaglandin-induced increase in mucus thickness (8, 10) but by prevention of vascular stasis (176).

REFERENCES

1. ALLEN, A. The structure and function of gastrointestinal mucus. In: *Basic Mechanisms of Gastrointestinal Mucosal Cell Injury and Protection*, edited by J. W. Harmon. Baltimore, MD: Williams & Wilkins, 1981, p. 351–357.
2. ALLEN, A. Structure and function of gastrointestinal mucus. In: *Physiology of the Gastrointestinal Tract* (1st ed.), edited by L. R. Johnson. New York: Raven, 1981, vol. I, p. 617–639.
3. ALLEN, A. Structure of gastrointestinal mucus glycoproteins and the viscous and gel-forming properties of mucus. *Br. Med. Bull.* 34: 28–33, 1978.

4. ALLEN, A., AND N. J. H. CARROLL. Adherent and soluble mucus in the stomach and duodenum. *Dig. Dis. Sci.* 30: 558–628, 1985.
5. ALLEN, A., N. J. H. CARROLL, AND B. H. HIRST. Gastric mucus in the anaesthetised rat: response to secretin, prostaglandin, ethanol and pepsin (Abstract). *J. Physiol. Lond.* 371: 135P, 1985.
6. ALLEN, A., G. FLEMSTRÖM, A. GARNER, W. SILEN, AND L. A. TURNBERG (editors). *Mechanisms of Mucosal Protection in the Upper Gastrointestinal Tract.* New York: Raven, 1983.
7. ALLEN, A., AND A. GARNER. Gastric mucus and bicarbonate secretion and their possible role in mucosal protection. *Gut* 21: 249–262, 1980.
8. ALLEN, A., A. GARNER, A. C. HUNTER, AND J. P. KEOGH. The gastroduodenal mucus barrier and the place of eicosanoids. In: *Eicosanoids and the Gastrointestinal Tract*, edited by K. Hillier. Lancaster, UK: MTP, 1988, p. 195–213.
9. ALLEN, A., AND L. C. HOSKINS. Colonic mucus in health and disease. In: *Diseases of the Rectum and Colon*, edited by J. B. Kirsner and R. G. Shorter. Baltimore, MD: Williams & Wilkins, 1988, p. 65–94.
10. ALLEN, A., D. A. HUTTON, A. J. LEONARD, J. P. PEARSON, AND L. A. SELLERS. The role of mucus in protection of the gastroduodenal mucosa. *Scand. J. Gastroenterol. Suppl.* 125: 71–77, 1986.
11. ALLEN, A., D. A. HUTTON, D. MANTLE, AND R. H. PAIN. Structure and gel formation in pig gastric mucus. *Trans. Biochem. Soc.* 12: 612–615, 1984.
12. ALLEN, A., D. HUTTON, S. MCQUEEN, AND A. GARNER. Dimensions of gastroduodenal surface pH gradients exceed those of adherent mucus gel layers. *Gastroenterology* 85: 463–466, 1983.
13. ALLEN, A., D. HUTTON, J. P. PEARSON, AND L. A. SELLERS. Mucus glycoprotein structure, gel formation and gastrointestinal mucus function. In: *Mucus and Mucosa*. London: Pitman, 1984, p. 137–156. (Ciba Found. Symp. 109.)
14. ALLEN, A., R. H. PAIN, AND T. ROBSON. Model for the structure of gastric mucus gel. *Nature Lond.* 264: 88–89, 1976.
15. ALLEN, A., R. WARD, W. J. CUNLIFFE, D. A. HUTTON, J. P. PEARSON, AND C. W. VENABLES. Changes in adherent mucus gel and pepsinolysis in peptic ulcer patients (Abstract). *Dig. Dis. Sci.* 30: 365, 1985.
16. ANDRE, C., R. LAMBERT, AND F. DESCOS. Stimulation of gastric mucous secretions in man by secretin. *Digestion* 7: 284–293, 1972.
17. BAGSHAW, P. F., D. J. MUNSTER, AND J. G. WILSON. Molecular weight of gastric mucus glycoprotein is a determinant of the degree of subsequent aspirin induced chronic gastric ulceration in the rat. *Gut* 28: 287–293, 1987.
18. BARONDES, S. H. Soluble lectins: a new class of extracellular proteins. *Science Wash. DC* 223: 1259–1264, 1984.
19. BELL, A. E., AND A. ALLEN. Gastrointestinal mucus, electrolytes and mucosal protection. In: *Electrolyte and Water Transport Across Gastrointestinal Epithelia*, edited by R. M. Case, A. Garner, L. Turnberg, and J. Young. New York: Raven, 1982, p. 253–255.
20. BELL, A. E., A. ALLEN, E. R. MORRIS, AND S. B. ROSS-MURPHY. Functional interactions of gastric mucus glycoprotein. *Int. J. Biol. Macromol.* 6: 309–315, 1984.
21. BELL, A. E., L. A. SELLERS, A. ALLEN, W. J. CUNLIFFE, E. R. MORRIS, AND S. B. ROSS-MURPHY. Properties of gastric and duodenal mucus: effect of proteolysis, disulfide reduction, bile, acid, ethanol and hypertonicity on mucus gel structure. *Gastroenterology* 88: 269–280, 1985.
22. BELL, R. G., L. S. ADAMS, AND R. W. OGDEN. Intestinal mucus trapping in the rapid expulsion of *Trichinella spiralis* by rats: induction and expression analyzed by quantitative worm recovery. *Infect. Immun.* 45: 267–272, 1984.
23. BELLA, A., JR., AND Y. S. KIM. Iron binding of gastric mucins. *Biochim. Biophys. Acta* 304: 580–585, 1973.
24. BENNETT, G., C. P. LEBLOND, AND A. HADDAD. Migration of glycoprotein from the Golgi apparatus to the surface of various cell types as shown by radioautography after labeled fucose injection into rats. *J. Cell Biol.* 60: 258–284, 1974.
25. BERSIMBAEV, R. I., M. M. TAIROV, AND R. I. SALGANIK. Biochemical mechanisms of regulation of mucus secretion by prostaglandin E_2 in rat gastric mucosa. *Eur. J. Pharmacol.* 115: 259–266, 1985.
26. BICKEL, M., AND G. L. KAUFFMAN, JR. Gastric gel mucus thickness: effect of distention, 16,16-dimethyl prostaglandin E_2, and carbenoxolone. *Gastroenterology* 80: 770–775, 1981.
27. BLACK, J. W., J. E. BRADBURY, AND J. H. WYLLIE. Stimulation of colonic mucus output in the rat. *Br. J. Pharmacol.* 66: 456–457, 1979.
28. BLASZCZYK, M., M. HERLYN, Z. STEPELEWSKI, AND H. KOPROWSKI. Monoclonal antibody localization of Lewis antigens in fixed tissue. *Lab. Invest.* 50: 394–400, 1984.
29. BLOOMFIELD, V. A. Hydrodynamic properties of mucus glycoproteins. *Biopolymers* 22: 2141–2154, 1983.
30. BOLAND, C. R., AND D. J. AHNEN. Binding of lectins to goblet cell mucin in malignant and premalignant colonic epithelium in the CF-1 mouse. *Gastroenterology* 89: 127–137, 1985.
31. BOLAND, C. R., C. K. MONTGOMERY, AND Y. S. KIM. Alterations in human colonic mucin occurring with cellular differentiation and malignant transformation. *Proc. Natl. Acad. Sci. USA* 79: 2051–2055, 1982.
32. BOLLARD, J. E., M. A. VANDERWEE, G. W. SMITH, C. TASMAN-JONES, J. B. GAVIN, AND S. P. LEE. Preservation of mucus in situ in rat colon. *Dig. Dis. Sci.* 31: 1338–1344, 1986.
33. BOLTON, J. P., D. PALMER, AND M. M. COHEN. Stimulation of mucus and non-parietal cell secretion by the E_2 prostaglandins. *Am. J. Dig. Dis.* 23: 359–364, 1978.
34. CARLSON, D. M. Structures and immunochemical properties of oligosaccharides isolated from pig submaxillary mucins. *J. Biol. Chem.* 243: 616–626, 1968.
35. CARLSON, D. M. Chemistry and biosynthesis of mucin glycoproteins. In: *Mucus in Health and Disease*, edited by M. Elstein and D. V. Parke. New York: Plenum, 1977, p. 251–273.
36. CARLSTEDT, I., AND J. K. SHEEHAN. Is the macromolecular architecture of cervical, respiratory and gastric mucins the same? *Trans. Biochem. Soc.* 12: 615–617, 1984.
37. CARLSTEDT, I., AND J. K. SHEEHAN. Macromolecular properties and polymeric structure of mucus glycoproteins. In: *Mucus and Mucosa*. London: Pitman, 1984, p. 157–172. (Ciba Found. Symp. 109.)
38. CARLSTEDT, I., J. K. SHEEHAN, A. P. CORFIELD, AND J. T. GALLAGHER. Mucous glycoproteins, a gel of a problem. *Essays Biochem.* 20: 40–76, 1985.
39. CHANTLER, E., J. B. ELDER, AND M. ELSTEIN. Mucus in health and disease. II. *Adv. Exp. Med. Biol.* 144: 1–441, 1982.
40. CLAMP, J. R., AND J. M. CREETH. Some non-mucin components of mucus and their possible biological roles. In: *Mucus and Mucosa*. London: Pitman, 1984, p. 121–131. (Ciba. Found. Symp. 109.)
41. CLAMP, J. R., AND L. REID. Mucus. *Br. Med. Bull.* 34: 1–96, 1978.
42. COHEN, P. S., R. ROSSOLL, V. J. CABELLI, S. L. YANG, AND D. C. LAUX. Relationship between the mouse colonizing ability of a human fecal *Escherichia coli* strain and its ability to bind a specific mouse colonic mucous gel protein. *Infect. Immun.* 40: 62–69, 1983.
43. CORFIELD, A. P., AND R. SCHAUER. Metabolism of sialic acids. In: *Sialic Acids: Chemistry, Metabolism, and Function*, edited by R. Schauer. New York: Springer-Verlag, 1982, p. 195–249.
44. CORFIELD, A. P., A. J. P. WILLIAMS, S. A. WAGNER, J. R. CLAMP, AND R. A. MOUNTFORD. Mucus glycoprotein degrading enzymes in inflammatory bowel disease detection of a novel sialic acid O acetyl esterase (Abstract). *Gut* 27: A1261, 1986.
45. CREETH, J. M. Constituents of mucus and their separation. *Br. Med. Bull.* 34: 17–24, 1978.
46. CREETH, J. M., AND B. COOPER. Studies on molecular weight distributions of two mucins. *Trans. Biochem. Soc.* 12: 618–621, 1974.

47. CREETH, J. M., AND M. A. DENBOROUGH. The use of equilibrium density gradient methods for preparation and characterization of blood group specific glycoproteins. *Biochem. J.* 117: 879–891, 1970.
48. CROSS, C. E., B. HALLIWELL, AND A. ALLEN. Antioxidant protection: a function of tracheobronchial and gastrointestinal mucus. *Lancet* 1: 1328–1330, 1984.
49. CROWTHER, R. S., C. MARRIOTT, AND S. L. JAMES. Cation induced changes in the rheological properties of purified mucus glycoproteins. *Biorheology* 21: 253–263, 1984.
50. CULLING, F. A., AND P. E. REID. Histochemistry of sialic acids. In: *Sialic Acids: Chemistry, Metabolism, and Function*, edited by R. Schauer. New York: Springer-Verlag, 1984, p. 173–191.
51. DAVENPORT, H. W. Physiological structure of the gastric mucosa. In: *Handbook of Physiology. Alimentary Canal*, edited by C. F. Code. Washington, DC: Am. Physiol. Soc., 1967, sect. 6, vol. II, chapt. 43, p. 759–779.
52. DAVENPORT, H. W. Mucosal barrier to hydrogen ion back diffusion: disrupting factors and mucosal response. In: *Peptic Ulcer Disease*, edited by R. Fisher. New York: Biomed. Information, 1979, p. 77–88.
53. DOMSCHKE, W., S. DOMSCHKE, D. HORNIG, AND L. DEMLING. Prostaglandin-stimulated gastric mucus secretion in man. *Acta Hepato-Gastroenterol.* 25: 292–294, 1979.
54. DONALD, A. S. R. The products of pronase digestion of purified blood group-specific glycoproteins. *Biochim. Biophys. Acta* 317: 420–436, 1973.
55. DOWNING, S. W., W. L. SALO, R. H. SPITZER, AND E. A. KOCH. The hagfish slime gland: a model system for studying the biology of mucus. *Science Wash. DC* 214: 1143–1145, 1981.
56. DUNSTONE, J. R., AND W. T. MORGAN. Further observations on the glycoprotein in human ovarian cyst fluids. *Biochim. Biophys. Acta* 101: 300–314, 1965.
57. FAHIM, E. F., G. G. FORSTNER, AND J. F. FORSTNER. Heterogeneity of rat goblet cell mucin before and after reduction. *Biochem. J.* 209: 117–124, 1983.
58. FEIZI, T. Demonstration by monoclonal antibodies that carbohydrate structures of glycoproteins and glycolipids are oncodevelopmental antigens. *Nature Lond.* 314: 53–57, 1985.
59. FERRY, J. D. *Viscoelastic Properties of Polymers* (3rd ed.). New York: Wiley, 1980.
60. FILIPE, M. I. Mucins in the human gastrointestinal epithelium: a review. *Invest. Cell Pathol.* 2: 195–216, 1979.
61. FILIPE, M. I., AND J. R. JASS. *Gastric Carcinoma*. Edinburgh: Churchill Livingstone, 1986.
62. FISHER, J., P. J. KLEIN, M. VIERBUCHEN, B. SKUTTA, G. UNLENBRUCK, AND R. FISCHER. Characterization of glycoconjugates of human gastrointestinal mucosa by lectins. *J. Histochem. Cytochem.* 32: 681–689, 1984.
63. FLEMSTRÖM, G. Gastric and duodenal mucosal bicarbonate secretion. In: *Physiology of the Gastrointestinal Tract* (2nd ed.), edited by L. R. Johnson. New York: Raven, 1987, vol. 2, p. 1011–1029.
64. FLEMSTRÖM, G., AND A. GARNER. Gastroduodenal HCO_3^- transport: characteristics and proposed role in acidity regulation and mucosal protection. *Am. J. Physiol.* 242 (*Gastrointest. Liver Physiol.* 5): G183–G193, 1982.
65. FLEMSTRÖM, G., AND E. KIVILAAKSO. Demonstration of a pH gradient at the luminal surface of rat duodenum in vivo and its dependence on mucosal alkaline secretion. *Gastroenterology* 84: 787–794, 1983.
66. FLEMSTRÖM, G., AND L. A. TURNBERG. Gastroduodenal defense mechanisms. *Clin. Gastroenterol.* 13: 327–355, 1984.
67. FLOREY, H. Mucin and the protection of the body. *Proc. R. Soc. Lond. B Biol. Sci.* 143: 144–148, 1955.
68. FORSTNER, J. F., AND G. G. FORSTNER. Calcium binding to intestinal goblet cell mucin. *Biochim. Biophys. Acta* 386: 283–292, 1975.
69. FORSTNER, J. F., I. JABBAL, B. P. FINDLAY, AND G. G. FORSTNER. Interaction of mucins with calcium, H^+ ion and albumin. *Mod. Probl. Paediatr.* 19: 54–65, 1976.
70. FORSTNER, J. F., I. JABBAL, AND G. G. FORSTNER. Intestinal goblet cell mucus. Chemical and physical characterization. *Can. J. Biochem.* 51: 1154–1166, 1973.
71. FORSTNER, J. F., F. OFOSU, AND G. G. FORSTNER. Radioimmunoassay of intestinal goblet cell mucin. *Anal. Biochem.* 83: 657–665, 1977.
72. FORSTNER, J. F., N. W. ROOMI, R. E. F. FAHIM, AND G. G. FORSTNER. Cholera toxin stimulates secretion of immunoreactive intestinal mucin. *Am. J. Physiol.* 240 (*Gastrointest. Liver Physiol.* 3): G10–G16, 1981.
73. FORSTNER, G., M. SHIH, AND B. LUKIE. Cyclic AMP and intestinal glycoprotein synthesis. The effect of β-adrenergic agents, theophylline and dibutyryl cyclic AMP. *Can. J. Physiol. Pharmacol.* 51: 122–129, 1973.
74. FRETER, R. Factors involved in the penetration of the mucous layer in experimental cholera. In: *Attachment of Organisms to the Gut Mucosa*, edited by E. C. Boedeker. Boca Raton, FL: CRC, 1984, vol. II, p. 43–50.
75. GIBBONS, R. A. Physico-chemical methods for determination of purity, molecular size and shape of glycoproteins. In: *Glycoproteins, Their Composition, Structure and Function* (2nd ed.), edited by A. Gottschalk. Amsterdam: Elsevier, 1972, p. 31–109.
76. GIBBONS, R. A. Mucus of the mammalian genital tract. *Br. Med. Bull.* 34: 34–38, 1978.
77. GIBBONS, R. J., AND J. V. QURESHI. Selective binding of blood group-reactive salivary mucins by *Streptococcus mutans* and other oral organisms. *Infect. Immun.* 22: 665–671, 1978.
78. GOLD, D. V., AND F. MILLER. Characterization of human colonic mucoprotein antigen. *Immunochemistry* 11: 369–375, 1974.
79. GOLD, D. V., AND F. MILLER. Comparison of human colonic mucoprotein antigen from normal and neoplastic mucosa. *Cancer Phila.* 38: 3204–3211, 1978.
80. GOTTSCHALK, A. (editor). *Glycoproteins: Their Composition, Structure and Function* (2nd ed.). Amsterdam: Elsevier, 1972.
81. GOTTSCHALK, A., A. S. BHARGAVA, AND V. L. N. MURTY. Submaxillary gland glycoprotein. In: *Glycoproteins: Their Composition, Structure and Function* (2nd ed.), edited by A. Gottschalk. Amsterdam: Elsevier, 1972, p. 810–829.
82. GREEN, K., AND T. OTORI. Direct measurements of membrane unstirred layers. *J. Physiol. Lond.* 207: 93–102, 1970.
83. HALLETT, P., A. J. ROWE, AND S. E. HARDING. A highly expanded spheroidal conformation for a mucin from a cystic fibrosis patient: new evidence from electron microscopy. *Trans. Biochem. Soc.* 12: 878–879, 1984.
84. HARDING, S. E., A. J. ROWE, AND J. M. CREETH. Further evidence for a flexible and highly expanded spheroidal model for mucus glycoproteins in solution. *Biochem. J.* 209: 893–896, 1983.
85. HARMON, J. W. (editor). *Basic Mechanisms of Gastrointestinal Mucosal Cell Injury and Protection*. Baltimore, MD: Williams & Wilkins, 1981.
86. HEATLEY, N. G. Mucosubstance as a barrier to diffusion. *Gastroenterology* 37: 313–318, 1959.
87. HILL, H. D., JR., J. A. REYNOLDS, AND R. L. HILL. Purification, composition, molecular weight, and subunit structure of ovine submaxillary mucin. *J. Biol. Chem.* 252: 3791–3798, 1976.
88. HILLS, B. A., B. D. BUTLER, AND I. M. LICHTENBERGER. Gastric mucosal barrier: hydrophobic lining to the lumen of the stomach. *Am. J. Physiol.* 244 (*Gastrointest. Liver Physiol.* 7): G561–G568, 1983.
89. HIRST, B. H., R. KAURA, AND A. ALLEN. Secretin stimulation of gastric mucus secretion in the cat: the viscosity of gastric juice in relation to glycoprotein structure and concentration. In: *Nutrition, Digestion, Metabolism*, edited by T. Gati, L. G. Szollar, and G. Ungvary. New York: Pergamon, 1980, vol. 12, p. 237–242. (Adv. Physiol. Sci. Ser.)
90. HOLLANDER, F. The two-component mucus barrier. *Arch. Intern. Med.* 93: 107–128, 1954.
91. HOLLANDER, F. The electrolyte patterns of gastric secretions: its implications for cystic fibrosis. *Ann. NY Acad. Sci.* 106:

298-310, 1963.
92. HOROWITZ, M. I., AND W. PIGMAN (editors). *The Glycoconjugates: Mammalian Glycoproteins and Glycolipids.* New York: Academic, 1977, vol. 1.
93. HOSKINS, L. C. Degradation of mucus glycoproteins in the gastrointestinal tract. In: *The Glycoconjugates,* edited by M. I. Horowitz and W. Pigman. New York: Academic, 1978, vol. 2, p. 235-250.
94. HOSKINS, L. C. Human enteric population ecology and degradation of gut mucins. *Dig. Dis. Sci.* 26: 769-772, 1981.
95. HOSKINS, L. C., M. AGUSTINES, W. B. MCKEE, E. T. BOULDING, M. KRIARIS, AND G. NIEDERMEYER. Mucin degradation in human colon ecosystems. Isolation and properties of fecal strains that degrade ABH blood group antigens and oligosaccharides from mucin glycoproteins. *J. Clin. Invest.* 75: 944-953, 1985.
96. HOSKINS, L. C., AND E. T. BOULDING. Degradation of blood group antigens in human colon ecosystems. *J. Clin. Invest.* 57: 63-73, 1976.
97. HOUNSELL, E. F., AND T. FEIZI. Gastrointestinal mucins. Structures and antigenicities of their carbohydrate chains in health and disease. *Med. Biol. Helsinki* 60: 227-236, 1982.
98. HUBBARD, S. C., AND R. J. IVATT. Synthesis and processing of asparagine-linked oligosaccharides. *Annu. Rev. Biochem.* 50: 555-583, 1981.
99. HUTTON, D. A., A. ALLEN, W. J. CUNLIFFE, AND J. P. PEARSON. Proteolytic degradation of mucus in the colon. *Trans. Biochem. Soc.* 15: 1074, 1988.
100. ITO, S. Functional gastric morphology. In: *Physiology of the Gastrointestinal Tract* (1st ed.), edited by L. R. Johnson. New York: Raven, 1981, vol. I, p. 517-550.
101. ITO, S., AND E. R. LACY. Morphology of rat gastric mucosal damage, defense, and restitution in the presence of luminal ethanol. *Gastroenterology* 88: 150-260, 1985.
102. IZHAR, M., Y. NUCHAMOWITZ, AND D. MIRELMAN. Adherence of *Shigella flexneri* to guinea pig intestinal cells is mediated by a mucosal adhesion. *Infect. Immun.* 35: 1110-1118, 1982.
103. JACOBS, L. R., AND P. W. HUBER. Regional distribution and alterations of lectin binding to colorectal mucin in mucosal biopsies from controls and subjects with inflammatory bowel diseases. *J. Clin. Invest.* 75: 112-118, 1985.
104. JAROWITZ, H. D., AND F. HOLLANDER. Viscosity of cell free canine gastric mucus. *Gastroenterology* 36: 582-594, 1954.
105. JENTJENS, T., A. L. SMITS, AND G. J. STROUS. 16,16-Dimethyl prostaglandin E_2 stimulates galactose and glucosamine but not serine incorporation in rat gastric mucous cells. *Gastroenterology* 87: 409-416, 1984.
106. JENTJENS, T., AND G. J. STROUS. Quantitative aspects of mucus glycoprotein biosynthesis in rat gastric mucosa. *Biochem. J.* 228: 227-232, 1985.
107. JENTJENS, T., A. VAN DE KAMP, R. SPEE-BRAND, AND G. J. STROUS. Biosynthesis, processing and secretion of mucus glycoprotein in the rat stomach. *Biochim. Biophys. Acta* 887: 133-141, 1986.
108. JOHANSSON, C., AND B. KOLLBERG. E_2 prostaglandins of human gastric mucus output. *Eur. J. Clin. Invest.* 9: 229-232, 1979.
109. KATZ, D., H. I. SIEGEL, AND G. B. J. GLASS. Acute gastric mucosal lesions produced by augmented histamine test. *Am. J. Dig. Dis.* 14: 447-455, 1969.
110. KENT, P. W., AND A. ALLEN. The biosynthesis of intestinal mucins. *Biochem. J.* 106: 645-658, 1968.
111. KERSS, S., A. ALLEN, AND A. GARNER. A simple method for measuring thickness of the mucus gel layer adherent to rat, frog and human gastric mucosa: influence of feeding, prostaglandin, N-acetylcysteine and other agents. *Clin. Sci. Lond.* 63: 187-195, 1982.
112. KHAN, M. A., D. P. WOLF, AND M. LITT. Effect of mucolytic agents on the rheological properties of tracheal mucus. *Biochim. Biophys. Acta* 444: 369-373, 1976.
113. KING, M. Viscoelastic properties of airway mucus. *Federation Proc.* 39: 3061-3080, 1980.
114. KIVILAAKSO, E., AND G. FLEMSTRÖM. Surface pH gradient in gastroduodenal mucosa. *Scand. J. Gastroenterol. Suppl.* 105: 50-52, 1984.
115. KIVILAAKSO, E., D. FROMM, AND W. SILEN. Effect of the acid secretory state on intramural pH of rabbit gastric mucosa. *Gastroenterology* 75: 641-648, 1978.
116. KOWALEWSKI, K., T. PACHKOWSKI, AND A. KOLODEJ. Effect of secretin on mucinous secretion by the isolated canine stomach perfused extracorporeally. *Pharmacology Basel* 16: 78-82, 1979.
117. KRAMER, M. F., J. J. GEUZE, AND G. J. A. M. STROUS. Site of synthesis, intracellular transport and secretion of glycoprotein in exocrine cells. In: *The Respiratory Tract Mucus.* Amsterdam: Elsevier, 1978, p. 25-51. (Ciba Found. Symp. 54.)
118. LACY, E. R. Gastric mucosal resistance to a repeated ethanol insult. *Scand. J. Gastroenterol. Suppl.* 10: 63-72, 1985.
119. LACY, E. R., AND S. ITO. Rapid epithelial restitution of the rat gastric mucosa after ethanol injury. *Lab. Invest.* 51: 573-583, 1984.
120. LAKE, A. M., K. J. BLOCH, M. R. NEUTRA, AND W. A. WALKER. Intestinal goblet cell mucus release. II. In vivo stimulation by antigen in the immunized rat. *J. Immunol.* 122: 834-837, 1979.
121. LAKE, A. M., K. J. BLOCH, K. J. SINCLAIR, AND W. A. WALKER. Anaphylactic release of goblet cell mucus. *Immunology* 39: 173-178, 1980.
122. LAMONT, J. T., AND A. S. VENTOLA. Stimulation of colonic glycoprotein synthesis by dibutyryl cyclic AMP and theophylline. *Gastroenterology* 72: 82-86, 1977.
123. LAMONT, J. T., AND A. S. VENTOLA. Purification and composition of colonic mucin. *Biochim. Biophys. Acta* 626: 234-244, 1980.
124. LAMONT, J. T., A. S. VENTOLA, E. A. MAULL, AND S. SZABO. Cysteamine and prostaglandin $F_{2\beta}$ stimulate rat gastric mucin release. *Gastroenterology* 84: 306-313, 1983.
125. LAUX, D. C., E. F. MCSWEEGAN, AND P. S. COHEN. Adhesion of enterotoxigenic *E. coli* to immobilized intestinal mucosal preparations: a model for adhesion to mucosal surface components. *J. Microbiol. Methods* 2: 27-39, 1984.
126. LEE, A., J. L. O'ROURKE, P. J. BARRINGTON, AND T. J. TRUST. Mucus colonization as a determinant in the pathogenicity in intestinal infection by *Campylobacter jejuni*: a mouse cecal model. *Infect. Immun.* 51: 536-546, 1986.
127. LEE, G. S., AND B. M. OGILVIE. The mucus layer in intestinal nematode infections. In: *The Mucosal Immune System in Health and Disease,* edited by P. L. Ogra and J. Bienstock. Columbus, OH: Ross Lab., 1981, p. 69-73.
128. LEONARD, A., AND A. ALLEN. Gastric mucosal damage by pepsin. *Gut* 27: A1236-A1237, 1986.
129. LEV, R. Histochemistry. In: *The Glycoconjugates: Mammalian Glycoproteins and Glycolipids,* edited by M. I. Horowitz and W. Pigman. New York: Academic, 1977, vol. 1, p. 35-49.
130. LEWIS, I. D., AND J. S. FORDTRAN. Effect of perfusion rate on absorption, surface area, unstirred water layer thickness, permeability, and intraluminal pressure in the rat ileum in vivo. *Gastroenterology* 68: 1509-1516, 1975.
131. LICHTENBURGER, L. M., L. A. GRAZIANI, E. J. DIAL, B. D. BUTLER, AND B. A. HILLS. Role of surface-active phospholipids in gastric glycoprotein. *Science Wash. DC* 219: 1327-1329, 1983.
132. LINDAHL, M., AND T. WADSTROM. K99 surface haemagglutinin of enterotoxigenic *E. coli* recognize terminal N-acetylgalactosamine sialic acid residues of glycophorin and other complex glycoconjugates. *Vet. Microbiol.* 9: 249-257, 1984.
133. LIST, S. J., B. P. FINDLAY, G. G. FORSTNER, AND J. F. FORSTNER. Enhancement of the viscosity of mucin by serum albumin. *Biochem. J.* 175: 565-567, 1978.
134. MACDERMOTT, R. P., R. M. DONALDSON, JR., AND J. S. TRIER. Glycoprotein synthesis and secretion by mucosal biopsies of rabbit colon and human rectum. *J. Clin. Invest.* 54: 545-554, 1974.
135. MACHADO, G., J. R. CLAMP, AND A. E. READ. Carbohydrate

content of endoscopic gastric biopsies in carcinoma of the stomach. *Gut* 18: 670–672, 1977.
136. MACHEN, T. E., AND A. M. PARADISO. Regulation of intracellular pH in the stomach. *Annu. Rev. Physiol.* 49: 21–25, 1987.
137. MAGNANI, J. L., Z. STEPELEWSKI, H. KOPROWSKI, AND V. GINSBURG. Identification of the gastrointestinal and pancreatic cancer-associated antigen detected by monoclonal antibody 19-9 in the sera of patients as a mucin. *Cancer Res.* 43: 5489–5492, 1983.
138. MANTLE, M., AND A. ALLEN. A colorimetric assay for glycoproteins based on the periodic acid/Schiff stain. *Biochem. Soc. Trans.* 6: 607–609, 1978.
139. MANTLE, M., AND A. ALLEN. Isolation and characterization of the native glycoprotein from pig small-intestinal mucus. *Biochem. J.* 195: 267–275, 1981.
140. MANTLE, M., G. G. FORSTNER, AND J. F. FORSTNER. Antigenic and structural features of goblet-cell mucin of human small intestine. *Biochem. J.* 217: 159–167, 1984.
141. MANTLE, M., G. G. FORSTNER, AND J. F. FORSTNER. Biochemical characterization of the component parts of intestinal mucin from patients with cystic fibrosis. *Biochem. J.* 224: 345–354, 1984.
142. MANTLE, M., AND J. F. FORSTNER. The effects of delipidation on the major antigenic determinant of purified human intestinal mucin. *Biochem. Cell Biol.* 64: 223–238, 1986.
143. MANTLE, M., D. MANTLE, AND A. ALLEN. Polymeric structure of pig small intestinal mucus glycoproteins: dissociation by proteolysis or by reduction of disulphide bridges. *Biochem. J.* 19: 277–285, 1981.
144. MARRIOTT, C., C. K. SHIH, AND M. LITT. Changes in the gel properties of tracheal mucus induced by divalent cations. *Biorheology* 16: 331–337, 1979.
145. MARSHALL, T., AND A. ALLEN. Isolation and characterization of the high molecular weight glycoproteins from pig colonic mucus. *Biochem. J.* 173: 569–578, 1978.
146. MARTIN, G. P., C. MARRIOTT, AND I. W. KELLAWAY. Direct effect of bile salts and phospholipids on the physical properties of mucus. *Gut* 19: 103–107, 1978.
147. MAWHINNEY, T. P., E. ADELSTEIN, D. A. MORRIS, A. M. MAWHINNEY, AND G. J. BARBERO. Structure determination of oligosaccharides derived from tracheobronchial mucus glycoproteins. *J. Biol. Chem.* 262: 2994–3001, 1987.
148. MCQUEEN, S., A. ALLEN, AND A. GARNER. Measurements of gastric and duodenal mucus gel thickness. In: *Mechanisms of Mucosal Protection in the Upper Gastrointestinal Tract*, edited by A. Allen, G. Flemström, A. Garner, W. Silen, and L. A. Turnberg. New York: Raven, 1984, p. 215–221.
149. MCQUEEN, S., D. HUTTON, A. ALLEN, AND A. GARNER. Gastric and duodenal surface mucus gel thickness in rat: effects of prostaglandins and damaging agents. *Am. J. Physiol.* 245 (*Gastrointest. Liver Physiol.* 8): G388–G393, 1983.
150. MEYER, F., AND A. SILBERBERG. The rheology and molecular organization of epithelial mucus. *Biorheology* 17: 163–168, 1980.
151. MIIKKELSEN, A., AND B. T. STOKKE. Flexibility of human bronchial mucin studied using low shear viscometry, direfringence relaxation analysis and electron microscopy. *Biopolymers* 24: 1683–1704, 1985.
152. MILLER, H. R. P., J. F. HUNTLEY, AND G. R. WALLACE. Immune exclusion and mucus trapping during the rapid expulsion of *Nippostrongylus brasiliensis* from primed rats. *Immunology* 44: 419–429, 1981.
153. MILLER, H. R. P., AND Y. NAWA. *Nippostrongylus brasiliensis*: intestinal goblet-cell response in adoptively immunized rats. *Exp. Parasitol.* 47: 81–90, 1979.
154. MILLER, R. S., AND L. C. HOSKINS. Mucin degradation in human colon ecosystems. *Gastroenterology* 81: 759–765, 1981.
155. MORRIS, E. R., AND D. A. REES. Principles of biopolymer gelation. *Br. Med. Bull.* 34: 49–53, 1978.
156. MORRIS, G. P. The myth of the mucus barrier. *Gastroenterol. Clin. Biol.* 9: 106–107, 1985.
157. MORRIS, G. P., R. J. HARDING, AND J. L. WALLACE. A functional model for extracellular gastric mucus in the rat. *Virchows Arch. B. Cell Pathol.* 46: 239–251, 1984.
158. MORRIS, G. P., AND J. L. WALLACE. The roles of ethanol and of acid in the production of gastric mucosal erosions in rats. *Virchows Arch. B Cell Pathol.* 38: 23–38, 1981.
159. MURTY, V. L. N., J. SAROSIEK, A. SLOMIANY, AND B. L. SLOMIANY. Effect of lipids and proteins on the viscosity of gastric mucus glycoprotein. *Biochem. Biophys. Res. Commun.* 1221: 521–529, 1984.
160. NEUTRA, M. R. The functional ultrastructure of mucous cells. *Chest* 81, Suppl.: 14S–19S, 1982.
161. NEUTRA, M. R., AND J. F. FORSTNER. Gastrointestinal mucus: synthesis, secretion, and function. In: *Physiology of the Gastrointestinal Tract* (2nd ed.), edited by L. R. Johnson. New York: Raven, 1987, vol. 2, p. 975–1009.
162. NEUTRA, M. R., R. J. GRAND, AND J. S. TRIER. Glycoprotein synthesis, transport and secretion by epithelial cells of human rectal mucosa. *Lab. Invest.* 36: 535–546, 1977.
163. NEUTRA, M. R., AND C. P. LEBLOND. Synthesis of the carbohydrate of mucus in the Golgi complex as shown by electron microscope radioautography of goblet cells from rats injected with glucose-H³. *J. Cell Biol.* 30: 119–136, 1966.
164. NEUTRA, M. R., AND C. P. LEBLOND. Radioautographic comparison of the uptake of galactose-H³ in the Golgi region of various cells secreting glycoproteins or mucopolysaccharides. *J. Cell Biol.* 30: 137–150, 1966.
165. NEUTRA, M. R., J. L. O'MALLEY, AND R. D. SPECIAN. Regulation of intestinal goblet cell secretion. II. A survey of potential secretagogues. *Am. J. Physiol.* 242 (*Gastrointest. Liver Physiol.* 5): G380–G387, 1982.
166. NEUTRA, M. R., T. H. PHILLIPS, AND T. E. PHILLIPS. Regulation of intestinal goblet cells in situ, in mucosal explants and in the isolated epithelium. In: *Mucus and Mucosa*. London: Pitman, 1984, p. 20–29. (Ciba Found. Symp. 109.)
167. NUGENT, J., AND M. O'CONNOR (editors). *Mucus and Mucosa*. London: Pitman, 1984. (Ciba Found. Symp. 109.)
168. OFOSU, F., J. F. FORSTNER, AND G. FORSTNER. Mucin degradation in the intestine. *Biochim. Biophys. Acta* 543: 476–483, 1978.
169. PEARSON, J. P., A. ALLEN, AND S. PARRY. A 70,000 molecular weight glycoprotein isolated from purified pig gastric mucus glycoprotein by reduction of disulphide bridges and its implication in the polymeric structure. *Biochem. J.* 197: 155–162, 1981.
170. PEARSON, J. P., A. ALLEN, AND C. W. VENABLES. Gastric mucus: isolation and polymeric structure of the undegraded glycoprotein: its breakdown by pepsin. *Gastroenterology* 78: 709–715, 1981.
171. PEARSON, J. P., R. KAURA, W. TAYLOR, AND A. ALLEN. The composition and polymeric structure of mucus glycoprotein from human gall bladder bile. *Biochim. Biophys. Acta* 706: 221–228, 1982.
172. PEARSON, J. P., R. WARD, A. ALLEN, N. B. ROBERTS, AND W. TAYLOR. Mucus degradation by pepsin: comparison of mucolytic activity of human pepsin 1 and pepsin 3: implications in peptic ulceration. *Gut* 27: 243–248, 1986.
173. PFEIFFER, C. J. Experimental analysis of hydrogen ion diffusion in gastrointestinal mucus glycoprotein. *Am. J. Physiol.* 240 (*Gastrointest. Liver Physiol.* 3): G176–G182, 1981.
174. PHILLIPS, T. E., T. H. PHILLIPS, AND M. R. NEUTRA. Regulation of intestinal goblet cell secretin. III. Isolated intestinal epithelium. *Am. J. Physiol.* 247 (*Gastrointest. Liver Physiol.* 10): G674–G681, 1984.
175. PHILLIPS, T. E., T. H. PHILLIPS, AND M. R. NEUTRA. Regulation of intestinal goblet cell secretion. IV. Electrical field stimulation in vitro. *Am. J. Physiol.* 247 (*Gastrointest. Liver Physiol.* 10): G682–G687, 1984.
176. PIHAN, G., D. MAJZOUBI, C. HAUDERISCHILD, J. S. TRIER, AND S. SZABO. Early microcirculatory stasis in acute gastric mucosal injury in the rat and prevention by 16,16-dimethyl prostaglandin E₂. *Gastroenterology* 91: 1415–1425, 1986.
177. PODOLSKY, D. K. Oligosaccharide structures of human colonic

mucin. *J. Biol. Chem.* 260: 8262–8266, 1985.
178. PODOLSKY, D. K. Oligosaccharide structures of isolated human colonic mucin species. *J. Biol. Chem.* 260: 15510–15516, 1985.
179. PODOLSKY, D. K., AND K. J. ISSELBACHER. Composition of human colonic mucin. Selective alteration in inflammatory bowel disease. *J. Clin. Invest.* 72: 142–153, 1983.
180. QUERESHI, R., G. G. FORSTNER, AND J. F. FORSTNER. Radioimmunoassay of human intestinal goblet cell mucin. *J. Clin. Invest.* 64: 1149–1156, 1979.
181. RAINFORD, K. D. The effects of aspirin and other non-steroid anti-inflammatory/analgesic drugs on gastrointestinal mucus. Glycoprotein synthesis in vivo: relationship to ulcerogenic actions. *Biochem. Pharmacol.* 27: 877–885, 1978.
182. RAMBOURG, A., W. HERNANDEZ, AND C. P. LEBLOND. Detection of complex carbohydrates in the Golgi apparatus of rat cells. *J. Cell Biol.* 40: 395–414, 1969.
183. RATHBONE, B. J., J. I. WYATT, AND R. V. HEATLEY. *Campylobacter pyloridis*—a new factor in peptic ulcer disease? *Gut* 27: 635–641, 1986.
184. RAVDIN, J. I., J. E. JOHN, L. I. JOHNSTON, D. J. INNES, AND R. L. GURRANT. Adherence of *Entamoeba histolytica* trophozoites to monolayers of human cells. *Infect. Immun.* 48: 292–297, 1985.
185. REID, P. E., C. F. A. CULLING, W. L. DUAN, C. W. RAMEY, A. B. MAGIL, AND M. G. CLAY. Differences between the O-acetylated sialic acids of the epithelial mucins of human colonic tumours and normal controls. *J. Histochem. Cytochem.* 28: 217–222, 1980.
186. ROBERT, A. Cytoprotection by prostaglandins. *Gastroenterology* 77: 761–767, 1979.
187. ROBERT, A., W. BOTTCHER, E. GOLANSKA, AND G. L. KAUFFMAN, JR. Lack of correlation between mucus gel thickness and gastric cytoprotection in rats. *Gastroenterology* 86: 670–674, 1984.
188. ROBERT, A., J. E. NEZAMIS, C. LANCASTER, AND A. J. HANCHAR. Cytoprotection by prostaglandins in rats. Prevention of gastric necrosis produced by alcohol, HCl, NaOH, hypertonic NaCl, and thermal injury. *Gastroenterology* 77: 433–443, 1979.
189. ROBERTS, G. P. Chemical aspects of respiratory mucus. *Br. Med. Bull.* 34: 34–38, 1978.
190. ROGERS, C. M., K. B. COOKE, AND M. I. FILIPE. Sialic acids of human large bowel mucosa: O-acylated variants in normal and malignant states. *Gut* 19: 587–592, 1978.
191. ROOMI, N., M. LABURTHE, N. FLEMING, R. CROWTHER, AND J. FORSTNER. Cholera-induced mucin secretion from rat intestine: lack of effect of cAMP, cycloheximide, VIP, and colchicine. *Am. J. Physiol.* 247 (*Gastrointest. Liver Physiol.* 10): G140–G148, 1984.
192. ROSS, I. N., H. M. M. BAHARI, AND L. A. TURNBERG. The pH gradient across mucus adherent to rat fundic mucosa in vivo and the effect of potential damaging agents. *Gastroenterology* 81: 713–718, 1981.
193. ROSS, I. N., AND L. A. TURNBERG. Studies of the "mucus-bicarbonate" barrier on rat fundic mucosa: the effects of luminal pH and a stable prostaglandin analogue. *Gut* 24: 1030–1033, 1983.
194. ROTH, J. Cytochemical localization of terminal N-acetyl-D-galactosamine residues in cellular compartments of intestinal goblet cells: implications for the topology of O-glycosylation. *J. Cell Biol.* 98: 399–406, 1984.
195. ROWE, P. H., R. LANGE, G. MARRONE, J. B. MATTHEWS, E. KASDON, AND W. SILEN. In vitro protection of amphibian gastric mucosa by nutrient HCO_3 against aspirin injury. *Gastroenterology* 89: 767–778, 1985.
196. ROZEE, K. R., D. COOPER, K. LAM, AND J. W. COSTERTON. Microbial flora of the mouse ileum mucous layer and epithelial surface. *Appl. Environ. Microbiol.* 43: 1451–1463, 1982.
197. SAKATA, T., AND W. V. ENGLEHART. Luminal mucin in the large intestine of mice, rats and guinea pigs. *Cell Tissue Res.* 219: 629–635, 1981.
198. SANDERS, M. J., A. AYALON, M. ROLL, AND A. H. SOLL. The apical surface of canine chief cell monolayers resists H^+ back diffusion. *Nature Lond.* 313: 82–84, 1985.
199. SANDOZ, D., G. NICOLAS, AND M. C. LAINE. Two mucous cell types revisited after quick-freezing and cryosubstitution. *Cell. Biol.* 54: 79–88, 1985.
200. SAROSIEK, J., A. SLOMIANY, A. TAKAGI, AND B. L. SLOMIANY. Hydrogen ion diffusion in dog gastric mucus glycoprotein: effect of associated lipids and covalently bound fatty acids. *Biochem. Biophys. Res. Commun.* 118: 523–531, 1984.
201. SCAWEN, M., AND A. ALLEN. The action of proteolytic enzymes on the glycoprotein from pig gastric mucus. *Biochem. J.* 163: 363–368, 1977.
202. SCHACHTER, H. Glycoprotein biosynthesis. In: *The Glycoconjugates. Mammalian Glycoproteins*, edited by M. I. Horowitz and W. Pigman. New York: Academic, 1978, vol. 2, p. 87–181.
203. SCHACHTER, H., S. NARASIMHAN, P. GLEESON, G. J. VELLA, AND I. BROCKHAUSEN. Oligosaccharide branching of glycoproteins: biosynthetic mechanisms and possible biological functions. *Philos. Trans. R. Soc. Lond. B Biol. Sci.* 300: 145–159, 1982.
204. SCHACHTER, H., AND S. ROSEMAN. Mammalian glycosyltransferases: their role in the synthesis and function of complex carbohydrates and glycolipids. In: *Biochemistry of Glycoproteins and Proteoglycans*, edited by W. J. Lennarz. New York: Plenum, 1980, p. 85–160.
205. SCHACHTER, H., AND D. WILLIAMS. Biosynthesis of mucus glycoproteins. In: *Mucus in Health and Disease*, edited by E. N. Chantler, J. B. Elder, and M. Elstein. New York: Plenum, 1982, vol. II, p. 3–28. (Adv. Exp. Med. Biol. Ser. 144.)
206. SCHAUER, R. (editor). *Sialic Acids: Chemistry, Metabolism, and Function.* New York: Springer-Verlag, 1982, p. 21–33.
207. SCHRAGER, J., AND M. D. G. OATES. The isolation and partial characterization of the principal gastric glycoprotein of "visible" mucus. *Digestion* 4: 1–12, 1971.
208. SELLERS, L. A., AND A. ALLEN. Studies on pig and sheep salivary mucins. *Biochem. Soc. Trans.* 12: 650, 1984.
209. SELLERS, L. A., A. ALLEN, AND M. K. BENNETT. Formation of a fibrin based gelatinous coat over repairing rat gastric epithelium following acute ethanol damage: interaction with adherent mucus. *Gut* 28: 835–843, 1987.
210. SELLERS, L. A., A. ALLEN, E. MORRIS, AND S. B. ROSS-MURPHY. Mechanical characterization and properties of gastrointestinal mucus gel. *Biorheology* 24: 615–623, 1987.
211. SELLERS, L. A., N. J. H. CARROLL, AND A. ALLEN. Misoprostol-induced increases in adherent gastric mucus thickness and luminal mucus output. *Dig. Dis. Sci.* 31, Suppl.: 91S–95S, 1986.
212. SHEEHAN, J., K. OAKES, AND I. CARLSTEDT. Electron microscopy of cervical, gastric and bronchial mucus glycoproteins. *Biochem. J.* 239: 147–153, 1986.
213. SHEFFNER, A. L. The reduction in vitro in viscosity of mucoprotein solutions by a new mucolytic agent, N-acetylcysteine. *Ann. NY Acad. Sci.* 106: 298–309, 1963.
214. SHORA, W., G. G. FORSTNER, AND J. F. FORSTNER. Stimulation of proteolytic digestion by intestinal goblet cell mucus. *Gastroenterology* 68: 470–479, 1975.
215. SILEN, W., AND S. ITO. Mechanisms for rapid re-epithelialization of the gastric mucosal surface. *Annu. Rev. Physiol.* 47: 217–229, 1985.
216. SLAYTER, H. S., A. G. COOPER, AND M. C. BROWN. Electron microscopy and physical parameters of human blood group i, A, B, and H antigens. *Biochemistry* 13: 3365–3371, 1974.
217. SLAYTER, H. S., G. LAMBLIN, A. LE TREUT, C. GALABERT, N. HOUDRET, P. DEGAND, AND P. ROUSSEL. Complex structure of human bronchial mucus glycoprotein. *Eur. J. Biochem.* 142: 209–218, 1984.
218. SLOMIANY, A., Z. JOZWIAK, A. TAKAGI, AND B. L. SLOMIANY. The role of covalently bound fatty acids in the degradation of human gastric mucus glycoprotein. *Arch. Biochem. Biophys.* 229: 560–566, 1984.
219. SLOMIANY, B. L., AND K. MEYER. Isolation and structural studies of sulphated glycoproteins of hog gastric mucosa. *J. Biol. Chem.* 247: 5062–5070, 1972.

220. SLOMIANY, B. L., V. L. MURTY, AND A. SLOMIANY. Isolation and characterization of oligosaccharides from rat colonic mucus glycoprotein. *J. Biol. Chem.* 255: 9719–9723, 1980.
221. SLOMIANY, A., B. L. SLOMIANY, H. WITAS, M. AONO, AND L. J. NEWMAN. Isolation of fatty acids covalently bound to the gastric mucus glycoprotein of normal and cystic fibrosis patients. *Biochem. Biophys. Res. Commun.* 113: 286–293, 1983.
222. SLOMIANY, A., H. WITAS, M. AONO, AND B. L. SLOMIANY. Covalently linked fatty acids in gastric mucus glycoprotein of cystic fibrosis patients. *J. Biol. Chem.* 258: 8535–8538, 1983.
223. SLOMIANY, B. L., E. ZDEBSKA, AND A. SLOMIANY. Structural characterization of neutral oligosaccharides of human H^+Le^{b+} gastric mucin. *J. Biol. Chem.* 259: 2863–2869, 1984.
224. SMITH, B. F., AND J. T. LAMONT. Hydrophobic binding properties of bovine gallbladder mucin. *J. Biol. Chem.* 259: 12170–12177, 1984.
225. SMITH, G. W., C. TASMAN-JONES, P. M. WIGGINS, AND S. P. LEE. Pig gastric mucus: a one-way barrier for H^+. *Gastroenterology* 89: 1313–1318, 1985.
226. SMITHSON, K. W., D. B. MILLAR, L. R. JACOBS, AND G. M. GRAY. Intestinal diffusion barrier: unstirred water layer or membrane surface mucous coat? *Science Wash. DC* 214: 1241–1244, 1981.
227. SNARY, D., A. ALLEN, AND R. H. PAIN. Structural studies on gastric mucoproteins. Lowering of molecular weight after reduction with 2-mercaptoethanol. *Biochem. Biophys. Res. Commun.* 40: 844–851, 1970.
228. SNARY, D., A. ALLEN, AND R. H. PAIN. Conformational changes in gastric mucoproteins induced by caesium chloride and guanidinium chloride. *Biochem. J.* 141: 641–646, 1974.
229. SPECIAN, R. D., AND M. R. NEUTRA. Goblet cells: membrane loss during rapid secretion (Abstract). *J. Cell Biol.* 83: 429, 1979.
230. SPECIAN, R. D., AND M. R. NEUTRA. Mechanisms of rapid mucus secretion in goblet cells stimulated by acetylcholine. *J. Cell Biol.* 85: 626–640, 1980.
231. SPECIAN, R. D., AND M. R. NEUTRA. Regulation of intestinal goblet cell secretion. I. Role of parasympathetic stimulation. *Am. J. Physiol.* 242 (*Gastrointest. Liver Physiol.* 5): G370–G379, 1982.
232. SPECIAN, R. D., AND M. R. NEUTRA. Cytoskeleton of intestinal goblet cells in rabbit and monkey. The theca. *Gastroenterology* 18: 1313–1325, 1984.
233. SPREE-BRAND, R., G. J. A. M. STROUS, AND M. F. KRAMER. Isolation and partial characterization of rat gastric mucus glycoprotein. *Biochem. Biophys. Acta* 621: 104–116, 1980.
234. STANLEY, R. A., S. P. RAM, R. K. WILKINSON, AND A. M. ROBERTON. Degradation of pig gastric and colonic mucins by bacteria isolated from the pig colon. *Appl. Environ. Microbiol.* 51: 1104–1109, 1986.
235. STARKEY, B. J., D. SNARY, AND A. ALLEN. Characterization of gastric mucoproteins isolated by equilibrium density-gradient centrifugation in caesium chloride. *Biochem. J.* 141: 633–639, 1974.
236. STROMBECK, D. R., AND D. HARROLD. Binding of cholera toxin to mucins and inhibition by gastric mucin. *Infect. Immun.* 10: 1266–1272, 1974.
237. SWEELY, C. C., AND H. A. NUNEZ. Structural analysis of glycoconjugates by mass spectrometry and nuclear magnetic resonance spectroscopy. *Annu. Rev. Biochem.* 54: 765–801, 1985.
238. TACHE, Y. Bombesin: central nervous system action to increase gastric mucus in rats. *Gastroenterology* 83: 75–80, 1982.
239. TAKEUCHI, K. D., D. MAGEE, J. CRITCHLOW, AND W. SILEN. Studies of the pH gradient and thickness of frog gastric mucus gel. *Gastroenterology* 84: 787–794, 1983.
240. TAO, P., AND D. E. WILSON. Effects of prostaglandin E_2, 16,16-dimethyl prostaglandin E_2 and a prostaglandin endoperoxide analogue (U46619) on gastric secretory volume, [H^+], and mucus synthesis and secretion in the rat. *Prostaglandins* 28: 353–365, 1984.
241. TAYLOR, W. H. Biochemistry and pathological physiology of pepsin 1. *Adv. Clin. Enzymol.* 2: 79–81, 1982.
242. TETTEMANTI, G., AND W. PIGMAN. Purification and characterization of bovine and ovine submaxillary mucins. *Arch. Biochem. Biophys.* 124: 41–50, 1969.
243. THOMSON, A. B. R. Unstirred water layers: a basic mechanism of gastrointestinal mucosal cell cytoprotection. In: *Basic Mechanisms of Gastrointestinal Mucosal Cell Injury and Cytoprotection*, edited by J. W. Harmon. Baltimore, MD: Williams & Wilkins, 1981, p. 327–350.
244. VADGAMA, P., AND K. G. M. M. ALBERTI. The possible role of bicarbonate in mucosal protection and peptic ulceration. *Digestion* 27: 203–213, 1983.
245. VAGNE, M., AND G. PERRETT. Regulation of gastric mucus secretion. *Scand. J. Gastroenterol.* 42: 63–74, 1976.
246. VAN HALBEEK, H. Structural analysis of the carbohydrate chains of mucin-type glycoproteins by high-resolution ^{1}H-n.m.r. spectroscopy. *Biochem. Soc. Trans.* 12: 601–605, 1984.
247. VAN HALBEEK, H., L. DORLAND, J. F. G. VLIEGENTHART, N. K. KOCHETKOV, N. P. ARBATSKY, AND V. A. DEREVITSKAYA. Characterization of the primary structure and the microheterogeneity of the carbohydrate chains of porcine blood-group H substance by 500-MHz ^{1}H-NMR spectroscopy. *Eur. J. Biochem.* 127: 21–29, 1982.
248. VARIYAM, E. P., AND L. C. HOSKINS. In vitro degradation of gastric mucin. Carbohydrate side chains protect polypeptide core from pancreatic proteases. *Gastroenterology* 84: 533–537, 1983.
249. VERDUGO, P. Hydration kinetics of exocytosed mucins in cultured secretory cells of the rabbit trachea: a new model. In: *Mucus and Mucosa*. London: Pittman, 1984, p. 212–225. (Ciba Found. Symp. 109.)
250. WALKER, W. A., M. WU, AND K. J. BLOCH. Stimulation by immune complexes of mucus release from goblet cells of the rat small intestine. *Science Wash. DC* 97: 370–371, 1977.
251. WALLACE, J. L., AND B. J. R. WHITTLE. Role of mucus in the repair of gastric epithelial damage in the rat. Inhibition of epithelial recovery by mucolytic agents. *Gastroenterology* 91: 603–611, 1986.
252. WARNER, R. R., AND J. R. COLEMAN. Electron probe analysis of calcium transport by small intestine. *J. Cell Biol.* 64: 54–74, 1975.
253. WATKINS, W. M. Genetic regulation of the structure of blood group specific glycoproteins. *Biochem. Soc. Symp.* 40: 125–146, 1974.
254. WATKINS, W. M. Biochemistry and genetics of the ABO, Lewis, and P blood group systems. *Adv. Hum. Genet.* 10: 1–136, 1980.
255. WESLEY, A. W., J. F. FORSTNER, AND G. G. FORSTNER. Structure of intestinal mucus glycoprotein from human postmortem on surgical tissue: inferences from correlation analyses of sugar and sulphate composition of individual mucins. *Carbohydrate Res.* 115: 151–163, 1983.
256. WESLEY, A. W., M. MANTLE, D. MANTLE, R. QUERESHI, G. FORSTNER, AND J. F. FORSTNER. Neutral and acidic species of human intestinal mucin. Evidence for different core peptides. *J. Biol. Chem.* 260: 7955–7959, 1985.
257. WILLIAMS, S. E., AND L. A. TURNBERG. Retardation of acid diffusion by pig gastric mucus: a potential role in mucosal protection. *Gastroenterology* 79: 299–304, 1980.
258. WILLIAMS, S. E., AND L. A. TURNBERG. Studies of the "protective" properties of gastric mucus: evidence for a mucus-bicarbonate barrier. *Gut* 22: 94–96, 1981.
259. WILSON, C. E., T. RAJAPAKSA, M. NOAR, E. QUADROS, AND A. ADAMS. Stimulation of gastric mucus secretion by misoprostol in humans. *Dig. Dis. Sci.*, 31 Suppl. 2: 126S–130S, 1986.
260. WITAS, H., J. SAROSIEK, M. AONO, V. L. N. MURTY, A. SLOMIANY, AND B. L. SLOMIANY. Lipids associated with rat small-intestinal mucus glycoprotein. *Carbohydrate Res.* 120: 67–76, 1983.
261. WITAS, H., B. L. SLOMIANY, E. ZDEBSKA, K. KOJIMA, Y. H. LIAU, AND A. SLOMIANY. Lipid associated with dog gastric

mucus glycoprotein. *J. Appl. Biochem.* 5: 16–24, 1983.
262. YOUNAN, F., J. P. PEARSON, AND A. ALLEN. Gastric mucus degradation in vivo in peptic ulcer patients and the effects of vagotomy. In: *Mucus in Health and Disease*, edited by E. N. Chantler, J. B. Elder, and M. Elstein. New York: Plenum, 1982, vol. II, p. 235–273. (Adv. Exp. Med. Biol. Ser. 144.)
263. YOUNAN, F., J. P. PEARSON, A. ALLEN, AND C. W. VENABLES. Changes in the structure of the mucous gel on the mucosal surface of the stomach in association with peptic ulcer disease. *Gastroenterology* 82: 827–831, 1982.
264. ZALEWSKY, C. A., AND F. G. MOODY. Mechanisms of mucus release in exposed canine gastric mucosa. *Gastroenterology* 77: 719–722, 1979.
265. ZALEWSKY, C. A., F. G. MOODY, M. ALLEN, AND E. K. DAVIS. Stimulation of canine gastric mucus secretion with intraarterial acetylcholine chloride. *Gastroenterology* 85: 1067–1075, 1983.

CHAPTER 20

Pancreatic secretion of electrolytes and water

R. M. CASE | Department of Physiological Sciences, University of Manchester, Manchester, United Kingdom

B. E. ARGENT | Department of Physiological Sciences, University of Newcastle upon Tyne, Newcastle upon Tyne, United Kingdom

CHAPTER CONTENTS

Pancreatic Structure
    Ductal tree
    Ductal cells
    Acinar cells
    Intercellular junctions
Methods of Study
    Intact gland preparations
    Ductal perfusion
    Micropuncture
    Isolated ductal tissue
Species-Dependent Patterns of Electrolyte Secretion
    Dog, cat, and human
    Rat and mouse
    Rabbit
    Pig
    Guinea pig and hamster
    Sheep, cow, and horse
    Primates
Sites of Electrolyte and Water Secretion
    Secretin-stimulated bicarbonate secretion
    Secretion evoked by cholecystokinin and vagal stimulation
Permeability of the Pancreatic Epithelium
    Nonelectrolytes
    Electrolytes
Cellular Mechanisms of Electrolyte Secretion From Acinar Cells
    Ionic requirements for secretion
    Electrophysiology and stimulus-secretion coupling
    Cellular models of acinar cell secretion
Cellular Mechanisms of Electrolyte Secretion From Duct Cells
    Stimulus-secretion coupling
    Ionic requirements and mechanisms of secretion
        Sodium and potassium
        Calcium and magnesium
        Chloride
        Bicarbonate
Cellular Models of Duct Cell Secretion
Future Progress

PANCREATIC JUICE IS THE PRODUCT of two distinct secretory processes: enzyme secretion and electrolyte secretion. Pancreatic enzymes constitute the most significant mechanism for food digestion. The functions of electrolyte secretion are less obvious but include acting as a vehicle for enzymes and being partly responsible for neutralization of gastric acid.

The synthesis, intracellular transport, storage, and ultimate discharge from the cell of pancreatic enzymes are considered in the chapters by Scheele and Kern in this *Handbook*. The purposes of this chapter are *1*) to describe the variable patterns of electrolyte secretion that are observed in different species, *2*) to explore the causes of these patterns on the basis of different control mechanisms (hormones and neurotransmitters) acting at different sites within the gland (acini and ducts), and *3*) to consider the transport mechanisms that may be involved in elaborating secretions at these different sites. Important structural features of the pancreas and methods for studying pancreatic electrolyte secretion are considered first.

PANCREATIC STRUCTURE

The pancreas develops from two endodermal outgrowths of the primitive gut, the dorsal and ventral pancreatic buds, which grow together, fuse, and give rise to the complex branched ductular mass characteristic of an exocrine gland [Fig. 1; (78, 124)].

Traditionally the blind endings of the terminal or intercalated ducts are described as being surrounded by pyramidal acinar cells grouped together to form an acinus (from Latin, berry or grape); duct cells lining an acinus are called centroacinar cells (Fig. 2). However, the organization of intercalated ducts and acini is more complex than this. From light-microscopic analysis of serial sections of rat, dog, and human pancreas and scanning electron-microscopic analysis of casts prepared by retrograde injection of silicone rubber into the duct system, Bockman and colleagues (1, 16–19) suggest that acinar tissue may be organized in an anastomosing tubular arrangement so that an intercalated duct does not necessarily end in an acinus but could emerge on the other side of it. This has been denied by Takahashi (213), who used scanning electron microscopy to study partially digested rat tissue. However, both agree that acini are not regular in size or shape; in some the lumen may be branched, which

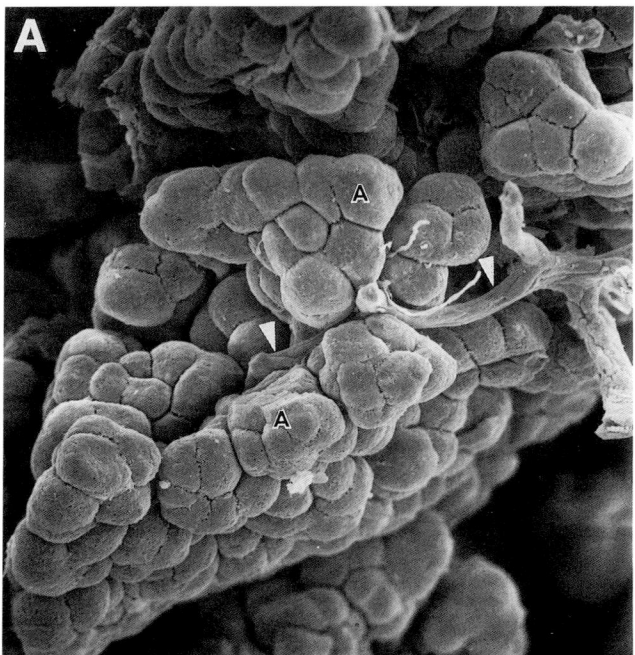 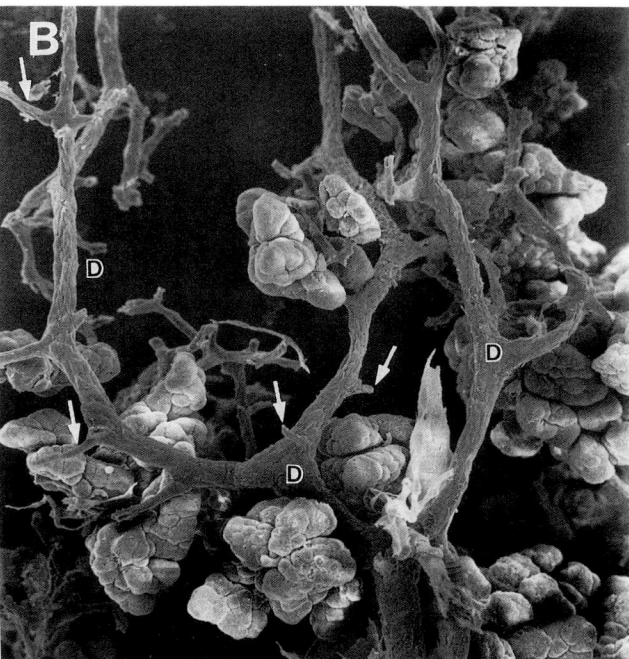

FIG. 1. *A*: scanning electron-microscopic view of rat pancreatic lobule treated with HCl to remove connective tissue. *Arrowheads*, long intercalated duct that connects with numerous acini (A). × 669. *B*: scanning electron-microscopic overview of ductal system of rat pancreas after removal of most acini by ultrasonic vibration. Excretory duct (D) branches dichotomously. *Arrows*, intercalated ducts. × 220. [From Takahashi (213).]

explains why acinar, centroacinar, and intralobular duct cells sometimes appear to be in contact with more than one lumen profile (Figs. 2 and 3*A*).

The architecture of the acini and intercalated ducts may be very important when considering how viscid secretions leave the acini and how pancreatic diseases develop. However, it is not crucial for this chapter, and we refer to enzyme-secreting cells as acinar cells and the remainder of the gland as ductal.

Ductal Tree

The ductal system can be divided into a number of parts. The intercalated ducts empty into intralobular ducts, which run within lobules of pancreatic tissue. These empty into interlobular ducts, which run in the connective tissue between lobules, and these in turn drain into the main pancreatic duct. These divisions are found in all common laboratory animals except the rat, mouse, and hamster, where a variable number of ducts open into the common bile-pancreatic duct from different lobes of the gland (214). The site at which the main pancreatic duct (or bile-pancreatic duct) empties into the duodenum varies between species. In the human, dog, cat, rat, mouse, and hamster it opens into the second part of the duodenum; in rabbit and guinea pig (and some other species) it enters in the third part of the duodenum, a long way from the pylorus (136). In some species, notably the dog, an accessory duct, which anastomoses with the main duct, opens into the duodenum as a vestige of the dorsal pancreatic bud.

In a comparative, qualitative study of the pancreatic ductal tree, Kodama (118) states that the ductal system is developed to a greater extent in carnivores and omnivores than in herbivores. Although morphometric analyses are necessary to confirm this assertion, it nevertheless accords with the species-dependent patterns of electrolyte secretion described later. In the rat, which according to Kodama (118) has a poorly developed ductal system, centroacinar and duct cells comprise ~11% of the total number of cells present (116). However, because of their small size they account for an even smaller proportion of gland cell volume, ~4% in the guinea pig (21).

In all species studied, unmyelinated nerve fibers travel in the lamina propria of all ducts (Fig. 4). Nerve endings are found near the basal surface of the ducts but are almost always separated from the ductal epithelium by the basal lamina (118, 141, 230). They also extend into the interacinar spaces where the basal lamina may or may not intervene between nerve and acinar cell (229, 230). These fibers are largely cholinergic, although some are adrenergic (116, 230). Peptidergic nerves containing vasoactive intestinal polypeptide (VIP) are also present, with the density in pig and chicken greater than in human, dog, and cat and much greater than in rat (209). Other peptidergic

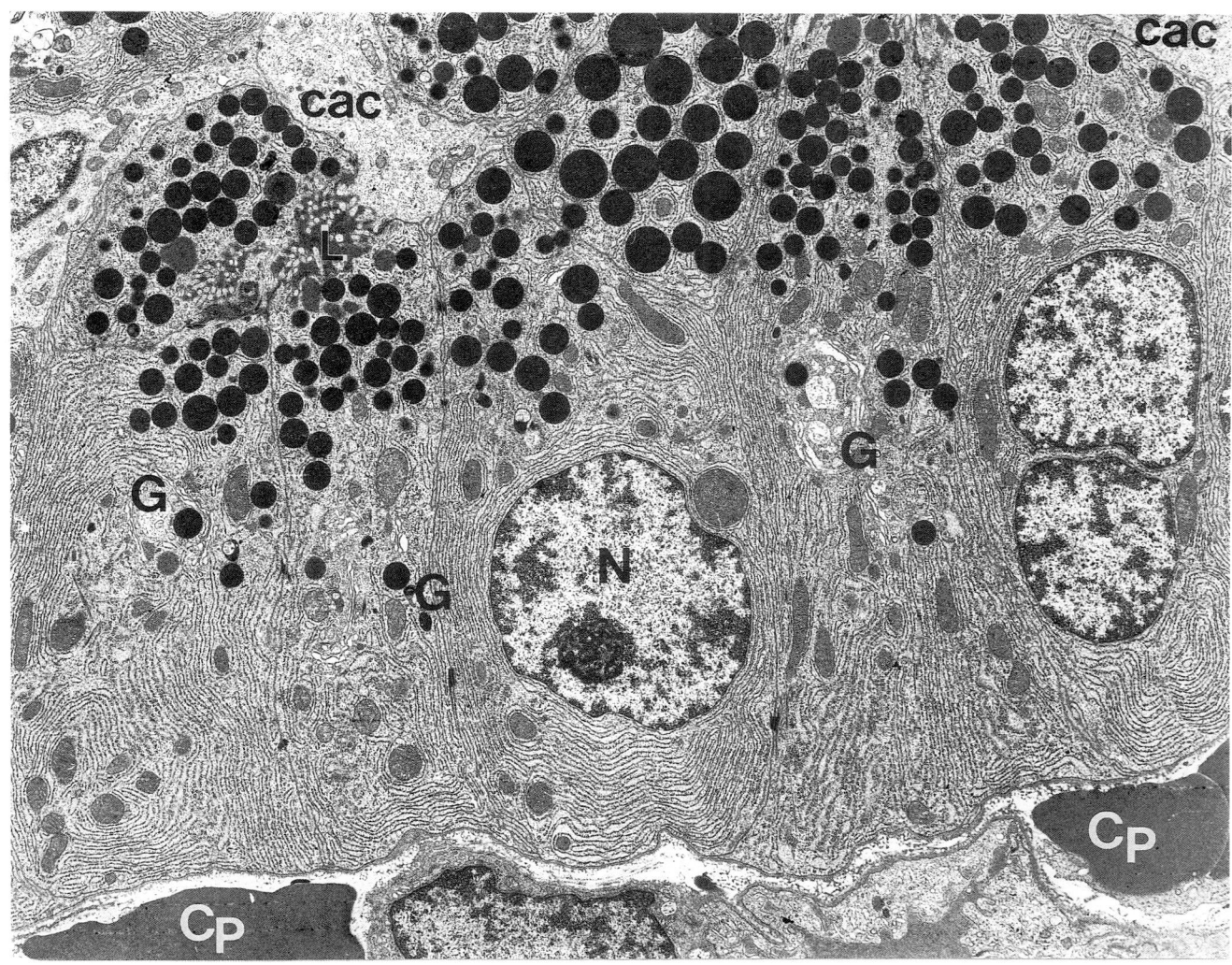

FIG. 2. Low-power micrograph of a segment from a human acinus. In acinar cells, polarized arrangement of rough endoplasmic reticulum, Golgi complex (*G*), and zymogen granules in relation to the nucleus (*N*) is evident. Parts of 2 centroacinar cells (*cac*) close to the lumen (*L*) are visible. *Cp*, capillary. × 6,000. [From Kern (117). In: *The Exocrine Pancreas: Biology, Pathobiology, and Diseases*, © 1986, Raven Press, New York.]

nerves have also been observed containing, for example, substance P, neurotensin, and bombesin-like immunoreactivity. Because intravenous bombesin and its mammalian counterpart, gastrin-releasing peptide, strongly stimulate pancreatic secretion in a number of species, including humans, it is tempting to suggest that bombesin-like peptides may be involved directly in pancreatic regulation.

Ductal Cells

The fine structure of the ductal epithelium has been described in detail in a number of species, most recently in the human (117, 118), starling (227, 234), and rat (213). These studies, which have been reviewed elsewhere (34), may be summarized as follows. Centroacinar and intercalated duct cells and those lining intralobular and small interlobular ducts are similar in structure (Figs. 2 and 3). The apparatus required for protein secretion (endoplasmic reticulum, the Golgi complex, and secretory vesicles) is poorly developed, whereas mitochondria are fairly prominent. Basolateral membranes of adjacent cells are interdigitated, a characteristic of striated duct cells in salivary glands and of renal proximal tubule cells. The luminal membrane bears some short microvilli and often a single cilium whose function is unknown; it might be involved in propelling viscid acinar secretion through the small ductules, although recent studies suggest it has a sensory function (20).

Cells lining the interlobular and main ducts are more columnar than those higher up the ductal tree.

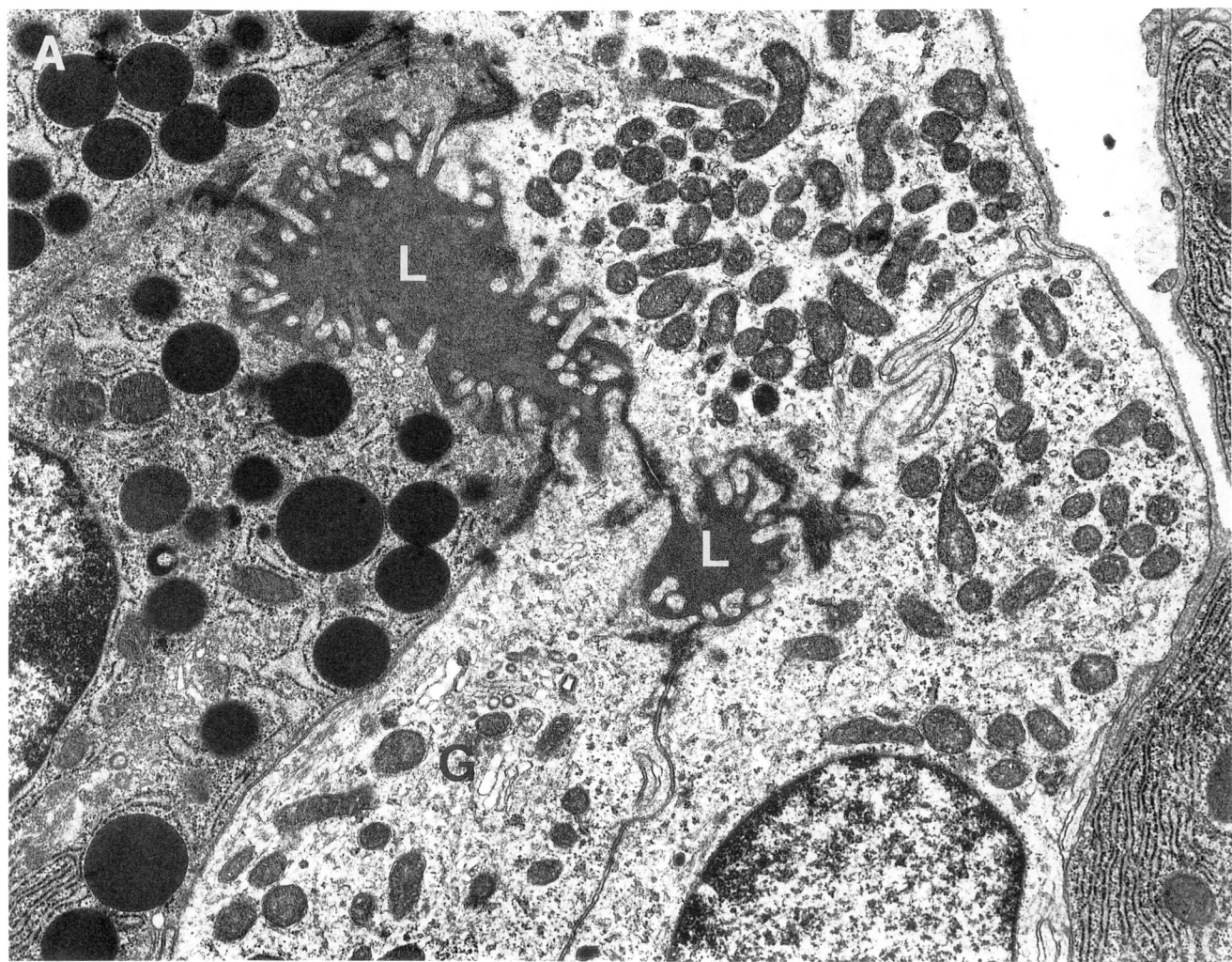

FIG. 3A. Fine structural characteristics of intralobular ducts in human pancreas. Large number of mitochondria, especially in vicinity of Golgi complex (G). L, duct lumen. × 12,000. [From Kern (117). In: *The Exocrine Pancreas: Biology, Pathobiology, and Diseases,* © 1986, Raven Press, New York.]

Rough endoplasmic reticulum, the Golgi complex, and secretory vesicles become more prominent, as do luminal microvilli (Fig. 4).

Acinar Cells

Acinar cells comprise 77% of the cells present in rat pancreas (116) or ~90% of cell volume in guinea pig pancreas (21). Their structure is typical of cells responsible for protein secretion and varies little among a wide variety of vertebrates (33). The nucleus occupies a central or basal position and is surrounded basally and laterally by extensive rough endoplasmic reticulum and apically by a prominent Golgi complex. The apical pole of the cell is filled with mature secretory vesicles (zymogen granules) or their immature precursors (condensing vacuoles). The luminal plasma membrane bears prominent microvilli (Figs. 2 and 3A), but the lateral and basal membranes are relatively straight and uncomplicated.

Intercellular Junctions

Adjacent pancreatic cells, whether acinar or ductal, are linked at their luminal margins by junctional complexes that thereby separate the acinar and ductal lumens from direct contact with the intercellular space (Fig. 3A). These complexes comprise three components: occluding junctions (zonulae occludentes) are closest to the lumen, beneath which are belt desmosomes (zonulae adherentes) and spot desmosomes (maculae adherentes) (117). Occluding junctions are usually referred to as tight junctions because they appear in thin sections as trilaminar structures resulting from the fusion of adjacent plasma membranes. However, in freeze-fracture replicas they are

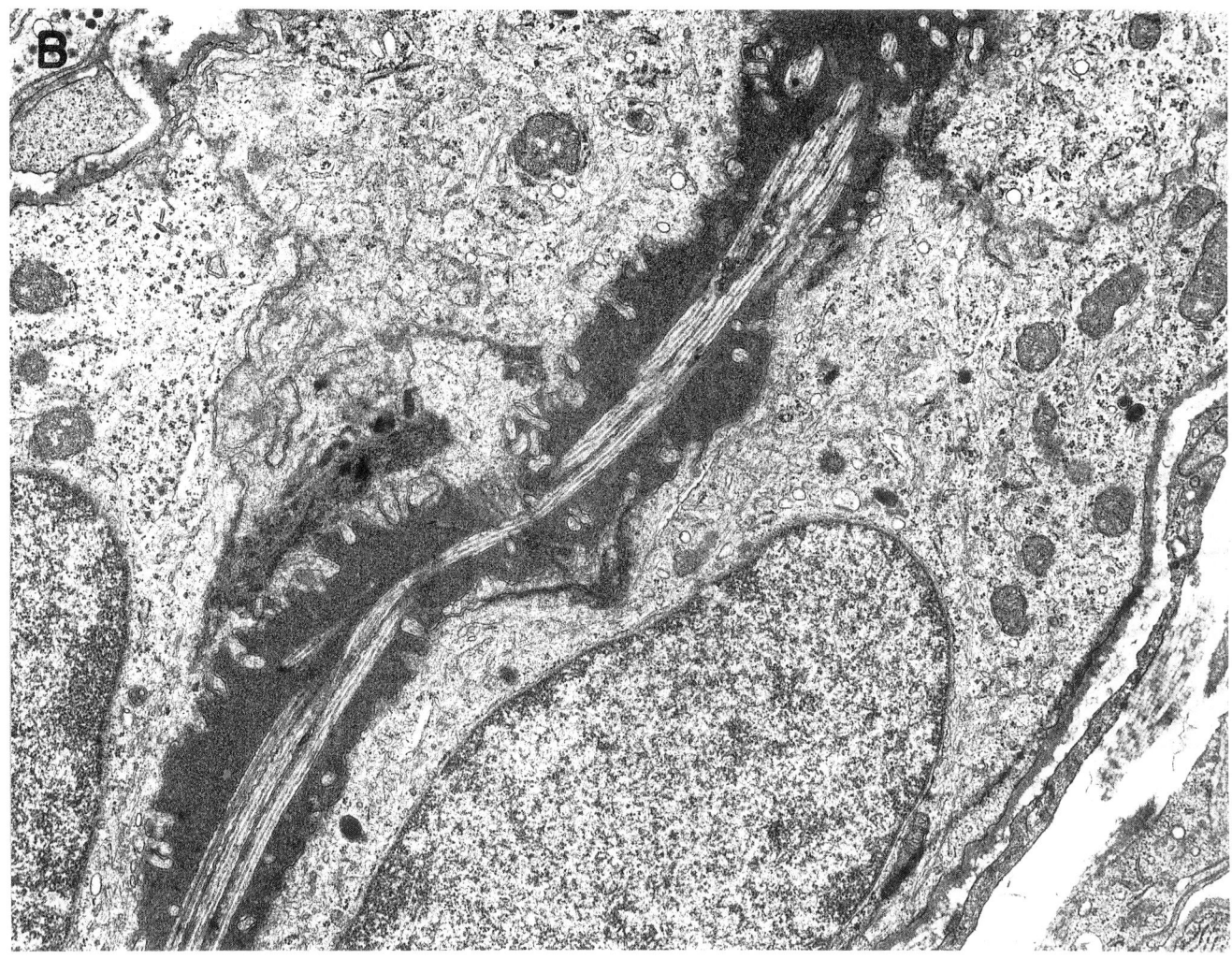

FIG. 3B. Fine structural characteristics of intralobular ducts in human pancreas. Interdigitations between lateral plasma membranes in adjacent cells, short microvilli on the luminal membrane, and cilia in the duct lumen. × 12,000. [From Kern (117). In: *The Exocrine Pancreas: Biology, Pathobiology, and Diseases,* © 1986, Raven Press, New York.]

seen to be composed of anastomosing strands of intramembranous particles. The name *tight junction* is unfortunate, because it implies a barrier to the passage of all solutes between the lumen and intercellular space. This is not so. Indeed, electrolyte secretion may depend on the fact that it is not. As their names imply, belt desmosomes form a continuous band around each cell, whereas spot desmosomes are patchlike areas of contact.

The lateral plasma membranes of neighboring acinar cells (there is no evidence in the case of duct cells) also become closely opposed at places other than junctional complexes, referred to as gap junctions since they are separated by a gap of ~2 nm. The intramembranous particles in gap junctional membranes form a pore structure across the membrane bilayer that, because of the symmetrical arrangement of the opposed membranes, allows the passage of ions, metabolites, and nucleotides, but not macromolecules, between adjacent cells (130). In other words, all the cells of an acinus (and perhaps more than one acinus) are in electrical and metabolic communication, allowing them to act as a unit (157).

METHODS OF STUDY

Until 1965 pratically all knowledge of pancreatic electrolyte secretion had been obtained in studies on conscious or anesthetized animals, principally dog, cat, and human. Investigations into the physiological control of secretion will continue to rely heavily on such studies. Similar but more recent studies in other species have contributed greatly to the understanding

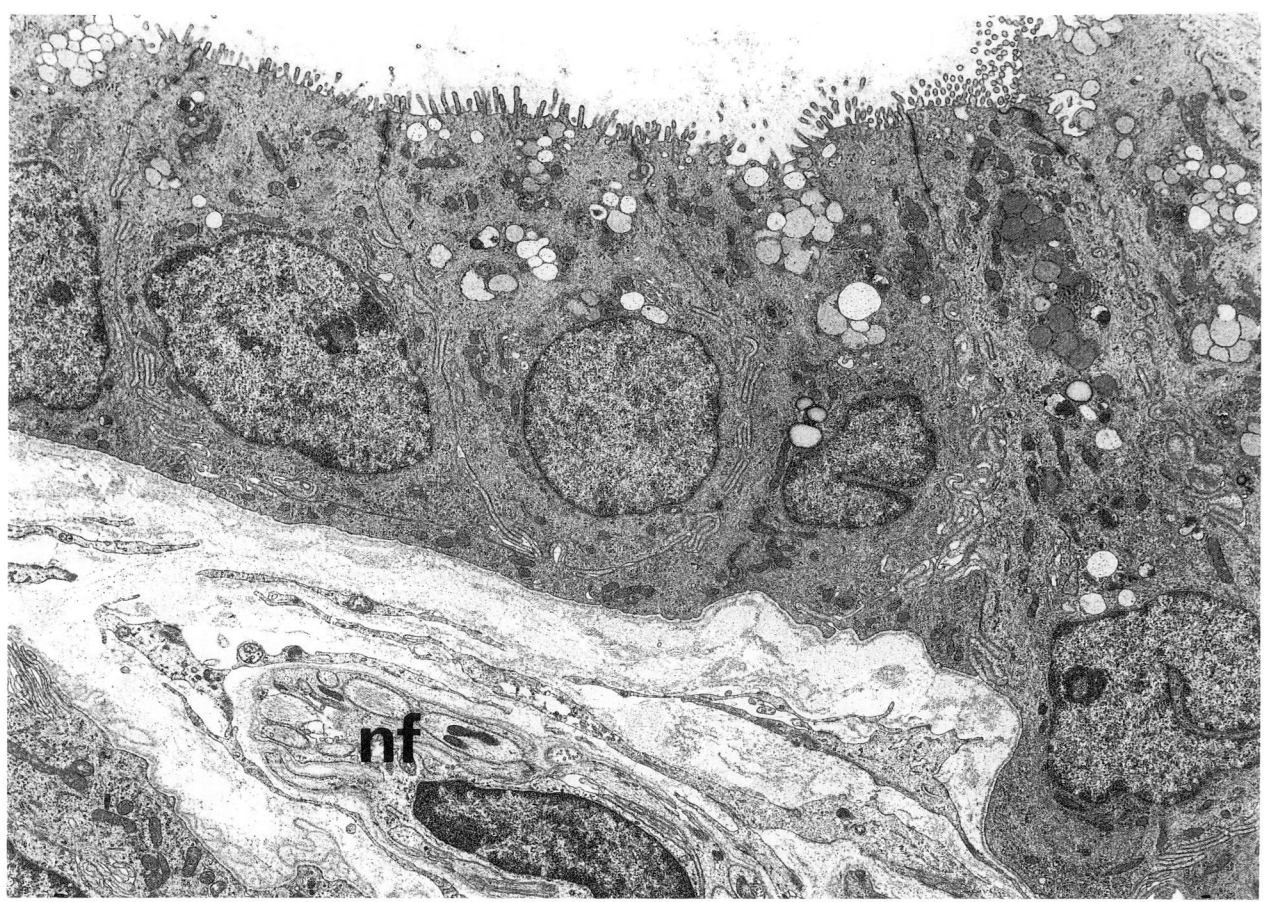

FIG. 4. Fine structure of an interlobular duct in the human pancreas. *nf*, Nerve fibers in lamina propria. × 5,700. [From Kern (117). In: *The Exocrine Pancreas: Biology, Pathobiology, and Diseases*, © 1986, Raven Press, New York.]

of control, as well as causing a little confusion, as we discuss next.

However, it is difficult to study cellular control mechanisms (stimulus-secretion coupling) in intact animals and almost impossible to study the mechanisms responsible for electrolyte and water transport. Isolated preparations of the pancreas are required for such studies. These preparations and the biophysical techniques used in conjunction with them have been described in detail elsewhere (34, 35). However, because they are fundamental to much that is discussed in this chapter, they are summarized here.

Intact Gland Preparations

Pancreatic enzyme secretion can be studied in a variety of in vitro preparations (slices, pieces, lobules, and dissociated acini and acinar cells) because the secretory products discharged into the bathing medium can be easily measured. This is not true of pancreatic electrolyte secretion. In this case the gland must be maintained sufficiently intact to allow direct collection of the secretory products from the pancreatic duct.

Two major techniques have been devised to solve this problem. Together they have produced a great deal of information about pancreatic electrolyte secretion and its cellular control. Taking advantage of the thinness of the pancreas in the rabbit, Rothman and Brooks (180) devised a method whereby the gland was suspended in a bath filled with oxygenated physiological salt solution (Fig. 5). This preparation has the advantage of being relatively easy to set up but the potential disadvantage that it has a high, spontaneous rate of secretion (higher than in vivo) and is therefore insensitive to exogenous stimuli.

The second technique, which is more conventional, involves isolation of the gland and perfusion through its arterial supply of either blood or, more usually, a physiological salt solution. Because the vascular supply of the pancreas and neighboring organs is complex (Fig. 6), isolation of the gland requires extensive surgery and is therefore time consuming. However, vascular perfusion permits rapid changes in extracellular milieu and precise regulation with exogenous stimuli. Perfusion studies have been performed on dog, cat, rat, pig, guinea pig, and hamster pancreas (for details see refs. 35 and 192).

Ductal Perfusion

Another perfusion technique that has provided some information about the permeability properties of the ductal epithelium consists of perfusing the main pancreatic duct, either in anesthetized animals or in the perfused gland. The main duct is cannulated at both the tail of the gland and at the point where it pierces the duodenal wall. Artificial "pancreatic juice" is perfused from the tail to the head of the gland with a motor-driven syringe.

The technique is particularly suited to those species in which there is no spontaneous secretion, such as the cat, where the method was first described (39). With the inclusion of a volume marker in the perfusion fluid it has also been used in other species, such as the rabbit (174).

Micropuncture

Micropuncture techniques, devised by renal physiologists to study transport in single renal tubules, can also be used in exocrine glands in vivo and in vitro to provide information about transport at specific sites in the ductal tree. This information can be gained in three ways. *1)* Micropuncture samples of pancreatic juice can be collected under free-flow conditions from different locations. *2)* A fluid of known composition can be used to split a stationary column of oil, a technique known as split-drop or stationary microperfusion. Changes in the composition of the injected fluid and the concentration of an impermeant marker allow conclusions to be drawn about water and ion fluxes in the ductal segment. *3)* Data on ion fluxes can be gained from stop-flow experiments, in which a sample of ductal fluid is collected under free-flow conditions, after which flow is stopped for a brief interval by occlusion of the duct, after which a second sample is taken.

Micropuncture of exocrine glands is difficult; it is particularly difficult to puncture high up the ductal tree and usually impossible to see the tip of the micro-

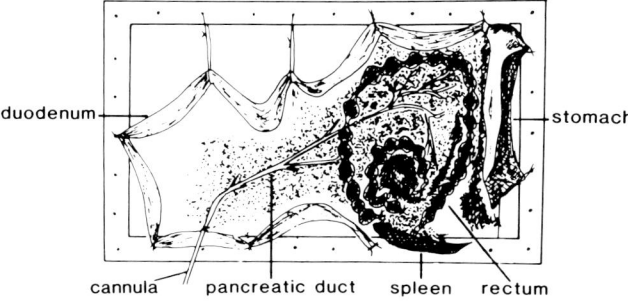

FIG. 5. Schematic diagram of isolated rabbit pancreas and attached duodenum mounted on a polyvinylchloride frame, which is then suspended in a chamber filled with a physiological buffer solution. *Stippled area*, pancreatic tissue. (Diagram courtesy of J. J. H. H. M. De Pont.)

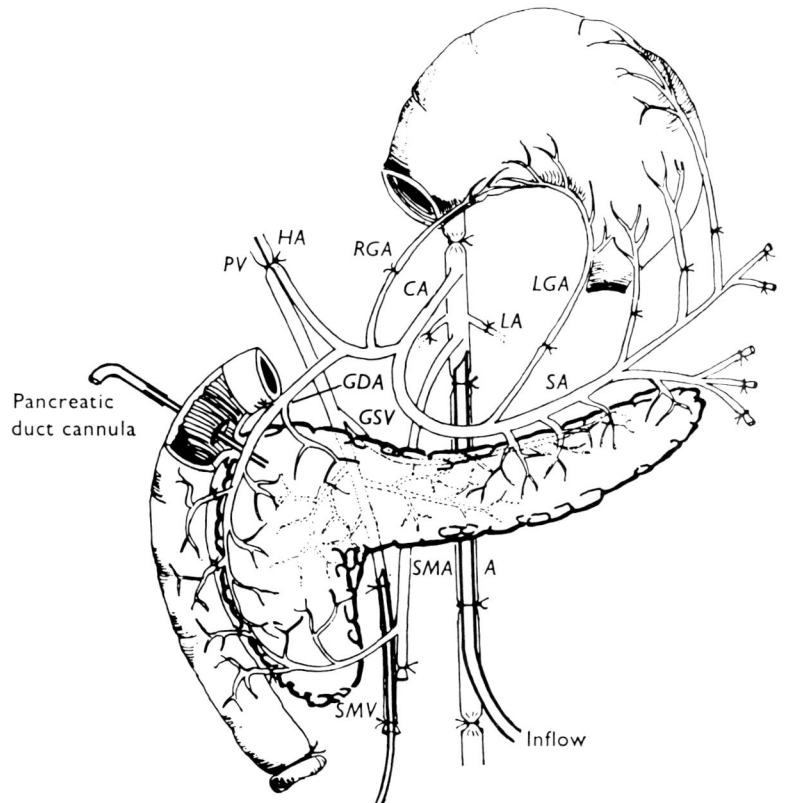

FIG. 6. Vascular supply of cat pancreas. Stomach is shown separated from the duodenum, displaced anteriorly, and turned through 180°. A, aorta; CA, celiac axis; GDA, gastroduodenal artery; GSV, gastrosplenic vein; HA, hepatic artery; LA, lumbar artery; LGA, left gastric artery; PV, portal vein; RGA, right gastric artery; SA, splenic artery; SMA, superior mesenteric artery; SMV, superior mesenteric vein. [From Case et al. (38).]

pipette within the duct lumen. In addition, most studies in the pancreas have been performed on the rat and rabbit, which, as we describe later, have complex patterns of electrolyte secretion. These studies are therefore difficult to interpret.

Isolated Ductal Tissue

Pancreatic tissue may be dissociated into acini and duct fragments by limited digestion with enzymes such as collagenase. However, although it is relatively simple to obtain quite pure preparations of acini from such digests, it is very difficult to separate ductal tissue, largely because, as mentioned previously, it comprises only ~4% of the gland mass. Some success has been achieved using a Beckman Elutriator rotor to separate duct cells from dissociated rat pancreas (186, 188). The problem would be much reduced if the mass of acinar tissue could be reduced. This can be done. Several treatments have been described that produce acinar cell atrophy, including regular administration of ethionine (52) and maintenance on a copper-deficient diet (67, 70, 202). The latter regimen is more precise and has allowed the isolation of duct fragments (71) and, more recently, lengths of intact duct (6). However, whereas ethionine acts in a variety of species, including dog, cat, rat, and monkey (52), copper deficiency has so far been shown to act only in the rat and cattle (66).

SPECIES-DEPENDENT PATTERNS OF ELECTROLYTE SECRETION

Species differences are often the last refuge of biologists unable to reconcile conflicting data. However, in the case of pancreatic electrolyte secretion, genuine and marked variations do exist between the common laboratory species.

The areas where variations occur are as follows:
1. The presence or absence of spontaneous, basal, or unstimulated secretion. (The term *basal* is best avoided, as it is also used in secretion studies to define the nonapical pole of the cell, and the term *unstimulated* implies the absence of any endogenous stimulation, which is always difficult to prove, especially in intact animals.) There is a marked species variation in the volume of spontaneous secretion (i.e., that which occurs in the absence of exogenous stimulation).
2. Physiological regulation of secretion. Variations in patterns of electrolyte secretion are inextricably linked to differences in control mechanisms. The pancreas receives neural and hormonal stimuli. Nervous control is mediated principally through vagal, cholinergic fibers, although the pancreas also receives a variable supply of vagal peptidergic (e.g., VIPergic) neurons and sympathetic adrenergic fibers from the splanchnic nerves. Hormonal control is achieved principally through secretin and cholecystokinin (CCK), which are released from the duodenal mucosa when it is exposed to acidic chyme. The response to CCK in particular shows marked species variation. Peptides with a structure similar to secretin (e.g., VIP) or sharing a common COOH-terminal tetrapeptide sequence with CCK (e.g., gastrins, caerulein) can have similar effects or (under appropriate conditions) may act as competitive inhibitors of secretion.
3. The electrolyte composition of secretion. Irrespective of the species, pancreatic juice is a clear, alkaline fluid that is always isosmotic with blood plasma and (except perhaps in the rat) contains sodium and potassium ions in concentrations slightly above those in plasma. However, the concentrations of the two major anions, chloride and bicarbonate, vary greatly: between species, in response to different stimuli, and at different flow rates.

Recognition of these variations has been delayed because until quite recently most experiments were carried out in dog, cat, and human, where the pattern is similar. However, many experimental studies are now performed in other species, where the pattern may vary greatly. We explore these variations in detail next.

Dog, Cat, and Human

The pattern of secretion in dog, cat, and human seems to be quite similar and represents the classic pattern that is described in most textbooks of physiology. There is almost no spontaneous secretion. Secretin causes a brisk flow of pancreatic juice that is poor in enzymes but contains bicarbonate in concentrations reaching 5–6 times that in plasma. Neither CCK nor vagal stimulation evokes much secretion.

Most early studies were performed in unanesthetized dogs prepared with a duodenal fistula opposite the main pancreatic duct, through which a cannula could be temporarily inserted into the pancreatic duct. In this way, Hart and Thomas (90) clearly described the relationship between bicarbonate concentration and flow rate that is characteristic of these three species: an asymptotic rise in juice bicarbonate concentration with increasing flow rate, reaching a plateau of ~145 mM, with a reciprocal fall in juice chloride concentration such that the sum of the two anions is constant at all secretory rates (Fig. 7).

Hart and Thomas (90) evoked secretion by injecting a variety of substances into the upper intestine (HCl, glutamic acid, peptones, and soaps) alone or in combination; occasionally they gave intravenous injections of secretin. Although such substances would clearly have evoked release of both secretin and CCK (and perhaps other peptides), the relationship between bicarbonate concentration and flow rate seemed not to be influenced by the nature of the stimulus. When pure hormones became available the situation was

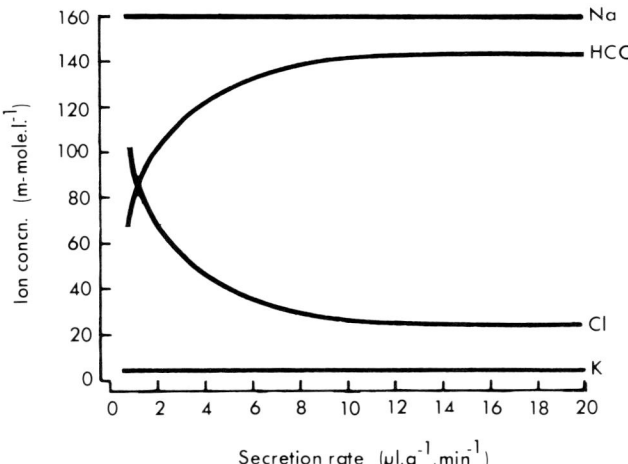

FIG. 7. Electrolyte composition of pancreatic juice in anesthetized cat stimulated to secrete at different flow rates by continuous infusion of different doses of secretin. Flow rate is expressed per gram gland weight (cf. Fig. 8) by assuming a gland weight of 10 g; this is only approximate. [From Case (32).]

clarified by Debas and Grossman (53), who studied the effect of pure CCK, CCK-octapeptide (CCK-8), caerulein, and secretin on secretion from chronic fistula dogs. The CCK analogues evoked ~10% of the volume of secretion evoked by secretin. At the highest flow rate common to all stimuli (which was rather low) the bicarbonate concentrations were, respectively, 58, 73, 57, and 118 mM, which suggests that the CCK analogues may not be acting in exactly the same way as secretin. Although CCK alone is a weak stimulant of bicarbonate secretion, it greatly augments the action of secretin (10, 53) so that, under physiological conditions, it may play a prominent role in evoking pancreatic electrolyte secretion.

In the cat the most comprehensive data on flow curves have been obtained from studies in anesthetized animals (39, 43, 127) and a perfused gland preparation (38) in which pancreatic juice was collected by direct cannulation of the main pancreatic duct after stimulation with secretin. Although the curves generated by secretin stimulation (Fig. 7) are very similar to those for the dog, the cat seems even less sensitive to CCK than the dog. Alone, CCK evokes almost no electrolyte secretion in anesthetized (26) or conscious (232) cats, although (as in the dog) it potentiates the action of secretin (26, 232). When a large dose of CCK is superimposed on a maximally effective dose of secretin, it evokes only a small additional component of secretion, which must contain a low-bicarbonate concentration since it slightly reduces juice bicarbonate concentration (39). In the perfused gland, large doses of CCK can evoke a fluid secretion of unknown composition whose volume reaches ~35% of that evoked by secretin (111).

Since the introduction of endoscopic retrograde cannulation of the pancreatic duct it has been possible to collect pure pancreatic juice in humans. In this way it has been confirmed that the pattern of secretion is similar to that in dog and cat, with maximum bicarbonate concentrations of ~135 mM achieved during secretin stimulation (58).

Although vagal stimulation is well known to stimulate pancreatic enzyme secretion in dog and cat (and probably human), its precise role in electrolyte secretion is less clear. Like CCK, intravenous methacholine and vagal stimulation potentiate the action of secretin on fluid secretion in cats (26, 125, 203), though perhaps not in dogs (10). If the cat duct is first flushed with saline, vagal stimulation in the absence of secretin can evoke a small secretion of unknown composition (126). However, this effect is not inhibited by atropine and is presumably therefore not evoked by acetylcholine. Certainly acetylcholine has little direct effect on fluid secretion in the perfused cat pancreas.

Early attempts to explain the characteristic pattern of pancreatic electrolyte secretion in these species gave rise to two hypotheses: *1*) the admixture, or two-component, hypothesis and *2*) the exchange hypothesis. Lim et al. (128) first suggested that pancreatic juice comprised two separate solutions containing either chloride or bicarbonate as sole anions. They further suggested that these two solutions were mixed in constant proportions, which proved to be wrong in view of the asymptotic flow curves described by Hart and Thomas (90). Hollander and Birnbaum (96) subsequently proposed that pancreatic juice was a mixture of four distinct components: *1*) mucus, derived from mucous epithelial cells in the walls of the main duct: *2*) enzymes, arising from the acinar cells; *3*) a chloride-containing fluid, similar in composition to interstitial fluid (and possibly identical with it); and *4*) a specific bicarbonate secretion, probably derived from centroacinar cells and terminal duct cells.

In the exchange hypothesis, first proposed by Dreiling et al. (60), pancreatic electrolyte secretion is assumed to consist of a single primary solution containing bicarbonate as the sole anion, some of which is subsequently exchanged by diffusion for chloride during its passage through the ducts.

A third hypothesis has been proposed largely on the basis of studies in the rabbit, where the pattern of secretion is different from that in dog, cat, and human. However, this hypothesis was hinted at in an early review of the literature on dog pancreas (204) and, for completeness, we mention it here. In this unicellular model, Rothman and Brooks (181) propose that both bicarbonate and chloride are secreted by one cell, with bicarbonate secretion dominating as secretion rate increases.

As reviewed by Makhlouf and Blum (132), it is not possible from flow curves to distinguish mathematically between these hypotheses. There is probably some truth in all three hypotheses because *1*) in the rat, admixture of two electrolyte secretions definitely takes place (see below); *2)* exchange diffusion of bicarbonate for chloride can take place in (at least) the main pancreatic duct (39, 43); and *3)* a unicellular model of secretion exists in the mandibular salivary gland, where either bicarbonate or chloride is secreted, depending on what extracellular anions are available (41). Thus elements of all three models may contribute to the pattern of pancreatic electrolyte secretion in dog, cat, and human, although admixture is probably least important, especially in the cat.

Rat and Mouse

Studies on the rat pancreas lagged behind those on the dog pancreas, perhaps because of the technical difficulty of obtaining bile-free pancreatic juice. However, early studies (listed in ref. 197) showed that pancreatic secretion in the rat differs from that in the dog in three important ways: *1)* there is a high spontaneous secretion, *2)* the response to secretin is rather poor, and *3)* there is a definite electrolyte secretion in response to CCK.

Pancreatic secretion in the rat varies markedly, depending on experimental conditions. Thus, although spontaneous secretion is quite significant in anesthetized animals (~ 2.5 $\mu l \cdot min^{-1} \cdot kg^{-1}$ body wt), it is 10-fold greater in conscious animals (147, 156, 197). Most studies of electrolyte secretion have been performed on anesthetized animals. Unfortunately the earlier of these studies produced confusing data. Mangos and colleagues (133, 135) claimed that during secretin stimulation, bicarbonate concentration *fell* with increasing flow rate from 79 to 35 mM, with a reciprocal rise in chloride concentration from 70 to 105 mM. A subsequent thorough study by Sewell and Young (197) clarified the matter. They showed that pure, natural, or synthetic secretin produced the expected rise in bicarbonate secretion as flow rate increased, whereas CCK-8 and caerulein evoked a fluid whose bicarbonate concentration remained constant at all flow rates, at ~ 30 mM (Fig. 8). The response to injecting pure secretin and caerulein together was the arithmetic sum of the individual responses (Fig. 8) and was almost identical with that evoked by a commercially available but impure preparation of secretin (Boots). In other words, contamination of this impure secretin by CCK causes increasing amounts of a chloride-rich secretion to be added to a bicarbonate-rich secretion as the dose of "secretin" is increased. This study, which was later confirmed (155), probably explains the data of Mangos et al. (133, 135), who used the Boots secretin.

Like CCK, pilocarpine (135), carbachol (102), and vagal stimulation (49) also evoke secretion of a chloride-rich fluid from rat pancreas. However, $\sim 30\%$ of the vagal response is resistant to blockade by atropine (and adrenoceptor blockers), suggesting participation of nonadrenergic, noncholinergic fibers in vagally evoked secretion (49).

Most of the data described here were obtained in anesthetized animals. Because of the high spontaneous secretion in conscious animals, it is perhaps not surprising that the effects of stimuli are less obvious under these conditions. This has caused Oates and Morgan (147) to question the validity of the anesthetized rat model. Unfortunately these authors used impure CCK in their study. Because the characteristics of pancreatic secretion observed in the anesthe-

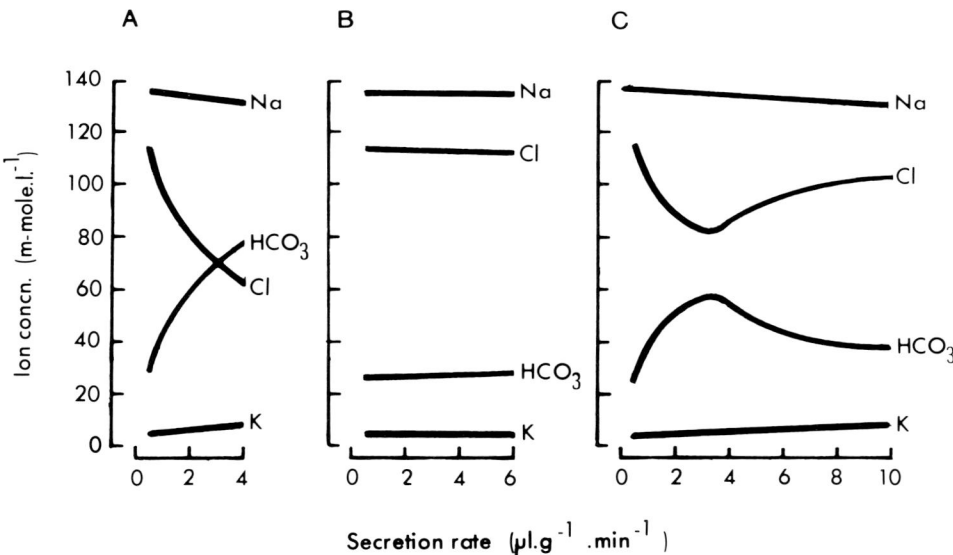

FIG. 8. Electrolyte composition of pancreatic juice in the anesthetized rat stimulated by secretin (*A*), caerulein (an analogue of cholecystokinin) (*B*), and a combination of secretin and caerulein (*C*). [From Case (32); data from Sewell and Young (197).]

tized rat, including the spontaneous secretion, have been observed in studies on isolated, perfused glands (113, 129, 164, 206, 207), they are presumably direct effects.

Although we have stated that sodium and potassium concentrations remain constant in pancreatic juice, in the rat this is not quite true. After stimulation with secretin (but not caerulein), potassium concentration doubles, from ~5 mM at rest to ~10 mM at maximum flow rates (155, 197). Presumably sodium concentration falls by the same amount, and although this was not observed in one study (197), it was in the other (155). This effect has not been observed in other species and remains enigmatic.

Only one study has been made of electrolyte secretion in the mouse, in which Mangos et al. (134) showed that both secretin and pilocarpine could evoke secretion. The composition of pilocarpine-stimulated secretion was the same as that in the rat (i.e., the bicarbonate concentration was ~30 mM and independent of flow rate). Strangely, however, the composition of secretion evoked by secretin was the same. Unfortunately the authors again used the Boots secretin in this study. Even so, it is difficult to explain why no rise in bicarbonate concentration was observed at any flow rate. It thus remains unsure whether secretin really evokes a low-bicarbonate secretion in mouse; if it does, this action is unique among species so far studied.

The uncertainties attached to the studies by Mangos and co-workers (133–135), which also include observations in the rabbit [see next section; (133)], are especially unfortunate because these studies comprise a substantial proportion of the total pancreatic micropuncture data, including all the data on rat and mouse.

In the rat, admixture of two distinct solutions plays a major role in determining the pattern of electrolyte secretion. However, because the concentration of bicarbonate evoked by CCK-8 can be increased by cycloheximide (by an unknown means) when flow rate is reduced (198), perhaps the composition of the primary secretion evoked by CCK can be changed, in accordance with the unicellular model.

Rabbit

The isolated, superfused rabbit pancreas preparation (see METHODS OF STUDY, p. 387) has been used by a number of groups for studying pancreatic electrolyte secretion. Micropuncture studies have also been performed in this preparation and in the rabbit pancreas in vivo. No comprehensive study of pancreatic electrolyte secretion in this species has been published, although an abstract has appeared (196a).

In anesthetized rabbits there is a substantial spontaneous secretion; the magnitude of this secretion varies greatly in different studies, from a low of 1 μl/min (180) to a high of 13 μl/min (205). Presumably these variations reflect differences in such factors as anesthesia and level of endogenous stimulation, although spontaneous secretion is still observed after extirpation of the whole gastrointestinal tract in decerebrate animals (7). In an average 2-kg rabbit a typical secretory rate would be ~2 μl·min^{-1}·kg^{-1} body wt, which is close to the values reported for the anesthetized rat (see ref. 197). The mean bicarbonate concentration in spontaneous secretion varied from a low of 30 mM (Fig. 9) to 108 mM (205), with bicarbonate concentration rising with flow rate.

The effect of secretin in the rabbit is even less impressive than in the rat. Typically a maximal dose doubles flow rate and increases bicarbonate concentration from 65 to 85 mM (30, 31).

Thus the bicarbonate concentration of both "spontaneous" and secretin-stimulated secretion seems to be directly related to flow rate (Fig. 9). In the micropuncture study by Mangos (133), complete flow curves were constructed for rabbits stimulated with the Boots secretin. These show the bicarbonate concentration to rise from ~50 mM at a flow rate of 4 μl·min^{-1}·g^{-1} tissue wet wt to 110 mM at 35 μl·min^{-1}·g^{-1} tissue wet wt, with a corresponding fall in chloride concentration. The Boots secretin is probably contaminated with CCK, which therefore becomes a significant stimulant at high flow rates (i.e., during stimulation with large doses). Therefore, with hindsight, we can predict that CCK might also stimulate bicarbonate secretion

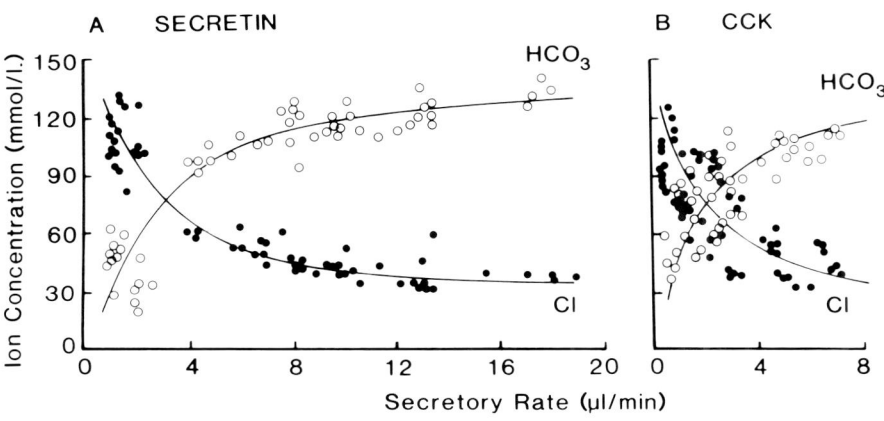

FIG. 9. Electrolyte composition of pancreatic juice in the anesthetized rabbit stimulated with secretin (*A*) and cholecystokinin (CCK) (*B*). [Data from Seow and Young (196a).]

in this species (if it did not, bicarbonate concentrations would decrease at high flow rates, as in the rat). This appears to be so. In the absence of secretin, pure CCK evokes a bicarbonate-rich secretion from the pancreas of the anesthetized rabbit (Fig. 9).

Vagal stimulation, pilocarpine, and acetylcholine accelerate the flow of pancreatic juice in the rabbit (8, 203). This secretion (at least that evoked by the cholinergic agonist carbachol) is also rich in bicarbonate (196a). Whether there is a noncholinergic component to vagal stimulation in the rabbit is untested.

The behavior of the rabbit pancreas in vitro is somewhat different from that in vivo. In vitro the rabbit pancreas has a very high rate of spontaneous secretion. Because of the different spontaneous rates observed in vivo (see above), the rate in vitro is variously reported as similar to that in vivo (205) or fivefold higher (180). In most studies it amounts to 5–10 μl/min (i.e., equivalent to ~2.5–5 μl·min^{-1}·kg^{-1} body wt), considerably more than the spontaneous secretion in the perfused rat pancreas (<1.0 μl·min^{-1}·kg^{-1} body wt, but dependent on perfusion flow rate). This rate is almost maximal and can only be marginally increased by secretin (177, 180), CCK (105), or carbachol (104), suggesting that the gland is already maximally stimulated. Consequently, bicarbonate concentration in pancreatic juice from the isolated rabbit pancreas is rather constant under all conditions at ~80–100 mM.

The data for rabbit pancreas can most easily be interpreted on the basis of the unicellular model, originally proposed by Rothman and Brooks (181) on the basis of experiments on the isolated rabbit pancreas.

Pig

The pig is the animal from which secretin and CCK have been prepared for use in other species. Pancreatic secretion in the pig shows some similarities with the classic pattern but also some marked differences.

In both anesthetized animals (94, 99) and the isolated, perfused pig pancreas (98, 106, 107) there is a small spontaneous secretion, ~0.4 μl·min^{-1}·kg^{-1} body wt. The gland is exquisitely sensitive to secretin, which increases the rate of secretion 100-fold (94, 99, 107). The resulting flow curves are very similar to the classic pattern, with bicarbonate concentration rising to a plateau value of ~150 mM and chloride falling to 20 mM (94).

The effects of CCK have not been extensively studied in the pig. In the perfused pig pancreas, CCK evokes a modest flow of pancreatic juice (~10% of the response to secretin) containing a low-bicarbonate concentration (<45 mM) (108).

It is the response to vagal stimulation that distinguishes the pig pancreas from those previously described. This response, a copious flow of bicarbonate-rich fluid almost identical to that evoked by secretin, was inexplicable when first described in detail by Hickson (94). Predictably, it was therefore ignored. More recent studies on the isolated, perfused pig pancreas have provided an explanation (65, 97, 107). In the pig, vagal innervation of the pancreas is rich in VIPergic neurons as well as the more conventional cholinergic neurons. On stimulation of the vagus nerve, these fibers release VIP, which evokes bicarbonate secretion, presumably in a manner analogous to secretin, although the potency of exogenous VIP is only ~1% of that of secretin (106, 107).

The effects of vagal stimulation are not inhibited by atropine (94, 99). This suggests that vagal cholinergic fibers have little or no direct effect on pancreatic electrolyte secretion in the pig. However, intra-arterial injections of acetylcholine in both the anesthetized animal (94, 99) and the perfused gland (98) do evoke a modest secretion, which can be blocked by atropine and contains modest amounts of bicarbonate.

Guinea Pig and Hamster

Although the guinea pig and hamster have been extensively used in pancreatic research, in neither has there been a comprehensive study of electrolyte and water secretion. From the available data on anesthetized guinea pigs (37a, 47, 51, 140) and an isolated, saline-perfused guinea pig pancreas preparation (140), we draw the following conclusions. There is a spontaneous secretion (variable, but ~3 μl·min^{-1}·kg^{-1} body wt). A copious, bicarbonate-rich fluid is evoked by secretin, reaching a maximum of ~150 μl·min^{-1}·kg^{-1} body wt (Fig. 10). As in the rabbit, CCK-8 also evokes a substantial bicarbonate-rich fluid (~100 μl·min^{-1}·kg^{-1} body wt) (Fig. 10).

The response to vagal stimulation is very similar to that in the pig: a copious secretion of a bicarbonate-rich fluid is evoked (51, 140). Because this effect is not sensitive to atropine, it is presumably brought about by excitation of VIPergic neurons, although this has not been proved. However, as in the pig, stimulation of the perfused gland with carbachol causes a modest secretion of fluid, the composition of which is unknown (140).

Only one study on electrolyte secretion by the Syrian golden hamster has been published (91, 92). Unfortunately this study is totally confusing, because the electrolyte concentrations of pancreatic juice varied widely. In one series of experiments the sum of anions was reported to be 250 mM, whereas in another the sum of cations was 43 mM. All other studies in all other species have shown pancreatic juice to be an isotonic secretion in which the sum of anions and cations is therefore constant at ~160 mM.

As a basis for studying pancreatitis in hamster Ali et al. (2) have obtained some information in this species. These experiments show that the anesthetized hamster has a spontaneous secretion (~12 μl·min^{-1}·kg^{-1} body wt), which may be increased fivefold by

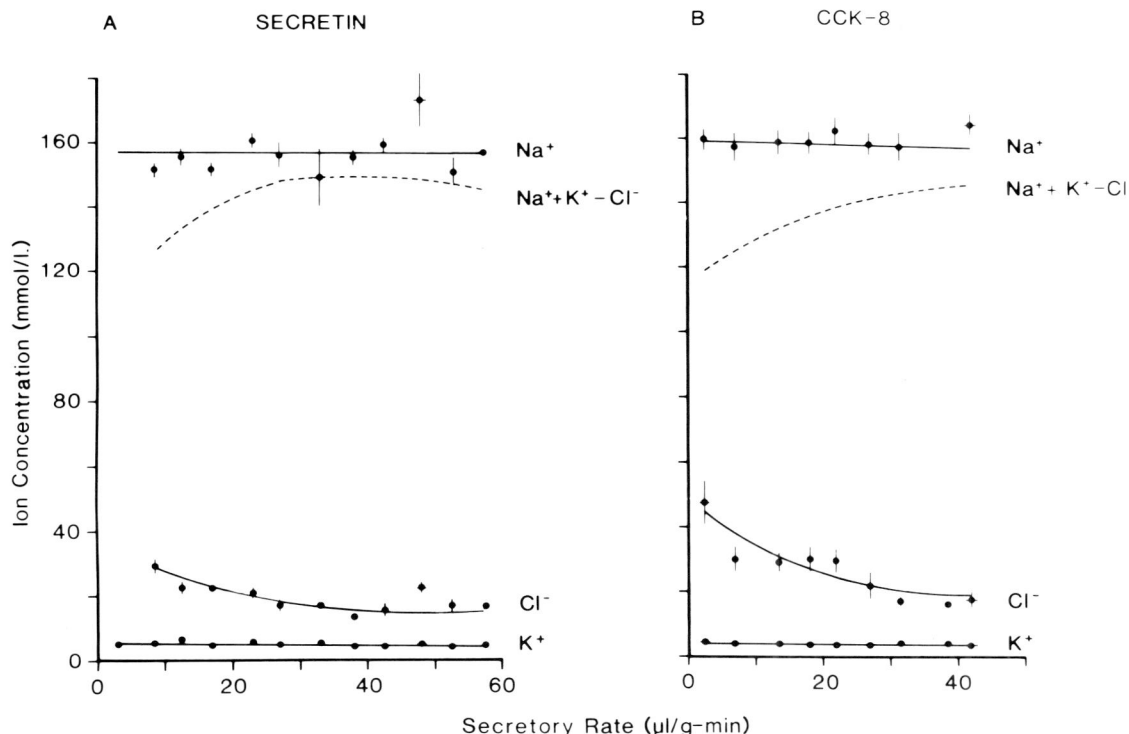

FIG. 10. Electrolyte composition of pancreatic juice in anesthetized guinea pig stimulated with secretin (A) and cholecystokinin-octapeptide (CCK-8) (B). Broken line, residual anions ($Na^+ + K^+ - Cl^-$), almost all bicarbonate. Flow rate is uncorrected for body weight or gland weight. [Data from Case et al. (37a).]

secretin, with the bicarbonate concentration approaching 150 mM (Fig. 11). Cholecystokinin-octapeptide increases secretion only slightly, with the bicarbonate concentration reaching ~50 mM. In other words, secretion in the hamster is quite close to the classic pattern observed in dog and human.

Sheep, Cow, and Horse

In conscious sheep, pancreatic juice is secreted continuously (~6.5 $\mu l \cdot min^{-1} \cdot kg^{-1}$ body wt) and shows little fluctuation over each 24-h period even when food is given. The main factor responsible for this pattern is probably the continual passage of chyme from abomasum to the duodenum. When abomasal contents are diverted, secretion is reduced 60%–70% (215). Even after a 48-h fast there is a spontaneous secretion (~3.0 $\mu l \cdot min^{-1} \cdot kg^{-1}$ body wt), which has a plasmalike electrolyte composition. Feeding doubles secretory rate and secretin causes a fourfold increase, with bicarbonate concentration reaching ~60 mM. Pilocarpine causes a two- to threefold increase in flow without changing bicarbonate concentration (131).

In anesthetized sheep the spontaneous secretion is reduced to <1.0 $\mu l \cdot min^{-1} \cdot kg^{-1}$ body wt (88, 175). Secretin, CCK, gastrin, carbachol, and pilocarpine all increase secretion to a greater or lesser extent (15, 88, 175, 215). Insulin hypoglycemia and direct vagal stimulation have a similar effect (88, 175, 215). These limited data suggest that pancreatic electrolyte secretion in the sheep most closely resembles that in the rat.

One brief communication on conscious and anesthetized horses indicates a substantial spontaneous secretion that is fivefold greater than in the conscious rat (25 $\mu l \cdot min^{-1} \cdot g^{-1}$ gland). This can be increased up to fivefold by either secretin or vagal stimulation. Chloride is the predominant anion, with bicarbonate concentrations not exceeding 70 mM (50). All that can be reliably said of the cow is that in 9-wk-old fed calves, the bicarbonate concentration increases from 20 to 90 mM as flow rate increases from 5 to 50 ml/h (216).

Primates

In conscious rhesus monkeys there is a marked spontaneous secretion (6 $\mu l \cdot min^{-1} \cdot kg^{-1}$ body wt) that contains 60–100 mM bicarbonate. A maximum dose of secretin increases secretion fourfold and elevates bicarbonate concentration to 140 mM, whereas CCK increases secretion by ~70% without changing bicarbonate concentration (76).

Some information has also been published on secretion in two species of anesthetized primates, *Erythrocebus patas* and *Papio mandrillus* (226). Spontaneous

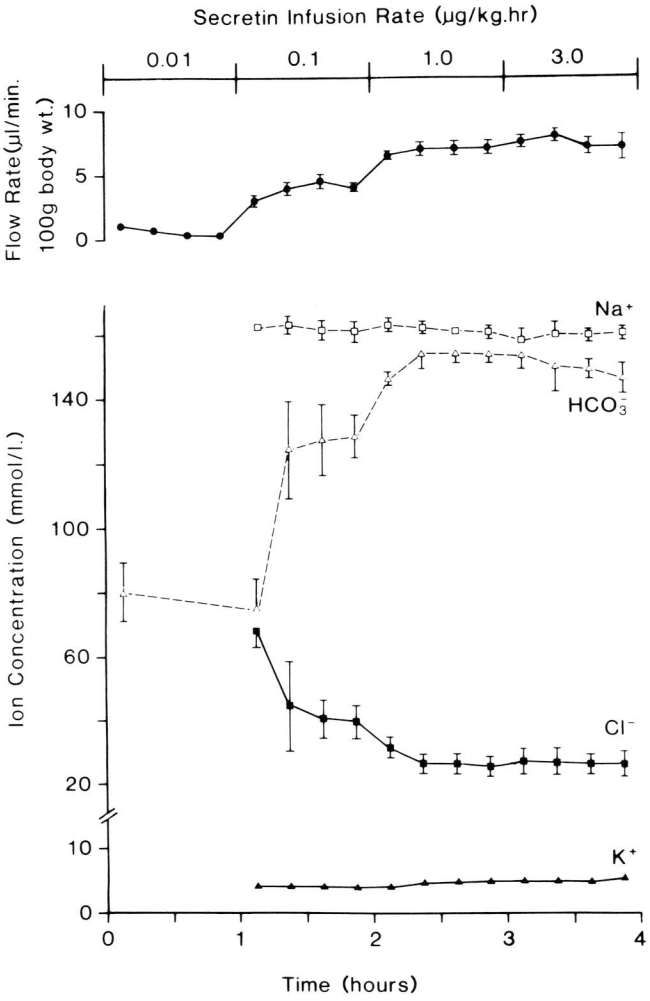

FIG. 11. Effect of increasing doses of secretin on flow rate and electrolyte composition of pancreatic juice in anesthetized Syrian golden hamster. (Unpublished data of A. E. Ali, R. M. Case, and S. C. B. Rutishauser.)

secretion was high in the latter (2 μl·min^{-1}·kg^{-1} body wt) but absent in the former. Secretin, CCK, and vagal stimulation increased secretion in both species to a variable extent. The response to vagal stimulation was reduced but not abolished by atropine. In response to secretin, the chloride concentration fell from ~100 to 20 mM (i.e., bicarbonate concentration rose from ~60 to 140 mM) as flow rate increased in E. patas. In P. mandrillus, chloride concentration was ~50 mM in spontaneous secretion (i.e., bicarbonate concentration was ~110 mM) and was apparently unchanged during stimulation (226). However, a pancreatic function test on conscious E. patas stated the bicarbonate concentration of duodenal fluid after stimulation with secretin plus CCK was only 65 mM (86).

These data suggest considerable variations in the patterns of secretion within primates. The pattern in E. patas most closely resembles that in the dog, that in P. mandrillus the rabbit, and that in the rhesus monkey is different from all others.

SITES OF ELECTROLYTE AND WATER SECRETION

The species-dependent patterns of electrolyte secretion just described pose a number of questions. What is the cause and origin of the spontaneous secretion? What is the origin of secretin-stimulated bicarbonate secretion? Do VIPergic neurons act at the same site as secretin? Where do CCK and acetylcholine act to evoke the chloride-rich secretion? Where does CCK act to evoke bicarbonate secretion? Definite answers exist to only two of these questions.

Secretin-Stimulated Bicarbonate Secretion

In the preceding description of the pattern of secretion in dog, cat, and human, reference was made to the suggestion of Hollander and Birnbaum (96) that pancreatic bicarbonate secretion was derived from centroacinar cells and terminal duct cells. The following early evidence favored this suggestion. *1*) In patients with pancreatitis, the volume of juice remained high while the enzyme content was reduced (96). *2*) Ethionine-induced destruction of acinar cells in cats, dogs, and monkeys caused the output of enzymes to be reduced more than that of bicarbonate (52, 109). *3*) Vacuolization of the ductal epithelium induced by alloxan in dogs decreased the response to secretin, whereas enzyme output was unaffected (85). *4*) Inhibition of carbonic anhydrase reduced bicarbonate but not enzyme secretion in humans (60), and carbonic anhydrase could be detected by histochemical means only in the ductal epithelium (9).

It is largely on the basis of this evidence that textbooks of physiology state that pancreatic bicarbonate secretion is derived from ductal elements within the gland. More recent evidence substantiates this point of view. *1*) As mentioned previously (see *Isolated Ductal Tissue*, p. 390), rats fed a copper-deficient diet develop acinar cell atrophy in the absence of changes in ductal tissue. In such animals secretin continues to evoke a bicarbonate-rich secretion (36, 70, 202). *2*) Carbonic anhydrase activity in ductal tissue prepared from such animals (36) or from normal animals (79) is enhanced sevenfold over that in whole-rat pancreas, and carbonic anhydrase C (the isoenzyme found in transporting epithelia) is located specifically in the ductal epithelium in human pancreas (122). *3*) With micropuncture techniques it has proved possible to obtain fluid samples from interlobular and intralobular ducts in the cat (127, 172) and also from the acini in rabbit (133, 173, 191, 211, 212), rat (133, 135), and mouse (134) glands. Although the results are somewhat difficult to interpret because of the different patterns of electrolyte secretion in these species, one observation seems irrefutable: secretin causes a bicarbonate-rich fluid to be secreted from the interlobular ducts (see ref. 34). On the other hand, fluid collected from the acini and intralobular ducts of the cat, rat, and mouse glands contains bicarbonate at concentrations similar to that in extracellular fluid and is virtually unaffected by stimulation with secretin (127, 133, 134). Fluid collected from the acini of the rabbit gland does contain bicarbonate at concentrations up to 4 times that in extracellular fluid (133, 191), but this is unaffected by secretin (191). These results suggest that in all species studied so far the interlobular ducts are the major site of secretin-stimulated bicarbonate and fluid transport. 4) Intact lengths of interlobular duct can now be isolated from the gland of copper-deficient rats [see FUTURE PROGRESS, p. 407; (6)]. When maintained in culture medium, the ends of the ducts seal to form closed vesicles. With micropuncture techniques such ducts have been observed to secrete in response to secretin but not caerulein (3). 5) As determined by microelectrode studies, secretin influences neither ion transport across the acinar cell plasma membrane (see *Electrophysiology and Stimulus-Secretion Coupling*, p. 399) nor oxygen consumption by acinar tissue (89) but does cause electrophysiological responses in duct cells (see Fig. 21). 6) Secretin stimulation depletes duct cells of cytoplasmic vesicles in the pig (26a).

These data allow the conclusion that an electrolyte secretion rich in bicarbonate is evoked by secretin from ductal elements within the pancreas of all species.

Secretion Evoked by Cholecystokinin and Vagal Stimulation

The existence of the chloride-rich component in pancreatic secretion proposed by Hollander and Birnbaum (96) in their admixture hypothesis was first clearly demonstrated in the experiments of Sewell and Young (197) on the anesthetized rat (see Fig. 8). Although the origin of this secretion was not addressed by Sewell and Young, the fact that it was evoked by CCK-8 and caerulein (i.e., by enzyme stimulants) led to the assumption that it was acinar in origin. The following evidence supports this assumption. 1) Neither CCK nor caerulein evoke secretion from the rat pancreas after acinar cell atrophy has been induced by feeding a copper-deficient diet (36, 70, 202). 2) Acinar micropuncture samples in the rat (but not rabbit) have a high-chloride concentration (133, 135). 3) In vitro studies in pancreatic acini show that CCK and acetylcholine evoke electrophysiological responses that are consistent with electrolyte secretion (see *Electrophysiology and Stimulus-Secretion Coupling*, p. 397).

Thus it seems likely that, in the rat at least, the chloride-rich component of pancreatic secretion evoked by CCK and acetylcholine is acinar in origin. However, in some species, notably rabbit and guinea pig, these agents evoke a secretion that is rich in bicarbonate (see SPECIES-DEPENDENT PATTERNS OF ELECTROLYTE SECRETION, p. 390). Very little of the evidence so far considered precludes the possibility that some bicarbonate is secreted by acini and some chloride is secreted by ductal cells or that secretin evokes some secretion from acini and CCK/acetylcholine some from ducts, especially in these species. Indeed experiments in dissociated guinea pig acinar cells show that sodium-pump activity is increased not only by carbachol and CCK-8 (by ~60%) but also by secretin (by ~40%), which suggests that secretin might evoked fluid secretion from acini in this species (100, 101). Specific experiments are required to answer these points.

We have also described how vagal VIPergic neurons can evoke a bicarbonate-rich secretion. Because VIP and secretin are similar in structure it is tempting to assume that in those species where VIPergic neurons play a role in secretion (especially pig and guinea pig but perhaps all species to a lesser extent) the released VIP acts on ductal cells. However, even though this seems likely, it may not be so. Thus acini prepared from one such species (guinea pig) possess two functionally distinct receptors for secretin and VIP. One receptor has a high affinity for secretin and a low affinity for VIP (secretin preferring); the other has reversed affinity (VIP preferring) (77). The number of the latter receptors in acini prepared from dog pancreas is reduced by 95% compared with that in guinea pig pancreas (123). If acinar VIP-preferring receptors are linked to electrolyte secretion, this difference in density could alone explain the greater importance of VIPergic neurons in guinea pig. However, it seems more likely that these acinar VIP-preferring receptors are linked to enzyme secretion (77) and that the secretin-preferring receptors modulate electrolyte secretion (100). Therefore those species-dependent differences in receptor density might also exist on ductal cells, which brings us back to the original suggestion that VIPergic neurons act on the ducts. It would be helpful to repeat the hormone-binding studies on duct cells to determine if they possess the same receptor types as acinar cells.

In conclusion, knowledge of secretion in dog, cat, and human on the one hand and rat on the other hand leads to the belief that duct cells are the source of bicarbonate in all species irrespective of the stimulus, whereas acini are the source of a variable quantity of chloride-rich fluid. However, there are very few data to support these beliefs.

PERMEABILITY OF THE PANCREATIC EPITHELIUM

Nonelectrolytes

The nonelectrolyte permeability of the isolated cat pancreas has been investigated by determining the apparent reflection coefficients for compounds of different molecular size added to the perfusion fluid (57). Substances with molecular volumes greater than ~0.45 cm^3/mol (i.e., sorbitol, xylose, and sucrose) are relatively impermeant, but below this value permeability increases in direct proportion to the reduction in molecular volume. These results suggest that the pancreas as a whole constitutes a leaky epithelium and that the overall permeability characteristics are largely determined by the properties of the paracellular pathway. However, studies on the intact gland cannot differentiate between the permeability characteristics of acinar and ductal epithelia.

Similar results have been obtained with the isolated rabbit pancreas (22, 104), although in this species the epithelium appears to be even more permeable than in the cat. Although this may indicate a very leaky paracellular pathway (104), an alternative explanation is uptake and secretion of large water-soluble molecules by the epithelial cells (142, 143). This latter effect may also explain the apparent increase in paracellular permeability that occurs when the rabbit gland is stimulated with acetylcholine, CCK, or other analogues (104).

Electrolytes

Altering the concentrations of sodium and potassium in fluid perfusing the cat pancreas produces equivalent alterations in the composition of the secreted juice (45). This would be expected for a leaky epithelium, and it seems likely that these ions (23), and also magnesium and calcium (183, 184), enter pancreatic juice via the paracellular pathway. Data on the absolute permeability of the pancreas to individual ions are not available, but there is some information on relative ion permeabilities. To calculate relative ion permeabilities it is necessary to record the alterations in transepithelial potential that occur when ion concentration gradients across the epithelium are changed. Such experiments require luminal perfusion and have only proved technically feasible on the main pancreatic duct.

When the lumen of the main duct in anesthetized cats (83, 231) and rabbits (143a, 174) is perfused with solutions containing bicarbonate at concentrations between 80 and 154 mM, the recorded transepithelial potentials are between +1.1 and −5 mV. These values are typical for leaky epithelia. The main duct of the cat (83) shows little selectivity between monovalent cations. The relative permeabilities are as follows: Li^+ = 1.08, Na^+ = 1.00, K^+ = 1.10, Rb^+ = 1.09, Cs^+ = 1.12. However, a definite permeability order exists for monovalent anions: F^- = 0.44, Cl^- = 1.08, Br^- = 1.38, I^- = 2.05. This is the Eisenmann sequence 1 (62, 237) and suggests that selectivity is due to weak positively charged sites in a highly hydrated channel, which is probably the junctional complex between the epithelial cells (221).

Secretin did not affect relative ion permeabilities but did increase lumen negativity by ~7 mV when there were no ion concentration gradients (and therefore no diffusion potentials) across the epithelium (231). This effect must result from either the transport of a negatively charged species from blood to lumen or a positively charged species from lumen to blood. Because the ductal epithelium actively transports bicarbonate ions when stimulated by secretin, the most likely candidates are either bicarbonate ions, hydroxyl ions, or protons (see CELLULAR MODELS OF DUCT CELL SECRETION, p. 405).

Transepithelial potentials in intralobular/interlobular duct segments, where secretin has a pronounced effect on bicarbonate and water transport, have occasionally been measured with microelectrodes and appear similar to those in the main duct (191, 211). No change in potential was observed after secretin stimulation (191), but in these "free-flow" experiments the duct was not perfused and thus transepithelial ion concentration gradients would have existed. This means that diffusion potentials could have masked any potentials generated as the result of active transport processes initiated by secretin. The availability of isolated interlobular ducts (3, 6) and thus the technical possibility of simultaneous luminal perfusion and transepithelial potential measurement should allow the precise ion permeability characteristics of these smaller and physiologically more interesting ducts to be determined.

Finally, a number of studies have examined the effects of luminal application of bile salts, alcohol, and drugs on the structure (151) and the permeability characteristics of the main pancreatic duct (171).

CELLULAR MECHANISMS OF ELECTROLYTE SECRETION FROM ACINAR CELLS

Any attempt to understand the cellular mechanisms of acinar electrolyte secretion requires the collation of data from the perfused rat pancreas and from ion flux and electrophysiological studies performed on isolated tissue from rat, mouse, pig, and human glands. Before we examine these data, it is worth recalling that the acini probably secrete an isotonic, sodium chloride-rich fluid and that the quantitative importance of this secretion, as opposed to ductal fluid secretion, varies greatly in these species (see SPECIES-DEPENDENT PATTERNS OF ELECTROLYTE SECRETION, p. 390).

Ionic Requirements for Secretion

This section refers to the rat pancreas, because this is the only species for which data are available.

Replacing extracellular sodium with Tris abolishes acetylcholine- and caerulein-stimulated electrolyte secretion from the perfused rat pancreas (164). When lithium is used to replace sodium, secretion is markedly inhibited but not abolished (164). Partial or total removal of extracellular potassium also inhibits fluid transport (110, 164). These results are usually interpreted as indicating that the sodium gradient, established across the basolateral membrane of the acinar cell by the Na^+-K^+-ATPase, plays an important role in the secretory process. In support of this idea, ouabain has an immediate inhibitory effect on caerulein-evoked fluid secretion (62a, 164).

Replacement of perfusate chloride by bromide has no significant effect on CCK-evoked secretion from the perfused rat pancreas (110, 112). However, replacement with acetate reduces the caerulein-evoked secretory response by ~70% (196). In these circumstances, acetate appears in the juice at the same concentration as chloride under normal conditions (196). Presumably acetate can support some fluid secretion, because the undissociated weak acid is lipophilic and therefore permeates cell membranes. When the impermeant substituents isethionate or sulfate are employed, fluid secretion is either substantially reduced (110, 112) or completely abolished (164, 196). These results indicate that acinar fluid transport is strongly dependent on a supply of anions. Under normal circumstances extracellular chloride is used, but other permeant anions (e.g., acetate) can partially substitute for chloride in the secretory mechanism.

Replacement of extracellular bicarbonate with either chloride or acetate causes little or no reduction in electrolyte secretion stimulated by acetylcholine, CCK, and caerulein (113, 164, 196). This is in marked contrast to the effects of bicarbonate replacement on secretin-stimulated electrolyte secretion. In the absence of chloride in the perfusate, bicarbonate ions can support some secretion from the acinar cells, because the gland responds to stimulation somewhat more vigorously when perfused with acetate and bicarbonate than with acetate alone (196).

Perfusing the rat gland with a calcium-free medium causes a rapid, marked inhibition of fluid secretion stimulated by either CCK or caerulein (112, 113, 219). Under these conditions the response to secretin is maintained (112, 219), indicating that the calcium sensitivity of ductal and acinar electrolyte secretory mechanisms must be different.

Electrophysiology and Stimulus-Secretion Coupling

Information on the electrophysiology of pancreatic acinar cells has come largely from the extensive investigations of Petersen and his co-workers [see the chapter by Petersen and Maruyama in this *Handbook*; (157–161)]. These studies have revealed important differences in the membrane permeability characteristics of acinar cells in pig and human as compared with those in rat and mouse. Consequently it is not possible to produce a common model to explain the cellular mechanisms of acinar fluid secretion in all these species.

Conventional microelectrode studies have shown that the resting membrane potential of acinar cells in the rat and mouse pancreas is about −40 mV (inside negative) (157). Acetylcholine, CCK-like peptides, and bombesin all cause a dose-related depolarization, which has a reversal potential of −15 to −20 mV, and a reduction in input resistance (increase in membrane conductance). These effects result from a stimulant-induced increase in permeability to chloride, sodium, and potassium ions; the relative permeability changes are $P_{Cl^-}:P_{Na^+}:P_{K^+} = 5.0:1.2:1.0$ (see ref. 157). This suggests that stimulation is associated with an uptake of sodium and chloride into the acinar cell with a slight loss of potassium. Small increases in intracellular chloride and sodium activity have been detected after stimulation with acetylcholine (150), although there appears to be no measurable decrease in intracellular potassium (165).

Neither secretin nor dibutyryl cAMP, even at very high concentrations, has any effect on the electrophysiological properties of pancreatic acinar cells (157). However, secretin does activate adenylate cyclase in membrane fractions, presumably largely acinar in origin, prepared from the whole pancreas (34) and at high (nanomolar) concentrations causes a small increase in amylase secretion from the perfused rat pancreas (112, 164). As discussed previously (see SITES OF ELECTROLYTE AND WATER SECRETION, p. 396), whether secretin has an effect on electrolyte secretion from acinar cells remains unanswered. In the rat, adrenergic agonists also evoke pancreatic enzyme secretion via the adenylate cyclase system but do not produce detectable electrophysiological effects in acinar cells (154).

Pig pancreatic acinar cells have a resting membrane potential of about −30 mV (inside negative) (152). However, in contrast to the mouse and rat, stimulation of the pig acinar cell with acetylcholine, pentagastrin, or bombesin results in hyperpolarization and a reduction in input resistance (152). Conventional microelectrode data are not available for the human pancreas.

In all species the permeability changes that occur after stimulation result from the opening of ion channels in the plasma membrane that are activated by a secretagogue-induced rise in intracellular calcium concentration. The electrophysiological effects of stimulants can be broadly mimicked either by exposing acinar cells to the calcium ionophore A23187 or by intracellular injection of calcium ions (see ref. 157), and stimulant-induced rises in intracellular calcium can be detected with a variety of techniques (148, 149). The source of this activating calcium is intracellular, at least when the stimulation period is short or during the early part of a prolonged exposure to secretagogues. Mobilization of this intracellular calcium store, which is probably located in the endoplasmic reticulum (or perhaps a specialized part of it), is triggered by the rise in inositol 1,4,5-trisphosphate that occurs after stimulation (see the chapter in this *Handbook* by Schulz). There is some evidence to suggest that during prolonged stimulation the electrophysiological events triggered by secretagogues are dependent on the presence of extracellular calcium and thus presumably on calcium influx into the acinar cell (112, 157, 158). The mechanism of this stimulant-induced calcium uptake is unknown; however, it appears to be controlled by inositol 1,4,5-trisphosphate and its phosphorylated metabolite, inositol 1,3,4,5-tetrakisphosphate (145).

The presence of calcium-activated ion channels in the basolateral membrane of pancreatic acinar cells has been confirmed with the patch-clamp technique [see the chapter in this *Handbook* by Petersen and Maruyama; (158, 159, 161)]. The ion selectivity of the channels, their voltage dependence, and their single-channel conductances vary between species. In the pig and humans the channels are potassium selective and voltage dependent (opened by depolarization of the membrane and vice versa), with a conductance of ~200 pS (161). About 50 of these so-called large potassium channels are present on each acinar cell (139, 161). Their activation after stimulation explains the hyperpolarization of pig acinar cells observed in conventional microelectrode studies (152). The human acinar cell also possesses a small potassium channel that has a conductance of ~50 pS (160).

In contrast to the acinar cells of pig and human, those of mouse and rat possess a voltage-independent, nonselective, monovalent-cation channel with a conductance of 30–50 pS (161). This channel, ~500 of which exist on each cell (159), is almost equally permeable to sodium and potassium ions but is virtually impermeable to chloride. After stimulation, sodium influx predominates, because the resting membrane potential is much closer to the potassium than to the sodium equilibrium potential, and thus the opening of this channel would explain the depolarization observed in rat and mouse acini. However, the mechanisms underlying the marked increase in chloride permeability (see above) that occurs after stimulation of rat and mouse acini remain unclear. Presumably this must be a conductive pathway, because alterations in extracellular chloride concentration have marked effects on the electrophysiological response to acetylcholine measured with conventional microelectrodes (157). The most likely explanation is that the chloride channels, presumably calcium activated, are either present on the luminal membrane of the acinar cell or are inserted into the luminal membrane during exocytosis of the zymogen granules (54). Their opening is detected in conventional microelectrode studies, in which the basolateral membrane is impaled, because of paracellular current flow via the junctional complexes (162).

Cellular Models of Acinar Cell Secretion

It is not yet possible to fit the experimental data into a single model that adequately describes acinar cell electrolyte secretion in all species. We are therefore obliged to describe three different models, one for the pig and human, another for the mouse, and a third for the rat. The three models have been devised on the basis of the data outlined in the two previous sections, together with studies on potassium fluxes in the gland, and the effects of pharmacological inhibitors on fluid secretion. The models possess one common and well-established feature: receptor activation releases calcium from an intracellular store, and the resulting increase in cytosolic calcium concentration opens ion channels on the basolateral membrane. Beyond this point the models are very tentative, rely heavily on analogies with other secretory epithelia, and are no more than working hypotheses at this time.

1. Pig and human (Fig. 12A). This model has been proposed on the basis of data obtained in the pig [see the chapter by Petersen and Maruyama in this *Handbook*; (152, 158, 159, 161)] but may equally apply to the human. In these species receptor activation causes the opening of a calcium-activated, voltage-sensitive, large potassium channel on the basolateral membrane, leading to potassium efflux and acinar cell hyperpolarization. It is envisaged that the sustained elevation of intracellular calcium concentration during stimulation would keep the channel in the open state and offset its tendency to close because of the membrane hyperpolarization. Released potassium is then reaccumulated into the acinar cell by the Na^+-K^+-ATPase and also by an electroneutral, diuretic-sensitive, Na^+-$2Cl^-$-K^+ cotransporter localized on the basolateral membrane. The cotransporter acts as a chloride pump, using the energy of the sodium gradient to increase the intracellular chloride concentration above electrochemical equilibrium. The rate at which the cotransporter moves chloride depends on the availability of extracellular potassium, which is controlled by the opening of the large potassium channel, and the chemical gradients for sodium and potassium across the basolateral membrane. If the accumulated chloride ions then enter the lumen through an anion channel on the apical membrane, the negative transepithelial potential generated by this event would draw sodium through the paracellular pathway.

The presence of a Na^+-$2Cl^-$-K^+ cotransporter has not been confirmed in the acinar cells of pig and human pancreas, and this aspect of the model relies on an analogy with salivary acini, where the presence of the cotransporter is better established [see the chapter by Cook and Young in this *Handbook*; (64, 163)]. In addition, there is no direct electrophysiological evidence for a stimulant-induced increase in chloride permeability in the pig pancreas (159), although whole-cell current recording has revealed a small in-

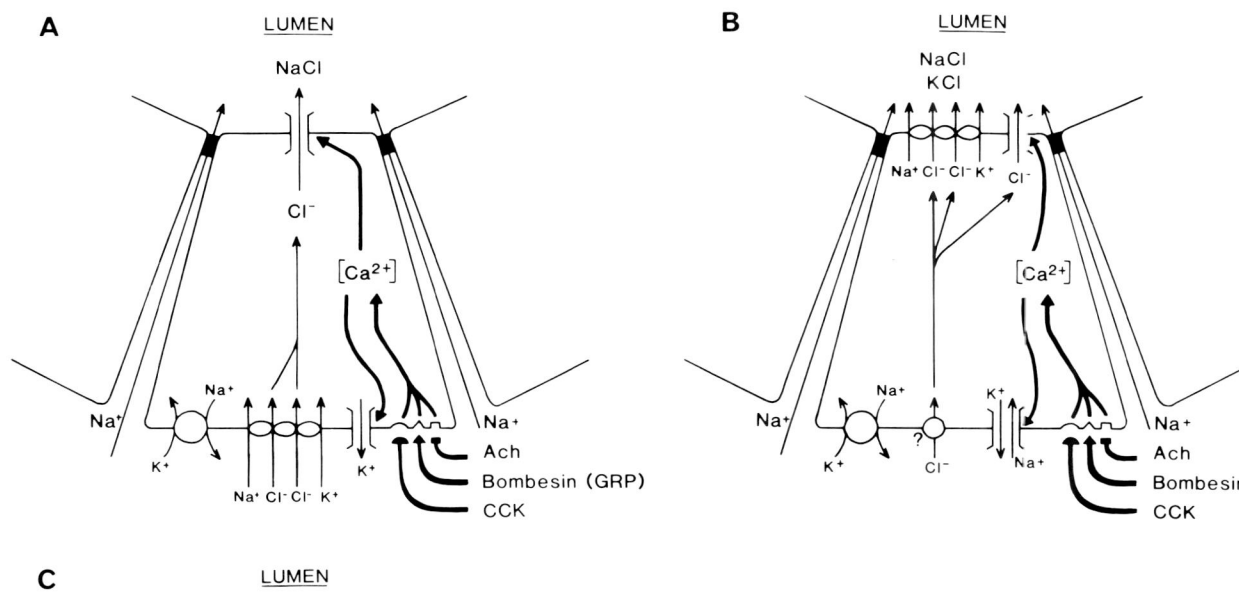

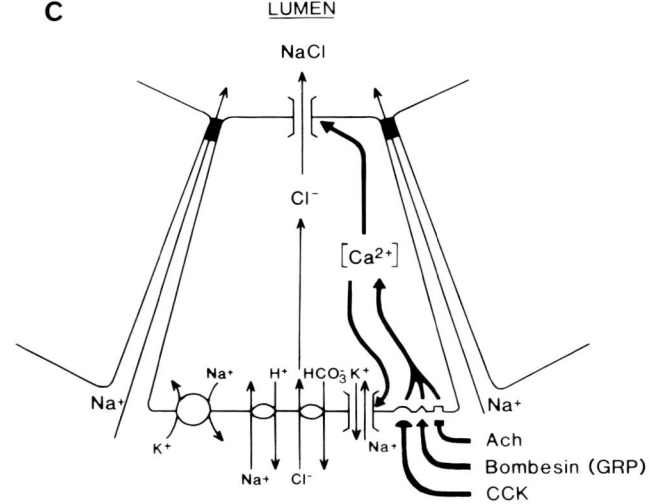

FIG. 12. Tentative models for cellular mechanisms of NaCl secretion by acinar cells of pig (A) and perhaps human, mouse (B), and rat (C). Ach, acetylcholine; CCK, cholecystokinin; GRP, gastrin-releasing peptide.

ward current that might be carried by chloride ions (158). There is, however, good evidence for the presence of chloride channels in the luminal membranes of other NaCl-secreting epithelia (13, 68, 71a, 72, 84, 137a, 138, 223). In similarity with the general concepts established in other epithelia, it is assumed that water moves into the acini lumen down osmotic gradients established by the ion-transport processes.

2. Mouse (Fig. 12B). In this species, receptor activation leads to the opening of calcium-activated, voltage-insensitive, nonselective cation channels on the basolateral membrane. This results in an uptake of sodium and a slight loss of potassium; the overall effect of these ion movements causes depolarization of the acinar cell. There is evidence for the presence of a diuretic-sensitive Na^+-$2Cl^-$-K^+ cotransporter in mouse pancreatic acinar cells (163, 200), but it appears to move ions from the intracellular to the extracellular compartment, i.e., in the opposite direction to the same cotransporter in salivary acini (64, 161). This means that if the cotransporter plays a role in secretion, it must be located on the luminal membrane (163). A basolateral location for the cotransporter might also seem unlikely on mechanistic grounds, because the increase in intracellular sodium caused by opening of the nonselective cation channel would lower the gradient for sodium across the basolateral membrane and thus tend to reduce the driving force available for accumulation of chloride. How chloride is pumped across the basolateral membrane in this species is unknown. However, in the absence of an Na^+-$2Cl^-$-K^+ cotransporter, the most likely alternative is a combination of Na^+-H^+ and Cl^--HCO_3^- exchange (37, 41, 218). The same mechanistic problem concerning the sodium gradient also applies to this double antiport scheme, unless one assumes that the antiporters are themselves regulated. Any model of the mouse acinar cell must also incorporate the clear

stimulant-induced increase in chloride permeability observed in conventional microelectrode studies. Because the patch-clamp technique has not revealed chloride channels on the basolateral membrane of mouse acini, they must be located on either the luminal or zymogen granule membranes. Thus this model incorporates two exit pathways for chloride across the luminal membrane, which seems a little unlikely.

3. Rat (Fig. 12C). In addition to providing information on the ionic dependency of acinar secretion, the perfused rat pancreas has also been used to study the effects of various transport inhibitors on the secretory process. At a concentration of 10^{-4} M, amiloride, SITS, furosemide, and methazolamide, which block Na^+-H^+ exchange, HCO_3^--Cl^- exchange, Na^+-$2Cl^-$-K^+ cotransport, and carbonic anhydrase, respectively, all reduced caerulein-evoked fluid secretion by 70%–80% (196). On the basis of these results it was suggested that chloride accumulation across the basolateral membrane in this species could best be explained by a combination of Na^+-H^+ and Cl^--HCO_3^- exchange rather than by an Na^+-$2Cl^-$-K^+ cotransporter (196). Recently, coupled Na^+-H^+ exchange has been directly demonstrated in rat (93) and guinea pig (61) acini. The inhibitory effect of methazolamide on secretion was ascribed to a fall in the rate of hydration of intracellular carbon dioxide and thus a reduction in the supply of bicarbonate for the Cl^--HCO_3^- exchanger. It was suggested that furosemide also acted in the same way, because it can inhibit carbonic anhydrase (137) and its effect on secretion was no greater than that observed with methazolamide (196). Tetraethylammonium (10^{-2} M) and barium (3×10^{-3} M), agents that block Ca^{2+}-activated potassium channels (see the chapter in this *Handbook* by Petersen and Maruyama), have also been shown to have an inhibitory effect on acinar secretion (62a).

It is difficult to reconcile these findings on the perfused gland with the presence of a calcium-activated, nonselective cation channel in the basolateral membrane of rat acini. Tetraethylammonium and barium have not been reported to block these channels, and any increase in intracellular sodium concentration that results when these channels open after stimulation reduces the chemical gradient for sodium and thus the net force driving chloride movement across the basolateral membrane. It is presumed that chloride leaves the acinar cell via a channel in the luminal membrane, although the presence of a stimulant-induced increase in chloride permeability has not been investigated in rat acini. The ability of acetate to partially substitute for chloride in the secretory mechanism (196) probably results from diffusion of the undissociated acid into the acinar cell. Once inside the cell one must speculate that dissociation occurs, producing the anion and a proton, and that the proton is transported back out of the cell on the Na^+-H^+ exchanger. Exactly how the charged anion crosses the apical membrane to enter the acinar lumen remains unclear.

CELLULAR MECHANISMS OF ELECTROLYTE SECRETION FROM DUCT CELLS

Stimulus-Secretion Coupling

Although most data have come from studies on the whole pancreas rather than on ductal tissue, there seems little doubt that the effects of secretin and VIP on duct cells are mediated by cAMP. The evidence can be summarized as follows:

1. Secretin activates adenylate cyclase in plasma membranes derived from isolated rat pancreatic duct cells; the dose required for half-maximal activation is $\sim 10^{-9}$ M (186). This is $\sim$10-fold lower than the dose required for half-maximal activation of adenylate cyclase in plasma membranes derived from acinar cells (186).

2. Secretin and VIP cause a dose-dependent increase in pancreatic cAMP content both in vivo and in various in vitro preparations derived from the cat (42), guinea pig (12, 153), rat (56, 115, 179, 225), and dog (59, 95). Studies using duct fragments isolated from copper-deficient rats (71) confirm that both hormones cause a dose-dependent elevation of cAMP content in duct cells, with secretin 10-fold more potent than VIP. However, both hormones also elevate the cAMP content of dissociated acinar cells (77).

3. Phosphodiesterase inhibitors markedly potentiate submaximal doses of secretin and at high concentrations can initiate pancreatic secretion in the cat (44) and dog (59, 217).

4. Dibutyryl cAMP stimulates bicarbonate secretion from the isolated cat pancreas (44).

5. Cholera toxin evokes electrolyte secretion when injected into the in vivo rat gland (114) and also when added to the fluid perfusing an isolated cat pancreas (201). In both species there is a simultaneous elevation in tissue cAMP content resulting from activation of adenylate cyclase.

It also seems likely that the actions of isoproterenol (48, 129, 144) and dopamine (74, 75, 144) on duct cells may be mediated by cAMP. Both compounds elevate the cAMP concentration in rat pancreas pieces (154, 225).

Ionic Requirements and Mechanisms of Secretion

Most information on the ionic requirements and mechanisms of secretion has come from studies on the vascular perfused cat pancreas (38) and the superfused rabbit pancreas (180), although some data on the perfused rat gland have also been published (113, 164, 195). Despite the different patterns of electrolyte secretion in these species (see SPECIES-DEPENDENT PATTERNS OF ELECTROLYTE SECRETION, p. 390) and the fact that the isolated rabbit gland secretes at maximum rates in the absence of secretin stimulation, essentially similar conclusions have been reached from studies on all preparations.

SODIUM AND POTASSIUM. Although it seems likely that sodium ions enter pancreatic juice by passive diffusion along the paracellular pathway, the presence of this cation on the blood side of the epithelium is an absolute requirement for ductal electrolyte secretion. When all sodium in the perfusate is replaced by lithium, an ion that has the same permeability as sodium in the main duct (83), there is an 80%–90% inhibition of secretion (45). As would be expected, the concentration of lithium in the juice under these conditions is approximately equal to that in the perfusion fluid (45). When a relatively impermeant substituent such as sucrose is employed, only half the sodium chloride needs to be osmotically replaced to produce a similar inhibitory effect on secretion rate (4, 23). Under these conditions isosmolality of juice and perfusion fluid is maintained by the development of a twofold sodium concentration gradient across the epithelium (23, 38). The presence of this gradient may indicate active sodium transport under these conditions, because the transepithelial potential, at least in the main duct of the rabbit gland, remains unchanged (23).

Like sodium, potassium probably enters the juice by passive diffusion along the paracellular pathway (23). However, complete removal of potassium ions from perfusion fluid does cause a 65% inhibition of secretion rate (45). Rubidium, but not cesium, can act as a complete substitute for potassium in supporting secretion (45).

Overall, these effects of sodium and potassium replacement probably indicate that the sodium gradient across the basolateral membrane of the duct cell plays an important role in bicarbonate secretion. One possibility is that the passive diffusion of sodium ions into the duct cell down their electrochemical gradient is coupled to the active transport of protons in the opposite direction (see CELLULAR MODELS OF DUCT CELL SECRETION, p. 405). Thus reducing the chemical gradient for sodium (sodium replacement with sucrose) or inhibiting the activity of Na^+-K^+-ATPase (27) at the inside face (sodium replacement with lithium) or outside face (potassium removal or replacement with cesium) of the basolateral membrane inhibits secretion. In support of this idea, ouabain is a potent inhibitor of spontaneous and secretin-stimulated electrolyte secretion from the gland (45, 121). Furthermore the dose-response curve for the inhibition of secretion is similar to that for inhibition of Na^+-K^+-ATPase activity in gland homogenates (178).

CALCIUM AND MAGNESIUM. Prolonged perfusion of the cat gland with a calcium-free solution causes a progressive and only partially reversible inhibition of secretin-stimulated fluid secretion (5, 103). This effect is probably associated with an increased permeability of the epithelium since, on returning to a calcium-containing perfusate, the juice calcium concentration was increased above normal. Alterations in the perfusate concentrations of magnesium have no effect on secretin-stimulated fluid secretion (5). These results are usually interpreted as indicating that calcium and magnesium play no direct role in the mechanisms of bicarbonate secretion.

CHLORIDE. Because the chloride concentration in secretin-stimulated pancreatic juice is always less than in blood or perfusion fluid, active transport of this ion seems unlikely. However, complete replacement of chloride on the blood side of the epithelium with isethionate or methyl sulfate inhibits ductal electrolyte secretion by ~60% (40, 120, 195). The ability of various anions to maintain secretion is as follows: chloride = bromide > nitrate > iodide > sulfate > methyl sulfate > isethionate (40). This order probably reflects their ability to cross the pancreatic epithelium and suggests that a supply of permeant anion is necessary for optimal secretion rates to be achieved. However, alterations in extracellular chloride concentration may change intracellular pH (224), and this parameter could affect bicarbonate transport and thus fluid secretion (see next section). Attempts to implicate chloride ions directly in the mechanism of bicarbonate secretion using pharmacological inhibitors, e.g., piretanide, ammonium ions, and SITS, which are known to affect chloride transport in other tissues, have not been convincing (34, 121, 193).

BICARBONATE. The bicarbonate concentration in secretin-stimulated pancreatic juice is 3–5 times higher than in blood or perfusion fluid. Given the low value of the transepithelial potential, this must indicate active transport of the ion. About 93% of secreted bicarbonate is derived from the blood and only ~7% is derived from metabolic sources (46).

Active bicarbonate transport could be explained by primary movement of any one of the constituent buffer components: bicarbonate ions, hydroxyl ions, protons, or carbon dioxide. Altering the concentration of these buffer components in blood (169, 170) and perfusion fluid (2a, 46, 191, 212) has shown that both fluid and bicarbonate secretion are directly related to extracellular pH and/or bicarbonate concentration. This suggests that either protons or bicarbonate ions are the primary transported species and means that two mechanisms could account for ductal bicarbonate secretion: either forward (blood-to-lumen) transport of bicarbonate or backward (lumen-to-blood) transport of protons. If the protons were derived from carbonic acid, the net effect of the latter mechanism would be forward transport of bicarbonate. From the mechanistic point of view, backward transport of protons could also be described as forward transport of hydroxyl ions, followed by combination with carbon dioxide to form bicarbonate (Fig. 13).

Forward bicarbonate movement was initially sup-

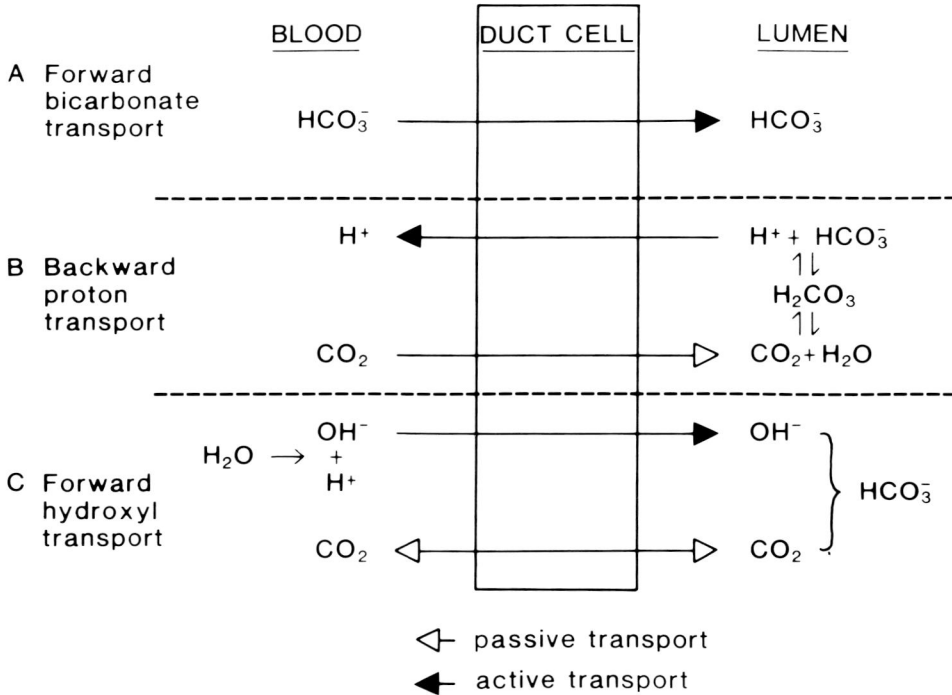

FIG. 13. Possible general mechanisms of active bicarbonate transport by pancreatic duct cell. Experimental evidence argues strongly in favor of either model B or C, which cannot be distinguished mechanistically. Both models could be written in many different ways. For example, hydration steps shown occurring in the lumen in B could equally take place inside the cell. In this case the bicarbonate ions generated would have to move across the apical membrane and into the lumen, possibly by diffusion down an electrochemical gradient. Bicarbonate ions might also enter the cell on an anion carrier located on the basolateral membrane. Once inside the cell they could then combine with protons being pumped back from the lumen to form carbon dioxide, which then diffuses across the apical membrane.

ported by the discovery of a bicarbonate-stimulated ATPase in pancreatic plasma membranes (235). However, it was later realized that its presence resulted from mitochondrial contamination of the membrane fractions (222). The best evidence for backward proton transport comes from the fact that the pancreas actively secretes buffer anions other than bicarbonate. These include the organic anions, acetate, butyrate, formate, and propionate (40, 195, 212), and also the weak organic acids sulfamerazine and glycodiazine (185, 189). The most likely explanation for their active secretion is the movement of protons, the common buffer component, since it seems highly improbable that a single ion pump could handle anions of such differing structure. As a general mechanism it could be envisaged that these compounds diffuse across the epithelium in their lipid soluble undissociated forms and then ionize in the lumen to form the appropriate anion and a proton. The latter is then actively transported back to the blood, leaving the anion trapped in the lumen. Thus it is reasonable to predict that the concentration of the undissociated buffer component in perfusion fluid is the factor determining secretion rate. This does seem to be the case for sulfamerazine (189). However, the situation is more complex for the other substituent compounds; for them the secretion rate is directly proportional to the perfusate pH and anion concentration (40, 185, 212). This indicates a similarity with bicarbonate secretion but also suggests that an anion carrier must exist on the basolateral membrane of the duct cell. Once inside the cell it is assumed that the anions interact with protons and then diffuse across the apical membrane into the lumen (Fig. 14). Finally, although backward proton transport appears the most plausible mechanism for active bicarbonate secretion, a note of caution should be added. If proton transport is the primary event, then one proton must enter the blood for every bicarbonate ion secreted. As would be predicted from this mechanism, the pH and partial pressure of CO_2 (P_{CO_2}) of venous effluent draining the perfused gland decrease and increase, respectively, after secretin stimulation (46). However, the changes are quite small (0.1 pH unit and 5 mmHg, respectively) and quantitatively too small to account for the amount of bicarbonate secreted (see discussion following ref. 193).

Because bicarbonate secretion is an active process, the transport route must be cellular and at least one active transport step must exist either at the basolateral or the luminal membrane of the duct cell. The

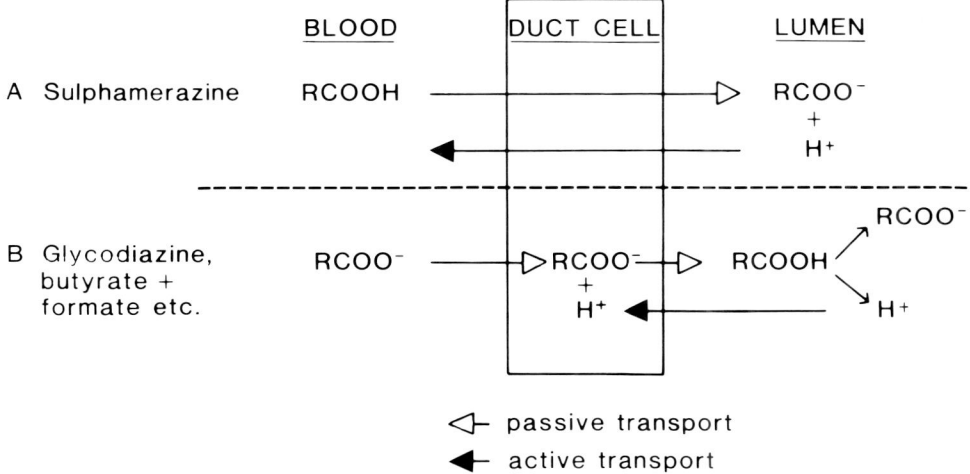

FIG. 14. Possible models for transport of weak organic acids and organic anions by pancreatic ductal epithelium. Such compounds can replace bicarbonate ions and support secretion. *A*: sulfamerazine. Secretion rates obtained with this compound depend on concentration of the lipid-soluble undissociated buffer component on blood side of the cell. This indicates that the weak acid delivers protons to the lumen by diffusing across the epithelium. *B*: glycodiazine, actetate, butyrate, formate, and propionate. Secretion rates obtained with these compounds depend on the concentration of the buffer anion on the blood side of the cell. Presumably this means that the anions cross the basolateral membrane of the duct cell on a carrier. Once inside the cell they could combine with protons moving back from the lumen and then diffuse across the apical membrane as undissociated buffer components.

location can only be established by determining the electrochemical gradient for the transported species across each cell membrane. Unfortunately this has not been achieved satisfactorily, largely because of the formidable technical problems associated with the direct measurement of intracellular pH in the small duct cells. For the moment only indirect observations are available from which we can make some simple calculations.

The electrical gradients across apical and basolateral membranes of the duct cell can be derived from microelectrode determinations of intracellular (−50 mV) and transepithelial (−5 mV) potentials (185). If blood and duct bicarbonate concentrations of 28 and 100 mM, respectively, are assumed, the Nernst equation can be employed to calculate the electrochemical gradients for bicarbonate across each membrane (185). These calculations suggest that the active transport step must be located at the luminal membrane if the intracellular bicarbonate concentration is < 4.2 mM (pH < 6.65) but at the basolateral membrane if the intracellular bicarbonate concentration is > 18 mM (pH > 7.27). The decision rests upon the accurate determination of intracellular pH. Reported values include "far lower than 7" (bromothymol blue staining of intact ducts) (185) and 6.4 ([^{3}H]nicotine bitartrate distribution in isolated duct cells) (186), both of which favor active transport at the luminal membrane, and 7.25 (2-[^{14}C]5′5′-dimethyloxazolidine-2,4-dione distribution in whole pancreas) (210, 211), which suggests active transport at the basolateral membrane. Because the location of the active transport step is undecided, the validity of the cellular models of duct cell secretion is still questionable.

CELLULAR MODELS OF DUCT CELL SECRETION

Based on the data outlined in the previous section, together with the effects of pharmacological inhibitors [many of which are of questionable specificity (34)], three different models have been proposed to account for duct cell electrolyte secretion. All employ the sodium gradient across the basolateral membrane, established by the Na$^+$-K$^+$-ATPase, as an energy source for active bicarbonate transport. Two of the models incorporate the idea of backward proton transport.

Swanson and Solomon (212) proposed the model shown in Figure 15. It is based on proton electrochemical gradients calculated from indirect intracellular pH measurements made on the whole gland. The initial step in bicarbonate secretion is viewed as diffusion of carbon dioxide across the epithelium, followed by hydration in the lumen to form bicarbonate ions and protons. The protons are then pumped back to the blood by a mechanism that requires active transport at both cell membranes. At the apical membrane an Na$^+$-H$^+$-ATPase (primary active transport) was proposed. This enzyme has never been detected in the pancreas, although Schulz and Terreros-Aranguren (190) have found an Mg^{2+}-ATPase in duct cell membranes that might act as a proton pump. The presence of a luminal proton pump has also been suggested on the basis of in vivo studies (167–170). At the basolateral membrane Swanson and Solomon proposed an Na$^+$-H$^+$ antiport that utilizes the energy of the sodium gradient (secondary active transport). There is some evidence for Na$^+$-H$^+$ exchange in isolated duct cells, but the activity of the exchanger was not affected by secretin (187). The basic Swanson-

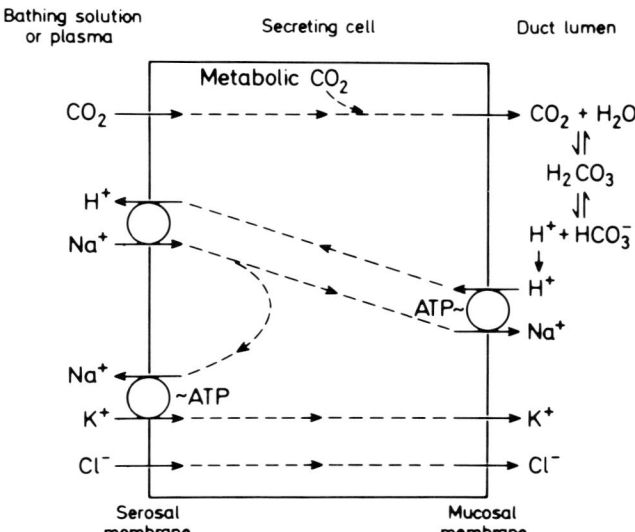

FIG. 15. Cellular model for duct cell electrolyte secretion. Forward (blood-to-lumen) transport of bicarbonate results from backward (lumen-to-blood) transport of protons. Primary (Na^+-H^+-ATPase) and secondary (Na^+-H^+ exchange) active transport steps are located at luminal and basolateral membranes, respectively. [From Swanson and Solomon (212).]

Solomon model has been modified by Scratcherd et al. (194) to include an HCO_3^--Cl^- antiport on the luminal membrane. It has been suggested that up to a third of pancreatic juice bicarbonate could be secreted by exchange with luminal chloride (193). Anion exchange plays an important role in bicarbonate secretion by the duodenal mucosa (69). Finally, in the Swanson-Solomon model all secreted ions are viewed as crossing the epithelium by the cellular route, a proposal that ignores the fact that the pancreas is a leaky epithelium (see PERMEABILITY OF THE PANCREATIC EPITHELIUM, p. 397).

On the assumption that the majority of sodium and potassium enters pancreatic juice via the paracellular pathway (23, 57, 121), De Pont et al. (55) have proposed the model shown in Figure 16, which is essentially the same as the one proposed for bicarbonate transport by the proximal tubule (28, 73). In this model the cellular hydration of carbon dioxide leads to intracellular accumulation of bicarbonate ions, which then diffuse down an electrochemical gradient into the lumen. Carbonic anhydrase, which catalyzes the hydration of carbon dioxide, is certainly localized in the ductal epithelium (36, 79, 122), and inhibitors of this enzyme (e.g., acetazolamide) reduce secretin-stimulated electrolyte secretion (14, 39, 40, 46, 176, 193, 212).

One useful prediction that arises from these models is that secretion should be reduced if the Na^+-H^+ antiport on the basolateral membrane is blocked. Amiloride, which at millimolar concentrations inhibits Na^+-H^+ exchange in various epithelial and nonepithelial tissues (11), does reduce secretin-stimulated electrolyte secretion from the in vitro cat (236) and rat glands (63) but has no effect on spontaneous secretion from the in vitro rabbit pancreas (121). This latter observation on the rabbit gland, together with the effects of other pharmacological inhibitors, has led Kuijpers et al. (121) to propose that the Na^+-H^+ antiport be replaced with an electroneutral Na^+-$2HCO_3^-$-Cl^- carrier (Fig. 17). Such a carrier is thought to play a role in the regulation of intracellular pH in squid axons, snail neurons, and barnacle muscle (24). One disadvantage of this model is that it does not include a specific role for carbonic anhydrase. It is also difficult to see how this scheme could account for the secretion of weak organic acids and anions in the absence of proton transport, unless the Na^+-$2HCO_3^-$-Cl^- carrier can also move protons (24).

Now that the models for ductal electrolyte secretion have been examined, can the stimulatory effect of secretin on this process be explained? Unfortunately the only certain piece of information about the cellular actions of this hormone is that it utilizes cAMP as a second messenger (see *Stimulus-Secretion Coupling*, p. 402). Presumably this cyclic nucleotide exerts its effect by stimulating the phosphorylation of cellular proteins that control transmembrane ion movements (199). Until recently (see FUTURE PROGRESS, p. 407), the only clue about the character and possible location of these biochemical events came from one electrophysiological study of the mouse pancreas. Random impalements of the gland identified a population of cells, presumed to be duct cells, that had resting potentials of about −23 mV and that hyperpolarized to about −41 mV in the presence of secretin (82). Interpretation of cellular electrophysiological data obtained from leaky epithelia is complicated because of

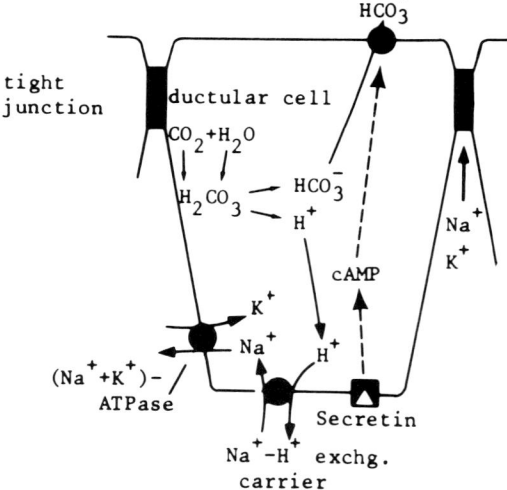

FIG. 16. Cellular model for duct cell electrolyte secretion. Forward (blood-to-cell) transport of bicarbonate results from backward (cell-to-blood) transport of protons. Secondary active transport step (Na^+-H^+ exchange) is located at the basolateral membrane, and bicarbonate leaves the cell down an electrochemical gradient across luminal membrane. [From De Pont et al. (55). In: *Electrolyte and Water Transport Across Gastrointestinal Epithelia*, © 1982, Raven Press, New York.]

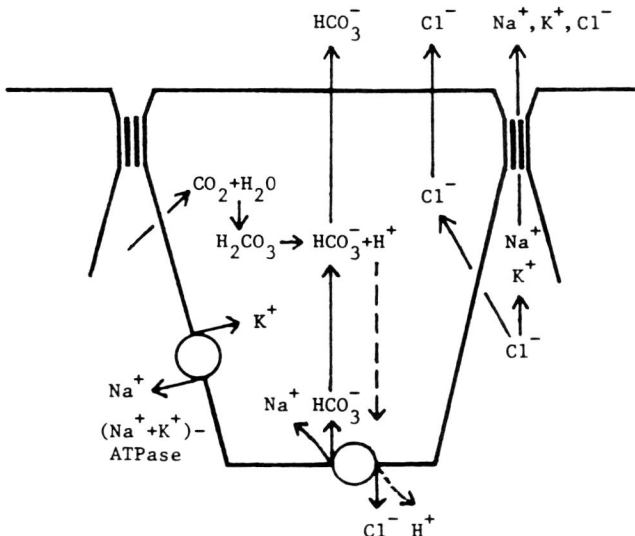

FIG. 17. Cellular model for duct cell electrolyte secretion. Forward (blood-to-cell) transport of bicarbonate is driven by a secondary active transport process (Na^+-$2HCO_3^-$-Cl^- exchange) on basolateral membrane, and bicarbonate leaves the cell down an electrochemical gradient across the luminal membrane. [From Kuijpers et al. (121).]

paracellular current flow (162). However, these results probably indicate an effect of secretin on ion movements across the basolateral membrane of the duct cell. Cultured cells derived from a pancreatic ductal carcinoma also hyperpolarize when stimulated with dibutyryl cAMP (220). At least three possibilities exist to explain these hyperpolarizing responses: 1) influx of bicarbonate ions, 2) efflux of potassium ions, and 3) increased activity of an electrogenic ion pump. The first possibility would require the presence of a conductive bicarbonate permeability on the basolateral membrane, which does not feature in any of the current models. There are no data that allow distinction between the other alternatives. However, an increase in potassium permeability would be consistent with the transient increase in perfusate potassium concentration that occurs after stimulation with secretin (39), and regulated potassium channels occur in the basolateral membranes of other epithelial cells (see Electrophysiology and Stimulus-Secretion Coupling, p. 399) and also in nerve cells (199). On the other hand, electrogenic Na^+-K^+ pumps are a well-documented feature of gland cells (157).

FUTURE PROGRESS

Because of the uncertainties regarding the mechanisms of duct cell electrolyte secretion, this area offers considerable opportunities for research. This is all the more important because the functioning of the ductal epithelium may be affected in diseases such as cystic fibrosis (119) and pancreatitis (25). More biophysical data are required before the cellular models of secretion proposed from whole-gland studies (Figs. 15–17) can be evaluated. The major difficulty with collection of such data has been technical, because the small duct cells are relatively scarce and also inaccessible in the intact gland (see PANCREATIC STRUCTURE, p. 383). Two alternative approaches are possible: 1) the isolation of interlobular ducts in lengths that allow luminal perfusion and electrophysiological studies and 2) the culture of duct cells as monolayers. These techniques have already been used to study electrolyte transport in other epithelia (29, 87, 166, 233), and recently some progress has been made with their application to the pancreas.

In 1980 Githens et al. (79) reported that millimeter lengths of interlobular ducts could be isolated from the rat pancreas with an enzymatic dissociation technique. Unfortunately the preparation shows poor morphological preservation [B. E. Argent and S. Arkle, unpublished observations; (79)], and although these changes appear to be reversed if the ducts are maintained in culture (80), the use of this preparation for physiological studies is questionable. In the same year Fölsch et al. (71) also described a method that involves enzymatic dissociation, followed by centrifugation, for the isolation of duct fragments from the pancreas of copper-deficient rats (see Isolated Ductal Tissue, p. 390). As a starting point for duct isolation this preparation has two advantages. 1) The proportion of duct cells in the gland is increased. 2) The content of potentially harmful digestive enzymes is markedly reduced. The isolated fragments appeared viable because they incorporated amino acids into proteins and also increased their cAMP content when stimulated with secretin (71). However, the fragments contained only 50–200 epithelial cells (71) and were therefore quite short (<50 μm), probably too short for biophysical experiments that would require luminal perfusion.

Recently we have been able to microdissect millimeter lengths of small interlobular ducts (Fig. 18), which show excellent morphological and biochemical preservation, from the glands of copper-deficient rats (6). When these ducts are cultured on polycarbonate filter rafts their ends seal within 8 h (3). Ducts isolated from the pancreas of copper-replete rats also seal in agarose matrix culture (80, 81), but this process takes 2–4 days, presumably reflecting their poor morphological preservation on isolation. In both preparations sealing is accompanied by luminal dilatation, flattening of the epithelium against the surrounding connective tissue layer, and an overall swelling of the duct. Puncturing the ducts causes a reduction in their size but increases the height of the epithelium (3). This confirms that the morphological changes occurring during culture result from fluid secretion into the closed luminal space. The epithelial cells within the cultured ducts retain secretin receptors and increase their cAMP content when exposed to this hormone (3).

Fluid secretion by cultured interlobular ducts has been directly measured using micropuncture techniques [Fig. 19; (3)]. A 14-fold increase in secretion

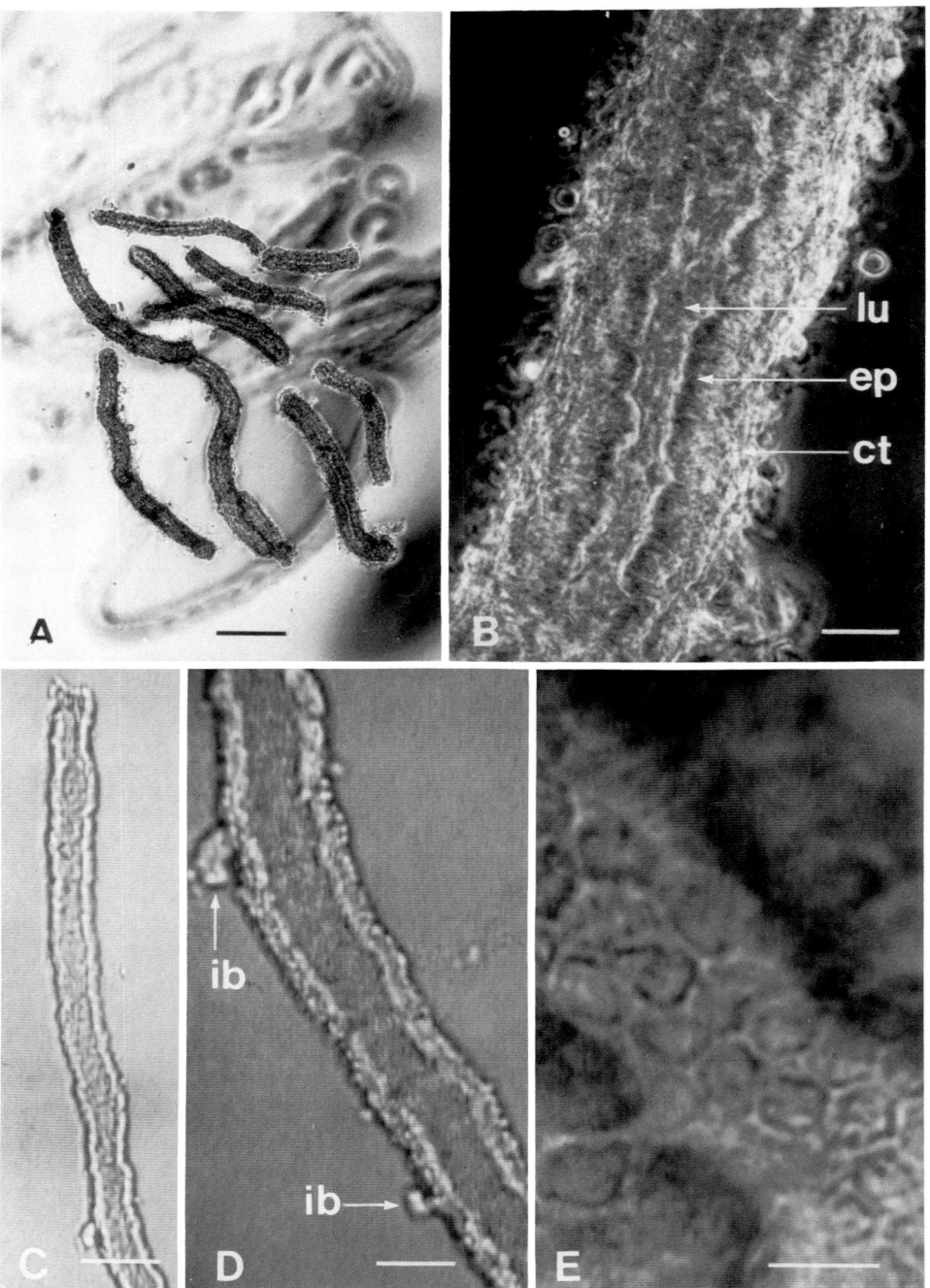

FIG. 18. Small interlobular ducts dissected from pancreas of a copper-deficient rat. *A*: collection of 10 ducts. Phase contrast. Bar, 250 μm. *B*: higher magnification of interlobular duct. Lumen (*lu*), epithelium (*ep*), and connective tissue layer (*ct*) are visible. Phase contrast. Bar, 25 μm. *C*: interlobular duct from which connective tissue layer has been microdissected. This low-magnification view shows that duct was ~1 mm in length. Bar, 100 μm. *D*: small projections visible on *left side* of duct are intralobular branches (*ib*) that have fractured close to their site of origin. Bar, 50 μm. *E*: apical region of individual cells can be seen clearly when the objective lens is focused on luminal membrane of epithelium. Bar, 10 μm. Bright-field optics. Plates *C–E* obtained from a videotape by photographing monitor screen. [From Arkle et al. (6).]

rate was caused by 10^{-8} M secretin, with a hormone concentration of $\sim 2 \times 10^{-11}$ M required for a half-maximal response (Fig. 20). This effect of secretin was abolished when bicarbonate in the perifusion buffer was replaced with *N*-2-hydroxyethylpiperazine-*N'*-2-ethanesulfonic acid (HEPES) and was mimicked by dibutyryl cAMP. These responses are characteristic of ductal electrolyte secretion from the isolated per-

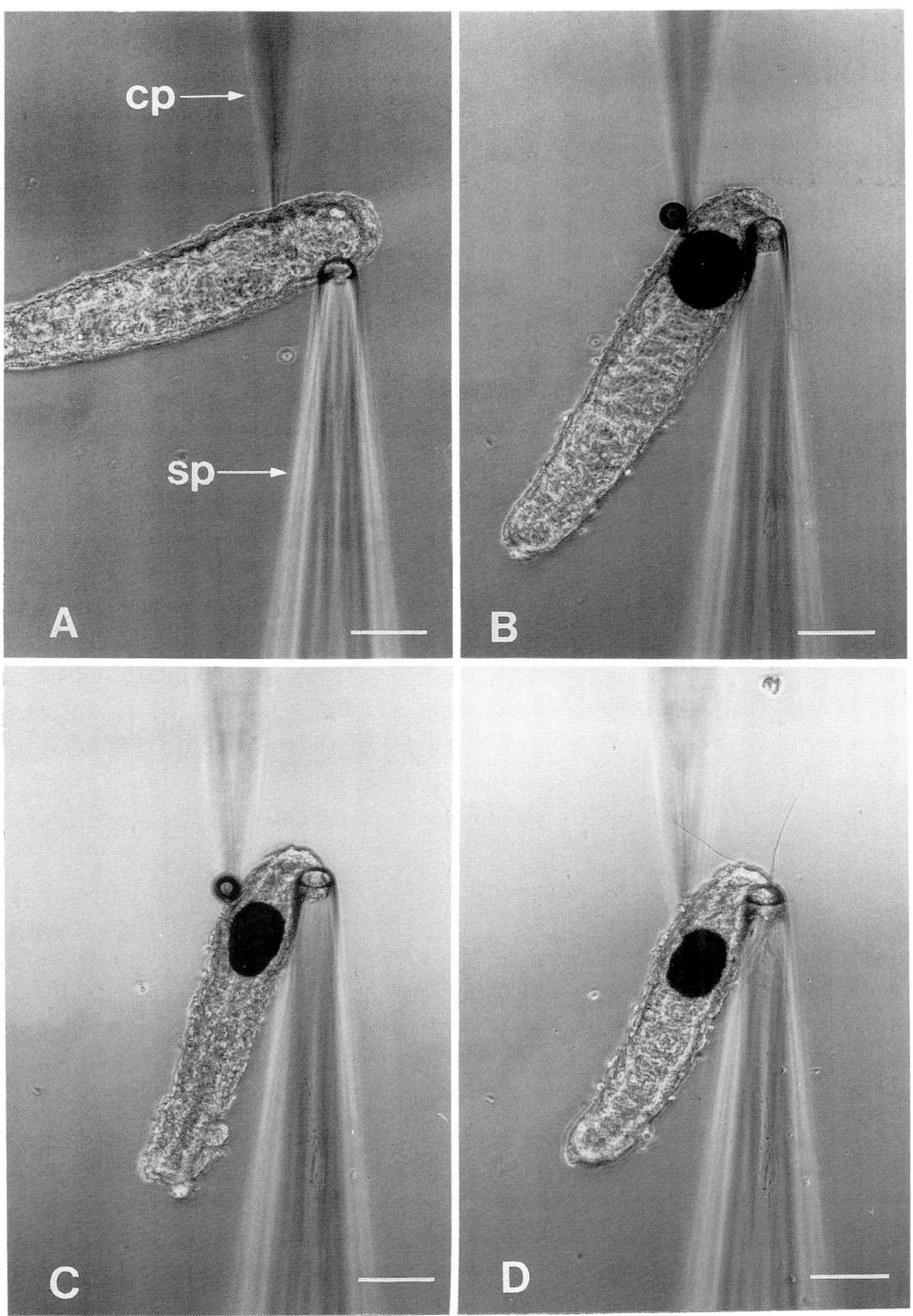

FIG. 19. Measurement of fluid secretion from isolated pancreatic duct using micropuncture techniques. Interlobular duct had been maintained in culture for 24 h, during which time the ends of the duct had sealed. Subsequent accumulation of fluid in closed luminal space has caused a dilatation of the lumen, a flattening of the epithelium against the connective tissue layer, and overall swelling of the duct. *A*: duct is first immobilized by applying a suction pipette (*sp*) to its outer connective tissue layer and then micropunctured using a beveled, oil-filled collection pipette (*cp*). *B*: success is confirmed by injection of a small volume of colored oil into the lumen. *C*: duct is deflated by aspirating luminal fluid into the collection pipette. Pipette is then withdrawn from the lumen, the fluid ejected to waste in the tissue bath, and the duct immediately repunctured along the same entry track. Usually the collection period is then started by application of subatmospheric pressure to the pipette. *D*: if suction is not applied to the collection pipette, secreted fluid accumulates within the closed lumen of the duct, causing it to dilate. Photograph was taken 40 min after *C*, during which time the duct was perifused with Krebs-Ringer bicarbonate buffer at 37°C. Phase contrast. Bars, 200 μm. [From Argent et al. (3).]

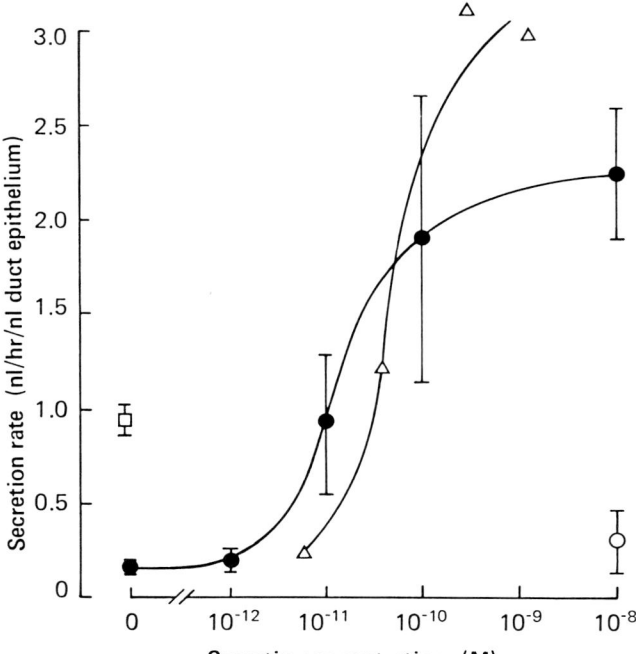

FIG. 20. Effects of secretin, bicarbonate ions, and dibutyryl cAMP on fluid secretion from isolated interlobular ducts that had been maintained in culture for 16–52 h. Individual ducts were micropunctured as shown in Fig. 19 and secreted fluid was collected. At the end of the collection period the dimensions of the epithelium were measured and epithelial volume was calculated. Secretion rates, plotted as mean ± SE, are expressed as $nl \cdot h^{-1} \cdot nl^{-1}$ duct epithelium. ●——●, Dose-response curve for effect of secretin on fluid secretion from isolated ducts. Each point is mean of 4–12 observations on different ducts. △——△, For comparison the dose-response curve for secretin-stimulated fluid secretion from the perfused copper-replete rat pancreas is also shown. Redrawn from data in ref. 113 assuming that 1 clinical unit of secretin = 7.5×10^{-11} mol (129). ○, Effect of replacing bicarbonate ions in perifusion buffer with HEPES on secretin-stimulated fluid secretion from isolated ducts. Mean of 6 observations on different ducts. □, Effect of 2×10^{-4} M dibutyryl cAMP on fluid secretion from isolated ducts. Mean of 7 observations on different ducts. [Data from Argent et al. (3).]

fused rat pancreas (113, 164). It can also be calculated that epithelial cells within isolated ducts secrete fluid at comparable rates to duct cells within the intact gland (3).

Recent electrophysiological studies on isolated ducts microdissected from copper-deficient (3a, 5a, 5b, 81a) and copper-replete rats (145a, 146) provide support for the cellular model of bicarbonate secretion shown in Figure 16. Conventional microelectrode experiments on perfused ducts (145a, 146) have confirmed the presence of an electrogenic Na^+-K^+-ATPase and an amiloride-sensitive Na^+-H^+ antiporter on the basolateral membrane (see Fig. 16) and also identified a barium-sensitive potassium conductance at the same location. Single-channel current recording has shown that this potassium conductance results from the presence of a large (237 pS), calcium-activated, voltage-dependent potassium channel on the basolateral side of the duct cell (3a).

In contrast to previous observations on the mouse pancreas [see CELLULAR MODELS OF DUCT CELL SECRETION, p. 405; (82)], secretin causes a reversible, long-lasting depolarization of epithelial cells in isolated rat pancreatic ducts [Fig. 21; (81a)]. At a hormone concentration of 10^{-8} M the depolarization has a rapid onset, measures 12.2 ± 2 mV (mean ± SE; $n = 11$) and is associated with a 28% ± 5% fall in input resistance. With 10^{-9} M secretin the depolarization measures 6 ± 2 mV ($n = 4$) and has a slower onset. On removal of secretin the membrane potential slowly repolarizes back to the resting level, but about 10 min were required for full recovery (Fig. 21). Often there was an increase in input resistance during the recovery phase (Fig. 21), which may have resulted from electrical uncoupling of the epithelial cells because of swelling of the duct as a consequence of increased fluid secretion into the closed luminal space (see Fig. 21 legend).

The depolarizing action of secretin and the associated fall in input resistance suggest that bicarbonate transport is electrogenic and probably associated with an increase in the ion permeability of the duct cell plasma membrane. These ideas are supported by the recent identification of a secretin-regulated, small-conductance (5 pS) chloride channel on the apical plasma membrane of the duct cell (5a, 81a). The open-state probability of this channel is increased about fourfold after stimulation with secretin (5a, 81a), and similar effects are observed with dibutyryl cAMP and forskolin, suggesting that the channel is regulated by a protein kinase A–mediated phosphorylation event. An increase in apical chloride conductance would explain the depolarizing action of secretin on duct cell membrane potential and also the associated fall in input resistance (Fig. 21). Bicarbonate secretion across the apical membrane could be achieved either by coupling this chloride channel to a chloride-bicarbonate exchanger or by direct movement of bicarbonate or a buffer component through the channel (81a).

Because both the proposed bicarbonate transport mechanisms at the apical membrane are rheogenic, there must be current flow across the basolateral membrane during secretion. This is provided by 1) the electrogenic Na^+-K^+ pump and 2) potassium efflux through the large potassium channel. Because the open-state probability of the potassium channel is voltage dependent, the depolarization that follows secretin stimulation would favor channel opening. However, one problem for this model at the moment is that tetraethylammonium and barium, agents that block Ca^{2+}-activated, voltage-dependent potassium channels (see the chapter in this Handbook by Petersen and Maruyama) do not inhibit secretin-stimulated fluid secretion from the rat pancreas (62a).

Monolayer cultures, with morphological characteristics of duct cells, have been obtained from explants of fetal rat pancreas (228) and from epithelial cells isolated from the main pancreatic duct (182, 208). It should be possible to apply these techniques to epithe-

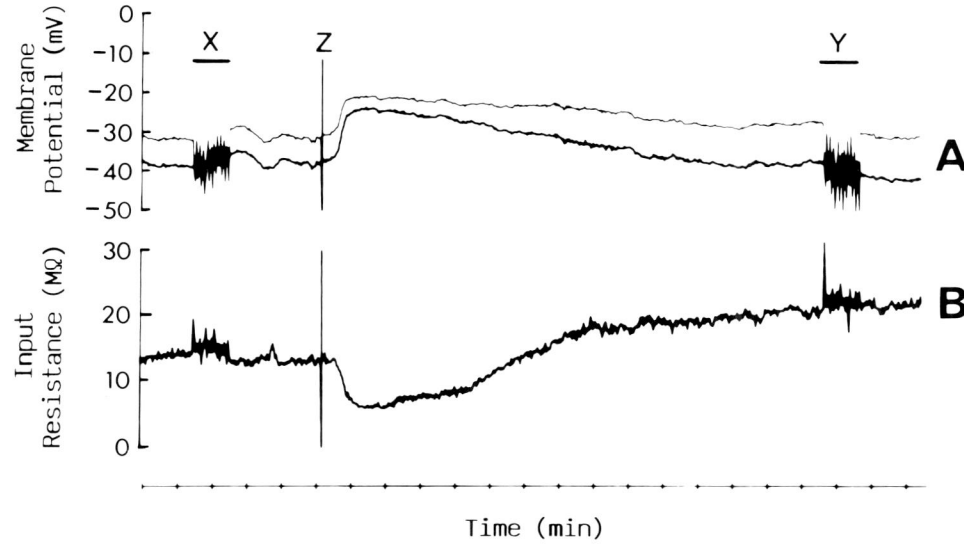

FIG. 21. Effects of secretin on membrane potential and input resistance of rat pancreatic duct cell. An isolated interlobular duct that had been maintained in culture for 24 h was first deflated by sectioning an end. Deflated duct was then placed in a tissue bath and held steady by drawing each end into a micropipette. Because the lumen was not cannulated and drained in these experiments, accumulation of secreted fluid often caused an obvious dilatation of ducts. Bath (volume 1 ml) was perfused with a Krebs-Ringer bicarbonate buffer (3 ml/min) at 37°C. Potassium acetate-filled microelectrode was then used to impale a duct cell across its basolateral membrane. A, lower of 2 traces shows duct cell membrane potential. Every 2 s a depolarizing square-wave current pulse (0.5 nA; 100 ms) was applied to the microelectrode. Upper trace, maximum change in membrane potential associated with these current pulses. X and Y, magnitude of injected current was varied to construct a current-voltage plot. Z, single dose of secretin was injected into the bath to give an instantaneous hormone concentration of 10^{-8} M. B: input resistance calculated from deflections in membrane potential caused by injected current pulses. [From Gray et al. (81a).]

lial cells within the small interlobular ducts that respond to secretin. Recently it has also proved possible to grow epithelial cells from explants of the human fetal pancreas that possess morphological, biochemical, and immunocytochemical characteristics typical of duct cells (89a). Such preparations will be very useful for biophysical, especially Ussing chamber, studies.

We thank Dr. Hiromi Takahasi-Iwanaga and Professor Horst Kern for providing us with the micrographs reproduced in Figures 1, 2, 3, and 4.

We also thank the following colleagues who have allowed us to present unpublished material: Professor Jan Joep de Pont (Fig. 5), Professor John Young and Dr. Francis Seow (Fig. 9), Dr. Andrew Garner and Dr. Philip Padfield (Fig. 10), Dr. Sigrid Rutishauser and Dr. Amir Ali (Fig. 11), and Dr. John Greenwell (Fig. 21).

We thank Sheila Millward for her drawings and Sheilah Long and Irene Littleford for preparing the manuscript.

REFERENCES

1. AKAO, S., D. E. BOCKMAN, P. LECHENE DE LA PORTE, AND H. SARLES. Three-dimensional pattern of ductuloacinar associations in normal and pathological human pancreas. *Gastroenterology* 90: 661–668, 1986.
2. ALI, A. E., R. M. CASE, AND S. C. B. RUTISHAUSER. Factors affecting the secretion of pancreatic and bile secretion in Syrian golden hamster (Abstract). *Digestion* 35: 3, 1986.
2a. AMMAR, E. M., D. HUTSON, AND T. SCRATCHERD. Absence of a relationship between arterial pH and pancreatic bicarbonate secretion in the isolated perfused cat pancreas. *J. Physiol. Lond.* 388: 495–504, 1987.
3. ARGENT, B. E., S. ARKLE, M. J. CULLEN, AND R. GREEN. Morphological, biochemical and secretory studies on rat pancreatic ducts maintained in tissue culture. *Q. J. Exp. Physiol. Cogn. Med. Sci.* 71: 633–648, 1986.
3a. ARGENT, B. E., S. ARKLE, M. A. GRAY, AND J. R. GREENWELL. Two types of calcium sensitive cation channels in isolated rat pancreatic duct cells (Abstract). *J. Physiol. Lond.* 386: 82P, 1987.
4. ARGENT, B. E., R. M. CASE, AND T. SCRATCHERD. Stimulation of amylase secretion from the perfused cat pancreas by potassium and other alkali metal ions. *J. Physiol. Lond.* 216: 611–624, 1971.
5. ARGENT, B. E., R. M. CASE, AND T. SCRATCHERD. Amylase secretion by the perfused cat pancreas in relation to the secretion of calcium and other electrolytes and as influenced by the external ionic environment. *J. Physiol. Lond.* 230: 575–593, 1973.
5a. ARGENT, B. E., M. A. GRAY, AND J. R. GREENWELL. Secretin-regulated anion channel on the apical membrane of rat pancreatic duct cells in vitro (Abstract). *J. Physiol. Lond.* 391: 33P, 1987.
5b. ARGENT, B. E., M. A. GRAY, AND J. R. GREENWELL. Characteristics of a large-conductance anion channel in membrane patches excised from rat pancreatic duct cells in vitro (Abstract). *J. Physiol. Lond.* 394: 146P, 1987.
6. ARKLE, S., C. M. LEE, M. J. CULLEN, AND B. E. ARGENT. Isolation of ducts from the pancreas of copper-deficient rats. *Q. J. Exp. Physiol. Cogn. Med. Sci.* 71: 249–265, 1986.
7. BAXTER, S. G. Continuous pancreatic secretion in the rabbit.

Am. J. Physiol. 96: 343-348, 1931.
8. BAXTER, S. G. Nervous control of the pancreatic secretion in the rabbit. *Am. J. Physiol.* 96: 349-355, 1931.
9. BECKER, V. Histochemistry of the exocrine pancreas. In: *The Exocrine Pancreas*, edited by A. V. S. de Reuck and M. P. Cameron. London: Churchill, 1962, p. 56-63. (Ciba Found. Symp.)
10. BEGLINGER, C., M. I. GROSSMAN, AND T. E. SOLOMON. Interaction between stimulants of exocrine pancreatic secretion in dogs. *Am. J. Physiol.* 246 (*Gastrointest. Liver Physiol.* 9): G173-G179, 1984.
11. BENOS, D. J. Amiloride: a molecular probe of sodium transport in tissues and cells. *Am. J. Physiol.* 242 (*Cell Physiol.* 11): C131-C145, 1982.
12. BENZ, J., B. ECKSTEIN, E. K. MATTHEWS, AND J. A. WILLIAMS. Control of amylase release in vitro: effects of ions, cyclic AMP, and colchicine. *Br. J. Pharmacol.* 46: 66-77, 1972.
13. BERRIDGE, M. J., B. D. LINDLEY, AND W. T. PRINCE. Membrane permeability changes during stimulation of isolated salivary glands of *Calliphora* by 5-hydroxytryptamine. *J. Physiol. Lond.* 244: 549-567, 1975.
14. BIRNBAUM, D., AND F. HOLLANDER. Inhibition of pancreatic secretion by the carbonic anhydrase inhibitor 2-acetylamino-1,3,4-thiadiazol-5-sulfonamide, Diamox (#6063). *Am. J. Physiol.* 174: 191-195, 1953.
15. BLOMFIELD, J., P. J. SETTREE, H. M. ALLARS, AND A. R. RUSH. Ultrastructural changes in the sheep pancreas stimulated in vivo by secretin, cholecystokinin and carbachol. *Exp. Mol. Pathol.* 36: 204-216, 1982.
16. BOCKMAN, D. E. Anastomosing tubular arrangement of the exocrine pancreas. *Am. J. Anat.* 147: 113-118, 1976.
17. BOCKMAN, D. E. Anastomosing tubular arrangement of dog exocrine pancreas. *Cell Tissue Res.* 189: 497-500, 1978.
18. BOCKMAN, D. E. Architecture of normal pancreas as revealed by retrograde injection. *Cell Tissue Res.* 205: 445-451, 1980.
19. BOCKMAN, D. E., W. R. BOYDSTON, AND I. PARSA. Architecture of human pancreas: implications for early changes in pancreatic disease. *Gastroenterology* 83: 55-61, 1983.
20. BOCKMAN, D. E., M. BÜCHLER, AND H. G. BEGER. Structure and function of specialized cilia in the exocrine pancreas. *Int. J. Pancreatol.* 1: 21-28, 1986.
21. BOLENDER, R. P. Sterological analysis of the guinea pig pancreas. I. Analytical model and quantitative description of nonstimulated pancreatic exocrine cells. *J. Cell Biol.* 61: 269-287, 1974.
22. BONTING, S. L., J. J. H. H. M. DE PONT, A. M. M. FLEUREN-JAKOBS, AND J. W. C. M. JANSEN. The reflexion coefficient as a measure of transepithelial permeability in the isolated rabbit pancreas. *J. Physiol. Lond.* 309: 547-555, 1980.
23. BONTING, S. L., J. J. H. H. M. DE PONT, AND J. W. C. M. JANSEN. The role of sodium ions in pancreatic fluid secretion in the rabbit. *J. Physiol. Lond.* 309: 533-546, 1980.
24. BORON, W. F. Transport of H^+ and of ionic weak acids and bases. *J. Membr. Biol.* 72: 1-16, 1983.
25. BROOKS, F. P. *Diseases of the Exocrine Pancreas*. Philadelphia, PA: Saunders, 1980.
26. BROWN, J. C., A. A. HARPER, AND T. SCRATCHERD. Potentiation of secretin stimulation of the pancreas. *J. Physiol. Lond.* 190: 519-530, 1967.
26a. BUANES, T., T. GROTMOL, T. LANDSVERK, AND M. G. RAEDER. Ultrastructure of pancreatic duct cells at secretory rest and during secretin-dependent $NaHCO_3$ secretion. *Acta Physiol. Scand.* 131: 55-62, 1987.
27. BUNDGAARD, M., M. MØLLER, AND J. H. POULSEN. Localisation of sodium pump sites in cat pancreas. *J. Physiol. Lond.* 313: 405-414, 1981.
28. BURCHARDT, B. C., K. SATO, AND E. FRÖMTER. Electrophysiological analysis of bicarbonate permeation across the peritubular cell membrane of rat kidney proximal tubule. I. Basic observations. *Pfluegers Arch.* 401: 34-42, 1984.
29. BURG, M. B. Introduction: background and development of microperfusion technique. *Kidney Int.* 22: 417-424, 1982.
30. CAFLISCH, C. R., S. SOLOMON, AND W. R. GALEY. Exocrine ductal pCO_2 in the rabbit pancreas. *Pfluegers Arch.* 380: 121-125, 1979.
31. CAFLISCH, C. R., S. SOLOMON, AND W. R. GALEY. In situ micropuncture study of pancreatic duct pH. *Am. J. Physiol.* 238 (*Gastrointest. Liver Physiol.* 1): G263-G268, 1980.
32. CASE, R. M. Pancreatic secretion: cellular aspects. In: *Scientific Basis of Gastroenterology*, edited by H. L. Duthie and K. G. Wormsley. London: Churchill Livingstone, 1978, p. 163-198.
33. CASE, R. M. Synthesis, intracellular transport and discharge of exportable protein in the pancreatic acinar cell and other cells. *Biol. Rev. Camb. Philos. Soc.* 53: 211-354, 1978.
34. CASE, R. M., AND B. E. ARGENT. Bicarbonate secretion by pancreatic duct cells: mechanisms and control. In: *The Exocrine Pancreas: Biology, Pathobiology, and Diseases*, edited by V. L. W. Go, J. D. Gardner, F. P. Brooks, E. Lebenthal, E. P. DiMagno, and G. A. Scheele. New York: Raven, 1986, p. 213-243.
35. CASE, R. M., AND B. E. ARGENT. Studies on the pancreas in vivo and during perfusion. In: *Biomembranes: Biological Transport*, edited by S. Fleischer and B. Fleischer. Orlando, FL: Academic, in press.
36. CASE, R. M., M. CHARLTON, P. A. SMITH, AND T. SCRATCHERD. Electrolyte secretory processes in exocrine pancreas and their intracellular control. In: *Biology of Normal and Cancerous Exocrine Pancreatic Cells*, edited by A. Ribet, L. Pradayrol, and C. Susini. Amsterdam: Elsevier/North-Holland, 1980, p. 41-54. (INSERM Symp. 15.)
37. CASE, R. M., A. D. CONIGRAVE, E. J. FAVALORO, I. NOVAK, C. H. THOMPSON, AND J. A. YOUNG. The role of buffer anions and protons in secretion by the rabbit mandibular salivary gland. *J. Physiol. Lond.* 322: 273-286, 1982.
37a. CASE, R. M., A. GARNER, AND P. J. PADFIELD. Pancreatic electrolyte and amylase secretion in the anaesthetized guinea-pig stimulated with either secretin, cholecystokinin octapeptide (CCK-8) or bombesin (Abstract). *J. Physiol. Lond.* 391: 104P, 1987.
38. CASE, R. M., A. A. HARPER, AND T. SCRATCHERD. Water and electrolyte secretion by the perfused pancreas of the cat. *J. Physiol. Lond.* 196: 133-149, 1968.
39. CASE, R. M., A. A. HARPER, AND T. SCRATCHERD. The secretion of electrolytes and enzymes by the pancreas of the anesthetized cat. *J. Physiol. Lond.* 201: 335-348, 1969.
40. CASE, R. M., J. HOTZ, D. HUTSON, T. SCRATCHERD, AND R. D. A. WYNNE. Electrolyte secretion by the isolated cat pancreas during replacement of extracellular bicarbonate by organic anions and chloride by inorganic anions. *J. Physiol. Lond.* 286: 563-576, 1979.
41. CASE, R. M., M. HUNTER, I. NOVAK, AND J. A. YOUNG. The anionic basis of fluid secretion by the rabbit mandibular salivary gland. *J. Physiol. Lond.* 349: 619-630, 1984.
42. CASE, R. M., M. JOHNSON, T. SCRATCHERD, AND H. S. A. SHERRATT. Cyclic adenosine $3',5'$-monophosphate concentration in the pancreas following stimulation by secretin, cholecystokinin-pancreozymin and acetylcholine. *J. Physiol. Lond.* 223: 669-684, 1972.
43. CASE, R. M., AND T. SCRATCHERD. On the permeability of the pancreatic duct membrane. *Biochim. Biophys. Acta* 219: 493-495, 1970.
44. CASE, R. M., AND T. SCRATCHERD. The actions of dibutyryl cyclic adenosine $3',5'$-monophosphate and methyl xanthines on pancreatic exocrine secretion. *J. Physiol. Lond.* 223: 649-667, 1972.
45. CASE, R. M., AND T. SCRATCHERD. The secretion of alkali metal ions by the perfused cat pancreas as influenced by the composition and osmolality of the external environment and by inhibitors of metabolism and Na^+,K^+-ATPase activity. *J. Physiol. Lond.* 242: 415-428, 1974.
46. CASE, R. M., T. SCRATCHERD, AND R. D. WYNNE. The origin

and secretion of pancreatic juice bicarbonate. *J. Physiol. Lond.* 210: 1–15, 1970.
47. CECCARELLI, B., F. CLEMENTE, AND J. MELDOLESI. Secretion of calcium in pancreatic juice. *J. Physiol. Lond.* 245: 617–638, 1975.
48. CHARIOT, J., C. ROZÉ, J. DE LA TOUR, M. SOUCHARD, AND C. VAILLE. Modulation of stimulated pancreatic secretion by sympathomimetic amines in the rat. *Pharmacology* 26: 313–323, 1983.
49. CHARIOT, J., C. ROZÉ, J. DE LA TOUR, AND C. VAILLE. Stimulation vagale non cholinergique de la sécrétion pancréatique externe chez le rat. *Gastroenterol. Clin. Biol.* 6: 371–378, 1982.
50. COMLINE, R. S., L. W. HALL, J. C. D. HICKSON, A. MURILLO, AND R. G. WALKER. Pancreatic secretion in the horse (Abstract). *J. Physiol. Lond.* 204: 10P–11P, 1969.
51. DAVISON, J. S., AND V. DICKSON. Vagal nonadrenergic, noncholinergic (NANC) nerves control pancreatic exocrine secretion in the guinea pig. In: *Secretion: Mechanisms and Control*, edited by R. M. Case, J. M. Lingard, and J. A. Young. Manchester, UK: Manchester Univ. Press, 1984, p. 225–230.
52. DE ALMEIDA, A. L., AND M. I. GROSSMAN. Experimental production of pancreatitis with ethionine. *Gastroenterology* 20: 554–577, 1952.
53. DEBAS, H. T., AND M. I. GROSSMAN. Pure cholecystokinin: pancreatic protein and bicarbonate response. *Digestion* 9: 469–481, 1973.
54. DELISLE, R. C., AND U. HOPFER. Electrolyte permeabilities of pancreatic zymogen granules (Abstract). *J. Cell Biol.* 97: 171a, 1983.
55. DE PONT, J. J. H. H. M., J. W. C. M. JANSEN, G. A. J. KUIJPERS, AND S. L. BONTING. A model for pancreatic fluid secretion. In: *Electrolyte and Water Transport Across Gastrointestinal Epithelia*, edited by R. M. Case, A. Garner, L. A. Turnberg, and J. A. Young. New York: Raven, 1982, p. 11–20.
56. DESCHODT-LANCKMAN, M., P. ROBBERECHT, P. DE NEEF, F. LABRIE, AND J. CHRISTOPHE. In vitro interactions of gastrointestinal hormones on cyclic adenosine 3′,5′-monophosphate levels and amylase output in the rat pancreas. *Gastroenterology* 68: 318–325, 1975.
57. DEWHURST, D. G., N. A. HADI, D. HUTSON, AND T. SCRATCHERD. The permeability of the secretin stimulated exocrine pancreas to non-electrolytes. *J. Physiol. Lond.* 277: 103–114, 1978.
58. DOMSCHKE, S., W. DOMSCHKE, W. RÖSCH, S. J. KONTUREK, E. WÜNSCH, AND L. DEMLING. Bicarbonate and cyclic AMP content of pure human pancreatic juice in response to graded doses of synthetic secretin. *Gastroenterology* 70: 533–536, 1976.
59. DOMSCHKE, S., S. J. KONTUREK, W. DOMSCHKE, A. DEMBIŃSKI, P. THOR, R. KRÖL, AND L. DEMLING. Cyclic-AMP and pancreatic bicarbonate secretion in response to secretin in dogs. *Proc. Soc. Exp. Biol. Med.* 150: 773–779, 1975.
60. DREILING, D. A., H. D. JANOWITZ, AND M. HALPERN. The effect of a carbonic anhydrase inhibitor, Diamox, on human pancreatic secretion. *Gastroenterology* 29: 262–279, 1955.
61. DUFRESNE, N., M.-J. BASTIE, N. VAYSSE, Y. CREACH, E. HOLLANDE, AND A. RIBET. The amiloride sensitive Na^+/H^+ antiport in guinea pig pancreatic acini. Characterization and stimulation by caerulein. *FEBS Lett.* 187: 126–130, 1985.
62. EISENMANN, G. Some elementary factors involved in specific ion permeation. In: *Proc. 23rd Int. Congr. Physiol. Sci., Tokyo*. Amsterdam: Excerpta Med., 1965, p. 489–506.
62a.EVANS, L. A. R., D. PIRANI, D. I. COOK, AND J. A. YOUNG. Intraepithelial current flow in rat pancreatic secretory epithelia. *Pfluegers Arch.* 407: S107–S111, 1986.
63. EVANS, L. A. R., AND J. A. YOUNG. The effect of transport blockers on pancreatic ductal secretion (Abstract). *Proc. Aust. Physiol. Pharmacol. Soc.* 16: 98P, 1985.
64. EXLEY, P. M., C. M. FULLER, AND D. V. GALLACHER. Potassium uptake in the mouse submandibular gland is dependent on chloride and sodium and abolished by piretanide. *J. Physiol. Lond.* 378: 97–108, 1986.
65. FAHRENKRUG, J. Role of VIPergic neurones in pancreatic bicarbonate secretion. *Biomed. Res.* 1: 84–87, 1980.
66. FELL, B. F., L. J. FARMER, C. FARQUHARSON, I. BREMNER, AND D. S. GRACA. Observations on the pancreas of cattle deficient in copper. *J. Comp. Pathol.* 95: 573–590, 1985.
67. FELL, B. F., T. P. KING, AND N. T. DAVIES. Pancreatic atrophy in copper-deficient rats: histochemical and ultrastructural evidence of a selective effect on acinar cells. *Histochem. J.* 14: 665–680, 1982.
68. FINDLAY, I., AND O. H. PETERSEN. Acetylcholine stimulates a Ca^{2+}-dependent Cl^- conductance in mouse lacrimal acinar cells. *Pfluegers Arch.* 403: 328–330, 1985.
69. FLEMSTRÖM, G., AND A. GARNER. Gastroduodenal HCO_3^- transport: characteristics and proposed role in acidity regulation and mucosal protection. *Am. J. Physiol.* 242 (*Gastrointest. Liver Physiol.* 5): G183–G193, 1982.
70. FÖLSCH, U. R., AND W. CREUTZFELDT. Pancreatic duct cells in rats: secretory studies in response to secretin, cholecystokinin-pancreozymin, and gastrin. *Gastroenterology* 73: 1053–1059, 1977.
71. FÖLSCH, U. R., H. FISCHER, H. D. SÖLING, AND W. CREUTZFELDT. Effects of gastrointestinal hormones and carbamylcholine on cAMP accumulation in isolated pancreatic duct fragments from the rat. *Digestion* 20: 277–292, 1980.
71a.FRIZZELL, R. A. Cystic fibrosis: a disease of ion channels? *Trends Neurosci.* 10: 190–193, 1987.
72. FRIZZELL, R. A., M. FIELD, AND S. C. SCHULTZ. Sodium-coupled chloride transport by epithelial tissues. *Am. J. Physiol.* 236 (*Renal Fluid Electrolyte Physiol.* 5): F1–F8, 1979.
73. FRÖMTER, E. Kidney proximal tubule—part of a workshop discussion. In: *Secretion: Mechanisms and Control*, edited by R. M. Case, J. M. Lingard, and J. A. Young. Manchester, UK: Manchester Univ. Press, 1984, p. 37–39.
74. FURUTA, Y., K. HASHIMOTO, AND M. WASHIZAKI. β-Adrenoceptor stimulation of exocrine secretion from the rat pancreas. *Br. J. Pharmacol.* 62: 25–29, 1978.
75. FURUTA, Y., K. IWATSUKI, O. TAKEUCHI, AND K. HASHIMOTO. Secretin-like activity of dopamine on canine pancreatic secretion. *Tohoku J. Exp. Med.* 108: 353–360, 1972.
76. GARDINER, B. N., AND D. M. SMALL. Simultaneous measurement of the pancreatic and biliary response to CCK and secretin. Primate biliary physiology. XIII. *Gastroenterology* 70: 403–407, 1976.
77. GARDNER, J. D., AND R. T. JENSEN. Receptors mediating the actions of secretagogues on pancreatic acinar cells. In: *The Exocrine Pancreas: Biology, Pathobiology, and Diseases*, edited by V. L. W. Go, J. D. Gardner, F. P. Brooks, E. Lebenthal, E. P. DiMagno, and G. A. Scheele. New York: Raven, 1986, p. 109–122.
78. GITHENS, S. Differentiation and development of the exocrine pancreas in animals. In: *The Exocrine Pancreas: Biology, Pathobiology, and Diseases*, edited by V. L. W. Go, J. D. Gardner, F. P. Brooks, E. Lebenthal, E. P. DiMagno, and G. A. Scheele. New York: Raven, 1986, p. 21–32.
79. GITHENS, S., III, D. R. G. HOLMQUIST, J. F. WHELAN, AND J. R. RUBY. Characterization of ducts isolated from the pancreas of the rat. *J. Cell Biol.* 85: 122–135, 1980.
80. GITHENS, S., III, D. R. G. HOLMQUIST, J. F. WHELAN, AND J. R. RUBY. Ducts of the rat pancreas in agarose matrix culture. *In Vitro Rockville* 16: 797–808, 1980.
81. GITHENS, S., III, D. R. G. HOLMQUIST, J. F. WHELAN, AND J. R. RUBY. Morphologic and biochemical characteristics of isolated and cultured pancreatic ducts. *Cancer Phila.* 47: 1505–1512, 1981.
81a.GRAY, M. A., J. R. GREENWELL, AND B. E. ARGENT. Ion channels in pancreatic duct cells: characterization and role in bicarbonate secretion. In: *Cellular and Molecular Basis of Cystic Fibrosis*, edited by G. Mastella and P. M. Quinton. San Francisco, CA: San Francisco Press, 1988, p. 205–221.
82. GREENWELL, J. R. The effects of cholecystokinin-pancreo-

zymin, acetylcholine and secretin on the membrane potentials of mouse pancreatic cells in vitro. *Pfluegers Arch.* 353: 159–170, 1975.
83. GREENWELL, J. R. The selective permeability of the pancreatic duct of the cat to monovalent ions. *Pfluegers Arch.* 367: 265–270, 1977.
84. GREGER, R., E. SCHLATTER, AND H. GOGELEIN. Cl^--channels in the apical cell membrane of the rectal gland "induced" by cAMP. *Pfluegers Arch.* 403: 446–448, 1985.
85. GROSSMAN, M. I., AND A. C. IVY. Effect of alloxan upon secretion of the pancreas. *Proc. Soc. Exp. Biol. Med.* 63: 62–63, 1946.
86. GYR, K., R. H. WOLF, AND O. FELSENFELD. Exocrine pancreatic function tests in patas monkeys (*Erythrocebus patas*). *Am. J. Vet. Res.* 35: 1361–1364, 1974.
87. HANDLER, J. S. Transport in cultured epithelia. *Curr. Eye Res.* 4: 317–322, 1985.
88. HARADA, E., K. NAKAGAWA, AND S. KATO. Characteristic secretory response of the exocrine pancreas in various mammalian and avian species. *Comp. Biochem. Physiol.* 73A: 447–453, 1982.
89. HARPER, S. L., V. H. PITTS, D. N. GRANGER, AND P. R. KVIETYS. Pancreatic tissue oxygenation during secretory stimulation. *Am. J. Physiol.* 250 (*Gastrointest. Liver Physiol.* 13): G316–G322, 1986.
89a. HARRIS, A., AND L. COLEMAN. Establishment of a tissue culture system for epithelial cells derived from human pancreas: a model for the study of cystic fibrosis. *J. Cell Sci.* 87: 695–703, 1987.
90. HART, W. M., AND J. E. THOMAS. Bicarbonate and chloride of pancreatic juice secreted in response to various stimuli. *Gastroenterology* 4: 409–420, 1945.
91. HELGESON, A. S., P. POUR, T. LAWSON, AND C. J. GRANDJEAN. Exocrine pancreatic secretion in the Syrian golden hamster *Mesocricetus auratus*. I. Basic values. *Comp. Biochem. Physiol. A Comp. Physiol.* 66: 473–477, 1980.
92. HELGESON, A. S., P. POUR, T. LAWSON, AND C. J. GRANDJEAN. Exocrine pancreatic secretion in the Syrian golden hamster *Mesocricetus auratus*. II. Effect of secretin and pancreozymin. *Comp. Biochem. Physiol. A Comp. Physiol.* 66: 479–483, 1980.
93. HELLMESSEN, W., A. L. CHRISTIAN, H. FASOLD, AND I. SCHULZ. Coupled Na^+-H^+ exchange in isolated acinar cells from rat exocrine pancreas. *Am. J. Physiol.* 249 (*Gastrointest. Liver Physiol.* 12): G125–G136, 1985.
94. HICKSON, J. C. D. The secretion of pancreatic juice in response to stimulation of the vagus nerves in the pig. *J. Physiol. Lond.* 206: 275–297, 1970.
95. HOLIAN, O., P. E. DONAHUE, L. M. NYHUS, AND C. T. BOMBECK. Effect of hormones on cyclic AMP in dog pancreatic tissue and secretion. *J. Surg. Res.* 32: 51–56, 1982.
96. HOLLANDER, F., AND D. BIRNBAUM. The role of carbonic anhydrase in pancreatic secretion. *Trans. NY Acad. Sci.* 15: 56–58, 1952.
97. HOLST, J. J., J. FAHRENKRUG, S. KNUHTSEN, S. L. JENSEN, S. S. POULSEN, AND O. V. NIELSEN. Vasoactive intestinal polypeptide (VIP) in the pig pancreas: role of VIPergic nerves in control of fluid and bicarbonate secretion. *Regul. Pept.* 8: 245–259, 1984.
98. HOLST, J. J., S. L. JENSEN, O. V. NIELSEN, AND T. W. SCHWARTZ. Oxygen supply, oxygen consumption, and endocrine and exocrine secretions of the isolated, perfused, porcine pancreas. *Acta Physiol. Scand.* 109: 7–13, 1980.
99. HOLST, J. J., O. B. SCHAFFALITZKY DE MUCKADELL, AND J. FAHRENKRUG. Nervous control of pancreatic exocrine secretion in pigs. *Acta Physiol. Scand.* 105: 33–51, 1979.
100. HOOTMAN, S. R., S. A. ERNST, AND J. A. WILLIAMS. Secretagogue regulation of Na^+-K^+ pump activity in pancreatic acinar cells. *Am. J. Physiol.* 245 (*Gastrointest. Liver Physiol.* 8): G339–G346, 1983.
101. HOOTMAN, S. R., D. L. OCHS, AND J. A. WILLIAMS. Intracellular mediators of Na^+-K^+ pump activity in guinea pig pancreatic acinar cells. *Am. J. Physiol.* 249 (*Gastrointest. Liver Physiol.* 12): G470–G478, 1985.
102. HOTZ, J., M. ZWICKER, H. MINNE, AND R. ZIEGLER. Pancreatic enzyme secretion in the conscious rat. *Pfluegers Arch.* 353: 171–189, 1975.
103. HUNTER, M., P. A. SMITH, AND R. M. CASE. The dependence of fluid secretion by mandibular salivary gland and pancreas on extracellular calcium. *Cell Calcium* 4: 307–317, 1983.
104. JANSEN, J. W. C. M., J. J. H. H. M. DE PONT, AND S. L. BONTING. Transepithelial permeability in the rabbit pancreas. *Biochim. Biophys. Acta* 551: 95–108, 1979.
105. JANSEN, J. W. C. M., A. M. M. FLEUREN-JAKOBS, J. J. H. H. M. DE PONT, AND S. L. BONTING. Blocking by 2,4,6-triaminopyrimidine of increased tight junction permeability induced by acetylcholine in the pancreas. *Biochim. Biophys. Acta* 598: 115–126, 1980.
106. JENSEN, S. L., J. FAHRENKRUG, J. J. HOLST, C. KÜHL, O. V. NIELSEN, AND O. B. SCHAFFALITZKY DE MUCKADELL. Secretory effects of secretin on isolated perfused porcine pancreas. *Am. J. Physiol.* 235 (*Endocrinol. Metab. Gastrointest. Physiol.* 4): E381–E386, 1978.
107. JENSEN, S. L., J. FAHRENKRUG, J. J. HOLST, O. V. NIELSEN, AND O. B. SCHAFFALITZKY DE MUCKADELL. Secretory effects of VIP on isolated perfused porcine pancreas. *Am. J. Physiol.* 235 (*Endocrinol. Metab. Gastrointest. Physiol.* 4): E387–E391, 1978.
108. JENSEN, S. L., J. F. REHFELD, J. J. HOLST, O. V. NIELSEN, J. FAHRENKRUG, AND O. B. SCHAFFALITZKY DE MUCKADELL. Secretory effects of cholecystokinins on the isolated perfused porcine pancreas. *Acta Physiol. Scand.* 111: 225–231, 1981.
109. KALSER, M. H., AND M. I. GROSSMAN. Pancreatic secretion in dogs with ethionine-induced pancreatitis. *Gastroenterology* 26: 189–197, 1954.
110. KANNO, T., AND A. SAITO. Influence of external potassium concentration on secretory responses to cholecystokinin-pancreozymin and ionophore A23187 in the pancreatic acinar cell. *J. Physiol. Lond.* 278: 251–263, 1978.
111. KANNO, T., AND A. SAITO. A comparison of secretory actions of VIP, secretin and CCK-PZ in the isolated and perfused kitten pancreas. *Endocrinol. Jpn.* 27, Suppl.: 51–57, 1980.
112. KANNO, T., I. SHIBUYA, AND N. ASADA. The role of extracellular calcium and other ions in the secretory responses of the exocrine pancreas to cholecystokinin. *Biomed. Res.* 4: 295–302, 1983.
113. KANNO, T., AND M. YAMAMOTO. Differentiation between the calcium-dependent effects of cholecystokinin-pancreozymin and the bicarbonate-dependent effects of secretin in exocrine secretion of the rat pancreas. *J. Physiol. Lond.* 264: 787–799, 1977.
114. KEMPEN, H. J. M., J. J. H. H. M. DE PONT, AND S. L. BONTING. Rat pancreas adenylate cyclase. III. Its role in pancreatic secretion assessed by means of cholera toxin. *Biochim. Biophys. Acta* 392: 276–287, 1975.
115. KEMPEN, H. J. M., J. J. H. H. M. DE PONT, AND S. L. BONTING. Pat pancreas adenylate cyclase. IV. Effect of hormones and other agents on cyclic AMP level and enzyme release. *Biochim. Biophys. Acta* 496: 65–76, 1977.
116. KEMPEN, H. J. M., J. J. H. H. M. DE PONT, AND S. L. BONTING. Rat pancreas adenylate cyclase. V. Its presence in isolated rat pancreatic acinar cells. *Biochim. Biophys. Acta* 496: 521–531, 1977.
117. KERN, H. F. Fine structure of the human exocrine pancreas. In: *The Exocrine Pancreas: Biology, Pathobiology, and Diseases*, edited by V. L. W. Go, J. D. Gardner, F. P. Brooks, E. Lebenthal, E. P. DiMagno, and G. A. Scheele. New York: Raven, 1986, p. 9–19.
118. KODAMA, T. A light and electron microscopic study on the pancreatic ductal system. *Acta Pathol. Jpn.* 33: 297–321, 1983.
119. KOPELMAN, H., P. DURIE, K. GASKIN, Z. WEIZMAN, AND G. FORSTNER. Pancreatic fluid secretion and protein hypercon-

centration in cystic fibrosis. *N. Engl. J. Med.* 312: 329–334, 1985.
120. KUIJPERS, G. A. J., I. G. P. VAN NOOY, J. J. H. H. M. DE PONT, AND S. L. BONTING. Anion secretion by the isolated rabbit pancreas. *Biochim. Biophys. Acta* 774: 269–276, 1984.
121. KUIJPERS, G. A. J., I. G. P. VAN NOOY, J. J. H. H. M. DE PONT, AND S. L. BONTING. The mechanism of fluid secretion in the rabbit pancreas studied by means of various inhibitors. *Biochim. Biophys. Acta* 778: 324–331, 1984.
122. KUMPULAINEN, T., AND P. JALOVAARA. Immunohistochemical localization of carbonic anhydrase isoenzymes in the human pancreas. *Gastroenterology* 80: 796–799, 1981.
123. LAVAL, J., G. BOMMELAER, C. SENARENS, N. VAYSSE, AND A. RIBET. Control of cyclic AMP accumulation in dog acini: its relation to amylase release. *Digestion* 28: 82–89, 1983.
124. LEBENTHAL, E., R. LEV, AND P. C. LEE. Prenatal and postnatal development of the human exocrine pancreas. In: *The Exocrine Pancreas: Biology, Pathobiology, and Diseases*, edited by V. L. W. Go, J. D. Gardner, F. P. Brooks, E. Lebenthal, E. P. DiMagno, and G. A. Scheele. New York: Raven, 1986, p. 33–43.
125. LENNINGER, S. Effects of parasympathomimetic agents and vagal stimulation on the flow in the pancreatic duct of the cat. *Acta Physiol. Scand.* 82: 345–353, 1971.
126. LENNINGER, S., AND P. OHLIN. The flow of juice from the pancreatic gland of the cat in response to vagal stimulation. *J. Physiol. Lond.* 216: 303–318, 1971.
127. LIGHTWOOD, R., AND H. A. REBER. Micropuncture study of pancreatic secretion in the cat. *Gastroenterology* 72: 61–66, 1977.
128. LIM, R. K. S., S. M. LING, A. C. LIU, AND I. C. YUAN. Quantitative relationships between the basic and other components of the pancreatic secretion. *Chin. J. Physiol.* 10: 475–492, 1936.
129. LINDGARD, J. M., AND J. A. YOUNG. β-Adrenergic control of exocrine secretion by perfused rat pancreas in vitro. *Am. J. Physiol.* 245 (*Gastrointest. Liver Physiol.* 8): G690–G696, 1983.
130. LOEWENSTEIN, W. R. Junctional intracellular communication: the cell-to-cell membrane channel. *Physiol. Rev.* 61: 829–913, 1981.
131. MAGEE, D. F. An investigation into the external secretion of the pancreas in sheep. *J. Physiol. Lond.* 158: 132–143, 1961.
132. MAKHLOUF, G. M., AND A. L. BLUM. An assessment of models for pancreatic secretion. *Gastroenterology* 59: 896–908, 1970.
133. MANGOS, J. A., AND N. R. MCSHERRY. Micropuncture study of excretion of water and electrolytes by the pancreas. *Am. J. Physiol.* 221: 496–503, 1971.
134. MANGOS, J. A., N. R. MCSHERRY, S. NOUSIA-ARVANITAKIS, AND K. IRWIN. Secretion and transductal fluxes of ions in exocrine glands of the mouse. *Am. J. Physiol.* 225: 18–24, 1973.
135. MANGOS, J. A., N. R. MCSHERRY, S. NOUSIA-ARVANITAKIS, AND R. F. SCHILLING. Transductal fluxes of anions in the rat pancreas. *Proc. Soc. Exp. Biol. Med.* 146: 321–328, 1974.
136. MANN, F. C., J. P. FOSTER, AND S. D. BRIMHALL. The relation of the common bile duct to the pancreatic duct in common domestic and laboratory animals. *J. Lab. Clin. Med.* 5: 203–206, 1920.
137. MAREN, T. H. Discussion following a paper by J. R. Martinez and N. Cassity. In: *Secretion: Mechanisms and Control*, edited by R. M. Case, J. M. Lindgard, and J. A. Young. Manchester, UK: Manchester Univ. Press, 1984, p. 93.
137a. MARTY, A. Control of ionic currents and fluid secretion by muscarinic agonists in exocrine glands. *Trends Neurosci.* 10: 373–377, 1987.
138. MARTY, A., Y. P. TAN, AND A. TRAUTMANN. Three types of calcium-dependent channel in rat lacrimal glands. *J. Physiol. Lond.* 357: 293–325, 1984.
139. MARUYAMA, Y., O. H. PETERSEN, P. FLANAGAN, AND G. T. PEARSON. Quantification of Ca^{2+}-activated K^+ channels under hormonal control in pig pancreas acinar cells. *Nature Lond.* 305: 228–232, 1983.

140. MATSUMOTO, T., AND T. KANNO. Potentiation of cholecystokinin-induced exocrine secretion by either electrical stimulation of the vagus nerve or exogenous VIP administration in the guinea pig pancreas. *Peptides Fayetteville* 5: 285–289, 1984.
141. MCALLISTER, R. M. R., AND M. D. KENDALL. The nerves of the accessory pancreatic ducts of the common starling (*Sturnus vulgaris*): an ultrastructural and light microscopic study. *J. Anat.* 139: 437–447, 1984.
142. MÉLÈSE, T., S. S. ROTHMAN. Distribution of three hexose derivatives across the pancreatic epithelium: paracellular shunts or cellular passage? *Biochim. Biophys. Acta* 763: 212–219, 1983.
143. MÉLÈSE, T., AND S. S. ROTHMAN. Pancreatic epithelium is permeable to sucrose and inulin across secretory cells. *Proc. Natl. Acad. Sci. USA* 80: 4870–4874, 1983.
143a. MOQTADERI, F., H. S. HIMAL, J. RUDICK, AND D. A. DREILING. Pancreatic transductal electrolyte flux. *Am. J. Gastroenterol.* 58: 177–184, 1972.
144. MORI, J., H. SATOH, Y. SATOH, AND F. HONDA. Amines and the rat exocrine pancreas. 3. Effects of amines on pancreatic secretion. *Jpn. J. Pharmacol.* 29: 923–933, 1979.
145. MORRIS, A. P., D. V. GALLACHER, R. F. IRVINE, AND O. H. PETERSEN. Synergism of inositol trisphosphate and tetrakisphosphate in activating Ca^{2+}-dependent K^+ channels. *Nature Lond.* 330: 653–655, 1987.
145a. NOVAK, I., AND R. GREGER. Cellular mechanisms of bicarbonate transport in isolated perfused pancreatic ducts (Abstract). *Acta Physiol. Scand.* 129: 13A, 1987.
146. NOVAK, I., E. SCHLATTER, AND R. GREGER. Electrophysiological study of HCO_3^- transport in isolated perfused small ducts of rat pancreas (Abstract). *Digestion* 35: 44, 1986.
147. OATES, P. S., AND R. G. H. MORGAN. Pancreatic response of anaesthetized and conscious rats to bolus injection of cholecystokinin-pancreozymin. *Aust. J. Biol. Sci.* 34: 283–293, 1981.
148. OCHS, D. L., J. I. KORENBROT, AND J. A. WILLIAMS. Intracellular free calcium concentrations in isolated pancreatic acini: effects of secretagogues. *Biochem. Biophys. Res. Commun.* 117: 122–128, 1983.
149. O'DOHERTY, J., AND R. J. STARK. Stimulation of pancreatic acinar secretion: increases in cytosolic calcium and sodium. *Am. J. Physiol.* 242 (*Gastrointest. Liver Physiol.* 5): G513–G521, 1982.
150. O'DOHERTY, J., AND R. J. STARK. A transcellular route for Na-coupled Cl transport in secreting pancreatic acinar cells. *Am. J. Physiol.* 245 (*Gastrointest. Liver Physiol.* 8): G499–G503, 1983.
151. O'LEARY, J. F., J. W. BORNER, W. J. RUNGE, L. P. DEHNER, AND R. L. GOODALE. Hyperplasia of pancreatic duct epithelium produced by exposure to sodium deoxycholate. *Am. J. Surg.* 147: 72–77, 1984.
152. PEARSON, G. T., P. M. FLANAGAN, AND O. H. PETERSEN. Neural and hormonal control of membrane conductance in the pig pancreatic acinar cell. *Am. J. Physiol.* 247 (*Gastrointest. Liver Physiol.* 10): G520–G526, 1984.
153. PEARSON, G. T., J. SINGH, M. S. DAOUD, J. S. DAVISON, AND O. H. PETERSEN. Control of pancreatic cyclic nucleotide levels and amylase secretion by noncholinergic, nonadrenergic nerves. A study employing electrical field stimulation of guinea pig segments. *J. Biol. Chem.* 256: 11025–11031, 1981.
154. PEARSON, G. T., J. SINGH, AND O. H. PETERSEN. Adrenergic nervous control of cAMP-mediated amylase secretion in the rat pancreas. *Am. J. Physiol.* 246 (*Gastrointest. Liver Physiol.* 9): G563–G573, 1984.
155. PERLMUTTER, J., AND J. R. MARTINEZ. The chronically reserpinized rat as a possible model for cystic fibrosis. VII. Alterations in the secretory response to cholecystokinin and to secretin from pancreas in vivo. *Pediatr. Res.* 12: 188–194, 1978.
156. PETERSEN, H., AND M. I. GROSSMAN. Pancreatic exocrine secretion in anesthetized and conscious rats. *Am. J. Physiol.* 233 (*Endocrinol. Metab. Gastrointest. Physiol.* 2): E530–E536, 1977.

157. PETERSEN, O. H. *The Electrophysiology of Gland Cells*. London: Academic, 1980.
158. PETERSEN, O. H. Calcium-activated potassium channels and fluid secretion by exocrine glands. *Am. J. Physiol.* 251 (*Gastrointest. Liver Physiol.* 14): G1–G13, 1986.
159. PETERSEN, O. H. Electrophysiology of acinar cells. In: *The Exocrine Pancreas: Biology, Pathobiology, and Diseases*, edited by V. L. W. Go, J. D. Gardner, F. P. Brooks, E. Lebenthal, E. P. DiMagno, and G. A. Scheele. New York: Raven, 1986, p. 141–161.
160. PETERSEN, O. H., I. FINDLAY, N. IWATSUKI, J. SINGH, D. V. GALLACHER, C. M. FULLER, G. T. PEARSON, M. J. DUNNE, AND A. P. MORRIS. Human pancreatic acinar cells: studies of stimulus secretion coupling. *Gastroenterology* 89: 109–117, 1985.
161. PETERSEN, O. H., AND Y. MARUYAMA. Calcium-activated potassium channels and their role in secretion. *Nature Lond.* 307: 693–696, 1984.
162. PETERSEN, O. H., Y. MARUYAMA, J. GRAF, R. LAUGIER, A. NISHIYAMA, AND G. T. PEARSON. Ionic currents across pancreatic acinar cell membranes and their role in fluid secretion. *Phil. Trans. R. Soc. Lond. B Biol. Sci.* 296: 151–166, 1981.
163. PETERSEN, O. H., AND J. SINGH. Acetylcholine-evoked potassium release in the mouse pancreas. *J. Physiol. Lond.* 365: 319–329, 1985.
164. PETERSEN, O. H., AND N. UEDA. Secretion of fluid and amylase in the perfused rat pancreas. *J. Physiol. Lond.* 264: 819–835, 1977.
165. POULSEN, J. H., AND B. OAKLEY. Intracellular potassium ion activity in resting and stimulated mouse pancreas and submandibular gland. *Proc. R. Soc. Lond. B Biol. Sci.* 204: 99–104, 1979.
166. QUINTON, P. M. Chloride impermeability in cystic fibrosis. *Nature Lond.* 301: 421–422, 1983.
167. RAEDER, M., AND Ø. MATHISEN. Plasma Na^+-ion concentration or pH as regulator of pancreatic HCO_3^- secretion. *Acta Physiol. Scand.* 112: 19–26, 1981.
168. RAEDER, M., AND Ø. MATHISEN. Abolished relationship between pancreatic HCO_3^- secretion and arterial pH during carbonic anhydrase inhibition. *Acta Physiol. Scand.* 114: 97–102, 1982.
169. RAEDER, M., A. MO, AND S. AUNE. Effect of plasma H^+-ion concentration on pancreatic HCO_3^- secretion. *Acta Physiol. Scand.* 105: 420–427, 1979.
170. RAEDER, M., A. MO, S. AUNE, AND Ø. MATHISEN. Relationship between plasma pH and pancreatic HCO_3^- secretion at different intravenous secretin infusion rates. *Acta Physiol. Scand.* 109: 187–191, 1980.
171. REBER, H. A., G. ADLER, AND K. R. WEDGWOOD. Studies in the perfused pancreatic duct in the cat. In: *The Exocrine Pancreas: Biology, Pathobiology, and Diseases*, edited by V. L. W. Go, J. D. Gardner, F. P. Brooks, E. Lebenthal, E. P. DiMagno, and G. A. Scheele. New York: Raven, 1986, p. 255–273.
172. REBER, H. A., AND R. LIGHTWOOD. Microcannulation—a new micropuncture technique application in cat pancreas secretion. *J. Appl. Physiol.* 40: 984–986, 1976.
173. REBER, H. A., AND C. J. WOLF. Micropuncture study of pancreatic electrolyte secretion. *Am. J. Physiol.* 215: 34–40, 1968.
174. REBER, H. A., C. J. WOLF, AND S. P. LEE. Role of the main duct in pancreatic electrolyte secretion. *Surg. Forum* 20: 382–384, 1969.
175. REYNOLDS, J., AND T. HEATH. Non-parallel secretion of pancreatic enzymes in sheep following hormonal or vagal stimulation. *Comp. Biochem. Physiol. A Comp. Physiol.* 68: 495–500, 1981.
176. RIDDERSTAP, A. S. The additive effect of acetazolamide and ouabain on pancreatic secretion in vitro. *Pfluegers Arch.* 311: 199–204, 1969.
177. RIDDERSTAP, A. S., AND S. L. BONTING. Enzyme secretion by the isolated rabbit pancreas: absence of a relation with the Na-K activated ATPase. *Pfluegers Arch.* 313: 53–61, 1969.
178. RIDDERSTAP, A. S., AND S. L. BONTING. Na-K-activated adenosine triphosphatase and pancreatic secretion in the dog. *Am. J. Physiol.* 216: 547–553, 1969.
179. ROBBERECHT, P., M. DESCHODT-LANCKMAN, P. DE NEEF, P. BORGEAT, AND J. CHRISTOPHE. In vivo effects of pancreozymin, secretin, vasoactive intestinal polypeptide and pilocarpine on the levels of cyclic AMP and cyclic GMP in the rat pancreas. *FEBS Lett.* 43: 139–143, 1974.
180. ROTHMAN, S. S., AND F. P. BROOKS. Electrolyte secretion from rabbit pancreas in vitro. *Am. J. Physiol.* 208: 1171–1176, 1965.
181. ROTHMAN, S. S., AND F. P. BROOKS. Pancreatic secretion in vitro in "Cl^--free," "CO_2-free," and low-Na^+ environment. *Am. J. Physiol.* 209: 790–796, 1965.
182. SATO, T., M. SATO, E. A. HUDSON, AND R. T. JONES. Characterization of bovine pancreatic ductal cells isolated by a perfusion-digestion technique. *In Vitro Rockville* 19: 651–660, 1983.
183. SCHREURS, V. V. A. M., H. G. P. SWARTS, J. J. H. H. M. DE PONT, AND S. L. BONTING. Role of calcium in exocrine pancreatic secretion. I. Calcium movements in the rabbit pancreas. *Biochim. Biophys. Acta* 404: 257–267, 1975.
184. SCHREURS, V. V. A. M., H. G. P. SWARTS, J. J. H. H. M. DE PONT, AND S. L. BONTING. Role of calcium in exocrine pancreatic secretion. III. Comparison of calcium and magnesium movements in rabbit pancreas. *Biochim. Biophys. Acta* 436: 664–674, 1976.
185. SCHULZ, I. Influence of bicarbonate-CO_2- and glycodiazine buffer on the secretion of the isolated cat's pancreas. *Pfluegers Arch.* 329: 283–306, 1971.
186. SCHULZ, I. Bicarbonate transport in the exocrine pancreas. *Ann. NY Acad. Sci.* 341: 191–209, 1980.
187. SCHULZ, I. Pancreas—part of a workshop discussion. In: *Secretion: Mechanisms and Control*, edited by R. M. Case, J. M. Lingard, and J. A. Young. Manchester, UK: Manchester Univ. Press, 1984, p. 34–36.
188. SCHULZ, I., K. HEIL, A. KRIBBEN, G. SACHS, AND W. HAASE. Isolation and functional characterisation of cells from the exocrine pancreas. In: *Biology of Normal and Cancerous Exocrine Pancreatic Cells*, edited by A. Ribet, L. Pradayrol, and C. Susini. Amsterdam: Elsevier/North-Holland, 1980, p. 3–18. (INSERM Symp. 15.)
189. SCHULZ, I., F. STRÖVER, AND K. J. ULLRICH. Lipid soluble weak organic acid buffers as "substrate" for pancreatic secretion. *Pfluegers Arch.* 323: 121–140, 1971.
190. SCHULZ, I., AND D. TERREROS-ARANGUREN. Sidedness of transport steps involved in pancreatic HCO_3^- secretion. In: *Electrolyte and Water Transport Across Gastrointestinal Epithelia*, edited by R. M. Case, A. Garner, L. A. Turnberg, and J. A. Young. New York: Raven, 1982, p. 143–156.
191. SCHULZ, I., A. YAMAGATA, AND M. WESKE. Micropuncture studies on the pancreas of the rabbit. *Pfluegers Arch.* 308: 277–290, 1969.
192. SCRATCHERD, T. The isolated perfused pancreas. In: *The Exocrine Pancreas: Biology, Pathobiology, and Disease*, edited by V. L. W. Go, J. D. Gardner, F. P. Brooks, E. Lebenthal, E. P. DiMagno, and G. A. Scheele. New York: Raven, 1986, p. 245–253.
193. SCRATCHERD, T., AND D. HUTSON. The role of chloride in pancreatic secretion. In: *Electrolyte and Water Transport Across Gastrointestinal Epithelia*, edited by R. M. Case, A. Garner, L. A. Turnberg, and J. A. Young. New York: Raven, 1982, p. 61–72.
194. SCRATCHERD, T., D. HUTSON, AND R. M. CASE. Ionic transport mechanisms underlying fluid secretion by the pancreas. *Phil. Trans. R. Soc. Lond. B Biol. Sci.* 296: 167–178, 1981.
195. SEOW, K. T. F. *Pancreatic Fluid and Electrolyte Secretion*. Sydney, Australia: Univ. of Sydney, 1986. PhD thesis.
196. SEOW, K. T. F., J. M. LINGARD, AND J. A. YOUNG. Anionic basis of fluid secretion by rat pancreatic acini in vitro. *Am. J. Physiol.* 250 (*Gastrointest. Liver Physiol.* 13): G140–G148, 1986.
196a. SEOW, F., AND J. A. YOUNG. Cholecystokinin evokes secretion

of a bicarbonate-rich juice by the rabbit pancreas (Abstract). *Proc. Aust. Physiol. Pharmacol. Soc.* 17: 199P, 1986.
197. SEWELL, W. A., AND J. A. YOUNG. Secretion of electrolytes by the pancreas of the anaesthetized rat. *J. Physiol. Lond.* 252: 379–396, 1975.
198. SEWELL, W. A., AND J. A. YOUNG. The effect of cycloheximide on cholecystokinin-evoked pancreatic juice of the anaesthetised rat. *Aust. J. Exp. Biol. Med. Sci.* 56: 385–394, 1978.
199. SIEGELBAUM, S. A., AND R. W. TSIEN. Modulation of gated ion channels as a mode of transmitter action. *Trends Neurosci.* 6: 307–313, 1983.
200. SINGH, J. Effects of acetylcholine and caerulein on $^{86}Rb^+$ efflux in the mouse pancreas. Evidence for a sodium-potassium-chloride cotransport system. *Biochim. Biophys. Acta* 775: 77–85, 1984.
201. SMITH, P. A., AND R. M. CASE. Effects of cholera toxin on cyclic adenosine 3′,5′-monophosphate concentration and secretory processes in the exocrine pancreas. *Biochim. Biophys. Acta* 399: 277–290, 1975.
202. SMITH, P. A., J. P. SUNTER, AND R. M. CASE. Progressive atrophy of pancreatic acinar tissue in rats fed a copper-deficient diet supplemented with D-penicillamine or triethylene tetramine: morphological and physiological studies. *Digestion* 23: 16–30, 1982.
203. SOLBERG, L. I., AND F. P. BROOKS. Cholinergic control of rabbit pancreatic secretion. *Am. J. Dig. Dis.* 14: 782–787, 1969.
204. SOLOMON, A. K. Electrolyte secretion in the pancreas. *Federation Proc.* 11: 722–731, 1952.
205. SOLOMON, T. E., N. SOLOMON, L. L. SHANBOUR, AND E. D. JACOBSON. Direct and indirect effect of nicotine on rabbit pancreatic secretion. *Gastroenterology* 67: 276–283, 1974.
206. SOMMER, H., AND H. KASPER. Effect of acetylcholine, gastrin, and glucagon alone and in combination with secretin and cholecystokinin on the secretion of the isolated perfused rat pancreas. *Res. Exp. Med.* 179: 239–247, 1981.
207. SOMMER, H., AND H. KASPER. The action of synthetic secretin, cholecystokinin-octapeptide and combinations of these hormones on the secretion of the isolated perfused rat pancreas. *Hepato-Gastroenterology* 28: 311–315, 1981.
208. STONER, G. D., C. C. HARRIS, D. G. BOSTWICK, R. T. JONES, B. F. TRUMP, E. W. KINGSBURY, E. FINEMAN, AND C. NEWKIRK. Isolation and characterization of epithelial cells from bovine pancreatic duct. *In Vitro Rockville* 14: 581–590, 1978.
209. SUNDLER, F., J. ALUMETS, R. HÅKANSON, J. FAHRENKRUG, AND O. SCHAFFALITZKY DE MUCKADELL. Peptidergic (VIP) nerves in the pancreas. *Histochemistry* 55: 173–176, 1978.
210. SWANSON, C. H., AND A. K. SOLOMON. Evidence for Na-H exchange in the rabbit pancreas. *Nature Lond.* 236: 183–184, 1972.
211. SWANSON, C. H., AND A. K. SOLOMON. A micropuncture investigation of the whole tissue mechanism of electrolyte secretion by the in vitro rabbit pancreas. *J. Gen. Physiol.* 62: 407–429, 1973.
212. SWANSON, C. H., AND A. K. SOLOMON. Micropuncture analysis of the cellular mechanisms of electrolyte secretion by the in vitro rabbit pancreas. *J. Gen. Physiol.* 65: 22–45, 1975.
213. TAKAHASHI, H. Scanning electron microscopy of the rat exocrine pancreas. *Arch. Histol. Jpn.* 47: 387–404, 1984.
214. TAKAHASKI, M., P. POUR, J. ALTHOFF, AND T. DONNELLY. The pancreas of the Syrian hamster (*Mesocricetus auratus*). I. Anatomical study. *Lab. Anim. Sci.* 27: 336–342, 1977.
215. TAYLOR, R. B. Pancreatic secretion in the sheep. *Res. Vet. Sci.* 3: 63–77, 1962.
216. TERNOUTH, J. H., AND H. L. BUTTLE. Concurrent studies on the flow of digesta in the duodenum and of exocrine pancreatic secretion of calves. The collection of the exocrine pancreatic secretion from a duodenal cannula. *Br. J. Nutr.* 29: 387–397, 1973.
217. TOMPKINS, R. K., AND S. L. KUCHENBECKER. Relationship of cyclic AMP to secretin stimulation of the pancreas. *J. Surg. Res.* 14: 172–176, 1973.
218. TURNBERG, L. A., F. A. BIEBERDORF, S. G. MORAWSKI, AND J. S. FORDTRAN. Interrelationships of chloride, bicarbonate, sodium and hydrogen transport in the human ileum. *J. Clin. Invest.* 49: 557–567, 1970.
219. UEDA, N., AND O. H. PETERSEN. The dependence of caerulein-evoked pancreatic fluid secretion on the extracellular calcium concentrations. *Pfluegers Arch.* 370: 179–183, 1977.
220. UEDA, N., Y. SUZUKI, M. UTSUMI, T. OBARA, K. OKAMURA, AND M. NAMIKI. Electrophysiological studies on the cultured cells obtained from transplantable pancreatic carcinoma in Syrian golden hamsters. *Peptides Fayetteville* 5: 423–428, 1984.
221. USSING, H. H., D. ERLIJ, AND U. LASSEN. Transport pathways in biological membranes. *Annu. Rev. Physiol.* 36: 17–49, 1974.
222. VAN AMELSVOORT, J. M. M., J. W. C. M. JANSEN, J. J. H. H. M. DE PONT, AND S. L. BONTING. Is there a plasma membrane located anion-sensitive ATPase? IV. Distribution of the enzyme in rat pancreas. *Biochim. Biophys. Acta* 512: 296–308, 1978.
223. VAN DRIESSCHE, W., AND W. ZEISKE. Ionic channels in epithelial cell membranes. *Physiol. Rev.* 65: 833–903, 1985.
224. VAUGHAN-JONES, R. D. Chloride activity and its control in skeletal and cardiac muscle. *Phil. Trans. R. Soc. Lond. B Biol. Sci.* 299: 537–548, 1982.
225. VAYSSE, N., J. P. ESTEVE, B. BRENAC, J. P. MOATTI, AND J. P. PASCAL. Action of dopamine on cyclic AMP-tissue level in the rat pancreas. Interaction with secretin. *Biomedicine Paris* 28: 342–347, 1978.
226. VEGA, D. F., E. MARTINEZ-VICTORIA, A. ESTELLER, AND A. MURILLO. Secretion of pancreatic juice in response to nervous and hormonal stimulation in the anaesthetized monkey (*Erythrocebus patas* and *Papio mandrillus*). *Comp. Biochem. Physiol. A Comp. Physiol.* 58: 259–264, 1977.
227. VINNICOMBE, S. J., AND M. D. KENDALL. The accessory pancreatic ducts of the starling *Sturnus vulgaris*: an ultrastructural and light microscopic study. *J. Anat.* 137: 341–355, 1983.
228. WALLACE, D. H., AND O. D. HEGRE. Development in vitro of epithelial-cell monolayers derived from fetal rat pancreas. *In Vitro Rockville* 15: 270–277, 1979.
229. WATANABE, T., AND M. YASUDA. Electron microscopic study on the innervation of the pancreas of the domestic fowl. *Cell Tissue Res.* 180: 453–465, 1977.
230. WATARI, N. Fine structure of nervous elements in the pancreas of some vertebrates. *Z. Zellforsch. Mikrosk. Anat.* 85: 291–314, 1968.
231. WAY, L. W., AND J. M. DIAMOND. The effect of secretin on electrical potential differences in the pancreatic duct. *Biochim. Biophys. Acta* 203: 298–307, 1970.
232. WAY, L. W., AND M. I. GROSSMAN. Pancreatic stimulation by duodenal acid and exogenous hormones in conscious cats. *Am. J. Physiol.* 219: 449–454, 1970.
233. WIDDICOMBE, J. H., M. J. WELSH, AND W. E. FINKBEINER. Cystic fibrosis decreases the apical membrane chloride permeability of monolayers cultured from cells of tracheal epithelium. *Proc. Natl. Acad. Sci. USA* 82: 6167–6171, 1985.
234. WILLIAMS, D. W., AND M. D. KENDALL. The ultrastructure of the centroacinar cells within the pancreas of the starling (*Sturnus vulgaris*). *J. Anat.* 135: 173–181, 1982.
235. WIZEMANN, V., A.-L. CHRISTIAN, J. WIECHMANN, AND I. SCHULZ. The distribution of membrane bound enzymes in the acini and ducts of the cat pancreas. *Pfluegers Arch.* 347: 39–47, 1974.
236. WIZEMANN, V., AND I. SCHULZ. Influence of amphotericin, amiloride, ionophores and 2,4-dinitrophenol on the secretion of the isolated cat's pancreas. *Pfluegers Arch.* 339: 317–338, 1973.
237. WRIGHT, E. M., AND J. M. DIAMOND. Anion selectivity in biological membranes. *Physiol. Rev.* 57: 109–156, 1977.

CHAPTER 21

Cellular regulation of pancreatic secretion

JOHN A. WILLIAMS | Department of Physiology, University of Michigan, Ann Arbor, Michigan

DANIEL B. BURNHAM | Department of Pharmacology, Smith Kline & French Laboratories, Swedeland, Pennsylvania

SETH R. HOOTMAN | Department of Physiology, Michigan State University, East Lansing, Michigan

CHAPTER CONTENTS

Receptors
    Functional characterization
    Receptor binding characteristics
    Molecular characterization of membrane receptors
    Regulation of pancreatic receptors
Intracellular Messengers
    Calcium-mediated secretagogues
        Source of calcium mobilization and release
        Signal for calcium mobilization
        Involvement of guanine nucleotide–binding proteins in coupling receptors to phospholipase C
        Diacylglycerol and activation of protein kinase C
    Cyclic AMP-mediated secretagogues
    Other possible intracellular messengers
Effectors
    Intracellular receptors for calcium, diacylglycerol, and cyclic AMP
    Protein phosphorylation as an effector system
        Protein kinases and phosphatases
        Secretagogue-induced changes in cellular protein phosphorylation
Exocytosis

A VARIETY OF DIGESTIVE ENZYMES, as well as an NaCl-rich pancreatic juice, are synthesized and secreted by pancreatic acinar cells. Physiologically secretion is activated by the gut hormones cholecystokinin (CCK) and secretin and by the neurotransmitters acetylcholine (ACh) and vasoactive intestinal polypeptide (VIP). Although their physiological relevance is less clear, other peptides such as bombesin, substance P, and peptide histidine isoleucine (PHI) also can act as secretagogues. Additional regulatory agents that may modulate secretion include insulin, epidermal growth factor (EGF), and somatostatin. These agents interact initially with receptors on the plasma membrane of the acinar cell. The steps after receptor occupancy that lead to secretion of either proteins or ions are generally referred to as *stimulus-secretion coupling*, after the pioneering work of Douglas (35). Although originally the term applied to the role of Ca^{2+} in mediating secretion by chromaffin and other cells, it is now used in the broader sense.

A consideration of stimulus-secretion coupling can be divided into consideration of receptors, intracellular messengers, and effectors. The term *intracellular messenger* is preferable to *second messenger*, because it avoids semantic arguments over the sequence of events that conveys a message from the plasma membrane to the inside of the cell or from the basolateral to the luminal side of a polarized cell such as the pancreatic acinar cell. Intracellular messengers in this definition are informational molecules and include substances such as ions, cyclic nucleotides, and phospholipids. Effectors are the systems activated by intracellular messengers that bring about the secretory response. Examples include enzymes, such as protein kinases, phosphatases, and adenosine triphosphatases (Ca^{2+}- and Na^+-K^+-activated), and specific membrane ion channels. The directed fusion of the zymogen granule with the luminal plasma membrane is the end result of protein secretion. The end result of fluid secretion is the net translocation of Na^+, Cl^-, and other osmotically active solutes into the luminal compartment of the acinus.

RECEPTORS

Functional Characterization

Given the definition of regulatory molecules as informational units, a receptor and presumably a specific receptor protein that initiates the cellular response to each regulatory molecule must exist. Neurohormonal regulators of distinct chemical structure generally act on distinct receptors, since the receptor displays a recognition site that is specific for a unique portion of the regulator. Regulatory molecules of similar structure may act via unique, related, or the same receptor, a distinction that is frequently difficult to make. Neurohumoral agents and hence their receptors can be characterized by their biological effects and more specifically by their mechanisms of action. Regulatory molecules acting on pancreatic acinar cells fall into four general categories (Table 1). In general, the Ca^{2+}-diacylglycerol–mediated agents are the main physiological regulators activating digestive enzyme secretion, while the cyclic AMP (cAMP)–mediated agents both activate fluid secretion and potentiate digestive enzyme secretion (for earlier reviews, see refs. 46, 47, 181). The tyrosine kinase–activating regulators have most of their effects on acinar cell metabolism and biosynthetic events but also potentiate digestive enzyme secretion (85, 183). In contrast to the first two classes of regulators, the tyrosine kinase–activating agents have been studied in other tissues and the gene structure coding for their receptors has been elucidated (170, 171). In general the properties of these receptors in the pancreas are similar to those in other tissues and are therefore not discussed in detail. Steroid hormones also can affect growth, differentiation, and gene expression in pancreatic acinar cells (see the chapter by Logsdon in this *Handbook*) but do not directly affect secretion by acinar cells and are not considered here.

A number of additional regulatory molecules may affect pancreatic acinar cell secretion in vivo or in the intact pancreas in vitro. Some stimulate or potentiate secretion (neurotensin, dopamine, epinephrine); others are inhibitory (pancreatic polypeptide, peptide YY, neuropeptide Y). Most of these agents, however, have not been shown to act directly on acinar cells, and specific acinar cell receptors have not been detected (88). Therefore their sites as well as mechanisms of action are unknown. Two possible indirect mechanisms of action are the influencing of either the release of islet hormones or the release of neurotransmitters from intrapancreatic nerve endings.

Receptor Binding Characteristics

Although the presence of a specific receptor can be inferred from studies of the action of agonists and antagonists on physiological responses, more direct information has been obtained from ligand-binding studies. Except for the muscarinic-cholinergic receptor, where labeled antagonists such as quinuclidinyl benzilate (QNB) or *N*-methylscopolamine (NMS) have been used, most pancreatic acinar receptors have been characterized primarily with radiolabeled peptide agonists (Table 2). In most cases these peptides have been labeled with radioiodine by oxidative labeling of tyrosine residues. In some peptides without an appropriate tyrosine, synthetic analogues have been used, such as Tyr^4-bombesin and Tyr^{11}-somatostatin, which behave very similarly to the native peptides. In cases where a ligand such as CCK contains sensitive methionine residues, conjugation with preiodinated moieties has been utilized (144).

TABLE 1. *Function Characterization of Regulatory Molecules Acting on Pancreatic Acinar Cells*

Calcium-diacylglycerol mediated
  Cholecystokinin, gastrin, caerulein
  Muscarinic-cholinergic
  Gastrin-releasing peptide, bombesin
  Substance P, physalaemin
Cyclic AMP mediated
  Secretin
  Vasoactive intestinal polypeptide
  Peptide histidine isoleucine
  Calcitonin gene–related peptide
  Somatostatin
Tyrosine kinase activating
  Insulin
  Insulin-like growth factor
  Epidermal growth factor
Steroids
  Glucocorticoids
  Estrogens

Characterization of pancreatic receptors by ligand binding has been carried out with isolated acinar cells or acini or with crude or purified cellular membrane fractions. In most instances studies that use intact cells are easier to relate to control of acinar cell function, as receptor occupancy and stimulation of function can be carried out under similar conditions. Studies of isolated membranes, however, allow more direct biochemical and structural characterization.

The major types of information obtained from ligand-binding studies include receptor affinity, number of receptors, and specificity for agonists and antagonists. Most of this information is obtained from competition-inhibition paradigms, where binding of a small fixed amount of labeled ligand is carried out in the presence of increasing quantities of unlabeled ligand, analogues, or inhibitors (Fig. 1). The data can then be fit graphically or by computer to derive binding constants. Table 2 lists the ligand-binding characteristics of receptors for the major regulators of pancreatic secretion. Most of the data listed are derived from studies on intact acini or cells that were carried out under physiological conditions. Some of the data on affinity and number of receptors, however, may have been affected by ligand internalization.

Several features may be noted. First, the estimated affinities for agonists of most hormone receptors are in the nanomolar or lower range, and the estimated number of receptors per cell varies from 500 to 200,000, with most ranging from 10,000 to 100,000 per cell. Because most gastrointestinal hormones such as

TABLE 2. Ligand-Binding Characteristics of Pancreatic Receptors Regulating Secretion

Regulator	Ligand	Affinity	Number/Cell	Agonist Specificity	Antagonists	Ref.
CCK	^{125}I-CCK-33 ^{125}I-CCK-8 [^{3}H]CCK-8	H20–50 pM L1–20 nM	H2,000* L1,200,000* 9,000†	Caerulein > CCK-8 > CCK-33 ≫ ds CCK-8 > gastrin > CCK-4	Bt$_2$cGMP, proglumide, CCK-27–32 NH$_2$, asperlicin	22, 47, 74, 145, 146, 157
ACh	[^{3}H]QNB [^{3}H]NMS	H0.2–1 μM‡ L10–30 μM‡	25,000§ 5,000†	ACh > carbamylcholine > methylcholine**	Atropine ≫ pirenzepine, QNB, NMS	3, 29, 63, 82, 83
Bombesin	[^{125}I-Tyr^4]bombesin	2 nM	5,400†	Bombesin > GRF > litorin	[D-Arg^1, D-Pro^2, D-Trp^{7,9}, Leu^{11}]substance P	72, 75
Substance P	^{125}I-physalaemin ^{125}I-substance P	0.4–2 nM	500† 2,500†	Physalaemin > substance P > eledoisin > kassinin	[D-Pro^2, D-Trp^{7,9}]substance P	71, 73, 151
VIP	^{125}I-VIP	0.5–1 nM†	9,000†	VIP > GRF > PHI > secretin	Secretin COOH-terminal fragments, secretin-5–27, secretin-14–27	10, 24, 76, 116
Secretin	^{125}I-secretin ^{125}I-VIP	11 nM† 1 nM§	93,000†	Secretin ≫ VIP = PHI	Secretin-5–27	10, 69, 110
Somatostatin	[^{125}I-Tyr^{11}]somatostatin	0.3–0.5 nM	14,000	Somatostatin-14 > somatostatin-28	None	38, 94, 142, 156

CCK, cholecystokinin; H, high affinity; L, low affinity; ds CCK-8, desulfated cholecystokinin octapeptide; ACh, acetylcholine; QNB, quinuclidinyl benzilate; NMS, N-methylscopolamine; VIP, vasoactive intestinal polypeptide; GRF, growth hormone–releasing factor; PHI, peptide histidine isoleucine. * Rat, mouse. † Guinea pig. ‡ High and low affinity for carbamylcholine. § Rat. ** High- and low-affinity agonist binding.

secretin and somatostatin are present in the circulation at concentrations of only 1–20 pM, only a small percentage of total receptors will normally be occupied. In the case of neurotransmitters such as ACh, the local concentration is unknown. Muscarinic-cholinergic receptors, however, have affinities for agonists in the micromolar range. In most cases the secretory dose-response curve is to the left of the receptor-occupancy curve. Thus target cells are able to respond to low concentrations of hormones by having an excess number of receptors or by linking receptor occupancy to generation of intracellular effectors in such a way that maximal physiological responses are generated when only a small percentage of the receptor population is activated (signal amplification). Some of the gastrointestinal hormones such as CCK have two classes of receptors, one of which is of much higher affinity (20–60 pM). That affinity constant is still greater than circulating levels of the hormone [1–6 pM (84)].

In the case of CCK and of muscarinic-cholinergic agonists, for which high- and low-affinity classes of receptors also exist, it is not always clear which class of receptor mediates which biological response, as different dose-response curves are seen for different actions. For CCK it is postulated that occupancy of high-affinity receptors activates enzyme secretion, while occupancy of low-affinity receptors inhibits enzyme release, enhances glucose transport, and activates adenylate cyclase (145). In the case of VIP and secretin, which are structurally related, there is overlapping agonist specificity, but in guinea pig acini occupancy of VIP receptors appears to stimulate enzyme release (47), whereas occupancy of secretin receptors activates transcellular ion transport (64). As both receptors are mediated by cAMP as an intracellular messenger, this may imply a compartmentalization of receptors or second messengers.

By the additional technique of autoradiography, the sites of radiolabeled ligand binding (i.e., functional receptors) can be determined. Autoradiographic analysis can be carried out at the light-microscopic level either on prelabeled and then embedded and sectioned material or by binding ligands directly to slide-mounted frozen and dried tissue sections. These techniques localize the cellular sites of ligand binding; subcellular localization usually requires extension to the electron-microscopic level (Fig. 2). Such studies have demonstrated that the initial site of insulin (7, 52, 143), CCK (134, 187), and insulin-like growth factor II (IGF-II) (102) binding in acinar cells is to the basolateral membrane domain. Immunofluorescence localization of an insulin antireceptor antibody substantiates the localization of the insulin receptor to the basolateral as opposed to the luminal membrane domain (R. De Lisle and J. A. Williams, unpublished observations).

The second major conclusion from autoradiographic studies is that at physiological temperatures, most

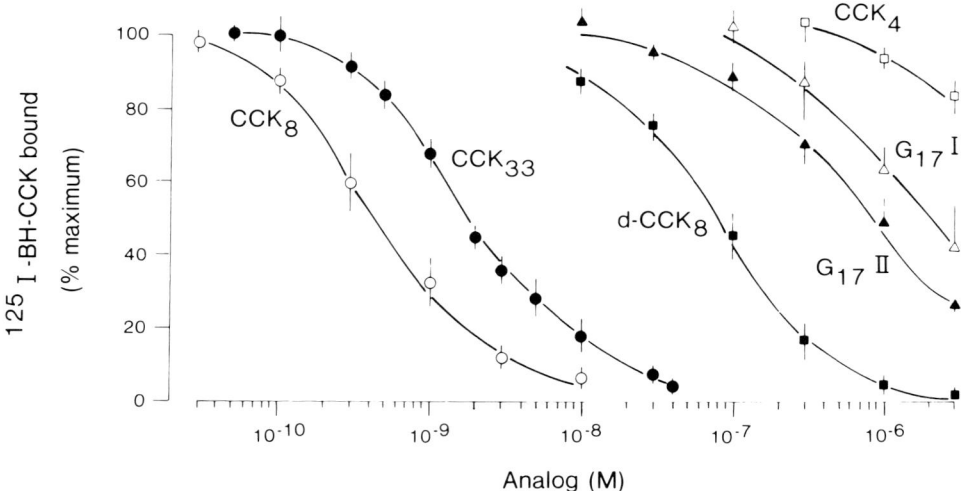

FIG. 1. Inhibition of ^{125}I–CCK-33 binding by cholecystokinin (CCK) analogues. Pancreatic membrane particles were incubated with 25 pM ^{125}I-CCK for 120 min in the presence or absence of various CCK analogues. Nonsaturable binding was subtracted from total binding, and saturable binding was expressed as the percent of maximal saturable binding. d-CCK-8, unsulfated COOH-terminal octapeptide of CCK; G-17-II, sulfated gastrin-17; G-17-I, unsulfated gastrin-17; CCK-4, COOH-terminal tetrapeptide of CCK. [From Steigerwalt and Williams (157).]

bound peptide ligands are subsequently internalized (Fig. 2). Internalization can occur within a few minutes and in most cases leads to ligand concentration within specific intracellular compartments, particularly the Golgi complex and lysosomes. In contrast to cell types such as fibroblasts, coated pits and vesicles are not prominent features of internalization in acinar cells; in autoradiograms, many silver grains overlie labeled ligand molecules that appear to be free in the cytoplasm and in rough endoplasmic reticulum, which occupies the largest portion of the cell. The fact that different ligands localize differently (102), along with the large amount of extralysosmal ligand, suggests that this mechanism is more than just a nonspecific degradative pathway, although no direct evidence yet links internalized ligand to specific cell functions. Less precise demonstration of peptide hormone internalization has been obtained using acid washing or trypsinization to selectively remove surface-bound ligand from acinar cells. Such studies are consistent with internalization by acinar cells of insulin, IGF-II, EGF, CCK, and somatostatin (78, 86, 101, 175) and, along with the autoradiographic data, indicate that all peptide regulators are internalized, although to varying extents.

These results, which show ligand internalization, complicate the interpretation of binding data obtained using intact cells at physiological temperature. Thus most of the current estimates of the number of functional receptors per acinar cell are probably overstatements. A single class of CCK-binding sites is observed on isolated plasma membranes (157); an additional class of higher affinity sites, whose presence is dependent on physiological temperatures and cellular metabolic energy, is observed on intact cells (145, 146). By contrast, similar binding of other ligands such as insulin and somatostatin is observed to intact cells and plasma membranes.

Molecular Characterization of Membrane Receptors

Although ligand-binding data give functional information about the receptor, these data alone provide little direct information about the physical composition of the receptor. Such information is provided by covalent attachment of labeled ligands to receptors, purification of receptors, determination of their amino acid and coding nucleotide sequences, and the use of specific antireceptor antibodies. These analyses have recently provided detailed compositional information for several tyrosine kinase receptors (insulin and EGF) and for some neurotransmitter receptors. However, little detailed information is available concerning receptors for the gut peptides, which regulate pancreatic function.

Preliminary structural information has been obtained recently by the use of bifunctional cross-linking agents to covalently label proteins at or near the binding site of several pancreatic receptors (Table 3). Evaluation on sodium dodecyl sulfate (SDS) polyacrylamide gels with and without reducing agents reveals the size and disulfide bond–linked subunit composition (Fig. 3), while solubilization of covalently labeled proteins in nonionic detergents followed by endoglycosidase treatment or lectin chromatography provides information as to the carbohydrate content of the receptor. Such studies have shown the receptors

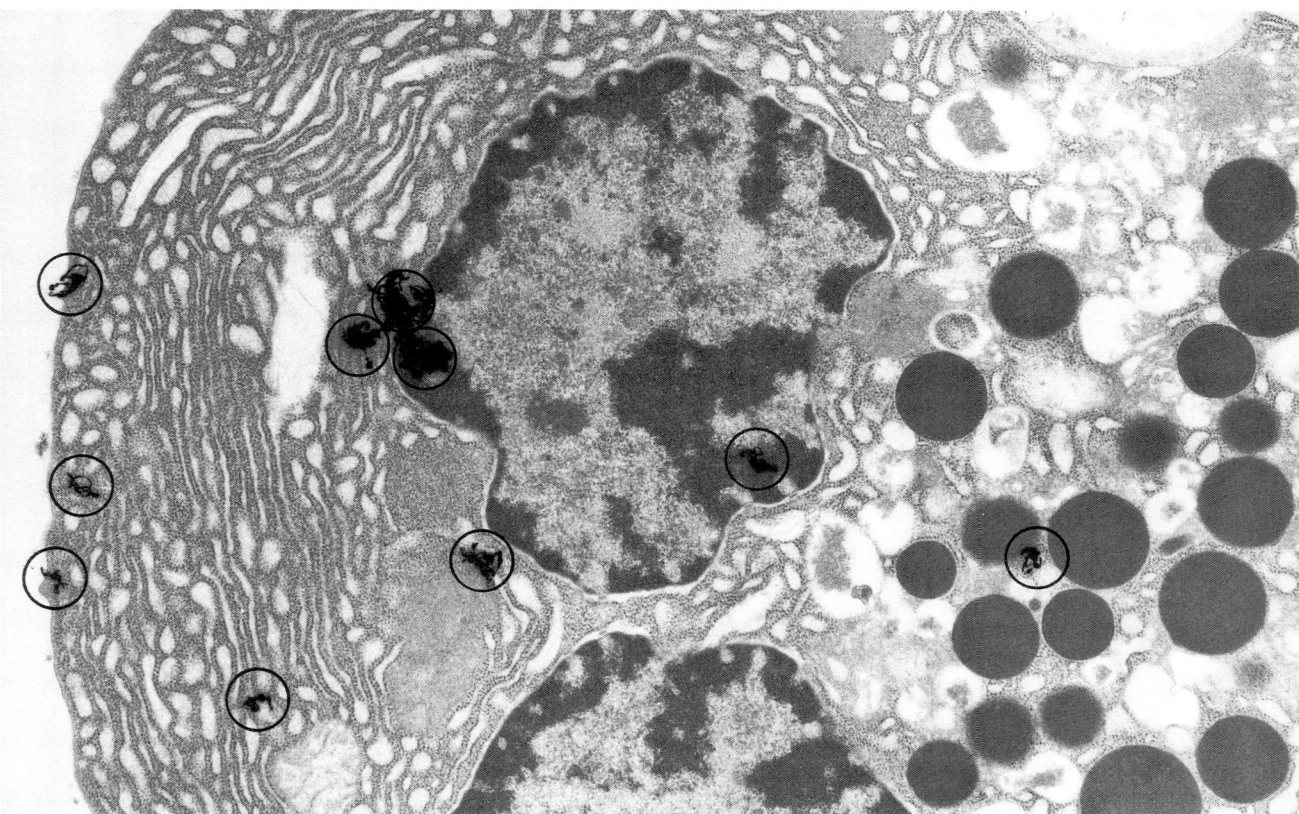

FIG. 2. Electron-microscopic autoradiograph of rat acini incubated for 1 h with ^{125}I–CCK-33 at 37°C. *Circles*, silver grains overlying ^{125}I-CCK molecules. Note grains both on basolateral membrane and intracellularly.

TABLE 3. *Physical Characterization of Pancreatic Receptors*

Regulator	Binding Protein Subunit, M_r	Subunits	Glycoprotein	Solubilized	Ref.
Cholecystokinin	75,000–80,000	Yes, disulfide linked	Yes	Yes, native	91, 133, 134, 140, 141, 166, 167
Somatostatin	92,000	No	Yes	Yes, cross-linked only	142, 165
Muscarinic-cholinergic	118,000	No	Yes*	Yes, native	66
Insulin	135,000	Yes, disulfide linked	Yes	Yes, native	52, 100

* Not yet carried out on pancreas but presumed from studies on other cells.

for CCK, somatostatin, and insulin to be glycoproteins, with binding subunits ranging in size from 75,000 to 125,000 M_r. The CCK receptor appears to contain additional disulfide-linked subunits, whereas the receptor for somatostatin appears to contain only one subunit. The muscarinic-cholinergic receptor has been covalently labeled with the specific alkylating agent [^{3}H]propylbenzilylcholine mustard and shows some tissue-specific differences in size when pancreas is compared with other tissues but contains no disulfide-linked subunits (66). As in other cells, the insulin receptor appears to be a tetramer, with two ligand-binding 130,000-M_r α-subunits and two 95,000-M_r β-subunits, which in other cells span the membrane and possess tyrosine kinase activity (52). The available information indicates that pancreatic receptors are quite similar to all peripheral insulin receptors. The development of anti-insulin–receptor antibodies and receptor cDNA probes is now allowing the biosynthesis and regulation of pancreatic insulin receptors to be studied.

More complete characterization can be provided by detergent solubilization and purification of functional receptors. This requires that solubilization not compromise the ligand-binding properties of the native receptor, so that ligand binding can be used to follow purification. Solubilization of pancreatic insulin (100) and muscarinic-cholinergic receptors (Y. Habara and S. Hootman, unpublished observations) has been carried out using techniques worked out on better studied tissues. For gastrointestinal hormone receptors, solubilization has been carried out only for CCK (167).

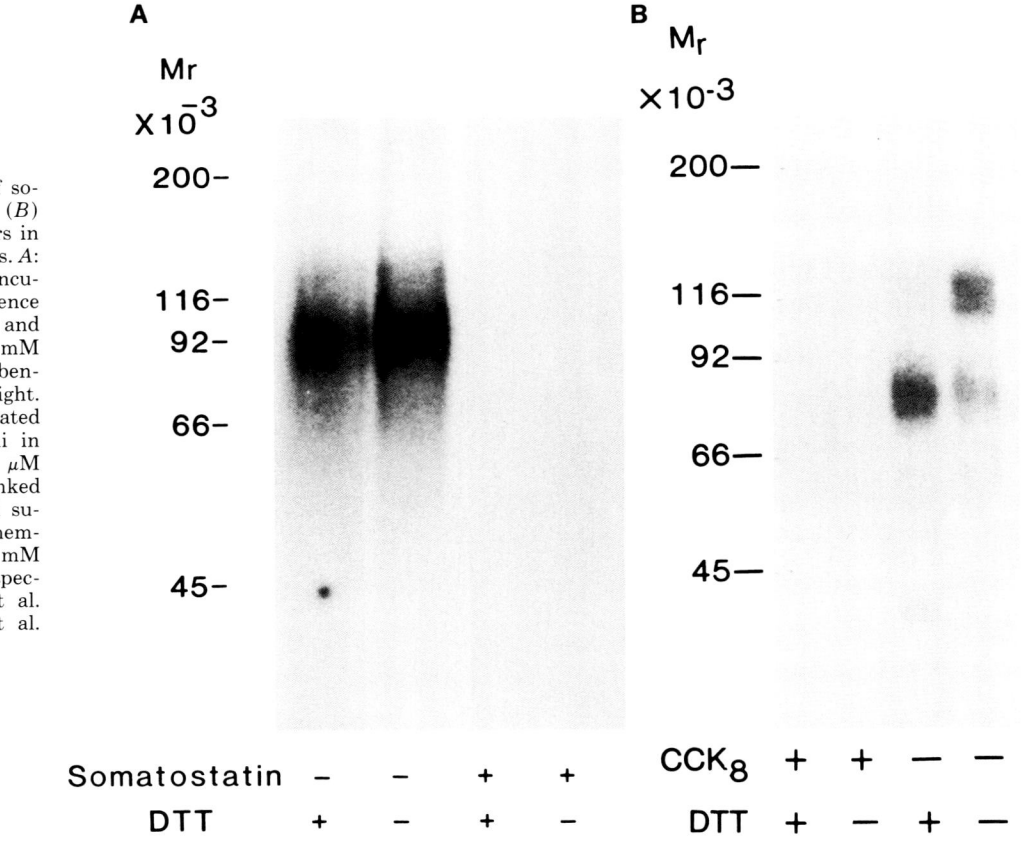

FIG. 3. Autoradiograms of somatostatin (A) and CCK (B) cross-linked to their receptors in pancreatic plasma membranes. A: ^{125}I-Tyr^1 somatostatin was incubated in the presence or absence of 1 µM cyclic somatostatin and then cross-linked with 0.1 mM n-hydroxysuccinimide azidobenzoate and exposure to UV light. B: ^{125}I-CCK-33 was incubated with mouse pancreatic acini in presence or absence of 0.1 µM CCK-8 and then cross-linked with 0.1 mM disuccinimidyl suberate. In both A and B, membranes were treated with 50 mM dithiothreitol (DTT) where specified. [A from Sakamoto et al. (142); B from Sakamoto et al. (140).]

Techniques such as lectin chromatography and ligand-affinity chromatography that have been used to purify other receptors can be applied to solubilized pancreatic receptors and have recently shown promise for purification of the CCK receptor (J. Szecowka, J. A. Williams, and I. D. Goldfine, unpublished observations).

Regulation of Pancreatic Receptors

The number of cellular receptors for a particular regulator is not fixed but can increase or decrease due to altered rates of receptor synthesis or degradation. The most common form of such regulation is downregulation of a receptor by its ligand. In this phenomenon, receptor occupancy leads to receptor internalization and degradation, with a subsequent reduced ability of the cell to respond to the regulator as a result of the decreased number of functional receptors. Thus the response to prolonged stimulation can be attenuated.

Several in vivo studies indicate that at least some pancreatic receptors are subject to such regulation. Larose et al. (83) demonstrated that chronic bethanechol injection induced a 43% decrease in pancreatic muscarinic receptor density in rats after 14 days. In acini prepared from bethanechol-treated rats, a fourfold decrease in the potency of carbachol to induce amylase secretion was noted, while the secretory response to caerulein, a CCK analogue, was relatively unaltered. The effect of increased somatostatin levels on pancreatic somatostatin-receptor levels has been studied in rats by injecting streptozotocin, which increases circulating somatostatin. The density of receptors on pancreatic membranes was reduced by 61% after 1 wk (156). Conversely, streptozotocin decreases circulating insulin levels and has been shown to upregulate or increase the number of pancreatic insulin receptors (100). The effects of injecting CCK in rats are more complex. When acinar CCK receptors were studied 2 h after injection, high-affinity CCK binding was increased (112). After 1 wk of chronic CCK injections, with attendant pancreatic hypertrophy and hyperplasia, the number of receptors per cell was normal, although the total number of receptors per pancreas was increased (113). This lack of downregulation may be important in permitting CCK-induced pancreatic growth.

In vitro studies permit a more detailed evaluation of receptor dynamics, but information from such investigations is only beginning to accumulate. Exposure of pancreatic acini to insulin in vitro reduced the number of insulin receptors, measured both on intact and solubilized acini (100). Thus the total number of receptors was reduced. In a study of guinea pig pancreatic muscarinic receptors, Hootman et al. (63) found a time- and concentration-dependent loss in [^{3}H]NMS binding in acini cultured with carbachol

(Fig. 4). On removal of carbachol from the culture medium, [³H]NMS binding slowly recovered to near control levels with a half time for recovery of 20–24 h. Restoration of cell surface receptors was blocked by the protein synthesis inhibitor cycloheximide. Thus chronic activation of muscarinic receptors appears to lead to a selective large-scale removal of muscarinic receptors, the reversal of which is slow and apparently requires de novo synthesis of receptor protein.

In addition to modulation of receptor number, agonists may also desensitize the response without a reduction in receptor number. In guinea pig and rat acini, exposure to cholinergic analogues for only 30–60 min leads to a three- to sixfold decrease in agonist potency (4). Over this short term, [³H]NMS binding is not decreased proportionately. It has been difficult to study similar effects of peptide secretagogues, due to the much higher affinity of their receptors and slow dissociation. Thus similar experiments with a short exposure to CCK have documented increased basal secretion when the agonist is removed, a phenomenon termed *residual stimulation* (27). In addition, however, CCK appears to induce desensitization at a postreceptor level, in that amylase release induced by all Ca^{2+}-diacylglycerol–mediated secretagogues is blunted (1).

Heterologous receptor regulation, where occupancy of a specific receptor can affect the binding properties of another distinct class of receptors, also has been documented. On acinar cells, CCK and other Ca^{2+}-diacylglycerol–mediated agents decrease the binding of EGF, IGF-II, substance P, and somatostatin (38, 78, 86, 94, 101, 152). Some of this effect is probably mediated by activation of protein kinase C and may involve receptor phosphorylation (86, 94). By contrast, insulin increases the binding of IGF-II to its receptor (101). The functional consequences of this heterologous receptor regulation are not clear.

INTRACELLULAR MESSENGERS

Secretagogues that elicit digestive enzyme secretion from the pancreatic acinus can be separated into two principal classes based on the intracellular coupling mechanisms that they activate (for earlier reviews, see refs. 46, 164, 181). Cholecystokinin and caerulein, bombesin and related peptides, substance P, and cholinergic agonists all appear to utilize Ca^{2+} and diacylglycerol as intracellular messengers (see Table 1). Secretin and VIP, on the other hand, appear to regulate acinar secretory activity through alterations in cellular cAMP levels. The mechanisms by which these two classes of agents cause changes in acinar cell Ca^{2+} and cAMP metabolism share some common elements and also exhibit very distinctive features.

Calcium-Mediated Secretagogues

The involvement of Ca^{2+} in stimulus-response coupling in the exocrine pancreas was first indicated by

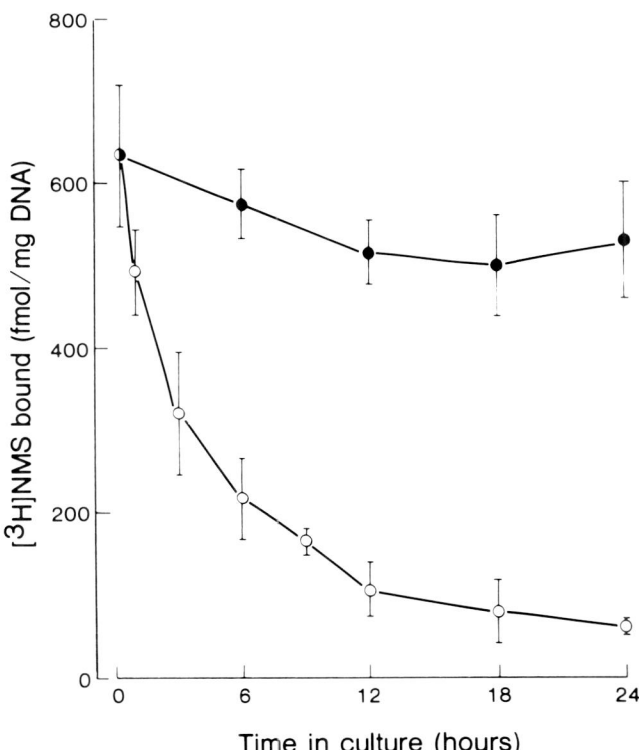

FIG. 4. Time dependence of decrease in specific [³H]N-methylscopolamine (NMS) binding to guinea pig pancreatic acini cultured with 0.1 mM carbachol (*open circles*) or in control media (*filled circles*). At each indicated time, acini were rinsed and incubated for 60 min with 0.5 nM [³H]NMS. [From Hootman et al. (63).]

studies that showed that incubation of pancreatic slices or fragments in Ca^{2+}-free media inhibited enzyme secretion in response to ACh or CCK (6, 19, 59). Additional evidence for the importance of Ca^{2+} as an intracellular messenger in the pancreatic acinar cell was provided by the observation that the Ca^{2+} ionophore A23187 provoked amylase release from pancreatic slices (37, 185) or isolated acini (121) incubated in the presence of extracellular Ca^{2+}. Later studies with isolated pancreatic lobules or acini demonstrated that while sustained release of digestive enzymes in response to physiological secretagogues depended on the presence of extracellular Ca^{2+}, a brief period of essentially normal secretion could be induced in its absence [Fig. 5; (46, 147, 179)]. However, depletion of acinar intracellular stores by combined use of extracellular chelator and a calcium ionophore (121) or intracellular addition of a chelator (32, 108) profoundly depresses secretion. These results indicated that intracellular Ca^{2+} reserves might play some role in initiating the secretory response. This supposition was supported by the observation that CCK and cholinergic agonists caused a dramatic but transient release of $^{45}Ca^{2+}$ from pancreatic acinar cells previously loaded with the radionuclide (19, 95, 182). Moreover this enhanced $^{45}Ca^{2+}$ efflux was not affected by removing Ca^{2+} from the medium. However, when un-

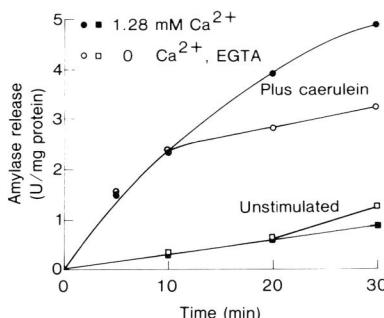

FIG. 5. Effect of Ca^{2+} removal on basal and caerulein-stimulated amylase release from mouse pancreatic acini. EGTA, ethylene glycol-bis(β-aminoethylether)-N,N'-tetraacetic acid. [From Williams (179).]

loaded acinar cells were stimulated in a $^{45}Ca^{2+}$-containing medium, a substantial enhanced uptake of the tracer was also observed (33, 77). Measurements of total exchangeable Ca^{2+} using $^{45}Ca^{2+}$ as a tracer showed that the efflux component initially predominates and that ~25% of total acinar cell Ca^{2+} is rapidly lost after exposure to CCK or cholinergic analogue but that prestimulus levels are regained within 1 h (33, 44, 128). These data are clearly consistent with an initial phase of intracellular Ca^{2+} mobilization and efflux followed by a latter period of Ca^{2+} influx.

Earlier data using Ca^{2+}-sensitive microelectrodes (109) and the Ca^{2+}-activated photoprotein aequorin (31) were consistent with a secretagogue-induced rise in acinar cytosolic free intracellular Ca^{2+} ($[Ca^{2+}]_i$). More sensitive and direct measurement of $[Ca^{2+}]_i$ in pancreatic acinar cells has been made possible recently by the development of a family of fluorescent Ca^{2+} chelators by Tsien et al. (169) that can be noninvasively loaded into cells. These probes, in the form of their acetoxymethyl esters, readily penetrate the plasma membranes of intact cells and are subsequently converted by cellular esterases to nonpermeant forms, which are thus trapped in the cytosol. The most widely used of these compounds is quin 2, which has a dissociation constant (K_d) for Ca^{2+} of ~120 nM, making it suitable for measuring Ca^{2+} in the range of 0.05–1.0 μM. One caveat in the use of quin 2 is that it acts as a buffer of intracellular Ca^{2+} and thus may dampen secretagogue-induced changes in $[Ca^{2+}]_i$ to some extent. The recent synthesis of a second generation of Ca^{2+} chelators with enhanced fluorescence properties (e.g., fura 2 and indo 1) has to a large extent alleviated this problem (54). Using quin 2, Ochs et al. (107) initially estimated a resting $[Ca^{2+}]_i$ of 180 nM in mouse pancreatic acini, a value that was reduced to 90 nM in a subsequent report (108) when a more exact determination of the K_d for quin 2 chelation of Ca^{2+} became available. Similar values have been obtained for mouse (123), guinea pig (65, 116), and rat pancreatic acinar cells (12, 96). Carbachol and CCK in all species induced rapid rises in $[Ca^{2+}]_i$ to 500–800 nM. In mouse acini, the rise in $[Ca^{2+}]_i$ was observed after a lag time following secretagogue addition of <2 s and was maximal by 5–6 s (Fig. 6). In Ca^{2+}-containing medium, $[Ca^{2+}]_i$ remained elevated by two- to threefold over resting levels for at least 15 min after exposure of acini to 1 μM carbachol (108). In the absence of extracellular Ca^{2+}, carbachol induced in mouse acini a similar initial elevation of Ca^{2+}, although the rate of rise to this level was slightly slowed and Ca^{2+} returned to resting levels within 4 min after agonist exposure. This is consistent with and provides an explanation for the brief period of normal secretion observed in the absence of extracellular Ca^{2+}. Pandol et al. (116) demonstrated using guinea pig pancreatic acini that carbachol, CCK, and bombesin all accessed the same source of intracellular Ca^{2+}, similar to results of earlier studies with $^{45}Ca^{2+}$ (159). Secretin, VIP, and other agents that elevate intracellular cAMP as well as the phorbol ester 12-O-tetradecanoylphorbol-13-acetate (TPA) had little effect on quin 2 fluorescence (65, 116). The Ca^{2+} ionophore ionomycin induced in mouse acini a rise in $[Ca^{2+}]_i$ similar to that elicited by carbachol, although it failed to comparably stimulate amylase release, suggesting that enzyme secretion is not solely regulated by Ca^{2+} (Fig. 7). Recent work by Bruzzone et al. (12) suggests that digestive enzyme release can be stimulated at resting levels of $[Ca^{2+}]_i$, and these authors suggest that protein kinase C may play an important role.

SOURCE OF CALCIUM MOBILIZATION AND RELEASE. Although sustained secretion after exposure to CCK or cholinergic agonists clearly depends on extracellular sources of Ca^{2+}, the initial rise in $[Ca^{2+}]_i$ appeared due to intracellular release. Current evidence implicates both the plasma membrane and rough endoplasmic reticulum as sites of secretagogue-responsive Ca^{2+} sequestration. The cytochemical studies of Haase et al. (56) revealed the presence of Ca^{2+} deposits along the plasma membranes of rat pan-

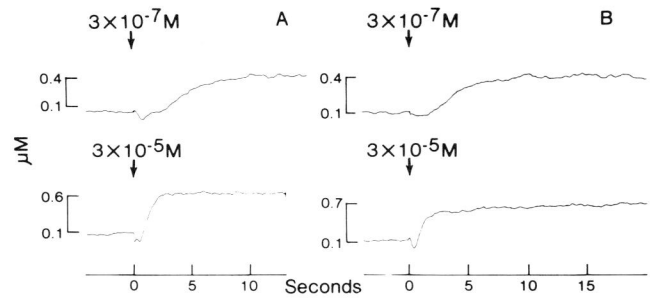

FIG. 6. Kinetics of initial changes in $[Ca^{2+}]$ of pancreatic acini measured with quin 2 after stimulation by carbachol. Acini were suspended in medium containing 1.28 mM Ca^{2+} (A) or in medium containing 0.1 mM EGTA and no added Ca^{2+} (B). Concentration of carbachol shown was added at the arrow. [From Ochs et al. (108).]

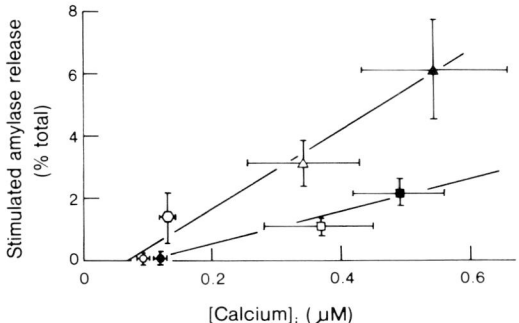

FIG. 7. Relationship between amount of amylase released from pancreatic acini and $[Ca^{2+}]_i$. Release of amylase and rise in $[Ca^{2+}]_i$ of quin 2-loaded pancreatic acini were measured after addition of various concentrations of either carbachol (0–1 µM) or inomycin (0–1 µM). [From Ochs et al. (108).]

creatic acinar cells that were reduced by >80% when cells were pretreated with carbachol or CCK. Exposure of acinar cells to atropine after carbachol caused reappearance of the deposits. Despite this and other suggestions of plasma membrane involvement in Ca^{2+} regulation, the magnitude of secretagogue-induced release (25% of total cell Ca^{2+}) makes it unlikely to be the sole reservoir for rapid Ca^{2+} mobilization in pancreatic acinar cells. Chandler and Williams (21), using the fluorescent probe chlorotetracycline, obtained evidence that cholinergic agonists release Ca^{2+} from intracellular sites that have a punctate distribution and that therefore clearly could not correspond to the plasma membrane. The probable identity of these sites was suggested by the studies of Dormer and Williams (34), who fractionated pancreatic acini after carbachol exposure and demonstrated a significant net loss of Ca^{2+} from a microsomal fraction composed predominantly of cisternae of rough endoplasmic reticulum. The rough endoplasmic reticulum also has been implicated in studies of Ca^{2+} sequestration by digitonin- or saponin-permeabilized pancreatic acinar cells (89, 176). In permeabilized cells, two separate Ca^{2+} pools could be distinguished, one in which Ca^{2+} uptake was supported by ATP and succinate and was inhibited by ruthenium red and azide and a second in which uptake also was supported by ATP, but which was insensitive to these two inhibitors (160). Oxalate markedly stimulated Ca^{2+} uptake into this second pool, which demonstrated substantial uptake at submicromolar $[Ca^{2+}]_i$. These characteristics of the first and second Ca^{2+} pools suggest that they can be identified with mitochondria and the rough endoplasmic reticulum, respectively. Subcellular fractionation studies have confirmed the existence of a distinct Ca^{2+}-uptake system in pancreatic rough endoplasmic reticulum (5, 124).

The second phase of the Ca^{2+} response to cholinergic agonists and CCK, reuptake and recharging of cellular stores, is dependent on influx of Ca^{2+} from outside the acinar cell. However, the mechanisms by which Ca^{2+} permeability of the acinar cell plasma membrane is regulated during this compensatory phase of the response to secretagogues are not known; secretagogue-regulated Ca^{2+}-selective channels have not been demonstrated in the pancreatic acinar cell.

SIGNAL FOR CALCIUM MOBILIZATION. Implication of the rough endoplasmic reticulum as the principal locus for Ca^{2+} release after stimulation of pancreatic acinar cells with CCK, ACh, or their analogues suggested that a message of some sort must be generated at the plasma membrane by receptor activation capable of bridging the gap of cytosol between the cell surface and the sequestered Ca^{2+} pool. The identity of this "missing messenger" has recently been established as inositol 1,4,5-trisphosphate (IP_3), one of two principal products of the hydrolysis of the plasma membrane phospholipid, phosphatidylinositol 4,5-bisphosphate (PIP_2) (see the chapter by Schulz in this Handbook). Phosphatidylinositol (PI) and its polyphosphate derivatives, phosphatidylinositol 4-phosphate (PIP) and PIP_2, together comprise only ~10% of cellular phospholipids, and the two polyphosphoinositides constitute <10% of this total (99). The involvement of this class of phospholipids in receptor-mediated events was first suggested by Hokin and Hokin (60), who demonstrated ACh stimulation of $^{32}P_i$ incorporation into PI in slices of pigeon pancreas. Michell and co-workers subsequently examined this response in several cell types and noted that it accompanied the action of a number of Ca^{2+}-mediated secretagogues (98). Evidence now suggests that the primary event elicited by agonist occupancy of either CCK or cholinergic receptors in the pancreatic acinar cell is increased hydrolysis of PIP_2 in the plasma membrane by an associated phospholipase C, which attacks the phosphoester bond between the fatty acid tail and sugar headgroup of the phospholipid (8). The products of this reaction are the lipophilic 1,2-diacylglycerol and the water-soluble IP_3. The earlier noted incorporation of $^{32}P_i$ into PI appears to reflect steps in the resynthesis and recharging of the secretagogue-responsive phosphoinositide pool (9).

The onset of PIP_2 hydrolysis after CCK- or cholinergic-receptor occupancy is very rapid, producing in rat pancreatic acini a maximal decrease of 30%–50% in $^{32}P_i$-labeled PIP_2 within 15 s after agonist exposure (111, 126), a decrease that was maintained for at least 10 min. By contrast, cellular content of labeled PIP did not vary appreciably over this time period, while labeling of phosphatidic acid, the principal phospholipid precursor of PI, slowly increased. Labeling of PI was unchanged for 5 min and then rose sharply. Simultaneously with the breakdown of PIP_2 there is also an increase in cellular IP_3 (137).

The rapidity with which hydrolysis of PIP_2 is stimulated indicates that its enzymatic breakdown either precedes or is contemporaneous with Ca^{2+} mobilization. Incubation of pancreatic acini in Ca^{2+}-free media containing ethylene glycol-bis(β-aminoethylether)-N-N'-tetraacetic acid (EGTA) or preloading acini with quin 2 to ablate the expected rise in [Ca^{2+}]$_i$ has little effect on the decrease in PIP_2 elicited by CCK or carbachol, and A23187 treatment does not provoke PIP_2 breakdown (111, 118).

Two recent studies have indicated further complexities in the effects of secretagogues on cellular metabolism of phosphatidylinositides. Farese et al. (40, 41) have identified secretagogue-responsive PI hydrolysis in rat pancreatic acini that is inhibited by incubation in Ca^{2+}-free media containing EGTA. This response is induced by secretin, VIP, and insulin, as well as by CCK and carbachol, and is therefore separable from the rapid hydrolysis of PIP_2 evoked by the latter two agents. The role of this response in initiating or amplifying the intracellular signaling process remains to be determined. Another complexity is the observation that a second isomer of IP_3, inositol 1,3,4-trisphosphate, is also liberated on stimulation of pancreatic acinar cells with caerulein (13). Unlike the release of IP_3, a latency period of up to 20 s after agonist exposure was seen before the cellular content of the newly discovered isomer began to rise. By several minutes after exposure of acini to caerulein, however, the 1,3,4 isomer accounted for 80% of total acinar cell IP_3. The physiological role of this newly discovered phospholipid metabolite is unknown, although it is believed to be produced by phosphorylation of IP_3 at position 3, followed by removal of the phosphate on position 5 (125).

The very rapid generation of IP_3 from PIP_2 after exposure of pancreatic and other cell types to secretagogues that raise [Ca^{2+}]$_i$ suggested that it might be the looked-for messenger between the plasma membrane and endoplasmic reticulum. Convincing evidence for this possibility was obtained by Streb et al. (161), using permeabilized acinar cells from rat pancreas. In this system, exogenously applied IP_3 stimulated efflux of Ca^{2+} from the cells, an efflux that was not affected by mitochondrial inhibitors (Fig. 8). Carbachol also elicited release of Ca^{2+} from permeabilized acinar cells, although the effect of the agonist was abolished in the presence of IP_3, suggesting that both agents access the same pool of intracellular Ca^{2+}. Moreover, IP_3 was able to release Ca^{2+} from pancreatic subcellular fractions enriched in endoplasmic reticulum (160). Similar results have been reported for several other cell types, indicating the common role of IP_3 in mediating stimulation of Ca^{2+} release evoked by a wide variety of secretagogues (125). The precise mechanism by which IP_3 causes Ca^{2+} efflux from the cisternal space of the endoplasmic reticulum is unknown, although recent studies have provided some insights. Low temperatures do not strongly inhibit IP_3-activated Ca^{2+} release from the endoplasmic reticulum in cultured vascular smooth muscle cells (154), an observation that suggests that a channel rather

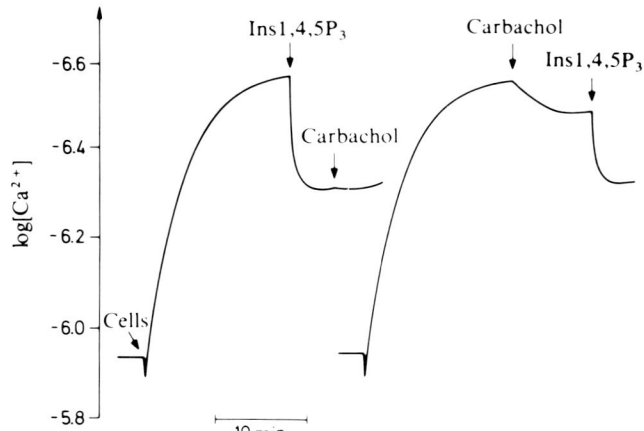

FIG. 8. Effect of inositol 1,4,5-trisphosphate (IP_3) on intracellular Ca^{2+} stores of leaky acinar cells. Calcium in the medium was measured with an ion-selective electrode. *Initial upward trace*, fall in medium Ca^{2+} as it is taken up by acinar organelles; *abrupt downward deflection*, Ca^{2+} release induced by IP_3. [From Streb et al. (161).]

than a carrier mediates the response. Spät et al. (155) reported specific binding of ^{32}P-labeled IP_3 to a saturable class of sites in permeabilized hepatocytes and neutrophils. These studies support the view that the rough endoplasmic reticulum in cells that are regulated by Ca^{2+}-mediated neurohumoral agents contains a finite population of Ca^{2+}-selective channels, the opening of which is modulated by binding of IP_3 to sites on or near the cytosolic aspect of the channel protein.

INVOLVEMENT OF GUANINE NUCLEOTIDE–BINDING PROTEINS IN COUPLING RECEPTORS TO PHOSPHOLIPASE C. Adenylate cyclase activity in mammalian cells is regulated by the combined action of two guanine nucleotide–binding proteins, one that stimulates the cyclase (N_s), and a second that is inhibitory (N_i). The regulatory interactions among these components of the adenylate cyclase complex are now well characterized (51). Recent evidence has also implicated an as yet uncharacterized G protein in the coupling of receptors for Ca^{2+}-mediated secretagogues to phospholipase C (97). In rat pancreatic acinar cells permeabilized by brief exposure to an intense electric field (97), as well as other types of cells (for review see ref. 125), the GTP analogues, GTPγS and Gpp(NH)p, each stimulated accumulation of IP_3 and potentiated the effects of hormones on PIP_2 breakdown. Pretreatment of acinar cells with cholera or pertussis toxins had no effect on these responses, indicating that neither N_i nor N_s was involved. An indication of the mechanism by which the novel GTP-binding protein activates PIP_2 hydrolysis has been provided by Smith et al. (153), who showed that activation of a guanine nucleotide–regulatory protein in human polymorphonuclear leukocyte plasma membranes reduces the Ca^{2+} requirement for phospholipase C from superphysiological to normal intracellular concentrations. The agonist-induced alterations in receptor conformation that allow interaction with this newly discovered species of regulatory protein remain to be determined.

DIACYLGLYCEROL AND ACTIVATION OF PROTEIN KINASE C. The second product of PIP_2 hydrolysis, diacylglycerol, acts as an intracellular messenger also, activating a Ca^{2+}- and phospholipid-dependent protein kinase, protein kinase C (105), which is present in the pancreatic acinar cell cytosol (18, 106, 188). Diacylglycerol and its analogues appear to activate the kinase at least in part by decreasing the concentration of Ca^{2+} required for its activation. Assessment of the role of protein kinase C in pancreatic stimulus-response coupling has been facilitated by studies utilizing tumor-promoting phorbol esters such as TPA, which substitute for diacylglycerol in sensitizing the kinase to Ca^{2+} (20). Mouse and guinea pig pancreatic acini secrete amylase in response to TPA, and the response is synergistically increased by the Ca^{2+} ionophore A23187 (Fig. 9; 15, 36, 96, 149). The membrane-permeant synthetic analogue of diacylglycerol, 1-oleoyl-2-acetolyl-sn,3-glycerol, and TPA also have been shown to stimulate activity of the plasma membrane enzyme Na^+-K^+-ATPase in dispersed guinea pig pancreatic acinar cells (62). These observations argue for an important role of protein kinase C in regulating ACh- or CCK-induced secretion of both digestive enzymes and electrolytes in the exocrine pancreas. The precise mechanisms by which these processes are effected by protein kinase C in the pancreatic acinar cell are not known, although kinase-mediated alterations in phosphorylation of particular proteins appear to be involved.

Cyclic AMP–Mediated Secretagogues

Along with Ca^{2+}, cAMP appears to play an important role in regulating acinar secretory activity in the exocrine pancreas. This view is based on the following evidence. *1*) Several neurotransmitters or hormones that stimulate secretion from acini also augment cellular cAMP levels. *2*) Synthetic analogues of cAMP such as dibutyryl cAMP (Bt_2cAMP) and 8-brcAMP stimulate secretion by acinar cells. *3*) Agents that stimulate adenylate cyclase or inhibit cyclic nucleotide phosphodiesterase activity either stimulate acinar cell secretion or potentiate the stimulation of secretion by other agonists that increase cAMP.

Several polypeptides in the secretin-VIP-glucagon family augment acinar secretory activity via modulation of cAMP levels (Fig. 10). Both secretin and VIP increase cAMP content of pancreatic acinar cells by 8- to 30-fold (79, 129). In the presence of the phosphodiesterase inhibitor isobutylmethylxanthine

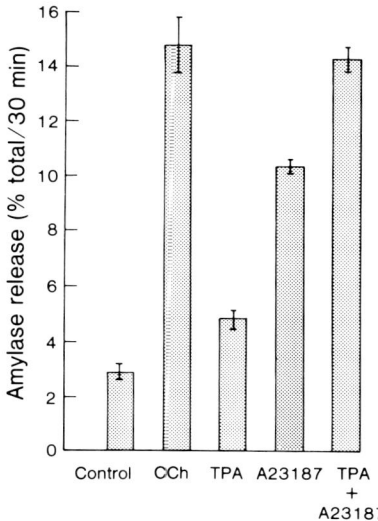

FIG. 9. Synergism of amylase release from mouse pancreatic acini induced by the calcium ionophore A23187 (2 μM) and the phorbol ester 12-O-tetradecanoylphorbol-13-acetate (TPA) (10^{-6} M). Together the two are able to reproduce effect of maximal concentration of carbachol (CCh).

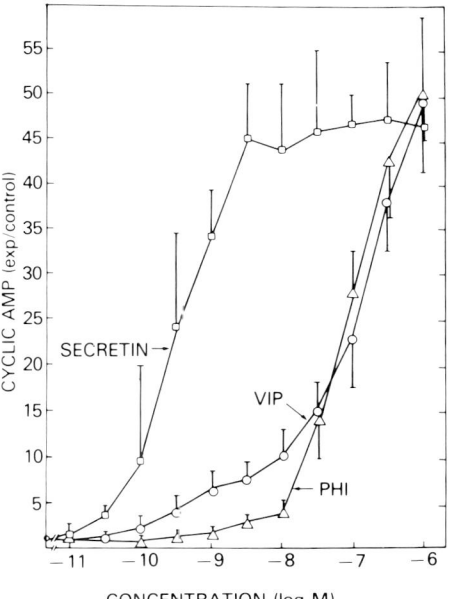

FIG. 10. Ability of secretin, vasoactive intestinal polypeptide (VIP), and peptide histidine isoleucine (PHI) to increase cAMP in dispersed acini from guinea pig pancreas. [From Jensen et al. (76).]

(IBMX), increases as large as 320-fold have been observed (48). In general, secretin appears to be roughly 10 times as potent but equally as effective as VIP in raising acinar cell cAMP levels (130). Recently other neurohormonal factors have been shown to modulate cAMP levels in the pancreatic acinus. The 27–amino acid brain-gut peptide PHI (76), as well as the

43-amino acid growth hormone–releasing factor from rat hypothalamus (117), both of which are structurally homologous to VIP, increase cAMP levels in guinea pig pancreatic acini. Somatostatin, which inhibits amylase release from the exocrine pancrease in vivo, inhibits the increase in cAMP induced in guinea pig acini by either secretin or VIP (39). The significance of these observations is not clear, however, because somatostatin does not inhibit VIP-induced amylase release by pancreatic acini in vitro [J. A. Williams, unpublished observations; (150)].

Both biochemical and cytochemical studies have established the presence of hormone-sensitive adenylate cyclase activity in the plasma membrane of the pancreatic acinar cell (87, 131, 139, 190). Besides VIP and secretin, CCK at high concentrations also can activate the pancreatic adenylate cyclase, although this effect is not believed to be physiologically relevant (46). Cholera toxin, which activates adenylate cyclase by covalent modification of its associated N_s, increased cAMP in both guinea pig (49) and rat (115) pancreatic acini and stimulated amylase release. The N_i is also present in acinar plasma membranes, as indicated by NAD ribosylation after exposure to pertussis toxin (81, 97); it presumably mediates the action of somatostatin to lower cAMP levels. Forskolin, a diterpene that directly activates the catalytic subunit of the cyclase complex, elicited a 30-fold elevation of cAMP in rat pancreatic acinar cells and induced amylase release (58).

Phosphodiesterase inhibitors such as IBMX and theophylline alone have only minor effects on acinar cell cAMP levels, although, as previously noted for IBMX, they dramatically increase the rise induced by secretin or VIP. Some specificity has been noted in their potentiating effects on the changes induced by the two secretagogues. In guinea pig pancreatic acini, IBMX equally augments the effects of both secretin and VIP, while theophylline is more potent as a potentiating agent for the secretin-induced response (48). These observations suggest that secretin and VIP may modulate distinct hormonally responsive cAMP pools in the acinar cell. This view is supported by the observation that VIP is a more effective stimulator of digestive enzyme release (46), while secretin is both more potent and effective in activating electrolyte secretion (64).

Other Possible Intracellular Messengers

A number of additional phenomena that have been proposed as intracellular regulatory processes occur coincident with stimulus-secretion coupling in the pancreatic acinar cell.

Intracellular levels of cyclic GMP (cGMP) are increased by CCK, cholinergic agonists, and A23187 in pancreatic acinar cells (25). However, elevation of intracellular cGMP by exposure of acinar cells to the membrane-permeant analogues, 8-brcGMP or Bt$_2$cGMP, or by activation of endogenous guanylate cyclase by nitrosourea or nitroprusside has no effect on Ca^{2+} fluxes or amylase release (50, 55). The observed rise in cGMP after exposure of acinar cells to secretagogues therefore appears to be a secondary consequence of the induced elevation of [Ca^{2+}]$_i$. The function, if any, of cGMP in pancreatic stimulus-response coupling is unknown.

Pancreatic secretagogues also enhance the release of arachidonic acid by activation of both phospholipase C and phospholipase A$_2$ (57). This metabolite supplies both the cyclooxygenase and lipoxygenase pathways, resulting in increased synthesis of prostaglandins and leukotrienes (93). However, exogenously applied prostaglandins do not stimulate pancreatic protein secretion nor do cyclooxygenase inhibitors such as indomethacin inhibit secretagogue action (23, 158). Inhibition of the lipoxygenase pathway also has no effect on secretion of amylase by rat pancreatic acini (138). Thus prostaglandins and leukotrienes do not appear to play intermediary roles in pancreatic stimulus-response coupling.

The changes in membrane potential of acinar cells that follow exposure to secretagogues also have been suggested as a mechanism of transmembrane signaling. However, in isolated acinar cell suspensions, elevation of medium K$^+$ depolarizes the plasma membrane without stimulating amylase release (122). Removal of Na$^+$ from the extracellular medium also does not block amylase release (180). In addition the direction of the secretagogue-induced change in acinar cell membrane potential appears to be species specific (see the chapter by Petersen and Maruyama in this *Handbook*). Thus changes in monovalent ion fluxes and resulting alterations in the transmembrane potential appear to be results rather than causes of the observed rise in Ca^{2+} and seem rather to represent a central event in the effector mechanism that subserves fluid and electrolyte secretion.

EFFECTORS

Intracellular Receptors for Calcium, Diacylglycerol, and Cyclic AMP

The term *effector* was defined earlier as the cellular machinery, usually consisting of structural and regulatory proteins, that is activated by intracellular messengers. The action of intracellular messengers is initiated by binding to a specific intracellular receptor. In the case of cAMP, all of its known effects are mediated by cAMP-activated protein kinase (80). Binding of cAMP to the regulatory subunit of the kinase results in dissociation and activation of the catalytic subunit (80). Cyclic AMP–activated protein kinase activity, associated with specific cAMP binding, has been demonstrated in a supernatant fraction of guinea pig pancreas (70).

Diacylglycerol, transiently generated from the breakdown of plasma membrane PIP_2, is a specific activator of a unique, Ca^{2+}- and phospholipid-dependent protein kinase, protein kinase C (105, 168). In the combined presence of diacylglycerol, phosphatidylserine, and phosphatidylethanolamine, the enzyme is active at submicromolar Ca^{2+} concentrations. Thus protein kinase C may be activated in the range of Ca^{2+} concentrations believed to exist in the stimulated, and possibly resting, acinar cell. It is also established that protein kinase C is the intracellular receptor for phorbol esters that can act as analogues of diacylglycerol (20).

The third and best-studied pancreatic intracellular messenger is Ca^{2+}. Although it is possible for Ca^{2+} to directly bind to and maintain activity of an enzyme such as mitochondrial pyruvate dehydrogenase phosphatase (127), most actions of Ca^{2+} are initiated by binding to proteins with no intrinsic enzyme activity. This latter type of Ca^{2+}-regulated protein is typified by calmodulin, the major Ca^{2+}-receptor protein in eucaryotic cells (11). Calmodulin is a ubiquitous 19,000-M_r acidic protein that binds 4 mol Ca^{2+}/1 mol protein. Upon binding Ca^{2+}, the Ca^{2+}-calmodulin complex undergoes a conformational change that promotes its association with target enzymes (in most cases) and subsequent enzyme activation (11). To activate an enzyme, at least three of the four Ca^{2+}-binding sites on calmodulin must be occupied, although whether binding to these sites is sequential or simultaneous, positively cooperative, or noncooperative, is unclear (127). The initial affinity of calmodulin for Ca^{2+} is generally believed to be of the order of a few micromolar, so in the resting cell most of the Ca^{2+}-binding sites are unoccupied. Calmodulin is responsible for activating several enzymes in pancreas, including one or more protein kinases, a protein phosphatase (see PROTEIN KINASES AND PHOSPHATASES, this page), and cyclic nucleotide phosphodiesterase (172). Because the binding domain of the Ca^{2+}-calmodulin complex is hydrophobic, several hydrophobic interacting drugs, including phenothiazines such as trifluoperazine, block enzyme activation. Although these agents inhibit pancreatic enzyme secretion, they have multiple effects on acinar cell function, which makes such results difficult to interpret (186).

In addition to calmodulin, several other Ca^{2+}-binding proteins have been characterized, although most appear to be limited to nonexocrine tissues (173). A notable exception is a class of Ca^{2+}-binding proteins purified from adrenal medulla [synexin (28)] and other tissues, including pancreas [calelectrins (162)], that bind secretory granule membranes at micromolar Ca^{2+} concentrations and promote membrane aggregation and, under certain conditions, membrane fusion. Because of these properties, these proteins have been suggested as mediators of the attachment of secretory granules to the plasma membrane, or granules to one another, in the process of compound exocytosis. However, the physiological significance of these proteins remains to be established, since promotion of membrane aggregation and fusion generally requires supraphysiological Ca^{2+} concentrations on the order of 10–100 μM.

Protein Phosphorylation as an Effector System

PROTEIN KINASES AND PHOSPHATASES. Cyclic nucleotide–activated protein kinase activity has been described in rat, guinea pig, and mouse pancreas (18, 70, 92). Jensen and Gardner (70) were able to separate distinct cAMP- and cGMP-activated kinase activities, each associated with appropriate cyclic nucleotide-binding activity. Cyclic AMP- but not cGMP-activated kinase activity was inhibited by the protein kinase inhibitor protein from bovine heart. The cAMP kinase activity is active in phosphorylating exogenous type II and III histone with half-maximal activation at 20–50 nM cAMP. Cyclic AMP kinase activity also phosphorylates soluble and particulate endogenous proteins, some of which are also phosphorylated by Ca^{2+}-activated kinases. Two isoforms of cAMP-activated kinase (types I and II), which differ in their regulatory subunits, exist in other tissues. Although type II cAMP kinase activity has been partially purified from rat pancreas (92), the relative amounts of the two isoforms in pancreas are unknown.

In situ activation of cAMP kinase in response to hormones can be estimated by homogenizing cells and measuring kinase activity without added cAMP. This activity can then be related to the maximal cAMP-activated activity. These studies have shown that secretin and VIP but not CCK activate the kinase in guinea pig and rat pancreas (61, 70). There is a reasonably good correlation between the ability of VIP and secretin to increase cellular cAMP and protein kinase activity. Because CCK, a Ca^{2+}-mediated secretagogue, does not alter the action of VIP or secretin on protein kinase activity, the potentiating action of these two classes of secretagogues is not at the level of the cAMP-activated protein kinase. Thus the fundamental properties of the cAMP-activated protein kinase in pancreas are similar to those of other tissues. Although this kinase is activated by cAMP-mediated secretagogues, its physiological substrates are largely unknown. Even less is known about the separate cGMP-activated kinase in acinar cells (174).

In contrast to a single ubiquitous cAMP-activated kinase, a number of distinct Ca^{2+}-activated kinases exist that differ in their substrate specificities. Phosphorylase kinase, myosin light-chain kinase, glycogen synthetase kinase, and synapsin kinase are specific kinases present in various tissues (26). All of these kinases are activated by Ca^{2+} and calmodulin. Distinct from the calmodulin-dependent kinases is the phospholipid-dependent protein kinase, protein kinase C. Although there is some overlap, the substrate specificities of the phospholipid-dependent and calmodulin-dependent kinases are generally different. Activation of both the phospholipid-dependent kinase and a calmodulin-mediated Ca^{2+} kinase is required in blood platelets to elicit a full secretory response (105). Pancreatic acinar cells possess both calmodulin- and phospholipid-dependent protein kinases that become active on addition of Ca^{2+} (Fig. 11). When Ca^{2+} is added to a high-speed supernatant fraction of pancreas, kinase activity on endogenous proteins can be demonstrated (18). This activity is calmodulin dependent, since the activity is lost when endogenous calmodulin is removed by extraction with phenothiazine coupled to a solid matrix. The activity is recovered by adding exogenous calmodulin, is particularly active on a protein substrate with a reported 92,000 to 100,000 M_r, but also acts on a 50,000- to 52,000-M_r protein. In the presence of calmodulin this activity is activated half maximally by 1 μM Ca^{2+} and maximally by 20 μM Ca^{2+} and is inhibited by trifluoperazine. It is not, however, active on exogenous substrates such as histones, α-casein, or rabbit muscle phosphorylase. Gorelick et al. (53) have purified a specific calmodulin-activated kinase from rat pancreas based on its ability to phosphorylate ribosomal protein S6. The 500,000-M_r holoenzyme contains a 51,000-M_r subunit that binds calmodulin and undergoes apparent autophosphorylation. Thus it is similar to glycogen synthetase kinase isolated from liver and the multifunctional calmodulin kinase II isolated from brain (148). Whether it accounts for part or all of the calmodulin-dependent kinase activity in pancreas, as well as its physiological substrates, is not known.

The possibility that more than one form of calmodulin-dependent kinase exists in pancreas was raised recently by Nairn et al. (104). These authors partially purified from rat pancreas cytosol a calmodulin-dependent enzyme that, based on its purification properties and substrate specificity, was distinct from previously identified calmodulin-dependent protein kinases. The enzyme, which they call calmodulin kinase III, was highly selective for a 100,000-M_r protein substrate purified from pancreatic cytosol.

Pancreatic cytosol also contains phospholipid-dependent kinase activity (protein kinase C) that can be separated from other kinases by ion-exchange chromatography and whose dissociation constant for Ca^{2+} is reduced to the micromolar level by diacylglycerol (18, 106, 188). In one study of rat pancreas this kinase phosphorylated endogenous proteins of 38,000, 30,000, 22,000, and 15,000 M_r (188). In a study of mouse pancreas the major proteins phosphorylated were 66,000- and 40,000-M_r proteins (18). In both studies the kinase was also shown to phosphorylate type III (lysine-rich) histone and was half maximally activated by 1–10 μM Ca^{2+}. It was inhibited by trifluoperazine but at higher concentrations than necessary to inhibit calmodulin-dependent protein kinase activity (18). In pancreas as well as other cells, phorbol esters such as

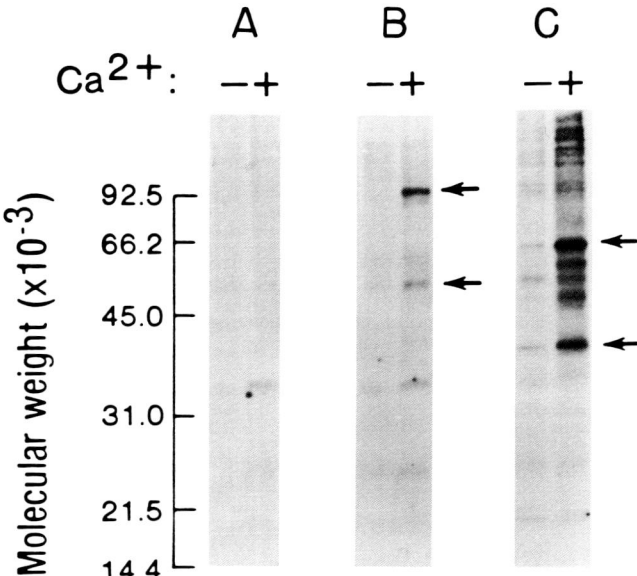

FIG. 11. Demonstration of separate calmodulin- and phospholipid-dependent, Ca^{2+}-activated kinase activity in mouse pancreas cytosol. Acinar cytosol was initially treated with phenothiazine coupled to a solid matrix to remove endogenous calmodulin. Cytosol was incubated with [γ-^{32}P]ATP in the absence ($-$) or presence ($+$) of Ca^{2+} alone (A), with added calmodulin (B), or added phosphatidylserine (C) and phosphorylated proteins resolved by polyacrylamide gel electrophoresis and autoradiography. Ca^{2+}-activated kinase activity is present in B and C and shown to act on different endogenous substrates. [From Burnham and Williams (18).]

TPA cause a rapid translocation of the protein kinase C; in addition 18,000-M_r and 22,000-M_r substrate proteins move from soluble to particulate fractions (178).

Although the previously discussed studies have been of cytosolic Ca^{2+}-activated kinases, kinase activity also can be observed in particulate fractions. Burnham and Williams (18) described calmodulin-dependent kinase activity in particulate fractions of mouse pancreas treated with ethylenediaminetetraacetic acid (EDTA), and Gorelick et al. (53) described ribosomal protein S6 kinase in a rat pancreatic microsomal fraction. Recently Ca^{2+}-activated as well as phospholipid-dependent Ca^{2+}-activated kinase activity has been described on zymogen granules purified from rat pancreas (16, 189). Here the major substrate is a 13,000- to 18,000-M_r protein present in the zymogen granule membrane. The kinase can be removed by high salt or EDTA, whereas the substrate is integral to the granule membrane (Fig. 12). Zymogen granules also possess cAMP-activated kinase activity, but this is active only on exogenous proteins (16, 90).

Regulation of cellular phosphoproteins also involves protein phosphatases, and the state of protein phosphorylation represents a balance between kinase and phosphatase activity (26, 67). Specific calmodulin-dependent Ca^{2+}-activated phosphatases have been described in skeletal muscle (type IIB phosphatase) and brain (calcineurin) (26). Pancreatic cytosol contains

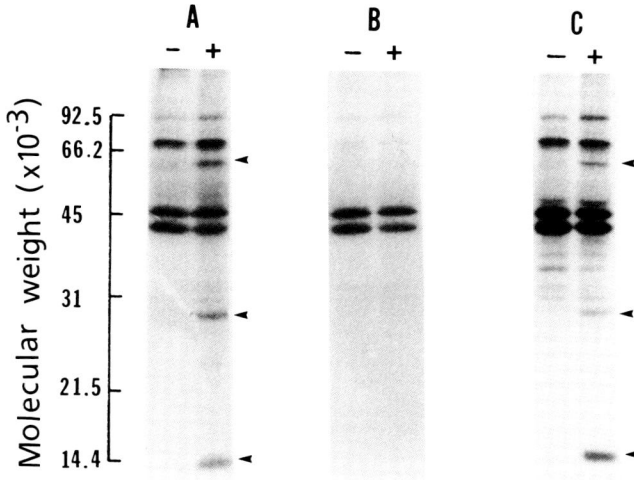

FIG. 12. Autoradiographs of polyacrylamide gels of purified zymogen granules incubated with [^{32}P]ATP and phosphatidylserine, in the presence (+) or absence (−) of Ca^{2+}. A: intact purified granules. B: granules extracted with KCl prior to ^{32}P labeling. C: granules extracted with KCl after ^{32}P labeling. [From Burnham et al. (16).]

Ca^{2+}-activated phosphatase activity that can be assayed with phosphorylated phosphorylase kinase or α-casein as substrates (Fig. 13). This activity has been partially purified and possesses similar properties to calcineurin (14). It is not known whether this activity accounts for the dephosphorylation induced by Ca^{2+}-mediated secretagogues of specific proteins in situ.

SECRETAGOGUE-INDUCED CHANGES IN CELLULAR PROTEIN PHOSPHORYLATION. Considerable data exist for changes in the phosphorylation of regulatory proteins involved in mediating the effects of hormones and neurotransmitters in diverse tissues and physiological processes (for review, see ref. 26). The description of the cascade of phosphorylation events involved in activation of hepatic and muscle glycogenolysis by cAMP-mediated hormones such as glucagon and epinephrine began this work. It was extended by the discovery that the activation of hepatic glycogenolysis by Ca^{2+}-mediated agonists such as α-adrenergic agents, angiotensin, and vasopressin also involved changes in phosphorylation of the enzymes phosphorylase and glycogen synthetase. Most of the specific enzymes or regulatory proteins involved in exocytosis or ion transport in secretory cells such as the pancreas have not been identified. Therefore the more general approach has been to determine if secretagogues induce changes in phosphorylation of any proteins.

In pancreatic acini and lobules prelabeled with $^{32}P_i$, secretagogues influence the phosphorylation of a number of proteins, as judged by one-dimensional SDS gel electrophoresis. In rat and mouse a major regulated 29,000- to 32,500-M_r phosphoprotein is present in a total particulate fraction (17, 43, 68). Phosphorylation of this protein is increased by CCK, secretin, carbachol, or Bt_2cAMP at concentrations similar to those that elicit secretion. In rat pancreas this effect was seen as early as 30 s, was maintained for 30 min, and was reversed when a receptor blocker was added (42). In mouse acini, however, this phosphorylation was only slowly reversible and was also induced by insulin, which by itself is not a secretagogue (17). This phosphoprotein has been identified by two-dimensional electrophoresis as the ribosomal S6 protein (43, 68) and therefore is probably not involved in mediation of acute secretory events. Another regulated phosphoprotein identified in exocrine pancreas is the 56,000-M_r vitamin D–binding protein (177). This protein, which was identified immunologically in cytosol of rat pancreatic acini, was a major substrate of Ca^{2+}-phospholipid–dependent protein kinase activity in broken-cell extracts of acini stimulated with carbachol. Although the role of this protein in pancreatic function is unknown, in B lymphocytes it appears to be involved in the linkage between membrane immunoglobulins and actin and therefore may be involved in connecting membrane glycoproteins to the cytoskeleton (119). Recently the regulated phosphorylation of myosin light chain was demonstrated in mouse acini (16a).

Pancreatic secretagogues alter phosphorylation of other, as yet unidentified, proteins. Freedman and Jamieson (42) reported that phosphorylation of a number of proteins was increased after 30-s exposure to secretagogues, but the effect was no longer seen at 30 min. This may be because in their experiments $^{32}P_i$ was removed from the medium at the time of stimu-

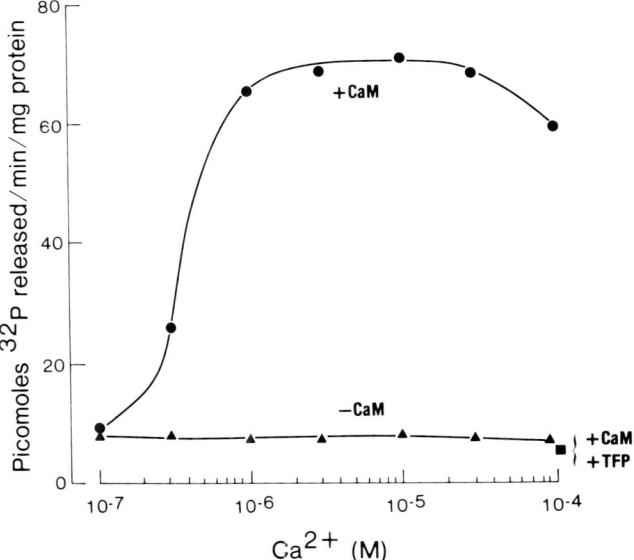

FIG. 13. Ca^{2+} sensitivity of Ca^{2+}-activated protein phosphatase activity purified from pancreatic cytosol. [^{32}P]casein was used as a substrate. Phosphatase activity is totally dependent on presence of calmodulin (CaM). TFP, trifluoperazine. [From Burnham (14).]

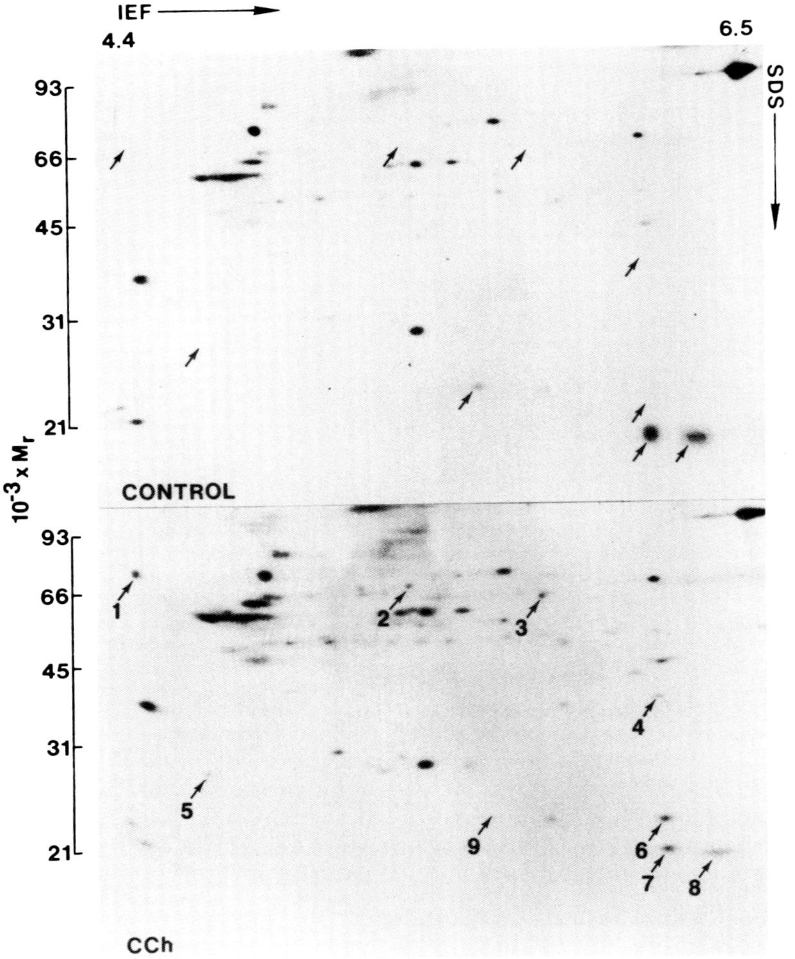

FIG. 14. Autoradiographs of soluble proteins obtained from mouse acini that were incubated with [^{32}P]phosphate for 60 min and stimulated with 3 µM carbachol (CCh) for 5 min. Soluble proteins were subjected to 2-dimensional polyacrylamide gel electrophoresis; proteins that undergo an alteration in phosphorylation in response to carbachol are numbered and indicated by *arrows*. IEF, isoelectric focusing dimension; SDS, sodium dodecyl sulfate. [From Burnham et al. (15).]

lation. Burnham and Williams (17) left ^{32}P$_i$ in the medium during stimulation and reported altered phosphorylation of four soluble proteins in response to carbachol, CCK, or A23187. Phosphorylation was increased for 16,000- and 23,000-M_r proteins and decreased for 20,500- and 21,000-M_r proteins. Most rapid changes were observed for the dephosphorylations, which were maximal after 60 s. After a receptor blocker was added, these dephosphorylations were rapidly reversible and were not induced by insulin, although insulin increased the phosphorylation of the 16,000- and 23,000-M_r soluble proteins. Dephosphorylation of the 20,500- and 21,000-M_r proteins correlated best with the onset of secretion. Application of two-dimensional gel electrophoresis has shown these proteins to have a pI = 6.2–6.4 (15), but their further identity and function remain to be determined.

In addition to the four aforementioned proteins resolved on one-dimensional SDS gels, at least six other soluble phosphoproteins have been identified by two-dimensional gel electrophoresis to undergo changes in phosphorylation in response to Ca^{2+}-mediated secretagogues (Fig. 14). In mouse and guinea pig acini, the phorbol ester TPA and Ca^{2+} ionophore A23187 altered the phosphorylation of distinct subsets of these proteins. The phosphorylation of four proteins was increased by TPA, whereas the ionophore increased the phosphorylation of two and decreased the phosphorylation of three other proteins. Administered together, these agents reproduced all of the changes in phosphorylation elicited by carbachol or CCK (15). Thus it is apparent that carbachol and CCK act via both diacylglycerol and Ca^{2+} to regulate pancreatic protein phosphorylation, presumably through activation of protein kinase C and various calmodulin-dependent protein kinases and phosphatases.

Studies of the effects of cAMP-mediated secretagogues on phosphorylation have generated less clear-cut results. Phosphorylation of ribosomal protein S6 in rat pancreatic lobules is increased in response to secretin (43). Recent studies in mouse acini using two-dimensional gel electrophoresis showed that VIP and 8-brcAMP altered the phosphorylation of a unique 56,000-M_r soluble protein as well as several of the previously mentioned proteins affected by carbachol and CCK (D. B. Burnham, C. Sung, and J. A. Williams, unpublished observations). Roberts and

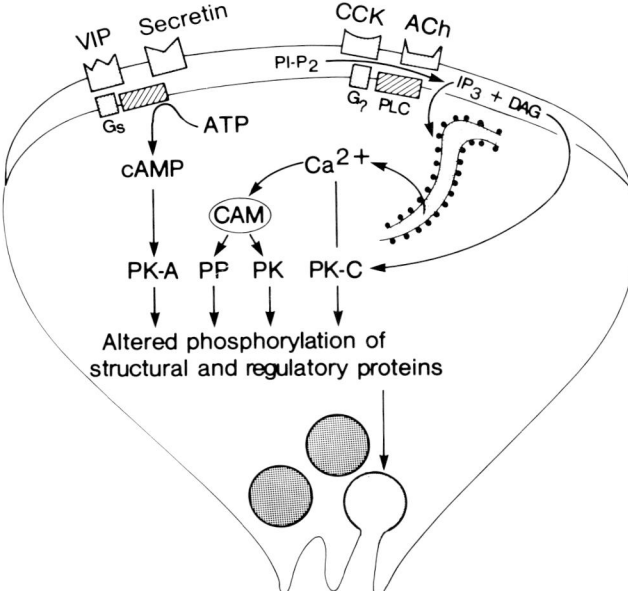

FIG. 15. Schematic diagram of stimulus-secretion coupling of pancreatic acinar cell protein secretion. ACh, acetylcholine; CAM, calmodulin; CCK, cholecystokinin; DAG, diacylglycerol; G_s, stimulatory guanine nucleotide–binding protein; $G_?$, unknown guanine nucleotide–regulated protein; IP_3, inositol 1,4,5-trisphosphate; PIP_2, phosphatidylinositol 4,5-bisphosphate; PK, protein kinase; PK-A, cAMP-activated protein kinase; PK-C, phospholipid-dependent protein kinase; PLC, phospholipase C; PP, protein phosphatase; VIP, vasoactive intestinal polypeptide.

Butcher (132), using minced mouse pancreas, reported that Bt_2cAMP induced the phosphorylation of 95,000-, 32,000-, and 20,000-M_r proteins. Thus although there appears to be some similarity in the phosphoprotein substrates affected by Ca^{2+}- and cAMP-mediated secretagogues, particularly ribosomal protein S6, these secretagogues also exert different patterns of phosphorylation. It remains to be established whether potentiation of enzyme secretion by Ca^{2+} and cAMP occurs at one or more shared and/or distinct phosphoprotein substrates.

The role of intracellular messengers and effectors in pancreatic enzyme secretion is summarized in Figure 15. Stimulation of secretion normally involves a synergistic interaction between intracellular messenger pathways, although in some species, such as the guinea pig, a sizable secretory response can be elicited via a single pathway. Although altered protein phosphorylation is currently a favored mode of messenger action in pancreatic stimulus-secretion coupling, much work is needed to link changes in phosphorylation of specific proteins with the regulation of acinar cell function.

Exocytosis

A terminal step in pancreatic stimulus-secretion coupling is the release of digestive enzymes stored in zymogen granules. Because the various enzymes are distributed in all granules, only a single mechanism is required to control the release of granule contents, and this is believed to occur by the process of exocytosis (114). Such a scheme requires all enzymes to be released in parallel in proportion to their relative amounts in the acinar cell granules. Clear data exist, however, documenting nonparallel secretion in various situations (135). This has led to the promulgation of theories other than exocytosis, most notably the equilibrium model, in which proteins exit from the cell by diffusional processes (136). Adelson and Miller (2), however, recently reconciled these views by showing that nonparallel secretion was compatible with a secretory model involving exocytosis. Moreover, different granules from the same pancreas differ markedly in protein composition (103).

The basic scheme outlined by Palade (114) is still adequate to describe the overall morphological sequence of events in exocytosis. This sequence includes *1*) translocation of the zymogen granule from its point of formation, usually near the center of the cell, to a position near the luminal plasma membrane; *2*) close apposition of the granule and the plasma membrane; *3*) fusion of granule and plasma membranes; and *4*) release of granule content and retrieval of membrane. Translocation of secretory granules to the luminal membrane may involve microtubules and microfilaments. Based on studies in other systems, microtubules may be involved in longer movements, whereas microfilaments are involved in short-range positioning (120). Because in acinar cells exocytosis is limited to one membrane domain, either the luminal membrane must possess unique release sites or the intracellular transport system channels granules specifically to this membrane. Whether this translocating system is influenced by secretagogues is unclear, although the reversible phosphorylation of pancreatic myosin in response to CCK (16a) certainly raises this possibility.

How second messengers and effectors are involved in promoting membrane fusion is a question just beginning to be addressed (30). In acinar cells the major components are the zymogen granule and the luminal plasma membranes, which must interact in a milieu of cytoplasmic proteins and ions. Calcium itself may promote fusion by neutralizing net negative surface charge on the granule and plasma membrane. However, this effect requires supraphysiological concentrations and would not explain fusion in cAMP- or diacylglycerol-mediated secretion. The various work reviewed in this chapter suggests that the second messengers act by altering protein phosphorylation. Cytosolic granule-binding proteins such as synexin may also facilitate fusion. In chromaffin cells one granule-binding protein has been shown to be a substrate for protein kinase C (163). Thus reversible phosphorylation of a cytosolic protein could regulate the ability to promote fusion. Similarly, either a cytosolic- or granule-associated kinase or phosphatase could alter the phosphorylation state of a granule or plasma membrane protein, rendering it fusion competent. In other systems, in vitro reconstitution is being used to define the minimal components necessary for regulated secretion and explain how second messengers and effectors bring this about at the molecular level (30). How these data relate to pancreatic acinar secretion remains to be determined.

REFERENCES

1. ABDELMOUMENE, S., AND J. D. GARDNER. Cholecystokinin-induced desensitization of enzyme secretion in dispersed acini from guinea pig pancreas. *Am. J. Physiol.* 239 (*Gastrointest. Liver Physiol.* 2): G272–G279, 1980.
2. ADELSON, J. W., AND P. E. MILLER. Pancreatic secretion by nonparallel exocytosis: potential resolution of a long controversy. *Science Wash. DC* 228: 993–996, 1985.
3. APPERT, H. E., T. H. CHIU, G. C. BUDD, A. J. LEONARDI, AND J. M. HOWARD. ^{3}H-methyl scopolamine binding to dispersed pancreatic acini. *Cell Tissue Res.* 220: 673–684, 1981.
4. ASSELIN, J., L. LAROSE, AND J. MORISSET. Short-term cholinergic desensitization of the rat pancreatic secretory response. *Am. J. Physiol.* 252 (*Gastrointest. Liver Physiol.* 15): G392–G397, 1987.
5. BAYERDÖRFFER, E., H. STREB, L. ECKHARDT, W. HAASE, AND I. SCHULZ. Characterization of calcium uptake into rough endoplasmic reticulum of rat pancreas. *J. Membr. Biol.* 81: 69–82, 1984.
6. BENZ, L., B. ECKSTEIN, E. K. MATTHEWS, AND J. A. WILLIAMS. Control of pancreatic amylase release in vitro: effects of ions, cyclic AMP and colchicine. *Br. J. Pharmacol.* 46: 66–77, 1972.
7. BERGERON, J. J. M., R. RACHUBINSKI, N. SEARLE, R. SIKSTROM, D. BORTS, P. BASTIAN, AND B. I. POSNER. Radioautographic visualization of in vivo insulin binding to the exocrine pancreas. *Endocrinology* 107: 1069–1080, 1980.
8. BERRIDGE, M. J. Rapid accumulation of inositol trisphosphate reveals that agonists hydrolyze polyphosphoinositides instead of phosphatidylinositol. *Biochem. J.* 212: 849–858, 1983.
9. BERRIDGE, M. J. Inositol trisphosphate and diacylglycerol as second messengers. *Biochem. J.* 220: 345–360, 1984.
10. BISSONNETTE, B. M., M. J. COLLEN, H. ADACHI, R. T. JENSEN, AND J. D. GARDNER. Receptors for vasoactive intestinal peptide and secretin on rat pancreatic acini. *Am. J. Physiol.* 246 (*Gastrointest. Liver Physiol.* 9): G710–G717, 1984.
11. BROSTROM, C. O., AND D. J. WOLFF. Properties and functions of calmodulin. *Biochem. Pharmacol.* 30: 1395–1405, 1981.
12. BRUZZONE, R., T. POZZAN, AND C. B. WOLLHEIM. Caerulein and carbamoylcholine stimulate pancreatic amylase release at resting cytosolic free Ca^{2+}. *Biochem. J.* 235: 139–143, 1986.
13. BURGESS, G. M., J. S. MCKINNEY, R. F. IRVINE, AND J. W. PUTNEY, JR. Inositol 1,4,5-trisphosphate and inositol 1,3,4-trisphosphate formation in Ca^{2+}-mobilizing-hormone-activated cells. *Biochem. J.* 232: 237–243, 1985.
14. BURNHAM, D. B. Characterization of Ca^{2+}-activated protein phosphatase activity in exocrine pancreas. *Biochem. J.* 231: 335–342, 1985.
15. BURNHAM, D. B., P. MUNOWITZ, S. R. HOOTMAN, AND J. A. WILLIAMS. Regulation of protein phosphorylation in pancreatic acini: distinct effects of Ca^{2+} ionophore A23187 and 12-O-tetradecanoylphorbol 13 acetate. *Biochem. J.* 235: 125–131, 1986.
16. BURNHAM, D. B., P. MUNOWITZ, N. THORN, AND J. A. WILLIAMS. Protein kinase activity associated with pancreatic zymogen granules. *Biochem. J.* 227: 743–751, 1985.
16a. BURNHAM, D. B., H.-D. SOLING, AND J. A. WILLIAMS. Evaluation of myosin light chain phosphorylation in isolated pancreatic acini. *Am. J. Physiol.* 254 (*Gastrointest. Liver Physiol.* 17): G130–G134, 1988.
17. BURNHAM, D. B., AND J. A. WILLIAMS. Effects of carbachol, cholecystokinin and insulin on protein phosphorylation in isolated pancreatic acini. *J. Biol. Chem.* 257: 10523–10528, 1982.
18. BURNHAM, D. B., AND J. A. WILLIAMS. Activation of protein kinase activity in pancreatic acini by calcium and cAMP. *Am. J. Physiol.* 246 (*Gastrointest. Liver Physiol.* 9): G500–G508, 1984.
19. CASE, R. M., AND T. CLAUSEN. The relationship between calcium exchange and enzyme secretion in the isolated rat pancreas. *J. Physiol. Lond.* 235: 75–102, 1973.
20. CASTAGNA, M., Y. TAKAI, K. KAIBUCHI, K. SANO, V. KIKKAWA, AND Y. NISHIZUKA. Direct activation of calcium activated, phospholipid-dependent protein kinase by tumor-promoting phorbol ester. *J. Biol. Chem.* 257: 7847–7851, 1982.
21. CHANDLER, D. E., AND J. A. WILLIAMS. Intracellular divalent cation release in pancreatic acinar cells during stimulus-secretion coupling. II. Subcellular localization of the fluorescent probe chlorotetracycline. *J. Cell Biol.* 76: 389–399, 1978.
22. CHANG, R. S. L., V. J. LOTTI, R. L. MONAGHAN, J. BIRNBAUM, E. O. STAPLEY, M. A. GOETZ, G. ALBERS-SCHONBERG, AND A. A. PATCHETT. A potent nonpeptide cholecystokinin antagonist selective for peripheral tissues isolated from *Aspergillus alliaceus*. *Science Wash. DC* 230: 177–180, 1985.
23. CHAUVELOT, L., S. HEISLER, J. HUET, AND D. GANGON. Prostaglandins and enzyme secretion from dispersed rat pancreatic acinar cells. *Life Sci.* 25: 913–920, 1979.
24. CHRISTOPHE, J. P., T. P. CONLON, AND J. D. GARDNER. Interaction of porcine vasoactive intestinal peptide with dispersed pancreatic acinar cells from the guinea pig. *J. Biol. Chem.* 251: 4629–4634, 1976.
25. CHRISTOPHE, J. P., E. K. FRANDSEN, T. P. CONLON, G. KRISHNA, AND J. D. GARDNER. Action of cholecystokinin, cholinergic agents and A23187 on accumulation of guanosine 3',5'-monophosphate in dispersed guinea pig pancreatic acinar cells. *J. Biol. Chem.* 251: 4640–4645, 1976.
26. COHEN, P. The role of protein phosphorylation in neural and hormonal control of cellular activity. *Nature Lond.* 296: 613–620, 1982.
27. COLLINS, S. M., S. ABDELMOUMENE, R. T. JENSEN, AND J. D. GARDNER. Cholecystokinin-induced persistent stimulation of enzyme secretion from pancreatic acini. *Am. J. Physiol.* 240 (*Gastrointest. Liver Physiol.* 3): G459–G465, 1981.
28. CREUTZ, C. E., AND D. C. STERNER. Calcium dependence of the binding of synexin to isolated chromaffin granules. *Biochem. Biophys. Res. Commun.* 114: 355–364, 1983.
29. DEHAYE, J. P., J. WINAND, P. POLOCZEK, AND J. CHRISTOPHE. Characterization of muscarinic cholinergic receptors on rat pancreatic acini by N-[^{3}H]methylscopolamine binding. Their relationship with calcium 45 efflux and amylase secretion. *J. Biol. Chem.* 259: 294–300, 1984.
30. DE LISLE, R. C., AND J. A. WILLIAMS. Regulation of membrane fusion in secretory exocytosis. *Annu. Rev. Physiol.* 48: 225–238, 1986.
31. DORMER, R. L. Direct demonstration of increases in cytosolic free Ca^{2+} during stimulation of pancreatic enzyme secretion. *Biosci. Rep.* 3: 233–240, 1983.
32. DORMER, R. L. Introduction of calcium chelators into isolated rat pancreatic acini inhibits amylase release in response to carbamylcholine. *Biochem. Biophys. Res. Commun.* 119: 876–883, 1984.
33. DORMER, R. L., J. H. POULSEN, V. LICKO, AND J. A. WILLIAMS. Calcium fluxes in isolated pancreatic acini: effects of secretagogues. *Am. J. Physiol.* 240 (*Gastrointest. Liver Physiol.* 3): G38–G49, 1981.
34. DORMER, R. L., AND J. A. WILLIAMS. Secretagogue-induced changes in subcellular Ca^{2+} distribution in isolated pancreatic

acini. *Am. J. Physiol.* 240 (*Gastrointest. Liver Physiol.* 3): G130–G140, 1981.
35. DOUGLAS, W. W. Stimulus-secretion coupling: the concept and clues from chromaffin and other cells. *Br. J. Pharmacol.* 34: 451–474, 1968.
36. DUPONT, J. J. H. H. M., AND A. M. M. FLEUREN-JAKOBS. Synergistic effect of A23187 and a phorbol ester on amylase secretion from rabbit pancreatic acini. *FEBS Lett.* 170: 64–68, 1984.
37. EIMERL, S., N. SAVION, O. HEICHAL, AND Z. SELINGER. Induction of enzyme secretion in rat pancreatic slices using the ionophore A23187 and calcium. An experimental bypass of the hormone receptor pathway. *J. Biol. Chem.* 249: 3991–3993, 1974.
38. ESTEVE, J. P., C. SUSINI, N. VAYSSE, H. ANTONIOTTI, E. WUNSCH, G. BERTHON, AND A. RIBET. Binding of somatostatin to pancreatic acinar cells. *Am. J. Physiol.* 247 (*Gastrointest. Liver Physiol.* 10): G62–G69, 1984.
39. ESTEVE, J. P., N. VAYSSE, C. SUSINI, J. M. KUNSCH, D. FOURMY, L. PRADAYROL, E. WUNSCH, L. MORODER, AND A. RIBET. Bimodal regulation of pancreatic exocrine function in vitro by somatostatin-28. *Am. J. Physiol.* 245 (*Gastrointest. Liver Physiol.* 8): G208–G216, 1983.
40. FARESE, R. V., R. E. LARSON, AND M. A. SABIR. Ca^{2+}-dependent and Ca^{2+}-independent effects of pancreatic secretagogues on phosphatidylinositol metabolism. *Biochim. Biophys. Acta* 710: 391–399, 1982.
41. FARESE, R. V., J. L. ORCHARD, R. E. LARSON, M. A. SABIR, AND J. S. DAVIS. Phosphatidylinositol hydrolysis and phosphatidylinositol 4',5'-diphosphate hydrolysis are separable responses during secretagogue action in the rat pancreas. *Biochim. Biophys. Acta* 846: 296–304, 1985.
42. FREEDMAN, S. D., AND J. D. JAMIESON. Hormone-induced protein phosphorylation. I. Relation between secretagogue action and endogenous protein phosphorylation in intact cells from exocrine pancreas and parotid. *J. Cell Biol.* 95: 903–908, 1982.
43. FREEDMAN, S. D., AND J. D. JAMIESON. Hormone-induced protein phosphorylation. II. Localization to the microsomal fraction from rat exocrine pancreas and parotid of a 29,000-dalton protein phosphorylated in situ in response to secretagogues. *J. Cell Biol.* 95: 909–917, 1982.
44. GARDNER, J. D., T. P. CONLON, H. L. KLAEVEMAN, T. D. ADAMS, AND M. A. ONDETTI. Action of cholecystokinin and cholinergic agents on calcium transport in isolated pancreatic acinar cells. *J. Clin. Invest.* 56: 366–375, 1975.
45. GARDNER, J. D., C. L. COSTENBADER, AND E. R. UHLEMANN. Effect of extracellular calcium on amylase release from dispersed pancreatic acini. *Am. J. Physiol.* 236 (*Endocrinol. Metab. Gastrointest. Physiol.* 5): E754–E762, 1979.
46. GARDNER, J. D., AND R. T. JENSEN. Regulation of pancreatic enzyme secretion in vitro. In: *Physiology of the Digestive Tract* (1st ed.), edited by L. R. Johnson. New York: Raven, 1981, p. 831–871.
47. GARDNER, J. D., AND R. T. JENSEN. Receptors mediating the actions of secretagogues on pancreatic acinar cells. In: *The Exocrine Pancreas: Biology, Pathobiology, and Diseases*, edited by V. L. W. Go, J. D. Gardner, F. P. Brooks, E. Lebenthal, E. P. DiMagno, and G. A. Scheele. New York: Raven, 1986, p. 109–122.
48. GARDNER, J. D., L. Y. KORMAN, M. D. WALKER, AND V. E. SUTLIFF. Effects of inhibitors of cyclic nucleotide phosphodiesterase on the actions of vasoactive intestinal peptide and secretin on pancreatic acini. *Am. J. Physiol.* 242 (*Gastrointest. Liver Physiol.* 5): G547–G551, 1982.
49. GARDNER, J. D., AND A. J. ROTTMAN. Action of cholera toxin on dispersed acini from guinea pig pancreas. *Biochim. Biophys. Acta* 585: 250–265, 1979.
50. GARDNER, J. D., AND A. J. ROTTMAN. Evidence against cyclic GMP as a mediator of the actions of secretagogues on amylase release from guinea-pig pancreas. *Biochim. Biophys. Acta* 627: 230–243, 1980.
51. GILMAN, A. G. Guanine nucleotide-binding regulatory proteins and dual control of adenylate cyclase. *J. Clin. Invest.* 73: 1–4, 1984.
52. GOLDFINE, I. D., AND J. A. WILLIAMS. Receptors for insulin and CCK in the acinar pancreas: relationship to hormone action. *Int. Rev. Cytol.* 85: 1–38, 1983.
53. GORELICK, F. S., J. A. COHN, S. D. FREEDMAN, N. G. DELAHUNT, J. M. GERSHONI, AND J. D. JAMIESON. Calmodulin-stimulated protein kinase activity from rat pancreas. *J. Cell Biol.* 97: 1294–1298, 1983.
54. GRYNKIEWICZ, G., M. POENIE, AND R. Y. TSIEN. A new generation of Ca^{2+} indicators with greatly improved fluorescence properties. *J. Biol. Chem.* 260: 3440–3450, 1985.
55. GUNTHER, G. R., AND J. D. JAMIESON. Increased cyclic GMP does not correlate with protein discharge from pancreatic acinar cells. *Nature Lond.* 280: 318–320, 1979.
56. HAASE, W., W. FRIESE, AND K. HEITMAN. Electron-microscopic demonstration of the distribution of calcium deposits in the exocrine pancreas of the rat after application of carbachol, atropine, cholecystokinin, and procaine. *Cell Tissue Res.* 235: 683–690, 1984.
57. HALENDA, S. P., AND R. P. RUBIN. Phospholipid turnover in isolated rat pancreatic acini. *Biochem. J.* 208: 713–721, 1982.
58. HEISLER, S. Forskolin potentiates calcium-dependent amylase secretion from rat pancreatic acinar cells. *Can. J. Physiol. Pharmacol.* 61: 1168–1176, 1983.
59. HOKIN, L. E. Effects of calcium omission on acetylcholine-stimulated amylase secretion and phospholipid synthesis in pigeon pancreas slices. *Biochim. Biophys. Acta* 115; 219–221, 1966.
60. HOKIN, M. R., AND L. E. HOKIN. Enzyme secretion and the incorporation of P^{32} into phospholipids of pancreas slices. *J. Biol. Chem.* 203: 967–977, 1953.
61. HOLIAN, O., C. T. BOMBECK, AND L. M. NYHUS. Hormonal stimulation of cyclic AMP-dependent protein kinase in rat pancreas. *Biochem. Biophys. Res. Commun.* 95: 553–561, 1980.
62. HOOTMAN, S. R., M. E. BROWN, AND J. A. WILLIAMS. Phorbol esters and A23187 regulate Na^+-K^+ pump activity in pancreatic acinar cells. *Am. J. Physiol.* 252 (*Gastrointest. Liver Physiol.* 15): G499–G505, 1987.
63. HOOTMAN, S. R., M. E. BROWN, J. A. WILLIAMS, AND C. D. LOGSDON. Regulation of muscarinic acetylcholine receptors in cultured guinea pig pancreatic acini. *Am. J. Physiol.* 251 (*Gastrointest. Liver Physiol.* 14): G75–G83, 1986.
64. HOOTMAN, S. R., S. A. ERNST, AND J. A. WILLIAMS. Secretagogue regulation of Na^+-K^+ pump activity in pancreatic acinar cells. *Am. J. Physiol.* 245 (*Gastrointest. Liver Physiol.* 8): G339–G346, 1983.
65. HOOTMAN, S. R., D. L. OCHS, AND J. A. WILLIAMS. Intracellular mediators of Na^+-K^+ pump activity in guinea pig pancreatic acinar cells. *Am. J. Physiol.* 249 (*Gastrointest. Liver Physiol.* 12): G470–G478, 1985.
66. HOOTMAN, S. R., T. M. PICADO-LEONARD, AND D. B. BURNHAM. Muscarinic acetylcholine receptor structure in acinar cells of mammalian exocrine glands. *J. Biol. Chem.* 260: 4186–4194, 1985.
66a. HOOTMAN, S. R., AND J. A. WILLIAMS. Stimulus-secretion coupling in the pancreatic acinus. In: *Physiology of the Gastrointestinal Tract* (2nd ed.), edited by L. R. Johnson. New York: Raven, 1987, vol. 2, p. 1129–1146.
67. INGEBRITSEN, T. S., AND P. COHEN. Protein phosphatases: properties and role in cellular regulation. *Science Wash. DC* 221: 331–338, 1983.
68. JAHN, R., AND H. D. SOLING. Phosphorylation of ribosomal protein S6 in response to secretagogues in the guinea pig exocrine pancreas, parotid and lacrimal gland. *FEBS Lett.* 153: 71–80, 1983.
69. JENSEN, R. T., C. G. CHARLTON, H. ADACHI, S. W. JONES, T. L. O'DONOHUE, AND J. D. GARDNER. Use of ^{125}I-secretin to identify and characterize high-affinity secretin receptors on pancreatic acini. *Am. J. Physiol.* 245 (*Gastrointest. Liver Physiol.* 8): G186–G195, 1983.

70. JENSEN, R. T., AND J. D. GARDNER. Cyclic nucleotide-dependent protein kinase activity in acinar cells from guinea pig pancreas. *Gastroenterology* 75: 806–817, 1978.
71. JENSEN, R. T., AND J. D. GARDNER. Interaction of physalaemin, substance P, and eledoisin with specific membrane receptors on pancreatic acinar cells. *Proc. Natl. Acad. Sci. USA* 76: 5679–5683, 1979.
72. JENSEN, R. T., S. W. JONES, K. FOLKERS, AND J. D. GARDNER. A synthetic peptide that is a bombesin receptor antagonist. *Nature Lond.* 309: 61–63, 1984.
73. JENSEN, R. T., S. W. JONES, Y. A. LU, M. C. XU, K. FOLKERS, AND J. D. GARDNER. Interaction of substance P antagonists with substance P receptors on dispersed pancreatic acini. *Biochim. Biophys. Acta* 804: 181–191, 1984.
74. JENSEN, R. T., G. F. LEMP, AND J. D. GARDNER. Interaction of cholecystokinin with specific membrane receptors on pancreatic acinar cells. *Proc. Natl. Acad. Sci. USA* 77: 2079–2083, 1980.
75. JENSEN, R. T., T. MOODY, C. PERT, J. E. RIVIER, AND J. D. GARDNER. Interaction of bombesin and litorin with specific membrane receptors on pancreatic acinar cells. *Proc. Natl. Acad. Sci. USA* 75: 6139–6143, 1978.
76. JENSEN, R. T., K. TATEMOTO, V. MUTT, G. F. LEMP, AND J. D. GARDNER. Actions of a newly isolated intestinal peptide PHI on pancreatic acini. *Am. J. Physiol.* 241 (*Gastrointest. Liver Physiol.* 4): G498–G502, 1981.
77. KONDO, S., AND I. SCHULZ. Calcium ion uptake in isolated pancreas cells induced by secretagogues. *Biochim. Biophys. Acta* 419: 76–92, 1976.
78. KORC, M., L. M. MATRISIAN, AND B. E. MAGUN. Cytosolic calcium regulates epidermal growth factor endocytosis in rat pancreas and cultured fibroblasts. *Proc. Natl. Acad. Sci. USA* 81: 461–465, 1984.
79. KORMAN, L. Y., M. D. WALKER, AND J. D. GARDNER. Action of theophylline on secretagogue-stimulated amylase release from dispersed pancreatic acini. *Am. J. Physiol.* 239 (*Gastrointest. Liver Physiol.* 2): G324–G333, 1980.
80. KREBS, E. G., AND J. A. BEAVO. Phosphorylation-dephosphorylation of enzymes. *Annu. Rev. Biochem.* 48: 923–959, 1979.
81. LAMBERT, M., M. SVOBODA, J. FURNELLE, AND J. CHRISTOPHE. Solubilization from rat pancreatic plasma membranes of a cholecystokinin (CCK) agonist-receptor complex interacting with guanine nucleotide regulatory proteins coexisting in the same macromolecular system. *Eur. J. Biochem.* 147: 611–617, 1985.
82. LAROSE, L., Y. DUMONT, J. ASSELIN, J. MORISSET, AND G. G. POIRIER. Muscarinic receptor of rat pancreatic acini: [^{3}H]QNB binding and amylase secretion. *Eur. J. Pharmacol.* 76: 247–254, 1981.
83. LAROSE, L., G. G. POIRIER, Y. DUMONT, C. FREGEAU, L. BLANCHARD, AND J. MORISSET. Modulation of rat pancreatic amylase secretion and muscarinic receptor populations by chronic bethanechol treatment. *Eur. J. Pharmacol.* 95: 213–223, 1983.
84. LIDDLE, R. A., I. D. GOLDFINE, AND J. A. WILLIAMS. Bioassay of plasma cholecystokinin in rats: effects of food, trypsin inhibitor, and alcohol. *Gastroenterology* 87: 542–549, 1984.
85. LOGSDON, C. D., AND J. A. WILLIAMS. Epidermal growth factor binding and biologic effects on mouse pancreatic acini. *Gastroenterology* 85: 339–345, 1983.
86. LOGSDON, C. D., AND J. A. WILLIAMS. Intracellular Ca^{2+} and phorbol esters synergistically inhibit internalization of epidermal growth factor in pancreatic acini. *Biochem. J.* 223: 893–900, 1984.
87. LONG, B. W., AND J. D. GARDNER. Effects of cholecystokinin on adenylate cyclase activity in dispersed pancreatic acinar cells. *Gastroenterology* 73: 1008–1014, 1977.
88. LOUIE, D. S., J. A. WILLIAMS, AND C. OWYANG. Action of pancreatic polypeptide on rat pancreatic secretion: in vivo and in vitro. *Am. J. Physiol.* 249 (*Gastrointest. Liver Physiol.* 12): G489–G495, 1985.
89. LUCAS, M., A. GALVAN, P. SOLANO, AND R. GOBERNA. Compartmentation of calcium in digitonin-disrupted guinea pig pancreatic acinar cells. *Biochim. Biophys. Acta* 731: 129–136, 1983.
90. MACDONALD, R. J., AND R. A. RONZIO. Phosphorylation of a zymogen granule membrane polypeptide from rat pancreas. *FEBS Lett.* 40: 203–206, 1974.
91. MADISON, L. D., S. A. ROSENZWEIG, AND J. D. JAMIESON. Use of the heterobifunctional cross-linker m-maleimidobenzoyl N-hydroxysuccinimide ester to affinity label cholecystokinin binding proteins on rat pancreatic plasma membranes. *J. Biol. Chem.* 259: 14818–14823, 1984.
92. MANGEAT, P. H., H. CHAHINIAN, AND G. J. MARCHIS-MOUREN. Characterization of the cyclic AMP-dependent protein kinase from rat pancreas, further purification of the catalytic subunit, substrate specificity, effect of the pancreatic heat-stable inhibitor. *Biochimie Paris* 60: 777–785, 1978.
93. MARSHALL, P. J., D. E. BOATMAN, AND L. E. HOKIN. Direct demonstration of the formation of prostaglandin E_2 due to phosphatidylinositol breakdown associated with stimulation of enzyme secretion in the pancreas. *J. Biol. Chem.* 256: 844–847, 1981.
94. MATOZAKI, T., C. SAKAMOTO, M. NAGAO, AND S. BABA. Phorbol ester or diacylglycerol modulates somatostatin binding to its receptors on rat pancreatic acinar cell membranes. *J. Biol. Chem.* 261: 1414–1420, 1986.
95. MATTHEWS, E. K., O. H. PETERSEN, AND J. A. WILLIAMS. Pancreatic acinar cells: acetylcholine-induced membrane depolarization, calcium efflux and amylase release. *J. Physiol. Lond.* 234: 689–701, 1973.
96. MERRITT, J. E., AND R. P. RUBIN. Pancreatic amylase secretion and cytoplasmic free calcium. *Biochem. J.* 230: 151–159, 1985.
97. MERRITT, J. E., C. W. TAYLOR, R. P. RUBIN, AND J. W. PUTNEY, JR. Evidence suggesting that a novel guanine nucleotide regulatory protein couples receptors to phospholipase C in exocrine pancreas. *Biochem. J.* 236: 337–343, 1986.
98. MICHELL, R. H. Inositol phospholipids and cell surface receptor function. *Biochim. Biophys. Acta* 415: 81–147, 1975.
99. MICHELL, R. H., J. N. HAWTHORNE, R. COLEMAN, AND M. L. KARNOVSKY. Extraction of polyphosphoinositides with neutral and acidified solvents. *Biochim. Biophys. Acta* 210: 86–91, 1970.
100. MÖSSNER, J., C. D. LOGSDON, I. D. GOLDFINE, AND J. A. WILLIAMS. Regulation of pancreatic acinar cell insulin receptors by insulin. *Am. J. Physiol.* 247 (*Gastrointest. Liver Physiol.* 10): G155–G160, 1984.
101. MÖSSNER, J., C. D. LOGSDON, N. POTAU, J. A. WILLIAMS, AND I. D. GOLDFINE. Effect of intracellular Ca^{2+} on insulin-like growth factor II internalization into pancreatic acini. Roles of insulin and cholecystokinin. *J. Biol. Chem.* 259: 12350–12356, 1984.
102. MÖSSNER, J., E. ROACH, I. D. GOLDFINE, AND J. A. WILLIAMS. Autoradiographic analysis of ^{125}I-insulin-like growth factor II internalization into pancreatic acini. *Diabetes Res. Clin. Pract.* 2: 75–82, 1986.
103. MROZ, E. A., AND C. LECHENE. Pancreatic zymogen granules differ markedly in protein composition. *Science Wash. DC* 232: 871–873, 1986.
104. NAIRN, A. C., B. BHAGAT, AND H. C. PALFREY. Identification of calmodulin-dependent protein kinase III and its major M_r 100,000 substrate in mammalian tissues. *Proc. Natl. Acad. Sci. USA* 82: 7939–7943, 1985.
105. NISHIZUKA, Y. The role of protein kinase C in cell surface signal transduction and tumor promotion. *Nature Lond.* 308: 693–698, 1984.
106. NOGUCHI, M., H. ADACHI, J. D. GARDNER, AND R. T. JENSEN. Calcium-activated, phospholipid-dependent protein kinase in pancreatic acinar cells. *Am. J. Physiol.* 248 (*Gastrointest. Liver Physiol.* 11): G692–G701, 1985.
107. OCHS, D. L., J. I. KORENBROT, AND J. A. WILLIAMS. Intra-

cellular free calcium concentrations in isolated pancreatic acini; effects of secretagogues. *Biochem. Biophys. Res. Commun.* 117: 122–128, 1983.
108. OCHS, D. L., J. I. KORENBROT, AND J. A. WILLIAMS. Relation between free cytosolic calcium and amylase release by pancreatic acini. *Am. J. Physiol.* 249 (*Gastrointest. Liver Physiol.* 12): G389–G398, 1985.
109. O'DOHERTY, J., AND R. J. STARK. Stimulation of pancreatic acinar secretion: increases in cytosolic calcium and sodium. *Am. J. Physiol.* 242 (*Gastrointest. Liver Physiol.* 5): G513–G521, 1982.
110. OKUMURA, K., S. IWAKAWA, K. I. INUI, AND R. HORI. Specific secretin binding sites in rat pancreas. *Biochem. Pharmacol.* 32: 2689–2695, 1983.
111. ORCHARD, J. L., J. S. DAVIS, R. E. LARSON, AND R. V. FARESE. Effects of carbachol and pancreozymin (cholecystokinin-octapeptide) on polyphosphoinositide metabolism in the rat pancreas in vitro. *Biochem. J.* 217: 281–287, 1984.
112. OTSUKI, M., Y. OKABAYASHI, A. OHKI, S. R. HOOTMAN, S. BABA, AND J. A. WILLIAMS. Amylase secretion by isolated pancreatic acini after acute cholecystokinin treatment in vivo. *Am. J. Physiol.* 246 (*Gastrointest. Liver Physiol.* 9): G419–G425, 1984.
113. OTSUKI, M., AND J. A. WILLIAMS. Amylase secretion by isolated pancreatic acini after chronic cholecystokinin treatment in vivo. *Am. J. Physiol.* 244 (*Gastrointest. Liver Physiol.* 7): G683–G688, 1983.
114. PALADE, G. E. Intracellular aspects of protein synthesis. *Science Wash. DC* 189: 347–358, 1975.
115. PAN, G. Z., M. J. COLLEN, AND J. D. GARDNER. Action of cholera toxin on dispersed acini from rat pancreas. Post-receptor modulation involving cyclic AMP and calcium. *Biochim. Biophys. Acta* 720: 338–345, 1982.
116. PANDOL, S. J., M. S. SCHOEFFIELD, G. SACHS, AND S. MUALLEM. Role of free cytosolic calcium in secretagogue-stimulated amylase release from dispersed acini from guinea pig pancreas. *J. Biol. Chem.* 260: 10081–10086, 1985.
117. PANDOL, S. J., H. SEIFERT, M. W. THOMAS, J. RIVIER, AND W. VALE. Growth hormone-releasing factor stimulates pancreatic enzyme secretion. *Science Wash. DC* 225: 326–328, 1984.
118. PANDOL, S. J., M. W. THOMAS, M. S. SCHOEFFIELD, G. SACHS, AND S. MUALLEM. Role of calcium in cholecystokinin-stimulated phosphoinositide breakdown in exocrine pancreas. *Am. J. Physiol.* 248 (*Gastrointest. Liver Physiol.* 11): G551–G560, 1985.
119. PETRINI, M., D. L. EMERSON, AND R. M. GALBRAITH. Linkage between surface immunoglobulin and cytoskeleton of B lymphocytes may involve Gc protein. *Nature Lond.* 306: 73–74, 1983.
120. PLATTNER, H., C. WESTPHAL, AND R. TIGGEMANN. Cytoskeleton-secretory vesicle interactions during the docking of secretory vesicles at the cell membrane in *Paramecium tetraurelia* cells. *J. Cell Biol.* 92: 368–377, 1982.
121. PONAPPA, B. C., AND J. A. WILLIAMS. Effects of ionophore A23187 on calcium flux and amylase release in isolated mouse pancreatic acini. *Cell Calcium* 1: 267–278, 1980.
122. POULSEN, J. H., AND J. A. WILLIAMS. Effect of extracellular K$^+$ concentration on resting potential, caerulein-induced depolarization and amylase release from mouse pancreatic acinar cells. *Pfluegers Arch.* 370: 173–177, 1977.
123. POWERS, R. E., P. C. JOHNSON, M. J. HOULIHAN, A. K. SALUJA, AND M. L. STEER. Intracellular Ca^{2+} levels and amylase secretion in Quin 2-loaded mouse pancreatic acini. *Am. J. Physiol.* 248 (*Gastrointest. Liver Physiol.* 17): C535–C541, 1985.
124. PREISSLER, M., AND J. A. WILLIAMS. Localization of ATP-dependent calcium transport activity in mouse pancreatic microsomes. *J. Membr. Biol.* 73: 137–144, 1983.
125. PUTNEY, J. W., JR. Formation and actions of calcium-mobilizing messenger, inositol 1,4,5-trisphosphate. *Am. J. Physiol.* 252 (*Gastrointest. Liver Physiol.* 15): G149–G157, 1987.
126. PUTNEY, J. W., JR., G. M. BURGESS, S. P. HALENDA, J. S. MCKINNEY, AND R. P. RUBIN. Effects of secretagogues on [^{32}P]phosphatidylinositol 4,5-biphosphate metabolism in the exocrine pancreas. *Biochem. J.* 212: 483–488, 1983.
127. RASMUSSEN, H., AND P. Q. BARRETT. Calcium messenger system: an integrated view. *Physiol. Rev.* 64: 938–984, 1984.
128. RENCKENS, B. A. M., J. J. SCHRIJEN, H. G. P. SWARTS, J. J. H. H. M. DEPONT, AND S. L. BONTING. Role of calcium in exocrine pancreatic secretion. IV. Calcium movements in isolated acinar cells of rabbit pancreas. *Biochim. Biophys. Acta* 544: 338–350, 1978.
129. ROBBERECHT, P., T. P. CONLON, AND J. D. GARDNER. Interaction of porcine vasoactive intestinal peptide with dispersed pancreatic acinar cells from guinea pig. *J. Biol. Chem.* 251: 4635–4639, 1976.
130. ROBBERECHT, P., M. DESCHODT-LANCKMAN, M. LAMMENS, P. DENEEF, AND J. CHRISTOPHE. In vitro effects of secretin and vasoactive intestinal peptide on hydrolase secretion and cyclic AMP levels in the pancreas of five animal species. A comparison with caerulein. *Gastroenterol. Clin. Biol.* 1: 519–525, 1977.
131. ROBBERECHT, P., M. WAELBROECK, M. NOYER, P. CHATELAIN, P. DENEEF, W. KONIG, AND J. CHRISTOPHE. Characterization of secretin and vasoactive intestinal peptide receptors in rat pancreatic plasma membranes using the native peptides, secretin-(7-27) and five secretin analogues. *Digestion* 23: 201–210, 1982.
132. ROBERTS, M. L., AND F. R. BUTCHER. The involvement of protein phosphorylation in stimulus-secretion coupling in mouse exocrine pancreas. *Biochem. J.* 210: 353–359, 1983.
133. ROSENZWEIG, S. A., L. D. MADISON, AND J. D. JAMIESON. Analysis of cholecystokinin-binding proteins using endo-β-N-acetylglucosaminidase F. *J. Cell Biol.* 99: 1110–1116, 1984.
134. ROSENZWEIG, S. A., L. J. MILLER, AND J. D. JAMIESON. Identification and localization of cholecystokinin-binding sites on rat pancreatic plasma membranes and acinar cells: a biochemical and autoradiographic study. *J. Cell Biol.* 96: 1288–1297, 1983.
135. ROTHMAN, S. S. The digestive enzymes of the pancreas: a mixture of inconstant proportions. *Annu. Rev. Physiol.* 39: 373–389, 1977.
136. ROTHMAN, S. S. Passage of proteins through membranes—old assumptions and new perspectives. *Am. J. Physiol.* 238 (*Gastrointest. Liver Physiol.* 1): G391–G402, 1980.
137. RUBIN, R. P., P. P. GODFREY, D. A CHAPMAN, AND J. W. PUTNEY, JR. Secretagogue-induced formation of inositol phosphates in rat exocrine pancreas: implications for a messenger role for inositol trisphosphate. *Biochem. J.* 219: 655–659, 1984.
138. RUBIN, R. P., K. L. KELLY, S. P. HALENDA, AND S. G. LAYCHOCK. Arachidonic acid metabolism in rat pancreatic acinar cells: calcium-mediated stimulation of the lipoxygenase system. *Prostaglandins* 24: 179–193, 1982.
139. RUTTEN, W. J., J. J. H. H. M. DEPONT, AND S. L. BONTING. Adenylate cyclase in the rat pancreas. Properties and stimulation by hormones. *Biochim. Biophys. Acta* 274: 201–213, 1972.
140. SAKAMOTO, C., I. D. GOLDFINE, AND J. A. WILLIAMS. Characterization of cholecystokinin receptor subunits on pancreatic plasma membranes. *J. Biol. Chem.* 258: 12707–12711, 1983.
141. SAKAMOTO, C., I. D. GOLDFINE, AND J. A. WILLIAMS. Pancreatic CCK receptors: characterization of covalently labeled subunits. *Biochem. Biophys. Res. Commun.* 118: 623–628, 1984.
142. SAKAMOTO, C., I. D. GOLDFINE, AND J. A. WILLIAMS. The somatostatin receptor on isolated pancreatic acinar cell plasma membranes. *J. Biol. Chem.* 259: 9623–9627, 1984.
143. SAKAMOTO, C., J. A. WILLIAMS, E. ROACH, AND I. D. GOLDFINE. In vivo localization of insulin binding to cells of the rat pancreas. *Proc. Soc. Exp. Biol. Med.* 175: 497–502, 1984.
144. SANKARAN, H., C. W. DEVENEY, I. D. GOLDFINE, AND J. A. WILLIAMS. Preparation of biologically active radioiodinated

cholecystokinin for radioreceptor assay and radioimmunoassay. *J. Biol. Chem.* 254: 9349-9351, 1979.
145. SANKARAN, H., I. D. GOLDFINE, A. BAILEY, V. LICKO, AND J. A. WILLIAMS. Relationship of cholecystokinin receptor binding to regulation of biological functions in pancreatic acini. *Am. J. Physiol.* 242 (*Gastrointest. Liver Physiol.* 5): G250-G257, 1982.
146. SANKARAN, H., I. D. GOLDFINE, C. W. DEVENEY, K. Y. WONG, AND J. A. WILLIAMS. Binding of cholecystokinin to high affinity receptors on isolated rat pancreatic acini. *J. Biol. Chem.* 255: 1849-1853, 1980.
147. SCHEELE, G., AND A. HAYMOVITS. Cholinergic and peptide-stimulated discharge of secretory protein in guinea pig pancreatic lobules. *J. Biol. Chem.* 254: 10346-10353, 1979.
148. SCHULMAN, H., J. KURET, A. B. JEFFERSON, P. S. NOSE, AND K. H. SPITZER. Ca^{2+}/calmodulin-dependent microtubule-associated protein 2 kinase: broad substrate specificity and multifunctional potential in diverse tissues. *Biochemistry* 24: 5320-5326, 1985.
149. SINGH, J. Phorbol ester (TPA) potentiates noradrenaline and acetylcholine-evoked amylase secretion in the rat pancreas. *FEBS Lett.* 180: 191-195, 1985.
150. SINGH, M. Effect of somatostatin on amylase secretion from in vivo and in vitro rat pancreas. *Dig. Dis. Sci.* 31: 506-512, 1986.
151. SJÖDIN, L. Binding and internalization of ^{125}I-Bolton-Hunter-substance-P by pancreatic acinar cells. *Biochem. Biophys. Res. Commun.* 124: 578-584, 1984.
152. SJÖDIN, L. Cholecystokinin inhibits binding of substance P to pancreatic acinar cells. *Acta Physiol. Scand.* 124: 471-474, 1985.
153. SMITH, C. D., C. C. COX, AND R. SNYDERMAN. Receptor-coupled activation of phosphoinositide-specific phospholipase C by an N protein. *Science Wash. DC* 232: 97-100, 1986.
154. SMITH, J. B., L. SMITH, AND B. L. HIGGINS. Temperature and nucleotide dependence of calcium release by myo-inositol 1,4,5-trisphosphate in cultured vascular smooth muscle cells. *J. Biol. Chem.* 260: 14413-14416, 1985.
155. SPÄT, A., P. G. BRADFORD, J. S. MCKINNEY, R. P. RUBIN, AND J. W. PUTNEY, JR. A saturable receptor for ^{32}P-inositol-1,4,5-trisphosphate in hepatocytes and neutrophils. *Nature Lond.* 319: 514-516, 1986.
156. SRIKANT, C. B., AND Y. C. PATEL. Somatostatin receptors on rat pancreatic acinar cells. *J. Biol. Chem.* 261: 7690-7696, 1986.
157. STEIGERWALT, R. W., AND J. A. WILLIAMS. Characterization of cholecystokinin receptors on rat pancreatic membranes. *Endocrinology* 109: 1746-1753, 1981.
158. STENSON, W. F., AND E. LOBOS. Metabolism of arachidonic acid by pancreatic acini: relation to amylase secretion. *Am. J. Physiol.* 242 (*Gastrointest. Liver Physiol.* 5): G493-G497, 1982.
159. STOLZE, H., AND I. SCHULZ. Effect of atropine, ouabain, antimycin A, and A23187 on "trigger Ca^{2+} pool" in exocrine pancreas. *Am. J. Physiol.* 238 (*Gastrointest. Liver Physiol.* 1): G338-G348, 1980.
160. STREB, H., E. BAYERDÖRFFER, W. HAASE, R. F. IRVINE, AND I. SHULZ. Effect of inositol-1,4,5-trisphosphate on isolated subcellular fractions of rat pancreas. *J. Membr. Biol.* 81: 241-253, 1984.
161. STREB, H., R. F. IRVINE, M. J. BERRIDGE, AND I. SCHULZ. Release of Ca^{2+} from a nonmitochondrial intracellular store in pancreatic acinar cells by inositol-1,4,5-trisphosphate. *Nature Lond.* 306: 67-69, 1983.
162. SÜDHOF, T. C. Calelectrins are a ubiquitous family of Ca^{2+}-binding proteins purified by Ca^{2+}-dependent hydrophobic affinity chromatography by a mechanism distinct from that of calmodulin. *Biochem. Biophys. Res. Commun.* 123: 100-107, 1984.
163. SUMMERS, T. A., AND C. E. CREUTZ. Phosphorylation of a chromaffin granule binding protein by protein kinase C. *J. Biol. Chem.* 260: 2437-2443, 1985.
164. SUNG, C. K., AND J. A. WILLIAMS. The role of calcium in pancreatic acinar cell secretion. *Miner. Electrolyte Meta.* 14: 71-77, 1988.
165. SUSINI, C., A. BAILEY, J. SZECOWKA, AND J. A. WILLIAMS. Characterization of covalently crosslinked pancreatic somatostatin receptors. *J. Biol. Chem.* 261: 16738-16743, 1986.
166. SVOBODA, M., M. LAMBERT, J. FURNELLE, AND J. CHRISTOPHE. Specific photoaffinity crosslinking of [^{125}I]cholecystokinin to pancreatic plasma membranes. Evidence for a disulfide-linked M_r 76,000 peptide in cholecystokinin receptors. *Regul. Pept.* 4: 163-172, 1982.
167. SZECOWKA, J., I. D. GOLDFINE, AND J. A. WILLIAMS. Solubilization and characterization of CCK receptors from mouse pancreas. *Regul. Pept.* 10: 71-83, 1985.
168. TAKAI, Y., K. KAIBUCHI, T. TSUDA, AND M. HOSHIJIMA. Role of protein kinase C in transmembrane signaling. *J. Cell. Biochem.* 29: 143-155, 1985.
169. TSIEN, R. Y., T. POZZAN, AND T. J. RINK. Calcium homeostasis in intact lymphocytes: cytoplasmic free calcium monitored with a new intracellular trapped fluorescent indicator. *J. Cell Biol.* 94: 325-334, 1981.
170. ULLRICH, A., J. R. BELL, E. Y. CHEN, R. HERRERA, L. M. PETRUZZELLI, T. J. DULL, A. GRAY, L. COUSSENS, Y. C. LIAO, M. TSUBOKAWA, A. MASON, P. H. SEEBURG, C. GRUNFELD, O. M. ROSEN, AND J. RAMACHANDRAN. Human insulin receptor and its relationship to the tyrosine kinase family of oncogenes. *Nature Lond.* 313: 756-761, 1985.
171. ULLRICH, A., L. COUSSENS, J. S. HAYFLICK, T. J. DULL, A. GRAY, A. W. TAM, J. LEE, Y. YARDEN, T. A. LIBERMANN, J. SCHLESSINGER, J. DOWNWARD, E. L. V. MAYES, N. WHITTLE, M. D. WATERFIELD, AND P. H. SEEBURG. Human epidermal growth factor receptor cDNA sequence and aberrant expression of the amplified gene in A431 epidermoid carcinoma cells. *Nature Lond.* 309: 418-425, 1984.
172. VANDERMEERS, A., M. C. VANDERMEERS-PIRET, J. RATHE, AND J. CHRISTOPHE. A calcium-dependent protein activator of guanosine-3',5'-monophosphate phosphodiesterase in bovine and rat pancreas. *Eur. J. Biochem.* 81: 377-386, 1977.
173. VAN ELDIK, L. J., J. G. ZENDEGUI, D. R. MARSHAK, AND D. M. WATTERSON. Calcium-binding proteins and the molecular basis of calcium action. *Int. Rev. Cytol.* 77: 1-61, 1982.
174. VAN LEEMPUT-COUTREZ, M., J. CAMUS, AND J. CHRISTOPHE. Cyclic nucleotide-dependent protein kinases of the rat pancreas. *Biochem. Biophys. Res. Commun.* 54: 182-190, 1973.
175. VIGUERIE, J. P., J. P. ESTÈVE, C. SUSINI, N. VAYSSE, AND A. RIBET. Processing of receptor-bound somatostatin: internalization and degradation by pancreatic acini. *Am. J. Physiol.* 252 (*Gastrointest. Liver Physiol.* 15): G535-G542, 1987.
176. WAKASUGI, H., T. KIMURA, W. HAASE, A. KRIBBEN, R. KAUFMANN, AND I. SCHULZ. Calcium uptake into acini from rat pancreas: evidence for intracellular ATP-dependent calcium sequestration. *J. Membr. Biol.* 65: 205-220, 1982.
177. WHOOTEN, M. W., A. E. NEL, P. J. GOLDSCHMIDT-CLERMONT, R. M. GALBRAITH, AND R. W. WRENN. Identification of the major endogenous substrate for the phospholipid/Ca^{2+}-dependent protein kinase in pancreatic acini as Gc (vitamin-D-binding protein). *FEBS Lett.* 191: 97-101, 1985.
178. WHOOTEN, M. W., AND R. W. WRENN. Redistribution of phospholipid/calcium-dependent protein kinase and altered phosphorylation of its soluble and particulate substrated proteins in phorbol ester-treated rat pancreatic acini. *Cancer Res.* 45: 3912-3917, 1985.
179. WILLIAMS, J. A. Regulation of pancreatic acinar cell function by intracellular calcium. *Am. J. Physiol.* 238 (*Gastrointest. Liver Physiol.* 1): G269-G279, 1980.
180. WILLIAMS, J. A. Multiple effect of Na^+ removal on pancreatic secretion in vitro. *Cell Tissue Res.* 210: 295-303, 1980.
181. WILLIAMS, J. A. Regulatory mechanisms in pancreas and salivary acini. *Annu. Rev. Physiol.* 46: 361-375, 1984.
182. WILLIAMS, J. A., AND D. CHANDLER. Ca^{2+} and pancreatic amylase release. *Am. J. Physiol.* 228: 1729-1732, 1975.

183. WILLIAMS, J. A., AND I. D. GOLDFINE. The insulin-pancreatic acinar axis. *Diabetes* 34: 980–986, 1985.
184. WILLIAMS, J. A., AND S. R. HOOTMAN. Stimulus-secretion coupling in pancreatic acinar cells. In: *The Exocrine Pancreas: Biology, Pathobiology, and Diseases*, edited by V. L. W. Go, J. D. Gardner, F. P. Brooks, E. Lebenthal, E. P. DiMagno, and G. A. Scheele. New York: Raven, 1986, p. 123–139.
185. WILLIAMS, J. A., AND M. LEE. Pancreatic acinar cells: use of a calcium ionophore to separate enzyme release from the earlier steps in stimulus-secretion coupling. *Biochem. Biophys. Res. Commun.* 60: 542–548, 1974.
186. WILLIAMS, J. A., J. H. POULSEN, AND M. LEE. Effects of membrane stabilizers on pancreatic amylase release. *J. Membr. Biol.* 33: 185–195, 1977.
187. WILLIAMS, J. A., H. SANKARAN, E. ROACH, AND I. D. GOLDFINE. Quantitative electron microscope autoradiographs of ^{125}I-cholecystokinin in pancreatic acini. *Am. J. Physiol.* 243 (*Gastrointest. Liver Physiol.* 6): G291–G296, 1982.
188. WRENN, R. W. Phospholipid-sensitive calcium-dependent protein kinase and its endogenous substrate proteins in rat pancreatic acinar cells. *Life Sci.* 32: 2385–2392, 1983.
189. WRENN, R. W. Phosphorylation of a pancreatic zymogen granule membrane protein by endogenous calcium/phospholipid-dependent protein kinase. *Biochim. Biophys. Acta* 775: 1–6, 1984.
190. ZAJIC, G., AND J. SCHACHT. Cytochemical demonstration of adenylate cyclase with strontium chloride in the rat pancreas. *J. Histochem. Cytochem.* 31: 25–28, 1983.

CHAPTER 22

Signaling transduction in hormone- and neurotransmitter-induced enzyme secretion from the exocrine pancreas

IRENE SCHULZ | *Max-Planck Institut für Biophysik, Frankfurt am Main, Federal Republic of Germany*

CHAPTER CONTENTS

In Vitro Preparations for Studying Cellular Mechanisms
   Isolated acini and acinar cells
   Isolation of intracellular organelles
Function of Pancreatic Cell Membranes
   Transport systems in plasma membrane
      ATP-driven ion pumps and carrier-mediated transport mechanisms
   Role of plasma membranes in hormonal stimulation of pancreatic secretion
      Adenylate cyclase–cyclic AMP system
      Phospholipase C system
      Activation of both messenger systems—adenylate cyclase and phospholipase C—by the same hormone
      Receptor-operated and intracellular messenger-operated ion channels
      Tyrosine kinase system
Role of Intracellular Membranes in Pancreatic Cell Function
   Passive ion-transport mechanisms in endoplasmic reticulum
      GTP-induced calcium release
   Active ion-transport mechanisms in endoplasmic reticulum
      MgATP-dependent calcium uptake into endoplasmic reticulum
      Ion gradient–coupled calcium uptake into endoplasmic reticulum
Ion Transport Involved in Exocytosis
   Is a chloride conductance in the membrane of pancreatic zymogen granules involved in exocytosis?

PHYSIOLOGICAL FUNCTION of many cell types is regulated by hormones and neurotransmitters that are recognized by the cell via specific receptors. These external signals are translated by the cell into intracellular signals. Intracellular messengers are formed very similarly in different cell types; however, the answers of the different cells to these intracellular messengers can be quite different and involve cell-specific responses such as metabolism, secretion, contraction, phototransduction, and cell growth. Three main pathways can be distinguished by which hormones induce intracellular signals that finally result in the physiological response. The most important pathway in pancreatic enzyme secretion employs receptor-mediated activation of phospholipase C and consequent breakdown of plasma membrane–bound phosphatidylinositol 4,5-bisphosphate (PIP_2) (12). This leads to generation of two messengers, inositol 1,4,5-trisphosphate (IP_3) (12, 211) and 1,2-diacylglycerol (217). Another pathway is mediated by stimulation of the plasma membrane–bound adenylate cyclase and production of the second messenger cyclic adenosine 3′,5′-monophosphate (cAMP) (213). A third pathway involves tyrosine kinase–activating receptors to which growth factors bind, such as epidermal growth factor, platelet-derived growth factor, and insulin. The latter pathway regulates cell metabolism and cell growth but also potentiates enzyme secretion from the exocrine pancreas (137, 236).

Hormones and neurotransmitters such as cholecystokinin (CCK) and acetylcholine can be considered main regulators of enzyme secretion. Release of Ca^{2+} by their intracellular messenger IP_3 leads to an increase in cytosolic free Ca^{2+} concentration ($[Ca^{2+}]_{cyt}$), from ~100 to ~800 nM (163, 212), and this increase activates several further events related to stimulus-secretion coupling. One consequence of raised $[Ca^{2+}]_{cyt}$ is activation of calmodulin-dependent protein kinases and phosphatases, which leads to altered phosphorylation of proteins assumed to regulate one or more final steps in exocytosis (29, 30). Another process is closely related to the action of 1,2-diacylglycerol, the other hydrolysis product of PIP_2 breakdown. Increased $[Ca^{2+}]_{cyt}$ leads to association of cytosolic protein kinase C with the plasma membrane (85, 88, 122, 237), which then can be activated by plasma membrane–located 1,2-diacylglycerol. Stimulated protein kinase C phosphorylates different regulatory proteins, some of which might be positively involved in stimulus-secretion coupling (20, 30, 113), whereas others lead to a negative feedback regulation. Thus a major function of protein kinase C is phosphorylation of receptors, with a consequent decrease in binding affinity to agonists (25, 26, 41, 49, 69, 114, 130, 131, 140). Negative feedback control by protein kinase C has also been proposed for intracellular processes such as smooth muscle contraction (159), secretion (3, 6, 168), and regulation of intracellular pH (15, 31, 154, 155).

Cyclic AMP is the main intracellular mediator for secretion- and vasoactive intestinal polypeptide (VIP)-induced secretion of $NaHCO_3$ and fluid from duct cells (36). These hormones also activate adenylate cyclase in acinar cells and increase cAMP levels in pancreatic cells. The degree of cAMP-mediated secretion, however, differs in different animal species. Although guinea pig responds to activators of adenylate cyclase with high-protein release, in other animal species, such as dog, cat, rat, and mouse, response to these hormones is rather low (186). Cyclic AMP activates protein kinase A, another protein-phosphorylating enzyme.

Although a variety of proteins on zymogen granule membranes and microsomal membranes have been shown to be substrates for Ca^{2+}- and cAMP-dependent protein kinases (29, 30, 204), direct involvement of these proteins in exocytosis has not been demonstrated.

In this chapter some aspects of intracellular events that are the consequence of hormonal stimulation are discussed. Because cellular regulation of pancreatic secretion in its main steps (i.e., receptor binding, intracellular messenger action, and cellular protein phosphorylation) are reviewed in the chapter by Williams, Burnham, and Hootman in this *Handbook*, these events are discussed only briefly in this chapter. The functional properties of cellular membranes, studied on both isolated acinar cells and membrane vesicles in vitro, are emphasized.

IN VITRO PREPARATIONS FOR STUDYING CELLULAR MECHANISMS

Isolated cells and purified subcellular components have become useful tools to study membrane-associated events such as primary and secondary active ion-transport mechanisms, coupling of receptors to membrane-bound enzymes (such as to adenylate cyclase and phospholipase C by GTP-binding proteins), opening of ion channels by intracellular messengers, and fusion of membranes such as zymogen granules with plasma membranes.

Depending on the problem to be studied, isolated acinar cells or acini (containing groups of 4-5 connected acinar cells) as well as isolated vesicles from plasma membranes or from endoplasmic reticulum or isolated zymogen granules have been used.

Isolated Acini and Acinar Cells

Isolated acini are obtained by digestion of pancreatic tissue with collagenase and separation of acini from other cells and cell debris by centrifugation in an albumin gradient. Acini are useful to study enzyme secretion because they respond to secretagogues much better than isolated acinar cells (70), probably because of the presence of intact gap junctions in acini as compared with acinar cells. Evidence suggests that gap junctions play a role in hormone-stimulated cell response (27, 100, 147, 174).

In order to introduce substances into cells and to study cellular responses to hormonal messengers such as cAMP, IP_3, or 1,2-diacylglycerol, which do not easily penetrate intact cells, the plasma membrane of acinar cells can be permeabilized by different means. Both saponin and digitonin bind to cholesterol in the plasma membrane, thereby forming micelles that have pore diameters of ~80-100 Å (5, 60, 72). Permeabilization by electric shock has also been applied to pancreas cells (117). Furthermore, washing of acinar cells in a nominal Ca^{2+}-free buffer that contains contaminating $[Ca^{2+}]$ of ~2-3 μM renders them leaky (212). The latter method works for isolated acinar cells but not for acini, which contain groups of 4-5 connected acinar cells. It therefore seems probable that in a nominal Ca^{2+}-free buffer, gap junctions of isolated cells are opened and substances that do not penetrate plasma membranes, such as ATP, IP_3, or ethylene glycol-bis(β-aminoethylether)-N,N'-tetraacetic acid (EGTA), can enter the cell by this pathway.

Isolation of Intracellular Organelles

Pancreatic tissue or isolated cells have been fractionated into their subcellular compartments, such as plasma membranes, endoplasmic reticulum, zymogen granules, or mitochondria, to study hormone binding, signal transfer through membranes, ion transport, and solute transport independently of cell metabolism or effects of test substances on other cell organelles.

Zymogen granules and mitochondria are easily separated from other organelles by differential centrifugation (148). Further purification can be obtained by density-gradient centrifugation with sucrose or other compounds giving high densities or viscosities at lower osmolarities than sucrose, such as Ficoll (87, 152, 232), metrizamide (48), or Percoll (162).

With Percoll gradient centrifugation, stable zymogen granules that do not lyse in electrolyte solutions can be separated from unstable granules (56).

Because pancreatic acinar cells have a highly developed endoplasmic reticulum, isolation of these membranes is relatively easy. Although a homogenate prepared from pancreatic tissue has also been used as starting material for the isolation of different types of organelles (152), it has proven useful to isolate and purify the cells first from which membranes of interest should be separated. With this method, relatively pure endoplasmic reticulum can be separated by differential centrifugation in two or three centrifugation steps, including centrifugation at 1,000 g that pellets down zymogen granules. The corresponding supernatant is centrifuged at 11,000 g, which brings down most of the mitochondria. Further centrifugation of the resulting supernatant at 27,000 g yields two- to threefold enrichment in rough endoplasmic reticulum, as compared with the starting material (11). For many studies this purification is sufficient and has the advantage that membranes can be obtained within 3–4 h after killing the animal. If degradation of receptor, enzyme, or transport functions of the membrane occurs rapidly during the separation procedure, the advantage of retained membrane function more than compensates for the lack of absolute purity.

Further purification of endoplasmic reticulum obtained in the 27,000-g pellet can be achieved by density centrifugation, e.g., in a Percoll gradient (11). This step yields further purification by approximately twofold. Because endoplasmic reticulum makes up ~20%–30% of cellular protein in acinar cells (152), a three- to fourfold enrichment of endoplasmic reticulum, as compared with homogenate, is the most that is obtainable. This is different for plasma membranes, which only make up 1%–2% of cell protein.

Although centrifugation in density gradients such as sucrose, sucrose and Ficoll, or Percoll has been used to purify plasma membranes (11, 152), the most efficient method to obtain a good yield of plasma membrane protein with high purity within a reasonable time of preparation seems to be a method that employs precipitation of membranes and cell organelles other than plasma membranes from a homogenate of pancreatic acinar cells with $MgCl_2$. This method, which has been used in the preparation of brush-border membranes from kidney proximal tubule (20) and small intestine (76), yields pancreatic plasma membranes with enrichment of their typical marker enzymes, such as Na^+-K^+-ATPase and hormone-stimulated adenylate cyclase, by 43- and 42-fold, respectively (9). These enrichment factors indicate a purity greater than has been described so far for plasma membranes from pancreatic acinar cells (123, 148, 152, 219).

FUNCTION OF PANCREATIC CELL MEMBRANES

Transport Systems in Plasma Membrane

One of the main functions of the plasma membrane is to separate the interior cell milieu from the surrounding milieu and to keep it in a state that allows maintenance of cell metabolism at rest and at stimulation. At rest transport systems such as channels, pumps, and carriers mainly serve to keep intracellular ion and nutrient concentrations constant. At stimulation these transport systems are activated and provide the basis for specific cell responses, such as excitability, osmoregulation, secretion, contraction, or cell growth.

ATP-DRIVEN ION PUMPS AND CARRIER-MEDIATED TRANSPORT MECHANISMS. As in other cells, intracellular ion concentrations are regulated by ATPase-transport and carrier-mediated transport systems in the plasma membrane. A schematic representation of primary active (MgATP driven) and secondary active (electrochemical ion-gradient driven) as well as passive ion-transport processes in the plasma membrane is given in Figure 1. In addition to Ca^{2+}-transport processes in the plasma membrane, regulation of $[Ca^{2+}]_{cyt}$ by intracellular organelles is indicated.

Intracellular Na^+ and K^+ concentrations are adjusted by Na^+-K^+-Mg^{2+}-ATPase at ~9 and 140 meq/l, respectively (177), and $[Ca^{2+}]$ by Ca^{2+}-Mg^{2+}-ATPase at ~100–200 nM (163, 197). Both the Na^+ and the K^+ gradients established by Na^+-K^+-ATPase over the plasma membrane are used for many secondary active transport processes, such as for Na^+-coupled uptake of amino acids (99, 100, 103, 104, 129, 178, 224), for regulation of intracellular pH by Na^+-H^+ countertransport (78), for extrusion of Ca^{2+} in Na^+-Ca^{2+} countertransport (10), or for uptake of Cl^- in coupled Na^+-$2Cl^-$-K^+ cotransport (172).

Activation of Na^+-$2Cl^-$-K^+ cotransport that leads to NaCl and fluid secretion from acinar cells is not a direct process but the consequence of several related steps following hormonal stimulation of plasma membrane–bound phospholipase C, generation of IP_3, increase in $[Ca^{2+}]_{cyt}$, and consequent activation of K^+ channels. Increased K^+ efflux stimulates the Na^+-$2Cl^-$-K^+ cotransporter that causes K^+ reuptake so that K^+ circulates over the plasma membrane (172). Increased Na^+ and Cl^- cotransport leads to a net uptake and a consequent disequilibrium of these ions, which is a prerequisite for secretion of NaCl and fluid over the luminal membrane. Although there is evidence for a regulated Cl^- conductance in the luminal membrane of exocrine acinar cells (67a), the luminal exit step for Na^+ during stimulation is not quite clear, and much of Na^+ that is taken up via the Na^+-$2Cl^-$-K^+ cotransporter is probably extruded at the contraluminal cell side by Na^+-K^+-ATPase.

The Na^+-H^+ countertransporter in the plasma membrane is an important mechanism in the regulation of intracellular pH (78, 153, 229) and probably also of cell volume, as in other tissues (32, 63, 73). It had been shown that hormones that activate phospholipase C also stimulate Na^+-H^+ exchange (15, 200). Because phorbol ester, a strong activator of protein kinase C, also stimulates Na^+-H^+ countertransport, it is assumed that the Na^+-H^+ countertransporter is activated by protein kinase C–mediated phosphorylation (15).

It was a hypothesis that Na^+-H^+ exchange is involved in $NaHCO_3$ secretion from pancreatic duct cells. The model is based on the observation that the underlying mechanism for HCO_3^- secretion into the duct lumen is H^+ transport from the lumen to the interstitial side (195, 198). Because HCO^- secretion is

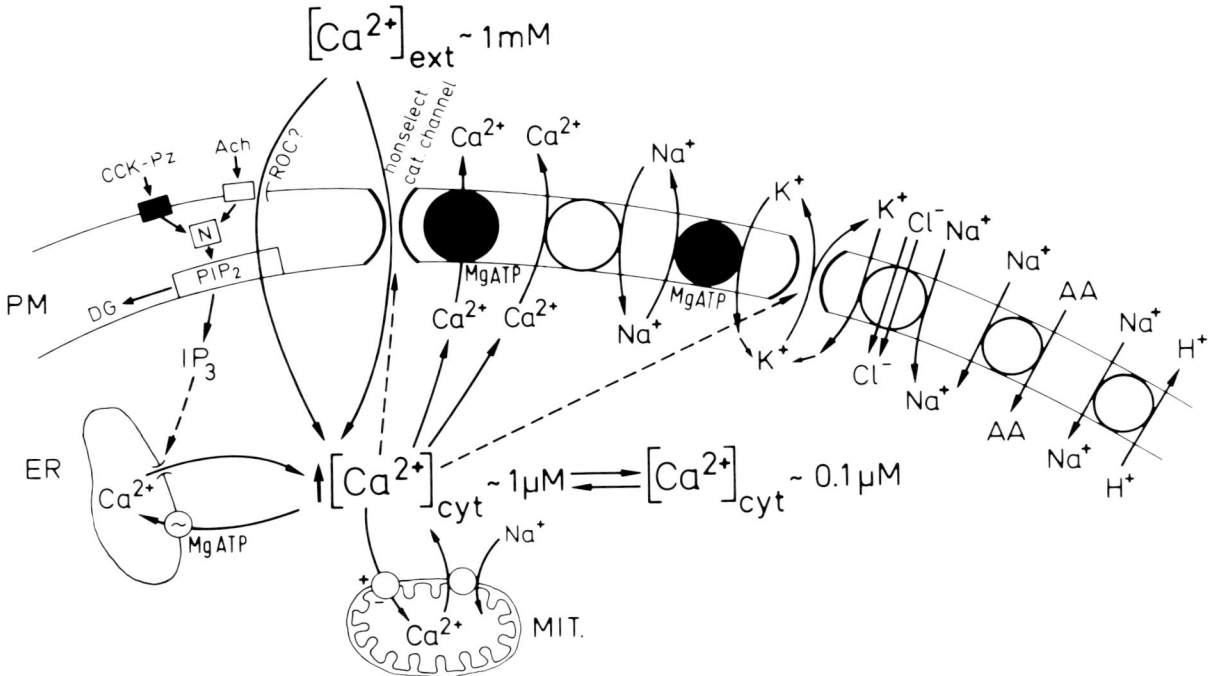

FIG. 1. Schematic representation of primary and secondary active transport processes in the plasma membrane of pancreatic acinar cells and of Ca^{2+}-transport processes in intracellular organelles. AA, amino acid; ACh, acetylcholine; $[Ca^{2+}]_{cyt}$, cytosolic free Ca^{2+} concentration; $[Ca^{2+}]_{ext}$, external free Ca^{2+} concentration; CCK-P$_z$ cholecystokinin-pancreozymin; DG, diacylglycerol; ER, endoplasmic reticulum; IP$_3$, inositol 1,4,5-trisphosphate; MIT, mitochondria; N, GTP-binding protein; PIP$_2$, phosphatidylinositol 4,5-bisphosphate; PM, plasma membrane; ROC, receptor-operated Ca^{2+} channel. *Filled circles*, primary active transport (MgATP driven); *open circles*, secondary active transport (electrochemical ion-gradient driven).

dependent on Na$^+$ (35), localization of both Na$^+$-K$^+$-ATPase and Na$^+$-H$^+$ countertransport at the contraluminal cell side combined with a passive exit step of OH$^-$ or HCO$_3^-$ at the luminal side seemed to be a reasonable assumption (28, 124, 196). The recent data of Novak (161) suggest that a luminal anion-conductance pathway is opened up by cAMP. These measurements of basolateral membrane potential differences in cells of isolated ducts allow the conclusion that there is a Ba^{2+}-sensitive K$^+$ conductance, an Na$^+$-K$^+$-ATPase and an Na$^+$-H$^+$ carrier in the basolateral membrane, and a rheogenic Cl$^-$-conductance pathway in the luminal membrane (161). These results are compatible with the model shown in Figure 2, in which Cl$^-$-HCO$_3^-$ exchange and Cl$^-$ exit are present (step 2). Because existence of an HCO$_3^-$-conductance pathway on the luminal membrane cannot be excluded by Novak (161), step 1 in Figure 2 is an alternative explanation for her data.

Stimulation of ion-transport processes is usually the consequence of increased passive ion flux down the electrochemical gradient for this ion, and the transport system readjusts the prestimulation level. This occurs during stimulation of enzyme secretion, when both increased Ca^{2+} release from intracellular stores and Ca^{2+} influx over the plasma membrane is followed by increased Ca^{2+}-ATPase and Ca^{2+}-Na$^+$ exchange, which leads to Ca^{2+} extrusion from the cell. Stimulation of these Ca^{2+}-extrusion mechanisms can be explained by increased $[Ca^{2+}]_{cyt}$ near to the K_m (Michaelis-Menten constant) values of these Ca^{2+}-transport systems and does not necessarily require activation of regulatory proteins associated with the Ca^{2+} transporter, as had been described for the Ca^{2+} pump in sarcoplasmic reticulum of heart that is activated by cAMP-dependent phosphorylation of phospholamban (115). In pancreatic plasma membranes the Ca^{2+}-binding protein calmodulin increases the Ca^{2+} sensitivity of the transport enzyme by about two orders of magnitude, which makes this Ca^{2+} pump a fine regulator of $[Ca^{2+}]_{cyt}$ (4).

Role of Plasma Membranes in Hormonal Stimulation of Pancreatic Secretion

Transfer of stimulant signals into the cell is promoted by transducing components located in the plasma membrane. Interaction of hormones or neurotransmitters with their specific plasma membrane-located receptors initiates signal-transferring reactions, which can be divided into four main groups: *1)* activation of adenylate cyclase, *2)* activation of phos-

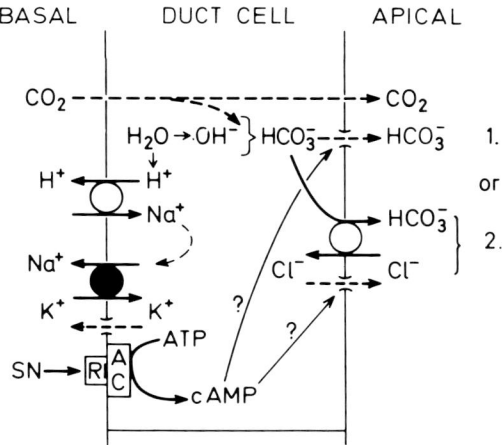

FIG. 2. Model for the mechanism of pancreatic electrolyte and fluid secretion. Na^+-H^+ antiport is located at the basal cell side coupled to the Na^+ gradient that is maintained by the Na^+-K^+-ATPase. On the luminal cell side either an HCO_3^- conductive pathway (1) or a Cl^- conductive pathway combined with a Cl^--HCO_3^- exchanger (2) is present. Secretin (SN) binds to its receptor (R) and stimulates adenylate cyclase (AC) to produce cAMP that opens up conductive pathways for HCO_3^- (1) or Cl^- (2). [Model based on data from Bungaard et al. (28), Kuijpers et al. (124), Novak (161), and Schulz (196).]

pholipase C, 3) opening of ion channels, and 4) activation of tyrosine kinase.

ADENYLATE CYCLASE–CYCLIC AMP SYSTEM. Binding of hormones such as secretin or VIP to their receptors leads to activation of adenylate cyclase and an increase in cAMP in the cell (186). Cyclic AMP in turn stimulates protein kinase A, which phosphorylates proteins assumed to be involved in secretion (29). Transducing components in the plasma membrane that couple receptors to adenylate cyclase are GTP-binding proteins (N or G proteins) (187). A schematic representation for hormonal activation of adenylate cyclase in general is shown in Figure 3. Whereas stimulatory N proteins (N_s) couple stimulatory receptors to adenylate cyclase, inhibitory N proteins (N_i) couple inhibitory receptors to this enzyme (102). Binding of a stimulatory agonist (H_s) to its receptor is assumed to lead to a change of receptor conformation, by which interaction of receptor with the α_s-subunit of the N protein is initiated. This leads to liberation of GDP, binding of GTP to the α_s-subunit, and consequent activation of the N protein, which probably is accompanied by dissociation of α_s-subunit from the $\beta\gamma$-complex of N_s protein (84). The α_s-subunit to which GTP is bound interacts with the catalytic subunit of adenylate cyclase (180). Stimulation of adenylate cyclase is terminated by the intrinsic GTPase activity of the N_s protein, which leads to hydrolysis of GTP and inactivation of N_s, presumably followed by reassociation of α- and $\beta\gamma$-subunits (38). The GTPase activity is inhibited if the α_s-subunit of N_s is ADP-ribosylated by cholera toxin. This results in permanent activation of N_s (39, 40).

CHAPTER 22: SIGNALING TRANSDUCTION 447

In hormonal inhibition of adenylate cyclase an N_i is interpolated (102). It is assumed that an inhibitory agonist (H_i) that causes a decrease in cellular cAMP content leads to exchange of GDP for GTP at the α_i-subunit of N_i, followed by dissociation of α_i-GTP from the $\beta\gamma$-complex. The mechanism of decrease in adenylate cyclase activity by N_i has not been clarified. Liberated α_i-subunit could inhibit the catalytic subunit of adenylate cyclase (Fig. 3). Evidence for direct inhibition of the catalytic subunit of adenylate cyclase by $\beta\gamma$-subunits liberated from N_i has also been obtained (18, 62, 108, 110). Similar to N_s, activated N_i is also inactivated when GTP is hydrolyzed by GTPase activity in N_i (45) and reassociation of α_i- and $\beta\gamma$-subunits occurs (1). The N_i can be ADP-ribosylated by an ectotoxin of Bordetella pertussis, also called islet-activating protein (111). This results in reduced ability of agonists to activate N_i and therefore in elimination of hormonal inhibition of adenylate cyclase.

In the pancreas the presence of both N_i and N_s has been demonstrated (127, 215). Stimulatory hormones of adenylate cyclase activity such as secretion or VIP act on enzyme and $NaHCO_3$ secretion and fluid secretion by mediation of cAMP. Cholecystokinin stimulates adenylate cyclase in broken-cell preparations (138, 188, 216). However, in intact acinar cells no CCK-induced increase in cellular cAMP has been observed (71, 183, 234). When cells were pretreated

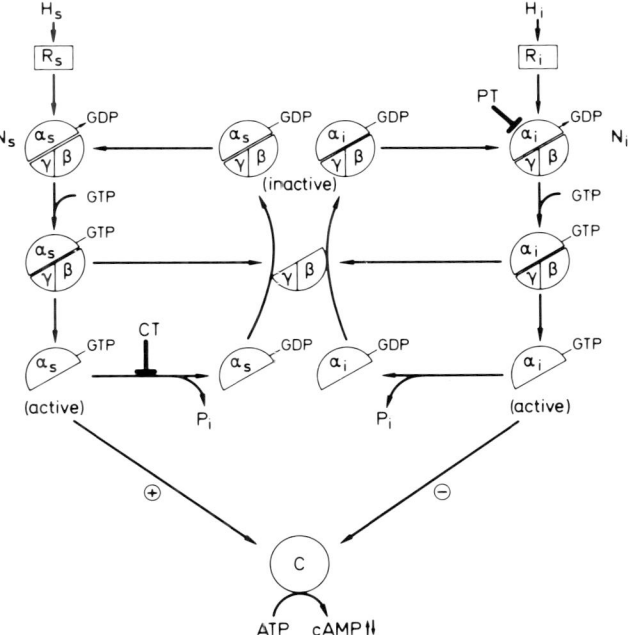

FIG. 3. Schematic representation for regulation of adenylate cyclase activity. C, catalytic subunit of adenylate cyclase; CT, cholera toxin; H_i, inhibitory ligands of adenylate cyclase; H_s, stimulatory ligands of adenylate cyclase; N_i, inhibitory N protein of adenylate cyclase; N_s, stimulatory N protein of adenylate cyclase; PT, pertussis toxin; R_i, inhibitory receptor; R_s, stimulatory receptor; α,β,γ, α,β,γ-complex of the respective N protein.

with pertussis toxin that inactivates N_i of the adenylate cyclase system, cAMP levels increased in acinar cells in response to cholecystokinin octapeptide (CCK-8) and the stimulatory effect of CCK-8 on amylase secretion was potentiated (235). These findings indicate that in intact pancreatic acinar cells N_i inhibition of the adenylate cyclase may be responsible for preventing CCK-8 from increasing cAMP (235). Inhibition of adenylate cyclase by α_2-receptor agonists, opioids, adenosine, prostaglandins E_1 and E_2, or somatostatin (which is known to act by activation of N_i in other cell types) (194) has not been shown for exocrine pancreas cells. Although inhibition of pancreatic HCO_3^- and fluid secretion by norepinephrine has been shown to be mediated by α-receptors (90), the mechanism of inhibition is not known. Similarly, how prostaglandins affect pancreatic secretion is not clear. Prostaglandins of the E series, which inhibit pancreatic electrolyte secretion in vivo (37), probably because of a fall in arterial blood pressure and consequent reduction in pancreatic blood flow, stimulate electrolyte secretion from the perfused cat gland (37). This effect is probably due to elevation of cellular cAMP, because the response is potentiated by theophylline. Somatostatin, which inhibited pancreatic fluid secretion when injected into intact animals (17, 135), was not inhibitory in an isolated, perfused preparation of the pancreas (2), suggesting an indirect effect on the gland in vivo. However, in fragments of rat pancreas, amylase release was stimulated by high concentrations (10^{-6} M) of somatostatin and this effect was accompanied by an increase in the tissue level of cAMP (185), which is not normally an effect of somatostatin. Studies with guinea pig acini showed that secretin- and VIP-stimulated cAMP production was inhibited by somatostatin (64). It thus appears that pancreatic adenylate cyclase is regulated by both N_s and N_i, as in other systems, and that close interactions between stimulatory and inhibitory pancreatic hormones occur. Furthermore, high concentrations of somatostatin-28 but not somatostatin-14 stimulated amylase release via the CCK receptors. Saturable binding of somatostatin to plasma membrane preparations of guinea pig, mouse, and rat pancreas was observed (189, 218), and CCK reduced somatostatin binding via the CCK receptor (189, 206).

PHOSPHOLIPASE C SYSTEM. Stimulated phosphoinositide metabolism was discovered by Hokin and Hokin (86), who showed that incorporation of ^{32}P into phospholipids of pancreatic tissue was stimulated by acetylcholine. The authors speculated that this effect might be involved in secretion. It later became clear that many different cells respond to different agonists with increased phospholipid turnover (12, 61, 151). Subsequently Michell (151) recognized that all agonists that act by mobilization of intracellular Ca^{2+} show this phospholipid effect, suggesting that phosphoinositide metabolism and Ca^{2+} mobilization are closely linked processes. It was not clear, however, if this reaction was a consequence of the Ca^{2+} signal or its cause.

Evidence now suggests that receptor-mediated activation of a plasma membrane–bound phospholipase C is not dependent on increased $[Ca^{2+}]_{cyt}$ (93, 158, 201, 210). Agonist-induced stimulation of the enzyme leads to breakdown of PIP_2 to both IP_3 and 1,2-diacylglycerol. A summary of the major pathways for biosynthesis and degradation of phosphoinositides and the fate of both second messengers, IP_3 and 1,2-diacylglycerol, are given in Figure 4.

The IP_3 releases Ca^{2+} from a nonmitochondrial Ca^{2+} pool that is likely to be part of the endoplasmic

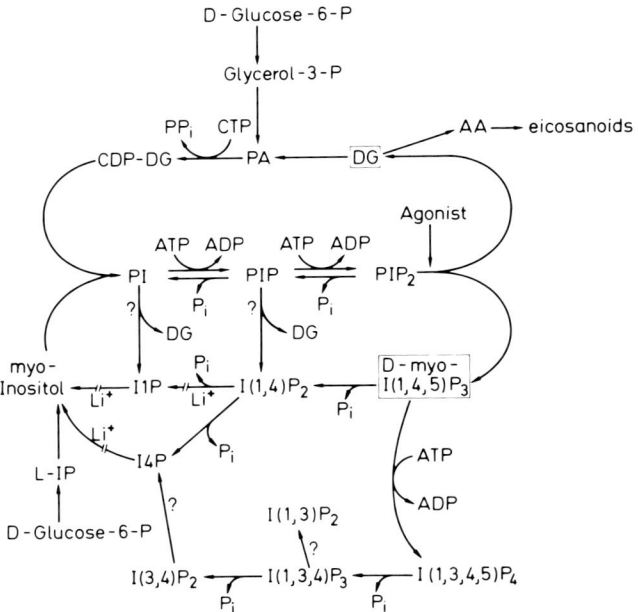

FIG. 4. Major pathways in agonist-dependent phosphoinositide metabolism and sites of Li^+ action. Both phosphatidic acid (PA) and myo-inositol are synthesized from D-glucose. Cytidine diphosphodiacylglycerol (CDP-DG), also named cytidine monophosphorylphosphatidate (CMPPA; CMP-PA), is formed from phosphatidic acid and CTP in the presence of CTP-PA cytidyltransferase. Biosynthesis of phosphatidylinositol (PI) from CDP-DG and myo-inositol via PI-synthetase occurs at the plasma membrane. Phosphorylation of PI by ATP in the presence of specific kinases to phosphatidylinositol 4-phosphate (PIP) and phosphatidylinositol 4,5-bisphosphate (PIP_2) occurs at the plasma membrane. Specific phosphatases in the plasma membrane can dephosphorylate PIP_2 to PIP and to PI. Agonists stimulate plasma membrane–bound phospholipase C, which leads to breakdown of phosphoinositides into diacylglycerol (DG) and water-soluble inositol phosphates. Diacylglycerol is either metabolized via lipases to release arachidonic acid (AA) for eicosanoid biosynthesis or is phosphorylated by ATP in the presence of the plasma membrane enzyme diacylglycerol kinase to regenerate PA. The other product of phospholipase C-mediated PIP_2 hydrolysis, inositol 1,4,5-trisphosphate $[I(1,4,5)P_3]$, is degraded via a specific phosphatase to inositol 1,4-bisphosphate $[I(1,4)P_2]$ and inositol 1-phosphate (I1P), and it is also phosphorylated via a specific kinase to inositol 1,3,4,5-tetrakisphosphate $[I(1,3,4,5)P_4]$. Inositol 1,3,4,5-tetrakisphosphate is dephosphorylated to inositol 1,3,4-trisphosphate $[I(1,3,4)P_3]$ and to inositol 3,4-bisphosphate $[I(3,4)P_2]$. Another pathway might involve degradation to inositol 1,3-bisphosphate $[I(1,3)P_2]$. [Model partly based on work by Shears et al. (199) and Irvine et al. (96).]

reticulum (209, 211); 1,2-diacylglycerol stimulates protein kinase C, an enzyme that is also activated by phorbol esters (160). Inositol 1,3,4,5-tetrakisphosphate (IP$_4$), the phosphorylation product of IP$_3$, can probably be considered an intracellular, third messenger from metabolism of phosphoinositides that might stimulate Ca^{2+} entry into the cell from the extracellular environment [see RECEPTOR-OPERATED AND INTRACELLULAR MESSENGER-OPERATED ION CHANNELS, p. 450; (97, 155a).]

In pancreatic acinar cells, Ca^{2+} is necessary for stimulation of enzyme secretion. Receptor-mediated stimulation of phospholipase C and generation of IP$_3$ lead to release of Ca^{2+} from endoplasmic reticulum and rise of [Ca^{2+}]$_{cyt}$ (211). As has been shown in permeabilized acinar cells, Ca^{2+} is not the only (113), probably not even the most important, messenger for secretion. Activation of protein kinase C by 12-O-tetradecanoylphorbol-13-acetate (TPA), however, leads to Ca^{2+}-dependent enzyme release in the range that can be also elicited by secretagogues (58, 113, 117).

Role of GTP-binding proteins in activation of phospholipase C. Evidence suggests that, like the adenylate cyclase system, one or more GTP-binding proteins also couple receptors to phospholipase C. In permeabilized thrombocytes, guanine nucleotides act synergistically with hormones on polyphosphoinositide breakdown (74, 75). A direct demonstration that guanine nucleotides regulate the activity of the membrane-bound phospholipase C came from studies with neutrophils (44) and blowfly salivary glands (136). In the latter study it was shown that serotonin could stimulate phospholipase C in the presence of GTP, indicating coupling between the receptor and the phospholipase C by a GTP-binding protein. The effect of hormones, GTPγS, and Ca^{2+} on stimulation of phospholipase C and breakdown of PIP$_2$ has also been demonstrated in isolated plasma membranes (42, 139, 142, 149, 164, 226). Thus the current working model is that receptors couple to the phospholipase C via a distinct guanine nucleotide regulatory protein (N$_p$) (Fig. 5). In permeabilized pancreatic acinar cells, GTPγS, a weakly hydrolyzable GTP analogue, stimulates phospholipase C and leads to a rise in IP$_3$ that then releases Ca^{2+} (150, 196a). The effect of additions of hormones subsequently to GTPγS on phospholipase C stimulation is reduced, probably because of a decrease in receptor affinity for the hormone in the presence of GTP or its analogues (14, 120, 132, 203). Some cells, such as leukocytes, respond to treatment with pertussis toxin with inhibition of hormonal stimulation of phospholipase C (21, 165, 227). This indicates that ADP ribosylation of the coupling N protein causes its functional elimination similar to the elimination of N$_i$ of the adenylate cyclase system. In other cell types, including pancreatic acinar cells, pertussis toxin has no effect on phospholipase C (150, 196a). However, treatment with cholera toxin reduces hormonal stimulation of phospholipase C in these cells (196a). Because cAMP had no effect on hormonal activation of phospholipase C in the pancreas (196a), the interpretation of these results considers direct cholera toxin–mediated ADP ribosylation of the hypothetical N protein (N$_p$) that couples receptor to phospholipase C. Another possible interpretation of the data is the assumption that free $\beta\gamma$, which is liberated by activation of N$_s$ protein in the presence of cholera toxin, could associate with the α_p-subunit of hormonally activated N$_p$ protein, and thus conversion to the inactive, fully associated form could take place (108, 110). Kinetic calculations, however, make it unlikely that during the on-off cycle a complete α- to $\beta\gamma$-separation takes place. It is more reasonable to assume a conformational transition that involves all three subunits but does not necessarily cause complete subunit separation (133). Alternatively, direct inhibition of phospholipase C by cholera toxin–activated N$_s$ or its α_s- or $\beta\gamma$-subunits could be possible.

Because inhibitory effects of cholera toxin on binding of chemoattractants to membranes from leukocytes have been described (228) and ADP ribosylation of an N protein that is also substrate for pertussis toxin and different from N$_s$ or N$_i$ has been shown (166, 228), the assumption that in some systems cholera toxin can directly ADP-ribosylate N$_p$, thereby altering its function, obtains further support.

Feedback regulation of phospholipase C. As mentioned in PHOSPHOLIPASE C SYSTEM, p. 448, 1,2-diacylglycerol, one of the products of receptor-mediated

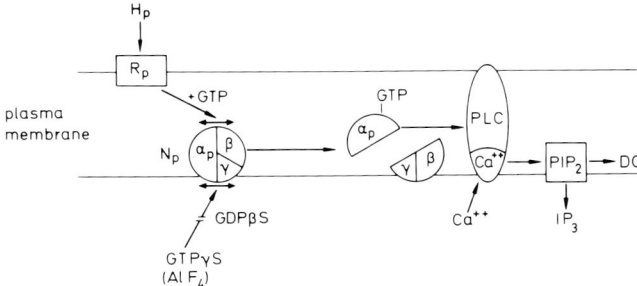

FIG. 5. Model for coupling of receptor to phospholipase C (PLC). When an agonist binds to its receptor, bound GDP is exchanged for GTP on the α-subunit of the N$_p$ protein. This leads to dissociation of the heterotrimer to α- and $\beta\gamma$-subunits. Dissociation can also occur in the presence of GTPγS or (AlF$_4$)$^-$ in the absence of an agonist. GDPβS can inhibit subunit dissociation. The dissociated GTP-α_p complex then activates phospholipase C, which catalyzes the hydrolysis of phosphatidylinositol 4,5-bisphosphate (PIP$_2$) to inositol 1,4,5-trisphosphate (IP$_3$) and diacylglycerol (DG). IP$_3$ diffuses into the cytosol and releases Ca^{2+} from the endoplasmic reticulum, while diacylglycerol remains in the plasma membrane and activates protein kinase C. The α-subunit possesses GTPase activity and so hydrolyzes the bound GTP, thus terminating the activation of phospholipase C. In the presence of GTPγS and (AlF$_4$)$^-$, inactivation of the α-subunit does not occur. Phospholipase C possesses a high-affinity site for Ca^{2+}, which has to be occupied for α-subunit activation. The minimum Ca^{2+} requirement is 100 nM, the concentration found in unstimulated cells. However, Ca^{2+} in the range of 1–500 μM (depending on cell type and conditions of assay) can also stimulate phospholipase C directly without involving the N$_p$ protein. H$_p$, stimulatory hormone of phospholipase C; R$_p$, receptor of phospholipase C. [Adapted from Cockcroft (43).]

stimulation of phospholipase C, stimulates protein kinase C, of which some target proteins reside in the plasma membrane. Thus feedback inhibition of phospholipase C (57, 107, 118, 126, 128, 134, 179, 233, 240) and desensitization of plasma membrane–bound β-receptor (134) have been described.

In a study on the role of regulation of pancreatic enzyme secretion, Pandol and Schoeffield (168) found an increase in cellular 1,2-diacylglycerol only with supramaximal concentrations of secretagogues that has an inhibitory effect on stimulated amylase release, as compared with a maximal dose of secretagogue. The protein kinase C inhibitor 1-(5-isoquinolinesulfonyl)-2-methylpiperazine augmented amylase release from acini stimulated not only by CCK-8 or carbachol but also by TPA. The authors concluded that protein kinase C does not stimulate but inhibits pancreatic stimulus-secretion coupling and suggested that the stimulatory effect of TPA on enzyme release was independent of its effect on protein kinase C (168). In studies on isolated pancreas cells, TPA stimulated amylase release (3, 58, 113); however, carbachol-induced increase in $[Ca^{2+}]_{cyt}$ was inhibited by this phorbol ester (3), which could indicate a feedback inhibition on the phospholipase C system. There is evidence that short-term incubation with TPA stimulates enzyme secretion, whereas preincubation of cells for 10–30 min inhibits carbachol-induced enzyme secretion. (R. Machala and I. Schulz, unpublished observations). It could be that N_p, the hypothetical coupling protein of receptors to phospholipase C, is phosphorylated by protein kinase C and thereby inactivated similarly to N_i of the adenylate cyclase system (8, 101, 109). Treatment of rabbit neutrophils with phorbol esters results in increased ADP ribosylation of a 41-kDa protein catalyzed by pertussis toxin and inhibition of f-Met-Leu-Phe–activated GTPase (145). This indicates that either N_i or a closely analogous protein such as the hypothetical N_p is inactivated by protein kinase C–induced phosphorylation.

ACTIVATION OF BOTH MESSENGER SYSTEMS—ADENYLATE CYCLASE AND PHOSPHOLIPASE C—BY THE SAME HORMONE. Cholecystokinin is a peptide hormone that stimulates enzyme secretion from the exocrine pancreas by activating phospholipase C (210). This hormone also activates adenylate cyclase in broken-cell membrane preparations (188), but no rise of cAMP levels occurred in intact pancreatic cells in response to CCK (71, 183, 234). Recently it has been shown that the peptide hormone secretin can activate both messenger systems in pancreatic acinar cells (221). Secretin at low concentrations (10^{-10}–10^{-8} M) stimulates cAMP accumulation and at concentrations $>10^{-8}$ M it increases IP_3 production, $[Ca^{2+}]_{cyt}$, and 1,2-diacylglycerol content in pancreatic acini, indicating that secretin stimulates cAMP and IP_3 production in rat pancreatic acinar tissue by two different mechanisms. Furthermore these authors (222) have shown that cholera toxin treatment causes an increase in cellular cAMP content and also results in reduction of secretin-stimulated IP_3 production. Because the time courses of these effects were quite different, the authors concluded that different cholera toxin substrates are involved. Similarly in a study on IP_3 production and Ca^{2+} release in permeabilized pancreatic acinar cells, preincubation of cells with cholera toxin inhibited the subsequent effect of CCK-8 on IP_3 production and Ca^{2+} release (196a). Because cAMP itself did not inhibit CCK-8–induced IP_3 production, the authors concluded that cholera toxin inhibited coupling of the CCK receptor to phospholipase C, most likely by directly affecting a GTP-binding protein (N_p) of phospholipase C. Conversely the opinion that activation of N_s can inhibit phospholipase C has also been put forward (231). Thus Wakelam et al. (231) have shown that liver cells possess two distinct receptors for the peptide hormone glucagon, one type coupled to phospholipase C and the other one coupled to adenylate cyclase. Glucagon-stimulated production of IP_3 was biphasic. Inhibition occurred at higher glucagon concentrations, when adenylate cyclase activity was maximally stimulated. The authors suggested that this inhibition of phospholipase C activity at higher glucagon concentrations occurred through activation of N_s of the adenylate cyclase system. On the other hand, treatment of intact hepatocytes with a glucagon analogue, (1-N-α-trinitrophenylhistidine,12-homoarginine)glucagon (TH-glucagon), which does not activate adenylate cyclase but stimulates production of IP_3 (231), could elicit desensitization of glucagon-stimulated adenylate cyclase, a process mimicked by phorbol esters (82). The authors assumed that stimulation of protein kinase C as a consequence of 1,2-diacylglycerol formation in the process of phospholipase C activation could evoke desensitization of glucagon-stimulated adenylate cyclase (82). A schematic diagram of the two signal-transduction systems [Fig. 6; (173)] illustrates mutual interrelations between both systems. The possibility that protein kinase C activation could evoke desensitization of adenylate cyclase could explain the old finding in pancreatic acinar cells that no cAMP accumulation occurs in response to CCK. However, Willems et al. (235) recently observed that CCK-8 did increase cAMP accumulation if cells were pretreated with pertussis toxin, an inhibitor of N_i activation. These results indicate that in the action of CCK on adenylate cyclase direct activation of N_i by CCK binding to an inhibitory receptor could be involved (235) and that this effect could explain the missing cAMP accumulation in pancreatic acinar cells in response to CCK.

RECEPTOR-OPERATED AND INTRACELLULAR MESSENGER-OPERATED ION CHANNELS. During stimulation with agonists, ion channels are opened in the plasma membrane, which leads to influx into the cell of ions such as Na^+ and Cl^- and efflux of K^+ (172).

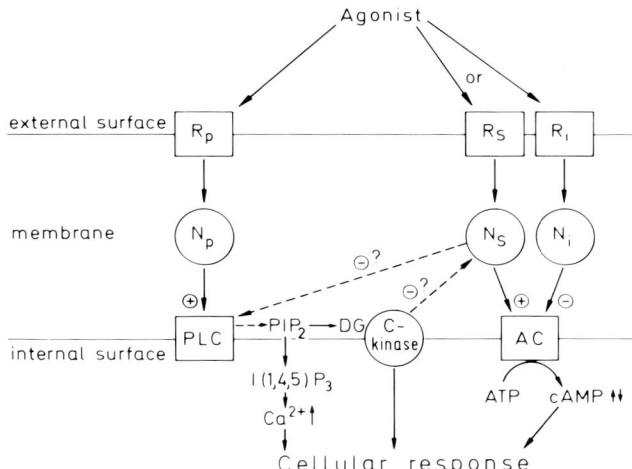

FIG. 6. Schematic diagram of the 2 signal-transduction systems. AC, adenylate cyclase; C-kinase, protein kinase C; DG, diacylglycerol; $I(1,4,5)P_3$, inositol 1,4,5-trisphosphate; N, GTP-binding protein; N_i, inhibitory N proteins; N_p, N protein of phospholipase C; N_s, stimulatory N proteins; PIP_2, phosphatidylinositol 4,5-bisphosphate; PLC, phospholipase C; R, receptor; R_p, receptor of phospholipase C; R_i, inhibitory receptor of adenylate cyclase; R_s, stimulatory receptor of adenylate cyclase.

How Ca^{2+} influx into pancreatic acinar cells is stimulated (119) is not clear. Voltage-dependent Ca^{2+} channels, present in excitable cells (146, 184), have not been found in pancreatic acinar cells (175). Whether Ca^{2+} influx occurs through so-called receptor-operated channels or by mediation of an intracellular messenger is also not decided. Although hormone-stimulated Ca^{2+} influx has been detected by tracer methods (119), electrical studies have not revealed any information about a possible Ca^{2+}-conductance pathway.

Influx of Ca^{2+} through nonspecific Ca^{2+}-dependent cation channels (172) is a possibility but has not been proven and might be hard to detect if Ca^{2+} fluxes are small. Whereas in heart, neuronal, and muscle cells, voltage-dependent Ca^{2+} channels are present (146, 184), no evidence for such channels exists for pancreatic acinar cells (175). One of the main questions that has intrigued cell physiologists is whether receptors are directly coupled to Ca^{2+} channels, i.e., whether Ca^{2+} channels can be opened by binding of an agonist to its receptor without mediation of a second messenger. In cardiac cells, both muscarinic receptors and β-adrenergic receptors, which activate an inwardly rectifying K^+ current and a slow Ca^{2+} current, respectively, can be functionally uncoupled from ion channels by GTP analogues. Intracellular application of the weakly hydrolyzable 5'-guanylylimidodiphosphate (GppNHp) brought about an agonist-induced, antagonist-resistant, persistent stimulation of K^+ and Ca^{2+} currents (24). In visceral smooth muscle, α_2-adrenergic stimulation caused $^{45}Ca^{2+}$ influx into the cell, which was not accompanied by depolarization (77), and in human granulocytes N-f-Met-Leu-Phe caused $^{45}Ca^{2+}$ influx into cells through a Ca^{2+} channel that could not be opened by depolarization of cells. In the latter two experiments, hormone-induced Ca^{2+} influx was inhibited by treatment with pertussis toxin. These findings indicate that an N protein might be involved in the opening of a Ca^{2+} channel. More direct evidence for regulation of a voltage-dependent Ca^{2+} channel by the GTP-binding protein G_o was obtained in neuroblastoma × glioma hybrid cells, suggesting that G_o is involved in the functional coupling of opiate receptors to neuronal voltage-dependent Ca^{2+} channels (80).

By measuring patch-clamp single-channel and whole-cell currents in isolated pancreatic acinar cells, the group of O. H. Petersen gave evidence for internal messenger-activated Ca^{2+} influx into the cell (214). The Ca^{2+}-activated K^+-channel currents were recorded from electrically isolated cell-attached membrane patches that contained Ca^{2+} at the outside of the patch (in the patch pipette) but to which the hormone added to the bath medium had no access (Fig. 7). Because the experimental procedure did not allow penetration of intracellular Ca^{2+} released by the hormone into the patch area, the authors concluded that Ca^{2+} influx into the cell is not directly linked to hormone receptor interaction. It was assumed that IP_3 or another intracellular messenger (perhaps IP_4, the phosphorylation product of IP_3) generated by the hormone that could diffuse more easily than Ca^{2+} into the patch area could perhaps mediate the hormone-evoked Ca^{2+} influx, raising $[Ca^{2+}]_{cyt}$ locally under the patch area so that Ca^{2+}-controlled opening of K^+ channels was possible. Because the authors excluded Ca^{2+}-induced Ca^{2+} influx, as had been postulated for human neutrophils (230), the type of this intracellular messenger remains unknown. Direct evidence for IP_3-activated Ca^{2+} channels was obtained in human T lymphocytes (125). Another candidate is IP_4, a novel inositol phosphate, which is produced rapidly on hormonal stimulation of cells (7, 167) by conversion of IP_3 by a specific ATP-dependent kinase (94) that is

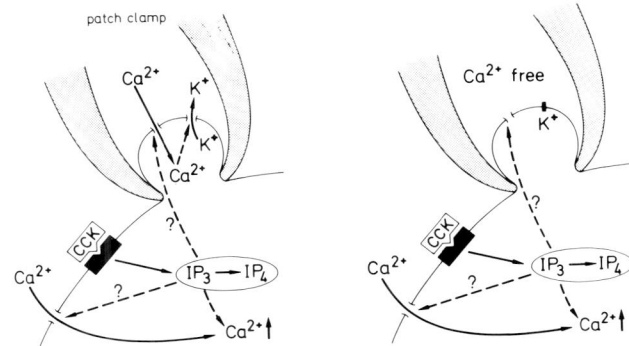

FIG. 7. Schematic diagram to explain patch-clamp experiments, which give evidence for internal messenger-activated Ca^{2+} influx into the cell. CCK, cholecystokinin; IP_3, inositol 1,4,5-trisphosphate; IP_4, inositol 1,3,4,5-tetrakisphosphate. [Adapted from Suzuki et al. (214).]

Ca^{2+} and calmodulin dependent (15a, 16). Phosphatase action then yields inositol 1,3,4-trisphosphate (see Fig. 4). Inositol 1,3,4,5-tetrakisphosphate cannot mobilize Ca^{2+} from intracellular stores (95, 238), but Irvine and Moor (97) have provided indirect evidence that suggests that IP_4 together with IP_3 acts as a second messenger by promoting extracellular Ca^{2+} entry into the cell.

Furthermore, intracellular injections of inositol 1,3,4-trisphosphate or IP_4 into NG108-15 hybrid cells caused an increase in nonspecific cation conductance. Because these derivatives were unable to activate the Ca^{2+}-dependent K^+ current in these cells, an effect that could be obtained by injection of IP_3, it was assumed that they did not release sufficient Ca^{2+} to activate K^+ current but generated an inward membrane current independent of Ca^{2+} release (83). It was not clear, however, if IP_4 had a direct effect on cation conductance. The nature and significance of this observation therefore may warrant further investigation. In a patch-clamp current study recently, Morris et al. (155a) measured transmembrane Ca^{2+}-activated K^+ current in single lacrimal acinar cells. They have shown that neither IP_3 alone nor IP_4 alone is able to activate this current, whereas in combination IP_3 and IP_4 evoke a sustained increase in Ca^{2+}-activated K^+ current, which is dependent on external Ca^{2+}. These data provide important criteria for a second messenger function for IP_4.

A further candidate for a messenger that opens Ca^{2+} channels might be 1,2-diacylglycerol, an activator of protein kinase C that could open channels by phosphorylation. Activators of protein kinase C induce an increase in voltage-dependent Ca^{2+} currents in vertebrate and invertebrate neurons (106), and phosphorylation of the α-subunit of the purified voltage-sensitive Na^+ channel from rat brain by protein kinase C has been demonstrated (50).

TYROSINE KINASE SYSTEM. The main signal-transducing systems in stimulus-secretion coupling from the exocrine pancreas are those previously described (see ADENYLATE CYCLASE–CYCLIC AMP SYSTEM, p. 447, and PHOSPHOLIPASE C SYSTEM, p. 448), including both adenylate cyclase– and phospholipase C–mediated pathways. In addition to receptor-mediated generation of intracellular messengers such as cAMP as well as IP_3 and 1,2-diacylglycerol, respectively, a third pathway regulated by growth factors is present in pancreatic acinar cells. A detailed description of growth regulation is given in the chapter by Logsdon in this Handbook; this subject is only briefly discussed here. As in other cells studied, growth factors, such as epidermal growth factor, and insulin activate specific receptors that possess intrinsic tyrosine-specific protein kinase activity (91). Growth factor binding induces a rapid stimulation of receptor kinase, resulting in autophosphorylation of the receptor itself as well as phosphorylation of various substrate proteins involved in the action of growth factors, such as DNA synthesis and cell division. It has been shown for different cell systems that in addition to activation of tyrosine kinase, growth factors such as epidermal growth factor and platelet-derived growth factor also stimulate the breakdown of phosphatidylinositol 4,5-bisphosphate with consequent generation of IP_3 and 1,2-diacylglycerol (12–14, 192). Whereas IP_3-induced Ca^{2+} release and 1,2-diacylglycerol-stimulated protein kinase C activity are important messenger functions in pancreatic enzyme secretion, they also play a role in cell growth. Thus an early rise in intracellular pH in response to growth factors, assumed to be involved in DNA synthesis, is mediated by an Na^+–H^+–exchange mechanism in the plasma membrane that is turned on by protein kinase C (154). Furthermore the rise in $[Ca^{2+}]_{cyt}$ induced by IP_3 could function as a second messenger that mediates the early transcriptional effects of growth factors. Of particular relevance to mitogen action is the finding that artificially raising $[Ca^{2+}]_{cyt}$ with the Ca^{2+} ionophore A23187 can mimic the effect of epidermal growth and platelet-derived growth factor on the rapid induction of the c-*fos* and c-*myc* protoncogenes (23, 223).

In exocrine pancreas, cell growth factors regulate biosynthesis and cell metabolism, as in other cells, but they also potentiate digestive enzyme secretion (66, 137, 236). Thus, in isolated superfused mouse pancreas, insulin alone had no effect on amylase secretion but it potentiated acetylcholine-evoked amylase secretion. The mechanism of this potentiating action remains largely unexplained. In other cell systems, growth factors have been shown to stimulate breakdown of phosphoinositides (13, 154). However, controversial findings exist concerning this effect of insulin (65–67, 171). Insulin stimulates precursor incorporation into phospholipids of the phosphoinositide cycle (59, 68, 143, 208), probably by insulin receptor–associated tyrosine kinase, which acts as a phosphatidylinositol kinase (141). Although not yet demonstrated for the exocrine pancreas, this would increase the phosphoinositide pool available for generation of second messengers IP_3 and 1,2-diacylglycerol by activation of phospholipase C. Another possible way that insulin may exert its potentiating action on secretagogue-evoked amylase output from pancreatic acinar cells is by elevating the intracellular level of cAMP (202), which has a potentiating effect on the Ca^{2+}-mediated pathway in stimulus-secretion coupling (70, 113). Recently it has been proposed that insulin might activate a phosphatidylinositol-specific phospholipase C, which is able to hydrolyze a novel glycolipid, releasing sugar-containing mediators together with 1,2-diacylglycerol with distinct acyl chain composition (89, 190, 191). Whether these mediators have actions similar to both hydrolysis products of PIP_2, i.e., IP_3 and 1,2-diacylglycerol, which release Ca^{2+} and activate protein kinase C, respectively, remains to be investigated.

ROLE OF INTRACELLULAR MEMBRANES IN PANCREATIC CELL FUNCTION

Signals that are generated in the plasma membrane are transferred to intracellular organelles via second, third, or multiple messengers. The mechanisms that are activated in intracellular membranes by intracellular messengers are in principle similar to those that are developed in the plasma membrane when a hormone binds to its receptor. There are also special binding sites for intracellular messengers that put subsequent processes into operation, such as opening of channels or activation of enzymes. The main function of intracellular compartments is uptake and storage of substances that can be made available if necessary. Further functions are uptake of precursors that are components for synthesis of other substances, such as amino acids for protein synthesis.

In the exocrine pancreas, synthesis and storage of proteins play a dominant role; the rough endoplasmic reticulum is therefore well developed. Along with these long-term functions, short-term regulatory mechanisms are also located in endoplasmic reticulum. Some of these short-term mechanisms are discussed next.

Passive Ion-Transport Mechanisms in Endoplasmic Reticulum

INOSITOL 1,4,5-TRISPHOSPHATE–INDUCED CALCIUM RELEASE. One of the most important passive ion-transport mechanisms in stimulus-secretion coupling is IP_3-induced Ca^{2+} release that leads to an increase in $[Ca^{2+}]_{cyt}$, which triggers many further processes in the cascade of events leading to exocytosis. Although the detailed mechanism by which IP_3 releases Ca^{2+} is not clear, evidence suggests that a Ca^{2+} channel is opened in the endoplasmic reticulum membrane by IP_3, allowing Ca^{2+} efflux from the intravesicular compartment into the cytosol (156). Main arguments that favor this conclusion are based on the observations that in isolated liver microsomal vesicles and in permeabilized pancreatic acinar cells IP_3 only releases Ca^{2+} if K^+ is present in the medium (156). The authors therefore concluded that IP_3 acts either by stimulating a coupled Ca^{2+}-K^+ exchange or a Ca^{2+} channel when K^+ enters the endoplasmic reticulum through a K^+ channel. The existence of the latter channel was demonstrated by dilution of K^+-gluconate–loaded endoplasmic reticulum vesicles into K^+-free $^{86}Rb^+$-containing medium. This experimental condition results in a K^+ gradient across the vesicle membrane whose inside is negative relative to the outside if a K^+ conductance exists in the endoplasmic reticulum membrane. Under these conditions the K^+ marker $^{86}Rb^+$ was taken up by the vesicles and the effect was blocked by the K^+-channel blocker tetraethylammonium. That the K^+ gradient can provide the driving force for Ca^{2+} uptake was shown by loading the vesicles with K^+ and resuspending them in $^{45}Ca^{2+}$-containing K^+-free medium. The IP_3-stimulated $^{45}Ca^{2+}$ uptake was inhibited by tetraethylammonium. This effect could be restored by addition of the K^+ ionophore valinomycin. It thus appears that IP_3 activates a Ca^{2+} channel in the endoplasmic reticulum membrane and that a K^+ flux across the membrane provides charge compensation for IP_3-induced Ca^{2+} release. The fact that IP_3-induced Ca^{2+} release is largely unaffected by different temperatures lets one also assume that a Ca^{2+} channel rather than a carrier-mediated process is involved (22, 157).

GTP-INDUCED CALCIUM RELEASE. Guanine nucleotides are also capable of mobilizing Ca^{2+} from certain preparations of endoplasmic reticulum and permeabilized cells (52, 54, 79, 225). The action, in contrast to that of IP_3, is competitively inhibited by GDP. The GTP-induced Ca^{2+} mobilization seems to require hydrolysis of the nucleotide, since the weakly hydrolyzable analogue of GTP, GTPγS, did not show such an effect. The GTP effect is more temperature dependent than that of IP_3, and the effect is potentiated by treatment with polyethylene glycol. In some preparations of endoplasmic reticulum, such as in liver microsomes, IP_3 was only capable of releasing Ca^{2+} if both GTP and polyethylene glycol (PEG) were present (52).

These data were interpreted to mean that a yet unknown loosely associated cytosolic factor that is necessary for the action of GTP could be aggregated to the endoplasmic reticulum membrane in the presence of polyethylene glycol. In permeabilized acinar cells from exocrine pancreas the ability of IP_3 to release Ca^{2+} was not dependent on the presence of GTP. However, inhibition of IP_3-induced Ca^{2+} release was seen in the presence of GTPγS (196a). In a preparation such as permeabilized liver cells the nucleotide increased IP_3-releasable Ca^{2+}. The GTP alone did not promote Ca^{2+} mobilization, but this effect could be brought about by washing the cells, which probably removes a cytosolic factor (220).

When endoplasmic reticulum was fractionated by Percoll gradient centrifugation into different subfractions, it could be clearly demonstrated that both IP_3 and GTP act on different Ca^{2+} pools (79). However, the experiments in which IP_3-induced Ca^{2+} release was enhanced in the presence of GTP (52, 54) suggest that GTP may act on the same Ca^{2+} store as IP_3. In an electron-microscopic investigation, Dawson (53) observed that addition of GTP to rat liver microsomes caused the appearance of very large vesicles. Formation of these vesicles was blocked by the omission of PEG or by the addition of GTPγS. The author concluded that GTP induced vesicle fusion and that this could explain the effects of GTP on microsomal Ca^{2+} release. Such fusion would be expected to 1) cause Ca^{2+} release due to leakage of vesicle contents during the fusion process and 2) greatly enhance IP_3-dependent Ca^{2+} release by increasing the pool size of Ca^{2+} available to a given IP_3-sensitive channel. For microsomal preparations that show large Ca^{2+} release in response to IP_3 in the absence of GTP (e.g., N1E-

115 cells, parotid gland, pancreas), vesicle fusion would not be expected to enhance the effect much further, since a great part of the intravesicular Ca^{2+} is accessible via IP_3-sensitive sites initially. However, GTP could still cause Ca^{2+} release due to leakage during vesicle fusion (53). Because neither the exact locations of both IP_3 and GTP-sensitive Ca^{2+} stores nor the detailed mechanisms of Ca^{2+} release are known, the effect of GTP on Ca^{2+} release is hard to explain.

Active Ion-Transport Mechanisms in Endoplasmic Reticulum

After Ca^{2+} release from intracellular stores, uptake of Ca^{2+} to the prestimulation level is promoted by intracellular active Ca^{2+}-transport processes. Whereas endoplasmic reticulum can be viewed as the site of rapid and fine regulation of $[Ca^{2+}]_{cyt}$, mitochondria only possess low-affinity systems and must therefore be regarded as long-term Ca^{2+}-buffering compartments. Normal cells usually do not retain much Ca^{2+} in mitochondria (205), which only play a significant role under conditions of pathological cellular Ca^{2+} overload. In a comparison of areas and activities of Ca^{2+}-transporting membranes in a heart muscle cell, Carafoli and Penniston (34) calculated that at $[Ca^{2+}]_{cyt}$ of 10^{-7} M at rest and 10^{-6} M at contraction the Ca^{2+}-pumping ATPase in sarcoplasmic reticulum accounts for most Ca^{2+} removal from the cytosol. At rest ~70% of the total is taken up into sarcoplasmic reticulum, whereas ~30% is extruded by Ca^{2+}-pump mechanisms in the plasma membrane and none is taken up by mitochondria. In the contracting cell at ~10^{-6} M $[Ca^{2+}]_{cyt}$, ~90% of total Ca^{2+} uptake occurs into sarcoplasmic reticulum, 3%-4% is extruded over the plasma membrane, and ~6% is taken up by mitochondria. When the $[Ca^{2+}]_{cyt}$ reaches 10^{-5} M, the mitochondria become dominant and take up ~51% of total Ca^{2+} uptake (~47% goes into sarcoplasmic reticulum and ~2% over the plasma membrane). These values are reflected in the affinities of the various transport systems for Ca^{2+}, which is $<10^{-6}$ M for the sarcoplasmic or endoplasmic reticulum and ~1.5×10^{-5} M and $1-2 \times 10^{-5}$ M for the electrophoretic uniporter and the Na^+-Ca^{2+} exchanger in mitochondria, respectively. The plasma membrane Ca^{2+}-ATPase has a K_m of ~4×10^{-7} M; the plasma membrane Na^+-Ca^{2+} exchanger has a K_m of ~$1-20 \times 10^{-6}$ M (33). These data are comparable for the pancreas. The apparent K_m value for Ca^{2+}-Mg^{2+}-ATPase activity in plasma membranes has been found to be ~3×10^{-7} M (123). The same K_m value has been found for Ca^{2+}-Mg^{2+}-ATPase activity in endoplasmic reticulum (11). Accordingly, fine regulation of $[Ca^{2+}]_{cyt}$ in pancreatic acinar cells is achieved by Ca^{2+}-transport systems in the endoplasmic reticulum (11, 212) and in the plasma membrane (9) but not by those in mitochondria (212).

MgATP-DEPENDENT CALCIUM UPTAKE INTO ENDOPLASMIC RETICULUM. The Ca^{2+} transport in isolated vesicles from pancreatic endoplasmic reticulum showed the same properties as Ca^{2+} transport in permeabilized acinar cells. Concerning K_m values for Ca^{2+}, Mg^{2+}, and ATP, Ca^{2+} uptake was dependent on cations in the order $Rb^+ > K^+ > Na^+ > Li^+ >$ choline$^+$ and dependent on anions in the order $Cl^- > Br^- > NO_3^- > SO_4^{3-} > SCN^- =$ cyclamate (11). Both Ca^{2+}-Mg^{2+}-ATPase activity and its phosphorylated intermediate were also dependent on cations and anions. The Ca^{2+}-Mg^{2+}-ATPase was stimulated in an order that was similar to that for Ca^{2+} uptake (92, 112). However, incorporation of $^{32}P_i$ from $[^{32}P]$ATP into the intermediate of the Ca^{2+}-Mg^{2+}-ATPase was highest with choline$^+$ and lowest with K^+ ions (92). These results were interpreted to mean that K^+ is necessary for the dephosphorylation step in the turnover cycle. The influence of anions on Ca^{2+}-Mg^{2+}-ATPase activity was quite different. In closed membrane vesicles from endoplasmic reticulum both Ca^{2+}-Mg^{2+}-ATPase activity and $^{32}P_i$ incorporation were increased in the presence of Cl^- and other permeant anions such as Br^- and NO_3^-, as compared with the weakly permeant anion cyclamate (112). However, when vesicles were opened by treatment with a low-detergent concentration that breaks up the membrane without changing Ca^{2+}-Mg^{2+}-ATPase activity, the anion sequence was abolished and Ca^{2+}-Mg^{2+}-ATPase activity as measured in the presence of cyclamate was increased to the same degree as in the presence of Cl^-. These results indicate that MgATP-driven Ca^{2+} transport into pancreatic endoplasmic reticulum is electrogenic or that during Ca^{2+} uptake electrogenic transport of another cation to which Ca^{2+} is coupled takes place: the charge is compensated by permeant anions. When vesicles were broken, the permeability barrier for anions was abolished. Therefore also limitation of Ca^{2+}-transport and Ca^{2+}-Mg^{2+}-ATPase activity due to the electrochemical potential difference that is built up by positive charge transfer during electrogenic Ca^{2+} transport was abolished. Electrogenicity during Ca^{2+} transport had been further confirmed by experiments in which an electrical potential difference was established over the vesicle membrane by use of different K^+ concentrations at both sides of the membrane in the presence of the K^+ ionophore valinomycin. If the K^+ gradient was directed from the outside to the inside of the vesicle, creating an inside-positive membrane potential, Ca^{2+}-Mg^{2+}-ATPase activity was decreased as compared with conditions without a membrane potential difference. If the K^+ gradient was directed from inside to outside, the resulting inside-negative membrane potential increased Ca^{2+}-Mg^{2+}-ATPase activity (112).

ION GRADIENT–COUPLED CALCIUM UPTAKE INTO ENDOPLASMIC RETICULUM. In the discussion on active ion transport an observation should be mentioned that appears like an active MgATP-promoted Ca^{2+} uptake. This Ca^{2+} uptake, however, is not directly coupled to ATP hydrolysis. It is driven by an H^+ gradient directed from the inside of the endoplasmic reticulum to the cytosol. The H$^+$ gradient is established by an H$^+$ pump located in the same compartment as the electrogenic Ca^{2+} pump. When MgATP-driven Ca^{2+} uptake into isolated permeabilized rat parotid cells was measured, ~10% of total ^{45}Ca^{2+} uptake could not be inhibited by vanadate, which in pancreas cells completely blocks nonmitochondrial MgATP-driven Ca^{2+} uptake (11, 212), Ca^{2+}-ATPase activity, and ^{32}P incorporation from [^{32}P]ATP into the 100-kDa intermediate of the Ca^{2+}-ATPase (92). This residual vanadate-insensitive Ca^{2+} uptake in parotid cells could be blocked, however, by protonophores such as carbonyl cyanide m-chlorophenylhydrazone (CCCP) and nigericin as well as by H$^+$-ATPase inhibitors (198a). In isolated endoplasmic reticulum vesicles from both rat parotid and pancreas it was further shown that ^{45}Ca^{2+} uptake could be induced in the presence of an H$^+$ gradient across the vesicle membrane. When vesicles were preincubated in the presence of MgATP, of vanadate, and with Ca^{2+} chelators to abolish MgATP-driven Ca^{2+} uptake, MgATP-driven H$^+$ uptake (as measured by the acridine orange method) still took place. When vesicles were then transferred into a medium that contained ^{45}Ca^{2+} and vanadate but did not contain any ATP, ^{45}Ca^{2+} uptake took place. This ^{45}Ca^{2+} uptake was not seen if the vesicles were preincubated in the presence of protonophores or in the presence of an H$^+$-ATPase inhibitor such as chloronitrobenzoxadiazole or if ATP was lacking in the preincubation medium, indicating that the presence of an H$^+$ gradient was involved in ^{45}Ca^{2+} uptake. It was concluded that Ca^{2+} uptake into parotid and pancreatic endoplasmic reticulum can occur via an MgATP-independent Ca^{2+}-H$^+$ exchange (198a). Whether both MgATP-dependent Ca^{2+} uptake and Ca^{2+}-H$^+$ exchange are promoted by the same transport protein remains to be investigated.

ION TRANSPORT INVOLVED IN EXOCYTOSIS

The process by which fusion of zymogen granules with the luminal plasma membrane of acinar cells is initiated and the mechanisms that are involved in exocytosis of zymogens into the lumen are not clearly understood. Specific protein kinases that are activated by intracellular messengers of exocytosis such as Ca^{2+}, 1,2-diacylglycerol, and cAMP are probably involved in fusion of zymogen granules with the luminal plasma membrane. It appears that both phosphorylation and dephosphorylation of target proteins in the plasma membrane, in the zymogen granule membrane, and in the cytosol play an important role in these processes [see the chapter by Williams, Burnham, and Hootman in this *Handbook*; (30, 47, 51, 239)]. The function of phosphorylated and dephosphorylated proteins in exocytosis is not clear. Previous work on both model systems and biological systems has suggested that there are two crucial steps that must occur to cause exocytosis of granular contents. *1)* Both granular and plasma membranes must be brought into close apposition (81, 169, 181, 242). *2)* The osmolarity of the granular content must increase enough to cause granule swelling and subsequent fusion of the two membranes (46, 181, 241, 242). Simultaneous electrical and optical measurements showed, however, that membrane fusion preceded secretory granule swelling during exocytosis of mouse mast cells (243).

In a chemosmotic model for exocytosis of proteins from secretory cells, Pollard and co-workers (170, 182) suggested that the granule membrane might have an anion-transport site. When fusion of granules with the plasma membrane is induced by Ca^{2+}, the anion-transport site becomes exposed to the external medium. Release of granular content involves a sequence of events, including a granule inwardly directed proton pump at the cytoplasmic side that results in an inside-positive membrane potential. Chloride can then enter the granule from the external medium down its electrochemical gradient by the specific pathway. The osmotic content of the granule is directly increased by influx of electrolytes, and water simultaneously enters the granule, causing it to swell. Increased granule volume finally results in lysis or fission and release of granule content into the extracellular milieu (170, 182). Because a proton pump has not been found in pancreatic acinar cells, this model is not applicable to the pancreas.

In studies on permeabilized adrenal medullary cells, Knight and Baker (116) found that high concentrations of Cl$^-$ (160 mM) could promote release of both catecholamines and dopamine β-hydroxylase in the absence of Ca^{2+}. The Ca^{2+}-dependent release, however, was inhibited by Cl$^-$. Of a variety of anions examined, their order of effectiveness at supporting Ca^{2+}-dependent exocytosis was glutamate$^-$ > acetate$^-$ > Cl$^-$ > Br$^-$ > SCN$^-$ (116). These data therefore are not consistent with the suggestion of Pollard et al. (182) that entry of Cl$^-$ into the granules promotes exocytosis. A model by Stanley and Ehrenstein (207) that carries the role of Ca^{2+} in fusion and secretory exocytosis a little further proposes Ca^{2+}-activated opening of K$^+$ channels present in granule membranes. The opening of these channels coupled with anion transport across the granular membrane would result in an influx of K$^+$ and anions, increasing the osmotic pressure of the granular content. For those granules close to the plasma membrane this would lead to fusion with the membrane and exocytosis of the granular contents. The model also requires a mechanism to pump or just to keep K$^+$ out of the granule (207).

Is a Chloride Conductance in the Membrane of Pancreatic Zymogen Granules Involved in Exocytosis?

Our recent findings suggest that secretion is Cl$^-$ dependent in permeabilized cells from the exocrine pancreas. In the presence of Cl$^-$, enzyme secretion is stimulated two- to threefold with secretagogues, whereas in the presence of weakly permeant anions such as cyclamate, enzyme secretion cannot be stimulated by secretagogues. When Cl^- is replaced by a lipophilic anion such as SCN^-, however, enzyme secretion is as high as with Cl^- in the presence of secretagogues. Further stimulation by secretagogues, however, is not possible. Only Cl^--dependent secretion can be inhibited by the Cl^--channel blocker 4,4'-diisothiocyanatostilbene-2,2'-disulfonic acid (DIDS) (C. M. Fuller, L. Eckhardt, and I. Schulz, unpublished observations). A DIDS-sensitive Cl^- pathway has been found in the granular membrane of isolated zymogen granules (55). The question then arises of whether such a Cl^- pathway is also present in the membrane of zymogen granules in vivo. A Cl^- channel could also be opened by stimulatory events, such as protein phosphorylation. Another possibility is opening of a Cl^- channel in the zymogen granule membrane as a consequence of fusion of the granule with the luminal plasma membrane. In this case the Cl^- channel would become an integral part of the luminal membrane. If in addition to an anion pathway the granular or luminal membrane also contains a cation conductance or if cations take the paracellular pathway (19), both cations and anions could move into the granule and the lumen, respectively. This would result in an increase in the osmotic pressure of the vesicles and would be followed by water flux that could flush out proteins from the granules into the lumen. The driving force for ion transport is increased $[Cl^-]_{cyt}$ above the electrochemical equilibrium during stimulation. This increase in $[Cl^-]_{cyt}$ results from a number of related steps, for each of which experimental evidence is given. Receptor-mediated stimulation of phospholipase C leads to breakdown of $PIP_2 \rightarrow$ generation of $IP_3 \rightarrow$ mobilization of $Ca^{2+} \rightarrow$ increase in $[Ca^{2+}]_{cyt} \rightarrow$ activation of Ca^{2+}-dependent K^+ channels in the basolateral plasma membrane $\rightarrow K^+$ efflux and thereby to activation of a coupled Na^+-$2Cl^-$-K^+ cotransport over the basolateral membrane with increased Cl^- influx into the cell (176). If a Cl^- conductance or Cl^- channel is present in zymogen granule membranes, this would become part of the luminal plasma membrane of the acinar cell, when granules fuse with this membrane, and Cl^- could move into the lumen followed by Na^+ and water. Indirect evidence for an increase in luminal Cl^- conductance during stimulation was obtained from electrophysiological studies (172, 176).

A schematic model of events proposed to be involved in stimulus-secretion coupling is shown in Figure 8. Because stimulation of protein release by Ca^{2+}-mediated secretagogues is accompanied by NaCl and fluid secretion from pancreatic acinar cells (176), although

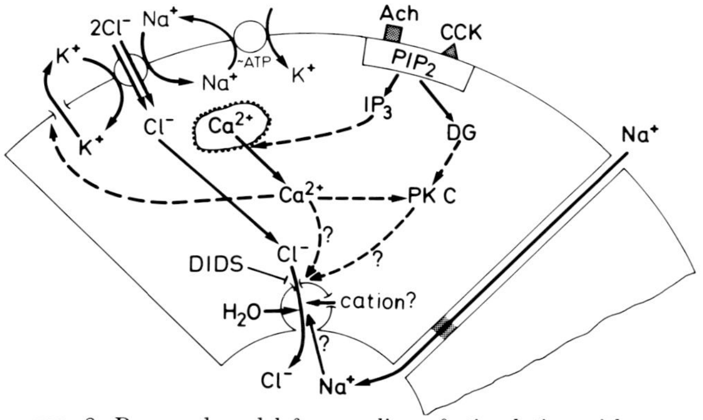

FIG. 8. Proposed model for coupling of stimulation with secretion of enzymes, NaCl, and fluid in pancreatic acinar cells. ACh, acetylcholine; CCK, cholecystokinin; DG, diacylglycerol; DIDS, 4,4'-diisothiocyanatostilbene-2,2'-disulfonic acid; IP_3, inositol 1,4,5-trisphosphate; PIP_2, phosphatidylinositol bisphosphate; PK C, protein kinase C.

to variable extent in different animal species, it is possible that both protein secretion and NaCl and fluid secretion are based on this common mechanism in the granular membrane. Whether this is true also for other species, such as the cat, which does not show Ca^{2+}-dependent NaCl and fluid secretion but cAMP-dependent $NaHCO_3$ and fluid secretion, or for other glands, like the parotid gland, which responds to cAMP with protein secretion and to Ca^{2+} with NaCl and fluid secretion from acinar cells, remains to be investigated.

In a study using quantitative electron-microprobe analysis, elemental dry-weight concentrations in cytoplasm, secretory granules, and nuclei of resting and pilocarpine-stimulated rat parotid gland acinar cells were measured (105). Secretory granules in resting cells had lower concentrations of Na^+, Cl^-, and K^+ and higher Ca^{2+} concentrations than cytoplasm. In pilocarpine-stimulated cells, however, the concentrations of Na^+ and Cl^- increased in secretory granules, while both Na^+ and Ca^{2+} concentrations increased and K^+ concentration decreased in cytoplasm. It is not known if several different observations, including those on protein phosphorylation of zymogen granular membranes (29, 30), on Ca^{2+}-induced cell capacitance fluctuation indicating Ca^{2+}-dependent exocytosis (144), and on anion dependency of secretion (C. M. Fuller, L. Eckhardt, and I. Schulz, unpublished observations) are related processes and part of the final step in stimulus-secretion coupling that leads to exocytosis. Because highly developed techniques to study membrane functions in isolated and reconstituted systems are available, it can be assumed that these questions will be solved in future work.

REFERENCES

1. AKTORIES, K., G. SCHULTZ, AND K. H. JAKOBS. Somatostatin-induced stimulation of a high-affinity GTPase in membranes of S49 lymphoma cyc⁻ and H21a variants. *Mol. Pharmacol.* 24: 183–188, 1983.

2. ALBINUS, M., E. L. BLAIR, R. M. CASE, D. H. COY, A. GOMEZ-PAN, B. H. HIRST, J. D. REED, A. V. SCHALLY, B. SHAW, P. A. SMITH, AND J. R. SMY. Comparison of the effect of somatostatin on gastrointestinal function in the conscious and anaesthetized cat and on the isolated cat pancreas. *J. Physiol. Lond.* 269: 77–91, 1977.

3. ANSAH, T.-A., S. DHO, AND R. M. CASE. Calcium concentration and amylase secretion in guinea pig pancreatic acini: interactions between carbachol, cholecystokinin octapeptide and the phorbol ester, 12-O-tetradecanoylphorbol 13-acetate. *Biochim. Biophys. Acta* 889: 326–333, 1986.

4. ANSAH, T.-A., A. MOLLA, AND S. KATZ. Ca^{2+}-ATPase activity in pancreatic acinar plasma membranes. Regulation by calmodulin and acidic phospholipids. *J. Biol. Chem.* 259: 13442–13450, 1984.

5. BANGHAM, A. D., AND R. W. HORNE. Action of saponin on biological cell membranes. *Nature Lond.* 196: 952–953, 1962.

6. BARROWMAN, M. M., S. COCKCROFT, AND B. D. GOMPERTS. Potentiation and inhibition of secretion from neutrophils by phorbol ester. *FEBS Lett.* 201: 137–142, 1986.

7. BATTY, I. R., S. R. NAHORSKI, AND R. F. IRVINE. Rapid formation of inositol 1,3,4,5-tetrakisphosphate following muscarinic receptor stimulation of rat cerebral cortical slices. *Biochem. J.* 232: 211–215, 1985.

8. BAUER, S., AND K. H. JAKOBS. Phorbol ester treatment impairs hormone- but not stable GTP analog-induced inhibition of adenylate cyclase. *FEBS Lett.* 198: 43–46, 1986.

9. BAYERDÖRFFER, E., L. ECKHARDT, W. HAASE, AND I. SCHULZ. Electrogenic calcium transport in plasma membrane of rat pancreatic acinar cells. *J. Membr. Biol.* 84: 45–60, 1985.

10. BAYERDÖRFFER, E., W. HAASE, AND I. SCHULZ. Na^+/Ca^{2+} countertransport in plasma membrane of rat pancreatic acinar cells. *J. Membr. Biol.* 87: 107–119, 1985.

11. BAYERDÖRFFER, E., H. STREB, L. ECKHARDT, W. HAASE, AND I. SCHULZ. Characterization of calcium uptake into rough endoplasmic reticulum of rat pancreas. *J. Membr. Biol.* 81: 69–82, 1984.

12. BERRIDGE, M. J. Inositol trisphosphate and diacylglycerol as second messengers. *Biochem. J.* 220: 345–360, 1984.

13. BERRIDGE, M. J., J. P. HESLOP, R. F. IRVINE, AND K. D. BROWN. Inositol trisphosphate formation and calcium mobilization in Swiss 3T3 cells in response to platelet-derived growth factor. *Biochem. J.* 222: 195–201, 1984.

14. BERRIDGE, M. J., AND F. R. IRVINE. Inositol trisphosphate, a novel second messenger in cellular signal transduction. *Nature Lond.* 312: 315–321, 1984.

15. BESTERMAN, J. M., W. S. MAY, JR., H. LEVINE III, E. J. CRAGOE, JR., AND P. CUATRECASAS. Amiloride inhibits phorbol ester-stimulated Na^+/H^+ exchange and protein kinase C. An amiloride analog selectively inhibits Na^+/H^+ exchange. *J. Biol. Chem.* 260: 1155–1159, 1985.

15a.BIDEN, T. J., M. COMTE, J. A. COX, AND C. B. WOLLHEIM. Calcium-calmodulin stimulates inositol 1,4,5-trisphosphate kinase activity from insulin-secreting RINm5F cells. *J. Biol. Chem.* 262: 9437–9440, 1987.

16. BIDEN, T. J., AND C. B. WOLLHEIM. Ca^{2+} regulates the inositol Tris/tetrakisphosphate pathway in intact and broken preparations of insulin-secreting RINm5F cells. *J. Biol. Chem.* 261: 11931–11934, 1986.

17. BODEN, G., M. C. SIVITZ, O. E. OWEN, W. ESSA-KOUMAR, AND J. H. LANDOR. Somatostatin suppresses secretin and pancreatic exocrine secretion. *Science Wash. DC* 190: 163–165, 1975.

18. BOKOCH, G. M. The presence of free G protein β/γ subunits in human neutrophils results in suppression of adenylate cyclase activity. *J. Biol. Chem.* 262: 589–594, 1987.

19. BONTING, S. L., J. J. H. H. M. DE PONT, AND J. W. C. M. JANSEN. The role of sodium ions in pancreatic fluid secretion in the rabbit. *J. Physiol. Lond.* 309: 533–546, 1980.

20. BOOTH, A. G., AND A. J. KENNY. A rapid method for the preparation of microvilli from rabbit kidney. *Biochem. J.* 142: 575–581, 1974.

21. BRADFORD, P. G., AND R. P. RUBIN. Pertussis toxin inhibits chemotactic factor-induced phospholipase C stimulation and lysosomal enzyme secretion in rabbit neutrophils. *FEBS Lett.* 183: 317–320, 1985.

22. BRASS, L. F., AND S. K. JOSEPH. A role for inositol trisphosphate in intracellular Ca^{2+}-mobilization and granule secretion in platelets. *J. Biol. Chem.* 260: 15172–15179, 1985.

23. BRAVO, R., J. BURCKHARDT, T. CURRAN, AND R. MÜLLER. Stimulation and inhibition of growth by EGF in different A431 cell clones is accompanied by the rapid induction of c-fos and c-myc proto-oncogenes. *EMBO J.* 4: 1193–1197, 1985.

24. BREITWIESER, G. E., AND G. SZABO. Uncoupling of cardiac muscarinic and β-adrenergic receptors from ion channels by a guanine nucleotide analogue. *Nature Lond.* 317: 538–540, 1985.

25. BROCK, T. A., S. E. RITTENHOUSE, C. W. POWERS, L. S. EKSTEIN, M. A. GIMBRONE, JR., AND R. W. ALEXANDER. Phorbol ester and 1-oleoyl-2-acetylglycerol inhibit angiotensin activation of phospholipase C in cultured vascular smooth muscle cells. *J. Biol. Chem.* 260: 14158–14162, 1985.

26. BROWN, K. D., J. BLAY, R. F. IRVINE, J. P. HESLOP, AND M. J. BERRIDGE. Reduction of epidermal growth factor receptor affinity by heterologous ligands: evidence for a mechanism involving the breakdown of phosphoinositides and the activation of protein kinase C. *Biochem. Biophys. Res. Commun.* 123: 377–384, 1984.

27. BRUZZONE, R., E. R. TRIMBLE, A. GJINOVCI, O. TRAUB, K. WILLECKE, AND P. MEDA. Regulation of pancreatic exocrine function. A role for cell-to-cell communication? *Pancreas* 2: 262–271, 1987.

28. BUNDGAARD, M., M. MØLLER, AND J. H. POULSEN. Localization of sodium and potassium activated ATPase in the pancreas. *J. Physiol. Lond.* 313: 405–414, 1981.

29. BURNHAM, D. B., P. MUNOWITZ, N. THORN, AND J. A. WILLIAMS. Protein kinase activity associated with pancreatic zymogen granules. *Biochem. J.* 227: 743–751, 1985.

30. BURNHAM, D. B., AND J. A. WILLIAMS. Activation of protein kinase activity in pancreatic acini by calcium and cAMP. *Am. J. Physiol.* 246 (*Gastrointest. Liver Physiol.* 9): G500–G508, 1984.

31. BURNS, C. P., AND E. ROZENGURT. Serum, platelet-derived growth factor, vasopressin and phorbol esters increase intracellular pH in Swiss 3T3 cells. *Biochem. Biophys. Res. Commun.* 116: 931–938, 1983.

32. CALA, P. M. Volume regulation by *Amphiuma* red blood cells. The membrane potential and its implications regarding the nature of the ion-flux pathways. *J. Gen. Physiol.* 76: 683–708, 1980.

33. CARAFOLI, E. Membrane transport in the messenger function of calcium (Ca). In: *Advances in Immunopharmacology 3: Proc. 3rd Int. Conf. Immunopharmacology, Florence, 6–9 May 1985*, edited by L. Chedid, J. W. Hadden, F. Spreafico, P. Dukor, and D. Willoughby. Oxford: Pergamon, 1986, p. 101–107.

34. CARAFOLI, E., AND J. T. PENNISTON. The calcium signal. *Sci. Am.* 253: 70–78, 1985.

35. CASE, R. M., A. A. HARPER, AND T. SCRATCHERD. Water and electrolyte secretion by the perfused pancreas of the cat. *J. Physiol. Lond.* 196: 133–149, 1968.

36. CASE, R. M., AND T. SCRATCHERD. The actions of dibutyryl cyclic adenosine 3′,5′-monophosphate and methyl xanthines on pancreatic exocrine secretion. *J. Physiol. Lond.* 223: 649–667, 1972.

37. CASE, R. M., AND T. SCRATCHERD. Prostaglandin action on pancreatic blood flow and on electrolyte and enzyme secretion by exocrine pancreas in vivo and in vitro. *J. Physiol. Lond.* 226: 393–405, 1972.

38. CASSEL, D., H. LEVKOVITZ, AND Z. SELINGER. The regulatory GTPase cycle of turkey erythrocyte adenylate cyclase. *J. Cyclic Nucleotide Res.* 3: 393–406, 1977.

39. CASSEL, D., AND T. PFEUFFER. Mechanism of cholera toxin action: covalent modification of the guanyl nucleotide-binding protein of the adenylate cyclase system. *Proc. Natl. Acad. Sci. USA* 75: 2669–2673, 1978.

40. CASSEL, D., AND Z. SELINGER. Mechanism of adenylate cyclase activation by cholera toxin: inhibition of GTP hydrolysis at the regulatory site. *Proc. Natl. Acad. Sci. USA* 74: 3307–3311, 1977.
41. COCHET, C., G. N. GILL, J. MEISENHELDER, J. A. COOPER, AND T. HUNTER. C-kinase phosphorylates the epidermal growth factor receptor and reduces its epidermal growth factor-stimulated tyrosine protein kinase activity. *J. Biol. Chem.* 259: 2553–2558, 1984.
42. COCKCROFT, S. The dependence on Ca^{2+} of the guanine-nucleotide-activated polyphosphoinositide phosphodiesterase in neutrophil plasma membranes. *Biochem. J.* 240: 503–507, 1986.
43. COCKCROFT, S. Polyphosphoinositide phosphodiesterase: regulation by a novel guanine nucleotide binding protein, G_p. *Trends Biochem. Sci.* 12: 75–78, 1987.
44. COCKCROFT, S., AND B. D. GOMPERTS. Role of guanine nucleotide binding protein in the activation of polyphosphoinositide phosphodiesterase. *Nature Lond.* 314: 534–536, 1985.
45. CODINA, J., J. D. HILDEBRANDT, L. BIRNBAUMER AND R. D. SEKURA. Effects of guanine nucleotides and Mg on human erythrocyte N_i and N_s, the regulatory components of adenylyl cyclase. *J. Biol. Chem.* 259: 11408–11418, 1984.
46. COHEN, F. S., M. H. AKABAS, AND A. FINKELSTEIN. Osmotic swelling of phospholipid vesicles causes them to fuse with a planar phospholipid bilayer membrane. *Science Wash. DC* 217: 458–460, 1982.
47. COHEN, P. The role of protein phosphorylation in neural and hormonal control of cellular activity. *Nature Lond.* 296: 613–620, 1982.
48. COLLOT, M., S. WATTIAUX-DECONICK, AND R. WATTIAUX. Isopycnic centrifugation of rat liver subcellular particles in sucrose and in metrizamide. In: *Biological Separation in Iodinated Density Gradient Media*, edited by D. Rickwood. Washington, DC: Information Retrieval Limited, 1976, p. 89–96.
49. CORVERA, S., AND J. A. GARCÍA-SÁINZ. Phorbol esters inhibit $alpha_1$ adrenergic stimulation of glycogenolysis in isolated rat hepatocytes. *Biochem. Biophys. Res. Commun.* 119: 1128–1133, 1984.
50. COSTA, M. R. C., AND W. A. CATTERALL. Phosphorylation of the α subunit of the sodium channel by protein kinase C. *Cell. Mol. Neurobiol.* 4: 291–297, 1984.
51. CREUTZ, C. E., L. G. DOWLNG, J. J. SANDRO, C. VILLAR-SANDO, J. H. WHIPPLE, AND W. J. ZAKS. Characterization of the chromobindins. *J. Biol. Chem.* 258: 14664–14674, 1983.
52. DAWSON, A. P. GTP enhances inositol trisphosphate-stimulated Ca^{2+} release from rat liver microsomes. *FEBS Lett.* 185: 147–150, 1985.
53. DAWSON, A. P. The mechanism of action of GTP in promoting inositol 1,4,5 trisphosphate-stimulated Ca^{2+} release from rat liver microsomes. In: *Table Ronde Roussel UCLAF No. 58, Polyphosphoinositides in Cell Biology, Paris, Feb. 16–17, 1987*, p. 17–18.
54. DAWSON, A. P., J. G. COMERFORD, AND D. V. FULTON. The effect of GTP on inositol 1,4,5-trisphosphate-stimulated Ca^{2+} efflux from a rat liver microsomal fraction. *Biochem. J.* 234: 311–315, 1986.
55. DE LISLE, R. C., AND U. HOPFER. Electrolyte permeabilities of pancreatic zymogen granules: implications for pancreatic secretion. *Am. J. Physiol.* 250 (*Gastrointest. Liver Physiol.* 13): G489–G496, 1986.
56. DE LISLE, R. C., I. SCHULZ, T. TYRAKOWSKI, W. HAASE, AND U. HOPFER. Isolation of stable pancreatic zymogen granules. *Am. J. Physiol.* 246 (*Gastrointest. Liver Physiol.* 9): G411–G418, 1984.
57. DELLA BIANCA, V., M. GRZESKOWIAK, M. A. CASSATELLA, L. ZENI, AND F. ROSSI. Phorbol 12,myristate 13,acetate potentiates the respiratory burst while inhibits phosphoinositide hydrolysis and calcium mobilization by formyl-methionyl-leucyl-phenylalanine in human neutrophils. *Biochem. Biophys. Res. Commun.* 135: 556–565, 1986.
58. DE PONT, J. J. H. H. M., AND A. M. M. FLEUREN-JAKOBS. Synergistic effect of A23187 and a phorbol ester or amylase secretion from rabbit pancreatic acini. *FEBS Lett.* 170: 64–68, 1984.
59. DE TORRONTEGUI, G., AND J. BERTHET. The action of insulin on the incorporation of [^{32}P]phosphate in the phospholipids of rat adipose tissue. *Biochim. Biophys. Acta* 116: 477–481, 1966.
60. DOURMASHKIN, R. R., R. M. DOUGHERTY, AND R. J. C. HARRIS. Electron microscopic observations on Rous sarcoma virus and cell membranes. *Nature Lond.* 194: 1116–1119, 1962.
61. DOWNES, P., AND R. H. MICHELL. Phosphatidylinositol 4-phosphate and phosphatidylinositol 4,5-bisphosphate: lipids in search of a function. *Cell Calcium* 3: 467–502, 1982.
62. ENOMOTO, K., AND T. ASAKAWA. Inhibition of catalytic unit of adenylate cyclase and activation of GTPase of N_i protein by $\beta\gamma$-subunits of GTP-binding proteins. *FEBS Lett.* 202: 63–68, 1986.
63. ERICSON, A.-C., AND K. R. SPRING. Volume regulation by *Necturus* gallbladder: apical Na^+-H^+ and Cl^--HCO_3^- exchange. *Am. J. Physiol.* 243 (*Cell Physiol.* 12): C146–C150, 1982.
64. ESTEVE, J. P., N. VAYSSE, C. SUSINI, J. M. KUNSCH, D. FOURMY, L. PRADAYROL, E. WUNSCH, L. MORODER, AND A. RIBET. Bimodal regulation of pancreatic exocrine function in vitro by somatostatin-28. *Am. J. Physiol.* 245 (*Gastrointest. Liver Physiol.* 8): G208–G216, 1983.
65. FARESE, R. V., J. S. DAVIS, D. E. BARNES, M. L. STANDAERT, J. S. BABISCHKIN, R. HOCK, N. K. ROSIC, AND R. J. POLLET. The de novo phospholipid effect of insulin is associated with increases in diacylglycerol, but not inositol phosphates or cytosolic Ca^{2+}. *Biochem. J.* 231: 269–278, 1985.
66. FARESE, R. V., R. E. LARSON, AND M. A. SABIR. Insulin and its secretagogues activate Ca^{2+} dependent phosphatidylinositol breakdown and amylase secretion in rat pancreas in vitro. *Diabetes* 30: 396–401, 1981.
67. FARESE, R. V., J. L. ORCHARD, R. E. LARSON, M. A. SABIR, AND J. S. DAVIS. Phosphatidylinositol hydrolysis and phosphatidylinositol 4′,5′-diphosphate hydrolysis are separable responses during secretagogue action in the rat pancreas. *Biochem. Biophys Acta* 846: 296–304, 1985.
67a. FINDLEY, J., AND O. H. PETERSEN. Acetylcholine stimulates a Ca^{2+}-dependent Cl^- conductance in mouse lacrimal acinar cells. *Pfluegers Arch.* 403: 328–330, 1985.
68. GARCÍA-SÁINZ, J. A., AND J. N. FAIN. Effect of insulin, catecholamines and calcium ions on phospholipid metabolism in isolated white fat-cells. *Biochem. J.* 186: 781–789, 1980.
69. GARCÍA-SÁINZ, J. A., F. MENDLOVIC, AND M. A. MARTINÉZ-OLMEDO. Effects of phorbol esters on α_1-adrenergic-mediated and glucagon-mediated actions in isolated rat hepatocytes. *Biochem. J.* 228: 277–280, 1985.
70. GARDNER, J. D., AND R. T. JENSEN. Secretagogue receptors on pancreatic acinar cells. In: *Physiology of the Gastrointestinal Tract* (2nd ed.), edited by L. R. Johnson. New York: Raven, 1987, vol. 2, p. 1109–1127.
71. GARDNER, J. D., V. E. SUTLIFF, M. D. WALKER, AND R. T. JENSEN. Effects of inhibitors of cyclic nucleotide phosphodiesterase on actions of cholecystokinin, bombesin, and carbachol on pancreatic acini. *Am. J. Physiol.* 245 (*Gastrointest. Liver Physiol.* 8): G676–G680, 1983.
72. GLAUERT, A. M., J. T. DINGLE, AND J. A. LUCY. Action of saponin on biological cell membranes. *Nature Lond.* 196: 953–955, 1962.
73. GRINSTEIN, S., C. A. CLARK, AND A. ROTHSTEIN. Activation of Na^+/H^+ exchange in lymphocytes by osmotically induced volume changes and by cytoplasmic acidification. *J. Gen. Physiol.* 82: 619–638, 1983.
74. HASLAM, R. J., AND M. M. L. DAVIDSON. Guanine nucleotides decrease the free [Ca^{2+}] required for secretion of serotonin from permeabilized blood platelets. Evidence of a role for a GTP-binding protein in platelet activation. *FEBS Lett.* 174: 90–95, 1984.
75. HASLAM, R. J., AND M. M. L. DAVIDSON. Receptor-induced diacylglycerol formation in permeabilized platelets; possible

role for a GTP-binding protein. *J. Recept. Res.* 4: 605–629, 1984.
76. HAUSER, H., K. HOWELL, R. M. C. DAWSON, AND D. E. BOWYER. Rabbit small intestine brush border membrane preparation and lipid composition. *Biochim. Biophys. Acta* 602: 567–577, 1980.
77. HÄUSLER, G., J.-E. DE PEYER, M. YAJIMA, AND G. SCHULTZ. Vascular smooth muscle: availability of calcium through α-adrenoceptor stimulation. *J. Cardiovasc. Pharmacol.* 8, Suppl. 8: 107–110, 1986.
78. HELLMESSEN, W., A. L. CHRISTIAN, H. FASOLD, AND I. SCHULZ. Coupled Na^+-H^+ exchange in isolated acinar cells from rat exocrine pancreas. *Am. J. Physiol.* 249 (*Gastrointest. Liver Physiol.* 12): G125–G136, 1985.
79. HENNE, V., AND H.-D. SÖLING. Guanosine 5'-triphosphate releases calcium from rat liver and guinea pig parotid gland endoplasmic reticulum independently of inositol 1,4,5-trisphosphate. *FEBS Lett.* 202: 267–273, 1986.
80. HESCHELER, J., W. ROSENTHAL, W. TRAUTWEIN, AND G. SCHULTZ. The GTP-binding protein, G_o, regulates neuronal calcium channels. *Nature Lond.* 325: 445–447, 1987.
81. HEUSER, J. E. Synaptic vesicle exocytosis revealed in quick-frozen frog neuromuscular junctions treated with 4-aminopyridine and given a single electrical shock. In: *Approaches to the Cell Biology of Neurons*, edited by W. M. Cowan and J. A. Ferrendelli. Bethesda, MD: Soc. Neurosci., 1977, p. 215–239. (Soc. Neurosci. Symp. II.)
82. HEYWORTH, C. M., S. P. WILSON, D. J. GAWLER, AND M. D. HOUSLAY. The phorbol ester TPA prevents the expression of both glucagon desensitisation and the glucagon-mediated block of insulin stimulation of the peripheral plasma membrane cyclic AMP phosphodiesterase in rat hepatocytes. *FEBS Lett.* 187: 196–200, 1985.
83. HIGASHIDA, H., AND D. A. BROWN. Membrane current responses to intracellular injections of inositol 1,3,4,5-tetrakisphosphate and inositol 1,3,4-trisphosphate in NG108-15 hybrid cells. *FEBS Lett.* 208: 283–286, 1986.
84. HILDEBRANDT, J. D., J. CODINA, W. ROSENTHAL, T. SUNYER, R. IYENGAR, AND L. BIRNBAUMER. Properties of human erythrocyte N_s and N_i, the regulatory components of adenylate cyclase, as purified without regulatory ligands. *Adv. Cyclic Nucleotide Protein Phosphorylation Res.* 19: 87–101, 1985.
85. HIROTA, K., T. HIROTA, G. AGUILERA, AND K. J. CATT. Hormone-induced redistribution of calcium-activated phospholipid-dependent protein kinase in pituitary gonadotrophs. *J. Biol. Chem.* 260: 3243–3246, 1985.
86. HOKIN, M. R., AND L. E. HOKIN. Enzyme secretion and the incorporation of ^{32}P into phospholipids of pancreas slices. *J. Biol. Chem.* 203: 967–977, 1953.
87. HONDA, S. I., T. HONGLADAROM, AND G. G. LATIES. A new isolation medium for plant organelles. *J. Exp. Bot.* 17: 460–472, 1966.
88. HORN, W., AND M. L. KARNOVSKY. Features of the translocation of protein kinase C in neutrophils stimulated with the chemotactic peptide f-Met-Leu-Phe. *Biochem. Biophys. Res. Commun.* 139: 1169–1175, 1986.
89. HOUSLAY, M. D., M. J. O. WAKELAM, AND N. J. PYNE. The mediator is the message: is it part of the answer of insulin's action? *Trends Biochem. Sci.* 11: 393–394, 1986.
90. HUBEL, K. A. Response of rabbit pancreas in vitro to adrenergic agonists and antagonists. *Am. J. Physiol.* 219: 1590–1594, 1970.
91. HUNTER, T., AND J. A. COOPER. Protein-tyrosine kinases. *Annu. Rev. Biochem.* 54: 897–930, 1985.
92. IMAMURA, K., AND I. SCHULZ. Phosphorylated intermediate of $(Ca^{2+} + K^+)$-stimulated Mg^{2+}-dependent transport ATPase in endoplasmic reticulum from rat pancreatic acinar cells. *J. Biol. Chem.* 260: 11339–11347, 1985.
93. IRVINE, R. F., A. J. LETCHER, AND R. M. C. DAWSON. Phosphatidylinositol-4,5-bisphosphate phosphodiesterase and phosphomonoesterase activities of rat brain. Some properties and possible control mechanisms. *Biochem. J.* 218: 177–185, 1984.
94. IRVINE, R. F., A. J. LETCHER, J. P. HESLOP, AND M. J. BERRIDGE. The inositol tris/tetrakisphosphate pathway—demonstration of $Ins(1,4,5)P_3$ 3-kinase activity in animal tissues. *Nature Lond.* 320: 631–634, 1986.
95. IRVINE, R. F., A. J. LETCHER, D. J. LANDER, AND M. J. BERRIDGE. Specificity of inositol phosphate-stimulated Ca^{2+} mobilization from Swiss-mouse 3T3 cells. *Biochem. J.* 240: 301–304, 1986.
96. IRVINE, R. F., A. J. LETCHER, D. J. LANDER, J. P. HESLOP, AND M. J. BERRIDGE. Inositol(3,4)bisphosphate and inositol(1,3)bisphosphate in GH_4 cells—evidence for complex breakdown of inositol(1,3,4)trisphosphate. *Biochem. Biophys. Res. Commun.* 143: 353–359, 1987.
97. IRVINE, R. F., AND R. M. MOOR. Micro-injection of inositol 1,3,4,5-tetrakisphosphate activates sea urchin eggs by a mechanism dependent on external Ca^{2+}. *Biochem. J.* 240: 917–920, 1986.
98. IWATSUKI, N., AND O. H. PETERSEN. Pancreatic acinar cells: acetylcholine-evoked electrical uncoupling and its ionic dependency. *J. Physiol. Lond.* 274: 81–96, 1978.
99. IWATSUKI, N., AND O. H. PETERSEN. Amino acids evoke short-latency membrane conductance increase in pancreatic acinar cells. *Nature Lond.* 283: 492–494, 1980.
100. IWATSUKI, N., AND O. H. PETERSEN. Amino acid-evoked membrane potential and resistance changes in pancreatic acinar cells. *Pfluegers Arch.* 386: 153–159, 1980.
101. JAKOBS, K. H., S. BAUER, AND Y. WATANABE. Modulation of adenylate cyclase of human platelets by phorbol ester. Impairment of the hormone-sensitive inhibitory pathway. *Eur. J. Biochem.* 151: 425–430, 1985.
102. JAKOBS, K. H., AND G. SCHULTZ. Occurrence of a hormone-sensitive inhibitory coupling component of the adenylate cyclase in S49 lymphoma cyc$^-$ variants. *Proc. Natl. Acad. Sci. USA* 80: 3899–3902, 1983.
103. JAUCH, P., AND P. LÄUGER. Electrogenic properties of the sodium-alanine cotransporter in pancreatic acinar cells. II. Comparison with transport models. *J. Membr. Biol.* 94: 117–127, 1986.
104. JAUCH, P., O. H. PETERSEN, AND P. LÄUGER. Electrogenic properties of the sodium-alanine cotransporter in pancreatic acinar cells. I. Tight-seal whole-cell recordings. *J. Membr. Biol.* 94: 99–115, 1986.
105. JZUTSU, K. T., AND D. E. JOHNSON. Changes in elemental concentrations of rat parotid acinar cells following pilocarpine stimulation. *J. Physiol. Lond.* 381: 297–309, 1986.
106. KACZMAREK, L. K. Phorbol esters, protein phosphorylation and the regulation of neuronal ion channels. *J. Exp. Biol.* 124: 375–392, 1986.
107. KANBA, S., K. S. KANBA, AND E. RICHELSON. The protein kinase C activator, 12-O-tetradecanoylphorbol-13-acetate (TPA), inhibits muscarinic (M_1) receptor-mediated inositol phosphate release and cyclic GMP formation in murine neuroblastoma cells (clone N1E-115). *Eur. J. Pharmacol.* 125: 155–156, 1986.
108. KATADA, T., G. M. BOKOCH, M. D. SMIGEL, M. UI, AND A. G. GILMAN. The inhibitory guanine nucleotide-binding regulatory component of adenylate cyclase. Subunit dissociation and the inhibition of adenylate cyclase in S49 lymphoma cyc$^-$ and wild type membranes. *J. Biol. Chem.* 259: 3586–3595, 1984.
109. KATADA, T., A. G. GILMAN, Y. WATANABE, S. BAUER, AND K. H. JAKOBS. Protein kinase C phosphorylates the inhibitory guanine-nucleotide-binding regulatory component and apparently suppresses its function in hormonal inhibition of adenylate cyclase. *Eur. J. Biochem.* 151: 431–437, 1985.
110. KATADA, T., J. K. NORTHUP, G. M. BOKOCH, M. UI, AND A. G. GILMAN. The inhibitory guanine nucleotide-binding regulatory component of adenylate cyclase. Subunit dissociation and guanine nucleotide-dependent hormonal inhibition *J. Biol. Chem.* 259: 3578–3585, 1984.
111. KATADA, T., AND M. UI. Direct modification of the membrane adenylate cyclase system by islet-activating protein due to

ADP-ribosylation of a membrane protein. *Proc. Natl. Acad. Sci. USA* 79: 3129–3133, 1982.
112. KEMMER, T. P., E. BAYERDÖRFFER, H. WILL, AND I. SCHULZ. Anion dependence of Ca^{2+} transport and $(Ca^{2+} + K^+)$-stimulated Mg^{2+}-dependent transport ATPase in rat pancreatic endoplasmic reticulum. *J. Biol. Chem.* 262: 13758–13734, 1987.
113. KIMURA, T., K. IMAMURA, L. ECKHARDT, AND I. SCHULZ. Ca^{2+}-, phorbol ester-, and cAMP-stimulated enzyme secretion from permeabilized rat pancreatic acini. *Am. J. Physiol.* 250 (*Gastrointest. Liver Physiol.* 13): G698–G708, 1986.
114. KING, A. C., AND P. CUATRECASAS. Resolution of high and low affinity epidermal growth factor receptors. Inhibition of high affinity component by low temperature, cycloheximide, and phorbol esters. *J. Biol. Chem.* 257: 3053–3060, 1982.
115. KIRCHBERGER, M. A., M. TADA, AND A. M. KATZ. Phospholamban: a regulatory protein of the cardiac sarcoplasmic reticulum. In: *Recent Advances in Studies on Cardiac Structure and Metabolism*, edited by A. Fleckenstein and N. S. Dhalla. Baltimore, MD: University Park Press, 1975, vol. 5, p. 103–115.
116. KNIGHT, D. E., AND P. F. BAKER. Calcium-dependence of catecholamine release from bovine adrenal medullary cells after exposure to intense electric fields. *J. Membr. Biol.* 68: 107–140, 1982.
117. KNIGHT, D. E., AND E. KOH. Ca^{2+} and cyclic nucleotide dependence of amylase release from isolated rat pancreatic acinar cells rendered permeable by intense electric fields. *Cell Calcium* 5: 401–418, 1984.
118. KOJIMA, I., H. SHIBATA, AND E. OGATA. Phorbol ester inhibits angiotensin-induced activation of phospholipase C in adrenal glomerulosa cells. Its implication in the sustained action of angiotensin. *Biochem. J.* 237: 253–258, 1986.
119. KONDO, S., AND I. SCHULZ. Ca^{++} uptake in isolated pancreas cells induced by secretagogues. *Biochim. Biophys. Acta* 419: 76–92, 1976.
120. KOO, C., R. J. LEFKOWITZ, AND R. SNYDERMAN. Guanine nucleotides modulate the binding affinity of the oligopeptide chemoattractant receptor on human polymorphonuclear leukocytes. *J. Clin. Invest.* 72: 748–753, 1983.
121. KORC, M., H. SANKARAN, K. Y. WONG, J. A. WILLIAMS, AND I. D. GOLDFINE. Insulin receptors in isolated mouse pancreatic acini. *Biochem. Biophys. Res. Commun.* 84: 293–299, 1978.
122. KRAFT, A. S., AND W. B. ANDERSON. Phorbol esters increase the amount of Ca^{2+}, phospholipid-dependent protein kinase associated with plasma membrane. *Nature Lond.* 301: 621–623, 1983.
123. KRIBBEN, A., T. TYRAKOWSKI, AND I. SCHULZ. Characterization of Mg-ATP-dependent Ca^{2+} transport in cat pancreatic microsomes. *Am. J. Physiol.* 244 (*Gastrointest. Liver Physiol.* 7): G480–G490, 1983.
124. KUIJPERS, G. A. J., I. G. P. VAN NOOY, J. J. H. H. M. DE PONT, AND S. L. BONTING. The mechanism of fluid secretion in the rabbit pancreas studied by means of various inhibitors. *Biochim. Biophys. Acta* 778: 324–331, 1984.
125. KUNO, M., AND P. GARDNER. Ion channels activated by inositol 1,4,5-trisphosphate in plasma membrane of human T-lymphocytes. *Nature Lond.* 326: 301–304, 1987.
126. LABARCA, R., A. JANOWSKY, J. PATEL, AND S. M. PAUL. Phorbol esters inhibit agonist-induced [^{3}H]inositol-1-phosphate accumulation in rat hippocampal slices. *Biochem. Biophys. Res. Commun.* 123: 703–709, 1984.
127. LAMBERT, M., M. SVOBODA, J. FURNELLE, AND J CHRISTOPHE. Solubilization from rat pancreatic plasma membranes of a cholecystokinin (CCK) agonist-receptor complex interacting with guanine nucleotide regulatory proteins coexisting in the same macromolecular system. *Eur. J. Biochem.* 147: 611–617, 1985.
128. LASSEGUE, B., AND J.-C. STOCLET. Phorbol ester inhibition of vasopressin-induced calcium efflux from cultured rat aortic myocytes. *FEBS Lett.* 205: 251–254, 1986.
129. LAUGIER, R., AND O. H. PETERSEN. Two different types of electrogenic amino acid action on pancreatic acinar cells. *Biochim. Biophys. Acta* 641: 216–221, 1981.
130. LEE, L.-S., AND I. B. WEINSTEIN. Epidermal growth factor, like phorbol esters, induces plasminogen activator in HeLa cells. *Nature Lond.* 274: 696–697, 1978.
131. LEEB-LUNDBERG, L. M. F., S. COTECCHIA, J. W. LOMASNEY, J. F. DEBERNARDIS, R. F. LEFKOWITZ, AND M. G. CARON. Phorbol esters promote α_1-adrenergic receptor phosphorylation and receptor uncoupling from inositol phospholipid metabolism. *Proc. Natl. Acad. Sci. USA* 82: 5651–5655, 1985.
132. LEVITZKI, A. Receptor to effector coupling in the receptor-dependent adenylate cyclase system. *J. Recept. Res.* 4: 399–409, 1984.
133. LEVITZKI, A. Hypothesis. Regulation of adenylate cyclase by hormones and G-proteins. *FEBS Lett.* 211: 113–118, 1987.
134. LIMAS, C. J., AND C. LIMAS. Phorbol ester- and diacylglycerol-mediated desensitization of cardiac β-adrenergic receptors. *Circ. Res.* 57: 443–449, 1985.
135. LIN, T.-M., D. C. EVANS, C. J. SHAAR, AND M. A. ROOT. Action of somatostatin on stomach, pancreas, gastric mucosal blood flow, and hormones. *Am. J. Physiol.* 244 (*Gastrointest. Liver Physiol.* 7): G40–G45, 1983.
136. LITOSCH, I., C. WALLIS, AND J. N. FAIN. 5-Hydroxytryptamine stimulates inositol phosphate production in a cell-free system from blowfly salivary glands. Evidence for a role of GTP in coupling receptor activation to phosphoinositide breakdown. *J. Biol. Chem.* 260: 5464–5471, 1985.
137. LOGSDON, C. D., AND J. A. WILLIAMS. Epidermal growth factor binding and biologic effects on mouse pancreatic acini. *Gastroenterology* 85: 339–345, 1983.
138. LONG, B. W., AND J. D. GARDNER. Effects of cholecystokinin on adenylate cyclase activity in dispersed pancreatic acinar cells. *Gastroenterology* 73: 1008–1014, 1977.
139. LUCAS, D. O., S. M. BAJJALIEH, J. A. KOWALCHYK, AND T. F. J. MARTIN. Direct stimulation by thyrotropin-releasing hormone (TRH) of polyphosphoinositide hydrolysis in GH_3 cell membranes by a guanine nucleotide-modulated mechanism. *Biochem. Biophys. Res. Commun.* 132: 721–728, 1985.
140. LYNCH, C. J., R. CHAREST, S. B. BOCCKINO, J. H. EXTON, AND P. F. BLACKMORE. Inhibition of hepatic α_1-adrenergic effects and binding by phorbol myristate acetate. *J. Biol. Chem.* 260: 2844–2851, 1985.
141. MACHICAO, E., AND O. H. WIELAND. Evidence that the insulin receptor-associated protein kinase acts as a phosphatidylinositol kinase. *FEBS Lett.* 175: 113–116, 1984.
142. MAGNALDO, I., H. TALWAR, W. B. ANDERSON, AND J. POUYSSÉGUR. Evidence for a GTP-binding protein coupling thrombin receptor to PIP_2-phospholipase C in membranes of hamster fibroblasts. *FEBS Lett.* 210: 6–10, 1987.
143. MANCHESTER, K. L. Stimulation by insulin of incorporation of [^{32}P]phosphate and ^{14}C from acetate into lipid and protein of isolated rat diaphragm. *Biochim. Biophys. Acta* 70: 208–210, 1963.
144. MARUYAMA, Y. Ca^{2+}-induced excess capacitance fluctuation studied by phase-sensitive detection method in exocrine pancreatic acinar cells. *Pfluegers Arch.* 407: 561–563, 1986.
145. MATSUMOTO, T., T. F. P. MOLSKI, M. VOLPI, C. PELZ, Y. KANAHO, E. L. BECKER, M. B. FEINSTEIN, P. H. NACCACHE, AND R. I. SHA'AFI. Treatment of rabbit neutrophils with phorbol esters results in increased ADP-ribosylation catalyzed by pertussis toxin and inhibition of the GTPase stimulated by fMet-Leu-Phe. *FEBS Lett.* 198: 295–300, 1986.
146. MCCLESKEY, E. W., A. P. FOX, D. FELDMAN, AND R. W. TSIEN. Different types of calcium channels. *J. Exp. Biol.* 124: 177–190, 1986.
147. MEDA, P., R. BRUZZONE, S. KNODEL, AND L. ORCI. Blockage of cell-to-cell communication within pancreatic acini is associated with increased basal release of amylase. *J. Cell Biol.* 103: 475–483, 1986.
148. MELDOLESI, J., J. D. JAMIESON, AND G. E. PALADE. Composition of cellular membranes in the pancreas of the guinea pig. I. Isolation of membrane fraction *J. Cell Biol.* 49: 109–129, 171.
149. MELIN, P.-M., R. SUNDLER, AND G. JERGIL. Phospholipase C

in rat liver plasma membranes. Phosphoinositide specificity and regulation by guanine nucleotides and calcium. *FEBS Lett.* 198: 85–88, 1986.
150. MERRIT, J. E., C. W. TAYLOR, R. P. RUBIN, AND J. W. PUTNEY. JR. Evidence suggesting that a novel guanine nucleotide regulatory protein couples receptors to phospholipase C in exocrine pancreas. *Biochem. J.* 236: 337–343, 1986.
151. MICHELL, R. H. Inositol phospholipids and cell surface receptor function. *Biochim. Biophys. Acta* 415: 81–147, 1975.
152. MILUTINOVIĆ, S., G. SACHS, W. HAASE, AND I. SCHULZ. Studies on isolated subcellular components of cat pancreas. I. Isolation and enzymatic characterization. *J. Membr. Biol.* 36: 253–279, 1977.
153. MOOLENAAR, W. H., J. BOONSTRA, P. T. VAN DER SAAG, AND S. W. DE LAAT. Sodium/proton exchange in mouse neuroblastoma cells. *J. Biol. Chem.* 256: 12883–12887, 1981.
154. MOOLENAAR, W. H., L. H. K. DEFIZE, AND S. W. DE LAAT. Ionic signalling by growth factor receptors. *J. Exp. Biol.* 124: 359–373, 1986.
155. MOOLENAAR, W. H., L. G. J. TERTOOLEN, AND S. W. DE LAAT. Phorbol ester and diacylglycerol mimic growth factors in raising cytoplasmic pH. *Nature Lond.* 312: 371–374, 1984.
155a. MORRIS, A. P., D. V. GALLACHER, R. F. IRVINE, AND O. H. PETERSEN. Synergism of Ins(1,4,5)P_3 with Ins(1,3,4,5)P_4 in activating Ca^{2+}-dependent K^+ channels. *Nature Lond.* In press.
156. MUALLEM, S., M. SCHOEFFIELD, S. PANDOL, AND G. SACHS. Inositol trisphosphate modification of ion transport in rough endoplasmic reticulum. *Proc. Natl. Acad. Sci. USA* 82: 4433–4437, 1985.
157. NAKAMURA, T., AND M. UI. Simultaneous inhibition of inositol phospholipid breakdown, arachidonic acid release, and histamine secretion in mast cells by islet-activating protein, pertussis toxin. A possible involvement of the toxin-specific substrate in the Ca^{2+}-mobilizing receptor-mediated biosignaling system. *J. Biol. Chem.* 260: 3584–3593, 1985.
158. NAKANISHI, H., H. NOMURA, U. KIKKAWA, A. KISHIMOTO, AND Y. NISHIZUKA. Rat brain and liver soluble phospholipase C: resolution of two forms with different requirements for calcium. *Biochem. Biophys. Res. Commun.* 132: 582–590, 1985.
159. NISHIKAWA, M., H. HIDAKA, AND R. S. ADELSTEIN. Phosphorylation of smooth muscle heavy meromyosin by calcium-activated, phospholipid-dependent protein kinase. The effect on actin-activated MgATPase activity. *J. Biol. Chem.* 258: 14069–14072, 1983.
160. NISHIZUKA, Y. The role of protein kinase C in cell surface signal transduction and tumor promotion. *Nature Lond.* 308: 693–698, 1984.
161. NOVAK, I. Pancreatic bicarbonate secretion. In: *pH Homeostasis: Mechanism and Control*, edited by D. Häussinger. London: Academic, 1988, p. 447–470.
162. ÖBRINK, B., B. WÄRMEGÅRD, AND H. PERTOFT. Specific binding of rat liver plasma membranes by rat liver cells. *Biochem. Biophys. Res. Commun.* 77: 665–670, 1977.
163. OCHS, D. L., J. I. KORENBROT, AND J. A. WILLIAMS. Intracellular free calcium concentrations in isolated pancreatic acini; effects of secretagogues. *Biochem. Biophys. Res. Commun.* 117: 122–128, 1983.
164. OKAJIMA, F., T. KATADA, AND M. UI. Coupling of guanine nucleotide regulatory protein to chemotactic peptide receptors in neutrophil membranes and its uncoupling by islet-activating protein, pertussis toxin. A possible role of the toxin substrate in Ca^{2+}-mobilizing receptor-mediated signal transduction. *J. Biol. Chem.* 260: 6761–6768, 1985.
165. OKAJIMA, F., AND M. UI. ADP-ribosylation of the specific membrane protein by islet-activating protein, pertussis toxin, associated with inhibition of a chemotactic peptide-induced arachidonate release in neutrophils. A possible role of the toxin substrate in Ca^{2+}-mobilizing biosignaling. *J. Biol. Chem.* 259: 13863–13871, 1984.
166. OWENS, J. R., L. T. FRAME, M. UI, AND D. M. F. COOPER. Cholera toxin ADP-ribosylates the islet-activating protein substrate in adipocyte membranes and alters its function. *J. Biol. Chem.* 260: 15946–15952, 1985.
167. PALMER, S., P. T. HAWKINS, R. H. MICHELL, AND C. J. KIRK. The labelling of polyphosphoinositides with [^{32}P]P_i and the accumulation of inositol phosphates in vasopressin-stimulated hepatocytes. *Biochem. J.* 238: 491–499, 1986.
168. PANDOL, S. J., AND M. S. SCHOEFFIELD. 1,2-Diacylglycerol, protein kinase C, and pancreatic enzyme secretion. *J. Biol. Chem.* 261: 4438–4444, 1986.
169. PARSEGIAN, V. A. Considerations in determining the mode of influence of calcium on vesicle-membrane interaction In: *Approaches to the Cell Biology of Neurons*, edited by W. M. Cowan and J. A. Ferrendelli. Bethesda, MD: Soc. Neurosci., 1977, p. 161–171. (Soc. Neurosci. Symp. II.)
170. PAZOLES, C. J., AND H. B. POLLARD. Evidence for stimulation of anion transport in ATP-evoked transmitter release from isolated secretory vesicles. *J. Biol. Chem.* 253: 3962–3969, 1978.
171. PENNINGTON, S. R., AND B. R. MARTIN. Insulin-stimulated phosphoinositide metabolism in isolated fat cells. *J. Biol. Chem.* 260: 11039–11045, 1985.
172. PETERSEN, O. H. Electrophysiology of acinar cells. In: *The Exocrine Pancreas: Biology, Pathobiology, and Diseases*, edited by V. L. W. Go, J. D. Gardner, F. P. Brooks, E. Lebenthal, E. P. DiMagno, and G. A. Scheele. New York: Raven, 1986, p. 141–161.
173. PETERSEN, O. H., AND C. BEAR. Two glucagon transducing systems. *Nature Lond.* 323: 18, 1986.
174. PETERSEN, O. H., AND N. IWATSUKI. Hormonal control of cell to cell coupling in the exocrine pancreas. In: *Hormone Receptors in Digestion and Nutrition*, edited by G. Rosselin, P. Fromageot, and S. Bonfils. Amsterdam: Elsevier/North-Holland, 1979, p. 191–202.
175. PETERSEN, O. H., AND Y. MARUYAMA. What is the mechanism of the calcium influx to pancreatic acinar cells evoked by secretagogues? *Pfluegers Arch.* 396: 82–84, 1983.
176. PETERSEN, O. H., AND Y. MARUYAMA. Calcium-activated potassium channels and their role in secretion. *Nature Lond.* 307: 693–696, 1984.
177. PETERSEN, O. H., Y. MARUYAMA, J. GRAF, R. LAUGIER, A. NISHIYAMA, AND G. T. PEARSON. Ionic currents across pancreatic acinar cell membranes and their role in fluid secretion. *Philos. Trans. R. Soc. Lond. B Biol. Sci.* 296: 151–166, 1981.
178. PETERSEN, O. H., AND J. SINGH. Amino acid-evoked membrane current in voltage clamped mouse pancreatic acini. *J. Physiol. Lond.* 319: 99P–100P, 1981.
179. PFEILSCHIFTER, J. Tumour promotor 12-O-tetradecanoylphorbol 13-acetate inhibits angiotensin II-induced inositol phosphate production and cytosolic Ca^{2+} rise in rat renal mesangial cells. *FEBS Lett.* 203: 262–266, 1986.
180. PFEUFFER, E., R.-M. DREHER, H. METZGER, AND T. PFEUFFER. Catalytic unit of adenylate cyclase: purification and identification by affinity crosslinking. *Proc. Natl. Acad. Sci. USA* 82: 3086–3090, 1985.
181. POLLARD, H. B., C. J. PAZOLES, C. E. CREUTZ, A. RAMU, C. A. STROTT, P. RAY, E. M. BROWN, G. D. AURBACH, K. M. TACK-GOLDMAN, AND N. R. SHULMAN. A role for anion transport in the regulation of release from chromaffin granules and exocytosis from cells. *J. Supramol. Struct.* 7: 277–285, 1977.
182. POLLARD, H. B., C. J. PAZOLES, C. E. CREUTZ, AND O. ZINDER. The chromaffin granule and possible mechanisms of exocytosis. *Int. Rev. Cytol.* 58: 159–197, 1979.
183. RENCKENS, B. A. M., S. E. VAN EMST-DE VRIES, J. J. H. H. M. DE PONT, AND S. L. BONTING. Rat pancreas adenylate cyclase. VII. Effect of extracellular calcium on pancreozymin-induced cyclic AMP formation. *Biochim. Biophys. Acta* 630: 511–518, 1980.
184. REUTER, H., S. KOKUBUN, AND B. PROD'HOM. Properties and modulation of cardiac calcium channels. *J. Exp. Biol.* 124: 191–201, 1986.
185. ROBBERECHT, P., M. DESCHODT-LANCKMAN, P. DE NEEF, AND J. CHRISTOPHE. Effects of somatostatin on pancreatic

exocrine function. Interaction with secretin. *Biochem. Biophys. Res. Commun.* 67: 315–323, 1975.

186. ROBBERECHT, P., M. DESCHODT-LANCKMAN, M. LAMMENS, P. DE NEEF, AND J. CHRISTOPHE. "In vitro" effects of secretin and vasoactive intestinal polypeptide on hydrolase secretion and cyclic AMP levels in the pancreas of five animal species. A comparison with caerulein. *Gastroenterol. Clin. Biol.* 1: 519–525, 1977.

187. RODBELL, M., L. BIRNBAUMER, S. L. POHL, AND H. M. J. KRANS. The glucagon-sensitive adenyl cyclase system in plasma membranes of rat liver. V. An obligatory role of guanyl nucleotides in glucagon action. *J. Biol. Chem.* 246: 1877–1882, 1971.

188. RUTTEN, W. J., J. J. H. H. M. DE PONT, AND S. L. BONTING. Adenylate cyclase in the rat pancreas. Properties and stimulation by hormones. *Biochim. Biophys. Acta* 274: 201–213, 1972.

189. SAKAMOTO, C., I. D. GOLDFINE, AND J. A. WILLAMS. The somatostatin receptor on isolated acinar cell plasma membranes. *J. Biol. Chem.* 259: 9623–9627, 1984.

190. SALTIEL, A. R., AND P. CUATRECASAS. Insulin stimulates the generation from hepatic plasma membranes of modulators derived from an inositol glycolipid. *Proc. Natl. Acad. Sci. USA* 83: 5793–5797, 1986.

191. SALTIEL, A. R., J. A. FOX, P. SHERLINE, AND P. CUATRECASAS. Insulin-stimulated hydrolysis of a novel glycolipid generates modulators of cAMP phosphodiesterase. *Science Wash. DC* 233: 967–972, 1986.

192. SAWYER, S. T., AND S. COHEN. Enhancement of calcium uptake and phosphatidylinositol turnover by EGF in A431 cells. *Biochemistry* 20: 6280–6286, 1981.

194. SCHULTZ, G., AND W. ROSENTHAL. Prinzipien der transmembranären Signalumsetzung bei der Wirkung von Hormonen und Neurotransmittern. *Arzneim.-Forsch.* 35(II) 12a: 1879–1885, 1985.

195. SCHULZ, I. Influence of bicarbonate CO_2^- and glycodiazine buffer on the secretion of the isolated cat pancreas. *Pfluegers Arch.* 329: 283–306, 1971.

196. SCHULZ, I. Electrolyte and fluid secretion in the exocrine pancreas. In: *Physiology of the Gastrointestinal Tract* (2nd ed.), edited by L. R. Johnson. New York: Raven, 1987, vol. 2, p. 1147–1171.

196a. SCHULZ, I., S. SCHNEFEL, H. BANFIĆ, AND L. ECKHARDT. Ca^{2+} signalling in exocrine glands in comparison to that in vascular smooth muscle cells. In: *Membrane Pathology*, edited by G. Bianchi, E. Carafoli, and A. Scarpa. New York: NY Acad. Sci., 1986, vol. 488, p. 240–251.

197. SCHULZ, I., H. STREB, E. BAYERDÖRFFER, AND K. IMAMURA. Hormonal and neurotransmitter regulation of Ca^{2+} movements in pancreatic acinar cells. In: *Hormones and Cell Regulation*, edited by J. E. Dumont, B. Hamprecht, and J. Nunez. Amsterdam: Elsevier, 1985, vol. 9, p. 325–342.

198. SCHULZ, I., F. STRÖVER, AND K. J. ULLRICH. Lipid soluble weak organic acid buffer as "substrate" for pancreatic secretion. *Pfluegers Arch.* 323: 121–140, 1971.

198a. SCHULZ, I., F. THÉVENOD, T. P. KEMMER, AND H. P. STREB. Proton-dependent Ca^{2+} uptake into inositol 1,4,5-trisphosphate-sensitive endoplasmic reticulum from red salivary glands. In: *Molecular Mechanisms in Secretion*, edited by N. A. Thorn, M. Treiman, O. H. Petersen, and J. H. Thaysen. Copenhagen: Munksgaard, in press. (Alfred Benzon Symp. 25.)

199. SHEARS, S. B., D. J. STOREY, A. J. MORRIS, A. B. CUBITT, J. B. PARRY, R. H. MICHELL, AND C. J. KIRK. Dephosphorylation of myo-inositol 1,4,5-trisphosphate and myo-inositol 1,3,4-trisphosphate. *Biochem. J.* 242: 393–402, 1987.

200. SIFFERT, W., G. SIFFERT, AND P. SCHEID. Activation of Na^+/H^+ exchange in human platelets stimulated by thrombin and a phorbol ester. *Biochem. J.* 241: 301–303, 1987.

201. SIMON, M.-F., H. CHAP, AND L. DOUSTE-BLAZY. Activation of phospholipase C in thrombin-stimulated platelets does not depend on cytoplasmic free calcium concentration. *FEBS Lett.* 170: 43–48, 1984.

202. SINGH, J. Mechanism of action of insulin on acetylcholine-evoked amylase secretion in the mouse pancreas. *J. Physiol. Lond.* 358: 469–482, 1985.

203. SOKOLOVSKY, M., D. GURWITZ, AND R. GALRON. Muscarinic receptor binding in mouse brain: regulation by guanine nucleotides. *Biochem. Biophys. Res. Commun.* 94: 487–492, 1980.

204. SÖLING, H.-D., U. PADEL, R. JAHN, G. THIEL, P. KRICKE, AND W. FEST. Regulation of protein kinases in exocrine secretory cells during agonist-induced exocytosis. In: *Advances in Enzyme Regulation*, edited by G. Weber. Oxford: Pergamon, 1985, vol. 23, p. 141–156.

205. SOMLYO, A. P., A. V. SOMLYO, H. SHUMAN, A. SCARPA, M. ENDO, AND G. INESI. Mitochondria do not accumulate significant Ca concentrations in normal cells. In: *Calcium and Phosphate Transport Across Biomembranes*, edited by F. Bronner and M. Peterlik. New York: Academic, 1981, p. 87–98.

206. SRIKANT, C. B., AND Y. C. PATEL. Somatostatin receptors on rat pancreatic acinar cells. Pharmacological and structural characterization and demonstration of down-regulation in streptozotocin diabetes. *J. Biol. Chem.* 261: 7690–7696, 1986.

207. STANLEY, E. F., AND G. EHRENSTEIN. A model for exocytosis based on the opening of calcium-activated potassium channels in vesicles. *Life Sci.* 37: 1985–1995, 1985.

208. STEIN, J. M., AND C. N. HALES. The effect of insulin on $^{32}P_i$ incorporation into rat fat cell phospholipids. *Biochim. Biophys. Acta* 337: 41–49, 1974.

209. STREB, H., E. BAYERDÖRFFER, W. HAASE, R. F. IRVINE, AND I. SCHULZ. Effect of inositol-1,4,5-trisphosphate on isolated subcellular fractions of rat pancreas. *J. Membr. Biol.* 81: 241–253, 1984.

210. STREB, H., J. P. HESLOP, R. F. IRVINE, I. SCHULZ, AND M. J. BERRIDGE. Relationship between secretagogue-induced Ca^{2+} release and inositol polyphosphate production in permeabilized pancreatic acinar cells. *J. Biol. Chem.* 260: 7309–7315, 1985.

211. STREB, H., R. F. IRVINE, M. J. BERRIDGE, AND I. SCHULZ. Release of Ca^{2+} from a nonmitochondrial intracellular store in pancreatic acinar cells by inositol-1,4,5-trisphosphate. *Nature Lond.* 306: 67–69, 1983.

212. STREB, H., AND I. SCHULZ. Regulation of cytosolic free Ca^{2+} concentration in acinar cells of rat pancreas. *Am. J. Physiol.* 245 (*Gastrointest. Liver Physiol.* 8): G347–G357, 1983.

213. SUTHERLAND, E. W., G. A. ROBISON, AND R. W. BUTCHER. Some aspects of the biological role of adenosine 3′,5′-monophosphate (cyclic AMP). *Circulation* 37: 279–306, 1968.

214. SUZUKI, K., C. C. H. PETERSEN, AND O. H. PETERSEN. Hormonal activation of single K^+ channels via internal messenger in isolated pancreatic acinar cells. *FEBS Lett.* 192: 307–312, 1985.

215. SVOBODA, M., J. FURNELLE, F. ECKSTEIN, AND J. CHRISTOPHE. Guanosine 5′-O-(2-thiodiphosphate) as a competitive inhibitor of GTP in hormone or cholera toxin-stimulated pancreatic adenylate cyclase. *FEBS Lett.* 109: 275–279, 1980.

216. SVOBODA, M., M. LAMBERT, AND J. CHRISTOPHE. Distinct effects of the C-terminal octapeptide of cholecystokinin and of a cholera toxin pretreatment on the kinetics of rat pancreatic adenylate cyclase activity. *Biochim. Biophys. Acta* 675: 46–61, 1981.

217. TAKAI, Y., A. KISHIMOTO, Y. IWASA, Y. KAWAHARA, T. MORI, AND Y. NISHIZUKA. Calcium-dependent activation of a multifunctional protein kinase by membrane phospholipids. *J. Biol. Chem.* 254: 3692–3695, 1979.

218. TAPAREL, D., J. P. ESTEVE, C. SUSINI, N. VAYSSE, D. BALAS, G. BERTHON, E. WUNSCH, AND A. RIBET. Binding of somatostatin to guinea-pig pancreatic membranes: regulation by ions. *Biochem. Biophys. Res. Commun.* 115: 827–833, 1983.

219. TARTAKOFF, A. M., AND J. D. JAMIESON. Subcellular fractionation of the pancreas. *Methods Enzymol.* 31: 41–59, 1974.

220. THOMAS, A. P. Modulation by GTP of the inositol 1,4,5-trisphosphate-activated calcium channel. In: *Inositol Lipids in Cellular Signaling. Current Communications in Molecular Bi-*

ology, edited by R. H. Michell and J. W. Putney, Jr. New York: Cold Spring Harbor, 1987, p. 133–139.
221. TRIMBLE, E. R., R. BRUZZONE, T. J. BIDEN, AND R. V. FARESE. Secretin induces rapid increases in inositol trisphosphate, cytosolic Ca^{2+} and diacylglycerol as well as cyclic AMP in rat pancreatic acini. *Biochem. J.* 239: 257–261, 1986.
222. TRIMBLE, E. R., R. BRUZZONE, T. J. BIDEN, C. J. MEEHAN, D. ANDREU, AND R. B. MERRIFIELD. Secretin stimulates cyclic AMP and inositol trisphosphate production in rat pancreatic acinar tissue by two fully independent mechanisms. *Proc. Natl. Acad. Sci. USA* 84: 3146–3150, 1987.
223. TSUDA, T., K. KAIBUCHI, B. WEST, AND Y. TAKAI. Involvement of Ca^{2+} in platelet-derived growth factor-induced expression of c-myc oncogene in Swiss 3T3 fibroblasts. *FEBS Lett.* 187: 43–46, 1985.
224. TYRAKOWSKI, T., S. MILUTINOVIĆ, AND I. SCHULZ. Studies on isolated subcellular components of cat pancreas. III. Alanine-sodium cotransport in isolated plasma membrane vesicles. *J. Membr. Biol.* 38: 333–346, 1978.
225. UEDA, T., S. H. CHURCH, M. W. NOEL, AND D. L. GILL. Influence of inositol 1,4,5-trisphosphate and guanine nucleotides on intracellular calcium release within the N1E-115 neuronal cell line. *J. Biol. Chem.* 261: 3184–3192, 1986.
226. UHING, R. J., H. JIANG, V. PRPIC, AND J. H. EXTON. Regulation of a liver plasma membrane phosphoinositide phosphodiesterase by guanine nucleotides and calcium. *FEBS Lett.* 188: 317–320, 1985.
227. VERGHESE, M. W., C. D. SMITH, AND R. SNYDERMAN. Potential role for a guanine nucleotide regulatory protein in chemoattractant receptor mediated polyphosphoinositide metabolism, Ca^{++} mobilization and cellular responses by leukocytes. *Biochem. Biophys. Res. Commun.* 127: 450–457, 1985.
228. VERGHESE, M., R. J. UHING, AND R. SNYDERMAN. A pertussis/choleratoxin-sensitive N protein may mediate chemoattractant receptor signal transduction. *Biochem. Biophys. Res. Commun.* 138: 887–894, 1986.
229. VIGNE, P., C. FRELIN, AND M. LAZDUNSKI. The amiloride-sensitive Na^+/H^+ exchange system in skeletal muscle cells in culture. *J. Biol. Chem.* 257: 9394–9400, 1982.
230. VON TSCHARNER, V., B. PROD'HOM, M. BAGGIOLINI, AND H. REUTER. Ion channels in human neutrophils activated by a rise in free cytosolic calcium concentration. *Nature Lond.* 324: 369–372, 1986.
231. WAKELAM, M. J. O., G. J. MURPHY, V. J. HRUBY, AND M. D. HOUSLAY. Activation of two signal-transduction systems in hepatocytes by glucagon. *Nature Lond.* 323: 68–71, 1986.
232. WALDER, I. A., AND J. B. LUNSETH. A technic for separation of the cells of the gastric mucosa. *Proc. Soc. Exp. Biol. Med.* 112: 494–496, 1963.
233. WATSON, S. P., AND E. G. LAPETINA. 1,2-Diacylglycerol and phorbol ester inhibit agonist-induced formation of inositol phosphates in human platelets: possible implications for negative feedback regulation of inositol phospholipid hydrolysis. *Proc. Natl. Acad. Sci. USA* 82: 2623–2626, 1985.
234. WILLEMS, P. H. G. M., A. M. M. FLEUREN-JAKOBS, J. J. H. H. M. DE PONT, AND S. L. BONTING. Potentiating role of cyclic AMP in pancreatic enzyme secretion, demonstrated by means of forskolin. *Biochim. Biophys. Acta* 802: 209–214, 1984.
235. WILLEMS, P. H. G. M., R. H. J. TILLY, AND J. J. H. H. M. DE PONT. Pertussis toxin stimulates cholecystokinin-induced cyclic AMP formation but is without effect on secretagogue-induced calcium mobilization in exocrine pancreas. *Biochim. Biophys. Acta* 928: 179–185, 1987.
236. WILLIAMS, J. A., AND I. D. GOLDFINE. The insulin-pancreatic acinar axis. *Diabetes* 34: 980–986, 1985.
237. WOLF, M., H. LEVINE, III, W. S. MAY, JR., P. CUATRECASAS, AND N. SAHYOUN. A model for intracellular translocation of protein kinase C involving synergism between Ca^{2+} and phorbol esters. *Nature Lond.* 317: 546–549, 1985.
238. WOLLHEIM, C. B., AND T. J. BIDEN. Signal transduction in insulin secretion: comparison between fuel stimuli and receptor agonists. *Ann. NY Acad. Sci.* 488: 317–333, 1986.
239. WRENN, R. W. Phosphorylation of a pancreatic zymogen membrane protein by endogenous calcium/phospholipid-dependent protein kinase. *Biochim. Biophys. Acta* 775: 1–6, 1984.
240. ZAVOICO, G. B., S. P. HALENDA, R. I. SHA'AFI, AND M. B. FEINSTEIN. Phorbol myristate acetate inhibits thrombin-stimulated Ca^{2+} mobilization and phosphatidylinositol 4,5-bisphosphate hydrolysis in human platelets. *Proc. Natl. Acad. Sci. USA* 82: 3859–3862, 1985.
241. ZIMMERBERG, J., F. S. COHEN, AND A. FINKELSTEIN. Fusion of phospholipid vesicles with planar phospholipid bilayer membranes. I. Discharge of vesicular contents across the planar membrane. *J. Gen. Physiol.* 75: 241–250, 1980.
242. ZIMMERBERG, J., F. S. COHEN, AND A. FINKELSTEIN. Micromolecular Ca^{2+} stimulates fusion of lipid vesicles with planar bilayers containing a calcium-binding protein. *Nature Lond.* 210: 906–908, 1980.
243. ZIMMERBERG, J., M. CURRAN, F. S. COHEN, AND M. BRODWICK. Simultaneous electrical and optical measurements show that membrane fusion precedes secretory granule swelling during exocytosis of beige mouse mast cells. *Proc. Natl. Acad. Sci. USA* 84: 1585–1589, 1987.

CHAPTER 23

Regulation of digestive reactions by the pancreas

STEPHEN S. ROTHMAN | Departments of Physiology and Stomatology, University of California, San Francisco, California

CHAPTER CONTENTS

Regulation of Digestion as a Function of Bulk
Regulation of Specific Digestive Reactions
    Effect of diet on digestive enzyme content of the pancreas
    Parallel secretion
    Adaptation versus regulation
    Nonparallel transport
Properties of Regulation
    Sensory input
    Modes of action
    Regulatory outcomes
Mechanism of Secretion
End Products and Zymogen Granules
Conclusion

TWO MAJOR CHEMICAL REACTION SEQUENCES, digestion and metabolism, provide the chemical foundation for assimilation, a central life function in animals. Metabolic reactions occur in the form of highly organized, regulated sequences or cycles that permit the organism to accommodate to its changing needs for one particular product or another. This process of regulation is in great part internal, an aspect of self-organization that results from the variable presence of reaction substrates, end products, and enzyme catalysts. Until recently the regulation of digestive reactions has been regarded quite differently. In this view, regulation is solely a function of the amount (or "bulk") of food in need of digestion and does not involve regulating the rates of specific digestive reactions, analogous to metabolism.

REGULATION OF DIGESTION AS A FUNCTION OF BULK

The idea that digestive reactions are regulated as a function of the bulk of food is largely derived from the work of the great nineteenth century physiologists Bernard, Pavlov, and Heidenhain. Bernard's studies on the autonomic nervous system demonstrated that parasympathetic discharge activates most gastrointestinal functions; i.e., it provides a general activation of "vegetative" processes. This gave rise to the idea that digestion is regulated by a general activation of the parasympathetic nervous system. Digestion-related stimuli, because they are able to produce a sensory or afferent discharge, elicit a parasympathetic efferent outflow that is graded in relation to the bulk of food and operates more or less independently of the particular nature of the foodstuffs in need of digestion.

Although many stimuli that enhance gastrointestinal function have been discovered, their quantitative relation to bulk digestive need remains unclear. Effective sensory stimuli often elicit responses that appear quite unrelated to digestive need or are related to it in indirect and uncertain ways. For example, stimuli premonitory to ingestion or digestion (those of Pavlov's cephalic or psychic phase) produce a general activation of alimentary function that may be quite unrelated in magnitude to the need for digestion. Certainly the smell of food or its sweet taste may be poor indicators of digestive need and yet may be very effective activators of gastrointestinal functions (9, 14). Gastric stretch receptors (9, 14) may also provide an erroneous signal if quantitative regulation is the goal. Although a full stomach certainly suggests the need to activate digestive functions, the volume of material in the stomach may not provide a particularly accurate indication of the quantitative need for enzymatic digestion. Of course, a small meal may require more substantial enzymatic digestion than a larger meal in terms of volume. Chemical, food-related stimuli, such as polypeptides, amino acids, and fatty acids in the stomach and intestine (9, 14, 17, 19, 22, 42, 60), present similar problems. For example, at least in some species, neither starch nor its end products in the stomach or the small bowel appear to be effective in activating gastrointestinal functions (22, 44, 60), whereas certain amino acids and polypeptides are quite effective (17, 19, 22, 42, 60). In such a circumstance, the body perceives the presence of amino acids in the stomach or intestine as a more effective signal for the digestion of starch than the presence of starch itself.

It is not clear how an accurate "general" signal of digestive need, indicating the overall or bulk (quantitative) requirements for digestion independent of the particular polymeric bonds in need of breakdown, is provided. Given such regulation, it is not clear what aspect of the meal is monitored: for example, its volume, its protein content, or the rate of liberation of amino acids.

REGULATION OF SPECIFIC DIGESTIVE REACTIONS

However bulk regulation may be accomplished, it has been the belief for most of the twentieth century that digestive reactions themselves are not regulated. Thus, although all chemical reactions that occur after digestion have evolved into highly and precisely regulated sequences (metabolism), and complex differentiated processes (including higher nervous function) have evolved to provide food for digestion, digestion itself remains an unregulated series of reactions.

Pavlov disagreed with this view. He believed that the body regulated digestive processes with the utmost care and precision and that in its service the reactions of digestion themselves were regulated. Pavlov and his student Walther proposed that regulation occurred by the variable presence of the different enzyme catalysts in the gut (38, 59). The particular needs of a meal undergoing digestion would be met by appropriately varying the rates of secretion of different digestive enzymes, principally the numerous enzymes secreted by the pancreas. This view, put forth at the turn of the last century, was widely discounted until recently, primarily because it seemed that all of the digestive enzymes were secreted at constant relative rates (or in parallel) regardless of the nature of the digestive load (6, 9, 14, 45). Thus only secretion of the overall amount of enzyme could be varied, not secretion of individual enzymes concerned with particular digestive reactions and as such, these variations could only be related to the bulk need for digestion, not to specific reactive situations.

Effect of Diet on Digestive Enzyme Content of the Pancreas

In the early 1940s, experiments by Ivy and his students, Grossman and Greengard (26, 27), provided the first substantial experimental evidence in support of Pavlov's theory. They found that varying the composition of diets fed to rats led to variations in the relative enzyme content of pancreatic tissue, increasing the proportion of one enzyme relative to another. Thus the animal responded to the chemical content of the diet, altering the proportions, not merely the quantity, of the various digestive enzymes in the pancreas. These observations have been corroborated numerous times during the more than 40 years since the original report (10, 15, 21, 51, 55). However, it cannot be stated that, for example, a high-carbohydrate or high-protein diet always produces the same alteration in tissue content. Differences, as well as similarities, have been reported in response to diets of the same general type in different studies.

Measuring amylase, lipase, and trypsin (proteolytic activity), Ivy and co-workers (26) reported that a high-carbohydrate diet produced a pronounced increase in amylase, a decrease in trypsin, and no change in the lipase content of the tissue. A high-protein diet led to increased tryptic activity, a lesser increase in lipase content, and no change in amylase. Finally, a high-fat diet did not alter either lipase or trypsin but did decrease the amylase content of the tissue. Some of these results suggest positive feedback by food substrates; for example, an increase in the carbohydrate content of the diet led to an increase in tissue amylase content. A cross-reactive negative feedback was also observed; for example, the high-carbohydrate diet decreased tissue content of trypsin. Effects that appeared contradictory, at least on a superficial level, were also seen. For example, a high-protein diet increased tissue lipase content. Finally, one major dietary change did not alter the tissue content of the enzyme centrally involved; i.e., a high-fat diet did not alter tissue lipase content, although it reduced amylase content and thereby increased lipase content relative to amylase, although not relative to trypsin.

Desnuelle, Reboud, and Ben Abdeljlil (15), repeating the study some 20 years later, were unable to demonstrate an increase in the amylase content of tissue as a result of a high-carbohydrate diet, and although they reported almost a doubling in chymotrypsinogen content with a high-protein diet, they did not observe an increase in the trypsinogen content of the same tissue. In addition, unlike the Ivy group, they found that a high-protein diet, as well as a high-fat diet, greatly depressed the amylase content of tissue. A summary comparison of the two studies is given in Table 1.

In other experiments, Ivy and co-workers (27) fed rats dextrose rather than starch and substituted a casein hydrolysate for casein in the diet. In the former case, they found an increase in the amylase content of pancreatic tissue, whereas in the latter, trypsin content was reduced. That is, the same type of manipulation for the two different foodstuffs led to opposite effects. Furthermore they observed that insulin injection reduced tissue amylase content. Desnuelle and co-workers (7), on the other hand, found that both glucose and insulin injection increased the rate of amylase synthesis.

Desnuelle et al. (15) thought that the differences in the results of the two groups, if not the apparent internal inconsistencies, might be attributable to

TABLE 1. *Effect of Diet on Pancreatic Digestive Enzyme Content*

Diet	% Change Compared With Mixed-Diet Control						
	Grossman et al. (26)			Desnuelle et al. (15)			
	Amylase	Lipase	"Trypsin"	Amylase	Lipase	ChTg	Tg
High CHO	+41	(−7)	−55	(+9)	−30	(−3)	(−19)
High protein	(−19)	+30	+200	−83	−60	+79	(−6)
High fat	−75	(−12)	(+20)	−72	(−2)	(−20)	(−15)

Values in parentheses, changes in the mean (20% or less) not reported as statistically distinguishable from controls. ChTg, chymotrypsinogen. Tg, trypsinogen. "Trypsin," proteolytic activity. Values from Grossman et al. are based on content per gland. Values from Desnuelle et al. are based on "specific activity" (activity per mg tissue protein).

measurement techniques. They noted that the means for estimating enzymes had been greatly improved since the original work of Grossman et al. (26, 27), and perhaps their method of measurement was at fault. Perhaps some of the differences also were attributable to minor experimental variations, such as the age, sex, or strain of animals, duration of the diet, or feeding pattern. Despite uncertainties, from these and other similar observations two things seem clear. Diet alters the enzyme composition of pancreatic tissue, and such variations indicate that the animal is responsive to changes in diet and specific digestive need. However, putting the possibility of measurement artifact aside, the alterations seem to be variable, even with essentially the same nutrient diet.

Parallel Secretion

Some of the confusion may stem from the view, held by Grossman et al. (26, 27), that the tissue content of enzymes is the appropriate parameter to assess the effect of diet. Secretion of specific enzymes is the only relevant parameter, regardless of how tissue enzyme content may change. The appropriateness of the tissue measurement was based on the belief that secretion was parallel. The concept of parallel secretion was proposed by Babkin in the 1930s. In this view the secretion of the various digestive enzymes by the pancreas occurred without regard to kind and thus in parallel with each other (6). Variations between enzymes that might be observed would merely reflect random statistical differences. Accordingly, the mixture of enzymes in secreted fluid would be identical to or parallel to the mixture in tissue, and measuring tissue content would in essence be a simple means of measuring the enzyme contents of secreted fluid.

Thus observations that supported Pavlov's ideas (diet-induced changes in the relative proportions of different digestive enzymes in the pancreas) and the major observation that had been thought to disprove them (parallel secretion) were joined. Changes were produced by diet, in accordance with Pavlov's theory, but such changes reflected changes in protein synthesis and only dependently reflected changes in the secretion or transport of these molecules out of the cell; synthesis—not transport—was nonparallel.

Adaptation Versus Regulation

At least in some species, the secretion of enzyme during a normal digestive period appears to be accounted for in great part by material manufactured prior to ingestion of the meal (47). Thus, if parallel secretion applies, the mixture of enzymes secreted would be stereotyped by its previous construction and could not provide a regulatory response to the particular contents of a given meal. This among other reasons led Grossman et al. (26, 27) to propose that what they were observing was a long-term adaptive reaction of the organism to a chronic change in its diet and not the regulation of specific digestive reactions during the digestion of single meals or the ongoing regulation of these reactions in general (26, 27, 48). Therefore dietary effects were viewed as being adaptive but in this sense not regulatory. An animal faced with long-term changes in diet might adapt by varying the rates of synthesis of the various enzymes accordingly. However, no short-term adaptation would occur during the digestion of individual meals. Only with continuing reinforcement could such effects be seen.

Nonparallel Transport

It now seems clear that the pancreas is capable of secreting its digestive enzymes in varying, or nonparallel, proportions, one enzyme relative to the other, because of differences in rates of transport, no matter to what extent synthesis may be affected by a particular experimental condition. Almost 20 years after publication of the first modern experimental evidence (41), the range of published observations amounts to ~50 separate examples, involving many different enzymes, animal species, and experimental circumstances (45, 47, 49).

If nonparallel secretion exists, then the measurement of tissue content is inadequate to assess the effect of a particular diet on secretion. For example, depending on the particular experimental circumstances, a positive feedback effect of starch on amylase secretion might be reflected in an increase or a decrease in tissue content, or no change at all.

The ability to vary the secretion rates of different

enzymes in the short run, i.e., within minutes, makes a distinction between adaptive and regulatory effects unnecessary. The regulation of digestive reactions can occur both during the course of a given meal and over the long run in response to need as part and parcel of the same regulated system. Short-run effects primarily reflect changes in the relative transport rates of different enzymes, whereas in the long run, synthesis also comes into play.

The report of nonparallel transport also raised questions about the adequacy of the exocytosis model for secretion. In exocytosis it was proposed that granules fused randomly with the appropriate portion of the cell membrane and as a result were emptied of their contents (37, 47). In this case, secreted fluid should contain the digestive enzymes in the same proportions found in granules within the cell and in constant proportions relative to each other. This should be true regardless of the rate of secretion (rate of fusion of granule and cell membrane) or the nature of the stimulant. (I consider nonparallel transport in terms of potential mechanisms briefly in MECHANISM OF SECRETION, p. 472.)

PROPERTIES OF REGULATION

Commonly, nonparallel transport involves alterations in the ratio of enzymes in fluid secreted by the pancreas in relative, not absolute, terms. The rates of secretion of different enzymes may increase or decrease together but not proportionately or the proportions of specific enzymes in the mixture may be altered for a given level of secretory activity.

Analysis of secretion data comparing average or mean values is often relatively insensitive and may give the erroneous impression that an effective treatment has been ineffective. Even large nonparallel effects, severalfold changes in the mix of proteins, may easily be missed. For example, in Figure 1, glucose

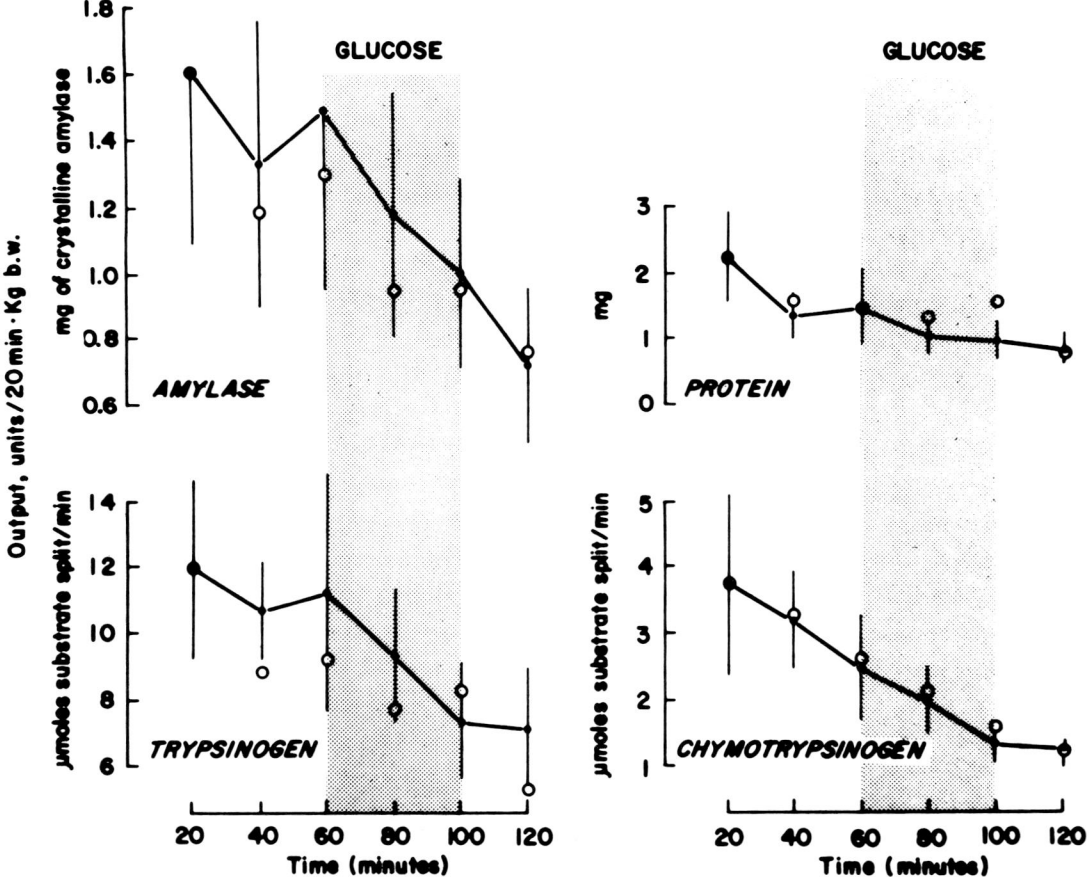

FIG. 1. Secretion of amylase, trypsinogen, chymotrypsinogen, and total protein into cannulated duct of rabbit pancreas in situ as a function of time. Duodenum was perfused for all 6 20-min periods (time on abscissa), either in absence (●; $n = 9$, where $n =$ the number of rabbits; values are means ± SE) or presence (○; $n = 6$) of glucose. When 5 mM glucose was added, it was added at 60 min (period 3) and removed at 100 min (period 5). For the glucose-positive series only mean values are presented. None of the glucose-positive points (○) for different times or measures was significantly different from time-paired controls (●). [From Rothman (44).]

perfused through the duodenum of rabbits does not alter either the overall rate of digestive enzyme secretion or the rate of secretion of individual enzyme species when secretion rate or output is averaged (44). Thus, in terms of mean values, treatment appears to have had no effect on the relative rates of secretion of different enzymes (44). However, when the data are replotted to assess the secretion of one enzyme versus that of another in a particular sample, glucose was found to have produced clear changes in the relative rates of secretion of the three proteins that were followed [Figs. 2 and 3; (44)].

Furthermore, when a particular agent does not enhance overall protein secretion, effects may be easily missed because of difficulty in distinguishing slopes of functions near the origin when there is a substantial variance in the placement of data points. Adding a general stimulant of protein secretion along with the agent of interest, by moving the data away from the origin, may help to disclose effects that might otherwise be missed. In one example (25), in the rat, differential effects of glucose and lysine on the proportion of secreted amylase and trypsinogen were clearly observed when a stimulant of overall protein secretion, the gastrointestinal hormone cholecystokinin (CCK),

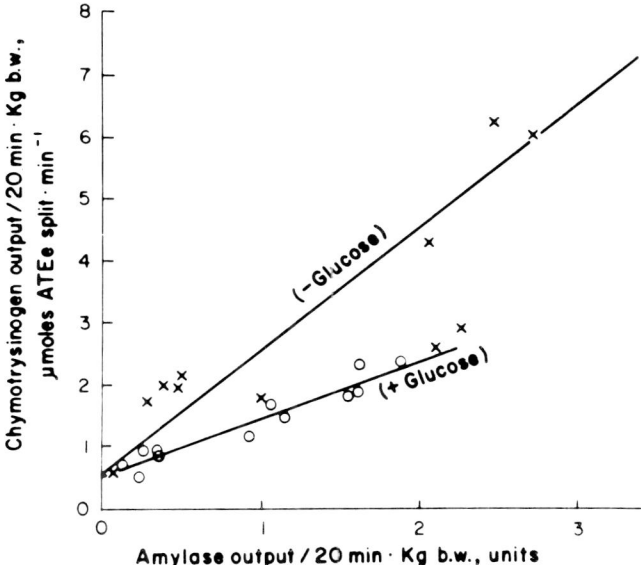

FIG. 3. Plot of secretion of amylase-chymotrypsinogen enzyme pair into cannulated duct of rabbit pancreas in situ before (×; periods 1 and 2 in Fig. 1) and after (○; periods 5 and 6) glucose was added to perfusion medium. *Lines* are calculated linear regressions. The glucose-positive intercept is significantly different from the origin. [From Rothman (44).]

was injected along with the digestive end products (Fig. 4).

Plotting the rate of secretion of one protein relative to another for individual samples of secreted fluid to look for changes in their relative proportions is particularly important. This process was first applied in work (the results of which were published in the early 1970s) in which regulatory, as opposed to merely nonparallel, secretion was seen (43). The amino acid lysine, an end product of protein digestion, was tested for its ability to alter the secretion of the digestive enzyme trypsinogen, which is centrally involved in its liberation from the peptide polymer, and the results were compared with its effect on the secretion of the functionally distinct enzyme, chymotrypsinogen. The amino acid was administered in three ways: *1*) intravenously, *2*) perfused through the duodenum, and *3*) added to medium in which whole rabbit pancreas was cultured in vitro. In all three cases, lysine at effective doses altered the relative proportions of the two enzymes in secretion.

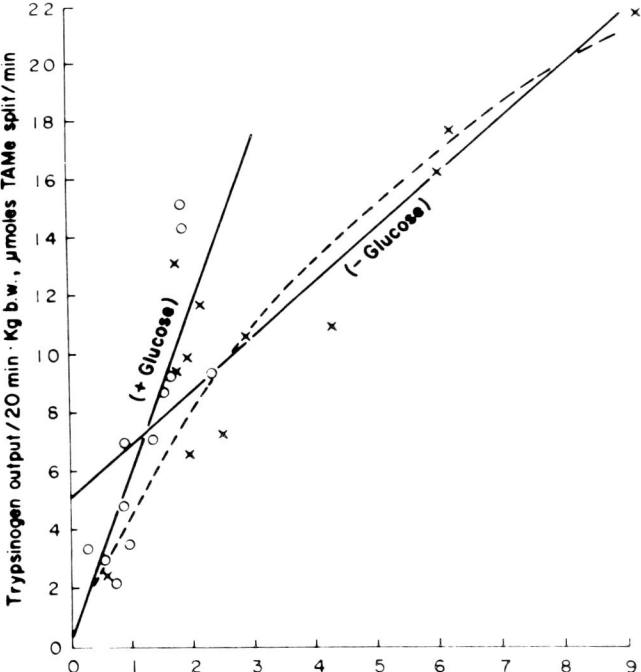

FIG. 2. Plot of secretion of chymotrypsinogen-trypsinogen enzyme pair into cannulated duct of rabbit pancreas in situ before (×; periods 1 and 2 in Fig. 1) and after (○; periods 5 and 6) glucose was added to perfusion medium. *Solid lines* are calculated by linear regression analysis for both groups. *Broken line* shows that the glucose-negative data fit a nonlinear function about as well as a linear function. (b.w., body weight; TAMe, p-toluenesulfonyl-L-arginine methyl ester·HCl; ATEe, N-acetyl-L-tyrosine ethyl ester·H$_2$O.) [From Rothman (44).]

Since these observations were made, there have been numerous studies that indicate that digestive reactions are regulated. Various digestive end products (25, 36, 44, 56), known hormones such as insulin and glucagon (58), as well as putative peptide hormones such as chymodenin (1–5) and duodenal extracts prepared after feeding particular meals (16, 20, 57) have been shown to cause short-term changes in the mixture of digestive enzymes in fluid secreted by the pancreas. In addition the effects of ingested trypsin inhibitors on pancreatic secretion (48, 51), the inhi-

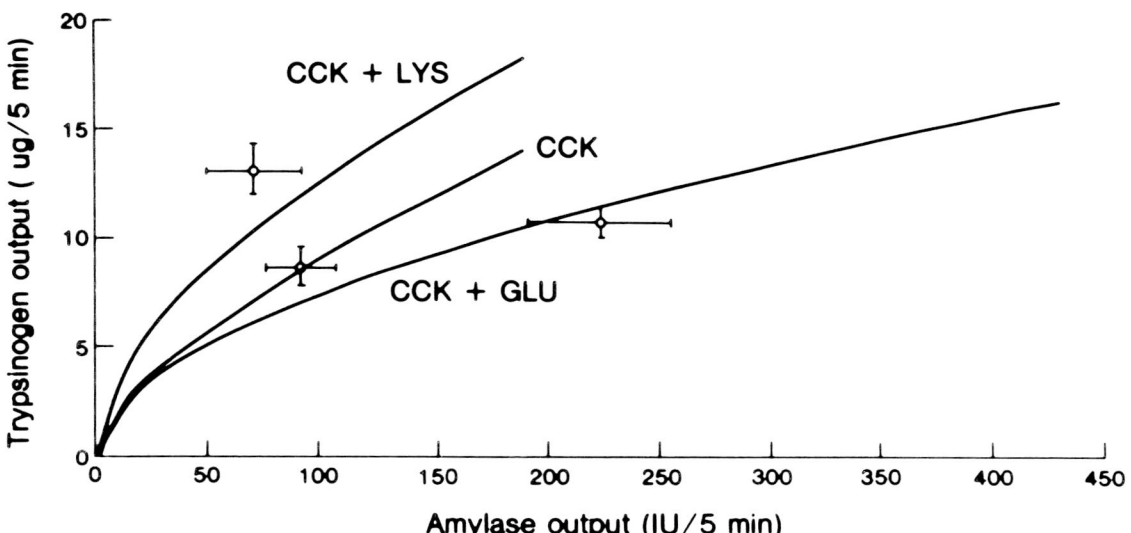

FIG. 4. Regression lines calculated from scatter of individual pairs of enzyme measurements for all data points over a 50-min period after intrapancreatic injection of cholecystokinin (CCK), CCK + Glucose (Glu), or CCK + Lysine (Lys). Each regression line is extended to upper limit of range of values observed for amylase output. In each case, best fit was using an equation $y = ax^b$. For CCK, $y = 0.44\ x^{0.66}$, $n = 87$, where n = the number of independent data points, $r = 0.87$, $P < 0.01$; for CCK + Glu, $y = 0.66\ x^{0.53}$, $n = 97$, $r = 0.91$, $P < 0.01$; and for CCK + Lys, $y = 0.97\ x^{0.53}$, $n = 97$, $r = 0.70$, $P < 0.01$. Differences among regressions were assessed by log-log transformation to approximate linear functions. Each regression differed from the others at $P < 0.001$. Open circles and error bars represent average peak output (mean ± SE) for condition indicated by adjacent regression line. [From Grendell et al. (25).]

bition of enzyme secretion by intestinal proteolytic enzymes (23), and the abrupt induction of intestinal brush-border enzyme in the presence of an appropriate digestive substrate (11, 40), among other observations, point to the existence of reaction-level regulation of digestive processes.

Despite the realization that it occurs, our knowledge of the regulation of digestive reactions is nevertheless still quite limited. However, we can develop some beginning notions, perhaps more a program for experimentation than a finished model.

Sensory Input

Because the purpose of such regulation would be to control reaction rate, the potential sensory input would of course be the stuff of the digestive reactions: the substrates, end products, and enzymes. In the past, our perception of whether a particular one of these substances in the intestine or blood had an effect on pancreatic enzyme secretion was determined by whether the substance augmented overall protein secretion by the gland. This was based on the ingrained concept of parallel transport, and when the evidence was negative for a particular concentration, means of administration, or experimental circumstance, the conclusion was drawn that the substance was generally without effect (9, 14, 17, 19, 22, 60). An increase in overall protein secretion was assumed to reflect a parallel response for all the different enzymes, even though only protein or one enzyme, most commonly amylase, was measured. Moreover, regardless of the nature of the stimulatory agent added to the intestinal lumen, if it was effective in augmenting protein secretion, it was assumed to do so through the release of a common substance, the duodenal hormone CCK (42, 60).

However, it cannot be assumed that all such effects are necessarily attributable to CCK, because other substances are known to augment overall protein secretion [e.g., vasoactive intestinal polypeptide (VIP), gastrin-inhibiting polypeptide (GIP), secretin, chymodenin, and histamine]. Furthermore, parallel secretion cannot be assumed either. Even in the absence of an effect on overall protein secretion, the response may involve altering the mixture of enzymes secreted by the gland. Indeed, the sine qua non of a response, if one is concerned with the regulation of digestive reactions, is to alter the mix of enzyme in secretion, not the overall rate of secretion.

From studies in which overall protein secretion was followed, as well as from those in which various enzymes were measured, it seems clear that some members of all three classes (end products, substrates, and enzymes) are able to alter digestive enzyme secretion by the gland in one way or another (9, 14, 17, 19, 22, 42, 60). Thus far the effect of substances on the mix of the enzymes secreted by the pancreas has primarily been studied with end products.

Glucose and lysine alter the relative rates of secretion of different digestive enzymes by the pancreas, both in vitro and in situ, when their concentrations in blood and extracellular fluid, as well as in the intestine, are varied (25, 43, 44). This may occur with or without changes in overall protein secretion. In addition, the intravenous injection of a mixture of fats has been reported to increase lipase secretion relative to trypsin in humans (36). To this point then, non-parallel secretion has been observed in the presence of end products of the three major classes of foodstuffs: proteins, lipids, and carbohydrates. Some of these effects have been seen at physiological concentrations, i.e., at concentrations similar to those that occur in extracellular fluid or blood during the digestion of a meal or that might normally occur otherwise.

Modes of Action

Direct regulation can be defined as regulation involving effects between reaction constituents, such as the binding of end product to enzyme to alter catalytic rate, whereas "indirect" regulation would involve processes outside the proximate reaction or sequence of reactions (e.g., varying the availability of enzyme by means of its genetic expression). For digestive reactions, all known regulatory effects are indirect and involve actions remote from ongoing enzyme catalysis (i.e., they alter the rate of secretion of different enzymes). Regulation does not appear to involve direct interactions between reaction constituents already present in the gut, although studies to clearly test for a direct regulation among reaction constituents have not been performed.

The actions of digestive substances can also be direct or indirect in another sense. Digestive stimuli capable of modifying the secretory rate of particular digestive enzymes may act directly at the gland or indirectly by means of hormonal or nervous intermediaries. Indirect effects may occur either within the intestine or at other tissues or organs. For end products, there is evidence of both direct (pancreatic) and indirect (intestinal and other) modes of action. As noted, effective end products are able to directly alter enzyme transport (25, 36, 43, 44), as well as synthesis (7), at the pancreas. End products also appear able to act indirectly, causing the release of hormones from secretory cells in the duodenum and elsewhere that in turn produce enzyme-selective secretion from the pancreas.

Two lines of evidence support a role for duodenal hormones. First, extracts of duodenum have been found to elicit the secretion of a specific enzyme after the duodenum has been exposed to a particular end product of digestion (16, 20, 57). For example, injection of duodenal extracts, collected after feeding a glucose meal to one rat, into the intrapancreatic circulation of another rat, favors amylase secretion by the second rat [although these results have been questioned (30, 57)]. Second, a polypeptide previously purified from duodenal mucosa has been shown to alter the relative proportions of digestive enzymes secreted by the pancreas. The particular effect was the selective secretion of chymotrypsinogen relative to lipase by the rabbit pancreas, both in situ and in vitro, in the presence of a peptide subsequently named *chymodenin*, a chymotrypsinogen-favoring peptide from the duodenum (1–5).

There is also evidence that end products can affect the secretion of specific enzymes through the release of nonintestinal humoral substances. The islet hormones insulin and glucagon are released by circulating glucose and amino acids into an intrapancreatic portal circulation at high concentrations relative to those found in the systemic circulation. A variety of studies have reported modest effects of these islet hormones on amylase secretion, although not always consistently (13, 18, 28, 32, 53, 54, 61), and have demonstrated that both peptides can be biologically active in regard to protein secretion by the pancreas. Insulin is capable of modestly enhancing pancreatic protein secretion in general, and glucagon may inhibit this secretion somewhat. In a recent study by Tseng et al. (58), both hormones were found to produce substantial enzyme-selective effects when injected into the intrapancreatic circulation along with a general stimulant of protein secretion (CCK) (Fig. 5). On the average, insulin augmented trypsinogen secretion, leaving amylase secretion unaltered, whereas glucagon inhibited amylase secretion but did not alter trypsinogen secretion (58).

Regulatory Outcomes

If one had to make a single generalization about the regulation of digestive reactions based on current knowledge, it would probably be that certain digestive end products produce positive feedback effects on the secretion of the enzymes involved in their liberation

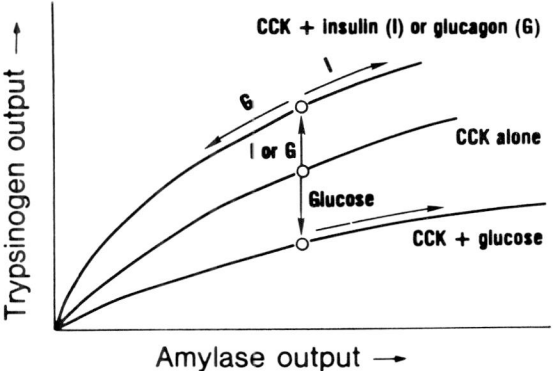

FIG. 5. Idealized schema showing effect of insulin (I), glucagon (G), and glucose on relative rates of amylase and trypsinogen secretion by rat pancreas in presence of cholecystokinin (CCK). CCK augments output of both enzymes, increasingly favoring amylase secretion as overall enzyme output is increased. Both glucagon and insulin shift CCK curve (CCK alone) toward a more trypsinogen-dominant function; however, glucagon points (G) tend to fall closer to origin, whereas insulin (I) points tend to be further away. Glucose shifts CCK curve toward a more amylase-dominant function, and points are found further from the origin. [From Tseng et al. (58).]

from polymeric substrates. This has been shown for glucose, amino acids, and fats and for direct pancreatic, as well as intestinal sites of action, in situ, in vitro, as well as in cell-free systems. This is the opposite of the apparent dominance of negative feedback of end product seen in metabolic reactions.

For positive feedback to provide effective regulation, there must be a means of limiting the cycle so that rate does not accelerate continously. This is accomplished in three ways. First, the presentation of substrate to the site of the reaction can be limited, thereby limiting reaction rate. This condition is elicited by limiting food intake and by delaying gastric emptying. Second, the digestive stimuli that lead to enzyme secretion (i.e., cause positive feedback) are removed from the intestine and blood, taken up by cells throughout the body, utilized, stored, and converted to other substances. In this way their positive-feedback actions are mitigated over time. Third, the positive-feedback drive is limited by the responsiveness of the pancreas to the stimulant. The gland is only able to secrete so much protein per unit of time, and thus secretory capability may be rate limiting.

Positive feedback is also balanced by negative-feedback effects. Several types of negative-feedback influence have been observed. The amino acid lysine acts on trypsinogen secretion by positive feedback, i.e., by increasing its proportionate presence in the mixture of enzymes secreted (43). However, when the concentration of lysine is raised beyond its maximum effective level (1–3 mM), it produces a negative-feedback effect, inhibiting trypsinogen secretion, at least relative to chymotrypsinogen (24, 43), as well as inhibiting protein secretion overall (42, 43). Thus lysine can produce both positive- and negative-feedback effects, depending on its concentration.

Insulin and glucagon provide another negative-feedback influence; they inhibit amylase secretion when glucose levels in blood are elevated. This occurs in the face of other positive-feedback effects of glucose directly at the acinar cell and indirectly via intestinal modalities (16, 20, 24, 25, 36, 57). The two hormones achieve their effects in different ways. Insulin augments overall secretion somewhat, but in the process spares amylase, decreasing its relative concentration in secreted fluid. Glucagon, on the other hand, reduces secretion overall, preferentially inhibiting amylase secretion (see Fig. 5). Thus both hormones lead to the same outcome in relative terms, a decrease in amylase secretion, but accomplish this goal in different ways.

MECHANISM OF SECRETION

In the first modern report of nonparallel transport in 1967 (41), three potential mechanisms were suggested to explain the ability of the cell to secrete its different proteins in varying proportions. First, protein contents of intracellular secretion granules could be ordered so that different granules (or cells) contained different individual digestive enzymes or groups of enzymes, and exocytic release is not random but selective. Second, the proteins might be secreted independently of a vesicle transport system by their direct movement across the relevant membranes, notably granule and cell membranes. This model, known as the equilibrium theory, describes protein transport in traditional membrane transport terms, as a process in which individual molecules move through permeable membranes. In this case, variations in the rates of secretion of different proteins would result from differences in the permeability of granule and cell membrane to them, as well as differences in their concentration gradients across these membranes. Finally, nonparallel transport might result from the variable mixing of secretory products derived from both types of process (exocytosis and membrane transport).

The evidence then available lent no support to the notion of intergranular ordering. At the electron-microscopic level, granules looked the same, within each cell and from cell to cell, and were recovered by differential centrifugation as a single apparently homogeneous sediment; immunocytochemistry at the light-microscopic level indicated that each cell appeared to contain the same variety of enzymes (62).

Although membrane transport had the advantage that it did not require an ordering of secretion granules, the lipid bilayer model of the biological membrane, as then envisioned (the Davson-Danielli model), prohibited or at least appeared to prohibit the passage of individual protein molecules through intact membranes. According to this view of membrane structure, which was widely held at the time, proteins could not penetrate, much less cross, the continuous lipid core of the membrane.

These ideas posed a difficulty that was responsible in great part for the prolonged controversy that followed. If both no intergranular order and membrane transport were correct, how could nonparallel transport occur? As the evidence has accumulated in the intervening years, it has become harder and harder to argue that nonparallel transport is not a real or even a common phenomenon, as the observations presented earlier amply demonstrate. If this is the case, then either intergranular order exists or these proteins are secreted by membrane transport processes, individually crossing membranes in the process.

With respect to the former hypothesis, electron-microscopic immunocytochemistry has demonstrated that "every" enzyme (~10 have been examined) is found in every granule in every cell (8, 29), and the pancreatic secretion granule appears to be a single anatomical and physical entity of homogeneous appearance when collected from tissue after homogenization (all cells), as well as from within individual cells in situ (47, 49). Furthermore, with the exception of one report of minor differences in the proportions of several enzymes in peri-insular acinar cells (~1%

of the total acinar mass) (31), a difference that could not account for a significant nonparallel effect, there is no evidence of regional ordering of enzyme content within the tissue.

Although the possibility can be excluded that nonparallel transport can be accounted for by the exocytosis of granules containing a single enzyme species, the possibility has not been excluded that there may be different proportions of enzymes in granules keyed to undergo exocytosis selectively; indeed, the idea has had a recent renaissance (3, 34). Adelson and Miller (3) claim to provide proof of "nonparallel exocytosis": the presence of granules or secretory cells that contain particular mixes of digestive enzymes (intergranular order) that can be called forth in a differential fashion by stimuli (selective release). They observed [reproducing earlier observations (4, 5)] that the duodenal peptide chymodenin causes nonparallel secretion and at the same time greatly increases covariance in the secretion of certain enzymes, particularly lipase and chymotrypsinogen. This observation, they conclude, proves the existence of nonparallel exocytosis. They propose that a tight correlation in the secretion of two enzymes demonstrates their "linked" transport (i.e., exocytosis), whereas a loose correlation would support their independent movement via membrane transport. Because secretion was nonparallel and the covariance high, transport occurred by nonparallel exocytosis. However, a tight correlation between two variables is not proof of linkage. Many independent events, including the secretion of substances from different cells, even different glands, are tightly covariant, responding similarly to a common external variable or set of variables. If a tight covariance were always observed (i.e., under all conditions of study), the hypothesis of linked transport certainly would be the first-order hypothesis. However, this is not the case for lipase and chymotrypsinogen or more generally for the secretion of digestive enzymes by the pancreas. The degree of covariance is itself variable. Indeed, the correlation between the secretion of lipase and chymotrypsinogen is quite weak in the absence of the peptide. In addition, if nonparallel exocytosis accounts for nonparallel transport from granules of mixed enzyme composition, then all enzyme pairs (e.g., amylase and trypsinogen) should display the same covariance in the presence of any agent that produces this effect, indicating their linked transport. This is not the case either for chymodenin (3) or more generally (44, 47).

Mroz and Lechene (34) have used micromanipulation and microfluorometric methods to measure the enzyme content of individual granules. They found that the relative proportions of amylase and chymotrypsin varied widely from granule to granule (34) and pointed out that such differences are prerequisite for selective or nonparallel exocytosis. Although this is true, differences in the enzyme content of different granules may merely reflect random variations, not ordered differences. Moreover, whatever differences in content may occur among granules, without the ability to select among them exocytosis would still produce only parallel secretion. The observed distribution of enzyme ratios (as plotted in their Fig. 2a (34)) appears roughly normal, suggesting that differences in granule content are not ordered by type but are distributed randomly, with a defined variance, around a single mean ratio. If indeed substantial variation exists from granule to granule among all 20 or so digestive enzymes, then a multitude of granule types would exist, with each granule differing from every other in terms of secretory proteins. To select among so many different granules, a remarkable system (one beyond our understanding of the coupling of stimulus to response) would be needed. This is not to say that such a system does not exist but rather that evidence demonstrating its existence is not available.

However, much positive evidence has been obtained for the membrane transport of these proteins. The concept of membrane transport for proteins is no longer the anathema that it was 20 years ago; it is even proposed in a restricted context by supporters of the exocytosis model of protein secretion in the signal hypothesis. The Davson-Danielli concept was incorrect. Proteins are embedded within biological membranes, transect them, and are transported across them. The posttranslational transport of proteins across membranes has been established, and formed proteins are known to cross a great variety of membranes, including those of bacteria, mitochondria, and nuclei, as well as the zymogen granule membrane and plasma membrane of the acinar cell. Two recent books (47, 49), and two review articles (46, 50) deal much more extensively with these ideas.

END PRODUCTS AND ZYMOGEN GRANULES

One cell-free experimental approach that bears on the question of the mechanism of secretion also seems to provide substantial information about the nature of regulation. It allows the assessment of effects of end products on the secretion of digestive enzymes by the pancreas, by studying their actions on isolated cellular organelles, most importantly on the zymogen granule of the acinar cell. Certain digestive end products produce the selective release or solubilization of a major portion of a particular digestive enzyme in a tissue homogenate at end-product concentrations normally found in plasma (24, 35). Release appears to be related to reaction level regulation in that end products of particular digestive reactions lead to the solubilization of the enzymes involved in those reactions. For example, lysine and arginine, amino acids in the peptide chain at the site where tryptic cleavage occurs, cause the release of trypsinogen but of neither chymotrypsinogen, which acts at aromatic amino acid–containing peptide bonds, nor the starch-splitting enzyme amylase [Fig. 6; (24, 35)]. Glucose, apparently

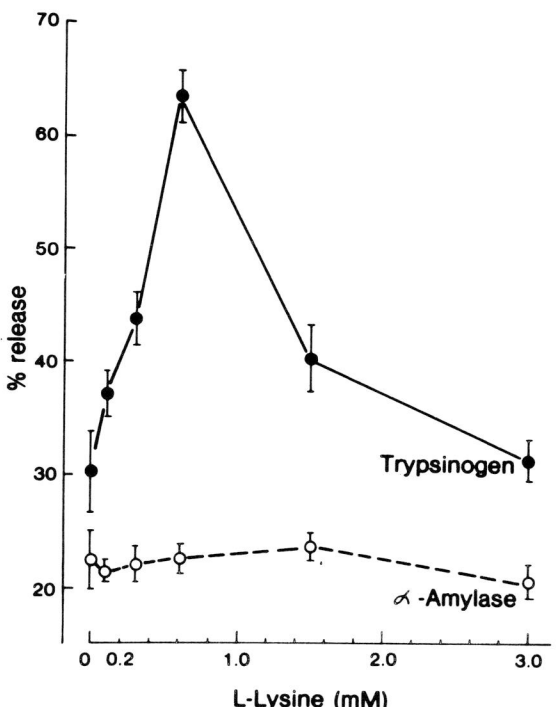

FIG. 6. Percent release of α-amylase (*open circles, broken line*) and trypsinogen (*closed circles, solid line*) in experiments with tissue homogenates, as related to concentration of L-lysine. Both enzymes are measured in the same experiment. Points, means; error bars, SE; $n = 3$, where $n =$ the number of tissue homogenates. [From Grendell and Rothman (24).]

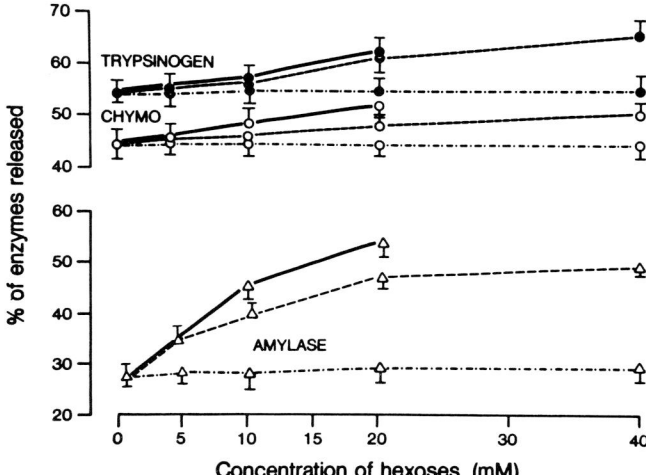

FIG. 7. Percent release of α-amylase, trypsinogen, and chymotrypsinogen (CHYMO) due to hexoses in experiments with tissue homogenates. *Solid lines*, experiments with glucose 1,6-diphosphate; *broken lines*, experiments with glucose 1-phosphate; and *dotted broken lines*, experiments with glucose 6-phosphate. Data are means ± SE ($n = 8$–18 values from 4–6 separate rat preparations). Glucose 1,6-diphosphate and glucose 1-phosphate produced substantial release of amylase (between 1.5 and 2 times control values at 10–20 mM) but only slight release of chymotrypsinogen and trypsinogen (up to ~10% above control values). Glucose 6-phosphate was ineffective in causing release of any of the enzymes. [From Niederau et al. (35).]

only if phosphorylated at its first carbon atom (C1), leads to the release of amylase but neither proteolytic enzyme [Fig. 7; (35)]. Phenylalanine and tryptophan, two aromatic amino acids, cause release of chymotrypsinogen, as well as trypsinogen, but not amylase (35).

These effects reflect positive feedback of end product, producing release of the enzyme or enzymes that accelerate its evolution from polymeric sources. This is consistent with the effects of end products observed in situ, as discussed earlier. Positive feedback is limited in that at a specified concentration release reaches maximal values for each particular substance. Thus far only lysine appears to provide an exception to this rule. At a concentration above that maximally effective, lysine has a negative-feedback influence, reducing trypsinogen release (Fig. 6). This is consistent with observations with this amino acid in situ (43).

The release of specific enzymes from isolated zymogen granules in the presence of particular end products of digestion is a property that one might well expect if the membrane of the zymogen granule were permeable to the variety of secreted proteins contained within it. It is difficult to account for this within a construct of secretion exclusively by exocytosis.

CONCLUSION

Pavlov believed that digestive reactions were regulated. He felt that, in its wisdom, the body would do no less. The evidence discussed shows that he was in all likelihood correct. However, obtaining comprehensive knowledge of the regulation of digestive reactions is a substantial task that in great part lies ahead.

REFERENCES

1. ADELSON, J. W. Un nouveau polypeptide, extrait du duodenum porcin, possedant une activité biologique chez le rat. *Biol. Gastroenterol.* 4: 355, 1971.
2. ADELSON, J. W. Chymodenin, an overview. In: *Gastrointestinal Hormones*, edited by J. C. Thompson. Austin: Univ. of Texas Press, 1975, p. 563–574.
3. ADELSON, J. W., AND P. E. MILLER. Pancreatic secretion by nonparallel exocytosis: potential resolution of a long controversy. *Science Wash. DC* 228: 993–996, 1985.
4. ADELSON, J. W., AND S. S. ROTHMAN. Selective pancreatic enzyme secretion due to a new peptide called chymodenin. *Science Wash. DC* 183: 1087–1089, 1974.
5. ADELSON, J. W., AND S. S. ROTHMAN. Chymodenin, a duodenal peptide: specific stimulation of chymotrypsinogen secretion. *Am. J. Physiol.* 229: 1680–1686, 1975.
6. BABKIN, B. P. *Secretory Mechanisms of the Digestive Glands*. New York: Hoeber, 1950.
7. BEN ABDELJLIL, A., J. C. PALLA, AND P. DESNUELLE. Effect of insulin on pancreatic amylase and chymotrypsinogen. *Biochem. Biophys. Res. Commun.* 18: 71–75, 1965.

8. BENDAYAN, M., J. ROTH, A. PERRELET, AND L. ORCI. Quantitative immunocytochemical localization of pancreatic secretory proteins in subcellular compartments in the rat acinar cell. *J. Histochem. Cytochem.* 28: 149–160, 1980.
9. BROOKS, F. P. *Control of Gastrointestinal Function: An Introduction to the Physiology of the Gastrointestinal Tract.* New York: Macmillan, 1970.
10. BUČKO, A., AND Z. KOPEC. Adaptation of enzymes activity of the rat pancreas on altered food intake. *Nutr. Dieta* 10: 276–287, 1968.
11. CEZARD, J.-P., J.-P. BROYART, P. CUISINIER-GLEIZES, AND H. MATHIEU. Sucrase-isomaltase regulation by dietary sucrose in the rat. *Gastroenterology* 84: 18–25, 1983.
12. DAGORN, J. C. Non-parallel enzyme secretion from rat pancreas: in vivo studies. *J. Physiol. Lond.* 280: 435–448, 1978.
13. DANIELSSON, A. Effects of glucose, insulin and glucagon on amylase secretion from incubated mouse pancreas. *Pfluegers Arch.* 348: 333–342, 1974.
14. DAVENPORT, H. W. *Physiology of the Digestive Tract* (5th ed.). Chicago, IL: Year Book, 1982.
15. DESNUELLE, P., J. P. REBOUD, AND A. BEN ABDELJLIL. Influence of the composition of the diet on the enzyme content of rat pancreas. In: *The Exocrine Pancreas: Normal and Abnormal Functions*, edited by A. V. S. DeReuck and M. P. Cameron. Boston, MA: Little, Brown, 1962, p. 90–107.
16. DICK, J., AND J. FELBER. Specific hormonal regulation, by food, of the pancreas enzymatic (amylase and trypsin) secretion. *Horm. Metab. Res.* 7: 161–166, 1975.
17. DIMAGNO, E. P., V. L. W. GO, AND W. H. J. SUMMERSKILL. Intraluminal and postabsorptive effects of amino acids on pancreatic enzyme secretion. *J. Lab. Clin. Med.* 82: 241–248, 1973.
18. DYCK, W. P., E. C. TEXTER, JR., J. M. LASATER, AND N. C. HIGHTOWER, JR. Influence of glucagon on pancreatic exocrine secretion in man. *Gastroenterology* 58: 532–539, 1970.
19. ERTAN, A., F. P. BROOKS, J. D. OSTROW, D. A. ARBAU, C. N. WILLIAMS, AND J. C. CERDA. Effect of jejunal amino acid perfusion and exogenous cholecystokinin on the exocrine pancreatic and biliary secretions in man. *Gastroenterology* 61: 686–692, 1971.
20. FELBER, J. P., A. ZERMATTEN, AND J. DICK. Modulation, by food, of hormonal system regulating rat pancreatic secretion. *Lancet* 2: 185–188, 1974.
21. GIDEZ, L. I. Effect of dietary fat on pancreatic lipase level in the rat. *J. Lipid Res.* 14: 169–177, 1973.
22. GO, V. L. W., A. F. HOFMAN, AND W. H. J. SUMMERSKILL. Pancreozymin bioassay in man based on pancreatic enzyme secretion: potency of specific amino acids and other digestive products. *J. Clin. Invest.* 49: 1558–1564, 1970.
23. GREEN, G. M., AND R. L. LYMAN. Feedback regulation of pancreatic enzyme secretion as a mechanism for trypsin inhibitor-induced hypersecretion in rats. *Proc. Soc. Exp. Biol. Med.* 140: 6–12, 1972.
24. GRENDELL, J. H., AND S. S. ROTHMAN. Digestive end products mobilize secretory proteins from subcellular stores in the pancreas. *Am. J. Physiol.* 241 (*Gastrointest. Liver Physiol.* 4): G67–G73, 1981.
25. GRENDELL, J. H., H. C. TSENG, AND S. S. ROTHMAN. Regulation of digestion. I. Effects of glucose and lysine on pancreatic secretion. *Am. J. Physiol.* 246 (*Gastrointest. Liver Physiol.* 9): G445–G450, 1984.
26. GROSSMAN, M. I., H. GREENGARD, AND A. C. IVY. The effect of dietary composition on pancreatic enzymes. *Am. J. Physiol.* 138: 676–682, 1943.
27. GROSSMAN, M. I., H. GREENGARD, AND A. C. IVY. On the mechanism of the adaptation of pancreatic enzymes to dietary composition. *Am. J. Physiol.* 141: 38–41, 1944.
28. KANNO, T., AND A. SAITO. The potentiating influences of insulin on pancreozymin-induced hyperpolarization and amylase release in the pancreatic acinar cell. *J. Physiol. Lond.* 261: 505–521, 1976.
29. KRAEHENBUHL, J. P., L. RACINE, AND J. D. JAMIESON. Immunocytochemical localization of secretory proteins in bovine pancreatic exocrine cells. *J. Cell Biol.* 72: 406–423, 1977.
30. LA BELLA, A., R. G. LAHAIE, H. SARLES, AND J. C. DAGORN. Pancreatic enzyme secretion: nonspecific stimulation by duodenal mucosal extracts. *Am. J. Physiol.* 240 (*Gastrointest. Liver Physiol.* 3): G109–G113, 1981.
31. MALAISSE-LAGAE, F., M. RAVAZZOLA, P. ROBBERECHT, A. VANDERMEERS, W. J. MALAISSE, AND L. ORCI. Exocrine pancreas: evidence for topographic partition of secretory function. *Science Wash. DC* 190: 795–797, 1975.
32. MANABE, T., AND M. L. STEER. Effects of glucagon on pancreatic content and secretion of amylase in mice. *Proc. Soc. Exp. Biol. Med.* 161: 538–542, 1979.
33. MARCHIS-MOUREN, G., L. PASERO, AND P. DESNUELLE. Further studies of amylase biosynthesis by pancreas of rats fed on a starch-rich or a casein-rich diet. *Biochem. Biophys. Res. Commun.* 13: 262–266, 1963.
34. MROZ, E. A., AND C. LECHENE. Pancreatic zymogen granules differ markedly in protein composition. *Science Wash. DC* 232: 871–873, 1986.
35. NIEDERAU, C., J. H. GRENDELL, AND S. S. ROTHMAN. Digestive end products release enzymes from particulate cellular pools, including zymogen granules. *Biochim. Biophys. Acta* 881: 281–291, 1986.
36. NIEDERAU, C., A. SONNENBERG, AND J. ERCKENBRECHT. Effects of intravenous infusion of amino acids, fat, or glucose on unstimulated pancreatic secretion in healthy humans. *Dig. Dis. Sci.* 30: 445–455, 1985.
37. PALADE, G. Intracellular aspects of the process of protein synthesis. *Science Wash. DC* 189: 347–358, 1975.
38. PAVLOV, I. P. *The Work of the Digestive Glands.* London: Griffith, 1910.
39. REBOUD, J. P., G. MARCHIS-MOUREN, A. COZZONE, AND P. DESNUELLE. Variations in the biosynthesis rate of pancreatic amylase and chymotrypsinogen in response to a starch-rich or a protein-rich diet. *Biochem. Biophys. Res. Commun.* 22: 94–99, 1966.
40. REISENAUER, A. M., AND G. M. GRAY. Abrupt induction of a membrane digestive enzyme by its intraintestinal substrate. *Science Wash. DC* 227: 70–72, 1985.
41. ROTHMAN, S. S. "Non-parallel transport" of enzyme protein by the pancreas. *Nature Lond.* 213: 460–462, 1967.
42. ROTHMAN, S. S. The release of cholecystokinin-pancreozymin: the effects of duodenal perfusion of various amino acids. In: *Frontiers in Gastrointestinal Hormone Research*, edited by S. Andersson. Stockholm: Almqvist & Wiksell, 1973, p. 185–190. (Nobel Symp. 16.)
43. ROTHMAN, S. S. Molecular regulation of digestion: short-term and bond specific. *Am. J. Physiol.* 226: 77–83, 1974.
44. ROTHMAN, S. S. Independent secretion of different digestive enzymes by the pancreas. *Am. J. Physiol.* 231: 1847–1851, 1976.
45. ROTHMAN, S. S. The digestive enzymes of the pancreas: a mixture of inconstant proportions. *Annu. Rev. Physiol.* 39: 373–389, 1977.
46. ROTHMAN, S. S. Passage of protein through membranes—old assumptions and new perspectives. *Am. J. Physiol.* 238 (*Gastrointest. Liver Physiol.* 1): G391–G402, 1980.
47. ROTHMAN, S. S. *Protein Secretion: A Critical Analysis of the Vesicle Model.* New York: Wiley, 1985.
48. ROTHMAN, S. S. The biological functions and physiological effects of ingested inhibitors of digestive reactions. In: *The Nutritional and Toxicological Significance of Enzyme Inhibitors in Foods*, edited by M. Friedman. New York: Plenum, 1986, p. 19–31.
49. ROTHMAN, S. S., AND J. J. L. HO. *Nonvesicular Transport.* New York: Wiley, 1985.
50. ROTHMAN, S. S., AND C. LIEBOW. Permeability of zymogen granule membrane to protein. *Am. J. Physiol.* 248 (*Gastrointest. Liver Physiol.* 11): G385–G392, 1985.
51. ROTHMAN, S. S., AND H. WELLS. Selective effects of dietary egg white trypsin inhibitor on pancreatic enzyme secretion. *Am. J. Physiol.* 216: 504–507, 1969.
52. ROTHMAN, S. S., AND H. WILKING. Differential rates of diges-

tive enzyme transport in the presence of cholecystokinin-pancreozymin. *J. Biol. Chem.* 253: 3543–3549, 1978.
53. SAITO, A., J. A. WILLIAMS, AND T. KANNO. Potentiation of cholecystokinin-induced exocrine secretion by both exogenous and endogenous insulin in isolated and perfused rat pancreata. *J. Clin. Invest.* 65: 777–782, 1980.
54. SINGH, M. Effect of glucagon on digestive enzyme synthesis, transport and secretion in mouse pancreatic acinar cells. *J. Physiol. Lond.* 306: 307–322, 1980.
55. SNOOK, J. T. Dietary regulation of pancreatic enzymes in the rat with emphasis on carbohydrate. *Am. J. Physiol.* 221: 1383–1387, 1971.
56. STASTNÁ, R., I. SKÁLA, F. HRUBÁ, G. K. SHLYGIN, AND Y. SYSOEV. The effect of some N solutions for parenteral nutrition on gastric and pancreatic secretion. *Nutr. Metab.* 23: 349–356, 1979.
57. TSENG, H. C., J. H. GRENDELL, AND S. S. ROTHMAN. Food, duodenal extracts, and enzyme secretion by the pancreas. *Am. J. Physiol.* 243 (*Gastrointest. Liver Physiol.* 6): G304–G312, 1982.
58. TSENG, H. C., J. H. GRENDELL, AND S. S. ROTHMAN. Regulation of digestion. II. Effects of insulin and glucagon on pancreatic secretion. *Am. J. Physiol.* 246 (*Gastrointest. Liver Physiol.* 9): G451–G456, 1984.
59. WALTHER, A. A. The Secretory Work of the Pancreatic Gland. Saint Petersburg, Russia: University of Saint Petersburg, 1897. PhD thesis.
60. WANG, C. C., AND M. I. GROSSMAN. Physiological determination of release of secretin and pancreozymin from intestine of dogs with transplanted pancreas. *Am. J. Physiol.* 164: 527–545, 1951.
61. WIZEMAN, V., P. WEPPLER, AND R. MAHRT. Effect of glucagon and insulin on the isolated exocrine pancreas. *Digestion* 11: 432–435, 1974.
62. YASUDA, K., AND A. H. COONS. Localization by immunofluorescence of amylase, trypsinogen, and chymotrypsinogen in the acinar cells of the pig pancreas. *J. Histochem. Cytochem.* 14: 303–313, 1966.

CHAPTER 24

Cellular compartmentation and protein processing in the exocrine pancreas

GEORGE A. SCHEELE | Laboratory of Cell and Molecular Biology, The Rockefeller University, New York, New York

HORST F. KERN | Department of Cell Biology, Philipps University, Marburg, Federal Republic of Germany

CHAPTER CONTENTS

Translocation of Proteins Across the Rough Endoplasmic Reticulum Membrane
Translocation Receptors
Asymmetric Integration of Membrane Proteins into the Lipid Bilayer
Intracisternal Processing and Cellular Sorting of Proteins
    Covalent intracisternal processing
    Golgi compartmentation
    Molecular sorting of proteins
    Importance of cisternal pH in the cellular sorting of proteins
    Intracisternal transport and processing of secretory proteins

THE EXOCRINE PANCREAS contains two types of epithelial cells with dissimilar but complementary functions: acinar cells, which secrete large amounts of protein, and duct cells, which secrete fluid and bicarbonate ions. These two cells are organized in an acinar-ductular complex that defines a common ductular space. Within this space the secretory products of these two cells are mixed and ultimately delivered to the intestinal lumen. Figure 1 demonstrates with an electron micrograph the differences observed in the subcellular structures of acinar and ductular cells. The presence of distinct subcellular compartments within the acinar cell was first described more than 100 years ago by Heidenhain (44) with the light microscope. For the past 30 years Palade (84) and his associates have used electron microscopy, tissue autoradiography, and cell fractionation to describe the intracisternal pathway by which secretory proteins are transported to the cell surface. This pathway comprises the sequential movement of exportable products through four intracellular compartments: rough endoplasmic reticulum (RER) → Golgi complex → condensing vacuoles → zymogen granules. In this chapter we focus on the biochemical mechanisms by which cells maintain an array of intracellular organelles and utilize these organelles for the chemical processing and cellular distribution of protein products, derived both from endogenous and exogenous sources. Within acinar cells such as those observed in the exocrine pancreas, these processes are organized in a specialized manner to effect the differentiated functions of this cell type: the preparation and storage of large quantities of digestive (pro)enzymes and, with peptidergic and cholinergic stimulation, their release into the digestive milieu of the intestinal lumen. We demonstrate in this chapter that the specific steps involved in protein processing are directed by peptide signals contained within individual gene products.

TRANSLOCATION OF PROTEINS ACROSS THE RER MEMBRANE

Earlier studies, which defined the secretory pathway in the guinea pig pancreas, indicated that secretory proteins are transported across only a single membrane, that of the RER (34, 101). Secretory proteins that were synthesized on ribosomes attached to the cytoplasmic leaflet of the RER membrane were vectorially transported to the membrane-bound cisternal space of this organelle. To define biochemical mechanisms by which secretory proteins are translocated across this membrane, Scheele and co-workers (103, 110) developed a highly efficient system for reconstitution of rough microsomes, using dog pancreas microsomal membranes that were stripped of endogenous mRNA by micrococcal nuclease treatment and a reticulocyte lysate translation system supplemented with bovine liver tRNA, human placental ribonuclease inhibitor (107), and dog pancreas poly(A)$^+$ mRNA, which codes for 14 well-characterized secretory products (see the other chapter by Scheele and Kern in this Handbook). In the absence of microsomal membranes the products of cell-free translation were in all

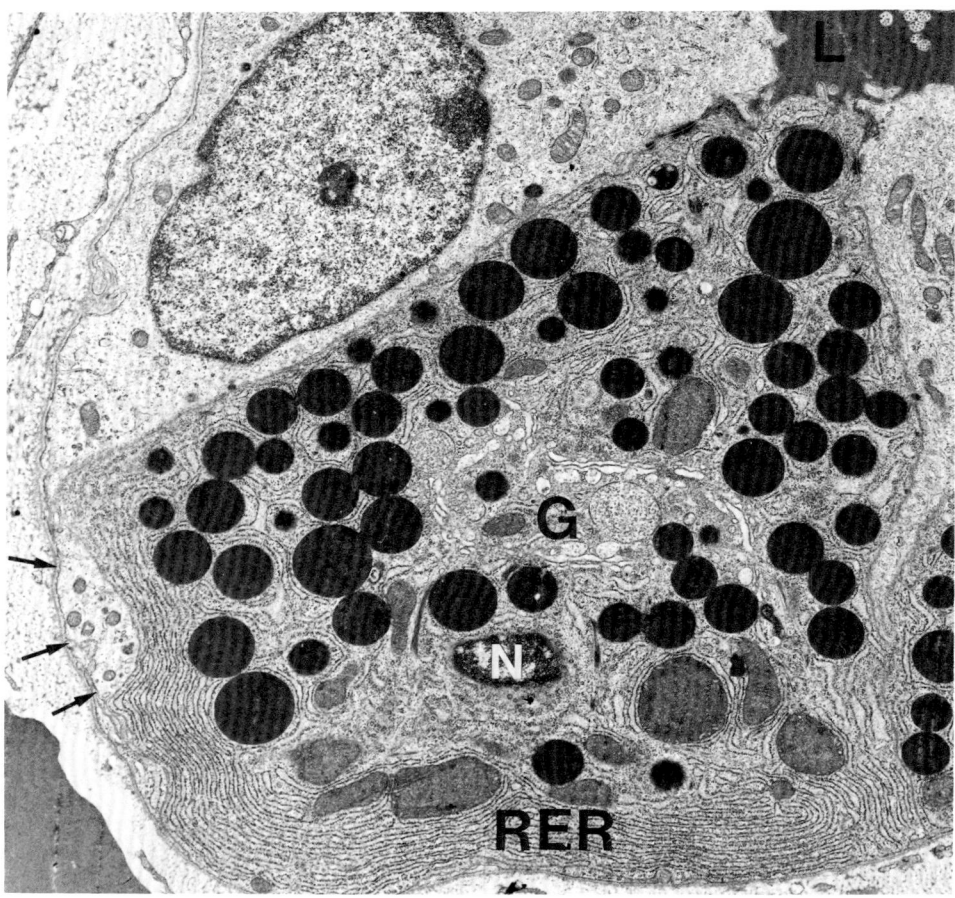

FIG. 1. Comparison of fine structure of an acinar and intralobular duct cell in human pancreas. In the acinar cell, membrane-bound compartments are distributed in a polar fashion with the rough endoplasmic reticulum (RER) located in the infra- and paranuclear regions (N, tangential section of nucleus); Golgi complex (G) and zymogen granules are located in the apical part of the cell. Cytoplasmic compartmentation is less elaborate in the adjacent duct cell, which largely contains free polysomes and minimal RER elements. L, luminal space. Arrows near the basal surface of the acinar cell point to a nerve ending. × 7,000.

cases 1,500–3,000 Da larger than authentic secreted proteins obtained from in vivo studies. In the presence of microsomal membranes, each of the 14 pancreatic presecretory proteins studied was correctly processed to mature secretory proteins by microsomal membranes, as judged by both one-dimensional and two-dimensional gel electrophoresis. Earlier studies that measured the topological location of nascent polypeptide chains (intracisternal vs. extracisternal) by introduction of proteases into the extracisternal medium during the posttranslational period were limited by the disruptive effect of proteases themselves on the microsomal membranes. Scheele and co-workers (103, 110) showed that this disruptive effect could be prevented by the treatment of membranes with the amphiphilic agent tetracaine prior to the introduction of proteases. When this improved assay was used, all secretory translation products were observed to be segregated within microsomal vesicles, as judged by their complete resistance to the effect of proteases added after completion of protein synthesis. Segregation of nascent secretory proteins was irreversible, since radioactive amylase, as well as the other labeled secretory proteins, remained quantitatively sequestered in microsomal vesicles during 90 min of incubation at 22°C after the cessation of protein synthesis (110).

These findings indicated that pancreatic proteins destined for secretion are synthesized as larger polypeptide chains (presecretory proteins) containing additional peptide extensions of 15–20 amino acid residues. Earlier studies (11, 76) on the in vitro synthesis of immunoglobulin chains agree with our conclusion that these peptide extensions served as peptide signals that mediate the translocation of nascent secretory polypeptide chains across the RER membrane. The extent of processing and segregation observed with pancreatic secretory proteins, complete for each of the 14 proteins studied, allowed us to rule out, for the first time, alternative mechanisms of protein secretion

TABLE 1. *Signal Translocation Sequences for Pancreatic (Pro)enzymes*

−20	−15	−10	−5	−1	(Pro)enzyme	pI	Ref.
	Met Ala Phe Leu	Trp Leu Leu Ser	Cys Phe Ala	Leu Leu Gly Thr Ala Phe Gly	Dog chymotrypsinogen 2	7.1	19, 86
	Met Ala Phe Leu	Trp Leu Val Ser	Cys Phe Ala	Leu Val Gly Ala Thr Phe Gly	Rat chymotrypsinogen B	4.8	105
		Met Lys Thr Phe Ile	Phe Leu Ala	Leu Leu Gly Ala Thr Val Ala	Dog trypsinogen 2	8.1	19, 87
		Met Asn Pro Leu Leu Ile	Leu Ala Phe	Leu Gly Ala Ala Val Ala	Dog trypsinogen 1	4.7	19, 87
		Met Ser Ala Leu Leu Ile	Leu Ala Leu	Val Gly Ala Ala Val Ala	Rat trypsinogen I	4.3	64
	Met - Phe - Val	- Val Leu Ala	Phe Leu Leu Ala	Tyr Ala -	Dog proelastase 1	5.0	19
	Leu Arg Phe Leu Val	Phe Ala Ser	Leu Val Leu Tyr	Gly His Ser	Rat proelastase I		120
	Met Ile Arg Thr Leu	Leu Leu Ser Ala	Phe Val Ala	Gly Ala Leu Ser	Rat proelastase II		120
		Met Lys Leu Ile	Leu Val Phe Gly Ala	Leu Leu Gly His Ile Tyr Cys	Dog procarboxypeptidase A1	5.0	*
		Met Lys Arg Leu Leu Ile	Leu Ser Leu	Leu Leu Glu Ala Val Cys Gly	Rat procarboxypeptidase A		89
		Met Lys Phe Phe Leu Leu Leu Ser	Val Ile	Gly Phe Cys Trp Ala	Dog amylase	6.0	†
		Met Lys Phe Val	Leu Leu Leu Ser Leu Ile	Gly Phe Cys Trp Ala	Rat amylase		64
		Met Lys Phe Leu Val Leu Ala	Ala Leu Leu	Thr Val Ala Ala Ala	Dog prophospholipase A₂	7.3	54
	Met Val Ser Ile	Trp Thr Ile	Ala Leu Phe Leu Leu Gly	Ala Ala Lys Ala	Dog lipase	5.9	54
Ser Leu Phe Leu Phe Ser	Leu Leu Val Leu Val	Leu Gly Trp Val	Gln Pro Ser	Leu Gly Lys Glu Leu Gly Met	Rat ribonuclease		65

pI, isoelectric point.   * S. Pinsky and G. Scheele, unpublished observations.   † K. S. LaForge and G. Scheele, unpublished observations.

(98). Studies that employed synchronized protein synthesis in the in vitro system and delayed addition of microsomal membranes (110) indicated that the peptide extensions resided at the NH$_2$ termini of polypeptide chains and verified that they mediated the cotranslational binding of presecretory proteins to microsomal membranes and the transport of nascent secretory proteins to the vesicular space. Accordingly these peptide extensions have been designated *signal-translocation sequences*.

Structural studies at the level of NH$_2$-terminal amino acid sequencing or deduction of amino acid sequences from nucleotide sequences representing pancreatic genes have provided the primary structures for signal-translocation sequences associated with exocrine proteins from both the dog and rat pancreas (Table 1). Pancreatic NH$_2$-terminal transport peptides generally vary between 15 and 18 residues, although the sequence for ribonuclease in the rat contains 25 residues. A comparison of these peptides indicates that sequence identity is low (<50%). Among enzymes that are closely related in evolution, sequence identity between signal peptides is higher. Thus signal peptides for dog and rat anionic trypsinogen show 73.3% identity.

Although the signal peptide is necessary for targeting and translocation of nascent secretory polypeptide chains across the microsomal membrane, Scheele and Jacoby (108) have shown that proteolytic removal of this peptide extension is required for the conformational maturation of biologically active proteins. Presecretory proteins synthesized in the absence of microsomal membranes showed conformational insta-

bility due to the absence or incorrect formation of disulfide bonds and nonspecific protein-protein interactions among nascent polypeptide chains, including the intermolecular formation of disulfide bonds, and the formation of oligomeric structures with little or no biological activity. Secretory proteins synthesized in the presence of microsomal membranes containing signal peptidase activity and an optimal oxidation-reduction potential showed native conformations, including formation of the correct sets of disulfide bonds. In contrast to presecretory proteins, processed and conformationally mature translocation products represent stable secretory molecules soluble in an aqueous environment. As judged by *1*) the binding of nascent amylase to its substrate, glycogen, *2*) the binding of nascent trypsinogen 1, trypsinogen 2 + 3, and chymotrypsinogen 1 to Sepharose-bound soybean trypsin inhibitor, and *3*) the activation of nascent trypsinogen by porcine enterokinase, conformationally mature translocation products showed biological activities similar to those of authentic secretory proteins synthesized in vivo (109). These findings have indicated not only that *1*) the translocation mechanisms studied in vitro closely mimic those that occur in vivo but also that *2*) the proteolytic removal of the transport peptide is a required step in the synthesis of biologically active pancreatic proteins and *3*) the development of these conformationally mature and biologically active exportable (pro)enzymes is coupled to their translocation across the RER membrane. The latter two points are of considerable physiological importance in the maintenance of cellular integrity. In the event that small quantities of presecretory proteins are released directly into the cytosolic space, their conformational instability would preclude the deleterious effects of enzyme activity (RNase, lipase, and amylase activity on intracellular RNA, triglyceride, and glycogen, respectively) and in all likelihood would lead to their breakdown by intracellular protein-degradation mechanisms.

TRANSLOCATION RECEPTORS

The introduction by Scheele (103) of canine pancreatic microsomal membranes, which are highly efficient in the translocation of secretory proteins across the RER membrane, has allowed the purification of membrane components involved in the translocation process.

In 1978 Warren and Dobberstein (130) reported that a high-salt wash of rough microsomal membranes removed a factor necessary for nascent protein translocation across the membrane. Recombination of incompetent membranes with the high-salt extract resulted in reconstitution of translocation activity. Stabilization of this factor in the presence of low concentrations of Nikkol, a nonionic detergent, allowed Walter and Blobel (127) to purify it by hydrophobic chromatography on aminopentyl agarose and sucrose gradient sedimentation. The factor was subsequently shown to be an 11S ribonucleoprotein particle that contained a 7S RNA (129) and proteins with molecular weights of 72,000, 68,000, 54,000, 19,000, 14,000, and 9,000 (127). The 7S RNA contains ~300 nucleotides and was shown to be identical to cytoplasmic 7SL RNA, which has been highly conserved throughout evolution. The 11S ribonucleoprotein complex was termed the *signal-recognition particle* (SRP) based on its high-affinity binding to ribosomes actively engaged in the synthesis of secretory proteins containing bona fide signal sequences [dissociation constant $(K_d) < 8 \times 10^{-9}$]. Alteration of the signal sequence, through the replacement of leucine with hydroxyleucine during amino acid incorporation, resulted in a sharp reduction in SRP binding ($K_d < 5 \times 10^{-5}$). Kurzchalia et al. (58) recently showed that the signal sequence of nascent preprolactin interacts with the 54,000 polypeptide of the SRP.

Addition of purified SRP to an in vitro wheat germ translation mix arrested the translation of nascent proteins containing NH_2-terminal signal sequences (128). Synchronized protein-synthesis studies indicated that the length of arrested chains approximated 8,000 Da or 70 amino acid residues. The molecular basis of inhibition of chain elongation in the wheat germ translation system is unknown. Within individual ribosomes the nascent chain exit site and the peptidyltransferase locus are separated by 16 nm. However, electron-microscopic analysis by negative-staining, dark-field imaging of unstained specimens and platinum shadowing has shown that the SRP is a rod-shaped particle, 5×23 nm (3). It is therefore conceivable that SRP bound to the signal sequence may also block the aminoacyl tRNA binding site. Signal-recognition particle devoid of Alu-like RNA sequences and the 9,000 and 14,000 heterodimer after micrococcal nuclease digestion continued to promote secretory protein translocation but no longer caused an elongation arrest (115). In addition, SRP-dependent translation arrest has not been observed in reticulocyte lysate, a mammalian in vitro translocation system (71). It is unlikely that translation arrest is a physiologically important step, as originally proposed (128), but rather is due to the heterologous mixture of SRP derived from animal sources and translation components specific for protein synthesis in plant systems.

Meyer, Dobberstein, and collaborators demonstrated that treatment of rough microsomes with low doses of elastase or trypsin released, in the presence of high salt, a 60,000-Da fragment (72) that was later shown to be the endoplasmic domain of a 72,000-Da integral membrane protein associated with the RER membrane, designated the *docking protein* (72, 73). Functional reconstitution of such membranes in regard to protein translocation activity could only be achieved by the addition of both SRP, removed by high salt, and the 60,000-Da fragment, removed by proteolysis. With the use of affinity chromatography with SRP agarose, Gilmore et al. (37) isolated from detergent-treated microsomal membranes a similar 72,000-Da integral membrane protein, which they termed the *SRP receptor*. In their study, arrest of translocation of preprolactin mRNA in the presence of purified SRP could only be achieved by the addition of unproteolyzed rough microsomes containing intact SRP receptors or a combination of the 60,000-Da fragment and protease-treated membranes. Lauffer et al. (60) recently cloned the SRP receptor, deduced the complete amino acid sequence from the nucleotide structure, and established that the NH_2 terminus is anchored to the RER membrane. Because the stoichiometric amounts of SRP and SRP receptor are small in comparison with ribosomes associated with microsomal membranes (37), these signal peptide receptors may function only transiently to direct presecretory proteins to the RER membrane and in this way facilitate the insertion of the transport peptide into the hydrophobic layer of the membrane.

It can be expected that additional components will be identified as part of the translocation machinery of the RER membrane. Ribophorins I and II, with molecular weights of 65,000 and 63,000, respectively, appear in stoichiometric amounts with membrane-bound ribosomes and stabilize the attachment of ribosomes to the RER membrane (94). Other protein components have been suggested by the following observation: microsomal membranes treated with high doses of proteases lost their ability to bind ribosomes despite supplementation of translation mixtures with

SRP and the 60,000-Da membrane fragment and the finding that ribophorins I and II were undegraded (46). In addition, recent studies that have begun to look at the mechanism of polypeptide chain transfer in the absence of peptide chain elongation have indicated that tight binding of the nascent chain to the microsomal membrane required a GTP-binding protein (21). Subsequent polypeptide chain translocation after their release by puromycin did not require either ATP or GTP (21), a finding that agrees with earlier predictions by Carne and Scheele (20).

Figure 2 shows in schematic fashion the molecular events that are believed to be involved in the selective compartmentation of secretory proteins into the cisternal space of the RER. An important feature of this vectorial transport mechanism is that protein translocation occurs across the lipid bilayer in the form of polypeptide chains containing only primary or minimal secondary structure. Native tertiary structure is believed to be exclusively developed within the intracisternal space of the RER. The presence of structure (even minimal structure) in the polypeptide chain at the cisternal leaflet of the membrane and the absence of structure (extended chain) within the channel of the large ribosomal subunit at the cytoplasmic leaflet of the membrane ensure that translocation of the polypeptide chain occurs in a vectorial manner across the RER membrane and into the cisternal space in the absence of additional energy (20). Furthermore, as secondary and tertiary structure accumulate in the polypeptide chain within the cisternal space, the motive force for vectorial transfer correspondingly increases.

Carne and Scheele (20) have analyzed the primary and secondary structural features of 53 NH_2-terminal transport peptides from both eucaryotic and procaryotic sources. These sequences varied between 15 and 31 residues. Little or no homology is observed in the

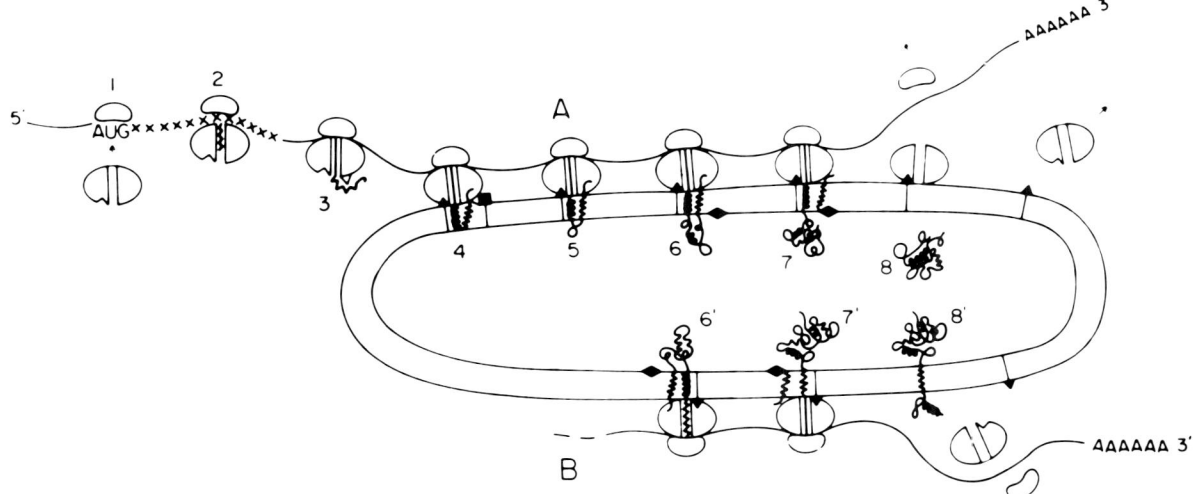

FIG. 2. Biochemical events believed to be involved in translocation of secretory proteins across (A) and integration of membrane proteins into (B) the RER membrane. 1, Initiation of protein synthesis begins at the AUG initiation codon either on free ribosomes or ribosomes indirectly attached to the microsomal membrane via the mRNA strand. 2, Following the AUG initiation codon, mRNAs for secretory proteins contain a characteristic set of codons (xxx) that result in translation of the signal-translocation sequence containing a core region (ww) rich in hydrophobic amino acids, particularly those with large side chains, Phe, Leu, Ile, Val. Messenger RNAs for nonsecretory proteins do not contain this sequence of codons. 3, The signal peptide when emerged from the channel in the large ribosomal subunit binds to the signal-recognition particle (SRP; ▽). 4, Formation of a functional ribosome-membrane junction: interaction of SRP with the docking protein associated with the RER membrane (▼) allows polypeptide chain translocation across the membrane to proceed. Formation of a functional ribosome-membrane junction also involves binding of ribosomes to ribophorins I and II located in the RER membrane (▲). The mechanism by which the signal peptide is inserted into the membrane is unknown. 5 and 6, Transfer of the polypeptide chain to the cisternal space of the RER. 7, Prior to chain termination, the signal peptide is cleaved from the nascent protein by a protease (♦) associated with the cisternal leaflet of the RER membrane. The fate of the signal peptide after cleavage is unknown. 8, In the absence of a stop-transport sequence, chain termination occurs and the completed polypeptide is released to the intravesicular space. Synthesis and translocation of the nascent protein is now complete. The small ribosomal subunit and the mRNA dissociate from the large ribosomal subunit, which remains attached to the membrane for a short time through electrostatic linkages and then is released. B: when a stop-transport sequence follows the start-transport signal, the nascent protein remains associated with the lipid bilayer in a transmembrane configuration, as depicted in steps 6', 7', and 8'. When the signal-transport sequence is not cleaved, the nascent polypeptide chain may also remain associated with the membrane, as depicted in Fig. 4B1. [From Scheele (105).]

primary structure of these sequences. However, based on regions of similarity, transport peptides can be divided into three domains. The NH_2-terminal domain, which is hydrophilic, contains a partially charged NH_2-terminal residue (eucaryotes) and, in many instances, additional charged residues. Immediately following the hydrophilic terminus there appears a hydrophobic domain consisting of a core region of hydrophobic amino acids, 9–18 residues in length. Within this hydrophobic region there exist clusters of hydrophobic amino acids with large side chains (Leu, Ile, Phe, Val). These clusters are observed toward the NH_2-terminal end of the hydrophobic segment and are 2–8 residues in length. Because codons for each of these four amino acid residues contain uridine as the middle base, the core region of the transport peptide is characterized by a high degree of structure-function conservation during the evolutionary process. The hydrophobic segment at this site in the polypeptide chain appears to be responsible for binding of SRP and insertion of the nascent chain into the lipid bilayer of the RER membrane. The third domain represents the region in which the signal peptide is proteolytically cleaved by a membrane-associated signal protease. Proteolytic changes generally occur after a small uncharged residue (Ala > Gly > Ser = Cys and in single cases each Thr and Trp). In addition, predicted β-turns (chain reversals) are frequently found at or near the cleavage site. The presence of a hydrophilic NH_2 terminus and a β-turn after the hydrophobic domain can be expected to facilitate the insertion of the transport peptide into the membrane in a loop configuration (20, 101, 110), as shown in Figure 2.

Insertion of the hydrophobic signal sequence into the membrane in a loop configuration can be expected to provide sufficient energy to insert an adjacent peptide of roughly equal length into the lipid bilayer. The presence of polar and charged residues in this adjacent peptide destabilizes the peptide's presence in the hydrophobic layer of the membrane (20). Such unfavorable energy conditions result in the transfer of this peptide, which represents the NH_2 terminus of the authentic protein, to the cisternal space of the RER. Once the chain is transferred, the energy requirements for continued chain translocation are minimized under equilibrium conditions because the free energy required for the entrance of new residues into the lipid bilayer is offset by the release of free energy associated with the exit of residues from the lipid bilayer. To minimize free-energy requirements in the transfer of free hydrogen-bond donors and acceptors across the hydrophobic layer of the membrane, polypeptide chain transfer can be expected to occur largely in alpha or 3_{10} helical structures (29).

ASYMMETRIC INTEGRATION OF MEMBRANE PROTEINS INTO THE LIPID BILAYER

Integral membrane proteins may contain one or more hydrophobic domains that span the lipid bilayer. To understand the role of these hydrophobic domains in the topological orientation of membrane proteins, it is instructive to compare start-translocation sequences, which initiate chain translocation across the lipid bilayer, and stop-translocation sequences, which halt such transfer. Figure 3 compares the start-translocation signals for five representative secretory proteins with the stop-translocation signals for five representative transmembrane proteins. Both types of sequences contain a hydrophobic domain in juxtaposition to a hydrophilic domain. However, the important differences between start and stop signals are their orientation and location within the polypeptide chain. Start-transport sequences are usually located at the NH_2 terminus, and a hydrophilic domain precedes the hydrophobic domain. In contrast, stop-transport sequences are located within the interior of the sequence and the hydrophobic domain precedes the hydrophilic domain. In addition, both the hydrophobic and hydrophilic domains in stop signals appear to be more extensively developed than their counterparts in start signals. For example, the hydrophobic sequence in stop signals varies between 15 and 26 residues, whereas the comparable sequence in start signals varies between 9 and 18 residues. Multiple charged amino acids, invariably including Arg and Lys residues, occur within the hydrophilic domains of stop-translocation sequences. The apparent requirement for positively rather than negatively charged residues relates to the higher levels of free energy required to insert the positively charged residues into the lipid bilayer of the membrane in comparison with negatively charged residues (122). In addition, positive charges in the polypeptide chain at this point can be expected to form electrostatic interactions with the negative charges associated with the polar head groups of membrane phospholipids (105). Clusters of negatively charged residues following hydrophobic sequences do not result in interruption of chain transfer across the lipid bilayer, as evidenced by the ability of trypsinogen, which contains four contiguous Asp residues immediately following the NH_2-terminal hydrophobic signal sequence, to translocate across the RER membrane as a secretory protein.

From a knowledge of the relative positions of start- and stop-translocation signals in nascent polypeptide chains, the topological relationships of secretory and membrane proteins with respect to the RER membrane can be predicted. Because of the absence of structural information on membrane proteins in the exocrine pancreas, examples that demonstrate the topological relationships are taken from a wide variety of cellular proteins. The general concepts that have been developed can be expected to pertain to membrane proteins in all eucaryotic cells, including the exocrine pancreas.

Figure 4A shows the potential relationships when the NH_2-terminal start-transport signal, which initiates chain translocation, is cleaved from the parent molecule. In the absence of a subsequent stop-trans-

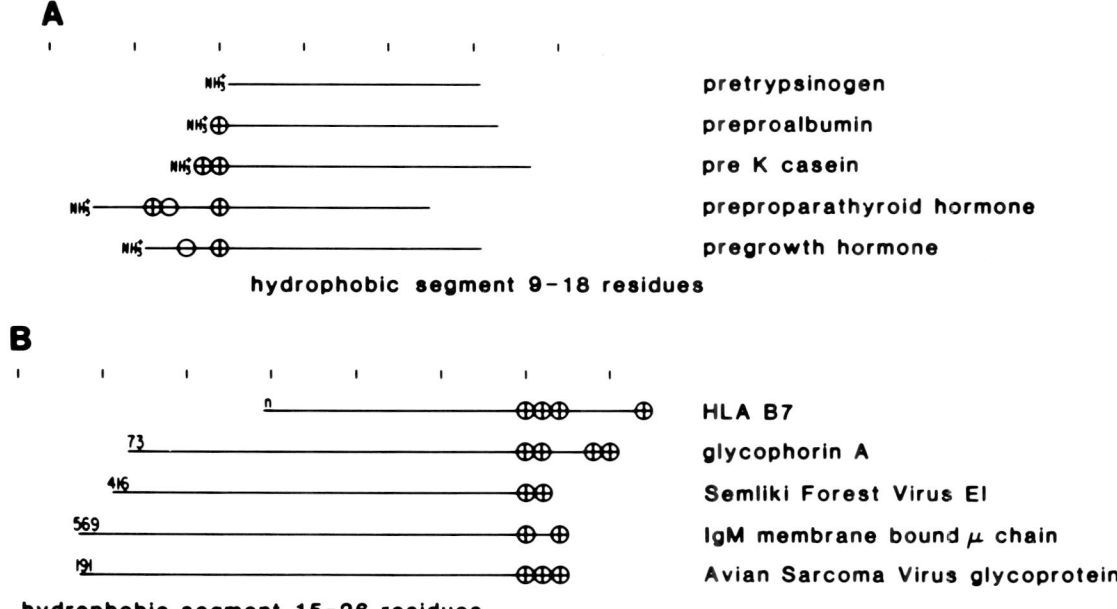

FIG. 3. Schematic comparison of start-translocation (A) and stop-translocation (B) peptide signals. Sequences are lined up at the interface between hydrophilic and hydrophobic domains. Hydrophilic domains contain charged residues shown with the appropriate sign enclosed by a *circle* and partially charged NH$_2$-terminal residues. *Continuous horizontal lines* represent hydrophobic domains. *Vertical lines* (top of each panel) indicate intervals of 5 amino acid residues. Numbers introducing stop-transport sequences indicate the internal position of these signals (residues from NH$_2$ terminus). *Right column*, proteins from which peptide signals were taken. Sequences were selected from the literature as follows: anionic pretrypsinogen (87), preproalbumin (116), pre K casein (70), preproparathyroid hormone (42), pregrowth hormone (114), histocompatibility antigen B7 (88), glycophorin A (125), Semliki forest virus protein E1 (35), membrane-bound form of IgM μ-chain (92), and avian sarcoma virus glycoprotein (22). [From Scheele (105).]

port sequence, the nascent polypeptide is completely transferred across the membrane and exists as either a soluble protein in the cisternal medium (Fig. 4A1) or a peripheral membrane protein associated with the cisternal leaflet of the RER membrane. As described in the following sections, soluble proteins may be routed from the cisternal secretory pathway into secondary and tertiary compartments or, in the case of pancreatic exocrine proteins, secreted into the extracellular medium. In the presence of a subsequent stop-transport signal, the nascent polypeptide chain assumes a transmembrane bitopic configuration with the NH$_2$ terminus directed into the cisternal space (Fig. 4A2). This orientation has been observed in glycophorin, a membrane protein in the erythrocyte (125), the membrane form of the heavy chain of histocompatibility antigens (88), and several viral membrane proteins (50, 93). In the presence of a stop-transport signal followed by an uncleaved start-transport signal, the nascent polypeptide chain assumes a transmembrane tritopic configuration with both the NH$_2$ terminus and COOH terminus directed into the cisternal space (Fig. 4A3). The alternating presence of stop signals and uncleaved start signals gives rise to higher ordered structures representing polytopic membrane proteins. 3-Hydroxy-3-methylglutaryl coenzyme A (HMG CoA) reductase, a smooth endoplasmic reticulum (SER) membrane protein in liver involved in cholesterol synthesis, shows the latter orientation but contains seven hydrophobic helical segments (3 uncleaved start signals and 4 stop signals) that span the lipid bilayer (61).

If the NH$_2$-terminal start signal is uncleaved and remains both attached to the parent molecule and inserted into the lipid bilayer, opposite orientations may be obtained (Fig. 4B). In the absence of a subsequent stop signal, the nascent polypeptide chain assumes a transmembrane bitopic configuration, with the NH$_2$ terminus directed into the cytosolic space (Fig. 4B1). This orientation is characteristic of membrane hydrolases that are observed in the intestinal brush border: aminopeptidase N (33), dipeptidyl peptidase (66), and isomaltase (43). In the presence of a subsequent stop-transport signal, the nascent chain assumes a transmembrane tritopic configuration, with both the NH$_2$ terminus and COOH terminus directed to the cytosolic space (Fig. 4B2). An example of this configuration may be observed in the signal-recognition particle receptor, exclusively confined to the RER membrane (60).

An important variation in this latter configuration is depicted in Figure 4B3, which shows an uncleaved hydrophobic segment near the NH$_2$ terminus that serves as both a start-transport and stop-transport

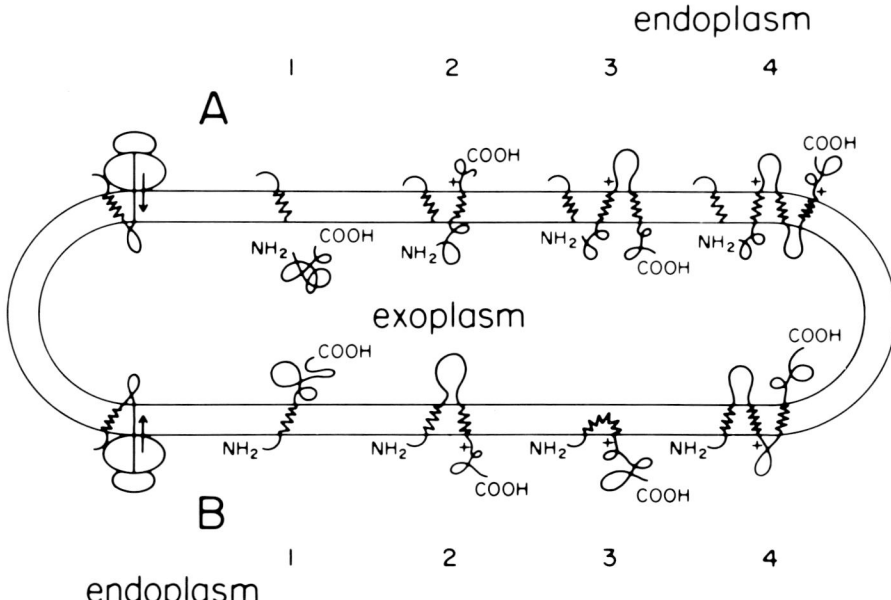

FIG. 4. Asymmetric insertion of membrane proteins. A: potential structures of secretory (A1) and membrane (A2–4) proteins derived after proteolytic cleavage of the start-translocation signal sequence. B: potential structures of membrane proteins derived in the absence of cleavage of the initial start signal. Ribosomes containing nascent polypeptide chains in the process of loop insertion and chain translocation are shown for reference to topological relationships and biogenetic mechanisms. NH$_2$ and COOH termini are shown in relationship to cytosolic (endoplasmic) and cisternal (exoplasmic) compartments. ᗢ, Hydrophobic amino acid residues in start-transport and stop-transport signal sequences. +, High-density regions of basic residues, which in combination with preceding hydrophobic sequences constitute stop-translocation signals. Thermodynamic considerations predict that hydrophobic portions of stop-transport sequences exist in helical configurations within the lipid bilayer. Configurations of start-transport sequences in the lipid bilayer are unknown and depend on whether such sequences interact with integral transmembrane proteins. [From Scheele (105).]

signal. The presence of this combination signal is characteristic of several membrane proteins residing in the endoplasmic reticulum membrane of the liver: cytochrome P-450, NADPH cytochrome P-450 reductase, and epoxide hydrolase (57). Alternating uncleaved start signals and stop signals give rise to higher ordered structures representing polytopic transmembrane proteins with NH$_2$ termini directed into the cytosolic space and the direction of COOH termini dependent on the last signal, either start or stop, in the polypeptide chain. A variant of this last configuration is observed in the anion transport protein (band III) of the erythrocyte membrane (13). The presence of a 40,000-Da domain on the cytoplasmic face of the erythrocyte membrane suggests that the start-transport signal is internal, residing ~0.5 units from the NH$_2$ terminus. Although the orientation of the COOH terminus is unknown, the molecule is believed to span the membrane two or more times.

In the case of ovalbumin, synthesized and secreted by cells in the chick oviduct without cleavage of a peptide signal (85), translocation across the RER membrane may occur by a mechanism related to the bitopic structure shown in Figure 4B1. If the energy of chain folding provides a motive force that exceeds the energy of retention of the transport sequence in the lipid bilayer, then the entire protein is released from the membrane and, on further conformational maturation, may assume a stable structure soluble in an aqueous environment. A similar mechanism resulting in "pull through" of the NH$_2$-terminal start-transport sequence appears to be operative in the case of two transmembrane proteins, p62 of the Semliki forest virus (35) and rhodopsin of the bovine retina (99).

Opsin is of particular interest because, with seven hydrophobic domains that span the lipid bilayer, it has served as a useful model for the study of these integration mechanisms. Friedlander and Blobel (34) used genetic engineering techniques to verify that two of the hydrophobic domains represent bona fide start-translocation signal sequences.

The thermodynamic forces associated with insertion of membrane proteins into the lipid bilayer are of considerable strength. Once full insertion has been achieved within the RER membrane, the general orientation of a membrane protein with respect to endoplasmic and exoplasmic compartments remains fixed. For example, in the case of membrane proteins that are transported to the cell surface, the exoplasmic segment becomes exposed to the external environment.

In contrast to membrane proteins that appear to be

integrated in a cotranslational manner into membranes via NH_2-terminal or internal signal sequences, two membrane proteins, cytochrome-b_5 reductase (12, 74, 80) and cytochrome b_5 (24, 80), which are part of an electron-transport chain involved in the desaturation of fatty acids, appear to be integrated via hydrophobic sequences residing within 25–30 residues of their COOH termini. Each of these proteins is synthesized on free polysomes, is inserted into the endoplasmic surfaces of a variety of cellular membranes in a posttranslational manner, and shows spontaneous insertion into phospholipid membranes. Because 35–40 amino acid residues within the nascent polypeptide chain are occluded by the large ribosomal subunit (68), it is not surprising that chain termination is required to permit these sequences to bind to intracellular membranes.

INTRACISTERNAL PROCESSING AND CELLULAR SORTING OF PROTEINS

The molecular signals that determine the distribution of proteins between the cisternal space of the RER, the extracisternal cytosolic space, or the membrane that separates these two compartments have been discussed in the first part of this chapter. The cisternal RER space is interconnected to other cisternal compartments of the cell (Golgi complex, secretory granules, and lysosomes) largely via membrane fusion and fission events. Soluble proteins and membrane proteins associated with these secondary and tertiary cisternal compartments are believed to be synthesized on membrane-bound ribosomes of the RER. After these proteins are integrated into the membrane or translocated to the cisternal space, a number of intracisternal covalent modifications may be observed that result in structural alterations necessary to determine the ultimate destination and function of these proteins.

Covalent Intracisternal Processing

Exoprocessing events include modification of chain termini, either acylation or cyclization at the NH_2 terminus, or amide formation at the COOH terminus (132). Endoprocessing events include the covalent modification of amino acid side chains or peptide bonds. Within the lumen of the RER, endoprocessing events include proteolytic cleavage of the signal-transport sequence by a membrane-associated complex of six proteins (30) and disulfide bond formation catalyzed by glutathione and a protein disulfide isomerase (75). Secretory proteins in the exocrine pancreas contain an abundance of disulfide bonds.

Certain proteins translocated to cisternal compartments have recently been found to associate with cellular membranes through the attachment of glycophospholipid moieties. For example, a glycophospholipid containing phosphatidylinositol, mannose, and glucosamine residues is attached to alkaline phosphatase, acetylcholinesterase, and the interleukin receptor through an ethanolamine-peptide linkage (32, 63). Without the glycophospholipid moiety these proteins are released from the cellular membrane and secreted into the extracellular medium.

Although the majority of pancreatic secretory products do not contain attached carbohydrate, N-linked glycosylation events in which lipid-linked core oligosaccharides (2 GlcNAc, 9 Man, and 3 Glc residues) are transferred to Asn residues represent important modifications particularly for membrane proteins. Core oligosaccharide transfer occurs cotranslationally, prior to polypeptide chain termination. The oligosaccharide transfer to Asn is specific for the amino acid sequence Asn-X-Ser(Thr) (119). However, only one-third of potential sites are glycosylated and these are generally observed at β-turns accessible at the molecular surface. Oligosaccharide transfer is believed to occur after initiation of protein domain folding in the nascent polypeptide chain. With completion of the peptide chain, further folding may occur that takes into account long-range intramolecular interactions. Addition of carbohydrate to the chain may influence these later folding pathways. The presence of carbohydrate moieties at the surface of proteins may not only stabilize the folded structure of the protein but may also prevent nonspecific protein-protein interactions, not only between adjacent proteins in the same membrane but also between proteins residing in opposing membranes associated with adjacent cells. Because glycosylation mechanisms are largely confined to the cisternal space of the RER and Golgi [addition of single GlcNAc residues in O-linkages to cytosolic proteins and proteins attached to nuclear pore complexes are exceptions (45)], carbohydrate moieties attached to protein segments appear only in the exoplasmic compartments of the cell (cisternal compartments of intracellular organelles or the external compartment of the cell).

Golgi Compartmentation

Figure 5 and Table 2 indicate that modification of core oligosaccharide chains occurs sequentially within discrete subcompartments of the Golgi complex. Earlier electron-microscopic studies showed that the Golgi consists of a parallel series of lamellae with cis elements facing the RER and trans elements facing secretory compartments (condensing vacuoles, secretory granules) or centriolar structures (Figs. 6A and C). Recent electron-microscopic studies following the movement of vesicular stomatitis virus through the cell have shown clearly that movement through the Golgi complex occurs from cis to trans (8). Orci et al. (81) demonstrated an increasing gradient of cholesterol across the Golgi, from cis to trans. This feature allowed Dunphy et al. (27) to successfully fractionate

Golgi elements according to their density characteristics by sucrose-gradient sedimentation. This work demonstrated that individual glycosidases and transferases reside in elements with different densities and allowed them to define three functionally distinct compartments of the Golgi: *1*) cis elements (high density), where lysosomal enzymes are phosphorylated; *2*) mid elements (intermediate density), where Man residues are removed and GlcNAc residues are added; and *3*) trans elements (low density) where Gal, sialic acid, and fucose residues are added.

Functionally distinct subcompartments of the Golgi complex have also been demonstrated by in situ localization using the electron microscope. By the use of monospecific antibodies, GlcNAc transferase 1 has been localized to mid-Golgi elements (26) and galactosyltransferase to trans elements (95). Localization of sugar residues by specific lectins has indicated that concanavalin A, specific for Man residues, binds to cis elements (124); wheat germ agglutinin, specific for GlcNAc and sialic acid residues, binds to both mid and trans elements (124); and ricin, specific for Gal residues, binds to trans elements (41).

Movement of nascent products from the RER to the Golgi complex and between the subcompartments of the Golgi is believed to take place by the fission and fusion of vesicular carriers, predominantly at the dilated rims of the Golgi saccules (31). Balch and co-workers (4, 5) have recently shown the dissociative transfer, via 60- to 80-nm vesicles, of proteins from donor (cis-) to acceptor (mid-) Golgi fractions and have demonstrated that this process depends on ATP and cytosolic components.

In the Golgi complex, O-linked glycosylation events of two general types are also observed. In the first type, GalNAc is linked to Ser or Thr residues and the oligosaccharide chain is extended via the alternating attachment of Gal and GalNAc residues (28). The second type is observed in the synthesis of proteoglycans, where xylose is linked to Ser or Thr residues

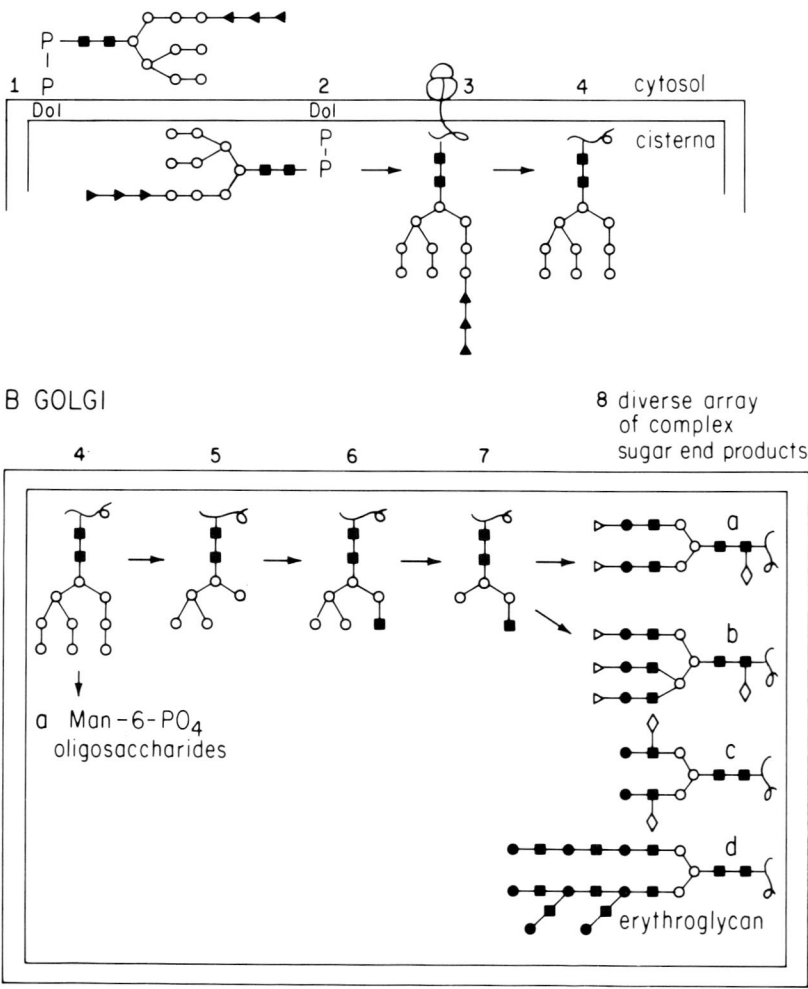

TABLE 2. *Compartmental Localization of Cisternal-Oriented Protein- and Saccharide-Processing Enzymes and Sorting Receptors*

Cellular Compartments	Enzymes and Receptors
(Pre)lysosomal	ATP-dependent proton pump Proteolytic processing of prolysosomal enzymes
Condensing vacuole	ATP-dependent proton pump Proteolytic processing of prosecretory proteins Sulfotransferase
Trans Golgi	ATP-dependent proton pump Sulfotransferase Sialyltransferase Fucosyltransferase Galactosyltransferase
Mid Golgi	GlcNAc transferases I, II, and III Mannosidases I and II
Cis Golgi	GlcNAc-1-phosphotransferase Phosphodiester glycosidase Man-6-PO_4 receptors Fatty acid acylase Xylosyltransferase
Rough endoplasmic reticulum	Xylosyltransferase Mannosidase (removes 1 residue) Glucosidases I, II Core oligosaccharide transferase -KDEL receptor Protein disulfide isomerase Translocation signal protease

and the oligosaccharide chain is extended via disaccharide units consisting of GalNAc and glucuronic acid to form glycosaminoglycan (GAG) structures (25). Glycosaminoglycans are polyanions containing not only the COOH groups of glucuronic acid but also N- and O-linked sulfate groups, normally one per disaccharide unit. Highly sulfated glycosaminoglycans containing 2–3 sulfate groups per disaccharide unit have also been observed. Further work is needed to determine the precise Golgi location of glycosyltransferases participating in O-linked glycosylation. However, the sulfotransferase is known to reside in trans-Golgi and condensing vacuole compartments.

Molecular Sorting of Proteins

Secretory, lysosomal, transmembrane, and cisternal membrane–associated proteins (CMAP) are synthesized by ribosomes bound to the RER membrane. As described earlier in this chapter, these proteins are synthesized in precursor form containing NH_2-terminal transport or signal peptides (prepeptides) that mediate their translocation across or integration into the RER membrane. The cell has evolved a number of mechanisms responsible for the subsequent distribution or sorting of these proteins to specific topological surfaces (membrane domains) or into specific cellular compartments. One relatively simple sorting mechanism is the cisternal retention of membrane-

FIG. 5. Formation of glycoproteins through the attachment of N-linked oligosaccharides. Oligosaccharides of this type occur in 2 general forms, high mannose and complex. Although differing in their final structure, these 2 oligosaccharides have a common biosynthetic origin, being derived from a lipid-linked oligosaccharide precursor. The lipid carrier is a dolichopyrophosphate (Dol-P-P), the lipid moiety of which contains an α-saturated polyisoprenoid moiety containing 20 isoprene units (together constituting an 80-carbon chain with branched methyl groups and 20 unsaturated bonds). *A*: Dol-P is synthesized in the outer mitochondrial membrane and, by an unknown mechanism, is transferred to the RER membrane. Oligosaccharide precursor is first assembled onto the Dol-P by enzymes believed to reside on the cytosolic leaflet of the RER membrane (*step 1*). Assembly involves attachment of 2 *N*-acetylglucosamine (GlcNAc), 9 mannose (Man), and 3 glucose (Glc) residues. Linkage patterns vary in a highly specific and reproducible manner. By an unknown mechanism, the Dol-P-P-oligosaccharide precursor then flips in the membrane bilayer, transferring the oligosaccharide to the cisternal surface of the RER (*step 2*). After transfer of the oligosaccharyl precursor to the protein acceptor (*step 3*), 3 glucosyl residues are removed by glucosidases I and II, which reside as integral membrane proteins at the cisternal surface of the rough and smooth endoplasmic reticulum (*step 4*). *B*: further processing of oligosaccharide chains occurs in the Golgi complex along divergent pathways. Within cis-Golgi elements, oligosaccharides attached to lysosomal enzymes are phosphorylated at position 6 (*step 4a*) by a modification process involving the activity of 2 enzymes, GlcNAc-1-phosphotransferase and phosphodiester glycosidase. Within mid- and trans-Golgi elements, glycoproteins proceed along a pathway that results in formation of a diverse array of complex oligosaccharides. This latter pathway begins with the removal of mannosyl residues from unbranched saccharides (mannosidase I; *step 5*). An addition of a GlcNAc residue (*step 6*) signals the further removal of mannosyl residues from branched saccharides (*step 7*). Then begins the progressive addition of GlcNAc, galactose (Gal), sialic acid, and fucose (Fuc) residues in a variety of combinations, several of which are depicted in *steps 8a–d*. GlcNAc and Gal form the core of these terminal additions. Heterogeneity is generated in the number of branch points (biantennary, *step 8a*; triantennary, *step 8b*), the fucose addition to the proximal GlcNAc residue, and the type of residue that terminates the oligosaccharide chain (sialic acid or fucose, in a mutually exclusive manner). Repeating disaccharides of the Gal-β-1,4-GlcNAc and GlcNAc-β-1,3-Gal types give rise to erythroglycan, a recently discovered proteoglycan associated with the erythrocyte plasma membrane and to a lesser extent with plasma membranes of other cells. Consistent with these cell biological concepts, glycoproteins associated exclusively with the RER contain only the high-mannose oligosaccharide. [Adapted from Hubbard and Ivatt (47) and Scheele (105).]

associated proteins in the RER through their association with membrane receptors residing specifically in this organelle. It has been recently shown that the presence of a COOH-terminal tetrapeptide, -KDEL (-Lys-Asp-Glu-Leu), constitutes a recognition determinant responsible for this association (78). Among

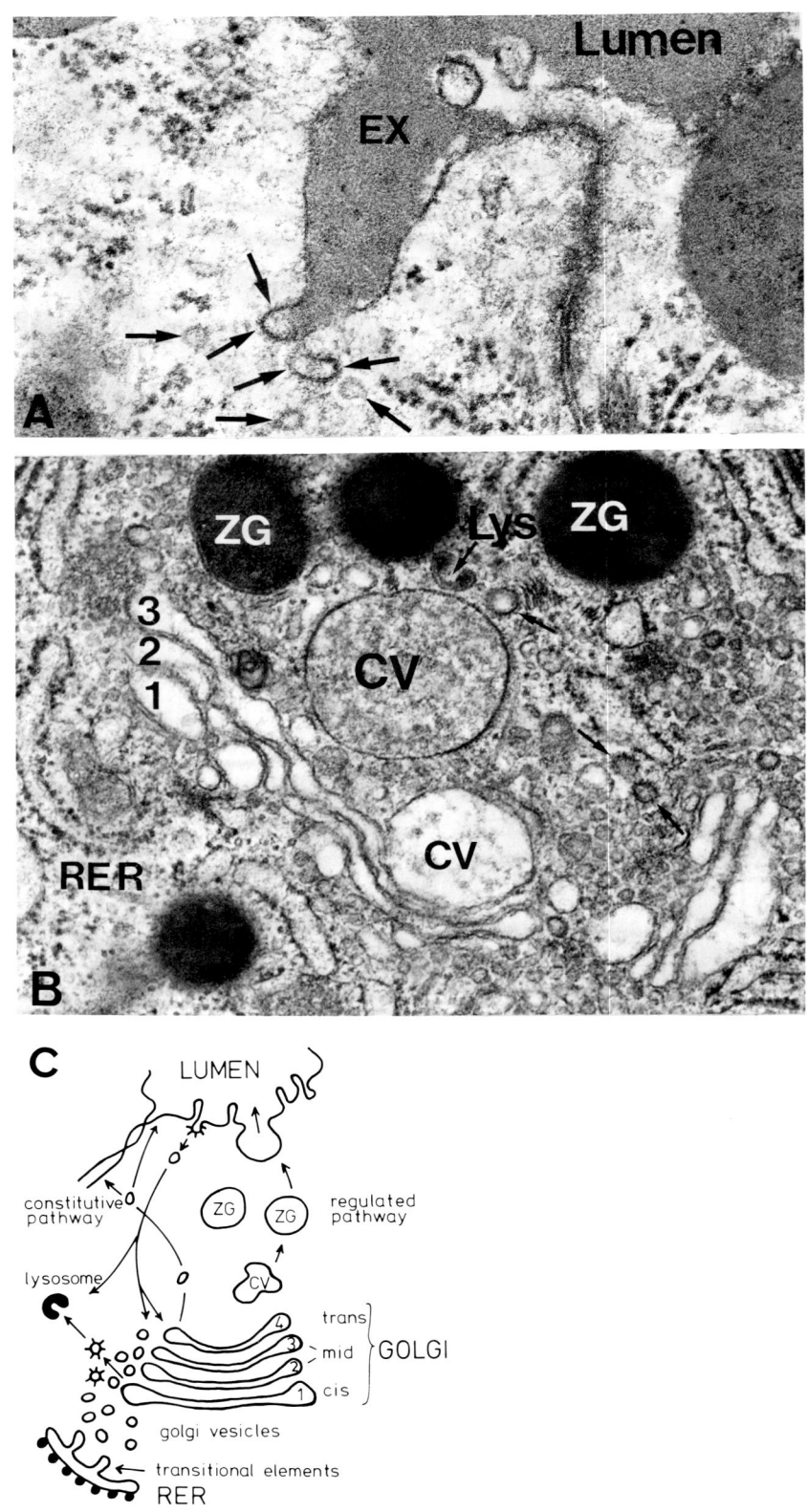

translocation products that exit the RER (secretory, lysosomal, and non-RER CMAPs), additional mechanisms are required for further sorting.

Whereas pancreatic secretory proteins (a mixture of ~20 neutral hydrolases) are transported to the extracellular space via the Golgi complex and secretory granules, lysosomal enzymes (a mixture of ~75 acid hydrolases) are efficiently directed from the Golgi into the lysosomal compartment. This sorting mechanism involves the phosphorylation of lysosomal enzymes in the cis elements of the Golgi by the concerted action of two enzymes: 1) GlcNAc-1-phosphotransferase attaches GlcNAc-1-PO$_4$ to Man residues associated with lysosomal enzymes (126) (5 of the 8 Man residues can be potentially modified), and 2) phosphodiester glycosidase removes the GlcNAc ring structure, leaving the phosphate group attached to the Man residue at position 6 (39). Because isolated high-mannose oligosaccharides are poor substrates for the enzymes and the high-mannose chains of lysosomal enzymes are the same as those observed in glycoproteins that are not phosphorylated and therefore are directed to other areas of the cell, it is believed that the intact protein portion of the lysosomal enzyme contains the recognition determinants for enzymes that phosphorylate these proteins (62).

Lysosomal enzymes containing Man-6-PO$_4$ residues are believed to be transported from cis elements of the Golgi via coated vesicles to endosomal and lysosomal structures. Immunocytochemical localization studies of Man-6-PO$_4$ receptors have been particularly revealing. In many normal cells, including the exocrine pancreas, these receptors are observed in 1–2 cisternae on the cis side of the Golgi, neighboring coated vesicles, lysosomes, and endosomes (15, 16). After tunicamycin treatment, which blocks core glycosylation of these enzymes and results in their secretion, Man-6-PO$_4$ receptors are confined to the cis Golgi and coated vesicles (14). After chloroquine alkalinization of endosomal and lysosomal compartments, receptors are confined to vacuolated endosomes and lysosomes (14). These findings suggest not only that ligand binding triggers movement of receptors to endosomes but also that ligand dissociation, at the low pH of these compartments, triggers their movement back to the cis Golgi.

Some, but not all, of the lysosomal enzymes are sequestered within the cisternal space of the RER, cis Golgi, and coated transport vesicles as proenzyme forms (18). Although the function of their propeptides is unknown, they could be involved in the inhibition of potential enzyme activity or in stabilization of the protein structures prior to their introduction into the acidic environment of the lysosomal system.

Cocompartmentation of acid (lysosomal) and neutral (secretory) hydrolases within the cisternal space of the RER of the exocrine pancreas presents no difficulty, because the pH of this compartment is neutral and each of the neutral proteases is synthesized in inactive precursor or zymogen form. It is clear, however, that molecular sorting mechanisms are not complete. Histochemical studies have indicated that low levels of lysosomal hydrolases, e.g., acid phosphatase, are observed in condensing vacuoles and zymogen granules, usually at the peripheral margins of the granule matrix just below the limiting membrane (79). In addition, Rinderknecht et al. (91) have demonstrated the presence of several acid hydrolases at low levels in human pancreatic juice obtained by cannulation of the main pancreatic duct. The evidence suggests that several of these lysosomal enzymes are copackaged into zymogen granules, because stimulation with cholecystokinin (CCK) resulted in a marked increase in secretion of N-acetyl-β-D-glucosaminidase and arylsulfatase similar to that observed for trypsinogen and amylase.

Sorting of protein products also occurs at the trans face of the Golgi. Certain cell types (plasma, liver

---

FIG. 6. Electron micrographs (A and B) and schematic drawing (C) of apical compartments in the pancreatic acinar cell. Rough endoplasmic reticulum (RER), Golgi, and granules are organized in polarized fashion. RER contains membrane-bound ribosomes and shows transitional elements from which bud smooth-surfaced vesicles (B). The Golgi complex is composed of 3–4 flattened cisternae in stacked array. Numerous 60- to 80-nm vesicles are concentrated on the cis face and at the lateral rims of the Golgi complex. Some vesicles reveal projections on the outer surface of the membrane (coated vesicles indicated by *arrows* in B). The Golgi complex serves as a central membrane compartment to which secretory, lysosomal, and membrane proteins are transported for further processing and targeting to their final destinations. The dominant pathway of protein processing and sorting in the acinar cell is represented by the storage, in granule form, and by the release of large quantities of digestive (pro)enzymes into the extracellular ductular space mediated through a secretagogue-regulated exocytotic pathway (A). Membrane material is recycled (*arrows* in A) to a large extent to the Golgi complex, where it may be utilized again for the secretory process. Membrane vesicles derived from various types of endocytosis are directed to prelysosomal and lysosomal compartments largely for utilization and/or digestion of receptor-recruited ligands from the cellular environment. Thus the numerous vesicles observed around the Golgi complex may constitute a variety of functional subgroups involved in a number of transport processes: *a*) transport of secretory, lysosomal, and membrane proteins from RER to Golgi; *b*) transport of lysosomal (pro)enzymes to (pre)lysosomal compartments; *c*) inter-Golgi transport; *d*) targeting of membrane proteins to polar surfaces of the cell; and *e*) internalization of membrane domains from the cell surface. These cellular pathways are schematically depicted in C. CV, condensing vacuole; EX, exocytosis; LYS, lysosome; ZG, zymogen granule. A: × 44,000; B: × 29,000.

cells) secrete constitutively, whereas other cell types (exocrine, endocrine cells) exhibit a secretory process regulated by hormones and neurotransmitters. Moore et al. (77) demonstrated that AtT-20 cells (endocrine cells derived from a mouse pituitary tumor) in tissue culture display both constitutive and regulated (dibutyryl cAMP-released) secretory pathways. A membrane glycoprotein derived from murine leukemia virus, the precursor to ACTH, and laminin were secreted constitutively; ACTH, a sulfated proteoglycan, and growth hormone or trypsinogen (the last two introduced by gene-transfer techniques) were secreted via the regulated pathway (17). However, the molecular signals involved in direction of secretory products into the constitutive versus regulated pathways are unknown.

Trans-Golgi protein sorting regulates the distribution of membrane proteins to separate plasma membrane domains in polarized epithelial cells where apical (luminal) membranes are separated from basolateral membranes by tight intercellular junctions. At the experimental level, Madin-Darby canine kidney (MDCK) cells cultured on Millipore filters to render both apical and basolateral membranes accessible have proved to be a valuable model for the study of these cellular processes (69). Early studies with this system have indicated that membrane proteins are delivered directly to appropriate surface domains. These findings suggest that plasma membrane proteins exhibit signals, perhaps contained in cytosolic domains, that determine to which surface domain they are transported.

Protein sorting is also involved in the separation of ligands and receptors after receptor-mediated endocytosis. In this process, clathrin cage assembly at the cytoplasmic surface of coated pits (96) is involved in the vesicular internalization of ligands (polypeptide hormones, transferrin, low-density lipoproteins, α_2-macroglobulin, asialoglycoproteins) bound to their cell surface receptors. Once internalized the clathrin coat is enzymatically removed and the vesicle is fused to an endosomal compartment, where a cisternal pH of ~6.5 releases the ligand from its receptor. The endosomal compartment, or CURL (compartment for uncoupling of receptor and ligand), then appears to mature into two structures (36). The vesicular portion containing the dissociated ligand is transformed into a secondary lysosome (see preceding discussion for delivery of lysosomal enzymes to the endosomal compartment). The tubular portion, containing receptors, recycles to the cell surface. Receptor-mediated endocytic mechanisms explain the physiological turnover of polypeptide hormones and could explain, in part, the downregulation of hormone receptors if a portion of internalized receptors are directed into secondary lysosomes. Cellular compartments involved in internalization pathways are less prominent in pancreatic acinar cells than in cells that are specialized for endocytic or phagocytic functions.

Importance of Cisternal pH in the Cellular Sorting of Proteins

A number of cellular compartments contain an ATP-dependent electrogenic H^+ pump (in parallel with a Cl^- conductance), which reduces the pH of the respective cisternal compartments (6, 38). The pharmacological characteristics of this pump distinguish it both from the F_0-F_1 ATPase of mitochondria and the H^+-ATPase of the gastric mucosa. The cisternal pH of trans-Golgi elements and condensing vacuoles is unknown, but sulfotransferase activity and proteolytic processing enzymes show pH optima near 6.5. Endosomes show a cisternal pH of 5.5–6.5 and lysosomes show a pH of ~4.5. Both lysosomotropic agents with weak base characteristics (chloroquine and NH_4Cl) and monovalent cation ionophores that interfere with the H^+ pump (monensin and nigericin) result in alkalinization of these compartments. The structural consequences of these agents are the formation of large vacuoles in Golgi and endosomal compartments. The functional consequences are several. Receptors, both cell surface and Man-6-PO_4, become trapped in endosomal compartments unable to release their ligands at neutral pH, a process that prevents subsequent sorting processes from taking place. Intracellular transport of nascent secretory proteins, proteoglycans, and plasma membrane glycoproteins are markedly inhibited (123). In AtT-20 cells, chloroquine diverted ACTH from the regulated to the constitutive secretory pathway (77). These last findings suggest that sorting of exportable proteins to the regulated pathway may be associated with acidic trapping or precipitation of secretory products.

Condensation of secretory proteins allows the storage, in aggregate form, of large quantities of secretory protein within a minimum of intracellular granule space. Intermediate stages of this process are easily visualized within condensing vacuoles with the electron microscope. The nonosmotic nature of zymogen granules (average diam, 1 μm) indicates that at completion of the concentration process there is extensive formation of heterologous multimetric complexes by the complement of exocrine proteins. Reggio and Dagorn (90) have observed aggregation among pancreatic exocrine proteins, which is dependent on pH and ionic strength and, under optimal conditions, is strong enough to induce precipitation.

Intracisternal Transport and Processing of Secretory Proteins

Jamieson and Palade (49) determined through a series of electron-microscopic radioautography studies the route and time course of intracellular transport of proteins in the exocrine pancreas. By the use of guinea pig pancreatic slices and an in vitro pulse-chase protocol, they determined that proteins pulse-labeled with [^{3}H]leucine first appeared in the area of the cell occupied by the RER. Transport of radiolabeled proteins from one intracisternal compartment to another (RER → Golgi → condensing vacuoles → zymogen granules) did not require ongoing protein synthesis. Transport of pulse-labeled proteins from transitional elements of the RER to small vesicles (60–80 nm) on the cis face of the Golgi complex and from zymogen granules to the extracellular, ductular space required ATP, findings that suggested that membrane fusion-fission events occurred at these two loci.

Scheele, Palade, and Tartakoff (112) employed a pulse-chase protocol followed by an analysis of cell fractions by two-dimensional gel electrophoresis (100) to study the kinetics of intracellular transport of ~20 individual secretory (pro)enzymes. By use of a double-radiolabel protocol, these authors were able to calculate 1) the extent of leakage of secretory proteins from membrane-enclosed compartments along the secretory pathway and 2) the extent of redistribution of leaked molecules to the surfaces of membrane fractions by nonspecific absorption. Correction of the data for leakage and relocation artifacts resulted in intracellular transport kinetics that indicated, in agreement with the radioautographic data, that secretory proteins are transported successively from the RER to the Golgi complex and from there to condensing vacuoles and zymogen granules. These studies indicate that the appearance of negatively charged exportable proteins in the postmicrosomal supernatant fraction could be accounted for by the uniform leakage of proteins during tissue homogenization and the selective reabsorption of positively charged molecules to the negatively charged surfaces of membrane-bound organelles. As such these studies provide strong support for the hypothesis that secretory proteins remain confined to membrane-bound organelles from their time of translocation into the cisternal space of the RER to their time of release, by exocytosis, from the cell.

The kinetic curves, which describe the movement of pulse-labeled proteins through the intermediate (Golgi, condensing vacuole) and late (zymogen granule) compartments of the cell, were broad and showed relatively low peak heights. Scheele and Tartakoff (113) showed that the kinetic dispersion observed during intracellular transport is due to a considerable asynchrony observed in the movement of individual proteins from compartment to compartment along the secretory pathway. Asynchronous transport was manifest on exit from the RER, and the relative kinetic order remained largely unchanged up to the moment of discharge from the cell. In the presence of carbachol or CCK stimulation, the times required for 40% discharge of labeled chymotrypsinogen 2, trypsinogen, amylase, and procarboxypeptidase B were 98, 102, 148, and 180 min, respectively. Transport rates did not correlate with isoelectric point, molecular weight, or the presence of carbohydrate. Recent studies by Keim and Rohr (53) have shown a similar asynchrony in the intracellular transport of proteins in the in vivo rat pancreas and demonstrated that the relative kinetic order of movement of individual proteins was unchanged by secretagogue stimulation. The molecular basis for asynchrony in the exit of proteins from the RER remains to be elucidated.

During cellular conditions of rest or stimulation by secretagogues, the contents of zymogen granules (average diam, 1 µm) are secreted from the acinar cell by exocytosis (83, 102). This process involves the fusion of the granule membrane with the luminal plasmalemma, followed by the formation of an opening within the area of fusion and discharge in bulk of the secretory granule contents (see Fig. 6A). With transmission electron microscopy, relatively few exocytotic images have been observed, suggesting that the exocytotic events occur rapidly. Analysis by freeze fracture (which provides an advantage over transmission electron microscopy in the study of membrane events), however, has revealed considerably increased numbers of exocytotic images in both the mouse and the guinea pig pancreas (121).

Kern and co-workers (55) have demonstrated that each of the steps in the secretory process is modulated by secretagogue stimulation. Optimal caerulein stimulation in the in vivo rat pancreas resulted in a marked depletion in tissue stores of zymogen granules within 2–3 h. After this period the total rate of protein synthesis increased progressively to levels at 24 h that were 160%–220% compared with control levels (9). With a pulse-chase protocol it was observed that in vivo hormonal stimulation accelerated intracellular transport of newly synthesized proteins from RER to Golgi and their packaging into granules (10). Under control conditions the minimal transit time through the sequential cellular compartments was ~30 min; in vivo stimulation for 24 h shortened this period to 10 min (Fig. 7A). This prolonged stimulation resulted in pronounced changes in the configuration of the Golgi complex (marked enlargement of the complex and appearance of proteinaceous material in cisternae) and in a decrease in size of condensing vacuoles and zymogen granules to one-third of their regular size, which is below the resolution of the light microscope. There was no indication that prolonged hormonal stimulation led to a rerouting of secretory proteins or to an omission of one of the regular steps in intracellular transport, as demonstrated by autoradiographic studies at the electron-microscopic level (Fig. 7B). The observed increase in protein synthesis, combined with the observed acceleration in intracellular transport, packaging, and granule discharge, enabled a sustained increase in the rate of exocrine secretion over the period of hormone stimulation (55). These findings provide an explanation for the apparent paradox observed previously when the stimulated pancreas continued to secrete high levels of amylase despite the disappearance of zymogen granules visualized at the light-microscopic level (97).

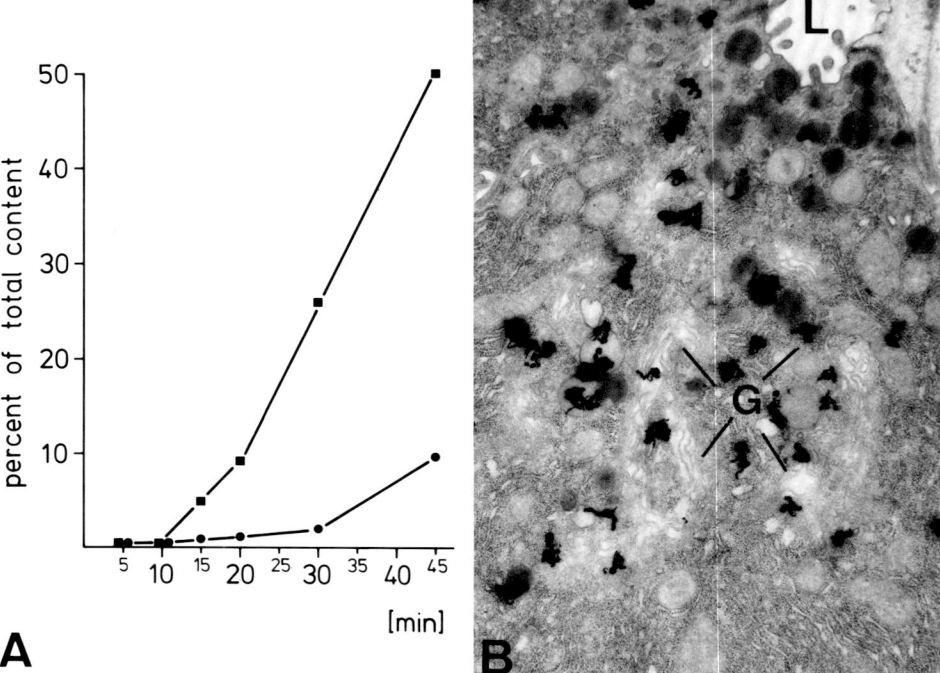

FIG. 7. *A*: kinetics of in vitro discharge of pulse-labeled proteins from rat pancreatic lobules prepared from control animals (●) and animals infused with 0.25 $\mu g \cdot kg^{-1} \cdot h^{-1}$ caerulein for 24 h (■). Values are expressed as percentage of total content released at each time point studied over 45 min. Minimal transit time is 30 min in control animals and 10 min in stimulated animals. *B*: autoradiographic demonstration of pancreatic proteins pulse-labeled for 3 min and chased for 7 min in lobules from an animal stimulated for 24 h, as in *A*. Autoradiographic grains appear over Golgi cisternae (*G*), condensing vacuoles, and granules in proximity to the acinar lumen (*L*). At the same time period in the absence of in vivo hormone stimulation, the majority of grains appear over elements of the RER. × 9,000. [*A* adapted from Bieger et al. (10).]

Under conditions of supramaximal secretagogue stimulation, both in vitro and in vivo, we have observed a redirection in the final step of the secretory pathway (Fig. 8). Exocytotic images disappeared at the luminal plasma membrane, and the luminal space appeared markedly contracted. Within acinar cells, fusion occurred between individual zymogen granules, large vacuolar structures appeared in the vicinity of the Golgi apparatus, and exocytotic images were observed at the lateral plasma membrane (106). The extensive alterations observed in the configurations of cellular compartments were accompanied by activation of lysosomal elements, as evidenced by the appearance of multivesicular bodies and the appearance of lysosomal enzymes in Golgi vacuolar structures (131). These morphological findings were consistent with the near cessation of enzyme secretion into the pancreatic duct and the increase observed in pancreatic enzymes observed in the blood circulation. Such cellular changes were accompanied by inflammation and edema within the pancreatic tissue, which together led to the development of acute edematous pancreatitis (1, 2, 59).

Analysis of the normal exocrine pancreas by immunocytochemical methods at the electron-microscopic level has not revealed qualitative specialization between individual acinar cells or between individual membrane-bound compartments (RER, Golgi, secretory granules) within cells with respect to secretory products (7, 56). The most comprehensive of these studies is that conducted by Bendayan et al. (7), who studied the tissue distribution of nine enzymes. Antibody localization for all nine enzymes was observed in each intracellular compartment and showed increasing concentrations along the RER → Golgi → granule pathway.

The concept of bulk discharge of exocrine proteins by exocytosis has been supported not only by immunolocalization studies but also by biochemical studies. Greene et al. (40) separated the exocrine pancreatic proteins of the cow pancreas by column chromatographic methods and showed that the mass proportions of these proteins in resting pancreatic juice and zymogen granule fractions were identical. Scheele and Palade (111) developed an in vitro system of guinea pig pancreatic lobules and studied the kinetics of discharge of three enzymes and four protease zymogens into the incubation medium under basal conditions and during stimulation with various secretagogues, carbachol, caerulein, and CCK. For each ex-

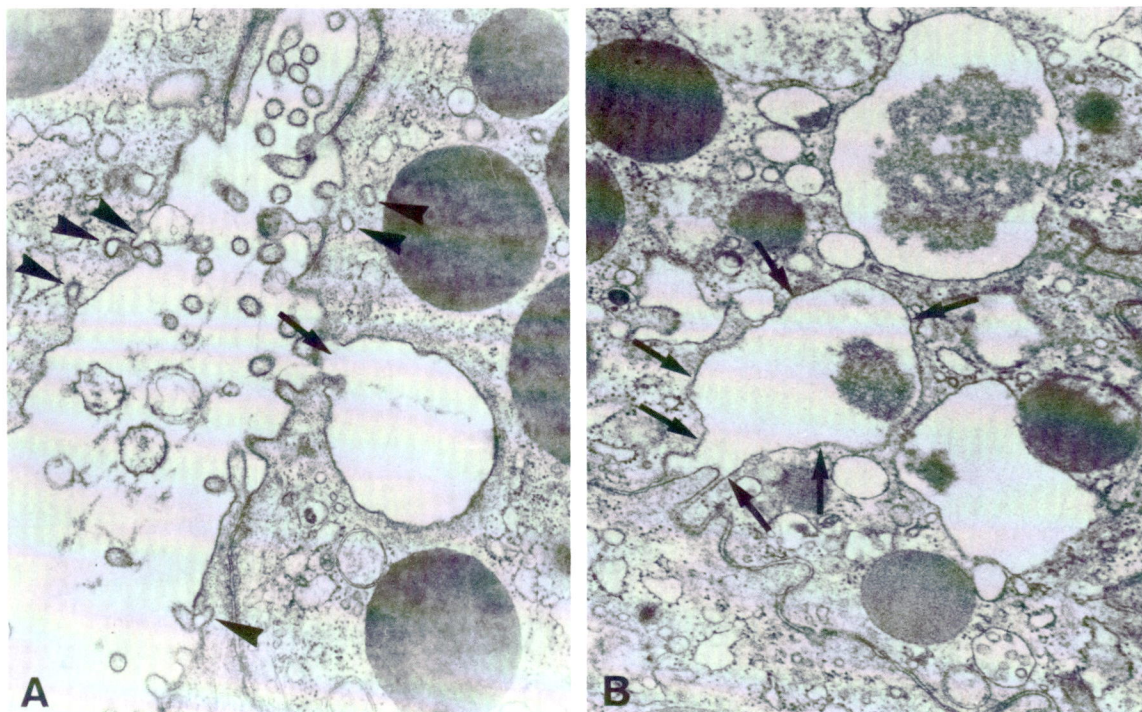

FIG. 8. Redirection of the final step in the secretory pathway during supraoptimal secretagogue stimulation. Under normal conditions (basal and stimulatory) exocytosis occurs exclusively at the luminal plasma membrane as depicted by the micrograph (A), where pancreatic lobules were fixed after stimulation for 15 min with optimal concentrations of caerulein (10^{-9} M). Functional activity at the luminal membrane is indicated by the exocytotic image (arrow) and by the presence of numerous coated pits and vesicles (arrowheads) at or near the membrane surface, respectively. Under conditions of supramaximal stimulation for 15 min with 10^{-7} M caerulein (B), exocytotic images are absent at the luminal membrane but appear in limited numbers at the lateral plasma membrane (opposing arrows). × 25,000. [Adapted from Scheele et al. (106).]

perimental condition investigated, all seven activities, estimated to represent ~89% of the secretory output of the guinea pig pancreas (100), were discharged in constant proportions and were released from the tissue to the same proportional extent (Fig. 9). Steer and co-workers (117, 118) studied the appearance of two zymogens and one enzyme in pancreatic juice from the stimulated in vitro rabbit pancreas. With methacholine or CCK stimulation, the discharges of amylase, trypsinogen, and chymotrypsinogen were parallel during the two 1-h collections before stimulation (basal secretion) and the two 1-h collections after stimulation. Dagorn (23) studied the discharge of proteins from the in vivo rat pancreas in response to CCK, pilocarpine, and caerulein stimulation. Each of the three secretagogues studied discharged the same mixture of amylase, lipase, and chymotrypsinogen, and this mixture was identical to that found in the pancreatic tissue itself prior to stimulation.

In vivo studies in the rat, however, have indicated that the proportions of secretory proteins were changed when basal secretion after an overnight fast was compared with stimulated secretion. For example, Dagorn (23) found that the amylase–chymotrypsinogen ratio decreased ~70% when the pancreas in the anesthetized rat was stimulated with CCK, caerulein, or pilocarpine. Nonparallel findings were most pronounced when basal secretion rates were minimal. Dagorn calculated that the pool of secretory protein responsible for this effect represented 0.35% of the exportable content of the gland. In two recent studies, Keim has demonstrated (51, 52) that the nonparallel findings of Dagorn were due, in part, to the effects of anesthetic agents. Conscious rats were studied after surgical procedures were performed to cannulate the pancreatic duct and recycle the pancreatic juice to the intestinal lumen. Rats were administered isocaloric diets in which carbohydrate was varied inversely with protein for 12 days. Secreted proteins were then analyzed by nondenaturing gel electrophoresis under basal conditions and during intervals of 10 min (through 3 h) after endogenous stimulation (pancreatic juice diversion) or exogenous stimulation (intravenous caerulein, 0.2 $\mu g \cdot kg^{-1} \cdot h^{-1}$, or intraduodenal administration of amino acids, lipid, or 10% sucrose). Although the expected changes were observed in the ratio of enzymes after administration for 12 days of the varied diets (see the other chapter by Scheele and

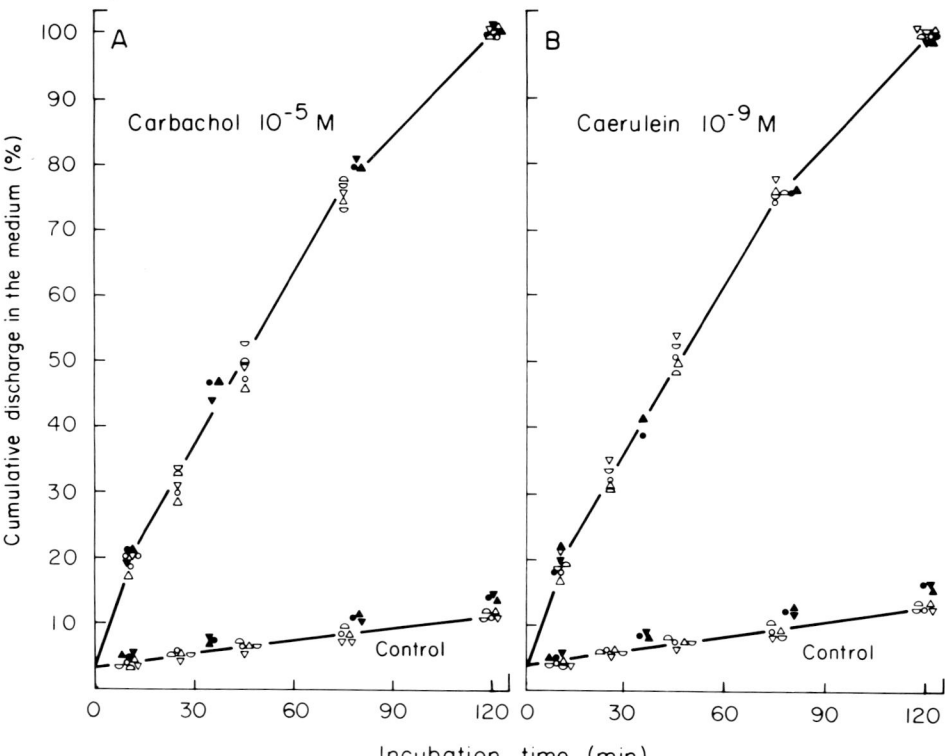

FIG. 9. Kinetics of discharge of 7 secretory proteins from guinea pig pancreatic lobules incubated in the presence and absence of 10^{-5} M carbachol (A) and 10^{-9} M caerulein (B). In a first experiment, discharge of amylase (○) was compared with discharges of chymotrypsinogen (△), trypsinogen (▽), procarboxypeptidase A (⌒), and procarboxypeptidase B (⌣). In a second experiment, discharge of amylase (●) was compared with that of lipase (▲) and ribonuclease (▼). Results are expressed as percentage of enzyme activities released into the medium at each time point relative to the sum of activities retained in tissue and discharged into medium at the end of the incubation period. Values given for the 4 zymogens represent corresponding enzyme activities assayed after enzyme activation. Data are normalized to stimulated values at 2 h. Under these conditions the 7 enzymes were discharged in synchrony and in constant proportions during control or secretagogue stimulation. Enzyme activities reflect equilibrium conditions in the cell. Consequently the asynchrony, which occurs during the intracellular transport of exocrine proteins and reflects the nonequilibrium conditions associated with a pulse-chase protocol, is not observed. Similar findings of parallel discharge of exocrine enzymes have been recently observed under a variety of dietary and hormonal conditions in vivo in the unanesthetized rat (51, 52). [From Scheele and Palade (111).]

Kern in this *Handbook*), no further changes were observed after acute pancreatic stimulation. The in vitro studies in the guinea pig, rat, and rabbit pancreas and the in vivo studies in the rat and cow pancreas indicate that, under physiological conditions and within the time period of a single meal (3 h), secretory proteins are largely discharged from the exocrine pancreas in parallel and in constant proportions.

Under conditions of secretagogue stimulation, the observation of anticoordinate changes at the level of protein synthesis (see the other chapter by Scheele and Kern in this *Handbook*) might appear to be at variance with the general observation of parallel discharge of exocrine proteins under physiological conditions. These two observations, however, provide important insight into the evolutionary development of mechanisms that regulate the secretory process in the exocrine pancreas. To rapidly deliver large quantities of digestive enzymes into the intestinal lumen during food ingestion, the secretory process in the exocrine pancreas has developed an intracellular storage compartment consisting of zymogen granules, the contents of which are released via the actions of physiological secretagogues. Within short periods of time (minutes to hours) the capacity of this storage compartment far exceeds the biosynthetic capacity of acinar cells. Thus the changes that occur at the level of protein synthesis are muted by the presence of large numbers of preformed zymogen granules and the apparent random nature in the selection of zymogen granules during the exocytotic process. This random process is due to the efficient mixing within the cytosolic space of zymogen granules synthesized at different points in time [autoradiography studies at the electron-microscopic level indicate that nascent (labeled) granules mix with preformed (unlabeled) granules in the exocrine pancreas (48)] and within the pancreatic ductal system of secretory products derived from different regions [peri- and teloinsular (67)] of the pancreatic gland. When perturbation of acinar cells by secretagogues or nutritional substrates persists for a longer period of time (hours to days), existing zymogen granule stores are replaced by granules containing secretory proteins synthesized under new conditions, and the composition of secreted proteins reflects the changes observed at the level of protein synthesis. Under these conditions the secreted products are determined by the adaptive changes observed at the level of protein synthesis.

REFERENCES

1. ADLER, G., T. H. HUPP, AND H. F. KERN. Course and spontaneous regression of acute pancreatitis in the rat. *Virchows Arch. A Pathol. Anat. Histol.* 382: 31–47, 1979.
2. ADLER, G., G. ROHR, AND H. F. KERN. Alterations of membrane fusion as a cause of acute pancreatitis in the rat. *Dig. Dis. Sci.* 27: 993–1002, 1982.
3. ANDREWS, D. W., P. WALTER, AND F. P. OTTENSMEYER. Structure of the signal recognition particle by electron microscopy. *Proc. Natl. Acad. Sci. USA* 82: 785–789, 1985.
4. BALCH, W. E., W. G. DUNPHY, W. A. BRAELL, AND J. E. ROTHMAN. Reconstitution of the transport of protein between successive compartments of the Golgi measured by the coupled incorporation of N-acetylglucosamine. *Cell* 39: 405–416, 1984.
5. BALCH, W. E., B. S. GLICK, AND J. E. ROTHMAN. Sequential intermediates in the pathway of intercompartmental transport in a cell-free system. *Cell* 39: 525–536, 1984.
6. BARR, R., K. SAFRANSKI, I. L. SUN, F. L. CRANE, AND D. J. MORRE. An electrogenic proton pump associated with the Golgi apparatus of mouse liver driven by NADH and ATP. *J. Biol. Chem.* 259: 14064–14067, 1984.
7. BENDAYAN, M., J. ROTH, A. PERRELET, AND L. ORCI. Quantitative immunocytochemical localization of pancreatic secretory proteins in subcellular compartments of the rat acinar cell. *J. Histochem. Cytochem.* 28: 149–160, 1980.
8. BERGMANN, J. E., AND S. J. SINGER. Immunoelectron microscopic studies of the intracellular transport of the membrane glycoprotein (G) of vesicular stomatitis virus in infected Chinese hamster ovary cells. *J. Cell Biol.* 97: 1777–1787, 1983.
9. BIEGER, W., A. MARTIN-ACHARD, M. BASSLER, AND H. KERN. Studies on intracellular transport of secretory proteins in the rat exocrine pancreas. IV. Stimulation by in vivo infusion of caerulein. *Cell Tissue Res.* 165: 435–453, 1976.
10. BIEGER, W., J. SEYBOLD, AND H. KERN. Studies on intracellular transport of secretory proteins in the rat exocrine pancreas. V. Kinetic studies on accelerated transport following caerulein infusion in vivo. *Cell Tissue Res.* 170: 203–219, 1976.
11. BLOBEL, G., AND B. DOBBERSTEIN. Transfer of proteins across membranes. I. Presence of proteolytically processed and unprocessed nascent immunoglobulin light chains on membrane-bound ribosomes of murine myeloma. *J. Cell Biol.* 67: 835–851, 1975.
12. BORGESE, N., G. PIETRINI, AND J. MELDOLESI. Localization and biosynthesis of NADH-cytochrome b_5 reductase, an integral membrane protein in rat liver cells. III. Evidence for the independent insertion and turnover of the enzyme in various subcellular compartments. *J. Cell Biol.* 86: 38–45, 1980.
13. BRAELL, W. A., AND H. F. LODISH. The erythrocyte anion transport protein is cotranslationally inserted into microsomes. *Cell* 28: 23–31, 1982.
14. BROWN, W. J., E. CONSTANTINESCU, AND M. G. FARQUHAR. Redistribution of mannose-6-phosphate receptors induced by tunicamycin and chloroquine. *J. Cell Biol.* 99: 320–326, 1984.
15. BROWN, W. J., AND M. G. FARQUHAR. The mannose-6-phosphate receptor for lysosomal enzymes is concentrated in cis Golgi cisternae. *Cell* 36: 295–307, 1984.
16. BROWN, W. J., AND M. G. FARQUHAR. Accumulation of coated vesicles bearing mannose 6-phosphate receptors for lysosomal enzymes in the Golgi region of I-cell fibroblasts. *Proc. Natl. Acad. Sci. USA* 81: 5135–5139, 1984.
17. BURGESS, T. L., C. S. CRAIK, AND R. B. KELLY. The exocrine protein trypsinogen is targeted into the secretory granules of an endocrine cell line: studies by gene transfer. *J. Cell Biol.* 101: 639–645, 1985.
18. CAMPBELL, C. H., AND L. H. ROME. Coated vesicles from rat liver and calf brain contain lysosomal enzymes bound to mannose-6-phosphate receptors. *J. Biol. Chem.* 258: 13347–13352, 1983.
19. CARNE, T., AND G. SCHEELE. Amino acid sequences of transport peptides associated with canine exocrine pancreatic proteins. *J. Biol. Chem.* 257: 4133–4140, 1982.
20. CARNE, T., AND G. SCHEELE. The role of presecretory proteins in the secretory process. In: *The Secretory Process*, edited by M. Cantin. Basel: Karger, 1983, p. 73–101.
21. CONNOLLY, T., AND R. GILMORE. Formation of functional ribosome-membrane junction during translocation requires the participation of a GTP-binding protein. *J. Cell Biol.* 103: 2253–2261, 1986.
22. CZERNILOFSKY, A. P., A. D. LEVINSON, H. E. VARMUS, J. M. BISHOP, E. TISCHER, AND H. M. GOODMAN. Nucleotide sequence of an avian sarcoma virus oncogene (src) and proposed amino acid sequence for the gene product. *Nature Lond.* 287: 198–203, 1980.
23. DAGORN, J. C. Nonparallel enzyme secretion from rat pancreas: in vivo studies. *J. Physiol. Lond.* 280: 435–448, 1978.
24. DAILEY, H. A., AND P. STRITTMATTER. Orientation of the carboxyl and NH_2 termini of the membrane-binding segment of cytochrome b_5 on the same side of phospholipid bilayers. *J. Biol. Chem.* 256: 3951–3955, 1981.
25. DORFMAN, A. Proteoglycan biosynthesis. In: *Cell Biology of Extracellular Matrix*, edited by E. D. Hay. New York: Plenum, 1981, p. 115–138.
26. DUNPHY, W. G., R. BRANDS, AND J. E. ROTHMAN. Attachment of terminal N-acetylglucosamine to asparagine-linked oligosaccharides occurs in central cisternae of the Golgi stack. *Cell* 40: 463–472, 1985.
27. DUNPHY, W. G., E. FRIES, L. J. URBANI, AND J. E. ROTHMAN. Early and late functions associated with the Golgi apparatus reside in distinct compartments. *Proc. Natl. Acad. Sci. USA* 78: 7453–7457, 1981.
28. ELHAMMER, A., AND S. KORNFELD. Two enzymes involved in the synthesis of O-linked oligosaccharides are localized on membranes of different densities in mouse lymphoma BW5147 cells. *J. Cell Biol.* 99: 327–331, 1984.
29. ENGELMAN, D. M., AND T. A. STEITZ. The spontaneous insertion of proteins into and across membranes. *Cell* 23: 411–422, 1981.
30. EVANS, E. A., R. GILMORE, AND G. BLOBEL. Purification of microsomal signal peptidase as a complex. *Proc. Natl. Acad. Sci. USA* 83: 581–585, 1986.
31. FARQUHAR, M. G. Progress in unraveling pathways of Golgi traffic. *Annu. Rev. Cell. Biol.* 1: 447–488, 1985.
32. FATEMI, S. H., R. HAAS, N. JENTOFT, T. L. ROSENBERRY, AND A. M. TARTAKOFF. The glycophospholipid anchor of Thy-1. Biosynthetic labeling experiments with wild type and class E Thy-1 negative lymphomas. *J. Biol. Chem.* 262: 4728–4732, 1987.
33. FERACCI, H., S. MAROUX, J. BONICEL, AND P. DESNUELLE. The amino acid sequence of the hydrophobic anchor of rabbit

intestinal brush border aminopeptidase N. *Biochim. Biophys. Acta* 684: 133–136, 1982.
34. FRIEDLANDER, M., AND G. BLOBEL. Bovine opsin has more than one signal sequence. *Nature Lond.* 318: 338–343, 1985.
35. GAROFF, H., A. FRISCHAUF, K. SIMONS, H. LEHRACH, AND H. DELIUS. Nucleotide sequence of cDNA coding for Semliki forest virus membrane glycoproteins. *Nature Lond.* 288: 236–241, 1980.
36. GEUZE, H. J., J. W. SLOT, G. J. A. M. STROUS, H. LODISH, AND A. L. SCHWARTZ. Intracellular site of asialoglycoprotein receptor-ligand uncoupling: double-label immunoelectron microscopy during receptor-mediated endocytosis. *Cell* 32: 277–287, 1983.
37. GILMORE, R., R. WALTER, AND G. BLOBEL. Protein translocation across the endoplasmic reticulum. Isolation and characterization of the signal recognition particle receptor. *J. Cell Biol.* 95: 470–477, 1982.
38. GLICKMAN, J., K. CROEN, S. KELLY, AND Q. AL-AWQATI. Golgi membranes contain an electrogenic H^+ pump in parallel to a chloride conductance. *J. Cell Biol.* 97: 1303–1308, 1983.
39. GOLDBERG, D. E., AND S. KORNFELD. Evidence for extensive subcellular organization of asparagine-linked oligosaccharide processing and lysosomal enzyme phosphorylation. *J. Biol. Chem.* 258: 3159–3165, 1983.
40. GREENE, L. J., C. H. W. HIRS, AND G. PALADE. On the protein composition of bovine pancreatic zymogen granules. *J. Biol. Chem.* 238: 2054–2070, 1963.
41. GRIFFITHS, G., R. BRANDS, B. BURKE, D. LOUVARD, AND G. WARREN. Viral membrane proteins acquire galactose in trans Golgi cisternae during intracellular transport. *J. Cell Biol.* 95: 781–792, 1982.
42. HABENER, J. F., M. ROSENBLATT, B. KEMPER, H. M. KRONENBERG, A. RICH, AND J. T. POTTS. Pre-proparathyroid hormone: amino acid sequence, chemical synthesis and some biological studies of the precursor region. *Proc. Natl. Acad. Sci. USA* 75: 2616–2620, 1978.
43. HAURI, H.-P., A. QUARONI, AND K. J. ISSELBACHER. Biogenesis of intestinal plasma membrane: posttranslational route and cleavage of sucrase-isomaltose. *Proc. Natl. Acad. Sci. USA* 76: 2616–2620, 1979.
44. HEIDENHAIN, R. Beitraege zur Kenntniss des Pankreas. *Pfluegers Arch. Gesamte Physiol. Menschen Tiere* 10: 557–632, 1875.
45. HOLT, G. D., C. M. SNOW, L. GERACE, AND G. W. HART. Localization of glycosylation to the cytosolic faces of the nuclear pore complex (Abstract). *J. Cell Biol.* 103: 320a, 1986.
46. HORTSCH, M., D. AVOSSA, AND D. I. MEYER. Characterization of secretory protein translocation: ribosome-membrane interaction in endoplasmic reticulum. *J. Cell Biol.* 103: 241–253, 1986.
47. HUBBARD, S. C., AND R. J. IVATT. Synthesis and processing of asparagine-linked oligosaccharides. *Annu. Rev. Biochem.* 50: 555–583, 1981.
48. JAMIESON, J. D., AND G. E. PALADE. Intracellular transport of secretory proteins in the pancreatic exocrine cell. II. Transport to condensing vacuoles and zymogen granules. *J. Cell Biol.* 34: 597–615, 1967.
49. JAMIESON, J. D., AND G. E. PALADE. Production of secretory proteins in animal cells. In: *International Cell Biology, 1976–1977*, edited by B. R. Brinkley and K. R. Porter. New York: Rockefeller Univ. Press, 1977, p. 308–317.
50. JOU, W. M., M. VERHOEYEN, R. DEVOS, E. SAMAN, R. FANG, D. HUYLEBROECK, M. CAREY, AND S. EMTAGE. Complete structure of the hemagglutinin gene from the human influenza A/Victoria/3/75 (H3N2) strain as determined from cloned DNA. *Cell* 19: 683–696, 1980.
51. KEIM, V. Rapid adaptation of pancreatic enzyme secretion in the conscious rat. I. Influence of endogenous and exogenous stimulation. *Ann. Nutr. Metab.* 30: 104–112, 1986.
52. KEIM, V. Rapid adaptation of pancreatic enzyme secretion in the conscious rat. II. Effects of fasting and dietary modulation. *Ann. Nutr. Metab.* 30: 113–119, 1986.
53. KEIM, V., AND G. ROHR. Influence of secretagogues on asynchronous secretion of newly synthesized pancreatic proteins in the conscious rat. *Pancreas* 2: 562–567, 1987.
54. KERFELEC, R., A. PUIGSERVER, K. S. LAFORGE, AND G. SCHEELE. Primary structures of canine pancreatic lipase and phospholipase A2 messenger RNA's. *Pancreas* 1: 430–437, 1986.
55. KERN, H., G. ADLER, AND G. SCHEELE. Structural and biochemical characterization of maximal and supramaximal hormonal stimulation in the rat exocrine pancreas. *Scand. J. Gastroenterol.* 20: 20–29, 1985.
56. KRAEHENBUHL, J. P., L. RACINE, AND J. D. JAMIESON. Immunocytochemical localization of secretory proteins in bovine pancreatic exocrine cells. *J. Cell Biol.* 72: 406–423, 1977.
57. KREIBICH, G., D. SABATINI, AND M. ADESNICK. Biosynthesis of hepatocyte endoplasmic reticulum proteins. *Methods Enzymol.* 96: 530–542, 1983.
58. KURZCHALIA, T. V., M. WIEDMANN, A. S. GIRSHOVICH, E. S. BOCHKAREVA, H. BIELKA, AND T. A. RAPOPORT. The signal sequence of nascent preprolactin interacts with the 54K polypeptide of the signal recognition particle. *Nature Lond.* 320: 634–636, 1986.
59. LAMPEL, M., AND H. F. KERN. Acute interstitial pancreatitis in the rat induced by excessive doses of pancreatic secretagogues. *Virchows Arch. A Pathol. Anat. Histol.* 373: 97–117, 1977.
60. LAUFFER, L., P. D. GARCIA, R. N. HARKINS, L. COUSSENS, A. ULLRICH, AND P. WALTER. Topology of signal recognition particle receptor in endoplasmic reticulum membrane. *Nature Lond.* 318: 334–338, 1985.
61. LISCUM, L., J. FINER-MOORE, R. M. STROUD, K. L. LUSKEY, M. S. BROWN, AND J. L. GOLDSTEIN. Domain structure of 3-hydroxy-3-methylglutaryl coenzyme A reductase, a glycoprotein of the endoplasmic reticulum. *J. Biol. Chem.* 260: 522–530, 1985.
62. LONG, L., M. REITMAN, J. TANG, R. M. ROBERTS, AND S. KORNFELD. Lysosomal enzyme phosphorylation. Recognition of a protein-dependent determinant allows specific phosphorylation of oligosaccharides present on lysosomal enzymes. *J. Biol. Chem.* 259: 14663–14671, 1984.
63. LOW, M. Biochemistry of the glycosyl phospholipid inositol membrane protein anchor. *Biochem. J.* 244: 1–13, 1987.
64. MACDONALD, R. J., S. J. STARY, AND G. H. SWIFT. Two similar but nonallelic rat pancreatic trypsinogens. Nucleotide sequences of the cloned cDNAs. *J. Biol. Chem.* 257: 9724–9732, 1982.
65. MACDONALD, R. J., S. J. STARY, AND G. H. SWIFT. Rat pancreatic ribonuclease messenger RNA. The nucleotide sequence of the entire mRNA and the derived amino acid sequence of the pre-enzyme. *J. Biol. Chem.* 257: 14582–14585, 1982.
66. MACNAIR, D. C., AND A. J. KENNEY. Proteins of the kidney microvillar membrane. The amphipathic form of dipeptidyl peptidase IV. *Biochem. J.* 179: 379–395, 1979.
67. MALAISSE-LAGAE, F., M. RAVAZZOLA, P. ROBBERECHT, A. VANDERMEERS, W. J. MALAISSE, AND L. ORCI. Exocrine pancreas: evidence for topographic partition of secretory function. *Science Wash. DC* 190: 795–797, 1975.
68. MALKIN, L. I., AND A. RICH. Partial resistance of nascent chains to proteolytic digestion due to ribosomal shielding. *J. Mol. Biol.* 26: 329–346, 1967.
69. MATLIN, K. S., AND K. SIMONS. Sorting of an apical plasma membrane glycoprotein occurs before it reaches the cell surface in cultured epithelial cells. *J. Cell Biol.* 99: 2131–2139, 1984.
70. MERCIER, J. C., G. HAZE, P. GAYE, AND D. HUE. Amino terminal sequence of the precursor of ovine β-lactoglobulin. *Biochem. Biophys. Res. Commun.* 82: 1236–1245, 1978.
71. MEYER, D. I. Signal recognition particle (SRP) does not mediate a translational arrest of nascent secretory proteins in mammalian cell-free systems. *EMBO J.* 4: 2031–2033, 1985.
72. MEYER, D. I., AND B. DOBBERSTEIN. Identification and characterization of a membrane component essential for the translocation of nascent proteins across the membrane of the en-

doplasmic reticulum. *J. Cell Biol.* 87: 503–508, 1980.
73. MEYER, D. I., E. KRAUSE, AND B. DOBBERSTEIN. Secretory protein translocation across membranes. The role of the "docking protein." *Nature Lond.* 297: 647–650, 1982.
74. MIHARA, K., R. SATO, R. SAKAKIBARA, AND H. WADA. Reduced nicotinamide adenine dinucleotide-cytochrome b_5 reductase: location of the hydrophobic membrane binding region at the carboxy-terminal end and the masked amino terminus. *Biochemistry* 17: 2829–2834, 1978.
75. MILLS, E. N. C., N. LAMBERT, AND R. B. FREEDMAN. Identification of protein disulphide-isomerase as a major acidic polypeptide in rat liver microsomal membranes. *Biochem. J.* 213: 245–248, 1983.
76. MILSTEIN, C., G. G. BROWNLEE, T. M. HARRISON, AND M. B. MATHEWS. A possible precursor of immunoglobulin light chains. *Nature New Biol.* 239: 117–120, 1972.
77. MOORE, H.-P., B. GUMBINER, AND R. B. KELLY. Chloroquine diverts ACTH from a regulated to a constitutive secretory pathway in AtT-20 cells. *Nature Lond.* 302: 434–436, 1983.
78. MUNRO, S., AND H. B. R. PELHAM. A C-terminal signal prevents secretion of luminal ER proteins. *Cell* 48: 899–907, 1987.
79. NOVIKOFF, A. B., E. ESSNER, S. GOLDFISCHER, AND M. HAUS. Nucleosidephosphate activities of cytomembranes. In: *The Interpretation of Ultrastructure*, edited by R. J. C. Harris. New York: Academic, 1962, p. 149–192.
80. OKADA, Y., A. B. FREY, T. M. GUENTHER, F. OESCH, D. D. SABATINI, AND G. KREIBICH. Studies on the biosynthesis of microsomal membrane proteins. Site of synthesis and mode of insertion of cytochrome b_5, cytochrome b_5 reductase, cytochrome P-450 and epoxide hydratase. *Eur. J. Biochem.* 122: 393–402, 1982.
81. ORCI, L., R. MONTESANO, P. MEDA, F. MALAISSE-LAGAE, D. BROWN, A. PERRELET, AND P. VASSALLI. Heterogenous distribution of filipin-cholesterol complexes across the cisternae of the Golgi apparatus. *Proc. Natl. Acad. Sci. USA* 78: 293–297, 1981.
82. ORCI, L., M. RAVAZZOLA, AND R. G. W. ANDERSON. The condensing vacuole of exocrine cells is more acidic than the mature secretory vesicle. *Nature Lond.* 326: 77–79, 1987.
83. PALADE, G. E. Functional changes in the structure of cell components. In: *Subcellular Particles*, edited by T. Hayashi. New York: Ronald, 1959, p. 64–83.
84. PALADE, G. Intracellular aspects of the process of protein synthesis. *Science Wash. DC* 189: 347–358, 1975.
85. PALMITER, R., D. J. GAGNON, AND K. A. WALSH. Ovalbumin: a secreted protein without a transient hydrophobic leader sequence. *Proc. Natl. Acad. Sci. USA* 75: 94–98, 1978.
86. PINSKY, S., S. LAFORGE, V. LUC, AND G. SCHEELE. Identification of cDNA clones encoding specific secretory isoenzyme forms: determination of the primary structure of dog pancreatic prechymotrypsinogen 2 mRNA. *Proc. Natl. Acad. Sci. USA* 80: 7486–7490, 1983.
87. PINSKY, S., S. LAFORGE, AND G. SCHEELE. Differential regulation of trypsinogen mRNA translation: full-length mRNA sequences encoding two oppositely charged trypsinogen isoenzymes in the dog pancreas. *Mol. Cell. Biol.* 5: 2669–2676, 1985.
88. PLOEGH, H. L., H. T. ORR, AND J. L. STROMINGER. Molecular cloning of a human histocompatibility antigen cDNA fragment. *Proc. Natl. Acad. Sci. USA* 77: 6081–6085, 1980.
89. QUINTO, C., M. QUIROGA, W. F. SWAIN, W. C. NIKOVITS, D. N. STANDRING, R. L. PICTET, P. VENEZUELA, AND W. J. RUTTER. Rat preprocarboxypeptidase A: cDNA sequence and preliminary characterization of the gene. *Proc. Natl. Acad. Sci. USA* 79: 31–35, 1982.
90. REGGIO, H., AND J. C. DAGORN. Packaging of pancreas secretory proteins in the condensing vacuoles of the Golgi complex. In: *Biology of Normal and Cancerous Exocrine Pancreatic Cells*, edited by A. Ribet, I. Pradayrol, and C. Susini. New York: Elsevier/North-Holland, 1980, p. 229–244. (INSERM Symp. 15.)

91. RINDERKNECHT, H., I. G. RENNER, AND H. H. KOYAMA. Lysosomal enzymes in pure pancreatic juice from normal healthy volunteers and chronic alcoholics. *Dig. Dis. Sci.* 24: 180–186, 1979.
92. ROGERS, J., P. EARLY, C. CARTER, K. CALAME, M. BOND, L. HOOD, AND R. WALL. Two mRNAs with different 3' ends encode membrane-bound and secreted forms of immunoglobulin μ chain. *Cell* 22: 303–312, 1980.
93. ROSE, J., W. WELCH, B. SEFTON, F. ESCH, AND N. LING. Vesicular stomatitis virus glycoprotein is anchored in the viral membrane by a hydrophobic domain near the COOH terminus. *Proc. Natl. Acad. Sci. USA* 77: 3884–3888, 1980.
94. ROSENFELD, M. G., E. E. MARCANTONIO, J. HAKIMI, V. M. ORT, P. H. ATKINSON, D. D. SABATINI, AND G. KREIBICH. Biosynthesis and processing of ribophorins in the endoplasmic reticulum. *J. Cell Biol.* 99: 1076–1082, 1984.
95. ROTH, J., J. M. LUCOCQ, E. G. BERGER, J. C. PAULSON, AND W. M. WATKINS. Terminal glycosylation is compartmentalized in the Golgi apparatus (Abstract). *J. Cell Biol.* 99: 229a, 1984.
96. ROTHMAN, J. E., AND S. L. SCHMID. Enzymatic recycling of clathrin from coated vesicles. *Cell* 46: 5–9, 1986.
97. ROTHMAN, S. S. Enzyme secretion in the absence of zymogen granules. *Am. J. Physiol.* 228: 1828–1834, 1975.
98. ROTHMAN, S. S., AND L. D. ISENMAN. Secretion of digestive enzyme derived from two parallel intracellular pools. *Am. J. Physiol.* 226: 1082–1087, 1974.
99. SCHECTER, I., Y. BURNSTEIN, R. ZEMELL, E. ZIV, F. KANTOR, AND D. PAPERMASTER. Messenger RNA of opsin from bovine retina: isolation and partial sequence of the in vitro translation product. *Proc. Natl. Acad. Sci. USA* 76: 2654–2658, 1979.
100. SCHEELE, G. A. Two dimensional gel analysis of soluble proteins; characterization of guinea pig exocrine pancreatic proteins. *J. Biol. Chem.* 250: 5375–5385, 1975.
101. SCHEELE, G. Biosynthesis, segregation, and secretion of exportable proteins by the exocrine pancreas. *Am. J. Physiol.* 238 (*Gastrointest. Liver Physiol.* 1): G467–G477, 1980.
102. SCHEELE, G. Pancreatic zymogen granules. In: *The Secretory Granule*, edited by A. M. Poisner and J. M. Trifaro. New York: Elsevier, 1982, p. 213–246.
103. SCHEELE, G. Methods for the study of protein translocation across the RER membrane using the reticulocyte lysate translation system and canine pancreatic microsomal membranes. *Methods Enzymol.* 96: 94–111, 1983.
104. SCHEELE, G. Pancreatic lobules in the in vivo study of pancreatic acinar cell function. *Methods Enzymol.* 98: 17–28, 1983.
105. SCHEELE, G. Cellular processing of proteins in the exocrine pancreas. In: *The Exocrine Pancreas: Biology, Pathobiology, and Diseases*, edited by V. L. W. Go, J. D. Gardner, F. P. Brooks, E. Lebenthal, E. P. DiMagno, and G. A. Scheele. New York: Raven, 1986, p. 69–85.
106. SCHEELE, G. A., G. ADLER, AND H. F. KERN. Exocytosis occurs at the lateral plasma membrane of the pancreatic acinar cell during supramaximal secretagogue stimulation. *Gastroenterology* 92: 245–253, 1987.
107. SCHEELE, G., AND P. BLACKBURN. Role of the mammalian ribonuclease inhibitor in cell-free protein synthesis. *Proc. Natl. Acad. Sci. USA* 76: 4898–4902, 1979.
108. SCHEELE, G., AND R. JACOBY. Conformational changes associated with proteolytic processing of presecretory proteins allow glutathione-catalyzed formation of native disulfide bonds. *J. Biol. Chem.* 257: 12277–12282, 1982.
109. SCHEELE, G., AND R. JACOBY. Proteolytic processing of presecretory proteins is required for development of biological activities in pancreatic exocrine proteins. *J. Biol. Chem.* 258: 2005–2009, 1982.
110. SCHEELE, G., R. JACOBY, AND T. CARNE. Mechanism of compartmentation of secretory proteins. Transport of exocrine pancreatic proteins across the microsomal membrane. *J. Cell Biol.* 87: 611–628, 1980.
111. SCHEELE, G., AND G. PALADE. Studies on the pancreas of the guinea pig. Parallel discharge of exocrine enzyme activities. *J.*

Biol. Chem. 250: 2660–2670, 1975.
112. SCHEELE, G. A., G. E. PALADE, AND A. M. TARTAKOFF. Cell fractionation studies on the guinea pig pancreas. Redistribution of exocrine proteins during tissue homogenization. *J. Cell Biol.* 78: 110–130, 1978.
113. SCHEELE, G., AND T. TARTAKOFF. Exit of nonglycosylated secretory proteins from the RER is asynchronous in the exocrine pancreas. *J. Biol. Chem.* 260: 926–931, 1985.
114. SEEBURG, P. H., J. SHINE, J. A. MARTIAL, J. D. BAXTER, AND H. M. GOODMAN. Nucleotide sequence and amplification of structural gene for rat growth hormone. *Nature Lond.* 240: 486–494, 1977.
115. SIEGEL, V., AND P. WALTER. Removal of the Alu structural domain from signal recognition particle leaves its protein translocation activity intact. *Nature Lond.* 320: 81–84, 1986.
116. STRAUS, A. W., C. D. BENNETT, A. M. DONOHUE, J. A. RODKEY, AND A. W. ALBERTS. Rat liver preproalbumin. Complete amino acid sequence of the prepiece. Analysis of the direct translation product of albumin messenger RNA. *J. Biol. Chem.* 252: 6846–6855, 1977.
117. STEER, M. L., AND G. GLAZER. Parallel secretion of digestive enzymes by the in vitro rabbit pancreas. *Am. J. Physiol.* 231: 1860–1865, 1976.
118. STEER, M. L., AND T. MANABE. Cholecystokinin-pancreozymin induces the parallel discharge of digestive enzymes from the in vitro rabbit pancreas. *J. Biol. Chem.* 254: 7228–7229, 1979.
119. STRUCK, D. K., AND W. J. LENNARZ. In: *The Biochemistry of Glycoproteins and Proteoglycans*, edited by W. J. Lennarz. New York: Plenum, 1980, p. 35–83.
120. SWIFT, G., C. CRAIK, S. STARY, C. QUINTO, R. LAHAIE, W. RUTTER, AND R. MACDONALD. Structure of the two related elastase genes expressed in the rat pancreas. *J. Biol. Chem.* 259: 14271–14278, 1984.
121. TANAKA, Y., P. DECAMILLI, AND J. MELDOLESI. Membrane interactions between secretion granules and plasmalemma in three exocrine glands. *J. Cell Biol.* 84: 438–453, 1980.
122. TANFORD, C. *The Hydrophobic Effect: Formation of Micelles and Biological Membranes* (2nd ed.). New York: Wiley, 1980.
123. TARTAKOFF, A. Perturbation of vesicular traffic with the carboxylic ionophore monensin. *Cell* 32: 1026–1028, 1983.
124. TARTAKOFF, A. M., AND P. VASSALLI. Lectin binding sites as markers of Golgi subcompartments: proximal to distal maturation of oligosaccharides. *J. Cell Biol.* 97: 1243–1248, 1983.
125. TOMITA, M., AND V. T. MARCHESI. Amino acid sequences and oligosaccharide attachment sites of human erythrocyte glycophorin. *Proc. Natl. Acad. Sci. USA* 72: 2964–2968, 1975.
126. WAHEED, A., A. HASLIK, AND K. VON FIGURA. UDP-N-acetylglucosamine: lysosomal enzyme precursor N-acetylglucosamine-1-phosphotransferase. Partial purification and characterization of the rat liver Golgi enzyme. *J. Biol. Chem.* 257: 12322–12331, 1982.
127. WALTER, P., AND P. BLOBEL. Purification of a membrane-associated protein complex required for protein translocation across the endoplasmic reticulum. *Proc. Natl. Acad. Sci. USA* 77: 7112–7116, 1980.
128. WALTER, P., AND G. BLOBEL. Translocation of proteins across the endoplasmic reticulum. III. Signal recognition protein (SRP) causes signal sequence-dependent and site-specific arrest of chain elongation that is released by microsomal membranes. *J. Cell Biol.* 91: 557–561, 1981.
129. WALTER, P., AND G. BLOBEL. Signal recognition particle contains a 7S RNA essential for protein translocation across the endoplasmic reticulum. *Nature Lond.* 299: 691–698, 1982.
130. WARREN, G., AND B. DOBBERSTEIN. Protein transfer across microsomal membranes reassembled from separated membrane components. *Nature Lond.* 273: 569–571, 1978.
131. WATANABE, O., F. M. BACCINO, M. L. STEER, AND J. MELDOLESI. Supramaximal caerulein stimulation and ultrastructure of rat pancreatic acinar cell: early morphological changes during development of experimental pancreatitis. *Am. J. Physiol.* 246 (*Gastrointest. Liver Physiol.* 9): G457–G467, 1984.
132. WOLD, F. In vivo chemical modification of proteins (posttranslational modification). *Annu. Rev. Biochem.* 50: 783–814, 1981.

CHAPTER 25

Selective regulation of gene expression in the exocrine pancreas

GEORGE A. SCHEELE | Laboratory of Cell and Molecular Biology, The Rockefeller University, New York, New York

HORST F. KERN | Department of Cell Biology, Philipps University, Marburg, Federal Republic of Germany

CHAPTER CONTENTS

Differential Regulation of Gene Expression
   Separation of gene products by two-dimensional gel electrophoresis
   Adaptation to inverse changes in nutritional substrates in the diet
   Response to hormonal stimulation
      Cholecystokinin-like hormones
      Secretin
      Insulin
      Glucocorticoids
   Pavlov's dietary adaptation is mediated by specific hormones
   Nutritional shock
      Protein deficiency
      Glucose deprivation
Gene Expression is Regulated at Multiple Levels
   Prenatal and postnatal development
   Nutritional substrates and hormones
Mechanisms in Selective Regulation of Gene Transcription
   Processing signals in eucaryotic genes
   Promoter elements
   Enhancer elements
      Tissue-specific enhancers
      Enhancer elements responsive to stimuli in the cellular environment
Selective Regulation of mRNA Translation Efficiency: Possible Mechanisms

IN THE EARLY PHASE of embryonic development, the pancreatic rudiment exists in a protodifferentiated state. The presence of small quantities of secretory proteins (50) and their corresponding mRNAs (28) indicates a low level of corresponding gene expression during this period. In the transition from the protodifferentiated to the early differentiated state, cytodifferentiation results in the appearance of intracellular organelles required for high levels of synthesis and storage of exportable products within acinar cells (50). Over the same period there is a dramatic increase in the activity of the genes that code for these differentiated products. Gene activation continues progressively, resulting in 3- to 4-log increases in specific mRNAs and their corresponding protein products by birth. This increase in tissue-specific genetic activity during the embryonic period is believed to occur mainly in response to intrinsic genetic programs whose regulatory mechanisms are largely unknown.

After birth the organism becomes increasingly dependent on its own abilities to adapt to changing nutritional conditions. The appearance of polypeptide hormone receptors in the basolateral membrane of the acinar cell within two days after birth in the rat allows the pancreas to respond to hormonal signals from the digestive tract (11). Among the many hormones liberated from the gut mucosa and pancreatic islets, cholecystokinin (CCK), secretin, and insulin are each released by specific nutritional substrates, protein, fat, and carbohydrate, respectively. These three polypeptide hormones not only modulate enzyme release from the pancreas but exert specific regulatory effects on the genes that code for functional groups of enzymes (proteolytic, lipolytic, and glycolytic enzymes, respectively). These recent findings largely explain the ability of the pancreas to adapt to dietary changes. In this chapter we describe the pattern of changes in gene expression observed in the pancreas in response to hormones and nutritional substrates. We also summarize what is known regarding the molecular mechanisms by which gene expression is regulated at both the transcriptional and translational levels in eucaryotic organisms. Repeated stimulation of such mechanisms by the periodic ingestion of nutritional substrates leads to changing patterns of gene expression in the exocrine pancreas. These processes appear to be mediated by specific hormones.

DIFFERENTIAL REGULATION OF GENE EXPRESSION

Separation of Gene Products by Two-Dimensional Gel Electrophoresis

Two-dimensional gel electrophoresis, combining gel isoelectric focusing in the first dimension and gradient gel electrophoresis in sodium dodecyl sulfate (SDS) in the second dimension, was introduced in 1975 by Scheele (63) for the analysis of complex mixtures of proteins, such as those synthesized and secreted by the exocrine pancreas. This procedure separates proteins according to two physical parameters: isoelectric point (IEP) in the first dimension and molecular weight (M_r) in the second dimension. Over the past 10 years this technique has been used to study the differential regulation of gene expression in the rat pancreas by hormones (55, 70) and nutritional substrates (71, 91, 92).

Figure 1 shows the exocrine proteins of the Wistar rat separated by two-dimensional isoelectric focusing and SDS gel electrophoresis. Proteins are labeled by abbreviations that are identified in Table 1. Twenty discrete proteins were observed. Multiple forms of individual enzymes were found, including three forms of trypsinogen and two forms each for chymotrypsinogen, proelastase, and amylase (70). Single forms were found for lipase and ribonuclease. Several forms of procarboxypeptidase A and B showed similar isoelectric point and molecular weight values and could not be completely separated by the two-dimensional gel procedure. They were accordingly analyzed as a group. Four proteins (labeled 1 to 4) could not be identified by enzymatic or proenzymatic activity. The protein P23 (IEP = 6.2; M_r = 23,000) was observed by Coomassie blue stain in proteins extracted from pancreatic tissue under conditions of hormonal stimulation but not under basal conditions. Table 1 lists each

TABLE 1. *Wistar Rat Exocrine Pancreatic Proteins*

Label	(Pro)enzyme	IEP	M_r	Distributed* Mean, %	SD
A1	Amylase 1	8.6	55,000	10.6	1.5
A2	Amylase 2	8.9	53,000	15.9	2.4
L	Lipase	6.8	50,000	4.5	0.5
T1	Trypsinogen 1	4.3	21,000	13.2	1.4
T2	Trypsinogen 2	4.4	21,000	1.6	0.3
T3	Trypsinogen 3	8.0	22,500	6.9	1.0
C1	Chymotrypsinogen 1	4.8	25,000	11.9	1.5
C2	Chymotrypsinogen 2	9.0	25,000	4.4	0.7
PE1	Proelastase 1	4.9	25,000	2.5	0.4
PE2	Proelastase 2	9.2	28,500	3.6	0.6
PCA	Procarboxypeptidase A	4.4	49,000	5.8	0.4
B1	Procarboxypeptidase B1	4.3	47,000	5.2	0.4
B2	Procarboxypeptidase B2	4.5	47,000	3.7	0.3
B3	Procarboxypeptidase B3	4.6	47,000	1.7	0.3
R	Ribonuclease	9.2	14,000	0.6	0.2
1	Unidentified	4.7	77,500	2.4	0.6
2	Unidentified	5.4	56,000	1.6	0.5
3	Unidentified	5.4	54,000	0.9	0.2
4	Unidentified	5.0	13,000	1.8	0.7

Isoenzymatic forms are labeled numerically with increasing isoelectric points in agreement with the IUPAC-IUB Commission on Biochemical Nomenclature for the identification of multiple forms of enzymes separated by gel electrophoresis. IEP, isoelectric point. M_r, apparent molecular weight of nonreduced proteins. * Distribution of proteins as quantified after incorporation of a mixture of 15 ^{14}C-labeled amino acids into pancreatic lobules incubated in vitro for 2 h and separation of proteins by 2-dimensional isoelectric focusing and sodium dodecyl sulfate gel electrophoresis. Data are expressed as percent relative to radioactivity contained in all exocrine protein spots. Mean and standard deviation values were derived from 20 animals fed a diet containing 22% protein, 63% carbohydrate, and 3% fat. [From Schick et al (70).]

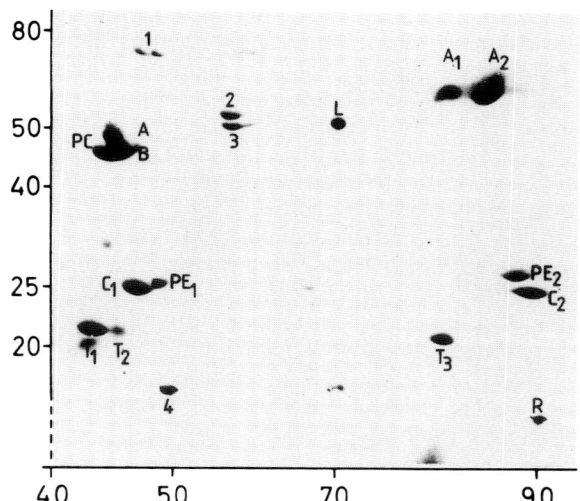

FIG. 1. Separation of gene products in the rat pancreas by 2-dimensional gel electrophoresis. Proteins that are identified by actual or potential enzyme activity are labeled by abbreviations that are identified in Table 1. *Numbers 1–4*, unidentified proteins. P23, present after hormonal stimulation (70), does not appear by Coomassie blue stain among proteins collected under basal conditions. *Ordinate*, apparent $M_r \times 10^{-3}$. *Abscissa*, isoelectric points. Enzymes and zymogens are numbered consecutively from anode to cathode, following the recommendations of the IUPAC-IUB Commission on the biochemical nomenclature of multiple forms of enzymes separated by polyacrylamide gel electrophoresis (2).

of the observed proteins, along with their isoelectric points, apparent molecular weights, and quantitative distribution, as judged by the percent incorporation of a mixture of 15 ^{14}C-labeled amino acids into individual proteins relative to the incorporation into all exocrine proteins. Pulse-chase studies suggest that each of the proteins in Figure 1 is coded by separate cytoplasmic mRNA species, because the appearance of none of the two-dimensional spots can be attributed to posttranslational modifications. Of particular interest is the observation that a complete set of serine protease zymogens (trypsinogen, chymotrypsinogen, and proelastase) appears in both anionic and cationic form. This pattern is generally observed among pancreatic proteins synthesized in mammals (66). As indicated next, anionic and cationic forms of both trypsinogen and chymotrypsinogen are independently regulated by hormones and nutritional factors.

Adaptation to Inverse Changes in Nutritional Substrates in the Diet

Pavlov first recognized that the ferment in pancreatic juice was altered by changes in the nutritional substrates in the diet (48). Investigations over the past

40 years have confirmed and extended these findings. Studies by Grossman et al. (26), Desnuelle et al. (16), Snook (75), and Dagorn and Lahaie (14) indicated that when protein levels >10% (by weight) were present in the diet for 6–10 days, concentrations of individual enzymes and proenzymes were regulated in direct proportion to carbohydrate and protein in the diet. These studies measured the level of tissue content of (pro)enzymes and rates of amino acid incorporation into exocrine proteins separated by column chromatography (16) or by isoelectric focusing (14). In the latter studies changes in synthetic rates preceded the changes observed in tissue levels of enzymes. Recently, two-dimensional gel electrophoresis has been used to follow the response of 16 individual exocrine proteins in the rat pancreas to inverse changes in carbohydrate and protein in isocaloric diets (71). Figure 2 shows the findings for four of the proteins, amylase forms 1 and 2 and chymotrypsinogen forms 1 and 2. During adaptation to diets containing normal protein (22%) or increased levels of protein (30%, 45%, 64%, 82%) and correspondingly decreased levels of carbohydrate, the two amylase forms and the two chymotrypsinogen forms were synthesized in direct proportion to nutritional substrates in the diet. With the exception of trypsinogen 3 and proelastase 2, other endoproteases and exoproteases followed the pattern shown for the two chymotrypsinogens (71).

The ability of the rat pancreas to adapt to changes in lipid in the diet was studied by administration of isocaloric and isonitrogenous diets containing increasing amounts of lipid (0%–30%) and decreasing amounts of carbohydrate (68.7%–1.25%). In response to diets containing increased unsaturated lipid (5%, 10%, 20%, 25%, and 30% sunflower oil), pancreatic lipase and colipase showed progressive increases to levels that were 2.5- and 2.0-fold higher, respectively, than those observed during administration of a normal diet with 3% lipid (91). In an independent study (61), increases in either saturated or unsaturated lipid in the diet led to increases in lipase levels in pancreatic tissue.

In previous studies that have sought to define pancreatic adaptation to changes in nutritional substrates in the diet, inverse changes in dietary components have been used to maintain a constant number of calories per gram of food. Because individual substrates have different and specific effects on the release of gastrointestinal hormones (see next section), an alternative approach, in which changes in single dietary substrates are studied, may need to be included in the analysis. Associated changes in calorie consumption may lead to less ambiguity in the interpretation of data than that which occurs when the concentrations of two dissimilar substrates are varied inversely in the diet.

Response to Hormonal Stimulation

In earlier studies that measured protein synthesis by incorporation of radioactive amino acids into trichloroacetic acid–insoluble protein, little change was observed in response to hormonal stimulation in vivo (for a review, see ref. 70). In recent studies that have followed rates of synthesis of individual proteins by two-dimensional gel electrophoresis, dramatic changes in protein synthesis have been observed with hormone stimulation. In these studies caerulein (an analogue of cholecystokinin) or secretin was infused into conscious rats over varying periods of time, up to 24 h. No attempt was made to mimic serum levels of hormones during normal alimentation. Instead infusions were conducted with optimal doses of caerulein or secretin to observe the limits of biochemical response to hormonal stimulation. After sacrifice of animals, pancreatic lobules were prepared and incorporation of amino acids into cellular proteins, measured in vitro in the absence of hormone, indicated persistent changes in the synthesis of exocrine proteins. Both coordinate and anticoordinate changes were observed in protein synthesis in response to administration of hormones.

CHOLECYSTOKININ-LIKE HORMONES. Figure 3 shows the changes observed in the synthesis of five selected

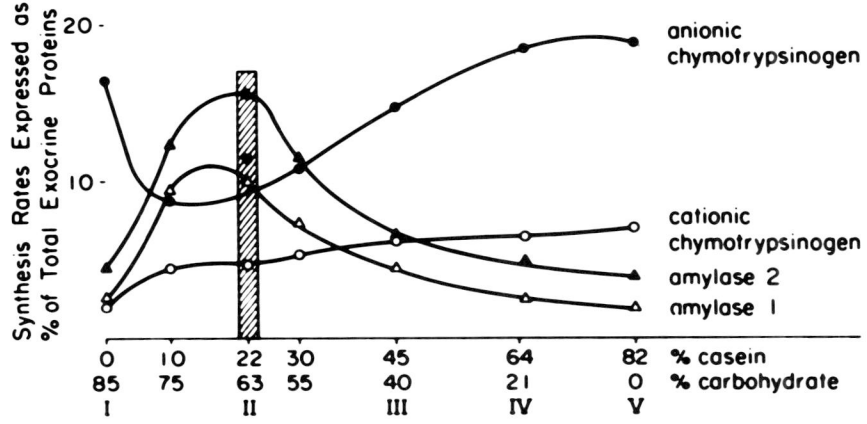

FIG. 2. Differential changes in synthesis of exocrine pancreatic proteins in response to reciprocal changes in carbohydrate and protein in the diet. After administration of isocaloric diets for 12 days, animals were sacrificed and rates of protein synthesis were measured by in vitro incorporation of a mixture of 15 ^{14}C-labeled amino acids into pancreatic lobules followed by analysis of individual proteins after their separation by 2-dimensional isoelectric focusing and sodium dodecyl sulfate gel electrophoresis. [From Scheele (65).]

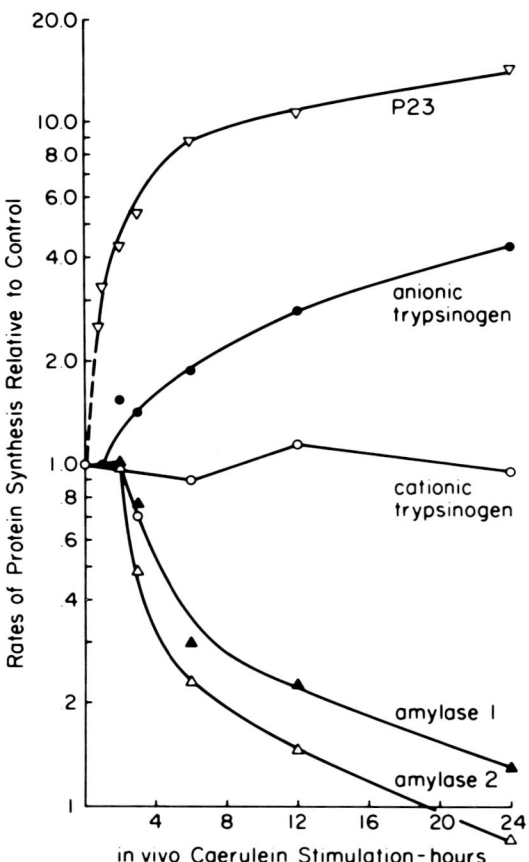

FIG. 3. Differential changes in the synthesis of exocrine proteins in response to hormonal stimulation in the rat. Caerulein, a peptide analogue of CCK, was infused into the tail vein of conscious rats at a rate of 0.25 $\mu g \cdot kg^{-1} \cdot h^{-1}$ for up to 24 h. Animals were sacrificed at indicated times, and protein synthesis rates were measured in absence of caerulein (see Fig. 2). [From Scheele (65).]

exocrine proteins in response to infusion with caerulein, a decapeptide that has the same COOH-terminal pentapeptide amide as CCK. Values for protein synthesis at 24 h relative to control levels showed that P23 synthesis increased 14.2-fold, anionic trypsinogen forms 1 and 2 increased 4.3-fold, cationic trypsinogen showed no change, and amylase forms 1 and 2 decreased to levels 0.14 and 0.07, respectively, of the control values obtained in the absence of hormone stimulation (70). Less dramatic changes or no changes were observed in the synthesis of other exocrine proteins, including lipase. Coordinate changes are defined as those that are similar in latency, kinetics, and extent (compare amylase forms 1 and 2); anticoordinate changes are defined as those that are dissimilar in these respects and in certain cases are opposite in direction of response (compare P23, anionic trypsinogen, cationic trypsinogen, and amylase). Coordinate changes were observed for the following isoenzymatic pairs: trypsinogen forms 1 and 2, chymotrypsinogen forms 1 and 2, and amylase forms 1 and 2. In other instances the hormone-induced changes appeared to be isoenzyme specific (e.g., the case of anionic trypsinogen versus cationic trypsinogen, as previously mentioned). In addition, synthesis of proelastase 1 but not proelastase 2 showed a response to caerulein stimulation (70).

For the proteins shown in Figure 3, the period of latency for the appearance of persistent changes in protein synthesis varied. Changes in the synthetic rates of amylase, anionic trypsinogen, and P23 occurred after 2, 1, and 0.5 h, respectively. Such variations in the latency period presumably reflect differences in the molecular mechanisms by which each of these three proteins is regulated. Studies conducted in vivo by stimulation of the rat pancreas with CCK (15) and in vitro by stimulation of guinea pig pancreatic lobules with carbamylcholine (64) have shown that similar but less extensive anticoordinate changes in protein synthesis occur within 15 min of secretagogue stimulation.

SECRETIN. In contrast to changes in levels of total protein synthesis, where secretin and caerulein effects were largely indistinguishable (2-fold increases in both cases), secretin infusion resulted in stimulation of the synthesis of a different set of exocrine proteins than was observed with caerulein infusion [Fig. 4; (54, 55)]. In response to secretin administration, lipase synthesis increased to the greatest extent: 2.8-, 5.8-, and 4.8-fold, at 6, 12, and 24 h, respectively. Proelastase 2 synthesis increased less: 3.4-fold at 24 h (54). Combined stimulation with optimal secretin and ca-

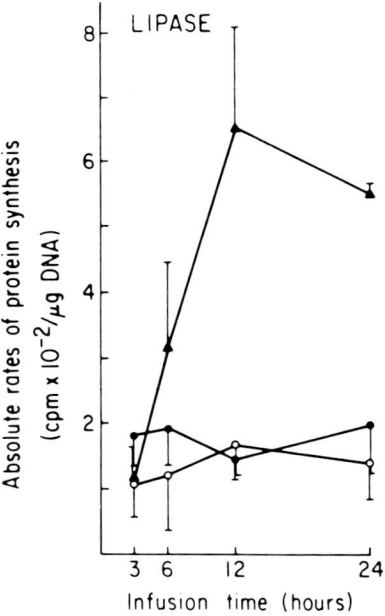

FIG. 4. Changes in the synthesis of pancreatic lipase in response to secretin stimulation in the rat. Secretin at 16 $CU \cdot kg^{-1} \cdot h^{-1}$ (*solid triangles*), caerulein at 0.25 $\mu g \cdot kg^{-1} \cdot h^{-1}$ (*closed circles*), and saline (*open circles*) were infused for up to 24 h. Animals were sacrificed at indicated times, and protein synthesis rates were measured (see Fig. 2). [From Rausch et al. (55).]

erulein did not augment increases in total or individual rates of protein synthesis (54).

INSULIN. In animals rendered diabetic by chemical destruction of β-cells, amylase levels in the exocrine pancreas decreased to 10% compared with controls (1, 30, 47, 76). After administration of insulin, normal levels were observed in amylase mRNA (30), amylase biosynthesis (47, 76), and levels of amylase in pancreatic tissue (1). In obese Zucker rats, which show diabetes due to insulin resistance (insulin levels 5-fold higher than in normal rats), glucose metabolism was reduced by 50%, tissue levels of amylase and its corresponding mRNA were reduced by 60%, and tissue lipase was increased twofold compared with controls (84, 85). On reversal of insulin resistance by ciglitazone therapy, insulin levels decreased to twice the normal level and tissue amylase, lipase, and glucose metabolism returned to normal values. However, in normal animals increases in serum insulin in response to glucose infusion did not modulate amylase levels, even in animals containing low concentrations of pancreatic amylase in response to a high-protein, low-carbohydrate diet (25a; H. Kern, unpublished observations). Thus insulin appears to exert a permissive effect on the expression of the amylase gene.

Cholecystokinin, secretin, and insulin are polypeptide hormones that interact with specific receptors on the plasma membranes of the pancreatic acinar cell. The cellular effects of these hormones are associated with activation of separate intracellular second-messenger pathways (see the chapter by Williams, Burnham, and Hootman in this *Handbook*). It will therefore be of interest to determine if second-messenger pathways are responsible for the hormone effects on gene expression.

GLUCOCORTICOIDS. Recent studies have indicated that levels of pancreatic amylase are regulated by glucocorticoid levels. Adrenalectomy resulted in a 50%–60% decrease in amylase levels, and during subsequent glucocorticoid administration, tissue levels of amylase varied proportionately with serum glucocorticoid levels (25a; R. Bruzzone and G. A. Scheele, unpublished observations). Work is needed to determine if glucocorticoids modulate the synthesis of other pancreatic enzymes.

Pavlov's Dietary Adaptation is Mediated by Specific Hormones

Table 2 correlates the changes observed in the synthesis of individual exocrine proteins in the rat pancreas in response to polypeptide hormones with those observed in response to nutritional substrates. The correlations suggest that differential regulation of functional groups of digestive enzymes in the rat exocrine pancreas by changes in nutritional substrates in the diet is controlled largely through the effects of specific hormones. This hypothesis is supported by studies that have measured changes in the levels of circulating hormones in the rat in response to food components delivered to the intestinal tract. Protein (but not amino acids, fatty acids, or glucose) releases CCK (35) and unsaturated lipid or fatty acids release secretin (19, 44, 62) from the intestinal mucosa. Thus positive feedback mechanisms appear to operate in the rat between nutritional substrates in the diet and regulation of the synthesis of pancreatic enzymes that are required for digestion of these substrates. Two of the feedback mechanisms appear to be mediated by specific gut hormones, CCK and secretin (Table 2). The mechanism(s) by which starch in the diet of normal animals leads to regulation of the synthesis of pancreatic amylase is less well understood. The lack of effect of insulin on amylase levels in normal animals fed a diet low in carbohydrate and high in protein suggests either that the effect of CCK dominates the effects of insulin in nondiabetic animals or that a different gut hormone mediates amylase synthesis in response to changes in dietary carbohydrate. The effects of other gut and islet hormones on the synthesis of individual proteins in the exocrine pancreas need to be determined.

Further study is also required to determine if digestive end products and their metabolic derivatives are directly involved in the regulation of pancreatic gene expression. In preliminary experiments (L. Trimble,

TABLE 2. *Regulation of Protein Synthesis in Rat Exocrine Pancreas: Correlation Between Nutritional Substrates and Specific Hormones*

Nutritional Substrates	Digestive End Products	Hormones	(Pro)enzyme Synthesis		
			Proteases	Lipases	Amylases
Protein	Amino acids	Cholecystokinin	Trypsinogen 1 and 2, ↑ Chymotrypsinogen 1 and 2, ↑ Proelastase 1, ↑ Procarboxypeptidase A and B, ↑		Amylase 1 and 2, ↓
Triglyceride phospholipid	Unsaturated fatty acids	Secretin	Proelastase 2, ↑	Lipase, ↑	
Starch	Glucose	Insulin*			Amylase 1 and 2, ↑

↑, Increase; ↓, decrease.   * Insulin effect observed only in diabetic animals.

unpublished observations) conducted on AR42J cells (a tissue culture line derived from a rat acinar cell carcinoma), addition of glucose to 200 mg/dl in the medium increased cellular levels of amylase by 50%. Addition of oleic acid to 2 mM increased lipase levels by 70%. Thus digestive end products appear to directly affect gene expression in pancreatic tissue, but their effects appear to be small in comparison with those achieved with hormone stimulation.

Nutritional Shock

PROTEIN DEFICIENCY. During adaptation to diets containing normal (22%) or decreased (0%–10%) levels of protein with corresponding increases in carbohydrate, amylase and the subgroup of anionic protease zymogens were largely synthesized in inverse proportion to nutritional substrates in the diet (71). During administration of a protein-free diet, changes in synthetic rates were related to isoelectric points of exocrine isoenzymes: anionic proteins (including trypsinogen forms 1 and 2, chymotrypsinogen 1, proelastase 1, and procarboxypeptidases A and B) increased 1.4- to 2.8-fold; neutral and cationic proteins (including lipase, trypsinogen 3, amylase forms 1 and 2, chymotrypsinogen 2, proelastase 2, and ribonuclease) uniformly decreased to levels 0.23–0.57 times control levels (Table 3). Figure 2 indicates that, under conditions of a protein-free diet, the relative synthesis of anionic chymotrypsinogen increased, whereas the relative synthesis of the cationic forms, chymotrypsinogen and amylase forms 1 and 2, decreased. Synthesis of anionic proteins, representing 45.6% of exocrine proteins synthesized under normal laboratory dietary conditions, increased to 95% after a 12-day administration of a protein-free diet. Under conditions of protein deprivation, the acinar cell redirects its protein synthetic activity toward production of a group of anionic protease zymogens (trypsinogen forms 1 and 2, chymotrypsinogen 1, proelastase 1, and procarboxypeptidases A and B). Under these conditions, a select group of pancreatic endo- and exoproteases continues to be synthesized and secreted by the acinar cell. The continued presence of these pancreatic proteases in the digestive tract allows for the digestion of newly found sources of protein, a process that is necessary, at this point, for survival of the organism. Further work is needed to determine the mechanisms by which anticoordinate changes occur in the synthesis of anionic compared with neutral and cationic proteins during administration of a zero-protein diet.

GLUCOSE DEPRIVATION. In AR42J cells, propagated in tissue culture, glucose starvation for 20–30 h increased trypsinogen mRNA levels while decreasing levels for amylase, chymotrypsinogen, and procarboxypeptidase A mRNA (77). In other cellular systems, e.g., Chinese hamster ovary (CHO) cells, glucose-regulated proteins appeared to be identical to stress proteins (90). However, the implications of these results for pancreatic physiology are not clear.

TABLE 3. *Changes in Fractional Protein Synthesis Rates in Response to High- and Low-Protein Diets Relative to Control Diet*

	IEP	Diet Composition*	
		High protein (82%)	Low protein (0%)
Trypsinogen 1	4.3	1.29	1.86
Trypsinogen 2	4.4	1.75	2.81
Procarboxypeptidase A and B	4.3–4.6	1.31	1.28
Chymotrypsinogen 1	4.8	1.62	1.41
Proelastase 1	4.9	2.76	2.20
Lipase	6.8	0.74	0.23
Trypsinogen 3	8.0	1.06	0.36
Amylase 1	8.6	0.36	0.41
Amylase 2	8.9	0.13	0.19
Proelastase 2	9.1	0.56	0.36
Chymotrypsinogen 2	9.2	1.64	0.57
Ribonuclease	9.3	0.90	0.47

IEP, isoelectric point. * Fractional rates of synthesis (FRS) of indicated proteins after 12 days on the indicated diet divided by corresponding FRS after control (22% protein) diet. [From Schick et al. (71).]

GENE EXPRESSION IS REGULATED AT MULTIPLE LEVELS

Prenatal and Postnatal Development

Han et al. (28), using cDNA probes, have recently shown that the large increases observed in the synthesis of pancreatic secretory proteins during embryonic differentiation in the rat are due to corresponding increases in mRNA concentrations. Depending on the individual mRNA, levels in the rat embryonic pancreas increased from 6–150 molecules per cell on day 14 to 500–100,000 molecules per cell on day 20. In both tissue culture (36, 36a, 53; B. Swarovsky, W. Steinhilber, and H. F. Kern, unpublished observations) and in vivo experiments (53), glucocorticoids accelerated cytodifferentiation in the pancreatic acinar cell during the early embryonic period. This was manifested by an increase in the content of enzymes, which varied among the individual proteins. In other systems responsive to glucocorticoids, formation of a hormone-receptor complex in the nucleus leads to its interaction with defined nucleotide regions and stimulates gene transcription (cf. Table 5).

In the pig, where prenatal (115 days) and postnatal (30 days) development are prolonged, the results indicate a considerable asynchrony in the appearance of exocrine proteins (46) and presumably in the activation of their corresponding genes. A group of proteins, including chymotrypsinogens A and B, proelastase II, procarboxypeptidase A, and amylase, appear at ~65 days. At 75 days anionic trypsinogen emerges. At about day 15 of the postnatal weaning period another group of (pro)enzymes, cationic trypsinogen, chymotrypsinogen C, and protease zymogen E, appears. Finally, 5 days after weaning (day 35) proelastase I appears. Further studies are needed to determine the mechanisms by which genes coding for isoenzymatic forms are activated at widely different times during the pre- and postnatal development periods in the pig.

Nutritional Substrates and Hormones

Wicker et al. (92) first studied the mechanisms by which pancreatic acinar cells regulate protein synthesis in response to changes in nutritional substrates in the diet. To determine whether changes in protein synthesis are mediated by alterations in mRNA levels, the relative levels of mRNA coding for pancreatic amylase, lipase, procarboxypeptidases A and B, and the family of serine protease zymogens were determined by the ability of isolated RNA to direct the synthesis of these products in a high-fidelity reticulocyte lysate system (92). As shown in Figure 5, translation products synthesized in vitro correlated with products synthesized in intact cells in pancreatic lobules. These findings indicated that changes in mRNA levels mediated the observed alterations in protein synthesis in response to dietary changes in carbohydrate and protein. These observations have been confirmed by Giorgi et al. (21, 22) with the use of cDNA probes.

In contrast to results observed during embryonic differentiation and in adult animals after changes in diets, the initial effects on protein synthesis achieved with a single period of caerulein stimulation are mediated at the level of efficiency of mRNA translation (Fig. 6). Despite changes in protein synthesis within 1–2 h of hormone stimulation, absolute levels of mRNA, as measured with cDNA probes, showed no changes over the 24-h period studied. However, administration of repeated periods of hormone stimulation (twice daily for 7 days) led to changes in mRNA levels (58). Under both single and multiple periods of caerulein stimulation, synthesis of anionic trypsinogen and amylase is increased and decreased, respectively. Thus the sequential phases of biological response, which can be expected to depend on the integration of divergent mechanisms activated over different periods of time, result in an increase in biochemical commitment with repeated hormone stimulation (65). Further work is needed to determine the time course and mechanisms by which repeated hormone stimulation alters levels of mRNA and the extent to which translational and peritranslational (pre- and posttranslational) control are integrated in the response to hormones and nutritional substrates.

MECHANISMS IN SELECTIVE REGULATION OF GENE TRANSCRIPTION

Processing Signals in Eucaryotic Genes

Figure 7 presents in schematic fashion the general organization of eucaryotic genes and the nucleotide signals believed to regulate, in part, gene transcription and mRNA translation. Gene transcription occurs after DNA strand separation and functional insertion of RNA polymerase near the promoter sequence, TATAAA (Goldberg-Hogness box). Transcription of

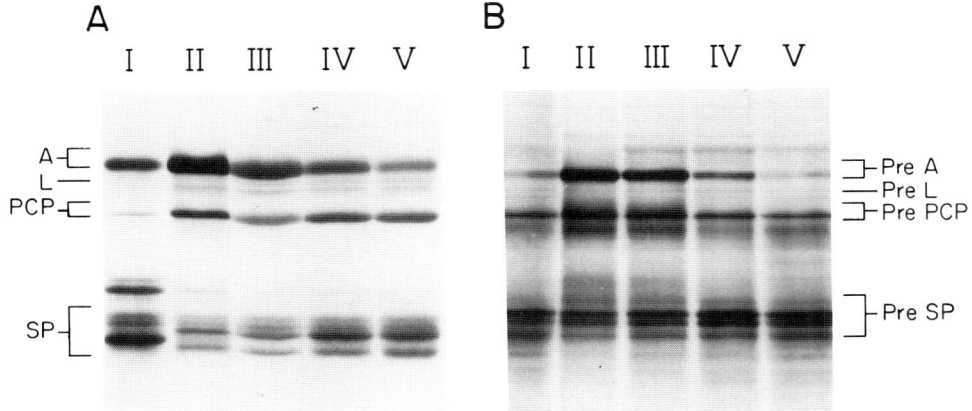

FIG. 5. Adaptation in the synthesis of exocrine proteins to dietary changes mediated by alterations in mRNA levels. Diets I–V (with compositions indicated in Fig. 2) were administered for 8 days. *A*: fluorogram of exocrine protein products synthesized in intact pancreatic cells. *B*: fluorogram of protein products synthesized by isolated mRNA in a rabbit reticulocyte in vitro translation system, a study that measures functional levels of mRNA. Synthesis of individual proteins in intact cells (*A*) is proportional to functional levels of mRNA (*B*). A, amylase; L, lipase; PCP, procarboxypeptidases A and B; SP, serine proteases (including trypsinogen, chymotrypsinogen, and proelastase). Presecretory proteins demonstrating small increases in molecular-weight values are synthesized during in vitro translation in the absence of microsomal membranes. [From Wicker et al. (92).]

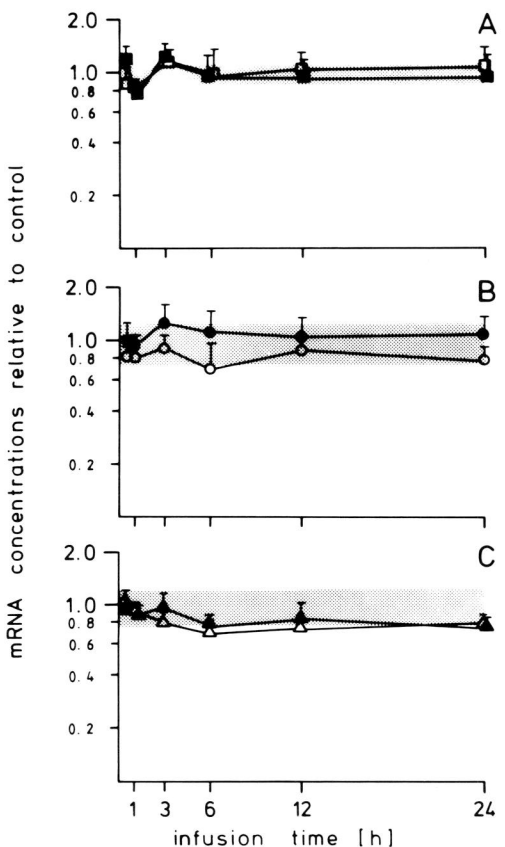

FIG. 6. Absence of kinetic changes in concentrations of mRNA coding for rat pancreatic proteins, anionic trypsinogen, lipase, and amylase, during a single period of caerulein infusion. Hormone was infused at 0.25 $\mu g \cdot kg^{-1} \cdot h^{-1}$ for up to 24 h. Messenger RNA concentrations were measured with nick-translated cDNA probes on nitrocellulose dot blots. Measurements were made on pancreatic tissue taken from hormone-infused (*closed symbols*) or saline-infused (*open symbols*) animals and normalized to a standard sample of pancreatic RNA obtained from a group of noninfused animals; 3–4 animals were studied at each time point. *Vertical bars*, standard deviation values. *Hatched areas*, 1 standard deviation obtained in the measurement of the standard sample of RNA. *Ordinate*, values are given on a logarithmic scale. (W. Steinhilber, J. Poensgen, U. Rausch, H. Kern, and G. A. Scheele, unpublished observations.)

the DNA coding sequence into an RNA intermediate begins at a purine residue ~30 base pairs downstream from the initial T in the TATA box and proceeds to a point at least 20 bp (base pairs) downstream from the polyadenylation signal, AATAAA. The molecular details of transcription termination have yet to be determined but appear to involve further transcription to downstream site(s) followed by endonuclease cleavage at the point of poly(A) addition (43). The nucleotide at which transcription begins is designated as the cap site, because it codes for the 5′-terminal residue of pre-mRNA, which is covalently linked (capped) by an inverted guanosine residue methylated in the N-7 position (^{7}m guanosine). Bases downstream from the cap site are given in positive integers and include the gene-coding sequence and its 3′ flanking sequence. Bases upstream from the cap site are given in negative integers and represent the 5′ flanking sequence of the gene. In addition to the promoter, which includes the TATA box and upstream promoter elements, there are now believed to exist enhancer sequences, usually further upstream but in a few cases within introns, that confer tissue-specific and hormone-modulated expression of structural genes.

In 1977 evidence (8, 12) was first provided that eucaryotic genes are organized along the chromosome in a segmented manner, with exons containing sequences found in mature mRNA separated by intervening sequences (introns). All exocrine pancreatic genes studied to date, including five rat protease zymogens (4, 13, 80), dog prophospholipase A$_2$ (B. Kerfelec, K. S. LaForge, and G. A. Scheele, unpublished observations) and lipase (F. Weidenbach, B. Ohlsson, and G. A. Scheele, unpublished observations), and mouse amylase (52) contain introns. Intron sequences invariably contain the dinucleotide GU at their 5′ end and the dinucleotide AG at their 3′ end (41). The consensus sequences at the 5′ and 3′ intron junctions are as follows: 5′^A_C AG/GURAGU.....(Y)$_{11}$NYAG/G 3′ where the / lines delimit the intron, R is purine, Y is pyrimidine, and N is any nucleotide. Although the precise mechanism of splicing is unknown, four small

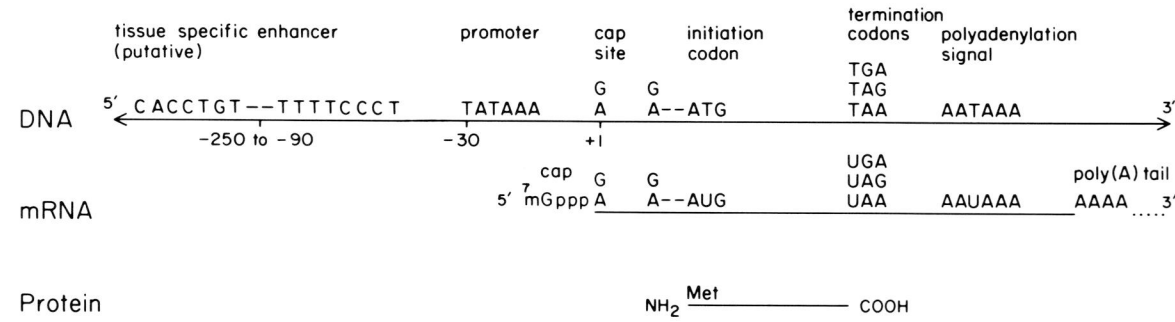

FIG. 7. Nucleotide signals involved in gene transcription and mRNA translation in the exocrine pancreas. Schematic depiction of a pancreatic gene shown without introns. Nomenclature for nucleotide signals is given directly above each sequence. *Dashes*, absence of nucleotide preference in consensus sequences. Putative tissue-specific enhancer is taken from ref. 9.

TABLE 4. *Transcriptional Elements in Eucaryotic Promoters*

	A. Upstream Promoter Elements			Ref.
Simian virus 40 early promoter	......ACTCCGCCCA......ACTCCGCCCA.....ACTCCGCCCA......TATA... ......TGAGGCGGGT......TGAGGCGGGT.....TGAGGCGGGT............ ←———— ←——/—— ←———— −106 −97 −86 −76 −56 −47 ——→			17, 40
Herpes simplex virus Thymidine kinase promoter	......ACCCCGCCCA......ATTGGCGAATT....GGGGCGGCGC......TATA... ......TGGGGCGGGT......TAACCGCTTAA.....CCCCGCCGCG............ ←———— ←—— −95 −87 −79 −72 ——→ −30			40
β-Globin promoter	......GCCACACCC......GGCCAATC......................TATA... ......CGGTGTGGG......CCGGTTAG..............................			42

B. Types of Promoter Elements and Binding Proteins			Ref.
Promoter element	Abbrev.	Binding protein	
TATA	TATA	II D	57
GGGGCGGGGC TA TAAT	GC	Sp I	17
CCAAT	CAAT	CTF	*
GCCACACCC	CACCC		

* R. Tjian, unpublished observations.

nuclear ribonuclear proteins (RNPs) and possibly a number of heterogeneous ribonuclear proteins (hn-RNPs) appear to be involved (56). The studies indicate that exon sequences also play an important role in distinguishing between normal splice sites and the many nearly identical cryptic splice sites. Pairing of the correct 5′ and 3′ splice sites appears to be due to a balance between differential splice-site strength (affinity for splice-mediating proteins) and the proximity of 5′ and 3′ splice sites.

The mature mRNA contains a coding region defined by the AUG initiation codon and the downstream sequence extending to an in-frame termination codon. In each mRNA the coding region is flanked by 5′ and 3′ nontranslated regions that contain chemical constituents involved in the stability and function of mRNA. The inverted ^{7}m guanosine cap binds a 24,000-M_r cap-binding protein that organizes a cap-binding protein complex believed to be involved in the binding of functional mRNA to the small ribosomal subunit during formation of the initiation complex in protein synthesis (82). The cap structure may also protect mRNA from degradation by 5′ exonucleases in the cell cytoplasm. The 3′ nontranslated region contains the polyadenylation signal and a poly(A) tract of variable length, 25–250 residues, which may be involved in regulation of mRNA turnover (29). With translation of the mRNA, the coding region directs the synthesis of the protein product, beginning with the initiator methionine at the amino terminus.

Promoter Elements

All eucaryotic genes contain a TATA box (or a close variant), which acts as the promoter element responsible for accurate (site-specific) initiation of transcription ~30 bp downstream from the first T in the TATA sequence. Alone, the TATA sequence acts only as a weak promoter. Recent studies on eucaryote viral and cellular genes have shown that promoter elements upstream of the TATA sequence determine the strength of the promoter region. Upstream promoter elements have been mapped only for a few eucaryotic genes. Analysis of the upstream promoter elements for the three genes shown in Table 4 is instructive. In each case the elements are confined to the region (−105 to −30) immediately preceding the TATA sequence. For each of the three genes, multiple promoter elements are present in this regulatory region. In addition to the TATA box, three discrete promoter elements regulate the transcription of these genes: the "GC" box (40), the "CAAT" box (7, 18, 40, 42), and the "CACCC" box (42). Similar transcriptional elements are found in the promoter regions of different genes. For example, GC promoter elements are observed in both simian virus 40 (SV40) early and thymidine kinase (tk) genes. The CAAT promoter elements are present in both the tk and β-globin genes. In contrast, the CACCC element has been observed only in β-globin promoters. As shown in Table 4, both the GC box and CAAT box may function in a bidirectional manner. For example, the CAAT box is in the same and opposite orientation to the TATA sequence in the β-globin and tk promoters, respectively. Within the tk promoter region the two "GC" transcriptional elements are in opposite orientation. Promoter strength is believed to be determined by the strength (affinity for DNA-binding proteins), number, and arrangements of these types of transcriptional elements (40).

Studies in the past two years have begun to identify

DNA-binding proteins (*trans*-acting factors) that interact with specific promoter elements (*cis*-acting factors). Examples of these specific DNA-binding proteins are presented in Table 4. In in vitro experiments, proteins that bind to the transcriptional elements shown in Table 4 usually protect, from DNase cleavage, a DNA sequence of up to ~20 bp, representing two complete turns of the DNA helix. Protein-base interactions are believed to occur in the major groove of the helix. The strength of a given promoter element appears to be determined by the affinity with which a protein binds to the nucleotide recognition sequence. Small variations in the sequence of promoter elements (e.g., those observed in the three "GC" elements in the SV40 early promoter) result in differences in affinity for the binding protein and therefore its individual promoter strength (40).

Although the sequences of promoter elements for parotid and liver amylase (*Amy-1a*) and pancreatic amylase (*Amy-2a*) have not been defined, the chromosomal positions and strengths of the three promoters have been determined (52). In A/J mice the *Amy-1a* locus, which encodes parotid and liver amylase (1 gene copy), is tightly coupled on chromosome 3 to the *Amy-2a* locus, which encodes pancreatic amylase (4 gene copies encoding identical mRNA species). The RNA transcripts from the *Amy-1a* and *Amy-2a* loci show 10% sequence heterogeneity.

Tissue-specific differences in splicing account for the differential expression of the *Amy-1a* locus in liver compared with salivary gland in the mouse (68, 93). The *Amy-1a* gene has two transcriptional promoters upstream from the structural amylase gene. The strong promoter, 7.5 kb upstream (kb, kilobase), is exclusively selected in the parotid gland. The weak promoter, 4.5 kb upstream, is selected not only in the liver but also in the parotid gland and pancreas (27). The "parotid" promoter is 30–40 times more efficient in transcription than the "liver" promoter (69). Together, promoters for the four copies of the *Amy-2a* locus encoding pancreatic amylase in A/J mice are 550-fold more active than the liver promoters (27). Nucleotide sequences representing upstream promoter elements have yet to be defined for pancreas-specific genes.

Enhancer Elements

TISSUE-SPECIFIC ENHANCERS. Studies carried out on eucaryotic viruses have indicated the presence of transcriptional enhancer sequences, typically 100 to 300 bp upstream from the transcription start site (89). In viral genes the presence of such regulatory elements is required for transcription under in vivo conditions. In a manner similar to upstream promoter elements, the function of enhancer sequences is orientation independent. Unlike promoter elements, they may remain functional when located at sites distant to the TATA sequence: within sequences 2–3 kb upstream of the cap site, within introns, or when inserted downstream from the structural gene. Viral enhancer sequences that show the consensus core sequence $GTGG^{AAA}_{TTT}G$ show relative but not absolute tissue specificity.

Two laboratories have recently obtained evidence that enhancer elements are also present in the 5' flanking sequences of pancreatic genes and that these sequences are responsible for tissue-specific expression in the exocrine pancreas. Walker et al. (88) constructed a hybrid gene containing the 5' flanking sequence of the rat chymotrypsin gene, presumed to contain the signals for transcriptional activity, and the coding sequence for bacterial chloramphenicol acetyltransferase (CAT). This gene construct was transformed into a variety of tissue culture cells, and transcription activity present in the flanking sequence derived from the chymotrypsin gene was followed by measuring cellular levels of CAT activity. Transcription activity was measured in three cell lines derived from different tissues: Chinese hamster ovary (CHO) fibroblast cells; HIT cells, representing transformed hamster pancreatic endocrine cells that produce insulin at levels 2%–5% of normal islet cells; and AR42J cells, representing a rat exocrine pancreatic tumor line that contains trypsin and chymotrypsin mRNAs at levels 0.9% and 18%, respectively, of normal acinar cells. The transcriptional activity of the 5' flanking region of the chymotrypsin gene was 50- to 200-fold higher in AR42J cells than in CHO and HIT cells. In studies that deleted increasing segments in the flanking region from the 5' terminus, transcriptional enhancer activity was mapped to the region between bases -274 and -192.

In an independent approach, Swift et al. (81) introduced the entire rat pancreatic elastase I gene containing 7 kb of 5' flanking sequence and 5 kb of 3' flanking sequence into fertilized mouse eggs and analyzed for tissue-specific expression of rat elastase in resulting postnatal mice. In 20% of cases, one or more elastase gene copies were incorporated into the germ line DNA. In five cases, postnatal tissues were examined in detail for expression of the rat elastase gene, as judged by the presence of rat pancreatic elastase mRNA sequences measured by hybridization to a specific 3' noncoding sequence that could distinguish the rat elastase mRNA from the endogenous mouse elastase mRNA. In four cases the expression of rat elastase within the mouse pancreas was three orders of magnitude higher than that observed in a variety of other tissues. These studies suggest that tissue-specific enhancer elements are associated with the flanking sequences of the elastase I gene.

Because the evidence pointed to the presence of tissue-specific enhancer elements in the 5' flanking regions of both chymotrypsin and elastase genes in the rat pancreas, these investigators analyzed the 5' flanking nucleotide sequences of other rat pancreatic genes for homologous regions that might define potential sites representing such enhancer elements (9, 80). From these comparisons a 20 bp consensus sequence (..GTCACCTGTGCTTTTCCCTG..) has been judged to represent the enhancer element responsible for tissue-specific expression of genes in the rat exocrine pancreas (9). Regions showing homology to this consensus sequence have been identified between −233 and −89 in the 5′ flanking regions of these genes. Developmentally coordinated proteins synthesized in the pancreatic acinar cell may interact with such enhancer sequences and activate genes in a tissue-specific manner during embryogenesis.

ENHANCER ELEMENTS RESPONSIVE TO STIMULI IN THE CELLULAR ENVIRONMENT. Table 5 lists the known eucaryotic enhancer or regulatory elements that mediate genetic responses to critical changes in the cellular environment. These enhancer elements mediate the long-term response to changes in the cell's environment by selectively altering concentrations of mRNA in the cell. For example, toxic concentrations of the heavy metals cadmium or zinc stimulate metal regulatory elements, which activate the expression of metallothionein, a protein that chelates heavy metals and thus protects the cell from their toxic effects (72, 78). Viral infection results in the activation of the β-interferon gene through stimulation of the interferon regulator element. The interferon element contains regions that are modulated separately by positive and negative effectors (23, 94, 95). Under basal conditions the negative effector is bound to the interferon regulatory element. After viral infection the negative effector dissociates and a positive effector binds and activates transcription of β-interferon mRNA. Enhancer elements activated by steroid hormone-receptor complexes have been defined for glucocorticoids, progesterone, and estrogen. For each of the elements listed in Table 5, nucleotide sequences have been defined and activation of these elements is believed to occur through their interaction with DNA-binding proteins. Modulation of protein-DNA interactions may occur as a result of second-messenger–driven modifications to these DNA-binding proteins.

From knowledge obtained in these other regulatory systems, we can predict that pancreatic genes will be found associated with multiple enhancer elements that respond not only to tissue-specific factors but also to a variety of factors, both positive and negative, that are modulated by nutritional, hormonal, and other changes that occur in the environment of the acinar cell. The precise levels of expression of each of the pancreatic genes can be expected to reflect the sum of the effects of these multiple elements. This pattern of multiple regulatory elements responsible for the multifactorial control of gene expression is consistent with information derived from the first detailed studies of other eucaryotic genes (see *Promoter Elements*, p. 507).

SELECTIVE REGULATION OF MRNA TRANSLATION EFFICIENCY: POSSIBLE MECHANISMS

A large body of work over the last two decades has established many of the factors involved in the initiation, elongation, and termination of mRNA translation. Initiation, the rate-limiting step of mRNA translation, depends on the functional collaboration of three types of elements: ribosomal subunits, an array of initiation (protein) factors, and mRNA containing a methylated 5′ cap structure (31, 45, 73). Much has been learned regarding the sequence of events by which small (40S) ribosomal subunits and mRNA are separately primed through the binding of specific protein factors. The 40S ribosomal subunit is primed through the association of a ternary complex [GTP·Met-tRNA$_i$·eIF-2 (tRNA$_i$, initiator tRNA; eIF, eucaryotic initiator factor); (6, 45)] and eIF-3 [a complex of ~10 proteins; (5)]. Messenger RNA is believed to be primed largely through the binding of several proteins to the 5′ methylated cap structure, a process that leads to the formation of a cap-binding protein complex (CBPB) (25). This protein complex contains eIF-4E or the cap-binding protein (24 kDa), eIF-4A (48 kDa), eIF-4B (80 kDa), and possibly other proteins yet to be discovered. The binding of eIF-4A and eIF-4B depends on the presence of ATP (24).

The ultimate goal of the initiation process is to pair the AUG initiation codon with the UAC anticodon of initiator tRNA associated with the small ribosomal subunit so that the large ribosomal subunit may attach, which then allows mRNA-directed polymerization of amino acids to proceed. The sequence immediately flanking AUG codons in mRNA sequences is important in the process of selection of AUG initiation codons. In animal mRNAs, structural studies (33) indicate that the consensus sequence that determines the site of translational initiation is CACCAUG. Studies both in vivo (32, 34) and in vitro (37) indicate that

TABLE 5. *Eucaryotic Enhancer Elements Responsive to Environmental Stimuli*

Regulatory Elements	Structural Genes	Ref.
Metal (Cd, Zn) (MRE)	Metallothionein	72, 78
Interferon (IRE)	β-Interferon	23, 95
Steroid hormone (SRE)		
Glucocorticoid (GRE)	Uteroglobin	79
	Lysozyme	59
	Human growth hormone	74
	Mouse mammary tumor virus	67
	Metallothionein	60
Progesterone (PRE)	Lysozyme	59
	Uteroglobin	79
	Mouse mammary tumor virus	10
Estrogen (ERE)	Lysozyme	59
Heat shock	Heat shock proteins	49

the base in position −3 relative to the A (position +1) of the AUG initiation codon plays a dominant role in determining the efficiency of mRNA translation. A decreasing order of translational efficiency is observed with A > G ≫ U > C in position −3. Further work is needed to define the mechanisms by which the cell distinguishes AUG initiation codons from other (cryptic) AUG codons present in either upstream or downstream mRNA sequences; AUG initiation codon selection factors, possibly proteins, may be involved in this process. The affinity of interaction of such factors with AUG codons may be determined by the composition of the surrounding sequence. The ability of the translational machinery to distinguish an authentic AUG start site from cryptic AUG codons may depend on the affinity of such an interaction together with the proximity of the AUG codon to the 5' end of the message (in 90% of the cases initiation of translation in eucaryotic mRNAs begins at the first AUG codon downstream from the 5' end).

Amounts of proteins synthesized within cells depend both on concentrations of mRNA and the efficiency with which individual mRNAs are translated. Modulation of protein synthesis in the exocrine pancreas in response to caerulein stimulation depends initially on alterations in the efficiency of translation of individual mRNAs. Changes in translation of this type are probably due to interactions of specific regulatory proteins with the 5' nontranslated regions (or 5' ends) of individual mRNAs. The binding of these proteins may be regulated by second-messenger–activated pathways. Consistent with this possibility is the finding that the 5' ends of mRNAs for both amylase and anionic trypsinogen are significantly conserved between rat and dog [Fig. 8; (51)]. The presence of this degree of conservation at the 5' ends of these two mRNAs represents a dramatic finding when compared with the considerable evolutionary drift observed in the 3' noncoding regions of these same mRNAs, which reflects more accurately the estimated evolutionary distance (70 million years) between rat and dog.

The 5' conserved sequences in amylase and anionic

```
                              -10    -5   +1
rat anionic trypsinogen       CCUUCUGCCACCAUG
dog anionic trypsinogen    AUACUUCUGCCAUCAUG

                        -15    -10    -5   +1   +5    +10
rat amylase             GACAACUUCAAAGCAAAAUGAAGUUCGUU
mouse amylase           GACAACUUCAAAGCAAAAUGAAGUUCGUU
dog amylase             GACAACU-CAAAGCAAAAUGAAGUUCGUU
```

FIG. 8. Conserved sequences at 5' termini of mRNAs encoding pancreatic anionic trypsinogen and amylase. *Vertical lines*, identities. *Dash* in dog amylase mRNA sequence is introduced as a deletion to maximize sequence alignments. AUG initiation codons are underlined. *Brackets*, complementary sequences that may base-pair to form stem-loop structures. Anionic trypsinogen mRNA sequences from the rat (39) and dog (51); amylase mRNA sequences from the mouse (69), rat (38), and dog. (K. S. LaForge and G. A. Scheele, unpublished observations.)

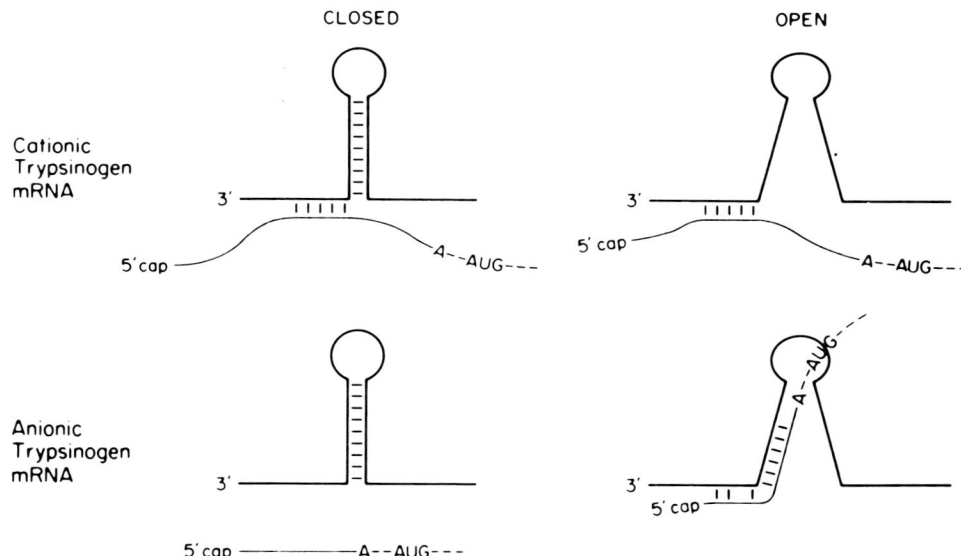

FIG. 9. Potential function of the stem-loop structure at the 3' end of 18S ribosomal RNA (rRNA) as a molecular switch responsible for differential regulation of trypsinogen mRNA translation (51, 65). The 5' nontranslated region of cationic trypsinogen mRNA contains 5 contiguous nucleotides that show potential base-pairing with the single-stranded 3'-terminal sequence of 18S rRNA. In contrast, the 5' nontranslated region of anionic trypsinogen mRNA contains a conserved region in which 9 out of 10 nucleotides show potential hybridization largely to the left arm of the double-stranded rRNA (stem) structure. Development of an open configuration in the stem-loop sequence would have little effect on potential interactions with cationic trypsinogen mRNA but would allow extensive interactions with anionic trypsinogen mRNA, which could promote the alignment of mRNA during initiation of mRNA translation.

trypsinogen mRNAs have already suggested mechanisms by which the efficiency of translation of these two mRNAs may be modulated. The conserved 5' sequences found in the rat and dog anionic trypsinogen mRNAs, but not in the mRNA sequence for cationic trypsinogen, correlates with the observed regulation in the translation of anionic but not cationic trypsinogen mRNA (51). In this case modulation of translation efficiency through the binding of a regulatory protein(s) to the 5' nontranslated mRNA sequence could be involved. In this mechanism the binding of a regulatory protein may promote the efficiency of translation of the cognate mRNA. In amylase mRNAs from rat and mouse, potential hybridization occurs over a stretch of seven contiguous bases (positions -8 to -14 paired with positions $+2$ to $+8$). The high degree of identity observed at the 5' end of the three amylase mRNAs and the potential secondary structure involving the AUG initiation codon suggest that regulation of secondary structure could be involved in the translational control of amylase mRNA translation.

Finally, RNA-RNA interactions between mRNA and ribosomal RNA (rRNA) may also be involved in regulation of mRNA translation under certain cellular states. The conserved 5' nontranslated region of anionic trypsinogen mRNA is complementary to a sequence near the 3' end of 18S rRNA, which normally exists in a double-stranded form in the small ribosomal subunit [Fig. 9; (86)]. It is possible, however, that the stem-loop structure in the rRNA may be opened under specific conditions (e.g., hormonal stimulation, nutritional shock, or viral infection). A change in the conformation of 18S rRNA could explain the differential regulation observed in the translation of mRNAs that code for trypsinogen isoenzymes (51). The open configuration of this stem-loop structure would have little effect on the potential interaction between cationic trypsinogen mRNA and the 18S rRNA. However, base pairing between anionic trypsinogen mRNA and 18S rRNA could only occur with the open structure depicted in Figure 9. Interaction in this manner may facilitate the functional binding of anionic trypsinogen mRNA in proximity to and in optimal alignment with the initiator UAC anticodon [alignment theory for the differential regulation of trypsinogen mRNA translation (65)]. Under such conditions the efficiency of translation of anionic but not cationic trypsinogen mRNA may increase, resulting in an increased synthesis of anionic but not cationic trypsinogen, a finding observed during caerulein stimulation (70) and during nutritional shock with a zero-protein diet (71). Further work is needed to demonstrate that the stem-loop structure near the 3' end of rRNA may be destabilized and opened under physiological conditions. Conceivably a small cytoplasmic RNA could perform this function by forming a more stable interaction with the right arm of the double-stranded rRNA structure.

REFERENCES

1. ADLER, G., AND H. F. KERN. Regulation of exocrine pancreatic secretory process by insulin in vivo. *Horm. Metab. Res.* 7: 290–296, 1975.
2. ANONYMOUS. Nomenclature of multiple forms of enzymes. *Eur. J. Biochem.* 82: 1–3, 1978.
3. BANERJI, J., L. OLSON, AND W. SCHAFFNER. A lymphocyte-specific cellular enhancer is located downstream of the joining region in immunoglobulin heavy chain genes. *Cell* 33: 729–740, 1983.
4. BELL, G., C. QUINTO, M. QUIROGA, P. VALENZUELA, C. CRAIK, AND W. RUTTER. Isolation and sequence of a rat chymotrypsinogen B gene. *J. Biol. Chem.* 259: 14265–14270, 1984.
5. BENNE, R., AND J. W. B. HERSHEY. Purification and characterization of initiation factor IF-E3 from rabbit reticulocytes. *Proc. Natl. Acad. Sci. USA* 73: 3005–3009, 1976.
6. BENNE, R., C. WONG, M. LUEDI, AND J. W. B. HERSHEY. Purification and characterization of initiation factor IF-E2 from rabbit reticulocytes. *J. Biol. Chem.* 251: 7675–7681, 1976.
7. BENOIST, C., K. O'HARE, R. BREATHNACH, AND P. CHAMBON. The ovalbumin gene-sequence of putative control regions. *Nucleic Acids Res.* 8: 127–142, 1980.
8. BERGET, S., C. MOORE, AND P. A. SHARP. Spliced segments at the 5' terminus of adenovirus 2 late mRNA. *Proc. Natl. Acad. Sci. USA* 74: 3171–3175, 1977.
9. BOULET, A. M., C. R. ERWIN, AND W. J. RUTTER. Cell-specific enhancers in the rat exocrine pancreas. *Proc. Natl. Acad. Sci. USA* 83: 3599–3603, 1986.
10. CATO, A. C., R. MIKSICEK, G. SCHÜTZ, J. ARNEMANN, AND M. BEATO. The hormone regulatory element of mouse mammary tumor virus mediates progesterone induction. *EMBO J.* 5: 2237–2240, 1986.
11. CHANG, A., AND J. D. JAMIESON. Stimulus-secretion coupling in the developing exocrine pancreas: secretory responsiveness to cholecystokinin. *J. Cell Biol.* 103: 2353–2365, 1986.
12. CHOW, L. T., R. E. GELINAS, T. R. BROKER, AND R. J. ROBERTS. An amazing sequence arrangement at the 5' ends of adenovirus 2 messenger RNA. *Cell* 12: 1–8, 1977.
13. CRAIK, C., Q. CHOO, G. SWIFT, C. QUINTO, R. MACDONALD, AND W. RUTTER. Structure of two related rat pancreatic trypsin genes. *J. Biol. Chem.* 259: 14255–14264, 1984.
14. DAGORN, J. C., AND R. G. LAHAIE. Dietary regulation of pancreatic protein synthesis. I. Rapid specific modulation of enzyme synthesis by changes in dietary composition. *Biochim. Biophys. Acta* 654: 111–118, 1981.
15. DAGORN, J. C., AND R. MONGEAU. Different action of hormonal stimulation on the biosynthesis of three pancreatic enzymes. *Biochim. Biophys. Acta* 498: 76–82, 1977.
16. DESNUELLE, P., J. P. REBOUD, AND A. BEN ABDELJLIL. Influence of the composition of the diet on the enzyme content of rat pancreas. In: *The Exocrine Pancreas*, edited by A. V. S. de Reuck and M. P. Cameron. Boston, MA: Little, Brown, 1962, p. 90–114. (Ciba Found. Symp.)
17. DYNAN, W. S., AND R. TJIAN. The promoter-specific transcription factor SpI binds to upstream sequences in the SV40 early promoter. *Cell* 35: 79–87, 1983.
18. EFSTRATIADIS, A., J. W. POSAKONY, T. MANIATIS, R. M. LAWN, C. O'CONNELL, R. A. SPRITZ, J. K. DERIEL, B. G. FORGET, S. M. WEISSMAN, J. L. SLIGHTOM, A. E. BLECHL, O. SMITHIES, F. E. BARALLE, C. C. SHOULDERS, AND N. J. PROUDFOOT. The structure and evolution of the human beta-globin gene family. *Cell* 21: 653–668, 1980.
19. FAICHNEY, A., W. Y. CHEY, Y. C. KIM, K. Y. LEE, AND M. S. KIM. Effect of sodium oleate on plasma secretin concentration and pancreatic secretion in dog. *Gastroenterology* 81: 458–462,

1981.
20. GILLIES, S. D., S. L. MORRISON, V. T. OI, AND S. TONEGAWA. A tissue-specific transcription enhancer element is located in the major intron of a rearranged immunoglobulin heavy chain gene. *Cell* 33: 717–728, 1983.
21. GIORGI, D., J. P. BERNARD, R. LAPOINTE, AND J. C. DAGORN. Regulation of amylase messenger RNA concentration in rat pancreas by total food content. *EMBO J.* 3: 1521–1524, 1984.
22. GIORGI, D., W. RENAUD, J. P. BERNARD, AND J. C. DAGORN. Regulation of proteolytic enzyme activities and mRNA concentrations in rat pancreas by food content. *Biochem. Biophys. Res. Commun.* 127: 937–942, 1985.
23. GOODBOURN, S., H. BURNSTEIN, AND T. MANIATIS. The β-interferon gene enhancer is under negative control. *Cell* 45: 601–610, 1986.
24. GRIFO, J. A., S. M. TAHARA, J. P. LEIS, M. A. MORGAN, A. J. SHATKIN, AND W. C. MERRICK. Characterization of eukaryotic initiation factor 4A, a protein involved in ATP-dependent binding of globin mRNA. *J. Biol. Chem.* 257: 5246–5252, 1982.
25. GRIFO, J. A., S. M. TAHARA, M. A. MORGAN, A. J. SHATKIN, AND W. C. MERRICK. New initiation factor activity eIF4F required for globin mRNA translation. *J. Biol. Chem.* 258: 5804–5810, 1983.
25a. GROSSMAN, A., A. M. BOCTOR, P. BAND, AND B. LANE. Role of steroids in secretions—modulating effect of triamcinolone and estradiol on protein synthesis and secretion from the rat exocrine pancreas. *J. Steroid Biochem.* 19: 1069–1081, 1983.
26. GROSSMAN, M. I., H. GREENGARD, AND A. C. IVY. The effect of dietary composition of pancreatic enzymes. *Am. J. Physiol.* 138: 676–682, 1943.
27. HAGENBÜCHLE, O., U. SCHIBLER, S. PETRUCCO, G. C. VAN TUYLE, AND P. K. WELLAUER. Expression of mouse Amy-2a alpha-amylase genes is regulated by strong pancreas-specific promoters. *J. Mol. Biol.* 185: 285–293, 1985.
28. HAN, J. H., L. RALL, AND W. J. RUTTER. Selective expression of rat pancreatic genes during embryonic development. *Proc. Natl. Acad. Sci. USA* 83: 110–114, 1986.
29. HUEZ, G., C. BRUCK, AND Y. CLEUTER. Translational stability of native and deadenylated rabbit globin mRNA injected into HeLa cells. *Proc. Natl. Acad. Sci. USA* 78: 908–911, 1981.
30. KORC, M., D. OWERBACH, C. QUINTO, AND W. J. RUTTER. Pancreatic islet-acinar cell interaction: amylase messenger RNA levels are determined by insulin. *Science Wash. DC* 213: 351–353, 1981.
31. KOZAK, M. Comparison of initiation of protein synthesis in prokaryotes, eukaryotes and organelles. *Microbiol. Rev.* 47: 1–45, 1983.
32. KOZAK, M. Point mutations close to the AUG initiation codon affect the efficiency of translation of rat preproinsulin in vivo. *Nature Lond.* 308: 241–246, 1984.
33. KOZAK, M. Compilation and analysis of sequences upstream from the translational start site in eukaryotic mRNAs. *Nucleic Acids Res.* 12: 857–872, 1984.
34. KOZAK, M. Point mutations define a sequence flanking the AUG initiator codon that modulates translation by eukaryotic ribosomes. *Cell* 44: 283–292, 1986.
35. LIDDLE, R. A., G. M. GREEN, C. K. CONRAD, AND J. A. WILLIAMS. Proteins but not amino acids, carbohydrates, or fats stimulate cholecystokinin secretion in the rat. *Am. J. Physiol.* 251 (*Gastrointest. Liver Physiol.* 14): G243–G248, 1986.
36. LOGSDON, C. D., J. MOESSNER, J. A. WILLIAMS, AND I. D. GOLDFINE. Glucocorticoids increase amylase mRNA levels, secretory organelles, and secretion in pancreatic acinar AR42J cells. *J. Cell Biol.* 100: 1200–1208, 1985.
36a. LOGSDON, C. D., K. J. PEROT, AND A. R. MCDONALD. Mechanism of glucocorticoid-induced increase in pancreatic amylase gene transcription. *J. Biol. Chem.* 262: 15765–15769, 1987.
37. LÜTCKE, H., K. C. CHOW, F. S. MICKEL, K. A. MOSS, H. F. KERN, AND G. A. SCHEELE. Selection of AUG initiation codons differs in plants and animals. *EMBO J.* 6: 43–48, 1987.
38. MACDONALD, R. J., M. CRERAR, W. SWAIN, R. PICTET, G. THOMAS, AND W. RUTTER. Structure of a family of rat amylase genes. *Nature Lond.* 287: 117–122, 1980.
39. MACDONALD, R. J., S. STARY, AND G. H. SWIFT. Two similar but nonallelic rat pancreatic trypsinogens. *J. Biol. Chem.* 257: 9724–9732, 1982.
40. MCKNIGHT, S., AND R. TJIAN. Transcriptional selectivity of viral genes in mammalian cells. *Cell* 36: 795–805, 1986.
41. MOUNT, S. M. A catalogue of splice junction sequences. *Nucleic Acids Res.* 10: 459–472, 1982.
42. MYERS, R. M., K. TILLY, AND T. MANIATIS. Fine structure genetic analysis of a β-globin promoter. *Science Wash. DC* 232: 613–618, 1986.
43. NEVINS, J. R. The pathway of eukaryotic mRNA formation. *Annu. Rev. Biochem.* 52: 441–466, 1983.
44. NIEBEL, W., M. V. SINGER, L. E. HANSSEN, AND H. GOEBELL. Effect of atropine on pancreatic bicarbonate output and plasma concentrations of immunoreactive secretin in response to intraduodenal stimulants. *Scand. J. Gastroenterol.* 18: 803–808, 1983.
45. OCHOA, S. Regulation of protein synthesis initiation in eukaryotes. *Arch. Biochem. Biophys.* 223: 325–349, 1983.
46. OHLSSON, B. Characterization and Physiological Function of Proteinases of Pancreatic Origin and Proteinase Inhibitors in the Developing Pig. Univ. of Lund, Sweden, 1987. Thesis.
47. PALLA, J., A. BEN ABDELJLIL, AND P. DESNUELLE. Action de l'insuline sur le biosynthese se l'amylase et de quelques autres enzymes du pancreas de rat. *Biochim. Biophys. Acta* 158: 25–35, 1968.
48. PAVLOV, I. P. *The Digestive Glands.* Birmingham, AL: Classics Med. Library, 1897, p. 41–42.
49. PELHAM, H. R. B., AND M. BIENZ. A synthetic heat shock promoter element confers heat inducibility on the herpes simplex virus thymidine kinase gene. *EMBO J.* 1: 1473–1477, 1982.
50. PICTET, R., AND W. J. RUTTER. Development of the embryonic endocrine pancreas. In: *Handbook of Physiology. Endocrinology. Endocrine Pancreas*, edited by D. F. Steiner and N. Freinkel. Washington, DC: Am. Physiol. Soc., 1972, sect. 7, vol. I, chapt. 2, p. 25–66.
51. PINSKY, S., S. LAFORGE, AND G. SCHEELE. Differential regulation of trypsinogen mRNA translation: full-length mRNA sequences encoding two oppositely charged trypsinogen isoenzymes in the dog pancreas. *Mol. Cell. Biol.* 5: 2669–2676, 1985.
52. PITTET, A.-C., AND U. SCHIBLER. Mouse α-amylase loci, Amy-1a and Amy-2a, are closely linked. *J. Mol. Biol.* 182: 359–365, 1985.
53. RALL, L., R. PICTET, S. GITHENS, AND W. J. RUTTER. Glucocorticoids modulate the in vitro development of the embryonic rat pancreas. *J. Cell Biol.* 75: 398–409, 1977.
54. RAUSCH, U., P. VASILOUDES, K. RÜDIGER, AND H. F. KERN. In-vivo stimulation of rat pancreatic acinar cells by infusion of secretin. II. Changes in individual rates of enzyme and isoenzyme biosynthesis. *Cell Tissue Res.* 242: 641–644, 1985.
55. RAUSCH, U., P. VASILOUDES, K. RÜDIGER, H. KERN, AND G. SCHEELE. Synthesis of lipase in the rat pancreas is regulated by secretin. *Pancreas* 1: 522–528, 1986.
56. REED, R., AND T. MANIATIS. A role for exon sequences and splice-site proximity in splice-site selection. *Cell* 46: 681–690, 1986.
57. REINBERG, D., M. HORIKOSHI, AND R. G. ROEDER. Factors involved in specific transcription in mammalian RNA polymerase II. Functional analysis of initiation factors IIA and IID and identification of a new factor operating at sequences downstream of the initiation site. *J. Biol. Chem.* 262: 3322–3330, 1987.
58. RENAUD, W., D. GIORGI, J. JOVANNA, AND J. C. DAGORN. Regulation of concentrations of mRNA for amylase, trypsinogen I and chymotrypsinogen B in rat pancreas by secretagogues. *Biochem. J.* 235: 305–308, 1986.
59. RENKAWITZ, R., G. SCHÜTZ, D. VON DER AHE, AND M. BEATO. Sequences in the promoter region of the chicken lysosome gene required for steroid regulation and receptor binding. *Cell* 37: 503–510, 1984.
60. RICHARDS, R. I., A. HEGUY, AND M. KARIN. Structural and functional analysis of the human metallothionein-IA gene: dif-

ferential induction by metal ions and glucocorticoids. *Cell* 37: 263–272, 1984.
61. SAAB, J. E., P. GODFREY, AND P. M. BRANNON. Adaptive response of rat pancreatic lipase to dietary fat: effects of amount and type of fat. *J. Nutr.* 116: 892–899, 1986.
62. SCHAFFALITSKY DE MUCKADELL, O. B., O. OLSEN, P. CANTOR, AND E. MAGID. Concentrations of secretin and CCK in plasma and pancreaticobiliary secretion in response to intraduodenal acid and fat. *Pancreas* 1: 536–543, 1986.
63. SCHEELE, G. A. Two-dimensional gel analysis of soluble proteins. Characterization of guinea pig exocrine pancreatic proteins. *J. Biol. Chem.* 250: 5375–5385, 1975.
64. SCHEELE, G. A. Analysis of the secretory process in the exocrine pancreas by two-dimensional isoelectric focusing/sodium dodecyl sulfate gel electrophoresis. *Methods Cell Biol.* 23: 345–358, 1981.
65. SCHEELE, G. A. Regulation of gene expression in the exocrine pancreas. In: *The Exocrine Pancreas: Biology, Pathobiology, and Diseases*, edited by V. L. W. Go, F. P. Brooks, E. P. DiMagno, J. D. Gardner, E. Lebenthal, E. P. DiMagno, and G. A. Scheele. New York: Raven, 1986, p. 55–67.
66. SCHEELE, G. A. Two dimensional electrophoresis in the analysis of exocrine pancreatic proteins. In: *The Exocrine Pancreas: Biology, Pathobiology, and Diseases*, edited by V. L. W. Go, F. P. Brooks, E. P. DiMagno, J. D. Gardner, E. Lebenthal, and G. A. Scheele. New York: Raven, 1986, p. 185–192.
67. SCHEIDEREIT, C., AND M. BEATO. Contacts between hormone receptor and DNA double helix within a glucocorticoid regulatory element of mouse mammary tumor virus. *Proc. Natl. Acad. Sci. USA* 81: 3029–3033, 1984.
68. SCHIBLER, U., O. HAGENBÜCHLE, P. K. WELLAUER, AND A. C. PITTET. Two promoters of different strengths control the transcription of the mouse alpha-amylase gene Amy-1a in the parotid gland and the liver. *Cell* 33: 501–508, 1983.
69. SCHIBLER, U., A. C. PITTET, R. YOUNG, O. HAGENBÜCHLE, M. TOSI, S. GELLMAN, AND P. WELLAUER. The mouse alpha-amylase multigene family: sequence organization of members expressed in the pancreas, salivary gland and liver. *J. Mol. Biol.* 155: 247–266, 1982.
70. SCHICK, J., H. KERN, AND G. SCHEELE. Hormonal stimulation in the exocrine pancreas results in coordinate and anticoordinate regulation of protein synthesis. *J. Cell Biol.* 99: 1559–1564, 1984.
71. SCHICK, J., R. VERSPOHL, H. KERN, AND G. SCHEELE. Two distinct adaptive responses in the synthesis of exocrine pancreatic enzymes to inverse changes in protein and carbohydrate in the diet. *Am. J. Physiol.* 247 (*Gastrointest. Liver Physiol.* 10): G611–G616, 1984.
72. SEARLE, P. F., G. W. STUART, AND R. D. PALMITER. Building a metal-responsive promoter with synthetic regulatory elements. *Mol. Cell. Biol.* 5: 1480–1489, 1985.
73. SHATKIN, A. J. Cap structure at the 5' end of eukaryotic cellular and viral mRNAs. *Cell* 9: 645–652, 1976.
74. SLATER, E. P., O. RABENAU, M. KARIN, J. D. BAXTER, AND M. BEATO. Glucocorticoid receptor binding and activation of a heterologous promoter by dexamethasone by the first intron of the human growth hormone gene. *Mol. Cell. Biol.* 5: 2984–2992, 1985.
75. SNOOK, J. T. Dietary regulation of pancreatic enzymes in the rat with emphasis on carbohydrate. *Am. J. Physiol.* 221: 1383–1387, 1971.
76. SÖLING, H. D., AND K. O. UNGER. The role of insulin in the regulation of alpha-amylase synthesis in the rat pancreas. *Eur. J. Clin. Invest.* 2: 199–212, 1972.
77. STRATOWA, C., AND W. J. RUTTER. Selective regulation of trypsin gene expression by calcium and by glucose starvation in a rat exocrine pancreas cell line. *Proc. Natl. Acad. Sci. USA* 83: 4292–4296, 1986.
78. STUART, G. W., P. F. SEARLE, AND R. D. PALMITER. Identification of multiple metal regulatory elements in mouse metallothionein-I promoter by assaying synthetic sequences. *Nature Lond.* 317: 828–831, 1985.
79. SUSKE, G., M. WENZ, A. C. CATO, AND M. BEATO. The uteroglobin gene region: hormonal regulation, repetitive elements and complete nucleotide sequence of the gene. *Nucleic Acids Res.* 11: 2257–2271, 1983.
80. SWIFT, G., C. CRAIK, S. STARY, C. QUINTO, R. LAHAIE, W. RUTTER, AND R. MACDONALD. Structure of the two related elastase genes expressed in the rat pancreas. *J. Biol. Chem.* 259: 14271–14278, 1984.
81. SWIFT, G. H., R. E. HAMMER, R. J. MACDONALD, AND R. L. BRINSTER. Tissue specific expression of rat pancreatic elastase I gene in transgenic mice. *Cell* 38: 639–646, 1984.
82. TAHARA, S. M., M. A. MORGAN, AND A. J. SHATKIN. Two forms of purified ^{7}mG-cap binding protein with different effects on capped mRNA translation in extracts of uninfected and poliovirus-infected HeLa cells. *J. Biol. Chem.* 256: 7691–7694, 1981.
83. TREISMAN, R. Transient accumulation of c-fos RNA following serum stimulation requires a conserved 5' element and c-fos 3' sequences. *Cell* 42: 889–902, 1985.
84. TRIMBLE, E. R., R. BRUZZONE, AND D. BELIN. Insulin resistance is accompanied by impairment of amylase gene expression in the exocrine pancreas of the obese Zucker rat. *Biochem. J.* 237: 807–812, 1986.
85. TRIMBLE, E. R., U. RAUSCH, AND H. F. KERN. Changes in individual rates of pancreatic enzyme and isoenzyme biosynthesis in the obese Zucker rat. *Biochem. J.* 248: 771–777, 1987.
86. VAN KNIPPENBERG, P. H., J. M. A. VAN KIMMENADE, AND H. A. HEUS. Phylogeny of the conserved 3' terminal structure of the RNA of small ribosomal subunits. *Nucleic Acids Res.* 12: 2595–2604, 1984.
87. VON DER AHE, D., S. JANICH, C. SCHEIDEREIT, R. RENKAWITZ, G. SCHÜTZ, AND M. BEATO. Glucocorticoid and progesterone receptors bind to the same sites in two hormonally regulated promoters. *Nature Lond.* 313: 706–709, 1985.
88. WALKER, M., T. EDLUND, A. BOULET, AND W. RUTTER. Cell specific expression controlled by the 5' flanking region of insulin and chymotrypsin genes. *Nature Lond.* 306: 557–561, 1983.
89. WEIHER, H., M. KÖNIG, AND P. GRUSS. Multiple point mutations affecting the simian virus 40 enhancer. *Science Wash. DC* 219: 626–631, 1983.
90. WELCH, J. W., J. I. GARRELS, G. P. THOMAS, J. J.-C. LIN, AND J. R. FERAMISCO. Biochemical characterization of the mammalian stress proteins and identification of two stress proteins as glucose- and calcium-ionophore regulated proteins. *J. Biol. Chem.* 258: 7102–7111, 1983.
91. WICKER, C., AND A. PUIGSERVER. Effects of inverse changes in dietary lipid and carbohydrate on the synthesis of some pancreatic secretory proteins. *Eur. J. Biochem.* 162: 25–30, 1987.
92. WICKER, C., A. PUIGSERVER, AND G. A. SCHEELE. Dietary regulation of levels of active mRNA coding for amylase and serine protease zymogens in the rat pancreas. *Eur. J. Biochem.* 139: 381–387, 1984.
93. YOUNG, R. A., O. HAGENBÜCHLE, AND U. SCHIBLER. A single mouse alpha-amylase gene specifies two different tissue-specific mRNAs. *Cell* 23: 451–458, 1981.
94. ZINN, K., D. DIMAIO, AND J. DEMAEYER-GUIGNARD. Identification of two distinct regulatory regions adjacent to the human beta-interferon gene. *Cell* 34: 865–879, 1983.
95. ZINN, K., AND T. MANIATIS. Detection of factors that interact with the human beta-interferon regulatory region in vivo by DNase I footprinting. *Cell* 45: 611–618, 1986.

Long-term regulation of pancreatic function studied in vitro

CRAIG D. LOGSDON | Laboratory of Cell Biology, Mount Zion Hospital and Medical Center, and Department of Physiology, University of California, San Francisco, California

CHAPTER CONTENTS

Long-Term In Vitro Models of the Exocrine Pancreas
  Organ explant cultures
    Explants of fetal pancreas
    Explants of adult pancreas
  Isolated Cell Cultures
    Acinar cells in suspension cultures
    Acinar cells in attached cultures
    Duct cells
  Pancreatic tumors and cell lines
    Acinar carcinomas
    Ductal cell lines
Physiological Regulation of Functions Studied in Long-Term In Vitro Systems
  Regulation of differentiation: effects of factors on the synthesis of secretory enzymes
  Regulation of growth
  Regulation of secretion
  Regulation of hormone receptors
Conclusions

MUCH OF THE CURRENT understanding of the neurohormonal regulation of the exocrine pancreas has been derived from investigations carried out with intact animals. These studies have demonstrated an array of complex interactions between the pancreas and the gut that involves multiple feedback pathways that are largely unknown in detail. In addition, they have suggested the existence of important interactions between the endocrine and exocrine portions of the pancreas (110). The multiplicity of complex interactions in vivo complicates the interpretation of experimental data; thus understanding of direct interactions and molecular mechanisms of neurohormonal regulation of the exocrine pancreas requires use of in vitro systems.

The development of acute in vitro preparations of pancreatic exocrine cells played a major role in advancing investigation of the regulation of the pancreas. Early in vitro studies of pancreatic function used either gland slices or fragments. Although significant data were obtained, major problems existed in terms of tissue viability, representational sampling, diffusional barriers, and the presence of multiple cell types. To circumvent these problems, Amsterdam and Jamieson (1) developed a procedure for isolating pancreatic acinar cells. The basis of this method was the use of collagenase and other proteases to digest the connective tissue stroma and of Ca^{2+} chelators and mechanical disruption to free individual cells (1, 2). Acinar cells isolated in this manner have been used effectively to investigate receptors, second messengers, and ion fluxes. By contrast, secretagogue-elicited secretion of digestive enzymes is reduced in the isolated cell preparation when compared with the in situ pancreas (1, 2). To overcome this problem, a technique was devised that used purified collagenase and eliminated the Ca^{2+} chelation step to produce intact acini that are composed of several acinar cells connected to a central lumina and that retain their normal intercellular junctions and membrane polarity (2, 111). Preparations of intact acini respond to secretagogues to a degree comparable with that of the original pancreas. Isolated acini have proven to be the system of choice for the investigation of acute acinar cell–hormone interactions. However, the viability of isolated acini maintained in physiological Ringer's solution decreases with time. Thus this preparation is not useful for studies of hormone action that occur over periods of more than a few hours.

The study of long-term effects of hormones on the pancreas is important because growth, differentiation, and dietary adaptation are likely under hormonal regulation. Moreover, long-term in vitro preparations would be useful for studying various pathologies; chronic stimulation with supramaximal doses of secretagogues results in a destructive effect closely resembling acute pancreatitis. In vitro model systems would also have great value as a tool for investigation of pancreatic cancer.

LONG-TERM IN VITRO MODELS OF
THE EXOCRINE PANCREAS

It is now established that the maintenance of differentiated cellular functions in vitro is influenced by a number of factors, including nutrients, vitamins, hormones, extracellular matrix components, and cell-cell interactions (3, 19). The requirements of different cell types vary greatly so that development of a successful culture system for a particular cell type is an empirical process. The endocrine components of the pancreas are maintained well in culture under conditions in which the exocrine components rapidly degenerate, a fact that has been used to enrich the endocrine component (17, 59–64, 73). Consequently, much is known concerning the in vitro requirements of the endocrine pancreas, but much less is known concerning the factors required for the maintenance of the exocrine pancreas in culture. Various types of cultures and what they have indicated about the in vitro requirements of the exocrine pancreas are described next.

Organ Explant Cultures

In organ explant culture a portion of the intact organ is removed from the animal and placed directly into culture. Of the various methods of culturing that have been applied to the pancreas, organ explant culture was the first, is the simplest, and is the most often attempted. Carrel and Lindbergh (10) reported placing cat pancreatic explants in culture in 1938. Organ explant cultures have advantages: *1*) cell-cell interactions remain intact, *2*) the extracellular matrix is preserved, and *3*) hormones or autocrine factors present in the whole organ remain present in the culture. Organ explant cultures also have disadvantages: *1*) diffusion of nutrients and gases toward, and metabolic wastes away from, the cells is difficult, and organ slices often show rapid central necrosis; and *2*) the culture retains a complex mixture of cells, complicating elucidation of direct interactions of hormonal regulators and their molecular mechanisms.

The majority of pancreatic organ explant culture studies have been conducted to study either normal pancreatic embryological development or the pathological alterations that occur in pancreatic cancer. In general, these studies have been based on qualitative descriptions of alterations in structure. Functional parameters have rarely been measured. Explant cultures have not been used to study the regulation of physiological functions of the pancreas. However, these studies have provided important information concerning the conditions necessary for the long-term maintenance of pancreatic cells in vitro. The variables that have been identified that affect the success of pancreatic explant culture are *1*) age of the animal from which the explant is taken, *2*) the growth substrate, *3*) interactions with mesenchymal tissue, and *4*) the addition of protease inhibitors and glucocorticoids to the medium.

EXPLANTS OF FETAL PANCREAS. Age is the most important consideration in the survival of pancreatic explants in vitro. Conditions that maintain fetal tissues are inadequate for maintenance of adult tissues. McEvoy et al. (61) studied the age question directly using fetal rat pancreas explants and found that survival of acinar components was inversely proportional to the age of the donor animal. In general, developmentally undifferentiated pancreas (younger than 13-day-old fetal tissue) shows parallel in vivo and in vitro differentiation (42, 76, 78, 94), whereas developmentally more mature pancreas (17 days postcoitus and older) shows a selective development of the endocrine pancreas at the expense of the exocrine component [Table 1; (62, 63)].

The substrate on which explants are cultured is critical to their survival, growth, and differentiation. Various substrata have been employed for fetal explant cultures, including a silk grid (60–64), tissue culture plastic (28), Gelfoam sponge (87–89), Millipore filter strips (22, 26, 42, 75–81), irradiated fibroblasts (59), and irradiated pigskin (31). Pancreatic explants are poorly maintained on plastic or glass. Best results have been obtained with substrates that float at or near the surface of the medium. The advantage of these substrates may be that they help to overcome the problem of gas diffusion into the mass of the explant. For this same reason, perfusion culture, in which the cells are constantly exposed to fresh, aerated medium, has also been reported to be useful for culturing pancreatic explants (70, 90).

The importance of an epitheliomesenchymal interaction in pancreatic morphogenesis was originally suggested by Golosow and Grobstein (26). They showed that intact 11-day-old dorsal pancreatic rudiment, cultured in vitro, developed to a stage of differentiation corresponding to the pancreas of a 15-day-old embryo, whereas isolated epithelium stripped of mesenchyme did not develop. When the isolated epithelium was recombined with pancreatic mesenchyme or with a number of heterogeneous mesenchymes, even when

TABLE 1. *Comparison of Absolute Changes in Pancreatic Tissue Component Masses*

Tissue	Percent Acinar	
	In vivo	4-Day-old explant
16-Day-old fetal pancreas	0	25.6±1.9
18-Day-old fetal pancreas	4.5±0.8	22.9±2.9
22-Day-old fetal pancreas	67.5±0.2	0
4-Day-old neonatal pancreas	57.8±2.1	1.0±0.5

Pancreases were removed from animals of the indicated ages and quantitated as to the percentage of total tissue mass, which was judged to be acinar either before or after 4 days of culture. Quantitation was conducted by using morphometric analysis based on a linear scan method. [Adapted from McEvoy et al. (62).]

separated by a filter, it continued to undergo morphogenesis resembling that of the intact rudiment. These findings were confirmed in greater ultrastructural detail by Kallman and Grobstein (42). Wessells (108) utilized an autoradiographic approach and demonstrated that the presence of salivary mesenchyme increased the number of embryonic pancreatic explant nuclei that synthesized DNA after 24 h in culture. Rutter et al. (93) further showed that the presence of mesenchyme was required for the dramatic increase in the specific activity of the exocrine proteins that normally occurs during development. These experiments demonstrated that differentiation of pancreatic epithelial cells is influenced by some transmissible factor present in the mesoderm. Later it was determined that the mesodermal requirement could be replaced by addition of a crude embryo extract or a particulate fraction of the extract (45, 91). The active component appeared to act at the cell surface (45) and is likely a protein, because it was sensitive to proteases, but no further identification has been made.

In contrast to these studies, which suggested the importance of a mesodermally derived factor for development of the pancreas, Parsa and Marsh (76) developed a chemically defined medium without serum that supported the growth and differentiation of 13-day-old postcoital fetal rat pancreas explants on Millipore filters for 10 wk (78, 79, 80). Using this system, they showed a pattern of development that paralleled that of in utero development and that had previously only been seen in the presence of mesenchyme (Table 2). Electron-microscopic observation of 10-wk-old explants indicated the presence of acinar cells, centroacinar cells, islet cells containing typical B-cell granules, and duct cells (78). These studies seemed to rule out the necessity for mesenchyme, serum, or chick embryo extract in acinar cell differentiation. However, it is possible that the mesenchyme present in the explant itself may have had significant effects on differentiation. This culture system has been used extensively as an in vitro model for studies of pancreatic carcinogenesis (75, 80, 81). Physiological studies have not been attempted with this system, and it is not known whether the acinar cells that differentiate in these cultures are responsive to secretagogues or whether hormones influence their growth or digestive enzyme content.

Other medium components are also important for maintenance of pancreatic explants. The concentration and type of nutrients in the medium is a major determinant (66). Also, although pancreatic digestive enzymes are normally stored in an inactive form, they are apparently activated by unknown mechanisms in the culture medium. Therefore, protease inhibitors are required to prevent autodigestion of the explants by their own digestive enzymes (31). Adrenocortical steroids are also important medium constituents for pancreatic explant culture. Acinar cell cytodifferentiation was more fully maintained in fetal pancreatic explants when the cells were grown in parabiotic organ culture with fetal adrenal tissue (64). In particular, glucocorticoid hormones have been found to both increase the rate of differentiation in early embryonic explants (16, 27, 60, 82, 103, 113) and to maintain the differentiation of later fetal explants (11, 61, 62). Gonadal steroids were not able to preserve the pancreas in vitro (60).

EXPLANTS OF ADULT PANCREAS. Adult pancreas has been more difficult to culture than fetal pancreas. Typically, adult pancreatic acinar cells begin a pattern of degeneration and dedifferentiation within 24 h of culture. The dedifferentiation involves a loss of zymogen granules and a consequent decrease in digestive enzyme contents. Because most studies have been morphological, little else is known about the characteristics of the dedifferentiated cells. Explant organ culture systems of adult pancreas have been described for humans (80, 81), hamsters (87–89), and lizards (90).

For long-term maintenance of adult pancreatic tissue from the Syrian golden hamster (87–89), explants were placed in tissue culture dishes on Gelfoam sponge rafts. These rafts float the tissue near the air-medium interface. These cultures were reported to be viable for up to 70 days, with some zymogen granule–containing cells characteristic of acinar cells present throughout the entire culture period. However, most acinar cells underwent the typical dedifferentiation. The cellular processes involved in dedifferentiation of pancreatic acinar cells were studied in detail with this culture system (89). Degeneration was found to occur by the processes of crinophagy and autophagia, which involve lysosomal digestion of cell structures. Degeneration was prevented by low temperature or the inclusion in the medium of cycloheximide, a protein-synthesis inhibitor, indicating that newly synthesized

TABLE 2. *Comparison of Enzyme Activities in Pancreases of Embryos and Organ Cultures*

Day		Amylase		Chymotrypsin		Lipase	
Gestation	Culture	Embryo	Culture	Embryo	Culture	Embryo	Culture
13	0	trace	0.44	0.00	0.48	0.00	0.20
16	3	1.22	1.81	trace	1.57	0.00	1.40
19	6	3.02	2.50	4.00	3.62	3.62	2.72
22	9	3.62	3.11	4.54	4.25	4.05	3.04

Pancreases from 13-day-old fetuses were cultured for indicated times, and enzyme content of these explants was compared with that found in pancreases of animals of similar ages. Values are logarithms of specific enzyme activities. [Adapted from Parsa et al. (78).]

proteins may be involved in regulating these cellular activities. Degeneration was not prevented by inclusion in the medium of a secretogogue, as has been reported for cultures of isolated acinar cells (72). This model has also proven useful for the study of the effects of chemical carcinogens on the structure of the cells (87).

Hamster pancreas has also been maintained in explant culture on layers of irradiated fibroblastic cells (59), again a substrate that floats near the air-medium interface. Some acinar cells were identifiable after 17 days, although again the majority of cells underwent the usual pattern of dedifferentiation.

Adult human pancreas also has been studied in organ explant culture. The hormonally defined medium that successfully allowed differentiation of embryonic pancreas was used to attempt to maintain adult human pancreas (80, 81). Degeneration of acinar cells, ranging from vacuolization to loss of apical cytoplasm, began within 24 h in culture. By the end of the second week of culture, few clusters of acini were viable, although intact acini were reported to be more abundant immediately adjacent to major ducts. Further necrosis of the acinar cells progressed throughout the 6-wk culture period. These experiments again underscore the difference between fetal and adult pancreatic tissues in terms of maintenance in vitro.

Isolated Cell Cultures

ACINAR CELLS IN SUSPENSION CULTURES. Hay (29) placed isolated guinea pig pancreatic acinar cells in suspension with gyration and found that cell aggregates formed, which he then cultured on plastic substrate, although the cells rapidly dedifferentiated. Ruoff and Hay (92) conducted studies using this system, which documented the loss of cell function over time.

Oliver (72) described a culture system for isolated rat pancreatic lacrimal and parotid acinar cells in which the cells were maintained as colonial aggregates in suspension for up to 1 mo. The author suggested that the inclusion in the culture medium of a secretagogue was the most critical factor for long-term survival. Without the secretagogue, acinar cells degenerated rapidly. In the presence of secretagogue, the cells were reported to have remained morphologically differentiated and to have continued to synthesize protein at rates similar to those of freshly isolated cells for up to 4 wk. No attempt was made to measure the secretory ability of the pancreatic acinar cells cultured in this manner. These results have not been reproduced by others.

More recently, Logsdon and Williams (53) attempted to systematically define the optimal conditions for maintenance of differentiated functions of pancreatic acini in short-term suspension cultures. Mouse acini were cultured for 24 h in the presence of a variety of hormones and regulatory molecules. Digestive enzyme release in response to secretagogues was then measured. Acini were chosen for this study because, as just described, previous research with acute preparations showed that acini were more functionally intact than isolated cells (111). This study was the first to investigate the effects of short-term culture on subsequent secretory capabilities. Secretion of digestive enzymes is a complex, highly differentiated response that requires the interaction of several cellular systems and therefore is a particularly good indication of the functional integrity of the acinar cell. Although the length of the culture period was short, this study demonstrated for the first time that epidermal growth factor (EGF) increased protein synthesis and secretory abilities of the acinar cells. This study also demonstrated that high concentrations of corticosterone directly inhibited protein synthesis and digestive enzyme secretion in the cultured cells. These effects of EGF and corticosterone on acinar cells were not expressed during the time afforded by typical acute in vitro preparations. This short-term culture system, and modifications of it, has now been utilized for a variety of studies (32, 51, 52, 54, 68). It seems particularly useful for investigations of the regulation of hormone receptor homeostasis.

In a recent study, Brannon et al. (9) cultured pancreatic acinar cells for up to 4 days in serum-free medium and reported effects of hormones on physiological functions. Although the acinar cells in this preparation retained only 10% of their initial amylase content, they remained responsive to the cholinergic secretagogue carbamyl choline in terms of the ability to secrete amylase. Furthermore, the cells cultured in a defined medium with 1 μg/ml insulin contained 36% more amylase than did control cells. Because insulin is required for maintenance of normal levels of pancreatic amylase in vivo (98), these data were taken to support a role for insulin in the regulation of amylase. However, it was not determined whether insulin had a similar effect on other enzymes or whether the effect of insulin was expressed on the synthesis or degradation of amylase.

ACINAR CELLS IN ATTACHED CULTURES. To study the effects of regulatory molecules on the growth and differentiation of adult pancreatic acinar cells, Logsdon and Williams (55) cultured adult mouse acini on collagen gels. Cultures were examined biochemically and morphologically over a 2-wk period. The cells underwent a period of adaptation, during which there was the typical dedifferentiation similar to that reported in other cultures, including a decrease in the normal contents of digestive enzymes and a loss of zymogen granules. However, in these cultures the cells then underwent a period of growth, during which the number of cells increased and the cells spread to form a monolayer. There then appeared to be a period of redifferentiation, such that after 2 wk the cultures consisted of monolayers predominantly of a single

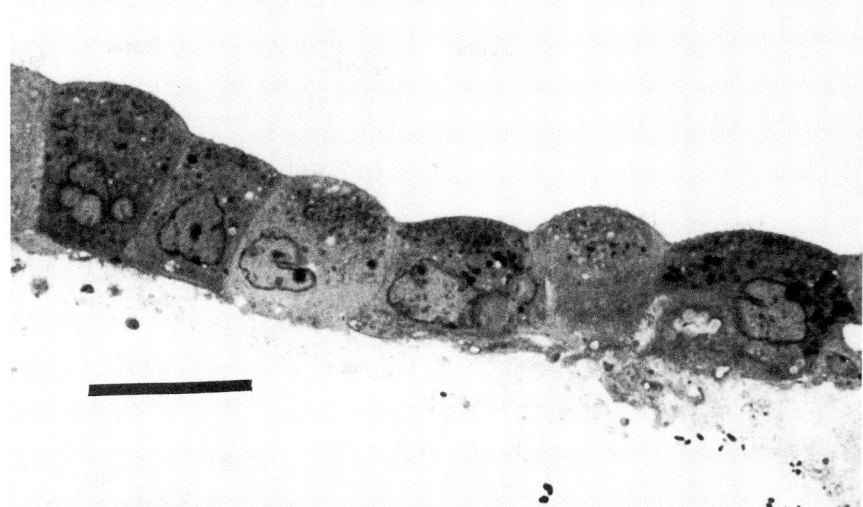

FIG. 1. Light micrograph of cross section of 14-day-old pancreatic cultures on collagen gels. Cells form a monolayer of cuboidal cells with obvious polarity. Bar, 20 μm. × 1,100. [From Logsdon and Williams (55).]

type of cuboidal cell (Fig. 1). The monolayers showed a polarity typical of secretory cells. Nuclei were located basally and the basal surface was relatively flat. The apical surface was rounded and the plasma membrane was covered with small microvilli; this portion of the cell was filled with dark-staining granules (Fig. 1). Under these culture conditions, fibroblastic overgrowth did not occur and endocrine cells were not observed. However, occasional cells were noted that displayed a ductlike appearance (not shown).

Further examination of the cultures at the ultrastructural level revealed characteristics of secretory cells (Fig. 2). The apical cytoplasm was filled with granules of varying sizes and shapes. Many of the granules were heterogeneous, containing small regions of higher electron density. The granules formed a gradient from larger to smaller from the cell interior toward the apex. Also evident in the cell cytoplasm were some large dark bodies that appeared to contain myelin whorls and were probably lysosomes. Rough endoplasmic reticulum was reduced in the cultured cells, but cisternae of rough endoplasmic reticulum containing electron-dense material, indicative of new protein synthesis, were evident. This preparation, although lacking the normal complement of secretory enzymes, responds to trophic factors and has proven useful for studies of the regulation of pancreatic cell growth (49).

Chen et al. (12) recently reported a different method of culturing pancreatic acinar cells on a solid support. These authors plated isolated rat pancreatic cells onto plastic that was then overlaid with a collagen gel. In these studies the cells formed three-dimensional organoid structures with well-defined epithelial lumina. Chen et al. suggested that these cells appeared ductlike on the basis of a lack of prominent zymogen granules, although whether these cells arise from ductal or acinar cell populations (or both) was not established.

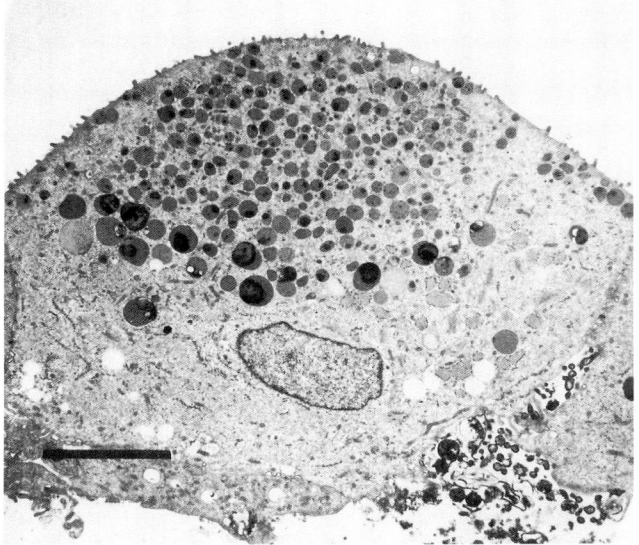

FIG. 2. Electron micrograph of a typical cell from a 14-day-old pancreatic culture in cross section. Bar, 5 μm. × 3,600. [From Logsdon and Williams (55).]

The behavior of the cells in these cultures indicates a considerable amount of cell-cell and cell-matrix interaction. The signals and mechanisms involved in these interactions are not understood.

DUCT CELLS. In organ explant cultures, pancreatic duct cells are often reported to survive more readily than acinar cells. However, it is not known whether the duct cells retain differentiated characteristics, because they have no obvious morphological or biochemical markers of differentiation, and therefore dedifferentiation may be more difficult to detect. The major difficulty in culturing the pancreatic duct cells is their low abundance; they represent only ±4% of the volume of the pancreas (8).

Cells of the main and largest interlobular ducts are the most accessible duct cells, although probably not the most interesting functionally. These duct cells have been cultured and studied in a variety of ways by Jones and co-workers (39–41). In one approach, Jones et al. removed bovine (39, 40) and human (39, 41) pancreatic ducts, cut them open, and cultured them as planar explants for up to 12 wk. The explants consisted of epithelium and underlying connective tissue, with the latter in contact with the culture dish. Although no functional studies were conducted, the cells appeared to be well maintained morphologically.

Isolated duct cells have also been cultured. Cells were isolated from the large ducts by scraping (99) or by enzymatic digestion (41, 95). Sato et al. (95) tested three different techniques for isolating bovine pancreatic duct cells and found that the digestion technique provided the greatest number of viable cells. In the digestion technique, the main pancreatic duct is perfused within the organ with a mixture of proteases. With this technique, 2×10^6 viable cells were obtained per bovine pancreas. These cells were shown to survive for as long as 20 passages with the cells maintaining epithelial features. Isokinetic Ficoll gradient separation provided poor yields of viable duct cells, possibly because of the length of time the cells were maintained under anoxic conditions.

To study the cells of the interlobular and intralobular ducts, Githens et al. (23–25) used a modification of the enzymatic digestion technique originally described for pancreatic islet isolation. The critical feature of this procedure was the optimization of the extent of digestion, such that the ducts were freed from associated tissues but were not themselves digested. When placed into agarose matrix culture, the cut ends of most interlobular ducts isolated from the rat sealed to create enclosed lumina. Some ducts retained their original cylindrical organization; others enlarged to varying degrees, resulting in structures that ranged from cylindrical to spherical in shape. This enlargement of the cultured ducts suggested the continuation of normal processes of ion and fluid secretion.

Pancreatic Tumors and Cell Lines

Another approach to the development of in vitro models for the study of the long-term regulation of the pancreas is the use of transplantable tumors and cell lines. Both acinar and ductal pancreatic cell lines are available (Table 3), and some have proven useful for understanding normal pancreatic acinar physiology. Cell lines have at least two advantages for the investigation of the long-term regulation of pancreatic exocrine cells. First, unlike primary cultures that undergo rapid changes in culture, the cell lines maintain stable characteristics. Second, cell lines allow the use of clonal cell selection techniques to develop cells with particular characteristics. The disadvantage of

TABLE 3. *Pancreatic Tumor Cells*

Species	Cell Name	Cell Type	Growth	Ref.
Rat		Acinar	TT	85
		Acinar	CL	84
	AR42J*	Acinar	CL	38
	ARIP*	Acinar	CL	38
Hamster	HT-2	Ductal	CL	101
		Ductal	TT	96
Human	COLO 357	Acinar	CL	69
	PANC-1*	Ductal	CL	46
	MIA PaCa-2*	Unknown	CL	114
		Unknown	CL	74
	Capan-1*	Ductal	CL	44
	Capan-2*	Ductal		
	ASPC-1*	Unknown	CL	13
	Hs 766T*	Unknown	CL	97
	RWP-1	Ductal	CL	18
	RWP-2	Ductal		
	T3M-4	Unknown	CL	71
	BxPC-3*	Unknown	CL	59

Pancreatic tumor cells are maintained either as transplantable tumors (TT) or as cell lines in vitro (CL). * Cell lines that are currently available through the American Type Culture Collection.

cell lines is that the cells may be altered in such a way as to no longer represent an accurate model of the normal cell. In general, primary cultures are superior for identifying regulatory factors because they more closely represent the normal cell, whereas cell lines are superior for studying mechanisms of action because they are stable and more easily manipulated. Transplantable tumors are less useful in that they must be maintained in vivo, with the subsequent complications.

ACINAR CARCINOMAS. The recent development of several animal models of pancreatic carcinogenesis indicates that certain chemical agents preferentially induce neoplastic lesions of either the ductal or the acinar cell type. Of particular interest are nafenopin and azaserine, which preferentially induce acinar cell carcinomas of the pancreas (56, 57, 85).

Rao and Reddy (84, 85) developed a transplantable pancreatic acinar carcinoma from a nafenopin-induced tumor in rat. This tumor underwent multiple passages in rats without obvious changes in structure and growth proportions, except that the gross metastatic lesions found with the primary tumor were no longer seen. This tumor has been extensively utilized for the investigation of cytodifferentiation, growth, and neoplasia (4, 14, 33, 35–37, 86, 105, 106). Although the size, shape, and distribution of secretory granules in these cells are abnormal, these acinar tumor cells are clearly capable of synthesizing, packaging, and discharging secretory proteins in response to a variety of secretagogues. Immunocytochemical demonstration of 10 enzymes in all secretory granules and the Golgi apparatus of these cells, similar to that seen in normal pancreas (5), further suggests that these well-differentiated neoplastic cells retain the ability to express pancreatic-specific genes.

Several types of evidence indicate that the extracellular matrix may play a role in the organization of the cell polarity in cells of the transplantable carcinoma. A morphological study by Ingber et al. (33) found that the cells of the parenchyma of the tumor do not organize into acinar structure and are randomly oriented with respect to each other, whereas those tumor cells that abut on blood vessels undergo polarization and form oriented epithelial layers. Indirect immunofluorescence revealed reorientation of the tumor cells only along linearly deposited laminin and type IV collagen. These data suggest that this nonmetastatic tumor lost the ability to produce or maintain a completed basal lamina within its disorganized parenchyma, although its cells retain the capacity to produce and reorganize along linear basal lamina when in contact with vascular adventitia. Thus it has been suggested that the failure to maintain a complete basal lamina may be involved in the disorganization of normal tissue architecture in the tumor.

Recently, these cells have been adapted to in vitro growth by culturing them on rat testicular seminiferous tubular basement membrane (106) or human amnion basement membrane (34). The cells were successfully maintained in vitro on these basement membranes for at least 7 days without significant deterioration in the morphological characteristics of the cells. The pattern of radiolabeled secretory proteins discharged into the medium by the cells maintained on seminiferous tubular basement membrane was similar to that seen in the transplantable tumor. The cells raised on amnion basement membrane showed a polarization not seen in cells grown on amnion stroma, thus demonstrating a role for basement membrane as a spatial organizer of pancreatic exocrine cells (34). The use of basement membranes could potentially overcome the limitations associated with the necessity of growing these cells in vivo as transplantable tumors. The cells showed no ability to divide or grow when cultured on these basement membranes; therefore investigation of the regulation of growth may not be possible. However, studies on the hormonal regulation of other functions, such as protein synthesis, gene expression, and secretion, could be conducted after placing the cells into the in vitro environment. Other pancreatic cell lines have shown preferential attachment to type IV collagen (65), further indicating the importance of the extracellular matrix to pancreatic cell function.

Another interesting carcinoma, the AR42J cell line, was developed from an azaserine-induced tumor in the rat pancreas (38). These cells contain significant amounts of amylase, are stable, consist of a uniform population of cells, and can be maintained in continuous culture. These characteristics make the AR42J cell one of the most useful model systems currently available for studying the long-term regulation of the pancreatic acinar cells.

AR42J cells have been used as a model system to study tissue-specific and hormonal regulation of gene expression. For the study of tissue-specific gene expression, genetic constructs, including the 5' regulatory sequences of pancreatic genes, were spliced in front of reporter molecules, such as chloramphenicol acetyltransferase. These constructs show high levels of expression when transfected into AR42J cells but not when transfected into fibroblasts or other nonpancreatic acinar cell lines. AR42J cells have been used in this fashion to delineate the DNA sequences important for the tissue-specific expression of chymotrypsin (8, 104) and amylase (8). For the study of hormonal regulation of gene expression, AR42J cells were treated with hormones or intracellular messengers and the expression of digestive enzyme genes measured.

AR42J cells have also been used to study the regulation of hormone receptors. These cells appear to have a complete complement of acinar cell hormone receptors, some of which, including those for insulin (86), cholecystokinin (CCK) (48), and substance P (112), have been well characterized. AR42J cells secrete amylase in response to secretagogues and may be useful for studies of stimulus-secretion coupling. The growth and differentiation of AR42J cells are regulated by hormones, and the cells have been used as a model system to study these processes as well.

DUCTAL CELL LINES. The majority of pancreatic carcinomas are either ductal in origin or are undifferentiated and resemble duct cells. Therefore, several pancreatic ductal cell lines exist (see Table 2). Reports on the establishment of transplantable tumor lines in inbred strains of Syrian hamsters have been published (96). Binding studies using ^{125}I-labeled secretin and vasoactive intestinal peptide (VIP) in pancreatic ductal cell lines of hamsters and humans suggest the presence of secretin receptors on the membrane of some of these cells (20, 21, 97). However, in other ductal cancer cell lines, neither secretin nor VIP had any apparent effect on cell ion-transport properties (102). A systematic assessment of the functional characteristics of the various ductal cell lines has not been carried out but could potentially yield a cell line of enormous value for understanding of regulation of ductal secretion in the intact pancreas.

PHYSIOLOGICAL REGULATION OF FUNCTIONS STUDIED IN LONG-TERM IN VITRO SYSTEMS

Regulation of Differentiation: Effects of Factors on the Synthesis of Secretory Enzymes

The primary differentiated function of pancreatic acinar cells is the synthesis and secretion of digestive enzymes. Therefore, the presence and/or secretion of enzymes is a useful and appropriate measure of pancreatic acinar cell differentiation. Furthermore, the relative content of various digestive enzymes is affected by the diet, and this dietary adaptation is likely the result of hormonal regulation. Consequently, it is of interest to investigate the factors that regulate the levels of digestive enzymes in the in vitro culture systems.

Glucocorticoids regulate the differentiation of acinar cells in fetal pancreatic organ explants (15, 16, 103, 113). A precocious rise in de novo synthesis of embryonic chick pancreatic enzymes was produced in ovo (15) and in vitro (16, 113) in response to hydrocortisone. However, islet cell hormones are also important regulators of the acinar pancreas (110), and glucocorticoids inhibit both the secretion of insulin and the growth of islet cells (60, 64). Thus, in studies conducted with organ cultures of embryonic pancreas, the effects of glucocorticoids noted on acinar cell function could be due to direct effects on endocrine cells. However, recently it has been shown that glucocorticoids have direct effects in vitro on AR42J cells (48, 50). Glucocorticoids were found to elicit a pleotropic effect on the differentiation of AR42J cells, including increases in amylase content (Fig. 3), synthesis, mRNA levels, and secretion (Fig. 4) and in numbers of secretory organelles (Fig. 5) and CCK receptors (Fig. 6).

Insulin is another important hormonal regulator of pancreatic acinar cells. In rats made diabetic with either streptozocin or alloxan, pancreatic weight is reduced and the contents of pancreatic enzymes are altered (98). In particular, levels of amylase and amylase mRNA are markedly diminished (43, 98). All of these defects are reversible after treatment with insulin. To investigate the direct effects of insulin on the exocrine pancreas, AR42J cells have been studied (69). Insulin increased the amylase activity of AR42J

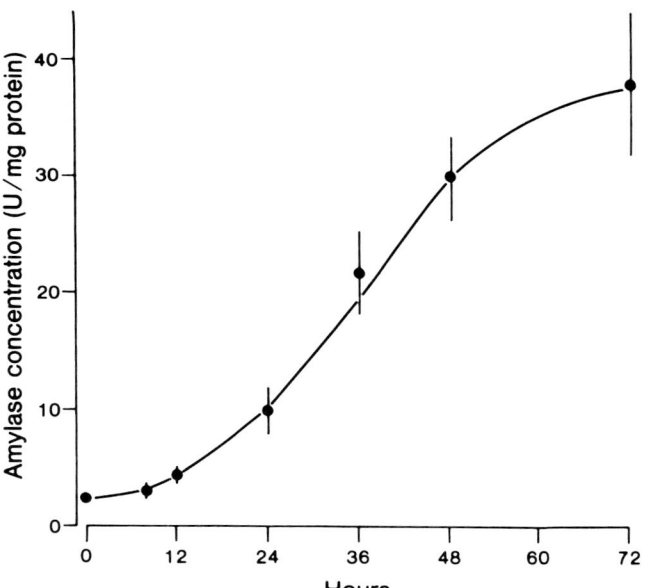

FIG. 3. Time course of the dexamethasone-induced increase in amylase concentration in AR42J cells. Dexamethasone (10 nM) was incubated with AR42J cells for up to 72 h. Cells were then analyzed for contents of amylase and protein. Values are units of amylase per milligrams of total protein and are means ± SD. [From Logsdon et al. (50), by copyright permission of the Rockefeller University Press.]

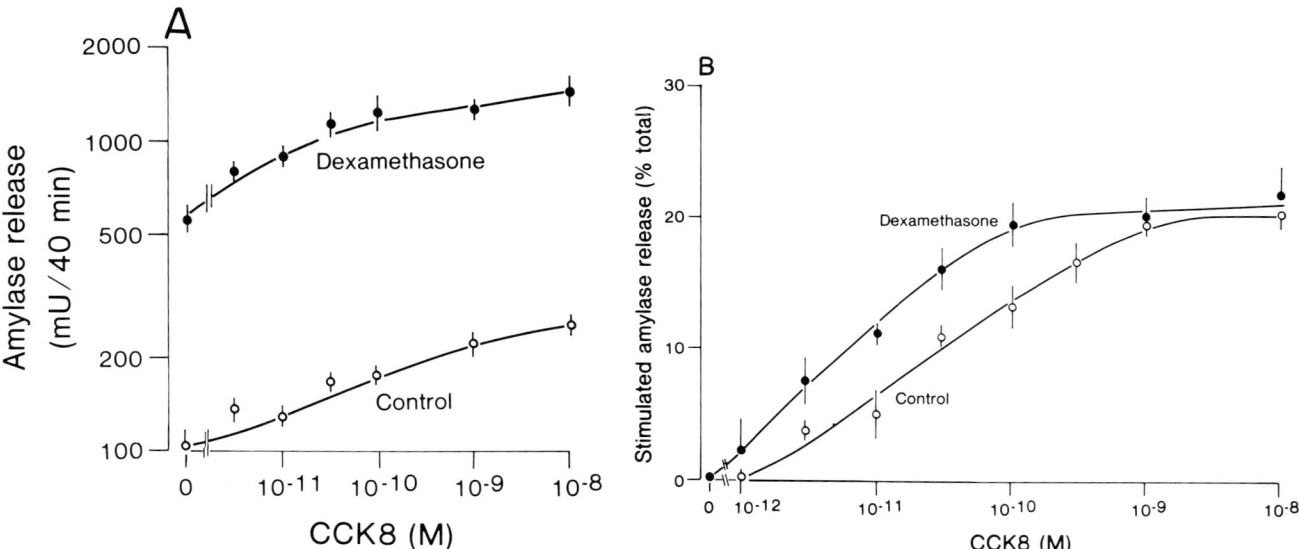

FIG. 4. Concentration dependence of amylase release stimulated by cholecystokinin octapeptide (CCK-8) for AR42J cells raised for 48 h in the absence (*open circles*) or presence (*filled circles*) of 10 nM dexamethasone. Amylase release of 40 min is plotted as a function of concentration of CCK-8 in the medium. *A*: amylase activity released into the medium expressed per culture. *B*: amylase release expressed as percentage of total amylase activity initially present in the cultures. [From Logsdon (48).]

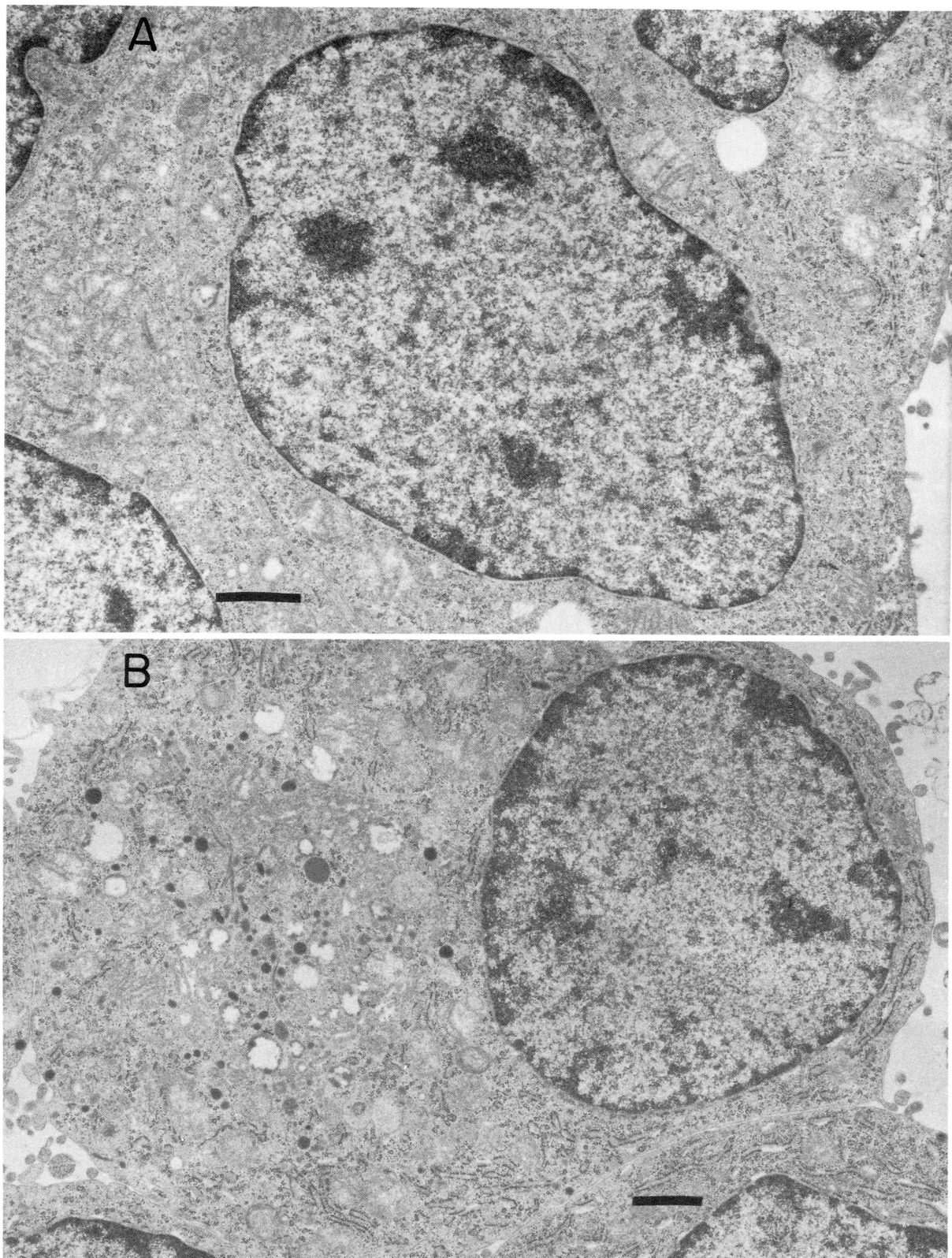

FIG. 5. Electron micrographs of AR42J cells. *A*: electron micrograph of a typical AR42J cell cultured in the absence of dexamethasone. Cell displays an undifferentiated appearance. Structural specializations for secretion are not present. *Bar*, 10 μm. × 10,400. *B*: electron micrograph of a typical AR42J cell cultured for 48 h in the presence of 10 nM dexamethasone. Cell has secretory granules and abundant rough endoplasmic reticulum. *Bar*, 10 μm. × 11,600. [From Logsdon et al. (50), by copyright permission of the Rockefeller University Press.]

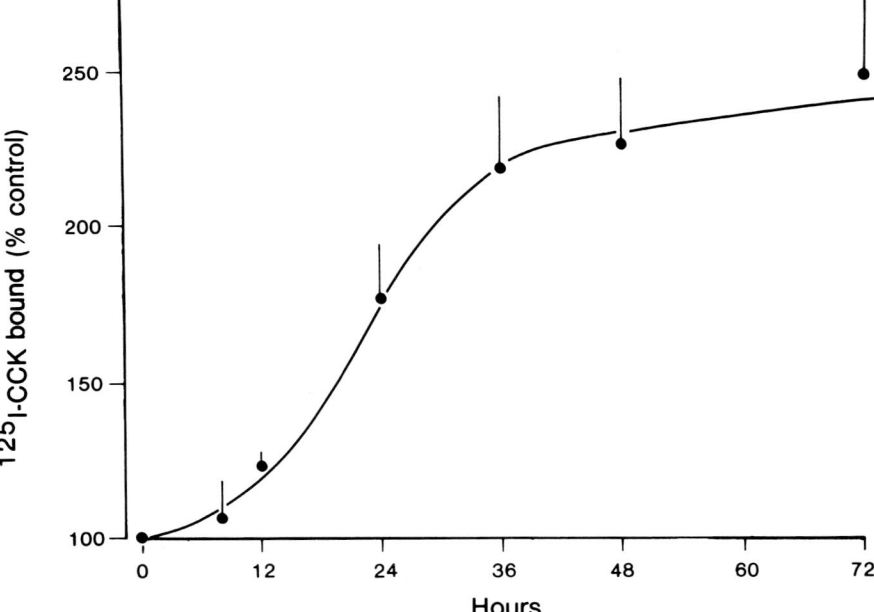

FIG. 6. Time course of dexamethasone-induced increase in ^{125}I-labeled CCK binding to AR42J cells. AR42J cells were cultured in the presence of dexamethasone (10 nM for up to 72 h) and were then incubated with 30 pM ^{125}I-labeled CCK for 1 h at 22°C. Values are percent total ^{125}I-labeled CCK bound per milligram of protein and are means ± SE. [From Logsdon (48).]

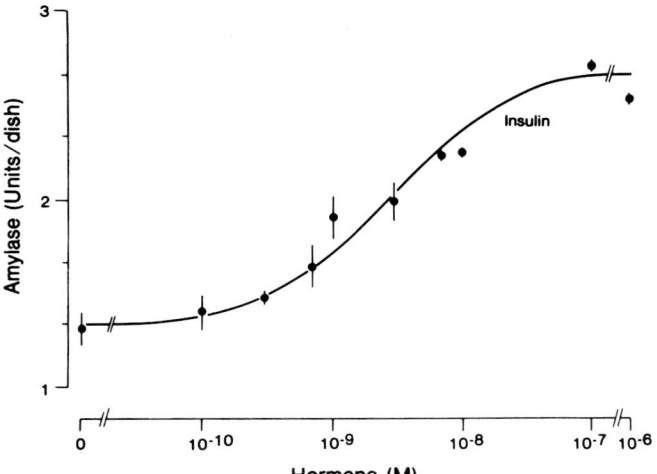

FIG. 7. Effect of increasing insulin concentrations on amylase levels in AR42J cells. Cells were plated onto 35-mm dishes at a density of 4–6 × 10^5 cells/dish. After 24 h, media were removed and fresh medium either without or with increasing concentrations of insulin was added. After a further 24 h, medium and hormone concentrations were replaced, and amylase levels were determined the following day. [From Mössner et al. (69).]

cells at concentrations as low as 1 nM and had maximal effects at 100 nM (Fig. 7). Immunoprecipitation of radiolabeled proteins revealed that insulin induced a selective increase of amylase synthesis over general protein synthesis. As noted previously, insulin also increased amylase content in a suspension culture of rat acinar cells (9).

Recently, Stratowa and Rutter (100) have studied the effects of pharmacological secretagogues on the expression of digestive enzyme genes in AR42J cells. They found that the Ca^{2+} ionophore A23187, which increases intracellular Ca^{2+}, and phorbol esters, which activate protein kinase C, selectively and synergistically increased trypsin gene expression. Glucose starvation similarly elicited a selective increase in trypsin mRNA. Furthermore, chimeras formed from the 5' flanking region of the trypsin gene linked to the coding sequence of the bacterial chloramphenical acetyltransferase gene were also induced by the same treatments, thus localizing the control sequences. The physiological significance of these findings is unclear. Naturally occurring secretagogues that act via Ca^{2+} as an intracellular second messenger, such as CCK, were not tested. In vivo treatment with caerulein, a CCK analogue, increases the mRNA levels of several proteases, including, but not exclusively, trysin (109). Also, the effect of caerulein treatment in vivo on trypsin mRNA is very small compared with the effects of the pharmacological secretagogues on trypsin mRNA in AR42J cells.

Regulation of Growth

It is now known that rapid and relatively large increases in pancreatic growth can occur under certain circumstances. These adaptive and regenerative responses of the pancreas are likely regulated by the same hormones and factors that regulate other pancreatic functions such as secretion, because a common property of gastrointestinal hormones is the ability to stimulate growth of their target tissues. Several secretagogues have been examined for effects on pancreatic growth in vivo. Trophic effects have been clearly demonstrated for CCK and its analogue caerulein. However, it is less clear whether other potential regulatory agents such as secretin and VIP or cholinergic analogues induce trophic effects. This is due partly to the complexity of the in vivo situation. Several potential growth regulators either increase or inhibit the release of other regulatory factors and/or have general systemic effects. These interactions in vivo complicate interpretation of experimental data. It therefore remains uncertain which factors have direct effects to initiate the growth of pancreatic acinar cells.

The monolayer culture system described by Logsdon and Williams (55) has proven useful for elucidation of the factors affecting the growth of the exocrine pancreas. This culture system provided the first demonstration of direct effects of CCK on pancreatic acinar cell growth. Cells were cultured in a medium containing 2.5% fetal bovine serum with or without caerulein. Caerulein led to a 152% increase in DNA and an 89% increase in protein. A caerulein concentration of 0.1–10 nM induced a dose-dependent increase in [^{3}H]thymidine incorporation (Fig. 8) and nuclear-labeling index with a maximal threefold increase.

More recently this in vitro culture system was utilized to identify other regulatory factors and to elucidate the mechanisms involved in initiation and control of growth and regenerative processes in the pancreas (49). Factors that regulate pancreatic growth, digestive enzyme synthesis, and secretion can be divided into four general categories: those that activate the turnover of phosphotidylinositides and thereby increase intracellular Ca^{2+}; those that affect intracellular cAMP; those whose receptors are tyrosine kinases; and steroid hormones. Factors from each of these categories were examined for their effects on the incorporation of [^{3}H]thymidine into DNA in pancreatic

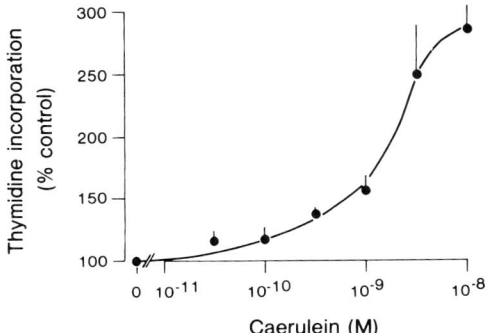

FIG. 8. Dose response for effects of caerulein on incorporation of [^{3}H]thymidine into pancreatic monolayer cultures. Cells were plated for 24 h in basal medium, then changed to basal medium plus indicated concentrations of caerulein for 3 days, the last 24 h of which included 0.1 μCi of [^{3}H]thymidine. Values are percent incorporation measured in cultures maintained in basal media and are means ± SE of means for 3–6 experiments. [From Logsdon and Williams (55).]

acinar cells in vitro. The CCK analogue caerulein and CCK-8 each led to threefold increases in [^{3}H]thymidine incorporation. Gastrin, which interacts weakly with the CCK receptor, stimulated DNA synthesis, but only at much higher concentrations. In contrast, other secretagogues that utilize Ca^{2+} as an intracellular messenger, including carbachol, bombesin, substance P, and the ionophore A23187, did not induce trophic responses. Substances that affect intracellular cAMP concentration, such as secretin, somatostatin, VIP, dibutyryl cAMP, and forskolin, did not increase DNA synthesis in cultured pancreatic cells. Insulin and EGF induced twofold and threefold increases in [^{3}H]thymidine incorporation, respectively. The effects of insulin were mediated via insulin-like growth factor I receptors. Steroid hormones had little effect on pancreatic acinar cell DNA synthesis. The stimulatory effects of CCK, insulin, and EGF were additive. The combination of caerulein, EGF, and insulin in a hormonally defined medium led to a 10-fold increase in the incorporation of [^{3}H]thymidine into DNA. These data indicate that CCK, EGF, and insulin directly increase DNA synthesis in pancreatic acinar cells.

Pancreatic cancer cells also show growth regulation by hormones in vitro. The AR42J cell line responds to insulin by increased growth (68). Several lines of evidence indicate that this effect of insulin on AR42J cell growth is mediated by an interaction with insulin receptors and not with insulin-like growth factor receptors. The reason for this difference between AR42J cells and primary cultures of acinar cells is unknown. AR42J cell growth is inhibited by glucocorticoids (6, 50). The growth of MIA PaCa-2 cells, an undifferentiated pancreatic cell line, was stimulated by EGF and inhibited by somatostatin (47). These results are particulary interesting because it has also been shown that EGF stimulates the phosphorylation of a tyrosine kinase in these cells and that somatostatin causes its dephosphorylation (30).

Regulation of Secretion

Secretion of digestive enzymes involves multiple cellular mechanisms, including those of receptor-effector coupling, protein synthesis and transport, and exocytosis. Brannon et al. (9) found that rat acinar cells cultured for 4 days in a hormonally defined medium still secreted amylase in response to carbachol to the same extent as freshly prepared cells, even though the content of amylase in the cultured cells was reduced by 90%.

Secretory ability also is well maintained in mouse pancreatic acini during short culture periods. Logsdon and Williams (53) showed that mouse acini cultured for 24 h maintained their secretory ability, and that inclusion of EGF, insulin, or carbachol in the culture medium increased the responsiveness of the cells after culturing. Utilizing this information to formulate a hormonally defined medium, Hootman et al. (32) improved this culture system and demonstrated that maximal release of amylase evoked by CCK or carbachol was not significantly reduced in cultured guinea pig pancreatic acini relative to that elicited in freshly dispersed acini. This system was used to investigate the effects of chronic treatment with carbachol on secretory responsiveness and muscarinic receptor density.

Cells isolated from the transplantable tumor cell line of Reddy and Rao (85) also retain the ability to secrete amylase in response to a number of secretagogues, including carbachol (36, 105), caerulein, bombesin, secretin, and VIP (35). The magnitude of the secretory response of the tumor cells (2- to 3-fold) was somewhat less than that of pancreatic lobules. Although detailed dose-response data were not shown, Iwanij and Jamieson (36) stated that the concentrations required for a maximal secretory response in this cell line were similar for most secretagogues to that required in normal acinar cells except for secretin, which was 10-fold less potent.

AR42J cells also secrete amylase in response to CCK (48, 50). In these cells, pretreatment with 10 nM dexamethasone for 48 h increased basal and CCK-stimulated amylase secretion five- to sixfold. This effect was primarily due to an increased amylase content. On closer examination, when amylase release was studied as a function of the concentration of CCK-8, the same maximal 2.6-fold increase was seen by using either control cells or cells that had been pretreated with dexamethasone, even though the basal amylase release was only 18% as great in control cells. When the data for amylase release were normalized based on the initial content in the cells, control cells showed the same maximal responsiveness to CCK-8 as dexamethasone-treated cells. However, they were less sensitive in terms of the threshold and one-half maximally effective concentrations of CCK-8 (Fig. 4). The concentration of CCK-8 required for one-half maximal stimulation of amylase secretion was decreased threefold after pretreated with dexamethasone. These findings indicate that treatment with glucocorticoids decreased the concentration of CCK required for stimulation of amylase release. This effect most likely is due to a dexamethasone-stimulated increase in CCK receptors in the AR42J cells.

AR42J cells also express a high density of substance P receptors and have been shown to secrete amylase in response to stimulation with this hormone (112). Other membrane and intracellular events elicited by exposure of AR42J cells to substance P have also been examined. Intracellular recordings from AR42J cells revealed a resting membrane potential of −40 to −65 mV, and application of substance P evoked prolonged depolarization of 20–40 mV. The intracellular free Ca^{2+} concentration in AR42J cells, measured with quin 2, increased transiently on stimulation with substance P. Thus substance P receptors on AR42J cells seem to be functionally coupled to membrane, intracellular, and secretory responses normally observed with pancreatic acinar cells. AR42J cells should provide a suitable cellular system for examining the properties and regulation of the substance P receptor.

Regulation of Hormone Receptors

Hormone receptors both regulate target cell functions and are themselves regulated. In many cell types, chronic exposure of a receptor to its ligand leads to altered cellular functions resulting from either changes in hormone receptor number and/or affinity or from postreceptor desensitization. Hormone receptors may also be affected by heterologous receptor interactions. In this case the binding of one ligand to its receptor on a target cell alters the association of another ligand to its own receptor. Investigation of aspects of receptor regulation requires longer time periods than are allowed by acute in vitro systems. Moreover it is important to understand the direct interactions of hormones with receptors in the absence of changes in the levels of other hormones. Therefore long-term in vitro systems have proved extremely valuable in this regard.

The islets of Langerhans are dispersed throughout the pancreas. Pancreatic acini are therefore exposed in vivo to very high concentrations of insulin. By measuring the affinity and capacity of acinar cells taken from normal and diabetic mouse pancreas, it was determined that high normal concentrations of insulin resulted in a downregulation of insulin receptors on the pancreatic acinar cells (68). To further study the regulation of acinar cell insulin receptors, acini from diabetic mice were placed in suspension culture for 24 h. Addition of 1 μM insulin during the culture period led to a 30% decrease in subsequent ^{125}I-labeled insulin binding; the presence or absence of either EGF or carbachol was without effect on insulin binding. Computer analysis of competition-inhibition curves showed a decrease in the number of insulin receptors in acini treated with insulin when compared with untreated acini from diabetic mice, although the receptor affinity remained unchanged. Solubilization of acini in 1% Triton X-100, which allows access to intracellular receptors, showed that this insulin-induced decrease was due to a change in the total number of cellular insulin receptors and not just to a decrease in receptors on the cell surface. This study indicated that insulin can directly regulate its own receptor on pancreatic acini and that in vivo insulin receptors in normal pancreatic cells are downregulated, presumably because of high ambient insulin concentrations.

The effects of glucocorticoids on the receptor binding and biological effects of CCK were investigated in AR42J cells (49). AR42J cells were found to possess

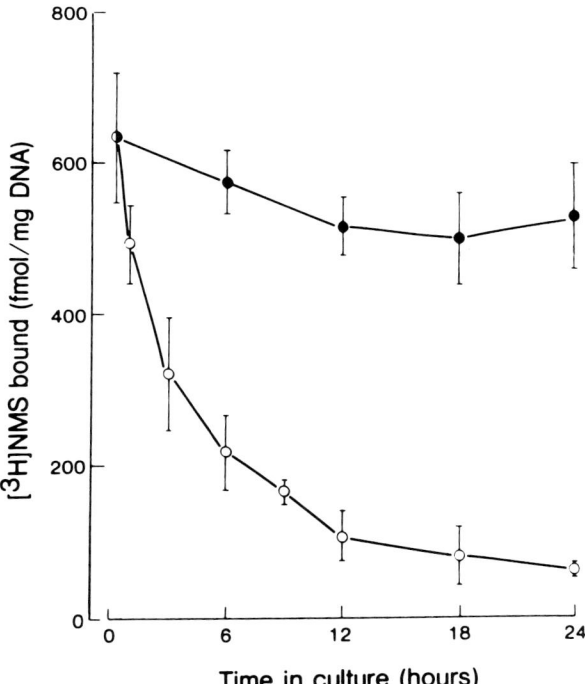

FIG. 9. Time dependence of decrease in specific [N-methyl-^{3}H]-scopolamine ([^{3}H]NMS) binding to guinea pig pancreatic acini cultured with 0.1 mM carbachol (CCh) (open circles). Acini cultured in hormonally defined medium without CCh (filled circles). At each time indicated, acini were rinsed with N-2-hydroxyethylpiperazine-N'-2-ethanesulfonic acid (HEPES)-buffered Ringer's solution and incubated for 60 min at 37°C in HEPES-buffered Ringer's solution containing 0.5 nM [^{3}H]NMS. Results represent means ± SE of 3 experiments. [From Hootman et al. (32).]

specific, high-affinity receptors for CCK. Competitive-inhibition experiments indicated that the CCK receptor on the AR42J cell has a specificity of CCK analogues that is similar to the CCK receptor on the normal rat pancreatic acinar cell. Treatment with 10 nM dexamethasone for 48 h increased the specific binding of ^{125}I-labeled CCK (Fig. 6). This increase in binding was time dependent, with maximal effects occurring after 48 h, and dose dependent, with a one-half maximal effect elicited by 1 nM dexamethasone. Other steroid analogues were also effective, and their potencies paralleled their relative effectiveness as glucocorticoids. Analyses of competitive-binding experiments conducted at 4°C to minimize hormone internalization and degradation revealed the presence of a single class of CCK binding sites with a dissociation constant of ±6 nM, and indicated that dexamethasone treatment nearly tripled the number of CCK receptors per cell with little change in receptor affinity. This dexamethasone-induced increase in the number of CCK receptors likely accounts for the increased sensitivity of the cells to CCK.

Long-term regulation of muscarinic receptors in cultured guinea pig pancreatic acini has also been investigated by assessing the effects of chronic cholinergic agonist exposure on binding of the muscarinic antagonist [N-methyl-^{3}H]scopolamine ([^{3}H]NMS) and on amylase release (32). Culture of acini with 0.1 mM carbachol decreased [^{3}H]NMS binding by 50% at 3-4 h and 85%-90% at 24 h. This decrease was attributable primarily to a reduction in receptor number rather than affinity. The binding also was decreased to a similar extent by the cholinergic agonists bethanechol and methacholine but not by other secretagogues. The decrease in antagonist binding induced by carbachol was dose dependent, with the 50% inhibitory concentration (IC$_{50}$) approximating the 50% effective concentration (EC$_{50}$) for amylase release. Culture of acini for 24 h with carbachol lead to a decrease in [^{3}H]NMS binding (Fig. 9) and abolished subsequent amylase release in response to carbachol but not to CCK-8. When carbachol was removed from the culture medium after 24 h and acini recultured in its absence, [^{3}H]NMS binding increased with a half time for recovery of 20-24 h. This recovery was blocked by cycloheximide. These results indicated that muscarinic receptor turnover in the pancreatic acinus is regulated by receptor activation and that a decrease in receptor number follows prolonged cholinergic agonist exposure.

CONCLUSIONS

Long-term in vitro preparations of pancreatic cells have already provided important insights into several aspects of the neurohormonal regulation of pancreatic physiology. Several different preparations are now available and will likely be utilized to a greater extent in the future. Of particular importance will be the development and exploitation of pancreatic cell lines with novel characteristics by using clonal cell selection techniques and somatic cell genetics. Useful characteristics would include an overabundance of hormone receptors or alterations in secretory mechanisms. Improvements in tissue culture conditions, including the use of complex biomatricies, also should lead to improvements in the ability to maintain primary cultures of untransformed acinar and duct cells.

REFERENCES

1. AMSTERDAM, A., AND J. D. JAMIESON. Structural and functional characterization of isolated pancreatic exocrine cells. *J. Cell Biol.* 19: 3028-3032, 1973.

2. AMSTERDAM, A., T. E. SOLOMON, AND J. D. JAMIESON. Sequential dissociation of the exocrine pancreas into lobules, acini, and individual cells. *Methods Cell Biol.* 20: 361-378, 1978.

3. BARNES, D., AND G. SATO. Serum-free cell culture: a unifying approach. *Cell* 22: 649–655, 1980.
4. BECICH, M. J., M. BENDAYAN, AND J. K. REDDY. Intracellular transport and storage of secretory proteins in relation to cytodifferentiation in neoplastic pancreatic acinar cells. *J. Cell Biol.* 96: 949–960, 1983.
5. BENDAYAN, M., M. J. BECICH, AND J. K. REDDY. Immunocytochemical localization of exocrine enzymes in rat pancreatic acinar carcinoma. *Eur. J. Cell Biol.* 34: 151–158, 1984.
6. BENZ, C., C. HOLLANDER, AND B. MILLER. Endocrine-responsive pancreatic carcinoma: steroid binding and cytotoxicity studies in human tumor cell lines. *Cancer Res.* 46: 2276–2281, 1986.
7. BOLENDER, R. P. Stereological analysis of the guinea pig pancreas. I. Analytical model and quantitative description of nonstimulated pancreatic exocrine cells. *J. Cell Biol.* 61: 269–287, 1974.
8. BOULET, A. M., C. R. ERWIN, AND W. J. RUTTER. Cell-specific enhancers in the rat exocrine pancreas. *Proc. Natl. Acad. Sci. USA* 83: 3599–3603, 1986.
9. BRANNON, P. M., B. M. ORRISON, AND N. KRETCHMER. Primary cultures of rat pancreatic acinar cells in serum-free medium. *In Vitro Cell. Dev. Biol.* 21: 6–14, 1985.
10. CARREL, A., AND C. A. LINDBERGH. *The Culture of Organs.* New York: Hoeber, 1938, p. 151.
11. CHASE, H. P., I. OCRANT, AND D. W. TALMAGE. The effects of different conditions of organ culture on the survival of the mouse pancreas. *Diabetes* 28: 990–993, 1979.
12. CHEN, J., E. C. STUCKEY, AND C. L. BERRY. Three-dimensional culture of rat exocrine pancreatic cells using collagen gels. *Br. J. Exp. Pathol.* 66: 551–559, 1985.
13. CHEN, W. H., J. S. HOROSZEWICS, S. S. LEONG, T. SHIMANO, R. PENETRANTE, W. H. SANDERS, R. BERJIAN, H. O. DOUGLASS, E. W. MARTIN, AND T. M. CHU. Human pancreatic adenocarcinoma: in vitro and in vivo morphology of a new tumor line established from ascites. *In Vitro* 18: 24–34, 1982.
14. CHIEN, J. L., AND J. R. WARREN. Differentiation of muscarinic cholinergic receptors in acinar carcinoma of rat pancreas. *Cancer Res.* 45: 4858–4863, 1985.
15. COHEN, A., H. HELLER, AND R. G. KULKA. The effect of hydrocortisone on exportable enzyme accumulation in the differentiating chick pancreas. *Dev. Biol.* 29: 293–306, 1972.
16. COHEN, A., AND R. G. KULKA. Relationship of steroid structure to induction of chymotrypsinogen in embryonic chick pancreas in vitro. *Endocrinology* 97: 475–478, 1975.
17. COLLIER, S., T. E. MANDEL, L. HOFFMAN, AND G. CARUSO. Organ culture of fetal mouse pancreas. The effect of culture conditions on insulin and glucagon secretion. *Diabetes* 30: 804–812, 1981.
18. DEXTER, D. L., G. M. MATOOK, P. A. MEITNER, H. A. BOGAARS, G. A. JOLLY, M. D. TURNER, AND P. CALABRESI. Establishment and characterization of two human pancreatic cancer cell lines tumorigenic in athymic mice. *Cancer Res.* 42: 2705–2714, 1982.
19. EHMANN, U. K., W. D. PETERSON, AND D. S. MISFELDT. To grow mouse mammary epithelial cells in culture. *J. Cell Biol.* 98: 1026–1032, 1984.
20. ESTIVAL, A., F. CLEMENTE, AND A. RIBET. Adenocarcinoma of the human exocrine pancreas: presence of secretin and caerulein receptors. *Biochem. Biophys. Res. Commun.* 102: 1336–1341, 1984.
21. ESTIVAL, A., P. MOUNIELOU, V. TROCHERIS, J. L. SCEMAMA, F. CLEMENTE, E. HOLLANDE, AND A. RIBET. Presence of VIP receptors in a human pancreatic adenocarcinoma cell line. Modulation of the cAMP response during cell proliferation. *Biochem. Biophys. Res. Commun.* 111: 958–963, 1983.
22. FELL, P. E., AND C. GROBSTEIN. The influence of extra-epithelial factors on the growth of embryonic mouse pancreatic epithelium. *Exp. Cell Res.* 53: 301–304, 1968.
23. GITHENS, S., III, D. R. HOLMQUIST, J. F. WHELAN, AND J. R. RUBY. Characterization of ducts isolated from the pancreas of the rat. *J. Cell Biol.* 85: 122–135, 1980.
24. GITHENS, S., III, D. R. HOLMQUIST, J. F. WHELAN, AND J. R. RUBY. Ducts of the rat pancreas in a agarose matrix culture. *In Vitro* 16: 797–808, 1980.
25. GITHENS, S., III, D. R. HOLMQUIST, J. F. WHELAN, AND J. R. RUBY. Morphologic and biochemical characteristics of isolated and cultured pancreatic ducts. *Cancer Phila.* 47: 1505–1512, 1981.
26. GOLOSOW, N., AND C. GROBSTEIN. Epitheliomesenchymal interaction in pancreatic morphogenesis. *Dev. Biol.* 4: 242–255, 1962.
27. HARDING, J. D., A. E. PRZYBYLA, R. J. MACDONALD, R. L. PICTET, AND W. J. RUTTER. Effects of dexamethasone and 5-bromodeoxyuridine on the synthesis of amylase mRNA during pancreatic development in vitro. *J. Biol. Chem.* 253: 7531–7537, 1978.
28. HAY, R. J. Pancreatic epithelial cells in vitro: a possible model system for studies in carcinogenesis. *Cancer Res.* 35: 2289–2291, 1975.
29. HAY, R. J. Isolation and maintenance of pancreatic epithelial cells as colonial aggregates in culture. *Tissue Culture Association Manual.* Rockville, MD: Tissue Culture Assoc., 1978, p. 809–811.
30. HIEROWSKI, M. T., C. LIEWBOW, K. DU SAPIN, AND A. V. SCHALLY. Stimulation by somatostatin of dephosphorylation of membrane proteins in pancreatic cancer MIA PaCa-2 cell line. *FEBS Lett.* 179: 252–256, 1985.
31. HIRATA, K., T. OKU, AND A. E. FREEMAN. Duct, exocrine, and endocrine components of cultured fetal mouse pancreas. *In Vitro* 18: 789–799, 1982.
32. HOOTMAN, S. R., M. E. BROWN, J. A. WILLIAMS, AND C. D. LOGSDON. Regulation of muscarinic acetylcholine receptors in cultured guinea pig pancreatic acini. *Am. J. Physiol.* 251 (*Gastrointest. Liver Physiol.* 14): G75–G83, 1986.
33. INGBER, D. E., J. A. MADRI, AND J. D. JAMIESON. Role of basal lamina in neoplastic disorganization of tissue arachitecture. *Proc. Natl. Acad. Sci. USA* 78: 3901–3905, 1981.
34. INGBER, D. E., J. A. MADRI, AND J. D. JAMIESON. Basement membrane as a spatial organizer of polarized epithelia. Exogenous basement membrane reorients pancreatic epithelial tumor cells in vitro. *Am. J. Pathol.* 122: 129–139, 1986.
35. IWANIJ, V., B. E. HULL, AND J. D. JAMIESON. Structural characterization of a rat acinar cell tumor. *J. Cell Biol.* 95: 727–733, 1982.
36. IWANIJ, V., AND J. D. JAMIESON. Biochemical analysis of secretory proteins synthesized by normal rat pancreas and by pancreatic acinar tumor cells. *J. Cell Biol.* 95: 734–741, 1982.
37. IWANIJ, V., AND J. D. JAMIESON. Comparison of secretory protein profiles in developing rat pancreatic rudiment and rat acinar tumor cells. *J. Cell Biol.* 95: 742–746, 1982.
38. JESSOP, N. W., AND R. J. HAY. Characteristics of two rat pancreatic exocrine cell lines derived from transplantable tumors (Abstract). *In Vitro* 16: 212, 1980.
39. JONES, R. T., L. A. BARRETT, C. VAN HAAFTEN, C. C. HARRIS, AND B. F. TRUMP. Carcinogenesis in the pancreas. I. Long-term explant culture of human and bovine pancreatic ducts. *J. Natl. Cancer Inst.* 58: 557–565, 1977.
40. JONES, R. T., E. A. HUYDSON, AND J. H. RESAU. A review of in vitro and in vivo culture techniques for the study of pancreatic carcinogenesis. *Cancer Phila.* 47: 1490–1496, 1981.
41. JONES, R. T., B. F. TRUMP, AND G. D. STONER. Culture of human pancreatic ducts. *Methods Cell Biol.* 21: 429–439, 1980.
42. KALLMAN, F., AND C. GROBSTEIN. Fine structure of differentiating mouse pancreatic exocrine cells in transfilter culture. *J. Cell Biol.* 20: 399–413, 1964.
43. KORC, M., D. OWERBACH, C. QUINTO, AND W. J. RUTTER. Pancreatic islet-acinar cell interaction: amylase messenger RNA levels are determined by insulin. *Science Wash. DC* 213: 351–353, 1981.
44. KYRIAZIA, A. P., A. A. DYRIAZIS, D. G. SCARPELLI, J. FOGH, M. S. RAO, AND R. LEPERA. Human pancreatic adenocarcinoma line Capan-1 in tissue culture and the nude mouse:

morphologic, biologic, and biochemical characteristics. *Am. J. Pathol.* 106: 250–260, 1982.
45. LEVINE, S., R. PICTET, AND W. J. RUTTER. Control of cell proliferation and cytodifferentiation by a factor reacting with the cell surface. *Nature Lond.* 246: 49–52, 1973.
46. LIEBER, M., J. A. MAZZETTA, W. NELSON-REES, M. KAPLAN, AND G. TODARO. Establishment of a continuous tumor-cell line (PANC-1) from a human carcinoma of the exocrine pancreas. *Int. J. Cancer* 15: 741–747, 1975.
47. LIEBOW, C., M. HIEROWSKI, AND K. DU SAPIN. Hormonal control of pancreatic cancer growth. *Pancreas* 1: 44–48, 1986.
48. LOGSDON, C. D. Glucocorticoids increase cholecystokinin receptors and amylase secretion in pancreatic acinar AR42J cells. *J. Biol. Chem.* 261: 2096–2101, 1986.
49. LOGSDON, C. D. Stimulation of pancreatic acinar cell growth by CCK, epidermal growth factor, and insulin in vitro. *Am. J. Physiol.* 251 (*Gastrointest. Liver Physiol.* 14): G487–G494, 1986.
50. LOGSDON, C. D., J. MÖSSNER, J. A. WILLIAMS, AND I. D. GOLDFINE. Glucocorticoids determine the differentiation of pancreatic AR42J cells: effects on morphology, secretion, and amylase mRNA. *J. Cell Biol.* 100: 1200–1208, 1985.
51. LOGSDON, C. D., AND J. A. WILLIAMS. Epidermal growth factor binding and biologic effects on mouse pancreatic acini. *Gastroenterology* 85: 339–345, 1983.
52. LOGSDON, C. D., AND J. A. WILLIAMS. Epidermal growth factor: intracellular Ca^{2+} inhibits its association with pancreatic acini and A431 cells. *FEBS Lett.* 164: 335–339, 1983.
53. LOGSDON, C. D., AND J. A. WILLIAMS. Pancreatic acini in short-term culture: regulation by EGF, carbachol, insulin, and corticosterone. *Am. J. Physiol.* 244 (*Gastrointest. Liver Physiol.* 7): G675–G682, 1983.
54. LOGSDON, C. D., AND J. A. WILLIAMS. Intracellular Ca^{2+} and phorbol esters synergistically inhibit EGF internalization. *Biochem. J.* 223: 893–900, 1984.
55. LOGSDON, C. D., AND J. A. WILLIAMS. Pancreatic acinar cells in monolayer culture: direct trophic effects of caerulein in vitro. *Am. J. Physiol.* 250 (*Gastrointest. Liver Physiol.* 13): G440–G447, 1986.
56. LONGNECKER, D. S., AND T. J. CURPHEY. Adenocarcinoma of the pancreas in azaserine-treated rats. *Cancer Res.* 35: 2249–2258, 1975.
57. LONGNECKER, D. S., H. S. LILJA, J. FRENCH, E. KUHLMANN, AND W. NOLL. Transplantation of azaserine-induced carcinomas of pancreas in rats. *Cancer Lett.* 7: 197–202, 1979.
58. LOOR, R., N. J. NOWAK, M. L. MANZO, H. O. DOUGLASS, AND T. M. CHU. Use of pancreas-specific antigen in immunodiagnosis of pancreatic cancer. *Clin. Lab. Med.* 2: 567–578, 1982.
59. MALICK, L. E., A. TOMPA, C. KUSZYNSKI, P. POUR, AND R. LANGENBACH. Maintenance of adult hamster pancreas cells on fibroblastic cells. *In Vitro* 17: 947–955, 1981.
60. MCEVOY, R. C., AND O. D. HEGRE. Foetal rat pancreas in organ culture: effects of media supplementation with various steroid hormones in the acinar and islet components. *Differentiation* 6: 105–111, 1976.
61. MCEVOY, R. C., O. D. HEGRE, AND A. LAZAROW. Fetal and neonatal rat pancreas in organ culture: age-related effects of corticosterone on the acinar cell component. *Am. J. Anat.* 146: 133–150, 1976.
62. MCEVOY, R. C., O. D. HEGRE, AND A. LAZAROW. Foetal rat pancreas in organ culture: effect of corticosterone concentrations on the acinar and islet cell components. *Differentiation* 6: 17–26, 1976.
63. MCEVOY, R. C., O. D. HEGRE, R. J. LEONARD, AND A. LAZAROW. Fetal rat pancreas: differentiation of the acinar cell component in vivo and in vitro. *Diabetes* 22: 584–589, 1973.
64. MCEVOY, R. C., A. LAZAROW, AND O. D. HEGRE. Organ culture of foetal rat pancreas. Effects of parabiotic culture with foetal adrenal glands. *Differentiation* 3: 69–77, 1975.
65. MCINTYRE, L. J., H. K. KLEINMAN, G. R. MARTIN, AND Y. S. KIM. Attachment of human pancreatic tumor cell lines to collagen in vitro. *Cancer Res.* 41: 3296–3299, 1981.
66. MILNER, G. R., M. GASPARO, R. KAY, AND R. D. MILNER. Effects of glucose and amino acids in insulin, glucagon and zymogen granule size of foetal rat pancreas grown in organ culture. *J. Endocrinol.* 82: 179–189, 1979.
67. MORGAN, R. T., L. K. WOODS, G. E. MOORE, L. A. QUINN, L. MCGAVRAN, AND S. G. GORDON. Human cell line (COLO-357) of metastatic pancreatic adenocarcinoma. *Int. J. Cancer* 25: 591–598, 1980.
68. MÖSSNER, J., C. D. LOGSDON, I. D. GOLDFINE, AND J. A. WILLIAMS. Regulation of pancreatic acinar cell insulin receptors by insulin. *Am. J. Physiol.* 247 (*Gastrointest. Liver Physiol.* 10): G155–G160, 1984.
69. MÖSSNER, J., C. D. LOGSDON, J. A. WILLIAMS, AND I. D. GOLDFINE. Insulin, via its own receptor, regulates growth and amylase synthesis in pancreatic acinar AR42J cells. *Diabetes* 34: 891–897, 1985.
70. MURELL, L. R., K. H. GERMAIN, AND D. M. LYNCH. Survival of functional pancreatic acinar tissue in circumfusion organ culture enhanced by chemically defined medium with hydrocortisone. *Cancer Res.* 35: 2286–2288, 1975.
71. OKABE, T., N. YAMAGUCHI, AND N. OHSAWA. Establishment and characterization of a carcinoembryonic antigen (CEA)-producing cell line from a human carcinoma of the exocrine pancreas. *Cancer Phila.* 51: 662–668, 1983.
72. OLIVER, C. Isolation and maintenance of differentiated exocrine glands in vitro. *In Vitro* 16: 290–305, 1980.
73. ORCI, L., A. A. LIKE, M. AMHERDT, B. BLONDEL, Y. KANAZAWA, E. B. MARLISS, A. E. LAMBERT, C. B. WOLLHEIM, AND A. E. RENOLD. Monolayer cell culture of neonatal rat pancreas: an ultrastructural and biochemical study of functioning endocrine cells. *J. Ultrastruct. Res.* 43: 270–297, 1973.
74. OWENS, R. B., H. S. SMITH, W. A. NELSON-REES, AND E. L. SPRINGER. Epithelial cell cultures from normal and cancerous human tissues. *J. Natl. Cancer Inst.* 56: 843–849, 1976.
75. PARSA, I., R. D. BLOOMFIELD, C. A. FOYE, AND A. L. SUTTON. Methylnitrosouria induced carcinoma in organ-cultured fetal human pancreas. *Cancer Res.* 44: 3530–3538, 1984.
76. PARSA, I., AND W. H. MARSH. Long-term organ culture of embryonic rat pancreas in a chemically defined medium. *Am. J. Pathol.* 82: 119–128, 1976.
77. PARSA, I., W. H. MARSH, AND P. J. FITZGERALD. Chemically defined medium for organ culture differentiation of rat pancreas anlage. *Exp. Cell Res.* 59: 171–175, 1969.
78. PARSA, I., W. H. MARSH, AND P. J. FITZGERALD. Pancreas acinar cell differentiation. I. Morphology and enzymatic comparisons of embryonic rat pancreas and pancreatic anlage grown in organ culture. *Am. J. Pathol.* 57: 457–487, 1969.
79. PARSA, I., W. H. MARSH, AND P. J. FITZGERALD. Pancreas acinar cell differentiation. II. Comparative DNA and protein synthesis of the embryonic rat pancreas and the pancreatic anlage growth in organ culture. *Am. J. Pathol.* 57: 489–521, 1969.
80. PARSA, I., W. H. MARSH, AND A. L. SUTTON. An in vitro model of human pancreas carcinogenesis: effects of nitroso compounds. *Cancer Phila.* 47: 1543–1551, 1981.
81. PARSA, I., W. H. MARSH, AND K. M. H. BUTT. Effects of dimethylnitrosamine on organ-cultured adult human pancreas. *Am. J. Pathol.* 102: 403–411, 1981.
82. RALL, L., R. PICTET, S. GITHENS, AND W. J. RUTTER. Glucocorticoids modulate the in vitro development of the embryonic rat pancreas. *J. Cell Biol.* 75: 398–409, 1977.
83. RAO, K. N., S. TAKAHASHI, AND H. SHINOZUKA. Acinar cell carcinoma of the rat pancreas grown in cell culture and in nude mice. *Cancer Res.* 40: 592–597, 1980.
84. RAO, M. S., AND J. K. REDDY. Transplantable acinar cell carcinoma of the rat pancreas. *Am. J. Pathol.* 94: 333–348, 1979.
85. REDDY, J. K., AND M. S. RAO. Transplantable pancreatic carcinoma of the rat. *Science Wash. DC* 198: 78–80, 1977.
86. REDDY, J. K., M. S. RAO, J. R. WARREN, S. QURESHI, AND E. I. CHRISTENSEN. Differentiation and DNA synthesis in pancreatic acinar carcinoma of rat. *Cancer Res.* 40: 3443–3454, 1980.

87. RESAU, J. H., J. R. COTTRELL, E. A. HUDSON, B. F. TUMP, AND R. T. JONES. Studies on the mechanisms of altered exocrine acinar cell differentiation and ductal metaplasia following nitrosamine exposure using hamster pancreatic explants. *Carcinogenesis Lond.* 6: 29–35, 1985.
88. RESAU, J. H., E. A. HUDSON, AND R. T. JONES. Organ explant culture of adult Syrian golden hamster pancreas. *In Vitro* 19: 315–325, 1983.
89. RESAU, J. H., L. MARZERLLA, B. F. TRUMP, AND R. T. JONES. Degradation of zymogen granules by lysosomes in cultured pancreatic explants. *Am. J. Pathol.* 115: 139–150, 1984.
90. RHOTEN, W. H. Continuous-perfusion tissue culture of fetal and adult pancreas of the lizard *Anulis carolinensis*. *Anat. Rec.* 203: 165–173, 1982.
91. RONZIO, R. A., AND W. J. RUTTER. Effects of a partially purified factor from chick embryos on macromolecular synthesis of embryonic pancreatic epithelia. *Dev. Biol.* 30: 307–320, 1973.
92. RUOFF, N. M., AND R. J. HAY. Metabolic and temporal studies on pancreatic exocrine cells in culture. *Cell Tissue Res.* 204: 245–252, 1979.
93. RUTTER, W. J., N. K. WESSELLS, AND C. GROBSTEIN. Control of specific synthesis in the developing pancreas. *Natl. Cancer Inst. Monogr.* 13: 51–65, 1964.
94. SANDERS, T. G., AND W. J. RUTTER. The developmental regulation of amylolytic and proteolytic enzymes in the embryonic rat pancreas. *J. Biol. Chem.* 249: 3500–3509, 1974.
95. SATO, T., M. SATO, E. A. HUDSON, AND R. T. JONES. Characterization of bovine pancreatic ductal cells isolated by a perfusion-digestion technique. *In Vitro* 19: 651–660, 1983.
96. SCARPELLI, D. G., AND M. S. RAO. Transplantable ductal adenocarcinoma of the Syrian hamster pancreas. *Cancer Res.* 39: 452–458, 1979.
97. SMITH, H. S. In vitro properties of epithelial cell lines established from human carcinomas and nonmalignant tissue. *J. Natl. Cancer Inst.* 62: 225–230, 1979.
98. SÖLING, H. D., AND K. O. UNGER. The role of insulin in the regulation of α-amylase synthesis in the rat pancreas. *Eur. J. Clin. Invest.* 2: 199–212, 1971.
99. STONER, G. D., C. C. HARRIS, D. G. BOSTWICK, R. T. JONES, B. F. TRUMP, E. W. KINGSBURY, E. FINEMAN, AND C. NEWKIRK. Isolation and characterization of epithelial cells from bovine pancreatic duct. *In Vitro* 14: 581–590, 1978.
100. STRATOWA, C., AND W. J. RUTTER. Selective regulation of trypsin gene expression by calcium and by glucose starvation in a rat exocrine pancreas cell line. *Proc. Natl. Acad. Sci. USA* 83: 4292–4296, 1986.
101. TOWNSEND, C. M., R. B. FRANKLIN, F. B. GELDER, E. GLASS, AND J. C. THOMPSON. Development of a transplantable model of pancreatic duct adenocarcinoma. *Surgery St. Louis* 92: 72–78, 1982.
102. UEDA, N., Y. SUZUKI, M. UTSUMI, T. OBARA, K. OKAMURA, AND M. NAMIKI. Electrophysiological studies on the cultured cells obtained from transplantable pancreatic carcinoma in Syrian golden hamsters. *Peptides Fayetteville* 5: 423–428, 1984.
103. VAN NEST, G., R. K. RAMAN, AND W. J. RUTTER. Effects of dexamethasone and 5-bromodeoxyuridine on protein synthesis and secretion during in vitro pancreatic development. *Dev. Biol.* 98: 295–303, 1983.
104. WALKER, M. D., T. EDLUND, A. M. BOULET, AND W. J. RUTTER. Cell-specific expression controlled by the 5'-flanking region of insulin and chymotrypsin genes. *Nature Lond.* 306: 557–561, 1983.
105. WARREN, J. R., M. J. TRUMP, J. K. REDDY, AND M. J. BECICH. Carbamylcholine stimulation of protein secretion in pancreatic acinar carcinoma of rat. *Cancer Lett.* 15: 245–253, 1982.
106. WATANABE, T. K., L. J. HANSEN, N. K. REDDY, Y. S. KANWAR, AND J. K. REDDY. Differentiation of pancreatic acinar carcinoma cells cultured on rat testicular semeniferous tubular basement membranes. *Cancer Res.* 44: 5361–5368, 1984.
107. WECKER, C., A. PUIGSERVER, G. RAUSCH, G. SCHEELE, AND H. KERN. Multiple-level caerulein control of the gene expression of secreteory proteins in the rat. *Eur. J. Biochem.* 151: 461–466, 1985.
108. WESSELLS, N. K. DNA synthesis, mitosis and differentiation in pancreatic acinar cells in vitro. *J. Cell Biol.* 20: 415–433, 1964.
109. WESSELLS, N. K., AND J. H. COHEN. Effects of collagenase on developing epithelia in vitro: lung, ureteric bud, and pancreas. *Dev. Biol.* 18: 294–309, 1968.
110. WILLIAMS, J. A., AND I. D. GOLDFINE. The insulin-pancreatic acinar axis. *Diabetes* 34: 980–986, 1985.
111. WILLIAMS, J. A., M. KORC, AND R. L. DORMER. Action of secretagogues on a new preparation of functionally intact, isolated pancreatic acini. *Am. J. Physiol.* 235 (*Endocrinol. Metab. Gastrointest. Physiol.* 4): E517–E524, 1978.
112. WOMACK, M. D., M. R. HANLEY, AND T. M. JESSELL. Functional substance P receptors on a rat pancreatic acinar cell line. *J. Neurosci.* 12: 3370–3378, 1985.
113. YALOVSKY, U., H. HELLER, AND R. G. KULKA. Accumulation of amylase by chick pancreas in organ culture. Effect of hydrocortisone. *Exp. Cell Res.* 80: 322–328, 1973.
114. YUNIS, A. A., G. K. ARIMURA, AND D. J. RUSSIN. Human pancreatic carcinoma (MIA PaCa-2) in continuous culture: sensitivity to asparaginase. *Int. J. Cancer* 19: 128–135, 1977.

CHAPTER 27

Structural and secretory polarity in the pancreatic acinar cell

AMY CHANG
JAMES D. JAMIESON

Department of Cell Biology, Yale University School of Medicine, New Haven, Connecticut

CHAPTER CONTENTS

Development of Structural Polarity
    The developing pancreas
    Tight junctions
    Cell-substrate interactions
Polarized Secretion in the Pancreas
Relationship Between Structural and Functional Polarity
Ontogeny of Pancreatic Secretion
Distal Events in Stimulus-Secretion Coupling
    Participation of the cytoskeleton
    Regulatory events
    Compensatory endocytosis
Regulated and Constitutive Secretion
Biogenesis of Membrane and Secretory Polarity

POLARITY OF EPITHELIAL CELLS exhibits several levels of increasing complexity: *1*) structural polarity involving the organization of cellular organelles; *2*) membrane polarity, the asymmetric distribution of plasma membrane components between apical and basolateral domains; and *3*) the functional manifestation of structural and membrane polarity as a physiological activity that occurs only at one plasmalemmal domain, e.g., polarized secretion. The exocrine pancreas has served as a paradigm for a structurally and functionally polarized epithelium. The pancreatic acinar cell exhibits structural polarity with rough endoplasmic reticulum and nucleus localized at the base of the cell, Golgi cisternae in a supranuclear position, and zymogen granules at the cell apex. Moreover the plasma membrane is divided by tight junctions into two discrete regions: an apical domain that delineates the glandular lumen and represents ~5% of the total plasma membrane area (1, 11) and a basolateral domain that rests on an investing basal lamina. The functional polarity of the acinar cell reflects the structural polarity of the plasma membrane: discharge of secretory granule content occurs only at the apical membrane, whereas receptors for peptide hormones and cholinergic agents that regulate pancreatic secretion appear localized at the basolateral plasmalemma. Therefore stimulus-secretion coupling in the pancreatic acinar cell and other polarized secretory cells has the additional complexity that the secretory response generated by a neurohumoral signal occurs at a site distinct and spatially separated from that of hormone binding.

In this chapter we introduce the developing pancreas as a model system for studying stimulus-secretion coupling in polarized epithelia. Structural and membrane polarity and secretory capability during development serve as an approach to understand the relationship between structural and secretory polarity in the pancreatic acinar cell. This chapter summarizes the current information that has been obtained through morphogenetic studies as well as in vitro experiments on the reestablishment of polarity in pancreatic acinar cells and other epithelial cells. We discuss studies of secretory responsiveness in the developing exocrine pancreas in relation to functional polarity in the fetal gland as well as possible mechanisms by which an epithelial cell maintains its structural and functional integrity in the face of extensive membrane intermixing that occurs during stimulated secretion.

DEVELOPMENT OF STRUCTURAL POLARITY

The Developing Pancreas

Morphogenesis of the pancreas begins with the evagination of the primitive foregut epithelium into the surrounding mesenchyme at 11 days of gestation in the rat [for review see Pictet et al. (93)]. Even at this early stage (days 11–15) the epithelial cells exhibit a polarized organization with junctional complexes separating microvilli at the apical plasmalemma from the lateral and basal plasma membrane. The cells rest on a continuous basement membrane. During this period epithelial growth and morphogenesis depend on the presence of mesenchymal tissue (119). Cytodifferentiation proceeds as low but significant levels of secretory enzymes/proenzymes are synthesized. By

16–17 days of gestation, acinar structures with recognizable lumina are formed in the rat; the amount of rough endoplasmic reticulum within acinar cells increases rapidly, and zymogen granules are formed. Concomitant with the accumulation (i.e., an increase in volume density) of the component organelles of the cell secretory machinery, the biosynthesis of digestive enzymes/proenzymes increases dramatically (~1,000-fold) beginning approximately at day 17 (46, 106). At this time cellular organelles begin to exhibit a polarized distribution: Golgi cisternae are predominantly localized in supranuclear regions, the flattened cisternae of the rough endoplasmic reticulum are oriented in parallel stacks in the basal region of the cell, and a few zymogen granules are typically distributed at the apical region (76, 89, 93). The size and number of zymogen granules continue to increase so that by 22 days of gestation (the day of birth) granules occupy the greater part (~56%) of the acinar cell volume [see Fig. 2; (125)] and the apparent polarity in granule distribution is lost.

The distribution of cell surface glycoproteins and glycolipids during plasmalemmal differentiation of the rat exocrine pancreas has been analyzed with a battery of lectins (71, 76, 84). Specific binding of wheat germ agglutinin (recognizing N-acetyl-D-glucosamine and sialic acid residues) and *Ricinus communis* agglutinin II (recognizing N-acetyl-D-galactosamine and galactose moieties) to the apical domain of the acinar cell plasma membrane is observed as early as day 13 of gestation (71). However, maturation of the apical membrane, as detected by binding of *Ulex* and *Lotus* lectins to fucosyl residues, does not begin until near the time of birth in some acinar cells and is not complete in all cells until 2–3 wk postpartum (76). Although the biochemical and functional identities of the lectin-binding sites are not known, the differential binding of lectins to the apical and basolateral membranes of the pancreatic acinar cell is consistent with observations made in other epithelia that distinct plasmalemmal domains have unique protein and lipid constituents (28, 114). In addition, freeze-fracture study has shown a heterogeneous distribution of intramembranous particles between the apical and basolateral plasma membrane in the acinar cell (22). Cytochemical labeling of cholesterol by filipin indicates that the luminal plasma membrane is cholesterol poor in comparison with the basolateral domain (86). Specific proteins that demonstrate polarized plasma membrane distribution in pancreatic acinar cells include apical leucine aminopeptidase (53), basolateral Na^+-K^+-ATPase (118, 133), and basolateral cholecystokinin (CCK) receptors (102). The pancreas-specific glycoprotein 2 (108) and δ-glutamyl transpeptidase (18) are predominantly apically distributed. The developmental expression of these markers for apical and basolateral membrane domains has not been examined.

Tight Junctions

The mechanism by which epithelial cell polarity is formed and maintained appears exceedingly complex and probably involves multiple interactions between intracellular (e.g., protein synthetic and cytoskeletal) machinery and the extracellular environment. Stabilization of cell polarity appears to involve junctional complexes, specifically the tight junctions that are characteristic of epithelial cells. Work on both the morphogenetic formation of junctional complexes and their disassembly/reassembly supports the idea that the tight junction may act as a barrier, limiting the diffusion of apical and basolateral membrane proteins and lipids (30, 114). During epithelial differentiation of fetal colon, development of distinct phosphatase activities that are segregated to apical and basolateral membrane domains occurs subsequent to the establishment of tight junctions (21). Freeze-fracture studies of junctional complexes during morphogenesis of fetal small intestine and liver have not examined concomitantly the development of membrane polarity (70, 79). Disruption of tight junctions in adult pancreatic lobules by incubation in Ca^{2+}-free medium results in an apparent loss of discrete plasmalemmal domains, detected by freeze-fracture analysis as the loss of an asymmetric distribution of intramembranous particles (77). Similar studies in other epithelia on the redistribution of plasma membrane macromolecules after cell dissociation and junctional complex disassembly also suggest that tight junctions maintain cell surface specialization (25, 28, 94, 127, 132). Nevertheless such experiments have not established a direct relationship between tight junction integrity and membrane polarity because Ca^{2+} chelation may disrupt a number of cell functions. Moreover the developmental studies only make temporal correlations between tight junction formation and expression of plasma membrane domain markers.

Some recent studies indicate that cell surface polarity can exist in the absence of tight junctions. Madden and Sarras (71) monitored the development of tight junctions and the formation of an apical membrane domain in fetal pancreas by freeze-fracture analysis and lectin binding, respectively. Although binding of wheat germ agglutinin and *Ricinus communis* agglutinin II was restricted to the luminal plasma membrane as early as day 13 of gestation in the rat, incompletely formed tight junctions did not appear until day 14, and tight junctions containing continuous sealing strands were formed by day 15. These results suggest that the lateral diffusion of apical membrane glycoproteins and glycolipids is restricted when complete tight junctional complexes are not present during the development of the rat pancreas. Similar observations have been made in other systems. In primary cultures of aortic endothelial cells, a polarized distribution of plasmalemmal proteins is established before the formation of a confluent monolayer (82). The polarity of a cell surface marker remains the same in Madin-Darby canine kidney (MDCK) cells with tight junctions that maintain different transepithelial resistances (36). Cells such as the osteoclast (5), the differentiating red blood cell (116), and sperm (33) have structurally and immunologically distinct plasma membrane domains, although they lack tight junctions. Plasma membrane microdomains, e.g., the tips of the postsynaptic folds at the neuromuscular junction with clusters of acetylcholine receptors (105), and enzymatic and antigenic sites concentrated at the base and tips of microvilli in the brush border of the kidney proximal tubule (61) are also maintained in the absence of junctional complexes. Therefore it appears that tight junctions are not necessary in establishing polarity. Nevertheless these structures may stabilize the polarized state, once formed: to perpetuate the polarity of epithelial cells during mitosis, junctional relationships with neighboring cells are not disturbed and cleavage of the two daughter cells occurs around a newly formed tight junction (93).

Cell-Substrate Interactions

It is clear from a variety of studies that the extracellular matrix participates in triggering the formation of the polarized state, probably through interaction with the cytoskeleton. The preferential budding of viruses from either the apical or basolateral membranes of a confluent monolayer of the kidney epithelial cell line MDCK reflects the cells' membrane polarity. Virus budding in dissociated MDCK cells infected in suspension is not asymmetric; polarized budding is restored when the cells are attached and spread on collagen gels, even though tight junctions are absent or incomplete in the sparsely plated cells (100). The repolarization of MDCK cells that is triggered by cell-substrate contact is a gradual process (4, 49) involving the removal and reinsertion of membrane proteins (74); it appears that complex correction mechanisms also participate in generating and maintaining polarity. The biogenesis of plasma membrane domains is discussed in BIOGENESIS OF MEMBRANE AND SECRETORY POLARITY, p. 543. The notion that cell attachment to a substratum acts as an early trigger of cell polarization is supported by work from our laboratory on a rat pancreatic acinar tumor. The acinar cells of the pancreatic carcinoma have lost their normal polarized structure along with their ability to maintain a normal basal lamina (53). However, when these cells are plated on basement membrane derived from amnion or on laminin-coated dishes, they attach, spread, and the polarized orientation of intracellular organelles characteristic of the normal epithelium is restored (54). Repolarization of these pancreatic tumor cells is accompanied by a change in the organization of cytoskeletal actin. Similarly, the cell line AR42J derived from a rat pancreatic tumor forms a structurally polarized monolayer when plated on laminin (E. Sachs and J. D. Jamieson, unpublished observation). In isolated thyroid cells and mammary gland cells cultured on or within collagen gels, cell-substrate interactions induce the formation of a basal pole (19, 82). When mammary epithelial cells in culture reach confluence and become polarized, actin filaments at the base of the cells are bundled into highly organized arrays; it has been suggested that the rearrangement of the cytoskeleton is induced by extracellular matrix components via cell surface proteoglycans that act as matrix receptors (96).

In some cases (MDCK and thyroid cells), the structural polarity induced by cell-substrate interaction is accompanied by the segregation of plasma membrane constituents into apical and basolateral domains (31, 49). However, plating the pancreatic acinar tumor on exogenous basement membrane is not sufficient to reestablish plasma membrane polarity, as assayed by the distribution of an apical membrane marker (54). It seems then that structural, membrane, and functional polarity, although interrelated, are not always tightly coupled. Obviously the kind of experimental analysis used to measure cell polarity should temper the conclusions. For instance, polarized viral budding is not a stringent measure of membrane polarity, because budding does not occur unless a critical concentration of viral envelope proteins has been reached (35).

The asymmetry of the extracellular milieu is responsible, at least in part, for initiating the formation of a polarized orientation. The induction of polarity in mouse 8-cell blastomeres is dependent on the points of cell contact between the individual cells; no polarity develops in cells that are completely surrounded by other cells (59). The asymmetric environment may also participate in stabilizing cell polarity, because exposure of a cell's preformed apical pole to a substratum causes the cell to dismantle the domain and reorganize a new apical surface that is free of the adhesive surface (19, 41, 82). By restricting intermixing of components between the luminal space and the interstitium, tight junctions maintain the anisotropic environment.

The epithelial cell helps to form its own polarized environment by secreting extracellular matrix components from the basolateral surface. Lee et al. (68) and Parry et al. (88) have suggested, based on their observations of mammary epithelial cells in culture, that the cell substratum may influence the polarity of glycosaminoglycan secretion (68, 88). Similarly it has been observed that a transformed phenotype and the loss of a polarized structure in endothelial cell cultures grown in the absence of growth factor occur concomitantly with the secretion of basal lamina in both apical and basolateral directions (42). It is possible that the interaction of basement membrane constituents with cell surface receptors initiates changes in the directed delivery of membrane components (and secretory products) to a specific domain. The participation of basal lamina in establishing acinar cell polarity during pancreatic development is not clear; although a continuous basal lamina underlies epithelial cells even before the formation of the pancreatic bud (day 10 of gestation in the rat), the deposition of specific components of the basement membrane with respect to formation of acinar cell polarity has not been examined.

POLARIZED SECRETION IN THE PANCREAS

Protein secretion in a polarized fashion is a functional manifestation of the structural polarity of an epithelium. In the pancreatic acinar cell, digestive enzymes/proenzymes packaged in secretory granules are released into the glandular lumen by exocytosis. Although this event occurs rapidly, exocytotic images in the pancreas have been captured by transmission electron microscopy (87) and by freeze-fracture technique (123). Exocytosis does not depend on new protein synthesis but requires energy and Ca^{2+} and is contingent on the directed movement of the secretory granule to apical plasma membrane and the specific recognition of the two partners (56, 87). Although the secretory enzyme α-amylase has been detected in serum (126), it is believed that fusion of the secretory granule membrane with the basolateral plasmalemma does not normally occur. Although it is possible that leakage of pancreatic luminal content into the interstitial space may occur, the secretion of hepatic α-amylase into the sinusoidal space can account for the serum α-amylase activity (109). During neurohumoral stimulation the rate of granule exocytosis increases; in addition, once a zymogen granule has fused with the apical membrane, other zymogen granules can then fuse with it to form an interconnected series of granules opening onto the lumen in a process called *compound exocytosis* (52). The stimulated insertion of secretory granule membrane into the luminal plasma membrane represents the substantial addition of ~900 μm^2 of granule membrane surface area per cell to ~30 μm^2 of apical plasma membrane out of a total plasmalemmal area of ~600 μm^2 (1, 11). Nevertheless the surface area of the plasma membrane and intracellular membranes remains constant during stimulation, although there is an initial transient enlargement of the apical membrane (58). Membrane homeostasis during exocytosis is maintained by membrane retrieval and recycling (for discussion see DISTAL EVENTS IN STIMULUS-SECRETION COUPLING, p. 536).

Analysis of apically secreted proteins by cannulation of the main pancreatic duct reveals the same mixture of digestive enzymes/proenzymes in the pancreatic juice as that packaged in granules (43). Synthesis of the digestive enzymes/proenzymes represents >90% of total protein synthesis in the pancreatic acinar cell. After translation these proteins are transported vectorially from the endoplasmic reticulum, sequentially through the cisternae of the Golgi apparatus, and delivered to condensing vacuoles; precipitation of secretory proteins within the condensing vacuoles results in the mature storage granules at the apex of the cell (57). In the fetal pancreas, secretory proteins are vectorially transported with approximately the same kinetics as that reported for the adult pancreas (see Fig. 4). In addition to the discharge of proteins packaged in granules from the apical surface of the cell, it is believed that protein is also secreted from the basolateral domain. Certainly epithelial cells secrete their own basement membrane from the basolateral surface (10, 17). The basal lamina components (type IV collagen, laminin, fibronectin, and heparan sulfate proteoglycan) have been localized at the base of the pancreatic acinar cell (53), although the secretion of these proteins by the epithelium has not been examined. It has been reported recently that protein species that are not observed in granule content comprise a minor component of secretion from pancreatic acini incubated in vitro. Some of these proteins may represent basolaterally secreted species since they are not found in cannulated pancreatic juice (2). Thus it appears that a small population of secretory proteins is not packaged into granules but is secreted apically and/or basolaterally.

During protein biosynthesis in the pancreas, secretory proteins destined for storage in granules are sorted from the subpopulation of proteins that are not packaged but released constitutively from the apical or basolateral surface (60, 124). Of the secretory proteins that are not stored in granules, those destined for release from the apical cell surface must be sorted from proteins headed toward the basolateral surface. The transport vesicles containing these secretory proteins must recognize and deliver their content to the topologically correct surface in the same way that a zymogen granule specifically recognizes the luminal plasmalemma. Although the general pathway for delivery of proteins into storage granules has been defined (56, 57), it is not known in which intracellular organelle(s) selective deviation from this pathway takes place. No experimental system is currently available in which it is possible to dissociate the intracellular pathways involved in transport of secretory proteins to storage granules or directly to the luminal surface or the basolateral surface (see BIOGENESIS OF MEMBRANE AND SECRETORY POLARITY, p. 543).

RELATIONSHIP BETWEEN STRUCTURAL AND FUNCTIONAL POLARITY

It appears, based on studies of a variety of cell types, that the relationship between structural and membrane polarity and functional polarity is a dynamic one that is dependent on the physiological role of the cell. For instance, mast cells, neutrophils, and cytotoxic lymphocytes do not have any apparent structural polarity (65, 67). However, when these cells are stimulated, the release of secretory granules from mast cells and neutrophils and the secretion of cytotoxic elements into the membrane of target cells by cytotoxic lymphocytes occur exclusively in the direction of the stimulus. Local perturbation of the mast cell surface by beads coated with concanavalin A can induce granule exocytosis (67). Presumably, polarized secretion in the mast cell is closely associated with a structural polarity that is locally and temporarily formed in the membrane. Similarly, environmental stimuli can reverse the functional polarity in kidney epithelial cells, i.e., acid loading of the animal results in a change in the direction of transepithelial transport, endocytosis, and exocytosis (110); concomitantly, changes in the plasma membrane must occur to allow the recognition of exocytic and endocytic vesicles (containing transport proteins) at the appropriate pole of the cell.

It has been reported for mammary and thyroid epithelial cells and the kidney cell line A6 derived from toad that the expression of structural polarity does not necessarily predict the expression of physiological functions (19, 45, 66). For example, no hormonal stimulation of A6 cells is detected when the cells are grown on plastic culture dishes, although the cells form domes (indicative of tight junctions and transepithelial transport activity) and express functional catalytic subunit of adenylate cyclase (66). Hormonal sensitivity develops when the cells are grown on filters that permit exposure of the basolateral cell surface to a physiological environment, inducing either the expression of receptors at the basolateral membrane or coupling of receptor binding to adenylate cyclase. Interestingly A6 cells continue to express hormonal sensitivity for some time after monolayer dissociation and the loss of structural polarity. Similarly in mammary and thyroid cells, functional polarity, in contrast to structural polarity, is greatly influenced by culture conditions (19, 45, 112). In these cells, the accessibility of the basolateral surface to nutrients as well as the cell shape determine whether prolactin can induce the expression, packaging, and vectorial discharge of milk proteins (46).

To study the relationship between secretory polarity and structural polarity in the pancreatic acinar cell, we have focused on the developing pancreas (3, 20). It has previously been observed that although some plasma membrane glycoproteins and/or glycolipids of the acinar cell are distributed in a polarized manner as early as day 13 of gestation (71), the cell surface continues to undergo modulation with the appearance of novel lectin-binding sites at day 19 of gestation and changing patterns of lectin binding until 2–3 wk after birth (76). Secretory granules first appear at the apex of the acinar cell at ~16–17 days of gestation; it has been suggested, based on protein release observed in vitro, that resting (unstimulated) secretory activity begins at this time (128). However, stimulated secretion does not occur until after birth (20, 24, 30). These observations suggest that structural polarization or the maturation of the acinar cell surface phenotype are not strictly coupled to the onset of secretory activity in the developing pancreas. We investigated this idea further by studying stimulated secretion and resting secretion in the developing pancreas (3, 20).

ONTOGENY OF PANCREATIC SECRETION

Shortly before birth, pancreatic acinar cells in fetal rats are packed with secretory granules; however, addition of secretagogues does not increase the rate of release of digestive enzymes/proenzymes from pancreatic lobules incubated in vitro (20, 24, 129), whereas the peptide hormone CCK dramatically increases the rate of release of the secretory protein α-amylase from glands of neonatal rats 1 day after birth (Fig. 1). To determine whether the secretory responsiveness of the neonatal pancreas is associated with the cell surface expression of CCK receptors, we measured radiolabeled CCK binding to pancreatic lobules from fetal (1 day before birth) and neonatal rats by light-microscopic autoradiography (Fig. 2). There are no apparent differences in the plasmalemmal binding sites for CCK in fetal or neonatal glands. Experiments in which binding and affinity cross-linking of CCK to membrane preparations were examined indicate further that there are no differences in the affinity or structure of the receptors in pancreas at the two ages (20).

The plasma membrane receptors for CCK are functional in the fetal pancreas because hormone binding is coupled to an increase in the cytosolic free Ca^{2+} concentration, as assayed by stimulated $^{45}Ca^{2+}$ efflux (20) and aequorin measurements (A. Chang and W. Apfeldorf, unpublished observations). Presumably, the release of Ca^{2+} from intracellular stores in response to stimulation occurs through the hydrolysis of phosphatidylinositol at the plasma membrane. (For discussion of signal transduction events in the pancreas see chapters by Schulz and by Williams, Burnham, and Hootman in this Handbook). The CCK-stimulated increase in the cytosolic Ca^{2+} concentration in the fetal pancreas appears quantitatively the same as that observed in the neonatal pancreas. Nevertheless the rate of secretion is not increased in the fetal gland in response to an elevated intracellular Ca^{2+} concentration generated either by hormone binding or by the addition of the Ca^{2+} ionophore A23187 to in vitro incubations of pancreatic lobules. Furthermore there is no stimulated secretory responses after the addition of phorbol dibutyrate or dibutyryl cAMP, which suggests that the secretory machinery of the fetal gland is not sensitive to activation of the Ca^{2+}-phospholipid–dependent protein kinase C or to stim-

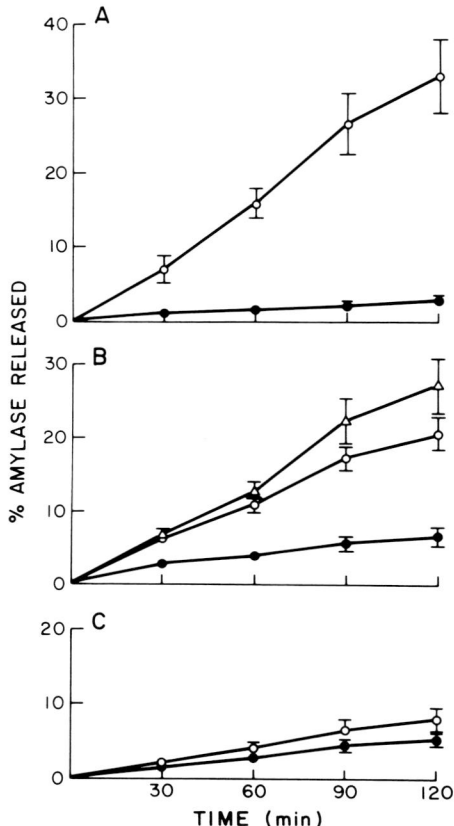

FIG. 1. Kinetics of amylase secretion from pancreatic lobules incubated in vitro. Pancreatic lobules were isolated from fetal (1 day before birth), neonatal (1 day after birth), and adult rats and preincubated in oxygenated medium at 37°C. Secretion assay was initiated in fresh medium in presence (○, △) or absence (●) of an optimal dose of cholecystokinin COOH-terminal octapeptide (CCK-8). Amylase activity was assayed in medium aliquots removed at 30-min intervals and in tissue homogenized at the end of 2-h incubation period. Secretion from adult (A), neonatal (B), and fetal (C) pancreas is expressed as percent of total tissue amylase released into medium and is approx. linear with time for up to 2 h. Maximal secretory response from neonatal and adult pancreas occurs in response to 10 nM (△) and 1 nM (○) CCK-8, respectively, and rates of secretion from glands at the 2 ages are comparable. In presence of a range of CCK-8 doses, there is no significant increase in rate of amylase secretion from fetal pancreas. Rate of CCK-8–stimulated secretion from neonatal pancreas (minus resting secretion) at 10 nM CCK-8 [0.23 ± 0.03%/min (mean ± SE)] is 8-fold greater than that from fetal gland at same dose of secretagogue [0.03 ± 0.01%/min (mean ± SE)].

ulation of the cAMP-dependent protein kinase. We concluded that some events in stimulus-secretion coupling are not yet mature in fetal pancreas but become functional after birth; these events may occur distal to receptor-mediated second messenger production during signal transduction.

Analogous coupling of receptor-mediated Ca^{2+} mobilization from secretory response has been observed in a rat mast cell line 2H3 undergoing mitosis (51) or treated with metalloendoprotease inhibitors or Zn^{2+} (8). Based on several indirect observations, it was proposed that secretory response as well as vesicular traffic are inhibited in these mitotic cells because membrane fusion events in general are inhibited (51). Similarly, we have considered the idea that in the fetal pancreas there is an interruption in exocytosis itself. At present, we are investigating the relationship between the maturation of secretory responsiveness and biochemical and structural events occurring during exocytosis, although we cannot rule out the involvement of other, as yet undefined, events occurring subsequent to Ca^{2+} mobilization.

DISTAL EVENTS IN STIMULUS-SECRETION COUPLING

The sequence of structural events occurring during exocytosis were outlined by Palade in 1975 (87) as: *1*) the movement of the zymogen granule to the apical membrane followed by *2*) recognition of the two partners and then *3*) fusion (and fission) of the membranes in an energy-dependent manner. Many hypotheses and some experimental work later, the biochemical and structural events that are involved in exocytosis remain obscure. In this section, we review current hypotheses on the regulation of exocytosis. Also, we suggest the developing pancreas as a possible model system in which to study distal events in stimulus-secretion coupling.

Participation of the Cytoskeleton

One of the first events occurring during exocytosis, the translocation of the secretory granule to the apical pole, may require propulsive forces. Many studies have been based on the hypothesis that such forces are generated by microtubules and/or the contractile interactions between actin and myosin (i.e., microfilaments). Most of the studies of the relationship between secretion and the cytoskeleton have used pharmacological agents that disrupt cytoskeletal elements. These experiments have been difficult to interpret because the drugs have a number of secondary effects. For instance, because microtubule-dissociating drugs also insert into membranes (111), direct interactions with the plasma membrane or secretory granule membrane could account for any effect of these drugs on secretion, independently of the status of microtubule assembly. At present, however, the consensus is that microtubules are not directly involved in the regulation of secretion. In isolated mouse pancreatic acini, colchicine-induced disassembly of microtubules has no effect on amylase secretion but does affect intracellular transport of newly synthesized proteins (130).

Although microtubule-dissociating drugs may not affect epithelial cell secretion per se, it has been suggested that they may disrupt polarized secretion and polarized delivery of membrane to the cell surface (65, 101). Microtubule depolymerization in natural killer lymphoid cells strongly inhibits cytotoxic activity. Be-

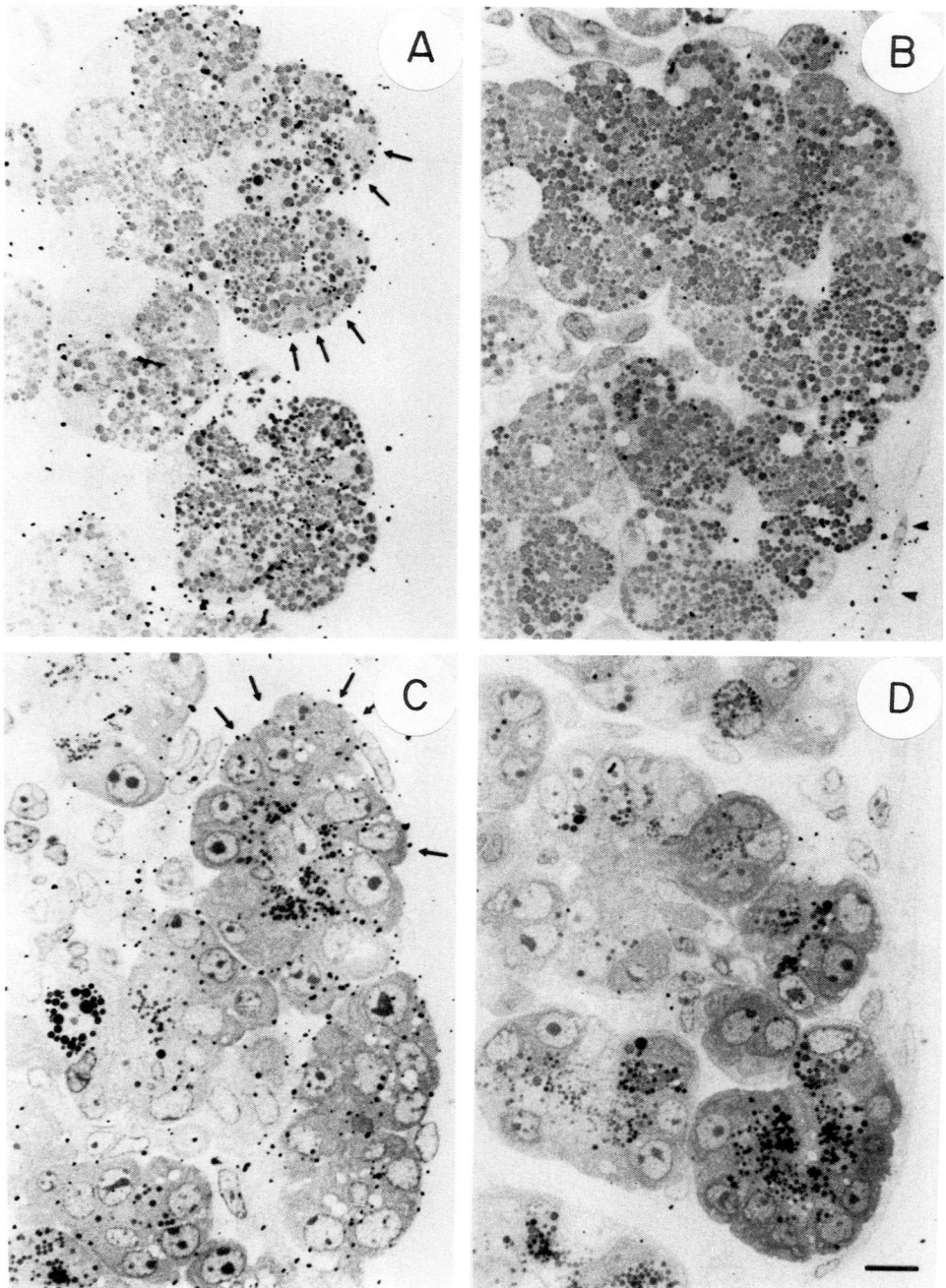

FIG. 2. Light-microscopic autoradiography of pancreatic lobules labeled with radiolabeled cholecystokinin (CCK). Pancreatic lobules were incubated in buffer containing ^{125}I-labeled cholecystokinin triacontatriapeptide (^{125}I-CCK-33) for 5 min at 23°C before being fixed and processed for light-microscopic autoradiography. *A*: fetal pancreas labeled with ^{125}I-CCK-33; *B*: control preparation of fetal pancreas incubated with radiolabeled CCK and 200-fold excess unlabeled hormone; *C*: neonatal pancreas labeled with ^{125}I-CCK-33; *D*: control preparation of neonatal pancreas incubated with radioligand in presence of unlabeled hormone. *Arrows,* autoradiographic grains localized around the periphery of acinar cells; *arrowheads,* in *B,* autoradiographic grains nonspecifically associated with mesenchymal tissue. *Bar,* 10 µm. Autoradiographic labeling of acinar cells in both fetal and neonatal pancreas is specific (*A* and *C*), because a very low level of nonspecific radioactivity is randomly associated with cells when ^{125}I-CCK-33 labeling occurs in presence of unlabeled CCK (*B* and *D*). Cell surface expression of CCK-binding sites is similar in fetal and neonatal pancreas, supporting the idea that events distal to hormone binding account for difference in secretory responsiveness at the 2 ages. Structure of acinar cells of fetal and neonatal pancreas is strikingly different. Acinar cells of fetal pancreas are packed with zymogen granules, whereas secretory granules of neonatal pancreas are smaller and restricted to cell apical region. [From Chang and Jamieson (20).]

cause disassembly of microtubules concomitantly disrupts the orientation of the Golgi apparatus and microtubule-organizing center in the killer cell toward the bound target cell, it is thought that the intracellular dispersion of Golgi elements results in the loss of polarized secretion of toxic components into the target cell and, consequently, the inhibition of cell lysis (65). In the exocrine pancreas, microtubule disruption results in structural alterations of the Golgi region; the fusion of transport vesicles, formation of Golgi cisternae, and organization of cisternae into integral stacks appear impaired (90). It is not known whether disorganization of the Golgi apparatus in the pancreatic acinar cell disrupts the polarity of secretion. Microtubule dissociation could also affect polarized secretion if one postulates that microtubules act as guidelines that direct secretory granules to the cell apex in a manner analogous to axonal transport (58). Although transport of secretory proteins may occur by the same or similar mechanism as the polarized delivery of membrane proteins to the cell surface (see BIOGENESIS OF MEMBRANE AND SECRETORY POLARITY, p. 543), there are conflicting reports on the effect of microtubule disruption on the polarized expression of viral proteins at the plasma membrane (101, 104).

In the pancreatic acinar cell, contractile proteins such as actin and myosin are closely associated with the membranes of the Golgi transcisternae and immature and mature secretory granules. The terminal web under the apical plasma membrane is composed of the cytoskeletal proteins villin and tropomyosin, in addition to actin and myosin (9, 26). Based on morphological observations, Jamieson (56) and others (6, 26) have suggested that during stimulated secretion the filamentous network under the luminal membrane is dissipated, allowing the zymogen granules to approach the luminal plasma membrane. In support of the idea, it has been reported that cytochalasin B, which interferes with the polymerization of actin, increases the level of nonstimulated secretion from pancreatic lobules at low dosages, whereas at high dosages it reversibly inhibits secretagogue-stimulated secretion (6). It is not clear whether the putative participation of microfilamentous proteins in exocytosis of storage granules is an active one, involving actin-myosin interaction to generate a force that drives granules to the apical membrane. Alternatively microfilamentous proteins may participate passively by simply allowing or guiding the approximation of granules with the cell apex during secretion.

Regulatory Events

The specific recognition of the secretory granule and the apical plasmalemma is thought to lie in the unique protein and lipid constituents of the two membranes. By analogy with the entry of animal viruses into cells, it is possible that specific polypeptides or specific polypeptide domains mediate membrane binding and that membrane fusion occurs on activation of the fusogen. Shared functions of storage granule membranes of many secretory cells are suggested by the observation that there is a subset of membrane proteins that is common to these granules (15). Furthermore work in an in vitro model for exocytosis indicates that an N-ethylmaleimide and trypsin-sensitive protein(s) are involved in Ca^{2+}-sensitive membrane fusion in sea urchin eggs (55, 107). It was postulated based on in vitro studies of the interaction of chromaffin granules with the plasma membrane that soluble and membrane proteins in the adrenal medulla act as recognition and/or fusogenic proteins (63, 78, 95). Because metalloendoprotease inhibitors block stimulus-secretion coupling in mast cells and adrenal chromaffin cells at a step distal to Ca^{2+} mobilization, it has been suggested that a protease associated with the plasma membrane is required for exocytosis; perhaps the proteolysis of a membrane protein is required to remove charge or steric restraints to membrane fusion or to generate a fusogenic peptide (83). The observation that immature secretory granules from pancreatic and parotid acinar cells appear not to fuse with the luminal plasma membrane during secretagogue stimulation is of relevance for the understanding of the mechanism of exocytosis (47, 64, 113). Perhaps the immature granule membrane lacks a recognition/fusogenic component that it will acquire during granule maturation.

In addition to the analysis of the compositional characteristics of the two fusion partners, another approach to understanding the recognition and fusion of granule and luminal membrane is to study the regulation of exocytosis. It is thought that Ca^{2+}-dependent protein phosphorylation/dephosphorylation is involved in the regulation of the secretory response to neurohumoral stimulation (23, 44, 97). There is little direct evidence to support the hypothesis. However, injection of Ca^{2+}-calmodulin–dependent protein kinase into the presynaptic terminal of the giant squid axon stimulates neurotransmitter release, suggesting that phosphorylation reactions catalyzed by the kinase may control synaptic vesicle release (69). In addition, it has been reported that anti-calmodulin antibody inhibits cortical granule exocytosis in sea urchin eggs (120), perhaps by modulating the activity of a Ca^{2+}-calmodulin–dependent protein kinase or phosphatase. In the pancreas, a Ca^{2+}-calmodulin–dependent protein kinase (38) and protein kinase C mediate changes in the phosphorylation states of both soluble and membrane proteins during hormone stimulation (14, 32, 91). Whether any of the phosphorylation events are involved in the enhanced secretion has not been established. For instance, it is unlikely that the stimulated phosphorylation of the ribosomal protein S6 is directly involved in the secretory response (32). On the assumption that Ca^{2+}-dependent phosphorylation might regulate the activity of a recognition protein or fusogen, several studies have focused on phosphoproteins in the zymogen granule membrane. Several proteins in the granule membrane serve as in vitro substrates for Ca^{2+}-calmodulin–dependent protein kinases and/or the Ca^{2+}-phospholipid–dependent protein kinase C (14, 91, 131). When acinar cells are labeled with [^{32}P]orthophosphate, a membrane protein of secretory granules becomes phosphorylated (91). However, it is uncertain whether the phosphorylation state of granule membrane proteins is modulated by neurohumoral stimulation in vivo.

We have been studying secretagogue-stimulated phosphorylation/dephosphorylation events in the developing pancreas to detect phosphoproteins that may be involved in the regulation of secretory response. The activity of the Ca^{2+}-calmodulin–dependent protein kinase measured in vitro as well as the apparent molecular mass of the kinase increase during pancreatic development (38). Whether the developmental changes in the Ca^{2+}-calmodulin–dependent protein kinase are related to the postnatal onset of secretory responsiveness is not known. However, in situ phosphorylation patterns induced by Ca^{2+}-mobilizing secretagogues in the neonatal pancreas (in comparison with the fetal gland) might reveal whether specific substrates become phosphorylated during stimulated exocytosis.

Compensatory Endocytosis

After exocytosis the excess membrane that is added to the apical plasma membrane is selectively retrieved by endocytosis and reused for packaging secretion anew (29, 87). Membrane retrieval and reutilization ensures that the amount of membrane of the secretory granules and plasma membrane remain relatively constant and that each compartment retains its molecular identity. The results of morphological studies with electron-dense tracers to map endocytic pathways have varied considerably, depending on the tracer and cell type (29, 121). In the exocrine pancreas, it appears that uptake of the tracers from the apical domain in smooth-surfaced vesicles can occur by more than one distinct pathway (50). Dextran, a fluid-phase marker, as well as cationized ferritin, which adsorbs to membranes, can enter compartments of the secretory pathway (Golgi cisternae, immature and mature granules) from the pancreatic duct, in concordance with the idea that membrane is retrieved from the cell surface and reutilized in packaging secretory products (50). The bulk-phase tracer (horseradish peroxidase) traverses different pathways, depending on whether it is internalized from the apical or basolateral surfaces of the acinar cell (85). Endocytosis from the apical cell surface occurs via a route separate from the endocytic pathway at the basolateral surface. Thus endocytosis, like polarized secretion, is a functional manifestation of the structural polarity of the pancreatic acinar cell.

REGULATED AND CONSTITUTIVE SECRETION

As first suggested by Tartakoff and Vassalli (124), protein secretion falls into two classes: *1*) regulated secretion of proteins stored in secretory granules that is activated on stimulation by neurohumoral agents and *2*) constitutive secretion of newly synthesized proteins that are not packaged within conspicuous storage compartments and whose discharge is not stimulated by secretagogues. A cell that exhibits regulated secretion also secretes some of its proteins constitutively, albeit by a distinct intracellular pathway (60). During biosynthesis, constitutively secreted proteins are sorted from proteins destined for packaging and storage within secretory granules. It has been suggested that constitutively secreted proteins are released from the cell surface via small vesicles (60). In a polarized secretory cell such as the pancreatic acinar cell, protein sorting takes on an additional level of complexity, because it is necessary to segregate proteins destined for secretion from the apical pole (e.g., digestive enzymes/proenzymes) from those aimed for release from the basal surface (e.g., basement membrane components).

The intracellular pathway involving packaging and storing secretory proteins in the exocrine pancreas has been well characterized (56, 57, 87). Recent work on the pancreas has focused on defining the intracellular pathway(s) taken by constitutively secreted proteins in the pancreatic acinar cell (2, 3). A number of characteristics of the exocrine pancreas from fetal rats prompted us to make our initial studies in the developing gland: between days 17 and 21 of gestation, pancreatic acinar cells accumulate zymogen granules, so that by day 1 before birth the cells contain granules in numbers far greater than that found in cells of the adult gland (27, 125). The resting secretion from the fetal pancreas is at least as high as that of the adult gland but it is not stimulated by secretagogues or permeant second messengers, i.e., it is constitutive. Although it has been presumed previously that resting secretion of pancreatic enzymes/proenzymes is a result of zymogen granule exocytosis (58, 87), we tested the hypothesis that the secretion from the fetal gland occurs in the absence of zymogen granule exocytosis. Therefore we examined the in vitro secretion of newly synthesized proteins from the fetal pancreas.

Pancreatic lobules from fetal rats 1 day before birth secrete newly synthesized proteins in a biphasic manner, which suggests that constitutive secretion in the fetal pancreas occurs via two distinct pathways [Figs. 3 and 4; (3)]. The digestive enzymes/proenzymes that are exclusively released during the second phase of secretion are also the predominant protein species

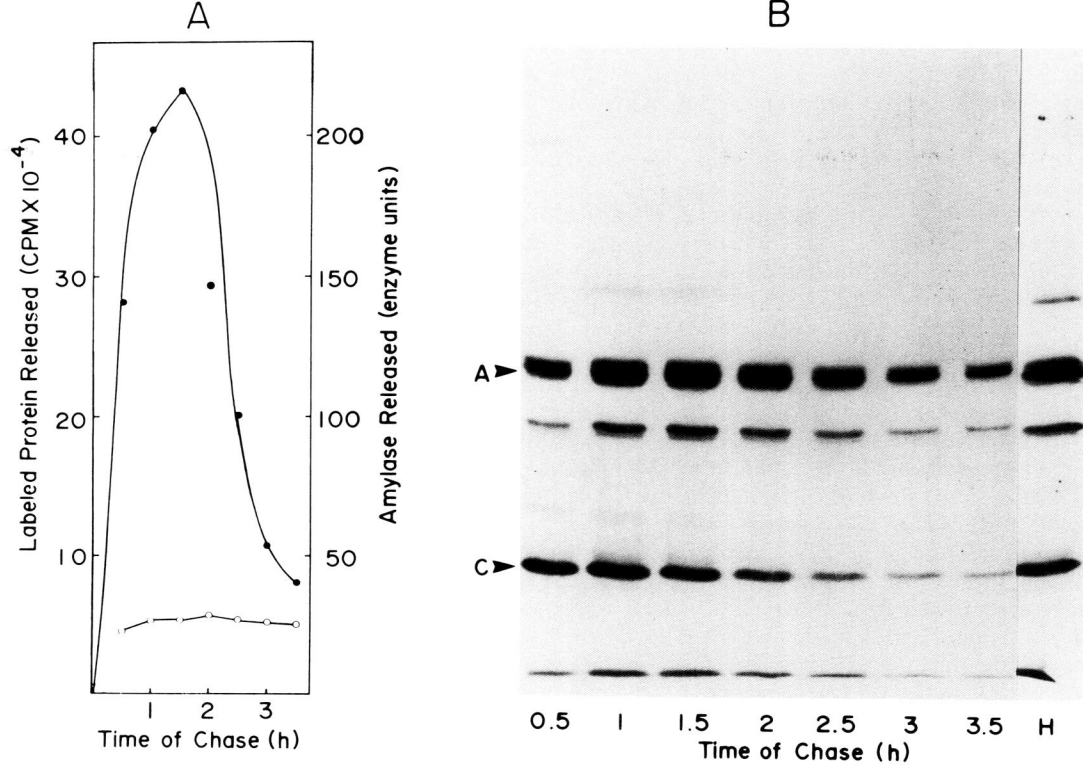
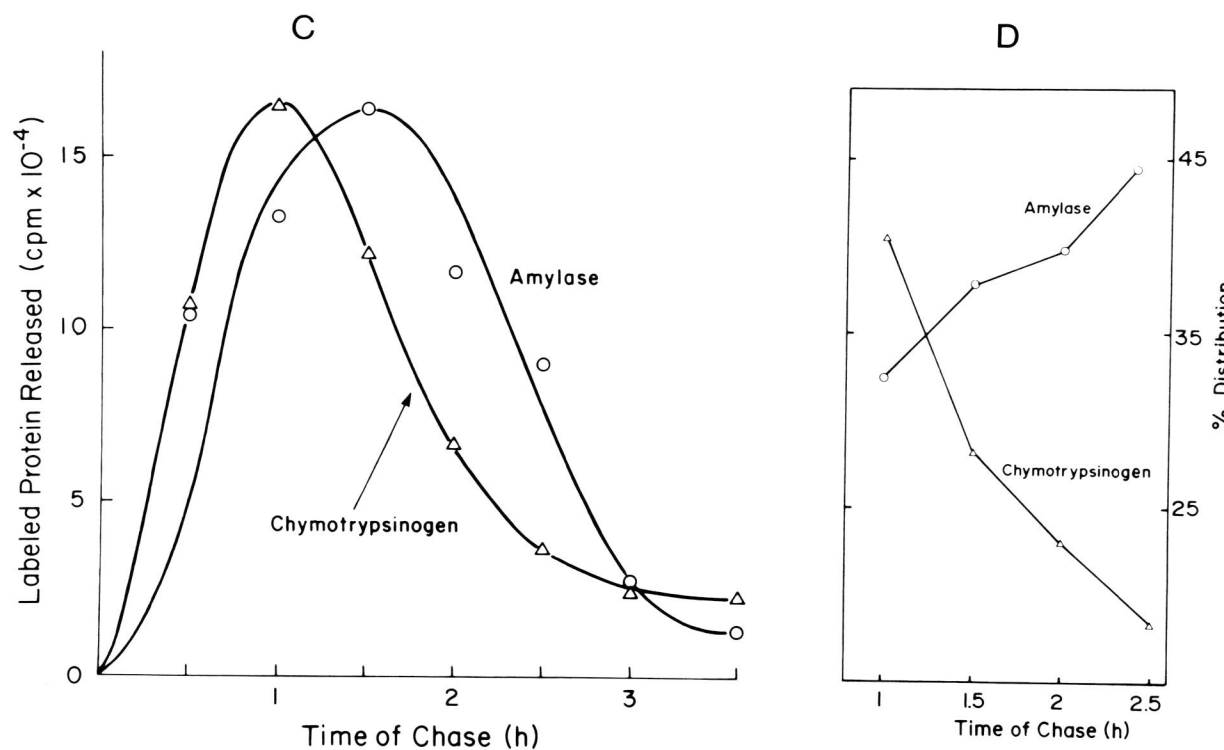

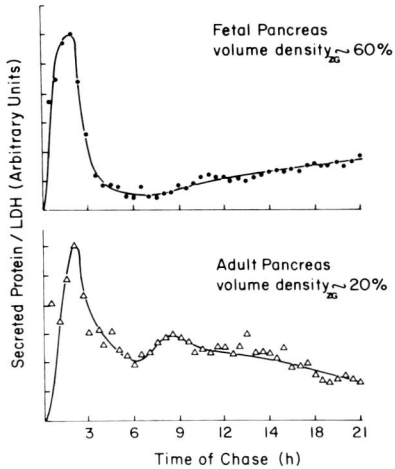

FIG. 4. Release of newly synthesized proteins from fetal pancreas in comparison with that of adult pancreas. Pancreatic lobules were pulse labeled with [^{35}S]L-methionine for 2 min and chased for sequential 30-min periods, as described in Figure 3. Release of radiolabeled proteins from fetal pancreas, normalized for leakage of lactate dehydrogenase (●), occurs in 2 distinct phases: the 1st phase peaks by 2 h of chase, is complete by 6.5 h, and comprises ~12% of total incorporated radioactivity; the 2nd phase is not yet maximal at 21 h of chase. In adult gland (△), a discrete peak in the 2nd secretory phase is observed at 9–10 h of chase. Because stored population of preformed zymogen granules (26) in adult gland (volume density ~20%) is small, relative to fetal gland (volume density ~60%), it is possible that rate of exocytosis of newly formed granules depends on number of preformed granules. [From Arvan and Chang (3; *top panel*) and Arvan and Castle (2; *bottom panel*).]

released during the first secretory phase. The kinetics of intracellular protein transport were analyzed by electron-microscopic autoradiography of pulse-labeled pancreas to determine the compartments of the secretory pathway from which the two phases of secretion derive. Figure 5 shows that radiolabeled proteins are vectorially transported in the fetal pancreas from the rough endoplasmic reticulum to Golgi cisternae to immature granules/condensing vacuoles and mature zymogen granules, as previously described in the adult pancreas (57). The first secretory phase appears to represent constitutive release of small vesicles that bud from immature granules/condensing vacuoles, because maximal secretion during the first phase corresponds kinetically with maximal autoradiographic labeling of the immature granules/condensing vacuoles. The second phase of secretion appears to derive from exocytosis of zymogen granules. Thus secretion from the fetal pancreas is constitutive because it is not stimulated by secretagogues and both a granule and a nongranule pathway are involved in the constitutive secretion. The pancreatic digestive enzymes/proenzymes traverse both pathways.

Secretion of newly synthesized proteins from the adult pancreas in vitro also occurs in a biphasic manner [Fig. 4; (2)]. The first phase of secretion in the adult pancreas, as in the fetal gland, may represent constitutive vesicular discharge and initially it is not changed in the presence of hormonal stimulation. The second phase of secretion from the adult pancreas is secretagogue-responsive, comprised exclusively of digestive enzymes/proenzymes, and likely represents resting exocytosis of mature granules. In both secretory phases, the digestive enzymes/proenzymes are detected in cannulated pancreatic juice and thus appear to be released from the apical surface in vivo. In addition to the pancreatic zymogens, a number of minor proteins are released during the first phase of secretion in both the adult and fetal pancreas. Because these proteins are not detected in the apical secretion in vivo, it has been inferred that the proteins are released from the basolateral surface of the acinar cell (2). The constitutive release of a fraction of newly synthesized digestive enzymes is compatible with previous suggestions (7, 115, 117) that there are two distinct pools of secretory proteins in the acinar cell.

FIG. 3. Nonlinear secretion of newly synthesized proteins from fetal pancreas. Pancreatic lobules were pulse labeled with [^{35}S]L-methionine for 10 min and chased for sequential 30-min periods with a complete change of medium at each interval. At the conclusion of experiment, lobules were homogenized and medium samples were cleared of particulates by centrifugation. Aliquots of homogenate and medium were divided for analysis by acid precipitation, sodium dodecyl sulfate–polyacrylamide gel electrophoresis (SDS-PAGE), and enzyme assays. *A*: amylase enzyme activity (○) is released into medium with linear kinetics during sequential and discrete chase intervals, whereas radiolabeled protein (●) is released nonlinearly during same period. *B*: SDS-PAGE and fluorography of proteins secreted into the medium during sequential 30-min chase intervals. Identical amount of unlabeled amylase activity was loaded into each lane. Results indicate that each of major protein species follows roughly same nonlinear kinetic pattern as that observed for overall protein radioactivity in *A*. Well-characterized digestive enzymes/proenzymes of pancreatic acinar cell comprise majority of secreted proteins. *Lane H*, pattern of labeled proteins remaining in tissue at conclusion of experiment, after 3.5 h: tissue still contains >80% of labeled protein, which is nearly exclusively in form of digestive enzymes/proenzymes. *C*: release of newly synthesized amylase and chymotrypsinogen was quantitated by liquid scintillation counting radioactivity contained within gel *bands A* and *C*, respectively, on fluorogram in *B*. The appearance in medium of amylase was delayed in comparison with that of chymotrypsinogen; this probably reflects asynchronous exit of these proteins from rough endoplasmic reticulum, as previously suggested by Scheele and Tartakoff (107a). *D*: asynchronous release of amylase and chymotrypsinogen is expressed as a time-related change in the fraction of total radioactivity secreted during each chase interval. [From Arvan and Chang (3).]

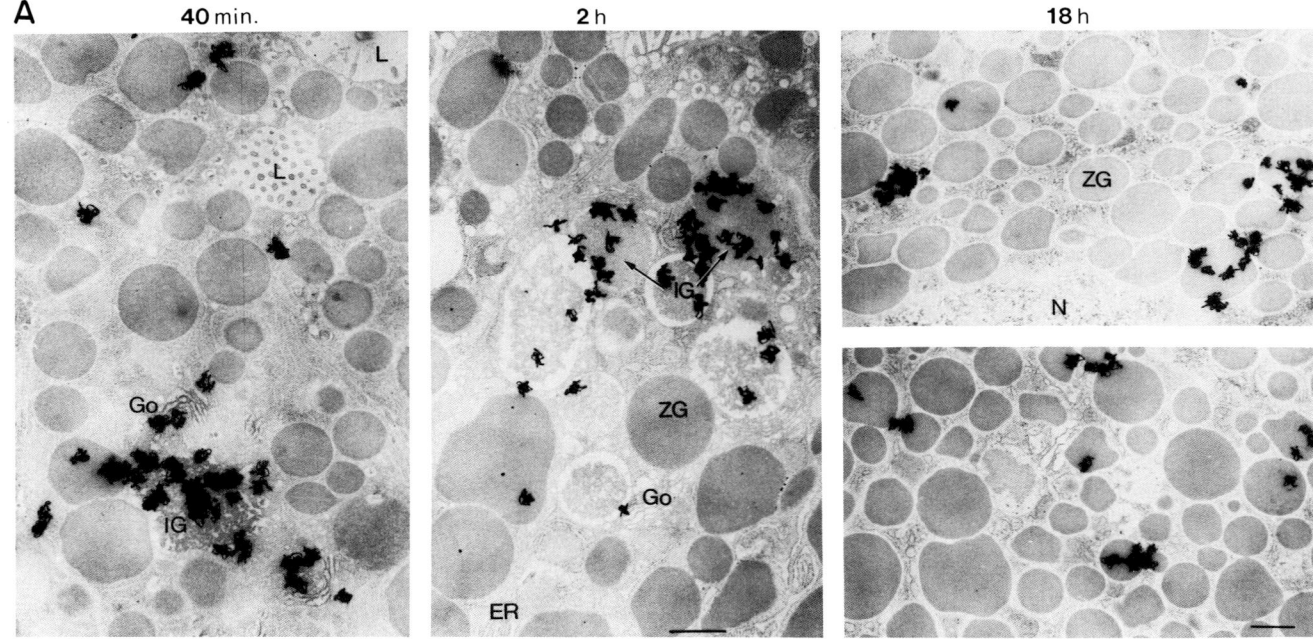

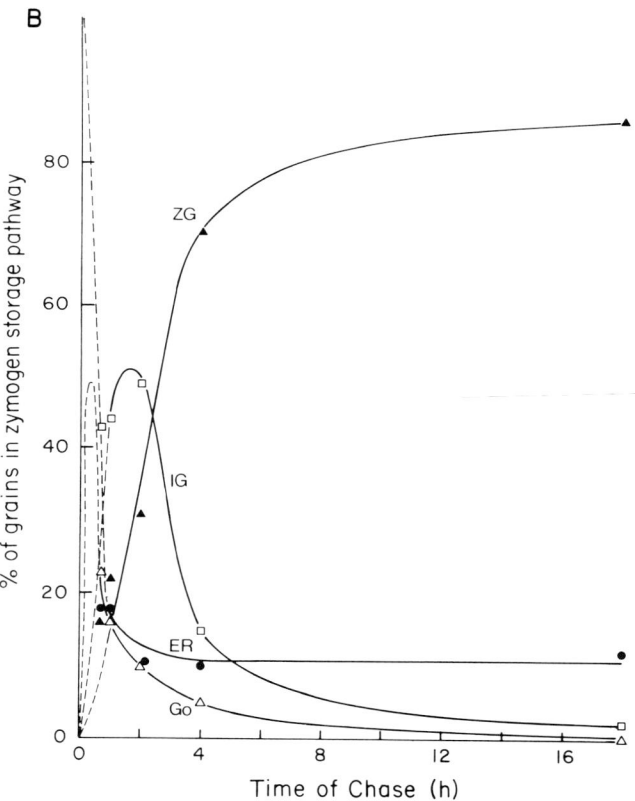

FIG. 5. Kinetics of autoradiographic labeling of the zymogen secretory pathway in the fetal pancreas. Pancreatic lobules were pulse labeled with [^{3}H]amino acids for 2 min and chased for 18 h with a complete change of medium every 30 min. At 40 min, 1 h, 2 h, 4 h, and 18 h lobules were removed from incubation, fixed, and processed for electron-microscopic autoradiography. A: pulse-labeled tissue after 40 min, 2 h (near peak of 1st phase of secretion of newly synthesized protein), and 18 h of chase (during the 2nd phase of secretion) (see Fig. 4). Compartments of the zymogen storage pathway: ER, rough endoplasmic reticulum; GO, Golgi apparatus; IG, immature granule/condensing vacuole; ZG, zymogen granule; N, portion of a cell nucleus; L, acinar lumena. Note that majority of autoradiographic grains overlie immature granules/condensing vacuoles (arrows) after 2 h of chase, and grains are concentrated over ZG after 18 h of chase. Bars, 1 μm. B: quantification of autoradiographic grains over established intermediates in zymogen storage pathway. Sum of grains over these structures is set at 100% at each time point. During chase periods corresponding to 1st phase of secretion (<6.5 h), several compartments in storage pathway contain radiolabel. Labeling of ZG is rising sharply when 1st phase of secretion of newly synthesized proteins is declining (2–6 h). Immature granules/condensing vacuoles are maximally labeled at 2 h of chase. Kinetics of autoradiographic labeling indicates that mature ZG do not serve as origin of 1st secretory phase. Data suggest that immature granules/condensing vacuoles could play such a role. [From Arvan and Chang (3).]

In contrast to the second phase of secretion from the fetal pancreas, which continued to increase at 21 h of chase, the second phase from the adult gland reached a maximum at 9–10 h and declined thereafter (Fig. 4). The volume density of storage granules in the fetal pancreas (~60%) is ~3 times greater than that of the adult gland (~20%) (27, 125) and could account for the ostensible delay in the peak of the second secretory phase in the fetal pancreas. We suggest that the release of newly synthesized proteins from zymogen granules by resting exocytosis depends on the population of preformed granules; i.e., there is limited intermixing of old and new granules (2, 3). Similarly, Sharoni et al. (113) and Singh (115) have reported that under resting conditions older proteins are preferentially secreted in comparison with newly synthesized proteins and during secretagogue stimulation both old and new granules are able to fuse with the apical membrane. It is not known how resting exocytosis differs from stimulated exocytosis, aside from the rate at which the two events occur. It is possible that the resting exocytosis of granules is ordered because of spatial constraints at the apical region of the cell, and resulting in limited access of the granules to the luminal plasma membrane. Stimulated exocytosis could represent freedom from such restrictions. Constitutive secretion from the fetal pancreas suggests a possible mechanistic difference between resting and stimulated exocytosis of zymogen granules.

Moore and Kelly (81) have proposed that a sorting receptor recognizes and directs newly synthesized proteins destined for packaging into secretory granules; proteins that are constitutively secreted are those that are not recognized by the receptor-mediated sorting mechanism. Because autoradiographic evidence indicates that the first phase of secretion in the adult and fetal pancreas derives from immature granules/condensing vacuoles, it is likely that the constitutive secretion of digestive enzymes/proenzymes occurs after a receptor-mediated sorting of these proteins into condensing vacuoles. Constitutive secretion of digestive enzymes/proenzymes from the fetal pancreas is therefore not synonymous with an unsorted pathway.

BIOGENESIS OF MEMBRANE AND SECRETORY POLARITY

Recent information on the biogenesis of structural and functional polarity in epithelial cells (for reviews see Refs. 73, 99, 114) has come principally from studies on MDCK cells. The cells form a highly polarized monolayer when grown on filters, providing ready experimental access to the apical and basolateral surface of the cells. Based on experiments on the MDCK cells it appears that newly synthesized plasma membrane proteins [e.g., basolateral Na^+-K^+-ATPase (16), basolaterally expressed G protein of the vesicular stomatitus virus (92), and apical hemagglutinin of the influenza virus (75)] are delivered directly to the correct membrane domain. Therefore sorting, targeting, and delivery of apical and basolateral membrane proteins must occur intracellularly. During biosynthesis, the hemagglutinin and vesicular stomatitus virus G protein have been colocalized in the Golgi transcisternae, suggesting that newly synthesized proteins directed to the apical and basolateral poles of the cell have in common part of the biosynthetic pathway (34, 98). Plasma membrane proteins and constitutively secreted serum proteins have been found within the same post-Golgi cytoplasmic vesicles (122), suggesting the possibility that some secretory proteins and membrane proteins may be delivered to the correct cell surface domain via the same route. Although the acidic environment of the endosome is involved in the sorting of internalized proteins (48), the sorting of membrane proteins is apparently not dependent on an acidic compartment; weak bases, which raise the pH of intracellular acidic compartments, do not affect the direct delivery of membrane marker proteins [e.g., influenza virus hemagglutinin (72) and Na^+-K^+-ATPase (16)] to the correct cell surface domain. The mechanisms responsible for the targeting of plasmalemmal proteins to their correct domains remain unknown.

In the case of several secretory proteins and lysosomal hydrolases, it appears that receptor-mediated processes are responsible for their sorting and targeting to the correct organelle. The best characterized example is that of lysosomal enzymes that are delivered to a prelysosomal compartment after interaction of the recognition marker on the hydrolase with the mannose-6-phosphate receptor (103). Weak bases, such as ammonium chloride, interfere with the correct delivery of lysosomal enzymes, ostensibly by disrupting the acid environment that is necessary for the dissociation of the ligand from the mannose-6-phosphate receptor (12, 37). It seems that the delivery of secretory proteins into storage granules also involves an acid-mediated sorting mechanism. Thus in the nonpolar AtT-20 cells derived from a pituitary tumor the weak base chloroquine diverts secretory proteins from being packaged into secretory granules and causes their release in a constitutive, nonstimulatable manner (60, 80). It is possible that a similar mechanism involving a pH-dependent interaction with a sorting receptor is involved in the polarized delivery and release of some secretory proteins. Caplan et al. have recently found that ammonium chloride perturbs the secretion of the basement membrane components, laminin, and heparan sulfate proteoglycan from the basolateral surface; instead, the proteins are released from both the apical and basolateral cell surfaces (17).

One approach to dissecting the mechanism by which proteins are directed to the apical or basolateral surface has been to transfect MDCK cells with genes coding for exogenous secretory proteins (40, 62). In this system, however, it appears that the sorting mechanism in MDCK cells may not recognize sorting signals on foreign proteins: the liver protein, α_2-microglobulin, is released from both the apical and basolateral surfaces in MDCK cells, although it is exclusively secretory from the sinusoidal (basolateral) surface of the hepatocyte (40). Nevertheless the secretion of exogenous proteins in both apical and basolateral directions implies that the polarized secretion of endogenous proteins is the result of sorting of proteins specifically to the apical or specifically to the basolateral surface.

Protein traffic in the pancreas is necessarily more complex than that of the MDCK cell, because the acinar cell packages secretory proteins into zymogen granules as well as releases them constitutively in an apical or basolateral direction. Pancreatic digestive enzymes/proenzymes are released constitutively from the apical pole of the cell after they have entered the storage granule pathway (2, 3). In addition, it is possible that proteins other than the digestive enzymes/proenzymes are constitutively secreted from the apical surface of the acinar cell, perhaps by a route distinct from that taken by the constitutively released zymogens. Future research must focus on the molecular mechanisms responsible for the polarized delivery of secretory and membrane proteins to the cell surface in polar, regulated epithelial cells such as the pancreatic acinar cell.

REFERENCES

1. AMSTERDAM, A., AND J. D. JAMIESON. Studies on dispersed pancreatic exocrine cells. I. Dissociation technique and morphological characteristics of separated cells. *J. Cell Biol.* 63: 1037–1056, 1974.
2. ARVAN, P., AND J. D. CASTLE. Phasic release of newly synthesized secretory proteins in the unstimulated rat exocrine pancreas. *J. Cell Biol.* 104: 243–252, 1987.
3. ARVAN, P., AND A. CHANG. Constitutive protein secretion from the fetal exocrine pancreas. *J. Biol. Chem.* 262: 3886–3890, 1987.
4. BALCAROVA-STANDER, J., S. E. PFEIFFER, S. D. FULLER, AND K. SIMONS. Development of cell surface polarity in the epithelial Madin-Darby canine kidney (MDCK) cell line. *EMBO J.* 3: 2687–2694, 1984.
5. BARON, R., L. NEFF, D. LOUVARD, AND P. J. COURTOY. Cell-mediated extracellular acidification and bone resorption: evidence for a low pH in resorbing lacunae and localization of a 100-kD lysosomal membrane protein at the osteoclast ruffled border. *J. Cell Biol.* 101: 2210–2222, 1985.
6. BAUDUIN, H., C. STOCK, D. VINCENT, AND J. F. GRENIER. Microfilamentous system and secretion of enzyme in the exocrine pancreas. *J. Cell Biol.* 66: 165–181, 1975.
7. BEAUDOIN, A. R., A. VACHEREAU, AND P. ST.-JEAN. Evidence that amylase is released from two distinct pools of secretory proteins in the pancreas. *Biochim. Biophys. Acta* 757: 302–305, 1983.
8. BEAVEN, M. A., J. P. MOORE, G. A. SMITH, T. R. HESKETH, AND J. C. METCALFE. The calcium signal and phosphatidylinositol breakdown in 2H3 cells. *J. Biol. Chem.* 259: 7137–7142, 1984.
9. BENDAYAN, M. Ultrastructural localization of cytoskeletal proteins in pancreatic secretory cells. *Can. J. Biochem. Cell. Biol.* 63: 680–690, 1985.
10. BERNFIELD, M. R., S. D. BANERJEE, AND R. H. COHN. Dependence of salivary epithelial morphology and branching morphogenesis upon acid mucopolysaccharide-protein (proteoglycan) at the epithelial surface. *J. Cell Biol.* 52: 674–689, 1972.
11. BOLENDER, R. P. Stereological analysis of the guinea pig pancreas. I. Analytical model and quantitative description of nonstimulated pancreatic exocrine cells. *J. Cell Biol.* 61: 269–287, 1974.
12. BROWN, W. J., E. CONSTANTINESCU, AND M. G. FARQUHAR. Redistribution of mannose-6-phosphate receptors induced by tunicamycin and chloroquine. *J. Cell Biol.* 99: 320–326, 1984.
13. BUNDGAARD, M., M. MØLLER, AND J. H. POULSEN. Localization of sodium pump sites in cat pancreas. *J. Physiol. Lond.* 313: 405–414, 1981.
14. BURNHAM, D. B., P. MUNOWITZ, N. THORN, AND J. A. WILLIAMS. Protein kinase activity associated with pancreatic zymogen granules. *Biochem. J.* 227: 743–751, 1985.
15. CAMERON, R. S., P. L. CAMERON, AND J. D. CASTLE. A common spectrum of polypeptides occurs in secretion granule membranes of different exocrine glands. *J. Cell Biol.* 103: 1299–1313.
16. CAPLAN, M. J., H. C. ANDERSON, G. E. PALADE, AND J. D. JAMIESON. Intracellular sorting and polarized cell surface delivery of (Na$^+$,K$^+$)ATPase, and endogenous component of MDCK cell basolateral plasma membranes. *Cell* 46: 623–631, 1986.
17. CAPLAN, M. J., J. L. STOW, A. P. NEWMAN, J. A. MADRI, H. C. ANDERSON, M. G. FARQUHAR, G. E. PALADE, AND J. D. JAMIESON. Sorting of newly synthesized secretory and lysosomal proteins in polarized MDCK cells (Abstract). *J. Cell Biol.* 103: 8a, 1986.
18. CASTLE, J. D., R. S. CAMERON, P. L. PATTERSON, AND A. K. MA. Identification of high molecular weight antigens structurally related to gamma-glutamyl transferase in epithelial tissues. *J. Membr. Biol.* 87: 13–26, 1985.
19. CHAMBARD, M., J. GABRION, AND J. MAUCHAMP. Influence of collagen gel on the orientation of epithelial cell polarity: follicle formation from isolated thyroid cells and from preformed monolayers. *J. Cell Biol.* 91: 157–166, 1981.
20. CHANG, A., AND J. D. JAMIESON. Stimulus-secretion coupling in the developing and exocrine pancreas: secretory responsiveness to cholecystokinin. *J. Cell Biol.* 103: 2353–2365, 1986.
21. COLONY, P. C., AND M. R. NEUTRA. Epithelial differentiation in the fetal rat colon. *Dev. Biol.* 97: 349–363, 1983.
22. DECAMILLI, P., D. PELUCHETTI, AND J. MELDOLESI. Structural difference between lumenal and lateral plasmalemma in pancreatic acinar cells. *Nature Lond.* 248: 245–246, 1974.
23. DE LISLE, R. C., AND J. A. WILLIAMS. Regulation of membrane fusion in secretory exocytosis. *Annu. Rev. Physiol.* 48: 225–238, 1986.
24. DOYLE, C. M., AND J. D. JAMIESON. Development of secretagogue response in rat pancreatic acinar cells. *Dev. Biol.* 65: 11–27, 1978.
25. DRAGSTEN, P. R., R. BLUMENTHAL, AND J. S. HANDLER. Membrane asymmetry in epithelia: is the tight junction a barrier to diffusion in the plasma membrane. *Nature Lond.* 294: 718–722, 1981.
26. DRENCKHAHN, D., AND H. G. MANNHERZ. Distribution of actin and the actin-associated proteins myosin, tropomyosin, alpha-actinin, vinculin, and villin in rat and bovine exocrine glands. *Eur. J. Cell Biol.* 30: 167–176, 1983.
27. ERMAK, T. H., AND S. S. ROTHMAN. Large decrease in zymogen granule size in the postnatal rat pancreas. *J. Ultrastruct. Res.* 70: 242–256, 1980.
28. EVANS, W. H. A biochemical dissection of the functional polarity of the plasma membrane of the hepatocyte. *Biochim. Biophys. Acta* 604: 27–64, 1980.
29. FARQUHAR, M. G. Membrane recycling in secretory cells: implications for traffic of products and specialized membranes within the Golgi complex. *Methods Cell Biol.* 23: 399–427, 1981.
30. FARQUHAR, M. G., AND G. E. PALADE. Junctional complexes in various epithelia. *J. Cell Biol.* 17: 375–412, 1963.
31. FERACCI, H., A. BERNADAC, S. HOVSEPIAN, G. FAYET, AND S. MAROUX. Aminopeptidase N is a marker for the apical pole of porcine thyroid epithelial cells in vivo and in culture. *Cell*

32. FREEDMAN, S. D., AND J. D. JAMIESON. Hormone-induced protein phosphorylation. III. Regulation of the phosphorylation of the secretagogue-responsive 29,000-dalton protein by both Ca^{++} and cAMP in vitro. *J. Cell Biol.* 95: 918–923, 1982.
33. FRIEND, D. S. Plasma-membrane diversity in a highly polarized cell. *J. Cell Biol.* 93: 243–249, 1982.
34. FULLER, S. D., R. BRAVO, AND K. SIMONS. An enzymatic assay reveals that proteins destined for the apical or basolateral domains of an epithelial cell line share the same late Golgi compartments. *EMBO J.* 4: 297–307, 1985.
35. FULLER, S., C.-H. VON BONSDORFF, AND K. SIMONS. Vesicular stomatitis virus infects and matures only through the basolateral surface of the polarized epithelial cell line, MDCK. *Cell* 38: 65–77, 1984.
36. FULLER, S. D., AND K. SIMONS. Transferrin receptor polarity and recycling accuracy in "tight" and "leaky" strains of Madin-Darby canine kidney cells. *J. Cell Biol.* 103: 1767–1779, 1986.
37. GONZALEZ-NORIEGA, A., J. H. GRUBB, V. TALKAD, AND W. S. SLY. Chloroquine inhibits lysosomal enzyme pinocytosis and enhances lysosomal enzyme secretion by impairing receptor recycling. *J. Cell Biol.* 85: 839–852, 1980.
38. GORELICK, F. S., A. CHANG, AND J. D. JAMIESON. Calcium-calmodulin stimulated protein kinase in developing pancreas. *Am. J. Physiol.* 253 (*Gastrointest. Liver Physiol.* 16): G469–G476, 1987.
39. GORELICK, F. S., J. A. COHN, S. D. FREEDMAN, N. G. DELAHUNT, J. M. GERSHONI, AND J. D. JAMIESON. Calmodulin-stimulated protein kinase activity from rat pancreas. *J. Cell Biol.* 97: 1294–1298, 1983.
40. GOTTLIEB, T. A., G. BEAUDRY, L. RIZZOLO, A. COLMAN, M. RINDLER, M. ADESNIK, AND D. D. SABATINI. Secretion of endogenous and exogenous proteins from polarized MDCK cell monolayers. *Proc. Natl. Acad. Sci. USA* 83: 2100–2104, 1986.
41. GREENBURG, G., AND E. D. HAY. Epithelia suspended in collagen gels can lose polarity and express characteristics of migrating mesenchymal cells. *J. Cell Biol.* 95: 333–339, 1982.
42. GREENBURG, G., I. VLODAVSKY, J.-M. FOIDART, AND D. GOSPODAROWICZ. Conditioned medium from epithelial cell cultures can restore the normal phenotypic expression of vascular endothelium maintained in vitro in the absence of fibroblast growth factor. *J. Cell. Physiol.* 103: 333–347, 1980.
43. GREENE, L. J., C. H. W. HIRS, AND G. E. PALADE. On the protein composition of bovine pancreatic zymogen granules. *J. Biol. Chem.* 238: 2054–2070, 1963.
44. GREENGARD, P. Phosphorylated proteins as physiological effectors. *Science Wash. DC* 199: 146–152, 1978.
45. HAEUPTLE, M.-T., Y. L. M. SUARD, E. BOGENMANN, H. REGGIO, L. RACINE, AND J.-P. KRAEHENBUHL. Effect of cell shape change on the function and differentiation of rabbit mammary cells in culture. *J. Cell Biol.* 96: 1425–1434, 1983.
46. HAN, J. H., L. RALL, AND W. J. RUTTER. Selective expression of rat pancreatic genes during embryonic development. *Proc. Natl. Acad. Sci. USA* 83: 110–114, 1986.
47. HAND, A. R., AND C. OLIVER. Effects of secretory stimulation on the Golgi apparatus and GERL of rat parotid acinar cells. *J. Histochem. Cytochem.* 32: 403–412, 1984.
48. HELENIUS, A., I. MELLMAN, D. WALL, AND A. HUBBARD. Endosomes. *Trends Biochem. Sci.* 8: 245–250, 1983.
49. HERZLINGER, D. A., AND G. K. OJAKIAN. Studies on the development and maintenance of epithelial cell surface polarity with monoclonal antibodies. *J. Cell Biol.* 98: 1777–1787, 1984.
50. HERZOG, V., AND H. REGGIO. Pathways of endocytosis from luminal plasma membrane in rat exocrine pancreas. *Eur. J. Cell Biol.* 21: 141–150, 1985.
51. HESKETH, T. R., M. A. BEAVEN, J. ROGERS, B. BURKE, AND G. B. WARREN. Stimulated release of histamine by a rat mast cell line is inhibited during mitosis. *J. Cell Biol.* 98: 2250–2254, 1984.
52. ICHIKAWA, A. Fine structural changes in response to hormonal stimulation of the perfused canine pancreas. *J. Cell Biol.* 24: 3369–3385, 1965.
53. INGBER, D. E., J. A. MADRI, AND J. D. JAMIESON. Neoplastic disorganization of pancreatic epithelial cell-cell relations. *Am. J. Pathol.* 121: 248–260, 1985.
54. INGBER, D. E., J. A. MADRI, AND J. D. JAMIESON. Basement membrane as a spatial organizer of polarized epithelia. *Am. J. Pathol.* 122: 129–139, 1986.
55. JACKSON, R. C., K. K. WARD, AND J. G. HAGGERTY. Mild proteolytic digestion restores exocytotic activity to N-ethylmaleimide-inactivated cell surface complex from sea urchin eggs. *J. Cell Biol.* 101: 6–11, 1985.
56. JAMIESON, J. D. Summary and perspectives. *Methods Cell Biol.* 23: 547–558, 1981.
57. JAMIESON, J. D., AND G. E. PALADE. Intracellular transport of secretory proteins in the pancreatic exocrine cell. I. Role of the peripheral elements of the Golgi complex. II. Transport to condensing vacuoles and zymogen granules. *J. Cell Biol.* 34: 577–596, 597–615, 1967.
58. JAMIESON, J. D., AND G. E. PALADE. Synthesis, intracellular transport, and discharge of secretory proteins in stimulated pancreatic exocrine cells. *J. Cell Biol.* 50: 135–158, 1971.
59. JOHNSON, M. H., AND C. A. ZIOMEK. Induction of polarity in mouse 8-cell blastomeres: specificity, geometry, and stability. *J. Cell Biol.* 91: 303–308, 1981.
60. KELLY, R. B. Pathways of protein secretion in eukaryotes. *Science Wash. DC* 230: 25–32, 1985.
61. KERJASCHKI, D., L. NORONHA-BLOB, B. SACKTOR, AND M. G. FARQUHAR. Microdomains of distinctive glycoprotein composition in the kidney proximal tubule brush border. *J. Cell Biol.* 98: 1505–1513, 1984.
62. KONDOR-KOCH, C., R. BRAVO, S. D. FULLER, D. CUTLER, AND H. GAROFF. Exocytotic pathways exist to both the apical and the basolateral cell surface of the polarized epithelial cell MDCK. *Cell* 43: 297–306, 1985.
63. KONINGS, F., AND W. DEPOTTER. A role for sialic acid containing substrates in the exocytosis-like in vitro interaction between adrenal medullary plasma membranes and chromaffin granules. *Biochem. Biophys. Res. Commun.* 106: 1191–1195, 1982.
64. KRAMER, M. F., AND C. POORT. Unstimulated secretion of protein from rat exocrine pancreas cells. *J. Cell Biol.* 52: 147–158, 1972.
65. KUPFER, A., G. DENNERT, AND S. J. SINGER. Polarization of the Golgi apparatus and the microtubule-organizing center within cloned natural killer cells bound to their targets. *Proc. Natl. Acad. Sci. USA* 80: 7224–7228, 1983.
66. LANG, M. A., J. MULLER, A. S. PRESTON, AND J. S. HANDLER. Complete response to vasopressin requires epithelial organization in A6 cells in culture. *Am. J. Physiol.* 250 (*Cell Physiol.* 19): C138–C145, 1986.
67. LAWSON, E., C. FEWTRELL, AND M. C. RAFF. Localized mast cell degranulation induced by concanavalin A-sepharose beads. *J. Cell Biol.* 79: 394–400, 1978.
68. LEE, E. Y.-H., G. PARRY, AND M. J. BISSELL. Modulation of secreted proteins of mouse mammary epithelial cells by the collagenous substrata. *J. Cell Biol.* 98: 146–155, 1984.
69. LLINAS, R., T. L. MCGUINNESS, C. S. LEONARD, M. SUGIMORI, AND P. GREENGARD. Intraterminal injection of synapsin I or calcium/calmodulin-dependent protein kinase II alters neurotransmitter release at the squid giant synapse. *Proc. Natl. Acad. Sci.* 82: 3035–3039, 1985.
70. MADARA, J. L., M. R. NEUTRA, AND J. S. TRIER. Junctional complexes in fetal rat small intestine during morphogenesis. *Dev. Biol.* 86: 170–178, 1981.
71. MADDEN, M. E., AND M. P. SARRAS. Development of an apical plasma membrane domain and tight junctions during histogenesis of the mammalian pancreas. *Dev. Biol.* 112: 427–442, 1985.
72. MATLIN, K. S. Ammonium chloride slows transport of the influenza virus hemagglutinin but does not cause mis-sorting in a polarized epithelial cell line. *J. Biol. Chem.* 261: 15172–

73. MATLIN, K. S. The sorting of proteins to the plasma membrane in epithelial cells. *J. Cell Biol.* 103: 2565–2568, 1986.
74. MATLIN, K. S., D. F. BAINTON, M. PESONEN, D. LOUVARD, N. GENTY, AND K. SIMONS. Transepithelial transport of a viral membrane glycoprotein implanted into the apical plasma membrane of MDCK cells. I. Morphological evidence. II. Immunological quantitation. *J. Cell Biol.* 97: 627–637, 638–643, 1983.
75. MATLIN, K. S., AND K. SIMONS. Sorting of an apical plasma membrane glycoprotein occurs before it reaches the cell surface in cultured epithelial cells. *J. Cell Biol.* 99: 2131–2139, 1984.
76. MAYLIE-PFENNINGER, M.-F., AND J. D. JAMIESON. Development of cell surface saccharides on embryonic pancreatic cells. *J. Cell Biol.* 86: 96–103, 1980.
77. MELDOLESI, J., G. CASTIGLIONI, R. PARMA, N. NASSIVERA, AND P. DECAMILLI. Ca^{++}-dependent disassembly and reassembly of occluding junctions in pancreatic acinar cells. Effect of drugs. *J. Cell Biol.* 79: 156–172, 1978.
78. MEYER, D. I., AND M. M. BURGER. Isolation of a protein from the plasma membrane of adrenal medulla which binds to secretory vesicles. *J. Biol. Chem.* 254: 9854–9859, 1979.
79. MONTESANO, R., D. S. FRIEND, A. PERRELET, AND L. ORCI. In vivo assembly of tight junctions in fetal rat liver. *J. Cell Biol.* 67: 310–319, 1975.
80. MOORE, H.-P., B. GUMBINER, AND R. B. KELLY. Chloroquine diverts ACTH from a regulated to a constitutive secretory pathway in AtT-20 cells. *Nature Lond.* 302: 434–436, 1983.
81. MOORE, H.-P. H., AND R. B. KELLY. Rerouting of a secretory protein by fusion with human growth hormone sequences. *Nature Lond.* 321: 443–446, 1986.
82. MULLER, W. A., AND M. A. GIMBRONE, JR. Plasmalemmal proteins of cultured vascular endothelial cells exhibit apical-basal polarity: analysis by surface-selective iodination. *J. Cell Biol.* 103: 2389–2402, 1986.
83. MUNDY, D. I., AND W. J. STRITTMATTER. Requirement for metalloendoprotease in exocytosis: evidence in mast cells and adrenal chromaffin cells. *Cell* 40: 645–656, 1985.
84. MURESAN, V., M. P. SARRAS, JR., AND J. D. JAMIESON. Distribution of sialoglycoconjugates on acinar cells of the mammalian pancreas. *J. Histochem. Cytochem.* 330: 947–955, 1982.
85. OLIVER, C. Endocytic pathways at the lateral and basal cell surfaces of exocrine acinar cells. *J. Cell Biol.* 95: 154–161, 1982.
86. ORCI, L., A. PERRELET, AND R. MONTESANO. Differential filipin labeling of the luminal membranes lining the pancreatic acinus. *J. Histochem. Cytochem.* 31: 952–955, 1983.
87. PALADE, G. Intracellular aspects of the process of protein synthesis. *Science Wash. DC* 189: 347–358, 1975.
88. PARRY, G., E. Y.-H. LEE, D. FARSON, M. KOVAL, AND M. J. BISSELL. Collagenous substrata regulate the nature and distribution of glycosaminoglycans produced by differentiated cultures of mouse mammary epithelial cells. *Exp. Cell Res.* 156: 487–499, 1985.
89. PARSA, I., W. H. MARSH, AND P. J. FITZGERALD. Pancreas acinar cell differentiation. *Am. J. Pathol.* 57: 457–487, 1969.
90. PAVELKA, M., AND A. ELLINGER. Effect of colchicine on the Golgi complex of rat pancreatic acinar cells. *J. Cell Biol.* 97: 737–748, 1983.
91. PEIFFER, A., C. GAGNON, AND S. HEISLER. Phosphorylation of zymogen granule membrane proteins in intact rat pancreatic acinar cells. *Biochem. Biophys. Res. Commun.* 122: 413–419, 1984.
92. PFEIFFER, S., S. D. FULLER, AND K. SIMONS. Intracellular sorting and basolateral appearance of the G protein of vesicular stomatitis virus in MDCK cells. *J. Cell Biol.* 101: 470–476, 1985.
93. PICTET, R. L., W. R. CLARK, R. H. WILLIAMS, AND W. J. RUTTER. An ultrastructural analysis of the developing embryonic pancreas. *Dev. Biol.* 29: 436–467, 1972.
94. PISAM, M., AND P. RIPOCHE. Distribution of surface macromolecules in dissociated epithelial cells. *J. Cell Biol.* 71: 907–920, 1976.
95. POLLARD, H. B., C. E. CREUTZ, V. FOWLER, J. SCOTT, AND C. J. PAZOLES. Calcium-dependent regulation of chromaffin granule movement, membrane contact, and fusion during exocytosis. *Cold Spring Harbor Symp. Quant. Biol.* 46: 819–834, 1981.
96. RAPRAEGER, A., M. JALKANEN, AND M. BERNFIELD. Cell surface proteoglycan associates with the cytoskeleton at the basolateral cell surface of mouse mammary epithelial cells. *J. Cell Biol.* 103: 2683–2696, 1986.
97. RASMUSSEN, H., I. KOJIMA, K. KOJIMA, W. ZAWALICH, AND W. APFELDORF. Calcium as intracellular messenger: sensitivity modulation, C-kinase pathway, and sustained cellular response. *Adv. Cyclic Nucleotide Res.* 18: 159–193, 1984.
98. RINDLER, M. J., I. E. IVANOV, H. PLESKEN, E. RODRIGUEZ-BOULAN, AND D. D. SABATINI. Viral glycoproteins destined for apical or basolateral plasma membrane domains traverse the same Golgi apparatus during their intracellular transport in doubly infected Madin-Darby canine kidney cells. *J. Cell Biol.* 98: 1304–1319, 1984.
99. RODRIGUEZ-BOULAN, E., D. E. MISEK, D. V. DE SALAS, P. J. I. SALAS, AND E. BARD. Protein sorting in the secretory pathway. *Curr. Top. Membr. Transp.* 24: 251–294, 1985.
100. RODRIGEUZ-BOULAN, E., K. T. PASKIET, AND D. D. SABATINI. Assembly of enveloped viruses in Madin-Darby canine kidney cells: polarized budding from single attached cells and from clusters of cells in suspension. *J. Cell Biol* 96: 866–874, 1983.
101. ROGALSKI, A. A., J. E. BERGMANN, AND S. J. SINGER. Effect of microtubule assembly status on the intracellular processing and surface expression of an integral membrane protein of the plasma membrane. *J. Cell Biol.* 99: 1101–1109, 1984.
102. ROSENZWEIG, S. A., L. J. MILLER, AND J. D. JAMIESON. Identification and localization of cholecystokinin-binding sites on rat pancreatic plasma membranes and acinar cells: a biochemical and autoradiographic study. *J. Cell Biol.* 96: 1288–1297, 1983.
103. SAHAGIAN, G. G. The mannose-6-phosphate receptor: function, biosynthesis, and translocation. *Biol. Cell* 51: 207–214, 1984.
104. SALAS, P. J. I., D. E. MISEK, VEGA-SALAS, D. GUNDERSEN, M. CEREIJIDO, AND E. RODRIGUEZ-BOULAN. Microtubules and actin filaments are not critically involved in the biogenesis of epithelial cell surface polarity. *J. Cell Biol.* 102: 1853–1867, 1986.
105. SALPETER, M. M., C. D. SMITH, AND J. A. MATTHEWS-BELLINGER. Acetylcholine receptor at neuromuscular junctions by EM autoradiography using mask analysis and linear sources. *J. Electron Microsc. Tech.* 1: 63–81, 1984.
106. SANDERS, T. G., AND W. J. RUTTER. The developmental regulation of amylolytic and proteolytic enzymes in the embryonic rat pancreas. *J. Biol. Chem.* 249: 3500–3509, 1974.
107. SASAKI, H. Modulation of calcium sensitivity by a specific cortical protein during sea urchin egg cortical vesicle exocytosis. *Dev. Biol.* 101: 125–135, 1984.
107a. SCHEELE, G., AND A. TARTAKOFF. Exit of nonglycosylated secretory proteins from the rough endoplasmic reticulum is asynchronous in the exocrine pancreas. *J. Biol. Chem.* 260: 926–931, 1985.
108. SCHEFFER, R. C. T., C. POORT, AND J. W. SLOT. Fate of the major zymogen granule membrane associated glycoproteins from rat pancreas. A biochemical and immunocytochemical study. *Eur. J. Cell Biol.* 23: 122–128, 1980.
109. SCHIBLER, V., O. HAGENBUCHLE, P. K. WELLAVER, AND A. C. PITTET. Two promoters of different strengths control the transcript of the mouse alpha-amylase gene Amy-1a in the parotid gland and the liver. *Cell* 33: 501–508, 1983.
110. SCHWARTZ, G. J., J. BARASCH, AND Q. AL-AWQATI. Plasticity of functional epithelial polarity. *Nature Lond.* 318: 368–371, 1985.
111. SEEMAN, P., M. CHAU-WONG, AND S. MOYYEN. Membrane expansion by vinblastin and strychnine. *Nature New Biol.* 241: 22, 1973.

112. SHANNON, J. M., AND D. R. PITELKA. The influence of cell shape on the induction of functional differentiation in mouse mammary cells in vitro. *In Vitro Rockville* 17: 1016–1028, 1981.
113. SHARONI, Y., S. EIMERL, AND M. SCHRAMM. Secretion of old versus new exportable protein in rat parotid slices. *J. Cell Biol.* 71: 107–122, 1976.
114. SIMONS, K., AND S. D. FULLER. Cell surface polarity in epithelia. *Annu. Rev. Cell Biol.* 1: 295–340, 1985.
115. SINGH, M. Nonparallel transport of exportable proteins in rat pancreas in vitro. *Can. J. Physiol. Pharmacol.* 60: 597–603, 1982.
116. SKUTELSKY, E., AND M. G. FARQUHAR. Variation in distribution of con A receptor sites and anionic groups during red blood cell differentiation in the rat. *J. Cell Biol.* 71: 218–231, 1976.
117. SLABY, F., AND J. BRYAN. High uptake of myo-inositol by rat pancreatic tissue in vitro stimulates secretion. *J. Biol. Chem.* 251: 5078–5086, 1976.
118. SMITH, Z. D. J., M. J. CAPLAN, B. FORBUSH III, AND J. D. JAMIESON. Monoclonal antibody localization of Na^+-K^+-ATPase in the exocrine pancreas and parotid of the dog. *Am. J. Physiol.* 253 (*Gastrointest. Liver Physiol.* 16): G99–G109, 1987.
119. SPOONER, B. S., H. I. COHEN, AND J. FAUBION. Development of the embryonic mammalian pancreas: the relationship between morphogenesis and cytodifferentiation. *Dev. Biol.* 61: 119–130, 1977.
120. STEINHARDT, R. A., AND J. M. ALDERTON. Calmodulin confers calcium sensitivity on secretory exocytosis. *Nature Lond.* 295: 154–155, 1982.
121. STEINMAN, R. M., I. S. MELLMAN, W. A. MULLER, AND Z. A. COHN. Endocytosis and the recycling of plasma membrane. *J. Cell Biol.* 96: 1–27, 1983.
122. STROUS, G. J. A. M., R. WILLEMSEN, P. VAN KERKHOF, J. W. SLOT, H. J. GEUZE, AND H. F. LODISH. Vesicular stomatitus virus glycoprotein, albumin, and transferring are transported to the cell surface via the same Golgi vesicles. *J. Cell Biol.* 97: 1815–1822, 1983.
123. TANAKA, Y., P. DECAMILLI, AND J. MELDOLESI. Membrane interactions between secretion granules and plasmalemma in three exocrine glands. *J. Cell Biol.* 84: 438–453, 1980.
124. TARTAKOFF, A., AND P. VASSALLI. Comparative studies of intracellular transport of secretory proteins. *J. Cell Biol.* 79: 694–707, 1978.
125. UCHIYAMA, Y., AND M. WATANABE. A morphometric study of developing pancreatic acinar cells of rats during prenatal life. *Cell Tissue Res.* 237: 117–122, 1984.
126. UCHIYAMA, Y., AND M. WATANABE. Morphometric and fine-structural studies of rat pancreatic acinar cells during early postnatal life. *Cell Tissue Res.* 237: 123–129, 1984.
127. VAN MEER, G., AND K. SIMONS. The function of tight junctions in maintaining differences in lipid composition between the apical and the basolateral cell surface domains of MDCK cells. *EMBO J.* 5: 1455–1464, 1986.
128. VAN NEST, G. A., R. J. MACDONALD, R. K. RAMAN, AND W. J. RUTTER. Proteins synthesized and secreted during rat pancreatic development. *J. Cell Biol.* 86: 784–794, 1980.
129. WERLIN, S. L., AND R. J. GRAND. Development of secretory mechanisms in rat pancreas. *Am. J. Physiol.* 236 (*Endocrinol. Metab. Gastrointest. Physiol.* 5): E446–E450, 1979.
130. WILLIAMS, J. A. Effects of antimitotic agents on ultrastructure and intracellular transport of protein in pancreatic acini. *Methods Cell Biol.* 23: 247–258, 1981.
131. WRENN, R. W. Phosphorylation of a pancreatic zymogen granule membrane protein by endogenous calcium/phospholipid-dependent protein kinase. *Biochim. Biophys. Acta* 775: 1–6, 1984.
132. ZIOMEK, C. A., S. SCHULMAN, AND M. EDIDIN. Redistribution of membrane proteins in isolated mouse intestinal epithelial cells. *J. Cell Biol.* 86: 849–857, 1980.

CHAPTER 28

Overview of bile secretion

ALAN F. HOFMANN | Division of Gastroenterology, Department of Medicine, University of California at San Diego, La Jolla, California

CHAPTER CONTENTS

Biological Overview: Significance of Bile Formation in Vertebrates
    Major biliary components and major functions of bile
    Differences between hepatic excretory function and renal excretory function
Anatomy and Circulation of Biliary Tract
    Functional anatomy of biliary tract
    Circulation of liver and biliary tract
    Imaging of biliary tract
Methods for Studying Physiology of Bile
    Biliary fistula methods
    Perfused liver preparations
    Isolated hepatocytes or cultured hepatocytes
    Hepatic cell doublets
    Isolated membrane fractions
Overview of Biliary Secretion
    Bile as resultant of secretory, hydrolytic, and absorptive processes
    Canalicular bile formation
    Ductular modification of canalicular bile
    Gallbladder modification of hepatic bile
    Spontaneous changes in bile composition
    Markers of bile flow
Classification of Biliary Constituents
Organic Constituents of Bile
    Biliary lipids
        Bile acids
        Phospholipids
        Cholesterol
    Minor organic components
        Bilirubin
        Fatty acids
    Trace organic components
Inorganic Components of Bile
    Physiological determinants of concentration of common ions
    Biliary calcium
    Significance of biliary electrolyte composition
    Minor and trace metals
Secretory and Excretory Functions of Bile
    Secretory function
        Water
        Phospholipids
        Cholesterol
        Bile acids
    Excretory function
        Features of biliary excretion
        Criterion for enterohepatic cycling
Cholestasis

THIS CHAPTER PRESENTS an overview of bile formation and events in the biliary tree. It serves as an introduction to the following seven chapters in which most of the topics discussed in it are covered in more detail. The first chapter, authored by me, reviews the circulation of bile acid molecules—molecules that induce bile flow and solubilize biliary and dietary lipids. Such solubilization is a physicochemical process, and its principles are reviewed by Cabral and Small. Bile also serves as a route for immunoglobulin secretion and for the secretion of other proteins that may, for example, transport heavy metals; current knowledge is summarized by Jones and Burwen.

Organic ions that are bound to albumin can only be eliminated in bile, and canalicular transport involves a variety of molecules other than bile acids. Hepatic biotransformation of such ions influences bile flow since addition of a conjugating group is often necessary for efficient canalicular transport and because the addition of a conjugating group prevents passive absorption in the biliary tree. The general subject of hepatic transport and biotransformation is an evolving discipline with one foot in physiology and the other in pharmacology. Meijer summarizes current viewpoints. Forker reviews current concepts of hepatic uptake of organic solutes emphasizing kinetic aspects of the process. Hepatocyte lysosomes process the proteins and lipids that enter the hepatocyte by receptor mediated endocytosis, and some lysosomal enzymes and products enter bile. LaRusso covers this subject, which is of great clinical importance because of the prevalence of lysosomal storage diseases. Three topics are not covered in this *Handbook* and are discussed briefly in my introductory chapter. The first is the function of the biliary epithelium; the second is vesicular secretion of biliary lipids by the hepatocyte; the final is the absorptive function of the gallbladder (89).

BIOLOGICAL OVERVIEW: SIGNIFICANCE OF
BILE FORMATION IN VERTEBRATES

*Major Biliary Components and Major
Functions of Bile*

In vertebrates the liver secretes a concentrated yellow or green solution into the small intestine. This solution, termed *bile,* is the only excretory pathway for the water-soluble catabolic products of two key body constituents—cholesterol and hemoglobin. Cholesterol is converted into a group of amphipathic molecules termed *bile salts,* and the heme moiety of hemoglobin is converted to an open tetrapyrole biliverdin (bilo = bile, verdi = green), which is excreted as such or in the form of bilirubin (rubin = red), its reduced derivative. Bile is also a secretion, since its major organic component, the bile acids, are amphipathic molecules that promote lipid absorption. In addition, in some species, bile contains high concentrations of bicarbonate, which neutralizes acidic gastric contents in the duodenum, and may also contain immunoglobulin A.

Bile is not a simple ionic solution. Hepatic bile contains vesicles of hydrated lipid bilayers of phospholipid and cholesterol, and these vesicles are rapidly transformed into mixed micelles by the amphipathic properties of bile salts. The mixed phospholipid–bile salt micelle has a surface that provides an amphipathic environment for other amphipathic molecules but whose interior is quite hydrophobic and that can serve as a vehicle for solubilizing and transporting lipids. It is believed that bile also serves as an excretory route for amphipathic derivatives of steroid hormones, of drugs, and of other xenobiotics. Bile also serves as an excretory route for antigen antibody complexes, but the overall significance of this pathway is not known.

Bile serves an excretory route for heavy metals such as copper (57). The precise physical form of transport of heavy metals in bile has not been defined, but the mixed micelles present in bile possess a charged surface that can bind multivalent cations nonspecifically. In addition, the primary bile acid anions bind calcium, iron, and probably other polyvalent cations at submicellar concentrations (70).

Bile thus serves as a pathway for the elimination of metabolites, antigens, and heavy metals. Amphipathic molecules are not readily eliminated in urine because their lipophilic regions enhance albumin binding, which decreases their filtration in the glomerulus; these lipophilic domains may also contribute to efficient passive tubular reabsorption. Obviously, large antigens are not filtered. Heavy metals are generally extensively protein bound in plasma, which also decreases their filtration in the glomerulus.

As noted, bile is also a secretory route, in that a number of substances in bile have important functions in the biliary tree or small intestine. Probably the most important of these is biliary secretion of fluid and electrolytes per se, which helps to prevent precipitation or sludging in the biliary tree. Another protective secretion is biliary phospholipid, which probably protects the epithelium of the biliary tree from damage by the bile salt molecules. Secretion of bile salts is a vital process, because bile acids have an essential function in the small intestinal lumen where they serve to solubilize digestion products of ingested triglyceride as well as to solubilize fat-soluble vitamins; bile salts may influence colonic motility and defecation. Another important secretion, at least in some species, is immunoglobulin A, which probably protects the biliary tree and the small intestine from invasion by bacteria or parasites (22). In certain species such as the guinea pig and the hog, bile is rich in bicarbonate. In these species, as well as several others (65), the common bile duct enters the second portion of the duodenum well above the entrance of the pancreatic duct. One may speculate that in such a species biliary bicarbonate may serve an important role in neutralizing gastric acid that enters the duodenum (116).

Differences Between Hepatic Excretory Function and Renal Excretory Function

There are a number of key differences between bile formation and function and urine formation and function, and in these days of specialization it may be useful to summarize these briefly. Bile is rather fixed in composition, and variations in biliary output, at least for the major constituents, are generally accompanied by changes in flow. Hepatic excretory function, as evidenced by bile composition, plays little role in water, electrolyte, acid-base, or nitrogen balance, which are key renal functions. Bile is formed in response to the osmotic activity of actively secreted anions (bile acid and/or bicarbonate anions), whereas urine is formed by hydrostatic filtration. The molecular weight of major solutes in bile other than immunoglobulins is low, under 1,000, whereas urinary solutes may have a molecular weight of up to 40,000, depending on the pore size of the glomerulus. Bile is concentrated and acidified in the gallbladder, whereas urine probably undergoes little change in composition during storage in the urinary bladder.

Biliary obstruction has two consequences. First, it leads to retention of biliary constituents in the liver, causing destruction of functional liver cell mass and eventually fatal hepatic failure. Obstruction to urinary flow causes uremia, which eventually is fatal because of renal insufficiency. Second, in growing children the deficiency of biliary secretion in the small intestine, in the absence of therapy, causes severe malabsorption leading to growth retardation. In contrast, the constituents of urine may be considered waste products, and urinary products are secreted for regulatory purposes such as acid-base homeostasis. Finally, the nonexcretory functions of the liver are far more important than its excretory functions; consequently, hepatic failure causes death because of the deficiency of hepatic biosynthetic and metabolic functions, rather than the accumulation of biliary constituents. In contrast, the nonexcretory functions of the kidney such as blood pressure regulation or erythropoietin synthesis are of less importance and life can be well sustained by dialysis procedures.

Considerable progress in understanding the physiology of bile has been made during the past two decades, but progress has been slow because the problems involved in the physiology of bile are difficult and multidisciplinary. Furthermore, in an era of spectacular advances in molecular biology, the worker in this rather old-fashioned field of organ physiology often feels as if his work is no longer in the exciting mainstream of science, and the funding fathers do not act in a way to dissuade this impression. However, in fact, biliary physiology has a lofty historical heritage (100), and great advances have and are being made. Not only is the physiology of great interest, but it has profound clinical implications on the understanding, diagnosis, and treatment of biliary disease, which is an important medical problem throughout the world. This chapter is meant to serve as an introduction to the detailed considerations of individual topics in subsequent chapters.

ANATOMY AND CIRCULATION OF BILIARY TREE

Functional Anatomy of Biliary Tract

The biliary canaliculus is a cylindrically shaped space that begins in the hepatocytes located in the central venous area and continues toward the portal triad (54). There is extensive branching of the tubes that constitute the canaliculus, so that the canaliculi in fact form a network like chicken wire but with frequent blind ends where the canaliculus begins (75). The membrane of the canaliculus is the membrane of the hepatocyte, so that canalicular bile may be considered extracellular. The canaliculus is separated from the plasma present in the space of Disse by the tight junctional complexes connecting the hepatocytes. These are readily permeable to low-molecular-weight solutes. Thus the canaliculus is separated from the hepatocyte by the canalicular membrane and from the plasma space by the paracellular junctional complexes.

Bile formation begins at the central region of the hepatocyte, and bile flows in the opposite direction to sinusoidal blood flow. The spectacular time-lapse microphotographs of Phillips and co-workers (29, 86) have shown clearly that the canaliculi contract, so that bile is presumably pumped toward the biliary ductules by active contraction of the microfilaments surrounding the canaliculi. The canaliculi are connected by the canals of Hering to the bile ductules, which are lined by a well-developed columnar epithelium. This epithelium is probably both absorptive and secretory, as is discussed in *Ductular Modification of Canalicular Bile,* p. 554. The ductules merge to form bile ducts, which in turn merge to form hepatic ducts. The largest hepatic ducts leaving the liver join to form the common hepatic duct. In animals with gallbladders, the cystic duct connects the gallbladder to the common hepatic duct, after which the major bile duct is termed the common bile duct; this enters the duodenum. A sphincter (the sphincter of Oddi) is present at the junction of the common duct and small intestine. In humans and many other vertebrates the common bile duct is joined by the pancreatic duct close to its entry into the small intestine, but, as noted, in some species the pancreatic duct may enter the duodenum more distally (65).

Bile that reaches the bile ducts is termed *hepatic bile*. A variable fraction of hepatic bile is partitioned into the gallbladder according to the pressure relationship in the biliary tree; when the sphincter of Oddi contracts, bile enters the gallbladder, since its pressure is considerably below the hydrostatic pressure generated by canalicular bile (37, 113).

The major function of the gallbladder is to concentrate bile by absorption of sodium chloride (and sometimes also sodium bicarbonate) and water in isosmotic proportions. Gallbladder bile becomes progressively concentrated but because of the micellar concentration of bile salts it remains isotonic. Thus the gallbladder acts to store the bile salts secreted by the liver in a cumulative manner, and the size of the gallbladder can remain relatively constant despite an increased content of bile acids because of the gallbladder's ability to selectively remove the low-molecular-weight anions (49). Precipitation of calcium carbonate is prevented by simultaneous secretion of hydrogen ions, which converts carbonate to bicarbonate (and bicarbonate to carbon dioxide), and this changes bile from a solution supersaturated in calcium carbonate to a solution that is grossly unsaturated (89).

Circulation of Liver and Biliary Tract

Detailed consideration of the hepatic circulation is available in numerous articles (35, 87), but a few of its features merit mentioning in the context of bile secretion. The liver is a highly vascular organ receiving blood from both the portal vein and hepatic artery. The venous blood streams down the sinusoids, and plasma constituents enter the space of Disse through its fenestrations of the endothelial cells. Spurts of arterial blood periodically jet from the arteriolar sphincters at the beginning of the sinusoid causing local turbulence and decreasing laminar flow. Before reaching the sinusoids, the hepatic artery gives off branches that form a capillary plexus around the ductular cells, which also drains to the sinusoids. The sinusoids thus receive blood from three sources: *1)* portal blood, which delivers blood from the spleen (with its rich reticuloendothelial system), the small intestine (rich in nutrients as well as enteric and pancreatic hormones), and the large intestine (containing bacterial metabolites); *2)* the hepatic artery (rich in oxygen and tissue metabolic products as well as free fatty acids from peripheral adipose tissue stores); and *3)* capillary blood from the biliary ductules that contains materials removed by the ductular cells from bile.

The first-pass extraction of substances reaching the liver varies from nearly 90% for high first-pass uptake substances such as bile acids (said to be blood flow limited) to essentially no uptake for large hydrophilic substances such as inulin. There is continuous endocytosis by the hepatocyte membrane of plasma so that every circulating substance probably passes into the hepatocyte to some extent (see the chapter by Van Dyke, Lake, and Scharschmidt) (60). There are a great variety of receptors for circulating proteins that induce protein uptake by receptor-mediated endocytosis (105). Substances that are not extracted by the liver return to the liver via arterial blood, but since only a fraction of the cardiac output goes to the liver, plasma clearance will be lower. For a substance such as the bile acid cholyltaurine, which has a first-pass fractional extraction of 90%, the consequence of complete diversion of portal blood into the systemic circulation causes about a fourfold increase in peripheral plasma levels.

Imaging of Biliary Tract

Direct (invasive) imaging of the biliary tree has become a conventional diagnostic technique in clinical medicine and has revolutionized the care of the jaundiced patient (17, 18, 75, 76, 97, 98). The spectacular advances in design and construction of fiber-optic endoscopes led to successful duodenoscopy, which was soon followed by successful passage of a catheter through the sphincter of Oddi; injection of radiopaque dye visualized the biliary tree (119). The technique of endoscopic retrograde cholangiography is now widely practiced and is a standard and key technique in the diagnostic evaluation of the patient with presumed obstructive jaundice (19). At about the same time, Okuda and colleagues (76) at Chiba University described the use of a long, flexible needle (the Chiba needle) that could be inserted safely and rapidly into the biliary tract by a percutaneous approach (transhepatic cholangiography). When a radiopaque dye was injected, the biliary tree was opacified (76). This technique is also widely used to distinguish "medical" jaundice, in which hyperbilirubinemia is caused by defective bile formation, from "surgical" jaundice, in which the flow of bile is blocked by a space occupying lesion. Endoscopic placement of catheters to drain the obstructed biliary tree is common clinical practice (17, 18, 31, 98).

Noninvasive imaging techniques have improved greatly in power and are occasionally useful. The most widely used features molecules that have a hydrophobic backbone affixed to an iminodiacetic side chain; such molecules chelate ^{99}Tc, an isotope that is ideal for imaging, and are selectively excreted into bile (114). Such a technique does not give high-resolution imaging of the biliary tree, but it does permit the presence of gallbladder filling to be evaluated. This has some clinical utility, since a gallbladder that visualizes well with the radionuclide is unlikely to be acutely inflamed because of obstruction by a stone in the cystic duct.

Bile salts are rapidly extracted by the liver and are highly concentrated in bile. The bile acid molecule is not readily iodinated, but the side chain has been modified to incorporate a gamma-emitting label such as ^{11}C or ^{75}Se (cf. ref. 74).

The most commonly used method of gallbladder imaging is real-time ultrasound. The technique of oral cholecystography is still of value. Iodinated benzoic acid derivatives that are secreted preferentially in bile are used. During gallbladder concentration, the dye becomes more concentrated causing gallbladder opacification by X ray (10).

METHODS FOR STUDYING PHYSIOLOGY OF BILE

Biliary Fistula Methods

The traditional method of studying biliary secretion is to fashion an acute or chronic biliary fistula in an experimental animal or to perfuse the isolated liver; such methods, although quite primitive in principle, continue to be used for lack of better methods. The problem with these methods is that what is recovered in bile is the algebraic sum of both canalicular and ductular events. There is an urgent need for a methodological breakthrough.

In animal studies, bile is sampled at the level of the common hepatic duct or common bile duct (after cholecystectomy). A variety of techniques have been described for exteriorizing the enterohepatic circulation in chronic animal preparations, and stream-splitting devices that permit a controlled interruption of the flow of bile have been extensively used in primates (23) and even in humans (111). It is possible to use a balloon-occludable T tube to divert bile to the outside, as desired (101). In some studies, the gallbladder is fixed just beneath the skin, so that it may be aspirated at will; this technique is called cholecystopexy (21). Whipple and colleagues popularized the use of the cholecystopyelostomy in which bile drains into the renal pelvis and thence into a reservoir such as a hemisected urinary bladder (56, 123). This technique is still used and is quite useful (84, 85). Thomas (110) described a fistula preparation in which an opening is made in the second portion of the duodenum just opposite the sphincter of Oddi. The opening is closed with a removable "bung," permitting insertion of a small catheter into the common duct followed by studies of biliary secretion in the unanesthetized animal. In humans, bile may be sampled from the duodenum, provided gallbladder contraction is maintained by a continuous intravenous infusion of cholecystokinin or by stimulating the continuous release of endogenous cholecystokinin by a constant infusion of a liquid meal into the intestine (25, 71–73, 93).

Stop-flow studies have been reported on occasion (16) but have aroused little interest. It is also possible to perfuse isolated segments of bile ducts (77, 78), but it is not yet clear that the absorption transport properties of these larger ducts are identical to those of the smaller biliary ductules.

The gallbladder may be studied in situ or ex vivo either in its natural configuration or everted. The gallbladder mucosa may be stripped and mounted in an Ussing chamber for conventional ion transport studies (121). In animals such as the *Necturus,* the entire gallbladder wall may be used for transport studies, since the muscle layer is so thin.

Perfused Liver Preparations

For experiments aimed at gaining insight into bile flow or composition, it is desirable to know the composition of sinusoidal plasma. This is most conveniently done using an isolated perfused liver preparation; this technique is now widely used and is considered to provide useful information (32). One possible limitation of this technique is that blood enters the liver only via the portal vein. Thus it was originally argued that in the conventional portally perfused liver there is no ductular function. This seems not to be the case, however, and ductular function may be well preserved even though blood is perfused solely by the portal vein (109). The technique is most commonly carried out using the rat liver, but ex vivo liver perfusion has also been carried out with livers from both mammals and nonmammals. The dog liver has been perfused in situ with blood provided by a living donor animal (42). In principle, liver from any species can be used. The liver may be used with a continuously recycling perfusate or a single-pass perfusion may be performed. More recently liver perfusion is carried out in a retrograde manner, by attaching the entry cannula to the central vein and the exit cannula to the portal vein (20, 81). With retrograde perfusion, the pericentral cells are exposed to the highest concentration of solute in the perfusate, in contrast to the prograde perfusion in which periportal cells are exposed to the highest concentration of solute in the perfusate. In some studies, blood has been supplied by both arterial and portal venous catheters (2).

Isolated Hepatocytes or Cultured Hepatocytes

If the liver is perfused with collagenase-elastase solutions, the hepatocytes may be dislodged and a preparation of isolated hepatocytes prepared. This is a powerful technique, especially for uptake or biotransformation studies (4, 80).

In the past few years, there has been a spectacular increase in the use of primary liver cell cultures or cell hybrids for studying aspects of endobiotic or xenobiotic uptake or biotransformation (cf. ref. 118). To date, however, none of these techniques has proved useful for studying hepatic canalicular secretion.

Hepatic Cell Doublets

To date, micropuncture of the biliary canaliculus has not been successful. Phillips and colleagues (86) showed that when isolated hepatocytes were incubated, they self-aggregated to form doublets that appear to secrete bile into canaliculi between them. Such apparent canaliculi are visualized with fluorescent dyes, but the dye rapidly leaks into the medium, presumably because there is direction to bile flow. Recently, Boyer and colleagues (34) have modified this technique and developed optical techniques to obtain information on the rate of canalicular bile formation with these doublets (30). However, since the volume of fluid collecting in the special quasi-canalicular space is too small to be sampled or analyzed, the chemical driving forces for bile flow cannot be quantified.

Isolated Membrane Fractions

For studying membrane transport, membrane fractions are prepared by conventional or novel techniques and their purity assessed by morphological and marker enzyme analysis. In the past decade, membrane fractions containing predominantly sinusoidal membranes or predominantly canalicular membranes have been used to study ion or bile acid transport (6, 67–69). Carrier proteins are being isolated, and reconstitution experiments are in progress.

OVERVIEW OF BILIARY SECRETION

Bile as Resultant of Secretory, Hydrolytic, and Absorptive Processes

Bile is secreted into the canaliculus and flows through the canals of Hering and into the biliary ductules. Hydrolysis by brush-border enzymes on the biliary ductular cells, and active and passive absorption and/or secretion of solutes and water by the ductular cells modify the composition of canalicular bile to form hepatic bile. The osmotic effects of actively absorbed or secreted solutes cause the flow of water and accompanying filtrable solutes from blood into bile (if ductular secretion occurs) and from bile into blood (if ductular absorption occurs). Thus hepatic bile is the resultant of canalicular secretion and ductular modification. A portion of hepatic bile enters the gallbladder where hepatic bile is transformed into gallbladder bile by active electrolyte absorption and secretion by the gallbladder as well as passive lipid absorption. In addition, the gallbladder mucosa, at least in some species, secretes immunoglobulins and possible other substances. The result of these multiple processes is the formation of gallbladder bile, which is a highly concentrated yet isosmotic micellar solution.

Canalicular Bile Formation

Canalicular bile formation results from the osmotic effects of actively secreted solutes. The most important of these are the bile acid anions (12, 83, 103), and the flow induced by these bile acid anions is called bile acid–dependent bile flow. The canalicular flow that is apparently induced by other solutes is called bile acid–independent bile flow. The substances that are actively secreted and generate bile acid–independent flow have not been clearly identified, but prime candidates are the tripeptide glutathione and bicarbonate ion (33). Those biliary constituents that are actively secreted by the hepatocyte into the canaliculus may be defined as primary solutes and are the prime movers for bile flow. In addition, there is probably a continuing exocytosis of vesicles into the canaliculus. These are an important determinant of the lipid and protein composition of bile but probably make a negligible contribution to bile volume.

Details of hepatocyte transport of bile acids are discussed in the chapter by Meijer. Amidated bile acids enter the hepatocyte by a sodium-coupled transport system that is secondary, since the primary energy is considered to originate from the Na^+-K^+-ATPase (52, 95). Unconjugated hydrophilic bile acids appear to enter by a BA^-/OH^- anion exchange system, and unconjugated lipophilic bile acids appear to enter passively, presumably by partitioning into the lipid domains of the sinusoidal membranes of the hepatocyte. Canalicular secretion of bile acids is remarkably concentrative; intracellular concentrations of 20–100 μM are likely, and the monomeric concentration of bile acid anions in the canaliculus is >1,000 μM. The transcanalicular electrical gradient favors bile acid translocation into the canaliculus, but the magnitude of the electrical gradient is too small to explain the concentration gradient, and an active transport system is probable. The capacity of this canalicular transport system (V_{max} of ~30 $\mu mol \cdot kg^{-1} \cdot min^{-1}$ in rodents) is considered to be below that of the sinusoidal uptake system (83).

The natural bile acids induce canalicular secretion of phospholipids [mostly lecithin (phosphatidylcholines)] and cholesterol in a coupled manner, in that the induced output of these lipids is directly proportional to the amount of bile acid traversing the hepatocyte (36). The mechanism by which bile acids induce biliary lipid secretion is not understood. It is likely to involve the stimulation of vesicle formation in the hepatocyte with migration of the vesicle to the canaliculus and exocytosis into bile (63).

A number of substances, such as conjugates of bilirubin, fat-soluble vitamins, and hormonal steroids, are also secreted into canalicular bile. Although biliary secretion may be the major route of elimination of such substances, the rate of biliary secretion of these substances is so low that it is unlikely that they play any important role in canalicular bile formation. Vesicles containing phospholipid and cholesterol (and possibly bile acids and also other constituents in smaller proportions) are also secreted into the canaliculus; however, it is generally believed that the osmotic effects of these secreted vesicles is relatively modest, since their molecular weight is quite high. [For a general summary of the relationship between chemical structure and secretion into bile, the monograph of Smith (99) and the recent review of Klaassen and Watkins (58) are invaluable.] In general, compounds that are concentrated into bile of a molecular weight 300–1,000 are lipophilic and are tightly bound by albumin. Obviously they must be transported by the canalicular membrane, although transport for anions could involve passive movement down the electrical gradients.

The osmotic pressure induced by the active secretion of primary solutes into the canaliculus induces the flow of plasma water and accompanying filtrable solutes into the canaliculus through the junctional complexes between the hepatocytes (see the chapter by Van Dyke, Lake, and Scharschmidt). These substances may be termed secondary (canalicular) solutes. Canalicular bile, thus containing a mixture of primary and secondary solutes, flows through the canals of Hering and enters the biliary tree.

Ductular Modification of Canalicular Bile

The ductular cells (cholangiocytes) actively absorb glucose (37, 78) and presumably amino acids, so that the concentration of these substances in ductular bile falls. Glutathione is hydrolyzed in the biliary ductules to its constituent amino acids, which are presumably actively absorbed (7). In some species, ductular cells actively secrete a bicarbonate-rich fluid in response to secretin and actively absorb a bicarbonate-rich fluid in response to somatostatin (90). The osmotic effect of actively absorbed or secreted solutes causes the flow of water and accompanying filtrable solutes from bile into blood (when ductular absorption occurs) and from blood into bile (when ductular secretion occurs). In addition, based on studies with bile acids, it seems highly likely that weak acids or weak bases, if sufficiently lipophilic (in uncharged form), will be absorbed passively by the cholangiocyte (125). Some species such as the pig and the guinea pig have extremely high bicarbonate concentrations and extremely low bile acid concentrations. In such species, it is possible that the majority of bile flow is driven by ductular bicarbonate secretion.

Gallbladder Modification of Hepatic Bile

Depending on pressure relationships within the biliary tree, as noted, a fraction of hepatic bile enters the gallbladder. The major change in hepatic bile is removal of water that is driven by active salt (or sodium chloride) transport by the gallbladder mucosal cells (123). The luminal membranes contain a hydrogen-sodium exchange system and a chloride-bicarbonate exchange system, so that sodium and chloride enter the mucosal cells and carbon dioxide is formed in the gallbladder lumen, from which it can readily diffuse passively. As a consequence of these transport processes, the concentration of bicarbonate and chloride falls, that of sodium increases, pH falls, and the concentration of bile acid anions increases markedly. Vesicles (containing phospholipid and cholesterol) are transformed into mixed micelles. The micelles bind counterions, and both micelles and bile acid monomers are too large to be absorbed across the junctional complexes of the gallbladder mucosa. The final concentration of cations, present both in the form of micellar counterions as well as in simple solution, is determined by the principles of Donnan equilibrium, since bile acid anions are too large and too charged to be absorbed passively by the gallbladder epithelium, and electrical neutrality must be preserved (88). Some cholesterol is probably absorbed passively from the micelles. In some species, immunoglobulin A is believed to be secreted by the gallbladder epithelium (22). Mucus (containing mucoprotein) is also secreted into the gallbladder lumen (61). Thus the end result of gallbladder modification of hepatic bile is to form gallbladder bile, which is a concentrated, yet isotonic micellar solution containing chiefly biliary lipids (as well as mucus and immunoglobulins).

Spontaneous Changes in Bile Composition

Certain chemical and physical changes occur spontaneously in bile, i.e., they result solely from the physicochemical interaction of canalicular constituents. The anatomical site at which these changes occur in the biliary tract depends on the rate of the process in relation to the rate of bile flow. A major physical change is the transformation of vesicles, which appear to be the dominant form of lipid in canalicular bile to micelles, which is the dominant physical form of lipid in gallbladder bile. This is considered to be a relatively rapid change and is likely to occur during the flow of bile in the canaliculus and biliary ductules, so that micelles are already present in the bile entering the gallbladder (102). The one chemical change that has been postulated to occur spontaneously is hydrolysis of bilirubin monoglucuronide to form unconjugated bilirubin. In these experiments, which have been reported only in preliminary form, the spontaneous hydrolysis of bilirubin monoglucuronide (at pH 7.8) was ~4%/h (104).

Markers of Bile Flow

Wheeler (122) and Forker (28) independently explored the permeability of the hepatocyte to nonmetabolizable, low-molecular-weight polyols such as erythritol and found that the hepatocyte was readily permeable to molecules up to the size of about mannitol. They introduced the idea of erythritol clearance, which as for renal clearance is biliary output/plasma concentration per anatomical unit. The assumption of the erythritol clearance method was that it could be used to measure canalicular bile flow. Accordingly agents such as bile acids that caused an increase in bile flow associated with an increase in erythritol clearance were considered to induce the formation of canalicular bile, whereas agents such as the hormone secretin, which increased bile flow despite no change in erythritol clearance, were considered to be acting at a ductular level.

The clearance idea is attractive in principle but has several problems in practice. First, the biliary ductules, at least in some species, possess some permeability to erythritol (108). The consequence of this is that erythritol clearance can be influenced by ductular bile flow. Second, if a larger molecule such as sucrose is used, it does not equilibrate with hepatocyte water for several hours. Consequently, during the early part of the experiment, its concentration in bile may reflect only paracellular solvent drag; only after it has reached equilibrium in hepatocyte water will its movement into bile reflect total canalicular bile flow. Finally, there are some molecules that move rapidly into the hepatocyte but are only slowly excreted into bile (5). Such molecules will be rapidly cleared from plasma, yet have a low biliary clearance.

In principle, there are molecules that are sufficiently large that they can be used to measure solely paracellular movement. Polyethylene glycol 900 has been proposed as such a paracellular pathway marker, but this molecule has an anomalously high concentration in bile, and it is sufficiently hydrophobic that it may well enter the hepatocyte by passing directly through lipid domains of the cell membrane. Another potential paracellular marker is the calcium ion, which appears to enter bile rapidly via the paracellular route (19).

A second kind of marker is a biliary recovery marker, which is a substance that is secreted exclusively into bile and that does not reabsorb from the small intestine. Indocyanine green (ICG) has been shown to fulfill the requirements of a biliary recovery marker (9).

CLASSIFICATION OF BILIARY CONSTITUENTS

In the past decade, it has been common practice to define the three major organic constituents of bile (bile acids, phospholipids, cholesterol) as the sole components of the biliary lipid fraction. This practice stems from the utility of expressing biliary cholesterol saturation in terms of the relative proportion of the three chemical constituents, as proposed first by Isaksson (53) and subsequently by Admirand and Small (1). Conjugated bile salts are not lipids in the usual sense, since they are insoluble in organic solvents. Conversely, unconjugated bilirubin and unes-

TABLE 1. *Classification of Biliary Constituents*

I. Organic constituents
   A. Major lipids
      1. Bile acids
      2. Phospholipids
      3. Cholesterol
   B. Minor lipids
      1. Bilirubins
      2. Fatty acids
   C. Trace constituents
      1. Steroids
      2. Vitamins and vitamin analogues
         a. Vitamin B_{12} and vitamin B_{12} analogues
         b. Fat-soluble vitamins (A, D, E, and K)
      3. Plasma solutes (sugars, purines, pyrimidines, amino acids, etc.)
      4. Miscellaneous dietary lipids
   D. Proteins and peptides
      1. Proteins
         a. Plasma proteins
         b. Immunoglobulins
         c. Bile-specific proteins secreted by the hepatocyte
         d. Mucoproteins
      2. Peptides
         a. Glutathione
         b. Other
      3. Amino acids
   E. Xenobiotics or dietary constituents
      1. Enterolactone and enterodiol
      2. Drugs

II. Inorganic constituents
   A. Actively secreted anions
      1. Bicarbonate
      2. Bile salts (acids)
   B. Passively secreted ions
      1. Anions
         a. Chloride
         b. Phosphate
         c. Sulfate
         d. Other
      2. Cations
         a. Major cations
            1) Sodium
            2) Potassium
            3) Calcium
            4) Magnesium
         b. Minor cations
            1) Iron
            2) Copper
         c. Trace metals

There are many semantic and scientific problems in this categorization. Bile acids (salts) are listed as both organic constituents and as actively secreted anions. Bicarbonate, and even phosphate and sulfate, could be classified as organic solutes. Cations are classified as major or minor according to their concentration in bile.

terified fatty acids that are always present in low concentrations in bile are, in fact, lipids since they are soluble in organic solvents; yet, by tradition, they are not included in the biliary lipid fraction. Conjugated bilirubin does not have the physical properties of a lipid and is not included in the lipid fraction by definition.

A simple and logical classification is the division of biliary constituents into an organic fraction, i.e., those consisting of the organic elements (carbon, hydrogen, oxygen, nitrogen, and sulfur) and an inorganic fraction. Table 1 shows a simple classification of biliary constituents using this categorization, but it must be noted that such definitions have not been accepted by the community of biliary physiologists.

The solids in bile are composed of the organic and inorganic solutes. The percent solids of bile, i.e., the weight of solids per 100 ml of bile, may be determined by ashing and weighing (15). The composition of the solid fraction varies greatly from species to species. In some species most of the solids are composed of biliary lipids. In other species the solids are composed of biliary proteins or other organic constituents. The measurement of percent solids can be considered to be of historic interest only, and in general it has been replaced by more specific analytical methods.

The major lipids in bile are the bile acids, phospholipids, and cholesterol. Minor lipids may be classified as those substances that are present in easily measurable concentrations in bile and that precipitate in bile, forming gallstones; these are conjugated bilirubin (or similar bile pigments) and fatty acids. Trace organic constituents may be classified as substances that are present at extremely low concentrations and whose presence in bile has no significance for the physiology or pathophysiology of bile. These are conjugates of fat-soluble vitamins, steroid hormones, and other dietary constituents such as sterols and saponins. Bile may contain appreciable concentrations of enterolactone or esterodiol, phenolic derivatives present in fiber. Finally, bile contains low concentrations of the secondary solutes (filtrated from plasma) that have not been reabsorbed by the biliary ductular epithelium.

Biliary proteins are quite heterogeneous. Some originate from canalicular secretion and probably others from leakage of plasma proteins through patulous junctional complexes (59, 91). The composition and origin of biliary proteins are discussed in the chapter by Jones and Burwen. In principle, bile may also contain formed elements from sloughed epithelial cells. The fate of sloughed epithelial cells in bile has not been determined and will not be considered further.

Bile is also a major route of elimination of many drugs and their metabolites (58, 99).

ORGANIC CONSTITUENTS OF BILE

Biliary Lipids

The chemistry of bile acids is discussed in detail in the chapter by Hofmann on the enterohepatic circulation, and that of phospholipids and cholesterol is discussed in the chapter by Cabral and Small on the physical chemistry of bile.

BILE ACIDS. The major functional constituents of bile in all vertebrates is a group of compounds termed bile salts because historically they were isolated from bile in the form of an insoluble (bile) salt (40, 41). Bile

TABLE 2. *Names and Structures of Common Bile Acids in Mammalian Bile*

Name	Position and Orientation of Hydroxyl Groups					Type	Comment
	Nucleus				Side Chain		
	3	6	7	12	23		
Lithocholic	α-OH					S	BDP of 8 and 9 (and possibly others)
Hyodeoxycholic	α-OH	α-OH				S	BDP of 3 and 6
Hyocholic	α-OH	α-OH	α-OH			P	PBA in pig
α-Muricholic	α-OH	β-OH	α-OH			P or S	BA of mouse and rat
β-Muricholic	α-OH	β-OH	β-OH			P or S	BA of mouse and rat
ω-Muricholic	α-OH	α-OH	β-OH			P or S	BA of mouse and rat
Murideoxycholic	α-OH	β-OH				S	BDP of 4 and 5
Chenodeoxycholic	α-OH		α-OH			P	PBA of many mammals
Ursodeoxycholic	α-OH		β-OH			P or S	HBEP of 8
Cholic	α-OH		α-OH	α-OH		P	PBA of many mammals
Ursocholic	α-OH		β-OH	α-OH		P or S	HBEP of 10
Deoxycholic	α-OH			α-OH		S	BDP of 10 and 11
Phocacholic	α-OH		α-OH	α-OH	−OH (R)*	P	PBA of seal and walrus
Phocadeoxycholic	α-OH			α-OH	−OH (R)	S	BDP of 13

Secondary; BDP, bacterial dehydroxylation product; P, primary; BA, bile acid; HBEP, hepatic or bacterial epimerization product. Bile acids shown have cholane nucleus with C_5 side chain and 5β (A/B cis) ring junction. Molecular weights of protonated acids are as follows: monohydroxy, 376.6; dihydroxy, 392.6; trihydroxy, 408.6 (for monosodium salt, add 22.0). These bile acids will be present in bile as the N-acyl glycine or taurine amidates with exception of lithocholic acid, which will be sulfated (in part) at 3 position (at least in humans). Ursocholic acid and muricholic acid are not present in major proportions in biliary bile acids of any known species. * The (R) denotes the steric configuration of the hydroxy group at the C_{23} position.

salts are anions in bile that are water-soluble products of cholesterol metabolism. In most bile salts, cholesterol has been converted to a saturated steroid in which the A and B rings are in cis configuration. (The hydrogen atom at C_5 is in the β-configuration). The hydroxyl group at the 3 position has been epimerized to the 3β-configuration. Additional hydroxyl groups have been added, usually at the 7 but in some species at the 1, 6, 12, or 16 positions. In lower vertebrates the side chain is hydroxylated, forming bile alcohols. In higher vertebrates, the side chain is further oxidized to form an acidic group to form bile acids. The cholesterol side chain has 8 carbons, so that in these vertebrates (such as reptiles) the structure of the bile acid side chain is that of isoctanoic acid. In still higher vertebrates, especially mammals, 3 carbons are removed from the side chain to form a C_5 side chain acid, having the structure of isopentanoic acid (40, 41).

In the biotransformation of cholesterol to bile alcohols or bile acids the added hydroxyl groups are all on one side of the molecule, so that the result is a molecule possessing a hydrophilic side, as well as a hydrophobic side, thus bestowing the properties of amphipathicity (94). For bile acids, the key aspect of amphipathicity is the ability to form mixed micelles with phospholipid and cholesterol in bile and with fatty acids and monoglycerides and fat soluble vitamins in the intestine. (For review of the physical chemistry of bile acids as they pertain to biological systems, see ref. 14; see also ref. 48.)

In most vertebrates, the bile acids formed in the liver, which are primary bile acids, contain two or three hydroxyl groups, although in lower vertebrates, bile acids with additional hydroxyl groups on the side chain are present. During their enterohepatic circulation, the nuclear hydroxyl groups are modified by bacteria to form new bile acids called secondary bile acids (13, 44, 64). Conversion of a primary bile acid to a secondary bile acid involves dehydroxylation, dehydrogenation (to convert a hydroxyl group to an oxo group), or epimerization. The bile acids in bile are thus a mixture of primary bile acids formed in the liver and secondary bile acids formed by intestinal bacteria. In some animals, e.g., humans and dogs, the primary bile acids predominate in bile. In the rabbit, secondary bile acids predominate. The names and structures of the common bile acids present in mammals are summarized in Table 2. Details of the enterohepatic circulation of bile acids are given in the chapter by Hofmann.

After their formation bile alcohols are sulfated, so

that bile alcohol sulfates are the major amphipathic substances in the bile of primitive vertebrates. In mammals, the side chain of bile acids is amidated with glycine or taurine to form conjugated bile acids that are N-acyl amidates. The bile acids present in human bile are virtually all amidated. The amide bond of glycine and taurine conjugated bile acids is a remarkable bond in that it is resistant to pancreatic carboxypeptidase and to other proteolytic enzymes. If bile acids are synthesized with amino acids other than glycine or taurine, e.g., leucine or lysine, such unnatural conjugated bile acids are rapidly hydrolyzed by the carboxypeptidases present in pancreatic juice (51). Conjugated bile acids circulate in their conjugated form until they encounter organisms possessing enzymes that hydrolyze the amide bonds. Such enzymes are termed cholyl amidases or cholyl glycine hydrolases, and their substrate specificities are being defined (8, 51).

Conjugation influences both the physicochemical and physiological properties of bile acids. Conjugation makes bile acids much stronger acids, which in turn decreases their tendency to precipitate from solution under acidic conditions such as might be encountered in the duodenum (14, 48). From a physiological standpoint, conjugation greatly retards passive absorption in the biliary tract so that all bile acids secreted into the canaliculus reach the duodenum.

We have recently examined the biliary bile acid composition from a large number of vertebrate species (L. R. Hagey, S. Rossi, C. D. Schteingart, and A. F. Hofmann, unpublished observations). It seems that in the evolution of species, a number of different biochemical modifications of the bile acid molecule have evolved—all resulting in the formation of molecules that are so hydrophilic that they undergo little passive absorption in the biliary tree and small intestine. In general, dihydroxy bile acids are readily absorbed passively whereas trihydroxy bile acids are not. Thus in some species, e.g., birds such as the vulture, bile acids are present entirely in unconjugated form but are predominantly trihydroxy bile acids whose three polar groups or the nucleus is sufficient to prevent passive absorption. (In the Australian opossum, there is a third hydroxy group on the 1 position.) When only two dihydroxy bile acids are present on the nucleus, additional polarity is required on the side chain. This is commonly accomplished by amidation with glycine or taurine, but in the manatee, the side chain ends in an ester sulfate (thus a bile salt is present as a sulfate, analogous to dodecyl sulfate). Finally, in some species (walrus) there is hydroxylation of the 23 carbon, which enhances the ionization of the terminal carboxyl group lowering its pK_a from 5 to 4 and thus having virtually the same effect on acidic strength as amidation with glycine. This observation leads to the prediction that such bile acids need not be amidated, since they are pseudo-conjugates. It will be of interest to find out whether this prediction will be proved correct.

PHOSPHOLIPIDS. In most vertebrates studied to date, the main phospholipid of bile is phosphatidylcholine with a rather well-defined fatty acid composition (see the chapter by Cabral and Small). Lysolecithin, sphingomyelin, or plasmologens are not present in appreciable amounts. Neutral lipids, such as diglycerides or triglycerides, and acidic lipids, such as fatty acids, are not present in bile in appreciable proportions in those vertebrates whose biliary lipid composition has been measured.

CHOLESTEROL. The major sterol in bile is cholesterol, which is present in unesterified form. Trace amounts of other sterols are present in bile reflecting endogenous formation as well as dietary constituents.

Minor Organic Components

BILIRUBIN. Bilirubin is a complex molecule, being a linear tetrapyrole of the biladiene-ac type, which consists of two asymmetrical dipyromethanones connected by a methylene bridge. A monograph on bilirubin that contains a detailed discussion of its chemistry, metabolism, and biliary secretion has been published recently (79). The natural IXα-Z isomer predominates in vivo and each dipyrolic half molecule carries a proprionic side chain that may be partly or completely ionized, depending on its local environment and the extent of internal hydrogen bonding between the carobxyl group(s) and the hydroxyl groups on the end of the bilirubin molecule. Bilirubin is formed in the reticuloendothelial cells of the spleen and the liver, transported in albumin-bound form to the hepatocyte, where it is taken up; hepatic uptake probably involves a carrier molecule. In the cell the bilirubin is bound to ligandin and delivered to the microsomes, where it is conjugated with glucuronic acid by the microsomal enzyme bilirubin [uridine diphosphate (UDP)] glucuronyl transferase. The majority of bilirubin in human bile is present as the diglucuronide, the glucuronic acid being esterified in ester linkage at its 1 position. The proportion of diglucuronide, the two isomeric monoglucuronides (the 9 or the 12), and that of unconjugated bilirubin may be determined with accuracy by high-pressure liquid chromatography. (In a few species, bilirubin is esterified with other monosaccharides such as xylose.) The consensus of recent studies is that conjugated bilirubin is not associated with a micelle (92) but that the monoglucuronide could partition somewhat into the micelle. Unconjugated bilirubin probably is strongly associated with the surface of the micelle, and there is increasing in vitro evidence that some bilirubin may be present in bile as a heterodimer with (monomeric) bile acid molecules.

FATTY ACIDS. Fatty acids in unesterified form are not present in bile in an appreciable portion. Trace amounts are present, but it is not known whether they are secreted as such or are formed during the flow of bile by hydrolysis of biliary phospholipid. Calcium salts of saturated fatty acids are a major component of bile duct stones formed in patients with infection or inflammation in the biliary tract. Presumably, leukocytes enter the biliary tract and release phospholipases or lysophospholipases.

Trace Organic Components

The trace constituents are unimportant in the context of biliary secretion, although for a given substance, bile may be an important or even sole route of elimination. Conjugates of steroid hormones are present, as are vitamin B_{12} analogues termed corrinoids (55). As noted, bile serves as the sole pathway for excretion of many amphipathic substances ingested in the diet or as drugs.

INORGANIC COMPONENTS OF BILE

The major cation of bile is sodium, and all other cations are present at much lower concentrations.

Physiological Determinants of Concentration of Common Ions

In most species, the driving force for canalicular bile formation is the active secretion of bile acid anions and in addition those ions that contribute to bile acid–independent flow. Accordingly, bile is enriched in these anions and accompanying counterions. For example, it is conceivable that bile acid anions are accompanied during their active canalicular secretion by potassium counterions that enter bile via a potassium channel in the canalicular membrane. The remainder of the ions entering canalicular bile should have a composition similar to a plasma filtrate, although the solute patterns will be distorted by differing reflection coefficients as ions are pulled through the junctional complexes by solvent drag. Canalicular bile has not been sampled because of technical difficulty as noted, but under conditions of rapid bile flow it is likely that hepatic bile does not differ greatly from canalicular bile. As would be anticipated, the composition of biliary electrolytes resembles that of plasma, although in general the bicarbonate-chloride ratio is greater than that present in plasma.

Ductular secretion is rich in bicarbonate, and under circumstances where ductular secretion contributes importantly to total biliary secretion, the concentration of bicarbonate ion should be increased. In an animal such as the guinea pig or pig whose bile is invariably greatly enriched in bicarbonate, it seems likely that secretion of bicarbonate is the major driving force for bile formation and is quantitatively far more important than canalicular secretion of bile acids.

Biliary Calcium

The concentration of calcium in hepatic bile is similar to that of ultrafiltrable plasma calcium, and this value is at least three orders of magnitude higher than the estimated concentration of calcium inside the hepatocyte (19). Thus although the uphill movement of bile acids into the canaliculus causes perhaps a 20-fold enrichment in bile acid concentration, the concentration of calcium in canalicular bile is 10,000 times that of the hepatocyte. The cell vigorously defends its calcium homeostasis, and the need to pump calcium out of the cell might explain the localization of a calcium ATPase pump on the canalicular membrane (6).

Significance of Biliary Electrolyte Composition

The concentration of inorganic electrolytes in bile would not be of any physiological interest were it not for the frequent precipitation of calcium carbonate to form gallstones.

Hepatic bile is supersaturated in calcium carbonate because the transport of unbound calcium ions from plasma exceeds the binding capacity of the bile acid monomers and vesicles present in hepatic bile (in the dog) (89). In hepatic bile, the fraction of calcium bound is ~20%. In the gallbladder, the saturation with respect to calcium carbonate is strikingly reduced by a series of simultaneous events. First, calcium and bicarbonate are absorbed rapidly from the gallbladder diminishing their concentration. The gallbladder mucosa secretes hydrogen ions, converting carbonate to bicarbonate and hence increasing calcium solubility. Finally, vesicles are converted to micelles, and the micelles have a greater binding capacity for calcium.

Minor and Trace Metals

Animal studies suggest that bile is an important excretory route for trace metals (58). The manner by which metals are transported in bile is not well established. Possibilities include simple ionic solution, complexation with proteins such as metallothionein, or complexation with glutathione. Harvey et al. (39) recently used inductively coupled plasma atomic absorption spectrometry to measure the concentration of major, minor, and trace metals in human gallbladder bile obtained at laparotomy. The values reported by these authors for samples from healthy subjects are as follows (mM, means ± SD): sodium, 210 ± 12; potassium, 12.7 ± 3.5; calcium, 7.4 ± 2.9; magnesium, 6.9 ± 2.2; copper, 0.09 ± 0.05; iron, 0.02 ± 0.01; zinc,

0.02 ± 0.006; manganese, 0.02 ± 0.02, and molybdenum, 0.02 ± 0.01. Values in bile samples obtained from gallstone patients did not differ significantly.

SECRETORY AND EXCRETORY FUNCTIONS OF BILE

Secretory Function

Bile has two major functions: secretory and excretory. The secretory function of bile involves bile acids, water, phospholipids, and immunoglobins. Its excretory function is the removal of lipids (especially cholesterol and bilirubin) and heavy metals and amphipathic metabolites of protein-bound substances such as hormonal steroids and fat soluble vitamins, as well as corrinoids, vitamin B_{12} congeners that are present in the diet.

WATER. Major functions of water (in bile) are to keep the biliary channels open and to prevent precipitation of insoluble salts. Water contribution from bile to the gastrointestinal secretory pool is small relative to gastric and small intestinal secretions. Thus the function of biliary water is self-protection, i.e., to prevent the precipitation or adsorption of biliary constituents to the biliary epithelium so that flow is obstructed. The location in the biliary tract where most rapid flow is needed is in the fine radicles of the biliary tract where bile is supersaturated with calcium carbonate.

A second major function of the aqueous phase of bile is to act as the liquid vehicle for the vesicles and micelles in bile to transport them from the canaliculus to the intestine and thus permit their excretion. Finally the aqueous phase transports the most important constituents of bile—the bile acids—to the intestine.

PHOSPHOLIPIDS. The phospholipids in bile serve to provide the lipid core of the mixed micelle, which is essential for its ability to solubilize cholesterol. In the absence of phospholipid secretion, bile would be grossly supersaturated with cholesterol, and cholesterol precipitation would be likely. Indeed, if one analyzes biliary lipid composition in gallstone patients, one can identify a group of patients with supersaturated bile because of an apparent deficiency of phospholipid (45). Phospholipid may conceivably protect the gallbladder mucosa from damage by the high concentrations of bile acids present in the gallbladder and some very low phospholipid/bile acid proportions have been observed in patients with chronic cholecystitis. Biliary phospholipid may also play a role in emulsification of dietary phospholipid. However, the functional importance of biliary phospholipid may remain unclear until a disease state is recognized in which biliary phospholipid secretion is defective or absent. It is possible that a clue to the function of biliary phospholipid will be provided by a comparison of biliary lipid secretion in different species.

CHOLESTEROL. The function of biliary cholesterol excretion is homeostatic when conversion to bile acids is insufficient. Whether the cholesterol in bile protects the cells of the biliary tract from damage by bile is unknown. In some species such as the dog, the concentration (and proportion in biliary lipids) of cholesterol is so low as to make this unlikely.

BILE ACIDS. Bill acids have multiple functions in bile and in the small and large intestine. First, bile acids are the water-soluble end product of cholesterol metabolism, so that their biosynthesis can be considered as cholesterol degradation, and bile formation serves to maintain cholesterol homeostasis. Second, they are actively secreted into bile and induce bile acid–dependent flow as discussed. Third, they induce the movement of vesicles containing phospholipid and cholesterol into bile, thus simultaneously promoting cholesterol excretion, as well as the secretion of the phospholipid, which is needed to form mixed micelles. In addition, bile acids bind calcium ions both as monomers and micelles, decreasing calcium activity and preventing calcium precipitation in bile. The mixed micelle in bile that results from the interaction of bile acid molecules and vesicles occurs as a "sink" for any lipids either secreted into bile or formed as bile moves down the biliary tree. For example, if unesterified fatty acids are formed (from phospholipids) or if unconjugated bilirubin is formed, such lipids can dissolve in the mixed micelles present in bile.

The property of bile acids to form micelles permits gallbladder bile to become highly concentrated, yet remain isotonic. As a consequence, the gallbladder can remain small (protected by the liver and rib cage), and when it contracts it delivers a jet of highly concentrated detergent solution to the small intestine.

Bile is discharged into the intestine when a meal is ingested. In the proximal small intestinal lumen, bile acids facilitate the digestion and absorption of dietary lipids as well as keep heavy metals such as iron and calcium in solution, probably promoting their absorption.

A detailed discussion of fat digestion and absorption is beyond the scope of this chapter. These topics will be reviewed in the volume on intestinal transport in preparation for this *Handbook* section of the gastrointestinal system. Excellent reviews have been published (82, 111). Bile acids form a compound with lipase and colipase at the oil/water interface of emulsified dietary lipid and promote the movement of lipolysis products from the surface of the oil droplet into the aqueous phase. Initially the lipid products are large liquid crystalline aggregates that may be formed vesicles. Then the vesicles are gradually transformed into mixed micelles, where the lipid core is largely fatty acid and monoglyceride; these are termed primary micellar solutes. The mixed micelle in the small intestinal lipid solubilizes fat-soluble vitamins that, in the absence of micelles, are not absorbed.

The effect of micellar solubilization is to accelerate diffusion through the unstirred layer coating the intestinal mucosa. The diffusive flux is the product of concentration times the diffusion coefficient. The micelle greatly increased aqueous concentration (by up to 1,000-fold) of lipolytic products such as long-chain fatty acids, yet the diffusion coefficient of the micelle falls by only a factor of seven. The end result is a great acceleration of the diffusive flux through the unstirred layer. For lipid absorption this is a critical event, since their passive membrane permeation is so high that diffusion is rate limiting in overall absorption. Thus the faster the diffusion rate, the faster the absorption.

Bile acids themselves are not absorbed to any great extent in the proximal small intestine because of their size and charge. They are actively absorbed in the terminal ileum and recycled in the enterohepatic circulation, as is fully discussed in the next chapter by Hofmann. Between meals, the majority of the bile acid pool is stored in the gallbladder.

Excretory Function

FEATURES OF BILIARY EXCRETION. Obviously the requirement for biliary excretion is secretion into the biliary canaliculus. Direct secretion of molecules by the biliary ductular cells into hepatic bile has been sought for many years, but no convincing evidence has as yet been advanced. The most attractive candidate molecules would be cationic molecules in those species in which there is an acidic microclimate at the luminal side of the biliary ductular wall if such exists.

Thus the current view is that all organic constituents of bile are secreted into the canaliculus either actively or passively. At least two canalicular transport systems have been tentatively identified by physiological methods—one for the common conjugated bile acids and one for the organic strongly charged molecules such as bilirubin diglucuronide and bile acid sulfates (24). The bile acid transport system may also transport ester glucuronides of endobiotics such as bilirubin or xenobiotics. The system that transports bilirubin diglucuronide may also transport the glutathione conjugates of molecules such as bromosulfophthalein. (A defect in such a transport mechanism is likely to be the key defect in the Dubin-Johnson syndrome.) It is of interest that bile acids with cationic, zwitterionic, or neutral side chains show virtually no secretion into canalicular bile (5).

For a compound once secreted into canalicular bile to reach the small intestine, it must not be completely reabsorbed in the biliary tract. Passive absorption is related to biliary pH (and/or the pH present in the microenvironment at the biliary epithelium/lumen interface and the lipophilicity of the molecule). The lipophilicity is determined by the steric arrangement of the polar and hydrophobic parts of the molecule. Polarity is obviously determined by the presence of any ionizing groups that greatly inhibit passive absorption as well as the presence of weaker polar groups such as hydroxyl groups. The effects of these various hydrophobic regions and polar groups, which in turn determine the overall lipophilicity of the molecule, can be estimated at least semiquantitatively by determining the octanol/water position coefficient at a pH similar to that expected to be present in the biliary tract (cf. ref 117). For weak acids, if the evidence from work with unconjugated bile acids applies, the presence of two or less polar groups will mean appreciable passive ductular absorption, except in species such as the guinea pig that have extremely alkaline bile and whose biliary epithelium presumably has an alkaline microclimate at the luminal surface (125). For weak bases, no information is available. In general, however, the biliary excretion rates of bases is likely to be considerably less than that of bile acids, which under physiological conditions is in the range of 0.2–2 μmol/kg·body wt^{-1}·min^{-1}.

If a molecule is not passively absorbed during passage in the biliary tract, it then is stored in the gallbladder or secreted into the duodenum. In the duodenum and proximal small intestine, the pH slowly increases from about pH 5 to 6.5, but there are occasional brief drops in the pH to as low as pH 4, because jets of acidic gastric content are not well mixed and neutralized by pancreatic and duodenal bicarbonate secretions (66). Under these circumstances, weak acids should be absorbed. In the more distal small intestine, the pH gradually rises and becomes similar to that present in the biliary tract, which suggests that little additional passive absorption occurs.

For a compound that enters the small intestine to reach the large intestine, it must neither be destroyed in nor absorbed from the small intestine. The small intestinal lumen is in a sense a hydrolytic environment, and it thus differs from the biliary tract where luminal enzyme activity is considered to be minimal. Thus a second requirement for effective excretion of compounds secreted into bile is that the linkage between the conjugating group and the xeno- or endobiotic be not susceptible to pancreatic or brush-border enzymes. The two most common groups—glucuronate and sulfate—both appear to be completely resistant to the intrinsic digestive enzymes. An important question is whether the bacterial flora present in the distal small intestine contains glucuronidase and sulfatase activities. Since these enzymes release the unconjugated drug into solution, reabsorption would be favored. What should reduce absorption in the case of weak acids is the relatively alkaline pH of the distal ileum.

If conjugates escape hydrolysis in the distal ileum, they pass into the colon where the density of the luminal flora increases by three to six orders of magnitude. Here, probably all endobiotic or xenobiotic conjugates are hydrolyzed. Absorption of the liberated unconjugated moiety is considered to be of limited magnitude because of *1*) binding of the liberated (amphipathic or lipophilic) compound to bacteria or dietary residue and/or *2*) the relatively small surface-to-volume ratio of the colon. Thus efficient conjugation with glucuronide or sulfate prevents passive absorption in the biliary tract or small intestine and digestion by pancreatic or mucosal enzymes. Efficient translocation to the colon occurs, and overall absorption from the colon is poor because of intraluminal binding that contains extremely low concentration of dissolved molecules. As a result of this low concentration of dissolved species and the limited absorptive area of the colon, products are incorporated into feces and pass to the exterior during defecation.

For bile acids the situation is completely different. An active transport system in the distant ileum "recognizes" the conjugated bile acid anions that are efficiently transported from the intestinal lumen into the portal blood. Bile acids return to the liver, are efficiently extracted by the hepatocyte, induce biliary lipid secretion by the hepatocyte, and are secreted into bile, inducing bile acid–dependent flow.

CRITERION FOR ENTEROHEPATIC CYCLING. The criterion for enterohepatic cycling is that biliary secretion exceeds hepatic biosynthesis and return of a substance or its precursors to the liver from peripheral sources.

$$\text{enterohepatic amplification} = \frac{\text{biliary secretion}}{\text{hepatic production or return of new molecules to the liver}}$$

The two limiting cases are bilirubin, where secretion is equal to the production rate, and bile acids, where secretion rate exceeds production rate by a factor of 10–30.

For both endobiotics and xenobiotics, a key question is the partition between renal excretion and biliary excretion. This may be expressed as a simple fraction, defining the fraction that is excreted in bile.

$$f_b = \frac{\text{biliary excretion}}{\text{total excretion (biliary + urinary)}}$$

Despite an enormous amount of work on bile acid metabolism, relatively few values are available because careful study of this problem using bile sampling methods that did not perturb the enterohepatic circulation have seldom been performed. It would seem helpful to define a fraction:

$$f_u = \frac{\text{urinary excretion}}{\text{total excretion}}$$

and

$$f_b = 1 - f_u$$

In general, it has been the tradition of pharmacology to equate fecal excretion of a drug with its biliary excretion. Thus the assumption has been made that no absorption of the drug occurs in the biliary tract or small or large intestine. For bile acids this is clearly not the case, as deoxycholic acid and lithocholic acid are both absorbed from the colon.

It has been proposed that it is possible to estimate the enterohepatic cycling by infusing bile from a donor animal to the small intestine of a recipient animal (115) or even to use several animals in a cascade fashion (107). Such a method is valid, although in the ideal circumstance the donor animal should be in a steady state with respect to biliary secretion. A very complete physiological, pharmacokinetic model for the enterohepatic circulation of bile acids has been proposed and used to simulate bile acid metabolism satisfactorily (47). This model has not been used for drugs since as yet drugs with active intestinal reabsorption have not been identified.

CHOLESTASIS

Cessation of bile flow, whether caused by mechanical obstruction of the biliary tract or absence of bile formation by the liver, is termed *cholestasis*. Cholestasis causes retention of biliary constituents, increased levels of these constituents in the blood and presumably in the liver cell, and increased elimination of the water soluble constituents in urine. In addition, cholestasis is associated with abolition of bile acid secretion into the small intestine, so that the intraluminal concentration of bile acids is greatly reduced.

Each of these alterations in the normal concentration/secretion pattern of biliary constituents leads to a number of secondary events, which will be summarized briefly.

Bile acid deficiency in the small intestine causes the absence of mixed micelle formation with the digestive products of dietary lipids (for review, see ref. 44). Fat-soluble vitamins are not absorbed, leading to a true deficiency state for vitamins E, D, K, and A. Fatty acids are absorbed as molecules; their diffusive flux through the aqueous boundary layer coating the intestinal epithelium is far slower than that of mixed micelles. Fat absorption occurs through the small intestine and is incomplete. Unsaturated fatty acids are absorbed more efficiently than are saturated fatty acids, presumably because they have greater aqueous solubility at the pH value present in the small intestine. Unabsorbed fatty acids pass into the colon, where they damage the epithelial cells causing inhibition of sodium and water absorption (3). Colonic permeability is increased. In the small intestine, the absence of bile acids leads to increased calcium activity, since bile acids bind calcium in both monomeric and micellar form. The calcium ions form insoluble calcium soaps. Dietary oxalate, usually precipitated as its insoluble calcium salt in the jejunal lumen, now remains in solution and may be hyperabsorbed from the colon, causing hyperoxaluria, which in turn may lead to nephrolithiasis (24, 46).

Other biliary lipids increase in blood. Phospholipids and cholesterol form mixed-disk aggregates that associate loosely with albumin and other serum lipoproteins to form a complex termed lipoprotein-X (27, 95). This complex may be formed in vitro by mixing bile and blood. A new steady state for cholesterol metabolism occurs, characterized by very low rates of body cholesterol synthesis and no cholesterol absorption. The only loss from the cholesterol and bile acid pool is by urinary elimination of bile acids.

Cholestasis probably also causes retention of biliary constituents in the hepatocyte. When the concentration of bile acids is sufficiently high, bile acids undergo extensive novel conjugation biotransformations. The resulting (amidated) sulfates and glucuronides are probably less tightly bound to serum albumin than are their corresponding nonsulfated derivatives and there is increased filtration in the glomerulus (106). The sulfates and glucuronides are poorly reabsorbed from the renal tubule, and there is a marked increase in the renal clearance of bile acids. Since the amount of bile acids formed is considerably less than that in health, and since there is little tubular reabsorption of bile acids, renal excretion of bile acids in cholestasis is far less than biliary secretion in health.

The anatomical path and biochemical mechanism by which bile acid conjugates move from hepatocyte or biliary ductular lumen to plasma are not known.

Bilirubin conjugates—particularly the monoglucuronide—increase markedly in plasma. A steady state is reached in which increased renal excretion of bilirubin becomes equal to the synthesis rate. A small fraction of the bilirubin conjugates becomes covalently coupled to serum albumin and cannot be excreted in urine (120).

Copper, iron, and other trace metal constituents of bile accumulate in the hepatocyte. The accumulation of trace metals and the accumulation of bile acids in the hepatocytes, as well as other as yet unidentified factors, lead in time to hepatocyte death. White cells migrate into the hepatic lobule; inflammation becomes evident microscopically; a marked increase in collagen deposition occurs; and portal fibrosis results, causing secondary biliary cirrhosis. The bundles of collagen obstruct venous inflow to the liver causing portal hypertension. With time, portal hypertension leads to esophageal varices and hemorrhage, which is often difficult to stem because of the loss of clotting factors. The result of intractable hemorrhage in the face of decreased functional cell mass leads to hepatic failure and death.

The work described in this article from this laboratory was supported by NIH Grants DK-21506 and DK-32130, as well as grants-in-aid from the Falk Foundation and Gipharmex S.p.A. The author acknowledges helpful discussions with L. R. Hagey, D. Gurantz, R. H. Dowling, and E. W. Moore. V. Huebner edited and prepared the typescript. The author apologizes to his colleagues whose publications were not cited in the interests of limiting the number of references.

REFERENCES

1. ADMIRAND, W. H., AND D. M. SMALL. The physiochemical basis of cholesterol gallstone formation in man. *J. Clin. Invest.* 47: 1043–1052, 1968.
2. AHMAD, A. B., P. N. BENNETT, AND M. ROWLAND. Influence of route of hepatic administration on drug availability. *J. Pharmacol. Exp. Ther.* 230: 718–725, 1984.
3. AMMON, H., AND S. F. PHILLIPS. Inhibition of colonic water and electrolyte absorption by fatty acids in man. *Gastroenterology* 65: 744–749, 1973.
4. ANWER, M. S. Effect of organic anions on bile acid uptake by isolated rat hepatocytes. *Hoppe-Seyler's Z. Physiol. Chem.* 359: 1027–1030, 1978.
5. ANWER, M. S., E. R. L. O'MAILLE, A. F. HOFMANN, R. A. DIPIETRO, AND E. MICHELOTTI. Influence of side-chain charge on hepatic transport of bile acids and bile acid analogues. *Am. J. Physiol.* 249 (*Gastrointest. Liver Physiol.* 12): G479–G488, 1985.
6. BACHS, O., K. S. FAMULSKI, F. MIRABELLI, AND E. CARAFOLI. ATP-dependent Ca^{2+} transport in vesicles isolated from the bile canalicular region of the hepatocyte plasma membrane. *Eur. J. Biochem.* 147: 1–7, 1985.
7. BALLATORI, N., R. JACOB, AND J. L. BOYER. Intrabiliary glutathion hydrolysis—a source of glutamate, cyst(e)ine, and glycine in bile (Abstract). *Hepatology Baltimore* 5: 951, 1985.
8. BATTA, A. K., AND G. SALEN. Substrate specificity of cholylglycine hydrolase for the hydrolysis of bile acid conjugates. *J. Biol. Chem.* 259: 15035–15039, 1984.
9. BERGE HENEGOUWEN, G. P. VAN, AND A. F. HOFMANN. Nocturnal gallbladder storage and emptying in gallstone patients and healthy subjects. *Gastroenterology* 75: 879–885, 1978.
10. BERK, P., J. T. FERRUCCI, JR., AND G. R. LEOPOLD. *Radiology of the Gallbladder and Bile Ducts.* Philadelphia, PA: Saunders, 1983.
11. BJORKHEM, I. Mechanism of bile acid biosynthesis in mammalian liver. In: *Sterols and Bile Acids*, edited by H. Danielsson and J. Sjovall. Amsterdam: Elsevier, 1985, p. 231–278.
12. BOYER, J. L. New concepts of mechanisms of hepatocyte bile formation. *Physiol. Rev.* 60: 303–326, 1980.
13. CAREY, M. C. The enterohepatic circulation. In: *The Liver: Biology and Pathobiology*, edited by I. Arias, H. Popper, D. Schachter, and D. A. Shafritz. New York: Raven, 1982, p. 429–465.
14. CAREY, M. C. Physico-chemical properties of bile acids and their salts. In: *Sterols and Bile Acids*, edited by H. Danielsson and J. Sjovall. Amsterdam: Elsevier, 1985, p. 345–403.
15. CAREY, M. C., AND D. M. SMALL. The physical chemistry of cholesterol solubility in bile. *J. Clin. Invest.* 61: 998–1026, 1978.
16. CHENDEROVITCH, J. Bile duct function in biliary system. In: *The Hepatobiliary System*, edited by W. Taylor. New York: Plenum, 1976, p. 267–284.
17. CLASSEN, M., J. GEENEN, AND K. KAWAI. *Nonsurgical Biliary Drainage.* New York: Springer-Verlag, 1984.
18. COTTON, P. B., J. GEENEN, AND B. HELM. Direct cholegraphy and related diagnostic methods (ERC, PTC). *Clin. Gastroenterol.* 12: 101–123, 1983.
19. CUMMINGS, S. A., AND A. F. HOFMANN. Physiological determinants of biliary calcium secretion in the dog. *Gastroenterology* 87: 664–673, 1984.
20. DAWSON, J. R., J. G. WEITERING, G. J. MULDER, R. N. STILLWELL, AND K. S. PANG. Alteration of transit time and

direction of flow to probe the heterogeneous distribution of conjugating activities for harmol in the perfused rat liver preparation. *J. Pharmacol. Exp. Ther.* 234: 691–699, 1985.
21. DEITRICK, J. E., C. K. MCSHERRY, B. THORBJARNARSON, AND F. GLENN. The study of bile salt kinetics in the experimental animal using a new technique. *J. Surg. Res.* 16: 559–563, 1974.
22. DELACROIX, D. L., H. J. F. HODGSON, A. MCPHERSON, C. DIRE, AND J. P. VAERMAN. Selective transport of polymeric immunoglobulin A in bile. Quantitative relationships of monomeric and polymeric immunoglobulin A, immunoglobulin M and other proteins in serum, bile and saliva. *J. Clin. Invest.* 70: 230–241, 1982.
23. DOWLING, R. H., E. MACK, J. PICOTT, J. BERGER, AND D. M. SMALL. Experimental model for the study of the enterohepatic circulation of bile in the rhesus monkey. *J. Lab. Clin. Med.* 72: 169–176, 1968.
24. EARNEST, D. L. Enteric hyperoxaluria. *Adv. Intern. Med.* 24: 407–427, 1979.
25. ENG, C., AND N. B. JAVITT. Chenodeoxycholic acid-3-sulfate. Metabolism and excretion in the rat and hamster and effects on hepatic transport systems. *Biochem. Pharmacol.* 32: 3555–3558, 1983.
26. EVERSON, G. T., M. J. LAWSON, AND C. MCKINLEY. Gallbladder and small intestinal regulation of biliary lipid secretion during intraduodenal infusion of standard stimuli. *J. Clin. Invest.* 71: 596–603, 1983.
27. FELLIN, R., G. BALDO, AND S. ZOTTI. Alterations of plasma lipoproteins in cholestasis. In: *Liver and Lipid Metabolism*, edited by S. Calandra, N. Carulli, and G. Salvioli. Amsterdam: Excerpta Med., 1984, p. 71–81.
28. FORKER, E. L. Two sites of bile formation as determined by mannitol and erythritol clearance in the guinea pig. *J. Clin. Invest.* 46: 1189–1195, 1967.
29. FRENCH, S. W. Role of canalicular contraction in bile flow. *Lab. Invest.* 53: 245–249, 1985.
30. GAUTAM, A., D. SCARAMUZZA, AND J. L. BOYER. Quantitative assessment of primary canalicular secretion in isolated rat hepatocyte couplets (IRHC) by optical planimetry (Abstract). *Gastroenterology* 90: 1727, 1986.
31. GOLDBERG, H. I. Percutaneous transhepatic cholangiography and biliary drainage. In: *Gastrointestinal Disease: Pathophysiology, Diagnosis, and Management*, edited by M. H. Sleisenger and J. S. Fordtran. Philadelphia, PA: Saunders, 1983, p. 1745–1760.
32. GORES, G. J., L. J. KOST, AND N. F. LARUSSO. The isolated perfused rat liver. Conceptual and practical considerations. *Hepatology Baltimore* 6: 511–517, 1986.
33. GRAF, J. Canalicular bile salt-independent bile formation: concepts and clues from electrolyte transport in rat liver. *Am. J. Physiol.* 244 (*Gastrointest. Liver Physiol.* 7): G233–G246, 1983.
34. GRAF, J., A. GAUTAM, AND J. L. BOYER. Isolated rat hepatocyte couplets: a primary secretory unit for electrophysiologic studies of bile secretory function. *Proc. Natl. Acad. Sci. USA* 81: 6516–6520, 1984.
35. GREENWAY, C. V., AND R. D. STARK. Hepatic vascular bed. *Physiol. Rev.* 51: 23–65, 1971.
36. GURANTZ, D., AND A. F. HOFMANN. Influence of bile acid structure on bile flow and biliary lipid secretion in the hamster. *Am. J. Physiol.* 247 (*Gastrointest. Liver Physiol.* 7): G736–G748, 1984.
37. GUZELIAN, P., AND J. L. BOYER. Glucose reabsorption from bile—evidence for a biliohepatic circulation. *J. Clin. Invest.* 53: 526–535, 1974.
38. HALLENBECK, G. A. Biliary and pancreatic intraductal pressures. In: *Handbook of Physiology. Alimentary Canal. Secretion*, edited by C. F. Code. Washington, DC: Am. Physiol. Soc., 1967, sect. 6, vol. II, chapt. 57, p. 1007–1025.
39. HARVEY, R. C., D. TAYLOR, C. N. PETRUNKA, A. D. MURRAY, AND S. M. STRASBERG. Quantitative analysis of major, minor and trace elements in gallbladder bile of patients with and without gallstones. *Hepatology Baltimore* 5: 129–132, 1985.
40. HASLEWOOD, G. A. D. *Bile Salts*. London: Methuen, 1967.
41. HASLEWOOD, G. A. D. *The Biological Importance of Bile Salts*. Amsterdam: North-Holland, 1978.
42. HOFFMAN, N. E., D. E. DONALD, AND A. F. HOFMANN. Effect of primary bile acids on bile lipid secretion from perfused dog liver. *Am. J. Physiol.* 229: 714–720, 1975.
43. HOFMANN, A. F. The enterohepatic circulation of bile acids in man. *Clin. Gastroenterol.* 6: 3–24, 1977.
44. HOFMANN, A. F. The bile loss syndrome: a doubtful entity. In: *Nonsurgical Biliary Drainage*, edited by M. Classen, J. Geenen, and K. Kawai. New York: Springer-Verlag, 1984, p. 120–126.
45. HOFMANN, A. F., S. M. GRUNDY, J. M. LACHIN, S.-P. LAN, R. A. BAUM, R. F. HANSON, T. HERSH, N. C. HIGHTOWER, JR., J. W. MARKS, H. MEKHJIAN, R. A. SCHAEFER, R. D. SOLOWAY, J. L. THISTLE, F. B. THOMAS, M. P. TYOR, AND THE NATIONAL COOPERATIVE GALLSTONE STUDY GROUP. Pretreatment biliary lipid composition in white patients with radiolucent gallstones in the National Cooperative Gallstone Study. *Gastroenterology* 83: 738–752, 1982.
46. HOFMANN, A. F., M. F. LAKER, K. DHARMSATHAPHORN, H. P. SHERR, AND D. LORENZO. Complex pathogenesis of hyperoxaluria after jejunoileal bypass surgery: oxalogenic substances in diet contribute to urinary oxalate. *Gastroenterology* 84: 293–300, 1983.
47. HOFMANN, A. F., G. MOLINO, M. MILANESE, AND G. BELFORTE. Description and simulation of a physiological pharmacokinetic model for the metabolism and enterohepatic circulation of bile acids in man. Cholic acid in healthy man. *J. Clin. Invest.* 71: 1003–1022, 1983.
48. HOFMANN, A. F., AND A. RODA. Physiochemical properties of bile acids and their relationship to biological properties: an overview of the problem. *J. Lipid Res.* 25: 1477–1489, 1984.
49. HOLZBACH, R. T. Effects of gallbladder function on human bile: compositional and structural changes. *Hepatology Baltimore* 4, Suppl.: 57S–60S, 1984.
50. HUIJGHEBAERT, S. M., AND A. F. HOFMANN. Pancreatic carboxypeptidase hydrolysis of bile acid-amino acid conjugates: selective resistance of glycine and taurine amidates. *Gastroenterology* 90: 306–315, 1986.
51. HUIJGHEBAERT, S. M., AND A. F. HOFMANN. Influence of the amino acid moiety on deconjugation of bile acid amidates by cholylglycine hydrolase or human fecal cultures. *J. Lipid Res.* 27: 742–752, 1986.
52. INOUE, M., R. KINNE, T. TRAN, AND I. M. ARIAS. Taurocholate transport by rat liver canalicular membrane vesicles. Evidence for the presence of an Na^+-independent transport system. *J. Clin. Invest.* 73: 659–663, 1984.
53. ISAKSSON, B. On the lipid constituents of bile from human gallbladder containing cholesterol gallstones. A comparison with normal human bladder bile. *Acta Soc. Med. Ups.* 59: 277, 1953–1954.
54. JONES, A. L. Anatomy of the normal liver. In: *Hepatology: A Textbook of Liver Disease*, edited by D. Zakim and T. D. Boyer. Philadelphia, PA: Saunders, 1982, p. 3–31.
55. KANAZAWA, S., V. HERBERT, B. HERZLICH, G. DRIVAS, AND C. MANUSSELIS. Removal of cobalamin analogue in bile by enterohepatic circulation of vitamine B_{12}. *Lancet* 1: 707–708, 1983.
56. KAPSINOW, R. The experimental production of duodenal ulcer by exclusion of bile from the intestine. *Ann. Surg.* 83: 614–617, 1926.
57. KLAASSEN, C. D. Biliary excretion of metals. *Drug Met. Rev.* 5:165–196, 1976.
58. KLAASSEN, C. D., AND J. B. WATKINS III. Mechanisms of bile formation, hepatic uptake and biliary excretion. *Pharmacol. Rev.* 36: 1–67, 1984.
59. KLOPPEL, T. M., W. R. BROWN, AND J. REICHEN. Mechanisms of secretion of proteins into bile: studies in the perfused rat liver. *Hepatology Baltimore* 6: 587–594, 1986.
60. LAKE, J. R., V. LICKO, R. W. VAN DYKE, AND B. SCHARSCHMIDT. Biliary secretion of fluid-phase markers by the isolated perfused rat liver. Role of transcellular vesicular trans-

port. *J. Clin. Invest.* 76: 676–684, 1985.
61. LAMONT, J. T., B. F. SMITH, AND J. R. L. MOORE. Role of gallbladder mucin in pathophysiology of gallstones. *Hepatology Baltimore* 4: 51S–56S, 1984.
62. LEISS, O., AND K. VON BERGMANN. Comparison of biliary lipid secretion in non-obese cholesterol gallstone patients with normal, young, male volunteers. *Klin. Wochenschr.* 63: 1163–1169, 1985.
63. LOWE, P. J., S. G. BARNWELL, AND R. COLEMAN. Rapid kinetic analysis of the bile-salt-dependent secretion of phospholipid, cholesterol and a plasma-membrane enzyme into bile. *Biochem. J.* 222: 631–637, 1984.
64. MACDONALD, I. A., V. D. BOKKENHEUSER, J. WINTER, A. M. MCLERNON, AND E. H. MOSBACH. Degradation of steroids in the human gut. *J. Lipid Res.* 24: 675–700, 1983.
65. MANN, F. C., J. P. FOSTER, AND S. D. BRIMHALL. The relation of the common bile duct to the pancreatic duct in common domestic and laboratory animals. *J. Lab. Clin. Med.* 5: 203–206, 1920.
66. MCCLOY, R. F., R. G. GREENBERG, AND J. H. BARON. Duodenal pH in health and duodenal ulcer disease: effect of a meal. *Gut* 25: 386–392, 1984.
67. MEIER, P. J., R. KNICKELBEIN, R. H. MOSELEY, J. W. DOBBINS, AND J. L. BOYER. Evidence for carrier-mediated chloride/bicarbonate exchange in canalicular rat liver plasma membrane vesicles. *J. Clin. Invest.* 75: 1256–1263, 1985.
68. MEIER, P. J., A. ST. MEIER-ABT, C. BARRETT, AND J. L. BOYER. Mechanisms of taurocholate transport in canalicular and basolateral rat liver plasma membrane vesicles. Evidence for an electrogenic canalicular organic anion carrier. *J. Biol. Chem.* 259: 10614–10622, 1984.
69. MEIER, P. J., E. S. SZTUL, A. REUBEN, AND J. L. BOYER. Structural and functional polarity of canalicular and basolateral plasma membrane vesicles isolated in high yield from rat liver. *J. Cell Biol.* 98: 991–1000, 1984.
70. MOORE, E. W. The role of calcium in the pathogenesis of gallstones: Ca^{++} electrode studies of model bile salt solutions and other biologic systems. *Hepatology Baltimore* 4, Suppl.: 228S–243S, 1984.
71. NILSELL, K., B. ANGELIN, B. LEIJD, AND K. EINARSSON. Comparative effects of ursodeoxycholic acid and chenodeoxycholic acid on bile acid kinetics and biliary lipid secretion in humans. *Gastroenterology* 85: 1248–1256, 1983.
72. NORTHFIELD, T. C., AND A. F. HOFMANN. Biliary lipid output during three meals and an overnight fast. I. Relationship to bile acid pool size and cholesterol saturation of bile in gallstone and control subjects. *Gut* 16: 1–11, 1975.
73. NORTHFIELD, T. C., N. F. LARUSSO, A. F. HOFMANN, AND J. L. THISTLE. Biliary lipid output during three meals and an overnight fast. II. Effect of chenodeoxycholic acid treatment in gallstone subjects. *Gut* 16: 12–17, 1975.
74. NYHLIN, H., M. V. MERRICK, M. A. EASTWOOD, AND W. G. BRYDON. Evaluation of ileal function using 23-selena-25-homotaurocholate, a gamma-labeled conjugated bile acid. *Gastroenterology* 84: 63–68, 1983.
75. ODA, M., V. M. PRICE, M. M. FISHER, AND M. J. PHILLIPS. Ultrastructure of bile canaliculi with special reference to the surface coat of the pericanalicular web. *Lab. Invest.* 31: 314–323, 1974.
76. OKUDA, K., K. TANIKAWA, T. EMUA, S. KURATOMI, S. JINNOUCHI, K. URABE, T. SUMIKOSHI, Y. KANDA, Y. FUKAYAMA, H. MUSHA, H. MORI, Y. SHIMOKOWA, F. YAKUSHIJI, AND Y. MATSUURA. Nonsurgical percutaneous transhepatic cholangiography—diagnostic significance in medical problems of the liver. *Am. J. Dig. Dis.* 19: 21–36, 1974.
77. OLSON, J. R., AND J. M. FUJIMOTO. Evaluation of hepatobiliary function by the segmented retrograde intrabiliary injection technique. *Biochem. Pharmacol.* 29: 205–211, 1980.
78. OLSON, J. R., AND J. M. FUJIMOTO. Demonstration of a D-glucose transport system in the biliary tree of the rat by use of the segmented retrograde intrabiliary injection technique. *Biochem. Pharmacol.* 29: 213–219, 1980.
79. OSTROW, J. D. *Bile Pigments and Jaundice.* New York: Dekker, 1986.
80. PANG, K. S., P. KONG, J. A. TERRELL, AND R. E. BILLINGS. Metabolism of acetaminophen and phenacetin by isolated rat hepatocytes. A system in which the spatial organization inherent in the liver is disrupted. *Drug. Metab. Dispos.* 13: 42–50, 1985.
81. PANG, K. S., H. KOSTER, I. C. M. HALSEMA, E. SCHOLTENS, G. J. MULDER, AND R. N. STILLWELL. Normal and retrograde perfusion to probe the zonal distribution of sulfation and glucuronidation activities of harmol in the perfused rat liver preparation. *J. Pharmacol. Exp. Ther.* 224: 647–653, 1983.
82. PATTON, J. S. Gastrointestinal lipid digestion. In: *Physiology of the Gastrointestinal Tract,* edited by L. R. Johnson. New York: Raven, 1981, p. 1147–1220.
83. PAUMGARTNER, G., AND T. SAUERBRUCH. Secretion, composition and flow of bile. *Clin. Gastroenterol.* 12: 3–23, 1983.
84. PERTSEMLIDIS, D., E. H. KIRCHMAN, AND E. H. AHRENS, JR. Regulation of cholesterol metabolism in the dog. I. Effects of complete bile diversion and of cholesterol feeding on absorption, synthesis, accumulation and excretion rates measured during life. *J. Clin. Invest.* 52: 2353–2367, 1973.
85. PERTSEMLIDIS, D., E. H. KIRCHMAN, AND E. H. AHRENS, JR. Regulation of cholesterol metabolism in the dog. II. Effects of complete bile diversion and of cholesterol feeding on pool size of tissue cholesterol measured at autopsy. *J. Clin. Invest.* 52: 2368–2378, 1973.
86. PHILLIPS, M. J., C. OSHIO, M. MIYAIRI, H. KATZ, AND C. R. SMITH. A study of bile canalicular contractions in isolated hepatocytes. *Hepatology Baltimore* 2: 763–768, 1982.
87. RAPPAPORT, A. M. Hepatic blood flow: morphologic aspects and physiologic regulation. *Int. Rev. Physiol.* 21: 1–63, 1980.
88. REGE, R. V., M.-J. LEE, AND E. W. MOORE. The Gibbs-Donnan equilibrium: its effect on distribution of calcium (Ca^{++}) across rabbit gallbladder (GB) epithelium (Abstract). *Gastroenterology* 90: 1761, 1986.
89. REGE, R. V., AND E. W. MOORE. Pathogenesis of calcium-containing gallstones. Canine ductular bile, but not gallbladder bile, is supersaturated with calcium carbonate. *J. Clin. Invest.* 77: 21–26, 1986.
90. RENE, E., R. G. DANZINGER, A. F. HOFMANN, AND M. NAKAGAKI. Pharmacologic effect of somatostatin on bile formation in the dog: enhanced ductular reabsorption as the major mechanism of anticholeresis. *Gastroenterology* 84: 120–129, 1983.
91. REUBEN, A. Biliary proteins. *Hepatology Baltimore* 4, Suppl.: 46S–50S, 1984.
92. REUBEN, A., K. E. HOWELL, AND J. L. BOYER. Effects of taurocholate on the size of mixed lipid micelles and their associations with pigment and proteins in rat bile. *J. Lipid Res.* 23: 1039–1052, 1982.
93. REUBEN, A., P. N. MATON, G. M. MURPHY, AND R. H. DOWLING. Bile lipid secretion in obese and non-obese individuals with and without gallstones. *Clin. Sci.* 69: 71–79, 1985.
94. RODA, A., A. F. HOFMANN, AND K. J. MYSELS. The influence of bile salt structure on self-association in aqueous solutions. *J. Biol. Chem.* 258: 6362–6370, 1983.
95. SABESIN, S. M. Cholestatic lipoproteins—their pathogenesis and significance. *Gastroenterology* 83: 704–709, 1982.
96. SCHARSCHMIDT, B. F., AND R. W. VAN DYKE. Mechanisms of hepatic electrolyte transport. *Gastroenterology* 85: 1199–1214, 1983.
97. SETCHELL, K. Hepatic conjugation of enterolactone and enterodiol: the first mammalian lignans. In: *Advances in Glucuronide Conjugation,* edited by S. Matern, K. W. Bock, and W. Gerok. Lancaster: MTP, 1985, p. 287–291.
98. SILVIS, S. *Therapeutic Gastrointestinal Endoscopy.* New York: Igaku Shoin, 1985.
99. SMITH, R. L. *The Excretory Function of Bile: The Elimination of Drugs and Toxic Substances in Bile.* London: Chapman & Hall, 1973.
100. SOBOTKA, H. *Physiological Chemistry of the Bile.* Baltimore,

MD: Williams & Wilkins, 1937.
101. SOLOWAY, R. D., H. C. CARLSON, AND L. J. SCHOENFIELD. A balloon-occludable T-tube for cholangiography and quantitative collection and reinfusion of bile in man. *J. Lab. Clin. Med.* 79: 500–504, 1972.
102. SOMJEN, G. J., AND T. GILAT. Contribution of vesicular and micellar carriers to cholesterol transport in human bile. *J. Lipid Res.* 26: 699–704, 1985.
103. SPERBER, I. Secretion of organic anions in the formation of urine and bile. *Pharmacol. Rev.* 11:109, 1951.
104. SPIVAK, W., W. YUEY, AND D. DIVENUTO. Role of bilirubin-mono- and diglucuronide (BMG, BDG) in the formation of unconjugated bilirubin (UCB) in model bile and native bile systems (Abstract). *Gastroenterology* 90: 1771, 1986.
105. STEER, C. J., AND R. D. KLAUSNER. Clathrin-coated pits and coated vesicles: functional and structural studies. *Hepatology* 3: 437–454, 1983.
106. STEIHL, A. Disturbances of bile acid metabolism in cholestasis. *Clin. Gastroenterol.* 6: 45–67, 1977.
107. SUWELACK, D., AND H. WEBER. Assessment of enterohepatic circulation of radioactivity following a single dose of [^{14}C]-Nimodipine in the rat. *Eur. J. Drug Metab. Pharmacokinet.* 10: 231–239, 1985.
108. TAVOLONI, N. Permeation patterns of polar nonelectrolytes across the guinea pig biliary tree. *Am. J. Physiol.* 247 (*Gastrointest. Liver Physiol.* 10): G527–G536, 1984.
109. TAVOLONI, N., AND F. SCHAFFNER. The intrahepatic biliary epithelium in the guinea pig: is hepatic artery blood flow essential in maintaining its function and structure? *Hepatology Baltimore* 5: 666–672, 1985.
110. THOMAS, P. E. An improved cannula for gastric and intestinal fistulas. *Proc. Soc. Exp. Biol. Med.* 46: 260–265, 1941.
111. THOMPSON, A. B. R., AND J. M. DIETSCHY. Intestinal lipid absorption. In: *Physiology of the Gastrointestinal Tract*, edited by L. R. Johnson. New York: Raven, 1981, p. 1147–1220.
112. THUREBORN, E. Human hepatic bile: composition changes due to altered enterohepatic circulation. *Acta Chir. Scand. Suppl.* 303: 1–63, 1962.
113. TOOULI, J., J. E. GEENEN, W. J. HOGAN, W. J. DODDS, AND R. C. ARNDORFER. Sphincter of Oddi motor activity: a comparison between patients with common bile duct stones and controls. *Gastroenterology* 82: 111–117, 1982.
114. TRIGER, D. R. Evaluation of liver disease by radionuclide scanning. In: *Hepatology: A Textbook of Liver Disease*, edited by D. Zakim and T. D. Boyer. Philadelphia, PA: Saunders, 1982, p. 633–646.
115. TSE, F. L. S., F. BALLARD, AND J. SKINN. Estimating the fraction reabsorbed in drugs undergoing enterohepatic circulation. *J. Pharmacokinet. Biopharm.* 10: 455–461, 1982.
116. UDDIN, K. K., AND R. M. CASE. Patterns of bicarbonate secretion in the upper gastrointestinal tract and their relation to gastric acid secretion. In: *Mechanisms of Mucosal Protection in the Upper Gastrointestinal Tract*, edited by A. Allen, G. Flemström, A. Garner, W. Silen, and L. A. Turnberg. New York: Raven, 1984, p. 129–134.
117. VADNERE, M., AND S. LINDENBAUM. Distribution of bile salts between 1-octanol and aqueous buffer. *J. Pharm. Sci.* 71: 875–880, 1982.
118. VAN DYKE, R. W., J. E. STEPHENS, AND B. F. SCHARSCHMIDT. Bile acid transport in cultured rat hepatocytes. *Am. J. Physiol.* 243 (*Gastrointest. Liver Physiol.* 6): G484–G492, 1982.
119. VENNES, J. A., AND S. E. SILVIS. Endoscopic retrograde cholangiopancreatography. In: *Gastrointestinal Disease: Pathophysiology, Diagnosis, and Management*, edited by M. H. Sleisenger and J. S. Fordtran. Philadelphia, PA: Saunders, 1983, p. 1727–1742.
120. WEISS, J. S., A. GAUTAM, J. J. LAUFF, M. W. SUNDBERG, P. JATLOW, J. L. BOYER, AND D. SELIGSON. The clinical importance of a protein-bound fraction of serum bilirubin in patients with hyperbilirubinemia. *N. Engl. J. Med.* 309: 147–150, 1983.
121. WHEELER, H. O. Concentrating function of the gallbladder. *Am. J. Med.* 51: 588–595, 1971.
122. WHEELER, H. O., E. D. ROSS, AND S. E. BRADLEY. Canalicular bile production in dogs. *Am. J. Physiol.* 214: 866–874, 1968.
123. WHIPPLE, G. H., AND W. B. HAWKINS. Bile fistulas and related abnormalities. Bleeding, osteoporosis, cholelithiasis, and duodenal ulcers. *J. Exp. Med.* 62: 599–620, 1935.
124. WOOD, J. R., AND J. SVANVIK. Gall-bladder water and electrolyte transport and its regulation. *Gut* 24: 579–593, 1983.
125. YOON, Y. B., L. R. HAGEY, A. F. HOFMANN, D. GURANTZ, E. L. MICHELOTTI, AND J. STEINBACH. Effect of side-chain shortening on the physiological properties of bile acids: hepatic transport and effect on biliary secretion of 23-nor-ursodeoxycholate in rodents. *Gastroenterology* 90: 837–852, 1986.

CHAPTER 29

Enterohepatic circulation of bile acids

ALAN F. HOFMANN | *Division of Gastroenterology, Department of Medicine, University of California, San Diego, California*

CHAPTER CONTENTS

Historical Aspects
General Features of Enterohepatic Circulation in Vertebrates
  Anatomical distribution
  Overall balance
  Description methods
  Determinants of composition and distribution
  Dynamic aspects of enterohepatic circulation in humans
    General description
    Bile acid concentrations
    Chemical constituents
  Bile acid biotransformations
    Biotransformations by hepatic enzymes
    Biotransformations by bacterial enzymes
  Movements of bile acids in enterohepatic circulation
    Transcellular bile acid transport
    Flow of bile acids between organs
Application of General Principles in Humans
  Cholic acid and deoxycholic acid
  Chenodeoxycholic acid and lithocholic acid
Determinants of Steady-State Biliary Bile Acid Composition in Humans
  Steroid moiety
  Amino acid moiety
Methods for Characterizing Enterohepatic Circulation of Bile Acids
  Bile acid pool size
  Bile acid biosynthesis or input
    Biosynthesis rates of primary bile acids
    Input rates of secondary bile acids
  Measurement of bile acid fluxes
    Fluxes due to interorgan flows
    Fluxes due to cellular transport
    Fluxes due to biotransformation
  Regulation of enterohepatic circulation
    Regulation of bile acid synthesis: negative feedback hypothesis
    Regulation of bile acid pool size
Perturbations of Enterohepatic Circulation
  Alterations in input
    Decreased input
    Increased input
  Alterations in interorgan flow
    Portal-systemic shunting
    Gallbladder motility and storage disturbances
  Alterations in transport
    Defective hepatic transport
    Defective enteral transport
  Alterations in biotransformation
    Increased deconjugation because of intestinal bacteria proliferation
  Alteration of amidation pattern by glycine or taurine administration
  Influence of conjugation pattern on enterohepatic circulation of bile acids
Epilogue: Comparison of Bile Acids and Conventional Drugs

THIS CHAPTER REVIEWS the current state of knowledge about the enterohepatic circulation (EHC) of bile acids. General principles are developed that should apply to the EHC in all vertebrates, although most of these principles have been derived from studies in humans. As noted in the chapter by Van Dyke, Lake, and Scharschmidt in this *Handbook*, the transport of bile acids through the hepatocyte is similar to that of many other xenobiotics. However, bile acids differ from many other substances in showing extremely concentrative transport into bile, and, in contrast to all other compounds that are secreted in conjugated form into bile, bile acids are actively reabsorbed from the terminal ileum.

The functions of bile acids are reviewed in the preceding chapter by Hofmann in this *Handbook*; only the movements of the bile acid molecules themselves are considered in this chapter. Several reviews on the EHC of bile acids have been published in the past two decades (35, 120, 171, 235); an older monograph (103) and a collection of reviews (181–181b) on the chemistry and biology of bile acids are still useful. The proceedings of a symposium on the physical chemistry of bile (124) and other symposia (15, 16, 34, 172, 202–205) dealing with various aspects of bile acid metabolism in health and disease have recently been published. An entire volume in the *Comprehensive Biochemistry* series deals with bile acids and cholesterol (46).

The physiological and pharmacological aspects of the EHC are described in this chapter, and the common disturbances of the EHC occurring in human disease are broadly outlined. In the past decade, bile acids have been introduced as drugs to alter hepatic cholesterol secretion and thus induce the dissolution of preexisting cholesterol gallstones. Some differences between the metabolism of bile acids and conventional drugs are noted.

HISTORICAL ASPECTS

In 1855 Moritz Schiff, working in Florence, measured the rate of bile secretion in the bile fistula dog after instilling varying amounts of bile into the small intestine (220). Schiff found that the amount of bile secreted was directly related to the amount of bile placed into the intestine. He concluded that bile must be reabsorbed from the intestine, pass to the liver in portal blood, and immediately be secreted into the biliary tract. He proposed the term *enterohepatic circulation*. Schiff's seminal observation was subsequently reinterpreted (143), since it became clear that of the major biliary constituents, only bile acids are efficiently reabsorbed and reexcreted in bile. Charming reviews of the history of the development of current ideas about the EHC are available (102, 112).

Some of the important advances in understanding the EHC of bile acids may be classified as follows:

I. Chemical advances
   A. Elucidation of chemical structure of major bile alcohols and bile acids and development of chemical methods for their interconversion (74)
   B. Adaptation of peptide synthesis methods for preparation of glycine and taurine amidates of bile acids (154, 188, 262), bile acid sulfates (87, 201, 263), and glucuronides (13, 248)
   C. Development of methods for synthesis of bile acids tagged with nonexchanging, stable, or radioactive isotopes in defined positions (26, 41, 137, 261)
   D. Development of chromatographic (232, 243) and spectroscopic (29, 51, 65, 223, 233, 240, 275) methods for bile acid separation, identification, and analysis; subsequent development of sensitive radioimmunoassay (212) and bioluminescence techniques (214)
   E. Description of bile alcohols and bile acids present in living vertebrates (101, 140)

II. Physicochemical advances
   A. Recognition that bile acids are amphipathic molecules and description of their lipid solubilizing properties by principles of colloid chemistry (1, 36, 37, 123, 135, 234)

III. Biochemical advances
   A. A benchmark experiment showing that bile acids were formed from cholesterol (32)
   B. Discovery of difference between primary bile acids (biosynthesized from cholesterol in the liver) and secondary bile acids (formed by bacterial modification of the substituents of primary bile acids) (24)

IV. Physiological advances
   A. Measurement of exchangeable bile acid pool size by isotope dilution (163)
   B. Measurement of bile acid secretion rates in primates (57, 58) and in humans (71, 90, 91, 158, 210, 259)
   C. Discovery of negative feedback of bile acid biosynthesis (25, 180)
   D. Discovery of active transport of bile acids by terminal ileum (254, 278, 280)
   E. Discovery of concept of deconjugation and reconjugation during enterohepatic cycling (105–107, 189), which, together with other experimental data, has permitted pharmacokinetic modeling of EHC of bile acids (115, 116, 132, 178)

V. Pathophysiological advances
   A. Discovery of inborn errors of bile acid metabolism (summarized in the preceding chapter by Hofmann in this *Handbook*; see also ref. 28) and characterization of disturbances of bile acid metabolism in hepatic and intestinal disease (80, 121, 241, 273)

VI. Therapeutic advances
   A. Discovery of ability of two bile acids, chenodeoxycholic acid (CDCA) and ursodeoxycholic acid (UDCA), to decrease proportion of cholesterol in biliary lipids (256) and their use to induce dissolution of cholesterol gallstones (12, 55, 168, 255)

GENERAL FEATURES OF ENTEROHEPATIC CIRCULATION IN VERTEBRATES

Anatomical Distribution

The tissue distribution of bile acids is restricted predominantly to the liver, biliary tract, intestine, and circulatory system. This has been shown experimentally by determining the tissue distribution of radiolabeled bile acids (cf. ref. 185). The major factor responsible for the selective localization of bile acids is the efficient uptake of bile acids by the liver (50%–90% first-pass fractional extraction) from the sinusoidal blood. The tight binding of bile acids to serum albumin, which diminishes diffusion of bile acids into other tissues as well as passage into the glomerular filtrate, makes a minor contribution in addition; however, bile acid metabolism does not appear to be markedly different in the analbuminemic rat (250). The majority of bile acids in the EHC are always located in the biliary tract or intestine, since the only "capacitance" in the EHC is in the lumina of these organs, where bile acids are present in extremely high aqueous concentrations.

Overall Balance

The efficient conservation of the secreted bile acids by the active and passive transport systems of the intestine results in the accumulation of a large mass of bile acids in the EHC. This mass remains relatively constant in size throughout the day and night and can be measured by the conventional technique of isotope dilution (for reviews, see refs. 128, 130). Such a measurement gives the mass of bile acids in the exchangeable bile acid pool.

The input of primary bile acids into the pool originates by biosynthesis from cholesterol in the liver, and a schematic depiction of overall cholesterol balance is shown in Figure 1. If bile acids such as CDCA and UDCA are administered, then input into the bile acid pool has two sources: oral intake of exogenous bile acids and biosynthesis of endogenous bile acids from cholesterol. In the steady state, input of bile acids into the EHC from synthesis (and/or oral ingestion) must be balanced by loss. In the healthy human, bile acids are lost from the EHC almost solely by fecal excretion, and there is no degradation of the steroid moiety during intestinal transit. The absence of appreciable urinary loss has three explanations: *1*) bile acids have low concentrations in peripheral blood because of efficient hepatic extraction; *2*) the bile acids entering the kidney are protein bound [75%–99%, depending on the bile acid; (213)], which diminishes their loss into the glomerular filtrate; and *3*) the natural bile acids that enter the glomerular filtrate are reabsorbed in the tubules by both active and passive transport.

Description Methods

The EHC of bile acids is composed of multiple EHCs, since each individual bile acid has its own EHC. The aggregation of these individual EHCs constitutes the overall EHC of bile acids in humans. Although the EHC of each bile acid is broadly similar, there are important quantitative differences between the EHCs of individual bile acids; these differences are responsible for the differing bile acid composition of serum bile acids and biliary bile acids, as well as the differing effects of individual bile acids on biliary bile acid composition when they are administered to humans.

The quantitative description of the movement of any molecule through the body in the blood circulation, its tissue distribution and concentration, its biotransformation into other compounds, and its excretion lie in the domain of pharmacokinetics. Traditionally, pharmacokinetics used simple "black-box" models featuring one, two, or three compartments, which, although often adequate to describe drug metabolism, have little or no physiological meaning (20, 52, 216). More recently, much more realistic, so-called physiological pharmacokinetic models have been described (27, 111). Such models are based on anatomical, physiological, pharmacological, and/or pathological considerations. (For a useful nonmathematical review of the principles of multicompartmental modeling, see ref. 11; for more mathematical treatments, see refs. 75, 85, 161, 211, 228.)

Bile acid metabolism can also be described using pharmacokinetic principles. The EHC of bile acids can be described satisfactorily by constructing a linear multicompartmental model consisting of well-mixed compartments of individual bile acid species. This chapter does not describe the EHC of any given bile acid in detailed pharmacokinetic terms, but it is possible to describe the metabolism of any individual bile acid with this approach. A realistic pharmacokinetic model has been described for cholic acid (CA) and its two conjugates, cholylglycine and cholyltaurine (132). [In this chapter the term *cholyltaurine* (which is more correct chemically) is used rather than the traditional but misleading term *taurocholate* (or taurocholic acid). See ref. 122 for a discussion of this practice.] More recently this pharmacokinetic approach has been extended to CDCA and its glycine and taurine amidates (178). The utility of such an approach is that it permits the EHC to be described quite quantitatively and even permits simulation of the movements of bile acids in health and disease.

Determinants of Composition and Distribution

The overall composition of the bile acids within the EHC is determined by the balance between input and loss for each of its constituent primary and secondary bile acids. Input is by hepatic biosynthesis (from cholesterol) or from intestinal absorption of secondary bile acids newly formed by bacterial enzymes from primary bile acids. Loss is by fecal excretion or biotransformation into another bile acid.

The distribution of individual bile acids in the EHC is determined by the balance between input and loss for each of the anatomical regions of the EHC. In

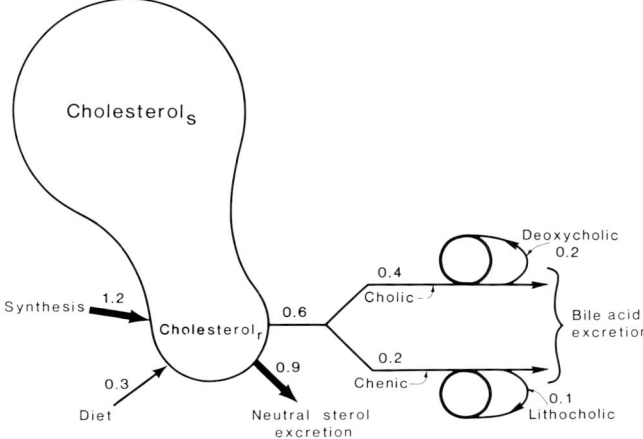

FIG. 1. Schematic depiction of overall cholesterol metabolism in humans. Enclosed area for cholesterol indicates 2 exchangeable cholesterol pools. Cholesterol$_s$, slowly exchangeable pool. Cholesterol$_r$, rapidly exchangeable pool. Numbers are within range of values obtained in healthy adults (g/day). Only the enteral loop of enterohepatic circulation (EHC) of bile acids is shown. In humans, excretion of cholesterol as neutral sterol exceeds acidic sterol excretion (bile acid excretion), whereas in many animals the majority of cholesterol is eliminated as bile acids rather than as cholesterol.

discussing the EHC, it is useful to subdivide the EHC into anatomical systems: the circulatory system, the hepatobiliary system, and the enteral system. Each of these systems may be considered to be composed of spaces, as shown in Figure 2. Figure 3 shows a more schematic depiction of the EHC; the continuous flow depicted in Figure 3 can easily be rearranged to show the spaces and interspace flux shown in Figure 2. In the model of the EHC developed by our laboratory, the amount of each individual bile acid in a space is termed a *compartment*, and the sum of all the bile acids in all compartments constitutes the bile acid in a space. The amount of a given bile acid in any space represents the balance between input and loss for that individual bile acid. For each bile acid, input is by de novo synthesis, by biotransformation from another bile acid, or by translocation (either by transport or flow) from another space. Loss is by biotransformation to another bile acid or by translocation (by flow or transport) to another space; such movements may be termed *interorgan flow*. This classification of bile acid inputs and outputs is used in this chapter.

Dynamic Aspects of Enterohepatic Circulation in Humans

GENERAL DESCRIPTION. During fasting the majority of the bile acid pool is in the gallbladder. Ingestion of a meal induces gallbladder contraction and discharge of bile as well as pancreatic enzyme secretion into the duodenum, initiating digestion. In turn, the liberated digestion products release cholecystokinin and possibly other hormones, which stimulate continual gallbladder contraction and pancreatic enzyme secretion. The bile acids that have entered the small intestine are propelled distally. A small fraction of bile acids is absorbed passively in the proximal small intestine, but most bile acids pass distally to the terminal ileum, where they are actively absorbed.

The bile acids that are absorbed from the intestine pass into the portal venous blood and are bound to albumin. They enter the sinusoids and are efficiently extracted by the hepatocytes. They are then actively secreted into the biliary canaliculus by a second ion-transport system. The bile acid molecules induce bile flow and proceed (as biliary solutes) down the biliary tract to be either stored in the gallbladder or secreted in the duodenum. If the sphincter of Oddi is closed, the gallbladder fills and bile acids are stored in the gallbladder until it contracts. If the sphincter of Oddi is open, bile acids pass into the duodenum. During the fasting state, about one-half of the bile acids secreted by the liver bypass the gallbladder (22). Although only

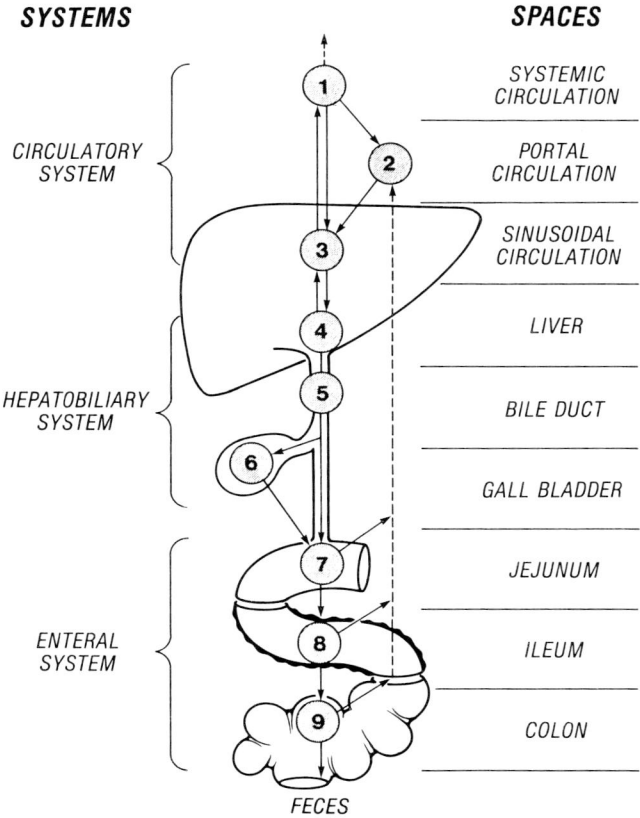

FIG. 2. Representation of EHC of bile acids as linear multicompartmental model. *Arrow* connecting systemic circulation (*space 1*) with sinusoidal circulation (*space 3*) corresponds to hepatic arterial blood flow; *arrow* connecting systemic circulation to portal circulation (*space 2*) corresponds to mesenteric circulation. In the pharmacokinetic model that has been developed, each space is further subdivided into 3 compartments that correspond to the glycine conjugate, the taurine conjugate, and the unconjugated species for any bile acid. [From Hofmann et al. (132), by copyright permission of the American Society for Clinical Investigation.]

FIG. 3. Schematic depiction of EHC shown as a continuing movement of molecules into bile, then into the intestine, then as reabsorption from the intestine, with spillover past the liver into the systemic circulation. This model can be transformed to that shown in Figure 2 without difficulty; both depict EHC of bile acids as consisting of an enterohepatic circle and a circulatory circle.

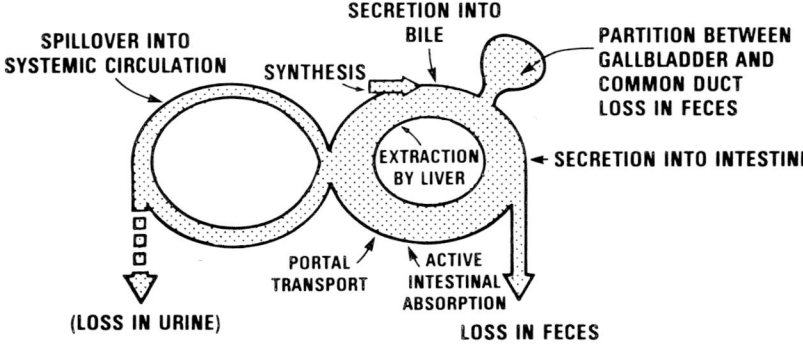

one-half of the bile flow enters the gallbladder, the gallbladder bile is continuously concentrated. During overnight fasting, the proportion of the bile acid pool that is present in the gallbladder continues to increase, since hepatic bile is continuously diverted into the gallbladder.

The movement of bile acids in the EHC fluctuates, accelerating with meals and decelerating during fasting. The movement of the EHC is largely determined by the activity of its mechanical pump, the small intestine, whose motility is governed by neurohormonal factors initiated by entrance of food into the small intestine from the stomach. Thus for all practical purposes the "rhythm" of the EHC is determined by the duration of gastric emptying, which in turn is determined by the eating pattern and dietary content. Contraction of the gallbladder is essential for mobilizing the stored bile acids but is not critical for further movement of bile acids, which depends on intestinal propulsion.

The increase in bile acid secretion occurring with meals is responsible for the postprandial elevation of serum bile acids (159). The fractional hepatic extraction of each bile acid remains rather constant in animals (162, 209) and in humans (8, 72), since the increased load presented to the liver during digestion is still far below the maximal hepatic capacity for bile acid uptake. Consequently a constant proportion of the bile acids returning to the liver spills over into the systemic circulation, causing an increase in the postprandial level of serum bile acids. Provided the fractional hepatic extraction rate remains constant, the magnitude of the postprandial increase is directly proportional to the rate at which bile acids are absorbed from the intestine. The serum bile acid level may thus be compared to a "flow meter" that gives a continuous readout of the flux of bile acids returning from the intestine and spilling past the liver (23, 84).

Bile Acid Concentrations

The aqueous concentration of a bile acid in a tissue (or compartment in the model) depends on three factors: *1*) the bile acid mass, which is determined by its relative rates of influx and efflux; *2*) the volume of water in the compartment; and *3*) binding or trapping by intracellular macromolecules, membranes, or organelles. The total bile acid concentrations are low in portal venous blood (20–50 μM) and peripheral blood (2–8 μM), since the flux of bile acids into and out of these compartments is rapid because of rapid blood flow. The concentration of total bile acids in the hepatocyte is also quite low (<50 μM), since active secretion of bile acids into the canaliculus is quite rapid in relation to hepatic uptake. (For approximate concentrations of bile acids in different tissues, see tables in refs. 132 and 178). On the other hand, the concentration of bile acids in hepatic bile is extremely high (20,000–40,000 μM), since bile acid transport is quite rapid relative to water transport. Because most of the water transported into the canaliculus is induced by the osmotic activity of bile acids (bile acid–dependent flow), the concentration of bile acids in hepatic bile remains rather constant (unless there is extensive ductular modification of canalicular bile). In the gallbladder there is little or no flux of bile acids across the gallbladder mucosa, but there is considerable water absorption. Consequently the concentration of bile acids increases to extremely high levels (50,000–200,000 μM). In the small intestine the flow of intestinal content is relatively low (1–5 ml/min) so that the concentration of bile acids remains high (5,000–20,000 μM). The concentration of bile acids in the cells of the terminal ileum is probably low, since transport of bile acids into portal blood is rapid. Thus bile acid concentrations are high where the bile acids are forming micelles that transport lipids, and low where the bile acids are merely being transported as single molecules.

Chemical Constituents

In humans, two primary bile acids, CA and CDCA, are synthesized from cholesterol in the liver. These bile acids are not interconvertible, as CDCA is never 12 hydroxylated by hepatic enzymes (to form cholate) (28). One factor influencing the ratio of CA to CDCA synthesis is believed to be determined by the degree of 12 hydroxylation of an early sterol precursor (3-oxo-7α-hydroxy-cholest-4-one) in bile acid biosynthesis from cholesterol (28).

Both CA and CDCA are modified by bacteria during their enterohepatic cycling. The secondary bile acids that are formed are reabsorbed to a varying extent and may or may not be biotransformed as they pass through the liver. Most secondary bile acids that reach the liver are secreted into bile; a tiny fraction regurgitates into plasma and is excreted in urine. Thus both bile and urine contain two families of bile acids—one originates from CA and the other from CDCA. The amount of bile acids secreted in bile exceeds that in urine by at least 50,000 times.

Bile Acid Biotransformations

It is useful to divide biotransformations into two types: those catalyzed by hepatic enzymes and those catalyzed by bacterial enzymes.

BIOTRANSFORMATIONS BY HEPATIC ENZYMES. It is currently believed that the liver is the only site of tissue biotransformation of the common, natural bile acids. [The term *common naturally occurring bile acids* refers to the 5 bile acids predominating in human bile: CA and its bacterial metabolite deoxycholic acid (DCA), and CDCA and its bacterial metabolites lithocholic acid (LCA) and UDCA. These are also the predominant bile acids in most other land mammals, with the exception of mouse and rat, which contain high concentrations of muricholates, and the pig, which contains hyocholate and hyodeoxycholate.] The major biotransformation that occurs in the liver is bile acid conjugation with glycine or taurine. This type of conjugation may be termed *amidation* to distinguish it from *glucuronidation* or *sulfation*.

All of the common naturally occurring bile acids are efficiently amidated with glycine or taurine during hepatic passage. Most of the bile acid conjugation in the liver involves reconjugation of bile acids that had been deconjugated during intestinal transit (116, 132, 178). In addition, newly synthesized bile acids are also conjugated with glycine or taurine before secretion into bile. The efficiency of hepatic amidation of bile acids is not known, since canalicular bile cannot be sampled, but it has generally been assumed that conjugation is complete, since unconjugated bile acids are present in only trace proportions in gallbladder bile. It now seems likely that unconjugated bile acids are secreted in bile in small proportions; however, if they are sufficiently lipophilic, they will be reabsorbed to some extent in the biliary ductules so that their proportion in the bile acids of hepatic bile is well below those of canalicular bile (133, 198, 282). Thus hepatic conjugation, although efficient, is not as complete as previously believed. The results of recent animal studies indicate that in some animals (birds, Australian opossum) the majority of bile acids present in bile are in unconjugated form (L. R. Hagey, S. Rossi, and A. F. Hofmann, unpublished observations).

Bile acid conjugation with glycine or taurine involves two enzymes. The first enzyme, coenzyme A (CoA) ligase, converts the bile acid to its CoA derivative. This enzyme is a cytosolic enzyme, and the conjugation step is fully reversible (145, 270). The second enzyme, which is microsomal, converts a bile acid CoA derivative irreversibly to a glycine or taurine amidate, which is then secreted in bile (44, 45, 146, 269). The enzyme involved in bile acid conjugation, the cholyl-CoA amino acid transferase, uses taurine in preference to glycine. The rate of bile acid CoA formation is influenced greatly by the structure of the bile acid side chain, as C_{23} nor–bile acids do not form the CoA derivative (44).

The taurine content of the hepatocyte is determined by the relative rates of taurine input resulting from active transport of taurine into the cell from plasma (as well as biosynthesis of taurine in the hepatocyte in some species) in relation to the rate of hepatic amidation of bile acids (and in some species other xenobiotics) with taurine. In humans taurine is synthesized to only a limited extent, and consequently diet is the predominant source of taurine for bile acid conjugation. The half-life of hepatocyte taurine is extremely short, so that the fraction of bile acids that is conjugated with glycine or taurine may vary from moment to moment. When the cells become deficient in taurine, conjugation occurs with glycine (cf. ref. 92). [This is not true in all species. The dog is an obligate taurine "amidator." When the cells become depleted in taurine, at least under acute conditions when a large bile acid load is presented to the liver, unconjugated bile acids are secreted into bile (cf. ref. 195).]

The taurine and glycine conjugates are fully ionized at intracellular pH, since the pK_a of the glycine-conjugated bile acids is ~3.9 and that of taurine-conjugated bile acids is <1 (135). Thus amidation prevents passive reflux back across the sinusoidal membrane into sinusoidal plasma (39); in addition, amidation may decrease the partition of bile acids into the lipid membranes of microsomes where glucuronidation and sulfation are believed to occur. A glucuronyl transferase for bile acids exists in the hepatocyte (148, 173), but it does not use conjugated bile acids as a substrate, except when they are present at very high concentrations. The natural bile acids are not glucu-

TABLE 1. *Trivial Names and Suggested Abbreviations for Major Conjugated Bile Acids in Human Bile*

Steroid Moiety	Conjugating Moiety	Traditional Name	Preferred Name
Cholic	Glycine	Glycocholate (GC)	Cholylglycine (CG, chl-gly)
	Taurine	Taurocholate (TC)	Cholyltaurine (CT, chl-tau)
Chenodeoxycholic	Glycine	Glycochenodeoxycholate (GCDC)	Chenodeoxycholylglycine (CDCG, cdc-gly)
	Taurine	Taurochenodeoxycholate (TCDC)	Chenodeoxycholyltaurine (TDCG, cdc-tau)
Deoxycholic	Glycine	Glycodeoxycholate (GDC)	Deoxycholylglycine (DCG, dex-gly)
	Taurine	Taurodeoxycholate (TDC)	Deoxycholyltaurine (DCT, dex-tau)
Lithocholic	Glycine	Glycolithocholate (GLC)	Lithocholylglycine (LCG, lit-gly)
	Taurine	Taurolithocholate (TLC)	Lithocholyltaurine (LCT, lit-tau)
Ursodeoxycholic	Glycine	Glycoursodeoxycholate (GUDC)	Ursodeoxycholylglycine (UDCG, udc-gly)
	Taurine	Tauroursodeoxycholate (TUDC)	Ursodeoxycholyltaurine (UDCT, udc-tau)
Ursocholic	Glycine	Glycoursocholate (GUC)	Ursocholylglycine (UCG, urs-gly)
	Taurine	Tauroursocholate (TUC)	Ursocholyltaurine (UCT, urs-tau)
Lithocholylglycine	Sulfate	Glycolithocholate sulfate (GLCS)	Sulfolithocholylglycine (SLCG, sul-lit-gly)
Lithocholyltaurine	Sulfate	Taurolithocholate sulfate (TLCS)	Sulfolithocholyltaurine (SLCT, sul-lit-tau)

Chenic has been suggested as a useful curtailed term for *chenodeoxycholic acid*, and the combining form *chenyl-* has been used as a curtailed form for *chenodeoxycholyl-*. Abbreviations C, CDC, DC, LC, and UDC were agreed on by an ad hoc committee that met in 1975 in Aalborg, Denmark (32). The term *cholylglycine* is more accurate than the term *glycocholate* (33).

ronidated (249) or sulfated (167, 242) to any appreciable extent, with the exception of LCA, which, after amidation with glycine or taurine, is sulfated extensively in the liver (42). The names of the common conjugated bile acids present in human bile are shown in Table 1, and the chemical structures of some of these bile acids are shown in Figure 4.

A number of other biotransformations of bile acids may occur in the hepatocyte, but the magnitude of these biotransformations is much less than that of amidation. These include (reversible) oxidation and stereospecific reduction at the 3 or 7 positions (30, 79, 183, 227); in addition, additional hydroxylations of the nucleus (cf. refs. 33, 182) may also occur but to an extremely limited extent in humans. The rat and hamster, however, have an active 7-hydroxylation system for bile acids (66; see also ref. 184). In rodents, deconjugation of bile acid amidates has been shown (94); whether this occurs in humans is unknown.

BIOTRANSFORMATIONS BY BACTERIAL ENZYMES. In health, bile acids are almost completely biotransformed by intestinal bacterial enzymes before they are excreted from the body (253). These bacterial biotransformations, which are reviewed by Macdonald et al. (166), may be indicated by the appearance of the bacterial metabolites (or their subsequent hepatic metabolites) in bile. Many minor bacterial biotransformations occur (166, 252, 253) but are not detectable from analysis of biliary bile acids, either because the metabolite is not absorbed or, if absorbed, is not secreted in bile but in urine.

Bacterial biotransformations begin when the bile acids reach the bacteria-rich regions of the intestine. In humans the bacterial flora begins to increase in number and diversity in the midileum, with a marked increase in the cecum. Bile acid biotransformations occurring in the ileum are mediated by aerobic or facultative anaerobic bacteria, whereas those occurring in the cecum are probably largely mediated by strictly anaerobic bacteria. The major bacterial biotransformation occurring in both ileum and cecum is deconjugation, which results in the formation of an unconjugated bile acid and glycine or taurine. This biotransformation is catalyzed by a large number of bacteria, such as *Bacteroides*. The enzyme involved is termed a *cholylglycine hydrolase* and has recently been characterized in some detail (17, 141).

The second major biotransformation is hydrogenation at the 3 (or 7) position, i.e., oxidation of the 3-hydroxy group to a 3-oxo group (or a 7-hydroxy group to a 7-oxo group). Bacterial dehydrogenases are widely distributed. However, the extent to which bile acids

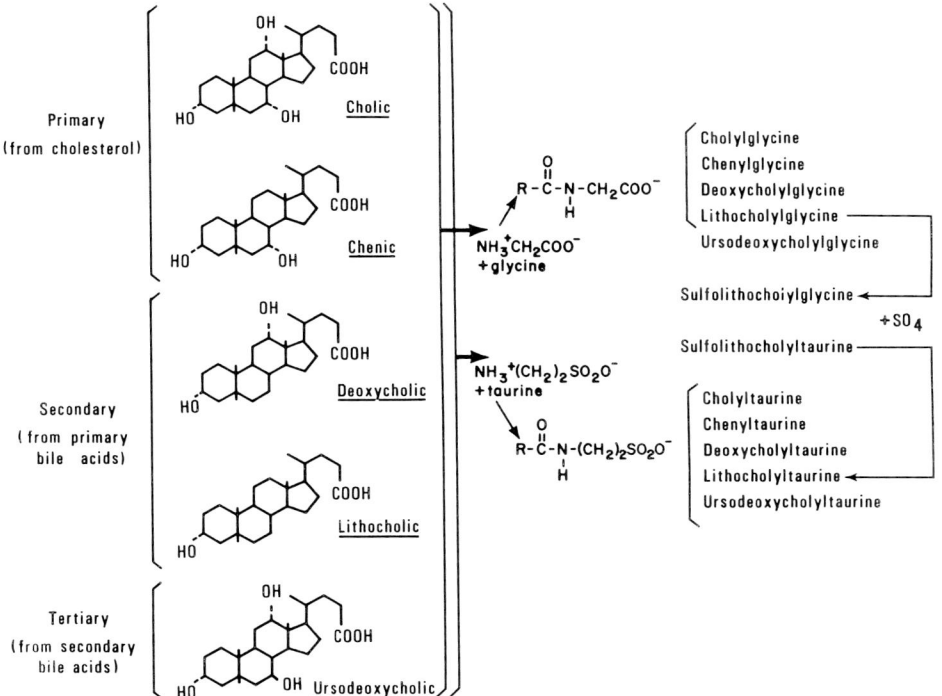

FIG. 4. Chemical structure of 5 major bile acids present in human bile. Bile acids are present as their glycine or taurine amidates; in addition, glycine and taurine amidates of lithocholic acid are sulfated. Term *chenic* is used as a curtailed name for chenodeoxycholic acid (CDCA). Hexagons denote saturated 6-membered rings of carbon atoms. *Solid lines* above the juncture of A and B rings and C and D rings correspond to methyl groups, whereas that at the bottom of the A-B ring juncture indicates a 5β-hydrogen group and that at the A and B rings are in a *cis* configuration. R, bile acid (minus its carboxyl group). The 14 bile acids in this figure comprise at least 95% of biliary bile acids in most individuals.

are dehydrogenated during enterohepatic cycling in the small intestine is not well characterized, since such oxo bile acids, although probably well absorbed from the intestine, undergo reduction during hepatic passage (30, 79, 183). The 3-oxo group of oxo bile acids may be further reduced by bacterial enzymes to a 3β-hydroxy group to form an "iso" bile acid, e.g., isodeoxycholic acid (3β,12α-dihydroxy-5β-cholanoic acid). Iso bile acids are absorbed from the intestine and are present in serum bile acids (222). They are taken up by the liver but are reepimerized to 3α-hydroxy bile acids during hepatic transport (227), presumably through a 3-oxo intermediate (251). This conversion of an iso bile acid to a natural bile acid may be considered useful, since the iso bile acids have high critical micellization concentration (CMC) values (215) and are presumably less efficient than natural bile acids in solubilizing dietary lipids.

The final major biotransformation mediated by bacteria, and one that occurs largely in the cecum and under strict anaerobic conditions, is 7 dehydroxylation to form the two most important secondary bile acids in bile. Cholic acid is 7 dehydroxylated to form DCA (3α,12α-dihydroxy), and CDCA is 7 dehydroxylated to form LCA (3α-hydroxy); DCA and LCA are the major fecal bile acids in humans (224, 253). This step is believed to proceed via a Δ^6 intermediate, one of which has recently been isolated from in vitro incubations (169), in which the conversion of CDCA to LCA was examined in detail.

Other minor biotransformations may occur, as evidenced by the great variety of bile acids found in urine (5) and feces (67, 68, 166, 224, 252, 253). Such changes include degradation of the side chain to form C_{23} nor-bile acids, dehydrogenation of the 7- or 12-hydroxy groups to form their respective 7-oxo and 12-oxo derivatives, additional nuclear or side-chain hydroxylation, and desaturation of the A, B, or C rings (for a review see refs. 166, 224). The most common biotransformations are summarized in Figure 5.

Movements of Bile Acids in Enterohepatic Circulation

Bile acids move from one space to another space in the EHC by transport across cellular membranes or by movement between organs by the flow of body fluids, such as bile, intestinal content, or blood.

TRANSCELLULAR BILE ACID TRANSPORT. Active bile acid transport by the hepatocyte and ileal enterocyte are each likely to be two-stage processes involving two carrier systems. The properties of the carriers involved in bile acid transport are discussed in detail in the chapter by Van Dyke, Lake, and Scharschmidt in this *Handbook*. Canalicular transport is extremely concentrative, as the concentration of bile acid anions in the hepatocyte is not above 50 μM, whereas the monomeric concentration of bile acids in bile is at least 1,000 μM, even if micelle formation occurs. Micelle formation is quite unnecessary for efficient canalicular bile acid transport (194).

A similar active transport system appears to be present in the terminal ileum but is less well characterized than the hepatocyte transport system (278, 280). Two distinct transport systems are likely: the first, which is present in the microvillus membrane, transports bile acids into the enterocyte and is an Na^+-coupled cotransport system; the second, which transports bile acids from the enterocyte into blood, appears to be an ion-exchange system (277). Bile acids bind to serum albumin (213) and are not sufficiently nonpolar to partition into the chylomicrons; there is no appreciable transport of bile acids from the intestine in chylomicrons by the intestinal lymphatics (73).

Passive bile acid absorption probably also contributes importantly to bile acid fluxes during enterohepatic cycling of bile acids. The bile acid molecule is too large to penetrate the tight junctions either between the hepatocytes or between the enterocytes. Accordingly, passive bile acid absorption must involve dissolution of the protonated bile acid molecule in the

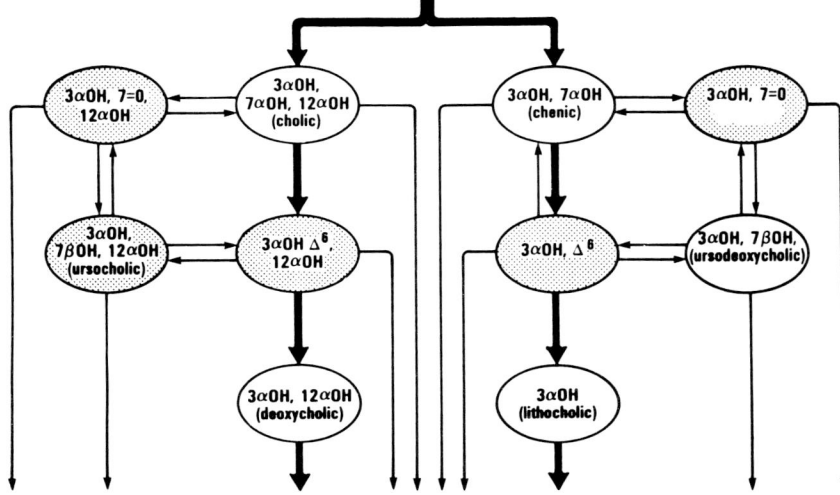

FIG. 5. Major pathways of bacterial biotransformation of cholic acid (CA) and CDCA in humans. 12-Dehydroxylation is believed not to occur. Figure does not show dehydrogenation at the 3 position or 12 position to form 3-oxo and 12-oxo bile acids, respectively. Both occur, since 3-oxo, 7-oxo, and 12-oxo bile acids are present in fecal bile acids (224).

lipid domains of the hepatocyte (266) or enterocyte (for a review, see ref. 178). However, a fatty acid–binding protein has recently been identified in liver and intestinal mucosa, and it has been proposed that this protein plays an important role in fatty acid transport (244). Because unconjugated monohydroxy and dihydroxy bile acids are quite hydrophobic and resemble fatty acids in being diffusion limited in their rate of absorption, this protein may also play a role in the passive uptake of unconjugated bile acids in the liver and small intestine (61). Blitzer et al. (31) have described a bile acid–hydroxide (bicarbonate) exchanger present in hepatocytes. The specificity of this system appears to be for unconjugated but hydrophilic bile acids, e.g., cholate. Such a transport system may also be present in the small intestine (92).

The rate of passive absorption into a cell membrane of any large uncharged molecule is directly proportional to its "hydrophobicity," which correlates highly with its distribution between an aqueous phase of the appropriate pH and a suitable immiscible, polar solvent such as n-octanol. The factors influencing n-octanol–water partition coefficients are well characterized (53, 238). Because a molecule can dissolve in the lipid part of a membrane only in its protonated form, the pH of the aqueous phase must be sufficiently acidic for *some* protonated species to exist. However, the aqueous phase pH can be as great as 2 or 3 pH units above the pK_a of a weak acid, provided the distribution ratio of the protonated compound between the oil phase and the aqueous phase is sufficiently high. For example, for an unconjugated bile acid with a pK_a of 5, at pH 8, 0.1% of the bile acid would be present in the protonated form (3). However, if the distribution ratio between n-octanol and water of this protonated bile acid were 1,000, then one would observe a distribution ratio of 1 at pH 8. Experimentally, unconjugated bile acids are readily extracted into organic solvents at pH 7, despite their pK_a value of 5.0 (264).

In addition to the requirement that some bile acids be present in protonated form, the bile acid must also not have too many polar groups, as these diminish its hydrophobicity. In experiments with perfused rat small intestinal segments, unconjugated dihydroxy and monohydroxy bile acids are absorbed quite rapidly, whereas trihydroxy bile acids, such as CA or ursocholic acid (UCA), are absorbed more slowly (21, 61, 219, 274). The amide bond of conjugated bile acids is also a polar group so that one would expect a glycine-conjugated bile acid to be absorbed more slowly than its corresponding unconjugated derivative, even for a given degree of ionization (53, 238). Finally, there are steric considerations in that changing the 7-hydroxy group from an α- to a β-configuration increases the polarity of a molecule and causes a decreased rate of passive absorption. Glycine-conjugated dihydroxy bile acids are absorbed extremely slowly at pH 7.5, but such bile acids should be absorbed passively below pH 6. The situation with respect to taurine-conjugated monohydroxy bile acids, such as lithocholyltaurine, is still unclear. The side chain of the molecule is extremely hydrophilic because of the sulfonic acid group, but the nuclear moiety is quite hydrophobic, making it difficult to predict how such molecules will behave.

These principles can now be applied to the enteral loop of the EHC. When bile acids are discharged into the small intestine, they arrive in the duodenum, where postprandial pH frequently shows spikes of acidic pH (174). The glycine conjugates of dihydroxy and monohydroxy bile acids are likely to be absorbed passively. Such absorption in the proximal small intestine is believed to be responsible for the early postprandial increase in the serum level of conjugates of CDCA (for a collation of experimental data, see ref. 178). The magnitude of passive absorption of conjugated bile acids in the proximal small intestine is not known but is probably 15%–30% of total bile acid absorption (178). The glycine conjugates of CDCA (and DCA), which are not absorbed, together with the remaining bile acid conjugates, then pass down the small intestine and are not absorbed passively because the intraluminal pH is too alkaline. They are actively absorbed in the terminal ileum. In addition, in the mid- and distal ileum, bacterial deconjugation also begins (192). Unconjugated dihydroxy and trihydroxy bile acids are absorbed passively.

FLOW OF BILE ACIDS BETWEEN ORGANS. Bile acids move between organs in the EHC because of bile flow in the biliary system, the flow of luminal content in the enteral system, and blood flow in the circulatory system (129, 178). The fate of bile acids in the intestinal lumen is dependent on small intestinal motility, since conjugated trihydroxy bile acids must be transported to the terminal ileum before absorption can occur. The postprandial peak of cholyl conjugates occurs ~2 h after a meal is ingested (157); this interval corresponds to the time required for bile acids to pass from the gallbladder into the duodenum (a rapid process) and then along the small intestine to the terminal ileum (a much slower process).

APPLICATION OF GENERAL PRINCIPLES IN HUMANS

Cholic Acid and Deoxycholic Acid

Cholic acid is formed in the hepatocyte from cholesterol and conjugated with glycine or taurine to form cholylglycine (C-Gly) and cholyltaurine (C-Tau). These amidated derivatives are then secreted into the canaliculus and enter the biliary tree. After variable storage time in the gallbladder, they enter the duodenum. Both conjugates are too polar to be absorbed in the duodenum; they are transported along the jejunum and into the ileum without appreciable absorption. During their passage through the midileum, some bacterial deconjugation occurs, forming unconjugated

CA. Results of human experiments (105, 107) suggest that C-Gly is deconjugated to a much greater extent than C-Tau. Most of the CA that is formed is absorbed. The C-Gly and C-Tau molecules that are not deconjugated are reabsorbed by active transport from the terminal ileum. The absorbed C-Gly and C-Tau molecules pass via the portal venous blood to the liver, where they are efficiently extracted (80%–90% fractional hepatic extraction). The small fraction (10%–20%) that spills over into the systemic circulation causes a postprandial increase in the serum level of cholyl conjugates ~2 h after a meal. The unconjugated CA absorbed from the intestine also passes to the liver, where the majority is removed (fractional hepatic extraction ~70%), and the remainder spills over into the circulation. Within a short time, however, all of the CA molecules are rapidly extracted by the hepatocyte, reconjugated with glycine or taurine, and secreted into bile.

The C-Tau and C-Gly that escape absorption from the terminal ileum pass into the cecum. Here both molecules are fully deconjugated, and the unconjugated CA formed in this manner plus the unabsorbed CA that entered from the ileum are fully 7 dehydroxylated by anaerobic bacteria to form DCA.

About one-third to one-half of the DCA that is newly formed is absorbed passively from the colon (106, 265). The DCA passes to the liver, where it is partially extracted (fractional hepatic extraction is probably 30%–50%). The remainder spills over into the systemic circulatory compartment, causing an elevation in the serum level of unconjugated DCA. The DCA that is extracted by the liver is conjugated with glycine or taurine to form deoxycholylglycine (DC-Gly) and deoxycholyltaurine (DC-Tau), which are secreted into bile. Conjugation with glycine or taurine in the hepatocyte may not be complete, and if unconjugated DCA enters the canaliculus, it is likely to be passively reabsorbed during its passage down the biliary tree (133, 184, 198, 282). The DC-Gly and DC-Tau are secreted into bile and pass down the biliary tract into the intestine. Some passive absorption of DC-Gly from the duodenum and jejunum may occur during spikes of acidic pH, and this absorption causes a transient, early postprandial increase in the serum level of DC-Gly. (The DC-Tau is thought not to be absorbed because it is present fully in its charged form, but this has not been proved experimentally.) The DC-Gly and the DC-Tau molecules that are not absorbed pass down the small intestine. Some deconjugation occurs in the ileum, but the majority of molecules are absorbed by active ileal transport to return to the liver and recycle repeatedly in the EHC. The degree of deconjugation of DC-Gly (106, 127) is estimated to be ~30% per meal, but that of its taurine amidates is probably lower (107). The DC-Gly, DC-Tau, and DCA molecules not absorbed from the terminal ileum pass into the cecum, where some additional passive absorption of unconjugated DCA occurs. Eventually the input of newly formed DCA from the large intestine to the liver must be balanced by fecal excretion of previously formed (conjugated) DCA molecules lost during enterohepatic cycling. These fluxes are shown in Figure 6.

Chenodeoxycholic Acid and Lithocholic Acid

Chenodeoxycholic acid is also formed in the hepatocyte from cholesterol and is conjugated with glycine or taurine to form chenodeoxycholylglycine (CDC-Gly) and chenodeoxycholyltaurine (CDC-Tau). These amidated derivatives are then secreted into the canaliculus and undergo the same general pattern of flow, transport, and biotransformation as conjugates of CA. There are, however, several minor differences between the metabolism of CDCA and CA. The CDC-Gly is absorbed passively to some extent from the duodenum or proximal jejunum during its brief moments of acid-

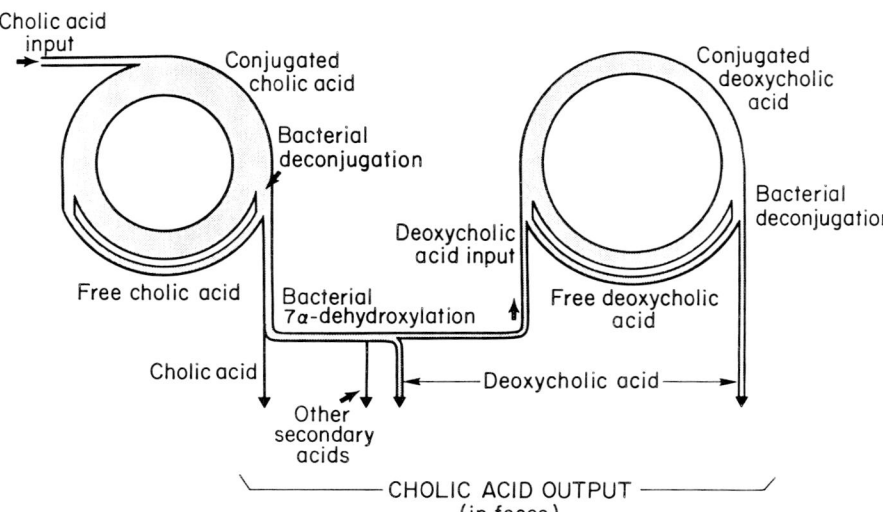

FIG. 6. Enterohepatic circulation of steroid moiety of CA and deoxycholic acid (DCA) in humans. Only the enterohepatic circle is shown. *Lower arc*, unconjugated bile acid that is formed in distal intestine, absorbed, and returned to the liver for reconjugation. CA passing into the colon is completely 7-dehydroxylated, but only a fraction of the DCA that is formed is reabsorbed from the large intestine. After its reabsorption, DCA is conjugated with glycine or taurine and conjugates then join the EHC of primary bile acids.

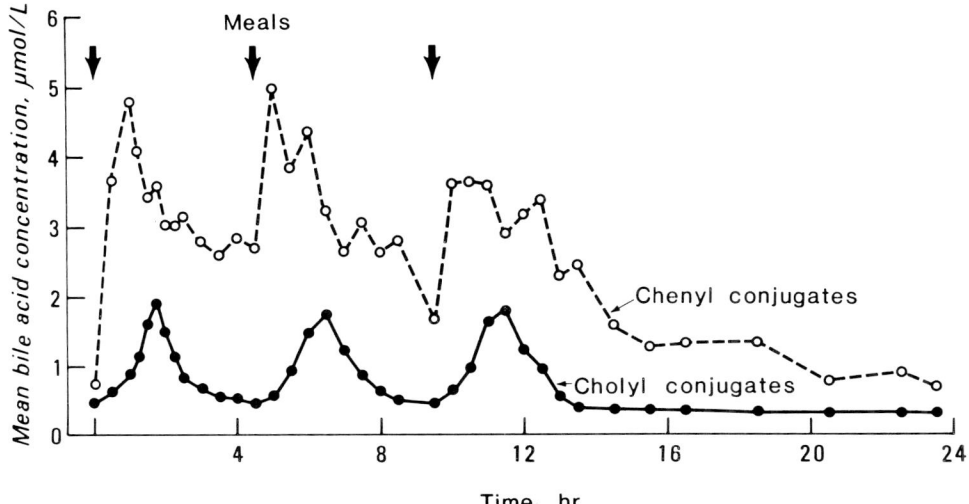

FIG. 7. Immunoreactive level of CA conjugates and CDCA conjugates (chenyl conjugates) after 3 equicaloric liquid meals in healthy volunteers. Level of CDCA conjugates rises sooner than that of CA conjugates, reflecting their earlier absorption from proximal small intestine. Delay of 2 h between meal ingestion and peak serum level of CA conjugates reflects transit time required for these bile acids to pass from the duodenum to the terminal ileum, where they are actively absorbed.

ification, whereas C-Gly is not (178). The hepatic fractional extraction of both conjugated and unconjugated CDCA is less than that of their corresponding CA derivatives, so that the systemic circulation is always relatively enriched in conjugated and unconjugated CDCA, as compared with the corresponding forms of CA (8, 72, 178). As a consequence of the earlier absorption of CDCA conjugates from the small intestine, their postprandial peak appears sooner than those of CA conjugates, as shown in Figure 7. As a consequence of lower fractional hepatic extraction of CDCA conjugates, their concentration is always greater than that of CA conjugates in the systemic circulation, as shown in Figure 8.

The small fraction of CDCA conjugates (and possibly some unconjugated CDCA formed by deconjugation) that escapes active absorption by the terminal ileum (10%/meal; 30%/day) passes into the large intestine. Here they undergo rapid deconjugation and 7 dehydroxylation to form LCA. At body temperature (135) LCA is insoluble and is also hydrophobic; it adsorbs to dietary residue as well as to the colonic bacteria, with the result that its aqueous concentration remains quite low. Probably about one-fifth of newly formed LCA is absorbed from the colon in humans (4). The absorbed LCA enters the mesenteric venous blood and becomes tightly bound to albumin. It is transported to the liver and is extracted by the hepatocyte by an unknown mechanism.

The hepatic metabolism of LCA, the major bacterial metabolite of CDCA, is unique among the natural bile acids: this monohydroxy bile acid is not only amidated with glycine or taurine during its hepatic transport but is also sulfated at the 3 position to form two

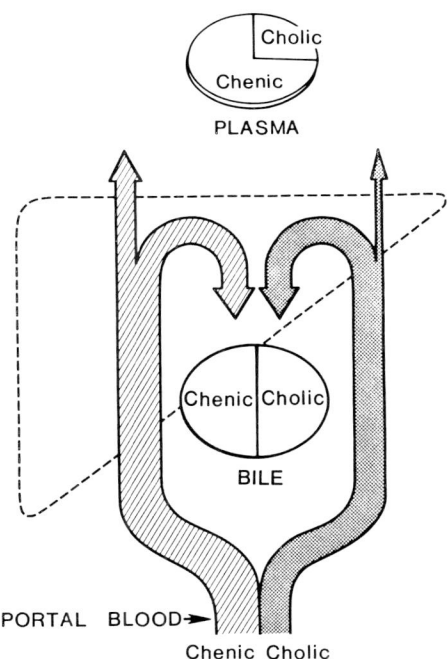

FIG. 8. Schematic depiction of hepatic uptake showing greater spillover of CDCA conjugates into systemic circulation as compared with CA conjugates. Unconjugated bile acids have a lower fractional hepatic extraction than their corresponding conjugates, so that systemic circulation is also enriched in unconjugated bile acids, as compared with their corresponding conjugated derivatives.

double conjugates, sulfolithocholylglycine and sulfolithocholyltaurine (4, 42, 199). These double conjugates are believed to be poorly absorbed from the small intestine (43), so that the sulfated lithocholyl amidates

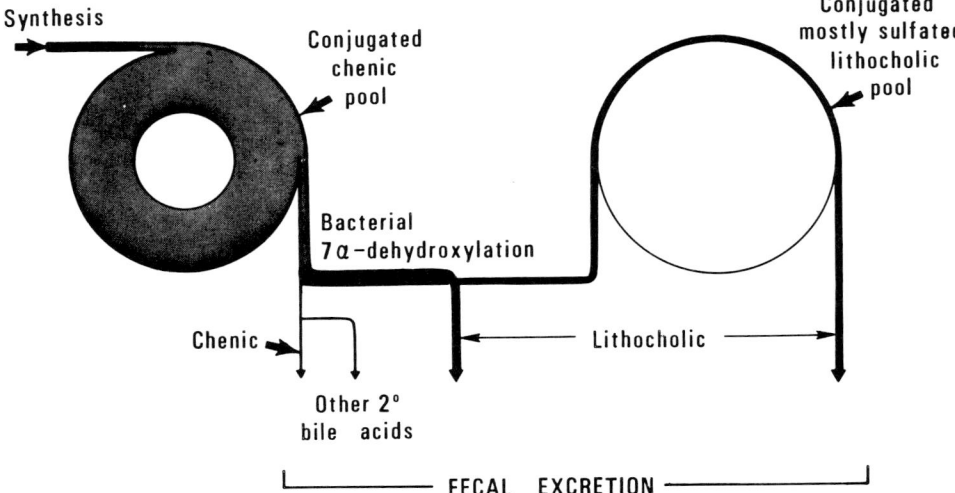

FIG. 9. Schematic depiction of EHC of CDCA (termed *chenic*) and lithocholic acid (LCA) in humans. Sulfolithocholyl conjugates are poorly absorbed from the small intestine. Figure does not show a small fraction of the amidated sulfated lithocholates that are fully hydrolyzed during colonic transit, resulting in reabsorption of a small fraction of unconjugated LCA.

that are secreted in bile are rapidly eliminated from the body. The EHCs of CDCA and LCA are summarized in Figure 9.

Other bacterial biotransformations may occur, such as dehydrogenation or epimerization at the 3 and/or 7 position. The 3- or 7-oxo derivatives, if formed and absorbed, are reduced stereospecifically during hepatic transport to form the α-hydroxy epimer. [Reduction is believed to be quite stereospecific at the 3 position (>90%) (227), but at the 7 position the reduction is less specific, with 10%–20% of the β-epimer being formed (79, 183).] If 3β-hydroxy forms are absorbed from the intestine, they are completely epimerized during hepatic transport; if 7β-derivatives are absorbed, they are believed to pass through the hepatocyte without epimerization. Thus human bile usually contains a small proportion of UDCA, which may be formed either in the liver (by hydrogenation of the 7-oxo precursor) or in the intestine. The 7β-epimer of CA, UCA, has only rarely been detected in bile (239). It may well be formed in the colon but is presumably too hydrophilic to undergo appreciable passive absorption from the limited surface area of the colonic epithelium. A simple depiction of the chemical structure of the major primary and secondary bile acids is shown in Figure 10.

DETERMINANTS OF STEADY-STATE BILIARY BILE ACID COMPOSITION IN HUMANS

The principles outlined above, which have emphasized dynamic aspects of the EHC of bile acids, may now be used to explain the steady-state biliary bile acid composition in humans. The preceding discussion has emphasized that the total EHC of bile acids is composed largely of the four individual EHCs of the two major primary and two major secondary bile acids.

Each EHC has a mass or pool whose steady-state size reflects the balance between input and loss. Each of the individual bile acid species [e.g., C-Gly (as distinguished from C-Tau)] may be considered to constitute an individual bile acid pool. Therefore it is useful to discuss the composition of biliary bile acids, considering first the steroid moiety and second the amino acid moiety.

Steroid Moiety

The two primary bile acids, CA and CDCA, have long been considered to be present in bile in about equal proportions, although more recent analyses suggest that CDCA is the most common biliary bile acid in Caucasians (for summary, see ref. 129). The synthesis rate of CDCA is about half that of CA, but its intestinal conservation is greater, as reflected in its slower turnover rate; the experimental data on relative synthesis rates of CDCA and CA have been tabulated (128, 130). The greater intestinal conservation of CDCA conjugates is probably explained by their passive absorption from the proximal small intestine.

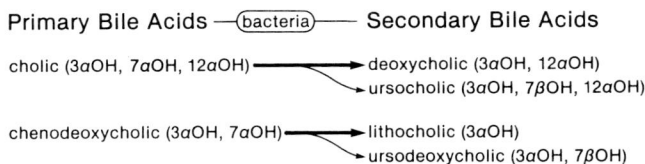

FIG. 10. Schematic depiction of major biotransformations in humans that influence biliary bile acid composition. Ursocholic acid, although probably formed to a considerable extent in the large intestine, is so hydrophilic that its proportion in bile is extremely low and usually it is not detectable.

The third most abundant biliary bile acid is DCA, whose proportion varies considerably but averages ~20%. The input of newly formed DCA into the EHC from the colon is ~30%–50% of that which is formed (106, 265). [The maximal amount that can be formed is approximately equal to the amount of CA (and conjugates of CA) that is lost from the EHC.] The small intestinal conservation of the amidates of DCA is nearly identical to that of CDCA, presumably because the physical properties of conjugates of DCA and conjugates of CDCA are quite similar. Because the intestinal conservation of conjugates of DCA is greater than that of the conjugates of CA, its proportion in bile may even exceed that of CA, its precursor (116).

In bile, LCA is present in extremely small proportions. The percent of LCA in biliary bile acids is much less than that of DCA because of at least three factors: *1*) less LCA is formed than DCA because the synthesis of its precursor, CDCA, is only 50% that of CA, the precursor of DCA; *2*) the fraction of LCA formed that is absorbed from the colon (~20%) is less than that of DCA (30%–50%) (4); and *3*) the sulfated lithocholyl conjugates formed in the liver are poorly conserved from the small intestine (43). Thus, when compared with the other natural bile acids, LCA has a smaller input and a much greater turnover rate. A depiction of the enteral loop of the four major bile acids is shown in Figure 11.

Amino Acid Moiety

For a given steroid moiety the proportion of bile acids conjugated with glycine (or taurine) also reflects the balance between input (from hepatic amidation) and loss (by bacterial deconjugation and also by lack of intestinal absorption). If bacterial deconjugation of one group of conjugates is less (e.g., taurine-conjugated bile acids), that class of amidates tends to accumulate in the EHC. The studies of Hepner et al. (105–107) suggested that glycine conjugates of all of the major biliary bile acids are deconjugated much more rapidly during enterohepatic cycling than the corresponding taurine conjugates. This observation should mean that the EHC is relatively enriched in taurine conjugates. Because the normal proportion of bile acids conjugated with glycine is 60%–90%, it would seem that hepatic conjugation favors glycine to an even greater extent, i.e., 75%–95%, since the amino acid moiety of taurine conjugates, when formed, remains in the EHC

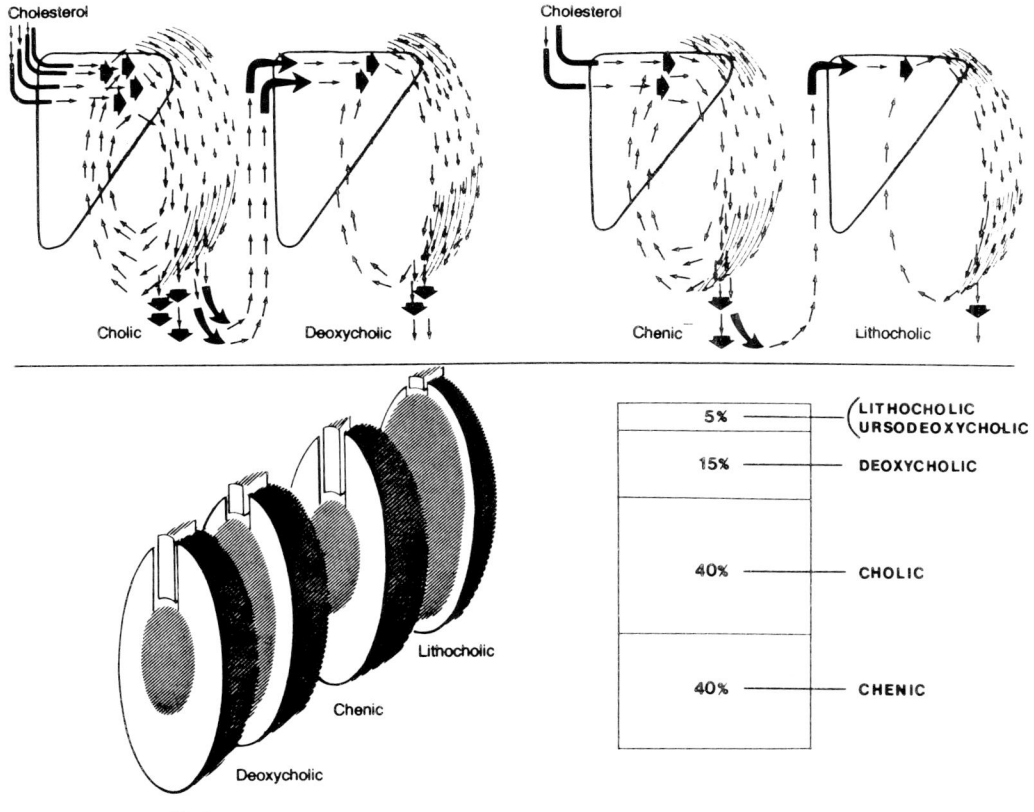

FIG. 11. Schematic depiction of EHC of 4 main bile acids present in human bile. Steady-state composition of biliary bile acids represents relative balance between input and intestinal conservation for each bile acid.

for a longer period than that of glycine-conjugated bile acids (107, 115, 116). The conjugation, deconjugation, and reconjugation of the three major bile acids of human bile are shown in Figure 6.

There is considerable intersubject variation in the proportion of bile acids conjugated with taurine, and it is likely that this proportion does not even remain constant throughout the day. Two factors contribute to this variability. *1*) Whether the liver conjugates bile acids with glycine or taurine depends on the momentary availability of taurine in the hepatocyte (95, 96). *2*) The degree of bacterial deconjugation may vary from person to person and for a person from meal to meal. In humans, taurine cannot be appreciably synthesized, and the diet is the predominant source of blood taurine to be transported into the hepatocyte. Accordingly the ingestion of a meal containing taurine (meat) or taurine supplementation in the diet causes an instantaneous increase in the proportion of bile acids conjugated with taurine (95).

Some animals have exclusively taurine-conjugated bile acids in bile (many carnivores but also some herbivores, including birds), whereas other animals have exclusively glycine-conjugated bile acids (101); still other animals have a mixture of glycine and taurine amidates. The transferase, whenever studied, conjugates bile acids preferentially with taurine, if it is available (146, 268, 270). Thus whether an animal conjugates its bile acids with glycine or taurine depends on the availability of taurine to the hepatic amino acid transferase, which in turn is probably largely determined by dietary taurine (245).

METHODS FOR CHARACTERIZING ENTEROHEPATIC CIRCULATION OF BILE ACIDS

The principles of the most common methods for characterizing the EHC of bile acids are discussed briefly, since most have been reviewed in detail elsewhere (for a tabulation, see ref. 121).

Bile Acid Pool Size

The mass in the exchangeable bile acid pool may be measured by the conventional technique of isotope dilution (163). In this technique a tracer dose of radioactive bile acid is administered (usually intravenously but occasionally orally) and the specific activity of the administered bile acid in the bile acid pool is determined repeatedly after the injected bile acid has had adequate time to mix with the bile acid pool. The method involves repeated sampling of bile (163), serum (51, 240), or intestinal content (117). Principles of this method, pitfalls, and utility have been thoroughly reviewed (60, 128, 130). An improvement in the technique by Vantrappen et al. (267) permits determination of pool size and fractional turnover rate from a single bile or serum sample. The pool size may also be determined by a washout technique, in which the pool is equated with the amount of bile acids recovered in a short time after an acute biliary fistula is inserted (57).

Bile Acid Biosynthesis or Input

BIOSYNTHESIS RATES OF PRIMARY BILE ACIDS. In the steady state, fecal bile acid excretion is essentially equal to bile acid synthesis from cholesterol, since the diet usually contains only trace amounts of bile acids. Consequently, bile acid synthesis can be determined by measuring fecal bile acids chemically (cf. refs. 88, 224, 253).

There are three great methodological problems, however. *1*) It is difficult to collect stools quantitatively, and fecal recovery markers (50) must be used to correct for incomplete collections. *2*) In the subsequent analysis of fecal homogenates, it is difficult to be certain that any extraction procedure used will provide full recovery of all bile acids. *3*) There is no simple analytical method for measuring fecal bile acids. The only accurate method is capillary gas chromatography coupled to mass spectrometry for compound identification; this technique requires extraordinarily expensive instrumentation and is immensely tedious (223, 224). If such a method is used, each fecal bile acid can be identified. The peaks can then be assigned to the CA family or the CDCA family, and the daily rate of synthesis of each of these bile acids determined.

Total bile acid synthesis may also be measured by using a suitable ^{3}H-labeled precursor whose ^{3}H is lost into body water when the compound is converted into a bile acid. N. Carulli and his colleagues (26a) have used [7β-^{3}H]cholesterol and Davidson et al. (48) have used [24,25-^{3}H]cholesterol. Bile acid synthesis may also be measured by labeling the cholesterol pool with [26-^{14}C]cholesterol and measuring $^{14}CO_2$ in breath, which is formed when the cholesterol side chain is oxidized to form bile acids (207).

INPUT RATES OF SECONDARY BILE ACIDS. The absorption of newly formed secondary bile acids from the large intestine corresponds to an "input" and is similar to the de novo synthesis of primary bile acids. In humans the metabolism of DCA may be described with a single pool model, so that it is possible to measure the exchangeable DCA pool size and the input of DCA into this pool with the conventional Lindstedt method (63, 127). The value obtained represents the input of newly formed DCA, which is some fraction of the DCA that is formed in the large intestine (115). The decay in the specific activity curve of DCA is caused by the input of newly formed DCA molecules that have been formed in the distal intestine by bacterial dehydroxylation of CA. If labeled CA is given, the label appears in DCA in a conventional product-precursor relationship [Fig. 12; (131)]. It is useful to define the fraction, $F_{dehydroxy}$, which is defined as the input of DCA divided by the maximum amount of DCA that could be formed (115, 265).

When the exchangeable pools of CA, CDCA, and DCA are labeled with a tracer, each behaves as a single well-mixed compartment; LCA, however, does not, because the specific activity decay curve is biexponential. The input of LCA can be calculated by a complex mathematical procedure (4).

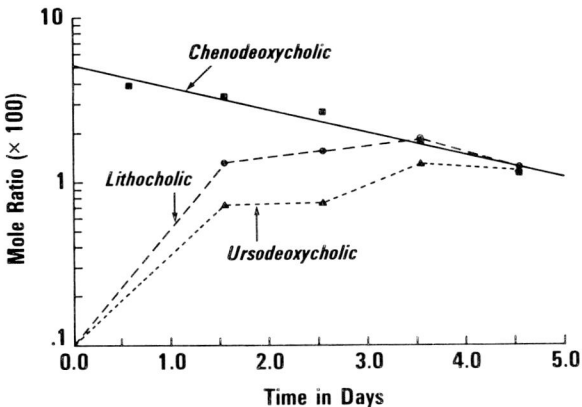

FIG. 12. Atoms percent excess (unit for stable isotopes that is comparable with specific activity for radioactive isotopes) of biliary bile acids after administration of [24-^{13}C]chenodeoxycholic acid to a healthy volunteer. LCA and ursodeoxycholic acid (UDCA), which are formed from CDCA, slowly appear in bile in labeled form. Atoms percent excess of CDCA declines exponentially, indicating that CDCA metabolism (as that of CA and DCA) can be described by a single-pool model.

Measurement of Bile Acid Fluxes

FLUXES DUE TO INTERORGAN FLOWS. Fluxes of bile acids due to flow of bile, intestinal content, or blood can be measured with indicator-dilution techniques (119, 162). Another method for measuring biliary bile acid secretion involves sampling (by controlled interruption) of the EHC, in such a manner as to cause minimal disturbance of the flux of bile acids in the EHC (56, 153, 236, 259).

Methods for measuring gallbladder contraction involve imaging techniques, such as real-time ultrasound (70, 139), or the use of the γ-camera together with administration of ^{99}Tc-labeled compounds that are stored in the gallbladder (151). It is also possible to measure the fraction of hepatic bile secretion that is stored in the gallbladder by infusing a biliary recovery marker (indocyanine green) at a constant rate intravenously and simultaneously measuring output into the duodenum [Fig. 13; (22)]. A duodenal output less than the intravenous infusion rate indicates gallbladder storage; when output into the duodenum exceeds the intravenous infusion rate, gallbladder emptying is occurring. This technique is being refined to give absolute rates of gallbladder storage and emptying. A combination of the biliary recovery marker method and radionuclide imaging may be used to give absolute flow rates into and out of the gallbladder (156).

The discharge of bile acids into the duodenum is also measured by indicator-dilution techniques. The most widely used method features measurement of

FIG. 13. Use of biliary recovery marker to measure gallbladder storage and emptying in humans. Indocyanine green, a compound quantitatively secreted into bile, is infused intravenously at a constant rate. *Left panel*: in absence of gallbladder storage, duodenal output (O_d) equals parenteral input (I_p). *Center panel*: when duodenal output is less than parenteral input, gallbladder storage is occurring. *Right panel*: when duodenal output exceeds parenteral input, gallbladder emptying is occurring. [From Berge Henegouwen and Hofmann (22).]

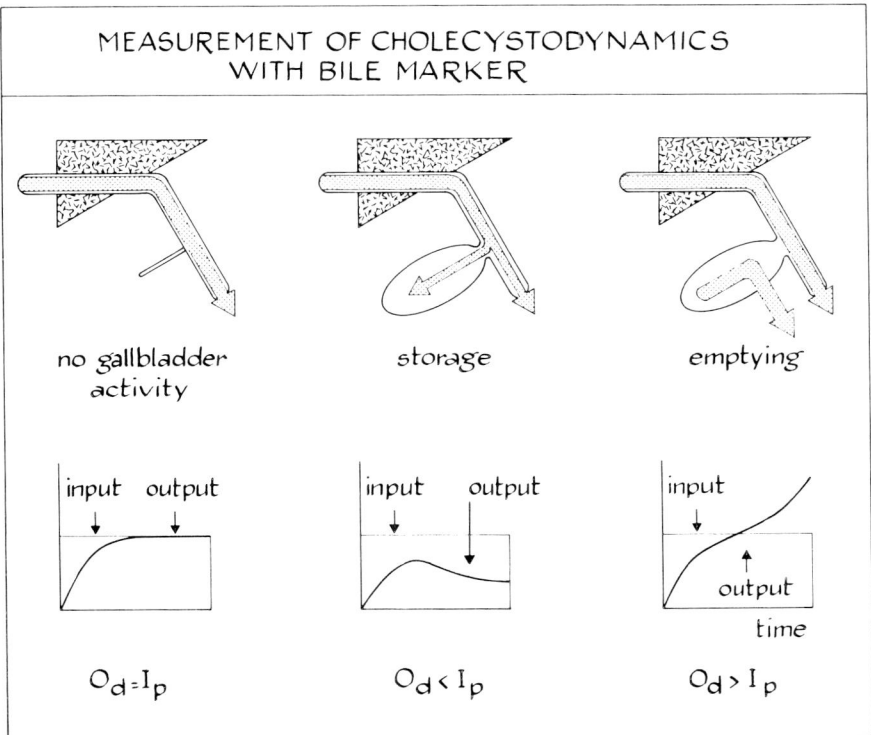

biliary secretion of bile acids into the duodenum during steady-state conditions, which are achieved by the continuous infusion of a liquid test meal into the small intestine (71, 91, 210). The digestive products of the meal stimulate cholecystokinin release so that the gallbladder remains nearly completely contracted throughout the procedure; this permits the duodenal input of bile acids to be equated with hepatic secretion. The technique has also been used under non-steady-state conditions to measure bile acid output during a normal eating pattern and overnight fasting (191). Both methods give similar results, but neither method is very precise.

Historically the bile acid pool size was measured by isotope dilution, an assumption was made about the number of times that this pool cycled per day, and the product was calculated to give a figure for bile acid secretion. When indicator-dilution methods were developed for measuring bile acid secretion, investigators divided bile acid secretion by the pool size to arrive at cycling frequency. Whether the unit *cycling frequency* has physiological meaning continues to be debated. If an analogy with blood circulation is made, cycling frequency is similar to the cardiac output divided by the blood volume, a term that is never used.

FLUXES DUE TO CELLULAR TRANSPORT. To measure bile acid fluxes from the intestinal lumen into the enterocyte in humans, intestinal perfusion studies (cf. refs. 150, 176) that use a reference marker are useful. Investigators must decide whether they wish to trace the absorption of endogenous bile acids during the fasting state or digestion or to measure the absorption of bile acids from a solution initially having a fixed concentration and pH. Such studies give useful relative values but may not accurately estimate true membrane permeability, since the absorption rate is influenced not only by the thickness of the unstirred layer but also by the mucosal surface area, neither of which can be accurately measured (for a discussion of the problem, see ref. 113). It is possible to use this technique to measure relative rates of absorption, and it may even be possible to measure approximate membrane permeability coefficients by including in the perfusate a substance for which absorption rate is diffusion controlled (cf. refs. 61, 279). The problems in quantitating absorption are well covered by Ho et al. (113), and a rather theoretical treatment has been presented by Amidon et al. (7).

To measure the flux of bile acids into the colon, a multilumen tube can be positioned so that an aspiration port is located close to the ileocecal valve. A marker may be infused proximally, and with suitable calculations the amount of bile acids passing into the colon can be measured (83). If fecal bile acids are measured at the same time, an estimate of colonic absorption can be made.

For measuring hepatic uptake of bile acids in humans, two techniques have been used. One technique used frequently is to measure the average hepatic uptake over the duration of absorption by comparing the integral of the plasma concentration time curve (AUC, area under the curve) in peripheral blood after intravenous administration with that after oral administration (21). (The following equation is used: first-pass extraction = $[1 - (AUC_{oral}/AUC_{iv})] \times 100$. This equation was derived by making a number of assumptions, not all of which may be true for bile acids. The extraction of bile acids is so rapid that hepatic uptake may occur before complete mixing in the plasma space occurs. However, the magnitude of this error is probably small.) Data obtained using this technique are illustrated in Figure 14. A second approach, which is much more difficult technically, is to measure the concentration of bile acids in portal venous and hepatic venous blood so that moment-by-moment hepatic uptake can be measured. Valuable studies using this methodology have been performed by Angelin et al. (8) and Ewerth et al. (72). Results obtained by the two methods are in satisfactory agreement.

For measuring bile acid uptake by the liver in animals, it is customary to use the indicator-dilution technique described by Goresky (86), in which the test substance and a nonabsorbable marker are injected simultaneously into the portal blood and samples are taken rapidly from the hepatic vein. From the concentration profile of test substance and marker, unidirectional first-pass uptake can be calculated. This procedure is best done by achieving a steady state of uptake of the test substance and then superimposing a tracer dose on this.

Measurement of unidirectional uptake and unidirectional reflux may also be determined using the single-pass perfusion methods (200). Such techniques may be performed in the conventional direction (prograde) or retrograde. In the latter case the pericentral cells are exposed to the highest concentration of test substance, and the pattern of hepatic secretion and/or biotransformation of bile acids may be changed (cf. ref. 18).

Net hepatic uptake may be measured using the isolated perfused liver with or without the presence of albumin in the perfusate. Luxon et al. (165) have proposed that the recycling perfusate system can be used provided the test substance is infused into the perfusate at a rate equal to the removal rate to maintain a relatively constant concentration gradient down the sinusoid.

FLUXES DUE TO BIOTRANSFORMATION. To define hepatic biotransformation of bile acids, the bile acid must be prepared in radioactive form, either with ^{14}C or ^{3}H. If a ^{3}H label is used, it must be located at a position that is stable during hepatic transport. The labeled bile acid is presented to the isolated perfused liver at a physiological dose. Radioactive bile acids are isolated from bile, and their identity is inferred by

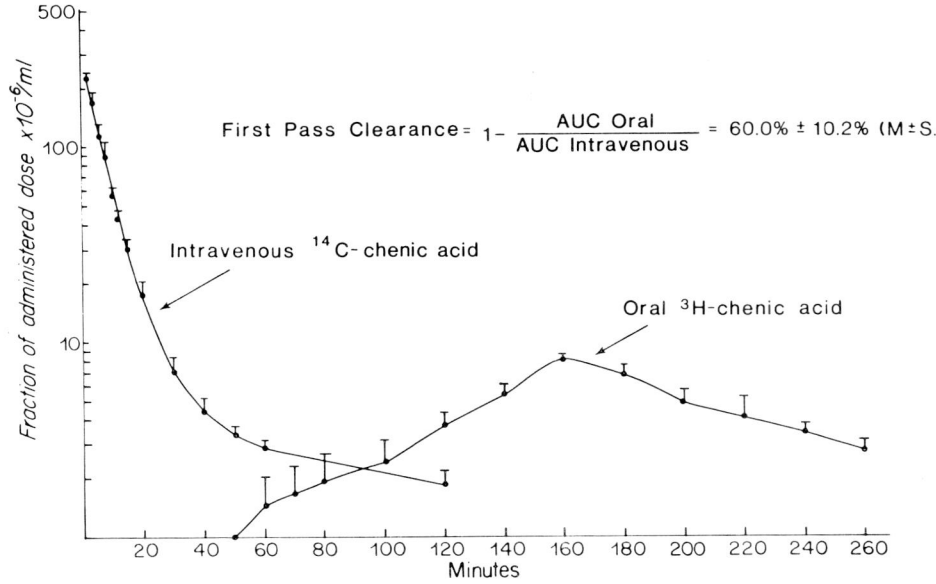

FIG. 14. Plasma disappearance of intravenously administered [^{14}C]CDCA (chenic acid) and plasma appearance of orally administered [^{3}H]CDCA in a healthy volunteer. Area under the curve after oral administration is only 40% of that after intravenous administration, permitting calculation of first-pass clearance to be ~60%. Figure has been confirmed by direct venous sampling (72).

chromatographic methods such as thin-layer chromatography, gas-liquid chromatography, and high-pressure liquid chromatography (cf. ref. 182). Tentative structural identification can be made by use of group-specific chemical reactions after chromatographic separation, but definitive identification usually requires mass spectrometry (65, 233) together with other spectroscopic techniques such as nuclear magnetic resonance spectroscopy (275). Bacterial biotransformation is usually assessed by in vitro incubations of radioactive bile acids with isolation of the radioactive products. Nonradioactive bile acids can be used, especially with pure bacterial strains (196, 197).

To characterize the biotransformation in individual cell fractions, conventional biochemical and pharmacological techniques are used (155).

All of these techniques indicate what can occur but do not necessarily indicate what does occur in the intact animal. To obtain such information, one can administer the compound to the intact animal and sample bile (in a nonperturbing manner) or intestinal content or feces. Hepatic biotransformation of bile acids has been defined in humans by giving the bile acid intravenously and measuring the chemical form appearing in duodenal bile under experimental conditions in which gallbladder storage is prevented by continuous cholecystokinin administration (42). Cholecystectomized subjects (or animals without a gallbladder) may also be used. To estimate the flux of bile acids due to biotransformation, deductions based on models of the EHC are used with analytical measurements (cf. refs. 105–107).

Regulation of Enterohepatic Circulation

REGULATION OF BILE ACID SYNTHESIS: NEGATIVE-FEEDBACK HYPOTHESIS. It has been known for about 30 years that when an acute biliary fistula is fashioned in an animal, a predictable sequence of events occurs, as first described by Eriksson (69). First, the bile acid pool containing its constituent primary and secondary bile acids is eliminated. When the pool is drained, the only bile acids then secreted are those bile acids currently being synthesized, which by definition are primary bile acids. With time, however, there is a gradual and striking increase in synthesis, so that by 2 or 3 days after creation of the bile fistula, the bile acid output will increase 4–10 times, and only primary bile acids are secreted. In adult humans, maximal synthesis rate is probably 2–6 g/day (5–12 mmol/day) (134), and usual synthesis rates are ~0.5–1 mmol/day.

Bile acid biosynthesis was also markedly increased in patients with bile acid malabsorption because of ileal dysfunction (89, 134) and severalfold increased in patients ingesting cholestyramine (82). In animals the infusion of conjugated bile acids returns the elevated biosynthesis rates toward normal (25, 180). The feeding of primary bile acids causes a decrease in primary bile acid biosynthesis in the rodent (180, 226) and in humans (47, 72, 158, 160). These observations led to the "negative-feedback hypothesis," namely, that the rate of return of bile acids to the liver had a negative-feedback effect on bile acid biosynthesis.

The entire problem of negative feedback of bile acid biosynthesis of bile acids by their transhepatic flux is being actively investigated and abounds with controversy. In cell cultures, added bile acids do not suppress cholesterol or bile acid biosynthesis (49, 152). Attempts to repeat older in vivo experiments have not always succeeded, and there are reports that in biliary fistula animals, the infusion of bile acids may not suppress bile acid biosynthesis (109, 110). A search for a hormone regulating bile acid biosynthesis gave negative results (114).

REGULATION OF BILE ACID POOL SIZE. The size of the bile acid pool reflects the balance between input and intestinal conservation. The major factor regulating the pool size is active ileal absorption, and ileal transport appears to be nearly saturated even in healthy individuals. Bile acid ingestion does not cause a consistent increase in total bile acid pool size, because ileal transport cannot be increased by bile acid feeding (cf. refs. 158, 160, 186). In contrast, loss of ileal transport function leads to a compensatory increase in synthesis; if this compensatory increase is sufficient to compensate for decreased intestinal return, there is no change in the fasting-state bile acid pool size, even though it decreases during the day and is restored during overnight fasting (116, 157). If intestinal transit is increased by drugs, the pool size decreases, cycling frequency increases, and secretion rate (the product of the two) remains relatively constant (59, 100). The one remarkable defect in regulation of the bile acid pool size appears to occur in some patients with cholesterol gallstones. In such patients, the pool size is decreased, bile acid secretion is probably decreased, yet there is no compensatory increase in bile acid biosynthesis (210, 225, 271).

PERTURBATIONS OF ENTEROHEPATIC CIRCULATION

Disturbances of the EHC usually involve *1*) alterations in input, *2*) alterations in interorgan flow, *3*) alterations in cellular transport, or *4*) alterations in biotransformation. Combinations may occur, especially in liver disease.

Alterations in Input

DECREASED INPUT. *Defective or decreased biosynthesis of primary bile acids or genetic defects in bile acid.* Alterations in input of bile acids into the EHC may be qualitative or quantitative. Qualitative changes are inborn errors of bile acid biosynthesis. It seems likely that the total inability to synthesize a water-soluble derivative of cholesterol is a fatal mutation. This reasoning is based on the fact that the currently recognized genetic defects of bile acid biosynthesis involve a defect in synthesis of only one of the two primary bile acids. Genetic defects in CDCA synthesis or CA synthesis are summarized in the preceding chapter by Hofmann in this *Handbook*.

Subtle quantitative defects in bile acid biosynthesis must occur in a variety of disease conditions and may be important. For example, in many patients with cholesterol gallstones, bile acid synthesis appears to be inexplicably decreased (210, 225, 271). Cholic acid synthesis is preferentially decreased in patients with cirrhosis (272).

Decreased input of secondary bile acids. In the germ-free animal, secondary bile acids are not formed. The major secondary bile acids, DCA and LCA, are less water soluble and more hydrophobic than their precursors; these changes in physicochemical properties result in a marked fall in the aqueous concentration of bile acids in the large intestinal lumen (in the healthy human) (118, 175). In contrast, in the germ-free animal there is an increase in the aqueous concentration of luminal bile acids (93); such animals have diarrhea that responds to the administration of cholestyramine, a bile acid–binding resin (10).

Decreased input of secondary bile acids may result from decreased formation of secondary bile acids or decreased absorption of secondary bile acids. Decreased dehydroxylation of bile acids may be induced by cecectomy (193, 280), the site of dehydroxylation, and probably by antibiotic administration, at least in some species (99, 126). Decreased absorption of newly formed DCA may be caused by intraluminal binding, for example, to fiber or resins (170, cf. 207), or by decreased colonic mucosal surface.

In humans there is considerable circumstantial evidence to suggest that the absorption of DCA is undesirable, since enrichment of biliary bile acids may at least under some conditions cause bile to become supersaturated in cholesterol (38), and there is an association between an increased proportion of DCA in bile and the presence of cholesterol cholelithiasis (170). Were humans to have an efficient 7-rehydroxylation system in the liver (as has been noted in the hamster), DCA would not accumulate in the EHC. In animals, little is known about the significance of the input of secondary bile acids. The rabbit appears unique among common mammals in that most of its circulating bile acids are secondary, and indeed the EHC of secondary bile acids is likely to inhibit the biosynthesis of primary bile acids from cholesterol. In the rhesus monkey (276), baboon (179), and rabbit (40, 76), administration of CDCA [or UDCA in the rhesus monkey (218) and rabbit (40)] causes the accumulation of LCA in the EHC and consequent hepatic inflammation and cirrhosis. Such does not occur in humans when CDCA or UDCA is administered because of the ability of the human liver to sulfate LCA and thus detoxify it (4, 77, 138).

INCREASED INPUT. *Increased biosynthesis of bile acids.* Increased biosynthesis of bile acids occurs in type IV hyperlipidemia but is not considered to have any clinical significance from the standpoint of bile acids (62).

Bile acid administration. The effect of unconjugated bile acid feeding on the steady-state composition of bile acids on the EHC has been studied rather carefully in humans and to some extent in animals. In every instance of administration of the major natural biliary bile acids in humans, the bile acid pool has become enriched in the administered bile acid or its principal biotransformation products or both. The ratio of the administered bile acid to that of its biotransformation product depends on the relative rates of input and efficiency of intestinal conservation. In addition there are interactions with endogenous bile acids. Expansion of the bile acid pool by bile acid feeding may suppress hepatic synthesis of primary bile acids either selectively or nonselectively, and the secretion of large amounts of exogenous bile acids into the intestine results in competition for active intestinal absorption between the administered bile acid and endogenous bile acids, since ileal transport is close to "saturation" during digestion. Consequently the pool size of the administered bile acid and its metabolites may expand and the pool size of endogenous bile acids may shrink (2, 47, 158, 160, 186). However, interpretation of the mechanism of expansion or shrinkage of the pool size of any bile acid cannot be made with certainty solely from an examination of biliary bile acid composition. To define the mechanism by which bile acid administration alters the proportion of endogenous bile acids in the biliary bile acids, it is necessary to perform a formal kinetic study of endogenous bile acid metabolism.

The effects of feeding individual bile acids in humans are summarized briefly. When DCA is fed, the bile acid pool becomes predominantly DCA, not only because DCA is well absorbed and well conserved but also because the synthesis of CA and CDCA is suppressed (2, 160). When CA is fed, biliary bile acids become composed of about equal parts of CA and DCA, although in some patients the proportion of DCA becomes larger than that of CA (9, 38, 158). This change in biliary bile acid composition is explained by efficient absorption of CA, as well as efficient absorption of DCA, its bacterial metabolite. There is greater intestinal conservation of the conjugates of DCA than those of CA; in addition, endogenous CDCA synthesis is suppressed. Thus DCA may become the predominant biliary bile acid. The increase in DCA in biliary bile acids obtained during CA feeding may be diminished by the simultaneous administration of antibiotics such as ampicillin (38).

When CDCA is fed to humans, there is a dose-related enrichment in the proportion of CDCA in biliary bile acids. At doses >40 μmol·kg^{-1}·day^{-1} (15 mg·kg^{-1}·day^{-1}), the proportion of CDCA in biliary bile acids may exceed 90% (47, 138). There are several reasons for this remarkable change in biliary bile acid composition: *1*) CDCA is efficiently absorbed, so that probably all of the administered dose is absorbed from the small intestine; *2*) CDCA administration suppresses the endogenous synthesis of CA; *3*) CDCA administration causes a more rapid turnover of the CA pool, presumably because of competition for active ileal transport; and *4*) although CDCA administration probably causes increased input of LCA into the EHC from the colon, the LCA is efficiently sulfated and does not accumulate in the EHC. When UDCA is fed, the biliary bile acids become enriched in UDCA but the enrichment, which with increasing dosage plateaus after about half of the biliary bile acids are replaced by UDCA, is clearly less than that obtained when feeding a similar dose of CDCA. The lower enrichment of the biliary bile acid pattern by administered UDCA has several explanations: *1*) UDCA does not suppress primary bile acid biosynthesis to the same extent as CDCA, *2*) UDCA conjugates may not compete as effectively for active ileal transport as CDCA conjugates, *3*) UDCA at high doses is probably not fully absorbed because of its physicochemical properties, and *4*) UDCA may be converted to CDCA by intestinal bacteria.

In animals the same principles apply, but there are some complexities. *1*) Bile acid feeding may induce anorexia so that pair feeding (or gavage) is necessary to be certain that the dose intended to be ingested is actually ingested. *2*) The administered bile acid or its bacterial biotransformation product may be hepatotoxic so that liver injury confounds the results; the most remarkable example of this is the accumulation of LCA in rabbits, rhesus monkeys, and baboons that were fed CDCA. *3*) Certain species are capable of efficient 7 hydroxylation of administered bile acids or their actual metabolites. Thus administration of UDCA to the hamster results in increased CDCA in bile, presumably because the administered UDCA is converted to LCA, which is 7α-hydroxylated to CDCA in the liver (237). The effect of bile acid administration in experimental animals on either endogenous bile acid synthesis or intestinal conservation of circulating bile acids has received relatively little attention.

Under some circumstances an administered bile acid may not accumulate in the EHC. Thistle and Schoenfield (256) noted that administered hyodeoxycholic acid did not accumulate in biliary bile acids when administered to women with gallstones. An explanation for this puzzling finding was not provided until many years later when Sacquet et al. (217) found that hyodeoxycholic acid is efficiently glucuronidated in humans; the glucuronide is not conserved by ileal absorption and there is also a considerable urinary loss (217). Cohen et al. (40) observed that nor-UDCA, the C_{23} homologue of UDCA, did not accumulate in biliary bile acids when administered to the rabbit. They showed that nor-UDCA was conjugated with glucuronate in the liver and not amidated; presumably the glucuronide was excreted in bile and not reabsorbed from the small intestine.

Alterations in Interorgan Flow

Two common perturbations in the EHC result from changes in interorgan flow. Both are anatomical. In portal-systemic shunting, portal blood passes directly into the systemic circulation, bypassing the liver. In defects of gallbladder storage and emptying, the gallbladder has a defect in delivering its content of bile acids; in the limiting case, it is absent.

PORTAL-SYSTEMIC SHUNTING. A portal-systemic shunt may be fashioned surgically either for research purposes or in the clinical setting to relieve portal hypertension. In cirrhosis a portal-systemic shunt is usually associated with impaired hepatic function. The effect of portal-systemic shunting per se on the EHC is merely to increase the level of bile acids in the systemic circulation. With normal hepatic function, the effect is not marked based on a computer simulation. If there is a compensatory increase in arterial blood flow so that total hepatic blood flow is unchanged, the increase in peripheral plasma levels is about fourfold (C. Cravetto, A. F. Hofmann, G. Molino, G. Belforte, and B. Bona, unpublished observations). Obviously the increase in systemic plasma levels will be greater if *1*) hepatic arterial blood flow is not increased in a compensatory manner and *2*) there is associated hepatic dysfunction. Whether such an increase in peripheral levels of bile acids has important pathophysiological consequences, e.g., increased red cell hemolysis, is not known.

GALLBLADDER MOTILITY AND STORAGE DISTURBANCES. Gallbladder contractility may be impaired by intrinsic gallbladder muscle disease (or scarring) or loss of the neurohormonal stimuli. Gallbladder compliance may be impaired by gallbladder muscle disease (or scarring). Removal of the gallbladder is a common surgical procedure; some humans are born with congenital absence of the gallbladder; a number of animal species, such as the rat, horse, and whale, do not have a gallbladder (221).

In humans, loss of the hormonal stimulus to gallbladder contraction occurs in patients with celiac sprue, in whom there is flattening of the intestinal villi. Cholecystokinin release is impaired, and the gallbladder contracts poorly after a meal is ingested (54). The exchangeable bile acid pool enlarges greatly, and the increased mass of bile acids is stored in the sluggish gallbladder (164). Impaired gallbladder motility has also been claimed to be present in gallstone patients (78).

In the absence of a gallbladder, bile acids must be stored in the small intestinal lumen. In cholecystectomized patients, the pattern of response of serum bile acid to a meal differs consistently but only slightly from those of control subjects. Fasting levels are slightly higher, and after meals the level takes longer to return to its premeal level (157). Thus cholecystectomized patients appear to be able to store their bile acid pool in the small intestine without difficulty.

There have been relatively few studies on the effect of cholecystectomy on the EHC in experimental animals, but in the prairie dog, cholecystectomy causes a decrease in the size of the bile acid pool and an increase in the proportion of secondary bile acids in the EHC (187), changes similar to those reported to occur in humans after cholecystectomy (6, 104, 206).

Alterations in Transport

DEFECTIVE HEPATIC TRANSPORT. Alterations in hepatic transport involve defective uptake and defective canalicular secretion. These could occur as genetic or developmental defects or be induced by drugs or disease. No genetic abnormalities of bile acid transport have been identified. A genetic defect in canalicular secretion of bile acids should present as intractable cholestasis. In the newborn infant and the developing rat, systemic levels of bile acids are high, suggesting inefficient hepatic uptake (14).

Bile acid analogues can be shown to inhibit bile acid uptake by using isolated hepatocytes (97), but elevated bile acid levels due to pharmacological inhibition of bile acid uptake have not been identified. Defective canalicular transport of bile acids is evident for the monohydroxy bile acid LCA (177) or its taurine amidate LCATau (142), but it is not clear whether the resulting cholestasis is caused by a biochemical inhibition of bile acid–independent flow (147) or frank swelling of the periportal cells that occludes the canaliculi, obstructing bile acid–independent flow that may originate in the pericentral cells (19). Obviously bile acid uptake and canalicular secretion of bile acids are impaired in severe liver disease. However, cell necrosis may also occur so that bile leaks from dam-

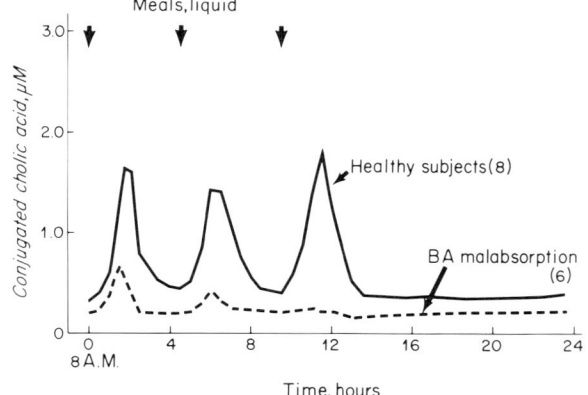

FIG. 15. Diurnal levels in serum of conjugates of CA, as determined by radioimmunoassay on samples collected at intervals of 15 to 30 min, except during the night when samples were taken at hourly intervals. Healthy subjects and patients with bile acid (BA) malabsorption because of ileal resection (164) were studied. In patients with bile acid malabsorption, peak of postprandial level declines progressively during the day, indicating progressive depletion of bile acid pool. Increased synthesis occurs throughout the day; during interval between supper and breakfast, increased synthesis restores the pool in part, so that postprandial peak after breakfast is largest of 3 postprandial peaks.

aged canaliculi. Thus the elevated systemic level of bile acids observed in patients with severe liver disease is likely to have multiple explanations.

DEFECTIVE ENTERAL TRANSPORT. Defective ileal transport is the most common transport defect of bile acids. This may occur because of ileal disease or resection or be a genetic transport defect (108). It is easily detected by a lower postprandial increase in the serum bile acid concentration (Fig. 15). A net decrease in ileal transport also occurs if bile acids are bound in the lumen by resins and possibly by dietary constituents, e.g., saponins (229). The clinical consequences of bile acid malabsorption are reviewed in the preceding chapter by Hofmann in this *Handbook*. When bile acid malabsorption is present, there is a lower postprandial elevation in serum bile acid levels, as well as increased fecal loss of bile acids.

Alterations in Biotransformation

INCREASED DECONJUGATION BECAUSE OF INTESTINAL BACTERIA PROLIFERATION. The major physiological factor responsible for controlling the proliferation of bacteria in the small intestine is intestinal motility, which continuously sweeps small intestine content into the large intestine. When stasis occurs because of diseases of intestinal motility or surgical misadventure, bacteria proliferate (230). Such proliferation is generally associated with increased bile acid deconjugation (190). The bile acids are absorbed passively and undergo the same fate as administered oral unconjugated bile acids (which is discussed extensively in *Alterations in Input*, p. 584). In the past, unconjugated bile acids were considered to have an important pathophysiological role in the malabsorption frequently observed in such patients, but the current view is that unconjugated bile acids, whether present in small intestine content or the systemic circulation, serve as a "marker" for increased deconjugation (247).

Bile acid deconjugation occurs normally during the enterohepatic cycling of bile acids in humans, but there is little information about the degree of deconjugation of bile acids during enterohepatic cycling in animals. It has been shown to occur in rats and can be greatly increased by surgical fashioning of a stagnant loop (268). Methods for quantifying bile acid deconjugation involve determining the turnover rates of the steroid and amino acid moieties of bile acids using the isotope dilution technique. Glycine-conjugated bile acids tagged with $[1\text{-}^{14}C]$glycine can be administered; the rate of $^{14}CO_2$ production is highly correlated with the turnover of the glycine moiety (105–107).

Because there is always deconjugation during enterohepatic cycling in humans, the systemic circulation always contains unconjugated bile acids. In the multicompartmental model for the EHC of bile acids in humans proposed by this laboratory, deconjugation and reconjugation of both CA and CDCA have been fully described and simulated. The first-pass clearance of unconjugated bile acids is less than that of conjugated bile acids. Consequently the systemic circulation is always enriched in unconjugated bile acids.

ALTERATION OF AMIDATION PATTERN BY GLYCINE OR TAURINE ADMINISTRATION. The enzyme involved in bile acid conjugation has a strong preference for taurine, and the rate of bile acid conjugation is linearly related to the concentration of intracellular taurine. Because taurine is mainly derived from diet and because the rate of active transport of taurine into the hepatocyte should be related to the plama level of taurine, the greater the dietary intake of taurine, the greater should be the fraction of bile acids conjugated with taurine. This has been shown to be true repeatedly in humans: supplementation of the diet with taurine causes a marked increase in the proportion of bile acids conjugated with taurine (81, 98, 107, 231). The data of Hardison (95) indicate that an oral taurine load immediately shifts the direction of hepatic amidation toward taurine and away from glycine. [In animals, taurine administration also increases the proportion of bile acids amidated with taurine, even in animals such as the guinea pig, which normally conjugate bile acids almost solely with glycine (144).]

The reverse is not true. Glycine administration does not alter the proportion of bile acids conjugated with glycine, since it does not influence taurine metabolism (231). In animals, administration of taurine analogues such as β-alanine (149) or guanidoethanesulfonic acid (50), which competitively inhibit taurine uptake by the hepatocyte but are not used for bile acid amidation, decreases the proportion of bile acids amidated with taurine.

INFLUENCE OF CONJUGATION PATTERN ON ENTEROHEPATIC CIRCULATION OF BILE ACIDS. The only known difference in the metabolism of glycine-conjugated and taurine-conjugated bile acids is that passive absorption of taurine-conjugated dihydroxy bile acids (DC-Tau and CDC-Tau) from the duodenum is unlikely to occur because these taurine-conjugated bile acids have an extremely low pK_a and therefore cannot be absorbed passively. The administration of taurine should increase the ratio of taurine-conjugated dihydroxy bile acids to glycine-conjugated dihydroxy bile acids in bile, and the bile acid pool should diminish modestly since the taurine-conjugated dihydroxy bile acid conjugates should undergo less passive absorption in the duodenum. In fact, taurine administration to healthy volunteers has been shown to cause a reduction in the size of the exchangeable CDCA pool size (98).

EPILOGUE: COMPARISON OF BILE ACIDS
AND CONVENTIONAL DRUGS

As a result of an intense multidisciplinary effort, the overall principles of the EHC in humans have

TABLE 2. *Properties of CDCA and UDCA, the First Enterohepatic Drugs, Versus Properties of Conventional Systemic Drugs*

	CDCA and UDCA	Conventional Systemic Drugs
Absorption	Passive	Passive
Albumin-binding	High	Variable
Hepatic extraction	50%–60%	As low as possible
Half time for plasma disappearance	5–10 min	Variable, usually longer
Hepatic biotransformation	Amidation	Glucuronidation, sulfation, others
Biliary secretion of metabolite	Complete	Variable
Daily biliary secretion, multiple of dose	10–40	Probably <1
Fate of biliary metabolites	Reabsorbed	Not reabsorbed
Site and mechanism of absorption of metabolites	Ileum, active	Colon, passive
Half time, days for body pool	2–3	Usually <1
Route of excretion	Fecal	Mostly urinary
Desired plasma level	Low	High
Target tissue	Gallbladder bile	Variable
Tissue distribution	Gut, portal and systemic circulation; liver and biliary tract	Variable; many tissues
Degree of bacterial degradation of compound	Extensive	Usually extremely low
Absorption of bacterial biotransformation product	Yes	Rarely
Toxicity in animals	Severe in nonsulfating species	Usually none
Monitoring of compliance	Bile levels	Blood levels

CDCA, chenodeoxycholic acid; UDCA, ursodeoxycholic acid.

been described and a fairly satisfactory model developed. In the coming decade this model will be improved, and it should be possible to apply it to both health and disease. The model should also apply, at least in broad outline, to the EHC of bile acids in animals. The model may in time be useful for the description of drug metabolism, but some of the fundamental qualities of bile acids are not present in most drugs for the simple reason that they are considered undesirable properties. To specify, ideal drugs have low first-pass clearance, high blood levels, minimal biliary excretion, and maximal urinary excretion (125). Nonetheless, some drugs have been identified in which the EHC is important, and improved methods for measuring the importance of the EHC in drug metabolism are being developed (246, 260). Evolution during recent millennia has resulted in improvement in the "design" of bile acids, yet the very properties that have evolved are considered undesirable for drugs. Differences in the properties of the natural bile acids and common drugs are summarized in Table 2. Figure 16 compares schematically the fate of drugs that are secreted in bile, not actively reabsorbed, and rapidly eliminated with that of bile acids that are secreted in bile, actively reabsorbed from the ileum, and persist in the EHC. The pharmaceutical industry will eventually seek to develop enterohepatic drugs if suitable target diseases can be identified. Such diseases should be where bile acids have their highest concentrations, namely the biliary tract and the intestinal tract.

The criticisms of R. Hermon Dowling have greatly improved the coherence of this chapter. The structure of the chapter is a result of the modeling efforts between our laboratory and that of the Torino group, and the many valuable discussions with G. Molino, C. Cravetto, M. Milanese, G. Belforte, and B. Bona are gratefully acknowledged. Initial modeling efforts were led by N. E. Hoffman,

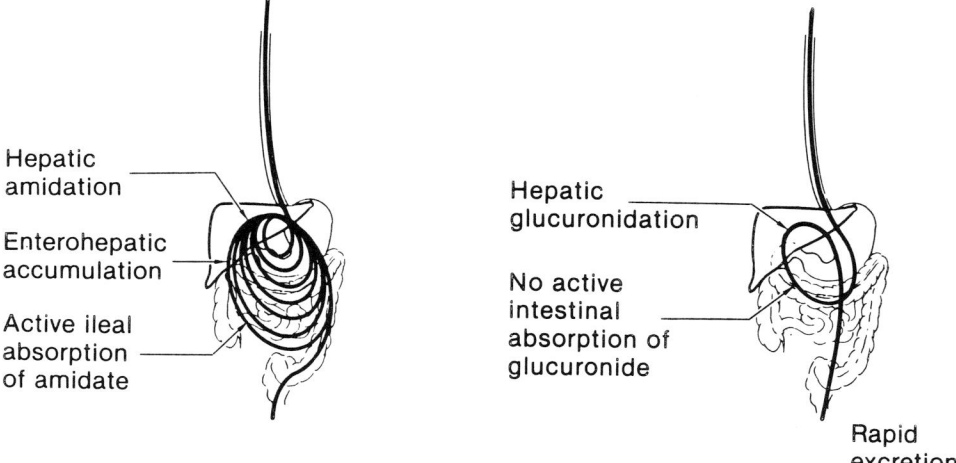

FIG. 16. Comparison of EHC of natural bile acids (*left*) and a drug that is excreted into bile as its glucuronide (*right*). In the colon, hydrolysis of glucuronide conjugates can occur and reabsorption of the aglycone can occur but is not shown.

whose insights were most valuable. V. Huebner prepared the manuscript, with the assistance of S. Davis.

The author's work is supported by National Institutes of Health Grants DK-21506 and DK-32130, as well as grants from The Falk Foundation e.V. (Freiburg, Federal Republic of Germany) and Gipharmex S.p.A. (Milan, Italy).

REFERENCES

1. ADMIRAND, W. H., AND D. M. SMALL. The physiochemical basis of cholesterol gallstone formation in man. *J. Clin. Invest.* 47: 1043–1052, 1968.
2. AHLBERG, J., B. ANGELIN, K. EINARSSON, K. HELLSTRÖM, AND B. LEIJD. Influence of deoxycholic acid on biliary lipids in man. *Clin. Sci. Mol. Med.* 53: 249–256, 1977.
3. ALBERT, A., AND E. P. SERJEANT. *Ionization Constants of Acids and Bases.* London: Methuen, 1962.
4. ALLAN, R. N., J. L. THISTLE, AND A. F. HOFMANN. Lithocholate metabolism during chenotherapy for gallstone dissolution. II. Absorption and sulphation. *Gut* 17: 413–419, 1976.
5. ALME, B., A. BREMMELGAARD, J. SJÖVALL, AND P. THOMASSEN. Analysis of metabolic profiles of bile acids in urine using a lipophilic anion exchanger and computerized gas-liquid chromatography-mass spectrometry. *J. Lipid Res.* 18: 339–362, 1977.
6. ALMOND, H. R., Z. R. VLAHCEVIC, C. C. BELL, JR., D. H. GREGORY, AND L. SWELL. Bile acid pools, kinetics and biliary lipid composition before and after cholecystectomy. *N. Engl. J. Med.* 289: 1213–1216, 1973.
7. AMIDON, G. L., J. H. KOU, R. L. ELLIOTT, AND E. N. LIGHTFOOT. Analysis of models for determining intestinal wall permeabilities. *J. Pharm. Sci.* 69: 1369–1373, 1969.
8. ANGELIN, B., I. BJÖRKHEM, K. EINARSSON, AND S. EWERTH. Hepatic uptake of bile acids in man—fasting and postprandial concentrations of individual bile acids in portal venous and systemic serum. *J. Clin. Invest.* 70: 724–731, 1982.
9. ANGELIN, B., AND B. LEIJD. Effects of cholic acid on the metabolism of endogenous plasma triglyceride and on biliary lipid composition in hyperlipoproteinemia. *J. Lipid Res.* 21: 1–9, 1980.
10. ASANO, T. Modification of cecal size in germfree rats by long-term feeding of anion exchange resin. *Am. J. Physiol.* 217: 911–918, 1969.
11. ATKINS, G. L. *Multicompartment Models for Biological Systems.* London: Methuen, 1969.
12. BACHRACH, W. H., AND A. F. HOFMANN. Ursodeoxycholic acid in the treatment of cholesterol cholelithiasis: a review. *Dig. Dis. Sci.* 27: 737–761, 833–856, 1982.
13. BACK, P., AND D. BOWEN. Chemical synthesis and characterization of glucuronic acid coupled mono-, di- and tri-hydroxy bile acids. *Hoppe Seylers Z. Physiol. Chem.* 357: 219–222, 1976.
14. BALISTRERI, W. F., W. M. BELKNAP, F. J. SUCHY, AND L. ZIMMER-NECHEMIAS. Immaturity of the enterohepatic circulation of bile acids in early life: factors responsible for increased peripheral serum bile acid concentrations. In: *Enterohepatic Circulation of Bile Acids and Sterol Metabolism*, edited by G. Paumgartner, A. Stiehl, and W. Gerok. Lancaster, UK: MTP, 1985, p. 87–93.
15. BARBARA, L., R. H. DOWLING, A. F. HOFMANN, AND E. RODA (editors). *Bile Acids in Gastroenterology.* Lancaster, UK: MTP, 1983.
16. BARBARA, L., R. H. DOWLING, A. F. HOFMANN, AND E. RODA (editors). *Recent Advances in Bile Acid Research.* New York: Raven, 1985.
17. BATTA, A. K., G. SALEN, AND S. SHEFER. Substrate specificity of cholylglycine hydrolysis of bile acid conjugates. *J. Biol. Chem.* 259: 15035–15039, 1984.
18. BAUMGARTNER, U., K. MIYAI, AND W. G. M. HARDISON. Greater taurodeoxycholate biotransformation during backward perfusion of rat liver. *Am. J. Physiol.* 251 (*Gastrointest. Liver Physiol.* 14): G431–G435, 1986.
19. BAUMGARTNER, U., K. MIYAI, AND W. G. HARDISON. Taurolithocholate cholestasis: physiologically and morphologically a canalicular disorder (Abstract). *Gastroenterology* 90: 1711, 1986.
20. BENET, L. Z., AND N. MASSOUND. Pharmacokinetics. In: *Pharmacokinetic Basis for Drug Treatment*, edited by L. Z. Benet, N. Massoud, and J. G. Gambertoglio. New York: Raven, 1984, p. 1–28.
21. BERGE HENEGOUWEN, G. P. VAN, AND A. F. HOFMANN. Pharmacology of chenodeoxycholic acid. II. Absorption and metabolism. *Gastroenterology* 73: 300–309, 1977.
22. BERGE HENEGOUWEN, G. P. VAN, AND A. F. HOFMANN. Nocturnal gallbladder storage and emptying in gallstone patients and healthy subjects. *Gastroenterology* 75: 879–885, 1978.
23. BERGE HENEGOUWEN, G. P. VAN, AND A. F. HOFMANN. Systemic spill-over of bile acids. *Eur. J. Clin. Invest.* 13: 433–437, 1983.
24. BERGSTRÖM, S. Metabolism of bile acids. *Federation Proc.* 20: 121–126, 1961.
25. BERGSTRÖM, S., AND H. DANIELSSON. On the regulation of bile acid formation in the rat liver. *Acta Physiol. Scand.* 43: 1–7, 1958.
26. BERGSTRÖM, S., M. ROTTENBERG, AND J. VOLTZ. The preparation of some carboxyl-labelled bile acids. *Acta Chem. Scand.* 7: 481–484, 1953.
26a. BERTOLOTTI, M., N. CARULLI, D. MENOZZI, F. ZIRONI, A. DIGRISOLO, A. PINETTI, AND M. GRAZIA BELDINIA. In vivo evaluation of cholesterol 7δ-hydroxylation in humans: effect of disease and drug treatment. *J. Lipid Res.* 27: 1278–1286, 1986.
27. BISCHOFF, K. B., R. L. DEDRICK, D. S. ZAHARKO, AND J. A. LONGSTRETH. Methotrexate pharmacokinetics. *J. Pharm. Sci.* 60: 1128–1133, 1971.
28. BJÖRKHEM, I. Mechanism of bile acid biosynthesis in mammalian liver. In: *Sterols and Bile Acids: New Comprehensive Biochemistry*, edited by H. Danielsson and J. Sjövall. Amsterdam: Elsevier, 1985, vol. 12, p. 231–278.
29. BJÖRKHEM, I., AND O. FALK. Assay of the major bile acids in serum by isotope dilution-mass spectrometry. *Scand. J. Clin. Lab. Invest.* 43: 163–170, 1983.
30. BJÖRKHEM, I., L. LILLEQUIST, K. NILSELL, AND K. EINARSSON. Oxidoreduction of different hydroxyl groups in bile acids during their enterohepatic circulation. *J. Lipid Res.* 27: 177–182, 1986.
31. BLITZER, B. L., C. TERZAKIS, AND K. A. SCOTT. Hydroxyl-bile acid exchange: a new mechanism for the uphill transport of cholate by basolateral liver plasma membrane vesicles. *J. Biol. Chem.* 261: 12042–12046, 1986.
32. BLOCH, K., B. N. BERG, AND D. RITTENBERG. The biological conversion of cholesterol to cholic acid. *J. Biol. Chem.* 149: 511–517, 1943.
33. BREMMELGAARD, A., AND J. SJÖVALL. Hydroxylation of cholic, chenodeoxycholic, and deoxycholic acids in patients with intrahepatic cholestasis. *J. Lipid Res.* 21: 1072–1081, 1980.
34. CALANDRA, S., N. CARULLI, AND G. SALVIOLI (editors). *Liver and Lipid Metabolism.* Amsterdam: Excerpta Med., 1984.
35. CAREY, M. C. The enterohepatic circulation. In: *The Liver: Biology and Pathobiology*, edited by I. Arias, H. Popper, D. Schachter, and D. A. Shafritz. New York: Raven, 1982, p. 429–465.
36. CAREY, M. C. Physico-chemical properties of bile acids and their salts. In: *Sterols and Bile Acids: New Comprehensive Biochemistry*, edited by H. Danielsson and J. Sjövall. Amsterdam: Elsevier, 1985, vol. 12, p. 345–403.

37. CAREY, M. C., AND D. M. SMALL. The physical chemistry of cholesterol solubility in bile. *J. Clin. Invest.* 61: 998–1026, 1978.
38. CARULLI, N., M. PONZ DE LEON, P. LORIA, R. IORI, A. ROSI, AND M. ROMANI. Effect of the selective expansion of the cholic acid pool on bile lipid composition: possible mechanism of bile acid induced biliary cholesterol desaturation. *Gastroenterology* 81: 539–546, 1981.
39. CLAYTON, L. M., D. GURANTZ, L. R. HAGEY, C. SCHTEINGART, M. S. ANWER, AND A. F. HOFMANN. Lack of conjugation greatly increases reflux of dihydroxy bile acids from the hepatocyte (Abstract). *Gastroenterology* 90: 1786, 1986.
40. COHEN, B. I., A. F. HOFMANN, E. H. MOSBACH, R. J. STENGER, M. A. ROTHSCHILD, L. R. HAGEY, AND Y. B. YOON. Differing effects of nor-ursodeoxycholic or ursodeoxycholic acid on hepatic histology and bile acid metabolism in the rabbit. *Gastroenterology* 91: 189–197, 1986.
41. COWEN, A. E., A. F. HOFMANN, D. L. HACHEY, P. J. THOMAS, D. T. E. BELOBABA, P. D. KLEIN, AND L. TOKES. Synthesis of 11,12-^{2}H$_2$- and 11,12-^{3}H$_2$-labeled chenodeoxycholic and lithocholic acids. *J. Lipid Res.* 17: 231–238, 1976.
42. COWEN, A. E., M. G. KORMAN, A. F. HOFMANN, AND O. W. CASS. Metabolism of lithocholate in healthy man. I. Biotransformation and biliary excretion of intravenously administered lithocholate, lithocholylglycine, and their sulfates. *Gastroenterology* 69: 59–66, 1975.
43. COWEN, A. E., M. G. KORMAN, A. F. HOFMANN, O. W. CASS, AND S. B. COFFIN. Metabolism of lithocholate in healthy man. II. Enterohepatic circulation. *Gastroenterology* 69: 67–76, 1975.
44. CZUBA, B., AND D. A. VESSEY. The effect of bile acid structure on the activity of bile acid-CoA-glycine/taurine-N-acyltransferase. *J. Biol. Chem.* 257: 8761–8765, 1982.
45. CZUBA, B., AND D. A. VESSEY. Structural characterization of cholyl-CoA:glycine/taurine N-acyl-transferase and a covalent substrate intermediate. *J. Biol. Chem.* 261: 6260–6263, 1986.
46. DANIELSSON, H., AND J. SJÖVALL (editors). *Sterols and Bile Acids: New Comprehensive Biochemistry.* Amsterdam: Elsevier, 1985, vol. 12.
47. DANZINGER, R. G., A. F. HOFMANN, J. L. THISTLE, AND L. J. SCHOENFIELD. Effect of oral chenodeoxycholic acid on bile acid kinetics and biliary lipid composition in women with cholelithiasis. *J. Clin. Invest.* 52: 2809–2821, 1973.
48. DAVIDSON, N. O., H. L. BRADLOW, E. H. AHRENS, JR., R. S. ROSENFELD, AND C. S. SCHWARTZ. Bile acid production in human subjects: rate of oxidation of (24,25-^{3}H)cholesterol compared to fecal bile acid excretion. *J. Lipid Res.* 27: 183–195, 1986.
49. DAVIS, R. A., W. E. HIGHSMITH, M. M. MCNEAL, J. A. SCHEXNAYDER, AND J. C. W. KUAN. Bile acid synthesis by cultured hepatocytes. Inhibition by mevinolin, but not by bile acids. *J. Biol. Chem.* 258: 4079–4082, 1983.
50. DE LA ROSA, J., AND M. H. STIPANUK. The effect of taurine depletion with guanidinoethanesulfonate on bile acid metabolism in the rat. *Life Sci.* 36: 1347–1351, 1985.
51. DEMARK, B. R., G. T. EVERSON, P. D. KLEIN, R. B. SHOWALTER, AND F. KERN. A method for the accurate measurement of isotope ratios of chenodeoxycholic and cholic acids in serum. *J. Lipid Res.* 23: 204–210, 1982.
52. DETTLI, L., AND P. SPRING. Pharmacokinetics as a basic medical problem. In: *Physico-Chemical Aspects of Drug Action*, edited by E. J. Ariens. Oxford, UK: Pergamon, 1968, p. 5–32.
53. DIAMOND, J. M., AND E. M. WRIGHT. Biological membranes: the physical basis of ion and non-electrolyte selectivity. *Annu. Rev. Physiol.* 31: 581–646, 1969.
54. DIMAGNO, E. P., V. L. W. GO, AND W. H. J. SUMMERSKILL. Impaired cholecystokinin-pancreozymin secretion, intraluminal dilution, and maldigestion of fat in sprue. *Gastroenterology* 63: 25–32, 1972.
55. DOWLING, R. H. Chenodeoxycholic acid therapy of gallstones. *Clin. Gastroenterol.* 6: 141–163, 1977.
56. DOWLING, R. H., E. MACK, J. PICOTT, J. BERGER, AND D. M. SMALL. Experimental model for the study of the enterohepatic circulation of bile in the rhesus monkey. *J. Lab. Clin. Med.* 72: 169–176, 1968.
57. DOWLING, R. H., E. MACK, AND D. M. SMALL. Effects of controlled interruption of the enterhepatic circulation of bile salts by biliary diversion and by ileal resection on bile salt secretion, synthesis, and pool size in the rhesus monkey. *J. Clin. Invest.* 49: 232–242, 1970.
58. DOWLING, R. H., E. MACK, AND D. M. SMALL. Primate biliary physiology. IV. Biliary lipid secretion and bile composition after acute and chronic interruption of the enterohepatic circulation in the rhesus monkey. *J. Clin. Invest.* 50: 1917–1926, 1971.
59. DUANE, W. C. Simulation of the defect of bile acid metabolism associated with cholesterol cholelithiasis by sorbitol ingestion in man. *J. Lab. Clin. Med.* 91: 969–978, 1978.
60. DUANE, W. C., D. E. HOLLOWAY, S. W. HUTTON, P. J. CORCORAN, AND N. A. HAAS. Comparison of bile acid synthesis determined by isotope dilution versus fecal acidic sterol output in human subjects. *Lipids* 17: 345–348, 1982.
61. DUPAS, J.-L., AND A. F. HOFMANN. Passive jejunal absorption of bile acids in vivo: structure-activity relationships and rate limiting steps (Abstract). *Gastroenterology* 86: 1067, 1984.
62. EINARSSON, K., AND K. HELLSTRÖM. The formation of bile acids in patients with three types of hyperlipoproteinemia. *Eur. J. Clin. Invest.* 2: 225–230, 1972.
63. EINARSSON, K., AND K. HELLSTRÖM. The formation of deoxycholic and chenodeoxycholic acid in man. *Clin. Sci. Mol. Med.* 46: 183–190, 1974.
64. EINARSSON, K., K. HELLSTRÖM, AND M. KALLNER. Feedback and regulation of bile acid formation in man. *Metabolism* 22: 1477–1483, 1973.
65. ELLIOTT, W. H. Mass spectra of bile acids. In: *Biochemistry of Applied Mass Spectrometry*, edited by G. R. Waller and O. C. Dermer. New York: Wiley, 1980, p. 229–253.
66. ELLIOTT, W. H. Metabolism of bile acids in liver and extrahepatic tissues. In: *Sterols and Bile Acids: New Comprehensive Biochemistry*, edited by H. Danielsson and J. Sjövall. Amsterdam: Elsevier, 1985, vol. 12, p. 303–329.
67. ENEROTH, P., B. GORDON, R. RYHAGE, AND J. SJÖVALL. Identification of some mono- and dihydroxy bile acids in human feces by gas-liquid chromatography and mass spectrometry. *J. Lipid Res.* 7: 511–523, 1966.
68. ENEROTH, P., B. GORDON, AND J. SJÖVALL. Characterization of trisubstituted cholanoic acids in human feces. *J. Lipid Res.* 7: 524–530, 1966.
69. ERIKSSON, S. Biliary excretion of bile acids and cholesterol in bile fistula rats. *Proc. Soc. Exp. Biol. Med.* 94: 578–583, 1957.
70. EVERSON, G. T., D. Z. BRAVERMAN, M. L. JOHNSON, AND F. KERN. A critical evaluation of real-time ultrasonography for the study of a gallbladder volume and contraction. *Gastroenterology* 79: 40–46, 1980.
71. EVERSON, G. T., M. J. LAWSON, C. MCKINLEY, R. SHOWALTER, AND F. KERN, JR. Gallbladder and small intestinal regulation of biliary lipid secretion during intraduodenal infusion of standard stimuli. *J. Clin. Invest.* 71: 596–603, 1983.
72. EWERTH, S., B. ANGELIN, K. EINARSSON, K. NILSELL, AND I. BJÖRKHEM. Serum concentrations of ursodeoxycholic acid in portal venous and systemic venous blood of fasting humans as determined by isotope dilution-mass spectrometry. *Gastroenterology* 88: 126–133, 1985.
73. EWERTH, S., I. BJÖRKHEM, K. EINARSSON, AND L. OST. Lymphatic transport of bile acids in man. *J. Lipid Res.* 23: 1183–1186, 1982.
74. FIESER, L. F., AND M. FIESER. *Steroids.* New York: Reinhold, 1959.
75. FINKELSTEIN, L., AND E. R. CARSON. *Mathematical Modelling of Dynamic Biological Systems.* New York: Wiley, 1985.
76. FISCHER, C. D., N. S. COOPER, M. A. ROTHSCHILD, AND E. H. MOSBACH. Effect of dietary chenodeoxycholic acid and lithocholic acid in the rabbit. *Am. J. Dig. Dis.* 18: 877–886, 1974.
77. FISHER, R. L., A. F. HOFMANN, S. ROSSI, J. L. CONVERSE,

AND S. P. LAN. Pathogenesis of morphological damage and major serum aminotransferase (AT) elevations in NCGS patients: enhanced liver cell sensitivity—not defective lithocholate metabolism (Abstract). *Hepatology Baltimore* 6: 1169, 1968.
78. FORGACS, I. C., M. N. MAISEY, G. M. MURPHY, AND R. H. DOWLING. Influence of gallstones and ursodeoxycholic acid therapy on gallbladder emptying. *Gastroenterology* 87: 299–307, 1984.
79. FROMM, H., G. L. CARLSON, A. F. HOFMANN, S. FARIVAR, AND P. AMIN. Metabolism in man of 7-ketolithocholic acid: precursor of cheno- and ursodeoxycholic acids. *Am. J. Physiol.* 239 (*Gastrointest. Liver Physiol.* 2): G161–G166, 1980.
80. FROMM, H., AND M. MALAVOLTI. Bile acid-induced diarrhoea. *Clin. Gastroenterol.* 15: 567–582, 1986.
81. GARBUTT, J. T., K. W. HEATON, L. LACK, AND M. P. TYOR. Increased ratio of glycine to taurine-conjugated bile salts in patients with ileal disorders. *Gastroenterology* 56: 711–720, 1969.
82. GARBUTT, J. T., AND T. J. KENNEY. Effect of cholestyramine on bile acid metabolism in normal men. *J. Clin. Invest.* 51: 2781–2789, 1972.
83. GILLER, J., AND S. F. PHILLIPS. The contribution of the colon to electrolyte and water conservation in man. *J. Lab. Clin. Med.* 81: 733–746, 1973.
84. GILMORE, I. T., AND A. F. HOFMANN. Altered drug metabolism and elevated serum bile acids in liver disease: a unified pharmacokinetic explanation. *Gastroenterology* 78: 177–179, 1980.
85. GODFREY, K. *Compartmental Models and Their Applications.* London: Academic, 1983.
86. GORESKY, C. A. Kinetic interpretation of hepatic multiple-indicator dilution studies. *Am. J. Physiol.* 245 (*Gastrointest. Liver Physiol.* 8): G1–G12, 1983.
87. GOTO, J., H. KATO, K. KANEKO, AND T. NAMBARA. Studies on steroids. CLXIII. Synthesis of monosulfates of cholic acid derivatives. *Chem. Pharm. Bull. Tokyo* 28: 3389–3394, 1980.
88. GRUNDY, S. M., E. H. AHRENS, JR., AND T. A. MIETTINEN. Quantitative isolation and gas-liquid chromatographic analysis of total fecal bile acids. *J. Lipid Res.* 6: 397–410, 1965.
89. GRUNDY, S. M., E. H. AHRENS, JR., AND G. SALEN. Interruption of the enterohepatic circulation of bile acids in man: comparative effects of cholestyramine and ileal exclusion on cholesterol metabolism. *J. Lab. Clin. Med.* 78: 94–121, 1971.
90. GRUNDY, S. M., W. C. DUANE, R. D. ADLER, J. M. ARON, AND A. L. METZGER. Biliary lipid output in young women with cholesterol gallstones. *Metabolism* 23: 67–73, 1974.
91. GRUNDY, S. M., AND A. L. METZGER. A physiological method for estimation of hepatic secretion in man. *Gastroenterology* 62: 1200–1217, 1972.
92. GURANTZ, D., AND A. F. HOFMANN. Influence of bile acid structure on bile flow and biliary lipid secretion in the hamster. *Am. J. Physiol.* 247 (*Gastrointest. Liver Physiol.* 10): G736–G748, 1984.
93. GUSTAFSSON, B. E., AND A. NORMAN. Physical state of bile acids in intestinal contents of germ-free and conventional rats. *Scand. J. Gastroenterol.* 3: 625–631, 1968.
94. HAGEY, L. R., J. P. NEOPTOLEMOS, S. S. ROSSI, H.-T. TON-NU, A. F. HOFMANN, AND J. O. WHITNEY. Deamidation and ester glucuronidation: new side chain biotransformations of bile acids occurring in man (Abstract). *Hepatology Baltimore* 5: 1023, 1985.
95. HARDISON, W. G. M. Hepatic taurine concentration and dietary taurine as regulators of bile acid conjugation with taurine. *Gastroenterology* 75: 71–75, 1978.
96. HARDISON, W. G. M. Relation of hepatic taurine pool size to bile-acid conjugation in man and animals. In: *Sulfur Amino Acids: Biochemical and Clinical Aspects*, edited by K. Kuriyama, R. J. Huxtable, and H. Iwata. New York: Liss, 1983, p. 407–418.
97. HARDISON, W. G. M., S. BELLANTANI, V. HEASLEY, AND D. SHELLHAMER. Specificity of an Na^+-dependent taurocholate transport site in isolated rat hepatocytes. *Am. J. Physiol.* 246 (*Gastrointest. Liver Physiol.* 9): G477–G483, 1984.
98. HARDISON, W. G. M., AND S. M. GRUNDY. Effect of bile acid conjugation pattern on bile acid metabolism in normal humans. *Gastroenterology* 84: 617–620, 1983.
99. HARDISON, W. G. M., AND H. I. ROSENBERG. The effect of neomycin on bile salt metabolism and fat digestion in man. *J. Lab. Clin. Med.* 74: 564–573, 1969.
100. HARDISON, W. G. M., N. TOMASZEWSKI, AND S. M. GRUNDY. Effect of acute alterations in small bowel transit time upon the biliary excretion rate of bile acids. *Gastroenterology* 76: 568–574, 1979.
101. HASLEWOOD, G. A. D. *The Biological Importance of Bile Salts.* Amsterdam: North-Holland, 1978.
102. HEATON, K. W. Bitter humour: the development of ideas about bile salts. *J. R. Coll. Phys.* 6: 281–286, 1971.
103. HEATON, K. W. *Bile Salts in Health and Disease.* Edinburgh: Churchill Livingstone, 1972.
104. HEPNER, G. W., A. F. HOFMANN, J.-R. MALAGELADA, P. A. SZCZEPANIK, AND P. D. KLEIN. Increased bacterial degradation of bile acids in cholecystectomized patients. *Gastroenterology* 66: 556–564, 1974.
105. HEPNER, G. W., A. F. HOFMANN, AND P. J. THOMAS. Metabolism of steroid and amino acid moieties of conjugated bile acids in man. I. Cholylglycine. *J. Clin. Invest.* 51: 1889–1897, 1972.
106. HEPNER, G. W., A. F. HOFMANN, AND P. J. THOMAS. Metabolism of steroid and amino acid moieties of conjugated bile acids in man. II. Glycine-conjugated dihydroxy bile acids. *J. Clin. Invest.* 51: 1898–1905, 1972.
107. HEPNER, G. W., J. A. STURMAN, A. F. HOFMANN, AND P. J. THOMAS. Metabolism of steroid and amino acid moieties of conjugated bile acids in man. III. Cholyltaurine (taurocholic acid). *J. Clin. Invest.* 52: 433–440, 1973.
108. HEUBI, J. E., W. F. BALISTRERI, J. D. FONDACARO, J. C. PARTIN, AND W. K. SCHUBERT. Primary bile acid malabsorption: defective in vitro ileal active bile acid transport. *Gastroenterology* 83: 804–811, 1982.
109. HEUMAN, D. M., C. R. HERNANDEZ, W. M. KUBASKA, M. D. LAW, P. B. HYLEMON, AND Z. R. VLAHCEVIC. Failure of tauroursodeoxycholate (TUDCA) to suppress rat bile acid synthesis in vivo (Abstract). *Hepatology Baltimore* 5: 1047, 1985.
110. HEUMAN, D. M., M. D. LAW, C. HARTMEN, W. M. KUBASKA, P. B. HYLEMON, AND Z. R. VLAHCEVIC. Effect of taurocholate infusion (TCI) on bile acid synthesis (BAS) and cholesterol 7α-hydroxylase ($C7\alpha H$) activity (Abstract). *Gastroenterology* 90: 1734, 1986.
111. HIMMELSTEIN, K. J., AND R. J. LUTZ. A review of the applications of physiologically based pharmacokinetic modeling. *J. Pharmacokinet. Biopharm.* 7: 127–145, 1979.
112. HISLOP, I. G. The absorption and entero-hepatic circulation of bile salts. An historical review. *Med. J. Aust.* 1: 1223–1226, 1970.
113. HO, N. F. H., H. P. MERKLE, AND W. I. HIGUCHI. Quantitative, mechanistic and physiologically realistic approach to the biopharmaceutical design of oral drug delivery system. *Drug Dev. Ind. Pharm.* 9: 1111–1184, 1983.
114. HOFFMAN, N. E., D. E. DONALD, AND A. F. HOFMANN. Increased bile acid synthesis after interruption of the enterohepatic circulation: evidence against hormonal mediation. *Proc. Soc. Exp. Biol. Med.* 154: 45–52, 1977.
115. HOFFMAN, N. E., AND A. F. HOFMANN. Metabolism of steroid and amino acid moieties of conjugated bile acids in man. IV. Description and validation of a multicompartmental model. *Gastroenterology* 67: 887–897, 1974.
116. HOFFMAN, N. E., AND A. F. HOFMANN. Metabolism of steroid and amino acid moieties of conjugated bile acids in man. V. Equations for the perturbed enterohepatic circulation and their application. *Gastroenterology* 72: 141–148, 1977.
117. HOFFMAN, N. E., N. F. LARUSSO, AND A. F. HOFMANN. Sampling intestinal content with a sequestering capsule: a noninvasive technique for determining bile acid kinetics in

man. *Mayo Clin. Proc.* 51: 171–175, 1976.
118. HOFMANN, A. F. Bile acids, diarrhea, and antibiotics: data, speculation and a unifying hypothesis. *J. Infect. Dis.* 135: S126–S132, 1977.
119. HOFMANN, A. F. Biliary lipid secretion in man. In: *Liver and Bile*, edited by L. Bianchi, W. Gerok, and K. Sickinger. Lancaster, UK: MTP, 1977, p. 101–118.
120. HOFMANN, A. F. The enterohepatic circulation of bile acids in man. *Clin. Gastroenterol.* 6: 3–24, 1977.
121. HOFMANN, A. F. The enterohepatic circulation of bile acids in health and disease. In: *Gastrointestinal Disease: Pathophysiology, Diagnosis, Management* (3rd ed.), edited by M. H. Sleisenger and J. S. Fordtran. Philadelphia, PA: Saunders, 1983, p. 115–131.
122. HOFMANN, A. F. Chemistry and enterohepatic circulation of bile acids. *Hepatology Baltimore* 4: 4S–14S, 1984.
123. HOFMANN, A. F. (editor). *Hepatology Baltimore* 4, Suppl.: 1984. (Kroc Workshop Proc.)
124. HOFMANN, A. F. (editor). The physical chemistry of bile in health and disease. *Hepatology Baltimore* 4: 1S–252S, 1984.
125. HOFMANN, A. F. Targeting drugs to the enterohepatic circulation: lessons from bile acids and other endobiotics. *J. Control. Rel.* 2: 2–11, 1985.
126. HOFMANN, A. F., V. BOKKENHEUSER, R. L. HIRSCH, AND E. H. MOSBACH. Experimental cholelithiasis in the rabbit induced by cholestanol feeding: effect of neomycin treatment on bile composition and gallstone formation. *J. Lipid Res.* 9: 244–253, 1968.
127. HOFMANN, A. F., C. CRAVETTO, G. MOLINO, G. BELFORTE, AND B. BONA. Simulation of the metabolism and enterohepatic circulation of endogenous deoxycholic acid in man using a physiological pharmacokinetic model for bile acid metabolism. *Gastroenterology*. In press.
128. HOFMANN, A. F., AND S. A. CUMMINGS. Measurement of bile acid and cholesterol kinetics in man by isotope dilution: principles and applications. In: *Bile Acids in Gastroenterology*, edited by L. Barbara, R. H. Dowling, A. F. Hofmann, and E. Roda. Lancaster, UK: MTP, 1983, p. 75–117.
129. HOFMANN, A. F., S. M. GRUNDY, J. M. LACHIN, S.-P. LAN, R. A. BAUM, R. F. HANSON, T. HERSH, N. C. HIGHTOWER, JR., J. W. MARKS, H. MEKHJIAN, R. A. SHAEFER, R. D. SOLOWAY, J. L. THISTLE, F. B. THOMAS, M. P. TYOR, AND THE NATIONAL COOPERATIVE GALLSTONE STUDY GROUP. Pretreatment biliary lipid composition in white patients with radiolucent gallstones in the National Cooperative Gallstone Study. *Gastroenterology* 83: 738–752, 1982.
130. HOFMANN, A. F., AND N. E. HOFFMAN. Measurement of bile acid kinetics by isotope dilution in man. *Gastroenterology* 67: 314–323, 1974.
131. HOFMANN, A. F., AND P. D. KLEIN. Characterization of bile acid metabolism in man using bile acids labeled with stable isotopes. In: *Stable Isotopes*, edited by T. A. Baillie. Baltimore, MD: University Park, 1978, p. 189–204.
132. HOFMANN, A. F., G. MOLINO, M. MILANESE, AND G. BELFORTE. Description and simulation of a physiological pharmacokinetic model for the metabolism and enterohepatic circulation of bile acids in man. Cholic acid in healthy man. *J. Clin. Invest.* 71: 1003–1022, 1983.
133. HOFMANN, A. F., K. R. PALMER, Y. B. YOON, L. R. HAGEY, D. GURANTZ, S. HUIJGHEBAERT, J. L. CONVERSE, S. CECCHETTI, AND E. MICHELOTTI. The biological utility of bile acid conjugation with glycine or taurine. In: *Advances in Glucuronide Conjugation*, edited by S. Matern, K. W. Bock, and W. Gerok. Lancaster, UK: MTP, 1985, p. 245–264.
134. HOFMANN, A. F., AND J. R. POLEY. Role of bile acid malabsorption in pathogenesis of diarrhea and steatorrhea in patients with ileal resection. I. Response to cholestyramine or replacement of dietary long chain triglyceride by medium chain triglyceride. *Gastroenterology* 62: 918–934, 1972.
135. HOFMANN, A. F., AND A. RODA. Physicochemical properties of bile acids and their relationship to biological properties: an overview of the problem. *J. Lipid Res.* 25: 1477–1489, 1984.
136. HOFMANN, A. F., AND D. M. SMALL. Detergent properties of bile salts: correlation with physiological function. *Annu. Rev. Med.* 18: 333–376, 1967.
137. HOFMANN, A. F., P. A. SZCZEPANIK, AND P. D. KLEIN. Rapid preparation of tritium labeled bile acids by enolic exchange on basic alumina containing tritiated water. *J. Lipid Res.* 9: 707–713, 1968.
138. HOFMANN, A. F., J. L. THISTLE, P. D. KLEIN, P. A. SZCZEPANIK, AND P. Y. S. YU. Chenotherapy for gallstones. II. Induced changes in bile composition and gallstone response. *J. Am. Med. Assoc.* 239: 1138–1144, 1978.
139. HOPMAN, P. M., F. M. BROUWER, G. ROSENBUSCH, J. B. M. J. JANSEN, AND C. B. H. W. LAMERS. A computerized method for rapid quantification of gallbladder volume from real-time sonograms. *Radiology* 154: 236–237, 1985.
140. HOSHITA, T. Bile alcohols and primitive bile acids. In: *Sterols and Bile Acids: New Comprehensive Biochemistry*, edited by H. Danielsson and J. Sjövall. Amsterdam: Elsevier, 1985, vol. 12, p. 279–302.
141. HUIJGHEBAERT, S. M., AND A. F. HOFMANN. Influence of amino acid moiety on deconjugation of bile acid amidates by cholylglycine hydrolase or human fecal cultures. *J. Lipid Res.* 27: 742–752, 1986.
142. JAVITT, N. B. Cholestasis in rats induced by taurolithocholate. *Nature Lond.* 210: 1262–1263, 1966.
143. JOSEPHSON, B. The circulation of the bile acids in connection with their production, conjugation and excretion. *Physiol. Rev.* 21: 463–486, 1941.
144. KIBE, A., C. WAKE, T. KURAMOTO, AND T. HOSHITA. Effect of dietary taurine on bile acid metabolism in guinea pigs. *Lipids* 15: 224–229, 1979.
145. KILLENBERG, P. G. Measurement and subcellular distribution of cholyl-CoA synthetase and bile acid-CoA:amino acid N-acyltransferase activities in rat liver. *J. Lipid Res.* 19: 24–31, 1978.
146. KILLENBERG, P. G., AND J. T. JORDAN. Purification and characterization of bile acid-CoA:amino acid N-acyltransferase from rat liver. *J. Biol. Chem.* 253: 1005–1009, 1978.
147. KING, J. E., AND L. J. SCHOENFIELD. Cholestasis induced by sodium taurolithocholate in isolated hamster liver. *J. Clin. Invest.* 50: 2305–2312, 1971.
148. KIRKPATRICK, R. B., M. D. GREEN, L. R. HAGEY, A. F. HOFMANN, AND T. R. TEPHLY. Effect of side chain length on bile acid conjugation: glucuronidation, sulfation, and CoA formation of nor-bile acids and their natural C_{24} homologues by human rat liver fractions. *Hepatology Baltimore*. In press.
149. KITANI, K., AND S. KANAI. Ursodeoxycholate-induced choleresis in taurine-deprived and taurine-supplemented rats. *Jpn. J. Physiol.* 35: 443–462, 1985.
150. KRAG, E., AND S. F. PHILLIPS. Active and passive bile acid absorption in man. Perfusion studies of the ileum and jejunum. *J. Clin. Invest.* 53: 1686–1694, 1974.
151. KRISHNAMURTHY, G. I. Radionuclide ejection fraction: a new technique for quantitative analysis of motor function of the human gallbladder. *Gastroenterology* 80: 482–490, 1981.
152. KUBASKA, W. M., E. C. GURLEY, P. B. HYLEMON, P. S. GUZELIAN, AND Z. R. VLAHCEVIC. Absence of negative feedback control of bile acid biosynthesis in cultured rat hepatocytes. *J. Biol. Chem.* 260: 13459–13463, 1985.
153. KUIPERS, F., R. HAVINGA, H. BOSSCHIETER, G. P. TOOROP, F. R. HINDRIKS, AND R. J. VONK. Enterohepatic circulation in the rat. *Gastroenterology* 88: 403–411, 1985.
154. LACK, L., F. O. DARRITY, T. WALKER, AND G. D. SINGLETARY. Synthesis of conjugated bile acids by means of a coupling agent. *J. Lipid Res.* 14: 367–370, 1973.
155. LA DU, B. N., H. G. MANDEL, AND E. L. WAY (editors). *Fundamentals of Drug Metabolism and Drug Disposition*. Baltimore, MD: Williams & Wilkins, 1971.
156. LANZINI, A., R. P. JAZRAWI, AND T. C. NORTHFIELD. Simultaneous quantitative measurements of hepatic secretion and of absolute gallbladder storage and emptying during fasting and eating in man. *Gastroenterology* 92: 852–861, 1987.

157. LARUSSO, N. F., N. E. HOFFMAN, AND A. F. HOFMANN. Dynamics of the enterohepatic circulation of bile acids. *N. Engl. J. Med.* 291: 689–692, 1974.
158. LARUSSO, N. F., N. E. HOFFMAN, A. F. HOFMANN, T. C. NORTHFIELD, AND J. L. THISTLE. Effect of primary bile acid ingestion on bile acid metabolism and biliary lipid secretion in gallstone patients. *Gastroenterology* 69: 1301–1314, 1975.
159. LARUSSO, N. F., N. E. HOFFMAN, M. G. KORMAN, A. F. HOFMANN, AND A. E. COWEN. Determinants of fasting and postprandial serum bile acid levels in healthy man. *Am. J. Dig. Dis.* 23: 385–391, 1978.
160. LARUSSO, N. F., P. A. SZCZEPANIK, A. F. HOFMANN, AND S. B. COFFIN. Effect of deoxycholic acid ingestion on bile acid metabolism and biliary lipid secretion in normal subjects. *Gastroenterology* 72: 132–140, 1977.
161. LASSEN, N. A., AND W. PERL. *Tracer Kinetic Methods in Medical Physiology.* New York: Raven, 1979.
162. LEGRAND-DEFRETIN, V., C. JUSTE, T. CORRING, AND A. RERAT. Enterohepatic circulation of bile acids in pigs: diurnal pattern and effect of a reentrant biliary fistula. *Am. J. Physiol.* 250 (*Gastrointest. Liver Physiol.* 13): G295–G301, 1986.
163. LINDSTEDT, S. The turnover of cholic acid in man. *Acta Physiol. Scand.* 40: 1–9, 1957.
164. LOW-BEER, T. S., K. W. HEATON, E. W. POMARE, AND A. E. READ. The effect of coeliac disease upon bile salts. *Gut* 14: 204–208, 1973.
165. LUXON, B. A., P. D. KING, AND E. L. FORKER. How to measure first-order hepatic transfer coefficients by distributed modeling of a recirculating rat liver perfusion system. *Am. J. Physiol.* 243 (*Gastrointest. Liver Physiol.* 6): G518–G531, 1982.
166. MACDONALD, I. A., V. D. BOKKENHEUSER, J. WINTER, A. M. MCLERNON, AND E. H. MOSBACH. Degradation of steroids in the human gut. *J. Lipid Res.* 24: 675–700, 1983.
167. MAKINO, I., H. HASHIMOTO, K. SHINOZAKI, K. YOSHINO, AND S. MALAGAWA. Sulfated and nonsulfated bile acids in urine, serum and bile of patients with hepatobiliary diseases. *Gastroenterology* 68: 545–553, 1975.
168. MAKINO, I., K. SHINOZAKI, K. YOSHINO, AND S. NAKAGAWA. Dissolution of cholesterol gallstones by ursodeoxycholic acid. *Jpn. J. Gastroenterol.* 72: 690–702, 1975.
169. MALAVOLTI, M., H. FROMM, B. COHEN, AND S. CERYAK. Isolation and identification of Δ^6-lithocholenic acid (3α-hydroxy-5β-6-cholen-24-oic acid) as an intestinal bacterial metabolite of chenodeoxycholic acid in man. *J. Biol. Chem.* 260: 11011–11015, 1985.
170. MARCUS, S. N., AND K. W. HEATON. Intestinal transit, deoxycholic acid and the cholesterol saturation in bile—three interrelated factors. *Gut* 27: 550–558, 1986.
171. MATERN, S., AND W. GEROK. Pathophysiology of the enterohepatic circulation of bile acids. *Rev. Physiol. Biochem. Pharmacol.* 85: 126–159, 1979.
172. MATERN, S., J. HACKENSCHMIDT, P. BACK, AND W. GEROK (editors). *Advances in Bile Acid Research.* Stuttgart, FRG: Schattauer Verlag, 1975.
173. MATERN, S., H. MATERN, E. H. FARTHMANN, AND W. GEROK. Hepatic and extrahepatic glucuronidation of bile acids in man. Characterization of bile acid uridine 5'-diphosphate-glucuronosyl transferase in hepatic, renal, and intestinal microsomes. *J. Clin. Invest.* 74: 402–410, 1984.
174. MCCLOY, R. F., R. G. GREENBERG, AND J. H. BARON. Duodenal pH in health and duodenal ulcer disease: effect of a meal. *Gut* 25: 386–392, 1984.
175. MCJUNKIN, B., H. FROMM, R. P. SARVA, AND P. AMIN. Factors in the mechanism of diarrhea in bile acid malabsorption: fecal pH—a key determinant. *Gastroenterology* 80: 1454–1464, 1981.
176. MEKHJIAN, H. S., S. F. PHILLIPS, AND A. F. HOFMANN. Colonic absorption of unconjugated bile acids: perfusion studies in man. *Dig. Dis. Sci.* 24: 545–550, 1979.
177. MIYAI, K., A. L. RICHARDSON, W. LMAYR, AND N. B. JAVITT. Subcellular pathology of rat liver in cholestasis and choleresis induced by bile salts. I. Effects of lithocholic, 3β-hydroxy-5-cholenoic, cholic, and dehydrocholic acids. *Lab. Invest.* 36: 249–258, 1977.
178. MOLINO, G., A. F. HOFMANN, C. CRAVETTO, G. BELFORTE, AND B. BONA. Simulation of the metabolism and enterohepatic circulation of endogenous chenodeoxycholic acid in man using a physiological pharmacokinetic model. *Eur. J. Clin. Invest.* 16: 397–414, 1986.
179. MORRISSEY, K. P., C. K. MCSHERRY, R. L. SWARM, W. H. NIEMAN, AND J. E. DEITRICK. Toxicity of chenodeoxycholic acid in the nonhuman primate. *Surgery St. Louis* 77: 851–860, 1975.
180. MOSBACH, E. H. Hepatic synthesis of bile acids. Biochemical steps and mechanism of rate control. *Arch. Intern. Med.* 130: 478–487, 1972.
181. NAIR, P. P., AND D. KRITCHEVSKY (editors). *The Bile Acids: Chemistry, Physiology, and Metabolism. Chemistry.* New York: Plenum, 1971, vol. 1.
181a. NAIR, P. P., AND D. KRITCHEVSKY (editors). *The Bile Acids: Chemistry, Physiology, and Metabolism. Physiology and Metabolism.* New York: Plenum, 1973, vol. 2.
181b. NAIR, P. P., AND D. KRITCHEVSKY (editors). *The Bile Acids: Chemistry, Physiology, and Metabolism. Pathophysiology.* New York: Plenum, 1976, vol. 3.
182. NAKAGAKI, M., R. G. DANZINGER, A. F. HOFMANN, AND R. A. DIPIETRO. Biliary secretion and hepatic metabolism of taurine conjugated 7α-hydroxy and 7β-hydroxy bile acids in the dog: defective hepatic transport and bile hyposecretion. *Gastroenterology* 87: 647–659, 1984.
183. NAKAGAKI, M., R. G. DANZINGER, A. F. HOFMANN, AND A. RODA. Hepatic biotransformation of two hydroxy-7-oxotaurine-conjugated bile acids in the dog. *Am. J. Physiol.* 245 (*Gastrointest. Liver Physiol.* 8): G411–G417, 1983.
184. NEOPTOLEMOS, J. P., N. MARASSI, D. GURANTZ, L. R. HAGEY, AND A. F. HOFMANN. The unusual physiological properties and unique hepatic biotransformation of nor-deoxycholate (Abstract). *Gastroenterology* 88: 1682, 1985.
185. NG, P. Y., AND A. F. HOFMANN. Tissue distribution of [1^{14}C] cholylglycine in rats and hamsters with a bile fistula or bile duct ligation. *Proc. Soc. Exp. Biol. Med.* 154: 134–137, 1977.
186. NILSELL, K., B. ANGELIN, B. LEIJD, AND K. EINARSSON. Comparative effects of ursodeoxycholic acid and chenodeoxycholic acid on bile acid kinetics and biliary lipid secretion in humans. Evidence for different modes of action on bile acid synthesis. *Gastroenterology* 85: 1248–1256, 1983.
187. NILSSON, L. O., T. A. STEIN, G. P. BURNS, G. ORANGIO, AND L. WISE. The effects of ileal resection with cholecystectomy on bile salt metabolism in the prairie dog. *J. Surg. Res.* 37: 304–308, 1984.
188. NORMAN, A. Preparation of conjugated bile acids using mixed carboxylic acid anhydrides. *Ark. Kemi* 8: 331–342, 1955.
189. NORMAN, A. Metabolism of glycocholic acid in man. *Scand. J. Gastroenterol.* 5: 231–236, 1970.
190. NORTHFIELD, T. C., B. S. DRASAR, AND J. T. WRIGHT. Value of small intestinal bile acid analysis in the diagnosis of the stagnant loop syndrome. *Gut* 14: 341–347, 1973.
191. NORTHFIELD, T. C., AND A. F. HOFMANN. Biliary lipid output during three meals and an overnight fast. I. Relationship to bile acid pool size and cholesterol saturation of bile in gallstone and control subjects. *Gut* 16: 1–11, 1975.
192. NORTHFIELD, T. C., AND I. MCCOLL. Postprandial concentrations of free and conjugated bile acids down the length of the normal human small intestine. *Gut* 14: 513–518, 1973.
193. OKHUYSEN-YOUNG, C., AND T. F. KELLOGG. The effect of cecectomy on fecal bile acid and neutral steroid excretion of the rat. *Comp. Biochem. Physiol. B Comp. Biochem.* 70B: 345–347, 1981.
194. O'MAILLE, E. R. L., AND A. F. HOFMANN. Relatively high biliary secretory maximum for non-micelle-forming bile acid: possible significance for mechanism of secretion. *Q. J. Exp. Physiol.* 71: 475–482, 1986.
195. O'MAILLE, E. R. L., T. G. RICHARDS, AND A. H. SHORT. Acute taurine depletion and maximal rates of hepatic conjugation

and secretion of cholic acid in the dog. *J. Physiol. Lond.* 180: 67–79, 1965.
196. OWEN, R. W. Biotransformation of bile acids by clostridia. *J. Med. Microbiol.* 20: 233–238, 1985.
197. OWEN, R. W., M. J. HILL, AND R. F. BILTON. Biotransformation of chenodeoxycholic acid by *Pseudomonas* species NCIB 10590 under anaerobic conditions. *J. Lipid Res.* 24: 1109–1118, 1983.
198. PALMER, K. R., D. GURANTZ, A. F. HOFMANN, L. M. CLAYTON, L. R. HAGEY, AND S. CECCHETTI. Hypercholeresis induced by norchenodeoxycholate in biliary fistula rodent. *Am. J. Physiol.* 252 (*Gastrointest Liver Physiol.* 15): G219–G228, 1987.
199. PALMER, R. H., AND M. G. BOLT. Bile acid sulfates. I. Synthesis of lithocholic acid sulfates and their identification in human bile. *J. Lipid Res.* 12: 671–679, 1971.
200. PANG, K. S., AND J. A. TERRELL. Retrograde perfusion to probe the heterogenous distributions of hepatic drug metabolizing enzymes in rats. *J. Pharmacol. Exp. Ther.* 2216: 339–346, 1981.
201. PARMENTIER, G., AND H. EYSSEN. Synthesis and characteristics of the specific monosulfates of chenodeoxycholate, deoxycholate, and their taurine or glycine conjugates. *Steroids* 30: 583–590, 1977.
202. PAUMGARTNER, G., AND A. STIEHL (editors). *Bile Acid Metabolism in Health and Disease.* Lancaster, UK: MTP, 1977.
203. PAUMGARTNER, G., A. STIEHL, AND W. GEROK (editors). *Biological Effects of Bile Acids.* Lancaster, UK: MTP, 1979.
204. PAUMGARTNER, G., A. STIEHL, AND W. GEROK (editors). *Bile Acids and Lipids.* Lancaster, UK: MTP, 1981.
205. PAUMGARTNER, G., A. STIEHL, AND W. GEROK (editors). *Bile Acids and Cholesterol in Health and Disease.* Lancaster, UK: MTP, 1983.
206. POMARE, E. W., AND K. W. HEATON. The effect of cholecystectomy on bile salt metabolism. *Gut* 14: 753–762, 1973.
207. POMARE, E. W., K. W. HEATON, T. S. LOW-BEER, AND H. J. ESPINER. The effect of wheat bran upon bile salt metabolism and upon the lipid composition of bile in gallstone patients. *Am. J. Dig. Dis.* 21: 521–526, 1976.
208. POOLER, P., AND W. DUANE. Effects of bile acid administration on bile acid synthesis and its circadian rhythm in man (Abstract). *Gastroenterology* 92: 1765, 1987.
209. PRIES, J. M., C. A. SHERMAN, G. C. WILLIAMS, AND R. F. HANSON. Hepatic extraction of bile salts in conscious dogs. *Am. J. Physiol.* 236 (*Endocrinol. Metab. Gastrointest. Physiol.* 5): E191–E197, 1979.
210. REUBEN, A., P. N. MATON, G. M. MURPHY, AND R. H. DOWLING. Bile lipid secretion in obese and non-obese individuals with and without gallstones. *Clin. Sci. Lond.* 69: 71–79, 1985.
211. ROBERTSON, J. S. *Compartmental Distribution of Radiotracers.* Boca Raton, FL: CRC, 1983.
212. RODA, A. Sensitive methods for serum bile acid analysis. In: *Bile Acids in Gastroenterology*, edited by L. Barbara, R. H. Dowling, A. F. Hofmann, and E. Roda. Lancaster, UK: MTP, 1983, p. 57–68.
213. RODA, A., G. CAPPELLERI, R. ALDINI, E. RODA, AND L. BARBARA. Quantitative aspects of the interaction of bile acids with human serum albumin. *J. Lipid Res.* 23: 490–495, 1982.
214. RODA, A., S. GIROTTI, G. CARREA, L. J. KRICKA, M. DELUCA, AND A. F. HOFMANN. Luminometric methods. In: *Methods of Enzymatic Analysis. Metabolites 3: Lipids, Amino Acids and Related Compounds*, edited by H. U. Bergmeyer, J. Bergmeyer, and M. Grassl. Weinheim, FRG: VCH, 1985, vol. VIII, p. 304–316.
215. RODA, A., A. F. HOFMANN, AND K. J. MYSELS. The influence of bile salt structure on self-association in aqueous solutions. *J. Biol. Chem.* 258: 6362–6370, 1983.
216. ROWLAND, M., AND T. N. TOZER. *Clinical Pharmacokinetics: Concepts and Applications.* Philadelphia, PA: Lea & Febiger, 1980.
217. SACQUET, E., M. PARQUET, M. RIOTTOT, A. RAIZMAN, P. JARRIGE, C. HUGUET, AND R. INFANTE. Intestinal absorption, excretion, and biotransformation of hyodeoxycholic acid in man. *J. Lipid Res.* 24: 604–613, 1983.
218. SARVA, R. P., H. FROMM, S. FARIVAR, R. F. SEMBRAT, H. MENDELOW, H. SHINOZUKA, AND S. K. WOLFSON. Comparison of the effects between ursodeoxycholic and chenodeoxycholic acid on liver function and structure and on bile acid composition in the rhesus monkey. *Gastroenterology* 79: 629–636, 1980.
219. SCHIFF, E. R., N. C. SMALL, AND J. M. DIETSCHY. Characterization of the kinetics of the passive and active transport mechanisms for bile acid absorption in the small intestine and colon of the rat. *J. Clin. Invest.* 51: 1351–1362, 1972.
220. SCHIFF, M. Gallenbildung, abhangig von der Aufsaugung der Gallenstoffe. *Pflueger Arch. Gesamte Physiol. Menschen Tiere* 3: 598–613, 1870.
221. SCHMIDT, C. R., AND A. C. IVY. The general function of the gallbladder. Do species lacking a gallbladder possess its functional equivalent? *J. Cell Comp. Physiol.* 10: 365–383, 1937.
222. SETCHELL, K. D. R., A. M. LAWSON, E. J. BLACKSTOCK, AND G. M. MURPHY. Diurnal changes in serum unconjugated bile acids in normal man. *Gut* 23: 637–642, 1983.
223. SETCHELL, K. D. R., A. M. LAWSON, N. TANIDA, AND J. SJÖVALL. General methods for the analysis of metabolic profiles of bile acids and related compounds in feces. *J. Lipid Res.* 24: 1085–1100, 1983.
224. SETCHELL, K. D. R., J. M. STREET, AND J. SJÖVALL. Fecal bile acids. In: *The Bile Acids. Methods and Applications*, edited by K. D. R. Setchell, D. Kritchevsky, and P. P. Nair. New York: Plenum, vol. 4, in press.
225. SHAFFER, E. A., AND D. M. SMALL. Biliary lipid secretion in cholesterol gallstone disease. The effect of cholecystectomy and obesity. *J. Clin. Invest.* 59: 828–840, 1977.
226. SHEFER, S., S. HAUSER, V. LAPAR, AND E. H. MOSBACH. Regulatory effects of sterols and bile acids on hepatic 3-hydroxy-3-methylglutaryl CoA reductase and cholesterol 7α-hydroxylase in the rat. *J. Lipid Res.* 14: 573–580, 1973.
227. SHEFER, S., G. SALEN, S. HAUSER, B. DAYAL, AND A. K. BATTA. Metabolism of iso-bile acids in the rat. *J. Biol. Chem.* 257: 1401–1406, 1982.
228. SHIPLEY, R. A., AND R. E. CLARK. *Tracer Methods for In Vivo Kinetics. Theory and Applications.* New York: Academic, 1972.
229. SIDHU, G. S., AND D. G. OAKENFULL. A new mechanism for the hypocholesterolaemic of saponins. *Br. J. Nutr.* 55: 643–649, 1986.
230. SIMON, G. L., AND S. H. GORBACH. Intestinal flora in health and disease. In: *Physiology of the Gastrointestinal Tract*, edited by L. R. Johnson. New York: Raven, 1981, p. 1361–1380.
231. SJÖVALL, J. Dietary glycine and taurine on bile acid conjugation in man. *Proc. Soc. Exp. Biol. Med.* 100: 676–678, 1959.
232. SJÖVALL, J. Separation and determination of bile acids. In: *Methods of Biochemical Analysis*, edited by D. Glick. New York: Wiley, 1964, vol. 12, p. 97–141.
233. SJÖVALL, J. Gas chromatography-mass spectrometry in studies of steroids, bile acids and bile alcohols. *Proc. Jpn. Soc. Med. Mass Spec.* 8: 29–46, 1983.
234. SMALL, D. M. The physical chemistry of the cholanic acids. In: *The Bile Acids: Chemistry, Physiology, and Metabolism. Chemistry*, edited by P. P. Nair and D. Kritchevsky. New York: Plenum, 1971, vol. 1, p. 249–356.
235. SMALL, D. M., R. H. DOWLING, AND R. N. REDINGER. The enterohepatic circulation of bile salts. *Arch. Intern. Med.* 130: 552–573, 1972.
236. SOLOWAY, R. D., H. C. CARLSON, AND L. J. SCHOENFIELD. A balloon-occludable T-tube for cholangiography and quantitative collection and reinfusion of bile in man. *J. Lab. Clin. Med.* 79: 500–504, 1972.
237. SPADY, D. K., E. F. STANGE, L. E. BILHARTZ, AND J. M. DIETSCHY. Bile acids regulate hepatic low density lipoprotein receptor activity in the hamster by altering cholesterol flux across the liver. *Proc. Natl. Acad. Sci. USA* 83: 1916–1920, 1986.

238. STEIN, W. D. *Transport and Diffusion Across Cell Membranes.* Orlando, FL: Academic, 1986.
239. STELLAARD, F., P. D. KLEIN, A. F. HOFMANN, AND J. M. LACHIN. Mass spectrometry identification of biliary bile acids in bile from gallstone patients before and during treatment with chenodeoxycholic acid. An ancillary study of the National Cooperative Gallstone Study (NCGS). *J. Lab. Clin. Med.* 105: 504–513, 1985.
240. STELLAARD, F., M. SACKMANN, T. SAUERBRUCH, AND G. PAUMGARTNER. Simultaneous determination of cholic acid and chenodeoxycholic acid pool sizes and fractional turnover rates in human serum using ^{13}C-labeled bile acids. *J. Lipid Res.* 23: 1313–1319, 1984.
241. STIEHL, A. Disturbances of bile acid metabolism in cholestasis. *Clin. Gastroenterol.* 6: 45–67, 1977.
242. STIEHL, A., R. RAEDSCH, G. RUDOLPH, U. GUNDERT-REMY, AND M. SENN. Biliary and urinary excretion of sulfated, glucuronidated and tetrahydroxylated bile acid in cirrhotic patients. *Hepatology Baltimore* 5: 492–495, 1985.
243. STREET, J. M., D. J. H. TRAFFORD, AND H. L. J. MAKIN. The quantitative estimation of bile acids and their conjugates in human biological fluids. *J. Lipid Res.* 24: 491–511, 1983.
244. STREMMEL, W., G. STROHMEYER, AND P. D. BERK. Hepatocellular uptake of oleate is energy dependent, sodium linked, and inhibited by an antibody to a hepatocyte plasma membrane fatty acid binding protein. *Proc. Natl. Acad. Sci. USA* 83: 3584–3588, 1986.
245. STURMAN, J. A., G. W. HEPNER, A. F. HOFMANN, AND P. J. THOMAS. Metabolism of ^{35}S taurine in man. *J. Nutr.* 105: 1206–1214, 1975.
246. SUWELACK, D., AND H. WEBER. Assessment of enterohepatic circulation of radioactivity following a single dose of [^{14}C]-nimodipine in the rat. *Eur. J. Drug Metab. Pharmacokinet.* 10: 231–239, 1985.
247. TABAQCHALI, S. The pathophysiological role of small intestinal bacterial flora. *Scand. J. Gastroenterol. Suppl.* 6: 139–163, 1970.
248. TAKIKAWA, H., H. OTSUKA, T. BEPPU, Y. SEYAMA, AND T. YAMAKAWA. Quantitative determination of bile acid glucuronides in serum by mass fragmentography. *J. Biochem. Tokyo* 92: 985–998, 1982.
249. TAKIKAWA, H., H. OTSUKA, T. BEPPU, Y. SEYAMA, AND T. YAMAKAWA. Serum concentrations of bile acid glucuronides in hepatobiliary diseases. *Digestion* 27: 189–195, 1983.
250. TAKIKAWA, H., Y. SEYAMA, Y. SUGIYAMA, AND S. NAGASE. Bile acid profiles in analbuminemia rats. *J. Biochem. Tokyo* 97: 199–203, 1985.
251. TAKIKAWA, H., A. STOLZ, AND N. KAPLOWITZ. Redox cycling of bile acids (BA) catalyzed by 3α-hydroxysteroid dehydrogenase ($3\alpha HSD$) (Abstract). *Hepatology Baltimore* 6: 1225, 1986.
252. TANDON, R., M. AXELSON, AND J. SJÖVALL. Selective liquid chromatographic isolation and gas chromatographic-mass spectrometric analysis of ketonic bile acids in faeces. *J. Chromatogr.* 302: 1–14, 1984.
253. TANIDA, N., Y. HIKASA, T. SHIMOYAMA, AND K. D. R. SETCHELL. Comparison of faecal bile acid profiles between patients with adenomatous polyps of the large bowel and healthy subjects in Japan. *Gut* 25: 824–832, 1984.
254. TAPPEINER, A. J. F. H. Ueber die Aufsaugung der Gallensaeuren alkalien im Dunndarme. *Wien. Akad. Sitzber.* 77: 281–304, 1878.
255. THISTLE, J. L., AND A. F. HOFMANN. Efficacy and specificity of chenodeoxycholic acid therapy for dissolving gallstones. *N. Engl. J. Med.* 289: 655–659, 1973.
256. THISTLE, J. L., AND L. J. SCHOENFIELD. Induced alterations in composition of bile of persons having cholelithiasis. *Gastroenterology* 61: 488–496, 1971.
257. THOMAS, G. P., M. NAKAGAKI, A. F. HOFMANN, AND S. MUNDLOS. A systemic marker for the rapid detection of experimental bacterial overgrowth: the ureolysis breath test (Abstract). *Gastroenterology* 82: 1197, 1982.
258. THOMSON, A. B. R., AND J. M. DIETSCHY. Intestinal kinetic parameters: effects of unstirred layers and transport preparation. *Am. J. Physiol.* 239 (*Gastrointest. Liver Physiol.* 2): G372–G377, 1980.
259. THUREBORN, E. Human hepatic bile. *Acta Chir. Scand. Suppl.* 303: 7–63, 1962.
260. TSE, F. L. S., F. BALLARD, AND J. SKINN. Estimating the fraction reabsorbed in drugs undergoing enterohepatic circulation. *J. Pharmacokinet. Biopharm.* 10: 455–461, 1982.
261. TSERNG, K.-Y., D. L. HACHEY, AND P. D. KLEIN. An improved synthesis of 24-^{13}C-labeled bile acids using formyl esters and modified lead tetra-acetate procedure. *J. Lipid Res.* 18: 400–403, 1977.
262. TSERNG, K.-Y., D. L. HACHEY, AND P. D. KLEIN. An improved procedure for the synthesis of glycine and taurine conjugates of bile acids. *J. Lipid Res.* 18: 404–407, 1977.
263. TSERNG, K.-Y., AND P. D. KLEIN. Bile acid sulfates. III. Synthesis of 7- and 12-monosulfates of bile acids and their conjugates using a sulfur trioxide-triethylamine complex. *Steroids* 33: 167–182, 1979.
264. VADNERE, M., AND S. LINDENBAUM. Distribution of bile salts between 1-octanol and aqueous buffer. *J. Pharm. Sci.* 71: 875–881, 1982.
265. VAN DER WERF, S. D. J., A. W. M. HUIJBREGTS, H. L. M. LAMERS, G. P. VAN BERGE HENEGOUWEN, AND J. H. M. VAN TONGEREN. Age dependent differences in human bile acid metabolism and 7α-dehydroxylation. *Eur. J. Clin. Invest.* 11: 425–431, 1981.
266. VAN DYKE, R. W., J. E. STEPHENS, AND B. F. SCHARSCHMIDT. Bile acid transport in cultured rat hepatocytes. *Am. J. Physiol.* 243 (*Gastrointest. Liver Physiol.* 6): G484–G492, 1982.
267. VANTRAPPEN, G., P. RUTGEERTS, AND Y. GHOOS. A new method for the measurement of bile acid turnover and pool size by a double-label, single intubation technique. *J. Lipid Res.* 22: 528–531, 1981.
268. VESSEY, D. A. The biochemical basis for the conjugation of bile acids with either glycine or taurine. *Biochem. J.* 174: 621–626, 1978.
269. VESSEY, D. A. The co-purification and common identity of cholyl CoA:glycine- and cholyl CoA:taurine-N-acyltransferase activities from bovine liver. *J. Biol. Chem.* 254: 2059–2063, 1979.
270. VESSEY, D. A., M. H. CRISSEY, AND D. ZAKIM. Kinetic studies on the enzymes conjugating bile acids with taurine and glycine in bovine liver. *Biochem. J.* 163: 181–183, 1977.
271. VLAHCEVIC, Z. R., C. C. BELL, JR., I. BUHAC, J. T. FARRAR, AND L. SWELL. Diminished bile acid pool size in patients with gallstones. *Gastroenterology* 59: 165–173, 1970.
272. VLAHCEVIC, Z. R., P. JUTTIJUDATA, C. C. BELL, JR., AND L. SWELL. Bile acid metabolism in patients with cirrhosis. II. Cholic and chenodeoxycholic acid metabolism. *Gastroenterology* 62: 1174–1181, 1972.
273. VLAHCEVIC, Z. R., M. F. PRUGH, D. H. GREGORY, AND L. SWELL. Disturbances of bile acid metabolism in parenchymal liver cell disease. *Clin. Gastroenterol.* 6: 25–43, 1977.
274. WALKER, S., A. STIEHL, R. RAEDSCH, P. KLOTERS, AND B. KOMMERELL. Absorption of ursodeoxycholic and chenodeoxycholic acid and their taurine and glycine conjugates in rat jejunum, ileum, and colon. *Digestion* 32: 47–52, 1985.
275. WATERHOUS, D. V., S. BARNES, AND D. D. MUCCIO. Nuclear magnetic resonance spectroscopy of bile acids. Development of two-dimensional NMR methods for the elucidation of proton resonance assignments for five common hydroxylated bile acids, and their parent bile acid, 5β-cholanoic acid. *J. Lipid Res.* 26: 1068–1078, 1985.
276. WEBSTER, K. H., M. C. LANCASTER, A. F. HOFMANN, D. F. WEASE, AND A. H. BAGGENSTOSS. Influence on primary bile acid feeding on cholesterol metabolism and hepatic function in the rhesus monkey. *Mayo Clin. Proc.* 50: 134–138, 1975.
277. WEINBERG, S. L., G. BURCKHARDT, AND F. A. WILSON. Taurocholate transport by rat intestinal basolateral membrane vesicles. Evidence for the presence of an anion exchange transport system. *J. Clin. Invest.* 78: 44–50, 1986.

278. WEINER, I. M., AND L. LACK. Bile salt absorption: enterohepatic circulation. In: *Handbook of Physiology. Alimentary Canal*, edited by C. F. Code. Washington, DC: Am. Physiol. Soc., 1968, sect. 6, vol. III, chapt. 73, p. 1439–1455.
279. WESTERGAARD, H., K. H. HOLTERMÜLLER, AND J. M. DIETSCHY. Measurement of resistance of barriers to solute transport in vivo in rat jejunum. *Am. J. Physiol.* 250 (*Gastrointest. Liver Physiol.* 13): G727–G735, 1986.
280. WILSON, F. A. Intestinal transport of bile acids. *Am. J. Physiol.* 241 (*Gastrointest. Liver Physiol.* 4): G83–G92, 1981.
281. YAHIRO, K., T. SETOGUCHI, AND T. KATSUKI. Effect of cecum and appendix on 7α-dehydroxylation and 7β-epimerization of chenodeoxycholic acid in the rabbit. *J. Lipid Res.* 21: 215–222, 1980.
282. YOON, Y. B., L. R. HAGEY, A. F. HOFMANN, D. GURANTZ, E. L. MICHELOTTI, AND J. H. STEINBACH. Effect of side-chain shortening on the physiological properties of bile acids: hepatic transport and effect on biliary secretion of 23-nor-ursodeoxycholate in rodents. *Gastroenterology* 90: 837–852, 1986.

CHAPTER 30

Cellular mechanisms of hepatic fluid and electrolyte transport

REBECCA W. VAN DYKE
JOHN R. LAKE
BRUCE F. SCHARSCHMIDT

Department of Medicine and Liver Center,
University of California, San Francisco, California

CHAPTER CONTENTS

Microanatomy of the Bile Secretory Unit
Cellular Mechanisms of Hepatocyte Electrolyte and
 Solute Transport
 Membrane potential
 Sinusoidal-lateral membrane transport mechanisms
  Na^+-K^+-ATPase and sodium-coupled solute transport
  Calcium transport
  Potassium channels
 Canalicular membrane transport mechanisms
  Chloride-bicarbonate exchange
  Chloride conductance
  Bile acid transport
 Organelle ion-transport mechanisms
  H^+-ATPase
  Chloride conductance
Canalicular Bile Formation
 Introduction and definition of terms
 Pathways of water and solute excretion
  Water and inorganic electrolyte movement into
   the canaliculus
  Use of inert solutes as fluid-phase markers
 Bile acid–stimulated canalicular bile formation
 Bile acid–independent bile formation
  Sodium-potassium-chloride cotransport
  Bicarbonate secretion
 Intracellular bile acid transport and bile acid–lipid coupling
 Regulation of canalicular water and solute secretion
Summary

THIS CHAPTER DEALS with the cellular mechanisms by which hepatocytes transport inorganic electrolytes, bile acids, and other selected compounds, with a particular focus on those processes believed to be important in the formation of canalicular bile. A brief review of the microanatomy of canalicular bile formation sets the stage for subsequent discussions of the cellular mechanisms involved in solute transport and canalicular bile formation and the routes of water and solute movement between blood and bile. Recent advances, newly emerging concepts, and important areas of persisting uncertainty are emphasized. Other chapters in this *Handbook* deal with events occurring "downstream" from the canaliculus as well as the biotransformation and enterohepatic recycling of certain organic solutes such as bile acids.

MICROANATOMY OF THE BILE SECRETORY UNIT

Hepatocytes are arranged in the form of single-cell-thick plates surrounded on either side by sinusoidal blood [Fig. 1; (108)]. One study with the scanning electron microscope reported a gradual increase in average diameter of sinusoids in the periportal (~4 μm) versus the pericentral regions (~5.5 μm); larger estimates (≥15 μm) of sinusoidal diameter have also been reported. Sinusoids are lined by a fenestrated endothelium. The fenestrations also increase gradually in number from the periportal to the pericentral ends of the sinusoid and are large enough (avg 0.1 μm diam) and numerous enough (9–14/μm^2 or ~5%–7% of endothelial surface area) to permit molecules such as albumin and lipoproteins free access to the space of Disse, where they can come into direct contact with the hepatocyte plasma membrane (108, 157, 158). Thus these fenestrations undoubtedly facilitate the bidirectional exchange of macromolecules between blood and hepatocytes as well as the uptake by hepatocytes of albumin-bound substances such as bile acids and other organic solutes. Because of continuous removal of solutes and oxygen from blood as it flows from portal tract to central vein, periportal and pericentral hepatocytes are exposed to different environments. These differences in the environment, as well as possibly intrinsic differences, contribute to specialization in the function of periportal and pericentral cells (93, 94). In this chapter the transport function of hepatocytes is discussed without regard to their location along the length of the sinusoid.

Hepatic bile is secreted into canaliculi, which are bounded by specialized portions of the plasma mem-

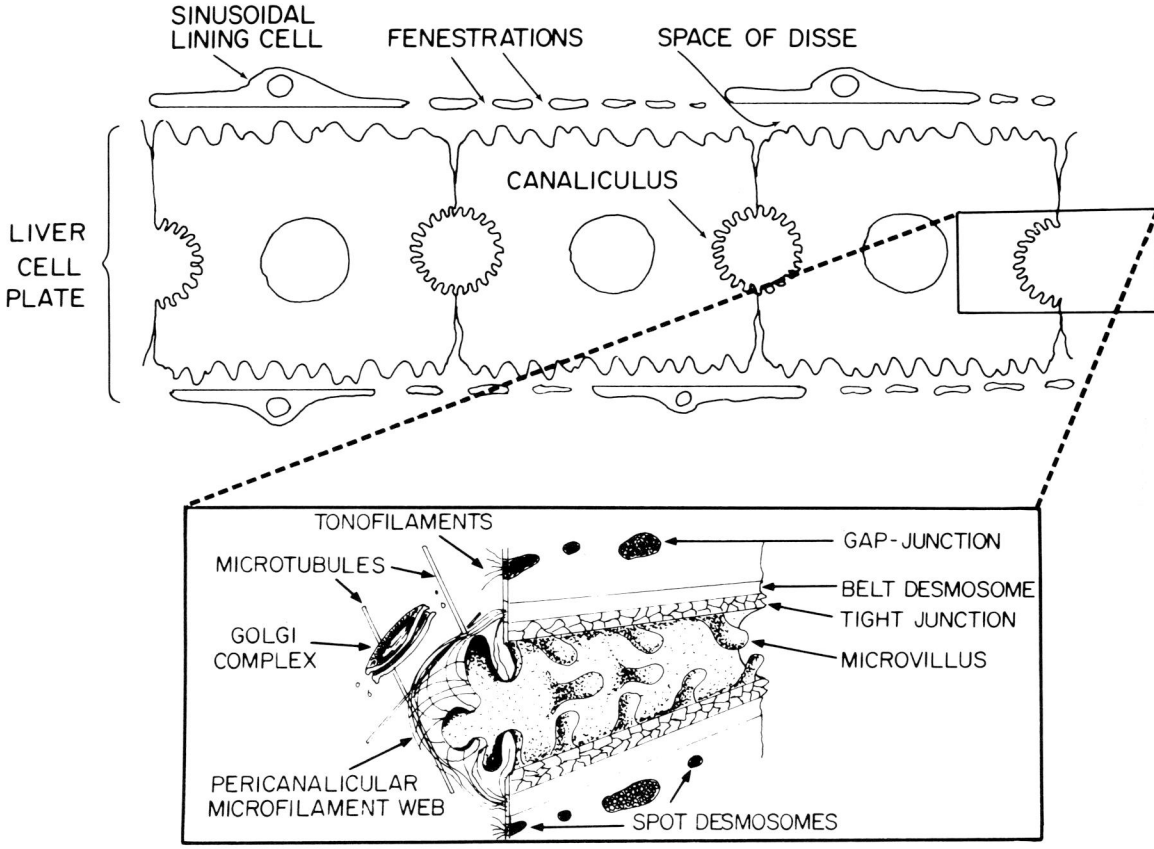

FIG. 1. Schematic illustration of a liver cell plate. Single-cell-thick liver plate (*top panel*) is separated from sinusoidal blood on either side by sinusoidal lining cells. Lining cells have large fenestrations in the cytoplasm, which allow free access of plasma to space of Disse. Each hepatocyte surface membrane consists of 3 distinct areas: the sinusoidal surface, the lateral surface, and the canalicular surface. *Bottom panel*, enlargement of the hemicanaliculus shown above. Canalicular membranes of adjacent hepatocytes are joined by tight junctions (shown here en face), which seal the canalicular space from the lateral intercellular space and sinusoidal space. Belt desmosomes adjacent to the tight junction help to hold cells together and serve as a site of attachment of pericanalicular microfilaments. Spot desmosomes can be thought of as "spot welds" between adjacent liver cells. They also help to hold cells together and are the sites of insertion of intracellular tonofilaments, which transmit passive forces to the membrane surface. Gap junctions allow direct passage of certain molecules between the cytoplasms of adjacent cells. Golgi complex, microtubules, and microfilaments are characteristically found in the pericanalicular area and may play a role in bile secretion. [From Scharschmidt (189).]

brane of two or more adjacent hepatocytes. These canaliculi interconnect to form a chicken wire–like network that surrounds individual hepatocytes within each plate and empties into interlobular ducts in the portal tracts that are lined by cuboidal epithelium (108).

The plasma membrane of hepatocytes, like that of other epithelial cells, consists of specialized domains. The sinusoidal and lateral domains account for 37% and 50%, respectively, of the cell surface and correspond to the basolateral membrane of other epithelial cells (243). The canalicular or apical membrane constitutes an estimated 13% of the cell surface (243). The three surface domains are not static entities; for example, measurements of fluid-phase endocytosis in cultured hepatocytes suggest that an amount of membrane equivalent to several times the total plasma membrane surface area is internalized each hour via endocytosis (191).

Hepatocytes are joined by junctional complexes, which separate the blood space (intercellular space) from the bile space (canaliculus). The tight junction between hepatocytes exhibits an intermediate number of sealing strands, as compared with that of other epithelia (53, 132). This fact, plus the observation that bile is isotonic with respect to plasma and rapidly mirrors changes in plasma osmolality, is consistent with the current classification of liver as a moderately leaky epithelium.

Contractile microfilaments in hepatocytes are concentrated around the canaliculus, where they form a pericanalicular web. Cinephotomicrographic studies of isolated hepatocyte couplets suggest that coordinated contraction of this web throughout the length of the canaliculus, possibly involving changes in cytosolic Ca^{2+} concentration, produces a peristaltic wave that "massages" bile into the larger collecting structures (162, 206, 242). Conversely, experimental agents (e.g., cytochalasins) and possibly drugs that impair microfilament function produce cholestasis that is accompanied by canalicular dilatation (64). Other elements of the cytoskeleton, such as microtubules, as well as intracellular organelles that may play a role in canalicular bile formation, are discussed later in this chapter.

CELLULAR MECHANISMS OF HEPATOCYTE ELECTROLYTE AND SOLUTE TRANSPORT

In this section a wide variety of plasma membrane–associated electrolyte transport mechanisms are discussed, a number of which may play a role in bile formation. Because many of these transport mechanisms occur in other cell types, this chapter emphasizes evidence bearing on the existence, presumed function, and peculiar features (if any) of these mechanisms in the liver. This discussion is organized according to the cellular location of each transport mechanism: sinusoidal-lateral plasma membrane, canalicular plasma membrane, and organelle membrane (Figs. 2, 3).

Membrane Potential

The cell membrane potential (E_m), an important determinant for electrolyte and solute transport, has been measured in rat and mouse hepatocytes in vivo or in perfused liver. Reported mean values range from -33 to -43 mV (51, 69, 90, 164, 247, 250). The hepatocyte E_m appears to be accounted for by the operation of an electrogenic Na^+-K^+ pump and conductance of the hepatocyte plasma membrane to ions (52) [($P_{Cl} > P_K > P_{Na}$ in rat liver (52, 90) and $g_K > g_{Cl} > g_{Na}$ for the sinusoidal membrane of isolated rat hepatocyte couplets, where P is permeability and g is conductance (88)]. This membrane potential is not fixed but rather varies considerably in response to a number of stimuli such as fasting (depolarization) (69, 164), exposure to glucagon and cAMP (hyperpolarization) (69, 166), exposure to alanine (transient depolarization followed by hyperpolarization) (69), and partial hepatectomy (hyperpolarization) (164). Isolated rat hepatocytes in primary culture exhibit similar properties with a resting E_m of -10 to -45 mV (88, 208, 250) that depolarizes after exposure to insulin (249).

Recent microelectrodes have been placed in the apparently closed bile canaliculus between couplets of

Space of Disse

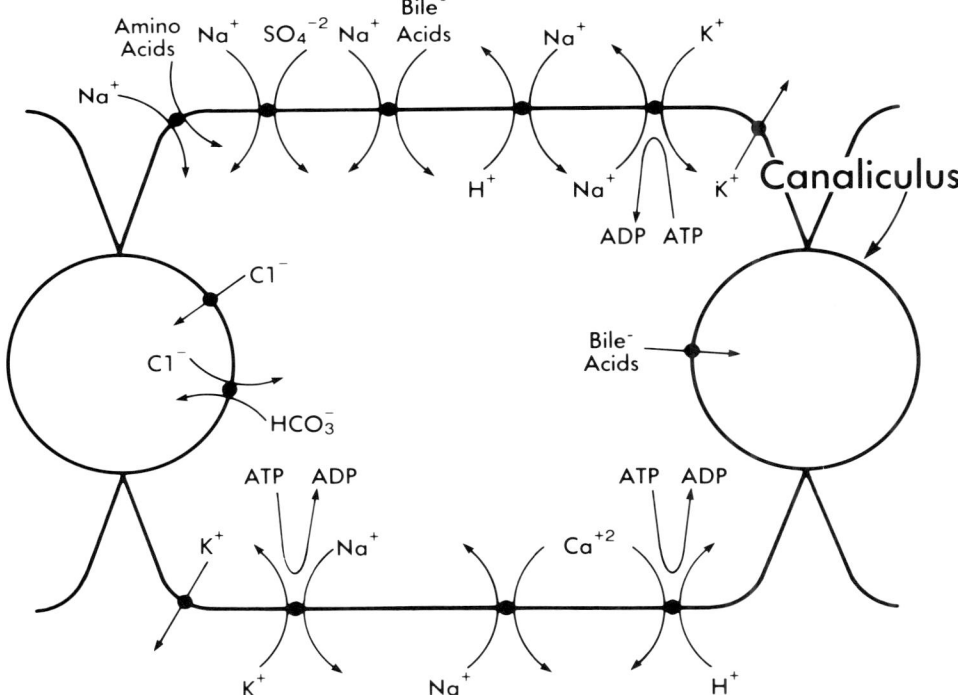

FIG. 2. Plasma membrane ion-transport mechanisms thought to be present on hepatocytes. For those transport processes (e.g., Na^+-Ca^{2+} exchange and Ca^{2+}-ATPase) that are poorly characterized in liver, the mechanism is depicted as it is known to occur in other cell types.

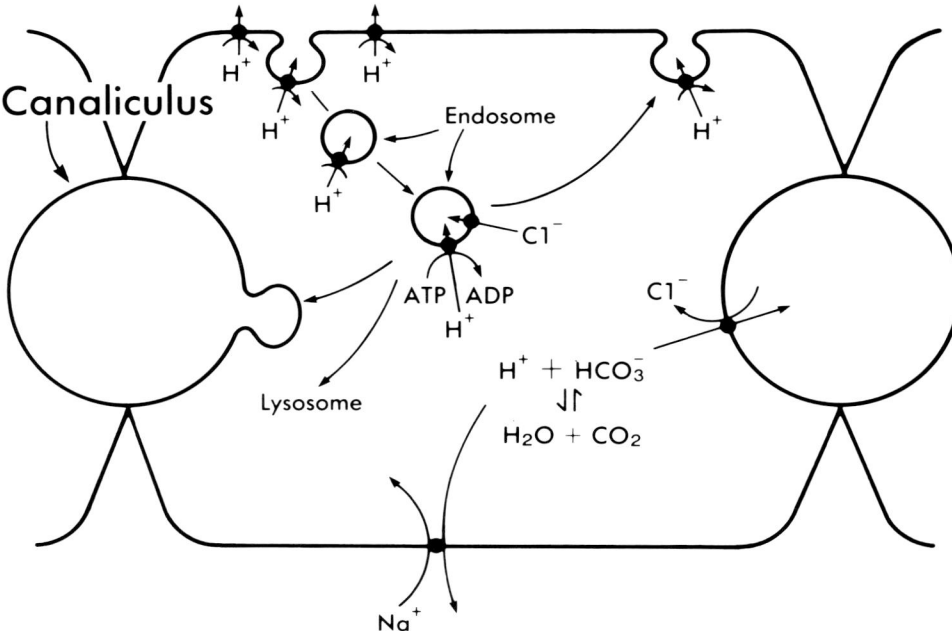

FIG. 3. Relationships between hepatocyte proton transport and endocytosis. Hepatocyte proton transport, mediated via a primary proton pump or Na^+-H^+ exchange, is postulated to play an important role both in biliary HCO_3^- secretion and bile formation and in endocytosis.

freshly isolated rat hepatocytes, and an electrical potential difference of -5.9 ± 3.3 mV was recorded between the canalicular space and the incubation medium (87). This is similar to the bile-to-peritoneal electrical potential difference of -4.4 mV recorded in intact rats (30) and is consistent with the classification of liver as a leaky epithelium (see MICROANATOMY OF THE BILE SECRETORY UNIT, p. 597).

Sinusoidal-Lateral Membrane Transport Mechanisms

Na^+-K^+-ATPASE AND SODIUM-COUPLED SOLUTE TRANSPORT. *Localization of Na^+-K^+-ATPase.* The ubiquitous Na^+ pump found in all epithelial cells, Na^+-K^+-ATPase, has been localized cytochemically, by the ouabain-sensitive and K^+-dependent nitrophenylphosphatase technique, to the sinusoidal and lateral plasma membrane in rat hepatocytes (31, 136, 251), a location analogous to that in virtually all other epithelial cells.

Immunofluorescence studies, however, utilizing both polyclonal (222) and monoclonal (141, 195) antibodies (raised against the α-subunit of canine or rat kidney Na^+-K^+-ATPase) that inhibit Na^+-K^+-ATPase activity in liver plasma membrane preparations (141, 222), have shown staining of both canalicular and sinusoidal plasma membranes of canine and rat hepatocytes. It is estimated that the density of antibody-binding sites is 2.5-fold higher on the canalicular membrane than on the sinusoidal and lateral plasma membrane (222). A number of issues, including antibody specificity, remain to be answered before the apparently disparate results with histochemical and immunofluorescent techniques can be reconciled.

Regulation of Na^+-K^+-ATPase. It is estimated that rat hepatocytes contain $\sim 2.4 \times 10^5$ pump units per cell with a maximum turnover rate of 9,000 ATP molecules hydrolyzed per pump per minute (195). In the intact cell, however, Na^+-K^+-ATPase probably operates at less than maximum velocity (V_{max}), and factors involved in the regulation of hepatic Na^+-K^+-ATPase have been studied by investigators interested in the relationships between Na^+-K^+-ATPase and solute transport, bile formation, or hepatic regeneration. In these studies, Na^+-K^+-ATPase activity was assessed in one of two ways: *1*) ouabain-sensitive Na^+-K^+–dependent ATP hydrolysis by liver plasma membranes, which measures the V_{max} of Na^+-K^+-ATPase in broken-cell preparations, or *2*) ouabain-sensitive ^{86}Rb uptake into hepatocytes, which assesses the actual rate of cation pumping by intact cells.

A variety of agents are reported to alter hepatic Na^+-K^+-ATPase, and the best characterized of these are summarized in Table 1. The mechanisms by which these agents alter Na^+-K^+-ATPase activity are not fully understood but may include substrate availability (especially Na^+), altered plasma membrane fluidity, altered synthesis of pump units, and changes in intra-

TABLE 1. Regulation of Hepatic Na^+-K^+-ATPase

Agent	Preparation Studied	Time Course	Effect	Mechanism	Ref.
Intracellular sodium	Cultured hepatocytes	Seconds to minutes	Increase	Substrate availability	101, 232
Insulin Epidermal growth factor	Isolated hepatocytes	Seconds to minutes	Increase	Increased Na^+-H^+ exchange and intracellular Na^+	66, 67
α-Agonists Vasopressin Angiotensin	Isolated hepatocytes	Seconds to minutes	Increase	Unknown mechanism related to intracellular Ca^{2+} concentration	21, 28
Glucagon	Isolated hepatocytes	Minutes	Increase	Unknown	101
	Cultured hepatocytes	Hours	Increase	? Synthesis of new pump units	67
Ethinyl estradiol	Liver plasma membranes Isolated hepatocytes	Days	Decrease	Decreased membrane fluidity	57, 116
Thyroxine	Liver plasma membranes Cultured hepatocytes	Days	Increase	? Synthesis of new pump units	116
				? Increased membrane fluidity	104
Glucocorticoids	Liver plasma membranes	Days	Increase	? Synthesis of new pump units	116, 155
				? Increased membrane fluidity	
Phenobarbital	Liver plasma membranes	Days	Increase	? Synthesis of new pump units	203
Chlorpromazine	Liver plasma membranes	Seconds to minutes	Decrease	? Direct inhibition of pump	115, 188
	Cultured hepatocytes			? Altered membrane fluidity	233
Bile salts	Liver plasma membranes	Seconds to minutes	Decrease	Unknown	190
		Days	Increase	Unknown	204, 241
Endotoxin	Liver plasma membranes	Seconds to minutes	Decrease	Unknown	227
Ethanol	Liver plasma membranes	Seconds to minutes	Increase	? Increased membrane fluidity	184
		Days	Decrease	? Decreased membrane fluidity	85

cellular Ca^{2+}. There is good evidence that small increases in intracellular Na^+ concentration (~12 mM normally in cultured hepatocytes) caused, for example, by enhanced Na^+-coupled uptake of bile acids or amino acids, are associated with an increase in Na^+-K^+-ATPase–mediated cation pumping (232). In contrast to such secondary changes in Na^+-K^+-ATPase activity, which presumably represent a compensatory mechanism designed to preserve electrochemical driving forces, little is known regarding the effects of primary changes in Na^+-K^+-ATPase activity and the electrochemical Na^+ gradient on Na^+ and solute transport. In three studies that have addressed this issue, cultured or freshly isolated hepatocytes were exposed to glucagon, chlorpromazine, or ethinyl estradiol. Membrane hyperpolarization produced by glucagon was associated with enhanced Na^+-coupled transport of taurocholate (63), whereas inhibition of Na^+-K^+-ATPase–mediated cation transport by chlorpromazine (233) and ethinyl estradiol (25) was associated with decreased (25, 233) transport of taurocholate and alanine. The overall importance of this regulatory mechanism for hepatic (and epithelial) physiology remains to be determined.

Sodium-coupled solute transport. Sodium-coupled "secondary-active" transport mechanisms driven by the electrochemical Na^+ gradient attributed to Na^+-K^+-ATPase accounts for the transport of a wide variety of solutes into and out of nonpolarized cells as well as epithelial cells (13). A number of these Na^+-coupled solute-transport processes that have been identified in the sinusoidal-lateral membrane of liver cells are described next. The role of some of these transporters in hepatic secretion and bile formation is discussed in CANALICULAR BILE FORMATION, p. 606.

Sodium-coupled amino acid transport. Amino acid metabolism takes place primarily in the liver, and amino acid transport may be an important rate-limiting step for metabolic processes such as gluconeogenesis. At least eight different amino acid–transport systems (six of which exhibit Na^+ dependency) have been described in the liver (50). The Na^+-dependent systems include system A, which transports most small neutral amino acids, can utilize Li^+ in place of Na^+, and is subject to adaptive and hormonal regulation; system ASC, which transports neutral amino acids, particularly those with an –OH or –SH group; an anionic system, system Z, which transports dicarboxylic amino acids; system Gly, which transports glycine and scarosine; and system N, which mediates uptake of histidine, glutamine, and asparagine and is stimulated by fasting (119).

Further characterization of Na^+-coupled alanine uptake (system A primarily) in isolated hepatocytes and plasma membrane vesicles has demonstrated that alanine uptake is saturable [K_m ~2–10 mM; (129, 205,

228, 229)], concentrative (intracellular:extracellular ratio ~10; (129)], electrogenic (127, 205, 228, 229), stereospecific, and localized to the basolateral plasma membrane (228).

Sodium-coupled transport of alanine (and other A and N system amino acids), which may be rate limiting for gluconeogenesis, is increased by fasting [adaptive regulation; (49)] through changes in intracellular amino acid concentration (246) and derepression of synthesis of system A transport proteins, resulting in an increase in the V_{max} for high-affinity amino acid transport (68, 117, 119, 129). System A amino acid transport in intact hepatocytes is also stimulated by insulin (67, 117, 118), glucagon (62, 117, 118), glucocorticoids (45, 117, 118), dibutyryl cAMP (187), and catecholamines (45, 118); it increases during growth of neonatal rats (22) and hepatic regeneration (249). Stimulation by hormones may occur by synthesis of new transporters, possibly through changes in cAMP or intracellular Ca^{2+} (118).

In summary, hepatocytes appear to possess a complex set of amino acid–transport mechanisms, some of which are Na^+ dependent and are responsive to amino acid availability or hormones important in regulating amino acid and glucose metabolism.

Sodium-coupled bile acid transport. Bile acids are synthesized exclusively in the liver and cycle between liver and intestine in the enterohepatic circulation (see the chapter by Hofmann in this *Handbook*). Specific bile acid–transport mechanisms have been described in hepatocytes, terminal ileal cells, and renal tubule cells. Studies utilizing intact liver, isolated and cultured hepatocytes, and liver plasma membrane vesicles have all demonstrated that certain bile acids (e.g., taurocholate) enter hepatocytes via Na^+ coupling (7, 60, 102, 192).

Sodium-dependent taurocholate uptake is saturable (with estimates of K_m ranging from 15 to 250 μM) (7, 60, 102, 192, 236) and concentrative (with an intracellular-extracellular ratio of up to 50:1) (192, 236). In both intact cells and membrane vesicles, taurocholate uptake is inhibited by other bile acids (7, 60, 95, 102), as well as by furosemide and bumetanide (35). Using purified membrane vesicles, Inoue et al. (102) and Meier et al. (152) localized Na^+-dependent uptake to the basolateral (sinusoidal) plasma membrane. Developmental changes in hepatic taurocholate transport in rats also have been demonstrated; the V_{max} for Na^+-coupled taurocholate uptake increases from 7 to 56 days of age (219).

The stoichiometry of Na^+-coupled bile acid uptake remains uncertain. Studies in isolated or cultured hepatocytes have suggested equimolar stoichiometry (192) but also binding of more than one Na^+ to the carrier (7, 192) and a dependence of taurocholate uptake on membrane potential (63). Studies examining the potential dependence of taurocholate transport by membrane vesicles have also been interpreted as consistent with electroneutral (60, 152) as well as electrogenic (102) transport. Finally, electrophysiological studies of rat liver in vivo (69) indicate that hepatic uptake of taurocholate is associated with depolarization of the membrane potential similar to that observed with alanine, the uptake of which is known to be electrogenic. Although these latter findings suggest an electrogenic uptake mechanism (i.e., Na^+:taurocholate ratio $> 1:1$), this conclusion must be considered tentative.

There exists considerable heterogeneity among physiological bile acids with respect to their dependence on Na^+ coupling for hepatic uptake (236). Relatively hydrophobic unconjugated di- and monohydroxy bile acids enter isolated or cultured hepatocytes more rapidly than do conjugated trihydroxy bile acids such as taurocholate, and much of their uptake is nonsaturable, Na^+ independent, and correlates closely with their decane-water partition coefficient (236), suggesting that hydrophobic bile acids readily penetrate or cross the membrane bilayer without benefit of a specialized Na^+-coupled carrier system. These findings do not necessarily indicate that di- and monohydroxy bile acids are unsuitable substrates for the Na^+-coupled carrier; many of these hydrophobic bile acids competitively inhibit Na^+-coupled taurocholate transport (95), indicating considerable affinity for the carrier. These characteristics (2, 95) are very similar to those of bile acid transport in the distal ileum (131), suggesting that a fundamentally similar transport mechanism is responsible for bile acid transport at both the hepatic and intestinal "poles" of the enterohepatic circulation.

Many investigators have tried to identify and/or isolate the hepatic bile acid carriers. Sodium-independent cholic acid–binding sites on liver plasma membranes have been identified (1, 9) and have been shown to increase with chronic administration of cholic acid to rats (204). In addition, a photoreactive taurocholate analogue (7-ADTC) was found to label a 54-kDa protein in hepatocyte membranes (238), while three other photoreactive taurocholate derivatives were observed to label a total of five different polypeptides (20–67 kDa) in both basolateral and canalicular plasma membrane preparations (124). The functions of these proteins remain unclear.

In summary, hepatocytes have a sinusoidal membrane bile acid–transport mechanism that is saturable, Na^+ dependent, intrinsically concentrative, and exhibits considerable substrate selectivity. Trihydroxy conjugated bile acids are most heavily dependent on this system for hepatic uptake. Bile acid–transport capacity and carrier density appear to vary as a function of age and may be increased by chronic bile acid loading.

Sodium-hydrogen exchange. The Na^+-H^+ antiporters are widely distributed in both procaryotic and eucaryotic membranes, and numerous biologic roles have been attributed to them, including regulation of intracellular pH (185), urinary acidification (12), and initiation of cell growth and differentiation (130). Hepatocytes, like other cells (185), maintain an intracellular pH that is more alkaline than expected [pH 6.85–7.18; (54, 169)] if protons were passively distributed across the plasma membrane, and a decrease in intracellular pH is noted after removal of Na^+ (169). These observations indicate the presence of an Na^+-dependent mechanism for active proton extrusion. Studies with rat liver membrane vesicles have directly demonstrated Na^+-H^+ exchange in sinusoidal (11, 156) but not canalicular (156) membranes. A population of vesicles (presumably with an "inside-out" orientation) containing both Na^+-H^+ exchange and Na^+-K^+-ATPase has been identified (11).

In many cell types, cell proliferation and differentiation are preceded by an influx of Na^+ and cytoplasmic alkalinization (105, 185), apparently mediated via accelerated Na^+-H^+ exchange. Similarly, proliferation of cultured rat hepatocytes, induced by insulin, glucagon, and epidermal growth factor (EGF), is accompanied by an increased flux of Na^+ (intracellular pH has not been measured) and proliferation is blocked by amiloride (122, 139). Because amiloride also inhibits hepatocyte protein synthesis (140), an etiologic role for Na^+-H^+ exchange in hepatic regeneration remains to be clarified. The role of Na^+-H^+ exchange in biliary HCO_3^- secretion and bile flow is discussed in CANALICULAR BILE FORMATION, p. 606.

Sodium-calcium exchange. Like Na^+-H^+ exchange, Na^+-Ca^{2+} exchange has been described in the plasma membrane of a variety of cell types. Sodium-calcium exchange, which exchanges 3 Na^+ for 1 Ca^{2+}, is best characterized in excitable tissues, such as the cardiac cell sarcolemma, where it is thought to extrude Ca^{2+} from the cell during repolarization and to maintain a low resting cytoplasmic Ca^{2+} concentration [see CALCIUM TRANSPORT, this page; (46, 168)].

Sodium-calcium exchange in hepatocytes is poorly characterized, perhaps partly because of difficulties in distinguishing Ca^{2+} efflux due to Na^+-Ca^{2+} exchange from that due to Ca^{2+}-ATPase. Two groups have been unable to show Na^+-dependent Ca^{2+} efflux from intact hepatocytes (24, 121). After reversal of the Na^+ gradient, however, Na^+-dependent Ca^{2+} uptake is demonstrable (24). The observation that ATP-dependent Ca^{2+} uptake into liver plasma membrane vesicles (mediated by Ca^{2+}-ATPase) is partially released by the addition of extravesicular Na^+ is further evidence for the existence of Na^+-Ca^{2+} exchange in the hepatocyte plasma membrane vesicles (125). The physiological role of this transporter in hepatocytes is not clear.

Sodium-coupled chloride uptake. Sodium-coupled Cl^- uptake, via either Na^+-Cl^- symport, Na^+-K^+-$2Cl^-$ symport, or coupled Na^+-H^+, Cl^--HCO_3^--exchangers, has been described in many epithelial tissues, notably kidney, intestine, trachea, and salt gland (58, 163), where replacement of Na^+ (or K^+ or both for Na^+-K^+-Cl^- symport) reduces Cl^- uptake; replacement of Cl^- reduces Na^+ uptake; and intracellular Cl^- activity is higher than predicted for passive distribution (4, 78, 163).

Studies using cultured rat hepatocytes show that Cl^- uptake is saturable ($K_m \sim 117$ mM) (192) and is not significantly affected by Na^+ replacement or by preincubation with ouabain (194) and that Cl^- replacement fails to significantly decrease Na^+-entry rate (194). In the intact rat liver, in cultured hepatocytes, and in the perfused rat liver, apparent steady-state exchangeable intracellular Cl^- concentrations do not differ from those predicted based on cell membrane potential (51, 52, 194, 247). Sodium-dependent Cl^- transport has not been demonstrated in either basolateral or canalicular membrane vesicles prepared from rat liver (151), and recent studies with Cl^--selective microelectrodes suggest that Cl^- is passively distributed in rat hepatocytes (70). Thus current evidence does not support the existence of Na^+-coupled Cl^- transport in hepatocytes.

Other sodium-coupled transport mechanisms. Sodium-coupled transport of a variety of other inorganic and organic anions and neutral compounds has been described in other epithelia. In liver cells, sulfate, and perhaps HCO_3^-, uptake occurs via an Na^+-dependent process (239). Hepatic transport of other solutes, especially Krebs cycle intermediates and other metabolically important molecules such as lactate, have not been studied.

Interactions between solutes transported by sodium-coupled mechanisms. Because uptake of Na^+, and often net positive charge, accompanies Na^+-coupled solute transport, it has been suggested that transport of one solute might decrease the electrochemical Na^+ gradient and thus impair transport of a second solute. In liver plasma membrane vesicles, for example, alanine clearly inhibits taurocholate uptake by rapidly dissipating the Na^+ gradient (33). In intact cells, however, interactions between the uptake of amino acids and sugars or bile acids occur only at high concentrations (34, 160), and these interactions do not parallel the amount of Na^+ carried into the cell (3). Because hepatic alanine transport actually hyperpolarizes hepatocytes (69, 126) and stimulates Na^+-K^+-ATPase (232), relatively little change in the overall electrochemical Na^+ gradient may occur under physiological conditions (69).

CALCIUM TRANSPORT. Hepatocytes, like many other cells, maintain a very low concentration of intracellular free (cytoplasmic) Ca^{2+} [<200 nM; (19, 109)] and utilize intracellular Ca^{2+} as a second messenger for a variety of hormones (27, 248). Maintaining this low concentration of intracellular Ca^{2+} requires transport mechanisms for removal of Ca^{2+} from the cytoplasm to storage pools in mitochondria and endoplasmic reticulum as well as to the extracellular space (42, 44, 110). Furthermore, Ca^{2+} is also excreted by hepatocytes into bile (56), although the route or mechanism of Ca^{2+} secretion is not identified. Three plasma membrane Ca^{2+}-transport mechanisms have been described in other cell types: *1*) Ca^{2+}-ATPase, a primary Ca^{2+} pump; *2*) Na^+-Ca^{2+} exchange (see *Sodium-calcium exchange*, p. 603); and *3*) Ca^{2+} channels.

Ca^{2+}-ATPase. A Ca^{2+}-stimulated ATPase activity has been identified (146) in liver plasma membrane vesicles that appears to correlate with ATP-dependent Ca^{2+} transport (15, 65, 146, 147) and presumably represents an ATP-driven Ca^{2+} pump similar to that identified in the plasma membrane of a wide variety of cells (46). In liver this ATPase activity does not require free Mg^{2+}, has a high affinity for Ca^{2+} [13 nM; (146)], is calmodulin independent (146), appears to be uncoupled by other metal ions [Fe^{2+}, Mn^{2+}, Co^{2+}; (147, 165)], and may be inhibited in vitro by glucagon (147). The ATP-dependent Ca^{2+} uptake into vesicles prepared from liver plasma membrane has also been demonstrated; however, contamination by mitochondrial or endoplasmic reticulum Ca^{2+} pumps remains an issue (5, 47, 65, 125, 172). This problem has been addressed (125) by using an inhibitor of mitochondrial Ca^{2+} transport (ruthenium red) and by demonstrating very different pH optimum for Ca^{2+} transport by plasma membranes (pH 8.0) and endoplasmic reticulum (pH 6.8). Whether the liver plasma membrane Ca^{2+}-stimulated ATPase actually represents this Ca^{2+} transporter remains controversial; one investigator has reportedly purified two different Ca^{2+}-stimulated ATPases from rat liver plasma membrane, only one of which mediates Ca^{2+} transport (143, 144).

Studies exploring the regulation of Ca^{2+}-ATPase have demonstrated that *1*) glucagon apparently inhibits the pump when added directly to membrane vesicles in vitro (147), *2*) free sulfhydryl groups may be important for normal pump function (23), and *3*) in vivo perfusion of liver with the intracellular Ca^{2+}-mobilizing hormones vasopressin, angiotensin II, and α_1-agonists (but not glucagon) inhibits ATP-dependent Ca^{2+} uptake into liver plasma membrane vesicles (172). These latter findings suggest that transient inhibition of Ca^{2+}-ATPase may prolong the elevation of cytosolic Ca^{2+} concentration and thus the duration of action of some hormones.

Calcium channels. Calcium channels have been described in a variety of excitable tissues where they participate in the rapid influx of Ca^{2+} during action potentials or muscle contraction (100). The observation that stimulation of hepatocytes by Ca^{2+}-mobilizing hormones increases plasma membrane Ca^{2+} permeability suggests the presence of Ca^{2+} channels in hepatocytes (28, 109, 248). However, the function of such putative Ca^{2+} channels in hepatocytes, as in other epithelial cells (248), is poorly understood.

POTASSIUM CHANNELS. A variety of K^+ channels, often regulated by Ca^{2+} and/or by voltage, have been described in epithelial tissues. These channels appear to mediate efflux of K^+ taken up by Na^+-K^+-ATPase and to regulate cell volume and membrane potential (167). Glucagon- and α-$_1$-agonist–stimulated K^+ loss from hepatocytes has been observed by a number of investigators (42, 106) and in guinea pig (but not rat) hepatocytes appears to be Ca^{2+} dependent and blocked by quinine, apamin, and tetraethylammonium (TEA) (42, 43, 55), findings consistent with the presence of Ca^{2+}-activated K^+ channels. In rat hepatocytes, both uptake of alanine and cell swelling (17, 29, 126, 128) are associated with an increase in K^+ permeability and efflux (17, 29, 128), as well as a decrease in cell volume (17).

Canalicular Membrane Transport Mechanisms

Transport across the apical pole of the hepatocyte, the canalicular plasma membrane, has been difficult to study directly. The recent development of techniques for preparation of relatively pure canalicular membrane vesicles has permitted questions regarding canalicular transport mechanisms to be addressed (107, 153).

CHLORIDE-BICARBONATE EXCHANGE. Chloride-bicarbonate exchange in epithelial cells may function either in concert with Na^+-H^+ exchange to effect net uptake of NaCl ("double exchangers") or, in cells engaged in proton secretion, to facilitate HCO_3^- efflux across the opposite cell membrane (173). Chloride-bicarbonate exchange appears to be an electroneutral process and is inhibited by stilbene compounds such as DIDS (4,4′-diisothiocyano-2,2′-disulfonic acid stilbene). Recent studies have demonstrated the presence in canalicular plasma membrane vesicles, but not basolateral (sinusoidal) plasma membrane vesicles, of electroneutral, HCO_3^-, and pH gradient–sensitive Cl^- uptake that is inhibited by DIDS and probably represents Cl^--HCO_3^- exchange (151). This exchange mechanism may facilitate biliary HCO_3^- excretion.

CHLORIDE CONDUCTANCE. Chloride conductances (or channels) in salt-transporting epithelia are believed to function in concert with an Na^+-dependent Cl^- transporter on the opposite cell membrane to mediate net transepithelial Cl^- absorption or secretion (123, 202).

Studies using purified rat liver canalicular plasma membrane vesicles (151) have demonstrated voltage-dependent, DIDS-insensitive Cl^- transport that is consistent with the presence of a Cl^- channel. Basolateral plasma membrane vesicles exhibit no such Cl^- transport (151). Studies with isolated or cultured hepatocytes are also consistent with a transport mechanism for Cl^- in the hepatocyte plasma membrane (20, 192), and a Cl^--conductance pathway has been described (230) in rat liver endocytic vesicles, presumably derived from the basolateral plasma membrane. Although these studies are consistent with the presence of a Cl^--conductance pathway in canalicular (and possibly basolateral) membranes, their physiological role is uncertain. In contrast to salt-transporting epithelia, neither Na^+-coupled Cl^- transport nor active Cl^- secretion has been observed in hepatocytes, and Cl^- appears to be passively distributed across the hepatocyte plasma membrane.

BILE ACID TRANSPORT. Liver in vivo exhibits a maximal secretory rate for bile acids, and different bile acids appear to compete for biliary secretion. Moreover, most studies have suggested that the uptake rate for bile acids considerably exceeds the maximal secretory rate. Thus characteristics of bile acid transport in vivo are consistent with the presence of a carrier-mediated, rate-limiting mechanism for bile acid secretion across the canalicular membrane. Studies using two different canalicular membrane preparations have further demonstrated saturable, electrical potential-sensitive, Na^+-independent taurocholate transport, which exhibits transstimulation (103, 152). Collectively these studies indicate the presence of a canalicular "carrier" for bile acids. They further suggest that membrane potential (cell interior $\sim$30–40 mV negative relative to blood and probably canaliculus) represents an important driving force for the bile acid secretion, and it is tempting to assume that the characteristics of the putative carrier are the major determinants of overall rates of hepatic bile acid transport.

However, bile acid secretion by intact liver presents a more complex picture than these data imply. *1*) The maximum secretory rate for bile acids does not remain constant in the face of increasing rates of bile acid delivery. Rather, bile acid secretion rate reaches a maximum and then declines with bile flow at increasing rates of bile acid infusion. This decline appears to represent a toxic effect of high concentrations of bile acid on the liver, and the maximum secretory rate for a particular bile acid may reflect its toxicity as well as intrinsic limitations in membrane carrier density or kinetics (96). *2*) At least for certain unconjugated bile acids, the maximal rate of secretion appears to reflect limitations in conjugation rather than secretion (252). *3*) The relationship between bile acid transport across the canalicular membrane via carriers and the possible role of vesicles in transcellular bile acid transport is uncertain (see *Bile Acid–Stimulated Canalicular Bile Formation*, p. 609).

Organelle Ion-Transport Mechanisms

Hepatocyte endocytic vesicles (derived from the plasma membrane) exhibit several mechanisms for ion transport, notably a primary proton pump and Cl^- conductance. Because these transport mechanisms might also occur in the plasma membrane and play important roles in cell pH regulation and bile formation, they are included in this discussion.

H^+-ATPase. Proton transport that is ATP dependent has been demonstrated in a variety of subcellular organelles from liver, including lysosomes (98), clathrin-coated vesicles (235), endosomes (186), Golgi (81), endoplasmic reticulum (174), and multivesicular bodies (231), as well as in organelles from other cell types (71, 107, 114, 214). The characteristics of proton transport in these organelles, where it has been studied, appear quite similar. Vesicle acidification is ATP dependent and inhibited by N-ethylmaleimide but not by ouabain, vanadate, or oligomycin (231, 234, 235). Acidification is electrogenic (i.e., the pump appears to transport only H^+) and requires the presence of Cl^- for maximal rates. Acidification is relatively insensitive to cation replacement; anions other than halides support acidification poorly, if at all (114, 231, 234, 235), but instead result in development of a large, interior-positive membrane potential (231, 234, 235). Thus Cl^- appears to serve as a permeable charge-compensating anion. In addition, experiments performed under voltage-clamp conditions with renal endosomes suggest that Cl^- may also directly activate the proton pump (114).

Morphological studies of cultured fibroblasts and hepatoma cells have confirmed that a wide variety of intracellular organelles are acidified, probably by this same proton pump (6, 198). Furthermore, antibodies raised against a 100-kDa lysosomal membrane protein (that may represent the proton pump) label a variety of acidic intracellular compartments as well as the plasma membrane of rat liver (175), suggesting the presence of this pump also in the hepatocyte plasma membrane.

A physiological role for this H^+-ATPase (proton pump) in receptor-mediated endocytosis of proteins by liver and other tissues is well established. Acidification appears to allow separation of receptor from ligand after endocytosis, because the binding affinity of ligand and receptor is pH sensitive (218). Separation is followed by sorting of ligand (destined for lysosomal degradation) from receptor (recycled back to the plasma membrane and reutilized) (41, 240).

This H^+-ATPase may have other functions as well. An apparently similar proton pump, located on the apical plasma membrane of urinary epithelial cells, is thought to mediate acidification by turtle bladder (5) and to play a role in acid secretion by mammalian kidney (12, 120, 199). In these tissues a proton pump seems to be present in plasma membrane and in intracellular vesicles and to shuttle between these two sites via fusion of these vesicles with the plasma membrane (5, 83, 120). Thus in liver a primary proton pump might play a role in transepithelial H^+ transport and bile formation, in intracellular pH regulation, and in receptor-mediated endocytosis.

CHLORIDE CONDUCTANCE. Endocytic vesicles, including clathrin-coated vesicles, and multivesicular bodies, appear to contain a Cl^- conductance in parallel with the proton pump just discussed. In these vesicles, Cl^- uptake is markedly enhanced by the membrane potential established by the proton pump (81, 215, 230) or, in the absence of ATP, by an interior-positive

K⁺ plus valinomycin diffusion potential (81, 215). This Cl⁻ channel presumably allows rapid movement of Cl⁻ and thus maximal vesicle acidification. Furthermore it has been suggested that this Cl⁻ conductance is linked to, and regulated by, the proton pump (114).

CANALICULAR BILE FORMATION

Introduction and Definition of Terms

Bile secretory pressure in the isolated perfused rat liver has been repeatedly demonstrated to be relatively independent of and (depending on experimental conditions) to exceed perfusion pressure (223). Thus the formation of canalicular bile is not attributable to ultrafiltration. Rather it depends on the active secretion by hepatocytes of inorganic electrolytes and organic solutes such as bile acids. This section builds on the preceding discussion of hepatocellular transport mechanisms and focuses specifically on those mechanisms currently believed to play a role in canalicular bile formation, as well as the pathways of water on solute excretion.

Certain general areas of uncertainty pertain to canalicular bile formation. Because of the technical inability to micropuncture the canaliculus (0.5–1.0 μm diam) of intact liver, it is not currently possible to measure either the concentration of various solutes in canalicular bile or the bile-to-blood electrical potential difference. [Studies in isolated hepatocyte couplets (87) suggest a blood-to-bile potential difference of 2–4 mV, bile negative.] Thus liver physiologists are currently unable to employ certain conventional electrophysiological techniques used to study the transport of charged solutes by other epithelia. A second consequence of the inability to sample bile proximal to the major ducts is that the relative contribution in vivo to bile flow, via secretion of absorption, of canaliculi and larger ducts and ductules cannot be directly measured. Instead the rate of canalicular bile formation is inferred from the rate of clearance of inert markers such as erythritol, which were originally believed to enter bile at the level of the canaliculus only. The hepatic handling of fluid-phase markers such as erythritol and sucrose, including current information on the routes of transport of these markers from blood to bile and the attendant implications regarding their use in experimental studies of bile formation, is described next.

The formation of canalicular bile is often divided operationally into two components or processes: *1)* bile acid–dependent bile formation (BADBF) and *2)* bile acid–independent bile formation (BAIBF). Conventionally BADBF is attributed to the osmotic effect of actively secreted bile acids, and operationally it is defined by the slope of the line relating canalicular bile flow and bile acid output [Fig. 4; (32, 36, 189, 193)]. By contrast, BAIBF is operationally defined as

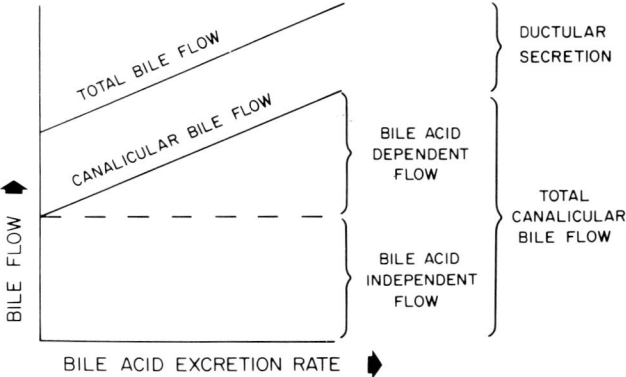

FIG. 4. Schematic representation of components of bile flow. Total bile flow consists of a theoretically constant ductular secretion, a theoretically constant bile acid–independent canalicular secretion, and a bile-dependent canalicular secretion that varies directly with bile acid output. In humans these components have been estimated by Boyer and Bloomer (37) to be ~180 ml/day, 225 ml/day, and 200 ml/day, respectively. [From Scharschmidt (189).]

the extrapolated "Y" intercept (Fig. 4) of the line relating canalicular bile flow to bile acid output and is conventionally attributed to the active secretion of inorganic electrolytes (36, 193). However, these generalizations may be incorrect and the terms *BADBF* and *BAIBF* may imply a greater degree of functional separation between these two processes than is justified by current information. Although these conventional terms are in common use and are alluded to in this chapter, we emphasize their inherent limitations (189).

The final section of this chapter is divided into several parts. First the current concepts regarding pathways of water and solute transport by liver and the mechanisms responsible for BAIBF and BADBF are discussed. The final two sections of this chapter deal with intracellular bile acid transport, bile acid–lipid coupling, and regulation of canalicular water and solute secretion.

Pathways of Water and Solute Excretion

The pathways by which water and solutes enter the biliary tree are of considerable interest and importance for our understanding of the mechanisms and regulation of bile formation. It has not been possible to sample fluid or measure electrical driving forces at the level of either the canaliculus or the proximal ducts and ductules in intact liver. Thus many fundamental questions regarding the locus or mechanism(s) of water and solute entry (canaliculi versus ductules; paracellular versus transcellular) are unresolved. This situation has generated, perhaps paradoxically, considerable interest in the study of certain inert, fluid-phase markers that have traditionally been viewed as entering bile solely at the level of the canaliculus. (The term *fluid-phase marker* is used here to refer to substances that, generally speaking, are readily soluble in water, are not known to be transported via specific pumps or carriers or to bind to receptors, and are metabolically inert. The movement of such solutes thus is assumed to reflect the movement of surrounding fluid.) Conventional studies of bile formation routinely employ such markers to address two different issues. *1*) The first issue addressed is the separation of water movement at the canaliculus from that occurring "downstream" [e.g., erythritol, mannitol; (19, 37, 72–74, 217, 245], much like inulin clearance is used to measure the rate of glomerular filtration. *2*) Because certain markers [e.g., sucrose, inulin; (75, 76)] have been generally viewed as moving via the paracellular pathway, the concentration of such markers in bile versus plasma has often been interpreted as reflecting the integrity or tightness of hepatocyte tight junctions (64, 177, 178).

This section focuses on several specific areas related to pathways of water and solute movement. *1*) The limited evidence on pathways of water and inorganic ion movement in liver is briefly summarized. *2*) A considerably larger body of evidence on the movement of inert fluid-phase markers is reviewed, both from the standpoint of the level in the biliary tree at which they enter bile and the microanatomic locus (paracellular versus transcellular) of entry.

The movement of organic solutes such as bile acids or bilirubin, which obviously enter bile at the level of the canaliculus and pass through hepatocytes, is not discussed. For a general discussion of the kinetics of organic solute transport, see the chapter by Forker in this *Handbook*.

WATER AND INORGANIC ELECTROLYTE MOVEMENT INTO THE CANALICULUS. Current evidence indicates that liver is similar to other leaky epithelia, which are characterized by high ion conductivity across their tight junctions (52, 59, 79, 148, 197). This conclusion is based on the following observations: *1*) bile is isotonic with respect to plasma and, in isolated liver preparations, changes in perfusate osmolality produce identical changes in bile osmolality (36, 48, 52, 244); *2*) the number of sealing strands in hepatocyte tight junctions is less than that in tight epithelia (77, 86, 132); *3*) the electrical potential difference between blood and common duct bile is low [≤ 5 mV; (30)]; and *4*) recently a low-resistance paracellular pathway has been directly demonstrated in isolated hepatocyte couplets (87, 88). These observations are consistent with the view that passive movement of inorganic electrolytes into the canaliculus occurs readily via a paracellular pathway. Studies involving the use of radionuclide tracers in perfused liver, in which the ionic composition of perfusate was rapidly changed and radionuclide washout in bile followed, suggested that the paracellular pathway accounts for much or most of the movement of Na^+ (95%), Cl^- (73%), and K^+ (74%) from perfusate to bile (89). These studies further suggest a canalicular permeability sequence for cations of $Li^+ > Na^+ > K^+ >$ Tris > choline and for anions of $NO_3^- > Cl^- >$ acetate > sulfate.

The pathways of water movement by epithelia, in general, are not fully understood and, for technical reasons, have been little investigated in liver. Paracellular movement of water has been inferred from the study of electron-dense substances such as ionic lanthanum (La^{3+}). Lanthanum can be seen by electron microscopy to penetrate hepatocyte tight junctions under basal conditions, and enhanced junctional penetration can be observed under choleretic conditions (38, 138). Ultrastructural studies of rat liver have also identified balloonlike outpouchings in the basolateral membrane adjacent to the tight junction in association with bile acid–induced choleresis (138). By analogy with other epithelia (148, 226), these studies collectively have been interpreted as evidence that bile acid–stimulated bile flow is associated with enhanced paracellular water movement.

Recent studies, however, suggest that the hydraulic conductivity of the plasma membranes of epithelial cells in other "leaky" tissues, such as gallbladder and proximal tubule, is sufficiently great that paracellular water flow is unlikely to account for >5% of total water flux and that transcellular water movement via membrane pores probably accounts for most transepithelial water flux (26, 171, 210). Any conclusions regarding the microanatomic loci of water movement in liver are highly inferential.

USE OF INERT SOLUTES AS FLUID-PHASE MARKERS. *Level of entry into bile.* Early studies using inert solutes such as erythritol and mannitol suggested that the biliary clearance of these markers varied directly with experimental maneuvers believed to selectively alter canalicular water flow (e.g., bile acid–induced choleresis) but not ductular water flow (e.g., secretin infusion) (19, 73, 76). This led to the conclusion that such markers entered bile solely at the canaliculus, and they subsequently have been widely used as markers of canalicular water flow (19, 73, 217).

Recently, however, it has become clear that these conclusions are species dependent and not wholly correct. Studies in dogs (in contrast to earlier studies in guinea pigs) have demonstrated that secretin choleresis is associated with an increase in erythritol clearance, albeit a proportionally smaller (60%–70% smaller) increase in erythritol clearance than that produced by taurocholate choleresis (19). This observation suggests that erythritol may enter bile in part via ducts and ductules (the site in the biliary tree on which secretin is assumed to act) or that secretin acts partly to stimulate canalicular water secretion; neither explanation can be excluded. More direct evidence for a ductular contribution to the movement of these markers has been provided by studies of in situ preparations of guinea pig and rat common bile ducts (207,

224). At physiological rates of isolated bile duct perfusion, intravenously injected marker entered the ductular lumen in significant quantities and the perfusate-to-bile ratio of erythritol averaged 0.32. Moreover, recent preliminary studies suggest that considerable water exchange occurs across the ducts and ductules in isolated liver perfused dually via both the hepatic artery (which supplies the biliary system) and the portal vein (179). These observations indicate that the current use of markers to quantitate canalicular bile flow as separate from ductular bile flow may be subject to error. The magnitude of this error is likely to be species dependent, and the degree to which such sources of error occur in humans is uncertain (170). In dogs, evidence suggests that up to 30% of net secretion or absorption occurring distal to the canaliculus may be incorrectly assumed to be of canalicular origin, based on measurement of erythritol clearance (19). Unfortunately an alternate approach to measuring a ductular or canalicular flow is not available.

Transcellular versus paracellular pathway of entry. It has generally been assumed that markers equal in size to sucrose or larger move from blood to bile predominantly or exclusively via a paracellular pathway (36). This conclusion rests primarily on the observation that sucrose enters hepatocytes slowly, if at all, yet rapidly reaches a steady-state concentration in bile much larger than that present in hepatocellular water (75). Sucrose and inulin, both of which are commonly used in studies of bile formation, are also routinely employed as markers of extracellular water in a variety of cell systems based on their lack of movement across biologic membranes (169). The movement of markers such as sucrose and inulin from blood to bile under steady-state conditions can be satisfactorily represented mathematically with equations that describe solute movement via pore-restricted diffusion and convection (72, 73, 76, 180). Based on the presumption that such pores or their functional equivalents reside in the tight junction, changes in the biliary transport and bile-plasma ratio of inert markers such as sucrose are often discussed in terms of changes in osmotic permeability (diffusion) or sieving coefficients (convection) and thus assumed to reflect altered permeability or leakiness of the hepatocyte tight junctions (64, 177, 178).

The use of inert marker movement as an indicator of junctional permeability, however, has yielded apparently conflicting findings. For example, an increase in the sucrose bile-plasma ratio has been seen under both choleretic and cholestatic conditions. The increase in the sucrose bile-plasma ratio under choleretic conditions (e.g., addition of bile acids) has been interpreted as enhanced permeability of the tight junction to fluid movement (137). Under cholestatic conditions (e.g., Ca^{2+} deprivation), an enhanced bile-plasma ratio has been interpreted as showing a loss of tight-junction integrity, allowing reflux of biliary content back into the space of Disse with a consequent loss of osmotic driving forces (176). These inconsistent observations have raised important questions regarding the utility and/or validity of these markers as indicators of tight-junction permeability and thus have not permitted the formulation of any general hypotheses regarding the role of altered junctional permeability in accelerated or impaired formation of canalicular bile.

A second pathway that might account for the biliary secretion of these markers is transcellular movement via vesicles, or transcytosis. Vesicles mediate the uptake and intracellular transport of a variety of macromolecules in hepatocytes (such as low-density lipoproteins, asialoglycoproteins, and immunoglobulin A), which enter hepatocytes via receptor-mediated endocytosis (14, 84, 133, 182, 196). Most such substances are transported to lysosomes, where they are catabolized (14, 133, 196), with only a small proportion appearing intact in bile. Ligands such as IgA, however, appear to be transported, in large part, directly to bile without passing through the lysosomal compartment (182, 196).

As endocytic vesicles pinch off from the plasma membrane (see Fig. 3), they trap within them extracellular fluid and solutes contained within that fluid. Thus inert fluid-phase markers, as well as ligands, should be transported to bile and lysosomes. Although the transcellular vesicular transport of certain fluid-phase markers (e.g., horseradish peroxidase) has been recognized for some time (111, 149), it is only recently that the contribution of transcytosis to biliary secretion has been systematically examined (134). In the perfused rat liver, studies with inert fluid-phase markers of varying molecular weights yielded the following results. *1*) Under steady-state conditions, even large markers (e.g., dextran, 70 kDa) were secreted in bile in appreciable amounts (bile-plasma ratio = 0.09), and there was no apparent ordering of marker bile-plasma ratios based on size (1.8–70 kDa). *2*) Under non-steady-state conditions in which the appearance in or disappearance from bile of selected markers was studied after their abrupt addition to or removal from perfusate (Fig. 5), erythritol reached a bile-plasma ratio of 1 within 2 min. Dextran appeared in bile only after a lag period of ~12 min and then slowly approached maximal values, whereas sucrose exhibited kinetically intermediate behavior. A similar pattern was observed after removal of >95% of the markers from the perfusate. Erythritol rapidly reapproached a bile-plasma ratio of 1, whereas the bile-plasma ratio for sucrose and dextran fell much more slowly and exceeded 1 for a full 30 min after perfusate washout (Fig. 5).

Biliary transport of dextran (Fig. 5) was kinetically incompatible with simple movement between two compartments (perfusate and bile) via a pathway described as a single rate constant but rather required the presence of multiple interposed delay compartments compatible with transcellular vesicle transport.

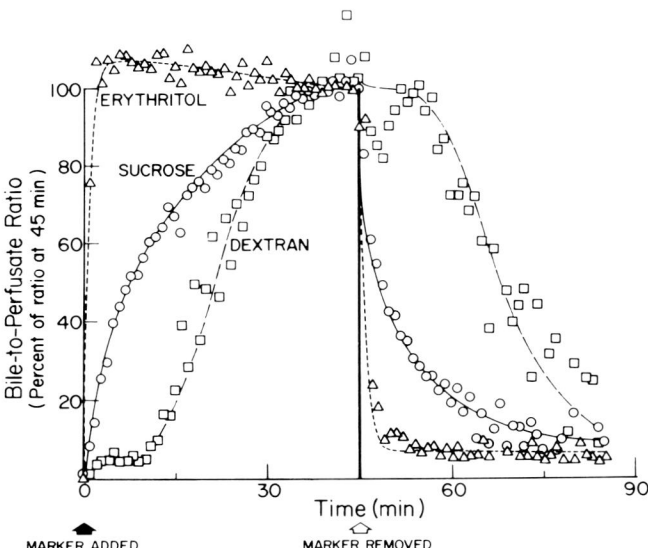

FIG. 5. Appearance-disappearance curves for biliary erythritol (120 Da), sucrose (342 Da), and dextran (70 kDa avg) after their abrupt addition to and removal from perfusate. [From Lake et al. (134).]

Sucrose entry into bile (Fig. 5), which was intermediate between that of dextran and erythritol, was mathematically described by a combination of the dextran pathway and a second pathway kinetically represented by a single-delay compartment, possibly representing paracellular entry of sucrose into bile (134).

Additional evidence supports the vesicular nature of dextran transport. Colchicine, which impairs vesicle traffic by disrupting microtubules, markedly decreased the biliary secretion of dextran without altering its time course of appearance in bile (133). Moreover, electron microscopy and fluorescence microscopy of cultured hepatocytes demonstrated the presence of horseradish peroxidase and fluorescein dextran in intracellular vesicles, and fractionation of perfused liver homogenates revealed that at least 35%–80% of sucrose, inulin, and dextran was associated with subcellular organelles (134, 145).

Collectively these observations are most compatible with a transcytosis pathway that contributes minimally to the secretion of very small solutes such as erythritol but accounts for a substantial fraction of sucrose secretion and virtually all (>95%) of the blood-to-bile transport of large solutes such as dextran. Moreover, these results also suggest that the possible contribution of such a mechanism to bulk fluid movement merits further evaluation. Finally, from these studies it is apparent that conclusions regarding canalicular permeability and/or tight-junction integrity based on changes in the bile-plasma ratio of solutes such as sucrose after experimental manipulation are clearly hazardous. For example, an agent that inhibits active solute transport and osmotically obligated water flux but that has little effect on vesicular movement might be expected to increase biliary marker concentration independent of any changes in effective canalicular pore size or tight-junction integrity. This would pertain to markers transported wholly (e.g., dextran) or in part (e.g., sucrose) via a vesicular mechanism. The effects of colchicine on sucrose movement into bile illustrate this point (133). Administration of colchicine, which produced little or no change in bile flow, increased the steady-state sucrose bile-plasma and accelerated the appearance of sucrose in bile. Kinetic analysis of these results indicated that the increase in sucrose output was due entirely to nonvesicular transport and that sucrose transport via the vesicular pathway decreased. Thus, even under conditions in which total bile flow is unchanged, changes in the secretion and bile-plasma ratio of inert markers may occur and reflect changes in vesicular and/or nonvesicular transport.

Although these observations provide considerable evidence for a role of transcytosis in the blood-to-bile transport of some macromolecules and may represent an important source of biliary proteins (159), certain areas of persisting uncertainty merit emphasis. *1)* Species differences in the existence or importance of this pathway may exist but have not been explored. *2)* It is unclear what type(s) of vesicles mediate this process. Specifically it is unclear whether this process involves vesicles that mediate receptor-mediated endocytosis, separate "pinosomes," or both. It is also unclear whether such vesicles are acidic, as has been reported for clathrin-coated vesicles and endosomes (231, 235). *3)* Although molecular weight has not appeared to be an important determinant of vesicular transport via this mechanism, the importance of other factors such as charge (40) and shape have not been determined. Finally, while the present studies provide quantitative estimates of the contribution of transcytosis to the blood-to-bile transport of fluid-phase markers, the implications with respect to bulk fluid movement via transcytosis are uncertain. Loss or accumulation of fluid by endocytic vesicles after their internalization would not be accompanied by corresponding changes in marker content. Thus, in this situation, biliary secretion of a fluid-phase marker may not accurately reflect transcellular bulk fluid movement. In addition, biliary secretion of fluid-phase marker does not provide a measure of the total amount of fluid internalized by hepatocytes via endocytosis, much of which may be recycled back to the space of Disse (191). The perfusate-to-bile transport of markers such as dextran seems to provide only a minimal estimate [$\sim 0.1\ \mu l \cdot g^{-1} \cdot min^{-1}$; (134)] of total fluid internalization via vesicles at the sinusoidal cell surface.

Bile Acid–Stimulated Canalicular Bile Formation

Bile acid–dependent or, perhaps more appropriately, bile acid–stimulated bile flow is generally attributed to the osmotic effect of actively secreted bile acids. This general formulation is based primarily on two observations. *1*) In all mammals thus far studied (including humans, dogs, rabbits, and rats) a nearly linear relationship has been observed between canalicular bile flow and bile acid output (36, 193). *2*) Certain non-micelle-forming (and therefore presumably more osmotically active) bile acids exhibit a greater choleretic potency (volume of bile output per unit bile acid secreted) than do certain micelle-forming bile acids (36).

A variety of observations, however, suggest that bile flow stimulated by at least certain bile acids may not be wholly attributable to the osmotic effects of these bile acids, per se. *1*) Striking differences, unexplained by micelle-forming properties but directly related to differences in bile acid–induced secretion of inorganic electrolytes (particularly HCO_3^-), exist in the choleretic efficiency of different bile acids given to the same animal species and of the same bile acid given to different animal species (161, 200, 216). Moreover the line relating canalicular bile flow to bile acid output differs up to an order magnitude between species, that is, 7 ml/mmol bile acid in humans versus 70 ml/mmol bile acid in the rabbit (36, 193). *2*) Ursodeoxycholic acid and some other bile acids elicit a dramatic (3- to 5-fold) increase in bile flow or "hypercholeresis" that is accompanied by a selective increase in biliary HCO_3^- concentration to 2–3 times that in plasma (61). *3*) The biliary output of certain bile acids can be experimentally dissociated from their associated choleresis. For example, perfusion of the isolated rat liver with an Na^+-free solution virtually abolishes the normal HCO_3^--rich choleresis produced by ursodeoxycholic acid while decreasing biliary ursodeoxycholic acid output by only ~30% (135), and exposure to amiloride (an inhibitor of Na^+-H^+ exchange) or amiloride analogues decreases by 30%–50% the amount of bile produced per unit of ursodeoxycholic acid or taurocholic acid secreted (135, 181). These observations provide compelling evidence that the stimulation of canalicular bile flow by certain bile acids is attributable, at least partly, to the stimulation of active electrolyte secretion, in particular the secretion of HCO_3^- (193).

Active HCO_3^- transport by other epithelia as well as liver is currently believed to result from the active transport of H^+ (or OH^-) ions (e.g., ref. 97). This raises the possibility that bile acids may stimulate H^+ transport via Na^+-H^+ exchange or a proton-translocating ATPase (as is found in endocytic vesicles), thereby increasing intracellular pH and HCO_3^- activity. This would be expected to enhance HCO_3^- efflux, and selective HCO_3^- secretion into the canaliculus may be mediated via the Cl^--HCO_3^- exchange mechanism recently identified in canalicular but not basolateral hepatocyte plasma membrane (151). These observations suggest that the HCO_3^--rich hypercholeresis produced by certain bile acids such as ursodeoxycholic acid requires an intact Na^+-H^+ exchange mechanism (135, 181).

An alternative mechanism for stimulation of HCO_3^- secretion by bile acids might involve uptake of bile acid anion across the basolateral membrane, possibly via Na^+ coupling, followed by intracellular protonation in the comparatively acidic cytoplasm [pH 7.1; (169)], efflux of the hydrophobic protonated species into the space of Disse, loss of the proton, and reuptake of the bile acid anion. This recycling of bile acid could represent Na^+-H^+ exchange. A conceptually similar type of recycling occurring downstream at the level of the ducts and/or ductules might also occur. The possible mechanisms involved in bile acid–stimulated bile formation and HCO_3^- secretion are also summarized in Figure 3. The mechanism(s) by which various bile acids stimulate bile formation represent an active and important area of investigation.

Bile Acid–Independent Bile Formation

Because the intact liver always synthesizes and secretes bile acids, BAIBF cannot be precisely measured. Moreover, because the relationship between bile flow and bile acid output is inconstant and nonlinear under certain circumstances (16, 18), the conventional definition of BAIBF as the extrapolated Y intercept of this line is subject to error. Despite these uncertainties, certain findings suggest that a substantial fraction of canalicular bile formation is attributable to the active transport of inorganic electrolytes. These findings include the observations that *1*) bile acid–free perfusion of the perfused rat liver results in a >95% fall in bile acid secretion but only an ~50% fall in canalicular bile flow, *2*) bile secretory pressure by the perfused rat liver during bile salt depletion exceeds perfusion pressure but decreases after administration of metabolic inhibitors, and *3*) certain drugs and hormones increase canalicular bile flow without altering bile acid output (38, 193).

The cellular mechanism(s) responsible for BAIBF are largely unknown, and concepts regarding this process have undergone considerable evolution. It was initially proposed that BAIBF was directly attributable to the activity of Na^+-K^+-ATPase, which (based on comparatively crude cell-fractionation techniques) was postulated to reside on the canalicular membrane (189). This formulation was based on the observations that inhibitors of Na^+-K^+-ATPase such as ouabain also appeared to inhibit bile formation in intact animals and the perfused liver and that certain drugs or hormones that increased (e.g., phenobarbital, taurocholate, thyroid hormone) or decreased (ethinyl estradiol, rose bengal, chlorpromazine) hepatic Na^+-K^+-ATPase activity produced apparently parallel changes in BAIBF (189). Subsequently each of the observations on which this hypothesis rested has been questioned or refuted. Histochemical studies have localized Na^+-K^+-ATPase to the sinusoidal-lateral membrane (although localization using monoclonal antibodies has yielded conflicting results; see *Localization of Na^+-K^+-ATPase*, p. 600). The effect of ouabain on BAIBF is stimulatory rather than inhibitory (201), and the apparent parallelism between BAIBF and Na^+-K^+-ATPase activity has been refuted (116). Other mechanisms for BAIBF have also been explored and are summarized here.

SODIUM-POTASSIUM-CHLORIDE COTRANSPORT. Sodium-potassium-chloride cotransport has been postulated to play a role in BAIBF (8). Localized to the sinusoidal-lateral membrane, Na^+-K^+-Cl^- cotransport would mediate intracellular accumulation of Cl^- above electrochemical equilibrium and, in junction with an asymmetric distribution of Cl^--conductance pathways or carriers (151), net canalicular Cl^- secretion (see Figs. 2, 3). However, studies in cultured hepatocytes have not provided evidence of coupling between the uptake of Na^+ and Cl^-, BAIBF by perfused rat liver is minimally altered when Cl^- is replaced by certain other permeant anions (194), and recent studies suggest that Cl^- is passively distributed across the hepatocyte plasma membrane (70).

BICARBONATE SECRETION. There is a moderate amount of evidence (albeit circumstantial) that HCO_3^- secretion is important in BAIBF. There are two important observations in this regard. *1*) Several investigators have demonstrated that removal of HCO_3^- from the perfusate of the isolated rat liver decreased BAIBF (8, 97, 237). These results are in sharp contrast to the lack of effects of replacing Na^+ with Li^+ or Cl^- with other anions (134, 194), and they suggest that HCO_3^- is necessary for BAIBF. *2*) The increase in bile flow produced by certain bile acid–independent choleretics (e.g., secretin, salicylates) appears to be at least partly attributable to selective enhancement of canalicular HCO_3^- secretion (19).

The secretion of HCO_3^- by hepatocytes is presumably attributable to the active secretion of H^+ ions via Na^+-H^+ exchange, a proton-translocating ATPase, or possibly other mechanisms. The observations that basal HCO_3^- secretion by isolated rat liver is minimally affected by removal of perfusate Na^+ or exposure to amiloride (135, 181, 237) raise the possibility that an H^+-translocating ATPase, as identified in endocytic vesicles and possibly residing also on the hepatocyte plasma membrane, may play a role in basal BAIBF and HCO_3^- secretion in a manner analogous to the putative role of such an H^+-ATPase in acid-alkali transport by turtle bladder and distal nephron (82, 83). The presence of an H^+ pump in endocytic vesicles that recycle to and from the plasma membrane also raises the possibility that H^+ extrusion from hepatocytes, intracellular pH, and possibly biliary HCO_3^- secretion may be mediated partly by insertion into or removal from the basolateral membrane of an H^+-ATPase via a vesicular shuttle mechanism (83, 199, 211). The mechanism(s) responsible for BAIBF represent an important and currently active area of investigation. Moreover the possible importance of HCO_3^- transport in this process suggests that fundamentally similar transport mechanisms may operate in both BADBF and BAIBF.

Intracellular Bile Acid Transport and Bile Acid–Lipid Coupling

In addition to stimulating the secretion of water and electrolytes, certain bile acids also stimulate biliary lipid secretion. As discussed more thoroughly in the chapter by Hofmann in this *Handbook*, biliary bile acid secretion is coupled in a nonlinear fashion with the output of biliary lipids. By comparison with hepatic uptake, however, relatively little is known about the mechanism(s) involved in intrahepatic bile acid transport and the coupling of bile acid and biliary lipid secretion. This section briefly summarizes current information and uncertainties regarding these related processes.

The major lipids in bile are nonesterified cholesterol and the phospholipid phosphatidylcholine (lecithin). Increased bile acid flux through the perfused rat liver increases the incorporation of radiolabeled lecithin precursors into the lecithins of both the endoplasmic reticulum and plasma membrane, and the specific activity of biliary lecithin exceeds that of either membrane fraction (92). This suggests the existence of a subpool of newly synthesized lecithin destined for biliary excretion. Animal studies similarly suggest the existence in hepatocytes of a labile pool of free cholesterol, the size of which depends on relative rates of cholesterol synthesis and esterification, that contributes importantly to biliary cholesterol secretion (213). The cellular location of these kinetically defined precursor pools and the mechanisms for transport to the canalicular membrane of cholesterol and phospholipid and assembly into mixed micelles are not known. Two possible mechanisms for intracellular transport of bile acids and/or lipid involving, respectively, vesicles and cytosolic binding protein are discussed next.

Studies of both native bile and model systems indicate that, in addition to micelles, mixed lipid vesicles containing cholesterol and phospholipids also represent an important vehicle for biliary lipid secretion (150, 209). These findings raise the possibility that such vesicles also represent the intracellular precursor of biliary mixed micelles. Bile acids are surface-active agents known to interact with biologic membranes in vitro (190). It is tempting to postulate that similar physical interaction occurs in vivo and that solubilization of membrane lipids by bile acids, perhaps to form vesicles, represents an early event in bile acid–lipid coupling and biliary lipid secretion. A role for vesicles in the intracellular transport of lipids and, by implication, bile acids are further supported by the prominence of vesicular structures in pericanalicular cytoplasm (39), the inhibitory effect of colchicine on biliary lipid secretion (91), and recent studies from nonpolarized cells that implicate a vesicular mechanism for transport of newly synthesized cholesterol from microsomes to the plasma membrane (113). Moreover, bile acid–lipid coupling is altered by enhanced conjugation and secretion of organic anions such as sulfobromophthalein and bilirubin, possibly at an intracellular locus (10). These observations suggest that the bile acid–stimulated biliary secretion of cholesterol and lecithin in the form of vesicles or micelles begins at some intracellular locus and is consistent with a role for vesicles in intracellular transport.

There is also evidence that cytosolic binding proteins are involved in intracellular bile acid transport. Hepatic cytoplasmic proteins that bind bile acids have been identified, although their bile acid transport remains uncertain. Glutathione S-transferases, which account for ~5% of rat hepatic cytosolic protein, have been demonstrated to bind bile acids (99), and these proteins bind lithocholic acids and bilirubin at the same nonsubstrate ligand-binding site (220). A bile acid–binding protein distinct from the glutathione S-transferases has also recently been identified in rat liver (where it accounts for <0.5% of cytosolic proteins) and in human hepatic cytosol (221). Preliminary observations suggest that this latter protein may also exhibit bile acid sulfotransferase activity (212). Although such proteins may participate in intracellular bile acid transport and "shuttle" bile acids between the sinusoidal membrane and putative canalicular membrane carriers, the existence or quantitative importance of such an intracellular pathway and the relationships of such a pathway to bile acid–lipid coupling or bile acid transport via vesicles is uncertain.

Regulation of Canalicular Water and Solute Secretion

Bile flow is slowest during periods of fasting, when bile acids are largely sequestered in the gallbladder and the flux of bile acids through the enterohepatic circulation and hence BADBF is at its nadir. With ingestion of a meal, hormonal (and possibly neural) stimulation of gallbladder contraction causes release into the intestine of gallbladder bile, thus priming the enterohepatic circulation with a bolus of bile acids that may recycle two or more times each meal (see the chapter by Hofmann in this *Handbook*).

In contrast to BADBF, gallbladder contraction, and bile ductular secretion, which are regulated largely by events that occur with feeding and fasting, relatively little is known regarding the normal regulation of BAIBF. Corticosteroids and thyroid hormone stimulate canalicular BAIBF in animals, and estrogens have the opposite effect (116, 189). Secretin, which is known to stimulate secretion by ducts and/or ductules, may also stimulate bile acid–independent secretion at the level of the canaliculus (19). Like secretin, glucose and insulin also appear to stimulate secretion by hepatocytes (80, 217), whereas somatostatin has the opposite effect (154, 183). The presence and quantitative significance of such hormonal effects in humans are uncertain.

The intracellular events and possible second messengers responsible for mediating such hormonal effects on bile formation are also largely unknown. Dibutyryl cAMP increases canalicular bile flow in the portal vein–perfused isolated rat liver (225). The degree to which hepatocytes and/or ductular cells contribute to this choleresis is unknown. If ducts and/or ductules are inactive in this preparation, these findings would suggest that cAMP may play a role in mediating the hepatocellular effects of certain hormones such as glucagon (225). The effects of dibutyryl cAMP on bile formation, however, are highly species dependent (112, 225). In humans, secretin-induced choleresis is associated with an increase in biliary secretion of cAMP; theophylline and dibutyryl cAMP both increase total bile flow two- to threefold (142). Secretin also produces a choleresis in the baboon that is associated with an increase in the cAMP content of bile duct tissue (142). These observations suggest that cAMP helps mediate the choleretic effect of secretin on the bile ducts in primates; less is known about the possible role of cAMP in mediating changes in hepatocellular secretion in humans. The possible roles of cytosolic Ca^{2+} and/or calmodulin and protein phosphorylation in regulating canalicular bile formation are largely unexplored in any animal species.

SUMMARY

In recent years, study of a variety of experimental models (isolated perfused liver, isolated and cultured hepatocytes, plasma membrane vesicles) has considerably increased our understanding of hepatic solute transport and bile formation. It is now established that hepatocytes represent a polarized transporting epithelium, which, like kidney and intestine, is actively involved in cellular and transepithelial transport of ions, organic solutes, proteins, and inert large-molecular-weight solutes. Despite the growing body of knowledge in this field, many important areas remain unresolved, including *1*) quantitation of electrical potential gradients at the level of the canaliculus and ductules and their role in solute transport and bile formation, *2*) delineation of the factors that regulate hepatic water and solute transport, *3*) quantitation of paracellular and transcellular movement of solutes and water, *4*) definition of the active solute-transport mechanism responsible for BADBF and BAIBF.

This study was supported by National Institutes of Health Grants AM-26270, AM-26743, AM-01254, and AM-07453 and an American Liver Foundation Fellowship Award.

REFERENCES

1. ACCATINO, L., AND F. R. SIMON. Identification and characterization of a bile acid receptor in isolated liver surface membranes. *J. Clin. Invest.* 57: 496–508, 1976.
2. ALDINI, R., A. RODA, A. M. MORSELLI LABATE, G. CAPPELLERI, E. RODA, AND L. BARBARA. Hepatic bile acid uptake: effect of conjugation, hydroxyl and keto groups, and albumin binding. *J. Lipid Res.* 23: 1167–1173, 1982.
3. ALVARADO, F., AND J. W. ROBINSON. A kinetic study of the interactions between amino acids and monosaccharides at the intestinal brush-border membrane. *J. Physiol. Lond.* 295: 457–475, 1979.
4. ALVO, M., J. CALAMIA, AND J. EVELOFF. Lack of potassium effect on Na-Cl cotransport in the medullary thick ascending limb. *Am. J. Physiol.* 249 (*Renal Fluid Electrolyte Physiol.* 18): F34–F39, 1985.
5. ANDERSEN, O. S., J. E. N. SILVEIRA, AND P. R. STEINMETZ. Intrinsic characteristics of the proton pump in the luminal membrane of a tight urinary epithelium. The relationship between transport rate and $\Delta\bar{\mu}_H$. *J. Gen. Physiol.* 86: 215–234, 1985.
6. ANDERSON, R. G. W., J. R. FALCK, J. L. GOLDSTEIN, AND M. S. BROWN. Visualization of acidic organelles in intact cells by electron microscopy. *Proc. Natl. Acad. Sci. USA* 81: 4838–4842, 1984.
7. ANWER, M. S., AND D. HEGNER. Effect of Na^+ on bile acid uptake by isolated rat hepatocytes. *Hoppe-Seyler's Z. Physiol. Chem.* 359: 181–192, 1978.
8. ANWER, M. S., AND D. HEGNER. Role of inorganic electrolytes in bile acid-independent canalicular bile formation. *Am. J. Physiol.* 244 (*Gastrointest. Liver Physiol.* 7): G116–G124, 1983.
9. ANWER, M. S., R. KROKER, D. HEGNER, AND A. PETTER. Cholic acid binding to isolated rat liver plasma membranes. *Hoppe-Seyler's Z. Physiol. Chem.* 358: 543–553, 1977.
10. APSTEIN, M. D. Inhibition of biliary phospholipid and cholesterol secretion by bilirubin in the Sprague-Dawley and Gunn rat. *Gastroenterology* 87: 634–638, 1984.
11. ARIAS, I. M., AND M. FORGAC. The sinusoidal domain of the plasma membrane of rat hepatocytes contains an amiloride-sensitive Na^+/H^+ antiport. *J. Biol. Chem.* 259: 5406–5408, 1984.
12. ARONSON, P. S. Mechanisms of active H^+ secretion in the proximal tubule. *Am. J. Physiol.* 245 (*Renal Fluid Electrolyte Physiol.* 14): F647–F659, 1983.
13. ARONSON, P. S. Electrochemical driving forces for secondary active transport: energetics and kinetics of Na^+-H^+ exchange and Na^+-glucose cotransport. In: *Electrogenic Transport: Fundamental Principles and Physiological Implications*, edited by M. P. Blaustein and M. Lieberman. New York: Raven, 1984, p. 49–70.
14. ASHWELL, G., AND J. HARFORD. Carbohydrate-specific receptors of the liver. *Annu. Rev. Biochem.* 51: 531–554, 1982.
15. BACHS, O., K. S. FAMULSKI, F. MIRABELLI, AND E. CARAFOLI. ATP-dependent Ca^{2+} transport in vesicles isolated from the bile canalicular region of the hepatocyte plasma membrane. *Eur. J. Biochem.* 147: 1–7, 1985.
16. BAKER, A. L., R. A. B. WOOD, A. R. MOOSSA, AND J. L. BOYER. Sodium taurocholate modifies the "bile acid independent" fraction of canalicular bile flow in the rhesus monkey. *J. Clin. Invest.* 64: 312–320, 1979.
17. BAKKER-GRUNWALD, T. Potassium permeability and volume control in isolated rat hepatocytes. *Biochim. Biophys. Acta* 731: 239–242, 1983.
18. BALABAUD, C., K. KRON, AND J. J. GUMUCIO. The assessment of the bile salt-nondependent fraction of canalicular bile water in the rat. *J. Lab. Clin. Med.* 89: 393–399, 1977.
19. BARNHART, J. L., AND B. COMBES. Erythritol and mannitol clearances with taurocholate and secretin-induced choleresis. *Am. J. Physiol.* 234 (*Endocrinol. Metab. Gastrointest. Physiol.* 3): E146–E156, 1978.
20. BEAR, C. E., C. N. PETRUNKA, AND S. M. STRASBERG. Evidence for a channel for the electrogenic transport of chloride ion in the rat hepatocyte. *Hepatology Baltimore* 5: 383–391, 1985.
21. BECKER, J., AND A. JAKOB. α-Adrenergic stimulation of glycolysis and Na^+,K^+-transport in perfused rat liver. *Eur. J. Biochem.* 128: 293–296, 1982.
22. BELLEMANN, P. Enhanced amino acid transport in cultured hepatocytes during liver development. *J. Biochem.* 90: 1821–1824, 1981.
23. BELLOMO, G., F. MIRABELLI, P. RICHELMI, AND S. ORRENIUS. Critical role of sulfhydryl group(s) in ATP-dependent Ca^{2+} sequestration by the plasma membrane fraction from rat liver. *FEBS Lett.* 163: 136–139, 1983.
24. BERNSTEIN, J., AND G. SANTACANA. The Na^+/Ca^{2+} exchange system of the liver cell. *Res. Commun. Chem. Pathol. Pharmacol.* 47: 3–34, 1985.
25. BERR, F., F. R. SIMON, AND J. REICHEN. Ethynylestradiol impairs bile salt uptake and Na-K pump function of rat hepatocytes. *Am. J. Physiol.* 247 (*Gastrointest. Liver Physiol.* 10): G437–G443, 1984.
26. BERRY, C. A. Characteristics of water diffusion in the rabbit proximal convoluted tubule. *Am. J. Physiol.* 249 (*Renal Fluid Electrolyte Physiol.* 18): F729–F738, 1985.
27. BERTHON, B., A. BINET, J.-P. MAUGER, AND M. CLARET. Cytosolic free Ca^{2+} in isolated rat hepatocytes as measured by quin2. *FEBS Lett.* 167: 19–24, 1984.
28. BERTHON, B., T. CAPIOD, AND M. CLARET. Effects of noradrenalin, vasopressin and angiotensin on the Na-K pump in rat isolated liver cells. *Br. J. Pharmacol.* 86: 151–161, 1985.
29. BERTHON, B., M. CLARET, J. L. MAZET, AND J. POGGIOLI. Volume- and temperature-dependent permeabilities in isolated rat liver cells. *J. Physiol. Lond.* 305: 267–277, 1980.
30. BINDER, H. J., AND J. L. BOYER. Bile salts: a determinant of the bile-peritoneal electrical potential difference in the rat. *Gastroenterology* 65: 943–948, 1973.
31. BLITZER, B. L., AND J. L. BOYER. Cytochemical localization of Na^+,K^+-ATPase in the rat hepatocyte. *J. Clin. Invest.* 62: 1104–1108, 1978.
32. BLITZER, B. L., AND J. L. BOYER. Cellular mechanisms of bile formation. *Gastroenterology* 82: 346–357, 1982.
33. BLITZER, B. L., AND R. L. BUELER. Kinetic and energetic aspects of the inhibition of taurocholate uptake by Na^+-dependent amino acids: studies in rat liver plasma membrane vesicles. *Am. J. Physiol.* 249 (*Gastrointest. Liver Physiol.* 12): G120–G124, 1985.
34. BLITZER, B. L., S. L. RATOOSH, AND C. B. DONOVAN. Amino acid inhibition of bile acid uptake by isolated rat hepatocytes: relationship to dissipation of transmembrane Na^+ gradient. *Am. J. Physiol.* 245 (*Gastrointest. Liver Physiol.* 8): G399–G403, 1983.
35. BLITZER, B. L., S. L. RATOOSH, C. B. DONOVAN, AND J. L. BOYER. Effects of inhibitors of Na^+-coupled ion transport on bile acid uptake by isolated rat hepatocytes. *Am. J. Physiol.* 243 (*Gastrointest. Liver Physiol.* 6): G48–G53, 1982.
36. BOYER, J. L. New concepts of mechanisms of hepatocyte bile formation. *Physiol. Rev.* 60: 303–326, 1980.
37. BOYER, J. L., AND J. R. BLOOMER. Canalicular bile secretion in man: studies utilizing the biliary clearance of ^{14}C-mannitol. *J. Clin. Invest.* 54: 773–781, 1974.
38. BOYER, J. L., E. ELIAS, AND T. J. LAYDEN. The paracellular pathway and bile formation. *Yale J. Biol. Med.* 52: 61–67, 1979.
39. BOYER, J. L., M. ITABASHI, AND Z. HRUBAN. Formation of pericanalicular vacuoles during sodium dehydrocholate choleresis—a mechanism for bile acid transport. In: *The Liver. Quantitative Aspects of Structure and Function*, edited by R. Preisig and J. Bircher. Aulendorf, Switzerland: Editio Cantor, 1979, p. 163–178.
40. BRADLEY, S. E., AND R. MERZ. Permselectivity of biliary canalicular membrane in rats: clearance probe analysis. *Am.*

J. Physiol. 235 (*Endocrinol. Metab. Gastrointest. Physiol.* 4): E570–E576, 1978.

41. BROWN, M. S., R. G. W. ANDERSON, AND J. L. GOLDSTEIN. Recycling receptors: the round-trip itinerary of migrant membrane proteins. *Cell* 32: 663–667, 1983.
42. BURGESS, G. M., M. CLARET, AND D. H. JENKINSON. Effects of catecholamines, ATP and ionophore A23187 on potassium and calcium movements in isolated hepatocytes. *Nature Lond.* 279: 544–546, 1979.
43. BURGESS, G. M., M. CLARET, AND D. H. JENKINSON. Effects of quinine and apamin on the calcium-dependent potassium permeability of mammalian hepatocytes and red cells. *J. Physiol. Lond.* 317: 67–90, 1981.
44. BURGESS, G. M., P. P. GODFREY, J. S. MCKINNEY, M. J. BERRIDGE, R. F. IRVINE, AND J. W. PUTNEY, JR. The second messenger linking receptor activation to internal Ca release in liver. *Nature Lond.* 309: 63–66, 1984.
45. CANIVET, B., M. FEHLMANN, AND P. FREYCHET. Glucocorticoid and catecholamine stimulation of amino acid transport in rat hepatocytes. *Mol. Cell. Endocrinol.* 19: 253–261, 1980.
46. CARAFOLI, E. Calcium-transporting systems of plasma membranes, with special attention to their regulation. *Adv. Cyclic Nucleotide Protein Phosphorylation Res.* 17: 543–549, 1984.
47. CHAN, K.-M., AND K. D. JUNGER. Calcium transport and phosphorylated intermediate of (Ca^{2+} + Mg^{2+})-ATPase in plasma membranes of rat liver. *J. Biol. Chem.* 258: 4404–4410, 1983.
48. CHENDEROVITCH, J., E. PHOCAS, AND M. MATUREAU. Effects of hypertonic solutions on bile formation. *Am. J. Physiol.* 205: 863–867, 1963.
49. CHRISTENSEN, H. N. Hypothesis: control of hepatic utilization of alanine by membrane transport or by cellular metabolism? *Biosci. Rep.* 3: 905–913, 1983.
50. CHRISTENSEN, H. N. On the strategy of kinetic discrimination of amino acid transport systems. *J. Membr. Biol.* 84: 97–103, 1985.
51. CLARET, B., M. CLARET, AND J. L. MAZET. Ionic transport and membrane potential of rat liver cells in normal and low-chloride solutions. *J. Physiol. Lond.* 230: 87–101, 1973.
52. CLARET, M., AND J. L. MAZET. Ion fluxes and permeability of cell membranes in rat liver. *J. Physiol. Lond.* 223: 279–295, 1972.
53. CLAUDE, P., AND D. A. GOODENOUGH. Fracture faces of zonulae occludentes from "tight" and "leaky" epithelia. *J. Cell Biol.* 58: 390–400, 1975.
54. COHEN, R. D., R. M. HENDERSON, R. A. ILES, AND J. A. SMITH. Metabolic inter-relationships of intracellular pH measured by double-barrelled micro-electrodes in perfused rat liver. *J. Physiol. Lond.* 330: 69–80, 1982.
55. COOK, N. S., AND D. G. HAYLETT. Effects of apamine, quinine and neuromuscular blockers on calcium-activated potassium channels in guinea-pig hepatocytes. *J. Physiol. Lond.* 358: 373–394, 1985.
56. CUMMINGS, S. A., AND A. F. HOFMANN. Physiologic determinants of biliary calcium secretion in the dog. *Gastroenterology* 87: 664–673, 1984.
57. DAVIS, R. A., F. KERN, JR., R. SHOWALTER, E. SUTHERLAND, M. SINENSKY, AND F. R. SIMON. Alterations of hepatic Na^+,K^+-ATPase and bile flow by estrogen: effects on liver surface membrane lipid structure and function. *Proc. Natl. Acad. Sci. USA* 75: 4130–4134, 1978.
58. DHARMSATHAPHORM, K., K. G. MANDEL, H. MASUI, AND J. A. MCROBERTS. Vasoactive intestinal polypeptide-induced chloride secretion by a colonic epithelial cell line. Direct participation of a basolaterally localized Na^+, K^+, Cl^- cotransport system. *J. Clin. Invest.* 75: 462–471, 1985.
59. DIAMOND, J. M. Tight and leaky junctions of epithelia: a prospective on kisses in the dark. *Federation Proc.* 33: 2220–2224, 1974.
60. DUFFY, M. C., B. L. BLITZER, AND J. L. BOYER. Direct determination of the driving forces for taurocholate uptake into rat liver plasma membrane vesicles. *J. Clin. Invest.* 72: 1470–1481, 1983.
61. DUMONT, M., S. ERLINGER, AND S. UCHMAN. Hypercholeresis induced by ursodeoxycholic acid and 7-ketolithocholic acid in the rat: possible role of bicarbonate transport. *Gastroenterology* 79: 82–89, 1980.
62. EDMONDSON, J. W., AND L. LUMENG. Biphasic stimulation of amino acid uptake by glucagon in hepatocytes. *Biochem. Biophys. Res. Commun.* 96: 61–68, 1980.
63. EDMONDSON, J. W., B. A. MILLER, AND L. LUMENG. Effect of glucagon on hepatic taurocholate uptake: relationship to membrane potential. *Am. J. Physiol.* 249 (*Gastrointest. Liver Physiol.* 12): G427–G433, 1985.
64. ELIAS, E., Z. HRUBAN, J. B. WADE, AND J. L. BOYER. Phalloidin-induced cholestasis: a microfilament-mediated change in junctional complex permeability. *Proc. Natl. Acad. Sci. USA* 77: 2229–2233, 1980.
65. EPPING, R. J., AND F. L. BYGRAVE. A procedure for the rapid isolation from rat liver of plasma membrane vesicles exhibiting Ca^{2+}-transport and Ca^{2+}-ATPase activities. *Biochem. J.* 223: 733–745, 1984.
66. FEHLMANN, M., B. CANIVET, AND P. FREYCHET. Epidermal growth factor stimulates monovalent cation transport in isolated rat hepatocytes. *Biochem. Biophys. Res. Commun.* 100: 254–260, 1981.
67. FEHLMANN, M., AND P. FREYCHET. Insulin and glucagon stimulation of (Na^+-K^+)-ATPase transport activity in isolated rat hepatocytes. *J. Biol. Chem.* 256: 7449–7453, 1981.
68. FEHLMANN, M., A. LE CAM, P. KITABGI, J.-F. REY, AND P. FREYCHET. Regulation of amino acid transport in the liver. Emergence of a high affinity transport system in isolated hepatocytes from fasting rats. *J. Biol. Chem.* 254: 401–407, 1979.
69. FITZ, J. G., AND B. F. SCHARSCHMIDT. Regulation of transmembrane electrical potential gradient in rat hepatocytes in situ. *Am. J. Physiol.* 252 (*Gastrointest. Liver Physiol.* 15): G56–G64, 1987.
70. FITZ, J. G., AND B. F. SCHARSCHMIDT. Intracellular chloride activity in intact rat liver: relationship to membrane potential and bile flow. *Am. J. Physiol.* 252 (*Gastrointest. Liver Physiol.* 15): G699–G706, 1987.
71. FORGAC, M., L. CANTLEY, B. WIEDENMANN, L. ALTSTIEL, AND D. BRANTON. Clathrin-coated vesicles contain an ATP-dependent proton pump. *Proc. Natl. Acad. Sci. USA* 80: 1300–1303, 1983.
72. FORKER, E. L. Two sites of bile formation as determined by mannitol and erythritol clearance in the guinea pig. *J. Clin. Invest.* 46: 1189–1195, 1967.
73. FORKER, E. L. Bile formation in guinea pigs: analysis with inert solutes of graded molecular radius. *Am. J. Physiol.* 215: 56–62, 1968.
74. FORKER, E. L. The effect of estrogen on bile formation in the rat. *J. Clin. Invest.* 48: 654–663, 1969.
75. FORKER, E. L. Hepatocellular uptake of inulin, sucrose, and mannitol in rats. *Am. J. Physiol.* 219: 1568–1573, 1970.
76. FORKER, E. L. Mechanisms of hepatic bile formation. *Annu. Rev. Physiol.* 39: 323–347, 1977.
77. FRIEND, D. S., AND N. B. GILULA. Variations in tight gap junction in mammalian tissues. *J. Cell Biol.* 53: 758–776, 1972.
78. FRIZZELL, R. A., AND M. E. DUFFEY. Chloride activities in epithelia. *Federation Proc.* 39: 2860–2864, 1980.
79. FROMTER, E., AND J. DIAMOND. Route of passive ion permeation in epithelia. *Nature Lond.* 235: 9–13, 1972.
80. GARBEROGLIO, C. A., H. M. RICHTER III, A HENAREJOS, A. R. MOOSSA, AND A. L. BAKER. Pharmacological and physiological doses of insulin and determinants of bile flow in dogs. *Am. J. Physiol.* 245 (*Gastrointest. Liver Physiol.* 8): G157–G163, 1983.
81. GLICKMAN, J., K. CROEN, S. KELLY, AND Q. AL-AWQATI. Golgi membranes contain an electrogenic H^+ pump in parallel to a chloride conductance. *J. Cell Biol.* 97: 1303–1308, 1983.
82. GLUCK, S., AND Q. AL-AWQATI. An electrogenic proton-translocating adenosine triphosphatase from bovine kidney medulla. *J. Clin. Invest.* 73: 1704–1710, 1984.

83. GLUCK, S., C. CANNON, AND Q. AL-AWQATI. Exocytosis regulates urinary acidification in turtle bladder by rapid insertion of H^+ pumps into the luminal membrane. *Proc. Natl. Acad. Sci. USA* 79: 4327–4331, 1982.
84. GOLDSTEIN, J. L., R. G. W. ANDERSON, AND M. S. BROWN. Coated pits, coated vesicles, and receptor-mediated endocytosis. *Nature Lond.* 279: 679–685, 1979.
85. GONZALEZ-CALVIN, J. L., J. B. SAUNDERS, AND R. WILLIAMS. Effects of ethanol and acetaldehyde on hepatic plasma membrane ATPases. *Biochem. Pharmacol.* 32: 1723–1728, 1983.
86. GOODENOUGH, D. A., AND J. P. REVEL. A fine structural analysis of intercellular junctions in the mouse liver. *J. Cell Biol.* 45: 272–290, 1970.
87. GRAF, J., A. GAUTAM, AND J. L. BOYER. Isolated rat hepatocyte couplets: a primary secretory unit for electrophysiologic studies of bile secretory function. *Proc. Natl. Acad. Sci. USA* 81: 6516–6520, 1984.
88. GRAF, J., R. H. HENDERSON, B. KRUMPHOLZ, AND J. L. BOYER. Cell membrane and transepithelial voltages and resistances in isolated rat hepatocyte couplets (IRHC). *J. Membr. Biol.* 95: 241–254, 1987.
89. GRAF, J., AND M. PETERLIK. Mechanism of transport of inorganic ions into bile. In: *The Hepatobiliary System—Fundamental and Pathological Mechanisms*, edited by W. Taylor. New York: Plenum, 1975, p. 43–58.
90. GRAF, J., AND O. H. PETERSEN. Cell membrane potential and resistance in liver. *J. Physiol. Lond.* 284: 105–126, 1978.
91. GREGORY, D. H., Z. R. VLAHCEVIC, M. F. PRUGH, AND L. SWELL. Mechanism of secretion of biliary lipids: role of a microtubular system in hepatocellular transport of biliary lipids in the rat. *Gastroenterology* 74: 93–100, 1978.
92. GREGORY, D. H., Z. R. VLAHCEVIC, P. SCHATZKI, AND L. SWELL. Mechanism of secretion of biliary lipids. I. Role of bile canalicular and microsomal membranes in the synthesis and transport of biliary lecithin and cholesterol. *J. Clin. Invest.* 55: 105–114, 1975.
93. GROOTHUIS, G. M. M., J. G. WEITERING, M. J. HARDONK, AND D. K. F. MEIJER. Heterogeneity of rat hepatocytes in transport and hepatic binding of asialoalkaline phosphatase studied after induction of selective acinar damage by N-hydroxy-2-acetylaminofluorene and carbon tetrachloride. *Biochem. Pharmacol.* 32: 2721–2727, 1983.
94. GUMUCIO, J. J., C. BALABAUD, D. L. MILLER, L. F. DEMASON, H. D. APPELMAN, T. J. STOCCHU, AND D. R. FRANZBLAU. Bile secretion and liver cell heterogeneity in the rat. *J. Lab. Clin. Med.* 91: 350–362, 1978.
95. HARDISON, W. G. M., S. BELLENTANI, V. HEASLEY, AND D. SHELLHAMER. Specificity of an Na^+-dependent taurocholate transport site in isolated rat hepatocytes. *Am. J. Physiol.* 246 (*Gastrointest. Liver Physiol.* 9): G477–G483, 1984.
96. HARDISON, W. G. M., D. E. HATOFF, K. MIYAI, AND R. G. WEINER. Nature of bile acid maximum secretory rate in the rat. *Am. J. Physiol.* 241 (*Gastrointest. Liver Physiol.* 4): G337–G343, 1981.
97. HARDISON, W. G. M., AND C. A. WOOD. Importance of bicarbonate in bile salt independent fraction of bile flow. *Am. J. Physiol.* 235 (*Endocrinol. Metab. Gastrointest. Physiol.* 4): E158–E164, 1978.
98. HARIKUMAR, P., AND J. P. REEVES. The lysosomal proton pump is electrogenic. *J. Biol. Chem.* 258: 10403–10410, 1983.
99. HAYES, J. D., R. C. STRANGE, AND I. W. PERCY-ROBB. Cholic acid binding by gluatathione S-transferase from rat liver cytosol. *Biochem. J.* 185: 83–87, 1980.
100. HESS, P., AND R. W. TSIEN. Mechanism of ion permeation through calcium channels. *Nature Lond.* 309: 453–456, 1984.
101. IHLENFELDT, M. J. A. Stimulation of Rb^+ transport by glucagon in isolated rat hepatocytes. *J. Biol. Chem.* 256: 2213–2218, 1981.
102. INOUE, M., R. KINNE, T. TRAU, AND I. M. ARIAS. Taurocholate transport by rat liver sinusoidal membrane vesicles: evidence of sodium cotransport. *Hepatology Baltimore* 2: 572–579, 1982.
103. INOUE, M., R. KINNE, T. TRAU, AND I. M. ARIAS. Taurocholate transport by rat liver canalicular membrane vesicles. Evidence for the presence of an Na^+-independent transport system. *J. Clin. Invest.* 73: 659–663, 1984.
104. ISMAIL-BEIGI, F., D. M. BISSELL, AND I. S. EDELMAN. Thyroid thermogenesis in primary monolayer culture. *J. Gen. Physiol.* 73: 369–383, 1979.
105. IVES, H. E. Na^+-H^+ exchange, oncogenes and growth regulation in normal and tumor cells. *West. J. Med.* 143: 365–370, 1985.
106. JAKOB, A., AND S. DIEM. Metabolic responses of perfused rat livers to alpha- and beta-adrenergic agonists, glucagon and cyclic AMP. *Biochim. Biophys. Acta* 404: 57–66, 1975.
107. JOHNSON, R. G., M. F. BEERS, AND A. SCARPA. H^+ ATPase of chromaffin granules. *J. Biol. Chem.* 257: 10701–10707, 1982.
108. JONES, A. L., D. L. SCHMUCKER, R. H. RENSTON, AND T. MURAKAMI. The architecture of bile secretion. *Dig. Dis. Sci.* 25: 609–629, 1980.
109. JOSEPH, S. K., K. E. COLL, A. P. THOMAS, R. RUBIN, AND J. R. WILLIAMSON. The role of extracellular Ca^{2+} in the response of the hepatocyte to Ca^{2+}-dependent hormones. *J. Biol. Chem.* 260: 12508–12515, 1985.
110. JOSEPH, S. K., AND J. R. WILLIAMSON. The origin, quantitation, and kinetics of intracellular calcium mobilization by vasopressin and phenylephrine in hepatocytes. *J. Biol. Chem.* 258: 10425–10432, 1983.
111. KACICH, R. L., R. H. RENSTON, AND A. L. JONES. Effects of cytochalasin D and colchicine on the uptake, translocation, and biliary secretion of horseradish peroxidase and [^{14}C]taurocholate in the rat. *Gastroenterology* 85: 385–394, 1983.
112. KAMINSKI, D. L., W. H. BROWN, AND Y. G. DESHPANDE. Effect of glucagon on bile cAMP secretion. *Am. J. Physiol.* 238 (*Gastrointest. Liver Physiol.* 1): G119–G123, 1980.
113. KAPLAN, M. R., AND R. D. SIMONI. Transport of cholesterol from the endoplasmic reticulum to the plasma membrane. *J. Cell Biol.* 101: 446–453, 1985.
114. KAUNITZ, J. D., R. D. GUNTHER, AND G. SACHS. Characterization of an electrogenic ATP and chloride-dependent proton translocating pump from rat renal medulla. *J. Biol. Chem.* 260: 11567–11573, 1985.
115. KEEFFE, E. B., N. M. BLANKENSHIP, AND B. F. SCHARSCHMIDT. Alteration of rat liver plasma membrane fluidity and ATPase activity by chlorpromazine hydrochloride and its metabolites. *Gastroenterology* 79: 222–231, 1980.
116. KEEFFE, E. B., B. F. SCHARSCHMIDT, N. M. BLANKENSHIP, AND R. K. OCKNER. Studies of relationships among bile flow, liver plasma membrane Na,K-ATPase, and microviscosity in the rat. *J. Clin. Invest.* 64: 1590–1598, 1979.
117. KELLEY, D. S., H. A. CAMPBELL, AND V. R. POTTER. Effects of hormones and amino acid depletion on the kinetic parameters of amino acid uptake in monolayer cultures of rat hepatocytes. *J. Cell. Physiol.* 112: 67–75, 1982.
118. KELLEY, D. S., T. EVANSON, AND V. R. POTTER. Calcium-dependent hormonal regulation of amino acid transport and cyclic AMP accumulation in rat hepatocyte monolayer cultures. *Proc. Natl. Acad. Sci. USA* 77: 5953–5957, 1980.
119. KILBERG, M. S. Amino acid transport in isolated rat hepatocytes. *J. Membr. Biol.* 69: 1–12, 1982.
120. KINNE-SAFFRAN, E., R. BEAUWENS, AND R. KINNE. An ATP-driven pump in brush-border membranes from rat renal cortex. *J. Membr. Biol.* 64: 67–76, 1982.
121. KOCH, K. S., R. GROSSE, H. SKELLY, AND H. L. LEFFERT. Initiation of cultured rat hepatocyte proliferation does not involve Na^+-dependent plasma membrane Ca^{2+} fluxes. *Cell Biol. Int. Rep.* 8: 309–316, 1984.
122. KOCH, K. S., AND H. L. LEFFERT. Growth control of differentiated adult rat hepatocytes in primary culture. *Ann. NY Acad. Sci.* 349: 111–127, 1980.
123. KOLB, H. A., C. D. A. BROWN, AND H. MURER. Identification of voltage-dependent anion channel in the apical membrane of a Cl^--secretory epithelium (MDCK). *Pfluegers Arch.* 403: 262–265, 1985.

124. KRAMER, W., U. BICKEL, H.-P. BUSCHER, W. GEROK, AND G. KURZ. Bile-salt-binding polypeptides in plasma membranes of hepatocytes revealed by photoaffinity labelling. *Eur. J. Biochem.* 129: 13-24, 1982.

125. KRAUS-FRIEDMANN, N., J. BIBER, H. MURER, AND E. CARAFOLI. Calcium uptake in isolated hepatic plasma-membrane vesicles. *Eur. J. Biochem.* 129: 7-12, 1982.

126. KRISTENSEN, L. O. Energization of alanine transport in isolated rat hepatocytes. *J. Biol. Chem.* 255: 5236-5243, 1980.

127. KRISTENSEN, L. O., AND M. FOLKE. Coupling ratio of electrogenic Na^+-alanine cotransport in isolated rat hepatocytes. *Biochem. J.* 210: 621-624, 1983.

128. KRISTENSEN, L. O., AND M. FOLKE. Volume-regulatory K^+ efflux during concentrative uptake of alanine in isolated rat hepatocytes. *Biochem. J.* 221: 265-268, 1984.

129. KRISTENSEN, L. O., L. SESTOFT, AND M. FOLKE. Concentrative uptake of alanine in hepatocytes from fed and fasted rats. *Am. J. Physiol.* 244 (*Gastrointest. Liver Physiol.* 7): G491-G500, 1983.

130. KRULWICH, T. A. Na^+/H^+ antiporters. *Biochim. Biophys. Acta* 726: 245-264, 1983.

131. LACK, L. Properties and biological significance of the ileal bile salt transport system. *Environ. Health Perspect.* 33: 79-90, 1979.

132. LAGARDE, S., E. ELIAS, J. B. WADE, AND J. L. BOYER. Structural heterogeneity of hepatocyte "tight" junctions: a qualitative analysis. *Hepatology Baltimore* 1: 193-203, 1981.

133. LAKE, J., P. GEORGE, V. LICKO, AND B. F. SCHARSCHMIDT. Vesicular transport of fluid phase markers (FPM) and ligands by liver: effects of colchicine and chloroquine (Abstract). *Clin. Res.* 33: 322, 1985.

134. LAKE, J. R., V. LICKO, R. W. VAN DYKE, AND B. F. SCHARSCHMIDT. Biliary secretion of fluid-phase markers by the isolated perfused rat liver: role of transcellular vesicular transport. *J. Clin. Invest.* 76: 676-684, 1985.

135. LAKE, J. R., R. W. VAN DYKE, AND B. F. SCHARSCHMIDT. Effects of Na^+ replacement and amiloride on ursodeoxycholic acid-stimulated choleresis and biliary bicarbonate secretion. *Am. J. Physiol.* 252 (*Gastrointest. Liver Physiol.* 15): G163-G169, 1987.

136. LATHAM, P. S., AND M. KASHGARIAN. The ultrastructural localization of transport ATPase in the rat liver at non-bile canalicular plasma membranes. *Gastroenterology* 76: 988-996, 1979.

137. LAYDEN, T. J., E. ELIAS, AND J. L. BOYER. Bile formation in the rat—the role of the paracellular shunt pathway. *J. Clin. Invest.* 62: 1375-1385, 1978.

138. LAYDEN, T. J., J. SCHWARZ, AND J. L. BOYER. Scanning electron microscopy of the rat liver: studies of the effect of taurolithocholate and other models of cholestasis. *Gastroenterology* 69: 724-738, 1975.

139. LEFFERT, H. L., AND K. S. KOCH. Ionic events at the membrane initiate rat liver regeneration. *Ann. NY Acad. Sci.* 339: 201-215, 1980.

140. LEFFERT, H. L., K. S. KOCH, M. FEHLMAN, W. HEISER, P. J. LAD, AND H. SKELLY. Amiloride blocks cell-free protein synthesis at levels attained inside cultured rat hepatocytes. *Biochem. Biophys. Res. Commun.* 108: 738-745, 1982.

141. LEFFERT, H. L., D. B. SCHENK, J. J. HUBERT, H. SKELLY, M. SCHUMACHER, R. ARIYASU, M. ELLISMAN, K. S. KOCH, AND G. A. KELLER. Hepatic (Na^+,K^+)-ATPase: a current view of its structure, function and localization in rat liver as revealed by studies with monoclonal antibodies. *Hepatology Baltimore* 5: 501-507, 1985.

142. LEVINE, R. A., AND R. C. HALE. Cyclic AMP in secretin choleresis: evidence for a regulatory role in man and baboons but not in dogs. *Gastroenterology* 70: 537-544, 1976.

143. LIN, S.-H. Novel ATP-dependent calcium transport component from rat liver plasma membranes. *J. Biol. Chem.* 260: 7850-7856, 1985.

144. LIN, S.-H. The rat liver plasma membrane high affinity (Ca^{2+}-Mg^{2+})-ATPase is not a calcium pump. *J. Biol. Chem.* 260: 10976-10980, 1985.

145. LORENZINI, P., P. ILTER, P. MEIER, AND J. L. BOYER. Taurochenodeoxycholic acid stimulates hepatic uptake of ^{3}H-methoxyinulin (^{3}HMI) into membrane-bound compartments (Abstract). *Hepatology Baltimore* 2: 737, 1982.

146. LOTERSZTAJN, S., J. HANOUNE, AND F. PECKER. A high affinity calcium-stimulated magnesium-dependent ATPase in rat liver plasma membranes. *J. Biol. Chem.* 256: 11209-11215, 1981.

147. LOTERSZTAJN, S., A. MALLAT, C. PAVOINE, AND F. PECKER. The inhibitor of liver plasma membrane (Ca^{2+}-Mg^{2+})-ATPase. *J. Biol. Chem.* 260: 9692-9698, 1985.

148. MACHEN, T. E., E. ERLIJ, AND F. B. P. WOODING. Permeable junctional complexes. The movement of lanthanum across rabbit gallbladder and intestine. *J. Cell Biol.* 54: 302-312, 1972.

149. MATTER, A., L. ORCI, AND C. ROVILLER. A study on the permeability barriers between Disse's space and the bile canaliculus. *J. Ultrastruct. Res. Suppl.* 11: 5-71, 1969.

150. MAZER, N. A., AND M. C. CAREY. Quasi-elastic light-scattering studies of aqueous biliary lipid systems. Cholesterol solubilization and precipitation in model bile solutions. *Biochemistry* 22: 426-442, 1983.

151. MEIER, P. J., R. KNICKELBEIN, R. H. MOSELEY, J. W. DOBBINS, AND J. L. BOYER. Evidence for carrier-mediated chloride/bicarbonate exchange in canalicular rat liver plasma membrane vesicles. *J. Clin. Invest.* 75: 1256-1263, 1985.

152. MEIER, P. J., A. S. MEIER-ABT, C. BARRETT, AND J. L. BOYER. Mechanisms of taurocholate transport in canalicular and basolateral rat liver plasma membrane vesicles. Evidence for an electrogenic canalicular organic anion carrier. *J. Biol. Chem.* 259: 10614-10622, 1984.

153. MEIER, P. J., E. S. SZTUL, A. RUEBEN, AND J. L. BOYER. Structural and functional polarity of canalicular and basolateral plasma membrane vesicles isolated in high yield from rat liver. *J. Cell Biol.* 98: 991-1000, 1984.

154. MEYERS, W. C., J. B. HANKS, AND R. S. JONES. Inhibition of basal and meal-stimulated choleresis by somatostatin. *Surgery St. Louis* 86: 301-306, 1979.

155. MINER, P. B., JR., E. SUTHERLAND, AND F. R. SIMON. Regulation of hepatic sodium plus potassium-activated adenosine triphosphatase activity by glucocorticoids in the rat. *Gastroenterology* 79: 212-221, 1980.

156. MOSELEY, R. H., P. J. MEIER, P. S. ARONSON, AND J. L. BOYER. Na^+-H exchange in rat liver basolateral but not canalicular membrane vesicles. *Am. J. Physiol.* 250 (*Gastrointest. Liver Physiol* 13): G35-G43, 1986.

157. MOTTA, P. A scanning electron microscopic study of the rat liver sinusoid: endothelial and Kupffer cells. *Cell Tissue Res.* 164: 371-385, 1975.

158. MOTTA, P., AND K. R. PORTER. Structure of rat liver sinusoids and associated tissue spaces as revealed by scanning electron microscopy. *Cell Tissue Res.* 148: 111-125, 1974.

159. MULLOCK, B. M., M. DOBRATA, AND R. H. HINTON. Sources of the proteins of rat bile. *Biochim. Biophys. Acta* 543: 497-507, 1978.

160. MURER, H., K. SIGRIST-NELSON, AND U. HOPPER. On the mechanism of sugar and amino acid interaction in intestinal transport. *J. Biol. Chem.* 250: 7392-7396, 1975.

161. O'MAILLE, E. R. L. The influence of micelle formation on bile salt secretion. *J. Physiol. Lond.* 302: 107-120, 1980.

162. OSHIO, C., AND M. J. PHILLIPS. Contractility of the bile canaliculi: implication for liver function. *Science Wash. DC* 212: 1041-1042, 1981.

163. PALFREY, H. C., AND M. C. RAO. Na/K/Cl co-transport and its regulation. *J. Exp. Biol.* 106: 43-54, 1983.

164. PALOHEIMO, M., J. LINKOLA, M. LEMPINEN, AND M. FOLKE. Time-courses of hepatocellular hyperpolarization and cyclic adenosine 3′,5′-monophosphate accumulation after partial hepatectomy in the rat. *Gastroenterology* 87: 639-646, 1984.

165. PECKER, F., AND S. LOTERSZTAJN. Fe^{2+} and other divalent metal ions uncouple Ca^{2+} transport from (Ca^{2+}-Mg^{2+})-ATPase in rat liver plasma membranes. *J. Biol. Chem.* 260: 731-735,

1985.
166. PETERSEN, O. H. The effect of glucagon on the liver cell membrane potential. *J. Physiol. Lond.* 239: 647–656, 1974.
167. PETERSEN, O. H., AND Y. MARUYAMA. Calcium-activated potassium channels and their role in secretion. *Nature Lond.* 307: 693–696, 1984.
168. PHILIPSON, K. D. Sodium-calcium exchange in plasma membrane vesicles. *Annu. Rev. Physiol.* 47: 561–571, 1985.
169. POLLOCK, A. S. Intracellular pH of hepatocytes in primary monolayer culture. *Am. J. Physiol.* 246 (*Renal Fluid Electrolyte Physiol.* 15): F738–F744, 1984.
170. PRANDI, D., S. ERLINGER, J.-C. GLASINOVIC, AND M. DUMONT. Canalicular bile production in man. *Eur. J. Clin. Invest.* 5: 1–6, 1975.
171. PREISIG, P. A., AND C. A. BERRY. Evidence for transcellular osmotic water flow in rat proximal tubules. *Am. J. Physiol.* 249 (*Renal Fluid Electrolyte Physiol.* 18): F124–F131, 1985.
172. PRPIC, V., K. C. GREEN, P. F. BLACKMORE, AND J. H. EXTON. Vasopressin-, angiotensin II-, and α_1-adrenergic-induced inhibition of Ca^{2+} transport by rat liver plasma membrane vesicles. *J. Biol. Chem.* 259: 1382–1385, 1984.
173. RABON, E., J. CUPPOLETTI, D. MALINOWSKA, A. SMOLKA, H. F. HELANDER, J. MENDLEIN, AND G. SACHS. Proton secretion by the gastric parietal cell. *J. Exp. Biol.* 106: 119–133, 1983.
174. REES-JONES, R., AND Q. AL-AWQATI. Proton-translocating adenosinetriphosphatase in rough and smooth microsomes from rat liver. *Biochemistry* 23: 2236–2240, 1984.
175. REGGIO, H., D. BAINTON, E. HARMS, E. COUDRIER, AND D. LOUVARD. Antibodies against lysosomal membranes reveal a 100,000-mol-wt protein that cross-reacts with purified H^+,K^+ ATPase from gastric mucosa. *J. Cell Biol.* 99: 1511–1526, 1984.
176. REICHEN, J., F. BERR, M. LE, AND G. H. WARREN. Characterization of calcium deprivation-induced cholestasis in the perfused rat liver. *Am. J. Physiol.* 249 (*Gastrointest. Liver Physiol.* 12): G48–G57, 1985.
177. REICHEN, J., AND M. LE. Taurocholate, but not taurodehydrocholate, increases biliary permeability to sucrose. *Am. J. Physiol.* 245 (*Gastrointest. Liver Physiol.* 8): G651–G655, 1983.
178. REICHEN, J., AND M. LE. Effect of taurocholate on place and rate of entry of sucrose into the bile canaliculus (Abstract). *Gastroenterology* 84: 1283, 1983.
179. REICHEN, J., AND H. SAEGESSEN. Role of the hepatic artery in bile formation in the rat (Abstract). *Hepatology Baltimore* 5: 987, 1985.
180. RENKIN, E. M. Filtration, diffusion and molecular sieving through porous membranes. *J. Gen. Physiol.* 38: 225–243, 1954.
181. RENNER, E. L., J. R. LAKE, E. J. CRAGOE, JR., R. W. VAN DYKE, AND B. F. SCHARSCHMIDT. Ursodeoxycholic acid choleresis: relationship to biliary HCO_3^- and effects Na^+-H^+ exchange inhibitors. *Am. J. Physiol.* 254 (*Gastrointest. Liver Physiol.* 17): G232–G241, 1988.
182. RENSTON, R. H., A. L. JONES, W. D. CHRISTIANSEN, AND G. T. HRADEK. Evidence for a vesicular transport mechanism in hepatocytes for biliary secretion of immunoglobulin A. *Science Wash. DC* 208: 1276–1278, 1980.
183. RICCI, G. L., AND J. FEVERY. Quantitative aspects of the effect of somatostatin on bile flow in the rat. *Biochem. Soc. Trans.* 8: 53–54, 1980.
184. RICCI, R. L., S. S. CRAWFORD, AND P. B. MINER, JR. The effect of ethanol on hepatic sodium plus potassium activated adenosine triphosphatase activity in the rat. *Gastroenterology* 80: 1445–1450, 1981.
185. ROOS, A., AND W. F. BORON. Intracellular pH. *Physiol. Rev.* 61: 296–434, 1981.
186. SAERMARK, T., N. FLINT, AND W. H. EVANS. Hepatic endosome fractions contain an ATP-driven proton pump. *Biochem. J.* 225: 51–58, 1985.
187. SAMSON, M., AND M. FEHLMANN. Plasma membrane vesicles from isolated hepatocytes retain the increase of amino acid transport induced by dibutyryl cyclic AMP in intact cells. *Biochim. Biophys. Acta* 687: 35–41, 1982.
188. SAMUELS, A. M., AND M. C. CAREY. Effects of chlorpromazine hydrochloride and its metabolites on Mg^{2+}- and Na^+,K^+- ATPase activities of canalicular-enriched rat liver plasma membranes. *Gastroenterology* 74: 1183–1190, 1978.
189. SCHARSCHMIDT, B. F. Bile formation and cholestasis, metabolism and enterohepatic circulation of bile acid, and gallstone formation. In: *Hepatology. A Textbook of Liver Diseases*, edited by D. Zakim and T. D. Boyer. Philadelphia, PA: Saunders, 1982, p. 297–351.
190. SCHARSCHMIDT, B. F., E. B. KEEFFE, D. A. VESSEY, N. M. BLANKENSHIP, AND R. K. OCKNER. In vitro effect of bile salts on rat liver plasma membrane lipid fluidity and ATPase activity. *Hepatology Baltimore* 1: 137–145, 1981.
191. SCHARSCHMIDT, B. F., J. R. LAKE, E. L. RENNER, V. LICKO, AND R. W. VAN DYKE. Fluid phase endocytosis by cultured rat hepatocytes and perfused rat liver. *Proc. Natl. Acad. Sci. USA* 83: 9488–9492, 1986.
192. SCHARSCHMIDT, B. F., AND J. F. STEPHENS. Transport of sodium, chloride and taurocholate by cultured rat hepatocytes. *Proc. Natl. Acad. Sci. USA* 78: 986–990, 1981.
193. SCHARSCHMIDT, B. F., AND R. W. VAN DYKE. Mechanisms of hepatic electrolyte transport. *Gastroenterology* 85: 1199–1214, 1983.
194. SCHARSCHMIDT, B. F., R. W. VAN DYKE, AND J. E. STEPHENS. Chloride transport by intact rat liver and cultured rat hepatocytes. *Am. J. Physiol.* 242 (*Gastrointest. Liver Physiol.* 5): G628–G633, 1982.
195. SCHENK, D. B., J. J. HUBERT, AND H. L. LEFFERT. Use of a monoclonal antibody to quantify (Na^+,K^+)-ATPase activity and sites in normal and regenerating rat liver. *J. Biol. Chem.* 259: 14941–14951, 1984.
196. SCHIFF, J. M., M. M. FISHER, AND B. J. UNDERDOWN. Receptor-mediated biliary transport of immunoglobulin A and asialoglycoprotein; sorting and missorting of ligands revealed by two radiolabeling methods. *J. Cell Biol.* 98: 79–89, 1984.
197. SCHULZ, S. G. Transport across epithelia: some basic principles. *Kidney Int.* 9: 65–75, 1976.
198. SCHWARTZ, A. L., G. J. A. M. STROUS, J. W. SLOT, AND H. J. GEUZE. Immunoelectron microscopic localization of acidic intracellular compartments in hepatoma cells. *EMBO J.* 4: 899–904, 1985.
199. SCHWARTZ, G. J., AND Q. AL-AWQATI. Carbon dioxide causes exocytosis of vesicles containing H^+ pumps in isolated perfused proximal and collecting tubules. *J. Clin. Invest.* 75: 1638–1644, 1985.
200. SEWELL, R. B., N. E. HOFFMAN, R. A. SMALLWOOD, AND S. COCKBAIN. Bile acid structure and bile formation: a comparison of hydroxy and keto bile acids. *Am. J. Physiol.* 238 (*Gastrointest. Liver Physiol.* 1): G10–G17, 1980.
201. SHAW, H. M., AND T. J. HEATH. Regulation of bile formation in rabbits and guinea pigs. *Q. J. Exp. Physiol. Cogn. Med. Sci.* 59: 93–102, 1974.
202. SHOROFSKY, S. R., M. FIELD, AND H. A. FOZZARD. The cellular mechanism of active chloride secretion in vertebrate epithelia: studies in intestine and trachea. *Philos. Trans. R. Soc. Lond. B Biol. Sci.* 299: 597–607, 1982.
203. SIMON, F. R., E. SUTHERLAND, AND L. ACCATINO. Stimulation of hepatic sodium and potassium-activated adenosine triphosphatase activity by phenobarbital. *J. Clin. Invest.* 58: 849–861, 1977.
204. SIMON, F. R., E. M. SUTHERLAND, AND M. GONZALEZ. Regulation of bile salt transport in rat liver. *J. Clin. Invest.* 70: 402–411, 1982.
205. SIPS, H. J., AND K. VAN DAM. Amino acid-dependent sodium transport in plasma membrane vesicles from rat liver. *J. Membr. Biol.* 62: 231–237, 1981.
206. SMITH, C. R., C. OSHO, M. MIYAURI, H. KATZ, AND M. J. PHILIPS. Coordination of the contractile activity of bile canaliculi: evidence from spontaneous contractions in vitro. *Lab. Invest.* 53: 270–274, 1985.
207. SMITH, N. D., AND J. L. BOYER. Permeability characteristics of bile duct in the rat. *Am. J. Physiol.* 242 (*Gastrointest. Liver Physiol.* 5): G52–G57, 1982.

208. SMOCK, T. K., R. W. VAN DYKE, D. M. BISSELL, AND B. F. SCHARSCHMIDT. Membrane potential, membrane resistance, and cation fluxes in cultured rat hepatocytes (Abstract). *Gastroenterology* 82: 1246, 1982.
209. SOMJEN, G. J., AND T. GILAT. A non-micellar mode of cholesterol transport in human bile. *FEBS Lett.* 156: 265–268, 1983.
210. SPRING, K. R. Fluid transport in gallbladder epithelium. *J. Exp. Biol.* 106: 181–194, 1983.
211. STANTON, B. A. Regulation of ion tranpsort in epithelia: role of membrane recruitment from cytoplasmic vesicles. *Lab. Invest.* 51: 255–257, 1984.
212. STOLZ, A., Y. SUGIYAMA, J. KUHLENKAMP, AND N. KAPLOWITZ. Identification and purification of a 36 kDa bile acid binding protein in human hepatic cytosol. *FEBS Lett.* 177: 31–35, 1984.
213. STONE, B. G., S. K. ERICKSON, AND A. D. COOPER. Regulation of rat biliary cholesterol secretion by agents that alter intrahepatic cholesterol metabolism: evidence for a distinct biliary precursor pool. *J. Clin. Invest.* 76: 1773–1781, 1985.
214. STONE, D. K., X.-S. XIE, AND E. RACKER. An ATP-driven proton pump in clathrin-coated vesicles. *J. Biol. Chem.* 258: 4059–4062, 1983.
215. STONE, D. K., X.-S. XIE, AND E. RACKER. Inhibition of clathrin-coated vesicle acidification by duramycin. *J. Biol. Chem.* 259: 2701–2703, 1984.
216. STRASBERG, S. M., R. G. ILSON, AND J. E. PALOHEIMO. Bile salt-associated electrolyte secretion and the effect of sodium taurocholate on bile flow. *J. Lab. Clin. Med.* 101: 317–326, 1983.
217. STRASBERG, S. M., C. N. PETRUNKA, R. G. ILSON, AND J. E. PALOHEIMO. Characteristics of inert solute clearance by the monkey liver. *Gastroenterology* 67: 259–266, 1979.
218. STROUS, G. J., A. DU MAINE, J. E. ZIJDERHAND-BLEEKEMOLEN, J. W. SLOT, AND A. L. SCHWARTZ. Effect of lysosomotropic amines on the secretory pathway and on the recycling of the asialoglycoprotein receptor in human hepatoma cells. *J. Cell Biol.* 101: 531–539, 1985.
219. SUCHY, F. J., S. M. COURCHENE, AND B. L. BLITZER. Taurocholate transport by basolateral plasma membrane vesicles isolated from developing rat liver. *Am. J. Physiol.* 248 (*Gastrointest. Liver Physiol.* 11): G648–G654, 1985.
220. SUGIYAMA, Y., A. STOLZ, M. SUGIMOTO, AND N. KAPLOWITZ. Evidence for a common affinity binding site on glutathione S-transferase B for lithocholic acid and bilirubin. *J. Lipid Res.* 25: 1177–1183, 1984.
221. SUGIYAMA, Y., T. YAMADA, AND N. KAPLOWITZ. Newly identified bile acid binders in rat liver cytosol: purification and comparison with glutathione S-transferases. *J. Biol. Chem.* 258: 3602–3607, 1983.
222. TAKEMURA, S., K. OMORI, K. TANAKA, K. OMORI, S. MATSUURA, AND Y. TASHIRO. Quantitative immunoferritin localization of [Na$^+$,K$^+$]ATPase on canine hepatocyte cell surface. *J. Cell Biol.* 99: 1502–1510, 1984.
223. TAVOLONI, N., J. S. REED, AND J. L. BOYER. Hemodynamic effects on determinants of bile secretion in isolated rat liver. *Am. J. Physiol.* 234 (*Endocrinol. Metab. Gastrointest. Physiol.* 3): E584–E592, 1978.
224. TAVOLONI, N., H. R. WYSSBROD, AND J. T. JONES. Permeability characteristics of the guinea pig biliary apparatus (Abstract). *Gastroenterology* 88: 1700, 1985.
225. THOMSEN, O. Ø. Mechanism and regulation of hepatic bile production: with special reference to the bile acid-independent canalicular bile production. *Scand. J. Gastroenterol. Suppl.* 97: 1–52, 1984.
226. TISHER, C. C., AND W. E. YARGER. Lanthanum permeability of the tight junction (zonula occludens) in the renal tubule of the rat. *Kidney Int.* 3: 238–250, 1973.
227. UTILI, R., C. O. ABERNATHY, AND H. J. ZIMMERMAN. Inhibition of Na$^+$,K$^+$-adenosine triphosphatase by endotoxin: a possible mechanism for endotoxin-induced cholestasis. *J. Infect. Dis.* 136: 583–587, 1977.
228. VAN AMELSVOORT, J. M. M., H. J. SIPS, M. E. A. APITULE, AND K. VAN DAM. Heterogeneous distribution of the sodium-dependent alanine transport activity in the rat hepatocyte plasma membrane. *Biochim. Biophys. Acta* 600: 950–960, 1980.
229. VAN AMELSVOORT, J. M. M., H. J. SIPS, AND K. VAN DAM. Sodium-dependent alanine transport in plasma-membrane vesicles from rat liver. *Biochem. J.* 174: 1083–1086, 1978.
230. VAN DYKE, R. W. Anion inhibition of the proton pump in rat liver multivesicular bodies. *J. Biol. Chem.* 261: 15941–15948, 1986.
231. VAN DYKE, R. W., C. A. HORNICK, J. BELCHER, B. F. SCHARSCHMIDT, AND R. J. HAVEL. Identification and characterization of ATP-dependent proton transport by rat liver multivesicular bodies. *J. Biol. Chem.* 260: 11021–11026, 1985.
232. VAN DYKE, R. W., AND B. F. SCHARSCHMIDT. (Na,K)-ATPase-mediated cation pumping in cultured rat hepatocytes. *J. Biol. Chem.* 258: 12912–12919, 1983.
233. VAN DYKE, R. W., AND B. F. SCHARSCHMIDT. Effects of chlorpromazine on Na$^+$-K$^+$-ATPase pumping and solute transport in rat hepatocytes. *Am. J. Physiol.* 253 (*Gastrointest. Liver Physiol.* 16): G613–G621, 1987.
234. VAN DYKE, R. W., B. F. SCHARSCHMIDT, AND C. J. STEER. ATP-dependent proton transport by isolated brain clathrin-coated vesicles. Role of clathrin and other determinants of acidification. *Biochim. Biophys. Acta* 812: 423–436, 1985.
235. VAN DYKE, R. W., C. J. STEER, AND B. F. SCHARSCHMIDT. Clathrin-coated vesicles from rat liver: enzymatic profile and characterization of ATP-dependent proton transport. *Proc. Natl. Acad. Sci. USA* 81: 3108–3112, 1984.
236. VAN DYKE, R. W., J. E. STEPHENS, AND B. F. SCHARSCHMIDT. Bile acid transport in cultured rat hepatocytes. *Am. J. Physiol.* 243 (*Gastrointest. Liver Physiol.* 6): G484–G492, 1982.
237. VAN DYKE, R. W., J. E. STEPHENS, AND B. F. SCHARSCHMIDT. Effect of ion substitution on bile acid-dependent and bile acid-independent bile formation by the isolated perfused rat liver. *J. Clin. Invest.* 70: 505–517, 1982.
238. VAN DIPPE, P., P. DRAIN, AND D. LEVY. Synthesis and transport characteristics of photoaffinity probes for the hepatocyte bile acid transport system. *J. Biol. Chem.* 258: 8890–8895, 1983.
239. VAN DIPPE, P., AND D. LEVY. Analysis of the transport system for inorganic anions in normal and transformed hepatocytes. *J. Biol. Chem.* 257: 4381–4385, 1982.
240. WALL, D. A., AND T. MAACK. Endocytic uptake, transport, and catabolism of proteins by epithelial cells. *Am. J. Physiol.* 248 (*Cell Physiol.* 17): C12–C20, 1985.
241. WANNAGAT, R. J., R. D. ADLER, AND R. K. OCKNER. Bile acid-induced increase in bile acid-independent flow and plasma membrane NaK-ATPase activity in rat liver. *J. Clin. Invest.* 61: 297–307, 1978.
242. WATANABE, S., C. R. SMITH, AND M. J. PHILLIPS. Coordination of the contractile activity of bile canaliculi: evidence from calcium microinjection of triplet hepatocytes. *Lab. Invest.* 53: 275–279, 1985.
243. WEIBEL, E. R., W. STÄUBLI, H. R. GNÄGI, AND F. A. HESS. Correlated morphometric and biochemical studies on the liver cell. I. Morphometric model, stereologic methods, and normal morphometric data for rat liver. *J. Cell Biol.* 42: 68–91, 1969.
244. WHEELER, H. O. Water and electrolytes in bile. In: *Handbook of Physiology. Alimentary Canal*, edited by C. F. Code. Washington, DC; Am. Physiol. Soc., 1968, sect. 6, vol. V, chapt. 113, p. 2409–2431.
245. WHEELER, H. O., E. D. ROSS, AND S. E. BRADLEY. Canalicular bile production in dogs. *Am. J. Physiol.* 214: 866–874, 1968.
246. WHITE, M. F., AND H. N. CHRISTENSEN. Simultaneous regulation of amino acid influx and efflux by system A in the hepatoma cell HTC. *J. Biol. Chem.* 258: 8028–8038, 1983.
247. WILLIAMS, J. A., C. D. WITHROW, AND D. M. WOODBURY. Effects of nephrectomy and KCl on transmembrane potentials, intracellular electrolytes, and cell pH of rat muscle and liver in vivo. *J. Physiol. Lond.* 212: 117–128, 1971.

248. WILLIAMSON, J. R., R. H. COOPER, S. K. JOSEPH, AND A. P. THOMAS. Inositol trisphosphate and diacylglycerol as intracellular second messengers in liver. *Am. J. Physiol.* 248 (*Cell Physiol.* 17): C203–C216, 1985.
249. WONDERGEM, R. Insulin depolarization of rat hepatocytes in primary monolayer culture. *Am. J. Physiol.* 244 (*Cell Physiol.* 13): C17–C23, 1983.
250. WONDERGEM, R., AND D. R. HARDER. Transmembrane potential and amino acid transport in rat hepatocytes in primary monolayer culture. *J. Cell. Physiol.* 104: 53–60, 1980.
251. YAMAMOTO, K., H. MAYAHARA, AND K. OGAWA. Cytochemical localization of ouabain-sensitive, K-dependent *p*-nitrophenylphosphatase in the rat hepatocyte. *Acta Histochem. Cytochem.* 17: 23–35, 1984.
252. ZOUBOULIS-VAFIADIS, I., M. DUMONT, AND S. ERLINGER. Conjugation is rate limiting in hepatic transport of ursodeoxycholate in the rat. *Am. J. Physiol.* 243 (*Gastrointest. Liver Physiol.* 6): G208–G213, 1982.

CHAPTER 31

Physical chemistry of bile

DONNA J. CABRAL
DONALD M. SMALL

Biophysics Institute, Housman Medical Research Center, Departments of Medicine and Biochemistry, Boston University School of Medicine, Boston, Massachusetts

CHAPTER CONTENTS

Structural and Surface Properties of Bile Salts
  Molecular structure
  Crystal structure
  Surface chemistry
    Bile acids
    Bile salts
  Hydrophobic-hydrophilic balance
Aqueous Solution Properties of Bile Salts
  Solubility in water
    Bile acids
    Bile salts
  Solubility in organic solvents
  Ionization behavior
  Aggregation behavior
    Critical micelle concentration
    Effect of hydroxyl constituents and conjugation state
    Effects of temperature
    Effects of added sodium ion
    Effects of side-chain length
    Micelle size and shape
    Bile salt–calcium interactions
Bile Salt–Lecithin Interactions
  Composition of lecithins in bile
  Bile salt–phospholipid–water systems
    Phase rule
    Egg yolk lecithin–sodium cholate–water system: general characteristics
    Structures of isotropic (micellar) zone
    "Micelle-to-vesicle transition"
    Bilayered aqueous two-phase region (region 7)
    Effects of bile salts on phospholipid vesicles
    Kinetics of multilamellar phosphatidylcholine liposomes to mixed micelle formation
    Kinetics of transbilayer movement of bile acids
Cholesterol
  Solubility in aqueous media
  Cholesterol–bile salt interactions
  Cholesterol-phospholipid interactions
  Quaternary system: bile salt–phospholipid–cholesterol–water
Bile
  Normal bile
  Metastable bile
  Cholesterol crystal formation in bile
  Secretion of bile from the hepatocyte

BILE IS AN AQUEOUS SECRETION produced in the canaliculi of the hepatocytes of the liver (84). It passes from the canaliculi to the bile ducts and in some species into the gallbladder, where it is concentrated. It is discharged into the intestine, where a major constituent, bile salts, aids in the digestion and absorption of fat and fat-soluble vitamins (24). The bile salts are reabsorbed by an active process in the ileum (61) and returned to the liver. The principal constituents of bile include the bile salts, lecithins (10) (phosphatidylcholines), cholesterol, bile pigments, inorganic ions, and proteins. This chapter covers the physical properties of bile salts, their interactions with lecithins and cholesterol in aqueous systems, and the relation of model systems of bile salts, lecithin, and cholesterol to native bile.

First we describe the structure and surface properties of bile salts (see next section), noting the different ways in which molecules of bile salts may pack together in a crystalline lattice and how bile acids concentrate at aqueous interfaces. A key to understanding the behavior of bile salts is recognition that they are polar amphipathic molecules; that is, they have a hydrocarbon side that tends to partition with other hydrocarbon surfaces and a hydrophilic side that interacts strongly with water. The amphipathic properties of bile salts allow them to aggregate into micelles in aqueous solutions, as described in AQUEOUS SOLUTION PROPERTIES OF BILE SALTS, p. 628. We characterize micelle formation by bile salts and illustrate how this is affected by changes in, for example, the chemical structure, temperature, and added counterions. Moving to more complicated systems, we elucidate the interaction of bile salts in aqueous systems with lecithin, the second most abundant lipid in bile (see BILE SALT–LECITHIN INTERACTIONS, p. 637). Here we introduce the use of the Gibbs phase rule and phase diagrams to describe the equilibrium structures (e.g., micelles, liquid crystals) that are formed by aqueous mixtures of lecithin and bile salts. In CHOLESTEROL, p. 649, we show how micelles and liquid crystals may interact with cholesterol, the third important lipid in bile, to bring this rather insoluble molecule into a micellar solution or liquid crystalline suspension. Having described the characteristics of the bile salt–phospholipid–cholesterol–water systems in detail, we relate

BILE ACID	R_1	R_2	R_3
LCA	αOH	H	H
DCA	αOH	H	αOH
CDCA	αOH	αOH	H
UDCA	αOH	βOH	H
CA	αOH	αOH	αOH
UCA	αOH	βOH	αOH

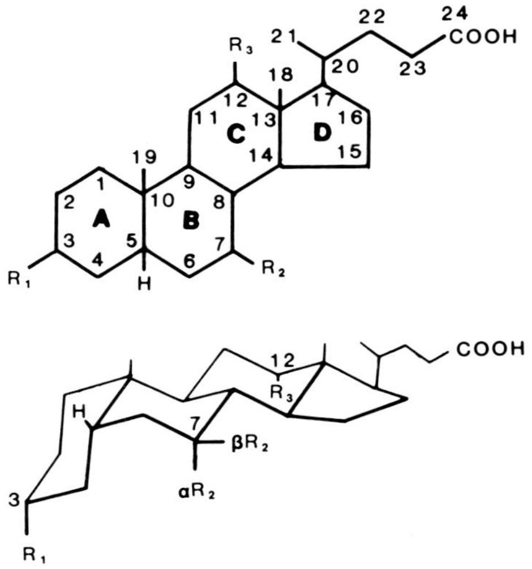

FIG. 1. Molecular structure of common bile acids showing common steroid ring and side-chain structure. Hydroxyl group(s) location and orientation are given for each bile acid.

this to the composition and state of native bile (see BILE, p. 652). The similarities of the bile salt–phospholipid–cholesterol–water system to the compositionally more complex native bile are discussed and related to metastable states (i.e., states not strictly at equilibrium) and the formation of cholesterol crystals in bile.

The literature on bile acid physical chemistry up to 1971 has been extensively reviewed (113). A concise review of the more recent work (20) and an entire supplement to the journal *Hepatology* (47a), which summarizes the more recent work of many investigators in this field, are valuable resources.

STRUCTURAL AND SURFACE PROPERTIES OF BILE SALTS

The physical properties of molecules arise from their structure and conformational flexibility. In this section we describe the molecular structure of the different common bile salts and bile acids. In all bile salts discussed in this chapter the ring system is the same; however, the number and position of hydroxyl groups and the presence or absence of conjugation to amino acids bring about important differences in the structure and consequent physical properties. Subtle changes, such as the addition of one hydroxyl group or the change from α- to β-configuration of a hydroxyl group may give very different crystalline packing, solubility, interfacial behavior, and behavior in aqueous systems. Ultimately different physical properties of individual bile salts may be important in determining the characteristics of biles rich in certain kinds of bile salts.

Molecular Structure

The common bile acids are synthesized from cholesterol in the liver and contain a saturated ring system and a five-carbon side chain terminating in a carboxyl group. The carboxyl is usually conjugated in the liver to taurine or glycine. Figure 1 shows the molecular structure of the C_{24} bile acids, and Table 1 lists physical constants for each acid to be discussed.

TABLE 1. *Common Bile Acids*

	MW	mp, °C
Lithocholic acid (LCA)	376.56	184–186
Chenodeoxycholic acid (CDCA)	392.56	143
Deoxycholic acid (DCA)	392.56	176–178
Ursodeoxycholic acid (UDCA)	392.56	203
Cholic acid (CA)	408.56	198
Ursocholic acid (UCA)	408.56	128*

MW, molecular weight; mp, melting point.   * Anhydrous.
[Data from Carey (20) and Small (113).]

The naturally occurring bile acids in humans are cholic acid (CA), chenodeoxycholic acid (CDCA), deoxycholic acid (DCA), and lithocholic acid (LCA). They are heterocyclic steroidal structures with a cis A-B ring juncture. The unconjugated bile acid molecules are ~20–21 Å long and nearly circular in cross section. Lithocholic acid has one α-hydroxyl group at carbon number 3 (C-3); DCA has α-hydroxyls at C-3 and C-12; CDCA has α-hydroxyl groups at C-3 and C-7; CA, a trihydroxy bile acid, has α-hydroxyls at C-3, C-7, and C-12. The α-hydroxyl groups all lie on one side of the ring (Fig. 1) and give the molecule amphipathic character with a polar and a nonpolar face responsible for its solubilizing properties (96, 113). Ursodeoxycholic acid (UDCA), another dihydroxy bile acid, has a C-3 α-hydroxyl and a C-7 β-hydroxyl group. Ursocholic acid (UCA) has a β-hydroxyl at C-7 and α-hydroxyls at C-3 and C-12. The change in hydroxyl orientation at C-7 makes UDCA and UCA more hydrophilic than the corresponding C-7 α-hydroxyl bile acids (55). A summary of physical constants for the

Na⁺ salts of the bile acids in aqueous solution is given in Table 2. In general bile salts are more dense than water, have molecular volumes of 508–539 Å³/molecule, which are somewhat less than the 623 Å³/molecule of their parent molecule cholesterol (117), and have low coefficients of expansion.

The infrared (IR) spectra for the bile acids have been summarized in detail in reference 113. The main IR features involve the hydroxyl and carboxyl groups. The predominant intermolecular interaction is hydrogen bonding between the carboxyl groups. The 3α- and 7α-hydroxyls, but not 12α-hydroxyl, are involved in hydrogen bonds in the crystalline lattice. However, in the amorphous phase all hydroxyls engage in a random pattern of hydrogen bonding. The ^{13}C and some ^{1}H nuclear magnetic resonance (NMR) assignments were determined for micellar solutions of sodium cholate (NaC), sodium deoxycholate (NaDC), and sodium chenodeoxycholate (NaCDC) (6, 65, 122). Recently Waterhous et al. (133), using two-dimensional NMR spectroscopy correlating the ^{1}H and ^{13}C spectra, completely assigned the ^{1}H spectra for the common bile salts and clarified discrepancies in previously published ^{13}C assignments. A comprehensive listing of ^{1}H and ^{13}C resonance assignments for LCA, CDCA, UDCA, DCA, CA, and cholanic acid is given in Table 3. The ^{1}H and ^{13}C chemical shifts and linewidths of some, but not all, resonances change with increases in concentration (79, 88). These effects have been related to micellar structure and are discussed in *Aggregation Behavior*, p. 631.

Crystal Structure

Elucidation of the single-crystal structure of the common bile acids has progressed greatly in the last 15 years (113). Crystal structures of most of the unconjugated bile acids, the Na⁺ and Rb⁺ salts of deoxycholate and the Na⁺ and Ca²⁺ salts of cholate have been determined and are summarized in Table 4. Lithocholic acid crystallized from acetic acid in an anhydrous state yielded a crystal of dimensions 0.4 × 0.3 × 0.15 mm (4). Lithocholic acid has an orthorhombic structure, with space group P2₁2₁2 with four molecules per unit cell (Fig. 2A). The structure is one of monolayers of alternately directed LCA molecules. All oxygen atoms in the molecule are involved in hydrogen bonds. Rings A, B, and C have a normal chair conformation common to all bile acids, whereas ring D of LCA has a β-α-envelope conformation. A high melting polymorph of CDCA forms prismatic crystals when recrystallized from acetonitrile (68). The anhydrous crystal structure is monoclinic with space group P2 and two molecules in the asymmetric unit (Fig. 2B). The structure is not layered. The D rings of the two molecules are in different conformations, one in a half-chair conformation and the other between a half-chair and β-envelope conformation. The molecules are involved in a network of hydrogen bonding involving all hydroxyl and carboxyl groups, which form a head-to-tail helix parallel to the β-axis (Fig. 2B).

Crystals of 2:3 DCA:H₂O have a tetragonal structure with space group P4₁2₁2 and **Z** = 16 (128). The single crystal with dimensions 0.3 × 0.3 × 0.4 mm was formed by slow evaporation in a 95%-methanol solution. The DCA molecules are in hydrogen-bonded layers, which form channels filled by chains of H₂O molecules (Fig. 2C). The H₂O molecules form hydrogen bonds with each other and with DCA. Crystals of UCDA are orthorhombic, with space group P2₁2₁2 and **Z** = 8 (80).

The monohydrate Rb⁺ salt of deoxycholate forms a monoclinic crystal with space group P2 and **Z** = 2 [Fig. 2D; (30)]. The molecules pack to form an assembly of wavy bilayers without channels. Each monolayer contains DCA molecules in head-to-tail fashion joined by hydrogen bonds to H₂O molecules. The crystal structure is stabilized by hydrogen bonding with H₂O and by ion-ion and ion-dipole interactions between Rb⁺ and deoxycholate.

Monohydrate crystals of NaC with crystal size 0.35 × 0.45 × 0.35 mm, as well as the heptahydrate Ca²⁺ salt with size 0.04 × 0.11 × 0.78 mm, are monoclinic with space group P2 and **Z** = 2 (29, 48a). In both, the cholate molecules form a planar bilayer structure with

TABLE 2. *Physical Constants of Bile Salts*

	LCA	DCA	CDCA	UDCA	CA	Ref.
Density						
By flotation	1.163	1.174	1.160		1.156	80
By pyknometry in H₂O	1.16	1.19	1.20		1.39	112
Volume						
Partial specific, cm³/g Na⁺ salt*		0.780	0.765		0.77	113
Molecular, Å³/molecule Na⁺ salt*	508	538	526		539	113
Molal, 25°C, cm³/mol						
Monomer by extrapolation to infinite dilution		316.0	320.0	323.5	314.4	112
Micellar at 0.08 M		327.9	326.2	325.0	319.5	112
Coefficient of expansion, cm³/°C Na⁺ salt†		0	0.345	0.390	0.305	132

* Calculated from density in aqueous solution at 22°C.   † Calculated by plotting molal volume vs. temperature at infinite dilution.
For definitions of abbreviations, see Table 1.

TABLE 3. ^{13}C and ^1H Resonance Assignments for Bile Acids

Carbon No.	5β-Cholanoic Acid				Lithocholic Acid				Chenodeoxycholic Acid			
	Type	Carbon	Proton α	Proton β	Type	Carbon	Proton α	Proton β	Type	Carbon	Proton α	Proton β
1	CH₂	37.6	1.74	0.88	CH₂	35.3	1.75	0.94	CH₂	36.5	1.83	0.99
2	CH₂	21.4	1.34 (2)		CH₂	30.3	1.29	1.60	CH₂	31.3	1.36	1.59
3	CH₂	27.1	1.74	1.18	CH	70.5		3.51	CH	72.9		3.37
4	CH₂	27.3	1.72	1.23	CH₂	36.3	1.71	1.45	CH₂	40.4	2.25	1.66
5	CH₂	43.8		1.27	CH	41.9		1.35	CH	43.1		1.36
6	CH₂	27.6	1.20	1.86	CH₂	27.1	1.23	1.83	CH₂	35.9	1.52	1.98
7	CH₂	26.6	1.08	1.37	CH₂	26.3	1.09	1.39	CH	69.1		3.80
8	CH	36.0		1.39	CH	35.6		1.38	CH	40.2		1.50
9	CH	40.6	1.39		CH	40.2	1.41		CH	34.0	1.87	
10	C	35.3			C	34.2			C	36.2		
11	CH₂	20.9	1.39	1.24	CH₂	20.6	1.38	1.23	CH₂	21.8	1.48	1.35
12	CH₂	40.3	1.12	1.92	CH₂	40.0	1.14	1.96	CH₂	41.0	1.21	2.00
13	C	42.8			C	42.4			C	43.7		
14	CH	56.7	1.07		CH	56.3	1.05		CH	51.5	1.48	
15	CH₂	24.3	1.02	1.59	CH₂	24.0	1.04	1.56	CH₂	24.6	1.09	1.74
16	CH₂	28.2	1.88	1.29	CH₂	28.0	1.85	1.27	CH₂	29.2	1.90	1.32
17	CH	56.1	1.10		CH	55.8	1.10		CH	57.2	1.18	
18	CH₃	12.1	0.65		CH₃	11.9	0.64		CH₃	12.2	0.69	
19	CH₃	24.3	0.92		CH₃	23.3	0.91		CH₃	24.6	0.93	
20	CH	35.4	1.43		CH	35.1	1.41		CH	36.7	1.45	
21	CH₃	18.3	0.95		CH₃	18.2	0.92		CH₃	18.8	0.96	
22	CH₂	30.9	1.79	1.34	CH₂	30.9	1.75	1.29	CH₂	32.4	1.79	1.31
23	CH₂	31.1	2.38	2.24	CH₂	30.9	2.32	2.12	CH₂	32.3	2.31	2.24
24	C	180.2			C	178.1			C	178.2		

Carbon No.	Ursodeoxycholic Acid				Deoxycholic Acid				Cholic Acid			
	Type	Carbon	Proton α	Proton β	Type	Carbon	Proton α	Proton β	Type	Carbon	Proton α	Proton β
1	CH₂	36.1	1.81	1.03	CH₂	36.3	1.77	0.98	CH₂	36.5	1.81	0.99
2	CH₂	31.0	1.28	1.62	CH₂	30.9	1.44	1.59	CH₂	31.2	1.45	1.59
3	CH	71.9		3.47	CH	72.5		3.54	CH	72.9		3.37
4	CH₂	38.6	1.81	1.55	CH₂	37.0	1.79	1.48	CH₂	40.5	2.29	1.66
5	CH	42.4		1.47	CH	43.5		1.39	CH	43.2		1.38
6	CH₂	38.0	1.60 (2)		CH₂	28.3	1.26	1.89	CH₂	35.9	1.53	1.95
7	CH	72.1	3.49		CH₂	27.3	1.19	1.42	CH	69.1		3.80
8	CH	43.4		1.45	CH	37.3		1.46	CH	41.0		1.55
9	CH	40.7	1.48		CH	34.6	1.89		CH	27.9	2.25	
10	C	35.1			C	35.3			C	35.9		
11	CH₂	22.4	1.47	1.34	CH₂	29.8	1.53 (2)		CH₂	29.6	1.58 (2)	
12	CH₂	41.5	1.19	2.03	CH	74.0		3.98	CH	74.0		3.97
13	C	44.8			C	44.8			C	47.6		
14	CH	56.5	1.09		CH	49.1	1.62		CH	43.0	2.00	
15	CH₂	27.9	1.46	1.90	CH₂	24.8	1.09	1.62	CH₂	24.2	1.12	1.76
16	CH₂	29.6	1.86	1.30	CH₂	28.5	1.87	1.29	CH₂	28.7	1.90	1.32
17	CH	57.4	1.25		CH	48.0	1.83		CH	48.1	1.86	
18	CH₃	12.7	0.71		CH₃	13.2	0.71		CH₃	13.0	0.72	
19	CH₃	24.0	0.94		CH₃	23.7	0.93		CH₃	23.2	0.92	
20	CH	36.6	1.44		CH	36.6	1.42		CH	37.8	1.43	
21	CH₃	19.0	0.96		CH₃	17.5	1.01		CH₃	17.7	1.02	
22	CH₂	32.3	1.80	1.32	CH₂	32.2	1.78	1.35	CH₂	32.4	1.79	1.35
23	CH₂	32.0	2.35	2.21	CH₂	32.0	2.38	2.23	CH₂	32.0	2.37	2.21
24	C	178.2			C	178.0			C	178.3		

Numbers in parentheses indicate number of protons. [From Waterhous et al. (133).]

the hydrophobic rings alternating with layers of polar groups (Fig. 2E). The Na^+ or Ca^{2+} ions and H_2O molecules interact with the polar portions of cholate through ionic and hydrogen bonds to form the polar region (Fig. 2F, G). Each Na^+ and Ca^{2+} is associated with 5 oxygen atoms. This bilayer structure for salts of CA is similar to that for rubidium deoxycholate (RbDC) (Fig. 2H). There is a different side-chain conformation for each salt that has been proposed to be important in accommodating different cation coordination schemes (48a). Thus both bile acids and salts form crystals that are often layered structures.

TABLE 4. *Crystal Data*

Molecule	Crystal Structure	Space Group	Unit Cell Dimensions	z	Unit Cell Volume	Density, g/cm^3	Ref.
LCA ($C_{24}H_{40}O_3$)	Orthorhombic	$P2_12_12_1$	a = 6.807 b = 12.178 c = 26.779	4	2,219.9	1.89 × 10^{-3}	4
CDCA ($C_{24}H_{40}O_4$)	Monoclinic	$P2_1$	a = 18.785 b = 8.120 c = 14.889 β = 99.10°	4	2,242.5	1.78 × 10^{-3}	68
DCA:H$_2$O (2:3) ($C_{24}H_{40}O_4$ 3/2 H$_2$O)	Tetragonal	$P4_12_12_1$	a = b = 13.999 c = 48.903	16	9,583.6	1.67 × 10^{-3}	128
NaDC:H$_2$O (1:4) ($C_{24}H_{39}O_4$ Na$^+$ 4H$_2$O)	Hexagonal	$P6_1$ or $P6_5$	a = b = 34.56 c = 11.75 γ = 120° α = β = 90°	18	36,461.9	1.2	17
RbDC:H$_2$O (1:1) ($C_{24}H_{39}O_4$ Rb† H$_2$O)	Monoclinic	$P2_1$	a = 13.118 b = 7.817 c = 11.785 β = 97.74°	2	1,197.5	1.67 × 10^{-3}	30
NaC:H$_2$O (1:1) ($C_{24}H_{39}O_5$ Na$^+$ H$_2$O)	Monoclinic	$P2_1$	a = 12.197 b = 8.214 c = 12.559 β = 108.07°	2	1,196.2	1.67 × 10^{-3}	29
CaCCl:H$_2$O (1:1) ($C_{24}H_{39}O_5$ Na$^+$ H$_2$O)	Monoclinic	$P2_1$	a = 11.918 b = 8.636 c = 15.302 β = 97.93°	2	1,545.0	1.29 × 10^{-3}	48a
UDCA	Orthorhombic	$P22_12_1$	a = 13.530 b = 26.737 c = 12.37	8	4,474.9	1.79 × 10^{-3}	80
NaC	Monoclinic	$P2_1$	a = 12.593 b = 8.215 c = 12.196 β = 107.86°	2	1,143.0	1.75 × 10^{-3}	80
CA:4H$_2$O	Monoclinic	$P2_1$	a = 14.043 b = 7.849 c = 13.697 β = 113.53°	2	1,269.1	1.58 × 10^{-3}	80

NaC, sodium cholate; NaDC, sodium deoxycholate; RbDC, rubidium deoxycholate. For other abbreviations see Table 1.

The oxygens are extensively hydrogen bonded to adjacent molecules and to H$_2$O if present. The salts of Rb$^+$, Na$^+$, and Ca^{2+} all crystallize with H$_2$O, which is highly hydrogen bonded. The metal ions are coordinated to several oxygens through a combination of ionic, dipolar, and hydrogen bonds.

Surface Chemistry

The surface chemistry of the unconjugated bile acids was extensively studied during the 1960s, was reviewed in 1971 by Small (113), and is summarized next. The reader is referred to the review by Small and references cited therein for a more detailed discussion.

BILE ACIDS. The bile acids have been classified as insoluble amphiphiles (113), but this is not strictly true. Certainly CA, UCA, and the C-23 and C-22 bile acids have appreciable solubility (37). The surface chemistry of some of the usual bile acids is summarized in Figure 3. When spread on H$_2$O the bile acids form gaseous or liquid monolayers. The surface pressure at film collapse varies with pH and salt concentration. The area of the acids at the surface increases from a minimum of 40 Å^2/molecule for cholanic acid to a maximum of 190 Å^2/molecule for CA at pH 2 (3 M NaCl). The nonhydroxylated cholanic acid stands up in the monolayer and occupies an area equal to the cross-sectional area of the steroid nucleus. The hydroxylated acids lie flat on the surface with the carboxyl and hydroxyl functional groups in the aqueous phase (Fig. 3). Ionizing LCA (high pH) allows the carboxyl to adhere to the surface more strongly, and 3 M NaCl salts out the 3α-hydroxyl, which allows it to be expelled from the surface so that lithocholate (LC) stands upright with only the carboxylate in the aqueous solvent (113).

In mixed systems of bile acid and lecithin at low pH, the collapse pressure increases with increasing amounts of lecithin for mono- and dihydroxy bile acids (113). At certain pressures there are breaks in the pressure-area curve where the bile acid is pushed out from the monolayer into a bulk phase, but some bile acid remains in the monolayer even at high pressure (LCA > DCA). These pressures are above the collapse

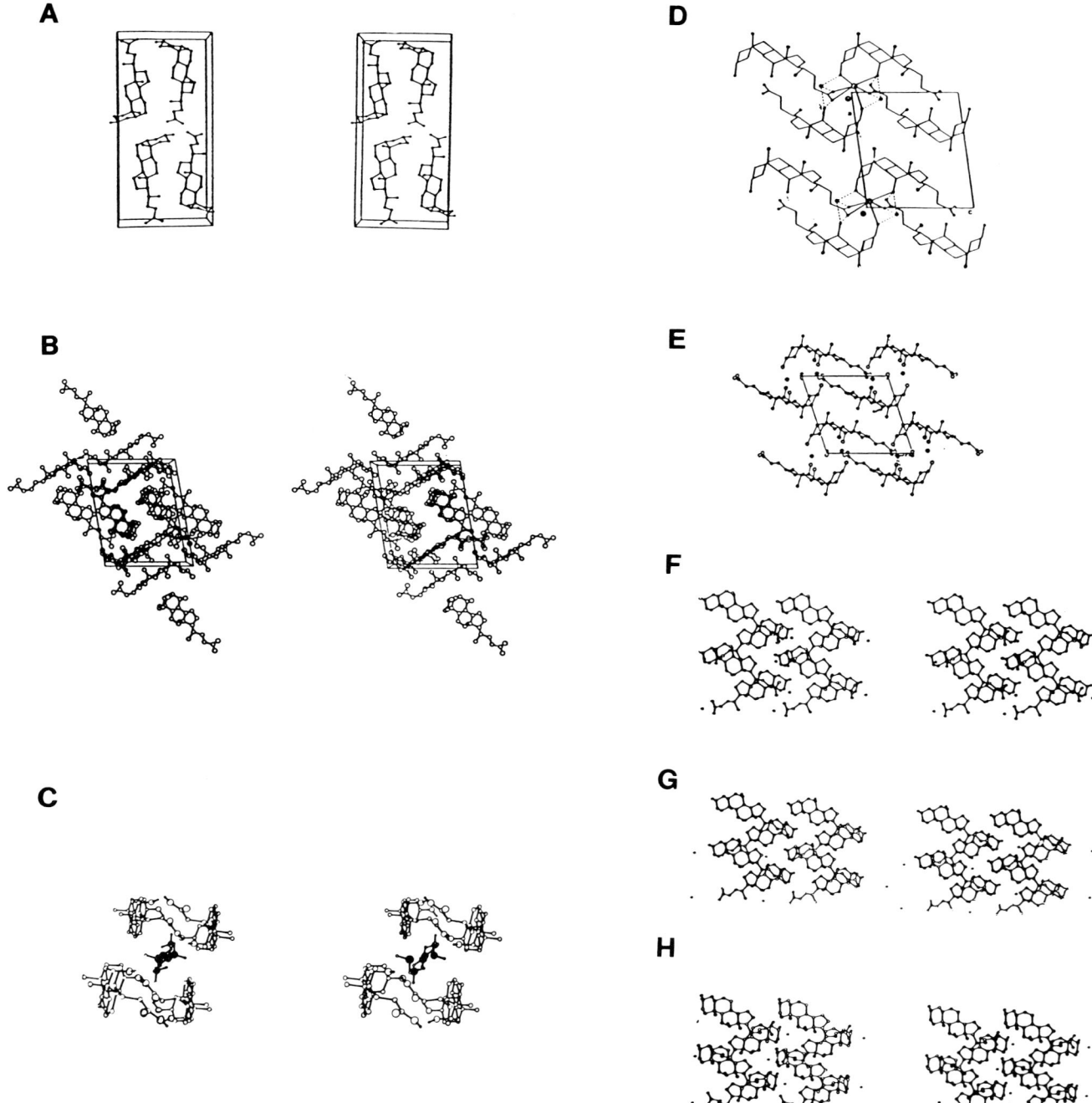

FIG. 2. Crystal structure of bile acids and salts. *A*: stereoview of the cell of lithocholic acid: *a*-axis projection with *c*-axis vertical. Oxygen molecules are blackened. *B*: stereoview of chenodeoxycholic acid; *b*-axis projection. *C*: stereoview of unit cell of deoxycholic acid: H_2O, showing channel filled with H_2O molecules. Oxygens of H_2O molecules are blackened. *D*: crystal packing of DCA:Rb viewed along *b*-axis, showing planar wavy bilayer patterns. *Broken lines*, hydrogen bonds; *large closed circles*, Rb^+; *smaller closed circles*, oxygen from H_2O; *small open circles*, oxygen from the bile acid. *E*: crystal packing of unit cell of Na-cholate-monohydrate *b*-axis projection. Wavy bilayer pattern is illustrated. *Broken lines*, hydrogen bonds; *larger closed circles*, Na^+; *smaller closed circles*, carbon atoms; *open circles*, oxygen atoms. *F*: stereoview of bilayer packing of calcium cholate chloride heptahydrate. *Closed circles* of cholate molecules represent oxygen atoms of hydroxyl and carboxyl groups. *Closed circles* between cholate molecules represent Ca^{2+}. *G*: stereoview of bilayer packing of sodium cholate monohydrate. *Closed circles* of cholate molecules represent oxygen atoms of hydroxyl and carboxyl groups. *Closed circles* between molecules represent H_2O. *H*: stereoview of bilayer packing of rubidium deoxycholate. *Closed circles* of deoxycholate molecules represent oxygen atoms of hydroxyl and carboxyl groups. *Closed circles* between molecules represent H_2O. [*A* from Arora et al. (4); *B* from Lindley et al. (68); *C* from Tang et al. (128); *D* from Coiro et al. (30); *E* from Cobbledick and Einstein (29); *F–H* from Hogan et al. (48a).]

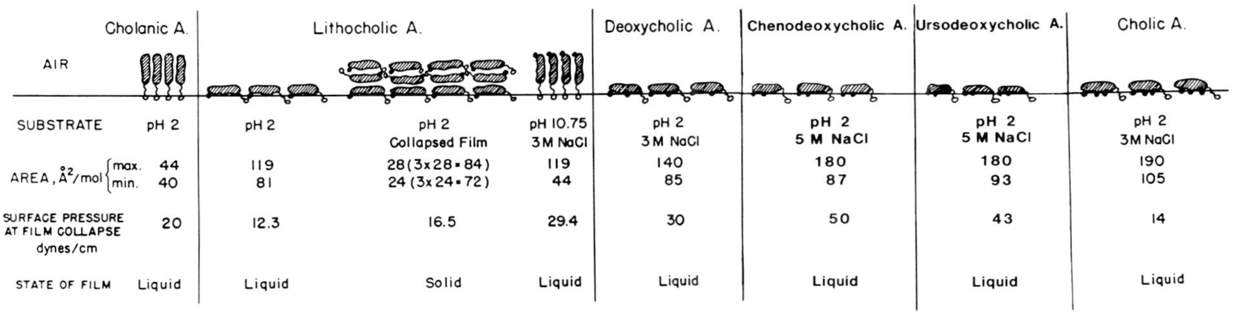

	Cholanic A.	Lithocholic A.			Deoxycholic A.	Chenodeoxycholic A.	Ursodeoxycholic A.	Cholic A.
AIR								
SUBSTRATE	pH 2	pH 2	pH 2 Collapsed Film	pH 10.75 3M NaCl	pH 2 3M NaCl	pH 2 5 M NaCl	pH 2 5 M NaCl	pH 2 3M NaCl
AREA, Å²/mol {max. / min.}	44 / 40	119 / 81	28(3×28·84) / 24(3×24·72)	119 / 44	140 / 85	180 / 87	180 / 93	190 / 105
SURFACE PRESSURE AT FILM COLLAPSE dynes/cm	20	12.3	16.5	29.4	30	50	43	14
STATE OF FILM	Liquid	Liquid	Solid	Liquid	Liquid	Liquid	Liquid	Liquid

FIG. 3. Schematic of surface configuration of cholanic acid and hydroxylate derivatives. Maximum area per molecule was taken from first inflection point of compression isotherms; minimum area was taken from collapse point. Pressure at film collapse and state of film are given. Collapse pressure for chenodeoxycholic acid in 3 M NaCl low pH was determined to be ~27 dyn/cm, similar to that of deoxycholic acid. [Data from Carey (20) and Small (113) and references therein.]

pressure of simple monolayers of the bile acid, indicating an interaction between the lecithin and bile acid. At low mole ratios of lecithin to CA (1:4) the collapse pressure does not increase. When the ratio approaches 1:1, the collapse pressure increases slightly. There is no CA left in the monolayer at very high pressures. From plots of area/molecule versus mole % lecithin, i.e., 100 × [moles lecithin/(moles lecithin + moles bile acid)], it was found that LCA and DCA appear to be held upright in the lecithin monolayer, whereas CA lies flat on the interface surface (113).

BILE SALTS. The majority of studies of the surface behavior of the alkaline salts of bile acids measured surface tension as a function of concentration to establish critical micelle concentrations (CMCs). These results have been included in a summary table of CMC's (see Table 7). Using a simplified Gibbs adsorption isotherm, Small (113) calculated the area of the salt. For sodium glycodeoxycholate (NaGDC) and sodium taurodeoxycholate (NaTDC) this surface area is ~90 Å²/molecule.

Hydrophobic-Hydrophilic Balance

Bile acids undergo modifications during their enterohepatic circulation. The primary bile acids excreted from the liver (CDCA, CA) are conjugated through amide bonds to glycine or taurine. Conjugation, especially with taurine, makes the bile acids resistant to precipitation at acid pH. The secondary bile acids (DCA, LCA) are formed by bacterial 7α-dehydroxylation from the primary bile acids in the intestine. Some of the bile acids are also deconjugated before absorption occurs. Deconjugation and dehydroxylation result in major changes in the hydrophilic character of the bile acids, essentially making the secondary bile acids less polar than their conjugated primary precursors. The physical chemistry of bile acids in terms of the relative contribution of the hydrophobic and hydrophilic faces of the molecules to the overall polarity (more commonly referred to as the hydrophobic-hydrophilic balance) has been shown to be correlated with physiological functions (3, 100).

Reverse-phase high-performance liquid chromatography (HPLC) takes advantage of the different degrees of polarity to provide a simple separation method for bile acids in synthetic mixtures and native bile (3, 9, 85, 104, 105). The stationary phase is usually porous silica beads covalently bonded to a linear long-chain hydrocarbon (9, 85) to which the hydrophobic side of the bile acid can associate. A mobile phase of aqueous methanol (85%–100%) at acidic pH [pH 2 (85); pH 5 (3)] gives one-step separation of a class of bile acids (tauryl, glycyl, or unconjugated). The HPLC mobility is dependent on both number and position of hydroxyl groups and on conjugation. A tracing (UV absorbance at 210 nm) of the HPLC separation of mixtures of monomer concentrations of bile acids is shown in Figure 4 (3). An increase in number of hydroxyl groups corresponds with a decrease in retention time. For the dihydroxy bile acids the 7α-position is more polar than the 12α-position. A change from 7α-hydroxyl to 7β-hydroxyl gives an even more dramatic effect, with UDCA more mobile than the tri-α-hydroxy CA. Thus, for a given conjugation state, mobility at pH 5 decreases in the order UCA > UDCA > CA > CDCA > DCA > LCA (3, 9, 85). The sulfated taurine group is more polar than the amino glycyl group, which is more polar than the free carboxyl group. For a given bile acid, mobilities are of the order taurine conjugated > glycine conjugated > unconjugated. The HPLC mobilities are in the reverse order of the hydrophobic-hydrophilic balance of the bile acids and conjugation state. Armstrong and Carey (3) found a reverse correlation between HPLC mobility and the cholesterol solubilization capacity of simple micellar solutions of bile acids. A more detailed discussion of the relationship between the physical chemistry of bile acids and cholesterol solubilization is given in BILE SALT–LECITHIN INTERACTIONS, p. 637.

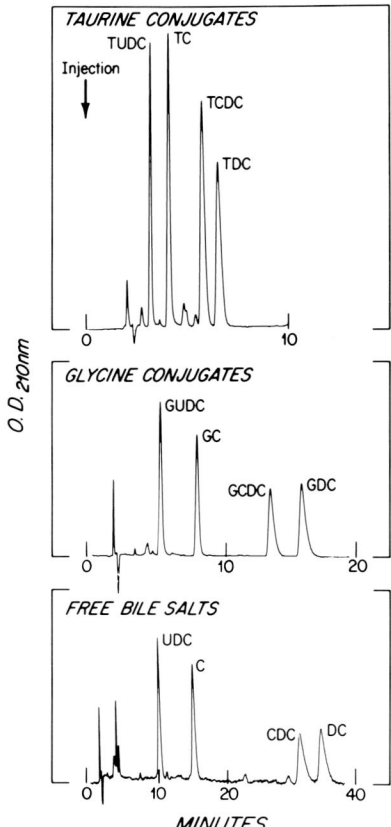

FIG. 4. Reverse-phase high-performance liquid chromatography (HPLC) chromatographs of mixture of bile salts. For a given conjugation state, mobility decreases in the order UDCA > CA > CDCA > DCA. Column used was an Ultrasphere ODS 250 × 4.6 mm, with a mobile phase of 75% MeOH–25% 0.005 M KH_2PO_4/H_3PO_3, pH 5.0. TUDC, tauroursodexoycholate; TC, taurocholate; TCDC, taurochenodeoxycholate; TDC, taurodeoxycholate; GUDC, glycoursodeoxycholate; GC, glycocholate; GCDC, glycochenodeoxycholate; GDC, glycodeoxycholate; UDC, ursodeoxycholate; C, cholate; CDC, chenodeoxycholate; DC, deoxycholate. OD, optical density. [From Armstrong and Carey (3).]

AQUEOUS SOLUTION PROPERTIES OF BILE SALTS

The prevention of cholesterol precipitation and subsequent stone formation in bile necessitates an understanding of how cholesterol is solubilized in model systems. Bile salts act as detergents, helping to keep phospholipids and cholesterol in solution. To determine the details of the interactions of these molecules, the physicochemical behavior of each and of simple mixtures has been extensively studied. The aqueous solution properties of bile salts are presented here, followed by the behavior of mixtures of bile salts and phospholipids, and bile salts and cholesterol to provide information needed to help understand the more complicated in vivo system.

Solubility in Water

BILE ACIDS. The solubility of the common bile acids and salts in H_2O and several organic solvents is given in Table 5. Early studies carried out by Ekwall et al. in the 1950s (35) gravimetrically estimated the aqueous solubility of DCA and CA. Ekwall found the solubilities at 37°C to be 0.0043 g/100 ml and 0.0078 g/ml, respectively. Igimi and Carey (55) determined solubilities of five ^3H-labeled bile acids in H_2O. These values agree with those found by Small (113) by microbalance techniques (Table 5) and are similar to those of Ekwall within the latter's experimental error. In the same study, Igimi and Carey also determined the solubility of UDCA and CDCA as a function of NaCl concentration (0–1.0 M NaCl) and temperature (20–60°C). The solubility of CDCA increased slightly from 0 to 0.3 M NaCl, then decreased to 80% maximum value at 1.0 M, while UDCA solubility increased from 0 to 0.6 M and only decreased slightly at higher NaCl concentrations. As the temperature increased, the solubility of the bile acids studied increased but to varying degrees.

TABLE 5. *Solubilities of Bile Acids*

	Cholanic Acid	LCA	UDCA	DCA	CDCA	CA	UCA	Ref.
Water								
37°C				0.0043		0.0078		35
				(0.114)		(0.190)		
	insol	0.0004	0.0021	0.0045	0.0100	0.0188		55
		(0.011)	(0.056)	(0.120)	(0.266)	(0.460)		
25°C + 3°C	insol	insol		0.0045	insol	0.0120		113
				(0.120)		(0.293)		
35°C	insol	0.0003	0.00045	<0.00114	0.00118	0.0132	0.0699	37
		(0.008)	(0.012)	(0.030)	(0.031)	(0.323)	(1.711)	
Alcohol	0.6	1.1	sol	22.07	3.4	3.06		
Acetone	0.3	0.1	0	1.05	2.2	2.82		
Ether	0.9	0.2		0.12	0.3	0.12		
$CHCl_3$	6.6*	0.7*	sl sol	0.29	0.31*	0.59		
Benzene	1.7*	0.1*		0.04	<0.04*	0.1		
Acetic acid	sol	sol	sol	0.81	6.2	15.2		

Solubilities in g/100 ml; numbers in parentheses are solubilities in mM. insol, Insoluble; sol, soluble; sl sol, slightly soluble. For other abbreviations see Table 1. * Data from Small (113).

A careful solubility temperature-dependence study was done by Fini et al (37). Solubilities of nine bile acids varying in hydroxyl number, location, and orientation were determined at nine temperatures, from 10°C to 50°C. Values at 35°C are given in Table 5 to compare with those at 37°C (55, 113). These values are somewhat lower, and the difference cannot be accounted for from the 2°C temperature difference or pH difference (all studies were done from pH 2.4 to 3.0 pH, where the unconjugated bile acids are fully protonated). It is possible that some of the discrepancy is due to different methods of separating nonsolubilized bile acid from solution. Igimi and Carey (55) centrifuged their sample and determined radioactivity in the visually clear supernatant, whereas Fini et al. (37) filtered the bile acid solution through 0.22-μm Millipore disks. Solubility increased to 2 or 3 times the lowest value between 10°C and 50°C for all bile acids except deoxycholate, which increased by one-third. In general, for bile acids at low concentration in H_2O, solubility increases with hydroxyl number and temperature and is dependent on hydroxyl position. The order of solubility from highest to lowest is UCA > CA > CDCA > DCA > UDCA > LCA. Note that UDCA eluted before CA on HPLC. This indicates that the 7β-hydroxyl interferes with adsorption to the hydrocarbon chains of the stationary phase. The thermodynamics of aqueous bile salt solutions has been discussed (92).

BILE SALTS. Fully ionized di- and trihydroxy bile salts are quite water soluble. The solubility limits for NaDC and NaC are >33.3 g/100 ml (0.803 M) and >56.9 g/100 ml (1.09 M) (113). The Na^+ and Ca^{2+} salts of LCA are only weakly soluble in water (20). Bile salts above their CMC can solubilize bile acid above the aqueous solubility of the acid (113). The amount of solubilized acid is dependent on bile salt species, concentration, temperature, and conjugation state. The pH of the bile acid–bile salt solution determines the ratio of a given bile salt to bile acid and hence the amount of acid that can be solubilized (113).

Solubility in Organic Solvents

Table 5 lists the solubility of unconjugated bile acids in common solvents. The values were taken largely from reference 113. Solubility in the less polar organic solvents (benzene and ether) increases as the number of hydroxyl groups decreases. Cholanic acid has the highest solubility. In more polar solvents (alcohol and acetone) the solubility increases with increasing number of hydroxyls. Deoxycholic acid in alcohol and CA in acetic acid appear to have unpredictably high solubilities. At higher concentrations "reverse micelles" of di- and trihydroxy bile acids may form in organic solvents. These micelles are characterized by hydrophilic associations (e.g., hydrogen bonding) between the bile acid molecules via hydroxyl and carboxyl groups and an association between the hydrophobic face of the bile acid and the solvent (7, 131).

Ionization Behavior

Bile acids encounter pH values ranging from 2 to 8.5 during their enterohepatic circulation. Their solubility in H_2O changes greatly with ionization state. Also, the interactions of bile salts with phospholipids and cholesterol and the ability to maintain cholesterol in solution varies between ionized and protonated forms. It is therefore important to comprehend the ionization behavior of the bile acids in aqueous solution and in the different molecular environments in which they can occur in bile (e.g., bound to vesicles or proteins, in micelles, or as monomers).

The effect of pH on the solution behavior of bile acids and salts has been extensively studied by titrating aqueous samples of bile salts in simple and in mixed systems (30, 33–36, 55, 113). A characteristic titration curve for a simple bile salt solution is shown in Figure 5 (113). The initial part of the curve represents titration of excess NaOH. At the inflection point W the bile salt begins to titrate until point X. At this point the mixed bile salt micelle cannot solubilize any additional bile acid and the bile acid precipitates with each addition of HCl (pH ~6.5–7 for unconjugated bile acids). There is no change in bulk pH during precipitation, and there are two phases present (liquid and crystalline bile acid). Near the end of the titration the pH changes and protonation is complete. The curve changes slope at an inflection point Z, the final equivalence point. Point Y represents the maximum solubility of bile acid in the bile salt micelle before supersaturation. Supersaturation occurs between X and Y. From these titration curves one can determine pH of precipitation, the moles of bile salt needed to solubilize one mole of bile acid, and the apparent pK_a (113). Glycine-conjugated bile acids precipitate from solution at lower pH values (pH ~4.5) and have lower apparent pK_a values (3.8–4.3) (55, 113). Taurine conjugates do not precipitate from solution even at pH 1.5 and have apparent pK_a values in the range 1.8–2.0. Small amounts of unconjugated bile acids, ~10 wt%, i.e., 100 × [wt bile acid/(wt bile acid + wt taurine conjugate)], can be bound to taurine-conjugated micelles, where they will remain solubilized at low pH values (113). In general, as the percentage of conjugated bile salts increases, the ratio of bile salt to bile acid needed to solubilize the acid decreases, and hence the pH at which the acid precipitates also decreases [see Small (113) for further discussion].

Ekwall et al. (35) titrated NaC and NaDC at various concentrations. An increase was found in the pKa from a constant 4.98 (CA) or 5.15 (DCA) below the CMC to 5.48 and 6.35, respectively, at concentrations well above the CMC. Concentration-dependent pK_a behavior was confirmed for CA and DCA and also found for CDCA (113). Igimi and Carey (55) generated

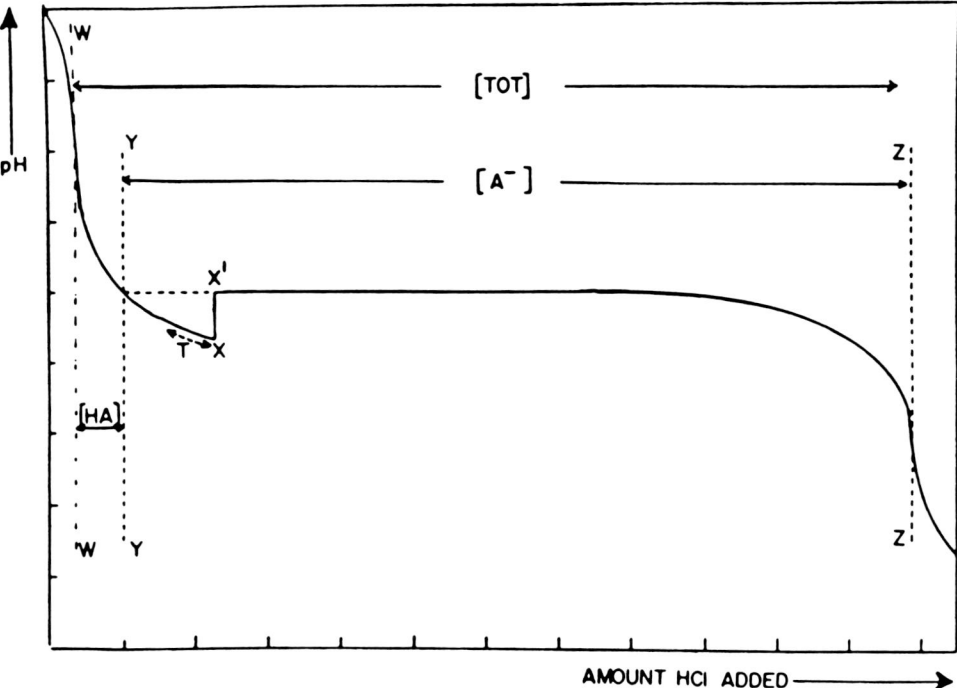

FIG. 5. Hypothetical titration curve for solutions of free bile salts or for glycine conjugates. W, first equivalence point where titration of bile salt with hydrochloric acid commences. Y, last point where bile salt solution is in thermodynamic equilibrium as a single aqueous phase. T, Tyndall effect noted in this region of titration curve. X, point where precipitation of bile acid crystals commences. X′, equilibrium pH at point of bile acid precipitation. Z, second equivalence point where titration of bile salt with hydrochloric acid is complete. TOT, total amount of acid required to complete the titration. HA, amount of acid added from first equivalence point W to point Y, which represents maximum solubility of bile acid (HA) in bile salt solution (A-). [From Small (113).]

similar titration curves for CDCA, UDCA, and their glycine conjugates. Table 6 gives the pK_a values for four bile acids at different concentrations, conjugation states, and aqueous environments.

DeMaria et al. (33) and Fini et al. (36) titrated unconjugated bile salts in aqueous methanol. The molecular interactions between bile salt and solvent in alcohol are such that micelles probably do not form (113). Therefore it may be possible to find monomeric pK_a values in alcohol solvents. Bile salts were titrated in solutions with varying mole fractions of methanol (36). A linear relationship was found when plotting apparent pK_a versus mole fraction of methanol. Apparent pK_a values for monomeric bile acids in pure H_2O were determined by extrapolating to zero-methanol concentration. All of the bile acids studied (LCA, DCA, CDCA, UDCA, and CA) had a similar pK_a of ~5.06. This pK_a value is similar to those determined by Ekwall et al. (35) for DCA and CA below the CMC but differs somewhat from values presented for monomeric bile acids given in Table 6. Note that the monomeric pK_a is lowest for CA (4.6) and highest for CDCA (5.88). Why all bile acids extrapolate to the same pK_a at zero-methanol concentration (36) but give different pK_a values when titrated in H_2O is not clear. It is probable, however, that the methanol in the solvent interacts with the bile salt in a manner that alters the activity of the carboxyl group for protonation or alters the H^+ activity of H_2O and may not provide an accurate measurement of the subtle effects of structure on the aqueous monomeric pK_a of the bile acids.

An alternative and nonperturbing method to potentiometric titration for monitoring changes in the ionization state with changes in pH is NMR spectroscopy. The chemical shift of the carboxyl carbon is quite different for the protonated and ionized state. As the relative population of each state changes during titration, a weighted average chemical shift of the two populations is found. The apparent pK_a is found by plotting chemical shift versus pH (pK_a equals the pH at one-half the maximum chemical shift change) (27). Carbon-13 (^{13}C) NMR spectroscopy has been used to determine the pK_a of CA, DCA, and CDCA in several aqueous environments (15, 121). The chemical shift of the carboxyl carbon (C-24) for the unconjugated bile acids is well resolved downfield from the other peaks in the spectrum (see Table 3). Changes in the ionization state of the bile acid can be directly monitored even when other titratable groups are present

TABLE 6. *Apparent pK_a Values of Bile Acids*

	Aqueous Medium, mM	Apparent pK_a	Ref.
Bile acid, unconjugated			
DCA	2	5.3	113
	23	6.21	113
	80	6.3	113
	12 mM DCA in 86 mM NaTDC	6.1	15
	4.5 mM DCA in 78 mM EYL vesicles	6.5	15
	DCA/BSA complex (2:1 mol/mol)	4.9	15
CDCA	2.2	5.88	113
	2.3	6.18	113
	101	6.53	113
	12 mM CDCA in 86 mM NaTCDC	6.3	15
	4.5 mM CDCA in 78 mM EYL vesicles	6.6	15
	CDCA/BSA complex (2:1 mol/mol)	4.2	15
UDCA	2.5	5.51	55
	20.0	5.83	55
	100.0	6.25	55
CA	0.2	4.6	113
	2	4.98	113
	22	5.21	113
	80	5.5	113
	11.5 mM CA in 83.5 mM NaTC	5.3	15
	4.4 mM CA in 78 mM EYL vesicles	6.8	15
	CA/BSA complex (2:1 mol/mol)	4.5	15
Glycine conjugates			
GDCA	1.5	4.23	113
	7.6	4.34	113
	76.0	4.20	113
GCDCA	5.0	4.21	55
	20.0	4.61	55
	40.0	4.54	55
GUDCA	5.0	4.00	55
	20.0	4.89	55
	50.0	5.12	55
GCA	1.8	3.95	113
	20	3.80	113
	86	4.09	113
Taurine conjugates			
TDCA	19.0	1.93	55
	38.0	1.95	55
TCA	18.5	1.85	55
	37.0	1.85	55

GDCA, glycodeoxycholic acid; GCDCA, glycochenodeoxycholic acid; GUDCA, glycoursodeoxycholic acid; GCA, glycocholic acid; TDCA, taurodeoxycholic acid; TCA, taurocholic acid; NaTDC sodium taurodeoxycholate; NaTCDC, sodium taurochenodeoxycholate; EYL, egg yolk lecithin; BSA, bovine serum albumin. For other abbreviations see Table 1.

(e.g., bile acid bound to albumin). By ^{13}C-enriching the carboxyl carbon at C-24, the chemical shift of this carbon can be observed and titrated when very low concentrations of bile salts are present in vesicles, mixed micelles, albumin complexes, or monomers (15). Figure 6 shows the titration curves for CA. A unique titration curve with a different apparent pK_a was found for CA in each environment. The apparent pK_a values ranged from 4.6 for 0.2 mM CA to 6.8 for CA in the outer monolayer of egg yolk lecithin (EYL) vesicles. The apparent pK_a values for the three bile acids studied are included in Table 6. The pK_a values for the trihydroxy and two dihydroxy bile acids are not significantly different from each other when bound to bovine serum albumin or when incorporated in EYL vesicles. A difference was found when the bile acids were in mixed micelles with their taurine conjugates (1:9 wt/wt unconjugated/conjugated). The taurine conjugates did not titrate in the pH range studied, so that only the titration of the free carboxyl is obtained. It is important, therefore, to consider the molecular environment as well as the bulk pH to accurately predict the ionization state of bile acids in model and native systems.

Aggregation Behavior

CRITICAL MICELLE CONCENTRATION. Bile salts associate in solution to form micelles of a characteristic

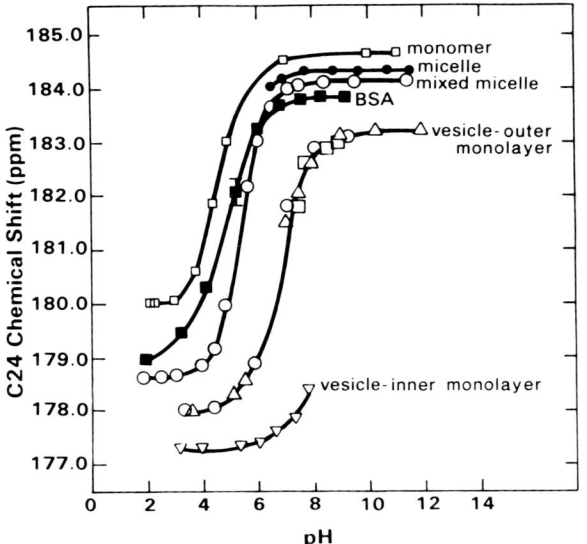

FIG. 6. Titration curves for cholic acid (CA) in several molecular environments. The pH is changed by adding HCl or NaOH, and chemical shift of carboxyl carbon (C-24) is monitored by ^{13}C NMR. Aqueous monomeric CA (0.2 mM below its CMC and solubility limit). Aqueous micellar CA (116 mM). CA-sodium taurocholate mixed micelles (1:7 mol/mol). CA-BSA (bovine serum albumin) complexes (2:1 mol/mol). CA-egg yolk lecithin vesicles (1:20 mol/mol), major peak from CA in outer monolayer of vesicle, minor peak from CA in inner monolayer of vesicle. [Adapted from Cabral et al. (15).]

shape and size dependent on, for example, bile salt concentration, ionic strength, temperature, and cation bound. The concentration at and above which spontaneous aggregation occurs is generally referred to as the critical micelle concentration (CMC), which is a range of concentrations but is often reduced to a single value. The CMC may be determined by a number of methods, each giving somewhat different values because they measure different physical properties or perturb the system differently. The methods currently used for studying aggregation properties of bile salt solutions include surface tension (81, 96, 98, 129), dye solubilization (96), light scattering (26, 60, 72, 107), kinetic dialysis (62), NMR spectroscopy (31, 57, 79), X-ray scattering (31), spectral shifting (37, 96), and photon correlation spectroscopy (97). Even the results of a stated method may differ because the techniques are used differently by different investigators. For instance, Roda et al. (96) used the maximum bubble surface tension method, whereas others use a Wilhelmy blade surface tension method (113). The former method is not strictly an equilibrium method and may give higher values. With these caveats the CMCs of the physiological bile salts and several analogues are summarized in Table 7. The method used, temperature, pH, and solvent are given where available. Where many measurements are made for a given bile salt we have also given a consensus value for the CMC that is a rounded-off average of all values excluding outliers. The actual CMCs are different for each bile salt, but some trends are observed. In general, for a given bile salt the CMC decreases with the addition of Na^+. The CMC increases with increasing number of hydroxyl groups, with a change in orientation from 7α-hydroxyl to 7β-hydroxyl, with increasing temperature, and with a shortening of the side chain.

EFFECT OF HYDROXYL CONSTITUENTS AND CONJUGATION STATE. Both number and position of the hydroxyl groups on the bile salt influence its CMC. Sodium cholate has higher CMC values in all media than the two all α-hydroxyl dihydroxy salts, NaDC and NaCDC. A 7β-hydroxyl greatly increases the CMC. Sodium ursodeoxycholate (NaUDC) has CMCs comparable to NaC (~12 mM in H_2O), whereas sodium ursocholate (NaUC) is much higher [37 mM in H_2O (96)]. This CMC effect of hydroxyl number and position correlates with the hydrophobic-hydrophilic balance found by HPLC separation for the unconjugated bile salts (see *Hydrophobic-Hydrophilic Balance*, p. 627). The CMCs of glycine conjugates of NaCDC and NaDC have slightly lower CMCs than the unconjugated state, whereas glycine conjugates of NaC and NaUDC do not. The taurine conjugates of the bile salts generally have lower CMCs (~1-3 mM) than either the unconjugated or glycine-conjugated salts. This is probably explained by the extra —CH_2— group that taurine has, which would be expected to lower the CMC (117).

EFFECTS OF TEMPERATURE. It is difficult to establish a relationship between temperature and CMC by comparing results from different investigators using varied methods and conditions. Temperature dependence was found in a systematic study by Carey and Small (22, 113) for sodium taurocholate (NaTC), NaTDC, and mixtures of the two. The CMC fell between 10°C and 20°C and remained constant between 20°C and 40°C. Above 40°C the CMC continually increased with a rise in temperature. For example, the CMC of NaTC in 0.15 M NaCl was 4.1, 2.7, 4.0, and 6.0 mM at 10°C, 20°C, 40°C, and 80°C, respectively. The exact CMC and temperature dependence of the rate of increase varied, but the trend was similar for both bile salts and the mixtures in H_2O and NaCl solutions. Thorough studies of temperature effects have not been reported for the other bile salts.

EFFECTS OF ADDED SODIUM ION. For all bile salts studied, addition of Na^+ (especially in the form of NaCl) decreases the CMC (see Table 7 and references cited therein). The change is less for NaC <0.7 M NaCl (113). For NaDC, NaGDC, NaTDC, NaCDC, and NaUDC different concentrations of NaCl were used, ranging from 0.01 M to 0.5 M NaCl (96, 98, 113, 129). The CMC was inversely related to NaCl concentration. For example, the CMC of NaDC was 6.4, 2,

TABLE 7. *Critical Micellular Concentration of Bile Salts*

Bile Salt	Method	T, °C	pH	Medium	CMC, mM	Ref.
NaC	Light scattering	20	6.6–7.6	0.1 N NaCl	20	DeMoerloose[a]
	Photon correlation spectroscopy	25	10	0.15 M NaCl	12.5	97
	Solubilization	20	NS	H_2O	13	Ekwall[a]
	1-Anilino-8-naphthalene sulfate	20	NS	H_2O	10	88
	Azulene	25	NS	0.15 M NaCl	16	96
	Dansylcadaverine	20	NS	H_2O	9.9	88
	Methylcholanthrene		NS	H_2O	12	Norman[a]
	Orange-OT	25	NS	0.15 M NaCl	11	96
	Spectral shift	25	10–11	H_2O	6.6	83
	Surface tension	22	9.0	H_2O	12	Small[a]
		22	7.4	M/15 Na phosphate[a]	4.9	Small[a]
		22	7.4	M/5 Na phosphate[a]	3.25	Small[a]
		25	NS	H_2O	13	96
		25	NS	0.15 M NaCl	14	98
		37	7	0.01 M Na_2HPO_4	7.5	81
Consensus values[b]		*20–25*		*H_2O*	*11*	
		25		*0.15 M NaCl*	*13*	
NaC[d]	Surface tension	25	NS	H_2O	30	96
		25	NS	0.15 M NaCl	21	96
NaC[e]	Surface tension	25	NS	H_2O	68	96
		25	NS	0.15 M NaCl	38	96
NaDC	Kinetic dialysis	20	7.4	0.01 M Na_2HPO_4	1.26	81
	Photon correlation spectroscopy	25	10	0.15 M NaCl	3.5	97
	Solubilization	20	NS	H_2O	5	Ekwall[a]
	1-Anilino-8-naphthalene sulfate	20	NS	0.02 M Na_2HPO_4	1.4	62
					4	
	Dansylcadaverine	20	NS	H_2O	2.5	88
	Methylcholanthracine	NS	NS	H_2O	5	Norman[a]
	Spectral shift	25	10–11	0.50 M NaCl	0.9	129
	Surface tension	20	6.6–7.6	H_2O	2	DeMoerloose[a]
		20	6.6–7.6	0.1 N NaCl	1	DeMoerloose[a]
		20	NS	NS	5	Miyake[a]
		25	NS	H_2O	10	96
		25	NS	0.15 M NaCl	3	96
		25	NS	0.001 M NaOH	6.4	98
		25	NS	0.45 M NaCl	2.8	98
		25	12	H_2O	6.4	129
		25	12	0.15 M NaCl	2	129
		25	12	0.30 M NaCl	1.1	129
		37	7	H_2O	7	83
Consensus values[b]		*20–25*		*H_2O*	*5*[c]	
		25		*0.15 M NaCl*	*3*	
NaDC[d]	Surface tension	25	NS	H_2O	23	96
		25	NS	0.15 M NaCl	10	96
NaGC	Solubilization	20	NS	H_2O	10	Ekwall[a]
	Azobenzene	37	6.3	0.15 M Na^+	8.0	Hofmann[a]
	Azulene	25	NS	0.15 M NaCl	10	96
	1-Monoolein	37	6.3	0.15 M Na^+	4.2	Hofmann[a]
	Orange-OT	25	NS	0.15 M NaCl	9	96
	Surface tension	25	NS	H_2O	12	96
		25	NS	0.15 M NaCl	10	96
		37	5.47	0.03 M $NaOOCCH_3$	1.5	81
Consensus values[b]		*20–25*		*H_2O*	*11*	
		25		*0.15 M NaCl*	*10*	
NaLC	Spectral shift	60	8	H_2O	0.2	Small[a]
NaTC	Diffusion	25	NS	H_2O	6.7	Woodford[a]
	Electron spin resonance	33	6.8	H_2O	6.0	Carey et al.[a]
	Kinetic dialysis					
	At 5 mM ATC	20	7.4	0.02 M Na_2HPO_4	2.1	
	At 10 mM ATC	20	7.4	0.02 M Na_2HPO_4	2.8	
	Self-diffusion	37	7.4	0.15 M NaCl	8.5	67
	Solubilization	20	NS	H_2O	10	Ekwall[a]
	Azobenzene	37	6.3	0.15 M Na^+	10.0	Hofmann[3]
	Azulene	25	NS	0.15 M NaCl	4	96
	Griseofulvin	37	5.4–6.6	H_2O	8.0	Bates[a]
	Hexestrol	37	5.4–6.6	H_2O	14.0	Bates[a]
	20-Methylcholanthrene	RT	NS	H_2O	12.0	Norman[a]
	1-Monoolein	37	6.3	0.15 M Na^+	4.2	Hofmann[a]
	Orange-OT	25	NS	0.15 M NaCl	9	96
	Spectral shift					
	Rhodamine 6G	25	6.8	H_2O	4.5	Carey et al.[a]
	Surface tension	25	NS	H_2O	10	96
		25	NS	0.15 M NaCl	6	96
		37	NS	H_2O	7.4	81
		37	4.09	0.03 M $NaOOCCH_3$	33	62
Consensus values[b]		*20–25*		*H_2O*	*8*[c]	
		20–25		*0.15 M NaCl*	*5*[c]	

TABLE 7–Continued

Bile Salt	Method	T, °C	pH	Medium	CMC, mM	Ref.
NaCDC	Photon correlation spectroscopy	25	10	0.15 M NaCl	4.5	97
	Solubilization					
	Azulene	25	NS	H_2O	10	96
		25	NS	0.15 M NaCl	5	96
	Methylcholanthrene	NS	NS	H_2O	6	Norman[a]
	Orange-OT	25	NS	H_2O	10	96
		25	NS	0.15 M NaCl	4	96
	Surface tension	25	NS	H_2O	9	96
		25	NS	0.15 M NaCl	4	96
		25	NS	0.01 M NaCl	5.7	98
		25	NS	0.15 M NaCl	2.7	98
		37	7	0.01 M NaCl	3.7	81
Consensus values[b]		25		*H_2O*	*9*	
		25		*0.15 M NaCl*	*4*	
NaCDC[d]	Surface tension	25	NS	H_2O	40	96
		25	NS	0.15 M NaCl	15	96
NaCDC[e]	Surface tension	25	NS	H_2O	65	96
		25	NS	0.15 M NaCl	40	96
NaGDC	Light scattering	20	6	0.15 M NaCl	1.1	Kratohvil and Dellicolli[a]
		20	6	0.50 M NaCl	0.74	Kratohvil and Dellicolli[a]
	Solubilization					
	And other methods	20	NS	H_2O	4	Ekwall[a]
	Azobenzene	37	6.3	0.15 M Na^+	1.9	Hofmann[a]
	Azulene	25	NS	0.15 M NaCl	3	96
	1-Monoolein	37	6.3	0.15 M Na^+	0.6	Hofmann[a]
	Orange-OT	25	NS	0.15 M NaCl	2.2	96
	Surface tension	20	6	H_2O	2.12	Kratohvil and Dellicolli[a]
		20	6	0.015 M NaCl	1.7	Kratohvil and Dellicolli[a]
		20	6	0.05 M NaCl	1.3	Kratohvil and Dellicolli[a]
		20	6	0.15 M NaCl	1.1	Kratohvil and Dellicolli[a]
		20	6	0.30 M NaCl	0.9	Kratohvil and Dellicolli[a]
		20	6	0.50 M NaCl	0.73	Kratohvil and Dellicolli[a]
		25	NS	H_2O	6	96
		25	NS	0.15 M NaCl	2	96
Consensus values[b]		20–25		*H_2O*	*4^c*	
		20–25		*0.15 M NaCl*	*2^c*	
NaGUC	Surface tension	25	NS	H_2O	35	96
		25	NS	0.15 M NaCl	30	96
NaTDC	Electron spin resonance	33	6.8	H_2O	3.5	Carey et al.,[a] Small[a]
	Light scattering	25	8.0	H_2O	1.5	Small and Carey[a]
		25	8.0	0.1 M NaCl	1.4	Small and Carey[a]
		25	8.0	0.3 M NaCl	1.1	Small and Carey[a]
		25	8.0	0.5 M NaCl	1.0	Small and Carey[a]
		25	6.0	H_2O	2.9	Kratohvil and Dellicolli[a]
		25	6.0	0.15 M NaCl	1.8	Kratohvil and Dellicolli[a]
		25	6.0	0.50 M NaCl	1.3	Kratohvil and Dellicolli[a]
		20	7.4	0.02 M Na_2HPO_4	0.8	62
		25	NS	0.15 M NaCl	1.67	60
		25	NS	0.6 M NaCl	0.96	60
	Spectral shift					
	Rhodamine 6G	25	6.8	H_2O	1.5	Carey et al.,[a] Small[a]
	Solubilization					
	And other methods	20	NS	H_2O	4	Ekwall[a]
	Azobenzene	37	6.3	0.15 M NaCl	1.9	Hofmann[a]
	Azulene	25	NS	0.15 M NaCl	6	96
	20-Methylcholanthrene	RT	NS	H_2O	5.0	Norman[a]
	Orange-OT	25	NS	0.15 M NaCl	3	96
	Spectral shift	25	NS	0.15 M NaCl	3	96
	Surface tension	25	6.0	H_2O	3.1	Kratohvil and Dellicolli[a]
		25	6.0	0.015 M NaCl	2.4	Kratohvil and Dellicolli[a]
		25	6.0	0.050 M NaCl	1.9	Kratohvil and Dellicolli[a]
		25	6.0	0.15 M NaCl	1.7	Kratohvil and Dellicolli[a]
		25	6.0	0.30 M NaCl	1.2	Kratohvil and Dellicolli[a]
		25	6.0	0.50 M NaCl	1.2	Kratohvil and Dellicolli[a]
		25	NS	H_2O	6	96
		25	NS	0.15 M NaCl	2.4	96
		37	4.09	0.03 M $NaOOCCH_2$	2.4	81
Consensus values[b]		20–25		*H_2O*	*3^c*	
		25		*0.15 M NaCl*	*2*	
NaTUC	Surface tension	25	NS	H_2O	52	96
		25	NS	0.15 M NaCl	40	96
NaUDC	Photon correlation spectroscopy	25	10	0.15 M NaCl	7.5	97
	Solubilization					
	Azulene	25	NS	H_2O	15	96
		25	NS	0.15 M NaCl	5	96
	Orange-OT	25	NS	H_2O	25	96
		25	NS	0.15 M NaCl	5	96

TABLE 7–*Continued*

Bile Salt	Method	T, °C	pH	Medium	CMC, mM	Ref.
	Surface tension	25	NS	H_2O	19	96
		25	NS	H_2O	37	96
		25	NS	0.15 M NaCl	21	96
		25	NS	0.15 M NaCl	7	96
		25	NS	0.01 M NaCl	11.5	98
		25	NS	0.15 M NaCl	6.3	98
NaGCDC	Solubilization					
	Azobenzene	37	6.3	0.15 M Na^+	2.4	Hofmann[a]
	Azulene	25	NS	H_2O	7	96
		25	NS	0.15 M NaCl	2	96
	1-Monoolein	37	6.3	0.15 M Na^+	0.8	Hofmann[a]
	Orange-OT	25	NS	H_2O	7	96
	Surface tension	25	NS	H_2O	6	96
		25	NS	0.15 M NaCl	1.8	96
		37	4.37	0.03 M $NaOOCCH_3$	1.2	81
Consensus values[b]		25		*H_2O*	*7*	
		25		*0.15 M NaCl*	*2*	
NaGUDC	Solubilization					
	Azulene	25	NS	H_2O	15	96
	Orange-OT	25	NS	H_2O	10	96
		25	NS	0.15 M NaCl	3	96
	Surface tension	25	NS	H_2O	12	96
		25	NS	0.15 M NaCl	4	96
NaTCDC	Self-diffusion	37	NS	H_2O	3.4	67
	Open-end capillary tube method	37	7.4	0.15 M NaCl	2.2	67
	Solubilization					
	Azobenzene	37	6.3	0.15 M Na^+	2.5	Hofmann[a]
	Azulene	25	NS	0.15 M NaCl	7	96
	1-Monoolein	37	6.3	0.15 M Na^+	0.8	Hofmann[a]
	Orange-OT	25	NS	0.15 M NaCl	4	96
	Surface tension	25	NS	H_2O	7	96
		25	NS	0.15 M NaCl	3	96
		37	4.09	0.03 M $NaOOCCH_3$	0.96	81
NaTUDC	Self-diffusion	37	7.4	0.15 M NaCl	3.1	67
		37	NS	H_2O	4.0	67
	Solubilization					
	Azulene	25	NS	0.15 M NaCl	1.6	96
	Orange-OT	25	NS	0.15 M NaCl	2.3	96
	Surface tension	25	NS	H_2O	8	96
		25	NS	0.15 M NaCl	2.2	96
Mixture of conjugated bile salts	Solubilization					
	Surface tension	22	5.3/7.5	M/15 Na phosphate[a]	0.7	Small[a]

T, temperature; CMC, critical micelle concentration; NS, not stated; RT, room temperature. NaC, sodium cholate; NaDC, sodium deoxycholate; NaGC, sodium glycocholate; NaLC, sodium lithocholate; NaTC, sodium taurocholate; NaCDC, sodium chenodeoxycholate; NaGDC, sodium glycodeoxycholate; NaGUC, sodium glycoursocholate; NaTDC, sodium taurodeoxycholate; NaTUC, sodium tauroursocholate; NaUDC, sodium ursodeoxycholate; NaGCDC, sodium glycochenodeoxycholate; NaGUDC, sodium glycoursodeoxychlolate; NaTCDC, sodium taurochenodeoxycholate; NaTUDC, sodium tauroursodeoxycholate. [a] Cited in Small (113). [b] Mean rounded-off value, excluding outliers and techniques that use monoolein solubilization. [c] Large variability with different methods. [d] C-23 analogue. [e] C-22 analogue.

1.1, and 0.9 mM in 0.0, 0.15, 0.30, and 0.50 M NaCl, respectively.

EFFECTS OF SIDE-CHAIN LENGTH. The effect of side-chain length on the common α-hydroxyl bile salts (NaC, NaDC, and NaCDC) were studied by Roda et al. (96). The CMC increased exponentially as the chain became shorter; for example, for NaCDC in H_2O the CMC was 9, 40, and 65 mM for a C_5, C_4, and C_3 side chain, respectively. This increase reflects in part the decreased hydrophobicity of the bile salts as —CH_2— groups are removed from the side chain. For aliphatic detergents the effect of chain length may be described by the equation log CMC = $A - Bn$, where A and B are characteristics of a particular detergent and n is the number of —CH_2— groups (117). The term A is related to the polar part of the molecule, and B is a function of the chain length. Although A differs from detergent to detergent, all single-chain aliphatic detergents have a value of B (i.e., the slope when log CMC is plotted against N, the number of carbons) of about log 2 (0.3). This relates directly to the energy necessary to move one —CH_2— group from a hydrophobic environment (in the micelle) to an aqueous environment (as a monomer). Thus the longer the chain the lower the monomer concentration. A plot for log CMC versus length of the side chain (96) gives values of 0.37–0.6, all higher than usual for aliphatic detergents. Nevertheless the general trend is similar to aliphatic detergents. That bile salts apparently deviate from the log CMC = $A - Bn$ rule could be due to a change in molecular packing in the micelle brought about by chain shortening. Further work is required.

MICELLE SIZE AND SHAPE. Aggregate size varies with concentration, number of hydroxyl groups, conjugation state, ionic strength, and temperature (8, 22, 26, 59, 60, 79, 98, 107, 112, 129). Free and conjugated deoxycholates have been extensively studied. The aggregation numbers ranged from two to thousands (112). The lowest value was found in H_2O (pH ~9) at low NaDC concentrations. In 0.053–1.0 M NaCl the aggregation jumps to 20–50. At high salt and low pH (4.9), micelles with >2,000 NaGDC molecules were found. Aggregate size increased only slightly (from 1.5 to 3.2) when the temperature was increased from 10°C to 80°C (113). Sodium taurodeoxycholate in 0.8 M NaCl doubled its aggregate size from 50 to 104 monomers when its concentration was increased from 0.2 to 4.95 g/dl (107). The micelle size remained constant up to concentrations of ~7 g/dl, then decreased slightly (to ~80 monomers) at higher concentrations. Glycine and taurine conjugates have similar aggregation numbers (6–50) that are somewhat less variable than unconjugated NaDC.

The pattern of changes in aggregation number with conditions is the same for NaCDC and NaUDC (98), but the actual micelle size is smaller, 7–30 monomers (NaCDC > NaUDC). Sodium cholate and its conjugates have the lowest aggregation numbers [2–73 for NaC, 2–9 for NaTC and NaGC); (98, 112)]. Light scattering (60) and self-diffusion studies (69) indicate a progressive aggregation for NaC and NaTC and association even at very low concentrations. A clearly defined CMC was not found, especially when compared with deoxycholate solutions. The micelle size does change with ionic strength, temperature, and concentration but to a lesser extent than the dihydroxy bile salts.

The structure of smaller micelles (aggregation number 2–10), referred to as primary micelles, was proposed to be a back-to-back association between the hydrophobic steroid ring portion, with the hydrophilic faces exposed to the aqueous medium [Fig. 7; (112)]. Larger aggregates, secondary micelles, were thought to arise from association of primary micelles at the polar faces stabilized by hydrogen bonding via hydroxyl groups. The primary micelle structure has been supported by ^1H and ^{13}C NMR data. Linewidths of C_{18} and C_{19} protons increase with increasing concentration, while C_{21} does not (79). This indicates greater immobilization of the ring methyls consistent with hydrophobic association with another bile salt molecule. Murata et al. (79) found differential changes in ^{13}C chemical shift with increasing concentration for NaDC. Ring methyl carbon resonances shifted downfield, but C_{21} was not affected. Hydroxyl carbons exhibited upfield shifts, as did some methylene carbons. This was interpreted as evidence for back-to-back association of primary micelles and for an ordered structure formed by hydrogen bonding between hydroxyl and carboxyl groups consistent with the proposed structures for secondary micelles. The authors

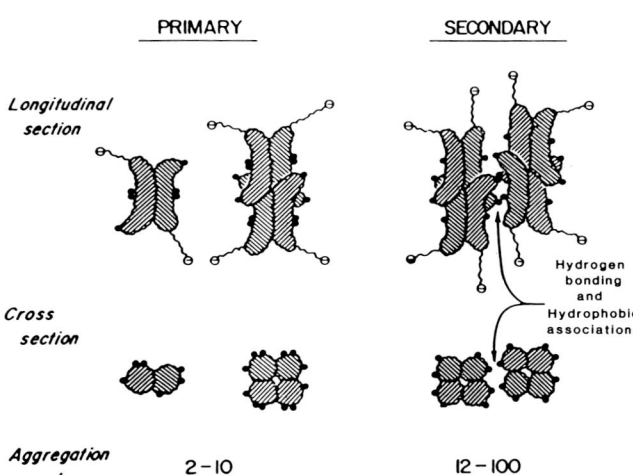

FIG. 7. Schematic of structure of bile salt micelles. Molecules of primary micelles associate via hydrophobic regions with aggregation numbers from 2 to 10. At higher concentrations some bile salts form larger aggregates, secondary micelles, stabilized by hydrophobic associations and hydrogen bonding. [From Small (112), printed with permission from *Advances in Chemistry Series*, copyright 1968, American Chemical Society.]

have confirmed the ^{13}C chemical shift changes for varying concentrations of NaDC and NaC. However, because the causes for the upfield and downfield changes cannot be unambiguously assigned, micelle structure cannot be determined from this evidence alone. Light scattering (107) of NaTDC also indicated formation and growth of elongated structures with increasing concentration above the CMC. Discontinuities in the NaTDC surface-tension curves and conductance at certain concentrations (81, 113) are also consistent with a primary-to-secondary micelle transition for the dihydroxy bile salt.

BILE SALT–CALCIUM INTERACTIONS. The centers of cholesterol gallstones often have a core of pigment that contains high levels of precipitated Ca^{2+} salts (76, 77, 134). These Ca^{2+} salts can be of several types (e.g., carbonate, phosphate, palmitate, or bilirubinate), and it has been proposed by Moore (76) that cholesterol stone formation is a two-stage process. First bile becomes supersaturated with a Ca^{2+} salt that precipitates; then cholesterol precipitates from cholesterol-supersaturated bile onto the Ca^{2+}-salt nidus. If bile is not supersaturated with cholesterol the process will stop with the precipitation of the Ca^{2+} salt. If these precipitates enlarge, then calcium stones (e.g., calcium bilirubinate) result.

Bile salts in both monomeric and micellar form bind free calcium (76, 78, 90, 133b, 134) and are thought to be important in regulating the activity of free calcium in bile and to aid in preventing the potentially insoluble Ca^{2+} salts from precipitating. Two techniques have been used to study binding, NMR (78) and calcium-sensitive electrodes (76, 77). The calcium electrodes measure the activity of calcium in solution, while the NMR technique measures displacement of bile-bound dysprosium (a paramagnetic ion that binds like calcium) by added calcium. The NMR technique has the added advantage that it can identify the specific atoms of the bile salt that take part in the binding. At pH 5, monomeric glycocholate binds Ca^{2+} to the carboxylate (—COO^-) group to form a 1:1 —$COOCa^+$ complex with a dissociation constant $K_m = 9.5 \pm 1.5$ mM. Monomeric taurocholate binds Ca^{2+} more weakly ($K_m = 58.9 \pm 0.2$) to the sulfonate group. No Ca^{2+} binding to any hydroxyl was seen by NMR (78), although such complexes have been suggested (76) and could be present in deoxycholates, since they have not been studied by the more site-specific NMR techniques. Calcium ions bind to glycocholate micelles in a reversible manner, with K_m ~26 mM (133b). Thus monomers bind Ca^{2+} more tightly but have a low capacity, whereas micelles bind less strongly but have high capacity (76). Taurocholate will also bind Ca^{2+} as monomers or micelles (77). Monomers formed the 1:1 "free acid–salt" ($CaTC^+$) with high affinity, whereas in micelles two taurocholate anions associated with one Ca^{2+} to form a neutral complex [$Ca(TC)_2$] with lower affinity. If Ca^{2+} concentration is greatly increased, then 1:1 complexes can occur in micelles (76).

The effect of hydroxyl number and position and the difference between glycine- and taurine-conjugated micellar bile salts was determined by Rajagopalan and Lindenbaum (90). Like many other bile salt properties, Ca^{2+} binding increased with increasing hydrophobicity. For a given conjugation state, NaGDC bound more Ca^{2+} per mole bile acid than sodium glycochenodeoxycholate (NaGCDC), which bound more than sodium glycocholate (NaGC) (158 vs. 148 vs. 56 mmol, respectively). For a given hydroxyl state, glycine conjugates bound more Ca^{2+} than taurine conjugates (56 vs. 33 mmol for cholate). For glycine conjugates, Ca^{2+} binding increased with the addition of lecithin. Thus the effect may be due to binding of the cation to the phosphate headgroup of lecithin. It seems likely, therefore, that bile salts and lecithin and other substituents in bile (e.g., proteins), which contain anionic groups capable of binding Ca^{2+}, act as buffers to keep potentially insoluble Ca^{2+}-salts from precipitating and growing into pigment stones or acting as nidi for cholesterol stones. Several animal models have been found that are currently being used to investigate the role of calcium in stone formation (48).

BILE SALT–LECITHIN INTERACTIONS

Composition of Lecithins in Bile

Bile salts are the most abundant lipid in bile; phospholipids are the second most abundant. The phospholipid composition of bile varies somewhat from species to species but in general is almost completely made up of lecithins (phosphatidylcholines). The composition of lecithins in human and rat bile and in egg yolk is given in Table 8 (86, 95). Bile and EYL are rich in sn-1 16:0 lecithins. The major sn-2 fatty acids are 18:1, 18:2, and other polyunsaturates. The fact that sn-1 18:0 lecithins are sparse in bile may reflect their very slow exchange rate compared with sn-1 16:0 lecithins (87). Certainly the in vitro dissolution rate of liquid crystals of lecithin by bile salts is more rapid if the sn-1 fatty acid is shorter (2). The sn-1 fatty acid extends three to four carbons more deeply into the bilayer and may serve as the main anchor, and the increased length increases the activation energy necessary to dislodge it from the membrane.

The crystalline and solution properties of bile salts have been examined, and the interaction of lecithin with bile salts is now discussed. Lecithins are zwitterionic phospholipids that swell in H_2O to form lamellar liquid crystalline phases (117). These lamellae consist of bilayers of lipids interspaced with layers of H_2O. The aliphatic chains may be in a liquid state or at lower temperatures frozen in a "gel" state. The lecithins found in bile (Table 8) contain appreciable quantities of unsaturated fatty acids that lower this gel-to-liquid chain transition to temperatures well below body temperature. Thus the state of the chains in the biliary lecithins is always liquid, and the liquid crystalline phases formed are the lamellar liquid crystals. Most model systems of bile salts and unsaturated lecithins have been carried out with EYL. The composition of EYLs varies somewhat, but they are usually rich in sn-1 palmitoyl, sn-2 oleyl, and linoleyl lecithins and therefore are similar to human biliary lecithins (Table 8). In the studies where the interac-

TABLE 8. *Specific Phosphatidylcholines (Lecithins) of Bile and Egg Yolk*

Phosphatidylcholine, mol%		Egg Yolk	Rat Bile	Human Gallbladder Bile	
sn-1	sn-2	*	†	‡	†	§
16:0	16:1	1.1			1.3	4.2
16:0	18:1	33.8	39.0	8.8	17.8	21.8
16:0	18:2	24.7	17.8	40.9	28.7	54.9
16:0	20:4	1.0		16.1	11.3	9.6
16:0	20:5				1.9	
16:0	22:5				5.5	
16:0	22:6	0.1		4.3	7.2	
18:0	18:1	21.7	8.8	0.4		2.0
18:0	18:2	5.2	8.2	7.8	3.7	1.7
18:0	20:4	3.8		5.5	1.8	1.7
18:0	22:6			1.3	4.1	
18:1	18:1	1.9				0.3
18:1	18:2	0.7	2.8	2.4	5.3	2.9
18:2	18:2	0.1			1.2	
Others		6.0	25.4	12.4	10.2	0.9

sn-1 and sn-2, position of fatty acids on the phosphatidylcholine glycerol backbone. * Data from S. Bennett-Clark and E. Belur (unpublished observations). † Data from Cantafora et al. (18). ‡ Data from Patton et al. (86). § Data from Robins et al. (95).

tions of bile salts with human biliary and EYL have been compared, the results have been quite comparable (75). In some of the studies of interactions of bile salts with phospholipids, saturated lecithins in which the chain-melting transition is considerably higher have been utilized.

Bile Salt–Phospholipid–Water Systems

To understand as completely as possible the interactions of bile salts with phospholipid in aqueous systems, the phase diagram of that system should be understood. Two such phase diagrams have been constructed by use of techniques of observation, direct and polarizing microscopy, and X-ray diffraction. The two systems are an EYL-NaC-H_2O system and a similar system in which a mixture of naturally occurring bile salts (taurine conjugates, glycine conjugates, di- and trihydroxy bile salts) was substituted for NaC (118–120). No complete diagram exists for a pure dihydroxy bile salt, but the phases formed would probably be the same, although the phase boundaries would

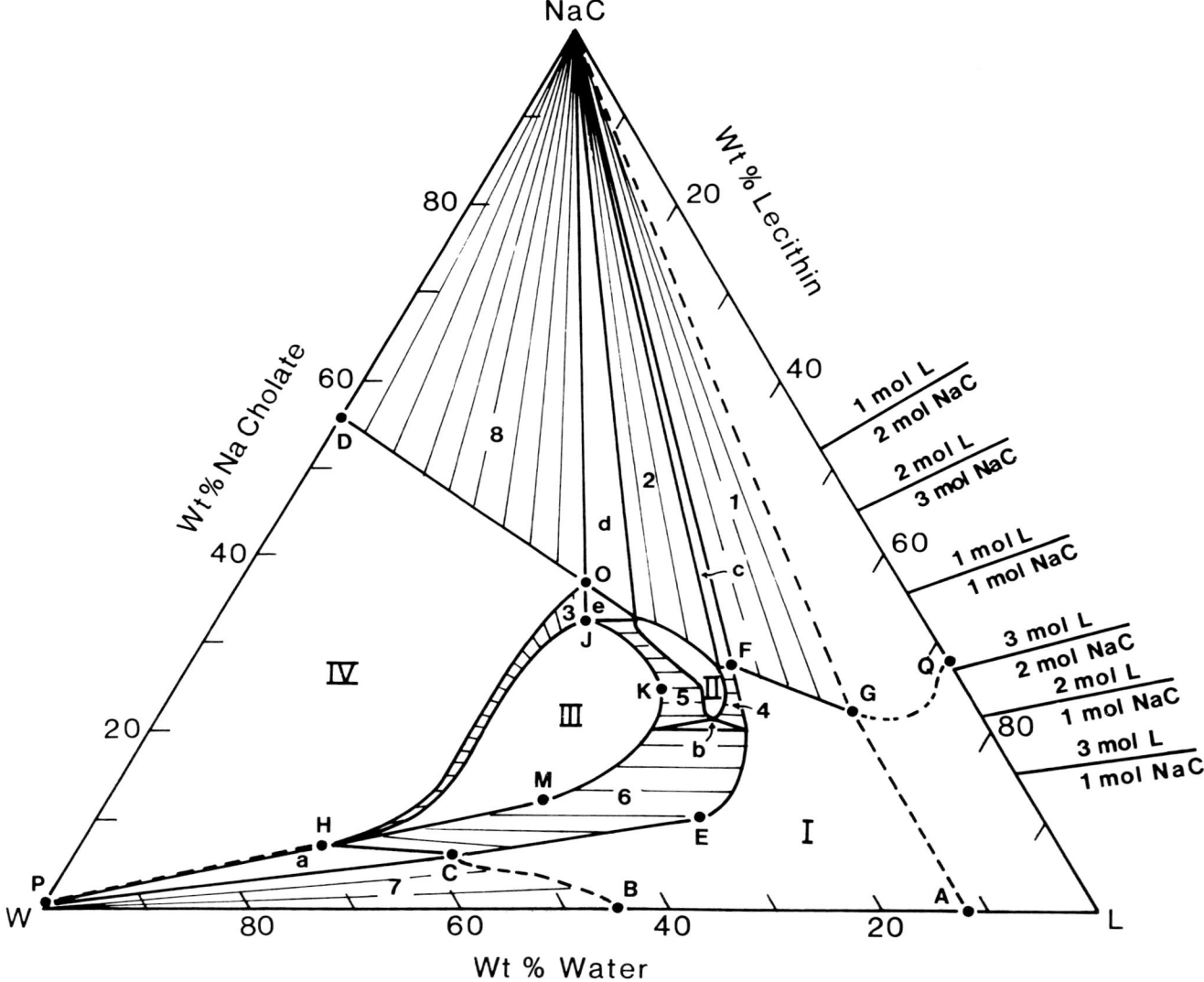

FIG. 8. Lecithin–sodium cholate–water diagram. W, H_2O; L, lecithin; NaC, sodium cholate. Along *side W-L*, numbers indicate H_2O wt percentage; along *side L-NaC*, numbers indicate lecithin wt percentage; along *side NaC-W*, numbers indicate NaC wt percentage. *Solid lines*, well-defined boundaries (±2%); *broken lines*, less well-defined frontiers. *I*, Zone of neat phase. *II*, Zone of cubic phase. *III*, Zone of middle phase. *IV*, Zone of isotropic phase. *1*, Zone of separation of neat phase and NaC crystals. *2*, Zone of separation of cubic phase and NaC crystals. *3*, Zone of separation of middle phase and isotropic solution. *4*, Zone of separation of neat phase and cubic phase. *5*, Zone of separation of cubic phase and middle phase. *6*, Zone of separation of neat phase and middle phase. *7*, Zone of separation of neat phase and isotropic phase. *8*, Zone of separation of isotropic solution and NaC crystals. *a–e*, Zones of separation of 3 phases whose compositions for each zone are indicated by apices of the triangles. *Points A–H, J, K, M, O–Q* are discussed in text. *Right side of diagram*, molar ratios of lecithin to NaC.

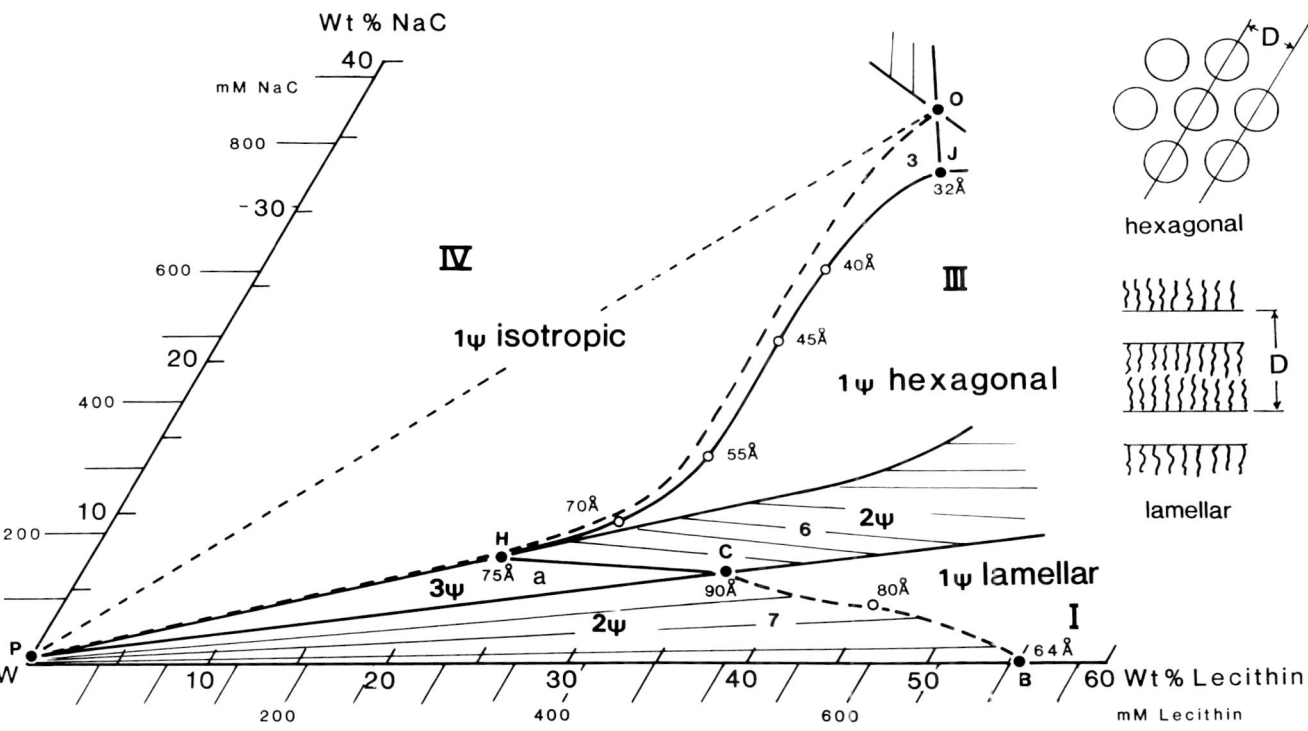

FIG. 9. Lecithin–sodium cholate–water diagram at high-H_2O concentrations. NaC, sodium cholate; W, H_2O. *Left*: *larger numbers*, wt%, *smaller numbers*, concentrations (mM). Parts of single-phase zones I, III, and IV are shown, as well as 2-phase regions 3, 6, and 7. Approximate tie lines in regions 6 and 7 and 3-phase region are illustrated. *Right*, Bragg spacings (D) from lamellar and hexagonal lattices. These lattice parameters are given in angstroms along high-H_2O phase boundaries of lamellar phase and hexagonal phase. *Dashed line* from P (critical micellar concentration of NaC in H_2O) to O, separates micelles formed by dilution of hexagonal phase (*below*) from micelles formed by dissolution of NaC crystals (*above*).

be different. The EYL-NaC-H_2O system at 25°C is shown in Figures 8–10. The original diagram (118–120) has been slightly modified in the high-H_2O region to be consistent with more recent data (108). The diagram was constructed from gross, microscopic, and X-ray examination of 160 separate mixtures. The results indicate that there are four zones comprised of a single homogeneous phase, eight regions of two coexisting phases, and five triangular regions of three coexisting phases.

PHASE RULE. At equilibrium the phase rule states that in a system at constant temperature and pressure, the number of degrees of freedom (F) is equal to the number of components (C) minus the number of coexisting phases (P); i.e. (F = C − P). Because this system contains three components, EYL, NaC, and H_2O, the phase diagram reduces to F = 3 − P. The implications are important. In any zone containing a single phase (I, II, III, or IV) there are F = 3 − 1 = 2 degrees of freedom. This means that the composition of two of the components in the zone must be fixed before this system is invariant. In other words, each point within a single-phase zone has different physical characteristics. This should be kept in mind when discussing the properties of the "micellar phase" or any of the other one-phase zones. In regions 1 to 8, in which two phases coexist, there is one degree of freedom (F = 3 − 2 = 1). Thus for any mixture in a two-phase region the composition of one of the phases must be determined. From the composition of one of the phases and the composition of the mixture, the composition of the other phase can be deduced. A straight line connecting the composition of the known phase to the composition of the mixture can be extended to the phase boundary of the second phase. The intersection of this line with the phase boundary gives the composition of the second phase. Such lines are called tie lines. A tie line connects the composition of the two phases in equilibrium. The approximate positions of some of the tie lines in two-phase regions are shown in Figures 8–10. Finally, in the triangular three-phase regions there are no degrees of freedom (F = 3 − 3 = 0); the system is invariant. All mixtures in three-phase regions separate into three phases whose compositions are given by the corners of the triangular regions.

EGG YOLK LECITHIN–SODIUM CHOLATE–WATER SYSTEM: GENERAL CHARACTERISTICS. The LW border in Figure 8 represents the binary lecithin-H_2O system. At H_2O contents <12% this system is heterogeneous and not well defined. Between 12% and 45% H_2O (AB), only the lamellar liquid crystal phase is

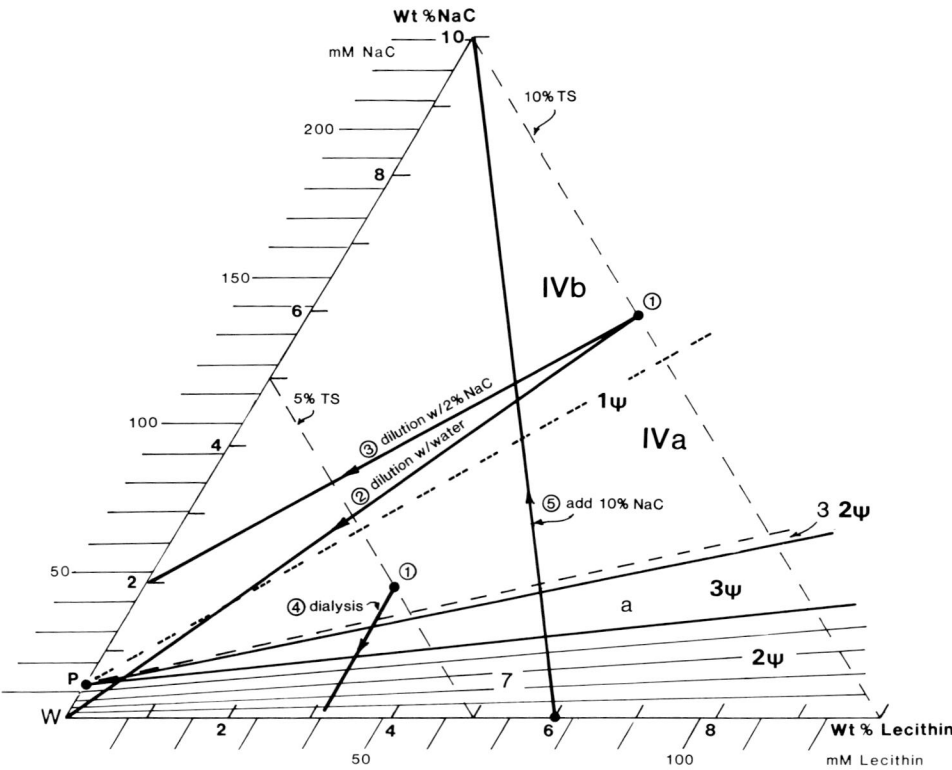

FIG. 10. Lecithin–sodium cholate system at >90% H₂O, illustrating different techniques for preparing mixtures. NaC, sodium cholate; W, H₂O. *NaC-W axis*: *larger numbers*, wt% NaC from 0% to 10%; *smaller numbers*, concentration (mM). *Point P*, critical micelle concentration of NaC in H₂O (0.52% or ~12 mM). *Lecithin axis*: *large numbers*, wt% lecithin; *small numbers*, concentrations (mM). Zone IV has been divided into 2 regions: *IVa* and *IVb*. *Line* separating 2 regions is a straight line running from *point P* to *point O* (see Fig. 9) and indicates a change in micellar structure. Narrow 2-phase *region 3* and larger 2-phase *region 7* are speareted by 3-phase *region a*. Heavy lines, various methods of preparing mixtures. Any mixture may be made up by coprecipitation method (e.g., *points* ① *on 10% solid line* and *5% solid line*; TS, total solutes). A mixture made up by coprecipitation may then be diluted with H₂O or buffer and will follow *line* ②. If a coprecipitated mixture is diluted with a specific concentration of bile salt, e.g., 2%, it will follow *line* ③. A coprecipitated mixture such as that plotted on *5% line* at 40 mM lecithin and 45 mM NaC may be placed in a dialysis bag and dialyzed against H₂O, buffer, or a bile salt solution. Because lecithin does not escape from the dialysis bag and its concentration remains unchanged, the composition will move toward *W-lecithin side* along a constant 40-mM line, *line* ④. Concentration of dialysate (not shown) will be found along W-NaC axis. Finally, any concentration of lecithin, eg., a 6% suspension (either coarse liposomes or unilamellar vesicles), may be diluted with a specific concentration of bile salt, for instance at 10%, as shown on *line* ⑤. Compositions will fall on *line* ⑤, depending on amount of 10% bile salt solution added.

present. As H₂O increases from 12% to 45%, the periodicity from one lamellae to the next (Bragg or d spacing) increases from 51 Å to 64 Å, while the calculated thickness of the bilayer decreases from ~47 Å to 32 Å. This is directly related to an increase in the surface area swept out by each EYL molecule at the plane of the bilayer. The surface area per EYL molecule increases from 60 Å² to ~70 Å² at 45% H₂O (109). For >45% H₂O, two phases exist: H₂O and a lamellar phase having a fixed d spacing at 64 Å. At 50% H₂O excess H₂O pockets can be identified within the liquid crystal phase, while at 80% H₂O the liquid crystal phase, seen as myelin figures and anisotropic droplets, floats in a continuous aqueous phase. On the W-NaC side, WD represents the range of solubility of NaC in H₂O at 20°C and pH 10. Saturation occurs at 56% NaC (~1.3 M). Point P represents the approximate CMC of NaC (0.52% = 12 mM). The third side of the diagram corresponds to dry mixtures of EYL and NaC.

All the points inside the diagram represent mixtures containing the three components in different proportions. Such mixtures may form one homogeneous phase or separate into two or three phases. There are four separate single homogeneous phase zones, eight areas of two coexisting phases, and five triangular areas of three phases. All mixtures in zone I are clear, rather fluid, and give microscopic texture and X-ray spacings of the lamellar liquid crystal structure. The finite limits of this phase are enclosed by ABCEFG. Line BC gives the maximum swelling of the lamellar phase, i.e., the maximum H_2O incorporated into the liquid crystal. As NaC is incorporated into the lamellar phase, H_2O content also increases and the periodicity between layers increases (Fig. 9) to 80 Å at ~52% H_2O, 44% EYL, and 4% NaC. An increase in H_2O content beyond BC (i.e., toward W) produces two phases, the lamellar phase in equilibrium with an isotropic aqueous phase. Line CE is straight and coincides with the line of constant composition for 85% EYL and 15% NaC (~3 EYL/1 NaC). It represents the maximum NaC that can be incorporated into the lamellar liquid crystalline lattice when the mixture contains >30% H_2O. This maximum is ~1 molecule of NaC to 3 molecules of EYL. Point C is a triple point (36% lecithin, 6% NaC, 58% H_2O), which marks the maximum incorporation of H_2O and NaC into the lamellar phase at high-H_2O concentration. The lamellar repeat at point C is ~90 Å. Point F is another triple point (52% lecithin, 27% NaC, 21% H_2O), which marks the maximum amount of NaC that the lamellar lattice can hold at any H_2O concentration. It occurs at a mole ratio of near 1:1 and has a lamellar repeat of only 38 Å. As the amount of H_2O is decreased along FG from 21% to 12%, the amount of NaC incorporated into the lattice falls from nearly 1:1 to ~1:2 (G).

Mixtures in zone II are clear, isotropic, and stiff. Bubbles are angular, and X-ray analysis shows that this phase has a face-centered cubic lattice with a d_{111} of 40–45 Å (118). This phase can only form within H_2O limits of 22%–28% and within the limits of 2:3–3:2 EYL/NaC. The arrangement of the molecules within the cubic lattice has not been determined.

Mixtures in zone III are nearly clear but more viscous than those in zone I. They have characteristic polarizing microscopic texture of hexagonally packed cylinders (119). The zone is bounded by the continuous line HJKM. The frontier HJ limits the maximum swelling that this hexagonal phase can undergo without breaking up to form an isotropic solution. This varies from ~70% H_2O, 7% NaC, and 23% lecithin (H) to ~32% H_2O, 33% NaC, and 35% lecithin (J). Point K shows the minimum H_2O (28.5%) compatible with the formation of this phase. Line MH has a fixed proportion of ~2 EYL to 1 NaC and represents the minimum NaC necessary to form the hexagonal phase. Each different mixture within zone III is composed of cylinders packed in hexagonal array. Because in a single-phase region the number of degrees of freedom is 2, each mixture will have distinct characteristics dependent on the composition. Thus the X-ray parameters, optical properties, viscosity, and density vary with composition (118, 119). Maximum d spacings in the hexagonal phase along the boundary HJ are given in Figure 9. As the amount of NaC increases (H to J) the amount of H_2O incorporated into the phase decreases and the d spacings also decrease. This also means that the distance between cylinder centers decreases with increased bile salt concentration. These findings have been interpreted (113) to indicate a structure for the hexagonal phase in which discoidal micelles stacked in rods pack into a hexagonal lattice, giving rise to the X-ray spacings In this model the diameter of the cylinder (the discoidal micelle) would decrease from ~75 Å at point H to ~32 Å at point J. Based on NMR evidence, Ulmius et al. (130a) have suggested that the cylinders are continuous rods rather than stacked micelles. The absolute location of the molecules within the hexagonal lattice is not known.

Zone IV is an isotropic liquid phase. It is bounded by WPDO (see expansions, Fig. 9). Point P is at 0.52% NaC, the CMC of NaC in H_2O (12 mM). (In 0.15 M NaCl the CMC is ~4 mM.) Along WP are low monomeric concentrations of NaC in H_2O. Point D is the maximum solubility of NaC in H_2O, and O is a triple point at a minimum amount of H_2O that can form the isotropic solution (30% H_2O, 33.5% EYL, 36.5% bile salt). The proportion of bile salt to EYL is 52% to 48%, and the mole ratio is ~2 EYL to 3 NaC. Line OP (Fig. 8) is a border of the isotropic phase. Frontier DO is straight and its extrapolation passes through the EYL apex. This represents the solubility of NaC in H_2O in the presence of low but increasing amounts of EYL, and it indicates that the solubility of NaC in H_2O is independent of the EYL present in this zone.

Between these one-phase zones are regions in which mixtures separate into the bordering two or three phases. There are eight two-phase regions numbered 1–8 in Figure 8 and five triangular three-phase regions lettered a–e. Regions 1, 2, and 8 are regions of separation of NaC crystals from phases I (lamellar), II (cubic), and IV (isotropic), respectively. Characteristic X-ray and microscopic findings define these phases. A narrow region 3 separates isotropic from hexagonal phases. Microscopically it is characterized by islands of the hexagonal phase appearing as fanlike units or cloudy streaks suspended in an isotropic liquid. This separated hexagonal phase could be clearly observed out to >80% H_2O but was extremely difficult to identify in very dilute systems. However, the phase rule dictates that it must extend over to the W-NaC border (see Figs. 9 and 10). In region 4 (cubic and lamellar) lamellar liquid crystals (I) are observed encircling rigid blocks of the isotropic phase (II). X-ray analysis gives spacings of both cubic and lamellar lattices. Region 5 is marked by isotropic areas (II) circling islands of anisotropic hexagonal phase (III). Mixtures in region 6 are characterized by islands of hexagonal phase (III) surrounded by the lamellar phase (I). X-ray analysis gives two sets of spacings, one corresponding to the hexagonal phase and the other to the lamellar phase. Mixtures in zone 7 are turbid. Close to the lamellar-phase boundary (BC) pockets of H_2O are entrapped in the lamellar phase. In water concentrations >80% the aqueous phase becomes continuous and myelin figures and isotropic droplets are suspended in the aqueous phase. The five invariant triangular zones of three coexisting phases are difficult to place with great accuracy. The phases present in mixtures falling in each triangle are given by the compositions at each apex. For instance, in the three-phase zone marked a, the phases present are a lamellar phase of composition C, a hexagonal phase of composition H, and an aqueous phase of composition P. The other three-phase zones are listed in Figure 8 and have been discussed in reference 119.

STRUCTURES OF ISOTROPIC (MICELLAR) ZONE. In the past 20 years most of the investigations have centered on the structure and characteristics of the "micellar zone." The compositions studied vary widely but as a rule have been dilute systems containing 90% or more H_2O and therefore would have compositions plotted in Figure 10. Each composition in zone IV will have different physical characteristics (e.g., size, shape) because there are 2 degrees of freedom in this large isotropic zone. Not only have the compositions varied widely but the techniques for the preparation of the solutions have varied. These include *1*) a coprecipitation technique in which bile salts and EYL are dried from an organic solvent and then brought up in an aqueous solution to a certain concentration (118–120); *2*) coprecipitation followed by dilution; *3*) coprecipitation followed by dilution with a bile salt solution; *4*) coprecipitation followed by dialysis against H_2O, buffer, or bile salt solutions; and *5*) interactions of multilamellar liposomes or unilamellar vesicles (usually produced by sonication) with bile salt solutions. These different methods of sample preparation are illustrated in Figure 10. Furthermore the time and temperature of equilibration have often differed, salt and/or buffers are often added, and proof of equilibrium has not always been sought.

These diverse systems have been studied by one or more methods, including direct observation, turbidity measurements, quasi-elastic light scattering for estimation of diffusion of aggregates from which an effective hydrodynamic radius (R_H) can be calculated, X-ray–scattering experiments from which a radius of gyration (R_G) can be calculated, NMR experiments (including ^1H, ^2H, ^{31}P, ^{13}C) in which molecular motions can be estimated, fluorescence polarization measurements, differential scanning calorimetry measurements, dialysis measurements, solubilization measurements, surface chemistry, ultracentrifugal measurements, and even column chromatographic methods. We have selected from the literature a few examples that illustrate both the consensus and divergence of opinion concerning the variable nature of zone IV.

Early conventional light-scattering studies on 10% solutions of NaC and EYL indicated that the system was made up of aggregates whose weight increased appreciably as the ratio of EYL to NaC increased (103, 120). Equilibrium ultracentrifugation and sedimentation and diffusion studies confirmed this (113). These early studies indicated that the aggregate weight increased sharply as the proportion of EYL was increased. A discoidal model was proposed in which a small bilayered EYL disk was surrounded by a perimeter of bile salts (110, 113). Support for the model came from the X-ray–diffraction studies on the hexagonal liquid crystalline phase (Fig. 9, zone III). The hexagonal liquid crystalline phase was thought to be composed of indefinitely long cylinders packed in a compact hexagonal lattice. The Bragg spacing (d) times $\sqrt{3/2}$ gives the distance from the center of one cylinder to another (D_T). This includes the total diameter of the cylinder (D_C) and the thickness of the H_2O layer between cylinders (D_W). In most detergent systems the dimensions of the hexagonal phase are very similar to the dimensions of the micellar phase formed when a small excess of H_2O is added to the hexagonal phase (117). It is thought that the cylinder of the hexagonal phase breaks up into rods of variable but finite length when excess H_2O is added. Thus the diameter of the cylinder is very similar to the cross-sectional diameter of the rod-shaped micelle produced from the hexagonal phase. Using the X-ray–diffraction data at the phase boundary of the hexagonal phase (Fig. 9, zone III), the distance between cylinders was calculated and plotted against the molar ratio of EYL to NaC. This plot indicated that D_T increases in a linear fashion from ~48.5 Å to 80 Å as the mole ratio increased from 0.7 to 1.75 (113). Using the simple disk model with all bile salts at the perimeter, the calculated disk diameter over the same mole ratio range increased from 40 Å to 71 Å and had the same slope. It was thus proposed that lipid cylinders were composed of stacks of disk-shaped micelles. Each cylinder in the hexagonal lattice was separated by ~9 Å H_2O space (3 H_2O molecules) at the phase boundary.

Although the simple discoidal micelle served as a working model, the complexity of micellar sizes and structures in zone IV became apparent as newer techniques were applied. First classic light scattering on 10% solutions or 10% solutions diluted with NaC at ~CMC (103) indicated that the micellar weight became significantly greater than that predicted for simple discoidal micelles (113). Next bile salt–lecithin systems of varying mole ratios were prepared as 10% solutions, then serially diluted to different concentrations, and the mean R_H estimated by quasi-elastic light scattering (70). Figure 11A shows that as the mole ratio of EYL to NaTC changes, R_H increases precipitously. Furthermore, the R_H values obtained are much larger than those calculated for the simple discoidal model (dashed line). However, note that as the *total* concentration increases, the divergence from the discoidal model becomes progressively less, e.g., compare 1.25 g/dl with 10 g/dl. Plotting R_H at a given mole ratio as a function of the total composition of lipid (Fig. 11B) shows that R_H approaches that of the simple discoidal model at high-lipid concentration.

9), Mazer et al. postulated that the discoidal micelles might contain bile salt within the disk. They calculated from the R_H that the total bile salt concentration was distributed into three separate species in each micellar system. *1*) An intermicellar species (not associated with EYL) was estimated for NaTC at 3.2–4.2 mM at 20°C in zone IVa. The lower value 3.2–4.2 mM is essentially the same as the CMC at 20°C for NaTC in 0.15 M NaCl (22). Bile salts in the mixed micelle were divided into *2*) those within the disk and *3*) those around the perimeter. Their calculations indicated that the bilayer of the disk contained ~1 bile salt per 2 lecithins and that the rest of the bile salt made up the perimeter of the disk. Thus at these more dilute concentrations within the region marked on Figure 10 as zone IVa, the size varies with the mole ratio and the total concentrations, as indicated in Figure 11.

In the region high in bile salts (Fig. 10, zone IVb), their results showed a more marked polydispersity of the radius and a sharp increase in non-EYL-associated NaTC. They suggested that two types of micelles coexist in zone IVb: small discoidal micelles of R_H ~30–35 Å and small taurocholate micelles of R_H ~10 Å. It had been suggested (118–120) that the micellar zone high in NaC (above the straight line PO in Fig. 9) was different than that rich in lecithin. The rationale for this was that micelles in zone IVa originate from dilution of the hexagonal phase, whereas those in zone IVb originate from dilution of systems containing crystalline NaC. Early studies showed a marked increase in conductivity and the action of snake venom phospholipase on lecithin above ~50 wt% NaC (82), and these authors suggested that there was a marked change in the micellar solution properties on crossing from zone IVa to IVb. Recent NMR studies in this region support the contention that zone IVb consists of two coexisting micelle types (106).

Using two coprecipitated systems, one having 1:1 EYL/bile salt (zone IVa) and the other having 1:3 EYL/bile salt (zone IVb), Muller (78a) studied the systems at 90% H_2O and more dilute by X-ray scattering. The micelles having 1:1 EYL/bile salt gave strong evidence of a bilayer ~50 Å thick. The R_G (corresponding to the largest radius swept out by a tumbling disk) increased with dilution compatible with the light-scattering experiments in this same region of the diagram. Thus the X-ray-scattering studies clearly indicate a mixed discoidal micelle in this region of zone IVa. However, the 1:3 EYL/bile salt mixture appeared to give, at all dilutions from 10% to 1.25%, a single micellar R_G and calculated diameter of ~62 Å. This diameter is about the same as that of the larger mixed micelle proposed from light-scattering studies (70). These authors believe that this is consistent with a mixed small EYL–bile salt micelle containing ~12 lecithins and 36 bile salts packed in a pseudoglobular structure. Because this composition is similar to the bile salt–lecithin ratio in

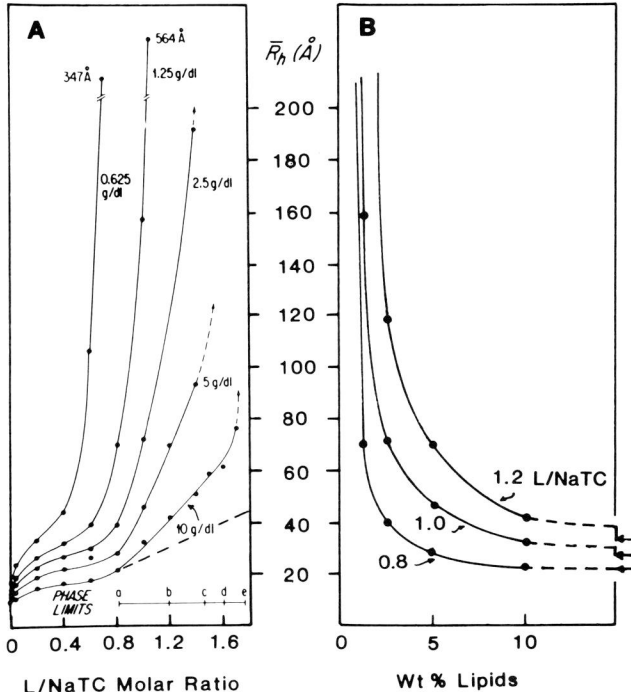

FIG. 11. *A*: mean hydrodynamic radii ($\bar{R}_h$) of sodium taurocholate–lecithin solutions (L/NaTC) at various total lipid concentrations (0.625%–10% at 24°C in 0.15 M NaCl). $\bar{R}_h$ diverges as lecithin–bile salt ratio approaches micellar-phase limits, and divergence limits are very concentration dependent. *Dashed line*, radius of simple discoid micelle. Phase limits shown at *bottom* of each panel represent appropriate lecithin–bile salt ratio for total lipid concentrations of *a*, 0.0625%; *b*, 1.25%; *c*, 2.5%; *d*, 5.0%; and *e*, 10.0%. $\bar{R}_h$ values indicated at tops of certain curves represent size of bilayer vesicles formed in supersaturated systems. *B*: $\bar{R}_h$ at 3 L-NaTC ratios as a function of the total wt% of lipids. As the total percentage of lipids increases, $\bar{R}_h$ decreases markedly and approaches the dimensions of the simple discoidal micelle (*arrows* at *right*) at high lipid concentrations. [*A* reprinted with permission from Mazer et al. (70). Copyright 1980 American Chemical Society.]

Thus we believe that at high-lipid and low H_2O concentrations, i.e., those approaching the phase boundary HJ (Figs. 8 and 9) of the hexagonal phase (zone III), the micelles are probably simple disks, with all or nearly all the bile salt at the perimeter. However, as the composition becomes more dilute the micelles enlarge and cannot fit a simple disk model (70). Mazer et al. (70) use these data to derive another model for the lecithin–bile salt micelle.

First, the asymptote to each of the curves in Figure 11*A* indicated a very large micelle size, and this was taken as the limit of the micellar phase at that particular lipid composition. These boundaries of the EYL-NaTC–0.15 M NaCl system have been plotted on Figure 12 on both rectangular (70) and triangular coordinates. For comparison the estimated phase boundaries of NaTC-EYL–0.15 M NaCl system are also shown. Second, realizing that zone I of the lamellar phase contains some bile salt (see Figs. 8 and

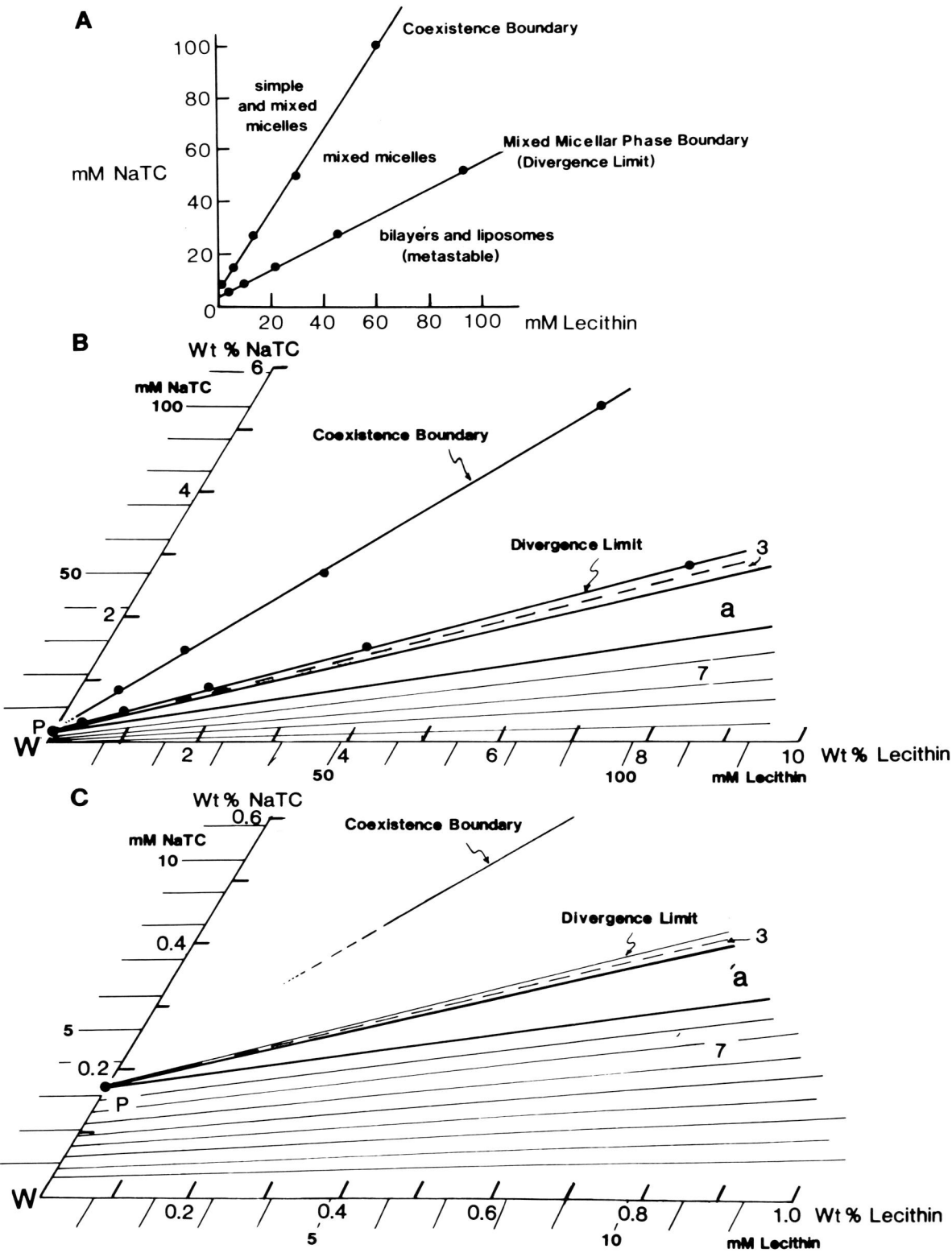

FIG. 12. Lecithin–sodium taurocholate (NaTC) system at high H$_2$O (W) content. *A*: divergence limit (see Fig. 11*A*) and coexistence boundary plotted on rectangular coordinates. *B–C*: system plotted on triangular coordinates (note difference in scale). *Dashed line*, phase boundary between micellar phase and *zone 3*. Phase boundary is virtually same as divergence limit line. *Point P* (critical micellar concentration of NaTC in H$_2$O) has a value of 3.2 mM; 2-phase *regions 3* and *7* are shown. Tie lines in *region 7* and the 3-phase *region a* are also shown. [*A* from Mazer et al. (70); *B–C*, coexistence and divergence boundaries from Mazer et al. (70).]

human bile, it is not purely an academic argument as to whether there are two species of micelles, pure bile salts and small disks (70), or one micelle consisting of a variable number of EYLs and bile salts similar to that of the starting mixture (78a).

Although there is little doubt that bile salts cover the perimeter of simple and mixed discoidal micelles, a question arises as to the way in which bile salts are incorporated into the bilayer part of the disk. Our initial X-ray–diffraction studies on zone I indicated that the calculated lipid bilayer thickness containing 10% NaC was ~40 Å at maximum H_2O content, whereas the pure EYL was only ~34 Å (119). Similar increases have been noted with cholesterol in EYL bilayers (11, 64). Because cholesterol packs parallel to the lecithin acyl chains in bilayers, it was proposed that bile salts did likewise (111). With the knowledge that methyl esters of CA can form hydrogen-bonded aggregates in organic solvents (7), Small (111, 113, 122) suggested that the bile acids dimerized in the bilayer with the axis of the ring systems parallel to the lecithin hydrocarbon chains. This picture appears to have been accepted in the literature. For NaC or other cholates the proof of this conformation in the bilayer is still lacking. Recently, using specifically deuterated dihydroxy (NaDC, NaCDC) bile acids, at a mole ratio 1:4 in multibilayers of EYL, ^2H NMR indicated that the DB ring axis (see Fig. 1) of the bile salt was parallel to the phospholipid chains, that is, perpendicular to the plane of the bilayer. This suggests that the dimerized-interdigitated conformation may be correct, at least for dihydroxy bile salts at the 1:4 ratio (100). On the other hand it is possible that bile salt monomers (especially trihydroxy bile salts or bile salts at very low bile salt–lecithin ratios) sit in the upper part of the bilayer surface and are not self-associated by hydrogen bonds.

With the use of saturated lecithins such as dipalmitoyl phosphatidylcholine (DPPC), which has a gel-to-lamellar liquid crystal transition that can be monitored by differential scanning calorimetry (DSC) or spectroscopy (117), mixed micelles of varying molar ratios were studied as 10% solutions (28). The solutions were made up at temperatures above the DPPC chain transition (42°C) and then cooling DSC was performed. At low bile salt/EYL ratios (1:1.5), a chain-freezing peak at ~34°C was present. This was accompanied by a marked change in the specific heat above and below this transition. The chain-freezing peak suggests that a relatively large cooperative unit (>100 molecules) can undergo the transition within the micelle. This raises questions as to the position of bile salts within the DPPC bilayer of the micelle. If bile salts were present in the disk at one for every two lecithins, as suggested by EYL systems (70), then they would disrupt the cooperative units in DPPC and prevent chain freezing. Perhaps bile salt is excluded from the bilayer of DPPC mixed micelles. A high-sensitivity DSC study was carried out on much more dilute solutions of DPPC and NaTDC or NaTC (124). In these systems the micellar solutions were first cooled and then heated in the calorimeter. Those rich in DPPC clearly showed a sharp chain-melting transition. Unfortunately the behavior was extremely complex and depended on the molar ratio, the bile salt present, and the time stored below the transition. This last fact indicates that these studies were carried out under nonequilibrium conditions, and thus the results cannot be interpreted in terms of the micellar (equilibrium) system. When DPPC–bile salt micelles are stored below the transition temperature of DPPC, phase transitions occur that probably partially dissociate bile salt from DPPC and allow it to crystallize into one or more of its solid phases (117).

A number of NMR experiments have been carried out (25, 66, 122, 126, 127). All of these studies are consistent with micellar-sized aggregates and are also consistent with a bilayered simple or mixed discoidal structure of a micelle. However, none of the studies have been particularly helpful in determining the finer structure of the micelles.

"MICELLE-TO-VESICLE TRANSITION." The change in composition from a micellar phase (zone IV) to a region where vesicles or lamellar liquid crystals and an aqueous phase are in equilibrium (Figs. 8 and 9, region 7) is called the micelle-vesicle transition. The phase diagram, however, indicates that in passing from zone IV to region 7 one must pass through a very narrow two-phase region (region 3) in which the aqueous phase is in equilibrium with the hexagonal phase, then through a region of invariant three-phase composition (region a) in which an aqueous phase at composition P, a hexagonal phase at composition H, and a lamellar phase at composition C are in equilibrium (Figs. 8 and 9) before reaching region 7. Diluting a micellar system follows a line toward the W apex and must cross these boundaries as this system becomes more dilute (see Fig. 10, line 2). A careful quasi-elastic light study, using both dilution and dialysis of 5% micellar solutions, has considerably extended our knowledge in this region (108). We discuss the NaGC-EYL system, although NaGDC was also studied. The R_H and the polydispersity index V are plotted as a function of dilution of an initial 5% solution (Fig. 13); V is an estimate of how diverse the particle sizes are. The micellar-phase limit is indicated by the dotted line. As micelles are diluted (left of the dotted line) the micelles grow in size, as noted earlier (see Fig. 11B). They suddenly become very large at the micellar-phase limit. At very high dilutions vesicles are formed. The greatest polydispersity occurs just at the micellar-phase boundary. This corresponds to the regions where the two phases (see Figs. 9 and 10, region 3) or three phases (see Figs. 9 and 10, region a) would be present. Thus the marked polydispersity is probably related to the precipitation of some of the hexagonal phase.

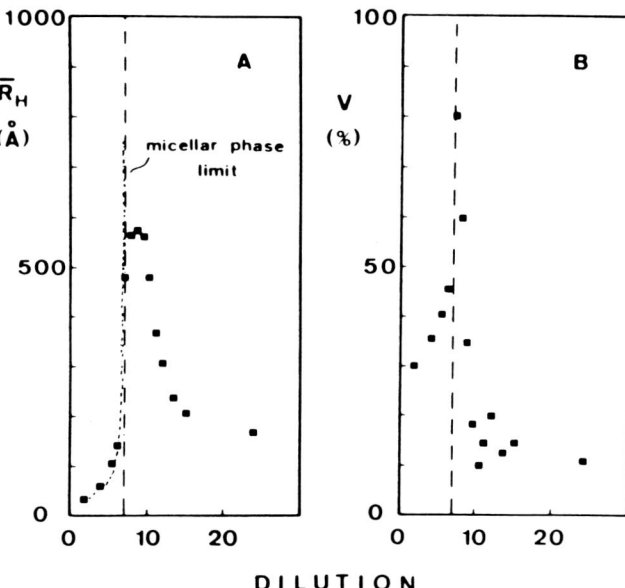

FIG. 13. A: mean hydrodynamic radius $\bar{R}_H$. B: polydispersity index V of a mixed micellar stock solution sodium glycocholate-lecithin [lecithin/sodium glycocholate = 0.85; total concentration (C_{tot}) = 5 g/dl; temperature = 20°C] as a function of dilution. Dilution factor of 1 corresponds to stock solution C_{tot}, and dilution factor of n corresponds to C_{tot} divided by n. [From Schurtenberger et al. (108). Reprinted with permission from *Journal of Physical Chemistry*, copyright 1985, American Chemical Society.]

After dilution the light scattering changes continuously for ~24 h and then steadies. Thus it takes ~24 h to move toward equilibrium after dilution. Nuclear magnetic resonance studies on a very similar system (125) also indicate that equilibrium is reached only slowly. After dilution into region 7, two different phospholipid environments were identified by NMR. One could be shifted with a lanthanide reagent (Pr^{+3}); the other could not. From the R_H of the particles and the presence of a protected environment, it was concluded that the particles were unilamellar vesicles. When the vesicles were further diluted with buffer, a very small decrease in size was noted. This decrease in size can probably be attributed to bile salts leaving the vesicle. Therefore the marked changes in size caused by the first dilution occur as they pass through the transition zones (Figs. 9 and 10, regions 3 and a). If adequate bile salt concentrations are added back to the vesicles they are reversed to micelles.

The mechanism of disk assembly into vesicles has been approached thermodynamically (38). Specifically the formation of a vesicle from a disk depends directly on the radius of the disk (R_d) and the surface tension of the edge γ_m and inversely on 4 times the elastic bending modulus K_c. Fromhertz (38) has defined a vesiculation factor, $V_E = R_d \gamma_m (\frac{1}{4} K_c)$. Thus, if the disk bilayer is rather rigid (i.e., K_c is high) or if the edge tension and diameter are small, then the disk will be stable. However, as the radius increases or as the surface tension increases or the stiffness decreases, disk closure into a vesicle becomes possible. A naked lecithin disk would have a high edge tension, like alkane:H_2O (~52 dyn/cm) (117), and some energy would be required to bend the planar disk into a vesicle. When pure EYL is sonicated, disks are probably produced first but are converted toward vesicle shape by cavitation energy and therefore close rapidly because of high edge surface tension. If the surface tension of the edge is lowered by adding a small amount of an edge-active agent such as a bile salt, then the γ_m is decreased and the process should slow appreciably (39). Mixtures of 0.8 mM NaTDC and 2.5 mM lecithin that fall in region 7 (see Fig. 12) were dried from solvent and hydrated. The lamellar liquid crystals that formed on hydration were sonicated 20–40 min to produce disks. At varying times after sonication the mixtures were negatively stained for electron microscopy. At 30 min after sonication the overall picture was one of stacks of large disks, whereas 3 h after sonication vesicles predominated. If each disk closed to form a vesicle, then the mean EYL content of starting disks should be the same as the resulting vesicles. The actual calculated mass of EYL in the discoidal particles was somewhat less than that of the vesicles, indicating that not only does closure take place but fusion of discoidal particles may also occur during the process.

BILAYERED AQUEOUS TWO-PHASE REGION (REGION 7). Region 7 contains bilayers (vesicles or liposomes) and an aqueous phase. The true equilibrium structure is probably a hydrated planar bilayer liquid crystalline phase in equilibrium with an aqueous phase. Bile salts distribute into both phases. An important experiment with equilibrium dialysis indicated the composition of these phases and the distribution of bile salts between them (108). Aqueous EYL-NaGC micellar solutions (1 ml) previously equilibrated were dialyzed at 20°C against different volumes of buffer (1–5 ml). Three hours were required to reach equilibrium (108). The concentrations of NaGC and EYL in the dialysis bag and in the dialysate were determined. The sizes of the beginning mixture were measured by light scattering. Because all of the lecithin remained in the dialysis bag and bile salts could move freely back and forth, the intermicellar concentration (IMC) was given by the dialysate. For the micellar system, the IMC decreased from ~6 mM to ~4.5 mM as the micelles became lecithin rich, enlarged, and approached the phase boundary. In the vesicle region (region 7) the concentration of bile salts in equilibrium with vesicles was less than the IMC and decreased with decreasing amounts of bile salts present. These compositions are plotted on the phase diagram in Figure 14. At constant EYL concentration, as NaGC concentration decreases in the dialysis bag, the IMC decreases. If the volume of the aqueous phase in the dialysis bag is known, one can calculate the amount of bile salt bound to the

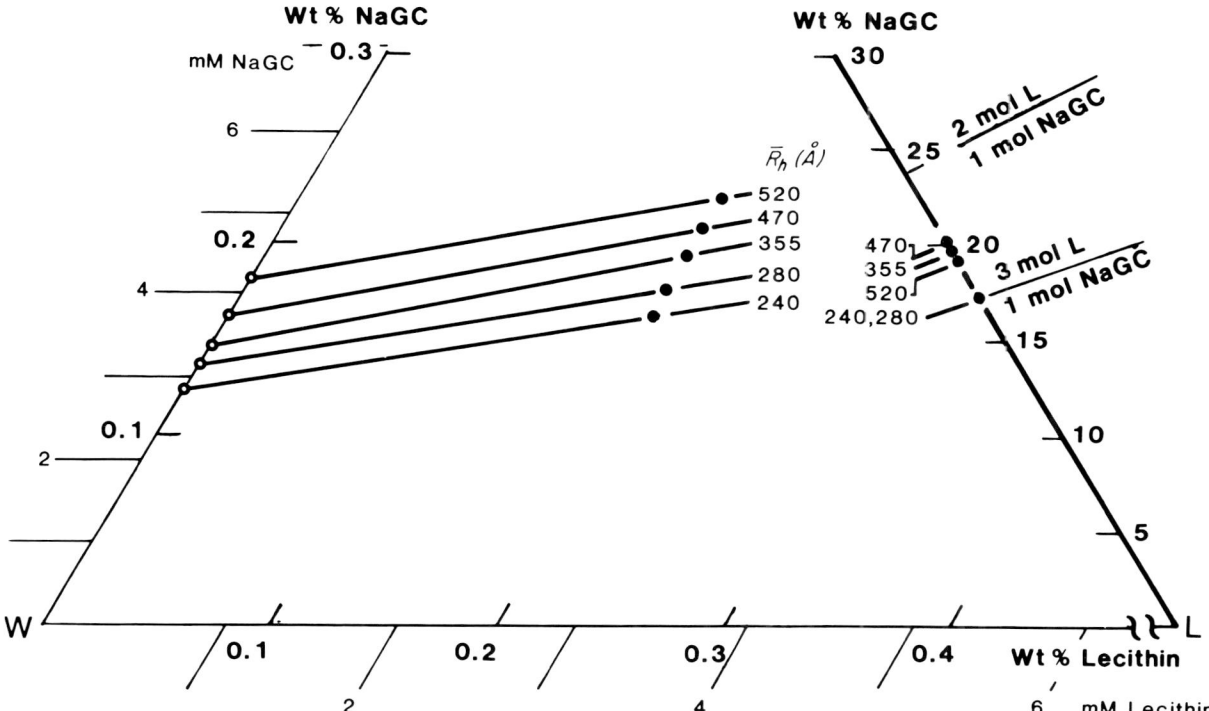

FIG. 14. Lecithin–sodium glycocholate–water system in dilute region of *zone 7*. L, lecithin; NaGC, sodium glycocholate; W, water. *Closed circles*, composition of fluid within dialysis bag; *open circles*, corresponding compositions (connected by a *line*) of dialysate fluid. *Lines* connect aqueous phase (intermicellar concentration or dialysate) with mixture composition (*closed circles*) for a series of vesicles of different sizes [$\bar{R}_h$ (Å)]. Extensions of these lines intersect L-NaGC border (as shown by *closed circles* on *right*) and give ratio of NaGC bound to lecithin vesicle. R_h, mean hydrodynamic radii.

vesicle. This is plotted as an EYL/NaGC ratio along the EYL-NaGC border (Fig. 14, *right*). In general, as NaGC IMC decreases, less is bound to the vesicle bilayer. Within experimental limits the ratio of bile salt bound to bilayer per gram of bilayer to bile salt distributed in the H$_2$O per gram of H$_2$O was constant at ~100. This is actually a distribution coefficient K_{L/H_2O} and indicates that there are 100 molecules of bile salt per gram of phospholipid to 1 molecule per gram of H$_2$O over a fair range of region 7. These lines between IMC and dialysis bag composition (two-phase mixture) are analogous to tie lines for region 7. The actual compositions of the lamellar phase could be determined from their intersection with boundary BC (see Figs. 8 and 9). From the position in the two-phase region and K_{L/H_2O}, the mass of bile salt associated with the vesicle and with the aqueous fraction can be calculated.

EFFECTS OF BILE SALTS ON PHOSPHOLIPID VESICLES. Several studies have indicated that bile salts bound to phospholipid vesicles enhance the movement of phospholipids. Less than 1 mol bile salt per 100 mol phospholipid enhances the off-rate of phospholipid from the vesicle in the presence of phosphatidylcholine-exchange protein from the liver (58). Thus bile salts appear to loosen the outer surface. The chemical shift difference between inner and outer phospholipid carbonyls merges when the ratio of bile salt to phospholipid in the vesicle approaches 7%. This may indicate that the phospholipid flip-flop rate becomes rapid on an NMR time scale. Using Pr^{+3} as a shift reagent, Hunt (53) and Hunt and Jawaharlal (54) followed the shift of the inner choline peak as a function of time to measure the flux of Pr^{+3} across the bilayer of EYL vesicles with added NaTC. They found that the rate constants for movement of Pr^{+3} were a linear function of the log of the bile salt concentration and that the slopes of the curves were all 4, suggesting that four bile salts bind one lanthanide ion and shuttle it as a complex across the bilayer. The rate of movement is enhanced by 10 mol% cholesterol, i.e., 100 × [mol cholesterol/(mol cholesterol + mol EYL + mol bile salt)], whereas 40 mol% blocked the shuttling of Pr^{+3}. A critical concentration of bile salts in the bilayer appeared to be necessary for any movement, and this was estimated to be ~50 molecules of bile salt per vesicle (~1 bile salt/100 EYL). When bile salts were cosonicated with phospholipids to produce vesicles rather than simply added to the outside of vesicles, the rate was only about half as great at a given bile salt concentration. This appeared to indicate that bile salts on the outside are critical for mediating the transport of lanthanides from outside to inside (25).

KINETICS OF MULTILAMELLAR PHOSPHATIDYLCHOLINE LIPOSOMES TO MIXED MICELLE FORMATION. Using a fixed concentration of multilamellar liposomes (1.5 mM), Rajagopalan and Lindenbaum (91) added equal volumes of different concentrations of bile salts (~1–40 mM), and followed turbidity (see Fig. 10, line 5, for path). The turbidity fell with time, and an estimated rate constant (K) and time for half the change in turbidity ($t_{1/2}$) to occur could be calculated (Fig. 15). Sodium chenodeoxycholate and NaDC cause far more rapid dissolution of liposomes into micelles than NaC; NaUDC is almost inactive.

KINETICS OF TRANSBILAYER MOVEMENT OF BILE ACIDS. The passive reabsorption of bile acids (especially unconjugated bile acids) in the small intestine and colon and the movement of bile acids from plasma into hepatocytes probably necessitate transbilayer movement (or flip-flop) of the bile acids in the phospholipid bilayer. Estimated rates have recently been determined by Cabral et al. (15, 16). Small amounts of [^{13}C]carboxyl-enriched bile acids were incorporated in small EYL vesicles (~1:22 mol/mol, no added salt). The rate of flip-flop was found by obtaining ^{13}C NMR spectra of the vesicle system at various temperatures. When the rate is slow (<70/s) separate resonances are observed for bile acid on the outer and inner monolayers of the vesicle. As the rate of flip-flop increases, these two peaks merge into a single, relatively broad peak and sharpen to a narrow peak at fast exchange rates (more than ~100/s). The first-order rate constant of flip-flop can be estimated from the change in linewidth of the peak(s) and peak separation. Rates were found for unconjugated CA, DCA, and CDCA. In the ionized state (pH 10.0) the half-life in the monolayer at 37°C was >24 h for all three bile salts. In the fully protonated form (pH 3.6), the flip rate was much higher and was different for each bile acid: CA flip rate at 37°C ~5/s ($t_{1/2}$ 139 ms), CDCA flip rate ~100/s ($t_{1/2}$ 6.9 ms), and DCA flip rate ~150/s ($t_{1/2}$ 4.6 ms). The rate of transbilayer movement at 37°C is therefore dependent on the ionization state (protonated ≫ ionized) and on the hydrophobicity of the bile acids, with the more hydrophobic bile acids flipping faster (DCA > CDCA > CA). The presence of cholesterol in the bilayer slows the flip rate of all bile acids (16a). To test the hypothesis that hydrogen-bonded dimers or tetramers of bile acids flip together, Cabral et al. (16a) studied vesicles with both CA and DCA or CA and methyl cholate. Flip rates were independent of the presence of another species, and it was concluded that flipping does not require dimers to be present in these bilayers.

Thus, in summary, the structures in zone IV are complex. Micelles in concentrated solutions near the phase boundary of the hexagonal zone are probably relatively small discoidal micelles consisting of simple EYL disks covered with a perimeter of bile salts. In more dilute systems, 10% or less in a region rich in EYL (that is, a region that corresponds to the molar ratios encompassed by the hexagonal phase zone, i.e., Fig. 10, zone IVa), the larger mixed-disk model appears to prevail. At a given molar bile salt/EYL ratio, decreasing the H$_2$O concentration leads to smaller micelles, whereas at a constant H$_2$O concentration, increasing the NaC leads to smaller micelles. In the zone having high–bile salt concentration, smaller, possibly more polydisperse micelles are present (Fig. 10, zone IVb). Whether these are of a single type (i.e., mixed EYL–bile salt micelles) or whether two separate micellar species coexist (pure bile salt and small disks) is not resolved. The micelle-vesicle phase boundary is really more complex; it involves crossing a two-phase zone (Fig. 10, zone 3), a three-phase zone (Fig. 10, zone a), and going back into a two-phase zone (Fig. 10, zone 7) in which the stable system is a lamellar

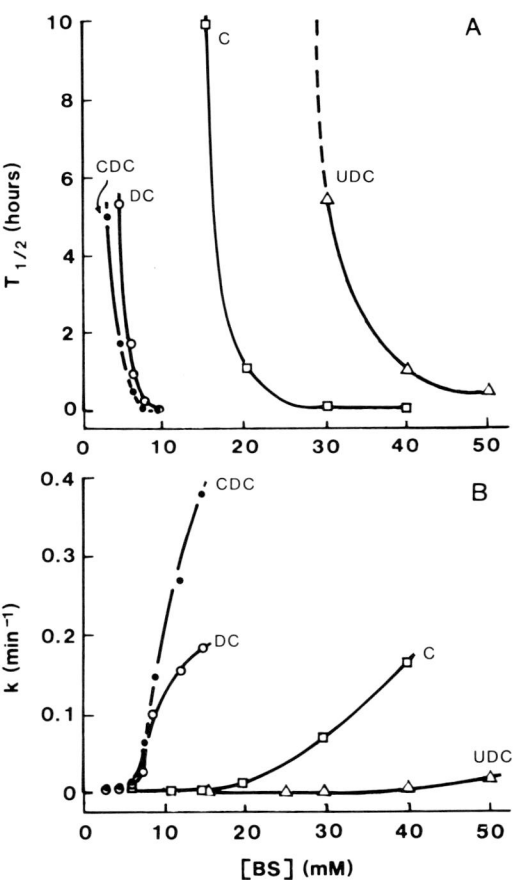

FIG. 15. Rates of dissolution of multilamellar liposomes by bile salts (BS) of different concentration. Lecithin concentrations of 1.5 mM were diluted with an equal volume of bile salts at concentrations on *horizontal axis*. *A*: time for half dissolution (T$_{1/2}$) is plotted as a function of bile salt concentration; *B*: complex rate constant k_1 (91) for the process is plotted. Order of dissolution rates, chenodeoxycholate (CDC) > deoxycholate (DC) > cholate (C) > ursodeoxycholate (UDC).

liquid crystalline phase containing some bile salt in equilibrium with a low concentration of monomeric bile salt. The apparent distribution coefficient for this is ~1/100, H_2O/phospholipid. The presence of bile salts in vesicles perturbs the structure of the vesicle, allowing phospholipids to flip-flop more rapidly and permeability changes to various ions to increase appreciably. The flip-flop of bile salts within vesicles is a function of the bile salt type (dihydroxys flip more rapidly than trihydroxys), temperature (flip rate increases as temperature increases), and ionization state (protonated bile acids flip much faster than ionized salts). Finally, bile acids appear to flip as monomers, not as hydrogen-bonded dimers.

CHOLESTEROL

Cholesterol is a relatively insoluble amphiphile (117). Its solubility is $\sim 20-30 \times 10^{-9}$ M in H_2O, but in gallbladder bile $20-30 \times 10^{-3}$ M cholesterol can be maintained in solution. To what is this 10^6 increase in solubility due? This section deals with the interaction of cholesterol with bile salts and lecithin and explains how these substances solubilize cholesterol in mixed micelles, thus immensely increasing its apparent aqueous solubility.

Solubility in Aqueous Media

Saad and Higuchi (99) found the maximum solubility of cholesterol to be 26 nM. Haberland and Reynolds (43) listed a maximum solubility of 4.7 μM. They found that cholesterol self-associated above 25–40 nM to "micellar" structures of variable size. The lower value was confirmed by Renshaw et al. (94), but cholesterol monohydrate microcrystals were detected at all higher concentrations when supersaturated solutions were filtered through 1.5-nm filters that prevented submicroscopic cholesterol monohydrate crystals from passing through. No evidence for a micellar cholesterol phase was seen. The lower value, 26 nM, is the more accurate and now generally accepted solubility limit. The "micellar aggregation" found earlier (43) was most probably microcrystals of cholesterol.

Cholesterol–Bile Salt Interactions

Bile salt micelles will incorporate small amounts of cholesterol (120). The amount solubilized depends on the hydrophobic-hydrophilic balance of the bile salt [hydrophobic bile salts > hydrophilic bile salts (21)] and on the total concentration. There is an inverse relationship between cholesterol solubilization and order of elution on reverse-phase HPLC columns (3). For a given concentration of bile salt, the amount of cholesterol solubilized decreased in the order NaDC > NaCDC > NaC ≫ NaUDC and unconjugated > glycine conjugated > taurine conjugated (37°C, 0.15 M Na^+, pH 7.0 for taurine conjugates, pH 10.0 for glycine conjugates and unconjugated bile salts). Expressed as the number of moles of bile salt needed to solubilize 1 mol of cholesterol, representative values are 324 for NaUDC, 30 for NaC, 17 for NaCDC, and 14 for NaDC. Because the association at the hydrophobic interior of the micelles should be similar for the various bile salts (especially the all α-hydroxyl conformation), the differences in cholesterol solubilization were thought to be consistent with some cholesterol binding at the hydrophilic side of the micelles (3).

Self-diffusion measurements indicate a change in micelle size when cholesterol is added to NaTC solution [0.15 M NaCl, 37°C (135)]. The aggregation number increases from 5 for simple NaTC micelles to 26 for NaTC-cholesterol micelles containing 25 molecules of NaTC and 1 molecule of cholesterol. The precise location of cholesterol in the micelle is not known. At concentrations >6.7 mM NaTC, the measured NaTC/cholesterol ratio was 50:1. This will occur if the mixed micelles coexist with simple NaTC micelles and monomers, such that five simple NaTC micelles are present for each NaTC-cholesterol micelle.

Cholesterol-Phospholipid Interactions

Surface-chemistry studies of mixed monolayers of EYL and cholesterol suggested that the orientation at the interface was such that the polar portions of the molecules were in the aqueous media (choline headgroup and glycerol backbone of EYL, and —OH group of cholesterol) with the acyl chains and steroid ring parallel and extended in the air. The phase behavior of the EYL-cholesterol-H_2O system as determined by polarizing microscopy and by X-ray diffraction was reported (11). Regions of one, two, or three phases were found, depending on the concentration of the constituents. A lamellar liquid crystalline phase is present up to mole ratios of 1:1 cholesterol/EYL and up to 35%–45% H_2O. The lamellar phase consists of bilayers of EYL and cholesterol separated by layers of H_2O. At higher H_2O levels, excess H_2O is in equilibrium with the lamellar phase. When the cholesterol/EYL ratio exceeds 1:1, crystals of cholesterol are present with one or both of the other phases. The thickness of the EYL bilayer increases by ~6 Å with the addition of cholesterol. The increase is not enough to accommodate all the cholesterol molecules lying perpendicular to the acyl chains between monolayers (11, 64). These data, plus the calculated areas per molecule, indicate that the stable form is a lamellar structure with cholesterol molecules interdigitating between the EYL fatty acyl chains, with the hydroxyl group facing the aqueous surface. This is analogous to the surface chemistry behavior of cholesterol-EYL monolayers. Plank et al. (89) determined the effect of added cholesterol on the surface charge density of mixtures of multilamellar liposomes of EYL and egg yolk phosphatidylglycerol (PG) by electrophoretic light scattering. Electrophoretic mobilities were consistent with those of the pure phospholipid multilayers up to 50 mol% cholesterol (<3% change in spacing of charged groups at the surface). This suggests that cholesterol molecules associate with the acyl chains of EYL and PG with little displacement of the phospholipid headgroup. The bilayer fluidity is decreased by the addition of cholesterol. Transbilayer movement of NaTC-Pr^{3+} in phosphatidylcholine (PC) vesicles (summarized in EFFECTS OF BILE SALTS ON PHOSPHOLIPID VESICLES, p. 647) is decreased with the addition of 40 mol% cholesterol (54).

The exchange or transfer of phospholipid and cholesterol between sonicated bilayer vesicles has been found to occur by diffusion of monomers through the aqueous phase (5, 13, 74). The rate was not dependent on acceptor concentration, and exchange was measurable when the donor and acceptor vesicles were separated by barriers that would allow monomeric cholesterol to pass through but not small vesicles. The rate of exchange was first order, with half times of 2.3 h for cholesterol and 48 h for phospholipids. The rate-limiting step was proposed to be dissociation of molecules from the outer monolayer of the vesicle into the aqueous media (74).

Quaternary System: Bile Salt–Phospholipid–Cholesterol–Water

To gain insight into the solubility of cholesterol in gallbladder bile, the simpler bile salt–phospholipid–cholesterol–H_2O systems at neutral pH were studied in detail (1, 12, 23, 118, 120). The results were expressed in a series of phase diagrams. The phase diagram of a four-component system can be represented as a regular tetrahedron, with each component at the corners at 100% concentration. By holding one component at a constant concentration and varying the other three, the system can be represented by a triangular phase diagram, as discussed for bile salt–phospholipid–H_2O system.

A series of phase diagrams was constructed, keeping the cholesterol content fixed at concentrations of 0%–25% for NaC-EYL-H_2O-cholesterol (12, 120). The 0% cholesterol case equals the three-component system NaC-EYL-H_2O of Figure 8. The addition of cholesterol does not change the basic structure of any of the phases present. Cholesterol does increase the X-ray long spacings and the thickness of the lipid layers and diameters of the cylinders. As the cholesterol concentration increases, the size of the one-phase regions (micellar, hexagonal, cubic, and lamellar) decreases. The maximum amount of cholesterol that can be incorporated into each phase is found from the concentration at which each one-phase region disappears (12, 120). The micellar phase can incorporate up to 5 wt% cholesterol. The maximum solubility for the cubic and hexagonal phases is <7% cholesterol. The lamellar liquid crystalline phase can incorporate much more cholesterol and remains a small one-phase region at 25% cholesterol.

Admirand and Small (1) presented the diagram as the three-lipid component system: bile salt, lecithin, and cholesterol at a constant 90% H_2O. They defined an isotropic micellar region that set the limits for cholesterol solubility. Hegardt and Dam (47), Holzbach et al. (52), and Carey and Small (23) showed that the original isotropic region (1) was somewhat too

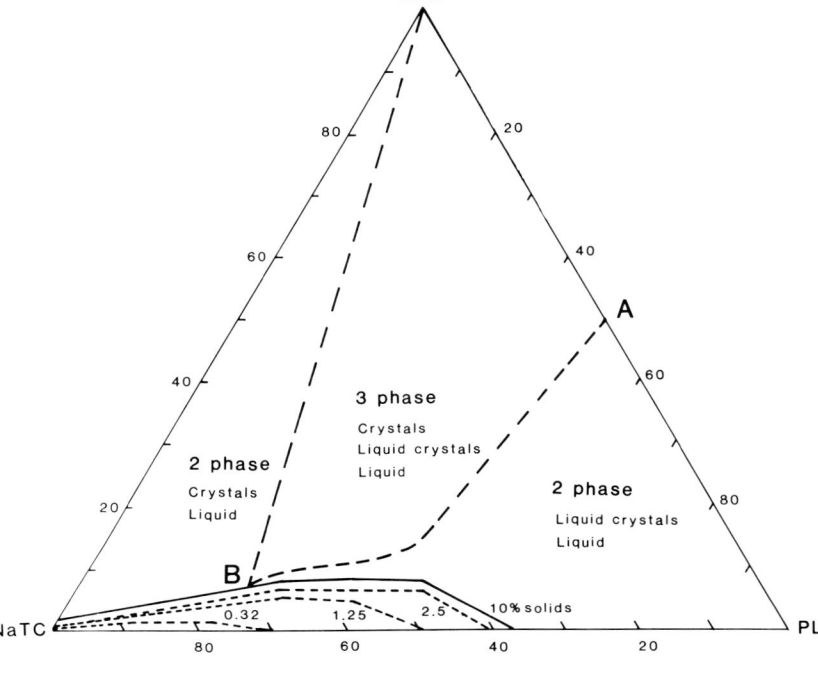

FIG. 16. Sodium taurocholate–lecithin–cholesterol phase diagram. Chol, cholesterol; NaTC, sodium taurocholate; PL, phospholipid. At 10% total lipid, micellar phase is bounded by *solid line*. *Dotted lines*, micellar boundaries with decreasing concentrations of total lipids. Liquid crystal phase is a phospholipid phase with varying amounts of cholesterol and small amounts of bile salts. Crystalline phase is cholesterol monohydrate. We suggest that a 3-phase region cannot directly abut on a 1-phase region, except at a point where the 2-phase region on the *right* extends to *point B*. The 3-phase region consists of *1*) cholesterol crystals, *2*) liquid crystals of phospholipid, cholesterol, and bile salt, and *3*) a micellar solution of composition B. [Adapted from Carey et al. (24).]

large. Realizing that gallbladder bile is not always 90% H_2O (10% lipids) and particularly that hepatic bile is considerably more dilute, Carey and Small (23) described the equilibrium solubility of cholesterol at H_2O concentrations from 80% to 95.7%. The general aspects of this phase diagram using NaTC are given in Figure 16. The micellar zone at the *bottom* consists of mixed micelles of bile salt, phospholipid, and cholesterol in equilibrium with a monomeric concentration of bile salts. As previously predicted (112a), less cholesterol is solubilized as the total lipid concentration decreases (23). *Above* and to the *right* of the micellar zone there are regions where other phases separate. To the *right*, toward the phospholipid side of the diagram, the liquid crystals (composed of phospholipid and cholesterol with small amounts of bile salts) are in equilibrium with the micellar solution. To the *left*, cholesterol crystals separate from the micellar solution. In the *middle* there is a three-phase region in which both cholesterol crystals and liquid crystals are in apparent equilibrium with a micellar solution at composition about B. If the solution is more than saturated with cholesterol and has a bile salt–lecithin ratio of ≥7:3, it will precipitate cholesterol monohydrate crystals when nucleated. On the other hand, if a solution is slightly richer in lecithin, that is, the ratio is <7:3, it will first precipitate a liquid crystalline phase and only at higher degrees of supersaturation (i.e., above line BA, whose absolute position is not clearly known and depends on the total concentration of lipids) will cholesterol crystals precipitate. Of substantial importance is the nature of the liquid crystalline phase precipitated. This phase, made up of bilayers of lecithin, cholesterol, and a small amount of bile salt, may be present as large multilamellar liposomes or as small unilamellar vesicles ~500 Å diam. Strictly speaking the unilamellar vesicle is a form of the liquid crystalline phase reduced to its smallest particulate dimensions. The vesicles are a suspension of a liquid crystalline phase dispersed in an aqueous phase. From a practical point of view they are so small that they act as "solubilizing" agents. Thus, from the standpoint of cholesterol crystal precipitation, the solubilization of cholesterol may be enhanced by the presence of a liquid crystalline phase so that it approaches the horizontal part of the dashed line BA. In the very dilute systems such as hepatic bile, the zone in which vesicles can occur is much larger and yet bile may appear grossly free of precipitate. Strictly speaking the vesicles are insoluble, but from a practical standpoint they act as solubilizers of cholesterol and prevent its crystallization. These comments relate to the finding of vesicles in bile that can carry suspended free cholesterol without it precipitating into cholesterol monohydrate crystals (123, 123a).

The bile salt composition of human bile consists mainly of CA and CDCA conjugates of glycine and taurine with smaller amounts of DCA conjugates and ~1% of LCA conjugates (Table 9). The glycine-taurine

TABLE 9. *Composition of Normal Human Bile*

	Hepatic Bile	Gallbladder Bile	Ref.
Specific gravity	1.009–1.013	1.026–1.032	26a
pH	7.1–8.5	5.5–7.7	
Total solids, %	1–3.5	4–17	
Total base, meq/l*	150–180		
Chloride, meq/l	75–110	15–30	
Lipids, % BA + L + Chol			93
Bile acids†	71.3	77.5 ± 4.7	
Phospholipids	21.1	15.6 ± 4.8	
Cholesterol	7.6	6.9 ± 2.0	
Lipids, μmol/ml			102
Bile acids		148.2 ± 25.1	
Phospholipids		38.3 ± 6.5	
Cholesterol		13.2 ± 3.1	
Proteins, μg/ml			
Total protein		97.0 ± 12.9	102
Albumin	155–1,485 (405)		32
Transferrin	11.4–160 (36.3)		32
α_2-Macroglobin	2.7–1C0 (13.5)		32
Immunoglobulin G	32–480 (88.8)		32
Immunoglubulin M	2.2–60 (19.6)		32
Apoprotein AI	2.9 ± 0.5	19.1 ± 2.2	102
Apoprotein AII	1.5 ± 0.4	10.4 ± 1.1	102
Apoprotein CI	12.4 ± 5.5	7.8 ± 1.4	102
Apoprotein CII	3.4 ± 1.1	3.9 ± 0.8	102
Apoprotein B	10.6 ± 5.2	38.5 ± 4.7	102
Elements, mM			46
Ca		7.38 ± 2.92	
Cu ($\times 10^2$)		9.58 ± 5.43	
Fe ($\times 10^2$)		1.59 ± 1.32	
K		12.68 ± 3.49	
Mg		6.91 ± 2.23	
Mn ($\times 10^2$)		1.18 ± 2.13	
Mo		2.09 ± 1.04	
Na		210.14 ± 12.13	
P		54.18 ± 15.27	
Sr		0.10 ± 0.10	
Zn ($\times 10^2$)		1.77 ± 0.63	

Values are means ± SE or range (mean). BA, bile acid; PL, phospholipid; Chol, cholesterol. * Total base in gallbladder bile must be similar to that in hepatic bile. A variable component of total base is bicarbonate, which can be as high as 60 meq/l in hepatic bile but is usually quite low (1–5 meq/l) in gallbladder bile. † The major bile acids in human bile are the primary bile acids, cholic acid and chenodeoxycholic acid, which account for ~40% each of the bile acids. The secondary bile acids, deoxycholic acid (~20%) and lithocholic acid (~1%), account for the rest. Bile acids are conjugated mainly with glycine (~60%) and with taurine (~40%).

ratio is ~6:4, although this varies with diet. Furthermore, in the treatment of gallstones, UDCA has been used and can become an appreciable component of the bile when used in gallstone therapy. Thus Salvioli et al. (101) studied the phase behavior of three taurine-conjugated bile salts [tauroursodeoxycholate (TUDC), taurocholate (TC), taurochenodeoxycholate (TCDC)] at a fixed H_2O concentration (10 g total lipid/dl, in 0.2 M NaCl). Each was studied separately, as were 1:1 TCDC-TUDC and glycochenodeoxycholate-glycoursodeoxycholate (GCDC-GUDC) mixtures (pH 7.4 for the taurine conjugates and pH 8–9 for the glycine conjugates). The phase diagrams for the individual bile salt systems are shown in Figure 17 (101). There is a central three-phase region surrounded by two two-

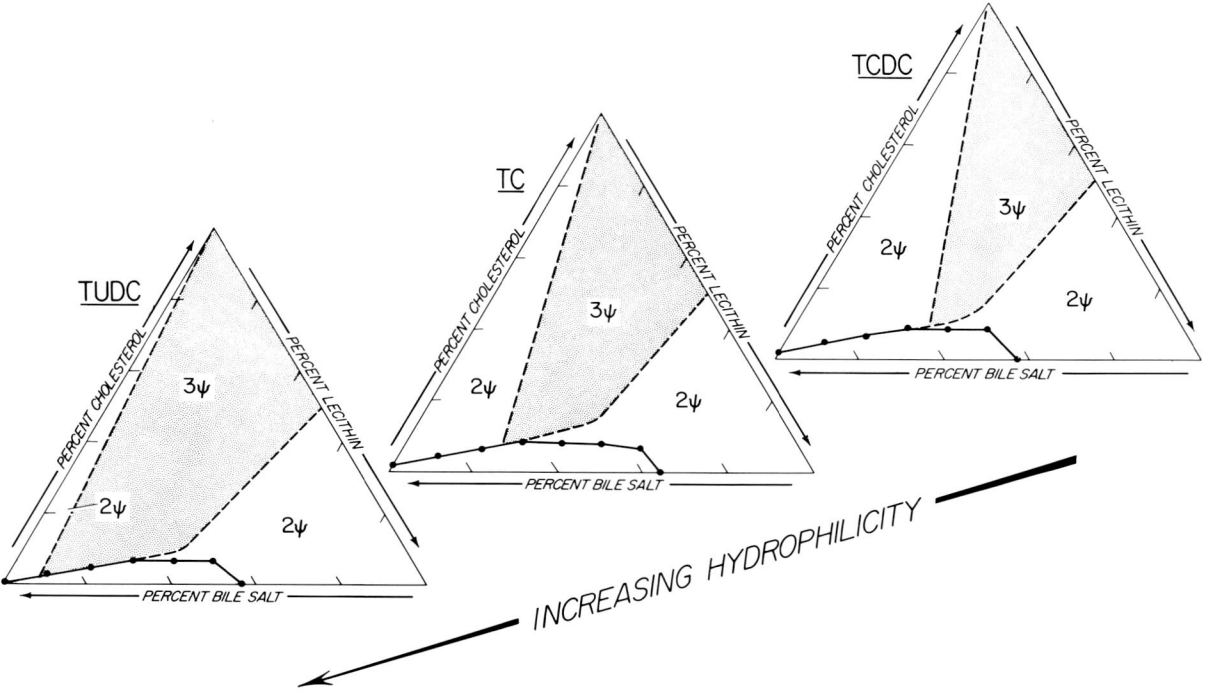

FIG. 17. Phase diagrams of aqueous taurine-conjugated ursodeoxycholic acid (TUDC)–, cholic acid (TC)–, or chenodeoxycholic acid (TCDC)–lecithin–cholesterol monohydrate at a total lipid concentration of 10 g/100 ml (0.20 M Na$^+$, pH 7.4, 37°C). Axes are expressed in mol%. Although phase diagrams show 3-phase region abutting 1-phase region, there must be a 2-phase region, however small, separating them. [From Salvioli et al. (101).]

phase regions and a one-phase region. The number and location of the phase regions are similar for all three bile salts, but the boundaries change. A one-phase micellar region (not labeled) exists at >35%–45% bile salt and up to ~8% cholesterol. The two-phase region at high-lecithin concentration (>50%) is composed of liquid crystals and micelles. The boundaries of this area are the same for the three taurine-conjugated bile salts. The two-phase region at the *left* (low-lecithin content) contains micelles and crystals of cholesterol. The size of this region varies with the hydrophobicity of the bile salt. As the hydrophobicity decreases (see *Hydrophobic-Hydrophilic Balance*, p. 627), the two-phase region increases, TCDC > TC > TUDC, with a loss in the three-phase region area. The three-phase region contains micelles, liquid crystals, and cholesterol crystals. The phase diagrams for the 1:1 TCDC-TUDC mixtures showed that the phase regions are the same as those of the individual bile salts, but the boundaries are different. The boundary between the left two-phase region and the three-phase region is intermediate to those of TCDC and TUDC in Figure 17. In fact, it looks about the same as the NaTC diagram in Figure 16. Studies of the phase behavior of a mixture of naturally occurring bovine (12) or human bile salts (75) and egg or human biliary lecithin showed that cholesterol solubilization was quantitatively nearly the same as in model systems of pure cholates or deoxycholates.

BILE

Normal Bile

The in vitro systems used to model human bile have concentrated on the three main lipid components: bile salts, lecithin, and cholesterol. Only recently have some of the other components in bile (i.e., proteins and Ca^{2+}) been studied for their effects on cholesterol-solubilizing capacity. The major lipids, proteins, and elements found in human hepatic and gallbladder bile are summarized in Table 9. Albumin appears to be the most abundant protein in bile. Other proteins that have been identified but not quantitated in bile include haptoglobin, insulin, epidermal growth factor, cholecystokinin, amylase, and lysosomal hydrolases (63 and refs. cited therein). Some of these proteins are known to bind bile salts and cholesterol (albumin and apoprotein AI) and have been implicated in affecting cholesterol precipitation in supersaturated bile (51). Calcium ions bind to bile salt micelles and may also affect cholesterol precipitation and Ca^{2+} anion salt precipitation (76, 90), as noted in BILE SALT–CALCIUM INTERACTIONS, p. 636.

Stable bile exists when the cholesterol concentration is below saturation. This occurs in the micellar region of the bile salt–lecithin–cholesterol phase diagram (see Fig. 16). It has not been conclusively shown that any of the inorganic or organic minor components listed in Table 9 change maximum cholesterol solubility. This solubility limit depends not only on the phospholipid–bile salt ratio but on the total lipid concentration in bile (24). Table 10 may be used to calculate the saturating level (as mol%) of cholesterol at different total and relative lipid concentrations of bile (19, 24). Cholesterol concentrations for intermediate lipid concentration can be interpolated from the values given; for more precise values the reader is referred to the complete critical tables of cholesterol saturation (19). The values may also be used to calculate the cholesterol saturation index (CSI). Biles having cholesterol in excess of the maximum equilibrium solubility have CSI > 1. At cholesterol concentrations equal to (CSI = 1) or less than saturating cholesterol concentrations (CSI < 1), there will be no spontaneous cholesterol precipitation from solution.

Metastable Bile

Above the unsaturated micellar zone in Figure 18 is a metastable region. The solution is supersaturated with cholesterol, but crystals of cholesterol monohydrate or multilamellar liquid crystals do not form readily (24). Because the upper boundary of this region (~3%–12% cholesterol and more than ~47% bile salt) was determined in in vitro mixtures of the components that might have contained heterogeneous nucleating agents (e.g., dust) the true limit for the metastable region is probably much higher (perhaps up to 13 times higher than cholesterol saturation limit) (114–116). Dilute solutions can contain large excesses of cholesterol without spontaneous crystal formation for very long periods of time. It is not known if such solutions contain supersaturated mixed micelles or vesicles or both, nor are the kinetics of micelle-to-vesicle transformations in dilute bile known, although in model systems it can be rather slow (see *Bile Salt–Phospholipid–Water Systems*, p. 638). Bile samples from individuals without evidence of cholesterol stones often have cholesterol concentrations that fall in or even above the metastable region of the phase diagrams (52, 101a). This has led many investigators to propose the presence of nucleating factors (factors that promote cholesterol crystallization) in the bile of patients with gallstones.

Cholesterol Crystal Formation in Bile

Cholesterol gallstone formation was proposed to occur in stages (116). The initial stage is supersaturation of bile, followed by precipitation of microcrystals of cholesterol monohydrate and finally growth into macroscopic crystals and stones. It is the initial formation of cholesterol crystals from supersaturated bile that is the key rate-limiting step in gallstone development. This process is generally called nucleation (114–116). Liquid crystals or vesicles (50, 71) may be an intermediate, especially if the bile composition is

TABLE 10. *Values for Maximum Cholesterol Solubility (mol% Cholesterol) in Bile*

N	Total Lipid Concentration*							
	0.50	1.00	2.00	3.00	4.00	5.00	10.00	20.00
0.085	1.546	2.182	2.661	2.981	3.189	3.323	3.753	4.338
0.135	2.146	2.781	3.444	3.886	4.169	4.352	4.929	5.629
0.185	2.378	3.496	4.315	4.830	5.173	5.395	6.105	6.907
0.235	2.839	4.497	5.213	5.733	6.102	6.346	7.147	8.028
0.285	3.336	5.113	5.939	6.430	6.792	7.055	7.912	8.850
0.335		5.034	6.263	6.773	7.109	7.400	8.301	9.274
0.385	2.182	4.023	6.047	6.733	7.038	7.367	8.311	9.289
0.425		2.624	5.518	6.537	6.828	7.173	8.136	9.086

To use this table, first estimate the total lipid concentration (g/dl) in the bile sample and find the column closest to this value. Next calculate the N value [mol ratio lecithin/(lecithin + bile salt)] of your sample and find the row with the closest value. Read the maximum cholesterol solubility in mol% cholesterol; i.e., 100 × [mol cholesterol/(mol cholesterol + mol lecithin + mol bile salt)]. For example, for a total lipid concentration of 2.0 g/dl and a lecithin mol ratio of 0.185, the maximum cholesterol solubility is 4.315 mol%. Intermediate values may be extrapolated. To convert total bile salts, phospholipids, and cholesterol from mol/l to g/dl, divide by 50, 77, and 38.7, respectively. * Lecithin + bile salt + cholesterol in g/dl. [Data from Carey (19) and Carey and Small (23).]

richer in lecithin than point B in Figure 16. Clearly, point B depends on the bile salt composition, as shown in Figure 17, and may also depend on the concentration of bile lipids. This, however, needs to be studied in more detail.

Quasi-elastic light-scattering studies (42, 123) showed that freshly secreted human bile, in contrast to dog bile (73), contains some vesicles. Using video-enhanced contrast microscopy, Halpern et al. (44) showed that the vesicles in normal bile were stable for many days, but in cholesterol stone patients, vesicles were noted to first increase in size, then aggregate, and finally nucleate cholesterol crystals.

The difference in nucleation time of native bile samples for normal and gallstone patients was studied by Hogan et al. (49). Crystal-free fractions of bile were incubated at 37°C and examined daily by polarizing microscopy to detect microcrystals. Although 68% of the normal patients had supersaturated biles (mean CSI = 1.42 ± 0.68), their mean nucleation time was 15 days compared with only 3 days for cholesterol gallstone patients (mean CSI = 1.80 ± 1.02). There was a correlation between CSI and nucleation time for normal patients but not for gallstone patients.

Using supersaturated model bile solutions containing cholesterol, NaC, and soybean lecithins (0.15 M NaCl, 10% lipid, pH 8.2, 37°C), Whiting and Watts (133a) found spontaneous crystal formation depended on the region of supersaturation. Thus nucleation time increased from ~2 days to ~8 days with decreasing amounts of cholesterol. There was an increase in nucleation time when H_2O was used with no added salt, but no change when KCl was substituted for NaCl. The nucleation time doubled when the incubation temperature was reduced from 37°C to 22°C. This observation possibly relates to increased kinetic mo-

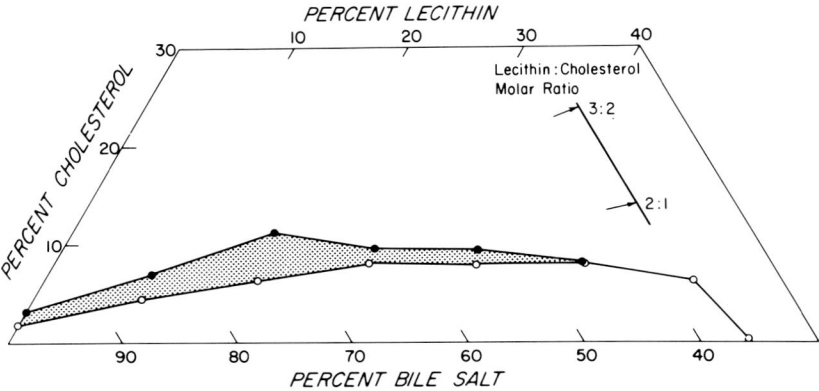

FIG. 18. Partial phase diagram for aqueous system sodium taurocholate–lecithin–cholesterol (20 g/100 ml, 0.15 M NaCl, 24°C). *Shaded area*, metastable region; *open circles*, maximum equilibrium cholesterol solubility; *closed circles*, metastable-labile limit. [From Carey et al. (24).]

tion, since cholesterol solubility is actually less at lower temperatures (24).

Using electron microscopy, quasi-elastic light scattering, and video-enhanced microscopy, Kibe et al. (56) looked at the effects of dilution, CSI, Ca^{2+} concentration, and bile salt–lecithin ratio on nucleation time in model bile. They found formation of large numbers of unilamellar vesicles on dilution of the samples when the bile salt–lecithin ratio was reduced to ≤1.9 (20% total lipid). A few vesicles were also found in more concentrated solutions with high CSI. Nucleation time at 37°C increased with dilution and with lowered bile salt–lecithin ratios and decreased with increasing CSI. In suspensions with vesicles the addition of Ca^{2+} resulted in crystal formation. They proposed that vesicles can serve as a nonmicellar carrier of cholesterol that may account for the metastability of supersaturated bile solutions.

Precipitated Ca^{2+} salts can promote cholesterol crystal formation in bile and are considered to be nucleating factors. Calcium binding to bile salt micelles, especially taurocholate micelles, was determined by Moore et al. (76, 77), who proposed that bile salts act as buffers in gallbladder bile to prevent calcium precipitation and that other factors are necessary to promote calcium precipitation from bile (see BILE SALT–CALCIUM INTERACTIONS, p. 636). Thus bile dilution, a decreased CSI, low-salt concentration, absence of calcium, decreased temperature, and a decrease in bile salt–phospholipid ratio all inhibit nucleation and prolong nucleation time. On the other hand, concentration, an increased CSI, higher salt and calcium, increased temperature, and a high bile salt–phospholipid ratio enhance nucleation.

In contrast to the required presence of nucleating factors in patients with cholesterol gallstones, some investigators have proposed that antinucleating factors (agents that prevent microcrystal formation) occur in normal bile and are reduced in abnormal bile (51) and that these agents inhibit supersaturated biles from normal patients from precipitating cholesterol. However, the importance of nucleating agents in human gallbladder bile from gallstone patients was most convincingly shown by Strasberg's group (14, 40, 40a, 41, 45). After eliminating cholesterol microcrystals as potential nucleating agents by filtration, they confirmed that supersaturated bile from normal patients had a much greater nucleation time than similarly supersaturated bile from gallstone patients. When they mixed equal proportions of normal and gallstone patient bile, nucleation occurred just as it had with the gallstone bile, indicating that nucleating agents were in gallstone bile and that any antinucleating agents present in normal bile were not adequate to overcome the gallstone patient's nucleating factors. These factors appeared quite potent, since small quantities of gallstone patient's bile were capable of nucleating large amounts of supersaturated normal bile. They attempted to determine what component in the bile of human gallstone patients was the nucleating agent. First they showed that the mucus concentration was not correlated with the nucleation time (45). Second, when mucus was removed by filtration the nucleation time did not change. They proposed that mucus was not the principal nucleating factor in human bile (41). Third, the addition of ethylenediaminetetraacetic acid (EDTA) to bile to remove all free calcium did not affect cholesterol precipitation but did prevent calcium bilirubinate precipitation. One can conclude that neither calcium nor calcium bilirubinate precipitates served as nuclei in these human gallbladder biles. They noted that heating to 90°C appeared to destroy the nucleation factor. Adding a small amount of filtered, nonheated, human gallstone patient's bile restored the nucleation (41). Finally, they show that isolated proteins from patients with gallstones accelerate nucleation, whereas proteins from control patients do not (40a). Thus, while certain protein molecules such as apoprotein AI may serve as antinucleating agents (51), they do not overcome the nucleation by another factor, possibly a heat-labile protein present in human gallstone patient's bile.

We advise caution in assuming that a nucleating agent found in the bile of patients with gallstones was the same nucleating agent responsible for the original precipitation of cholesterol from the patients' lithogenic bile before gallstones were present. Lithogenic agents found in gallstone bile may be the product of the gallstones rather than the initiating force for the first precipitation. Thus it will be important in both animal (64a, 64b, 69a) and human studies to look for nucleating agents in lithogenic bile that has not yet precipitated cholesterol. Such patients are now available in groups that have had their stones dissolved by bile acid therapy and who are then taken off therapy. Such patients develop supersaturated bile within a few weeks, and at least 50% go on to form gallstones at a later date (55a, 98a). A careful search for nucleating agents in the bile of such patients should be made.

Although it has not been unequivocally proven as a nucleating agent in human bile (41), mucin probably does play an important role in the nucleation of cholesterol from bile of prairie dogs (64a, 64b) and ground squirrels (69a). When such rodents are fed a high-cholesterol diet they develop gallstones within a few weeks. MacPherson et al. (69a) showed that the CSI increased within 12 h of feeding a cholesterol-enriched diet. By 18 h a globular material suggested to be mucus was seen on the surface of the gallbladder epithelium. At 24 h there was a sharp decrease in the absolute concentration of bile salts and phospholipids and an increase of cholesterol concentration. Simultaneously the first cholesterol monohydrate crystals were observed. Thus, although the true nucleating agents were not identified in this case, it appears that the sequence of events is supersaturation, followed by putative mucous formation, followed by nucleation. Aggregation of monohydrate crystals and growth into stone-size aggregates occur over the next few weeks (69a).

In Figure 19 we speculate concerning the formation of crystals from gallbladder bile. On the *left* is the phase diagram for very dilute hepatic bile and six separate hypothetical biles with different compositions (*a–f*). On the *right* is the phase diagram for gallbladder bile (as estimated from ref. 24) with the same compositions plotted. It is assumed that heterogeneous nucleation is necessary for the formation of cholesterol crystals; however, we do not have any particular choice of nucleating agents (e.g., specific proteins, inorganic Ca^{2+} salts, mucins). Furthermore we do not know the exact form in which bile is secreted. We speculate on the mechanisms and the recent findings of vesicles in the canalicular space (130) in *Secretion of Bile From the Hepatocyte*, p. 657.

Bile *a* has a composition like dog or rat bile. It is probably secreted as micelles, although it could be secreted as vesicles and rapidly converted to micelles by the action of the relatively high bile salt composition (see *Secretion of Bile From the Hepatocyte*, p. 657). It falls within the micellar phase on the hepatic

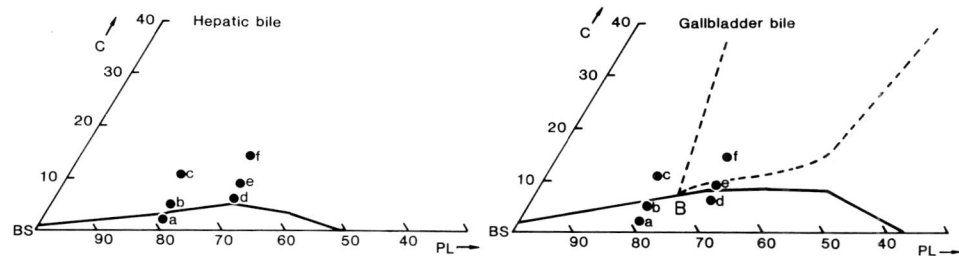

FIG. 19. Behavior of hypothetical biles of compositions *a–f* as they transform from dilute hepatic bile to concentrated gallbladder bile. Total lipids for hepatic bile are ~1.5% and for gallbladder bile 10%. Phase boundaries are from Carey et al. (24). BS, bile salt. *Point B* is the same as that discussed in Fig. 16. For further explanation, see text.

Secreted into Hepatic Bile as
  a. Unsaturated micelles.
  b. Supersaturated micelles.
  c. Vesicles with high ratio of cholesterol (C) to phospholipid (PL) (C/PL) and perhaps a few micelles.
  d. Vesicles with low C/PL and micelles.
  e. Vesicles with moderate C/PL (<1:1) and perhaps a few micelles.
  f. Vesicles with saturated 1:1 C/PL and perhaps a few micelles.

Transformed in Gallbladder to
  a. Unsaturated micelles.
  b. Unsaturated micelles (probably fairly rapid).
  c. Supersaturated micelles, which readily nucleate cholesterol monohydrate crystals.
  d. Micelles (kinetics unknown).
  e. Some vesicles are converted to saturated micelles (kinetics unknown); other vesicles remain. Cholesterol cannot precipitate because C/PL vesicle is less than saturated (<1:1). Vesicles may aggregate and fuse to produce multilamellar liquid crystals (kinetics unknown).
  f. Converts some vesicles to micelles, leaving vesicles with high C/PL supersaturated with cholesterol. These may remain as "metastable" vesicles, aggregate, and fuse into metastable multilamellar liquid crystals or precipitate cholesterol crystals from lamellar liquid crystals if a strong nucleating agent is present (kinetics fairly slow).

bile side, and when it is concentrated in the gallbladder (Fig. 19, *right*) it remains as micelles that are highly unsaturated with cholesterol.

Bile *b* is initially secreted in a supersaturated state, presumably as either vesicles with a fairly high ratio of cholesterol to phospholipid coexisting with bile salt micelles and/or monomers or as supersaturated micelles. Because of the concentration in the gallbladder, vesicles or supersaturated micelles are converted to unsaturated micelles. The kinetics for this reaction are presumably rapid, since the bile salt–lecithin ratio is high (24).

Bile *c* is secreted in a highly supersaturated state with a high ratio of bile salt to phospholipid. The ground squirrel secretes such a bile after 24 h of cholesterol feeding (69a). Either highly supersaturated micelles or vesicles with a very high ratio of cholesterol to phospholipid coexisting with micelles and/or monomeric bile salts are secreted. In the gallbladder any vesicles present would be converted to supersaturated micelles that, with a mild nucleating agent, could nucleate cholesterol monohydrate crystals. Crystals would then coexist with saturated micelles and bile salt monomers.

Bile *d* is presumably secreted as vesicles with a low ratio of cholesterol to phospholipid; perhaps some micelles and/or bile salt monomers are also present. In the gallbladder they should be converted to micelles, but the kinetics are unknown, so that if the conversion were slow the gallbladder might consist initially of both micelles and vesicle fractions. The vesicles would be converted to micelles over some finite period of time. No cholesterol precipitation takes place, because the final concentration is within the limits of solubility.

Bile *e* is presumably secreted as vesicles with a moderate ratio of cholesterol to phospholipid; perhaps a few micelles and/or monomers are also present. In the gallbladder some of the vesicles should be converted to saturated micelles (the kinetics of this process are also unknown) and some vesicles remain. Cholesterol cannot precipitate because the cholesterol-phospholipid mole ratio is less than saturated (i.e., <1). These vesicles may aggregate or fuse to produce multilamellar liquid crystals, which are sometimes seen in biles. The kinetics of this process are also not known. These biles may become cloudy because of the formation of multilamellar aggregates, but cholesterol crystals do not precipitate.

Bile *f* is highly supersaturated and is probably secreted largely as vesicles that are saturated with cholesterol, e.g., ~1:1 cholesterol–phospholipid mole ratio. Possibly a few micelles and/or monomers are also secreted. Because the bile is dilute the cholesterol is not readily precipitated from the vesicles. In the gallbladder some of the vesicles are converted to micelles

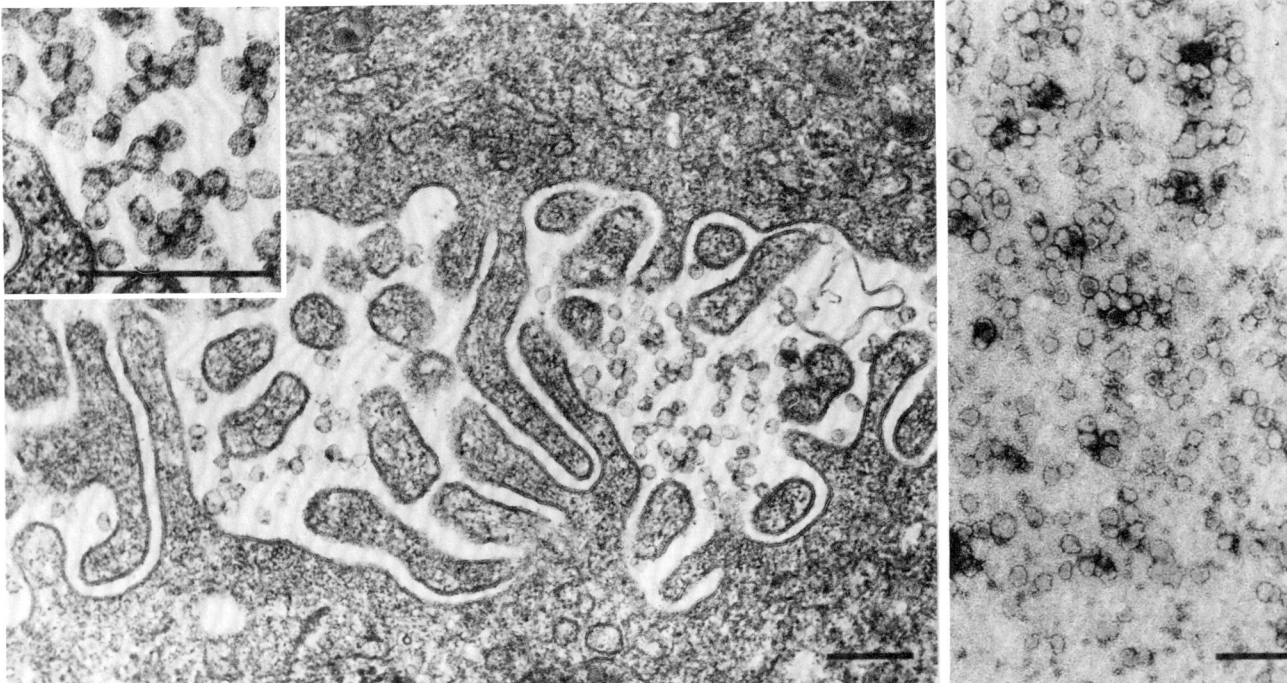

FIG. 20. Electron micrograph of a bile canaliculus of a bile salt–depleted rat liver that shows the presence of vesicular material within the lumen (*left*) (× 32,800). Higher magnification (*inset*) illustrates more clearly unilamellar structure of vesicle and can be compared with structure of canalicular membrane. Mean diameter of these vesicles is 56 nm (30–85 nm). Similar vesicles are also apparent in fresh bile specimen of same animal (*right*). Bars, 0.25 μm. [From Ulloa et al. (130), © by Am. Assoc. of the Study of Liver Diseases, 1987.]

because of the concentration effects. These micelles presumably have the composition at point B (Fig. 19) and leave vesicles with a high mole ratio of cholesterol to phospholipid (>1:1) supersaturated with cholesterol. These vesicles may remain as metastable vesicles containing cholesterol or may fuse to form metastable liquid crystals in bile. If a strong nucleating agent is present, crystalline cholesterol monohydrate may be nucleated and precipitated, moving the system toward its equilibrium state (see Fig. 16). Both normal biles and those from gallstone patients can have compositions similar to or falling between c and f. The presence of cholesterol crystallization in gallstone formation probably depends on the presence of potent nucleating factors in the case of f and moderate nucleating factors in the case of c. The exact position of point B and the other phase boundaries depends on the bile salt composition and total lipid composition, as shown in Figures 16 and 17.

Once cholesterol monohydrate microcrystal formation occurs, growth to macroscopic crystals and stones proceeds. Growth is believed to occur at the surfaces of the crystal in two directions (116) to quickly form large platelike cholesterol crystals. Aggregates of crystals and continued growth lead to stones that may result in clinical symptoms requiring treatment by either mechanically removing the stones or by administering chenodeoxycholates and ursodeoxycholates to attempt to dissolve the cholesterol stones (101).

Secretion of Bile From the Hepatocyte

The mechanisms of secretion of bile from the liver cell have been a topic of considerable discussion for nearly 20 years (112a). The secretion rate of bile salt is generally coupled to the secretion rate of phospholipids and cholesterol, although the cholesterol-phospholipid ratio may change under varying circumstances (33b). Furthermore the secretion rate of bile salts is related to the enterohepatic circulation and the synthesis of bile salts by hepatic cells (33a). The highest rates of secretion are found during high rates of return of bile salts from the intestine or nonphysiologically during the perfusion of high rates of bile salts into the duodenum or portal vein. The lowest rates of bile salt secretion, apart from partial biliary obstruction (127a), are found immediately after washout of the bile salt pool during a period when synthesis of endogenous bile salts has been depressed by rapid recirculation of bile salts (33a). The rate of bile salt synthesis could play a role in the particular mechanism of secretion.

Bile could be secreted by a number of different mechanisms. *1*) It could be secreted as preformed lecithin-cholesterol-bile salt micelles transported by a secretory body to the canalicular membrane and exocytosed into the canalicular space. *2*) It could be secreted as micelles formed by bile salts combining with parts of the canalicular membrane as they pass through it (112a). *3*) Bile salts could be independently secreted as monomers and then interact with the outer leaflet of the canalicular membrane to produce budding of lecithin to form vesicles. *4*) Bile salts could be secreted separately as monomers, and the phospholipid and cholesterol could be secreted independently as vesicles from secretory bodies exocytosing their contents into the canalicular space. This would require some intracellular coupling of two independent secretory pathways. Once the nascent bile was in the canaliculus, if the bile salt concentration were high enough to solubilize all the phospholipid in micelles, micelles might exist there. If, on the other hand, the bile salt concentrations were below the CMC or too low to solubilize the lipid present, vesicles might form and move toward an equilibrium with micellar and/or nonmicellar bile salts.

Although the specific mechanisms of secretion of bile are not precisely known, recent work indicates that some vesicles do form within the canaliculus during low-bile salt secretion rates and can be visualized by electron microscopy (130). In these experiments the bile salt pool was allowed to drain out through a bile fistula. At the nadir of bile secretion ($\sim$1 μl$\cdot$g$^{-1}\cdot$min^{-1}) the total lipid concentration was

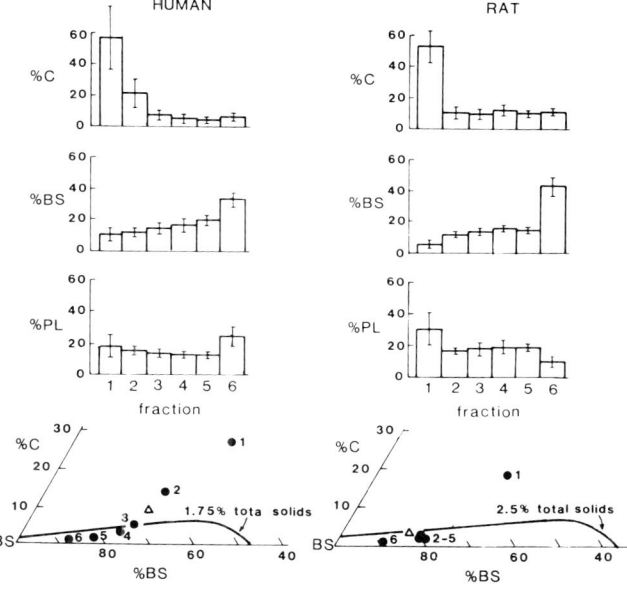

FIG. 21. Distribution and composition of biliary lipids in bile fractionated on a continuous metrizamide density gradient. *Left*, T-tube human hepatic bile; *right*, rat hepatic bile. Histograms are mean ± SD of percentage of each lipid in total sample of 7 human biles and 4 rat biles. Total lipid composition of rat biles was 2.4 g/dl; total composition of human T-tube biles was 1.7 g/dl. BS, bile salt; C, cholesterol; PL, phospholipid. Compositions of each of the density cuts are plotted below on triangular coordinates with appropriate line as determined by Carey et al. (24). *Open triangle*, mean composition of bile; *solid circle*, density gradient fraction. Human bile is supersaturated with respect to *1.75% total solids line*, whereas rat bile is not supersaturated. Least-dense fraction (fraction 1) of both human and rat bile contains vesicles and has a composition very different from other density cuts. [Data from Ulloa (130).]

only 0.25 g/100 dl bile and the bile salt concentration of 3.45 mM was near the CMC. This dilute bile was clearly supersaturated. Vesicles were seen by electron microscopy in both the bile and in the canaliculus space (Fig. 20).

Furthermore, in the same paper, fresh, supersaturated human hepatic bile from T-tube patients and fresh, unsaturated, rat hepatic bile, collected during pool washout, were spun within a few hours in a metrizamide gradient to separate fractions of different density (Fig. 21). In both human bile and surprisingly in unsaturated rat bile, the lightest fraction was extremely rich in phospholipid and cholesterol and poor in bile salt and contained vesicles. If the rat bile was allowed to incubate for 24 h the vesicle fraction disappeared and merged into the denser micellar fractions. When plotted on triangular coordinates (Fig. 21) the light fractions of both rat and human (fraction 1) bile fall well above the appropriate solubility line and consist of vesicles coexisting with some bile salts. In the unsaturated rat bile, all of the other fractions (fractions 2–6) fall in the micellar region and presumably consist of unsaturated cholesterol–phospholipid micelles. In the supersaturated human biles, fractions 2 and probably 3 also consist of vesicles and some bile salts, whereas fractions 4–6 fall in a micellar zone and presumably exist as micelles. The presence of vesicles in the unsaturated fresh hepatic bile of the rat suggest that the vesicles are either secreted into or formed within the dilute canalicular bile. It is tempting to speculate that the vesicle fraction was the last to be secreted during the washout period, and that at this time the bile salt secretion rate was very low. It has been known for nearly two decades that bile composition becomes progressively more supersaturated during acute bile diversion and pool washout (33b). It is at the end of the pool washout, where bile salt secretion is at a minimum, that Ulloa et al. (130) found vesicles in the canaliculus.

The mean total lipid composition of the rat bile collected at the bile duct was 2.4 g/100 dl. This composition is similar to the composition of point a in Figure 19. Its mean composition predicted micelles (Fig. 21), but it contained both vesicles and micelles. With time the vesicular fraction was converted into micelles. The supersaturated human bile, on the other hand, would have contained both vesicles and micelles. Its composition is rather like composition e in Figure 19. Once transported to the gallbladder, vesicles and micelles probably would continue to coexist in such a bile.

Although the mechanism of secretion of vesicles is not established, the failure to find secretory bodies within the liver filled with unilamellar liposomes like those found in bile (130) makes it likely that the vesicles are formed by budding of the canalicular membrane or on dilution of micelles within the canaliculus. Whether the same putative mechanisms occur at high–bile salt secretion rates must be explored.

We thank Anne M. Gibbons for preparation of the manuscript.
This work was supported by National Institutes of Health Grants HL-26335 and HL-07291.

REFERENCES

1. ADMIRAND, W. H., AND D. M. SMALL. The physico-chemical basis of cholesterol gallstone formation in man. *J. Clin. Invest.* 47: 1045–1052, 1968.
2. ANGELICO, M. Subselection of phospholipids during formation of native bile: physical-chemical coupling of bile salts and phospholipids (Abstract). *Clin. Res.* 33: 538A, 1985.
3. ARMSTRONG, M. C., AND M. C. CAREY. The hydrophobic-hydrophilic balance of bile salts. Inverse correlation between reverse phase high pressure liquid chromatographic mobilities and micellar cholesterol-solubilizing capacities. *J. Lipid Res.* 23: 70–80, 1982.
4. ARORA, S. K., G. GERMAIN, AND J. P. DECLERCQ. The crystal and molecular structure of lithocholic acid. *Acta Crystallogr. Sect. B. Struct. Crystallogr. Cryst. Chem.* 32: 415–419, 1982.
5. BACKER, J. M., AND E. A. DAWIDOWICZ. Mechanism of cholesterol exchange between phospholipid vesicles. *Biochemistry* 20: 3805–3810, 1981.
6. BARNES, S., AND J. M. GECKLE. High resolution nuclear magnetic resonance spectroscopy of bile salts: individual proton assignments for sodium cholate in aqueous solution at 400 MHz. *J. Lipid Res.* 23: 161–170, 1982.
7. BENNETT, W. S., G. ELLINGTON, AND S. KOVACK. Self-association of phenolics and bile acid derivatives. *Nature Lond.* 214: 776–780, 1967.
8. BIRDI, K. S. Aggregation of bile salts (Na-deoxycholate). *Finn. Chem. Lett.* 6-8: 142–146, 1982.
9. BLOCH, C. A., AND J. B. WATKINS. Determination of conjugated bile acids in human bile and duodenal fluid by reverse phase high pressure liquid chromatography. *J. Lipid Res.* 19: 510–513, 1978.
10. BORGSTROM, B. Studies of the phospholipids of human bile and small intestine. *Acta Chem. Scand.* 11: 749, 1957.
11. BOURGES, M., D. M. SMALL, AND D. G. DERVICHIAN. Biophysics of lipidic associations. II. The ternary systems. Cholesterol-lecithin-water. *Biochim. Biophys. Acta* 137: 157–167, 1967.
12. BOURGES, M., D. M. SMALL, AND D. G. DERVICHIAN. Biophysics of lipidic associations. III. The quaternary systems. Bile salt-lecithin-cholesterol-water. *Biochim. Biophys. Acta* 144: 189–202, 1967.
13. BRUCKDIRFER, K. R., AND M. K. SHERRY. The solubility of cholesterol and its exchange between membranes. *Biochim. Biophys. Acta* 769: 187–196, 1984.
14. BURNSTEIN, M. J., R. G. ILSON, C. N. PETRUNKA, R. D. TAYLOR, AND S. M. STRASBERG. Evidence of a potential nucleating factor in the gallbladder bile of patients with cholesterol gallstones. *Gastroenterology* 85: 801–807, 1983.
15. CABRAL, D. J., J. A. HAMILTON, AND D. M. SMALL. The ionization behavior of bile acids in different aqueous environments. *J. Lipid Res.* 27: 334–343, 1986.
16. CABRAL, D. J., D. M. SMALL, H. S. LILLY, AND J. A. HAMILTON. Transbilayer movement of bile acids in model membranes. *Biochemistry* 26: 1801–1804, 1987.
16a. CABRAL, D. J., D. M. SMALL, AND J. A. HAMILTON. Exchange behavior of bile acids in phosphatidylcholine vesicles (Abstract). *Biophys. J.* 51: 540a, 1987.
17. CAMPANELLI, A. R., D. FERRO, E. GIGLIO, P. IMPERATORI, AND V. PIACENTE. Thermal and x-ray study of sodium deox-

ycholate crystal and fiber. *Thermochim. Acta* 67: 223–232, 1983.
18. CANTAFORA, A., A. DIBIASE, D. ALVARO, M. ANGELICO, M. MARIN, AND A. F. ATTILI. High performance liquid chromatographic analysis of molecular species of phosphatidylcholine-development of quantitative assay and its application to human bile. *Clin. Chim. Acta* 134: 281–285, 1983.
19. CAREY, M. C. Critical tables for calculating the cholesterol saturation of native bile. *J. Lipid Res.* 19: 945–955, 1978.
20. CAREY, M. C. Physical-chemical properties of bile acids and their salts. In: *New Comprehensive Biochemistry: Sterols and Bile Acids*, edited by J. Sjovall and H. Danielson. Amsterdam: Elsevier, 1985, p. 345–403.
21. CAREY, M. C., M. J. ARMSTRONG, N. A. MAZER, H. IGIMI, AND G. SALVIOLI. Measurement of the hydrophilic-hydrophobic balance of bile salt: correlation with physical-chemical interactions. In: *Bile Acids and Cholesterol in Health and Disease*, edited by G. Paumgartner, A. Stiehl, and W. Gerok. Lancaster, UK: MTP, 1983, p. 31–42.
22. CAREY, M. C., AND D. M. SMALL. Micellar properties of dihydroxy and trihydroxy bile salts: effects of counterion and temperature. *J. Colloid Interface Sci.* 31: 382–396, 1969.
23. CAREY, M. C., AND D. M. SMALL. The physical chemistry of cholesterol solubility in bile: relationship to gallstone formation and dissolution in man. *J. Clin. Invest.* 61: 998–1026, 1978.
24. CAREY, M. D., D. M. SMALL, AND C. M. BLISS. Lipid digestion and absorption. *Annu. Rev. Physiol.* 45: 651–677, 1983.
25. CASTELLINO, F. J., AND B. N. VIOLAND. ^{31}P-nuclear magnetic resonance and ^{31}P[^{1}H] nuclear Overhauser effect analysis of mixed egg phosphatidylcholine-sodium taurocholate vesicles and micelles. *Arch. Biochem. Biophys.* 193: 543–550, 1979.
26. CHANG, Y., AND J. R. CARDINAL. Light scattering studies on bile acid salts. II. Pattern of self-association of sodium deoxycholate, sodium taurodeoxycholate and sodium glycodeoxycholate in aqueous electrolyte solutions. *J. Pharm. Sci.* 67: 994–999, 1978.
26a.CHRISTOPHER, F. *Textbook of Surgery* (9th ed.), edited by L. Davis. Philadelphia, PA: Saunders, 1968.
27. CISTOLA, D. P., D. M. SMALL, AND J. A. HAMILTON. Ionization behavior of aqueous short-chain carboxylic acids: a carbon-13 NMR study. *J. Lipid Res.* 23: 795–799, 1982.
28. CLAFFEY, W. J., AND R. T. HOLZBACH. Dimorphism in bile salt/lecithin mixed micelles. *Biochemistry* 20: 415–418, 1981.
29. COBBLEDICK, R. E., AND F. W. B. EINSTEIN. The structure of $3\alpha,7\alpha,12\alpha$-trihydroxy-5β-cholan-24-oate monohydrate (sodium cholate monohydrate). *Acta Crystallogr. Sect. B Struct. Crystallogr. Cryst. Chem.* 36: 287–292, 1980.
30. COIRO, Y. M., E. GIGLIO, S. MOROSETTE, AND A. PALLESCHI. A monoclinic phase of the deoxycholic acid rubidium salt. *Acta Crystallogr. Sect. B Struct. Crystallogr. Cryst. Chem.* 36: 1478–1480, 1980.
31. CONTE, G., R. DIBLASI, E. GIGLIO, A. PARRETT, AND N. V. PAVEL. Nuclear magnetic resonance and x-ray studies on micellar aggregates of sodium deoxycholate. *J. Phys. Chem.* 88: 5720–5724, 1984.
32. DELACROIX, D. L., H. J. F. HODGSON, A. MCPHERSON, C. DIVE, AND J. P. VAERMAN. Selective transport of polymeric immunoglobulin A in bile. *J. Clin. Invest.* 70: 230–241, 1982.
33. DEMARIA, P., A. FINI, AND A. RODA. Chemical properties of bile acids. I. Thermodynamic dissociation constants of some cholanic acid derivatives in 50 weight percent aqueous methanol. *Gazz. Chim. Ital.* 111: 95–97, 1981.
33a.DOWLING, R. H., E. MACK, AND D. M. SMALL. Effects of controlled interruption of the enterohepatic circulation of bile salts by biliary diversion and by ileal resection on bile salt secretion, synthesis and pool size in the Rhesus monkey. *J. Clin. Invest.* 49: 232–242, 1970.
33b.DOWLING, R. J., E. MACK, AND D. M. SMALL. Biliary lipid secretion and bile composition following acute and chronic interruption of the enterohepatic circulation in the Rhesus monkey. *J. Clin. Invest.* 50: 1917–1926, 1971.
33c.DOWLING, R. H., AND D. M. SMALL. The effect of pH on the solubility of varying mixtures of free and conjugated bile salts in solution. *Gastroenterology* 54: 1291, 1968.
34. EKWALL, P., T. ROSENDAHL, AND N. LOFMAN. Studies on bile acid salt solutions. *Acta Chem. Scand.* 11: 590–598, 1957.
35. EKWALL, P., T. ROSENDAHL, AND A. STEN. Studies on bile acid salt solutions. II. The solubility of cholic acid in sodium cholate solutions and that of deoxycholate in sodium deoxycholate solutions. *Acta Chem. Scand.* 12: 1622–1633, 1958.
36. FINI, A., A. RODA, AND P. DEMARIA. Chemical properties of bile acids. Part 2. pKa values in H_2O and aqueous methanol of some hydroxy bile acids. *Eur. J. Chem.* 17: 467–470, 1982.
37. FINI, A., A. RODA, R. FUGAZZA, AND B. GRIGOLO. Chemical properties of bile acids. III. Bile acid structure and solubility in water. *J. Solution Chem.* 14: 595–603, 1985.
38. FROMHERTZ, P. Lipid-vesicle structure: size control by edge-active agents. *Chem. Phys. Lipids* 94: 259–266, 1983.
39. FROMHERTZ, P., AND D. RUPPEL. Lipid vesicle formation: the transition from open disk to closed shells. *FEBS Lett.* 179: 155–159, 1985.
40. GALLINGER, S., P. R. C. HARVEY, C. N. PETRUNKA, AND S. M. STRASBERG. The effect of binding of ionized calcium on the in vitro nucleation of cholesterol and calcium bilirubinate in human gallbladder bile. *Gut* 27: 1382–1386, 1986.
40a.GALLINGER, S., P. ROBERT, C. HARVEY, C. N. PETRUNKA, R. G. ILSON, AND S. M. STRASBERG. Biliary proteins and the nucleation defect in cholesterol cholelithiasis. *Gastroenterology* 91: 867–875, 1987.
41. GALLINGER, S., R. D. TAYLOR, P. R. C. HARVEY, C. N. PETRUNKA, AND S. M. STRASBERG. Effect of mucous glycoprotein on nucleation time of human bile. *Gastroenterology* 89: 648–658, 1985.
42. GILAT, T., AND G. J. SÖMJEN. Cholesterol solubility in human bile. *J. Clin. Gastroenterol.* In press.
43. HABERLAND, M. E., AND J. A. REYNOLDS. Self-association of cholesterol in aqueous solution. *Proc. Natl. Acad. Sci. USA* 70: 2313–2316, 1973.
44. HALPERN, Z., M. A. DUDLEY, A. KIBE, M. P. LYNN, A. C. BREUER, AND R. T. HOLZBACH Rapid vesicle formation and aggregation in abnormal human biles. A time-lapse video-enhanced contrast microscopy study. *Gastroenterology* 90: 875–885, 1986.
45. HARVEY, P. R. C., C. A. RUPAR, S. GALLINGER, C. N. PETRUNKA, AND S. M. STRASBERG. Quantitative and qualitative comparison of gallbladder mucus glycoprotein from patients with and without gallstones. *Gut* 27: 374–381, 1986.
46. HARVEY, P. R. C., D. TAYLOR, C. N. PETRUNKA, A. D. MURRAY, AND S. M. STRASBERG. Quantitative analysis of major, minor and trace elements in gallbladder bile of patients with and without gallstones. *Hepatology Baltimore* 5: 129–132, 1985.
47. HEGARDT, F. G., AND H. DAM. The solubility of cholesterol in aqueous solutions of bile salts and lecithin. *Z. Ernaehrungswiss.* 10: 223–233, 1971.
47a.*Hepatology Baltimore* 4, Suppl.: 1S–252S, 1984.
48. HOFMANN, A. F. Animal models of calcium cholelithiasis. *Hepatology Baltimore* 4, Suppl.: 209S–211S, 1984.
48a.HOGAN, A., S. E. EALICK, C. E. BUGG, AND S. BARNES. Aggregation patterns of bile salts: crystal structure of calcium cholate chloride heptahydrate. *J. Lipid Res.* 25: 791–798, 1984.
49. HOLAN, K. R., R. T. HOLZBACH, R. E. HERMANN, A. C. COOPERMAN, AND W. J. CLAFFEY. Nucleation time: a key factor in the pathogenesis of cholesterol gallstone disease. *Gastroenterology* 77: 611–617, 1979.
50. HOLZBACH, R. T., AND C. CORBUSIER. Liquid crystals and cholesterol nucleation during equilibration in supersaturated bile analogues. *Biochim. Biophys. Acta* 528: 436–444, 1978.
51. HOLZBACH, R. T., A. KIBE, E. THIEL, J. H. HOWELL, M. MARSH, AND R. E. HERMANN. Biliary proteins: unique inhibitors of cholesterol crystal nucleation in human gallbladder bile. *J. Clin. Invest.* 73: 35–45, 1984.
52. HOLZBACH, R. T., M. MARSH, M. OLSZEWSKI, AND K. HOLAN. Cholesterol solubility in bile: evidence that supersatu-

rated bile is frequent in healthy man. *J. Clin. Invest.* 52: 1467–1479, 1973.
53. HUNT, G. R. A. A comparison of triton X-100 and the bile salt taurocholate as micellar ionophones or fusogenes in phospholipid vesicular membranes. A ^{1}H NMR method using the lanthanide probe ion Pr^{3+}. *FEBS Lett.* 119: 132–136, 1980.
54. HUNT, G. R. A., AND K. JAWAHARLAL. A ^{1}H-NMR investigation of the mechanism for the ionophore activity of the bile salts in phospholipid vesicular membranes and the effects of cholesterol. *Biochim. Biophys. Acta* 601: 678–684, 1980.
55. IGIMI, H., AND M. C. CAREY. pH-solubility relations of chenodeoxycholic and ursodeoxycholic acids: physical-chemical basis for dissimilar solution and membrane phenomena. *J. Lipid Res.* 21: 72–90, 1980.
55a.ISER, J. H., G. M. MURPHY, AND R. H. DOWLING. Speed of change in biliary lipids and bile acids with chenodeoxycholic acid—is intermittent therapy feasible? *Gut* 18: 7–15, 1977.
56. KIBE, A., M. A. DUDLEY, Z. HALPERN, M. P. LYNN, A. C. BREUER, AND R. T. HOLZBACH. Factors affecting cholesterol monohydrate crystal nucleation time in model systems of supersaturated bile. *J. Lipid Res.* 26: 1102–1111, 1985.
57. KOLEHMAINEN, E. Solubilization of aromatics in aqueous bile salts. I. Benzene and alkylbenzenes in sodium cholate: ^{1}H NMR study. *J. Colloid Interface Sci.* 105: 273–277, 1985.
58. KRAMER, R. M., H. J. HASSELBACH, AND G. SEMENZA. Rapid transmembrane movement of phosphatidylcholine in small unilamellar lipid vesicles formed by detergent removal. *Biochim. Biophys. Acta* 643: 233–242, 1985.
59. KRATOHVIL, J. P., AND H. T. DELLICOLLI. Micellar properties of bile salts. Sodium taurodeoxycholate and sodium glycodeoxycholate. *Can. J. Biochem.* 46: 945–952, 1968.
60. KRATOHVIL, J. P., W. P. HSU, M. A. JACOBS, T. M. AMINABHAVI, AND Y. MUKUNOKI. Concentration-dependent aggregation patterns of conjugated bile salts in aqueous sodium chloride solutions. *Colloid Polym. Sci.* 261: 781–785, 1983.
61. LACK, L., AND I. M. WEINER. Intestinal bile salt transport: structure-activity relationships and other properties. *Am. J. Physiol.* 210: 1142–1152, 1966.
62. LAKE, M., AND D. T. ORGANISCIAK. Determination of the composition of mixed micelles of bile salts by kinetic dialysis. *Lipids* 19: 553–557, 1984.
63. LARUSSO, N. F. Proteins in bile: how they get there and what they do. *Am. J. Physiol.* 247 (*Gastrointest. Liver Physiol.* 10): G199–G205, 1984.
64. LECUYER, H., AND D. G. DERVICHIAN. Structure of aqueous mixtures of lecithin and cholesterol. *J. Mol. Biol.* 45: 39–57, 1969.
64a.LEE, S. P., M. C. CAREY, AND J. T. LAMONT. Aspirin prevention of cholesterol gallstone formation in the prairie dog. *Science Wash. DC* 211: 1420, 1981.
64b.LEE, S. P., J. T. LAMONT, AND M. C. CAREY. Role of gallstone mucin hypersecretion in the evolution of cholesterol gallstones: studies in the prairie dog. *J. Clin. Invest.* 67: 1712–1723, 1981.
65. LEIBFRITZ, D., AND J. D. ROBERTS. Nuclear magnetic resonance spectroscopy. Carbon-13 spectra of cholic acids and hydrocarbons included in sodium deoxycholate solutions. *J. Am. Chem. Soc.* 95: 4996–5003, 1973.
66. LICHTENBERG, D., Y. ZILBERMAN, P. GREENZAID, AND S. ZAMIR. Structural and kinetic studies on the solubilization of lecithin by sodium deoxycholate. *Biochemistry* 18: 3517–3525, 1979.
67. LINDHEIMER, M., J. C. MONTET, J. MOLENAT, R. BONTEMPS, AND B. BRUN. Ionic self-diffusion of various bile salts. *J. Chim. Phys.* 78: 447–455, 1981.
68. LINDLEY, P. F., M. M. MAHMOUD, F. E. WATSON, AND W. A. JONES. The structure of chenodeoxycholic acid, $C_{24}H_{40}O_4$. *Acta Crystallogr. Sect. B Struct. Crystallogr. Cryst. Chem.* 36: 1893–1897, 1980.
69. LINDMAN, B., N. KAMENKA, H. FABRE, J. ULMIUS, AND T. WEILOCH. Aggregation, aggregate composition and dynamics in aqueous sodium cholate solutions. *J. Colloid Interface Sci.* 73: 556–565, 1980.
69a.MACPHERSON, B. R., R. S. PEMSINGH, AND G. W. SCOTT. Experimental cholelithiasis in the ground squirrel. *Lab. Invest.* 56: 138–145, 1987.
70. MAZER, N. A., G. B. BENEDEK, AND M. C. CAREY. Quasielastic light-scattering studies of aqueous biliary lipid systems. Mixed micelle formation in bile salt-lecithin solutions. *Biochemistry* 19: 601–615, 1980.
71. MAZER, N. A., AND M. C. CAREY. Quasi-elastic light scattering studies of aqueous biliary lipid systems. Cholesterol solubilization and precipitation in model bile solutions. *Biochemistry* 22: 426–442, 1983.
72. MAZER, N. A., M. C. CAREY, R. F. KWASNICK, AND G. B. BENEDEK. Quasi-elastic light scattering studies of aqueous biliary lipid systems. Size, shape and thermodynamics of bile salt micelles. *Biochemistry* 18: 3064–3075, 1979.
73. MAZER, N. A., P. SCHURTENBERGER, M. C. CAREY, R. PRESIG, K. WEIGAND, AND W. KANIZ. Quasi-elastic light scattering studies of native bile from the dog: comparison with aggregation behavior of model biliary lipid systems. *Biochemistry* 23: 1994–2005, 1984.
74. MCLEAN, L. R., AND M. C. PHILLIPS. Mechanism of cholesterol and phospholipid exchange or transfer between unilamellar vesicles. *Biochemistry* 20: 2893–2900, 1981.
75. MONTET, J. C., AND D. G. DERVICHIAN. Solubilisation micellaire du cholestérol par les sels biliaires et les lécithines extraits de la bile humaine. *Biochimi Paris* 53: 751–754, 1971.
76. MOORE, E. W. The role of calcium in the pathogenesis of gallstones: Ca^{++} electrode studies of model bile salt solutions and other biologic systems. *Hepatology Baltimore* 4, Suppl.: 228S–243S, 1984.
77. MOORE, E. W., L. CELIC, AND J. D. OSTROW. Interactions between ionized calcium and sodium cholate: bile salts are important buffers for prevention of calcium-containing gallstones. *Gastroenterology* 83: 1079–1089, 1982.
78. MUKIDJAM, E., S. BARNES, AND G. A. ELGAVISH. NMR studies of the binding of sodium and calcium ions to the bile salts glycocholate and taurocholate in dilute solution, as probed by the paramagnetic lanthanide dysprosium. *J. Am. Chem. Soc.* 108: 7082–7089, 1986.
78a.MULLER, K. Structural dimorphism of bile salt/lecithin mixed micelles. A possible regulatory mechanism for cholesterol solubility in bile? X-ray structure analysis. *Biochemistry* 20: 404–414, 1981.
79. MURATA, Y., G. SUGIHARA, K. FUKUSHIMA, AND M. TANAKA. Study of the micelle formation of sodium deoxycholate. Concentration dependence of carbon-13 nuclear magnetic resonance chemical shift. *J. Phys. Chem.* 86: 4690–4694, 1982.
80. NORTON, D. A., AND B. HANER. Crystal data (I) for some bile acid derivatives. *Acta Crystallogr. Sect. B. Struct. Crystallogr. Cryst. Chem.* 19: 477–479, 1965.
81. O'CONNOR, C. J., B. T. CH'NG, AND R. G. WALLACE. Studies in bile salt solutions. 1. Surface tension evidence for a stepwise aggregation model. *J. Colloid Interface Sci.* 95: 410–419, 1983.
82. OLIVE, J., AND D. G. DERVICHIAN. Action d'une phospholipase sur la lécithine a l'état micellaire. *Bull. Soc. Chim. Biol.* 50: 1409–1418, 1968.
83. PAL, S., A. R. DAS, AND S. P. MOULIK. Interaction of bile salts (sodium cholate and sodium deoxycholate) with a nonionic surfactant (Triton X-100) and polyethylene glycols. *Indian J. Biochem. Biophys.* 19: 295–300, 1982.
84. PALMER, R. H. Bile salts and the liver. *Prog. Liver Dis.* 7: 221–242, 1982.
85. PARRIS, N. A. Liquid chromatographic separation of bile acids. *J. Chromatogr.* 133: 273–279, 1977.
86. PATTON, G. M., S. B. CLARK, J. M. FASULO, AND S. J. ROBINS. Utilization of individual lecithins in intestinal lipoprotein formation in the rat. *J. Clin. Invest.* 73: 231–240, 1984.
87. PATTON, G. M., S. J. ROBINS, J. M. FASULO, AND S. B. CLARK. Influence of lecithin acyl chain composition on the kinetics of exchange between chylomicrons and high density lipoproteins. *J. Lipid Res.* 26: 1285–1293, 1985.

88. PAUL, R., M. K. MATHEW, R. NARAYANAN, AND P. BALARAM. Fluorescent probe and NMR studies of the aggregation of bile salts in aqueous solutions. *Chem. Phys. Lipids* 25: 345–356, 1979.
89. PLANK, L., C. E. DAHL, AND B. R. WARE. Effect of sterol incorporation on headgroup separation in liposomes. *Chem. Phys. Lipids* 36: 319–328, 1985.
90. RAJAGOPALAN, N., AND S. LINDENBAUM. The binding of Ca^{2+} to taurine and glycine-conjugated bile salt micelles. *Biochim. Biophys. Acta* 711: 66–74, 1982.
91. RAJAGOPALAN, N., AND S. LINDENBAUM. Kinetics and thermodynamics of the formation of mixed micelles of egg phosphatidylcholine and bile salt. *J. Lipid Res.* 25: 135–147, 1984.
92. RAJAGOPALAN, N., M. VADNERE, AND S. LINDENBAUM. Thermodynamics of aqueous bile salt solutions: heat capacity, enthalpy and entropy of dilution. *J. Solution Chem.* 10: 785–801, 1981.
93. REDINGER, R. N., AND D. M. SMALL. Bile composition, bile salt metabolism and gallstones. *Arch. Intern. Med.* 130: 618–630, 1972.
94. RENSHAW, P. F., A. S. JANOFF, AND K. W. MILLER. On the nature of dilute aqueous cholesterol suspensions. *J. Lipid Res.* 24: 47–51, 1983.
95. ROBINS, S. J., J. M. FASULO, AND G. M. PATTON. Lipids of pigment gallstones. *Biochim. Biophys. Acta* 712: 21–25, 1982.
96. RODA, A., A. F. HOFMANN, AND K. J. MYSELS. The influence of bile salt structure and aggregation in aqueous solutions. *J. Biol. Chem.* 258: 6362–6370, 1983.
97. ROE, J. M., AND B. W. BARRY. Measurement of critical micelle concentration by photon correlation spectroscopy. *J. Colloid Interface Sci.* 94: 580–583, 1983.
98. ROE, J. M., AND B. W. BARRY. Bile salt association (cholate, deoxycholate, chenodeoxycholate and ursodeoxycholate) and interactions with aromatic alcohols (benzyl, 2-phenylethanol, and 3-phenylpropanol). *J. Colloid Interface Sci.* 107: 398–404, 1985.
98a. RUPPIN, D. C., G. M. MURPHY, R. H. DOWLING, AND THE BRITISH GALLSTONE STUDY GROUP. Gallstone disease without gallstones—bile acid and bile lipid metabolism after complete gallstone dissolution. *Gut* 27: 559–566, 1986.
99. SAAD, H. Y., AND W. I. HIGUCHI. Water solubility of cholesterol. *J. Pharm. Sci.* 54: 1205–1206, 1965.
100. SAITO, H., Y. SUGIMOTO, R. TABETA, S. SUZUKI, G. IZUMI, M.KODAMA, S. TOYOSHIMA, AND C. NAGATA. Incorporation of bile acids of low concentration into model and biological membranes studied by ^{2}H and ^{31}P NMR. *J. Biochem. Tokyo* 94: 1877–1887, 1983.
101. SALVIOLI, G., H. IGIMI, AND M. C. CAREY. Cholesterol gallstone dissolution in bile. Dissolution kinetics of crystalline cholesterol monohydrate by conjugated chenodeoxycholate-lecithin and conjugated ursodeoxycholate-lecithin mixtures: dissimilar phase equilibria and dissolution mechanisms. *J. Lipid Res.* 24: 701–720, 1983.
101a. SEDAGHAT, S., AND S. M. GRUNDY. Cholesterol crystals and the formation of cholesterol gallstones. *N. Engl. J. Med.* 302: 1274–1277, 1980.
102. SEWELL, R. B., S. J. T. MAO, T. KAWAMOTO, AND N. F. LARUSSO. Apolipoproteins of high, low, and very low density lipoproteins in human bile. *J. Lipid Res.* 24: 391–401, 1983.
103. SHANKLAND, W. The equilibrium and structure of lecithin-cholate mixed micelles. *Chem. Phys. Lipids* 4: 109–130, 1970.
104. SHAW, R., AND W. H. ELLIOTT. Bile acids. XLVIII. Separation of conjugated bile acids by high-pressure liquid chromatography. *Anal. Biochem.* 74: 273–281, 1976.
105. SHAW, R., M. RIVETNA, AND W. H. ELLIOTT. Bile acids. LXIII. Relationship between the mobility on reverse-phase high performance liquid chromatography and the structure of bile acids. *J. Chromatogr.* 21: 347–361, 1980.
106. SHURTENBERGER, P., AND B. LINDMAN. Coexistence of simple and mixed bile salt micelles: an NMR self-diffusion study. *Biochemistry* 24: 7161–7165, 1985.
107. SHURTENBERGER, P., N. MAZER, AND W. KANZIG. Static and dynamic light scattering studies of micellar growth and interactions in bile salt solutions. *J. Phys. Chem.* 87: 308–315, 1983.
108. SCHURTENBERGER, P., N. MAZER, AND W. KANZIG. Micelle to vesicle transition in aqueous solutions of bile salt and lecithin. *J. Phys. Chem.* 89: 1042–1049, 1985.
109. SMALL, D. M. Phase equilibria and structure of dry and hydrated egg lecithin. *J. Lipid Res.* 8: 551–557, 1967.
110. SMALL, D. M. Physico-chemical studies of cholesterol gallstone formation. *Gastroenterology* 52: 607–610, 1967.
111. SMALL, D. M. A classification of biological lipids based upon their interaction in aqueous systems. *J. Am. Oil Chem. Soc.* 45: 108–119, 1968.
112. SMALL, D. M. Studies on the size and structure of bile salt micelles, influences of structure, concentration, counterion concentration, pH and temperature. *Adv. Chem. Ser.* 84: 31–52, 1968.
112a. SMALL, D. M. The formation of gallstones. *Adv. Intern. Med.* 16: 243–264, 1970.
113. SMALL, D. M. The physical chemistry of cholanic acids. In: *The Bile Acids: Chemistry, Physiology and Metabolism. Chemistry*, edited by P. P. Nair and D. Kritchevsky. New York: Plenum, 1971, vol. 1, p. 249–356.
114. SMALL, D. M. Cholesterol nucleation and growth in gallstone formation. *N. Engl. J. Med.* 302: 1305–1307, 1980.
115. SMALL, D. M. Nucleation and growth of cholesterol gallstones. *Med. Chir. Dig.* 9: 619–635, 1930.
116. SMALL, D. M. The staging of cholesterol gallstones with respect to nucleation and growth. (Workshop on dissolution of gallstones. Proc. 6th Bile Acid Meet., Freiburg, W. Germany, October 9–11, 1980.) In: *Bile Acids and Lipids*, edited by G. Paumgartner, A. Stiehl, and W. Gerok. Lancaster, UK: MTP, 1981, pt. 1, p. 291–300.
117. SMALL, D. M. The physical chemistry of lipids from alkanes to phospholipids. In: *Handbook of Lipid Research*, edited by D. Hanahan. New York: Plenum, 1986, vol. 4, p. 1–672.
118. SMALL, D. M., AND M. BOURGES. Lyotropic paracrystalline phases obtained with aqueous ternary systems of amphiphilic substances in water. *Mol. Cryst. Liq. Cryst.* 1: 541–561, 1966.
119. SMALL, D. M., M. BOURGES, AND D. G. DERVICHIAN. Biophysics of lipid associations. I. The ternary systems. Lecithin-bile salt-water. *Biochim. Biophys. Acta* 125: 563–580, 1966.
120. SMALL, D. M., M. BOURGES, AND D. G. DERVICHIAN. Ternary and quaternary aqueous systems containing bile salts, lecithin and cholesterol. *Nature Lond.* 211: 816–818, 1966.
121. SMALL, D. M., D. J. CABRAL, D. P. CISTOLA, J. S. PARKS, AND J. A. HAMILTON. The ionization behavior of fatty acids and bile acids in micelles and membranes. *Hepatology Baltimore* 4, Suppl.: 77S–79S, 1984.
122. SMALL, D. M., S. A. PENKETT AND D. CHAPMAN. Studies on simple and mixed bile salt micelles by nuclear resonance spectroscopy. *Biochim. Biophys. Acta* 176: 178–189, 1969.
123. SOMJEN, G. J., AND T. GILAT. A non-micellar mode of cholesterol transport in human bile. *FEBS Lett.* 156: 265–268, 1983.
123a. SOMJEN, G. J., Y. MARIKOVSKY, P. LELKES, AND T. GILAT. Cholesterol-phospholipid vesicles in human bile: an ultrastructural study. *Biochim. Biophys. Acta* 879: 14–21, 1986.
124. SPINK, C. H., K. MULLER, AND J. M. STURTEVANT. Precision scanning calorimetry of bile salt-phosphatidylcholine micelles. *Biochemistry* 26: 6598–6605, 1982.
125. STARK, R. E., G. J. GOSSELIN, J. M. DONOVAN, M. C. CAREY, AND M. F. ROBERTS. Influence of dilution on the physical state of model bile systems: NMR and quasi-elastic light-scattering investigations. *Biochemistry* 24: 5599–5605, 1985.
126. STARK, R. E., J. L. MANSTEIN, W. CURATOLO, AND B. SEARS. Deuterium nuclear magnetic resonance studies of bile salt/phosphatidylcholine mixed micelles. *Biochemistry* 22: 2486–2490, 1983.
127. STARK, R. E., AND M. R. ROBERTS. 500 MHz ^{1}H-NMR studies of bile salt-phosphatidylcholine vesicles. Evidence for differential motional restraint on bile salt and phosphatidylcholine resonances. *Biochim. Biophys. Acta* 770: 115–121, 1984.
127a. STRASBERG, S. M., R. N. REDINGER, D. M. SMALL, AND R.

H. Egdahl. The effect of elevated biliary tract pressure on biliary lipid metabolism and bile flow in nonhuman primates. *J. Lab. Clin. Med.* 99: 343-353, 1982.
128. Tang, C. P., R. Popovitz-Biro, M. Lahav, and L. Leiserowitz. The tetragonal crystal structure of 2:3 deoxycholate acid-water complex. *Isr. J. Chem.* 18: 385-389, 1979.
129. Thomas, D. C., and S. D. Christian. Micellar and surface behavior of sodium deoxycholate characterized by surface tension and ellipsometric methods. *J. Colloid Interface Sci.* 78: 466-478, 1980.
130. Ulloa, N., J. Garrido, and F. Nervi. Ultracentrifugal isolation of vesicular carriers of biliary cholesterol in native human and rat bile. *Hepatology Baltimore* 7: 235-244, 1987.
130a. Ulmius, J., G. Lindblom, H. Wennerstrom, L. B. A. Johansson, K. Fontell, O. Soderman, and G. Arvidson. Molecular organization in liquid-crystal phases of lecithin-sodium cholate-water systems studied by nuclear magnetic resonance. *Biochemistry* 21: 1553-1560, 1982.
131. Vadnere, M., and S. Lindenbaum. Association of deoxycholic acid in organic solvents. *J. Pharm. Sci.* 71: 881-883, 1982.
132. Vadnere, M., R. Natarajan, and S. Lindenbaum. Apparent molal volumes of bile salts in water and water-d_2 solution. *J. Phys. Chem.* 84: 1900-1903, 1980.
133. Waterhous, D. V., S. Barnes, and D. D. Muccio. Nuclear magnetic resonance spectroscopy of bile acids. Development of two-dimensional NMR methods for the elucidation of proton resonance assignments for five common hydroxylated bile acids and their parent bile acid, 5β-cholanoic acid. *J. Lipid Res.* 26: 1068-1078, 1985.
133a. Whiting, M. J., and J. M. Watts. Cholesterol crystal formation and growth in model bile solutions. *J. Lipid Res.* 24: 861-868, 1983.
133b. Williamson, B. W. A., and I. W. Percy-Robb. The interaction of calcium ions with glycocholate micelles in aqueous solution. *Biochem. J.* 181: 61-66, 1979.
134. Williamson, B. W. A., and I. W. Percy-Robb. Contribution of biliary lipids to calcium binding in bile. *Gastroenterology* 78: 696-702, 1980.
135. Woodford, F. P. Enlargement of taurocholate micelles by added cholesterol and monoolein: self-diffusion measurements. *J. Lipid Res.* 10: 539-545, 1969.

CHAPTER 32

Pathways and functions of biliary protein secretion

ALBERT L. JONES
SUSAN JO BURWEN

Cell Biology and Aging Section, Veterans Administration Medical Center and Departments of Anatomy, Medicine, and the Intestinal Immunology and Liver Centers, University of California, San Francisco, California

CHAPTER CONTENTS

Anatomy of Biliary Protein Secretion
  Liver
  Hepatocyte
Receptor-Mediated Endocytosis and Intracellular Transport
  Pathways
  Lysosomal degradation (degradative pathway)
  Secretion of intact proteins into bile (transcellular pathway)
  Utilization by hepatocytes
    Transferrin
    Epidermal growth factor
    Insulin
  Mechanisms of pathway regulation
Role of the Cytoskeleton in Protein Transport by Hepatocytes
Functional Significance of Hepatobiliary Protein Secretion
  Intestinal immune response
  Biliary cholesterol metabolism
  Pathophysiological conditions
    Functional significance of hepatobiliary transport of epidermal growth factor

BILE CONTAINS MANY INTACT PROTEINS and polypeptides. Only recently has there been any significant interest in characterizing the various proteins in bile and determining the function of biliary protein secretion.

The two major proteins in rat bile are albumin and secretory component [the receptor for immunoglobulin A (IgA); (40, 47, 50)]. Secretory component is found both free and linked to polymeric IgA by disulfide bonds. Secretory IgA is the major immunoglobulin in bile, although both rat and human bile contain small amounts of IgG and IgM as well. Albumin, which is produced by the liver, is secreted mainly into the bloodstream, although a significant amount is still found in bile. It is not clear how much of the albumin found in bile is derived from the plasma or contributed by de novo synthesis. In experiments with isolated perfused rat liver, very little albumin is removed from the perfusate by the liver (A. L. Jones, unpublished observations). These data suggest that most albumin found in bile represents newly synthesized protein that has been secreted into the biliary space instead of into the sinusoid. However, nonspecific fluid-phase endocytosis may account for the presence of some plasma-derived albumin in bile. In addition to these major biliary proteins, many trace proteins, such as insulin (58) and epidermal growth factor (EGF) (62), are found as well. There are also many unidentified peptides, probably representing degradation products from a wide variety of proteins that have been taken up from plasma by liver and catabolized.

There are only two sources of protein found in bile: proteins such as EGF, insulin, and IgA are derived from the plasma; proteins such as lysosomal enzymes, albumin, and secretory component are produced locally within the liver and transported directly to the bile. The apoproteins of lipoproteins represent a special case, because they are produced within the liver, secreted into the bloodstream, and later taken up by hepatocytes and catabolized (33). Small amounts of the sequestered apoproteins are also found in bile (67).

The structural and functional aspects of biliary protein secretion are the emphasis of this chapter.

ANATOMY OF BILIARY PROTEIN SECRETION

Liver

The liver is the largest organ in the body and has a dual blood supply, receiving portal vein blood from the gastrointestinal tract (~80% of the blood supply) and hepatic arterial blood from the hepatic artery (a branch of the celiac artery). Another unique feature of the liver vasculature, in addition to the dual blood supply, is the nature of the sinusoidal lining cells. Two types of cells, endothelial cells and Kupffer cells, occur in a ratio of ~2:1. There has been speculation that Kupffer cells, which are fixed macrophages, may participate in the secretion of biliary proteins by processing plasma-derived substances in such a way that they become attractive substrates for hepatocyte receptors. However, there is no definitive evidence for the role of Kupffer cells in hepatocyte protein uptake. The liver endothelial cells, like the endothelial cells in many endocrine glands, contain numerous fenestrae; however, unlike the endocrine endothelial cell fenestrae, the liver endothelial cell fenestrae do not contain diaphragms. Although type 4 collagen in the hepatic sinusoid can be detected by immunocytochemistry (31), no distinct membrane beneath the endothelial lining of liver sinusoids is revealed by electron microscopy (except in ruminants). Therefore there may be little, if any, barrier between the liver cell surface and blood plasma material smaller than 1,000 Å. It is not at all unusual for microvilli on the hepatocyte surface to protrude through the fenestrae directly into the sinusoidal lumen (Fig. 1).

Bile formed by hepatocytes first enters minute channels between the cells, called bile canaliculi, which conduct the bile to the periphery of the liver lobules. Bile enters progressively larger ducts that feed into the common bile duct and is eventually secreted into the intestinal lumen. The normal human liver consists of ~1 million of these little lobules, which are usually depicted as hexagonal structures, with the periphery of the lobule containing the bile ducts, portal vein, and hepatic artery (the triad) and the lymphatics. In the center of the lobule the central vein provides the main venous drainage site from the liver lobule (Fig. 2). Approximately 12–24 hepatocytes extend from the portal area to the central vein. Consequently, hepatocytes in the periphery of the lobule (periportal or zone 1 cells) are exposed to the portal blood supply well before cells in the central lobular region (zone 3). Zone 1 cells remove certain substances from portal blood with such great efficiency that the effective concentration of these substances to which zone 3 cells are exposed is virtually nil. As a result these substances exhibit a steep portal-to-central lobular concentration gradient. Lobular concentration gradients were first demonstrated to be formed by certain sugars (26) and have since been found for bile acids (34), EGF (62), and asialoglycoproteins (A. L. Jones, unpublished observations). However, many other substances, which are also removed from blood by a receptor-mediated process, do not form gradients, such as insulin (58) and certain apolipoproteins (10).

It is important when investigating the transport function of the liver to establish whether the particular substance being studied forms a lobular concentra-

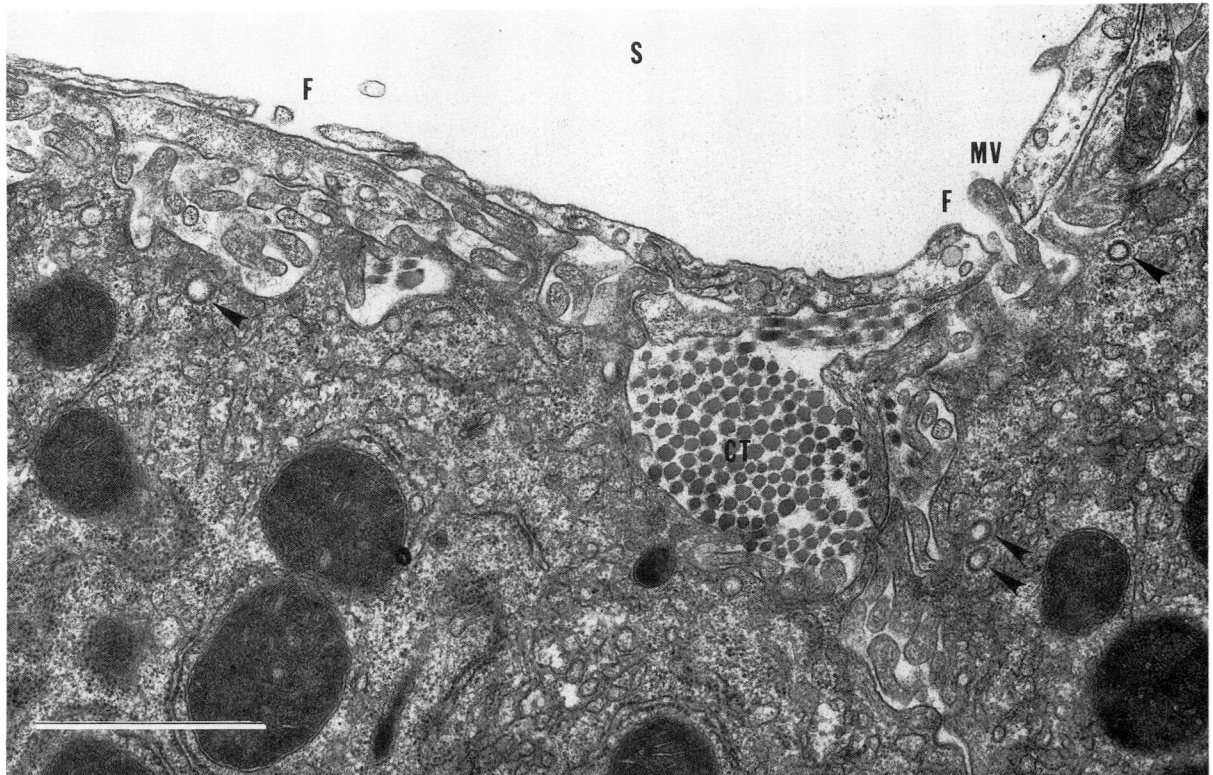

FIG. 1. Electron micrograph of normal rat hepatocyte. Microvillus (*MV*) is protruding through endothelial cell fenestrae (*F*) directly into the sinusoid (*S*). Numerous clathrin-coated pits and vesicles (*arrowheads*) are visible adjacent to sinusoidal plasma membrane. *CT*, connective tissue. *Bar*, 1 μm.

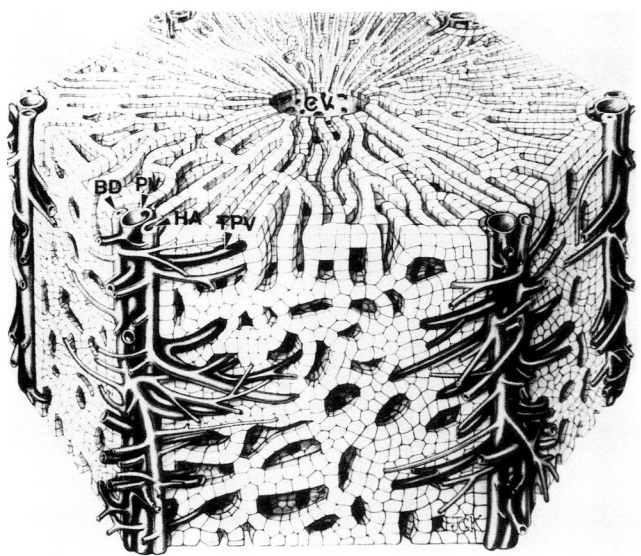

FIG. 2. Diagrammatic representation of structure of liver lobule. *BD*, bile duct; *PV*, portal vein; *HA*, hepatic artery; *TPV*, terminal portal venule; *CV*, central vein. [From Jones and Schmucker (38).]

tion gradient, because this might have a direct bearing on the interpretation of experimental results. For example, in the presence of a hepatotoxin that selectively injures only zone 3 cells, the hepatic clearance of substances that form a lobular concentration gradient would not be affected, even though substantial liver damage has occurred.

Hepatocyte

The hepatocyte accomplishes biliary protein transport by using subcellular structures common to many cells: membrane receptors, vesicles, Golgi, multivesicular bodies and lysosomes, and microtubules. The hepatocyte sinusoidal surface plasma membrane has many folds and microvilli, which enhance its surface area dimensions. This membrane contains so many receptors that it is difficult to imagine where there is space for any proteins or glycoproteins without receptor capabilities. Multiple invaginations, some of which contain clathrin coats, occur along this entire membrane (Fig. 1). Intracellular vesicles, usually without coats, are dispersed throughout the cytoplasm but are often more concentrated near the bile canaliculus. Multivesicular bodies and lysosomes are often associated with the Golgi-lysosome region of the cell, which is also generally located near the bile canaliculus. Microtubules are distributed randomly in hepatocytes in vivo, but in hepatocytes in primary monolayer culture, microtubules project radially from microtubule organizing centers (centrioles) near the nucleus [Fig. 3; (7, 24)]. Tight junctions between adjacent hepatocytes prevent regurgitation of biliary components from the canalicular lumen into blood. No paracellular route for proteins from canalicular lumen to blood has ever been demonstrated in normal liver.

The contribution of other liver cells, besides hepatocytes, to the protein composition of bile is unknown. The surfaces of parenchymal, ductule, duct, and gallbladder epithelial cells that abut the biliary space have numerous microvilli. In primates, duct, ductule, and gallbladder epithelial cells contain receptors for IgA (52), so these cells may be responsible for biliary IgA secretion in higher mammals. Besides having a biliary secretion function, these cells may modify bile composition by selectively taking up biliary components. The ducts are surrounded by blood capillary plexuses (Fig. 4), and the duct epithelial cells contain vesicles that might have transport functions. There is currently no method for sampling bile from canalicular spaces or ductules. Because only the final product contained in the common bile duct can be sampled, modifications in bile composition en route from canalicular spaces to the common bile duct cannot be followed. This makes interpretation of data unclear. For example, most of the insulin in bile collected via cannulation of the rat common bile duct is in the form of catabolized peptide fragments (58). Although intact insulin may have been initially secreted into bile canaliculi by hepatocytes, it may have been removed from bile by ductule cells, as a conservation measure, during its journey to the common bile duct. Until there is a way to sample bile canalicular contents, we can only speculate about what is actually secreted by hepatocytes.

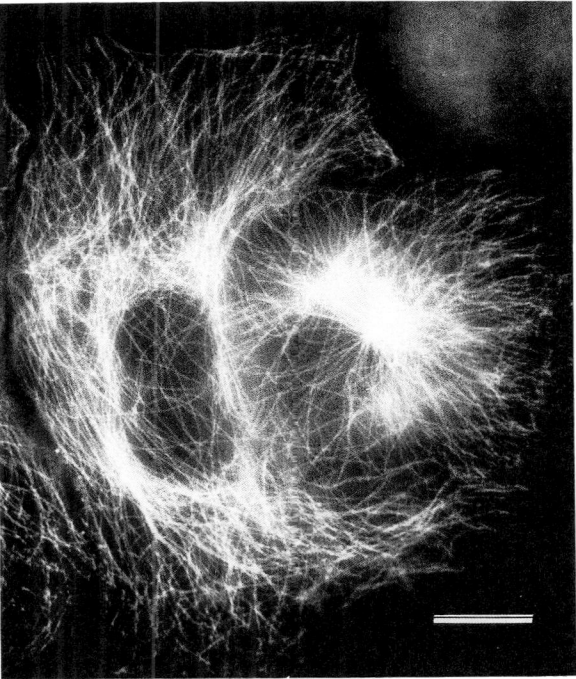

FIG. 3. Immunofluorescent staining of microtubules in rat hepatocytes in primary monolayer culture. Microtubules radiate from multiple microtubule organizing centers in perinuclear regions of cells. *Bar*, 10 μm. (Light micrograph courtesy of J. M. Caron.)

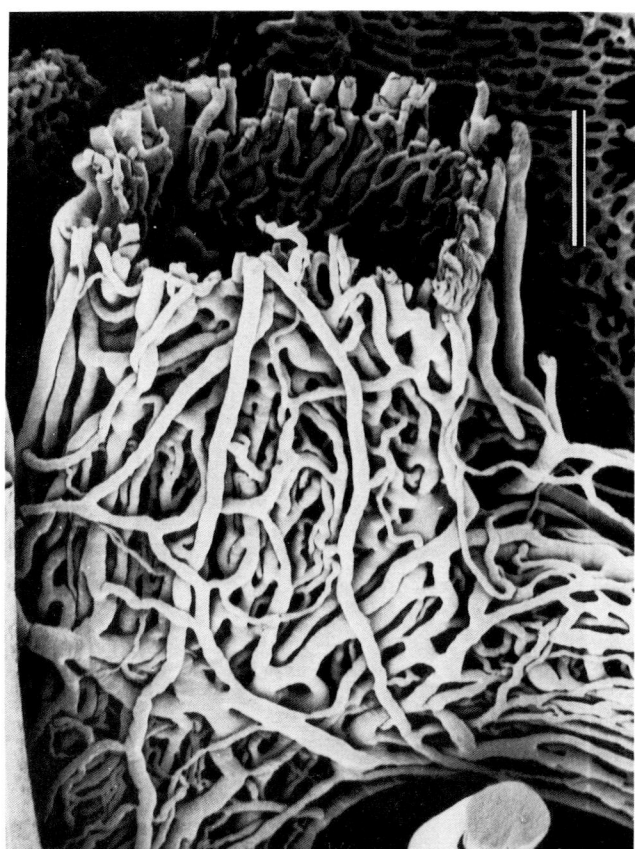

FIG. 4. Peribiliary plexus as visualized by methacrylate injection replica scanning electron microscopy. Plexus consists of an inner capillary network and an outer venous network. Bar, 200 μm. [From Ohtani (54).]

RECEPTOR-MEDIATED ENDOCYTOSIS
AND INTRACELLULAR TRANSPORT PATHWAYS

Plasma-derived proteins bind to specific receptors on the sinusoidal plasma membrane surface of hepatocytes, and the protein-receptor complexes are internalized into endocytic vesicles.

Proteins that are taken up by the liver are either degraded, secreted intact into bile, or utilized by the liver cell. Receptors on the hepatocyte sinusoidal plasma membrane determine which proteins are specifically taken up, but once endocytosis has occurred, the intracellular destinations of the proteins determine which of these three outcomes will result.

Our laboratory, among others, has provided much information on the intracellular transport pathways utilized by endocytosed proteins in hepatocytes. The proteins studied include human and rat IgA (35, 57), EGF (6, 62), insulin (58), asialoglycoproteins (30), low-density lipoproteins (10), and chylomicron and very-low-density lipoprotein remnants (29, 33). Several distinct pathways have emerged from these investigations. Once the proteins are internalized, they do not necessarily share a common fate with their receptors.

Lysosomal Degradation (Degradative Pathway)

Some proteins are primarily degraded by hepatocytes, such as asialoglycoproteins (30), low-density lipoproteins (10), and chylomicron and very-low-density lipoprotein remnants (29, 33). They are endocytosed in clathrin-coated pits and vesicles, the vesicles fuse to form endosomes, the endosomes mature into multivesicular bodies, and the multivesicular bodies, upon fusion with primary lysosomes, form secondary lysosomes (Fig. 5). Their contents are degraded by lysosomal enzymes, and the degradation products are secreted either back into the bloodstream or into bile.

In the presence of the acidic environment of the endosome compartment, these proteins dissociate from their receptors. From this point on the transport pathways for the proteins and their receptors diverge. The proteins proceed to the lysosomes, where they are degraded. The receptors are generally recycled; i.e., they are returned to the sinusoidal plasma membrane by an as yet unknown mechanism, where they are reutilized for receptor-mediated endocytosis [Fig. 5; (4)]. Vesicle acidification and receptor-ligand uncoupling are prerequisites for subsequent lysosomal processing of the protein and for receptor recycling.

Secretion of Intact Proteins into Bile (Transcellular Pathway)

Some proteins, such as IgA, are primarily secreted intact into bile (65). In this transport pathway, endocytic pits and vesicles may not be clathrin coated. The endocytic vesicles do not fuse but remain a constant size after formation (~1,000–1,500 Å) and travel directly to the bile canalicular membrane [Fig. 6; (57)]. They do not interact with other organelle compartments en route, and consequently there is no opportunity for lysosomal degradation. These endocytic "shuttle" vesicles empty their contents into the bile canalicular lumen by exocytosis. Immunoglobulin A is secreted still bound to a portion of its receptor, secretory component (46, 49). The part of the receptor that remains membrane associated may be recycled.

The transcellular pathway has also been found to operate in the reverse direction (Fig. 7). When polymeric or secretory IgA was perfused retrograde up the common bile duct, they were found to enter the bloodstream intact (36). They were taken up by hepatocytes at the bile canalicular surface and were carried across the cells to the sinusoidal surface, making use of shuttle vesicles of the same size and appearance as those seen in the normal transport direction (from sinusoidal to bile canalicular surface). There was no evidence for the involvement of any liver cells other than hepatocytes in this reverse transport process. The paracellular pathway between adjacent hepatocytes was also ruled out.

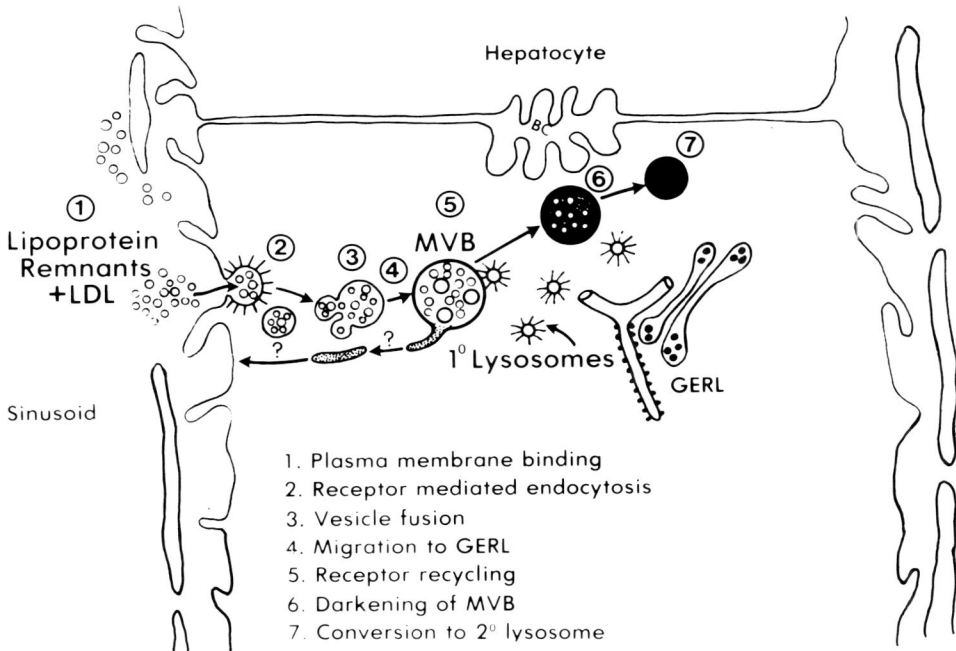

FIG. 5. Diagram depicts 7 major steps in macromolecular processing in degradative or lysosomal pathway. Although only lipoprotein remnants and low-density lipoproteins (LDL) are shown, steps are identical for all ligands entering this pathway. Taillike appendage that appears to be leaving the multivesicular body (MVB) in *step 5* may contain receptors that are in the process of recycling. Primary lysosomes derived from the Golgi, endoplasmic reticulum, lysosome (GERL) region of the cell fuse with MVB and release their acid-activated hydrolases into acidic interior of MVB. This results in degradation of MVB contents accompanied by a gradual condensation of MVB into a dense secondary lysosome or residual body. *BC*, bile canaliculus. [From Jones and Burwen (32).]

Mullock et al. (51) provided evidence that shuttle vesicles with membrane-associated secretory component are continuously formed at the sinusoidal surface and translocated to the bile canalicular surface even in the absence of internalized IgA. This evidence and the preceding observations indicate that this vesicular transport system operates continuously and can transport material in either direction.

Utilization by Hepatocytes

The uptake of certain endocytosed proteins, such as growth factors and transferrin, has direct metabolic consequences for hepatocytes. Their transport pathways are more specialized.

TRANSFERRIN. The transport pathway for transferrin was actually described in hepatoma cells (11), but because iron is stored in the liver, it is likely that the same mechanism for iron sequestration exists in normal hepatocytes. Ferric iron circulates in the blood bound to transferrin. At neutral pH, transferrin-Fe^{2+} binds to its receptor on the sinusoidal plasma membrane and is endocytosed as transferrin-Fe^{2+}-receptor complex. This complex enters the prelysosomal compartment, and in the acidic environment, Fe^{2+} dissociates from the complex. The transferrin, however, remains bound to its receptor at the acidic pH. The transferrin-receptor complex travels back to the plasma membrane, and on exposure to the neutral pH at the cell surface, the transferrin dissociates from its receptor and is released back into the blood. The receptor remains associated with the membrane. Meanwhile, inside the cell, upon dissociation from the transferrin, iron is released from the transport vesicle (by an unknown mechanism) and binds to ferritin for storage in the cytoplasm.

Under conditions of normal hepatic iron storage, biliary iron appears to be derived from a cytosolic pool (48). However, under pathological conditions of iron overload, both ferritin (3) and iron (48) accumulate in hepatocyte lysosomes and are excreted into bile, presumably via a lysosome-to-bile hepatic excretory pathway (48).

EPIDERMAL GROWTH FACTOR. A potent mitogen for cultured cells, including hepatocytes (63), EGF has been implicated as a hepatotrophic factor during liver regeneration (5, 13). Transport of EGF by rat hepatocytes in vivo has been well characterized (6, 62). In the rat, >90% of an intraportally injected dose of EGF is taken up by hepatocytes on its first pass through the liver, forming a steep lobular concentration gradient (62). Most of the endocytosed EGF is degraded in lysosomes. However, a small but significant portion is secreted intact into bile by the transcellular path-

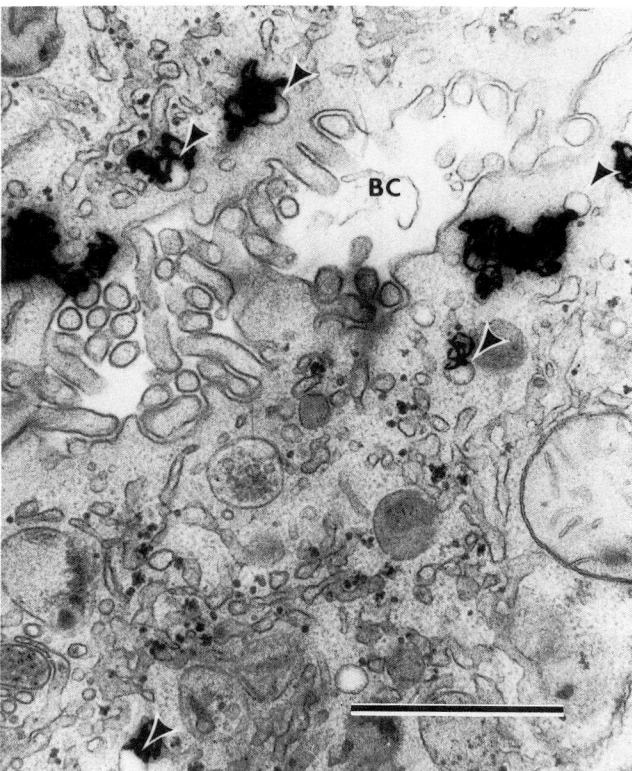

FIG. 6. Electron-microscopic autoradiograph showing association of silver grains with endocytic vesicles (*arrowheads*) in vicinity of bile canaliculi (*BC*) 30 min after injection of ^{125}I-labeled IgA into rat portal vein. *Bar*, 1 μm. [From Renston et al. (57), © 1980 by the American Association for the Advancement of Science.]

way, as indicated by the following data. *1*) Anti-EGF immunoprecipitable radioactivity peaks in bile at 20 min after intraportal injection of ^{125}I-labeled EGF, whereas radioactivity in bile associated with the degradative pathway peaks at 40 min postinjection (Fig. 8). *2*) In the presence of chloroquine (an inhibitor of lysosomal function), secretion of degraded EGF is significantly decreased, whereas the amount of intact EGF appearing in bile remains unchanged [Fig. 9; (6)].

The receptor for EGF is not recycled. Although it appears to uncouple from its ligand in the acidic environment of the prelysosomal compartment, the receptor apparently proceeds with its ligand to the lysosomal compartment, where both undergo degradation (70). The destruction of the receptor accounts for the phenomenon known as downregulation, whereby the amount of hormone taken up by a cell is self-limiting.

During the pre-S phase of liver regeneration (initiated by 70% hepatectomy), there is a marked change in the transport and metabolism of EGF by hepatocytes in vivo. Up to 27% of the EGF is translocated to the nucleus, accompanied by a concomitant decrease in lysosomal degradation [Table 1; (56)]. The nuclear-associated EGF is primarily intact, and a small but significant portion becomes covalently associated with a larger protein, whose molecular weight is consistent with the size of the EGF receptor.

INSULIN. The kinetics of insulin transport across liver cells is consistent with the utilization of a transcellular pathway (58). However, insulin collected from the common bile duct is in degraded form (58). Chloroquine, an inhibiter of lysosomal function, inhibits hepatic insulin degradation by only 25%. Nonlysosomal processing of insulin is also evident in hepatoma cells (69). Therefore insulin degradation may be due to the presence of insulinases in hepatocyte plasma membranes (72) or it may reflect a modification in bile protein composition subsequent to hepatocyte secretion, which cannot be detected due to the limitations of sampling only common bile duct contents (see *Hepatocyte*, p. 665). Nuclear accumulation of ^{125}I-labeled insulin has been demonstrated in several cultured cells (23, 68). This nuclear transport pathway for insulin may be similar to that observed for EGF in regenerating rat liver in vivo.

Insulin receptors, unlike EGF receptors, appear to be recycled by hepatocytes to some extent (9, 21, 45). Leupeptin and chloroquine do not affect the steady-state turnover of insulin receptors (45), suggesting that a nonlysosomal mechanism accounts for any degradation of receptors that does occur. In addition, hepatic insulin receptors are capable of forming oligomers, which affects their binding affinity and may influence their recycling or degradation as well (14).

Mechanisms of Pathway Regulation

Almost nothing is known of the factors or mechanisms that regulate the protein transport pathways of hepatocytes. It has been postulated that the receptor for each protein might contain the address for the intracellular destination of the endocytic vesicle in its cytoplasmic domain (4, 49). However, there is a growing body of evidence indicating that the receptor alone does not determine the transport pathway utilized by its ligand protein.

For proteins that are transported by more than one pathway, such as EGF, the selection of transport pathway at the level of the receptor could only take place if there were a subgroup of receptors for each pathway. There is no evidence for the existence of more than one class of receptor for EGF at the hepatocyte surface (18).

The association of clathrin with plasma membrane in areas where receptor-mediated endocytosis occurs suggests a possible regulatory role for this protein in the endocytic process. The conformational change in clathrin, from hexagonal to pentagonal arrays, appears to be instrumental in the pinching off of pits to form endocytic vesicles (28). In addition the transient presence of clathrin may affect the behavior of endo-

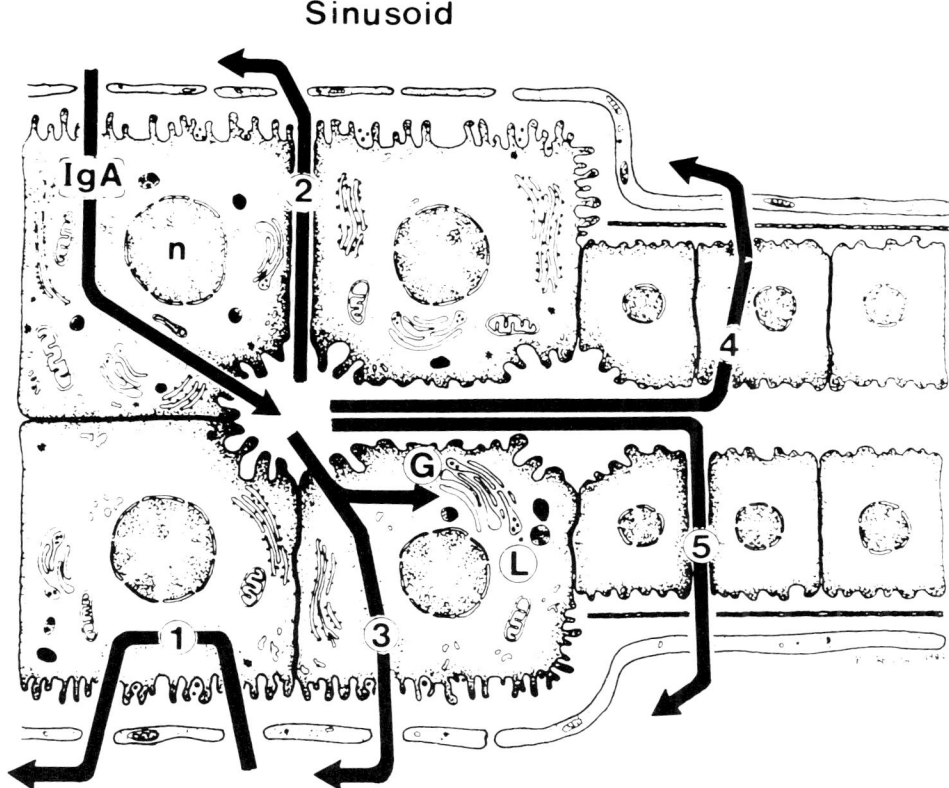

FIG. 7. Schematic diagram representing several possible pathways whereby IgA entering hepatocytes or biliary space may be regurgitated back into plasma. Evidence supports utilization of a direct transport pathway across parenchymal (3) cells. There is no experimental evidence for the paracellular route through tight junctional complexes (2 and 5). n, Nucleus; G, Golgi; L, lysosomes. [From Jones et al. (36), © 1984 by the American Association for the Study of Liver Diseases.]

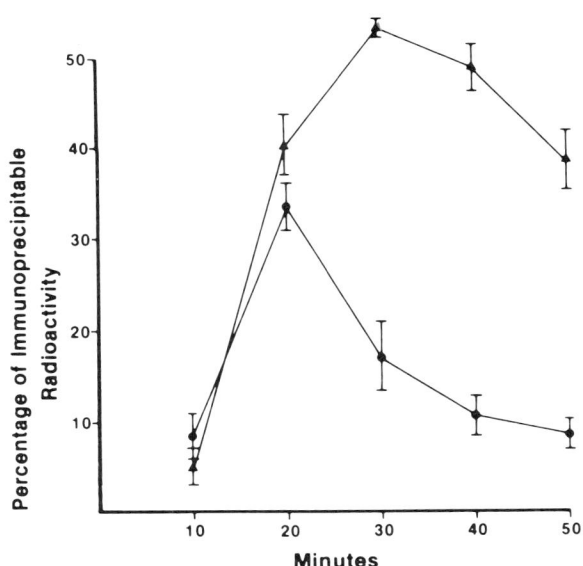

FIG. 8. Anti-EGF immunoprecipitable radioactivity secreted into bile, expressed as a percentage of total. ^{125}I-labeled EGF was injected into rat portal veins, and bile was collected via cannulae over 10-min intervals. Bile was immunoprecipitated with rabbit anti–mouse EGF antisera. ▲, Rats pretreated with chloroquine [number of rats (n) = 5]; ●, rats pretreated with saline (controls; n = 4). Error bars, standard error of mean. [From Burwen et al. (6).]

cytic vesicles, allowing them to fuse and enter into endosome formation. In the absence of clathrin, fusion may be prevented from occurring, inhibiting the formation of endosomes (hence inhibiting participation in the degradative pathway), and resulting instead in the formation of 1,000- to 1,500-Å shuttle vesicles (and a commitment to the transcellular pathway). This model requires that the segregation of receptors on the cell surface into clathrin-coated versus uncoated areas determines their entry into different pathways. It does not explain how one protein, and presumably one receptor, can have different pathways.

Some evidence suggests that the sorting event committing ligand to its ultimate intracellular destination can take place subsequent to internalization. Proteins that are known to bind to different receptors and utilize different pathways have been observed in the same endocytic vesicles. Transferrin and asialoglycoproteins have been localized in the same endocytic vesicles in hepatoma cells (53). Immunoglobulin A has also been observed in endocytic vesicles along with asialoglycoproteins (22). However, the results with IgA are difficult to interpret because the polymeric IgA species was not designated. As much as 30% of human IgA taken up by rat liver enters the lysosomal pathway in the same manner as asialoglycoproteins and lipoproteins (35), whereas rat dimeric IgA is transported exclusively by the transcellular route. Furthermore a proportion of human IgA, corresponding to the

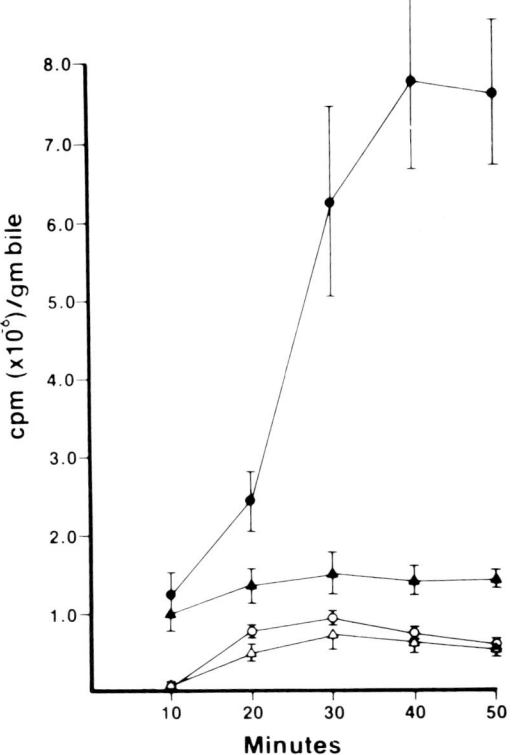

FIG. 9. Total and immunoprecipitable radioactivity in rat bile collected over a period of 50 min after intraportal injection of ^{125}I-labeled EGF. ●, Total radioactivity [control; number of rats (n) = 4]; ▲, total radioactivity, chloroquine pretreated (n = 5); ○, immunoprecipitable radioactivity (control; n = 4); △, immunoprecipitable radioactivity, chloroquine pretreated (n = 5). Error bars, standard error of mean. [From Burwen et al. (6).]

TABLE 1. *Intracellular Distribution of ^{125}I-EGF Autoradiographic Grains*

	HPX, % grains			SHAM, % grains		
	3 min	15 min	60 min	3 min	15 min	60 min
Nucleus	2.1	9.9	27.1	0.6	0.3	0.5
MVB-Lys	11.4	11.9	8.8	27.5	30.8	41.4

^{125}I-labeled epidermal growth factor (EGF) was injected into portal veins of rats 8 h after hepatectomy. At 3, 15, or 60 min after injection, livers were perfusion fixed and processed for electron-microscopic autoradiography. Autoradiographic grain distribution was quantitated by concentric circle analysis (58). Changes in volume density (71) of nuclear or MVB-Lys (multivesicular bodies and lysosomes) compartments cannot account for change in distribution of grains as a result of hepatectomy. HPX, 70% hepatectomy; SHAM, sham hepatectomy. [Adapted from Raper et al. (56).]

percentage associated with the lysosomal pathway, does not bind to secretory component but binds to a different receptor (probably the asialoglycoprotein receptor) for internalization (55, 64).

The nature of the binding between the receptor and its protein can influence which transport pathway is utilized. For example, when receptors are internalized by binding to antireceptor antibodies, instead of to their normal ligands, the normal transport pathways are perturbed. Upon internalization, transferrin receptor bound to antireceptor antibody gets degraded in lysosomes (44). Similarly, internalized secretory component bound to antisecretory component is also degraded in lysosomes (43). These two example, provided by transferrin receptor (normally recycled) and secretory component (normally secreted), demonstrate clearly that what the receptor binds can affect its intracellular destination and metabolic fate. The association between the receptors and their antibodies is polyvalent, instead of the usual monovalent association between these receptors and their normal ligands. The polyvalent nature of the binding may result in cross-linking at the cell surface; the resultant receptor aggregation may dramatically affect the intracellular destination after internalization (Fig. 10).

Some intrinsic property of the protein, beyond the nature of its binding to a particular receptor, may influence its transport pathway selection. For example, both EGF and transforming growth factor α (TGFα), which have a high degree of structural homology, utilize the same receptor on cell plasma membranes for binding and internalization (17). However, EGF stimulates normal growth and TGFα induces malignant transformation (16, 60), suggesting that once internalized their transport and processing diverge, resulting in a different growth response by the target cell. If this is the case, since the same receptor (and binding) is involved, a difference in transport pathway would have to be due to some other property of the protein inherent in its structure.

Microtubules may also be involved in the regulation of protein transport pathways. Their role in the vectorial movement of transport vesicles has been demonstrated (Fig. 11) and is discussed in more detail in the next section.

ROLE OF THE CYTOSKELETON IN PROTEIN TRANSPORT BY HEPATOCYTES

The liver contains a rich supply of cytoskeletal elements, including microtubules, microfilaments, and intermediate filaments. Because of their contractile properties, microfilaments were originally suspected of playing the major role in vesicular transport of proteins across liver cells. However, Kacich et al. (39) demonstrated that cytochalasin D, an agent that disrupts microfilaments, had no apparent effect on protein transport. On the other hand, the microtubule depolymerizing agent, colchicine, markedly disrupted the vesicular protein transport system in rat liver.

With regard to microtubules and the transcellular pathway, Goldman et al. (24) demonstrated that, although colchicine did not affect IgA uptake, internalization, and formation of endocytic shuttle vesicles, the subsequent vectorial movement of these IgA-containing vesicles was inhibited. In colchicine-treated

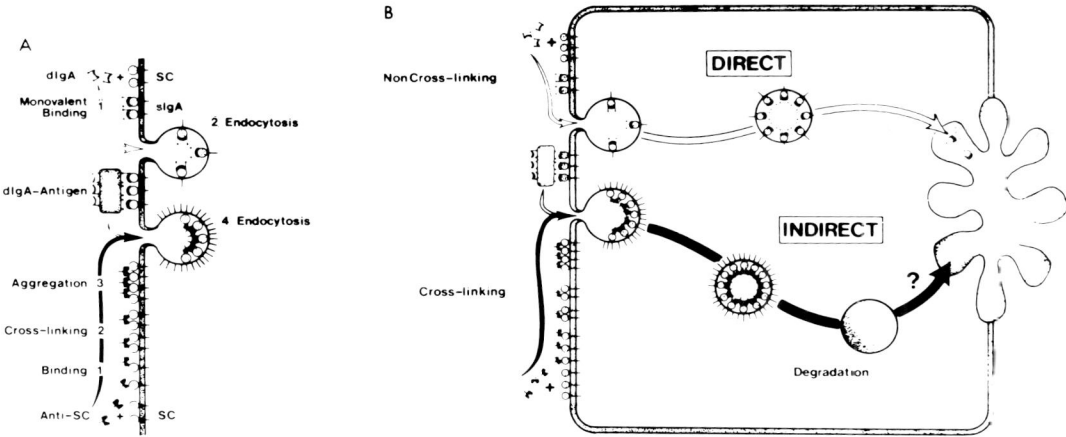

FIG. 10. *A*: dimeric IgA (dIgA) binds to its receptor, secretory component (SC), monovalently through disulfide linkages. However, IgG anti-SC antibody binds in a polyvalent fashion that may result in receptor cross-linking and aggregation. *B*: SC-receptor aggregation may result in degradation of both receptor and ligand via lysosomal (indirect) pathway, rather than their direct transport into bile intact. This phenomenon may be a physiological mechanism for selective degradation of circulating dIgA-antigen complexes. sIgA, secretory IgA. [From Kim et al. (43).]

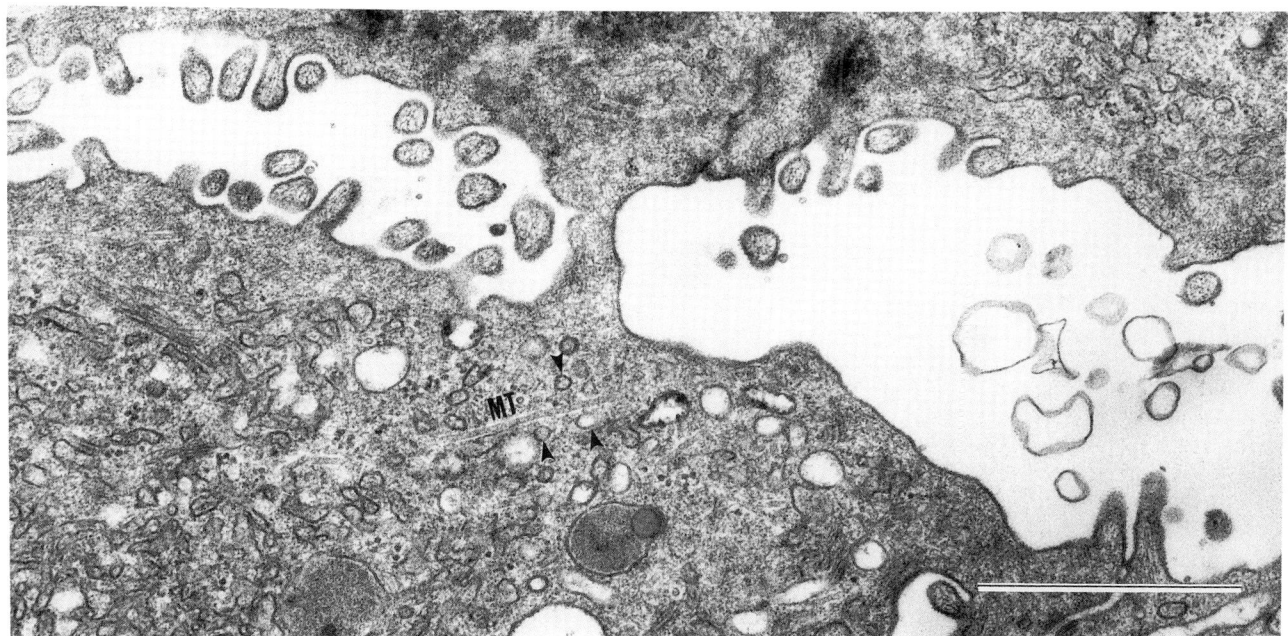

FIG. 11. Electron micrograph of bile canalicular region of hepatocyte from rat with extrahepatic cholestasis. Numerous endocytic vesicles (*arrowheads*) are in close association with microtubules (*MT*). Bar, 1 μm. [From Renston et al. (59), © 1983 by the American Association for the Study of Liver Diseases.]

primary cultures of rat hepatocytes, IgA-containing vesicles remained at the cell periphery, whereas in control cells not exposed to colchicine the vesicles were transported well into the cell interior.

Liver microtubules play a major role in directing vesicles to the lysosomal pathway, as well as to the transcellular pathway. In primary cultures of rat hepatocytes, the length of the microtubule that radiates out from the microtubule organizing center could be controlled by exposing the cells to graduated doses of Colcemid, whereby the length of the microtubules was inversely proportional to the Colcemid concentration (7). Furthermore, by measuring the rate of degradation of asialoglycoproteins (which are transported by the degradative pathway), it was possible to demonstrate a relationship between microtubule length and the amount of protein reaching the lysosomal compartment of the cell. The rate of lysosomal degradation of asialoglycoproteins was inversely proportional to the Colcemid concentration, as was microtubule length (7).

These findings raise important questions as to the mechanisms by which microtubules can direct vesicles. Are the same microtubules responsible for the vectorial movement of vesicles to lysosomes, the bile canaliculus, or the nucleus? There may be subtle differences in microtubule structure that allow them to specialize in different transport destinations. Eukaryotic cells contain multiple genes for both α- and β-tubulin (12). Furthermore, differential expression of tubulin genes occurs during rat brain development (2). Therefore there may be a structural basis for the heterogeneity in microtubule function. More will have to be learned about microtubule function before biliary protein secretion is completely understood.

FUNCTIONAL SIGNIFICANCE OF HEPATOBILIARY PROTEIN SECRETION

Intestinal Immune Response

The protein in bile with the most obvious functional significance is IgA. The importance of the biliary pathway of IgA secretion in primates is not clear. However, in rat, up to 90% of the specific IgA antibodies present in the intestinal lumen have reached this destination via hepatic secretion into bile (F. Koster, unpublished observations).

The liver appears to be the only source of IgA found in the intestinal lumen during the early phases of the intestine-initiated immune response. Beginning at 5 days after intestinal exposure to the antigen, cholera toxin, specific anti–cholera toxin antibody-producing cells are found only in liver (1). Lymphocytes in the lamina propria of the intestine become involved in the immune response only after boosting with additional antigen exposure. Peyer's patch lymphocytes "home" to the intestine and other secretory organs, where they mature into plasma cells and begin secreting specific IgA antibodies. At this point in the immune response these plasma cells become the major source of the IgA that appears in bile via hepatic transport.

The amount of biliary IgA secretion in primates, as compared with rats, is quite variable (46). In human bile, IgA is present as both secretory IgA and polymeric IgA without secretory component (46). Secretory component was originally thought to be present only on hepatocyte plasma membranes. However, evidence now indicates that in primates it is associated with bile ductule, duct, and gallbladder epithelial cells (52). Therefore biliary IgA in primates is probably secreted by these epithelial cells rather than by hepatocytes.

Squirrel monkeys secrete into their bile only minute amounts of exogenous human or rat polymeric IgA, or Fc_2 fragments derived from dimeric IgA, that has been injected into their portal veins (37). However, squirrel monkey liver takes up and processes asialoglycoproteins in a manner very similar to that of rat liver (65a). The sinusoidal endothelial cells of squirrel monkey liver are also morphologically similar to those of rat liver, having the same large fenestrae. This morphological feature, together with the evidence that asialoglycoproteins are transported by hepatocytes and Fc fragments are not, indicates that the lack of transport of IgA by squirrel monkey liver is not due to the inaccessibility of the hepatocyte sinusoidal surface to large immunoglobulins but is due instead to a selective barrier to IgA transport. This barrier may be the absence of the necessary receptor, secretory component, on the hepatocyte surface.

In rhesus monkeys with bile fistulae, exposure of the intestinal lumen to the antigen, cholera toxin, results in a 20-fold increase in endogenous IgA secretion into bile (41). The mechanism by which this IgA reaches the bile remains unknown.

Kupffer cells, with Fc receptors, clear antigen-antibody complexes from the blood. Recent data from Russell et al. (61) indicate that hepatocytes in mouse liver also play an important role in the clearance of immune complexes, and some complexes enter the bile intact. It is speculated that IgA immune complexes enter the hepatocyte via binding of the Fc portion of the IgA to secretory component. The IgA molecule is in turn bound to the antigen by its Fab portion.

Biliary Cholesterol Metabolism

Apolipoproteins A1 and A2, the major protein components of high-density lipoproteins, are present in bile as intact polypeptides (67). These apolipoproteins are carriers of cholesterol and phospholipid in plasma, and they may perform an analogous function in bile (42). Two observations suggest their involvement in biliary cholesterol metabolism: 1) they inhibit the formation of cholesterol monohydrate crystals in vitro, and 2) they are present in the chromatography fractions of biliary proteins that are capable of inhibiting cholesterol nucleation (42).

Therefore apolipoproteins A1 and A2 may help prevent gallstone formation in human bile, even under conditions of cholesterol supersaturation.

Pathophysiological Conditions

Under certain pathophysiological conditions the hepatobiliary secretion of proteins is dramatically altered. The transport pathways normally utilized are modified, resulting in metabolic consequences for the organism, as well as for the protein.

In studies using senescent rats, the hepatobiliary secretion of IgA declines sixfold as a function of age (66). This decline in transport function is due to a sharp decrease in the number of IgA receptors on hepatocyte plasma membranes in senescent rats (15).

However, there is no age-associated decline in the hepatobiliary secretion of asialoglycoproteins (C. K. Daniels, unpublished observations), indicating that the overall protein transport function of the liver does not necessarily decline as a function of age.

This age-dependent loss of IgA receptors in rat liver may be a specific, isolated event for IgA receptors or it may be indicative of age-related pathophysiological changes that tend to selectively affect the transcellular protein transport pathway. Changes in hepatocyte pathophysiology that also selectively affect the transcellular transport of IgA can be experimentally induced. In the presence of ethynylestradiol, an estrogen with cholestatic properties, the uptake and processing of IgA by rat liver is markedly inhibited (25). However, proteins utilizing the degradative pathway, such as asialoglycoproteins, continue to be taken up and transported. Although asialoglycoprotein degradation is somewhat inhibited by ethynylestradiol treatment, nonetheless the degradative pathway continues to be operative in the presence of ethynylestradiol.

Functional Significance of Hepatobiliary Transport of Epidermal Growth Factor

Epidermal growth factor secreted into bile reaches the intestinal lumen, where its mitogenic activity may contribute to the stimulation of intestinal epithelial cell proliferation. In the intestinal lumen, EGF also comes directly from the secretions of the two major glands that synthesize EGF: salivary and Brunner's glands.

Once EGF is in the blood it has access to all tissues in the organism that require it for normal growth. Circulating EGF also inhibits gastric acid secretion (see Ref. 63 for review). The function of the liver in taking up EGF may be to provide a homeostatic mechanism for regulating levels of circulating EGF.

Epidermal growth factor has also been implicated as a hepatotrophic factor during liver regeneration. *1)* In normal liver, EGF infused intraperitoneally stimulates DNA synthesis (5). *2)* Periportal hepatocytes take up >90% of an injected dose of ^{125}I-labeled EGF in the first pass through the liver (62); these cells demonstrate the highest rate of DNA synthesis and mitosis after partial hepatectomy (20, 27). *3)* The number of EGF receptors in regenerating liver is reduced dramatically several hours before DNA synthesis commences (19); such a downregulation of receptors is associated with EGF-stimulated mitosis (8). *4)* It has recently been reported that endogenous EGF concentration in the plasma of hepatectomized rats is significantly elevated from 6 to 18 h posthepatectomy, with the increase detectable as early as 2 h (13). Because liver has the capacity to regenerate, its ability to sequester and transport EGF may represent a latent function held in reserve, to be called on when needed for regeneration.

Recently we demonstrated that the regenerating rat liver handles EGF very differently than the normal adult rat liver (56). During the pre-S phase of liver regeneration, at 8 h after its initiation by 70% hepatectomy, the regenerating liver fragment retains the same percentage of an injected dose of ^{125}I-labeled EGF as a sham-hepatectomized control. Due to the smaller size of the regenerating liver fragment, the resulting amount of EGF retained per gram liver is almost 3 times greater for regenerating than for control liver.

In addition to the increased retention of EGF by regenerating liver, there is an accompanying marked decrease in the amount of degraded EGF secreted into bile. Quantitative electron-microscopic autoradiography was used to determine the intracellular distribution of ^{125}I-labeled EGF at various times after its intraportal injection into rats at 8 h after 70% hepatectomy. There is a time-dependent accumulation of autoradiographic grains in hepatocyte nuclei (Table 1). Up to 27% of the grains become associated with nuclei by 1 h after ^{125}I-labeled EGF injection (as compared with ~0.5% in sham-hepatectomized controls). The increase in nuclear-associated grains is accompanied by a substantial decrease in grains associated with multivesicular bodies and lysosomes, organelles of the degradative pathway.

This nuclear-associated radioactivity, when further analyzed by isolating liver nuclei from regenerating livers exposed to ^{125}I-labeled EGF, is ~70% immunoprecipitable with a specific anti-EGF antiserum. In addition, sodium dodecyl sulfate–polyacrylamide gel electrophoresis and autoradiography reveal the presence of a 180,000-Da radioactive protein band, which is a size consistent with EGF complexed to its receptor. Positive identification of this band has not been made.

In summary, the altered intracellular transport of EGF during liver regeneration suggests the EGF and/or its receptor may participate at the nuclear level in the initiation of DNA synthesis or alteration of gene expression.

REFERENCES

1. ALTORFER, J., S. J. HARDESTY, J. H. SCOTT, AND A. L. JONES. Specific antibody synthesis and biliary secretion by the rat liver after intestinal immunization with cholera toxin. *Gastroenterology* 93: 539–549, 1987.
2. BOND, J. F., AND S. R. FARMER. Regulation of tubulin and actin mRNA production in rat brain: expression of a new β-tubulin mRNA with development. *Mol. Cell. Biol.* 3: 1333–1342, 1983.
3. BRADFORD, W. D., J. G. ELCHLEPP, A. U. ARSTILA, B. R. TRUMP, AND T. D. KINNEY. Iron metabolism and cell membranes: relation between ferritin and hemosiderin in bile and biliary excretion of lysosomal contents. *Am. J. Pathol.* 56: 201–228, 1969.
4. BROWN, M. S., R. G. W. ANDERSON, AND J. L. GOLDSTEIN.

Recycling receptors: the round trip itinerary of migrant membrane proteins. *Cell* 32: 663–667, 1983.
5. BUCHER, N. L. R., U. PATEL, AND S. COHEN. Hormonal factors concerned with liver regeneration. In: *Hepatotrophic Factors*. Amsterdam: Elsevier, 1978, p. 95–110. (Ciba Found. Symp.).
6. BURWEN, S. J., M. E. BARKER, I. S. GOLDMAN, G. T. HRADEK, S. E. RAPER, AND A. L. JONES. Transport of epidermal growth factor by rat liver: evidence for a nonlysosomal pathway. *J. Cell Biol.* 99: 1259–1265, 1984.
7. CARON, J. M., A. L. JONES, AND M. W. KIRSCHNER. Autoregulation of tubulin synthesis in hepatocytes and fibroblasts. *J. Cell Biol.* 101: 1763–1772, 1985.
8. CARPENTER, G., AND S. COHEN. Epidermal growth factor. *Annu. Rev. Biochem.* 48: 193–216, 1979.
9. CARPENTIER, J.-L., H. GAZZANO, E. VAN OBBERGHEN, M. FEHLMAN, P. FREYCHET, AND L. ORCI. Intracellular pathway followed by the insulin receptor covalently coupled to ^{125}I-photoreactive insulin during internalization and recycling. *J. Cell Biol.* 102: 989–996, 1986.
10. CHAO, Y.-S., A. L. JONES, G. T. HRADEK, E. E. T. WINDLER, AND R. J. HAVEL. Autoradiographic localization of the sites of uptake, cellular transport, and catabolism of low density lipoproteins in the liver of normal and estrogen-treated rats. *Proc. Natl. Acad. Sci. USA* 78: 597–601, 1981.
11. CIECHANOVER, A., A. L. SCHWARTZ, A. DAUTRY-VARSAT, AND H. F. LODISH. Kinetics of internalization and recycling of transferrin and the transferrin receptor in a human hepatoma cell line. *J. Biol. Chem.* 258: 9681–9689, 1983.
12. CLEVELAND, D. W., M. A. LOPATA, R. J. MACDONALD, N. J. COWAN, W. J. RUTTER, AND M. W. KIRSCHNER. Number and evolutionary conservation of α- and β-tubulin and cytoplasmic β- and γ-actin genes using specific cloned cDNA probes. *Cell* 20: 95–105, 1980.
13. CORNELL, R. P. Gut-derived endotoxin elicits hepatotrophic factor secretion for liver regeneration. *Am. J. Physiol.* 249 (*Regulatory Integrative Comp. Physiol.* 18): R551–R562, 1985.
14. CRETTAZ, M., I. JIALEL, M. KASUGA, AND C. R. KAHN. Insulin receptor regulation and desensitization in rat hepatoma cells. *J. Biol. Chem.* 259: 11543–11549, 1984.
15. DANIELS, C. K., D. L. SCHMUCKER, AND A. L. JONES. Age-dependent loss of dimeric immunoglobulin A receptors in the liver of the Fischer 344 rat. *J. Immunol.* 134: 3855–3858, 1985.
16. DE LARCO, J. E., AND G. J. TODARO. Growth factors from murine sarcoma virus-transformed cells. *Proc. Natl. Acad. Sci. USA* 75: 4001–4005, 1978.
17. DERYNCK, R., A. B. ROBERTS, M. E. WINKLER, E. Y. CHEN, AND D. V. GOEDDEL. Human transforming growth factor-α: precursor structure and expression in *E. coli*. *Cell* 38: 287–297, 1984.
18. DUNN, W. A., T. P. CONNOLLY, AND A. L. HUBBARD. Receptor-mediated endocytosis of epidermal growth factor by rat hepatocytes: receptor pathway. *J. Cell Biol.* 102: 24–36, 1986.
19. EARP, H. S., AND E. J. O'KEEFE. Epidermal growth factor receptor number decreases during rat liver regeneration. *J. Clin. Invest.* 67: 1580–1583, 1981.
20. FABRIKANT, J. I. Kinetic analysis of hepatic regeneration. *Growth* 31: 311–315, 1967.
21. FEHLMANN, M., J.-L. CARPENTIER, E. VAN OBBERGHEN, P. FREYCHET, P. THAMM, D. SAUNDERS, D. BRANDENBURG, AND L. ORCI. Internalized insulin receptors are recycled to the cell surface in rat hepatocytes. *Proc. Natl. Acad. Sci. USA* 79: 5921–5925, 1982.
22. GEUZE, H. J., J. W. SLOT, G. J. A. M. STROUS, J. PEPPARD, K. VON FIGURA, A. HASILIK, AND A. L. SCHWARTZ. Intracellular receptor sorting during endocytosis: comparative immunoelectron microscopy of multiple receptors in rat liver. *Cell* 37: 195–204, 1984.
23. GOLDFINE, I. D., A. L. JONES, G. T. HRADEK, K. Y. WONG, AND J. S. MOONEY. Entry of insulin into human cultured lymphocytes: electron microscope autoradiographic analysis. *Science Wash. DC* 202: 760–763, 1978.

24. GOLDMAN, I. S., A. L. JONES, G. T. HRADEK, AND S. HULING. Hepatocyte handling of immunoglobulin A in the rat: the role of microtubules. *Gastroenterology* 85: 130–140, 1983.
25. GOLDSMITH, M. A., A. L. JONES, B. J. UNDERDOWN, AND J. M. SCHIFF. Multiple estrogen effects on hepatic processing of immunoglobulin A and asialoglycoprotein (Abstract). *J. Cell Biol.* 97: 158a, 1983.
26. GORESKY, C. A., G. G. BACH, AND B. E. NADEAU. On the uptake of materials by the intact liver: the transport and net removal of galactose. *J. Clin. Invest.* 52: 991–1009, 1973.
27. GRISHAM, J. W. A morphologic study of deoxyribonucleic acid synthesis and cell proliferation in regenerating rat livers: autoradiography with thymidine-H^3. *Cancer Res.* 22: 842–849, 1962.
28. HARRISON, S. C., AND T. KIRCHHAUSEN. Clathrin, cages, and coated vesicles. *Cell* 33: 650–652, 1983.
29. HORNICK, C. A., A. L. JONES, G. RENAUD, G. HRADEK, AND R. J. HAVEL. Effect of chloroquine on low-density lipoprotein catabolic pathway in rat hepatocytes. *Am. J. Physiol.* 246: (*Gastrointest. Liver Physiol.* 9): G187–G194, 1984.
30. HUBBARD, A. L., AND H. STUKENBROK. An electron microscope autoradiographic study of the carbohydrate recognition systems in rat liver. II. Intracellular fates of the ^{125}I-ligands. *J. Cell Biol.* 83: 65–81, 1979.
31. IRVING, M. G., F. J. ROLL, S. HUANG, AND D. M. BISSELL. Characterization and culture of sinusoidal endothelium from normal rat liver: lipoprotein uptake and collagen phenotype. *Gastroenterology* 87: 1233–1247, 1984.
32. JONES, A. L., AND S. J. BURWEN. Hepatic receptors and their ligands: problems of intracellular sorting and vectorial movement. *Semin. Liver Dis.* 5: 136–146, 1985.
33. JONES, A. L., G. T. HRADEK, C. HORNICK, G. RENAUD, E. E. T. WINDLER, AND R. J. HAVEL. Uptake and processing of remnants of chylomicrons and very low density lipoproteins by rat liver. *J. Lipid Res.* 25: 1151–1158, 1984.
34. JONES, A. L., G. T. HRADEK, R. H. RENSTON, K. Y. WONG, G. KARLAGANIS, AND G. PAUMGARTNER. Autoradiographic evidence for hepatic lobular concentration gradient of bile acid derivative. *Am. J. Physiol.* 238 (*Gastrointest. Liver Physiol.* 1): G233–G237, 1980.
35. JONES, A. L., G. T. HRADEK, AND D. S. SCHMUCKER. Intracellular processing of human vs. rat immunoglobulin A in the rat liver. *Hepatology Baltimore* 5: 1172–1178, 1985.
36. JONES, A. L., G. T. HRADEK, D. L. SCHMUCKER, AND B. J. UNDERDOWN. The fate of polymeric and secretory immunoglobulin A infusion into the common bile duct in rats. *Hepatology Baltimore* 4: 1173–1183, 1984.
37. JONES, A. L., S. HULING, B. J. UNDERDOWN, AND J. M. SCHIFF. Hepatic transport of IgA and asialoglycoprotein in non-human primates (Abstract). *Gastroenterology* 86: 1324, 1984.
38. JONES, A. L., AND D. L. SCHMUCKER. Current concepts of liver structure as related to function. *Gastroenterology* 73: 833–851, 1977.
39. KACICH, R. L., R. H. RENSTON, AND A. L. JONES. Effects of cytochalasin D and colchicine on the uptake, translocation, and biliary secretion of horseradish peroxidase and [^{14}C]sodium taurocholate in the rat. *Gastroenterology* 85: 385–394, 1983.
40. KAKIS, G., AND I. YOUSEF. Protein composition of rat bile. *Can. J. Biochem.* 56: 287–290, 1978.
41. KECLIK, M., R. H. WOLF, O. FELSENFELD, AND H. F. SMETANA. Immunoglobulins and antibodies in gallbladder bile. *Am. J. Gastroenterol.* 54: 19–29, 1970.
42. KIBE, A., R. T. HOLXBACH, N. F. LARUSSO, AND S. J. T. MAO. Inhibition of cholesterol crystal formation by apolipoproteins in supersaturated model bile. *Science Wash. DC* 225: 514–516, 1984.
43. KIM, E., G. T. HRADEK, AND A. L. JONES. Degradative intracellular transport of antisecretory component in cultured hepatocytes. *Gastroenterology* 88: 1791–1798, 1985.
44. KLAUSNER, R. D., G. ASHWELL, J. VAN RENSWONDE, J. B. HARFORD, AND K. R. BRIDGES. Binding of apotransferrin to

K562 cells: explanation of the transferrin cycle. *Proc. Natl. Acad. Sci. USA* 80: 2263-2266, 1983.

45. KRUPP, M. N., AND M. D. LANE. Evidence for different pathways for the degradation of insulin and insulin receptor in the chick liver cell. *J. Biol. Chem.* 257: 1372-1377, 1982.

46. KUTTEH, W. H., S. J. PRINCE, J. O. PHILLIPS, J. G. SPENCE, AND J. MESTECKY. Properties of immunoglobulin A in serum of individuals with liver diseases and in hepatic bile. *Gastroenterology* 82: 184-193, 1982.

47. LaRUSSO, N. F. Proteins in bile: how they get there and what they do. *Am. J. Physiol.* 247 (*Gastrointest. Liver Physiol.* 10): G199-G205, 1984.

48. LE SAGE, G. D., L. J. KOST, S. S. BARHAM, AND N. F. LaRUSSO. Biliary excretion of iron from hepatocyte lysosomes in the rat: a major excretory pathway in experimental iron overload. *J. Clin. Invest.* 77: 90-97, 1986.

49. MOSTOV, K. E., J.-P. KRAEHENBUHL, AND G. BLOBEL. Receptor-mediated transcellular transport of immunoglobulin: synthesis of secretory component as multiple and larger transmembrane forms. *Proc. Natl. Acad. Sci. USA* 77: 7257-7261, 1980.

50. MULLOCK, B. M., M. DOBROTA, AND R. HINTON. Sources of the protein in rat bile. *Biochim. Biophys. Acta* 543: 497-507, 1978.

51. MULLOCK, B. M., R. S. JONES, AND R. H. HINTON. Movement of endocytic shuttle vesicles from the sinusoidal to the bile canalicular face of hepatocytes does not depend on occupation of receptor sites. *FEBS Lett.* 113: 201-205, 1980.

52. NAGURA, H., P. D. SMITH, P. K. NAKANE, AND W. R. BROWN. IgA in human bile and liver. *J. Immunol.* 126: 587-595, 1981.

53. NEUTRA, M. R., H. F. LODISH, AND A. J. CIECHANOVER. Transferrin and asialoorosomucoid share the same endocytic vesicles during receptor-mediated endocytosis in hepatoma cells (Abstract). *J. Cell Biol.* 97: 167a, 1983.

54. OHTANI, O. The peribiliary portal system in the rabbit liver. *Arch. Histol. Jpn.* 42: 153-167, 1979.

55. PHILLIPS, J. O., M. W. RUSSELL, T. A. BROWN, AND J. MESTECKY. Selective hepatobiliary transport of human polymeric IgA in mice. *Mol. Immunol.* 21: 907-914, 1984.

56. RAPER, S. E., S. J. BURWEN, M. E. BARKER, AND A. L. JONES. Translocation of epidermal growth factor to the hepatocyte nucleus during rat liver regeneration. *Gastroenterology* 92: 1243-1250, 1987.

57. RENSTON, R. H., A. L. JONES, W. D. CHRISTIANSEN, G. T. HRADEK, AND B. J. UNDERDOWN. Evidence for a vesicular transport mechanism in hepatocytes for biliary secretion of immunoglobulin A. *Science Wash. DC* 208: 1276-1278, 1980.

58. RENSTON, R. H., D. G. MALONEY, A. L. JONES, G. T. HRADEK, K. Y. WONG, AND I. D. GOLDFINE. The bile secretory apparatus: evidence for a vesicular transport mechanism for proteins in the rat, using horseradish peroxidase and [^{125}I]insulin. *Gastroenterology* 78: 1373-1388, 1980.

59. RENSTON, R. H., G. ZSIGMOND, R. A. BERNHOFT, S. J. BURWEN, AND A. L. JONES. Vesicular transport of horseradish peroxidase during chronic bile duct obstruction in the rat. *Hepatology Baltimore* 3: 673-680, 1983.

60. ROBERTS, A. B., C. A. FROLIK, M. A. ANZANO, AND M. B. SPORN. Transforming growth factors from neoplastic and non-neoplastic tissues. *Federation Proc.* 42: 2621-2625, 1983.

61. RUSSELL, M. W., T. A. BROWN, J. L. CLAFLIN, K. SCHROER, AND J. MESTECKY. Immunoglobulin A-mediated hepatobiliary transport constitutes a natural pathway for disposing of bacterial antigens. *Infect. Immun.* 42: 1041-1048, 1983.

62. ST. HILAIRE, R. J., G. T. HRADEK, AND A. L. JONES. Hepatic sequestration and biliary secretion of epidermal growth factor: evidence for a high-capacity uptake system. *Proc. Natl. Acad. Sci. USA* 80: 3797-3801, 1983.

63. ST. HILAIRE, R. J., AND A. L. JONES. Epidermal growth factor: its biologic and metabolic effects with emphasis on the hepatocyte. *Hepatology Baltimore* 2: 601-613, 1982.

64. SCHIFF, J. M., M. M. FISHER, A. L. JONES, AND B. J. UNDERDOWN. Human IgA as a heterovalent ligand: switching from the asialoglycoprotein receptor to secretory component during transport across the rat hepatocyte. *J. Cell Biol.* 102: 920-931, 1986.

65. SCHIFF, J. M., M. M. FISHER, AND B. J. UNDERDOWN. Receptor-mediated biliary transport of immunoglobulin A and asialoglycoprotein: sorting and missorting of ligands revealed by two radiolabeling methods. *J. Cell Biol.* 98: 79-89, 1984.

65a. SCHIFF, J. M., S. L. HULING, AND A. L. JONES. Receptor-mediated uptake of asialoglycoprotein by the primate liver initiates both lysosomal and transcellular pathways. *Hepatology Baltimore* 6: 837-847, 1986.

66. SCHMUCKER, D. L., R. GILBERT, G. T. HRADEK, A. L. JONES, AND H. BAZIN. Effect of aging on the hepatobiliary transport of dimeric immunoglobulin A in the male Fischer rat. *Gastroenterology* 88: 436-443, 1985.

67. SEWELL, R. B., S. J. MAO, T. KAWAMOTO, AND N. F. LaRUSSO. Apoproteins of high, low, and very low density lipoproteins in human bile. *J. Lipid Res.* 24: 391-401, 1983.

68. SMITH, R. M., AND L. JARETT. Ultrastructural evidence for the accumulation of insulin in nuclei of intact 3T3-L1 adipocytes by an insulin-receptor mediated process. *Proc. Natl. Acad. Sci. USA* 84: 459-463, 1987.

69. SMITH, R. M., N. H. LAUDENSLAGER, N. SHAH, AND L. JARETT. Insulin binding and processing by H4IIEC3 hepatoma cells: ultrastructural and biochemical evidence for a unique route of internalization and processing. *J. Cell Physiol.* 130: 428-435, 1987.

70. STOSCHECK, C. M., AND G. CARPENTER. Down regulation of epidermal growth factor receptors: direct demonstration of receptor degradation in human fibroblasts. *J. Cell Biol.* 98: 1048-1053, 1984.

71. WEIBEL, E. R., W. STAUBLI, H. R. GNAGI, AND F. A. HESS. Correlated morphometric and biochemical studies on the liver cell. I. Morphometric model, stereologic methods, and normal morphometric data for rat liver. *J. Cell Biol.* 42: 68-91, 1969.

72. YOKONO, K., R. A. ROTH, AND S. BABA. Identification of insulin-degrading enzyme on the surface of cultured human lymphocytes, rat hepatoma cells, and primary cultures of rat hepatocytes. *Endocrinology* 111: 1102-1108, 1982.

CHAPTER 33

Hepatocyte lysosomes in intracellular digestion and biliary secretion

NICHOLAS F. LaRUSSO | *Gastroenterology Research Unit, Mayo Medical School, Clinic, and Foundation, Rochester, Minnesota*

CHAPTER CONTENTS

Lysosomes: Background and General Considerations
  Perspectives
  Definitions
    Biochemical
    Morphological
    Histochemical
    Functional
  Characteristics
    Biochemical
    Morphological
  Methods of study
    Isolation of lysosomes
    Assay of lysosomal hydrolases
    Electron microscopy, including histochemistry
Lysosomes: Formation and Function
  Enzyme biogenesis
  Endocytosis
  Functions
    Digestion
    Storage
    Transport
    Secretion
Summary

THE INDIVIDUAL EUKARYOTIC CELL possesses the necessary biochemical armamentarium to allow the complete degradation of both extracellular material and endogenous cellular constituents. These digestive processes largely depend on the availability of an array of hydrolytic enzymes that are capable of degrading virtually all proteins, lipids, and carbohydrates. These enzymes are localized largely within lysosomes, which are generally considered to be the most important organelles for cell digestion.

Although an extensive amount of data have been generated characterizing the role of lysosomes in intracellular digestion, only recently has the importance of the extracellular release or secretion of lysosomal contents been appreciated. Indeed, "the extracellular secretion of lysosomal enzymes is thought to be an integral part of the lysosomal activity of normal cells" (25). With this in mind, and given the well-established role of biliary secretion as a major excretory route for a variety of exogenous compounds and intracellular constituents, attention was directed to the possibility that exocytosis of lysosomal contents into biliary canaliculi might be a major excretory route for hepatocyte lysosomes (22, 60). Results from experiments to test this hypothesis support the existence of a lysosome-to-bile vesicular excretory pathway; the data are also compatible with the concept that this pathway may be a final common pathway whereby the hepatocyte disassembles macromolecules and secretes into bile some end products of partial or complete lysosomal hydrolysis.

The overall objectives of this chapter are *1*) to provide an in-depth review of the processes by which the eukaryotic cell engulfs, degrades, and disposes of exogenous extracellular material and endogenous intracellular constituents and *2*) to summarize evidence defining the physiological and biochemical bases for and the functional significance of the release of hepatocyte lysosomal contents into bile. In addition, techniques for studying lysosomes in the laboratory are briefly reviewed and, where possible, comments addressing disturbances of normal lysosomal digestion and secretion resulting in pathology are made.

LYSOSOMES: BACKGROUND AND GENERAL CONSIDERATIONS

Perspectives

The lysosome was discovered in 1955 by Christian de Duve in the laboratory of Physiological Chemistry at the University of Louvain, Belgium. "All we wanted to know was something about the localization of glucose-6-phosphatase, which we thought might provide a possible clue to the mechanism of action or lack of action of insulin on the liver cell," said de Duve on acceptance of the Nobel Prize for Physiology or Medicine, December 12, 1974, a prize he shared with Albert Claude and George Palade (21). A detailed review of the experiments leading to the discovery of this organelle is beyond the scope of this chapter and has been summarized elsewhere (4, 19).

Definitions

Lysosomes can be defined on several levels (16, 19, 22, 78). *1*) The literal definition of a lysosome is "lytic particle." *2*) The biochemical definition of a lysosome is a membrane-bound, subcellular organelle containing hydrolytic enzymes with acid pH optima that exhibit structure-linked latency. *3*) The morphological definition of a lysosome is a membrane-bound, round or oval organelle of heterogeneous size and electron density averaging ~0.5 µm in diameter. *4*) The histochemical definition of a lysosome is a membrane-bound organelle that stains positively for acid phosphatase reaction product. *5*) The functional definition of a lysosome is the digestive tract of the cell.

There are several pertinent points worth making regarding each of these levels of definition.

BIOCHEMICAL. Lysosomes contain >50 enzymes identified biochemically by their enzymatic activity (5, 6). Virtually all of these enzymes are glycoproteins, all exhibit pH optima in the acid range, and most are hydrolases that catalyze reactions of this general type

$$A—B + H_2O \rightarrow A—H + B—OH$$

The lysosomal hydrolases can be categorized into groups, depending on their general substrate specificities (e.g., lipases, nucleases, glycosidases, and so forth). Moreover the enzymes are sequestered within a lipoprotein membrane that is impermeable to the enzymes and their substrates and that prevents the escape of the hydrolases into the cell cytoplasm (Fig. 1). Because the lipoprotein membrane prevents physical contact between intralysosomal enzymes and extralysosomal substrate, lysosomal enzyme activity is "latent" (19). Specifically, no lysosomal enzyme activity is measurable enzymatically in biological samples unless the lipoprotein membrane is physically disrupted to allow exposure of enzyme to substrate (e.g., by addition of a detergent to the assay mix).

MORPHOLOGICAL. Lysosomes can generally be identified with moderate confidence on transmission electron microscopy because of their size (0.5–1.5 µm); shape (round or oval); electron density (heterogeneous, electron-dense center); single, limiting membrane; and occasionally, characteristic intracellular location (e.g., pericanalicular in hepatocytes) (74). In addition, lysosomes commonly display a "halo" that can be seen just below the limiting membrane surrounding the lysosomal matrix (Fig. 2). Morphometric data suggest that the rat hepatocyte contains ~15–20 lysosomes (107). However, definitive morphological identification of a lysosome requires techniques that specifically visualize the enzymes themselves (e.g.,

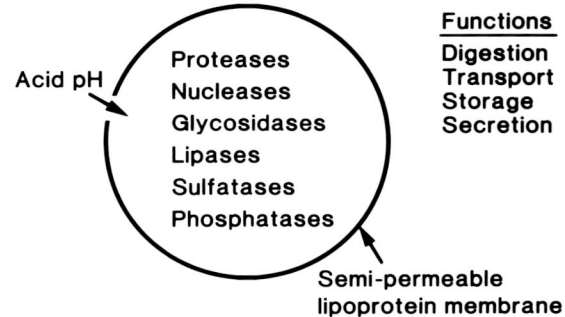

FIG. 1. The lysosome.

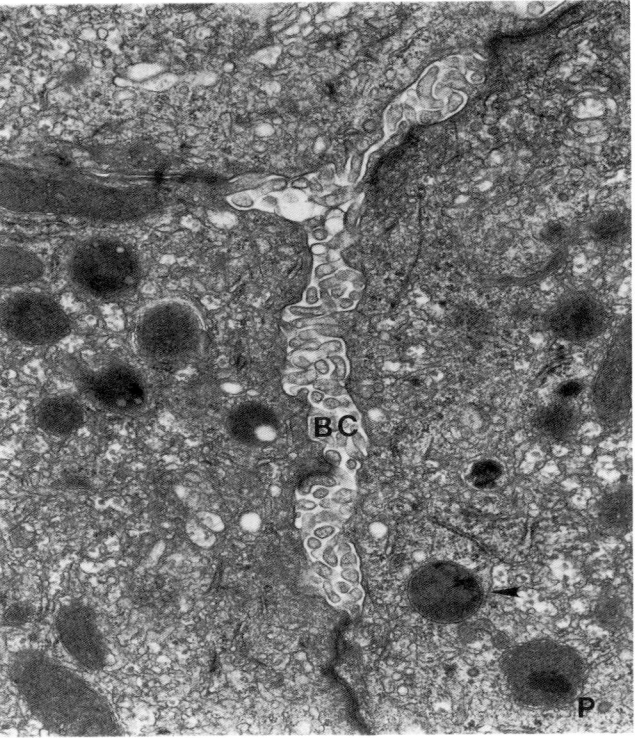

FIG. 2. Transmission electron microscopy of rat liver showing lysosome-like organelle (*arrowhead*) in region of bile canaliculus (*BC*) of hepatocyte. Peroxisome (*P*) is also seen (×23,273). [From LaRusso (59).]

immunohistochemistry) or reaction products of enzymes specific to lysosomes. In addition, certain exogenous compounds that are selectively sequestered in lysosomes (i.e., lysosomotropic) alter the appearance of lysosomes and allow confident morphological identification. Finally, the finding within a membrane-limited structure of membrane remnants resulting from the partial degradation of other organelles (e.g., mitochondria) strongly suggests that the membrane-limited structure is a lysosome.

HISTOCHEMICAL. Heavy metals, such as lead, produce products of enzyme reactions (e.g., lead phosphate) that are specifically localized to lysosomes if the substrate utilized (e.g., β-glycerophosphate) is hydrolyzed specifically by lysosomal hydrolases (75, 93, 95).

FUNCTIONAL. The functional categories of lysosomes include the following. *1*) Primary lysosomes are "virgin lysosomes" or lysosomes, likely newly synthesized, that contain only unused enzymes packaged for future use but not as yet having encountered material or substrates to be degraded (45). *2*) Secondary lysosomes are "used lysosomes" containing both hydrolytic enzymes and material to be degraded, being degraded, or already degraded.

Characteristics

BIOCHEMICAL. *Enzymology of hydrolases.* Table 1 provides a selective list of hydrolase activities thought to be present in lysosomes (7). The actual number of distinct enzymes (as opposed to activities) in lysosomes is not definitely known. Virtually all lysosomal enzymes are glycoproteins that often overlap in their substrate specificities, may occur in multiple "isozyme forms" differing perhaps mainly in their carbohydrate side chains, and are thought to have rather slow turnover rates (3.5–5 days) (51, 100). These glycoproteins are generally of high molecular weights (>100,000) and can differ in their intralysosomal localization (i.e., bound to the inner aspect of the lysosomal membrane or relatively free within the intralysosomal matrix) as manifested in their release after disruptive treatments (102). It is important to emphasize that there are also nonlysosomal intracellular hydrolases that are active at both acid and neutral pH.

Because lysosomal hydrolases have an acid pH optimum, the intralysosomal pH must be maintained in the acid range, generally around pH 5. Recent data indicate that the acid milieu within lysosomes is maintained by an ATP-driven proton pump in the lysosomal membrane (86).

Lysosomal membranes. The lysosomal membrane has a lipoprotein composition that is both similar to and different from the plasma membrane (65). The lipid composition is similar, containing a high content of cholesterol and sphingomyelin; the protein and carbohydrate components are likely different, but these have not been well characterized (11). Although available data are limited, it seems likely that only small molecules normally traverse the lysosomal membrane (83). In general, disaccharides and larger dipeptides fall above the apparent size limit for molecules that can enter or leave mammalian cell lysosomes readily, whereas many monosaccharides, dipeptides with mol wt <200, and amino acids fall within this limit. As one would expect, charged molecules have more difficulty traversing the membrane than do uncharged molecules of comparable molecular weights; in fact, this characteristic has been utilized to trap certain basic substances within lysosomes. Although the existence of active or facilitated membrane transport systems in the lysosomal membrane has been postulated, unequivocal evidence is difficult to obtain. Also, the precise mechanism whereby molecules actually leave the lysosome is unclear and may differ for different molecules. More specifically, whereas some end products of lysosomal hydrolysis may actually pass directly through the membrane itself (e.g., small, lipid-soluble molecules), others may be shuttled out of lysosomes in small vesicles formed by an evagination of a portion of the lysosomal membrane.

MORPHOLOGICAL. Because lysosomes are morphologically heterogeneous, the morphological criteria for the tentative identification of an organelle as a lysosome require emphasis; reasonable criteria should include the presence of cytochemically demonstrable acid hydrolase activity plus the presence of a delimiting membrane in an organelle of appropriate size, shape, appearance, and intracellular location. Without histochemical confirmation, morphological identification of a structure as a lysosome is tenuous, and properly speaking, such organelles should be referred to as lysosomelike (9, 85). The presence of membrane remnants within a vesicle or parallel biochemical studies can also serve to support the tentative identification of an organelle as a lysosome.

The most commonly used cytochemical method for identifying lysosomes is the Gomori method for identifying acid phosphatase reaction product (34). When fixed tissue is incubated at an acid pH in a mixture containing lead ions and β-glycerophosphate, the β-glycerophosphate is hydrolyzed by lysosomal acid phosphatase and the liberated phosphates precipitate rapidly in the form of lead phosphates; lead phosphates survive preparation of tissue for electron mi-

TABLE 1. *Major Enzymatic Activities of Lysosomes*

Substrate Class	Enzymatic Activities
Lipids	Acid lipase
	Phospholipase A_1, A_2, and C
	Sphingomyelin phosphodiesterase
Proteins	Acid carboxypeptidase
	Cathepsin A, B, C, and D
Polysaccharides, carbohydrate side chains of glycoproteins, and complex lipids	Aspartyl glucosylaminidase
	Fucosidase(s)
	α-Galactosidase
	β-Glucosidase
	β-Galactosidase
	β-Glucuronidase
	Hyaluronidase
	L-Iduronidase(s)
	α- and β-Mannosidases
	Neuraminidase
	Hexosaminidase A
Nucleic acids	Acid deoxyribonuclease
	Acid ribonuclease
Miscellaneous	Acid phosphatases and phosphodiesterases
	Acid esterases
	Acid sulfatases

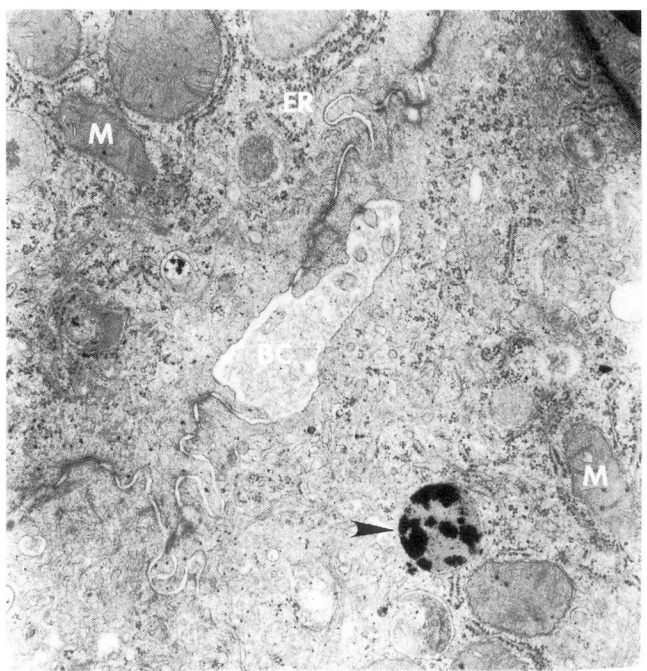

FIG. 3. Transmission electron microscopy of rat liver showing acid phosphatase reaction product in lysosome (*arrowhead*) in region of a bile canaliculus (*BC*) of hepatocyte. Mitochondria (*M*) and endoplasmic reticulum (*ER*) are also seen (×27,083).

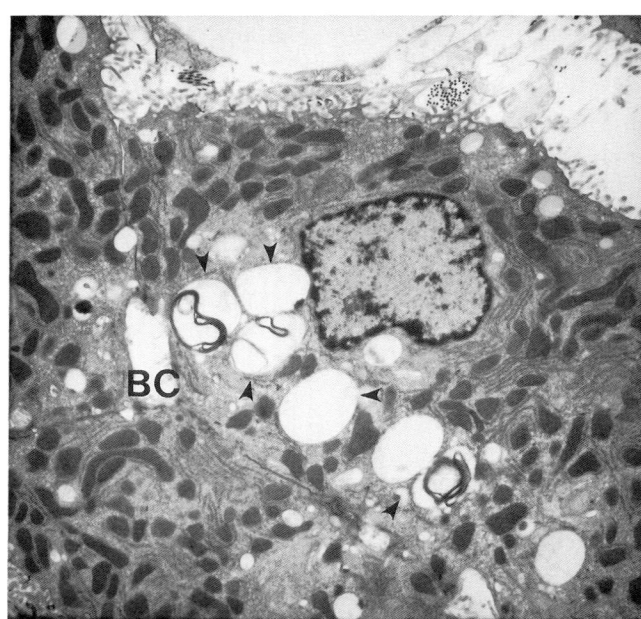

FIG. 4. Transmission electron microscopy of rat liver showing secondary lysosomes (*arrowheads*) containing Triton WR-1339 in vicinity of bile canaliculus (*BC*) (×2,857) [From LaRusso (59).]

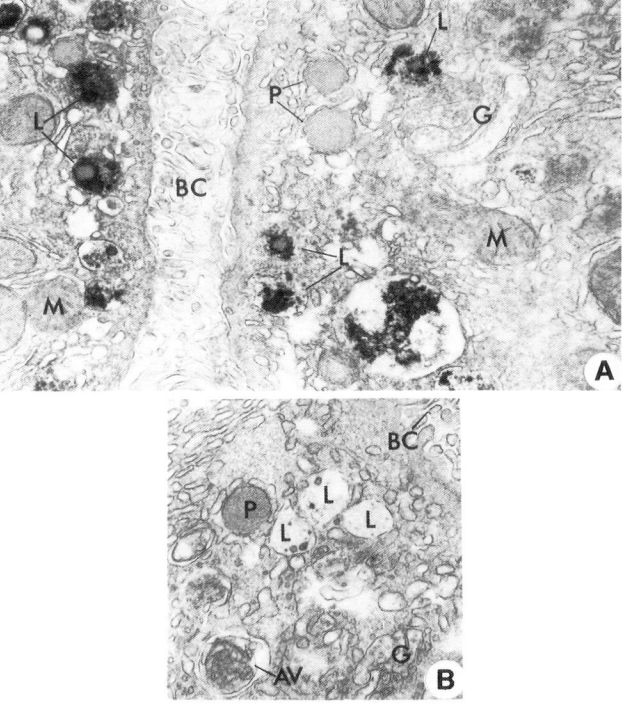

FIG. 5. Transmission electron microscopy of rat liver showing definitive identification of lysosomes (*L*) by immunocytochemical demonstration of β-galactosidase in rat hepatocytes (*Panel A*). *AV*, autophagic vacuole; *BC*, bile canaliculus. Note absence of reaction product in peroxisome (*P*), Golgi apparatus (*G*), and mitochondria (*M*). Panel B shows control section of hepatocyte exposed to nonimmune rabbit serum and absence of reaction product in lysosomes (L) (×27,234). [From Novikoff et al. (79), by copyright permission of the Rockefeller University Press.]

croscopy and are sufficiently electron dense or opaque to be readily visible (Fig. 3). Lysosomes can also be stained with several types of vital (neutral red) or fluorescent (acridine orange) dyes that can accumulate within minutes in lysosomes of isolated or cultured cells (1). In addition, macromolecules, such as Triton WR-1339 (a nonionic polymer detergent of mol wt >100,000), may be sequestered in lysosomes and alter the appearance (size, shape, electron density) of lysosomes (106), allowing confident morphological identification (Fig. 4). Finally, immunocytochemistry also can be used to identify an organelle as a lysosome. For example, an immunocytochemical method for directly visualizing lysosomal β-galactosidase in tissue has recently been described [Fig. 5; (79)].

Morphologists have given a variety of names to secondary lysosomes, depending on the appearance of the materials accumulating within them. Pinocytosed materials often accumulate in a morphologically distinctive type of secondary lysosome known as a multivesicular body, a structure that contains many small vesicles within a delimiting membrane. In addition, residual bodies are a type of secondary lysosome in which much of the content is residual matter not susceptible to or only slowly or partially susceptible to further degradation.

Methods of Study

ISOLATION OF LYSOSOMES. Tissue fractionation is the most commonly used preparative technique for isolating specific intracellular organelles, including lysosomes (17, 18, 20). A variety of specific protocols for various tissues have been published and have stood the test of time; their details are beyond the scope of this chapter (see refs. 8, 15).

Centrifugation has been the major isolation technique applied to the preparation of fractions of tissue homogenates enriched in lysosomes relative to the starting material. Differential centrifugation (which separates particles sedimenting at different rates) and isopycnic centrifugation (which separates particles equilibrating at different densities in a gradient) have both been employed, frequently in sequence (72). More recently, free-flow electrophoresis (38, 41) has been employed for the purification of lysosomes. Indeed, the apparent enrichment of a tissue fraction by this technique is considerably greater than can be achieved by centrifugation alone. Both biochemical and morphological techniques should be employed to assess the extent of enrichment of the particle fraction of interest.

Even under the best of circumstances, purification of unmodified lysosomes by differential or isopycnic centrifugation results in fractions heavily contaminated with mitochondria and peroxisomes. In fact, lysosomes isolated by differential centrifugation likely contain more peroxisomal than lysosomal protein. Also, the median equilibrium density of lysosomes (1.22) in sucrose gradients lies between that of mitochondria (1.19) and that of peroxisomes (1.23–1.25), giving rise to important contamination of lysosomes in all parts of the gradient.

A considerable improvement in the degree of enrichment of lysosomal fractions can be accomplished by loading the lysosomes in vivo with a compound that alters their physical properties in such a way as to modify their equilibrium density. Triton WR-1339 is sequestered in hepatocyte, kidney, and gastric cell lysosomes and causes a marked decrease in the density of these lysosomes from ~1.22 to 1.10 (the approximate density of Triton WR-1339 itself). This resulting decrease in density, which reflects the fact that the intralysosomal Triton WR-1339 is resistant to hydrolysis and remains localized to lysosomes, permits the separation of lysosomes from both mitochondria and peroxisomes. Other compounds such as dextran, Thorotrast (colloidal thorium hydroxide), and sucrose can be employed to alter the physical properties of lysosomes, permitting isolation of purer fractions (15). However, the main disadvantage of each of these techniques is that the physical and biochemical properties of the lysosomes themselves are altered in the purification process, thereby making the interpretation of any results from experiments on such altered lysosomes difficult. An alternative approach, in which the properties of mitochondria are altered by exposure to calcium (111), removes contaminating mitochondria without affecting the lysosomes themselves (Fig. 6).

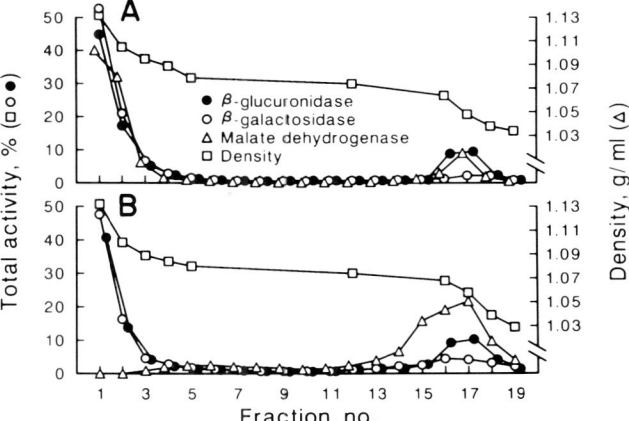

FIG. 6. Example of use of isopycnic centrifugation on gradient of Percoll to isolate rat liver lysosomes. In this experiment, fraction of rat liver containing predominately mitochondria and lysosomes (ML fraction) was initially prepared by homogenization and differtial centrifugation. After exposing (B) or not exposing (A) ML fraction to 1 mM CaCl$_2$, fractions were layered on nonlinear gradient ranging in density from 1.03 to 1.13 g/ml. Marker enzymes for lysosomes (i.e., β-glucuronidase and β-galactosidase) and for mitochondria (i.e., malate dehydrogenase) were measured. Note that in the absence of prior exposure to CaCl$_2$ (A), distribution in activities for all three enzymes is similar, with major peaks at density of 1.13 g/ml. After exposure to CaCl$_2$ (B), distribution patterns for lysosomal enzymes are unchanged. In contrast, peak in activity of malate dehydrogenase has now shifted up the gradient to density of 1.05 g/ml, reflecting CaCl$_2$-induced osmotic swelling, and hence lightening, of mitochondria.

ASSAY OF LYSOSOMAL HYDROLASES. Numerous methods have been published over the years; a detailed account of these techniques is beyond the scope of this chapter (see refs. 7, 102, 103). However, it does seem worthwhile to make several general points regarding lysosomal enzyme activity measurements.

Virtually all assays for lysosomal enzymes include the addition of a detergent (e.g., Triton X-100, deoxycholic acid) to the assay mix to assure complete disruption of the lysosomal membrane and guarantee complete access of enzyme to substrate. When lysosomal enzyme assays are being developed, kinetic studies such as activity versus time of incubation and activity versus concentration of enzyme should be performed; also, measurements of pH optima, enzyme stability versus storage time, and enzyme activity versus temperature should be explored. Such studies are absolutely essential in each tissue or biological fluid studied to identify valid assay conditions and assure quantitative estimates of enzyme activity.

Recently, fluorometric assays have become popular to measure lysosomal enzyme activities (82). This methodology is rapid, accurate, reproducible, and amenable to automation. Other methods for quantitating the actual amount, rather than the activity, of lysosomal enzymes are also available. For example, radioimmunoassays for measurement of two lysosomal glycosidases, β-galactosidase and β-glucuronidase, have recently been described (23). Such sensitive and specific methodology permits quantification of the mass (i.e., micrograms) of these lysosomal proteins in biological samples. When used in conjunction with activity measurements, knowledge about the actual amount or concentration of a protein in response to perturbations permits inferences about the mechanisms regulating alterations in enzyme activity.

ELECTRON MICROSCOPY, INCLUDING HISTOCHEMISTRY. A complete description of the methodology involved in the morphological study of the lysosome is beyond the scope of this chapter. However, it is worth emphasizing that considerable information concerning the functional aspects of the lysosomal system has been generated by extensions of the general technique of transmission electron microscopy. These methodologies include histochemistry, quantitative morphometry, and kinetic tracer studies. For example, the use of lysosomotropic agents permits a kinetic morphological analysis of lysosomal behavior. Radiolabeled compounds may be assessed by autoradiography or nonradiolabeled, electron-dense substances visualized by conventional transmission electron microscopy.

LYSOSOMES: FORMATION AND FUNCTION

Enzyme Biogenesis

It is generally felt that biogenesis of lysosomal enzymes begins in GERL, an acronym for the Golgi-associated Endoplasmic Reticulum involved in the formation of Lysosomes, originally described by Novikoff (76, 77, 80). Nascent lysosomal enzymes are equipped with signal peptides to facilitate their entry into the rough endoplasmic reticulum (RER) (39, 92). The enzymes are then glycosylated and phosphorylated within the RER (30, 63). Delivery of the newly synthesized, phosphorylated glycoproteins to lysosomes depends on the presence of receptors on the inner aspect of the membrane of the RER to which lysosomal enzymes with a mannose 6-phosphate residue bind (12, 91, 109). Thus, most newly synthesized lysosomal enzymes rely on the phosphomannosyl recognition marker for segregation from other products of the RER. Hydrolases that are bound to receptors on the inner aspect of the cisternae of the RER collect in coated vesicles that bud off the Golgi or GERL to become primary lysosomes. Before or when primary lysosomes become secondary lysosomes, a fall in pH allows pH-dependent release of enzyme from receptors and recycling of free receptors back to GERL.

Endocytosis

For extracellular, macromolecular materials to be digested by the lysosomal system, they must enter the cell and become incorporated into lysosomes (Fig. 7). Because materials of macromolecular dimensions cannot traverse the plasma membrane directly, these substances enter the cell in membrane-delimited compartments formed by the invagination of a portion of the plasma membrane. This process is called endocytosis (50) and is the major mechanism whereby extracellular material is routed to lysosomes. Generally, several types of endocytosis are defined, depending on either the size or nature of the material being endocytosed or the mechanism of endocytosis. For example, phagocytosis (the uptake of solids) is commonly differentiated from pinocytosis (the uptake of fluids, including soluble proteins) (99). From a mechanistic point of view, endocytosis can be subdivided into receptor-mediated (adsorptive) or bulk flow (fluid-phase) endocytosis (33, 81). The former involves the

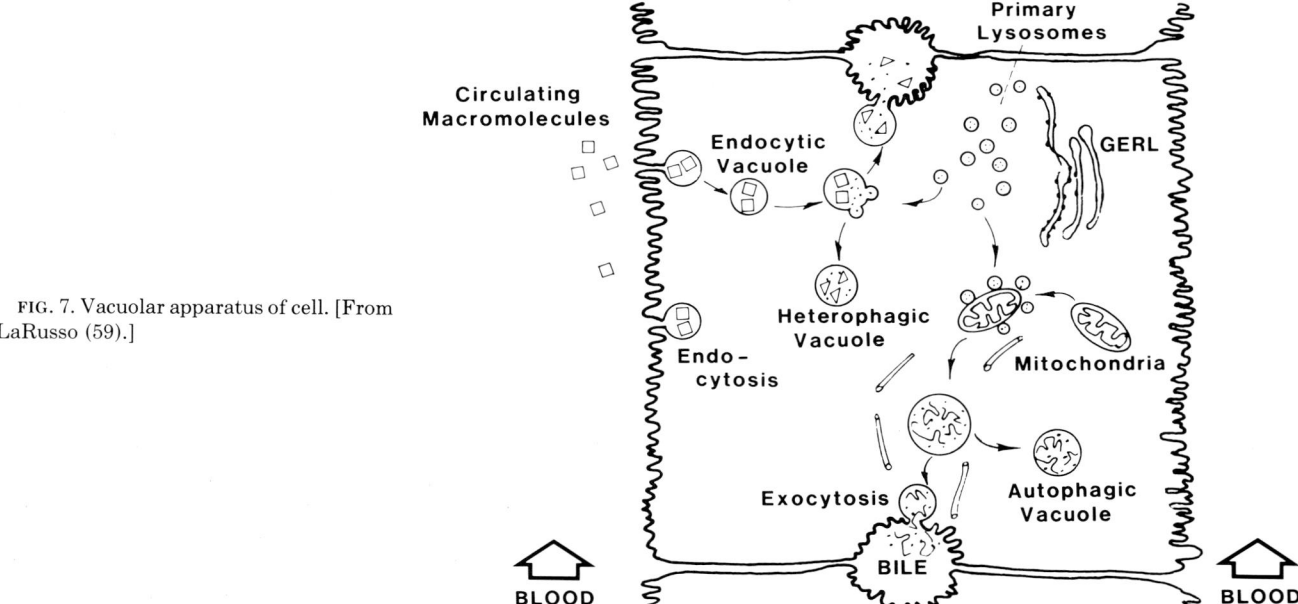

FIG. 7. Vacuolar apparatus of cell. [From LaRusso (59).]

binding of a ligand (e.g., transport proteins, such as low-density lipoprotein and transferrin; peptide hormones, such as insulin; and growth factors, such as epidermal and transforming growth factors) to a specific receptor within the plasma membrane but extending from the external surface of the cell. Frequently, the initial interaction of ligand and receptor provokes a clustering of receptors into clathrin-containing "coated pits" evident by electron microscopy (33). Internalization of ligand and receptor can then occur, leading to formation of an endocytic vesicle. Many proteins internalized in this fashion are delivered to lysosomes for degradation or disassembly, whereas others are not degraded but are delivered to cellular structures other than lysosomes (52).

Functions

DIGESTION. The major function of lysosomes is intracellular digestion. Macromolecules can be hydrolyzed to the level of fatty acids, oligopeptides, and individual amino acids within lysosomes. The fate of the products of digestion varies. If digestion is complete, the released materials are often small enough to leave the lysosome for reutilization or excretion by the cell. Also, intralysosomal digestive products remaining sequestered within lysosomes may be released directly from the cell by a process called exocytosis. Finally, some undigestible or poorly digestible materials may accumulate in lysosomes; in some cells, these materials may remain in the lysosomes over the life span of the cell forming lipofuscin or "aging pigment." In other cells, these materials may also be discharged from the cell by exocytosis.

Three types of lysosomal digestion can be differentiated, depending on the source and nature of the digested material.

Heterophagy. Heterophagy is the combined process whereby extracellular macromolecules enter the cell by endocytosis and are delivered to the lysosomal system (Fig. 7). It is believed that the endocytic vesicle fuses with primary lysosomes, which deliver their hydrolytic enzymes for digestion (59). The resulting vesicle may be referred to as a phagolysosome. Heterophagy is an important process whereby the organism protects itself [e.g., killing of bacteria by polymorphonuclear leukocytes (29) or macrophages (13)]. Also, heterophagy results in elimination of senescent cells (e.g., disposal of aged red blood cells by spleen macrophages) and catabolism of circulating glycoproteins [e.g., asialoglycoproteins are removed from the circulation by hepatocytes and sequestered in lysosomes (3, 98)], lipoproteins [low-density lipoprotein is processed in the lysosomal system of many cells (71)], and hormones [e.g., insulin, epidermal growth factor, and transforming growth factor-β (TGF-β) are processed in hepatocyte lysosomes (32, 55)].

Autophagy. Autophagy is the process whereby intracellular constituents (e.g., mitochondria) are degraded by lysosomes (Fig. 7). Although the mechanism(s) whereby endogenous cellular material becomes incorporated into secondary lysosomes is unclear, complete organelles or organelle fragments are apparent in lysosomes (i.e., autophagic vacuoles), particularly after administration of certain compounds such as chloroquine (Fig. 8). This process is important in differentiation, cell involution, and turnover of intracellular constituents (26).

Crinophagy. Crinophagy is the process whereby endocrine glands can degrade excess secretory material rather than discharge it from the cell (94). Presumably, secretion granules containing hormone not destined for extracellular release fuse with primary lysosomes that affect degradation of excess polypeptide.

STORAGE. Undigested or partially digested materials can accumulate within lysosomes for extended periods of time. Also, trace metals (copper, iron, lead) may be sequestered and stored in lysosomes of mammalian liver [Fig. 9; (96)].

TRANSPORT. Certainly lysosomes are involved in the intracellular movement of materials and, in that sense, serve a transport function. Moreover, this function becomes of considerable importance relative to the roles of the lysosome in digestion and secretion. For example, the primary lysosome carries its complement of hydrolytic enzymes to endocytic vesicles as well as to cellular constituents during heterophagy and autophagy, respectively. The secondary lysosome has the capacity to transport both enzymes and digestive res-

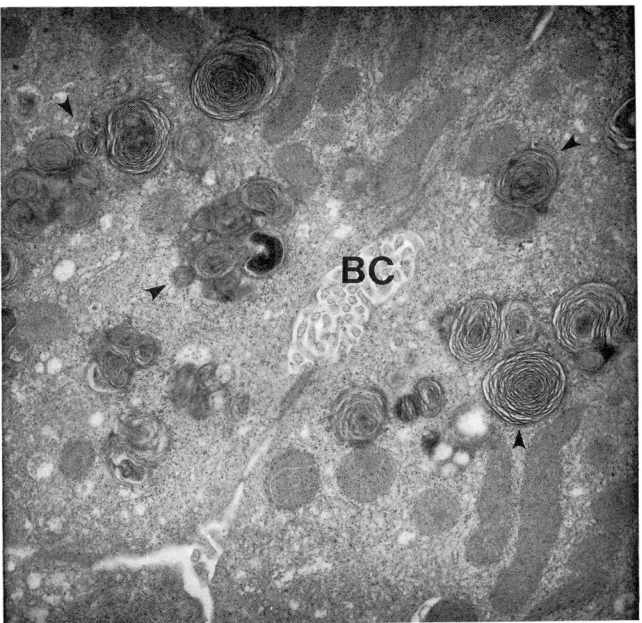

FIG. 8. Transmission electron micrograph of bile canalicular region in liver of chronic chloroquine-treated animal. Many pericanalicular autophagic vacuoles (*arrowheads*) surrounding bile canaliculus (*BC*) contain concentric arrays of membranous structures. Some autophagic vacuoles occur in clusters (×9,638). [From Sewell et al. (88), copyright 1983 by The American Gastroenterological Association.]

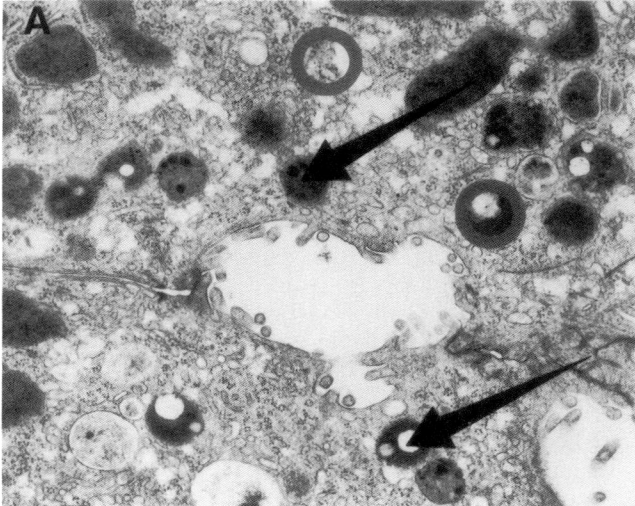
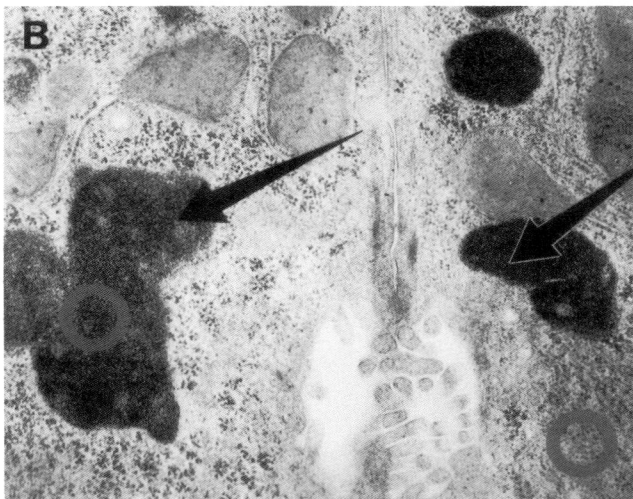

FIG. 9. Transmission electron micrograph of liver sample from normal (A) and iron-loaded (B) rat. Arrows indicate pericanalicular, lysosome-like structures in normal rat liver and electron-dense, membrane-bound, pericanalicular organelles in iron-loaded rat liver. [From LeSage et al. (62), by copyright permission of The American Society for Clinical Investigation.]

idues to the plasma membrane for extracellular release.

SECRETION. *Secretion by nonhepatic cells.* The extracellular release of lysosomal constituents is an integral part of the lysosomal activity of normal cells. Considerable in vitro data support the selective, extracellular release of lysosomal hydrolases by viable, nonhepatic, mammalian cells (14, 40, 42, 46, 57, 66, 87, 108). Most studies have employed isolated cells or cells in culture because these systems are free from hormonal influences and permit collection of all products secreted into the medium. Nonhepatic mammalian cells in which the selective and regulated release of lysosomal protein has been demonstrated include cultured cells from cartilage, bone, thyroid, and adrenal medulla;

cultured embryonic and human skin fibroblasts and isolated polymorphonuclear leukocytes have also been shown to secrete lysosomal enzymes.

Although the mechanism of the selected, extracellular release of lysosomal hydrolases from nonhepatic cells has not been clarified, de Duve and Wattiaux (22) have emphasized the importance of coalescence of membranes in the regulation of the vacuolar system. Such a phenomenon is likely important to lysosomal secretory processes because it would allow bulk extracellular release of lysosomal hydrolases without passage of these degradative enzymes into the cell cytoplasm. Alternatives to such an exocytic process compatible with the maintenance of cellular organization are difficult to envision. Therefore, the secretory process by which lysosomal enzymes are released in bulk from these cells likely involves movement of lysosomes to the cell periphery and fusion of the lysosomal membrane with the cytoplasmic surface of the plasma membrane. Coalescence of lysosomal and plasma membranes occurs and the lysosomal membrane becomes incorporated into the plasma membrane. Ultimately, this process results in the bulk, exocytic discharge from the cell of all lysosomal contents.

The physiochemical factors regulating both membrane fusion in general and the coalescence of lysosomal and plasma membranes during secretion are unclear. However, it seems likely that fusion if facilitated by high surface tension. Unstable membranes are characterized chemically by a high proportion of short-chain fatty acids, the incorporation of *cis* isomers, or the presence of polar groups in the hydrocarbon chain. These properties tend to increase the surface area occupied by each molecule, thereby reducing the van der Waals forces between them.

In addition to the intrinsic properties of the lysosomal and plasma membrane that are relevant to an exocytic secretory mechanism, other factors that likely influence the exocytic secretion of lysosomal enzymes include cyclic nucleotides, microtubules, and microfilaments.

The role of cyclic nucleotides in the extracellular release of lysosomal enzymes is currently under investigation. Considerable work has been done with human neutrophils, and evidence to date suggests that autonomic neurohormones and prostaglandins can modulate lysosomal enzyme release by virtue of their ability to affect intracellular levels of cyclic nucleotide (31). For example, prostaglandins are an example of compounds that inhibit release of lysosomal enzymes from a variety of nonhepatic, mammalian cell types, presumably by AMP-mediated mechanisms (48). It is unclear, however, what intracellular mechanisms are utilized by cyclic nucleotides to affect lysosomal enzyme secretion. One possibility is that cyclic nucleotides alter the physiochemical properties of lysosomal membranes and/or microtubules via certain nucleotide-dependent, protein kinases (49). In fact, data indicate that cAMP, in conjunction with calcium, directly affects the physical form and response to colchicine of microtubules (28). Additionally, intracellular concentrations of calcium correlate with the secretion of lysosomal enzymes induced by a calcium ionophore in polymorphonuclear leukocytes (47).

Finally, it appears that the intracellular movement of lysosomes depends on normal microtubule function (27, 68). In the presence of colchicine and other agents known to interfere with microtubule function, lysosomal movement and, in certain cell systems, the secretion of lysosomal hydrolases, appear to be inhibited. Indeed, morphological data indicate that a single intravenous injection of a microtubule-binding agent such as vinblastine leads to both the disappearance of microtubules and the accumulation of autophagic vacuoles (43).

Despite these data, the precise factors controlling the intracellular movement of primary and secondary lysosomes remain relatively obscure. Indeed, little true insight exists as to why a lysosome may fuse with one endocytic vesicle but not with another, with one organelle but not with another, and with one domain of the plasma membrane but not with another.

Secretion by hepatocytes: lysosome-to-bile hepatic excretory pathway. Because biliary secretion is a major excretory route for a variety of exogenous and endogenous constituents and given the accumulated data indicating that nonhepatic cells can selectively release their lysosomal contents by exocytosis in a regulated fashion, the possibility that a vesicular excretory route exists from hepatocyte lysosomes to bile seems plausible.

Evidence supporting its existence. Evidence supporting the existence of the lysosome-to-bile hepatic excretory pathway comes from four general categories of observations (Table 2).

First, with both enzymatic assay measurements and radioimmunoassays, it has been demonstrated that lysosomal enzymes are present in bile, generally in amounts greater than enzymes originating from other organelles (14, 16, 60, 73, 104). In addition, there are kinetic and immunological similarities between hepatic and biliary lysosomal enzymes, suggesting that the biliary lysosomal enzymes originate from hepatocytes (60, 73).

Second, under a variety of basal and perturbed conditions, data indicate coordinate or parallel release of several lysosomal enzymes into bile, suggesting bulk exocytic discharge of these proteins [Fig. 10; (60, 64, 88)]. Indeed, observations by transmission electron microscopy demonstrate the fusion of hepatocyte lysosomes with canaliculi (Fig. 11).

Third, biochemical and morphological data indicate that certain exogenous macromolecules can be sequestered in hepatocyte lysosomes and can also be released into bile in parallel with endogenous lysosomal enzymes (53, 61). For example, when [^{3}H]Triton WR-1339 was administered to rats, biochemical and morphological studies showed that hepatocyte lysosomes sequestered this large molecule (61). Tritium was also excreted into bile in bile fistula rats in parallel with

TABLE 2. *Evidence Supporting Existence of Lysosome-to-Bile Hepatic Excretory Pathway*

Lysosomal enzymes are present in bile.
Lysosomal enzymes are released coordinately into bile.
Exogenous compounds are sequestered in hepatocyte lysosomes and are released into bile in parallel with endogenous lysosomal enzymes.
Biliary secretion of lysosomal enzymes and hepatocyte lysosomal morphology can be modulated by microtubule binding agents, hormones, and lysosomotropic compounds.

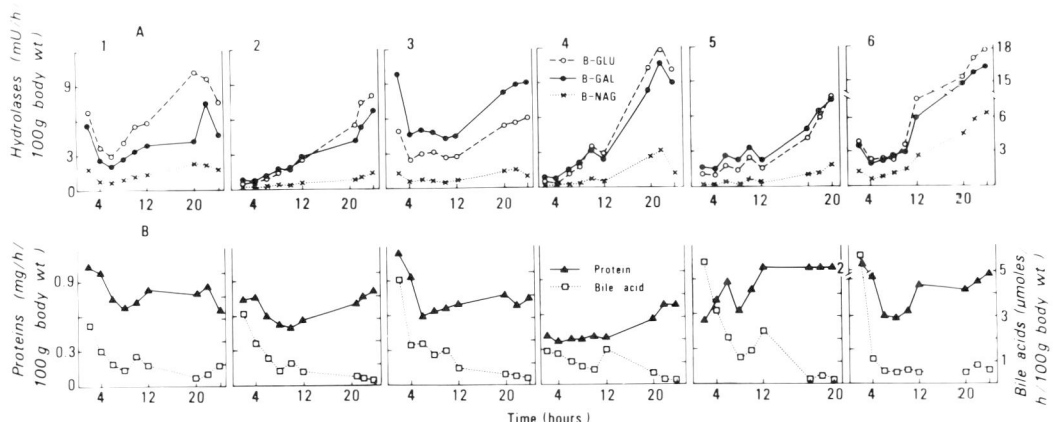

FIG. 10. Twenty-four-hour biliary excretion of acid hydrolases, total protein, and bile acids. *A*: β-glucuronidase (β-GLU), β-galactosidase (β-GAL), and *N*-acetyl-β-glucosaminidase (β-NAG) outputs are presented for each of 6 rats (1–6). *B*: total protein and bile acid outputs for same rats. Although excretory patterns for lysosomal enzymes varied among rats, in each rat, output patterns of the 3 lysosomal enzymes were parallel. No such parallelism existed between total protein and bile acid outputs. [From LaRusso and Fowler (60), by copyright permission of The American Society for Clinical Investigation.]

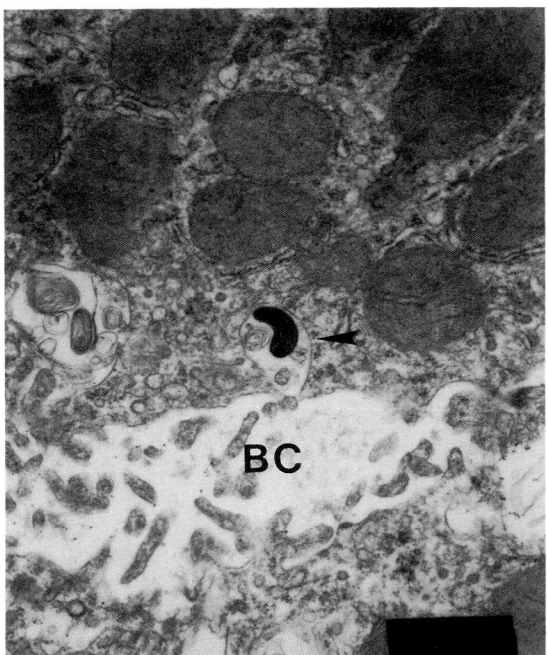

FIG. 11. Transmission electron micrograph of rat liver showing autophagic vacuole (*arrowhead*), a type of secondary lysosome, containing remnants of mitochondrion. Autophagic vacuole is fusing with bile canaliculus (*BC*) and is in the process of exocytosis (×62,500).

likely, the functional significance, if any, of this pathway is less apparent. It seems reasonable to postulate that this pathway may be important *1*) in the turnover of endogenous and exogenous lysosomal constituents, *2*) in the biliary excretion of metals and lipids, and *3*) in the hepatic processing of selected proteins.

1. Although studies have been performed assessing half lives of lysosomal enzymes (51, 100), the precise mechanism whereby senescent lysosomal enzymes are turned over by the cell remains completely unknown. Because lysosomal enzymes are glycoproteins, theoretically they should be susceptible to intralysosomal degradation by coexistent lysosomal proteases. Certainly intralysosomal modifications in pH or conformational changes in enzyme structure with age may provoke intralysosomal autodigestion of individual lysosomal enzymes and account for the turnover by a process confined to the interior of the lysosome itself. Alternatively, as has been proposed for a variety of nonhepatic cells systems, "the secretion of lysosomal three lysosomal enzymes. These observations support the existence of a pathway from blood to hepatocyte lysosomes to bile by which exogenous (e.g., [^{3}H]Triton WR-1339) and endogenous (e.g., lysosomal enzymes) materials located within lysosomes undergo biliary excretion.

Finally, with a variety of experimental models, it has been demonstrated that the biliary output of endogenous and exogenous lysosomal constituents can be modulated by microtubule binding agents (Fig. 12) [e.g., colchicine and vinblastine (89)], by hormones [e.g., insulin, glucagon, and ethinyl estradiol (64; R. B. Sewell, S. Grinpukel, A. R. Zinsmeiser, and N. F. LaRusso, unpublished observations)], and by lysosomotropic agents [e.g., chloroquine (88; R. B. Sewell, S. Grinpukel, A. R. Zinsmeiser, and N. F. LaRusso, unpublished observations)]. In many of these situations, concomitant drug-induced alterations in hepatic lysosomal structure and in biliary lysosomal enzyme output were also shown.

Taken together, results from all studies strongly indicate that the hepatocyte possess the ability to process (i.e., extract, metabolize, transport, and excrete) molecules endocytosed from the circulation and that hepatocyte lysosomes participate in the metabolic, transport, and excretory components of this processing pathway (58).

Functional significance. Although the existence of the lysosome-to-bile hepatic excretory pathway seems

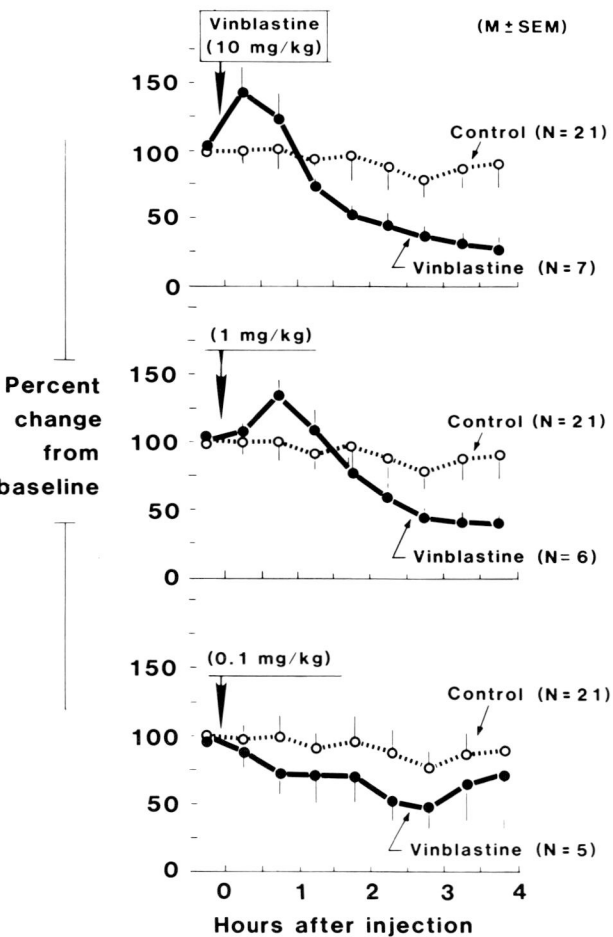

FIG. 12. Biliary output of lysosomal enzyme, *N*-acetyl-β-glucosaminidase before and after vinblastine in bile fistula rats. *Arrows* indicate time at which vinblastine was given. [From Sewell et al. (89).]

enzymes may be an important process to regulate lysosomal enzyme synthesis" (105). More specifically, the turnover of lysosomal enzymes in certain cells may be accounted for largely by extracellular release. Currently, however, the quantitative contribution of biliary excretion of lysosomal constituents to the turnover of individual lysosomal glycoproteins in the hepatocyte is unknown. Preliminary data suggest that biliary excretion probably accounts for a minimum of 25%–35% of the daily turnover of two lysosomal glycosidases (A. Nakano and N. F. LaRusso, unpublished observations). Also, the quantitative relationship between the digestive activity of hepatocyte lysosomes (via both heterophagy and autophagy) and the release of intralysosomal constituents into bile requires additional clarification.

2. It also seems likely that the lysosome-to-bile hepatic excretory pathway could be involved in biliary metal secretion (96). Bile is the major excretory route for copper in mammals, but little is known about the subcellular mechanisms accounting for biliary copper secretion (67). Moreover, excess hepatic copper is sequestered in hepatocyte lysosomes in genetic and acquired diseases of the liver associated with hepatic copper overload (e.g., Wilson's disease, primary biliary cirrhosis, primary sclerosing cholangitis) (24, 97, 110). The factors that regulate the sequestration of excess copper in lysosomes and the role of the lysosome-to-bile hepatic excretory pathway in biliary copper secretion are under investigation. Preliminary data in an animal model for hepatic copper overload indicate that excess copper is sequestered in hepatocyte lysosomes and, under these conditions, the sequestered copper is released into bile in parallel with endogenous lysosomal enzymes (37).

Similarly, the liver plays a major role in iron metabolism, and evidence suggests a primary role for lysosomes in the intracellular handling of iron (70, 96). Moreover, as with copper, excess hepatic iron is sequestered almost entirely into hepatocyte lysosomes in certain human diseases (e.g., genetic hemochromatosis) (2). Although biliary excretion has not been considered a major excretory route for iron, recent work is compatible with an important role for biliary excretion in overall hepatic iron metabolism. For example, in rats fed 2% carbonyl iron, there is a 45-fold increase in hepatic iron concentration compared with controls (62). Electron microscopy with quantitative morphometry and X-ray microanalysis show that this excess iron is sequestered in an increased number of lysosomes concentrated in the pericanalicular region of the hepatocyte (see Fig. 9). Iron loading was also demonstrated to be associated with a fourfold increase in biliary iron excretion (62). In iron-loaded rats, but not in controls, biliary iron excretion was tightly coupled to the release into bile of each of three lysosomal hydrolases. In contrast, no such relationships were found among biliary iron excretion and biliary outputs of either a plasma membrane marker enzyme or total protein. Finally, after administration of colchicine, an agent that binds microtubules, there were parallel alterations in the biliary excretion of iron and lysosomal enzymes in iron-loaded rats (62). This work demonstrates that experimental iron overload causes a dramatic increase in hepatic iron content, a sequestration of excess iron in increased numbers of hepatocyte lysosomes, and a marked increase in biliary iron excretion that is closely coupled to the biliary outputs of three lysosomal enzymes under basal and perturbed (i.e., colchicine treatment) conditions. The data strongly indicate that biliary iron excretion from hepatocyte lysosomes is a major excretory route for excess hepatic iron.

3. The lysosome-to-bile hepatic excretory pathway may likely have functional importance in the hepatic processing of selected proteins. For example, recent data indicate the importance of the liver in the processing of TGF-β, a recently discovered polypeptide that regulates growth and differentiation of normal and neoplastic cells. This growth-active polypeptide is detectable in liver and inhibits hepatic DNA synthesis induced by epidermal growth factor, a smaller growth-active protein (69). With biological active radiolabeled TGF-β and in vivo (bile fistula rat) and in vitro (isolated perfused rat liver) experimental models, over two-thirds of this radiolabeled peptide can be recovered in liver and bile after intrafemoral injection (55, 56). By 90 min, over four-fifths of the label removed by the liver has been slowly excreted into bile; most of the label in bile is soluble in trichloroacetic acid (TCA), indicating extensive degradation and raising the possibility that the peptide might be sequestered in hepatocyte lysosomes. Two lysosomotropic agents, chloroquine and leupeptin, double the amount of TCA-precipitable label in bile (55). Moreover, after fractionation of liver by differential or isopycnic centrifugation, the majority of the label associates with the fractions enriched in lysosomes as judged by comparison to marker enzymes (56). These data indicate rapid, extensive, and organ-selective extraction of TGF-β by the liver. After extraction, TGF-β associates with organelles that affect efficient transhepatic transport, extensive intracellular metabolism, and slow but complete biliary excretion. Use of pharmacologic agents and tissue fractionation indicates that these organelles include lysosomes. These data are compatible with the functional importance of the lysosome-to-bile excretory pathway in the hepatic processing of specific polypeptides.

One cannot assume that all circulating proteins that are cleared and metabolized by the liver and are excreted into bile in degraded forms are necessarily processed in the lysosomal system of the hepatocyte. For example, certain circulating, biologically active molecular forms of cholecystokinin, an important gastrointestinal peptide hormone, are efficiently cleared by the liver and extensively metabolized; the metabolites are then excreted into bile in nearly quantitative fashion (35, 36). However, based on published data using pharmacologic agents and describing the kinetics of biliary excretion of these metabolites (36) and other observations employing tissue fractionation (G. J. Gores and N. F. LaRusso, unpublished observations), the subcellular site of degradation of cholecystokinin appears not to be lysosomal. Thus there appear to exist other nonlysosomal pathways in the hepatocyte by which this cell can extract, metabolize, and excrete into bile circulating materials.

Pathophysiologic considerations. Currently few data are available that directly address the possibility that disturbances in the lysosome-to-bile hepatic excretory pathway might be relevant to disease. Nevertheless, given the likely functions of this pathway, one can infer the types of defects that could result from its malfunction. For example, congenital or acquired iron- and copper-overload diseases of the liver might involve primary or secondary defects in biliary lysosomal metal excretion. Indeed, given the data indicating that biliary iron excretion from hepatocyte lysosomes containing excess iron can be modulated by drugs, pharmacologic agents that accelerate the biliary excretion of hepatocyte lysosomal contents may have potential therapeutic benefit in patients with iron-overload diseases or other diseases associated with the lysosomal sequestration of cell toxins.

It is also plausible that alterations in the biliary excretion or activity of lysosomal β-glucuronidase could be involved in pigment gallstone formation because these stones are composed largely of unconjugated bilirubin; there are preliminary data that are consistent with this possibility (101). Finally, alterations in the physicochemical properties or the structure of specific protein constituents in bile resulting from disordered lysosomal processing could be related to the pathogenesis of certain metabolic hepatobiliary disorders, such as cholesterol cholelithiasis. Recent work has suggested that certain biliary proteins, some of which may result at least partly from the lysosomal disassembly of precursor lipoproteins, may act as nucleating or antinucleating factors in bile (54).

SUMMARY

Since the discovery of the lysosome by de Duve over 30 years ago, it has become clear that this organelle plays a pivotal role in the digestion and secretion of endogenous and exogenous cellular constituents by all eukaryotic cells. In addition to describing from biochemical, morphological, and functional viewpoints the properties, characteristics, and methods of isolation of lysosomes in general, this chapter has focused in particular on the ability of hepatocyte lysosomes to digest and secrete into bile both extracellular and intracellular components and has reviewed the functional and potential pathologic importance of this secretory pathway.

This work was supported by Grants DK-24031 and DK-34988 from the National Institutes of Health and by the Mayo Clinic. The author thanks Drs. G. J. Gores, L. J. Miller, and S. F. Phillips for reviewing the manuscript and Marilyn Breyer for typing it. The author also acknowledges the following postdoctoral research fellows who have worked in his laboratory and on whose work much of the material in this chapter is based: P. C. deGroen, G. J. Gores, J. B. Gross, Jr., G. D. LeSage, V. Lopez del Pino, D. M. Nagorney, A. Nakano, and R. B. Sewell.

REFERENCES

1. ALLISON, A. C., AND M. R. YOUNG. Vital staining in fluorescence microscopy of lysosomes. In: *Lysosomes in Biology and Pathology*, edited by J. T. Dingle and H. B. Fell. Amsterdam: Elsevier, 1969, vol. 2, p. 600–628.
2. ARBROGH, B. A., H. GLAUMANN, AND J. L. E. ERICSSON. Studies on iron loading of rat liver lysosomes: effects on the liver and distribution and fate of iron. *Lab. Invest.* 30: 664–673, 1974.
3. ASHWELL, G., AND A. G. MORELL. The role of surface carbohydrates in the hepatic recognition and transport of circulated glycoproteins. *Adv. Enzymol.* 41: 99–129, 1974.
4. BAINTON, D. F. The discovery of lysosomes. *J. Cell Biol.* 91: 66S–76S, 1981.
5. BARRETT, A. J. Properties of lysosomal enzymes. In: *Lysosomes in Biology and Pathology*, edited by J. T. Dingle and H. B. Fell. Amsterdam: Elsevier, 1969, vol. 2, p. 245–312.
6. BARRETT, A. J. *Tissue Proteinase*. Amsterdam: Elsevier, 1971.
7. BARRETT, A. J. Lysosomal enzymes. In: *Lysosomes: A Laboratory Handbook*, edited by J. T. Dingle. Amsterdam: Elsevier, 1972, p. 46–135.
8. BEAUFAY, H. Methods for the isolation of lysosomes. In: *Lysosomes in Biology and Pathology*, edited by J. T. Dingle and H. B. Fell. Amsterdam: Elsevier, 1969, vol. 2, p. 207–244.
9. BECK, F., AND J. B. LLOYD. Histochemistry and electron microscopy of lysosomes. In: *Lysosome in Biology and Pathology*, edited by J. T. Dingle and H. B. Fell. Amsterdam: Elsevier, 1969, vol. 2, p. 567–599.
10. BERGERON, J. J., R. SIKSTROM, A. R. HAND, AND B. I. POSNER. Binding and uptake of ^{125}I-insulin in rat hepatocytes and endothelium. *J. Cell Biol.* 80: 427–443, 1979.
11. BURNSIDE, J., AND D. L. SCHNEIDER. Characterization of the membrane proteins of rat liver lysosomes: composition, enzyme activities and turnover. *Biochem. J.* 204: 525–534, 1982.
12. CREEK, K. E., AND W. S. SLY. The role of the phosphomannosyl receptor in the transport of acid hydrolases to lysosomes. In: *Lysosomes in Biology and Pathology. Molecular and Cellular Aspects of Lysosomes*, edited by R. T. Dean, J. T. Dingle, and W. Sly. Amsterdam: Elsevier, 1984, vol. 7, p. 63–82.
13. D'ARCY HART, P. Phagosome-lysosome fusion in macrophages: a hinge in the intracellular fate of ingested microorganisms? In: *Liposomes in Biology and Pathology. Lysosomes and Applied Biology and Therapeutics*, edited by J. T. Dingle, P. J. Jacques, and I. H. Shaw. Amsterdam: Elsevier, 1979, vol. 6, p. 409–424.
14. DAVIES, P., AND A. C. ALLISON. The secretion of lysosomal enzymes. In: *Lysosomes in Biology and Pathology*, edited by J.

T. Dingle and R. T. Dean. Amsterdam: Elsevier, 1976, vol. 5, p. 61–98.
15. DEAN, R. T. Methods for the isolation of lysosomes. In: *Lysosomes: A Laboratory Handbook* (2nd ed.), edited by J. T. Dingle. Amsterdam: Elsevier, 1977, p. 1–18.
16. DE DUVE, C. The lysosome concept. In: *Ciba Foundation Symposium on Lysosomes*, edited by A. V. S. DeReuck and M. P. Cameron. Boston, MA: Little, Brown, 1963, p. 1–28.
17. DE DUVE, C. The separation and characterization of subcellular particles. *Harvey Lect.* 59: 49–87, 1965.
18. DE DUVE, C. General principles. In: *Enzyme Cytology*, edited by D. B. Roodyn. London: Academic, 1967, p. 1–26.
19. DE DUVE, C. The lysosome in retrospect. In: *Lysosomes in Biology and Pathology*, edited by J. T. Dingle and H. B. Fell. Amsterdam: Elsevier, 1969, vol. 1, p. 3–42.
20. DE DUVE, C. Tissue fractionation. Past and present. *J. Cell Biol.* 50: pp. 20D–55D, 1971.
21. DE DUVE, C. Exploring cells with a centrifuge. *Science Wash. DC* 189: 186–194, 1975.
22. DE DUVE, C., AND R. WATTIAUX. Functions of lysosomes. *Annu. Rev. Physiol.* 28: 435–492, 1966.
23. DEGROEN, P. C., G. D. LESAGE, AND N. F. LARUSSO. Development and initial application of specific radioimmunoassay for rat liver lysosomal hydrolases (Abstract). *Gastroenterology* 84: 1134, 1983.
24. DICKSON, E. R., C. R. FLEMING, AND J. LUDWIG. Primary biliary cirrhosis. *Prog. Liver Dis.* 6: 487–502, 1979.
25. DINGLE, J. T. The extracellular secretion of lysosomal enzymes. In: *Lysosomes in Biology and Pathology*, edited by J. T. Dingle and H. B. Fell. Amsterdam: Elsevier, 1969, vol. 2, p. 421–436.
26. ERICSSON, J. L. E. Mechanism of cellular autophagy. In: *Lysosomes in Biology and Pathology*, edited by J. T. Dingle and H. B. Fell. Amsterdam: Elsevier, 1969, vol. 2. p. 345–394.
27. FREED, J. J., AND M. M. LIEBOWITZ. The association of a class of saltatory movements with microtubules and cultured cells. *J. Cell Biol.* 45: 334–354, 1970.
28. FULLER, G. M., J. J. ELLISON, M. MCGILL, L. A. SORDAHL, AND B. R. BRINKLEY. Studies on the inhibitory role of calcium in the regulation of microtubular assembly in vitro and in vivo. In: *Microtubules and Microtubule Inhibitors*, edited by M. Borgers and M. De Brabander. Amsterdam: Elsevier, 1975, p. 379–390. (Proc. Int. Symp. Beerse, Belgium.)
29. GINSBURG, I. The role of lysosomal factors of leukocytes in the biodegradation and storage of microbial constituents in infectious granulomas. In: *Lysosomes in Biology and Pathology. Liposomes in Applied Biology and Therapeutics*, edited by J. T. Dingle, P. J. Jacques, and I. H. Shaw. Amsterdam: North-Holland, 1979, vol. 6, p. 327–408.
30. GOLDBERG, D., C. GABEL, AND S. KORNFELD. Processing of lysosomal enzyme oligosaccharide units. In: *Lysosomes in Biology and Pathology. Molecular and Cellular Aspects of Lysosomes*, edited by R. T. Dean, J. T. Dingle, and W. Sly. Amsterdam: Elsevier, 1984, vol. 7, p. 45–62.
31. GOLDBERG, N. D., M. K. HADDOX, D. K. HARTLE, AND J. W. HADDEN. Biological role of cyclic 3′,5′ guanosine monophosphate: In: *Pharmacology and the Future of Man. Cellular Mechanisms*, edited by R. A. Maxwell and G. H. Acheson. Basel: Karger, 1973, vol. 5, p. 146–169. (Proc. Int. Congr. Pharmacol., 5th, San Francisco, 1972.)
32. GOLDFINE, I. D., A. L. JONES, G. T. HRADEK, AND K. Y. WONG. Electron microscopic autoradiographic analysis of (^{125}I)iodoinsulin entry into adult rat hepatocytes in vivo: evidence for multiple sites of hormone localization. *Endocrinology* 108: 1821–1828, 1981.
33. GOLDSTEIN, J. L., R. ANDERSON, AND M. BROWN. Coated pits, coated vesicles in receptor-mediated endocytosis. *Nature Lond.* 279: 679–685, 1979.
34. GOMORI, G. *Microscopic Histochemistry: Principles in Practice.* Chicago: Univ. of Chicago Press, 1952.
35. GORES, G. J., N. F. LARUSSO, AND L. J. MILLER. Hepatic processing of cholecystokinin peptides. I. Structural specificity and mechanism of hepatic extraction. *Am. J. Physiol.* 250 (*Gastrointest. Liver Physiol.* 13): G344–G349, 1986.
36. GORES, G. J., L. J. MILLER, AND N. F. LARUSSO. Hepatic processing of cholecystokinin peptides. II. Cellular metabolism, transport, and biliary excretion. *Am. J. Physiol.* 250 (*Gastrointest. Liver Physiol.* 13): G350–G356, 1986.
37. GROSS, J. B., JR., L. KOST, P. TIETZ, S. BARHAM, J. MCCALL, AND N. F. LARUSSO. Mechanism of biliary copper excretion in experimental copper overload (Abstract). *Gastroenterology* 86: 1322, 1984.
38. HANNIG, K. Continuous free-flow electrophoresis as an analytical and preparative method in biology. *J. Chromatogr.* 159: 183–191, 1978.
39. HASILIK, A., AND K. VON FIGURA. Processing of lysosomal enzymes in fibroblasts. In: *Lysosomes in Biology and Pathology. Molecular and Cellular Aspects of Lysosomes*, edited by R. T. Dean, J. T. Dingle, and W. Sly. Amsterdam: Elsevier, 1984, vol. 7, p. 3–16.
40. HAWKINS, D. Neutrophilic leukocytes in immunologic reactions: evidence for the selective release of lysosomal constituents. *J. Immunol.* 108: 310–317, 1972.
41. HEIDRICH, H. G., AND M. E. DEW. Preparative free flow electrophoresis for the isolation of organelle and cell fractions from rabbit kidney cortex. *Curr. Probl. Clin. Biochem.* 6: 108–112, 1976.
42. HENSON, P. M. Secretion of lysosomal enzymes induced by immune complexes and complement. In: *Lysosomes in Biology and Pathology*, edited by J. T. Dingle and R. T. Dean. Amsterdam: Elsevier, 1976, vol. 5, p. 99–126.
43. HIRSIMAKI, P., B. F. TRUMP, AND A. U. ARSTILA. Studies on vinblastine-induced autophagocytosis in the mouse liver. *Virchows Arch. B Cell Pathol.* 22: 89–109, 1976.
44. HOLDSWORTH, G., AND R. COLEMAN. Enzyme profiles in mammalian bile. *Biochim. Biophys. Acta* 389: 47–50, 1975.
45. HOLTZMAN, E. Lysosomes: a survey. *Cell Biol. Monogr.* 3: 1–99, 1976.
46. HULTBERG, B., AND S. SJÖBLAD. Lysosomal enzymes in medium from cultured skin fibroblasts from normal individuals and patients with lysosomal diseases. *Clin. Chem. Acta* 80: 79–86, 1977.
47. IGNARRO, L. J. Non-phagocytic release of neutral protease and B-glucuronidase from human neutrophils. *Arthritis Rheum.* 17: 25–36, 1974.
48. IGNARRO, L. J., A. L. BORONSKY, AND R. J. PERPER. Effects of prostaglandins on release of enzymes from lysosomes of pancreas, spleen and kidney cortex. *Life Sci.* 12: 193–201, 1973.
49. IGNARRO, L. J., AND W. J. GEORGE. Hormonal control of lysosomal enzyme release from neutrophils: elevation of cyclic nucleotide levels by autonomic neural hormones. *Proc. Natl. Acad. Sci. USA* 71: 2027–2031, 1974.
50. JACQUES, P. J. Endocytosis. In: *Lysosomes in Biology and Pathology*, edited by J. T. Dingle and H. B. Fell. Amsterdam: Elsevier, 1969, vol. 2, p. 395–420.
51. JESSUP, W., J. L. BODMER, R. T. DEAN, V. A. GREENAWAY, AND P. LEONI. Intracellular turnover and secretion of lysosomal enzymes. *Biochem. Soc. Trans.* 12: 529–531, 1984.
52. JONES, A. L., R. H. RENSTON, AND S. J. BURWEN. Uptake and intracellular disposition of plasma-derived proteins by hepatocytes. *Prog. Liver Dis.* 7: 51–69, 1982.
53. KAGAWA, K., AND S. TOMIZAWA. Exocytic excretion of dextran sulfate from liver to bile. *Jpn. J. Pharmacol.* 30: 101–108, 1980.
54. KIBE, A., R. T. HOLZBACH, N. F. LARUSSO, AND S. J. T. MAO. Inhibition of cholesterol crystal formation by apoproteins A-I and A-II in model systems of supersaturated bile: implications for gallstone pathogenesis in man. *Science Wash. DC* 225: 514–516, 1984.
55. KOST, L. J., R. J. COFFEY, H. L. MOSES, AND N. F. LARUSSO. Hepatic extraction, metabolism and biliary excretion of transforming growth factor β (Abstract). *Gastroenterology* 90: 1739, 1986.

56. KOST, L. J., R. J. COFFEY, H. L. MOSES, AND N. F. LARUSSO. Hepatic handling of transforming growth factor β. *J. Cell Biol.* 103: 446a, 1986.
57. LARUSSO, N. F. Lysosomal enzymes in biological fluids: physiologic and pathophysiologic significance. *Dig. Dis. Sci.* 24: 177–179, 1979.
58. LARUSSO, N. F. Proteins in bile: how they get there and what they do. *Am. J. Physiol.* 247 (*Gastrointest. Liver Physiol.* 10): G199–G205, 1984.
59. LARUSSO, N. F. Lysosomes and the vacuolar apparatus: the gastrointestinal tract of the cell. In: *Viewpoints in Digestive Diseases*, edited by S. F. Phillips. Thorofare, NJ: Am. Gastroenterol. Assoc., 1986, vol. 18.
60. LARUSSO, N. F., AND S. FOWLER. Coordinate secretion of acid hydrolases in rat bile: hepatocye exocytosis of lysosomal protein? *J. Clin. Invest.* 64: 948–954, 1979.
61. LARUSSO, N. F., L. J. KOST, J. A. CARTER, AND S. S. BARHAM. Triton WR-1339, a lysosomotropic compound, is excreted into bile and alters the biliary excretion of lysosomal enzymes and lipids. *Hepatology* 2: 209–215, 1982.
62. LESAGE, G. D., L. J. KOST, S. S. BARHAM, AND N. F. LARUSSO. Biliary excretion of iron from hepatocyte lysosomes in the rat: a major excretory pathway in experimental iron overload. *J. Clin. Invest.* 77: 90–97, 1986.
63. LI, Y.-T., AND S.-C. LI. Activator proteins related to the hydrolases of glycosphingolipids catalyzed by lysosomal glycosidases. In: *Lysosomes in Biology and Pathology. Molecular and Cellular Aspects of Lysosomes*, edited by R. T. Dean, J. T. Dingle, and W. Sly. Amsterdam: Elsevier, 1984, vol. 7, p. 99–118.
64. LOPEZ DEL PINO, V., AND N. F. LARUSSO. Dissociation of bile flow and biliary lipid secretion from biliary lysosomal enzyme output in experimental cholestasis. *J. Lipid Res.* 22: 229–235, 1981.
65. LUCY, J. A. Lysosomal membranes. In: *Lysosomes in Biology and Pathology*, edited by J. T. Dingle and H. B. Fell. Amsterdam: Elsevier, 1969, vol. 2, p. 313–344.
66. LYSMAN, G., R. B. ZURIER, AND S. HOFFSTEIN. Leukocyte proteases and the immunologic release of lysosomal enzymes. *Am. J. Pathol.* 68: 539–559, 1972.
67. MAHONEY, J. R., J. A. BUSH, D. J. GUBLER, W. H. MORETZ, G. E. CARTWRIGHT, AND M. M. WINTROBE. Studies on copper metabolism. XV. The excretion of copper by animals. *J. Lab. Clin. Med.* 46: 702–708, 1955.
68. MOORE, P. L., H. L. BANK, N. T. BRISSIE, AND S. S. SPICER. Association of microfilament bundles with lysosomes in polymorphonuclear leukocytes. *J. Cell Biol.* 71: 659–666, 1976.
69. MOSES, H. L., E. L. BRANUM, J. A. PROPPER, AND R. A. ROBINSON. Transforming growth factor production by chemically transformed cells. *Cancer Res.* 41: 2842–2848, 1981.
70. MUNRO, H. N., AND M. C. LINDOR. Ferritin: structure, biosynthesis, and role in iron metabolism. *Physiol. Rev.* 58: 317–396, 1978.
71. MYANT, N. B. The catabolism of low-density lipoprotein by the LDL-receptor-lysosomal system. In: *Lysosomes in Biology and Pathology. Molecular and Cellular Aspects of Lysosomes*, edited by R. T. Dean, J. T. Dingle, and W. Sly. Amsterdam: Elsevier, 1984, vol. 7, p. 261–296.
72. NAGORNEY, D. M., N. F. LARUSSO, AND R. R. DOZOIS. Development and application of methodology for assessing the role of lysosomes in experimental ulcerogenesis in the guinea pig. *Gastroenterology* 85: 548–556, 1983.
73. NAKANO, A., P. D. DEGROEN, AND N. F. LARUSSO. Immunological characterization of lysosomal enzymes in rat bile: implications for biliary lysosomal enzyme secretion (Abstract). *Hepatology* 5: 1032, 1985.
74. NOVIKOFF, A. B. Lysosomes and related particles. In: *The Cell: Biochemistry, Physiology, Morphology*, edited by J. Brachet and A. E. Mirsky. New York: Academic, 1961, vol. 2, p. 423–488.
75. NOVIKOFF, A. B. Lysosomes and the physiology and pathology of cells: contributions of staining methods. In: *Ciba Foundation Symposium on Lysosomes*, edited by A. V. S. DeReuck and M. P. Cameron. Boston: Little, Brown, 1963, p. 36–71.
76. NOVIKOFF, A. B. Enzyme localization and ultrastructure of neurons. In: *The Neuron*, edited by H. Hyden. Amsterdam: Elsevier, 1967, p. 255–318.
77. NOVIKOFF, A. B. Lysosomes in nerve cells. In: *The Neuron*, edited by H. Hyden. Amsterdam: Elsevier, 1967, p. 319–377.
78. NOVIKOFF, A. B. Lysosomes: a personal account. In: *Lysosomes and Storage Diseases*, edited by H. G. Hers and F. Van Hoof. New York: Academic, 1973, p. 1–41.
79. NOVIKOFF, P. M., N. F. LARUSSO, A. B. NOVIKOFF, R. J. STOCKERT, A. YAM, AND G. D. LE SAGE. Immunocytochemical localization of lysosomal β-galactosidase in rat liver. *J. Cell Biol.* 97: 1559–1565, 1983.
80. NOVIKOFF, P. M., A. B. NOVIKOFF, N. QUINTANA, AND J. J. HAUW. Golgi apparatus, GERL and lysosomes within neurons in rat dorsal root ganglia studied by thick section and thin section cytochemistry. *J. Cell Biol.* 50: 859–886, 1971.
81. PASTAN, I. H., AND M. C. WILLINGHAM. Receptor-mediated endocytosis of hormones in cultured cells. *Annu. Rev. Physiol.* 43: 239–250, 1981.
82. PETERS, T. J., M. MULLER, AND C. DE DUVE. Lysosomes of the arterial wall. I. Isolation and subcellular fractionation of cells from normal rabbit aorta. *J. Exp. Med.* 136: 1117–1139, 1972.
83. REIGNGOUD, D.-J., AND J. M. TAGER. The permeability properties of the lysosomal membrane. *Biochim. Biophys. Acta* 472: 419–449, 1977.
84. ST. HILAIRE, R. J., G. T. HRADEK, AND A. L. JONES. Hepatic sequestration and biliary secretion of epidermal growth factor: evidence for a high capacity uptake system. *Proc. Natl. Acad. Sci. USA* 80: 3797–3801, 1983.
85. SCHELLENS, J. P. M., W. T. H. DAEMS, J. J. MEMEIS, P. BREDEROO, W. C. DEBRUIJN, AND WISSE. Electron microscopical identification of lysosomes. In: *Lysosomes: A Laboratory Handbook*, edited by J. T. Dingle. Amsterdam: Elsevier, 1972, p. 147–208.
86. SCHNEIDER, D. L. ATP-dependent acidification of membrane vesicles isolated from purified rat liver lysosomes: acidification activity requires phosphate. *J. Biol. Chem.* 258: 1833–1838, 1983.
87. SCHNEIDER, F. H. Observation on the release of lysosomal enzymes from isolated bovine adrenal gland. *Biochem. Pharmacol.* 17: 848–851, 1968.
88. SEWELL, R. B., S. S. BARHAM, AND N. F. LARUSSO. Effect of chloroquine on the form and function of hepatocyte lysosomes: morphologic modifications and physiologic alterations related to the biliary excretion of lipids and proteins. *Gastroenterology* 85: 1146–1153, 1983.
89. SEWELL, R. B., S. S. BARHAM, A. R. ZINSMEISTER, AND N. F. LARUSSO. Microtubule modulation of biliary excretion of endogenous and exogenous hepatic lysosomal constituents. *Am. J. Physiol.* 246 (*Gastrointest. Liver Physiol.* 9): G8–G15, 1984.
91. SHEPHERD, V. L., AND P. D. STAHL. Macrophage receptors for lysosomal enzymes. In: *Lysosomes in Biology and Pathology*, edited by J. T. Dingle, R. T. Dean, and W. Sly. Amsterdam: Elsevier, 1984, vol. 7, p. 83–98.
92. SKUDLAREK, M. D., E. K. NOVAK, AND R. T. SWANK. Processing of lysosomal enzymes in macrophages in kidneys. In: *Lysosomes in Biology and Pathology. Molecular and Cellular Aspects of Lysosomes*, edited by R. T. Dean, J. T. Dingle, and W. Sly. Amsterdam: Elsevier, 1984, vol. 7, p. 17–44.
93. SMITH, R. E., AND M. G. FARQUHAR. Preparation of nonfrozen sections for electron microscope cytochemistry. *RCA Sci. Instrum. News* 10: 13–17, 1965.
94. SMITH, R. E., AND M. G. FARQUHAR. Lysosome function in the regulation of the secretory process in cells of the anterior pituitary gland. *J. Cell Biol.* 31: 319–336, 1966.
95. SMITH, R. E., AND W. H. FISHMAN. P-(acetoxymercuric) aniline diazotate, a reagent for visualizing the naphthol AS-

BI product of acid hydrolase action at the level of the light and electron microscope. *J. Histochem. Cytochem.* 17: 1–22, 1969.
96. STERNLIEB, I., AND S. GOLDFISCHER. Heavy metals and lysosomes. In: *Lysosomes in Biology and Pathology*, edited by J. T. Dingle and R. T. Dean. Amsterdam: Elsevier, 1976, vol. 5, p. 185–202.
97. STERNLIEB, I., C. J. A. VAN DEN HAMER, A. G. MORELL, S. ALPERT, G. GREGORIADIS, AND I. H. SCHEINBERG. Lysosomal defect of hepatic copper excretion in Wilson's disease (hepatolenticular degeneration). *Gastroenterology* 64: 99–105, 1973.
98. STOCKERT, R. J., AND A. G. MORRELL. Hepatic binding protein: a galactose-specific receptor of mammalian hepatocytes. *Hepatology* 3: 750–757, 1983.
99. STOSSEL, T. P. Phagocytosis. *N. Engl. J. Med.* 290: 717–723, 774–780, 833–839, 1974.
100. STRAWSER, L. D., AND O. TOUSTER. The cellular processing of lysosomal enzymes and related proteins. *Rev. Physiol. Biochem. Pharmacol.* 87: 169–210, 1980.
101. TABATA, M., J. R. CORCORAN, AND J. D. OSTROW. Kinetics of β-glucuronidase (βGase) from bovine liver (Liv) and escherichia coli (EC) with bilirubin diglucuronide as substrate (Abstract). *Gastroenterology* 88: 1698, 1985.
102. TAPPEL, A. L. Lysosomal enzymes and other components. In: *Lysosomes in Biology and Pathology*, edited by J. T. Dingle and H. B. Fell. Amsterdam: Elsevier, 1969, vol. 2, p. 207–244.
103. TAPPEL, A. L. Methods for the study of lysosomal function. In: *Lysosomes in Biology and Pathology*, edited by J. T. Dingle and H. B. Fell. Amsterdam: Elsevier, 1969, vol. 2, p. 547–554.
104. TOYODA, S., Y. ETO, AND K. AOKI. Bile lysosomal enzymes: characteristics and pathological significance for various hepatobiliary disorders. *Clin. Chem. Acta* 79: 291–298, 1977.
105. WANG, C. C., AND O. TOUSTER. Turnover studies on proteins of rat liver lysosomes. *J. Biol. Chem.* 250: 4896–4902, 1975.
106. WATTIAUX, R., M. WIEBO, AND P. BAUDHUIN. Influence of the injection of Triton WR-1339 on the properties of rat-liver lysosomes. In: *Ciba Foundation Symposium on Lysosomes*, edited by A. V. S. DeReuck and M. P. Camerson. Boston: Little, Brown, 1963, p. 176–200.
107. WEIBEL, E. R., W. STAUBL, H. R. GNAGI, AND F. A. HESS. Correlated morphometric and biochemical studies on the liver cell. I. Morphometric model, sterologic methods, and normal morphometric data for rat liver. *J. Cell Biol.* 42: 68–91, 1969.
108. WEISSMANN, G., R. B. ZURIER, P. H. SPIELER, AND I. M. GOLDSTEIN. Mechanisms of lysosomal enzyme release from leukocytes exposed to immune complexes and other particles. *J. Exp. Med.* 134: 149–165, 1971.
109. WELLS, W. W., AND C. A. COLLINS. Phosphorylation of lysosomal membrane components as a possible regulatory mechanism. In: *Lysosomes in Biology and Pathology. Molecular and Cellular Aspects of Liposomes*, edited by R. T. Dean, J. T. Dingle, and W. Sly. Amsterdam: Elsevier, 1984, vol. 7, p. 119–140.
110. WIESNER, R. H., AND N. F. LaRUSSO. Clinicopathologic features of the syndrome of primary sclerosing cholangitis. *Gastroenterology* 79: 200–206, 1980.
111. YAMADA, H., H. HAYASHI, AND Y. NATORI. A simple procedure for the isolation of highly purified lysosomes from normal rat liver. *J. Biochem.* 95: 1155–1160, 1984.

CHAPTER 34

Hepatic transport of organic solutes

E. L. FORKER | Departments of Medicine and Physiology, University of Missouri School of Medicine, Columbia, Missouri

CHAPTER CONTENTS

Functional Anatomy
  Vascular organization of hepatic parenchyma
    Structural adaptations for efficient solute exchange between blood and tissue
  Liver cell plates, canaliculi, and tight junctions
  Intra-acinar heterogeneity
Experimental Methods
Transport Mechanisms
  Classification
  Carrier-mediated transport
    Facilitated diffusion, active transport, *trans* effects
    Countertransport and cotransport
    Simple diffusion and convection
  Compartmentation of solutes inside cell and in bile
  Effect of choleresis on canalicular anion transport
Mathematical Modeling
  Steady-state input-output relations
  Distribution of flow to multiple sinusoids
  Estimating uptake, efflux, and removal separately
    Single-pass indicator dilution curves
    Solute disappearance curves
Uptake of Albumin-Bound Organic Anions
  Historical perspective
  Dissociation-mediating sites on the cell surface?
  Specificity
Transport of Specific Solute Classes
  Bile acids
  Long-chain fatty acids
  Other organic anions
  Amino acids
  Organic cations
  Neutral compounds
    Cardiac glycosides
    Hexoses
  Intracellular binding proteins
  Putative membrane carriers
Conclusion

THE LIVER SERVES the rest of the body as a synthetic and catabolic center of exceptional diversity. It also serves via the bile as a major excretory pathway for organic wastes. Moreover, bile has an important digestive function, conveying bile acids to the intestinal lumen. These features bespeak a heavy traffic of organic solutes across the plasma membranes of liver cells. This chapter is a synthesis of current opinion about the mechanisms that mediate this traffic, giving particular attention to the liver's special anatomy, to the advantages and drawbacks of various experimental methods, and to the distinctive kinetic features of carrier-mediated transport. Additional sections review the power and pitfalls of mathematical models as well as postulated explanations for the effect of protein binding on hepatic uptake. The emphasis is on general principles rather than on specific solutes, but conspicuous transport features of several classes of solutes are briefly summarized. Extended discussions of vesicular transport and biotransformation appear in the chapters by Jones and Burwen, by LaRusso, and by Meijer and so are excluded here. The bibliographic citations are an inevitably arbitrary selection from a large, rapidly expanding literature. More comprehensive reference lists can be found in other reviews (7, 12, 62, 116).

FUNCTIONAL ANATOMY

Figures 1 and 2 are drawings of parenchymal microanatomy constructed to illustrate some of the special adaptations discussed in this section. A summary of useful morphometric estimates appears in Table 1.

Vascular Organization of Hepatic Parenchyma

Anatomists and physiologists have argued at length over the concept of functional parenchymal units. The terms *lobule* and *acinus* are in common parlance depending on whether one chooses to organize liver cells into groups drained by the same hepatic venule or into groups supplied by a pair of arterial and venous afferents. This controversy is somewhat artificial. Blood exiting the liver through a particular hepatic venule may arise from several afferent sources and blood entering the liver through a given afferent may leave via more than one hepatic venule. A useful alternative to the concept of functional units envisions the liver parenchyma as a continuous array of liver cell plates disposed around a matrix of more or less regularly spaced vascular sources and sinks. The sources are terminal branches of the portal vein and terminal branches of the hepatic artery. The arterial afferents

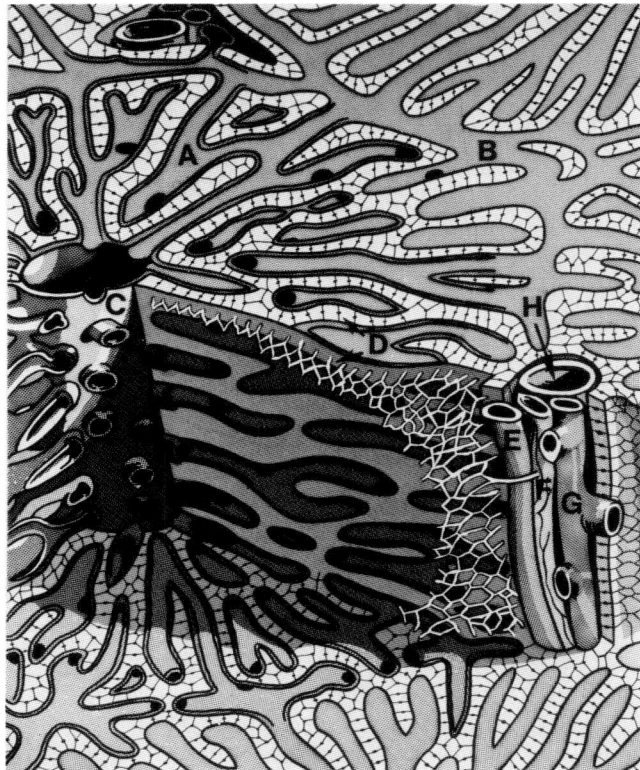

FIG. 1. Parenchymal architecture. *A*: sinusoid; *B*: liver cell plate; *C*: hepatic venule; *D*: canaliculi; *E*: hepatic arteriole; *F*: bile ductule with peribiliary capillary plexus; *G*: portal venule; *H*: portal vein. [Adapted from Leevy (71a).]

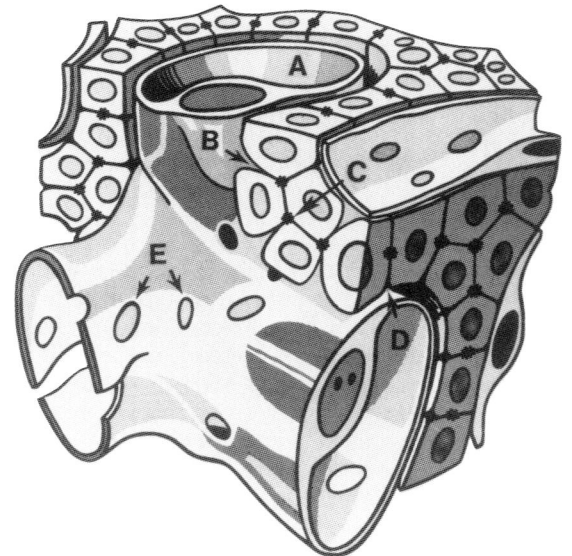

FIG. 2. Anatomy of sinusoidal solute exchange. *A*: sinusoid; *B*: lateral recess and junctional complex; *C*: canaliculus; *D*: Disse space; *E*: endothelial fenestrae. [Adapted from Leevy (71a).]

TABLE 1. *Hepatic Morphometry*

	Parenchymal Composition, % of total liver volume	Ref.		Dimension, μm	Ref.
Hepatocytes	77.8	11	Hepatocyte diam	21	78
Sinusoidal lumen	10.6	11	Sinusoidal diam	5	134
			Canalicular diam	1	46
Disse space	4.9	11	Disse space width	0.5*	4
Endothelial cells	2.8	11			
Kupffer cells	2.1	11			
Ito cells	1.4	11			
Bile canaliculi	0.4	11			

Because figures are averages for liver as a whole, zonal variations are suppressed. *Value given in ref. 4 agrees with that calculated from a mean sinusoidal diameter of 5 μm (134) and volume fractions of 10.6% and 4.9% for sinusoids and Disse space, respectively (11), on assumption that Disse space comprises a circular zone of constant width surrounding each sinusoid.

arise from the periductular capillary plexus and from hepatic arterioles, which together supply ~30% of the total blood flow. The sinks are terminal branches of the hepatic vein. To construct a useful analogy one may think of the liver as a sponge formed around the interdigitating branches of the afferent and efferent vascular trees. The tissue of the sponge corresponds to anastomosing plates of liver cells, while the holes in the sponge correspond to the sinusoids. Corrosion casts confirm that afferent blood supplies a more or less well-defined group of liver cells between a source and adjacent sinks (59). In this sense the flow pattern may be thought of as defining a functional unit, but the sponge analogy is appropriate in the sense that each sinusoid is connected to every other one within a given liver lobe. This view accounts for the remarkable plasticity of the sinusoidal circulation. Deformation of the liver surface leads to blanching of a surprisingly large area of subjacent tissue without discernible changes in portal flow or pressure—a reflection of the low sinusoidal pressures that govern flow in this system and the availability of multiple alternate channels.

Structural Adaptations for Efficient Solute Exchange Between Blood and Tissue

Liver sinusoids differ from the continuous or fenestrated capillaries of most other organs in two respects. The endothelial lining cells lack a basement membrane and their cytoplasm is penetrated by large holes or fenestrae that lack diaphragms. The liver sinusoids of sheep and other ruminants are said to be exceptional in having a well-defined basal lamina (42). For a contrary opinion see refs. 138, 139. The average diameter of the fenestrae is ~110 nm. The range of diameters is perhaps 75–300 nm, with a distribution that is approximately Gaussian. The aggregate area of the fenestrae comprises ~6% of the sinusoidal wall (134, 135). The importance of the fenestrae in accommodating transendothelial solute exchange can be appreciated by comparing their size to the Stokes-Einstein radii of representative plasma proteins: albumin

= 3.7 nm, immunoglobulin M = 12 nm, lipoproteins = 20–70 nm (123). It is clear from these dimensions that only the cellular elements and chylomicrons are constrained by the endothelial sieve. With the possible exception of the spleen and bone marrow (111), no other circulation in the body enjoys this remarkable specialization for transcapillary solute exchange.

The interstitial space of Disse comprises a thin layer of extracellular fluid between the sinusoidal endothelium and the surface of liver cells. It contains the fat storing cells of Ito (54), a few collagen fibers, and the proteoglycans and hyaluronidate of extracellular matrix as well as microvilli that project into it from the plasma membranes of subjacent hepatocytes. Scanning electron micrographs show that an occasional microvillus protrudes through the endothelial fenestrae and is thus in direct contact with blood (55, 93). Most of the basolateral hepatocyte surface lies within the Disse space, however, and thus has access to blood only insofar as diffusion and convection mediate the transfer of solutes across the intervening interstitial fluid. The role of convection in this process is generally ignored because diffusion alone is sufficient to erase radial concentration gradients that would otherwise develop between blood and the cell surface in the course of solute removal or secretion. The theoretical basis for this conclusion is that the radius of the sinusoidal lumen ($\sim$2.5 μm) and the thickness of the Disse layer ($\sim$0.5 μm) comprise so short a distance that diffusion equilibrium is predicted to occur rapidly even for large solutes such as albumin (6, 32). An observation consistent with the same conclusion is that albumin appears in liver lymph at virtually the same concentration as that in plasma. Small deviations from unity reported for the lymph-plasma concentration ratio probably reflect steric hindrance by the interstitial matrix and/or dilution of sinusoidal lymph by albumin-free fluid arising from the peribiliary capillary plexus (4). For a different reason axial diffusion (in the direction of blood flow) makes only a trivial contribution to the movement of solutes along the sinusoid. The explanation in this case is that axial movement is dominated by convection. Moreover the diameter of a typical sinusoid ensures that erythrocytes traverse the lumen in contact with the endothelium (134)—a feature that defeats the tendency that would otherwise occur for red cells to travel faster than plasma. Thus all the constituents of blood in a sinusoid are thought to move with the same velocity. The importance of these considerations to mathematical modeling of hepatic transport is discussed later.

Liver Cell Plates, Canaliculi, and Tight Junctions

Because adult liver cell plates are only one cell thick, typical hepatocytes face two sinusoids and so have two basolateral poles or surfaces. Each basolateral surface comprises a basal portion that lies close to the endothelial wall and a lateral domain composed of shallow clefts or recesses that occupy the boundary between adjacent cells. The intercellular recesses terminate in junctional complexes that include desmosomes, gap junctions, and tight junctions.

Bile canaliculi lie in the center of liver cell plates between adjacent hepatocytes whose plasma membranes form the canalicular wall. The junctional complexes form continuous seams along the canaliculi, sealing them on each side from what would otherwise be free communication with the Disse space by way of the intercellular clefts. The canalicular lumen is $\sim$1 μm in diameter. The large surface-to-volume ratio that this small diameter ensures is enlarged still further by microvillous extensions of the lining membrane. Because the bile capillaries are continuous along each cell-to-cell interface, they form a complex network resembling chicken wire. Like the sinusoids and the hepatocytes themselves, all the canaliculi within a liver lobe are at least potentially in communication with each other. The canalicular system is blind ended where liver cell plates abut a terminal hepatic venule, but they empty into interlobular bile ductules at the upstream end of the sinusoids. In this sense the flow of canalicular bile is typically countercurrent to the flow of sinusoidal blood.

Electron micrographs of hepatocyte tight junctions show that these structures are similar to those in so-called leaky epithelia such as the gallbladder and the small intestine. It is generally inferred on this basis and because bile is isosmotic with plasma that the tight junctions are more or less freely permeable to water and inorganic ions. Their penetration by ionic lanthanum during choleresis is consistent with this conclusion (71, 126), as is their low electrical resistance (40) and the rapid equilibration of Na^+ and K^+ between liver perfusate and bile (39). The importance of the tight junctions to blood-bile exchange of organic solutes is less clear. Small metabolically inert hydrophilic solutes such as sucrose and inulin achieve steady-state bile-plasma concentration ratios much faster than they fill their steady-state volumes of distribution in liver tissue (27). This finding originally led to a widely accepted proposal that the tight junctions are the principal pathway by which these and other fluid-phase markers achieve their unexpectedly high concentrations in bile. More recent studies (69) suggest that fluid-phase markers larger than sucrose enter bile by vesicular transcytosis. There is no consensus yet on the importance of transcytosis to the biliary excretion of smaller solutes such as erythritol and mannitol, nor is it known whether the vesicular pathway is indiscriminant or available only to certain solutes.

Intra-Acinar Heterogeneity

To this point the structural features subserving solute exchange have been presented as though they were uniform throughout the liver. In fact heterogeneities occur that further complicate an already complex anatomy. Parenchymal zones are a convenient reference frame for describing some of these variations (102). Zone 1 denotes hepatocytes adjacent to a vascular inlet. Zone 3 denotes the cells around an hepatic venous outlet. Zone 2 denotes an intervening region defined by exclusion instead of by landmarks of its own.

It is clear from the extensive arborization of the sinusoids that even if each channel had the same cross section and carried blood at the same velocity, the variation in path lengths would ensure a nonuniform distribution of sinusoidal transit times. In practice the situation is even more complicated because in vivo microscopy reveals that flow is faster in some sinusoids than in others (63) and morphometry shows that the sinusoids of zone 3 are on average larger than those in zone 1 (91). The increase in sinusoidal diameter from zone 1 to zone 3 implies a corresponding decrease in the surface area of the basolateral hepatocyte membrane per unit of vascular volume. The apparent uptake rate constant (defined as flux per mass of interstitial solute) thus becomes a function of location along a typical sinusoid even if the intrinsic transport capacity per unit area is regarded as constant. A similar dependence on location may apply to the rate constants for efflux from cytoplasm to interstitial fluid and for biliary excretion. In the case of efflux the potential for heterogeneity arises from the nonuniform ratio of cell volume to basolateral membrane area that in turn arises from the circumstance that more cells of uniform size are required to surround the sinusoids of zone 3 than of zone 1. Heterogeneity in the rate constant for excretion arises not only from the increase in cell volume per unit length of sinusoid from zone 1 to zone 3 but from the fact that canalicular diameter increases in the opposite direction (70). In addition to heterogeneities attributable to these geometric considerations, the intrinsic transport capacities of hepatocyte membranes may themselves depend on location along the vascular stream. Systematic zonal variations in the activities of hepatocellular enzymes are widely reported (43), and steady-state zonal concentration profiles of transported solutes have been visualized by autoradiography (37, 56) and by in vivo spectrometry (44). It has proved difficult, however, to relate any of these findings to quantitative changes in the transport kinetic parameters. Approaches to dealing with transport heterogeneity by determining average kinetic parameters for the liver as a whole are discussed in MATHEMATICAL MODELING, p. 701.

EXPERIMENTAL METHODS

Methodological approaches to the study of hepatic transport divide somewhat arbitrarily into two broad classes: those that employ morphologic techniques and those designed to measure the kinetics of solute movement.

Morphologic approaches include histochemistry, immunocytochemistry, autoradiography, morphometry, in vivo microscopy, and scanning electron microscopy. Such techniques have already supplied and should continue to provide important insights into the structural features that subserve the transport function as well as the intra-acinar and the intracellular location of transported solutes and mediating enzymes. Even dynamic aspects of the transport process may be elucidated. For example, electron microscopy has played a central role in exposing the vectorial features of endocytosis and exocytosis.

Morphologic techniques are not well suited to quantitative kinetic measurements, however. This information depends instead on measuring the time dependence of solute movement in living tissue or in isolated tissue components. Such systems include intact animals, artificially perfused livers, isolated hepatocytes in suspension or monolayer culture, membrane vesicles, and isolated hepatocyte couplets. Each of these approaches has its own advantages, but none offers the simplicity afforded by such tissues as gallbladder or intestinal mucosa that can be mounted in a flux chamber as an intact epithelial layer. Unlike the kidney, moreover, liver parenchyma does not lend itself to micropuncture or microperfusion of functional units in situ. These difficulties largely explain why the physiology of hepatic transport is less well understood than that of more accessible tissues.

The study of whole livers has the advantage of preserving the functional polarity of individual cells and of maintaining the spatial relationship of one cell to another. As explained in FUNCTIONAL ANATOMY, p. 693, this approach also minimizes diffusion artifacts that may otherwise arise from unstirred layers of extracellular fluid. The important disadvantage is that kinetic parameters (rate constants, clearances, affinities, and so forth) must be inferred from input-output relations obtained from the whole organ rather than from direct measurements of the flux rates and their conjugate driving forces at the level of individual cells. Heterogeneities in perfusate flow distribution and transport capacity further complicate the design of experiments and the construction of mathematical models to interpret the observations.

The use of isolated hepatocytes in suspension or monolayer culture simplifies these problems by allowing direct determination of flux rates and their extracellular driving concentrations under circumstances where the latter can be varied more or less at will. This advantage is costly, however. When hepatocytes are isolated by enzymatic and/or mechanical disruption of the liver, they lose the functional polarity they have as components of an intact liver cell plate. Moreover when cells are studied in suspension, it is impossible to distinguish fluxes occurring across the basolateral domain from those occurring at the canalicular pole even if the distinctive transport properties of these regions remain at their original locations. It is not known to what extent this difficulty is circumvented by culturing hepatocytes in confluent monolayers. Even if the surface of monolayers exposed to the bathing medium is considered as functionally equivalent to the basolateral domain, however, the best one can hope for is a preparation with the kinetic features of a liver with total biliary obstruction.

There are two other problems to consider in using isolated hepatocytes for kinetic work. First, it is difficult to reduce the unstirred layer of the bathing medium to a thickness that approaches to within an order of magnitude the radius of the sinusoid lumen and the width of the associated Disse space (24). If solute conductance across such layers becomes rate limiting, the resulting diffusion artifact may be substantial—not just quantitatively but qualitatively (5). (See, for example, the later discussion of the effect of protein binding on hepatic uptake.) Second, it is impossible to know whether the behavior of isolated hepatocytes truly represents the function of hepatocytes in the intact organ. This difficulty arises not just because membrane function may be altered during tissue disruption but because the isolation procedure may select a nonrepresentative subset of "viable" hepatocytes from the heterogeneous population of cells that operate in vivo.

The use of hepatocyte membrane vesicles has the potential to overcome the polarity problem and in principle this approach enjoys a number of other advantages. By careful density gradient centrifugations of liver homogenate, one can prepare vesicles whose membranes appear to come predominantly, if not quite exclusively, from the basolateral or the canalicular domain of hepatocytes (83). Vesicle volume can be modified osmotically to help make the important distinction between solute that is merely adsorbed to the vesicle surface and that which actually enters the interior. Ionic concentration differences can be imposed across the vesicle wall so that the kinetic effects of transmembrane differences in pH and electrical potential can be assessed and the role of Na^+ cotransport can be investigated.

Although these advantages are extraordinarily important, the vesicle method suffers from important ambiguities. Claims of membrane purity are generally defended by appealing to enrichment figures for so-called marker enzymes. Membrane-associated enzymes may move during membrane harvesting, however, and even their location in vivo may be disputed. For example, monoclonal antibodies to Na^+-K^+-ATPase, a widely accepted marker for basolateral membranes, have been used recently to claim (72) and to deny (121) that this enzyme is also at the canalicular pole of hepatocytes. Thus even elaborate purification schemes should not be expected to yield vesicles from one surface domain that are completely uncontaminated by components from the other surface domain or by intracellular membranes, to say nothing of contamination by other cell types. In typical preparations, only 70%–80% of membrane vesicles are right-side out—a feature that may cloud the quantitative interpretation of flux components. Finally, there is little reason to expect that vesicles prepared by even the gentlest of current methods will display the same quantitative behavior per unit area of membrane as do living cells. For the present, therefore, the best one can hope for is that transport kinetics observed with membrane vesicles will be qualitatively representative of an intact liver.

The closest approach yet to isolating intact functional units from the liver is provided by hepatocyte couplets, i.e., pairs of hepatocytes whose junctional complexes escape disruption during isolation and which contain a canalicular space between them. The canalicular vacuole seals shortly after isolation and thereafter one can observe cycles of expansion and collapse that presumably represent secretion of canalicular fluid and its subsequent escape—the latter occurring perhaps by intermittent disruption of the tight junctions. Undoubtedly the most important advantage of couplets is that the dilated canaliculus can be entered with a microelectrode. Limited experience with this preparation has already shown that couplets transport fluorescein from the bathing medium to the canalicular lumen, that the tight junctions are at least intermittently intact, and that the canalicular lumen is ~6 mV negative to the bathing medium, i.e., 25–35 mV positive with respect to the cell interior (40). Because bile secretion occurs into a closed space in this preparation and because this space is too small to allow the removal of fluid samples, it is unlikely that hepatocyte couplets can be used to study steady-state bile secretion or to dissect the unidirectional components of canalicular solute flux. Nevertheless couplets have considerable promise as a means to explore the electrical driving forces and the functional role of the tight junctions.

TRANSPORT MECHANISMS

Classification

A classification of transmembrane solute migration can be constructed from fundamental energetic considerations or mechanistic hypotheses. The energetic approach proceeds from thermodynamic axioms and is therefore unassailable, but it is concerned with only the direction or potential direction of the net flux and so has nothing to say about pathways or mechanisms. A mechanistic classification focuses on how the flux occurs rather than on simply whether it can occur. In considering the hepatic transport of organic solutes we are primarily concerned with this second approach, especially as it applies to carrier-mediated systems, but some elementary energetic considerations are appropriate first.

From the energetic point of view solute movement may be classified according to whether the net flux is driven solely by the difference in electrochemical potential of the solute in question or whether some other source of energy contributes to the driving force. In an isobaric, isothermal system the differences in electrochemical potential between two regions (e.g., between the inside and outside of a liver cell) is given by

$$\Delta \tilde{\mu} = RT \ln(a_1 c_1 / a_2 c_2) + zF(\psi_1 - \psi_2) \quad (1)$$

in which $\Delta \tilde{\mu}$ is the electrochemical potential difference of the solute in question, a and c are respectively the activity coefficient and the concentration of the solute, and ψ is the electric potential. The subscripts 1 and 2 denote the regions between which $\Delta \tilde{\mu}$ exists, and z is the charge, if any, on the solute. R is the gas constant, T is the absolute temperature, and F is the Faraday constant. Transport systems in which the net flux depends exclusively on $\Delta \tilde{\mu}$ are distinguished from those in which the net flux depends on one or more additional sources of energy. The former category includes simple diffusion and carrier-mediated diffusion. The latter includes solvent drag and active transport. In solvent drag systems the additional energy is supplied by convection. In active transport the extra energy is supplied directly or indirectly by coupling to an exergonic chemical reaction.

Carrier-Mediated Transport

Large organic solutes, including even lipophilic materials such as long-chain fatty acids, probably require carrier mediation or endocytosis for efficient hepatic extraction. A general model of carrier-mediated transport that accommodates most of our limited knowledge about how such systems work in the liver appears in Figure 3. For simplicity in illustrating the kinetic features, it is convenient to assume that concentrations are equivalent to activities and that transported solutes are electrically neutral. It should be emphasized, however, that these simplifications are rarely warranted in practice, because protein binding, steric exclusion, and micelle formation may easily result in activities that are very different from apparent concentrations and because many organic solutes of interest are weak electrolytes. It should also be recognized that although the scheme in Figure 3 envisions a mobile carrier that shuttles back and forth across the membrane, other kinetically equivalent formulations are plausible provided they incorporate the essential property that the carrier reacts with its ligand(s) alternately at each side of the membrane. This feature distinguishes carrier mediation from transport that proceeds via pores to which the solute has access at each side of the membrane simultaneously. For lucid discussions of this distinction and of carrier-mediated transport in general see refs. 2 and 109.

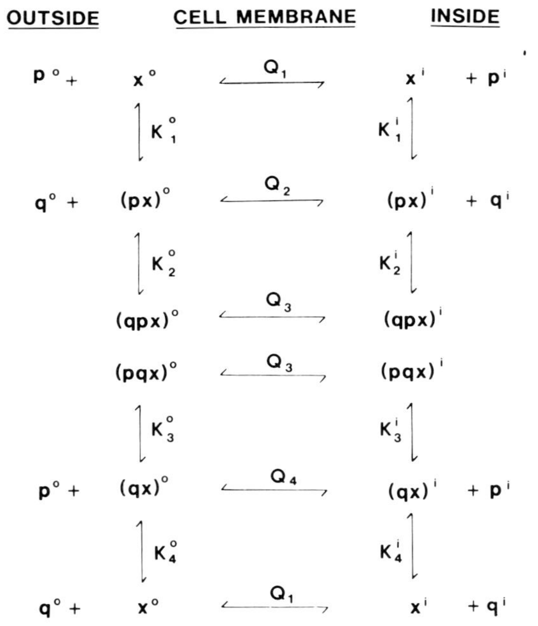

In Figure 3, p and q represent solutes that bind to a carrier x. The dissociation constants that govern these binding reactions are denoted by K_1, K_2, K_3, and K_4. The superscripts i and o denote the inside and outside of the membrane. Transmembrane migration of free carrier and its ligand complexes is governed by the mobilities Q_1, Q_2, Q_3, and Q_4. This process is regarded as rate limiting, moreover, so that the binding reactions at each membrane face are at equilibrium. The differential equations that describe even this simplified model are nonlinear and difficult. For present purposes, simpler, albeit still somewhat complicated, expressions can be constructed from the steady-state flux criterion for the carrier

$$\text{net carrier flux} = {}_xJ_{\text{net}} = {}_xJ_{\text{o}\to\text{i}} - {}_xJ_{\text{i}\to\text{o}} = 0 \quad (2)$$

and the carrier conservation constraint

$$\text{total carrier} = x_t = x^o + x^i + (xp)^o + (xp)^i + (xq)^o \\ + (xq)^i + (xpq)^o + (xpq)^i + (xqp)^o + (xqp)^i \quad (3)$$

The steady-state unidirectional flux of p derived from these considerations is

$$_pJ_{\text{o}\to\text{i}} = \frac{x_t S^i [S^o - Q_1 - Q_4(q^o/K_4^o)]}{S^i R^o + S^o R^i}$$

$$S^o = Q_1 + Q_2(p^o/K_1^o) + Q_3 p^o q^o [(1/K_1^o K_2^o) \\ + (1/K_3^o K_4^o)] + Q_4(q^o/K_4^o) \quad (4)$$

$$R^o = 1 + p^o/K_1^o + p^o q^o [(1/K_1^o K_2^o) \\ + (1/K_3^o K_4^o)] + q^o/K_4^o$$

S^i and R^i are identical to S^o and R^o except for the

FIG. 3. Reaction scheme for carrier-mediated transport.

indicated change in superscripts. Equations 4 display such characteristic kinetic features of a carrier-mediated unidirectional flux as saturation kinetics and analogue competition. Other kinetic features and the steady-state solute distribution implicit in Equations 4 are best illustrated by example.

FACILITATED DIFFUSION, ACTIVE TRANSPORT, *TRANS* EFFECTS. Suppose that $K_2^o = K_2^i = K_3^o = K_3^i = K_4^o = K_4^i = \infty$ so that only the complex xp can form. If $K_1^o = K_1^i$, the system mediates facilitated diffusion. Such systems are equilibrative in the sense that $p^o = p^i$ when $_pJ_{i \to o} = {_pJ_{o \to i}}$. If chemical work is supplied to make $K_1^i > K_1^o$, the system mediates primary active transport. In this case the movement of p is concentrative in the sense that when the net flux is zero, $p^i > p^o$. It is easy to show from Equations 4 that for either of these special cases the unidirectional flux is a function of solute concentration on both the *trans* and the *cis* side of the membrane if x and xp traverse the membrane with different mobilities. In other words, when $Q_1 \neq Q_2$, $_pJ_{i \to o}$ depends not only on p^i but also on p^o, the so-called *trans* effect. Otherwise $_pJ_{i \to o}$ depends on p^i alone. The potential for *trans* effects is a general property of Equations 4 whenever the solute-carrier complexes have mobilities that differ from each other or from the mobility of the free carrier.

COUNTERTRANSPORT AND COTRANSPORT. The model in Figure 3 accounts for two other kinetic features that are known to occur in the liver: countertransport and cotransport. These terms refer to the fact that the flux of one solute can be driven at a faster rate than would otherwise occur by the simultaneous flux of another solute that shares the same carrier. When the driven and driving fluxes are in opposite directions the phenomenon is called countertransport. When they are in the same direction the appropriate term is cotransport.

As an example of countertransport consider the situation in which the binary complexes xp and xq are transported but formation of the ternary complexes xpq and xqp is excluded. If the system is equilibrative, i.e., if $K_1^o = K_1^i$ and $K_4^o = K_4^i = \bar{K}_4$, it is easy to show from Equations 4 that when $_pJ_{net} = 0$

$$\frac{p^i}{p^o} = \frac{\bar{K}_4 Q_1 + Q_4 q^i}{\bar{K}_4 Q_1 + Q_4 q^o} \quad (5)$$

It follows that $p^i > p^o$ if $q^i > q^o$. To appreciate how this occurs suppose that in addition to the equilibrative carrier mechanism the system contains a pump that moves q from inside to outside. If the pump is switched on when $p^i = p^o$ and $q^i = q^o$, a new steady state will evolve in which $q^o > q^i$, $_pJ_{net} = 0$, and according to Equation 5, $p^o > p^i$. The energy for maintaining this disequilibrium is evidently supplied by the pump that moves q, but the coupling of this energy to the movement of p occurs because both solutes share the same facilitated diffusion system. In effect the facilitated diffusion of q down its concentration gradient supplies the energy for driving p up its concentration gradient in the opposite direction. Eliciting transient countertransport is a valuable experimental maneuver to investigate the presence of carrier mediation. If, for example, a bolus of q were added to one side of the system at equilibrium, a transient net flux of p would be observed toward that side as a result of having induced a transient net flux of q in the opposite direction. Recognizing that the prediction of Equation 5 does not depend on the transported species having different mobilities or dissociation constants, it should be clear that transient countertransport of a tracer may be induced by introducing a bolus of the unlabeled form of the same solute.

Cotransport is a feature of systems that transport ternary complexes. Suppose, for example, that the system in Figure 3 is equilibrative and that $Q_2 = Q_4 = 0$ so that only the free carrier and the ternary complexes xpq and xqp can cross the membrane. The potential for cotransport can be appreciated by recognizing that $_pJ_{o \to i} = {_qJ_{o \to i}}$ inasmuch as neither solute can cross without the other. It follows that if a bolus of q is introduced on the outside so that $q^o > q^i$, the facilitated diffusion of q in response to this concentration difference will be accompanied by a net flux of p in the same direction. Typically such systems feature Na^+ cotransport of an organic solute driven by a steady-state difference between extracellular and intracellular Na^+ activities. The term *secondary active transport* applies to this situation denoting that the energy for the uphill movement of the organic component is supplied indirectly through the action of Na^+-K^+-ATPase that maintains the driving transmembrane difference in Na^+ activity. When membrane vesicles are used to look for evidence of Na^+ cotransport, an inwardly directed Na^+ gradient may be imposed artificially. Under these circumstances the initial influx of organic solute may raise its intracellular concentration transiently above the equilibrium value until the accompanying inrush of Na^+ dissipates the Na^+ gradient. This "overshoot" effect is an important experimental clue to the presence of potentially concentrative Na^+ cotransport.

SIMPLE DIFFUSION AND CONVECTION. Diffusion and convection appear to play relatively minor roles in the hepatic transport of organic solutes except for those of small molecular weight such as ethanol, urea, and CCl_4. As discussed in FUNCTIONAL ANATOMY, p. 693, however, diffusion and solvent drag may be important in the paracellular movement of small hydrophilic solutes like erythritol and mannitol. Convection and diffusion through the tight junctions is also postulated as an explanation for bile to plasma reflux of larger biliary constituents during cholestasis (26). Compelling evidence that this mechanism is quantitatively important is still wanting, however.

Compartmentation of Solutes Inside Cell and in Bile

It should be clear from the foregoing discussion that understanding the physiology of kinetic observations depends on identifying the driving forces. For organic solutes traversing the plasma membranes of hepatocytes, this objective is elusive because activities have proved difficult to estimate and impossible to measure. Examples of some ambiguities that arise from this difficulty are discussed here with respect to cytosol and bile. The kinetic concomitants of protein binding in plasma are discussed separately in UPTAKE OF ALBUMIN-BOUND ORGANIC ANIONS, p. 707.

The distinction between facilitated diffusion and primary active transport rests on whether the system is fundamentally equilibrative or concentrative. For many solutes of interest, especially amphipaths, this issue remains unresolved owing to intracellular binding and/or to micellar sequestration in bile. The diagnostic dye bromosulfophthalein (BSP) provides an instructive example. Many reports have appeared of the apparently concentrative uptake and excretion of this dye, but a confident interpretation of the transport mechanism has not emerged. This difficulty is contributed to by the partial conjugation of BSP in transit through the cell, but the fundamental problem is the unknown activity of the dye in cytosol and bile. Uncertainty in deciding the functional importance of dye binding to cytosolic proteins and perhaps to other intracellular components is attributable to the dilution and physical disruption associated with preparing liver homogenates. The activities of BSP and its conjugates in bile are thought to be substantially less than their concentrations owing to an association with mixed micelles (106) and perhaps to protein as well, but confident estimates of the associated activity coefficients are unavailable. In the future, vesicles prepared from suitably purified basolateral and canalicular membranes may resolve some of these ambiguities and morphologic studies may clarify the role, if any, of exocytosis in the canalicular secretion of BSP. Meanwhile a different approach discussed presently suggests, although it does not establish, that the canalicular transport of BSP is effectively unidirectional and thus presumably active.

Intracellular compartmentation and/or binding of solutes to the cell surface may also lead to uncertainty in the interpretation of apparent countertransport. Suppose, for example, that the uptake of p^* (a tracer for p) occurs by simple diffusion and that p^* binds to intracellular protein. If the system is at steady state, an extracellular bolus of unlabeled p or of another transported solute that competes with p^* for the same cytosolic binding sites would result in a transient efflux of p^*, not because flux coupling occurs through a common transport carrier (countertransport), but because unlabeled solute on entering the cell would displace p^* from its binding protein and thus raise its intracellular activity. The distinction between this phenomenon and countertransport could be made if intracellular solute activities were accessible to measurement but not otherwise. A paper by Scharschmidt et al. (108) on the putative countertransport of BSP and other organic anions at the basolateral aspect of liver cells provides a good example of the ambiguity that may then prevail.

Effect of Choleresis on Canalicular Anion Transport

In 1966 O'Maille et al. (96) reported that taurocholate-induced choleresis increased the canalicular transport capacity for BSP. It was later confirmed that these observations applied to several other organic anions (106) and that the enhancing effect of taurocholate on the net canalicular flux of BSP occurred at dye levels well below those required for saturation (28). A number of postulates have been offered to explain this phenomenon. Boiled down to essentials they classify into two kinds of mechanisms: a bile acid–induced increase in the unidirectional flux of the dye from cell to bile or a decrease in the oppositely directed flux. The first group of possibilities includes an increase in the intracellular activity of BSP, incorporation of the dye into preformed bile acid micelles that then enter bile by exocytosis, and a bile acid–induced recruitment of additional transport sites in zone 3 of the liver lobule. The second group of possibilities includes a decrease in the biliary activity of the dye owing to the choleresis and/or to the formation of more or larger biliary micelles in the canalicular lumen or a reduction in the passive permeability of the canalicular membrane or its tight junctions.

The physiology of how taurocholate enhances the biliary excretion of BSP and of other cholephilic organic anions remains unresolved. Gibson and Forker (35) report an instructive analysis of the problem, however, suggesting that canalicular transport of BSP is effectively unidirectional. The thrust of Gibson's report is to show that a potent synthetic choleretic, SC-2644, produces a fourfold increment in canalicular bile production but has no effect on the excretion of taurocholate or BSP. To appreciate the interpretation of this finding, let J_n be the net canalicular flux of BSP comprising the component unidirectional fluxes J_1 and J_2. If F is the canalicular bile production rate and k is the transfer coefficient governing the back flux J_2, we have in the steady state

$$J_n = J_1 - J_2 = F B \quad (6)$$
$$J_2 = aBk$$

in which B is the dye concentration in bile and a is its activity coefficient. Equations 6 combine to yield

$$J_n = J_1[F/(F + ak)] \quad (7)$$

On differentiation regarding F Equation 7 becomes

$$\frac{dJ_n}{dF} = \frac{dJ_1}{dF}[F/(F+ak)] + J_1\frac{d}{dF}[F/(F+ak)] \quad (8)$$

Applying Equation 8 to the observation that taurocholate choleresis increases J_n, it is apparent that at least one of the derivatives on the right must be positive. Applying the same equation to the experiments with SC-2644, in which $dJ_n/dF = 0$ despite even larger increments in F, leads to two possible interpretations: *1*) the derivatives on the right are of opposite sign and of magnitudes that exactly nullify each other or *2*) both derivatives are zero. The first possibility is inherently improbable. The second could be realized if $ak = 0$ or if ak were exactly proportional to F over the fourfold increase in bile flow achieved with SC-2644. Because this last possibility is also inherently improbable the most plausible interpretation is that $ak = 0$. It follows that the effect of taurocholate on BSP excretion is not attributable to the associated increase in bile flow and that even in the absence of bile acid stimulation the canalicular transport of BSP is unidirectional and therefore presumably active.

In considering this argument it should be recognized that the model represented by Equations 6 is oversimplified to the extent that ductular modification of bile flow is ignored and canalicular bile is treated as a stirred compartment. In fact, canalicular bile production and the net solute fluxes are unknown functions of location within the lobule, bile flow is countercurrent to plasma flow, and bile production may be modified by extralobular ductules. Unfortunately equations that take account of these complicating factors are nonlinear integral-differential expressions that so far have not yielded important additional insights.

MATHEMATICAL MODELING

Compartmental analysis and related mathematical techniques are used widely in the study of hepatic transport to draw kinetic inferences from measurements made outside the liver. In its present state of development this approach is not well suited to dissecting transport mechanisms at the level of individual liver cells, but modeling has proved to be a powerful tool for characterizing the kinetic behavior of the liver as a whole and thus for studying the functional integration of multiple transport phenomena in a setting of intact liver cell plates with normal flows of blood and bile.

All of the models discussed here are concerned with estimating one or more of the apparent rate constants (or combinations of them) that govern uptake and efflux at the basolateral face of the cell as well as a removal rate constant that governs metabolism or biliary excretion. The removal process is considered unidirectional. As emphasized in the foregoing section this simplification may be warranted for organic anions such as BSP and bile acids, but its application to other classes of solutes has not been tested. All of the models involve several other simplifications that are important to understand. The highly arborized network of intrahepatic vascular channels is represented as an array of nonanastomosing pathways operating in parallel. Moreover, each sinusoid is assumed to be identical to all the others except insofar as it has a characteristic transit time. In particular the rate constants that govern transport in a given sinusoid do not depend on the transit time in that channel. At each point along a sinusoid a single intracellular pool is assumed to receive the uptake flux and to be the source for efflux and removal. Further, the rate of change of solute in this pool depends exclusively on the transmembrane fluxes. A simple example of intracellular compartmentation should make these constraints clear. Suppose that intracellular solute exists in two forms, free and bound. If the uptake flux enters the free pool and this pool is also the compartment of origin for efflux and removal, the models yield proper estimates of the transport rate constants provided that intracellular binding is fast enough to be considered at equilibrium. This last constraint is met automatically in steady-state experiments. It may be met in transient experiments as well, but this has not been verified. If efflux and removal arise from different compartments, the models discussed here are inappropriate even if the equilibrium binding constraint is met. Finally we ignore the possibility that a transported solute may diffuse from one cell to another via the gap junctions or that solute removed from one sinusoid may return to the circulation in an adjacent sinusoid. In effect, solute migration across gap junctions is assumed to be slow with respect to uptake, efflux, and removal, and half of each hepatocyte is assumed to exchange solutes only with its adjacent sinusoid. These compartmental assumptions appear reasonable and in any event are essential to writing tractable equations, but they have not been tested. The assumption of unbranched sinusoids is a simplification that obviously violates the anatomic facts. Its practical importance also awaits testing.

Steady-State Input-Output Relations

The simplest approach to modeling input-output relations is diagrammed in Figure 4A. If sinusoidal plasma and Disse fluid are considered as one compartment and all of the hepatocytes as another, the steady-state conservation relations are

$$\begin{aligned} Fu_o + k_2 zV_z &= Fu + k_1 uV_u \\ (k_2 + k_3)zV_z &= k_1 uV_u \end{aligned} \quad (9)$$

in which k_1, k_2, and k_3 are the rate constants for uptake, efflux, and removal, respectively, F is plasma flow, V_u is the extracellular volume, and zV_z is the amount of transported solute inside the cell. The symbols u_o and u denote the solute concentrations in

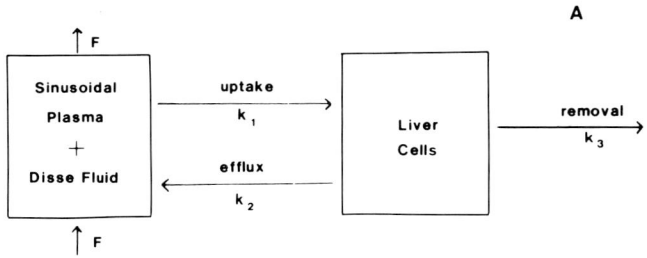

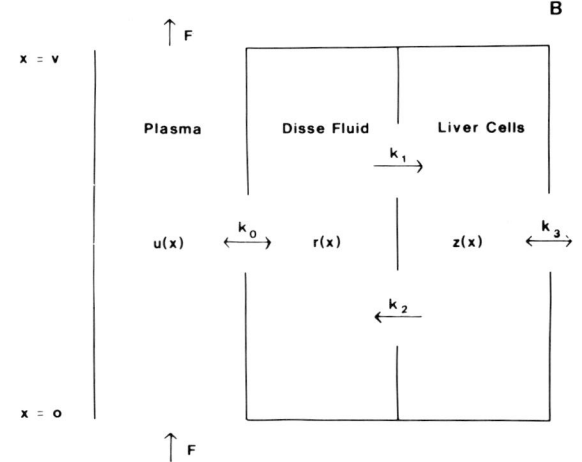

FIG. 4. *A*: entire liver lumped into two homogeneous compartments. *B*: distributed modeling of single sinusoid to account for changes in solute concentration with position.

afferent plasma and in extracellular fluid, respectively. Eliminating zV_z from this system yields

$$(u_o - u)/u_o = E = C/(F + C)$$
$$C = V_u k_1 k_3/(k_2 + k_3) = V_u K \quad (10)$$

in which E is the extraction fraction. C is frequently called the intrinsic clearance but is perhaps more easily understood as the product of V_u and an equivalent rate constant, $K [= k_1 k_3/(k_2 + k_3)]$, that governs the net removal of the solute from extracellular fluid.

The simplicity and popularity of Equations 10 notwithstanding, it should be emphasized that representing extracellular fluid and the entire population of liver cells as homogeneous compartments is a poor compromise with physiology. Solutes flowing along a liver sinusoid encounter individual hepatocytes in sequence. It follows that uptake and efflux by any particular cell influences the plasma concentration to which other cells are exposed downstream. So-called "lumped" models of the sort displayed in Figure 4A do not account for this phenomenon. Fortunately they can be replaced by more realistic "distributed" models.

Figure 4*B* illustrates distributed modeling of a single sinusoid. Let $u(x)$, $r(x)$, and $z(x)$ be the steady-state solute concentrations in plasma, in Disse fluid, and inside liver cells, respectively. The location variable x in these functions is the vascular volume contained between the sinusoidal inlet and any point downstream. If k_o is the rate constant for diffusional exchange between plasma and Disse fluid and if γ and θ are, respectively, the Disse volume and the intracellular volume per unit of vascular volume, the steady-state conservation relations are

$$-F(du/dx) + r(x)\gamma k_o = u(x) k_o$$
$$u(x) k_o + z(x)\theta k_2 = r(x)\gamma(k_1 + k_o) \quad (11)$$
$$r(x)\gamma k_1 = z(x)\theta(k_2 + k_3)$$

In writing these equations, diffusion in the vascular compartment is ignored as discussed in FUNCTIONAL ANATOMY, p. 693. Moreover the interstitial fluid layer is thin so that $k_o \gg k_1, k_2$ with the result that $u(x) = r(x)$. With this simplification and on eliminating $z(x)\theta$ the governing equations reduce to

$$-F(du/dx) = u\gamma k_1 k_3/(k_2 + k_3) \quad (12)$$

If γ and the rate constants are independent of x the solution is

$$[u(0) - u(v)]/u(0) = E = 1 - \exp(-C/F)$$
$$C = v\gamma k_1 k_3/(k_2 + k_3) = v\gamma K = -F \ln(1 - E) \quad (13)$$

in which v is the sinusoidal volume. An equivalent expression is

$$E = 1 - \exp(-\bar{t}\gamma K) \quad (14)$$

in which $\bar{t} (= v/F)$ is the sinusoidal transit time.

There are two important differences between Equations 13 and the lumped formulation in Equations 10. First, the lumped model refers the rate constant K to the whole of extracellular fluid, whereas the distributed model refers this rate constant to the volume of interstitial fluid. This difference in definition is an arbitrary but widely practiced convention. The numerical distinction is clarified by noting that $V_u = v(1 + \gamma)$. Second, for given observations of E and F, Equations 10 and 13 yield different values of C over and above that attributable to the difference between V_u and $v\gamma$. An exception occurs when E is close to zero, in which case the vascular concentration profile is flat so that large errors do not arise from the assumption that each liver cell is exposed to the same solute concentration.

In considering the distributed model it is important to recognize that if one or more of the rate constants depends on concentration, K becomes a function of u and thus of x. Recall also that the rate constants and/or γ may be functions of location owing to heterogeneities in geometry or transport capacity. Under these circumstances the value of C determined by Equations 13 is the average defined by

$$\overline{C} = \frac{1}{v} \int_0^v [v\gamma(x)K(x)]dx \quad (15)$$

The integral expression for $\overline{C}$ comes from solving Equation 12 with γ, k_1, k_2, and k_3 considered as unspecified functions of x. The average value of C defined in this way differs from that calculated by dividing the net flux for the whole sinusoid by the average solute concentration in Disse fluid. This distinction can be appreciated by noting that the equivalent rate constant $\gamma(x)K(x)$ is defined as the net flux per unit sinusoidal volume, $J(x)$, divided by the mass of Disse solute per unit sinusoidal volume, $M(x)$, and by recognizing that

$$\frac{1}{v} \int_0^v \gamma(x)K(x)dx = \frac{1}{v} \int_0^v \frac{J(x)}{M(x)} dx$$

$$\neq \frac{\int_0^v J(x)dx}{\int_0^v M(x)dx} \quad (16)$$

The integral of the ratio is the average value of γK implied by Equation 15. The ratio of integrals is a more conventional index of transport performance but is available from input-output relations only if $J(x)$ and $M(x)$ are known functions.

An instructive example of this distinction is illustrated by the case in which the net flux is a saturable function of the plasma concentration. The governing differential equation is

$$-F(du/dx) = u\gamma V/(u + K_m) \quad (17)$$

in which V is the maximum flux per unit of vascular volume and K_m is the value of u required to achieve half saturation. The solution is

$$J = \hat{u}\gamma V_{max}/(\hat{u} + K_m)$$
$$\hat{u} = [u(0) - u(v)]/\ln[u(0)/u(v)] \quad (18)$$

Here J is the steady-state flux for the whole sinusoid and V_{max} is its maximum value. The fact that J conforms to a Michaelis-Menten equation in $\hat{u}$ suggests that $\hat{u}$ is in some sense the average solute concentration along the sinusoid. It can be shown, however, that

$$J/\hat{u} = -F \ln(1 - E) = \overline{C} \quad (19)$$

Accordingly $J/\hat{u}$ is the average intrinsic clearance only in the special sense defined by Equation 15. Nevertheless fitting Equation 19 to a plot of J versus $\hat{u}$ yields formally correct estimates of γV_{max} and K_m. Moreover, because $\gamma V_{max} = \frac{1}{v}\int_0^v v\gamma(x)V(x)dx$, Equation 18 holds even if γ and V are unknown functions of location.

These considerations may be generalized as follows. In characterizing an unknown system the indicated procedure is to determine the steady-state extraction fraction at various values of u(0). A plot of $\overline{C}$ versus u(0) may then suggest the general form of the dependence, if any, between apparent intrinsic clearance and solute concentration. This dependency should then be modeled explicitly, as, for example, by Equation 17. This last step is a prerequisite for finding acceptable estimates of the transport parameters.

Distribution of Flow to Multiple Sinusoids

The liver has so far been considered as a single equivalent sinusoid. This amounts to assuming that each of the many sinusoids actually present has the same volume, the same distribution of transport rate constants, and carries the same flow. The last assumption can be removed by accounting for the flow distribution explicitly. Suppose there are N classes of sinusoids each receiving the same afferent solute concentration. If each member of the ith class receives flow f_i and the number of sinusoids in that class is n_i, the rate at which solute emerges from the ith class in the steady state is, according to Equation 14,

$$n_i f_i u(0) \exp(-\bar{t}_i \gamma K) \quad (20)$$

in which $\bar{t}_i$ is the vascular transit time. Adding the steady-state solute outflows from all N classes and rearranging yields

$$E = 1 - \sum_1^N (n_i f_i/F) \exp(-\bar{t}_i \gamma K) \quad (21)$$

which is Equation 14 weighted for the variation in sinusoidal transit times. Extending this reasoning to an indefinitely large number of flow classes yields

$$E = 1 - \int_0^\infty \psi(\lambda) \exp(-\lambda \gamma K) d\lambda \quad (22)$$

in which $\psi(\lambda)$ is the probability density function (PDF) of vascular transit times. Thus to estimate γK from Equation 22 it suffices to measure the steady-state extraction fraction and to define $\psi(\lambda)$. An approach to this latter objective is to record the transient appearance in the hepatic venous outflow of a vascular reference marker such as labeled erythrocytes. Figure 5A displays two such curves constructed to simulate the appearance of red cells in mixed hepatic venous blood after an impulse injection into the portal vein. If the ordinate is scaled to show the fraction of the dose emerging per unit time and these values are plotted against elapsed time after the injection, the resulting curve is the PDF of transit times between the injection and collection points. The frequency function of transit times recorded in this way is not exactly the desired PDF of sinusoidal transit times

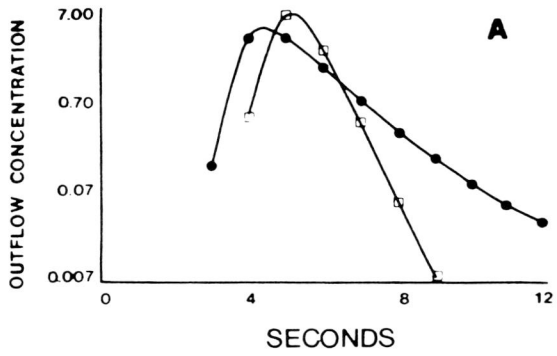

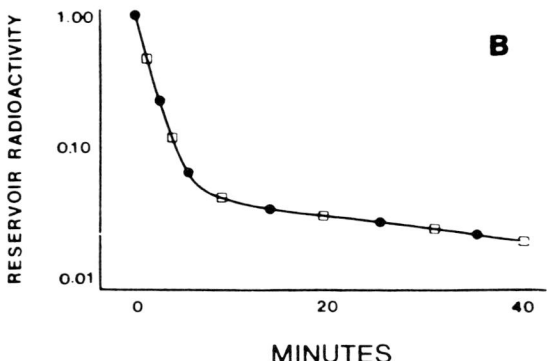

FIG. 5. *A*: hepatic venous outflow curves simulated for Gaussian flow distributions with different variances. *B*: perfusate disappearance curves simulated for same sinusoidal flow distributions as in *A*. [From Forker and Luxon (32a).]

because the record includes the population of transit times in the portal and hepatic veins (the so-called nonexchanging vessels) and in the collecting catheter. Goresky and Silverman (38) have discussed the deconvolution procedure required to correct such curves for catheter distortion. Unfortunately there is no remedy for the distortion imparted by the nonexchanging vasculature, because the transit time distribution in these vessels cannot be measured separately. The potential importance of this problem is considered next in a slightly different context.

Estimating Uptake, Efflux, and Removal Separately

The transport rate constants can be estimated individually by introducing time as a variable in the model equations. Because these equations arise from substantially more complicated derivations than the steady-state methods just discussed, the algebraic details are omitted in favor of references to the primary literature.

SINGLE-PASS INDICATOR DILUTION CURVES. Consider an experiment in which labeled erythrocytes and labeled albumin or some other nontransported extracellular marker are injected as an impulse into the portal vein. The time courses with which these materials appear in hepatic venous effluent define the PDF of

vascular transit times and the PDF of transit times in extracellular fluid. Because both markers emerge with the same flow, the difference in their mean transit times is the mean transit time in Disse fluid. In other words

$$\bar{t}_{ALB} - \bar{t}_{RBC} = \bar{t}_{NEV} + v(1 + \gamma)/F \\ - (\bar{t}_{NEV} + v/F) = v\gamma/F \quad (23)$$

Here $\bar{t}$ denotes mean transit time and the subscripts ALB, RBC, and NEV refer to albumin, red blood cells, and the nonexchanging vessels, respectively. Inasmuch as the cumulative outflow of each label must eventually equal the dose originally injected we also have

$$F \int_0^\infty u(v, t) dt = \text{dose} \quad (24)$$

in which $u(v, t)$ is the efferent concentration of either red cells or albumin. Thus if the dose is known, the area under the concentration-time curve serves to determine the perfusate flow F and thus from Equation 23 the Disse volume $v\gamma$. Notice that the sinusoidal volume v is not available by this approach because the PDF of erythrocyte transit times includes an unknown contribution from the nonexchanging vessels.

A widely adopted simplification to get around this difficulty is to assume that the nonexchanging vessels contribute nothing more than a transmission delay to the shape of an outflow curve. Under these circumstances every point on the red cell and albumin curves is delayed by $\bar{t}_{NEV}$, but the shapes of these curves are otherwise attributable to the distributions and extracellular fluid of sinusoidal transit times. Because the intralobular volume of distribution of albumin is $(1 + \gamma)$ times the intralobular volume of distribution of red cells, the albumin curve will be lower and later than the erythrocyte curve by just this factor. This relation and the correction for the delay in the nonexchanging vessels are expressed by

$$R(t - \bar{t}_{NEV}) = (1 + \gamma) A\left(\frac{t - \bar{t}_{NEV}}{1 + \gamma}\right) \quad (25)$$

in which $R(t)$ and $A(t)$ are the PDFs of outflow transit times for red cells and albumin, respectively. The parameters $(1 + \gamma)$ and $\bar{t}_{NEV}$ can now be determined from the adjustments in timing and amplitude of the albumin curve required to make it congruent with the erythrocyte curve. Once $(1 + \gamma)$ is determined in this way the sinusoidal volume v becomes available from Equation 23.

If the nonexchanging vessels contribute only a transmission delay, therefore, the vascular and extracellular reference curves suffice to determine v, γ, and F as well as the PDF of sinusoidal transit times. This information is sufficient to determine the transport rate constants from a simultaneously recorded outflow curve of a transported solute. The equations that govern this process in a single sinusoid are

$$F(\partial u/\partial x) + (1 + \gamma)(\partial u/\partial t) + k_1\gamma u - k_2\theta z = 0$$

$$\theta(\partial z/\partial t) + (k_2 + k_3)\theta z - k_1\gamma u = 0$$

$$u(0, t) = \delta(t - \bar{t}_{NEV}) \quad (26)$$

$$u(x, 0) = z(x, 0) = 0$$

in which the initial condition $u(0, t)$ defines an impulse arriving at the sinusoidal inlet ($x = 0$) after the delay, $\bar{t}_{NEV}$. It is convenient and mathematically valid to treat this delay as though it were all attributable to the afferent vasculature, although in fact it arises from both the portal and hepatic veins.

The procedure for extending the solution to Equations 26 to the case of many sinusoids is formally analogous to that employed to derive Equation 22. Specifically Equations 26 are solved for u(v, t) as a function of the unknowns k_1, k_2, and k_3 and this function is convoluted with the PDF of sinusoidal transit times. The resulting integral equation is fitted by an iterative least-squares procedure to find estimates of all three rate constants. Although the equations for a single sinusoid have an explicit analytical solution, the PDF of transit times is available only implicitly from the reference curves. The required convolution must therefore be carried out by numerical integration. This procedure and the most useful solution to Equations 26 are explained by Luxon and Forker (79). A logically equivalent but numerically less accurate procedure involving a different definition of the uptake parameter was originally suggested by Goresky et al. (37) and has been exploited by Goresky and others to study the transport of numerous organic solutes.

The complicated mathematical details notwithstanding, the rationale of the multiple indicator dilution approach is straightforward. The outflow curve of a transported solute differs from that of its extracellular reference marker according to three factors: the fraction of the dose that escapes uptake and thus emerges with the same time course as the extracellular reference marker, the fraction of the dose that enters liver cells but escapes removal and thus emerges in the venous effluent with a later time course, and the fraction of the dose that is irreversibly removed and thus does not appear in the outflow at all. These fractions are governed by the transport rate constants. The vascular and extracellular reference curves serve to determine plasma flow, its distribution among the sinusoids, and the interstitial volume of distribution. These measurements serve in turn to distinguish between the delays in solute outflow that are attributable to vascular and interstitial transit and those that arise from the transported solute's sojourn inside hepatocytes.

Errors in estimating the transport rate constants by this approach occur in two principal ways. First, the frequency function of outflow transit times is contaminated by an unknown contribution from the nonexchanging vessels. Second, it has proved infeasible to follow the tails of the outflow curves long enough to capture all the information they potentially contain. This latter difficulty (which is more important than the nonexchanging vessel artifact) occurs in preparations with an intact peripheral circulation because recirculation spoils the outflow data after ∼25 s. In nonrecirculating perfusion systems a similar, albeit less serious, problem arises from the difficulty of resolving three isotopes in late samples of low radioactivity. The effect in either case is to terminate the outflow data prematurely. The worst errors attributable to truncating the tail functions occur in estimating the rate constant for removal, because most of the information about this last step in the transport sequence is contained in the late portion of the solute outflow curve. Errors in k_3 attributable to this problem are generally so large and unpredictable as to render the method unreliable as a means to study the kinetics of removal. Smaller errors (on the order of 10%–20%) may also occur in estimating k_2, while estimates of k_1 that depend largely on the shape of the initial upstroke of the curve are available with substantially higher fidelity (79). A final theoretical objection whose practical importance has not been studied is the implicit assumption that γ and all three rate constants are independent of location along the sinusoids.

SOLUTE DISAPPEARANCE CURVES. Estimates of the transport rate constants are also available from solute disappearance curves. The simplest approach employs an extension of the lumped model in Figure 4A constructed by enlarging the extracellular compartment in the liver to include the entire extrahepatic volume of distribution. If a bolus of solute with exclusive hepatic removal is injected into this compartment at $t = 0$ the governing equations are

$$V_P(dP/dt) = zV_z k_2 - PV_P k_1$$

$$V_z(dz/dt) = PV_P k_1 - zV_z(k_2 + k_3) \quad (27)$$

$$V_P P(0) = \text{dose}$$

$$z(0) = 0$$

in which P and z are the concentrations in extracellular fluid and liver cells, respectively, and V_P and V_z are the corresponding compartment volumes. On eliminating zV_z from this system the solution is of the form

$$P(t) = A_1 \exp(-a_1 t) + A_2 \exp(-a_2 t) \quad (28)$$

in which the coefficients A_1, A_2, a_1, and a_2 are constants that depend on V_P and the transport rate constants. Fitting observations of $P(t)$ to the sum of two exponentials predicted by Equation 28 and using the relation $V_P(A_1 + A_2) = \text{dose}$ yields straightforward estimates of k_1, k_2, k_3, and V_P. The mathematical simplicity of this approach and its ready adaptability to human subjects are presumably responsible for its popularity, especially for pharmacokinetic work. Estimates of the rate constants obtained in this way are likely to be in serious error, however, because the underlying assumptions are greatly oversimplified. The error that arises from assuming that all liver cells are exposed to the same solute concentration has been emphasized in the previous discussion of steady-state input-output relations. The lumped approach to analyzing solute disappearance curves compounds this difficulty by assuming that the extrahepatic volume of distribution is homogeneous as well. Forker and Luxon (29) have provided a general analysis of this problem, including numerical examples of the errors that occur.

In pharmacokinetic work the focus of interest is frequently on the area under the disappearance curve rather than on its shape. The utility of estimating this area rests on a proof by Nosslin (95) that

$$\text{steady-state plasma clearance} = \frac{\text{dose}}{\int_0^\infty P(t)dt} \quad (29)$$

This relation holds for any compartmental model however complicated, provided the rate constants are independent of concentration and provided the dose is injected as an impulse into and subsequent samples are removed from the same compartment P. Equation 29 is thus free of the objections raised previously about lumped modeling. In practice, however, the integration to infinity required by Nosslin's theorem must proceed by extrapolation of the tail of the disappearance curve. Unfortunately substantial errors may occur if this extrapolation is based on the lumped assumption that the disappearance curve decays as the sum of two exponentials.

Luxon et al. (80) have suggested a hybrid approach to recording and analyzing solute disappearance curves that capitalizes on more realistic distributed modeling of intrahepatic events while avoiding the distortions that arise from an unstirred extrahepatic volume of distribution. This method employs an isolated perfused liver preparation in which the extrahepatic circuit comprises a mechanically stirred reservoir connected to the liver by short conduits with a well-defined distribution of transit times. In Luxon's system this conduit transfer function is a lagged exponential, i.e., the empirically determined PDF of conduit transit times conforms to that expected from a simple transmission delay and a stirred compartment operating in series. Because the aggregate volume of the conduits is much larger than morphometric estimates of the extralobular vascular volume, the nonexchanging vasculature contributes only a trivial distortion to the PDF of solute transit times between the reservoir and the sinusoids. Under these circumstances the governing equations for the one sinusoid version of the model are

$$V_P(dP/dt) = Fu(v, t) - FP$$
$$V_D(dD/dt) = FP(t - \bar{t}_c) - FD$$
$$F(\partial u/\partial x) + (1 + \gamma)(\partial u/\partial t) + k_1\gamma u - k_2\theta z = 0$$
$$\theta(\partial z/\partial t) + (k_2 + k_3)\theta z - k_1\gamma u = 0 \quad (30)$$
$$u(0, t) = D(t)$$
$$u(x, 0) = z(x, 0) = D(0) = 0$$
$$P(0) = \text{dose}/V_P$$

Here P and D are solute concentrations in the perfusate reservoir and at the sinusoidal inlet, respectively. V_P is the reservoir volume and V_D is the volume of the hypothetical compartment that accounts for solute dispersion in the conduits. The transit time $\bar{t}_c$ is the transmission delay. The total mean transit time between the reservoir and the sinusoids is thus $(V_D/F) + \bar{t}_c$ to within only a small error attributable to the unaccounted-for effect of the nonexchanging vasculature. The other parameters have the meanings previously defined by Equations 26, which it should be noted are identical to the sinusoidal portion of Equations 30. The solution constructed by Laplace transformation takes the form

$$P(t) = A_1\exp(-a_1t) + A_2\exp(-a_2t)$$
$$+ \sum_{i=3}^\infty A_i(\sin\theta_i t + \cos\theta_i t)\exp(-a_i t) \quad (31)$$
$$a_1 < a_2 < a_i, \quad i \geq 3$$

in which the subscripted coefficients A, a, and θ are constants that depend on the system parameters. The infinite sum of trigonometric functions appears because each recirculation is associated with an oscillation in $P(t)$. These oscillations are exponentially damped, however, and in the presence of many sinusoids with different flows they are out of phase with each other. For these reasons they are not usually apparent in the disappearance curve. Nevertheless they exert an important numerical effect on estimates of the rate constants. The oscillations are implicit in the Laplace transform of the solution, but they do not have an explicit closed representation in the time domain. To circumvent this difficulty $P(t)$ is fitted to the model equation through numerical inversion of the solution transform (80).

In practice, estimates of $v\gamma k_1$, k_2, and k_3 are obtained as follows. The perfusion system is set up with arbitrary predetermined values of F, V_P, and the conduit parameters V_D and $\bar{t}_c$. A constant infusion of unlabeled solute is provided to bring the system to steady state. Subsequently a tracer impulse of labeled solute is injected into the reservoir at $t = 0$ and its disappearance recorded until a final monoexponential decay is manifest in $P(t)$. Finally, $v(1 + \gamma)$ is estimated at the end of the experiment as the steady-state hepatic volume of distribution of labeled albumin or some other extracellular marker.

Simulation analysis of this approach has identified four features that recommend it over the analysis of single-pass outflow curves. First, the tail of a disappearance curve evolves as a mathematically explicit exponential defined by the first term in Equation 31. By contrast the tail of a single-pass outflow curve lacks this explicit asymptotic behavior and so cannot be extrapolated with confidence. Because estimates of the removal rate constant depend on the late part of the curve in both models, measurements of k_3 are made more reliably from disappearance curves than from single-pass outflow curves. Second, disappearance curves are insensitive to the PDF of transit times in the nonexchanging vasculature, whereas this unknown frequency function may make an important contribution to single-pass outflow patterns. Third, the shape of disappearance curves is virtually independent of the sinusoidal flow distribution, whereas this frequency function is crucial to the interpretation of outflow transients. This difference is illustrated in Figure 5 that displays pairs of simulated curves for Gaussian flow distributions with two different variances. In considering this illustration, notice that a symmetrical frequency function of flows corresponds to a skewed distribution of transit times, because flow and transit time are inversely related through a common sinusoidal volume. Figure 5A shows that the outflow patterns depend importantly on the variance. Figure 5B shows that the corresponding disappearance curves are virtually indistinguishable. In principle, this difference is immaterial to the precision with which one could recover the rate constants from the two kinds of procedures. In practice, however, single-pass outflow curves are distorted by an unknown contribution from the nonexchanging vessels, and the truncated tail functions cannot be extrapolated with confidence. These impediments to interpretation are largely absent from the corresponding disappearance curves. Finally, disappearance curves are remarkably insensitive to heterogeneities in transport capacity along the sinusoids (31)—a performance criterion as yet undefined for single-pass outflow curves.

The analytical advantages that disappearance curves enjoy over outflow curves is most easily understood in nonmathematical terms by recognizing that the former approach records the liver's transport performance over many recirculations, whereas useful data in the latter approach are rarely recordable beyond ~30 s. Outflow curves potentially contain detailed information about the PDF of solute transit times and hence about heterogeneities in flow distribution. However, this information is contaminated by an unwanted additional signal (the nonexchanging-vessel artifact), and for technical reasons it is difficult to record in its entirety (the tail-truncation problem). By contrast disappearance curves contain almost no information about the distribution of flow to individual sinusoids or about the variations in transport capacity along their length, but they depict the average kinetic behavior of the liver as a whole with remarkable fidelity.

The disappearance-curve method is fraught with the technical difficulty of maintaining stable liver performance and of keeping the reservoir volume constant for a substantially longer interval than is required to record a single-pass outflow curve. A less obvious aspect of this problem that warrants special emphasis is the requirement to keep the volume of recirculating perfusate small with respect to the plasma clearance. To the extent that this objective is not met, a disappearance curve declines relatively slowly, the difference between the exponential coefficients in Equation 31 becomes less well defined, and the information content of the data diminishes.

UPTAKE OF ALBUMIN-BOUND ORGANIC ANIONS

Organic solutes destined for hepatic disposal commonly arrive in afferent blood bound to protein. The effect of this binding on ligand transport has received concerted recent study—especially the effect of binding to albumin on the uptake of organic anions.

Historical Perspective

Early speculations (3, 9) to the contrary nonwithstanding, pharmacologists have taken it virtually as an article of faith that albumin-bound drugs are available to the liver only insofar as spontaneous dissociation of the albumin-ligand complex can generate the free or unbound form of the drug in sinusoidal plasma. Wide acceptance of this view appears to have been sustained by two considerations. First, the hepatic extraction fraction of albumin is trivial, suggesting that free ligands rather than albumin-ligand complexes are the ultimate substrate for hepatic uptake. Second, spontaneous dissociation could account, at least theoretically, for the otherwise surprising finding that solutes with small free fractions in afferent plasma may have high hepatic extraction fractions. The conventional interpretation runs as follows. If the uptake rate constant were large enough, liver cells at each point along a sinusoid could remove essentially all of the available free ligand from adjacent plasma. Mass action would then ensure the spontaneous dissociation of more free ligand that in turn would be available to neighboring hepatocytes downstream. The unexpectedly high extraction of tightly bound solutes can be reconciled in this way by postulating sufficiently large uptake rate constants and/or sufficiently long vascular transit times. Curiously this paradigm seems to have been examined critically only recently.

In 1981 two papers appeared that challenged the conventional teaching. Weisiger et al. (132) showed that the kinetics of oleate uptake by rat liver differed depending on whether various concentrations of the fatty acid were perfused in the presence of a fixed albumin concentration or whether oleate and albumin were perfused in a fixed molar ratio. It was recognized later (32) that this argument is not itself compelling, but Weisiger et al. also showed that the uptake rate correlated with the concentration of bound rather than of free oleate. Forker and Luxon (30) assumed that only spontaneously dissociating free taurocholate was available for uptake and showed that this premise led to contradictory estimates of the intrinsic clearance at two different albumin concentrations.

The design of the taurocholate experiment is instructive. Suppose that the binding reaction between albumin and taurocholate is fast compared with uptake so that taurocholate binding is effectively at equilibrium everywhere along the sinusoid. If free taurocholate is the only form available for removal, Forker and Luxon reasoned that uptake from a single sinusoid should be governed by

$$-F du/dx = \alpha \gamma K u \quad (32)$$

in which α is the equilibrium free fraction and the other notation conforms to the definitions specified by Equations 11. If α, γ, and the transport rate constant K are independent of total taurocholate concentration u, Equation 32 predicts

$$-(F/\alpha)\ln(1 - E) = \overline{C} \quad (33)$$

in which $\overline{C}$ is the average intrinsic clearance of free taurocholate in the sense specified by Equation 15. Because the stipulation that α and $\overline{C}$ be constants can be met experimentally by using tracer levels of the ligand, the prediction of Equation 33 is testable by measuring α and E at two different albumin concentrations. The experiments with taurocholate (equilibrium binding constant = 10^3 M^{-1}) showed that raising the perfusate albumin concentration by a factor of 10 led to an increase in $\overline{C}$ of 160%. To state the same conclusion differently, a 10-fold decrease in α led to only an 11% reduction in E—a far smaller change than predicted by Equation 33. Qualitatively similar but quantitatively even more striking discrepancies are reported for more tightly bound anions such as rose bengal (34).

Dissociation-Mediating Sites on the Cell Surface?

When the same tracer concentration of rose bengal (equilibrium binding constant = 6×10^5 M^{-1}) is used to explore its extraction by perfused rat liver over a wide range of bovine albumin concentrations, the data display two interesting features. First, the extraction fraction declines with each increment in perfusate albumin concentration. At each point, however, E is higher than that predicted by Equation 33. The higher the albumin concentration, and thus the more extensive the binding, the more pronounced this discrepancy becomes until at albumin concentrations approximating those of rat plasma, Equation 33 underestimates E by about two orders of magnitude. Second, even when sufficient albumin is already present to bind more than 99% of the dye, adding more free albumin reduces rose bengal extraction. This suggests, but does not prove, that free albumin competes with albumin-ligand complexes for access to the transport process. A graph of these kinetic relationships appears in Figure 6. Weisiger et al. (132) showed that bovine albumin binds reversibly to rat liver cells and that excess rat albumin displaces it. Fleischer et al. (25) showed that the affinity of bovine albumin for isolated hepatocytes is independent of whether the albumin is free or bound to rose bengal.

A model that accommodates these binding data as well as the kinetics of rose bengal removal appears in Figure 7. In this scheme free ligand and free albumin bind each other reversibly in the vascular lumen. Because the sinusoidal endothelium is freely permeable to even large solutes, all three species enter Disse fluid and are therefore free to diffuse to the cell surface

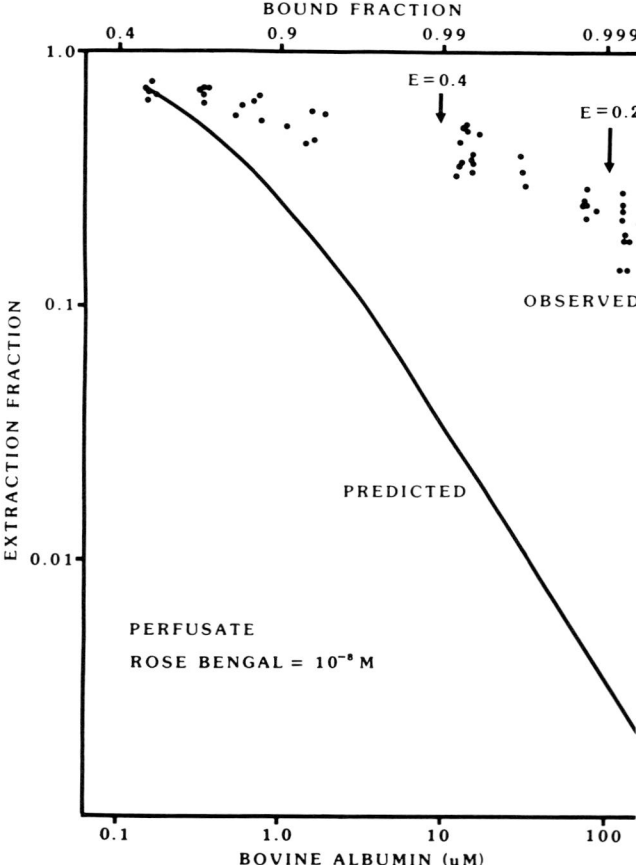

FIG. 6. Removal of tracer rose bengal by perfused rat liver at various concentrations of bovine albumin. Observations appear as discrete points. *Solid curve* depicts relation predicted by assuming that only free rose bengal is available for hepatic uptake.

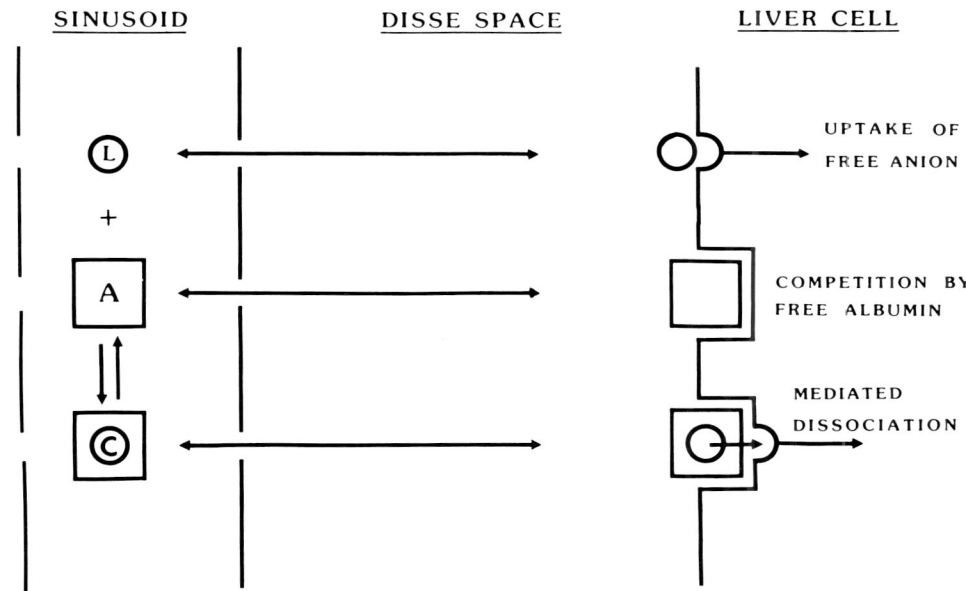

FIG. 7. Receptor model for surface-mediated dissociation of albumin-bound ligands. L, free ligand; A, free albumin; C, albumin-ligand complex.

(albeit at different rates). Free ligand engages the uptake carrier directly. In addition, however, sites are present on the liver cell surface that bind albumin-ligand complexes and promote their dissociation, presumably by inducing a transient conformational change in the albumin. Free ligand liberated in this process is subject to transport before it reenters the pool of free ligand in extracellular fluid, thus making more free ligand available for uptake than would otherwise occur. Because the cell surface sites respond to albumin, not to the ligand(s) that it carries, free albumin competes with albumin-ligand complexes for these sites and thus inhibits ligand uptake.

The simplest mathematical formulation of this concept assumes that the binding reaction in extracellular fluid is at equilibrium, that the diffusional resistance of the Disse layer is negligible, and that the rate constant governing uptake of free ligand from the Disse space is identical to the one that governs uptake of free ligand liberated by the debinding reaction at the cell surface. With these simplifications the solution to the model equation is

$$-(F/\alpha)\ln(1 - E) = \frac{\overline{C}\left[1 + \left(\frac{1-\alpha}{\alpha}\right)\lambda\right]}{1 - (1-\alpha)\lambda} \quad (34)$$

$$\lambda = S_t/(S_t + K_s + A)$$

A is the albumin concentration in extracellular fluid, α is the free fraction of ligand, S_t is the number of albumin-binding sites on the cell surface, and K_s is the equilibrium constant for the association of albumin with these sites. If this model is realistic it should fit the data for rose bengal extraction in Figure 6 and the fitting procedure should yield values of S_t and K_s that resemble independent estimates of the same parameters. As it turns out these criteria are met surprisingly well. In particular model independent estimates of S_t and K_s obtained with isolated liver cells (132) and the model-dependent estimates obtained from fitting Equations 34 to the rose bengal uptake data both yield $\sim 10^7$ sites per cell. The affinity of albumin for these sites is $\sim 10^4$ M^{-1} (33).

Some of the simplifying assumptions that constrain Equations 34 are not crucial to its qualitative predictions. In particular it can be shown that extending the model to account for a distribution of flows to many sinusoids does not change its qualitative behavior (34). Surface-mediated dissociation is also consistent with the effect of albumin on the initial uptake of palmitate by rat hepatocytes cultured as monolayers—experiments designed to ensure equilibrium binding in the bathing medium, to correct for diffusion artifacts arising from unstirred fluid layers, and to avoid the assumptions inherent in distributed modeling of an intact liver (24).

Specificity

Recent reports suggest that nonhepatic tissues, including the blood brain barrier (97–99) and cardiac muscle (48, 49), also display mediated dissociation. In the brain the dissociation-mediating surface is postulated to be the vascular endothelium, and proteins other than albumin are said to participate, including hormone-binding globulins. In perfused rat hearts the oxidation of long-chain fatty acids proceeds at substantially faster rates than can be accounted for by spontaneous dissociation of albumin–fatty acid complexes in plasma. It is not clear whether the dissociation-mediating mechanism resides at the capillary endothelium, the myocyte plasma membrane, or at some intervening location in interstitial fluid. Moreover the possibility that rate-limiting diffusion of bound ligand to the cell surface may account for the appearance of mediated dissociation has not been excluded in either of these nonhepatic tissues.

The way in which diffusion artifacts could be misleading has been explored by Luxon et al. (81), who studied the effect of albumin binding on the uptake of tracer taurocholate by perfused loops of rat distal ileum. The apparent intrinsic mucosal clearance of free taurocholate in this preparation is about three times higher in the presence of albumin than in its absence. This finding is reminiscent of the hepatic phenomenon, but when the diffusional resistance of the intervillous fluid layer is accounted for, the appearance of mediated dissociation vanishes. Modeling the unstirred layer on the assumption that mediated dissociation is absent leads to

$$-(F/\alpha)\ln(1 - E) = vK\phi/[\phi + \alpha K\delta r/2]$$
$$\phi = \alpha D_f + (1 - \alpha)D_b \quad (35)$$

in which K is the mucosal transfer coefficient of taurocholate and the left side is as before the apparent intrinsic clearance of free taurocholate. The geometric parameters v, r, and δ denote the volume of the intestinal lumen, its radius, and the thickness of the unstirred layer, respectively. The diffusion parameter ϕ is the weighted average of the aqueous diffusion coefficients for free (D_f) and bound (D_b) taurocholate. Equations 35 make it clear that the effect of diffusion is to reduce the apparent intrinsic clearance below its true value (vK) but that $-(F/\alpha)\ln(1 - E)$ is nevertheless a rising function of the albumin concentration. It is this latter feature of unstirred fluid layers that may be confused with mediated dissociation. Because Luxon's analysis of the taurocholate uptake data yields values of δ and K that agree with independent estimates of these parameters, there is no reason to invoke mediated dissociation in the ileal mucosa. In the liver, where the unstirred fluid layer is smaller by two to three orders of magnitude, this explanation is less appealing. The difference between the effect of albumin binding on bile acid uptake in the liver and in the ileum is especially intriguing because the Na$^+$ cotransport system for bile acids is similar in both organs.

There are two reports suggesting that mediated dissociation in the liver occurs with proteins other than albumin. Stollman et al. (114) report that the hepatic extraction of bilirubin is virtually independent not only of perfusate albumin concentration but of perfusate ligandin as well. Because both proteins display similarly high binding affinities for bilirubin, this finding calls into serious question the concept of a specific "albumin receptor," a term that appears in the literature frequently if prematurely. Jones et al. (57) report a similar independence of hepatic propranolol extraction from the binding of this drug to the sialated form of α_1-acid glycoprotein. In both cases the bound fraction of the ligand is high enough to make it unlikely that spontaneous dissociation could be the explanation, but neither study excludes this possibility rigorously.

It should be clear from these considerations that the model in Figure 7, though consistent with much of the kinetic data, is by no means established. Other possibilities not only remain to be studied but the specificity expected of a cell surface receptor has not been confirmed.

One alternative possibility invokes the negative electric field produced by anionic phospholipids in the plasma membrane. Charged proteins such as albumin should orient in this field, and because the pH should be lower in this region than elsewhere in extracellular fluid, the binding affinity of albumin for weak acids should also be lower. Such "microclimate" effects could lead to locally accelerated dissociation of albumin-ligand complexes and thus to the appearance that uptake is driven by the concentration of bound ligand as well as by the concentration of the free form. Another mechanism leading to the same effect can be visualized as follows. Albumin-anion complexes could offload their anions directly to the membrane proteins that mediate uptake. In this way bound anion would truly be a substrate for uptake even though only free anion would enter the cell. Yet a third possibility is that the concentration of albumin in extracellular fluid modulates the intrinsic transport capacity of hepatocytes, albeit by an unknown mechanism.

Finally, Bass and Pond (5a) have made a trenchant theoretical analysis of the diffusion phenomenon that should prompt careful reappraisal of experiments originally interpreted as excluding this explanation. Their treatment takes proper account of the concentration profiles of bound and free ligand in the unstirred layer, and they call attention to the binding disequilibrium that develops in this layer if only free ligand is subject to uptake. These details were ignored or treated simplistically in the experiments with rose bengal, taurocholate, and palmitate cited previously. A particularly intriguing kinetic feature of the diffusion phenomenon is that it can lend the appearance of saturation kinetics to the "albumin-mediated" uptake flux even though the concentration of total anion is far below the K_m of its membrane carrier.

TRANSPORT OF SPECIFIC SOLUTE CLASSES

This chapter concludes with a short summary of conspicuous kinetic features that characterize the hepatic transport of several classes of organic solutes together with brief discussions of putative membrane carriers and cytosolic binding proteins. More detailed (and controversial) information is available in recent reviews (7, 12, 62, 116).

Bile Acids

Mammalian bile acids are cholanic acid derivatives that differ according to the number and position of substituent hydroxyl groups and according to whether the carboxylic acid group is conjugated. The pK_a of unconjugated bile acids is ~6. Conjugation with glycine or taurine typically reduces this value to about 4 and 2, respectively (93a). Other physical properties of the bile acids such as aqueous solubility, binding affinity for albumin, and critical micellar concentration vary from one structure to another depending importantly on the number and position of hydroxyl groups on the ring structure (50).

All the naturally occurring bile acids enjoy efficient uptake from hepatic plasma followed by excretion in bile. Both steps display saturation kinetics and competitive inhibition by other bile acids. The canalicular step is generally rate-limiting in the transport sequence, including intervening metabolic steps such as hydroxylation and conjugation. The latter modification typically, though not invariably, precedes canalicular excretion.

The conspicuous feature of taurocholate uptake by isolated hepatocytes and by liver plasma membrane vesicles is Na^+ cotransport (1, 18, 51, 107, 129). In basolateral vesicle preparations with an imposed inwardly directed Na^+ gradient, taurocholate uptake displays a substantial concentration overshoot (8), but the stoichiometry of the cotransport system and whether it is neutral or electrogenic remains unsettled (1, 21, 51, 82, 107). Less hydrophilic bile acids such as unconjugated monohydroxy and dihydroxy derivatives of cholanic acid display uptake kinetics that suggest a partial role for simple diffusion. In particular uptake occurs in the absence of Na^+, is only partially saturable, and correlates with hydrophobicity as measured by decane-H_2O partition ratios (129). These more hydrophobic bile acids evidently share some affinity for the Na^+ cotransport system, however, because they competitively inhibit the uptake of taurocholate (45).

The mechanism of canalicular bile acid secretion remains poorly understood despite intensive study. Apical transport is apparently concentrative, but this inference is clouded by the unknown activities of bile acids inside the cell and in canalicular bile. To judge from the electrical potential difference measured in hepatocyte couplets (40), the positivity of the canalicular lumen over the cell interior (25–35 mV) could account for an equilibrium activity difference of threefold to fourfold in favor of bile. This difference is far smaller than the concentration differences actually observed, but confident corrections for intracellular compartmentation and sequestration in biliary micelles are not available. A conductive pathway for taurocholate in canalicular membrane vesicles is consistent with carrier mediation (53, 82) while morphologic evidence (13) and the inhibiting effects of such microtubular poisons as colchicine support the inference of a vesicular mechanism (41). Both ideas are consistent with the kinetic findings of saturation and analogue competition.

Long-Chain Fatty Acids

Recent findings by Stremmel, Berk, and co-workers (118–120) call into serious question the traditional view that hepatic uptake of long-chain fatty acids is mediated exclusively by simple diffusion. These authors report that oleate uptake by isolated hepatocytes displays saturation kinetics, competition by other long-chain fatty acids, Na^+ dependence, and inhibition by phloretin—all features suggestive of a carrier-mediated Na^+ cotransport system. The thrust of this inference is strengthened substantially by the isolation from hepatocyte plasma membranes of a fatty acid–binding protein and the finding that antibodies to this protein impair the uptake of oleate but not of bile acids or BSP.

Other Organic Anions

A diverse group of amphipathic organic anions including bilirubin and such familiar diagnostic agents as BSP, indocyanine green, rose bengal, iopanoic acid, and fluorescein resemble bile acids and long-chain fatty acids insofar as efficient hepatic extraction and tight binding to albumin are concerned, but their hepatic uptake is apparently mediated by yet a third transport system or systems. The salient features include saturation kinetics, mutual competitive inhibition, and independence from Na^+. Uptake is apparently concentrative, but because binding of these materials to albumin and to intracellular protein is extensive, the plasma-cytosolic activity differences and thus the driving forces that govern the uptake of these solutes are conjectural. These compartmentation effects obscure the interpretation of such epiphenomena as countertransport and transstimulation, both of which have nevertheless been reported (10, 36, 100, 101, 108). A popular and plausible, albeit still uncertain, interpretation is that this group of organic anions have their uptake mediated by facilitated diffusion systems that are similar if not identical. Berk and Stremmel (7) have contributed a balanced, comprehensive review of this question, including a discussion of the evidence that more than one transport system may be involved.

All members of this class are concentrated in bile and most (rose bengal and indocyanine green are notable exceptions) are conjugated with more polar compounds such as glucuronic acid or glutathione as a prerequisite for efficient biliary excretion. Other features of canalicular excretion, which for this group of anions is the rate-limiting step in the transport sequence, are saturation kinetics, mutual competition, and flux acceleration by micelle forming bile acids such as taurocholate. The inference that canalicular transport of BSP is unidirectional and thus presumably active has been discussed in an earlier section of this chapter. As for bile acids, however, it is still not clear whether canalicular excretion of bilirubin, BSP,

Amino Acids

Amino acids subject to hepatic uptake include at least eight classes, each with somewhat different kinetic features (16). Among the best characterized is a group of small neutral amino acids of which alanine is the prototype. Alanine uptake is mediated by a concentrative, rheogenic, Na^+ cotransport system that is stereospecific and apparently distinct from the systems that mediate the uptake of other organic solutes (61, 66, 67, 112, 127, 128).

Alanine does not accumulate in bile. Instead, bile contains surprisingly high concentrations of several other amino acids, among the more prominent and interesting of which are constituents of the tripeptide glutathione. Glutathione itself is also concentrated in bile predominantly as its oxidized form. Specific transport systems that mediate canalicular amino acid secretion have not been identified, but carrier-mediated transport of glutathione is reported to occur in canalicular membrane vesicles (52). An intriguing, as yet unconfirmed, suggestion is that glutathione and its amino acid constituents may contribute importantly to the osmotic driving force for basal canalicular H_2O flow (39).

Organic Cations

A variety of nitrogen-containing organic cations engage an efficient hepatic uptake mechanism that appears distinct from the systems that mediate the movement of organic anions and alanine (47, 94, 105). Intensively studied examples are procainamide ethobromide (20, 85, 89, 105) and d-tubocurarine (84, 86, 88–90). Hepatic uptake of these materials appears to be concentrative, mutually competitive, and Na^+ dependent, but little is known about the energetics or specificity of the system. Organic cations are also concentrated in bile but even less is known about the responsible mechanism.

Neutral Compounds

As discussed in FUNCTIONAL ANATOMY, p. 693, the biliary excretion of small metabolically inert hydrophilic solutes such as erythritol, mannitol, sucrose, and inulin probably occurs by nonspecific vesicular transcytosis and by diffusion and convection across the tight junctions. The specific transport of two other classes of neutral compounds, cardiac glycosides and the metabolically important hexoses, are discussed here.

CARDIAC GLYCOSIDES. Ouabain is taken up by hepatocytes and excreted in bile by transport systems that are saturable, concentrative, and inhibitable by other cardiac glycosides (19, 68, 104, 110). In general inhibition by organic anions and cations appears not to occur, although exceptions are reported (23, 87). Beyond the inference to be drawn for these data that the hepatic transport of cardiac glycosides is carrier mediated and active, little is known about the mechanisms of uptake or excretion, nor is it clear whether these systems are specific for the glycoside structure or available to a wider range of uncharged drugs.

HEXOSES. The hepatic uptake of glucose, galactose, and fructose displays the characteristic features of a common facilitated diffusion system: saturation kinetics, equilibrative energetics, analogue competition, and stereospecificity (17, 22, 37, 60, 133). The rate-limiting determinant of net sugar removal at normal rates of hepatic plasma flow is metabolic disposal via intracellular phosphorylation rather than saturation of the membrane transport system.

Intracellular Binding Proteins

All of the organic anions considered previously (bile acids, fatty acids, bilirubin, BSP, and so forth) bind to cytosolic proteins. At least four such proteins or groups of proteins have been identified: fatty acid binding or Z protein (14, 74, 92), ligandin (74), and two more recently reported bile acid binders (122). Ligandin displays enzymatic activity as glutathione S-transferase B, one of at least six forms of glutathione transferase in liver cytosol (58). The more recently discovered bile acid–binding proteins may have sulfotransferase activity (115). Apart from these catalytic activities the role of cytosolic anion binders is not known. Ligandin is abundant, comprising ~5% of hepatocyte cytosolic protein, and it binds numerous ligands such as bilirubin and indocyanine green that are not substrates for its catalytic activity. An early assumption that ligandin plays a pivotal role in the unidirectional flux of its ligands from plasma to cytosol (73, 75, 76) is not in accord with more recent estimates of the unidirectional transport rate constants (137). Whether ligandin and the other cytosolic binding proteins are important in protecting the cell from the toxic effects of free intracellular anions and/or whether they serve to promote the intracellular transport or metabolism of their ligands remains to be determined.

Putative Membrane Carriers

All three classes of organic anions also bind to liver cell plasma membranes. Some of this binding is class specific. This finding has led to an intensive search for corresponding membrane proteins with the hope of isolating transport carriers. Although this objective is not fully realized, substantial progress has been made in identifying candidate proteins and in a few instances the evidence for a carrier function is compelling.

In addition to the immunologic evidence of a transport role for membrane fatty acid–binding protein already discussed, several candidate bile acid carriers are reported (15, 64, 65, 130, 131), and one of these (77) is said to mediate Na^+-dependent taurocholate transport when incorporated into liposomes. Three groups of investigators have isolated membrane proteins that bind BSP and bilirubin but not bile acids or fatty acids. Reichen and Berk (103) and Stremmel et al. (117) have used affinity chromatography to isolate a binding protein of 55,000 daltons from rat liver plasma membranes. Antibody to this material inhibits the binding of BSP and bilirubin to liver cell membranes and inhibits the uptake of these anions by isolated rat hepatocytes. Wolkoff and Chung (136) report a similar protein with at least partial cross-reactivity to the same antibody. A European group (125) reports an apparently different membrane protein to which they give the engagingly optimistic name bilitranslocase. Bilitranslocase is said to mediate BSP transport when incorporated into vesicles (113).

CONCLUSION

Since the last edition of this *Handbook* nearly 20 years ago, more direct experimental techniques have evolved, more refined mathematical models have appeared, and the number of investigators pursuing problems in hepatic transport has increased greatly. These efforts have produced a large increase in information about hepatic structure and function and a corresponding growth in the number and sophistication of postulates. It is remarkable, nevertheless, how little we yet understand about the fundamental mechanisms of organic solute transport by the liver. As might be expected from the anatomic inaccessibility of the proximal biliary tree, this is especially true of canalicular transport. The forecast is for accelerating progress. This seems most likely to come from the purification of putative receptor and carrier proteins followed by confirmation of their functions by reconstitution experiments and from the development of techniques to measure the activities of transported solutes to replace their apparent concentrations. Meanwhile mathematical modeling should continue to be important as a means to measuring then understanding how the function of individual liver cells is integrated to accomplish the performance of the liver as a whole organ.

I am grateful for expert secretarial assistance by C. A. Ritter.
This work was supported by National Institutes of Health Grant DK-27623.

REFERENCES

1. ANWER, M. S., AND D. HEGNER. Role of inorganic electrolytes in bile acid-independent canalicular bile formation. *Am. J. Physiol.* 244 (*Gastrointest. Liver Physiol.* 7): G116–G124, 1983.
2. ARONSEN, P. S. Identifying secondary active solute transport in epithelia. *Am. J. Physiol.* 240 (*Renal Fluid Electrolyte Physiol.* 9): F1–F11, 1981.
3. BAKER, K. J., AND S. E. BRADLEY. Binding of sulfobromophthalein (BSP) sodium by plasma albumin: its role in hepatic BSP extraction. *J. Clin. Invest.* 45: 281–287, 1966.
4. BARROWMAN, J. D., AND D. N. GRANGER. Hepatic lymph. In: *Hepatic Circulation in Health and Disease*, edited by W. W. Lautt. New York: Raven, 1981, p. 137–150.
5. BARRY, P. H., AND J. M. DIAMOND. Effects of unstirred layers on membrane phenomena. *Physiol. Rev.* 64: 763–872, 1984.
5a. BASS, L., AND S. M. POND. The puzzle of rates of cellular uptake of protein-bound ligands. In: *Pharmokinetics, Mathematical and Statistical Approaches to Metabolism and Distribution of Chemicals and Drugs*, edited by A. Pecile and A. Rescigno. London: Plenum, 1988, p. 245–269.
6. BASSINGTHWAIGHTE, J. B., AND C. A. GORESKY. Modeling in the analysis of solute and water exchange in the microvasculature. In: *Handbook of Physiology. The Cardiovascular System. Microcirculation*, edited by E. M. Renkin and C. C. Michel. Bethesda, MD: Am. Physiol. Soc., 1984, sect. 2, vol. IV, pt. 1, chapt. 13, p. 549–626.
7. BERK, P. D., AND W. STREMMEL. Hepatocellular uptake of organic anions. In: *Progress in Liver Disease*, edited by H. Popper and F. Schaffner. New York: Grune & Stratton, 1986, vol. 8, p. 125–144.
8. BLITZER, B. L., AND C. B. DONOVAN. A new method for the rapid isolation of basolateral plasma membrane vesicles from rat liver. *J. Biol. Chem.* 259: 9295–9301, 1984.
9. BLOOMER, J. R., P. D. BERK, J. VERGALLA, AND N. I. BERLIN. Influence of albumin on the hepatic uptake of unconjugated bilirubin. *Clin. Sci. Mol. Med.* 45: 505–516, 1973.
10. BLOOMER, J. R., AND J. ZACCARIA. Effect of graded bilirubin loads on bilirubin transport by perfused rat liver. *Am. J. Physiol.* 230: 736–742, 1976.
11. BLOUIN, A., R. P. BOLENDER, AND E. R. WEIBEL. Distribution of organelles and membranes between hepatocytes and nonhepatocytes in rat liver parenchyma: a stereological study. *J. Cell Biol.* 72: 441–455, 1977.
12. BOYER, J. L. Mechanisms of bile secretion and hepatic transport. In: *Physiology of Membrane Disorders*, edited by T. E. Andreoli, J. F. Hoffman, D. D. Fenestil, and S. G. Schultz. New York: Plenum, 1986, p. 609–636.
13. BOYER, J. L., M. ITABASHI, AND Z. HRUBAN. Formation of pericanalicular vacuoles during sodium dehydrocholate choleresis—a mechanism for bile acid transport. In: *The Liver. Quantitative Aspects of Structure and Function*, edited by R. Preisig and J. Bircher. Aulendorf: Editio Cantor, 1979, p. 163–178. (Proc. 3rd Int. Gstaad Symp.)
14. BURNETT, D. A., N. LYSENKO, J. A. MANNING, AND R. K. OCKNER. Utilization of long chain fatty acids by rat liver: studies of the role of fatty acid binding protein. *Gastroenterology* 77: 241–249, 1979.
15. CHENG, S., AND D. LEVY. Characterization of the anion transport system in hepatocyte plasma membranes. *J. Biol. Chem.* 255: 2637–2640, 1980.
16. CHRISTENSEN, H. N. On the strategy of kinetic discrimination of amino acid transport systems. *J. Membr. Biol.* 84: 97–103, 1985.
17. CRAIK, J. D., AND K. R. ELLIOTT. Kinetics of 3-O-methyl-D-glucose transport in isolated rat hepatocytes. *Biochem. J.* 182: 503–508, 1979.
18. DUFFY, M. C., B. L. BLITZER, AND J. L. BOYER. Direct determination of the driving forces for taurocholate uptake into rat liver plasma membrane vesicles. *J. Clin. Invest.* 72: 1470–1481, 1983.
19. EATON, D. L., AND C. D. KLAASSEN. Carrier-mediated trans-

port of ouabain in isolated hepatocytes. *J. Pharmacol. Exp. Ther.* 205: 480–488, 1978.
20. EATON, D. L., AND C. D. KLAASSEN. Carrier-mediated transport of the organic cation procaine amide ethobromide by isolated rat liver parenchymal cells. *J. Pharmacol. Exp. Ther.* 206: 595–606, 1978.
21. EDMUNDSON, J. W., B. A. MILLER, AND L. LUMENG. Effect of glucagon on hepatic taurocholate uptake: relationship to membrane potential. *Am. J. Physiol.* 249 (*Gastrointest. Liver Physiol.* 12): G427–G433, 1985.
22. ELLIOTT, K. R., AND J. D. CRAIK. Sugar transport across the hepatocyte plasma membrane. *Biochem. Soc. Trans.* 10: 12–13, 1982.
23. ERTTMANN, R. R., AND K. H. DAMM. Influence of bile flow, theophylline and some organic anions on the biliary excretion of ^{3}H-ouabain in rats. *Arch. Int. Pharmacodyn. Ther.* 218: 290–298, 1975.
24. FLEISCHER, A. B., W. O. SHURMANTINE, B. A. LUXON, AND E. L. FORKER. Palmitate uptake by hepatocyte monolayers: the effect of albumin binding. *J. Clin. Invest.* 77: 964–970, 1986.
25. FLEISCHER, A. B., W. O. SHURMANTINE, F. L. THOMPSON, E. L. FORKER, AND B. A. LUXON. Effect of a transported ligand on the binding of albumin to rat liver cells. *J. Lab. Clin. Med.* 105: 185–189, 1985.
26. FORKER, E. L. The effect of estrogen on bile formation in the rat. *J. Clin. Invest.* 48: 654–663, 1969.
27. FORKER, E. L. Hepatocellular uptake of inulin, sucrose, and mannitol in rats. *Am. J. Physiol.* 219: 1568–1573, 1970.
28. FORKER, E. L., AND G. GIBSON. Interaction between sulfobromophthalein (BSP) and taurocholate: the kinetics of transport from liver cells to bile in rats. In: *The Liver: Quantitative Aspects of Structure and Function*, edited by G. Paumgartner and R. Preisig. Basel: Karger, 1973, p. 326–336.
29. FORKER, E. L., AND B. A. LUXON. Hepatic transport kinetics and plasma disappearance curves: distributed modeling versus conventional approach. *Am. J. Physiol.* 235 (*Endocrinol. Metab. Gastrointest. Physiol.* 4): E648–E660, 1978.
30. FORKER, E. L., AND B. A. LUXON. Albumin helps mediate removal of taurocholate by rat liver. *J. Clin. Invest.* 67: 1517–1522, 1981.
31. FORKER, E. L., AND B. A. LUXON. Hepatic transport kinetics: effect of anatomic and metabolic heterogeneity on estimates of the average transfer coefficients. *Am. J. Physiol.* 243 (*Gastrointest. Liver Physiol.* 6): G532–G540, 1982.
32. FORKER, E. L., AND B. A. LUXON. Albumin-mediated transport of rose bengal by perfused rat liver. Kinetics of the reaction at the cell surface. *J. Clin. Invest.* 72: 1764–1771, 1983.
32a. FORKER, E. L., AND B. A. LUXON. Analyzing tracing disappearance curves to study hepatic transport kinetics. *Am. J. Physiol.* 244 (*Gastrointest. Liver Physiol.* 7): G573–G577, 1983.
33. FORKER, E. L., AND B. A. LUXON. Effects of unstirred Disse fluid, nonequilibrium binding, and surface-mediated dissociation on hepatic removal of albumin-bound organic anions. *Am. J. Physiol.* 248 (*Gastrointest. Liver Physiol.* 11): G709–G717, 1985.
34. FORKER, E. L., B. A. LUXON, M. SNELL, AND N. SHURMANTINE. Effect of albumin binding on the hepatic transport of rose bengal: surface mediated dissociation of limited capacity. *J. Pharmacol. Exp. Ther.* 223: 342–347, 1982.
35. GIBSON, G. E., AND E. L. FORKER. Canalicular bile flow and bromosulfophthalein transport maximum: the effect of bile salt-independent choleretic SC-2644. *Gastroenterology* 66: 1046–1053, 1974.
36. GORESKY, C. A. The hepatic uptake and excretion of sulfobromophthalein and bilirubin. *Can. Med. Assoc. J.* 92: 851–857, 1965.
37. GORESKY, C. A., G. C. BACH, AND B. E. NADEAU. On the uptake of materials by the intact liver. The transport and net removal of galactose. *J. Clin. Invest.* 52: 991–1009, 1973.
38. GORESKY, C. A., AND M. SILVERMAN. Effect of correction of catheter distortion on calculated liver sinusoidal volumes. *Am. J. Physiol.* 207: 883–892, 1964.
39. GRAF, J. Canalicular bile salt-independent bile formation: concepts and clues from electrolyte transport in rat liver. *Am. J. Physiol.* 244 (*Gastrointest. Liver Physiol.* 7): 233–246, 1983.
40. GRAF, J., A. GANTAM, AND J. L. BOYER. Isolated rat hepatocyte couplets: a primary secretory unit for electrophysiologic studies of bile secretory function. *Proc. Natl. Acad. Sci. USA* 81: 6516–6520, 1984.
41. GREGORY, D. H., Z. R. VLAHCEVIC, M. F. PRUGH, AND L. SWELL. Mechanism of secretion of biliary lipids: role of a microtubular system in hepatocellular transport of biliary lipids in the rat. *Gastroenterology* 74: 93–100, 1978.
42. GRUBB, D. J., AND A. L. JONES. Ultrastructure of hepatic sinusoids in sheep. *Anat. Rec.* 170: 75–79, 1971.
43. GUMUCIO, J. J., AND D. L. MILLER. Functional implications of liver cell heterogeneity. *Gastroenterology* 80: 393–403, 1981.
44. GUMUCIO, J. J., D. L. MILLER, M. D. KRAUSS, AND C. S. CUTTER-ZANOLLI. The transport of fluorescent compounds into hepatocytes and the resultant zonal labeling of the hepatic acinus in the rat. *Gastroenterology* 80: 639–646, 1981.
45. HARDISON, W. G. M., S. BELLENTANI, V. HEASLEY, AND D. SHELLHAMER. Specificity of an Na$^+$-dependent taurocholate transport site in isolated rat hepatocytes. *Am. J. Physiol.* 246 (*Gastrointest. Liver Physiol.* 9): G477–G483, 1984.
46. HESS, F. A., E. R. WEIBEL, AND R. PREISIG. Morphometry of dog liver: normal base-line data. *Virchows Arch. B Cell Pathol.* 12: 303–317, 1973.
47. HIROM, P. C., R. D. HUGHES, AND P. MILLBURN. The physicochemical factor required for the biliary excretion of organic cations and anions. *Biochem. Soc. Trans.* 2: 327–330, 1974.
48. HIUTTER, J. F., H. M. PIPER, AND P. G. SPIECKERMANN. Kinetic analysis of myocardial fatty acid oxidation suggesting an albumin receptor mediated uptake process. *J. Mol. Cell. Cardiol.* 16: 219–226, 1984.
49. HIUTTER, J. F., H. M. PIPER, AND P. G. SPIECKERMANN. Myocardial fatty acid oxidation: evidence for an albumin-receptor-mediated membrane transfer of fatty acids. *Basic Res. Cardiol.* 79: 274–282, 1984.
50. HOFFMAN, A. F., AND A. RODA. Physicochemical properties of bile acids and their relationship to biological properties: an overview of the problem. *J. Lipid Res.* 25: 1477–1489, 1984.
51. INOUE, M., R. KINNE, T. TRAN, AND I. M. ARIAS. Taurocholate transport by rat liver sinusoidal membrane vesicles: evidence of sodium cotransport. *Hepatology* 2: 572–579, 1982.
52. INOUE, M., R. KINNE, T. TRAN, AND I. M. ARIAS. The mechanism of biliary secretion of reduced glutathione. Analysis of transport process in isolated rat liver canalicular membrane vesicles. *Eur. J. Biochem.* 134: 467–471, 1983.
53. INOUE, M., R. KINNE, T. TRAN, AND I. M. ARIAS. Taurocholate transport by rat liver canalicular membrane vesicles. Evidence for the presence of an Na$^+$-independent transport system. *J. Clin. Invest.* 73: 659–663, 1984.
54. ITO, T., AND S. SHIBASAKI. Electromicroscopic study on the hepatic sinusoidal wall and the fat-storing cells in the normal human liver. *Arch. Histol. Jpn.* 29: 137–192, 1968.
55. JONES, A. L. Anatomy of the normal liver. In: *Hepatology: A Textbook of Liver Disease*, edited by D. Zakim and T. D. Boyer. Philadelphia, PA: Saunders, 1982, p. 3–31.
56. JONES, A. L., G. T. HRADEK, R. H. RENSTON, K. Y. WONG, G. KARLAGANIS, AND G. PAUMGARTNER. Autoradiographic evidence for hepatic lobular concentration gradient of bile acid derivative. *Am. J. Physiol.* 238 (*Gastrointest. Liver Physiol.* 1): G233–G237, 1980.
57. JONES, D. B., D. J. MORGAN, AND G. W. MIHALY. Discrimination between the venous equilibrium and sinusoidal models of hepatic drug elimination in the isolated perfused rat liver by perturbation of propranolol protein binding. *J. Pharmacol. Exp. Ther.* 229: 522–526, 1984.
58. KAPLOWITZ, N. Physiological significance of glutathione S-transferases. *Am. J. Physiol.* 239 (*Gastrointest. Liver Physiol.* 2): G439–G444, 1980.

59. KARDON, R. H., AND R. G. KESSEL. Three-dimensional organization of the hepatic microcirculation in the rodent as observed by scanning electron microscopy of corrosion casts. *Gastroenterology* 79: 72–81, 1980.
60. KEIDING, S., S. JOHANSEN, AND K. WINKLER. Hepatic galactose elimination kinetics in the intact pig. *Scand. J. Clin. Lab. Invest.* 42: 253–259, 1982.
61. KILBERG, M. S. Amino acid transport in isolated rat hepatocytes. *J. Membr. Biol.* 69: 1–12, 1982.
62. KLAASSEN, C. D., AND J. B. WATKINS III. Mechanisms of bile formation, hepatic uptake, and biliary excretion. *Pharmacol. Rev.* 36: 1–67, 1984.
63. KOO, A., I. Y. S. LIANG, AND K. K. CHENG. The terminal hepatic microcirculation in the rat. *Q. J. Exp. Physiol.* 60: 261–266, 1975.
64. KRAMER, W., U. BICKEL, AND H. P. BUSCHER. Bile-salt-binding polypeptides in plasma membranes of hepatocytes revealed by photoaffinity labelling. *Eur. J. Biochem.* 129: 13–24, 1982.
65. KRAMER, W., U. BICKEL, H. P. BUSCHER, W. GEROK, AND G. KURZ. Binding proteins for bile acids in membranes of hepatocytes revealed by photo affinity labelling (Abstract). *Hoppe-Seyler's Z. Physiol. Chem.* 361: 1307, 1980.
66. KRISTENSEN, L. O., AND M. FOLKE. Coupling ratio of electrogenic Na^+-alanine cotransport in isolated rat hepatocytes. *Biochem. J.* 210: 621–624, 1983.
67. KRISTENSEN, L. O., L. SESTOFT, AND M. FOLKE. Concentrative uptake of alanine in hepatocytes from fed and fasted rats. *Am. J. Physiol.* 244 (*Gastrointest. Liver Physiol.* 7): G491–G500, 1983.
68. KUPFERBERG, H. J., AND L. S. SCHANKER. Biliary excretion of ouabain-^{3}H and its uptake by liver slices in the rat. *Am. J. Physiol.* 214: 1048–1053, 1968.
69. LAKE, J. R., V. LICKO, R. W. VAN DYKE, AND B. F. SCHARSCHMIDT. Biliary secretion of fluid-phase markers by the isolated perfused rat liver. Role of transcellular vesicular transport. *J. Clin. Invest.* 76: 676–684, 1985.
70. LAYDEN, T. J., AND J. L. BOYER. Influence of bile acids on bile canalicular membrane morphology and the lobular gradient in canalicular size. *Lab. Invest.* 39: 110–119, 1978.
71. LAYDEN, T. J., E. ELIAS, AND J. L. BOYER. Bile formation in the rat. The role of the paracellular shunt pathway. *J. Clin. Invest.* 62: 1375–1385, 1978.
71a. LEEVY, C. M. (editor). *Evaluation of Liver Function.* Indianapolis, IN: Eli Lilly Res. Lab., 1974.
72. LEFFERT, H. L., D. B. SCHENK, J. J. HUBERT, H. SKELLY, M. SCHUMACHER, R. ARIYASU, M. ELLISMAN, K. S. KOCH, AND G. A. KELLER. Hepatic (Na^+, K^+)-ATPase: a current view of its structure, function and localization in rat liver as revealed by studies with monoclonal antibodies. *Hepatology* 5: 501–507, 1985.
73. LEVI, A. J., Z. GATMAITAN, AND I. M. ARIAS. Deficiency of hepatic organic anion-binding protein as a possible cause of non-haemolytic unconjugated hyperbilirubinaemia in the newborn. *Lancet* 2: 139–140, 1969.
74. LEVI, A. J., Z. GATMAITAN, AND I. M. ARIAS. Two hepatic cytoplasmic protein fractions, Y and Z, and their possible role in the hepatic uptake of bilirubin, sulfobromophthalein, and other anions. *J. Clin. Invest.* 48: 2156–2167, 1969.
75. LEVI, A. J., Z. GATMAITAN, AND I. M. ARIAS. Deficiency of hepatic organic anion-binding protein, impaired organic anion uptake by liver and "physiologic" jaundice in newborn monkeys. *N. Engl. J. Med.* 283: 1136–1139, 1970.
76. LEVINE, R. I., H. REYES, A. J. LEVI, Z. GATMAITAN, AND I. M. ARIAS. Phylogenetic study of organic anion transfer from plasma into the liver. *Nat. New Biol.* 231: 277–279, 1971.
77. LEVY, D., AND P. VON DIPPE. Reconstitution of the bile acid transport system derived from hepatocyte sinusoidal membranes (Abstract). *Hepatology* 3: 837, 1983.
78. LOUD, A. V. A quantitative stereological description of the ultrastructure of normal rat parenchymal cells. *J. Cell Biol.* 37: 27–46, 1968.
79. LUXON, B. A., AND E. L. FORKER. Simulation and analysis of hepatic indicator dilution curves. *Am. J. Physiol.* 243 (*Gastrointest. Liver Physiol.* 6): G76–G89, 1982.
80. LUXON, B. A., P. D. KING, AND E. L. FORKER. How to measure first-order hepatic transfer coefficients by distributed modeling of a recirculating rat liver perfusion system. *Am. J. Physiol.* 243 (*Gastrointest. Liver Physiol.* 6): G518–G531, 1982.
81. LUXON, B. A., P. D. KING, AND E. L. FORKER. Only free bile acid drives ileal absorption of taurocholate. *Am. J. Physiol.* 250 (*Gastrointest. Liver Physiol.* 13): G648–G652, 1986.
82. MEIER, P. J., A. ST. MEIER-ABT, C. BARRETT, AND J. L. BOYER. Mechanisms of taurocholate transport in canalicular and basolateral rat liver plasma membrane vesicles. Evidence for an electrogenic canalicular organic anion carrier. *J. Biol. Chem.* 259: 10614–10622, 1984.
83. MEIER, P. J., E. S. SZTUL, A. REUBEN, AND J. L. BOYER. Structural and functional polarity of canalicular and basolateral plasma membrane vesicles isolated in high yield from rat liver. *J. Cell Biol.* 98: 991–1000, 1984.
84. MEIJER, D. K., J. W. ARENDS, AND J. G. WEITERING. The cardiac glycoside sensitive step in the hepatic transport of the bisquaternary ammonium compound, hexafluorenium. *Eur. J. Pharmacol.* 15: 245–251, 1971.
85. MEIJER, D. K., E. S. BOS, AND K. J. VAN DER LAAN. Hepatic transport of mono and bisquaternary ammonium compounds. *Eur. J. Pharmacol.* 11: 371–377, 1970.
86. MEIJER, D. K., G. A. VERMEER, AND G. KWANT. The excretion of hexafluorenium in man and rat. *Eur. J. Pharmacol.* 14: 280–285, 1971.
87. MEIJER, D. K., R. J. VONK, E. J. SCHOLTENS, AND W. G. LEVINE. The influence of dehydrocholate on hepatic uptake and biliary excretion of ^{3}H-taurocholate and ^{3}H-ouabain. *Drug Metab. Dispos.* 4: 1–7, 1976.
88. MEIJER, D. K., AND J. G. WEITERING. Curare-like agents: relation between lipid solubility and transport into bile in perfused rat liver. *Eur. J. Pharmacol.* 10: 283–289, 1970.
89. MEIJER, D. K., J. WESTER, AND M. GUNNINK. Distribution of quaternary ammonium compounds between particulate and soluble constituents of rat liver, in relation to their transport from plasma into bile. *Naunyn-Schmiedeberg's Arch. Pharmacol.* 273: 179–192, 1972.
90. MEYER, D. K., AND A. H. SCAF. Inhibition of the transport of d-tubo-curarine from blood to bile by K-strophantoside in the isolated perfused rat liver. *Eur. J. Pharmacol.* 4: 343–346, 1968.
91. MILLER, D. L., C. S. ZANOLLI, AND J. J. GUMUCIO. Quantitative morphology of the sinusoids of the hepatic acinus. Quantimet analysis of rat liver. *Gastroenterology* 76: 965–969, 1979.
92. MISHKIN, S., L. STEIN, AND Z. GATMAITAN. The binding of fatty acids to cytoplasmic proteins: binding to Z protein in liver and other tissues of the rat. *Biochem. Biophys. Res. Commun.* 47: 997–1003, 1972.
93. MOTTO, P., M. MUTO, AND T. FUJITA. *The Liver: An Atlas of Scanning Electronmicroscopy.* Tokyo: Igaku Shoin, 1978.
93a. NAIR, P. P., AND D. KRITCHEVSKY (editors). *The Bile Acids: Chemistry.* New York: Plenum, 1971, vol. 1.
94. NAYAK, P. K., AND L. S. SCHANKER. Active transport of tertiary amine compounds into bile. *Am. J. Physiol.* 217: 1639–1643, 1969.
95. NOSSLIN, B. Mathematical appendices. In: *Metabolism of Human Gamma Globulin*, edited by S. B. Andersen. Oxford, UK: Blackwell, 1964, p. 103–137.
96. O'MAILLE, E. R., T. G. RICHARDS, AND A. H. SHORT. Factors determining the maximal rate of organic anion secretion by the liver and further evidence on the hepatic site of action of the hormone secretin. *J. Physiol. Lond.* 186: 424–438, 1966.
97. PARDRIDGE, W. M. Transport of protein-bound hormones into tissues in vivo. *Endocrinol. Rev.* 2: 103–123, 1981.
98. PARDRIDGE, W. M., AND L. J. MIETUS. Transport of steroid hormones through the rat blood-brain barrier. Primary role of albumin-bound hormone. *J. Clin. Invest.* 64: 145–154, 1979.

99. PARDRIDGE, W. M., R. SAKIYAMA, AND H. L. JUDD. Protein-bound corticosteroid in human serum is selectively transported into rat brain and liver in vivo. *J. Clin. Endocrinol. Metab.* 57: 160–165, 1983.
100. PAUMGARTNER, G., P. PROBST, AND R. KRAINES. Kinetics of indocyanine green removal from the blood. *Ann. NY Acad. Sci.* 170: 134–147, 1970.
101. PAUMGARTNER, G., AND J. REICHEN. Kinetics of hepatic uptake of unconjugated bilirubin. *Clin. Sci. Mol. Med.* 51: 169–176, 1976.
102. RAPPAPORT, A. M. The structural and functional unit in the human liver (liver acinus). *Anat. Rec.* 130: 673–689, 1958.
103. REICHEN, J., AND P. D. BERK. Isolation of an organic anion binding protein from rat liver plasma membrane fractions by affinity chromatography. *Biochem. Biophys. Res. Commun.* 91: 484–489, 1979.
104. RUSSELL, J. Q., AND C. D. KLAASSEN. Species variation in the biliary excretion of ouabain. *J. Pharmacol. Exp. Ther.* 183: 513–519, 1972.
105. SCHANKER, L. S., AND H. M. SOLOMON. Active transport of quaternary ammonium compounds into bile. *Am. J. Physiol.* 204: 829–832, 1963.
106. SCHARSCHMIDT, B. F., AND R. SCHMID. The micellar sink: a quantitative assessment of the association of organic anions with mixed micelles and other macromolecular aggregates in rat bile. *J. Clin. Invest.* 62: 1122–1131, 1978.
107. SCHARSCHMIDT, B. F., AND J. F. STEPHENS. Transport of sodium, chloride and taurocholate by cultured rat hepatocytes. *Proc. Natl. Acad. Sci. USA* 78: 986–990, 1981.
108. SCHARSCHMIDT, B. F., J. G. WAGGONER, AND P. D. BERK. Hepatic organic anion uptake in the rat. *J. Clin. Invest.* 56: 1280–1292, 1975.
109. SCHULTZ, S. *Basic Principles of Membrane Transport.* Cambridge, UK: Cambridge Univ. Press, 1980.
110. SCHWENK, M., T. WIEDMANN, AND H. REMMER. Uptake, accumulation and release of ouabain by isolated rat hepatocytes. *Naunyn-Schmiedeberg's Arch. Pharmacol.* 316: 340–344, 1981.
111. SIMIONESCU, M., AND N. SIMIONESCU. Ultrastructure of the microvascular wall: functional correlations. In: *Handbook of Physiology. The Cardiovascular System. Microcirculation,* edited by E. M. Renkin and C. C. Michel. Bethesda, MD: Am. Physiol. Soc., 1984, sect. 2, vol. IV, pt. 1, chapt. 3, p. 41–102.
112. SIPS, H. J., AND K. VAN DAM. Amino acid-dependent sodium transport in plasma membrane vesicles from rat liver. *J. Membr. Biol.* 62: 231–237, 1981.
113. SOTTOCASA, G. L., G. BALDINI, AND G. SANDRI. Reconstitution in vitro of sulfobromophthalein transport by bilitranslocase. *Biochim. Biophys. Acta* 685: 123–128, 1982.
114. STOLLMAN, Y. R., U. GARTNER, L. THEILMANN, N. OHMI, AND A. W. WOLKOFF. Hepatic bilirubin uptake in the isolated perfused rat liver is not facilitated by albumin binding. *J. Clin. Invest.* 72: 718–723, 1982.
115. STOLZ, A., Y. SUGIYAMA, J. KUHLENKAMP, AND N. KAPLOWITZ. Identification and purification of a 36 kDa bile acid binders in human hepatic cytosol. *FEBS Lett.* 177: 31–35, 1984.
116. STRANGE, R. C. Hepatic bile flow. *Physiol. Rev.* 64: 1055–1102, 1984.
117. STREMMEL, W., M. A. GERBER, V. GLEZEROV, S. N. THUNG, S. KOCHWA, AND P. D. BERK. Physicochemical and immunohistological studies of a sulfobromophthalein-binding and bilirubin-binding protein from rat liver plasma membranes (Abstract). *Hepatology* 2: 717a, 1982.
118. STREMMEL, W., S. KOCHWA, AND P. D. BERK. Studies of oleate binding to rat liver plasma membranes. *Biochem. Biophys. Res. Commun.* 112: 88–95, 1983.
119. STREMMEL, W., G. STROHMEYER, AND P. D. BERK. Hepatocellular uptake of oleate is energy dependent, sodium linked, and inhibited by an antibody to a hepatocyte plasma membrane fatty acid binding protein. *Proc. Natl. Acad. Sci. USA* 83: 3584–3588, 1986.
120. STREMMEL, W., G. STROHMEYER, AND F. BORCHARD. Isolation and partial characterization of fatty acid binding protein in rat liver plasma membranes. *Proc. Natl. Acad. Sci. USA* 82: 4–8, 1985.
121. STUZL, E., M. CAPLAN, D. BIEMESDERFER, L. BARRETT, M. KASHGARIAN, AND J. L. BOYER. Localization of Na^+,K^+-ATPase α-subunit to the basolateral (BLPM) but not canalicular membrane (CLPM) of rat hepatocytes with poly- and monoclonal antibodies (Abstract). *Hepatology* 5: 1016, 1985.
122. SUGIYAMA, Y., T. YAMADA, AND N. KAPLOWITZ. Newly identified bile acid binders in rat liver cytosol. Purification and comparison with glutathione S-transferases. *J. Biol. Chem.* 258: 3602–3607, 1983.
123. TAYLOR, A. E., AND D. N. GRANGER. Exchange in macromolecules across the microcirculation. In: *Handbook of Physiology. The Cardiovascular System. Microcirculation,* edited by E. M. Renkin and C. C. Michel. Bethesda, MD: Am. Physiol. Soc., 1984, sect. 2, vol. IV, pt. 1, chapt. 11, p. 467–520.
125. TIRIBELLI, C., G. LUNAZZI, AND M. LUCIANI. Isolation of a sulfobromophthalein-binding protein from hepatocyte plasma membrane. *Biochim. Biophys. Acta* 532: 105–112, 1978.
126. TOYOTA, N., K. MIYAI, AND W. G. HARDISON. Effect of biliary pressure versus high bile acid flux on the permeability of hepatocellular tight junction. *Lab. Invest.* 50: 536–542, 1984.
127. VAN AMELSVOORT, J. M., H. J. SIPS, M. E. APITULE, AND K. VAN DAM. Heterogeneous distribution of the sodium-dependent alanine transport activity in the rat hepatocyte plasma membrane. *Biochim. Biophys. Acta* 600: 950–960, 1980.
128. VAN AMELSVOORT, J. M., H. J. SIPS, AND K. VAN DAM. Sodium-dependent alanine transport in plasma-membrane vesicles from rat liver. *Biochem. J.* 174: 1083–1086, 1978.
129. VAN DYKE, R. W., J. E. STEPHENS, AND B. F. SCHARSCHMIDT. Bile acid transport in cultured rat hepatocytes. *Am. J. Physiol.* 243 (*Gastrointest. Liver Physiol.* 6): G484–G492, 1982.
130. VON DIPPE, P., P. DRAIN, AND D. LEVY. Synthesis and transport characteristics of photoaffinity probes for the hepatocyte bile acid transport system. *J. Biol. Chem.* 258: 8890–8895, 1983.
131. VON DIPPE, P., AND D. LEVY. Characterization of the bile acid transport system in normal and transformed hepatocytes. Photoaffinity labelling of the taurocholate carrier protein. *J. Biol. Chem.* 258: 8896–8901, 1983.
132. WEISIGER, R., J. GOLLAN, AND R. OCKNER. Receptor for albumin on the liver cell surface may mediate uptake of fatty acids and other albumin-bound substances. *Science Wash. DC* 211: 1048–1051, 1981.
133. WILLIAMS, T. F., J. H. EXTON, C. R. PARK, AND D. M. REGEN. Stereospecific transport of glucose in the perfused rat liver. *Am. J. Physiol.* 215: 1200–1209, 1968.
134. WISSE, E., R. B. DE ZANGER, K. CHARELS, P. VAN DER SMISSEN, AND R. S. MCCUSKEY. The liver sieve: considerations concerning the structure and function of endothelial fenestrae, the sinusoidal wall and the space of Disse. *Hepatology* 5: 683–692, 1985.
135. WISSE, E., R. DE ZANGER, AND R. JACOBS. Lobular gradients in endothelial fenestrae and sinusoidal diameter favour centrolobular exchange processes: a scanning EM study. In: *Sinusoidal Liver Cells,* edited by D. L. Knook and E. Wisse. Amsterdam: Elsevier, 1982, p. 61–67.
136. WOLKOFF, A. W., AND C. T. CHUNG. Identification, purification and partial characterization of an organic anion binding protein from rat liver cell plasma membranes. *J. Clin. Invest.* 65: 1152–1161, 1980.
137. WOLKOFF, A. W., C. A. GORESKY, J. SELLIN, Z. GATMAITAN, AND I. M. ARIAS. Role of ligandin in transfer of bilirubin from plasma into liver. *Am. J. Physiol.* 236 (*Endocrinol. Metab. Gastrointest. Physiol.* 5): E638–E648, 1979.
138. WRIGHT, P. L., J. A. CLEMETT, K. F. SMITH, W. A. DAY, AND R. FRASER. Hepatic sinusoidal endothelium in goats. *Aust. J. Exp. Biol. Med. Sci.* 61: 739–741, 1983.
139. WRIGHT, P. L., K. F. SMITH, W. A. DAY, AND R. FRASER. Hepatic sinusoidal endothelium in sheep: an ultrastructural reinvestigation. *Anat. Rec.* 206: 385–390, 1983.

CHAPTER 35

Transport and metabolism in the hepatobiliary system

DIRK K. F. MEIJER | Department of Pharmacology and Therapeutics, University Center of Pharmacy, University of Groningen, Groningen, The Netherlands

CHAPTER CONTENTS

Methodology
Factors Influencing Clearance of Drugs by the Liver
   Chemical structure
   Protein binding
     Plasma protein binding
     Intracellular protein binding
   Bile flow
   Adaptive changes
   Biotransformation, detoxification, toxification
     General considerations
     Phase I and phase II biotransformation reactions
     Phase I reactions
     Phase II reactions
     Biotransformation of drugs leading to toxification
     Drug metabolism in liver disease
     Relationship between biliary excretion and metabolism
   Acinar heterogeneity
     Zonal heterogeneity in drug metabolism
     Zonal heterogeneity in drug transport
Mechanisms of Membrane Transport of Organic Compounds
   General considerations
     Multiplicity of transport mechanisms: model compounds
     Bidirectional character of transport
     Hepatobiliary concentration gradients
     Potential driving forces for hepatobiliary transport
   Organic anions
     Kinetic analysis: model compounds
     Mechanisms of hepatic uptake
     Mechanisms of biliary excretion
     Carrier proteins for anion transport
   Organic cations
     Kinetic analysis: model compounds
     Mechanisms for hepatic uptake of monovalent cations
     Mechanisms for hepatic uptake of bivalent cations
     Mechanisms for bile-canalicular transport
   Uncharged compounds
     Kinetic analysis: model compounds
     Mechanisms for hepatobiliary transport
Conclusion

THE LIVER IS A HETEROGENEOUS ORGAN that can perform a vast variety of functions, one of which is to clear endogenous and exogenous materials from the blood. The various cell types in the organ (152, 363) deal not only with nutrients and endogenous metabolic products from other parts of the body but also with drugs, toxins, immune complexes, denatured proteins, lipoid particles, and solid particles such as viruses, bacteria, parasites, and tumor cells. After primary uptake of material by the liver, a versatile metabolic apparatus for degradation and biotransformation is available in intracellular organelles such as lysosomes and endoplasmic reticulum. Further disposition may include secretion of the products from the organ via at least three pathways: *1*) the systemic circulation, *2*) the biliary system, and *3*) the lymphatic system. This active and flow-dynamic system is situated between the digestive tract and the rest of the body: not only nutrients but also pollutants, drugs, and other xenobiotic compounds absorbed in the gut have to pass this barrier before they can reach the general circulation.

Because the liver is a relatively large organ, constituting ~3% of body weight, exposure time during passage of the organ is relatively long compared with the kidneys. In addition, because of the unique endothelial lining (Fig. 1), with perforations of ~0.1 μm, organic compounds in the blood in the unbound as well as in the plasma protein bound form can be very efficiently exposed to the large villous plasma membrane of the hepatocytes (152, 363). The plasma membrane is a surprisingly heterogeneous modality with various domains with specific functions (80, 81). In addition to the sinusoidal part, with a range of discriminating receptors and carrier proteins (154), the lateral domain provides the cell contacts via desmosomes, gap junctions, and tight junctions, of which the latter constitute the barrier between the extracellular space and the bile canaliculi (45). The villous canicular membranes that have a special lipid composition and possess multiple-carrier systems form the primary channels of the biliary system. Biliary excretion does

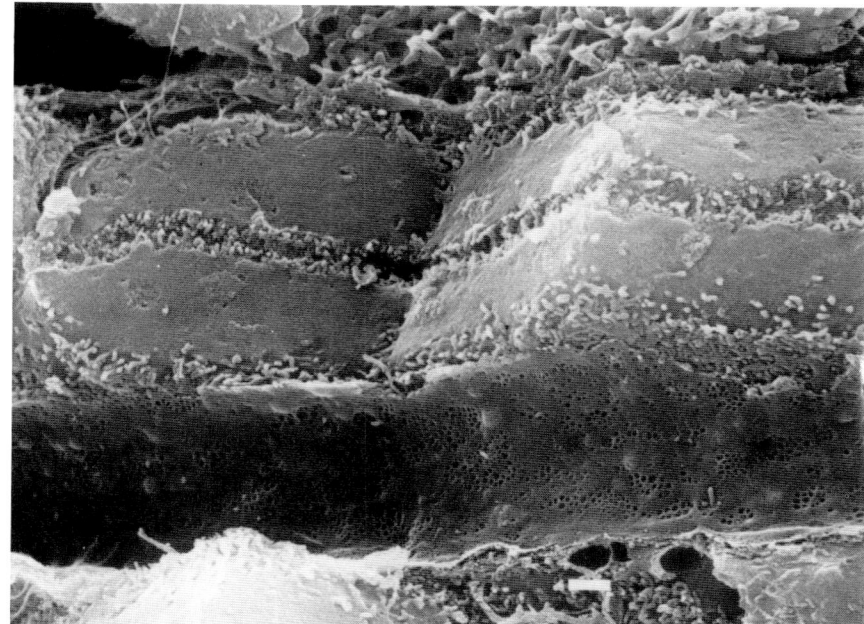

FIG. 1. Ultrastructure of the liver. Scanning electron micrograph (× 5,000) showing the sinusoidal wall with large fenestrae and a plate of parenchymal cells (hepatocytes) with microvilli at the sinusoidal and canalicular poles of the cell representing sites for carrier-mediated transport of drugs.

not necessarily imply removal from the body. In the gut many of the biotransformation processes that occur in the liver can be reversed, and reabsorption of the particular compounds produced may lead to enterohepatic cycling (55, 107, 125, 132, 172, 258). In individuals with a functioning gallbladder this is a discontinuous process and the homeostasis of this system is maintained by hormonal as well as nervous influences (97). This chapter focuses on the mechanisms involved in the hepatic disposition of drugs. There are many reviews of biliary physiology available (14, 36, 43–45, 75–78, 87, 94, 97, 103, 132, 149, 152, 270, 271, 276, 291, 295, 325, 354, 355). Comprehensive reviews on the biliary excretion of drugs are also available, some of them providing tabulated data on biliary excretion in animals and humans (29, 30, 94, 107, 171, 172, 182–184, 200, 209, 276, 278, 287, 288, 311, 312, 324).

How are organic solutes taken up by the liver and what processes are responsible for their removal? Many compounds reach their destination by passive permeation of the lipoid membranes of the various cell types. Others require more or less specific transport processes, such as membrane carrier transport or receptor-mediated endocytosis (154). Transport from blood to metabolic or secretory sites within the cells can be pictured as a sequential phenomenon: supply via the bloodstream, dissociation from circulating macromolecules in plasma, association with carrier molecules in the membrane, translocation across the membrane, dissociation from the carrier at the inner side of the membrane, vesicle transport and/or diffusion through the cytoplasm followed by similar processes in the entrance of organelles, or secretion from the cells (see Fig. 1 for the functional liver structure).

Drug transport processes in the liver are traditionally classified according to the charge of the compounds to be transported, i.e., transport processes for organic anions, organic cations, and uncharged compounds (287, 288). The classic approach to discriminate between carrier-mediated and non-carrier-mediated transport includes the demonstration of 1) saturation of membrane transport, 2) mutual competition in transport by structurally related compounds, 3) stereospecificity of membrane transport, 4) countertransport or accelerative exchange diffusion, 5) transport against a chemical or electrochemical gradient, and 6) metabolic energization of the transport. The latter two characteristics are relevant for a further differentiation between carrier-mediated facilitated diffusion and primary or secondary active transport. A definite demonstration of carrier-mediated transport along these lines must rely on experimentation with inherent limitations in the techniques employed and in the interpretation of the data produced. Saturation of transport often requires large concentrations of the drug, producing aspecific effects on cell metabolism and/or membrane conformation. Mutual competitive inhibition during hepatic transport is difficult to demonstrate, because often more than one process is involved in the transport of a substrate. Competition and saturation phenomena in the transport of hydrophobic organic ions have even been demonstrated in studies with protein-free artificial membranes (28). Countertransport can be shown in isolated membrane preparations under rather unphysiological conditions but is not easy to demonstrate in the intact organ or cells because of binding to sites other than the supposed carrier molecules. The unbound concentration of drugs in hepatocyte cytosol and primary bile in situ is as yet technically impossible to measure. Experiments with metabolic inhibitors or manipulation of such factors as the temperature raise questions about the specificity of these procedures.

Drug metabolites, including polar conjugates, are transported out of the hepatocytes by carrier-mediated processes at both poles of the cell. These processes not only determine the final elimination route (bile or urine) but also the local concentrations in the hepatobiliary system and the gut, with obvious implications for therapeutic actions and local toxicity (57, 65, 171). The quantitative measurement of these processes can provide crucial data in assessing the function of the organ under normal physiological as well as pathological conditions. Elucidation of the mechanism involved in hepatic uptake and removal of xenobiotic compounds may also improve the understanding of their therapeutic or toxic effects in the various parts of the organ (14, 110, 271) and may lead to a more rational design of probes to test the elimination function of the liver (243) or to development of organ-specific therapeutic agents (261).

METHODOLOGY

Liver function in the clearance of drugs through metabolism or excretion can be approached by using various in vivo and in vitro techniques (Table 1). In the intact organism, rates of plasma disappearance, biliary excretion, or production of metabolites can be evaluated by taking an adequate number of samples from plasma, urine, and/or bile (41, 53). In vivo techniques in animals with permanently implanted catheters in blood vessels, common bile duct, or urinary bladder enable long-term pharmacokinetic studies in mobile animals without the use of anesthetics (229). A compartmental analysis of the plasma disappearance and biliary appearance profiles can provide useful data on rate constants for distribution to and from the liver as well as for the excretion or metabolic process [Fig. 2; (41, 98, 204, 210)].

TABLE 1. *Techniques for Study of Hepatic Transport Function*

Intact liver in vivo
Isolated perfused liver
Liver slices and/or liver snips
Isolated hepatocytes
  Acute
  Cultured
  Couplets
Homogenate subfractions, organelles
Isolated membrane vesicles
Isolated carrier proteins
  Binding and/or photoaffinity labeling
  Antibodies
  Reconstitution
Carrier protein synthesis, regulation

Preparations are listed in order of decreasing structural organization.

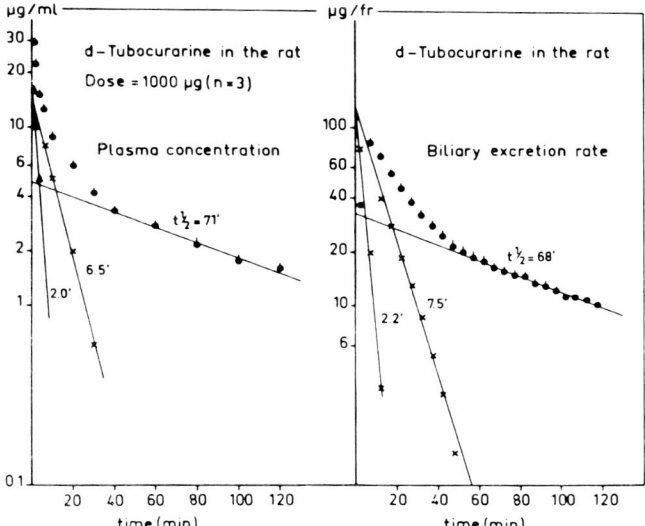

FIG. 2. Plasma disappearance and biliary excretion rate patterns after intravenous injection of 1,000 μg d-tubocurarine (organic cation). Curve stripping of triexponential curves reveals almost identical slopes in plasma and bile, enabling calculation of transport rate constants for hepatic uptake and biliary excretion using compartmental models (40) as well as plasma and biliary clearance [cf. Fig. 4; (242)].

Because of the complexity of the intact organism, a spectrum of in vitro techniques was developed to study liver function per se or to obtain more detailed kinetic data on the cellular mechanisms involved. Table 1 depicts these techniques in an order of decreasing structural organization of the preparations used. Each technique has special advantages but also limitations, and a proper interpretation of what is seen often requires the combined use of such techniques. The isolated perfused organ, as elaborated by Brauer et al. (46, 47) and Miller et al. (221), represents a versatile technique to study liver function per se (41, 208). Easy manipulation of the composition of the perfusion medium (solutes, proteins), perfusion rate, and perfusion direction [antegrade and retrograde; (109, 110, 248)]; the possibility of taking many samples both from inflowing and outflowing medium; and the intact architecture of the liver structure (sinusoidal lining, acinar microcirculation, intact polarity of hepatocytes, and biliary tree structure) yield the advantage of studying hepatic clearance function under approximately physiological conditions (46, 47, 208).

Liver perfusion is the starting point for the preparation of isolated hepatocytes by sequential perfusion with Ca^{2+}-free and collagenase-containing media and separation from other cell types by differential centrifugation, as developed by Berry and Friend (33). Despite these invasive procedures, acutely isolated hepatocytes retain most of their transport (41, 299) and metabolic capabilities (328) and can even be separately isolated from perivenous and periportal zones (119). Isolated hepatocyte preparations offer the advantages of homogeneous exposure and multiple identical samples for detailed kinetic studies. However, plasma membrane domain differentiation is lost and canalicular membrane components rapidly redistribute across the entire surface (96). Canalicular structure may reappear on long-term culturing of hepatocytes on plates (96, 108, 116, 247, 273) or may be inspected in pairs of hepatocytes still connected by tight junctions (couplets) (104). The latter techniques enabled cinematographic observations of microfilament-mediated canalicular contractions as well as concentrative transport of fluorescent dyes (96, 256, 257). Recently progress was made in introducing microelectrodes in the canalicular space of couplets to measure the transmembrane potential (104). The existence of a "sealed-off" space in couplets may eventually lead to an as yet unfulfilled option: micropuncture of primary canalicular bile.

Much progress in the study of hepatic transport was made by the preparation of closed membrane vesicles from the various domains of the hepatocyte plasma membrane [see *Organic Anions*, p. 742; (37, 38, 64, 80, 142–144, 197–199)]. If the origin (on the basis of marker enzymes) and orientation (right side in or out) of the membrane vesicles are properly defined, they offer the possibility of studying symport and antiport mechanisms at the membrane level, because, in contrast to intact cells, both the external and internal milieu can be manipulated. Also the electrogenic features of organic solute transport can be studied, because a membrane potential can be imposed by preloading and incubation of the vesicles in media with inorganic ions having unequal permeation rates, thereby inducing short-lasting diffusion potentials (133).

Very promising are recent approaches in the isolation and purification of potential carrier proteins (see *Organic Anions*, p. 742) by treating plasma membrane fractions with detergents and subsequent protein separation by electrophoretič, chromatographic, or affinity chromatographic methods, among others (268, 331, 365). Photoaffinity labeling can be used to identify and tag these proteins during the isolation procedures (51, 338, 371, 372). Relative binding affinity for transport model compounds, preparation of monospecific antibody proteins, and reconstitution of transport function in artificial membranes may lead to a better characterization of their carrier function (29).

The molecular features characterized in vitro should be put in perspective to the integrated physiological system of the intact organism. In the future such studies should be directed more to the excretory function in humans. A number of factors prevent real insight into the quantitative aspects of this process. Part of the problem is the collection of human bile continuously and completely under physiological conditions. Many studies only report concentration in incidentally taken duodenal aspirates, T-tube bile, gallbladder bile, or samples taken with percutaneous drainage techniques. Usually the total concentration of the drug, including its metabolites, is determined. On the basis of assumptions on average bile production in the human, the fraction of the dose eliminated via bile can only be roughly estimated. The use of the actual volume of T-tube bile collected may lead to gross underestimation of this fraction because of bile salt depletion. In addition, postcholecystectomy patients with T-tube bile drainage (who are often the object of study) may have had abnormal liver function and commonly receive many other drugs before and after surgery. Also the major pelvic operation procedures per se may depress bile flow for several weeks after surgery (126). Bile flow as well as elimination rate of dibromosulfophthalein (DBSP) was ~50% depressed in the postcholecystectomy patients (213).

In two reports interesting techniques were tested to quantify biliary excretion and enterohepatic cycling of drugs in normal subjects without disturbance of the physiological conditions. Balloon-occludable multilumen duodenal tubes (15) or triple-lumen tubes (66) provided reproducible methods to estimate the kinetics of biliary excretion and enterohepatic circulation of drugs administered parenterally or even orally. In four reviews the available data on biliary excretion of drugs in humans were tabulated (171, 183, 276, 278). Enterohepatic cycling of drugs has also partly been listed in these and other reports (258, 310). Examples of drugs excreted unchanged in human bile for >20% of the administered dose include the following: the more lipophilic antibiotics of the cephalosporin and penicillin groups as well as rifamide and other rifamycins (231); the antineoplastic agents doxorubicin (adriamycin) and vincristine; the cardiac glycoside digoxin; the β-blocking agent practolol; the peripheral muscle relaxants hexafluronium (276, 278) and vecuronium (27); the diagnostic dyes indocyanine green (ICG), rose bengal (276, 278), and DBSP (213); and various types of biliary contrast agents (31, 32). In most of these cases one or more metabolites are simultaneously detected in bile. Therefore very few model compounds are available for studying hepatobiliary transport per se in humans. There is also little information on interactions of drugs during hepatobiliary transport in the human. Probenecid lowers the biliary output of rifampicin and indomethacin (18, 160, 276). Rifamycins depress the biliary excretion of bromosulfophthalein (BSP) and ICG (2). Quinidine reduces biliary clearance of digoxin in patients in addition to influencing renal and intestinal secretion. Such effects explain the marked rise in steady-state plasma concentrations of the cardiac glycoside if it is combined with lipophilic cations like quinidine and certain Ca^{2+} antagonists (293). More systematic studies in the human are required to reveal drug interactions at the level of hepatobiliary clearance and enterohepatic circulation.

FACTORS INFLUENCING CLEARANCE OF DRUGS BY THE LIVER

Chemical Structure

Elimination of drugs from the body is a concerted action of kidneys, intestine, and liver. On the basis of the chemical structure and physicochemical features of drugs, such as the presence of certain functional groups, lipophilicity, and dissociation constant (pK) value of dissociable groups, a rough prediction can be made of the relative contribution of excretory and metabolic processes in the kidneys, intestine, and liver to the total body clearance (85, 186, 244, 303). The relative excretory activity of an organ can be estimated by measuring the fraction of an intravenous dose that is finally excreted via that particular pathway. The absolute capability of the organ to eliminate the drug is most adequately expressed as the organ blood clearance (volume/time).

How the chemical structure of a drug determines its elimination pattern via urine and bile has been studied by Hirom et al. (127–130), Hughes et al. (137, 138), and Smith (310–312). Central in their work is the "molecular-weight threshold hypothesis," which reflects the empirical finding that within a series of structurally related drugs, biliary excretion becomes appreciable only if a certain threshold in molecular weight is exceeded. Such thresholds are generally found in the molecular weight (MW) range of 200–600, depending on the charge of the compound, the number of charged groups, and the animal species studied. The molecular-weight threshold for organic anions in the rat, for instance, was proposed to be 325 ± 50 (130). Because the biliary secretion mechanism is fundamentally different from a filtration process, there is no obvious physiological basis for categorizing of drugs according to their molecular size. More likely, molecular weight indirectly reflects other physicochemical characteristics, such as lipophilicity and especially the balance between hydrophilic and hydrophobic properties. The lipophilic part of the molecule probably enables a hydrophobic interaction with the translocating membrane components, whereas hydrophilic groups such as SO_3^-, COO^-, and also sugar groups may provide the possibility for electrophilic or van der Waals' type of interactions.

Structure-transport correlation studies with a limited number of compounds have been performed mostly in rat for monovalent organic cations (127, 137, 241, 242), divalent organic cations (138, 200, 347), sulfonamides (61, 129), penicillins (284), organic anions in general (127, 130), iron chelates such as ferrioxamines (220), various hepatobiliary contrast agents [mostly radiopharmaceuticals; (243)], and bile acids (13). The conclusion from these studies is that not merely the presence of charged or hydrophilic groups nor the absolute lipophilicity (200) but rather the balance between these properties is a crucial factor in hepatic transport. In this respect one should realize that the extent of protein binding and metabolism may also increase with lipophilicity. These factors can oppose the promoting effect of lipophilicity on the extent of biliary excretion. Even relatively small structural factors may largely influence the interaction with carrier systems (311) and the distribution in the body, as has been shown for cardiac glycosides (194, 311, 312) and stereoisomers of drugs (312, 333).

A group of organic cations were studied with regard to the relative contributions of renal, intestinal, and hepatobiliary clearance that were correlated with lipophilicity, molecular weight, and protein binding (241, 242). The results supported the idea of earlier studies (127, 347) that biliary elimination becomes more important with increasing lipophilicity. It is striking, however, that variation in lipophilicity introduced a much more pronounced variation in the biliary than in the intestinal clearance. The modest biliary and intestinal clearance of the agents with relatively low lipid solubility is not due to a lack of penetration into the cells. The marked accumulation in hepatocytes and intestinal mucosa cells rather suggests that transport out of the cells to bile and gut lumen is deficient (242). Total clearance via the urine evidently correlated poorly with lipophilicity in this series of quaternary ammonium cations (Fig. 3). In a more biological context, compounds with sufficient hydrophobic features enjoy a high affinity for carrier systems, especially in the liver.

Protein Binding

Many drugs undergoing hepatic biotransformation or hepatobiliary transport are highly protein bound (279). This is partly because the amphipathic character that favors hepatobiliary transport and/or metabolism also may promote association with plasma proteins (303).

PLASMA PROTEIN BINDING. In the plasma mainly two different proteins are responsible for most of the binding: albumin and α_1-acid glycoprotein (orosomucoid). Roughly speaking, albumin, having a relatively high plasma concentration (4.5% or 0.6 mM), binds acidic (anionic) drugs, and α_1-acid glycoprotein, having a much lower concentration (0.1% or 1.5 μM), binds predominantly basic (cationic) drugs (35, 279, 332). Both proteins are synthesized in the liver, and, especially during chronic liver disease and/or renal disease, plasma concentration can be abnormally low (35, 332). During acute-phase reactions such as inflammation, tumors, and burns, concentration of orosomucoid in plasma may be elevated as a result of increased hepatic synthesis (332).

Drug binding to these proteins is a saturable phenomenon. The unbound fraction (f_u) of a drug is

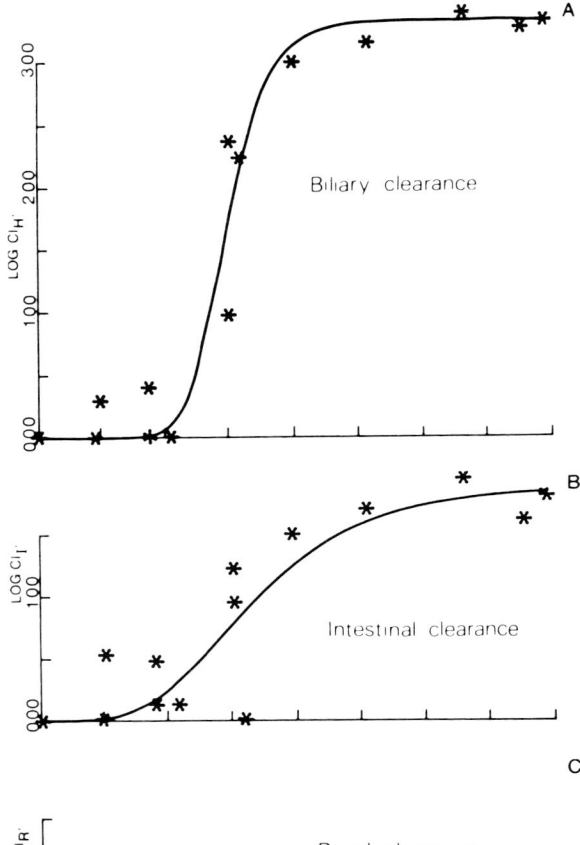

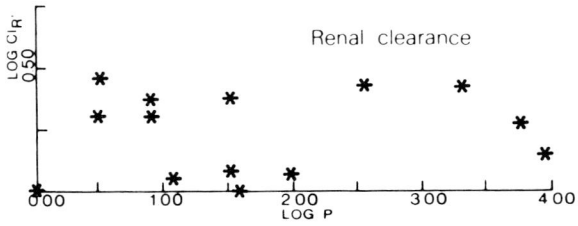

FIG. 3. Relation between lipophilicity (measured by partition between octanol and Krebs buffer and expressed relative to the value of tetraethyl methylammonium) and carrier-mediated clearance of 14 monovalent organic cations (drugs with a quaternary ammonium group) via biliary, intestinal, and renal routes in the rat. Clearance values are corrected for passive fluxes (e.g., glomerular filtration) and indicate carrier-mediated transport in the organs. Between log $P = 1$ and log $P = 2$, biliary clearance increases ~1,000 and intestinal clearance ~100 times in contrast to renal clearance, which shows no such correlative pattern. At log $P > 2.5$, clearance is partly blood-flow limited.

the estimation of liver function can be made. For instance, if albumin concentration is abnormally low due to liver disease, decreased intrinsic clearance function may be partly or even completely masked by an increased unbound fraction (35, 279, 332).

Hepatic clearance of anionic dyes was shown to be inversely related to the albumin concentration in humans (106), the intact animal (141), the isolated perfused rat liver [Fig. 4; (204, 210)], and isolated hepatocytes (40). Comparing uptake rate of BSP and its glutathione derivative BSP-GSH in isolated hepatocytes incubated without albumin suggested that BSP-GSH has a much lower uptake rate than BSP (297). In isolated perfused rat livers with albumin-containing perfusion medium (369), however, uptake of BSP-GSH in the liver occurred more rapidly than for BSP. This discrepancy is resolved if the large difference in protein binding is taken into account: BSP-GSH is bound to albumin 10–100 times less than BSP (20). Hepatic storage of drugs, apart from saturable membrane transport, is also determined by relative binding in the plasma and liver compartments. If albumin is added to a perfusate of a perfused liver preloaded with DBSP, redistribution from liver to the plasma occurs [Fig. 4; (204)].

Because the liver is able to extract compounds that are <1% unbound with an extraction fraction of up to 80%, it has been suggested that in addition to progressive spontaneous dissociation within the liver sinusoids and space of Disse (349, 350), direct interac-

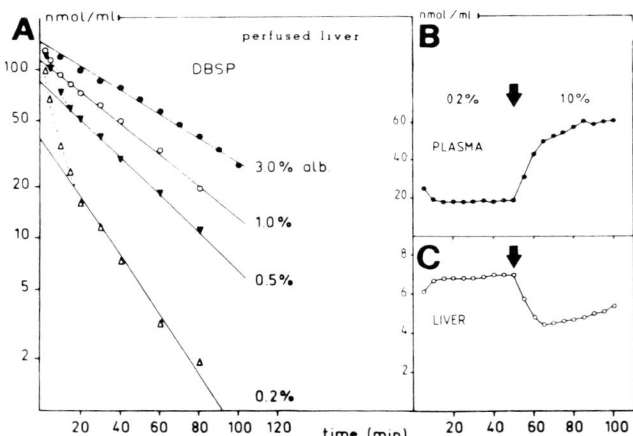

FIG. 4. Influence of albumin on clearance of dibromosulfophthalein (DBSP) in isolated perfused liver after single injection (A). Inverse relation is shown between albumin concentration and initial uptake rate. Lowering of albumin concentration increases unbound fraction and thereby the driving force for transport into the liver (hepatic storage) and liver to bile. B: influence of extracellular binding on hepatic storage of DBSP. Steady-state kinetics by constant infusion of 800 nmol/min of DBSP in isolated rat liver perfused with 0.2% albumin containing Krebs solution. At $t = 60$ min, albumin was added to the perfusion medium to a concentration of 1.0%. Before addition, constant concentrations in bile, liver, and perfusate were attained. Addition of albumin results in disturbance of steady state and efflux of DBSP from the liver (C) to the perfusate plasma (B).

determined by the total concentration of the drug and protein in plasma. If a drug is 99.9% bound, $f_u = 0.001\%$, and a slight displacement, for instance one that results in a new binding percentage of 99.0%, affects the unbound fraction 10-fold. Consequently, especially for drugs with binding >90%, displacement interactions or changes in protein concentration produce large relative changes in the unbound fraction of the drug in plasma, and this may also alter the clearance of drugs by the liver (35, 279, 332). If this aspect is not taken into account, major misinterpretations in

tions of the albumin-drug complex at the plasma membrane may facilitate dissociation of the complex (Fig. 5). The latter hypothesis (20, 42) underwent a revival through more recent observations that the hepatic uptake kinetics of taurocholate (89) and of fatty acids (350) in isolated perfused liver seemed to be better related to the bound rather than the unbound fraction. That the protein-bound fraction is not entirely inert with regard to transport was also concluded from studies in isolated hepatocytes (23) and isolated membrane vesicles (38) and raised extensive discussions (91, 224, 349) in relation to the classic concept that only the unbound fraction determines the rate of hepatic removal (279). The relative contribution of both mechanisms to the extraction of drugs by the liver may largely depend on the type of compound. Hydrophobic ligands with very high affinity [dissociation binding constant $(K_d) > 10^{-7}$ M] for plasma proteins such as fatty acids, BSP, iopanoate, and bilirubin may (partly) require some sort of protein-membrane surface interactions to provide sufficient unbound substrate. For other drugs with lower affinities, such as BSP-GSH (20, 349) and propranolol (153, 224), spontaneous dissociation due to progressive removal of the unbound drug may be the most prominent mechanism.

Drugs that are noncovalently linked to albumin dissociate from the protein before they are taken up with carrier-mediated mechanisms (20, 29, 336b). In contrast, drugs that are covalently linked to certain proteins can be taken up in various cell types of the liver by means of receptor-mediated endocytosis. For instance, glycoproteins, of which the cell type–specific endocytosis is determined by the nature of the terminal sugar group of the oligosaccharide chain (154), as well as transferrin, apolipoproteins, and certain immunoglobulins, can be used as potential carriers to deliver drugs more selectively to the liver and even to cell types within the organ (261). Examples are antiviral agents, antineoplastic and antiparasitic drugs, peptide toxins, and agents that modify cholesterol and/or bile salt metabolism (261). This "drug-targeting" concept assumes that at least part of the covalently linked drug, after endocytosis and intracellular migration to lysosomes, is released in an active form by proteolytic cleavage of the drug-protein bond and also that efficiency or cell specificity of the endocytotic processes is not affected by changes in net charge or in molecular conformation of the protein by the drug loading (261).

INTRACELLULAR PROTEIN BINDING. The intracellular binding of drugs is important for net hepatic uptake and prevention of intracellular toxic effects. For organic anions two classes of cytosolic binding proteins were discovered independently in the same period by three groups (162, 181, 223) studying bilirubin, cortisol metabolites, and carcinogens, respectively. Sephadex filtration of cytosol fractions to which these agents were added roughly showed two elution peaks, often called the Y and Z peaks (181). In the rat and human the Y peak contains the large family of glutathione S-transferases catalyzing the conjugation of GSH with a wide variety of ligands in the first step to mercapturic acid formation [see *Biotransformation, Detoxification, Toxification*, p. 726; (364)]. These proteins, with an MW of ~48,000, can be induced despite their abundant presence in the liver by compounds such as phenobarbital, styrene oxide, 3-methylcholantrene, spironolactone, and trans-stilbene oxide (172). Normally glutathione transferase B or ligandin constitutes 4%–5% of the cytosolic protein and is the major binding protein.

Another major binding protein is Z protein (24, 163), which is identical to hepatic fatty acid–binding protein (hFABP), aminoazo dye–binding protein, and sterol-carrier protein. It is a hydrophobic 12.5- to 14.2-kDa protein that in cytosol shows at least three forms that have different isoelectric points. Besides DBSP, BSP, and ICG, it binds organic anions such as bilirubin,

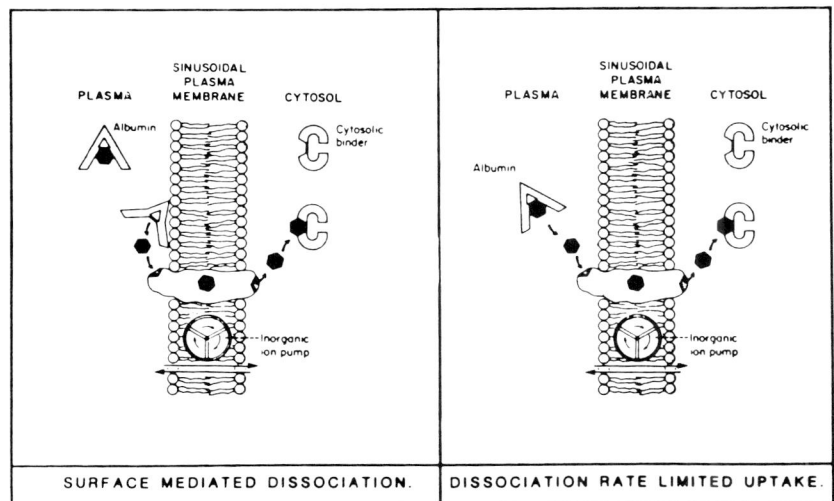

FIG. 5. Schematic representation of supposed mechanisms for organic anion transport at sinusoidal level. Net transport across the membrane is mediated by an undefined translocator (carrier or ion-selective channel protein) transporting the unbound substrate. At the plasma membrane surface, interaction of albumin may lead to facilitated dissociation of the organic anion (*left*). Alternatively, spontaneous dissociation by progressive removal of the unbound fraction may occur, in which dissociation from albumin may constitute the rate-limiting step (*right*). Within the cell the substrate is associated with cytosolic-binding proteins such as ligandin (glutathione transferase B) and Z protein (fatty acid–binding protein). [Adapted from Berk et al. (29).]

steroid sulfates, and also very hydrophobic uncharged compounds such as fatty acids, sex steroids, and hexachlorophene (163). Its concentration in cytosol, ~2% of cytosolic protein, is modulated by diurnal rhythm, sex steroid hormones, dietary fat, and hypolipidemic drugs (24, 25, 86, 163, 210, 327).

It has been advocated that ligandin and hFABP are involved in the primary uptake of organic anions from blood plasma into the liver (181, 219, 364). This idea was based on more indirect evidence from competitive, ontogenetic, and phylogenetic studies, as well as the influence of phenobarbital and other drug-metabolism inducers (181, 366). A facilitating influence could be imagined if dissociation from the supposed carrier sites at the inside of the cell would be rate limiting in the membrane translocating mechanism (219, 330). However, the apparent lack of correlation between levels of ligandin and uptake rate under various experimental conditions negated this idea (44, 160, 172, 210). Kinetic studies using two-compartment kinetic analysis (210) and multiple-indicator dilution techniques (366) in isolated perfused livers indicated an influence of the cytosolic protein on hepatic efflux rather than on influx of organic anions. A similar effect was suggested for Z protein as induced by the hypolipidemic agent nafenopin (210) after a relatively large dose of DBSP but not at much lower test doses of BSP and bilirubin (327). These findings agree with the observation that binding of organic anions to the Z fraction is only appreciable at a relatively high liver content (181, 210). The influence of intracellular proteins and extracellular plasma proteins on the efficiency of carrier-mediated transport deserves further study in experiments with isolated membrane vesicles or with liposomal systems in which transport can be reconstituted with purified carrier proteins (316).

Bile Flow

In many studies it was observed that the excretion rate of drugs may vary with bile flow. In particular, bile acids may stimulate the biliary output of cholephilic anions such as BSP (31, 60, 172, 182), and this effect is marked only at relatively high doses of the organic anion (343). Although these observations have not been adequately explained, a number of mechanisms have been proposed and partly tested, especially for organic anionic compounds.

1. Bile acids stimulate hepatic uptake of substances such as BSP (192).
2. Bile acids displace BSP from extracellular or intracellular binding sites and thereby, in non-steady-state conditions, increase the driving force for excretion (60).
3. Bile acids, by their choleretic effect, dilute drugs in the canalicular lumen and prevent backdiffusion from the canalicular lumen to the cell (increased net transport) or inhibit backdiffusion by competition for supposed carrier systems involved in lumen-to-cell transport (339).
4. Bile acids form mixed micelles (275) that bind amphipathic drugs and increase net transport across the canaliculus [biliary micellar sink theory; (290)].
5. Bile acids improve solubility of drugs in bile and prevent precipitation in the canalicular lumen (343) or decrease toxic influences of organic anions on mitochondrial metabolism (107) or canalicular membranes (212).
6. Bile acids in high doses may recruit pericentral hepatocytes in the excretion of drugs (101, 109).
7. Bile acids may change the function of the transport carriers, for instance by increasing the affinity for the carrier protein (allosteric effects) or influencing the membrane environment (88).

None of these mechanisms has been fully excluded. It is very likely that, depending on the dose (78, 339) and the physicochemical properties of the category of the drug under study as well as on the nature of the bile acid used (16, 74, 101, 343); more than one of these mechanisms may contribute to the observed stimulatory but variable effects of bile acids (107, 172, 339). The overlapping substrate specificity of the bile acid–carrier system and other carrier systems (see *Organic Anions*, p. 742) may partly explain the variable effects (Table 2) of different bile acids on the excretion rate of various organic compounds tested under similar experimental conditions (211, 239, 339–345). For some organic compounds bile salts are without any effect (32, 107, 218, 244) or even inhibit biliary output (339–344, 353). A lack of effect has also been reported for various potent non–bile salt choleretics (78, 88, 165, 205, 339).

TABLE 2. *Choleresis and Biliary Excretion of Drugs*

Drug	Taurocholate, 106 µmol/h	Dehydrocholate, 106 µmol/h	Biliary Micelle Binding
Uncharged compounds			
Ouabain	0	--	+
K-strophanthoside	0	0	+
Organic cations			
APAEB	±	0	+
d-Tubocurarine	0	0	+
Tributylmethyl-ammonium	+	+	+
Organic anions			
Indocyanine green	++	+	+
Dibromosulfophthalein	+	++	+
Rose bengal	+	+	+
Phenolphthalein glucuronide	0	0	+
Bile acids			
Taurocholate	*	0	+

APAEB, 4N-acetyl derivative of procaineamide ethobromide. 0, No effect; ±, variable effect; +, stimulation; ++, strong stimulation; --, strong inhibition.   * Not studied.

Competitive inhibitory effects of such choleretics and/or their acidic metabolites (78, 87, 205, 222) may partly mask a potential stimulatory effect on biliary excretion in the cases in which their effect on biliary excretion rate of anionic drugs was studied. Nevertheless the marked elevation of bile flow by such non–bile salt choleretics generally does not lead to any change in the biliary excretion rate of drugs, an observation that seems to rule out bile flow per se as a major determinant in stimulation of anion transport into bile. Other factors that may explain a lack of effect of choleresis on excretion rate include a relatively low canalicular concentration (339), as in the case of the toxic (low-dose) cardiac glycosides (345), or the fact that the hepatic uptake step rather than the biliary excretion step is rate limiting in hepatobiliary transport, as has been shown for some (342) but not all (239) organic cations.

The influence of bile acids on excretion of drugs may have practical consequences for elimination of drugs during bile salt depletion or diagnostic procedures in humans. For DBSP it was shown that excretion into bile is depressed in postcholecystectomy patients with bile drainage, compared with controls (213). In contrast to fed individuals (31), fasted volunteers did not show gallbladder opacification with iopanoic acid, which was suggested to be due to bile acid storage in the gallbladder preventing an effect of the bile salt on absorption and excretion of the cholecystographic agents.

Adaptive Changes

Hepatic clearance of drugs can be largely influenced by long-term changes in physiological conditions or by persistent exposure to pharmacological agents. The latter includes the influence of agents that induce hypertrophy and/or hyperplasia of liver tissue, such as microsomal enzyme inducers and certain hypolipidemic agents (Table 3). In addition partially hepatectomy, selective biliary obstruction, liver regeneration, and chronic exposure of the liver to cholephilic substrates can lead to adaptation of the hepatic transport system (172, 182–184). Such manipulations are inherently aspecific, and many potential rate-limiting factors in hepatic transport can be affected. For instance, enzyme inducers such as certain long-acting barbiturates may at the same time affect hepatic blood flow, body temperature, and nervous control and induce cytosolic binding proteins as well as drug-metabolizing enzymes involved in drug oxidation and conjugation. At the same time they can cause extra bile flow and also may stimulate membrane transport itself by promoting synthesis of carrier proteins (307). The time sequence of these various changes after exposure to such agents can be very different (172). Duration of pretreatment, the type and dose of the inducing agents, the dosage regimen, and the category of the transported drug also influence the outcome of the interaction studied. For example, within the group of organic anions, phenobarbital pretreatment exerts much more effect on hepatobiliary transport of BSP and DBSP than on that of ICG, phenol red, rose bengal, probenecid, taurocholate, and succinylsulfathiazole (107, 172), pointing to multiplicity in organic anion-transport mechanisms (see *General Considerations*, p. 739).

Despite the difficulties in comparing and interpreting the many studies involved, a number of possible mechanisms can be proposed.

1. All of the compounds listed in Table 3 increase liver weight. Many of the observed effects on hepatic uptake rate or biliary clearance of drugs are parallel to this change: transport expressed per unit of liver weight is often not affected. This observation, however, does not necessarily imply that the induced extra liver mass is functional with regard to drug transport.

2. Most of the compounds listed increase total hepatic blood flow but not blood flow per unit of liver weight (196). This increase in blood flow may explain many of the observed changes in initial plasma-disappearance rate, reflecting more rapid distribution to the liver. This is especially important for drugs with a high initial extraction ratio (196, 210).

3. Besides proliferation of endoplasmic reticulum or other cell organelles, these agents generally increase the cytosolic concentration of drug-binding proteins (Table 3) and thus hepatic storage of many anionic

TABLE 3. *Pharmacological Agents Affecting Hepatic Clearance*

	Dose, mg/kg	Liver Weight	Blood Flow	Bile Flow	Cytochrome P-450	UDPGT	Ligandin	Z Protein
Phenobarbital	50–90	+	+	++	+	+	++	0
3-Methylcholantrene	20–40	+	?	0	+	+	+	0
Spironolactone	75	+	?	+	+	+	+	0
Pregnolone-16α-carbonitril	75	+	?	++	+	+	+	+
Clofibrate	200	+	0	+	±	++	0	+
Nafenopin	200	++	0	++	0	0	0	++
Trans-stilbene oxide	200–400	+	?	+	0	++	+++	0

UDPGT, UDP-glucuronosyltransferase; 0, no effect; ±, slight effect; +, increased; ++, strong increase; +++, very strong increase; ?, not known.

model compounds, a factor that adds to the influence of the elevated liver volume (204, 210, 364).

4. The increased bile flow may increase the net flux of organic compounds into bile at high doses (see *Bile Flow*, p. 724). 3-Methylcholantrene, which induces no choleresis, has no effect on the biliary excretion of organic anions (172).

5. The inducing agents may stimulate synthesis or inhibit degradation of functional carrier protein or alternatively may directly or indirectly change the membrane environment of the carrier and improve its translocating function (203, 307). Such an effect may underlie the reported increase in bile acid transport by pretreatment with phenobarbital (304) and stimulation of ouabain excretion by pretreatment with pregnenolone-16α-carbonitril (PCN) (167, 168).

Two-thirds hepatectomy or selective biliary obstruction in the major liver lobes does not lead to a proportional decrease in hepatobiliary clearance of various organic anions (3, 166, 335). This may partly be due to the relatively higher supply of the bile acid pool to the remaining functional cells. If bile salts are depleted prior to the experiment, the increase in transport of organic anions on a liver weight basis is much less increased (335). Another factor may be the rapid cell reproduction in the periportal acinar zone that may have higher transport capacity (112, 335).

Adaptation of hepatobiliary transport after exposure to abnormally high concentrations of bile salts has been reported in studies with intraduodenal infusion (307), repeated oral administration (348), and selective biliary obstruction (3). Such phenomena were explained by an increased synthesis of bile acid–transport proteins (307), a phenomenon that can be blocked by cycloheximide. Such adaptive increases in carrier proteins have not been demonstrated for non–bile salt organic anions (83).

It may be concluded that the hepatic disposition of drugs may adapt in situations of extra demand and that this process is under the influence of many factors. Such adaptive mechanisms require further study, especially with regard to the molecular mechanisms at the level of protein synthesis and membrane transport.

Biotransformation, Detoxification, Toxification

GENERAL CONSIDERATIONS. Hepatic clearance, defined as a process for irreversible elimination of drugs from the body, may include both biliary excretion and metabolic conversion of a drug. The latter process is often called drug metabolism or biotransformation (Fig. 6). Although drug metabolism in the liver is predominantly localized in the hepatocytes, the nonparenchymal endothelial and Kupffer cells contain low but inducible amounts of drug-oxidation enzymes. Specific activity of drug conjugating enzymes in these sinusoidal cell types is only one-third of that in hep-

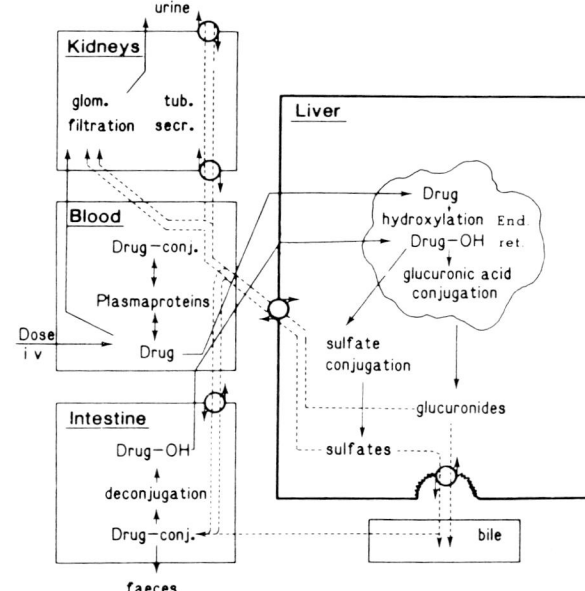

FIG. 6. Schematic representation of drug disposition by the liver. Hepatic uptake can occur passively or by carrier transport. Phase I and phase II biotransformation may produce polar conjugates that can be excreted into the general circulation or into bile, both via carrier-mediated transport. Conjugates or polar drugs may also be excreted via tubular secretion in the urine or via intestinal secretion. In the gut, conjugates can be deconjugated and reabsorbed, undergoing enterohepatic cycling.

atocytes (193). Besides the liver, the kidney, intestines, and lungs can significantly contribute to drug metabolism (267). Although their metabolic capacity is generally lower than for liver, the driving force for metabolic conversion can be relatively high because of local absorption or accumulation by active transport.

Biotransformation is generally seen as a detoxification process: less active (321) and less toxic compounds are formed. However, it has become clear in the last decades that drug biotransformation in some cases can result in formation of highly toxic products (115, 140, 321) and also that excretion of active drugs and metabolites in the bile may have serious consequences for local toxicity in the gastrointestinal tract (57, 65, 107).

PHASE I AND PHASE II BIOTRANSFORMATION REACTIONS. The metabolism of drugs can be viewed as a series of steps by which drugs are "prepared" for excretion. This may include oxidation, reduction, and hydrolytic reactions (phase I reactions), which often produce more reactive species of the drug, as well as synthetic reactions such as conjugation of the drug (259), or their phase I metabolites with glucuronic acid, sulfuric acid, or glutathione, by which hydrophilic groups are introduced (phase II reactions). Without such metabolic processes, lipophilic drugs would tend to have a very long residence time in the body. Only passive diffusion from blood across the gut mucosa and subsequent binding to intestinal contents (145) or aspiration of very volatile drugs via the lungs would be alternative elimination pathways.

Most of the phase I and some of the phase II enzymes are embedded in the lipid environment of the smooth endoplasmic reticulum (SER; see Fig. 9). Therefore the lipophilicity of the drug and its degree of ionization are important for access to the metabolic sites. Significant correlations have been reported for various classes of drugs between lipid-water partition coefficients of drugs and the rate of metabolic conversion. The most important biotransformation processes are depicted in Table 4, with special reference to the functional groups involved. The consequences of biotransformation of some selected drugs that are of special interest to gastroenterologists and hepatologists are discussed next.

PHASE I REACTIONS. *Drug oxidation by mixed-function oxidase.* Oxidation of drugs implies the loss of electrons at a carbon center of the drug molecule. Often a more electronegative O atom is introduced by which hydrophilicity (solubility in water) increases. This tends to reduce the distribution volume in the body and improve elimination by membrane transport or further metabolic attack. The monooxygenases cat-

TABLE 4. *Biotransformation Reactions of Xenobiotics*

Reaction	Enzymes Involved	Localization	Substrates/Functional Groups
Chemical transformations (phase I reactions)			
Oxidation			
Hydroxylation; N-, O-, and S-dealkylation; epoxidation; S- and N-oxidation; desulfuration; dehalogenation	Monooxygenases	Endoplasmic reticulum	Alkanes, alkenes, arenes, amines, thiones, ethers, alkylhalogenides, thioethers
Deamination	Monooxygenases, monoamine oxidases (MAO)	Endoplasmic reticulum, mitochondria	Amines
Alcohol oxidation	Monooxygenases, alcohol dehydrogenases	Endoplasmic reticulum, cytosol	Alcohols
Aldehyde oxidation	Aldehyde dehydrogenases, aldehyde oxidases	Cytosol	Aldehydes
Aromatization	Dehydrogenases	Mitochondria	Cyclohexanes
Reduction			
Nitro- and azoreduction, reductive dehalogenation, sulfoxide reduction	Monooxygenases, azoreductase, N-oxide reductases, sulfoxide reductases	Endoplasmic reticulum, mitochondria, Cytosol	Azo- and nitro- groups, arene oxides, alkyl halogenides, N-oxides, sulfoxides
Alcohol, ketone, and aldehyde reduction	Alcohol dehydrogenases, carbonyl reductases	Cytosol	Alcohols, ketones, aldehydes
Hydrolysis			
Cleavage of ester and amide groups	Esterases, amidases	Endoplasmic reticulum, mitochondria, cytosol	Esters, amides
Conjugations (phase II reactions)			
With H_2O	Epoxide hydrolases	Endoplasmic reticulum, cytosol	Epoxides
With glutathione	Glutathione S-transferases	Cytosol	Electrophiles
With glucuronic acid (UDPGA)*	UDP-glucuronosyltransferases†	Endoplasmic reticulum	OH, COOH, NH_2, NH, SH, aromatic OH, aromatic NH_2, same alcohols
With methyl group (SAM)*	S-, N-, and O-methyltransferases	Endoplasmic reticulum, cytosol	Aromatic OH, NH_2, NH, N, SH
With acetic acid (acetyl-CoA)*	N-acetyltransferases	Cytosol	Aromatic NH_2, some aliphatic NH_2, hydrazides, SO_2NH_2
With sulfuric acid (PAPS)*	Sulfotransferases	Cytosol	Aromatic OH, some aromatic NH_2
With amino acids (CoA)*	Acyl-CoA ligase	Mitochondria	Amino acids
Glycine	Glycine transferases	Endoplasmic reticulum, cytosol, lysosomes, mitochondria	Aromatic COOH, aromatic alkyl COOH
Glutamine‡, ornithine§, taurine	Amino acid acyltransferases	Mitochondria, possibly other cell compartments	COOH

* Abbreviations in parentheses are cosubstrates: UDPGA, uridine-5′-diphosphoglucuronic acid; PAPS, 3′-phosphoadenosine-5′-phosphosulfate; SAM, S-adenosylmethionine; CoA, coenzyme A.   † Deficient in pigs.   ‡ Humans and certain monkeys only.   § Certain birds and reptiles only.

alyze the following reaction: $XH + O_2 + NADPH + H^+ \rightarrow XOH + H_2O + NADP^+$. The cofactor nicotinamide adenine dinucleotide phosphate in its reduced form (NADPH) is produced in the cell via the mitochondrial citric acid cycle, fatty acid oxidation, and the pentose phosphate pathways. Under certain conditions the supply of NADPH may become rate limiting. Other cellular reactions that compete for the reduced cofactor such as fatty acid synthesis and reduction of oxidized glutathione, as well as the general oxidation-reduction state of the cell, influence this type of drug oxidation. NAD^+ is required for synthesis of activated glucuronic acid and ATP for the synthesis of activated sulfate. Thus both are cofactors in the conjugation of hydroxylated drugs formed by the cytochrome P-450 system. It is likely therefore that the various drug-metabolism steps are intimately linked to intermediary metabolism and in various zones of the liver acinus are under a coordinated control (328).

Oxidation of drugs via the mixed-function oxidation system (Fig. 7) occurs in a sequence of events: the oxidized cytochrome (Fe^{3+} form) binds the drug and a pair of electrons arises from the oxidation of the pyridine ring of NADPH. The flavoprotein NADPH cytochrome reductase transduces these electrons one at a time to the drug–cytochrome P-450 complex, by which the iron atom is reduced ($Fe^{3+} \rightarrow Fe^{2+}$), which form then binds a molecule of oxygen (O_2). Acceptance of another electron and rearrangement of charges leads to an intermediary complex with O_2^{2-} and Fe^{3+} atoms. Subsequently the O_2^{2-} is split and degradation of the cytochrome complex leads to transfer of one oxygen to the drug molecule and reduction of the other oxygen to water, using one proton and the H atom from NADPH. In this process the oxidized (Fe^{3+}) form of the cytochrome P-450 is released again and the cycle can restart (337). Thus a pair of electrons is supplied not to reduce the drug but to reduce molecular oxygen to H_2O, making one oxygen atom available for drug hydroxylation (Fig. 7). The particular enzyme system therefore is often called mixed-function oxidase, because at the same time it converts one oxygen to water (oxidase reaction) while the other is incorporated in the substrate (oxygenase). Another (trivial) name is the cytochrome P-450 system; however, cytochrome P-450 is only one of the enzymes involved.

The metalloprotein cytochrome P-450 is embedded in the SER combined with another hemoprotein cytochrome b_5 and a number of other enzymes, e.g., the flavoproteins NADPH–cytochrome b_5, NADPH–cytochrome P-450 reductase (involved in phase I reactions), as well as epoxide hydrolase and UDP-glucuronosyltransferase (17), representing some of the phase II metabolic processes. This series of proteins is supposed to represent an "assembly line" by which drugs are sequentially metabolized by the particular phase I and phase II enzymes without the intermediates leaving the proximity of this lipid complex. Removal of phase I products by phase II reactions very likely improves efficiency of the phase I metabolic processes such as drug hydroxylation. In addition to the oxygenase activity, the cytochrome P-450 can act as a (per)oxidase and as a (per)oxygenase. During the cycle of oxygen activation by the cytochrome P-450 system, instead of oxidizing a substrate, the enzyme may release the superoxide anion O_2^- or H_2O_2, both of which are very reactive species (see Fig. 7). However, under normal conditions the cellular levels of these toxic compounds are kept low by the protective enzymes superoxide dismutase, catalase, and glutathione (GSH) peroxidase. The latter enzyme during this detoxification reaction produces the oxidized (dimeric) form of GSH, which can be considered a glutathione conjugate of glutathione (GSSG). This product is efficiently secreted from hepatocytes into bile, an event that may reflect oxidative stress of the cells (305).

The hemoprotein cytochrome P-450, in the CO-reduced form, has a characteristic band at 450 nm, while in the oxidized form the absorption maximum is at 418 nm. The first step in metabolism (e.g., the binding of the drug to the protein) can be observed spectrally. By the technique of difference spectrophotometry, adding various drugs to suspended microsomes, the spectra between 360 and 460 nm reveal characteristic changes that reflect different spin states of the heme iron of the cytochrome P-450. Three fundamentally empirical types of spectra have been characterized. Type I spectra are produced by the binding to a hydrophobic region of the enzyme by drugs with mostly lipophilic moieties, which produces

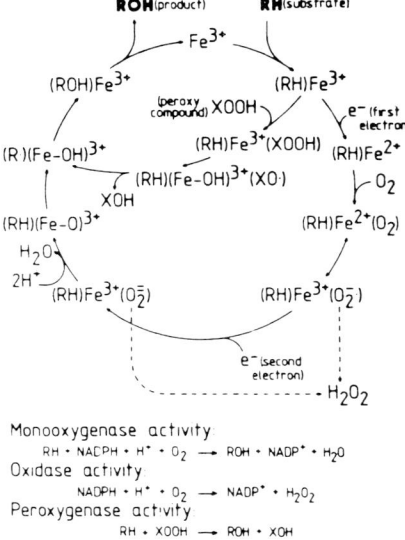

FIG. 7. Proposed scheme of the cytochrome P-450 oxidation-reduction cycle representing the monooxygenase, oxidase, and peroxygenase function. RH, xenobiotic compound; ROH, hydroxylated product. XOOH, formation of various peroxy compounds, including hydrogen peroxide.

a shift from 418 to 391 nm. Type II spectra are produced by drugs with an amine function such as cimetidine, showing a shift to longer wave lengths because of two-point attachment of the hydrophobic region and interaction of the lone pair of electrons of the N atom directly with the cytochrome iron atom. Type III spectra are a reversed type I spectrum due to interaction of an O atom in the drug with the heme iron directly. For instance, toluene, benzylamine, and benzyl alcohol produce type I, type II, and reversed type I spectra, respectively. Complex molecules possessing more than one of these features may give mixed types of spectra.

The activity of the cytochrome P-450 system can be pharmacologically reduced by various agents that in some way attack the heme iron. Cobalt, for instance, inhibits heme synthesis and also induces its degradation (337). Drugs with allyl or alkinyl groups such as secobarbital and 17α-ethinylestradiol form irreversible (covalent) adducts with cytochrome P-450 after oxidation of these side-chain groups and can be considered "suicide inhibitors." Amphetamines and other drugs with a tertiary amine function such as propoxyphene and the well-known experimental tool SKF 525A form reversible adducts with the enzyme after N de-ethylation. Other drugs such as cimetidine and metyrapone that have a remarkable affinity for cytochrome P-450 by themselves decrease the activity of cytochrome P-450 in drug oxidation by competitive binding. Not all cytochrome P-450 activity is necessarily affected by these reactions, because affinity of these compounds for isoforms of this enzyme can be different. Relative inhibition of the mixed-function oxidase system can also occur in vivo because of the presence of endogenous substrates that are converted by the mixed-function oxidase system. Examples are fatty acids, leukotrienes, prostaglandins, hormonal steroids, cholesterol, vitamin D_3, and thyroxine.

Data accumulated over the last decades clearly demonstrate that multiple forms of cytochrome P-450 exist. From livers of rats, mice, and rabbits at least 20 different forms have been isolated and from intestinal material at least 4 (114). These forms differ in substrate specificity, positional specificity, and stereospecificity (333) and have similar molecular weights but different amino acid sequences. This marked enzyme heterogeneity can be seen as a prerequisite to cope with the huge variety of xenobiotics to which the organism is exposed. Because of the subtle differences, purification to homogeneity of the isoforms is a difficult task. Very helpful in this regard is the possibility to induce the various types of cytochrome P-450 with different chemicals (238). For instance, phenobarbital, 3-methylcholantrene, β-naphthoflavone, tetrachlorodibenzo-p-dioxin (TCDD), PCN, cholestyramine, and rifampicin each induce distinct subspecies of cytochrome P-450 (Fig. 8). Enzyme induction implies an adaptive increase in the number of enzyme molecules

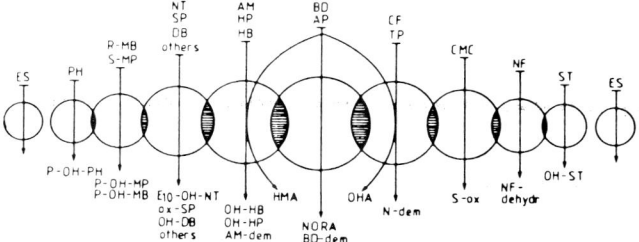

FIG. 8. Graphic representation of the different forms of cytochrome P-450 (circles) in the human with different, but probably overlapping, substrate and product selectivities. Arrows, single metabolic pathways. AM, aminopyrine; AP, antipyrine; BD, benzodiazepine; CF, caffeine; CMC, carboxymethylcysteine; DB, debrisoquine; ES, endogenous substrate; HB, hexobarbital; HP, heptabarbital; MB, methylphenobarbital; MP, mephenytoin; NF, nifedipine; NT, nortriptyline; PH, phenytoin; SP, sparteine; ST, steroid; TP, theophylline.

because of either an increase in the rate of enzyme synthesis or a decrease in its rate of degradation. Both may be related to gene expression during differentiation, development, and individual life as influenced by many physiological and pharmacological stimuli. Hormones such as glucocorticoids, thyroxine, and glucagon are the natural stimuli of many of the changes during fetal and postnatal development (237, 362). For such compounds specific receptors are present in various cell types. Glucocorticoids and polycyclic aromatic hydrocarbons, for instance, produce an accumulation of specific messenger RNA for the induced enzyme. This occurs via binding of the compounds to a cytoplasmic receptor, which is transferred to the nucleus, where binding to DNA occurs, whereas phenobarbital enhances mRNA transcription. The involvement of protein synthesis in these inductive mechanisms explains the blocking effect on induction of ethionine, cycloheximide, and actinomycin D, substances that inhibit protein synthesis at different levels (237, 238). Inducing agents themselves have a characteristically slow rate of elimination and may accumulate after multiple dosing or may persist in the body several weeks after exposure. This may not only lead to long-lasting induction but also may prevent binding of the endogenous (physiological) ligands previously mentioned.

Genetically determined or induced differences in the spectrum of the different isoforms of cytochrome P-450 may affect the metabolic profile of drugs, i.e., the relative amounts of various metabolites arising from one drug. Model substrates can be developed for which clearance is typical for the function of a certain form of cytochrome P-450 (monofunctional substrates) (Fig. 8). Relatively nontoxic substrates in the radioactive or stable isotope forms, such as aminopyrine and caffeine, which have minimal protein binding, negligible extrahepatic elimination, blood flow–independent clearance, and the possibility to measure clearance noninvasively via saliva or expired CO_2

(breath tests), are used to quantify the rate-limiting oxidative demethylation process. Caffeine demethylation, hydroxylation of zoxazolamine, and N-oxidation of sparteine are probably selectively mediated by cytochrome P-450c (also called P-448 and induced by 3-methylcholantrene), P-450b (phenobarbital inducible), and P-450h, respectively (Fig. 8). Antipyrine can be used as a polyfunctional substrate to characterize the different cytochrome P-450 isoenzymes involved in its metabolism by measuring the related metabolites excreted in the urine (48). In the human, cytochrome P-450 may even be more polymorph than in inbred strains of laboratory animals. In 5%–10% of the population certain metabolic pathways may be almost absent. This has been demonstrated for drugs such as sparteine, mephenytoin, debrisoquine, and nifedipine (48). Such drugs can be useful as model compounds to determine individual drug metabolism characteristics or to establish toxicity risk factors in drug use after exposure to pollutants, food products, and other environmental factors.

Other oxidative processes. In addition to the cytochrome P-450 mixed-function oxidase system, a number of other processes can mediate oxidation of substrates. The microsomal flavoprotein (FAD-containing) monooxygenase catalyzes hydroxylation of secondary amines and can convert tertiary amines to N-oxides. Mitochondrial monoamine oxidase (MAO) oxidizes primary amines such as adrenergic compounds. The latter enzyme is strongly inhibited by certain antidepressant drugs (MAO inhibitors). A serious side effect of such drugs was intoxication with tyramine-containing food. Cytosolic diamine oxidase breaks down histamine, whereas the cytosolic NAD^+-linked alcohol dehydrogenase catalyzes the oxidation of primary and secondary alcohols. Along with aldehyde oxidase, NAD^+-linked aldehyde dehydrogenase converts aldehyde to carboxylic acids. It is inhibited by disulfiram (Antabuse), a drug that is used to discourage ethanol consumption through accumulation of the toxic acetaldehyde.

Reduction of drugs. From the role of the flavoproteins NADPH–cytochrome P-450 reductase and NADH–cytochrome b_5 reductase as electron distributors it can be inferred that, along with various mitochondrial and cytosolic reductases (Table 4), these enzymes can also be potentially involved in reduction of certain functional groups in drug molecules. Such groups include nitro-, keto-, azo-, hydroxylamine, nitroso-, quinone, and alkylhalide groups. Examples are reduction of the NO_2 group in nitrofurantoin and chloramphenicol to an NH-OH group, reduction of the quinone structure of doxorubicin to a toxic intermediate, and removal of halogen atoms from halothane producing highly reactive radical species (Table 5). In most of these reactions cytochrome P-450 is also involved and oxidation-reduction cycles of drugs can therefore be imagined, depending on the O_2 concentration of the particular tissue involved. Especially in pericentral cells, the cytochrome system could operate in a reductive mode, because at a relatively low O_2 level the reducing equivalents are not used to activate O_2 but rather to reduce the drug.

Variability in phase I metabolism. Drug metabolism may be influenced by various factors, including liver mass and blood flow, genetic factors, sex, age, diet, autoinduction or use of other drugs, intake of alcohol, smoking habits, exposure to environmental pollutants, and pathological conditions (5, 337). Consequently variations of more than a factor of 10 may be present among individuals in clearance function for a certain drug, implying that at a fixed dosage regimen, average steady-state plasma concentration within a population of patients may vary with the same factor. Variations in cardiac output and changes in total hepatic blood flow are especially important if intrinsic drug clearance in the liver is high and blood flow is rate limiting. Certain β-blocking drugs consequently can affect their own clearance rate or the clearance of concomitantly given drugs with a high intrinsic clearance through a decrease in cardiac output (357). The carbohydrate-protein ratio of a diet as well as the total caloric intake can markedly influence the rate of drug oxidation, as shown for theophylline and antipyrine in animals and humans. Polycyclic aromatic substances can strongly induce drug oxidation locally in the mucosal cells of the respiratory and gastrointestinal tract. This may have major consequences not only for first-pass clearance in intestine and liver being one factor determining the bioavailability of drugs (134, 357) but also for local toxicity and carcinogenesis (57, 140).

Phase I and also phase II metabolism steps of a certain drug can be induced by the presence of other drugs, the drug itself, or, for instance, chronic alcohol consumption. Long-acting barbiturates, for example, induce the hepatic clearance of warfarin (5). A standard-dose regimen of the latter drugs can lead to underdosing so that the anticoagulant effect, as indicated by routine anticoagulant blood tests, is insufficient. In contrast, bleeding may occur if the dose requirement is adjusted but the intake of the particular barbiturate is discontinued. The anti-infective rifamycins may induce metabolism of estrogenic as well as progestogenic components of oral contraceptives, leading to relative underdosing and risk of unwanted pregnancy. Industrial chemicals such as polychlorinated hydrocarbons [e.g., polychlorinated biphenyls (PCBs)] may induce drug oxidation in liver and intestine, as may polycyclic hydrocarbons from tobacco smoke or charcoal-broiled meat (5). For drugs with a low margin between therapeutic effects and major side effects, such as the bronchodilatory drug theophylline, this has practical consequences. Induction of its metabolism may imply a 100% shortening of the elimination half-life and therefore the necessity for an impractical frequency of dosing. This kinetic pattern also implies a poor control of blood levels between the therapeutic and toxic ranges.

Inhibition of enzyme activity in the liver belongs to the most dangerous type of drug interaction. Impairment of hepatic drug metabolism by cimetidine, allopurinol, disulfiram, and nortriptyline are pertinent examples of this. For example, inhibition of demethylation or hydroxylation may affect the hepatic clearance of sedative agents, minor tranquilizers, and other psychoactive compounds. This leads to overdosing, and the resulting accumulation of such drugs may greatly affect intellectual performance and reaction time in, for example, traffic situations.

Biotransformation of drugs leads to detoxification, in the case that less active or even inactive products are formed. For example, hydroxylation of an aromatic ring yields products with higher hydrophilicity and often (but not always) less affinity for the particular pharmacological receptor. As previously discussed, the hydroxyl group is prone to further metabolism via phase II conjugation reactions.

PHASE II REACTIONS. Phase II biotransformation constitutes synthetic reactions in which the drug itself or its phase I metabolite(s) is conjugated. Such reactions are catalyzed by various transferases and require activated (mostly nucleotide) forms of endogenous cosubstrates such as glucuronic acid, sulfuric acid, acetic acid, and glucose or occur by prior activation of the substrate itself (see Fig. 9). These conjugation reactions can take place if the parent drug or its phase I metabolites contain a suitable functional group (52). Examples are phenolic, alcoholic, or carboxylic hydroxyl groups; nitrogen in cyclic structures or in amine groups; and thiol groups (see Table 4). The conjugation reactions can also be crucial in the detoxification of reactive metabolites generated by phase I reactions, such as carbonium ions and epoxides. If such activated exogenous compounds are formed, conjugation can even occur with nonactivated endogenous substrates such as glutathione or amino acids. Because both phase I and phase II reactions in principle have a saturable character, the capacity of the conjugation systems relative to the rate of formation of a reactive metabolite is a major factor in determining the harmful effect.

By conjugation, drugs and metabolites not only become less active but also are prepared for more efficient elimination. The greater rate of elimination of the more water-soluble drug conjugate compared with the drug itself can be related both to a smaller distribution volume and a more efficient clearance process. Conjugation with glucuronic or sulfuric acid, for instance, may lead to a higher affinity for the carrier-mediated anion-transport processes in kidney and liver. The relative contribution of the different conjugation reactions to the biotransformation of a drug depends on numerous factors: the affinity of the substrate for the particular transferase, the capacity of the system in relation to the dose of the drug administered, the acinar and intracellular localization of the process (329), the presence of competing endogenous substrates, the availability of the conjugating cosubstrates, and the extent of deconjugation that may eventually occur in the intestine and other parts of the body (52, 334).

The various types of drug-conjugating systems are briefly dealt with next. Detailed information can be found in recent reviews or monographs on glucuronidation (67, 195, 280), sulfation (52, 227, 228), acetylation, methylation and glycosylation (52), and glutathione and amino acid conjugation (158, 216, 217, 246).

Glucuronidation. Glucuronide formation is a quantitatively important and versatile conjugation reaction effected by UDP-glucuronosyltransferase (commonly abbreviated to glucuronyl transferase). The enzyme is associated with the endoplasmic reticulum and transfers glucuronic acid from the activated nucleotide UDP-α-1-glucuronic acid to nucleophilic atoms (O^-, S^-, C^-, N^-) to form β-1-glucuronides. Aryl acetic acid analgesics such as indomethacin are conjugated to acyl glucuronides in which the carboxylic hydroxyl of glucuronic acid is linked to the carboxylic acid group of the drug via an ester linkage. Such metabolites are highly unstable at pH > 6.0, and partial decomposition to the parent drug in alkaline urine may lead to gross overestimation of the fraction of the drug excreted unchanged (52). More common is the formation of ether glucuronides in which a hydroxyl group of the drug is linked to the hydroxyl group of glucuronic acid, forming quite stable conjugates that in the body can only be split by the hydrolyzing enzyme β-glucuronidase. The latter reaction is a major factor in deconjugation in the gut. Only primates may form N-glucuronides (e.g., from sulfonamides) and C-glucuronides (e.g., phenylbutazone).

Cats and related feloidea have a glucuronidation defect especially for relatively small water-soluble drugs. Also intraspecies differences are known, such as the major defect in Crigler-Najjar syndrome and relative defects in Gilbert disease in humans as well as the complete absence of glucuronidation in the well-known mutant animal model of the Gunn rat. For some substrates, such as 4-methylumbelliferone, the glucuronidation reaction is in competition with sulfation: the fraction of the dose converted to sulfate or glucuronide conjugates is therefore species dependent (67, 228). Glucuronidation can be considered a high-capacity low-affinity system, whereas sulfation is more easily saturated and is quantitatively commonly more important at low concentration (52, 227). The two systems also differ in intracellular localization (see Table 4; Fig. 9).

The cosubstrate UDP-glucuronic acid (UDPGA) is produced from UDP-glucose by the enzyme UDP-glucose dehydrogenase. This enzyme is strongly inhib-

732 HANDBOOK OF PHYSIOLOGY ~ THE GASTROINTESTINAL SYSTEM III

TABLE 5. *Reactive Intermediates of Drugs and Other Xenobiotics*

Compound	Structural Formula	Proposed Reactive Intermediate	Enzymes Involved	Toxicity
Acetaminophen			Cytochrome *P*-450	Hepatic necrosis, renal necrosis
Benzo[*a*]pyrene			Cytochrome *P*-450 and epoxide hydratase	Carcinogenesis
Bromobenzene			Cytochrome *P*-450	Hepatic necrosis, lung necrosis
Carbon disulfide			Cytochrome *P*-450	Hepatic necrosis
Carbon tetrachloride			Cytochrome *P*-450	Hepatic necrosis, renal necrosis
Chloramphenicol			Cytochrome *P*-450	Blood dyscrasias
Chloroform			Cytochrome *P*-450	Hepatic necrosis, renal necrosis
Dimethylnitrosoamine			Cytochrome *P*-450	Carcinogenesis
Furosemide			Cytochrome *P*-450	Hepatic necrosis, renal necrosis

Compound	Structure	Metabolite	Enzyme	Toxicity
Halothane	F H F—C—C—Cl F Br	F H F—C—C—Cl • F	Cytochrome *P*-450	Hepatic necrosis
Isoniazid	H₂N—NH—C(=O)—(pyridine)	O=C⊕—CH₃ (approx. CH₃—C⊕=O)	Transacetylase and cytochrome *P*-450	Hepatic necrosis
Iproniazid	CH₃ H—C—NH—NH—C(=O)—(pyridine) CH₃	CH₃ H—C⊕ CH₃	Cytochrome *P*-450	Hepatic necrosis
N-acetylaminofluorene	NH—C(=O)—CH₃ on fluorene	⊕N—C(=O)—CH₃ on fluorene	Cytochrome *P*-450 and sulfotransferase	Hepatotoxicity, carcinogenesis
Paraquat	CH₃—N⁺(pyridinium)—(pyridinium)—N⁺—CH₃	CH₃—N(pyridine)—(pyridyl)—N—CH₃ •	Cytochrome *P*-450 reductase	Lung toxicity
Parathion	(C₂H₅O)₂—P(=S)—O—(C₆H₄)—NO₂	:S:	Cytochrome *P*-450	Hepatic necrosis
Thioacetamide	S CH₃—C—NH₂	O O ⊖S⊖ CH₃—C—NH₂ ⊕	Cytochrome *P*-450	Hepatic necrosis, neoplasia
Xylene	H₃C—(C₆H₄)—CH₃	H₃C—(C₆H₄)—C(=O)H	Cytochrome *P*-450 and alcohol dehydrogenase	Lung toxicity

Adapted from Ingleman-Sundberg (140).

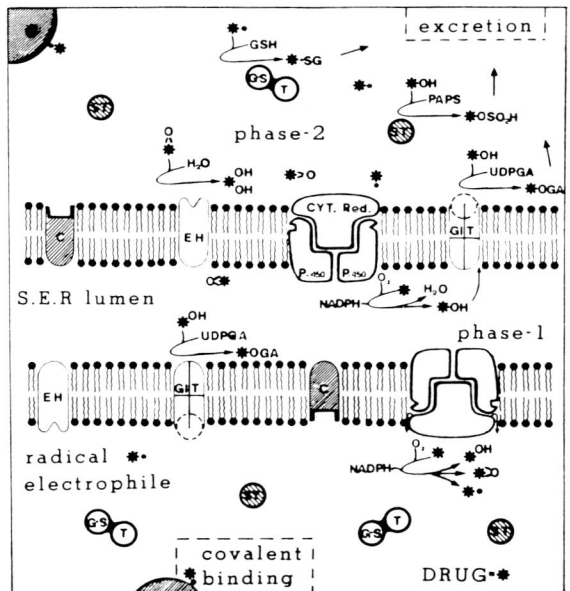

FIG. 9. Schematic representation of interrelation of phase I and phase II biotransformation reactions. Smooth endoplasmic reticulum (SER) membranes contain NADPH–cytochrome P-450 reductase (Cyt. Red.) closely associated cytochrome P-450, UDP-glucuronosyltransferase (GlT), epoxide hydratase (EH), and a supposed permease (carrier, C) for UDPGA transport. Products of the monooxygenase are hydroxylated drug, drug epoxides, and reactive drug metabolites (radicals and electrophiles). The latter can undergo covalent binding to macromolecules in cell organelles, cytoplasm, and plasma membranes. Phase II synthetic reactions of, for example, phase I products include conjugation with glutathione (GSH) via the various glutathione S-transferases (GST), sulfuric acid (SO$_3$H) via the sulfotransferase (ST) using 3-phosphoadenosine-5′-sulfate phosphate (PAPS), and glucuronic acid (GA) using UDPGA. Resulting polar conjugates are substrates for excretion mechanisms at sinusoidal and canalicular poles of the hepatocyte.

ited by one of its products, NADH (131). Other NADH-forming reactions, such as oxidation of ethanol, may therefore inhibit this enzyme, and this may explain the decrease of UDPGA levels after ethanol and acetaminophen administration (6, 131). It follows that glucuronidation is very dependent on carbohydrate reserves, and the rate of glucuronidation may vary manyfold in fasted and nonfasted states. The UDPGA concentration in the cell (0.25–0.40 mM) can be increased by pretreatment with microsomal enzyme inducers and decreased by D-galactosamine (135), other drugs that are glucuronidated, and diethylether (58, 63).

The glucuronosyltransferase is embedded in the membranes of the endoplasmic reticulum. The latency of the UDPGA-transferase in microsomal fractions in vitro can be manipulated mechanically or with detergents. This has been attributed to changes in the membrane conformation, which indirectly activate the enzymes, and to a possible internal localization of the active site in endoplasmic reticulum tubuli and microsomes (195, 280). The latter model requires a permease that transports UDPGA into the endoplasmic reticulum lumen (see Fig. 9). The endogenous UDP-N-acetylglucosamine may control the latency in vivo either by allosteric interactions or stimulation of the supposed permease. Selective inhibitors of glucuronosyltransferase are not known; salicylamide is a strong inhibitor; however, it also affects sulfation.

Several isoforms of UDP-glucuronosyltransferase (UDPGT) have been characterized: a group of smaller planar molecules, mostly phenols, is glucuronidated by an isoform that is induced by 3-methylcholanthrene and TCDD and develops in the late fetal period before birth (187, 361). Other isoforms are induced by phenobarbital and develop in the neonatal period (360). They catalyze the glucuronidation of more bulky substrates like bilirubin, 17-OH steroids, 3-OH androgens, and estrone (361). These groups of substrates are probably glucuronidated by different isoforms, although overlapping substrate specificity exists in that, for example, simple phenols are also substrates for the 17-OH steroid isoform (82). The subunit molecular masses of the different isoforms vary from 50 to 56 kDa, and also the amino acid composition varies, although significant homology exists (82). The holoenzymes probably are oligomers consisting of one to four subunits (251, 334).

Glucuronidation is not an exclusive hepatic function; among other tissues it also occurs in kidney and intestine. Glucuronides are excreted via carrier-mediated processes both at the hepatobiliary and renal tubular level. Glucuronides of relatively small hydrophilic compounds are also secreted at the sinusoidal pole of the hepatocyte to the general circulation (see Fig. 6). Deconjugation of drug-glucuronide compounds in the gut is important and may lead to enterohepatic circulation of the drug. Although glucuronidation is commonly a detoxification process, in some cases conjugation can lead to formation of toxic intermediates (228).

Conjugation with glucose is largely analogous to glucuronidation in cellular localization and type of cofactors. It has only incidentally been reported, for example, for bilirubin (295, 355).

Sulfation. Sulfate conjugation is important for relatively small hydrophilic drugs whose intracellular distribution is mainly cytoplasmic (see Fig. 9). This is related to the exclusive localization of the various sulfotransferases in that part of the cell (52). Phenols and alcohols are the major substrates. Sulfate conjugates are chemically stable, strongly ionized half-esters of sulfuric acid, although at pH < 6.0 they are more labile. The sulfate-donating cosubstrate is 3′-phosphoadenosine 5′-phosphosulfate (PAPS), synthesized from ATP and inorganic sulfate. Sulfation is much more sensitive to changes in the ATP/ADP ratio than glucuronidation; for example, uncoupling of oxidative phosphorylation depresses sulfation strongly.

The estimated cytosolic concentration of sulfate and the Michaelis-Menten constant (K_m) of overall sulfation is ~0.5–1.0 mM, implying that changes in serum sulfate may affect sulfation rate in vivo. However, if the sulfate plasma concentration (0.3–2.0 mM) decreases below normal levels, rapid protein catabolism and increased renal tubular reabsorption of sulfate tend to normalize the situation quite rapidly. For instance, depletion of the sulfate pool in vivo with large doses of substrates such as acetaminophen is transient and only moderate (227, 228). Cysteine and methionine are precursors for formation of inorganic sulfate as well as for production of glutathione and taurine. Sulfate absorption from the gut is an additional source. The latter is a carrier-mediated process, like uptake into hepatocytes, and can be inhibited by organic anions (136).

In addition to sulfotransferase activity, the relative PAPS concentration may determine the rate of sulfation in various tissues. Effective sulfation occurs in addition to the liver in intestine, kidney, lung, brain, and human platelets. Estrone and other steroids, bile acids, biogenic amines, and thyroid hormones are endogenous substrates and are probably converted via distinct forms of aryl sulfotransferase (52, 132, 227). Sulfation can be inhibited quite selectively with 2,6-dichloro-4-nitrophenol (DCNP) and pentachlorophenol (PCP) in vitro and in vivo (227, 228). The mechanism of inhibition of these agents is competitive. In contrast to glucuronidation, induction of sulfate conjugation by treatment with barbiturates and 3-methylcholantrene does not occur, although this could be different for membrane-bound sulfotransferases that catalyze sulfation of saccharides and proteins (52). The pig is deficient in sulfation of some phenolic compounds.

Net production of sulfate and also glucuronide conjugates of drugs is also influenced by simultaneous hydrolysis of the conjugates in the liver. The hydrolases sulfatase (both in endoplasmic reticulum and cytosol) and β-glucuronidase (lysosomal) are present in periportal and pericentral regions of the hepatic acinus and in combination with the transferases could provide regulatory futile cycles of conjugation-deconjugation processes. Drugs that stimulate net conjugation can in principle do so by inhibition of the hydrolases (225).

Epoxide hydration. Oxidation of aromatic rings can lead to formation of epoxides (also called oxiranes), binding one oxygen to two adjacent carbon atoms in a chemically reactive tricyclic ring structure. They may spontaneously rearrange to form diols or can undergo enzymatic hydration through catalysis with epoxide hydrolase (also called hydratase or hydrase) (337). Opening of the ring by nucleophilic attack of an H_2O molecule can be viewed as conjugation with H_2O (see Fig. 9), but some consider this a phase I reaction. The hydrolase is found in almost all tissues and especially in the pericentral zone of the hepatic acini. Activity is present both in the SER and cytoplasm. Multiple monomeric forms of 48–56 kDa have been isolated and show overlapping substrate specificity. Cyclohexene oxide is a strong inhibitor of these isoforms. Diol formation generally implies detoxification, but in some cases, as with polycyclic aromatic carcinogens, primarily formed diols are further metabolized to epoxides at adjacent carbon atoms that can only be slowly hydrated and consequently may readily form covalent adducts with proteins and DNA (337).

Acetylation. Acetylation is another important metabolic pathway for drugs with an NH_2 function in amines, amino acid, or sulfonamide groups. The acetyl group is provided by acetylcoenzyme A (acetyl-CoA) via intermediary metabolism, and the conjugation is catalyzed by N-acetyltransferases. The dog and related canine carnivores are unable to acetylate most of the drugs with an amine function. The N-acetyltransferase is a cytosolic 26,500-MW enzyme in humans. In the liver it is present both in parenchymal cells and cell types of the reticuloendothelial system, although this intrahepatic distribution is strongly species dependent. In humans, liver and gut contain most of the activity. The substrate specificity of the enzyme in the various tissues, however, may be different. Harmine is a potent inhibitor of the enzyme. The N-acetyltransferase shows genetic polymorphism in many species, including humans. In the United States the ratio of fast to slow acetylators is close to 1. This phenotype difference has been suggested to be related to the flexibility of the active site, resulting in differences of fitting of planar as opposed to bulky molecules at the active site of the enzyme (52). The polymorphism of the hepatic transferase may have major consequences for drug toxicity. For example, the tuberculostatic drug isoniazid may accumulate in slow acetylators, and this predisposes these individuals to neuropathy. Furthermore, lupus erythematosus may develop in such patients because of deficient acetylation of procainamide. Hepatotoxicity of isoniazid is related to formation of an acetyl derivative (acetylhydrazine) that after hydroxylation is converted to a very reactive intermediate. Rapid acetylators are more susceptible to these types of side effects (140, 337).

Glutathione conjugation. Various inducible forms of glutathione S-transferase catalyze the reaction of electrophilic agents with the free thiol group of the nucleophilic tripeptide (LGlu-LCys-Gly). The electrophilic substrates can in principle also react spontaneously with glutathione. This is especially true for epoxides and reactive N-oxidation products that are formed through phase I reactions. Other substrates are sufficiently electrophilic in themselves to undergo conjugation. Examples are haloalkanes and benzenes (such as 1-chloro-2,4-dinitrobenzene and sulfobromophthalein) or compounds with activated carbon-carbon double bonds (such as ethacrynic acid) and nitroalkanes (such as nitroglycerin).

The dimeric transferases have a cytosolic localization (Fig. 9) and possess binding sites for both glutathione and the electrophilic substrate, bringing both reactants closely together (14, 216, 217, 246, 364). Endogenous substrates include steroids and leukotrienes. The transferase activity can be strongly inhibited by nonsubstrate ligands such as bilirubin, ICG, and DBSP (161). After conjugation the glutamine and glycine moieties can be split off by peptidases such as γ-glutamyltransferase (γ-GT) in kidney and liver and the resulting cysteine group is acetylated, forming mercapturic acid (S-substituted N-acetylcysteines). The latter process is deficient in guinea pigs. Whereas glutathione conjugates are excreted via the bile and not in the final urine, mercapturic acid conjugates of drugs can be eliminated via both bile and urine.

The high concentration of GSH in the cytosol (in the millimolar range) protects cellular organelles from covalent binding of reactive drug metabolites to nucleophilic centers in proteins, DNA, and RNA and thereby prevents mutagenic, carcinogenic, or other harmful processes. This scavenger function of glutathione should also be seen in relation to its defensive role in removal of H_2O_2, which is continuously generated in the cell, through catalysis by glutathione peroxidase. The regulation of hepatic glutathione is a dynamic balance of synthesis, utilization, and transport (158). Liver glutathione content is altered by diurnal rhythm and fasting and probably has a heterogeneous acinar distribution (54, 161). The tripeptide can be depleted with agents such as diethylmaleate, resulting in a loss of protection and increased toxicity of chemicals. A dose of 650 mg/kg of this agent reduces hepatic GSH levels to ~10% within 2 h. The hepatocyte can export glutathione either across the sinusoidal or the canalicular domain of the plasma membrane by carrier-mediated processes (see *General Considerations*, p. 739) but cannot take up the substrate. Maintenance of intracellular glutathione in the liver therefore totally depends on in situ biosynthesis, and this largely determines the hepatotoxicity after exposure to drugs that are converted to reactive intermediates. For instance, high doses of acetaminophen saturate the normal glucuronide and sulfation detoxification pathways and lead, via cytochrome P-450 oxidation, to a reactive N-acetyl[p]benzoquinone (Table 5). Covalent binding of the positively charged intermediate to proteins, DNA, and RNA results in perturbation of cellular Ca^{2+}-homeostasis and finally cell death (246). These events can only be prevented as long as sufficient glutathione is available. Depletion of glutathione cannot be corrected by administration of the tripeptide itself but can be partly counteracted by supplementation of cysteine, methionine, or N-acetylcysteine, or even better by GSH monoesters (216).

Amino acid conjugation. Compounds with carboxylic acid groups (132) can be conjugated with the amino acids glycine and taurine. Also, conjugation with glutamine occurs in primate species (52). Formation of an amine bond between the α-amino group of the amino acid with the carboxyl moiety of the drug can only occur after activation of the carboxylic acid of the drug to a CoA thioester by acyl-CoA ligase, using ATP. Well-known examples are benzoic acid and salicylate, which are converted to hippuric acid and salicyluric acid, respectively, by conjugation with glycine. The amidation should be seen as competitive with glucuronidation, and the relative importance of these pathways depends on the respective affinities of the substrate for UDP-glucuronosyltransferases and the predominantly mitochondrial amino acid N-acyltransferase. Bile acid conjugation with glycine or taurine is catalyzed by enzymes distinct from those conjugating the types of drugs previously mentioned.

BIOTRANSFORMATION OF DRUGS LEADING TO TOXIFICATION. The dual role of drug-metabolizing systems in detoxification of drugs and in the production of very toxic drug metabolites is essential for understanding the phenomenon of hepatotoxicity. Acute cellular necrosis and mutagenic and carcinogenic effects of drugs can be the result of an interaction of nucleophilic cell macromolecules and reactive metabolic intermediates of drugs produced in the liver. These toxic intermediates include generators of oxygen intermediates, free radicals, electrophilic compounds such as certain epoxides, and even drug conjugates (Table 5).

Cytochrome P-450, alcohol dehydrogenase, epoxide hydrolase, sulfotransferases, and glutathione transferases, alone or in combination, can play a role in such bioactivation processes. For instance, the carcinogen 2-acetylaminofluorene is oxidized to an N-hydroxy derivative that is subsequently sulfated. Because the O-sulfate ester moiety is a good "leaving" group, a strongly electrophilic nitrenium ion is generated that can form adducts with DNA and RNA (228).

Other examples of phase II bioactivation include glucuronidation of N-hydroxy aromatic amines involved in bladder and colon cancer, glutathione conjugation of dihalogenated alkanes leading to mutagenic compounds, the previously mentioned epoxidation of diols formed from polycyclic aromatic compounds, and acetylation of isoniazid (140, 228). However, oxidative processes mediated by cytochrome P-450 and dehydrogenases form the most important toxification mechanisms. Well-known examples are the two-electron oxidation of acetaminophen leading to semiquinone radicals, epoxidation of the toxins styrene and vinyl chloride, and radical formation from halothane and carbon tetrachloride [Table 5; (115)].

The localization of the activation process within the cell, the intrinsic reactivity of the intermediates, and the nature and availability of the nucleophilic targets determine the site of covalent binding in the cell (Fig. 9). Export of the activated species to other cells within the organ where activation occurs or even to other tissues is also a potential possibility (57, 107, 115). It follows that local or general toxicity depends on the level of activating enzymes that can be induced by other xenobiotics and by the activity of phase II detoxification processes, some of which are also inducible. However, the final expression of covalent binding to nucleic acids or proteins in cellular damage or mutagenic and carcinogenic effect also depends on the presence of protecting pathways, for instance via DNA repair (140, 238). A number of relevant examples of drug-induced hepatotoxicity are given in Table 5.

DRUG METABOLISM IN LIVER DISEASE. Liver disease includes a wide variety of pathological conditions that, depending on the type and stage of the disease, may affect drug metabolism to a very variable extent. In addition to alterations in enzyme activity in the liver, changes in functional hepatic blood flow due to intra- and extrahepatic shunting (35, 357) and changes in the plasma levels of drug-binding proteins such as albumin (279) and α_1-acid glycoprotein (35) (orosomucoid) can be crucial factors. Concomitant renal disease, altered hormonal status, and accumulation of endogenous substrates or xenobiotic metabolites that are normally removed by the excretory organs also can play a role. Depending on the rate-limiting steps in biotransformation of the individual drugs, these multiple factors can lead to decreased, unchanged, and even increased metabolism. Cirrhosis, including the chronic alcoholic type, may depress phase I cytochrome P-450 activity much more than phase II glucuronidation. This is especially true if large portions of the liver are replaced by fibrous tissue. Either normal cells do not receive sufficient blood because of shunting (intact-cell hypothesis) or the function of individual cells is decreased despite normal blood supply (sick-cell hypothesis) (270, 271). The latter aspect is probably a major factor in chronic forms of hepatitis. In contrast, acute hepatitis can lead to increased hepatic blood flow and increased clearance, especially of drugs with a relatively high extraction ratio.

Acute ethanol exposure may competitively inhibit total cytochrome P-450 activity, although because of chronic intake it may have induced the hepatic content of the enzyme. Cancer of the liver parenchyma leads to formation of hepatoma tissue that by dedifferentiation can be almost devoid of drug metabolism activity. However, cytochrome P-450 activity can be increased in the tissue surrounding the tumor.

RELATIONSHIP BETWEEN BILIARY EXCRETION AND METABOLISM. Phase I and phase II biotransformation steps can be crucial in the rate of biliary elimination of drugs and their metabolites. These metabolic steps may improve the hydrophilicity-hydrophobicity balance necessary for interaction with the carrier systems in the canalicular membranes (see *Chemical Structure*, p. 721).

For a number of carcinogens it has been clearly shown that phase I and phase II metabolism is rate limiting in their biliary excretion (107, 183, 184). This explains the stimulatory and inhibitory effects of this process seen after pretreatment with the enzyme inducer 3-methylcholantrene and inhibitors of the mixed-function oxidase system, respectively (183). In contrast acetylation of the organic cation procaine amide ethobromide (PAEB) (139, 207, 235) or, for example, hydrolysis of another organic cation, methyldeptropine (180), retards rather than stimulates total biliary output (200). Induction of glucuronide conjugation in general stimulates biliary excretion of drug-glucuronide conjugates. However, for a number of phenolic compounds this was not the case. An increase of biliary output of the glucuronide conjugates was probably not observed, because for these substrates canalicular transport of the conjugates is rate limiting in the biliary output (183).

Depletion of glutathione by diethylmaleate only leads to a decrease in total BSP excretion if glutathione availability becomes rate limiting during infusion of large amounts of BSP but not after a single injection of the dye (107). Induction or inhibition of GSH transferases alters the excretion rate of total BSP or ethacrynic acid, which occur mainly in the form of GSH metabolites (107, 183). Altered compartmentalization of the conjugate within the hepatocytes may also play a role in the effect of inducing agents (93). The earlier observation that p-nitrophenol glucuronide formed in the liver is much more efficiently excreted into bile than the preformed conjugate also indicates that an intracellular sequestration may influence the secretory process (183).

Acinar Heterogeneity

It has become increasingly clear that hepatocytes in the various zones (Fig. 10) of the liver acinus [the microcirculatory unit of the liver (266)] are heterogeneous with respect to ultrastructure, micromilieu, and function. Two forms of heterogeneity in drug trans-

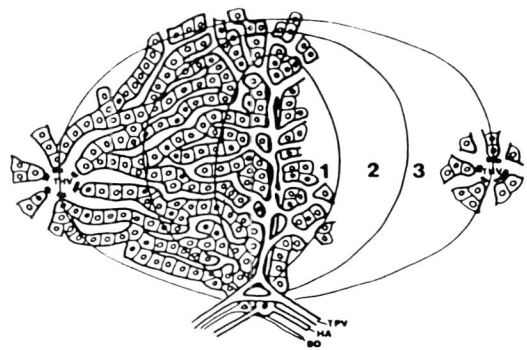

FIG. 10. Schematic representation of the microcirculatory unit of the liver: the hepatic acinus. An arbitrary division in 3 zones is indicated. BD, bile ductule; HA, hepatic artery; TPV, terminal portal venule.

port and metabolism can be envisioned by looking at the liver structure: *1*) a difference due to the localization of the cells in the bloodstream (i.e., zone 1 cells periportally and zone 3 cells perivenously), by which the cells are exposed to different concentrations of substrates and *2*) intrinsic cellular differences not directly related to concentration gradients of drugs along the sinusoids but related to an unequal cellular equipment (110, 117, 120). The latter differences may be secondary to acinar concentration gradients in O_2, CO_2, and hormones.

ZONAL HETEROGENEITY IN DRUG METABOLISM. Observations on selective zone 3 toxicity by CCl_4 and acetaminophen and the knowledge that liver injury is related to the production of reactive metabolites from these compounds suggested zonal differences in the rate of production of these toxic biotransformation products (117, 120). Microspectrophotometric, immunochemical, morphological, and microdissection studies of periportal and pericentral regions of rat liver demonstrated that zone 3 cells have a higher density of SER; higher concentrations of cytochrome P-450, cytochrome b_5, and glucuronosyltransferase; but a lower concentration of glutathione (156, 329). The phenobarbital-inducible subspecies of cytochrome P-450 seems to be localized preferentially in the perivenous zone (110, 119, 329). Selective damage of zone 1 and zone 3 (112, 117), as well as experiments in which the sequential rate of formation of drug metabolites after antegrade and retrograde single-pass perfusions of the liver was studied (248), indirectly suggests that the oxidative conversion predominantly occurs in zone 3 but that sulfate conjugation predominantly occurs in zone 1.

An interesting and more direct approach was taken by using microlight guides that can be placed on periportal and pericentral regions in the liver (124, 329). In this fiberoptic approach one excites and also collects fluorescence produced by fluorescent model compounds such as 7-hydroxycoumarin that is generated from nonfluorescent 7-ethoxycoumarin by mixed-function oxidase and is subsequently conjugated with glucuronic or sulfuric acid to nonfluorescent products. For these coumarins it appeared that monooxygenation and glucuronidation tend to be higher in the pericentral cells. Both processes in zone 3 can be specifically induced by phenobarbital but not by 3-methylcholantrene. In normal liver, sulfation appears to be predominantly localized in zone 1. These observations may explain the selective zone 1 toxicity of relatively low doses of *N*-hydroxy-2-acetylaminofluorene, for which sulfation is a determining factor (110, 112, 227).

In contrast to an earlier study with BSP (54), a recent study with nitrobenzene derivatives indicated a substrate-dependent zonal heterogeneity in glutathione conjugation that is not simply due to GSH availability or rate of adduct formation but also to unequal zonal substrate uptake and/or intracellular binding (124). Microdissection studies of human liver (225) confirm that UDP-glucuronosyltransferase and glutathione *S*-transferase are predominantly localized in pericentral regions, whereas sulfotransferase activity is greater in zone 1. In addition this study demonstrated an even distribution of the deconjugating enzymes glucuronidase and sulfatase, indicating that net production of drug conjugates is influenced by futile cycles of conjugation-deconjugation reactions in both zones in the liver.

It remains to be established what the signals are for unequal synthesis and/or turnover of drug-metabolizing enzymes and their cosubstrates in the various zones under normal and induced states.

ZONAL HETEROGENEITY IN DRUG TRANSPORT. Evidence is growing that, besides an unequal rate of metabolism, periportal and perivenous cells have unequal rates of uptake and biliary excretion of endogenous and exogenous compounds. Injection of tracer doses of [^{3}H]taurocholate into the portal vein of rats resulted in selective uptake of the label in zone 1 hepatocytes, as was detected by freeze autoradiography [Fig. 11; (109)]. Similar concentration gradients have been observed in the rat liver in vivo (109) and in perfused livers after administration of low doses of bile salt derivatives (151), BSP (117), chlorpromazine, propranolol (7), and some fluorescent compounds (110, 119, 121). A reversed gradient, with increasing concentrations going from zone 1 to zone 3, was seen less frequently, for instance when a high concentration of fluorescein isothiocyanate was infused (121). The existence of acinar gradients is not a general rule: for example, for the cardiac glycoside ouabain and [^{35}S]-BSP at physiological albumin concentration, no such pattern was observed (111, 112, 117, 118).

The concentration of the compound to be transported in the sinusoidal blood is likely to decrease from zone 1 to zone 3 as a consequence of uptake and further processing by the hepatocytes. These concentration gradients in the blood may thus lead to an unequal involvement of the hepatocytes in the blood-to-cell transfer of a particular compound and in steady state result in decreasing concentration gradients in the liver tissue, parallel to those in the sinusoidal blood. When the uptake process in zone 1 reaches saturation, higher exposition of zone 3 occurs and both the sinusoidal and the cellular gradient disappear. This phenomenon was shown for several substrates such as taurocholate (109), rhodamine B (121), and the organic anion BSP (117). In addition it was shown that when the sinusoidal concentration gradient was reversed by administering the label to a retrogradely perfused liver, the gradients in the liver tissue were also reversed, and virtually all label was observed in zone 3 (109).

Combination of autoradiographic and kinetic studies indicates that both periportal and perivenous cells

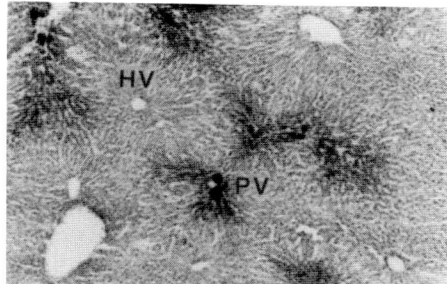

ANTEGRADE PERFUSION

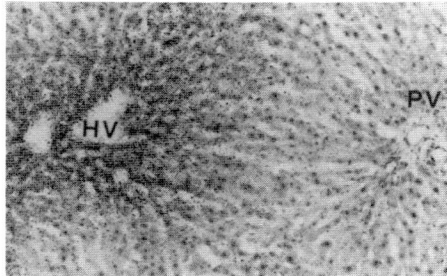

RETROGRADE PERFUSION

FIG. 11. Light-microscopic autoradiographs of rat liver 30 s after injection of [^{3}H]taurocholate in rat liver perfused in normal direction (antegrade) and retrograde, using freeze autoradiography to prevent diffusion of the water-soluble compound. Steep acinar gradients are observed in zone 1 and zone 3, respectively. *Large black dots*, material present in bile ducts at $t = 30$ s, which is only visible after antegrade perfusion. *HV*, hepatic venule; *PV*, portal venule.

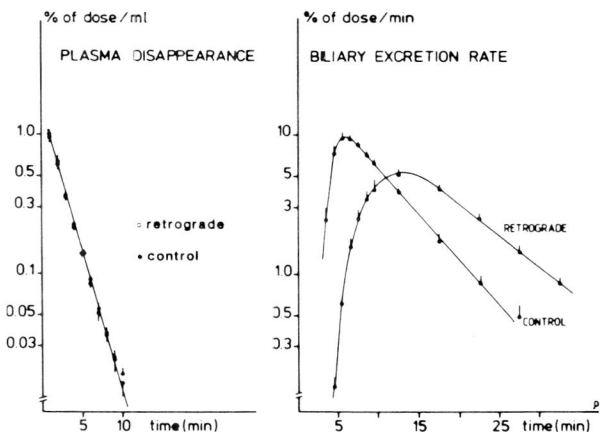

FIG. 12. Taurocholate kinetics in isolated perfused livers after antegrade and retrograde perfusion. Plasma disappearance rate (distribution to the liver, *left panel*) is independent of direction of flow, whereas biliary excretion rate (*right panel*) is more rapid at normal direction of flow. Because distribution to zone 1 is predominant at normal flow (cf. Fig. 12), zone 1 cells may be better equipped for bile secretion of taurocholate than zone 3 cells.

are well equipped to take up bile acids, as shown in Figure 11. The biliary excretion rate of the periportal cells, however, (as measured in normal perfusions in Fig. 12) is considerably higher than that of zone 3 cells [retrograde perfusions in Fig. 12; (109)]. This may either be due to a difference in canalicular membrane transport or to a higher intracellular binding of bile salts in the zone 3 cells. The predominant involvement of zone 1 cells in bile acid transport is supported by the finding that zone 3 damage did not affect bile acid transport to a significant degree (112, 117). The molecular basis for the observed heterogeneity in relation to number or function of the carrier proteins in the cells deserves further study.

MECHANISMS OF MEMBRANE TRANSPORT OF ORGANIC COMPOUNDS

General Considerations

The distribution of xenobiotics into the liver generally occurs according to the pH-partition hypothesis. This implies that the rate of passage from blood into the cells is linearly related to the amount of undissociated drug [as determined by the pH at the surface of the cells and pKa (dissociation constant of the acid form) value of the compound] and also linearly related to the lipid-water partition coefficient of the undissociated drug. The importance of these factors was demonstrated by Kurz (176) for a wide variety of compounds and indicated that permeation across the lipoid membranes by passive diffusion is the most probable mechanism for lipophilic drugs. Only very small hydrophilic molecules such as urea and monosaccharides can pass the liver cell membranes, probably via channel-like structures. In contrast, nonspecific fluid-phase endocytosis may lead to permeation of very large molecules that would normally not be able to penetrate. Although this process is relatively slow, a volume of 5%–25% of the cell may be endocytosed per hour, depending on the cell type (291). For the transport of very hydrophilic and/or charged organic molecules into the hepatocytes, or vice versa, specialized membrane processes are required that involve translocating carriers or channels for drugs and receptor-mediated endocytotic activity for various types of proteins (44, 154, 172, 183, 209, 287).

MULTIPLICITY OF TRANSPORT MECHANISMS: MODEL COMPOUNDS. The hepatic uptake mechanisms for drugs and xenobiotics are traditionally classified according to the type of charge of the compounds, i.e., transport pathways for organic anions, organic cations, and uncharged molecules [Fig. 13, (172, 183, 287)]. For several reasons this traditional scheme may require revision. For organic anions at least two (but probably more) different transport mechanisms exist. It is possible that this is related to the presence of various transport systems for naturally occurring anionic substrates such as bile salts, fatty acids, and amino acids, as well as bilirubin and its conjugates. On the basis of Na^+ dependency, one roughly differentiates in an Na^+-dependent transfer system and an

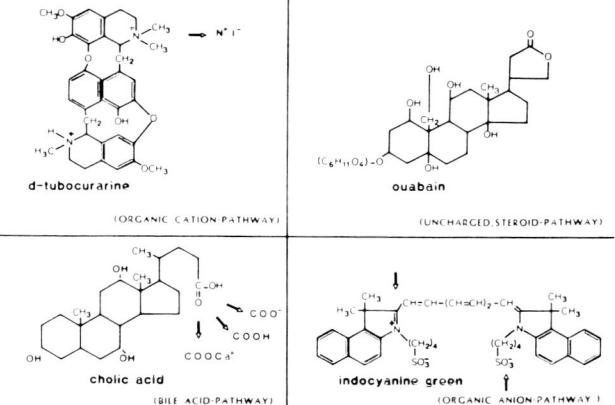

FIG. 13. Chemical structure of cholephilic compounds for which separate transport processes have been proposed. However, indocyanine green (organic anion) contains also a cationic group. Cholic acid can be present as organic anion, as an undissociated carboxylic steroidal compound, or complexed with Ca^{2+} with a net positive charge. The steroid structure of ouabain (uncharged compound) resembles that of bile acids, and the organic cation d-tubocurarine may form electroneutral ion pairs. These features enable multiple interactions and overlapping substrate specificity for transport systems.

Na^+-independent mechanism (92). The Na^+-dependent transport, probably related to the bile acid system, serves a group of substrates of surprising variety, such as methotrexate (159, 286), ethacrynic acid (250), iodipamide, iopodate, and even cyclopeptides such as antamanide, somatostatin, cyclosporin A, and phalloidin (92). The Na^+-independent uptake occurs for nonmetabolized organic anions such as DBSP, rose bengal, tartrazine, eosine, chlorothiazide, succinylsulfathiazole, and rifampicin, as well as numerous sulfate, glucuronide, or glutathione conjugates of drugs (172, 183).

Besides the Na^+ dependency as a discriminating parameter, other observations indicate that within the large group of anions a further differentiation should be made. Transport of morphine sulfate and morphine glucuronide in the rat in vivo is differently affected by phenobarbital treatment and temperature changes (252). Phenobarbital pretreatment and bile salts have markedly unequal influences on BSP and rose bengal transport (191). Fasting experiments and interaction studies with bile salts and rifampicin revealed at least two uptake mechanisms for BSP that are different from the one for rifampicin (177, 178). No competition in hepatobiliary transport was observed between a number of drug glucuronides on one hand and BSP and BSP-GSH on the other (336). Pretreatment of rats with the hypolipidemic agent nafenopin largely decreased hepatic transport of DBSP, enhanced transport of chlorothiazide, but left bile salt transport completely intact (183, 205). Partial hepatectomy increased hepatic transport of DBSP per unit of liver weight but decreased that for tartrazine (335). Even unconjugated BSP and BSP-GSH may differ with regard to biliary transport, as suggested by studies in mutant sheep (4) and observations in Dubin-Johnson patients (22).

For organic cations at least two (200) and for uncharged compounds three separate hepatic transport mechanisms were proposed (299). It is not easy therefore to anticipate interactions at the level of hepatobiliary transport only on the basis of chemical structure and net charge of the particular drugs. In this respect the following factors should be taken into account.

1. Organic anions like BSP are transported by a high-affinity/low-capacity as well as a low-affinity/high-capacity system (11, 178). The outcome of interactions on the level of hepatic transport is influenced by the respective substrate concentrations of the model compounds that are attained in the hepatobiliary system. Bile salts at high concentrations may mutually compete with BSP for hepatic uptake but certainly not in the physiological concentration range where the Na^+-dependent system is operating (11).

2. In a physiological milieu some organic compounds can be present in more than one form (Fig. 13). The most simple example is partial dissociation of acid or basic groups in the drug molecule. At a physiological pH, unconjugated bile salts such as cholic acid and dehydrocholic acid are partly present as undissociated and uncharged species. In that form they may fit the carrier system for uncharged steroids that is supposed to handle the cardiac glycosides, among others. Organic cations such as quaternary and tertiary amines may be present at the water-lipid membrane interphase complexed as ion pairs with endogenous inorganic or organic counterions and may present themselves to the liver partly as an electroneutral species (239, 240). Conversely it is not excluded that organic anions can in principle form complexes with Ca^{2+} or Mg^{2+} or other endogenous organic cations and may be partly transported as such (300). The model compounds depicted in Figure 10 are supposed to be transported in the liver via different mechanisms. However, marked inhibitory effects between these compounds were reported in the intact organ and in isolated hepatocytes (172, 200, 299, 344). Bile salts that (as indicated in Fig. 13) can be present in multiple forms cannot only strongly inhibit hepatic uptake of other organic anions (339, 341, 344) but also that of cations (344, 353) and steroids (68). Dehydrocholate inhibits the biliary excretion of cardiac glycosides (211) and organic anions (339). Indocyanine green is a zwitterion and was found to be the most potent inhibitor of uptake into hepatocytes of the organic cation PAEB (69), as well as the uncharged compound ouabain (68). Depending on their relative affinity for the various carrier processes in the plasma membrane, the multiple forms of these drugs consequently may induce unexpected inhibitory effects.

3. The alternative possibility should be considered that the liver possesses at least one general transport system for high-molecular-weight xenobiotics irrespective of their charge (51). Many of the interactions previously mentioned could be explained by involvement of such an aspecific carrier process. Recent studies using photoaffinity labeling techniques with bile acids, uncharged steroids, organic cations, and anionic dyes revealed only a very limited number of polypeptides as potential carrier molecules and also demonstrated mutual inhibitory effects of these agents (51).

BIDIRECTIONAL CHARACTER OF TRANSPORT. Although many studies have been performed on the hepatic uptake and biliary excretion systems, much less is known about the mechanisms for membrane transport from the hepatocytes into plasma or from the canalicular lumen back into the cells. If simple carrier-mediated facilitated diffusion is involved at both poles of the cell, in principle bidirectional transport should be anticipated at these levels. One feature of facilitated diffusion is accelerated exchange diffusion, which is identical to the phenomenon of countertransport and transstimulation. These phenomena are based on alteration of net flux of a substrate by disturbing the steady state of influx and efflux by competition with another compound added at one side of the membrane. For instance, if the liver is loaded with a radioactive-labeled organic anion, the addition of a competing anion (for instance the unlabeled substrate) to the plasma compartment would initially inhibit transport into the cell more than transport out of the cell and would therefore induce a transient efflux of label until the concentration of the unlabeled material in the cell is equilibrated with the plasma concentration.

Such countertransport at the hepatic uptake level was claimed to occur for bilirubin, BSP, and ICG in the rat in vivo (292). Competition for extracellular and intracellular binding sites (cytosolic proteins and cell organelles) might at least partly explain such a phenomenon (34, 204, 226, 230). However, various studies with organic anions in plasma membrane fractions enriched in canalicular or sinusoidal elements (142, 143, 197) demonstrated transstimulation effects, which might imply bidirectional facilitated diffusion as a principal equilibrative mechanism. Addition of albumin to the perfusion medium of isolated livers preloaded with DBSP (204) or agents that compete for intracellular binding sites clearly demonstrated efflux of organic anions from liver into plasma (34, 98, 204).

It is not known whether this sinusoidal efflux is carrier mediated or simply represents the counterpart of the hepatic uptake process. It has been speculated that the organic anion reflux process is related to a carrier system that exports glucuronide, sulfate, or glutathione conjugates of organic compounds to the bloodstream after their formation within the hepatocyte [see Fig. 6; (61a)]. The DBSP inhibits sinusoidal excretion of the sulfate conjugate of the phenolic drug harmol and thereby increases its excretion into bile (61a). Evidence for saturable, carrier-mediated transport to the bloodstream of GSH and GSH conjugates has been reported in other studies (158, 305). Transport of organic anions at the sinusoidal level is therefore very likely carrier mediated in both directions. Whether a single carrier system mediates both the plasma-to-cell and cell-to-plasma fluxes is an open question.

No conclusive evidence is available for reversed transport from canalicular lumen to the hepatocyte cytosol. Retrograde biliary injections of various compounds lead to partial loss of the injected compound from the biliary tree (94). The level at which this occurs, however, is debated (87, 94).

HEPATOBILIARY CONCENTRATION GRADIENTS. The transcellular transport of organic compounds to bile involves at least two membrane transport steps: sinusoidal uptake and canalicular excretion. The total biliary concentration of drugs often exceeds that of plasma manyfold, and the question is at which level hepatobiliary transport is intrinsically concentrative. Binding of many cholephilic drugs to proteins and biliary micelles, however, makes it difficult to assess whether real concentration gradients of unbound material exist. Also, such gradients are dose dependent and determined by relative rates of uptake and secretion. In the classic studies of Hanzon (123) in which fluorescein was injected in rats, fluorescence microscopic pictures of the liver structure gave the clear impression that a two-step concentrative process from plasma to bile was involved. Unequal binding to macromolecules in plasma, cell, and bile canaliculus cannot explain the observed gradients that were later also observed in cultured hepatocytes (96, 104, 257). In other studies attempts were made to calculate transmembrane concentration gradients of organic compounds by estimating unbound drug in plasma, liver cytosol, and bile. For the organic anions DBSP (204, 339), p-acetylaminohippuric acid (277), and taurocholate (10, 30, 44, 274, 296), for the organic cations PAEB and its acetylated form (139, 215), as well as for the uncharged ouabain (68, 175), such distribution studies indicate that a two-step concentrative process is involved: one from plasma to the cytosol and one from cytosol into bile. Figure 14 depicts the estimated unbound concentration gradients of some model compounds.

Transmembrane gradients should be seen in the light of the membrane potential across the barriers involved. Depending on the type of charge and the number of charged group, the inside negative electrical potential of the hepatocytes can potentially affect

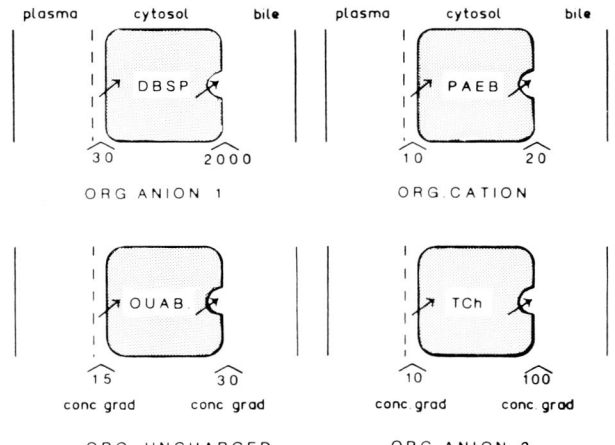

FIG. 14. Calculated concentration gradients of unbound drug across sinusoidal and canalicular membranes for the organic anion dibromosulfophthalein (DBSP), the organic cation procaineamide ethobromide (PAEB), the uncharged compound ouabain (ouab), and the bile salt taurocholate (TCh). Concentrations of the unbound drugs were determined in ultrafiltrates of plasma and cytosol fraction of liver homogenates. Cytosol concentrations were corrected for contamination with plasma and bile.

transmembrane distribution of charged molecules, if the membrane transport processes are bidirectional and thus equilibrative. Electrical potential differences between liver cell and plasma and between bile and plasma were estimated to be −40 mV and −4 mV, respectively (103, 104, 277). However, both for organic anions (204) and organic cations (200, 215) the gradients observed at the sinusoidal level significantly exceed the passive equilibrium value. Canalicular transport of organic cations from the negatively charged cells should be clearly uphill in the light of the very high bile-liver concentration ratios (215, 239), Distribution of the uncharged cardiac glycoside ouabain would not be anticipated to be influenced by the membrane potential, yet the calculated chemical gradients in the system clearly indicate concentrative transport, both at the sinusoidal and the canalicular level (68, 175).

POTENTIAL DRIVING FORCES FOR HEPATOBILIARY TRANSPORT. The general idea of uphill transport raises the question of the energization and driving forces for these processes. In the case of taurocholate, cotransport with Na^+ has been demonstrated in isolated perfused rat livers (62, 269), isolated hepatocytes (10, 39, 73, 84, 296, 338), and basolateral plasma membrane vesicles (64, 142, 197, 282, 306). Some of the studies with vesicles clearly indicate electrogenic transport of taurocholate with more than one Na^+, because inside-negative diffusion potentials highly stimulated the rate of vectorial transport into the vesicles (142, 282, 306). This is in line with the observation that hyperpolarization of hepatocytes by incubation with NO_3^- and SCN^- resulted in an increased rate of Na^+-dependent uptake of taurocholate and phalloidin (255). The reverse seems to be true for cholate (255) and iodipamide (155, 323). It was suggested that the latter organic anions can be transported electroneutrally with two Na^+ and one Cl^- ion. Phalloidin, an uncharged molecule, would be accompanied by one Na^+, rendering the transport system to be electrogenic and driven by a negative membrane potential (255). The likely mechanism for hepatic uptake of conjugated bile acids therefore is secondary active transport driven by concentration gradients of Na^+, which in its turn is maintained by the ATP-driven Na^+-K^+ pump. For the other categories of cholephilic agents such cofactors are much less clearly defined.

Recent studies with basolateral membrane vesicles indicate that the driving force for uptake of BSP-like compounds into hepatocytes could be the electrochemical gradient of hydroxyl ions produced by an Na^+-H^+ exchange mechanism that pumps protons out of the cell at the sinusoidal pole of the cell (136). In line with the results with cholate and iodipamide, a link with inwardly directed Cl^- gradients was suggested for BSP (320, 367). Countertransport of anions with inorganic anions could occur electroneutrally and in principle could explain the passage of the negative potential barrier at the sinusoidal level. The opportunities for both cotransport and countertransport would provide a very flexible and dynamic combination of transporting systems for organic compounds.

The driving forces for canalicular transport of organic anions are even more poorly defined. The negative membrane potential could drive electrogenic transport of anionic drugs. In addition countertransport with HCO_3^-(OH^-) has been claimed in studies with hepatocytes and isolated perfused livers (8), as well as in isolated canalicular membrane vesicles. Hepatobiliary transport of charged drugs may therefore largely depend on the transmembrane gradients of inorganic ions. The coupling between such gradients in the hepatobiliary system and drug transport is tentatively depicted in Figure 15.

Organic Anions

KINETIC ANALYSIS: MODEL COMPOUNDS. Organic anion transport in the liver has been more broadly studied than that of other types of compounds. This is apparently related to the physiological importance of hepatobiliary disposition of bile acids (132), bilirubin conjugates (295), and other bile pigments. Other factors in this respect are procedures to quantify hepatic transport functions with anionic dyes such as BSP, ICG, and rose bengal (53), as well as the methods that have been developed for imaging of the hepatobiliary system with cholecystographic contrast agents or radionuclide scanning agents of the acetanilidoiminodiacetic acid (IDA) group (243).

Anion transport has been often studied using the model compound BSP, which is a tetra-brom deriva-

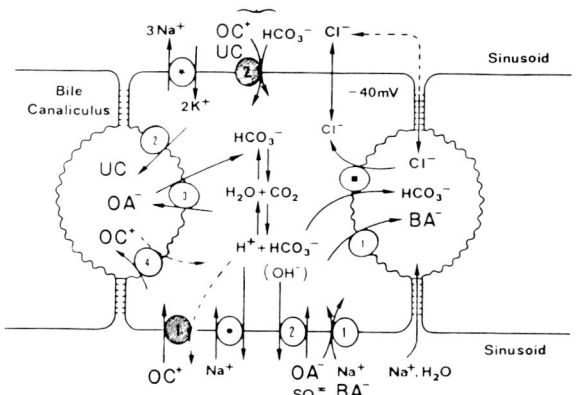

FIG. 15. Coupling between gradients of inorganic ions with carrier-mediated transport of organic anions (OA$^-$), bile acids (BA$^-$), organic cations (OC$^+$), and uncharged compounds (UC) in the hepatocyte. Inorganic ion pumps include electrogenic NA$^+$-K$^+$ exchange, Na$^+$-H$^+$, OH$^-$-SO$_4^-$ antiport (sinusoidal), and Cl$^-$-HCO$_3^-$ antiport (canalicular). Na$^+$-coupled bile acid transport (1) and OH$^-$-OA$^-$ antiport (2) at the sinusoidal level and HCO$_3^-$-OA$^-$ antiport at the canaliculi are anionic systems. Organic cations are taken up by system 1 and may share a transport system for uncharged compounds (2). In the latter system, ion-pair formation may play a role; at the canalicular level antiport of organic cations with protons is tentatively assumed. Transport processes at sinusoidal level have more overlapping substrate specificity than the projected 4 canalicular carrier systems. [Adapted from Hugentobler and Meier (136).]

tive. However, the compound undergoes conjugation with glutathione and is excreted predominantly in this form and further metabolized products (172, 183). These metabolites exhibit quite different protein binding and transport characteristics compared with BSP itself (20, 22, 297, 369). Because the commonly used bioanalytic methods do not distinguish between BSP and its metabolites, changes in the rate of metabolism can be misinterpreted to reflect altered hepatic transport. Use of the glutathione conjugate of BSP itself may only partly resolve this problem because it is not metabolically inert. The dibromo analogue (DBSP) is an alternative for studies in the rat, because it is not metabolically altered in this species (148, 172). In contrast, in humans partial conjugation of DBSP was reported (213). Other organic anions that are excreted predominantly unchanged in bile include ICG, iodipamide, Eosine, bromcresol green, p-aminohippuric acid, chlorothiazide, tartrazine, succinylsulfathiazole, cromoglycate, bromphenol blue, rose bengal (172, 183), and the IDA-type of scintiscanning agents (243, 245). Other organic anionic drugs that can be abundantly excreted in bile are various types of biliary contrast agents (31, 32, 312), sulfonamides (61, 312), penicillins (284), and cephalosporins (172, 183, 231, 312). The extent of biliary output of such agents may depend on the balance between hydrophobic and hydrophilic properties, as well as other structural factors (127, 312).

Cholephilic organic anions are commonly avidly bound to plasma proteins, which tends to limit the distribution in the body initially to the plasma compartment. After intravenous injection, at relatively low doses, mostly a biexponential plasma disappearance is observed, of which the first rapid phase is mainly due to uptake into the liver and the slope of the second phase reflects the biliary elimination rate and/or rate of metabolism. With appropriate models, the various processes involved can be quantitatively described in terms of rate constants for hepatic uptake, reflux, biliary excretion, and/or hepatic metabolism (41, 53, 98, 122, 210). Uptake of organic anions into the liver can be very rapid, leading to an initial hepatic extraction ratio of 0.40:0.60 (210, 213). The maximum velocity (V_{max}) and K_m values for uptake determined in isolated hepatocytes are fully compatible with such an effective uptake process (40, 41). In principle both the uptake and biliary excretion steps are saturable, and consequently the kinetic profiles of organic anions are dose dependent (53, 122, 202, 204, 249, 319, 358, 359). Also plasma protein binding may become saturated, and disproportionate changes in unbound fraction with the dose may lead to alterations in clearance (204, 210), extrahepatic distribution (169, 202, 204), and extrahepatic elimination (169, 202).

The carrier-mediated character of the hepatic uptake and biliary excretion processes is also reflected in competitive interactions between the anionic drugs during hepatic transport (292). The outcome of such mutual interactions in principle depends on the respective doses and affinities of the agents for the membrane transport process. However, simultaneous competition of the particular agents for plasma protein binding should be taken into account as a complicating factor (204). Another aspect mentioned before is that at least two different uptake systems are available for organic anions. System one may predominantly transport conjugated bile acids (19, 92), estrone sulfate (50, 301), iodipamide (92), methotrexate (159, 286) and probably also succinylsulfathiazole (1, 19), probenecid (113), and phenolphthalein glucuronide (335, 340). System two is supposed to preferentially transport DBSP and BSP (40), BSP-GSH (297), ICG (204, 292), unconjugated bile acids (10, 85), bilirubin (21, 29, 295), fusidic acid (9), rifamycins (177), penicillins (284, 331), and the scintiscanning IDA agents (245). The two systems very likely have overlapping substrate specificity.

The relative contribution of these transport processes to the hepatic uptake of a certain compound therefore is dose dependent and evidently influenced by the presence of competing endogenous substrates. It follows that specific inhibitors for the two systems cannot easily be developed. Rifamycins probably belong to the strongest inhibitors of organic anion transport in the liver. The antibiotics preferentially compete for uptake of BSP (177) and unconjugated bile acids but at higher concentration also affect carrier-mediated transport of conjugated bile acids and probably other type I compounds (12). Rifamycins interfere with BSP and bilirubin kinetics (2) as well as hepatobiliary transport of bile acids (95) in the human.

MECHANISMS OF HEPATIC UPTAKE. Uptake of organic anions like BSP, BSP-GSH, DBSP, and diisopropyl IDA in isolated hepatocytes has demonstrated temperature dependency, partial inhibition of metabolic inhibitors, and Na^+ independency of uptake (30, 40, 297, 300). The activation energy is generally a factor of 3–4 lower than that found for bile acids (297). Reduction of cellular ATP to <5% of controls only moderately affects uptake of BSP (294, 297, 300, 304), DBSP (40), and BSP-GSH (297) but in contrast strongly decreases uptake of iodipamide as well as cyclic peptides such as phalloidin and antamanide that are probably transported by an Na^+-dependent mechanism (92, 254). Nevertheless the analyses of Goresky (99, 100), Scharschmidt et al. (292), and Berk et al. (29) clearly indicate at least carrier-mediated facilitated diffusion for BSP. Uptake into hepatocytes of BSP (300) and DBSP (40) is competitively inhibited by ICG and that of BSP-GSH is competitively inhibited by BSP (297).

MECHANISMS OF BILIARY EXCRETION. Multiplicity of organic anion transport on the canalicular level is also evident. Observations in mutant sheep (4, 22) and a more recent study in rat (146, 147) showing that bile acid transport in the mutant species is normal whereas biliary transport of BSP and bilirubin is almost absent provided crucial information. Transport of BSP-GSH in these mutant species is much more affected than that of the parent compound, and cholic acid transport is also abnormal (22). Abnormality in the binding affinity for the canalicular carrier proteins may be involved in the mutant species.

That transport of some organic anions into the canaliculus is very likely carrier mediated is demonstrated in many interaction studies in vivo and was confirmed in recent studies with canalicular membrane vesicles (143, 144, 199, 326) showing transstimulation phenomena. Efflux of DBSP from the hepatocytes is inhibited by ICG (40), whereas the release of DBSP from the cells is not affected by taurocholate at concentrations of the bile acid that almost block the uptake into the hepatocytes (341). The mechanism of secretion from the cells therefore is principally different from that for uptake and might be related mainly to the biliary transport process in the intact organ (41). Efflux of DBSP from the cells (14 kcal/mol) is occurring down the electrochemical gradient, exhibits a lower activation energy than that for uptake (26 kcal/mol), and is in the same range as that for taurocholate (298). The release of DBSP (40) and of BSP-GSH (297) from hepatocytes is decreased by uncouplers of oxidative phosphorylation such as dinitrophenol and pentachlorophenol (102), respectively, an effect that was not due to competition with anionic conjugates of these inhibitors (102).

CARRIER PROTEINS FOR ANION TRANSPORT. Recently advances have been made in the study of the carrier proteins by using radioactive probes of various cholephilic model compounds provided with a photolabile group. The photochemical reaction with ultraviolet light makes it possible to label the carrier proteins covalently with the radioactive substrate, providing an important tool for further isolation, purification, and identification of the carrier protein involved. Differential photoaffinity labeling in which the photoreaction is performed in the absence and presence of other substrates may provide semiquantitative data on competitive effects that can be correlated with kinetic data obtained in in vitro liver preparations (51, 173, 185, 198, 268, 338, 356, 365, 368, 371, 372). Photoaffinity probes of bile acids and phalloidin, as well as BSP (which is photoreactive in itself), label at least five polypeptides with molecular mass of ~100, 68, 54, 48, and 43 kDa and one of ~30 kDa. The 68-kDa protein is albumin, the 30-kDa protein reaction is mitochondrial, and the 43-kDa component is related to the microfilament structures and probably represents actin (173).

Investigations with sinusoidal plasma membranes consistently revealed the 54- and 48-kDa polypeptides that are probably integral membrane proteins (30, 51). With [^{35}S]BSP, predominantly the 54-kDa species is labeled (368). The BSP less efficiently labels the 48-kDa component, which is clearly labeled by bile acid derivatives (173, 185, 356, 371, 372). The 54-kDa and 48-kDa polypeptides are also labeled by antamanide and phalloidin, which inhibit not only BSP transport but also bile acid transport in liver (51, 356, 371, 372). The 48-kDa protein might therefore represent the Na^+-coupled bile acid uptake system. Its distribution is restricted to liver and kidney, in contrast to the 54-kDa protein, which seems to be present in many other organs (56). Preliminary evidence indicates that with photoaffinity probes of organic cations and uncharged compounds, labeling occurs also on polypeptides in the region of 54 kDa and 48 kDa (51, 254). It is thought that these polypeptides represent rather broad aspecific hepatic uptake systems for organic compounds, irrespective of their charge. Alternatively a family of membrane proteins with approximately the same molecular mass could be involved (30).

Interesting attempts have been made to isolate, purify, and further characterize some of these transport proteins in the liver (30, 188, 268, 316, 317, 365). One important point is that the particular proteins not only bind the model compounds but are also involved in translocation across, or channeling through, the plasma membrane. Raising antibodies against the purified proteins provides a powerful tool to study body distribution and cellular localization, as well as the influence on transport. Reconstitution experiments were performed with a potential carrier protein of 105 kDa that was tentatively called bilitranslocase. The hydrophobic protein was included in monolaminar liposomal systems containing 150 mM KCl and BSP. Bilitranslocase but not albumin increased efflux of BSP from the liposomes, a phenomenon that was clearly accelerated by addition of valinomycin, which induces an (inside-negative) K^+-diffusion potential (315, 316). This suggests that transport of BSP in this system occurs in the anionic form and can be driven by a negative membrane potential, a process that could be imagined in relation to sinusoidal or canalicular secretion from the cells.

Recently a 100-kDa protein was isolated from purified canalicular membranes (198). Besides being present in hepatocytes it is also found in apical membranes of ileal and renal tubule epithelial cells but not in basolateral membranes in the three organs. The heavily glycosylated protein was demonstrated through photoaffinity labeling with bile acids and with monospecific antibodies (198). Binding of bile acids was inhibited by BSP and phalloidin. More detailed studies on the composition and behavior of these potential carrier proteins in lipid membranes should increase our understanding of the actual translocating mechanism. In addition it would be of great interest to correlate the amount of such carrier proteins with hepatic uptake rate in animal species or patients with abnormally low hepatic transport of organic anions. The study of regulation of the cellular synthesis and degradation of carrier proteins could then clarify their role under various physiological and pathological conditions.

Organic Cations

KINETIC ANALYSIS: MODEL COMPONENTS. This very large category of drugs mainly consists of compounds with one or more quaternary or tertiary amine groups. Quaternary ammonium groups with a nitrogen atom linked to four carbon atoms are permanently positively charged. The tertiary amines can also be considered as cationic because, depending on the pKa value of the strongly basic group, at physiological pH >90% of the molecules may be protonized (236, 285). Among the quaternary agents are anticholinergic and neuromuscular blocking agents. The tertiary amines represent a wide variety of drugs, ranging from peripherally acting local anesthetics, antihistaminics, β-blocking agents, and anticholinergic drugs to various centrally acting psychotropic drugs.

An important contribution to a better understanding of hepatic transport mechanisms for organic cations was made by Schanker and co-workers (139, 287, 289, 314). They showed that after intravenous injection of some of these compounds into the rat, concentrations were reached in the bile far exceeding the concomitant plasma levels. The biliary excretion was saturable and could be suppressed by structurally related compounds. Uptake into liver slices was blocked by anoxia and a number of metabolic inhibitors. These data provided clues for the existence of one or more carrier-mediated transport processes but do not necessarily imply primary active transport (200). Endogenous substrates for this hepatic transport system may include choline, acetylcholine, and thiamine, for which competition with quaternary drugs has been demonstrated (370). Another natural substrate could be nicotinamide, which is excreted in bile as nicotinamide riboside (189), a phenomenon that was suggested to be linked to the hepatic pyridine nucleotide cycle.

MECHANISMS FOR HEPATIC UPTAKE OF MONOVALENT CATIONS. Two compounds were most frequently used as model compounds in the study of hepatic transport: PAEB and its 4N-acetyl derivative (APAEB) (139, 207). Experiments with these compounds have been performed using several techniques, such as rats in vivo (139, 207, 215, 232, 235, 242, 289, 297), isolated perfused livers (207, 342), rat liver slices (139, 314), isolated hepatocytes (69, 341), and liver membrane vesicles (281). After intravenous administration in rat (207, 242), a rapid initial phase is seen in the disappearance curve that reflects distribution to various tissues, in particular the excretory organs. A second component in the decay curve is due to the sum of hepatic, renal, and intestinal clearance. The relative contribution of these elimination pathways depends on the lipophilicity-hydrophilicity balance of the individual compounds (see *Chemical Structure*, p. 721).

Transport of organic cations from plasma into bile seems to occur in at least two concentrative steps: from plasma into liver and from the cells into bile [see *General Considerations*, p. 739; Fig. 13; (139, 207, 215)]. After hepatic uptake, PAEB is partly acetylated and also excreted in this form in bile (139, 207). Because it is excreted unchanged in bile, APAEB was considered a more attractive model compound. However, the capacity of the saturable uptake process for APAEB is probably much lower than that of the biliary excretory step (234, 235). The relatively slow hepatic uptake of APAEB (200), which is quite atypical for this type of cations, was confirmed in studies with isolated hepatocytes (69) and may be related to the ionizable N-acetyl group or to steric hindrance of binding to the carrier sites.

Recently three structurally simple, aliphatic compounds were introduced as model compounds to characterize hepatobiliary transport of the monovalent cations: triethyl-, tripropyl-, and tri-n-butylmethyl ammonium (TEMA, TPMA, and TBuMA, respectively). In contrast to PAEB they are not protein bound in plasma and liver cytosol and are not metabolized (239). Partition coefficients of these agents vary with a factor of 10, and this parameter is linearly correlated with the bile-liver concentration ratio of these agents, confirming the idea that lipophilicity of organic cations is a crucial factor especially in determining the biliary excretion process. The tributyl derivative is extensively excreted in bile, very much resembling PAEB. All of the three-model compounds are efficiently taken up in the liver, but uptake of TBuMA and TPMA is saturable, whereas that of TEMA is not. The latter agent, of which biliary clearance is equal to bile flow, may enter the cells via channels or pores and may also pass into bile paracellularly (239).

Anionic compounds such as taurocholate and iodide ions increase net uptake of these aliphatic cations into the liver, probably because of ion-pair formation in the liver, but only taurocholate also increased biliary excretion rate (239, 240). Various metabolic inhibitors and lowering of temperature inhibited uptake of PAEB and/or APAEB in liver slices (139, 314) and isolated hepatocytes (69). Uptake is clearly Na^+ independent, not inhibited by ouabain, but strongly inhibited by sulfhydryl reagents such as ethacrynic acid and N-methylmaleimide (69, 139, 232). Other basic drugs like d-tubocurarine and tertiary amines such as morphine and quinidine also decreased uptake rate of PAEB (69). Sulfhydryl agents also affect hepatic uptake in vivo. Portal administration of sulfhydryl agents clearly lowered hepatic content of the cation with parachloromercury benzoate more potent than N-methylmaleimide at this level (232).

MECHANISMS FOR HEPATIC UPTAKE OF BIVALENT CATIONS. Drugs in this class have two positively charged centers: either two quaternary ammonium groups or one quaternary group combined with a protonized tertiary nitrogen group. If two or more quaternary groups are present, hepatic uptake and biliary excretion are generally poor (138, 200, 287). Hepatobiliary transport in that case is only appreciable if the two positive charges in the molecule are masked by the presence of large lipophilic ring structures, as is the case with hexafluoronium (201). If in the bis-onium compounds one group is tertiary, dissociation of the attached proton leads to a strongly amphipathic molecule, in which besides the remaining quaternary group a large uncharged moiety enables hydrophobic interactions with the supposed carrier sites. Examples are the classic drug d-tubocurarine (see Fig. 13) and the steroidal cation vecuronium. The importance of the lipophilicity-hydrophilicity balance is clearly reflected in the different hepatic disposition of d-tubocurarine and its less lipid-soluble trimethyl derivative metocurine (214), as well as in the very dissimilar kinetics of the more recently developed muscle relaxants vecuronium and pancuronium (26, 27, 262), which differ in only one methyl group at one of the amine centers. The fast uptake of vecuronium into the liver is probably the major factor in terminating its neuromuscular blocking action in animals (26) and humans (27).

Accumulation in the liver of these agents may, in addition to efficient membrane transport at the sinusoidal level, also be explained by extensive intracellular binding (200). After hepatic uptake, d-tubocurarine is stored in a deep compartment (41) that is only very slowly available for biliary excretion (Fig. 16). Liver subfractionation studies as well as electron microscopy of d-tubocurarine molybdate precipitates in liver sections strongly suggest association with lysosomes in hepatocytes (214, 352). The accumulation in these organelles can be partially inhibited by chloroquine (351). The time course of lysosomal uptake and biliary excretion of d-tubocurarine indicates that these processes are not directly related. Association of organic cations with lysosomes might explain the persistent hepatic storage of these compounds observed in other studies (59, 72). The cationic drug that entered the lysosomes may finally be very slowly excreted into the bile, along with endogenous lysosomal material, as has been demonstrated for other lysosomotropic agents (179). The mechanisms of lysosomal accumulation of this type of organic cations remain to be clarified. Carrier-mediated permeation of d-tubocurarine from the cytoplasm into the organelles, followed by intralysosomal protonation, might play a role in addition to proton antiport into this acidic compartment. Coendocytosis of the drug with asialoorosomucoid (α_1-acid glycoprotein), a glycoprotein that avidly binds cationic drugs (336a) and that is efficiently taken up into the liver by receptor-mediated endocytosis, was recently excluded as a possible mechanism of lysosomal accumulation (336b). Triggering

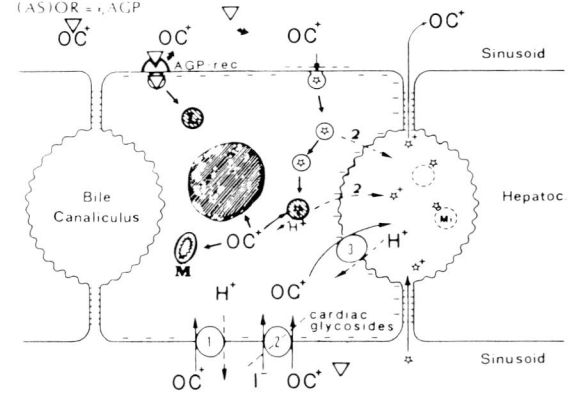

FIG. 16. Transport mechanisms for hepatobiliary transport of organic cations (OC^+) within organelles and canaliculi is indicated by ☆. Uptake can occur by 2 systems. System 2 is inhibitable by cardiac glycosides that may share a carrier process transporting high-molecular-weight (bulky) lipophilic organic cations, possibly as ion pairs with inorganic counteranions. System 1 may preferentially serve monovalent cations and may operate by proton antiport. Lipophilic organic cations bind in plasma to α_1-acid glycoprotein (α_1-AGP) [orosomucoid (OR)] but are not coendocytosed with asialoforms of this glycoprotein (ASOR). Accumulation in lysosomes (L) may occur via aspecific fluid-phase endocytosis at the plasma membrane or antiport with protons from the cytoplasm. Direct transport of drug from lysosomes to bile is not known. In addition to association with lysosomes, extensive binding can occur to mitochondria (M) and nuclei (N). Biliary excretion involves carrier-mediated transport possibly by antiport with protons. Binding to mixed biliary micelles (Mi) may facilitate net transport into bile canaliculi. AGP-rec., asialo-glycoprotein receptor.

of aspecific fluid-phase endocytosis by the organic cations at the surface of the cell (150, 157) and subsequent vesicular transport to the lysosomes might form an alternative explanation (Fig. 16). This mechanism has been proposed for renal accumulation of the strongly basic aminoglycosides (150, 157).

There is an apparent heterogeneity in the mechanisms for hepatic uptake of monovalent and bivalent organic cations. In the case of the bis-onium compounds, this process can be strongly inhibited with relatively low concentrations (1 μM) of cardiac glycosides in rat liver (200). The biliary excretion step, however, is not affected by these agents (201, 215). In contrast, uptake of the monoquaternary agents is not influenced by cardiac glycosides in these concentrations (201, 207, 215), indicating a separate uptake mechanism (Fig. 16). This idea is supported by the finding that large concentrations of PAEB, TBuMA, and related monovalent organic cations do not suppress the hepatic uptake of bis-onium compounds such as d-tubocurarine and vecuronium. On the contrary, such bis-onium compounds do inhibit hepatic uptake and also the biliary excretion of monovalent cations (207) and probably block the supposed carrier site for this transport process.

The actual mechanism by which the cardiac glycosides suppress the net hepatic uptake of bis-onium compounds is not completely understood. This effect is not directly linked to their well-known influence on the Na^+-K^+ pump (212). Digitoxin and K-strophanthoside are much more potent inhibitors than ouabain, yet their potency to inhibit this ATPase in rat liver is quite similar (212). It is more likely that binding of the two types of agents to a polyfunctional carrier system results in mutual competition (51, 253). The uptake of ouabain into hepatocytes is depressed by d-tubocurarine (69). This was confirmed by the observation of a competitive mutual inhibition (D. K. F. Meijer, unpublished observations). Photoaffinity labeling of proteins in hepatocyte plasma membrane with lipophilic organic cations such as N-(n-propyl)-21-deoxy-ajmalinium also can be strongly inhibited by cardiac glycosides, while these agents also kinetically interact during uptake in isolated hepatocytes (51). The particular studies thus might indicate the involvement of a multifunctional carrier process mediating hepatic uptake of amphipathic organic cations, cardiac glycosides, and probably also certain types of bile acids. In this respect it is important to note that bile salts, albeit at relatively high concentrations, inhibit hepatic uptake of various organic cations (344, 353) and also of ouabain (344, 345).

MECHANISMS FOR BILE-CANALICULAR TRANSPORT. In contrast to the multiplicit character of the hepatic uptake process of organic cations, the mono- and bivalent compounds probably share one biliary transport process. The canalicular transport step for the monovalent cations such as PAEB and TBuMA can be strongly suppressed by the more lipophilic bisonium compounds mentioned earlier (200). This interaction seems to be mutual, because PAEB depresses biliary excretion of d-tubocurarine (59). Retrograde biliary injection of sulfhydryl agents (233) strongly inhibits biliary excretion of some organic cations and affects hepatic uptake less so that hepatic content is increased. In this respect N-methylmaleimide was most potent. To leave the cell and enter the bile canaliculus, organic cations have to pass the barrier of the negative membrane potential, and it remains to be established what the driving force for this secretory process is. In addition to a partial association with biliary micelles, providing sink conditions in net transport (180, 190), electroneutral antiport with protons, as has been described in brush-border membranes of renal tubular cells (313), or transport as ion pairs with suitable counteranions should be considered as possible mechanisms. Nevertheless, only primary active transport, consuming energy-rich substrates such as ATP, or secondary active transport driven by gradients of other substrates could explain the considerable bile–cell water concentration gradients of unbound drugs that have been calculated in several studies with organic cations (137, 139, 200, 215, 239–242). The lipophilic organic anion tetraphenylborate, which stimulated net hepatic uptake of TBuMA, almost completely blocks its biliary excretion at an unchanged bile flow. The reason for this might be that either the compound forms intracellular ion pairs with TBuMA that are not transportable by the canalicular carrier system or that it is inserted in the canalicular membranes and facilitates electrogenic transport of TBuMA from bile to the cell cytoplasm (240), thereby preventing efficient biliary output. Such a membrane transport mechanism for organic cations was demonstrated in the presence of tetraphenylborate in isolated liver plasma membrane vesicles (281).

Attempts to identify potential carrier proteins for organic cations in apical and basolateral plasma membranes of the hepatocyte have a preliminary character. Photoaffinity labeling with photoreactive probes of the amphipathic cation N-(n-propyl)-21-deoxy-ajmalinium (51) (D. K. F. Meijer, unpublished observations) revealed quite specific labeling of 48- and 50-kDa polypeptides, whereas for a probe of PAEB showed incorporation in 48- and 72-kDa proteins. The involvement in organic cation transport of these various proteins and the relation with other drug-transporting systems remain to be established.

Uncharged Compounds

KINETIC ANALYSIS: MODEL COMPOUNDS. In addition to saccharides that reach concentration in bile close to the concomitant plasma concentration, two categories of uncharged drugs are highly excreted in bile in various species: the cardiac glycosides (310–312) and hormonal steroids (324). After primary hepatic uptake, especially the more lipophilic drugs in both groups are often metabolically altered and biliary excretion occurs predominantly in the form of various anionic metabolites (324). Some of the more hydrophilic steroidal compounds such as ouabain, however, are excreted unchanged and are used as model compounds to characterize the proposed hepatobiliary transport system for uncharged compounds. Uptake and secretion of ouabain have been characterized in the rat in vivo (41, 79, 164, 167, 170, 174, 175, 194, 211, 283), isolated perfused liver (41, 105, 212), and isolated rat hepatocytes (41, 68, 71, 253, 254, 302, 318) and kinetically been compared in vitro and in the intact organ (41). Affinity for the uptake system of the more lipophilic compounds digitoxin (68, 254, 302), K-strophanthoside (200, 202), and digoxin (68) is probably much higher than that of ouabain.

Ouabain can reach concentrations in isolated hepatocytes up to 100 times that in the medium. The drug is not appreciably bound to plasma proteins or to hepatic cytosolic proteins (68, 172). Cardiac glycosides have one or more hydrophilic sugar groups connected to the more hydrophobic steroid nucleus. The balance of these molecular features may favor excretion into bile. Digoxin and digitoxin undergo considerable metabolism and are only partly excreted in the unchanged form in various species. Ouabain is predominantly excreted via the bile in rats, but biliary excretion is much lower and urinary output more predominant in guinea pigs (194), dogs, and rabbits (283). Newborn rats also cannot concentrate ouabain in the liver. Development of this system occurs more slowly than that for taurocholate, as measured in vivo (170) and in isolated hepatocytes (318), suggesting principal differences in the two mechanisms involved.

In contrast to taurocholate at relatively low concentrations, uptake of ouabain is Na^+ independent (68, 253, 302). The compounds do not show mutual competitive inhibition in the uptake process (71). Also, interaction studies indicate dissimilar mechanisms: pretreatment with nafenopin depresses ouabain but not taurocholate uptake and excretion (206). The local anesthetic dibucaine blocks uptake of ouabain in addition to its choleretic activity in perfused liver but leaves taurocholate transport almost unchanged (105). Uptake of ouabain in hepatocytes does not show a pH optimum nor a preincubation stimulation, as observed for taurocholate (302). Hepatocytes isolated from rats pretreated with the enzyme inducer PCN have increased ouabain but unchanged taurocholate and PAEB uptake (70). However, there is probably overlapping substrate specificity with bile acids, and mutual interactions are possible. Taurocholate inhibits ouabain uptake into isolated hepatocytes (302). Inhibitors that affect uptake of cholate into hepatocytes also depress the uptake of ouabain in that system (253). Especially the unconjugated bile acids may accommodate the same Na^+-independent uptake as well as biliary transport systems that exist for ouabain and other cardiac glycosides and uncharged steroids (175). Major differences in affinity for Na^+-dependent and Na^+-independent systems for cholate and taurocholate were also inferred from other studies (9–11).

As for the biliary excretion step, there is also clear separation of ouabain and taurocholate transport, as has been clearly demonstrated recently in mutant rats with deficient ouabain excretion but normal taurocholate transport (146). Retrograde biliary injection of Triton X-100 markedly depressed ouabain excretion but left taurocholate excretion intact (322). Ouabain even increased the rate of secretion of taurocholate from isolated hepatocytes (298). Dehydrocholate but not taurocholate (211) depressed biliary excretion of ouabain.

MECHANISMS FOR HEPATOBILIARY TRANSPORT. Saturable carrier-mediated uptake of steroids other than ouabain has been reported in isolated hepatocytes for 17β-estradiol, testosterone, estrone (263), corticosterone (264), and cortisol (265). Not surprising for such lipophilic compounds, in addition to saturable also nonsaturable components are observed in the uptake kinetics. Interactions of dexamethasone and aldosterone in hepatocyte uptake studies have been reported (264). Some of these compounds also depress uptake of taurocholate and ouabain in liver slices (175) or isolated hepatocytes (68, 302). Estrone sulfate is a substrate for the Na^+-dependent taurocholate pathway (49, 301). In contrast to the sulfate conjugate, the glucuronides may be partly transported via the ouabain pathway (50), as well as by the organic anion pathway, as indicated by interactions with various model compounds out of this group (68, 79). It can be concluded that at least three pathways for hepatic uptake of steroidal compounds should be classified. One pathway is carrier mediated, in principle is open for all steroids, and is the major route for ouabain. Another is Na^+ dependent and preferentially transports conjugated bile acids and sulfate conjugates of hormonal steroids. The third pathway is simple membrane diffusion of the lipophilic steroids. That steroid glucuronides are partly transported by one of the anionic pathways awaits definite proof.

The carrier-mediated uptake of ouabain in hepatocytes is not correlated with binding to Na^+-K^+-ATPase, because the steroidal alkaloid cevadine (a tertiary amine) competitively inhibits uptake without any effect on binding of ouabain to this enzyme (253). Also uptake of ouabain in hepatoma cells is deficient, whereas the Na^+-K^+-ATPase is present and inhibitable by ouabain (253). The carrier system therefore seems to be a separate modality from the ATPase-pump system. In recent photoaffinity-label studies with a photoreactive probe of ouabain (254), predominant labeling of polypeptides in the 50-kDa region was found. The uptake transport processes of the cardiac glycosides are quite selectively induced by PCN (167), an effect that could also be demonstrated in newborn rats that have a largely deficient uptake system (168, 170). The specificity of this substrate induction may be related to the resemblance in structure of PCN and ouabain and the possibility that they share a common uptake system. Long-term exposure to PCN may lead to adaptive changes via increased synthesis or decreased degradation of the particular carrier proteins and should be an attractive object for further characterization of this transport system.

CONCLUSION

The process of hepatobiliary transport of drugs still retains many of its secrets. How many transport systems are involved? What is the relation between transport of endogenous and exogenous organic and inorganic compounds? To what extent do the sinusoidal and canalicular carrier systems differ? Are the transport processes homogeneously distributed within the liver acini? What is the explanation for some of the marked species differences that were observed? How do the transport systems adapt after substrate loading, hepatotoxic reactions, and during liver disease? Are genetically determined diseases with transport defects related to deficient carrier function? Are the supposed carriers mobile elements or ion-selective and/or membrane potential–sensitive channels?

Although in the last decades considerable progress was made in answering these questions, the study of transport of drugs in the liver remains to provide many challenges for hepatologists and pharmacologists.

REFERENCES

1. ABOU-EL-MAKAREM, M. M., P. MILLBURN, AND R. L. SMITH. Biliary excretion of [^{14}C]-succinylsulphathiazole in rat and rabbit. *Biochem. J.* 105: 1295–1299, 1967.
2. ACOCELLA, G., F. B. NICOLIS, AND L. T. TENCONI. The effect of an intravenous infusion of rifamycin SV on the excretion of bilirubin, bromsulphalein and indocyanine green in man. *Gastroenterology* 49: 521–525, 1965.
3. ADLER, R. D., F. J. WANNAGAT, AND R. K. OCKNER. Bile secretion in selective biliary obstruction: adaptation of taurocholate transport maximum to increased secretory load in the rat. *Gastroenterology* 73: 129–136, 1977.
4. ALPERT, S., M. MOSHER, A. SHANSKE, AND I. M. ARIAS. Multiplicity of hepatic excretory mechanisms for organic anions. *J. Gen. Physiol.* 53: 238–247, 1969.
5. ALVARES, A. P. Oxidative biotransformation of drugs. In: *The Liver: Biology and Pathobiology*, edited by I. Arias, H. Popper, D. Schachter, and D. A. Shafritz. New York: Raven, 1982, p. 265–280.
6. AN, T. Y., AND D. P. JONES. Intracellular inhibition of UDP-glucose dehydrogenase during ethanol oxidation. *Chem.-Biol. Interact.* 43: 283–288, 1983.
7. ANDERSON, J. H., R. C. ANDERSON, AND L. S. IBEN. Hepatic uptake of propranolol. *J. Pharmacol. Exp. Ther.* 206: 172–180, 1978.
8. ANWER, M. S. Biliary secretion of diisothiocyanostilbene disulfonate (DIDS) involves DIDS/HCO$_3$-(OH$^-$) exchange (Abstract). *Hepatology Baltimore* 6: 1213, 1986.
9. ANWER, M. S., AND D. HEGNER. Interaction of fusidates with bile acid uptake by isolated rat hepatocytes. *Naunyn-Schmiedeberg's Arch. Pharmacol.* 382: 329–332, 1978.
10. ANWER, M. S., AND D. HEGNER. Effect of Na$^+$ on bile acid uptake by isolated rat hepatocytes. *Hoppe-Seyler's Z. Physiol. Chem.* 359: 181–192, 1978.
11. ANWER, M. S., AND D. HEGNER. Effect of organic anions on bile acid uptake by isolated rat hepatocytes. *Hoppe-Seyler's Z. Physiol. Chem.* 359: 1027–1030, 1978.
12. ANWER, M. S., R. KROKER, AND D. HEGNER. Inhibition of hepatic uptake of bile acids by rifamycins. *Naunyn-Schmiedeberg's Arch. Pharmacol.* 302: 19–24, 1978.
13. ANWER, M. S., E. R. L. O'MAILLE, A. F. HOFMANN, R. A. DIPIETRO, AND E. MICHELOTTI. Influence of side-chain charge on hepatic transport of bile acids and bile acid analogues. *Am. J. Physiol.* 249 (*Gastrointest. Liver Physiol.* 12): G479–G488, 1985.
14. ARIAS, I. M., G. FLEISCHNER, I. LISTOWSKY, M. BHAGAVA, K. KAMISAKA, AND Z. GATMAITAN. Ligandin: structure and function. In: *Liver and Bile*, edited by L. Bianchi. Lancaster, UK: MTP, 1976, p. 157. (Falk Symp. 25.)
15. ASKIN, J. R., D. I. LYON, S. D. SHULL, C. I. WAGNER, AND R. D. SOLOWAY. Factors affecting delivery of bile to the duodenum in man. *Gastroenterology* 74: 560–565, 1978.
16. AVNER, D. L., AND M. M. BERENSON. Effect of choleretics on canalicular transport of protoporphyrin in the rat liver. *Am. J. Physiol.* 242 (*Gastrointest. Liver Physiol.* 5): G347–G353, 1982.
17. AXELROD, J. The discovery of the microsomal drug-metabolizing enzymes. In: *Drug Metabolism and Distribution*, edited by J. W. Lamble. Amsterdam: Elsevier, 1983, p. 1–11. (Curr. Rev. Biomed. Ser. 3.)
18. BABER, N., L. HALLIDAY, R. SIBEON, T. LITTLER, AND M. L. ORME. The interaction between indomethacin and probenecid. A clinical and pharmacokinetic study. *Clin. Pharmacol. Ther.* 24: 298–307, 1978.
19. BAILEY, D. G., AND G. S. JOHNSON. Multiple excretory mechanisms for organic anions. A study with succinylsulphathiazole and taurocholate in the rat. *Can. J. Physiol. Pharmacol.* 53: 97–103, 1975.
20. BAKER, K. J., AND S. E. BRADLEY. Binding of sulfobromophthalein (BSP) sodium by plasma albumin. Its role in hepatic BSP extraction. *J. Clin. Invest.* 45: 281–287, 1966.
21. BARNHART, J. L., AND R. CLARENBURG. Factors determining clearance of bilirubin in perfused rat liver. *Am. J. Physiol.* 225: 497–507, 1973.
22. BARNHART, J. L., R. R. GRONWALL, AND B. COMBES. Biliary excretion of sulfobromophthalein compounds in normal and mutant Corriedale sheep. Evidence for disproportionate transport defect for conjugated sulfobromophthalein. *Hepatology Baltimore* 1: 441–447, 1981.
23. BARNHART, J. L., B. L. WITT, W. G. HARDISON, AND R. N. BERK. Uptake of iopanoic acid by isolated rat hepatocytes in primary culture. *Am. J. Physiol.* 244 (*Gastrointest. Liver Physiol.* 7): G630–G636, 1983.
24. BASS, N. M. Function and regulation of hepatic and intestinal fatty acid binding proteins. *Chem. Phys. Lipids* 38: 95–114, 1985.
25. BASS, N. M., M. E. BARKER, AND A. L. JONES. Zonal expression of fatty acid binding protein (FABP) corresponds with fatty acid uptake by zones 1 and 2 of the hepatic acinus (Abstract). *Hepatology Baltimore* 5: 1011, 1985.
26. BENCINI, A. F., M. C. HOUWERTJES, AND S. AGOSTON. Effects of hepatic uptake of vecuronium bromide and its putative metabolites on their neuromuscular blocking actions in the

cat. *Br. J. Anaesth.* 57: 789–795, 1985.
27. BENCINI, A. F., A. H. J. SCAF, Y. J. SOHN, U. W. KERSTENKLEEF, AND S. AGOSTON. Hepatobiliary disposition of vecuronium bromide in man. *Anesthesiology* 58: 988–995, 1986.
28. BENZ, R. P., P. LAUGER, AND K. JANKO. Transport kinetics of hydrophobic ions in lipid bilayer membranes: charge-pulse relaxation studies. *Biochim. Biophys. Acta* 455: 701–720, 1976.
29. BERK, P. D., B. J. POTTER, AND W. STREMMEL. Role of plasma membrane ligand-binding proteins in the hepatocellular uptake of albumin-bound organic anions. *Hepatology Baltimore* 7: 165–176, 1987.
30. BERK, P. D., AND W. STREMMEL. Hepatocellular uptake of organic anions. *Prog. Liver Dis.* 8: 125–144, 1986.
31. BERK, R. N., J. L. BARNHART, AND L. E. GOLDBERGER. The enhancement of iopanoate excretion by taurocholate. *Invest. Radiol.* 15: S116–S121, 1980.
32. BERK, R. N., P. M. LOEB, A. COBO-FRENKEL, AND J. L. BARNHART. The biliary and urinary excretion of sodium tyropanoate and sodium ipodate in dogs: pharmacokinetics, influence of bile salts and choleretic effects with comparison to iopanoic acid. *Invest. Radiol.* 12: 85–95, 1977.
33. BERRY, M. N., AND D. S. FRIEND. High yield preparation of isolated rat liver parenchymal cells: a biochemical and fine structural study. *J. Cell Biol.* 43: 506–520, 1969.
34. BERTHELOT, P., AND B. H. BILLING. Effect of bunamiodyl on hepatic uptake of sulfobromophthalein in the rat. *Am. J. Physiol.* 211: 395–399, 1966.
35. BLASCHKE, T. F. Protein binding and kinetics of drugs in liver diseases. *Clin. Pharmacokinet.* 2: 32–44, 1977.
36. BLITZER, B. L., AND J. L. BOYER. Cellular mechanisms of bile formation. *Gastroenterology* 82: 346–357, 1982.
37. BLITZER, B. L., AND C. B. DONOVAN. A new method for the rapid isolation of basolateral plasma membrane vesicles from rat liver. Characterization, validation, and bile acid transport studies. *J. Biol. Chem.* 259: 9295–9301, 1984.
38. BLITZER, B. L., AND L. LYONS. Enhancement of Na^+-dependent bile acid uptake by albumin: direct demonstration in rat basolateral liver plasma membrane vesicles. *Am. J. Physiol.* 249 (*Gastrointest. Liver Physiol.* 12): G34–G38, 1985.
39. BLITZER, B. L., S. L. RATOOSH, C. B. DONOVAN, AND J. L. BOYER. Effects of inhibitors of Na^+-coupled ion transport on bile acid uptake by isolated rat hepatocytes. *Am. J. Physiol.* 243 (*Gastrointest. Liver Physiol.* 6): G48–G53, 1982.
40. BLOM, A., K. KEULEMANS, AND D. K. F. MEIJER. Transport of dibromosulphthalein by isolated rat hepatocytes. *Biochem. Pharmacol.* 30: 1809–1816, 1981.
41. BLOM, A., A. H. J. SCAF, AND D. K. F. MEIJER. Hepatic drug transport in the rat. A comparison between isolated hepatocytes, the isolated perfused liver and the liver in vivo. *Biochem. Pharmacol.* 31: 1553–1565, 1982.
42. BLOOMER, J. R., P. D. BERK, J. VERGALLA, AND N. I. BERLIN. Influence of albumin on the hepatic uptake of unconjugated bilirubin. *Clin. Sci. Mol. Med.* 45: 505–516, 1973.
43. BOYER, J. L. New concepts of mechanisms of hepatocyte bile formation. *Physiol. Rev.* 60: 303–326, 1980.
44. BOYER, J. L. Mechanisms of bile secretion and hepatic transport. In: *Physiology of Membrane Disorders* (2nd ed.), edited by T. E. Andreoli, J. F. Hoffman, D. D. Fanestil, and S. G. Schultz. New York: Plenum, 1986, p. 609–636.
45. BOYER, J. L., E. ELIAS, AND T. J. LAYDEN. The paracellular pathway and bile formation. *Yale J. Biol. Med.* 52: 61–67, 1979.
46. BRAUER, R. W., G. F. LEONG, AND R. J. HOLLOWAY. Mechanics of bile secretion: effect of perfusion pressure and temperature on bile flow and bile secretion pressure. *Am. J. Physiol.* 177: 103–112, 1954.
47. BRAUER, R. W., R. L. PESSOTTI, AND P. PIZZOLATO. Isolated rat liver preparation. Bile production and other basic properties. *Proc. Soc. Exp. Biol. Med.* 78: 174–181, 1951.
48. BREIMER, D. D., N. P. E. VERMEULEN, M. DANHOF, M. W. E. TEUNISSEN, R. P. JOERES, AND M. VAN DER GRAAFF. Assessment and prediction of in vivo oxidative drug metabolism activity. In: *Pharmacokinetics: A Modern View*, edited by L. Z. Benet, G. Levy, and B. L. Ferraiolo. New York: Plenum, 1984, p. 191–216.
49. BROCK, W. J., S. DURHAM, AND M. VORE. Characterization of the interaction between estrogen metabolites and taurocholate for uptake into isolated hepatocytes. Lack of correlation between cholestasis and inhibition of taurocholate uptake. *J. Steroid Biochem.* 20: 1181–1185, 1984.
50. BROCK, W. J., AND M. VORE. Characterization of uptake of steroid glucuronides into isolated male and female rat hepatocytes. *J. Pharmacol. Exp. Ther.* 229: 175–181, 1984.
51. BUSCHER, H. P., G. FRICKER, W. GEROK, W. KRAMER, G. KURZ, M. MULLER, AND S. SCHNEIDER. Membrane transport of amphiphilic compounds by hepatocytes. In: *Receptor-Mediated Uptake in the Liver*, edited by H. Greten, E. Windler, and U. Beisiegel. Heidelberg, FRG: Springer-Verlag, 1986, p. 189–199.
52. CALDWELL, J. Conjugation reactions in the metabolism of xenobiotics. In: *The Liver: Biology and Pathobiology*, edited by I. Arias, H. Popper, D. Schachter, and D. A. Shafritz. New York: Raven, 1982, p. 281–296.
53. CARSON, E. R., AND E. A. JONES. Use of kinetic analysis and mathematical modelling in the study of metabolic pathways in vivo. Application to hepatic organic anion metabolism. *N. Engl. J. Med.* 300: 1016–1027, 1979.
54. CHEN, E. H., J. J. GUMUCIO, N. H. HO, AND D. L. GUMUCIO. Hepatocytes of zones 1 and 3 conjugate sulfobromophthalein with glutathione. *Hepatology Baltimore* 4: 467–476, 1984.
55. CHEN, H. G., AND J. F. GROSS. Pharmacokinetics of drugs subject to enterohepatic circulation. *J. Pharm. Sci.* 68: 792–799, 1979.
56. CHENG, S., AND D. LEVY. Characterization of the anion transport system in hepatocyte plasma membrane. *J. Biol. Chem.* 255: 2637–2640, 1980.
57. CHIPMAN, J. K. Bile as a source of potential reactive metabolites. *Toxicology* 25: 99–111, 1982.
58. CHRISTENSSON, P. I., AND G. ERIKSSON. Effects of six anaesthetic agents on UDP-glucuronic acid and other nucleotides in rat liver. *Acta Anaesthesiol. Scand.* 29: 629–631, 1985.
59. COHEN, E. N., B. H. WINSTOW, AND D. SMITH. The metabolism and elimination of d-tubocurarine-^{3}H. *Anesthesiology* 28: 309–317, 1967.
60. DELAGE, Y., S. ERLINGER, M. DUVAL, AND J. P. BENHAMOU. Influence of dehydrocholate and taurocholate on bromosulphthalein uptake, storage, and excretion in the dog. *Gut* 16: 105–108, 1976.
61. DESPOPOULOS, A. Congruence of renal and hepatic excretory functions: sulfonic acid dyes. *Am. J. Physiol.* 220: 1755–1758, 1971.
61a. DE VRIES, M. H., G. M. M. GROOTHUIS, G. J. MULDER, H. NGUYEN, AND D. K. F. MEIJER. Secretion of the organic anion harmol sulfate from liver into blood. Evidence for a carrier-mediated mechanism. *Biochem. Pharmacol.* 34: 2129–2135, 1985.
62. DIETMAIER, A., R. GASSER, J. GRAF, AND M. PETERLIK. Investigations on the sodium dependence of bile acid fluxes in the isolated perfused rat liver. *Biochim. Biophys. Acta* 443: 81–91, 1976.
63. DILLS, R. L., AND C. D. KLAASSEN. Decreased glucuronidation of bilirubin by diethyl ether anesthesia. *Biochem. Pharmacol.* 33: 2813–2814, 1984.
64. DUFFY, M., B. L. BLITZER, AND J. L. BOYER. Direct determination of the driving forces for taurocholate uptake into rat liver plasma membrane vesicles. *J. Clin. Invest.* 72: 1470–1481, 1983.
65. DUGGAN, D. E., AND K. C. KWAN. Enterohepatic recirculation of drugs as a determinant of therapeutic ratio. *Drug Metab. Rev.* 9: 21–41, 1979.
66. DUJOVNE, C. A., J. H. GUSTAFSON, AND R. A. DICKEY. Quantification of biliary excretion of drugs in man. *Clin. Pharmacol. Ther.* 31: 187–194, 1982.

67. DUTTON, G. J.: *Glucuronidation of Drugs and Other Compounds.* Boca Raton, FL: CRC, 1980.
68. EATON, D. L., AND C. D. KLAASSEN. Carrier-mediated transport of ouabain in isolated hepatocytes. *J. Pharmacol. Exp. Ther.* 205: 480–488, 1975.
69. EATON, D. L., AND C. D. KLAASSEN. Carrier-mediated transport of the organic cation procaine amide ethobromide by isolated rat liver parenchymal cells. *J. Pharmacol. Exp. Ther.* 206: 595–606, 1978.
70. EATON, D. L., AND C. D. KLAASSEN. Effects of microsomal enzyme inducers on carrier-mediated transport systems in isolated rat hepatocytes. *J. Pharmacol. Exp. Ther.* 208: 381–385, 1979.
71. EATON, D. L., AND J. A. RICHARDS. Kinetic evaluation of carrier-mediated transport of ouabain and taurocholic acid in isolated hepatocytes. Evidence for independent transport system. *Biochem. Pharmacol.* 35: 2721–2725, 1986.
72. ECHIGOYA, Y., Y. MATSUMOTO, Y. NAKAGAWA, T. SUGA, AND S. NIINOBE. Metabolism of quaternary ammonium compound. I. Binding of tropane alkaloids to rat liver lysosomes. *Biochem. Pharmacol.* 21: 477–484, 1972.
73. EDMONDSON, J. W., B. A. MILLER, AND L. LUMENG. Effect of glucagon on hepatic taurocholate uptake: relationship to membrane potential. *Am. J. Physiol.* 249 (*Gastrointest. Liver Physiol.* 12): G427–G433, 1985.
74. ENGELKIND, L. R., R. GRONWALL, AND M. S. ANWER. Effect of dehydrocholic, chenodeoxycholic and taurocholic acids on the excretion of bilirubin. *Am. J. Vet. Res.* 41: 355–361, 1980.
75. ERLINGER, S. Hepatocyte bile secretion: current views and controversies. *Hepatology Baltimore* 1: 352–360, 1981.
76. ERLINGER, S. What is cholestasis in 1985? *J. Hepatol.* 1: 687–693, 1985.
77. ERLINGER, S. Bile flow. In: *The Liver: Biology and Pathobiology* (2nd ed.), edited by I. M. Arias, W. B. Jakoby, H. Popper, D. Schachter, and D. A. Shafritz. New York: Raven, 1988, p. 643–662.
78. ERLINGER, S., AND D. DHUMEAUX. Mechanisms and control of secretion of bile water and electrolytes. *Gastroenterology* 66: 281–304, 1974.
79. ERTMANN, R. R., AND K. H. DAMM. Influence on bile flow, theophylline and some organic anions on the biliary excretion of ^{3}H-ouabain in rats. *Arch. Int. Pharmacodyn. Ther.* 218: 290–298, 1975.
80. EVANS, W. H. A biochemical dissection of the functional polarity of the plasma membrane of the hepatocyte. *Biochim. Biophys. Acta* 604: 27–64, 1980.
81. EVANS, W. H. Membrane traffic at the hepatocytes sinusoidal and canalicular surface domains. *Hepatology Baltimore* 5: 452–458, 1981.
82. FALANY, C. M., AND T. R. TEPHLY. Separation, purification and characterization of three isoenzymes of UDP-glucuronyltransferase from rat liver microsomes. *Arch. Biochem. Biophys.* 227: 248–258, 1983.
83. FISCHER, E., AND F. VARGA. Effect of pretreatment with exogenous organic anions on biliary excretion in rats. *Arch. Int. Pharmacodyn. Ther.* 264: 135–143, 1983.
84. FITZ, J. G., AND B. F. SCHARSCHMIDT. Regulation of transmembrane electrical potential (E_m) of rat hepatocytes in situ (Abstract). *Hepatology Baltimore* 5: 1011, 1985.
85. FLECK, C., AND H. BRAUNLICH. Methods in testing interrelationships between excretion of drugs via urine and bile. *Pharmacol. Ther.* 25: 1–22, 1984.
86. FLEISCHER, G., D. K. F. MEIJER, W. G. LEVINE, Z. GATMAITAN, AND I. M. ARIAS. Effect of hypolipidemic drugs nafenopin and clofibrate on the concentration of ligandin and Z-protein in rat liver. *Biochem. Biophys. Res. Commun.* 67: 1401–1407, 1975.
87. FORKER, E. L. Mechanisms of hepatic bile formation. *Annu. Rev. Physiol.* 39: 323–347, 1977.
88. FORKER, E. L., AND G. GIBSON. Interaction between sulfobromophthalein (BSP) and taurocholate. The kinetics of transport from liver cells to bile in rats. In: *The Liver: Quantitative Aspects of Structure and Function*, edited by G. Paumgartner and R. Preisig. Basel: Karger, 1973, p. 326–336.
89. FORKER, E. L., AND B. A. LUXON. Albumin helps mediate removal of taurocholate by rat liver. *J. Clin. Invest.* 67: 1517–1522, 1981.
90. FORKER, E. L., AND B. A. LUXON. Analyzing tracer disappearance curves to study hepatic transport kinetics. *Am. J. Physiol.* 244 (*Gastrointest. Liver Physiol.* 7): G573–G577, 1983.
91. FORKER, E. L., AND B. LUXON. Lumpers vs. distributors. *Hepatology Baltimore* 5: 1236–1237, 1985.
92. FRIMMER, M. Organotropism by carrier-mediated transport. *Trends Pharmacol. Sci.* 3: 395–397, 1982.
93. FUHRMAN-LANE, C., AND J. M. FUJIMOTO. Trans-stilbene oxide administration increased hepatic glucuronidation of morphine but decreased biliary excretion of morphine glucuronide in rats. *J. Pharmacol. Exp. Ther.* 222: 526–533, 1982.
94. FUJIMOTO, J. M. Some in vivo methods for studying sites of toxicant action in relation to bile formation. In: *Toxicology of the Liver*, edited by G. Plaa and W. R. Hewitt. New York: Raven, 1982, p. 121–145.
95. GALEAZZI, R., I. LORENZINI, AND F. ORLANDI. Rifampicin-induced elevation of serum bile acids in man. *Dig. Dis. Sci.* 25: 108–112, 1980.
96. GEBHARDT, R. Use of cultured hepatocytes in studies on bile formation. In: *Research in Isolated and Cultured Hepatocytes*, edited by A. Guillouzo and C. Guguen-Guillouzo. Paris: INSERM, 1986, p. 353–376.
97. GEROLAMI, A., AND J. C. SARLES. Biliary secretion and motility. In: *Gastrointestinal Physiology II*, edited by R. K. Crane. Baltimore, MD: University Park, 1977, vol. 12, p. 224–256. (Int. Rev. Physiol. Ser.)
98. GOETZEE, A. E., T. G. RICHARDS, AND V. R. TINDALL. Experimental changes in liver function induced by probenecid. *Clin. Sci.* 19: 63–78, 1960.
99. GORESKY, C. A. Initial distribution and rate of uptake of sulfobromophthalein in the liver. *Am. J. Physiol.* 207: 13–26, 1964.
100. GORESKY, C. A. The processes of cellular uptake and exchange in the liver. *Federation Proc.* 41: 3033–3039, 1982.
101. GORESKY, C. A., H. H. HADDAD, W. S. KLUGER, B. E. NADEAU, AND G. G. BACH. The enhancement of maximal bilirubin excretion with taurocholate-induced increments in bile flow. *Can. J. Physiol. Pharmacol.* 52: 389–403, 1974.
102. GÖTZ, R., L. R. SCHWARZ, AND H. GREIM. Effect of pentachlorophenol and 2,4,6-trichlorophenol on the disposition of sulfobromophthalein and respiration of isolated liver cells. *Arch. Toxicol.* 44: 147–155, 1980.
103. GRAF, J. Canalicular bile salt-independent bile formation: concepts and clues from electrolyte transport in rat liver. *Am. J. Physiol.* 244 (*Gastrointest. Liver Physiol.* 7): G233–G246, 1983.
104. GRAF, J., A. GAUTAM, AND J. L. BOYER. Isolated rat hepatocyte couplets: a primary secretory unit for electrophysiologic studies of bile secretory function. *Proc. Natl. Acad. Sci. USA* 81: 6516–6520, 1984.
105. GRAF, J., AND M. PETERLIK. Ouabain-mediated sodium uptake and bile formation by isolated perfused rat liver. *Am. J. Physiol.* 230: 876–885, 1976.
106. GRAUSZ, H., AND R. SCHMID. Reciprocal relation between plasma albumin level and hepatic sulfobromophthalein removal. *N. Engl. J. Med.* 284: 1403–1406, 1971.
107. GREGUS, Z., AND C. D. KLAASSEN. Enterohepatic circulation of toxicants. In: *Gastrointestinal Toxicology*, edited by K. Roxman and O. Hanninen. Amsterdam: Elsevier, 1986, p. 57–118.
108. GRISHAM, J. W. Cell types in rat liver cultures: their identification and isolation. *Mol. Cell. Biochem.* 53/54: 23–33, 1983.
109. GROOTHUIS, G. M. M., M. J. HARDONK, K. P. T. KEULEMANS, P. NIEUWENHUIS, AND D. K. F. MEIJER. Autoradiographic and kinetic demonstration of acinar heterogeneity of taurocholate transport. *Am. J. Physiol.* 243 (*Gastrointest. Liver*

Physiol. 6): G455–G462, 1982.
110. GROOTHUIS, G. M. M., M. J. HARDONK, AND D. K. F. MEIJER. Hepatobiliary transport of drugs: do periportal and perivenous hepatocytes perform the same job? *Trends Pharmacol. Sci.* 6: 322–327, 1985.
111. GROOTHUIS, G. M. M., K. P. T. KEULEMANS, M. J. HARDONK, AND D. K. F. MEIJER. Acinar heterogeneity in hepatic transport of dibromosulfophthalein and ouabain studied by autoradiography, normal and retrograde perfusions and computer simulation. *Biochem. Pharmacol.* 32: 3069–3078, 1983.
112. GROOTHUIS, G. M. M., J. G. WEITERING, K. P. T. KEULEMANS, M. J. HARDONK, D. MULDER, AND D. K. F. MEIJER. Heterogeneity of rat hepatocytes in bile acid and DBSP transport studied after induction of selective acinar damage by N-hydroxy-2-acetylaminofluorene and carbon tetrachloride. *Naunyn-Schmiedeberg's Arch. Pharmacol.* 322: 310–318, 1983.
113. GUARINO, A. M., AND L. S. SCHANKER. Biliary excretion of probenecid and its glucuronide. *J. Pharmacol. Exp. Ther.* 164: 387–395, 1968.
114. GUENGERICH, F. P., G. A. DANNAN, S. T. WRIGHT, M. V. MARTIN, AND L. S. KAMINSKY. Purification and characterization of microsomal cytochrome P450s. *Xenobiotica* 12: 701–716, 1982.
115. GUENGERICH, F. P., AND D. C. LIEBLER. Enzymatic activation of chemicals to toxic metabolites. *Crit. Rev. Toxicol.* 14: 259–307, 1985.
116. GUGUEN-GUILLOUZO, C., AND A. GUILLOUZO. Modulation of functional activities in cultured rat hepatocytes. *Mol. Cell. Biochem.* 53/54: 35–56, 1983.
117. GUMUCIO, J. J. Functional and anatomic heterogeneity in the liver acinus: impact on transport. *Am. J. Physiol.* 244 (*Gastrointest. Liver Physiol.* 7): G578–G582, 1983.
118. GUMUCIO, D. L., J. J. GUMUCIO, J. A. P. WILSON, C. CUTTER, M. KRAUSS, R. CALDWELL, AND E. CHEN. Albumin influences sulfobromophthalein transport by hepatocytes of each acinar zone. *Am. J. Physiol.* 246 (*Gastrointest. Liver Physiol.* 9): G86–G95, 1984.
119. GUMUCIO, J. J., M. MAY, C. DVORAK, J. CHIANNALE, AND V. MASSEY. The isolation of functionally heterogeneous hepatocytes of the proximal and distal half of the liver acinus in the rat. *Hepatology Baltimore* 6: 932–944, 1986.
120. GUMUCIO, J. J., AND D. L. MILLER. Functional implications of liver cell heterogeneity. *Gastroenterology* 80: 393–403, 1981.
121. GUMUCIO, J. J., D. L. MILLER, M. D. KRAUSS, AND C. C. ZANOLLI. Transport of fluorescent compounds into hepatocytes and the resultant zonal labeling of the hepatic acinus in the rat. *Gastroenterology* 80: 639–645, 1981.
122. HACKI, W., J. BIRCHER, AND R. PREISIG. A new look at the plasma disappearance of sulfobromophthalein (BSP): correlation with BSP transport maximum and hepatic plasma flow in man. *J. Lab. Clin. Med.* 88: 1019–1029, 1976.
123. HANZON, V. Liver cell secretion under normal and pathologic conditions studied by fluorescence microscopy on living rats. *Acta Physiol. Scand. Suppl.* 101: 1–268, 1952.
124. HARRIS, C., AND R. G. THURMAN. A new method to study glutathione adduct formation in periportal and pericentral regions of the liver lobule by micro-reflectance spectrophotometry. *Mol. Pharmacol.* 29: 88–96, 1986.
125. HARRISON, L. I., AND M. GIBALDI. Influence of cholestasis on drug elimination: pharmacokinetics. *J. Pharm. Sci.* 65: 1346–1348, 1976.
126. HERMAN, R. H., R. N. REDINGER, AND D. M. SMALL. The effects of surgery on bile secretion and composition. *Surg. Forum* 22: 378–380, 1971.
127. HIROM, P. C., R. D. HUGHES, AND P. MILLBURN. The physicochemical factor required for the biliary excretion of organic cations and anions. *Biochem. Soc. Trans.* 2: 327–330, 1974.
128. HIROM, P. C., P. MILLBURN, AND R. L. SMITH. Bile and urine as complementary pathways for the excretion of foreign organic compounds. *Xenobiotica* 6: 55–64, 1976.
129. HIROM, P. C., P. MILLBURN, R. L. SMITH, AND R. T. WILLIAMS. Molecular weight and chemical structure as factors in the biliary excretion of sulphonamides in the rat. *Xenobiotica* 2: 205–214, 1972.
130. HIROM, P. C., P. MILLBURN, R. L. SMITH, AND R. T. WILLIAMS. Species variations in the threshold molecular-weight factor for the biliary excretion of organic anions. *Biochem. J.* 129: 1071–1077, 1972.
131. HJELLE, J. J. Hepatic UDP-glucuronic acid regulation during acetaminophen biotransformation in rats. *J. Pharmacol. Exp. Ther.* 237: 750–756, 1986.
132. HOFMANN, A. F. Chemistry and enterohepatic circulation of bile acids. *Hepatology Baltimore* 4: 4S–14S, 1984.
133. HOPFER, U. Isolated membrane vesicles as tools for analysis of epithelial transport. *Am. J. Physiol.* 233 (*Endocrinol. Metab. Gastrointest. Physiol.* 2): E445–E449, 1977.
134. HOUSTON, J. B. Kinetics of drug metabolism and disposition: physiological determinants. In: *Drug Metabolism and Disposition: Considerations in Clinical Pharmacology*, edited by G. R. Wilkinson and M. D. Rawlins. Lancaster, UK: MTP, 1985, p. 62–90.
135. HOWELL, S. R., G. A. HAZELTON, AND C. D. KLAASSEN. Depletion of hepatic UDP-glucuronic acid by drugs that are glucuronidated. *J. Pharmacol. Exp. Ther.* 236: 610–614, 1986.
136. HUGENTOBLER, G., AND P. J. MEIER. Multispecific anion exchange in basolateral (sinusoidal) rat liver plasma membrane vesicles. *Am. J. Physiol.* 251 (*Gastrointest. Liver Physiol.* 14): G656–G664, 1986.
137. HUGHES, R. D., P. MILLBURN, AND R. T. WILLIAMS. Molecular weight as a factor in the excretion of monoquaternary ammonium cations in the bile of rat, rabbit and guinea pig. *Biochem. J.* 136: 967–978, 1973.
138. HUGHES, R. D., P. MILLBURN, AND R. T. WILLIAMS. Biliary excretion of some diquaternary ammonium cations in the rat, guinea pig and rabbit. *Biochem. J.* 136: 979–984, 1973.
139. HWANG, S. W., AND L. S. SCHANKER. Hepatic uptake and biliary excretion of N-acetyl procaine amide ethobromide in the rat. *Am. J. Physiol.* 225: 1437–1443, 1973.
140. INGELMAN-SUNDBERG, M. Bioactivation or inactivation of toxic compounds. In: *Drug Metabolism and Distribution*, edited by J. W. Lamble. Amsterdam: Elsevier, 1983, p. 22–27. (Curr. Rev. Biomed. Ser. 3.)
141. INOUE, M., E. HIRATA, Y. MORINO, S. NAGASE, J. CHOWDHURY, N. R. CHOWDHURY, AND I, M. ARIAS. The role of albumin in the hepatic transport of bilirubin: studies in mutant analbuminemic rats. *J. Biochem. Tokyo* 97: 737–743, 1985.
142. INOUE, M., R. KINNE, T. TRAN, AND I. M. ARIAS. Taurocholate transport by rat liver sinusoidal membrane vesicles: evidence of sodium cotransport. *Hepatology Baltimore* 2: 572–579, 1982.
143. INOUE, M., R. KINNE, T. TRAN, AND I. M. ARIAS. Taurocholate transport by rat liver canalicular membrane vesicles. *J. Clin. Invest.* 73: 659–663, 1984.
144. INOUE, M., R. KINNE, T. TRAN, L. DIEMPICA, AND I. M. ARIAS. Rat liver canalicular membrane vesicles. Isolation and tropocological characterization. *J. Biol. Chem.* 258: 5183–5188, 1983.
145. ISRAILI, Z. H., AND P. G. DAYTON. Enhancement of xenobiotic elimination: role of intestinal excretion. *Drug Metab. Rev.* 15: 1123–1159, 1984.
146. JANSEN, P. L. M., G. M. M. GROOTHUIS, W. H. M. PETERS, AND D. K. F. MEIJER. Selective hepatobiliary transport defect for organic anions and neutral steroids in mutant rats with hereditary conjugated hyperbilirubinemia. *Hepatology Baltimore* 7: 71–76, 1987.
147. JANSEN, P. L. M., W. H. PETERS, AND W. M. LAMERS. Hereditary chronic conjugated hyperbilirubinemia in mutant rats caused by defective hepatic anion transport. *Hepatology Baltimore* 5: 573–579, 1985.
148. JAVITT, N. B. Phenol-3,6-dibromophthalein disulfonate, a new compound for the study of liver disease. *Proc. Soc. Exp. Biol. Med.* 117: 254–257, 1964.

149. JAVITT, N. B. Hepatic bile formation. *N. Engl. J. Med.* 295: 1464–1469, 1976.
150. JOHANSSON, P., J. O. JOSEFSOON, AND L. NASSBEYER. Induction and inhibition of pinocytosis by aminoglycoside antibiotics. *Br. J. Pharmacol.* 83: 615–623, 1984.
151. JONES, A. L., G. T. HRADEK, R. H. RENSTON, K. Y. WONG, G. KARLAGANIS, AND G. PAUMGARTNER. Autoradiographic evidence for hepatic lobular concentration gradient of bile acid derivative. *Am. J. Physiol.* 238 (*Gastrointest. Liver Physiol.* 1): G233–G237, 1980.
152. JONES, A. L., D. L. SCHMUCHER, R. H. RENSTON, AND T. MURAKAMI. The architecture of bile secretion. A morphological perspective of physiology. *Dig. Dis. Sci.* 25: 609–629, 1980.
153. JONES, D. B., M. S. CHING, R. SMALLWOOD, AND D. J. MORGAN. A carrier-protein receptor is not a prerequisite for avid hepatic elimination of highly bound compounds: a study of propranol elimination by the isolated perfused rat liver. *Hepatology Baltimore* 5: 590–593, 1985.
154. JONES, E. A., J. M. VIERLING, C. J. STEER, AND J. REICHEN. Cell surface receptors in the liver. *Prog. Liver Dis.* 6: 43–80, 1979.
155. JOPPEN, C., E. PETZINGER, AND M. FRIMMER. Properties of iodipamide uptake by isolated rat hepatocytes. *Naunyn-Schmiedeberg's Arch. Pharmacol.* 331: 393–397, 1985.
156. JUNGERMANN, K., AND N. KATZ. Functional hepatocellular heterogeneity. *Hepatology Baltimore* 3: 385–395, 1982.
157. KALOYANIDES, G. J. Renal pharmacology of aminoglycoside antibiotics. In: *Kidney, Small Proteins and Drugs*, edited by C. Bianchi, A. Bertelli, and C. G. Duarte. Basel: Karger, 1984, p. 148–167. (Contr. Nephrol. 42.)
158. KAPLOWITZ, N., T. Y. AW, AND M. OOKHTENS. The regulation of hepatic glutathione. *Annu. Rev. Pharmacol. Toxicol.* 25: 715–744, 1985.
159. KATES, R. E., AND T. N. TOZER. Biliary secretion of methotrexate in rats and its inhibition by probenecid. *J. Pharm. Sci.* 65: 1348–1351, 1976.
160. KENWRIGHT, S., AND A. J. LEVI. Impairment of hepatic uptake of rifamycin by probenecid and its therapeutic implications. *Lancet* 2: 1401–1405, 1973.
161. KETLEY, J. N., W. H. HABIG, AND W. B. JAKOBY. Binding of nonsubstrate ligands to the glutathione S-transferases. *J. Biol. Chem.* 250: 8670–8673, 1975.
162. KETTERER, B., P. ROSS-MANSELL, AND J. K. WHITEHEAD. The isolation of carcinogenic-binding protein from liver of rats given 4-dimethylaminoazobenzene. *Biochem. J.* 103: 316–324, 1967.
163. KETTERER, B., E. TIPPING, J. F. HACKNEY, AND D. BEALE. A low-molecular-weight protein that resembles ligandin in its binding properties. *Biochem. J.* 155: 511–521, 1976.
164. KITANI, K., S. KANAI, AND R. MIURA. Increased biliary excretion of ouabain induced by bucolome in the rat. *Clin. Exp. Pharmacol. Physiol.* 5: 117–124, 1978.
165. KITANI, K., R. MIURA, AND S. KANAI. Difference in the effect of bucolome on the hepatic transport maximum of sulfobromophthalein and indocyanine green. *Tohoku J. Exp. Med.* 126: 247–256, 1978.
166. KLAASSEN, C. D. Comparison of the effects of two-thirds hepatectomy and bile-duct ligation on hepatic excretory function. *J. Pharmacol. Exp. Ther.* 191: 25–31, 1974.
167. KLAASSEN, C. D. Effect of microsomal enzyme inducers on the biliary excretion of cardiac glycosides. *J. Pharmacol. Exp. Ther.* 191: 201–211, 1974.
168. KLAASSEN, C. D. Stimulation of the development of the hepatic excretory mechanism to ouabain in newborn rats with microsomal enzyme inducers. *J. Pharmacol. Exp. Ther.* 191: 212–218, 1974.
169. KLAASSEN, C. D. Extrahepatic distribution of sulfobromophthalein. *Can. J. Physiol. Pharmacol.* 53: 120–123, 1975.
170. KLAASSEN, C. D. Independence of bile acid and ouabain hepatic uptake: studies in the newborn rat. *Proc. Soc. Exp. Biol. Med.* 157: 66–69, 1978.
171. KLAASSEN, C. D., D. L. EATON, AND C. Z. CAGEN. Hepatobiliary disposition of xenobiotics. *Prog. Drug Metab.* 6: 1–75, 1981.
172. KLAASSEN, C. D., AND J. B. WATKINS. Mechanisms of bile formation, hepatic uptake and biliary excretion. *Pharmacol. Rev.* 36: 1–67, 1984.
173. KRAMER, W., U. BICKEL, H. P. BASCHER, W. GEROK, AND G. KURZ. Bile-salt binding polypeptides in plasma membranes of hepatocytes revealed by photoaffinity labeling. *Eur. J. Biochem.* 129: 13–24, 1982.
174. KUPFERBERG, H. J. Inhibition of ouabain-^{3}H uptake by liver slices and its excretion into the bile by compounds having a steroid nucleus. *Life Sci.* 8: 1179–1185, 1969.
175. KUPFERBERG, H. J., AND L. S. SCHANKER. Biliary secretion of ouabain-^{3}H and its uptake by liver slices in the rat. *Am. J. Physiol.* 214: 1048–1053, 1968.
176. KURZ, H. Die Permeation von Giften in die Leber; Eigenschaften der Zellmembran. *Naunyn-Schmiedebergs Arch. Pharmakol. Exp. Pathol.* 254: 33–44, 1966.
177. LAPERCHE, Y., C. GRAILLOT, J. ARONDEL, AND P. BERTHELOT. Uptake of rifampicin by isolated rat liver cells. Interaction with sulfobromophthalein uptake and evidence for separate carriers. *Biochem. Pharmacol.* 28: 2065–2069, 1979.
178. LAPERCHE, Y., A. M. PREAUX, G. FELDMANN, J. L. MAHU, AND P. BERTHELOT. Effect of fasting on organic anion uptake by isolated rat liver cells. *Hepatology Baltimore* 1: 617–621, 1981.
179. LaRUSSO, N. F., L. J. KOST, J. A. CARTER, AND S. S. BARHAM. Triton WR-1339, a lysosomotropic compound, is excreted into bile and alters the biliary excretion of lysosomal enzymes and lipids. *Hepatology Baltimore* 2: 209–215, 1982.
180. LAVY, U. I., W. HESPE, AND D. K. F. MEIJER. Uptake and excretion of the quaternary ammonium compound deptropine methiodide in the isolated perfused rat liver. *Naunyn-Schmiedeberg's Arch. Pharmacol.* 275: 183–192, 1972.
181. LEVI, A. J., Z. GATMAITAN, AND I. M. ARIAS. Two cytoplasmic protein fractions, Y and Z, and their possible role in hepatic uptake of bilirubin, sulfobromophthalein and other anions. *J. Clin. Invest.* 48: 2156–2167, 1969.
182. LEVINE, W. G. Biliary excretion of drugs and other xenobiotics. *Annu. Rev. Pharmacol. Toxicol.* 18: 81–96, 1978.
183. LEVINE, W. G. Biliary excretion of drugs and other xenobiotics. *Prog. Drug Res.* 25: 361–419, 1981.
184. LEVINE, W. G. Excretion mechanisms. In: *Biological Basis of Detoxification*, edited by J. Caldwell and W. B. Jakoby. New York: Academic, 1983, p. 251–285.
185. LEVY, D., AND S. CHENG. Photoaffinity labeling of anion transport components in hepatocyte plasma membranes. *Ann. NY Acad. Sci.* 346: 232–242, 1980.
186. LIEN, E. J. Structure-activity relationships and drug disposition. *Annu. Rev. Pharmacol. Toxicol.* 21: 31–61, 1981.
187. LILIENBLUM, W., A. K. WALLI, AND K. W. BOCK. Differential induction of rat liver microsomal UDP-glucuronosyltransferase activities by various inducing agents. *Biochem. Pharmacol.* 31: 907–913, 1982.
188. LUNAZZI, G., C. TIRIBELLI, B. GAZZIN, AND G. L. SOTTOCASA. Further studies on bilitranslocase, a plasma membrane protein involved in hepatic organic anion uptake. *Biochim. Biophys. Acta* 685: 117–122, 1982.
189. MACGREGOR, J. T., AND A. BURKHALTER. Biliary excretion of nicotinamide riboside. A possible role in the regulation of hepatic pyridine nucleotide dynamics. *Biochem. Pharmacol.* 22: 2645–2658, 1973.
190. MACGREGOR, J. T., AND T. W. CLARKSON. Metabolism and biliary excretion of phenanthridinium salts. II. Biliary excretion. *Biochem. Pharmacol.* 21: 1679–1696, 1972.
191. MAHU, J.-L., P. DUVALDESTIN, D. DHUMEAUX, AND P. BERTHELOT. Biliary transport of cholephilic dyes: evidence for two different pathways. *Am. J. Physiol.* 232 (*Endocrinol. Metab. Gastrointest. Physiol.* 1): E445–E450, 1977.
192. MARINOVIC, Y., J.-C. GLASINOVIC, B. SEMELLE, J.-F. BOIV-

192. IEUX, AND S. ERLINGER. Facilitation of hepatic uptake of phenol 3,6-dibromphthalein disulfonate by taurocholate. *Am. J. Physiol.* 232 (*Endocrinol. Metab. Gastrointest. Physiol.* 1): E560–E564, 1977.
193. MARK, W., H. GLATT, AND F. OESCH. Xenobiotic metabolizing enzymes of rat liver nonparenchymal cells. *Toxicol. Appl. Pharmacol.* 84: 500–511, 1986.
194. MARZO, A., AND P. GHIRARDI. Biliary and urinary excretion of five cardiac glycosides and its correlation with their physical and chemical properties. *Naunyn-Schmiedeberg's Arch. Pharmacol.* 298: 51–56, 1977.
195. MATERN, S., K. W. BOCK, AND W. GEROK (editors). *Advances in Glucuronide Conjugation.* Lancaster, UK: MTP, 1985. (Falk Symp. 40.)
196. MCDEVITT, D. G., A. S. NIES, AND G. R. WILKINSON. Influence of phenobarbital on factors responsible for hepatic clearance of indocyanine green in rat: relative contributions of induction and altered liver blood flow. *Biochem. Pharmacol.* 26: 1247–1250, 1977.
197. MEIER, P. J., A. S. MEIER-ABT, C. BARRETT, AND J. L. BOYER. Mechanisms of taurocholate transport in canalicular and basolateral rat liver plasma membrane vesicles. Evidence for an electrogenic canalicular organic anion carrier. *J. Biol. Chem.* 259: 10614–10622, 1984.
198. MEIER, P. J., S. RUETZ, G. FRICKER, AND L. LANDMAN. Identical bile acid (BA) transport systems are present in apical membranes of liver, ileum and kidney epithelial cells (Abstract). *Hepatology Baltimore* 6: 1134, 1986.
199. MEIER, P. J., E. S. SZTUL, A. REUBEN, AND J. L. BOYER. Structural and functional polarity of canalicular and basolateral plasma membrane vesicles isolated in high yield from rat liver. *J. Cell Biol.* 98: 991–1000, 1984.
200. MEIJER, D. K. F. The mechanisms for hepatic uptake and biliary excretion of organic cations. In: *Intestinal Permeation*, edited by M. Kramer and F. Lauterbach. Amsterdam: Excerpta Med., 1976, p. 196–207.
201. MEIJER, D. K. F., J. W. ARENDS, AND J. G. WEITERING. The cardiac glycoside sensitive step in the hepatic transport of the bisquaternary ammonium compound, hexafluorenium. *Eur. J. Pharmacol.* 15: 245–258, 1971.
202. MEIJER, D. K. F., AND A. BLOM. Hepatic transport of organic anions and organic cations in the rat in vivo, isolated perfused rat livers and isolated hepatocytes. *The Liver. Quantitative Aspects of Structure and Function*, edited by R. Preisig and J. Bircher. Aulendorf, Switzerland: Editio Cantor, 1979, p. 77–86. (Proc. Int. Gstaad Symp. 3.)
203. MEIJER, D. K. F., A. BLOM, AND J. G. WEITERING. The influence of phenobarbital on the subcellular distribution in liver and the transport rate in isolated hepatocytes of dibromosulfophthalein. *Biochem. Pharmacol.* 31: 2539–2542, 1982.
204. MEIJER, D. K. F., A. BLOM, J. G. WEITERING, AND R. HORNSVELD. Pharmacokinetics of the hepatic transport of organic anions: influence of extra- and intracellular binding on hepatic storage of dibromosulfophthalein and interactions with indocyanine green. *J. Pharmacokinet. Biopharm.* 12: 43–65, 1984.
205. MEIJER, D. K. F., J. BOGNACKI, AND W. G. LEVINE. Effect of nafenopin (SU-13,437) on liver function. Influence on the hepatic transport of organic anions. *Naunyn-Schmiedeberg's Arch. Pharmacol.* 290: 235–250, 1975.
206. MEIJER, D. K. F., J. BOGNACKI, AND W. G. LEVINE. Effect of nafenopin (SU-13,437) on liver function. Hepatic uptake and biliary excretion of ouabain in the rat. *Drug Metab. Dispos.* 3: 220–225, 1975.
207. MEIJER, D. K. F., E. S. BOS, AND K. J. VAN DER LAAN. Hepatic transport of mono and bisquaternary ammonium compounds. *Eur. J. Pharmacol.* 11: 371–377, 1970.
208. MEIJER, D. K. F., K. KEULEMANS, AND G. J. MULDER. Isolated perfused rat liver technique. *Methods Enzymol.* 77: 81–93, 1981.
209. MEIJER, D. K. F., C. NEEF, AND G. M. M. GROOTHUIS. Carrier-mediated transport of drug by the liver. In: *Topics in Pharmaceutical Sciences*, edited by D. D. Breimer and P. Speiser. Amsterdam: Elsevier, 1983, p. 167–189.
210. MEIJER, D. K. F., R. J. VONK, K. KEULEMANS, AND J. G. WEITERING. Hepatic uptake and biliary excretion of dibromosulphthalein. Albumin dependence, influence of phenobarbital and nafenopin pretreatment and the role of Y and Z protein. *J. Pharmacol. Exp. Ther.* 202: 8–21, 1977.
211. MEIJER, D. K. F., R. J. VONK, E. SCHOLTENS, AND W. G. LEVINE. The influence of dehydrocholate on hepatic uptake and biliary excretion of ^{3}H-taurocholate and ^{3}H-ouabain. *Drug Metab. Dispos.* 4: 1–7, 1976.
212. MEIJER, D. K. F., R. J. VONK, AND J. G. WEITERING. The influence of various bile salts and some cholephilic dyes on Na^+-, K^+- and Mg^{2+}-activated ATPase of rat liver in relation to cholestatic effects. *Toxicol. Appl. Pharmacol.* 43: 597–612, 1978.
213. MEIJER, D. K. F., J. G. WEITERING, B. L. BAJEMA, AND G. A. VERMEER. Pharmacokinetics of biliary excretion in man. V. Dibromosulfophthalein. *Eur. J. Clin. Pharmacol.* 24: 549–556, 1983.
214. MEIJER, D. K. F., J. G. WEITERING, AND R. J. VONK. Hepatic uptake and biliary excretion of d-tubocurarine and trimethyltubocurarine in the rat in vivo and in isolated perfused rat livers. *J. Pharmacol. Exp. Ther.* 198: 229–239, 1976.
215. MEIJER, D. K. F., J. WESTER, AND M. GUNNINK. Distribution of quaternary ammonium compounds between particulate and soluble constituents of rat liver in relation to their transport from plasma into bile. *Naunyn-Schmiedeberg's Arch. Pharmacol.* 273: 179–192, 1972.
216. MEISTER, A. Glutathione. In: *The Liver: Biology and Pathobiology* (2nd ed.), edited by I. M. Arias, W. B. Jakoby, H. Popper, D. Schachter, and D. A. Shafritz. New York: Raven, 1988, p. 401–417.
217. MEISTER, A., AND M. E. ANDERSON. Glutathione. *Annu. Rev. Biochem.* 52: 711–760, 1983.
218. MESA, V. A., J. FEVERY, AND J. DE GROOTE. The maximal biliary excretory rate (Tm) of ioglycamide in the rat. Effect of taurocholate. *J. Hepatol.* 1: 243–252, 1985.
219. MEUWISSEN, J. A. T. P., B. KETTERER, AND K. P. M. HEIRWEGH. Role of soluble binding proteins in overall hepatic transport of bilirubin. In: *Chemistry and Physiology of Bile Pigments*, edited by P. D. Berk and N. I. Berlin. Bethesda, MD: Natl. Inst. Health, 1977, p. 323–327.
220. MEYER-BRUNOT, H. G., AND H. KEBERLE. Biliary excretion of ferrioxamines of varying liposolubility in perfused rat liver. *Am. J. Physiol.* 214: 1193–1200, 1968.
221. MILLER, L. L., C. J. BLY, M. L. WATSON, AND W. F. DALE. The dominant role of the liver in plasma protein synthesis. *J. Exp. Med.* 94: 451–453, 1951.
222. MOHRI, K., T. UESUGI, AND K. KAMISAKA. Buculome N-glucuronide: purification and identification of a major metabolite of buculome in rat bile. *Xenobiotica* 15: 615–621, 1985.
223. MOREY, K. S., AND G. LITWACK. Isolation and properties of cortisol metabolite binding protein from livers of rats given 4-dimethyl-aminoazobenzene. *Biochem. J.* 103: 316–324, 1967.
224. MORGAN, D. J., D. B. JONES, AND R. A. SMALLWOOD. Modelling of substrate elimination by the liver: has the albumin receptor model superseded the well stirred model? *Hepatology Baltimore* 5: 1231–1235, 1985.
225. MOUELHI, M. E., AND F. C. KAUFFMAN. Sublobular distribution of transferases and hydrolases associated with glucuronide, sulfate and glutathione conjugation in human liver. *Hepatology Baltimore* 6: 450–456, 1986.
226. MUDGE, G. H., G. R. STIBITZ, M. S. ROBINSON, AND M. W. GEMBORYS. Competition for binding to multiple sites of human serum albumin for cholecystographic agents and sulfobromophthalein. *Drug Metab. Dispos.* 6: 440–452, 1978.
227. MULDER, G. J. Sulfation—metabolic aspects. In: *Progress in Drug Metabolism*, edited by J. W. Bridges and L. F. Chasseaud. London: Taylor & Francis, 1984, vol. 8, p. 35–100.
228. MULDER, G. J., J. H. N. MEERMAN, A. M. VAN DEN GOOR-

BERGH. Bioactivation of xenobiotics by conjugation. In: *Xenobiotic Conjugation Chemistry*, edited by G. D. Paulsen, J. Caldwell, D. H. Hutson, and J. J. Menn. Washington, DC: Am. Chem. Soc., 1986, p. 282–301.
229. MULDER, G. J., E. SCHOLTENS, AND D. K. F. MEIJER. Collection of metabolites in bile and urine from the rat. *Methods Enzymol.* 77: 21–30, 1981.
230. MULLER, W., AND A. E. STILLBAUER. Liver slice uptake of intravenous and oral biliary contrast media. *Arch. Int. Pharmacodyn. Ther.* 246: 187–204, 1980.
231. NAGAR, H., AND S. A. BERGER. The excretion of antibiotics by the biliary tract. *Surg. Gynecol. Obstet.* 158: 601–607, 1984.
232. NAKAE, H., Y. IUCHI, AND S. MURANISHI. Biopharmaceutical study of the hepato-biliary transport of drugs. VIII. Investigations of hepatic uptake of organic cations by portal infusion. *Chem. Pharm. Bull. Tokyo* 26: 88–95, 1978.
233. NAKAE, H., H. OKAMOTO, K. TAKADA, AND S. MURANISHI. Biopharmaceutical study of the hepato-biliary transport of drugs. VI. Inhibition of active biliary excretion of organic cations by retrograde infusion. *Chem. Pharm. Bull. Tokyo* 25: 427–433, 1977.
234. NAKAE, H., R. SAKATA, AND S. MURANISHI. Biopharmaceutical study of the hepato-biliary transport of drugs. V. Hepatic uptake and biliary excretion of organic cations. *Chem. Pharm. Bull. Tokyo* 24: 886–893, 1976.
235. NAKAE, H., K. TAKADA, S. ASADA, AND S. MURANISHI. Transport rates of hepatic uptake and biliary excretion of an organic cation, acetyl procainamide ethobromide. *Biochem. Pharmacol.* 29: 2573–2576, 1980.
236. NAYAK, P. K., AND L. S. SCHANKER. Active transport of tertiary amine compounds into bile. *Am. J. Physiol.* 217: 1639–1643, 1969.
237. NEBERT, D. W., H. J. EISEN, M. NEGISHI, M. A. LANG, L. M. HJELNULAND, AND A. B. OKEY. Genetic mechanisms controlling the induction of polysubstrate monooxygenase (P450) activities. *Annu. Rev. Pharmacol. Toxicol.* 21: 431–462, 1981.
238. NEBERT, D. W., AND F. J. GONZALEZ. Cytochrome P450 gene expression and regulation. *Trends Pharmacol. Sci.* 6: 160–164, 1985.
239. NEEF, C., K. T. P. KEULEMANS, AND D. K. F. MEIJER. Hepatic uptake and biliary excretion of organic cations. I. Characterization of three new model compounds. *Biochem. Pharmacol.* 33: 3977–3990, 1984.
240. NEEF, C., K. P. T. KEULEMANS, AND D. K. F. MEIJER. Hepatic uptake and biliary excretion of organic cations. II. The influence of ion pair formation. *Biochem. Pharmacol.* 33: 3991–4002, 1984.
241. NEEF, C., AND D. K. F. MEIJER. Structure-pharmacokinetics relationship of quaternary ammonium compounds. Correlation of physicochemical and pharmacokinetic parameters. *Naunyn-Schmiedeberg's Arch. Pharmacol.* 328: 111–118, 1984.
242. NEEF, C., R. OOSTING, AND D. K. F. MEIJER. Structure-pharmacokinetics relationship of quaternary ammonium compounds. Elimination and distribution characteristics. *Naunyn-Schmiedeberg's Arch. Pharmacol.* 328: 103–110, 1984.
243. NUNN, A. D., AND M. D. LOBERG. Hepatobiliary agents. In: *Radiopharmaceuticals: Structure-Activity Relationships*, edited by R. P. Spencer. New York: Grune & Stratton, 1981, p. 540–548.
244. OFFERHAUS, L. Drug interactions at excretory mechanisms. *Pharmacol. Ther.* 15: 69–78, 1981.
245. OKUDA, H., R. NUNES, S. VALLABHAJOSULA, A. STRASHUN, S. J. GOLDSMITH, AND P. D. BERK. Studies of the hepatocellular uptake of the hepatobiliary scinti-scanning agent ^{99m}Tc-DISIDA. *J. Hepatol.* 3: 251–259, 1986.
246. ORRENIUS, S., AND P. MOLDEUS. The multiple roles of glutathione in drug metabolism. *Trends Pharmacol. Sci.* 5: 432–435, 1984.
247. OSHIO, C., AND M. J. PHILLIPS. Contractility of bile canaliculi: implications for liver function. *Science Wash. DC* 212: 1041–1042, 1981.
248. PANG, K. S. The effect of intercellular distribution of drug-metabolizing enzymes on the kinetics of stable metabolite formation by liver: first pass effect. *Drug Metab. Rev.* 14: 61–76, 1983.
249. PAUMGARTNER, G., P. PROBST, R. KRAINES, AND C. M. LEEVY. Kinetics of indocyanine green removal from the blood. *Ann. NY Acad. Sci.* 170: 134–142, 1970.
250. PETERLIK, M., AND H. GAZDA. Sodium-linked transport of ethacrynic acid by rat liver. Possible significance for choleretic action. *Biochem. Pharmacol.* 29: 2733–2739, 1980.
251. PETERS, W. H. M., H. NAUTA, AND P. L. M. JANSEN. The molecular masses and molecular structure of UDP-glucuronyl transferase as determined by radiation-inactivation analysis. In: *Advances in Glucuronide Conjugation*, edited by S. Matern, K. W. Bock, and W. Gerok. Lancaster, UK: MTP, 1985, p. 235–244. (Falk Symp. 40.)
252. PETERSON, R. E., AND J. M. FUJIMOTO. Biliary excretion of morphine-3-glucuronide and morphine-3-ethereal sulfate by different pathways in the rat. *J. Pharmacol. Exp. Ther.* 184: 409–418, 1973.
253. PETZINGER, E., AND K. FISCHER. Transport function of the liver. Lack of correlation between hepatocellular ouabain uptake and binding to (Na$^+$,K$^+$)-ATPase. *Biochim. Biophys. Acta* 815: 334–340, 1985.
254. PETZINGER, E., K. FISCHER, AND H. FASOLD. Role of the bile acid transport system in hepatocellular ouabain uptake. In: *Cardiac Glycosides 1785–1985. Biochemistry, Pharmacology, Clinical Relevance*, edited by E. Erdmann, K. Greeff, and J. C. Skou. Darmstadt, FRG: Steinkopff-Verlag, 1986, p. 297–304.
255. PETZINGER, E., AND M. FRIMMER. Driving forces in hepatocellular uptake of phalloidin and cholate. *Biochim. Biophys. Acta* 778: 539–568, 1984.
256. PHILLIPS, M. J., M. ODA, E. MAK, M. M. FISHER, AND K. N. JEEJEEBHOY. Microfilament dysfunction as a possible cause of intrahepatic cholestasis. *Gastroenterology* 69: 48–58, 1975.
257. PHILLIPS, M. J., C. OSHIO, M. MIYAIRI, H. KATZ, AND C. R. SMITH. A study of bile canalicular contractions in isolated hepatocytes. *Hepatology Baltimore* 2: 763–768, 1982.
258. PLAA, G. L. The enterohepatic circulation. In: *Handbook of Experimental Pharmacology*, edited by J. R. Gillette and J. R. Mitchell. Berlin: Springer-Verlag, 1975, p. 130–149.
259. POLLARD, M. R., AND G. J. DUTTON. Liver snips. A simple, rapid and reproducible method for studying metabolism in small fragments of tissue, as applied to glucuronidation in rat liver. *Biochem. J.* 202: 469–473, 1982.
260. POTTER, B. J., B. BLADER, AND P. D. BERK. BSP uptake by rat liver sinusoidal membrane vesicles: further evidence for carrier mediated transport (Abstract). *Hepatology Baltimore* 5: 1042, 1985.
261. POZNANSKY, M. J., AND R. L. JULIANO. Biological approaches to the controlled delivery of drugs: a critical review. *Pharmacol. Rev.* 26: 277–336, 1984.
262. RAMZAN, M. I., A. A. SOMOGYI, J. S. WALKER, C. A. SHANKS, AND E. J. TRIGGS. Clinical pharmacokinetics of the nondepolarizing muscle relaxants. *Clin. Pharmacokinet.* 6: 25–60, 1981.
263. RAO, M. L., G. S. RAO, AND H. BREUR. Uptake of estrone, estradiol-17β and testosterone by isolated rat liver cells. *Biochem. Biophys. Res. Commun.* 77: 566–573, 1977.
264. RAO, M. L., G. S. RAO, J. ECKEL, AND H. BREUR. Factors involved in the uptake of corticosterone by rat liver cells. *Biochim. Biophys. Acta* 500: 322–332, 1977.
265. RAO, M. L., G. S. RAO, M. HOLLER, H. BREUR, P. J. SCHATTENBERG, AND W. D. STEIN. Uptake of cortisol by isolated rat liver cells. A phenomenon indicative of carrier-mediated and simple diffusion. *Hoppe-Seyler's Z. Physiol. Chem.* 357: 573–584, 1976.
266. RAPPAPORT, A. M. Hepatic blood flow: morphologic aspects and physiologic regulation. *Int. Rev. Physiol.* 21: 1–63, 1980.
267. RAWLINS, M. D. Extrahepatic drug metabolism. In: *Drug Metabolism and Disposition: Considerations in Clinical Phar-*

macology, edited by G. R. Wilkinson and M. D. Rawlins. Lancaster, UK: MTP, 1985, p. 21–35.
268. REICHEN, J., AND P. D. BERK. Isolation of an organic anion binding protein from rat liver plasma membrane fractions by affinity chromatography. *Biochem. Biophys. Res. Commun.* 91: 484–489, 1979.
269. REICHEN, J., AND G. PAUMGARTNER. Uptake of bile acids by perfused rat liver. *Am. J. Physiol.* 231: 734–742, 1976.
270. REICHEN, J., AND G. PAUMGARTNER. Excretory function of the liver. In: *Liver and Biliary Tract Physiology I*, edited by N. B. Javitt. Baltimore, MD: University Park, 1980, vol. 21, p. 103–150. (Int. Rev. Physiol. Ser.)
271. REICHEN, J., AND F. R. SIMON. Cholestasis: In: *The Liver: Biology and Pathobiology* (2nd ed.), edited by I. M. Arias, W. B. Jakoby, H. Popper, D. Schachter, and D. A. Shafritz. New York: Raven, 1988, p. 1105–1124.
272. REID, D. J. Cellular defense mechanisms against reactive metabolites. In: *Bioactivation of Foreign Compounds*, edited by M. W. Anders. New York: Academic, 1985, p. 71–108.
273. REID, L. M., AND D. M. JEFFERSON. Culturing hepatocytes and other differentiated cells. *Hepatology Baltimore* 4: 548–559, 1984.
274. REUBEN, A. Bile formation: sites and mechanisms. *Hepatology Baltimore* 4: 15S–24S, 1984.
275. REUBEN, A., K. E. HOWELL, AND J. L. BOYER. Effects of taurocholate on the size of mixed lipid micelles and their association with pigment and proteins in rat bile. *J. Lipid Res.* 23: 1039–1052, 1982.
276. ROLLINS, D. E. Pharmacokinetics of drug excretion in bile. In: *Pharmacokinetic Basis for Drug Treatment*, edited by L. Z. Benet, N. S. Massoud, and J. G. Gambertoglio. New York: Raven, 1984, p. 77–88.
277. ROLLINS, D. E., J. W. FRESTON, AND D. M. WOODBURRY. Transport of organic anions into liver cells and bile. *Biochem. Pharmacol.* 29: 1023–1028, 1980.
278. ROLLINS, D. E., AND C. D. KLAASSEN. Biliary excretion of drugs in man. *Clin. Pharmacokinet.* 4: 368–379, 1979.
279. ROWLAND, M. Protein binding and drug clearance. *Clin. Pharmacokinet.* 9, Suppl. 1: 10–17, 1984.
280. ROY-CHOWDHURY, J., N. ROY-CHOWDHURY, P. M. NOVIKOFF, A. B. NOVIKOFF, I. M. ARIAS. UDP-glucuronyl transferase: problems within a biological "family." In: *Advances in Glucuronide Conjugation*, edited by S. Matern, K. W. Bock, and W. Gerok. Lancaster, UK: MTP, 1985, p. 33–40. (Falk Symp. 40.)
281. RUIFROK, P. G. Uptake of quaternary ammonium compounds into rat liver plasma membrane vesicles. *Biochem. Pharmacol.* 31: 1431–1435, 1982.
282. RUIFROK, P. G., AND D. K. F. MEIJER. Sodium coupled uptake of taurocholate by rat liver plasma membrane vesicles. *Liver* 2: 28–34, 1982.
283. RUSSEL, J. Q., AND C. D. KLAASSEN. Species variation in the biliary excretion of ouabain. *J. Pharmacol. Exp. Ther.* 183: 513–519, 1972.
284. RYRFELDT, A. Biliary excretion of some penicillines, quaternary ammonium compounds and tertiary amines including aspects on mechanisms for their excretion. *Acta Pharmacol. Toxicol. Suppl.* 32: 1–23, 1973.
285. RYRFELDT, A., AND E. HANSSON. Biliary excretion of quaternary ammonium compounds and tertiary amines in the rat. *Acta Pharmacol. Toxicol.* 30: 59–68, 1971.
286. SAID, M. H., W. B. STRUM, AND D. HOLLANDER. Inhibitory effect of unconjugated bile acids on the enterohepatic circulation of methotrexate. *J. Pharmacol. Exp. Ther.* 213: 660–664, 1984.
287. SCHANKER, L. S. Secretion of organic compounds in bile. In: *Handbook of Physiology. Alimentary Canal*, edited by C. F. Code. Washington, DC: Am. Physiol. Soc., 1968, sect. 6, vol. V, chapt. 114, p. 2433–2499.
288. SCHANKER, L. S. Transport of drugs. In: *Metabolic Pathways. Metabolic Transport*, edited by L. E. Hokin. London: Academic, 1972, vol. 6, p. 543–579.
289. SCHANKER, L. S., AND H. M. SOLOMON. Active transport of quaternary ammonium compounds into bile. *Am. J. Physiol.* 204: 829–832, 1963.
290. SCHARSCHMIDT, B. F., AND R. SCHMID. The micellar sink. A quantitative assessment of the association of organic anions with mixed micelles and other macromolecular aggregates in rat bile. *J. Clin. Invest.* 62: 1122–1131, 1978.
291. SCHARSCHMIDT, B. F., AND R. W. VAN DYKE. Mechanisms of hepatic electrolyte transport. *Gastroenterology* 85: 1199–1214, 1983.
292. SCHARSCHMIDT, B. F., J. G. WAGGONER, AND P. D. BERK. Hepatic organic anion uptake in the rat. *J. Clin. Invest.* 56: 1280–1292, 1975.
293. SCHENCK-GUSTAFSSON. Quinidine induced reduction of the biliary excretion of digoxin in patients. In: *Cardiac Glycosides 1785–1985. Biochemistry-Pharmacology-Clinical Relevance*, edited by E. Erdmann, K. Greeff, and J. C. Skou. Darmstadt, FRG: Steinkopff-Verlag, 1986, p. 293–296.
294. SCHENKER, S., AND B. COMBES. Role of hepatic adenosine triphosphate in BSP transport and metabolism in vivo. *Am. J. Physiol.* 212: 295–300, 1967.
295. SCHMID, R. Bilirubin metabolism: state of the art. *Gastroenterology* 74: 1307–1312, 1978.
296. SCHWARZ, L. R., R. BURR, M. SCHWENK, E. PFAFF, AND H. GREIM. Uptake of taurocholic acid into isolated rat liver cells. *Eur. J. Biochem.* 55: 617–623, 1975.
297. SCHWARZ, L. R., R. GÖTZ, AND C. D. KLAASSEN. Uptake of sulfobromophthalein-glutathione conjugate by isolated hepatocytes. *Am. J. Physiol.* 239 (*Cell Physiol.* 8): C118–C123, 1980.
298. SCHWARZ, L. R., M. SCHWENK, E. PFAFF, AND H. GREIM. Excretion of taurocholate from isolated hepatocytes. *Eur. J. Biochem.* 71: 369–373, 1976.
299. SCHWENK, M. Transport systems of isolated hepatocytes. *Arch. Toxicol.* 44: 113–126, 1980.
300. SCHWENK, M., R. BURR, L. SCHWARZ, AND E. PFAFF. Uptake of bromosulfophthalein by isolated liver cells. *Eur. J. Biochem.* 64: 189–197, 1976.
301. SCHWENK, M., AND V. LÓPEZ DEL PINO. Uptake of estrone sulfate by isolated rat liver cells. *J. Steroid Biochem.* 13: 669–673, 1980.
302. SCHWENK, M., T. WIEDMANN, AND H. REMMER. Uptake, accumulation and release of ouabain by isolated rat hepatocytes. *Naunyn-Schmiedeberg's Arch. Pharmacol.* 316: 340–344, 1981.
303. SEYDEL, J. K., AND K. J. SCHAPER. Quantitative structure-pharmacokinetic relationships and drug design. *Pharmacol. Ther.* 15: 131–182, 1982.
304. SHOREY, J., S. SCHENKER, AND B. COMBES. Effect of acute hypoxia on hepatic excretory function. *Am. J. Physiol.* 216: 1441–1452, 1969.
305. SIES, H. Reduced and oxidized glutathione efflux from liver. In: *Glutathione: Storage, Transport and Turnover in Mammals*, edited by Y. Sakamoto, T. Higashi, and N. Taheiski. Tokyo: Japan Sci. Soc., 1983, p. 63–88.
306. SIMION, F. A., B. FLEISCHER, AND S. FLEISCHER. Ionic requirements for taurocholate transport in rat liver plasma membrane vesicles. *J. Bioenerg. Biomembr.* 16: 507–515, 1984.
307. SIMON, F. R., E. M. SUTHERLAND, AND M. GONZALEZ. Regulation of bile salt transport in rat liver. Evidence that increased maximum bile salt secretory capacity is due to increased cholic acid receptors. *J. Clin. Invest.* 70: 401–411, 1982.
310. SMITH, R. L. The biliary excretion and enterohepatic circulation of drugs and other organic compounds. *Prog. Drug Res.* 9: 299–360, 1966.
311. SMITH, R. L. Excretion of drugs in bile. In: *Handbook of Experimental Pharmacology*, edited by B. B. Brodie and J. G. Gilette. Berlin: Springer-Verlag, 1971, vol. 28, p. 354–389.
312. SMITH, R. L. *The Excretory Function of Bile. The Elimination of Drugs and Toxic Substances in Bile*. London: Chapman & Hall, 1973.

313. SOKOL, P. P., P. D. HOLOHAN, AND C. R. ROSS. Electroneutral transport of organic cations in canine renal brush border membrane vesicles. *J. Pharmacol. Exp. Ther.* 233: 694–699, 1985.
314. SOLOMON, H. M., AND L. S. SCHANKER. Hepatic transport of organic cations: active uptake of a quaternary ammonium compound procainamide ethobromide by rat slices. *Biochem. Pharmacol.* 12: 621–626, 1963.
315. SOTTOCASA, G. L., G. BALDINI, S. PASSAMONTI, AND G. C. LUNAZZI. Electrogenic sulfobromophthalein (BSP) movements mediated by bilitranslocase in liver plasma membrane vesicles (Abstract). *J. Hepatol.* (Suppl. 1): S133, 1985.
316. SOTTOCASA, G. L., G. BALDINI, G. SANDRI, G. LUNAZZI, AND C. TIRIBELLI. Reconstitutions in vitro of sulfobromophthalein transport by bilitranslocase. *Biochim. Biophys. Acta* 685: 123–128, 1982.
317. SOTTOCASA, G. L., C. TIRIBELLI, M. LURIANI, G. C. LUNAZZI, AND B. GAZZIN. Isolation and some properties of a protein molecule involved in hepatic bilirubin and other anion transport. In: *Functional and Molecular Aspects of Biomembrane Transport*, edited by E. Quagliariclo. Amsterdam: Elsevier/North-Holland, 1979, p. 451–458.
318. STACEY, N. H., AND C. D. KLAASSEN. Uptake of ouabain by isolated hepatocytes from livers of developing rats. *J. Pharmacol. Exp. Ther.* 211: 360–363, 1979.
319. STOECKEL, K., P. J. M. MCNAMARA, A. J. MCCLEAN, P. DU SOUICH, D. LALKA, AND M. GIBALDI. Nonlinear pharmacokinetics of indocyanine green in the rabbit and rat. *J. Pharmacokinet. Biopharm.* 8: 483–496, 1980.
320. STREMMEL, W., AND P. D. BERK. Hepatocellular BSP and bilirubin uptake is selectively inhibited by an antibody to the liver plasma membrane BSP/bilirubin binding protein (Abstract). *Hepatology Baltimore* 5: 1035, 1985.
321. SUTFIN, T., AND W. J. JUSKO. Compendium of active drug metabolites. In: *Drug Metabolism and Disposition: Considerations in Clinical Pharmacology*, edited by G. R. Wilkinson and M. D. Rawlins. Lancaster, UK: MTP, 1985, p. 91–159.
322. SWEENEY, E. F., AND J. M. FUJIMOTO. Effect of intrabiliary administration of Triton X-100 on biliary excretory function in the rat. *Biochem. Pharmacol.* 33: 2309–2313, 1984.
323. TAFLER, M., K. ZIEGLER, AND M. FRIMMER. Iodipamide uptake by rat liver plasma membrane vesicles enriched in the sinusoidal fraction: evidence for carrier mediated transport dependent on membrane potential. *Biochim. Biophys. Acta* 855: 157–168, 1986.
324. TAYLOR, W. The excretion of steroid hormone metabolites in bile and feces. *Vitam. Horm.* 29: 201–285, 1971.
325. THALHAMMER, T., R. FUCHS, M. PETERLIK, AND J. GRAF. Hepatocellular electrolyte transport and bile secretion. In: *Hepatology: Festschrift for Hans Popper*, edited by H. Brunner and H. Thaler. New York: Raven, 1985, p. 319–327.
326. THALHAMMER, T., G. HANSEL, AND J. GRAF. Analysis of hepatic uptake and biliary excretion of an anionic xenobiotic utilizing isolated plasmamembrane vesicles. In: *Pharmacochemistry Library VIII*, edited by M. Tichy. Amsterdam: Elsevier, 1985. (Proc. Symp. QSAR Toxicol. Xenobiochem.)
327. THEILMAN, L., Y. R. STOLLMAN, I. M. ARIAS, AND A. W. WOLKOFF. Does Z protein have a role in transport of bilirubin and bromosulfophthalein by isolated perfused rat liver. *Hepatology Baltimore* 4: 923–926, 1984.
328. THURMAN, R. G., AND F. C. KAUFFMAN. Factors regulating drug metabolism in intact hepatocytes. *Pharmacol. Rev.* 31: 229–251, 1980.
329. THURMAN, R. G., AND F. C. KAUFFMAN. Sublobular compartmentation of pharmacologic events (SCOPE): metabolic fluxes in periportal and pericentral regions of the liver lobule. *Hepatology Baltimore* 5: 144–151, 1985.
330. TIPPING, E., AND B. KETTERER. The influence of soluble binding proteins on lipophile transport and metabolism in hepatocytes. *Biochem. J.* 195: 441–452, 1981.
331. TIRIBELLI, C., G. C. LUNAZZI, AND G. L. SOTTOCASA. Mechanisms of hepatic uptake of organic anions. *Clin. Sci.* 71: 1–8, 1986.
332. TOZER, T. N. Implications of altered plasma protein binding in disease states. In: *Pharmacokinetic Basis for Drug Treatment*, edited by L. Z. Benet, N. Massoud, and J. G. Gambertoglio. New York: Raven, 1984, p. 173–193.
333. TRAGER, W. F., AND B. TESTA. Stereoselective drug disposition. In: *Drug Metabolism and Disposition: Considerations in Clinical Pharmacology*, edited by G. R. Wilkinson and M. D. Rawlins. Lancaster, UK: MTP, 1985, p. 35–61.
334. TUKEY, R. H., AND T. R. TEPHLY. Purification and properties of rabbit liver *p*-nitrophenol UDP-glucuronyltransferase. *Arch. Biochem. Biophys.* 209: 565–578, 1981.
335. UESUGI, T., J. BOGNACKI, AND W. G. LEVINE. Biliary excretion of drugs in the rat during liver regeneration. *Biochem. Pharmacol.* 25: 1187–1193, 1976.
336. UESUGI, T., AND M. IKEDA. Studies on biliary excretion mechanisms of drugs. IV. Inhibitory studies on sulfobromophthalein and glucuronides in the rat. *Biochem. Pharmacol.* 25: 1361–1368, 1976.
336a. VAN DER SLUIJS P., AND D. K. F. MEIJER. Binding of drugs with a quaternary ammonium group to alpha-1 acid glycoprotein and asialo alpha-1 acid glycoprotein. *J. Pharmacol. Exp. Ther.* 234: 703–707, 1985.
336b. VAN DER SLUIJS, P., H. H. SPANJER, AND D. K. F. MEIJER. Hepatic disposition of cationic drugs bound to asialo-orosomucoid: lack of co-endocytosis and evidence for intrahepatic dissociation. *J. Pharmacol. Exp. Ther.* 240: 668–673, 1987.
337. VESSEY, D. A. Hepatic metabolism of drugs and toxins. In: *Hepatology: A Textbook of Liver Disease*, edited by D. Zakim and T. D. Boyer. Philadelphia, PA: Saunders, 1982, p. 197–230.
338. VON DIPPE, P., AND D. LEVY. Characterization of the bile acid transport system in normal and transformed hepatocytes. Photoaffinity labeling of the taurocholate carrier protein. *J. Biol. Chem.* 258: 8896–8901, 1983.
339. VONK, R. J., M. DANHOF, T. COENRAADS, A. B. D. VAN DOORN, K. KEULEMANS, A. H. J. SCAF, AND D. K. F. MEIJER. Influence of bile salts on hepatic transport of dibromosulphthalein. *Am. J. Physiol.* 237 (*Endocrinol. Metab. Gastrointest. Physiol.* 6): E524–E534, 1979.
340. VONK, R. J., P. A. JEKEL, AND D. K. F. MEIJER. Choleresis and hepatic transport mechanisms. II. Influence of bile salt choleresis and biliary micelle binding on the biliary excretion of various organic anions. *Naunyn-Schmiedeberg's Arch. Pharmacol.* 290: 375–387, 1975.
341. VONK, R. J., P. A. JEKEL, D. K. F. MEIJER, AND M. J. HARDONK. Transport of drugs in isolated hepatocytes. The influences of bile salts. *Biochem. Pharmacol.* 27: 397–405, 1978.
342. VONK, R. J., E. SCHOLTENS, G. T. P. KEULEMANS, AND D. K. F. MEIJER. Choleresis and hepatic transport mechanisms. IV. Influence of bile salt choleresis on the hepatic transport of the organic cations, *d*-tubocurarine and N4-acetyl procainamide ethobromide. *Naunyn-Schmiedeberg's Arch. Pharmacol.* 302: 1–9, 1978.
343. VONK, R. J., H. VAN DER VEEN, G. PROP, AND D. K. F. MEIJER. The influence of taurocholate and dehydrocholate choleresis on plasma disappearance and biliary excretion of indocyanine green in the rat. *Naunyn-Schmiedeberg's Arch. Pharmacol.* 282: 401–410, 1974.
344. VONK, R. J., A. B. D. VAN DOORN, G. J. MULDER, AND D. K. F. MEIJER. The influence of bile salts on hepatocellular transport. *Bile Acid Meeting. V. Biological Effects of Bile Salts*, 1978, p. 121–126. (Falk Symp. 26.)
345. VONK, R. J., A. B. D. VAN DOORN, A. H. J. SCAF, AND D. K. F. MEIJER. Choleresis and hepatic transport mechanisms. III. Binding of ouabain and *K*-strophantoside to biliary micelles and influence of choleresis on their biliary excretion. *Naunyn-Schmiedeberg's Arch. Pharmacol.* 300: 173–177, 1977.
347. WASSERMAN, O. Influence of substituents on pharmacokinetics of bisquaternary ammonium compounds. *Naunyn-*

Schmiedeberg's Arch. Pharmacol. 270, Suppl.: R154, 1971.
348. WATKINS, J. B., AND C. D. KLAASSEN. Effect of repeated oral administration of taurocholate on hepatic excretory function in the rat. *J. Pharmacol. Exp. Ther.* 218: 182-187, 1981.
349. WEISIGER, R. A. Non-equilibrium drug binding and hepatic removal. In: *Protein Binding and Drug Transport*, edited by J. P. Tillement and E. Lindenlaut. New York: Schattauer, 1986.
350. WEISIGER, R., J. GOLLAN, AND R. OCKNER. Receptor for albumin on the liver cell surface may mediate uptake of fatty acids and other albumin-bound substances. *Science Wash. DC* 211: 1048-1051, 1981.
351. WEITERING, J. G., W. LAMMERS, D. K. F. MEIJER, AND G. J. MULDER. Localization of d-tubocurarine in rat liver lysosomes. Lysosomal uptake, biliary excretion and displacement by quinacrine in vivo. *Naunyn-Schmiedeberg's Arch. Pharmacol.* 299: 277-281, 1977.
352. WEITERING, J. G., G. J. MULDER, D. K. F. MEIJER, W. LAMMERS, M. VEENHUIS, AND S. E. WENDELAAR-BONGA. On the localisation of d-tubocurarine in rat liver lysosomes in vivo by electron microscopy and subcellular fractionation. *Naunyn-Schmiedeberg's Arch. Pharmacol.* 289: 251-256, 1975.
353. WESTRA, P., K. T. P. KEULMANS, M. C. HOUWERTJES, M. J. HARDONK, AND D. K. F. MEIJER. Mechanisms underlying the prolonged duration of action of muscle relaxants due to extrahepatic cholestasis. *Br. J. Anaesth.* 53: 217-227, 1981.
354. WHEELER, H. O. Secretion of bile acids by the liver and their role in the formation of hepatic bile. *Arch. Intern. Med.* 130: 533-541, 1972.
355. WHITMER, D. J., S. C. HAUSER, AND J. L. GOLLAN. Mechanisms of formation, hepatic transport and metabolism of bile pigments. In: *Intrahepatic Calculi*. New York: Liss, 1984, p. 29-52.
356. WIELAND, T., M. NASSAL, W. KRAMER, G. FRICKER, U. BICKEL, AND G. KURZ. Identity of hepatic membrane transport systems for bile salts, phalloidin, and antamidine by photoaffinity labeling. *Proc. Natl. Acad. Sci. USA* 81: 5232-5236, 1984.
357. WILKINSON, G. R., AND D. G. SHAND. A physiologic approach to hepatic drug clearance. *Clin. Pharmacol. Ther.* 18: 377-390, 1975.
358. WILLS, R. J., R. D. SMITH, AND G. J. YAKATAN. Dose dependent pharmacokinetics and biliary excretion of bromphenol blue in the rat. *J. Pharm. Sci.* 72: 1127-1131, 1983.
359. WINKLER, K., AND C. GRAM. Models for description of bromsulfalein elimination curves in man after single intravenous injection. *Acta Med. Scand.* 169: 263-272, 1961.
360. WISHART, G. J. Functional heterogeneity of UDP-glucuronosyltransferase as indicated by its differential development and inducibility by glucocorticoids. *Biochem. J.* 174: 485-489, 1978.
361. WISHART, G. J. Demonstration of functional heterogeneity of hepatic uridine diphosphase glucuronosyltransferase activities after administration of 3-methylcholanthrene and phenobarbital to rats. *Biochem. J.* 174: 671-672, 1978.
362. WISHART, G. J., AND G. J. DUTTON. Regulation of onset of development of UDP-glucuronosyltransferase activity towards O-aminophenol by glucocorticoids in late-foetal rat liver in utero. *Biochem. J.* 168: 507-511, 1977.
363. WISSE, E., R. B. DE ZANGER, K. CHARELS, P. VAN DER SMISSEN, AND R. S. MCCUSKEY. The liver sieve: considerations concerning the structure and function of endothelial fenestrae, the sinusoidal wall and the space of Disse. *Hepatology Baltimore* 5: 683-692, 1985.
364. WOLKOFF, A. W. The glutathione S-transferases: their role in the transport of organic anions from blood to bile. In: *Liver and Biliary Tracts Physiology I*, edited by N. B. Javitt. Baltimore, MD: University Park, 1980, vol. 21, p. 151-169. (Int. Rev. Physiol. Ser.)
365. WOLKOFF, A. W., AND C. T. CHUNG. Identification, purification and partial characterization of an organic anion binding protein from rat liver cell plasma membrane. *J. Clin. Invest.* 65: 1152-1161, 1980.
366. WOLKOFF, A. W., C. A. GORESKY, J. SELLIN, Z. GATMAITAN, AND I. M. ARIAS. Role of ligandin in transfer of bilirubin from plasma into liver. *Am. J. Physiol.* 236 (*Endocrinol. Metab. Gastrointest. Physiol.* 5): E638-E648, 1979.
367. WOLKOFF, A. W., R. NAKATA, K. L. JOHNSON, AND A. SOSIAK. Influence of Cl⁻ on organic anion transport in short term cultured rat hepatocytes and isolated perfused liver (Abstract). *Hepatology Baltimore* 6: 1199, 1986.
368. WOLKOFF, A. W., A. SOSIAK, H. C. GREENBLATT, J. VAN RENSWOUDE, AND R. J. STOCKERT. Immunological studies of an organic anion-binding protein isolated from rat liver cell plasma membrane. *J. Clin. Invest.* 76: 454-459, 1985.
369. YAM, J., M. REEVES, AND J. J. ROBERTS. Comparison of sulfobromophthalein (BSP) and sulfobromophthalein glutathione (BSP-SSH) disposition under conditions of altered liver function in the isolated perfused rat liver. *J. Lab. Clin. Med.* 87: 373-383, 1976.
370. YOSHIOKA, K., H. NISHIMURA, AND T. HASEGAWA. Effect of a phenyl group in quaternary ammonium compounds on thiamine uptake in isolated rat hepatocytes. *Biochim. Biophys. Acta* 819: 263-266, 1985.
371. ZIEGLER, K., M. FRIMMER, AND H. FOSOLD. Further characterization of membrane proteins involved in the transport of organic anions in hepatocyte. *Biochim. Biophys. Acta* 769: 117-129, 1984.
372. ZIEGLER, K., M. FRIMMER, S. MÜLLNER, AND H. FASOLD. 3′-Isothiocyanate-benzamdido[³H]cholate, a new photoaffinity label for hepatocellular membrane proteins responsible for the uptake of both bile acids and phalloidin. *Biochim. Biophys. Acta* 773: 11-22, 1984.

INDEX

Index

A23187 ionophore
  calcium and, 86
  differentiation, AR42J cells, 524
  mucin secretion, 83
  pancreatic secretion ontogeny, 535–536
  pepsinogen secretion, 274
Acetylation, hepatic drug clearance, 735
Acetylcholine
  acid secretion, 215
  acinar cell electrophysiology, 41–42
  gastric secretion, 138–139
  nonoxyntic cells, 215
  potassium conductance, basolateral membrane, 7
  salivary glands
    stimulus-permeability coupling, 52–54
N-Acetylcysteine, gastrointestinal mucus, 361
N-Acetylgalactosamine, gastrointestinal mucus, 364
Acid secretion
  gastric electrophysiology, 186–187
  oxyntic cell morphology, 207–211
Acinar cells
  carcinomas, 520–521
  defined, 1
  distal events—stimulus-secretion coupling, 536–539
    compensatory endocytosis, 539
    cytoskeleton participation, 536–537
    regulatory events, 538–539
  electrolyte secretion
    cellular mechanisms, 398–402
    ionic requirements, 398–399
    stimulus-secretion coupling, 399–400
  electrophysiology
    cell model, 45–46
    cell-to-cell communication, 28–29
    electrogenic pumps, 34–36
    fluid secretion, 44–46
    membrane effects of stimulants, 36–44
    nonselective cation channels and fluid secretion, 46
    potassium (K^+) channel
    protein secretion, 46–47
    research methods, 26–28
    resting membrane properties, 29–34
  hepatic drug clearance, 737–739
  hepatic transport, intra-acinar heterogeneity, 695–696
  historical background, 25–26
  isolated cell cultures
    attached cultures, 518–519
    suspension, 518
  membrane-secretory polarity biogenesis, 543–544
  pancreas protein processing, 477–478
  pancreatic secretion
    ontogeny, 535–536
    regulatory molecules, 419–420
  pancreatic structure, 386
  polarized pancreatic secretion, 534
  protein translocation, 477–478
  regulated and constitutive secretion, 539–543
  signal transduction, 444
  structural polarity, 531–535
    cell-substrate interactions, 533–534
    developing pancreas, 531–532
    tight junctions, 532–533
Actin, exocrine granule membranes, 117
Active transport, hepatic transport, 699
Adaptation
  hepatic drug clearance, 725–726
  vs. regulation of digestion, 467
Adenylate cyclase
  HCl acid secretion, parietal cells, 257
  oxyntic cell function, prostaglandin inhibition, 223–224
  pancreatic plasma membrane, signaling transduction, 447–448
  phospholipase C, CCK regulation, 450
  salivary gland differentiation, 100–101
Adrenergic receptors, pepsinogen secretion, 271
α-Adrenergic receptors
  gastric inhibition, 178
  parotid gland amylase release, 64
β-Adrenergic agonists
  amylase release
    calcium levels and, 68–69
    parotid gland, 65–66
  gastric inhibition, 178
  mucin secretion, signal transduction, 84
  parotid gland, 64
  salivary gland differentiation, 101–102
Adult pancreas, explant cultures, 517–518
Agonist-dependent phosphoinositide metabolism, signaling transduction, 448–449
Alanine
  hepatic transport, 712
  sodium-coupled transport, 601–602
L-Alanine, sodium-amino acid cotransport, 35–36
Albumin
  biliary secretion, 663
  hepatic drug clearance, 722–723
  hepatic transport, organic anions, 707–710
Alcohol, gastric mucosal barrier, 288–290
Amidation, bile acid biotransformation, 571–572
Amino acids
  hepatic drug clearance, 736
  hepatic transport, 712
  moiety, bile acid composition, 579–580
  sodium-coupled transport, 601–602
Amino acid sequences
  H^+-K^+-ATPase, 230–232
  pepsinogen secretion, 269–270
  protein translocation, 478–479
Aminoacyl tRNA, protein translocation, 480
[^{14}C]-Aminopyrine
  cAMP-dependent protein kinase, 258
  parietal cells, 256
Amylase release
  calcium involvement in, 66–70
    calcium-free medium, 66–67
    cAMP and, 65–66
    mechanism of action, 66
  diet and, 466–467
  digestive regulation, zymogen granules, 473–474
Anesthetics, gastric secretion, 132–133
Anion transport
  ductal electrolyte transport, 14
  hepatic drug clearance, 742–745

Antibodies
  gastrointestinal secretion, 327–331
    immunoaffinity chromatography, 331
    immunoassay, 329–330
    immunoblotting, 330
    immunocytochemistry, 328–329
    immunoprecipitation, 330–331
    molecular biology, 340
    monoclonal antibodies, 328
    polyclonal antisera, 328
Anticholinergics
  gastric inhibition, 167
  gastric secretion, 132
Antigens
  exocrine granule membranes, 111–112
  gastrointestinal mucus, 364–365
Anti-idiotypes, gastrointestinal secretion, 339
Antinucleating factors, cholesterol crystal formation, 654–655
Antral gastric mucosa, 281
Antral mucosa
  gastric inhibition, 165–167
  gastric secretion, 147
  vagotomy, 149–150
Antrectomy
  gastric inhibition, 160, 163
  gastric secretion, 147
Antrofundal interactions, gastric secretion, 147
Apical cell membrane
  gastric mucosa, 284–285
    electrophysiology, 186
    lipid constituents, 291–292
  HCl acid secretion, 219–220
  plasma membrane vesicles, 296
  proton impermeability, 295–296
Apolipoproteins, biliary protein secretion, 672
AR42J cells
  acinar carcinomas, 521
  differentiation, 522–524
  gene expression, glucose deprivation, 504
  pancreatic hormone regulation, 526–527
  pancreatic secretion regulation, 526
L-Arginine, sodium–amino acid cotransport, 36
Asialoglycoproteins, exocrine granule membranes, 119–120
Aspirin, gastric mucosal barrier, 286–287
ATP
  exocrine granule membranes, 115–116
    exocytosis, 118–119
  gastric microcirculation, 297–298
  pancreatic plasma membrane, 445
  pancreatic secretion, 427
  phosphorylation, 243
  salivary gland secretory mechanism, 6
ATP/ADP ratio
  hepatic drug clearance, sulfation, 734–735
  H^+-K^+-ATPase, 243–244
Atropine
  gastric secretion, 138
  stomach parietal cell stimulation, 347–348
Autophagy, lysosome function, 683
Autoradiographic labeling, pancreatic secretion, 541–542

Balloon-occludable multilumen duodenal tubes
  hepatic transport research, 720
  triple-lumen tubes, 720
Barium, acinar cell electrophysiology, 32
Basal acid secretion, stomach parietal cells, 346–347
Basement membrane, acinar carcinomas, 521
Basolateral maxi K^+ channel, 12–13
Basolateral membrane
  ductal electrolyte transport, 15
  gastric electrophysiology, 186
  HCl acid secretion, 220–221
  secretory mechanism, 5–9
    basolateral Na^+-K^+-$2Cl^-$ symport, 4–5, 7–8
    Na^+-H^+ and Cl^--HCO_3^- antiport, 8–9
    potassium conductance, 6–7
    sodium pump, 5–6
Benzimidazole derivatives, gastric inhibition, 167
Bethanechol, gastric secretion, 139
  gastrin release, 141
  vagotomy, 148–149
Bicarbonate secretion
  bile acid–independent bile formation, 611
  canalicular bile formation, 610
  duct cell electrolyte secretion, 403–405, 407–410
  duodenal secretion
    biochemical basis, 314
    gradient, 314
    measurement of, 311–312
    mucosal transport, 313–314
  gastric electrophysiology, 188–189
    cation and anion requirements, 187
  gastric microcirculation, 297
  gastroduodenal control, 312–317
    acid secretion and blood flow, 317
    humoral control, 314–315
    leakage, 313
    neural influence, 315–317
    prostaglandins, 317
  historical background, 309
  measurement techniques, 309–312
    duodenal secretion, 311–312
    gastric secretion, 310–311
    intragastric P_{CO_2}, 310–311
    osmolality, 310–311
  model, 5
  pancreatic electrolyte secretion, 396–397
  pancreatic plasma membrane, 446
  pharmacological modulation, 317–320
    alkaline secretion, 318–319
    cAMP, 319
    duodenal secretion by prostaglandins, 319–320
  protective role of, 320–321
  secretory mechanism, cytosolic activities, 10
Bile
  cholesterol crystal formation, 653–657
  defined, 621
  hepatocyte secretion of, 657–658
  hypothetical compositions a-f, 655–657
  metastable, 653–654
  normal human bile composition, 652–653
Bile acid–dependent bile formation (BADBF)
  canalicular bile formation, 606
  canalicular water and solute secretion, 612
Bile acid–independent bile formation (BAIBF)
  canalicular bile formation, 606
  cellular mechanisms, 610–611
Bile acid–lipid coupling, 611–612
Bile acids, 556–558
  biosynthesis, 580–581
    defector or decreased levels, 584
    increased input, 584–585
  canalicular membrane transport, 605
  conventional drugs and, 588–589
  crystal structure, 623–626
  flux measurements, 581–582
  gastric mucosal barrier, 287–288
  genetic defects, 584
  hepatic drug clearance, 724–725
  hydrophobic-hydrophilic balance, 627–628
  interorgan flow, 575
  ionization behavior, 629–631

molecular structure, 622–623
negative feedback hypothesis, 583–584
names and structures, 556–558
organic solvent solubility, 628–629
pK_a values, 630–631
pool size, 580, 584
^{13}C and ^{1}H resonance assignments, 623–625
secondary acid input, 584
secretory function, 560
sodium-coupled transport, 602
steady-state determinants, 578–580
    amino acid moiety, 579–580
    steroid moiety, 578–579
surface chemistry, 625–626
transbilayer movement kinetics, 648–649
transport mechanisms, 710–711
water solubility, 628–629
Bile salts
  aqueous solution properties, 628–637, 649
    aggregation behavior, 631–637
    calcium interactions, 636–637
    critical micelle concentration, 631–635
    hydroxyl constituents and conjugation state, 632
    ionization behavior, 629–631
    organic solvent solubility, 628–629
    side-chain length, 635–636
    sodium ion effects, 632, 635
    temperature effects, 632
    water solubility, 629
  cholesterol, 649–652
    aqueous solubility, 649
    phospholipid interactions, 649–650
    quaternary system, 650–652
  lecithin interactions, 637–649
    lecithin composition, 637–638
  normal human composition, 650–651
  phospholipid-water systems, 638–649
    bilayered aqueous two-phase region, 646–647
    egg yolk lecithin–sodium cholate–water system, 639–642
    isotropic (micellar) zone structure, 642–645
    micelle-to-vesicle transition, 645–646
    multilamellar liposome kinetics, 648
    phase rule, 639–640
    phospholipid vesicles, 647–648
    transbilayer movement kinetics, 648–649
  physical constants, 623
  structural and surface properties, 622–628
    crystal structure, 623–626
    hydrophobic-hydrophilic balance, 627–628
    molecular structure, 622–623
    surface chemistry, 627
Bile secretion
  biliary components and major functions, 549–550
  biliary lipids, 556–558
    bile acids, 556–558
    cholesterol, 558
    phospholipids, 558
  canalicular bile formation, 554
    ductal modification, 554
  cholestasis, 562–563
  classification, 555–556
    organic constituents, 556–559
  excretory function, 561–562
  flow markers, 555
  gallbladder modification, 554–555
  hepatic vs. renal excretion, 550–551
  inorganic components, 559–560
  microanatomy, 597–599
  minor organic compounds, 558–559
    bilirubin, 558
    fatty acids, 559

  physiology research, 552–553
  secretory function, 560–561
    hydrolytic and absorptive processes, 553–554
  spontaneous composition changes, 555
  trace organic components, 559
  vertebrates, 549–551
Biliary fistula methods, 552–553
Biliary protein secretion
  anatomy, 663–663
    hepatocyte, 665–666
    liver, 663–665
  cholesterol metabolism, 672
  hepatobiliary transport of EGF, 673
  hepatocyte transport
    cytoskeleton, 670–672
  intestinal immune response, 672
  intracellular transport pathways, 666–670
    hepatocyte utilization, 667–668
    lysosomal degradation, 666
    regulation mechanisms, 668–670
    transcellular pathways, 666–667
  pathophysiological conditions, 672–673
  receptor-mediated endocytosis, 666–670
Biliary tree
  anatomy and circulation, 551–552
  functional anatomy, 551
  imaging techniques, 552
  liver and biliary tract circulation, 551–552
Bilirubin
  biliary composition, 558
  hepatic transport, 711–712
Bilitranslocase, hepatic transport, 713
Biosynthesis
  H$^+$-K$^+$-ATPase, 248
  mucin, 81–82, 369–370
Biotransformation
  alterations, 587–588
  bile acid fluxes, 582–583
  hepatic drug clearance, 726–737
    amino acid conjugation, 736
    epoxide hydration, 735
    glucuronidation, 731, 734
    glutathione conjugation, 735–736
    metabolism, 737
    mixed-function oxidase, 727–731
    phase I and II reactions, 726–727
    phase I metabolism variability, 730–731
    phase II reactions, 731, 734–736
    reactive intermediates, 730, 732–733
    sulfation, 734–735
    toxification, 736–737
    xenobiotics, 726
Bivalent cations, hepatic drug clearance, 746–747
Blood flow, gastric bicarbonate secretion, 317
Bombesin, gastric inhibition, 173–174
Bovine serum albumin (BSA), gastrointestinal secretion, 329–330
Bromosulfophthalein (BSP), hepatic transport, 700
Brush-border enzymes, biliary secretion, 553–554
Bt$_2$cAMP. *See* Dibutyryl cAMP
Buccal glands, mucin secretion, 87–88
Bulbogastrone, intestinal gastric inhibition
  immunoneutralization, 169–170
  vs. secretin, 168
  vs. somatostatin, 168–169
Bulk digestion, regulation of, 465–466
Bulk discharge, intracisternal protein transport and processing, 492–494
Butanedione, H$^+$-K$^+$-ATPase inhibition, 247

Ca^{2+}-ATPase, sinusoidal-lateral membrane transport, 603–604
Cable analysis, gastric electrophysiology, 196–197

Caerulein
  gene expression and, 501–502
  pancreatic electrolyte secretion, 391, 424–425
  pancreatic growth regulation, 525
Calcitonin gene-related peptide (CGRP), gastric inhibition, 174
Calcium
  acinar cell electrophysiology, stimulant-evoked membrane changes, 43–44
  bile salt interactions, 636–637
  biliary concentration, 559
  cholesterol crystal formation, 654
  electrolyte secretion, duct cells, 403
  fluxes, amylase release, 68
  HCl acid secretion, 258–262
  ion gradient–coupled uptake, 454–455
  MgATP-dependent uptake, 454
  mucin secretion, 85–86
  oxyntic cell function, 224–225
  pancreatic cell function
    GTP-induced, 453–454
    IP_3-induced release, 453
  pancreatic secretion
    diacylglycerol and protein kinase C, 429
    guanine nucleotide–binding proteins, 428–429
    intracellular messengers, 425–429
    intracellular receptors, 430–431, 450–452
    mobilization and release, 426–427
    mobilization signal, 427–428
    ontogeny, 535–536
  parotid gland amylase release, 66–70
    β-agonists and calcium levels, 68–69
    calcium fluxes, 68
    calcium-free medium, 66–67
    intracellular calcium elevation, 67–68
    mechanism of action, 70
    medium of mobilization, 69–70
    receptors, 57–58
    source mobilization, 69
  pepsinogen secretion, intracellular mediators, 274
  phosphoinositides, intracellular release, 56–57
  potassium conductance, basolateral membrane, 7
  salivary gland
    receptors, 57
    signaling system, 51–58
  sinusoidal-lateral membrane transport, 603–604
  sodium-calcium exchange, 603
  sodium pump (Na^+-K^+-ATPase) regulation, 600–601
  stimulus-secretion coupling, regulatory events, 538–539
  voltage activation, acinar cell electrophysiology, 31–32
Calmodulin
  amylase release, 70
  exocrine granule membranes, 117–118
  HCl acid secretion, protein kinase, 258
  pancreatic secretion
    intracellular Ca^{2+} receptors, 431
    protein kinases and phosphatases, 431–432
Campylobacter jejuni, gastrointestinal mucus, 374
Canalicular anion transport, choresis, 700–701
Canalicular bile formation, 554, 606–612
  acid-independent formation, 610–611
  acid-stimulation, 609–610
  intracellular transport and lipid coupling, 611–612
  water and solute excretion pathways, 606–609
    inert fluid-phase markers, 607–609
    inorganic electrolyte movement, 607
    regulation, 612
Canalicular membrane transport, 604–605
Canaliculi
  hepatic drug clearance, 747
  hepatic transport, 695
Capacitance measurements, acinar cell electrophysiology, 27–28
Cap-binding protein complex (CBPB), 509–511

Carbachol
  amylase release, 67–68
  gastrointestinal mucus, 360, 371–72
  HCl acid secretion, calcium channels, 259–260
  pancreatic secretion
    calcium-mediated secretagogues, 426
    regulation, 424–425
  phosphorylation and amylase secretion, 71–72
  stomach parietal cell stimulation, 348
Carbohydrate, gastrointestinal mucus, 363–365
Carbon dioxide production, H^+-K^+-ATPase, 239–240
Carboxyl group reagents, H^+-K^+-ATPase inhibition, 247–248
Cardiac glycosides
  hepatic drug clearance, bivalent cations, 747
  hepatic transport, 712
Carrier-mediated transport, 698–699
Carrier proteins, anion transport, 744–745
Catecholamines, gastric inhibition, 177–178
Cation permeabilities, exocrine granule membranes, 115
Cats, pancreatic electrolyte secretion, 390–391
Cell-activation mechanisms, HCl acid secretion, 222–224
Cell separation, oxyntic cell function studies, 211–212
Cell-to-cell coupling, acinar cell electrophysiology, 28–29, 34
Cellular transport
  bile acids, flux measurements, 582
  in vitro preparations, 444–445
Central nervous system, gastric secretion, 127–129
Centrifugation, lysosome isolation, 681
Cephalic inhibition, gastric secretion, 159–163
Chenic (chenodeoxycholic) acid (CDCA)
  biotransformations, 573–574
Chenodeoxycholic acid, 576–578
  molecular structure, 622
Chief cells
  gastric electrophysiology, 190–191
  pepsinogen secretion, 268, 268, 270
  stomach, 348–350
    mucosal pepsinogen content, 349
    structure, 348
Chloride (Cl^-)
  canalicular membrane transport, 604–605
  ductal electrolyte transport
    basolateral membrane, 15
    duct cell structure, 403
  exocytosis, pancreatic zymogen granules, 455–456
  gastric electrophysiology
    acinar cells, 37–38
    cation and anion requirements, 187
    electrogenic transport, 199–200
    oxyntic cell secretion, 187–188
  organelle ion-transport mechanism, 605–606
  secretory mechanism, cytosol, 9–10
  sodium-coupled uptake, 603
Cl^--HCO_3^- antiport
  canalicular membrane transport, 604
  ductal electrolyte secretion, luminal membrane, 14–15
  HCl acid secretion, 220–221
Cholecystokinin (CCK)
  acinar cells
    carcinomas, 521
    electrophysiology, 39–40
  adenylate cyclase/phospholipase C regulation, 450
  digestive enzyme regulation, 469–470
    duodenal hormones, 471
    sensory input, 470–471
  gastric secretion, 129
  gastrointestinal mucus secretion, 371–372
  gene expression, 501–502
    protein synthesis, 503–504
  intestinal gastric inhibition, 169–170
  oxyntic cell function, 213–214
  pancreatic electrolyte secretion, 397

species differences, 390–396
pancreatic hormone regulation, 526–527
pancreatic secretion
  calcium-mediated
    mobilization and release, 426–427
  secretagogues, 426
  molecular characterization of receptors, 422–423
  ontogeny, 535–537
  receptor characteristics, 421–422, 272
  regulation, 424–425, 526
pepsinogen secretion, 271–272
Cholecystokinin-5 (CCK-5), acinar cell electrophysiology, 37
Cholecystokinin-8 (CCK-8)
  acinar cell electrophysiology, 39–40
  AR42J cell differentiation, 522–524
  HCl acid secretion, calcium channels, 259–260
  oxyntic cell function, 213–214
  pancreatic growth regulation, 525
  pancreatic plasma membrane, signaling transduction, 447–448
  pancreatic secretion
    electrolytes, 391
    regulation, 526
  pepsinogen secretion, 272
^{125}I-Cholecystokin-33 (CCK-33), pancreatic secretion, 421–423
Cholephilic organic anions, hepatic drug clearance, 743–744
Choleresis
  hepatic drug clearance, 724–725
  hepatic transport, 700–701
Cholestasis, biliary excretion, 562–563
Cholesterol
  bile acids, 558
    lipid coupling, 611–612
  bile salt interaction, 649–652
    aqueous media solubility, 649
    phospholipids, 649–650
  biliary secretion, 560
    apolipoproteins, 672
    proteins, 560
  crystal formation in bile, 653–657
  quaternary system, 650–652
  solubility values in bile, 652–653
Cholesterol-phospholipid ratio, gastric membrane composition, 291–292
Cholesterol saturation index (CSI)
  crystal formation in bile, 653–657
  solubility values, 653
Cholic acid (CA), 575–576
  molecular structure, 622
Cholinergic receptors
  gastrointestinal mucus secretion, 371–372
  oxyntic cell function, 213
  parotid gland amylase release, 64
  pepsinogen secretion, 270–271
  stomach parietal cell stimulation, 347
Cholylglycine hydrolase, biotransformations, 573–574
Cholyltaurine, bile acid metabolism, 569
Chromaffin granules, granule membranes, 114–115
Chromobindins, exocrine granule membranes, 117
CHYMO, digestive regulation, 474
Chymodenin, digestive enzyme regulation, 469–471
Cimetidine, parietal cell stimulation, 347–348
Cis configuration, bile acids, 556–557
Cisternal membrane-associated proteins (CMAP), 487–490
Cisternal pH, cellular protein sorting, 490
Cocompartmentation, molecular protein sorting, 489–490
Coenzyme A, bile acid conjugation, 572
Colchicine, canalicular bile formation, 609
Collagen, salivary gland morphogenesis, 97
Colonic secretion, 337–339
Common naturally occurring bile acids, 571–572
Compartmentation
  enterohepatic circulation, 570

hepatic transport, 700
pancreatic proteins, 485–487
Compensatory endocytosis, pancreatic function, 539
Complementary DNA (cDNA), gastrointestinal secretion, 340
Compound exocytosis, 534
Conjugation state
  bile acids, 558
  bile salts, 632
Conserved sequences, mRNA translation, 510–511
Constitutive pathways, exocrine secretory membranes, 107
Contractile proteins, stimulus-secretion coupling, 538
Convection, hepatic transport, 699
Corneal epithelium, luminal membrane, 11
Corticosterone, gastric development, 352–353
Corticotropin-releasing factor (CRF)
  duodenal bicarbonate secretion, 317
  gastric inhibition, 174
    intravenous administration, 177
Cotransport, hepatic transport, 699
Countertransport
  hepatic transport, 699
  liver ultrastructure, 718–719
Covalent intracisternal processing, pancreatic proteins, 485
Cow, pancreatic electrolyte secretion, 395
Crinophagy, lysosome function, 683
Critical micelle concentration (CMC)
  bile salts, 627
    aggregation behavior, 631–635
    calcium interactions, 636–637
    hydroxyl constituents and conjugation state, 632
    micelle size and shape, 636
    side-chain length, 635–636
    sodium ion effects, 632, 635
    temperature effects, 632
  biotransformations, 574
  gastric mucosal barrier, 287–288
Cultured hepatocytes, biliary function, 553
Cyclic adenosine 5'-monophosphate (cAMP)
  ductal transport control, 16
  duodenal mucosal alkaline secretion, 319
  gastric alkaline secretion, 319
  gastrointestinal mucus secretion, 372
  HCl acid secretion
    oxyntic cell activation, 222–223
    parietal cells, 255–257
  mucin secretion, 84–85
    sublingual gland, 86–87
  nonhepatic lysosome secretion, 684–685
  oxyntic cell function, histamine receptors, 212–213
  pancreatic plasma membrane, signaling transduction, 447–448
  pancreatic secretion, 429–430
    intracellular receptors, 430–431
    protein kinases and phosphatases, 431–433
  parotid gland amylase release, 65–66
  pepsinogen secretion, intracellular mediators, 273
  salivary gland differentiation, 100
    endogenous protein phosphorylation, 70–73
8-brcAMP, pancreatic secretion, 429–430
Cyclic guanosine monophosphate (cGMP)
  pancreatic secretion, 430
    protein kinases and phosphatases, 431–433
  salivary gland, stimulus-permeability coupling, 53–54
8-brcGMP
  pancreatic secretion, 430
  pepsinogen secretion, 273
8-brcIMP, pepsinogen secretion, 273
Cyclic nucleotides
  nonhepatic lysosome secretion, 684–685
  salivary glands, stimulus-permeability coupling, 53–54
Cytochalasin B, oxyntic cell cytoskeleton, 210–211
Cytochrome P-450
  hepatic drug clearance

Cytochrome P-450 (*continued*)
    biotransformation, 728–729
    biotransformation to toxification, 736–737
Cytodifferentiation, salivary gland, 97–100
Cytoplasmic proteins, exocrine granule membranes, 116–118
Cytoprotection, gastric mucosa, 299–302
Cytoskeleton
    biliary protein secretion, 670–672
    stimulus-secretion coupling, 536, 538
Cytosolic binding proteins
    hepatic drug clearance, 723–724
    hepatic transport, 712
    pancreatic secretion, protein kinases and phosphatases, 432–433
    secretory mechanism, 9–10

Davson-Danielli concept, secretion mechanism, 472
D cells
    gastrin release, 142–143
    neuroparacrine interactions, 143–144
Density separations, oxyntic cell function studies, 211–212
Deoxycholic acid (DCA), 575–576
    molecular structure, 622
2-Deoxy-D-glucose (2-DG), gastric secretion, 130–131
Detoxification, hepatic drug clearance, 726–737
Developmental regulation, salivary gland, 95–97
Dextran, canalicular bile formation, 608–609
Diacylglycerol (DAG), pancreatic secretion
    intracellular receptors, 430–431
    protein kinase C, 429
1,2-Diacylglycerol (DAG), pepsinogen secretion, 274
Dibutyryl cAMP (Bt_2cAMP)
    amylase release, 66–67
    canalicular water and solute secretion, 612
    duct cell electrolyte secretion, 407–410
    gastric secretion, 146
    gastrointestinal mucus secretion, 372
    HCl acid secretion, parietal cell, 258
    mucin secretion, 83
    oxyntic cell function
        histamine receptors, 212–213
        prostaglandin inhibition, 223–224
    pancreatic secretion, 429–430
Diet
    digestive enzyme content and, 466–467
    gastric development, 351–352
Diethyl pyrocarbonate (DEPC), H^+-K^+-ATPase inhibition, 247
Differentiation
    exocrine secretory membranes, 108
    secretory enzyme synthesis, 521–524
Diffusion, hepatic transport, 699
Digestion
    end products and zymogen granules, 473–474
    lysosome function, 683
    regulation
        vs. adaptation, 467
        bulk digestion, 465–466
        diet, 466–467
        modes of action, 471
        nonparallel transport, 467–468
        outcomes, 471–472
        parallel secretion, 467
        properties, 468–472
        sensory input, 470–471
    secretion mechanism, 472–473
Digestive enzymes, diet and, 466–467
Digitonin, gastric mucosal barrier, 288
Direct regulation of digestion, 471
Disk assembly, bile salt–phospholipid–water systems, 646
Dissociation-mediating sites, hepatic transport, 708–709
Distal events, stimulus-secretion coupling, 536–539

DNA synthesis
    gastric mucosal growth, 351
    gastric secretion, 146–147
Docking protein, protein translocation, 480
Dogs, pancreatic electrolyte secretion, 390–391
Donnan effect, exocrine granule membranes, 115
Dopamine
    gastric secretion, 132
    parotid gland
        amylase release, 64
        receptors, 64
Dorsomotor nucleus of the vagus (DMNV)
    enteric pathways, 137
    functional anaomy, 128–129
    gastric secretion, 127–129
Double antiport (Na^+-H^+ and Cl^--HCO_3^-), 5
Drug clearance
    hepatic transport
        acinar heterogeneity, 737–739
        adaptive changes, 725–726
        bidirectional character, 741
        bile flow, 724–725
        biotransformation, detoxification and toxification, 726–737
        chemical structure, 721
        hepatobiliary concentration gradients, 741–742
        metabolites, 719
        model compounds, 739–741
        organic anions, 742–745
        organic cations, 745–747
        protein binding, 721–724
        uncharged compounds, 747–749
Ductal electrolyte transport, 13–16
    absorptive mechanism elements, 13–15
        basolateral membrane, 15
        luminal membrane, 14–15
    control, 15–16
    model specifications, 16–17
    plasticity properties, 15
    water permeability, 15
Ductal perfusion, pancreatic electrolyte secretion, 389
Duct cells
    carcinomas, 521
    electrolyte secretion
        cellular mechanisms, 402–405
        ionic requirements, 402–403
        models, 405–407
        stimulus-secretion coupling, 402
    isolated cell cultures, 519–520
    modification of canalcular bile, 554
    pancreas protein processing, 477–478
    protein translocation, 477–478
Duodenal alkaline secretion
    cyclic AMP and, 319
    prostaglandins stimulation, 319–320
Duodenal bicarbonate secretion
    biochemical basis, 314
    gradient, 314
    measurement of, 311–312
    mucosal transport, 313–314
Duodenal hormones, digestive regulation, 471
Duodenal ulcer, gastric inhibition, 166

EC_{50}, HCl acid secretion, parietal cells, 256
E. coli, gastrointestinal mucus, 374
ED_{50}, gastric secretion, 138
EEDQ reagent, H^+-K^+-ATPase inhibition, 248
Effectors, pancreatic secretion, 430–436
    calcium, diacylglycerol and cAMP, 430–431
    exocytosis, 435–436
    protein phosphorylation, 431–435

Egg yolk lecithin–sodium cholate–water system
  isotropic (micellar) zone structures, 642–645
  phase diagrams, 638–642
EGTA, amylase release, calcium-free medium, 65–66
Electrodiffusion control of membrane potential, 29–31
Electrogenic pumps, acinar cell electrophysiology, 34–36
Electrolytes
  biliary concentration, 559
  canalicular bile formation, 606–608
  ductal transport, 13–16
  pancreatic permeability, 398
Embryology
  gene expression, 504–505
  salivary gland
    cytodifferentiation, 97–99
    morphogenesis, 94–95
Endocytosis
  compensatory, 539
  lysosome formation and function, 682–683
  protein sorting, 490
  receptor-mediated, biliary protein secretion, 666–670
Endoplasmic reticulum, pancreatic cell function, 453–455
Enhancer elements, gene transcription signals, 508–509
Entamoeba histolytica, 374
Enteral hepatic transport, 587
Enteric nervous system, gastric secretion, 136–138, 141–142
Enterochromaffin-like (ECL) cells, 215
Enterogastrone
  gastric secretion, 150–151
  intestinal gastric inhibition, 170–171
Enterohepatic circulation (EHC)
  anatomical distribution, 568
  bile acids, 562
    biotransformation, 571–574
    concentrations, 571
    conjugation pattern, 588
    metabolism, 569
    movements, 574–575
  chemical constituents, 571
  composition and distribution, 569–570
  description methods, 569
  dynamic aspects, 570–571
  historical aspects, 567–568
  overall balance, 568–569
  perturbations, 584–588
    biotransformation alterations, 587–588
    input alterations, 584–585
    interorgan flow alterations, 586
    transport alterations, 586–587
  regulation, 583–584
Enzyme-linked immunosorbent assay (ELISA), 329–330
Enzymes
  biotransformations, 573–574
  exocrine granule membranes, 111
  gastrointestinal mucus, 362–363
  lysosomal biogenesis, 682
  pancreatic secretion, 335–337
  second-messenger framework, 443–444
Eosin, H^+-K^+-ATPase conformation, 245
Epidermal growth factor (EGF)
  biliary protein secretion, 663
    functional significance, 673
    hepatocyte utilization, 667–670
    pathway regulation mechanisms, 668–670
  gastric inhibition, 172–173
Epinephrine
  acinar cell electrophysiology, 41–42
  salivary gland stimulus-permeability coupling, 52–54
Epithelial cell monolayers
  apical cell membranes, 295–296
  gastric mucosa, 284–285
Epithelial resistance, gastric electrophysiology, 186
Epoxide hydration, hepatic drug clearance, 735
Equivalent circuit analysis
  impedance measurements, 197–198
  microelectrodes, 194–197
Erythrocyte membrane composition, 291–292
Ester sulfate groups, gastrointestinal mucus, 365
Eucaryotic genes
  enhancer elements, 509
  transcription signals, 505–507
Excitation-contraction coupling, 116–117
Excytosis, acinar cell electrophysiology, 47
Exocrine pancreas. *See* Pancreas
Exocrine secretory membranes
  cytoplasmic protein interaction, 116–118
  exocytosis, 118–119
  granules
    formation and storage, 108–109, 114–116
    isolation and preparation, 109
    packaging, stability and osmotic behavior, 114–115
  hydrogen-ATPase activity, 116
  intragranular pH and buffering capacity, 115–116
  ion permeation, 115
  lipid composition, 109–110
  microfilamentous networks and organelle interactions, 116–117
  microtubules, 116
  plasmalemmal protein carrier, 120
  polypeptide binding, 117
  protein composition, 110–114
    antigens, 111–112
    enzyme activities, 111
    polypeptides, 111–114
  recycling, turnover and sorting, 119–121
  reinternalized membrane, 119–120
  sorting sites, 120–121
  stimulus-enhanced protein phosphorylation, 117–118
Exocytosis
  exocrine granule membranes, 118–119
  intracisternal protein transport and processing, 492–494
  ion transport, 454–455
  pancreatic secretion, 435–436
  pepsinogen secretion, 274–275
Extracellular matrix, acinar carcinomas, 521

Facilitated diffusion, hepatic transport, 699
Fat, intestinal gastric inhibition, 170–172
Fatty acids
  biliary composition, 559
  long-chain, hepatic transport, 711
Feedback
  pancreatic plasma membrane, 449–450
  See also Negative and Positive feedback
Fetal pancreas, explant cultures, 516–517
Fluid-phase markers
  canalicular bile formation, 607–609
  transcellular vs. paracellular pathway, 608
Fluorescein isothiocyanate (FITC), gastric membrane composition, 293–294
Fluorescence
  acinar cell electrophysiology, 29
  H^+-K^+-ATPase conformation, 245
Fundic gastric mucosa, 281
  gastric secretion, 146–147
Fundoantral reflexes, gastric secretion, 147
Fundus, gastric secretion, regional vagotomy, 149
Furosemide, basolateral Na^+-K^+-$2Cl^-$ symport, 8

G-17 (little gastrin), gastric secretion, 150–151
GABA (γ-aminobutyric acid)
  gastric secretion, 132
  glucoprivic stimulation, 132

Gallbladder
  bile acid flux measurements, 581–582
  canalicular water and solute secretion, 612
  functional anatomy, 551
  hepatic bile modification, 554–555
  imaging techniques, 552
  motility and storage disturbances, 586–587
GalNAc
  mucin biosynthesis, 369–370
  salivary mucins, 79–80
Gastric alkaline secretion inhibitors, 318
Gastric bicarbonate secretion
  direct titration, 311
  leakage, 313
  measurement of, 310–311
  origin and transport, 312–313
Gastric electrophysiology
  membrane permeability and transport properties, 186–200
    bicarbonate secretion, 188–189
    cation and anion requirements for acid secretion, 186–187
    chloride secretion—oxyntic cells, 187–188
    electrogenic chloride transport, 199–200
    electroneutral hydrogen transport, 199–200
    equivalent circuit analysis, 194–197, 197–198
    isolated oxyntic cells, 190–191
    oxyntic cells intact mucosa, 191–193
      stimulation changes, 193–195
    potassium transport, 188
    sodium absorption—surface cells, 187–188
    surface cells—intact, isolated mucosa, 189–190
    transport functions vs. associated shunt pathways, 198–199
  voltage, current, ionic flux and resistance, 185–186
Gastric glands
  dispersion and isolation techniques, 211
  pepsinogen secretion, 268
Gastric inhibition
  cephalic phase, 159–160
    pharmacological inhibition, 161–163
    somatostatin as vagogastrone, 161–162
    vagotomy and antrectomy, 160
    vagus-dependent inhibitor—vagogastrone, 160–161
  intestinal phase, 168–170
    acid, 168–170
    fat, 170–172
  neuropeptides, 173–177
    intracerebral administration, 173–174
    intravenous administration, 174–177
  stomach, 163–168
    antral mucosa, 165–167
    gastric phase, 163
    oxyntic mucosa, 163–165
    pharmacological inhibition, 167–168
  sympathetic nervous system, 177–178
  urogastrone and epidermal growth factor, 172–173
Gastric inhibitory peptide (GIP), gastric inhibition, 170–171
Gastric juice, electrolyte composition, 279–280
Gastric membrane composition, 291–294
  glycosubstances, 293–294
  lipid constituents, 291–293
Gastric mucosa
  apical membrane proton impermeability, 295–296
  barrier-breaking studies, 286–291
    alcohols, 288–290
    aspirin and weak acids, 286–287
    bile acids, 287–288
    pepsin, 290–291
  barrier constituents, 282
  cytoprotection, 299–302
    prostaglandins, 299–301
    sulfhydryl compounds, 302
  electrical characteristics, 280–281
  epithelial cell layer, 284–285
  intracellular pH regulation, 296–297
  microcirculation, 297–298
  mucus-bicarbonate layer, 282–284
  pepsinogen secretion, 268
  proton permeability, 281–282
  rapid reepithelialization, 298–300
  structure, 185
  surface hydrophobicity, 294–295
Gastric receptors, nonoxyntic cells, 215
Gastric secretion
  central controls, 127–136
    brain peptides, 129–130
    central vagal complex, 127–129
    glucoprivation excitation, 130–132
    non-glucose stimuli, 132–135
    vagal afferents, 135–136
  HCl acid, 332
  intrinsic factor, 334
  pepsinogen, 332–334
  peripheral controls, 136–151
    acetylcholine, 138–139
    antrofundal interactions, 147
    enteric nervous system, 136–138
    gastrin, 139–144
    histamine, 145–146
    intestinal phase, 150–151
    peptides, 144
    trophic effects, 146–147
    vagotomy, 147–150
Gastrin
  gastric secretion
    acid inhibition, 144–145
    enteric nervous system and paracrine regulation, 141–142
    gastric phase, 144
    neuroparacrine interactions, 143–144
    physiological effects, 139–140
    release, 140
    vagal and cholinergic regulation, 140–141
    vagotomy effects, 141
  gastrointestinal mucus secretion, 371–372
  pepsinogen secretion, 271–272
  receptors, 213–214
  serum levels, 350–351
  stomach, 350
  tissue levels, 350
^{125}I-labeled [Leu^{15}]Gastrin, oxyntic cell function, 213–214
Gastrin-releasing peptide (GRP), 142–143
G cell, neuroparacrine interactions, 143–144
Gel formation in mucins, 366–368
Gene expression
  differential regulation, 499–504
    adaptation to diet changes, 500–501
    diet adaptation and hormonal mediation, 503–504
    hormonal stimulation, 501–503
    nutritional shock, 504
    two-dimensional gel electrophoresis, 499–500
  mRNA translation mechanisms, 509–511
  multiple regulation, 504–505
    nutritional substrates and hormones, 505
    pre- and postnatal development, 504–505
  transcription mechanisms, 505–509
    enhancer elements, 508–509
    processing signals—eucaryotic genes, 505–507
    promoter elements, 507–508
Gland lumina resistance, gastric electrophysiology, 196
Glands of Blandin-Nuhn, mucin secretion, 87–88
Glucagon
  digestive enzyme regulation, 469–470, 472
  HCl acid secretion, 262
Glucocorticoids
  gastric development, 352–353
  gene expression, 503

pancreatic hormone regulation, 526–527
Glucoprivation, gastric secretion, 130–133
Glucose
  analogues, gastric secretion, 130–131
  deprivation, gene expression, 504
  digestive enzyme regulation, 468–469
  oxidation, oxyntic cell function studies, 212
  sensory input, 470–471
Glucuronidation
  bile acid biotransformation, 572
  hepatic drug clearance, 731, 734
Glucuronosyltransferase, hepatic drug clearance, 734
γ-Glutamyltransferase, exocrine granule membranes, 111, 120
Glutathione conjugation, hepatic drug clearance, 735–736
Glycine
  amidation pattern alteration, 587–588
  bile acid conjugation, 572–573
Glycoconjugate composition, gastric membrane composition, 293–294
Glycoprotein
  mucin structure, 81, 363–368
  pancreatic proteins, Golgi compartmentation, 485–487
Glycosaminoglycans, salivary gland morphogenesis, 95–97
Glycosubstances, gastric membrane composition, 293–294
O-Glycosylation, mucin biosynthesis, 81–82
Goblet cells, mucus secretion, 370
Goldberg-Hogness box, gene transcription signals, 505–507
Gold thioglucose, gastric secretion, 131–132
Golgi apparatus
  compartmentation, pancreatic proteins, 485–487
  exocrine granule membranes, 119–120
  fraction 1 (GF_1), 114
  parotid gland ultrastructure, 64
  salivary gland cytodifferentiation, 97–98
Golgi-associated endoplasmic reticulum (GERL), 682
Growth regulation, pancreatic function, 524–525
GTP-binding proteins, pancreatic plasma membrane
  calcium release, 453–454
  phospholipase C activation, 450
Guanine nucleotide-binding proteins
  pancreatic secretion, phospholipase C, 428–429
  oxyntic cell function, prostaglandin inhibition, 223–224
Guinea pig, pancreatic electrolyte secretion, 394–395

H_2 antagonists
  gastric secretion, 145–146
  gastric inhibition, 163, 167
  oxyntic cell function, 212–213
Hamster, pancreatic electrolyte secretion, 394–395
HCO_3^-. See Bicarbonate
Heavy metals, biliary excretion, 550
Heidenhain pouch, gastric inhibition, 163–164
Hepatic bile, 551
Hepatic cell doublets, 553
Hepatic enzyme biotransformations, 571–573
Hepatic fatty acid–binding protein (hFABP), 723–724
Hepatic fluid and electrolyte transport, 599–606
  canalicular membrane transport mechanisms, 604–605
  membrane potential, 599–600
  organelle ion-transport mechanisms, 605–606
  sinusoidal-lateral membrane transport mechanisms, 600–604
    calcium transport, 603–604
    sodium pump (Na^+-K^+-ATPase), 600–603
Hepatic parenchyma
  anatomy, 693–696
  blood-tissue solute exchange, 694–695
  intra-acinar heterogeneity, 695–696
  liver cell plates, canaliculi and tight junctions, 695
  vascular organization, 693–694
Hepatic transport
  albumin-bound anionic uptake, 707–710
  dissociation-mediating sites, 708–709
    historical background, 707–708
    specificity, 709–710
  choleresis and canalicular ion transport, 700–701
  drug clearance, 721–739
    acinar heterogeneity, 737–739
    adaptive changes, 752–726
    bile flow, 724–725
    biotransformation, detoxification and toxification, 726–737
    chemical structure, 721
    membrane transport mechanisms, 739–749
    protein binding, 721–724
  mathematical modeling, 701–707
    multiple sinusoid distributed flow, 703–704
    single-pass indicator dilution curves, 704–705
    solute disappearance curves, 705–707
    steady-state input-output relations, 701–703
    uptake, efflux and removal estimation, 704–707
  mechanisms, 697–701
    carrier-mediated transport, 698–699
    classification, 697–698
    countertransport and cotransport, 699
    diffusion and convection, 699
    facilitated diffusion, active transport, *trans* effects, 699
    secretion, 337
  research methodology, 696–697, 719–720
    liver perfusion, 719–720
  solute compartmentation, 700
  specific solute classes, 710–713
    amino acids, 712
    bile acids, 710–711
    intracellular binding proteins, 612
    long-chain fatty acids, 711
    neutral compounds, 712
    organic anions, 711–712
    organic cations, 712
    putative membrane carriers, 712–713
Hepatobiliary concentration gradients, drug clearance, 741–742
Hepatocytes
  bile secretion, 657–658
  isolation, transport research, 696–697
  protein secretion, 667–668
    anatomy, 663–666
    cytoskeleton, 670–672
  couplets, transport research, 697
  membrane vesicles, transport research, 697
Heterophagy, lysosome function, 683
Hexoses, hepatic transport, 712
Histamine
  acid secretion, fundic mucosa, 215
  gastric electrophysiology, 190
  gastric secretion, 145–146
    mucus secretion, 371–372
  HCl acid secretion, calcium channels, 261
  oxyntic cell function, receptor specificity, 212–213
  pepsinogen secretion, 271
  stomach parietal cell stimulation, 348
H^+-K^+-ATPase
  barrier-site model, 240–241
  biosynthesis, 248
  conformations, 245
  fluorescence, 245
  functions, 230
  gastric secretion, 332
  HCL acid secretion, 216–218
    apical cell membrane, 219–220
  inhibition, 245–248
    group-selective reagents, 247–248
    site specificity, 246–247
  kinetics, 240–245
    ATP-ADP exchange, 243–244
    biphasic effect of ATP, 242
    K^+-dependent phosphorylation, 244–245

H^+-K^+-ATPase (continued)
    phosphorylation—ATP, 243
        steady-state ATPase, 241–242
        steady-state phosphatase reaction, 242–243
    model, 248–249
    molecular weight, 233
    parietal cell acid secretion, 229–230
    structure, 230–233
        amino acid sequence, 230–232
        dimensions, 233–234
        protein composition, 232–233
    tissue distribution, 248
    transport, 233–240
        modes, 238–239
        parietal cell, 239–240
        resting vesicles, 233–238
        stimulated vesicles, 238
    tryptic digestion, 245
Hormones
    gene expression, 501–503
        pre- and postnatal development, 505
        protein synthesis, 503–504
    pancreatic regulation, 526–527
Horse, pancreatic electrolyte secretion, 395
Horseradish peroxidase (HRP), gastrointestinal secretion, 329
Humans, pancreatic electrolyte secretion, 390–391
Humoral control, duodenal bicarbonate secretion, 314–315
Hydrochloric acid secretion
    cell-activation mechanisms, 222–224
        calcium-dependent mechanisms, 224
        cAMP oxyntic cell activation, 222–223
        prostaglandin inhibition of oxyntic cell, 223–224
    cellular basis, 215–222
        apical cell membrane, 219–220
        basolateral membrane transport, 220–221
        H^+-K^+-ATPase, 216–217
        membrane changes, 216–219
        oxyntic cell transport model, 221–222
    gastric secretion, 332
    historical background, 255
    parietal cell
        calcium control, 258–262
        cAMP, 255–257
        protein kinases, 257–258
    oxyntic cell morphology, 207–211
        cytoskeleton, 210–211
        membrane turnover and recycling, 208–210
        ultrastructure, 208–209
    oxyntic cell receptors, 211–215
Hydrodynamic radii (R_h), EYL–sodium cholate–water system, 642–644
Hydrogen
    chief cell stimulation, 349–350
    fluxes, gastric mucosal barrier, 288
    gastric electrophysiology
        cation and anion requirements, 187
        electroneutrality, 199–200
    permeability coefficient, gastric mucosa, 281–282
    pump, gastric electrophysiology, 186
    transport
        H^+-K^+-ATPase, 237–238
        kinetics, 240–245
Hydrolase
    assay methods, 681–682
    enzymology, 679
Hydrophobicity
    bile acid movements, 574–575
    signal sequence, protein translocation, 482
Hydroxyapatite, salivary mucins, 80
Hydroxyl constituents
    bile acids, 556–557
    bile salts, 632
    calcium interactions, 637

Hyperosmolar solution, intestinal gastric inhibition, 172
Hyperpolarization
    acinar cell electrophysiology
        stimulant-evoked membrane potential changes, 41–42
Hypothalamus, gastric secretion, 128–129

Ileocolonic inhibition, peptide YY, 172
Immune response, biliary protein secretion, 672
Immunoaffinity chromatography
    gastrointestinal secretion, 331
    salivary secretion, 332
Immunoassay, gastrointestinal secretion, 329–330
Immunoblotting, gastrointestinal secretion, 330
Immunocytochemistry, gastrointestinal secretion, 328–329
Immunodissection, gastrointestinal secretion, 339
Immunoglobulin A (IgA)
    bile composition, 550
    biliary protein secretion, 666–668
        intestinal immune response, 672
        pathophysiological conditions, 672–673
    hepatobiliary secretion, 337
Immunoprecipitation, gastrointestinal secretion, 330–331
Impedance measurements, gastric electrophysiology, 197–198
Indirect regulation of digestion, 471
Indocyanine green, hepatic transport, 711–712
Initiation, mRNA translation, 509–511
[^{3}H]Inositol, salivary gland receptor mechanisms, 55–56
Inositol bisphosphate (IP_2), 55–56
Inositol hexakisphosphate (IP_6, phytic acid), 56
Inositol monophosphate (IP), 55–56
Inositol pentakisphosphate (IP_5), 56
Inositol tetrakisphosphate (IP_4)
    pancreatic plasma membrane, 452
    salivary gland receptor mechanisms, 56
Inositol trisphosphate (IP_3)
    HCl acid secretion, calcium channels, 260–261
    intracellular calcium release, 56–57
    pancreatic cell function, calcium release, 453
    pancreatic secretion, calcium mobilization, 427–428
    salivary gland receptor mechanisms, 55–56, 57
Insulin
    biliary protein secretion, 668
    digestive enzyme regulation, 469–470
    digestive regulation, 472
    gastrointestinal mucus secretion, 371–372
    gene expression, 503
        protein synthesis, 503–504
Intact gland preparations, pancreatic electrolyte secretion, 388–389
Integral membrane proteins, asymmetric integration into lipid bilayer, 482–485
Intercalated ducts, 1
Intercellular junctions, pancreatic structure, 386–387
Intergranular ordering, secretion mechanism, 472
Interlobular ducts, electrolyte secretion, 407–409
Interorgan flow
    bile acids, 586
    enterohepatic circulation, 570
    flux measurements, 581–582
Intestinal bacteria, biotransformation alterations, 587–588
Intestinal phase
    gastric inhibition, 168–170
    gastric secretion, 150–151
Intracellular ions
    gastric electrophysiology, 189–195
    isolated oxyntic cells, 190–191
Intracellular membranes, pancreatic cell function, 453–455
Intracellular messengers
    pancreatic plasma membrane ion channels, 450–452
    pancreatic secretion, 425–430
        calcium-mediated secretagogues, 425–429
        cAMP-mediated secretagogues, 429–430
    pepsinogen secretion, 273–274

Intracellular microelectrodes, acinar cell electrophysiology, 26–27
Intracellular organelles, signal transduction, 444–445
Intracellular protein binding
  hepatic drug clearance, 723–724
  hepatic transport, 712
Intracellular transport
  bile acid–lipid coupling, 611–612
  biliary protein secretion, 666–670
    hepatocyte utilization, 667–668
    lysosomal degradation pathway, 666
    regulation mechanisms, 668–670
    transcellular pathway, 666–667
  regulated and constitutive pancreatic secretion, 539–543
Intracisternal protein processing, 485–495
  covalent processing, 485
  Golgi compartmentation, 485–487
Intragranular packaging, exocrine granule membranes, 114–115
Intrinsic factor (IF)
  gastric secretion, 146, 334
  stomach parietal cell stimulation, 348
Intrinsic mechanism, 356
Ion channels
  exocrine granule membranes, 115
  exocytosis and, 455
  intracellular messengers, 450–452
  gastric electrophysiology, 200–201
  H^+-K^+-ATPase, 240
  pancreatic cell function, endoplasmic reticulum, 453–455
  See also specific channels, e.g., Calcium, Sodium
Ion concentration in bile, 559
Ion gradient–coupled calcium uptake, 454–455
Ionophores
  H^+-K^+-ATPase, 234–235
  See also specific ionophores
IR drop method, gastric electrophysiology, 196–197
Islets of Langerhans, pancreatic distribution, 526–527
Isobutylmethylxanthine (IBMX)
  oxyntic cell function, 214
  pancreatic secretion, 429–430
Isolated cell cultures
  acinar cells
    attached cultures, 518–519
    signal transduction, 444
    suspension cultures, 518
  biliary function, 553
  duct cells, 519–520
    pancreatic electrolyte secretion, 390
Isolated membrane fractions, biliary function, 553
Isoproterenol, phosphorylation and amylase secretion, 72–73

K^+-Cl^- transporter, HCl acid secretion, 217–219
K^+-H^+ antiport, ductal electrolyte transport, 14

Labial glands, mucin secretion, 87–88
Lanthanum, canalicular bile formation, 607
Lateral access, paracellular pathway, 11–12
Lecithin
  bile acid–lipid coupling, 611–612
  bile salt interactions, 637–649
    chemical composition, 637–638
    lecithin–sodium cholate–water diagrams, 638–640
Lecithin–sodium glycocholate–water system, 646–647
Lecithin–sodium taurocholate system, 643–644
Lectins, gastric membrane composition, 293–294
Ligand binding, pancreatic secretion, 420–421
Ligandin
  hepatic drug clearance, 724
  hepatic transport, 712
Lingual glands, mucin secretion, 87–88
Lipase, diet and, 466–467
Lipids
  bile constituents, 556–558
  exocrine granule membranes, 109–110
  gastric membrane composition, 291–293
  gastrointestinal mucus, 368
  gene expression adaptation, 501
Lipophilicity, hepatic drug clearance, 721–722
Lithocholic acid (LCA), 576–578
  molecular structure, 622
Liver
  cell plates, hepatic transport, 695
  drug clearance in disease states, 737
  excretory function vs. renal function, 550–551
  protein secretion, 663–665
  transport and metabolism, 717–719
  ultrastructure, 717–719
Long-chain fatty acids, hepatic transport, 711
Luminal acidification
  gastric inhibition, 163–164
  antral mucosa, 165–166
Luminal membrane
  ductal electrolyte transport, 14–15
  secretory mechanism, 10–11
Luminal Na^+ pump secretion model, 3
Lumped one-cell model, gastric electrophysiology, 198
Lysine, 470–471
L-Lysine, 473–474
Lysolecithin, gastric mucosal barrier, 288
Lysosomes
  biliary secretion
    historical background, 677–678
    protein secretion, 666–667
  biochemical definition, 678
  digestive function, 683
    autophagy, 683
    crinophagy, 683
    heterophagy, 683
    endocytosis, 682–683
  enzymatic activities, 679
  enzyme biogenesis, 682
  functional categories, 679
  histochemical definition, 678–679
  membrane composition, 679–680
  morphology, 678–679
    membrane structure, 679–680
  primary and secondary classification, 679
  research methodology, 680–682
    electron microscopy, 682
    hydrolase assay, 681–682
    isolation and tissue fractionation, 680–681
  secretion function, 684–685
    lysosome-to-bile hepatic excretory pathway, 685–688
  storage function, 683–684
  transport function, 683–684
Lysosome-to-bile hepatic excretory pathway, 685–688
  functional significance, 686–687
  metal secretion, 687
  pathophysiological conditions, 688
  TGF-beta processing, 687
Lysozyme, salivary secretion, 331–332

Macromolecules
  exchange, 107
  gastric secretion, afferent neurons, 135–136
Madin-Darby canine kidney (MDCK) cells
  membrane and secretory polarity, 543–544
  protein sorting, 490
Magnesium, ATP-dependent calcium uptake
  pancreatic cell function, 454
  electrolyte secretion, 403
  H^+-K^+-ATPase, 241
Manganese, H^+-K^+-ATPase, 241
Mast cells
  acid secretion, fundic mucosa, 215
  polarized secretion, 535

Mathematical modeling, hepatic transport, 701–707
Medial forebrain bundle, gastric secretion, 128–129
Membrane capacitance, acinar cell electrophysiology, 43
Membrane fusion
    HCl acid secretion, 216–219
    pancreatic secretion, exocytosis, 435–436
Membrane potentials
    gastric electrophysiology, 189–195
    hepatocyte electrolyte and solute transport, 599–600
    isolated oxyntic cells, 190–191
Membrane proteins, hepatic transport, 712–713
Membrane receptors
    mucin secretion, 83
    pancreatic secretion, 422–424
Membrane recycling hypothesis, oxyntic cell secretion, 208–210
Membrane resistance ratio (R_a/R_b), 194–197
Membrane transport
    hepatic drug clearance
        model compounds, 739–741
        organic compounds, 739–749
    secretion mechanism, 472
Membrane vesicles, hepatic transport research, 720
Mesenchyme, salivary gland morphogenesis, 96–97
Messenger RNA (mRNA)
    gastrointestinal secretion, 340
    gene transcription signals, 505–507
    H^+-K^+-ATPase biosynthesis, 248
    protein translocation, 477–480
    translation efficiency mechanisms, 509–511
Metabolism, hepatic drug clearance, 738
Metastable bile, 653
Met-enkephalin, gastric secretion, 132
3-O-Methylglucose (3-O-MG)
    gastric secretion, 128–129
    glucoprivation, 130–131
N-Methylscopolamine, oxyntic cell function, 213
N-Methyltransferase, gastric secretion, 146
Metrizamide density gradient, bile fractionation, 657–658
MG2 mucin cells, 80
Micelles
    bile acid secretory function, 560
    bile salts, size and shape configurations, 636
    cholestasis, 562–563
    egg yolk lecithin–sodium cholate–water system, 642–645
    gastric mucosal barrier, 287–288
    hepatocyte bile secretion, 657–658
    multilamellar phosphatidylcholine liposomes, 648
    See also Critical micelle concentration
Micelle-to-vesicle transition, 645–646
Microelectrodes, equivalent circuit analysis, 194–197
Microfilaments
    exocrine granule membranes, 116–117
    oxyntic cell cytoskeleton, 210–211
Micropuncture
    duct cell electrolyte secretion, 407, 409
    pancreatic electrolyte secretion, 389–390
    salivary secretion, 2–3
Microtubules
    biliary protein secretion cytoskeleton, 671–672
    exocrine granule membranes, 116–117
    hepatocyte protein secretion, 665
    oxyntic cell cytoskeleton, 210–211
Mitochondria
    amylase release, 69
    signal transduction, 444–445
Mitochondrial monoamine oxidase (MAO), 730
Mixed-function oxidase, 727–731
Molecular sorting of proteins, 487–490
Monoclonal antibodies
    gastrointestinal secretion, 328
    salivary secretion, 332
Monolayer cell culture, pancreatic growth regulation, 525

Monovalent cations, hepatic drug clearance, 745–746
Morphine, gastric secretion, 132
Mouse, pancreatic electrolyte secretion, 392–393
M_r proteins, 433–434
Mucin
    biosynthesis, 81–82
    chemical composition and structure, 79–80
    cholesterol nucleation, 655
    minor salivary glands, 87–88
    structural alteration, 81
    sublingual gland, 86–87
    submandibular gland secretion, 82–86
        calcium and, 85–86
        cellular mediators, 84–85
        membrane receptors, 83–84
        signal transduction, 84
Mucus
    adherent gel, 359–363
        physical and permeability properties, 362–363
        in vivo continuity and thickness, 359–362
    cellular biosynthesis and secretion, 368–370
    control of secretion, 371–372
    defined, 359
    degradation, 372–373
    function, 373–375
    gastroduodenal protection, 374–375
    mucin glycoprotein, 363–368
        carbohydrate chain structure, 363–365
        gel formation, 366–368
        lipid and protein components, 368
        protein core and macromolecular polymeric structure, 365–366
        technology research, 363
    secretion measurement, 370–371
Mucus-bicarbonate layer, gastric mucosa, 282–284
Multilamellar phosphatidylcholine liposomes, 648
Multiple cell types, gastric electrophysiology, 195–196
Multiple sinusoid flow distribution, hepatic transport, 703–704
Muscarinic-cholinergic receptors
    pancreatic secretion, 421–422
    molecular characterization, 423–424
Muscarinic receptors, pepsinogen secretion, 270–271
Myoepithelial cells, salivary gland cytodifferentiation, 98

N_3-ATP, H^+-K^+-ATPase inhibition, 246–247
Na^+-$2Cl^-$-K^+ cotransport
    electrolyte secretion, acinar cells, 400–402
    pancreatic plasma membrane, 445
NADPH, hepatic drug clearance, 728–729
Na^+-H^+ antiport
    ductal electrolyte transport
        basolateral membrane, 15
        luminal membrane, 14
    HCl acid secretion, 221
    membrane transport mechanisms, 602–603
    pancreatic plasma membrane, 445–446
Na^+-H^+-ATPase, duct cell electrolyte secretion, 405–407
Na^+-HCO_3^- cotransport, 192–193
Na^+-K^+-Cl^- cotransport, 611
Na^+-K^+-$2Cl^-$ symport
    cytosolic activities, 10
    secretory mechanism, 7–8
Negative feedback
    bile acid synthesis, 583–584
    digestive regulation, 472
Nerve cell stimulation, gastric secretion, 132
Neuropeptides
    gastric inhibition, 173–177
    intracerebral administration, 173–174
    intravenous administration, 174–177
Neurotensin, intestinal gastric inhibition, 171–172
Neurotransmitter receptors, parotid gland, 64
NH_2 terminal sequences

pepsinogen secretion, 269–270
protein translocation, 479–482
NH_3-terminal sequences, asymmetric membrane protein insertion, 482–484
Noise analysis, acinar cell electrophysiology, 27–28
Nonelectrolytes, pancreatic permeability, 397–398
Nonoxyntic cells, receptors, 214–215
Nonparallel transport, regulation of digestion and, 467–468
Nonselective cation channels
    acinar cell electrophysiology, 39–40
    fluid secretion, 46
N proteins, pancreatic plasma membrane, 447–448
Nucleating factors, cholesterol crystal formation, 654
Nucleus ambiguus (NA), gastric secretion, 127–129
Nucleus tractus solitarius (NTS), gastric secretion, 127–129
Nutritional shock, gene expression, 504
Nutritional substrates, gene expression
    adaptation, 500–501
    pre- and postnatal development, 505
    protein synthesis, 503–504

Oligosaccharide chains
    gastrointestinal mucus, 367–368
    mucin biosynthesis, 369–370
    pancreatic proteins, Golgi compartmentation, 485–487
    salivary mucins, 79–80
Omeprazole, H^+-K^+-ATPase inhibition, 246–247
"One linkage-one glycosyltransferase," mucin biosynthesis, 82
Opiates, gastric secretion, 132
Opioid peptides
    duodenal alkaline secretion, 320
    gastric inhibition, 174
    intravenous administration, 177
Opsin, translocation, 484
Organelles
    exocrine granule membranes, 116–117
    intracellular signal transduction, 444–445
    ion transport, 605–606
Organ explant cultures, exocrine pancreas, 516–518
Organic anions
    hepatic drug clearance, 742–745
        carrier proteins, 744–745
        excretion mechanisms, 744
        kinetic analysis, 742–744
        schematic, 723
        uptake mechanisms, 744
    hepatic transport, 711–712
Organic cations
    hepatic drug clearance
        bile-canalicular transport, 747
        bivalent cation uptake, 746–747
        kinetic analysis, 745
        monovalent cation uptake, 745–746
    hepatic transport, 712
Organic compounds
    hepatic drug clearance, 740–741
    hepatic transport, albumin-bound, 707–710
Osmolality
    exocrine granule membranes, 115
    gastric bicarbonate secretion, 310–311
Osmotic flow hypothesis, oxyntic cell secretion, 208–210
Ouabain, hepatic drug clearance, 748
Ovalbumin, translocation, 484
"Overshoot" effect, hepatic transport, 699
Oxygen consumption
    H^+-K^+-ATPase, 239–240
    oxyntic cell function studies, 212
Oxygen radical generation, gastric microcirculation, 298
Oxyntic cells
    calcium-dependent mechanisms, 224
    cytoskeleton, 210–211
    functional morphology, 207–211

gastric electrophysiology
    basolateral impalement, 191–192
    Cl^- secretion, 187–188
    intact mucosa, 191–193
    isolated cells, 190–191
    stimulation changes, 193–195
growth regulation, 351
HCl acid secretion transport model, 221–222
membrane turnover and recycling, 208–210
pepsinogen secretion, 270
receptors, 211–215
    specificity studies, 212–214
ultrastructure, 208–209
Oxyntic mucosa
    gastric inhibition, 163–165
    local inhibitors, 164–165

Pachymeter, gastrointestinal mucus, 359–360
Palatal glands, mucin secretion, 87–88
Pancreas
    cellular compartmentation, 477–478
    embryonic morphogenesis, 531–532
    end products and zymogen granules, 473–474
    epithelial permeability, 397–398
        electrolytes, 398
        nonelectrolytes, 397–398
    membrane and secretory polarity, 543–544
    polarity, 534–535
    protein processing
        cisternal pH, 490
        covalent intracisternal processing, 485
        Golgi compartmentation, 485–487
        intracisternal and cellular sorting, 485–495
        intracisternal transport of secretory proteins, 490–495
        molecular sorting, 487–490
    protein translocation
        integral membrane proteins, 482–485
        receptors, 480–482
        RER membrane, 477–480
    regulation of digestion
        vs. adaptation, 467
        diet and enzyme content, 466–467
        as function of bulk, 465–466
        modes of action, 471
        nonparallel transport, 467–468
        parallel secretion, 467
        properties, 468–472
        regulatory outcomes, 471–472
        sensory input, 470–471
    research methodology, 387–390
        ductal perfusion, 389
        intact gland preparations, 388–389
        isolated ductal tissue, 390
        micropuncture, 389–390
    structure, 383–384
        acinar cells, 386
        ductal cells, 385–387
        ductal tree, 384–385
        intercellular junctions, 386–387
    in vitro studies
        differentiation regulation, 521–524
        growth regulation, 524–525
        historical background, 515
        hormone receptor regulation, 526–527
        isolated cell culture, 518–520
        long-term models, 516–521
        organ explant cultures, 516–518
        physiological regulation of functions, 521–527
        secretion regulation, 525–526
        tumors and cell lines, 520–521
Pancreatic cell membranes
    exocytosis and ion transport, 455

Pancreatic cell membranes (continued)
  chloride conductance, 455–456
  hormonal stimulation of secretion, 446–447
    adenylate cyclase–cAMP system, 447–448
    phospholipase C system, 448–450
    receptor and intracellular messenger ion channels, 450–452
    single-hormone activation, 450
    tyrosine kinase system, 452
  intracellular membranes, 453–455
    active ion transport—endoplasmic reticulum, 454–455
    passive ion transport—endoplasmic reticulum, 453–454
  transport systems–plasma membrane, 445–446
Pancreatic proteins, gene expression, 500
Pancreatic secretion, 335–337
  cellular mechanisms, 398–402
    acinar cell models, 400–402
    bicarbonate, 403–405
    calcium and magnesium, 403
    chloride, 403
    duct cells, 402–405
    electrophysiology, 399–400
    ionic requirements, 398–399, 402–403
    sodium and potassium, 403
    stimulus-secretion coupling acinar cells, 399–400
      duct cells, 402
  duct cell models, 405–407
  effectors, 430–436
    exocytosis, 435–436
    intracellular receptors—calcium, cAMP, and DAG, 430–431
    protein phosphorylation, 431–435
  electrolyte secretion patterns, 390–396
    dog, cat, and human, 390–392
    guinea pig and hamster, 394–395
    pig, 394
    primates, 395–396
    rabbit, 393–394
    rat and mouse, 392–393
    sheep, cow, and horse, 395
  electrolyte secretion sites, 396–397
    cholecystokinin and vagal stimulation, 397
    secretin-stimulated bicarbonate secretion, 396–397
  intracellular messengers, 425–430
    calcium-mediated secretagogues, 425–429
    cAMP-mediated secretagogues, 429–430
  mechanism, 472–473
  ontogeny, 535–536
  receptors, 419–425
    binding characteristics, 420–422
    functional characterization, 419–420
    molecular characterization, 422–424
    regulation, 424–425
  regulatory agents, 419
Paracellular (shunt) pathway
  gastric mucosa, 281
  junction permeability, 198–199
  secretory mechanism, 11–12
Parallel secretion, regulation of digestion and, 467
Parietal cell
  gastric development, corticosterone, 353
  HCl acid secretion
    calcium control, 258–262
    cAMP, 255–257
    protein kinases, 257–258
    H^+-K^+-ATPase, 229–230
  stomach
    acid secretory response, 347–348
    basal acid secretion, 346–347
    structure, 345–346
Parotid gland
  amylase secretion, 63
    β-adrenergic mobilization of calcium, 69
    calcium, 66–70
    cAMP, 65–66
    future research trends, 73
    mechanism of action, 66
    saliva composition, 63–64
  capacitative calcium entry, 57–58
  cytodifferentiation, 97
  endogenous protein phosphorylation, 70–73
  neurotransmitter receptors, 64
  slice system, 65
  ultrastructure, 64
Partial pressure of carbon dioxide (P_{CO_2}), 310–311
Patch-clamp studies
  acinar cell electrophysiology, 26–28
  K^+ conductance, basolateral membrane, 7
Pentagastrin
  gastric secretion, 141–142
  pepsinogen secretion, 271–272
  stomach parietal cell stimulation, 347–348, 348
Pepsin
  gastric mucosal barrier, 290–291
  gastric secretion, 138
  mucus degradation, 372–373
Pepsinogen
  gastric secretion, 332–334
  secretion
    assay, 268–269
    basic model, 267–268
    biochemical properties, 269–270
    distribution, 270
    historical background, 267
    intracellular mediators, 273–274
      calcium, 274
      cAMP, 273–274
    mechanism, 274–275
    stimulants, 270–273
      adrenergic stimulation, 271
      cholinergic stimulation, 270–271
      gastrin/cholecystokinin, 271–272
      histamine, 271
      secretin and vasoactive intestinal peptide, 272–273
  stomach chief cell stimulation, 349
Pepsinogen I, gastric secretion, 333–334
Pepsinogen II, gastric secretion, 333–334
Peptides
  duodenal bicarbonate secretion, 315
  gastric secretion, 129–130
  gastrin release, 144
  pepsinogen secretion, 273
Peptide YY
  gastric secretion, 151
  ileocolonic inhibition, 172
Perfused liver preparation, biliary function, 553
Peribiliary plexus, protein secretion, 665–666
Permeability
  ductal transport properties, 15
  gastrointestinal mucus, 362–363
  pancreatic epithelium, 397–398
Peroxidase-antiperoxidase (PAP) system
  gastrointestinal secretion, 329
  mucin biosynthesis, 369–370
Pertussis toxin, oxyntic cell function, 223–224
pH
  bicarbonate secretion, surface gradient, 320–321
  bile acids, 287–288
  bile solubility, 629–631
  biliary excretion, 561–562
  cisternal, cellular protein sorting, 490
  exocrine granule membranes, 115–116
  gastrointestinal mucus, 360
  HCl acid secretion homeostasis, 221

H$^+$-K$^+$-ATPase, 235, 282
  intracellular regulation, 296–297
  lysosomal hydrolase, 679
  pepsinogen secretion, 269–270
  secretory mechanism, cytosolic activities, 10
Phase rule, bile salt–lecithin interactions, 638–640
Phosphatase
  pancreatic secretion, effector system, 431–433
  H$^+$-K$^+$-ATPase kinetics, 242–243
Phosphatidylcholine, gastric membrane composition, 294–295
Phosphatidylinositol (PI), 427–428
Phosphatidylinositol 4,5-bisphosphate (PIP$_2$)
  pancreatic secretion, 427–428
  salivary gland receptor mechanisms, 54–55
Phosphatidylinositol 4-phosphate (PIP)
  pancreatic secretion, 54–55
  salivary gland receptor mechanisms, 54–55
Phosphoinositides
  calcium entry, 57
  intracellular calcium release, 56–57
  1,4,5-IP$_3$ mechanisms, 57
  parotid gland entry, 57–58
  pathways, 54–56
  turnover, 54
Phospholipase C
  adenylate cyclase, CCK regulation, 450
  pancreatic plasma membrane
    feedback regulation, 449–450
    GTP-binding proteins, 449
    signaling transduction, 448–449
  pancreatic secretion, 428–429
  pepsinogen secretion, 275
Phospholipids
  bile acids, 558
  biliary secretory function, 560
  cholesterol interactions, 649–650
  monolayer hypothesis, 295
  vesicles, bile salt interactions, 647–648
Phosphorylation
  exocrine granule membranes, stimulus-enhanced, 117–118
  H$^+$-K$^+$-ATPase
    from ATP, 243
    K$^+$-dependent, 244–245
  pancreatic secretion
    protein, 431–435
    secretagogue-induced changes, 433–435
  stimulus-secretion coupling, 538–539
Pig, pancreatic electrolyte secretion, 394
Pirenzepine, oxyntic cell function, 213
Plasmalemma proteins, exocrine granule membranes, 120
Plasma protein binding, hepatic drug clearance, 721–723
Plasticity, ductal transport properties, 15
Polyclonal antisera, gastrointestinal secretion, 328
Polypeptides, granule membranes, 111–114, 117
Portal-systemic shunting, interorgan flow, 586
Positive feedback, digestive regulation, 472
Potassium
  acinar cell electrophysiology
    fluid secretion, 44–46
    protein secretion, 46–47
    resting membrane, 31–34
    stimulant-evoked membrane potential changes, 40–43
  basolateral membrane, 6–7
  dephosphorylation, H$^+$-K$^+$-ATPase, 244–245
  ductal electrolyte transport, 15
  electrolyte secretion, duct cells, 403
  gastric electrophysiology, 188
  pancreatic plasma membrane, intracellular messengers, 450–452
  secretory mechanism
    cytosol, 9–10
    model, 4
  single-channel current recording, 36–37
  sinusoidal-lateral membrane transport, 604
Potential carrier proteins, hepatic transport research, 720
Preproapomucin, 81
Primary fluid formation
  secretion models, 3–5
    basolateral Na$^+$-K$^+$-2Cl$^-$ model, 4–5
    biocarbonate model, 5
    double-antiport (Na$^+$-H$^+$ and Cl$^-$-HCO$_3$) model, 5
    K$^+$ secretion model, 4
    luminal Na$^+$ pump model, 3
Primates, pancreatic electrolyte secretion, 395–396
Probability density function (PDF)
  multiple sinusoid flow distribution, 703–704
  single-pass indicator dilution curves, 704–705
  solute disappearance curves, 705–707
(Pro)enzymes, protein translocation, 478–480
Promoter elements, gene transcription signals, 507–508
Prostaglandin E, gastric inhibition, 165
Prostaglandin E$_1$ (PGE$_1$), 256
Prostaglandin E$_2$
  analogues, gastric inhibition, 167–168
  duodenal bicarbonate secretion, 314
  gastric alkaline secretion, 318–319
  HCl acid secretion, parietal cells, 256
Prostaglandins
  duodenal bicarbonate secretion, 317
  gastric alkaline secretion, 318–319
  gastric inhibition
    antral mucosa, 166–167
    local inhibition, 165
  gastric mucosa cytoprotection, 299–301
  gastrointestinal mucus, 360, 372
  oxyntic cell function inhibition, 223–224
Protein
  amylase secretion, 70–73
  bile constituents, 556
  cytoplasmic, 116–118
  exocrine granule membranes, 110–114
  gastrointestinal mucus, 364–365, 368
    mucin and macromolecular structure, 365–366
  H$^+$-K$^+$-ATPase composition, 232–233
  pancreatic secretion
    phosphorylation and effector system, 431–435
    secretagogue-induced phosphorylation, 433–435
  translocation, RER membrane, 477–480
Protein A, gastrointestinal secretion, 329
Protein binding
  hepatic drug clearance, 721–724
  intracellular protein, 723–724
  plasma protein, 721–723
Protein deficiency, gene expression, 504
Protein kinase
  amylase release, 66
  HCl acid secretion, 257–258
  mucin secretion, 84–85
  pancreatic secretion effector system, 431–433
  parotid gland amylase release, 66
Protein kinase C
  amylase release, 70
  pancreatic secretion, 429
  pepsinogen secretion, 274
Protein secretion
  acinar cell electrophysiology, 46–47
  biliary
    anatomy, 663–666
    cholesterol metabolism, 672
    cytoskeleton and hepatocyte transport, 670–672
    epidermal growth factor (EGF), 673
    intestinal immune response, 672
    intracellular transport pathways, 666–670

Protein secretion (*continued*)
  pathophysiological conditions, 672–673
  receptor-mediated endocytosis, 666–670
Protein synthesis
  gene expression, 503–504
  intracisternal transport and processing, 494–495
  mRNA translation, 509–511
  regulated and constitutive pancreatic secretion, 539–541
  salivary gland cytodifferentiation, 99–100
Protein translocation
  schematic, 481–482
  start- and stop-translocation schematic, 482–485
Protonatable amines, H^+-K^+-ATPase inhibition, 246
Proton permeability
  apical cell membranes, 295–296
  gastric mucosa, 281–282
  mucus, 282–283
Proton pump (H^+-ATPase)
  exocrine granule membranes, 116
  gastric inhibition, 163
  organelle ion-transport mechanism, 605
  transport, 235
Pylorooxyntic reflex, gastric inhibition, 167

QNB ([^{3}H]quinuclidinyl benzilate), 213
Quaternary bile salt–phospholipid–cholesterol–water system, 650–652

Rabbit, pancreatic electrolyte secretion, 393–394
Radioimmunoassay
  gastrointestinal secretion, 329–330
  pepsinogen secretion, 268–269
Rat, pancreatic electrolyte secretion, 392–393
Receptors
  cell surface, 100–101
  endocytosis mediation, 666–670
  gastric inhibition, 161–163
  hepatic transport, 708–709
  mucin secretion, 83
  neurotransmitter, parotid gland, 64
  nonoxyntic cells, 214–215
  oxyntic cell, 211–215
    specificity studies, 212–214
  pancreatic plasma membrane, 450–452
  pancreatic secretion, 419–425
    binding characteristics, 420–422
    functional characterization, 419–420
    molecular characterization, 422–424
    regulation, 424–425, 526–527
  parotid gland amylase release, 64
  protein sorting, 487–490
  protein translocation, 480–482
    SRP receptor, 480
  salivary gland phosphoinositides, 54–58
Reepithelialization, gastric mucosa, 298–300
Regulated pathways, exocrine secretory membranes, 107–108
Residual stimulation, pancreatic secretion regulation, 425
Resting membrane properties, acinar cell electrophysiology, 29–31
Resting vesicles, H^+-K^+-ATPase, 233–236
Reverse hemolytic plaque assay (RHPA), 339–340
Ribonuclear proteins (RNPs), gene transcription signals, 505–507
Ribophorins I and II, protein translocation, 480–481
7S RNA, protein translocation, 480
7SL RNA, protein translocation, 480
40S RNA, mRNA translation, 509–511
Rose bengal, hepatic transport, 711–712
Rough endoplasmic reticulum (RER)
  gastric development, 353
  lysosomal biogenesis, 682
  pancreatic secretion, 335–336
  protein translocation, 477–480
  stomach chief cell stimulation, 350
rRNA (ribosomal RNA), mRNA translation, 510–511
Rubidium (^{86}Rb$^+$)
  exocrine granule membranes, 115
  HCl acid secretion, 217–218
  H^+-K^+-ATPase, 236
    stimulated vesicles, 237–238
  salivary glands, stimulus-permeability coupling, 52–53
Rubidium (Rb$^+$-Rb$^+$) exchange, proton pump, 236–237

S6 protein
  pancreatic secretion, 433–435
  phosphorylation and amylase secretion, 71–72
Salivary glands
  anatomical nomenclature, 1–2
  calcium signaling system, 51–52
    phosphoinositides and receptor mechanisms, 54–58
    stimulus-permeability coupling, 52–54
  cytodifferentiation, 97–100
    developmental regulation, 99–100
    embryology, 97–99
  differentiation, 93–102
  ductal electrolyte transport, 13–16
    absorptive mechanism, 13–15
    control, 15–16
    model, 16–17
    plasticity, 15
    water permeability, 15
  morphogenesis, 94–97
    developmental regulation, 95–97
    embryology, 94–95
  primary fluid formation, 3–13
  secretion, 331–332
    control, 12–13
    mechanism, 5–12
    models, 3–5
    two-stage hypothesis, 2–3
  stimulus-secretion coupling, 100–103
    cAMP, 100
    cell surface receptors, 100–101
    nerve coupling, 101–102
SCH 28080, H^+-K^+-ATPase inhibition, 246–247
SCN$^-$, HCl acid secretion, 216
Secondary active transport, 699
Second-messenger systems
  exocrine granule membranes, 117
  pancreatic enzyme secretion, 443–444
Secretagogues
  acinar cell electrophysiology, 29
    whole-cell current recordings, 37
  differentiation, AR42J cells, 524
  intracisternal protein transport and processing, 491–494
  mucin secretion, 83
  pancreatic secretion, 419
    calcium-mediated, 425–429
    cAMP-mediated, 429–430
    cellular protein phosphorylation, 433–435
  stomach chief cell stimulation, 349–350
Secretin
  duct cells
    carcinomas, 521
    electrolyte secretion, 406–407, 407–410
  gastrointestinal mucus, 360
    secretion mechanism, 371–372
  gene expression, 502–503
    protein synthesis, 503–504
  intestinal gastric inhibition
    vs. bulbogastrone, 168
    fat, 170
  pancreatic electrolyte secretion
    bicarbonate secretion, 396–397
    species differences, 390–396
  pepsinogen secretion, 272–273

Secretion
    digestive regulation, 472–473
    enzyme studies, 468–469
    lysosome function
        hepatocytes, 685–688
        nonhepatic cells, 684–685
    pancreatic function, 525–526
    regulated and constitutive, 539–543
    See also Secretion by specific organs
Secretory component, biliary secretion, 663
Secretory enzymes, differentiation and, 521–524
Secretory polarity, pancreatic cells, 534–535
Secretory proteins
    cellular sorting, cisternal pH and, 490
    intracisternal transport and processing, 490–495
Sensory input, digestive enzyme regulation, 470–471
Serous glands of von Ebner, 87–88
Sex and gastric development, 355–356
Sham feeding
    duodenal bicarbonate secretion, 315–316
    gastric secretion, cortical and cephalic phase, 133–134
Sheep, pancreatic electrolyte secretion, 395
Shigella flexneri, gastrointestinal mucus, 374
SH reagent, H^+-K^+-ATPase inhibition, 248
Shunt pathways, gastric electrophysiology, 196–199
Sialic acids, gastrointestinal mucus, 365
Side-chain length, critical micelle concentrations, 635–636
Signaling transduction
    cellular mechanism research, in vitro, 444–445
    hormone and neurotransmitter regulation, 443–444
    ion transport and exocytosis, 455–456
    mucin secretion, 84
    pancreatic cell membranes, 445–452
        adenylate cyclase-cAMP, 447–448
        intracellular membranes, 453–455
        ion-transport mechanisms, 453–455
        phospholipase C, 448–450
        plasma membranes and hormonal stimulation, 446–447
        receptor and intracellular messenger ion channels, 450–452
        single-hormone activation, 450
        transport systems, 445–446
        tyrosine kinase, 452
Signal-recognition particle, protein translocation, 480
Signal-translocation sequences, protein translocation, 478–479
Single-channel recordings
    acinar cell electrophysiology, 31–32
    chloride channels, 38
    intact cells, 32–33
    potassium channels, 36–37
Single-pass indicator dilution curves, 704–705
Sinusoidal-lateral membrane transport
    calcium transport, 603–604
    sodium pump (Na^+-K^+-ATPase), 600–603
        amino acid transport, 601–602
        calcium exchange, 603
        chloride uptake, 603
        hydrogen exchange, 602–603
        localization, 600
        regulation, 600–601
        solute transport, 601
SITS (4-acetamido-4′-isothiocyanostilbene-2-2′-disulfonic acid)
    double-antiport secretion model, 5, 9
    gastric electrophysiology, oxyntic cells, 192
SKF 525A, hepatic drug clearance, 729
Sodium
    bile acid transport, 711
    bile salts, critical micelle concentrations, 632, 635
    electrolyte secretion, duct cells, 14, 403
    gastric electrophysiology, 187–188
    gastric mucosal barrier, 288
    gastric secretion, enteric pathways, 137
    hepatic drug clearance, model compounds, 740–741

    secretory mechanism
        control, 12–13
        cytosol, 9–10
Sodium–amino acid cotransport, 35–36
Sodium-coupled solute transport, 601
Sodium pump (Na^+-K^+-ATPase)
    acinar cell electrophysiology, 34–35
    basolateral membrane, 5–6
    bile acid–independent bile formation, 610–611
    compared with H^+-K^+-ATPase, 230–232
    ductal electrolyte transport, 15
    hepatic drug clearance, 748–749
    hepatic transport, 699
    membrane and secretory polarity, 543–544
    salivary glands, stimulus-permeability coupling, 53
    sinusoidal-lateral membrane transport, 600–603
        amino acid transport, 601–602
        calcium exchange, 603
        chloride uptake, 603
        hydrogen exchange, 602–603
        localization, 600
        regulation, 600–601
        solute transport, 601
Sodium taurocholate–lecithin–cholesterol phase diagram, 650–651
Solute disappearance curves, hepatic transport modeling, 705–707
Somatostatin
    antral mucosa, 166–167
    vs. bulbogastrone, 168–169
    intravenous administration, 175–176
    local inhibition, 165
    SS-14, 175
    SS-28, 175
    as vagogastrone, 161–162
Somatostatin-like-immunoreactivity (SLI), 175–176
Soybean trypsin inhibitor (STI), 336–337
Space of Disse, hepatic transport, 695
Specificity, hepatic transport, 709–710
Steady-state kinetics
    ATPase, 241–242
    hepatic transport, 701–702
Steroid moiety, bile acid composition, 578–579
Stimulant-evoked membrane potential changes, 40–43
Stimulation-associated (SA) vesicles, 217–218
    HCl acid secretion, 223
    H^+-K^+-ATPase, 237–238
Stimuli, regulation of bulk digestion, 465–466
Stimulus-permeability coupling, 52–54
Stimulus-secretion coupling
    distal events
        compensatory endocytosis, 539
        cytoskeleton participation, 536, 538
        regulatory events, 538–539
    electrolyte secretion
        acinar cells, 399–400
        duct cells, 402
    exocrine granule membranes, 116–117
    oxyntic cell secretion, 209–210
    pancreatic secretion, 419–436
        electrolytes, 388
        model of events, 456
        schematic, 435
    salivary gland differentiation, 100–102
Stomach
    chief cells, 348–350
        mucosal pepsinogen content, 349
        pepsin secretory response, 349–350
        structure, 350
    gastric inhibition, 163–168
        antral mucosa, 165–167
        gastric phase, 163
        oxyntic mucosa, 163–165
        pharmacological inhibition, 167–168

Stomach (continued)
   gastric mucosal growth, 351
      corticosterone, 352–354
      diet or weaning, 351–352
      intrinsic mechanism, 356
      sexual influences, 355–356
      thyroxine, 354–355
   gastrin, 350
   parietal cells, 345–348
      acid secretory response, 347–348
      basal acid secretion, 346–347
      structure, 345–346
Storage granules
   formation, 108–109
   isolation and preparation, 109
Structural polarity, pancreatic cells, 534–535
Sublingual gland, mucin secretion, 86–87
Submandibular gland
   cytodifferentiation, 97
   mucin secretion, 82–86
      cellular mediators, 84–85
      membrane receptors, 83–84
      signal transduction, 84
Substance P, mucin secretion, 83–84
Succinate, pancreatic secretion, 427
Sugars
   gastric secretion, 132
   See also Glucose
Sulfation
   bile acid biotransformation, 572
   hepatic drug clearance, 734–735
Sulfhydryl compounds, gastric mucosa cytoprotection, 302
Surface epithelial cells (SEC)
   intact, isolated mucosa, 189–190
   sodium secretion, 187–188
Surface hydrophobicity, gastric membrane composition, 294–295
Sympathetic nervous system, gastric inhibition, 177–178
Synapsin, exocrine granule membranes, 117–118

Taurine
   amidation pattern alteration, 587–588
   bile acid conjugation, 572–573
Taurine-conjugated ursodeoxycholic acid phase diagram, 651–652
Taurocholate
   bile acid metabolism, 569
   hepatic transport, albumin-bound anions, 708
   uptake
      bile acid transport, 711
      sodium-coupled transport, 601–602
TCS protonophore, H^+-K^+-ATPase, 234–235
Telenzepine, oxyntic cell function, 213
Temperature
   bile salts
      critical micelle concentration, 632
5-Thioglucose, gastric secretion, 131–132
Thyrotropin-releasing hormone (TRH)
   gastric inhibition, 160
   gastric secretion, 129
   intravenous administration, 177
Thyroxine, gastric development, 354–355
Tight junctions
   gastric mucosa, 284
   hepatic transport, 695
   paracellular pathway, secretory mechanism, 12
Tissue acid-base balance, 297
Tissue fractionation, lysosome isolation, 680–681
Tissue oxygenation, gastric microcirculation, 297–298
Tissue-specific enhancers, gene transcription signals, 508–509
Toxification
   biotransformation, 736–737
   hepatic drug clearance, 726–737
Tracheal mucosa, luminal membrane, 11

Transcellular pathway
   bile acid transport, 574–575
   biliary protein secretion, 666–667
Transcription
   enhancer elements, 508–509
   processing signals, 505–507
   promoter elements, 507–508
   regulation mechanisms, 505–509
Transcytosis, canalicular bile formation, 608
Trans effect, hepatic transport, 699
Transepithelial resistance
   equivalent circuit analysis, 194–197
   gastric mucosa, 281
Transferrin, biliary protein secretion, 667
Transfer RNA (tRNA)
   mRNA translation, 509–511
   protein translocation, 477–480
Transforming growth factor (TGF)
   biliary protein secretion, 670
   lysosome-to-bile hepatic excretory pathway, 687
Trans-Golgi protein sorting, 489–490
Translation, mRNA, 509–511
Transport
   bile acids, 710–711
   classification, 697–698
      carrier-mediated, 698–699
      countertransport and cotransport, 699
      diffusion and convection, 699
      trans effects, 699
   defective hepatic transport, 586–587
   hepatic drug clearance
      bidirectionality, 741
      concentration gradients, 741–742
      driving forces, 742–743
      zonal heterogeneity, 738–739
   lysosome function, 683–684
   pancreatic plasma membrane, 445–446
Truncal vagotomy, gastric secretion, 147–150
Trypsin, diet and, 466–467
Tryptic digestion, H^+-K^+-ATPase conformation, 245
T-tube bile, hepatic transport research, 720
Tubulovesicle compartment, oxyntic cell secretion, 209–210
Two-component hypothesis, gastric juice composition, 280
Two-dimensional gel electrophoresis, gene expression, 499–500
Two-stage hypothesis, salivary secretion, 2–3
Tyrosine kinase, pancreatic plasma membrane, 452

UDP-glucuronic acid (UDPGA), hepatic drug clearance, 731, 734
UDP-glucuronosyltransferase (UDPGT), hepatic drug clearance, 734
Uncharged compounds, hepatic drug clearance, 747–749
Urogastrone, gastric inhibition, 172–173
Ursocholic acid (UCA), molecular structure, 622
Ursodeoxycholic acid (UDCA), molecular structure, 622

Vagal stimulation
   electrical stimulation, 135–136
   gastric secretion
      afferents stimulation, 135–136
      cortical and cephalic phase, 133–134
   pancreatic electrolyte secretion, 397
Vagogastrone
   gastric inhibition, 160–161
   gastric secretion, 148–149
   somatostatin as, 161–162
Vagotomy
   gastric inhibition, 160
   gastric secretion, 147–150
   stomach inhibition, 163
Vagovagal reflexes, gastric secretion, 135
L-Valine, sodium–amino acid cotransport, 35–36

Vanadate, H^+-K^+-ATPase inhibition, 246
Vasoactive intestinal peptide (VIP)
    duct cell carcinomas, 521
    gastric inhibition, 174
    intravenous administration, 176–177
    mucin secretion, 83–84
    pancreatic ductal tree, 384–385
    pancreatic secretion, 421–422
    pepsinogen secretion, 272–273
Ventromedian hypothalamus (VMH)
    gastric secretion, 128–129
    glucose analogues, 131–132
Vibrating-probe measurements, gastric electrophysiology, 197

Water permeability, ductal transport properties, 15
Weak acids
    gastric mucosal barrier, 286–287
    oxyntic cell function studies, 212
Weaning, gastric development, 351–352

Whole-cell current recording
    acinar cell electrophysiology, 33–34
        number of channels per cell, 34
    secretagogues, 37

Xenobiotics, hepatic drug clearance, 726–739
X-ray crystallography, pepsinogen secretion, 269–270

Zonal heterogeneity
    metabolism, 738
    transport, 738–739
Z protein
    hepatic drug clearance, 723–724
    hepatic transport, 712
Zymogen granules
    chloride conductance, 455–456
    digestive regulation, 473–474
    intracisternal protein transport and processing, 491–492
    pancreatic secretion, 435–436
    signal transduction, 444–445